STRUCTURE REPORTS

for 1984

Volume 51B

Part 1

STRUCTURE REPORTS

for 1984

Volume 51B (Part 1)

ORGANIC SECTION

General editor

G. Ferguson

Section editor

G. Ferguson

Published for the

INTERNATIONAL UNION OF CRYSTALLOGRAPHY

KLUWER ACADEMIC PUBLISHERS

DORDRECHT / BOSTON / LONDON

First published in 1992

ISSN 0166–7033

Printed on acid-free paper

ISBN 0–7923–1758–0

Printed in the Netherlands

TABLE OF CONTENTS

INTRODUCTION

The present volume continues the aim of Structure Reports to present critical accounts of all crystallographic structure determinations. Details of the arrangement in the volumes, symbols used etc. are given in previous volumes (e.g. 41B or 42A, pages vi–viii).

University of Guelph,
Guelph, Ontario, Canada

G. FERGUSON

26 January 1992

STRUCTURE REPORTS

SECTION III

ORGANIC COMPOUNDS

Edited by

G. Ferguson

with the assistance of

M. Currie
C. H. L. Kennard
A. L. Macdonald
D. F. Rendle
S. J. Rettig
S. N. Scrimgeour

To find a particular organic or organometallic compound the subject index or formula index at the end of Part 2 of this volume should be used. The general arrangement is: aliphatic or open chain compounds; open chains with N, S; benzene derivatives; cyclic hydrocarbons; condensed ring systems; heterocyclic compounds; carbohydrates; amino acids; natural products; molecular complexes; organometallic compounds - B, Si, P, As, Sb, groups IA, IIA, III, IV, VI; transition metal complexes - π-complexes, other ligands; inorganic ions which have organic counter ions. Only complete structure analyses are described.

TRIFLUOROACETIC ACID MONOHYDRATE
$C_2HF_3O_2,H_2O$ (I)

TRIFLUOROACETIC ACID TETRAHYDRATE
$C_2HF_3O_2,4(H_2O)$ (II)

D. MOOTZ and D. BOENIGK, 1984. Z. Naturforsch., B39, 298-304.

I. Tetragonal, $I\bar{4}2d$, a = 10.420, c = 18.120 Å, Z = 16. Mo radiation, R = 0.051 for 800 reflexions (at -63°C).

II. Triclinic, $P\bar{1}$, a = 6.353, b = 6.919, c = 9.649 Å, α = 79.63, β = 74.30, γ = 70.72°, Z = 2. Mo radiation, R = 0.057 for 1484 reflexions (at -40°C).

Compound I is shown in detail in Fig. 1 and is a true acid hydrate. The "tetrahydrate" II is a hydrated oxonium salt containing centrosymmetric hydrated $(H_5O_2)^+$ ions and centrosymmetric dimers $[(CF_3COO)_2H]^-$ (Fig. 2).

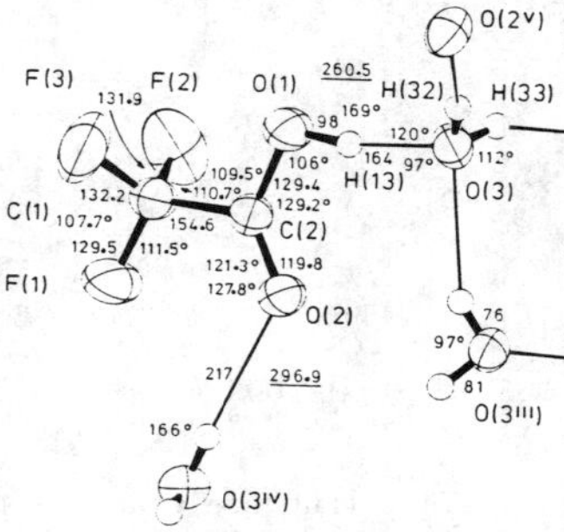

Fig. 1. The structure of $C_2HF_3O_2,H_2O$.

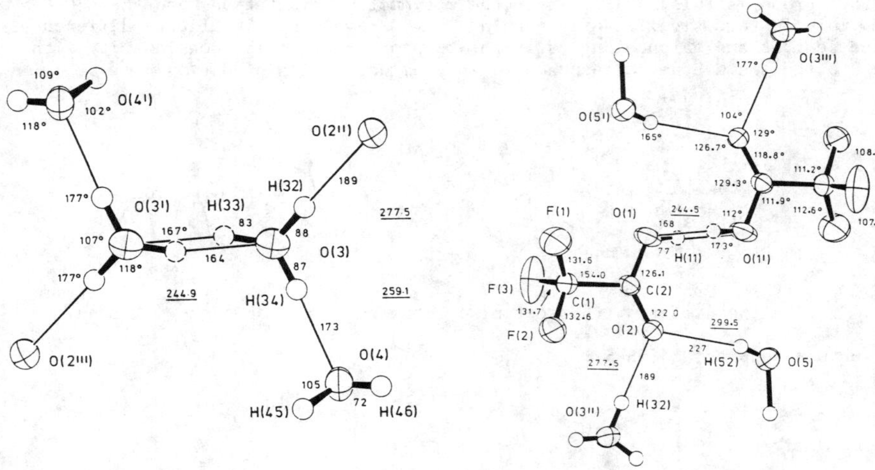

Fig. 2. The $(H_5O_2)^+$ and $[(CF_3COO)_2H]^-$ ions in (II). Atoms H33 and H11 are disordered over two sites.

CYANOFORMAMIDE
$C_2H_2N_2O$

U. DRUCK, W. BECKER and G. BECKER, 1984. Z. Kristallogr., 167, 131-134.

Orthorhombic, Pnma, a = 8.519, b = 5.427, c = 7.235 Å, Z = 4. Mo radiation, R = 0.059 for 363 reflexions.

All atoms of the molecule (Fig. 1) lie on a crystallographic mirror plane. Main bond lengths are: C1-N1 1.327(7), C1-O1 1.216(6), C1-C2 1.477(8), C2-N2 1.141(8) Å. Angles are: N1-C1-O1 126.8(5), N1-C1-C2 113.9(4), O1-C1-C2 119.3(5), C1-C2-N2 177.2(6)°. In the crystal structure there are hydrogen-bonded layers parallel to (010) with H...O 2.07(8) and H...N 2.21(7) Å.

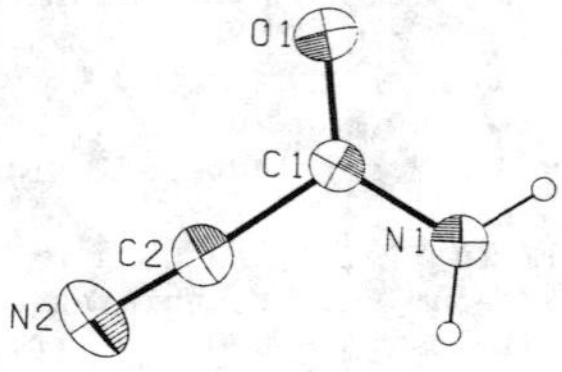

Fig. 1. The cyanoformamide molecule.

THIOACETAMIDE
C_2H_5NS

G.A. JEFFREY, J.R. RUBLE and J.H. YATES, 1984. J. Am. Chem. Soc., 106, 1571-1576.

Monoclinic, $P2_1/c$, a = 6.972, b = 9.873, c = 11.009 Å, β = 99.75°, Z = 8. Neutron radiation, R = 0.043 for 2048 reflexions (at 15 K).

For previous X-ray work see (1). The structure contains two symmetry-independent molecules with different orientations of the methyl groups (Fig. 1). One rotamer has close to m symmetry with planar S=C-C-N and C-C-NH_2 moieties. In the other rotamer, the methyl group is twisted 15.6° from the m symmetry orientation, and the S=C-C-N, C-C-NH_2 moieties are significantly nonplanar. There are no significant differences in the bond lengths and valence angles in the two conformers. The nonplanarity with respect to the sp^2 C atom corresponds to a pyramidalization of 0.6°.

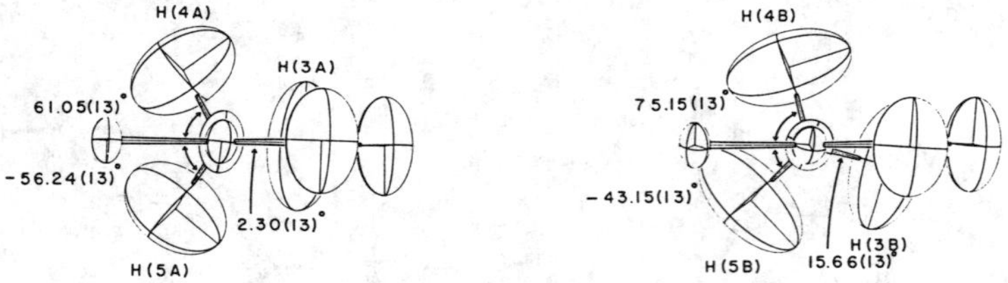

Fig. 1. The two different thioacetamide rotamers viewed down the C-C bond.

1. Structure Reports, 24, 545.

α-KETOGLUTARIC ACID
$C_5H_6O_5$ (I)

SODIUM HYDROGEN α-KETOGLUTARATE
$C_5H_5NaO_5$ (II)

POTASSIUM HYDROGEN α-KETOGLUTARATE
$C_5H_5KO_5$ (III)

T. LIS and J. MATUSZEWSKI, 1984. Acta Cryst., C40, 2016-2019.

I. Monoclinic, $P2_1/c$, a = 16.71, b = 6.375, c = 5.525 Å, β = 94.73°, D_m = 1.64, Z = 4. Mo radiation, R = 0.051 for 1338 reflexions.

II. Orthorhombic, Pbca, a = 6.113, b = 5.859, c = 34.76 Å, D_m = 1.79, Z = 8. Mo radiation, R = 0.040 for 1594 reflexions.

III. Monoclinic, $P2_1/c$, a = 6.517, b = 17.816, c = 8.050 Å, β = 131.79°, D_m = 1.74, Z = 4. Mo radiation, R = 0.033 for 1869 reflexions.

Views of the organic moieties in these structures are in Fig. 1. In (II) and (III) the carboxyl group bonded to the keto group is deprotonated. In all structures the α-ketoglutarate residue has a nearly planar zigzag carbon-chain conformation. All O atoms are almost in the plane through the C-atom chain except the carboxyl group in III which is perpendicular. In (II) each Na^+ ion is coordinated to six O atoms from five different glutarates, with Na...O 2.355(2) to 2.605(2) Å. In (III) each K^+ ion is surrounded by eight O atoms from seven different glutarates at distances varying from 2.747(2) to 3.013(2) Å.

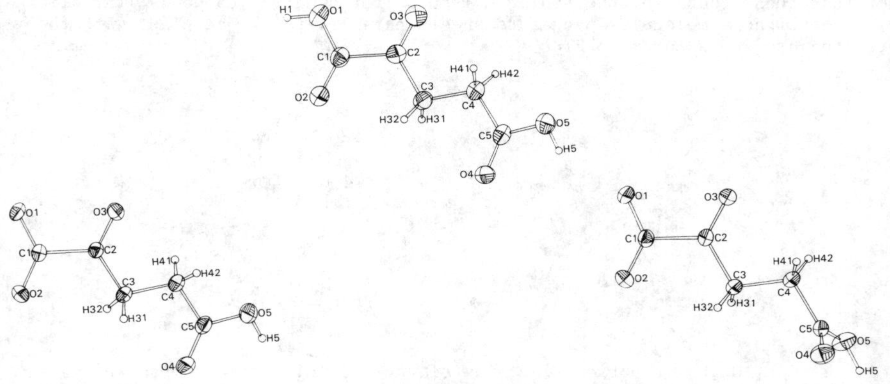

Fig. 1. Views of α-ketoglutaric acid (top) and the anions in the Na^+ (left) and K^+ (right) compounds.

(±)-2,3-DIHYDROXY-2,3-DIMETHYLBUTANEDIOIC ACID ((±)-DIMETHYLTARTARIC ACID)
$C_6H_{10}O_6$

E.N. DUESLER, M. MONDRAGON and R.E. TAPSCOTT, 1984. Acta Cryst., C40, 1887-1890.

Triclinic, $P\bar{1}$, a = 8.632, b = 10.064, c = 10.505 Å, α = 114.22, β = 106.71, γ = 92.92°, Z = 4. Mo radiation, R = 0.058 for 2086 reflexions.

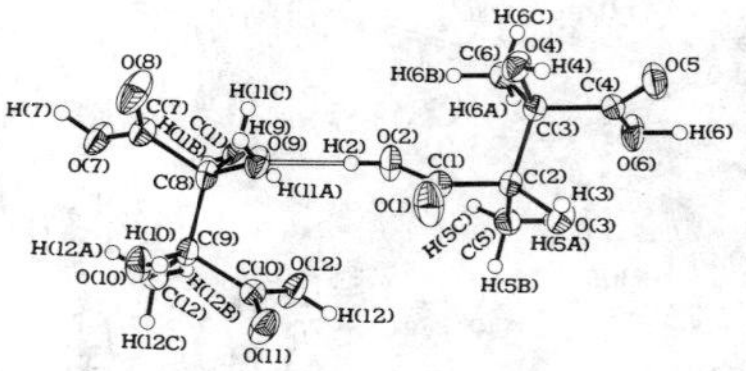

Fig. 1. The asymmetric unit of (±)-dimethyltartaric acid.

The molecular conformations are (-)-synclinal (R,R) and (+)-synclinal (S,S). The torsion angle deviation from perfect staggering is about 17° toward close contact for methyl substituents. The asymmetric unit consists of enantiomeric pairs of molecules linked by hydrogen bonding (Fig. 1). Pairs of symmetry equivalent molecules are joined by hydrogen bonding of carboxylic acid groups across centres of symmetry. These dimers are linked by further hydrogen bonding. C-C bond lengths are greater than those in tartaric acid e.g. C(2)-C(3) 1.574(5) and C(8)-C(9) 1.573(5) Å.

(R,S)-2,3-DIHYDROXY-2,3-DIMETHYLBUTANEDIOIC ACID MONOHYDRATE (meso-DIMETHYLTARTARIC ACID MONOHYDRATE)

$C_6H_{10}O_6,H_2O$

E.N. DUESLER, M. MONDRAGON and R.E. TAPSCOTT, 1984. Acta Cryst., C40, 1286-1288.

Monoclinic, C2/c, a = 12.199, b = 9.177, c = 7.589 Å, β = 98.57°, Z = 4. Mo radiation, R = 0.041 for 1013 reflexions.

The molecule (Fig. 1) has a centrosymmetric antiperiplanar conformation and has C-C bond lengths, viz. C(1)-C(2) 1.543(2), C(2)-C(2') 1.572(2) Å, significantly longer than those found for the non-methyl-substituted parent compound (tartaric acid). The organic molecules pack into sheets parallel to the ac plane with the sheets separated by water molecules.

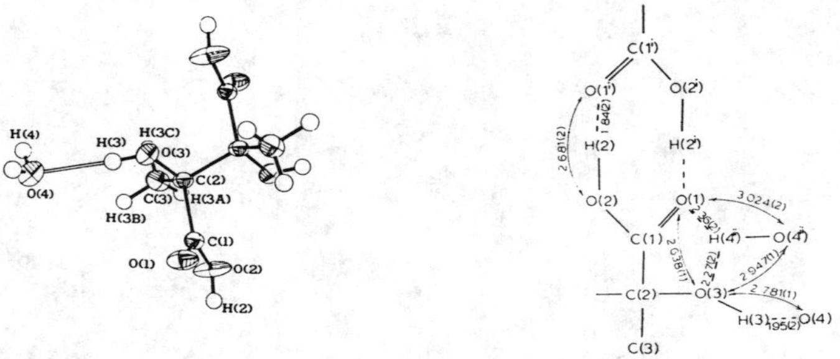

Fig. 1. $C_6H_{10}O_6,H_2O$: perspective view of the molecule and accompanying water molecule and a view showing hydrogen bonds (broken lines) with associated nonbonded contact distances (curved arrows).

(E)-(2RS,5RS)-2,4,5,6,6-PENTACHLORO-2-METHYL-6-NITROHEX-3-ENOIC ACID

$C_7H_6Cl_5NO_4$

M.P. HARTSHORN, R.J. MARTYN and J. VAUGHAN, 1984. Aust. J. Chem., 37, 1963-1968.

Triclinic, P$\bar{1}$, a = 7.985, b = 9.119, c = 10.799 Å, α = 107.47, β = 105.99, γ = 105.74°, D_m = 1.71, Z = 2. Mo radiation, R = 0.039 for 2660 reflexions.

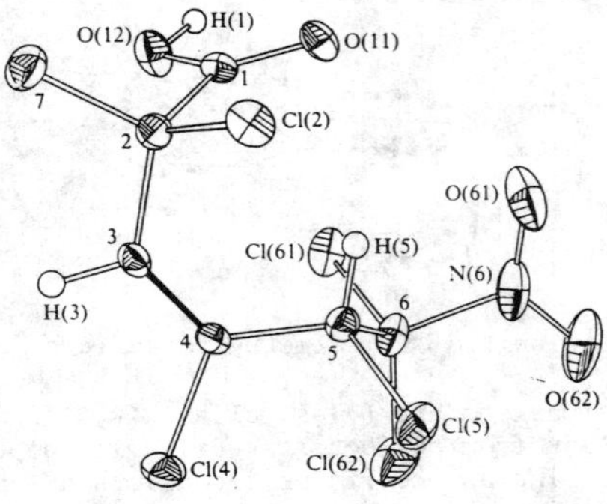

Fig. 1. The structure of $C_7H_6Cl_5NO_4$.

The molecule has different C-Cl distances, C(2)-Cl(2) 1.804(2), C(5)-Cl(5) 1.780(2), C(6)-Cl(61) 1.757(2), -Cl(62) 1.732(3), C(4)-Cl(4) 1.734(2) Å. Torsion angles defining the backbone of the molecule are C(1)-C(2)-C(3)-C(4) +62.5(3)°, C(2)-C(3)-C(4)-C(5) -0.2(4)°, C(3)-C(4)-C(5)-C(6) -120.1(3)°, C(4)-C(5)-C(6)-N(6) +166.0(2)° (Fig.1).

ETHYL 3-AMINO-2-CYANOCROTONATE
$C_7H_{10}N_2O_2$ (I)

ETHYL 3-AMINO-2-CYANODITHIOCROTONATE
$C_7H_{10}N_2S_2$ (II)

K. SZULZEWSKY, B. SCHULZ, S. KULPE and J. KREUTZMANN, 1984. Acta Cryst., C40, 280-283.

I. Monoclinic, C2/c, a = 9.437, b = 16.408, c = 10.453 Å, β = 90.68°, Z = 8. Mo radiation, R = 0.085 for 1442 reflexions.

II. Monoclinic, C2/c, a = 26.192, b = 5.038, c = 14.077 Å, β = 89.467°, Z = 8. Mo radiation, R = 0.063 for 2105 reflexions.

Both I and II form six-membered hydrogen-bonded rings (Fig. 1). It is suggested that the N-H...S bond is stronger than the N-H...O one. In both structures extended nets are produced by intermolecular hydrogen bonding.

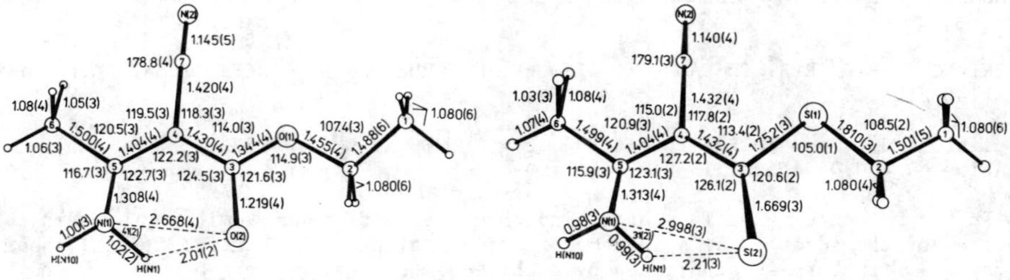

Fig. 1. Molecular dimensions for $C_7H_{10}N_2O_2$ (left) and $C_7H_{10}N_2S_2$ (right).

PHENYLPHOSPHINEDIACETIC ACID
$C_{10}H_{11}O_4P$ $C_6H_5P(CH_2COOH)_2$

L.M. SHKOL'NIKOVA, A.V. GASPARYAN, N.M. DYATLOVA, J. PODLAHA and J. PODLAHOVA, 1984. J. Struct. Chem., 25, 445-448 [Zh. Strukt. Khim., 25(3), 120-123].

Triclinic, P$\bar{1}$, a = 7.929, b = 8.314, c = 9.676 Å, α = 70.35, β = 76.11, γ = 64.62°, Z = 2. Mo radiation, R = 0.049 for 1295 reflexions.

The molecule has a dicarboxylic acid structure. The orientation of the phenyl ring and the two carboxyl groups with respect to the P-C bond corresponds to the crossed and gauche conformations, respectively. The active H atoms in the carboxylic acid groups are involved in typical carboxylic acid O-H...O hydrogen bonds about crystallographic centres of symmetry (O...O 2.680(6) and 2.664(5) Å), producing infinite chains of molecules along the [$\bar{1}$01] direction (Fig. 1).

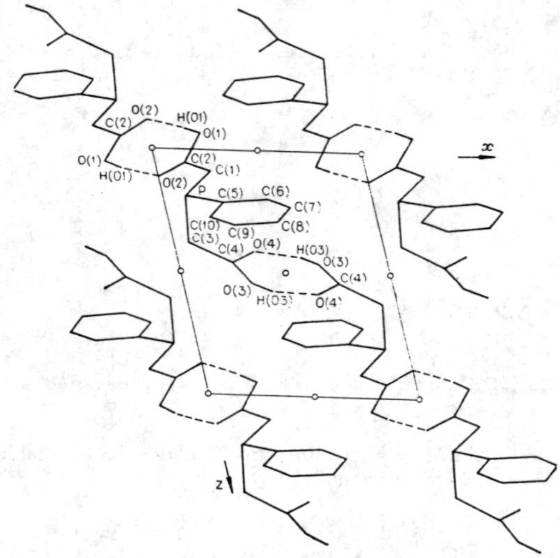

Fig. 1. The structure of phenylphosphinediacetic acid.

4-HYDROXY-3-PHENYLBUTANAMIDE MONOHYDRATE
$C_{10}H_{13}NO_2,2H_2O$

M. SORIANO-GARCÍA, R.A. TOSCANO, T. ALCOCER, R. CRUZ and F. GARCÍA, 1984. Acta Cryst., C40, 2106-2108.

Monoclinic, C2/c, a = 16.546, b = 6.445, c = 19.936 Å, β = 100.77°, Z = 8. Cu radiation, R = 0.039 for 1313 reflexions.

In the molecule (Fig. 1) steric hindrance from different chemical groups is minimized by the adoption of a staggered configuration at the central C(2)-C(3) bond. The crystal structure is stabilized by a three-dimensional network of O-H...O and N-H...O hydrogen bonds.

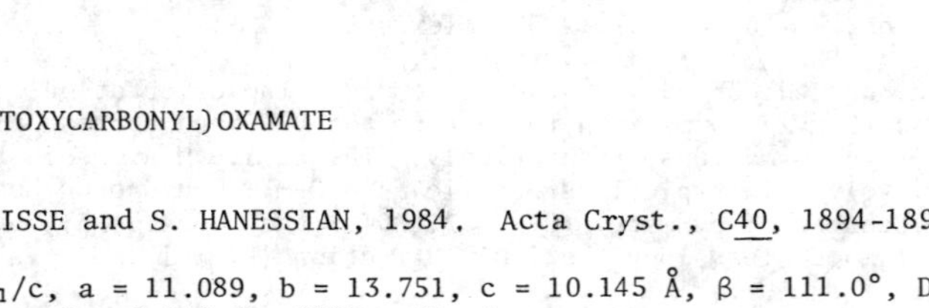

Fig. 1. A view of $C_{10}H_{13}NO_2,2H_2O$

BENZYL N-(t-BUTOXYCARBONYL)OXAMATE
$C_{14}H_{17}NO_5$

H. WYSS, F. BRISSE and S. HANESSIAN, 1984. Acta Cryst., C40, 1894-1897.

Monoclinic, $P2_1/c$, a = 11.089, b = 13.751, c = 10.145 Å, β = 111.0°, D_m = 1.23, Z = 4. Cu radiation, R = 0.040 for 2227 reflexions.

In this amide ester of oxalic acid, the nearly planar benzyl group is inclined at about 43° to the plane of the carboxylic group. The carboxylic plane is at

90(1)° to the amide group which is coplanar with the terminal butoxycarbonyl group. The amide-carboxylic C-C distance is 1.533(2) Å. Dimers are formed by hydrogen bonding between amide groups from adjacent molecules across centres of symmetry (Fig. 1); the N...O distance is 2.918(2) Å.

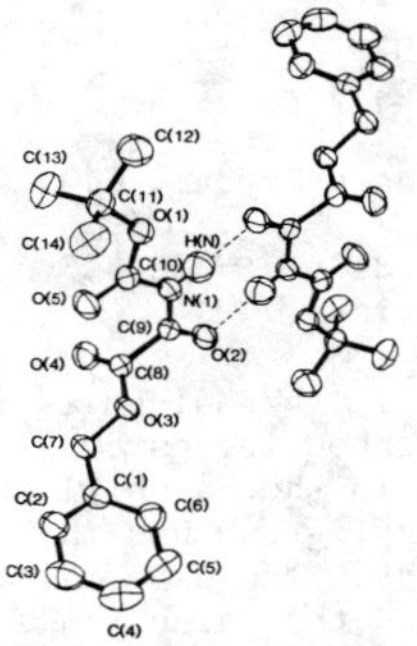

Fig. 1. The benzyl N-(t-butoxycarbonyl)oxamate dimer.

N-(DIPHENYLMETHYLENE)ACETAMIDE
$C_{15}H_{13}NO$ $CH_3CON{=}CPh_2$

R. ALLMANN, R. KUPFER, M. NAGEL and E.-U. WURTHWEIN, 1984. Chem. Ber., 117, 1597-1605.

Orthorhombic, $Pbc2_1$, a = 13.130, b = 11.054, c = 8.456 Å, Z = 4. Mo radiation, R = 0.072 for 696 reflexions.

The C=N-C=O torsion angle (73°) and the bond lengths (C=N 1.265(9), N-C 1.376(9), C=O 1.217(11) Å) indicate that there is only little interaction between the C=O and C=N groups in this molecule.

[4-(BENZYLOXY)PHENYL]ACETIC ACID
$C_{15}H_{14}O_3$

J.W. BATS and R. CANENBLEY, 1984. Acta Cryst., C40, 993-995.

Triclinic, $P\bar{1}$, a = 5.487, b = 8.121, c = 27.424 Å, α = 89.07, β = 86.65, γ = 88.98°, Z = 4. Mo radiation, R = 0.045 for 1798 reflexions.

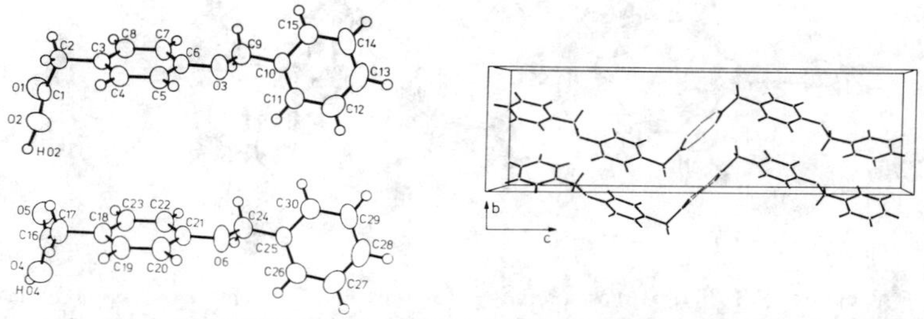

Fig. 1. $C_{15}H_{14}O_3$: perspective views of the two independent molecules and the view along a* of the molecular packing.

In the crystal structure (Fig. 1) both independent molecules form centrosymmetric dimers via double hydrogen bonds through the carboxyl groups with O(1)...O(2') 2.697(2) and O(4)...O(5') 2.639(2) Å. C-O distances in the carboxyl groups are C(1)-O(1) 1.237(2), C(1)-O(2) 1.273(2), C(16)-O(4) 1.258(2) and C(16)-O(5) 1.252(2) Å. The dihedral angles between the phenyl rings are 64.0(3) and 64.1(3)°. The two molecules differ in the relative orientations of the carboxyl groups, but are otherwise similar. The dimers are arranged in parallel stacks in the c direction.

[4-(PHENOXYMETHYL)PHENYL]ACETIC ACID
$C_{15}H_{14}O_3$

J.W. BATS and R. CANENBLEY, 1984. Acta Cryst., C40, 995-997.

Monoclinic, $P2_1/a$, a = 6.901, b = 6.812, c = 26.523 Å, β = 92.58°, Z = 4. Mo radiation, R = 0.047 for 1385 reflexions.

In the crystal structure (Fig. 1) the molecules form centrosymmetric dimers by hydrogen bonding between two neighbouring carboxyl groups with O...O 2.645(2) Å and O-H...O 173(2)°. C-O distances in the carboxyl group are 1.207(2) and 1.309(2) Å. The (phenoxymethyl)phenyl group has the Z conformation required to avoid steric hindrance by the phenyl groups. The dihedral angle between the phenyl groups is 71.1(2)°, of the same magnitude as the value found in the (benzyloxy)phenyl isomer (1).

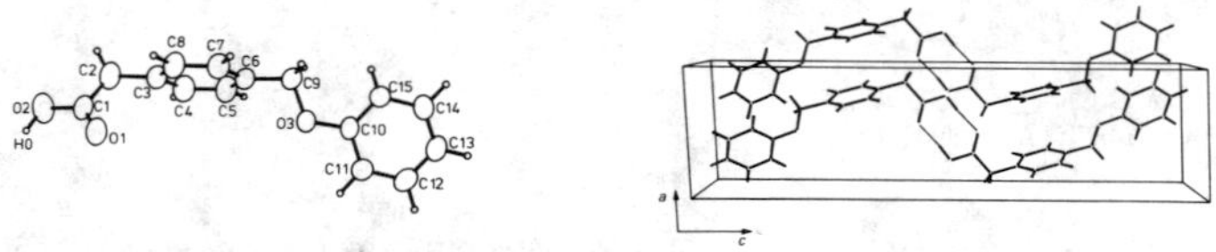

Fig. 1. $C_{15}H_{14}O_3$: perspective views of the molecule, and of the crystal structure along b.

1. Previous Report.

(2S,3R)-p-BROMOPHENACYL 3-ACETOXY-2-METHYL-4-OXOPENTANOATE
$C_{16}H_{17}BrO_6$

R. JANSEN, W.S. SHELDRICK and G. HOFLE, 1984. Justus Liebigs Ann. Chem., 78-84.

Monoclinic, $P2_1$, a = 24.718, b = 6.580, c = 5.291 Å, β = 95.33°, Z = 2. Cu radiation, R = 0.045 for 1506 reflexions.

The analysis establishes the absolute configuration as shown in Fig. 1.

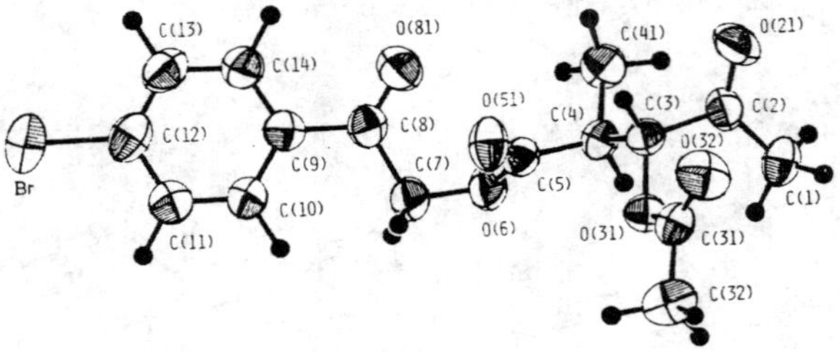

Fig. 1. A view of (2S,3R)-p-bromophenacyl 3-acetoxy-2-methyl-4-oxopentanoate.

1,2-BENZENEDIACETIC ACID α,α,α',α'-TETRAMETHYLDIMETHYLESTER
$C_{16}H_{22}O_4$ (I)

2,3-NAPHTHALENE DIACETIC ACID α,α,α',α'-TETRAMETHYLDIMETHYLESTER
$C_{20}H_{24}O_4$ (II)

L.R.C. BARCLAY, S. BROWNSTEIN, E.J. GABE and F.L. LEE, 1984. Can. J. Chem., 62, 1358-1362.

I. Monoclinic, A2/a, a = 10.665, b = 8.467, c = 17.198 Å, β = 105.94°, Z = 4. Mo radiation, R = 0.051 for 1244 reflexions.

II. Monoclinic, $P2_1/c$, a = 9.099, b = 18.849, c = 11.319 Å, β = 113.88°, Z = 4. Mo radiation, R = 0.048 for 1691 reflexions.

Molecular dimensions for I and II are in Fig. 1. Overcrowding causes the carbon atoms bonded to the aromatic rings to be displaced above and below the ring plane. The torsion angle between the 1,2-substituents in I is 12.1°; the analagous value in II is 8.9°.

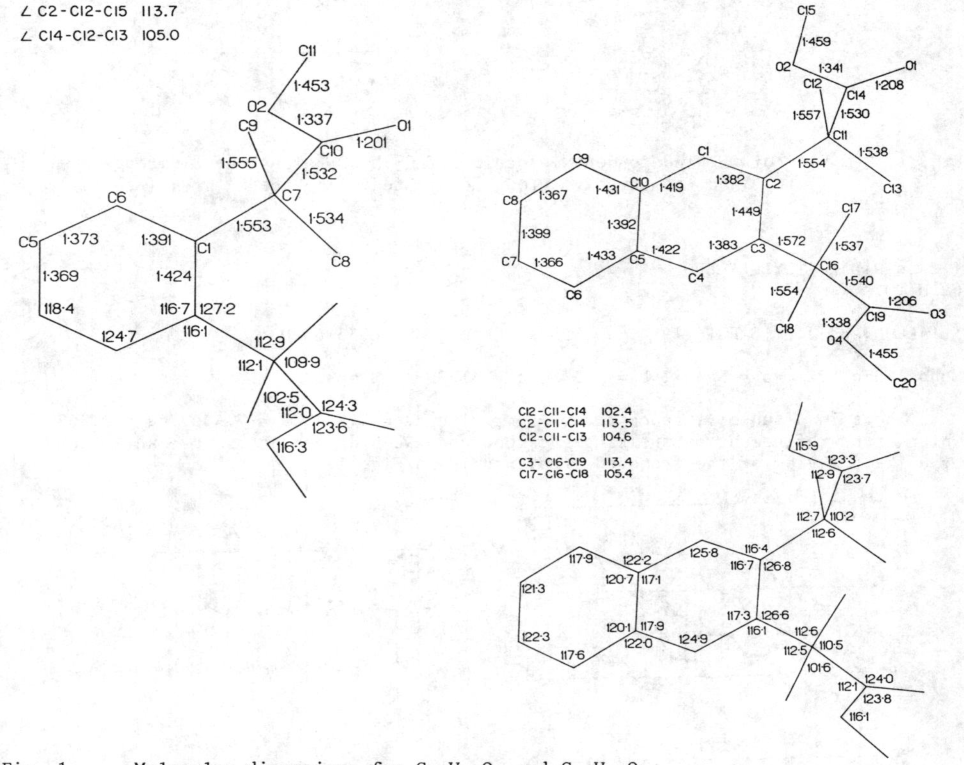

Fig. 1. Molecular dimensions for $C_{16}H_{22}O_4$ and $C_{20}H_{24}O_4$.

β'-HEPTADECANOIC ACID
$C_{17}H_{34}O_2$

M. GOTO and E. ASADA, 1984. Bull. Chem. Soc. Japan, 57, 1145-1146.

Triclinic, $P\bar{1}$, a = 5.561, b = 8.018, c = 47.90 Å, α = 114.18, β = 114.96, γ = 80.22°, Z = 4. Cu radiation, R = 0.077 for 1028 reflexions.

The β'-form of heptadecanoic acid has a characteristic layer structure and the molecules form all-trans zigzag chains (Fig. 1). The conformations of the two molecules in the asymmetric unit are similar to each other. These molecules are joined in pairs by hydrogen bonds between the carboxyl groups with O...O distances of 2.64 Å (molecule I-molecule I) and 2.62 Å (II-II). The dihedral angle between the planes containing the carboxyl groups is 11° (I-I) and 21° (II-II). The C-C distances average 1.53 Å with a range of 1.49 to 1.57 Å. The C(1)-O(1), C(1)-O(2), C(18)-O(3) and C(18)-C(4) distances are respectively 1.22, 1.31, 1.26 and 1.29 Å.

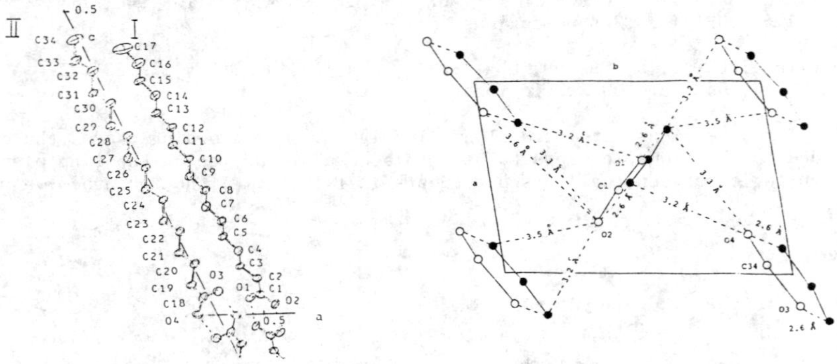

Fig. 1. View of the independent molecules of $C_{17}H_{34}O_2$ along the b-direction (left), and the carboxyl groups projected along the C* axis (right).

1,2-sn-DIPALMITOYLGLYCEROL
$C_{35}H_{68}O_5$

N. ALBON and J.F. BARET, 1984. J. Phys. Chem., 88, 6333-6340.

Monoclinic, $P2_1$, a = 5.498, b = 7.549, c = 43.39 Å, β = 93.62°.

There is a subcell, space group Pbnm, with a_s = 4.924, b = 7.549, c_s = 2.562 Å which after refinement of the single carbon and two hydrogen atoms gave an R-factor of 0.128. A view of the proposed structure is in Fig. 1.

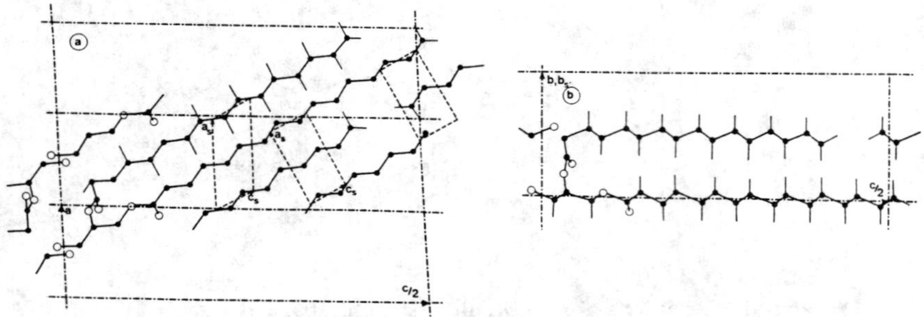

Fig. 1. Views of the model proposed for $C_{35}H_{68}O_5$ from X-ray and spectral data.

TRILAURIN (β FORM)
$C_{39}H_{74}O_6$

V. GIBON, P. BLANPAIN, B. NORBERG and F. DURANT, 1984. Bull. Soc. Chim. Belg., 93, 27-34.

Triclinic, $P\bar{1}$, a = 12.084, b = 31.617, c = 5.468 Å, α = 94.82, β = 100.45, γ = 96.41°, Z = 2. Cu radiation, R = 0.06 for 1850 reflexions.

The molecule (Fig. 1) adopts a chair-like conformation with two hydrocarbon chains pointing in one direction and the third chain in the opposite direction. The chains lie in parallel planes.

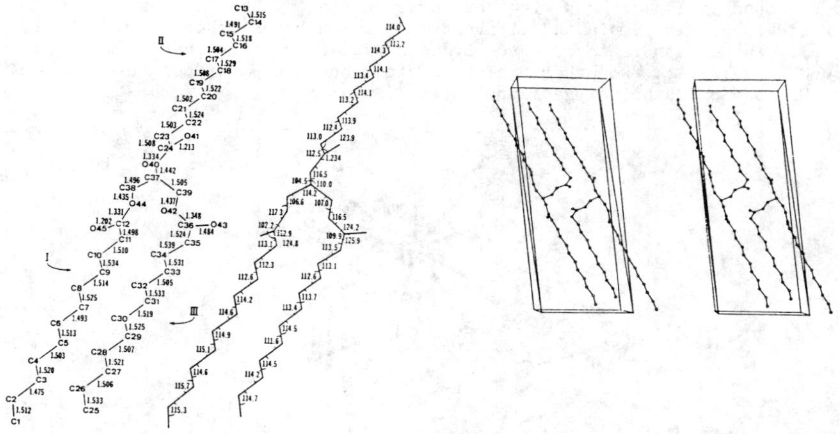

Fig. 1. Molecular dimensions and a stereoview of trilaurin.

SODIUM HYDROGEN OXALATE MONOHYDRATE

C_2HNaO_4,H_2O $NaHC_2O_4,H_2O$

R.G. DELAPLANE, R. TELLGREN and I. OLOVSSON, 1984. Acta Cryst., C40, 1800-1803.

Triclinic, P$\bar{1}$, a = 6.4235, b = 6.6580, c = 5.6941 Å, α = 85.048, β = 110.100, γ = 104.963, Z = 2. Neutron radiation (λ = 1.210 Å), R = 0.036 for 1122 reflexions (at 120K).

There is good agreement with the previously described structure at 295K (1). The O...O distances (Fig. 1) contract by 0.005(2), 0.012(2), 0.018(2) Å in order of hydrogen bond length (O...O 2.566(2), 2.794(1), 2.806(1) Å) with corresponding H...O decreases of 0.006(3), 0.014(3), 0.022(3) Å. The Na^+...O distances contract in the range 0.004(3) to 0.030(3) Å.

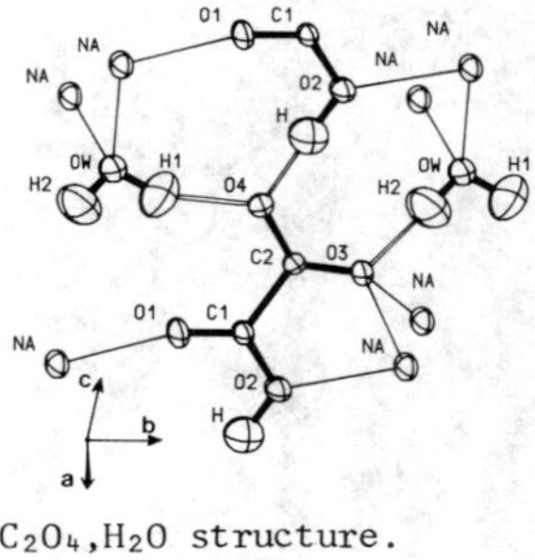

Fig. 1. A view of the $NaHC_2O_4,H_2O$ structure.

1. Structure Reports, 43B, 21.

CESIUM TETRAOXALATE DIHYDRATE

$C_4H_3CsO_8,2H_2O$ $C_2HO_4^-.C_2H_2O_4.Cs^+.2(H_2O)$

W. CLEGG, 1984. Z. Kristallogr., 169, 211-217.

Triclinic, P$\bar{1}$, a = 6.279, b = 7.440, c = 10.776 Å, α = 94.53, β = 99.49, γ = 95.33°, Z = 2. Mo radiation, R = 0.030 for 3358 reflexions.

The compound is isomorphous with the corresponding K^+ and NH_4^+ salts (1). The oxalate residues are in simple and bifurcated hydrogen bonds (shortest O...O 2.503(6) Å). The Cs ion is coordinated by five oxalate O atoms and one water molecule (Cs-O 3.060-3.195(2) Å); two other oxalate O atoms and one water molecule lie within 3.36 Å of Cs, and one other oxygen atom is at 3.56 Å.

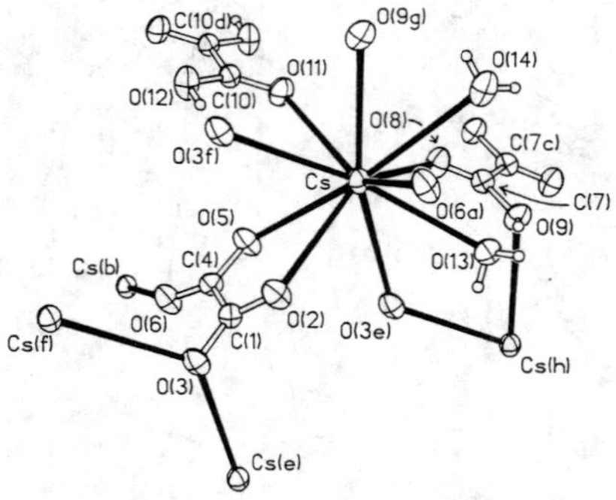

Fig. 1. The asymmetric unit of $C_4H_3CsO_8$ with symmetry related atoms to complete the coordination of Cs and O atoms.

1. Structure Reports, 32B, 23; 49B, 16.

SODIUM HYDROGEN MALEATE TRIHYDRATE

$C_4H_3NaO_4,3H_2O$ $NaH(C_4H_2O_4).3H_2O$

G. OLOVSSON, I. OLOVSSON and M.S. LEHMANN, 1984. Acta Cryst., C40, 1521-1526.

Triclinic, $P\bar{1}$, a = 5.927, b = 6.288, c = 11.222 Å, α = 104.00, β = 91.47, γ = 100.26°, Z = 2. Neutrons (λ = 0.8323 Å), R = 0.033 for 2508 reflexions (at 120K).

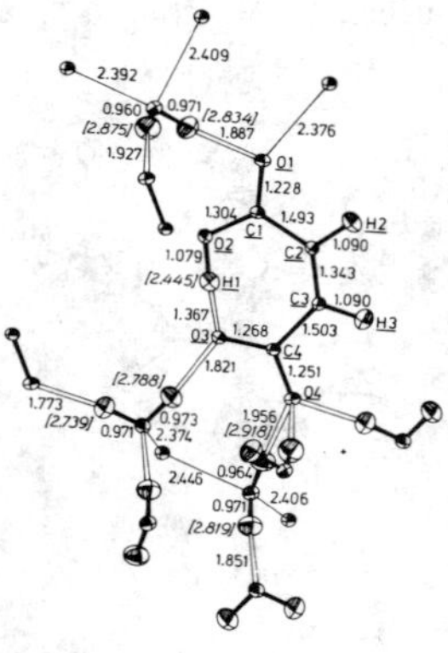

Fig. 1. Bond lengths within the sodium hydrogen maleate trihydrate structure.

In this previously determined structure (1), the hydrogen maleate ion has an almost planar ring structure (deviations of the seven heavy atoms from the C(1-4) plane are in the range 0.001-0.033 Å; 0.002-0.007 Å for the hydrogen atoms) containing a short, asymmetric, intramolecular hydrogen bond with O...O 2.445(1), O-H 1.079(2), H...O 1.367(2) Å (Fig. 1). These ions are linked through hydrogen bonds to water molecules forming double layers parallel to (010). These layers are linked through Na^+...O interactions (Na^+...O 2.374-2.446 Å). Each of the three independent water molecules has a distorted tetrahedral environment with all H atoms involved in hydrogen bonds (O...O 2.739-2.918 Å, O-H 0.960-0.973 Å, H-O-H 103.4-105.7°).

1. Structure Reports, 40B, 26.

CALCIUM meso-TARTRATE TRIHYDRATE
$C_4H_4CaO_6,3H_2O$ $Ca^{2+}.C_4H_4O_6^{2-}.3H_2O$

A.J. DE VRIES and J. KROON, 1984. Acta Cryst., C40, 1542-1544.

Monoclinic, $P2_1/c$, a = 8.921, b = 10.300, c = 9.881 Å, β = 91.78°, Z = 4. Cu radiation, R = 0.065 for 1817 reflexions.

The meso-tartrate dianion (Fig. 1) is dissymmetric with a synclinal conformation about the central C-C bond; the C(1)C(2)C(3)C(4) torsion angle is -77.0(2)°. The non-hydrogen skeleton comprises two almost planar halves with O(1)C(1)C(2)C(3) and O(5)C(4)C(3)O(4) torsion angles 4.1(3) and 13.0(3)°. The Ca^{2+} ion is coordinated by seven O atoms, including those of two water molecules, with Ca^{2+}...O 2.214(2)-2.419(2) Å. The structure is stabilized by extensive hydrogen bonding.

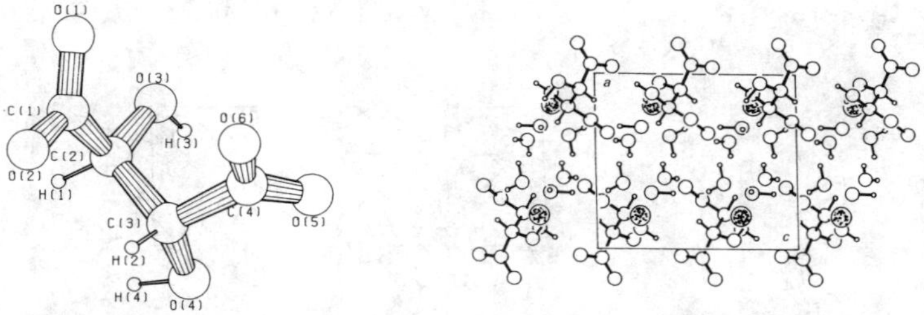

Fig. 1. The meso-tartrate dianion, and a view of the unit cell along the b-direction.

CALCIUM ACETATE MONOHYDRATE
$C_4H_6CaO_4,H_2O$ $Ca(C_2H_3O_2)_2,2H_2O$

E.A. KLOP, A. SCHOUTEN, P. VAN DER SLUIS and A.L. SPEK, 1984. Acta Cryst., C40, 51-53.

Triclinic, $P\bar{1}$, a = 6.750, b = 11.076, c = 11.782 Å, α = 116.49, β = 92.40, γ = 97.31°, Z = 4. Mo radiation, R = 0.037 for 2057 reflexions.

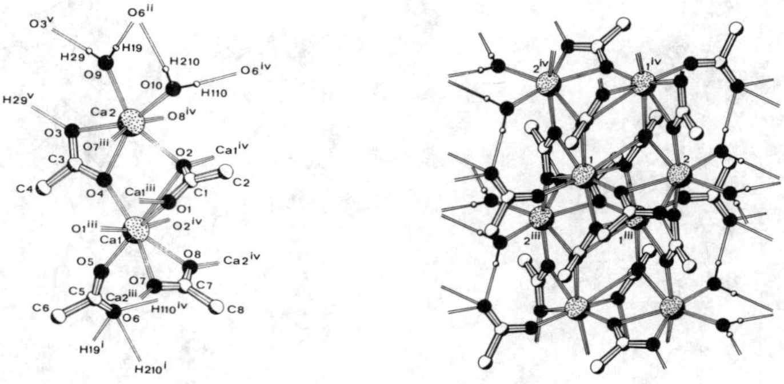

Fig. 1. The asymmetric unit of $Ca(C_2H_3O_2)_2,2H_2O$ and a view of a double-stranded Ca chain.

The structure (Fig. 1) consists of infinite multiple O-bridged double-stranded Ca chains that are cross-linked via an extensive hydrogen-bonded network. The Ca-O distances are in the range 2.270(4) to 2.933(3) Å and the O...O distances in the hydrogen bonded network are in the range 2.699(5) to 2.994(5) Å.

LITHIUM HYDROGEN ADIPATE

$C_6H_9LiO_4$ $LiH[O_2C(CH_2)_4CO_2]$

X. HE and B.M. CRAVEN, 1984. Acta Cryst., C40, 1157-1159.

Monoclinic, $P2_1/c$, a = 8.222, b = 7.448, c = 11.569 Å, β = 92.19°, D_m = 1.424, Z = 4. Cu radiation, R = 0.048 for 1329 reflexions.

In the crystal structure (Fig. 1) the molecules are arranged in layers parallel to (100) with hydrocarbon chains sandwiched between polar regions consisting of carboxylic acid, carboxylate and Li ions. The Li ion is tetrahedrally coordinated by O atoms, one of the Li...O distances being short (1.896(3) Å). The carboxylic acid and carboxylate groups cannot be distinguished from their C-O bond lengths. Each has one long (1.289, 1.290(2)) and one short (1.231, 1.233(2) Å) C-O bond. The O atoms of the long C-O bonds are bridged by a short asymmetric H-bond (O...O 2.486(2), H...O 1.47(3) Å, O-H...O 166(3)°).

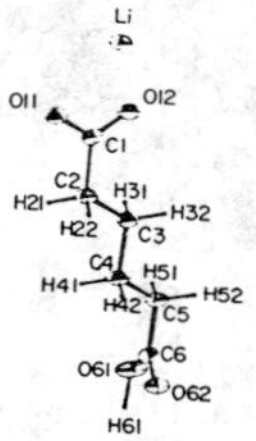

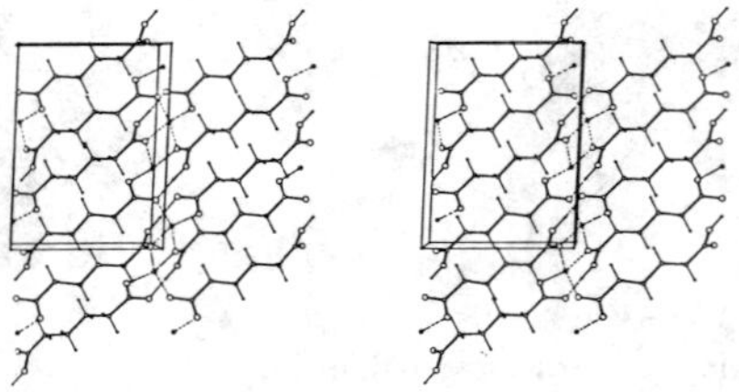

Fig. 1. $LiH[O_2C(CH_2)_4CO_2]$: molecular structure and stereoview along b of the crystal structure.

CALCIUM HYDROGEN TRIACETATE MONOHYDRATE

$C_6H_{10}CaO_6,H_2O$ $CaH(C_2H_3O_2)_3,H_2O$

E.A. KLOP and A.L. SPEK, 1984. Acta Cryst., C40, 1817-1819.

Triclinic, $P\bar{1}$, a = 6.858, b = 12.281, c = 12.520 Å, α = 83.86, β = 91.62, γ = 87.09°, Z = 4. Mo radiation, R = 0.046 for 2551 reflexions.

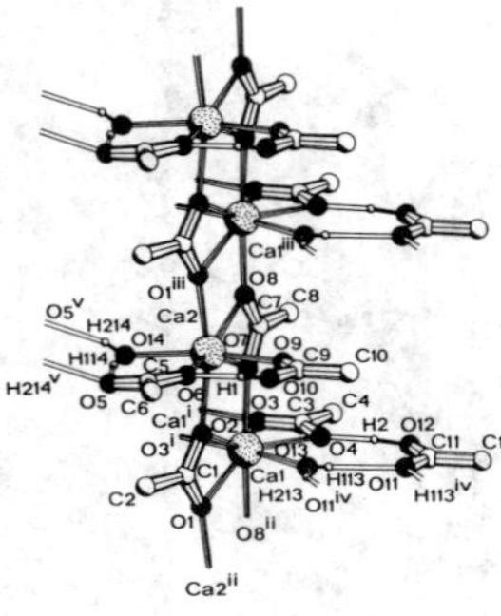

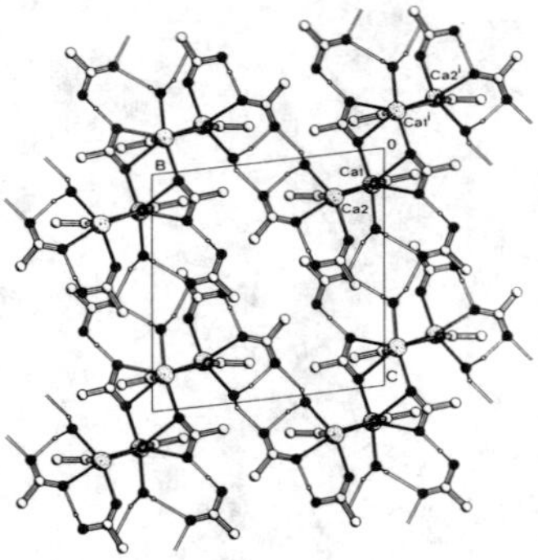

Fig. 1. Diagram showing the asymmetric unit in $CaH(C_2H_3O_2)_3,H_2O$, and a projection of the unit cell along the a-direction.

The structure contains infinite, double-stranded, multiple oxygen bridged Ca chains along a (Fig. 1). Ca(1) is eight coordinate (distorted dodecahedral) with Ca...O 2.347-2.594(3) Å while Ca(2) is seven coordinate (pentagonal bipyramidal) with Ca...O 2.352-2.524(3) Å. Three of the four OAc^- ions in the asymmetric unit act both as bidentate chelating and as O-bridging ligands; one OAc^- ion is monodentate as is a neutral AcOH molecule. The other AcOH molecule does not coordinate to Ca. The Ca chains are connected via hydrogen bonds involving the water molecules; O...O distances are in the range 2.548(5)...2.876(5) Å.

SODIUM D-GLUCONATE
$C_6H_{11}NaO_7$

T. LIS, 1984. Acta Cryst., C40, 376-378.

Monoclinic, $P2_1$, a = 8.560, b = 8.694, c = 5.654 Å, β = 104.59°, Z = 2. Mo radiation, R = 0.027 for 1119 reflexions.

The gluconate ion has a nearly planar zig-zag conformation with one intramolecular hydrogen bond (Fig. 1). The Na^+ ion is surrounded by six O atoms forming a distorted octahedron with Na-O 2.340-2.540(2) Å.

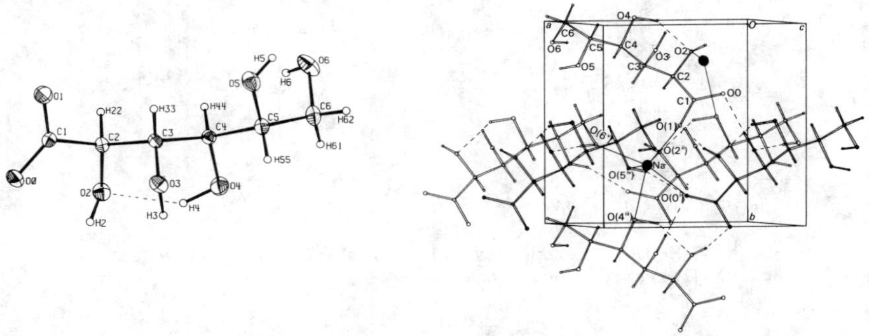

Fig. 1. A view of the D-gluconate anion, and the molecular packing and hydrogen bonding scheme.

MAGNESIUM BIS(HYDROGEN MALEATE) HEXAHYDRATE
$C_8H_6MgO_8,6H_2O$ $Mg(C_4H_3O_4)_2.6H_2O$

M.P. GUPTA, C. VAN ALSENOY and A.T.H. LENSTRA, 1984. Acta Cryst., C40, 1526-1529.

Monoclinic, $P2_1/a$, a = 6.742, b = 10.201, c = 11.816 Å, β = 104.06°, Z = 2. Mo radiation, R_w = 0.038 for 1365 reflexions.

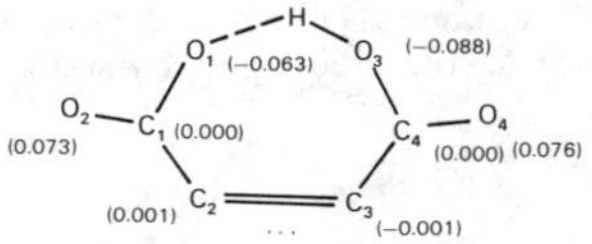

Fig. 1. The hydrogen maleate anion with displacements from the C(1),C(2),C(3),C(4) plane.

The Mg^{2+} cation lies at a centre of symmetry coordinated by a slightly distorted octahedron of O atoms all belonging to waters of crystallisation (Mg...O 2.036-2.059(1) Å). The cation is linked to the O atoms of the H-maleate anion by hydrogen bonds involving water molecules with O...O 2.732-2.900(1) Å; only O(3) is not involved. The anion is slightly non-planar (Fig. 1), the carboxyl groups being rotated about 3° and 4° with respect to the carbon skeleton and with the dihedral angle between the C(2),C(1),O(1),O(2) and C(3),C(4),O(3),O(4) planes being 6.0(1)°. The intramolecular hydrogen bond is unsymmetrical with O(3)-H 0.88(1), O(1)...H 1.55(1), O(3)...O(1) 2.410(1) and O-H...O 167(1)°.

(-)-α-METHYLBENZYLAMMONIUM HYDROGEN meso-TARTRATE

$C_{12}H_{17}NO_6$ $C_8H_{12}N^+,C_4H_5O_6^-$

J. KROON, A.J.M. DUISENBERG and A.F. PEERDEMAN, 1984. Acta Cryst., C40, 645-647.

Monoclinic, $P2_1$, a = 6.471, b = 13.74, c = 7.585 Å, β = 108.87°, Z = 2. Mo radiation, R = 0.036 for 2946 reflexions.

The meso-tartrate anion adopts a nearly centrosymmetric conformation (Fig. 1). The carboxyl groups are rotated from the C-C(α)-O(H) planes by 13 and 27° respectively. The molecules are linked in the crystal by a network of hydrogen bonds with O(3)...O(1') 2.508(1), O(5)...O(4') 2.808(1), O(6)...O(2') 2.801(1), N...O(5') 2.962(1), N...O(2') 2.918(1) and N...O(1') 2.863(1) Å.

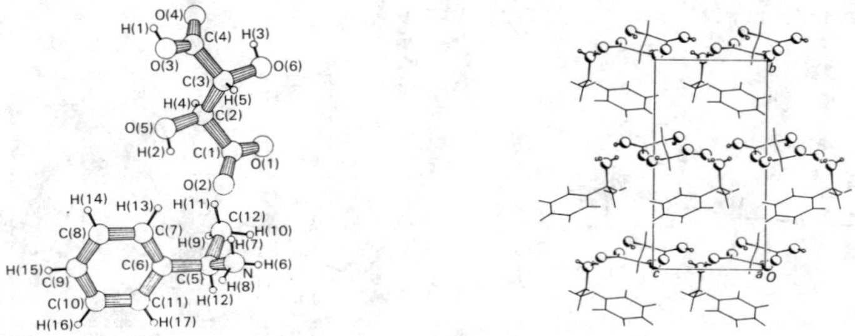

Fig. 1. $C_8H_{12}N^+,C_4H_5O_6^-$: perspective view of the complex molecules and the packing diagram projected down a.

CALCIUM BIS(2-FLUOROBENZOATE) DIHYDRATE

$C_{14}H_8CaF_2O_4,2H_2O$ $Ca(C_7H_4FO_2)_2,(H_2O)_2$

A. KARIPIDES and C. MILLER, 1984. J. Am. Chem. Soc., 106, 1494-1495.

Monoclinic, C2/c, a = 23.023, b = 11.146, c = 7.793 Å, β = 133.00°, D_m = 1.60, Z = 4. R = 0.034 for 1627 reflexions.

The Ca ion lies on a twofold crystallographic axis and the coordination is shown in Fig. 1. The CaO_8 coordination polyhedron resembles most closely a dodecahedron somewhat distorted towards a C_{2v}-bicapped trigonal prism, with Ca-O1a 2.360(1), Ca-O1 2.629(1), Ca-O2 2.463(1) Å. The carboxyl group is twisted 41° from the aryl ring plane. There is a short intermolecular hydrogen bond C2-F...H1-OW (F...O 2.994 Å).

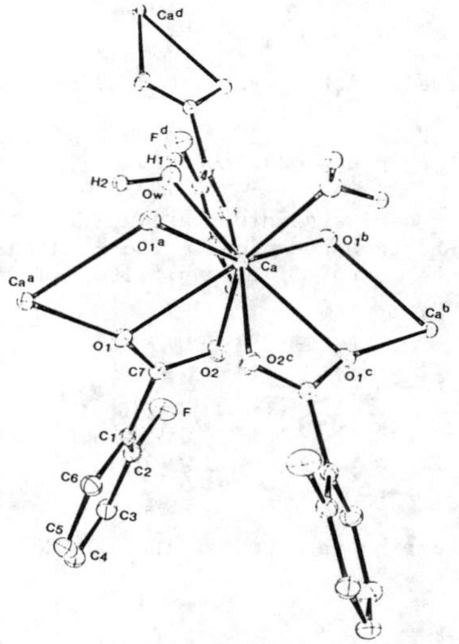

Fig. 1. A portion of the CaO_8 coordination in $C_{14}H_8CaF_2O_4,2H_2O$.

BARIUM BIS(p-HYDROXYBENZOATE) MONOHYDRATE
$C_{14}H_{10}BaO_6,H_2O$ $Ba(C_7H_5O_3)_2.H_2O$

A.N. SHNULIN, G.N. NADZHAFOV and Kh.S. MAMEDOV, 1984. J. Struct. Chem., 25, 588-592 [Zh. Strukt. Khim., 25(4), 89-94].

Orthorhombic, Pnam, a = 7.529, b = 5.559, c = 33.32 Å, Z = 4. Cu radiation, R = 0.053 for 821 reflexions.

A view of the structure along b is shown in Fig. 1. The Ba ions display undecahedral geometry, being coordinated by eight oxygen atoms from two water molecules and six different carboxylate groups. The Ba polyhedra are linked by a common face into columns extended along the a axis. The columns are linked into layers parallel to the mirror plane, which contains the Ba ions and the water molecules. The Ba-O distances range from 2.76(1) to 2.89(1) Å.

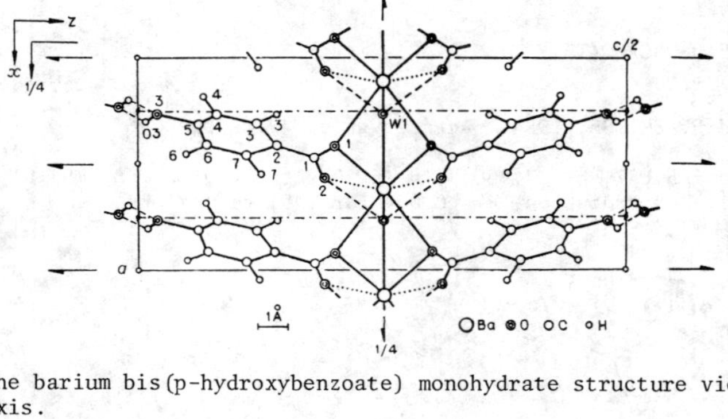

Fig. 1. The barium bis(p-hydroxybenzoate) monohydrate structure viewed down the b axis.

PIPERAZINIUM BIS(n-HEPTANOATE)

$C_{18}H_{38}N_2O_4$ $C_4H_{12}N_2^{2+}.2C_7H_{13}O_2^-$

J.-P. SANGIN and F. BRISSE, 1984. Acta Cryst., C40, 2091-2093.

Triclinic, $P\bar{1}$, a = 5.759, b = 7.549, c = 12.452 Å, α = 95.21, β = 95.25, γ = 100.21°, D_m = 1.06, Z = 1. Mo radiation, R = 0.044 for 1144 reflexions.

The structure consists of a piperazinium cation lying on an inversion centre and hydrogen bonded to the two n-heptanoate anions. These anions (Fig. 1) have a fully extended conformation, with the carboxylic group tilted 54° from the plane of the methylenic chain.

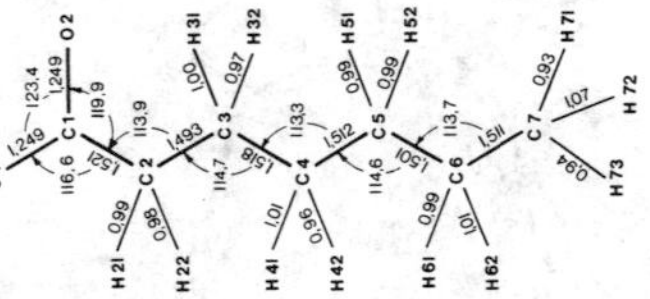

Fig. 1. Dimensions for the n-heptanoate ion in $C_{18}H_{38}N_2O_4$.

DICALCIUM BARIUM CYCLOPROPANECARBOXYLATE

$C_{24}H_{30}BaCa_2O_{12}$ $Ca_2Ba[(CH_2)_2CHCO_2]_6$

K. STADNICKA and A.M. GLAZER, 1984. Acta Cryst., B40, 139-145.

Cubic, Fd3m, a = 18.409 Å, D_m = 1.56, Z = 8. Mo radiation, R = 0.032 for 516 reflexions.

As in the isomorphous compound dicalcium barium propionate (1), the structure is disordered with the organic groups statistically in two sites related by a pseudo-mirror plane. X-ray diffuse scattering measurements show that there are correlated locally ordered regions in the form of platelets perpendicular to <100> and elongated along the <110> directions, corresponding to regions of opposite chiralities. The correlation lengths are found to be 23.4(5) Å projected along [110] and 4.1(1) Å projected along [001]. It has been found that at room temperature the crystal is normally orthorhombic, but that X-ray dosage stabilizes the prototype phase at this temperature. The orthorhombic-cubic transition in the absence of X-ray dosage is at about 308 K.

1. Structure Reports, 46B, 26.

PIPERAZINIUM BIS(n-DECANOATE)

$C_{24}H_{50}N_2O_4$ $C_4H_{12}N_2^{2+}.2C_{10}H_{19}O_2^-$

J.-P. SANGIN and F. BRISSE, 1984. Acta Cryst., C40, 2094-2096.

Triclinic, $P\bar{1}$, a = 5.687, b = 7.450, c = 16.208 Å, α = 93.09, β = 97.01, γ = 98.20°, D_m = 1.05, Z = 1. Cu radiation, R = 0.074 for 1401 reflexions.

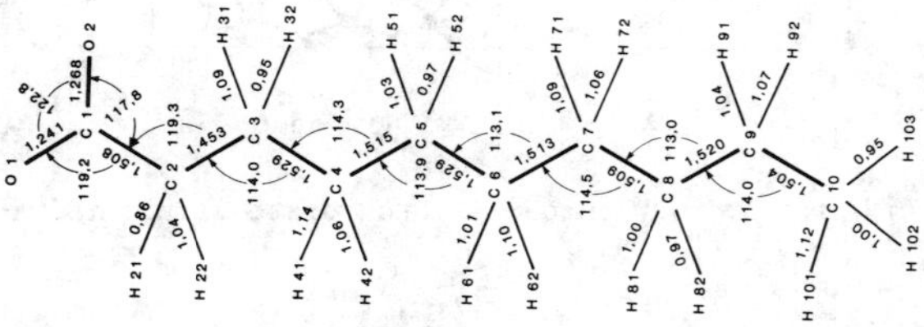

Fig. 1. Dimensions of the n-decanoate anion.

The structure consists of a centrosymmetric piperazinium cation in the chair conformation hydrogen bonded to two n-decanoate anions. The anions (Fig. 1) do not have the fully extended conformation since a torsion angle of 82.3(7)° is observed around the C(3)-C(4) bond. The dihedral angle between the carboxylic plane and the C(1) to C(4) plane is 17.3(5)°. The rest of the aliphatic chain, from C(3) to C(10), is nearly planar. This conformation is characteristic of anions with an even number of C atoms.

TRIMETHYLAMINE
C_3H_9N $N(CH_3)_3$

A.J. BLAKE, E.A.V. EBSWORTH and A.J. WELCH, 1984. Acta Cryst., C40, 413-415.

Trigonal, P$\bar{3}$, a = 6.1366, c = 6.8516, Z = 2. Mo radiation, R = 0.0559 for 384 reflexions (at 118 K).

The structure consists of discrete molecules (Fig. 1) arranged head-to-tail in columns which run alternately parallel and antiparallel to the c crystallographic axis. Individual molecules have effective C_{3v} symmetry with pyramidal geometry at N (C-N-C 110.40(7)°, C-N 1.4535(11) Å).

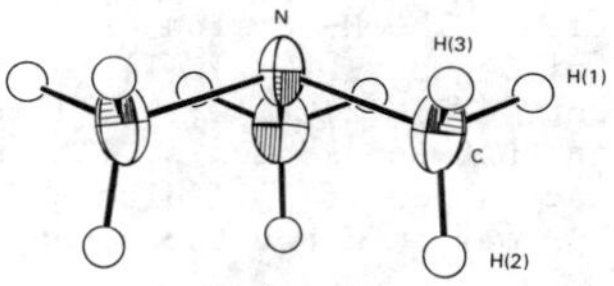

Fig. 1. A view of trimethylamine.

1,7-HEPTANEDIAMINE DIHYDROBROMIDE
$C_7H_{20}Br_2N_2$

J. BRISSON and F. BRISSE, 1984. Acta Cryst., C40, 1402-1404.

Monoclinic, $P2_1/c$, a = 4.766, b = 16.915, c = 14.856 Å, β = 97.54°, D_m = 1.53, Z = 4. Mo radiation, R = 0.043 for 1310 reflexions (at 173K).

The aliphatic chain has the all-trans conformation although it is markedly curved. Bond lengths and angles are normal. Each H of the $-NH_3^+$ groups is hydrogen bonded to a Br^- with N...Br in the range 3.28(1)-3.38(1) Å.

1,8-OCTANEDIAMINE DIHYDROBROMIDE
$C_8H_{22}Br_2N_2$

J. BRISSON and F. BRISSE, 1984. Acta Cryst., C40, 1405-1407.

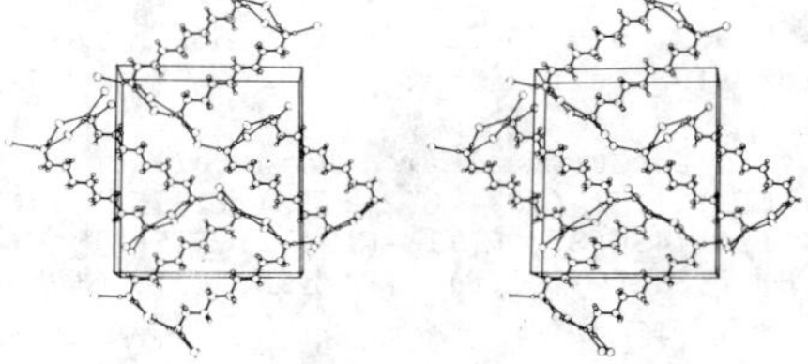

Fig. 1. Stereoview of the 1,8-octanediamine dihydrobromide structure.

Orthorhombic, Pba2, a = 18.396, b = 15.792, c = 4.588 Å, D_m = 1.49, Z = 4. Cu radiation, R = 0.037 for 1305 reflexions.

Layers of approximately planar diammonium cations in a fully extended conformation are sandwiched between layers of bromide ions (Fig. 1). Some of the torsion angles along the chain differ by up to 20° from the trans value of 180°. One $-NH_3^+$ group forms hydrogen bonds with three Br^- ions while the other is connected to four Br^-; N...Br distances are in the range 3.29(1) to 3.43(1) Å. Average values along the chain are C-C 1.497(7) Å, C-C-C 114.8(10)°.

N-[3-(3-AMMONIOPROPYLAMMONIO)PROPYL]SUCCINAMIC ACID SULPHATE
$C_{10}H_{23}N_3O_7S$ $C_{10}H_{23}N_3O_3^{2+}.SO_4^{2-}$

J.S. SUH, Z.B. XU, K. HOFMANN, C.S. YOO and M. SAX, 1984. Acta Cryst., C40, 131-134.

Monoclinic, $P2_1/c$, a = 17.693, b = 9.453, c = 8.958 Å, β = 90.42°, D_m = 1.460, Z = 4. Cu radiation, R = 0.042 for 2278 reflexions.

The backbone chain in the cation (Fig. 1) consists of four short segments which are in the extended conformation. The N-terminus segment is six atoms in length and it is followed by two five-atom segments [N(2)C(7)C(6)C(5)N(1) and C(5)N(1)C(4)C(3)-C(2)] and one four-atom segment [C(3)C(2)C(1)O(1)]. The chain changes direction abruptly at the three junctions between segments. The structure analysis proves unequivocally that selective acylation of 3,3'-diaminodipropylamine with succinic anhydride takes place at the primary and not the secondary amino group.

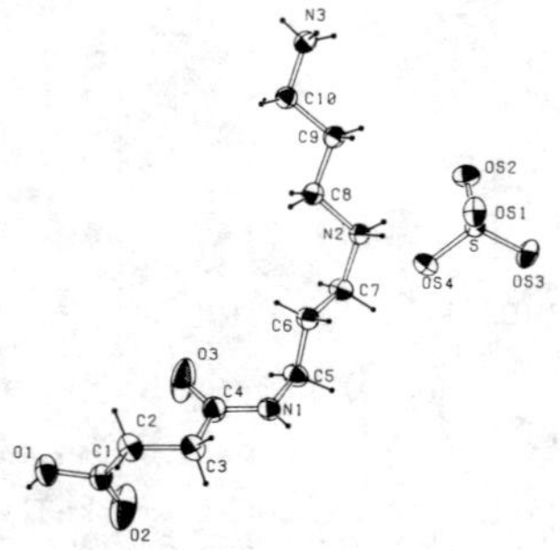

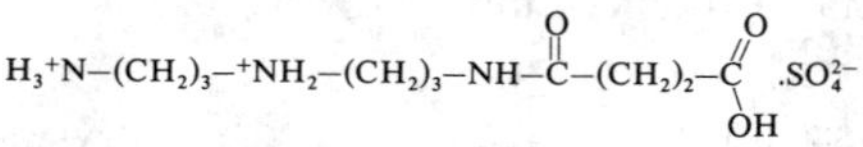

Fig. 1. A view of $C_{10}H_{23}N_3O_7S$.

N,N'-ETHYLENEBIS-(1,1,1-TRIFLUORO-4-IMINOPENTAN-2-ONE)
$C_{12}H_{14}F_6N_2O_2$

R. CEA-OLIVARES, I. RODRIGUEZ, M.J. ROSALES and A. TOSCANO, 1984. Aust. J. Chem., 37, 879-883.

Monoclinic, $P2_1/n$, a = 6.939, b = 20.529, c = 10.662 Å, β = 96.87°, Z = 4. Cu radiation, R = 0.073 for 1198 reflexions.

Discrete molecules in the crystal are characterised by a conjugated enamino ketone structure (C(1)-C(2) 1.524(10), C(2)-C(3) 1.373(8), C(3)-C(4) 1.387(8), C(4)-N(1) 1.327(8), N(1)-C(11) 1.456(8), C(11)-C(12) 1.515(8), C(12)-N(2) 1.443(7), N(2)-C(9) 1.319(8), C(9)-C(8) 1.382(9), C(8)-C(7) 1.389(9), C(7)-C(6) 1.503(10) Å (Fig. 1).

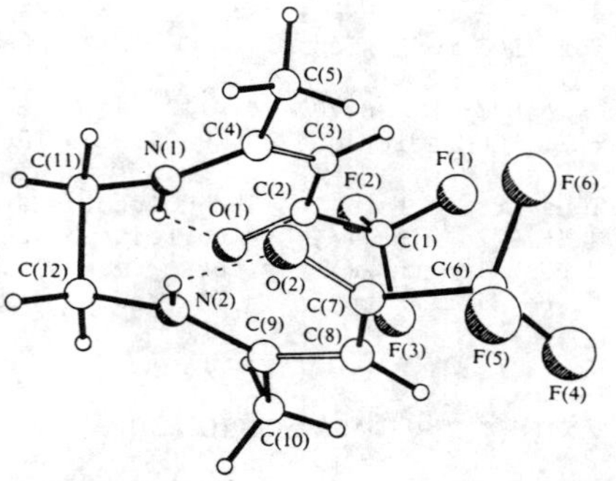

Fig. 1. The structure of $C_{12}H_{14}F_6N_2O_2$.

(2,5-DIMETHOXY-3-CYCLOPENTEN-1-YL)DIMETHYL(2-PROPENYL)AMMONIUM IODIDE
$C_{12}H_{22}INO_2$ $C_{12}H_{22}NO_2{}^+.I^-$

R. SWANSON, U.C. YOON and P.S. MARIANO, 1984. Acta Cryst., C40, 1967-1969.

Monoclinic, $P2_1/c$, a = 10.233, b = 14.005, c = 11.013 Å, β = 99.12°, Z = 4. Mo radiation, R = 0.031 for 1462 reflexions.

The molecule (Fig. 1) has a pseudo-plane of symmetry bisecting the ring π bond. Bond lengths and angles are normal.

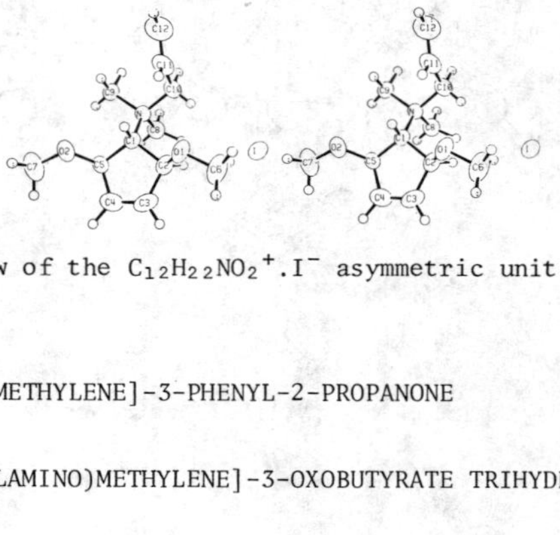

Fig. 1. Stereoview of the $C_{12}H_{22}NO_2{}^+.I^-$ asymmetric unit.

3-[BIS(DIMETHYLAMINO)METHYLENE]-3-PHENYL-2-PROPANONE
$C_{14}H_{20}N_2O$ (I)

METHYL 2-[BIS(DIMETHYLAMINO)METHYLENE]-3-OXOBUTYRATE TRIHYDRATE
$C_{10}H_{18}N_2O_3,3H_2O$ (II)

N.U. KAMATH and K. VENKATESAN, 1984. Acta Cryst., C40, 559-562.

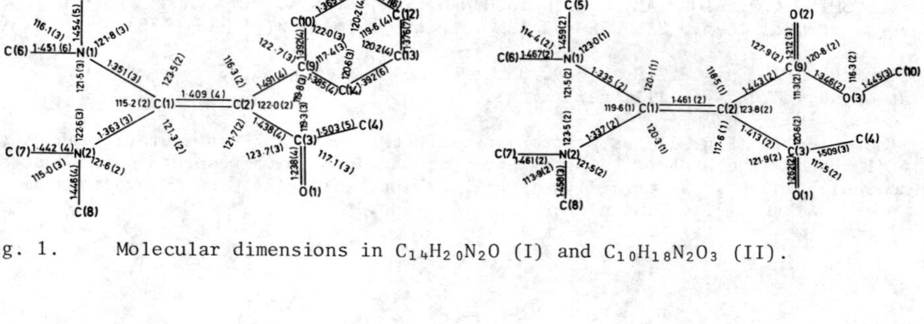

Fig. 1. Molecular dimensions in $C_{14}H_{20}N_2O$ (I) and $C_{10}H_{18}N_2O_3$ (II).

I. Orthorhombic, Pbca, a = 8.605, b = 14.845, c = 21.445 Å, D_m = 1.12, Z = 8. Cu radiation, R = 0.057 for 1563 reflexions.

II. Triclinic, P$\bar{1}$, a = 6.872, b = 8.707, c = 12.341 Å, α = 91.1, β = 92.8, γ = 100.5°, D_m = 1.22, Z = 2. Cu radiation, R = 0.055 for 1960 reflexions.

Molecular dimensions are in Fig. 1. The molecules are twisted about the C=C bond by 34.8(3)° in (I) and 56.9(2)° in (II). A correlation between rotational barriers (DNMR studies) and C=C bond lengths (X-ray measurements) is indicated. Rotational barriers are deduced to be 41.2 kJ mol^{-1} in (I) and < 34 kJ mol^{-1} for (II).

DIETHYLENETRIAMINE-N,N,N',N'',N''-PENTA-ACETIC ACID
$C_{14}H_{23}N_3O_{10}$

L.M. SHKOL'NIKOVA, G.V. POLYANCHUK, N.M. DYATLOVA and I.A. POLYAKOVA, 1984. J. Struct. Chem., 25, 264-272 [Zh. Strukt. Khim., 25(2), 103-113].

Monoclinic, P2_1/c, a = 17.912, b = 8.992, c = 10.794 Å, β = 105.28°, Z = 4. Mo radiation, R = 0.041 for 1422 reflexions.

The molecule (Fig. 1) exhibits a nonsymmetric betaine structure due to quaternization of two of the three neighbouring nitrogen atoms. The nonsymmetry of the molecule is ascribed to the formation of an intramolecular N-H...N hydrogen bond (N...N 2.830(3) Å). There are also intramolecular N-H...O hydrogen bonds and an extensive network of intermolecular O-H...O and C-H...O hydrogen bonds.

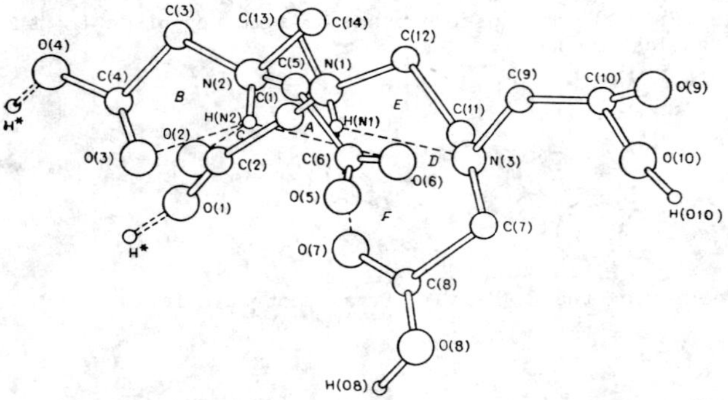

Fig. 1. The diethylenetriamine-N,N,N',N'',N''-penta-acetic acid molecule.

1-BENZOYL-2-(N-ACETYL-N-PHENYLAMINO)ETHYLENE
$C_{17}H_{15}NO_2$

S.M. ALDOSHIN, L.O. ATOVMYAN, O.A. D'YACHENKO, V.I. MINKIN, V.A. BREN' and G.D. PALUI, 1984. J. Struct. Chem., 25, 288-292 [Zh. Strukt. Khim., 25(2), 131-135].

Monoclinic, P2_1/b, a = 8.268, b = 9.703, c = 17.473 Å, γ = 83.33°, Z = 4. Cu radiation, R = 0.037 for 1110 reflexions.

The molecule (Fig. 1) has a ketoamine structure with a Z-configuration about the C(1)=C(9) bond. The N-phenyl substituent is rotated 76° with respect to the rest of the molecule, which is essentially planar. An intramolecular C-H...O hydrogen bond (H...O 2.23 Å) stabilizes the planar configuration.

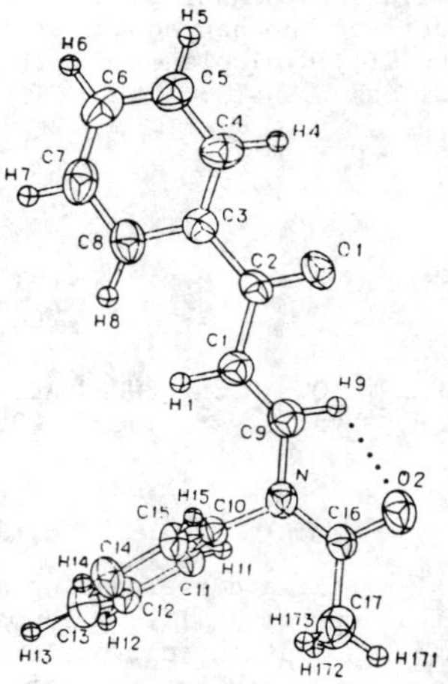

Fig. 1. The 1-benzoyl-2-(N-acetyl-N-phenylamino)ethylene molecule.

BENZOCTAMINE
$C_{18}H_{19}N$ (I)

N-METHYLBENZOCTAMINE
$C_{19}H_{21}N$ (II)

P. BROUANT, M. PIERROT, A. BALDY, J.C. SOYFER and J. BARBE, 1984. Acta Cryst., C40, 1590-1593.

I. Monoclinic, $P2_1/c$, a = 13.431, b = 7.358, c = 14.137 Å, β = 90.81°, Dm = 1.20, Z = 4. Mo radiation, R = 0.053 for 1433 reflexions.

II. Monoclinic, $P2_1/c$, a = 9.336, b = 28.489, c = 5.880 Å, β = 107.77°, Dm = 1.20, Z = 4. Mo radiation, R = 0.044 for 1574 reflexions.

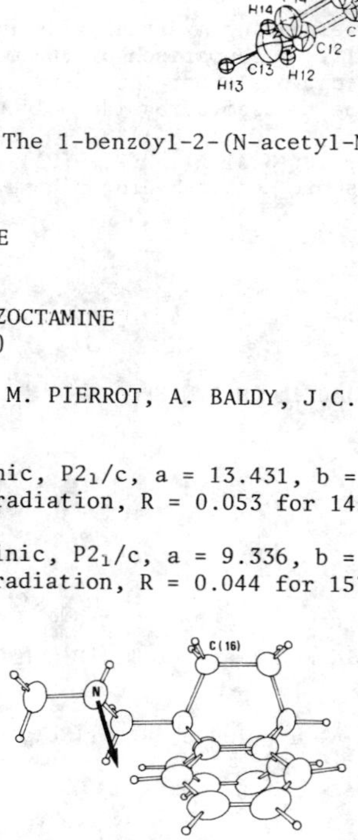

(I)

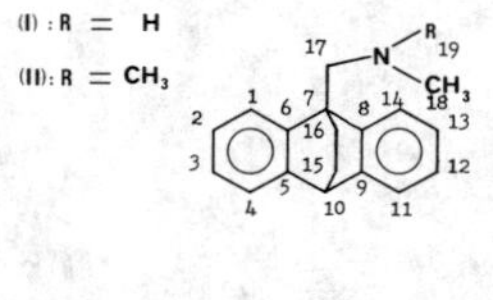

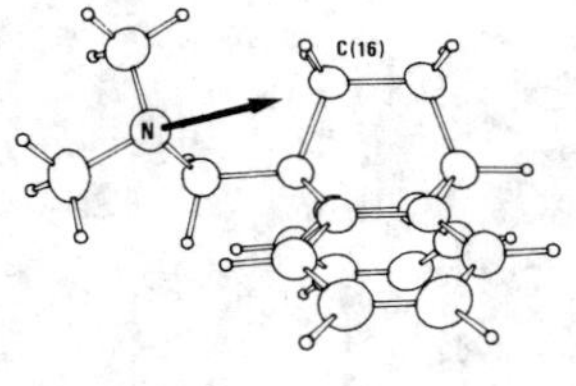

(II)

Fig. 1. The benzoctamine (I) and N-methylbenzoctamine (II) molecules. (The arrow indicates the orientation of the nitrogen lone pair.)

Both molecules have very similar conformations (Fig. 1) with aromatic rings and bridge atoms planar. The angle between the benzene rings is 118° (I) and 121° (II). The N atom is about 0.2 Å from the C(1)-C(6) plane in both molecules. Some bond lengths are: C(10)-C(15) 1.585(5) and 1.554(5), C(15)-C(16) 1.584(5) and 1.551(5), C(7)-C(16) 1.562(4) and 1.566(4) Å.

THIOSEMICARBAZIDE HYDROBROMIDE

CH_6BrN_3S $[NH_2CSNHNH_3]^+Br^-$

A.I. GUBIN, A. TASHENOV, N.N. NURAKHMETOV, M.Zh. BURANBAEV and V.A. BEREMZHANOV, 1984. Kristallografiya, 29, 163-165 [Sov. Phys. Crystallogr., 29, 97-98.

Orthorhombic, Pbca, a = 11.859, b = 7.863, c = 11.923 Å, Z = 8. Mo radiation, R = 0.055 for 343 reflexions.

The compound is isostructural with the corresponding hydrochloride (1). Although the hydrogen atoms were not located experimentally, a comparison of the molecular geometry of the two compounds suggests that the cations exist formed by protonation of the N(2) atom of the hydrazine fragment in the thiosemicarbazide molecule (Fig. 1). Bond distances are: C-S 1.68, C-N(1) 1.37, C-N(3) 1.30 and N(1)-N(2) 1.39 Å. A bifurcated hydrogen bond N(2)-H(23)...Br^- (N(2)...Br 3.31 Å), N(2)-H(23)...Br'^- (N(2)...Br' 3.30 Å) is present as opposed to a short single hydrogen bond N(2)-H(23)...Cl^- in the hydrochloride (N...Cl' 3.097 Å).

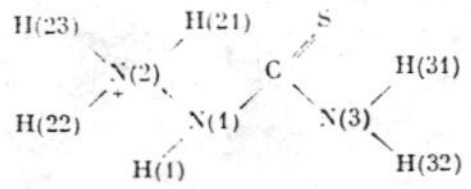

Fig. 1. Schematic representation of the cation in CH_6BrN_3S.

1. Structure Reports, 42B, 29.

THIOCARBONOHYDRAZIDE DIHYDROCHLORIDE HEMIHYDRATE

$CH_8Cl_2N_4S,0.5H_2O$ $CH_8N_4S^{2+},2Cl^-,0.5H_2O$

M. LANFRANCHI, A.M. MANOTTI LANFREDI, M.A. PELLINGHELLI and A. TIRIPICCHIO, 1984. Acta Cryst., C40, 1239-1242.

Triclinic, P$\bar{1}$, a = 9.936, b = 19.108, c = 9.421 Å, α = 104.5, β = 118.4, γ = 90.9°, Z = 8. Mo radiation, R = 0.0322 for 3884 reflexions.

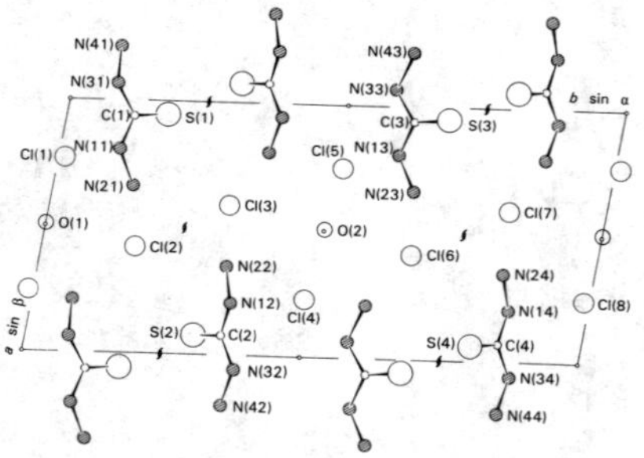

Fig. 1. $CH_8Cl_2N_4S,0.5H_2O$: projection of the structure along c showing the contents of a half-cell (monoclinic subcell).

There are four crystallographically independent units in the cell, approximately related by pseudo symmetry centres and twofold screw axes, so a monoclinic subcell can be recognised. The structure (Fig. 1) consists of deprotonated cations, in the expected cis,cis conformation, of Cl^- anions and of water molecules. The structural parameters of the four independent units are practically equal. Several >NH...Cl, $-NH_3^+$..Cl, $-NH_3^+$...O and OH...Cl hydrogen bonds determine the packing of the cations and anions in the crystal.

LITHIUM DIMETHYLDITHIOCARBAMATE TETRAHYDRATE
$C_3H_6LiNS_2,4H_2O$

I. YMEN, 1984. Acta Cryst., C40, 33-34.

Tetragonal, $P4_3$, a = 8.227, c = 15.260 Å, D_m = 1.289, Z = 4. Mo radiation, R = 0.038 for 1970 reflexions.

Pairs of distorted $[Li(H_2O)_4^+]$ units are linked by O-H...O hydrogen bonds to form chains along c (Fig. 1). A three-dimensional network is formed with channels of van der Waals packed dithiocarbamate ions linked to water molecules by O-H...S hydrogen bonds. The hydrogen bond lengths are O-H...O (O...O 2.823 and 2.912(4) Å), O-H...S (O...S 3.205 to 3.382(2) Å). The C-S distances in the thiocarbamate are 1.719 and 1.728(2) Å.

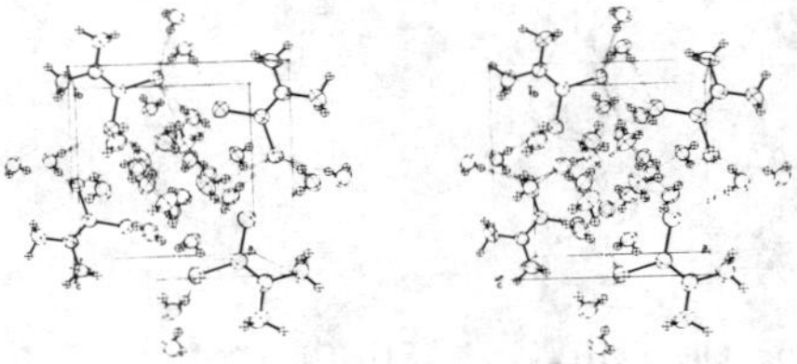

Fig. 1. A stereoview of $C_3H_6LiNS_2,4H_2O$.

3-HYDROXYIMINO-2-BUTANONE THIOSEMICARBAZONE
$C_5H_{10}N_4OS$

A.K. NANDI, S. CHAUDHURI, S.K. MAZUMDAR and S. GHOSH, 1984. Acta Cryst., C40, 1193-1196.

Monoclinic, $P2_1/c$, a = 11.090, b = 12.691, c = 6.112 Å, β = 104.51°, D_m = 1.370, Z = 4. Mo radiation, R = 0.035 for 1169 reflexions.

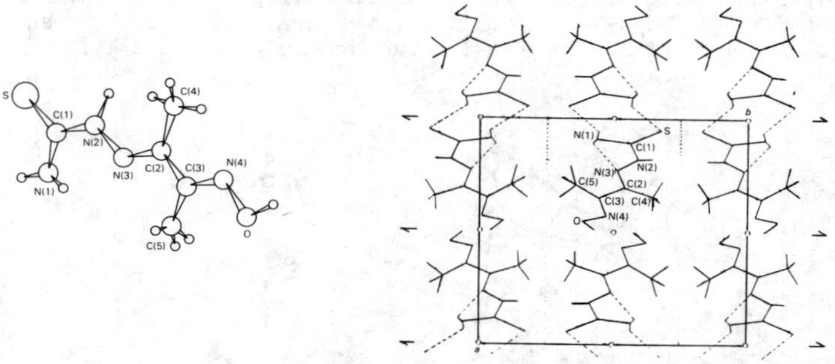

Fig. 1. $C_5H_{10}N_4OS$: perspective view of the molecule down the a axis and molecular packing; hydrogen bonds are indicated by dotted lines.

In the crystal structure (Fig. 1) the molecule is in the fully extended form. Electron delocalization in the thiosemicarbazide and oxime moieties is observed;

bond distances are S-C(1) 1.683(2), C(1)-N(2) 1.362(2), N(2)-N(3) 1.374(2) and N(3)-C(2) 1.287(2) Å. In packing, pairs of intermolecular N-H...S hydrogen bonds across centres of symmetry result in dimerisation of the molecules. Intramolecular N-H...N hydrogen bonds lend conformational stability to the molecules.

LITHIUM DIISOPROPYLDITHIOCARBAMATE TRIHYDRATE
$C_7H_{14}LiNS_2,3(H_2O)$

A. OSKARSSON and I. YMEN, 1984. Acta Cryst., C40, 30-32.

Triclinic, P$\bar{1}$, a = 6.143, b = 8.488, c = 12.796 Å, α = 95.71, β = 101.32, γ = 93.73°, D_m = 1.224, Z = 2. Cu radiation, R = 0.047 for 920 reflexions.

Pairs of distorted $[Li(H_2O)_4{}^+]_2$ tetrahedra (Fig. 1) are formed by edge sharing and these polyhedra are connected to form layers parallel to the ab plane by O-H...S hydrogen bonds (O...S 3.203 to 3.362(7) Å). There are no Li-S bonds. The Li-O distances are 1.870(13) to 2.031(12) Å. In the thiocarbamate, the C-S distances are 1.709 and 1.740(6) Å.

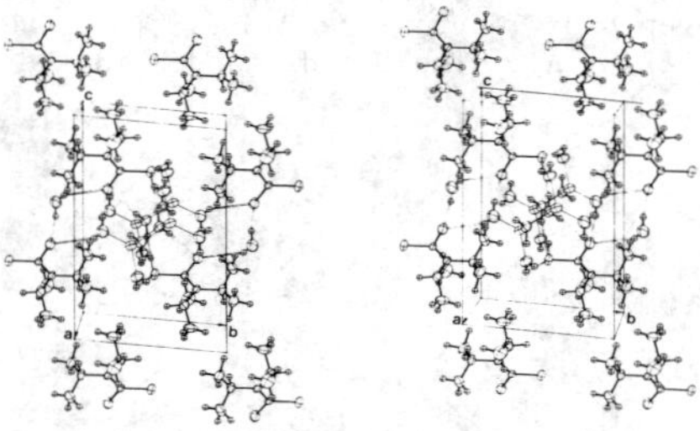

Fig. 1. A stereoview of the $C_7H_{14}LiNS_2,3(H_2O)$ structure.

3,3',5,5'-TETRAKIS(t-BUTYL)-2,2'-BIS(SULPHINYLAMINO)-1,1'-DIPHENYLDISULPHIDE
$C_{28}H_{40}N_2O_2S_4$

H.W. ROESKY, K.-L. WEBER, M. NOLTEMEYER and G.M. SHELDRICK, 1984. Z. Naturforsch., B39, 163-166.

Triclinic, P$\bar{1}$, a = 10.436, b = 10.888, c = 14.515 Å, α = 76.69, β = 80.67, γ = 77.91°, Z = 2. Mo radiation, R = 0.078 for 2162 reflexions.

The analysis establishes the structure shown in Fig. 1, with S(1)-S(1') 2.046(3), N-S 1.506(7) and 1.501(9), S-O 1.432(8) and 1.448(8) Å, N-S-O 120.6(4) and 119.9(4)°, C-S-S 104.0(3) and 103.8(3)°. The C-S-S-C torsion angle is -68.2(4)°.

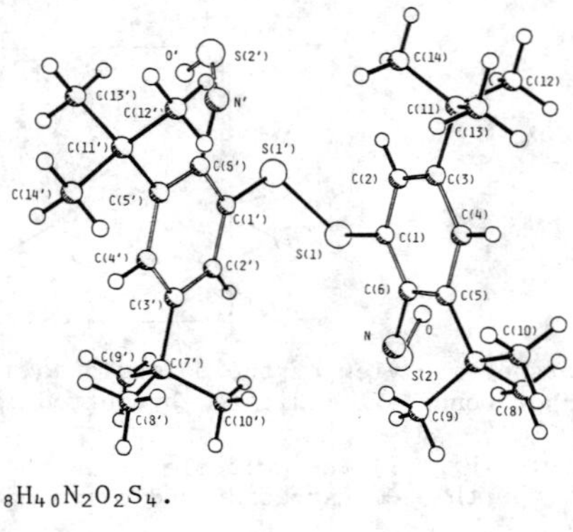

Fig. 1. A view of $C_{28}H_{40}N_2O_2S_4$.

METHYL t-BUTYL ETHER
$C_5H_{12}O$ (I)

DI-t-BUTYL-PEROXIDE
$C_8H_{18}O_2$ (II)

Yu.L. SLOVOKHOTOV, T.V. TIMOFEEVA, M.Yu. ANTIPIN and Yu.T. STRUCHKOV, 1984. J. Mol. Struct., 112, 127-140.

I. Orthorhombic, Pcab, a = 9.253, b = 9.742, c = 13.463 Å, Z = 8. Mo radiation, R = 0.032 for 812 reflexions (at -120°C).

II. Monoclinic, $P2_1/n$, a = 8.181, b = 11.833, c = 10.392 Å, β = 100.17°, Z = 4. Cu radiation, R = 0.063 for 1271 reflexions (at -80°C).

Molecules I and II are shown in Fig. 1. Main dimensions for I are C(Me)-O 1.412(2), C(Bu)-O 1.445(2), C-C 1.528(2) Å, C-O-C 116.2(1), C-C-O(ϕ) 103.3(1), C-C-O(χ) 110.7(1)°. In the peroxide (II) the C-O-O-C torsion angle is 164.0(4)°. Other dimensions include O-O 1.478(3), C-O 1.439(4), C-C 1.532(11) Å, C-O-O 106.7(2), C-C-O(ϕ) 100.7(3), C-C-O(χ) 110.8°. The results are compared with those from theoretical calculations and from gas-diffraction.

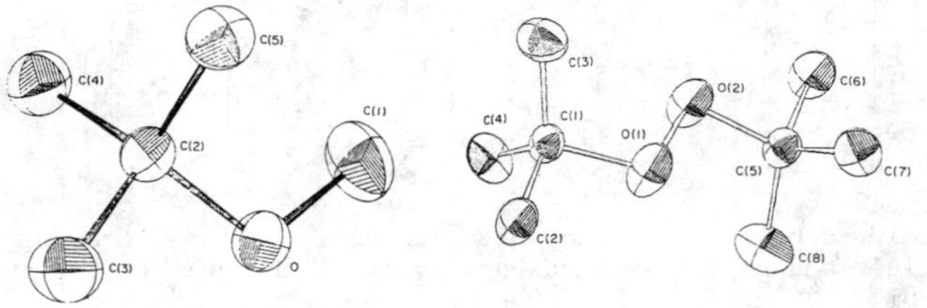

Fig. 1. The $C_5H_{12}O$ and $C_8H_{18}O_2$ molecules.

4,4'(1,2-ETHANEDIYLDIIMINO)BIS(1,1,1-TRICHLORO-3-PENTEN-2-ONE)
$C_{12}H_{14}Cl_6N_2O_2$

R. CEA-OLIVARES, I. RODRÍGUEZ, M. SORIANO-GARCÍA, R.A. TOSCANO and M. CÓRDOBA, 1984. Monatsh., 115, 485-491.

Monoclinic, P2/c, a = 15.915, b = 8.867, c = 12.517 Å, β = 94.92°, D_m = 1.65, Z = 4. Cu radiation, R = 0.061 for 2681 reflexions.

There are two independent molecules (A and B, Fig. 1) in the crystal structure, each lying on a crystallographic twofold axis. The crystal structure consists of alternate layers of A and B molecules, nearly perpendicular to the a axis. The relevant mean bond lengths are: $C(sp^3)-C(sp^3)$ 1.52, $C(sp^3)-C(sp^2)$ 1.501, $C(sp^2)-C(sp^2)$ 1.39 and C=O 1.25 Å. The geometry about the exocyclic C(11)-C(11') and C(12)-C(12') bonds is the commonly observed gauche conformation with torsion angles N(1)-C(11)-C(11')-N(1') and N(2)-C(12)-C(12')-N(2') of 57.3° and -52.6° respectively. There is a short N...N distance of 2.92 and 2.91 Å for the A and B molecules respectively. There are also intramolecular hydrogen bonds: N(1)...O(1) 2.738 Å in A with N(1)-H(1)...O(1) 165.2°; N(2)...O(2) 2.662 Å in B with N(2)-H(2)...O(2) 131.8°.

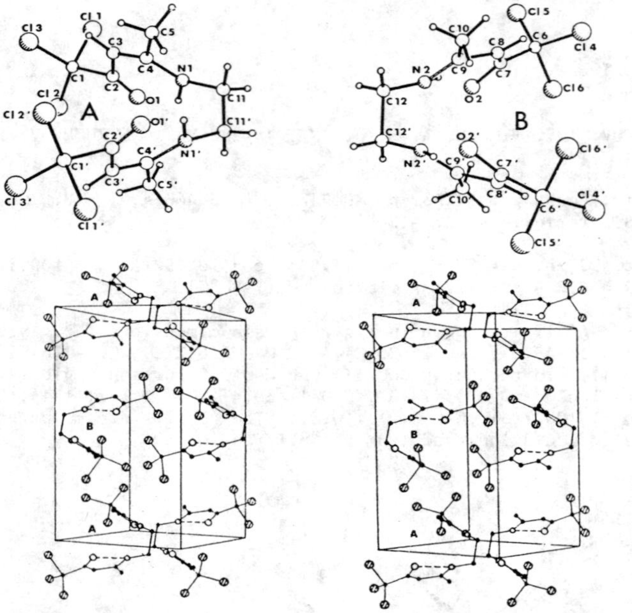

Fig. 1. Labelling scheme in the two molecular conformations of $C_{12}H_{14}Cl_6N_2O_2$ (top left and right), stereoscopic view of the unit cell contents (bottom).

TRIPHENYLBROMOMETHANE

$C_{19}H_{15}Br$

A. DUNAND and R. GERDIL, 1984. Acta Cryst., B40, 59-64.

Trigonal, $P\bar{3}$, a = 13.933, c = 13.438 Å, Dm = 1.42, Z = 6. Mo radiation, R = 0.049 for 976 reflexions.

Three independent molecules lie on special positions on crystallographic threefold axes. They are associated in pairs (Fig. 1) and form two distinct Ph_3C-Br ...Br-CPh_3 head-to-head arrangements characterized by long C-Br bonds (2.031, 2.016, 2.002(11) Å) and short Br...Br intermolecular contacts (Br(1)...Br'(1) 3.496(2), Br(2)...Br'(3) 3.203(2) Å).

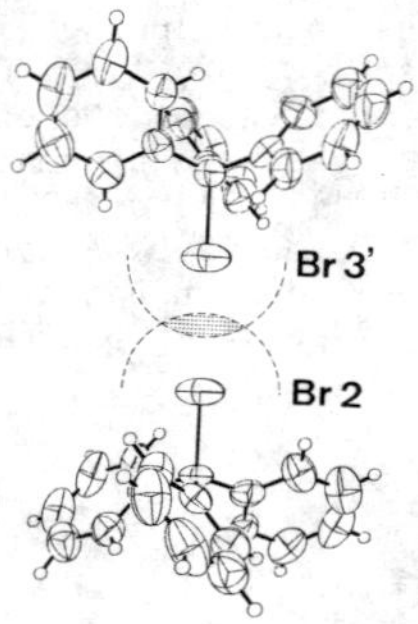

Fig. 1. Head-to-head pairing of two Ph_3CBr molecules. The dashed line outlines the shape of the Br atom with a van der Waals radius 1.85 Å.

HEXA-2,4-DIYNE-1,6-DIYLE BIS[(p-CHLOROPHENYL)CARBAMATE]
$C_{20}H_{14}Cl_2N_2O_4$

C. BROUTY, P. SPINAT and A. WHULER, 1984. Acta Cryst., C40, 1619-1624.

Triclinic, $P\bar{1}$, a = 12.115, b = 8.217, c = 4.856 Å, α = 101.5, β = 83.4, γ = 96.0°, Z = 1. Cu radiation, R = 0.09 for 824 reflexions (at 163K).

The polymer molecules grow inside the monomer crystal structure. The conformation of the monomer is as in the pure compound (1). Excluding the chlorophenyl group (which makes a 25° angle with the molecular mean plane) no atom is more than 0.2 Å from the mean plane. This plane is inclined at 25° to (110). The monomer units are linked by N-H...O hydrogen bonds (Fig. 1) which are lengthened from the pure compound (2.97 vs. 2.895 Å). The mean plane of the polymer molecule makes an angle of 12° with (110) with the chlorophenyl group at 39° to the mean plane. N is 0.4 Å from the mean plane. The polymer chain has an electronic structure characteristic of an acetylenic rather than a butatriene type.

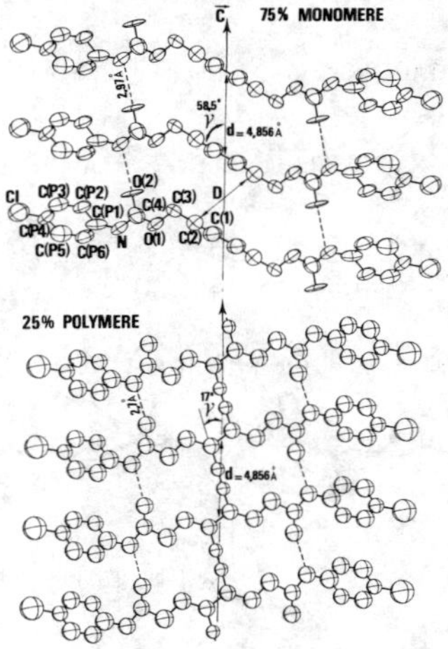

Fig. 1. Projection of the molecules on the (110) plane in the $C_{20}H_{14}Cl_2N_2O_4$ 75% monomer-25% polymer structure.

1. Structure Reports, 50B,

Z,E,Z-1,2,3,4,5-PENTAPHENYL-1,3,5-HEPTATRIENE
$C_{37}H_{30}$

N.F. WOOLSEY, L.J. RADONOVICH, F.M. SAAD and M. BROSTROM, 1984. J. Org. Chem., 49, 1937-1941.

Monoclinic, $P2_1/c$, a = 10.207, b = 29.86, c = 9.347 Å, β = 107.03°, D_m = 1.15, Z = 4. Mo radiation, R = 0.107 for 2370 reflexions.

The stereochemistry of the molecule is indicated in Fig. 1. Mean dimensions are C=C 1.347(7), Csp^2-Csp^2 1.494(7), Csp^2-C phenyl 1.487(8), Csp^2-CH_3 1.511(8) Å.

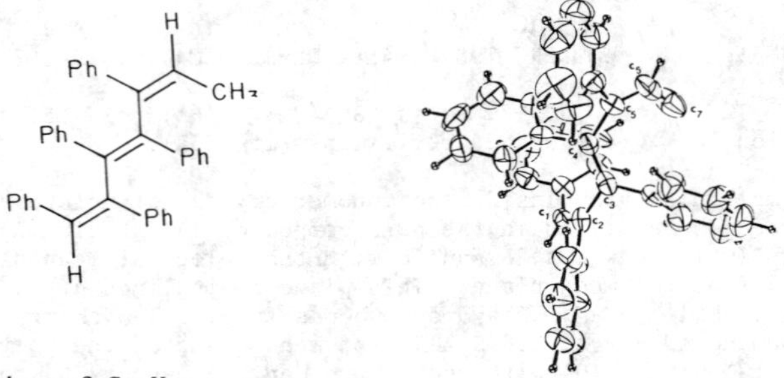

Fig. 1. A view of $C_{37}H_{30}$.

1,7-DIPHENYL-4-BUTYL-1,6-HEPTADIEN-3,5-DIONE
$C_{23}H_{24}O_2$

A. MOSTAD, U. PEDERSEN, P.B. RASMUSSEN and S.-O. LAWESSON, 1984. Acta Chem. Scand., B38, 479-484.

Orthorhombic, $Pca2_1$, a = 11.675, b = 21.399, c = 7.378 Å, D_m = 1.201, Z = 4. Mo radiation, R = 0.040 for 2357 reflexions (at 121K).

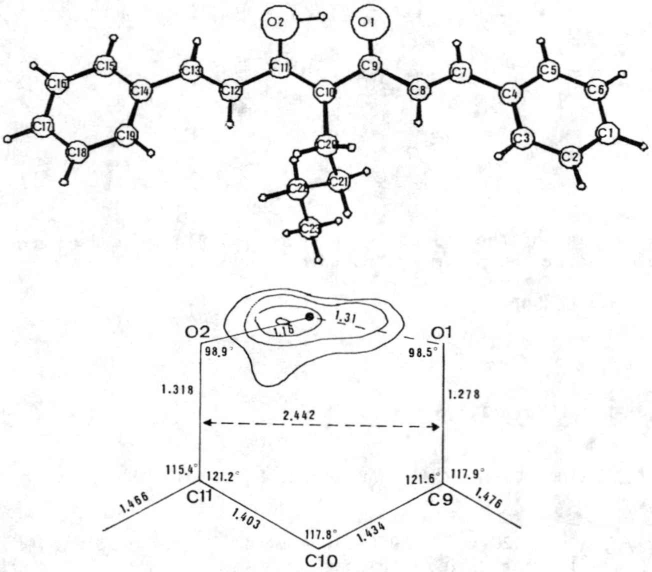

Fig. 1. Molecular structure of $C_{23}H_{24}O_2$ (top), and the geometry of the enol ring (below). Electron density contours are at 0.1, 0.2, 0.3 and 0.4 electrons $Å^{-3}$.

The central portion of the molecule (Fig. 1) appears to be reasonably planar, but the angle between the two phenyl ring planes is 22°. The molecule is quite symmetric, in terms of bond distances and angles, on either side of the enol ring. The enol ring itself (Fig. 1) is clearly not symmetric, the two C-O and C(9)-C(10) and C(10)-C(11) distances being significantly different. There are no intermolecular hydrogen bonds, and the molecules are held together by interactions between the enol rings (O(1)...C(1') 3.361, O(1)-H(1') 2.63, O(2)...C(17') 3.492, O(2)...H(C17') 2.72 Å).

3,3-BIS(METHYLTHIO)-2-NITRO-2-PROPENE-1-NITRILE
$C_5H_6N_2O_2S_2$

N.U. KAMATH and K. VENKATESAN, 1984. Acta Cryst., C40, 1211-1214.

Monoclinic, $P2_1/n$, a = 4.123, b = 13.096, c = 15.126 Å, β = 93.8°, D_m = 1.540, Z = 4. Cu radiation, R = 0.054 for 1119 reflexions.

In the molecule (Fig. 1) the C=C bond is longer than that found in ethylene whilst C(2)-N(2), C(2)-C(1), S(2)-C(3) and S(1)-C(3) are shorter than expected. These bond-length changes indicate that there is a small but significant push-pull effect in the molecule. The molecule is twisted about the C=C bond by 12.8(3)° and C(2) and C(3) have appreciable pyramidality.

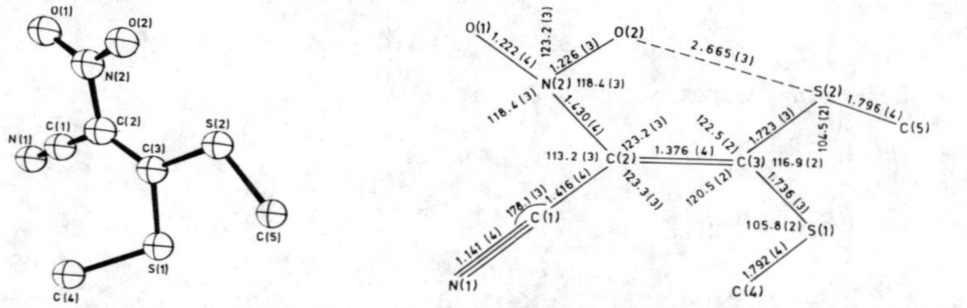

Fig. 1. $C_5H_6N_2O_2S_2$: perspective view of the molecule and bond lengths and angles.

TETRACYANOETHYLENE (MONOCLINIC FORM) (ETHYLENETETRACARBONITRILE)
$C_6H_4N_4$

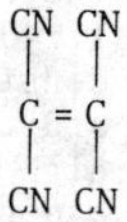

S.L. CHAPLOT, A. MIERZEJEWSKI and G.S. PAWLEY, 1984. Acta Cryst., C40, 663-666.

Monoclinic, $P2_1/n$, Z = 2, neutron powder diffraction data.
At 5K: a = 7.3313, b = 6.1108, c = 6.9315 Å, β = 97.043°, R = 0.074 for 10 reflexions.
At 150K: a = 7.3860, b = 6.1449, c = 6.9548 Å, β = 97.138°, R = 0.100 for 12 reflexions.
At 295K: a = 7.501, b = 6.218, c = 7.004 Å, β = 97.22°, R = 0.085 for 12 reflexions.

At 5K, 150K and 295K respectively average bond lengths are: C=C 1.38(3), 1.36(4) and 1.38(14); C-C 1.43(1), 1.42(2) and 1.45(8); C≡N 1.15(1), 1.15(2) and 1.12(7) Å. These compare with values of 1.317(9), 1.449(4) and 1.132(5) Å respectively determined in earlier room temperature work (1). Work has also been carried out on the cubic form of the title compound (2).

1. Structure Reports, 24, 560.
2. Ibid, 37B, 23.

TETRAFLUOROTEREPHTHALONITRILE
$C_8F_4N_2$

P. SEILER, W.B. SCHWEIZER and J.D. DUNITZ, 1984. Acta Cryst., B40, 319-327.

Orthorhombic, Cmca, a = 7.6848, b = 9.7350, c = 9.5549 Å, Z = 4. Mo radiation, R = 0.081 for 1406 reflexions (at 98 K).

The effect of selectively excluding weak reflections from least-squares refinements of crystal structures has been examined. A systematic bias in the intensities of weak reflections may be introduced by improper treatment of prescan measurements. Omission of reflections with weak or negative measured intensity from high-order ($\sin \theta/\lambda > 0.98$ Å^{-1}) refinements has practically no effect on the positional or vibrational parameters provided that this bias is eliminated. However, substantial errors in the parameters may be introduced by refining with only low-order reflections or even with full data sets that include low-order reflecions. Details of molecular geometry from one of the refinements are in Fig. 1.

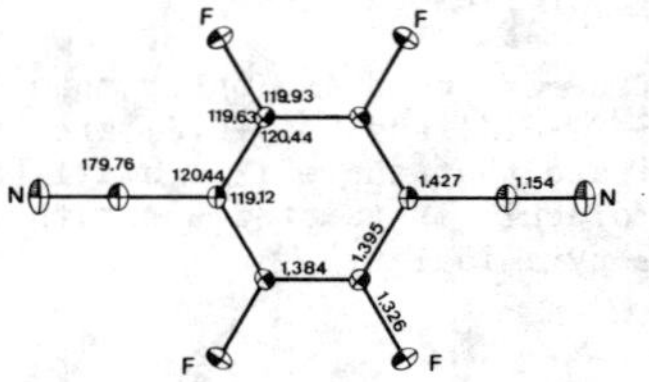

Fig. 1. Molecular geometry of $C_8F_4N_2$ at 98 K.

TETRAFLUOROTEREPHTHALONITRILE

$C_8F_4N_2$

F.L. HIRSHFELD, 1984. Acta Cryst., B40, 484-492.

The low-temperature X-ray data for TFT measured by Dunitz and coworkers (1) have led, via multipole refinement, to a static deformation density map. In support of the data and the model, final discrepancy indices are R = 0.036, $R_w(F^2)$ = 0.019, GOF = 1.140, the discrepancies ΔF show negligible systematic trends and yield a residual density showing mainly random noise, vibration parameters satisfy the rigid-bond test, and the total electron density is everywhere positive. Results accord well with the high-order refinement of Dunitz et al. (1). The deformation density closely resembles experimental and theoretical maps of related molecules. Both π-conjugative and σ-inductive effects seem responsible for unequal aromatic bond lengths and angles. Partitioned atomic moments for the C≡N and C-F fragments approximately match corresponding quantities from two other X-ray and six SCF studies. The small C-F bond peak does not provide the electrostatic binding required by the Hellmann-Feynman theorem; this must come mainly from an undetectable forward polarization in the core region around F. Details of molecular dimensions are in Fig. 1.

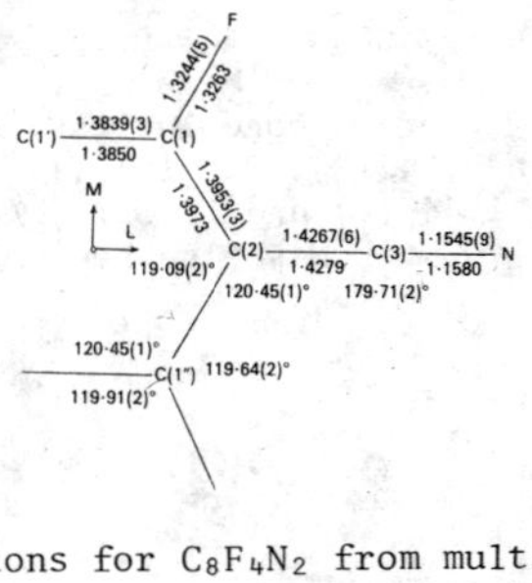

Fig. 1. Molecular dimensions for $C_8F_4N_2$ from multipole refinement.

1. Previous Report.

p-DICYANOBENZENE
$C_8H_4N_2$

M. COLAPIETRO, A. DOMENICANO, G. PORTALONE, G. SCHULTZ and I. HARGITTAI, 1984. J. Mol. Struct., 122, 141-157.

Triclinic, $P\bar{1}$, a = 3.843, b = 6.575, c = 7.313 Å, α = 114.50, β = 93.53, γ = 96.99°, D_m = 1.284, Z = 1. Mo radiation, R = 0.0426 for 842 reflexions.

A refinement with new data of a structure reported previously (1). Details of dimensions for this molecule (which lies about an inversion centre) are shown in Fig. 1 from electron diffraction, conventional refinement, and high-θ refinement (averaged where appropriate).

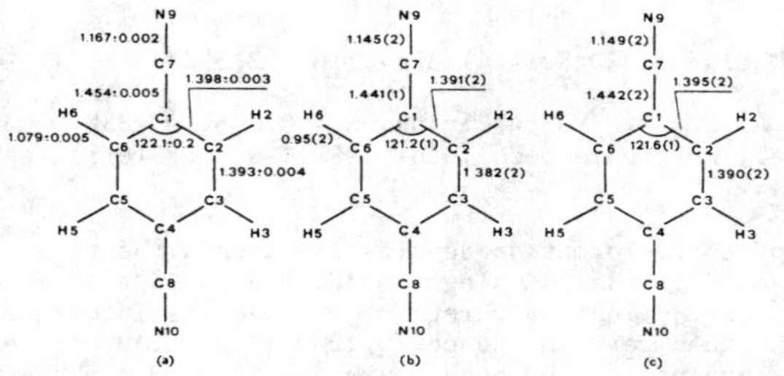

Fig. 1. Molecular geometry for p-dicyanobenzene (a) gas electron diffraction, (b) conventional X-ray, (c) high-θ refinement.

1. Structure Reports, 43B, 65; 44B, 46.

p-DIISOCYANOBENZENE
$C_8H_4N_2$

M. COLAPIETRO, A. DOMENICANO, G. PORTALONE, I. TORRINI, I. HARGITTAI and G. SCHULTZ, 1984. J. Mol. Struct., 125, 19-32.

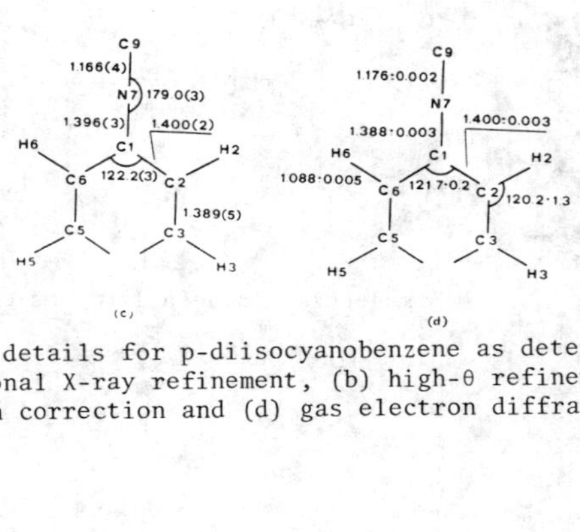

Fig. 1. Geometry details for p-diisocyanobenzene as determined by (a) conventional X-ray refinement, (b) high-θ refinement, (c) after libration correction and (d) gas electron diffraction.

Monoclinic, A2/m, a = 4.772, b = 7.232, c = 9.887 Å, β = 98.28°, Z = 2. Mo radiation, R = 0.0319 for 415 reflexions.

The molecules have crystallographic 2/m symmetry. The structure was refined conventionally and by the high-θ method and was also examined by gas electron diffraction. Dimensions are in Fig. 1 and are interpreted as indicating the contribution of a dipolar form is larger in the crystalline state than in the gas state. In the crystal, there are layers of parallel molecules; the molecular plane is close to the 102 plane.

HEXACYANOBENZENE

$C_{12}N_6$

U. DRUCK and A. KUTOGLU, 1984. Z. Kristallogr., 166, 233-244.

Cubic, Pa3, a = 10.781 Å, Z = 4. Mo radiation, R = 0.046, 0.035, 0.039 (high-angle data, charge-cloud model, multipole model) for 198, 495, 567 reflexions (at 120 K). For previous work see 1.

The peak heights of the deformation density lie within the range 0.25-0.46 $eÅ^{-3}$ in the ring, 0.26-0.48 $eÅ^{-3}$ in the C-C single bond, 0.33-0.74 $eÅ^{-3}$ in the C-N triple bond and 0.00-0.12 $eÅ^{-3}$ in the lone pair region, whereas the integrated charges in the deformation peaks show a good correspondence (0.12-0.15 e, 0.10-0.13 e, 0.18-0.29 e, 0.00 e). These charge contents of the bond peaks are in good agreement with theoretical values (C=C; 0.15 e. C-C; 0.09 e, C≡N; 0.26 e) computated for p-dicyanobenzene. The difference between the experimental and theoretical (0.19 e) values for the lone pair together with the high melting point (370°C) of hexacyanobenzene suggest a probability of an intermolecular charge compensation in the crystal.

1. U. DRUCK and A. KOTOGLU, 1983. Acta Cryst., C39, 638-640.

3-DIMETHYLAMINO-3-METHYLTHIO-2-PHENYLACRYLONITRILE

$C_{12}H_{14}N_2S$

N. SEN and K. VENKATESAN, 1984. Acta Cryst., C40, 1730-1733.

Orthorhombic, $P2_12_12_1$, a = 8.140, b = 10.434, c = 14.101 Å, Dm = 1.204, Z = 4. Mo radiation, R = 0.039 for 1169 reflexions.

Molecular dimensions and conformation are in Fig. 1. The molecule is twisted about the C=C bond by 22.2(3)°. Packing is controlled by van der Waals interactions.

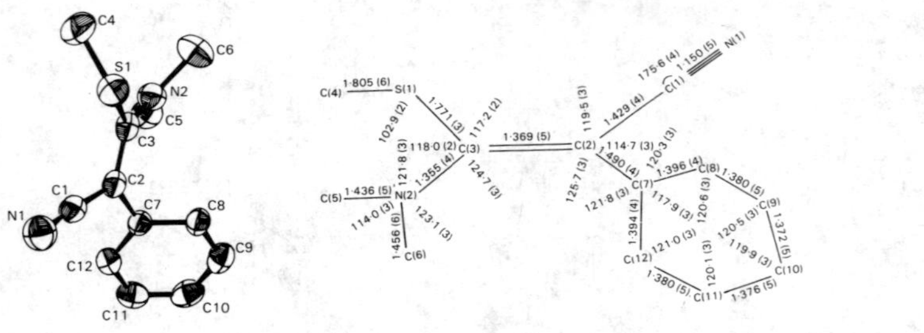

Fig. 1. A view of the $C_{12}H_{14}N_2S$ molecule and details of molecular dimensions.

3-CHLOROBIPHENYL-4-CARBONITRILE
$C_{13}H_8ClN$

H.H. SUTHERLAND and A. RAWAS, 1984. Acta Cryst., C40, 830-832.

Orthorhombic, Pbca, a = 24.05, b = 11.41, c = 7.70 Å, D_m = 1.32, Z = 8. Cu radiation, R = 0.087 for 1316 reflexions (photographic data).

The molecule is non-planar; the angle between the phenyl rings is 31.0(3)° (Fig. 1). The average C-C bond length in the phenyl rings is 1.383, the C-Cl bond is 1.723(10). In the C-C≡N moiety, C-C is 1.438(14) and C≡N is 1.140(16) Å, the C-C bond making an angle of 2.5(5)° with the phenyl plane.

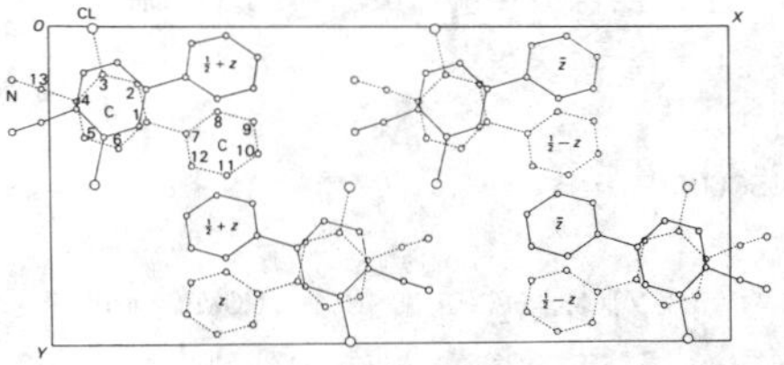

Fig. 1. The arrangement of the $C_{13}H_8ClN$ molecules in the unit cell, viewed along z.

UREA
CH_4N_2O

S. SWAMINATHAN, B.M. CRAVEN and R.K. McMULLAN, 1984. Acta Cryst., B40, 300-306.

Tetragonal, $P\bar{4}2_1/m$, a = 5.565, 5.565, 5.570, 5.576, 5.584, 5.590, 5.598, c = 4.684, 4.685, 4.688, 4.689, 4.689, 4.692, 4.694 Å (at 12, 30, 60, 90, 123, 150 and 173 K). Neutron radiation, R = 0.025, 0.025, 0.028 for 342 reflexions (at 12, 60 and 123 K).

The atomic anisotropic thermal parameters are consistent with overall rigid-body motion together with intramolecular librations of the NH_2 groups. With increasing temperature, the H-atom motion out of the molecular plane becomes dampened with respect to the overall rigid-body motion. The effect appears to be due to the H-bonding which couples the molecules in the crystal. After correction for thermal motion, including anharmonic N-H stretching, there is good agreement in the bond lengths: C-O 1.265(1), C-N 1.350(1), N-H 1.005-1.021(3) Å, N-C-N 117.2(1), O-C-N 121.4(1), C-N.H 119.2(1) and 120.5(1), H-N-H 120.2(2)°.

THIOUREA S,S-DIOXIDE
$CH_4N_2O_2S$

I.-C. CHEN and Y. WANG, 1984. Acta Cryst., C40, 1937-1938.

Orthorhombic, Pnma, a = 10.644, b = 10.106, c = 3.9348 Å, D_m = 1.66, Z = 4. Mo radiation, R = 0.046 for 562 reflexions.

The molecular geometry is similar to that from a previous photographic analysis (1). Bond lengths are C-S 1.8615(4), S-O 1.4890(4), C-N 1.2972(4) Å. The N...O hydrogen bond lengths are 2.8392(5) and 2.8429(9) Å.

1. Structure Reports, 27, 749.

GUANIDINIUM PERCHLORATE

$CH_6ClN_3O_4$ $CH_6N_3^+.ClO_4^-$

A.E. KOZIOL, 1984. Z. Kristallogr., 168, 313-315.

Rhombohedral, R3m, a = 7.606, c = 9.121 Å. Mo radiation, R = 0.030 for 139 reflexions.

A previous study (1) reported the space group as R3. This is now shown to be incorrect. Both guanidinium and perchlorate ions have C_{3v} symmetry. There are no changes to the ion geometries or interion distances.

1. Z. PAJAK, M. GROTTEL and A.E. KOZIOL, 1982. J. Chem. Soc. Faraday Trans., 2, 1529-1538.

N-METHYL-N'-NITRO-N-NITROSOGUANIDINE

$C_2H_5N_5O_3$

S. RICE, M.Y. CHENG, R.E. CRAMER, M. MANDEL, H.F. MOWER and K. SEFF, 1984. J. Am. Chem. Soc., 106, 239-243.

Monoclinic, $P2_1/c$, a = 11.673, b = 11.260, c = 9.892 Å, β = 114.56°, D_m = 1.64, Z = 8. Mo radiation, R = 0.075 for 1019 reflexions.

The molecules (Fig. 1) are nearly planar and contain intramolecular amino-to-nitro hydrogen bonds. The results are essentially the same as those reported previously (1).

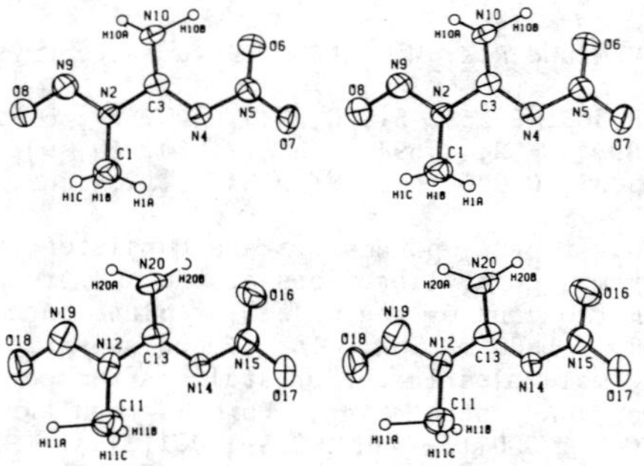

Fig. 1. The two independent $C_2H_5N_5O_3$ molecules.

1. Structure Reports, 48B, 51.

UREA - OXALIC ACID

$C_3H_6N_2O_5$ $CH_4N_2O.C_2H_2O_4$

S. HARKEMA, J.H.M. TER BRAKE and R.B. HELMHOLDT, 1984. Acta Cryst., C40, 1733-1734.

Monoclinic, C2/c, a = 13.0625, b = 6.6437, c = 6.8478 Å, β = 92.474°, Z = 4. Neutron radiation, R = 0.062 for 757 reflexions.

The crystal structure (Fig. 1) contains urea and oxalic acid molecules, held together by a two-dimensional hydrogen-bonding network. The neutron diffraction experiment provides a more accurate description of the geometry of the hydrogen bonds than the corresponding X-ray analysis (1).

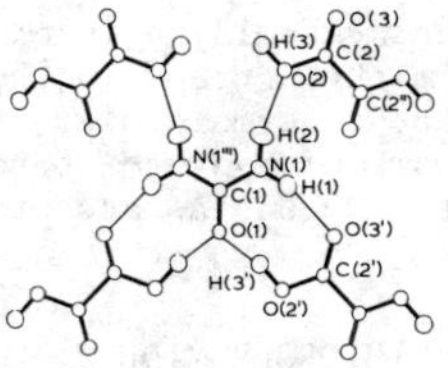

Fig. 1. Part of the crystal structure of $CH_4N_2O.C_2H_2O_4$.

1. Structure Reports, 45B, 44.

trans-GUANIDINODITHIOFORMIC ACID S-METHYL ESTER
$C_3H_7N_3S_2$

G. KIEL, G. GATTOW and W. EUL, 1984. Z. Anorg. Chem., 517, 198-206.

Monoclinic, $P2_1/c$, a = 8.701, b = 11.426, c = 8.221 Å, β = 125.7°, Dm = 1.493, Z = 4. Mo radiation, R = 0.032 for 799 reflexions.

The structure consists of isolated molecules which exist as E-conformers. Details of dimensions are in Fig. 1. The molecules have delocalized double bonds and are linked by S...H-N hydrogen bridges (shortest S...N contact 3.338(5), H...S 2.40(6) Å).

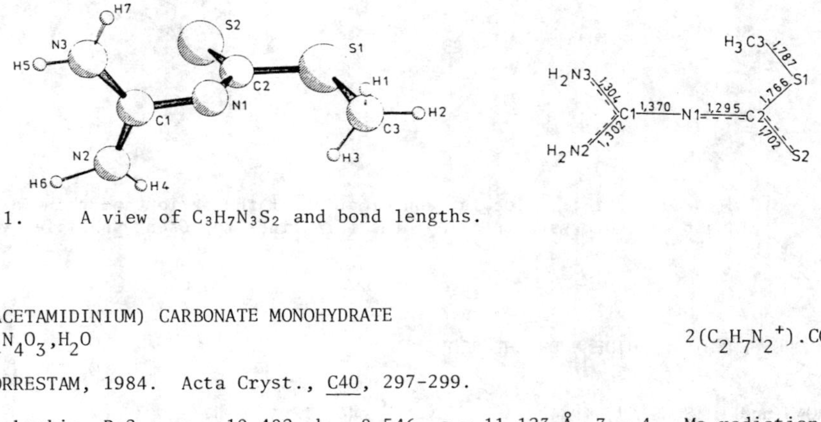

Fig. 1. A view of $C_3H_7N_3S_2$ and bond lengths.

BIS(ACETAMIDINIUM) CARBONATE MONOHYDRATE
$C_5H_{14}N_4O_3,H_2O$ $2(C_2H_7N_2^+).CO_3^{2-}.H_2O$

R. NORRESTAM, 1984. Acta Cryst., C40, 297-299.

Orthorhombic, $Pn2_1a$, a = 10.482, b = 8.546, c = 11.123 Å, Z = 4. Mo radiation, R = 0.051 for 1127 reflexions (at 108 K).

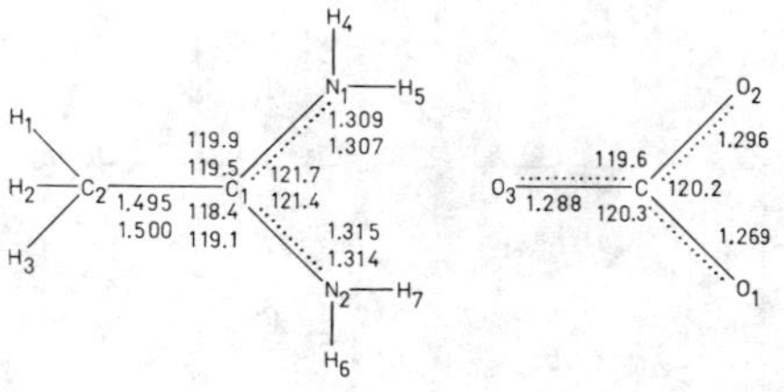

Fig. 1. Intramolecular distances and angles in bis(acetamidinium) carbonate monohydrate. Upper and lower values refer to the two independent cations. C-H and N-H distances are in the ranges 0.97-1.10 and 0.81-0.99(5) Å respectively.

The compound is obtained when acetamidine is exposed to air. Dimensions are in Fig. 1. The orientations of the methyl groups in the two independent cations differ, and this is attributed to different intermolecular interactions. The ions and water molecules are connected by a network of hydrogen bonds, with amino groups and water molecules as donors, and carbonate oxygens as acceptors.

BIS(2-GUANIDINOETHYL) DISULPHIDE DIHYDROBROMIDE

$C_6H_{18}Br_2N_6S_2$ $C_6H_{18}N_6S_2^{2+},2Br^-$

T. ISHIDA, M. INOUE, K. KITAMURA, A. WAKAHARA, K. TOMITA and M. SHINODA, 1984. Acta Cryst., C40, 660-663.

Orthorhombic, $Pca2_1$, a = 24.159, b = 4.904, c = 12.451 Å, D_m = 1.791, Z = 4. Mo radiation, R = 0.087 for 861 reflexions.

The molecule consists of two chemically equivalent ethyl-guanidine groups on a central S-S bond (1.96(2) Å) and has no conformational symmetry (Fig. 1). It is in the extended conformation and is elongated along the c axis; there is no indication of an intramolecular N...S interaction. The crystal packing is stabilised by the formation of intermolecular N-H...Br and N-H...S hydrogen bonds and by many short contacts.

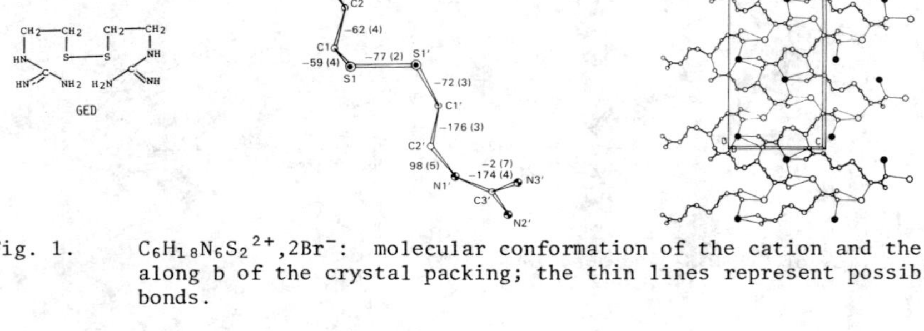

Fig. 1. $C_6H_{18}N_6S_2^{2+},2Br^-$: molecular conformation of the cation and the view along b of the crystal packing; the thin lines represent possible hydrogen bonds.

1-(p-CHLOROPHENYL)BIGUANIDE HYDROCHLORIDE

$C_8H_{11}Cl_2N_5$ $C_8H_{11}ClN_5^+,Cl^-$

C.J. BROWN and L. SENGIER, 1984. Acta Cryst., C40, 1294-1295.

Monoclinic, $P2_1/n$, a = 13.604, b = 5.719, c = 15.376 Å, β = 108.26°, D_m = 1.433, Z = 4. Mo radiation, R = 0.046 for 1843 reflexions.

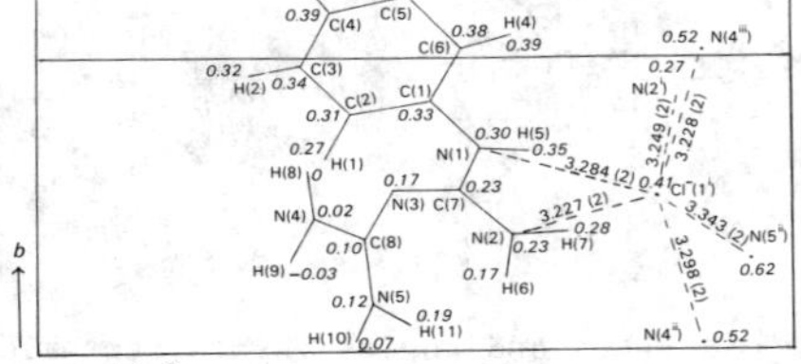

Fig. 1. $C_8H_{11}Cl_2N_5$: view of the molecule and Cl^- coordination.

In the crystal structure (Fig. 1) the Cl^- ion is coordinated octahedrally to six N atoms at a mean distance of 3.272(2) Å. The C-N bonds of the biguanide moiety are within the range 1.307(3)-1.352(3) Å; it consists of two planes at an angle of 50.72(5)° to each other.

N-CYANO-N'-METHYL-N''-(2-{[(5-METHYL-1H-IMIDAZOL-4-YL)METHYL]THIO}ETHYL)GUANIDINE (CIMETIDINE) (POLYMORPH Z)

$C_{10}H_{16}N_6S$

L. PÁRKÁNYI, A. KÁLMÁN, B. HEGEDÜS, K. HARSÁNYI and J. KREIDL, 1984. Acta Cryst., C40, 676-679.

Monoclinic, $P2_1/c$, a = 7.283, b = 10.808, c = 18.281 Å, β = 118.18°, Z = 4. Mo radiation, R = 0.034 for 1733 reflexions.

Bond distances and angles in polymorph Z of cimetidine agree well with those observed in polymorph A (1) (Fig. 1). However, due to the geometrical isomerisation of the guandidine moiety, their conformations differ markedly, e.g. the intramolecular bond distance N(2)...N(4) 2.881(2) Å in polymorph A, increases to 7.382(3) Å. Consequently, the hydrogen bonding of Z also differs significantly from that of A.

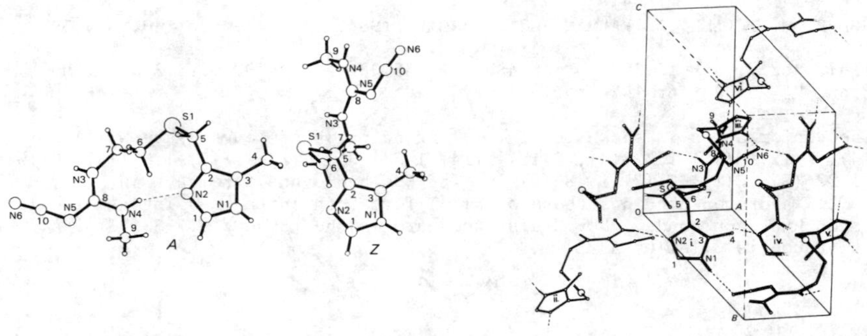

Fig. 1. $C_{10}H_{16}N_6S$: perspective views of polymorphs A and Z (left) and a view of intermolecular hydrogen bonding in polymorph Z (right).

1. Structure Reports, 44B, 192.

1-(p-DIMETHYLAMINOPHENYL)-3-(MERCAPTOACETYL)UREA

$C_{11}H_{15}N_3O_2S$

A. UCHIDA, Y. OHASHI, Y. SASADA, Y. KANEKO and T. ENDO, 1984. Acta Cryst., C40, 115-117.

Monoclinic, $P2_1/n$, a = 20.981, b = 8.1402, c = 7.2535 Å, β = 94.375°, D_m = 1.38, Z = 4. Cu radiation, R = 0.054 for 1779 reflexions.

Except for the SH group the molecule is approximately planar (Fig. 1); centrosymmetric hydrogen bonded dimers are formed by N-H...O hydrogen bonds (N...O 2.856(3) Å). The SH group makes no hydrogen bonds with neighbouring atoms. The acyl urea moiety, -NHCONHCO-, is planar within 0.017 Å and has an intramolecular hydrogen bond (N...O 2.681(3) Å). The acylurea group and benzene ring of an adjacent molecule overlap (Fig. 1).

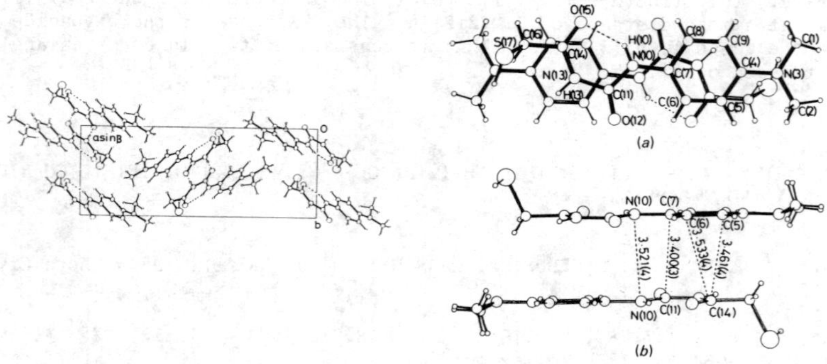

Fig. 1. The crystal structure of $C_{11}H_{15}N_3O_2S$ and views of the overlapping mode.

N-BENZOYL-N'-TOSYL-GUANIDINE
$C_{15}H_{15}N_3O_3S$

E. FISCHER, M. TELLER, A. KALMAN and G. ARGAY, 1984. Tetrahedron, 40, 385-390.

Monoclinic, C2/c, a = 15.395, b = 5.486, c = 35.086 Å, β = 92.01°, Z = 8. Mo radiation, R = 0.044 for 1527 reflexions.

The conformation is shown in Fig. 1. Principal bond lengths are N(11)-C(12) 1.301(4), C(12)-N(13) 1.328(4), C(12)-N(14) 1.381(5), N(14)-C(15) 1.381(5), C(15)-O(16) 1.226(5), C(15)-C(17) 1.486(5) Å. The N-H hydrogens take part in N-H...O and N-H...N hydrogen bonding; those on N(13) form both intra- and inter-molecular bifurcated hydrogen bonds. N...O distances are in the range 2.657-3.373(6) Å; the N...N distance is 3.131(6) Å.

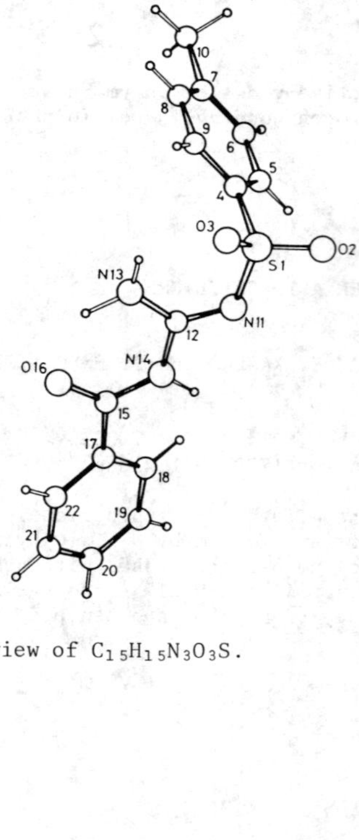

Fig. 1. A perspective view of $C_{15}H_{15}N_3O_3S$.

N,N'-BIS(2-METHYLPHENYL)GUANIDINE

$C_{15}H_{17}N_3$ $(CH_3C_6H_4\text{-}N)_2C\text{-}NH_3$

C.J. BROWN and D.J. GASH, 1984. Acta Cryst., C40, 562-564.

Orthorhombic, Pcab, a = 21.558, b = 13.588, c = 8.980 Å, D_m = 1.218, Z = 8. Cu radiation, R = 0.053 for 1245 reflexions.

The molecular packing is shown in Fig. 1. Difference maps show that all three H atoms of the guanidine residue are bonded to the terminal N(2) atom; C(8)-N(2) 1.382(4), C(8)-N(1) 1.364(4), C(8)-N(3) 1.365(4) Å. Molecules are linked in chains by N-H...N hydrogen bonds (N...N 3.024(4) Å).

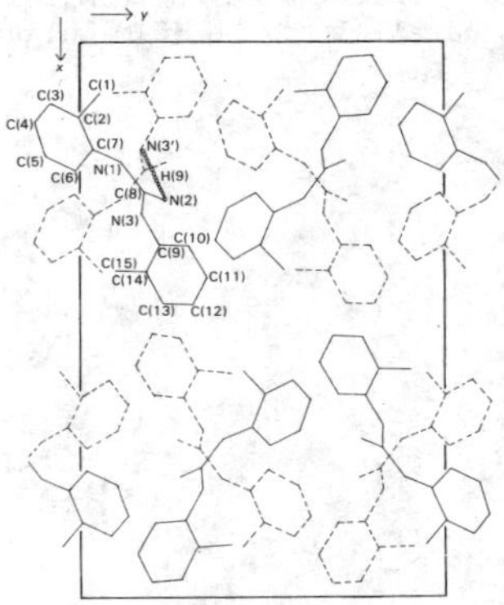

Fig. 1. Packing of $C_{15}H_{17}N_3$ in projection on (001).

1-(p-DIMETHYLAMINOPHENYL)-3-(PHENYLACETYL)UREA

$C_{17}H_{19}N_3O_2$

Y. OHASHI, A. UCHIDA, Y. SASADA, K. KINOSHITA and T. ENDO, 1984. Acta Cryst., C40, 117-119.

Monoclinic, $P2_1/a$, a = 12.878, b = 9.954, c = 12.245 Å, β = 94.92°, D_m = 1.26, Z = 4. Cu radiation, R = 0.049 for 2160 reflexions.

The benzyl group (Fig. 1) is approximately at right angles to the remainder of the molecule. Centrosymmetric dimers are formed by N-H...O hydrogen bonding (N...O 2.848(2) Å). The p-dimethylamino groups in two molecules around an inversion centre are stacked and the dimethylamino group of one molecule is capped by the benzyl group of the other.

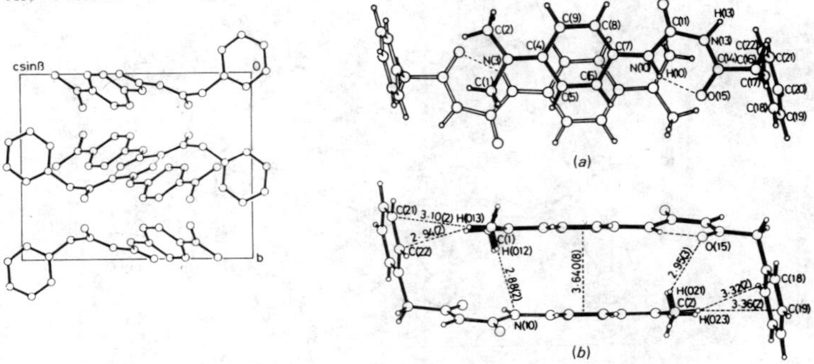

Fig. 1. A view of the crystal structure of $C_{17}H_{19}N_3O_2$ and diagrams showing the overlapping mode.

1-(p-DIMETHYLAMINOPHENYL)-3-(PHENYLACETYL)UREA - 1-ISOBUTYL-3-(p-NITROBENZOYL)UREA (1:1 COMPLEX)

$C_{29}H_{34}N_6O_6$ $C_{17}H_{19}N_3O_2,C_{12}H_{15}N_3O_4$

A. UCHIDA, Y. OHASHI, Y. SASADA, M. MORIYA and T. ENDO, 1984. Acta Cryst., C40, 120-122.

Triclinic, $P\bar{1}$, a = 11.413, b = 16.740, c = 8.569 Å, α = 91.76, β = 108.24, γ = 108.47°, D_m = 1.28, Z = 2. Mo radiation, R = 0.107 for 3541 reflexions.

The two component molecules are linked by pairs of N-H...O hydrogen bonds (Fig. 1) (N...O 2.820(4) and 2.895(4) Å). There are also intramolecular N-H...O hydrogen bonds (N...O 2.705 and 2.690(4) Å). The complex forms columns along the c-axis and is packed in sheets parallel to the $(2\bar{1}0)$ plane.

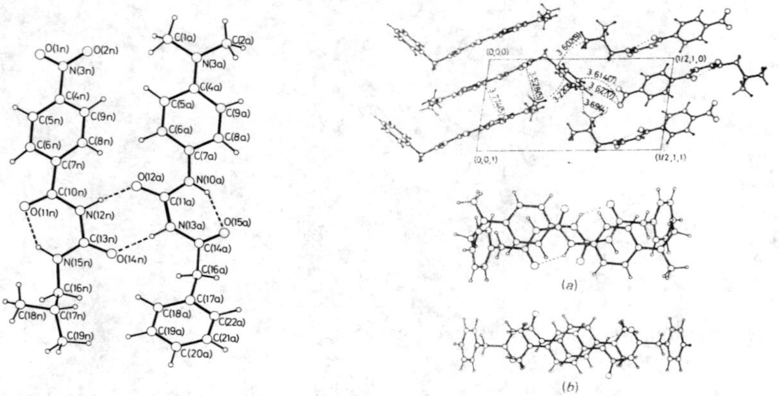

Fig. 1. The hydrogen-bonded complex and views of the molecular overlap.

DIPOTASSIUM 1,2-HYDRAZINEBIS(DITHIOCARBOXYLATE) METHANOL SOLVATE

$C_2H_2K_2N_2S_4,4(CH_4O)$

R. MATTES and B. FUSSER, 1984. Z. Naturforsch., B39, 1-5.

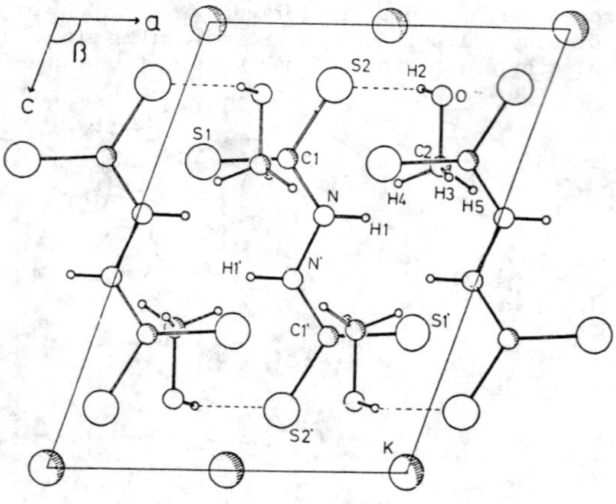

Fig. 1. A projection of the structure of $C_2H_2K_2N_2S_4,4(CH_4O)$.

Monoclinic, C2/m, a = 7.180, b = 13.293, c = 9.377 Å, β = 109.26°, Z = 2. Mo radiation, R = 0.021 for 915 reflexions (at -150°C).

The anion has 2/m crystallographic symmetry and the K^+ ion lies on a twofold axis. The anion is trans-planar (Fig. 1) with N-N 1.392(3), N-C1 1.340(3), C-S 1.691(2) and 1.725(2) Å. The K^+ ion is 3.412 and 3.426 Å from adjacent S atoms and 2.766 and 2.292 Å from methanol O atoms. The methanol is linked to S2 by an O-H...S hydrogen bond (O...S 3.332(1) Å).

2-PYRIDINE-ONN-AZOXYCYANIDE

$C_6H_4N_4O$

D. VITERBO, M. CALLERI, R. CALVINO and R. FRUTTERO, 1984. Acta Cryst., C40, 1728-1730.

Triclinic, P$\bar{1}$, a = 6.368, b = 7.856, c = 7.902 Å, α = 115.41, β = 75.32, γ = 105.42°, Z = 2. Mo radiation, R = 0.052 for 1173 reflexions.

The O atom of the azoxy group (Fig. 1) is bonded to the N atom adjacent to the pyridine ring; the cyano and pyridine moieties are in anti conformation. Some bond lengths are: N(2)-O 1.244(2), C(2)-N(2) 1.475(2), N(2)-N(3) 1.286(2), N(3)-C(7) 1.355(2), C(7)-N(4) 1.148(3) Å.

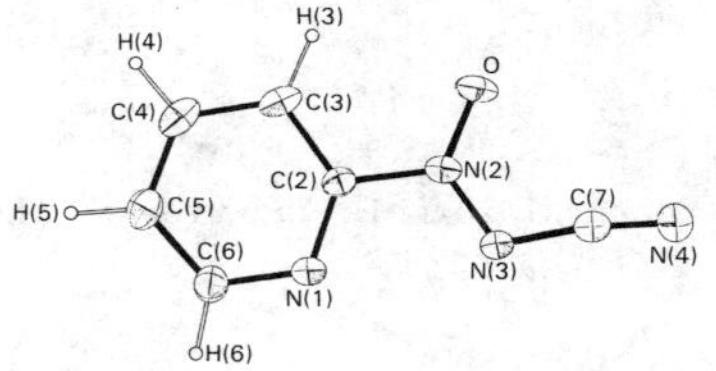

Fig. 1. A view of $C_6H_4N_4O$.

N-PHENYL-N'-SULPHINYLHYDRAZINE

$C_6H_6N_2OS$

F. SCHANDA, A. GIEREN and A. FILHOL, 1984. Acta Cryst., C40, 306-308.

Monoclinic, $P2_1/a$, a = 10.451, b = 15.561, c = 8.346 Å, β = 95.44°, Z = 8. Neutron radiation, R = 0.056 for 1789 reflexions (at 120 K).

There are two independent molecules in the asymmetric unit; mean dimensions are in Fig. 1.

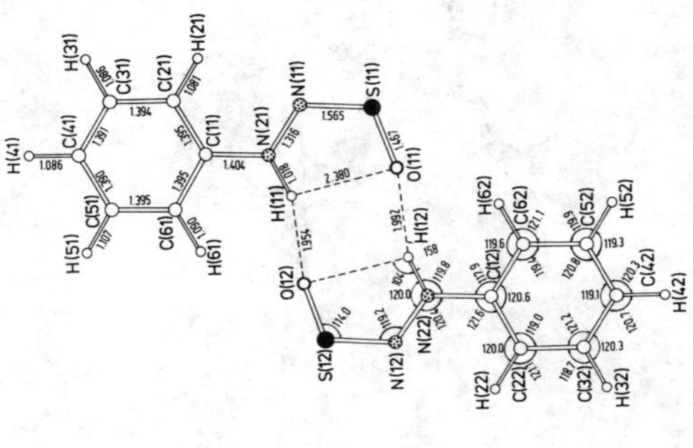

Fig. 1. Bond lengths and angles for $C_6H_6N_2OS$, averaged over the two independent molecules.

ACETONE O-[2-(1-HYDROXY-3-PHENYL-2-TRIAZENO)-2-PROPYL]OXIME
$C_{12}H_{18}N_4O_2$

K.R. ACHARYA, A.M. SHAIKH and NIZAMUDDIN, 1984. Acta Cryst., C40, 1199-1200.

Monoclinic, $P2_1/c$, a = 10.391, b = 13.257, c = 10.321 Å, β = 97.78°, Dm = 1.14, Z = 4. Cu radiation, R = 0.084 for 1181 reflexions.

In the molecule (Fig. 1) bond distances and angles in the phenyl ring are normal, mean C-C 1.377(11) Å. The N(2)-N(3) bond length of 1.274(7) Å suggests considerable double-bond character, as does the planarity of the bonding about N(3), which indicates sp^2 hybridisation. An intramolecular hydrogen bond and an N-H bond, N(1)-H(N1) 1.05(6) Å, support the existence of a resonance hybrid of two valence-bond contributors in the solid state. Marked electron delocalisation is present in the C-N-N-N chain.

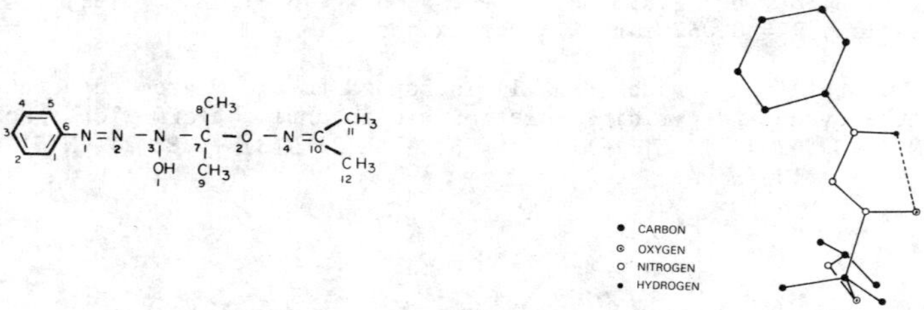

Fig. 1. $C_{12}H_{18}N_4O_2$: structural formula and view of the molecule down the b axis.

BENZALDEHYDE 2-NITROPHENYLHYDRAZONE
$C_{13}H_{11}N_3O_2$

M.G.B. DREW, B. VICKERY and G.R. WILLEY, 1984. Acta Cryst., C40, 304-306.

Monoclinic, $P2_1/n$, a = 16.71, b = 3.904, c = 17.71 Å, β = 99.7°, D_m = 1.41, Z = 4. Mo radiation, R = 0.082 for 711 reflexions.

The structure (Fig. 1) consists of centrosymmetric pairs of molecules linked by weak N-H...O hydrogen bonds (N...O 3.43(1) Å). There is also a strong intramolecular N-H...O hydrogen bond (N...O 2.61(2) Å). The N-N bond length is 1.347(15) Å and the molecule adopts the E-configuration.

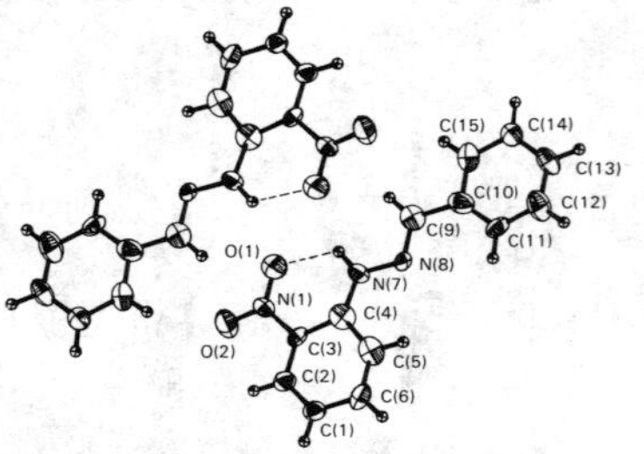

Fig. 1. A view of two molecules of $C_{13}H_{11}N_3O_2$.

DI-2-PYRIDYL KETONE 2-PYRIDYLHYDRAZONE
$C_{16}H_{13}N_5$

C.F. ISHAK, R.T. PFLAUM and N.C. BAENZIGER, 1984. Acta Cryst., C40, 2047-2049.

Orthorhombic, $P2_12_12_1$, a = 7.9213, b = 10.603, c = 16.499 Å, D_m = 1.304, Z = 4. Mo radiation, R = 0.039 for 1751 reflexions.

One pyridyl ring of the dipyridyl ketone group is coplanar with the 2-pyridylhydrazone moiety. The planarity results from the hydrogen bond formed between the N atom of one pyridyl ring from the dipyridyl ketone group and the hydrazone N atom which carries an H atom. The N...N distance in the N-H...N hydrogen bond is 2.607(2) Å, with N-H...N 129°. The N(1)-N(2) bond length is 1.349(2) Å.

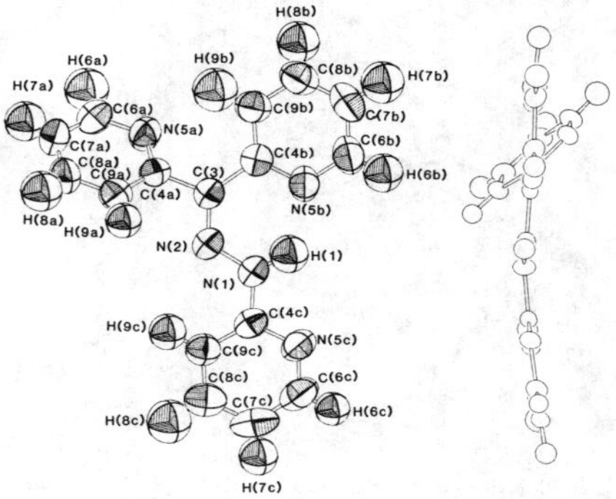

Fig. 1. Views of the $C_{16}H_{13}N_5$ molecule.

α-(4-METHYL-2-NITROPHENYLHYDRAZONO)ACETOACETANILIDE
$C_{17}H_{16}N_4O_4$

C.J. BROWN and H.R. YADAV, 1984. Acta Cryst., C40, 564-566.

Monoclinic, $P2_1/a$, a = 12.034, b = 20.388, c = 7.623 Å, β = 120.5°, D_m = 1.395, Z = 4. Cu radiation, R = 0.044 for 1156 reflexions.

The molecules (Fig. 1) are almost planar and stack in layers perpendicular to c* with three intramolecular hydrogen bonds per molecule and no strong intermolecular forces. The carbon atom C(10) is trigonal with its hydrogen atom transferred to the nitrogen adjacent to the toluene ring so that the pigment exists in the hydrazone tautomeric form. The N(2)-N(3) distance is 1.309(5) Å.

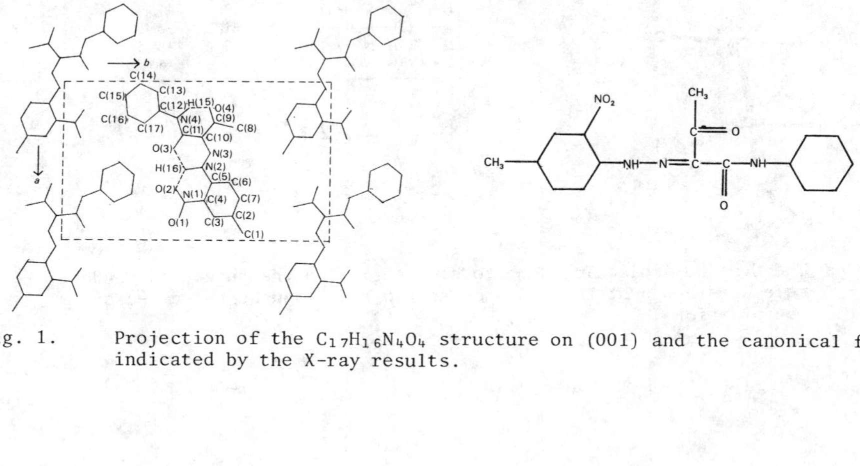

Fig. 1. Projection of the $C_{17}H_{16}N_4O_4$ structure on (001) and the canonical form indicated by the X-ray results.

1,2,3-PROPANETRIOL TRINITRATE (β-Form)
$C_3H_5N_3O_9$

A.A. ESPENBETOV, M. Yu ANTIPIN, Yu.T. STRUCHKOV, V.A. PHILIPPOV, V.G. TSIREL'SON, R.P. OZEROV and B.S. SVETLOV, 1984. Acta Cryst., C40, 2096-2098.

Orthorhombic, $Pna2_1$, a = 8.900, b = 13.608, c = 6.762 Å, Z = 4. Mo radiation, R = 0.026 for 783 reflexions (at 153 K).

Main bond lengths in the molecule (Fig. 1) are: O(C)-N 1.402-1.414(4), O-C 1.445-1.458(5) Å.

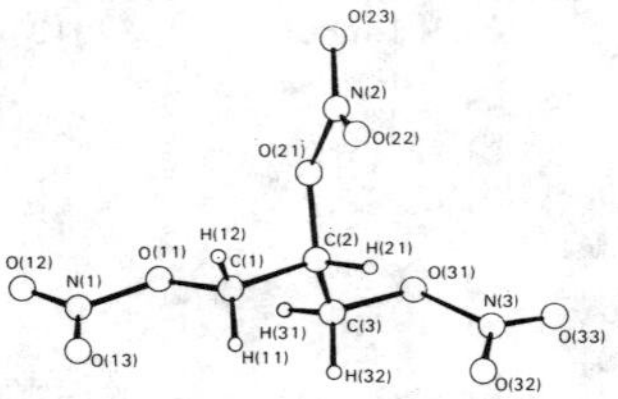

Fig. 1. A view of $C_3H_5N_3O_9$.

N,N-DIMETHYL ETHANOLAMINE N-OXIDE
$C_4H_{11}NO_2$

E.R. MAIA, A. PEGUY and S. PEREZ, 1984. Can. J. Chem., 62, 6-10.

Monoclinic, Cc, a = 25.725, b = 7.023, c = 9.483 Å, β = 101.16°, D_m = 1.21, Z = 12. Cu radiation, R = 0.063 for 575 reflexions.

There are three independent molecules in the asymmetric unit, one being pseudo-cyclic because of an intramolecular hydrogen bond (O...O 2.56 Å) leads to a cis-cis conformation. The other two molecules have trans-gauche conformations. Details of conformation are in Fig. 1. The trans-gauche molecules are linked by O-H...O-N hydrogen bonding (O...O 2.57 and 2.63(2) Å.

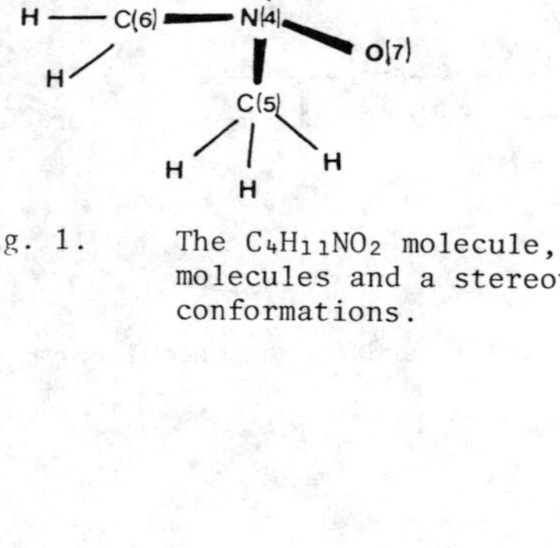

O(1)—C(2)—C(3)—N(4)	172(1)	175(1)	2(8)
C(2)—C(3)—N(4)—C(5)	61(2)	−65(2)	121(6)
C(2)—C(3)—N(4)—C(6)	−179(2)	176(2)	−124(6)
C(2)—C(3)—N(4)—O(7)	−59(2)	55(2)	0(7)

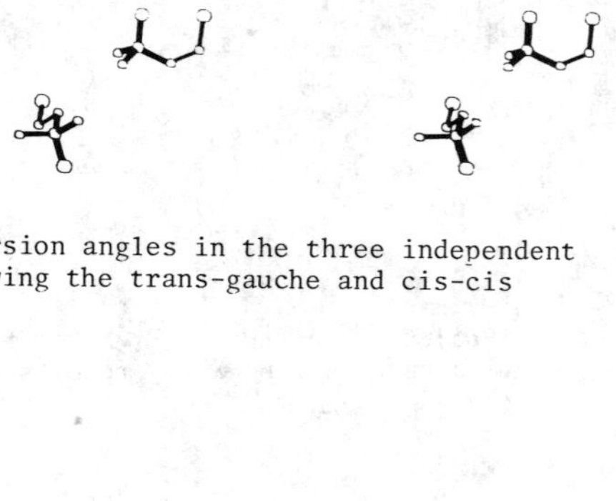

Fig. 1. The $C_4H_{11}NO_2$ molecule, some torsion angles in the three independent molecules and a stereoview showing the trans-gauche and cis-cis conformations.

N-(3-CYANOPHENYL)ACETOHYDROXAMIC ACID HYDRATE
$C_9H_8N_2O_2,0.3H_2O$

R. MOCHARLA, D.R. POWELL and D. VAN DER HELM, 1984. Acta Cryst., C40, 1369-1371.

Orthorhombic, $Pna2_1$, a = 25.202, b = 4.0433, c = 26.268 Å, Dm = 1.354, Z = 12. Cu radiation, R = 0.045 for 2149 reflexions.

The three independent hydroxamic acid molecules have the cyano groups anti to the hydroxyl O (Fig. 1). The hydroxamate groups are in the trans conformation with the non-hydrogen atoms slightly non-planar. Interplanar angles between phenyl rings and hydroxamate groups are 8.3(7), 2.4(6) and 5.0(7)°. Some average bond lengths are, C(1)-N(9) 1.417(5), N(9)-C(10) 1.352(5), N-O 1.405(5), C=O 1.240(5), C(7)-N(8) 1.140(6) Å. All active H atoms are involved in an extensive hydrogen bond network.

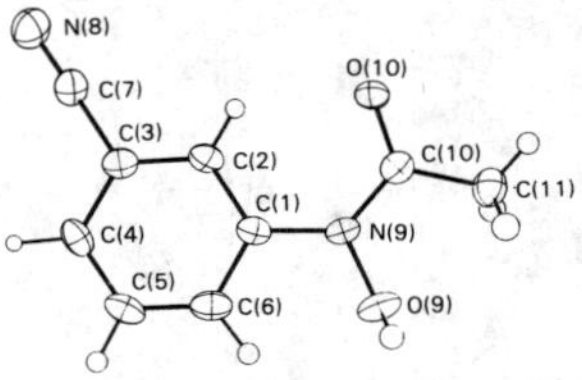

Fig. 1. The $C_9H_8N_2O_2$ molecule.

p-BROMOPHENYLAZOBIS(ACETOXIME)
$C_{12}H_{17}BrN_4O_2$

K.R. ACHARYA, A.M. SHAIKH and NIZAMUDDIN, 1984. Z. Kristallogr., 168, 25-30.

Monoclinic, $P2_1/n$, a = 11.270, b = 10.627, c = 12.631 Å, β = 97.50°, Dm = 1.41, Z = 4. Cu radiation, R = 0.046 for 1642 reflexions.

The molecule has the tautomeric form indicated in Fig. 1, with the H atom bonded to nitrogen atom N1. There is an intramolecular N(1)-H...O(1) hydrogen bond with N...O 2.497(5) Å (N-H...O 111.8°). The N2-N3 distance is 1.275(5) Å.

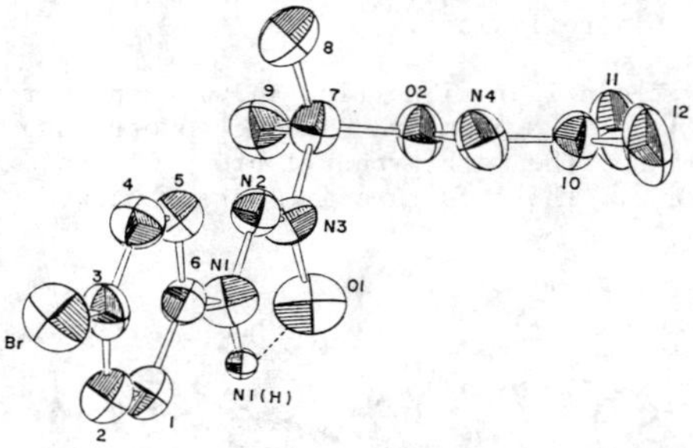

Fig. 1. A view of $C_{12}H_{17}BrN_4O_2$.

3,3'-(TRIMETHYLENEDIIMINO)BIS(3-METHYL-2-BUTANONE OXIME)
$C_{13}H_{28}N_4O_2$

M.S. HUSSAIN, MAZHAR-UL-HAQUE and J. AHMAD, 1984. Acta Cryst., C40, 813-816.

Tetragonal, $I4_1cd$, a = 15.458, c = 13.075 Å, Z = 8. Mo radiation, R = 0.048 for 749 reflexions.

In the molecule the two amine N atoms are twofold-related and are suitably oriented for chelation whereas the oxime N atoms are not suitably positioned for complexation (Fig. 1). The average N-O distance of 1.410(3) Å is significantly longer than the corresponding distance in the nickel complex (1) and is similar to the distance observed in 3,3'-(ethylenediimino)bis(3-methyl-2-butanone oxime) (2).

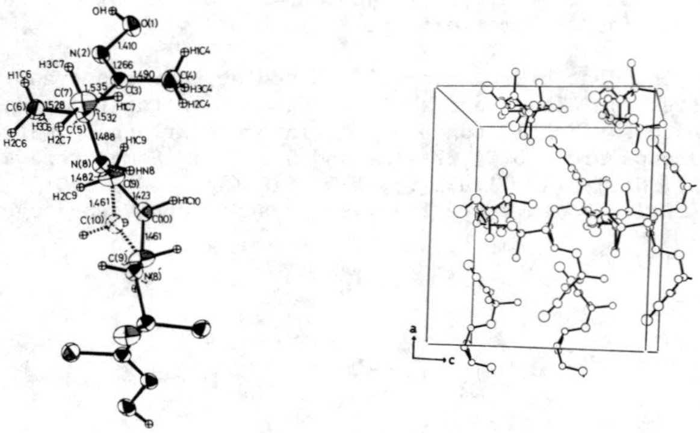

Fig. 1. $C_{13}H_{28}N_4O_2$: perspective view of the molecule, showing the disorder of C(10) and a view of the unit cell showing molecular packing.

1. Structure Reports, 45B, 1167, 1168.
2. Ibid, 50B,

(Z)-N-BENZYL-O-METHOXYBENZOTHIOHYDROXAMIC ACID
$C_{15}H_{15}NO_2S$

A.M. LOBO, S. PRABHAKAR, M.A. SANTOS, H.S. RZEPA and D.J. WILLIAMS, 1984. J. Chem. Soc. Perkin II, 1511-1516.

Monoclinic, $P2_1/n$, a = 8.534, b = 12.321, c = 13.704 Å, β = 99.12°, Z = 4. Cu radiation, R = 0.042 for 1614 reflexions.

The compound has the Z-configuration (Fig. 1) with an intramolecular OH...S hydrogen bond (2.822(3) Å). The N atom has planar geometry and the C-aryl group is orthogonal to the CNOH plane. The methoxyphenyl group is approximately normal (83°) to the thiohydroxamide plane. The structure is loosely packed with no intermolecular contacts of less than 3.5 Å.

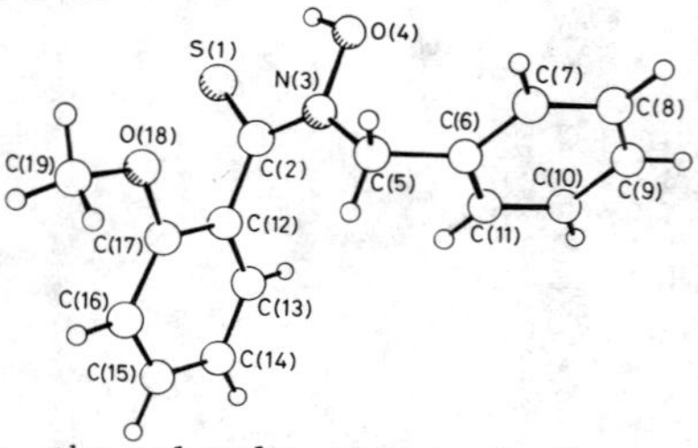

Fig. 1. $C_{15}H_{15}NO_2S$: the molecular structure.

METHYL (DIPHENYLMETHYLIMINO)ACETATE N-OXIDE
$C_{16}H_{15}NO_3$

Y. INOUYE, 1984. Acta Cryst., C40, 140-142.

Monoclinic, $P2_1/a$, a = 10.142, b = 13.851, c = 9.986 Å, β = 99.961°, D_m = 1.28, Z = 4. Mo radiation, R = 0.069 for 2399 reflexions.

The analysis establishes the configuration at the C=N bond to be Z (Fig. 1), and the molecular geometry conforms to that found in other nitrones. The N-O distance is 1.274(2) Å. The dihedral angle between the ester plane (O(2)-C(3)-O(4)) and the nitrone plane (C(5)-N(6)-O(7)) is 31.0(3)°. The dihedral angle between the two phenyl rings is 109.2(3)° and the conformation of the diphenylmethyl group is such that C(5) is eclipsed with H(4) and that C(9) and C(15) atoms are gauche to O(7).

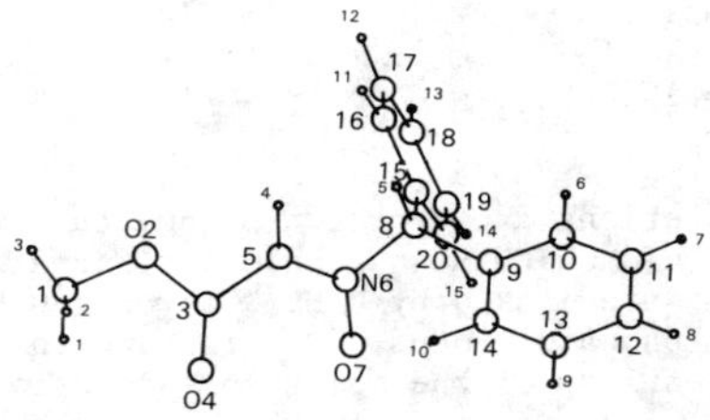

Fig. 1. A view of $C_{16}H_{15}NO_3$.

O-METHYLTHIOCARBAMATE
C_2H_5NOS $NH_2CS\text{-}O\text{-}CH_3$

LIN ZHENGYAN, R. MATTES, H. SCHNOCKEL, M. THUNEMANN, E. HUNTING, U. HOHNKE and C. MENDEL, 1984. J. Mol. Struct., 117, 117-127.

Monoclinic, $P2_1/c$, a = 4.139, b = 11.11, c = 9.39 Å, β = 103.74°, Z = 4. Mo radiation, R = 0.024 for 721 reflexions (at -150°C).

The molecule has the Z-conformation in the solid, with a nearly planar heavy-atom skeleton (Fig. 1). Main bond lengths are C-N 1.317(2), C=S 1.673(2), C-O 1.335(2), O-Me 1.447(2) Å, C-O-C 118.9(1)°. Molecules are linked by N-H...S hydrogen bonds.

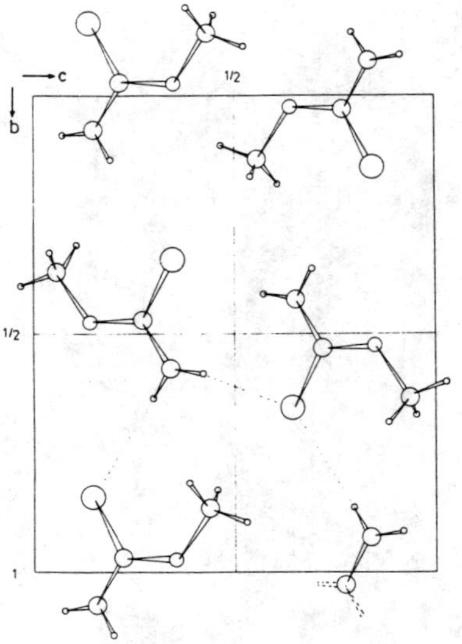

Fig. 1. A projection of the C_2H_5NOS crystal structure.

N-METHYLAMINOMETHANESULPHONIC ACID
$C_2H_7NO_3S$ (I)

DISODIUM N-METHYLIMINOBIS(METHANESULPHONATE) DIHYDRATE
$C_3H_7NNa_2O_6S_2,2(H_2O)$ (II)

T.S. CAMERON, W.J. CHUTE and O. KNOP, 1984. Can. J. Chem., 62, 540-548.

I. Orthorhombic, $P2_12_12_1$, a = 5.455, b = 7.791, c = 11.925 Å, Z = 4. Mo radiation, R = 0.0292 for 609 reflexions.

II. Orthorhombic, Pbca, a = 10.469, b = 6.039, c = 17.549 Å, Z = 4. Mo radiation, R = 0.0536 for 824 reflexions.

In I $^+MeNH_2CH_2SO_3^-$ zwitterions are hydrogen bonded to form infinite chains parallel to a. In II layers of $MeN(CH_2SO_3^-)_2$ anions alternate with layers of Na^+ ions parallel to (001); the anions are linked by O-H...O hydrogen bonds between water molecules and O(2) of the sulphonate groups. Deatils are in Fig. 1. There is extensive discussion and comparison of the molecular dimensions with numerous related acids and their salts.

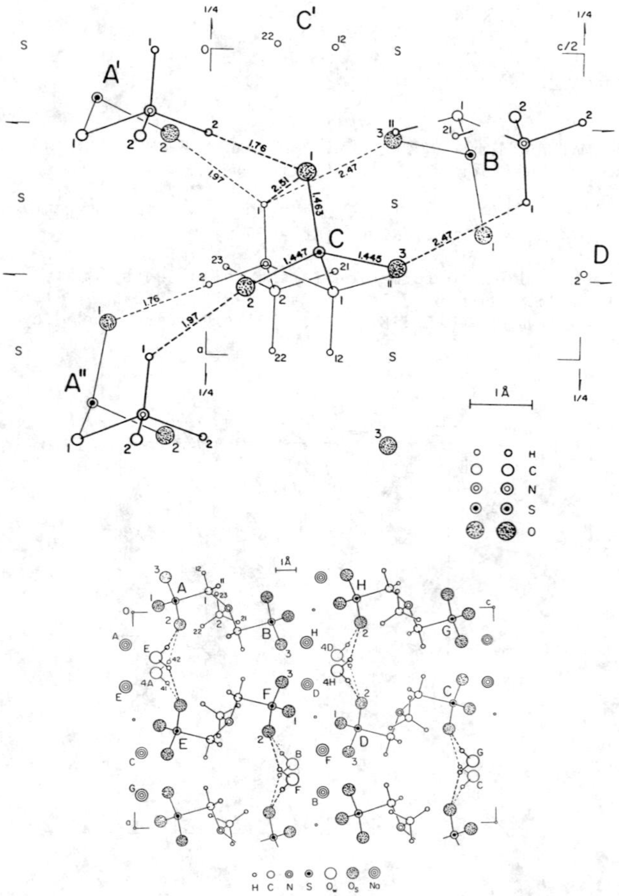

Fig. 1. Projections on (010) of the $C_2H_7NO_3S$ (top) and $C_3H_7NNa_2O_6S_2.2(H_2O)$ structure (bottom).

DISODIUM ETHYLENEBIS(DITHIOCARBAMATE) HEXAHYDRATE
$C_4H_6N_2Na_2S_4,2H_2O$ (I) $2Na^+.[S_2CNHCH_2^-]_2,6H_2O$

IMIDAZOLIDINE-2-THIONE (ETHYLENETHIOUREA)
$C_3H_6N_2S$ (II) $(CH_2NH)_2CS$

T.C.W. MAK, K.S. JASIM and C. CHIEH, 1984. Can. J. Chem., 62, 808-813.

I. Triclinic, $P\bar{1}$, a = 5.766, b = 7.212, c = 10.074 Å, α = 88.53, β = 98.41, γ = 105.50°, D_m = 1.518, Z = 1. Mo radiation, R = 0.022 for 2134 reflexions.

II. Monoclinic, $P2_1/c$, a = 5.800(1), b = 14.516, c = 5.780 Å, β = 101.33°, D_m = 1.422, Z = 4. Mo radiation, R = 0.031 for 975 reflexions.

In I (Fig. 1) the bisdithiocarbamate ion has a centre of symmetry between the two CH_2 groups, and therefore the two NCS_2^{2-} groups are separated as far apart as possible. Each Na^+ ion is octahedrally surrounded by five water molecules with Na...O distances ranging from 2.373 to 2.446 Å, and a sulphur atom of the thiocarbamate group with Na...S 3.060 Å. The two C-S bonds of the CS_2^- group are 1.710(1) and 1.727(1) Å, and SCS 121.9(1)°. Compound II has been reported previously (1). In this refinement with new data the principal distances are C-S 1.692(1), (S)C-N (mean) 1.328(2), (C)C-N (mean) 1.456(2), C-C 1.524(3) Å.

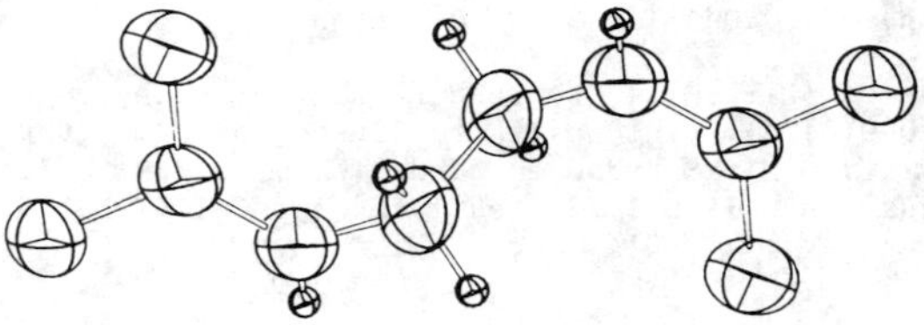

Fig. 1. The ethylenebisdithiocarbamate ion.

1. Structure Reports, 17, 745.

BIS[(METHYLTHIO)CARBONYL]TRISULPHANE
$C_4H_6O_2S_5$ (I)

BIS[(METHYLTHIO)CARBONYL]TETRASULPHANE
$C_4H_6O_2S_6$ (II)

A.W. MOTT and G. BARANY, 1984. J. Chem. Soc. Perkin I, 2615-2621.

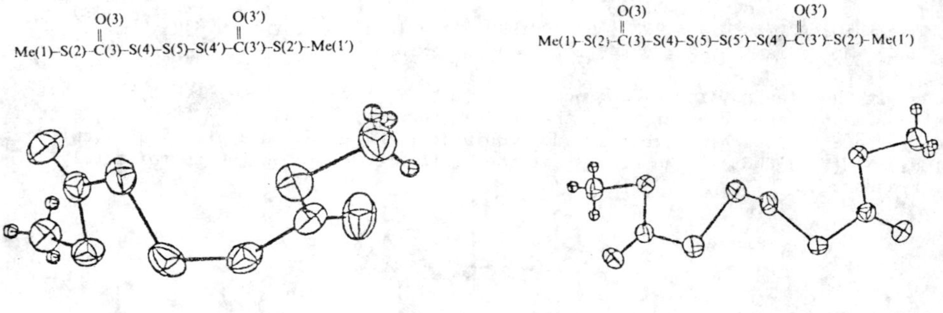

Fig. 1. The skeletons and views of the structures, $C_4H_6O_2S_5$ (top right) and $C_4H_6O_2S_6$ (bottom right).

I. Monoclinic, C2/c, a = 10.841, b = 4.323, c = 21.145 Å, β = 97.18°, Z = 4. Mo radiation, R = 0.034 for 891 reflexions (at -10°C).

II. Monoclinic, C2/c, a = 25.372, b = 5.563, c = 7.974 Å, β = 107.30°, Z = 4. Mo radiation, R = 0.033 for 1239 reflexions (at -10°C).

The X-ray analyses established the structures of the tri- and tetrasulphanes (Fig. 1). Bond lengths in I are: C(1)-S(2) 1.795(6), C(3)-S(2) 1.740(4), C(3)-O(3) 1.198(6), C(3)-S(4) 1.795(4) and S(4)-S(5) 2.040(2) Å. The C(3)-S(4)-S(5) and S(4)-S(5)-S(4') angles are 105.5(2) and 107.8(1)° respectively. In II bond lengths are: C(1)-S(2) 1.794(4), C(3)-S(2) 1.742(4), C(3)-O(3) 1.201(4), C(3)-S(4) 1.801(4), S(4)-S(5) 2.032(2), and S(5)-S(5') 2.055(2) Å. The C(3)-S(4)-S(5) and S(4)-S(5)-S(5') angles are 105.7(2) and 106.2(1)°. The S(4)-S(5)-S(4')-S(5') torsion angle is -78.8° and the structure is trans,trans about the two S(4)-S(5) bonds.

LITHIUM DIETHYLDITHIOCARBAMATE TRIHYDRATE

$C_5H_{10}LiNS_2,3H_2O$

I. YMEN, 1984. Acta Cryst., C40, 241-243.

Triclinic, P$\bar{1}$, a = 7.710, b = 8.784, c = 9.154 Å, α = 77.88, β = 82.67, γ = 64.93°, Z = 2. Mo radiation, R = 0.041 for 2573 reflexions.

Pairs of distorted $[Li(H_2O)_4]^+$ tetrahedra are formed by edge sharing (Li-O 1.895, 1.897, 2.019 and 2.049(3) Å). The pairs are linked to form a three-dimensional network (Fig. 1) by six O-H...S hydrogen bonds (O...S 3.229 to 3.452(2) Å). In the anion, main distances are C(1)-S 1.722 and 1.724(1), C(1)-N 1.336(2) Å.

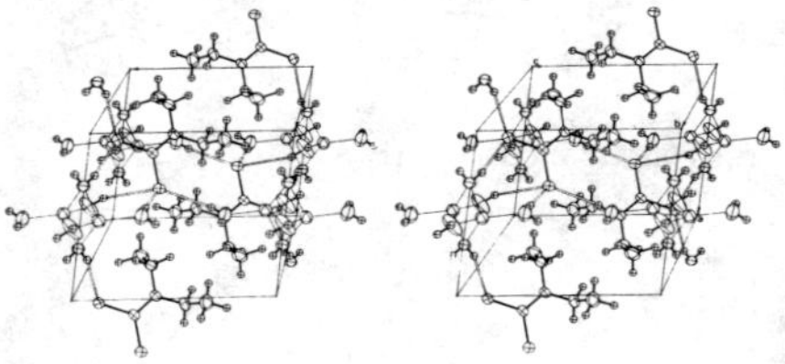

Fig. 1. A stereoview of the lithium diethyldithiocarbamate trihydrate.

N-(3-DIMETHYLAMINOPROPYL)DITHIOCARBAMIC ACID

$C_6H_{14}N_2S_2$

V. SCHRAMM, S.C. KOKKOU and P. KARAGIANNIDIS, 1984. Acta Cryst., C40, 149-151.

Monoclinic, $P2_1/n$, a = 6.9882, b = 10.9995, c = 12.3087 Å, γ = 96.830°, D_m = 1.250, Z = 4. Mo radiation, R = 0.048 for 1444 reflexions.

In the planar thioureide group $-NCS_2$ (Fig. 1), the C-N bond, 1.338(5) Å, has strong double-bond character and the two C-S bonds, 1.727(3) and 1.699(3) Å, with a difference in length of 6.6σ, are not equivalent. The molecule exists as a zwitterion $(CH_3)_2N^+H(CH_2)_3NHCS_2^-$, the bonding at the protonated N atom being approximately tetrahedral.

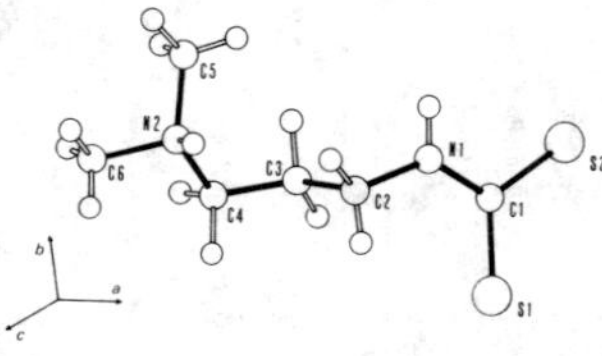

Fig. 1. A view of $C_6H_{14}N_2S_2$.

(3RS)-4,4-DIMETHYL-1-PENTYN-3-YL (SR)-METHANESULPHINATE
$C_8H_{14}O_2S$

B. KOJIĆ-PRODIĆ, A.L. SPEK, P. WIJKENS, G. TADEMA, C.J. ELSEVIER and P. VERMEER, 1984. Acta Cryst., C40, 1841-1843.

Monoclinic, $P2_1/n$, a = 6.0341, b = 11.8316, c = 14.410 Å, β = 93.88°, Z = 4. Cu radiation, R = 0.038 for 1433 reflexions.

The crystals contain a racemic mixture of the (3R,SS) and (3S,SR) enantiomers. The ethynyl and S=O groups are trans with respect to the plane through S(1)-O(2)-C(2) (Fig. 1). The dimethylpentyne moiety exhibits a staggered conformation about C(2)-C(5). There is a short intermolecular contact involving the sulphoxo group and the terminal ethynyl H atom (O(1)...H(4) 2.28(1) Å) suggesting a C-H...O-S interaction. Bond lengths and angles are normal.

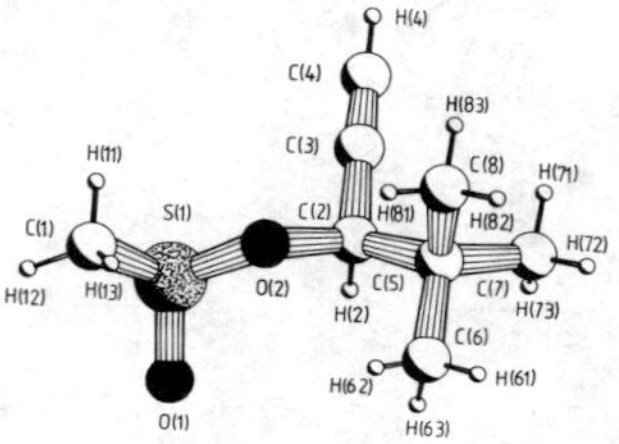

Fig. 1. The (3R,SS) enantiomer of $C_8H_{14}O_2S$.

1-(METHYLSULPHONYL)-2-(PHENYLTHIO)ETHANE
$C_9H_{12}O_2S_2$

F. MO, B.C. HAUBACK and S. WINTHER, 1984. Acta Cryst., B40, 288-294.

Orthorhombic, $P2_12_12_1$, a = 5.312, b = 8.599, c = 22.262 Å, Z = 4. Mo radiation, R = 0.025 for 1691 reflexions (at 86 K).

Both end groups of the molecule are oriented gauche (Fig. 1). Principal bond lengths are: S1-C1 1.757(2), S1-C2 1.776(2), S1-O1 1.446(1), S1-O2 1.450(1), S2-C3 1.810(2), S2-C4 1.774(2) Å. Structure data for a variety of crystalline molecules containing S in different oxidation states show that sulfides represent a structural discontinuity relative to the corresponding sulfoxides and sulfones.

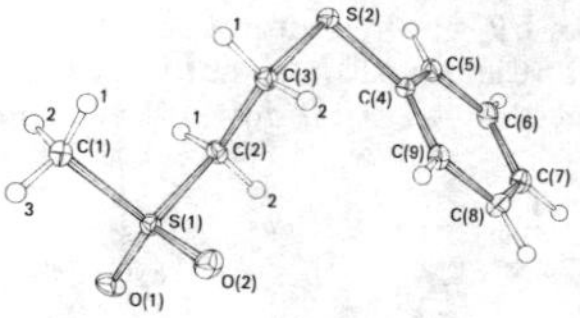

Fig. 1. A view of $C_9H_{12}O_2S_2$.

4-AMINO-4'-NITRODIPHENYLDISULPHIDE
$C_{12}H_{10}N_2O_2S_2$

J.D. KORP and I. BERNAL, 1984. J. Mol. Struct., 118, 157-164.

Monoclinic, $P2_1/c$, a = 15.360, b = 9.023, c = 9.069 Å, β = 98.95°, D_m = 1.47, Z = 4. Mo radiation, R = 0.031 for 1390 reflexions.

The molecular dimension data are consistent with there being a conjugation of a quinoidal form in both ring systems. The disulphide linkage (S-S 2.046(1) Å) shows little if any double bond character and the C-S-S-C torsion angle is 101.7° (Fig. 1), consistent with no communication between the two ring systems.

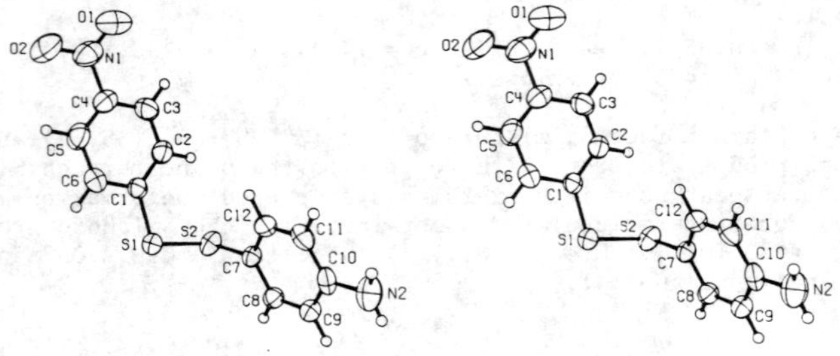

Fig. 1. A stereoview of $C_{12}H_{10}N_2O_2S_2$.

BENZENETHIOSULPHONIC ACID PHENYL ESTER
$C_{12}H_{10}O_2S_2$ (I) Ph-SO_2-S-Ph

p-TOLYLTHIOSULPHONIC ACID p-TOLYL ESTER
$C_{14}H_{14}O_2S_2$ (II) Me-C_6H_4-SO_2-S-C_6H_4-Me

p-CHLOROBENZENETHIOSULPHONIC ACID p-CHLOROPHENYL ESTER
$C_{12}H_8Cl_2O_2S_2$ (III) Cl-C_6H_4-SO_2-S-C_6H_4-Cl

R. CAPUTO, G. PALUMBO, M. NARDELLI and G. PELIZZI, 1984. Gazz. Chim. Ital., 114, 421-430.

I. Monoclinic, P2/n, a = 12.413, b = 8.537, c = 12.649 Å, β = 116.50°, Z = 4. Mo radiation, R = 0.0485 for 1033 reflexions.

II. Monoclinic, $P2_1/n$, a = 13.932, b = 12.076, c = 8.326 Å, β = 93.40°, Z = 4. Cu radiation, R = 0.0425 for 1084 reflexions.

III. Monoclinic, $P2_1/n$, a = 13.716, b = 11.932, c = 8.206 Å, β = 92.47°, Z = 4. Mo radiation, R = 0.0493 for 1040 reflexions.

The three structures (Fig. 1) are very similar with the conformation of the thiosulphonyl system determined by the tendency of non-bonding electron pairs of S(II) to assume synclinal and antiperiplanar conformation with respect to the S(VI)-O bonds. S-S distances corrected for rigid body libration are 2.093, 2.100 and 2.105 Å for I, II and III respectively.

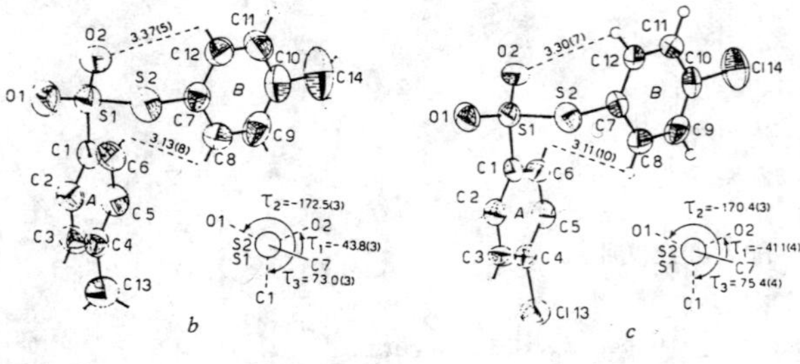

Fig. 1. Structures of the thiosulphonic esters II and III.

BIS(PHENYLSULPHONYL) TRISULPHIDE
$C_{12}H_{10}O_4S_5$

I.-C. CHEN and Y. WANG, 1984. Acta Cryst., C40, 1890-1892.

Tetragonal, $P4_12_12$, a = 7.758, c = 26.204 Å, D_m = 1.56, Z = 4. Mo radiation, R = 0.042 for 1146 reflexions.

The molecule (Fig. 1) contains a twofold axis through S(1). The S-O bond lengths are 1.422(2) and 1.428(2) with S-C 1.747(3) Å.

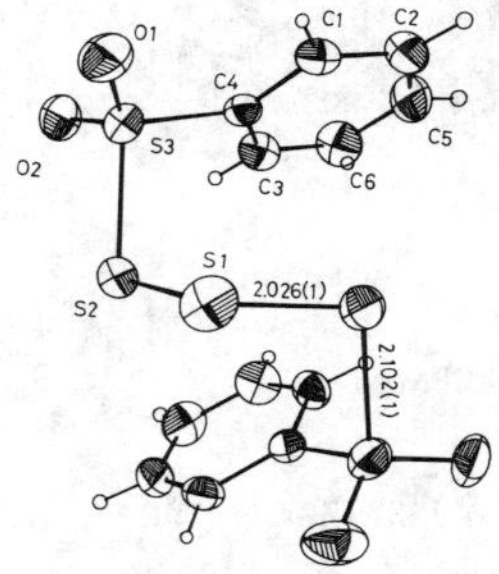

Fig. 1. The bis(phenylsulphonyl) trisulphide molecule with S-S bond lengths.

4-METHYL-4-NITRO-1-PHENYL-2-THIOXOPENTAN-1-ONE S-OXIDE
$C_{12}H_{13}NO_4S$

D.St.C. BLACK, G.D. FALLON, B.M. GATEHOUSE and J.D. WISHART, 1984. Aust. J. Chem., 37, 777-784.

Monoclinic, C2/c, a = 17.869, b = 5.920, c = 24.103 Å, β = 91.50°, D_m = 1.39, Z = 8. Cu radiation, R = 0.052 for 1054 reflexions.

The S-oxide oxygen group is on the side of the C=S bond opposite to the benzoyl group. Significant torsion angles are, O(1)-S-C(8)-C(9) -4.0(5), O(1)-S-C(8)-C(7) -173.4(4), S-C(8)-C(7)-O(2) +153.6(4), S-C(8)-C(7)-C(1) -25.6(6), S-C(8)-C(9)-C(10) +97.7(5)° (Fig. 1).

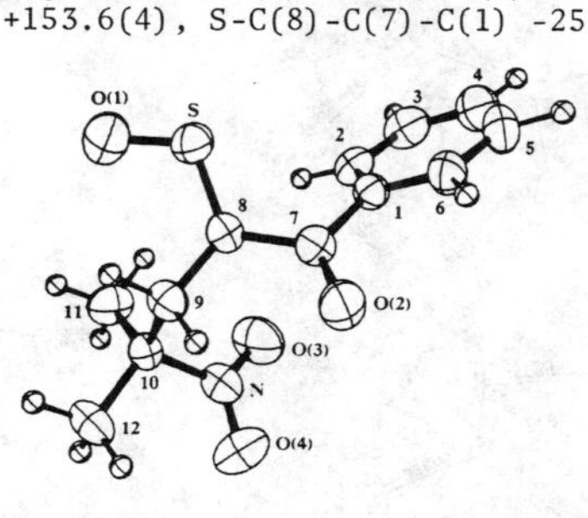

Fig. 1. The structure of $C_{12}H_{13}NO_4S$.

DI-p-TOLYL DISELENIDE
$C_{14}H_{14}Se_2$

G. VAN DEN BOSSCHE, M.R. SPIRLET, O. DIDEBERG and L. DUPONT, 1984. Acta Cryst., C40, 1979-1980.

Monoclinic, $P2_1/a$, a = 15.344, b = 6.189, c = 13.951 Å, β = 91.26°, D_m = 1.69, Z = 4. Mo radiation, R = 0.093 for 1710 reflexions.

The Se-Se distance is 2.328(1) Å with Se-C 1.91 and 1.92(1) Å. The dihedral angle between the planes C-Se-Se and Se-Se-C is 80.3(4)°. Se(1) and Se(2) (Fig. 1) are displaced from their respective phenyl planes by 0.050(1) and 0.068(1) Å.

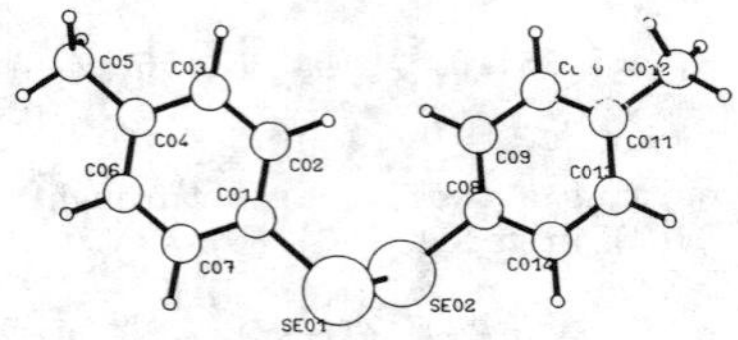

Fig. 1. The di-p-tolyl diselenide molecule.

4,4'-DIMETHOXY(THIOBENZOPHENONE)
$C_{15}H_{14}O_2S$ (I)

4,4'-BIS(DIMETHYLAMINO)THIOBENZOPHENONE
$C_{17}H_{20}N_2S$ (II)

P. ARJUNAN, V. RAMAMURTHY and K. VENKATESAN, 1984. Acta Cryst., C40, 552-555.

I. Triclinic, $P\bar{1}$, a = 9.810, b = 9.635, c = 15.015 Å, α = 79.11, β = 102.38, γ = 107.76°, Dm = 1.318, Z = 4. Cu radiation, R = 0.074 for 2663 reflexions.

II. Monoclinic, $P2_1/c$, a = 17.029, b = 6.706, c = 14.629 Å, β = 113.55°, Dm = 1.230, Z = 4. Mo radiation, R = 0.062 for 855 reflexions.

Fig. 1 shows dimensions for the two independent molecules of I and for the molecule of II. The C=S bond lengths are much shorter than normal. The orientation of the C=S chromophores in the crystal lattice and their reactivity are discussed.

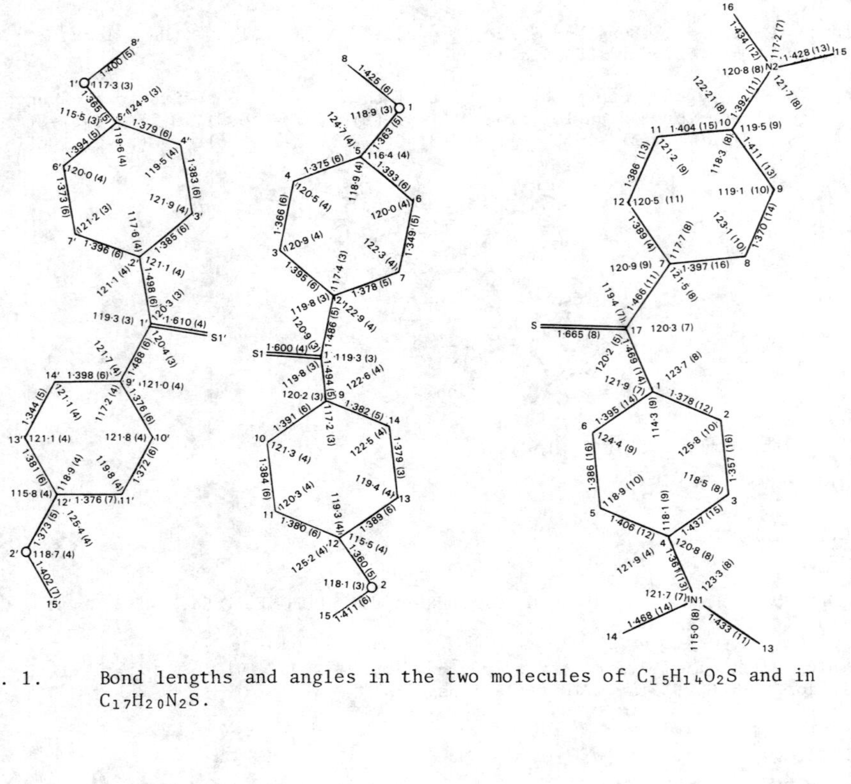

Fig. 1. Bond lengths and angles in the two molecules of $C_{15}H_{14}O_2S$ and in $C_{17}H_{20}N_2S$.

BIS(o-CHLOROPHENYLAMINOCARBONYLMETHYL) DISULPHIDE
$C_{16}H_{14}Cl_2N_2O_2S_2$

M. PIERROT, A. BALDY, J.C. MAIRE, R.C. MEHROTRA, T.S. KAPOOR and B.P. BACHLAS, 1984. Acta Cryst., C40, 1931-1934.

Monoclinic, $P2_1/c$, a = 9.394, b = 17.84, c = 10.576 Å, β = 91.9°, D_m = 1.5, Z = 4. Mo radiation, R = 0.031 for 1403 reflexions.

The molecules adopt a folded conformation (Fig. 1) with non-parallel phenyl rings (dihedral angle 151(1)°); axes through C(6)-C(3)-N(1) and C(14)-C(11)-N(2) are inclined at 35(1)°. The molecules are arranged in columns along the screw axes held together by N-H...O hydrogen bonds (N...O 2.85(2) and 2.88(2) Å). The S-S distance is 2.022(2) Å, with the S-S-C angles 103.0 and 102.8(2)°.

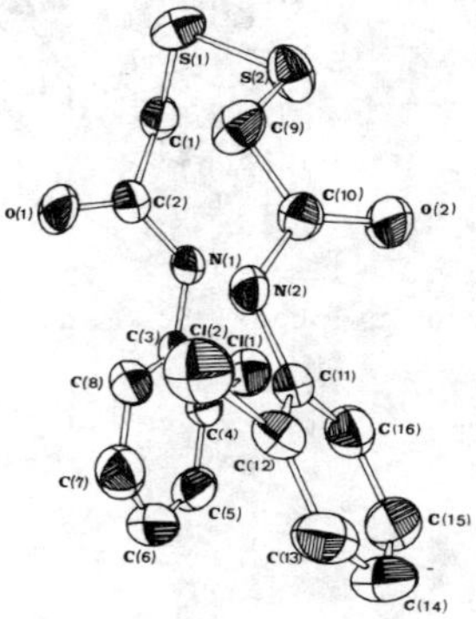

Fig. 1. The molecule of bis(o-chlorophenylaminocarbonylmethyl) disulphide.

1,8-DI-2-PYRIDYL-3,6-DITHIAOCTANE
$C_{16}H_{20}N_2S_2$

A. CASTINEIRAS, M.V. PAREDES and W. HILLER, 1984. Acta Cryst., C40, 2078-2079.

Monoclinic, $P2_1/c$, a = 15.189, b = 4.825, c = 11.833 Å, β = 107.64°, Z = 2. Mo radiation, R = 0.057 for 1032 reflexions.

The molecule (Fig. 1) lies about an inversion centre. S-C distances are both 1.809(5) Å and C-S-C is 100.0(2)°.

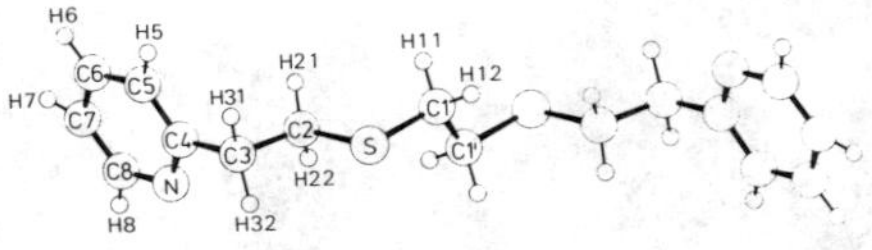

Fig. 1. A view of $C_{16}H_{20}N_2S_2$.

N-PHENYLSULPHONYL-Se,Se-DIPHENYLSELENIMIDE
$C_{18}H_{15}NO_2SSe$

N.G. FURMANOVA, R.G. GERR, V.P. KRASNOV and V.I. NADDAKA, 1984. J. Struct. Chem., 25, 821-825 [Zh. Strukt. Khim., 25(5), 166-169].

Monoclinic, $P2_1/c$, a = 10.570, b = 33.975, c = 10.229 Å, β = 112.53°, Z = 8. Cu radiation, R = 0.091 for 4680 reflexions (twinned crystal).

The structure and principal dimensions (averaged over the two crystallographically independent molecules) are shown in Fig. 1. The molecules are linked into chains in the solid state by weak Se...O interactions (Se-O 3.03 and 3.25 Å).

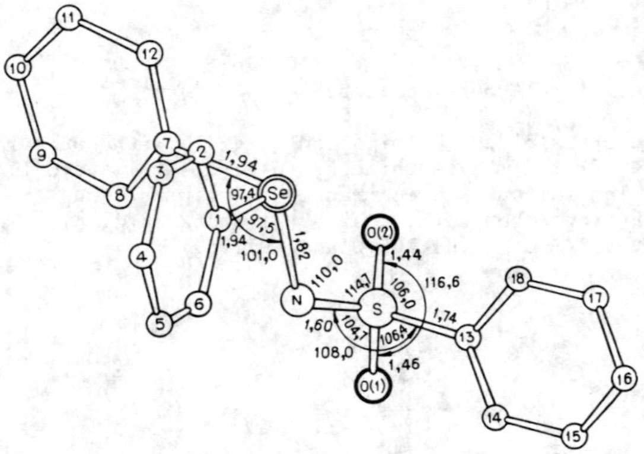

Fig. 1. The structure of N-phenylsulphonyl-Se,Se-diphenylselenimide.

4-BIPHENYL PHENYL THIOKETONE
$C_{19}H_{14}S$ (I)

1-NAPHTHYL PHENYL THIOKETONE
$C_{17}H_{12}S$ (II)

P. ARJUNAN, V. RAMAMURTHY and K. VENKATESAN, 1984. Acta Cryst., C40, 556-558.

I. Orthorhombic, Pbca, a = 7.443, b = 32.691, c = 11.828 Å, D_m = 1.216, Z = 8. Cu radiation, R = 0.068 for 1378 reflexions.

II. Orthorhombic, $P2_12_12_1$, a = 5.873, b = 13.677, c = 15.668 Å, D_m = 1.297, Z = 4. Cu radiation, R = 0.069 for 751 reflexions.

Both compounds undergo photo-oxidation in the solid state. The dihedral angle between the phenyl rings of the biphenyl moiety is 46(1)°. Dimensions are in Fig. 1. The correlation between molecular packing and reactivity is discussed.

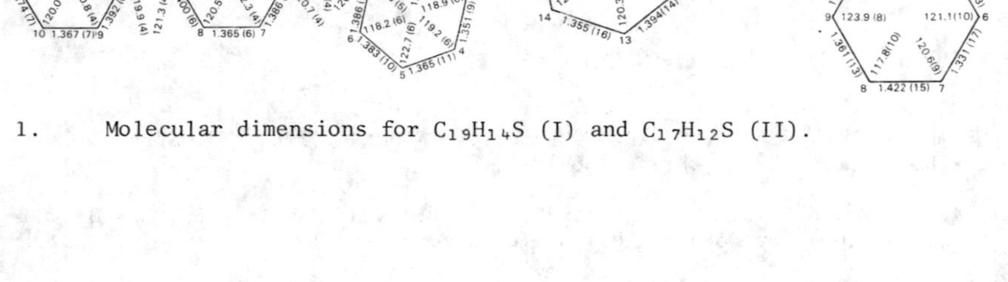

Fig. 1. Molecular dimensions for $C_{19}H_{14}S$ (I) and $C_{17}H_{12}S$ (II).

7,7'-DIPHENOXY-8,8'-DIQUINOLYLDISULPHIDE
$C_{20}H_{20}N_2O_2S_2$

O.G. MATYUKHINA, A.N. SOBOLEV, J.K. OZOLS, J.V. ASHAKS and Yu.A. BANKOVSKII, 1984. J. Struct. Chem., 25, 499-502 [Zh. Strukt. Khim., 25(3), 175-178].

Monoclinic, B2/b, a = 9.252, b = 16.452, c = 16.343, γ = 85.63°, Z = 4. Mo radiation, R = 0.0285 for 1088 reflexions.

The molecule (Fig. 1) has crystallographic twofold symmetry. The S-S distance is 2.074(1) Å, S-C 1.769(3) Å, and the C-S-S angle is 101.5(1)°, and the C-S-S-C torsion angle is 63.2°. The dihedral angle between normals to the quinoline ring mean planes is 13.7°. The aromatic rings are planar to within 0.04 Å.

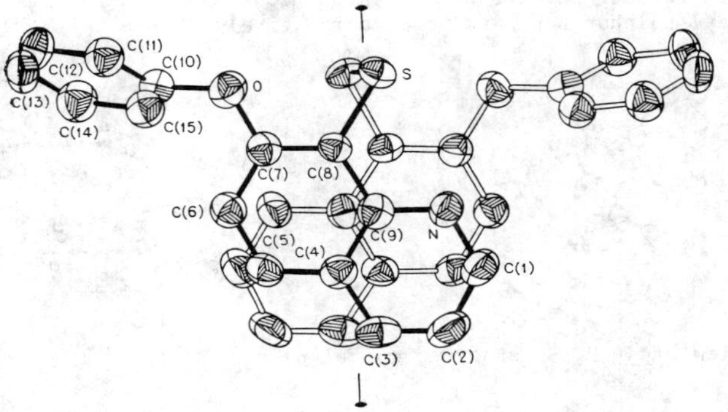

Fig. 1. The 7,7'-diphenoxy-8,8'-diquinolyldisulphide molecule.

S,S-[SULPHONYLBIS(1,4-PHENYLENE)]DI(THIOBENZOATE)
$C_{26}H_{18}O_4S_3$

J. GARBARCZYK, 1984. Makromol. Chem., 185, 2033-2041.

Monoclinic, $P2_1/a$, a = 14.050, b = 10.404, c = 16.734 Å, β = 110.57°, Z = 4. Cu radiation, R = 0.049 for 2466 reflexions.

Molecules of the title compound are asymmetric with the thiobenzoate substituents syn and anti with relation to the central sulphonyl group (Fig. 1). The thiocarbonyl groups do not lie in the planes of the phenylene rings. The torsional angles C(6)-C(7)-S(1)-C(8) and C(7)-S(1)-C(8)-C(9) are -171.55 and 102.40° respectively. Bond distances of interest are: S(1)-C(7) 1.785, S(1)-C(8) 1.771, S(2)-O(3) 1.444, S(2)-O(2) 1.441, S(2)-C(14) 1.761, S(2)-C(11) 1.778, S(3)-C(17) 1.769, S(3)-C(20) 1.780, C(7)-O(1) 1.199, and C(20)-O(4) 1.203 Å.

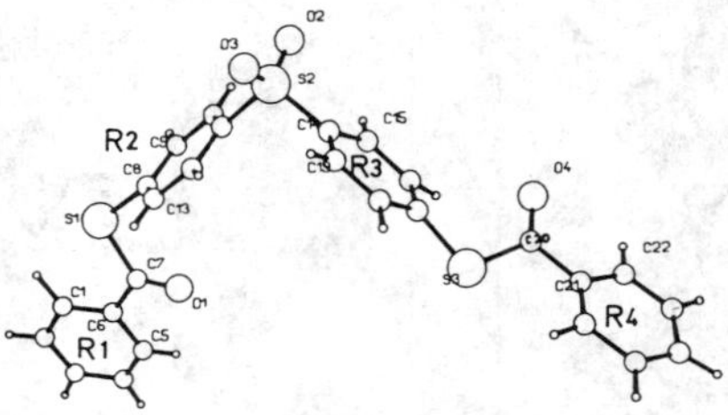

Fig. 1. Projection of $C_{26}H_{18}O_4S_3$ molecule in the [010] direction.

1,2,4,5-TETRAKIS(BENZYLTHIO)BENZENE

$C_{34}H_{30}S_4$

S.D. COX, C.W. DIRK, F. MORAES, D.E. WELLMAN, F. WUDL, M. SOLTIS and C. STROUSE, 1984. J. Am. Chem. Soc., 106, 7131-7133.

Triclinic, P$\bar{1}$, a = 5.358, b = 10.80, c = 14.77 Å, α = 106.02, β = 89.84, γ = 117.22°, Z = 1. Mo radiation, R = 0.047 for 1400 reflexions.

The molecule (Fig. 1) lies about an inversion centre; mean S-C(sp^2) and S-C(sp^3) distances are 1.771(4) and 1.830(5) Å respectively with C1-S1-C10 103.1 and C2-S2-C11 98.6(2)°. The material is photochromic and explanations are put forward to explain this involving intermolecular charge transfer between tetrathiobenzenes, intra-molecular charge transfer involving tetrathiobenzene and phenyl rings and homolytic cleavage of benzyl-sulphur bonds to produce relatively stable thiyl radicals.

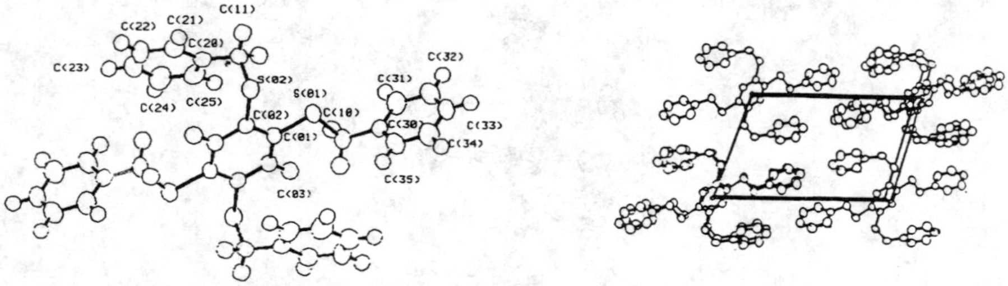

Fig. 1. A view of $C_{34}H_{30}S_4$ and the crystal packing.

POTASSIUM 4,5-DICHLORO-2-CYANO-3,6-DIOXO-1,4-CYCLOHEXADIENOLATE

$C_7Cl_2KNO_3$ $K^+.C_7Cl_2NO_3^-$

M. KONNO, 1984. Acta Cryst., C40, 236-238.

Tetragonal, $I4_1/a$, a = 20.611, c = 8.3990 Å, Z = 16. Mo radiation, R = 0.030 for 2313 reflexions.

The anions form centrosymmetric isolated dimers with a parallel interplanar distance of 3.206 Å (Fig. 1). The K^+ ion is coordinated by five oxygen atoms of three anions and one N atom of a cyano group (K...O 2.798(2) to 2.889(2), K...N 2.826(2) Å).

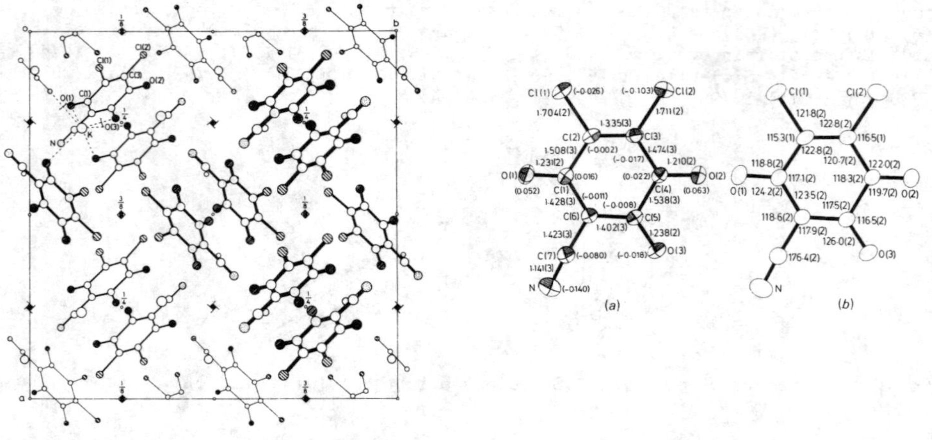

Fig. 1. A projection of the $K^+.C_7Cl_2NO_3^-$ structure and details of anion geometry.

CESIUM PENTACYANOPROPENIDE
C_8CsN_5

K.W. HIPPS, U. GEISER, U. MAZUR and R.D. WILLETT, 1984. J. Phys. Chem., 88, 2498-2504.

Monoclinic, $P2_1/c$, a = 8.285, b = 13.050, c = 9.000 Å, β = 95.44°, D_m = 2.2, Z = 4. Mo radiation, R = 0.047 for 2258 reflexions.

The anion (Fig. 1) is essentially planar with angle sums at C(6), C(7) and C(8) 360.0°. Some bond lengths are: C(6)-C(7) 1.395(7), C(7)-C(8) 1.398(7), C-C(N) 1.411-1.454(7), C-N 1.128-1.152(7) Å. The Cs^+ cation has nine nearest neighbours (at an average distance of 3.39 Å). The structure consists of sheets parallel to the b-axis.

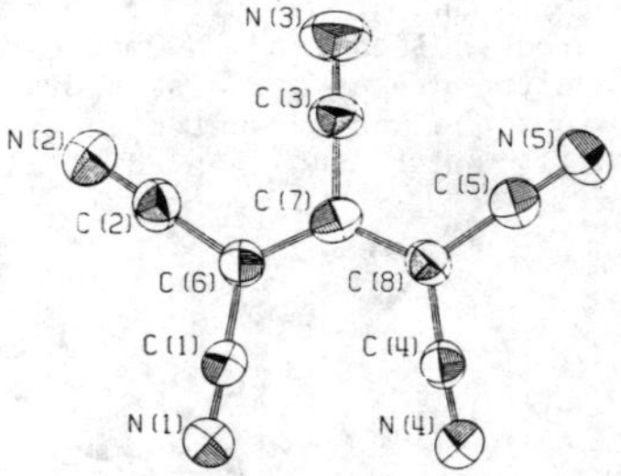

Fig. 1. The pentacyanopropenide anion.

DIETHYLMETHYLAMMONIUM TRIFLUOROACETYLMETHYLIDE MONOHYDRATE
$C_8H_{14}F_3NO,H_2O$

K. PETERS, E.-M. PETERS, H.G. VON SCHNERING, E. ZIEGLER and H. WITTMANN, 1984. Z. Kristallogr., 169, 305-308.

Monoclinic, $P2_1/n$, a = 15.770, b = 12.275, c = 12.224 Å, β = 109.39°, Z = 8. Mo radiation, R = 0.067 for three-dimensional data.

The two independent ylide molecules are linked by water molecules as shown in Fig. 1. The mean C2-C3 distance is 1.330(4) Å, with C3-O3 1.266(4), C2-N1 1.477(4), C3-C4 1.511(4) and C-F 1.311-1.315(4) Å.

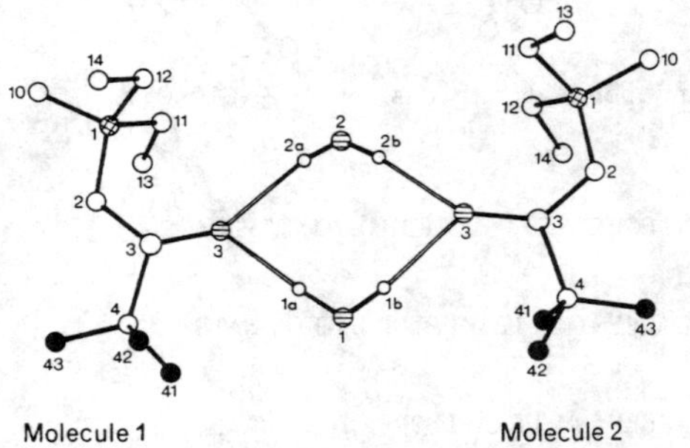

Fig. 1. A drawing of the two independent $C_8H_{14}F_3NO$ molecules linked by water molecules.

1,5-BIS(DIMETHYLAMINO)-2,4-DINITROPENTAMETHINIUM PERCHLORATE
$C_9H_{15}ClN_4O_8$ $[Me_2NCHC(NO_2)CHC(NO_2)CHNMe_2]^+ClO_4^-$

J. DALE, S. KRÜGER and C. RØMMING, 1984. Acta Chem. Scand., B38, 117-124.

Monoclinic, Cc, a = 11.812, b = 16.818, c = 15.805 Å, β = 113.04°, Z = 8. Mo radiation, R = 0.032 for 4570 reflexions (at -150°C).

The two crystallographically-independent cations (Fig. 1) have the same unsymmetric non-planar conformation, best described as a roughly planar delocalized 1,3-dinitro-syn, anti-allylic anion chromophore, substituted by two dimethyliminium cationic groups rotated out of the plane of the chromophore. The main carbon skeleton is nearly planar and the average bond lengths are 1.381 and 1.401 Å in molecules 1 and 2 respectively. The two outer C-C bonds are much longer (average 1.452 and 1.440 Å) and the C-N bonds to the dimethylamino groups are so short (average lengths 1.291 and 1.302 Å) that both are considered as localized CN double bonds carrying full positive charge. The C-N bonds to the nitro groups are also short (average lengths 1.448 and 1.429 Å). The perchlorate anions are uncoordinated in the crystal, with mean Cl-O distances of 1.460 Å around Cl(1) and 1.464 Å around Cl(2). The O-Cl-O angles range from 107.8 to 111.0°.

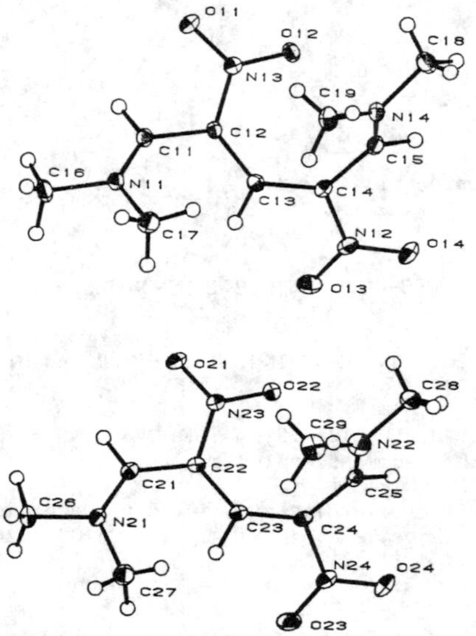

Fig. 1. The two independent cations in $C_9H_{15}ClN_4O_8$.

BIS(METHYLSELENO)BENZYLCARBENIUM HEXACHLOROANTIMONATE
$C_{10}H_{13}Cl_6SbSe_2$ (I)

BIS(METHYLTHIO)BENZYLCARBENIUM HEXACHLOROANTIMONATE
$C_{10}H_{13}Cl_6SbS_2$ (II)

BIS(METHYLTHIO)ETHYLCARBENIUM HEXAFLUOROANTIMONATE
$C_5H_{11}F_6S_2Sb$ (III)

L. HEVESI, S. DESAUVAGE, B. GEORGES, G. EVRARD, P. BLANPAIN, A. MICHEL, S. HARKEMA and G.J. VAN HUMMEL, 1984. J. Am. Chem. Soc., 106, 3784-3790.

I. Orthorhombic, Pbca, a = 16.372, b = 17.918, c = 13.093 Å, Z = 8. Mo radiation, R = 0.06 for 2813 reflexions (at 173 K).

II. Orthorhombic, Pbca, a = 16.276, b = 17.581, c = 13.012 Å, Z = 8. Mo radiation, R = 0.05 for 2060 reflexions (at 173 K).

III. Rhombohedral, R3c, a = 8.520 Å, α = 78.42°, Z = 2. Mo radiation, R = 0.06 for 303 reflexions (at 163 K).

I and II are isomorphous and both cations (Fig. 1) show trigonal planar geometry around the positively charged central carbon atom with mean C(1)-S 1.68(1), C(1)-Se 1.84(1), C(1)-C(4) 1.50 Å. Complex III has the cation disordered about a threefold axis and the "C(1)-S" distance is 1.67(1) Å.

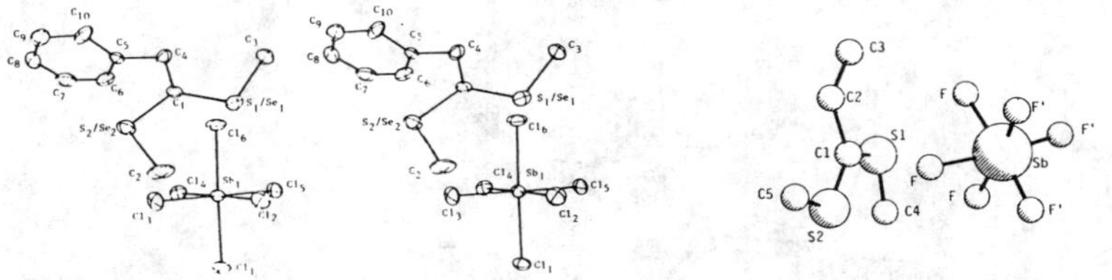

Fig. 1. A stereoview of $C_{10}H_{13}Cl_6SbSe_2$ and a view of the $C_5H_{11}F_6S_2Sb$ structure.

4-(4-BROMOPHENYL)-4-PHENYL-1-OXA-3-AZABUTATRIENIUM HEXACHLOROANTIMONATE
$C_{14}H_{19}BrCl_6NOSb$ (I) $C_{14}H_9BrNO^+.SbCl_6^-$

4,4-DIPHENYL-1-THIA-3-AZABUTATRIENIUM HEXACHLOROANTIMONATE
$C_{14}H_{10}Cl_6NSSb$ (II)

M. AL-TALIB, I. JIBRIL, J.C. JOCHIMS and G. HUTTNER, 1984. Chem. Ber., 117, 3211-3221.

I. Triclinic, P$\bar{1}$, a = 9.580, b = 10.391, c = 10.606 Å, α = 91.62, β = 97.48, γ = 72.65°, Z = 2. Mo radiation, R = 0.037 for 1905 reflexions.

II. Monoclinic, $P2_1/c$, a = 8.116, b = 12.686, c = 21.428 Å, β = 116.26°, Z = 4. Mo radiation, R = 0.056 for 1507 reflexions.

The asymmetric unit of I contains one complete cation (Fig. 1) and two independent halves of $SbCl_6^-$ ions, the Sb atoms lying on inversion centres. In both (I) and (II), the C-N-C moieties are bent (129° (I), 139° (II)).

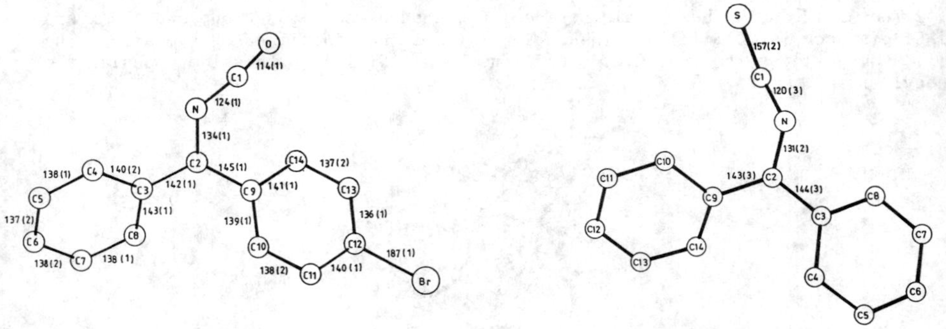

Fig. 1. Molecular diagrams and bond lengths (pm) for the butatrienium species (I) (left) and (II) (right).

1,1,3-TRI-t-BUTYL-3-ISOPROPYLAMINO-2-AZA-ALLENIUM HEXACHLOROANTIMONATE

$C_{17}H_{35}Cl_6N_2Sb$ $C_{17}H_{35}N_2^+.SbCl_6^-$

J.C. JOCHIMS, R. ABU-EL-HALAWA, I. JIBRIL and G. HUTTNER, 1984. Chem. Ber., 117, 1900-1912.

Triclinic, P$\bar{1}$, a = 10.319, b = 10.384, c = 13.880 Å, α = 103.85, β = 110.37, γ = 95.72°, Z = 2. Mo radiation, R = 0.067 for 2587 reflexions (at 228 K).

Cation dimensions are in Fig. 1. The C1-N1-C9 angle is large, 156(1)°, and the C1-N and C9-N distances are consistent with considerable double bond character.

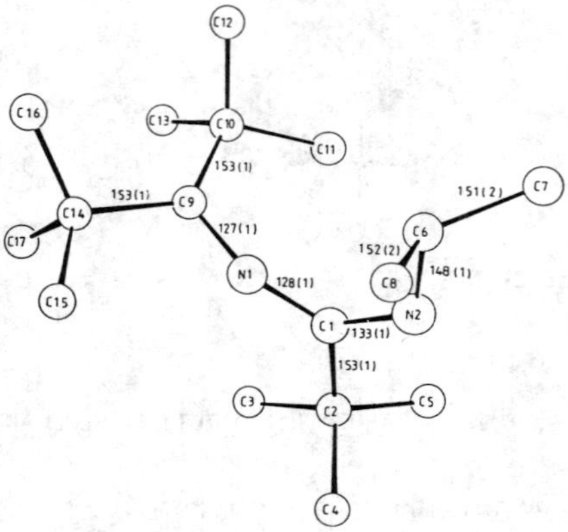

Fig. 1. Bond lengths in $C_{17}H_{35}N_2^+$.

1-PHENYL-1,2,3,4,5,6-HEXAMETHYLBENZENIUM TETRACHLOROALUMINATE

$C_{18}H_{23}AlCl_4$ $(C_{18}H_{23})^+(AlCl_4)^-$

G.I. BORODKIN, Sh.M. NAGI, I.Yu. BAGRYANSKAYA and Yu.V. GATILOV, 1984. J. Struct. Chem., 25, 440-445 [Zh. Strukt. Khim., 25(3), 114-119].

Orthorhombic, Pbca, a = 11.116, b = 16.923, c = 21.848 Å, Z = 8. Mo radiation, R = 0.088 for 2254 reflexions (at -110°C).

The $AlCl_4^-$ anion has an almost ideal tetrahedral configuration, and the average Al-Cl distance is 2.134(3) Å. The cation (Fig. 1) has idealized C_s symmetry. The C(1) atom deviates from the C(2)-C(6) plane by 0.105(7) Å to the side opposite the phenyl group.

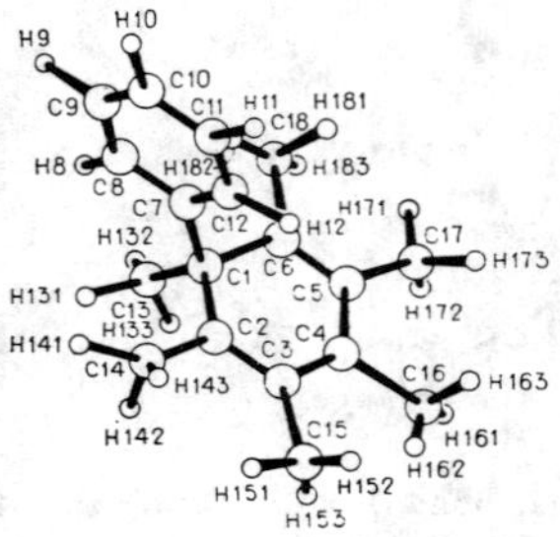

Fig. 1. The 1-phenyl-1,2,3,4,5,6-hexamethylbenzenium cation.

TETRAMETHYLAMMONIUM PENTAKIS(METHOXYCARBONYL)CYCLOPENTADIENOATE

$C_{19}H_{27}NO_{10}$ $[NMe_4][CH_5(CO_2Me)_5]$

M.I. BRUCE, P.A. HUMPHREY, B.W. SKELTON and A.H. WHITE, 1984. Aust. J. Chem., 37, 2441-2446.

Monoclinic, $P2_1/c$, a = 9.267, b = 16.249, c = 14.687 Å, β = 100.90°, D_m = 1.31, Z = 4. Mo radiation, R = 0.066 for 1610 reflexions.

The anion $[C_5(CO_2Me)_5]^-$ consists of a planar C_5 ring penta-substituted with CO_2Me which have random orientation with respect to the C_5 plane (Fig. 1).

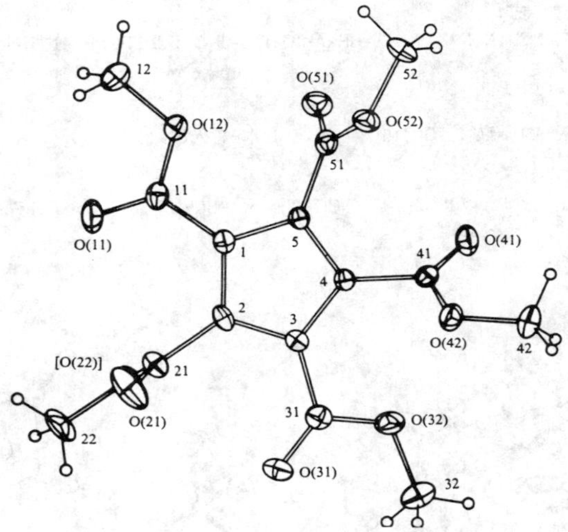

Fig. 1. The structure of $[NMe_4][CH_5(CO_2Me)_5]$.

BIS(DIMETHYLAMMONIUM) (3,4-BIS(DICYANOMETHYLENE)-1-CYCLOBUTEN-1,2-DIYL)BIS(DICYANO-METHANIDE)

$C_{20}H_{16}N_{10}$ $2(Me_2NH_2^+).C_{16}N_8^-$

B. GERECHT, T. KAMPCHEN, K. KOHLER, W. MASSA, G. OFFERMANN, R.E. SCHMIDT, G. SEITZ and R. SUTRISNO, 1984. Chem. Ber., 117, 2714-2729.

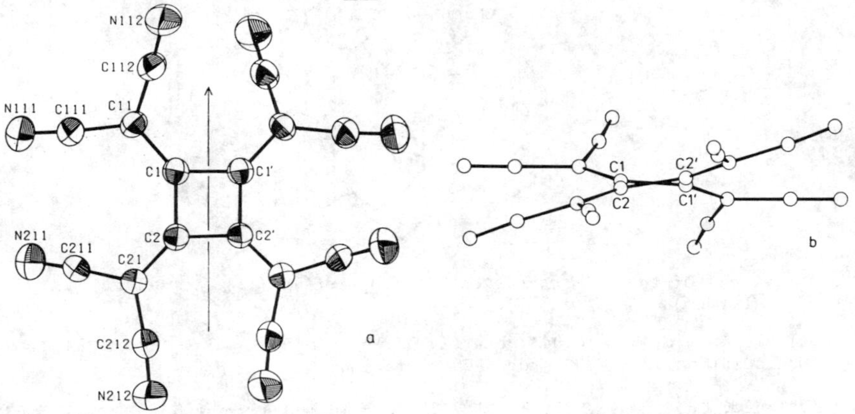

Fig. 1. Views of the $C_{16}N_8^-$ anion.

Monoclinic, C2/c, a = 15.704, b = 11.918, c = 12.257 Å, β = 113.02°, Z = 4. Mo radiation, R = 0.048 for 1322 reflexions.

The anion (Fig. 1) has twofold crystallographic symmetry and the four-membered ring is puckered (torsion angle C2-C1-C1'-C2' -12.5°). The C1-C2, C1-C2' and C2-C2' bond lengths are 1.451(4), 1.449(5) and 1.463(5) Å respectively.

N,N'-DIETHYL-N,N,N',N'-TETRAMETHYL-1,3-AZULENEDIMETHANAMINIUM DIIODIDE - TRIIODOMETHANE (1:2)
$C_{20}H_{32}I_2N_2,2(CHI_3)$ (I)

5-(1,1-DIMETHYLETHYL)-N,N,N,N',N',N'-HEXAMETHYL-1,3-BENZENEDIMETHANAMINIUM DIIODIDE - TRIIODOMETHANE (1:2)
$C_{18}H_{34}I_2N_2,2(CHI_3)$ (II)

N,N,N-TRIMETHYLBENZENEAMINIUM IODIDE - TRIIODOMETHANE (1:1)
$C_9H_{14}IN,CHI_3$ (III)

H.-G. LOHR, A. ENGEL, H.-P. JOSEL, F. VOGTLE, W. SCHUH and H. PUFF, 1984. J. Org. Chem., 49, 1621-1627.

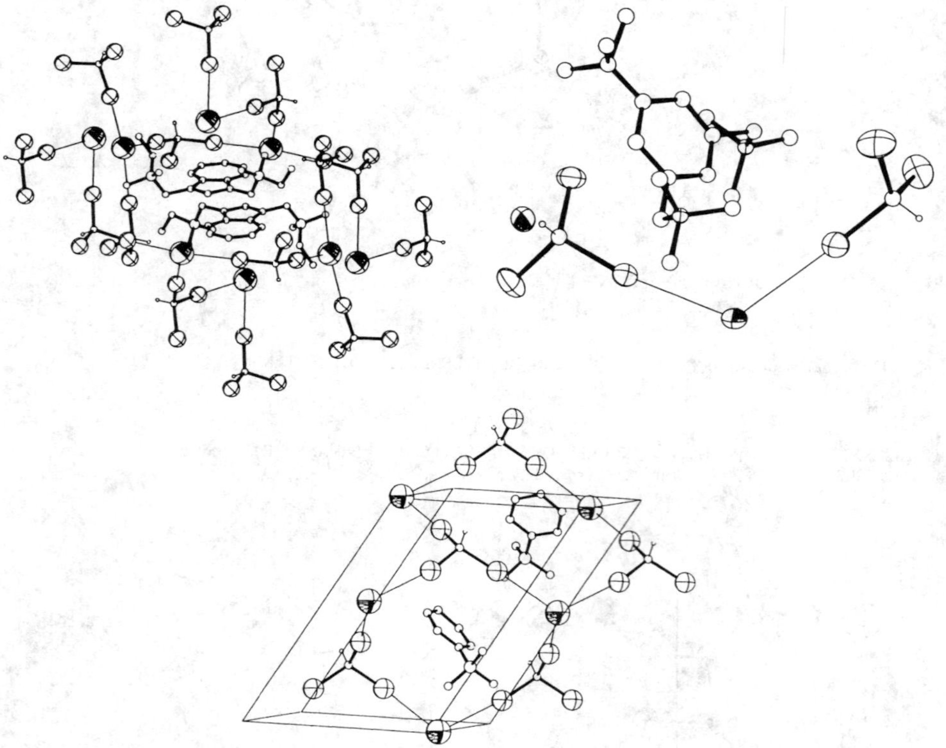

Fig. 1. Views of the CHI_3 complexes (I) (top left), (II) (top right) and (III) (bottom).

I. Monoclinic, $P2_1/c$, a = 16.043, b = 12.137, c = 18.360 Å, β = 100.06°, D_m = 2.53, Z = 4. Mo radiation, R = 0.063 for 3970 reflexions.

II. Monoclinic, Cc, a = 13.535, b = 16.425, c = 16.100 Å, β = 93.13°, D_m = 3.574, Z = 4. Mo radiation, R = 0.059 for 3053 reflexions.

III. Monoclinic, Pc, a = 9.908, b = 7.395, c = 13.808 Å, β = 123.38°, D_m = 2.48, Z = 2. Mo radiation, R = 0.057 for 1576 reflexions.

In all three structures (Fig. 1) the triodomethane molecules are connected by iodide ions. The presence of electron donor-acceptor bonds is indicated by the I...I distances (3.47-3.69 Å) being about 0.75 Å less than the sum of the van der Waals radii and C-I distances being slightly longer than in pure CHI_3. The I^-...I-C arrangements are always nearly linear. The I-C-I angles average 111.5°.

1,1-(2,2'-BIPHENYLENE)-3,3-DIPHENYL-2-AZA-ALLENIUM HEXACHLOROANTIMONATE

$C_{26}H_{18}Cl_6NSb$ $[C_{26}H_{18}N]^+ . [SbCl_6]^-$

M. AL-TALIB, I. JIBRIL, E.-U. WURTHWEIN, J.C. JOCHIMS and G. HUTTNER, 1984. Chem. Ber., 117, 3365-3373.

Monoclinic, $P2_1/c$, a = 11.61, b = 10.605, c = 25.79 Å, β = 121.93°, Z = 4. Mo radiation, R = 0.047 for 2618 reflexions.

The structure contains discrete cations (Fig. 1) and anions. The interplanar angle C15-C21-N/C3-C4-N is 85°. The compound is regarded as an allene and not as an azaallylium system.

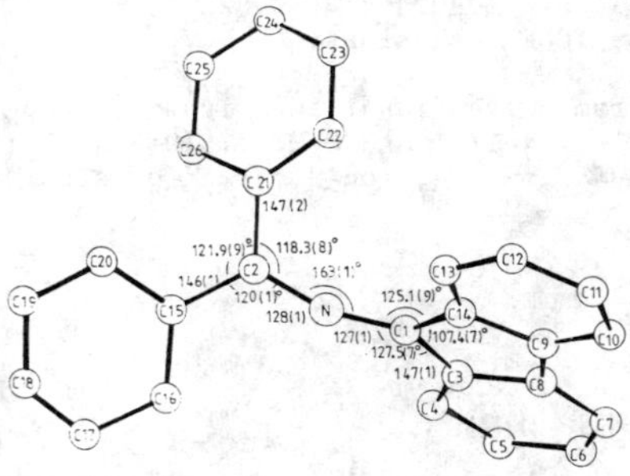

Fig. 1. A view of the $C_{26}H_{18}N^+$ cation with bond lengths (pm) and angles (°).

p-IODOBENZOIC ACID

$C_7H_5IO_2$

R.G. BAUGHMAN and J.E. NELSON, 1984. Acta Cryst., C40, 204-206.

Monoclinic, $P2_1/a$, a = 30.357, b = 6.0300, c = 4.1561 Å, β = 97.35°, Z = 4. Mo radiation, R = 0.072 for 1002 reflexions.

The molecules (Fig. 1) pack as centrosymmetric hydrogen bonded dimers, with O...O 2.61(1) Å.

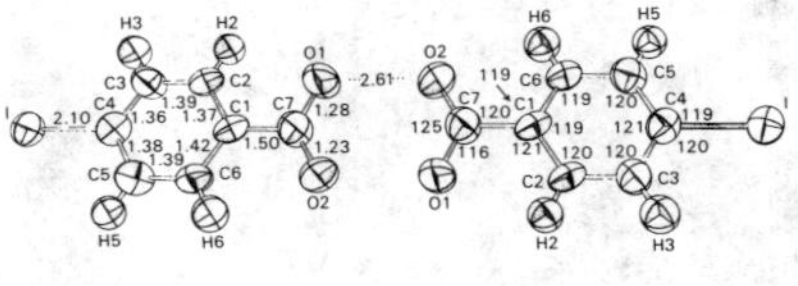

Fig. 1. A pair of inversion related p-iodobenzoic acid molecules, and dimensions (σ 0.02 Å. 1°).

p-METHYLSULPHONYLBENZOIC ACID

$C_8H_8O_4S$ CH_3-SO_2-C_6H_4-CO_2H

J. BRUNVOLL, M. COLAPIETRO, A. DOMENICANO, C. MARCIANTE, G. PORTALONE and I. HARGITTAI, 1984. Z. Naturforsch., B39, 607-609.

Triclinic, P$\bar{1}$, a = 5.614, b = 7.303, c = 11.056 Å, α = 80.26, β = 85.50, γ = 74.05°, D_m = 1.545, Z = 2. Mo radiation, R = 0.0334 for 2816 reflexions, (R = 0.0329 for 1751 high-θ reflexions).

The results of a high-θ refinement of the structure are found to agree with those of an electron-diffraction study of methylphenylsulphone. The acid crystallizes as centrosymmetric hydrogen-bonded dimers (O...O 2.629 Å). The $-SO_2Me$ conformation is the same in both studies with the C-S-C plane nearly perpendicular to the plane of the phenyl ring.

3,5-DIMETHYLBENZOIC ACID

$C_9H_{10}O_2$ $Me_2C_6H_3CO_2H$

M. COLAPIETRO, A. DOMENICANO, C. MARCIANTE and G. PORTALONE, 1984. Z. Naturforsch., B39, 1361-1367.

Monoclinic, $P2_1/n$, a = 8.165, b = 6.992, c = 14.422 Å, β = 93.21°, D_m = 1.211, Z = 4. Mo radiation, R = 0.0415 for 1662 reflexions.

Molecular dimensions from conventional and high-θ refinement are in Fig. 1. The mean aromatic C-C distance is 1.401(1) and C-Me 1.510(3) Å. The data are in excellent agreement with those obtained from electron-diffraction studies of related systems.

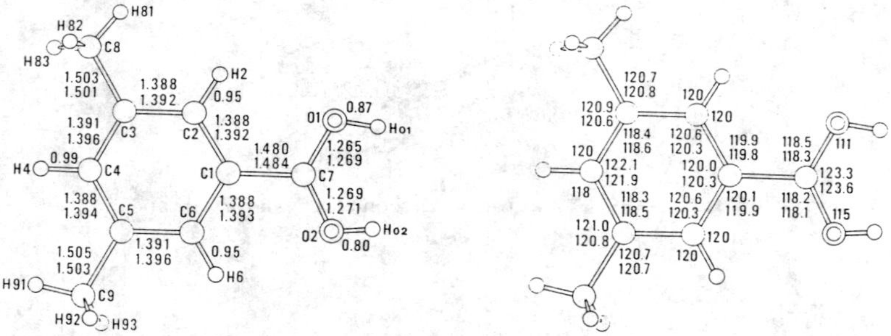

Fig. 1. Bond lengths and angles in 3,5-dimethylbenzoic acid (lower values are from the high-θ refinement). The carboxyl H atom is disordered.

METHYL 2-(METHYLTHIO)BENZOATE

$C_9H_{10}O_2S$ (I)

DIMETHYL 2,2'-DITHIODIBENZOATE

$C_{16}H_{14}O_4S_2$ (II)

DIMETHYL 2,2'-THIODIBENZOATE

$C_{16}H_{14}O_4S$ (III)

METHYL 2-(2-NITROPHENYLTHIO)BENZOATE

$C_{14}H_{11}NO_4S$ (IV)

2,2'-DINITRODIPHENYL SULPHIDE

$C_{12}H_8N_2O_4S$ (V)

METHYL 2-(2-NITROPHENYLTHIO)PHENYLACETATE
$C_{15}H_{13}NO_4S$ (VI)

A. KUCSMAN, I. KAPOVITS, L. PARKANYI, G. ARGAY and A. KALMAN, 1984. J. Mol. Struct., 125, 331-347.

I. Orthorhombic, $P2_12_12_1$, a = 5.445, b = 11.701, c = 14.144 Å, Z = 4. Mo radiation, R = 0.056 for 888 reflexions.

II. Monoclinic, $P2_1/n$, a = 13.564, b = 7.693, c = 15.128 Å, β = 97.15°, Z = 4. Mo radiation, R = 0.057 for 2395 reflexions.

III. Monoclinic, Cc, a = 10.531, b = 14.381, c = 9.650 Å, β = 99.00°, Z = 4. Mo radiation, R = 0.033 for 1232 reflexions.

IV. Monoclinic, $P2_1/n$, a = 12.035, b = 7.837, c = 15.551 Å, β = 112.02°, Z = 4. Mo radiation, R = 0.042 for 1629 reflexions.

V. Triclinic, $P\bar{1}$, a = 7.677, b = 7.840, c = 11.679 Å, α = 102.78, β = 88.95, γ = 118.94°, Z = 2. Mo radiation, R = 0.045 for 2033 reflexions.

VI. Monoclinic, $P2_1/c$, a = 13.227, b = 13.546, c = 7.910 Å, β = 96.11°, Z = 4. Mo radiation, R = 0.038 for 1744 reflexions.

The conformations of I - VI are shown in Fig. 1. Molecule I is planar, molecule II adopts a conformation in which both aryl rings are coplanar with the C(ar)-S-S plane. Compounds III and V assume a skew conformation in which one ring (A) remains coplanar with the C(ar)-S-C(ar) plane while the other ring becomes perpendicular to the C(ar)-S-C(ar) plane to minimise proximal-ortho group interaction. The unsymmetrical diaryl sulphides IV and VI also have a skew conformation; the phenyl ring with the stronger electron-withdrawing group occupying the planar ring position (A). Each of the compounds has a planar five-membered "ring" and a linear X-S...O arrangement. S...O contacts are in the range 2.619-3.077 Å.

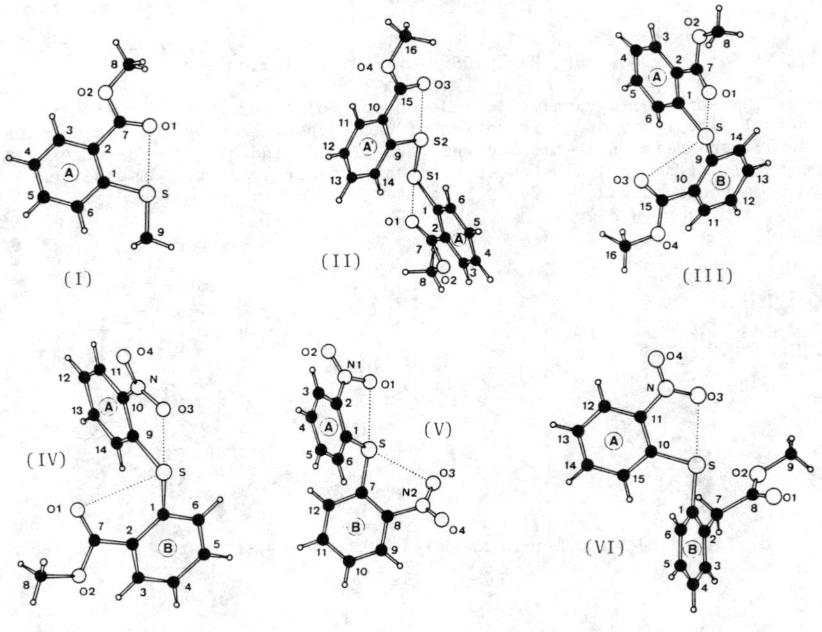

Fig. 1. Solid state conformations for sulphides I - VI.

2-METHOXYMETHOXYBENZOIC ACID
$C_9H_{10}O_4$

P.G. JONES, G.M. SHELDRICK, A.J. KIRBY and A.J. BRIGGS, 1984. Acta Cryst., C40, 545-547.

Triclinic, $P\bar{1}$, a = 4.964, b = 7.918, c = 11.777 Å, α = 99.99, β = 91.61, γ = 106.15°, Z = 2. Mo radiation, R = 0.062 for 1532 reflexions.

The crystal structure contains centrosymmetric hydrogen-bonded dimers. The compound has unusual reactivity towards acetal hydrolysis but the crystal structure does not reveal why. The conformation about the acetal centre (Fig. 1) is close to (+sc,+sc). The dihedral angle between the planes of the aromatic ring and the carboxylic acid group is 3(1)°. The hydrogen bonded O...O distance is 2.67(1) Å.

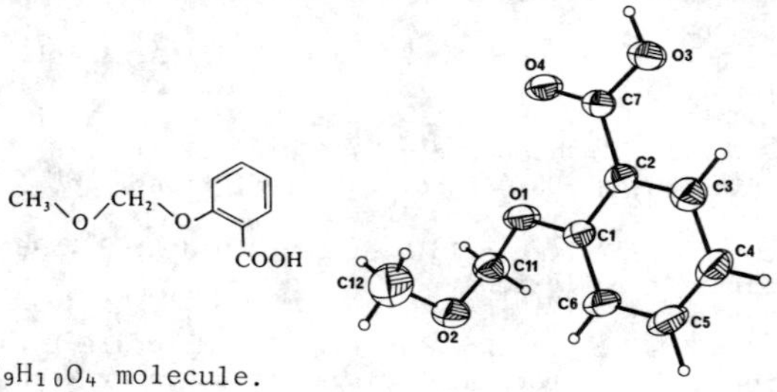

Fig. 1. The $C_9H_{10}O_4$ molecule.

3,5-DICHLORO-2-HYDROXY-4-METHOXY-6-n-PROPYLBENZOIC ACID (DIFFERANISOLE A)
$C_{11}H_{12}Cl_2O_4$

Y. IIMURA, T. SAKURAI, K. ASAHI, N. TAKAHASHI and H. OKA, 1984. Acta Cryst., C40, 2058-2061.

Triclinic, $P\bar{1}$, a = 10.242, b = 15.972, c = 7.797 Å, α = 91.93, β = 95.23, γ = 98.36°, D_m = 1.472, Z = 4. Mo radiation, R = 0.050 for 3501 reflexions.

There are two crystallographically independent molecules (Fig. 1) which are almost identical except for the orientation of the carboxylic groups. The dihedral angles between carboxylic and benzene planes are 11.6(2) and 8.4(1)° for molecules A and B respectively. The crystal structure contains centrosymmetric dimers linked by O-H...O hydrogen bonds (2.632(3), 2.679(3) Å). There is also an intramolecular hydrogen bond, as shown in Fig. 1.

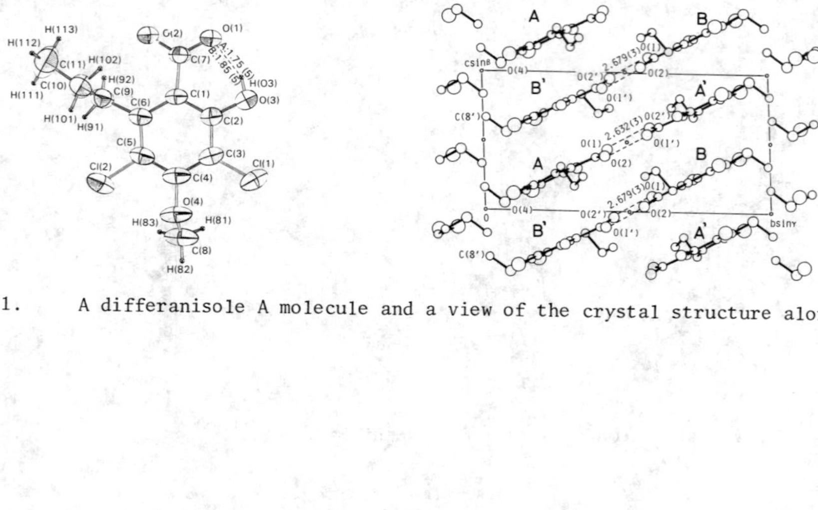

Fig. 1. A differanisole A molecule and a view of the crystal structure along a.

1-BENZOYLOXY-1-METHOXY-N-METHYLACETAMIDE
$C_{11}H_{13}NO_4$

T. NISHIO, N. NAKAJIMA, M. KONDO, Y. OMOTE and M. KAFTORY, 1984. J. Chem. Soc. Perkin I, 391-396.

Monoclinic, $P2_1/n$, a = 12.216, b = 19.604, c = 4.798 Å, β = 99.98°, Z = 4. Mo radiation, R = 0.048 for 1237 reflexions.

The atoms of the benzoyloxy group, C(1) and O(1) are all coplanar and the C(11)-O(1) bond is approximately perpendicular to this system (Fig. 1). The methyl amide moiety is also planar with a carbonyl distance of 1.227(2) and N(1)-C(2) distance of 1.375(3) Å. The benzoyloxy carbonyl distance is 1.205(3) Å. Molecules in the crystal are held together by hydrogen bonds formed between carbonyl O(3) of an amide and N(1) of an amide related by a translation along c (N(1)...O(3) 2.915 Å).

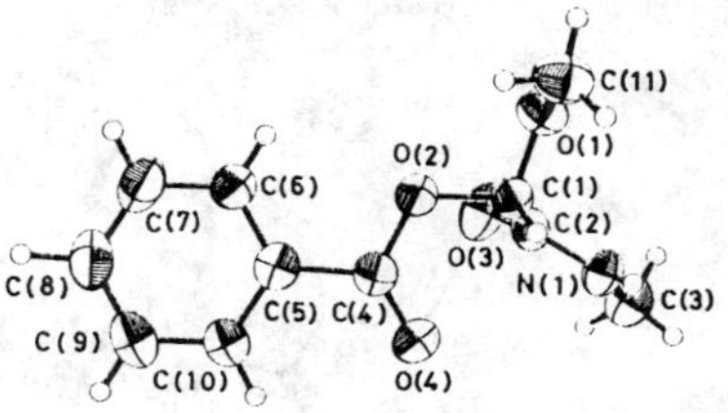

Fig. 1. $C_{11}H_{13}NO_4$: a perspective view of the molecule.

t-BUTYL 3-NITROPERBENZOATE
$C_{11}H_{13}NO_5$ (I)

t-BUTYL 4-NITROPERBENZOATE
$C_{11}H_{13}NO_5$ (II)

L. GOLIC and I. LEBAN, 1984. Acta Cryst., C40, 447-450.

I. Monoclinic, $P2_1/c$, a = 8.462, b = 17.408, c = 8.279 Å, β = 101.56°, D_m = 1.30, Z = 4. Mo radiation, R = 0.114 for 1069 reflexions.

II. Triclinic, $P\bar{1}$, a = 6.892, b = 8.077, c = 11.970 Å, α = 106.94, β = 102.11, γ = 93.37°, D_m = 1.27, Z = 2. Mo radiation, R = 0.096 for 535 reflexions.

Both crystals decayed during data collection. In both molecules, the peroxy-carbonyl groups are essentially planar, with a torsion angle about the O-O bond (Fig. 1) of 129(1)° for (I) and 120(2)° for (II).

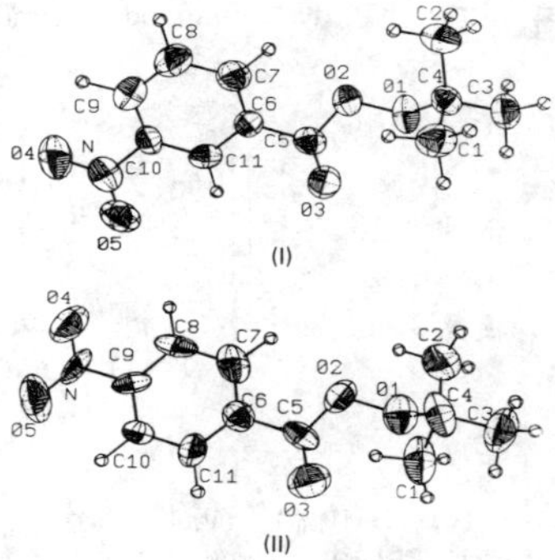

Fig. 1. Views of the two isomeric $C_{11}H_{13}NO_5$ molecules.

TRIMETHYL 4,5,6-TRICHLORO-1,2,3-BENZENETRICARBOXYLATE
$C_{12}H_9Cl_3O_6$

N. GALESIC, I. MATIJASIC and M. BRUVO, 1984. Acta Cryst., C40, 308-311.

Monoclinic, $P2_1/c$, a = 13.649, b = 5.819, c = 19.667 Å, β = 109.09°, Z = 4. Cu radiation, R = 0.059 for 2103 reflexions.

The molecular structure (Fig. 1) is strongly influenced by the three vicinal methoxycarbonyl groups as well by three vicinal Cl atoms attached to the benzene ring. The central ester group is twisted 34.9(1)° out of the best plane through the benzene ring. The other two groups are twisted 68.7(1) and 66.6(1)° out of this plane. The average C-Cl distance is 1.719(4) Å and the intramolecular Cl...Cl contacts are 3.099(2) and 3.098(2) Å.

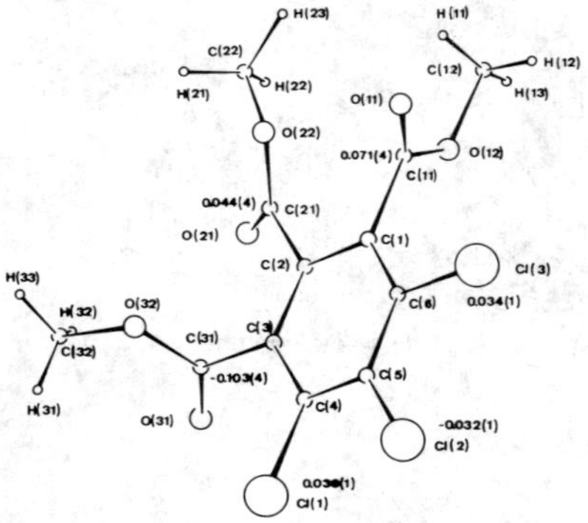

Fig. 1. The $C_{12}H_9Cl_3O_6$ molecule.

4-BIPHENYLCARBOXYLIC ACID
$C_{13}H_{10}O_2$

C.P. BROCK, J.R. BLACKBURN and K.L. HALLER, 1984. Acta Cryst., B40, 493-498.

Monoclinic, $P2_1/c$, a = 7.438, b = 26.907, c = 15.766 Å, β = 102.93°, D_m = 1.276, Z = 12. Mo radiation, R = 0.038 for 2848 reflexions.

The inter-ring dihedral angles for the three independent molecules are 28.7(2), 35.7(2), and 32.2(2)°. Averaged dimensions are in Fig. 1. The crystal packing unit is composed of six molecules, three dimers arranged around an inversion centre (Fig. 1).

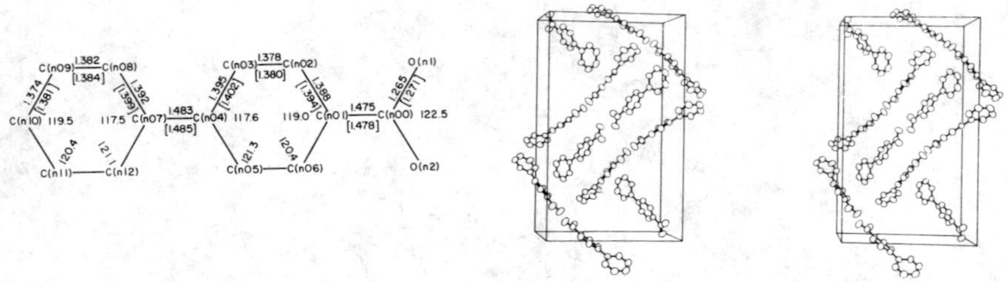

Fig. 1. Averaged dimensions for $C_{13}H_{10}O_2$ (values in square brackets corrected for rigid-body thermal motion), and a stereoview of the crystal structure.

3,3'-DICHLORODIBENZOYL PEROXIDE
$C_{14}H_8Cl_2O_4$

A. SYED, P. UMRIGAR, E.D. STEVENS, G.W. GRIFFIN and R.J. MAJESTE, 1984. Acta Cryst., C40, 1458-1460.

Monoclinic, C2/c, a = 14.655, b = 9.318, c = 10.373 Å, β = 109.67°, Z = 4. Mo radiation, R = 0.037 for 856 reflexions.

A twofold symmetry axis bisects the O-O peroxide bond which has O-O 1.453(2) Å. The C-Cl bond length is 1.743(2) Å. There is an antiparallel arrangement of the C=O and C-Cl bonds yielding a transoid disposition. The molecules form parallel chains through van der Waals attractions (Fig. 1).

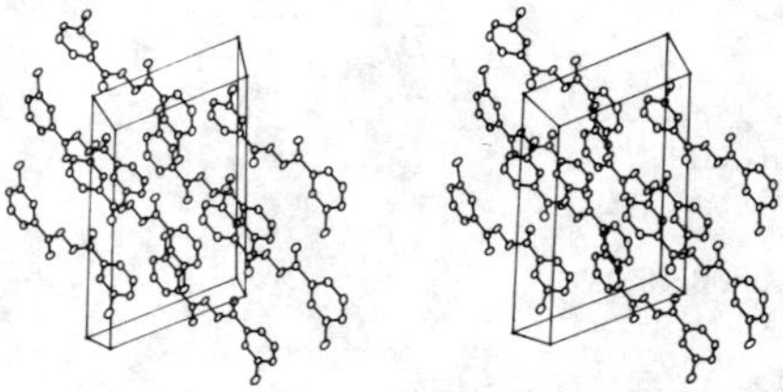

Fig. 1. Stereoview of the molecular packing in the structure of $C_{14}H_8Cl_2O_4$.

1,1,1-TRICHLORO-3,4-DIMETHYLPENT-4-EN-2-OL 3,5-DI-NITROBENZOATE
$C_{14}H_{13}Cl_3N_2O_6$

J.P. BENNER, G.B. GILL, S.J. PARROTT, B. WALLACE and M.J. BEGLEY, 1984. J. Chem. Soc. Perkin I, 315-329.

Monoclinic, $P2_1$, a = 9.453, b = 6.076, c = 15.593 Å, β = 96.36°, Z = 2. Cu radiation, R = 0.043 for 834 reflexions.

The molecular structure is shown in Fig. 1. Bond lengths and angles are of low precision but the values fall within expected ranges.

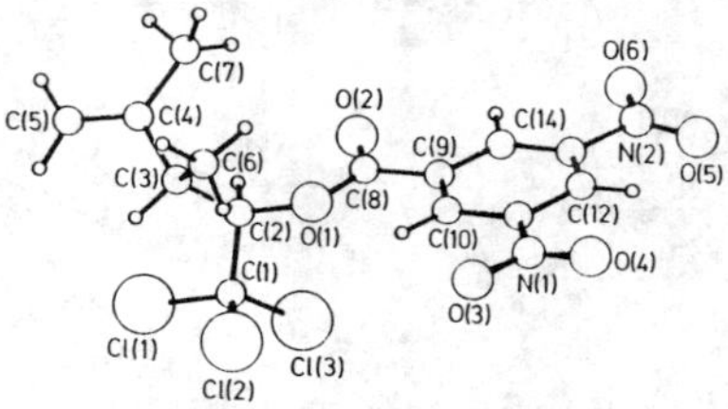

Fig. 1. $C_{14}H_{13}Cl_3N_2O_6$: the molecular structure.

2-[4-(ACETYLAMINOSULPHONYL)PHENYLCARBAMOYL]BENZOIC ACID (PHTHALYLSULFACETAMIDE)
$C_{16}H_{14}N_2O_6S$

T.P. SINGH, U. PATEL and M. HARIDAS, 1984. Acta Cryst., C40, 2088-2091.

Monoclinic, $P2_1/c$, a = 8.01, b = 13.00, c = 18.04 Å, β = 111.6°, Z = 4. Cu radiation, R = 0.092 for 1803 reflexions.

The two planar benzene rings in the structure (Fig. 1) make an angle of 62.1(5)° with each other. The molecule adopts a coiled conformation about the a axis. The molecules are arranged in the form of hydrogen-bonded loops. There are no intermolecular interactions in the interior of the loop.

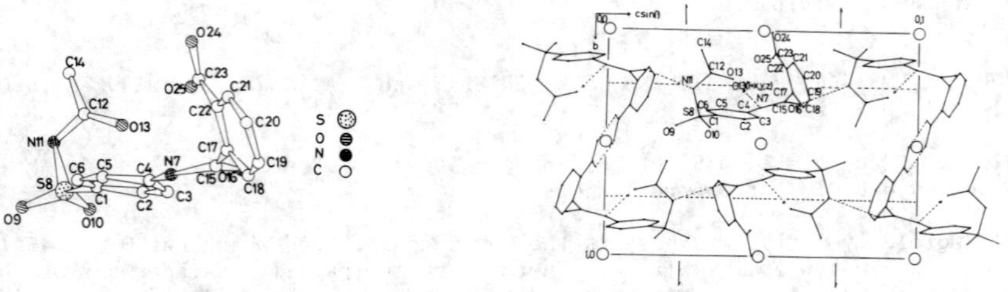

Fig. 1. A perspective view of $C_{16}H_{14}N_2O_6S$ and the crystal structure as viewed down the a-axis.

1,2-PROPANEDIYL BIS(p-CHLOROBENZOATE)
$C_{17}H_{14}Cl_2O_4$ (I)

2,3-BUTANEDIYL BIS(p-CHLOROBENZOATE)
$C_{18}H_{16}Cl_2O_4$ (II)

G. BOCELLI and M.F. GRENIER-LOUSTALOT, 1984. Acta Cryst., C40, 1394-1397.

I. Monoclinic, C2/c, a = 19.859, b = 8.837, c = 19.426 Å, β = 103.15°, Z = 8. Cu radiation, R = 0.041 for 1295 reflexions.

II. Monoclinic, $P2_1/c$, a = 14.679, b = 6.284, c = 19.638 Å, β = 93.24°, Z = 4. Cu radiation, R = 0.055 for 2920 reflexions.

The conformation of the centre of the chain of II is gauche (torsion angle O(2)-C(8)-C(10)-O(4) is 63.4°) while the methyl group in I shifts the chain from a trans arrangement at O(4) to O(2) (Fig. 1). The two rings are tilted by 11.9(2), 6.5(2)° in I; 6.7(1), 0.5(1)° in II with respect to the carboxylic groups and subtend dihedral angles of 10.0(1) and 65.4(1)° to each other in each molecule. There is one short C-C bond in II (C(8)-C(10) 1.471(9) Å); the C-Cl bonds are in the range 1.735(5)-1.745(4) Å.

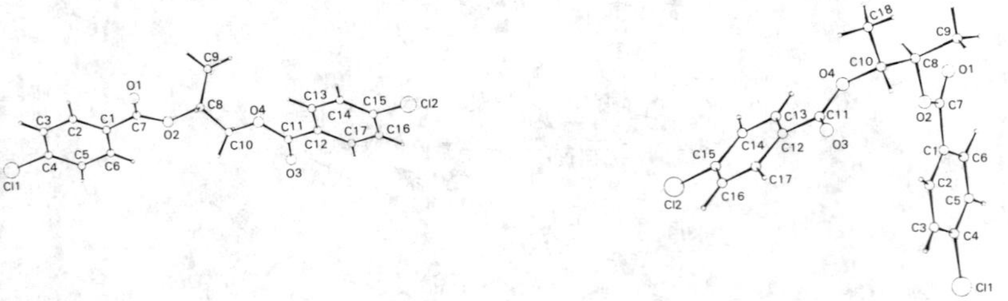

Fig. 1. The $C_{17}H_{14}Cl_2O_4$ and $C_{18}H_{16}Cl_2O_4$ molecules.

1,4-BUTANEDIYL BIS[o-CHLOROBENZOATE]
$C_{18}H_{16}Cl_2O_4$ (I)

1,4-BUTANEDIYL BIS[m-CHLOROBENZOATE]
$C_{18}H_{16}Cl_2O_4$ (II)

1,4-BUTANEDIYL BIS[p-CHLOROBENZOATE]
$C_{18}H_{16}Cl_2O_4$ (III)

G. BOCELLI and M.F. GRENIER-LOUSTALOT, 1984. Acta Cryst., C40, 679-683.

I. Monoclinic, $P2_1/n$, a = 8.858, b = 12.956, c = 7.472 Å, β = 97.08°, Z = 2. Cu radiation, R = 0.050 for 1183 reflexions.

II. Monoclinic, $P2_1/c$, a = 5.870, b = 19.270, c = 7.651 Å, β = 97.29°, Z = 2. Cu radiation, R = 0.049 for 1301 reflexions.

III. Triclinic, P1, a = 7.609, b = 10.318, c = 5.924 Å, α = 96.31, β = 98.35, γ = 111.05°, Z = 1. Cu radiation, R = 0.081 for 1006 reflexions.

The phenyl rings are planar in I and II, while some deformations are found in those of III. Mean bond lengths within the rings are as expected: 1.386(5) (I), 1.379(3) (II) and 1.410(11) (III). Values for C=O bond lengths are 1.142(10)-1.244(14) Å and for C-O 1.335(3)-1.466(3) Å, in good agreement with the usual findings. The usual tilt of the carboxylic groups with respect to their phenyl rings is observed and in the case of I the steric hindrance increases the value (31.2(1)°) noticeably. The conformation of the butanediyl chain is gauche-trans-gauche in I and all-trans in II and III. The crystal packing is determined by van der Waals forces only (Fig. 1).

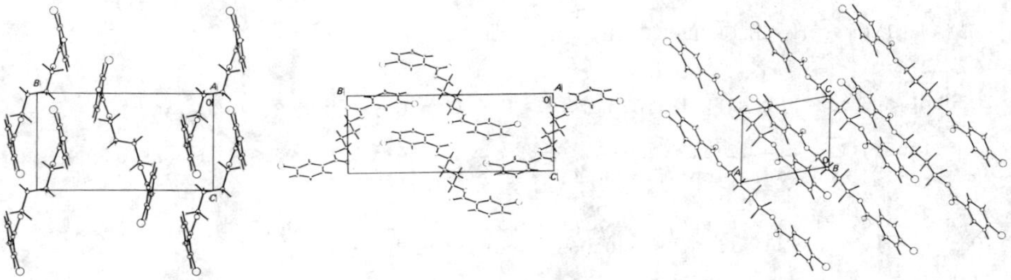

Fig. 1. $C_{18}H_{16}Cl_2O_4$: crystal packing along x in I, along x in II and along y in III.

1,6-HEXANEDIYL BIS[o-CHLOROBENZOATE]
$C_{20}H_{20}Cl_2O_4$ (I)

1,6-HEXANEDIYL BIS[m-CHLOROBENZOATE]
$C_{20}H_{20}Cl_2O_4$ (II)

1,6-HEXANEDIYL BIS[p-CHLOROBENZOATE]
$C_{20}H_{20}Cl_2O_4$ (III)

G. BOCELLI and M.F. GRENIER-LOUSTALOT, 1984. Acta Cryst., C40, 1391-1394.

I. Monoclinic, $P2_1/c$, a = 8.145, b = 20.359, c = 12.000 Å, β = 103.99°, Z = 4. Cu radiation, R = 0.060 for 2882 reflexions.

II. Triclinic, $P\bar{1}$, a = 8.354, b = 7.365, c = 8.431 Å, α = 71.21, β = 97.48, γ = 95.84°, Z = 1. Cu radiation, R = 0.045 for 1251 reflexions.

III. Triclinic, P1, a = 10.483, b = 7.760, c = 6.746 Å, α = 107.77, β = 98.32, γ = 70.43°, Z = 1. Cu radiation, R = 0.060 for 1455 reflexions.

All three molecules have the all-trans conformation in the aliphatic chains (Fig. 1). The tilt of the carboxylic groups with respect to the phenyl rings is greatest in the ortho compound (33.8 and 36.0° compared with 0.2-4.5°). Bond lengths and angles are normal; C-Cl distances are in the range 1.712-1.745(2) Å with Cl out-of-plane displacements 0.012-0.051 Å.

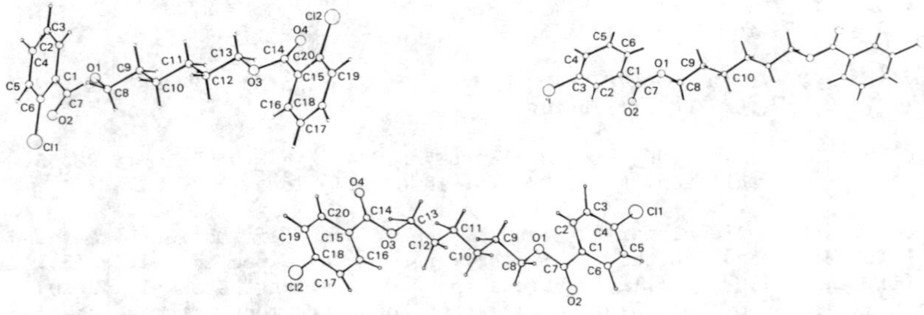

Fig. 1. Conformations of the three isomeric $C_{20}H_{20}Cl_2O_4$ molecules.

2,5-HEXANEDIYL BIS(p-CHLOROBENZOATE)
$C_{20}H_{20}Cl_2O_4$ (I)

2,5-HEXANEDIYL BIS(m-CHLOROBENZOATE)
$C_{20}H_{20}Cl_2O_4$ (II)

G. BOCELLI and M.F. GRENIER-LOUSTALOT, 1984. Acta Cryst., C40, 1400-1402.

I. Triclinic, $P\bar{1}$, a = 12.512, b = 12.333, c = 7.166 Å, α = 92.81, β = 85.84, γ = 63.84°, Z = 2. Cu radiation, R = 0.065 for 2608 reflexions.

II. Triclinic, $P\bar{1}$, a = 8.817, b = 8.084, c = 7.341 Å, α = 104.35, β = 79.06, γ = 87.70°, Z = 1. Cu radiation, R = 0.065 for 1637 reflexions.

The conformation of the tetramethylene chain is gauche-trans-gauche in II and is significantly shifted away from an all-trans arrangement in I (C(8)-C(10)-C(11)-C(12) torsion angle is 167.8(7)°). In I the aromatic rings are within 1.4° of being parallel and the carboxylic groups are tilted 0.4(2) and 2.3(2)° to their rings. In II the rings are parallel and are tilted at 6.0(1)° to their carboxylic moiety. In II the C(9) and C(13) methyl groups are trans to each other and axial with respect to the chain (Fig. 1); in I they lie on the same side with respect to the C(8),C(10),C(11),C(12) sequence and are 1.00 and 1.05 Å out of the plane.

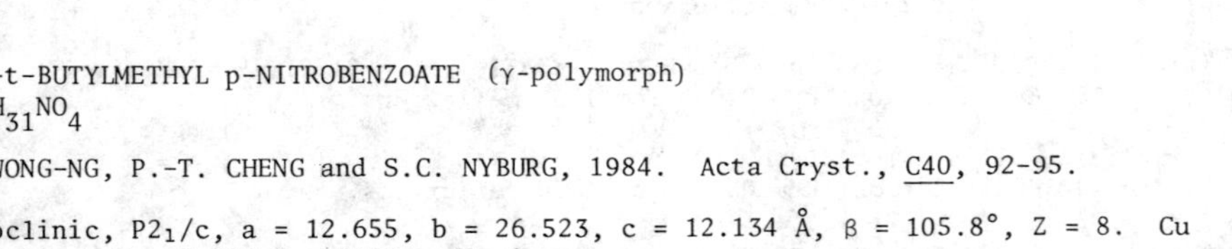

Fig. 1. The isomeric $C_{20}H_{20}Cl_2O_4$ molecules.

TRI-t-BUTYLMETHYL p-NITROBENZOATE (γ-polymorph)
$C_{20}H_{31}NO_4$

W. WONG-NG, P.-T. CHENG and S.C. NYBURG, 1984. Acta Cryst., C40, 92-95.

Monoclinic, $P2_1/c$, a = 12.655, b = 26.523, c = 12.134 Å, β = 105.8°, Z = 8. Cu radiation, R = 0.087 for 3453 reflexions.

The two independent molecules in the asymmetric unit have virtually identical structures (Fig. 1). They differ from the α-polymorph (1) in that the C(1)-C(7)-O(1)-C(8) torsion angles are -178.8 and 178.6° compared with 162.3° for the α-form. The three C(8)-C(methyl) distances are long averaging 1.617(6) Å (1.618(6) Å in α-form).

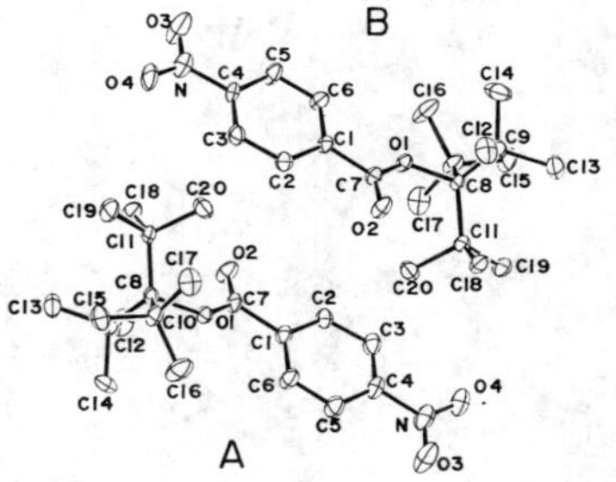

Fig. 1. A view of the two molecules of $C_{20}H_{31}NO_4$.

1. Structure Reports, 44B, 83.

1,10-DECANEDIYL BIS(o-CHLOROBENZOATE)
$C_{24}H_{28}Cl_2O_4$

G. BOCELLI and M.F. GRENIER-LOUSTALOT, 1984. Acta Cryst., C40, 1397-1399.

Triclinic, P$\bar{1}$, a = 13.665, b = 13.114, c = 7.213 Å, α = 103.53, β = 74.92, γ = 110.02°, Z = 2. Cu radiation, R = 0.074 for 3116 reflexions.

The carboxylic groups of the two independent, centrosymmetric molecules are tilted by 16.6(2) and 24.0(2) with respect to their phenyl rings. The torsion angles of the decane chains are near to 180° and the conformation is all trans. The two molecules are arranged almost perpendicularly to each other (Fig. 1). Bond lengths and angles are normal.

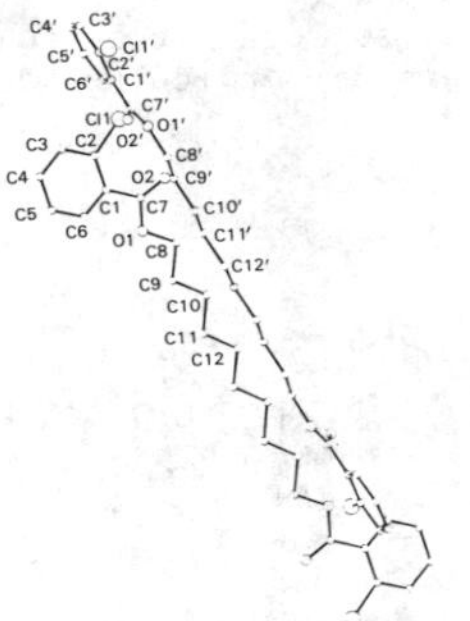

Fig. 1. The two independent $C_{24}H_{28}Cl_2O_4$ molecules.

RUBIDIUM ORTHOSULPHOBENZOATE
$C_7H_5O_5SRb$ $Rb^+[C_6H_4COOHSO_3^-]$

T.B. TEPLOVA, T.N. TURSKAYA, A.B. TOVBIS, V.E. ZAVODNIK, T.A. SHIBANOVA and G.S. BELIKOVA, 1984. Kristallografiya, 29, 726-730 [Sov. Phys. Crystallogr., 29, 429-431].

Orthorhombic, Pbca, a = 25.559, b = 10.492, c = 7.109 Å, Z = 8. Mo radiation, R = 0.037 for 1562 reflexions.

The crystal structure consists of Rb^+ cations and sulphobenzoate anions, with the cations forming corrugated layers parallel to the bc plane (Fig. 1). The Rb coordination polyhedron is a very distorted eight-pointed figure, pairs of which are linked to form dimers. The Rb...Rb distance in the pairs of polyhedra is 3.971 Å, and the Rb-O distances cover the range 2.875 to 3.142 Å. The Rb^+ ion is coordinated by eight O atoms, two of which belong to carbonyl groups and six to SO_3 groups. The network of Rb^+ ions is screened on both sides by anions and there is one hydrogen bond (O(2)-H(5)...O(5') O...O 2.647 Å).

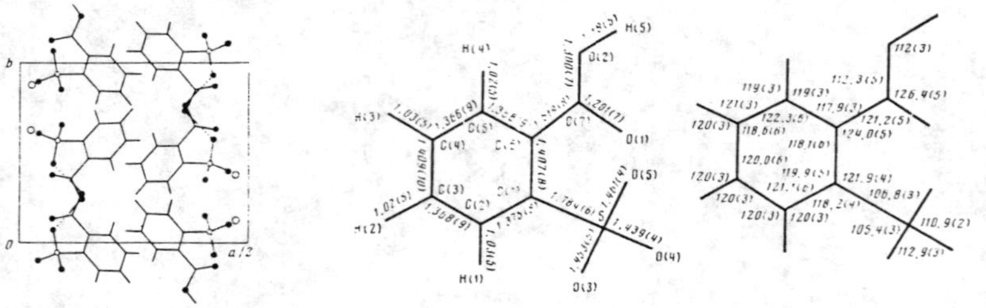

Fig. 1. Projection of the structure $C_7H_5O_5SRb$ on the ab plane, and bond distances (Å) and angles (°) in the anion.

HEXAMETHYLENEDIAMMONIUM PHTHALATE TRIHYDRATE

$C_{14}H_{22}N_2O_4,3H_2O$ $C_6H_{18}N_2^{2+}.C_8H_4O_4^{2-}.3H_2O$

N.R. JAGANNATHAN, E. SUBRAMANIAN, R. SRINIVASAN and J. TROTTER, 1984. Acta Cryst., C40, 1743-1744.

Triclinic, P$\bar{1}$, a = 8.171, b = 9.168, c = 12.217 Å, α = 83.04, β = 77.96, γ = 89.97°, D_m = 1.26, Z = 2. Mo radiation, R = 0.053 for 3083 reflexions.

The hexamethylenediammonium ions (Fig. 1) lie on centres of symmetry and adopt a planar zigzag configuration. The two carboxyl groups of the phthalate ion are twisted by 48.8(3)° in the same sense from the plane of the six-membered ring. The packing arrangement involves spatial segregation of polar and non-polar regions of the molecules into alternating layers perpendicular to the c axis.

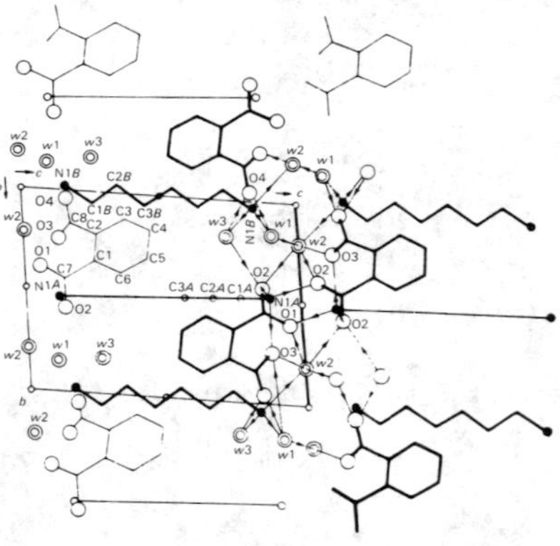

Fig. 1. A projection of the hexamethylenediammonium phthalate trihydrate down the a* axis.

4,4'-DINITROBIPHENYL 4-HYDROXYBIPHENYL (3:1)

$C_{48}H_{34}N_6O_{13}$ $3C_{12}H_8N_2O_4.C_{12}H_{10}O$

C.P. BROCK and K.L. HALLER, 1984. Acta Cryst., C40, 1387-1390.

Triclinic, P1, a = 11.034, b = 11.166, c = 9.431 Å, α = 90.15, β = 115.10, γ = 81.30°, Z = 1. Mo radiation, R = 0.062 for 1728 reflexions.

The structure is similar to that previously reported (1) except that the 4,4'-dinitrobiphenyl molecules are twisted by an average of 37.2(1)° about the central C-C bonds (instead of being mirror symmetrical). Infinite parallel stacks of 4,4'-dinitrobiphenyl molecules with intermolecular spacing 3.722 Å have almost planar 4-hydroxybiphenyl molecules ordered within the channels between the stacks (Fig. 1). The ordering is stabilised by a hydrogen bond between OH and the neighbouring NO_2 group (O...O 2.971(12) Å). The other O of the NO_2 group is 3.364(25) Å from the same OH group.

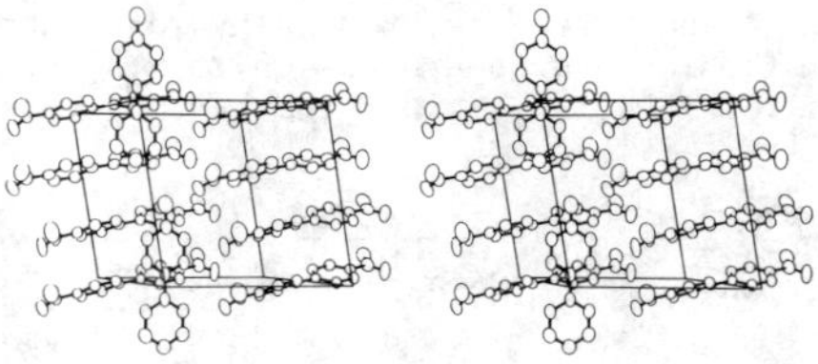

Fig. 1. Stereoview of the unit cell of the $3C_{12}H_8N_2O_4.C_{12}H_{10}O$ complex.

1. Structure Reports, 11, 669.

2,6-DINITROANILINE
$C_6H_5N_3O_4$

L. PARKANYI and A. KALMAN, 1984. J. Mol. Struct., 125, 315-320.

Monoclinic, $P2_1/n$, a = 3.782, b = 27.729, c = 7.152 Å, β = 95.41°, Z = 4. Cu radiation, R = 0.056 for 1338 reflexions.

The plane of the NH_2 group makes a dihedral angle of 24.08(4)° with the phenyl ring plane, and the nitro group are rotated -8.03(9)° and 5.25(14)° out of the phenyl plane (Fig. 1). The N(1)-C(1) bond is short 1.339(2) and the amino H atoms form intramolecular N-H...O hydrogen bonds to the vicinal O atoms (N(1)...O(1) 2.620, N(1)...O(3) 2.608(2) Å). Bond lengths in the phenyl ring are longer at C(1) (1.426 and 1.429(2)) compared with those at the other atoms (1.377-1.389(2) Å). The C-N(nitro) distances are 1.451 and 1.459(2) Å.

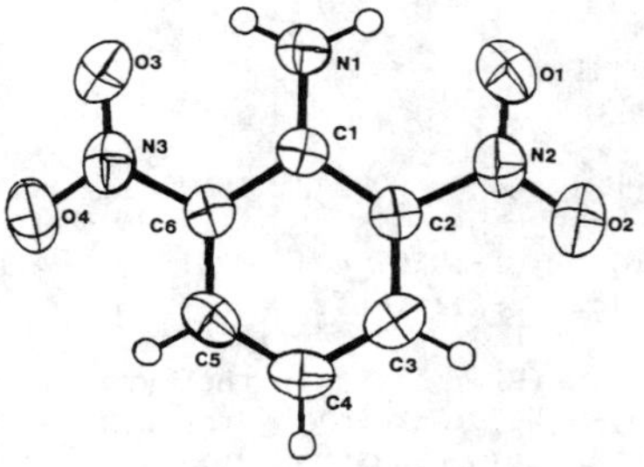

Fig. 1. 2,6-Dinitroaniline.

4-(4-CHLOROPHENYL)THIOSEMICARBAZIDE
$C_7H_8ClN_3S$

A.K. NANDI, S. CHAUDHURI, S.K. MAZUMDAR and S. GHOSH, 1984. J. Chem. Soc. Perkin II, 1729-1733.

Monoclinic, $P2_1/c$, a = 14.594, b = 10.495, c = 12.384 Å, β = 104.73°, Z = 8. Mo radiation, R = 0.035 for 1967 reflexions.

There are two molecules of the title compound which shows antibacterial activity in the asymmetric unit. In both molecules (Fig. 1) the hydrazine NH_2 and the S atom assume a trans-configuration. The S atom occurs as the keto form in both, C(3)-S 1.687(3) (A), 1.692(3) Å (B). Other distances (C(5)-N(4) 1.420(3), 1.422(4); C(3)-N(2) 1.327(3), 1.330(3); C(3)-N(4) 1.351(3), 1.332(4) Å for A, B respectively) indicate extensive delocalisation of the π-electrons in these regions of the molecules. The N(1)-N(2) distance is 1.403(4) (A) and 1.414(4) Å (B). The C(5),C(3),N(4) plane makes an angle of 46.43° with the phenyl ring plane in A and an angle of 80.40° with this plane in B. All N atoms, except N(2B), in both molecules take part in hydrogen bonding. The S atoms in A and B are involved in hydrogen bonding in different ways; S(B) accepts two intermolecular hydrogen bonds while S(A) accepts only one. The Cl atoms Cl (A) (x,y,z) and Cl (B) (x,y,1+z) of the two molecules approach each other to a distance of 3.443(1) Å.

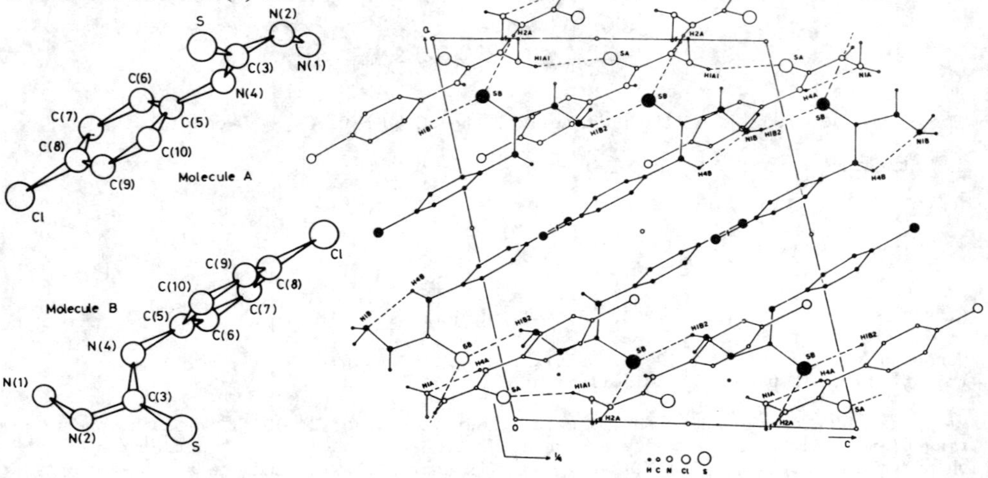

Fig. 1. $C_7H_8ClN_3S$: perspective views of the two molecules and a projection of the crystal structure.

α,α-DICHLORO-4'-NITROACETANILIDE
$C_8H_6Cl_2N_2O_3$

A. DE, A.K. BASAK, S. CHAUDHURI and G. PODDAR, 1984. Acta Cryst., C40, 2104-2106.

Monoclinic, $P2_1/c$, a = 7.966, b = 13.985, c = 9.301 Å, β = 92.55°, D_m = 1.611, Z = 4. Mo radiation, R = 0.039 for 1526 reflexions.

The acetanilide side chain (Fig. 1) is in the extended conformation. Intermolecular N(acetanilide)-H(N)...O(keto) hydrogen bonds between the two molecules related by a c-glide plane stabilize the structure (N...O 3.003(2) Å).

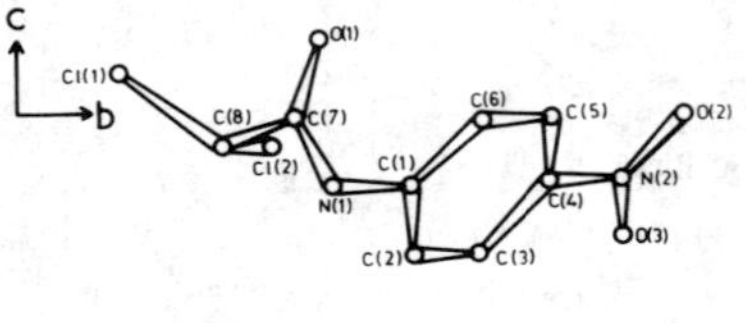

Fig. 1. $C_8H_6Cl_2N_2O_3$.

4'-AMINO-2,2,3'-TRICHLOROACETOPHENONE
$C_8H_6Cl_3NO$

A. DE, 1984. Acta Cryst., C40, 1972-1974.

Monoclinic, $P2_1/a$, a = 7.425, b = 9.089, c = 14.202 Å, β = 92.27°, D_m = 1.69, Z = 4. Cu radiation, R = 0.065 for 1279 reflexions.

The acetyl group is strongly conjugated with the phenyl ring; the dihedral angle between the two is 1.09° and the C(4)-C(7) bond length is 1.468(9) Å (Fig. 1). The C(5)...O(1) distance is 2.790(8) Å. Molecules are linked by an N-H...O(1) hydrogen bond (N...O 2.959(7) Å).

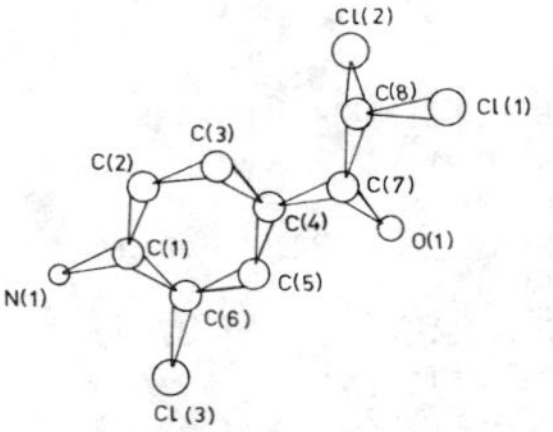

Fig. 1. The 4'-amino-2,2,3'-trichloroacetophenone molecule.

2-DIMETHYLAMINOBENZENEDIAZONIUM TETRAFLUOROBORATE
$C_8H_{10}BF_4N_3$

J.D. WALLIS and J.D. DUNITZ, 1984. Helv. Chim. Acta, 67, 1374-1378.

Monoclinic, $P2_1/n$, a = 7.198, b = 15.922, c = 9.278 Å, β = 96.96°, Z = 4. Mo radiation, R = 0.066 for 1549 reflexions.

The cation (Fig. 1) is very nearly planar; strong steric interactions between ortho-groups being diminished by in-plane angular distortions.

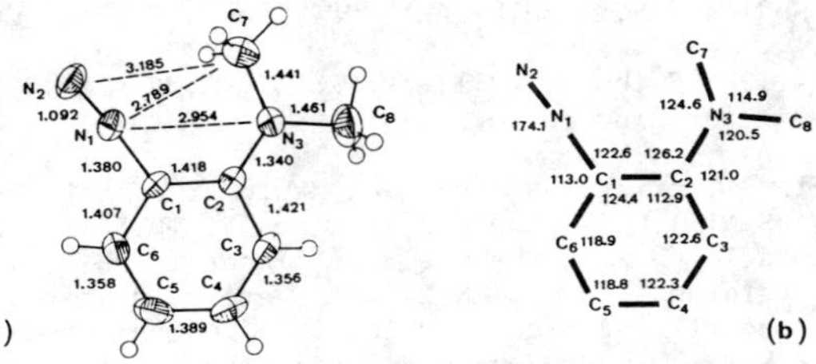

Fig. 1. A view of the $C_8H_{10}N_3^+$ cation with interatomic distances and angles (σ 0.005 Å, 0.4°).

3-METHYL-1-p-TOLYLTRIAZENE
$C_8H_{11}N_3$

A.J. RANDALL, C.H. SCHWALBE and K. VAUGHAN, 1984. J. Chem. Soc. Perkin II, 251-253.

Monoclinic, $P2_1/c$, a = 8.196, b = 8.545, c = 13.163 Å, β = 112.72°, D_m = 1.17, Z = 4. Mo radiation, R = 0.063 for 1088 reflexions.

The triazene unit in the molecule (Fig. 1) has a double bond between N(1) and N(2) (1.275(3) Å) with a hydrogen atom and a methyl group both carried by N(3). The N(2)-N(3) and N(3)-C(8) distances are 1.319(3) and 1.451(4) Å respectively. Bond angles C(1)-N(1)-N(2), N(1)-N(2)-N(3) and N(2)-N(3)-C(8) are 112.1(2), 115.4(2) and 123.5(2)° respectively. There is a 21-22° twist about the bond joining the phenyl ring to the triazene moiety. Molecules are linked by N(1)-H...N(3) hydrogen bonds (3.067(3) Å).

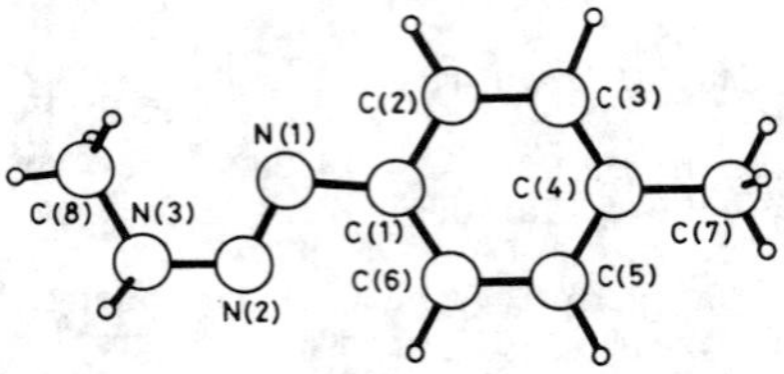

Fig. 1. $C_8H_{11}N_3$: the molecular structure.

4-METHYL-2-NITROACETANILIDE (AMBER POLYMORPH)

$C_9H_{10}N_2O_3$ $(CH_3)(NO_2)C_6H_3NHCOCH_3$

J.C. MOORE, A. YEADON and R.A. PALMER, 1984. J. Cryst. Spectrosc. Res., 14, 283-291.

Monoclinic, $P2_1/c$, a = 10.158, b = 11.635, c = 8.041 Å, β = 94.55°, D_m = 1.35, Z = 4. Cu radiation, R = 0.080 for 1027 reflexions.

The structure consists of nearly planar molecules, all approximately parallel to one another with their longitudinal axes parallel to b. The amide group forms an intramolecular hydrogen bond with the nitro group. Molecules related by c-glide are stacked along c in a very distinct columnar form. The columns are held together by dipole-dipole interactions between close antiparallel carbonyl groups or between close antiparallel nitro groups. For details of the other polymorphs, see (1).

1. Structure Reports, 50B,

4-AMINO-N-2-PYRIDYLBENZENESULPHONAMIDE (SULPHAPYRIDINE)

$C_{11}H_{11}N_3O_2S$

A.K. BASAK, S. CHAUDHURI and S.K. MAZUMDAR, 1984. Acta Cryst., C40, 1848-1851.

Monoclinic, C2/c, a = 12.807, b = 11.711, c = 15.379 Å, β = 94.07°, D_m = 1.441, Z = 8. Mo radiation, R = 0.041 for 2160 reflexions.

The molecule (Fig. 1) exists as the imido tautomer with H attached to the pyridine N atom. The S,N(1),C(7) plane is inclined at 10.5° to the pyridine ring plane. The dihedral angle between the two ring planes is 93.1(1)° with phenyl rings stacked along a and pyridine along c. N(2) and S are respectively -0.060 and -0.197(1) Å from the phenyl ring plane. Molecules are connected in pairs across a centre of symmetry by N-H...O hydrogen bonds; additional intra- and intermolecular N-H...O and N-H...N hydrogen bonds stabilise the structure. Some bond lengths and angles are S-N(1) 1.605(3), S-C(1) 1.749(2), N(1)-C(7) 1.336(3) Å, O-S-N(1) 111.6 and 105.8(1), C(11)-N(3)-C(7) 124.5(3)°.

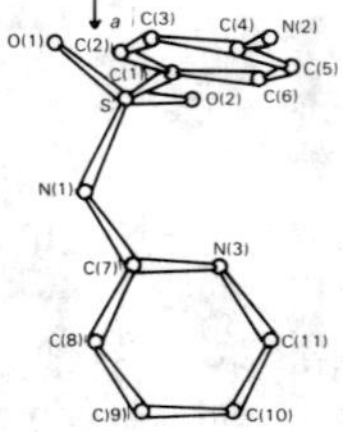

Fig. 1. The sulphapyridine molecule.

N[1]-PHENYLSULPHANILAMIDE (SULPHANILANILIDE)
$C_{12}H_{12}N_2O_2S$

C.J. BROWN, D.S. COOK and L. SENGIER, 1984. Acta Cryst., C40, 844-846.

Orthorhombic, Pcab, a = 24.230, b = 15.500, c = 6.034 Å, D_m = 1.460, Z = 8. Mo radiation, R = 0.035 for 814 reflexions.

Apart from an appreciable quinonoid character in the aniline ring caused by proton transfer on nitrogen, bond lengths and angles in the molecule have standard values: S-O 1.443(4) and 1.435(4), S-N 1.661(3) and S-C 1.775(4) Å (Fig. 1). The molecules are linked in the crystal along b by a system of three N-H...O hydrogen bonds per molecule through the unsubstituted amino group.

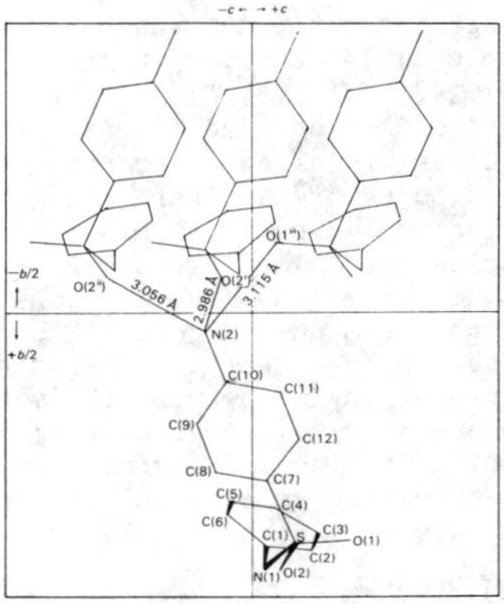

Fig. 1. $C_{12}H_{12}N_2O_2S$: projection of the structure on (100); hydrogen bond distances are shown.

2-AMINO-N-BENZOYLBENZAMIDINE
$C_{14}H_{13}N_3O$

D. KORBONITS, P. KISS, K. SIMON and P. KOLONITS, 1984. Chem. Ber., 117, 3183-3193.

Orthorhombic, Pbca, a = 7.352, b = 26.681, c = 24.729 Å, Z = 16. Mo radiation, R = 0.047 for 2280 reflexions.

The two independent molecules in the cell are linked by hydrogen bridges as shown in Fig. 1.

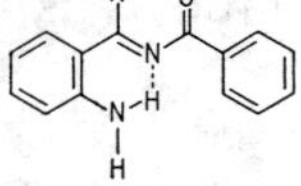

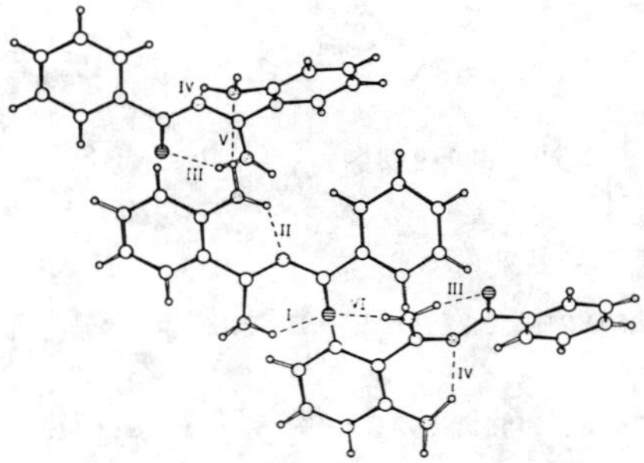

Fig. 1. Hydrogen bonding in $C_{14}H_{13}N_3O$.

S-METHYLDITHIZONE HYDROGEN TRIIODIDE
$C_{14}H_{15}I_3N_4S$ (I)

2,3-DIPHENYL-5-METHYLTHIOTETRAZOLIUM TRIIODIDE CHLOROFORM SOLVATE
$C_{14}H_{13}I_3N_4S,CHCl_3$ (II)

F.H. HERBSTEIN and W. SCHWOTZER, 1984. J. Chem. Soc. Perkin II, 1917-1921.

I. Triclinic, P$\bar{1}$, a = 12.367, b = 10.084, c = 8.451 Å, α = 109.0, β = 97.5, γ = 96.9°, D_m = 2.23, Z = 2. Mo radiation, R = 0.082 for 2438 reflexions.

II. Monoclinic, Cc, a = 15.923, b = 10.218, c = 16.118 Å, β = 109.91°, D_m = 2.11, Z = 4. Mo radiation, R = 0.051 for 2165 reflexions.

In I the cation (Fig. 1) has the anti, s-trans configuration. The formazan moiety is planar. The C(19)S(6)C(3)N(4) torsion angle is 96(1.5)°. The I_3^- anion is asymmetric with I(1)-I(2) 2.841(2), I(2)-I(3) 3.025(2) Å and I(1)-I(2)-I(3) 177.21(6)°. In the crystal, alternate sheets of cations and anions lie parallel to (010). In the cationic layer pairs of cations are arranged about a centre of symmetry such that respectively the C-S and S-CH_3 dipoles are antiparallel: these pairs are stacked to form columns along [001]. The cation layer is formed by such columns related by a translation along [100]. Interleaving anion layers contain oppositely aligned tri-iodide ions related by centres of inversion. The I_3^- ions lie approximately along [104] and are aligned parallel to the S-CH_3 vectors of the cations. In the cation of II, the tetrazolium ring is essentially planar and the methyl group is rotated out of this plane (-9(1.5)°). The dihedral angles between the mean planes of the phenyl rings and the tetrazolium ring are 58(2)° (phenyl C(7-12)) and 69(2)° (phenyl C(13-18)). The tri-iodide ion is fully ordered and symmetrical. The structure can be described in terms of C face-centred layers parallel to (007), with cations and anions segregated on alternate layers. Successive layers of any one type are related by the c-glide.

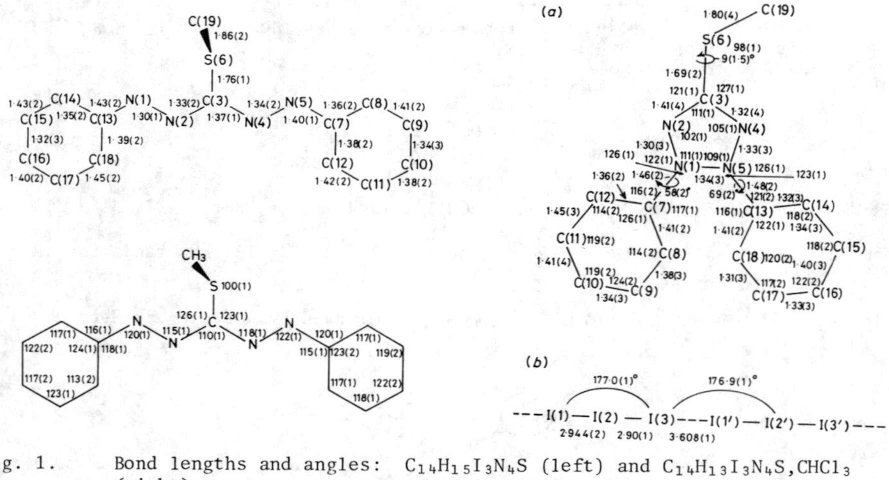

Fig. 1. Bond lengths and angles: $C_{14}H_{15}I_3N_4S$ (left) and $C_{14}H_{13}I_3N_4S,CHCl_3$ (right).

4-BROMO-N-(2-HYDROXYCINNAMYLIDENE)ANILINE
$C_{15}H_{12}BrNO$

K.Sh. KARAEV and N.G. FURMANOVA, 1984. Kristallografiya, 29, 462-465 [Sov. Phys. Crystallogr., 29, 277-279].

Monoclinic, $P2_1/n$, a = 11.718, b = 13.502, c = 8.931 Å, γ = 109.25°, Z = 4. Cu radiation, R = 0.072 for 1780 reflexions.

The conformation of the molecule is characterized by three approximately planar fragments: bromoaniline (Br, N, C(1)-C(6)), phenolate (O, C(11)-C(15)), and diene (N, C(4), C(7)-C(10)) (Fig. 1). The relative orientation of the fragments is determined by their rotation around the single bonds C(4)-N and C(9)-C(10). The dihedral angles between bromoaniline and diene, phenolate and diene, and bromoaniline and phenolate are 45.5, 11.9 and 49.6° respectively. The distribution of bond lengths suggests a large contribution from a resonance form containing the hydrogen atom transferred from the N atom to the O atom.

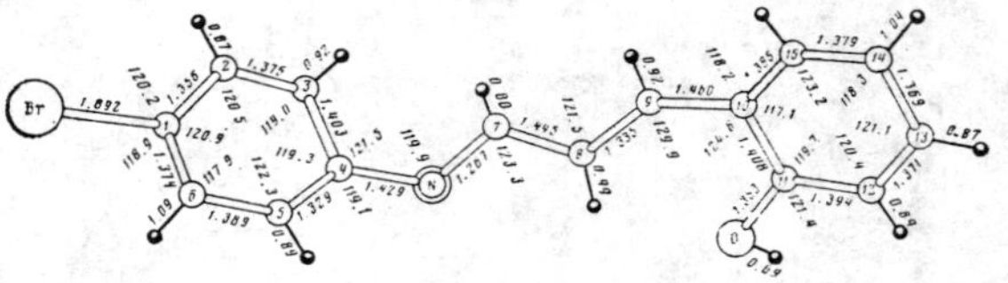

Fig. 1. Structure of $C_{15}H_{12}BrNO$ showing bond lengths and angles.

α-(4-METHYL-2-NITROPHENYLHYDRAZONO)ACETANILIDE
$C_{15}H_{14}N_4O_3$

C.J. BROWN and H.R. YADAV, 1984. Acta Cryst., C40, 95-96.

Monoclinic, $P2_1/a$, a = 23.344, b = 5.905, c = 12.164 Å, β = 121.07°, Z = 4. Cu radiation, R = 0.045 for 1183 reflexions.

The almost planar molecules exist as hydrogen bonded centrosymmetric dimers (Fig. 1) with N...O 3.119(4) Å. There is also an intramolecular N-H...O hydrogen bond (N...O 2.909(3) Å). Some bond lengths are N(2)-N(3) 1.334(4), N(3)-C(8) 1.284(5), C(8)-C(9) 1.448(5), C(9)-N(4) 1.380(5) Å.

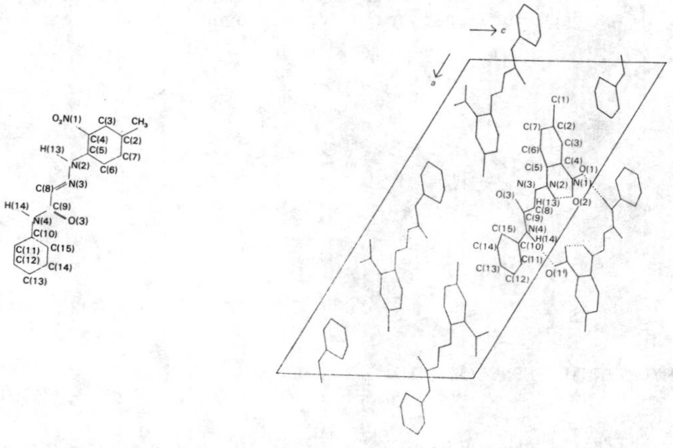

Fig. 1. The $C_{15}H_{14}N_4O_3$ molecule and a packing diagram.

2-PHENYLETHYNYL-N,N-DIMETHYLANILINE
$C_{16}H_{15}N$

J.D. WALLIS and J.D. DUNITZ, 1984. Helv. Chim. Acta, 67, 39-46.

Orthorhombic, Pbcn, a = 15.626, b = 7.107, c = 22.807 Å, Z = 8. Mo radiation, R = 0.043 for 1696 reflexions (at 98 K).

The dimethylamino group has pyramidal geometry and the nitrogen lone pair is not directed towards the triple bond; instead, one N-CH_3 bond lies almost in the plane of the disubstituted ring, directed away from the triple bond, while the other N-CH_3 bond and the lone pair are directed to opposite sites of the plane, nearer to the acetylene (Fig. 1). There is a remarkably short intramolecular contact (2.39 Å) between a methyl H- and an acetylenic C-atom. The Taft σ_I parameter of the arylethynyl substituent appears to be similar to that of a carboxylic ester, judging from the bond-angle deformation at the ipso-C-atom.

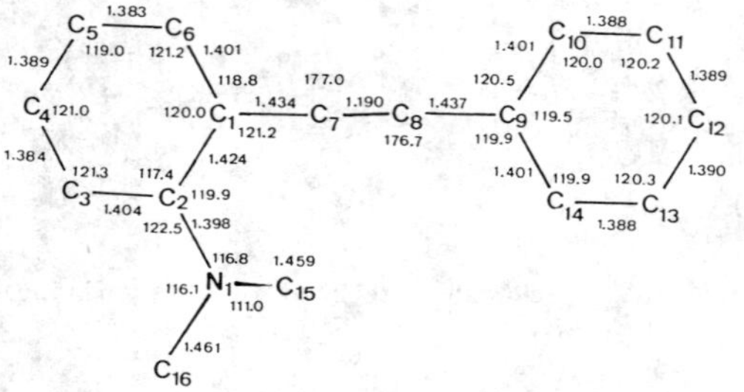

Fig. 1. Bond lengths and angles for $C_{16}H_{15}N$.

SULPHAPROXYLINE (p-ISOPROPOXY-N-SULPHANILYLBENZAMIDE)
$C_{16}H_{18}N_2O_4S$

A.K. BASAK, S.K. MAZUMDAR and S. CHAUDHURI, 1984. Acta Cryst., C40, 419-422.

Monoclinic, $P2_1/c$, a = 9.908, b = 17.263, c = 9.585 Å, β = 103.08°, Z = 4. Mo radiation, R = 0.043 for 1983 reflexions.

In the molecule (Fig. 1) the S-N(1)-C(7)-C(8) torsion angle of 161.4(2)° indicates that C(8) is (+)-antiperiplanar to the S atom. Molecules are linked in chains by N-H...O hydrogen bonds between amido nitrogen N(1) and keto oxygen O(3) (N...O 2.853(3) Å).

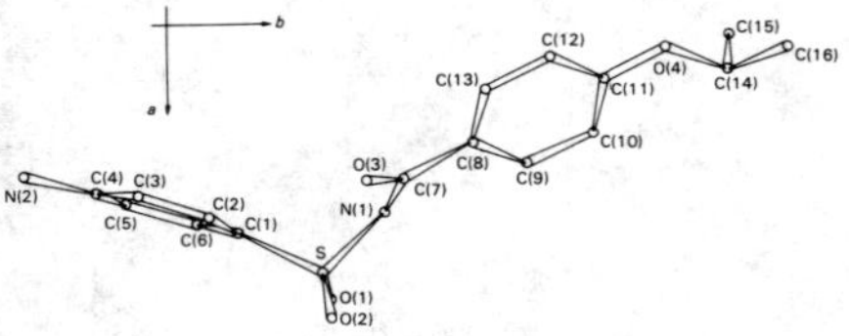

Fig. 1. A view of $C_{16}H_{18}N_2O_4S$.

4,4'-BIS(DIMETHYLAMINO)DIPHENYLAMMONIUMYL IODIDE
$C_{16}H_{21}IN_3$ $[(CH_3)_2NC_6H_4\dot{N}HC_6H_4N(CH_3)_2]^+,I^-$

K. TICHÝ and W. PETTER, 1984. Acta Cryst., C40, 1055-1057.

Monoclinic, C2/c, a = 13.114, b = 14.124, c = 9.633 Å, β = 109.914°, Z = 4. Mo radiation, R = 0.070 for 1741 reflexions and neutron diffraction, R = 0.261 for 467 reflexions.

The title compound has previously been studied by X-ray diffraction (1). In the molecule (Fig. 1) the mean C-H distance in the methyl group is 1.03 Å, in the benzene ring 1.06 Å. The hydrogen bridge N-H...I is linear (all three atoms are in special positions (0,y,0.25)) with N-H 1.53 and H...I 2.27 Å.

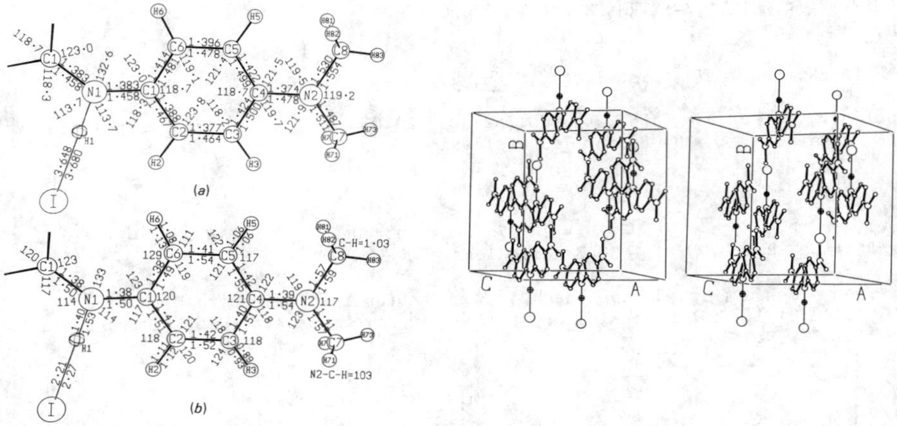

Fig. 1. $C_{16}H_{21}IN_3$: bond lengths and angles in the independent half of the molecule (a) X-ray measurement (upper: uncorrected bond lengths, lower: bond lengths corrected for thermal motion), (b) neutron measurement (left) and a stereoview of the molecular packing (right).

1. Structure Reports, 31B, 59.

α-(1-HYDROXYETHYLIDENE)-ACET-2-METHYL-4-CHLOROANILIDO-α-AZO-(4'-CHLORO-2'-NITROBENZENE)
$C_{17}H_{14}Cl_2N_4O_4$

A. WHITAKER, 1984. Z. Kristallogr., 167, 225-233.

Triclinic, P$\bar{1}$, a = 8.297, b = 10.551, c = 11.856 Å, α = 113.14, β = 67.84, γ = 107.88°, D_m = 1.562, Z = 2. Cu radiation, R = 0.060 for 2806 reflexions.

The molecule (Fig. 1) exists as the hydrazone tautomer and intramolecular H-bonds keep the molecule approximately planar. Molecules are packed in layers parallel to the (0$\bar{2}$3) plane and in columns parallel to b.

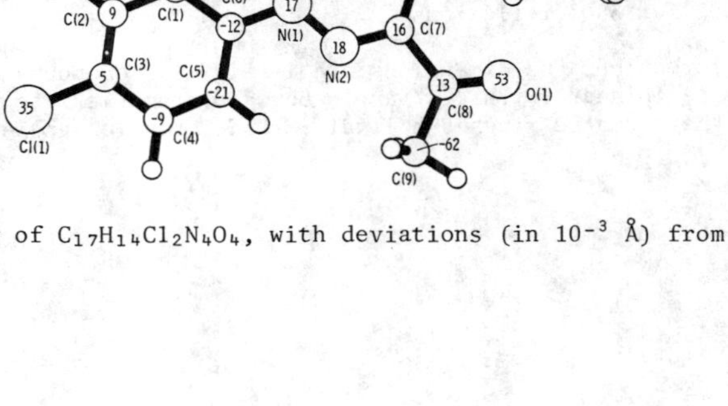

Fig. 1. A view of $C_{17}H_{14}Cl_2N_4O_4$, with deviations (in 10^{-3} Å) from the molecular plane.

N-PHENYL-2-(4-METHYL-2-NITRO-PHENYLHYDRAZONO)-3-OXO-BUTYRAMIDE
$C_{17}H_{16}N_4O_4$

A. WHITAKER, 1984. Z. Kristallogr., 166, 177-188.

Monoclinic, $P2_1/c$, a = 7.587, b = 20.362, c = 12.033 Å, β = 120.68°, D_m = 1.416, Z = 4. Cu radiation, R = 0.080 for 1696 reflexions.

E.F. PAULUS, 1984. Z. Kristallogr., 167, 65-72.

Monoclinic, $P2_1/n$, a = 7.598, b = 20.375, c = 10.440 Å, β = 98.18°, D_m = 1.41, Z = 4. Mo radiation, R = 0.092 for 3043 reflexions.

The molecule (Fig. 1) has the hydrazono tautomeric form, with N1-N2 1.327(2), N2-C7 1.312(2) Å. The molecules crystallize in layers separated by distances in the range 3.133-3.320 Å.

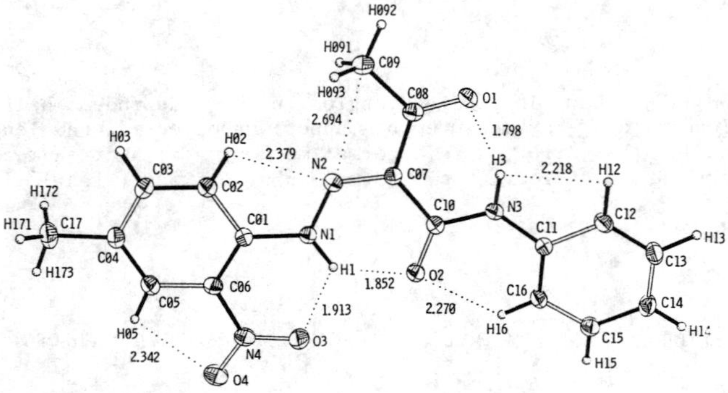

Fig. 1. The $C_{17}H_{16}N_4O_4$ molecule and hydrogen bond distances.

2-(p-DIMETHYLAMINOPHENYL)-3-PHENYLPROPANE-1-NITRILE
$C_{17}H_{18}N_2$ (I)

METHYL 2-(p-NITROPHENYL)-3-PHENYLBUTANOATE
$C_{17}H_{17}NO_4$ (II)

B. TINANT, R. TOUILLAUX, J.P. DECLERCQ, M. VAN MEERSSCHE, G. LEROY and J. WEILER, 1984. Bull. Soc. Chim. Belg., 93, 237-238.

I. Orthorhombic, $P2_12_12_1$, a = 12.339, b = 6.014, c = 19.463 Å, Z = 4. Mo radiation, R = 0.053 for 866 reflexions.

II. Monoclinic, Pn, a = 27.490, b = 12.434, c = 9.135 Å, β = 95.74°, Z = 8. Mo radiation, R = 0.092 for 3503 reflexions.

Molecular conformations and dimensions are in Fig. 1. Compound (I) has its aromatic rings antiperiplanar (with a 3° angle between aromatic ring planes); compound (II) has the aromatic rings synclinal (C-C-C-C torsion angle 53°).

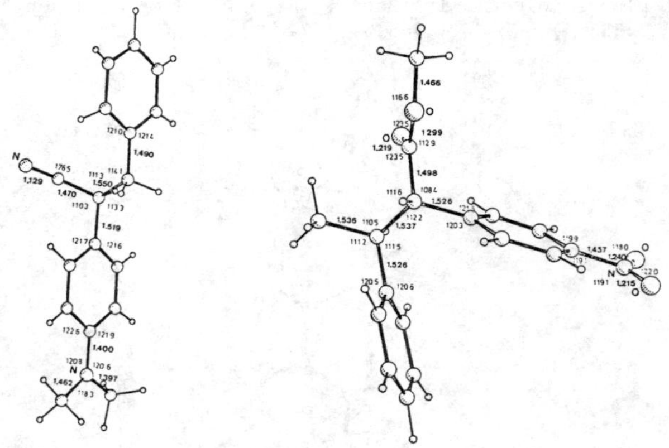

Fig. 1. Molecular conformations and dimensions for $C_{17}H_{18}N_2$ (I) (left) and $C_{17}H_{17}NO_4$ (II) (right) (averaged over the four independent molecules of (II) which have similar conformations).

3-OXO-5-PHENYLAMINOHEX-4-ENE ANILIDE
$C_{18}H_{18}N_2O_2$

P. BRIARD, A. DUBOURG, R. ROQUES, S. CASTILLO, V. HÉRAULT and J.-P. DECLERCQ, 1984. Acta Cryst., C40, 1374-1376.

Monoclinic, $P2_1/c$, a = 11.131, b = 15.778, c = 9.679 Å, β = 112.43°, D_m = 1.22, Z = 4. Mo radiation, R = 0.062 for 1470 reflexions.

The molecule (Fig. 1) is stabilised by an intramolecular hydrogen bond (H(6)... O(9) 1.974 Å). The atoms C(11)-C(16),N(6),C(5),C(4),C(3),C(2),O(9) are nearly coplanar as are C(17)-C(22),N(7),C(1). The angle between the two planes is 110°.

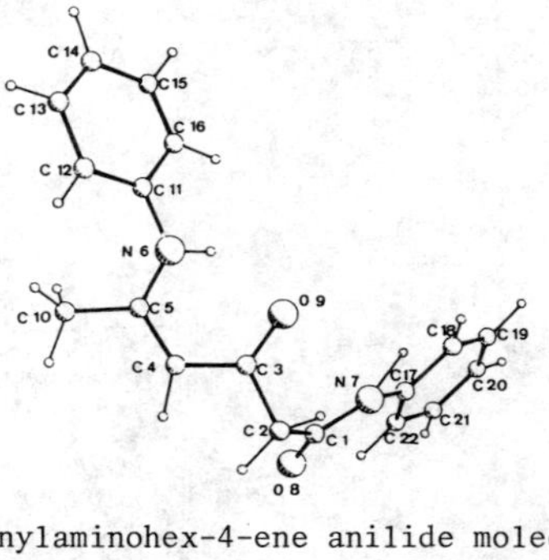

Fig. 1. The 3-oxo-5-phenylaminohex-4-ene anilide molecule.

BENZALDEHYDE-p-(TRICYANOVINYLIDENE)METHYLPHENYLHYDRAZONE
$C_{19}H_{13}N_5$

G. TOSI, P. BRUNI, L. CARDELLINI and G. BOCELLI, 1984. Gazz. Chim. Ital., 114, 111-115.

Orthorhombic, $Pna2_1$, a = 14.994, b = 16.011, c = 6.692 Å, Z = 4. Cu radiation, R = 0.048 for 1245 reflexions.

The almost planar molecule is shown in Fig. 1. Some bond lengths are: N1-N2 1.368(2), N1-C7 1.251(4), N2-C8 1.445(3), N2-C9 1.349(4) Å.

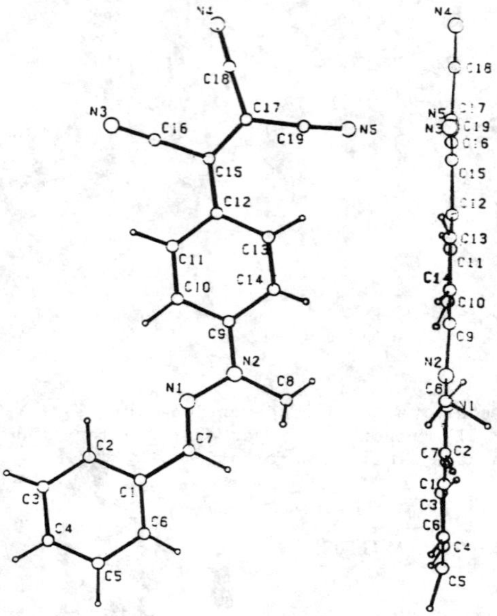

Fig. 1. Views of $C_{19}H_{13}N_5$.

HEXA-2,4-DIYNE-1,6-DIYL BIS(PHENYLCARBAMATE)
$C_{20}H_{16}N_2O_4$ $[C_6H_5,NHCOOCH_2C{\equiv}C]_2$

A. WHULER, P. SPINAT and C. BROUTY, 1984. Acta Cryst., C40, 693-696.

Monoclinic, C2/c, a = 23.54, b = 7.753, c = 18.66 Å, β = 94.48°, Z = 8. Cu radiation, R = 0.077 for 1513 reflexions (at 213K).

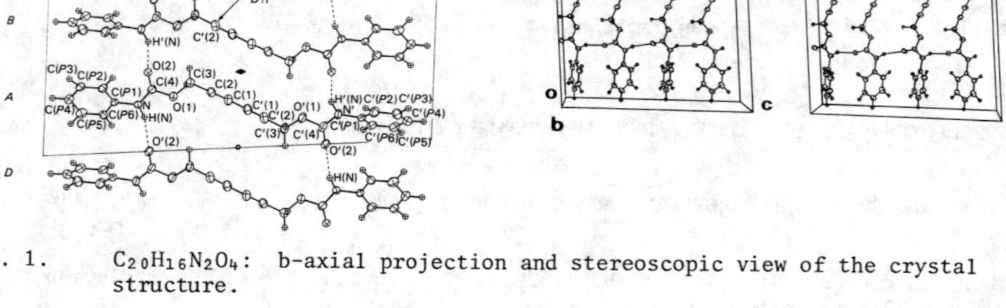

Fig. 1. $C_{20}H_{16}N_2O_4$: b-axial projection and stereoscopic view of the crystal structure.

The molecule (Fig. 1) is not centrosymmetric and the two phenyl groups are in different orientations with respect to the diacetylenic plane (33(2) and 11(2)° respectively). Some bond distances are: C(1)-C(1') 1.376(7), C(1)-C(2) 1.201(7) and C(1')-C(2') 1.191(7) Å. The molecules form parallel chains in the crystal and are linked by N-H...O hydrogen bonds with N...O 2.902(4) and 3.012(5) Å.

DIPHENYLKETENE-N-(2-METHYLPHENYL)IMINE
$C_{21}H_{17}N$ (I)

DIPHENYLKETENE-N-(4-BROMO-2,6-DIMETHYLPHENYL)IMINE
$C_{22}H_{18}BrN$ (II)

J.C. JOCHIMS, J. LAMBRECHT, U. BURKERT, L. ZSOLNAI and G. HUTTNER, 1984. Tetrahedron, 40, 893-903.

I. Orthorhombic, $P2_12_12_1$, a = 7.874, b = 17.07, c = 23.81 Å, Z = 8. Mo radiation, R = 0.081 for 1409 reflexions.

II. Monoclinic, $P2_1/c$, a = 4.183, b = 29.58, c = 14.19 Å, β = 88.93°, Z = 4. Mo radiation, R = 0.034 for 1480 reflexions (at 228 K).

There are two independent molecules in the asymmetric unit of I. There are no unusual bond lengths or angles in the molecules (Fig. 1). Both ketenimines I and II show bent C=C=N moieties (171(2), 174(1)° in I, 170.5(6)° in II). In I the N-phenyl ring is twisted through 6 and -2° from the C-N-C plane; in II the corresponding rotation is -63°.

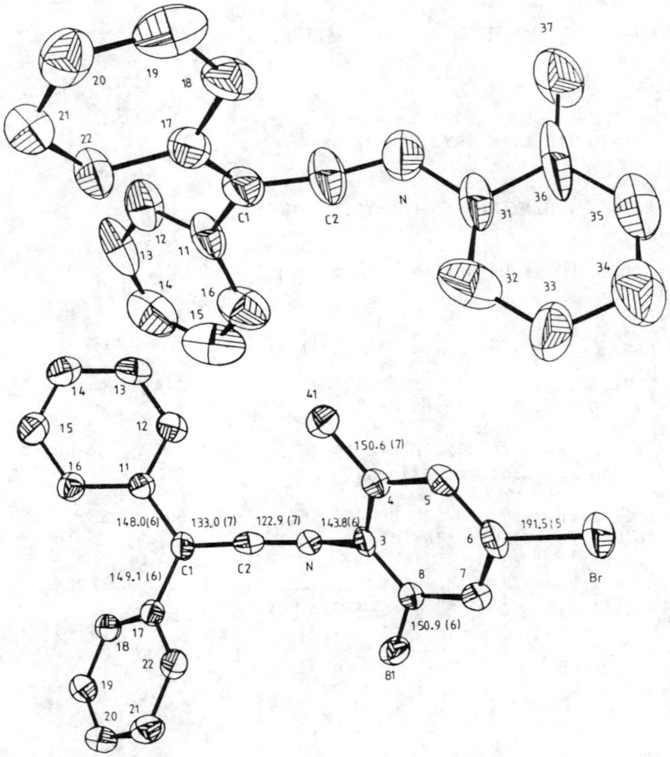

Fig. 1. Drawings of ketenimines I (top) and II (bottom) with some dimensions in pm.

N-BENZOYLOXY-N-(o-TOLYL)BENZAMIDE
$C_{21}H_{17}NO_3$

F. BAERT, J. LAMIOT, D. COUTURIER, D. ROUSSEL and G. RICART, 1984. Acta Cryst., C40, 1071-1072.

Triclinic, $P\bar{1}$, a = 11.268, b = 9.970, c = 8.801 Å, α = 104.30, β = 87.71, γ = 113.25°, Z = 2. Mo radiation, R = 0.052 for 2793 reflexions.

In the molecule (Fig. 1) the benzene ring of the anilide is exo (trans to the oxygen atom). There is no hydrogen bonding and no intermolecular distances between the molecules are less than the sum of the van der Waals radii, except one between C(19) and C(18) which is 3.385(10) Å.

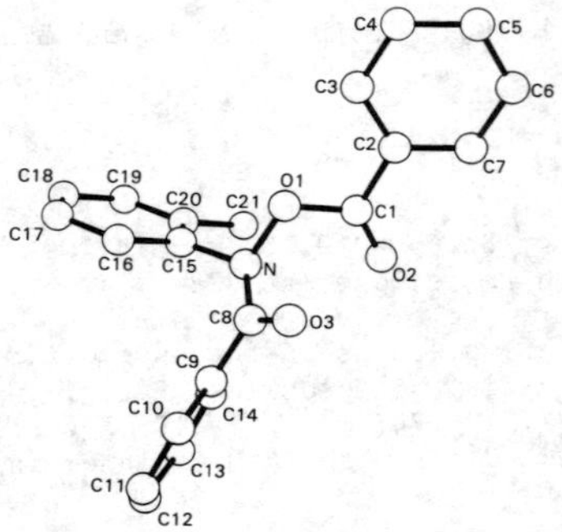

Fig. 1. $C_{21}H_{17}NO_3$: perspective view of the molecule.

1,3-PROPANEDIYLBIS(4-(PHENYLMETHYL)PHENYLCARBAMATE)
$C_{31}H_{30}N_2O_4$ (I)

1,5-PENTANEDIYLBIS(4-(PHENYLMETHYL)PHENYLCARBAMATE)
$C_{33}H_{34}N_2O_4$ (II)

1,6-HEXANEDIYLBIS(4-(PHENYLMETHYL)PHENYLCARBAMATE)
$C_{34}H_{36}N_2O_4$ (III)

L. BORN, 1984. Z. Kristallogr., 167, 145-151.

I. Monoclinic, $P2_1/b$, a = 9.2147, b = 11.301, c = 25.391 Å, α = 82.10°, Z = 4. Cu radiation, R = 0.058 for 2637 reflexions.

II. Monoclinic, C2/c, a = 25.423, b = 4.9991, c = 45.234 Å, β = 100.5°, Z = 8. Cu radiation, R = 0.038 for 2351 reflexions.

III. Monoclinic, $P2_1/b$, a = 5.0701, b = 42.561, c = 6.8604 Å, α = 80.88°, Z = 2. Cu radiation, R = 0.039 for 1455 reflexions.

The three diurethanes are shown in Fig. 1. All are linked by N-H...O=C hydrogen bonds. III lies about an inversion centre.

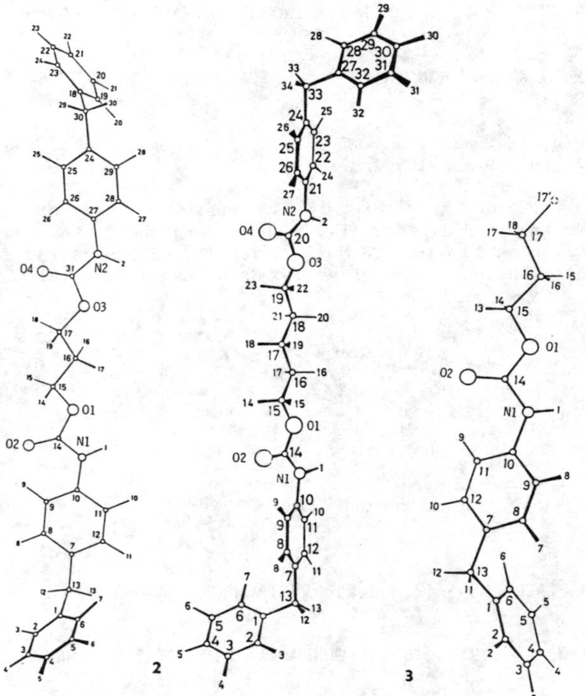

Fig. 1. Views of the diurethanes I - III (left to right).

(±)-NORPHENYLEPHRINE HYDROCHLORIDE
$C_8H_{12}ClNO_2$ $C_8H_{12}NO_2^+.Cl^-$

G. FERGUSON, M. PARVEZ, B.L. RUHL, J.F. MALONE and J.M. MIDGLEY, 1984. Acta Cryst., C40, 1436-1438.

Monoclinic, $P2_1/n$, a = 11.071, b = 5.879, c = 14.310 Å, β = 104.00°, D_m = 1.42, Z = 4. Mo radiation, R = 0.032 for 1293 reflexions.

The aminoethanol side chain is in the gauche conformation with CC-CN and OC-CN torsion angles of 60.9(3) and -64.1(3). The interplanar angle between the phenyl and C1C7C8 planes (Fig. 1) is 89.5(3)°. Each Cl^- ion is hydrogen bonded to three symmetry-related $-N^+H_3$ groups (N...Cl 3.219-3.349(3) Å) and to the M-OH group (O...Cl 3.053(3) Å). An O-H...O hydrogen bond (O...O 2.828(3) Å) is formed between the side chain OH and the M-OH of an adjacent molecule. The analysis indicates that 8% of the unreduced carbonyl precursor is present in the crystal.

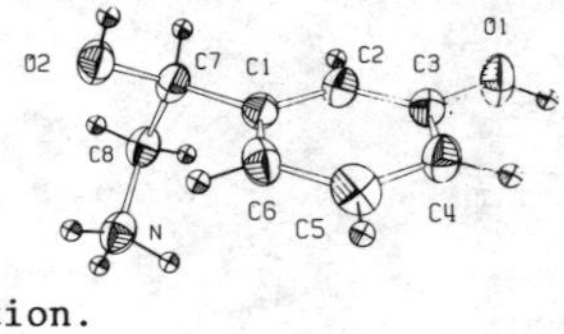

Fig. 1. The $C_8H_{12}NO_2^+$ cation.

DL-NORMETANEPHRINE HYDROCHLORIDE [α-(AMINOMETHYL)VANILLYL ALCOHOL HYDROCHLORIDE]
$C_9H_{14}ClNO_3$ $C_9H_{14}NO_3^+.Cl^-$

R.R. PATTANAYEK, J.K. DATTAGUPTA, S.C. BHATTACHARYYA and N.N. SAHA, 1984. Acta Cryst., C40, 294-297.

Monoclinic, $P2_1/c$, a = 5.218, b = 17.081, c = 12.260 Å, β = 91.50°, D_m = 1.340, Z = 4. Mo radiation, R = 0.069 for 1651 reflexions.

The ethylamine side chain (Fig. 1) is planar, maximally extended and approximately perpendicular to the phenyl ring. The N atom is gauche to the hydroxyl group at the side chain (torsion angle O3-C7-C8-N -55.5°). The crystal structure is stabilized by a three-dimensional network of hydrogen bonds of types N-H...Cl, O-H...Cl and N-H...O.

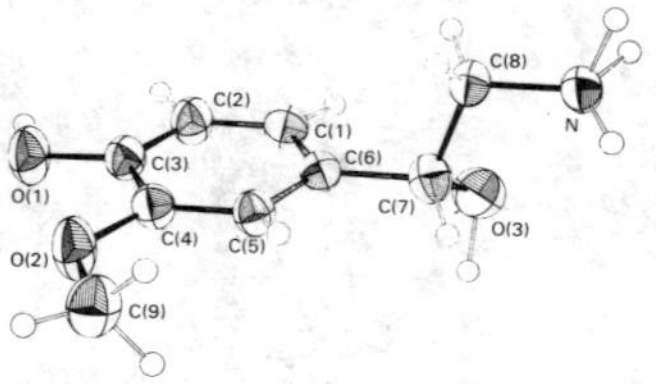

Fig. 1. A view of the cation $C_9H_{14}NO_3^+$.

2,4,4,5,5,6,6-HEPTACHLORO-3-(2,3,4,5,6-PENTACHLOROPHENOXY)-4-CYCLOHEXENONE
$C_{12}Cl_{12}O_2$

M.L. DEINZER, J.G. LAMBERTON, D.C. ROHRER and P.E. STRONG, 1984. J. Org. Chem., 49, 4074-4076.

Triclinic, $P\bar{1}$, a = 11.293, b = 12.601, c = 8.340 Å, α = 109.69, β = 107.44, γ = 108.67°, Z = 2. Mo radiation, R = 0.047 for 5109 reflexions.

The atoms of the phenoxy ring (Fig. 1) are planar and the 4-cyclohexenone ring has an envelope conformation with C2 0.68 Å from the plane of the remaining five atoms. The C13-C3-O3-C1' torsion angle is -172.7°.

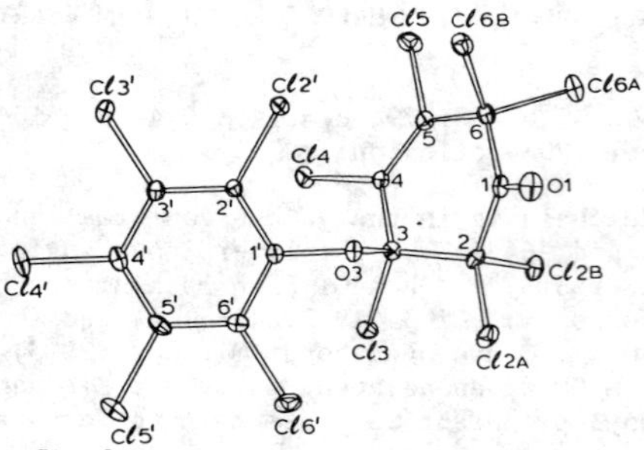

Fig. 1. A view of $C_{12}Cl_{12}O_2$.

4-HYDROXYBIPHENYL (Orthorhombic Form)
$C_{12}H_{10}O$ (I)

4-HYDROXYBIPHENYL (Monoclinic Form)
$C_{12}H_{10}O$ (II)

C.P. BROCK and K.L. HALLER, 1984. J. Phys. Chem., 88, 3570-3574.

I. Orthorhombic, $P2_12_12_1$, a = 15.472, b = 5.469, c = 20.600 Å, Z = 8. Mo radiation, R = 0.042 for 1171 reflexions.

II. Monoclinic, $P2_1/a$, a = 8.067, b = 5.449, c = 20.022 Å, β = 95.14°, Z = 4. Mo radiation, R = 0.042 for 570 reflexions.

Crystals grown from solution are orthorhombic and the asymmetric unit contains two nearly planar molecules (angles between phenyl ring pairs are 2.0 and 2.3(2)°). Mean dimensions are in Fig. 1. Crystals grown by sublimation are monoclinic and the unique molecule in the asymmetric unit is nearly planar (phenyl ring inter-planar angle 1.6(2)°). Mean dimensions are in Fig. 1. As in biphenyl, the structure seems to be complicated by subtle static and/or dynamic disorder.

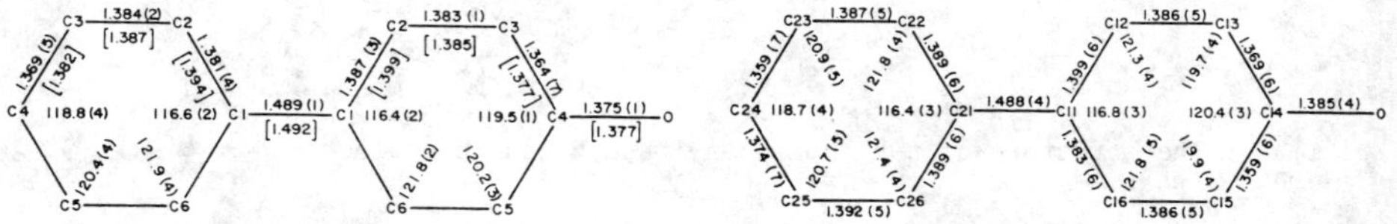

Fig. 1. Molecular dimensions averaged over the two independent molecules of $C_{12}H_{10}O$ for the orthorhombic form (I) (left) and for the monoclinic form. Bond lengths in square brackets are corrected for rigid-body thermal motion.

2-(4,6-DIMETHYL-2-PYRIDYLIMINOMETHYL)PHENOL
$C_{14}H_{14}N_2O$

C. ESCOBAR and M.T. GARLAND, 1984. Acta Cryst., C40, 889-891.

Orthorhombic, $P2_12_12_1$, a = 21.585, b = 8.8285, c = 6.2952 Å, Dm = 1.22, Z = 4. Mo radiation, R = 0.073 for 1054 reflexions.

The molecule is not planar; the pyridine-ring plane is inclined at 5.4(9)° to the salicylidene-ring plane (Fig. 1). The orientation of the pyridine ring is such that the exocyclic C atom is cis to the pyridine N. There is a strong intramolecular O-H...N hydrogen bond with O...N 2.60(1) Å. The crystal packing is effected only by van der Waals forces.

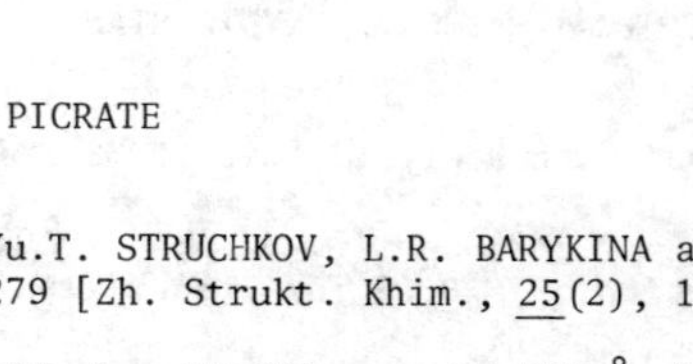

Fig. 1. Perspective view of the $C_{14}H_{14}N_2O$ molecule.

p-HYDROXYPHENYLTHIOPHANIUM PICRATE
$C_{16}H_{15}N_3O_8S$ $(C_{10}H_{13}OS)^+(C_6H_2N_3O_7)^-$

A.E. KALININ, D.S. YUFIT, Yu.T. STRUCHKOV, L.R. BARYKINA and E.N. KARAULOVA, 1984. J. Struct. Chem., 25, 273-279 [Zh. Strukt. Khim., 25(2), 114-121].

Monoclinic, $P2_1/c$, a = 7.5344, b = 12.690, c = 18.700 Å, β = 99.711°, Z = 4. Cu radiation, R = 0.057 for 2464 reflexions.

The structure consists of hydrogen bonded ion pairs (Fig. 1). The β-carbon atoms of the thiophanium ring were found to be twofold disordered. The S atoms have a pyramidal structure typical of sulphonium salts. The p-hydroxyphenyl substituent occupies a pseudoequatorial position of the C_4S ring. Bond lengths (Fig. 1) are generally as expected.

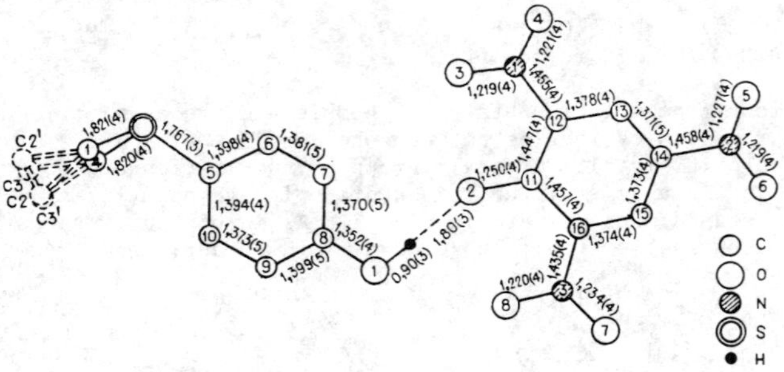

Fig. 1. The structure of p-hydroxyphenylthiophanium picrate.

(E)-1,2-BIS(2-METHOXYPHENYL)ETHENE
$C_{16}H_{16}O_2$

S.J. JUNGK, F.R. FRONCZEK and R.D. GANDOUR, 1984. Acta Cryst., C40, 1873-1875.

Orthorhombic, Pbca, a = 13.341, b = 14.412, c = 13.753 Å, Z = 8. Mo radiation, R = 0.044 for 946 reflexions.

The molecule (Fig. 1) deviates slightly from planarity; one phenyl ring is almost coplanar with the double bond (dihedral angle 2.7(3)°) while the other makes an angle of 15.0(3)° with it. The torsion angles C(1)-C(2)-O(1)-C(15) and C(9)-C(10)-O(2)-C(16) are 173.1(5) and 171.1(5)° while the C(7)-C(8) bond length is 1.320(4) Å.

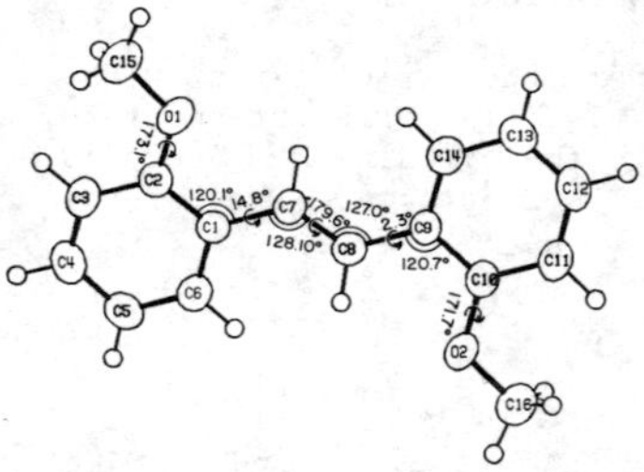

Fig. 1. The 1,2-bis(2-methoxyphenyl)ethene molecule. Some bond and torsion angles are indicated.

2,2,4,4-TETRAMETHYL-3-(2-METHYL-4-METHOXYPHENYL)PENTAN-3-OL
$C_{17}H_{28}O_2$

E. HOUGH and J.S. LOMAS, 1984. Acta Cryst., C40, 1938-1941.

Triclinic, $P\bar{1}$, a = 8.259, b = 7.888, c = 13.493 Å, α = 102.2, β = 96.8, γ = 110.1°, D_m = 1.12, Z = 2. Mo radiation, R = 0.05 for 1736 reflexions.

The molecule has the synperiplanar conformation with the OH group rotated 11.6° out of the phenyl ring plane (Fig. 1). Overcrowding at C(OH) is indicated by C-C distances of 1.604(7) and 1.618(7) Å and a C-C-C angle of 118.6(4)°; the C(1)-C(10)-O(11) angle is 105.6(3)°. The methoxy group is rotated 5.7° out of the phenyl ring plane. The crystal contains sheets of molecules parallel to (100) with the phenyl rings oriented at 45° to the sheets.

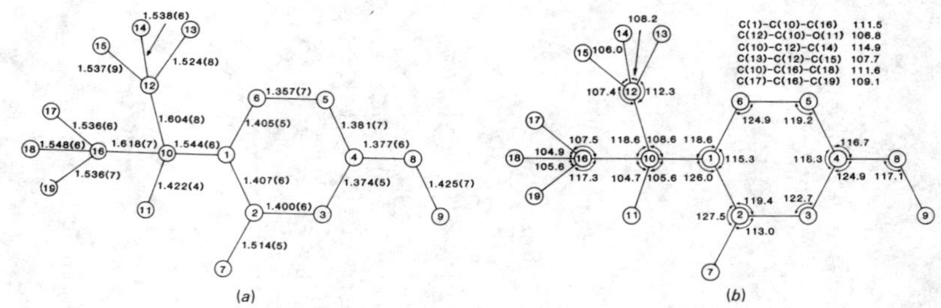

Fig. 1. Bond lengths and angles in the $C_{17}H_{28}O_2$ molecule.

(-)-3-(1-CYCLOPROPYLMETHYL-3-ISOBUTYL-3-PYRROLIDINYL)PHENOL HYDROBROMIDE

$C_{18}H_{28}BrNO$ $C_{18}H_{28}NO^+.Br^-$

F.R. AHMED, 1984. Acta Cryst., C40, 1423-1425.

Orthorhombic, $P2_12_12_1$, a = 14.551, b = 16.013, c = 7.797 Å, D_m = 1.294, Z = 4. Cu radiation, R = 0.025 for 2021 reflexions.

The five-membered pyrrolidine ring has the envelope conformation with C(3) -0.632 Å from the C(2),N(1),C(5),C(4) plane; the 3-isobutyl group is axial (Fig. 1). The 1-cyclopropylmethyl and 3-aryl groups are both equatorial and cis to each other. The absolute configuration of the molecule is 3S. Each Br is involved in two N-H...Br...H'-O' hydrogen bonds (N...Br 3.248(2), O...Br 3.213(3) Å) which bridge the molecules into zig-zag chains parallel to z. The three bonds in the cyclopropyl group and the adjacent C(6)-C(7) average 1.491(4) Å.

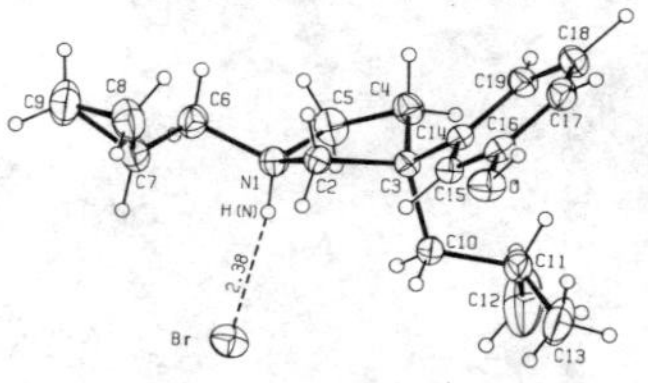

Fig. 1. The molecular structure of $C_{18}H_{28}NO^+.Br^-$.

1-{4-[2-(CYCLOPROPYLMETHOXY)ETHYL]PHENOXY}-3-ISOPROPYLAMINO-2-PROPANOL HYDROCHLORIDE (BETAXOLOL HYDROCHLORIDE)

$C_{18}H_{30}ClNO_3$ $C_{18}H_{30}NO_3{}^+.Cl^-$

C. PASCARD, E.T.H. DAU, P. MANOURY and B. MOMPON, 1984. Acta Cryst., C40, 1430-1432.

Triclinic, $P\bar{1}$, a = 16.646, b = 8.170, c = 7.631 Å, α = 93.01, β = 104.29, γ = 100.75°, Z = 2. Cu radiation, R = 0.14 for 1400 reflexions.

The structure contains partially disordered molecules in a fully extended conformation (Fig. 1). Each Cl^- ion is involved in hydrogen bonding to N and O atoms of the same molecule forming seven-membered rings with N...Cl 3.27(2) and O...Cl 3.04(2) Å. The second H atom, of the NH_2 group also links with another Cl^- ion at -x, -y, 1-2 with N...Cl 3.15(2) Å.

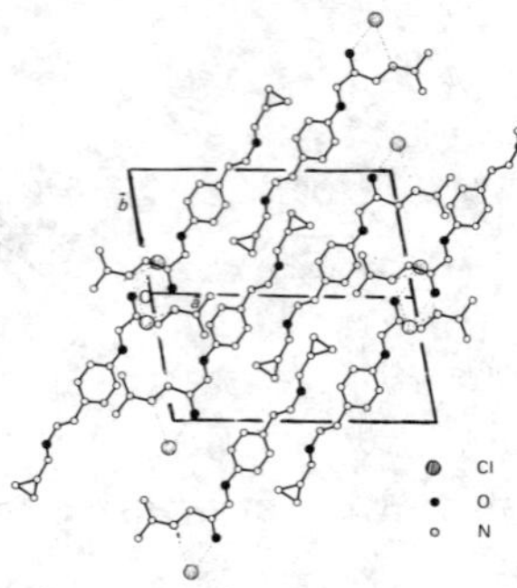

Fig. 1. The structure of $C_{18}H_{30}NO_3^+.Cl^-$ projected along c.

(R,R)-2-HYDROXY-5-[1-HYDROXY-2-(1-METHYL-3-PHENYLPROPYLAMINO)ETHYL]BENZAMIDE HYDROCHLORIDE ((R,R)-LABETALOL HYDROCHLORIDE)

$C_{19}H_{25}ClN_2O_3$ $C_{19}H_{25}N_2O_3^+,Cl^-$

P. MURRAY-RUST, J. MURRAY-RUST, D. HARTLEY, P. HALLETT and J. CLIFTON, 1984. Acta Cryst., C40, 825-828.

Orthorhombic, $P2_12_12_1$, a = 31.04, b = 19.21, c = 6.57 Å, Z = 8. Cu radiation, R = 0.075 for 3766 reflexions.

The crystal contains two crystallographically independent molecules which have different conformations in the propyl chain (Fig. 1). Both molecules adopt a conformation intermediate between 'perpendicular-trans' and 'planar-trans'. The molecules are linked in the crystal by O...O, O...N, N...O, N...Cl and O...Cl hydrogen bonds.

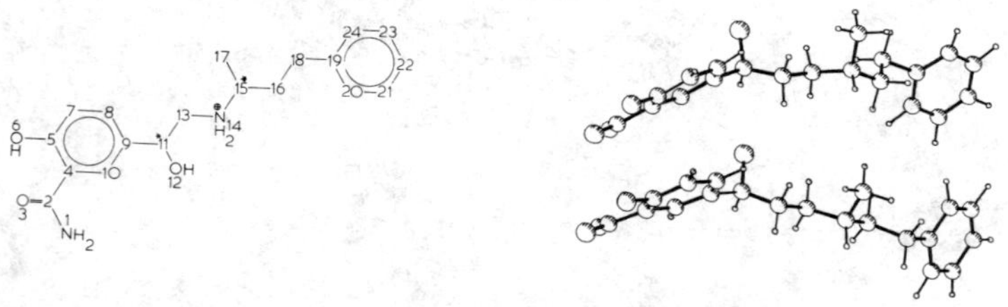

Fig. 1. $C_{19}H_{25}ClN_2O_3$: molecular structure of the cation and views of the two cations of the asymmetric unit roughly parallel to the plane of the substituted phenyl ring.

BIS[2-(o-METHOXYPHENOXY)ETHYL)ETHYL]ETHER - UREA

$C_{19}H_{26}N_2O_6$ $C_{18}H_{22}O_5.CH_4N_2O$

K.K. CHACKO, P. NARASIMHAN and W. SAENGER, 1984. Acta Cryst., C40, 160-163.

Monoclinic, C2, a = 26.084, b = 4.732, c = 7.612 Å, β = 101.75°, D_m = 1.32, Z = 2. Cu radiation, R = 0.052 for 774 reflexions.

The complex (Fig. 1) has twofold crystallographic symmetry. It exists in an 'S'-type conformation with each half of the 'S' loop bound to one half of two urea molecules through weak, bifurcated N-H...O hydrogen bonds to form an infinite chain.

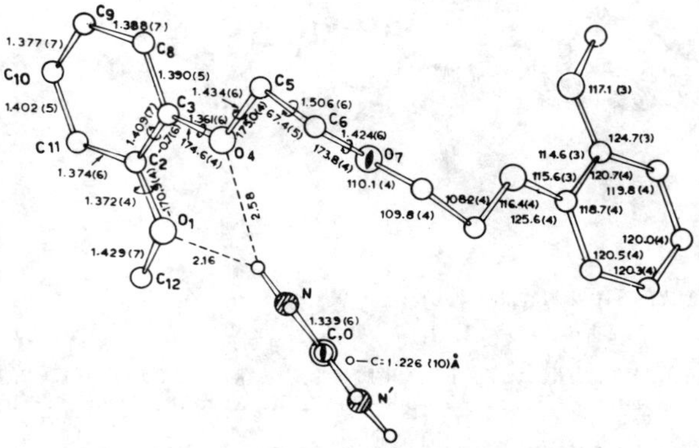

Fig. 1. Details of molecular geometry in the $C_{18}H_{22}O_5.CH_4N_2O$ complex.

2,2-BIS(3,5-DIBROMO-4-HYDROXYPHENYL)PROPANE DIGLYCIDYL ETHER
$C_{21}H_{20}Br_4O_4$

Yu.N. SAF'YANOV, V.P. GOLOVACHEV and E.A. KUZ'MIN, 1984. J. Struct. Chem., 25, 658-660 [Zh. Strukt. Khim., 25(4), 156-157].

Monoclinic, $P2_1/b$, a = 9.703, b = 25.35, c = 9.242 Å, γ = 91.0°, Z = 4. Mo radiation, R = 0.073 for 813 reflexions.

The molecule has an angular shape with a bending at the central carbon atom of the propane group (C(4)-C(1)-C(13) 105°). The Br and O atoms and the central atom of the propane group are located in the plane of the phenyl ring to which they are bonded. The angle between the planes of the phenyl rings is close to 90°. Bond distances and angles are shown in Fig. 1.

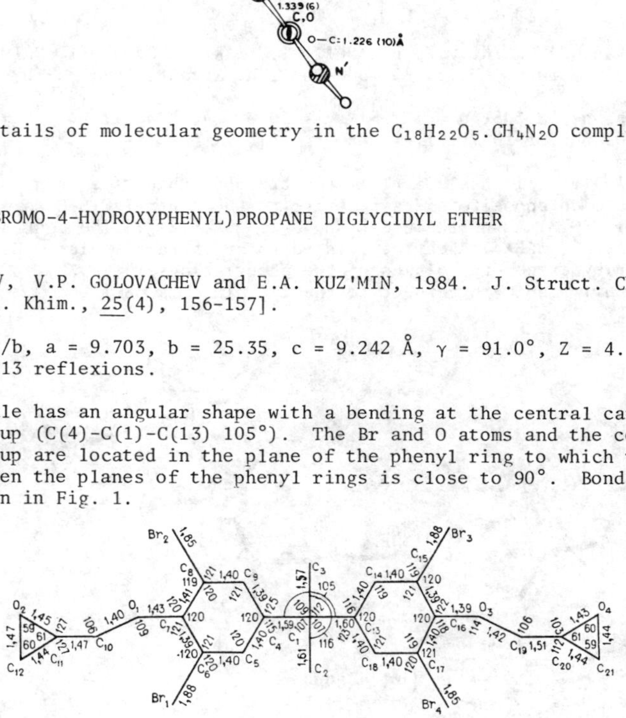

Fig. 1. Geometry of the 2,2-bis(3,5-dibromo-4-hydroxyphenyl)propane diglycidyl ether molecule.

4-(4'-ETHOXYPHENYLAZO)PHENYL HEPTANOATE
$C_{21}H_{26}N_2O_3$

A.M. SHAIKH, N.C. SHIVAPRAKASH, M.M.M. ABDOH and J.S. PRASAD, 1984. Z. Kristallogr., 169, 109-115.

Triclinic, $P\bar{1}$, a = 7.935, b = 8.568, c = 16.216 Å, α = 73.61, β = 75.01, γ = 71.26°, Z = 2. Mo radiation, R = 0.0457 for 1922 reflexions.

The molecule (Fig. 1) is maximally extended. The angle between the two phenyl ring planes is 15.0°. The crystal structure contains pairs of centrosymmetrically related molecules with partial overlap of the phenyl-azo-phenyl groups.

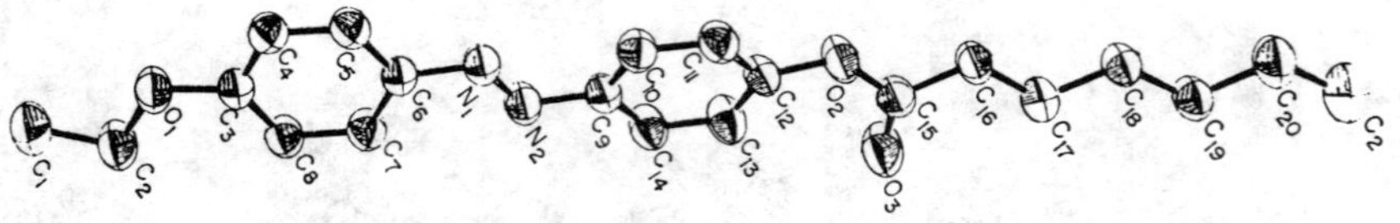

Fig. 1. A view of $C_{21}H_{26}N_2O_3$.

2,2'-THIOBIS(4-METHYL-6-t-BUTYLPHENOL)
$C_{22}H_{30}O_2S$

L.A. CHETKINA, V.E. ZAVODNIK. V.K. BEL'SKII, I.G. ARZAMANOVA, M.I. NAIMAN and Ya.A. GURVICH, 1984. J. Struct. Chem., 25, 940-943 [Zh. Strukt. Khim. 25(6), 114-117].

Monoclinic, A2/a, a = 8.845, b = 12.951, c = 19.352 Å, γ = 111.94°, Z = 4. Mo radiation, R = 0.036 for 1087 reflexions.

The molecule (Fig. 1) has exact C_2 symmetry and exhibits a trans structure: the hydroxy groups are on opposite sides relative to the central C-S-C moiety. The C-S bond length is 1.780(3) Å and the C-S-C angle is 105.5(2)°. The contraction of the C-OH bond length to 1.353(4) Å may be related to an intramolecular O-H...S interaction. The dihedral angle between the planes of the phenyl rings is 72.3°, and the twist angle of the phenyl ring with respect to the CSC plane is 68.1°.

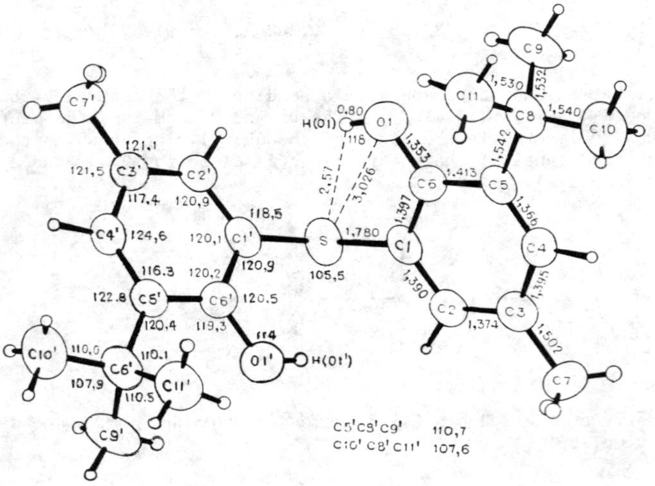

Fig. 1. The 2,2'-thiobis(4-methyl-6-t-butylphenol) molecule (σ = 0.003-0.005 Å and 0.2-0.3° for parameters involving non-hydrogen atoms).

2,2'-METHYLENEBIS(4-METHYL-6-t-BUTYLPHENOL)
$C_{23}H_{32}O_2$

L.A. CHETKINA, V.E. ZAVODNIK, V.K. BEL'SKII, I.G. ARZAMANOVA, M.I. NAIMAN and Ya.A. GURVICH, 1984. J. Struct. Chem., 25, 935-939 [Zh. Strukt. Khim., 25(6), 109-113].

Orthorhombic, $Pna2_1$, a = 16.111, b = 12.808, c = 10.159 Å, Z = 4. Mo radiation, R = 0.061 for 892 reflexions.

The conformation of the molecule (Fig. 1) is determined by an intramolecular hydrogen bond between the hydroxy groups, which are in a cis orientation with respect to the methylene bridge (O...O 2.76, O...H 2.1 Å). The dihedral angle between the planes of the phenyl rings is 62°, and the twist angles of the rings relative to the plane through the three central carbon atoms are 102 and 89°. The C-C-C angle at the central carbon is 113°. Intermolecular hydrogen bonding is prevented by the steric bulk of the t-butyl groups. Bond lengths and angles (σ = 0.015-0.020 Å and 0.9-1.1°) are as expected.

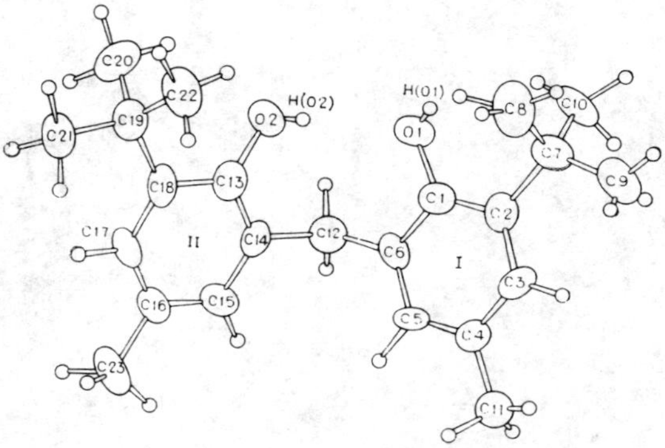

Fig. 1. The 2,2'-methylenebis(4-methyl-6-t-butylphenol) molecule.

4-(4-HYDROXY-3-IODOPHENOXY)-3,5-DIIODOPHENETHYLAMINE BIS(SALICYLATO)BORATE (1:1) SALT
$C_{28}H_{21}BI_3NO_8$ $C_{14}H_{13}I_3NO_2{}^+,C_{14}H_8BO_6{}^-$

V. CODY, 1984. Acta Cryst., C40, 1214-1217.

Monoclinic, $P2_1/c$, a = 14.886, b = 14.514, c = 15.870 Å, β = 108.92°, Z = 4. Mo radiation, R = 0.064 for 8316 reflexions.

In the salt (Fig. 1) the cation is cisoid, the 3'-iodine proximal, the protonated side chain extended, and the diphenyl ether conformation twist-skewed, whilst the anion has tetrahedral B coordination with the planes of each salicylate moiety bisecting one another. The protonated amine and 4'-OH of the cation are involved in an extended network of hydrogen bonds with the O atoms of the anion.

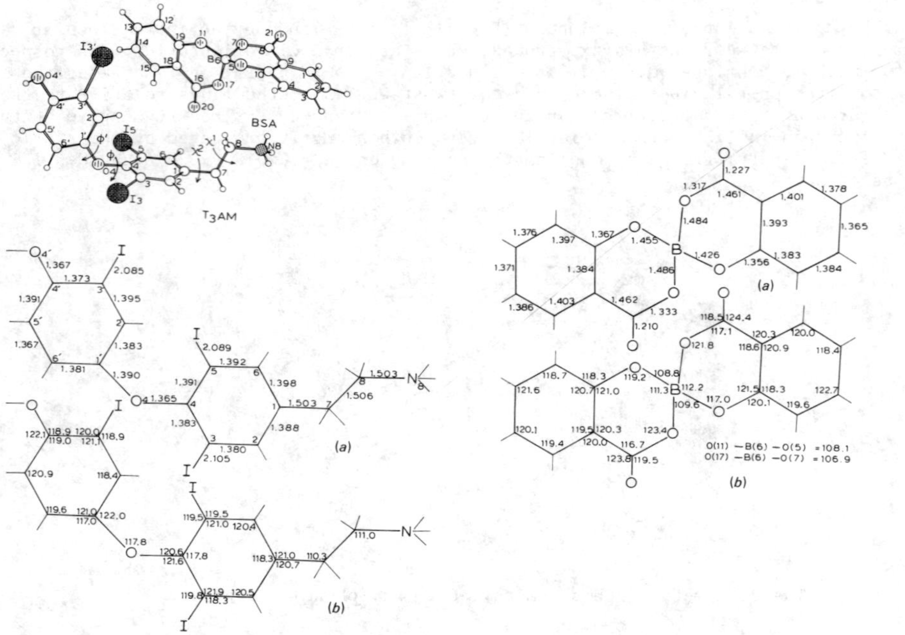

Fig. 1. $C_{14}H_{13}I_3NO_2^+,C_{14}H_8BO_6^-$: the structure and bond lengths and angles in the cation (left) and anion (right).

BIS[1-(4-METHOXYPHENYL)-2-PHENYL-2-[[TRIFLUOROMETHYLSULPHONYL]OXY]VINYL] ETHER
$C_{32}H_{24}F_6O_9S_2$

G. MAAS and W. LORENZ, 1984. J. Org. Chem., 49, 2273-2278.

Monoclinic, C2/c, a = 20.175, b = 9.475, c = 18.279 Å, β = 101.78°, Z = 4. Mo radiation, R = 0.045 for 2326 reflexions.

The structure of this reaction product is shown in Fig. 1. The molecule has twofold crystallographic symmetry.

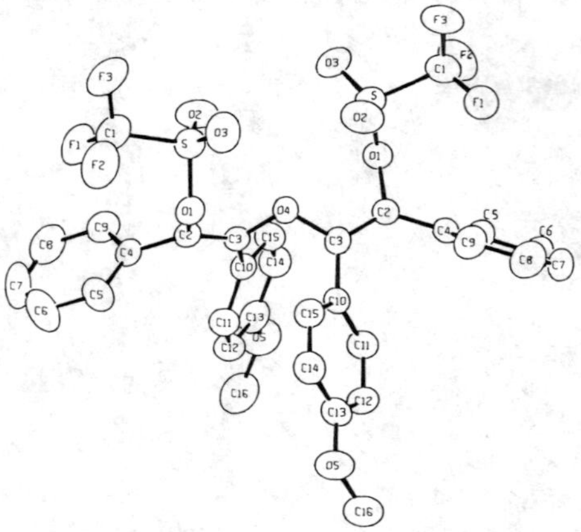

Fig. 1. A view of the $C_{32}H_{24}F_6O_9S_2$ molecule.

2-AMINO-3,5,6-TRICHLORO-1,4-BENZOQUINONE
$C_6H_2Cl_3NO_2$

H. ENDRES, E. ROSSATO and R. BALLES, 1984. Z. Naturforsch., B39, 445-448.

Orthorhombic, $P2_12_12_1$, a = 4.941, b = 10.725, c = 15.400 Å, Z = 4. Mo radiation, R = 0.035 for 780 reflexions.

Details of the molecular geometry of this intermediate in the preparation of 2,5-diamino-3,6-dichloro-1,4-benzoquinone are shown in Fig. 1. Molecules are linked by pairs of N-H...O hydrogen bonds (with H...O 2.20(6) and 2.15(6) Å).

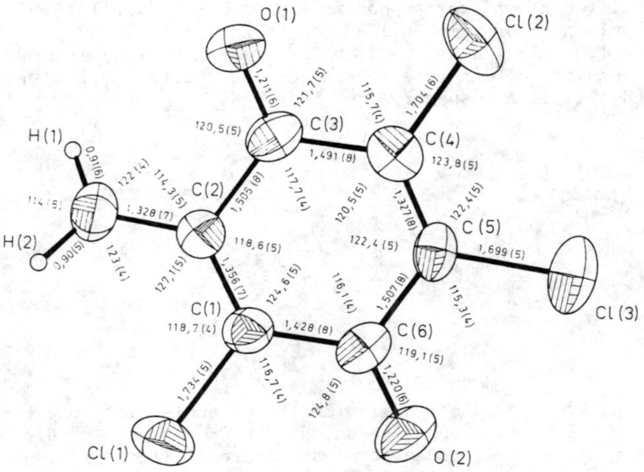

Fig. 1. Molecular geometry details for $C_6H_2Cl_3NO_2$.

2-METHOXY-6-PENTYL-1,4-BENZOQUINONE
$C_{12}H_{16}O_3$

H.W. SCHMALLE, O.H. JARCHOW, B.M. HAUSEN and K.H. SCHULZ, 1984. Acta Cryst., C40, 1084-1087.

Triclinic, $P\bar{1}$, a = 4.123, b = 9.759, c = 14.256 Å, α = 87.07, β = 87.38, γ = 79.15°, Z = 2. Cu radiation, R = 0.058 for 1518 reflexions.

In the molecule (Fig. 1) the angle between the quinonoid ring plane and the pentyl chain plane is 13.4(4)°. The average dimensions of the $C(sp^3)$-$C(sp^3)$ distances and corresponding angles of the aliphatic chain are 1.523(5) Å and 112.8(3)°, those for the quinone ring are C-C 1.477(4), C=C 1.338(5), C=O 1.225(4) Å, C-C-C 118.6(3), C=C-C 121.1(3), O=C-C 120.7(3)°, the shortest intermolecular C...O distance being 3.33(1) Å.

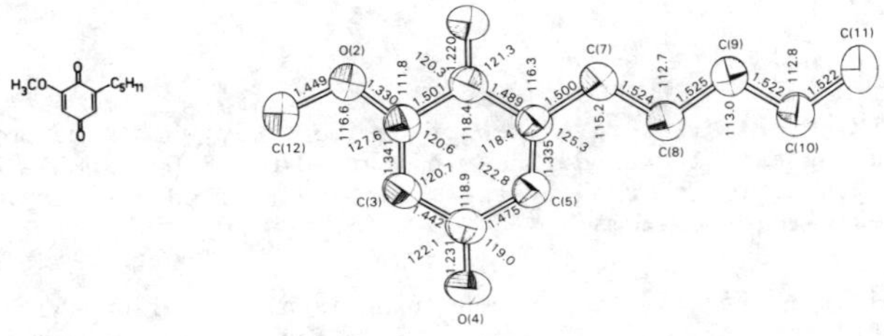

Fig. 1. $C_{12}H_{16}O_3$: bond distances and angles.

(R,S)-4-METHOXYDALBERGIONE
$C_{16}H_{14}O_3$

H.W. SCHMALLE, O.H. JARCHOW, B.M. HAUSEN and K.H. SCHULZ, 1984. Acta Cryst., C40, 1087-1090.

Monoclinic, $P2_1/n$, a = 6.701, b = 38.826, c = 5.365 Å, β = 109.56°, Z = 4. Cu radiation, R = 0.053 for 1449 reflexions.

In the molecule (Fig. 1) the quinone ring is almost planar with a maximum out-of-plane deviation of -0.031(3) Å for C(2). The average dimensions of the C(sp^3)-C(sp^2) distances and corresponding angles are 1.508(4) Å and 112.5(3)°; those for the quinone ring are C-C 1.477(4), C=C 1.333(4), C=O 1.220(4) Å, C-C-C 118.2(3), C=C-C 120.9(3) and O=C-C 121.0(3)°. The methylene distance is 1.279(5) Å. The structure consists of discrete molecules; the shortest O...O and C...O contacts are 3.28(1) and 3.41(1) Å respectively.

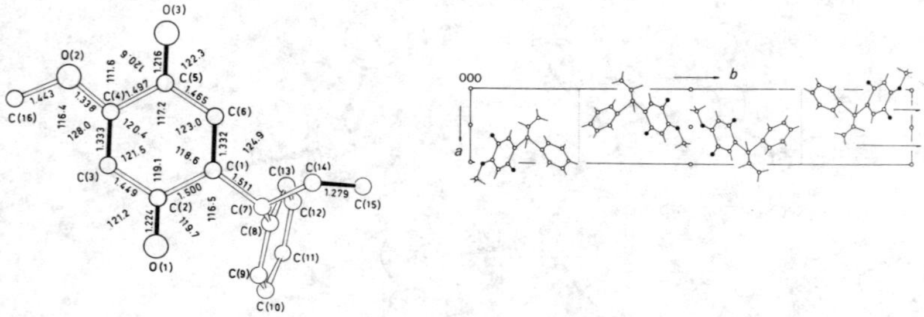

Fig. 1. $C_{16}H_{14}O_3$: perspective view of the molecule with some bond distances and angles (blackened lines indicate C=C and C=O double bonds) and the ab projection of the crystal structure (filled circles indicate O atoms and open circles C atoms).

(R)-3,4-DIMETHOXYDALBERGIONE
$C_{17}H_{16}O_4$

H.W. SCHMALLE, O.H. JARCHOW, B.M. HAUSEN and K.H. SCHULZ, 1984. Acta Cryst., C40, 1090-1092.

Orthorhombic, $P2_12_12_1$, a = 7.462, b = 7.961, c = 25.254 Å, Z = 4. Cu radiation, R = 0.095 for 885 reflexions.

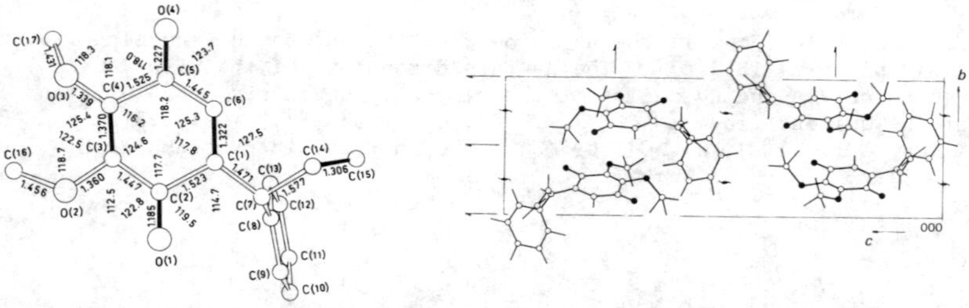

Fig. 1. $C_{17}H_{16}O_4$: perspective view of the molecule with some bond distances and angles (blackened lines indicate C=C and C=O double bonds) and the bc projection of the crystal structure (filled circles indicate O atoms, open circles C atoms).

In the molecule (Fig. 1) the quinone ring is almost planar with a maximum out-of-plane deviation of -0.02(1) Å for C(3). The average values of the C-C distances and corresponding angles of the chiral atom C(7) are 1.51(2) Å and

112.1(1.3)°. The mean dimensions of the quinone ring are C-C 1.485(11), C=C 1.346(16), C=O 1.206(14) Å and C-C-C 118(1), C=C-C 121(1), O=C-C 121(1)°. The methylene distance is 1.306(13) Å. The structure consists of discrete molecules.

(S)-4,4'-DIMETHOXYDALBERGIONE
$C_{17}H_{16}O_4$

H.W. SCHMALLE, O.H. JARCHOW, B.M. HAUSEN and K.H. SCHULZ, 1984. Acta Cryst., C40, 1092-1094.

Monoclinic, $P2_1$, a = 7.236, b = 6.479, c = 15.844 Å, β = 93.16°, Z = 2. Cu radiation, R = 0.052 for 1184 reflexions.

In the molecule (Fig. 1) the atoms of the quinone ring form an almost planar system with a maximum out-of-plane deviation of -0.022(5) Å for C(2). The average values of the $C(sp^3)$-$C(sp^2)$ bond lengths and corresponding angles are 1.519(7) Å and 112.7(5)°; the mean values of the quinone-ring dimensions are C-C 1.483(7), C=C 1.328(8), C=O 1.215(7) Å, C-C-C 117.5(5), C=C-C 121.5(5) and O=C-C 121.3(5)°. The methylene bond length is extremely short: 1.240(9) Å. With the exception of a very weak C-H...O interaction in the b direction the structure consists of discrete molecules.

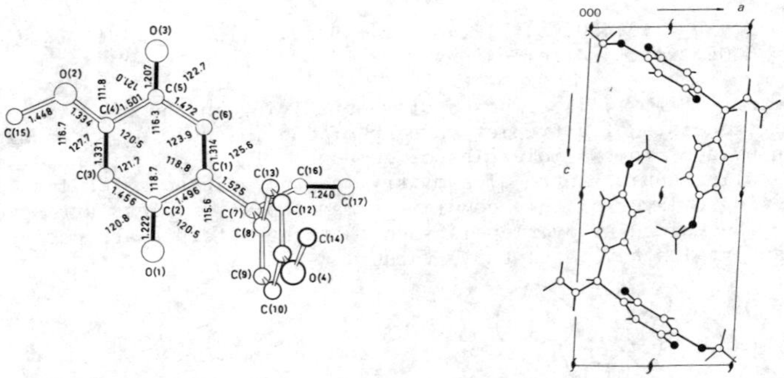

Fig. 1. $C_{17}H_{16}O_4$: perspective view of the molecule with some bond distances and angles (blackened lines indicate C=C and C=O double bonds) and the ac projection of the crystal structure (filled circles indicate O atoms, open circles C atoms).

PERCHLOROFUCHSONE CARBON TETRACHLORIDE SOLVATE
$C_{19}Cl_{14}O,CCl_4$

E. MOLINS, J. RIUS and C. MIRAVITLLES, 1984. Z. Kristallogr., 169, 149-158.

Triclinic, $P\bar{1}$, a = 7.256, b = 13.663, c = 16.341 Å, α = 75.22, β = 80.18, γ = 82.57°, Z = 2. Mo radiation, R = 0.047 for 4521 reflexions.

The molecule (Fig. 1) has a propeller conformation; the twist at the C=C moiety is 33.3(9)°.

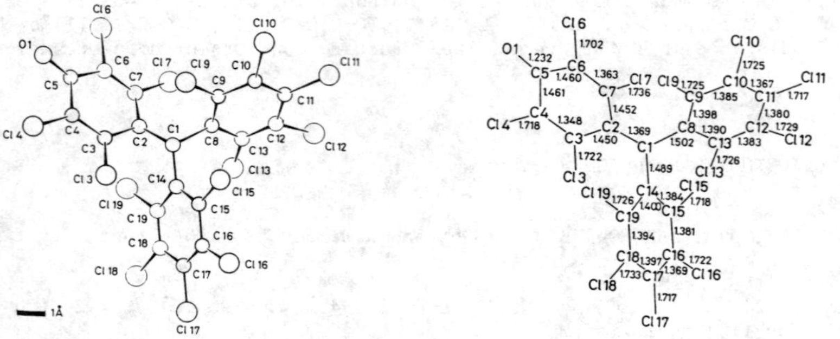

Fig. 1. A view of $C_{19}Cl_{14}O$ and bond lengths.

BIS(2,5-DIMETHYL-1,4-BENZOQUINONE) 2,5-DIMETHYL-1,4-DIHYDROXYBENZENE

$C_{24}H_{26}O_6$ $2(C_8H_8O_2),C_8H_{10}O_2$

A.O. PATIL, S.R. WILSON, D.Y. CURTIN and I.C. PAUL, 1984. J. Chem. Soc. Perkin II, 1107-1110.

Monoclinic, $P2_1/c$, a = 10.480, b = 12.344, c = 9.301 Å, β = 115.91°, Z = 2. Cu radiation, R = 0.051 for 1404 reflexions.

The structure (Fig. 1) is composed of sheets lying in the $\bar{3}02$ plane and the basic structural unit consists of a triplet made up of one molecule of hydroquinone which forms hydrogen bridges to two molecules of quinone. The hydroquinone of each triplet lies at a crystallographic centre of symmetry and the quinone molecules lie in general positions. A single layer of this complex is made up of triplets arranged in herring-bone fashion. The molecular overlap is such that the interplanar spacing is 3.30 Å. Bond lengths and angles have acceptable values.

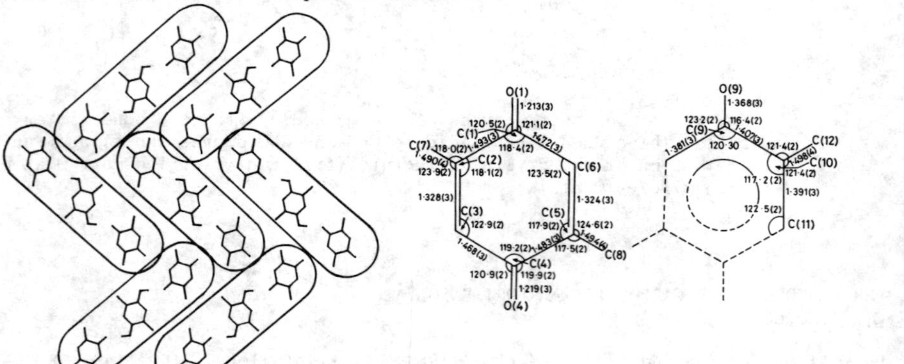

Fig. 1. $C_{24}H_{26}O_6$: a view of a sheet of molecules lying in the ($\bar{3}02$) plane showing the triplets and bond lengths and angles in the complex.

α-CHLOROACETOPHENONE

C_8H_7ClO (I) $PhCOCH_2Cl$

2-METHYLSULPHONYL-1-PHENYLETHANONE

$C_9H_{10}O_3S$ (II) $PhCOCH_2SO_2CH_3$

2-CHLORO-2-METHYLSULPHONYL-1-PHENYLETHANONE
$C_9H_9ClO_3S$ (III) $PhCOCHClSO_2CH_3$

CHLOROMETHYLMETHYLSULPHONE
$C_2H_5ClO_2S$ (IV) $ClCH_2SO_2CH_3$

J.S. GROSSERT, P.K. DUBEY, G.H. GILL, T.S. CAMERON and P.A. GARDNER, 1984. Can. J. Chem., 62, 798-807.

I. Orthorhombic, $P2_12_12_1$, a = 4.034, b = 9.54, c = 18.999 Å, Z = 4. Mo radiation, R = 0.096 for 436 reflexions (at -140°C).

II. Monoclinic, $P2_1$, a = 9.242, b = 5.231, c = 9.622 Å, β = 92.42°, Z = 2. Mo radiation, R = 0.038 for 811 reflexions.

III. Monoclinic, $P2_1/a$, a = 8.566, b = 8.354, c = 14.767 Å, β = 104.34°, Z = 4. Mo radiation, R = 0.031 for 812 reflexions.

IV. Triclinic, $P\bar{1}$, a = 5.162, b = 7.130, c = 7.151 Å, α = 101.59, β = 89.42, γ = 92.97°, Z = 2. Mo radiation, R = 0.046 for 376 reflexions.

The individual molecules are shown in Fig. 1. The S-C(α) distances in II, III and IV are 1.787(4), 1.815(4) and 1.781(9) Å respectively; the lengthening of the bond in III is a consequence of the substitution of hydrogen by the electron withdrawing chlorine. The S-Me distances in II, III and IV are 1.759(4), 1.760(5) and 1.765(9) Å respectively. In I, the conformation is defined by the Cl-C-C-Ph torsion angle of 177(2)°.

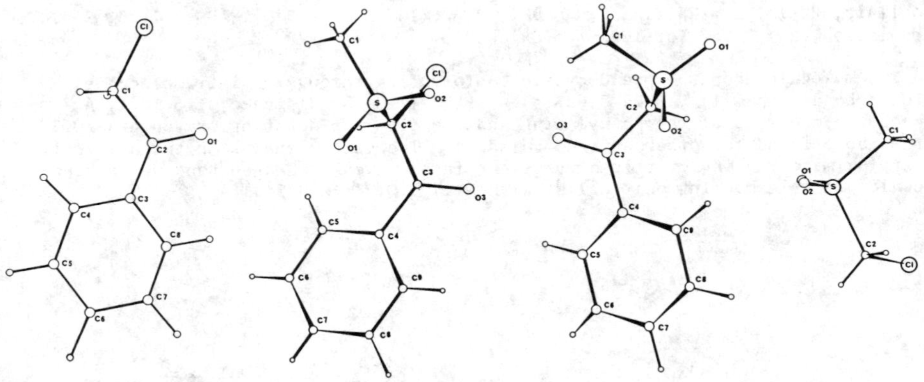

Fig. 1. Views of the compounds I, II, III and IV (left to right).

BIS(TRIFLUOROACETATO)(PHENYL)IODINE(III)
$C_{10}H_5F_6IO_4$

G.A. STERGIOUDIS, S.C. KOKKOU, A.P. BOZOPOULOS and P.J. RENTZEPERIS, 1984. Acta Cryst., C40, 877-879

Triclinic, $P\bar{1}$, a = 9.824, b = 9.111, c = 7.688 Å, α = 88.75, β = 99.80, γ = 90.91°, D_m = 2.076, Z = 2. Mo radiation, R = 0.052 for 1362 reflexions.

In the molecule iodine has characteristic T-shaped coordination with I-O 2.17(1) and 2.13(1) and I-C 2.08(1) Å (Fig. 1). This coordination is complimented by intra- and intermolecular I...O contacts of 3.00(1), 3.16(1) and 3.06(1) Å. O(3) is involved in both types of contact and this causes the formation of dimers between symmetry-equivalent molecules which, repeated along the b axis, form columns interconnected by O...H-C hydrogen bridging bonds.

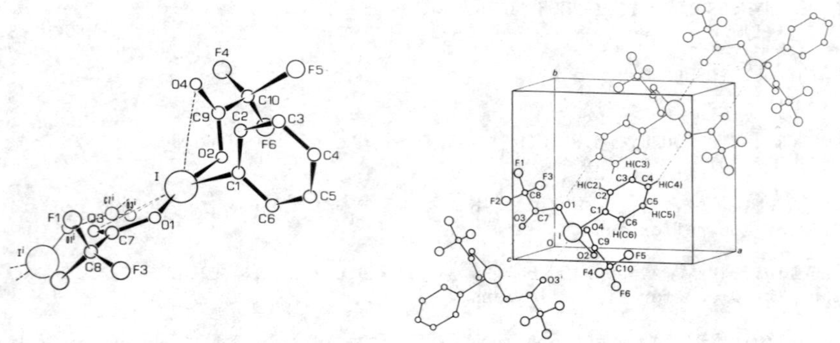

Fig. 1. Clinographic projections of the $C_{10}H_5F_6IO_4$ molecule and of the molecular packing.

4-HYDROXY-4-PHENYLPENTANAMIDE
$C_{11}H_{15}NO_2$

J. ZUKERMAN-SCHPECTOR, E.E. CASTELLANO, G. OLIVA and G. CARVAJAL, 1984. Acta Cryst., C40, 685-687.

Monoclinic, $P2_1/n$, a = 24.215, b = 6.981, c = 6.147 Å, β = 91.7°, D_m = 1.23, Z = 4. Cu radiation, R = 0.058 for 1091 reflexions.

The molecule adopts a pseudocyclic conformation through an intramolecular hydrogen bond with O(1)...O(2) 2.683(4) Å (Fig. 1). The distance of 6.10(1) Å between the centroid of the phenyl ring and the electron-donating oxygen atom O(2) is in the range found for other anti-convulsants. The carbon chains in the unit cell lie approximately along c; this arrangement favours weak hydrogen bonding between molecules in neighbouring unit cells, with N(1)...O(2') 2.925(5) Å.

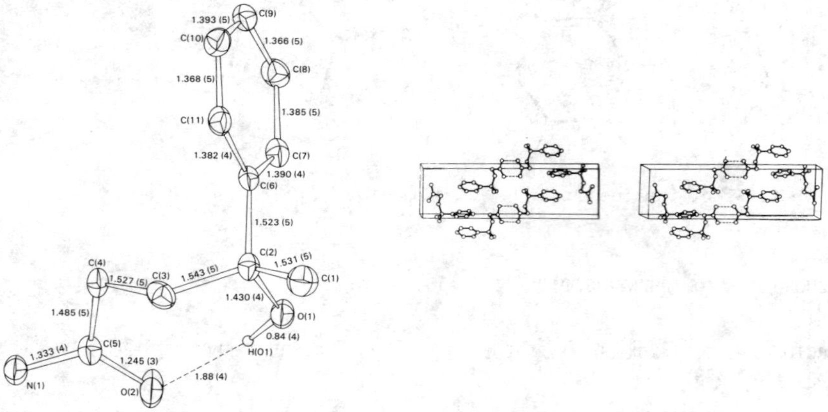

Fig. 1. $C_{11}H_{15}NO_2$: bond lengths in the molecule and a stereoview of the crystal structure; broken lines indicate hydrogen bonds.

N-[BIS(METHYLTHIO)METHYLENE]CINNAMAMIDE
$C_{12}H_{13}NOS_2$

N.U. KAMATH and K. VENKATESAN, 1984. Acta Cryst., C40, 1610-1612.

Monoclinic, $P2_1/n$, a = 14.626, b = 7.144, c = 11.996 Å, β = 90.03°, D_m = 1.326, Z = 4. Mo radiation, R = 0.035 for 1455 reflexions.

The molecule (Fig. 1) contains an intramolecular, attractive S...O interaction with S(1)...O(6) 2.658(3) Å. O(6) is 34.8° out of the C(15),S(1),C(3) plane and the C(15)-S(1)...O(6) angle is 175.8(3)°. Further bond lengths are, C(3)-N(4) 1.289(4), C(7)-C(8) 1.311(5), S(1)-C(3) 1.746(3), S(2)-C(3) 1.753(3) Å.

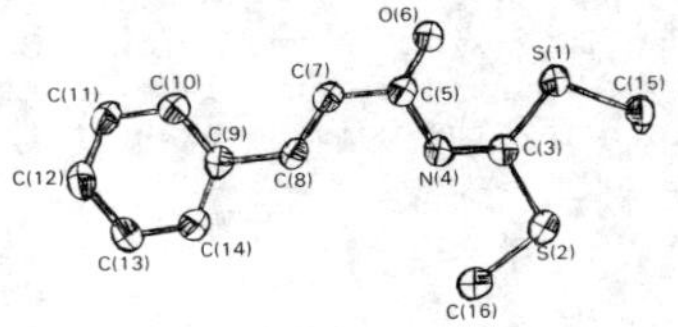

Fig. 1. The $C_{12}H_{13}NOS_2$ molecule.

1,4-DIBROMO-2,5-DIETHYL-3,6-DIMETHYLBENZENE
$C_{12}H_{16}Br_2$ (I)

1,3-DIBROMO-2,5-DIETHYL-4,6-DIMETHYLBENZENE
$C_{12}H_{16}Br_2$ (II)

R.A. WOOD, T.R. WELBERRY and M. PUZA, 1984. Acta Cryst., C40, 1255-1260.

I. Monoclinic, $P2_1$, a = 9.084, b = 4.459, c = 17.940 Å, β = 122.82°, Z = 2. Cu radiation, R = 0.064 for 661 reflexions.

II. Monoclinic, $P2_1$, a = 9.086, b = 4.442, c = 17.969 Å, β = 122.69°, Z = 2. Cu radiation, R = 0.063 for 933 reflexions.

In both isomers (Fig. 1) the 1,3,4,6 substituents show disorder of Br and CH_3. For I the 1,4 sites are occupied on average by 0.563(2) Br and 0.437(2) C and vice versa for 3,6, while for II the 1,3 sites are occupied on average by 0.506(1) Br and 0.494(1) C and vice versa for the 4,6 sites.

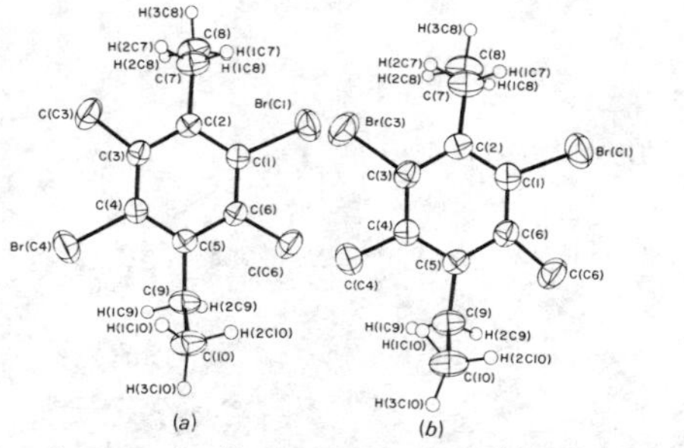

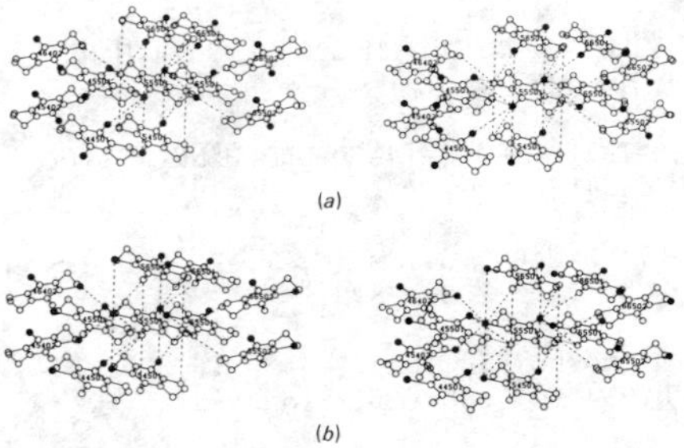

Fig. 1. $C_{12}H_{16}Br_2$: perspective views of the molecules and stereoviews of the molecular packing in (a) molecule I and (b) molecule II.

cis-DICHLOROSTILBENE
$C_{14}H_{10}Cl_2$ (I)

trans-DICHLOROSTILBENE
$C_{14}H_{10}Cl_2$ (II)

E. HEY, F. WELLER, K. DEHNICKE and G. MAIER, 1984. Z. Naturforsch., B39, 409-415.

I. Rhombohedral, $R\bar{3}$, a = 33.230, c = 6.006 Å, Z = 18. Mo radiation, R = 0.064 for 821 reflexions.

II. Monoclinic, $P2_1/n$, a = 5.719, b = 17.325, c = 6.406 Å, β = 111.09°, Z = 2. Mo radiation, R = 0.056 for 1138 reflexions.

In (I) there is one molecule per asymmetric unit (Fig. 1), with C=C 1.326(10) Å and the dihedral angles of the phenyl rings with olefin plane are 48 and 72°. In (II) the molecule lies about an inversion centre, with C=C 1.325(10) and the dihedral angle of the unique phenyl ring with the olefin plane is 71°. The analysis of (II) was complicated by disorder (3:7) of the molecules about the inversion centre as shown in Fig. 1.

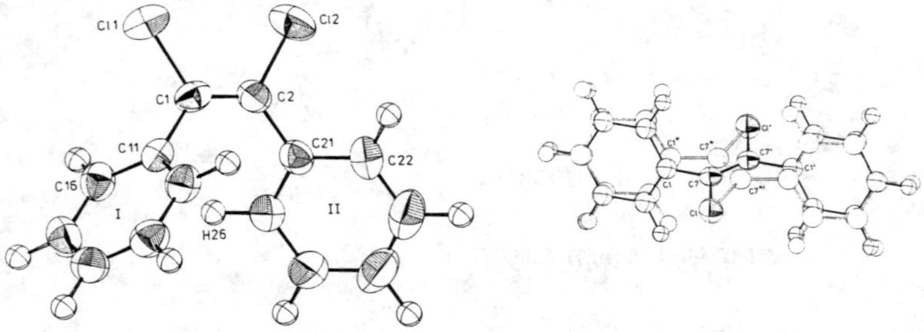

Fig. 1. Views of cis- and trans-dichlorostilbene.

1,1,1-TRICHLOROBIS(PHENYL)ETHANE
$C_{14}H_{11}Cl_3$ (I)

BIS(4-CHLOROPHENYL)METHANE
$C_{13}H_{10}Cl_2$ (II)

9-(TRICHLOROMETHYL)-9,10-DIHYDROANTHRACENE
$C_{15}H_{11}Cl_3$ (III)

1-CHLORO-2,2-BIS(4-CHLOROPHENYL)ETHANE
$C_{14}H_9Cl_3$ (IV)

1,1,1-TRICHLORO-2-(4-CHLOROPHENYL)-2-(2,4-DICHLOROPHENYL)ETHANE
$C_{14}H_8Cl$ (V)

C.H.L. KENNARD, G. SMITH, T.-B. PALM, S. HOVMOLLER and A. SJOGREN, 1984. J. Agric. Food Chem., 32, 886-895.

I. Orthorhombic, $Pbc2_1$, a = 16.593, b = 10.237, c = 7.955 Å, Z = 4. Mo radiation, R = 0.054 for 688 reflexions.

II. Monoclinic, $P2_1/a$, a = 26.620, b = 7.222, c = 5.988 Å, β = 91.40°, Z = 4. Mo radiation, R = 0.113 for 333 reflexions.

III. Orthorhombic, $Cmc2_1$, a = 15.732, b = 14.611, c = 5.928 Å, Z = 4. Cu radiation, R = 0.066 for 511 reflexions.

IV. Monoclinic, $P2_1$, a = 15.163, b = 5.824, c = 7.452 Å, β = 100.12°, Z = 2. Mo radiation, R = 0.038 for 1052 reflexions.

V. Monoclinic, $P2_1/c$, a = 11.621, b = 11.895, c = 12.455 Å, β = 115.60°, Z = 4. Mo radiation, R = 0.088 for 1759 reflexions.

These structures (Fig. 1) are of interest because of their relation to DDT. Of interest are the torsion angles τ_1 and τ_2 which describe the orientations of the phenyl rings. These are +131.8° and -92.9° for I, +64.9° and +87.5° for II. Compound III has crystallographic mirror symmetry (and equal $\tau_1 \cdot \tau_2$ values ±144.9°); the central ring has a distorted boat conformation. Compound IV is a DDT metabolite and has torsion angles C14-C13-C1-C6 +104°, C14-C13-C7-C8 -38°. Compound V has τ_1 and τ_2 +130.4° and -77.3°. A DDT mode of action theory is analyzed in terms of available crystallographic data.

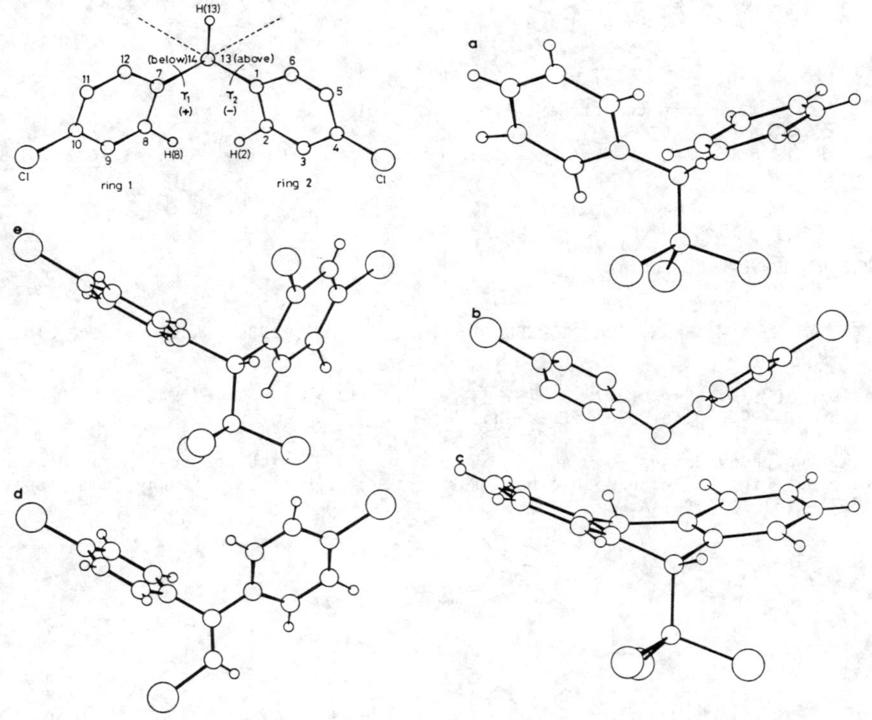

Fig. 1. The standard numbering scheme for DDT analogues with convention used for ring torsion angles, and diagrams (clockwise from top right) of the five DDT analogues I - V.

trans-STILBENE
$C_{14}H_{12}$

J.A. BOUWSTRA, A. SCHOUTEN and J. KROON, 1984. Acta Cryst., C40, 428-431.

Monoclinic, $P2_1/c$, a = 15.709, b = 5.723, c = 12.374 Å, β = 111.90, Z = 4. Mo radiation, R = 0.060 for 2276 reflexions.

For earlier work see (1). There are two independent molecules (A and B) lying about inversion centres. One molecule (B) is disordered over two sites (Fig. 1). The nature of the disorder and the degree of misorientation are very similar to those in trans-azobenzene.

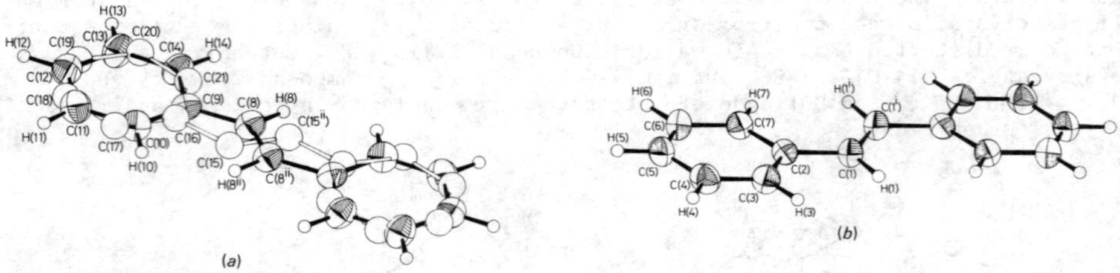

Fig. 1. Views of the two trans-stilbene molecules.

1. J.M. ROBERTSON and I. WOODWARD, 1937. Proc. R. Soc. London, Ser. A, 162, 568-583.

(Z)-2,3-DIPHENYL-2-BUTENE
$C_{16}H_{16}$

F.R. FRONCZEK, A.M. SWAN, J.A. CORKERN and R.D. GANDOUR, 1984. Acta Cryst., C40, 1875-1876.

Monoclinic, $P2_1/n$, a = 10.217, b = 8.703, c = 14.342 Å, β = 95.67°, Z = 4. Mo radiation, R = 0.050 for 998 reflexions.

The two phenyl rings form angles of 57.6 and 49.3° with the best plane of the double bond (Fig. 1). The double bond is twisted slightly (3.9°) away from planarity with C(7)-C(8) 1.343(4) Å.

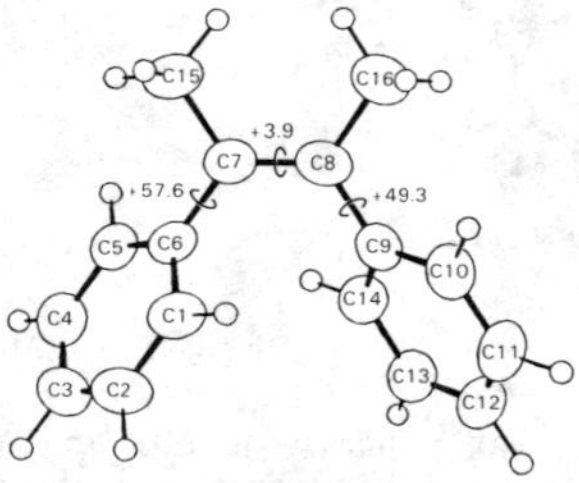

Fig. 1. The 2,3-diphenyl-2-butene molecule.

(1RS,2RS,3RS)-3,4-EPOXY-1,2-DIPHENYLBUTAN-1-OL
$C_{16}H_{16}O_2$

J.M. COXON, G.W. SIMPSON, P.J. STEEL and V.C. TRENERRY, 1984. Aust. J. Chem., 37, 65-72.

Monoclinic, C2/c, a = 27.965, b = 5.743, c = 17.668 Å, β = 116.01°, D_m = 1.245, Z = 8. Mo radiation, R = 0.035 for 1053 reflexions.

The phenyl groups are held in a gauche relationship (Fig. 1) (C(1)'-C(1)-C(2)-C(1)'' torsion angle 61.6°) by intermolecular hydrogen bonding between hydroxyl hydrogen and an epoxide oxide (O...H 1.92, O(H)...O 2.81 Å) between molecules related by a centre of inversion.

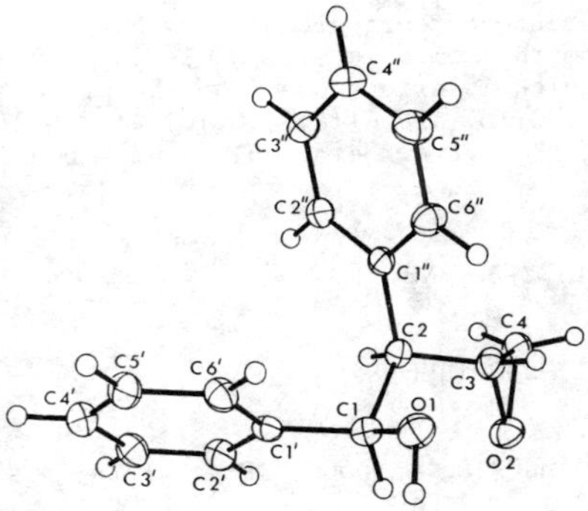

Fig. 1. The structure of $C_{16}H_{16}O_2$.

trans-CINNAMOYL PEROXIDE
$C_{18}H_{14}O_4$

D. BETHELL, D.J. CHADWICK, G.Q. MALING and M.M. HARDING, 1984. Acta Cryst., C40, 1628-1630.

Monoclinic, $P2_1$, a = 14.64, b = 9.10, c = 5.63 Å, β = 93.0°, Z = 2. Cu radiation, R = 0.069 for 604 reflexions.

The molecule (Fig. 1) has an extended conformation with its two acyloxy groups lying in nearly orthogonal planes. The α,β-unsaturated carbonyl moiety has s-cis stereochemistry with the C=O bond length 1.15(2) Å. The torsion angle about O-O is 87.1°.

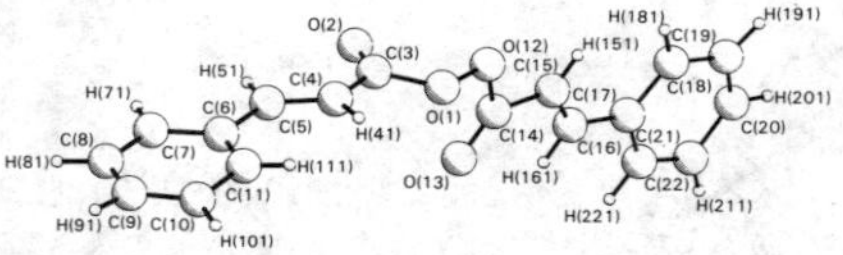

Fig. 1. The trans-cinnamoyl peroxide molecule.

(E)-1-(2-BROMO-4,5-DIMETHOXYPHENYL)-2-(2-HYDROXY-3,4-DIMETHOXYPHENYL)ETHYLENE
$C_{18}H_{19}BrO_5$

J.M. ARRIETA, E. DOMÍNGUEZ, E. LETE, M.J. VILLA and G. GERMAIN, 1984. Acta Cryst., C40, 859-861.

Monoclinic, $P2_1/c$, a = 12.971, b = 8.069, c = 16.158 Å, β = 95.04°, Z = 4. Mo radiation, R = 0.043 for 1713 reflexions.

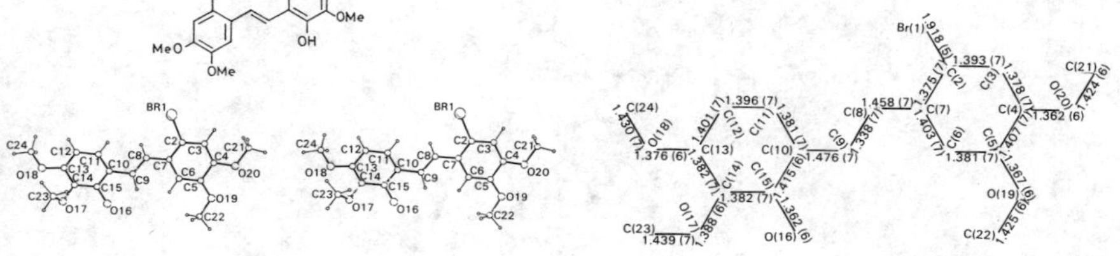

Fig. 1. $C_{18}H_{19}BrO_5$: stereoview of the molecule and bond lengths.

In the molecule both benzene rings are planar, but show a slight distortion (Fig. 1). The planarity of the molecule as a whole is only disturbed by the 3-methoxy group of the phenol ring which is oriented out of the plane of the ring. Intermolecular distances O(17)...O(16) and O(18)...O(16) are 2.72(1) and 2.93(1) Å, indicating hydrogen bonds, the latter being only a weak bond.

(E)-2,3-BIS(2-METHOXYPHENYL)-2-BUTENE
$C_{18}H_{20}O_2$ (I)

(Z)-2,3-BIS(2-METHOXYPHENYL)-2-BUTENE
$C_{18}H_{20}O_2$ (II)

F.R. FRONCZEK, M.A. OLIVER and R.D. GANDOUR, 1984. Acta Cryst., C40, 504-506.

I. Monoclinic, C2/c, a = 21.096, b = 7.473, c = 10.247 Å, β = 110.41°, Z = 4. Mo radiation, R = 0.042 for 461 reflexions.

II. Orthorhombic, $P2_12_12_1$, a = 8.307, b = 11.053, c = 16.854 Å, Z = 4. Mo radiation, R = 0.050 for 743 reflexions.

The (E)-isomer (I) has $\bar{1}$ symmetry (Fig. 1), with a dihedral angle of 68.6° between the aromatic plane and the plane of the double bond. The (Z)-isomer (II) has approximate twofold symmetry, with aromatic rings forming dihedral angles of 71.6 and 68.2° with the plane of the double bond.

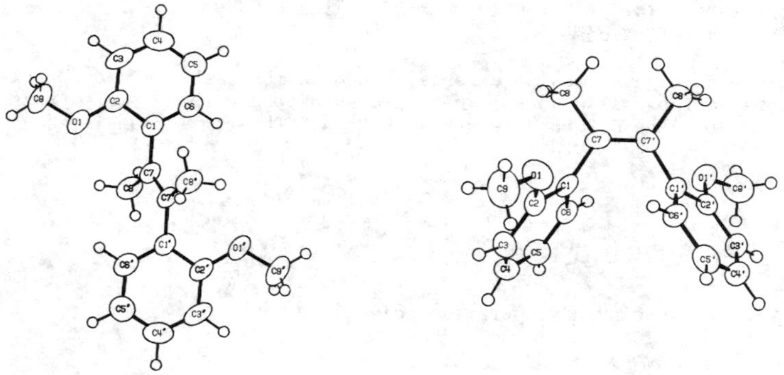

Fig. 1. Views of the (E)- (left) and (Z)- (right) isomers of $C_{18}H_{20}O_2$.

(R,R,S,S)-BIS[(trans-2-PHENYL-1-CYCLOPROPYL)CARBONYL] PEROXIDE
$C_{20}H_{18}O_4$

D. BETHELL, D.J. CHADWICK, M.M. HARDING and G.Q. MALING, 1984. Acta Cryst., C40, 1909-1911.

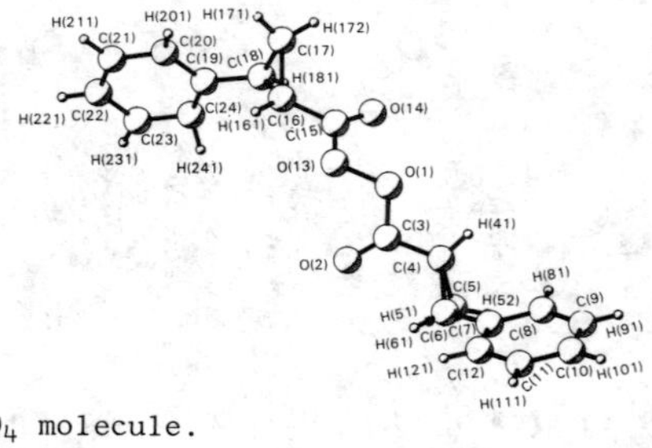

Fig. 1. The $C_{20}H_{18}O_4$ molecule.

Monoclinic, $P2_1/c$, a = 8.11, b = 18.50, c = 11.45 Å, β = 95.5°, Z = 4. Cu radiation, R = 0.053 for 1157 reflexions.

The molecules (Fig. 1) have an extended conformation; the torsion angle about the O-O bond is -121.8(7)°. The phenyl and carbonyl groups have the bisected conformation with respect to the cyclopropane ring. All O atoms have very anisotropic thermal motion leading to close contacts with neighbouring molecules (O...O 3.04 and 3.35, O...C 3.20-3.36 Å). The O(1)-O(13) bond distance is 1.452 Å.

ACETYLTRIPHENYLMETHANE
$C_{21}H_{18}O$

M.N. PONNUSWAMY and J. TROTTER, 1984. Acta Cryst., C40, 142-144.

Monoclinic, C2/c, a = 16.276, b = 7.067, c = 27.389 Å, β = 99.78°, D_m = 1.209, Z = 8. Mo radiation, R = 0.033 for 1086 reflexions.

In the molecule (Fig. 1) there are two distinct C(central)-C bond lengths; C-C(Ph) 1.538(2) Å and C-C(Ac) 1.555(3) Å, the lengthening of the latter being explicable in terms of the electronic structure of the molecule. For details of the Ge and Si analogues see (1).

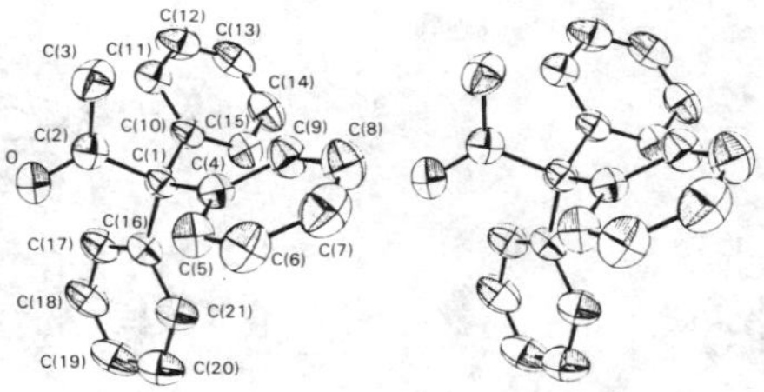

Fig. 1. A stereoview of $C_{21}H_{18}O$.

1. Structure Reports, 33B, 306; 34B, 373.

2,2'-DIPHENYL-1,1'-(1,3-PHENYLENE)DIETHANEDIONE
$C_{22}H_{14}O_4$

A. SYED, E.D. STEVENS and C.L. KLEIN, 1984. Acta Cryst., C40, 1271-1273.

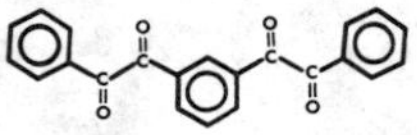

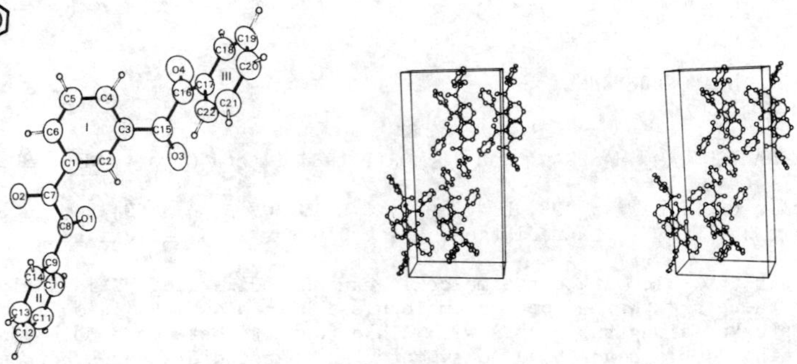

Fig. 1. $C_{22}H_{14}O_4$: perspective view of the molecule and stereoview of the molecular packing in the unit cell.

Orthorhombic, Pbca, a = 7.756, b = 14.864, c = 29.523 Å, D_m = 1.34, Z = 8. Mo radiation, R = 0.035 for 1217 reflexions.

In the molecule (Fig. 1) adjacent carbonyl groups are arranged in an antiparallel configuration. Three of the carbonyl groups are approximately planar with the central phenyl ring. The other two phenyl rings are twisted by 40.4(3) and 88.1(3)° with respect to the central phenyl ring. The molecules are held together by van der Waals forces.

2,2'-DIPHENYL-1,1'-(1,4-PHENYLENE)DIETHANEDIONE
$C_{22}H_{14}O_4$

A. SYED and E.D. STEVENS, 1984. Acta Cryst., C40, 1273-1276.

Triclinic, $P\bar{1}$, a = 4.466, b = 12.785, c = 16.104 Å, α = 68.46, β = 88.16, γ = 88.08°, Z = 2. Mo radiation, R = 0.035 for 1555 reflexions.

The molecular structure (Fig. 1) is closely related to that of 2,2'-diphenyl-1,1'-(1,3-phenylene)diethanedione (1) with twisted phenyl rings. The asymmetric unit, however, consists of two half-molecules at different centres of symmetry. Carbonyl groups are arranged in an antiparallel configuration and molecular packing is determined by van der Waals interactions.

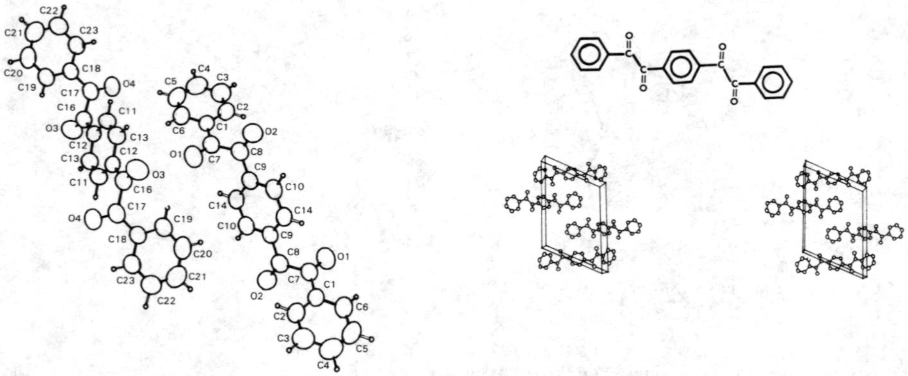

Fig. 1. $C_{22}H_{14}O_4$: perspective view of the molecules and stereoview of the molecular packing in the unit cell.

1. Previous Report.

1,4-DICINNAMOYLBENZENE
$C_{24}H_{18}O_2$

H. NAKANISHI, M. HASEGAWA and T. MORI, 1984. Acta Cryst., C40, 509-511.

Triclinic, $P\bar{1}$, a = 5.798, b = 7.923, c = 19.307 Å, α = 89.12, β = 82.12, γ = 88.67°, D_m = 1.28, Z = 2. Cu radiation, R = 0.050 for 2014 reflexions.

In the solid state the molecule can photodimerize. This analysis shows (Fig. 1) that there are two independent molecules, both of which lie on inversion centres, and are stacked along the b axis with close contacts between double bonds. Pairs of reacting double bonds have no symmetry so that the distances between photo-adductive olefin C atoms are 3.973(4), 4.086(4), 3.903(4) and 3.955(4) Å.

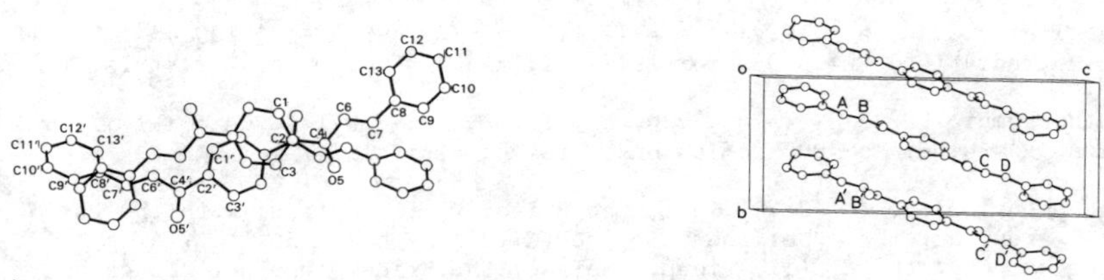

Fig. 1. A view showing overlapping of reactive molecules and the crystal structure projected onto the (100) plane.

3,3-DIPHENYL-1-(2,4,6-TRIMETHYLPHENYL)PROPANE-1-ONE
$C_{24}H_{24}O$

A.G. PINKUS, D.F. MULLICA, W.O. MILLIGAN, D.A. GROSSIE and P.W. HURD, 1984. Tetrahedron, 40, 4829-4835.

Monoclinic, $P2_1/n$, a = 10.875, b = 6.039, c = 29.015 Å, β = 91.29°, D_m = 1.14, Z = 4. Mo radiation, R = 0.066 for 1127 reflexions.

The dihedral angle between the mesityl ring and the C11,C1,O1,C2 plane is 114.4°. Molecular dimensions are as anticipated.

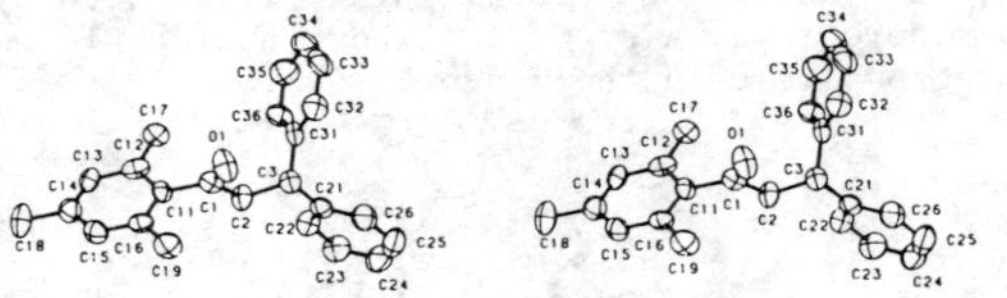

Fig. 1. A stereoview of $C_{24}H_{24}O$.

(E)-3-(N-BENZOYL-N-BENZYLAMINO)-4,4-DIMETHYL-1-PHENYLPENT-2-EN-1-ONE
$C_{27}H_{27}NO_2$ (I)

(Z)-3-(N-BENZOYL-N-BENZYLAMINO)-4,4-DIMETHYL-1-PHENYLPENT-2-EN-1-ONE
$C_{27}H_{27}NO_2$ (II)

J. CORNFORTH, V.A. PATRICK and A.H. WHITE, 1984. Aust. J. Chem., 37, 1453-1460.

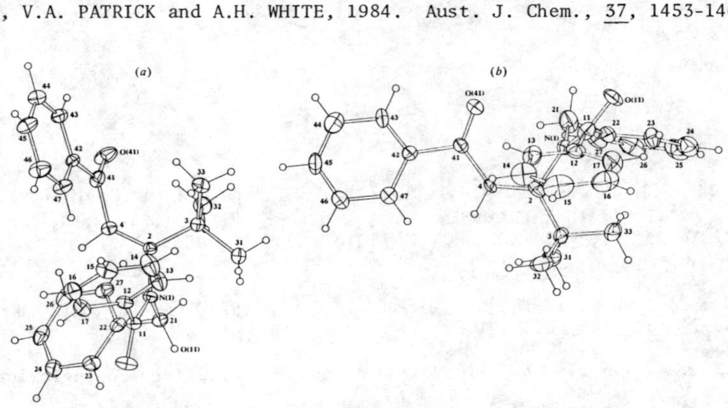

Fig. 1. The structures of (a) E and (b) Z isomers of $C_{27}H_{27}NO_2$.

I. Monoclinic, C2/c, a = 10.241, b = 17.199, c = 25.495 Å, β = 100.45°, D_m = 1.19, Z = 8. Mo radiation, R = 0.052 for 1947 reflexions. Colourless crystals, T = 295 K.

II. Orthorhombic, $P2_12_12_1$, a = 18.24, b = 15.60, c = 7.871 Å, D_m = 1.17, Z = 4. Mo radiation, R = 0.061 for 902 reflexions. Yellow crystals, T = 120 K.

The phenyl plane (C(42-7)) in I, the E isomer, is coplanar with the carbonyl (C,O(41)) group but the other phenyl plane (C(12-17)) has a 56.1° angle with its adjoining carboxyl the C,O(11) group. Neither the amide plane nor the carbonyl group attached to C(4) are coplanar with the ethylene (C(2,4)) plane. II is formed when aqueous ethanolic alkali is added to I. In the Z isomer (II), the phenyl (C(42-7)), carbonyl (C,O(41)) and ethylene (C(2,4)) segments are approximately coplanar. The amide group is also planar and forms a dihedral angle of 79.4° with the ethylene plane. On exposure to daylight, the yellow Z isomer is converted to E.

TRIMESITYLVINYL ACETATE
$C_{31}H_{30}O_2$

S.E. BIALI and Z. RAPPOPORT, 1984. J. Am. Chem. Soc., 106, 477-496.

Monoclinic, $P2_1/c$, a = 13.910, b = 11.599, c = 16.551 Å, β = 99.50°, Z = 4. Mo radiation, R = 0.131.

The molecule (Fig. 1) has a "vinyl propeller" structure. The acetate group is in an anti, anti conformation where the carbonyl plane is twisted by 54° from the ideally planar conformation. There is a twist of 10° about the double bond and the torsional angles of the mesityl groups are in the range 57-64°.

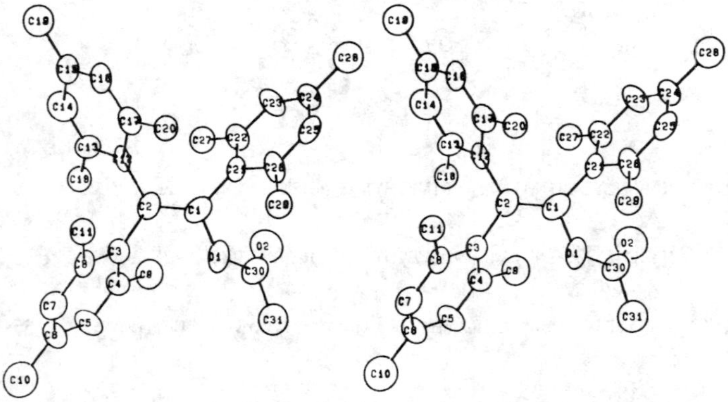

Fig. 1. Stereoview of the $C_{31}H_{30}O_2$ molecule.

31,32,33,34,35,36-HEXAFLUORO-4,9,14,19,24,29-HEXAMETHYLHEPTACYCLO[25.3.1.1.^{2,6}.$1^{7,11}$.-$1^{12,16}$.$1^{17,21}$.$1^{22,26}$]HEXATRIACONTA-1(31),2,4,6(36),7,9,11(35),12,14,16(34),17,19,-21(33),22,24,26(32),27,29-OCTADECAENE METHYLENE CHLORIDE SOLVATE
$C_{42}H_{30}F_6,CH_2Cl_2$ $(CH_3C_6H_2F)_6.CH_2Cl_2$

D.J. CRAM, S.B. BROWN, T. TAGUCHI, M. FEIGEL, E. MAVERICK and K.N. TRUEBLOOD, 1984. J. Am. Chem. Soc., 106, 695-701.

Cubic, Pa3, a = 15.621 Å, Z = 4. Mo radiation, R = 0.12 for 736 reflexions.

The molecule (Fig. 1) has exact $\bar{3}$ symmetry (approximately D_{3d}), with the F atoms in an octahedral arrangement. The dihedral angle between adjacent aryl mean planes is 44°. The aromatic rings are non-planar (5.7° fold) and the F atoms are displaced by 0.20 Å from the mean plane.

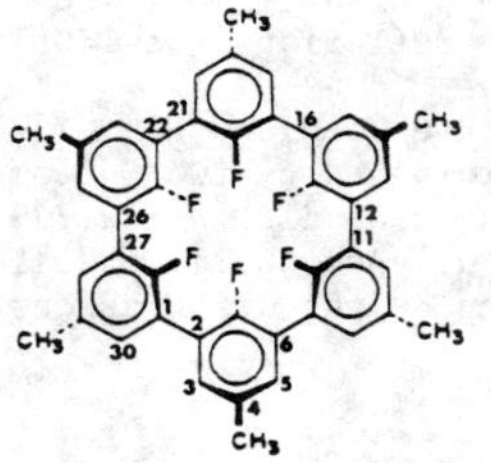

Fig. 1. The structure of $(CH_3C_6H_2F)_6$.

2,4,6-TRI(3',5'-DI-t-BUTYL-4'-HYDROXYBENZYL)MESITYLENE
$C_{54}H_{78}O_3$

L.A. CHETKINA, V.E. ZAVODNIK, A.N. SOBOLEV and V.K. BEL'SKII, 1984, Kristallografiya, 29, 389-391 [Sov. Phys. Crystallogr., 29, 234-236].

Orthorhombic, $P2_12_12_1$, a = 10.146, b = 17.145, c = 27.758 Å, Z = 4. R = 0.072 for 2598 reflexions.

The molecule (Fig. 1) consists of a fully substituted aromatic ring connected by methylene bridges to three external rings (B, C and D) of 3,5-ditert-butyl-4-hydroxybenzyl groups. The bond lengths in the central ring A appear to alternate: 1.384 and 1.411 Å, with angles of 118.8 and 121.1°. All the rings are planar, with rings B, C and D almost orthogonal to ring A (dihedral angles 79.7, 81.9 and 72.5° respectively) providing an overall "basket" shape to the molecule with ring A as the base. The average C-OH bond length is 1.381 Å and the bond angle at the methylene C atom is 112.6°.

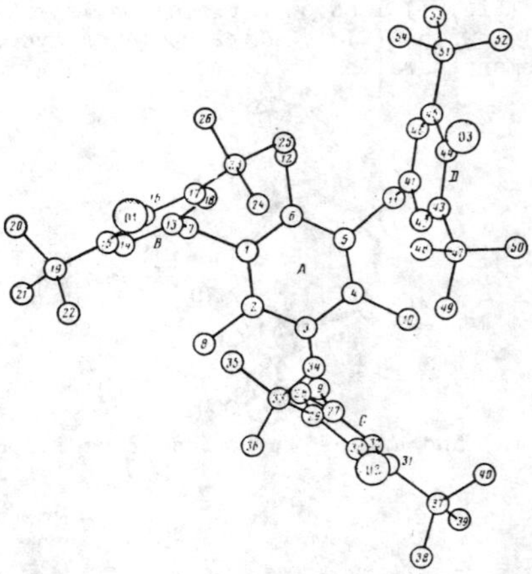

Fig. 1. Conformation of the molecule $C_{54}H_{78}O_3$.

DIPOTASSIUM SALT OF 3-HYDROXY-4-MERCAPTO-2-THIOXO-3-CYCLOBUTEN-1-ONE DIHYDRATE
(POTASSIUM 1,3-DITHIOSQUARATE DIHYDRATE)
$C_4K_2O_2S_2,2H_2O$ $K_2[C_4O_2S_2],2H_2O$

R. MATTES and G. JOHANN, 1984. Acta Cryst., C40, 740-742.

Monoclinic, $P2_1/c$, a = 7.309, b = 8.404, c = 8.120 Å, β = 112.6°, D_m = 1.82, Z = 2. Mo radiation, R = 0.026 for 945 reflexions (at 140K).

The planar 1,3-dithiosquarate dianion (Fig. 1) is situated on an inversion centre. Potassium is seven-coordinated by two S atoms at 3.469(1) and 3.445(1) Å respectively and five O atoms at mean distance 2.795(1) Å. In the crystal structure the planes of the dianions are aligned nearly parallel to the a axis, with an approximate angle of 45° between neighbouring anions.

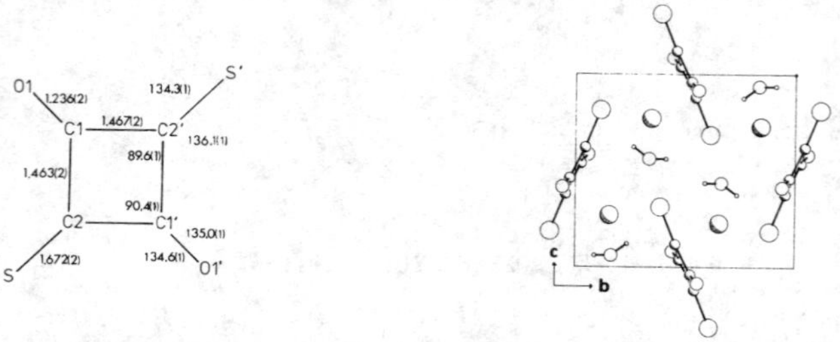

Fig. 1. $C_4K_2O_2S_2,2H_2O$: interatomic distances and angles and a projection of the unit-cell contents.

CYCLOPENTENE-1,2,3-TRIONE
$C_5H_2O_3$

H. IRNGARTINGER, M. NIXDORF, W. DOBLER and R. GLEITER, 1984. Acta Cryst., C40, 1481-1482.

Tetragonal, $P4_12_12$, a = b = 5.701, c = 14.113 Å, Z = 4. Mo radiation, R = 0.037 for 253 reflexions.

The planar molecule (Fig. 1) lies on a twofold axis. Close intermolecular interactions (O(2)...C(1)=O with O(2)...C(1) 2.946(4) Å and O...C=O angle 105.2(1)°) produce a helical arrangement parallel to c.

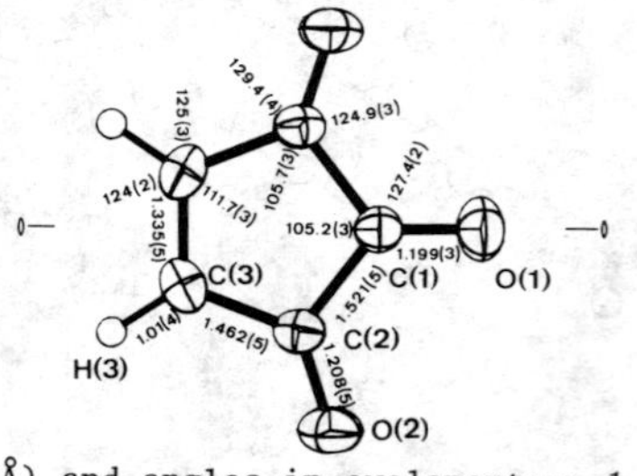

Fig. 1. Bond lengths (Å) and angles in cyclopentene-1,2,3-trione.

1,1'-BICYCLOPROPYLIDENE
C_6H_8

M. TRAETTEBERG, A. SIMON, E.-M. PETERS and A. DE MEIJERE, 1984. J. Mol. Struct., 118, 333-343.

Monoclinic, $P2_1/a$, a = 9.656, b = 5.107, c = 5.220 Å, β = 96.29°, Z = 2. Mo radiation, R = 0.095 for 417 reflexions (at -40°C).

The molecule (Fig. 1) lies about an inversion centre, with C=C 1.304(8), Csp^2-Csp^3 1.442 and 1.439(6), Csp^3-Csp^3 1.504(7) Å, C=C-C 147.9 and 149.2(6)°, ring C-C-C 58.4-62.9(3)°. An electron diffraction study yielded C=C 1.314(1), Csp^2-Csp^3

1.468(1), Csp^3-Csp^3 1.554(2) Å, C=C-C 148.1, ring C-C-C 58.1-63.8(1)°.

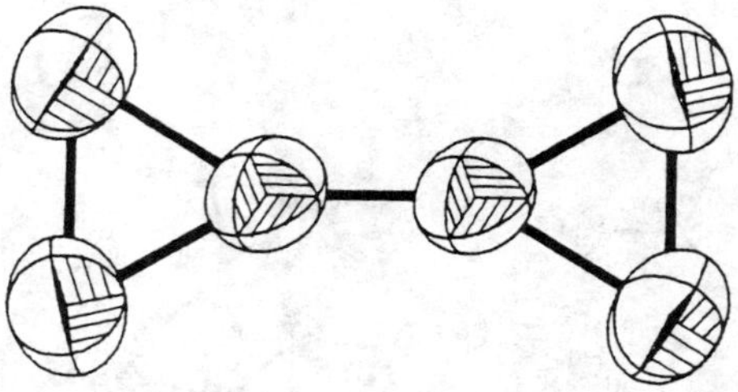

Fig. 1. The 1,1'-bicyclopropylidene molecule.

trans-3,4-DICHLORO-1,2,3,4-TETRAMETHYLCYCLOBUTENE
$C_8H_{12}Cl_2$

I.Yu. BAGRYANSKAYA, Yu.V. GATILOV and S.A. OSADCHII, 1984. J. Struct. Chem., 25, 502-503 [Zh. Strukt. Khim., 25(3), 178-180].

Monoclinic, C2/c, a = 10.013, b = 7.056, c = 13.566 Å, β = 107.91°, Z = 4. Mo radiation, R = 0.053 for 738 reflexions (at -65°C).

The molecule has exact C_2 symmetry, the twofold axis bisecting the C(1)-C(1)' and C(2)-C(2)' bonds. Bond lengths and angles are shown in Fig. 1.

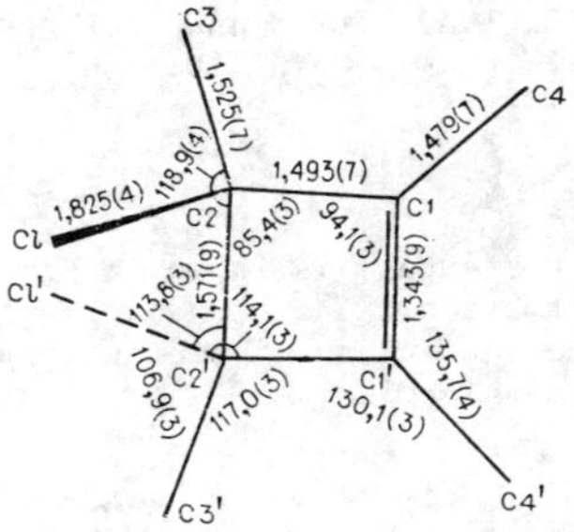

Fig. 1. Geometry of the trans-3,4-dichloro-1,2,3,4-tetramethylcyclobutene molecule.

DI(OXAMIDE DIOXIMINIUM) SQUARATE
$C_8H_{14}N_8O_8$ (I) $2C_2H_7N_4O_2^+.C_4O_4^{2-}$

OXAMIDE DIOXIMINIUM DI(HYDROGEN SQUARATE)
$C_{10}H_{10}N_4O_{10}$ $C_2H_8N_4O_2^{2+}.2C_4HO_4^-$

H. ENDRES and M. SCHENDZIELORZ, 1984. Acta Cryst., C40, 453-456.

I. Monoclinic, $P2_1/n$, a = 3.6797, b = 21.454, c = 8.807 Å, β = 96.93°, Z = 2. Mo radiation, R = 0.061 for 394 reflexions.

II. Monoclinic, C2/c, a = 13.248, b = 6.115, c = 17.918 Å, β = 113.69°, Z = 4. Mo radiation, R = 0.040 for 1540 reflexions.

In I (Fig. 1) the cation is present in the s-trans form, with one oxime N atom protonated. The squarate anion lies about a crystallographic inversion centre. In II, the dipositive cations lie on crystallographic twofold axes and the hydrogen squarate anion is in a general position. In the cation, both oxime N atoms are protonated and a gauche conformation is found.

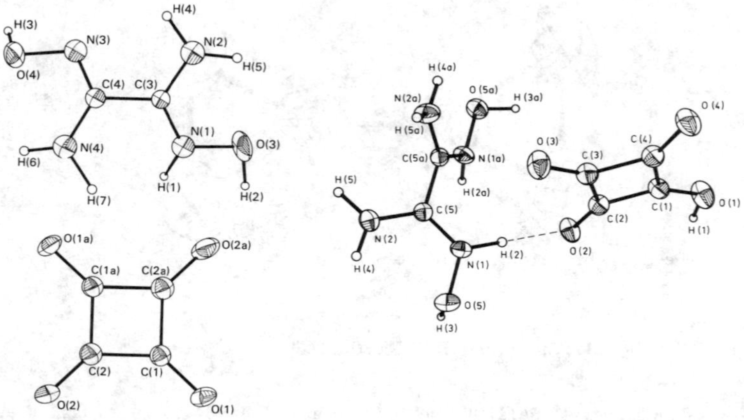

Fig. 1. Views of the ions in $C_8H_{14}N_8O_8$ (left) and $C_{10}H_{10}N_4O_{10}$ (II) (right).

2-(4'-CARBOXY-3'-HYDROXYBUTYL)CYCLOPENT-2-EN-1-ONE
$C_{10}H_{14}O_4$

A.F. MISHNEV, Ya.Ya. BLEIDELIS, I.A. MILMAN and Ya.F. FREIMANIS, 1984. J. Struct. Chem., 25, 815-817 [Zh. Strukt. Khim., 25(5) 161-163].

Monoclinic, $P2_1$, a = 5.814, b = 8.392, c = 10.980 Å, β = 104.12°, Z = 2. Cu radiation, R = 0.034 for 1492 reflexions.

The molecule (with bond distances) is shown in Fig. 1. The chiral centre C(3) was found to have the S conformation. The α chain is linear, and the atoms C(1-6) are located practically in the same plane. The five-membered ring is planar, with O(4) deviating by 0.044 Å from this plane.

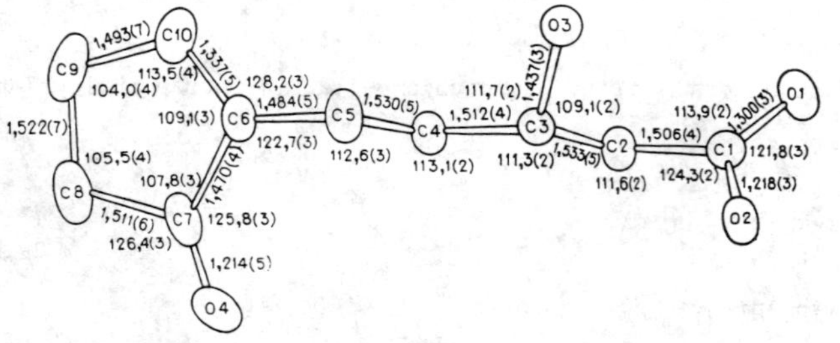

Fig. 1. The 2-(4'-carboxy-3'-hydroxybutyl)cyclopent-2-en-1-one molecule.

2-METHYLTHIO-1-r-2-t-DICYANO-1-PHENYLTHIOCYCLOPROPANE
$C_{12}H_{10}N_2S_2$

B. TINANT, J.P. DECLERCQ and M. VAN MEERSSCHE, 1984. Bull. Soc. Chim. Belg., 93, 921-922.

Orthorhombic, $Pc2_1n$, a = 7.757, b = 10.369, c = 15.265 Å, Z = 4. Mo radiation, R = 0.049 for 880 reflexions.

The ring bond length pattern for the molecule (Fig. 1) is C(1)-C(2) 1.535(12), C(2)-C(3) 1.506(12), C(3)-C(1) 1.501(11) Å. The C-S distances are C(1)-S(1) 1.793(8), C(2)-S(2) 1.762(8), C(5)-S(1) 1.769(7), C(12)-S(2) 1.793(8) Å.

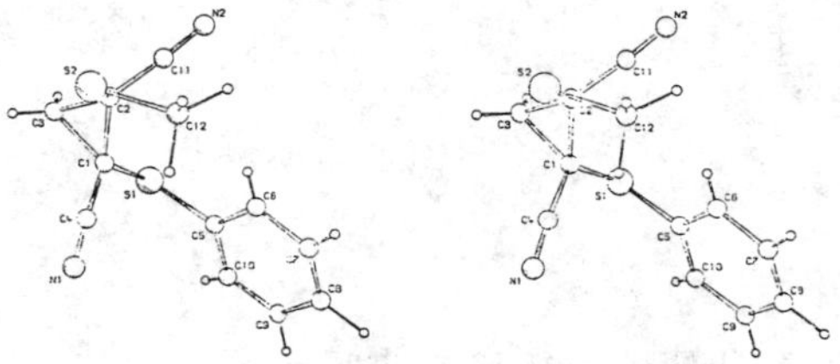

Fig. 1. $C_{12}H_{10}N_2S_2$: a stereoview.

METHYL 2-[2-(DIETHYLAMINO)ETHYL]AMINO-1-CYCLOPENTENECARBODITHIOATE
$C_{13}H_{24}N_2S_2$

T. RAY and S.P.S. GUPTA, 1984. Acta Cryst., C40, 1977-1978.

Monoclinic, Cc, a = 7.571, b = 17.772, c = 11.505 Å, β = 92.87°, Z = 4. Mo radiation, R = 0.061 for 1006 reflexions.

The molecules (Fig. 1) have intramolecular N-H...S hydrogen bonds (N...S 3.030(7) Å). There is partial delocalisation of the N lone pair in the N-C=C-C conjugated system. The five-membered ring has an envelope conformation with Δ = 38.7°.

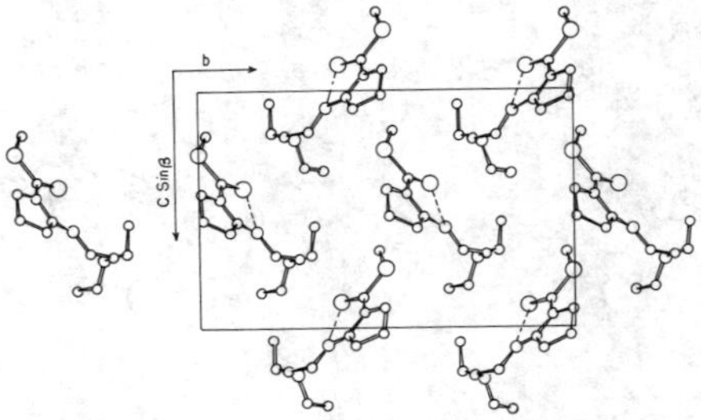

Fig. 1. Packing of the $C_{13}H_{24}N_2S_2$ molecules viewed along a.

DIMETHYL 2-(4-ACETOXY-5-ETHYL-3-METHOXY-2-METHOXYCARBONYLCYCLOPENT-2-ENYL)PROPANEDIOATE
$C_{17}H_{24}O_9$

R.T. BROWN, M.F. JONES and C.J. GILMORE, 1984. J. Chem. Res., (S) 294-295, (M) 2701-2737.

Triclinic, P$\bar{1}$, a = 8.092, b = 8.795, c = 15.026 Å, α = 104.55, β = 86.00, γ = 115.10°, Z = 2. Mo radiation, R = 0.041 for 2709 reflexions.

The structure and stereochemistry are in Fig. 1. The central five-membered ring has a somewhat flattened envelope conformation.

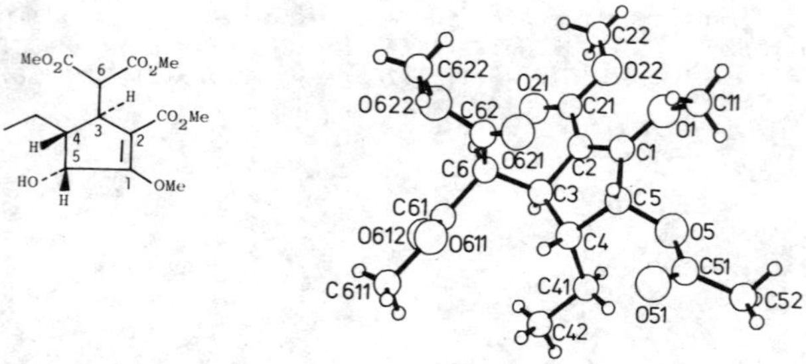

Fig. 1. $C_{17}H_{24}O_9$: structure and stereochemistry.

2-(2-HYDROXY-3-PHENYL-2-PROPEN-1-YLIDENE)-3,3,5,5-TETRAMETHYLCYCLOPENTANONE
$C_{18}H_{22}O_2$ (I)

1-HYDROXY-6,6,8,8-TETRAMETHYL-2-PHENYLBICYCLO[3.3.0]OCT-4-EN-3-ONE
$C_{18}H_{22}O_2$ (II)

T. SIMONEN and R. KIVEKÄS, 1984. Acta Chem. Scand., B38, 679-687.

I. Triclinic, P$\bar{1}$, a = 9.980, b = 14.061, c = 18.104 Å, α = 100.19, β = 101.06, γ = 96.35°, Z = 6. Mo radiation, R = 0.081 for 3214 reflexions.

II. Orthorhombic, $P2_12_12_1$, a = 7.888, b = 9.145, c = 22.161 Å, Z = 4. Mo radiation, R = 0.084 for 1174 reflexions.

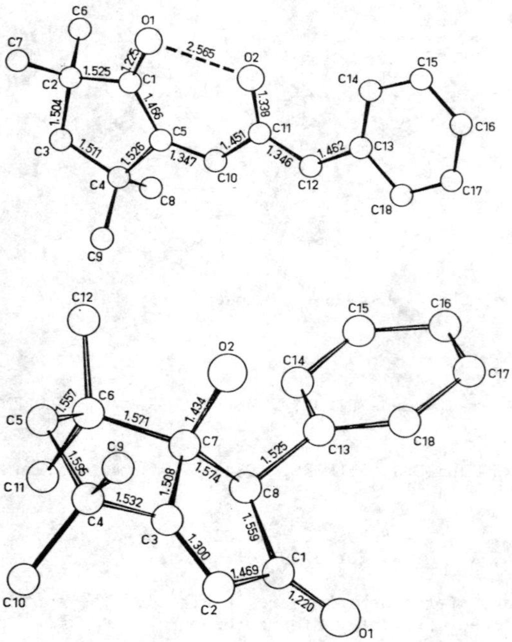

Fig. 1. Molecular structures of $C_{18}H_{22}O_2$ (I) showing mean values of selected bond lengths (top), and (II) showing selected bond lengths.

The vinyl alcohol derivative (I) contains three crystallographically-independent molecules in which the OH group is linked to the carbonyl O atom through a strong intramolecular hydrogen bond (O...O 2.565 Å (mean)) (Fig. 1). There are no significant differences in the geometry of the three independent molecules. They adopt a flattened conformation in which all but the methyl C atoms are nearly in a plane. In II, all bond lengths are as expected. The strain in the bicyclo[3.3.0]octenone system is revealed by the bond angles around C(7) and C(8), which range from 102.1 to 119.7°. There is an intermolecular hydrogen bond O(2)-H...O(1') with O...O 2.847 Å.

2,5-DIBENZYLIDENECYCLOPENTANONE
$C_{19}H_{16}O$ (I)

(+)-2,5-DIBENZYLIDENE-3-METHYLCYCLOPENTANONE
$C_{20}H_{18}O$ (II)

2,9-DIBENZYLIDENE-6,12-DIPHENYLDISPIRO[4.1.4.1]DODECANE-1,8-DIONE MONOHYDRATE
$C_{38}H_{32}O_2$ (III)

C.R. THEOCHARIS, W. JONES, J.M. THOMAS, M. MOTEVALLI, M.B. HURSTHOUSE, 1984. J. Chem. Soc. Perkin II, 71-76.

I. Orthorhombic, $C222_1$, a = 11.803, b = 5.698, c = 20.872 Å, Z = 4. Cu radiation, R = 0.051 for 597 reflexions.

II. Monoclinic, $P2_1$, a = 10.325, b = 7.723, c = 10.483 Å, β = 111.38°, Z = 2. Cu radiation, R = 0.046 for 1160 reflexions.

III. Tetragonal, $P4_12_12$, a = 10.860, c = 23.979 Å, Z = 4. Cu radiation, R = 0.051 for 1894 reflexions.

Compound III was prepared by solid state photodimerisation of I. II is photostable. The main interest in the structures was in the geometry which led to solid state reaction. In I O(1) and C(1) (Fig. 1) are at special positions located on the twofold axis parallel to [100], all other atoms being in general positions. Thus the asymmetric unit is half the molecule. Nearest neighbour molecules are related by translation along [010] such that C(2)...C(13') is 3.714(5) and C(6)...C(5') is 3.725(5) Å. The molecule as a whole is approximately planar. Consequently double bonds C(2)=C(6) and C(5')=C(13') lie in planes that are parallel to each other but the bonds themselves are not parallel (dihedral angle 56°). This geometry is not generally considered conducive to topochemical reaction. However the p_z orbital on C(2) is oriented in the direction of the corresponding orbital on C(13'). The formation of III within the crystal of I is presumed to result from reaction between C(2) and C(13') on one hand and C(6) and C(5') on the other. The cyclobutane ring of the dimer thus formed (III) is puckered as a result of dipole-dipole repulsion between carbonyl groups of the dimer. Molecules of II are planar with the exception of the methyl group which lies 1.65 Å above the plane of the molecule. Nearest-neighbour molecules are related by a twofold screw axis and the distance separating C(5)=C(13) and C(2')=C(6') is 3.871(5) Å. Although this distance is suitable for [2+2] topochemical reaction this crystal is photostable. It is suggested that this is due to the fact that the benzylidene groups are not parallel, preventing overlap of the necessary orbitals.

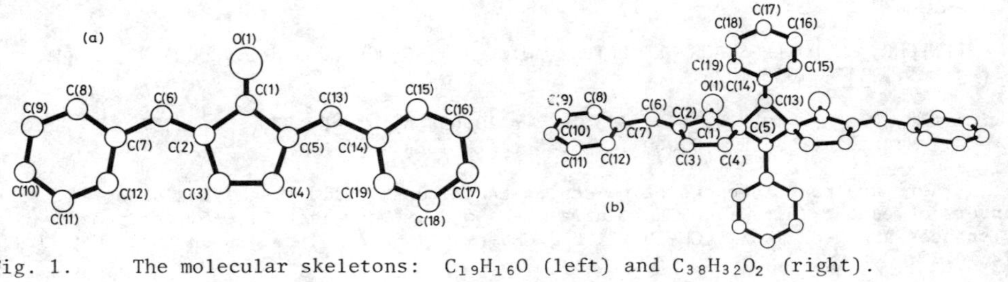

Fig. 1. The molecular skeletons: $C_{19}H_{16}O$ (left) and $C_{38}H_{32}O_2$ (right).

2-(p-METHYLBENZYL)-5-BENZYLIDENECYCLOPENTANONE 2-(p-CHLOROBENZYL)-5-BENZYLIDENE-CYCLOPENTANONE
$0.64(C_{20}H_{20}O),0.36(C_{19}H_{17}ClO)$ (I)

5-(p-BROMOBENZYLIDENE)-2-(p-METHYLBENZYL)CYCLOPENTANONE 5-(p-BROMOBENZYLIDENE)-2-(p-CHLOROBENZYL)CYCLOPENTANONE
$0.7(C_{20}H_{19}BrO),0.3(C_{19}H_{16}BrClO)$ (II)

2-(p-METHYLBENZYL)-5-BENZYLIDENECYCLOPENTANONE 2-(p-CHLOROBENZYL)-5-BENZYLIDENE-CYCLOPENTANONE (Irradiated Compound)
$0.2(C_{39}H_{35}ClO_2),0.51(C_{20}H_{20}O),0.29(C_{19}H_{17}ClO)$ (III)

C.R. THEOCHARIS, G.R. DESIRAJU and W. JONES, 1984. J. Am. Chem. Soc., 106, 3606-3609

I. Monoclinic, $P2_1/c$, a = 17.314, b = 10.667, c = 8.747 Å, β = 77.39°, Z = 4. R = 0.042 for 1103 reflexions.

II. Monoclinic, $P2_1/c$, a = 18.365, b = 11.152, c = 8.324 Å, β = 94.51°, Z = 4. R = 0.058 for 1185 reflexions.

III. Monoclinic, $P2_1/c$, a = 17.306, b = 10.667, c = 8.768 Å, β = 76.64°, Z = 4. R = 0.039 for 1107 reflexions.

The components of (I) have cell data very similar to those found for (I) and true solid solutions can be formed for all compositions (Fig. 1). The components of (II) have quite different cell data. The cell data for (II) are similar to those of the p-methyl component and the p-chloro component has been induced to adopt the crystal structure and conformation of the p-methyl compound. Compound (III) (Fig. 1) is obtained on irradiation of (I) and contains a photodimer and both component monomers.

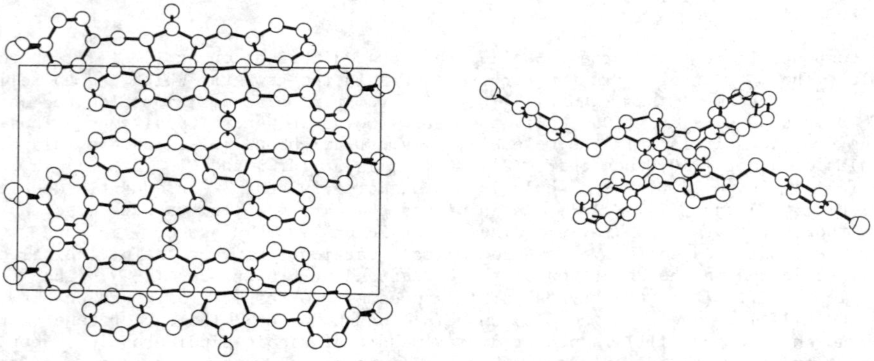

Fig. 1. The crystal structure of the mixed crystal of (I) (left) and (right) the dimer and incipient dimer obtained after irradiation of (I).

HEXACYCLOPROPYLETHANE
$C_{20}H_{30}$

W. BERNLOHR, H.-D. BECKHAUS, K. PETERS, H.G. VON SCHNERING and C. RUCHARDT, 1984. Chem. Ber., 117, 1013-1025.

Triclinic, $P\bar{1}$, a = 10.799, b = 14.347, c = 10.629 Å, α = 96.57, β = 90.57, γ = 96.74°, Z = 4. Mo radiation, R = 0.060 for 5389 reflexions.

There are four independent molecules in the unit cell, each lying about an inversion centre (Fig. 1). The central C-C bond is long (mean 1.636; other mean distances are C-CH 1.532, $CH-CH_2$ 1.507, CH_2-CH_2 1.490 Å. The torsion angle $C-C-CH-CH_2$ is 90.7°.

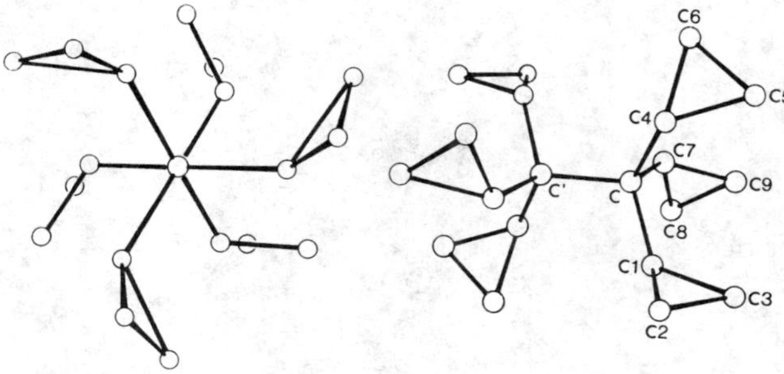

Fig. 1. Two views of $C_{20}H_{30}$; along the central C-C bond and normal to it.

4-HYDROXY-2,3,4-TRIPHENYL-2-CYCLOPENTEN-1-ONE BENZENE SOLVATE
$C_{23}H_{28}O_2$,0.5(C_6H_6)

K. VYAS and H. MANOHAR, 1984. Acta Cryst., C40, 134-136.

Triclinic, P$\bar{1}$, a = 6.068, b = 13.110, c = 13.527 Å, α = 110.63, β = 100.41, γ = 82.62°, D_m = 1.21, Z = 2. Cu radiation, R = 0.075 for 2558 reflexions.

The five-membered ring (Fig. 1) is puckered in a half-chair conformation (C(4) -0.153(4) Å, C(5) +0.175(4) Å from the plane of C(1),C(2),C(3)). The dihedral angles between the C(1),C(2),C(3) plane and phenyl rings A, B, and C are 44.7, 43.8 and 98.4(5)° respectively. Molecules are linked by O-H...O hydrogen bonds (O...O 2.885(4) Å).

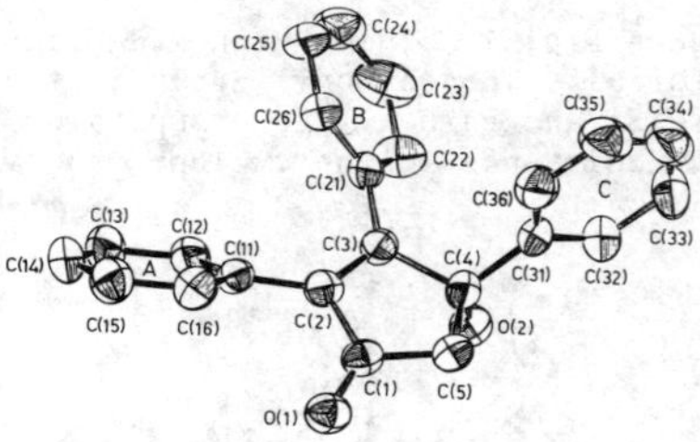

Fig. 1. A view of $C_{23}H_{28}O_2$.

ETHYL cis-1-(HYDROXYDIPHENYLMETHYL)-2-PHENYLCYCLOPROPYLCARBOXYLATE
$C_{25}H_{24}O_3$

B.A. FEIT, R. ELSER, U. MELAMED and I. GOLDBERG, 1984. Tetrahedron, 40, 5177-5180.

Monoclinic, P2_1/c, a = 11.935, b = 6.243, c = 27.837 Å, β = 101.82°, Z = 4. Cu radiation, R = 0.045 for 2558 reflexions.

In the molecule (Fig. 1) bond distances within the cyclopropane ring are CH_2-C(C_6H_5) 1.486 Å, CH_2-C(COOEt) 1.502 Å and C(C_2H_5)-C(COOEt) 1.538 Å.

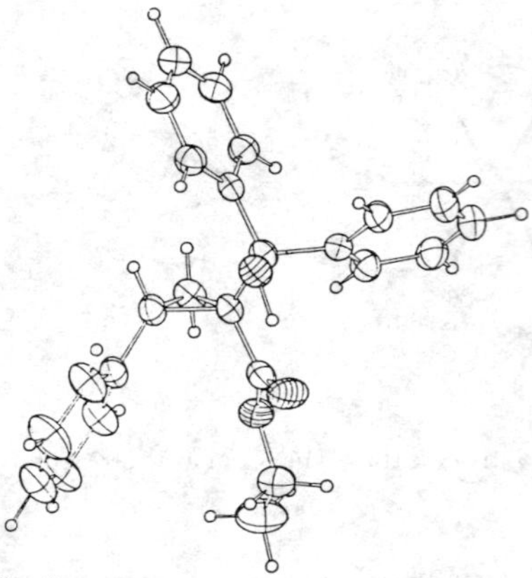

Fig. 1. A view of $C_{25}H_{24}O_3$.

L-MENTHYL (+)-1(S),3(S)-2,2-DIPHENYL-3-(TRIMETHYLSILYL)CYCLOPROPANECARBOXYLATE
$C_{29}H_{40}O_2Si$

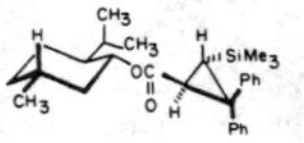

L.A. PAQUETTE, T. UCHIDA and J.C. GALLUCCI, 1984. J. Am. Chem. Soc., 106, 335-340.

Monoclinic, $P2_1$, a = 6.792, b = 18.599, c = 10.794 Å, β = 94.97°, Z = 2. Mo radiation, R = 0.064 for 1734 reflexions.

The structure is shown in Fig. 1. The cyclohexane ring has a chair conformation with all substituents equatorial. The carbonyl group is trans to the cyclopropyl ring hydrogen atom H(12). The Me_3Si and carboxylate groups are disposed on opposite sides of the C_3 ring. The mean C-C distance in the cyclopropane ring is 1.503(18) Å.

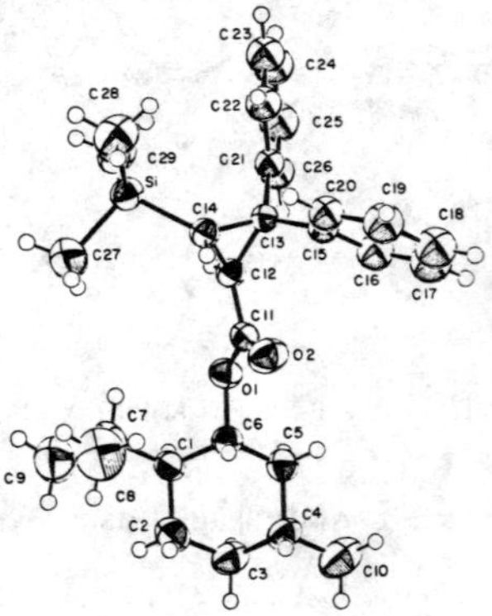

Fig. 1. The $C_{29}H_{40}O_2Si$ molecule.

trans-4-CYANO-2,3,4,5-TETRAPHENYL-2-CYCLOPENTEN-1-ONE
$C_{30}H_{21}NO$ (I)

cis-1,2,3,4-TETRAPHENYL-4-CYANO-5-METHYL-2-CYCLOPENTEN-1-ONE
$C_{31}H_{23}NO$ (II)

R.L. EAGEN, M.A. OGLIARUSO, B.H. ARISON and J.P. SPRINGER, 1984. J. Org. Chem., 49, 4248-4252.

I. Triclinic, $P\bar{1}$, a = 14.428, b = 15.954, c = 10.609 Å, α = 104.40, β = 109.48, γ = 84.16°, Z = 4. Cu radiation, R = 0.045 for 4970 reflexions.

II. Monoclinic, $P2_1/c$, a = 8.983, b = 16.254, c = 15.988 Å, β = 98.22°, Z = 4. Cu radiation, R = 0.039 for 2119 reflexions.

The trans- and cis-configurations for these molecules were established by X-ray analysis (Fig. 1).

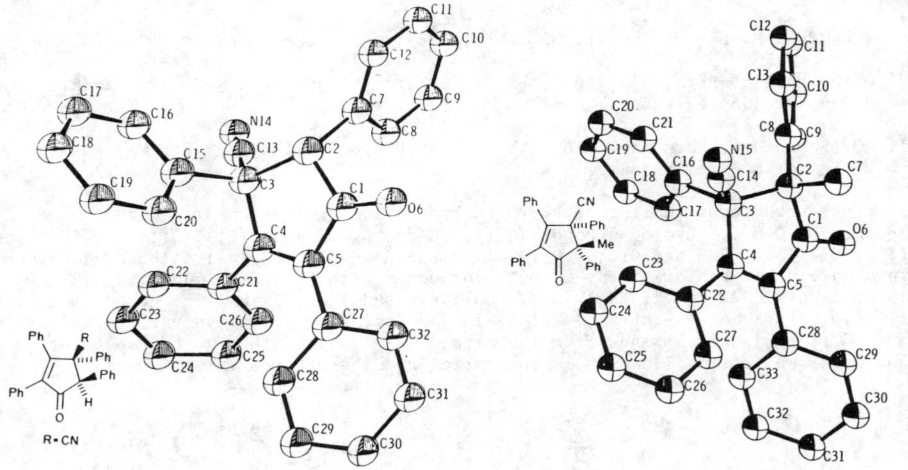

Fig. 1. Views of the cis (I) and trans (II) (right) 2-cyclopenten-1-one derivatives.

4-(1,3-DIPHENYL-1-PROPENYL)-2,3,5-TRIPHENYL-2-CYCLOPENTEN-1-ONE

$C_{38}H_{30}O$

L. CAPUANO, T. TRIESCH, V. SCHRAMM and W. HILLER, 1984. Chem. Ber., 117, 2785-2789.

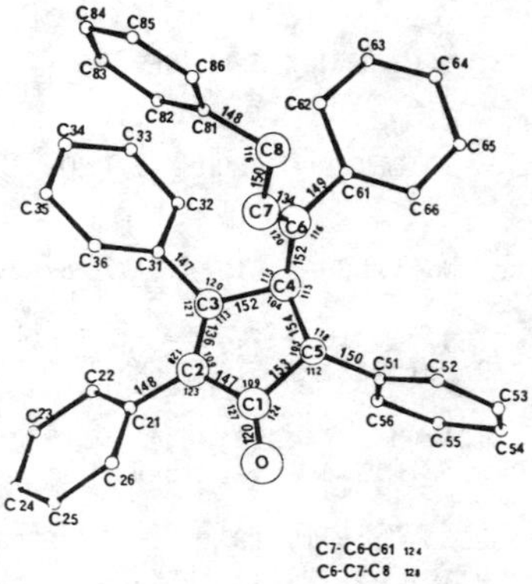

Fig. 1. The conformation of $C_{38}H_{30}O$ and dimensions (pm and degrees).

Orthorhombic, Pbca, a = 21.981, b = 25.216, c = 10.133 Å, Z = 8. Mo radiation, R = 0.063 for 947 reflexions.

A view of the molecule showing the conformation and some dimensions in Fig. 1.

(±)trans-1,2-CYCLOHEXANEDIOL
$C_6H_{12}O_2$ (I)

cis-1,2-CYCLOHEXANEDIOL
$C_6H_{12}O_2$ (II)

R. SILLANPÄÄ, M. LESKELÄ and L. HILTUNEN, 1984. Acta Chem. Scand., B38, 249-254.

I. Orthorhombic, Pbca, a = 7.885, b = 19.301, c = 8.498 Å, Z = 8. Mo radiation, R = 0.036 for 496 reflexions.

II. Orthorhombic, Pbca, a = 7.687, b = 19.696, c = 8.611 Å, Z = 8. Mo radiation, R = 0.034 for 804 reflexions.

Both structures consist of hydrogen-bonded 1,2-cyclohexanediol dimers held together by O-H...O hydrogen bonds (O(1)...O(2') 2.728 and O(2)...O(1'') 2.779 Å in I; O(1)...O(2''') 2.778 and O(2)...O(1'''') 2.758 Å in II) (Fig. 1). The cyclohexanediol rings have chair conformations in both compounds with C(1)-O(1) and C(2)-O(2) distances of 1.422 and 1.429 Å in I, and 1.428 and 1.426 Å in II. The non-bonded O(1)...O(2) distances (intramolecular) in I and II are respectively 2.820 and 2.834 Å. The C-C ring distances range from 1.503 to 1.521 Å. In I the two OH groups are equatorial whilst in II one OH is equatorial and the other axial.

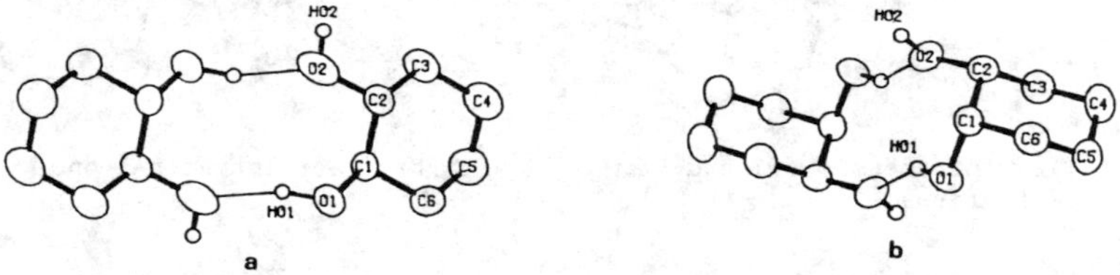

Fig. 1. The dimeric centrosymmetric structures of (a) (±)trans-$C_6H_{12}O_2$ and (b) cis-$C_6H_{12}O_2$.

trans-2,2,3,4,5,6-HEXACHLORO-6-METHYLCYCLOHEX-3-ENONE
$C_7H_4Cl_6O$ (I)

cis-2,4,5,6,6-PENTACHLORO-2-METHYLCYCLOHEX-3-ENONE
$C_7H_5Cl_5O$ (II)

r-2,4,t-5,c-6-TETRACHLORO-2,3,6-TRIMETHYLCYCLOHEX-3-ENONE
$C_9H_{10}Cl_4O$ (III)

M.P. HARTSHORN, R.J. MARTYN and J. VAUGHAN, 1984. Aust. J. Chem., 37, 795-808.

I. Monoclinic, $P2_1/c$, a = 8.410, b = 10.194, c = 13.329 Å, β = 98.04, D_m = 1.82, Z = 4. Mo radiation, R = 0.039 for 1336 reflexions.

II. Monoclinic, $P2_1/c$, a = 7.421, b = 14.319, c = 10.178 Å, β = 97.69°, D_m = 1.70, Z = 4. Mo radiation, R = 0.046 for 1585 reflexions.

III. Monoclinic, $P2_1/c$, a = 8.533, b = 10.187, c = 13.526 Å, β = 97.97°, D_m = 1.55, Z = 4. Mo radiation, R = 0.058 for 634 reflexions.

Structures I-III were obtained by chlorinating appropriate substituted phenols and anilines. These compounds have a twist-boat conformation for the cyclohex-3-

enone ring. The torsion angles C(1)-C(2)-C(3)-C(4) and C(3)-C(4)-C(5)-C(6) are +9.6(5), -29.0(5) for I: +6.7(8), -27.9(8) for II; and +11(1), -31(1)° for III respectively (Fig. 1).

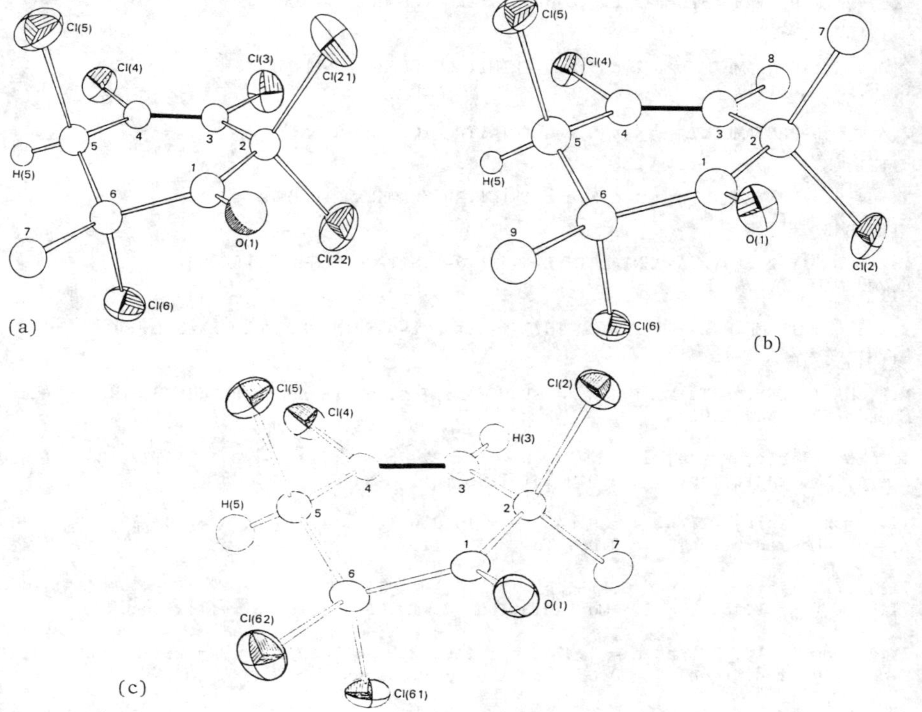

Fig. 1. The structures of (a) $C_7H_4Cl_6O$, (b) $C_7H_5Cl_5O$ and (c) $C_9H_{10}Cl_4O$.

cis-1,3-CYCLOHEXANEDICARBOXYLIC ACID
$C_8H_{12}O_4$ (I)

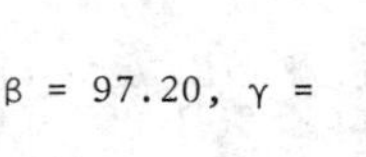

trans-1,3-CYCLOHEXANEDICARBOXYLIC ACID
$C_8H_{12}O_4$ (II)

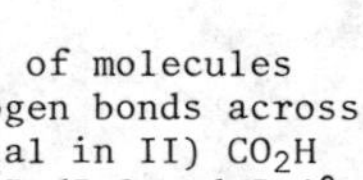

H. VAN KONINGSVELD, 1984. Acta Cryst., C40, 1857-1863.

I. Triclinic, P$\bar{1}$, a = 5.340, b = 6.558, c = 13.834 Å, α = 104.54, β = 96.36, γ = 69.64°, Z = 2. Mo radiation, R = 0.044 for 1634 reflexions.

II. Triclinic, P$\bar{1}$, a = 5.301, b = 6.192, c = 13.311 Å, α = 102.79, β = 97.20, γ = 90.50°, Z = 2. Mo radiation, R = 0.054 for 1671 reflexions.

The structures contain infinite chains along [201] ([111] in II) of molecules linked alternately by 2.664 and 2.648 (2.666 and 2.686 Å in II) hydrogen bonds across centres of symmetry. The carbonyl O atoms of the equatorial (and axial in II) CO_2H groups are twisted away from the C_β-eclipsed position by 21.3 and 15.5 (5.2 and 5.4° in II); i.e. both axial and equatorial groups adopt the synperiplanar C_β-C_α-C=O arrangement.

2,3,4,5,6-PENTAMETHYL-r-2,t-5,t-6-TRINITROCYCLOHEX-3-ENONE
$C_{11}H_{15}N_3O_7$ (I)

2,3,4,5,6-PENTAMETHYL-r-2,t-5,c-6-TRINITROCYCLOHEX-3-ENONE
$C_{11}H_{15}N_3O_7$ (II)

2,3,4,5,6-PENTAMETHYL-r-2,c-5,t-6-TRINITROCYCLOHEX-3-ENONE
$C_{11}H_{15}N_3O_7$ (III)

2,3,4,5,6-PENTAMETHYL-r-4,t-5,t-6-TRINITROCYCLOHEX-2-ENONE
$C_{11}H_{15}N_3O_7$ (IV)

2,3,4,5,6-PENTAMETHYL-r-4,c-5,c-6-TRINITROCYCLOHEX-2-ENONE
$C_{11}H_{15}N_3O_7$ (V)

r-2-HYDROXY-2,3,4,5,6-PENTAMETHYL-t-5,t-6-DINITROCYCLOHEX-3-ENONE
$C_{11}H_{16}N_2O_6$ (VI)

2,3,5,6-TETRAMETHYL-4-NITRATOMETHYL-r-2,t-5,t-6-TRINITROCYCLOHEX-3-ENONE
$C_{11}H_{14}N_4O_{10}$ (VII)

M.P. HARTSHORN, W.T. ROBINSON, J. VAUGHAN, J.M. WHITE and A.R. WHYTE, 1984. Aust. J. Chem., 37, 1489-1502.

I. Monoclinic, $P2_1/n$, a = 7.774, b = 16.427, c = 11.198 Å, β = 106.27, D_m = 1.46, Z = 4. Mo radiation, R = 0.049 for 1343 reflexions.

II. Monoclinic, $P2_1/c$, a = 9.045, b = 10.920, c = 14.429 Å, β = 103.26°, D_m = 1.43, Z = 4. Mo radiation, R = 0.057 for 806 reflexions.

III. Triclinic, $P\bar{1}$, a = 6.555, b = 8.710, c = 12.473 Å, α = 87.71, β = 81.31, γ = 71.96°, D_m = 1.48, Z = 2. Mo radiation, R = 0.057 for 1965 reflexions.

IV. Monoclinic, $P2_1/n$, a = 9.374, b = 11.248, c = 13.040 Å, β = 98.52°, D_m = 1.46, Z = 4. Mo radiation, R = 0.052 for 1758 reflexions.

V. Monoclinic, $P2_1/c$, a = 6.982, b = 14.973, c = 13.081 Å, β = 96.93°, D_m = 1.49, Z = 4. Mo radiation, R = 0.045 for 1320 reflexions.

VI. Monoclinic, $P2_1/c$, a = 6.119, b = 13.814, c = 16.621 Å, β = 100.84°, D_m = 1.39, Z = 4. Mo radiation, R = 0.069 for 1239 reflexions.

VII. Orthorhombic, $P2_12_12_1$, a = 11.105, b = 15.461, c = 8.774 Å, D_m = 1.59, Z = 4. Cu radiation, R = 0.042 for 1093 reflexions.

Nitration of pentamethylphenol with nitrogen dioxide in benzene gives a number of compounds including I-IV. Fuming nitric acid nitration also produces VII. All the cyclohex-3-enone rings are skewed boat except III and V which are a flattened half-chair (Fig. 1). Some torsion angles are in Table 1.

Table 1. Torsion angles (°)

	I	II	III	VI	VII
C(1)-C(2)-C(3)-C(4)	-11.2(5)	-15.9(9)	+12.9(3)	-6.1(7)	-11.9(4)
C(3)-C(4)-C(5)-C(6)	+31.3(4)	+35.7(9)	+19.7(3)	+26.6(6)	+32.5(4)

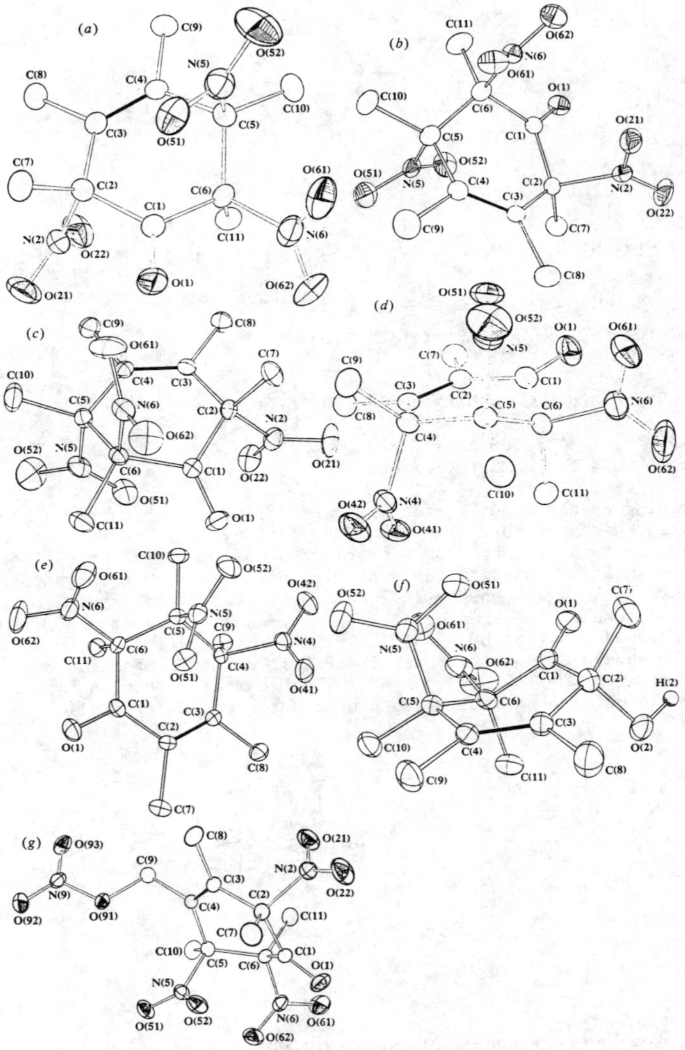

Fig. 1. The structures of I-VII (a to g) respectively.

trans-2-t-BUTYL-CYCLOHEX-3-ENE-1-CARBOXYLIC ACID
$C_{11}H_{18}O_2$

R. VIANI, J. LAPASSET and J.-P. AYCARD, 1984. Acta Cryst., C40, 2074-2076.

Monoclinic, $P2_1/c$, a = 6.478, b = 13.534, c = 12.457 Å, β = 92.58°, D_m = 1.10, Z = 4. Cu radiation, R = 0.058 for 1612 reflexions.

The ring conformation (Fig. 1) is intermediate between half-chair and sofa. Molecules are linked to form hydrogen bonded dimers about inversion centres.

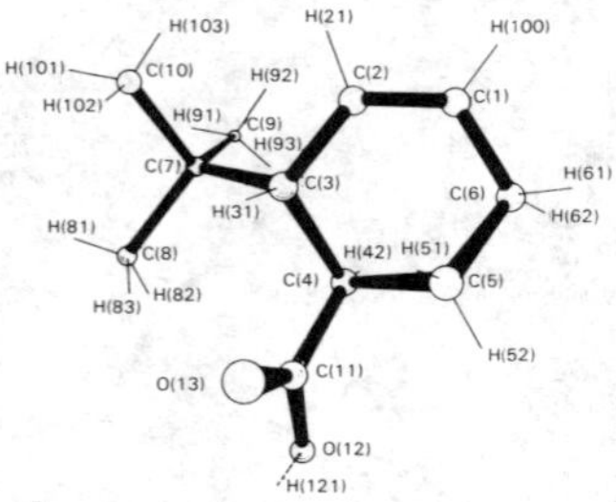

Fig. 1. A view of $C_{11}H_{18}O_2$.

trans-4-t-BUTYL-1-CYCLOHEXANECARBOXYLIC ACID
$C_{11}H_{20}O_2$

H. VAN KONINGSVELD and J.C. JANSEN, 1984. Acta Cryst., B40, 420-424.

Monoclinic, $P2_1/c$, a = 12.303, b = 7.843, c = 11.854 Å, β = 107.84°, Z = 4. Mo radiation, R = 0.044 for 2395 reflexions.

A flattening at the t-butyl side and a puckering at the carboxyl side of the ring are observed in this structure (Fig. 1). The carbonyl O atom of the equatorial carboxyl group is twisted away from the eclipsed position with an α,β bond in the ring by 20.2°, in agreement with the synperiplanar C_β-C_α-C=O arrangement observed for equatorial as well as axial carboxyl groups in several cyclohexanecarboxylic acids and related compounds. The equatorial t-butyl group is twisted away from the perfectly staggered position by about 4°. The off-staggering of the equatorial t-butyl group in several ring structures is described with a Gaussian distribution function G(ω,σ) with <ω> = 0° and σ ≈ 5°. The broad distribution implies a negligible barrier to rotation of the t-butyl group or a potential energy well with a 'flat' minimum.

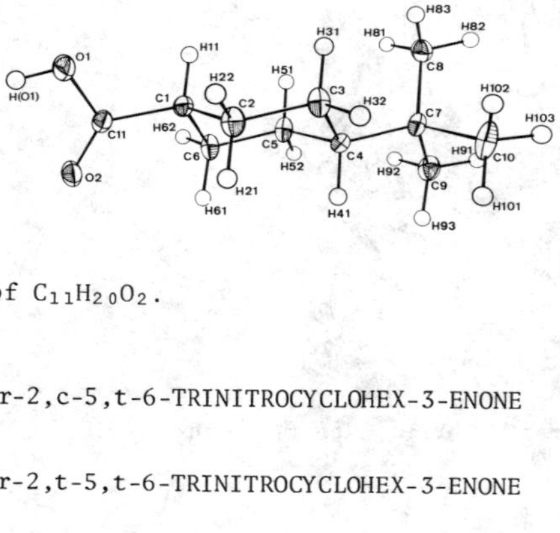

Fig. 1. A drawing of $C_{11}H_{20}O_2$.

4-t-BUTYL-2,6-DIMETHYL-r-2,c-5,t-6-TRINITROCYCLOHEX-3-ENONE
$C_{12}H_{17}N_3O_7$ (I)

4-t-BUTYL-2,6-DIMETHYL-r-2,t-5,t-6-TRINITROCYCLOHEX-3-ENONE
$C_{12}H_{17}N_3O_7$ (II)

4-t-BUTYL-2,6-DIMETHYL-r-2,c-5,c-6-TRINITROCYCLOHEX-3-ENONE
$C_{12}H_{17}N_3O_7$ (III)

4-t-BUTYL-c-6-HYDROXY-2,6-DIMETHYL-r-2,c-5-DINITROCYCLOHEX-3-ENONE
$C_{12}H_{18}N_2O_6$ (IV)

M.P. HARTSHORN, B.R. PENFOLD, K.H. SUTTON and J. VAUGHAN, 1984. Aust. J. Chem., 37, 809-810.

I. Monoclinic, $P2_1/c$, a = 6.846, b = 13.335, c = 16.612 Å, β = 94.06°, D_m = 1.38, Z = 4. Mo radiation, R = 0.048 for 1227 reflexions.

II. Monoclinic, $P2_1/n$, a = 7.338, b = 12.777, c = 16.379 Å, β = 100.46°, D_m = 1.40, Z = 4. Mo radiation, R = 0.042 for 1265 reflexions.

III. Orthorhombic, $P2_12_12_1$, a = 8.545, b = 9.655, c = 18.185 Å, D_m = 1.40, Z = 4. Mo radiation, R = 0.044 for 1946 reflexions.

IV. Monoclinic, $P2_1/n$, a = 6.677, b = 12.956, c = 17.185 Å, β = 95.46°, D_m = 1.28, Z = 4. Mo radiation, R = 0.052 for 1126 reflexions.

The reaction of 4-t-butyl-2,6-dimethylphenol with nitrogen dioxide in benzene yields 4-t-butyl-2,6-dimethyl-r-2,t-5,c-6-trinitrocyclohex-3-enone and I-IV. The cyclohex-3-enone ring in I-III exists in a flattened skewed boat configuration (C(1)-C(2)-C(3)-C(4), C(3)-C(4)-C(5)-C(6) +20.0(5), -34.8(4) for I; +6.7(5), -35.7(4) for II; +11.6(3), -38.4(3)° for III) and in IV as a flattened half-chair (-3.0(6), -28.5(5)° respectively) (Fig. 1).

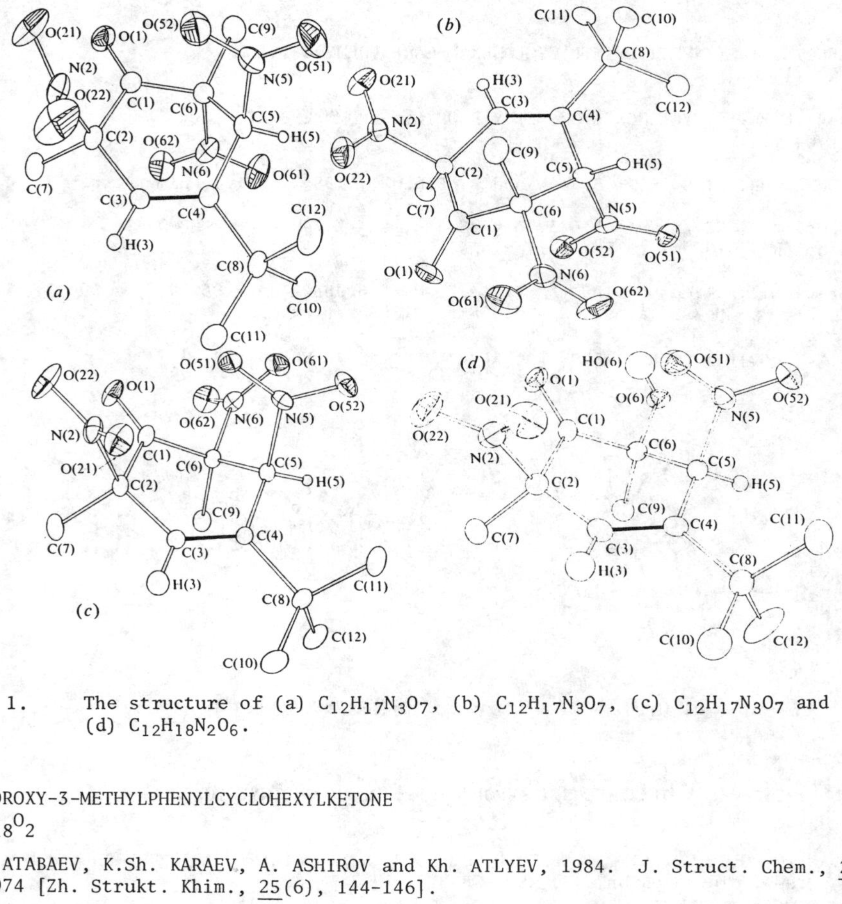

Fig. 1. The structure of (a) $C_{12}H_{17}N_3O_7$, (b) $C_{12}H_{17}N_3O_7$, (c) $C_{12}H_{17}N_3O_7$ and (d) $C_{12}H_{18}N_2O_6$.

4-HYDROXY-3-METHYLPHENYLCYCLOHEXYLKETONE
$C_{14}H_{18}O_2$

T.A. ATABAEV, K.Sh. KARAEV, A. ASHIROV and Kh. ATLYEV, 1984. J. Struct. Chem., 25, 972-974 [Zh. Strukt. Khim., 25(6), 144-146].

Orthorhombic, Pbca, a = 11.877, b = 8.616, c = 24.518 Å, D_m = 1.12, Z = 8. Mo radiation, R = 0.050 for 1664 reflexions.

Bond lengths and angles for the molecule are shown in Fig. 1 (σ = 0.002-0.004 Å and 0.2-0.3° for non-hydrogen atoms). The cyclohexane ring has the chair conformation with the keto substituent in an equatorial position. The dihedral angle between the mean planes of the phenyl and ketone groups is 12.3°.

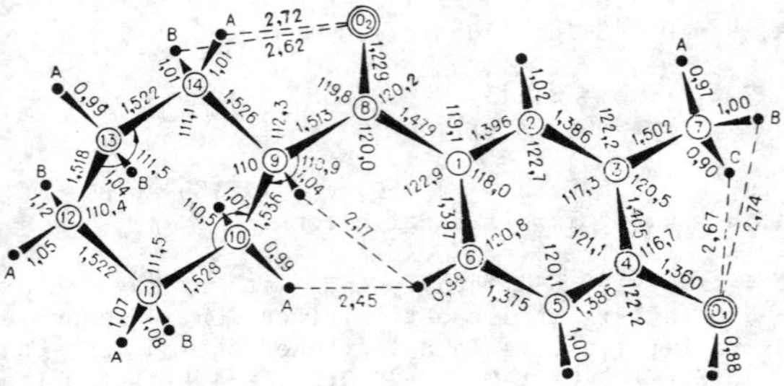

Fig. 1. The 4-hydroxy-3-methylphenylcyclohexylketone molecule.

4β-(t-BUTYL)-1α-(1-METHYLVINYL)CYCLOHEXANECARBALDEHYDE OXIME
$C_{14}H_{25}NO$ (I)

2-(cis-4-(t-BUTYL)CYCLOHEXYL)-2-METHYLPROPYL p-BROMOBENZOATE
$C_{21}H_{31}BrO_2$ (II)

P. SCHONHOLZER, D. SUSS, T.S. WAN and A. FISCHLI, 1984. Helv. Chim. Acta, 67, 669-683.

I. Monoclinic, $P2_1/n$, a = 6.623, b = 28.386, c = 7.742 Å, β = 98.07°, Z = 4. Cu radiation, R = 0.080 for 1593 reflexions.

II. Triclinic, $P\bar{1}$, a = 6.199, b = 10.457, c = 16.083 Å, α = 96.14, β = 90.31, γ = 95.02°, Z = 2. Mo radiation, R = 0.068 for 1338 reflexions.

X-ray analysis established the structures as shown in Fig. 1.

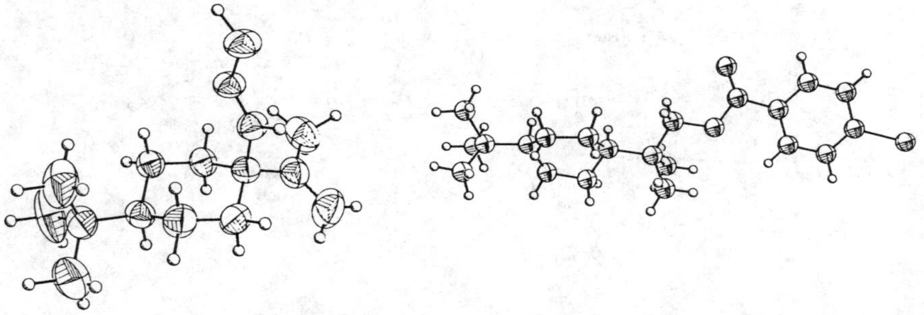

Fig. 1. Views of $C_{14}H_{25}NO$ (left) and of $C_{21}H_{31}BrO_2$ (right).

2-AMINO-1-CYANO-5,5-DIMETHYL-3-(2-PROPENYL)-6-DICYANOMETHYLCYCLOHEXA-1,3-DIENE
$C_{15}H_{16}N_4$

D.C. LANKIN, G.W. GRIFFIN, I. BERNAL, J. KORP, S.F. WATKINS, T.J. DELORD and N.S. BHACCA, 1984. Tetrahedron, 40, 2829-2836.

Triclinic, $P\bar{1}$, a = 9.043, b = 9.095, c = 9.934 Å, α = 78.76, β = 73.86, γ = 69.14°, Z = 2. Mo radiation, R = 0.035 for 1447 reflexions.

The molecule (Fig. 1) is in a half-chair conformation (ΔC_2 = 5.5°). The C5-C6 π-system does not interact with either the adjacent vinyl substituent or with the other endocyclic π system (C1-C2). Molecules are joined in pairs by N-H...N≡C hydrogen bonds.

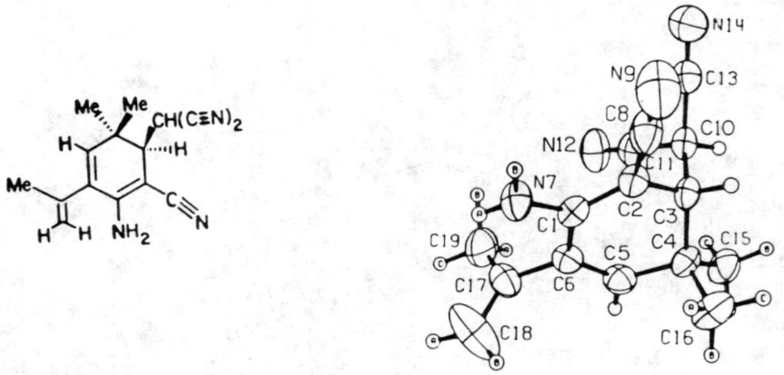

Fig. 1. The $C_{15}H_{16}N_4$ molecule.

3,5-DICYANO-6-(2-METHOXY-1,1,2-TRIMETHYLPROPYL)CYCLOHEXENE
$C_{15}H_{22}N_2O$ (I)

cis-1,3-DICYANO-4-(2-METHOXY-1,1,2-TRIMETHYLPROPYL)CYCLOHEXENE
$C_{15}H_{22}N_2O$ (II)

1,5-DICYANO-4-(2-METHOXY-1,1,2-TRIMETHYLPROPYL)CYCLOHEXENE
$C_{15}H_{22}N_2O$ (III)

R.M. BORG, D.R. ARNOLD and T.S. CAMERON, 1984. Can. J. Chem., 62, 1785-1802.

I. Triclinic, $P\bar{1}$, a = 6.558, b = 9.627, c = 11.967 Å, α = 101.584, β = 87.429, γ = 98.208°, Z = 2. Mo radiation, R = 0.034 for 1044 reflexions.

II. Orthorhombic, Pbca, a = 10.013, b = 11.541, c = 26.362 Å, Z = 8. Mo radiation, R = 0.036 for 943 reflexions.

III. Monoclinic, $P2_1/c$, a = 9.231, b = 6.884, c = 22.973 Å, β = 101.946°, Z = 4. Mo radiation, R = 0.033 for 981 reflexions.

X-ray crystallography was used to characterize these three cyclohexene derivatives.

1-HYDROXY-1-ETHYNYL-2-METHYL-2-(CYCLOHEX-1'-ENYL)CYCLOHEXANE
$C_{15}H_{22}O$

V. KRISHNASAMY, B.L. RAO and L.M. RAO, 1984. Z. Kristallogr., 169, 57-61.

Triclinic, $P\bar{1}$, a = 6.624, b = 9.315, c = 10.628 Å, α = 99.83, β = 101.09, γ = 101.36°, Z = 2. Mo radiation, R = 0.080 for 775 reflexions.

Molecular dimensions for this compound (a potential intermediate in steroid synthesis) are in Fig. 1.

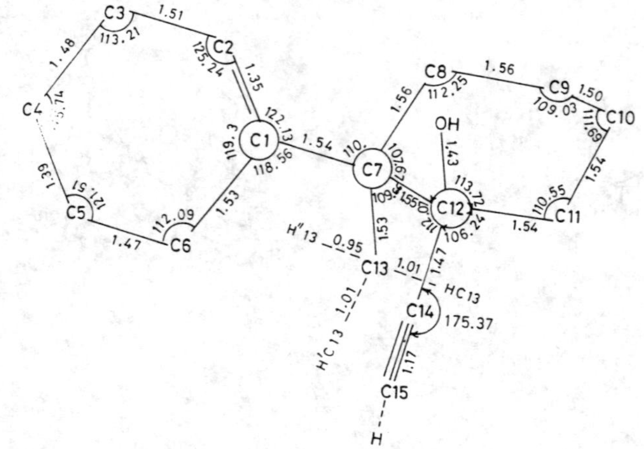

Fig. 1. Bond lengths and angles in $C_{15}H_{22}O$.

(2RS,4RS)-cis-2-[cis-1-CHLORO-4-(2-CHLORO-2-PROPYL)CYCLOHEXYL]-4-(2-CHLORO-2-PROPYL)-1-CYCLOHEXANONE
$C_{18}H_{29}Cl_3O$

J.M. COXON, G.J. HYDES and P.J. STEEL, 1984. Acta Cryst., C40, 1037-1038

Triclinic, $P\bar{1}$, a = 6.403, b = 10.616, c = 14.500 Å, α = 94.13, β = 97.85, γ = 96.86°, D_m = 1.26, Z = 2. Mo radiation, R = 0.065 for 1794 reflexions.

In the crystal structure the two chloropropyl side chains are each disordered between two rotameric forms (Fig. 1). The mean planes through the two cyclohexane chains are inclined by 64.2(4)° to one another, with a staggered conformation about the C(2)-C(1') bond and the Cl substituent axial; all other substituents are equatorial. The relatively long C(1')-Cl(2) bond length (1.862(4) Å) reflects the steric crowding at this centre and accounts for the ease with which the substrate loses HCl.

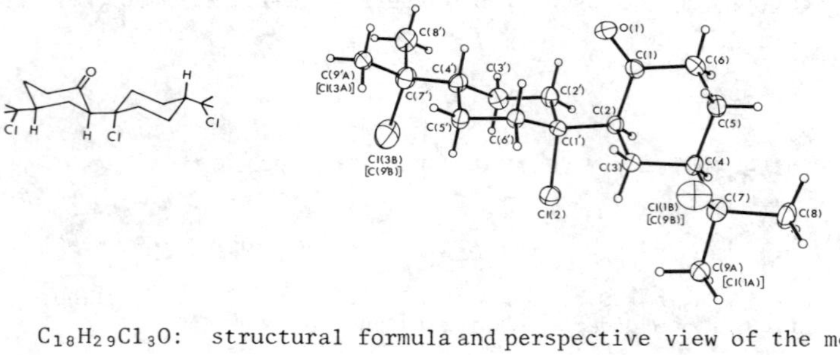

Fig. 1. $C_{18}H_{29}Cl_3O$: structural formula and perspective view of the molecule.

cis-(1R,3S)-3-t-BUTYLCYCLOHEXANYL (2R)-3,3,3-TRIFLUORO-2-METHOXY-2-PHENYLPROPIONATE
$C_{20}H_{27}F_3O_3$

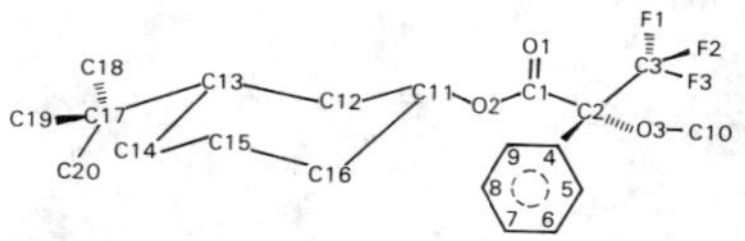

A.T.H. LENSTRA, M.P. GUPTA, L. VANHOECK and G. LEMIERE, 1984. Acta Cryst., C40, 566-569.

Orthorhombic, $P2_12_12_1$, a = 7.805, b = 15.150, c = 16.85 Å, Z = 4. Mo radiation, R = 0.046 for 1448 reflexions.

The cyclohexane ring has the chair conformation with the t-butyl and the propionate groups at equatorial positions. The anomeric effect plays a key role in determining the geometry of the propionate moiety which has C2-O3-C10 118.9(2)° and C2-O3 1.396(3), O3-C10 1.417(3) Å.

1,1'-BIS((1,3,5-TRIMETHYL-4-OXO-2,5-CYCLOHEXADIENYL)PEROXY)DIETHYL ETHER

$C_{22}H_{30}O_7$

C.W. JEFFORD, G. BERNARDINELLI and E.C. McGORAN, 1984. Helv. Chim. Acta, 67, 1952-1956.

Orthorhombic, Pccn, a = 8.237, b = 27.567, c = 10.375 Å, Z = 4. Mo radiation, R = 0.100 for 726 reflexions.

The molecule (Fig. 1) adopts a linear zig-zag conformation with twofold symmetry about O(4). Atom O(4) is disordered about the crystallographic twofold axis. The six-membered ring is almost perpendicular (87.9°) to the plane defined by atoms C(1),O(1),O(2). The molecule is the threo-diastereoisomer.

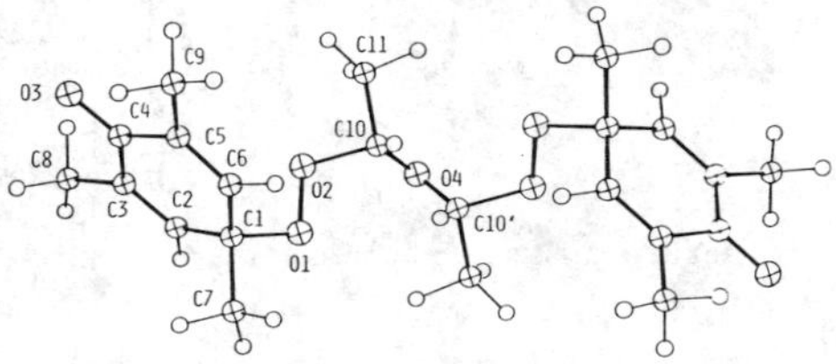

Fig. 1. A view of $C_{22}H_{30}O_7$.

4-t-BUTYL-r-2-(4'-t-BUTYL-2',6'-DIMETHYLPHENOXY)-t-6-HYDROXY-2,6-DIMETHYL-t-5-NITROCYCLOHEX-3-ENONE

$C_{24}H_{35}NO_5$ (I)

M.P. HARTSHORN, K.H. SUTTON and J. VAUGHAN, 1984. Aust. J. Chem., 37, 1369-1374.

Monoclinic, $P2_1/c$, a = 12.277, b = 9.683, c = 20.196 Å, β = 93.32°, D_m = 1.16, Z = 4. Mo radiation, R = 0.061 for 1763 reflexions.

There is an intra-molecular H-bond (HO(6)...O(1) 1.96 Å) with a flattened half-chair conformation for the alicyclic ring (C(4)-C(3)-C(2)-C(1) -2.9(6), C(3)-C(4)-C(5)-C(6) -29.6(5)°, Fig. 1).

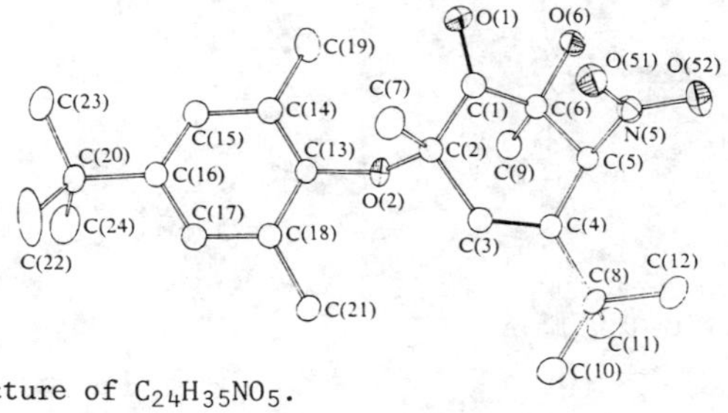

Fig. 1. The structure of $C_{24}H_{35}NO_5$.

2-BENZOYL-4-CYANO-4-ETHOXYCARBONYL-1,3,5-TRIPHENYLCYCLOHEXANOL
$C_{35}H_{31}NO_4$

P.S. MANCHAND, L.J. TODARO and R.S. MANCHAND, 1984. J. Chem. Res., (S) 116-117, (M) 1222-1228.

Triclinic, P$\bar{1}$, a = 10.673, b = 11.265, c = 24.581 Å, α = 94.79, β = 95.82, γ = 98.25°, Z = 4. Cu radiation, R = 0.053 for 6302 reflexions.

The analysis establishes that condensation of two equivalents of chalcone with ethyl cyanoacetate gives the cyclohexanol shown in Fig. 1.

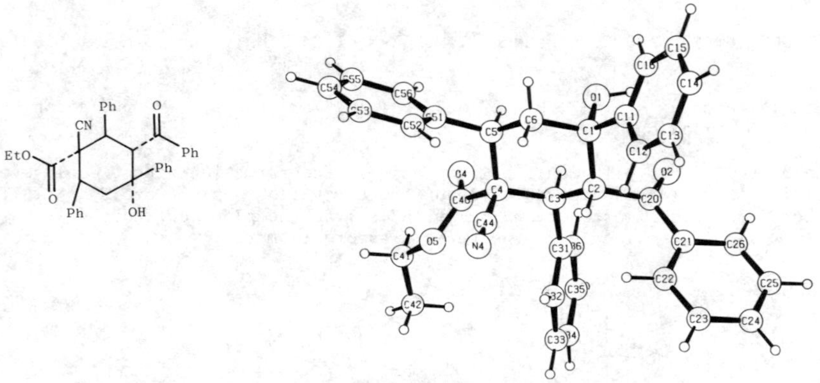

Fig. 1. $C_{35}H_{31}NO_4$: structure and stereochemistry.

2-MERCAPTOCYCLOHEPTATRIENETHIONE (DITHIOTROPOLONE)
$C_7H_6S_2$

B. KREBS, G. HENKEL and W. STUCKER, 1984. Z. Naturforsch., B39, 43-49.

Monoclinic, $P2_1/c$, a = 8.352, b = 13.806, c = 12.371 Å, β = 103.46°, Z = 8. Mo radiation, R = 0.046 for 2525 reflexions (at -130°C).

There are two independent molecules in the asymmetric unit, with similar geometries. A notable feature is the presence of S-H...S intramolecular hydrogen bonds (Fig. 1), with S...S 3.120(1) and 3.142(1), S-H 1.29(4) and 1.25(3), S...H 2.06(4) and 2.14(3) Å, S-H...S 137(3) and 134(2)°.

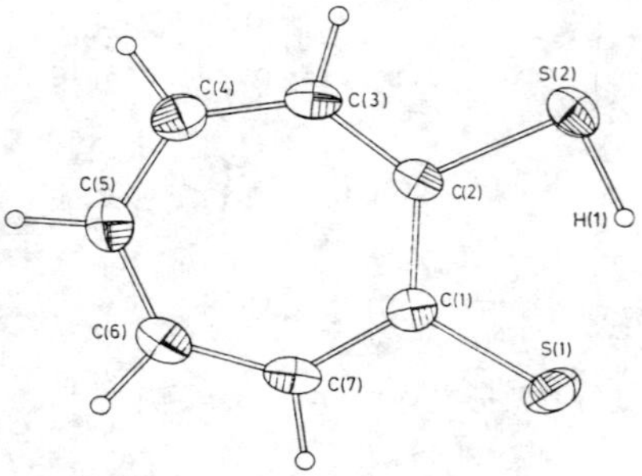

Fig. 1. One of the $C_7H_6S_2$ molecules.

(±)-N-BENZOYL-3-METHYLAMINOCYCLOHEPTANONE
$C_{15}H_{19}NO_2$ (I)

(+)-N-BENZOYL-4-METHYLAMINOCYCLOHEPTANONE
$C_{15}H_{19}NO_2$ (II)

A.T. McPHAIL and A.R. PINDER, 1984. Tetrahedron, 40, 1661-1666.

I. Monoclinic, $P2_1/c$, a = 12.943, b = 13.705, c = 7.557 Å, β = 102.59°, Z = 4. Cu radiation, R = 0.050 for 1442 reflexions.

II. Orthorhombic, $P2_12_12_1$, a = 10.225, b = 20.612, c = 6.423 Å, Z = 4. Cu radiation, R = 0.078 for 678 reflexions.

Structural details for both isomers are in Fig. 1. The seven-membered ring in I has a slightly skewed chair conformation with an approximate mirror plane passing through C(4) and the mid-point of the C(1)-C(7) bond. In II the seven-membered ring has approximate twofold symmetry, the twofold axis passing through C(1) and the mid-point of the C(4)-C(5) bond. In both compounds the orientation of the amide oxygen O(8') is such that it almost eclipses the H atom on the ring carbon bearing the substituent (C(3) in I, 9.7°; C(4) in II, 9.7°).

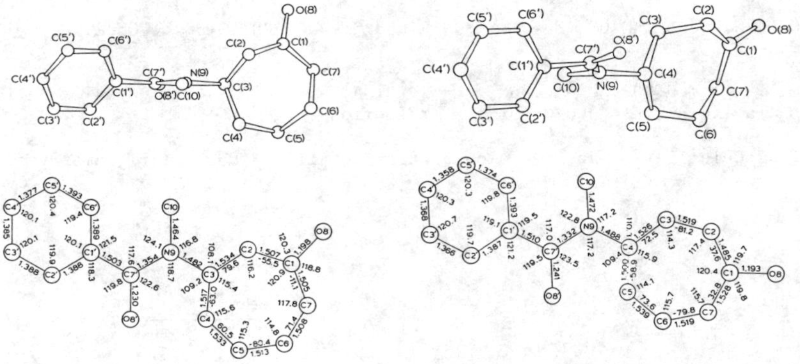

Fig. 1. Solid state conformation and dimensions (lengths (σ 0.009-0.018 Å), bond angles (σ 0.6-1.1°) and torsion angles (σ 0.8-0.9°)) in the two isomers of $C_{15}H_{19}NO_2$ (I), (left) and II, (right).

meso-(BI-2,4,6-CYCLOHEPTATRIEN-1-YL)-3,3'-DICARBOXYLIC ACID DIMETHYL ESTER
$C_{18}H_{18}O_4$

W. BAUER, J. DAUB, E. EIBLER, A. GIEREN, V. LAMM and H. LOTTER, 1984. Chem. Ber., 117, 809-826.

Monoclinic, $P2_1/b$, a = 7.650, b = 7.724, c = 13.451 Å, γ = 79.99°, D_m = 1.24, Z = 2. Cu radiation, R = 0.038 for 790 reflexions.

The molecule lies on an inversion centre (Fig. 1) and is the meso form.

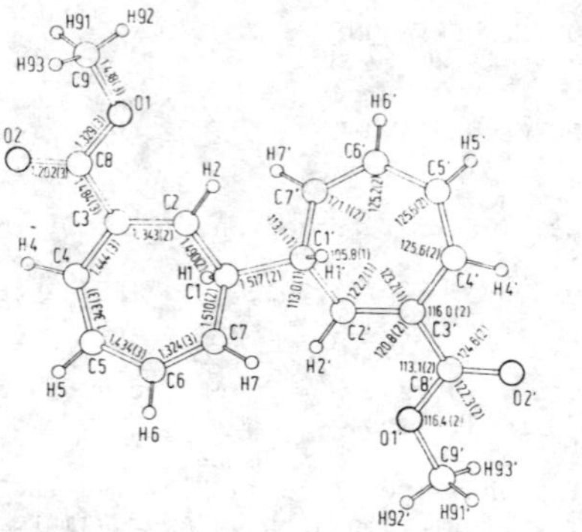

Fig. 1. Molecular dimensions and a view of $C_{18}H_{18}O_4$.

1,1'-DIPHENYL-1,1'-BICYCLOOCTYL
$C_{28}H_{38}$

W. BERNLOHR, H.-D. BECKHAUS, H.-J. LINDNER and C. RUCHARDT, 1984. Chem. Ber., 117, 3303-3319.

Triclinic, P$\bar{1}$, a = 9.802, b = 6.868, c = 9.201 Å, α = 80.94, β = 100.94, γ = 117.1°, Z = 1. Cu radiation, R = 0.076 for 1131 reflexions.

The molecule lies on an inversion centre; details of molecular geometry are in Fig. 1.

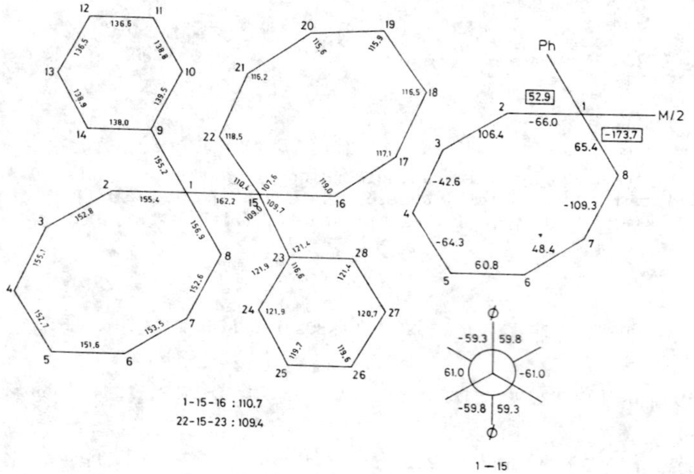

Fig. 1. Bond lengths (pm) and angles for $C_{28}H_{38}$.

1,4,5,8-TETRAKIS(p-TOLYLOXY)METHYL)-1,3,5,7-CYCLOOCTATETRAENE
$C_{40}H_{40}O_4$

R. DIERCKS, L. STAMP and H. TOM DIECK, 1984. Chem. Ber., 117, 1913-1919.

Triclinic, P$\bar{1}$, a = 9.410, b = 11.132, c = 15.887 Å, α = 100.04, β = 98.65, γ = 90.05°, Z = 2. Mo radiation, R = 0.119 for 3452 reflexions.

The analysis established the double bond and substituent positions as shown in Fig. 1 and defined in the title. The interplanar angles for the C1,C6,C7,C8-C2,C3,C4,C5 and C1,C2,C3,C8-C4,C5,C6,C7 moieties are 87.4 and 92.2° respectively.

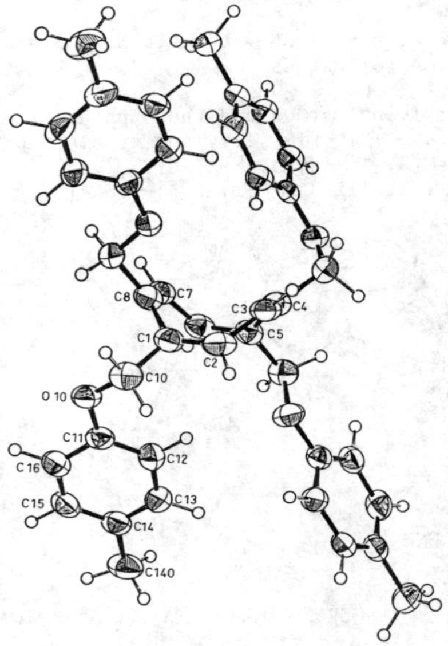

Fig. 1. A view of $C_{40}H_{40}O_4$.

1,5-DIBROMONAPHTHALENE
$C_{10}H_6Br_2$ (I)

1,8-DIBROMONAPHTHALENE
$C_{10}H_6Br_2$ (II)

R.C. HALTIWANGER, P.T. BEURSKENS, J.M.J. VANKAN and W.S. VEEMAN, 1984. J. Cryst. Spectrosc. Res., 14, 589-597.

I. Monoclinic, C2/c, a = 14.576, b = 4.049, c = 15.046 Å, β = 92.18°, Z = 4. Mo radiation, R = 0.039 for 554 reflexions.

II. Monoclinic, $P2_1/c$, a = 8.050, b = 10.898, c = 11.083 Å, β = 107.83°, Z = 4. Mo radiation, R = 0.11 for 2701 reflexions.

Molecule I lies about an inversion centre and is planar (Fig. 1), with C-Br 1.947(7) Å. Molecule II is disordered with two orientations related by a rotation of approximately 90°.

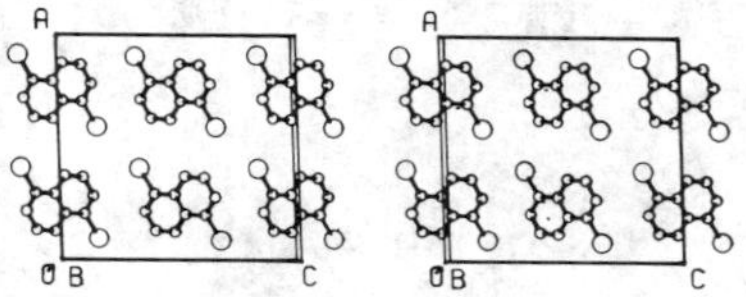

Fig. 1. A stereoview of 1,5-dibromonaphthalene

3-(N-METHYL-β-NAPHTHYLAMINO)ACROLEIN
$C_{14}H_{13}NO$

BAI CHUNLI, YU ZHONGHENG, FU HENG and TANG YOUQI, 1984. J. Struct. Chem., 3, 65-68.

Monoclinic, $P2_1/c$, a = 10.268, b = 11.581, c = 9.605 Å, β = 100.40°, Z = 4. Mo radiation, R = 0.041 for 1264 reflexions.

The planes of the formylvinyl group and the naphthyl ring are tilted with respect to the N(1),C(3),C(4),C(5) (Fig. 1) plane by 8.6 and 33.6° respectively. Main bond lengths are N(1)-C(3) 1.348(2), N(1)-C(5) 1.430(2), N(1)-C(4) 1.464(2), O(1)-C(1) 1.229(3), C(1)-C(2) 1.412(3), C(2)-C(3) 1.354(2) Å.

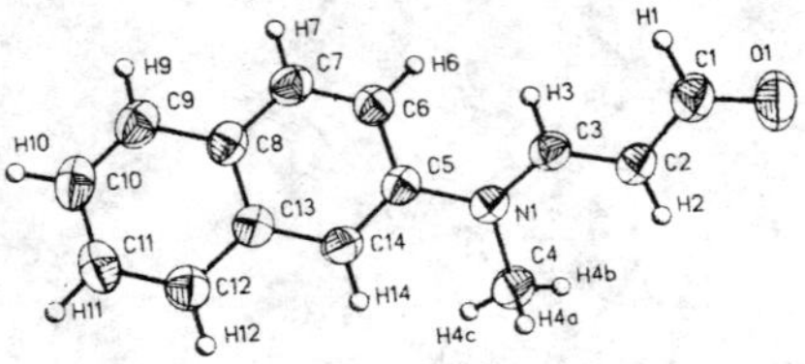

Fig. 1. A view of $C_{14}H_{13}NO$.

1,4,5,8-TETRAMETHYLNAPHTHALENE
$C_{14}H_{16}$

C.S. SHINER, J. NOORDIK, A.M. FISHER, D.M. ECKLEY, J. BODENHAMER and R.C. HALTIWANGER, 1984. Acta Cryst., C40, 540-542.

Monoclinic, $P2_1/n$, a = 7.962, b = 5.205, c = 12.922 Å, β = 104.98°, Z = 2. Mo radiation, R = 0.045 for 607 reflexions (at 300 K).

The molecule (Fig. 1) lies on an inversion centre. The carbon framework is planar; short intramolecular contacts between methyl groups are relieved by in-plane distortions (e.g. C(1)-C(2)-C(6) 125.8(2), C(1)-C(3)-C(7) 125.4(2)°).

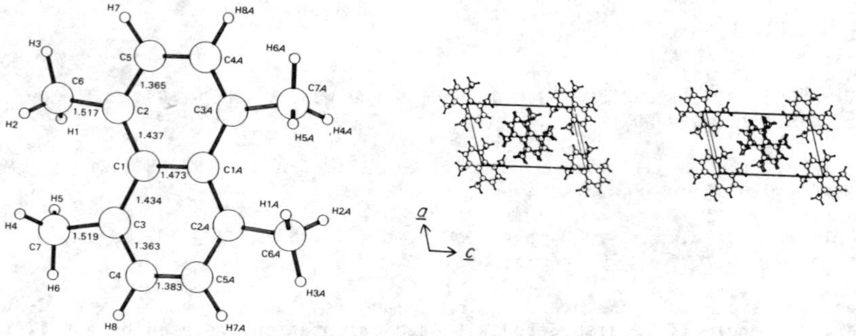

Fig. 1. The $C_{14}H_{16}$ molecule with bond lengths (e.s.d.'s are 0.003 Å) and a stereoview of the packing.

5-(N-METHYL-β-NAPHTHYLAMINO)PENTA-2,4-THIENAL
$C_{16}H_{15}NO$

BAI CHUNLI, YU ZHONGHENG, FU HENG and TANG YOUQI, 1984. J. Struct. Chem., 3, 69-72.

Orthorhombic, Pbca, a = 8.342, b = 15.675, c = 19.779, Z = 8. Mo radiation, R = 0.046 for 1178 reflexions.

The formylbutadienyl plane and the naphthyl ring plane (Fig. 1) form dihedral angles with the N(1),C(5),C(6),C(7) plane of 2.8 and 31.4° respectively. Main bond lengths are: N(1)-C(5) 1.357(3), N(1)-C(6) 1.456(4), N(1)-C(7) 1.423(3), C(1)-O(1) 1.220(3), C(1)-C(2) 1.425(4), C(2)-C(3) 1.349(4), C(3)-C(4) 1.422(4), C(4)-C(5) 1.349(4) Å.

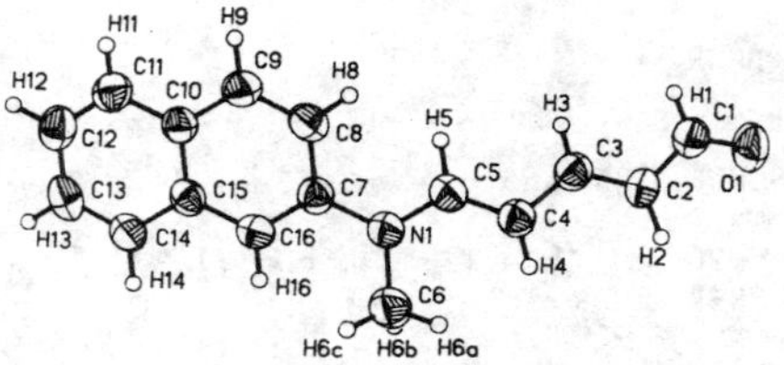

Fig. 1. A view of $C_{16}H_{15}NO$.

CALCIUM NAPHTHIONATE OCTAHYDRATE

$C_{20}H_{16}CaN_2O_6S_2,8H_2O$ $Ca^{2+}.2C_{10}H_8NO_3S^-,8H_2O$

C.J. BROWN, M. EHRENBERG and H.R. YADAV, 1984. Acta Cryst., C40, 58-60.

Monoclinic, $P2_1/n$, a = 23.605, b = 11.931, c = 9.708 Å, β = 94.0°, D_m = 1.528, Z = 4. Cu radiation, R = 0.029 for 4467 reflexions.

The Ca^{2+} ion is seven coordinated to water oxygens in a pentagonal bipyramidal array with mean Ca-O 2.411 Å. The water molecules are hydrogen bonded to sulphonate oxygens and amino nitrogens. All bond lengths and angles have standard values.

NAPHTHALENE - PERFLUOROBIPHENYL

$C_{22}H_8F_{10}$ $C_{10}H_8,C_{12}F_{10}$

L.I. FOSS, A. SYED, E.D. STEVENS and C.L. KLEIN, 1984. Acta Cryst., C40, 272-274.

Monoclinic, C2/c, a = 13.355, b = 6.472, c = 21.994 Å, β = 100.48°, D_m = 1.642, Z = 4. Mo radiation, R = 0.044 for 652 reflexions.

The naphthalene and perfluorobiphenyl molecules are separated by normal van der Waals distances. The naphthalene molecule lies on an inversion centre (Fig. 1) and the perfluorobiphenyl molecule lies on a twofold axis. The dihedral angle between the two rings of the perfluorobiphenyl is 55.3°. Each naphthalene molecule is placed between two perfluorobiphenyl rings; the closest contact between rings is 3.514(7) Å. Molecular dimensions are as expected.

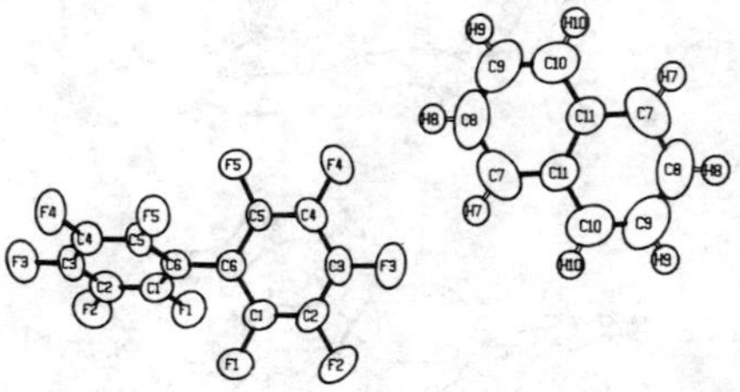

Fig. 1. A plot of the naphthalene and perfluorobiphenyl molecules.

cis-1,8-BIS(1-PYRENYL)NAPHTHALENE
$C_{42}H_{24}$ (I)

trans-1,8-BIS(1-PYRENYL)NAPHTHALENE
$C_{42}H_{24}$ (II)

1,8-BIS(2-PYRENYL)NAPHTHALENE
$C_{42}H_{24}$ (III)

P. WAHL, C. KRIEGER, D. SCHWEITZER and H.A. STAAB, 1984. Chem. Ber., <u>117</u>, 260-276.

I. Monoclinic, $P2_1/c$, a = 20.301, b = 11.414, c = 11.572 Å, β = 98.89°, Z = 4. Mo radiation, R = 0.040 for 3837 reflexions.

II. Monoclinic, $P2_1$, a = 8.581, b = 10.971, c = 14.655 Å, β = 92.50°, Z = 2. Mo radiation, R = 0.038 for 1824 reflexions.

III. Monoclinic, $P2_1/n$, a = 11.448, b = 10.790, c = 21.544 Å, = 93.50°, Z = 4. Mo radiation, R = 0.043 for 2309 reflexions.

The molecules are overcrowded (Fig. 1). Relief from crowding is obtained by in-plane and out-of-plane bond angle bending; details are in Fig. 1.

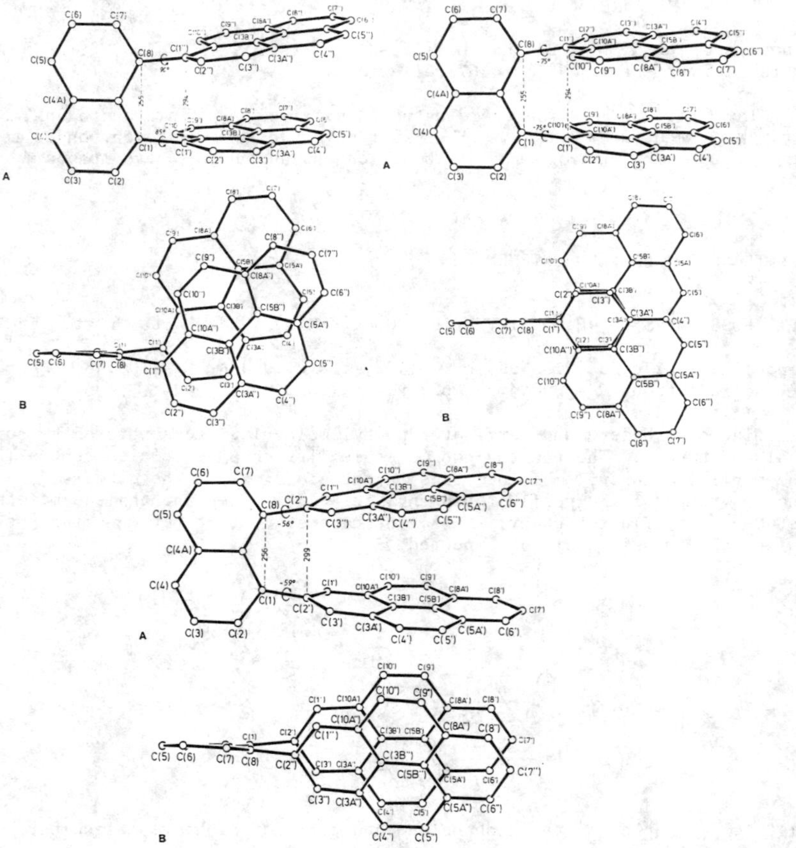

Fig. 1. The three $C_{42}H_{24}$ isomers viewed onto the naphthalene plane, and parallel to the naphthalene plane; I (top left), II (top right), III (lower left).

2,3-DIMETHYL-5,8-DIHYDROXY-1,4-NAPHTHOQUINONE (2,3-DIMETHYLNAPTHAZARIN)
$C_{12}H_{10}O_4$

J.G. RODRIGUEZ, P. SMITH-VERDIER, F. FLORENCIO and S. GARCIA-BLANCO, 1984. J. Mol. Struct., 112, 101-109.

Monoclinic, $P2_1/a$, a = 16.429, b = 6.524, c = 9.136 Å, β = 90.19°, D_m = 1.47, Z = 4. Mo radiation, R = 0.053 for 1278 reflexions.

Molecular dimensions are in Fig. 1; one ring has benzenoid character and the other is quinonoid. The molecules stack along the c-direction and overlap to form a charge transfer complex with 3.42 Å between overlapping planes.

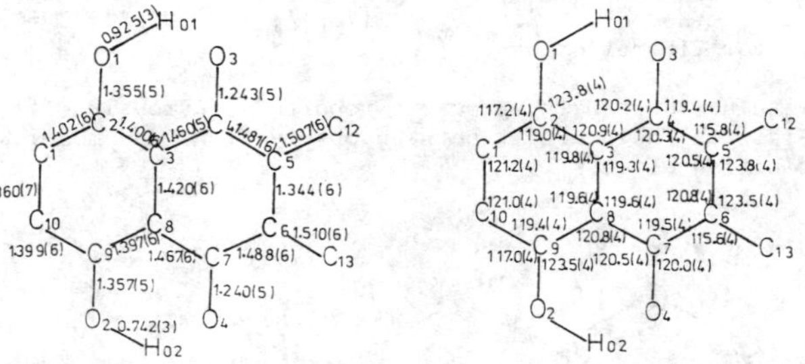

Fig. 1. $C_{12}H_{10}O_4$ with bond lengths and angles.

1,4-ANTHRAQUINONE
$C_{14}H_8O_2$

A.V. DZYABCHENKO and V.E. ZAVODNIK, 1984. J. Struct. Chem., 25, 159-161 [Zh. Strukt. Khim., 25(1), 177-179].

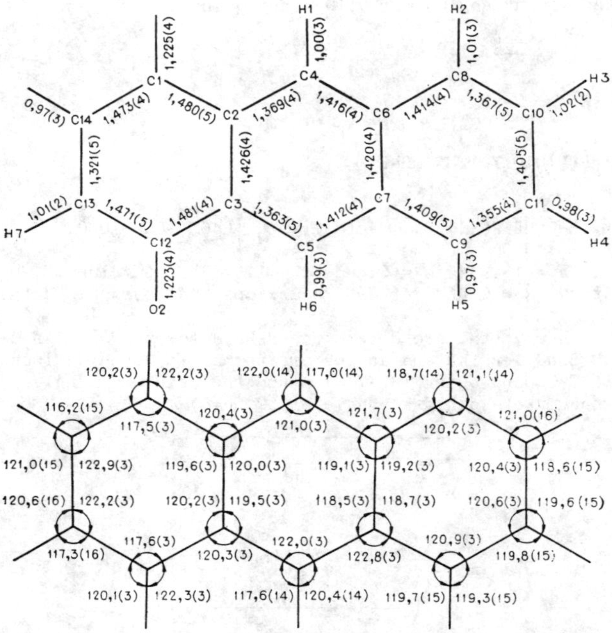

Fig. 1. Bond lengths and angles in the 1,4-anthraquinone molecule

Monoclinic, $P2_1/b$, a = 8.430, b = 19.846, c = 5.927 Å, γ = 100.21°, Z = 4. Mo radiation, R = 0.033 for 772 reflexions.

Bond lengths and angles are shown in Fig. 1. The carbonyl group C(1)-O(1) is displaced by 2.0° from the plane of the rest of the molecule.

9,10-ANTHRAQUINONE
$C_{14}H_8O_2$

A.T.H. LENSTRA and J.F.J. VAN LOOCK, 1984. Bull. Soc. Chim. Belg., 93, 1053-1055.

Monoclinic, $P2_1/c$, a = 3.956, b = 7.866, c = 15.766 Å, β = 102.7°, Z = 2. Mo radiation, R = 0.106 for 718 reflexions.

A refinement with new data of a structure reported previously (1). Details of dimensions are in Fig. 1. The poor fit between observations and model was shown not to be caused by extinction effects.

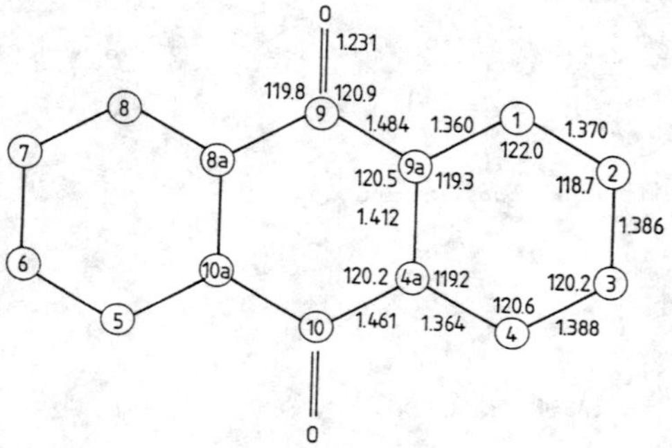

Fig. 1. Distances and angles for 9,10-anthraquinone.

1. Structure Reports, 32B, 129.

9,10-DIHYDRO-9,10-EPIDIOXYANTHRACENE
$C_{14}H_{10}O_2$

C.J. BROWN and M. EHRENBERG, 1984. Acta Cryst., C40, 1059-1060.

Monoclinic, $P2_1/a$, a = 15.944, b = 5.863, c = 11.432 Å, β = 108.2°, D_m = 1.356, Z = 4. Cu radiation, R = 0.062 for 798 reflexions (photographic data).

Molecules in the cyrstal are discrete and held together by van der Waals forces (Fig. 1). Bond lengths and angles are normal for anthraquinone-type compounds, the mean aromatic C-C bond being 1.377(8) Å and the mean C-C single bond 1.517(8) Å. The two benzo rings make a dihedral angle of 52.66(5)°.

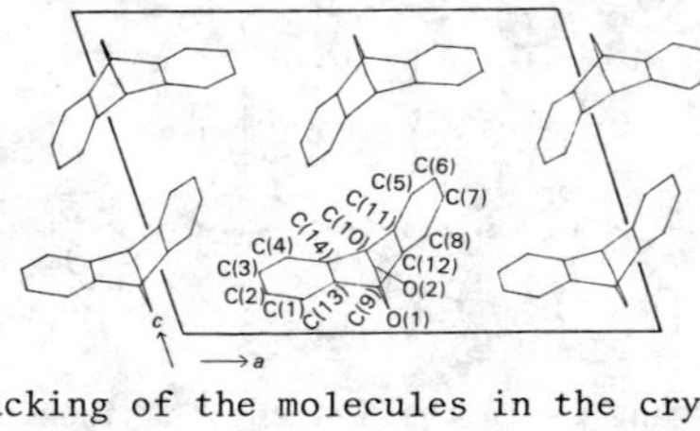

Fig. 1. $C_{14}H_{10}O_2$: packing of the molecules in the crystal.

9-ACETYLANTHRACENE
$C_{16}H_{12}O$

K. ANDERSSON, H.-D. BECKER, L.M. ENGELHARDT, L. HANSEN and A.H. WHITE, 1984. Aust. J. Chem., 37, 1337-1340.

Monoclinic, $P2_1/n$, $a = 9.419$, $b = 16.294$, $c = 8.315$ Å, $\beta = 113.99°$, $D_m = 1.26$, $Z = 4$. Mo radiation, $R = 0.054$ for 1440 reflexions.

The acetyl group is twisted out of the plane of the aromatic π-system by 88°. The compound is characterized by head to tail pairs with an interplanar distance of 3.42 Å between parallel aligned anthracene moieties (Fig. 1).

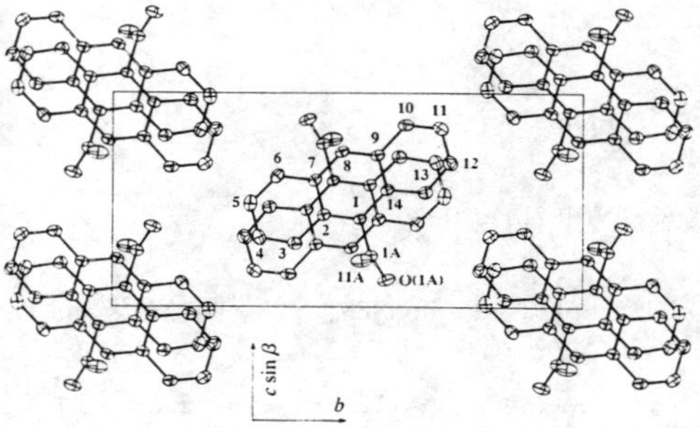

Fig. 1. The structure of 9-acetylanthracene.

(Z)-9-(β-STYRYL)ANTHRACENE
$C_{22}H_{16}$ (I)

(E)-9-(β-STYRYL)ANTHRACENE
$C_{22}H_{16}$ (II)

PHOTODIMER OF 9-PHENYLETHYNYLANTHRACENE
$C_{44}H_{28}$ (III)

H.-D. BECKER, V.A. PATRICK and A.H. WHITE, 1984. Aust. J. Chem., 37, 2215-2224.

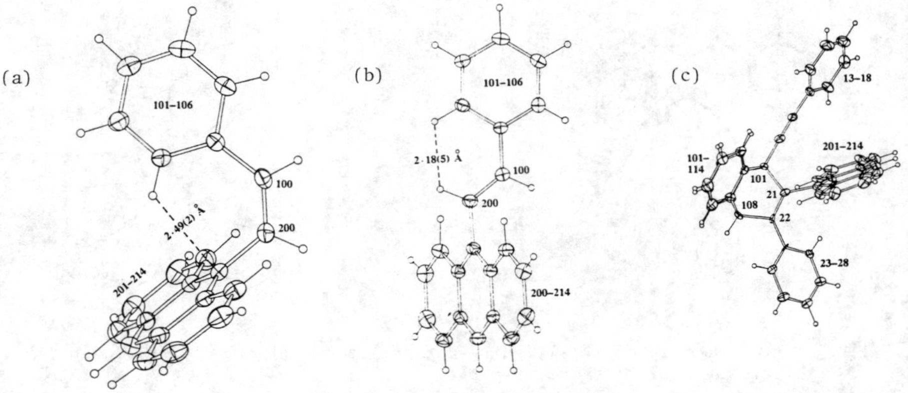

Fig. 1. The structures of (a) (Z)-9-(β-styryl)anthracene, (b) (E)-9-(β-styryl)-anthracene and (c) photodimer of 9-phenylethynylanthracene.

I. Monoclinic, $P2_1/c$, a = 5.507, b = 13.050, c = 21.425 Å, β = 95.73°, Z = 4. Mo radiation, R = 0.047 for 1604 reflexions.

II. Monoclinic, $P2_1/n$, a = 11.77, b = 24.35, c = 5.503 Å, β = 107.38°, Z = 4. Mo radiation, R = 0.050 for 1100 reflexions.

III. Monoclinic, $P2_1/c$, a = 10.719, b = 18.627, c = 15.509 Å, β = 101.10°, Z = 4. Mo radiation, R = 0.070 for 1055 reflexions.

There is an increase in the angle between the plane of the anthracene and the plane of the ethylenic double bond, 78.4, 65.5, 81.9° respectively in I, II, and III. (Fig. 1).

(E)-1,2-DI(9-ANTHRYL)ETHENE
$C_{30}H_{20}$ (I)

(E)-1,2-DI(10-BROMO-9-ANTHRYL)ETHENE
$C_{30}H_{18}Br_2$ (II)

H.-D. BECKER, L.M. ENGELHARDT, L. HANSEN, V.A. PATRICK and A.H. WHITE, 1984. Aust. J. Chem., 37, 1329-1335.

I. Monoclinic, C2/c, a = 13.42, b = 12.48, c = 12.87 Å, β = 118.07, Z = 4. Mo radiation, R = 0.049 for 1010 reflexions.

II. Triclinic, $P\bar{1}$, a = 9.299, b = 8.532, c = 7.474 Å, α = 111.47, β = 95.12, γ = 100.94°, D_m = 1.68, Z = 1. Mo radiation, R = 0.031 for 1519 reflexions.

Molecule (I) has 2-fold symmetry with a dihedral angle of 55.1° between the anthracene plane and the central ethene group, and 69.8° between two anthracene π-systems. II is at a centre of symmetry, dihedral angle 60.7°, (Fig. 1) and the two anthracene π-systems are aligned parallel to one another.

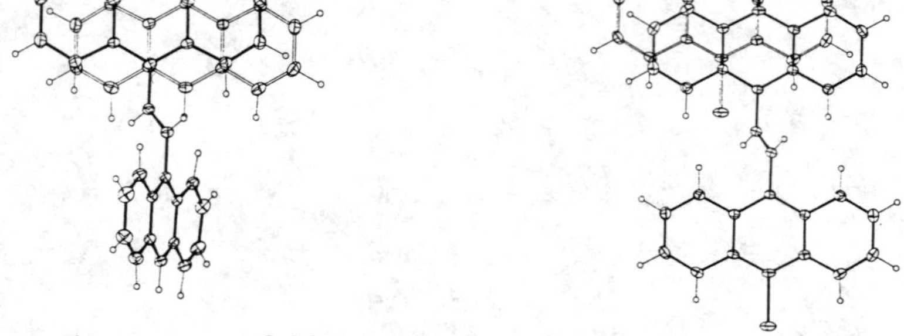

Fig. 1. The structure of (E)-1,2-di(9-anthryl)ethene (left) and (E)-1,2-di(10-bromo-9-anthryl)ethene (right).

2-NITRO-1,3-INDANDIOL
$C_9H_9NO_4$

S. NORDENSON, J. SKRAMSTAD and E. FLOTRA, 1984. Acta Chem. Scand., B38, 461-467.

Tetragonal, $I4_1/a$, a = 19.8402, c = 8.7315 Å, Z = 16. Mo radiation, R = 0.044 for 4799 reflexions (at 130K).

The conformation of the molecule is as expected with an almost planar carbon framework excepting C(2) which is out of the plane in such a way that the nitro group occupies an equatorial position (Fig. 1). The dihedral angle between the benzene ring plane and that formed by atoms C(1), C(4), C(9) and C(3) is 3°, and the angle between the C(1),C(2),C(3) plane and the benzene plane is 29°. The bond distances show almost mirror-like symmetry across the molecule, e.g. C(5)-C(6) 1.396, C(7)-C(8) 1.396, C(4)-C(5) 1.388, C(8)-C(9) 1.392, C(1)-C(9) 1.508, C(3)-C(4) 1.509 Å. Other distances are: C(1)-O(1) 1.420, C(3)-O(3) 1.427, C(1)-C(2) 1.549, C(2)-C(3) 1.535, C(2)-N(2) 1.497, N(2)-O(4) 1.223 and N(2)-O(5) 1.229 Å. The slight deviations from molecular mirror symmetry are caused by hydrogen bonding which comprises two O-H...O and two C-H...O bonds (Fig. 1).

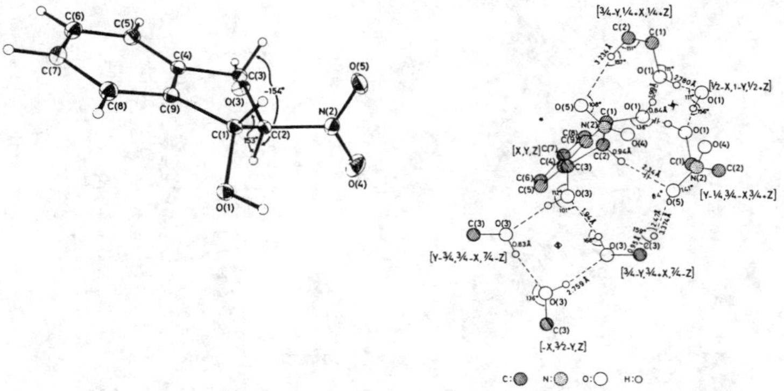

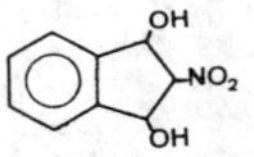

Fig. 1. Molecular structure of $C_9H_9NO_4$ (left), and hydrogen bonding in the crystal structure (right).

(±)-(1R*,2S*,3aR*,7aR*)-PERHYDRO-1,2-INDENEDIOL
$C_9H_{16}O_2$

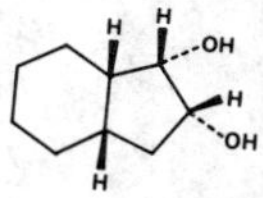

Y. MATOBA, Y. ISHII, M. OGAWA, H. NAKAI and M. SHIRO, 1984. Acta Cryst., C40, 1735-1736.

Triclinic, $P\bar{1}$, a = 11.368, b = 14.833, c = 5.319 Å, α = 95.76, β = 89.98, γ = 106.13°, Z = 4. Cu radiation, R = 0.046 for 2436 reflexions.

All hydroxyl groups form intermolecular hydrogen bonds but not intramolecular ones. The cyclohexane and cyclopentane rings adopt distorted chair and half-chair forms respectively. There are no unusual bond lengths or angles.

(1R*,2R*,3aS*,5S*,6S*,7aS*)-PERHYDRO-1,2,5,6-INDENETETROL
$C_9H_{16}O_4$ (I)

(±)-(1R*,2S*,3aS*,5R*,6S*,7aS*)-PERHYDRO-1,2,5,6-INDENETETROL
$C_9H_{16}O_4$ (II)

Y. MATOBA, Y. ISHII, M. OGAWA, H. NAKAI and M. SHIRO, 1984. Acta Cryst., C40, 691-693.

I. Monoclinic, $P2_1$, a = 6.989, b = 13.107, c = 5.293 Å, β = 100.41°, Z = 2. Cu radiation, R = 0.031 for 853 reflexions.

II. Orthorhombic, $Pna2_1$, a = 12.254, b = 5.177, c = 14.275 Å, Z = 4. Cu radiation, R = 0.025 for 779 reflexions.

The cyclohexane rings in I and II are in a slightly distorted chair form; the cyclopentane ring of I adopts the envelope form, while that of II is a half-chair (Fig. 1). The hydroxyl groups of O(12) and O(13) in I are equatorial, whereas in II they occupy axial and equatorial positions respectively. All the hydroxyl groups form intermolecular hydrogen bonds, but not intramolecular ones; O...O distances are in the ranges 2.671(4)-2.731(4) Å (I) and 2.682(2)-2.811(2) Å (II).

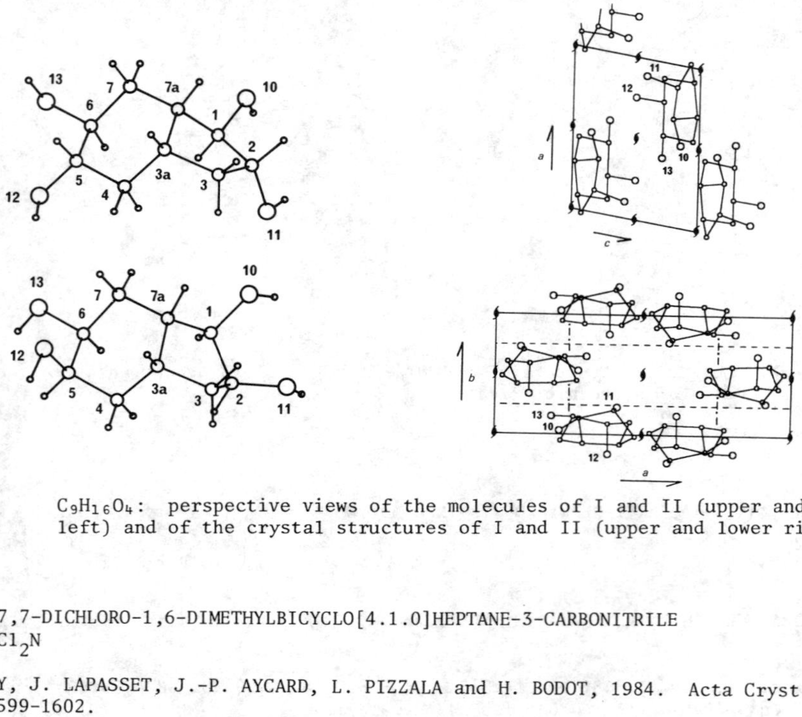

Fig. 1. $C_9H_{16}O_4$: perspective views of the molecules of I and II (upper and lower left) and of the crystal structures of I and II (upper and lower right).

trans-7,7-DICHLORO-1,6-DIMETHYLBICYCLO[4.1.0]HEPTANE-3-CARBONITRILE
$C_{10}H_{13}Cl_2N$

N. EBBY, J. LAPASSET, J.-P. AYCARD, L. PIZZALA and H. BODOT, 1984. Acta Cryst., C40, 1599-1602.

Orthorhombic, $Cmc2_1$, a = 7.336, b = 11.828, c = 11.868 Å, D_m = 1.36, Z = 4. Mo radiation, R = 0.051 for 1396 reflexions (at 109K).

Within the crystal both enantiomers occur, perfectly disordered, and occupy positions of mirror symmetry (Fig. 1). The molecule shows a tendency to be in a sofa conformation; C(1),C(2),C(5),C(6) are planar with C(3) and C(4) 0.66 and 0.11 Å respectively from the plane. The cyano group is axial and its dipole-dipole interactions with the C-Cl bonds stabilise this conformer.

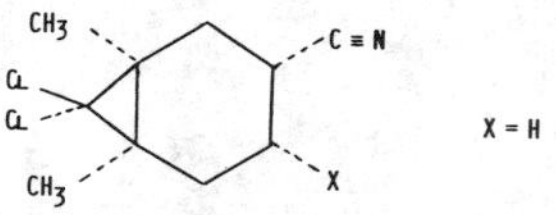

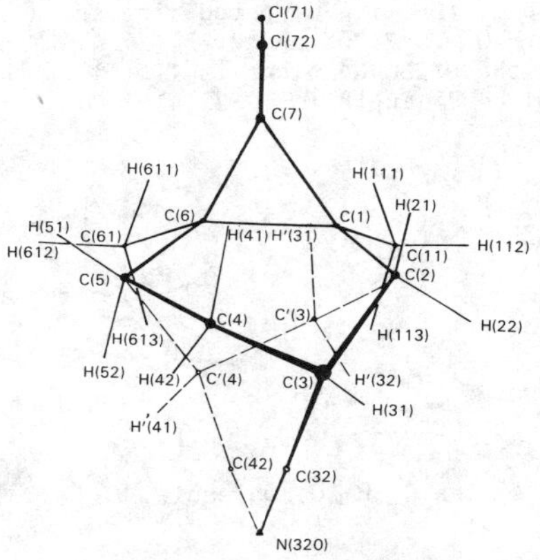

Fig. 1. The relative positions of the two enantiomeric forms of the $C_{10}H_{13}Cl_2N$ molecule within the structure.

2-AMINO-6,7-DIHYDROXYTETRALIN HYDROBROMIDE

$C_{10}H_{14}BrNO_2$ $C_{10}H_{14}NO_2^+.Br^-$

J.K. STALICK, C.R. HUBBARD, A.D. MIGHELL, J.R. RODGERS and A.S. HORN, 1984. Acta Cryst., C40, 317-320.

Monoclinic, $P2_1/c$, a = 11.061, b = 20.939, c = 9.420 Å, β = 92.81°, D_m = 1.54, Z = 8. Cu radiation, R = 0.039 for 2934 reflexions.

There are two independent formula units in the unit cell. Molecule A is essentially planar, except for two C atoms of the aliphatic ring which are displaced equally 0.33 Å above and below the plane of the molecule; the N atom also lies in the molecular plane defined by the aromatic ring. Molecule B exhibits partial disorder of the (+)- and (-)-enantiomers. The molecules pack in hydrogen-bonded layers.

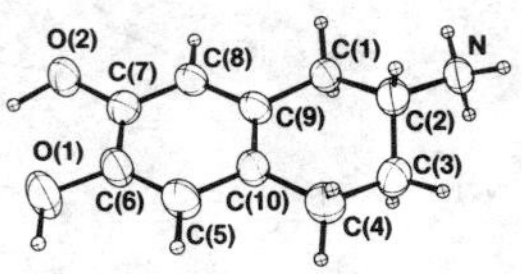

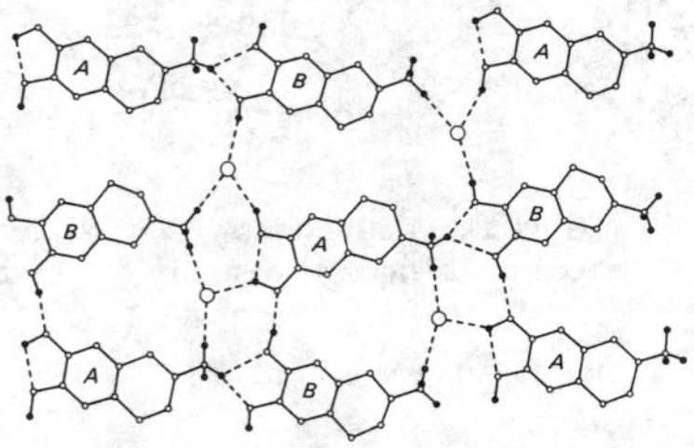

Fig. 1. The $C_{10}H_{14}NO_2^+$ structure and a hydrogen-bonded layer.

rel-(1S,3aR,4R,7S,7aR)-5,6,7,7a-TETRAHYDRO-4,7-METHANO-1H-INDENE-1,3a(4H)-DIOL

$C_{10}H_{14}O_2$

D.A. GROSSIE, W.H. WATSON, C. WU and P.D. BARTLETT, 1984. Acta Cryst., C40, 1416-1418.

Monoclinic, $P2_1/c$, a = 13.657, b = 6.212, c = 11.069 Å, β = 112.42°, Z = 4. Cu radiation, R = 0.043 for 1056 reflexions.

The planar cyclopentene ring (maximum deviation 0.011(7) Å) is cis fused to the norbornane moiety, interplanar angle 124(1)° (Fig. 1). The two OH groups lie on the same side of the five-membered ring and are involved in two intermolecular hydrogen bonds with O...O 2.703 and 2.737(2) Å. The interplanar angle between the cyclopentene ring and the H3C3C4H4 plane is 174(1)° with the H atoms bent towards the OH side. The C3-C4 bond length is 1.322(4) Å.

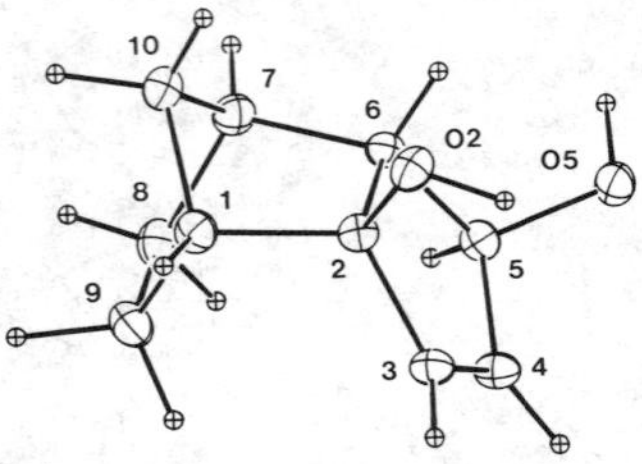

Fig. 1. The $C_{10}H_{14}O_2$ molecule.

trans-DECALIN-1,4-DIONE

$C_{10}H_{14}O_2$

S. ARIEL and J. TROTTER, 1984. Acta Cryst., C40, 1253-1255.

Monoclinic, $P2_1/a$, a = 10.030, b = 5.826, c = 16.191 Å, β = 103.06°, Z = 4. Cu radiation, R = 0.072 for 1025 reflexions.

In the molecule (Fig. 1) the chair cyclohexane ring is trans-fused to a twist-boat cyclohexane-1,4-dione ring; bond lengths and angles are normal. The molecule has an approximate twofold axis through the midpoints of the C(2)-C(3) and C(6)-C(7) bonds. The deviations from perfect C_2 symmetry may arise from the different lattice environments facing the two carbonyl groups.

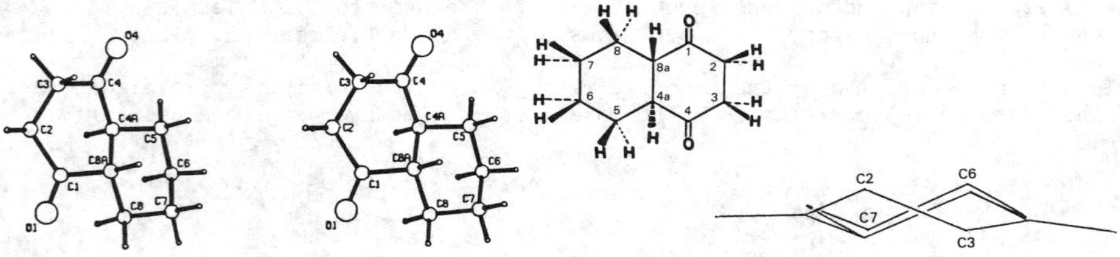

Fig. 1. $C_{10}H_{14}O_2$: stereoscopic view of the molecule and a view in a projection demonstrating the pseudo C_2 symmetry.

4aβ,5,8,8aβ-TETRAHYDRONAPHTHOQUIN-1α,4α-DIOL

$C_{10}H_{14}O_2$

S. ARIEL and J. TROTTER, 1984. Acta Cryst., C40, 100-102.

Orthorhombic, Pbca, a = 20.479, b = 8.101, c = 21.375 Å, Z = 16. Mo radiation, R = 0.048 for 1953 reflexions.

The two crystallographically independent molecules adopt similar conformations; a half-chair cyclohexene ring is cis-fused to a half-chair cyclohexendiol moiety. The hydroxyl configurations in both molecules are anti with respect to the bridgehead H atoms; one is pseudo-equatorial, the other is pseudo-axial. Molecules A and B are linked by O-H...O hydrogen bonds with O...O 2.729-2.909(3) Å.

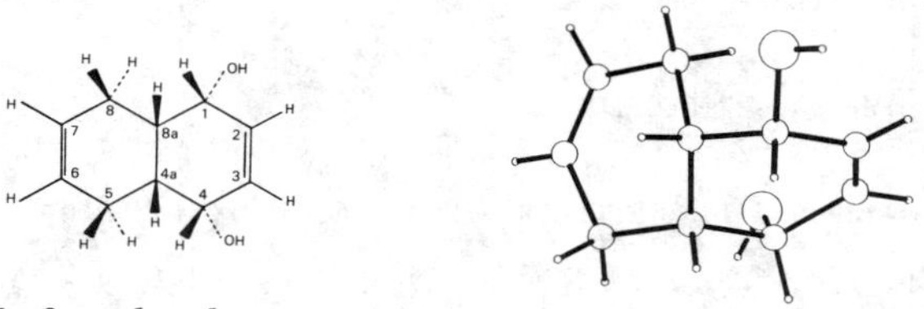

Fig. 1. A $C_{10}H_{14}O_2$ molecule.

5,7-DIMETHOXY-1-INDANONE
$C_{11}H_{12}O_3$

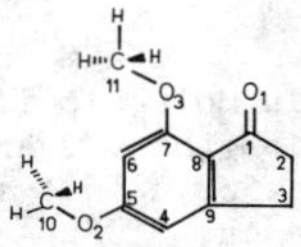

M.P. GUPTA, A.T.H. LENSTRA and H.J. GEISE, 1984. Acta Cryst., C40, 1846-1848.

Monoclinic, $P2_1/n$, a = 8.695, b = 9.782, c = 11.189 Å, β = 90.68°, D_m = 1.34, Z = 4. Mo radiation, R = 0.042 for 1417 reflexions.

The 5-membered ring has a slightly puckered C_s envelope conformation. C(10), O(2),C(11),O(3) are almost coplanar with the aromatic ring; the fragments H_3C-O-C(aryl) -C(aryl) have a staggered, synperiplanar conformation.

9-HYDROXY-10-METHYL-cis-2-DECALONE
$C_{11}H_{18}O_2$ (I)

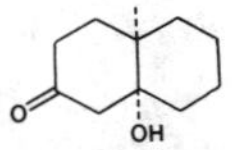

11-OXATRICYCLO[4.4.3.0^{1,6}]TRIDECANE-3,12-DIONE
$C_{12}H_{16}O_3$ (II)

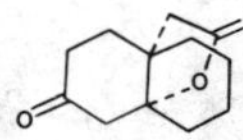

K.D. ONAN, B.J. MAYER and T.A. SPENCER, 1984. Acta Cryst., C40, 1041-1044.

I. Monoclinic, $P2_1/a$, a = 13.643, b = 5.989, c = 12.265 Å, β = 93.62°, D_m = 1.20, Z = 4. Cu radiation, R = 0.037 for 1181 reflexions.

II. Monoclinic, $P2_1$, a = 7.565, b = 10.585, c = 7.422 Å, β = 117.62°, D_m = 1.30, Z = 2. Cu radiation, R = 0.033 for 750 reflexions.

Both molecules adopt chair-chair conformations in which the oxygen ring substituent is axial in the ring containing the carboxyl group (Fig. 1). The presence of a γ-lactone which incorporates the atoms of the ring junction significantly flattens the entire decalone system of molecule II. All intermolecular separations for I and II are normal van der Waals contacts except for an O...O hydrogen bond of 2.837(2) Å in I.

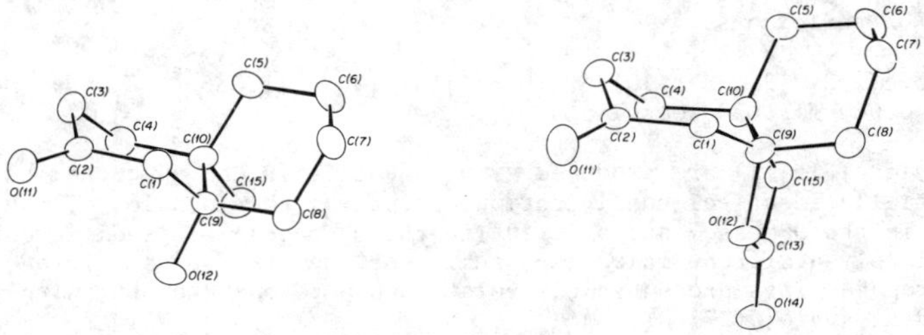

Fig. 1. Perspective views of the $C_{11}H_{18}O_2$ (I) and $C_{12}H_{16}O_3$ (II) molecules.

9-HYDROXY-10-METHYL-trans-2-DECALONE
$C_{11}H_{18}O_2$ (I)

9-ACETOXY-10-METHYL-trans-2-DECALONE
$C_{13}H_{20}O_3$ (II)

K.D. ONAN, B.J. MAYER and T.A. SPENCER, 1984. Acta Cryst., C40, 1044-1047.

I. Monoclinic, $P2_1/c$, a = 12.325, b = 6.357, c = 14.400 Å, β = 116.70°, D_m = 1.19, Z = 4. Cu radiation, R = 0.059 for 1199 reflexions.

II. Monoclinic, $P2_1/n$, a = 7.822, b = 10.731, c = 15.052 Å, β = 101.02°, D_m = 1.19, Z = 4. Cu radiation, R = 0.069 for 1424 reflexions.

Both molecules adopt relatively undistorted chair-chair conformations (Fig. 1). In molecule II the ester carbonyl group is synplanar to the $C(sp^3)$-O bond. Bond lengths and angles are generally normal, with the mean $C(sp^3)$-$C(sp^3)$ bond length 1.532 Å in I and 1.531 Å in II, though the C(9)-C(10) bond is slightly longer in each, 1.552 Å in I and 1.546 Å in II. All intermolecular separations appear to be normal van der Waals contacts apart from a weak O...O hydrogen bond of 2.938(3) Å in I.

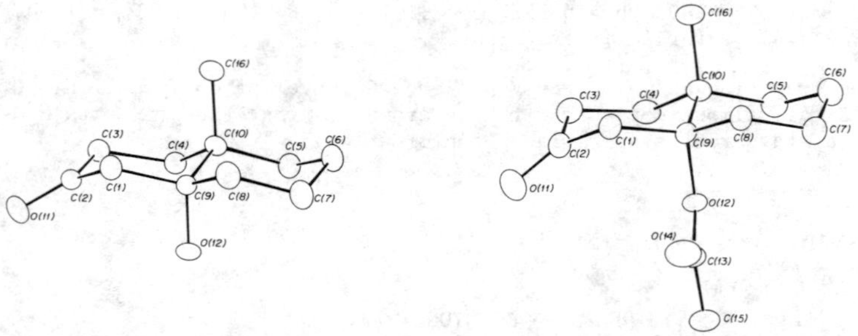

Fig. 1. Perspective views of the $C_{11}H_{18}O_2$ (I) and $C_{13}H_{20}O_3$ (II) molecules.

(1R,4R,6R,9S)-7,7-DICHLORO-1-METHYL-4-(2',2'-DICHLORO-1'-METHYLCYCLOPROPYL)BICYCLO-[4.1.0]HEPTAN-2-ONE
$C_{12}H_{14}Cl_4O$ (I)

(1R,4R,6R,9R)-7,7-DICHLORO-1-METHYL-4-(2',2'-DICHLORO-1'-METHYLCYCLOPROPYL)BICYCLO-[4.0.1]HEPTAN-2-ONE
$C_{12}H_{14}Cl_4O$ (II)

J. ZUKERMAN-SCHPECTOR, E.E. CASTELLANO, G. OLIVA, T.J. BROCKSOM and E.T. CANEVAROLO, 1984. Can. J. Chem., 62, 570-573.

I. Orthorhombic, $P2_12_12_1$, a = 7.982, b = 9.538, c = 18.785 Å, Z = 4. Mo radiation, R = 0.047 for 832 reflexions.

II. Orthorhombic, $P2_12_12_1$, a = 10.339, b = 11.510, c = 11.886 Å, Z = 4. Mo radiation, R = 0.060 for 620 reflexions.

The molecules (Fig. 1) are produced simultaneously in the same chemical reaction and have essentially identical configuration except for the location of the chlorine atoms at C-11 (in the 9S case) and at C-10 (in the 9R case). This difference gives rise to a completely different molecular packing of the structures. In each compound the two cyclopropane rings present a cis relationship to one another, with an identical dihedral angle of 77.0°.

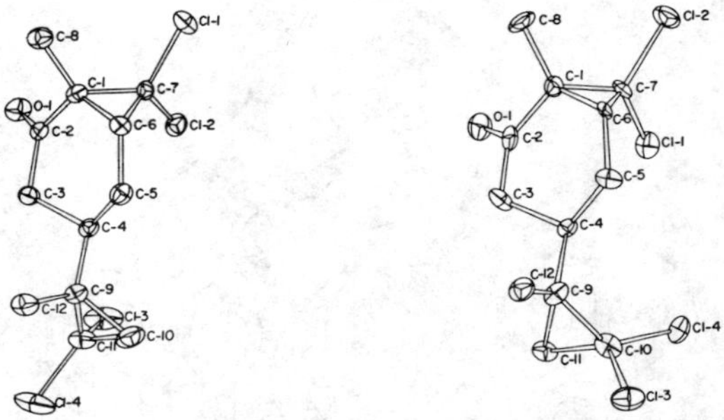

Fig. 1. Views of the 9S (I) (left) and 9R (II) (right) isomers of $C_{12}H_{14}Cl_4O$.

7H-BENZOCYCLONONEN-7-OL
$C_{13}H_{12}O$

MAZHAR-UL-HAQUE, J. AHMED and W. HORNE, 1984. Acta Cryst., C40, 193-195.

Triclinic, P$\bar{1}$, a = 5.688, b = 12.878, c = 14.138 Å, α = 80.85, β = 85.52, γ = 89.79°, Z = 4. Mo radiation, R = 0.038 for 2139 reflexions.

There are two independent molecules in the asymmetric unit with similar geometries. The 9-membered ring is in a tub conformation (Fig. 1) and the hydroxyl group is exo. The crystal structure has molecules linked into tetramers about an inversion centre by O-H...O hydrogen bonds O...O 2.714 and 2.740(5) Å.

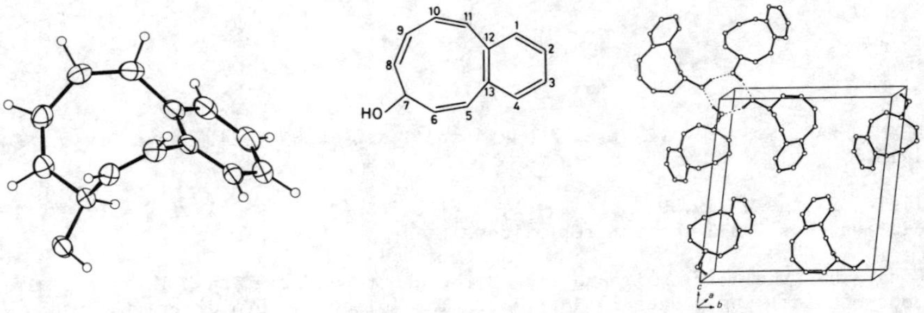

Fig. 1. A view of a $C_{13}H_{12}O$ molecule, and the packing arrangement.

4,6,8-TRIMETHYLAZULENE
$C_{13}H_{14}$

H.N.C. WONG, KI PUI SO and T.C.W. MAK, 1984. Z. Kristallogr., 169, 117-125.

Monoclinic, P2_1/c, a = 8.150, b = 11.999, c = 10.579 Å, β = 100.84°, Z = 4. Mo radiation, R = 0.074 for 1037 reflexions.

The carbon skeleton in this molecule is planar (Fig. 1) and the transannular C9-C10 bond length is 1.488(5) Å.

Fig. 1. The electron-density through the mean plane of $C_{13}H_{14}$. Contours are drawn at 1 eÅ^{-3} intervals starting at 0.5 eÅ^{-3}.

(1RS,7SR,8aRS)-4,7-DIMETHYL-1,2,3,5,6,7,8,8a-OCTAHYDRO-1-NAPHTHOIC ACID
$C_{13}H_{20}O_2$

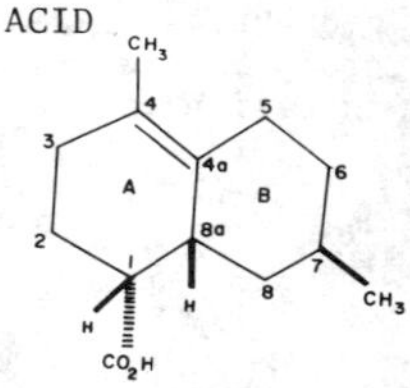

N.N. DHANESHWAR, S.S. TAVALE, K.R. ACHARYA and T.N.G. ROW, 1984. Acta Cryst., C40, 1914-1915.

Monoclinic, $P2_1/n$, a = 9.597, b = 9.461, c = 13.065 Å, β = 93.75°, D_m = 1.152, Z = 4. Mo radiation, R = 0.043 for 575 reflexions.

Ring A has the half-chair and ring B the chair conformation. The dihedral angle between the ring planes is 46.8(7)°. The molecules form dimers through hydrogen bonds linking the carboxyl groups across centres of symmetry (O...O 2.639(6) Å).

cis-2,3,4aβ,8aβ-TETRAMETHYL-4a,5,8,8a-TETRAHYDRO-1,4-NAPHTHOQUINONE
$C_{14}H_{18}O_2$ (I)

cis-2,3,4aβ,8aβ-TETRAMETHYL-4a,5,6,7,8,8a-HEXAHYDRO-1,4-NAPHTHOQUINONE
$C_{14}H_{20}O_2$ (II)

S. ARIEL and J. TROTTER, 1984. Acta Cryst., C40, 1032-1034.

I. Orthorhombic, Pbca, a = 26.255, b = 10.387, c = 8.966 Å, Z = 8. Mo radiation, R = 0.081 for 1905 reflexions.

II. Orthorhombic, Pbca, a = 26.699, b = 10.413, c = 9.003 Å, Z = 8. Mo radiation, R = 0.073 for 939 reflexions.

The crystals are isostructural. In both structures (Fig. 1) the cyclohexene (I) and cyclohexane (II) moieties are cis-fused to a half-chair cyclohex-2-ene-1,4-dione ring. In I the cyclohexene ring exists in a half-chair conformation and in II the cyclohexane ring adopts a chair conformation. All intermolecular interactions correspond to van der Waals contacts.

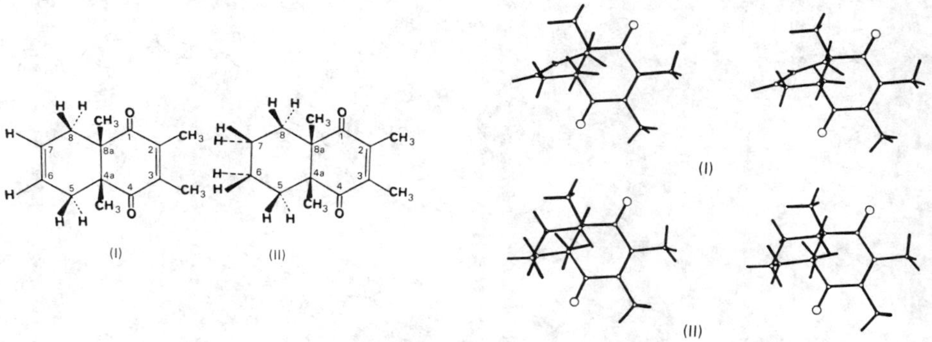

Fig. 1. $C_{14}H_{18}O_2$ (I) and $C_{14}H_{20}O_2$ (II): structural formulae and stereoscopic views of the molecules.

(+)-(1S,2S)-2-AMINO-5-METHOXYINDAN-1-OL (-)-(2S,3S)-TARTARIC ACID DIHYDRATE
$C_{14}H_{19}NO_8,2(H_2O)$ (I)

(+)-(6aR,11bR)-6,6a,7,11b-TETRAHYDRO-5H-INDENO[2,1-c]ISOQUINOLINE
$C_{16}H_{15}N$ (II)

(-)-(6aS,12aS)-5,6,6a,7,8,12a-HEXAHYDROBENZ[a]PHENANTHRIDINE
$C_{17}H_{17}N$ (III)

S. HAGISHITA, M. SHIRO and K. KURIYAMA, 1984. J. Chem. Soc. Perkin I, 1655-1669.

I. Orthorhombic, $P2_12_12_1$, a = 8.191, b = 28.760, c = 7.282 Å, Z = 4. Cu radiation, R = 0.043 for 1455 reflexions.

II. Orthorhombic, $P2_12_12_1$, a = 11.769, b = 13.439, c = 7.670 Å, Z = 4. Cu radiation, R = 0.041 for 1085 reflexions.

III. Monoclinic, $P2_1$, a = 16.125, b = 7.820, c = 10.421 Å, β = 105.91°, Z = 4. Cu radiation, R = 0.042 for 2195 reflexions.

X-ray analysis established the absolute configuration of the 2-amino-5-methoxy-indan-1-ol moiety in I (Fig. 1) to be 1S,2S. Likewise the C(6a)-C(11b) ring junction of II and the C(6a)-C(12b) ring junction of III were determined to be cis. The asymmetric unit of III contains two crystallographically independent molecules.

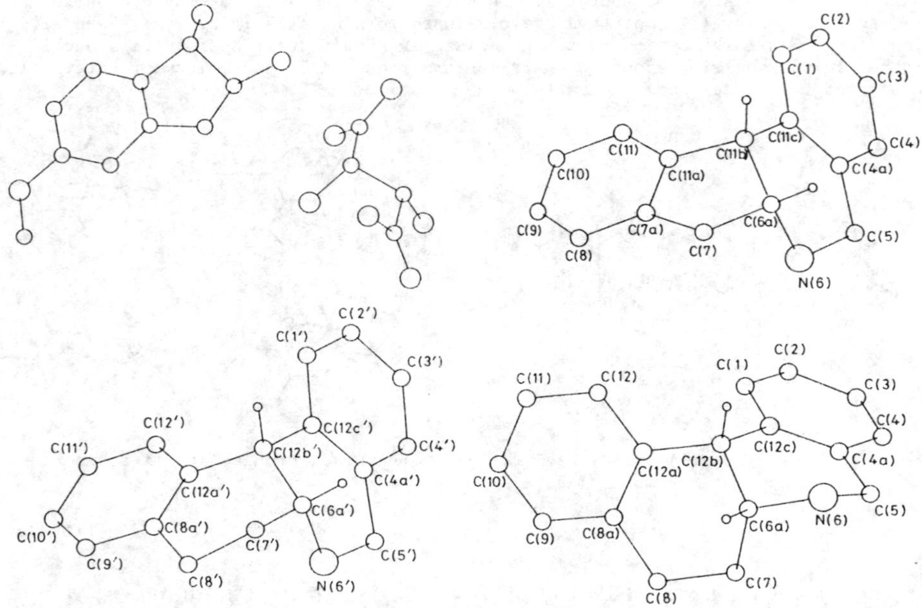

Fig. 1. The molecular structures: $C_{14}H_{19}NO_8,2(H_2O)$ (top left), $C_{16}H_{15}N$ (top right) and the two molecules of $C_{17}H_{17}N$ (bottom).

(1R,7R,8R)-7α-HYDROXY-8β-ISOPROPYL-1α-METHYLBICYCLO[5.3.0]DECANE-2,5-DIONE
$C_{14}H_{22}O_3$

C. PICCINNI-LEOPARDI, J.P. DECLERCQ, G. GERMAIN, M. VAN MEERSSCHE, F. AUDENAERT and M. VANDEWALLE, 1984. Bull. Soc. Chim. Belg., 93, 863-866.

Orthorhombic, $P2_12_12_1$, a = 6.779, b = 13.826, c = 14.130 Å, Z = 4. Mo radiation, R = 0.043 for 1045 reflexions.

In this molecule (Fig. 1) the cyclopentane ring adopts a half-chair conformation, with a symmetry axis passing through C(10). The seven-membered ring has a distorted twist-boat conformation, with a twofold axis bisecting C(4)-C(5).

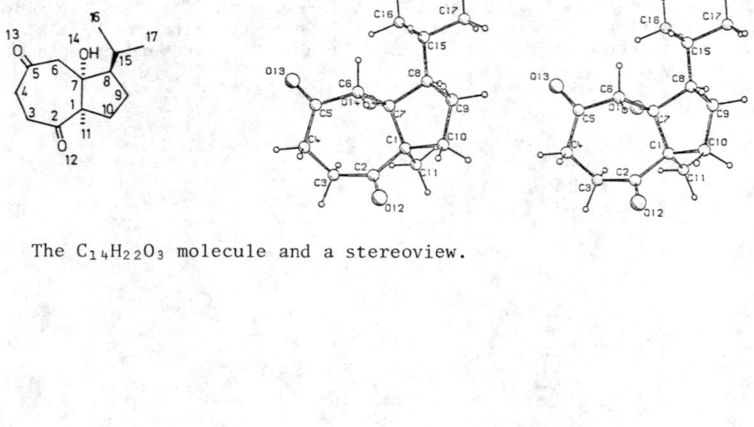

Fig. 1. The $C_{14}H_{22}O_3$ molecule and a stereoview.

4,4,8α-TRIMETHYL-8aβ-METHOXYCARBONYL-4aβ,5,8,8a-TETRAHYDRO-1(4H)-NAPHTHALENONE
$C_{15}H_{20}O_3$

S. ARIEL and J. TROTTER, 1984. Acta Cryst., C40, 2084-2086.

Triclinic, P$\bar{1}$, a = 8.308, b = 11.649, c = 8.233 Å, α = 98.16, β = 111.63, γ = 106.14°, Z = 2. Mo radiation, R = 0.065 for 1011 reflexions.

The half-chair cyclohexene ring (Fig. 1) is cis-fused to a half-chair cyclohexenone ring. Although the packing arrangement is ideally suited for [2+2] photodimerization, the molecule is photochemically inert in the solid state due to steric compression between potentially reacting molecules and their stationary lattice neighbours.

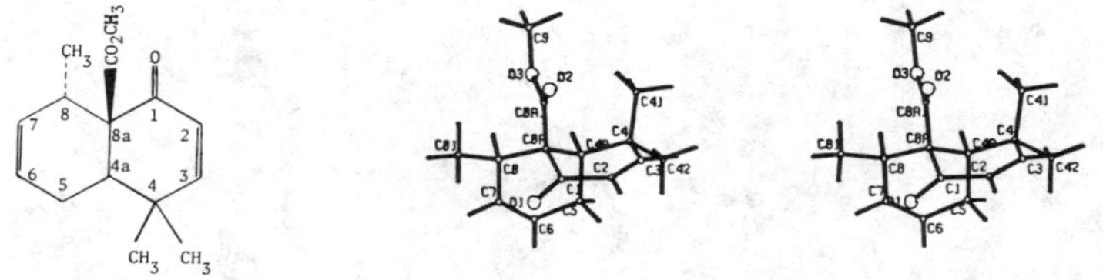

Fig. 1. $C_{15}H_{20}O_3$ and a stereoview.

HEXAHYDRO-4aβ-HYDROXY-5β,8aβ-DIMETHYL-3β-(1-METHYLETHENYL)-1,6(2H,5H)-NAPHTHALENEDIONE
$C_{15}H_{22}O_3$ (I)

HEXAHYDRO-4aβ-HYDROXY-5α,8aβ-DIMETHYL-3α-(1-METHYLETHENYL)-1,6(2H,5H)-NAPHTHALENEDIONE
$C_{15}H_{22}O_3$ (II)

C. AGAMI, F. MEYNIER, C. PUCHOT, J. GUILHEM and C. PASCARD, 1984. Tetrahedron, 40, 1031-1038.

I. Monoclinic, C2/c, a = 24.055, b = 8.075, c = 15.081 Å, β = 112.92°, Z = 8. Cu radiation, R = 0.044 for 1649 reflexions.

II. Monoclinic, P2_1/n, a = 19.532, b = 11.747, c = 12.130 Å, β = 97.61°, Z = 8. Cu radiation, R = 0.052 for 2135 reflexions.

The conformations of I and II are shown in Fig. 1.

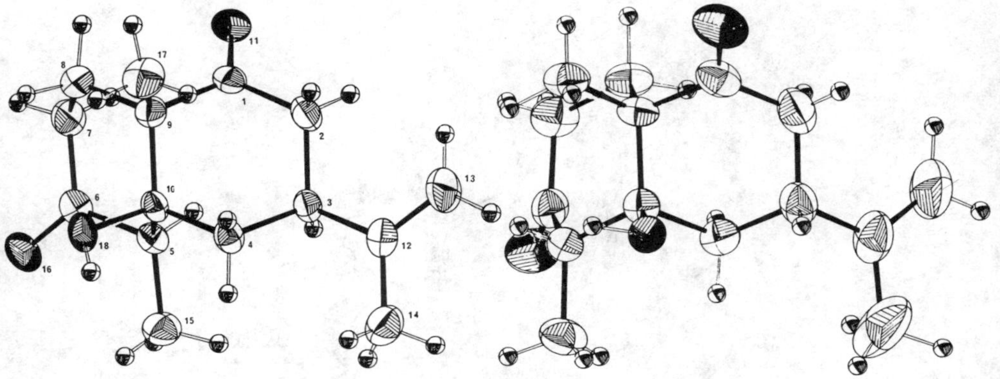

Fig. 1. The 3β (left) and 3α (right) isomer of $C_{15}H_{22}O_3$.

(1S,2R,6S)-1,3,7,7-TETRAMETHYLBICYCLO[4.4.0]DEC-3-ENE-2-METHANOL (DRIMENOL)
$C_{15}H_{26}O$

C. ESCOBAR and O. WITTKE, 1984. Acta Cryst., C40, 1469-1471.

Monoclinic, $P2_1$, a = 12.384, b = 22.615, c = 7.462 Å, β = 93.23°, Dm = 1.07, Z = 6. Mo radiation, R = 0.063 for 3177 reflexions.

The three independent molecules have the same conformation with the cyclohexane ring (chair) and the cyclohexene ring (half-chair) trans-fused. There is steric repulsion between the two axial methyl groups on the cyclohexane ring (C(11)...C(14) 3.359(6) Å). The molecules are linked into infinite chains parallel to c by O-H...O hydrogen bonds (O...O 2.744-2.774 Å).

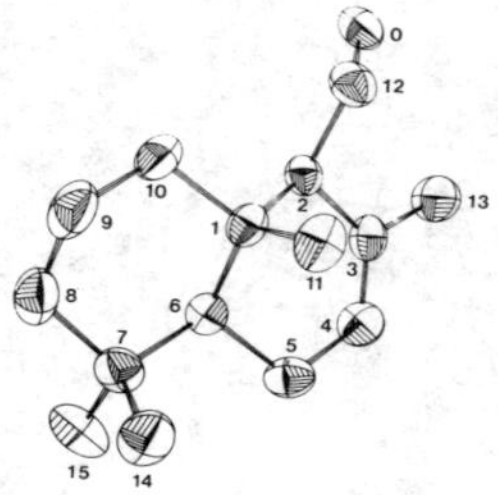

Fig. 1. The drimenol molecule.

2-(N-4-METHOXYPHENYLAMINOETHYLENE)-1,3-INDANEDIONE
$C_{17}H_{13}NO_3$

S.M. ALDOSHIN, L.O. ATOVMYAN, O.A. D'YACHENKO, V.I. MINKIN, V.A. BREN' and Zh.V. Bren', 1984. J. Struct. Chem., 25, 292-296 [Zh. Strukt. Khim., 25(2), 136-140].

Monoclinic, B2/b, a = 25.725, b = 27.890, c = 3.856 Å, γ = 73.80°, Z = 8. Cu radiation, R = 0.038 for 1016 reflexions.

The molecule (Fig. 1) has a planar structure stabilized by an intramolecular N-H...O hydrogen bond (H...N 2.08(2) Å). Bond lengths (σ = 0.002-0.004 Å) differ from those observed for other ketoamines.

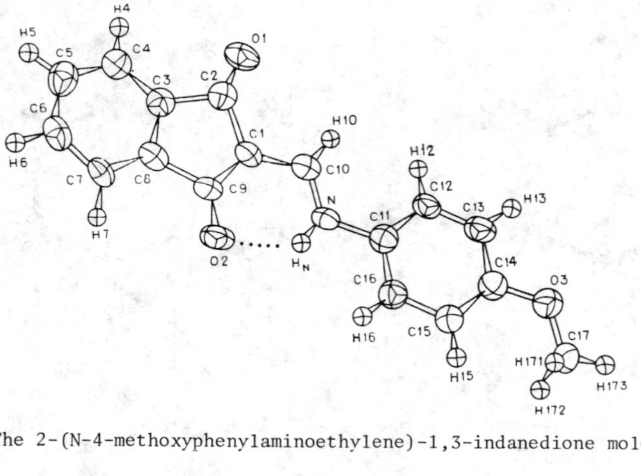

Fig. 1. The 2-(N-4-methoxyphenylaminoethylene)-1,3-indanedione molecule.

9-BENZOYL-2-DECALONE
$C_{17}H_{20}O_2$

N. SEURON, M. OUREVITCH, J. SEYDEN-PENNE and J. GUILHEM, 1984. Bull. Soc. Chim. Fr., II-124-128.

Monoclinic, $P2_1/n$, a = 10.344, b = 17.991, c = 7.769 Å, β = 108.14°, Z = 4. Cu radiation, R = 0.047 for 2122 reflexions.

The stereochemistry of this substituted decalone is shown in Fig. 1.

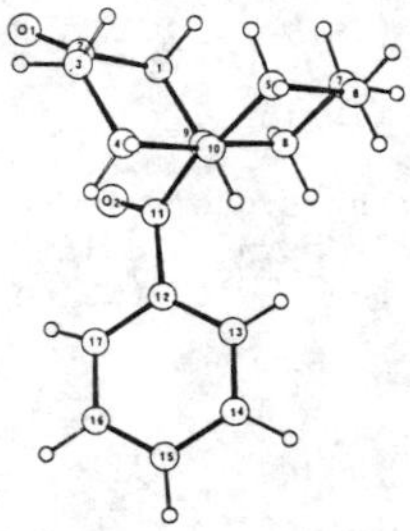

Fig. 1. A view of $C_{17}H_{20}O_2$.

(-)-cis,cis-DECAHYDRO-2-NAPHTHOL p-BROMOBENZOATE
$C_{17}H_{21}BrO_2$

T. ORITANI, K. YAMASHITA and C. KABUTO, 1984. J. Org. Chem., 49, 3689-3694.

Orthorhombic, $P2_12_12_1$, a = 16.980, b = 13.967, c = 6.517 Å, Z = 4. Cu radiation, R = 0.031 for 2703 reflexions (at 10°C).

The X-ray analysis confirmed the structure shown in Fig. 1 and established the (-) configuration by anomalous dispersion measurements. Both cyclohexane rings are in chair form; molecular dimensions are normal.

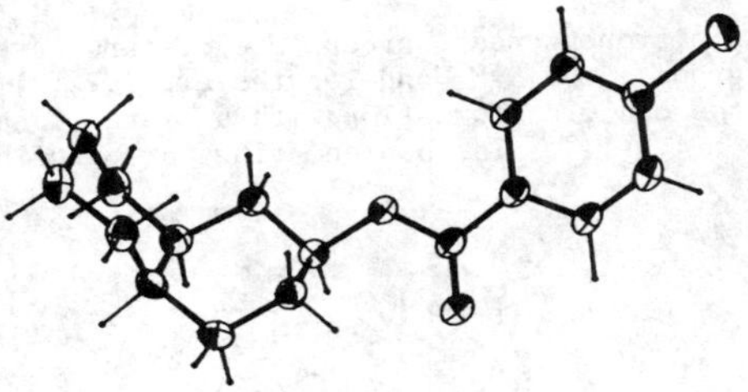

Fig. 1. A view of $C_{17}H_{21}BrO_2$.

2,2-DIMETHYL-5-t-BUTYL-7-ACETAMIDOINDANONE
$C_{17}H_{23}NO_2$

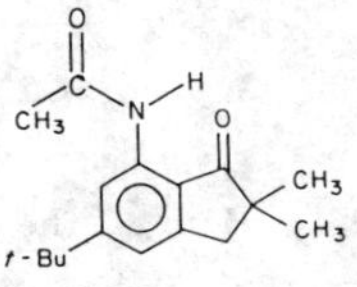

P. MASLAK, P.E. FANWICK and R.D. GUTHRIE, 1984. J. Org. Chem., 49, 655-659.

Orthorhombic, Pnma, a = 12.571, b = 7.302, c = 16.910 Å, D_m = 1.16, Z = 4. Mo radiation, R = 0.056 for 1224 reflexions.

The structure of this reaction product was established as that shown above.

3,4-DIHYDRO-2-[(p-HYDROXYPHENETHYL)AMINOETHYL]-1(2H)-NAPHTHALENONE HYDROCHLORIDE
$C_{19}H_{22}ClNO_2$

A. CARPY, J.-M. LEGER and J.-C. COLLETER, 1984. Acta Cryst., C40, 154-156.

Monoclinic, Cc, a = 17.759, b = 13.282, c = 7.521 Å, β = 107.09°, Z = 4. Cu radiation, R = 0.039 for 1178 reflexions.

The chain linking the two ring systems is fully extended (Fig. 1); the two aromatic ring systems are mutually perpendicular. The quaternary nitrogen and OH groups are involved in hydrogen bonds with the chloride ion.

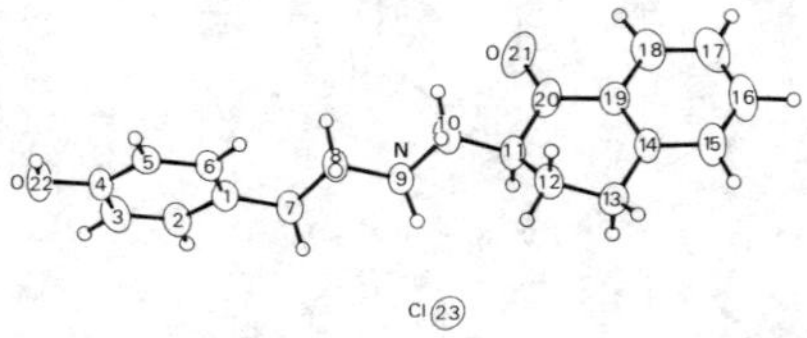

Fig. 1. A view of $C_{19}H_{22}ClNO_2$.

9,10-BIS(p-NITROBENZENEAZO)-cis-DECALIN
$C_{22}H_{24}N_6O_4$

C.A. KAVOUNIS and P.J. RENTZEPERIS, 1984. Acta Cryst., C40, 1434-1436.

Monoclinic, $P2_1/c$, a = 7.520, b = 20.180, c = 14.823 Å, β = 105.02°, D_m = 1.301, Z = 4. Mo radiation, R = 0.061 for 707 reflexions.

The almost planar p-nitrobenzeneazo groups are cis and have antiperiplanar geometry about the azo groups; the N=N bond lengths are 1.22(2) Å (Fig. 1). The cyclohexane rings have the chair conformation with an interplanar angle between the least-squares planes of 114.7(1)°. The benzene rings are inclined at 66.3(1)° to each other.

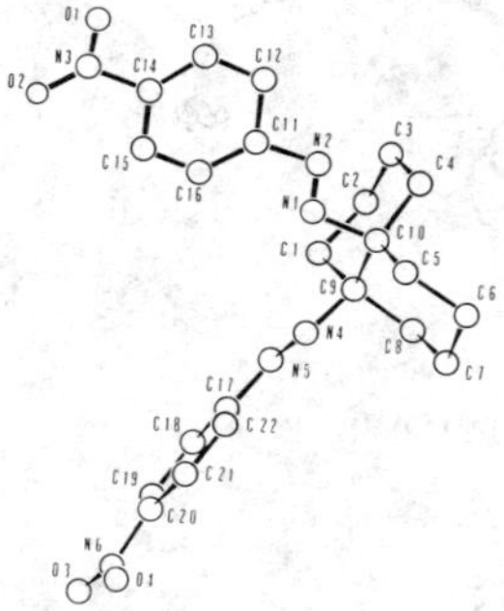

Fig. 1. The $C_{22}H_{24}N_6O_4$ molecule.

3,7-BIS(p-ACETYLPHENYL)BICYCLO[3.3.0]OCTANE
$C_{24}H_{26}O_2$ (I)

3,7-BIS(HYDROXYMETHYL)BICYCLO[3.3.0]OCTANE
$C_{10}H_{18}O_2$ (II)

D. GARDETTE, J.-C. GRAMAIN, J. LHOMME, C. PASCARD and T. PRANGE, 1984. Bull. Soc. Chim. Fr., II-404-408.

I. Monoclinic, $P2_1/c$, a = 18.287, b = 9.571, c = 11.719 Å, β = 107.22°, Z = 4. Cu radiation, R = 0.082 for 1724 reflexions.

II. Monoclinic, $P2_1/c$, a = 9.154, b = 12.145, c = 9.615 Å, β = 110.82°, Z = 4. Cu radiation, R = 0.048 for 1125 reflexions.

The two molecules are shown to have a "W" conformation (Fig. 1). Related NMR studies show that the solution conformations are identical to those found in the crystals.

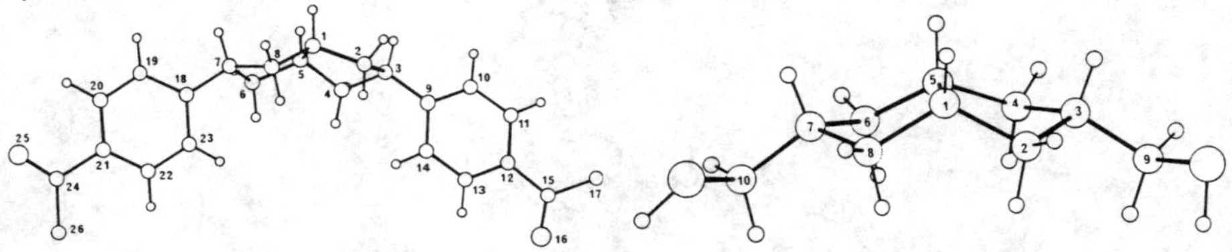

Fig. 1. Views of $C_{24}H_{26}O_2$ and $C_{10}H_{18}O_2$.

1,1,1',1',3,3,3',3'-OCTAMETHYL-2,2'-BIINDANYLIDENE
$C_{26}H_{32}$

T. PILATI and M. SIMONETTA, 1984. Acta Cryst., C40, 1407-1409.

Orthorhombic, Pbca, a = 21.093, b = 11.495, c = 8.235 Å, D_m = 1.131, Z = 4. Mo radiation, R = 0.045 for 1592 reflexions.

The molecules (Fig. 1) lie on inversion centres but have almost 2/m symmetry. The ethylenic part is planar within ±0.01 Å with the C(2)-C(2') distance 1.350(2) Å. The adjacent single bond lengths (1.584(1) Å), the angles involving the methyl groups (103.6-118.2(1)°) and the envelope conformation of the five-membered ring all reflect steric hindrance within the molecule. The shortest intramolecular C...C contacts are C(10)...C(13') 3.286(6), C(11)...C(13') 3.316(2), C(11)...C(12') 3.246(2) Å.

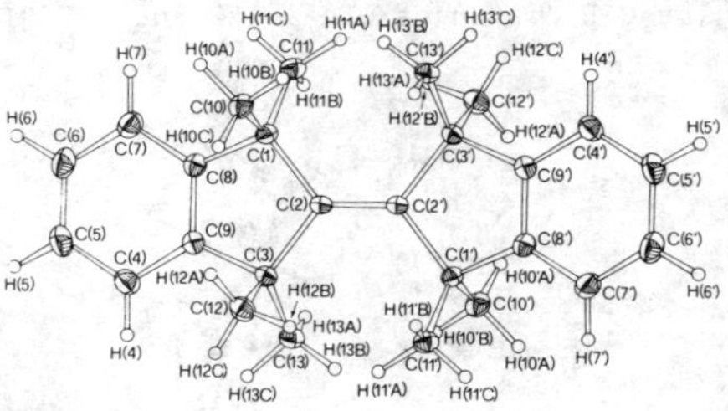

Fig. 1. The $C_{26}H_{32}$ molecule.

(7R)-7,8-EPOXY-9,10-SECO-5,10(19)-CHOLESTADIEN-3β-YL p-BROMOBENZOATE
$C_{34}H_{47}BrO_3$

K. NAKAYAMA, S. YAMADA, H. TAKAYAMA, Y. NAWATA and Y. IITAKA, 1984. J. Org. Chem., 49, 1537-1539.

Monoclinic, $P2_1$, a = 17.312, b = 6.407, c = 15.814 Å, β = 116.19°, D_m = 1.231, Z = 2. Cu radiation, R = 0.062 for 3484 reflexions.

The analysis establishes the stereochemistry shown in Fig. 1, with the oxygen of the epoxide ring oriented in the α-direction.

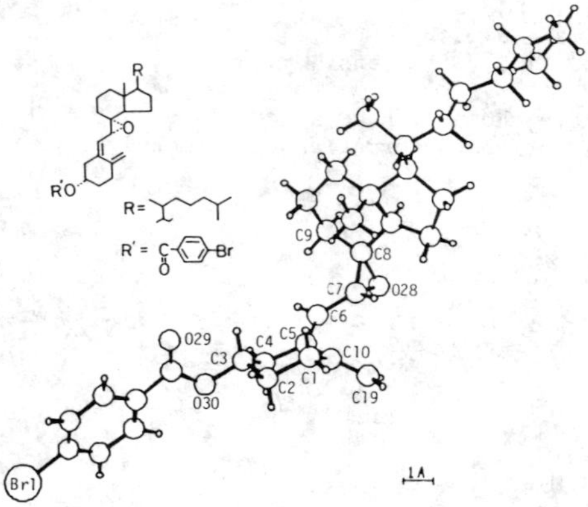

Fig. 1. A view of $C_{34}H_{47}BrO_3$.

anti-4,4,8,8-TETRAFLUOROTRICYCLO[$5.1.0.0^{3,5}$]OCTANE
$C_8H_8F_4$

W.R. DOLBIER, Jr., M. ODANIEC, E. GOMULKA, M. JASKOLSKI and H. KORONIAK, 1984. Tetrahedron, 40, 3945-3948.

Orthorhombic, Cmca, a = 7.2029, b = 8.6597, c = 11.9478 Å, Z = 4. Cu radiation, R = 0.036 for 266 reflexions.

The molecules (Fig. 1) are located at a site with crystallographic 2/m symmetry; the six-membered ring is exactly planar and the cyclopropane rings are anti. The C(1)-C(1)' distance in the cyclopropane ring is 1.552(4) Å, C(1)-C(2) is 1.503(2) and C(1)-C(3) 1.452(3) Å. The C-F distances are 1.352 and 1.359(3) Å.

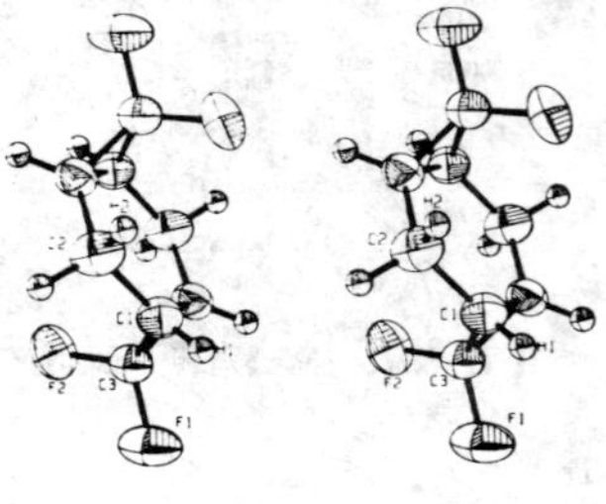

Fig. 1. A stereoview of $C_8H_8F_4$.

FLUORENE
$C_{13}H_{10}$

V.K. BELSKY, V.E. ZAVODNIK and V.M. VOZZHENNIKOV, 1984. Acta Cryst., C40, 1210-1211.

Orthorhombic, Pnma, a = 8.475, b = 18.917, c = 5.717 Å, Z = 4. Mo radiation, R = 0.056 for 375 reflexions.

The molecule (Fig. 1) is planar, with a maximum deviation of 0.013(6) Å for C(5), and possesses C_{2v} symmetry. Intermolecular contacts correspond to normal van der Waals interactions. For previous work see 1.

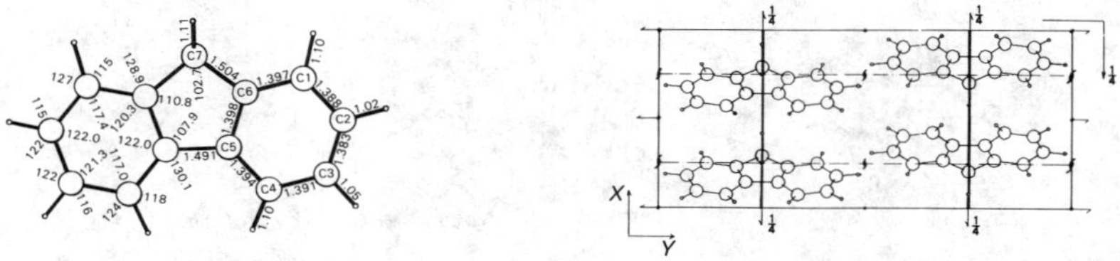

Fig. 1. $C_{13}H_{10}$: interatomic distances and bond angles, and the projection of the structure on to the (001) plane.

1. Structure Reports, 18, 718; 19, 583.

FLUORENE
$C_{13}H_{10}$

R.E. GERKIN, A.P. LUNDSTEDT and W.J. REPPART, 1984. Acta Cryst., C40, 1892-1894.

Orthorhombic, Pnma, a = 8.365, b = 18.745, c = 5.654 Å, Z = 4. Mo radiation, R = 0.043 for 1250 reflexions (at 159K).

The two planar, mirror-related halves of the molecule are inclined at 1.28(4)° (Fig. 1). There is no disorder in the structure, which is very similar to that reported at 295K (1) although the present authors disagree with the interpretation that the molecule is planar at room temperature.

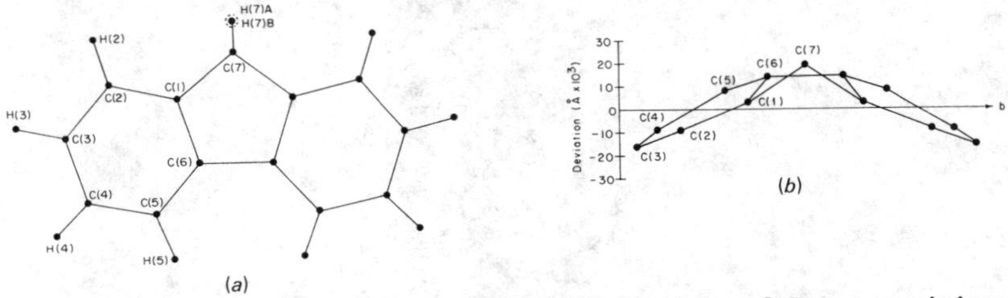

Fig. 1. (a) The fluorene molecule with (b) deviations of C-atom positions from the mean molecular plane.

1. Previous Report.

METHYL 8-OXOTRICYCLO[5.4.0.0^{2,6}]UNDECANE-1-CARBOXYLATE
$C_{13}H_{18}O_3$

G. FERGUSON, 1984. Acta Cryst., C40, 853-855.

Monoclinic, Cc, a = 8.399, b = 16.366, c = 9.310 Å, β = 108.19°, Z = 4. Mo radiation, R = 0.047 for 891 reflexions.

In the molecule the five- and six-membered rings are mutually trans-disposed about the central four-membered ring (Fig. 1). One of the atoms of the six-membered ring is disordered over two sites (0.75 : 0.25) leading to twist-boat and twist-chair conformations for this ring; the four-membered ring is puckered (torsion angles ±9.4(3)°) and the five-membered ring has an endo-envelope conformation. With the exception of those involving the disordered C(4) atom, the dimensions of the molecule are in accord with accepted values, with mean lengths $C(sp^3)-C(sp^3)$ 1.539(5), $C(sp^3)-C(sp^2)$ 1.499(5) and $C(sp^2)=O$ 1.199(5) Å. Intermolecular contacts correspond to van der Waals distances.

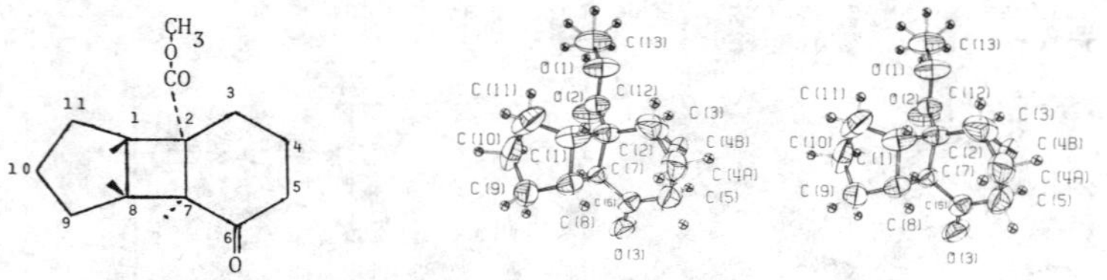

Fig. 1. $C_{13}H_{18}O_3$: structural formula and stereoview of the molecule.

endo-1-CARBAMOYL-exo-1-CARBOXY-1a-METHYL-1a,7a-DIHYDRO-1H-CYCLOPROPA[b]NAPHTHALENE-2,7-DIONE
$C_{14}H_{11}NO_5$

P.H. BOYLE, M.J. O'MAHONY and C.J. CARDIN, 1984. J. Chem. Soc. Perkin I, 593-601.

Triclinic, $P\bar{1}$, a = 8.120, b = 8.225, c = 9.603 Å, α = 105.5, β = 86.5, γ = 82.9°, D_m = 1.40, Z = 2. Mo radiation, R = 0.039 for 1186 reflexions.

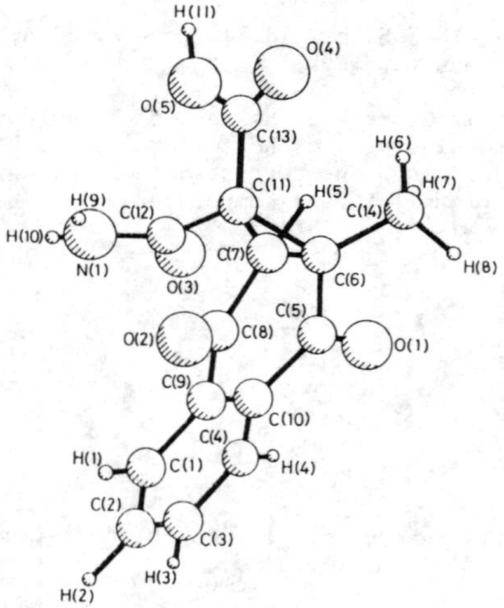

Fig. 1. $C_{14}H_{11}NO_5$: the molecular structure.

The molecule consists of a naphthaquinone skeleton fused to a substituted cyclopropane ring (Fig. 1). The fused cyclohexane ring adopts a shallow boat conformation. The benzene and cyclohexane ring planes are aligned at 6.5° to each other. The ring carbonyl C atoms, C(5),C(8), are both displaced 0.112 Å from the cyclohexane ring mean plane in the direction away from the cyclopropane ring. The normal to the cyclopropane ring makes an angle of 100° with the normal to the mean cyclohexane ring plane. The cyclopropane ring is a perfect equilateral triangle with a mean side of 1.528(4) Å. The amide group is not quite planar and the carboxy group, itself not strictly planar, is orthogonal to the amide group. In the crystal, both amide H atoms are intermolecularly hydrogen-bonded, H(9) to one of the ring carbonyl oxygen atoms, O(2), and H(10) to an adjacent amide O(3). The third bond is between carboxy hydrogen, H(9), and the other ring carbonyl O atom, O(1). Carboxy O atom O(4) is uncoordinated. The O(1)...O(5), N(1)...O(3) and N(1)...O(2) distances are 2.969, 2.975 and 2.919 Å, respectively.

1,2,3,4,5,6,7,8-OCTAHYDROANTHRACENE
$C_{14}H_{18}$

H. VAN KONINGSVELD and J.M.A. BAAS, 1984. Acta Cryst., C40, 311-313.

Monoclinic, $P2_1/c$, a = 9.354, b = 6.550, c = 9.622 Å, β = 116.99°, Z = 2. Mo radiation, R = 0.038 for 1593 reflexions.

The molecule (Fig. 1) lies on an inversion centre and also has approximate twofold symmetry in the oblong direction of the molecule. Dimensions (Fig. 1) are compared with those from force-field calculations.

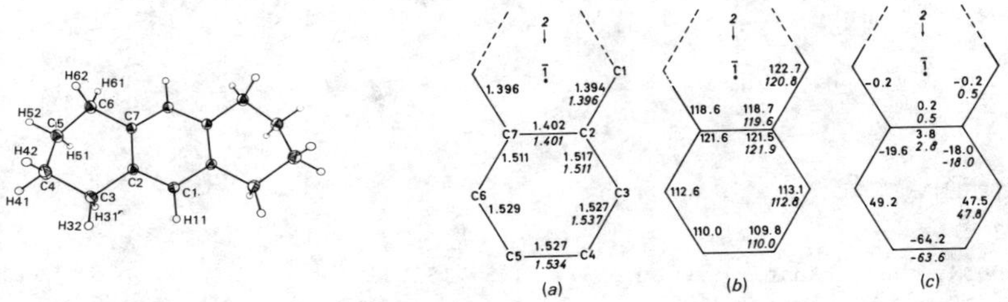

Fig. 1. The $C_{14}H_{18}$ molecule and a comparison of (a) bond lengths, (b) bond angles and (c) torsion angles as obtained by X-ray analysis (point group $\bar{1}$) and force-field calculations (values in italics, point group 2/m).

cis-cisoid-cis-PERHYDROANTHRACENE
$C_{14}H_{24}$

H. VAN KONINGSVELD, J.M.A. BAAS and B. VAN DE GRAAF, 1984. Acta Cryst., C40, 1463-1465.

Triclinic, $P\bar{1}$, a = 6.491, b = 9.442, c = 9.760 Å, α = 107.52, β = 94.37, γ = 82.89°, Z = 2. Mo radiation, R = 0.041 for 3150 reflexions (at 110K).

The molecule adopts the triple-chair conformation (Fig. 1), but is not mirror-symmetric about C(9),C(10). The short C...C distance (3.396 Å) between syn-axial methylene groups across the central ring causes serious deformations in some of the bond and torsion angles.

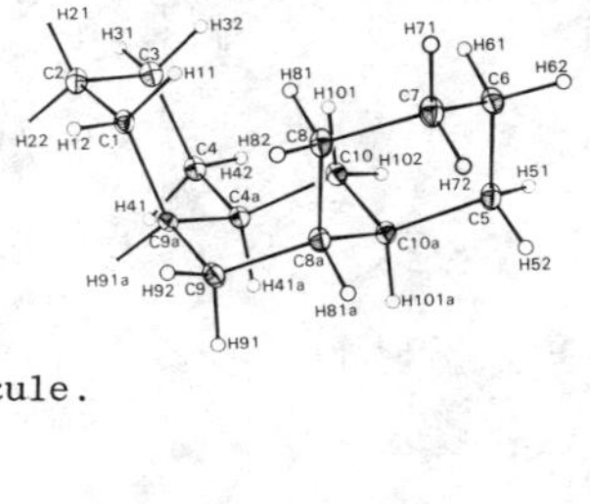

Fig. 1. The $C_{14}H_{24}$ molecule.

5H-DIBENZO[a,d]CYCLOHEPTEN-5-ONE
$C_{15}H_{10}O$

Y. ODDON, N. DARBON, J.P. REBOUL, B. CRISTAU, J.C. SOYFER and G. PEPE, 1984. Acta Cryst., C40, 524-526.

Monoclinic, Cc, a = 24.924, b = 3.930, c = 14.429 Å, β = 133.49°, Dm = 1.33, Z = 4. Mo radiation, R = 0.045 for 742 reflexions.

The molecule (Fig. 1) has a folded conformation; the dihedral angle between the aromatic rings is 142.8(6)°. The molecular geometry is compared with that found in other related tricyclic systems.

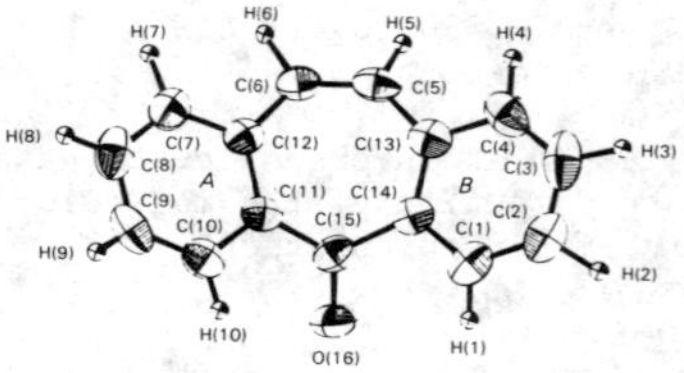

Fig. 1. A view of $C_{15}H_{10}O$.

9-BUTOXYPHENALENONE
$C_{17}H_{16}O_2$

R.C. HADDON, 1984. Aust. J. Chem., 37, 2145-2151.

Orthorhombic, $P2_12_12_1$, a = 9.798, b = 16.825, c = 8.328 Å, Z = 4. Cu radiation, R = 0.081 for 839 reflexions.

In the molecule, the butyl group adopts an S-trans conformation with respect to the carbonyl group with torsion angles C(8)-C(9)-O(2)-C(1') +7.2°, C(9)-O(2)-C(1')-C(2') +178.2° (Fig. 1).

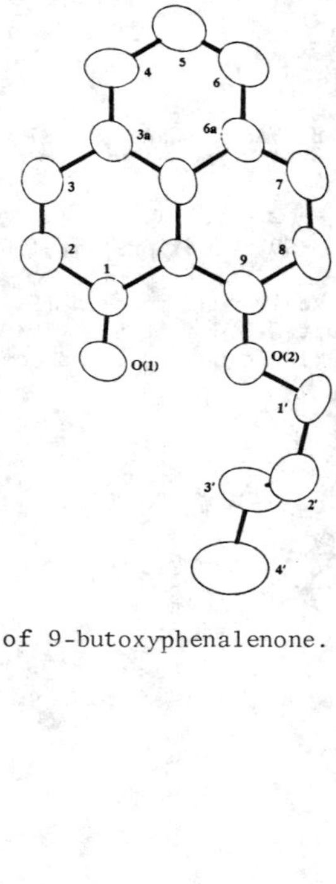

Fig. 1. The structure of 9-butoxyphenalenone.

METHYL ((2S*,4AR*,10AR*)-1,2,3,4,4a,9,10,10a-OCTAHYDRO-5,7-DIMETHOXY-4,9-DIOXO-8-HYDROXY-2-PHENANTHRYL)ACETATE
$C_{19}H_{22}O_7$

R.O. DUTHALER, P. MATHIES, W. PETTER, C. HEUBERGER and V. SCHERRER, 1984. Helv. Chim. Acta, 67, 1217-1221.

Triclinic, $P\bar{1}$, a = 8.458, b = 9.978, c = 11.733 Å, α = 90.25, β = 95.80, γ = 115.19°, Z = 2. Mo radiation, R = 0.039 for 2412 reflexions.

The analysis shows that the molecule has cis-fusion of the saturated rings (Fig. 1). The H-C(4a)-C(10a)-H torsion angle is 52° and the C(4)-C(4a)-C(4b)-C(5) value is 76°.

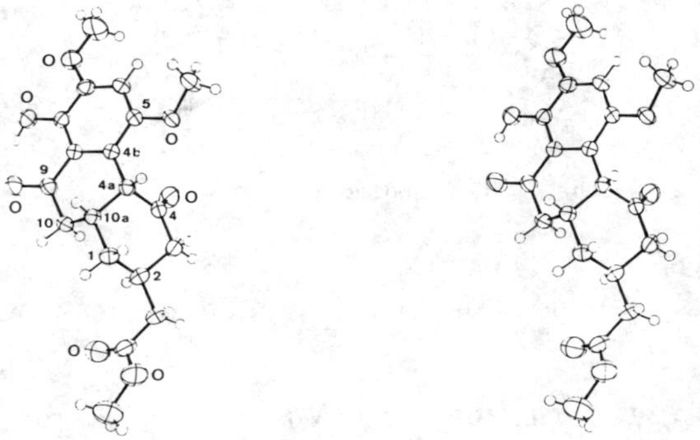

Fig. 1. A stereoview of $C_{19}H_{22}O_7$.

DIMETHYL 2-CYANO-3-[(9-FLUORENYLIDENE)HYDRAZINO]-(E)-BUT-2-ENEDIOATE
$C_{20}H_{15}N_3O_4$

W. BURGERT and D. REWICKI, 1984. Chem. Ber., 117, 2409-2421.

Monoclinic, $P2_1/c$, a = 13.871, b = 7.137, c = 19.828 Å, β = 112.95°, Z = 4. Cu radiation, R = 0.050 for 2609 reflexions.

The structure of this hydrazone was established by X-ray analysis. Geometry details are in Fig. 1.

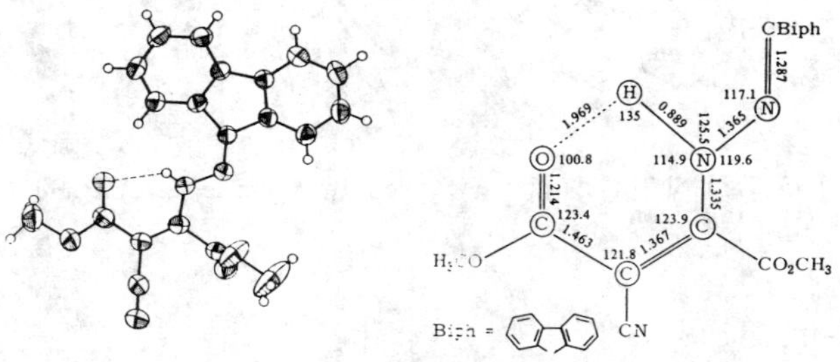

Fig. 1. A view of $C_{20}H_{15}N_3O_4$ and some dimensions.

ETHYL 1-ETHYL-7-METHOXY-3,4-DIOXO-2-PHENANTHRENECARBOXYLATE
$C_{20}H_{18}O_5$

D.F. MULLICA, W.O. MILLIGAN, J.S. BELEW, D.A. GROSSIE and E.L. SAPPENFIELD, 1984. Acta Cryst., C40, 1923-1925.

Monoclinic, $P2_1/c$, a = 14.815, b = 8.009, c = 13.911 Å, β = 90.90°, D_m = 1.35, Z = 4. Mo radiation, R = 0.075 for 851 reflexions.

The phenanthrene moiety (Fig. 1) is planar (mean deviation 0.034 Å) with the plane of the carboxylate group at 74.5° to it. The C(18)-C(19) group is almost perpendicular to the ring with the C(3) methoxy group coplanar. There is a short intramolecular contact between the H attached to C(1) and O(11) of 2.054 Å. The C(11)-C(12) bond length is 1.544(10) Å.

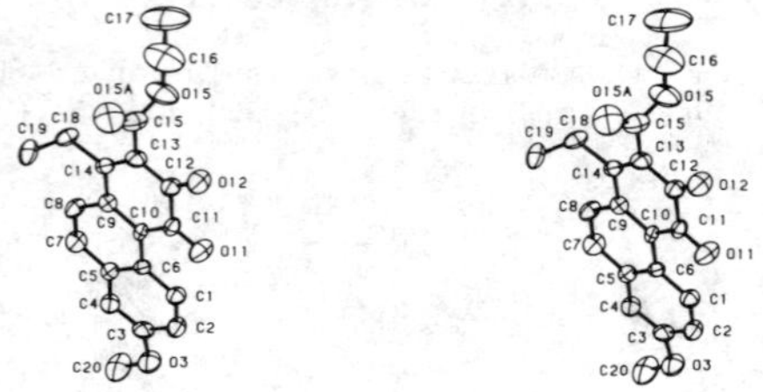

Fig. 1. Stereoview of the $C_{20}H_{18}O_5$ molecule.

BIS{1a,8b-DIHYDRO(1,1-$^2H_2$)-2H-BENZO[a]CYCLOPROPA[c]CYCLOHEPTEN-2-ONE}HYDROGEN(I) HEXACHLOROANTIMONATE

$C_{24}H_{17}D_4Cl_6O_2Sb$ $[H(C_{12}H_8D_2O)_2][SbCl_6]$

R.F. CHILDS, R. FAGGIANI, C.J.L. LOCK and A. VARADARAJAN, 1984. Acta Cryst., C40, 1291-1294.

Monoclinic, I2/c, a = 18.336, b = 8.056, c = 17.982 Å, β = 99.56°, Z = 4. Mo radiation, R = 0.059 for 2714 reflexions (at 238K).

In the crystal structure (Fig. 1) the cation consists of two ketones bridged by the proton giving a very strong hydrogen bond with O...O 2.44(1) Å. The hemi-protonation has not caused extensive delocalisation of the olefinic bond, 1.341(8) Å, nor changed the bonds, especially the bridging bond, 1.544(8) Å, in the cyclopropane group significantly. The side bond in the cyclopropane group, 1.515(9) Å, nearest the ketone group is longer than the other side bond, 1.471(9) Å, consistent with some conjugation with the hemi-protonated ketone group.

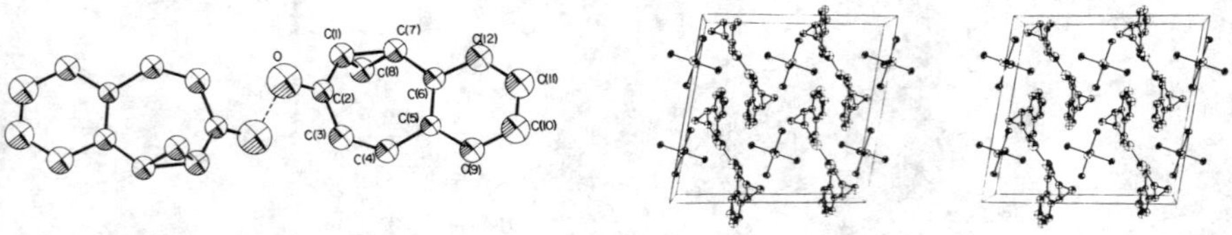

Fig. 1. $[H(C_{12}H_8D_2O)_2][SbCl_6]$: perspective view of the cation and a stereoview of the packing down b.

9-(2-METHYL-1-NAPHTHYL)FLUORENE
$C_{24}H_{18}$ (I)

1-METHYL-9-(2-METHYL-1-NAPHTHYL)FLUORENE
$C_{25}H_{20}$ (II)

1-FLUORO-9-(2-METHYL-1-NAPHTHYL)FLUORENE
$C_{24}H_{17}F$ (III)

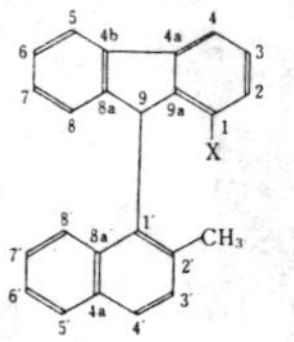

S. MURATA, T. MORI and M. ŌKI, 1984. Bull. Chem. Soc. Japan, 57, 1970-1975.

I. Monoclinic, $P2_1/n$, a = 9.264, b = 8.621, c = 21.083 Å, β = 91.53°, Z = 4. Cu radiation, R = 0.076 for 2011 reflexions.

II. Orthorhombic, Pbcn, a = 18.164, b = 7.414, c = 26.431 Å, Z = 8. Cu radiation, R = 0.065 for 2335 reflexions.

III. Monoclinic, $P2_1/c$, a = 9.438, b = 15.514, c = 12.032 Å, β = 107.79°, Z = 4. Cu radiation, R = 0.070 for 2352 reflexions.

One of the most interesting features in each of the three structures is that the naphthyl group does not bisect the C(8a)-C(9)-C(9a) angle and does not therefore eclipse the C(9)-H bond. The dihedral angles between the planes C(2')-C(1')-C(9) and C(1')-C(9)-C(8a) in I, II and III are respectively 71.7, 67.7 and 71.3°; similarly for planes C(2')-C(1')-C(9) and C(1')-C(9)-C(9a) (50.7, 54.3 and 47.1°). The molecular structures of I and II are quite normal except for the atoms in the vicinity of C(9). The C(8a)-C(9) and C(9)-C(9a) distances are long (1.505 Å in I and 1.527 Å in II; 1.531 Å in I and 1.517 Å in II) and the angle C(8a)-C(9)-C(9a) is small (102.0° in I and 102.4° in II). The corresponding distances and angles in III are 1.518 and 1.520 Å and 101.4°. In the fluoro-compound (III), the bond distance of C(1)-C(9a) is abnormally short (1.358 Å) compared to that in I and II (1.385 and 1.392 Å respectively). This is attributed to the presence of the F atom.

3,10-DIPHENYLDISPIRO[5.0.5.1]TRIDECANE
$C_{25}H_{30}$

H. REDLOF, F. VOGTLE, H. PUFF, H. REUTER and P. WOLLER, 1984. J. Chem. Res., (S) 314-315, (M) 2910-2929.

Monoclinic, C2/c, a = 33.402, b = 7.164, c = 18.601 Å, β = 115.71°, Z = 8. R = 0.0793 for 2604 reflexions.

The analysis establishes the compound to be the (Z)-isomer (with respect to the phenyl ring arrangement) (Fig. 1).

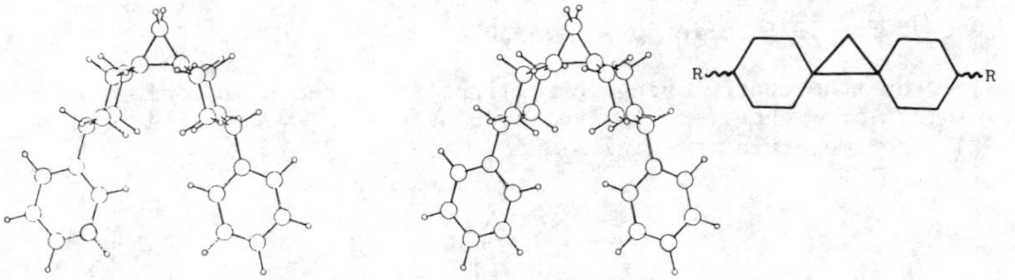

Fig. 1. A stereoview of $C_{25}H_{30}$ (R = Ph).

trans-1,2-DIHYDROXY-1,2-DI(p-TOLYL)ACENAPHTHENE
$C_{26}H_{22}O_2$

Yu.V. GATILOV, Sh.M. NAGI, T.V. RYBALOVA and G.I. BORODKIN, 1984. J. Struct. Chem., 25, 123-127 [Zh. Strukt. Khim., 25(1), 142-146].

Monoclinic, $P2_1/a$, a = 11.237, b = 14.678, c = 12.271 Å, β = 107.52°, Z = 4. Mo radiation, R = 0.053 for 2432 reflexions.

The molecule, along with bond lengths (σ = 0.005 Å) is shown in Fig. 1. The C1-C2 distance of 1.652(5) Å is the longest known for acenaphthene derivatives having normal geometry within the naphthalene core.

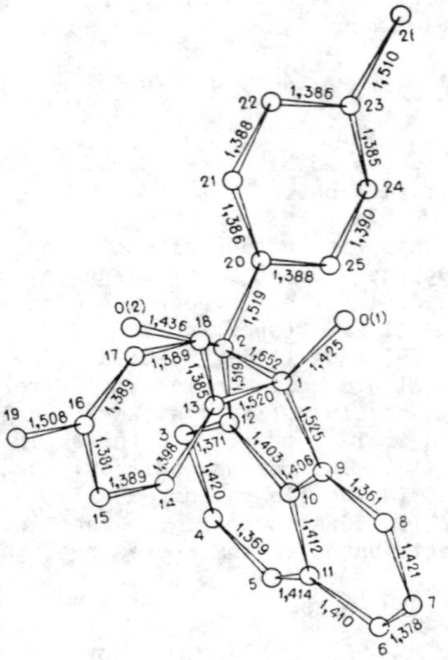

Fig. 1. The trans-1,2-dihydroxy-1,2-di(p-tolyl)acenaphthene molecule.

trans-syn-trans-PERHYDROPHENANTHRENE-9,10-DIOL DIBENZOATE
$C_{28}H_{32}O_4$

N.L. ALLINGER, H. HONIG, U. BURKERT, L. ZSOLNAI and G. HUTTNER, 1984. Tetrahedron, 40, 3449-3453.

Triclinic, $P\bar{1}$, a = 10.77, b = 11.73, c = 12.00 Å, α = 115.97, β = 87.50, γ = 116.66°, Z = 2. R = 0.0743 for 1925 reflexions (at -30°C).

This isomer has the central ring in a slightly twisted boat conformation as predicted by molecular mechanics. The twist about the C21...C22 direction is 19.1(9)° (18.4° from molecular mechanics).

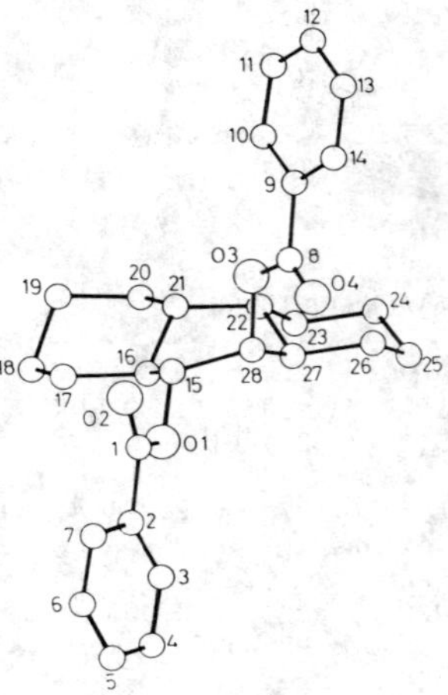

Fig. 1. A view of $C_{28}H_{32}O_4$.

1,3-DIBENZYLCYCLOBUTANE-2,4-BIS(2'-SPIRO-(5'-p-CHLORO)BENZYLCYCLOPENTANONE)
$C_{38}H_{34}Cl_2O_2$

W. JONES and C.R. THEOCHARIS, 1984. J. Cryst. Spectrosc. Res., 14, 447-455.

Monoclinic, $P2_1/c$, a = 16.688, b = 10.719, c = 8.702 Å, β = 103.57°, Z = 2. Cu radiation, R = 0.044 for 1551 reflexions.

The compound (Fig. 1) is the product of the photo-induced solid-state reaction of 2-(p-chloro)-benzyl-5-benzylidenecyclopentanone (1), a reaction which is shown to occur in a single-crystal to single-crystal fashion. The cell parameters of the dimer are very close to those of the monomer, with both structures belonging to the same space group. The compound has the molecular structure which is expected for the product of a topochemical reaction within the perfect monomer crystal. The packing of the dimer is compared with that of the monomer and found to be very similar. The molecule lies about an inversion centre (at the centre of the C_4 ring).

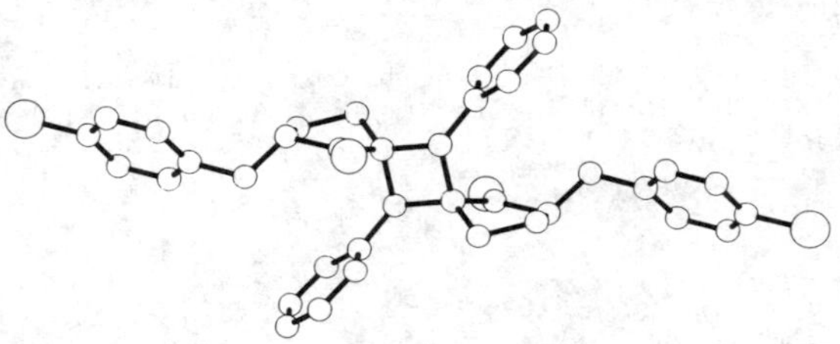

Fig. 1. A view of the $C_{38}H_{34}Cl_2O_2$ molecule.

9c-METHYL-9cH-CYCLOPENTA[jk]FLUORENE
$C_{16}H_{12}$ (I)

7b-METHYL-7bH-CYCLOPENT[cd]INDENE-1,2-DICARBALDEHYDE
$C_{14}H_{10}O_2$ (II)

R. McCAGUE, C.J. MOODY, C.W. REES and D.J. WILLIAMS, 1984. J. Chem. Soc. Perkin I, 909-914.

I. Orthorhombic, $P2_12_12_1$, a = 6.887, b = 9.297, c = 17.673 Å, Z = 4. Cu radiation, R = 0.032 for 896 reflexions.

II. Monoclinic, $P2_1/n$, a = 9.813, b = 8.940, c = 12.247 Å, β = 99.82°, Z = 4. Cu radiation, R = 0.039 for 1290 reflexions.

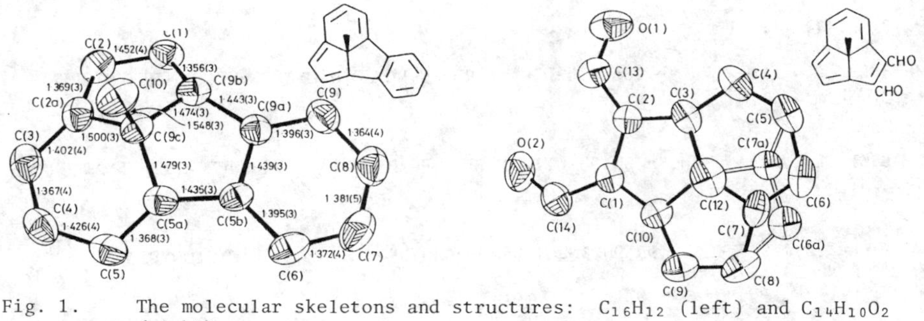

Fig. 1. The molecular skeletons and structures: $C_{16}H_{12}$ (left) and $C_{14}H_{10}O_2$ (right).

Molecules of I have a 'reversed umbrella' conformation with the central C atom tetrahedral. The peripheral bonds show departures from planarity of up to 39° for C(9a)-C(9b). This results in a distinct alternation in bond lengths around the periphery of the tricyclic [10]annulene portion (Fig. 1). Molecules of II are disordered and consist of diastereoisomers in which the five- and six-membered rings opposite the dialdehyde are reversed. The occupancies are 60 and 40%. This molecule also adopts a 'reversed umbrella' conformation. Bond lengths and angles have acceptable values except in the region of the disorder. The aldehyde groups are essentially coplanar, with a short intramolecular C-H...O bond 2.42 Å.

11b-AZIDO-5,6,7A,11b-TETRAHYDROBENZOCYCLOBUTA[1,2-f]-7H-BENZOCYCLOHEPTENE
$C_{17}H_{15}N_3$

G. ADAM, J. ANDRIEUX, M. PLAT, B. VIOSSAT and N. RODIER, 1984. Bull. Soc. Chim. Fr., II-101-108.

Orthorhombic, Pca2$_1$, a = 13.12, b = 13.86, c = 7.637 Å, Z = 4. Mo radiation, R = 0.050 for 1147 reflexions.

In the molecule (Fig. 1), the B/C ring junction is cis. The azido group is essentially linear (N-N-N 174.3(6)°) with N(1)-N(2) 1.229(6), N(2)-N(3) 1.120(6), N(1)-C(11b) 1.498(6) Å.

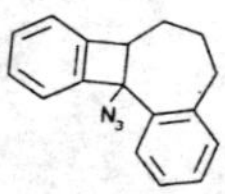

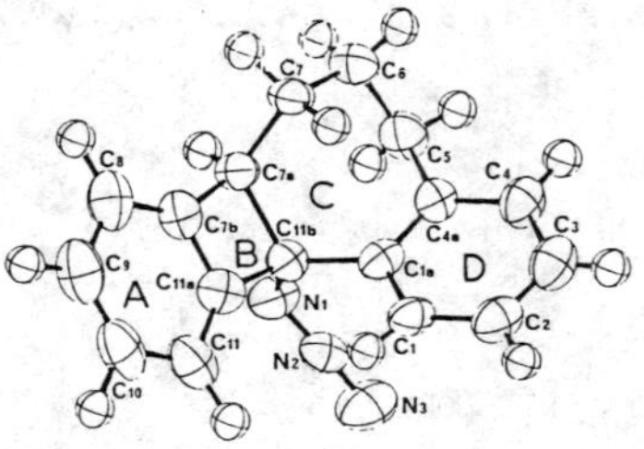

Fig. 1. A view of $C_{17}H_{15}N_3$.

6b,9-DIMETHYL-1,2,3,4,5,6,6a,6b,10a,10b-DECAHYDROBENZ[e]INDACENE-7,10-DIONE
$C_{18}H_{22}O_2$ (I)

8b,11-DIMETHYL-1,2,3,4,5,6,7,8,8a,8b,12a,12b-DODECAHYDRO-9,12-TRIPHENYLENEDIONE
$C_{20}H_{26}O_2$ (II)

7b,10-DIMETHYL-2,3,4,5,6,7,7a,7b,11a,11b-DECAHYDRO-1H-CYCLOPENTA[1]PHENANTHRENE-8,11-DIONE
$C_{19}H_{24}O_2$ (III)

7b-METHYL-2,3,4,5,6,7,7a,7b,13a,13b-DECAHYDRO-1H-BENZO[b]CYCLOPENTA[1]PHENANTHRENE-8,13-DIONE
$C_{22}H_{24}O_2$ (IV)

4b,7-DIMETHYL-2,3,4,4a,4b,8a,8b,9,10,11,12,13-DECAHYDRO-1H-CYCLOHEPTA[1]PHENANTHRENE-5,8-DIONE
$C_{21}H_{28}O_2$ (V)

15a-METHYL-2,3,4,5,6,7,8,9,9a,9b,15a,15b-DODECAHYDRO-1H-BENZO[b]CYCLOHEPTA[1]-PHENANTHRENE-10,15-DIONE
$C_{24}H_{28}O_2$ (VI)

2,4a-DIMETHYL-4a,4b,5,6,7,8,9,10,11,12,13,14,14a,14b-TETRADECAHYDRODICYCLOHEPTA[a,c]-NAPHTHALENE-1,4-DIONE
$C_{22}H_{30}O_2$ (VII)

10b-METHYL-1,2,3,4,5,6,7,8,9,10,10a,10b,16a.16b-TETRADECAHYDRODICYCLOHEPT[a,c]-ANTHRACENE-11,16-DIONE
$C_{25}H_{30}O_2$ (VIII)

1-HYDROXY-3,20-DIMETHYLPENTACYCLO[10.8.0.$0^{5,20}$.$0^{6.12}$.$0^{13,19}$]ICOSA-2,18-DIEN-4-ONE
$C_{22}H_{30}O_2$ (IX)

1,4-DIMETHYLHEXACYCLO[10.8.0.$0^{2,11}$.$0^{4,19}$.$0^{5,11}$.$0^{12,18}$]ICOSANE-3,20-DIONE
$C_{22}H_{30}O_2$ (X)

M. KAFTORY and A. WEISZ, 1984. Acta Cryst., C40, 456-464.

I. Monoclinic, $P2_1/c$, a = 8.491, b = 20.029, c = 8.650 Å, β = 96.48°, Z = 4. Mo radiation, R = 0.062 for 1985 reflexions.

II. Monoclinic, $P2_1/c$, a = 18.509, b = 9.250, c = 21.229 Å, β = 112.56°, Z = 8. Mo radiation, R = 0.079 for 3118 reflexions.

III. Orthorhombic, $Pna2_1$, a = 9.080, b = 15.706, c = 10.994 Å, Z = 4. Cu radiation, R = 0.038 for 1345 reflexions.

IV. Orthorhombic, $P2_12_12_1$, a = 34.069, b = 7.039, c = 7.093 Å, Z = 4. Cu radiation, R = 0.064 for 1631 reflexions.

V. Orthorhombic, $P2_12_12_1$, a = 20.091, b = 9.187, c = 9.331 Å, Z = 4. Mo radiation, R = 0.070 for 1230 reflexions.

VI. Monoclinic, $P2_1$, a = 19.254, b = 9.841, c = 9.985 Å, β = 93.74°, Z = 4. Cu radiation, R = 0.057 for 3145 reflexions.

VII. Monoclinic, $P2_1/n$, a = 18.389, b = 15.398, c = 6.370 Å, β = 90.22°, Z = 4. Mo radiation, R = 0.061 for 2191 reflexions.

VIII. Monoclinic, $P2_1/c$, a = 16.965, b = 10.122, c = 12.739 Å, β = 111.74°, Z = 4. Mo radiation, R = 0.081 for 1718 reflexions.

IX. Triclinic, $P\bar{1}$, a = 15.338, b = 9.268, c = 7.036 Å, α = 80.08, β = 103.41, γ = 102.70°, Z = 2. Mo radiation, R = 0.064 for 2560 reflexions.

X. Monoclinic, C2/c, a = 12.513, b = 14.103, c = 21.044 Å, β = 106.03°, Z = 8. Mo radiation, R = 0.054 for 2326 reflexions.

Stereoviews of the various molecules are shown in Fig. 1. In I the five-membered rings have different conformations; one adopts the envelope, the other the half-chair conformation. The first has a pseudo mirror plane passing through atom C(9) and the midpoint of C(6)-C(7), the second has a pseudo twofold axis passing through atom C(4) and the midpoint of C(1)-C(2). In II the unit cell contains two independent molecules in the asymmetric unit. The six-membered rings adopt the chair conformation; however, one is slightly flattened at the double-bond bridge. In III the five-membered ring adopts an envelope conformation with C(10) out of the plane. The six-membered ring adopts an almost perfect chair conformation. Compound IV is analogous to compound (III). The conformations of the five- and six-membered rings are identical to those found in (III). In V the six-membered ring adopts the chair conformation; the seven-membered ring adopts a slightly twisted chair conformation. Compound VI is analogous to V. There are two independent molecules in the asymmetric unit. The six- and seven-membered rings adopt the same conformations as in V. In VII both seven-membered rings adopt the same chair conformation. VIII is analogous to VII; the seven-membered ring conformations are identical to those found in VII. In IX the cycloheptene moiety is disordered. The disorder can be attributed to a flip of atoms C(3) and C(4) around the bond. The cycloheptene appears in two conformations due to the disorder; the main conformation is a chair, the minor is a boat. The unit-cell packing consists of molecules held by H bonding along the short c axis. The H bonding is between the keto O(2) atom and the alcohol OH group (O(1)... O(2) 2.788, O(2)...HO(1) 1.845 Å, O(1)-HO(1)...O(2) 171.1°). Compound X consists of a planar four-membered ring, two five-membered rings in envelope conformations with either C(15) or C(16) as flap, two six-membered rings in boat conformations and two seven-membered rings, both in twist-chair conformations.

The conformation of the rings in the ten adducts are compared and related to photochemical reactivity. In I through VI the molecular structures are controlled by the most stable conformations of the various ring systems. In VII and VIII the conformation of the central cyclohexene ring is the less stable one and is responsible for the photochemical rearrangement of VII.

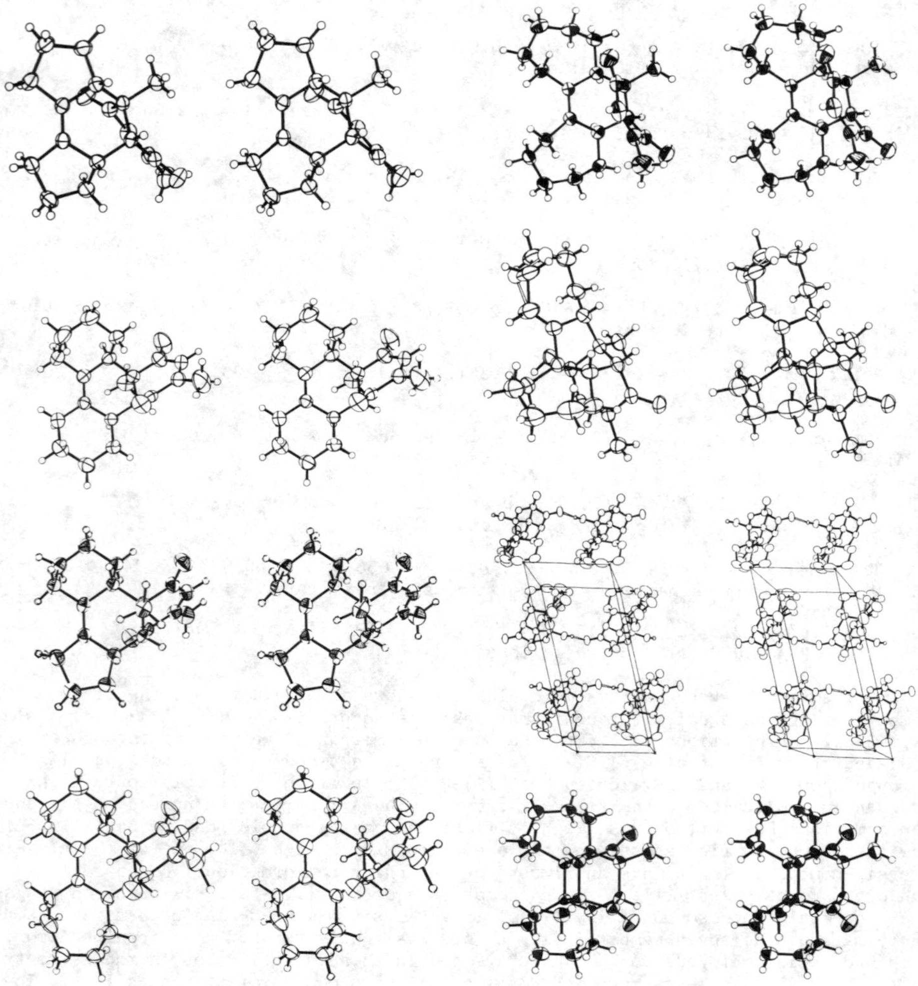

Fig. 1. Stereoviews of the adducts between diones and 1,1'-bicycloalkenyls. In the left column from top to bottom are molecules (I), (II), (III) and (V). In the right column from top to bottom are molecules (VII), (IX), a packing diagram for (IX), and molecule (X).

5-METHYLCHRYSENE
$C_{19}H_{14}$

S. KASHINO, D.E. ZACHARIAS, C.K. PROUT, H.L. CARRELL, J.P. GLUSKER, S. HECHT and R.G. HARVEY, 1984. Acta Cryst., C40, 536-540.

Monoclinic, $P2_1/c$, a = 21.680, b = 5.558, c = 23.876 Å, β = 116.07°, Z = 8. Cu radiation, R = 0.080 for 2636 reflexions.

There are two molecules in the asymmetric unit and one of these is disordered. The molecules (Fig. 1) are distorted because of steric hindrance between the 5-methyl group and the hydrogen atom on C(4); this distortion is evidenced by an out-of-plane displacement (0.2-0.4 Å) of the carbon atom of the methyl group and by increased angles (122-127°) in the 'bay region' between C(4) and C(5).

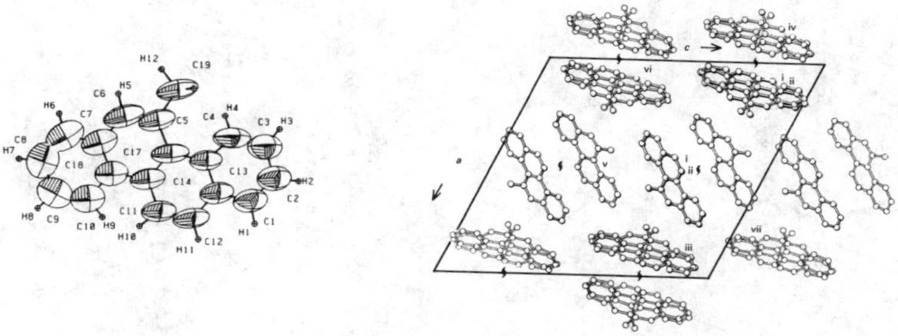

Fig. 1. The ordered chrysene molecule and a view of the crystal structure along b.

(4aR,8aS,13bS)-10,13-DIHYDROXY-2,12-DIMETHYL-6,7,8,8a-TETRAHYDRO-1,4,5,9-(13bH)BENZ-[e]ANTHRACENETETRAONE HEMI-TOLUENE SOLVATE
$C_{20}H_{18}O_6, 0.5C_7H_8$

Y. MIYAGI, K. MARUYAMA, N. TANAKA, M. SATO, T. TOMIZU, Y. ISOGAWA and H. KASHIWANO, 1984. Bull. Chem. Soc. Japan, 57, 791-795.

Monoclinic, $P2_1/a$, a = 17.537, b = 10.358, c = 10.588 Å, β = 91.942°, D_m = 1.378, Z = 4. Mo radiation, R = 0.088 for 2543 reflexions.

Photolysis of two acyl-1,4-benzoquinone moieties resulted in intramolecular cycloaddition and the structure of the product (Fig. 1) has been unequivocally established by single crystal diffraction studies.

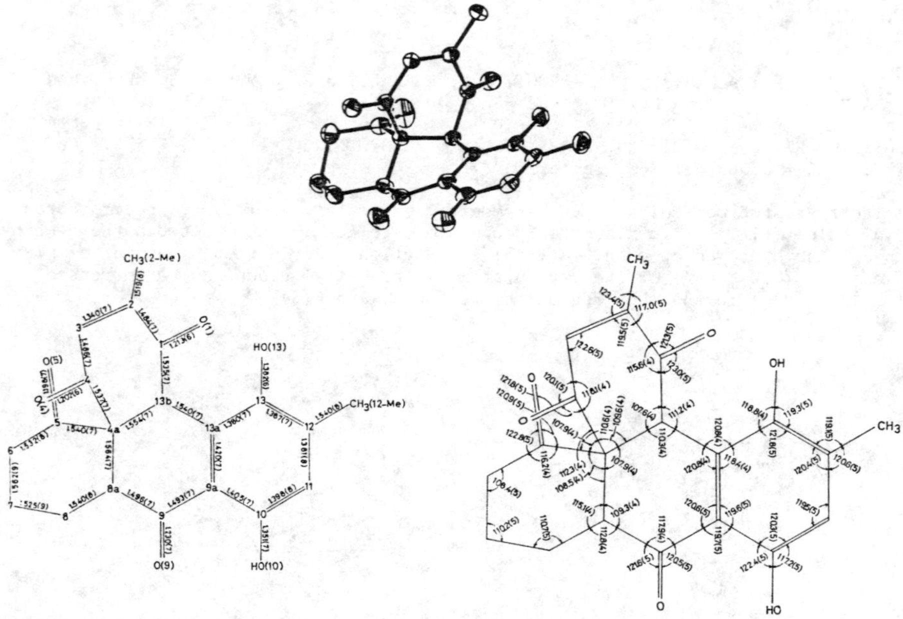

Fig. 1. Perspective view of $C_{20}H_{18}O_6$ (top), and bond distances (bottom left) and bond angles (bottom right).

trans,cis,cis-[10.4.3]TRIANNULANE-16,18-DIONE
$C_{20}H_{30}O_2$ (I)

trans,cis,cis-[10.4.4]TRIANNULANE-16,18-DIONE
$C_{21}H_{32}O_2$ (II)

J.A. MARSHALL, J.C. PETERSON and L. LEBIODA, 1984. J. Am. Chem. Soc., 106, 6006-6015.

I. Monoclinic, $P2_1/c$, a = 9.743, b = 21.439, c = 8.472 Å, β = 106.25°, Z = 4. Mo radiation, R = 0.045 for 1115 reflexions.

II. Monoclinic, $P2_1/a$, a = 18.483, b = 10.933, c = 18.373 Å, β = 97.31°, Z = 8. Mo radiation, R = 0.08 for 1974 reflexions.

The structures of (I) and (II) are shown in Fig. 1. In the case of (II) there are two independent molecules which were partially disordered in the asymmetric unit.

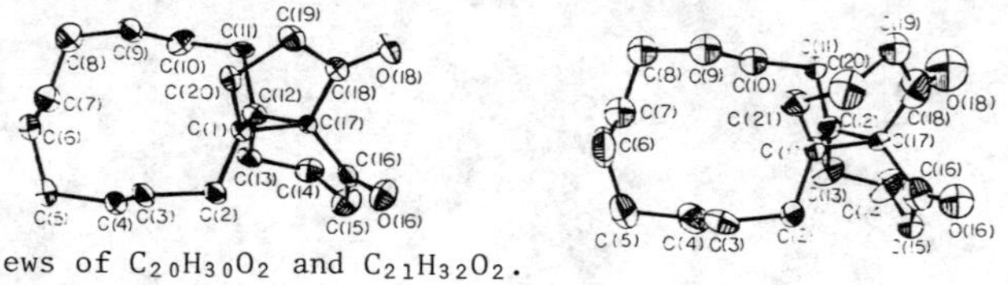

Fig. 1. Views of $C_{20}H_{30}O_2$ and $C_{21}H_{32}O_2$.

7-METHOXYMETHYL-12-METHYLBENZ[a]ANTHRACENE
$C_{21}H_{18}O$

C.J. GILMORE and T.R. RAE, 1984. Acta Cryst., C40, 1630-1632.

Monoclinic, $P2_1/a$, a = 7.899, b = 17.041, c = 11.388 Å, β = 99.44°, D_m = 1.2, Z = 4. Mo radiation, R = 0.037 for 1550 reflexions.

The planar anthracene skeleton is distorted around the attachment of the methyl group (Fig. 1) i.e. all ring torsion angles are less than 10° except C(1)-C(12b)-C(12a)-C(12) -24.5(2), C(4a)-C(12b)-C(12a)-C(6a) -17.6(2). C(12b)-C(12a)-C(6a)-C(6) 16.6(2), C(6a)-C(12a)-C(12)-C(11a) -13.5°. The C(5)-C(6) bond length is 1.364(3) with C(3)-C(4) 1.355(5) and C(8)-C(9) 1.377(4) Å. The shortest intramolecular contact distance is H(1)...H(132) at 2.05(4) Å.

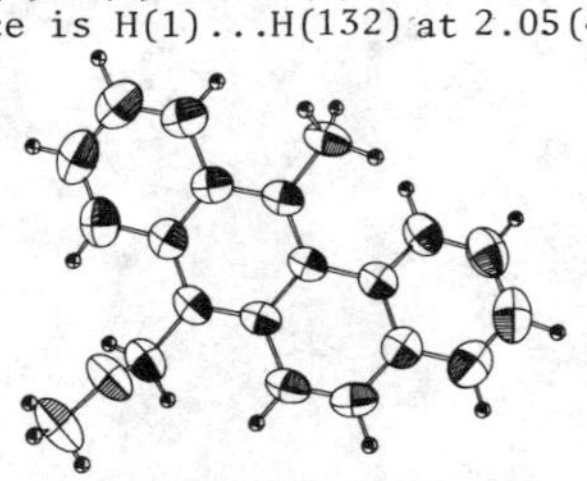

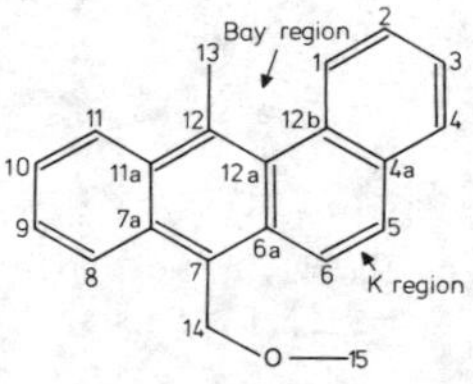

Fig. 1. The $C_{21}H_{18}O$ molecule.

2,8-DIMETHOXY-5,5,10b-TRIMETHYL-cis-4b,5,6,10b,11,12-HEXAHYDROCHRYSENE
$C_{23}H_{28}O_2$

D.J. COLLINS, J.D. CULLEN, G.D. FALLON and B.M. GATEHOUSE, 1984. Aust. J. Chem., 37, 2279-2294.

Triclinic, $P\bar{1}$, a = 11.621, b = 11.528, c = 7.911 Å, α = 114.4, β = 97.9, γ = 100.5°, D_m = 1.20, Z = 2. Cu radiation, R = 0.071 for 1127 reflexions.

The stereochemistry of the ring junction (Fig. 1) shows that its precursor ethyl (1SR,2RS)-2-methyl-2-[6'-methoxy-2'-(p-methoxyphenyl)-2'-methyl-1',2',3',4'-tetrahydronaphthalen-1'ξ-yl]propanoate has an isobutyrate group cis to the anisyl group.

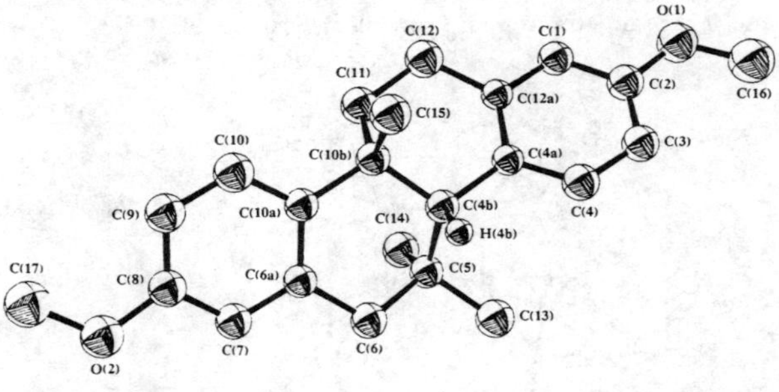

Fig. 1. The structure of $C_{23}H_{28}O_2$.

TRI-[b,e,h]-(4,5-DIMETHOXYBENZO)CYCLONONA-2,5,8-TRIEN-1-ONE
$C_{27}H_{28}O_7$

M.N. PONNUSWAMY and J. TROTTER, 1984. Acta Cryst., C40, 1420-1422.

Orthorhombic, $P2_12_12_1$, a = 16.735, b = 18.763, c = 7.805 Å, D_m = 1.253, Z = 4. Mo radiation, R = 0.054 for 1244 reflexions.

The molecule contains a nine-membered ring with a twist conformation fused to three planar six-membered rings. The molecule has approximately C_2 symmetry with the pseudo-twofold axis along the C=O bond; maximum deviations are 16.8° in torsion angle and 18.3° in dihedral angle from exactly symmetrical geometry. Bond lengths and angles are normal.

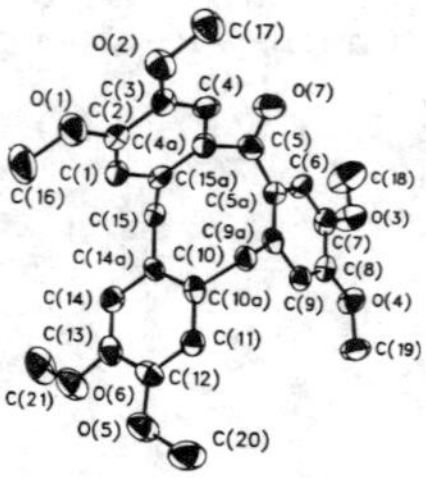

Fig. 1. The $C_{27}H_{27}O_7$ molecule.

[4]-ROTANE
$C_{12}H_{16}$ $(CCH_2CH_2)_4$

A. ALMENNINGEN, O. BASTIANSEN, B.N. CYVIN, S. CYVIN, L. FERNHOLT and C. RØMMING, 1984. Acta Chem. Scand., A38, 31-39.

Monoclinic, $P2_1/c$, a = 7.282, b = 7.318, c = 9.641 Å, β = 112.04°, Z = 2. Mo radiation, R = 0.030 for 1404 reflexions (at 115K).

The molecule (Fig. 1) has crystallographic inversion symmetry so the central cyclobutane ring is strictly planar. The cyclopropane rings are within 1° normal to the central ring and the plane of the cyclobutane ring bisects the bond between the outer C atoms of the cyclopropane rings. The important bond lengths (corrected for libration in parenthesis) are: C(1)-C(5) 1.496 (1.501), C(1)-C(6) 1.499 (1.503), C(5)-C(6) 1.520 (1.523), C(1)-C(2) 1.521 (1.526), C(1)-C(4) 1.520 (1.525), C(2)-C(7) 1.499 (1.503), C(2)-C(8) 1.497 (1.501), C(7)-C(8) 1.517 (1.520) and C-H (mean) 1.05 Å. The angles C(2)-C(1)-C(4) and C(1)-C(2)-C(3) are 90.03 and 89.97° respectively and the intra-cyclopropane ring angles range from 59.41 to 60.99°.

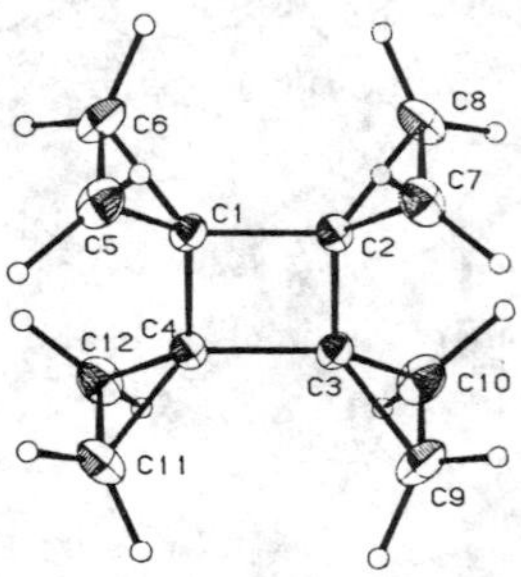

Fig. 1. Molecular structure of $C_{12}H_{16}$.

13H-DIBENZ[a,de]ANTHRACEN-13-ONE
$C_{21}H_{12}O$

I. OONISHI, S. FUJISAWA, J. AOKI, Y. OHASHI and Y. SASADA, 1984. Bull. Chem. Soc. Japan, 57, 3048-3051.

Orthorhombic, $Pn2_1a$, a = 14.227, b = 15.551, c = 12.436 Å, Z = 8. Cu radiation, R = 0.106 for 1964 reflexions.

The molecules are stacked in pairs (Fig. 1), and two pairs with different overlapping modes are randomly arranged with different populations. One of the two crystallographically-independent molecules takes a disordered structure, B and C, both situated very nearly on the same plane but with different orientations. The mean interplanar distances between A and B', A and C' are 3.47 and 3.58 Å respectively. The molecules are significantly distorted from a planar conformation due to overcrowding around the O atom. The repulsion between O and C(21)-H(21) is relieved by enlargement of the O-C(7)-C(16) angle (to 126° in A and 125° in B) and the twisting angle of O-C(7)...C(6)-C(21) (8.2° in A, and 6.3° in B). The esd's on bond lengths vary from 0.01 to 0.03 Å and preclude any detailed discussion of the distances.

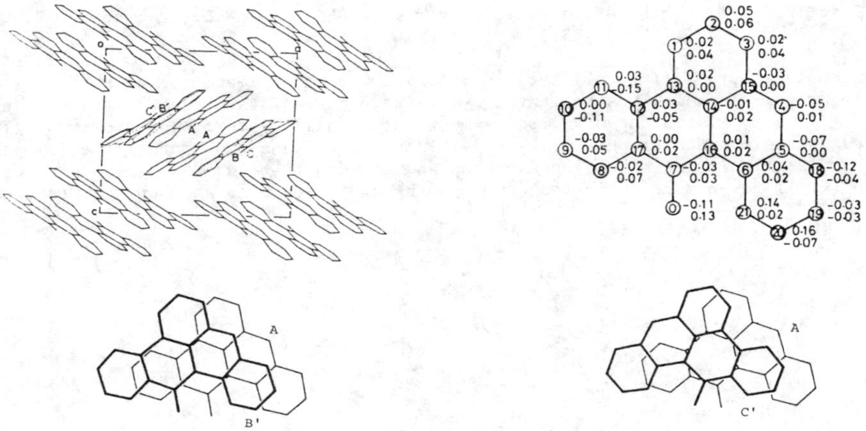

Fig. 1. Projection of the crystal structure of $C_{21}H_{12}O$ along b (top left), Numbering scheme and atomic displacement (Å) from molecule mean plane (upper values refer to molecule A) (top right), overlapping modes in dimer pairs A and B' (bottom left), and A and C' (bottom right).

cis-endo-DISPIRO[CYCLOBUTANONE-2,8'-(TETRACYCLO[5.5.1.0$^{2',6'}$.0$^{10',13'}$]TRIDECAN-4'-ONE ETHYLENE KETAL)-12',12''-CYCLOBUTANONE]
$C_{21}H_{26}O_4$ (I)

endo-HEXACYCLO[11.5.1.0^{2,6}.0^{7,19}.0^{8,18}.0^{14,18}]NONADEC-2(6)-ENE-3,10,17-TRIONE
$C_{19}H_{20}O_3$ (II)

endo-HEPTACYCLO[11.5.1.0^{2,6}.0^{2,15}.0^{7,19}.0^{8,12}.0^{14,18}]NONADECANE-5,9,16-TRIONE
$C_{19}H_{20}O_3$ (III)

endo-HEPTACYCLO[11.5.1.0^{2,6}.0^{2,15}.0^{7,19}.0^{8,12}.0^{14,18}]NONADEC-10-ENE-5,9,16-TRIONE
$C_{19}H_{18}O_3$ (IV)

HEPTACYCLO[11.5.1.0^{2,6}.0^{5,9}.0^{7,19}.0^{8,12}.0^{14,18}]NONADEC-14(18)-ENE-3,10,17-TRIONE
$C_{19}H_{18}O_3$ (V)

OCTACYCLO[11.5.1.0^{2,6}.0^{2,15}.0^{6,11}.0^{7,19}.0^{8,12}.0^{14,18}]NONADECANE-5,9,16-TRIONE
$C_{19}H_{18}O_3$ (VI)

J.E. BALDWIN, P.L.M. BECKWITH, J.D. WALLIS, A.P.K. ORRELL and K. PROUT, 1984. J. Chem. Soc. Perkin II, 53-61.

I. Monoclinic, $P2_1$, a = 8.139, b = 7.691, c = 13.842 Å, β = 95.61°, Z = 2. R =

0.058 for 864 reflexions.

II. Monoclinic, $P2_1/n$, a = 15.950, b = 13.246, c = 6.865 Å, β = 92.28°, Z = 4. R = 0.034 for 1172 reflexions.

III. Monoclinic, $P2_1/n$, a = 6.202, b = 15.827, c = 14.664 Å, β = 103.88°, Z = 4. R = 0.050 for 1333 reflexions.

IV. Monoclinic, $P2_1/n$, a = 8.869, b = 15.085, c = 10.532 Å, β = 92.29°, Z = 4. R = 0.049 for 1143 reflexions.

V. Monoclinic, $P2_1/a$, a = 13.526, b = 13.164, c = 16.574 Å, β = 110.82°, Z = 8. R = 0.056 for 1982 reflexions.

VI. Triclinic, $P\bar{1}$, a = 8.183, b = 8.826, c = 11.748 Å, α = 68.02, β = 66.59, γ = 65.18°, Z = 2. R = 0.045 for 2311 reflexions.

Compounds I to VI (Fig. 1) were isolated during an approach to the synthesis of dodecahedrane. The highly substituted C_{13}-tetraquinane (I) and the C_{19}-hexaquinanetrione (II) are twisted out of mirror symmetry to minimise steric interactions. The bridged C_{19}-hexaquinanetriones (III) and IV) and the C_{19}-heptaquinanetrione (V) show signs of strain in the molecular skeleton and the doubly-bridged C_{19}-hexaquinanetrione (VI) is even more distorted showing both bond stretching and bond compression.

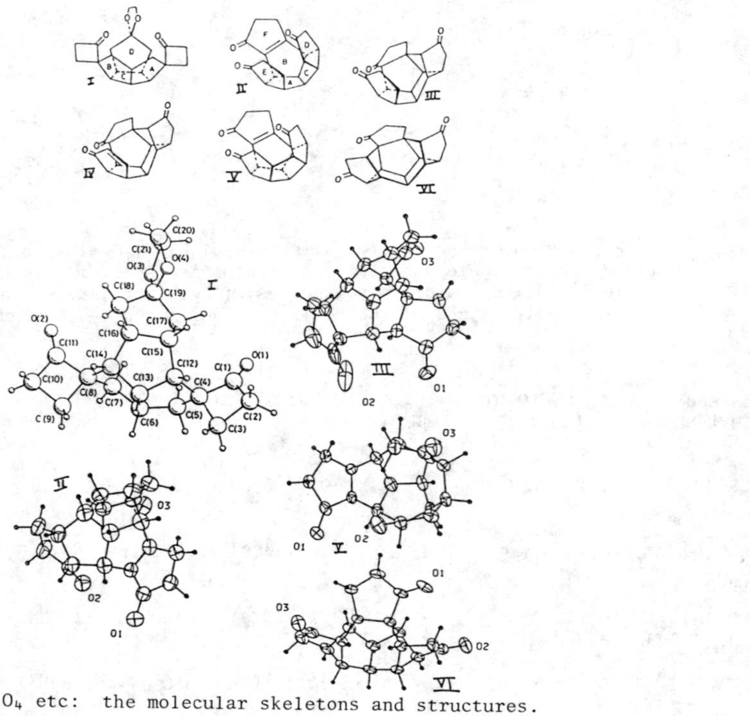

Fig. 1. $C_{21}H_{26}O_4$ etc: the molecular skeletons and structures.

HEXASPIRO[2.0.3.0.2.0.3.0.2.0.3.0]HENEICOSANE
$C_{21}H_{30}$

L. FITJER, U. KLAGES, W. KUHN, D.S. STEPHENSON, G. BINSCH, M. NOLTEMEYER, E. EGERT and G.M. SHELDRICK, 1984. Tetrahedron, 40, 4337-4349.

Monoclinic, $P2_1/m$, a = 7.190, b = 13.912, c = 8.094 Å, β = 96.74°, Z = 2. Mo radiation, R = 0.080 for 1542 reflexions.

The molecule (Fig. 1) has crystallographic mirror symmetry with the six-membered ring in chair conformation. Bond angles within the cyclohexane ring alternate by about 7°, the larger ones being at atoms which form part of cyclopropane rings. In the four-membered rings spiro C-C bond lengths are elongated (average 1.57 Å). One C_4 ring is puckered, the other forced to be planar by crystallographic symmetry.

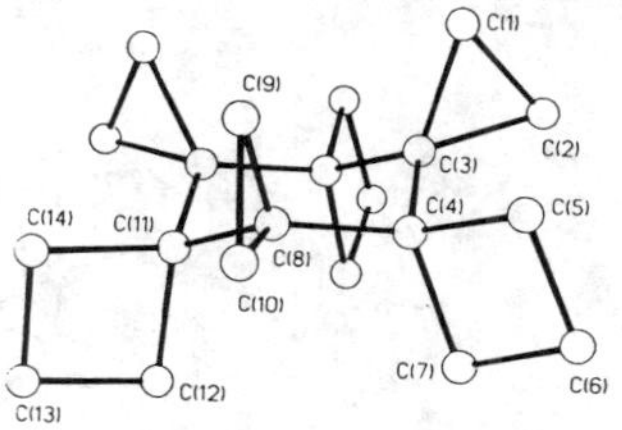

Fig. 1. A view of $C_{21}H_{30}$.

exo-(8bRS,8cSR,12aRS,12bSR)-8b,8c,9,10,12a,12b-HEXAHYDROBENZOCYCLOBUTA[1,2-e]PYRENE
$C_{22}H_{18}$

M. KIMURA, S. KASHINO, S. MOROSAWA and M. HAISA, 1984. Acta Cryst., C40, 1612-1614.

Monoclinic, $P2_1/a$, a = 21.698, b = 13.240, c = 5.1026 Å, β = 93.802°, D_m = 1.26, Z = 4. Cu radiation, R = 0.050 for 2044 reflexions.

The cyclobutane ring has a skewed conformation with c-c-c-c torsion angles in the ring of 15.3(1)-15.6(1)°; the hydrogen atoms are in the cis-anti configuration. Bond lengths at the ring junctions are C(18)-C(19) 1.572(3) and C(20)-C(21) 1.561(3) Å. The phenanthrene ring is planar within 0.109(2) Å while rings D and F (Fig. 1) have the half-chair conformation. The molecules form sheets parallel to (001) and are stacked along c.

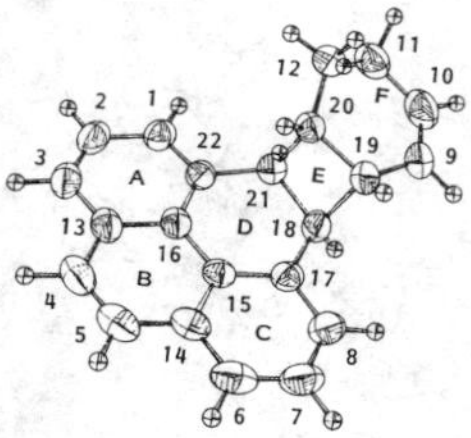

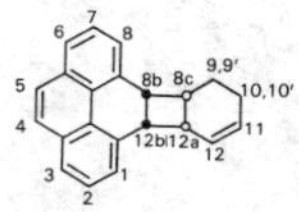

Fig. 1. The $C_{22}H_{18}$ molecule.

13-(3,5-DINITROBENZOYLOXY)HEXACYCLO[12.2.0.0^{1,4}.0^{4,7}.0^{7,10}.0^{10,13}]HEXADECANE
$C_{23}H_{24}N_2O_6$ (I)

11-(4-PHENYLAZOBENZOYLOXY)HEXACYCLO[12.2.0.0^{1,4}.0^{4,7}.0^{8,11}.0^{8,14}]HEXADECANE
$C_{29}H_{30}N_2O_2$ (II)

L. FITJER, D. WEHLE, M. NOLTEMEYER, E. EGERT and G.M. SHELDRICK, 1984. Chem. Ber., 117, 203-221.

I. Monoclinic, $P2_1/c$, a = 8.016, b = 7.516, c = 34.188 Å, β = 92.38°, Z = 4. Mo radiation, R = 0.067 for 2409 reflexions.

II. Monoclinic, $P2_1/c$, a = 12.797, b = 6.974, c = 28.498 Å, β = 112.97°, Z = 4. Mo radiation, R = 0.063 for 3248 reflexions.

The structures of these polyspiranes were established to be as shown in Fig. 1.

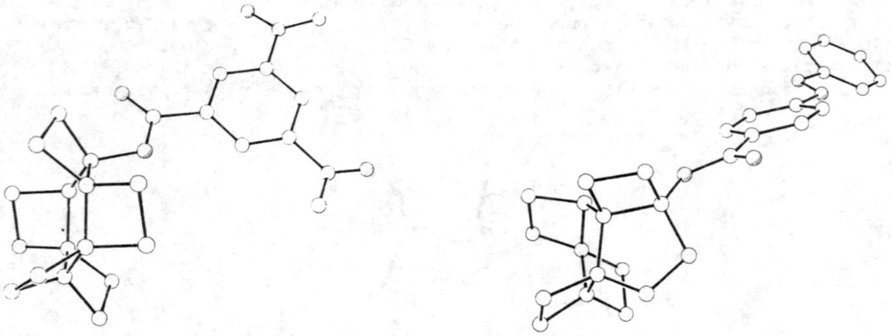

Fig. 1. Views of $C_{23}H_{24}N_2O_6$ (I) (left) and $C_{29}H_{30}N_2O_2$ (II) (right); nitrogen and oxygen atoms are partly shaded, hydrogens are omitted.

15bH-BENZO[cd](BENZO[6,7]CYCLOHEPTA)[1,2,3-fd]PYRENE
$C_{27}H_{16}$

M. VAN MEERSSCHE, J.P. DECLERCQ, G. GERMAIN and B. SOUBRIER-PAYEN, 1984. Bull. Soc. Chim. Belg., 93, 449-450.

Monoclinic, $P2_1$, a = 15.918, b = 10.864, c = 4.825 Å, β = 95.72°, Z = 2. Cu radiation, R = 0.034 for 1218 reflexions.

In rings I-IV, the geometry is similar to that in benzphenanthrene (1) and hexahelicene (2). Because of sp^3 atom C27, the molecule can be thought of in two planes; that of ring VI and carbons C(12) and C(27), and the second contains all the other carbons and C(27). The dihedral angle between these planes is 69°.

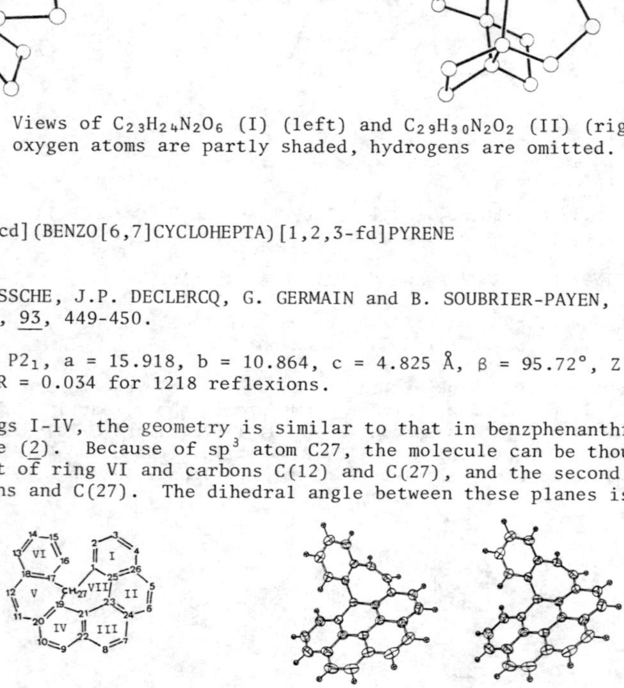

Fig. 1. The $C_{27}H_{16}$ molecule and a stereoview.

1. Structure Reports, 28, 444.
2. Ibid, 39B, 128.

1-(1-HYDROXYETHYL)HEXAHELICENE
$C_{28}H_{20}O$

M. VAN MEERSSCHE, J.P. DECLERCQ, G. GERMAIN, B. SOUBRIER-PAYEN, H.J. LINDNER and B. KITSCHKE, 1984. Bull. Soc. Chim. Belg., 93, 445-448.

Orthorhombic, $P2_12_12_1$, a = 8.541 (8.554), b = 14.100 (14.17), c = 15.812 (15.83) Å, Z = 4. Cu radiation, R = 0.049 (0.097) for 1032 (2434) reflexions. [Two independent determinations reported.]

The analyses show that the compound is the (dl) PR-MS isomer. The molecule has a pseudo twofold axis through C(21)-C(22).

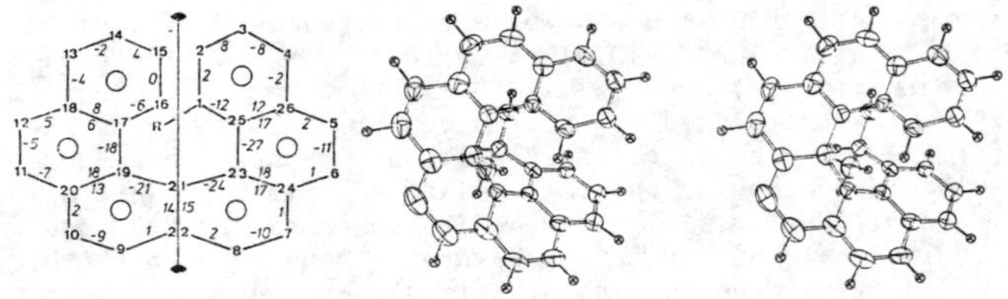

Fig. 1. The $C_{28}H_{20}O$ molecule and a stereoview.

5,7-BIS(3-DIMETHYLAMINOPROPYL)-12,13,15,16-TETRAHYDROTRIBENZO[1,2-d:1',2'-a:5',4'-a':1'',2''-d']DICYCLOHEPTENE-5,7-DIOL
$C_{34}H_{44}N_2O_2$ (I)

5,13-BIS(3-DIMETHYLAMINOPROPYL)-7,8,15,16-TETRAHYDROTRIBENZO[1,2-d:1',2'-a:4',5'-a':1",2"-d']DICYCLOHEPTENE-5,13-DIOL
$C_{34}H_{44}N_2O_2$ (II)

C. HORSBURGH, P. LINDLEY, F. STRIDE, Y. ASSCHER and I. AGRANAT, 1984. Acta Cryst., C40, 1296-1298.

I. Monoclinic, C2/c, a = 19.632, b = 7.726, c = 20.728 Å, β = 111.22°, Z = 4.
Cu radiation, R = 0.074 for 1754 reflexions.

II. Monoclinic, $P2_1/c$, a = 7.469, b = 11.206, c = 17.328 Å, β = 97.96°, Z = 2.
Cu radiation, R = 0.062 for 1842 reflexions.

Compound I is a racemate of molecules adopting the R,R and S,S configurations whereas in II the molecules have a meso-R,S configuration (Fig. 1). Both structures contain intramolecular hydrogen bonds between a hydroxyl group and the N atom of the aminopropyl moiety attached to the corresponding asymmetric C atom.

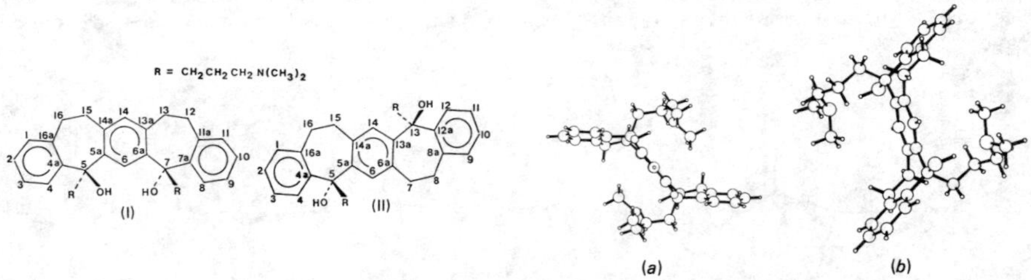

Fig. 1. $C_{34}H_{44}N_2O_2$: structural formulae of I and II and (a) I viewed along the b axis, (b) viewed along the c axis.

3-exo-CHLORO-6-exo-NITROBICYCLO[2.2.1]HEPTAN-2-exo-OL
$C_7H_{10}ClNO_3$ (I)

3-exo-CHLORO-6-exo-NITROBICYCLO[2.2.1]HEPTAN-2-endo-OL
$C_7H_{10}ClNO_3$ (II)

J.C.A. BOEYENS, L. DENNER and J.P. MICHAEL, 1984. J. Chem. Soc. Perkin II, 1569-1573.

I. Monoclinic, $P2_1/c$, a = 10.57, b = 7.76, c = 10.17 Å, β = 93.03°, D_m = 1.51, Z = 4.
Mo radiation, R = 0.042 for 1079 reflexions.

II. Monoclinic, $P2_1/n$, a = 6.91, b = 9.97, c = 12.11 Å, β = 92.36°, D_m = 1.54, Z = 4.
Mo radiation, R = 0.044 for 1229 reflexions.

In the structure of I (Fig. 1) there is a short intermolecular contact (2.10 Å) between the hydroxylic hydrogen atom of one molecule and O(2) of the nitro group of a neighbouring molecule. The H...O(3) distance is 2.92 Å. The H...O(2) hydrogen bond is almost linear with O(1)-H...O(2) 174°. In the crystal, molecules of a single enantiomeric series are linked by intermolecular hydrogen bonds along the b axis and each of the four molecules in the unit cell participates in a hydrogen-bonded assembly only with molecules occupying the same relative symmetry position in neighbouring unit cells. In II the H...O(2) and H...O(3) contacts are 2.26 and 2.42 Å respectively suggesting an intermolecular bifurcated hydrogen bond between the hydroxy and nitro groups. Two hydrogen-bonded chains of molecules per unit cell are propagated through the crystal lattice along the c-axis of the unit cell. The conformations of the nitro groups in I and II are such that the O(endo)-N-C(2)-C(3) dihedral angle is -70.6° for I and 10.4° for II.

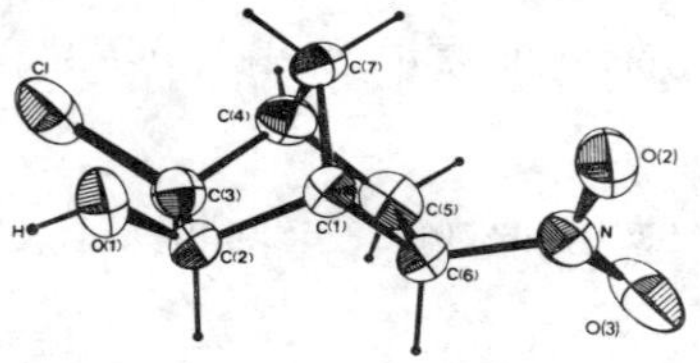

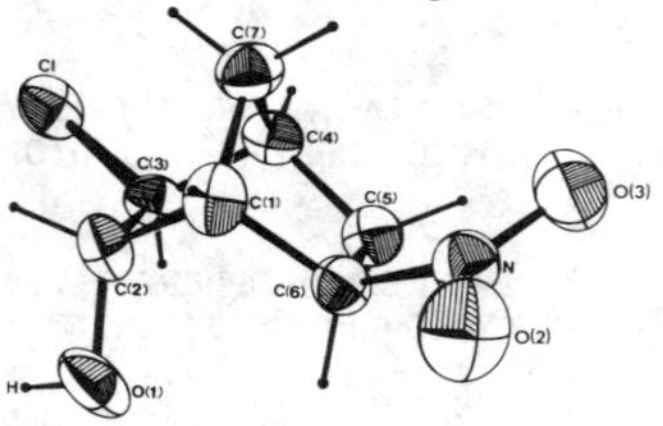

Fig. 1. $C_7H_{10}ClNO_3$: perspective views of the molecules, I (left) and II (right).

1,2,2,5,6,6-HEXAFLUOROTRICYCLO[3.3.0.0^{6,8}]OCTANE
$C_8H_6F_6$

D.C. ENGLAND, F.J. WEIGERT and J.C. CALABRESE, 1984. J. Org. Chem., 49, 4816-4819.

Orthorhombic, $Pna2_1$, a = 14.963, b = 5.671, c = 9.559 Å, Z = 4. Mo radiation, R = 0.049 for 704 reflexions.

The C-C bonds flanked by fluorine (Fig. 1) are slightly shorter than those on the opposite side of the molecule (compare C2-C3 1.498(8) Å vs. C4-C5 1.527(9) Å). The central ring C-C distances average 1.554 Å and the six 'external' C-C distances average 1.511 Å.

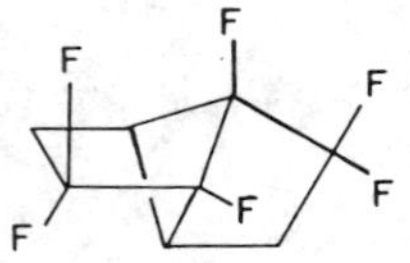

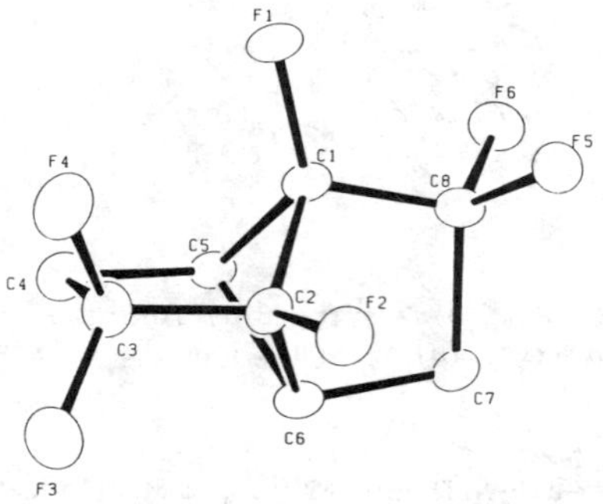

Fig. 1. A view of $C_8H_6F_6$.

3-exo,6-exo-DICHLORO-5-endo-HYDROXY-3-endo-NITRO-BICYCLO[2.2.1]HEPTANE-2-exo-CARBO-NITRILE
$C_8H_8Cl_2N_2O_3$

J.C.A. BOEYENS, L. DENNER and J.P. MICHAEL, 1984. J. Chem. Soc. Perkin II, 767-770.

Monoclinic, Cc, a = 6.62, b = 15.95, c = 9.90 Å, β = 92.94°, Dm = 1.60, Z = 4. Mo radiation, R = 0.038 for 714 reflexions.

The molecule (Fig. 1) possesses an intramolecular hydrogen bond between endo-nitro and hydroxy groups (O(1)-H...O(2) 2.04 Å). This is substantiated by IR and NMR data. Some bond distances and angles in the molecule are: C(3)-N(2) 1.52(1), C(5)-O(1) 1.42(1), C(3)-Cl(2) 1.787(7), C(2)-C(8) 1.47(1), C(8)-N 1.13(1), C(1)-C(7) 1.55(1), C(4)-C(7) 1.52(1) Å, C(1)-C(7)-C(4) 94.7(6)°.

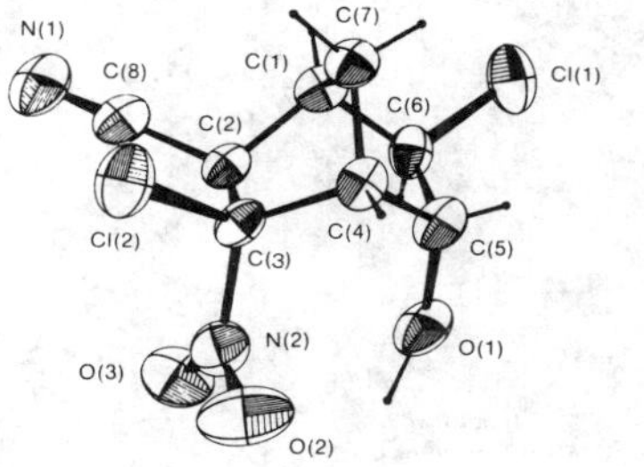

Fig. 1. $C_8H_8Cl_2N_2O_3$: the molecular structure.

4-HYDROXYBICYCLO[3.3.1]NON-3-ENE-2,9-DIONE
$C_9H_{10}O_3$ (I)

10-HYDROXYBICYCLO[5.3.1]UNDEC-9-ENE-8,11-DIONE
$C_{11}H_{14}O_3$ (II)

[9](2.4)-PHLOROGLUCINOPHANE MONOHYDRATE
$C_{15}H_{22}O_3,H_2O$ (III)

K.-H. SCHONWALDER, P. KOLLAT, J.J. STEZOWSKI and F. EFFENBERGER, 1984. Chem. Ber., 117, 3280-3296.

I. Monoclinic, $P2_1/c$, a = 6.6669, b = 11.7493, c = 20.4543 Å, β = 92.517°, Z = 8. Mo radiation, R = 0.04 for 6275 reflexions (at 120 K).

II. Monoclinic, C2/c, a = 13.3334, b = 6.6908, c = 22.3620 Å, β = 101.028°, Z = 8. Mo radiation, R = 0.059 for 2863 reflexions (at 120 K).

III. Monoclinic, C2/c, a = 28.901, b = 7.306, c = 13.723 Å, β = 98.50°, Z = 8. Mo radiation, R = 0.076 for 3754 reflexions (at 120 K).

Stereoviews of (I), (II) and (III) are in Fig. 1 along with principal dimensions. There are two independent molecules in (I) with similar geometry. The eight-membered ring in (II) is in chair-chair conformation.

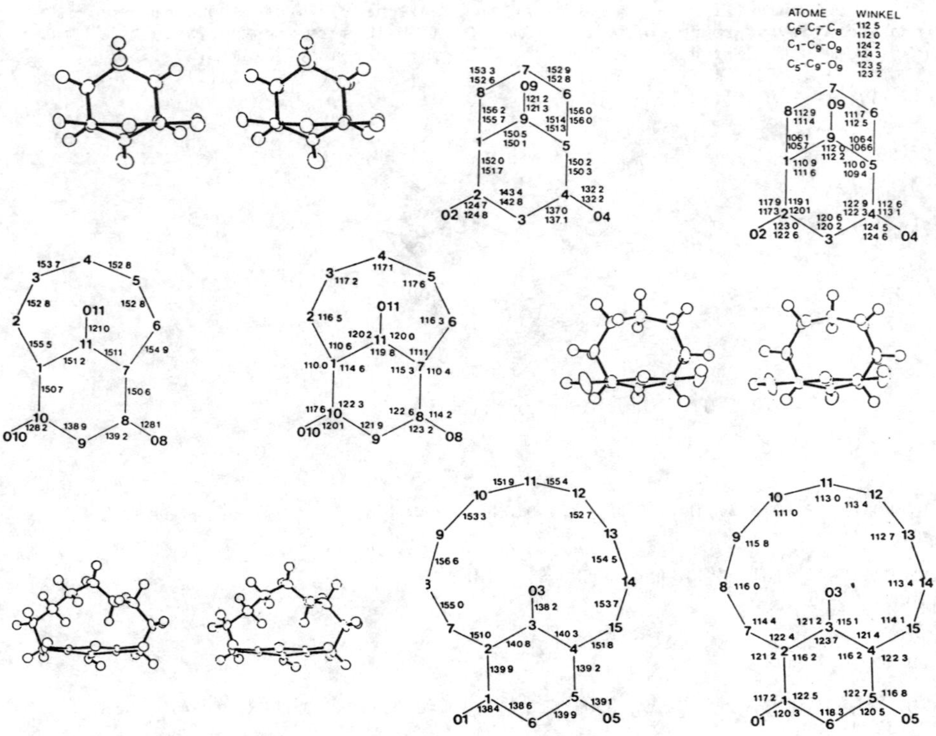

Fig. 1. Views and dimensions for $C_9H_{10}O_3$ (I) (top), $C_{11}H_{14}O_3$ (II) (centre) and $C_{15}H_{22}O_3$ (bottom) (bond lengths in pm, angles in degrees).

3-BROMOBICYCLO[3.3.1]NON-2-EN-4-ONE
$C_9H_{11}BrO$

A.S. BATSANOV and Yu.T. STRUCHKOV, 1984. J. Struct. Chem., 25, 653-655 [Zh. Strukt. Khim., 25(4), 150-152].

Triclinic, $P\bar{1}$, a = 6.73, b = 8.06, c = 8.43 Å, α = 89.2, β = 90.5, γ = 114.3°, D_m = 1.62 (at 20°C), Z = 2. Mo radiation, R = 0.084 for 1128 reflexions (at -120°C).

The structure is shown in Fig. 1. The oxygen atom was found to be disordered (2:1) over the O(1) and O(2) positions. The C(1-5) fragment is planar to within ±0.01 Å and the C(1,5-9) cyclohexane ring has the chair conformation. The Br-C distance is 1.92(1) Å.

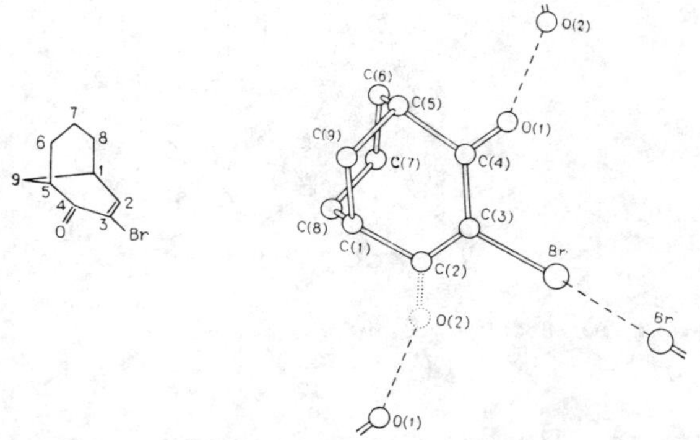

Fig. 1. The 3-bromobicyclo[3.3.1]non-2-en-4-one molecule. Short intermolecular distances are indicated by dotted lines.

PENTACYCLO[$4.3.0.0^{2,4}.0^{3,8}.0^{5,7}$]NONAN-4-ACETAMIDE

$C_{10}H_{11}NO$

R. FRÖHLICH, 1984. Acta Cryst., C40, 1289-1290.

Monoclinic, $P2_1/a$, a = 8.985, b = 7.454, c = 12.423 Å, β = 105.65°, Z = 4. Mo radiation, R = 0.042 for 858 reflexions.

The molecule (Fig. 1) contains two different cyclopropane units. The one bonded to the amide group shows the expected deviation from equilateral geometry. The bonds adjoining the carbon attached to the substituent are distinctly longer than the bond opposite. The other cyclopropane unit is exactly equilateral. The molecules form pairs, joined together by a N-H...O hydrogen bond.

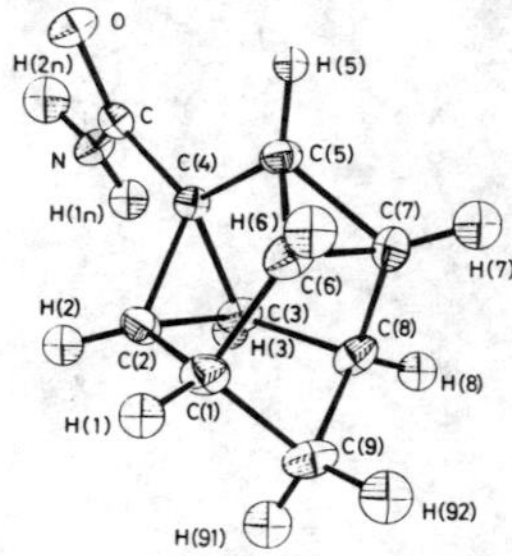

Fig. 1. $C_{10}H_{11}NO$: perspective view of the molecule.

1,3,5,7-TETRANITROADAMANTANE

$C_{10}H_{12}N_4O_8$

C. GEORGE and R. GILARDI, 1984. Acta Cryst., C40, 674-676.

Tetragonal, $P\bar{4}2_1c$, a = 7.874, b = 10.552 Å, Z = 2. Cu radiation, R = 0.027 for 296 reflexions.

The molecule (Fig. 1) lies on a $\bar{4}$ axis and the adamantane cage is only slightly distorted from ideal; the average C-C distance is 1.533 Å. The nitro groups are rotated approximately 15° from a symmetrical eclipsed position with the adjacent cage bonds to the C atoms on the $\bar{4}$ axis; the C-N distance is 1.529(3) Å and the O-N-O angle 124.1(3)°. The structure is compared with those of related adamantanes.

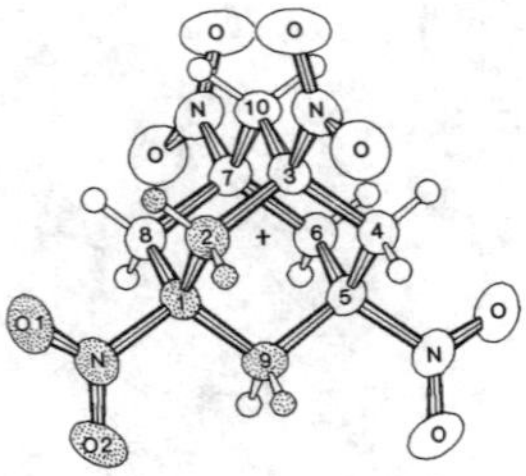

Fig. 1. $C_{10}H_{12}N_4O_8$: perspective view of the molecule.

2,5-DIBROMOTRICYCLO[6.2.1.0^{2,7}]UNDECA-4,9-DIENE-3,6-DIONE
$C_{11}H_8Br_2O_2$

A. SYED, P.P. UMRIGAR, G.W. GRIFFIN, E.D. STEVENS, R.J. MAJESTE, 1984. Acta Cryst., C40, 1916-1918.

Triclinic, P$\bar{1}$, a = 6.457, b = 7.668, c = 11.189 Å, α = 77.32, β = 81.40, γ = 71.19°, Z = 2. Mo radiation, R = 0.028 for 1482 reflexions (at 110K).

The molecule (Fig. 1) contains a half-chair cyclopentadiene ring cis-fused to a slightly puckered benzoquinone ring; the C(1),C(2),C(3),C(4) and C(1),C(6),C(5),C(4) interplanar angle is 10.1(4)°. The two double bonds are parallel with a 3.67(3) Å separation. The atoms C(7) to C(10) are planar while C(5) and C(6) deviate by -0.043 and +0.044(4) Å respectively from the C(7),C(6),C(5),C(10) plane. The angle at the C(7)-C(11)-C(10) bridgehead is 94.2(3)°. The C-Br distances are 1.979(3) for C(6)-Br(2) and 1.886(3) Å for C(3)-Br(1).

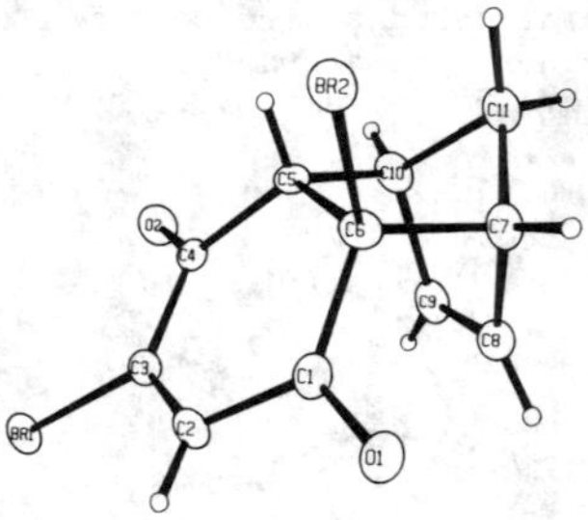

Fig. 1. The $C_{11}H_8Br_2O_2$ molecule.

2,6-DICYANOBICYCLO[3.3.1]NONA-2,6-DIENE
$C_{11}H_{10}N_2$ (I)

exo-4-BROMO-2,6-DICYANOBICYCLO[3.3.1]NONA-2,6-DIENE
$C_{11}H_9BrN_2$ (II)

exo,exo-4,8-DIBROMO-2,6-DICYANOBICYCLO[3.3.1]NONA-2,6-DIENE
$C_{11}H_{18}Br_2N_2$ (III)

4-BROMO-2,6-DICYANO-TRICYCLO[3.3.1.0^{2,8}]NONA-3,6-DIENE (4-BROMO-2,6-DICYANOBARBARALANE)
$C_{11}H_7BrN_2$ (IV)

H. QUAST, Y. GORLACH, J. STAWITZ, E.-M. PETERS, K. PETERS and H.G. VON SCHNERING, 1984. Chem. Ber., 117, 2745-2760.

I. Orthorhombic, Pccn, a = 7.134, b = 11.654, c = 10.961 Å, Z = 4. Mo radiation,

R = 0.084 for 760 reflexions.

II. Monoclinic, $P2_1/c$, a = 12.195, b = 7.911, c = 11.804 Å, β = 113.78°, Z = 4. Mo radiation, R = 0.065 for 1526 reflexions.

III. Orthorhombic, Pccn, a = 7.841, b = 13.215, c = 10.937 Å, Z = 4. Mo radiation, R = 0.065 for 845 reflexions.

IV. Monoclinic, $P2_1/n$, a = 9.428, b = 15.366, c = 7.027 Å, β = 95.85, Z = 4. Mo radiation, R = 0.055 for 1647 reflexions.

Details of molecular geometry are in Fig. 1 for I and II and in Fig. 2 for III and IV. The bromine atoms in II and III are shown to be exo.

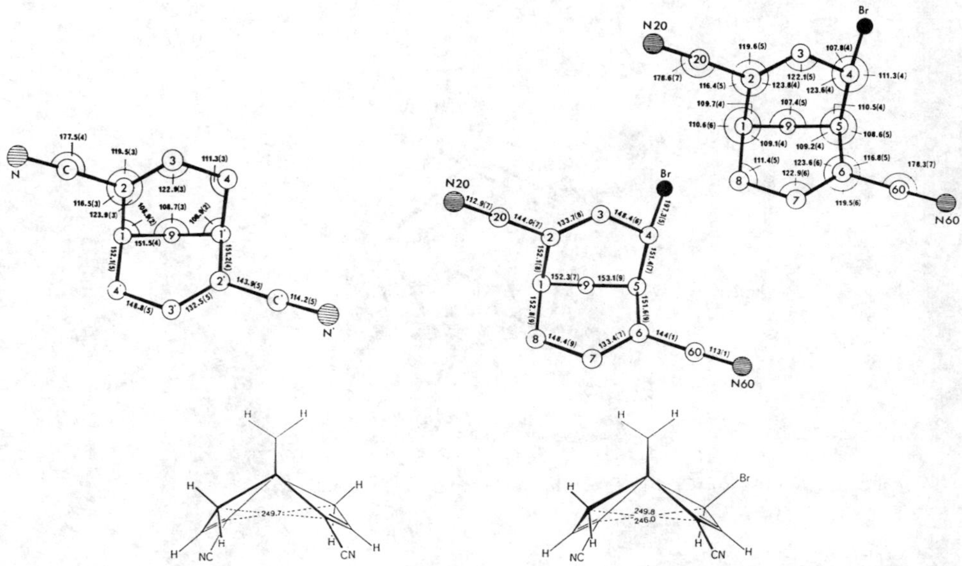

Fig. 1. Details of molecular geometry for $C_{11}H_{10}N_2$ (I) (left) and $C_{11}H_9BrN_2$ (II) (right).

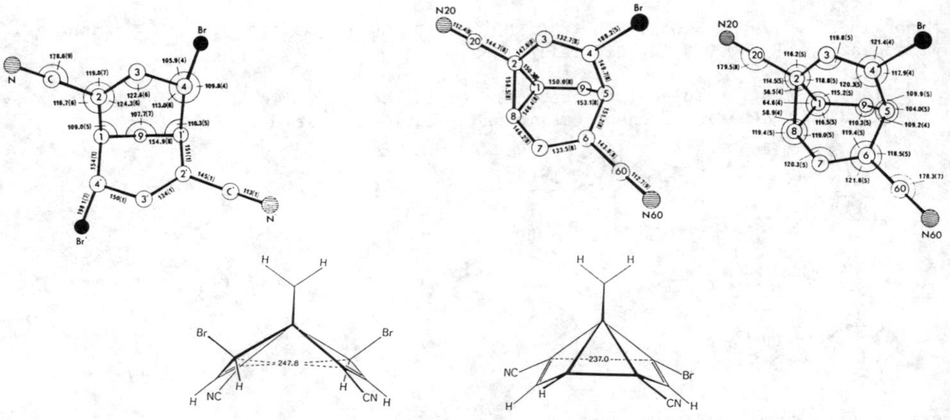

Fig. 2. Details of molecular geometry for $C_{11}H_{18}Br_2N_2$ (III) (left) and $C_{11}H_7BrN_2$ (IV) (right).

3-METHYLPENTACYCLO[5.4.0.0^{2,6}.0^{3,10}.0^{5,9}]UNDECANE-8,11-DIONE
$C_{12}H_{12}O_2$

A.P. MARCHAND, S.C. SURI, A.D. EARLYWINE, D.R. POWELL and D. VAN DER HELM, 1984. J. Org. Chem., 49, 670-675.

Monoclinic, $P2_1/n$, a = 9.8547, b = 11.4664, c = 7.8100 Å, β = 90.231°, D_m = 1.37, Z = 4. Cu radiation, R = 0.044 for 1813 reflexions at 138 K.

Much of the strain inherent in the molecule (Fig. 1) is accommodated by lengthening of the C(2)-C(7) and C(4)-C(5) bonds (both 1.589(2) Å). The two five-membered rings that contain C(11) have a nearly ideal envelope conformation. The five-membered rings with carbonyl groups are highly distorted. The four-membered ring is planar (maximum displacement 0.0011(12) Å).

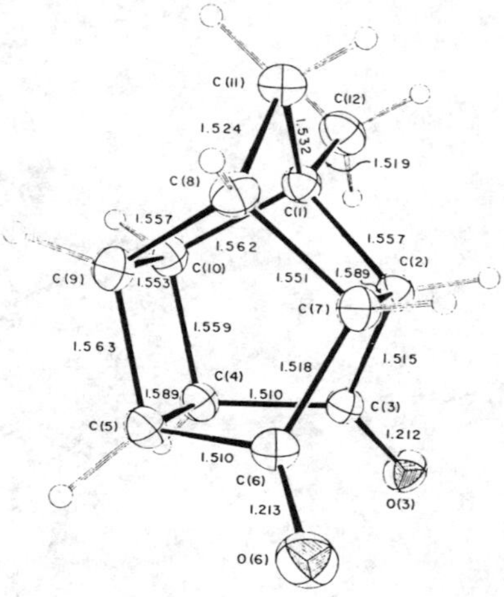

Fig. 1. A view of $C_{12}H_{12}O_2$ with bond lengths (σ 0.002 Å).

7-METHYL-1(2)a,1(6)a,3(4)a-TRIHOMOCUBANE-1(6)a,3(4)a-DIONE
$C_{12}H_{12}O_2$

G.S. MURTHY and K. VENKATESAN, 1984. Acta Cryst., C40, 1581-1584.

Monoclinic, $P2_1/n$, a = 6.219, b = 10.508, c = 7.339 Å, β = 107.64°, D_m = 1.360, Z = 2. Mo radiation, R = 0.058 for 614 reflexions.

The non-centric molecules occupy crystallographic centres of symmetry. This is explained on the basis of enantiomeric and rotational disorder (Fig. 1).

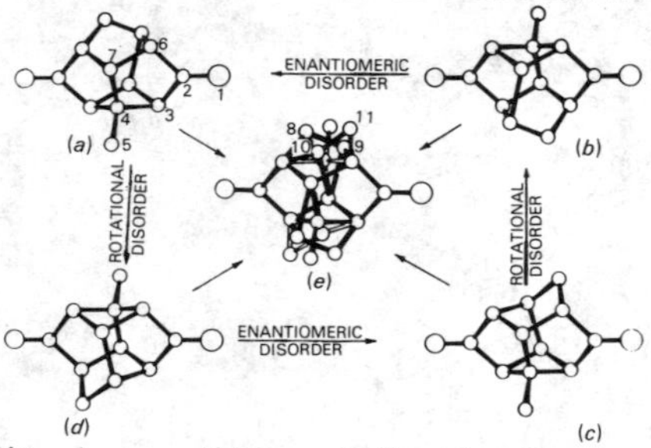

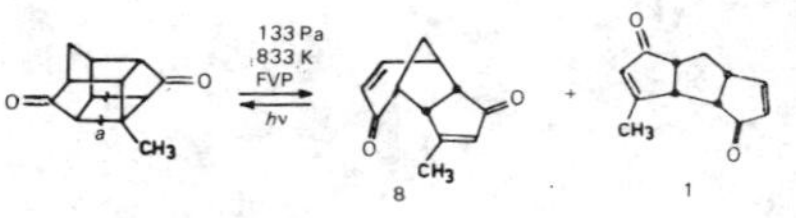

Fig. 1. Different disordered orientations of the $C_{12}H_{12}O_2$ molecule contributing to the final disordered crystal structure.

[4]PERISTYLANEDIONE
$C_{12}H_{12}O_2$

P. ENGEL, J.W. FISCHER and L.A. PAQUETTE, 1984. Z. Kristallogr., 166, 225-231.

Monoclinic, $P2_1/c$, a = 7.352, b = 11.349, c = 12.556 Å, β = 122.5°, Z = 4. Mo radiadiation, R = 0.047 for 1607 reflexions (at -2°C).

The molecule has almost perfect C_{2v} symmetry (Fig. 1). The four-membered ring and both five-membered rings bearing the keto groups are almost planar and the other two five-membered rings show half-chair conformation.

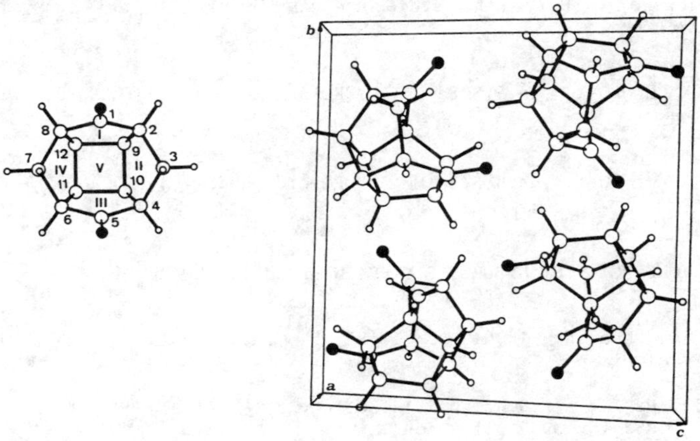

Fig. 1. The $C_{12}H_{12}O_2$ molecule and a view of the crystal structure.

TRICYCLO[4.2.2.2^{2,5}]DODECA-1,5-DIENE
$C_{12}H_{16}$ (I)

1,2,5,6-DIEPOXYTRICYCLO[4.2.2.2^{2,5}]DODECANE
$C_{12}H_{16}O_2$ (II)

1,5-DIBROMOTETRACYCLO[4.2.2.2^{2,5}.0^{2,6}]DODECANE
$C_{12}H_{16}Br_2$ (III)

K.B. WIBERG, R.D. ADAMS, P.J. OKARMA, M.G. MATTURRO and B. SEGMULLER, 1984. J. Am. Chem. Soc., 106, 2200-2206.

I. Monoclinic, C2/c, a = 13.942, b = 6.239, c = 14.198 Å, β = 133.91°, Z = 4. Mo radiation, R = 0.040 for 921 reflexions (at -25°C).

II. Orthorhombic, Pbca , a = 9.580, b = 9.258, c = 11.012 Å, Z = 4. Mo radiation, R = 0.043 for 932 reflexions.

III. Triclinic, $P\bar{1}$, a = 6.609, b = 7.318, c = 7.485 Å, α = 69.08, β = 62.24, γ = 64.72°, Z = 1. Mo radiation, R = 0.042 for 890 reflexions.

All three molecules have crystallographic inversion symmetry (Fig. 1). In I the bond length data are: C=C 1.354(1), Csp^3-Csp^2 1.516-1.521(1), Csp^3-Csp^3 1.595 and 1.596(1) Å. The two double bonds are separated by 2.395(1) Å and the olefinic C atoms are 0.40 Å out of the C3-C6 plane. The diepoxide derivative II has the C2-C3 and C5-C(6) distances 1.561 and 1.563(1) Å and the transannular distance between the carbon atoms of the epoxides is 2.459(1) Å. The dibromide III obtained from I has the C4-C4' bond elongated (1.570(5) Å) and the C1-C4 bond shortened (1.502(4) Å). Strain is evidenced by large exterior bond angles (C1-C4-C5' 123.7°, C1-C4-C6' 123.2(2)°).

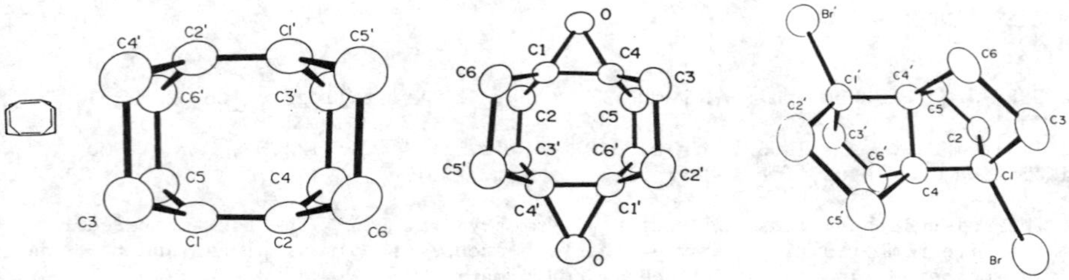

Fig. 1. Views of $C_{12}H_{16}$, and its diepoxy and dibromo derivatives.

5,6,7,8,9,9-HEXACHLORO-1,4,4a,5,8,8a-HEXAHYDRO-endo-endo-1,4-ETHANO-5,8-DIMETHANO-NAPHTHALENE

$C_{13}H_{10}Cl_6$ (I)

5,6,7,8,9,9-HEXACHLORO-1,4,4a,5,8,8a-HEXAHYDRO-endo-endo-1,4-PROPANO-5,8-DIMETHANO-NAPHTHALENE

$C_{14}H_{12}Cl_6$ (II)

D.C. DONG, W. WONG-NG, S.C. NYBURG, P.Y. SIEW and J.T. EDWARD, 1984. Can. J. Chem., 62, 452-456.

I. Orthorhombic, Pbca, a = 15.01, b = 13.27, c = 15.00 Å, Z = 8. Mo radiation, R = 0.057 for 1528 reflexions.

II. Monoclinic, $P2_1/c$, a = 9.42, b = 12.30, c = 27.84 Å, β = 97.27°, Z = 8. Mo radiation, R = 0.052 for 3945 reflexions.

These molecules (Fig. 1) are shown to exhibit decreasing across-space double-bond/double-bond repulsion which correlates with the observed decrease in the ease of cyclization in solution. In the absence of repulsion, it is calculated that the distance between double-bond mid-points would have been 2.37 and 2.67 Å (for I and II respectively). The actual values are 2.94 Å for I and 2.99 and 3.01 Å for the two independent molecules of II.

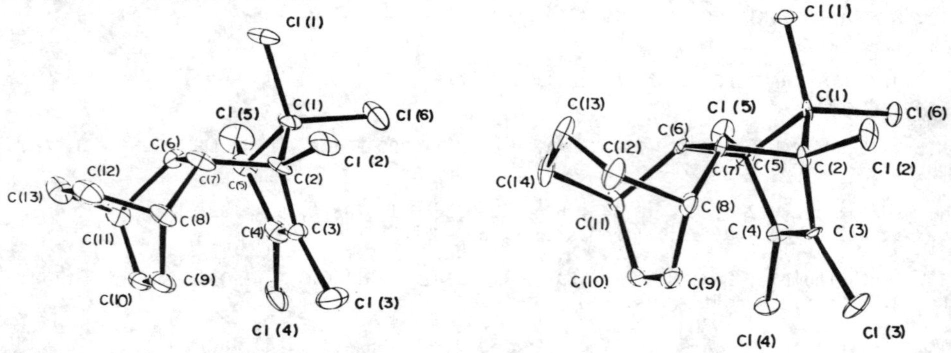

Fig. 1. Views of $C_{13}H_{10}Cl_6$ and one of the $C_{14}H_{12}Cl_6$ molecules.

5,6,7,8,9,9-HEXACHLORO-2,3-EPOXY-1,2,3,4,4a,5,8,8a-OCTAHYDRO-endo-5,8-endo-1,4-ETHANO-5,8-METHANONAPHTHALENE (HOMOENDRIN)

$C_{13}H_{10}Cl_6O$ (I)

5,6,7,8,9,9-HEXACHLORO-1,4,4a,5,8,8a-HEXAHYDRO-endo-5,8-exo-1,4-ETHANO-5,8-METHANO-NAPHTHALENE (HOMOISODRIN)

$C_{13}H_{10}Cl_6$ (II)

HOMOISODRIN ACETATE
$C_{15}H_{14}Cl_6O_2$ (III)

C.H.L. KENNARD and G. SMITH, 1984. J. Agric. Food Chem., 32, 895-900.

I. Monoclinic, $P2_1/n$, a = 11.541, b = 14.584, c = 8.888 Å, β = 92.88°, Z = 4. Mo radiation, R = 0.035 for 1309 reflexions.

II. Orthorhombic, Pbca, a = 14.988, b = 13.277, c = 14.999 Å, Z = 8. Mo radiation, R = 0.033 for 1127 reflexions.

III. Monoclinic, $P2_1/n$, a = 12.235, b = 11.538, c = 12.232 Å, β = 101.60°, Z = 4. Mo radiation, R = 0.037 for 1980 reflexions.

Views of these "cyclodiene" analogues are in Fig. 1. Structures for I and II are in accord with those previously proposed but III has its acetate group at C8 (rather than at C9). Comparative structural features are presented for a number of cyclodiene insecticides.

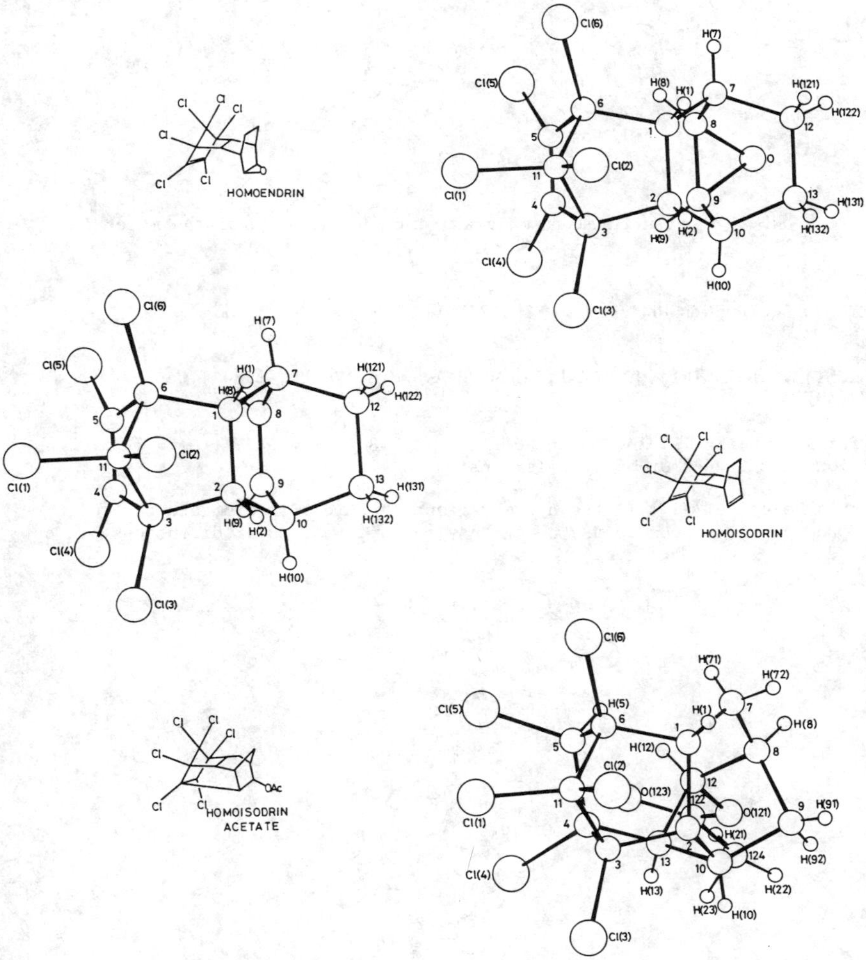

Fig. 1. Views of homoendrin (top), homoisodrin (centre) and homoisodrin acetate (bottom).

2',2'-DIMETHYLSPIRO[BICYCLO[2.2.1]HEPT-5-ENE-7,1'-CYCLOPROPANE]-2,3-DICARBONITRILE
$C_{13}H_{14}N_2$

N.W. ALCOCK, A.J.F. EDMUNDS and C.J. SAMUEL, 1984. Acta Cryst., C40, 1101-1102.

Monoclinic, $P2_1/c$, a = 10.973, b = 7.611, c = 13.842 Å, β = 91.27°, Z = 4. Mo radiation, R = 0.092 for 1285 reflexions.

In the molecule (Fig. 1) bond distances and angles are as expected, with C(7)-C(8) 1.518(10), C(8)-C(9) 1.540(10) and C(9)-C(7) 1.527(10) Å and C(7)-C(8)-C(9) 59.9(5), C(8)-C(9)-C(7) 59.3(5) and C(9)-C(7)-C(8) 60.7(5)° in the cyclopropane ring and C(12)-N(1) 1.128(10) and C(13)-N(2) 1.128(11) Å in the nitrile groups.

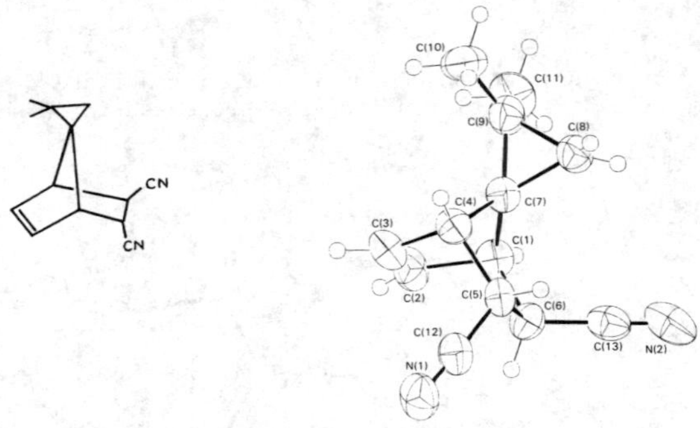

Fig. 1. $C_{13}H_{14}N_2$: structural formula and perspective view of the molecule.

3-HYDROXY-6-OXO-2-PHENYLSULPHONYLBICYCLO[2.2.1]HEPTANE
$C_{13}H_{14}O_4S$

O. DE LUCCHI, V. LUCCHINI, M. ZAMAI, G. MODENA and G. VALLE, 1984. Can. J. Chem., 62, 2487-2497.

Monoclinic, $P2_1/a$, a = 19.314, b = 6.001, c = 11.348 Å, β = 98.7°, D_m = 1.35, Z = 4. Mo radiation, R = 0.064 for 2005 reflexions.

In this molecule (Fig. 1) the hydroxy group (at C(6)) is exo, and the sulphonyl group is endo on the same side as the carbonyl group. Molecular dimensions are as anticipated.

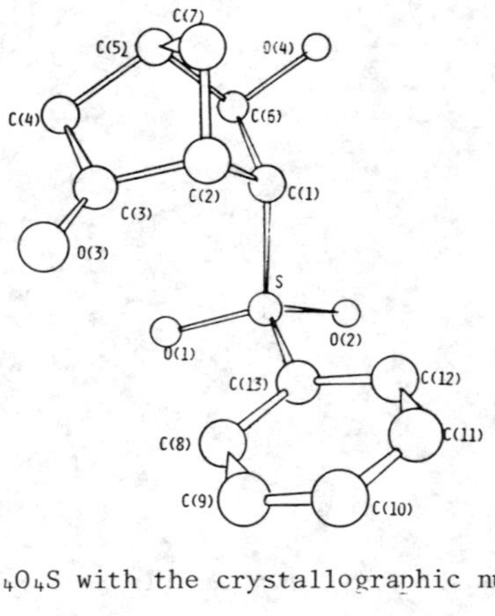

Fig. 1. A view of $C_{13}H_{14}O_4S$ with the crystallographic numbering scheme.

1-(TETRACYCLO[5.4.0.0^{3,8}.0^{4,8}]UNDEC-5-EN-9-YL)ETHANOL
$C_{13}H_{18}O$

K. CHANDRASEKHAR and J. HAUSER, 1984. Acta Cryst., C40, 283-285.

Rhombohedral, R$\bar{3}$, a = 26.371, c = 8.374 Å, D_m = 1.13, Z = 18. Mo radiation, R = 0.038 for 1160 reflexions.

The molecule (Fig. 1) has high internal strain at C(1); the exo-angle C(2)-C(1)-C(9) is 140.2(2)°. The cyclopentene ring C1-C5 has a flattened envelope conformation while the cyclopentane ring is puckered into a strongly bent half-chair. The ring C1,C9,C8,C7,C6,C5 has a distorted chair form and the seven-membered ring C1,C9,C8,C7,-C6,C11,C10 has a twisted boat conformation.

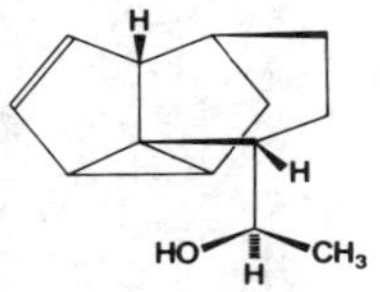

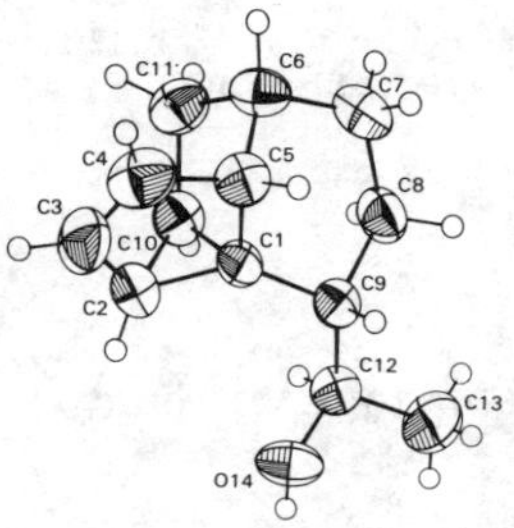

Fig. 1. The $C_{13}H_{18}O$ molecule.

rel-(1S,2S,3S,3aR,4R,7S,7aR)-1,2,3,4,5,6,7,7a-OCTAHYDRO-1,3a-ETHENO-4,7-METHANO-3aH-INDENE-2,3-DICARBOXYLIC ANHYDRIDE
$C_{14}H_{14}O_3$

W.H. WATSON, J. GALLOY, C. WU and P.D. BARTLETT, 1984. Acta Cryst., C40, 1418-1420.

Monoclinic, P2_1/a, a = 6.387, b = 10.216, c = 16.897 Å, β = 100.27°, Z = 4. Mo radiation, R = 0.055 for 1357 reflexions.

The six-membered ring of the norbornane system has a skew-boat conformation due to the strain imposed by ring fusions at C(1) and C(5) (Fig. 1). There are two short intramolecular contacts H(10)-endo...H(5) 2.20(4) and H(8)-endo...H(2) 2.17(5) Å. H(8)-endo may also interact with the C(2) π system being 2.41 Å above C(2). Some bond lengths are, C(2)-C(3) 1.338(3), C(13)-O(15) 1.397(3), C(14)-O(15) 1.387(3) Å.

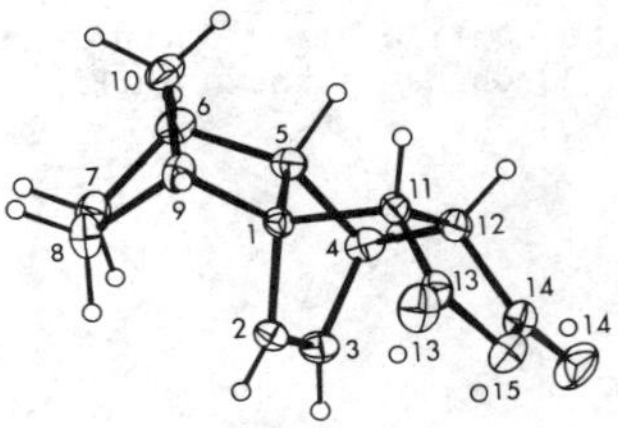

Fig. 1. The $C_{14}H_{14}O_3$ molecule.

(4aR*,6S*,7S*,9R*)-6-BROMO-3,4,6,7,8,9,10,11-OCTAHYDRO-7-HYDROXY-4a,9-METHANO-4aH-BENZOCYCLONONENE-2,12(5H)-DIONE
$C_{14}H_{17}BrO_3$ (I)

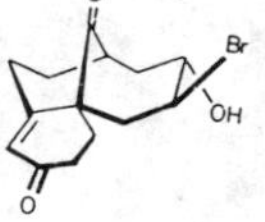

(4aR*, 6S*,7R*,9R*,12S*)-12-(t-BUTYLDIMETHYLSILOXY)-6,7-EPOXY-3,4,6,7,8,9,10,11-OCTAHYDRO-4a,9-METHANO-4aH-BENZOCYCLONONEN-2(5H)-ONE
$C_{20}H_{32}O_3Si$ (II)

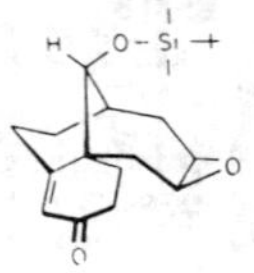

L.A. PAQUETTE, T.J. NITZ, R.J. ROSS, J.P. SPRINGER, 1984. J. Am. Chem. Soc., 106, 1446-1454.

I. Orthorhombic, $Pna2_1$, a = 11.907, b = 16.990, c = 6.376 Å, Z = 4. Cu radiation, R = 0.034 for 953 reflexions.

II. Orthorhombic, Pccn, a = 7.150, b = 38.433, c = 14.722 Å, Z = 8. Cu radiation, R = 0.057 for 1719 reflexions.

X-ray analysis established the structures of these intermediates as shown.

11,12-DIMETHYLBICYCLO[5.3.2]DODECA-1,11-DIEN-6-ONE
$C_{14}H_{20}O$

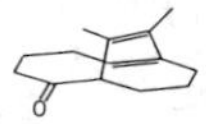

W. VON, E. DOERING and J.C. SCHMIDHAUSER, 1984. J. Am. Chem. Soc., 106, 5025-5026.

Monoclinic, $P2_1/n$, a = 8.122, b = 12.071, c = 12.574 Å, β = 105.21°, Z = 4. Mo radiation, R = 0.0872 for 1221 reflexions.

The analysis establishes the structure as shown above.

transoid-(1RS,3SR,6RS,9RS)-7-HYDROXYIMINO-9-METHYL-1-(2-METHYLPROPYL)TRICYCLO-[4.2.1.$0^{3,9}$]NONAN-8-ONE
$C_{14}H_{21}NO_2$

G.E. FREEMAN, D.C. FISHER and K. SEOK-HWAN, 1984. Acta Cryst., C40, 502-504.

Monoclinic, $P2_1/c$, a = 7.3367, b = 7.9791, c = 23.298 Å, β = 103.207°, D_m = 1.172, Z = 4. Mo radiation, R = 0.0647 for 1254 reflexions.

The analysis establishes the configuration of this intermediate in the synthesis of the sesquiterpene alkaloid dendrobine as shown in Fig. 1.

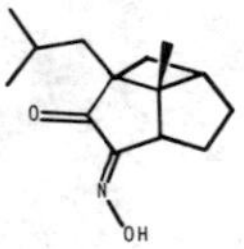

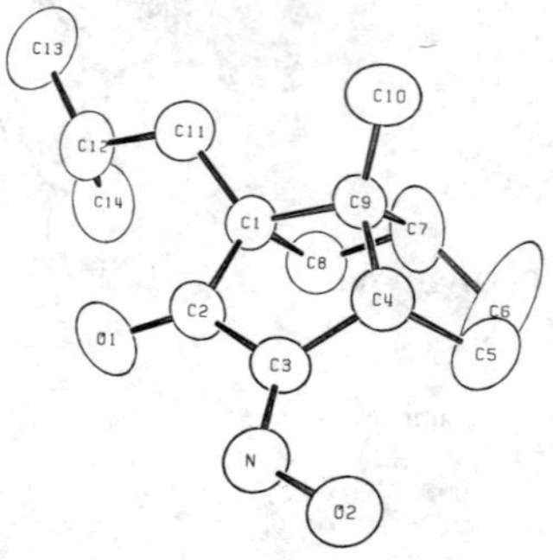

Fig. 1. A view of the intermediate $C_{14}H_{21}NO_2$.

2-HYDROXY-13-OXO-TRICYCLO[7.3.1.$0^{2,7}$]TRIDECANE
$C_{14}H_{22}O_2$

G. BUCHBAUER, A. FISCHLMAYR, E. HASLINGER, W. ROBIEN, H. VOLLENKLE and C. WASSMANN, 1984. Montash., 115, 739-748.

Monoclinic, $P2_1/n$, a = 8.518, b = 14.789, c = 19.321 Å, γ = 94.91°, Z = 8. R = 0.100 for 1719 reflexions.

There are two independent molecules in the asymmetric unit but their molecular geometries are similar within the limits of experimental error. The carbon framework can be considered as three cyclohexane rings, two forming trans-decalin and the third connected in an adamantane-like configuration (Fig. 1). The independent molecules are linked by two hydrogen bonds O(1)-H(1)...O(22) (O(1)...O(22) 2.897 Å) and O(2)...H(21)-O(21) (O(2)...O(21) 2.830 Å) in a pseudo-centrosymmetric dimeric arrangement. Bond lengths of interest are (for molecule I and II respectively): O(1)-C(2) 1.441, 1.435; O(2)-C(13) 1.233, 1.246 Å.

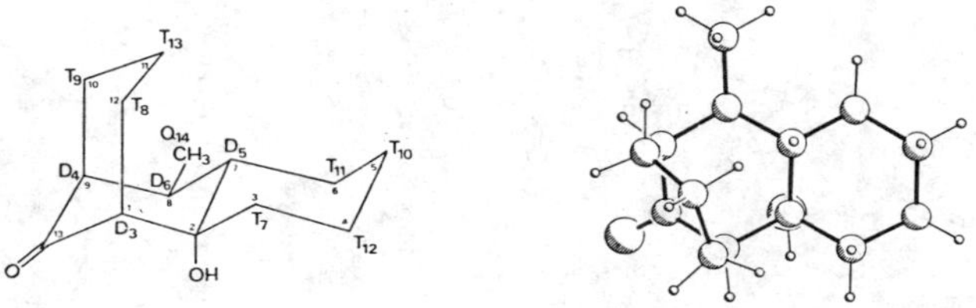

Fig. 1. Numbering scheme for $C_{14}H_{22}O_2$ and the molecular structure perpendicular to decalin plane.

2-exo-CHLORO-3-endo-(2-NITROPHENYLTHIO)-7,7-DIMETHYLBICYCLO[2.2.1]HEPTANE
$C_{15}H_{18}ClNO_2S$

M.V. SERGEYEVA, Yu.T. STRUCHKOV, E.N. KURKUTOVA, T.V. ZEMLYANUKHINA, N.K. SADOVAYA and N.S. ZEFIROV, 1984. Acta Cryst., C40, 1615-1616.

Triclinic, $P\bar{1}$, a = 7.3039, b = 7.9432, c = 13.4552 Å, α = 95.92, β = 98.66, γ = 99.74°, Z = 2. Mo radiation, R = 0.037 for 2430 reflexions.

The norbornane framework has a synchro-twist conformation with a difference between the C7C1C2C3 and C7C4C3C2 torsion angles of 13.0(2)°. The Cl and S atoms are displaced -1.198 and 1.251 Å respectively from the C1C2C3C4 mean plane and this plane makes an angle of 119.7° with the C(2)SC(10) plane which in turn makes an angle of 15.3° with the benzene ring plane. The S and N atoms lie 0.043 and 0.010 Å respectively from the benzene ring plane. There is a short (2.643(2) Å) S...O(2) interaction with the C(2)-S...O(2) angle 170.3(1)°.

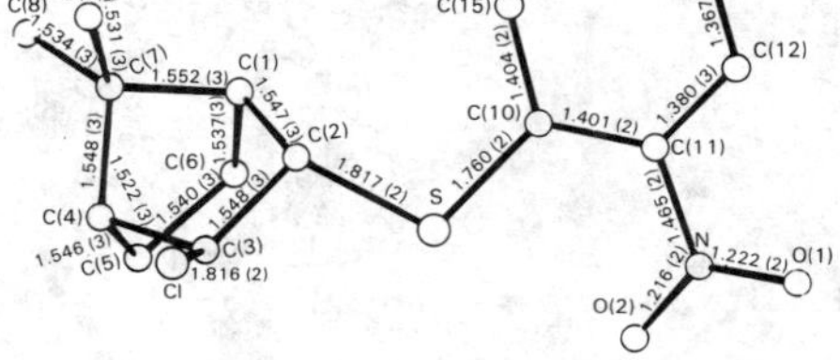

Fig. 1. The $C_{15}H_{18}ClNO_2S$ molecule.

3α,7α-BIS(2-HYDROXY-2-PROPYL)BICYCLO[3.3.1]NONANE
$C_{15}H_{28}O_2$

H. VAN KONINGSVELD, J.A. PETERS and J.C. JANSEN, 1984. Acta Cryst., C40, 158-160.

Orthorhombic, $P2_12_12$, a = 12.700, b = 10.474, c = 10.771 Å, Z = 4. Mo radiation, R = 0.043 for 1559 reflexions.

The molecule (Fig. 1) has a double twist-boat conformation. The two hydroxypropyl groups have different orientations with respect to the ring system; the group at C(11) is nearly antiperiplanar to H(1) (torsion angle C(11)-C(10)-C(1)-H(1) 177°), whereas in the other group O(2) and H(8) are nearly antiperiplanar (torsion angle O(2)-C(13)-C(8)-H(8) 171°). The molecules are linked by weak O-H...O hydrogen bonds (O...O 2.811 and 2.769 Å).

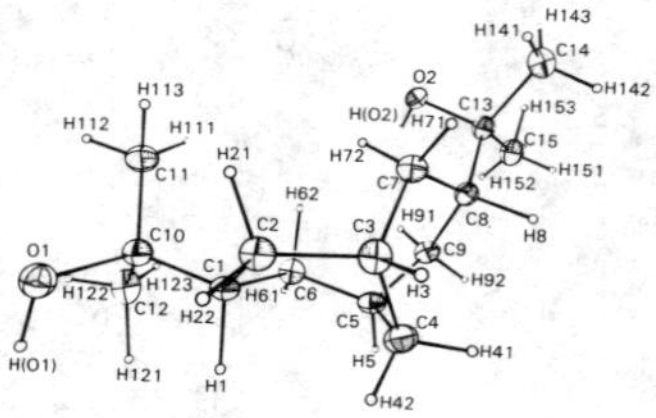

Fig. 1. A view of $C_{15}H_{28}O_2$.

1,2,2a,4a,5,6,7,8-OCTAHYDRO-5,8-METHANOCYCLOBUT[c]INDENE-1,1,2,2-TETRACARBONITRILE
$C_{16}H_{12}N_4$

R.P. KASHYAP, W.H. WATSON, C. WU and P.D. BARTLETT, 1984. Acta Cryst., C40, 861-863.

Monoclinic, $P2_1/n$, a = 11.415, b = 15.401, c = 7.751 Å, β = 96.90°, Z = 4. Cu radiation, R = 0.040 for 1508 reflexions.

The molecule contains a slightly puckered four-membered ring with three long C-C distances: C(1)-C(2) 1.581(3), C(2)-C(3) 1.596(3) and C(3)-C(4) 1.582(3) Å (Fig. 1). Both N-C-C-C-N moieties are planar and almost perpendicular to the four-membered ring and to each other. The four-membered ring makes an interplanar angle of 118.1(6)° with the almost planar five-membered ring. The cyclohexane ring is held in a boat conformation by the methylene bridge. Although there is little steric hindrance within the molecule, there is considerable angle strain as evidenced by the fourteen $C(sp^3)$ valence angles of less than 104°.

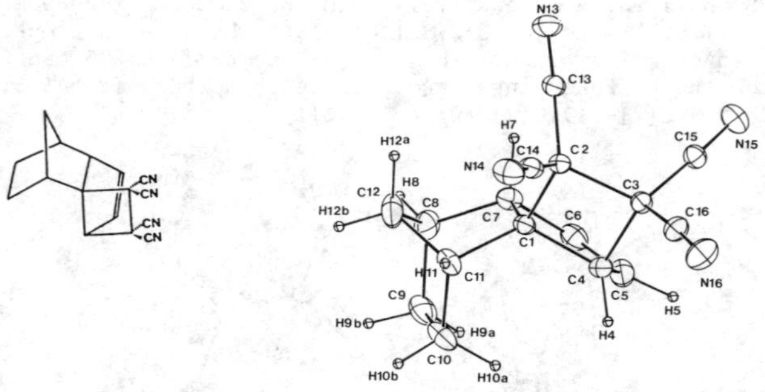

Fig. 1. Structural formula and perspective view of the $C_{16}H_{12}N_4$ molecule.

rel-(1R,4S,4aR,7R)-1,2,3,4,4a,5,6,7-OCTAHYDRO-1,4:4a,7-DIMETHANONAPHTHALENE-5,5,6,6-TETRACARBONITRILE
$C_{16}H_{12}N_4$ (I)

rel-(1R,4S,4aS,7S)-1,2,3,4,4a,5,6,7-OCTAHYDRO-1,4:4a,7-DIMETHANONAPHTHALENE-5,5,6,6-TETRACARBONITRILE
$C_{16}H_{12}N_4$ (II)

W.H. WATSON, R.P. KASHYAP, C. WU and P.D. BARTLETT, 1984. Acta Cryst., C40, 866-869.

I. Monoclinic, $P2_1/c$, a = 8.986, b = 9.661, c = 14.870 Å, β = 94.86°, Z = 4. Cu radiation, R = 0.0397 for 1494 reflexions.

II. Monoclinic, $P2_1/c$, a = 9.696, b = 8.925, c = 16.022 Å, β = 105.33°, Z = 4. Cu radiation, R = 0.0345 for 1447 reflexions.

The molecules can be considered to be composed of fused five-membered rings in envelope conformations or fused six-membered rings held in boat conformations by methylene bridges (Fig. 1). The sixteen $C(sp^3)$ valence angles of less than 105° imply considerable ring strain within the molecules. C-C bond lengths of 1.577(3) to 1.620(3) Å are associated with the tetracyanoethylene moiety. The bond elongation is attributed to the electron-withdrawing character of the cyano groups.

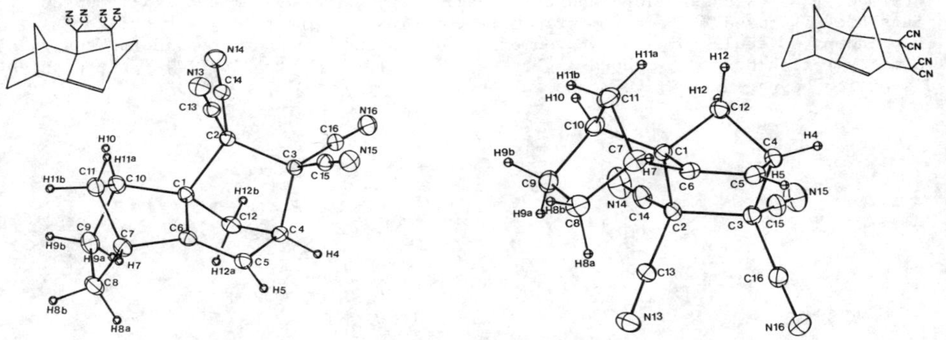

Fig. 1. $C_{16}H_{12}N_4$: perspective views of isomers I and II.

rel-(1R,4S,5S,8R)-1,2,3,4,5,6,7,8-OCTAHYDRO-1,4:5,8-DIMETHANONAPHTHALENE-2,2,3,3-TETRACARBONITRILE
$C_{16}H_{12}N_4$

W.H. WATSON, R.P. KASHYAP, P.D. BARTLETT and C. WU, 1984. Acta Cryst., C40, 863-866.

Monoclinic, $P2_1/c$, a = 18.123, b = 13.509, c = 12.073 Å, β = 109.35°, Z = 8. Cu radiation, R = 0.043 for 2497 reflexions.

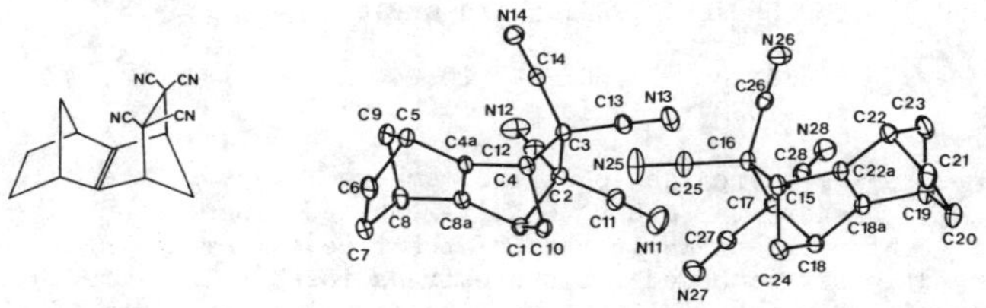

Fig. 1. Structural formula of the $C_{16}H_{12}N_4$ molecule and a perspective view of the two independent molecules.

The two independent molecules each contain a non-crystallographic mirror plane bisecting the double bond and passing through the two methylene bridges; there is excellent agreement between equivalent bond distances related by the mirror plane and between the two independent molecules (Fig. 1). The central π system deviates from planarity by ca. 4°; this is attributed primarily to an interaction between the central π bond and the π system of the cyano groups. The electron-withdrawing character of the cyano groups may account for the elongation of three C-C bonds in each molecule which range from 1.591(3) to 1.609(4) Å.

cis-3,4,4a,9,10,10a-HEXAHYDRO-1,4a-ETHANOPHENANTHRENE-2(1H),12-DIONE
$C_{16}H_{16}O_2$

A.K. MUKHERJEE and M. MUKHERJEE, 1984. Acta Cryst., C40, 981-983.

Orthorhombic, $Pca2_1$, a = 11.385, b = 11.938, c = 18.50 Å, Z = 8. Cu radiation, R = 0.044 for 1472 reflexions.

Of the three six-membered and one five-membered rings in the molecule (Fig. 1), the phenyl ring is substantially planar, the other two six-membered rings are in half-boat and distorted chair conformations, while the five-membered ring is intermediate between envelope and half-chair. The two non-aromatic six-membered rings are cis fused. The asymmetric unit contains two crystallographically independent molecules of similar geometry. The structure shows considerable strain; formally tetrahedral C-C-C bond angles vary from 100.9(3) to 116.6(4)°, the greatest distortion being in the five-membered ring.

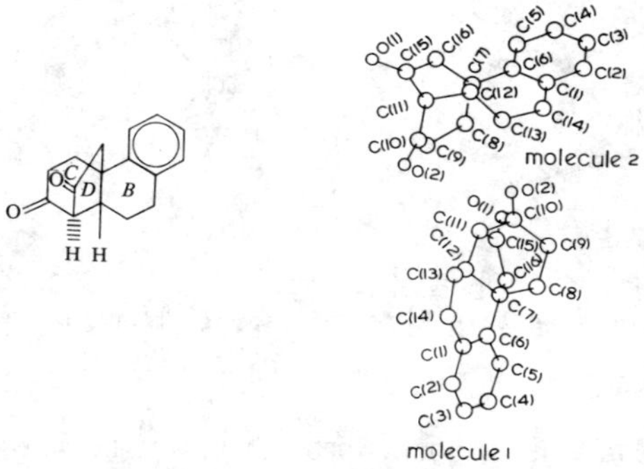

Fig. 1. $C_{16}H_{16}O_2$: perspective view of the asymmetric unit along b.

trans-3,4,4a,9,10,10a-HEXAHYDRO-3,4a-ETHANOPHENANTHRENE-2(1H),12-DIONE
$C_{16}H_{16}O_2$

M. MUKHERJEE and A.K. MUKHERJEE, 1984. Acta Cryst., C40, 983-985.

Monoclinic, $P2_1$, a = 8.638, b = 7.160, c = 10.554 Å, β = 104.08°, Z = 2. Mo radiation, R = 0.044 for 1080 reflexions.

Of the three six-membered and one five-membered rings in the molecule (Fig. 1) the phenyl ring is planar, the other two six-membered rings assume distorted half-chair and chair conformations while that of the five-membered ring is an envelope. The two non-aromatic six-membered rings are trans fused. The structure exhibits considerable strain, "tetrahedral" C-C-C valence angles vary from 102.0(4) to 115.1(4)°, the five-membered ring being distorted most.

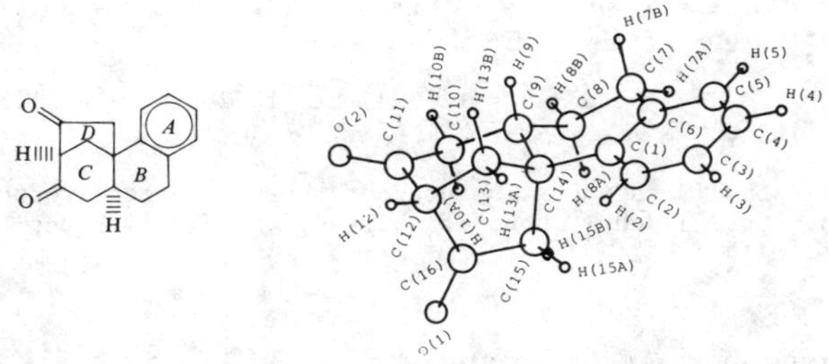

Fig. 1. $C_{16}H_{16}O_2$: perspective view of the molecule along b.

DIMETHYL rel-(1R,4S,4aR)-2,3,4,4a-TETRAHYDRO-1,4-METHANO-1H-BENZOCYCLOHEPTENE-5,6-DICARBOXYLATE

$C_{16}H_{18}O_4$

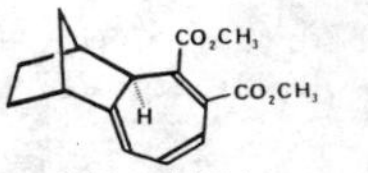

R.P. KASHYAP, W.H. WATSON, C. WU and P.D. BARTLETT, 1984. Acta Cryst., C40, 1854-1857.

Monoclinic, $P2_1/c$, a = 7.659, b = 10.833, c = 16.867 Å, β = 96.04°, Z = 4. Cu radiation, R = 0.044 for 1566 reflexions.

The cycloheptatriene ring is in a slightly distorted boat conformation with alternating double (C=C 1.327(4)-1.353(5) Å) and single (C-C 1.430(5), 1.457(4) Å) bonds. The two CCO_2CH_3 groups are planar with extended conformations and their planes intersect at 94.3(7)°. The norbornane system is slightly twisted because of fusion to the seven-membered ring.

trans-(BICYCLO[3.1.1]HEPT-6-YLIDENE)BICYCLO[3.1.1]HEPT-7-endo-YL ACETATE

$C_{16}H_{22}O_2$ (I)

3',3'-DICHLOROBIS(3'-exo-BICYCLO[3.1.1]HEPTANE-6-SPIRO)-1',2'-CYCLOPROPANE

$C_{15}H_{20}Cl_2$ (II)

R. ROMER, J. HARNISCH, A. RODER, A. SCHOFFER, G. SZEIMIES, G. GERMAIN and J.M. ARRIETA, 1984. Chem. Ber., 117, 925-947.

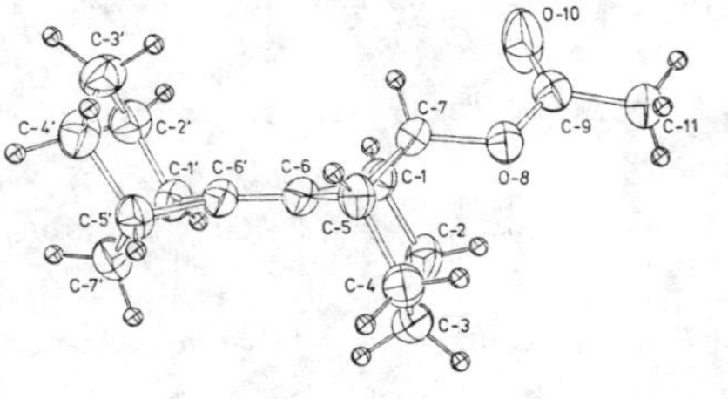

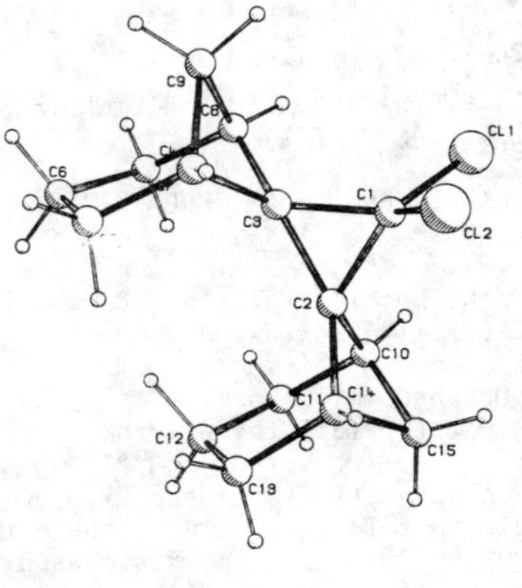

Fig. 1. Views of $C_{16}H_{22}O_2$ (I) (left) and $C_{15}H_{20}Cl_2$ (II) (right).

I. Monoclinic, $P2_1/a$, a = 9.998, b = 10.511, c = 13.435 Å, β = 93.49°, Z = 4. R = 0.045 for 1257 reflexions.

II. Monoclinic, $P2_1/c$, a = 6.387, b = 16.094, c = 13.284 Å, β = 91.33°, Z = 4. R = 0.081 for 1178 reflexions.

The analyses established the structures of these reaction products as shown in Fig. 1. In I, the central double bond is twisted, with torsion angles C1-C6-C6'-C1' 14.0 and C5-C6-C6'-C5' -12.9°.

2-(p-CHLOROBENZYLIDENE) (E)-(-)-3-BORNANONE
$C_{17}H_{19}ClO$

F. LABRUYERE, Ch. BERTRAND, C. METGE, A. DUBOURG, R. ROQUES, J.-P. DECLERCQ and G. GERMAIN, 1984. Acta Cryst., C40, 277-279.

Orthorhombic, $P2_12_12_1$, a = 10.703, b = 11.327, c = 12.004 Å, Z = 4. Mo radiation, R = 0.051 for 787 reflexions.

The analysis confirms the E configuration previously determined by other methods. The conjugated O=C-C=C system is slightly skewed in the sense of a left-handed helix (-4°).

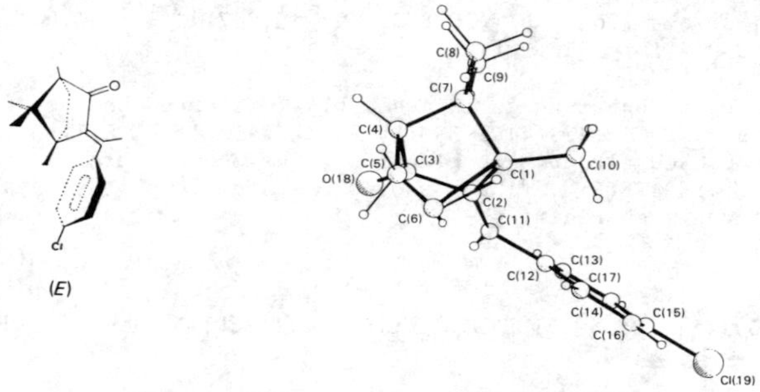

Fig. 1. The E configuration of $C_{17}H_{19}ClO$ and a view of the molecule.

1,3,5,7-TETRACARBOMETHOXYBICYCLO[3.3.1]NONA-2,6-DIENE-2,6-DIOL
$C_{17}H_{20}O_{10}$ (I)

1,3,5,7-TETRACARBOMETHOXYBICYCLO[3.3.1]NONA-2,6-DIENE-2,6-DIOL BENZENE SOLVATE
$C_{17}H_{20}O_{10}, 4C_6H_6$ (II)

M.D. RADCLIFFE, A. GUTIERREZ, J.F. BLOUNT and K. MISLOW, 1984. J. Am. Chem. Soc., 106, 682-687.

I. Orthorhombic, Pbcn, a = 8.235, b = 13.870, c = 15.963 Å, Z = 4. Cu radiation, R = 0.051 for 1173 reflexions.

II. Orthorhombic, Pnaa, a = 11.867, b = 12.667, c = 25.601 Å, Z = 4. Cu radiation, R = 0.086 for 1363 reflexions.

In both structures the $C_{17}H_{20}O_{10}$ molecules (Fig. 1) are located on crystallographic twofold axes. In II the ester molecules are sheathed in a channel of benzene molecules (Fig. 2). In this unusual inclusion compound, it is not clear whether I is the host or the guest molecule.

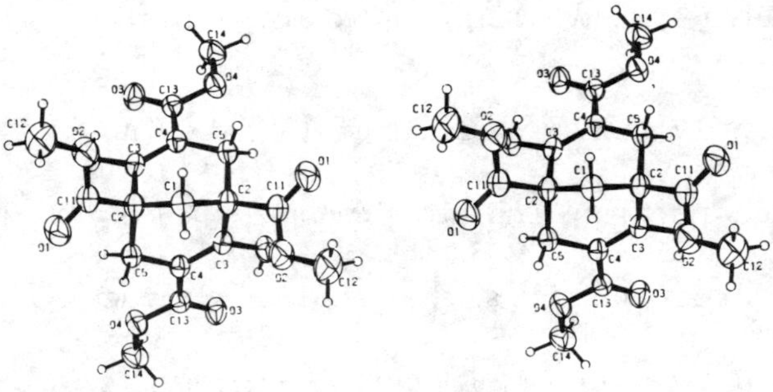

Fig. 1. Stereoview of $C_{17}H_{20}O_{10}$.

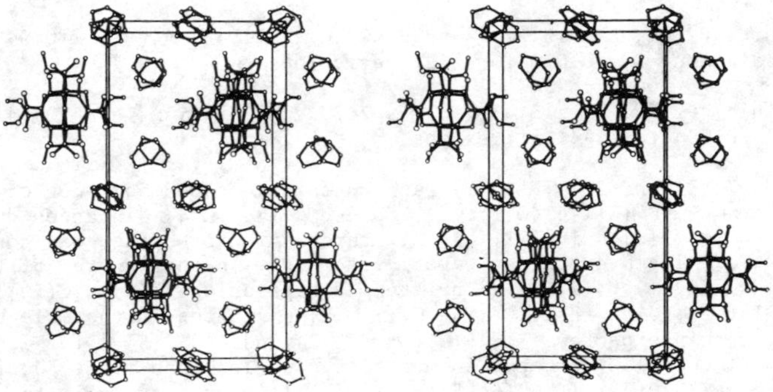

Fig. 2. Stereoview of $C_{17}H_{20}O_{10},4C_6H_6$.

11,12-DICHLORO-9,10-ETHANO-9,10-DIHYDRO-1-ANTHROIC ACID METHYL ESTER
$C_{18}H_{14}Cl_2O_2$

J.H. NOORDIK, 1984. Acta Cryst., C40, 123-125.

Triclinic, $P\bar{1}$, a = 9.782, b = 13.269, c = 12.942 Å, α = 88.77, β = 97.67, γ = 100.17°, Z = 4. Mo radiation, R = 0.042 for 2803 reflexions.

The two independent molecules in the asymmetric unit have the same overall geometry but differ in the orientation of the methoxycarbonyl group. The analysis establishes trans configuration of the Cl atoms in the -CHCl-CHCl- bridge (Fig. 1).

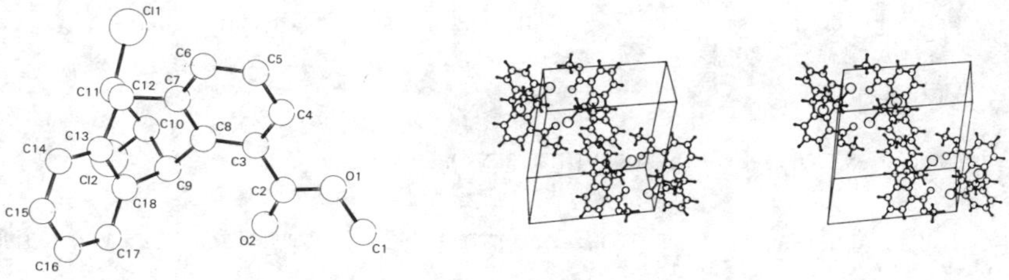

Fig. 1. A view of one molecule of $C_{18}H_{14}Cl_2O_2$ and a stereoview of the unit cell.

HEXAHYDRO-7-HYDROXY-1,2,4-ETHANYLYLIDENE-1H-CYCLOBUTA[c,d]PENTALEN-5(1aH)-ONE 4-NITROBENZOATE
$C_{18}H_{15}NO_5$ (I)

OCTAHYDRO-8-HYDROXY-1,3,6-ETHANYLYLIDENECYCLOBUT[cd]INDEN-2(1H)-ONE 4-NITROBENZOATE
$C_{19}H_{17}NO_5$ (II)

DECAHYDRO-9-HYDROXY-1,3,7-ETHANYLYLIDENE-2H-CYCLOBUT[cd]AZULEN-2-ONE 4-NITROBENZOATE
$C_{20}H_{19}NO_5$ (III)

R.V. CERNIK, G.-A. CRAZE, O.S. MILLS, I. WATT and S.N. WHITTLETON, 1984. J. Chem. Soc. Perkin II, 685-690.

I. Orthorhombic, Pccn, a = 10.140, b = 26.158, c = 11.608 Å, Z = 8. Mo radiation, R = 0.091 for 607 reflexions.

II. Monoclinic, $P2_1/c$, a = 15.313, b = 9.600, c = 11.439 Å, β = 110.04°, D_m = 1.40, Z = 4. Mo radiation, R = 0.029 for 1322 reflexions.

III. Monoclinic, $P2_1/c$, a = 15.433, b = 9.767, c = 11.461 Å, β = 107.41°, Z = 4. Mo radiation, R = 0.054 for 1444 reflexions.

Compounds I, II and III (Fig. 1) each contain imbedded boat-conformation 4-hydroxycyclohexanones (C(1-4),O(1),O(4)) whose C(1)...C(4) distances decrease from 2.67(3) in I to 2.479(6) Å in III. In all three structures the boat conformations are slightly twisted. The hydrocarbon cages differ only in the lengths of the remote methylene bridge C(7)...C(10). The bridgehead separation (C(7)...C(10)) increases from 2.25 in I to 2.762 Å in III. The C(1)-O(1) distances in the molecule are all slightly longer than expected - 1.45(2), 1.453(3) and 1.452(5) Å for I, II and III respectively. The C(2)-C(3) bond, common to the cyclobutane rings, decreases in length (1.62(2), 1.570(3), 1.560(6) Å in I, II, III respectively) with extension of the methylene bridge. The relationship of the structural variation to the barriers and reaction path for the 1,4-hydride transfer in the corresponding alkoxide anions is discussed.

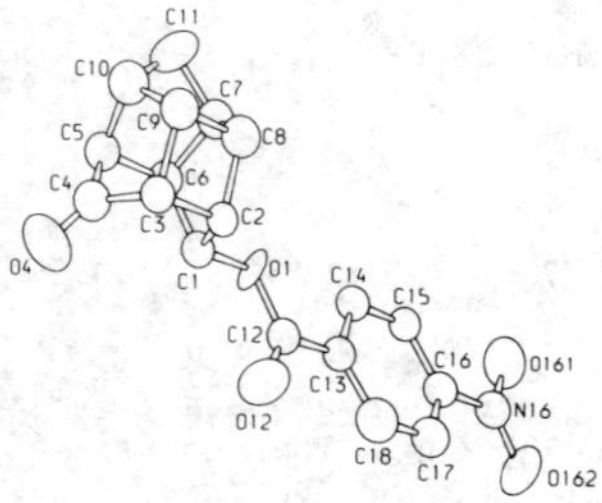

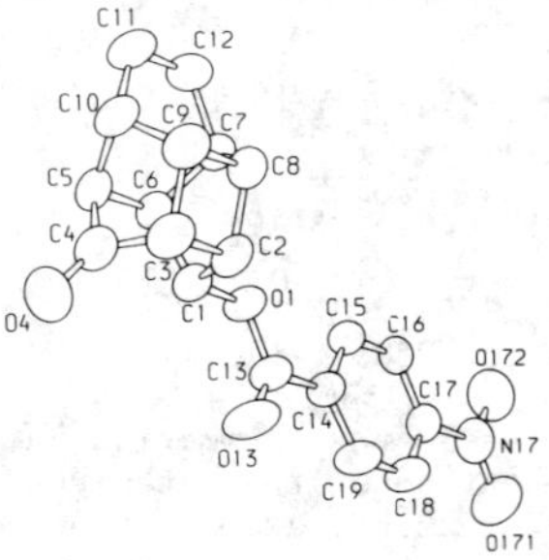

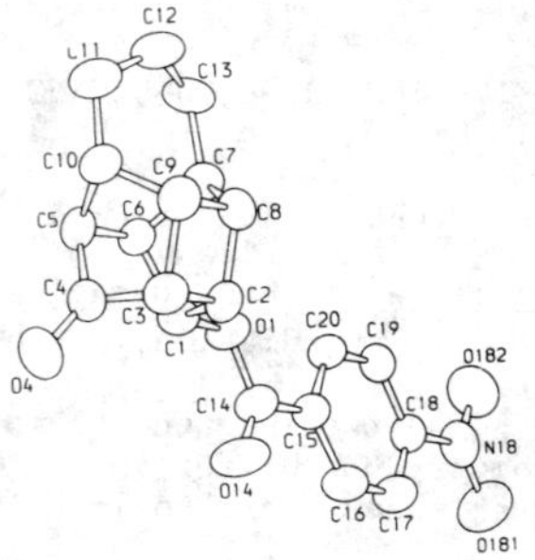

Fig. 1. The molecular structures: $C_{18}H_{15}NO_5$ (left), $C_{19}H_{17}NO_5$ (centre) and $C_{20}H_{19}NO_5$ (right).

2-endo-CHLORO-10-NITRO-3-exo-(p-TOLYLTHIO)BENZONORBORNENE
$C_{18}H_{16}ClNO_2S$

W. ADAM, N. CARBALLEIRA, D. SCHEUTZOW, K. PETERS, E.-M. PETERS and H.G. VON SCHNERING, 1984. Chem. Ber., 117, 1139-1152.

Monoclinic, $P2_1/a$, a = 20.540, b = 11.023, c = 7.521 Å, β = 90.09°, Z = 4. Mo radiation, R = 0.050 for 3244 reflexions.

The analysis established the structure and stereochemistry shown in Fig. 1. The site of nucleophilic attack by chloride ion on the precursor is established.

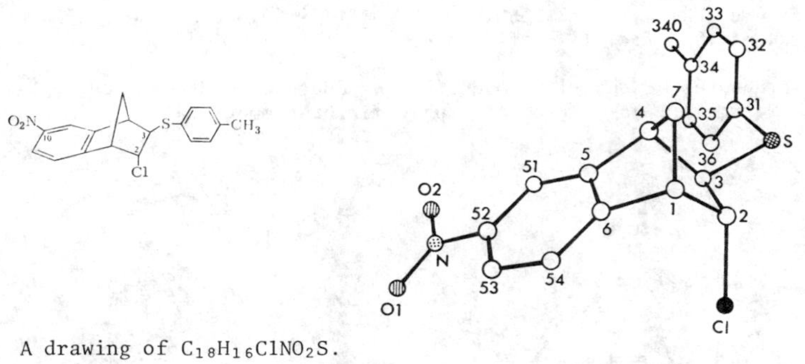

Fig. 1. A drawing of $C_{18}H_{16}ClNO_2S$.

1,4-DIBROMO-6-METHYL-5-NORDIISOPHORA-2(7)3,5-TRIEN-3-OL
$C_{18}H_{24}Br_2O$

F. KURZER, A.R. MORGAN and S.J. RETTIG, 1984. Monatsh., 115, 333-348.

Monoclinic, $P2_1/n$, a = 16.874, b = 8.734, c = 11.736 Å, β = 96.42°, Z = 4. Mo radiation, R = 0.031 for 1598 reflexions.

The fully-substituted aromatic ring A is significantly non-planar; and the distortion results in a boat-like conformation in which C(6) and C(8a) are displaced from the mean plane in a direction opposite to that of the remaining four atoms. The six-membered ring B has a distorted C(3)-sofa conformation (Fig. 1) whilst the fully saturated six-membered ring C has a distorted chair conformation flattened at one side as a result of steric interaction between C(15) and C(8a) (H(15)...C(8a) is 2.59 Å). Bond distances and angles are normal with the exception of C(8)-C(14) which is 1.563 Å in length. This is long when compared to the normal value of 1.511 Å for a C(sp^3)-C(sp^2) single bond.

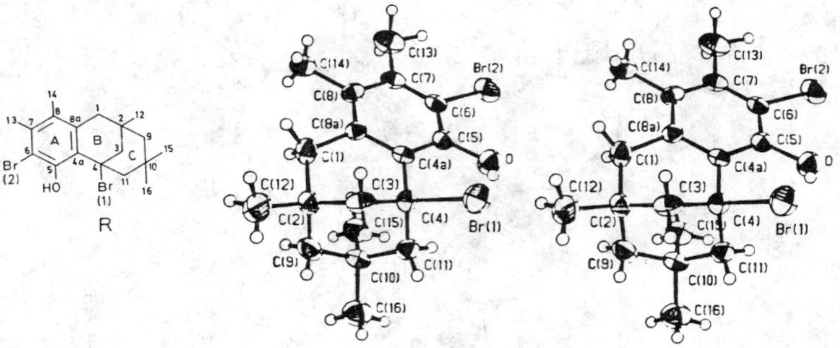

Fig. 1. Schematic drawing of $C_{18}H_{24}Br_2O$ and a stereoscopic view of the molecule.

(endo)-2,3-BIS(BENZENESULPHONYL)BICYCLO[2.2.1]HEPT-5-ENE
$C_{19}H_{18}O_4S_2$ (I)

8,9-BIS(BENZENESULPHONYL)TETRACYCLO[$4.3.0.0^{2,4}.0^{3,7}$]NONANE
$C_{21}H_{20}O_4S_2$

O. DE LUCCHI and G. VALLE, 1984. Z. Kristallogr., 167, 293-298.

I. Orthorhombic, $Ccm2_1$, a = 11.506, b = 16.759, c = 9.094 Å, Z = 4. Cu radiation, R = 0.039 for 397 reflexions.

II. Monoclinic, $P2_1/a$, a = 21.359, b = 8.064, c = 10.676 Å, β = 90.5°, Z = 4. Mo radiation, R = 0.054 for 2147 reflexions.

The structures of these cycloaddition products were determined by X-ray analysis (Fig. 1). Molecule I has crystallographic mirror symmetry.

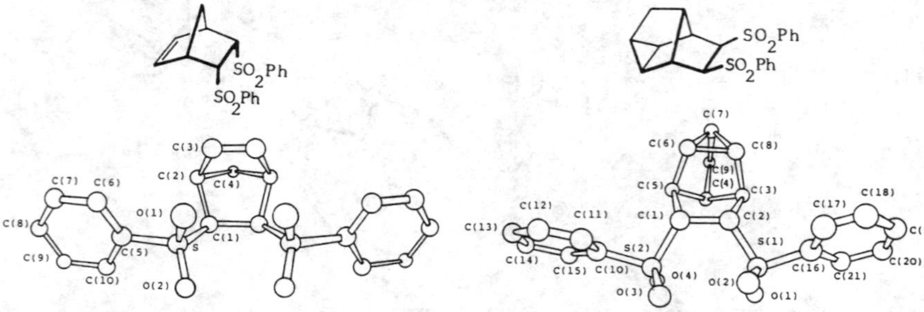

Fig. 1. Views of $C_{19}H_{18}O_4S_2$ (I) (left) and $C_{21}H_{20}O_4S_2$ (II) (right).

3,4,5,6-DIBENZO[6.2.2]PROPELLA-3,5,9,11-TETRAENE
$C_{20}H_{16}$

K. WEINGES, W. KASEL, U. HUBER-PATZ, H. RODEWALD and H. IRNGARTINGER, 1984. Chem. Ber., 117, 1868-1876.

Monoclinic, $P2_1/c$, a = 19.403, b = 10.301, c = 8.914 Å, β = 126.44°, Z = 4. Mo radiation, R = 0.036 for 2122 reflexions.

The structure and dimensions of this 1,4-bridged Dewar-benzene derivative are shown in Fig. 1. The angle between the C1,C8,C9,C10 and C1,C8,C11,C12 planes is 115.9° and both four-membered rings are individually planar.

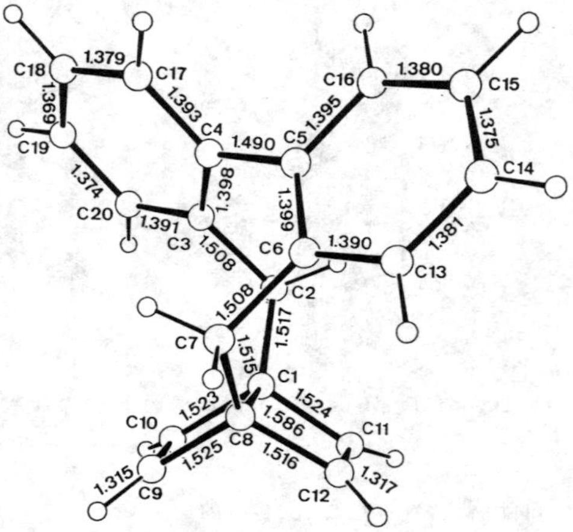

Fig. 1. $C_{20}H_{16}$ with bond lengths (σ 0.001-0.002 Å).

7,8-DIACETOXYBICYCLO[4.2.0]OCTA-2,4-DIENE 2,3-DIHYDRONAPHTHAZARIN ADDUCT
$C_{20}H_{20}O_8$

P. RUBIO, F. FLORENCIO, P. SMITH-VERDIER, S. GARCIA-BLANCO and J.G. RODRIGUEZ, 1984. Acta Cryst., C40, 450-452.

Monoclinic, $P2_1/c$, a = 7.744, b = 27.419, c = 9.449 Å, β = 113.16°, D_m = 1.50, Z = 4. Cu radiation, R = 0.052 for 2944 reflexions.

The analysis establishes the structure of the adduct as shown in Fig. 1. There is no charge-transfer interaction in the crystal lattice.

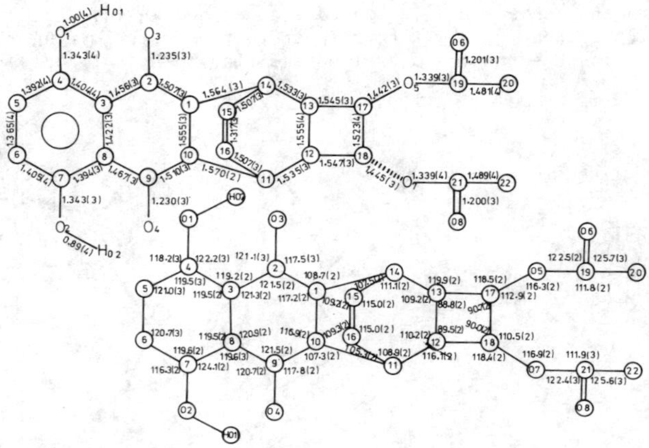

Fig. 1. Bond lengths and angles in $C_{20}H_{20}O_8$.

(9R*,2'S*)-2-CHLOROOXAPROTILINE HYDROCHLORIDE
$C_{20}H_{23}Cl_2NO$ $C_{20}H_{23}ClNO^+.Cl^-$

T. WINKLER, W. DIETERLE, P. HUG, H.-P. KRIEMLER and G. RIHS, 1984. Helv. Chim. Acta, 67, 2082-2086.

Monoclinic, $P2_1/c$, a = 7.810, b = 8.293, c = 29.221 Å, β = 98.38°, Z = 4. R = 0.064 for 3087 reflexions.

The analysis established the configuration of the cation as shown in Fig. 1.

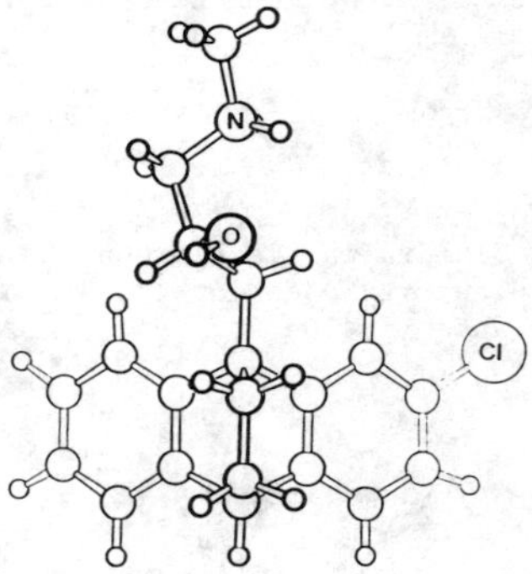

Fig. 1. A view of the $C_{20}H_{23}ClNO^+$ cation.

5-NITRO-8-(2-BICYCLO[2.2.1]HEPTYL)-1,2,3,4,4a,8b-HEXAHYDRO-1,4-METHANOBIPHENYLENE
$C_{20}H_{23}NO_2$

G. BOCELLI, M. CATELLANI, G.P. CHIUSOLI and S. LAROCCA, 1984. J. Organometal. Chem., 265, C9-C11.

Triclinic, $P\bar{1}$, a = 10.328, b = 9.028, c = 9.876 Å, α = 105.09, β = 97.32, γ = 107.72°, Z = 2. Cu radiation, R = 0.077 for 1368 reflexions.

The compound (Fig. 1) was prepared by reaction of p-nitrobromobenzene with two molecules of norbornene. The resulting structure has two bicycloheptyl rings, one linked directly to the aromatic ring and the other fused to a cyclobutene ring which in turn is fused to the benzene ring. Bond distances and angles are normal. The nitro group is slightly bent (7.3°) with respect to the aromatic ring which lies almost in the same plane as the fused cyclobutene ring (dihedral angle 2.9°).

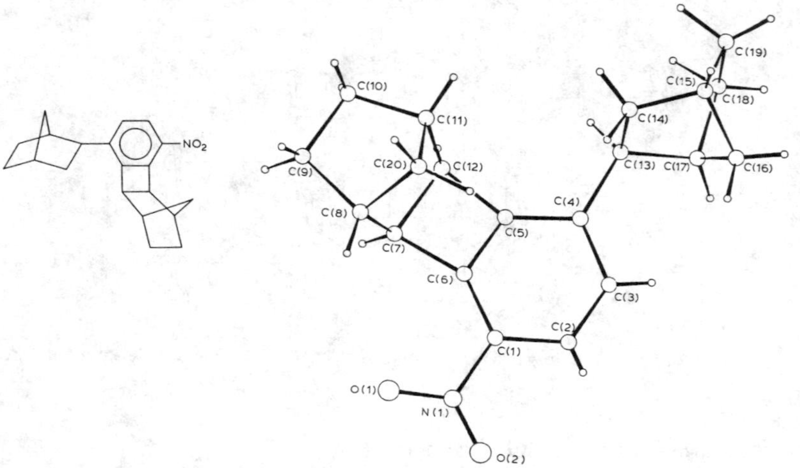

Fig. 1. Molecular structure of $C_{20}H_{23}NO_2$.

11,11-DIMETHYL-9,10-BIS(METHOXYCARBONYL)-12-SPIROCYCLOPROPYLTETRACYCLO[$6.2.1.1^{3,6}.0^{2,7}$]-DODECA-2(7),9-DIENE
$C_{20}H_{24}O_4$

L.A. PAQUETTE, K.E. GREEN, R. GLEITER, W. SCHAFER and J.C. GALLUCCI, 1984. J. Am. Chem. Soc., 106, 8232-8240.

Orthorhombic, $P2_12_12_1$, a = 9.608, b = 9.437, c = 19.157 Å, Z = 4. Mo radiation, R = 0.087 for 1554 reflexions (at -127°C).

The most notable feature of this molecule (Fig. 1) is the non-planarity of the double bond C4=C5 system. The dihedral angle between the C3,C4,C5,C6 and C10,C4,C5,C13 planes is 21.8(4)°. This non-planarity does not appear to be the result of steric interactions (the C8...C16 distance is 4.68 Å).

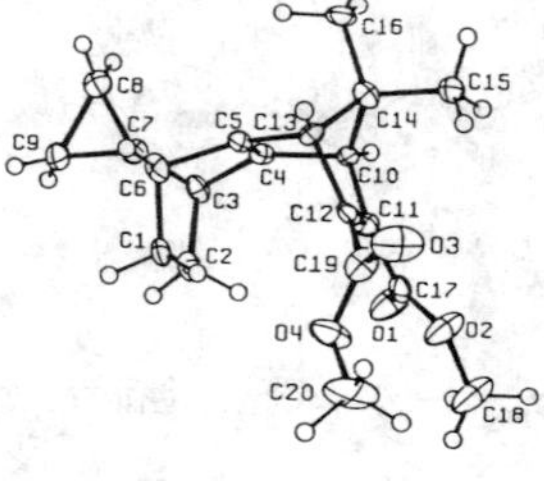

Fig. 1. A view of $C_{20}H_{24}O_4$.

(1aSR,3aRS,6aSR,6bRS,9aSR,9bRS,9cSR)-2,3,5,6,6b,7,8,9-OCTAHYDRO-6b-METHYL-1aH,4H-3a,6a:9a,9c-DIETHANOCYCLOPENT[1'2':1,2]ACENAPHTHO[2a,3-b]OXIRENE
$C_{20}H_{28}O$ (I)

(3aRS,6aSR,6bSR,9aRS,9bSR,9cSR)-3,5,6,6b,7,8,9b-OCTAHYDRO-6b-METHYL-4H-3a,6a:9a,9c-DIETHANOCYCLOPENT[a]ACENAPHTHYLEN-1(2H)-ONE
$C_{20}H_{28}O$ (II)

(1'RS,4'RS,6'SR,11'SR)-6'-(3,5-DINITROBENZOYLOXY)-1'-METHYLDISPIRO(CYCLOBUTAN-1,2'-[4,11]PROPANOTRICYCLO[$5.3.1.0^{4,11}$]UNDEC-7-ENE-3',1''-CYCLOPROPANE)
$C_{27}H_{30}N_2O_6$ (III)

L. FITJER, W. KUHN, U. KLAGES, E. EGERT, W. CLEGG, N. SCHORMANN and G.M. SHELDRICK, 1984. Chem. Ber., 117, 3075-3092.

I. Monoclinic, $P2_1/n$, a = 6.994, b = 16.811, c = 13.429 Å, β = 104.79°, Z = 4. Mo radiation, R = 0.050 for 1684 reflexions.

II. Orthorhombic, Pbca, a = 7.343, b = 16.679, c = 26.020 Å, Z = 8. Mo radiation, R = 0.079 for 1212 reflexions.

III. Monoclinic, $P2_1/n$, a = 14.305, b = 7.614, c = 22.938 Å, β = 111.88°, Z = 4. Mo radiation, R = 0.061 for 2777 reflexions.

The structures of these bridged ring species are shown in Fig. 1. Strain in (I) and (II) is shown by some long bond lengths (C6a-C6b 1.593, 1.570, C6b-C9a 1.592, 1.584 Å). In (III), two of the five-membered rings have a twist conformation and the one bonded to the dinitrobenzoate has four atoms in a plane and C(5') 0.42 Å from that plane.

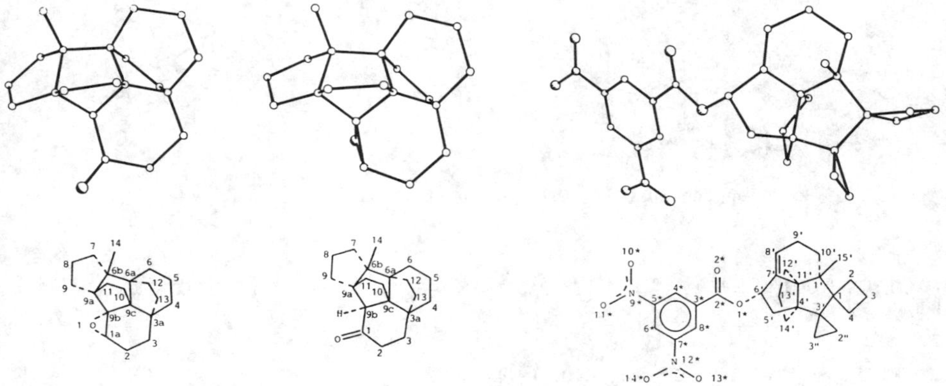

Fig. 1. Views of $C_{20}H_{28}O$ (I) (left), (II) centre and $C_{27}H_{30}N_2O_6$ (III) (right).

1-HYDROXY-5,11,15-TRIMETHYLPENTACYCLO[$9.5.1.0^{2,9}.0^{4,8}.0^{12,16}$]HEPTADEC-2(9)-EN-3-ONE
$C_{20}H_{28}O_2$

M. SORIANO-GARCIA, B. ORTIZ and R.A. TOSCANO, 1984. Acta Cryst., C40, 479-481.

Orthorhombic, $P2_12_12_1$, a = 10.677, b = 7.205, c = 22.749 Å, Z = 4. Cu radiation, R = 0.050 for 1095 reflexions.

The molecular structure (Fig. 1) determined from the X-ray data confirms the structure previously assigned on the basis of chemical and spectroscopic evidence. The molecule is stabilized by the formation of an intramolecular hydrogen bond between the hydroxyl and carbonyl groups [H(01)...O(2) 2.047(4) Å, O(1)-H(01)...O(2) 152.8(2)°]. Three five-membered rings have envelope conformations; ring B is planar.

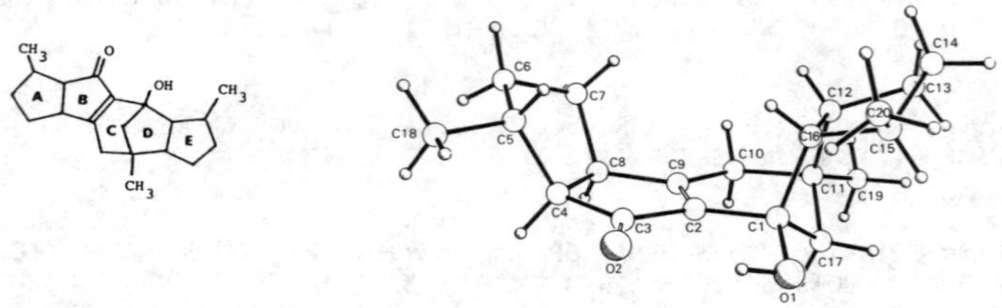

Fig. 1. The $C_{20}H_{28}O_2$ molecule.

(1R*,1'R*,2S*,2'S*,3R*,3'R*,5R*,5'R*,6S*,6'S*)-3-HYDROXY-3'-TRIMETHYLSILYLOXY[3,3'-BITRICYCLO[4.2.1.$0^{2,5}$]NON-7-ENYL]-4,4-DIONE
$C_{21}H_{26}O_4Si$

J.W. BATS, O. BELLINGER and W. RIED, 1984. Acta Cryst., C40, 184-186.

Triclinic, P$\bar{1}$, a = 9.041, b = 10.505, c = 11.546 Å, α = 76.59, β = 76.09, γ = 67.54°, Z = 2. Mo radiation, R = 0.045 for 3124 reflexions.

The structure (Fig. 1) consists of two endo-tricyclo[4.2.1.$0^{2,5}$]nona-3,7-diene groups linked by a single C-C bond of length 1.557(2) Å. The norbornene moieties have a bridging C-C-C angle of 93.6(2)°. The molecules form dimers by O-H...O hydrogen bonds.

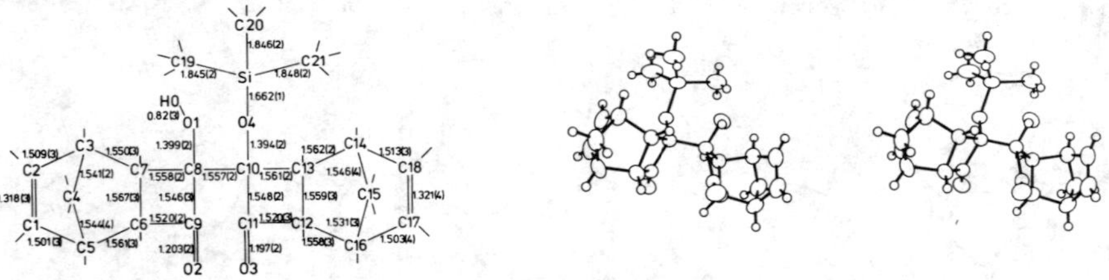

Fig. 1. Bond lengths in $C_{21}H_{26}O_4Si$ and a stereoview.

4-OXATETRACYCLO[9.2.1.$0^{2,10}$.$0^{3,8}$]TETRADEC-3(8)-ENE-5-SPIRO-4'-TRICYCLO[5.2.1.$0^{2',6'}$]-DECAN-3'-ONE
$C_{22}H_{28}O_2$

G. BOCELLI, 1984. Acta Cryst., C40, 1944-1945.

Monoclinic, $P2_1/c$, a = 13.410, b = 10.840, c = 12.444 Å, β = 98.42°, Z = 4. Cu radiation, R = 0.060 for 1822 reflexions.

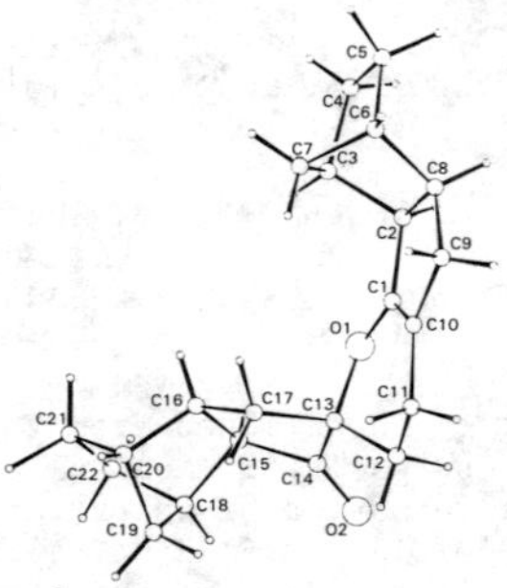

Fig. 1. The $C_{22}H_{28}O_2$ molecule.

The six-membered ring has a half-chair conformation (Fig. 1) with C(12) 0.66(5) Å out of the plane of the remaining atoms; the ring is rotated 72.2(2)° with respect to the spiro ring plane. The five-membered rings have envelope conformations. Some bond lengths are C(1)-O(1) 1.381(4), C(13)-O(1) 1.459(4), C(1)-C(10) 1.329(5) Å.

9-ACETOXY-2,10-DI-t-BUTYL-TETRACYCLO[6.3.2.0^{3,7}.0^{8,11}]TRIDECA-2,4,6,10,12-PENTAENE
$C_{23}H_{28}O_2$

B. SZECHNER, M. REY, A.S. DREIDING and R. GRIEB, 1984. Helv. Chim. Acta, 67, 1386-1396.

Monoclinic, $P2_1/b$, a = 11.85, b = 28.45, c = 6.10 Å, γ = 75.6°, Z = 4. R = 0.06 for 817 reflexions.

Molecular dimensions are in Fig. 1. There are three planes in the molecule: plane I (with 9 atoms, C1-C8 and C14), plane II (with 4 atoms, C1,C8,C12,C13) and plane III (with 6 atoms, C8-C11 and C1 and C18). The acute angle between planes I and II is 70°. Atom C1 is 0.38 Å from plane III.

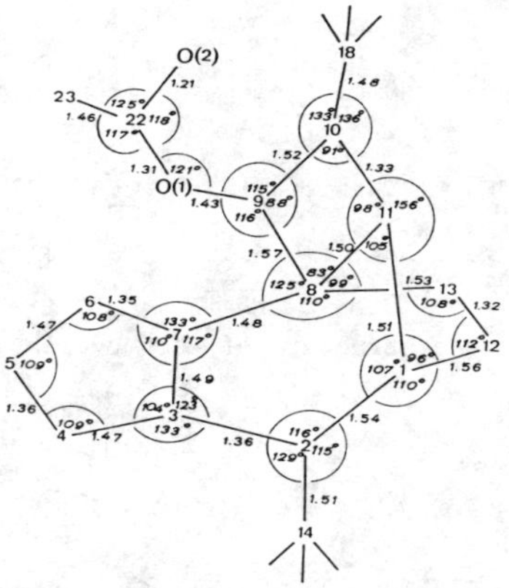

Fig. 1. Bond lengths and angles in $C_{23}H_{28}O_2$.

(+)-(7S,11S)-11-HYDROXY-10-METHYLBICYCLO[5.3.1]DEC-1(10)-EN-7-YLMETHYL (+)-10-CAMPHORSULPHONATE
$C_{23}H_{36}O_5S$

L. LEBIODA, J.A. MARSHALL and K.E. FLYNN, 1984. Acta Cryst., C40, 1188-1190.

Tetragonal, $P4_1$, a = 9.833, c = 23.46 Å, Z = 4. Mo radiation, R = 0.065 for 1168 reflexions.

The molecule consists of the camphorsulphate moiety and the bicyclo[5.3.1]decene moiety which, except for the covalent link, do not interact (Fig. 1). The shortest non-bonded contact is 3.11(1) Å between O(1) and C(106). The cyclooctane ring is in a conformation corresponding to that of minimum energy with an approximate D_2 symmetry. The best planes through it and the condensed cyclohexene form an angle of 101.7(6)°. Bond distances and angles are normal, as are intermolecular contacts. The short distance of 2.86(1) Å between O(11) and O(2) of a neighbouring molecule suggests a hydrogen bond.

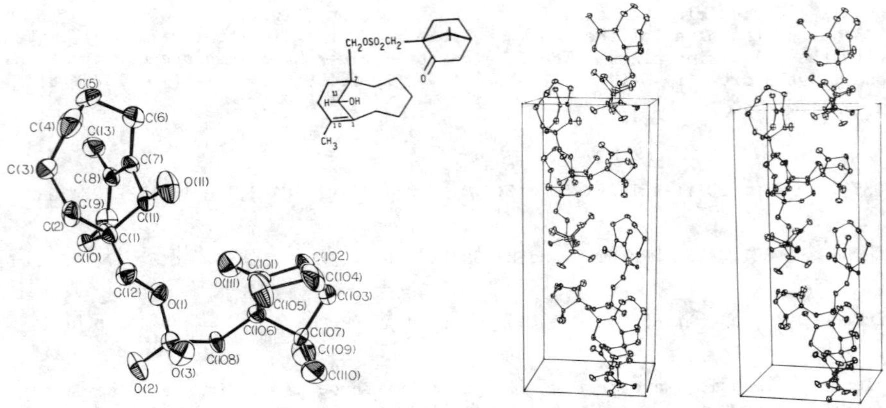

Fig. 1. $C_{23}H_{36}O_5S$: perspective view of the molecule and stereoview of the molecular packing.

2-OXO-5-exo-6-endo-BIS(PHENYLSULPHONYL)-7,8-BENZOBICYCLO[2.2.2]OCT-7-ENE
$C_{24}H_{20}O_5S_2$

O. DE LUCCHI and G. VALLE, 1984. Z. Kristallogr., 166, 307-310.

Triclinic, P$\bar{1}$, a = 12.747, b = 11.305, c = 8.163 Å, α = 110.4, β = 102.4, γ = 81.8°, Z = 2. Cu radiation, R = 0.053 for 1474 reflexions.

The analysis establishes the structure of this cycloadduct to be as shown in Fig. 1.

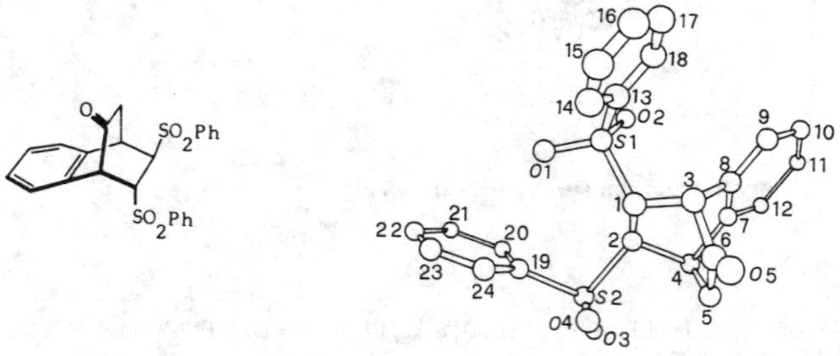

Fig. 1. A view of $C_{24}H_{20}O_5S_2$.

(1R,3R,4R)-(+)-3-BENZOYL-3-p-TOLYLAZO-2-BORNANONE
$C_{24}H_{26}N_2O_2$

A. DUBOURG, R. ROQUES, Ch. BERTRAND, C. METGE, F. LABRUYERE, J.-P. DECLERCQ and G. GERMAIN, 1984. Acta Cryst., C40, 274-277.

Orthorhombic, $P2_12_12_1$, a = 6.868, b = 11.247, c = 27.276 Å, Z = 4. Mo radiation, R = 0.089 for 950 reflexions.

The analysis establishes the configuration of C(3) in the molecule (Fig. 1).

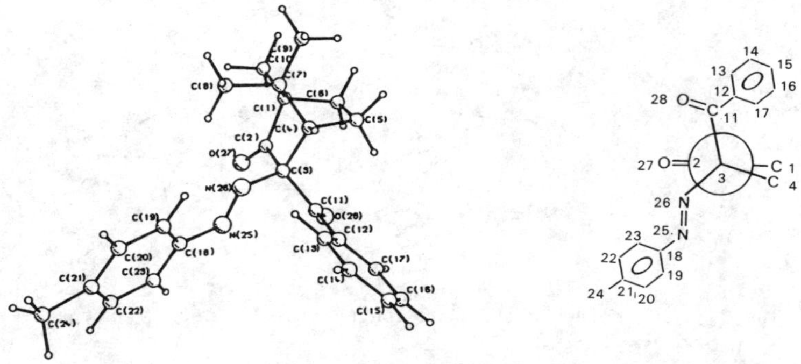

Fig. 1. A view of $C_{24}H_{26}N_2O_2$ and a Newman projection along C3-C2.

HEXACYCLO[7.5.1.$0^{3,13}$.$0^{5,12}$.$0^{7,11}$.$0^{10,14}$]PENTADECANE-2,6-DIONE-4-SPIRO-1'-4',5'-DIMETHYLCYCLOHEX-4'-ENE-2'-SPIRO-2''-CYCLOPENTANE-1'',3''-DIONE
$C_{26}H_{28}O_4$

P.E. EATON, W.H. BUNNELLE and P. ENGEL, 1984. Can. J. Chem., 62, 2612-2626.

Monoclinic, $P2_1/n$, a = 10.260, b = 11.583, c = 16.004 Å, β = 90.7°, Z = 4. Mo radiation, R = 0.048 for 1235 reflexions.

The analysis establishes the stereochemistry as shown in Fig. 1. The fused six-membered ring is in half-chair conformation and the cyclopentanedione unit is skewed to the (non-crystallographic) peristylane mirror plane.

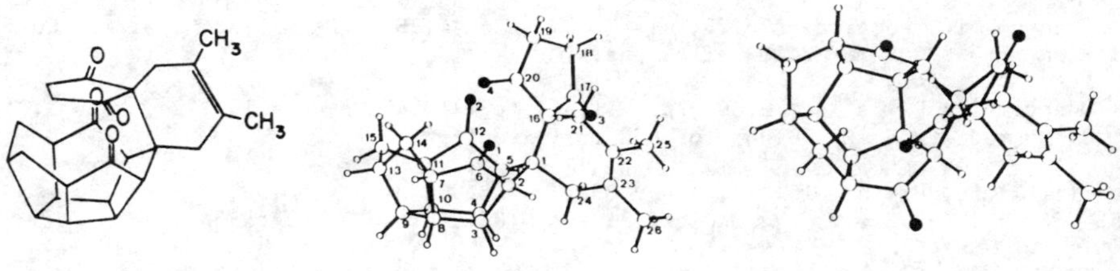

Fig. 1. The $C_{26}H_{28}O_4$ molecule.

14,15-BIS(2-METHOXYPHENYL)PENTACYCLO[9,2,1,$1^{4,7}$,$0^{2,10}$,$0^{3,8}$]PENTADECAN-9-ONE
$C_{29}H_{32}O_3$

A.P. MARCHAND, D.B. GOODWIN, M.B. HOSSAIN and D. VAN DER HELM, 1984. J. Org. Chem., 49, 2897-2900.

Monoclinic, $P2_1/c$, a = 9.466, b = 19.413, c = 13.095 Å, β = 106.82°, D_m = 1.238, Z = 4. Cu radiation, R = 0.053 for 4346 reflexions.

If the cyclopentanone ring is disregarded, the molecule (Fig. 1) has a pseudo centre of symmetry. The phenyl rings are nearly perpendicular to the central five-membered ring (interplanar angles 77 and 79°). The cyclopentanone ring is planar. The bond angles in the norbornyl moieties deviate significantly from the normal tetrahedral value because of the considerable strain in the molecule.

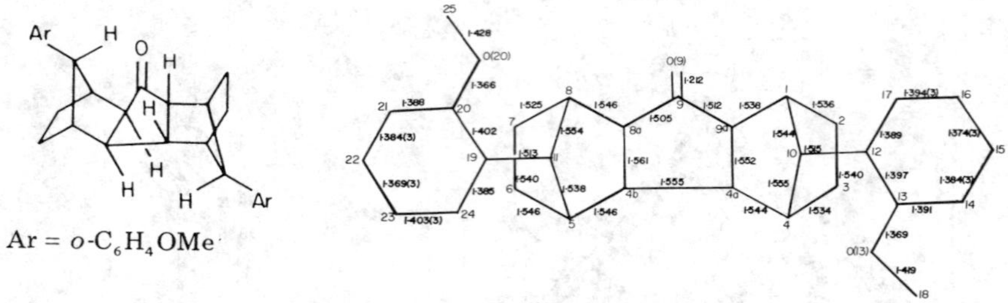

Fig. 1. The $C_{29}H_{32}O_3$ molecule and bond lengths (esd's 0.002 Å except where shown).

LEPIDOPTERENE (5,6,11,12-TETRAHYDRO-4b,12[1',2']:6,10b[1'',2'']DIBENZENOCHRYSENE)
$C_{30}H_{22}$ (I)

DIMETHYLLEPIDOPTERENE
$C_{32}H_{26}$ (II)

α-TETRACHLOROLEPIDOPTERENE
$C_{30}H_{18}Cl_4$ (III)

β-TETRACHLOROLEPIDOPTERENE
$C_{30}H_{18}Cl_4$ (IV)

H-D. BECKER, S.R. HALL, B.W. SKELTON and A.H. WHITE, 1984. Aust. J. Chem., 37, 1313-1327.

I. Orthorhombic, Pbcn, a = 36.75, b = 7.229, c = 44.85 Å, D_m = 1.28, Z = 24. Cu radiation, R = 0.065 for 2413 reflexions.

II. Monoclinic, $P2_1/c$, a = 9.607, b = 11.564, c = 10.576 Å, β = 107.72°, D_m = 1.21, Z = 2. Mo radiation, R = 0.041 for 632 reflexions.

III. Triclinic, $P\bar{1}$, a = 15.374, b = 9.858, c = 8.000 Å, α = 83.98, β = 83.73, γ = 71.53°, D_m = 1.49, Z = 2. Mo radiation, R = 0.037 for 1687 reflexions.

IV. Monoclinic, C2/c, a = 11.03, b = 13.67, c = 16.32 Å, β = 111.13°, D_m = 1.49, Z = 4. Mo radiation, R = 0.053 for 1005 reflexions.

A metastable phase(I) of lepidoterene whose gross molecular disposition is severely distorted, exhibits an absorption maximum at 580 nm rather than that of the stable form at 560 nm. II gives rise to emission at 540 nm. Dimerisation of 1,5-dichloro-9-anthrylmethyl yields metastable III, maxima at 590 nm, and IV, 575 nm. In all cases, the central C-C bond (C(8)-C(8)') is significantly longer (1.64-1.66 Å) than normal C-C bonds. Views of the molecules are in Fig. 1.

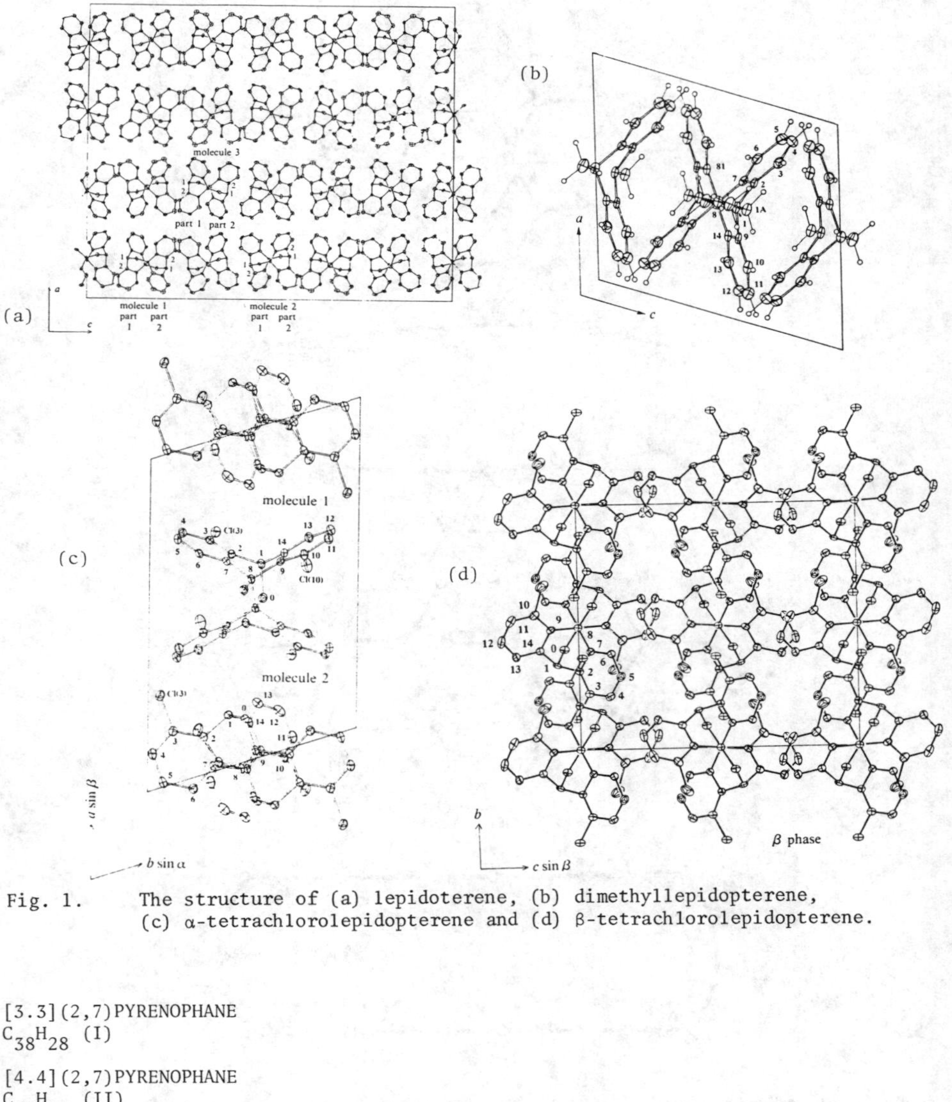

Fig. 1. The structure of (a) lepidoterene, (b) dimethyllepidopterene, (c) α-tetrachlorolepidopterene and (d) β-tetrachlorolepidopterene.

[3.3](2,7)PYRENOPHANE
$C_{38}H_{28}$ (I)

[4.4](2,7)PYRENOPHANE
$C_{40}H_{32}$ (II)

H.A. STAAB, N. RIEGLER, F. DIEDERICH, C. KRIEGER and D. SCHWEITZER, 1984. Chem. Ber., 117, 246-259.

I. Monoclinic, $P2_1/c$, a = 10.052, b = 10.137, c = 11.926 Å, β = 91.05°, Z = 2. Mo radiation, R = 0.043 for 1611 reflexions.

II. Monoclinic, $P2_1/c$, a = 10.468, b = 10.165, c = 12.552 Å, β = 96.29°, Z = 2. Mo radiation, R = 0.060 for 1462 reflexions.

Both molecules lie about crystallographic inversion centres (Fig. 1). In I there is considerable overlap of the two pyrene π-systems, and the pyrene units are bent. In II the pyrene units are almost ideally planar.

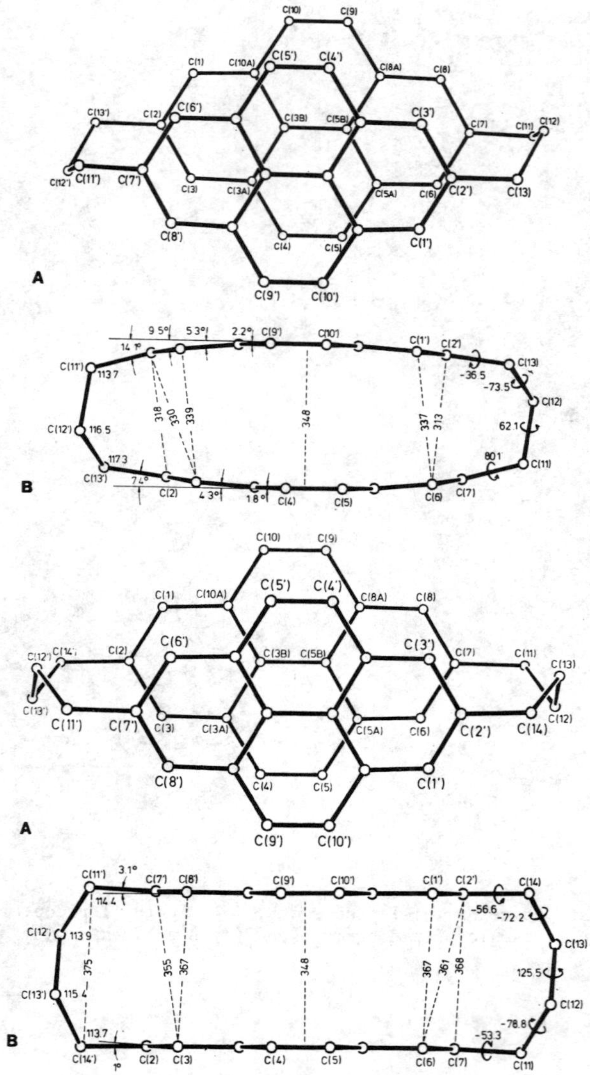

Fig. 1. Views and dimensions of $C_{38}H_{28}$ (I) (top) and $C_{40}H_{32}$ (II) (bottom).

BIS(9-TRIPTYCYL)ETHER BENZENE SOLVATE
$C_{40}H_{26}O,2(C_6H_6)$

H. IWAMURA, T. ITO, H. ITO, K. TORIUMI, Y. KAWADA, E. OSAWA, T. FUJIYOSHI and C. JAIME, 1984. J. Am. Chem. Soc., 106, 4712-4717.

Monoclinic, $P2_1/c$, a = 20.868, b = 10.405, c = 18.912 Å, β = 114.12°, D_m = 1.207, Z = 4. Cu radiation, R = 0.052 for 4494 reflexions.

The molecule (Fig. 1) adopts a beveled gear-shaped structure in which the ethereal C-O bond lengths are 1.407(2) and 1.418(3) Å with C-O-C 135.8(2)°.

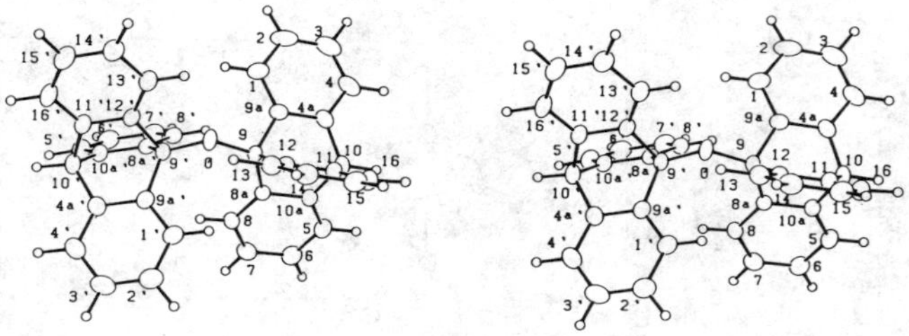

Fig. 1. A stereoview of $C_{40}H_{26}O$.

3-NITRO-1,2,4-TRIAZOLE

$C_2H_2N_4O_2$

G. EVRARD, F. DURANT, A. MICHEL, J.G. FRIPIAT, J.L. CLOSSET and A. COPIN, 1984. Bull. Soc. Chim. Belg., 93, 233-234.

Tetragonal, $P4_12_12$, a = 6.873, c = 19.220 Å, D_m = 1.65, Z = 8. Cu radiation, R = 0.043 for 598 reflexions.

The five-membered ring is planar and the angle between this plane and the nitro group is 2.94°. Molecular dimensions and formal charges are in Fig. 1.

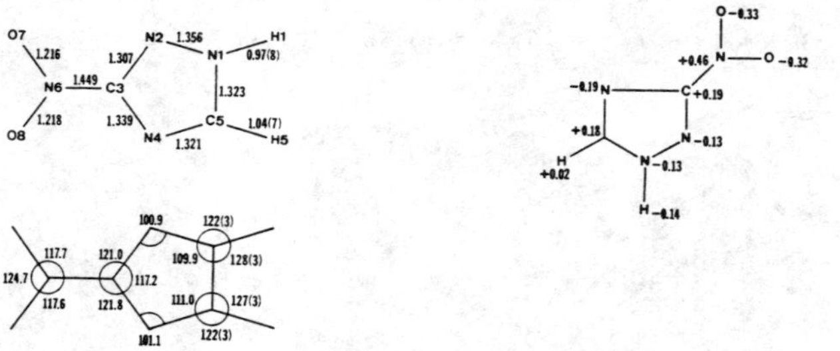

Fig. 1. Molecular dimensions and formal charges for $C_2H_2N_4O_2$.

AMMONIUM 5-CHLORODINITROMETHYLTETRAZOLE

$C_2H_4ClN_7O_4$ $(C_2ClN_6O_4)^-(NH_4)^+$

E.A. GOIKO, N.V. GRIGOR'EVA, G.V. MAKARENKO, N.V. MARGOLIS, A.A. MEL'NIKOV, T.K. STROCHKINA and I.V. TSELINSKII, 1984. J. Struct. Chem., 25, 336-338 [Zh. Strukt. Khim., 25(2), 180-182].

Monoclinic, $P2_1/c$, a = 5.75, b = 21.90, c = 7.22 Å, β = 109.83°, Z = 4. Cu radiation, R = 0.09 for 1011 reflexions.

The geometry of the anion (σ = 0.02 Å and 2°) is shown in Fig. 1. The tetrazole ring is planar to within 0.008 Å.

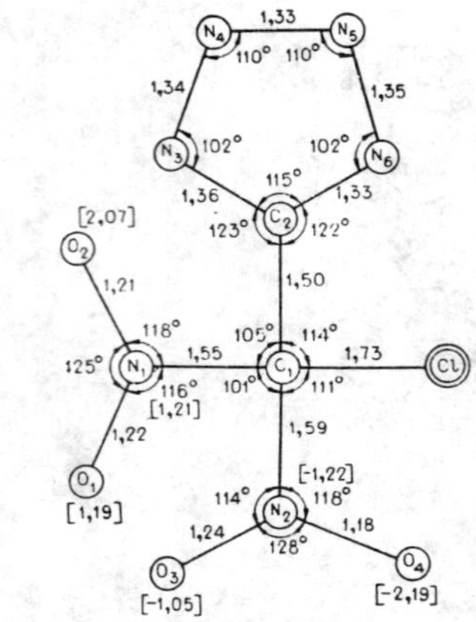

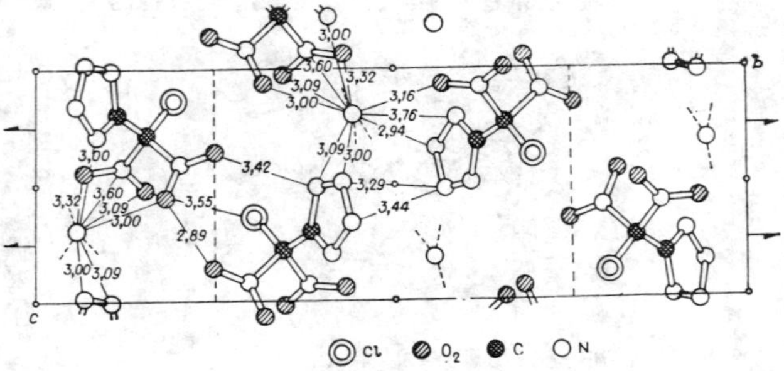

Fig. 1. The anion geometry (left, deviations from the tetrazole ring mean plane in square brackets) and a view of the structure of ammonium 5-chlorodinitromethyltetrazole perpendicular to the bc plane (right).

POTASSIUM 2-METHYL-5-DINITROMETHYLTETRAZOLE
$C_3H_3KN_6O_4$ $(C_3H_3N_6O_4)^-K^+$

E.A. GOIKO, N.V. GRIGOR'EVA, G.V. MAKARENKO, N.V. MARGOLIS, A.A. MEL'NIKOV, T.K. STROCHKINA and I.V. TSELINSKII, 1984. J. Struct. Chem., 25, 339-341 [Zh. Strukt. Khim., 25(2), 182-185].

Monoclinic, C2/c, a = 26.26, b = 8.16, c = 15.36 Å β = 96°, Z = 16. Cu radiation, R = 0.08 for 1348 reflexions.

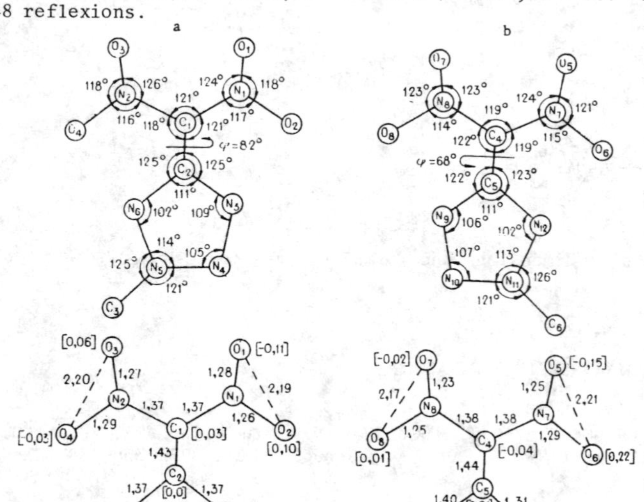

Fig. 1. Bond angles (top) and bond lengths and deviations from the ring mean planes (bottom) for the two independent 2-methyl-5-dinitromethyltetrazole anions.

The structure consists of six-coordinate potassium ions and $(C_3H_3N_6O_4)^-$ anions. The tetrazole rings in both of the two independent anions are planar to within experimental error and the nitro groups are twisted out of the tetrazole ring mean planes by 0-8°. Anion geometry (σ = 0.02 Å and 2°) is shown in Fig. 1.

AZOMYCIN (2-NITROIMIDAZOLE)
$C_3H_3N_3O_2$

I.K. LARSEN, 1984. Acta Cryst., C40, 285-287.

Monoclinic, $P2_1/c$, a = 7.319, b = 9.802, c = 6.754 Å, β = 115.70°, D_m = 1.646, Z = 4. Mo radiation, R = 0.040 for 1028 reflexions (at 105 K).

The molecule is planar (Fig. 1) and in the crystal molecules are linked by N-H...N hydrogen bonds (N...N 2.830(3) Å). Bond lengths are as expected.

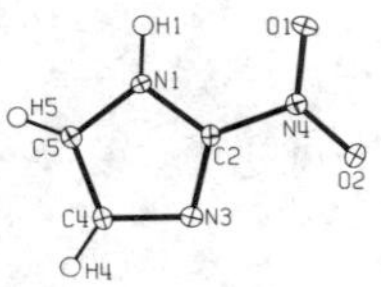

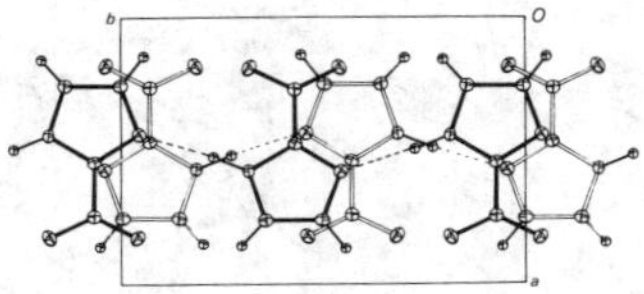

Fig. 1. The $C_3H_3N_3O_2$ molecule and a view of the crystal packing.

IMIDAZOLINE-2-THIONE HEMIHYDRATE
$C_3H_4N_2S,0.5H_2O$

E.S. RAPER, A.R.W. JACKSON and D.J. GARDINER, 1984. Inorg. Chim. Acta, 84, L1-L4.

Monoclinic, C2/c, a = 13.21, b = 5.105, c = 15.298 Å, β = 105.61°, D_m = 1.45, Z = 8. Mo radiation, R = 0.029 for 658 reflexions.

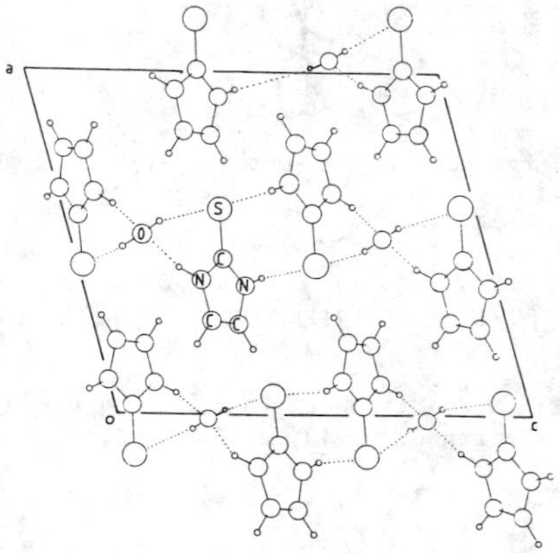

Fig. 1. Unit cell contents of $C_3H_4N_2S.0.5H_2O$ viewed down b.

The molecule is essentially planar with a maximum deviation of 0.011 Å for the S atom from the mean plane of the molecule (Fig. 1). Location of the H atoms attached to N(1) and N(3) confirms the thione character of the molecule. The C(2)-S(2) distance is 1.698 Å, the C(2)-N(1) and C(2)-N(3) distances 1.346 and 1.344 Å respectively, N(1)-C(5) 1.376 Å, N(3)-C(4) 1.377 Å, and C(4)-C(5) 1.339 Å. Thus canonical forms with a double bond at C(4)-C(5), single bonds at N(1)-C(5) and N(3)-C(4), double bonds at either C(2)-N(1) or C(2)-N(3) and a single bond at C(2)-S(2) would seem to make the major contribution. Two main hydrogen bonds occur, one involving the water molecule (N(1)-H(1)...O(1) with N...O 2.876 Å), and one the S atom (O(1)-H(6)...S(2') with O...S 3.208 Å).

N-METHYL-2-PYRROLIDONE-QUINOL
$C_5H_9NO,C_6H_6O_2$

P. BEHMEL and G. WEBER, 1984. J. Mol. Struct., 116, 271-277.

Monoclinic, $P2_1/c$, a = 7.826, b = 7.203, c = 19.522 Å, β = 98.71°, Z = 4. Mo radiation, R = 0.068 for 1240 reflexions.

In the crystal, the pyrrolidone is twofold disordered and is itself hydrogen bonded to chains of hydrogen-bonded quinols (Fig. 1) (O(1)-H...O(2) 2.690(3), O(2)-H...O(3) 2.575(4) Å.

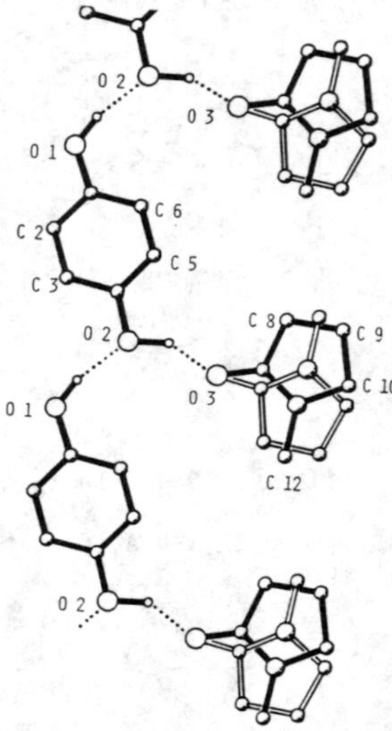

Fig. 1. A view of the 1:1 adduct of N-methyl-2-pyrrolidone with quinol.

(1S,2S,α-S)-1-α-CARBOXYETHYL-3,3-BIS(TRIFLUOROMETHYL)DIAZIRIDINE
$C_6H_6F_6N_2O_2$

G.V. SHUSTOV, A.B. ZOLOTOI, N.L. ZAICHENKO, O.A. DYACHENKO, L.O. ATOVMYAN and R.G. KOSTYANOVSKY, 1984. Tetrahedron, 40, 2151-2159.

Trigonal, $P3_1$, a = 10.023, c = 8.363 Å, Z = 3. Cu radiation, R = 0.040 for 632 reflexions.

The molecule (Fig. 1) has a conformation in which the H-atom, the smallest substituent at C4 is eclipsed by the diaziridine ring. Molecules are linked by O-H...O hydrogen bonds (O...O 2.698(6) Å) involving the carboxyl groups.

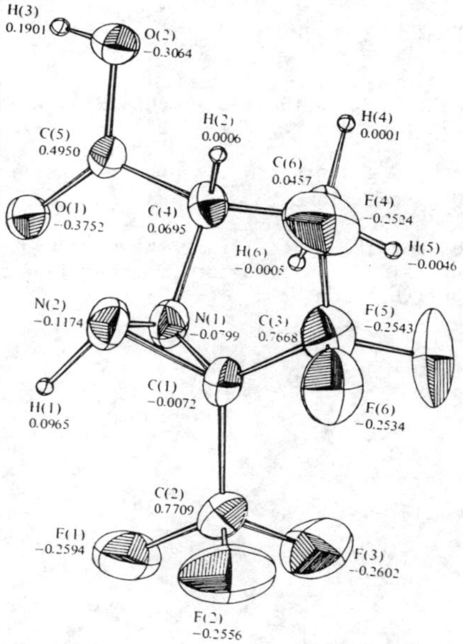

Fig. 1. A view of $C_6H_6F_6N_2O_2$; small numbers are atomic charges calculated by the INDO procedure.

(1-METHYL-4-NITRO-5-IMIDAZOLYL)-3-THIO-1H-1,2,4-TRIAZOLE
$C_6H_6N_6O_2S$

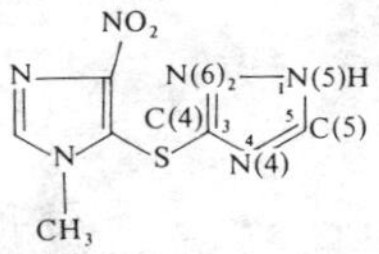

L. DUPONT, O. DIDEBERG and J.C. JAMOULLE, 1984. Acta Cryst., C40, 1269-1271.

Monoclinic, $P2_1/a$, a = 10.291, b = 14.485, c = 7.263 Å, β = 120.38°, Z = 4. Mo radiation, R = 0.071 for 1659 reflexions.

In the molecule (Fig. 1) the aromatic rings are planar and the S and methyl C are slightly out of the imidazole plane with the nitro group rotated by ~10° from this plane. Both the intermolecular hydrogen bond N-H...N between the imidazole and triazole rings and the difference-Fourier synthesis indicate that in crystals the 1H-1,2,4-triazole tautomer formula is more probable than the 2H one.

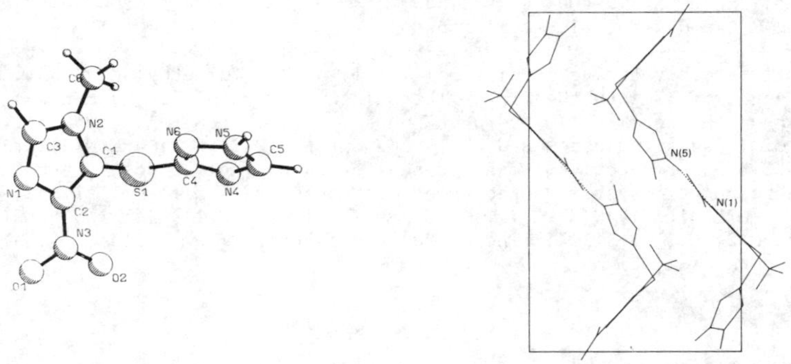

Fig. 1. $C_6H_6N_6O_2S$: view of the molecule and view along the c axis of the crystal structure.

Z,Z'-1-METHYL-2-(2-CARBAMOYLAZIRIDINO)GLYOXIME
$C_6H_{10}N_4O_3$

A.F. MISHNEV, M.F. BUNDULE, Ya. Ya. BLEIDELIS, V.G. ANDRIANOV and A.V. EREMEEV, 1984. J. Struct. Chem., 25, 769-772 [Zh. Strukt. Khim., 25(5), 117-121].

Monoclinic, $P2_1/c$, a = 13.050, b = 4.910, c = 14.099 Å, β = 111.2°, Z = 4. Cu radiation, R = 0.038 for 1075 reflexions.

Bond lengths and angles for the molecule are shown in Fig. 1. The configuration about the amidoxime fragment is Z. The glyoxime fragment is approximately planar and has an anti-s-trans-anti configuration.

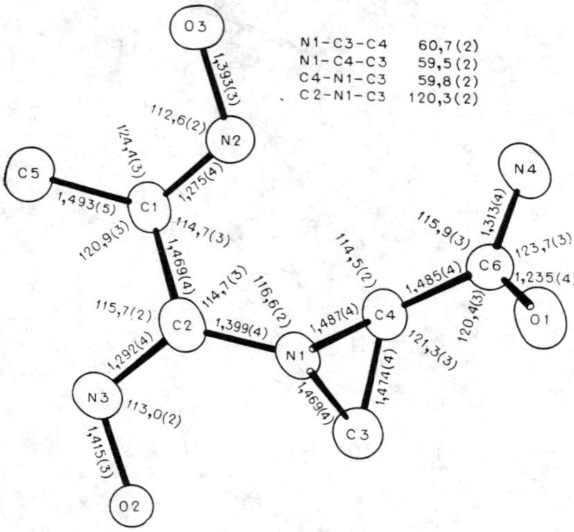

Fig. 1. Bond distances and angles for the Z,Z'-1-methyl-2-(2-carbamoylaziridino)-glyoxime molecule.

2,2-BIS(TRIFLUOROMETHYL)-4-DIMETHYLAMINO-1,3-DIAZOLONE
$C_7H_7F_6N_3O$ $(CH_3)_2NC_3N_2H(O)(CF_3)_2$

J.W. BATS, K. KELLER, A. THIEL and H.W. ROESKY, 1984. J. Fluor. Chem., 26, 313-319.

Monoclinic, $P2_1/n$, a = 9.469, b = 11.540, c = 10.205 Å, β = 106.02°, Z = 4. Ag radiation, R = 0.047 for 1168 reflexions.

The five-membered ring is planar (Fig. 1) and the dimethylamino group makes a dihedral angle of 3° with the ring. The molecules crystallize as centrosymmetric dimers linked through N-H...O hydrogen bonds (N(2)...O 2.817 Å). The N(1)-C(3) distance of 1.294 Å indicates a double bond but the N(2)-C(2) and N(3)-C(3) bonds, at 1.341 and 1.329 Å respectively, whilst shorter than single bonds, are not. They are part of a resonant π-system. Other bond distances are: N(1)-C(1) 1.438, N(2)-C(1) 1.436, N(3)-C(6) 1.431, N(3)-C(7) 1.443, O-C(2) 1.223, C(1)-C(4) 1.530, C(1)-C(5) 1.534, and C(2)-C(3) 1.513 Å. The C-F distances range from 1.313 to 1.337 Å with an average length of 1.321 Å.

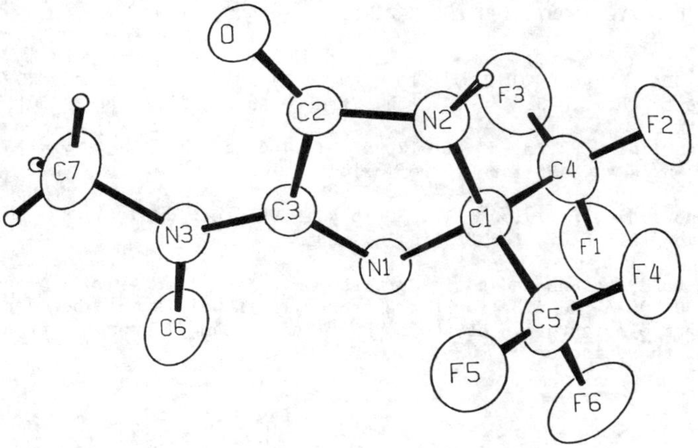

Fig. 1. Perspective drawing of the molecular structure of $C_7H_7F_6N_3O$.

3-AMINO-4-ACETYL-5-METHYLENE-Δ^3-PYRROLIN-2-ONE
$C_7H_8N_2O_2$ (I)

2-IMINO-3-AMINO-4-ACETYL-5-METHYLENE-2,5-DIHYDROFURAN
$C_7H_8N_2O_2$ (II)

M. BASATO, B. CORAIN, A.C. VERONESE, F. D'ANGELI, G. VALLE and G. ZANOTTI, 1984.
J. Org. Chem., 49, 4696-4700.

I. Triclinic, $P\bar{1}$, a = 7.898, b = 7.591, c = 7.393 Å, α = 116.9, β = 113.9, γ = 83.4°, Z = 2. Mo radiation, R = 0.060 for 970 reflexions.

II. Triclinic, $P\bar{1}$, a = 7.811, b = 8.028, c = 6.999 Å, α = 117.5, β = 81.2, γ = 111.3°, Z = 2. Mo radiation, R = 0.043 for 1157 reflexions.

The geometries of the two molecules are very similar and both are individually planar (Fig. 1). Some differences in bond lengths are C1-C2 1.476(5) in (I), 1,498(6) Å in (II), C3-C4 1.451(5) in (I), 1.473(6) Å in (II). In both structures, molecules are linked by hydrogen bonds and there are also intramolecular hydrogen bonds (N2...O2, N2...O1 in I, N2...O2, N2...N1 in II).

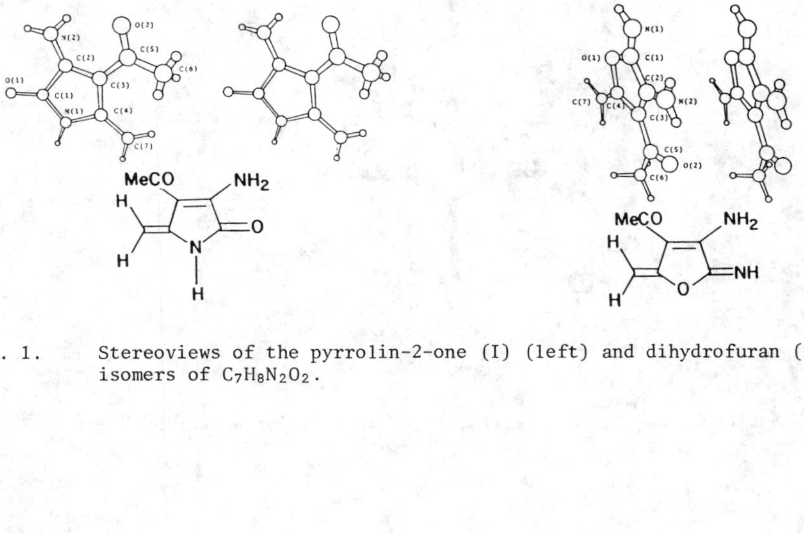

Fig. 1. Stereoviews of the pyrrolin-2-one (I) (left) and dihydrofuran (II) (right) isomers of $C_7H_8N_2O_2$.

2-CARBAMOYL-2-ETHOXYCARBONYL-1-METHOXYAZIRIDINE (α AND β ISOMERS)
$C_7H_{12}N_2O_4$

A.B. ZOLOTOI, O.A. D'YACHENKO, L.O. ATOVMYAN, V.F. RUDCHENKO and R.G. KOSTYANOVSKII, 1984. J. Struct. Chem., 25, 140-141 [Zh. Strukt. Khim., 25(1), 159-160].

α isomer: Monoclinic, $P2_1/b$, a = 6.017, b = 13.466, c = 12.460 Å, γ = 73.50°, Z = 4. Cu radiation, R = 0.053 for 925 reflexions.

β isomer: Monoclinic, $P2_1/a$, a = 12.530, b = 15.941, c = 4.904 Å, β = 103.90°, Z = 4. Cu radiation, R = 0.070 for 855 reflexions.

Selected bond lengths and angles are shown in Fig. 1 (σ = 0.003 Å and 0.2° for the α isomer, 0.005 Å and 0.3° for the β isomer). The difference between the isomers is in the relative orientation of the methoxy and carbamoyl groups: trans in the α isomer, cis in the β isomer.

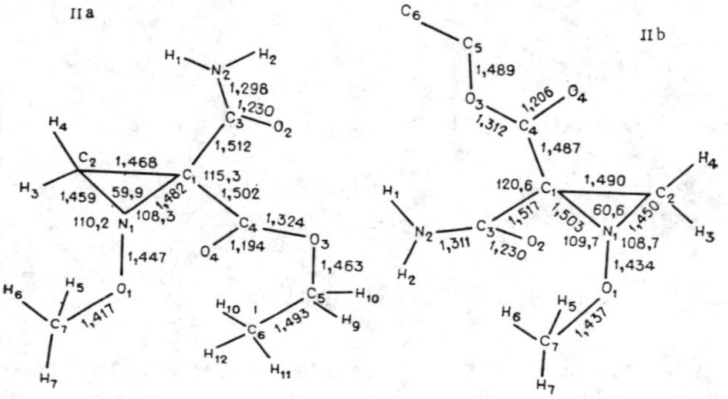

Fig. 1. The 2-carbamoyl-2-ethoxycarbonyl-1-methoxyaziridine molecules (α isomer, left, and β isomer, right).

1,1,3,3-TETRAMETHYLIMIDAZOLIDINIUM DIIODIDE METHYLENE DICHLORIDE SOLVATE
$C_7H_{18}I_2N_2,CH_2Cl_2$

M.A. KHAN, C. PEPPE and D.G. TUCK, 1984. Can. J. Chem., 62, 1662-1665.

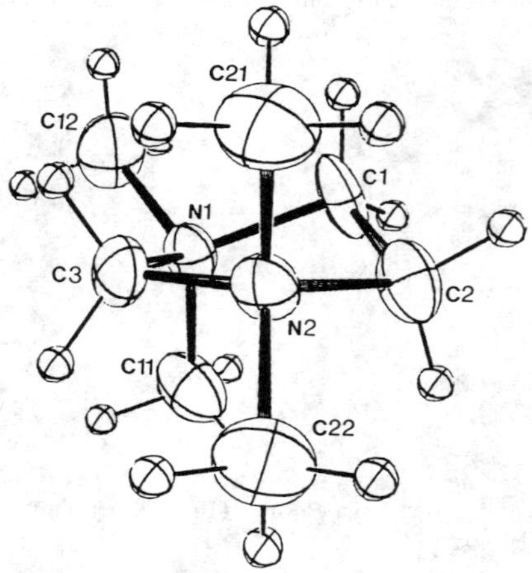

Fig. 1. A view of the 1,1,3,3-tetramethyl-imidazolidinium cation.

Monoclinic, $P2_1/n$, a = 8.109, b = 16.183, c = 11.968 Å, β = 93.45°, Z = 4. Mo radiation, R = 0.0415 for 1261 reflexions.

The structure consists of five-membered cyclic $C_7H_{18}N_2^{2+}$ cations, iodide anions, and methylene dichloride molecules which are held loosely in the lattice. The five-membered ring (Fig. 1) has an envelope conformation with C(1) 0.61(9) Å from the plane of N(1)N(2)C(2)C(3). The N-C distances are all in the 1.50-1.53(1) Å range; the C(1)-C(2) distance is 1.48(1) Å.

2-PHENYL-4-CHLORO-1,2,3-TRIAZOLE

$C_8H_6N_3Cl$ $C_6H_5N_3CHCCl$

I. SØTOFTE and K. NIELSEN, 1984. Acta Chem. Scand., A38, 497-499.

Monoclinic, $P2_1/n$, a = 5.742, b = 10.354, c = 13.680 Å, β = 94.17°, Z = 4. Mo radiation, R = 0.047 for 806 reflexions.

The phenyl and triazole groups are individually planar, deviations of the atoms from the least-squares planes through them being less than 0.004 Å. The Cl atom is 0.022 Å out of the triazole ring plane, and the angle between the planes of the phenyl and triazole rings is 1.6° (Fig. 1). Bond distances of note are: C(4)-Cl 1.703, C(4)-C(5) 1.400, C(5)-N(1) 1.331, N(1)-N(2) 1.342, N(2)-N(3) 1.329, N(3)-C(4) 1.315, N(2)-C(6) 1.432 Å, and those within the phenyl ring range from 1.372 to 1.390 Å. The deformations of the endocyclic angles in the phenyl group are symmetric around the C(6)-C(9) vector, with an angle of 122.2° at C(6) indicating a highly electronegative substituent bonded to C(6).

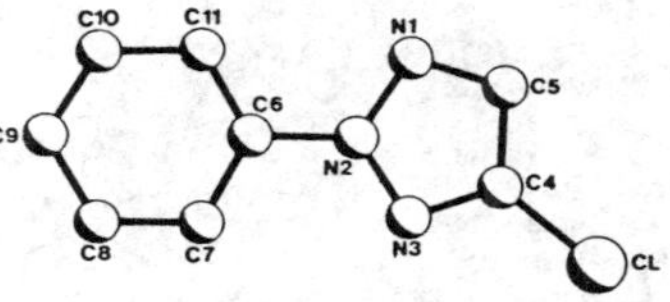

Fig. 1. The structure of $C_8H_6N_3Cl$ viewed perpendicular to the plane containing N(1), N(2) and N(3).

3-(2,6-DICHLOROBENZYLIDENEHYDRAZINO)-1,2,4-TRIAZOL-4-AMINE (NEBIDRAZINE)

$C_9H_8Cl_2N_6$

A. CARPY, M. SAUX and M. MONTAGUT, 1984. Acta Cryst., C40, 1265-1267.

Monoclinic, $P2_1/c$, a = 5.608, b = 8.486, c = 23.421 Å, β = 95.28°, Z = 4. Cu radiation, R = 0.034 for 952 reflexions.

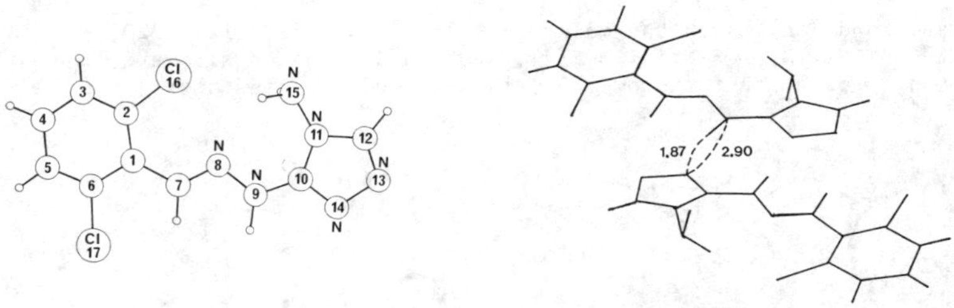

Fig. 1. $C_9H_8Cl_2N_6$: perspective view of the molecule and a view showing two molecules projected on (100) linked by a hydrogen bond.

In the molecule (Fig. 1) the phenyl and triazole rings are almost coplanar with each other. The shortest bond in the triazole ring is C(12)-N(13) 1.293(6) Å, cf. 1.265 Å for a pure C=N bond. The distance C(7)-N(8) 1.274(5) Å can be considered as typical of a pure double bond. The lengths of the four other C-N bonds are intermediate between those of a single and a double bond. The N-N bonds in the triazole ring are close to single bonds. Calculations indicate some π delocalisation in the bridge chain and in the triazole ring, with a strong tendency for double bonds to be localised at C(7)-N(8) and C(12)-N(13).

5-(3'-HYDROXYPYRIDYL-2'-THIO)-4-NITRO-1-METHYLIMIDAZOLE
$C_9H_8N_4O_3S$

S. PUIG-TORRES, G.E. MARTIN, S.B. LARSON and S.H. SIMONSEN, 1984. J. Heterocycl. Chem., 21, 155-159.

Monoclinic, $P2_1/c$, a = 7.573, b = 18.725, c = 8.189 Å, β = 113.34°, Z = 4. Mo radiation, R = 0.036 for 2565 reflexions (at 173 K).

The compound was obtained from a reaction which did not give the anticipated ring closure. The molecule and dimensions are in Fig. 1. It is suggested that the close approach of O(14) and S(1) (3.116 Å) plays a role in the orientation of the rings in this compound and that such a conformation could be a contributing factor in the failure of a precursor to undergo cyclization.

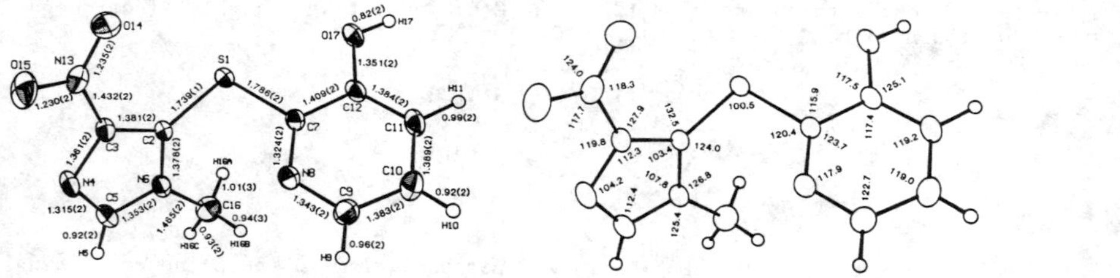

Fig. 1. Conformation and dimensions for $C_9H_8N_4O_3S$.

5-AMINO-4-PHENYL-3,4-DIHYDRO-2H-PYRROL-2-ONE
$C_{10}H_{10}N_2O$

P. GLUZINSKI, J. RAJEWSKI, J. LANGE, G.D. ANDREETTI and G. BOCELLI, 1984. Acta Cryst., C40, 108-110.

Monoclinic, $P2_1/a$, a = 12.770, b = 5.761, c = 12.001 Å, β = 97.94°, Z = 4. Cu radiation, R = 0.043 for 1085 reflexions.

Principal bond lengths in the molecule (Fig. 1) are C10-O12 1.220, N9-C10 1.374(4), N8-C9 1.324(3), C8-N13 1.302(4). Both hydrogens on N13 were located from a difference map, showing that tautomer(II) is obtained, not tautomer(I).

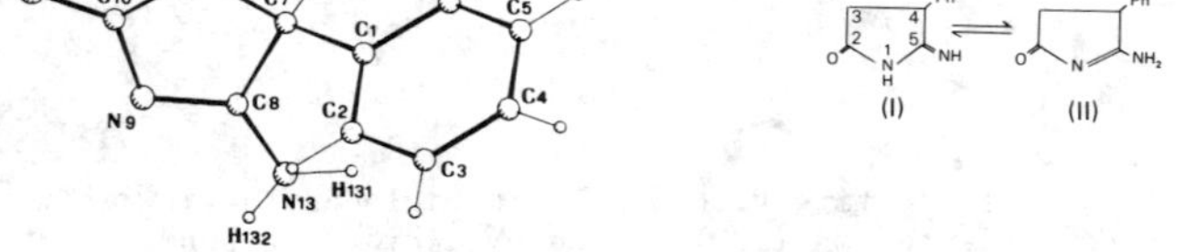

Fig. 1. The $C_{10}H_{10}N_2O$ molecule and the two possible tautomers.

1-(4',6'-DIMETHYLPYRIMIDIN-2'-YL)-3-METHYLPYRAZOL-5-OL
$C_{10}H_{12}N_4O$

P.J. STEEL and A.R. WHYTE, 1984. Aust. J. Chem., 37, 459-464.

Monoclinic, $P2_1/c$, a = 9.692, b = 7.371, c = 15.345 Å, β = 106.07°, Z = 4. Mo radiation, R = 0.072 for 574 reflexions.

The compound (Fig. 1) exists as the OH tautomer. Both the pyrazole and pyrimidine rings are planar. The short O(1)...N(1)' (2.60(1) Å) and H(10)...N(1)' (1.55 Å) indicate a strong intramolecular hydrogen bond, with the compound existing as the OH tautomer. Both the pyrazole and pyrimidine rings are planar.

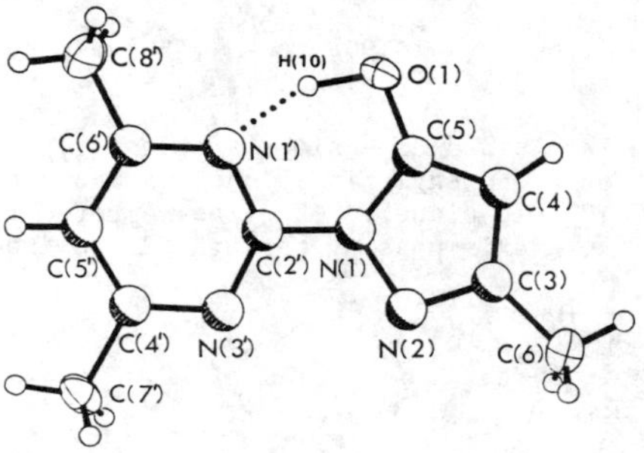

Fig. 1. The structure of $C_{10}H_{12}N_4O$.

1-[2-(2,6-DICHLOROPHENOXY)ETHYL]-3-METHYLTHIO-5-AMINO-1H-1,2,4-TRIAZOLE
$C_{11}H_{12}Cl_2N_4OS$ (I)

2-[2-(2,6-DICHLOROPHENOXY)ETHYL]-3-METHYLTHIO-5-AMINO-2H-1,2,4-TRIAZOLE
$C_{11}H_{12}Cl_2N_4OS$ (II)

A. KALMAN, L. PARKANYI and J. REITER, 1984. J. Mol. Struct., 118, 293-301.

I. Monoclinic, $P2_1/c$, a = 11.220, b = 10.005, c = 12.784 Å, β = 91.99°, Z = 4. Mo radiation, R = 0.060 for 1930 reflexions.

II. Orthorhombic, Pbca, a = 7.618, b = 14.289, c = 26.494 Å, Z = 8. Mo radiation, R = 0.061 for 1315 reflexions.

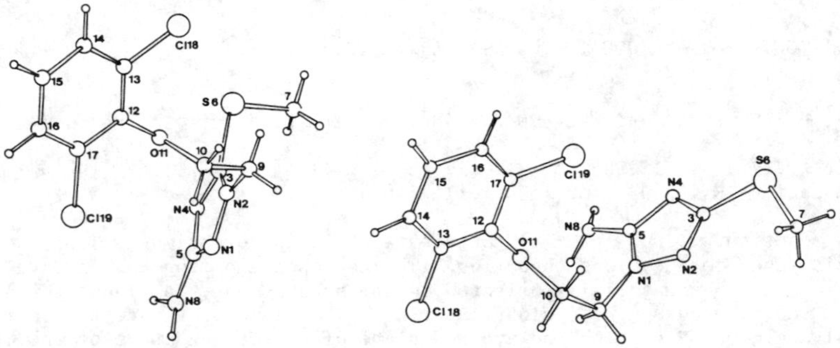

Fig. 1. Views of the isomeric $C_{11}H_{12}Cl_2N_4OS$ molecules (I) (right) and (II) (left).

Views of the molecules are in Fig. 1. There is a marked difference in the endocyclic bond angles at N(1) in I (C(5)-N(1)-C(9) 131.4, N(2)-N(1)-C(9) 119.6(5)°) and at N(2) in II (C(3)-N(2)-C(9) 130.0, N(1)-N(2)-C(9) 119.1(7)°). The S(6)-C(7) bond in (II) is twisted 79° out of the triazole ring plane. The N(1)-N(2) bond lengths are quite different in the two molecules (1.391(5) Å (I), 1.367(7) Å II).

N-(4-NITROPHENYL)-(L)-PROLINOL
$C_{11}H_{14}N_2O_3$

J. ZYSS, J.F. NICOUD and M. COQUILLAY, 1984. J. Chem. Phys., 81, 4160-4167.

Monoclinic, $P2_1$, a = 5.261, b = 14.908, c = 7.185 Å, β = 105.18°, Z = 2. Cu radiation, R = 0.050 for 1058 reflexions.

A view of the molecule and dimensions are in Fig. 1. The nitro group is rotated 5.8° about C1-N1 out of the phenyl ring plane. Amino nitrogen (N2) is pyramidal (N2 is 0.10 Å from the C4,C7,C10 plane). The five-membered ring has an approximate twofold symmetry (twofold axis passing through N2 and the C7-C8 midpoint).

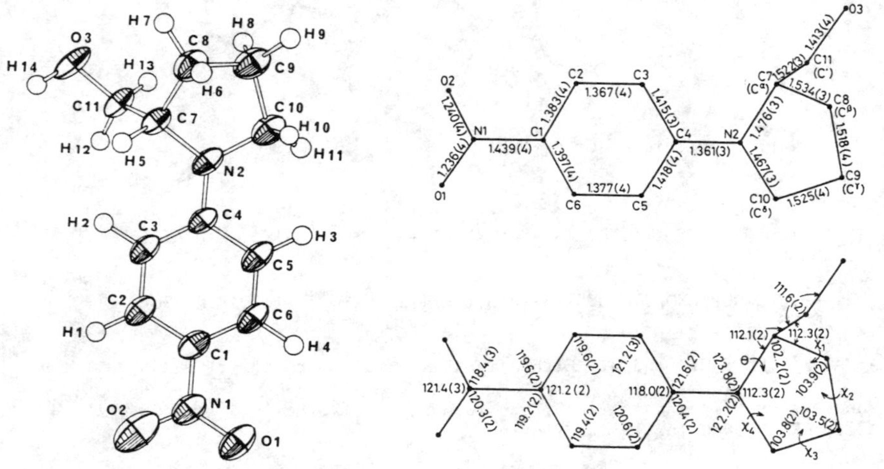

Fig. 1. The $C_{11}H_{14}N_2O_3$ molecule and dimensions.

5-METHYL-1-(OCTAHYDRO-1-IODE-1H-INDEN-4-YL)-1H-TETRAZOLE
$C_{11}H_{17}IN_4$

PUTTARAJA, D.S.S. GOWDA and W.L. DUAX, 1984. J. Cryst. Spectrosc. Res., 14, 13-20.

Monoclinic, $P2_1/a$, a = 11.586, b = 10.617, c = 10.417 Å, β = 94.567°, D_m = 1.70, Z = 4. Mo radiation, R = 0.031 for 1620 reflexions.

This reaction product (Fig. 1) contains a chair-conformation cyclohexane ring that is cis fused to a cyclopentane ring. The cyclopentane ring has an envelope conformation and an equatorially oriented iodine substituent. A planar tetrazole ring is axially substituted on the cyclohexane ring. The plane of the tetrazole ring makes a dihedral angle of 69.6° with the average plane of the carbon atoms of the cyclohexane ring.

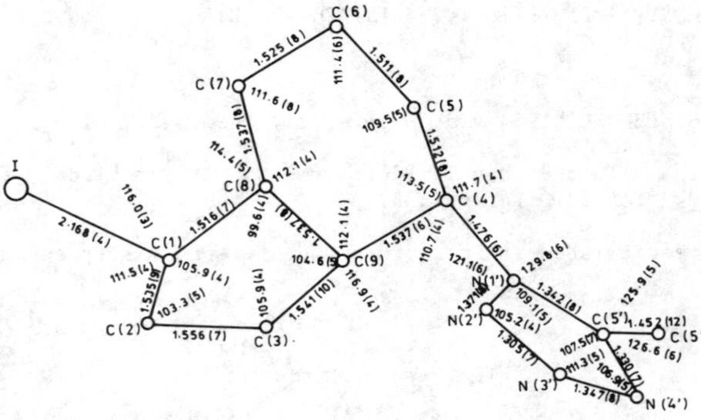

Fig. 1. Bond lengths and angles in $C_{11}H_{17}IN_4$.

3-ETHOXYCARBONYL-4-PHENYLPYRROLIDINE-2,5-DIONE
$C_{13}H_{13}NO_4$

H. ALPER, C.P. MAHATANTILA, F.W.B. EINSTEIN, A.C. WILLIS, 1984. J. Am. Chem. Soc., 106, 2708-2710.

Monoclinic, $P2_1/n$, a = 15.286, b = 5.377, c = 16.218 Å, β = 109.72°, Z = 4. R = 0.046.

The analysis establishes that the 3-ethoxy and 4-phenyl groups are in a trans relationship.

(1,3-DIMETHYL-2-IMIDAZOLIDINYLIDENE)PHENYLACETONITRILE
$C_{13}H_{15}N_3$

N. SEN, K. VENKATESAN, K.R. ACHARYA and T.N.G. ROW, 1984. Acta Cryst., C40, 2122-2124.

Orthorhombic, $P2_12_12_1$, a = 9.858, b = 10.951, c = 10.984 Å, D_m = 1.191, Z = 4. Mo radiation, R = 0.041 for 875 reflexions.

In the molecule (Fig. 1) the C=C bond length is 1.382(4) Å, significantly longer than that in ethylene, 1.336(2) Å. The molecule is twisted about the C=C bond by 23.1(4)° and the phenyl group is rotated out of the plane of this bond by 39.4(4)°. Molecular packing is determined by van der Waals forces.

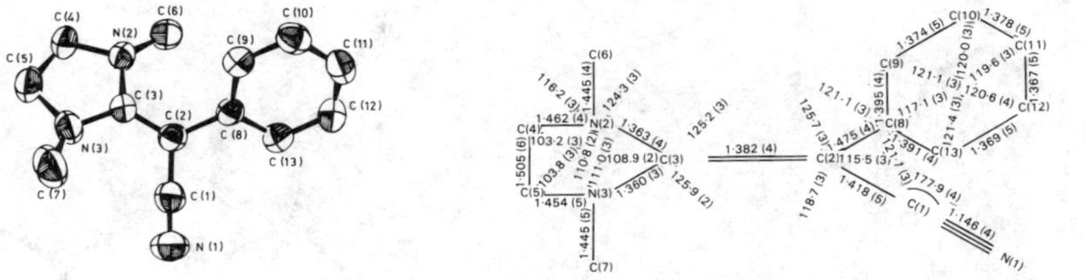

Fig. 1. $C_{13}H_{15}N_3$ and molecular dimensions.

METHYL 2-{[(3-METHOXY-1-METHYL-1H-1,2,4-TRIAZOL-5-YL)AMINOCARBONYL]AMINOSULPHONYL}-BENZOATE

$C_{13}H_{15}N_5O_6S$

T.P. SELBY and G.E. LEPONE, 1984. J. Heterocycl. Chem., 21, 61-64.

Monoclinic, $P2_1/c$, a = 13.690, b = 14.926, c = 7.955 Å, β = 91.85°, Z = 4. Mo radiation, R = 0.040 for 2279 reflexions.

The analysis established the structure and stereochemistry as shown in Fig. 1.

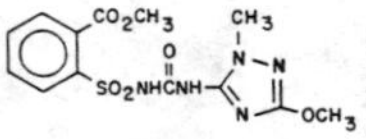

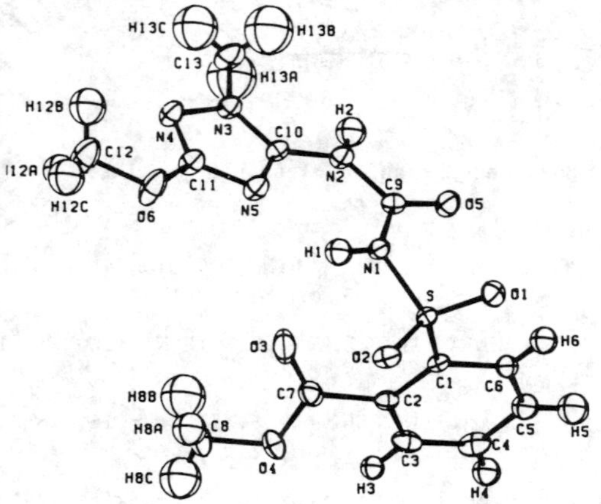

Fig. 1. A view of $C_{13}H_{15}N_5O_6S$.

1-BENZYL-2-(2-HYDROXYPROP-1-ENYL)-4,5-DIHYDROIMIDAZOLE

$C_{13}H_{16}N_2O$

M.W. ANDERSON, M.J. BEGLEY, R.C.F. JONES and J. SAUNDERS, 1984. J. Chem. Soc. Perkin I, 2599-2602.

Monoclinic, $P2_1/c$, a = 10.766, b = 5.575, c = 19.740 Å, β = 93.56°, Z = 4. Cu radiation, R = 0.068 for 513 reflexions.

The structure (Fig. 1) reveals the planar nature of the N(3)-C(2)-C(6)-C(7)-O system, the torsion angles N(3)-C(2)-C(6)-C(7) and C(2)-C(6)-C(7)-O(9) being -3.4 and 0.5° respectively. The C(2)-C(6) and C(6)-C(7) bond lengths are both 1.39(1) Å, intermediate between double and single bond lengths. The 1-benzyl aromatic ring is approximately perpendicular to the dihydroimidazole ring.

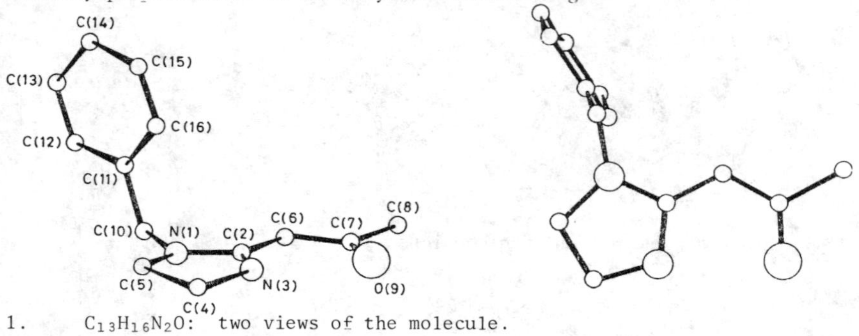

Fig. 1. $C_{13}H_{16}N_2O$: two views of the molecule.

5-AMINO-4-(2,6-DICHLOROPHENYL)-1-(2-NITROPHENYL)-1H-1,2,3-TRIAZOLE (Yellow form)
$C_{14}H_9Cl_2N_5O_2$ (I)

4-(2,6-DICHLOROPHENYL)-5-(2-NITROANILINO)-2H-1,2,3-TRIAZOLE (Red form)
$C_{14}H_9Cl_2N_5O_2$ (II)

N. SEN and K. VENKATESAN, 1984. Acta Cryst., C40, 1901-1905.

I. Monoclinic, $P2_1/n$, a = 9.525, b = 14.762, c = 11.268 Å, β = 107.82°, D_m = 1.539, Z = 4. Cu radiation, R = 0.088 for 2054 reflexions.

II. Triclinic, $P\bar{1}$, a = 9.796, b = 10.750, c = 7.421 Å, α = 95.29, β = 70.18, γ = 92.76°, D_m = 1.585, Z = 2. Cu radiation, R = 0.058 for 1866 reflexions.

The phenyl rings are inclined to the planar triazole ring at 81.7 and 86.1° (I) and at 71.7 and 23.8° (II). The nitro group/phenyl ring angles are 24.5 (I) and 1.1° (II) (Fig. 1).

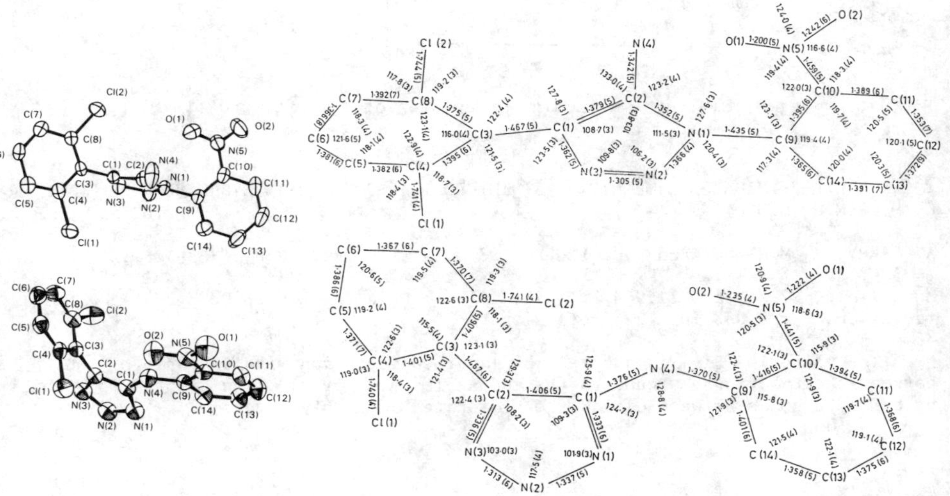

Fig. 1. The $C_{14}H_9Cl_2N_5O_2$ molecules and dimensions (I) (top), (II) (bottom).

(R)-(-)-1-CHLORO-2,2-DIPHENYLAZIRIDINE
$C_{14}H_{12}ClN$ (I)

(±)-1-CHLORO-2,2-DIPHENYLAZIRIDINE
$C_{14}H_{12}ClN$ (II)

A. FORNI, I. MORETTI, G. TORRE, S. BRÜCKNER, L. MALPEZZI and G. DI SILVESTRO, 1984. J. Chem. Soc. Perkin II, 791-797.

I. Monoclinic, $P2_1$, a = 21.370, b = 8.986, c = 13.522 Å, β = 108.40°, Z = 8. Cu radiation, R = 0.064 for 3681 reflexions.

II. Monoclinic, $P2_1/c$, a = 9.030, b = 11.330, c = 11.973 Å, β = 104.19°, Z = 4. Cu radiation, R = 0.057 for 1595 reflexions.

There are four independent molecules in the asymmetric unit of I. In both structures (Fig. 1) the N-C(2)-C(3) angle is significantly larger than the expected value of ca. 112° (I 117.3(7)-121.8(8), II 120.3(3)°). This is due to intramolecular interaction between one phenyl ring and Cl, C(3)...Cl 2.95 Å. This interaction may also be the cause of the narrow range of C(1)-C(2)-C(3)-C(8) torsion angles, 143-149°.

The second phenyl ring shows much more flexibility (30°) of orientation with respect to the aziridine ring. The N atom exhibits pyramidal geometry in all five molecules. The average C-C bond length in the three-membered ring is 1.49 Å, shorter than a single bond.

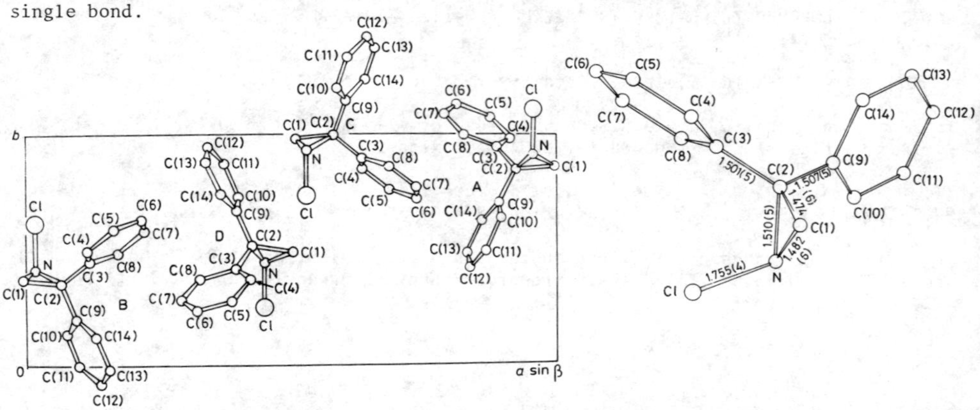

Fig. 1. $C_{14}H_{12}ClN$: the molecular structures of I (left) and II (right).

2-[(1-ETHYL-2-IMIDAZOLYL)METHYL]-2,3-DIHYDRO-1,4-BENZODIOXIN HYDROCHLORIDE MONOHYDRATE

$C_{14}H_{17}ClN_2O_2,H_2O$ $C_{14}H_{16}N_2O_2,HCl,H_2O$

A. CARPY, M. MONTAGUT and J.-M. LEGER, 1984. Acta Cryst., C40, 1267-1269.

Monoclinic, $P2_1/c$, a = 11.964, b = 13.410, c = 9.766 Å, β = 96.02°, Z = 4. Cu radiation, R = 0.038 for 1226 reflexions.

The molecule (Fig. 1) adopts a biplanar conformation with an angle of ∿60° between the two rings. Bond lengths and angles are normal. The crystal is stabilised by hydrogen bonds between Cl and the protonated imidazolyl N atom.

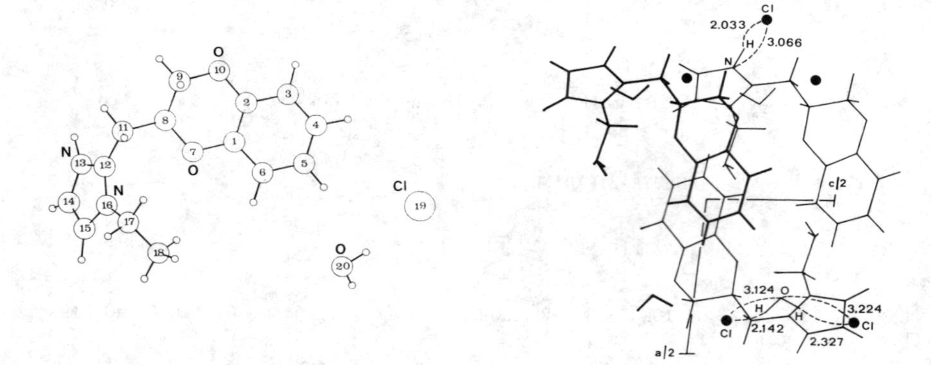

Fig. 1. $C_{14}H_{16}N_2O_2,HCl,H_2O$: perspective view of the molecule and a view of the molecular packing projected on (010) showing the hydrogen bonds.

(Z)-3,4-DIETHYL-5-(PHENYLMETHYLENE)-3-PYRROLIN-2-ONE

$C_{15}H_{17}NO$ (I)

(Z)(syn)-3,4-DIMETHYL-5-(2-PYRIDYLMETHYLENE)-3-PYRROLIN-2-ONE

$C_{12}H_{12}N_2O$ (II)

J.K. DATTAGUPTA, E.F. MEYER, JR., D.L. CULLEN, D.A. LIGHTNER, H. FALK, K. GRUBMAYR and I. BERNAL, 1984. Acta Cryst., C40, 88-92.

I. Monoclinic, $P2_1/n$, a = 11.392, b = 7.130, c = 15.789 Å, β = 97.76°, Z = 4. Cu radiation, R = 0.040 for 1057 reflexions.

II. Triclinic, $P\bar{1}$, a = 7.117, b = 8.223, c = 9.242 Å, α = 103.59, β = 92.08, γ = 92.41°, Z = 2. Mo radiation, R = 0.036 for 972 reflexions.

Both I and II (Fig. 1) show only small deviations from planarity. In I there is intermolecular hydrogen bonding so that the molecules form hydrogen bonded dimers with N...O 2.882(2) Å. Molecule II forms an intramolecular N-H...N hydrogen bond (N(1)...N(2) 2.771(3) Å).

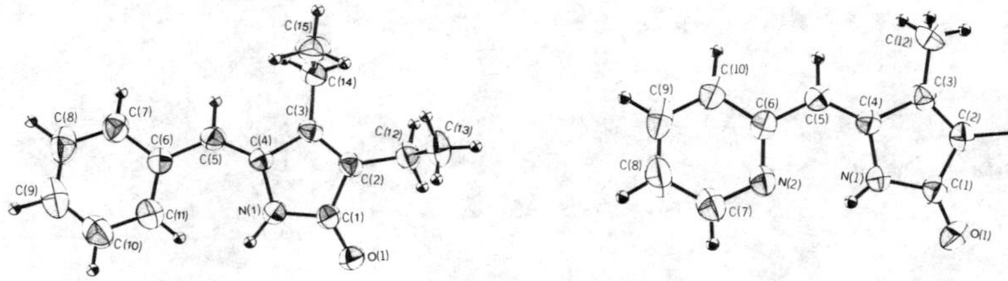

Fig. 1. Views of $C_{15}H_{17}NO$ (I) (left) and $C_{12}H_{12}N_2O$ (II) (right).

N-(1-ADAMANTYL)-2-(1-METHYLETHYLIDENE)AZIRIDINE
$C_{15}H_{23}N$

H. QUAST, R. JAKOB, K. PETERS, E.-M. PETERS and H.G. VON SCHNERING, 1984. Chem. Ber., 117, 840-849.

Monoclinic, $P2_1/n$, a = 18.612, b = 10.277, c = 6.721 Å, β = 93.72°, Z = 4. Mo radiation, R = 0.100 for 1594 reflexions.

A view of the molecule with dimensions is in Fig. 1.

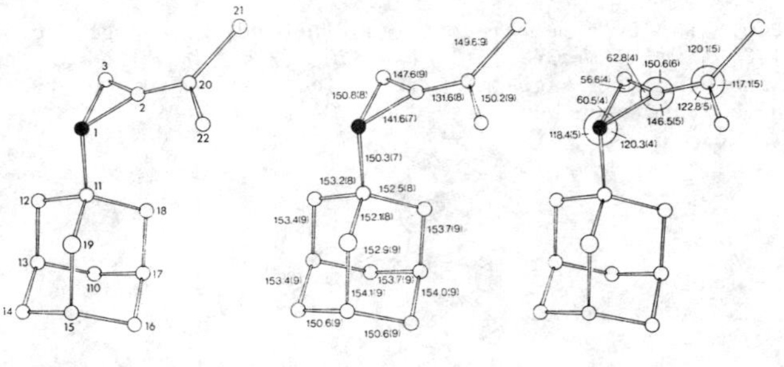

Fig. 1. The $C_{15}H_{23}N$ molecule with some bond lengths (pm) and angles (°).

1-PHENYL-3-METHYL-4-PHENYL-HYDRAZONO-PYRAZOLE-5-THIONE
$C_{16}H_{14}N_4S$

L.K. HANSEN, J. SIELER, R. RICHTER, G. HINSCHE and E. UHLEMANN, 1984. Acta Chem. Scand., A38, 331-334.

Monoclinic, C2/c, a = 27.642, b = 7.329, c = 14.528 Å, β = 93.87°, Dm = 1.35, Z = 8. Mo radiation, R = 0.039 for 2124 reflexions.

The structure analysis proves unambiguously that the molecule exists in the hydrazo form (Fig. 1). Atom H(1) was located in a difference map and it is bonded to N(4) with a distance of 1.14 Å. The six-membered ring so formed (S, C(3), C(2), N(3), N(4), H(1)) is almost planar and the intramolecular distance H(1)...S is 2.00 Å with the N(4)...S distance at 3.01 Å. The N(4)-H(1)...S angle is 145°. Other features of interest are bond lengths S-C(3) 1.678, N(1)-N(2) 1.411, N(1)-C(3) 1.368, N(1)-C(11) 1.419, N(2)-C(1) 1.295, N(3)-N(4) 1.308, N(3)-C(2) 1.326 and N(4)-C(5) 1.414 Å. The angles between the planes of the pyrazole ring and the phenyl rings (C(5) to C(10) and C(11) to C(16) are 7.4 and 11.9° respectively. The pyrazole ring and ring A are nearly coplanar (dihedral angle 1.6°).

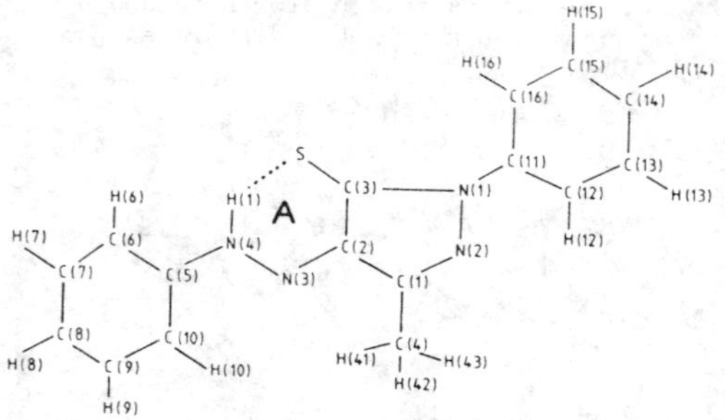

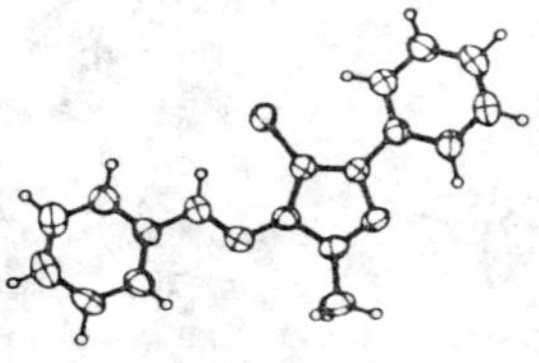

Fig. 1. Numbering scheme for $C_{16}H_{14}N_4S$ and a view of the molecule.

1-DIPHENYLMETHYL-3-HYDROXYAZETIDINIUM CHLORIDE

$C_{16}H_{18}ClNO$ $C_{16}H_{18}NO^+,Cl^-$

K.R. ACHARYA, S.S. TAVALE, T.N.G. ROW and K. VENKATESAN, 1984. Acta Cryst., C40, 846-848.

Monoclinic, $P2_1/a$, a = 12.356, b = 9.054, c = 14.043 Å, β = 100.34°, D_m = 1.14, Z = 4. Mo radiation, R = 0.053 for 1088 reflexions.

In the molecule the four-membered ring is buckled; the angle between the planes N(1),C(1),C(3) and C(1),C(2),C(3) is 13.0° (Fig. 1). The two phenyl rings are inclined to one another at an angle of 67.2°. The azetidinium moiety is linked to the Cl^- ion through a hydrogen bond with O-H...Cl 3.166(5) Å.

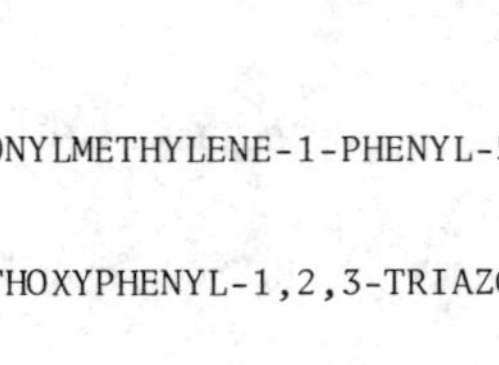

Fig. 1. $C_{16}H_{18}ClNO$: perspective view of the molecule.

5-METHOXYCARBONYL-4-METHOXYCARBONYLMETHYLENE-1-PHENYL-5-PROPYL-CΔ^2-TRIAZOLINE

$C_{16}H_{19}N_3O_4$ (I)

METHYL 4-METHOXYCARBONYL-1-p-METHOXYPHENYL-1,2,3-TRIAZOL-5-YL-ACETATE

$C_{14}H_{15}N_3O_5$ (II)

METHYL 5-METHOXYCARBONYL-1-p-METHOXYPHENYL-1,2,3-TRIAZOL-4-YL-ACETATE

$C_{14}H_{15}N_3O_5$ (III)

D. BARRACLOUGH, N.P. MOORHOUSE, E.I. ONWUYALI, F. SCHEINMANN, M.B. HURSTHOUSE and A.M.R. GALAS, 1984. J. Chem. Res., (S) 102-103, (M) 1101-1195.

I. Triclinic, $P\bar{1}$, a = 8.377, b = 9.512, c = 11.904 Å, α = 109.32, β = 101.63, γ = 102.11°, Z = 2. Cu radiation, R = 0.048 for 2064 reflexions.

II. Triclinic, $P\bar{1}$, a = 6.219, b = 14.105, c = 16.758 Å, α = 93.89, β = 91.32, γ = 93.04°, Z = 4. Cu radiation, R = 0.049 for 3503 reflexions.

III. Orthorhombic, Pbca, a = 7.859, b = 9.035, c = 42.261 Å, Z = 8. Cu radiation, R = 0.043 for 1367 reflexions.

Experiments in triazole synthesis led to these compounds (I-III) whose structures were determined to be as shown in Fig. 1.

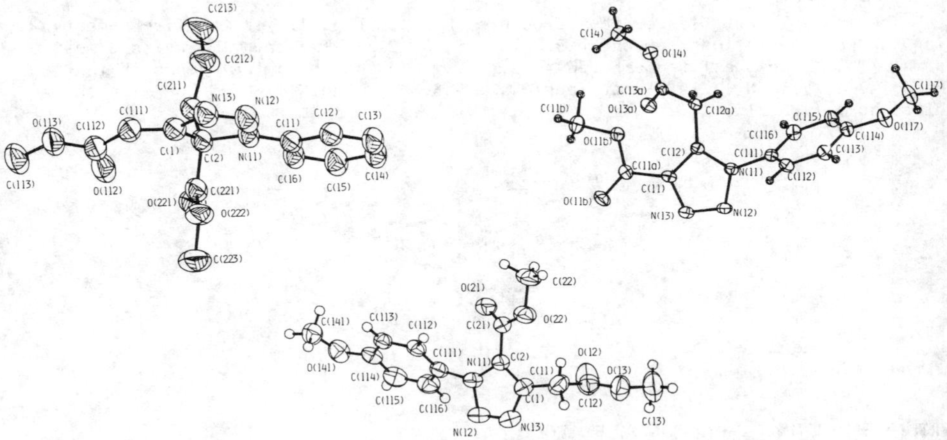

Fig. 1. Views of $C_{16}H_{19}N_3O_4$ (I) (left) and $C_{14}H_{15}N_3O_5$ (II) (top right) and (III) (lower right).

1'-(p-BROMOPHENYL)-3'-ETHYL-1',3',4',5'-TETRAHYDRO-1,2-DIDEOXY-D-glycero-D-gulo-HEPTOFURANOSO[1,2-d]IMIDAZOLE-2'-THIONE
$C_{16}H_{21}BrN_2O_5S$

M.D. ESTRADA, A. CONDE and R. MARQUEZ, 1984. Acta Cryst., C40, 898-901.

Monoclinic, $P2_1$, a = 6.924, b = 8.661, c = 15.039 Å, β = 96.92°, D_m = 1.598, Z = 2. Mo radiation, R = 0.039 for 2038 reflexions.

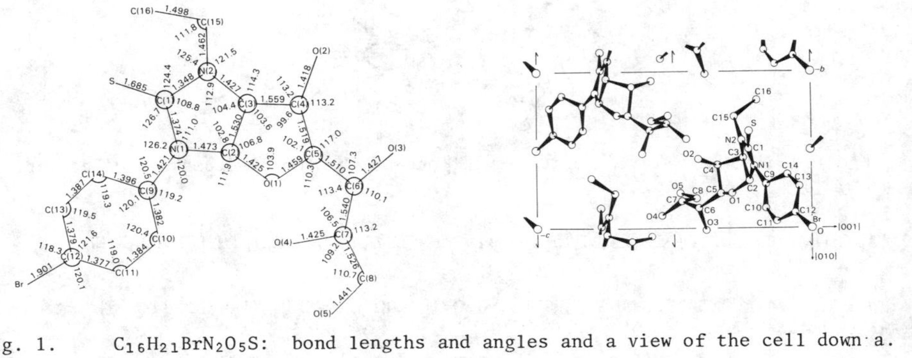

Fig. 1. $C_{16}H_{21}BrN_2O_5S$: bond lengths and angles and a view of the cell down a.

In the molecule the sugar ring adopts a ^{4}E conformation and the dihedral angle in the bicycle is 49.0(2)° (Fig. 1). Bond lengths and angles are generally as expected. Intermolecular hydrogen bonds link the molecules to form a three-dimensional network.

5-t-BUTYL-1,2,2-TRIMETHYL-3-(2,4,6-TRIBROMOPHENYL)-4-IMIDAZOLIDINONE
$C_{16}H_{21}Br_3N_2O$

D. MODERHACK, M. LORKE and D. SCHOMBURG, 1984. Justus Liebigs Ann. Chem., 1685-1695.

Orthorhombic, Pbca, a = 17.388, b = 26.449, c = 7.953 Å, Z = 8. Cu radiation, R = 0.058 for 3287 reflexions.

The analysis establishes the structure shown in Fig. 1. In the five-membered ring the mean deviation from the best plane is 0.12 Å. The phenyl ring is almost normal (81.2°) to the five-membered ring. Atom N(4) has pyramidal geometry (angle sum 337.7°) while N(1) is planar (angle sum 358.8°). The N(1)-C(2) bond length is quite short (1.352(5) Å).

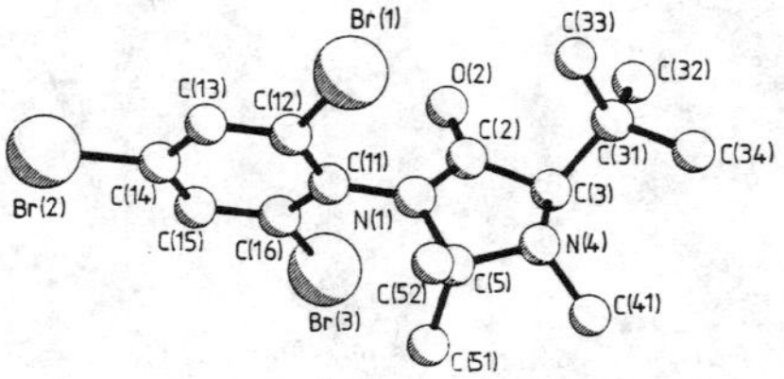

Fig. 1. The molecular structure of $C_{16}H_{21}Br_3N_2O$.

TRIS(3,5-DIMETHYL-1-PYRAZOLYL)METHANE
$C_{16}H_{22}N_6$

J.P. DECLERCQ and M. VAN MEERSSCHE, 1984. Acta Cryst., C40, 1098-1101.

Monoclinic, $P2_1/n$, a = 17.573, b = 23.000, c = 18.105 Å, β = 113.89°, Z = 16. Mo radiation, R = 0.108 for 4137 reflexions.

None of the four independent molecules (Fig. 1) shows threefold symmetry. In spite of the absence of crystallographic symmetry between these four molecules, their conformations and bond lengths and angles are very similar. There is an approximate inversion centre. The helical conformation is in good agreement with that of comparable molecules.

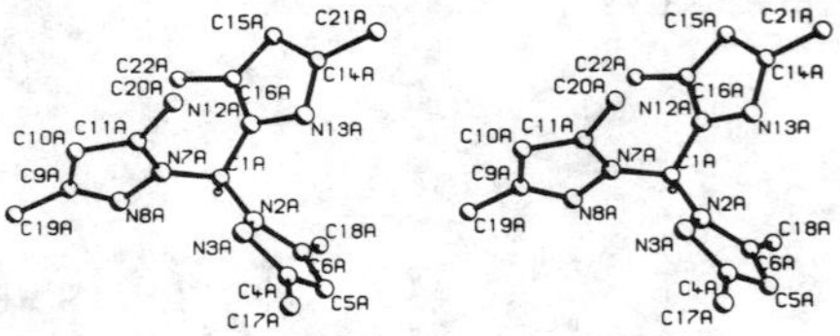

Fig. 1. $C_{16}H_{22}N_6$: stereoscopic view of one of the four independent molecules.

1,2-BIS(ETHOXYCARBONYL)-1-(2-HYDROXY-2-(1-PHENYLPYRAZOL-4-YL)ETHYL)HYDRAZINE
$C_{17}H_{22}N_4O_5$

R.A. JONES, B.A. GONZALEZ, J.S. ARQUES, J.Q. PARDO and T.J. KING, 1984. J. Chem. Soc. Perkin I, 1423-1425.

Triclinic, $P\bar{1}$, a = 5.639, b = 8.410, c = 20.238 Å, α = 84.32, β = 83.56, γ = 73.45°, Z = 2. R = 0.061 for 2275 reflexions [from a single twin].

The structure (Fig. 1) was disordered with two alternative sites for O(20) in the ratio 65:35. The disorder is due to inversion at the chiral centre C(6). Some bond distances are: N(1)-N(2) 1.359(5), N(2)-C(3) 1.315(6), N(1)-C(5) 1.351(5), N(1)-C(21) 1.414(5), C(4)-C(6) 1.493(6), C(6)-C(7) 1.499(7), C(6)-O(20) 1.319(7), C(7)-N(8) 1.458(5), N(8)-N(9) 1.396(4), N(8)-C(14) 1.365(5), N(9)-C(10) 1.350(6) Å.

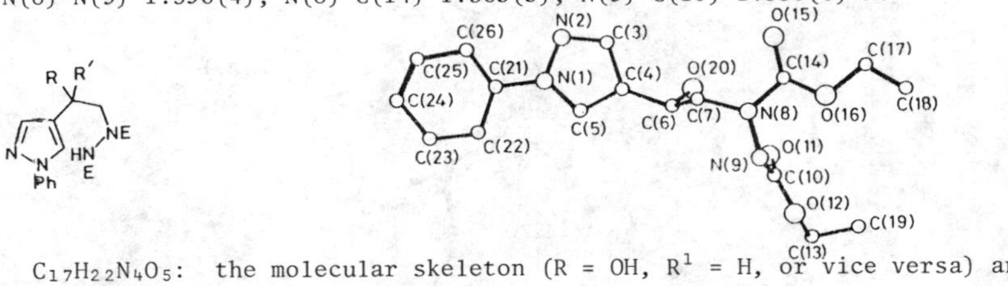

Fig. 1. $C_{17}H_{22}N_4O_5$: the molecular skeleton (R = OH, R^1 = H, or vice versa) and structure.

(±)-4β-(2-PYRROLYLCARBONYL)-1α-ETHYL-3aβ,4,5,7aα-TETRAHYDROINDAN-5β-METHANOL
$C_{17}H_{23}NO_2$ (I)

METHYL α(R),5(S)-DIMETHYL-6(R)-[1(S)-HYDROXYPROPYL]TETRAHYDROPYRAN-2(R)-ACETATE
$C_{13}H_{24}O_4$ (II)

M.P. EDWARDS, S.V. LEY, S.G. LISTER, B.D. PALMER and D.J. WILLIAMS, 1984. J. Org. Chem., 49, 3503-3516.

I. Monoclinic, P2/n, a = 14.546, b = 5.734, c = 19.187 Å, β = 105.60°, Z = 4. Cu radiation, R = 0.043 for 1573 reflexions.

II. Orthorhombic, $P2_12_12_1$, a = 5.006, b = 15.894, c = 17.448 Å, Z = 4. Cu radiation, R = 0.042 for 1092 reflexions.

X-ray analyses showed the structures of these intermediates in indanomycin synthesis to be as in Fig. 1. In (II), the substituents at C3 and C6 are axial.

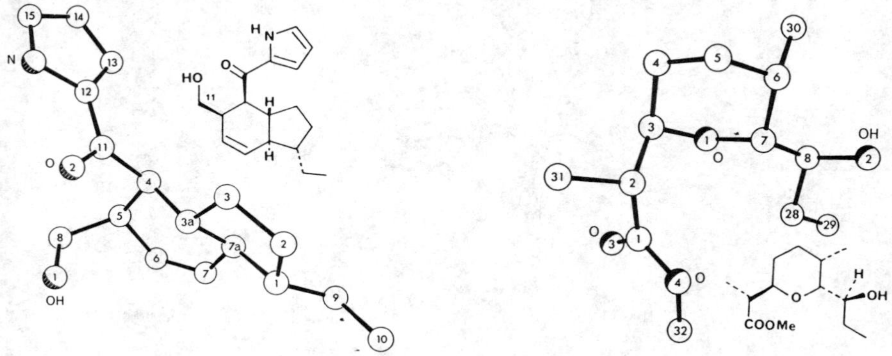

Fig. 1. Views of $C_{17}H_{23}NO_2$ (I) (left) and $C_{13}H_{24}O_4$ (II) (right).

DIETHYL 4,4',5,5'-TETRAMETHYL-3,3'-DIOXO-2,2'-BI(2,3-DIHYDRO)PYRROL-2,2'-DICARBOXYLATE
$C_{18}H_{24}N_2O_6$

R. GHAFFARI-TABRIZI, P. MARGARETHA and H.W. SCHMALLE, 1984. Helv. Chim. Acta, 67, 1957-1962.

Orthorhombic, $Pna2_1$, a = 11.353, b = 12.603, c = 13.422 Å, Z = 4. Cu radiation, R = 0.058 for 1287 reflexions.

The analysis establishes the structure shown in Fig. 1.

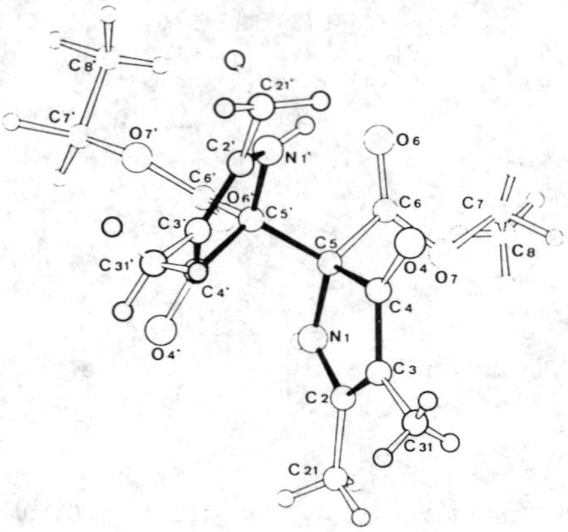

Fig. 1. A view of $C_{18}H_{24}N_2O_6$.

DIMETHYL 2-CHLORO-1,5-DIPHENYLPYRROLE-3,4-DICARBOXYLATE
$C_{20}H_{16}ClNO_4$

N. DE KIMPE, P. SULMON, L. DE BUYCK, R. VERHE, N. SCHAMP, J.-P. DECLERCQ and M. VAN MEERSSCHE, 1984. J. Chem. Res., (S) 82-83, (M) 0801-0833.

Triclinic, $P\bar{1}$, a = 10.588, b = 9.973, c = 11.316 Å, α = 119.11, β = 120.92, γ = 73.64°, Z = 2. Mo radiation, R = 0.043 for 2371 reflexions.

The analysis establishes the structure of this reaction product to be as shown in Fig. 1.

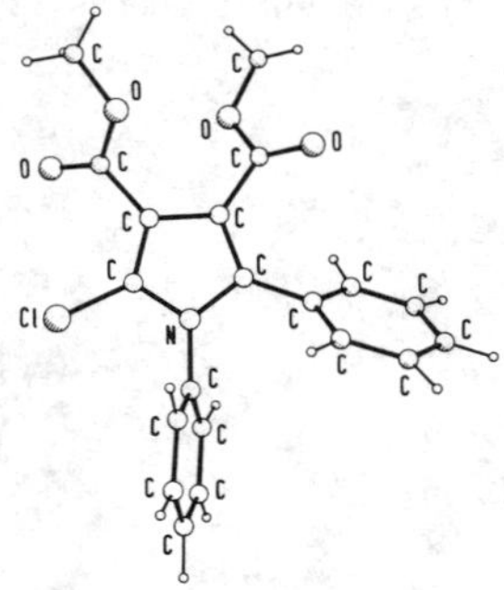

Fig. 1. A view of $C_{20}H_{16}ClNO_4$.

1-(2,6-DIMETHYLPHENYL)-(Z)-5-[(ETHOXYCARBONYL)IMINO]-2-PHENYL-4,4-BIS(TRIFLUORO-METHYL)-2-IMIDAZOLINE
$C_{22}H_{19}F_6N_3O_2$

A. GIEREN, T. HÜBNER, C.-P. KAERLEIN, K. BURGER, H. PARTSCHT and E. HUBER, 1984. Chem. Zeit., 108, 211-212.

Triclinic, $P\bar{1}$, a = 8.918, b = 10.269, c = 13.253 Å, α = 107.19, β = 99.71, γ = 102.45°, D_m = 1.41, Z = 2. Mo radiation, R = 0.043 for 3275 reflexions.

The heterocyclic five-membered ring is planar (Fig. 1) with the largest deviation from the best plane through the five atoms being 0.03 Å. Atoms C(11), C(21), N(3) and C(4) with respective deviations from this plane of -0.05, 0.14, 0.13 and 0.34 Å, are almost in the plane. Important bond lengths are: C(1)-N(1) 1.450, N(1)-C(2) 1.280, C(2)-N(2) 1.405, N(2)-C(3) 1.386, C(3)-C(1) 1.533, C(2)-C(11) 1.470, N(2)-C(21) 1.442, C(3)-N(3) 1.246, N(3)-C(4) 1.395 Å. The dihedral angles between the best plane through the five-membered ring and those through the phenyl ring, the 2,6-dimethylphenyl ring, and the ester grouping are respectively 43, 87 and 84°.

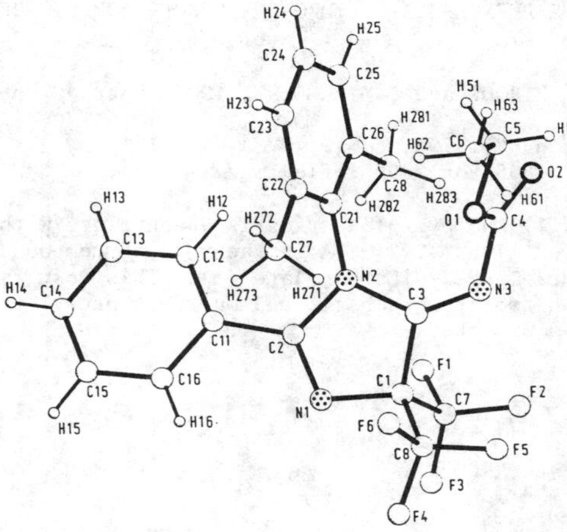

Fig. 1. Molecular structure of $C_{22}H_{19}F_6N_3O_2$.

DIETHYL 3,5-DIPHENYL-PYRROLE-1,2-DICARBOXYLATE
$C_{22}H_{21}NO_4$ (I)

DIETHYL 3-(2-CHLOROPHENYL)-5-PHENYL-PYRROLE-1,2-DICARBOXYLATE
$C_{22}H_{20}ClNO_4$ (II)

A. LAARIF, F. THÉOBALD, M. BIROUK and J.F. ROBERT, 1984. Acta Cryst., C40, 1278-1281.

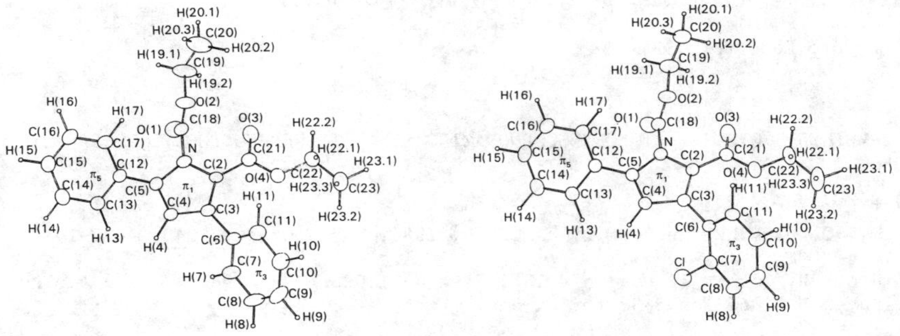

Fig. 1. Perspective views of molecules of $C_{22}H_{21}NO_4$ (I) and $C_{22}H_{20}ClNO_4$ (II).

I. Orthorhombic, Pbca, a = 17.213, b = 18.910, c = 11.968 Å, Z = 8. Cu radiation, R = 0.081 for 1762 reflexions.

II. Orthorhombic, Pbca, a = 16.955, b = 18.487, c = 13.048 Å, Z = 8. Cu radiation, R = 0.067 for 3122 reflexions.

In the molecules (Fig. 1) the modifications of the angles between the phenyl groups and the pyrrole ring are in agreement with the magnetic non-equivalence of the ethoxycarbonyl chains, which is more pronounced in II. The three aromatic rings are planar. The carbonyl groups are planar: that attached to C(2) is coplanar with the pyrrole ring plane but that attached to N is inclined to the ring plane by 72.4(7)° for I and 67.0(4)° for II.

DIETHYL 3-(2-NITROPHENYL)-5-PHENYL-3,4-DIHYDRO-2H-PYRROLE-2,2-DICARBOXYLATE
$C_{22}H_{22}N_2O_6$

A. LAARIF, F. THÉOBALD, M. BIROUK and J.F. ROBERT, 1984. Acta Cryst., C40, 1008-1011.

Triclinic, P$\bar{1}$, a = 11.201, b = 11.792, c = 9.137 Å, α = 112.13, β = 112.46, γ = 81.75°, Z = 2. Cu radiation, R = 0.068 for 3009 reflexions.

In the molecule (Fig. 1) the pyrroline ring is an envelope with C(4) out of plane by 0.420(4) Å compared with 0.292(5) Å in the related compound diethyl 3,5-diphenyl-3,4-dihydro-2H-pyrrole-2,2-dicarboxylate (1). This results from steric constraints caused by the nitro-group substituted on the 4-phenyl ring. Bond lengths and angles are generally as expected.

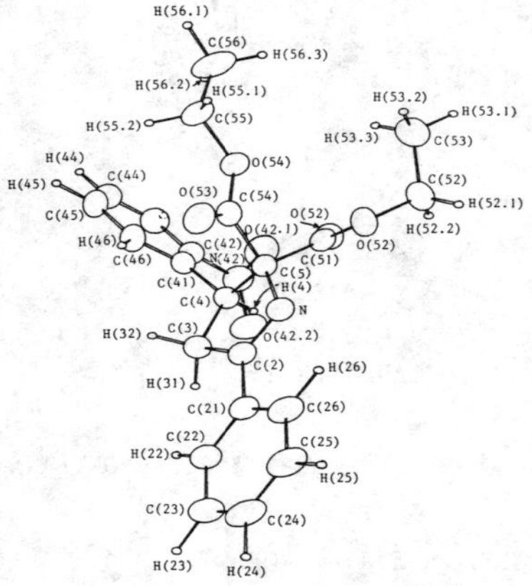

Fig. 1. Perspective view of the $C_{22}H_{22}N_2O_6$ molecule.

1. Structure Reports, 49B, 238.

ETHYL 2-METHYL-5-OXO-4-(p-TOLUENESULPHONAMIDO)-1-(p-TOLUENESULPHONYL)-2,5-DIHYDRO-2-PYRROLECARBOXYLATE
$C_{22}H_{24}N_2O_7S_2$

L. DUPONT, O. DIDEBERG and J.C.L. JAMOULLE, 1984. Acta Cryst., C40, 1472-1473.

Monoclinic, P2_1/c, a = 12.063, b = 18.545, c = 11.023 Å, β = 98.94°, Z = 4. Cu radiation, R = 0.065 for 3858 reflexions.

The two benzene rings are almost parallel (10.5° between normals) and 7.18 Å apart. S(1) and S(2) are 0.08(1) and 0.12(1) Å out of the planes of their attached benzene rings; S(1) is 0.19(1) Å out of the C(15)C(16)C(17)C(18)N(2) plane (Fig. 1). N(1)-H...O(6) hydrogen bonds (3.014(6) Å) link the molecules.

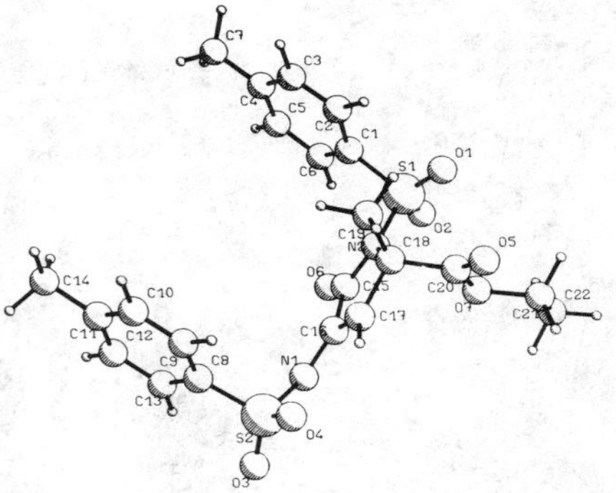

Fig. 1. The $C_{22}H_{24}N_2O_7S_2$ molecule.

(±)-1,2-DIPHENYL-4-[2-(PHENYLSULPHINYL)ETHYL]-3,5-PYRAZOLIDINEDIONE (SULPHINPYRAZONE)
$C_{23}H_{20}N_2O_3S$

K. GO and G. KARTHA, 1984. Acta Cryst., C40, 1869-1871.

Monoclinic, C2/c, a = 28.349, b = 5.696, c = 30.799 Å, β = 126.07°, Z = 8. Cu radiation, R = 0.064 for 3019 reflexions.

Each molecular location within the cell is occupied randomly by either one of two approximately superimposable diastereomers which differ only in the configuration around S (Fig. 1). The diastereomer whose S=O is gauche to the ethylpyrazolidinedione group predominates over the anti form (3:1). The two phenyl groups at the pyrazone make an angle of 86° with each other.

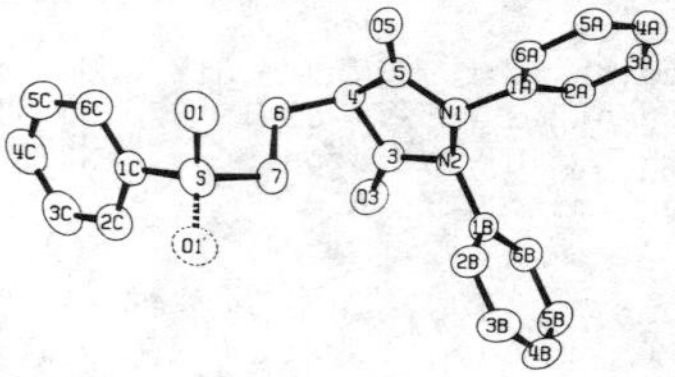

Fig. 1. The sulphinpyrazone molecule.

BIS(L-PHENYL-3,5-DIMETHYLPYRAZOLE) - OXALIC ACID
$C_{24}H_{26}N_4O_4$ (I) $2(C_{11}H_{12}N_2)C_2H_2O_4$

1-PHENYL-3,5-DIMETHYLPYRAZOLIUM PERCHLORATE
$C_{11}H_{13}N_2O_4Cl$ (II) $C_{11}H_{13}N_2^+ \cdot ClO_4^-$

J. ZUKERMAN-SCHPECTOR, E.E. CASTELLANO, G. OLIVA, A.C. MASSABNI and A.D. PINTO, 1984. Can. J. Chem., 62, 725-728.

I. Monoclinic, C2/c, a = 16.516, b = 11.685, c = 12.587 Å, β = 107.11°, Z = 4. Mo radiation, R = 0.057 for 906 reflexions.

II. Orthorhombic, a = 16.204, b = 15.329, c= 10.296 Å, Z = 8. Mo radiation, R = 0.071 for 2051 reflexions.

Both structures are stabilized by a hydrogen bond between N(2) of the pyrazole moiety and an oxygen of an acid. In I, the oxalic acid lies on an inversion centre and acts as a proton donor; in II, N(2) is protonated and acts as a donor to perchlorate (Fig. 1).

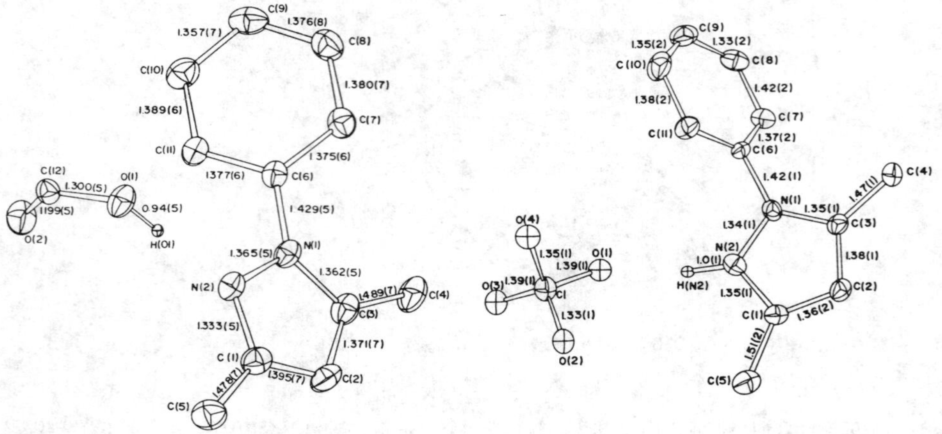

Fig. 1. Views with bond lengths of the asymmetric units of $C_{24}H_{26}N_4O_4$ (I) (left) and $C_{11}H_{13}N_2O_4Cl$ (II) (right).

BEPRIDIL HYDROCHLORIDE MONOHYDRATE

$C_{24}H_{35}ClN_2O,H_2O$ $(C_{24}H_{35}N_2O)^+Cl^-,H_2O$

J.P. DECLERCQ and M. VAN MEERSSCHE, 1984. Bull. Soc. Chim. Belg., 93, 235-236.

Monoclinic, $P2_1/a$, a = 30.106, b = 14.017, c = 11.725 Å, β = 100.78°, Z = 8. Mo radiation, R = 0.088 for 2083 reflexions.

The two independent molecules (Fig. 1) have different conformations in the crystal as shown by the torsion angles C6-C7-C8-C9 87 and -103° and C6-C7-C8-C16 -126 and 86°. Both molecules form N-H...Cl hydrogen bonds (3.13 and 3.07 Å). N^+-C distances are particularly long (mean 1.54 Å).

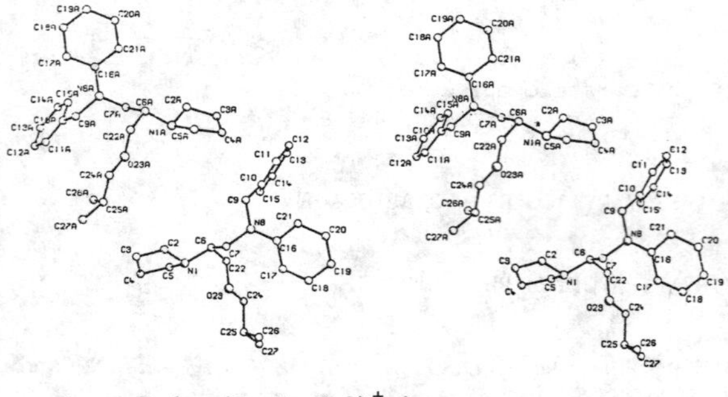

Fig. 1. A stereoview of the $(C_{24}H_{35}N_2O)^+$ ions.

1,1'-(1,2-BIS(4-METHOXYPHENYL)ETHANEDIYLIDENE)BIS(4-PHENYL-3,5-DIOXO-1,2,4-TRIAZOLIDINIUM) DIHYDROXIDE INNER SALT

$C_{32}H_{24}N_6O_6$

C.-C. CHENG, F.D. GREENE and J.F. BLOUNT, 1984. J. Org. Chem., 49, 2917-2922.

Monoclinic, A2/a, a = 22.393, b = 16.117, c = 16.614 Å, β = 102.12°, Z = 8. Cu radiation, R = 0.070 for 1436 reflexions.

The analysis shows that the molecule is a bis(azomethine imine) (Fig. 1). The p-methoxyphenyl group and attached heterocyclic ring are close to being individually planar, and these planes (the upper and lower halves of the molecule) are approximately perpendicular. Main bond lengths are in Fig. 1.

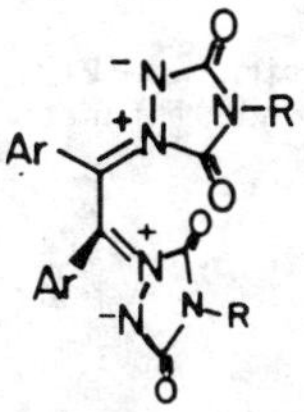

Ar = p-$CH_3OC_6H_4$

R = phenyl

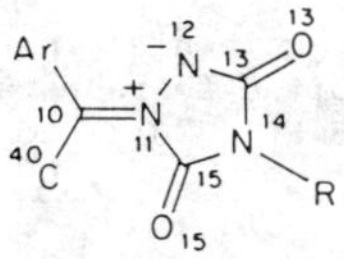

C_{10}–N_{11}	1.317	(1.346)	N_{14}–C_{15}	1.318	(1.335)
N_{11}–N_{12}	1.362	(1.362)	C_{15}–O_{15}	1.226	(1.212)
N_{12}–C_{13}	1.354	(1.350)	C_{15}–N_{11}	1.452	(1.469)
C_{13}–O_{13}	1.223	(1.199)	C_{10}–C_{40}	1.513	
C_{13}–N_{14}	1.426	(1.452)	C_{10}–Ar	1.428	(1.453)

Fig. 1. The $C_{32}H_{24}N_6O_6$ molecule and principal bond lengths (bond C10-C40 joins the two halves of the molecule; values in parentheses refer to the other half of the molecule).

cis-1,5-DIPHENYL-3-(DIPHENYLMETHYLENE)-4-MORPHOLINOCARBONYL-2-PYRROLIDINONE
$C_{34}H_{30}N_2O_3$

A.E. BAYDAR, G.V. BOYD, F. STRIDE and P.F. LINDLEY, 1984. Acta Cryst., C40, 1443-1445.

Orthorhombic, Pbca, a = 18.262, b = 23.285, c = 12.798 Å, Z = 8. Cu radiation, R = 0.056 for 2527 reflexions.

The morpholinocarbonyl (at C4) and the phenyl (at C5) substituents are cis-equatorial with respect to the γ-lactam ring plane; the C41-C4-C5-C51 torsion angle is 29.6(4)°. The pyrrolidine ring has the C4 envelope conformation (C4 is displaced -0.457 Å from the N1,C2,C3 plane) and the morpholine ring has the chair conformation. In the diphenylmethylene moiety at C3 the phenyl rings are inclined at 93.8(2)° while the phenyl rings at N1 and C5 are inclined at 25.9 and 62.2° with respect to the N1,C2,C3 plane.

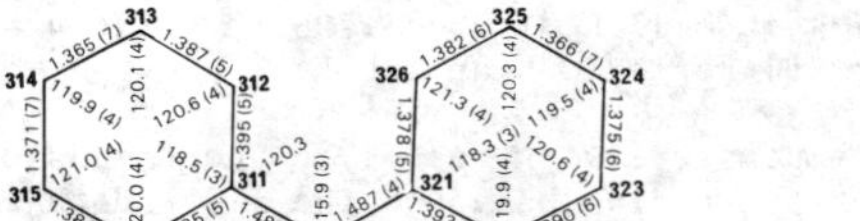

Fig. 1. Distances (Å) and angles (°) in the $C_{34}H_{30}N_2O_3$ molecule.

PYRIDINE BORON TRIFLUORIDE
$C_5H_5BF_3N$ C_5H_5N-BF_3

K. TÖPEL, K. HENSEN and J.W. BATS, 1984. Acta Cryst., C40, 828-830.

Monoclinic, $P2_1/c$, a = 17.679, b = 5.914, c = 14.917 Å, β= 118.90°, Z = 8. Mo radiation, R_w = 0.037 for 956 reflexions.

The two independent molecules in the asymmetric unit are similar (Fig. 1). The boron atoms have a distorted tetrahedral coordination. Some bond distances are: B-N 1.602(4) and 1.604(5), B-F 1.330(4) to 1.362(3) Å.

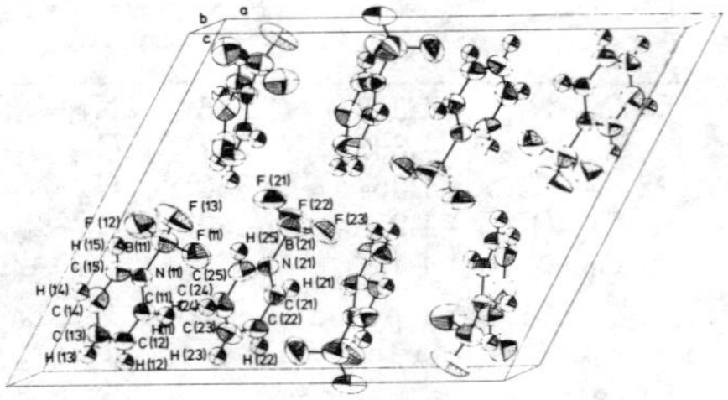

Fig. 1. $C_5H_5BF_3N$: view of the crystal structure.

2-PYRIDONE
C_5H_5NO

U. OHMS, H. GUTH, E. HELLNER, H. DANNOHL and A. SCHWEIG, 1984. Z. Kristallogr., 169, 185-200.

I. Orthorhombic, $P2_12_12_1$, a = 13.657, b = 5.912, c = 5.701 Å, Z = 4. Mo radiation, R = 0.034 for 647 reflexions.

II. Orthorhombic, $P2_12_12_1$, a = 13.645, b = 5.900, c = 5.692 Å, Z = 4. Neutron radiation, R = 0.041 for 625 reflexions.

III. Orthorhombic, $P2_12_12_1$, a = 13.564, b = 5.792, c = 5.598 Å, Z = 4. Mo radiation, R = 0.041 for 1174 reflexions at 120 K.

IV. Orthorhombic, $P2_12_12_1$, a = 13.564, b = 5.795, c = 5.604 Å, Z = 4. Neutron radiation, R = 0.041 for 981 reflexions at 120 K.

For previous room temperature work see (1). The dimensions at 120 K are shown in Fig. 1. Deformation density studies based on the 120 K data were performed using multipole and charge-cloud models.

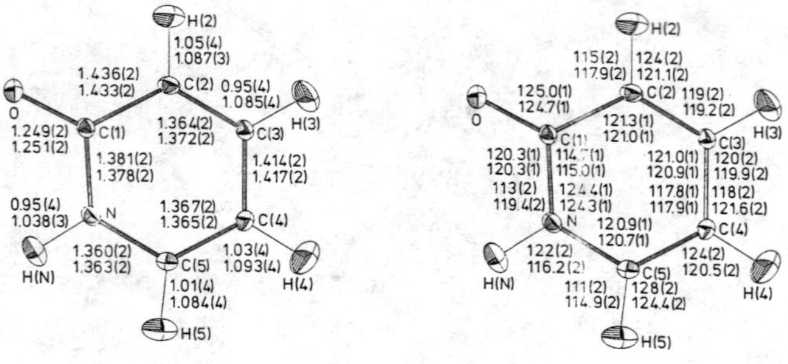

Fig. 1. Bond lengths and angles in 2-pyridone; upper values are X-ray, lower values are neutron results.

1. Structure Reports, 17, 748.

3-HYDROXY-4-(1'-CHLORO-1'-HYDROXIMINOMETHYL)PYRIDINIUM NITRATE
$C_6H_6ClN_3O_5$ (I)

FURO[2,3-c]PYRIDINIUM 3-HYDROXYPYRIDINE-4-NITROLATE
$C_{13}H_{10}N_4O_5$ (II)

3-(TRINITROMETHYL)-PYRIDIN-4-OL
$C_6H_4N_4O_7$ (III)

S. SHIOTANI, H. MORITA, M. INOUE, T. ISHIDA, Y. IITAKA and A, ITAI, 1984. J. Heterocycl. Chem., 21, 725-736.

I. Monoclinic, $P2_1/c$, a = 4.688, b = 23.161, c = 8.647 Å, β = 105.24°, D_m = 1.723, Z = 4. Cu radiation, R = 0.110 for 1541 reflexions.

II. Orthorhombic, Pbca, a = 7.394, b = 14.047, c = 26.206 Å, D_m = 1.474, Z = 8. Cu radiation, R = 0.107 for 2316 reflexions.

III. Monoclinic, $P2_1/n$, a = 9.896, b = 15.114, c = 6.560 Å, β = 104.60°, Z = 4. Cu radiation, R = 0.061 for 1714 reflexions.

The structures of molecules I - III were established by X-ray methods to be as shown in Fig. 1.

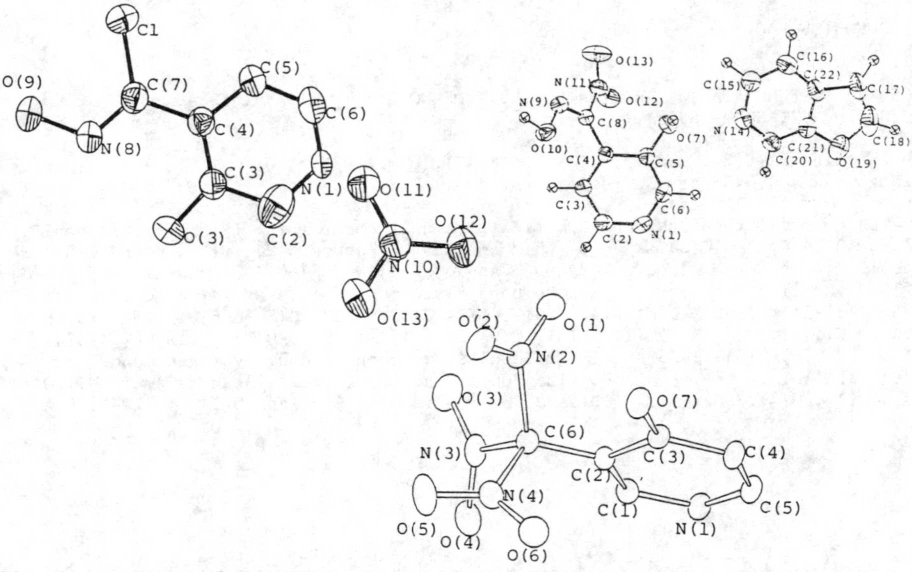

Fig. 1. Views of $C_6H_6ClN_3O_5$ (I), $C_{13}H_{10}N_4O_5$ (II) and $C_6H_4N_4O_7$ (III) (clockwise from top left).

DISODIUM 1,4-PIPERAZINEDICARBODITHIOATE HEXAHYDRATE
$C_6H_8N_2Na_2S_4,6H_2O$ $Na_2[S_2CN(CH_2)_4NCS_2],6H_2O$

J.-P. LEGROS, D. TROY and J. GALY, 1984. Acta Cryst., C40, 801-804.

Monoclinic, $P2_1/c$, a = 6.009, b = 12.185, c = 11.494 Å, β = 92.47°, D_m = 1.52, Z = 2. Mo radiation, R = 0.021 for 1917 reflexions.

In the crystal structure, sodium coordinates five water molecules and one sulphur atom forming a distorted octahedron (Fig. 1). These coordination polyhedra share edges to build parallel files running along a and linked together by 1,4-piperazinedecarbodithioate anions into layers parallel to the ac plane. The 1,4-piperazine-dicarbodithioate anion exhibits the chair conformation and is made up of two almost planar C_2NCS_2 moieties related by a centre of symmetry. The S-C-S bite angle is 119.6(1)°. The C-S distances are not equivalent: 1.740(1) and 1.713(1) Å; the shorter corresponds to the S...Na interaction and one S...H-O hydrogen bond while the longer corresponds to four S...H-O hydrogen bonds.

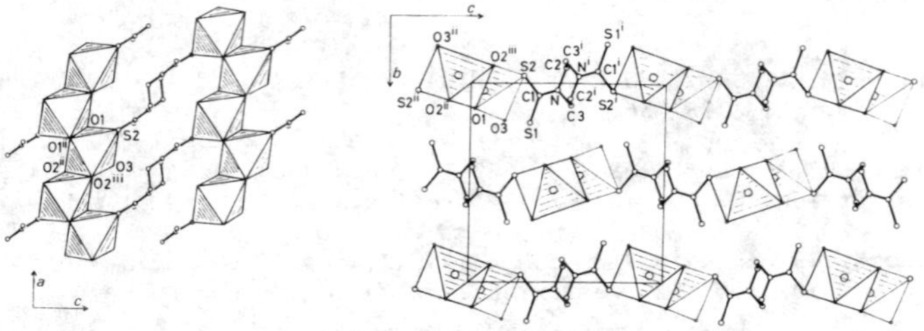

Fig. 1. $C_6H_8N_2Na_2S_4,6H_2O$: sodium coordination octahedra linked by piperazinebis-(dithiocarbamate) anions forming layers parallel to the ac plane and the projection of the crystal structure along a.

D-GLUCONO-1,5-LACTAM
$C_6H_{11}NO_5$

H. OGURA, K. FURUHATA, H. TAKAYANAGI, N. TSUZUNO and Y. IITAKA, 1984. Bull. Chem. Soc. Japan, 57, 2687-2688.

Orthorhombic, $P2_12_12_1$, a = 11.721, b = 13.471, c = 4.926 Å, D_m = 1.51, Z = 4. Cu radiation, R = 0.043 for 912 reflexions.

The molecule takes the conformation of a half-chair (Fig. 1). The torsional angle of C(2)-C(1)-N-C(5) is -4.8° which means that atoms C(2), C(1), N and C(5) are virtually in the same plane. Torsional angles C(1)-N-C(5)-C(4) and N-C(1)-C(2)-C(3) are 16.2 and 21.7° respectively. The bond distances of interest are: C(1)-C(2) 1.538, C(2)-C(3) 1.508, C(3)-C(4) 1.517, C(4)-C(5) 1.528, C(5)-C(6) 1.515, C(5)-N 1.463, N-C(1) 1.326, C(1)-O(1) 1.231, C(2)-O(2) 1.411, C(3)-O(3) 1.419, C(4)-O(4) 1.412, and C(6)-O(5) 1.436 Å. The molecules are connected by hydrogen bonds in layers parallel to the ab plane. The donor...acceptor atom distances are: N...O(3) 2.92, O(1)...O(4) 2.82, O(2)...O(5) 2.73, O(5)...O(1) 2.71 and O(3)...O(2) 2.70 Å.

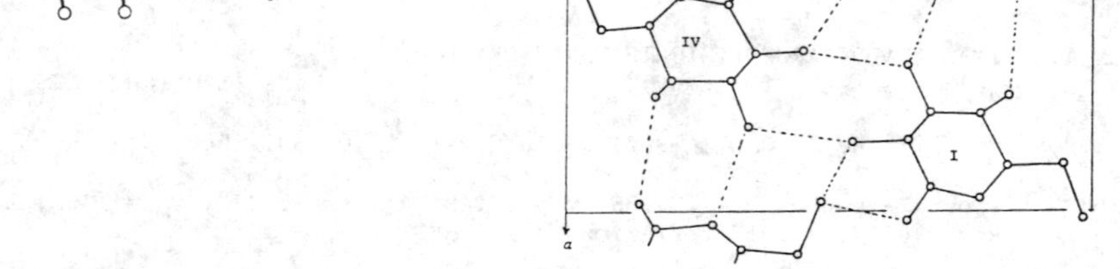

Fig. 1. Perspective view of the molecule $C_6H_{11}NO_5$, and projection of the crystal structure along the c axis with hydrogen bonds represented by dotted lines.

1,3,4,6-TETRAMETHYL-1,4-DIHYDRO-1,2,4,5-TETRAZINE
$C_6H_{12}N_4$

G.W. BEMIS, R.R. WHITTLE, S.L. MAYO and R.A. OLOFSON, 1984. Acta Cryst., C40, 2076-2078.

Monoclinic, $P2_1/n$, a = 10.612, b = 6.820, c = 10.975 Å, β = 95.31°, D_m = 1.13, Z = 4. Mo radiation, R = 0.077 for 704 reflexions.

This potentially antiaromatic or homoaromatic ring system (Fig. 1) has a flattened boat conformation with both N-methyl groups equatorial. Substantial delocalization of the electron pairs of N(1) and N(4) is found.

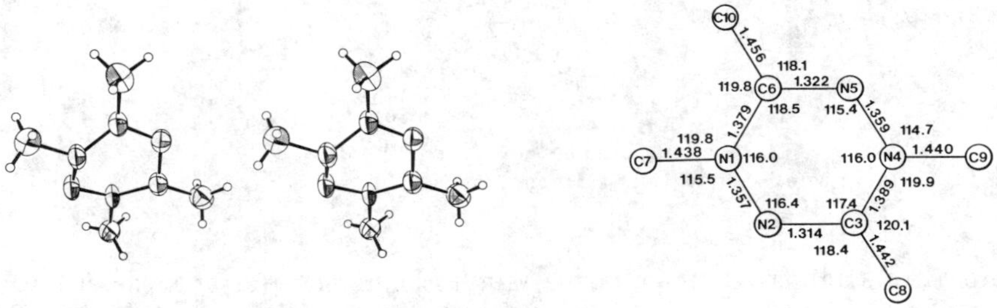

Fig. 1. A stereoview of $C_6H_{12}N_4$ and dimensions.

1-METHYL-2-THIOXO-1,2-DIHYDRONICOTINIC ACID
$C_7H_7NO_2S$

L. DUPONT, O. DIDEBERG and L. THUNUS, 1984. Acta Cryst., C40, 417-419.

Orthorhombic, $Pna2_1$, a = 14.368, b = 14.518, c = 7.169 Å, Z = 8. Cu radiation, R = 0.074 for 1167 reflexions.

The two independent molecules in the asymmetric unit (Fig. 1) are individually planar (neglecting methyl H atoms). The C-S distances are 1.687 and 1.688(7) Å. Although the hydroxyl hydrogen atoms were not located, the intramolecular S(1)...O(2) and S(11)...O(12) distances (2.856, 2.867 Å) are consistent with intramolecular O-H...S bonding.

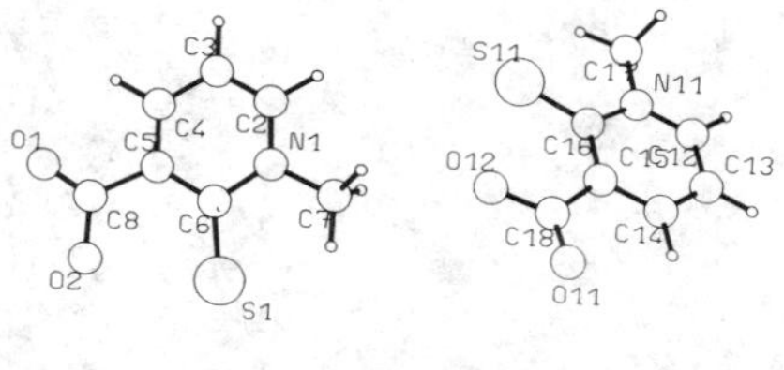

Fig. 1. The two independent $C_7H_7NO_2S$ molecules.

4-DIMETHYLAMINOPYRIDINE
$C_7H_{10}N_2$

U. OHMS and H. GUTH, 1984. Z. Kristallogr., 166, 213-217.

Monoclinic, $P2_1/n$, a = 14.813, b = 7.604, c = 6.103 Å, β = 93.2°, Z = 4. Mo radiation, R = 0.072 for 1137 reflexions.

Molecular dimensions are in Fig. 1. The dihedral angle between the pyridine ring plane and the dimethylamino group is 7(1)°.

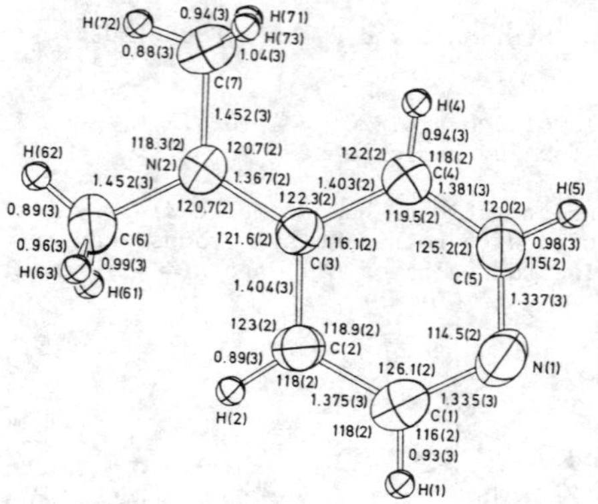

Fig. 1. A view of $C_7H_{10}N_2$ with dimensions.

HEXAHYDRO-1,3,5-TRINITRO-1,3,5-TRIAZINE TETRAHYDROTHIOPHENE 1,1-DIOXIDE 1:1 COMPLEX

$C_7H_{14}N_6O_8S$ $C_3H_6N_6O_6.C_4H_8O_2S$

T.M. HALLER, T.B. BRILL and A.L. RHEINGOLD, 1984. Acta Cryst., C40, 517-519.

Triclinic, $P\bar{1}$, a = 6.238, b = 10.241, c = 11.271 Å, α = 100.65, β = 100.50, γ = 94.193°, Z = 2. Mo radiation, R = 0.040 for 1605 reflexions.

The triazine system (Fig. 1) contains all of the three possible orientations of the NO_2 groups (axial, equatorial, and planar). The ring conformation is a chair. The tetrahydrothiophene 1,1-dioxide molecule is regular and normal; the crystal lattice contains discrete molecules which pack as shown in Fig. 1.

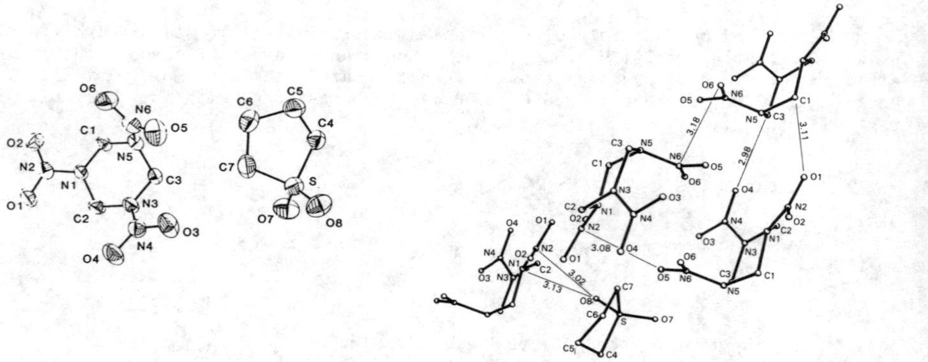

Fig. 1. The component molecules in $C_3H_6N_6O_6.C_4H_8O_2S$ and a diagram showing shortest intermolecular contacts.

2-[(METHYLSULPHONYL)ACETYL]PYRIDINE

$C_8H_9NO_3S$

T. SRIKRISHNAN, 1984. Acta Cryst., C40, 1441-1443.

Monoclinic, $P2_1/c$, a = 11.073, b = 8.352, c = 10.489 Å, β = 108.85°, D_m = 1.45, Z = 4. Cu radiation, R = 0.031 for 1282 reflexions.

In the molecule (Fig. 1), C(7) is almost in the ring plane with O(7) and the sulphonyl group twisted away from the ring. The O atoms of the sulphonyl group are involved in C-H...O interactions (e.g. C(4)...O(10) 3.357(3), C(12)...O(10) 3.349(3) Å). There is partial stacking of the pyridine rings; molecules related by a centre of symmetry lie on top of each other separated by 3.45 Å.

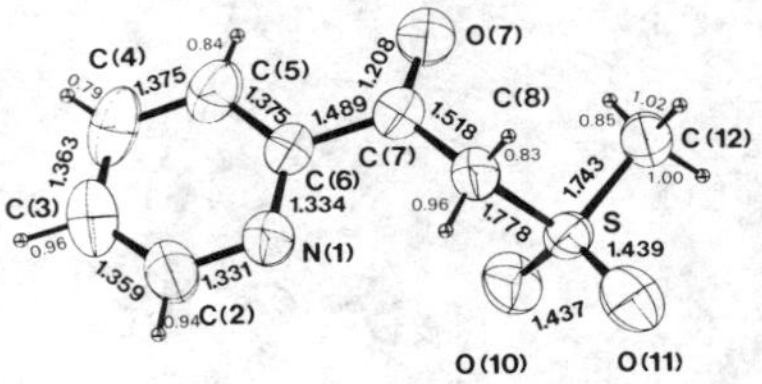

Fig. 1. Bond lengths (Å) in the $C_8H_9NO_3S$ molecule. (e.s.d.s. are 0.002-0.004 Å for heavier atoms.)

2,2,6,6-TETRAMETHYLPIPERIDINIUM BROMIDE

$C_9H_{20}BrN$ $C_9H_{20}N^+.Br^-$

M.D. WALKINSHAW, A.H. COWLEY and S.K. MEHROTRA, 1984. Acta Cryst., C40, 129-131.

Orthorhombic, $P2_12_12_1$, a = 8.885, b = 9.300, c = 13.089 Å, Z = 4. Mo radiation, R = 0.040 for 883 reflexions.

The piperidinium ring (Fig. 1) has a distorted chair conformation. The C7...C8 distance (between axial methyl groups) is only 3.205(11) Å. The C-N-C angle is 120.9(5)°.

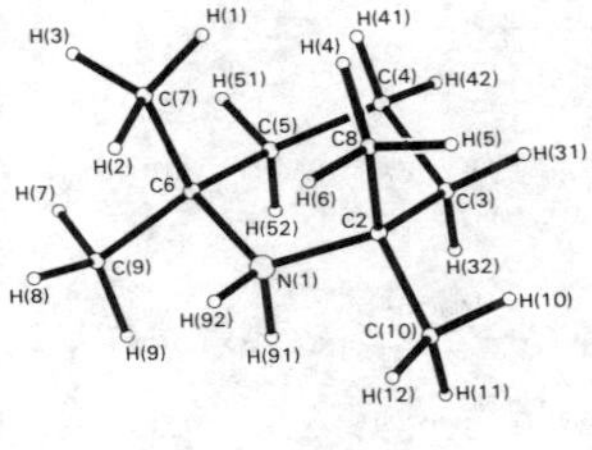

Fig. 1. A view of the $C_9H_{20}N^+$ cation.

2,3,3,5,6,6-HEXAMETHYL-3,6-DIHYDROPYRAZINE HEXAHYDRATE

$C_{10}H_{18}N_2.6(H_2O)$ (I)

3-AMINO-3-METHYL-2-BUTANONE HYDROCHLORIDE

$C_5H_{12}ClNO$ (II) $C_5H_{12}NO^+.Cl^-$

C.L. KLEIN, R.J. MAJESTE, A.E. LUEDTKE, W.J. RAY, E.D. STEVENS and J.W. TIMBERLAKE, 1984. J. Org. Chem., 49, 1208-1211.

I. Monoclinic, $P2_1/c$, a = 6.454, b = 6.462, c = 20.405 Å, β = 102.03°, D_m = 1.09, Z = 2. Cu radiation, R = 0.036 for 625 reflexions.

II. Orthorhombic, Pnma, a = 10.505, b = 21.936, c = 10.035 Å, D_m = 1.17, Z = 12. Mo radiation, R = 0.054 for 711 reflexions.

Molecule (I) (Fig. 1) has crystallographic inversion symmetry. In the crystal, molecules are sandwiched between layers of water molecules (all of which have their hydrogen atoms disordered). Some bond lengths are N1-C1 1.276(4), C1-C2 1.504(4), C(2)-N(1) 1.452(2) Å. In (II), the unit cell has eight cations in general positions and four on mirror planes. The cations (Fig. 1) are arranged in sheets which are bridged together by chloride ions hydrogen bonded to ammonium groups.

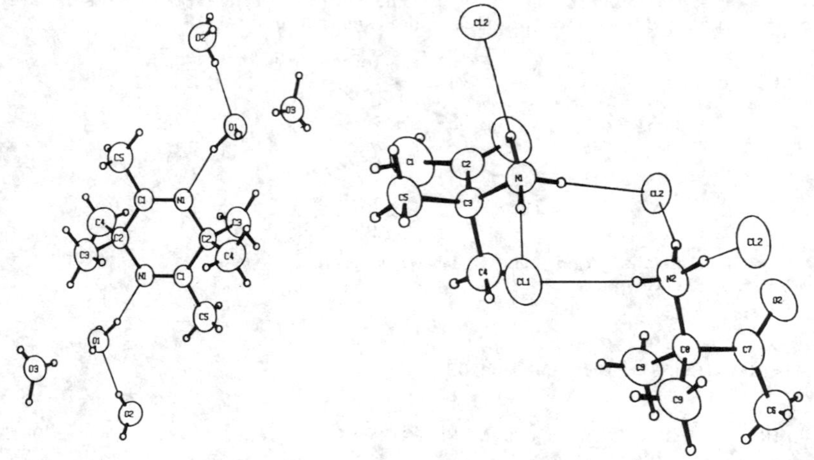

Fig. 1. Views of $C_{10}H_{18}N_2$,6(H_2O) (I) (left) and of $C_5H_{12}NO^+.Cl^-$ (II) (right).

4-(PHENYL-β-AZOXY)PYRIDINE N-OXIDE
$C_{11}H_9N_3O_2$ (I)

4-(PHENYL-α-AZOXY)PYRIDINIUM METHIODIDE
$C_{12}H_{12}IN_3O$ (II)

E. BUNCEL, S.R. KEUM, M. CYGLER, K.I. VARUGHESE and G.I. BIRNBAUM, 1984. Can. J. Chem., 62, 1628-1639.

I. Monoclinic, $P2_1/c$, a = 14.984, b = 5.627, b = 25.484 Å, β = 109.93°, D_m = 1.41, Z = 8. Cu radiation, R = 0.041 for 2484 reflexions.

II. Triclinic, $P\bar{1}$, a = 7.477, b = 9.939, c = 10.073 Å, α = 65.39, β = 70.38, γ = 79.92°, Z = 2. Cu radiation, R = 0.043 for 1718 reflexions.

In I there are two independent planar molecules in the asymmetric unit (Fig. 1), with N(2)-N(3) 1.274, N(2)-O(2) 1.265 Å. Molecule II (Fig. 1) is not planar, the pyridine ring being rotated 36.5° from the phenylazoxy plane; the N(2)-N(3) distance is 1.281(6) and N(3)-O(1) 1.255(6) Å. Evidence is presented for the existence of intramolecular C-H...O and C-H...N bonds in crystal structures of trans-azoxyarenes.

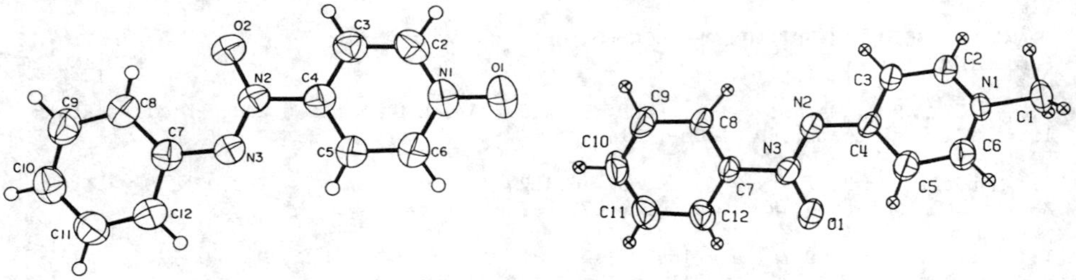

Fig. 1. Views of $C_{11}H_9N_3O_2$ (I) (right) and the cation in $C_{12}H_{12}IN_3O$ (II) (left).

4-(p-CHLOROBENZOYL)PYRIDINE
$C_{12}H_8ClNO$ $Cl-C_6H_4-CO-C_5H_5N$

A. SYED, L.I. FOSS and E.D. STEVENS, 1984. Acta Cryst., C40, 834-836.

Monoclinic, $P2_1$, a = 21.585, b = 5.762, c = 3.843 Å, β = 90.27°, Z = 2. Mo radiation, R = 0.0472 for 1853 reflexions (at 110K).

In the molecule the dihedral angle between the essentially planar aromatic rings is 52.2°; the repulsions between the non-bonded atoms are thus minimised. The carbonyl portion of the molecule exhibits a high degree of planarity and an increase in the endocyclic angle opposite the chlorine substituent is observed. The molecules are packed in the crystal in parallel stacks at normal van der Waals distances.

erythro-(5RS,7SR)-5-(α-HYDROXY-2-CHLOROBENZYL)-6-METHYL-4,5-DIHYDRO-2H-PYRIDAZIN-3-ONE
$C_{12}H_{13}ClN_2O_2$

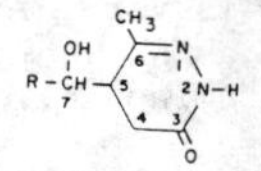

J. TAOUFIK, J.D. COUQUELET, J.M. COUQUELET and A. CARPY, 1984. J. Heterocycl. Chem., 21, 305-310.

Monoclinic, $P2_1/n$, a = 17.445, b = 7.079, c = 18.814 Å, β = 91.96°, Z = 8. Cu radiation, R = 0.041 for 2464 reflexions.

The analysis was used to establish the erythro-(5RS,7SR) configuration.

N-(2-PIPERIDINOETHYL)PIPERIDINIUM N'-OXIDE PERCHLORATE MONOHYDRATE
$C_{12}H_{25}ClN_2O_5,H_2O$ $(C_{12}H_{25}N_2O)^+,ClO_4^-,H_2O$

M. JASKÓLSKI and Z. KOSTURKIEWICZ, 1984. Acta Cryst., C40, 869-871.

Monoclinic, $P2_1/n$, a = 11.366, b = 6.540, c = 22.082 Å, β = 95.52°, Z = 4. Cu radiation, R = 0.062 for 1892 reflexions.

The $C_{12}H_{25}N_2O^+$ ion is cisoidal with a very short intramolecular O...H...N hydrogen bond with O...N 2.551(4) Å; the H atom in this heteroatomic bridge appears to be more or less central (Fig. 1). Water molecules and the cations are hydrogen-bonded to form infinite chains in which the H_2O molecules are double donors and the N-O groups double acceptors. The ClO_4^- ion is not involved in hydrogen bonding and its O atoms show relatively high thermal motion.

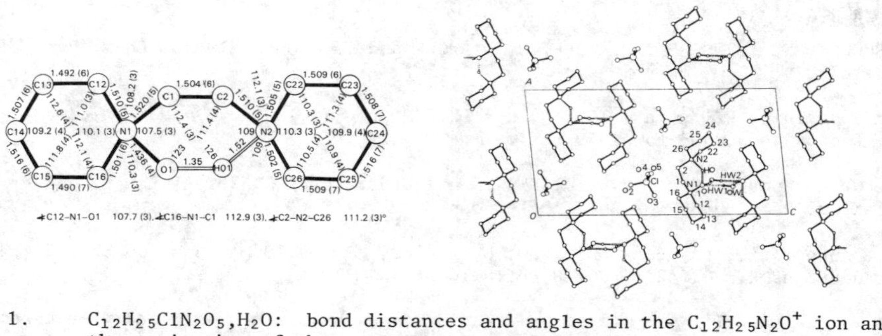

Fig. 1. $C_{12}H_{25}ClN_2O_5,H_2O$: bond distances and angles in the $C_{12}H_{25}N_2O^+$ ion and the projection of the crystal structure down b.

1-(2-FLUORO-2-PHENYLETHYL)PIPERIDINE HYDROCHLORIDE (ENANTIOMER)
$C_{13}H_{19}ClFN$ (I) $C_{13}H_{18}FN,HCl$

1-(2-FLUORO-2-PHENYLETHYL)PIPERIDINE HYDROCHLORIDE (RACEMIC) SOLID SOLUTION
$C_{13}H_{19}ClFN$ (II) $C_{13}H_{18}FN,HCl$

B. CHION, J. LAJZEROWICZ, C. CHARLON, D. LE BARS and C.L. DUC, 1984. Acta Cryst., C40, 709-712.

I. Orthorhombic, $P2_12_12_1$, a = 7.171, b = 8.020, c = 22.73 Å, Z = 4. Cu radiation, R = 0.055 for 716 reflexions.

II. Orthorhombic, Pnma, a = 7.190, b = 8.022, c = 22.81 Å, Z = 4. Cu radiation, R = 0.092 for 1116 reflexions.

Bond distances and angles for the enantiomer (I) are given in Fig. 1(a); the molecule has a pseudo-mirror plane. The orientation of the enantiomer molecule (I) is very close to one of the two possible orientations in the solid solution (II), as shown in Fig. 1(b); the molecule model for II is similar to those of (phenylethyl)-amines already known.

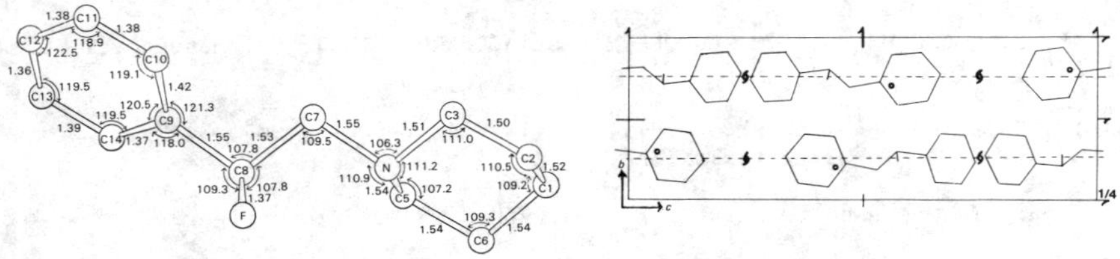

Fig. 1. $C_{13}H_{18}FN,HCl$: (a) bond distances and angles in the enantiomer (I), (b) projection of the enantiomer structure (I) down a; the solid-solution mirror is on the dashed line.

2-{(Z)-1-[1-(2,2-DIMETHYLHYDRAZINO)ETHYLIDENE]-2-OXOPROPYL}-(E)-BUTENEDICARBOXYLIC ACID DIMETHYL ESTER
$C_{13}H_{20}N_2O_5$ (I)

(3aα,4α,5β,5aα)-3-ACETYL-1-(DIMETHYLAMINO)-3a,4,5,5a-TETRAHYDRO-2-METHYL-1H-CYCLOBUTA-[b]PYRROLE-3a,4,5,5a-TETRACARBOXYLIC ACID TETRAMETHYL ESTER
$C_{19}H_{26}N_2O_9$ (II)

1-(DIMETHYLAMINO)-1,2-DIHYDRO-3-(METHOXYCARBONYL)-5-[(E)-2-(METHOXYCARBONYL)ETHENYL]-6-METHYL-2-PYRIDINEACETIC ACID METHYL ESTER
$C_{17}H_{24}N_2O_6$ (III)

W. SUCROW, W. TURNSCHEK, U. WOLF, H. D'AMOUR and C. KRUGER, 1984. Chem. Ber., 117, 1620-1635.

I. Monoclinic, $P2_1/n$, a = 10.704, b = 12.849, c = 11.159 Å, β = 93.45°, Z = 4. Mo radiation, R = 0.0486 for 1879 reflexions.

II. Orthorhombic, Pbca, a = 10.7910, b = 15.6021, c = 25.3231 Å, Z = 8. Cu radiation, R = 0.0485 for 3042 reflexions.

III. Monoclinic, $P2_1/c$, a = 10.430, b = 8.062, c = 22.705 Å, β = 100.52°, Z = 4. Mo radiation, R = 0.042 for 1138 reflexions.

Molecules I-III are formed on reaction of enehydrazines with acetylenecarboxylic esters. The structures (Fig. 1) were determined by X-ray analysis.

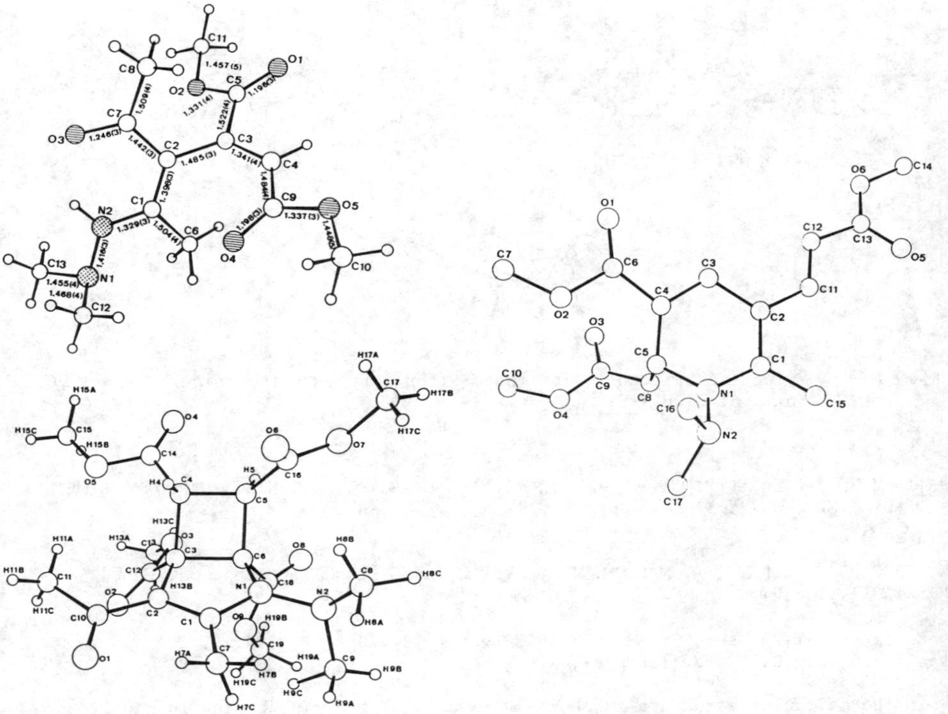

Fig. 1. Views of methyl esters I (top left), II (lower left) and III (right).

(±)-α-1,2,3-TRIMETHYL-4-PHENYL-4-PIPERIDINOL
$C_{14}H_{21}NO$ (I)

(±)-β-1,2,3-TRIMETHYL-4-PHENYL-4-PIPERIDINOL
$C_{14}H_{21}NO$ (II)

(±)-γ-1,2,3-TRIMETHYL-4-PHENYL-4-PIPERIDINOL HYDROCHLORIDE
$C_{14}H_{22}ClNO$ (III)

M. CYGLER and F.R. AHMED, 1984. Acta Cryst., B40, 436-440.

I. Orthorhombic, $Pna2_1$, a = 10.137, b = 11.446, c = 10.815 Å, D_m = 1.162, Z = 4. Cu radiation, R = 0.032 for 1294 reflexions.

II. Monoclinic, $P2_1/a$, a = 18.434, b = 9.391, c = 7.445 Å, β = 91.27°, D_m = 1.130, Z = 4. Cu radiation, R = 0.043 for 1617 reflexions.

III. Monoclinic, $P2_1/a$, a = 11.756, b = 16.299, c = 7.483 Å, β = 100.19°, D_m = 1.206, Z = 4. Cu radiation, R = 0.041 for 1733 reflexions.

The three isomers (Fig. 1) have the piperidine ring in the chair conformation, with the phenyl ring equatorial. Hydrogen bonding is present in all three structures. In both the α- and β-bases, intermolecular O-H...N' bonds link the molecules into chains along the a and b directions, respectively. In the γ-HCl, there are no direct links between the molecules, but each Cl accepts two hydrogen bonds to bridge the molecules into chains along a. The first is N-H...Cl and the second is O'-H'...Cl with an H...Cl...H' angle of 105(1)°.

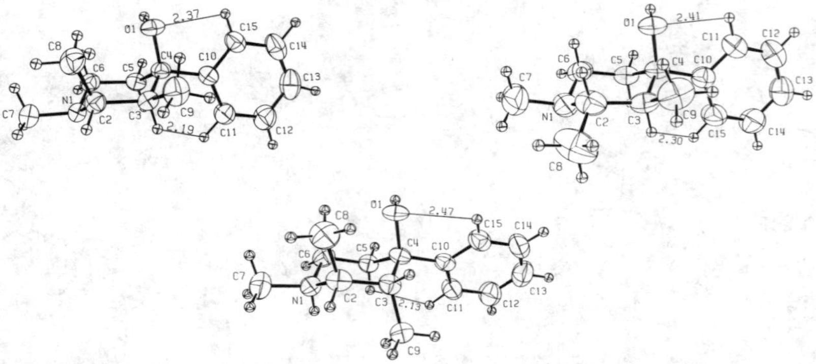

Fig. 1. Molecular structures for the α- (top left), β- (top right), and γ-HCl forms of $C_{14}H_{21}NO$.

2-(1-AMINO-2-ETHOXYCARBONYL-2-ACETYL)-VINYL-4-CARBOXAMIDO-5-ETHOXYCARBONYL-6-METHYL-1,3-DIAZINE
$C_{16}H_{20}N_4O_6$

M. BASOTO, B. CORAIN, A. MARCOMINI, G. VALLE and G. ZANOTTI, 1984. J. Chem. Soc. Perkin II, 965-973.

Monoclinic, $P2_1/n$, a = 13.334, b = 10.576, c = 12.679 Å, β = 90.5°, Z = 4. Mo radiation, R = 0.107 for 1289 reflexions.

The molecular structure is shown in Fig. 1. The pyridine ring is planar and coplanar with the N(4)-C(14)-O(4) group while it makes an angle of 56° with the C(6), C(7),N(1),C(3) moiety. N(1) makes a short (2.60 Å) intramolecular contact with the O(2) atom. The N(2)-C(7)-N(3) angle is large at 129(1)°. There is disorder in the crystal in the region of C(15).

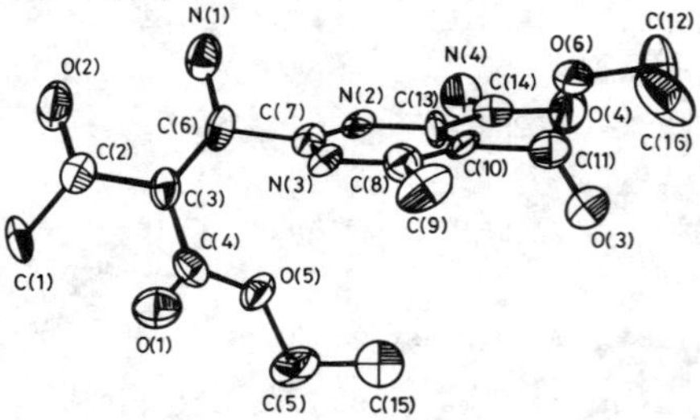

Fig. 1. $C_{16}H_{20}N_4O_6$: the molecular structure.

α(-)-1-ALLYL-2,3-DIMETHYL-3-ARYLPIPERIDINE HYDROBROMIDE
$C_{16}H_{24}BrNO$ $C_{16}H_{23}NO,HBr$

F.R. AHMED, 1984. Acta Cryst., C40, 1021-1023.

Monoclinic, $P2_1$, a = 9.068, b = 11.720, c = 7.693 Å, β = 94.44°, D_m = 1.329, Z = 2. Cu radiation, R = 0.031 for 1637 reflexions.

In the molecule the piperidine ring has the chair form with the 1-allyl and 3-aryl groups both equatorial, and the 2-methyl and 3-methyl groups both axial (Fig. 1). The N-C bonds of 1.502(6)-1.527(5) Å are slightly elongated as a result of protonation of the N atom. The $C(sp^3)$-$C(sp^3)$ bonds of 1.504(7)-1.559(5) Å and the C(sp^2)-C(sp^2) of 1.371(6)-1.398(6) Å are in the ranges commonly observed. The C(16)=C(17) double bond is only 1.299(8) Å, indicating a partial triple-bond character. The molecules are interlinked through N-H...Br...H'-O' hydrogen bonds into spirals parallel to b.

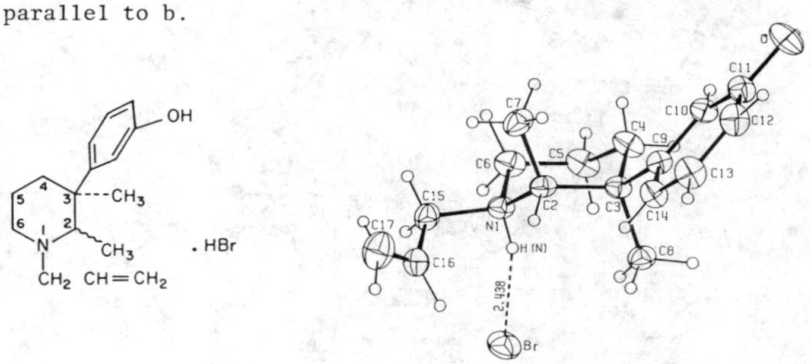

Fig. 1. $C_{16}H_{24}BrNO$: structural formula and molecular structure.

1-METHYL-5,6-DIPHENYLPYRAZIN-2-ONE
$C_{17}H_{14}N_2O$ (I, IV)

1-ETHYL-5,6-DIPHENYLPYRAZIN-2-ONE
$C_{18}H_{16}N_2O$ (II)

3,7-DIMETHYL-2,6,10,12-TETRAPHENYL-3,7,9,11-TETRAAZATRICYCLO[4.2.2.$2^{2,5}$]DODECA-9,11-DIENE-4,8-DIONE
$C_{34}H_{28}N_4O_2$ (III)

M. KAFTORY, 1984. J. Chem. Soc. Perkin II, 757-765.

I. Monoclinic, $P2_1$, a = 18.765, b = 6.584, c = 11.585 Å, β = 107.16°, Z = 4. Mo radiation R = 0.049 for 2207 reflexions. [Light sensitive modification.]

II. Triclinic, $P\bar{1}$, a = 14.994, b = 11.500, c = 8.754 Å, α = 87.13, β = 95.33, γ = 98.81°, Z = 4. Mo radiation, R = 0.066 for 3847 reflexions.

III. Tetragonal, $I4_1/a$, a = 22.93, c = 25.35 Å, Z = 16. Cu radiation, R = 0.11 for 3302 reflexions. [The crystal lost solvent during data collection and became opaque.]

IV. Orthorhombic, $P2_12_12_1$, a = 13.082, b = 10.374, c = 9.963 Å, Z = 4. Mo radiation, R = 0.051 for 1372 reflexions. [Light stable modification.]

Packing of the molecules in I and II is very similar. Molecules, arranged in parallel stacks, run along the shortest axis. The stacks are composed of two crystallographically independent molecules (Fig. 1) with their central portions parallel to each other. The angles between the phenyl and pyrazinone rings are 4.6 and 3.9° respectively. In I the independent molecules are related by a pseudo twofold axis while in II they are related by a pseudo-inversion centre. In I the proximity of the reaction centres ranges from 3.456 to 3.640 Å. In II there are two categories of distance - 4.295 and 4.356, and 5.538 and 5.656 Å. In IV, the light stable modification of I, the molecules are not packed with their central portion in a parallel fashion which is essential for cyclophoto dimerisation, the distances between possible reaction centres being >5.5 Å. In III, the photodimer, channels are formed along c in which solvent (ethyl acetate) molecules are occluded in a disordered fashion and are poorly resolved. The molecular structure is that of the syn-trans isomer. Bond lengths and angles in the five independent monomers are in good agreement.

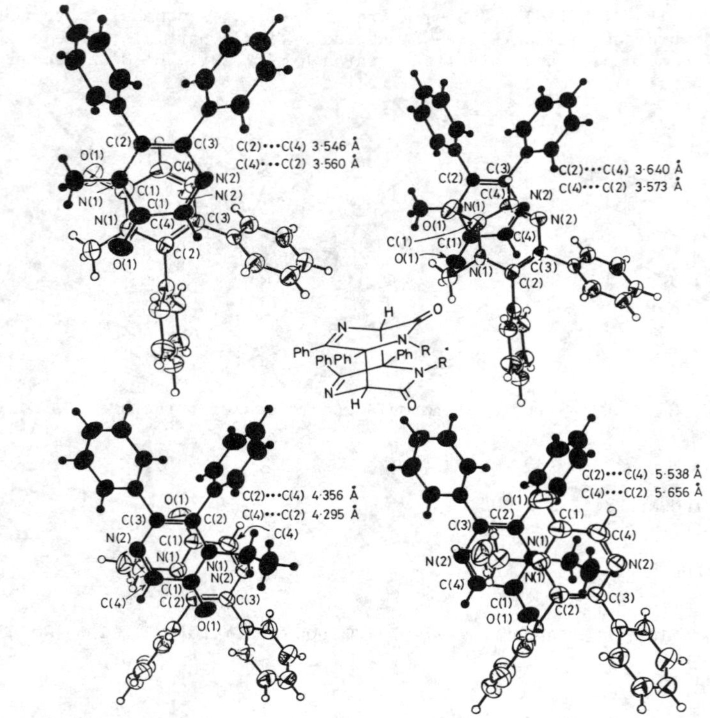

Fig. 1. Two types of overlap diagrams of the two independent molecules in $C_{17}H_{14}N_2O$ (I) (top) and $C_{18}H_{16}N_2O$ (bottom). The molecular skeleton for $C_{34}H_{28}N_4O_2$ is inset.

2-(1,2-DIHYDRO-1-METHYL-2-PYRIDYLIDENE)-3-(4-ETHOXYPHENYL)-3-OXOPROPANENITRILE
$C_{17}H_{16}N_2O_2$

K.R. ACHARYA, S.S. TAVALE and T.N.G. ROW, 1984. Acta Cryst., C40, 873-875.

Triclinic, P$\bar{1}$, a = 7.199, b = 8.473, c = 12.360 Å, α = 101.73, β = 94.37, γ = 100.10°, D_m = 1.24, Z = 2. Mo radiation, R = 0.052 for 1082 reflexions.

In the molecule the C=C distance is 1.438(6) Å, appreciably longer than in ethylene (1). The steric and electronic effects result in rotation about the C=C bond, the rotation angle being 40.7(6)° (Fig. 1). The crystal packing is solely through van der Waals interactions.

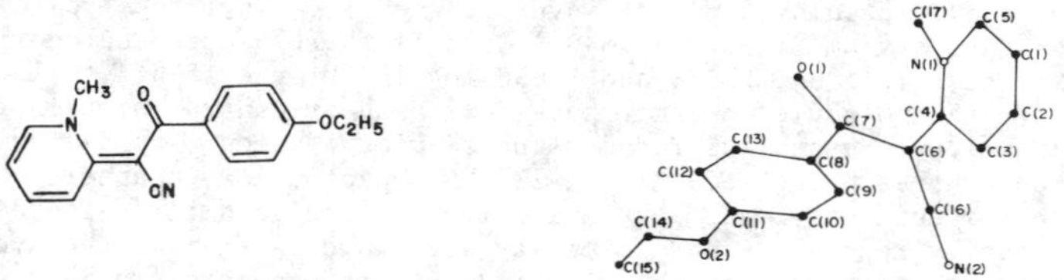

Fig. 1. $C_{17}H_{16}N_2O_2$: structural formula and perspective view of the molecule.

1. Structure Reports, 30B, 405.

(E)-2-(N-PHENYLBENZALDEHYDE HYDRAZONE-C[1])PYRIDINE
$C_{18}H_{15}N_3$

R.N. BUTLER and S.M. JOHNSON, 1984. J. Chem. Soc. Perkin I, 2109-2116.

Monoclinic, $P2_1/c$, a = 10.820, b = 9.335, c = 14.940 Å, β = 73.87°, Z = 4. Mo radiation, R = 0.059 for 1015 reflexions.

In the molecule (Fig. 1) the phenyl ring, which is cis- to the NHAr group, was twisted at an angle of 106.4° to the C(8)-C(1)-N(1) plane, Bond lengths include: C(1)-C(8) 1.487(6), C(1)-C(13) 1.488(7), C(1)-N(1) 1.283(6), N(1)-N(2) 1.352(6), and N(2)-C(2) 1.400(6) Å.

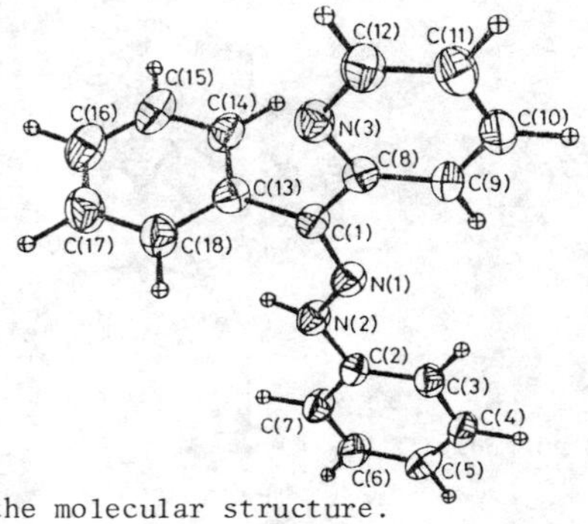

Fig. 1. $C_{18}H_{15}N_3$: the molecular structure.

1-(2,6-DIMETHYLPHENYL)-1,2-DIHYDRO-6-PHENYL-2,2,4-TRIS(TRIFLUOROMETHYL)-1,3,5-TRIAZINE
$C_{20}H_{14}F_9N_3$

A. GIEREN, T. HÜBNER, C.-P. KAERLEIN, K. BURGER, U. WASSMUTH and H. PARTSCHT, 1984. Chem. Zeit., 108, 207-208.

Monoclinic, $P2_1/n$, a = 14.513, b = 16.681, c = 16.996 Å, β = 102.84°, Z = 8. Mo radiation, R = 0.042 for 3169 reflexions.

There are few differences which are significant in the molecular geometry of the two independent molecules in the asymmetric unit. The maximum difference in bond angles is 1.6°, and 4° in torsion angles. The 1,2-dihydro-1,3,5-triazine ring is in a tub conformation (Fig. 1) with atoms N(1), N(3), C(4) and C(6) almost coplanar, and atoms C(2) and N(5) 0.43 and 0.16 Å from the plane. Bond lengths within that ring are: N(1)-C(2) 1.491, N(1)-C(6) 1.377, C(6)-N(5) 1.304, N(5)-C(4) 1.367, C(4)-N(3) 1.271 and N(3)-C(2) 1.453 Å. The phenyl ring at C(6) is twisted slightly out of the plane of N(1), N(3), C(4) and C(6) by 29° but the dimethylphenyl ring at N(1) is almost (87°) at right angles to this plane.

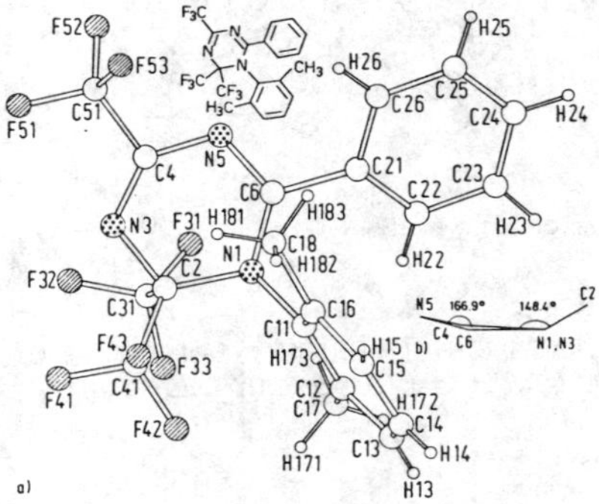

Fig. 1. Projection of $C_{20}H_{14}F_9N_3$ on the plane N(1), N(3) and C(4).

3-ACETONYLIDENE-5-BENZOYL-2-PHENYL-1,2,3,4-TETRAHYDROPYRAZINE
$C_{20}H_{18}N_2O_2$

B. VIOSSAT, N. RODIER, Y. METTEY and J.-M. VIERFOND, 1984. Acta Cryst., C40, 332-334.

Triclinic, $P\bar{1}$, a = 11.24, b = 10.97, c = 7.861 Å, α = 77.05, β = 106.27, γ = 119.47°, D_m = 1.35, Z = 2. Mo radiation, R = 0.046 for 1604 reflexions.

The dimensions are in accord with the molecular formulation shown in Fig. 1. The crystal structure consists of layers which spread out along the (001) planes. These layers are linked by N-H...O hydrogen bonds. There is also an intramolecular N-H...O hydrogen bond.

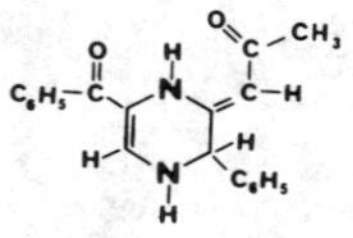

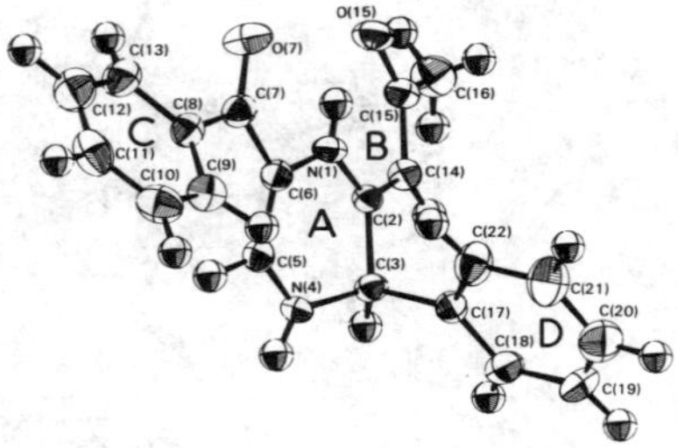

Fig. 1. The $C_{20}H_{18}N_2O_2$ molecule.

2,3-DIACETOXY-5,6-DIPHENYL-2,3-DIHYDROPYRAZINE
$C_{20}H_{18}N_2O_4$

A. OHTA, M. INOUE, J. YAMADA, Y. YAMADA, T. KURIHARA and T. HONDA, 1984. J. Heterocycl. Chem., 21, 103-105.

Triclinic, $P\bar{1}$, a = 10.206, b = 11.480, c = 9.197 Å, α = 106.96, β = 106.27, γ = 106.63°, Z = 2. Mo radiation, R = 0.059 for 2469 reflexions.

The analysis established that the acetoxy groups have a trans-arrangement (Fig. 1).

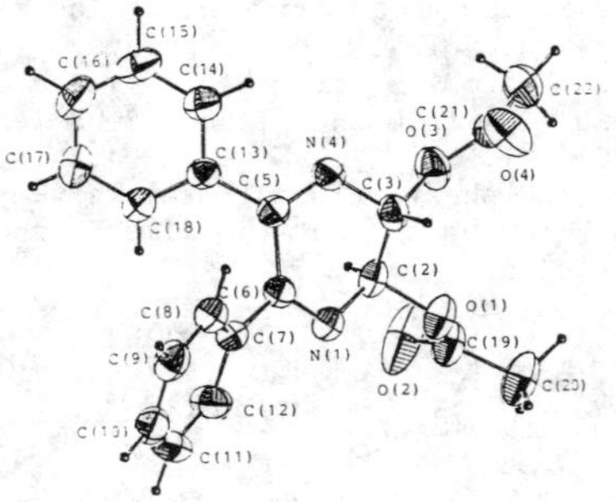

Fig. 1. A perspective view of $C_{20}H_{18}N_2O_4$.

(4α,5α,6β,(Z))-1,4,5,6-TETRAHYDRO-6-(3-METHOXY-2-METHYL-3-OXO-1-PHENYL-1-PROPENYL)-1-METHYL-4,5-PYRIDAZINEDICARBOXYLIC ACID DIMETHYL ESTER
$C_{20}H_{24}N_2O_6$ (I)

(1α,2β,3β)-3(2-FORMYL-1,2-DIMETHYLHYDRAZINO)-5-METHYL-6-OXO-4-PHENYL-4-CYCLOHEXENE-1,2-DICARBOXYLIC ACID DIMETHYL ESTER
$C_{20}H_{24}N_2O_6$ (II)

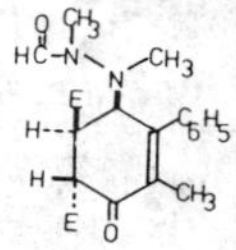

5,7a-DIHYDRO-1,6-DIMETHYL-5-METHYLENE-7-PHENYL-PYRANO[3,2-c]PYRAZOLE-3,3a(1H)-DICARBOXYLIC ACID DIMETHYL ESTER
$C_{19}H_{20}N_2O_5$ (III)

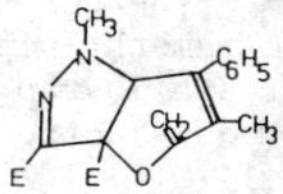

J.A. MOORE, O.S. ROTHENBERGER, W.C. FULTZ and A.L. RHEINGOLD, 1984. J. Org. Chem., 49, 1261-1265.

I. Monoclinic, $P2_1/c$, a = 10.198, b = 13.634, c = 15.132 Å, β = 104.97°, Z = 4. Mo radiation, R = 0.0747 for 1639 reflexions.

II. Monoclinic, $P2_1/n$, a = 19.855, b = 7.563, c = 26.820 Å, β = 104.91°, Z = 8. Mo radiation, R = 0.0802 for 3635 reflexions.

III. Monoclinic, C2/c, a = 29.82, b = 10.085, c = 12.377 Å, β = 106.34°, Z = 8. Mo radiation, R = 0.0640 for 1260 reflexions.

The structures of these rearrangement products were determined by X-ray methods.

(2S-(2α,3α,3aα,6aα))-HEXAHYDRO-2-METHYL-2-NITRO-3-PHENYL-6a-(1-PIPERIDINYL)-1(2H)-PENTALENONE
$C_{20}H_{26}N_2O_3$

G. BARBARELLA, S. BRUCKNER, G. PITACCO and E. VALENTIN, 1984. Tetrahedron, 40, 2441-2449.

Monoclinic, $P2_1/c$, a = 9.024, b = 16.754, c = 24.036 Å, β = 91.46°, Z = 8. Mo radiation, R = 0.054 for 5654 reflexions.

The analysis establishes the structure as shown in Fig. 1.

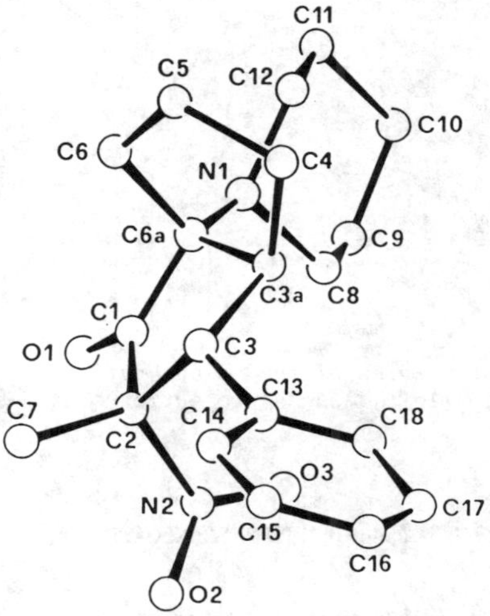

Fig. 1. A view of one of the two independent $C_{20}H_{26}N_2O_3$ molecules.

N,N,N',N'-TETRAMETHYLPIPERAZINEDIIUM BIS(2,4-DINITROPHENOLATE) DIHYDRATE
$C_{20}H_{26}N_6O_{10},2H_2O$ $(C_8H_{20}N_2)^{2+}2(C_6H_3N_2O_5)^-,2H_2O$

F. IWASAKI and K. MUTAI, 1984. Bull. Chem. Soc. Japan, 57, 1295-1298.

Triclinic, $P\bar{1}$, a = 11.504, b = 8.239, c = 7.119 Å, α = 106.22, β = 99.67, γ = 92.31°, Z = 1. Mo radiation, R = 0.067 for 2743 reflexions.

The compound consists of two 2,4-dinitrophenolate anions, one N,N,N',N'-tetramethylpiperazinediium cation and two water molecules (Fig. 1). The cation has crystallographic inversion symmetry and is connected to the two anions by hydrogen bonds via the water molecules forming a dimeric unit. Bond lengths in the anion are: C(1)-O(1) 1.254, C(1)-C(2) 1.440, C(2)-C(3) 1.379, C(3)-C(4) 1.373, C(4)-C(5) 1.402, C(5)-C(6) 1.351, C(6)-C(1) 1.432, C(2)-N(2) 1.444, N(2)-O(21) 1.203, N(2)-O(22) 1.196, C(4)-N(4) 1.436, N(4)-O(41) 1.218, N(4)-O(42) 1.236 Å, and in the cation: NP-CP(3) 1.507, NP-CP(1) 1.504, NP-CP(2) 1.506, NP-CP(4) 1.498, CP(1)-CP(2') 1.502 Å. The hydrogen bond distances are: O(1)...O(W) 2.764 and O(1)...O(W') 2.753 Å.

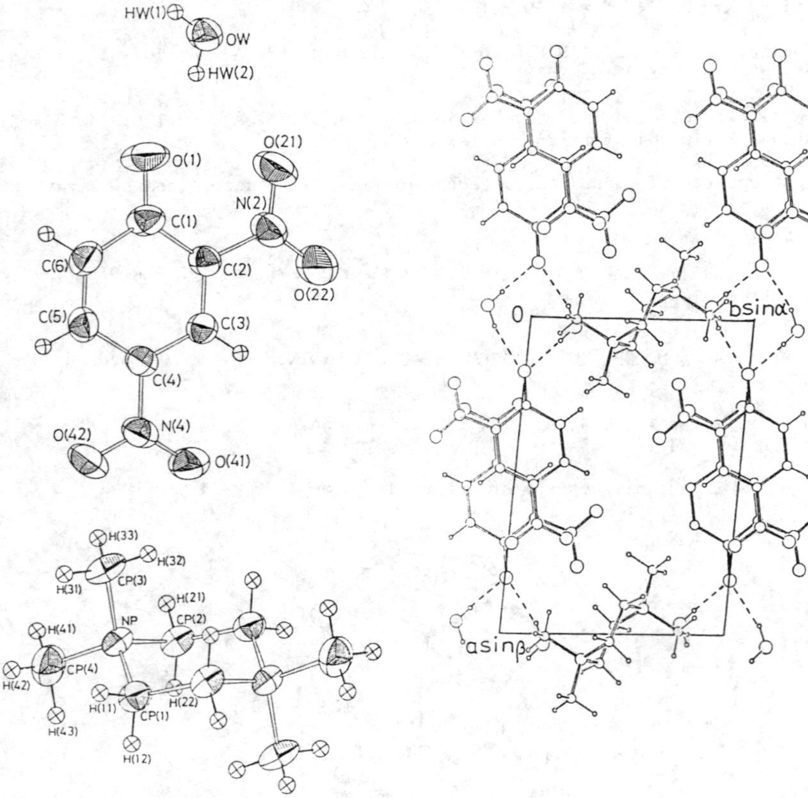

Fig. 1. The molecular structures of the anion, cation and water molecule in $C_{20}H_{26}N_6O_{10},2H_2O$ (left), and a projection of the structure along the c axis.

6-HEXYLAMINO-1,2,4,5-TETRAZIN-3-YL 4-PENTYLOXYBENZOATE
$C_{20}H_{29}N_5O_3$

H. HARTUNG, U. BAUMEISTER, F. THIELEMANN and M. JASKÓLSKI, 1984. Acta Cryst., C40, 482-484.

Triclinic, $P\bar{1}$, a = 11.646, b = 13.468, c = 7.321 Å, α = 103.51, β = 106.26, γ = 79.59°, D_m = 1.17, Z = 2. Cu radiation, R = 0.054 for 2293 reflexions.

The almost fully stretched molecule (Fig. 1) shows normal bond lengths and angles. The plane of the bridging carboxylic group is inclined to those of the benzene and tetrazine rings by 12.1 and 76.3°, respectively. Two of the hexyl-chain torsion angles are in the gauche region. The molecules are packed with their long axes parallel.

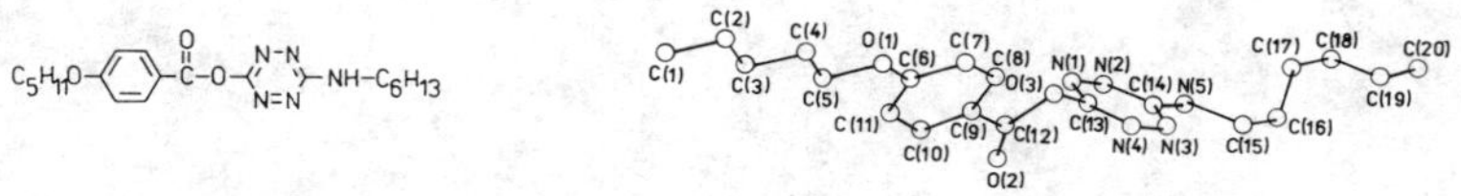

Fig. 1. The $C_{20}H_{29}N_5O_3$ molecule.

sym-TRIPHENYLTRIAZINE
$C_{21}H_{15}N_3$

S.V. LINDEMAN, V.E. SHKLOVER, Yu.T. STRUCHKOV, L.M. MITINA and V.A. PANKRATOV, 1984. J. Struct. Chem., 25, 162-164 [Zh. Strukt. Khim. 25(1), 180-181].

Monoclinic, $P2_1/c$, a = 10.958, b = 3.9113, c = 35.921 Å, β = 90.96°, Z = 4. Cu radiation, R = 0.041 for 2229 reflexions.

Bond lengths and angles (σ = 0.002 Å and 0.1°) are shown in Fig. 1. The phenyl rings are rotated out of the plane of the central C_3N_3 ring by 10.4, 6.6 and 7.5° for the A, B, and C rings, respectively.

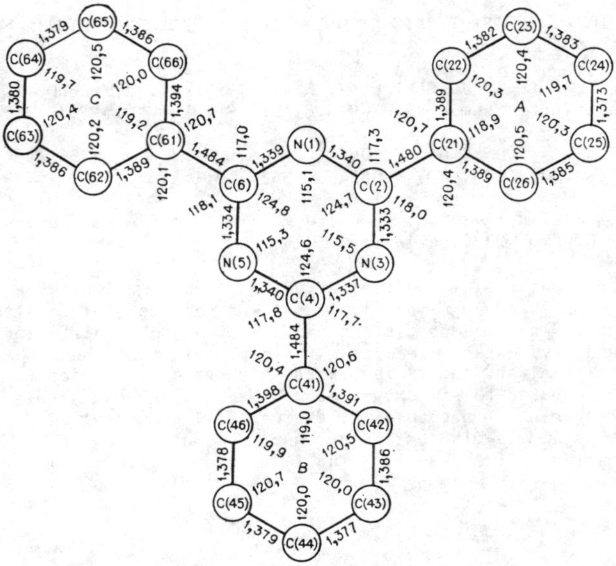

Fig. 1. Bond distances and angles in sym-triphenyltriazine.

4-p-CHLOROPHENYL-1-METHYL-2,6-DIPHENYLPYRIDINIUM IODIDE
$C_{24}H_{19}ClIN$ $C_{24}H_{19}ClN^+.I^-$

F. FLORENCIO and S. GARCIA-BLANCO, 1984. Acta Cryst., C40, 2086-2088.

Monoclinic, $P2_1/n$, a = 10.404, b = 10.475, c = 19.982 Å, β = 104.57°, Z = 4. Mo radiation, R = 0.073 for 2675 reflexions.

The cation (Fig. 1) has approximate m-symmetry. Bond lengths and angles are unexceptional. The positive charge is localized on the N atom. Packing is due to coulombic and van der Waals forces.

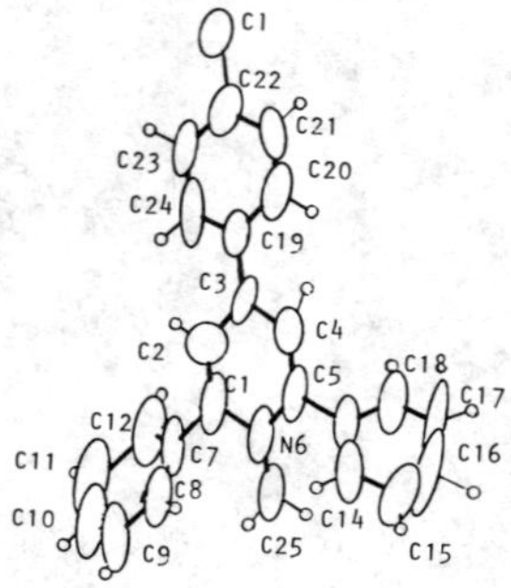

Fig. 1. The structure of $C_{24}H_{19}ClN^+$.

DISODIUM BIS[2,6-DIMETHYL-3,5-BIS(ETHOXYCARBONYL)-1,4-DIHYDROISONICOTINATE] PENTA-HYDRATE
$C_{28}H_{36}N_2Na_2O_{12}.5H_2O$ (I) $2(C_{14}H_{18}NO_6^-)2(Na^+).5H_2O$

SODIUM 2,6-DIMETHYL-3,5-BIS(ETHOXYCARBONYL)-1,4-DIHYDROISONICOTINATE MONOHYDRATE
$C_{14}H_{18}NNaO_6.H_2O$ (II)

A.A. KEMME, M.F. BUNDULE, J.J. BLEIDELIS, E.A. BISENIEKS, J.R. ULDRIKIS and G.J. DUBUR, 1984. J. Struct. Chem., 25, 309-316 [Zh. Strukt. Khim., 25(2), 155-162].

I. Triclinic, $P\bar{1}$, a = 9.036, b = 13.069, c = 15.609 Å, α = 98.35, β = 102.66, γ = 98.59°, Z = 2. Cu radiation, R = 0.063 for 3314 reflexions.

II. Monoclinic, C2/c, a = 18.243, b = 11.097, c = 16.026 Å, β = 95.24°, Z = 8. Cu radiation, R = 0.070 for 2276 reflexions.

The asymmetric units of I and II (including anion bond lengths and bond angles) are shown in Figures 1 and 2, respectively. The structure of I consists of two crystallographically independent cations and anions A and B, differing in their configuration, and five water molecules. The carbonyl groups have the s-trans-s-trans arrangement in anion A, and the s-cis-s-trans arrangement in anion B. The Na^+ cations in I are both coordinated to six oxygen atoms (Na-O 2.32-2.61 Å) in an edge-shared bioctahedral arrangement. In complex II the carbonyl groups and C=C bonds of the 1,4-dihydropyridine ring have s-cis configurations. The Na^+ cation in II is coordinated to five oxygen atoms (Na-O 2.29-2.46 Å). The 1,4-dihydro-pyridine rings in both structures have boat conformations.

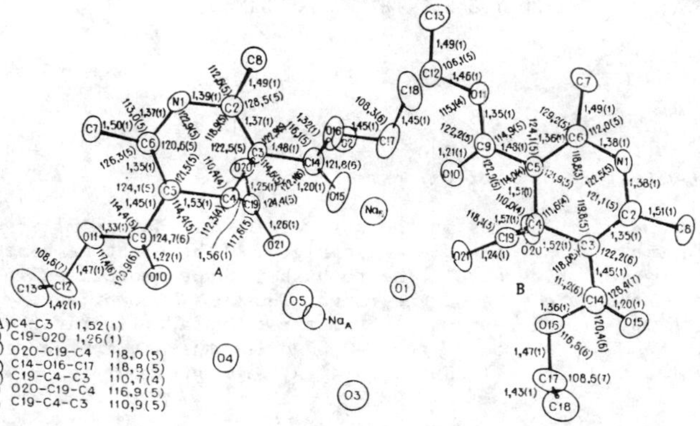

Fig. 1. The asymmetric unit of disodium bis(2,6-dimethyl-3,5-bis(ethoxycarbonyl)-1,4-dihydroisonicotinate) pentahydrate.

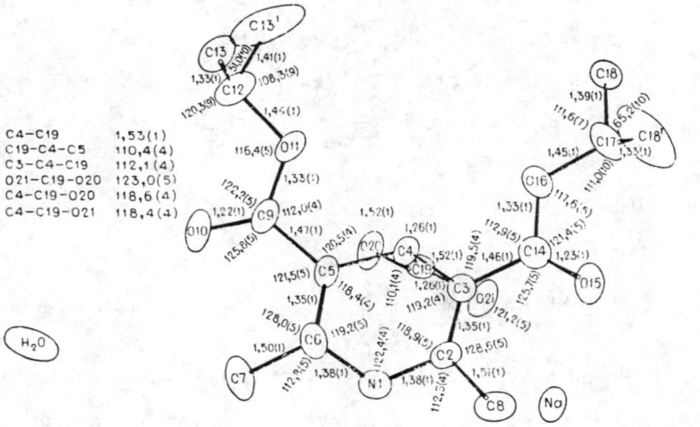

Fig. 1. The asymmetric unit of sodium 2,6-dimethyl-3,5-bis(ethoxycarbonyl)-1,4-dihydroisonicotinate monohydrate

1,3,5-TRIS(2,2,6,6-TETRAMETHYL-4-PIPERIDYL-N-OXIDE)-1,3,5-TRIAZACYCLOHEXANE MONOHYDRATE

$C_{30}H_{57}N_6O_3, H_2O$

A.A. SHEVYREV, R.M. LOBKOVSKAYA and R.P. SHIBAEVA, 1984. Kristallografiya, 29, 279-284 [Sov. Phys. Crystallogr., 29, 166-169].

Orthorhombic, Pbca, a = 34.501, b = 17.114, c = 11.469 Å, Z = 8. Cu radiation, R = 0.058 for 2122 reflexions.

All the six-membered hetero rings have distorted chair conformations. Relative to the central ring A, the hetero rings B and C occupy equatorial positions, and ring D is in an axial position. The N-O bond lengths are 1.298, 1.293 and 1.288 Å and are considered normal for nitroxyls. The distances between the paramagnetic centres of adjacent molecules in the crystal (N...N 5.43-5.86, 6.94 Å) are shorter than the intramolecular distances 9.91 Å, 8.50 Å and 8.62 Å. The water molecules link the triradicals into chains parallel to [001] by means of hydrogen bonds (3.005 Å, 2.924 Å).

1-(AZIDOMETHYL)-3,5,7-TRINITRO-1,3,5,7-TETRAAZACYCLOOCTANE
$C_5H_{10}N_{10}O_6$

T.B. BRILL, R.J. KARPOWICZ, T.M. HALLER and A.L. RHEINGOLD, 1984. J. Phys. Chem., 88, 4138-4143.

Monoclinic, $P2_1$, a = 12.994, b = 5.920, c = 15.697 Å, β = 96.45°, Z = 4. Mo radiation, R = 0.0366 for 1856 reflexions.

The two independent molecules in the asymmetric unit have similar conformations and geometries; a view of one molecule is in Fig. 1. The eight-membered ring has a chair-chair conformation which gives rise to basket shaped molecules in which all substituents are on the same side of the ring. The CH_2-N_3 bond is very long (1.515(5) Å) which may explain why HN_3 is the first product liberated during thermolysis. At the N-nitro groups, the N-O bond lengths are in the range 1.204-1.222(5) Å while the N-N bonds (1.359-1.389(4) Å) are intermediate between single and double bond lengths.

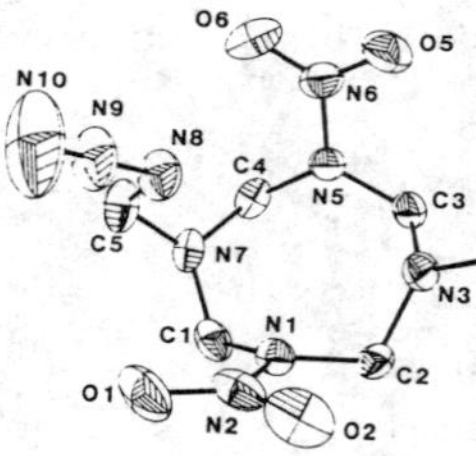

Fig. 1. A view of $C_5H_{10}N_{10}O_6$.

1H-AZEPINE-2,5-DIONE 5-(O-METHYLOXIME)
$C_7H_8N_2O_2$

D.T. CROMER, R.R. RYAN and D.S. WILBUR, 1984. Acta Cryst., C40, 301-303.

Triclinic, $P\bar{1}$, a = 3.877, b = 9.522, c = 10.674 Å, α = 107.99, β = 97.66, γ = 98.50°, Z = 2. Mo radiation, R = 0.030 for 825 reflexions.

The seven-membered ring is essentially planar although bond lengths (Fig. 1) indicate little π-delocalization. The methyloxime O atom is syn to the ring N atom. Molecules are linked to form centrosymmetric dimers by N-H...O hydrogen bonds (N...O 2.860(2) Å).

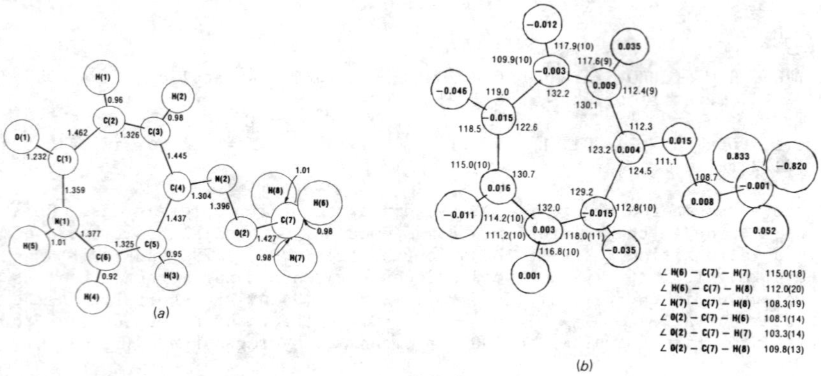

Fig. 1. Bond distances and angles in $C_7H_8N_2O_2$.

2,3-DIHYDRO-1,4-DIAZEPINIUM PICRATE
$C_{11}H_{11}N_5O_7$ $C_5H_9N_2^+.C_6H_2N_3O_7^-$

G. FERGUSON, B.L. RUHL, T. WIECKOWSKI, D. LLOYD and H. McNAB, 1984. Acta Cryst., C40, 1740-1742.

Monoclinic, P2₁/c, a = 11.042, b = 8.224, c = 15.113 Å, β = 92.40°, Z = 4. Mo radiation, R = 0.042 for 1336 reflexions.

The crystal structure contains discrete diazepinium cations and picrate anions linked in chains by N-H...O hydrogen bonds (2.898 and 2.777(3) Å). The cation (Fig. 1) has a five-membered delocalized 1,5-diazapentadienium chain with mean C-C 1.382(8), mean C-N 1.306(9) Å, in a helical conformation. The picrate ring plane is planar but the nitro groups are inclined at 37.5(3), 25.6(3)° (ortho) and 1.9(3)° (para) to the ring. Picrate dimensions are consistent with a significant contribution from a resonance form with an essentially normal C=O bond.

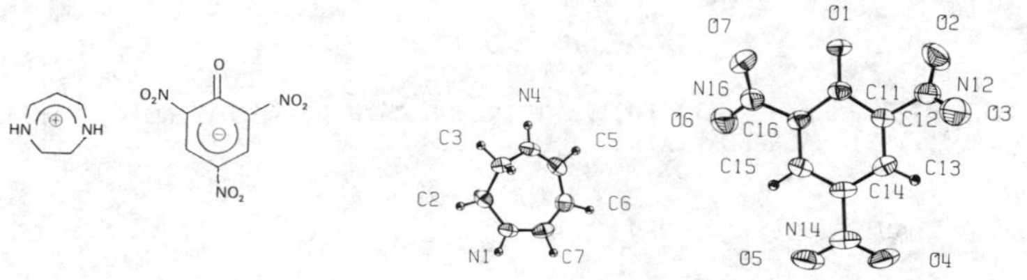

Fig. 1. The cation and anion in 2,3-dihydro-1,4-diazepinium picrate

cis-DIMETHYL 4-PHENYL-2,5,7-TRIMETHYL-4,5-DIHYDROAZEPINE-3,6-DICARBOXYLATE
$C_{19}H_{23}NO_4$ (I)

trans-DIMETHYL 4-BENZYL-2,5,7-TRIMETHYL-4,5-DIHYDROAZEPINE-3,6-DICARBOXYLATE
$C_{20}H_{25}NO_4$ (II)

D.A. CLAREMON, D.E. McCLURE, J.P. SPRINGER and J.J. BALDWIN, 1984. J. Org. Chem., 49, 3871-3874.

I. Monoclinic, P2₁/n, a = 14.181, b = 7.483, c = 16.643 Å, β = 90.24°, [Z = 4]. R = 0.050 for 2154 reflexions.

II. Triclinic, P$\bar{1}$, a = 12.175, b = 14.330, c = 11.779 Å, α = 95.26, β = 112.66, γ = 89.05°, Z = 4. R = 0.047 for 4358 reflexions.

The X-ray analyses established cis-geometry for (I) and trans-geometry for (II). The molecular skeletons are in Fig. 1.

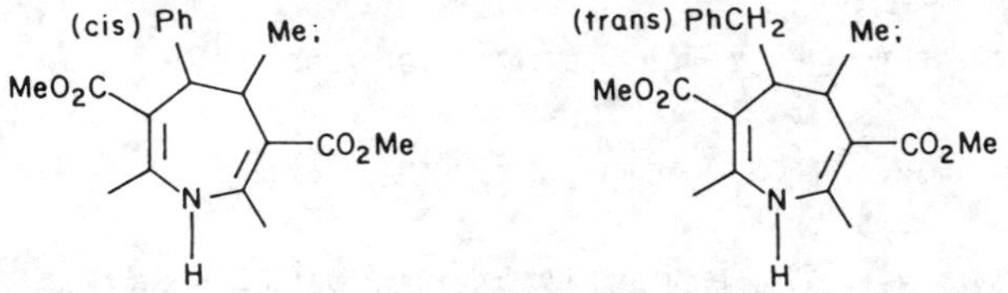

Fig. 1. The structures of $C_{19}H_{23}NO_4$ (I) (left), and $C_{20}H_{25}NO_4$ (II) (right).

2-(DIETHYLAMINO)-3-ISOPROPYL-5-PHENYL-3H-AZEPINE
$C_{19}H_{26}N_2$

J.W. STREEF, H.C. VAN DER PLAS, A. VAN VELDHUIZEN and K. GOUBITS, 1984. Rec. J.R. Neth. Chem. Soc., 103, 225-232.

Monoclinic, $P2_1/a$, a = 13.148, b = 12.641, c = 11.501 Å, β = 113.90°, Z = 4. Cu radiation, R = 0.067 for 2244 reflexions.

The analysis served to confirm the structure and stereochemistry shown in Fig. 1.

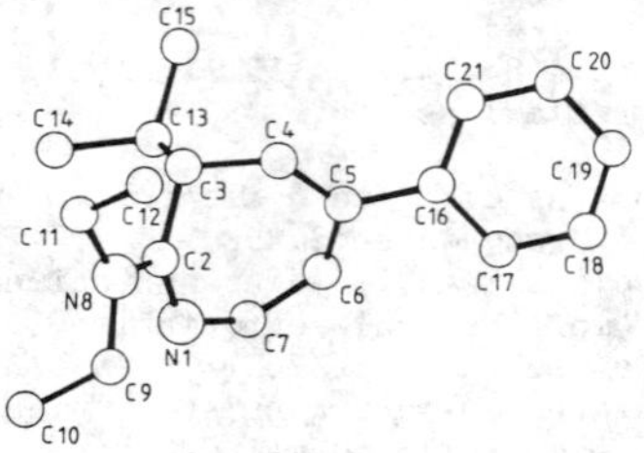

Fig. 1. A view of $C_{19}H_{26}N_2$ (H atoms omitted).

TETRAMETHYL 6-DIETHYLAMINO-3,7a,10,11b-TETRAHYDRO-7H-DIPYRIDAZINO[4,5-b:4',5'-d]-AZEPINE-1,4,8,11-TETRACARBOXYLATE
$C_{22}H_{28}N_6O_8$

M.R. BRYCE, Z. DAUTER, T. KÄMPCHEN, C.D. REYNOLDS and G. SEITZ, 1984. Acta Cryst., C40, 1934-1937.

Triclinic, $P\bar{1}$, a = 9.565, b = 13.699, c = 9.463 Å, α = 96.95, β = 97.12, γ = 97.05°, Z = 2. Cu radiation, R = 0.072 for 2469 reflexions.

The seven-membered azepine ring has a distorted boat conformation (Fig. 1) with the dihedral angle between the best planes through the two dihydropyridazino rings 88.5(4)°. The C(1),N(7),C(8),C(15) plane makes angles with the planes C(1),C(6),N(7) and C(8),C(9),C(10),C(15) of 50.7 and 51.5(4)° respectively and with the A and C rings of 75.9 and 46.1(3)° respectively. Molecules are linked by intermolecular hydrogen bonds; N(4)...N(12) 3.20(1) and N(13)...O(17) 2.87 Å.

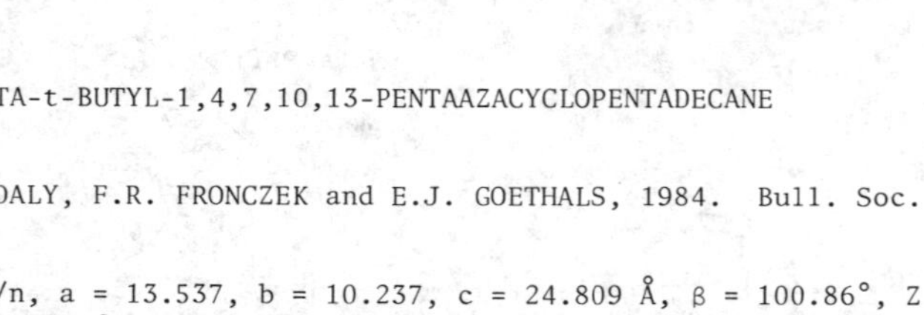

Fig. 1. Stereoview of the $C_{22}H_{28}N_6O_8$ molecule.

1,4,7,10,13-PENTA-t-BUTYL-1,4,7,10,13-PENTAAZACYCLOPENTADECANE
$C_{30}H_{65}N_5$

A. MUNIR, W.H. DALY, F.R. FRONCZEK and E.J. GOETHALS, 1984. Bull. Soc. Chim. Belg., 93, 133-138.

Monoclinic, $P2_1/n$, a = 13.537, b = 10.237, c = 24.809 Å, β = 100.86°, Z = 4. Mo radiation, R = 0.056 for 1969 reflexions.

The five N atoms of the cyclic pentamer (Fig. 1) lie within 0.85 Å of their common plane. The distortion from coplanarity is described as a twist towards a half-chair conformation, in which N1 and N5 lie on opposite sides of the plane formed by N2,N3,N4. The t-butyl groups are all exo-oriented. N-C (ring) bonds average 1.47(1) Å; N-C(t-butyl) average 1.49(1) Å. C-N-C angles within the ring average 110.9(7)° while exo C-N-C angles average 115(2)°. The N-C-C-N groups are all essentially anti, with torsion angles 169.0-179.4°. C-N-C-C torsion angles range from 67.3-121.0°.

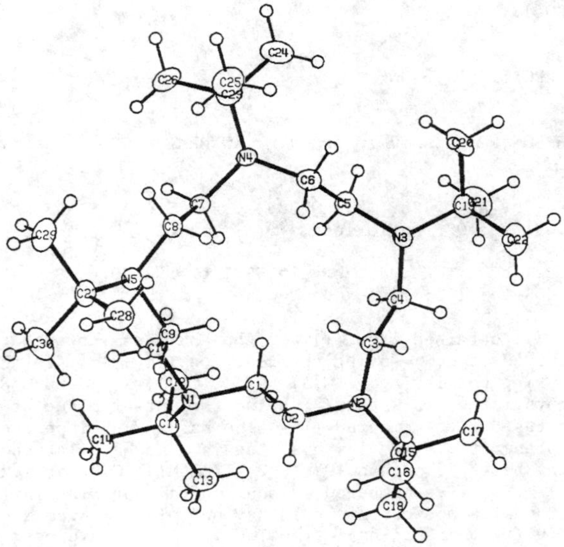

Fig. 1. A view of $C_{30}H_{65}N_5$.

α,ω-BIS(METHYLDODECA-1,12-DIYLAMMONIO)HEXANE DIBROMIDE
$C_{32}H_{66}Br_2N_2$

B.H. RUBIN, M. WILLIAMSON, M. TAKESHITA, F.M. MENGER, F.A.L. ANET, B. BACON and N.L. ALLINGER, 1984. J. Am. Chem. Soc., 106, 2088-2092.

Triclinic, $P\bar{1}$, a = 7.581, b = 7.315, c = 18.470 Å, α = 109.96, β = 99.23, γ = 107.67°, Dm = 1.19, Z = 1. Mo radiation, R = 0.077 for 1769 reflexions.

The molecule (Fig. 1) lies about an inversion centre and there is some disorder associated with C(9) and adjacent atoms. The conformations of 13-membered rings are discussed.

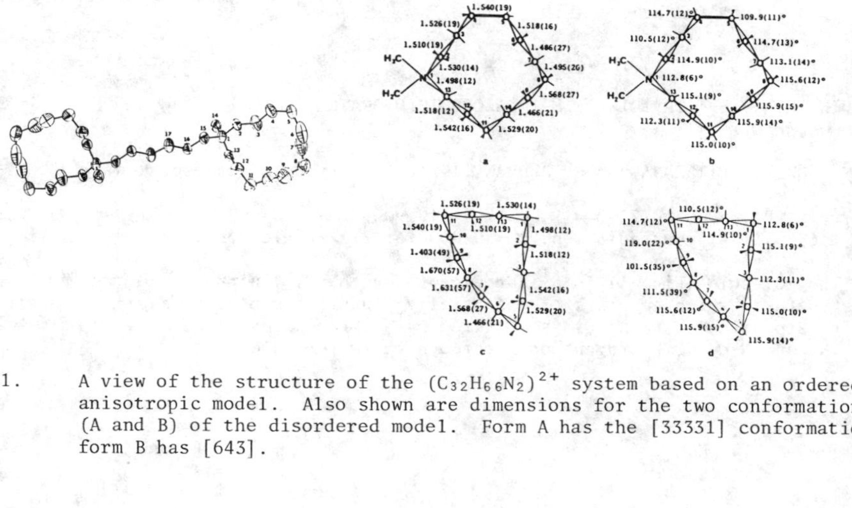

Fig. 1. A view of the structure of the $(C_{32}H_{66}N_2)^{2+}$ system based on an ordered anisotropic model. Also shown are dimensions for the two conformations (A and B) of the disordered model. Form A has the [33331] conformation; form B has [643].

(E)-5-HYDROXYPYRROLIZIDIN-3-ONE
$C_7H_{11}NO_2$ (I)

(Z)-5-THIOKETALPYRROLIZIDIN-3-ONE
$C_{12}H_{19}NOS_2$ (II)

J.L. FLIPPEN-ANDERSON, P. D'ANTONIO and J.H. KONNERT, 1984. J. Cryst. Spectrosc. Res., 14, 565-572.

I. Monoclinic, $P2_1/c$, a = 7.887, b = 9.788, c = 9.316 Å, β = 100.6°, Z = 4. Cu radiation, R = 0.051 for 926 reflexions.

II. Monoclinic, $P2_1/c$, a = 12.947, b = 10.553, c = 9.629 Å, β = 101.4°, Z = 4. Cu radiation, R = 0.085 for 1621 reflexions.

Molecule (II) is obtained from (I) and the analyses shows that a change in configuration occurred at C(5) during the reaction (Fig. 1). In the fused ring system the five-membered ring consisting of C(5), C(6), C(7), N(4) and C(8) has an envelope conformation in both molecules with C(7) being the out-of-plane atom in both cases. In (II) the other fused five-membered ring also has an envelope conformation with C(1) being the out-of-plane atom. In (I) this ring is somewhat flattened with two of the five torsion angles being less than 10° (C(2)-C(3)-N(4)-C(8) at 7.2° and C(3)-N(4)-C(8)-C(1) at 5.9°). The five-membered sulphur-containing ring in (II) is in a twisted conformation with no four atoms in any one plane. Atom N(4) is planar in both molecules. In (I) there is an intramolecular OH...O hydrogen bond (O(5)...O(3) 2.76 Å).

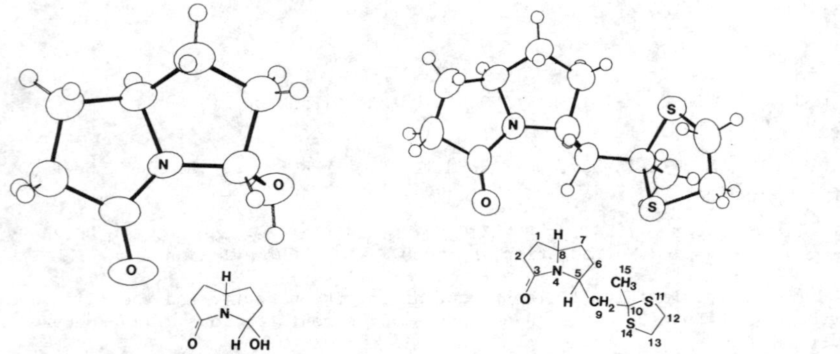

Fig. 1. Structures and stereochemistry for $C_7H_{11}NO_2$ (I) (left) and $C_{12}H_{19}NOS_2$ (II) (right).

4,6-DINITRO-1-(5-TETRAZOLYL)-1H-INDAZOLE TRIHYDRATE
$C_8H_4N_8O_4,3H_2O$

D.S.S. GOWDA, PUTTARAJA and R. RUDMAN, 1984. J. Cryst. Spectrosc. Res., 14, 359-368.

Triclinic, $P\bar{1}$, a = 7.022, b = 9.507, c = 10.906 Å, α = 84.99, β = 71.89, γ = 72.56°, D_m = 1.62, Z = 2. Cu radiation, R = 0.060 for 2112 reflexions.

The molecule (Fig. 1) is planar except for the N(7) nitro-group oxygen atoms. The structure, which consists of essentially planar molecules arranged in layers nearly perpendicular to a*, is stabilized by a three-dimensional network of O-H...O, O-H...N, and N-H...O hydrogen bonds through the water molecules.

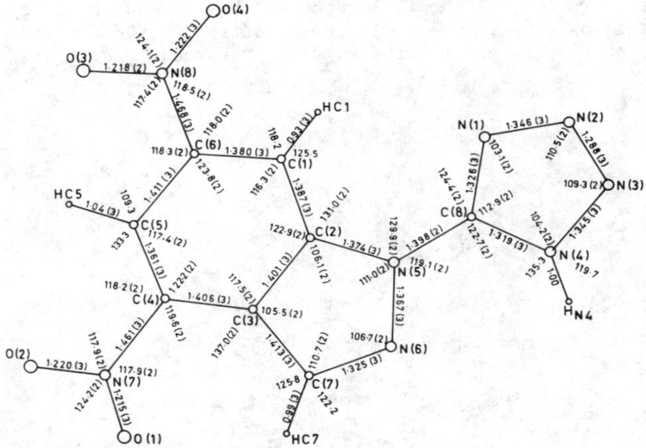

Fig. 1. Bond lengths and angles in $C_8H_4N_8O_4$.

4,8-DICHLORO-3-OXO-3H-OXAZOLO[3,4-a]AZEPINE
$C_8H_5Cl_2NO_2$ (I)

5,9-DICHLORO-4-OXO-3-OXA-PIPERIDINO[2,1-a]AZEPINE
$C_9H_7Cl_2NO_2$ (II)

J.L.M. DILLEN and O. METH-COHN, 1984. S. Afr. J. Chem., 37, 171-176.

I. Orthorhombic, $Pca2_1$, a = 11.881, b = 9.818, c = 7.440 Å, Z = 4. Mo radiation, R = 0.037 for 1221 reflexions.

II. Monoclinic, $P2_1/a$, a = 18.945, b = 10.405, c = 12.646 Å, β = 128.76°, Z = 8. Mo radiation, R = 0.050 for 2829 reflexions.

Molecule I is orange-red whereas compound II is colourless. Views of the molecules are in Fig. 1. Bond lengths are as anticipated with characteristic double and single bonds expected for a non-planar ring system with limited conjugation. Molecule I has a more planar azepine ring than found in II.

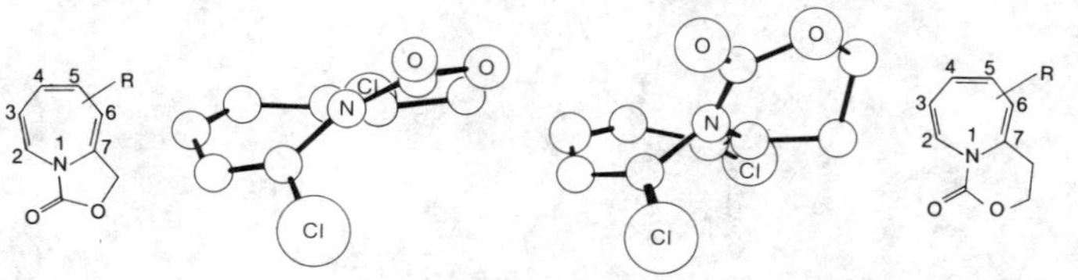

Fig. 1. Views of $C_8H_5Cl_2NO_2$ (I) (left) and $C_9H_7Cl_2NO_2$ (II) (right).

1,8-NAPHTHYRIDINE
$C_8H_6N_2$

P. DAPPORTO, C.A. GHILARDI, C. MEALLI, A. ORLANDINI and S. PACINOTTI, 1984. Acta Cryst., C40, 891-894.

Monoclinic, $P2_1/c$, a = 6.135, b = 10.407, c = 11.258 Å, β = 117.76°, Z = 4. Mo radiation, R = 0.049 for 1275 reflexions (at 163K).

The crystal structure of the title compound has already been reported elsewhere (1). The present low-temperature analysis suggests that the distortion of the molecule from planarity is due to crystal packing forces (Fig. 1).

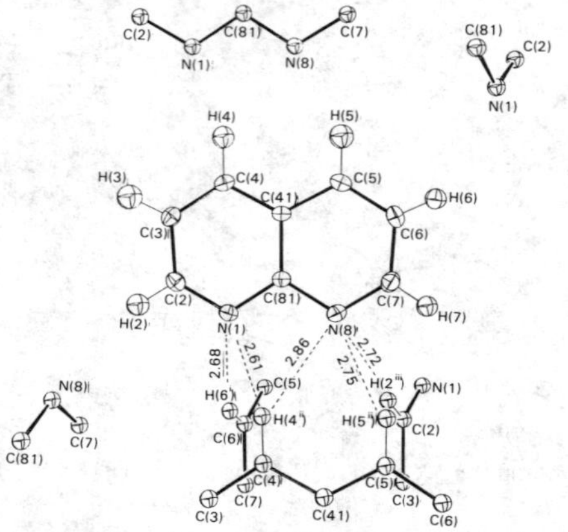

Fig. 1. A view of the $C_8H_6N_2$ molecule showing selected intermolecular contacts.

1. Structure Reports, 38B, 284.

7-(METHYLTHIO)-2,5,5-TRIMETHYL-1H,5H-[1,2,4]TRIAZOLO[1,2-a][1,2,4]TRIAZOLE-1,3-DIONE
$C_8H_{12}N_4O_2S$

Y. NAKAYAMA and Y. SANEMITSU, 1984. J. Org. Chem., 49, 1703-1707.

Monoclinic, C2/c, a = 22.106, b = 7.207, c = 15.821 Å, β = 11.795°, D_m = 1.362, Z = 8. Cu radiation, R = 0.038 for 1785 reflexions.

The analysis established the structure of this cyclization product to be as shown in Fig. 1.

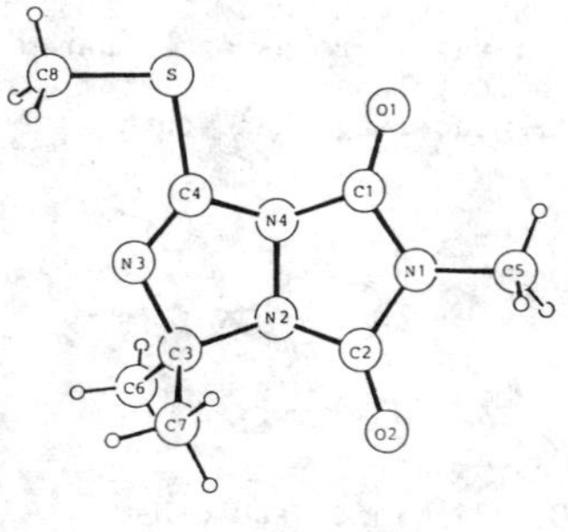

Fig. 1. A view of $C_8H_{12}N_4O_2S$.

trans-1,8-DINITROSO-1,8-DIAZADECALIN
$C_8H_{14}N_4O_2$

R.L. WILLER, C.K. LOWE-MA, D.W. MOORE and L.F. JOHNSON, 1984. J. Org. Chem., 49, 1481-1487.

Monoclinic, $P2_1/c$, a = 7.444, b = 9.546, c = 13.801 Å, β = 97.34°, Z = 4. Mo radiation, R = 0.074 for 1149 reflexions.

The molecule (Fig. 1) has the nitroso groups in a syn,anti orientation; the diazadecalin adopts a chair, twist-boat conformation. The N-N distances are 1.306(5) and 1.322(6) Å, and N-O 1.222(6) and 1.242(5) Å.

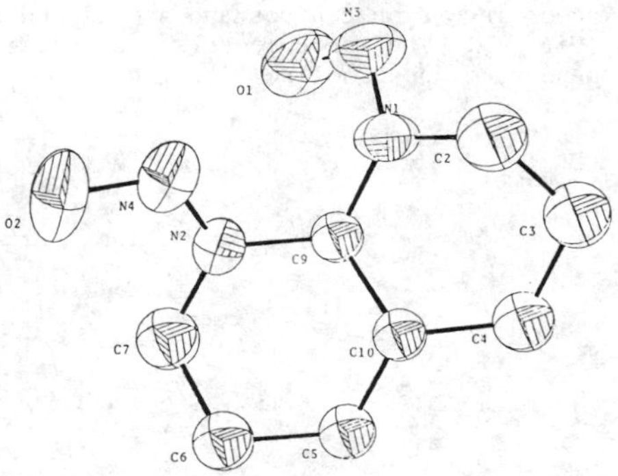

Fig. 1. A view of $C_8H_{14}N_4O_2$.

1H-INDOLE-3-CARBOXALDEHYDE
C_9H_7NO (I)

N,N-DIMETHYL-1H-INDOLE-3-METHANAMINE (GRAMINE)
$C_{11}H_{14}N_2$ (II)

S.N. GOLUBEV and Yu.D. KONDRASHEV, 1984. J. Struct. Chem., 25, 467-470 [Zh. Strukt. Khim., 25(3), 145-149].

I. Orthorhombic, $Pca2_1$, a = 14.145, b = 5.832, c = 8.714 Å, D_m = 1.342, Z = 4. Mo radiation, R = 0.047 for 1078 reflexions.

II. Orthorhombic, Pbca, a = 10.839, b = 9.292, c = 19.903 Å, D_m = 1.156, Z = 8. Mo radiation, R = 0.057 for 974 reflexions.

Bond lengths and angles (σ = 0.003-0.004 Å and 0.2° for non-hydrogen atoms, 0.03 Å and 1.5° involving hydrogen atoms) for the two molecules are shown in Fig. 1. Molecule I is planar and in II the plane of the side chain is perpendicular to that of the indole group. Hydrogen bonds form infinite chains in both structures.

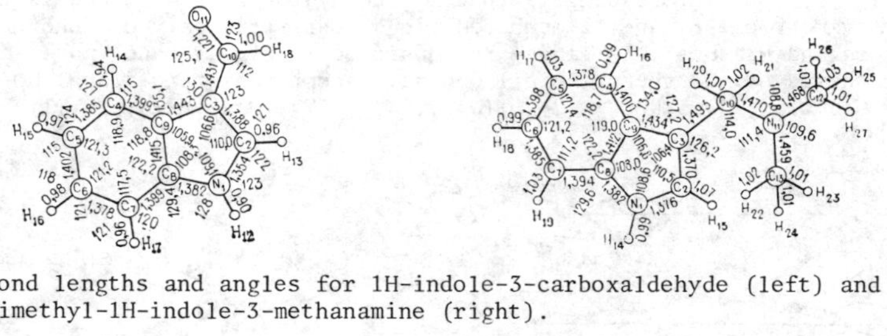

Fig. 1. Bond lengths and angles for 1H-indole-3-carboxaldehyde (left) and N,N-dimethyl-1H-indole-3-methanamine (right).

8-HYDROXY-5-QUINOLINESULPHONIC ACID DIHYDRATE
$C_9H_7NO_4S,2H_2O$

T. BANERJEE, A.K. BASAK, S.K. MAZUMDAR and S. CHAUDHURI, 1984. Acta Cryst., C40, 507-509.

Monoclinic, $P2_1/n$, a = 6.699, b = 16.711, c = 10.203 Å, β = 109.22°, D_m = 1.59, Z = 4. Mo radiation, R = 0.030 for 1585 reflexions.

The molecules pack about inversion centres and are linked by hydrogen bonds through water molecules (Fig. 1). The molecule exists as a zwitterion, with the sulphonic acid proton transferred to quinoline nitrogen.

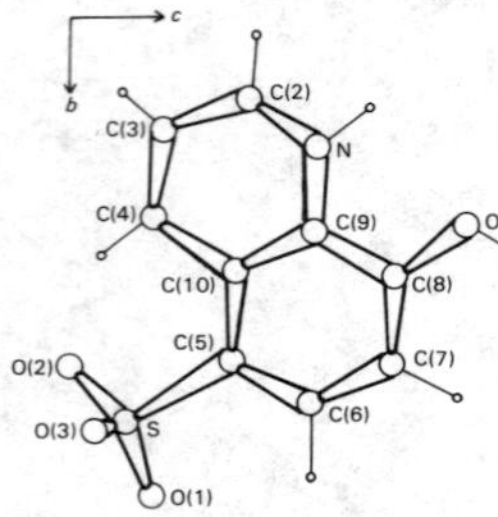

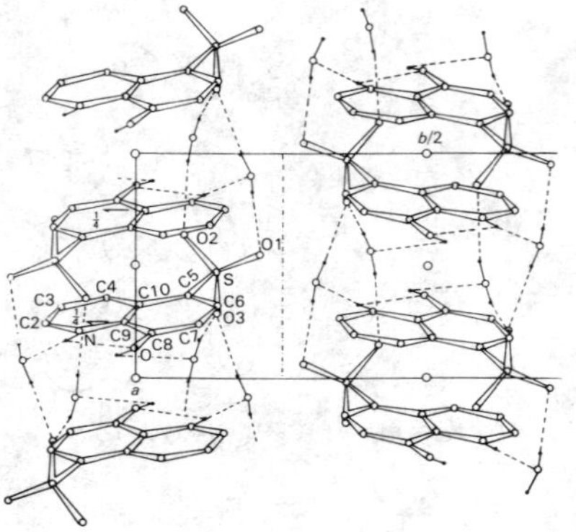

Fig. 1. The $C_9H_7NO_4S$ molecule and a view of the hydrogen bonding scheme.

2-AMINO-8-HYDROXYQUINOLINE
$C_9H_8N_2O$

I.N. POLYAKOVA, Z.A. STARIKOVA, B.V. PARUSNIKOV and I.A. KRASAVIN, 1984. Kristallografiya, 29, 923-927 [Sov. Phys. Crystallogr., 29, 544-547].

Monoclinic, $P2_1/c$, a = 11.847, b = 4.109, c = 15.427 Å, β = 90.47°, Z = 4. Mo radiation, R = 0.050 for 835 reflexions.

The molecule has a planar structure with no unusual bond lengths (C(1)-N(1) 1.321, C(1)-N(2) 1.377, C(1)-C(2) 1.425, C(2)-C(3) 1.354, C(3)-C(4) 1.410, C(4)-C(5) 1.427, C(4)-C(9) 1.414, C(5)-C(6) 1.359, C(6)-C(7) 1.409, C(7)-C(8) 1.373, C(8)-C(9) 1.419, C(8)-O 1.375 and C(9)-N(1) 1.361 Å). Hydrogen bonds N-H...N, bifurcated O-H...N and N-H...O form isolated regions within the layers of molecules parallel to the yz plane (Fig. 1). An interesting feature of the hydrogen bond scheme is the formation of centrosymmetrical 8- and 12-membered rings consisting only of donor and acceptor heteroatoms and H atoms; the rings are connected in zigzag bands parallel to the plane of the layer and the principal donor...acceptor distances are: O-H(0)...N(1) 2.717, O-H(0)...N(2') 3.008, N(2)-H1(N)...O'' 3.177, N(2)-H2(N)...N(1) 3.198 Å.

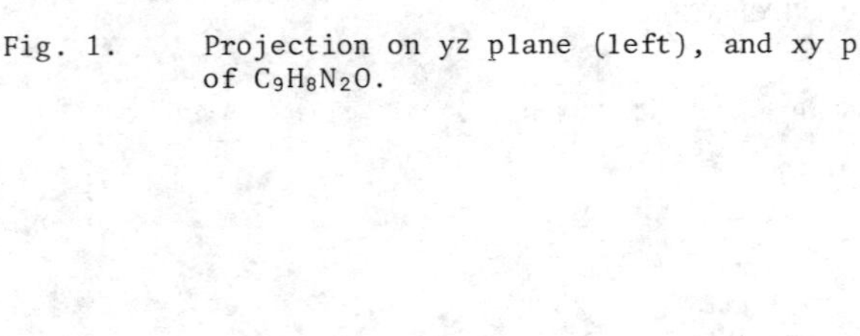

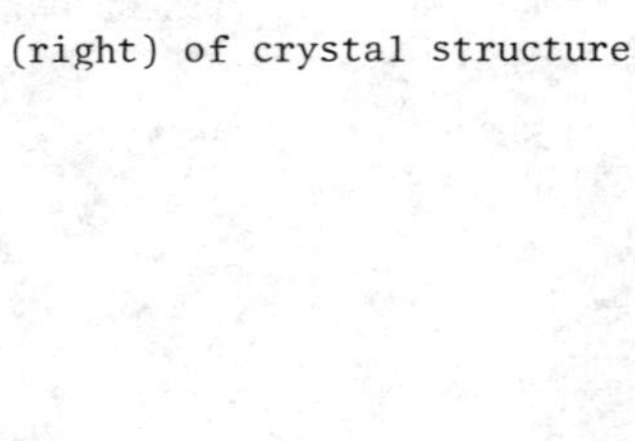

Fig. 1. Projection on yz plane (left), and xy plane (right) of crystal structure of $C_9H_8N_2O$.

1,2,3,4-TETRAHYDROQUINOLINIUM-7-SULPHONATE MONOHYDRATE
$C_9H_{11}NO_3S,H_2O$ (I)

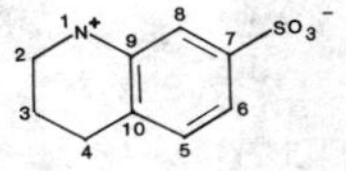

1,2,3,4-TETRAHYDRO-1-METHYLQUINOLINIUM-7-SULPHONATE
$C_{10}H_{13}NO_3S$ (II)

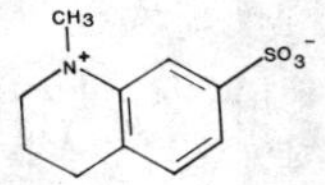

J.L. FLIPPEN-ANDERSON and R. GILARDI, 1984. Acta Cryst., C40, 1065-1068.

I. Monoclinic, $P2_1/c$, a = 7.979, b = 9.096, c = 14.492 Å, β = 96.03°, Z = 4. Mo radiation, R = 0.057 for 1374 reflexions.

II. Monoclinic, $P2_1/n$, a = 9.343, b = 14.235, c = 7.706 Å, β = 95.71°, Z = 4. Mo radiation, R = 0.069 for 1140 reflexions.

Bond distances and angles agree quite well in I and II and the conformations of both molecules are also essentially the same (Fig. 1). The sulphonyl group is in the meta position with respect to the quinoline moiety. The aromatic rings are planar and the saturated rings are in a half-chair conformation with C(2) and C(3) twisted out of the plane of the other four ring atoms. The packing of the molecules (both are zwitterions) in the crystal is influenced by hydrogen bonding, N-H...O and O-H...O in I and N-H...O in II.

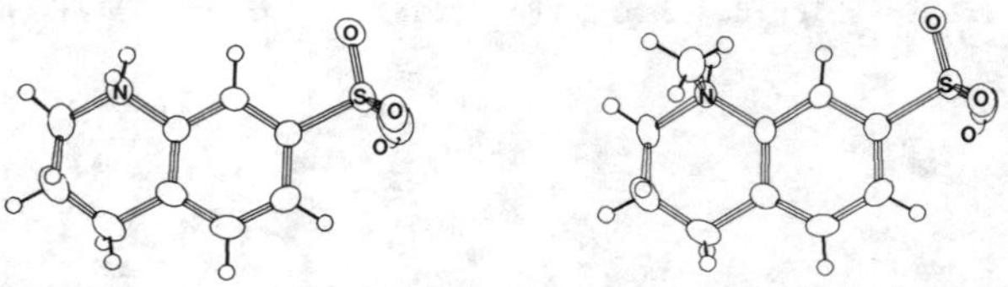

Fig. 1. Perspective views of the $C_9H_{11}NO_3S$ (I) and $C_{10}H_{13}NO_3S$ (II) molecules.

3-HYDROXY-1-OXO-2-PHENYL-1H-PYRAZOLO[1,2-a]PYRAZOL-4-IUM HYDROXIDE, INNER SALT
$C_{12}H_8N_2O_2$ (I)

C.L. RASTON and A.H. WHITE, 1984. Aust. J. Chem., 37, 2577-2582.

Monoclinic, $P2_1/a$, a = 14.375, b = 6.330, c = 10.780 Å, β = 102.52°, D_m = 1.47, Z = 4. Mo radiation, R = 0.050 for 1208 reflexions.

The molecule has a zwitterion skeleton (Fig. 1), substantially planar, with a delocalization pattern representative of a betaine comprised of a pyrazolinium cation and a delocalized β-diketone anion.

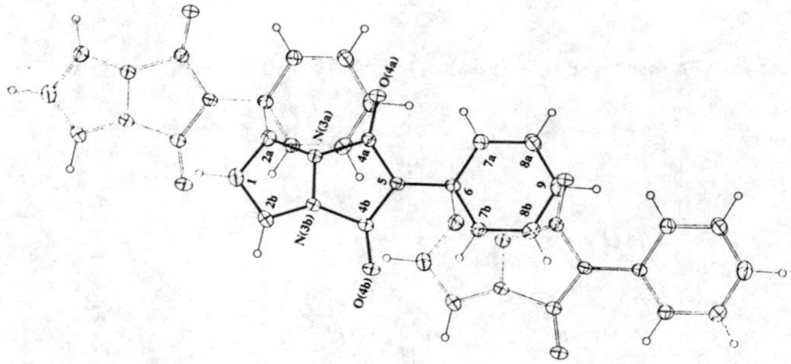

Fig. 1. The structure of $C_{12}H_8N_2O_2$.

α-AMINO-β-(3-INDOLYL)ACRYLIC ACID METHYL ESTER
$C_{12}H_{12}N_2O_2$ (I)

ACETYL-α-AMINO-β-(3-INDOLYL)ACRYLIC ACID METHYL ESTER
$C_{14}H_{14}N_2O_3$ (II)

TRIACETYL-α-AMINO-β-(3-INDOLYL)ACRYLIC ACID METHYL ESTER
$C_{18}H_{18}N_2O_5$ (III)

V.G. ANDRIANOV, Yu.T. STRUCHKOV, K.K. BABIEVSKII, N.I. CHERNOGLAZOVA and V.M. BELIKOV, 1984. J. Struct. Chem., 25, 758-768 [Zh. Strukt. Khim., 25(5), 105-116].

I. Monoclinic, $P2_1/a$, a = 9.781, b = 11.860, c = 9.882 Å, β = 109.140°, Z = 4. Cu radiation, R = 0.048 for 1365 reflexions.

II. Monoclinic, $P2_1/a$, a = 21.991, b = 8.5783, c = 7.204 Å, β = 109.35°, Z = 4. Cu radiation, R = 0.055 for 1980 reflexions.

III. Monoclinic, $P2_1/a$, a = 16.603, b = 14.294, c = 7.2361 Å, β = 101.33°, Z = 4. Cu radiation, R = 0.053 for 1594 reflexions.

Bond lengths and angles in molecules I, II, and III are shown in Figs. 1, 2 and 3 respectively. All three molecules have the Z configuration about the C(1)=C(2) double bond, the indolyl and amino substituents being located cis to one another. The steric and electronic effects which determine the actual conformations of the three molecules are discussed in detail.

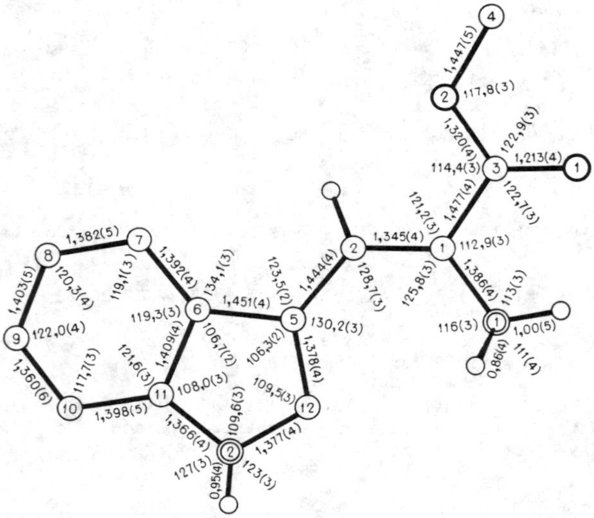

Fig. 1. Geometry of α-amino-β-(3-indolyl)acrylic acid methyl ester.

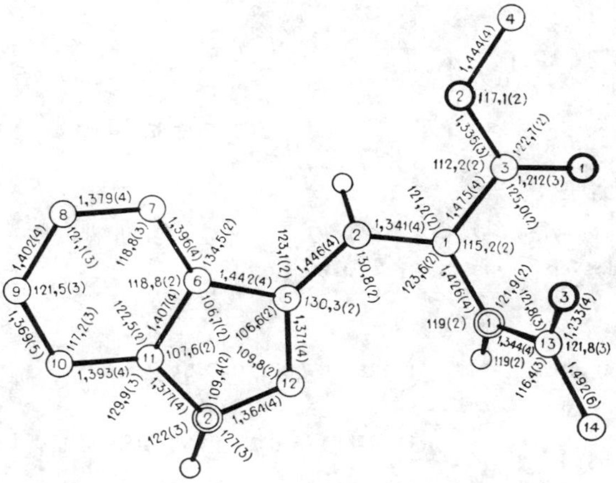

Fig. 2. Geometry of acetyl-α-amino-β-(3-indolyl)acrylic acid methyl ester.

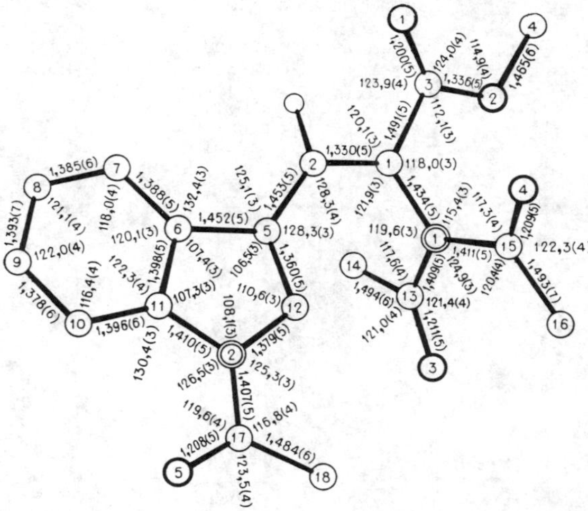

Fig. 3. Geometry of triacetyl-α-amino-β-(3-indolyl)acrylic acid methyl ester.

5-AMINO-4-METHYL-1-ISOQUINOLINECARBALDEHYDE THIOSEMICARBAZONE HYDROCHLORIDE

$C_{12}H_{14}ClN_5S$ $C_{12}H_{14}N_5S^+.Cl^-$

R. KURODA, S. NEIDLE and D.E.V. WILMAN, 1984. Acta Cryst., C40, 465-467.

Monoclinic, $P2_1/n$, a = 8.284, b = 13.906, c = 12.040 Å, β = 92.95°, D_m = 1.42, Z = 4. Cu radiation, R = 0.061 for 628 reflexions.

In the molecule (Fig. 1) the N(3)-N(4) bond (1.36(1) Å) is significantly shorter than anticipated for a normal N-N bond. The shortening can be explained by extended delocalisation involving not only the thiosemicarbazone chain but also aromatic heterocyclic rings. The short C(1)-C(12) distance, 1.43 Å, supports this hypothesis. The C-S bond length, 1.67 Å, is intermediate between single and double bond values.

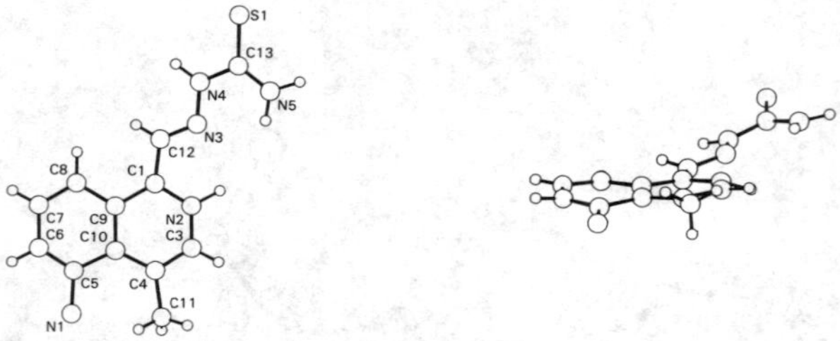

Fig. 1. Plan and side views of $C_{12}H_{14}ClN_5S$.

4-HYDROXY-N-METHYL-4-VINYL-1,2,3,4-TETRAHYDROQUINOLINE
$C_{12}H_{15}NO$

N. RODIER, L. PINELLI, G. ADAM, J. ANDRIEUX and M. PLAT, 1984. Bull. Soc. Chim. Fr., I-217-221.

Tetragonal, $I4_1/a$, a = 22.091, c = 9.099 Å, Z = 16. Mo radiation, R = 0.049 for 939 reflexions.

In the molecule (Fig. 1), ring A is planar and ring B is intermediate between sofa and half-chair conformations. Molecules are linked by O-H...O hydrogen bonds to form infinite spirals about a 4_1 axis (O...O 2.664(2) Å).

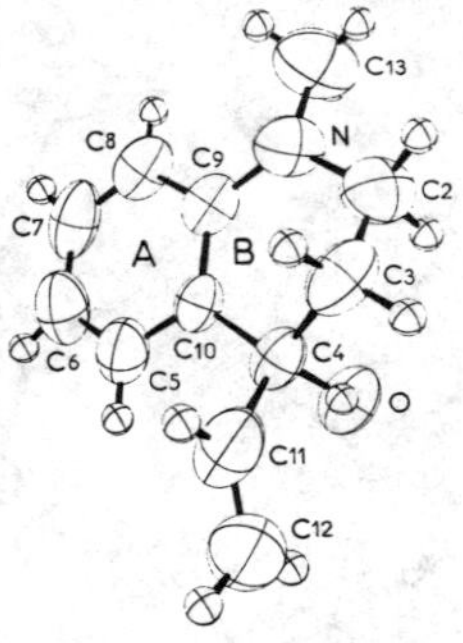

Fig. 1. A view of $C_{12}H_{15}NO$.

ETHYL 1-CYANO-1,2,3,4-TETRAHYDRO-2-ISOQUINOLINECARBOXYLATE
$C_{13}H_{14}N_2O_2$

A. GZELLA, M. JASKÓLSKI, U. RYCHLEWSKA and Z. KOSTURKIEWICZ, 1984. Acta Cryst., C40, 2098-2100.

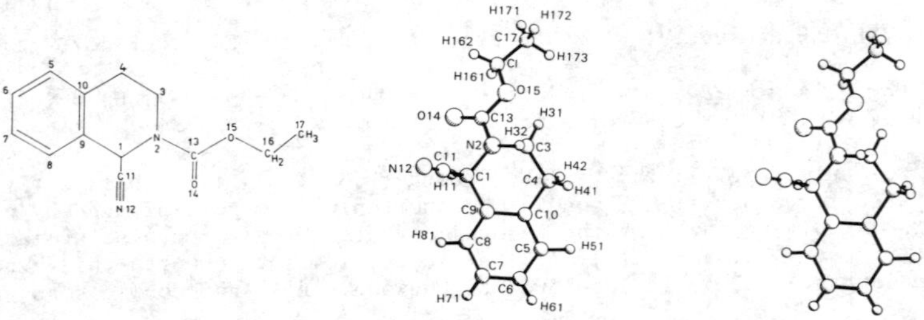

Fig. 1. $C_{13}H_{14}N_2O_2$ and a stereoview.

Triclinic, $P\bar{1}$, a = 8.868, b = 10.119, c = 8.2858 Å, α = 83.624, β = 119.147, γ = 111.051°, D_m = 1.28, Z = 2. Cu radiation, R = 0.053 for 1361 reflexions.

The hydrogenated heterocyclic ring (Fig. 1) has the conformation of a slightly deformed half-chair. The C(3)-N(2)-C(13)-O(14) torsion angle (-172.4(4)°) characterizes the amide bond as anti. The NCOO group is exactly planar and makes an angle of 31.2(4)° with the aromatic ring, minimizing steric interactions between the carbonyl group and neighbouring H atoms.

ETHYL 8-CYANO-5-HYDROXY-2-METHYLQUINOLINE-6-CARBOXYLATE
$C_{14}H_{12}N_2O_3$

I. HERMECZ, Z. MESZAROS, K. SIMON, L. SZABO and Z. PAL, 1984. J. Chem. Soc. Perkin I, 1795-1798.

Monoclinic, $P2_1/n$, a = 11.385, b = 11.102, c = 10.191 Å, β = 103.63°, Z = 4. Mo radiation, R = 0.043 for 1907 reflexions.

The X-ray analysis established the structure of the title compound (Fig. 1). The proton of the hydroxy group is involved in intramolecular hydrogen bonding with the ester oxygen.

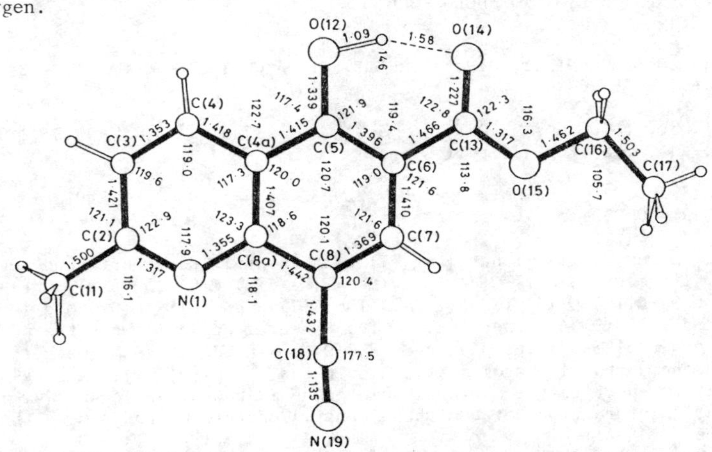

Fig. 1. $C_{14}H_{12}N_2O_3$: the molecular structure showing bond lengths (σ ≤ 0.003 Å) and angles (σ ≤ 0.5°).

trans-1,4,5,8-TETRAACETYL-1,4,5,8-TETRAAZADECALIN
$C_{14}H_{22}N_4O_4$ (I)

cis-1,4,5,8-TETRAACETYL-1,4,5,8-TETRAAZADECALIN
$C_{14}H_{22}N_4O_4$ (II)

F. BORREMANS, M. ANTEUNIS, U. SHMUELI, L. SCHLEIFER, H. SHVO and B. FUCHS, 1984. Tetrahedron, 40, 257-266.

I. Monoclinic, $P2_1/c$, a = 8.106, b = 11.115, c = 8.593 Å, β = 104.53°, Z = 2. Mo radiation, R = 0.041 for 955 reflexions.

II. Monoclinic, $P2_1/n$, a = 8.919, b = 15.157, c = 11.253 Å, β = 91.88°, Z = 4. Mo radiation, R = 0.038 for 2373 reflexions.

Molecule I (Fig. 1) lies on an inversion centre and consists of two fused centrosymmetrically related twist-boat rings. The cis-form II (Fig. 1) contains an asymmetrically flattened chair fused with a twist-boat.

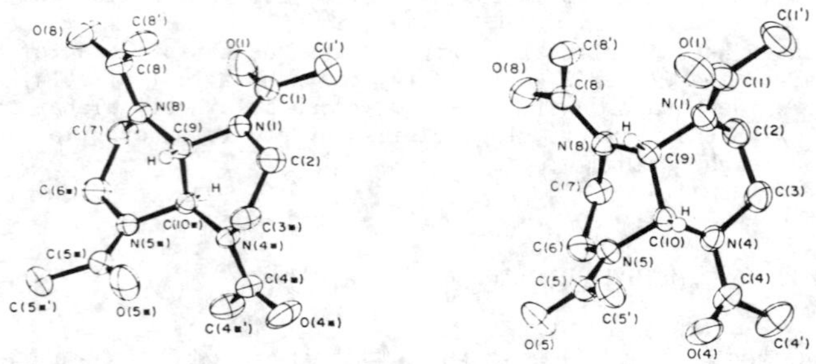

Fig. 1. The trans- and cis-forms of $C_{14}H_{22}N_4O_4$.

(Z)-BENZYLIDENE-1,3-DIHYDROINDOL-2-ONE
$C_{15}H_{11}NO$ (I)

(E)-BENZYLIDENE-1,3-DIHYDROINDOL-2-ONE
$C_{15}H_{11}NO$ (II)

A.C. CODA, A.G. INVERNIZZI, P.P. RIGHETTI, G. TACCONI and G. GATTI, 1984. J. Chem. Soc. Perkin II, 615-619.

I. Monoclinic, I2/c, a = 22.973, b = 5.915, c = 17.517 Å, β = 105.53°, Z = 8. Cu radiation, R = 0.083 for 426 reflexions.

II. Monoclinic, $P2_1/n$, a = 12.294, b = 22.232, c = 4.100 Å, β = 95.41°, Z = 4. Cu radiation, R = 0.052 for 761 reflexions.

The Z-isomer (I) is planar (Fig. 1) with strongly divergent bond angles of the O=C-C=C-Ar fragment (O-C(7)-C(8) 127.0(8), C(7)-C(8)-C(9) 130.8(8), C(8)-C(9)-C(10) 133.9(8), C(9)-C(10)-C(15) 129.5(9)°). The C(7)-O and C(8)-C(9) bond lengths are 1.224(9) and 1.350(9) Å respectively. The dihedral angle between the phenyl and 1,3-dihydro-2-oxoindol-3-ylidene planes is 8.0(9)°. The distance between the o-arylidene H atom and the carbonyl O atom is 2.1 Å. In the E-isomer (II), the aryl group is twisted with respect to the dihydroindole plane (dihedral angle 49.7(4)°) due to a strong steric interaction between the o-benzylidene proton and 4-H of the 1,3-dihydro-indol-2-one (2.48 Å).

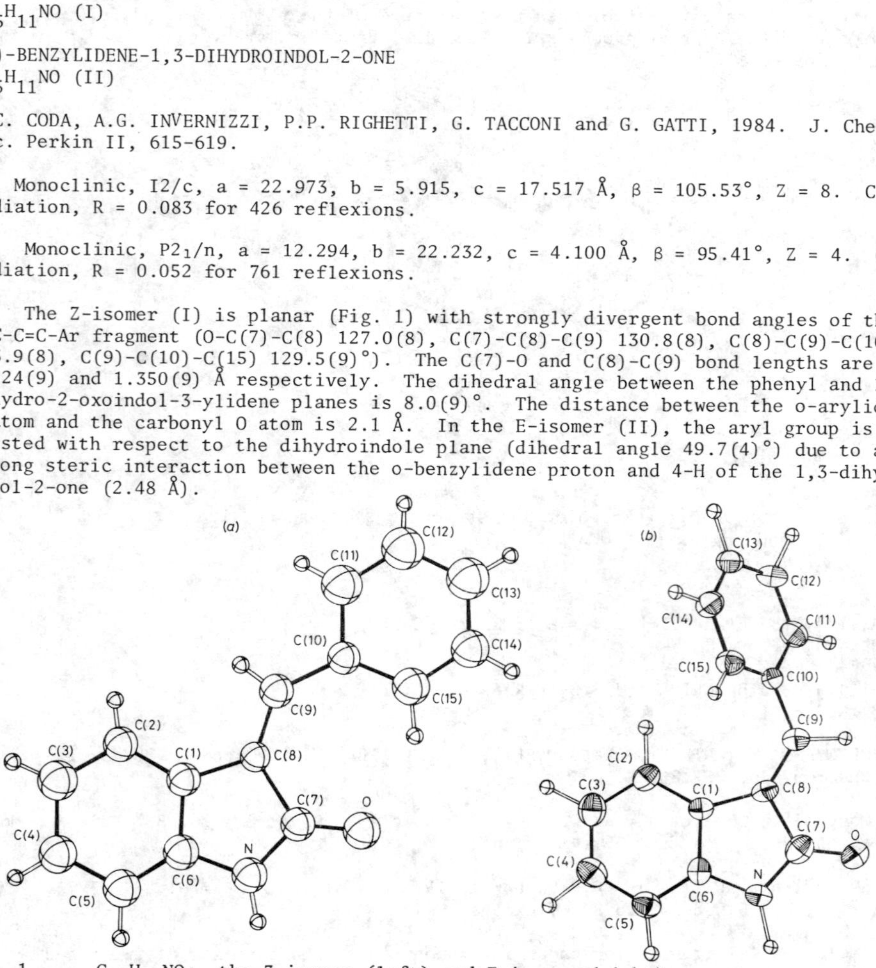

Fig. 1. $C_{15}H_{11}NO$: the Z-isomer (left) and E-isomer (right).

1-PHENYL-1,2-DIHYDRO-2-QUINOLINONE
$C_{15}H_{11}NO$

N.C. BAENZIGER and S. WAWZONEK, 1984. Acta Cryst., C40, 1576-1578.

Monoclinic, $P2_1/c$, a = 5.621, b = 16.098, c = 12.368 Å, β = 98.74°, Z = 4. Mo radiation, R = 0.042 for 1595 reflexions.

The phenyl ring at N(1) makes a dihedral angle of 71.9(1)° with the quinoline ring plane (non-hydrogen atoms planar to within ±0.010(2) Å). The C(3)-C(4) bond length (Fig. 1) is 1.333(3) Å indicating a double bond in that position.

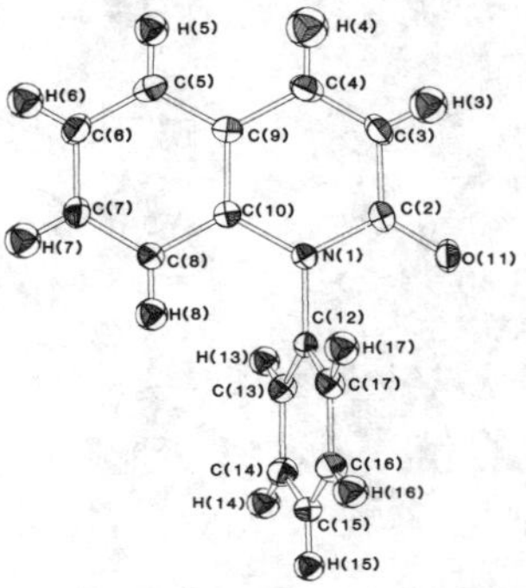

Fig. 1. The 1-phenyl-1,2-dihydro-2-quinolinone molecule.

N-{(1R*,5S*,7R*)-6-OXOBICYCLO[3.2.0]HEPT-2-EN-7-YL}PHTHALIMIDE
$C_{15}H_{11}NO_3$

B. SHELDRICK, D. AKRIGG, M.I. PAGE and D. AGATHOCLEOUS, 1984. Acta Cryst., C40, 1217-1219.

Orthorhombic, Pcab, a = 7.434, b = 13.163, c = 24.684 Å, Z = 8. Cu radiation, R = 0.082 for 1300 reflexions.

In the molecule (Fig. 1) the cyclobutanone and cyclopentene rings of the bicyclohept-2-en-6-one moiety are cis fused; the phthalimide system is essentially planar. Bond lengths and angles are normal.

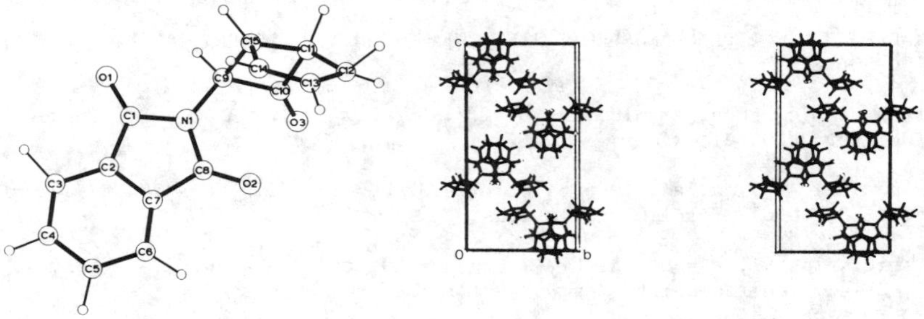

Fig. 1. $C_{15}H_{11}NO_3$: perspective view of the molecule and stereoview of the crystal structure along a.

6,9-DIHYDRO-6-METHOXY-9-(METHOXYCARBONYL)-3-PHENYLPYRROLO[1,2-a]-1,3,5-TRIAZINE-2,4-DIONE
$C_{15}H_{15}N_3O_5$

T. SHERADSKY and R. MOSHENBERG, 1984. J. Org. Chem., 49, 587-590.

Monoclinic, $P2_1/c$, a = 9.158, b = 14.214, c = 11.498 Å, β = 96.25°, Z = 4. Mo radiation, R = 0.052 for 1897 reflexions.

The X-ray analysis showed that the methoxy and methoxycarbonyl groups (bonds C3-C7 and C6-O5 in Fig. 1) are in a cis relationship.

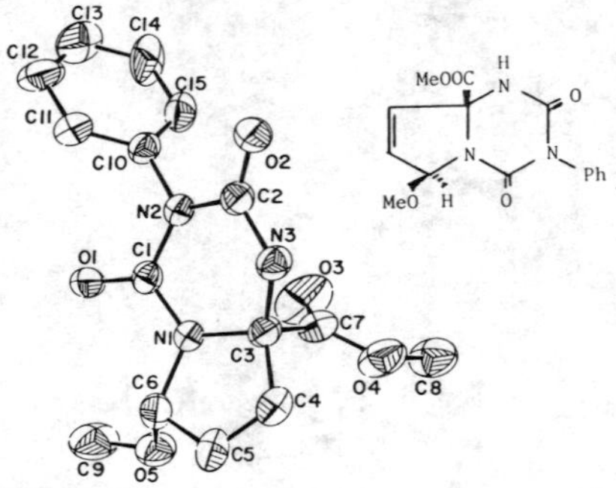

Fig. 1. A view of $C_{15}H_{15}N_3O_5$.

5,6,7,8-TETRAMETHOXY-1,2-DIMETHYL-1,2,3,4-TETRAHYDROISOQUINOLINE HYDROCHLORIDE (1-METHYLWEBERINE HYDROCHLORIDE)
$C_{15}H_{24}ClNO_4$ (I)

2-(2,3,4,5-TETRAMETHOXYPHENYL)ETHYLAMINE HYDROCHLORIDE
$C_{12}H_{20}ClNO_4$ (II)

5,6,7,8-TETRAMETHOXY-1-METHYLISOQUINOLINE HYDROCHLORIDE
$C_{14}H_{18}ClNO_4$ (III)

2,4-DIAMINO-5-(2,3,4,5-TETRAMETHOXYBENZYL)PYRIMIDINE (TETRAMETHOPRIM)
$C_{15}H_{20}N_4O_4$ (IV)

2,4-DIAMINO-5-(2,3,4,5,6-PENTAMETHOXYBENZYL)PYRIMIDINE (PENTAMETHOPRIM)
$C_{16}H_{22}N_4O_5$ (V)

I.L. KARLE, J.L. FLIPPEN-ANDERSON, J.F. CHIANG and A.H. LOWREY, 1984. Acta Cryst., B40, 500-506.

I. Orthorhombic, Pbca, a = 10.262, b = 17.865, c = 18.127 Å, Z = 8. Cu radiation, R = 0.068 for 2009 reflexions.

II. Monoclinic, $P2_1/a$, a = 10.646, b = 7.623, c = 18.862 Å, β = 104.47°, Z = 4. Cu radiation, R = 0.077 for 1105 reflexions.

III. Triclinic, $P\bar{1}$, a = 9.298, b = 9.679, c = 9.664 Å, α = 80.77, β = 63.44, γ = 70.36°, Z = 2. Cu radiation, R = 0.049 for 1787 reflexions.

IV. Monoclinic, $P2_1/n$, a = 12.335, b = 11.828, c = 12.511 Å, β = 119.00°, Z = 4. Mo radiation, R = 0.103, for 1075 reflexions.

V. Triclinic, $P\bar{1}$, a = 6.016, b = 8.319, c = 18.613 Å, α = 82.73, β = 83.58, γ = 89.29°, Z = 2. Mo radiation, R = 0.086 for 1531 reflexions.

Crystal and molecular structure analyses by X-ray diffraction have established the conformations of methoxy groups on phenyl rings when there are four or more adjacent methoxy groups. In three of the compounds studied, the methyl C atoms are placed alternately above and below the plane of the aryl group in a regular fashion, while in two very similar compounds (IV and V) there are unexpected irregularities in the rotations about the C(Ph)-O bonds and consequent crowding of adjacent OCH_3 groups. Views of the conformations are in Fig. 1.

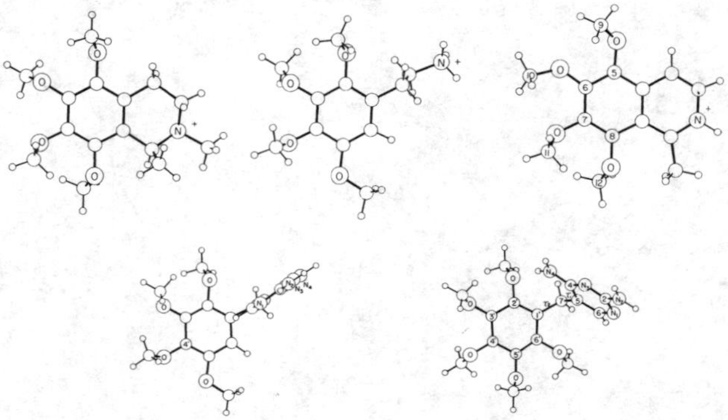

Fig. 1. Views of the tetramethoxy and pentamethoxy compounds I - V (I, II, III top, left to right, IV, V bottom left, right).

((3R*)-3-((1'R*)-1,2-BIS(METHOXYCARBONYL)ETHYL)-2-OXO-3H-IMIDAZO[1,2-a]PYRIDIN-3-YL)-ACETIC ACID METHYL ESTER
$C_{16}H_{18}N_2O_7$ (I)

((3R*)-3-((1'S*)-1,2-BIS(METHOXYCARBONYL)ETHYL)-2-OXO-3H-IMIDAZO[1,2-a]PYRIDIN-3-YL)-ACETIC ACID METHYL ESTER
$C_{16}H_{18}N_2O_7$ (II)

(2-OXO-3H-IMIDAZO[1,2-a]PYRIDIN-3-YL)ACETIC ACID
$C_9H_8N_2O_3$ (III)

3-(METHOXYCARBONYLMETHYL)-1H-IMIDAZO[1,2-a]PYRIDINIUM-2-OLATE
$C_{10}H_{10}N_2O_3, 2(H_2O)$ (IV)

M.E. BAUMANN, H. BOSSHARD, W. BREITENSTEIN, G. RIHS and T. WINKLER, 1984. Helv. Chim. Acta, 67, 1897-1905.

I. Triclinic, $P\bar{1}$, a = 6.719, b = 14.036, c = 10.103 Å, α = 110.93, β = 76.18, γ = 103.33°, Z = 2. R = 0.050 for 2379 reflexions.

II. Monoclinic, $P2_1/c$, a = 11.215, b = 12.444, c = 12.452 Å, β = 95.22°, Z = 4. R = 0.075 for 1395 reflexions.

III. Triclinic, P1, a = 9.437, b = 8.499, c = 5.170 Å, α = 100.71, β = 86.82, γ = 96.35°, Z = 2. R = 0.051 for 1614 reflexions.

IV. Monoclinic, C2/c, a = 23.181, b = 4.541, c = 22.842 Å, β = 102.24°, Z = 8. R = 0.060 for 1354 reflexions.

X-ray analysis established the structures as shown in Fig. 1. For III, both tautomeric forms are present, linked by hydrogen bonding.

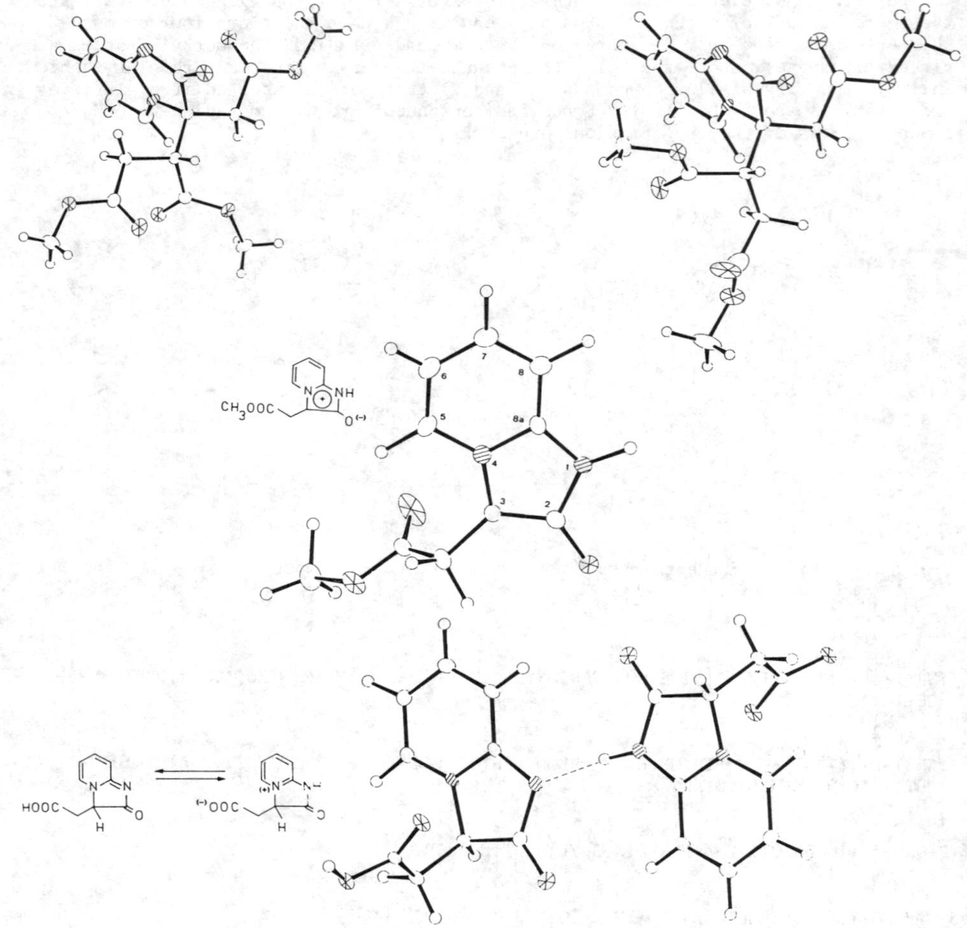

Fig. 1. Views of $C_{16}H_{18}N_2O_7$ (I and II, top left, right), $C_{10}H_{10}N_2O_3$ (centre) and $C_9H_8N_2O_3$ (bottom).

2-BENZOYL-1,2,3,4-TETRAHYDRO-1-ISOQUINOLINECARBONITRILE
$C_{17}H_{14}N_2O$

M. PŁYWACZYK, E. TYKARSKA, M. JASKÓLSKI and Z. KOSTURKIEWICZ, 1984. Acta Cryst., C40, 1107-1109.

Monoclinic, $P2_1/a$, a = 8.8155, b = 14.687, c = 10.6392 Å, β = 94.44°, D_m = 1.267, Z = 4. Cu radiation, R = 0.041 for 1532 reflexions.

The heterocyclic fragment of the molecule (Fig. 1) exhibits a flattened sofa conformation. The C(3)-N(2)-C(13)-O(14) torsion angle of -171.9(3)° characterizes the amide bond as anti. The amide group and the phenyl ring are deconjugated, the corresponding dihedral angle being 55.7(4)°. This twisting thus minimizes steric interactions between the carbonyl group and neighbouring H atoms. Bond distances and angles are generally as expected.

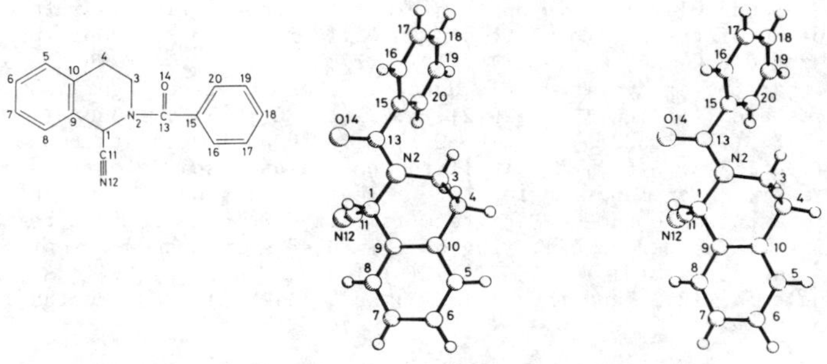

Fig. 1. $C_{17}H_{14}N_2O$: stereoview of the molecule.

N-(2,4,6-TRIMETHYLPHENYL)-PHTHALIMIDE
$C_{17}H_{15}NO_2$ (I)

N-(o-NITROPHENYL)-PHTHALIMIDE
$C_{14}H_8N_2O_4$ (II)

S. VOLIOTIS, J.M. ARRIETA and G. GERMAIN, 1984. Acta Cryst., C40, 1946-1948.

I. Triclinic, $P\bar{1}$, a = 11.260, b = 8.284, c = 8.114 Å, α = 108.69, β = 97.50, γ = 85.17°, Z = 2. Mo radiation, R = 0.043 for 1448 reflexions.

II. Triclinic, $P\bar{1}$, a = 12.578, b = 7.661, c = 7.324 Å, α = 61.53, β = 103.93, γ = 104.89°, Z = 2. Mo radiation, R = 0.037 for 1194 reflexions.

The angles between the normals to the two planes of the phthalimide moieties (Fig. 1) are 0.74 (I) and 2.28° (II). In I the phenyl ring is inclined at 71.0(2)° to the phthalimide moiety. This angle is 54.0(2)° in II.

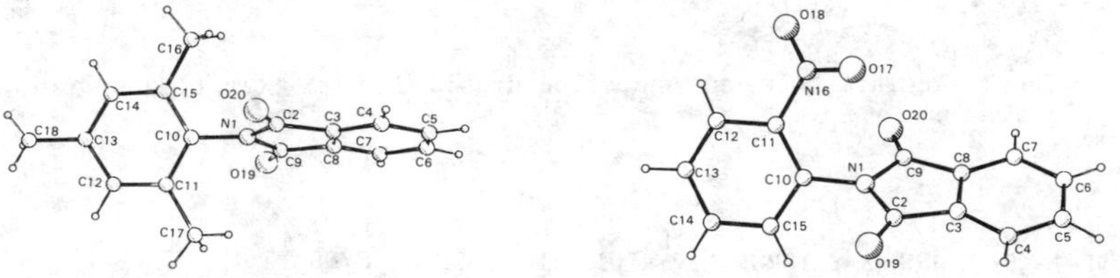

Fig. 1. The $C_{17}H_{15}NO_2$ (left) and $C_{14}H_8N_2O_4$ (right) molecules.

1-(N,N-DIMETHYLAMINOMETHYLIDENE)-3-PHENYLISOINDOLENINE
$C_{17}H_{16}N_2$ (I)

1-FORMYL-3-PHENYLISOINDOLE
$C_{15}H_{11}NO$ (II)

R. BONNETT, M.B. HURSTHOUSE, S.A. NORTH and J. TROTTER, 1984, J. Chem. Soc. Perkin II, 833-839.

I. Monoclinic, $P2_1/n$, a = 20.387, b = 7.161, c = 20.642 Å, β = 116.07°, Z = 8. Cu radiation, R = 0.051 for 3241 reflexions.

II. Monoclinic, $P2_1/c$, a = 14.956, b = 4.878, c = 16.951 Å, β = 117.3°, Z = 4. Cu radiation, R = 0.034 for 1784 reflexions.

Compound I contains two crystallographically independent molecules in the asymmetric unit which exhibit a difference in conformation about the phenylisoindolenine bond (Fig. 1). In one conformer the phenyl ring is rotated clockwise out of the plane of the heterocyclic system while in the other conformer this rotation is anticlockwise, C(9)-C(3)-C(10)-C(11) 29.2, -28.9°. The X-ray analysis established that the exocyclic double bond of I has the Z-configuration. Some C-N bond distances are: N(2)-C(3) 1.329(2), 1.327(2), C(1)-N(2) 1.386(3), 1.392(3), C(16)-N(17) 1.323(3), 1.322(3) Å. The C(1)-C(16) bond length has much single bond character, 1.386(3), 1.376(3) Å. In II the phenyl and isoindole rings are each planar but not coplanar, interplanar angle 29.2°. The formyl group is coplanar with the isoindole skeleton while the $CHNMe_2$ group is tilted 8° with respect to the plane of the fused rings. The carbocyclic ring of the fused system shows a marked degree of bond alternation, C(5)-C(6) 1.409, C(4)-C(5) 1.367, C(4)-C(9) 1.414 Å. The C(1)-N(2)-C(3) angle is 112.2° and the N(2)-C(3) and C(1)-N(2) bond lengths 1.348 and 1.376 Å respectively.

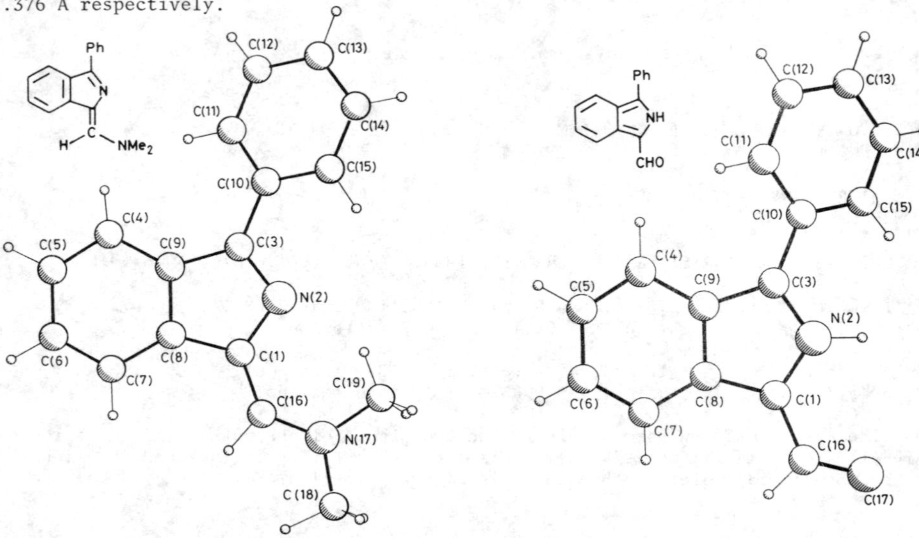

Fig. 1. The molecular skeletons and structures: $C_{17}H_{16}N_2$ (left) and $C_{15}H_{11}NO$ (right).

3-CARBAMOYL-1-(3-(INDOL-3-YL)PROPYL)PYRIDINIUM SULPHATE HYDRATE

$C_{17}H_{18}N_3O_3S_{0.5}$,3.5($H_2O$) $C_{17}H_{18}N_3O^+$,0.5(SO_4^{2-}),3.5(H_2O)

T. ISHIDA, S. IBE and M. INOUE, 1984. J. Chem. Soc. Perkin II, 297-304.

Triclinic, $P\bar{1}$, a = 9.692, b = 15.320, c = 7.117 Å, α = 92.93, β = 117.03, γ = 93.06°, D_m = 1.386, Z = 2. Cu radiation, R = 0.105 for 3198 reflexions.

The molecule (Fig. 1) is in the fully extended conformation and the indole and pyridinium rings are planar and parallel. The plane of the carbamoyl group makes an angle of 178.6(5)° with the plane of the pyridinium ring. In the crystal, the indole moiety is stacked with the pyridinium of a neighbouring molecule along the a-axis. The average interplanar spacings in the area of overlap are 3.407 and 3.480 Å. The carbamoyl group is linked with a centrosymmetrically related one by two hydrogen bonds, N(8)H(8)...O(9) 2.943(6) Å. N(8) is further hydrogen bonded to O(1)(water) (2.935(7) Å). O(3)(water) and O(5)(water) are disordered with occupancies of 0.5.

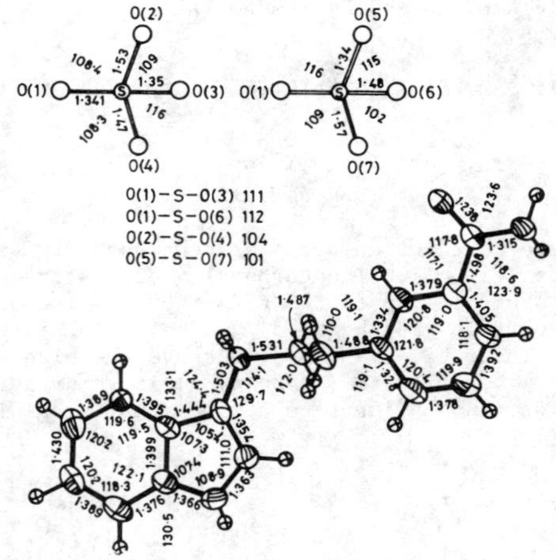

Fig. 1. $C_{17}H_{18}N_3O_3S_{0.5}$,3.5($H_2O$): bond lengths and angles.

N-1'-ADAMANTYLBENZOAZETINONE
$C_{17}H_{19}NO$

R.A. OLOFSON, R.K. VANDER MEER, D.H. HOSKIN, M.Y. BERNHEIM, S. STOURNAS and D.S. MORRISON, 1984. J. Org. Chem., 49, 3367-3372.

Triclinic, $P\bar{1}$, a = 10.594, b = 11.392, c = 6.555 Å, α = 95.28, β = 102.82, γ = 116.43°, D_m = 1.2, Z = 2. Mo radiation, R = 0.054 for 1072 reflexions.

The adamantyl moiety is disordered and principal dimensions for the benzoazetinone are in Fig. 1. Atom N1 is 0.54 Å from the plane of the C atoms to which it is bonded.

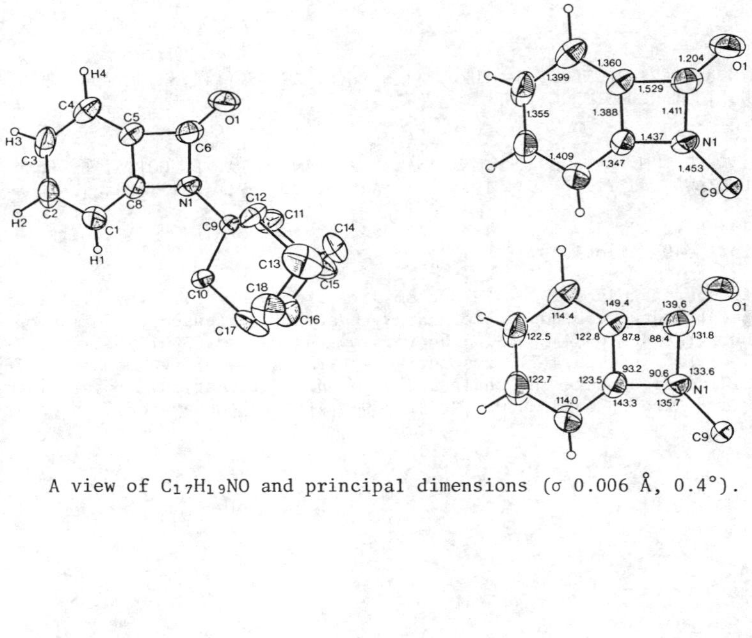

Fig. 1. A view of $C_{17}H_{19}NO$ and principal dimensions (σ 0.006 Å, 0.4°).

ETHYL 3-PHENYL-4,5,6,7-TETRAHYDROINDOLE-2-CARBOXYLATE
$C_{17}H_{19}NO_2$

K.W. LAW, T.-F. LAI, M.P. SAMMES, A.R. KATRITZKY and T.C.W. MAK, 1984. J. Chem. Soc. Perkin I, 111-118.

Triclinic, $P\bar{1}$, a = 8.024, b = 8.973, c = 11.293 Å, α = 103.32, β = 93.34, γ = 110.98°, Z = 2. Mo radiation, R = 0.049 for 1969 reflexions.

The crystal structure contains an array of centrosymmetrically-related hydrogen bonded pairs of molecules (Fig. 1). The carbonyl group is essentially coplanar with the pyrrole ring and is syn relative to the C(2)-N bond. The phenyl ring is rotated out of the plane of the pyrrole ring by 51(4)°. The cyclohexene moiety is puckered with C(6) and C(7) lying respectively below and above the plane defined by the pyrrole ring atoms. The C(6)-C(7) bond length (1.464(4) Å) is remarkably short. The atoms of the ten-membered ring hydrogen bonding site are remarkably co-planar. The N...O(1)' distance is 2.847(2) Å.

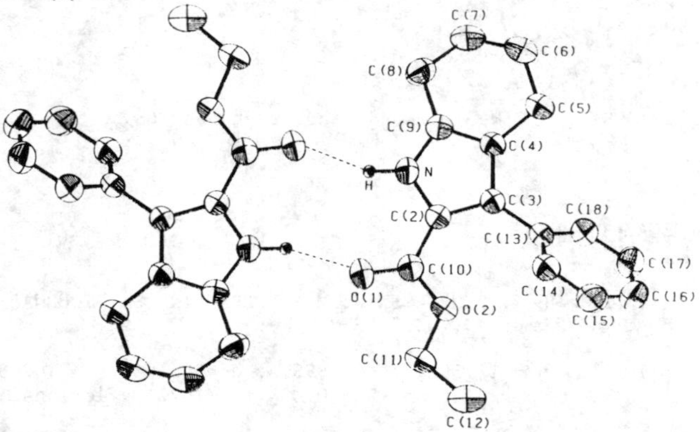

Fig. 1. $C_{17}H_{19}NO_2$: the centrosymmetric arrangement of a hydrogen-bonded pair of molecules.

(1S,2R,3S,7S,8S,8aR)-7-CHLORO-8-HYDROXY-1,2-METHYLENE-3-[(1E,3E,5E)-5-METHYL-1,3,5-HEPTATRIENYL]OCTAHYDROINDOLIZINE
$C_{17}H_{24}ClNO$

S. GOMI, D. IKEDA, H. NAKAMURA, H. NAGANAWA, F. YAMASHITA, K. HOTTA, S. KONDO, Y. OKAMI, H. UMEZAWA and Y. IITAKA, 1984. J. Antibiot., 37, 1491-1494.

Orthorhombic, $P2_12_12_1$, a = 15.329, b = 17.354, c = 6.260 Å, Z = 4. Cu radiation, R = 0.056 for 1740 reflexions.

The structure and absolute stereochemistry are in Fig. 1. The side-chain atoms of the conjugated double bonds are planar with a fully extended conformation which nearly bisects the angle N4-C3-C2. The torsion angles, N4-C3-C10-C11 and C2-C3-C10-C11 are -117.3(2)° and 125.4(2)° respectively. The six and five membered rings of the indolizidine group take the chair and envelope conformations, respectively. An intermolecular hydrogen bond, O-H...N4 of 2.903(4) Å was observed in the crystal structure.

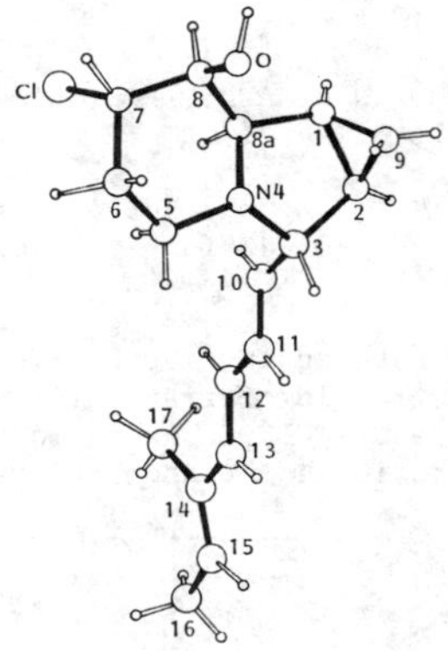

Fig. 1. A perspective view of $C_{17}H_{24}ClNO$.

8,8'-DIQUINOLYLDISULPHIDE-5,5'-DISULPHONIC ACID TRIHYDRATE
$C_{18}H_{12}N_2O_6S_4,3H_2O$

O.G. MATYUKHINA, A.D. OZOLA, Ya.K. OZOLS, A.N. SOBOLEV, Ya.V. ASHAKS and Yu.A. BANKOVSKII, 1984. Kristallografiya, 29, 741-744 [Sov. Phys. Crystallogr., 29, 437-439].

Monoclinic, P2/b, a = 9.263, b = 14.797, c = 8.258 Å, γ = 98.75°, Z = 2. Mo radiation, R = 0.025 for 1105 reflexions.

The molecule has crystallographic twofold symmetry, the rotation axis passing through the midpoint of the disulphide bond (Fig. 1). The disulphide bond is approximately perpendicular to the quinoline nuclei and the torsion angle of the S-S bond is -71.5°. The bond length S-S'is 2.062 Å and S-C(8) is 1.776 Å. The S'-S-C(8) angle is 98.8°. The small dihedral angle (21.3°) between the quinoline ring planes permits stacking of the molecules along the b axis with short intermolecular contacts between the O atoms of the sulpho groups and the S atoms of the disulphide bridge (O...S 2.85 Å). Between the stacks of molecules there are water molecules which involve the zwitterionic structure in a network of hydrogen bonds (N-H(N)...O(W1) 2.694, O(W1)-H(W1,1)...O(3') 2.939, O(W1)-H(W1,2)...O(W2) 2.766, O(W2)-H(W2)...O(2'') 2.726 Å).

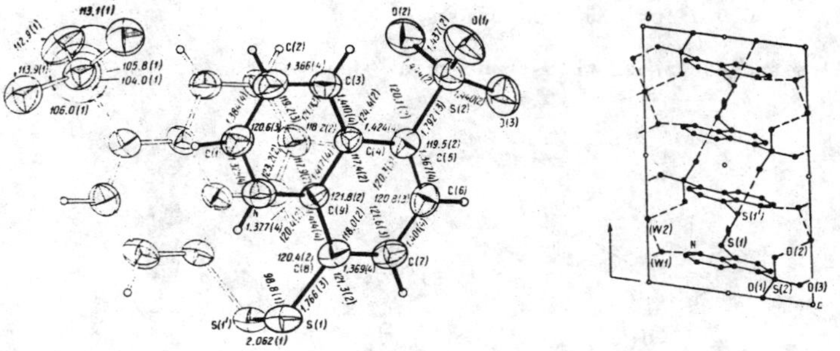

Fig. 1. Molecular structure of $C_{18}H_{12}N_2O_6S_4,3H_2O$ showing bond lengths and angles and xy projection of the crystal structure.

BIS(8-HYDROXYQUINOLINIUM) CHLORIDE TETRACHLOROFERRATE(III)

$C_{18}H_{16}Cl_5FeN_2O_2$ $[8hq]_2[FeCl_4]Cl$

G.A. BOTTOMLEY, A.M. CARTER, L.M. ENGELHARDT, F.J. LINCOLN, J.M. PATRICK and A.H. WHITE, 1984. Aust. J. Chem., 37, 871-877.

Triclinic, $P\bar{1}$, a = 14.243, b = 10.177, c = 7.583 Å, α = 84.71, β = 86.96, γ = 86.50°, D_m = 1.58, Z = 2. Mo radiation, R = 0.043 for 2415 reflexions, yellow crystals. T = 295 K.

This material releases HCl between 100-180°C. The cations are protonated at N (Fig. 1). The two anions are tetrachloroferrate(III) with a single negative charge delocalised over four Cl atoms, and a Cl^-. The cations and $FeCl_4^-$ are held together by hydrogen bonds. The Cl^- ions lie in a tunnel through the lattice down the c axis.

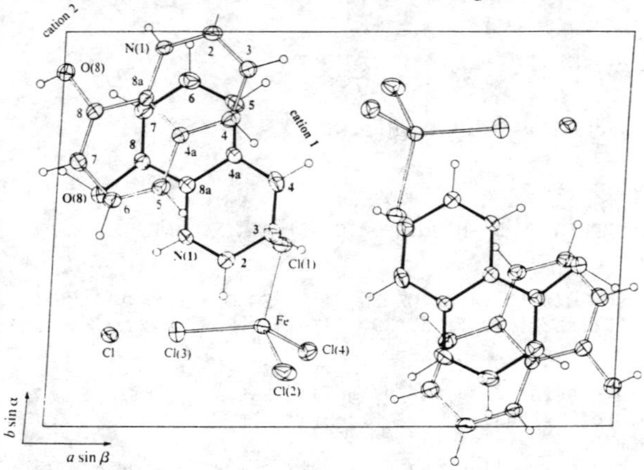

Fig. 1. The packing of $[8hq]_2[FeCl_4]Cl$.

2-(o-N-ACETYLAMINO-N'-ACETYLANILINOMETHYL)BENZIMIDAZOLE

$C_{18}H_{18}N_4O_2$

J. ARMAND, L. BOULARES, C. BELLEC, C. BOIS, M. PHILOCHE-LEVISALLES and J. PINSON, 1984. Can. J. Chem., 62, 1028-1030.

Monoclinic, $P2_1/c$, a = 12.660, b = 8.466, c = 15.172 Å, β = 94.21°, D_m = 1.32, Z = 4. Mo radiation, R = 0.057 for 1353 reflexions.

The analysis establishes the structure as shown in Fig. 1.

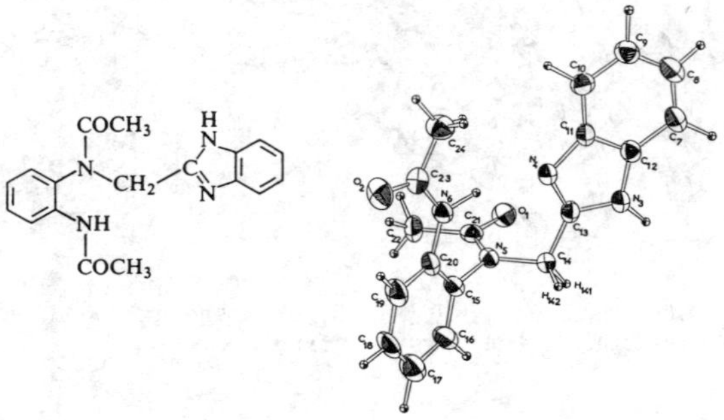

Fig. 1. The $C_{18}H_{18}N_4O_2$ molecule.

1-METHYLINDOL-3-YL 1,2,3,4-TETRAHYDRO-1-METHYL-2-METHOXYCARBONYLMETHYLENEPYRROL-3-YL KETONE
$C_{18}H_{20}N_2O_3$

I.K. AL-KHAWAJA, R.L. BEDDOES, D.I. BISHOP, R.J. CERNIK, J.A. JOULE and O.S. MILLS, 1984. J. Chem. Res., (S) 296-297, (M) 2738-2767.

Monoclinic, P2, a = 6.533, b = 9.392, c = 13.026 Å, β = 90.95°, Z = 2. Mo radiation, R = 0.0338 for 1051 reflexions.

X-ray analysis establishes the structure shown in Fig. 1 for this ketone.

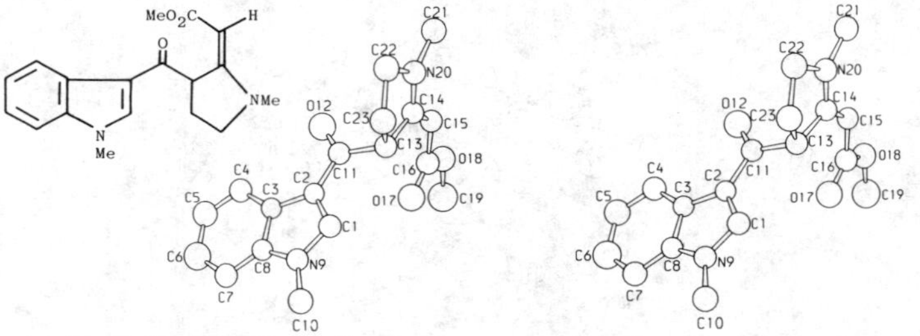

Fig. 1. A stereoview of $C_{18}H_{20}N_2O_3$.

ETHYL 1-p-BROMOBENZOYL-PERHYDROINDOLE-2-CARBOXYLATE MONOHYDRATE
$C_{18}H_{22}BrNO_3,H_2O$

F. ROBERT, Y. JEANNIN, M. VINCENT and M. LAUBIE, 1984. Acta Cryst., C40, 1219-1220.

Tetragonal, $P4_3$, a = 14.067, c = 9.654 Å, Z = 4. Mo radiation, R = 0.068 for 896 reflexions.

In the molecule (Fig. 1) the five-membered ring has an envelope conformation. Each water molecule is linked by three hydrogen bonds, two with other water molecules and one with the oxygen of the benzoyl group. The chain formed by the water molecules is wrapped around the fourfold axis.

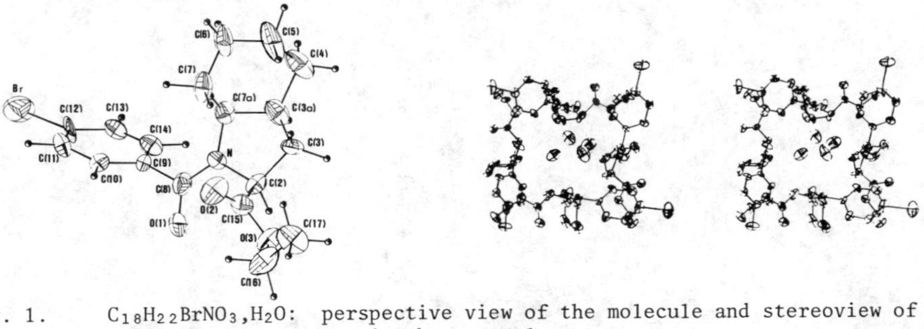

Fig. 1. $C_{18}H_{22}BrNO_3,H_2O$: perspective view of the molecule and stereoview of the molecular arrangement in the crystal.

3-BENZYLOXYMETHYL-1-(3-HYDROXYPROPYL)-6-METHYL-1,5,6,7-TETRAHYDRO-4H-INDAZOL-4-ONE
$C_{19}H_{24}N_2O_3$

W. SUCROW, R. BROCKMANN, H.-J. HAUPT and H. PREUT, 1984. Justus Liebigs Ann. Chem., 1711-1718.

Monoclinic, $P2_1/n$, a = 14.722, b = 7.990, c = 15.551 Å, β = 94.75°, Z = 4. Mo radiation, R = 0.097 for 2187 reflexions.

The analysis establishes the position of the hydroxypropyl group at N(1) (Fig. 1). Some bond lengths: O(1)-C(7) 1.220(5), N(1)-N(2) 1.370(5), N(1)-C(3) 1.332(5), N(2)-C(1) 1.319(5), C(1)-C(2) 1.403(6), C(2)-C(3) 1.369(5) Å.

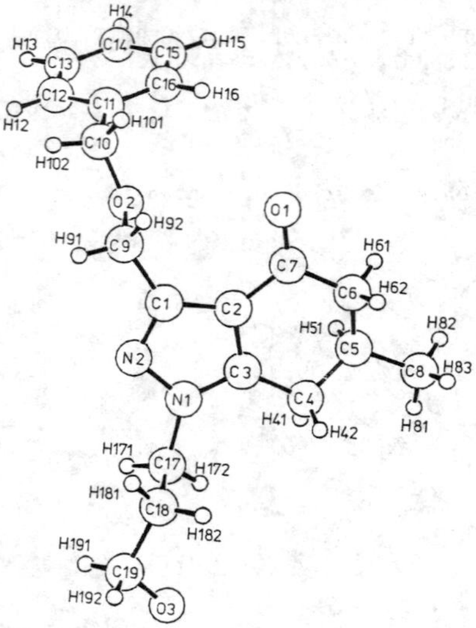

Fig. 1. A view of $C_{19}H_{24}N_2O_3$.

5,7'-DIMETHYL-8,8'-DIQUINOLYLDISULPHIDE

$C_{20}H_{16}N_2S_2$ $(CH_3C_9H_5N)SS(CH_3C_9H_5N)$

O.G. MATYUKHINA, A.N. SOBOLEV, A.P. STURIS, Ya.K. OZOLS and Yu.A. BANKOVSKII, 1984. Kristallografiya, 29, 610-612 [Sov. Phys. Crystallogr., 29, 363-364].

Monoclinic, $P2_1/b$, a = 12.772, b = 17.825, c = 7.532 Å, γ = 102.38°, Z = 4. Mo radiation, R = 0.047 for 965 reflexions.

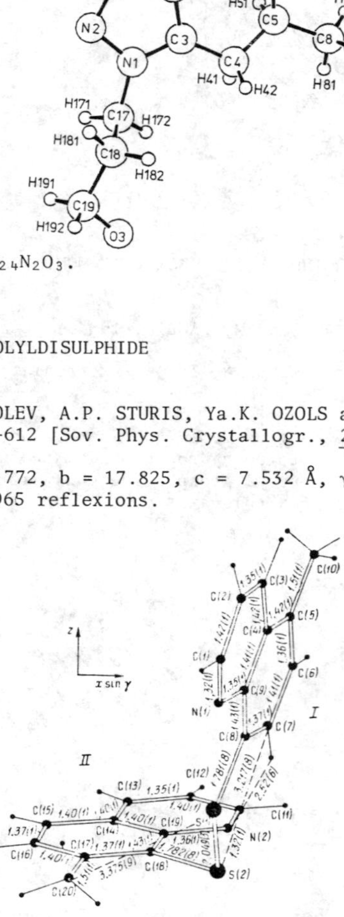

Fig. 1. Projection of the molecule $C_{20}H_{16}N_2S_2$ on the xz plane, showing bond distances (Å).

The fragment of the molecule including the S atoms and the atoms of the quinoline ring I has equatorial conformation (Fig. 1). The appropriate torsion angles are S(2), S(1), C(8), C(7) 1.0° and S(2), S(1), C(8), C(9) 179.2°. The S atoms deviate very little from plane I; S(1) by -0.082 and S(2) by -0.050 Å. In the other fragment of the molecule (the S atoms and the atoms of quinoline ring II) one would expect to find axial conformation but the torsion angles S(1), S(2), C(18), C(17) 61.7° and S(1), S(2), C(18), C(19) 122.8° indicate that it is distorted. The bond angles S(1)-C(8)-C(7) and S(1)-C(8)-C(9), respectively 125.6 and 114.3°, and the corresponding angles in fragment II of 123.3 and 116.4° give an idea of the steric repulsion between atoms C(7) and S(2) and C(20) and S(1). The angles S(2)-S(1)-C(8) and S(1)-S(2)-C(18) are 104.7 and 105.0° respectively.

2-(PHENYL-N-OXIDOIMINOMETHYL)-3-PHENYLAMINOINDOLE
$C_{21}H_{17}N_3O$

L. GRECI and P. SGARABOTTO, 1984. J. Chem. Soc. Perkin II, 1281-1284.

Triclinic, P$\bar{1}$. a = 11.982, b = 11.462, c = 6.253 Å, α = 83.4, β = 80.1, γ = 95.7°, Z = 2. Cu radiation, R = 0.045 for 1578 reflexions.

The fused two-ring system of the molecule (Fig. 1) is approximately planar, the dihedral angle between the mean planes of the two rings being 178.3°. The phenyl ring C(16-21) makes an angle of 64.7° with the plane of the indole ring. The N(2) atom shows some pyramidal character due to interaction between O(1) and N(1), O(1)...N(1) 2.578(4) Å. The N=C distance, N(2)-C(9), is 1.286(4) Å, The dihedral angles between the nitrone moiety and the phenyl ring C(10-15) and between the indole ring and the nitrone moiety are 35.3 and 14.8° respectively. Packing is determined by O...H-N hydrogen bonds (O(1)...N(1) 2.875(5), O(1)...N(3) 2.928(4) Å).

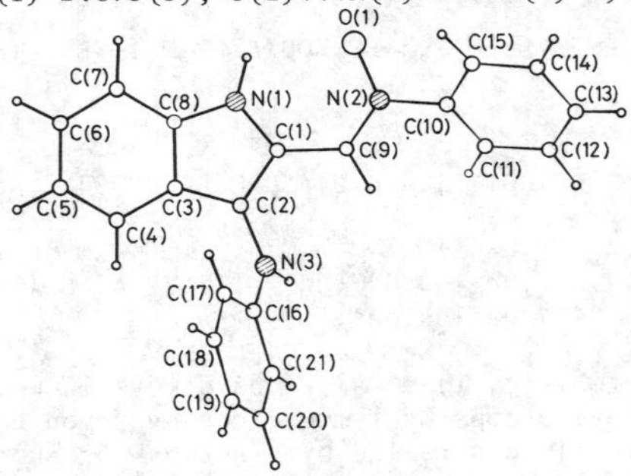

Fig. 1. $C_{21}H_{17}N_3O$: the molecular structure.

1-(p-ETHOXYPHENYL)-1,3-DIHYDRO-3-PHENYL-2H-BENZIMIDAZOLE-2-THIONE
$C_{21}H_{18}N_2OS$

A. CRIADO, A. CONDE and R. MARQUEZ, 1984. Acta Cryst., C40, 188-190.

Orthorhombic, $P2_12_12_1$, a = 7.600, b = 11.132, c = 20.767 Å, Z = 4. Cu radiation, R = 0.049 for 1531 reflexions.

The benzimidazole moiety (Fig. 1) is close to being planar (dihedral angle between the two fused rings 1.6(2)°). The orientations of the phenyl rings are defined by the dihedral angles which the unsubstituted (73.7(2)°) and substituted (61.4(2)°) rings make with the imidazole ring.

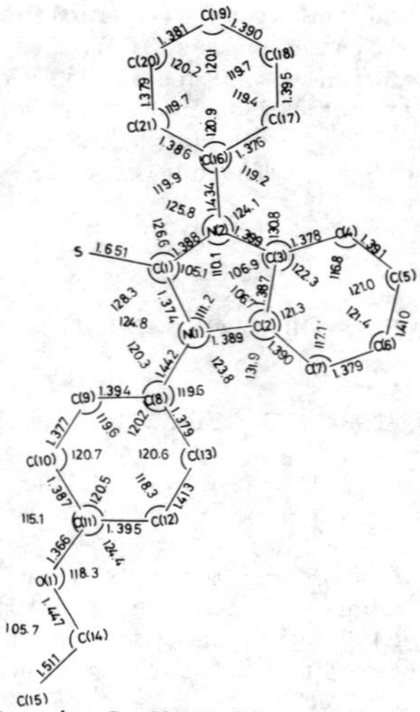

Fig. 1. Bond lengths and angles in $C_{21}H_{18}N_2OS$.

2-[7-CHLORO-2(3H)-OXO-5-PHENYL-1H-1,4-BENZODIAZEPIN-1-YL] ETHYL HYDROGEN SUCCINATE N^4-OXIDE
$C_{21}H_{19}ClN_2O_6$

V. FERRETTI, V. BERTOLASI, G. GILLI and P.A. BOREA, 1984. Acta Cryst., C40, 986-988.

Monoclinic, $P2_1/n$, a = 17.269, b = 9.823, c = 11.477 Å, β = 98.48°, Z = 4. Cu radiation, R = 0.039 for 2474 reflexions.

The crystal structure (Fig. 1) consists of dimers in which the two molecules are linked through the carbonyl groups by symmetrical hydrogen bonds, with O...O 2.684 Å. The crystal packing is mainly controlled by a network of short C-H...O contacts which could be considered weak hydrogen bonds. The diazepine ring adopts an almost perfect boat conformation. The molecule is chiral and both enantiomers are contained in the centrosymmetric space group. Bond distances and angles are generally as anticipated.

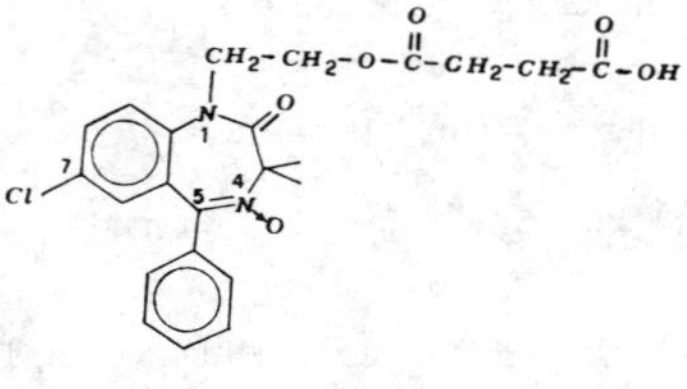

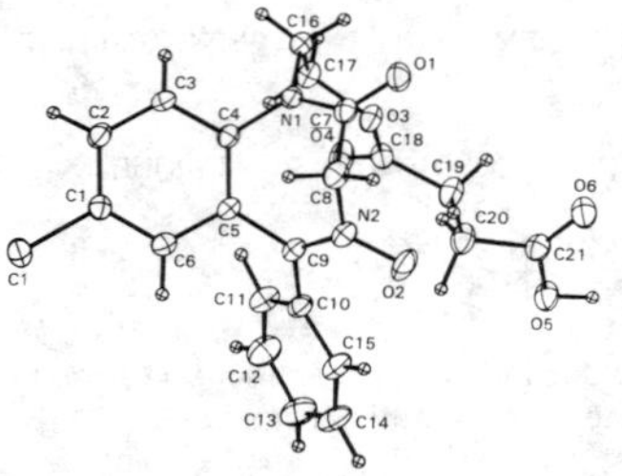

Fig. 1. $C_{21}H_{19}ClN_2O_6$: structural formula and perspective view of the molecule.

2-(2-QUINOLYL)CYCLOHEXANONE PHENYLHYDRAZONE
$C_{21}H_{21}N_3$

G. BOCELLI, G. TOSI and L. CARDELLINI, 1984. Acta Cryst., C40, 1952-1955.

Monoclinic, $P2_1/n$, a = 16.249, b = 9.058, c = 11.846 Å, β = 102.42°, Z = 4. Cu radiation, R = 0.041 for 1698 reflexions.

The quinoline moiety is slightly folded around the C(10)-C(11) bond (Fig. 1) with the two rings inclined at 1.7(1)° to each other. The cyclohexane ring is partially disordered and exhibits two different conformations, one midway between a boat and twist boat and the other close to a chair. The N(1)-N(2) distance is 1.379(3) Å.

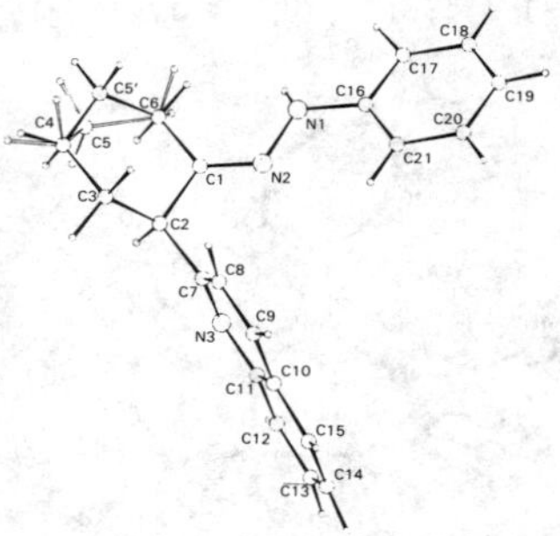

Fig. 1. The 2-(2-quinolyl)cyclohexanone phenylhydrazone molecule.

2-(2-HYDROXY-3-{[2-(1H-INDOL-3-YL)-1,1-DIMETHYLETHYL]AMINO}PROPOXY)BENZONITRILE (BUCINDOLOL)

$C_{22}H_{25}N_3O_2$

J.-M. LEǴER, M. GOURSOLLE and A. CARPY, 1984. Acta Cryst., C40, 706-708.

Triclinic, $P\bar{1}$, a = 8.801, b = 8.883, c = 13.662 Å, α = 99.79, β = 92.22, γ = 105.43°, Z = 2. Cu radiation, R = 0.035 for 1938 reflexions.

In the molecule the oxypropanolamine chain has an A conformation, as already defined (1) (Fig. 1). Bond lengths and angles have the expected values. The molecules are linked in the crystal by N-H...O and O-H...N hydrogen bonds.

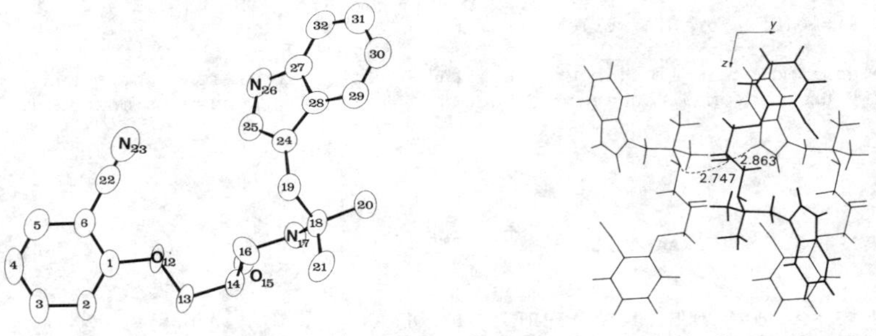

Fig. 1. $C_{22}H_{25}N_3O_2$: perspective view of the molecule and the projection along [100] of the molecular packing.

1. J.-M. LÉGER, M. GADRET and A. CARPY, 1980. Mol. Pharmacol. 17, 339.

7,7a-DIHYDRO-1,3-BIS(4-METHYLPHENYL)-4,7,7a-TRIPHENYL-1H-PYRROLO[2,3-d]PYRIMIDIN-2,5,6(3H)-TRIONE

$C_{38}H_{29}N_3O_3$

G. KOLLENZ, G. PENN, G. DOLENZ, Y. AKCAMUR, K. PETERS, E.-M. PETERS and H.G. VON SCHNERING, 1984. Chem. Ber., 117, 1299-1309.

Triclinic, $P\bar{1}$, a = 11.321, b = 14.584, c = 9.875 Å, α = 105.58, β = 99.87, γ = 98.88°, Z = 2. Mo radiation, R = 0.076 for 1644 reflexions.

X-ray analysis established the structure of product of cycloaddition reactions as shown in Fig. 1. Some bond lengths are: C1-N2 1.482(8), C1-N6 1.500(10), C1-N9 1.493(9), N2-C3 1.377(9), N2-C21 1.455(7), C3-N4 1.416(7), N4-C5 1.395(10), N4-C41 1.455(9), C8-N9 1.373(10), N9-C91 1.449(9) Å.

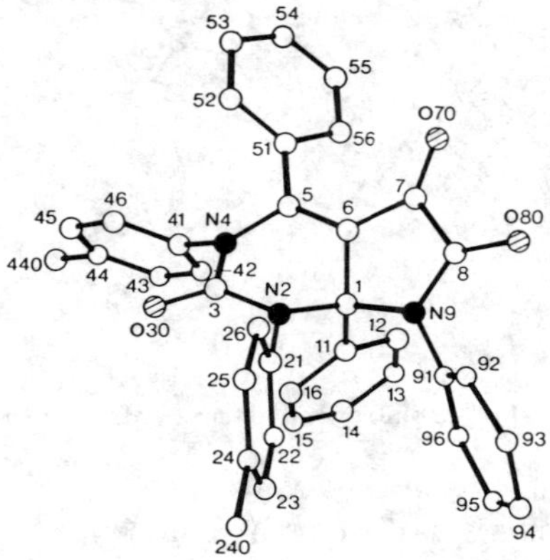

Fig. 1. A view of $C_{38}H_{29}N_3O_3$.

1,8-DIAZABIPHENYLENE
$C_{10}H_6N_2$

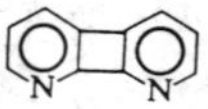

B.R. DEROSKI, J.S. RICCI Jr., J.A.H. MacBRIDE and J.H. MARKGRAF, 1984. Can. J. Chem., 62, 2235-2237.

Monoclinic, $P2_1/n$, a = 3.786, b = 8.532, c = 11.218 Å, β = 92.24°, Z = 2. Cu radiation, R = 0.059 for 614 reflexions.

The planar molecule lies about a crystallographic inversion centre and consequently the N atom sites are disordered. The distance between molecular planes is 3.52 Å.

2,5-DIMETHYL-1,3,4,6-TETRAAZACYCL[3.3.3]AZINE
$C_{10}H_9N_5$

G. LAURENT, J.-P. BOUTIQUE, J.J. VERBIST, F. DURANT and G.J. ASHWELL, 1984. Acta Cryst., C40, 2108-2109.

Monoclinic, P2/m, a = 4.022, b = 17.640, c = 6.568 Å, β = 96.02°, Z = 2. Cu radiation, R = 0.045 for 770 reflexions.

The molecule (Fig. 1) lies on a mirror plane (through C(2), N(1), C(8)), with a planar non-hydrogen skeleton. The bond pattern presents a strong atomatic character at the periphery, while bond lengths involving the central N atom are somewhat longer.

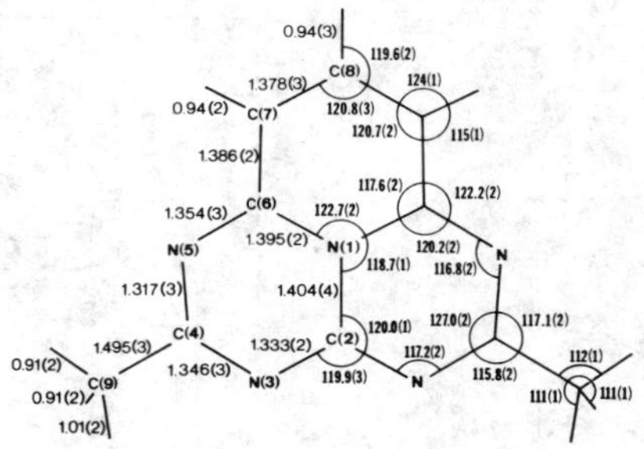

Fig. 1. Molecular dimensions for $C_{10}H_9N_5$.

NAPHTHO[2,1-e]TETRAZOLO[1,5-b]-as-TRIAZINE
$C_{11}H_{16}N_6$ (I)

NAPHTHO[1,2-e]TETRAZOLO[1,5-b]-as-TRIAZINE
$C_{11}H_{16}N_6$ (II)

Gy. HAJOS, A. MESSMER, A. NESZMELYI and L. PARKANYI, 1984. J. Org. Chem., 49, 3199-3203.

I. Monoclinic, $P2_1/n$, a = 15.355, b = 7.639, c = 17.223 Å, β = 104.22°, D_m = 1.50, Z = 8. Mo radiation, R = 0.134 for 1743 reflexions.

II. Monoclinic, $P2_1/c$, a = 15.467, b = 7.650, c = 20.108 Å, β = 124.40°, D_m = 1.51, Z = 8. Mo radiation, R = 0.049 for 2265 reflexions.

The analyses established the structures of these molecules to be "b-fused" tetrazoles (Fig. 1).

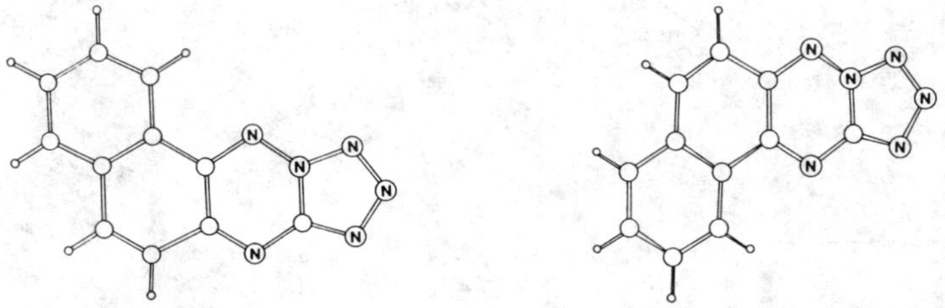

Fig. 1. The two isomeric $C_{11}H_{16}N_6$ molecules, (I) (left) and (II) (right).

PHENAZINE:IODINE
$C_{12}H_8I_2N_2$ $C_{12}H_8N_2.I_2$

T. UCHIDA and K. KIMURA, 1984. Acta Cryst., C40, 139-140.

Orthorhombic, Pbcn, a = 12.901, b = 11.602, c = 8.303 Å, D_m = 2.30, Z = 4. Mo radiation, R = 0.039 for 776 reflexions.

A refinement with new data of a structure reported previously (1). See Fig. 1 for further details.

Fig. 1.

1. Str

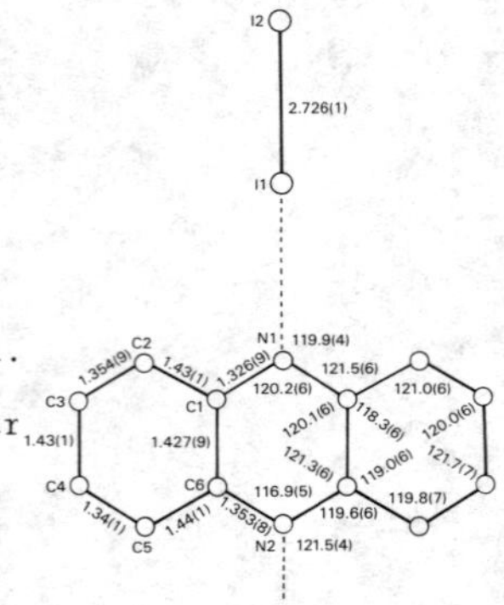

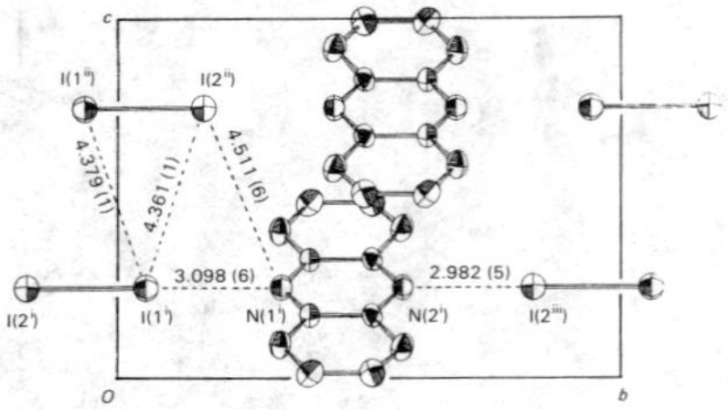

Fig. 1. Molecular dimensions and details of packing in $C_{12}H_8N_2.I_2$.

1. Structure Reports, 32B, 273.

POTASSIUM 1,3,3a,8a-TETRAHYDRO-3a,8a-DIHYDROXY-2,8-DIOXOINDENO[1,2-d]IMIDAZOLE-2-ETHANESULPHONATE
$C_{12}H_{11}KN_2O_7S$ $K^+.C_{12}H_{11}N_2O_7S^-$

D.D. BRAY, D.D. CLARKE, N. CHATTERJIE and B. SINHA, 1984. Acta Cryst., C40, 348-350.

Monoclinic, $P2_1/c$, a = 6.795, b = 17.701, c = 12.864 Å, β = 116.25°, D_m = 1.74, Z = 4. Cu radiation, R = 0.074 for 2091 reflexions.

The compound was prepared as shown in Fig. 1. The unsubstituted N atom of the ureidotaurine reacts with the central C of the ninhydrin.

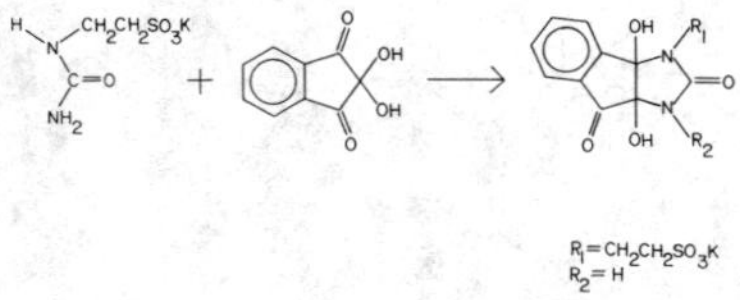

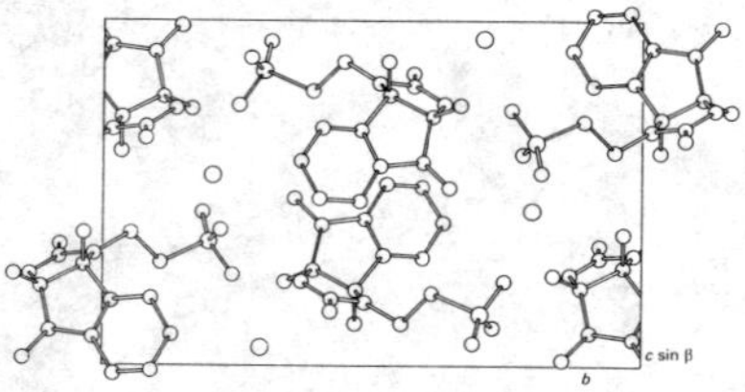

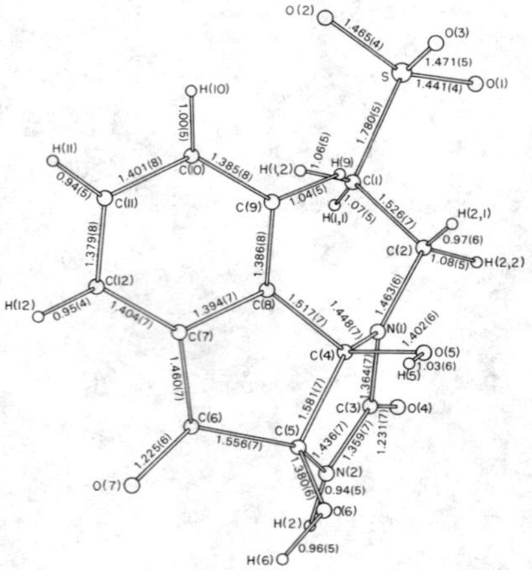

Fig. 1. The formation route for $K^+C_{12}H_{11}N_2O_7S^-$, details of bond lengths, and a packing diagram.

trans-1,9-DICHLORO-1,2,3,4,6,7,8,9-OCTAHYDROPHENAZINE 5-OXIDE
$C_{12}H_{14}Cl_2N_2O$

Z. GALDECKI, V. GROCHULSKI and Z. WAWRZAK, 1984. Acta Cryst., C40, 415-416.

Monoclinic, $P2_1/c$, a = 6.155, b = 11.390, c = 20.200 Å, β = 119.85°, D_m = 1.479, Z = 4. Cu radiation, R = 0.039 for 1671 reflexions.

In the molecule (Fig. 1) the central pyrazine ring is flat (maximum deviation 0.012(9) Å) and both unsaturated six-membered rings are in half-chair conformations with both C-Cl bonds in axial positions.

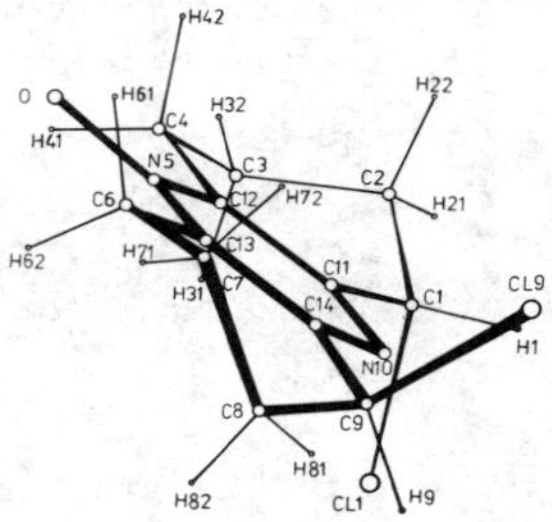

Fig. 1. A view of $C_{12}H_{14}Cl_2N_2O$.

3,5,9b-TRIMETHYL-4a,9b-DIHYDRO-3H,5H-[1,2,4]TRIOXINO[5,6-b]INDOLE
$C_{12}H_{15}NO_3$

C.W. JEFFORD, D. JAGGI, J. BOUKOUVALAS, S. KOHMOTO and G. BERNARDINELLI, 1984. Helv. Chim. Acta, 67, 1104-1111.

Monoclinic, $P2_1/c$, a = 7.687, b = 10.574, c = 14.908 Å, β = 104.26°, Z = 4. Mo radiation, R = 0.042 for 924 reflexions.

The analysis establishes that this particular molecule (of a pair of diastereoisomers) has the structure shown in Fig. 1. The trioxane ring has a chair form, cis-fused to the indole nucleus. The O(1)-O(2) distance is 1.466(5) Å and the C-O distances are O(1)-C(7) 1.450(6), O(2)-C(8) 1.431(7), O(3)-C(8) 1.429(6), O(3)-C(9) 1.446(6) Å.

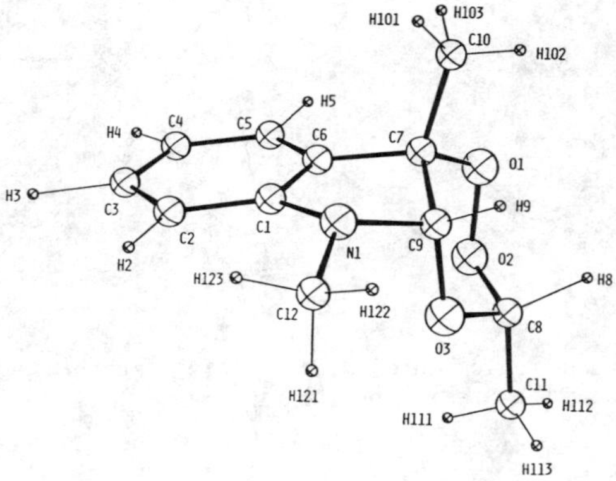

Fig. 1. A view of $C_{12}H_{15}NO_3$.

METHYL 9H-PYRIDO[3,4-b]INDOLE-3-CARBOXYLATE (METHYL β-CARBOLINE-3-CARBOXYLATE)
$C_{13}H_{10}N_2O_2$

V. BERTOLASI, V. FERRETTI, G. GILLI and P.A. BOREA, 1984. Acta Cryst., C40, 1981-1983.

Monoclinic, $P2_1/c$, a = 11.467, b = 5.803, c = 32.382 Å, β = 97.11°, Z = 8. Mo radiation, R = 0.058 for 1593 reflexions.

Both molecules in the asymmetric unit are planar and are linked by N(A1)-H... O(B1) hydrogen bonds (Fig. 1). The N...O distance is 2.825(5) Å with the N-H...O angle 140°. Ribbons of molecules are linked along a by further N-H...O hydrogen bonds (N...O 2.912(5) Å) into infinite chains. The structure is stabilised by short (3.388(6) and 3.095(7) Å) C-H...O and N...N (3.307(6) Å) contacts.

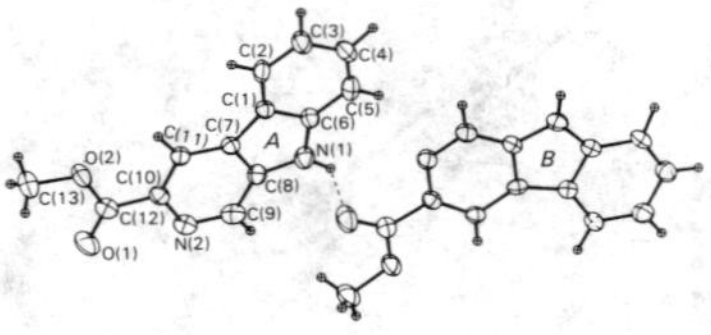

Fig. 1. The asymmetric unit in the methyl β-carboline-3-carboxylate structure.

(S,R)-5-METHYLENE-PYRROLIDINO[2,1-c][1,4]BENZODIAZEPIN-11-ONE
$C_{13}H_{14}N_2O$

S.N. GOLUBEV, Yu.D. KONDRASHEV and M.G. RYZHOV, 1984. Kristallografiya, 29, 736-740 [Sov. Phys. Crystallogr., 29, 435-437].

Triclinic, $P\bar{1}$, a = 6.950, b = 7.146, c = 11.859 Å, α = 95.70, β = 103.68, γ = 102.60°, Dm = 1.27, Z = 2. Mo radiation, R = 0.058 for 1333 reflexions.

The 1,4-diazepin ring consists of two planar fragments: a quasipeptide fragment (C(12),C(3),N(4),C(5),C(13),C(16)) and a peptide fragment (C(12),O(15),C(11),N(10),-C(14),H(10)) intersecting at the C(12) atom and closed on the other side by the C(13)-C(14) bond (Fig. 1). The angle between the planar fragments is 116°. The pyrrolidine ring has an envelope conformation with atom C(2) 0.529 Å from the plane through atoms C(1), C(3), N(4) and C(12). The molecules are linked into centro-symmetric dimers by hydrogen bonds (N(10)-H(10)...O(15') 2.884 Å).

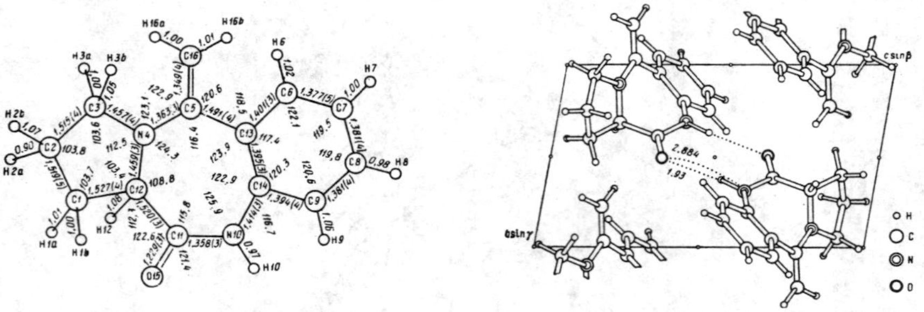

Fig. 1. Bond lengths and angles in $C_{13}H_{14}N_2O$ (left), and projection of the structure along the a axis (right).

4,9-DIHYDRO-7-METHOXY-1-METHYL-3H-PYRIDO[3,4-b]INDOLE (HARMALINE)
$C_{13}H_{14}N_2O$

W. REIMERS, H. GUTH and WANG ZU-TAO, 1984. Acta Cryst., C40, 977-978.

Orthorhombic, $P2_12_12_1$, a = 10.709, b = 13.576, c = 7.541 Å, Z = 4. Neutron diffraction, R = 0.076 for 976 reflexions.

The framework of the molecule (Fig. 1) is built up of three rings (benzene, pyrrole and fused dihydropyridine). The substituted dihydropyridine ring is rather puckered. The mean bond lengths involving H are: $C(sp^3)$-H 1.082(15), $C(sp^2)$-H 1.093(12) and N-H 1.023(10) Å. One molecule is connected to two other molecules by one N-H...N hydrogen bond each; the hydrogen-bond length is 1.989(12) Å and the bond angle is 170.3(14)°. For previous work see (1).

Fig. 1. $C_{13}H_{14}N_2O$: molecular structure and hydrogen bonding (dotted lines).

1. ATTA-UR-RAHMAN, E. FORESTI SERANTONI and L. RIVA DI SANSEVERINO, 1971. Pak. J. Sci. Ind. Res. 14, 490.

5,6-DIHYDRO[5,4-c]PYRIMIDINOCARBAZOLE
$C_{14}H_{11}N_3$

NGUYEN-HUY DUNG, B. VIOSSAT, J.-C. LANCELOT and M. ROBBA, 1984. Acta Cryst., C40, 997-999.

Monoclinic, $P2_1/c$, a = 9.127, b = 9.390, c = 14.903 Å, β = 119.30°, D_m = 1.28, Z = 4. Mo radiation, R = 0.055 for 1474 reflexions.

The crystal structure (Fig. 1) consists of chains of molecules linked by N-H...N hydrogen bonds with N...N 2.867(4) Å. The pyrimidine, pyrrole and benzene rings are planar and have normal bond lengths and angles. The cyclohexadiene ring has a half-chair conformation as is found in 1,3-cyclohexadiene although the bond lengths are different.

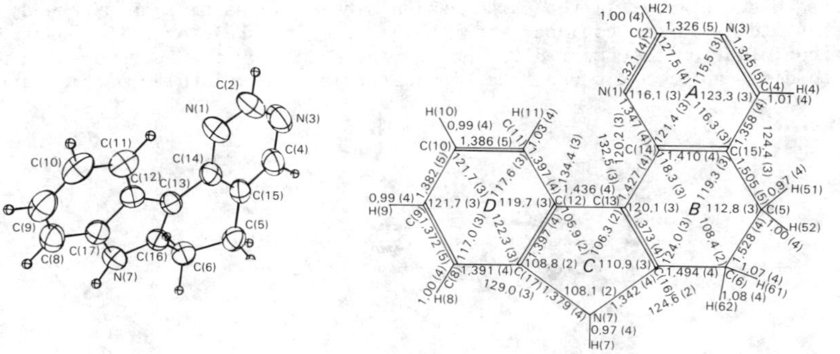

Fig. 1. $C_{14}H_{11}N_3$: conformation of the molecule and bond lengths and angles.

5,6-DIHYDRO[5,4-c]PYRIMIDINOCARBAZOLE (MONOCLINIC FORM II)
$C_{14}H_{11}N_3$

N.-H. DUNG, B. VIOSSAT, J.-C. LANCELOT and M. ROBBA, 1984. Acta Cryst., C40, 1602-1605.

Monoclinic, C2/c, a = 11.996, b = 12.219, c = 15.54 Å, β = 100.95°, D_m = 1.30, Z = 8. Mo radiation, R = 0.066 for 1411 reflexions.

The 1,3-cyclohexadiene ring (Fig. 1) is in a skew-chair conformation. Molecules are linked by N(7)-H...N(3') hydrogen bonds (N...N 2.882(4) Å) into infinite ribbons along c as in the monoclinic form I (1). Molecules related by inversion centres are stacked with their mean planes parallel with interplanar spacings of 3.51 and 3.95 Å.

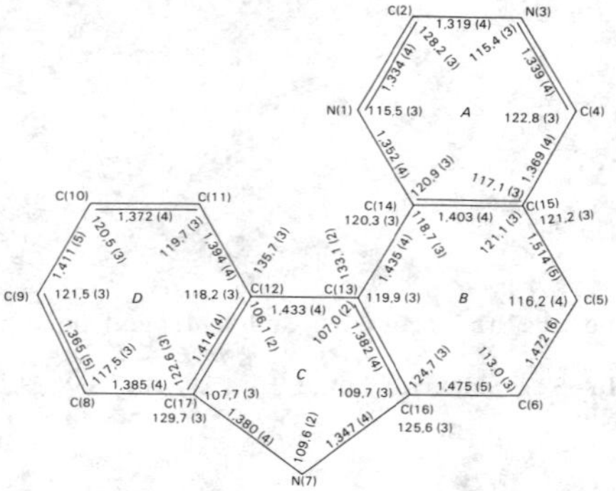

Fig. 1. Bond lengths (Å) and angles (°) in the $C_{14}H_{11}N_3$ molecule.

1. Previous Report.

N-ETHYL-3-CARBAMOYL-β-CARBOLINE
$C_{14}H_{13}N_3O$

A.K.S. MUIR and P.W. CODDING, 1984. Can. J. Chem., 62, 1803-1806.

Monoclinic, C2/c, a = 16.220, b = 7.728, c = 19.623 Å, β = 104.16°, Z = 4. Mo radiation, R = 0.032 for 1133 reflexions.

In the molecule (Fig. 1) the carboxyamide side chain assumes an extended conformation and is almost coplanar with the β-carboline skeleton. The observed molecular conformation is compared to the current model for the receptor binding site. Hydrogen bonding and aromatic ring stacking determine the molecular packing arrangement.

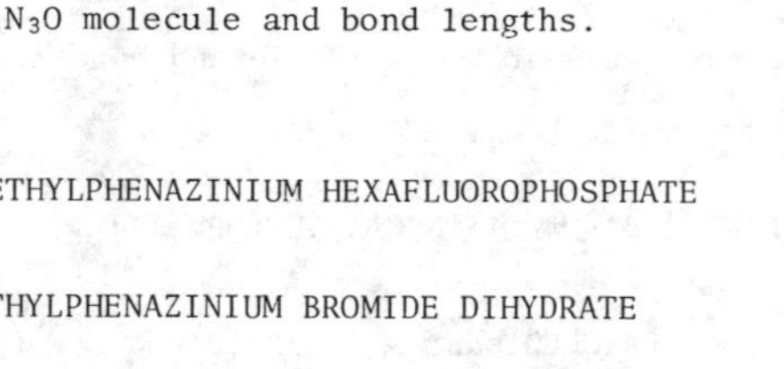

Fig. 1. The $C_{14}H_{13}N_3O$ molecule and bond lengths.

5,10-DIHYDRO-5,10-DIMETHYLPHENAZINIUM HEXAFLUOROPHOSPHATE
$C_{14}H_{14}F_6N_2P$ (I)

5,10-DIHYDRO-5,10-DIETHYLPHENAZINIUM BROMIDE DIHYDRATE
$C_{16}H_{18}BrN_2,2(H_2O)$ (II)

K. DIETZ, H.J. KELLER, D. NOTHE and D. WEHE, 1984. Z. Naturforsch., B39, 452-457.

I. Monoclinic, Cc, a = 14.544, b = 6.23, c = 18.542 Å, β = 121.32°, Z = 4. Mo radiation, R = 0.119 for 1064 reflexions.

II. Monoclinic, C2/c, a = 21.732, b = 5.346, c = 14.436 Å, β = 105.01°, Z = 4. Mo radiation, R = 0.045 for 1038 reflexions.

The cation in (I) is nearly planar being folded only 3.3° around the N...N line. In (II), the cation lies about an inversion centre and is planar. Details of geometry are in Fig. 1. The cations form segregated stacks along b with an interplanar distance of 3.47 Å. The Br^- ion lies on a twofold axis and is hydrogen bonded to the water molecules (H...Br 2.62 Å). The packing of (I) is described as containing weakly exchanging mixed stacks and is quite different from (II).

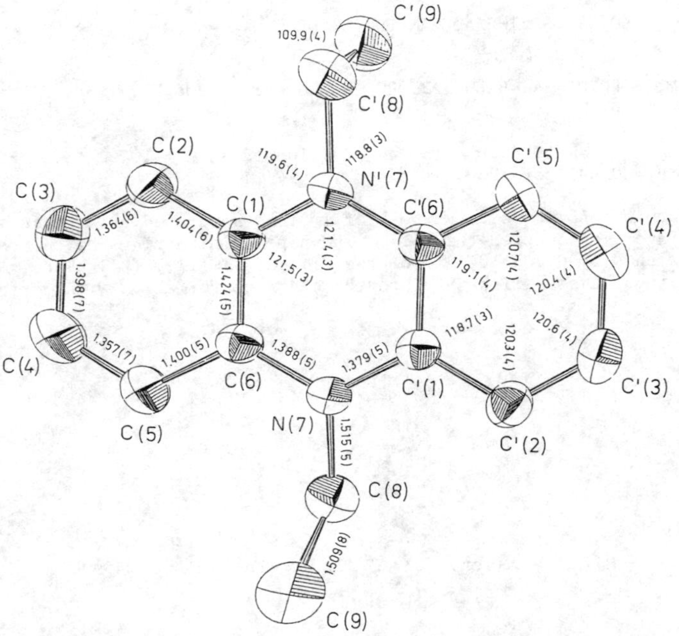

Fig. 1. Molecular dimensions for the cation in $C_{16}H_{18}BrN_2,2(H_2O)$ (II).

1,4,7,10,13,16-HEXAAZAPENTACYCLO[11.5.1.1^{4,10}.0^{7,20}.0^{16,19}]ICOSANE
$C_{14}H_{26}N_6$

M.M. OLMSTEAD, P.P. POWER and M. VIGGIANO, 1984. Acta Cryst., C40, 1439-1440.

Monoclinic, C2/c, a = 15.024, b = 6.275, c = 15.100 Å, β = 91.79°, Z = 4. Mo radiation, R = 0.040 for 1191 reflexions.

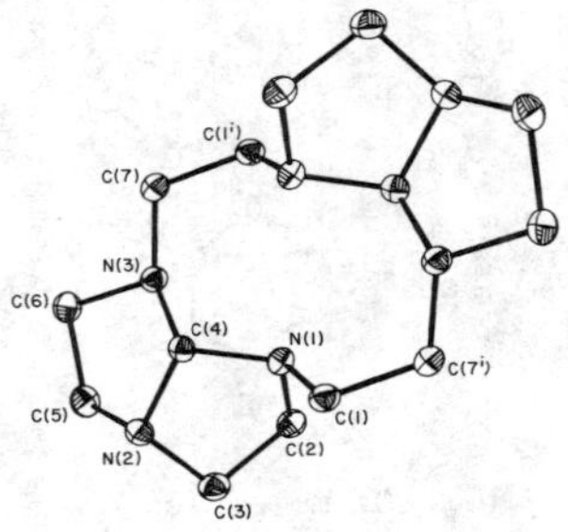

Fig. 1. The $C_{14}H_{26}N_6$ molecule.

The molecule (Fig. 1), which lies on a centre of symmetry, contains a tetraaza-cyclodecane ring within which the two methine carbons, each bridging three nitrogen atoms, have distorted geometry i.e. N(2)-C(4)-N(3) 104.5(1), N(1)-C(4)-N(2) 108.3(1), N(1)-C(4)-N(3) 112.8(1)°, N(2)-C(4) 1.499(2), N(3)-C(4) 1.461(2), N(1)-C(4) 1.461(2) Å. The shortest transannular distance is N(1)...N(1') 2.897(5) Å. The endocyclic angles of the five-membered rings are normal but the exocyclic angles are 113.2(1), 115.6(1) at N(3), 112.8 at C(4) and 112.2 and 113.1° at N(1). There are no short intermolecular contacts.

6,9-DIAZONIADISPIRO[5.2.5.2]HEXADECANE DIIODIDE

$C_{14}H_{28}I_2N_2$ $C_{14}H_{28}N_2^{2+}.2I^-$

J. ESTIENNE, M. PIERROT, A. BALDY, J. ROSENBERG and G. ROBERT, 1984. Acta Cryst., C40, 1478-1480.

Monoclinic, $P2_1/n$, a = 7.086, b = 13.144, c = 10.012 Å, β = 101.20°, D_m = 1.70, Z = 2. Mo radiation, R = 0.041 for 1236 reflexions.

The structure consists of a network of I^- ions separated by organic moieties which lie on inversion centres (Fig. 1). The piperazinium rings have a chair conformation with N(1) 0.665(10) Å from the C(1),C(2),C(1'),C(2') plane. The piperidinium rings have a slightly distorted chair conformation with dominant mirror symmetry.

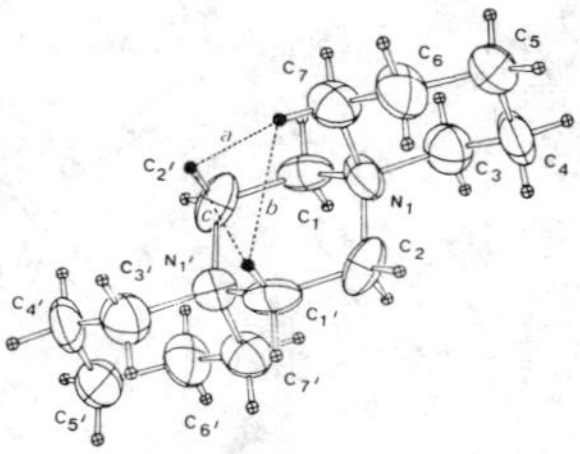

Fig. 1. The $C_{14}H_{28}N_2^{2+}$ cation, (a = 2.29, b = 2.21, c = 2.44 Å).

DIMETHYL PYRAZOLO[1,2-a]PYRAZOLO[1,2-b]BENZOTRIAZOLE-3,4-DICARBOXYLATE

$C_{15}H_{13}N_3O_4$

B. BOVIO, 1984. J. Cryst. Spectrosc. Res., 14, 369-382.

Triclinic, $P\bar{1}$, a = 9.268, b = 19.367, c = 8.821 Å, α = 96.76, β = 115.39, γ = 80.17°, D_m = 1.41, Z = 4. Mo radiation, R = 0.047 for 2653 reflexions.

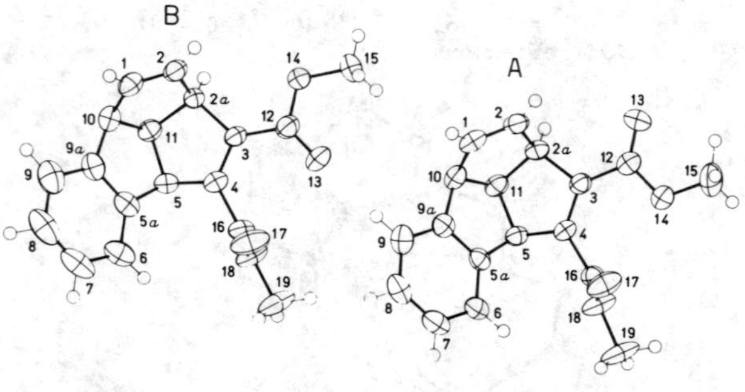

Fig. 1. The two independent $C_{15}H_{13}N_3O_4$ molecules.

The two independent molecules are almost identical in conformation (Fig. 1) except for the orientations of the 3-carbomethoxy groups. There is appreciable puckering of the condensed five-membered rings. Bond distances and angles are not all in accord with expected values and these are discussed in detail. The N5-N11 distances are 1.498(3) Å in both molecules, longer than that expected for a single N-N bond.

9-ISOPROPYLCARBAZOLE
$C_{15}H_{15}N$

O. SARAVARI, N. KITAMURA, S. TAZUKE, A. TAKENAKA and Y. SASADA, 1984. Acta Cryst., C40, 1617-1619.

Orthorhombic, Iba2, a = 16.8334, b = 17.9780, c = 7.9871 Å, Z = 8. Cu radiation, R = 0.069 for 890 reflexions.

Groups of four molecules are close-packed around the twofold axes with approximate 4_2 symmetry forming "polar arrows" along c (Fig. 1). Within the "arrows" the shortest distance between methyl groups related by the twofold axis is 3.53(2) Å; the other methyl group forms an electrostatic interaction with the N atom of a neighbouring molecule. There is no stacking between the planar aromatic moieties of the carbazole. The dihedral angle between the benzene ring planes is 1.8°

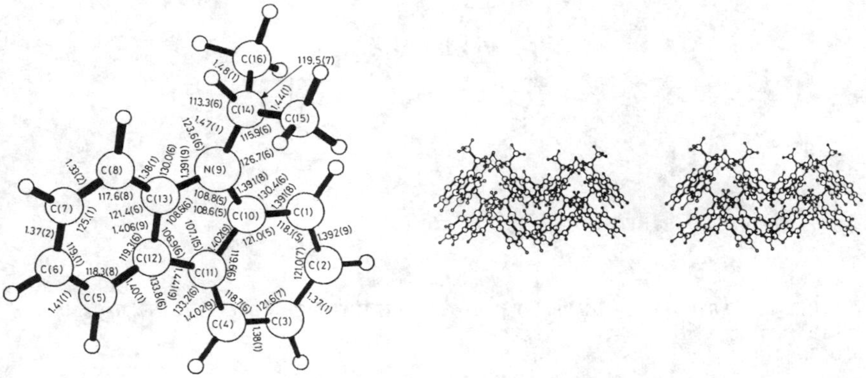

Fig. 1. Bond lengths (Å) and angles (°) in the 9-isopropylcarbazole molecule and a stereoview of the "polar arrows".

METHYL 8-ISOPROPYL-3,3a,8,8a-TETRAHYDROINDENO[2,1-c]PYRAZOLE-8a-CARBOXYLATE
$C_{15}H_{18}N_2O_2$

L. TOUPET and J.C. MESSAGER, 1984. Acta Cryst., C40, 330-331.

Triclinic, P$\bar{1}$, a = 5.791, b = 15.503, c = 15.954 Å, α = 82.24, β = 79.35, γ = 79.13°, D_m = 1.27, Z = 4. Mo radiation, R = 0.065 for 1518 reflexions.

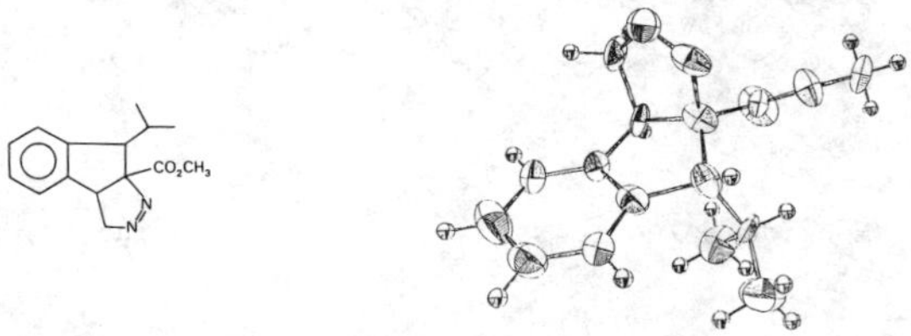

Fig. 1. A view of one of the $C_{15}H_{18}N_2O_2$ molecules.

There are two independent molecules in the asymmetric unit; the analysis establishes that both molecules have the isopropyl and methoxycarbonyl groups cis (Fig. 1).

3a-PHENYL-2,3,3a,4-TETRAHYDROPYRROLO[1,2-a]BENZIMIDAZOL-1-ONE
$C_{16}H_{14}N_2O$

A. CHIMIRRI, S. GRASSO, M. LONGERI, G. MENNITI, G. ROMEO and G. VALLE, 1984. J. Chem. Res., (S) 78-79, (M) 0601-0618.

Triclinic, P$\bar{1}$, a = 10.560, b = 9.182, c = 7.321 Å, α = 97.9, β = 88.6, γ = 113.3°, Z = 2. Cu radiation, R = 0.091 for three-dimensional data.

The analysis establishes the structure shown in Fig. 1. The imidazoline and fused benzene rings are nearly coplanar (dihedral angle 4.7°), while the pyrrolidone ring forms an angle of 35° with the benzimidazole system. The pyrrolidone ring has an envelope conformation, with C3 on the flap.

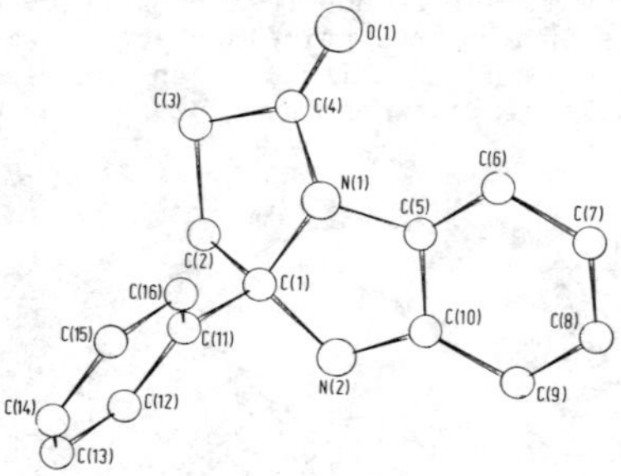

Fig. 1. A view of $C_{16}H_{14}N_2O$.

5,11-DIMETHYL-6H-PYRIDO[4,3-b]CARBAZOLE 2-OXIDE TRIHYDRATE (ELLIPTICINE N-OXIDE TRIHYDRATE)
$C_{17}H_{14}N_2O,3H_2O$

J.R. GARLICH, E.M. KAISER and E.O. SCHLEMPER, 1984. Acta Cryst., C40, 1871-1873.

Monoclinic, $P2_1/c$, a = 12.229, b = 6.962, c = 18.946 Å, β = 106.67°, D_m = 1.35, Z = 4. Mo radiation, R = 0.045 for 887 reflexions.

Except for the methyl H atoms, the molecule is nearly planar; the maximum deviation from planarity being 0.048(6) Å for a methyl C atom. The N-oxide O atom acts as acceptor for three hydrogen bonds (Fig. 1); the N-O distance is 1.347(4) Å.

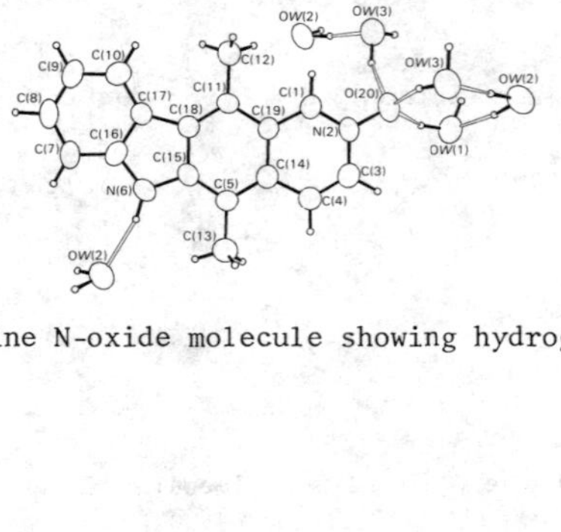

Fig. 1. The ellipticine N-oxide molecule showing hydrogen bonding to water molecules.

2-METHYL-1,2,3,4-TETRAHYDROBENZO[g]PYRIDAZINO[1,2-b]PHTHALAZINE-6,13-DIONE
$C_{17}H_{16}N_2O_2$ (I)

1,2,3,4-TETRAHYDROBENZO[g]PYRIDAZINO[1,2-b]PHTHALAZINE-6,13-DIONE
$C_{16}H_{14}N_2O_2$ (II)

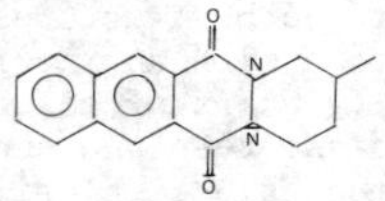

P.T. BEURSKENS, G. BEURSKENS, M.C. APREDA, C. FOCES-FOCES, F.H. CANO and S. GARCIA-BLANCO, 1984. Acta Cryst., C40, 1718-1721.

I. Triclinic, $P\bar{1}$, a = 8.0164, b = 10.6705, c = 17.5227 Å, α = 77.08, β = 101.24. γ = 99.28°, Z = 4. Cu radiation, R = 0.058 for 4952 reflexions.

II. Monoclinic, $P2_1$, a = 14.4453, b = 11.1088, c = 8.1103 Å, β = 99.59°, Z = 4. Cu radiation, R = 0.041 for 2294 reflexions.

In compound (I) there is a coupled disorder involving the methyl groups. The pyridazine rings have chair conformations in both compounds. There are no unusual bond distances or angles.

ACRIDINE ORANGE HYDROIODIDE
$C_{17}H_{20}IN_3$ (I)

ACRIDINE ORANGE HYDROCHLORIDE MONOHYDRATE
$C_{17}H_{20}ClN_3,H_2O$ (II)

C.A. MATTIA, L. MAZZARELLA, V. VITAGLIANO and R. PULITI, 1984. J. Cryst. Spectrosc. Res., 14, 71-87.

I. Monoclinic, $P2_1/c$, a = 7.440, b = 10.583, c = 21.115 Å, β = 96.34°, D_m = 1.60, Z = 4. Mo radiation, R = 0.038 for 2181 reflexions.

II. Monoclinic, $P2_1/c$, a = 8.548, b = 15.566, c = 14.545 Å, β = 120.71°, D_m = 1.28, Z = 4. Cu radiation, R = 0.058 for 1099 reflexions.

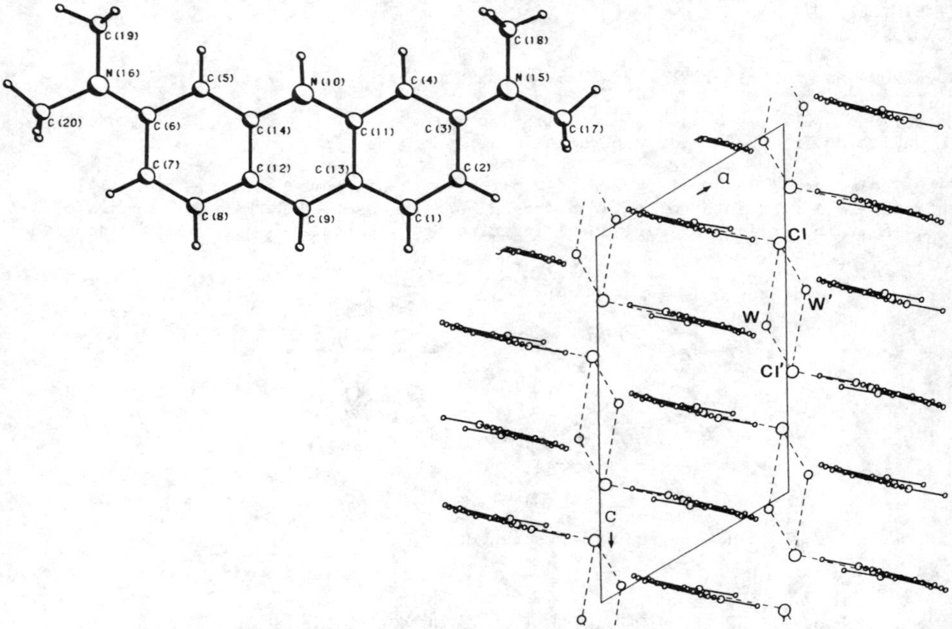

Fig. 1. A view of an acridine orange cation (left) and part of a packing diagram for the hydrochloride hydrate (II).

The acridine orange molecule is protonated at the central ring nitrogen (Fig. 1) in both structures. In the hydroiodide (I), the cations stack approximately parallel to a with an interplanar spacing 3.51 Å. The hydrochloride packing is shown in Fig. 1. Pairs of molecules related by an inversion centre, overlap each other with a plane-to-plane separation of 3.47 Å. There is hydrogen bonding involving the water molecules, chloride ions and cations.

N-[(2β,11bα)-1,3,4,6,7,11b-HEXAHYDRO-2H-BENZO[a]QUINOLIZIN-2-YL]-N-METHYL-1-PROPANE-SULPHONAMIDE HYDROCHLORIDE
$C_{17}H_{27}ClN_2O_2S$

A. CARPY, A. H'NAÏFI and J.-M. LÉGER, 1984. Acta Cryst., C40, 1965-1967.

Monoclinic, $P2_1/c$, a = 11.434, b = 14.761, c = 11.800 Å, β = 113.17°, Z = 4. Cu radiation, R = 0.048 for 1700 reflexions.

Ring A and atoms C(7) and C(10) are coplanar with C(8) and N(9) respectively -0.454(4) and 0.334(3) Å from the plane; ring C has the chair conformation. The C(11)-C(12)-N(15)-S(16) and S(16)-C(17)-C(18)-C(19) torsion angles are -131(1) and 177(1)°. A hydrogen bond (N...Cl 3.037(3) Å) links the protonated N(9) to the Cl^- anion. An important distance in the molecule is N(9)...S(16) which is 5.57(1) Å.

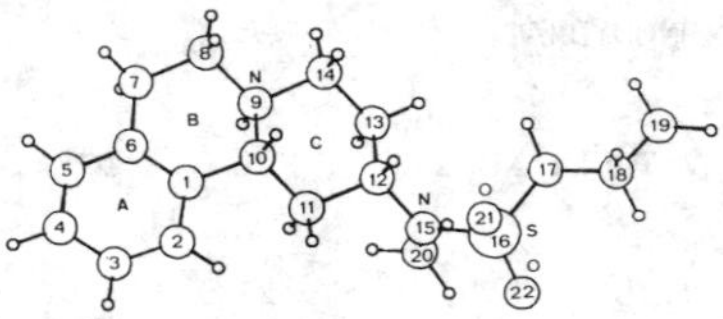

Fig. 1. The $C_{17}H_{27}ClN_2O_2S$ cation.

7-METHYLELLIPTICINE
$C_{18}H_{16}N_2$

R. KURODA and M. SAINSBURY, 1984. J. Chem. Soc. Perkin II, 1751-1753.

Monoclinic, $P2_1/n$, a = 4.042, b = 19.655, c = 16.343 Å, β = 94.48°, D_m = 1.34, Z = 4. Cu radiation, R = 0.048 for 806 reflexions.

The molecule (Fig. 1) is strictly planar with all three methyl C atoms lying on the plane. A weak intermolecular hydrogen bond is observed between N(2) and N(6) (3.14 Å). The ability of various ellipticines to intercalate with DNA is discussed.

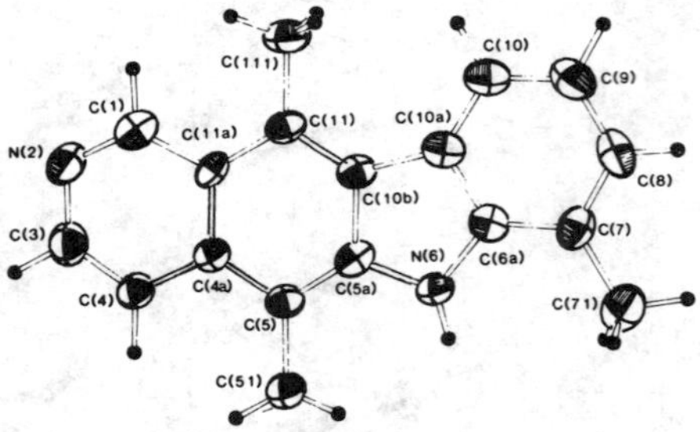

Fig. 1. $C_{18}H_{16}N_2$: the molecular structure.

(S)-(+)-4,5,6,7,11,12,13,14-OCTAHYDRO[1]BENZAZOCINO[7,6,5-efg]BENZAZOCINE (2,6'.2,6'-BIS(1-AZA-1,4-BUTANEDIYL)BIPHENYL)
$C_{18}H_{20}N_2$ (I)

(S)-(-)-2,2'-DIAMINO-6,6'-DIMETHYLBIPHENYL
$C_{14}H_{16}N_2$ (II)

(S)-(-)-(6R,7R)-5,6,7,8-TETRAHYDRO-1,6,7,12-TETRAMETHYLDIBENZO[e,g][1,4]DIAZOCINE DINITRATE (N,N'-(2,3-BUTANEDIYL)-2,2'-DIAMINO-6,6'-DIMETHYLBIPHENYL DINITRATE)
$C_{18}H_{24}N_4O_6$ (III)

K. SENO, S. HAGISHITA, T. SATO and K. KURIYAMA, 1984. J. Chem. Soc. Perkin I, 2013-2022.

I. Orthorhombic, $P2_12_12_1$, a = 14.901, b = 21.817, c = 8.744 Å, Z = 8. Cu radiation, R = 0.037 for 2392 reflexions.

II. Orthorhombic, $P2_12_12_1$, a = 7.976, b = 20.126, c = 7.559 Å, Z = 4. Mo radiation, R = 0.033 for 1034 reflexions.

III. Orthorhombic, $P2_12_12_1$, a = 8.094, b = 11.673, c = 10.304 Å, Z = 4. Cu radiation, R = 0.070 for 859 reflexions.

There are two crystallographically independent molecules (A, B) in the asymmetric unit of I (Fig. 1) but these do not differ significantly. Neither A nor B exhibits the C_2 symmetry expected from the molecular skeleton. Indeed one bridge involves an approximately planar N atom in both molecules while the other bridging N atom has pyramidal geometry. There is one intramolecular (C(1')...H-N(1') 2.43(4), 2.48(4) (A, B)) and two intermolecular (N(1')(A)...H-N(1')(B) 2.16(4); N(1')(B)...H-N(1)(A) 2.43(4) Å) short contacts in I. The angles between the phenyl mean planes are 118, 93 and 66° respectively for I, II and III. In III the cyclooctadiene ring has a twist-boat-chair conformation. The configuration of the diaminoethylene bridge C atoms is R,R. The cd spectra of the three compounds were discussed.

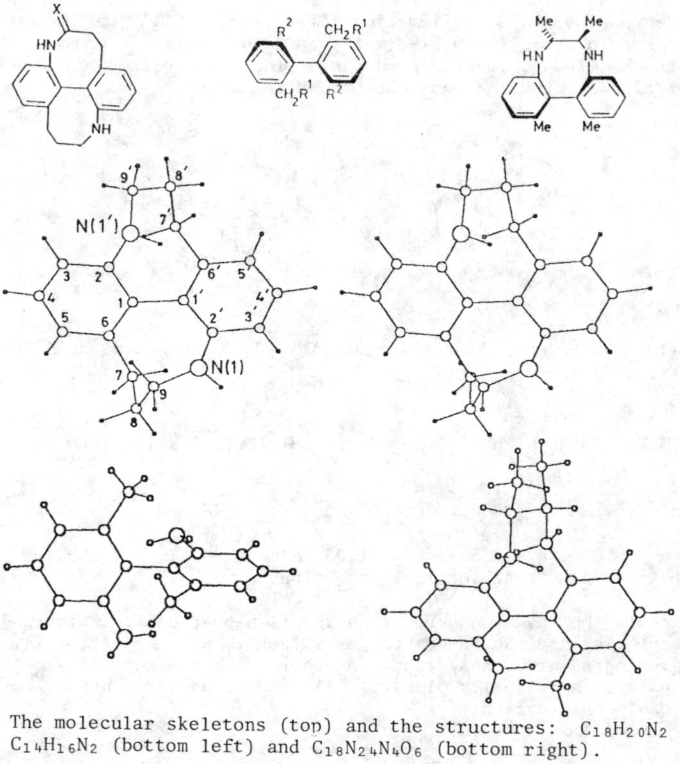

Fig. 1. The molecular skeletons (top) and the structures: $C_{18}H_{20}N_2$ (middle), $C_{14}H_{16}N_2$ (bottom left) and $C_{18}N_{24}N_4O_6$ (bottom right).

cis-2,3,9,9a-TETRAHYDRO-9-(PHENYLMETHYL)-1H-PYRROLO[1,2-a]INDOLE-9-CARBONITRILE
$C_{19}H_{18}N_2$

W. VERBOOM, D.N. REINHOUDT, R. VISSER and S. HARKEMA, 1984. J. Org. Chem., 49, 269-276.

Triclinic, $P\bar{1}$, a = 10.91, b = 9.570, c = 8.578 Å, α = 65.15, β = 66.94, γ = 71.24°, Z = 2. Mo radiation, R = 0.048 for 1934 reflexions.

The analysis established that the molecule was the cis-isomer (Fig. 1).

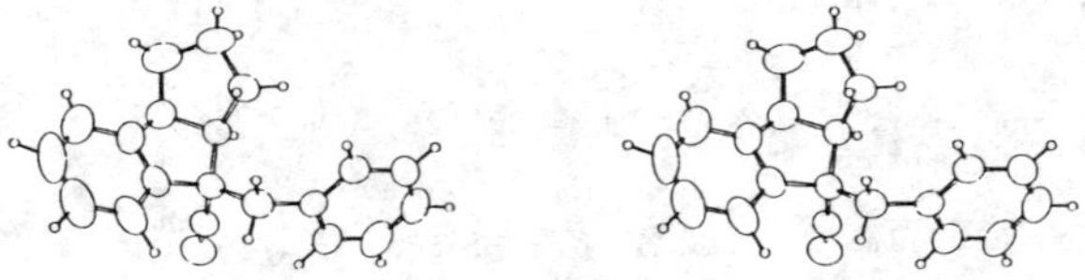

Fig. 1. A stereoview of $C_{19}H_{18}N_2$.

4-CHLOROIMIPRAMINE HYDROCHLORIDE
$C_{19}H_{24}Cl_2N_2$ $C_{19}H_{24}ClN_2^+.Cl^-$

A. HALLBERG, N.M. HINTERMEISTER, A.R. MARTIN, R.B. BATES and R.B. ORTEGA, 1984. Acta Cryst., C40, 2110-2112.

Monoclinic, $P2_1/c$, a = 16.972, b = 10.362, c = 11.018 Å, β = 107.40°, Z = 4. Mo radiation, R = 0.039 for 2013 reflexions.

This antidepressant has a confirmation (Fig. 1) with C(1') not in the synperiplanar position but in the C(4a)-N(5)-C(5a) plane and synclinal to avoid steric interaction with the Cl on the peri-fused ring. The amino nitrogen is protonated and the crystal contains alternating ionic and aromatic layers.

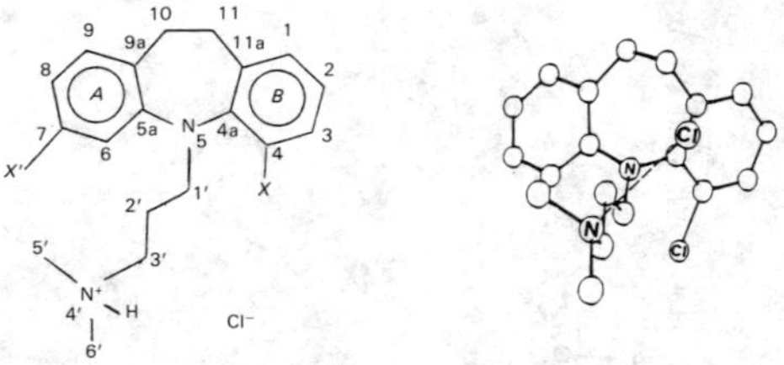

Fig. 1. 4-Chloroimipramine hydrochloride (X = Cl, X' = H) and an ORTEP plot.

8-CHLORO-1-[(DIMETHYLAMINO)METHYL]-6-PHENYL-4H-IMIDAZO[1,2-a][1,4]BENZODIAZEPINE
$C_{20}H_{19}ClN_4$

H.J. BUTCHER and T.A. HAMOR, 1984. Acta Cryst., C40, 848-850.

Triclinic, $P\bar{1}$, a = 9.036, b = 9.888, c = 11.035 Å, α = 94.62, β = 100.40, γ = 106.75°, Z = 2. Mo radiation, R = 0.042 for 1361 reflexions.

In the molecule the seven-membered ring adopts a boat conformation and the angle between the phenyl ring and the fused benzene moiety is 66.5(5)° (Fig. 1). Bond lengths and angles agree with those found in other 1,4-benzodiazepines. The benzo and imidazo rings are individually planar. All intermolecular contacts correspond to normal van der Waals interactions.

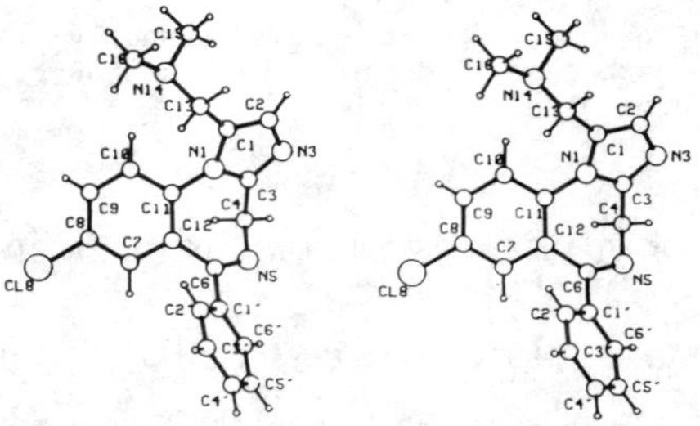

Fig. 1. $C_{20}H_{19}ClN_4$: stereoscopic view of the molecule in a direction perpendicular to the mean plane through atoms C(7)-C(12).

3,12-DIHYDRO-3,3,6,12-TETRAMETHYL-7H-PYRANO[2,3-c]ACRIDIN-7-ONE
$C_{20}H_{19}NO_2$

R.R. SMOLDERS, J. HANUISE, T. LEPOINT, N. VOGLET, B. TINANT, J.P. DECLERCQ and M. VAN MEERSSCHE, 1984. Tetrahedron, 40, 5181-5184.

Monoclinic, $P2_1$, a = 13.050, b = 15.011, c = 8.274 Å, β = 99.14°, Z = 4. Mo radiation, R = 0.065 for 1178 reflexions.

The two independent molecules in the asymmetric unit are related by a non-crystallographic inversion centre. There is intramolecular repulsion between methyl group C(21) and carbon C(1) (Fig. 1); the torsion angle C(21)-N(12)-C(16)-C(13) is -36° in one molecule and 25° in the other, and N(12)-C(16)-C(13)-C1 is -15° and +25°.

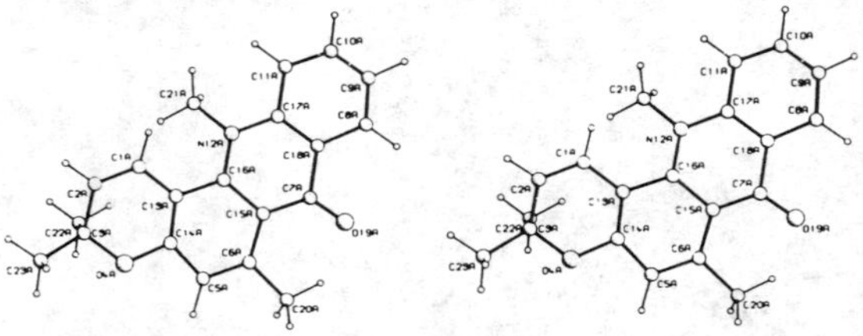

Fig. 1. A stereoview of $C_{20}H_{19}NO_2$ molecule A.

6-DIETHYLAMINO-3-METHOXY-11H-ISOINDOLO[2,1-a]INDOL-11-ONE
$C_{20}H_{20}N_2O_2$

R. KURODA, S.H. IMAM, M. HOOPER and J.R. BROWN, 1984. Acta Cryst., C40, 527-529.

Triclinic, $P\bar{1}$, a = 10.510, b = 12.906, c = 13.157 Å, α = 99.75, β = 107.45, γ = 96.06°, D_m = 1.29, Z = 4. Cu radiation, R = 0.048 for 3422 reflexions.

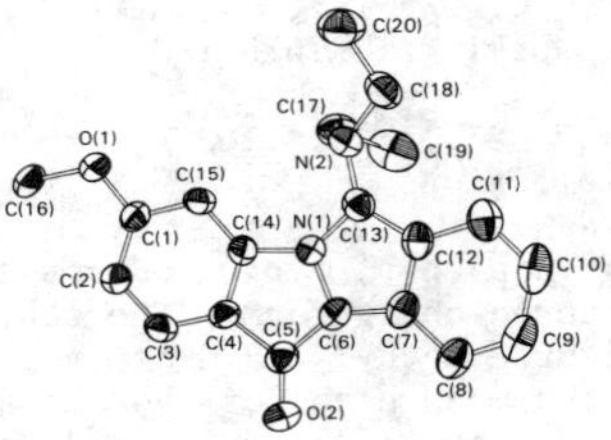

Fig. 1. A view of one molecule of $C_{20}H_{20}N_2O_2$.

Both molecules in the asymmetric unit are individually planar (except for one of the two N-ethyl groups in each case). One molecule is shown in Fig. 1; the methoxy groups lie in the chromophore planes.

4'-(9-ACRIDINYLAMINO)-3'-METHOXYMETHANESULPHONANILIDE (AMSACRINE)
$C_{21}H_{19}N_3O_3S$

J.S. BUCKLETON and T.N. WATERS, 1984. Acta Cryst., C40, 1587-1589.

Monoclinic, $P2_1/c$, a = 11.176, b = 21.142, c = 8.794 Å, β = 112.29°, Z = 4. Mo radiation, R = 0.041 for 1478 reflexions (at 256K).

In the amsacrine molecule (Fig. 1) the acridine group is non-planar, the four outer carbon atoms being displaced to the same side and the two central carbon atoms to the opposite side of the mean plane. The torsion angles C(8a),C(9)-N(11),C(12) and C(9),N(11)-C(12),C(17) are 129.0 and 19.1° respectively.

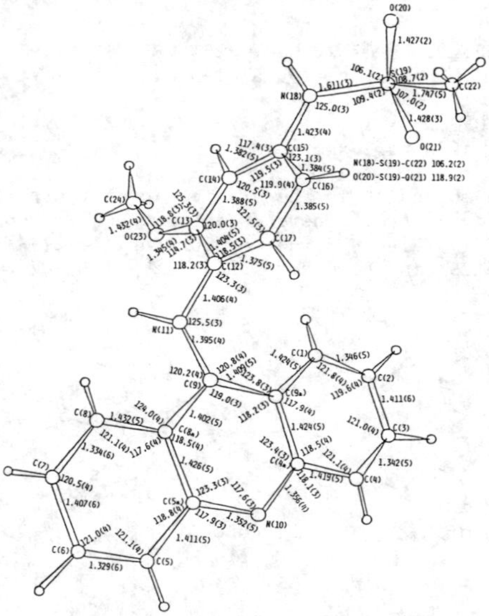

Fig. 1. Bond lengths (Å) and angles (°) in the amsacrine molecule.

(1S,12bS,2'R)-1α-ETHYL-1β-(2-HYDROXY-2-METHOXY-CARBONYLETHYL)-1,2,3,4,6,7,12,12bα-OCTAHYDROINDOLO[2,3-a]QUINOLIZINE
$C_{21}H_{28}N_2O_3$

L. SZABO, P. KOLONITS, G. KALAUS, C. SZANTAY, A. KALMAN, L. PARKANYI, M. KAJTAR and J. ANGYAN, 1984. J. Chem. Soc. Perkin II, 1629-1636.

Monoclinic, $P2_1$, a = 9.795, b = 7.387, c = 14.265 Å, β = 106.00°, Z = 2. Cu radiation, R = 0.044 for 2216 reflexions.

The indole moiety (Fig. 1) is planar, ring C adopts a twisted envelope E_6 (sofa) form and ring D a slightly flattened chair form. The ethyl group at C(1) is α-equatorial while the β-substituted methyl 2-hydroxypropionate which has a chiral centre at C(2') protrudes from the mean plane of the fused ring system. The substituents of C(2'), O(1), C(13) and C(1') are ranked in a clockwise direction. Consequently, the absolute configuration of C(2') is R.

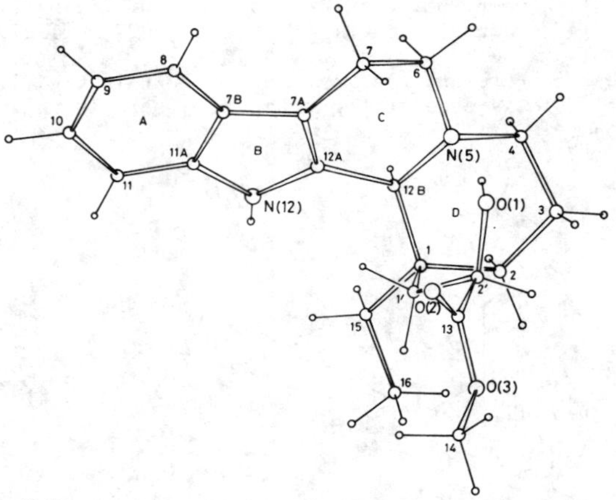

Fig. 1. $C_{21}H_{28}N_2O_3$: a perspective view of the molecule.

ACENAPHTE[9,10b]QUINOLIZINIUM-13-ETHOXYCARBONYL BROMIDE SESQUIHYDRATE
$C_{22}H_{16}BrNO_2,1.5(H_2O)$

F. FLORENCIO, P. SMITH-VERDIER and S. GARCIA-BLANCO, 1984. Z. Kristallogr., 167, 29-35.

Triclinic, P$\bar{1}$, a = 9.229, b = 10.568, c = 11.672 Å, α = 116.69, β = 70.32, γ = 100.68°, Z = 2. Mo radiation, R = 0.069 for 2939 reflexions.

The five rings in the molecule (Fig. 1) are very near to being coplanar (mean deviation 0.037 Å). The ethyl carboxylate group is in an extended conformation, with torsion angles C20-C21-O23-C24 176.2(8), C21-O23-C24-C25 95.2(9)°.

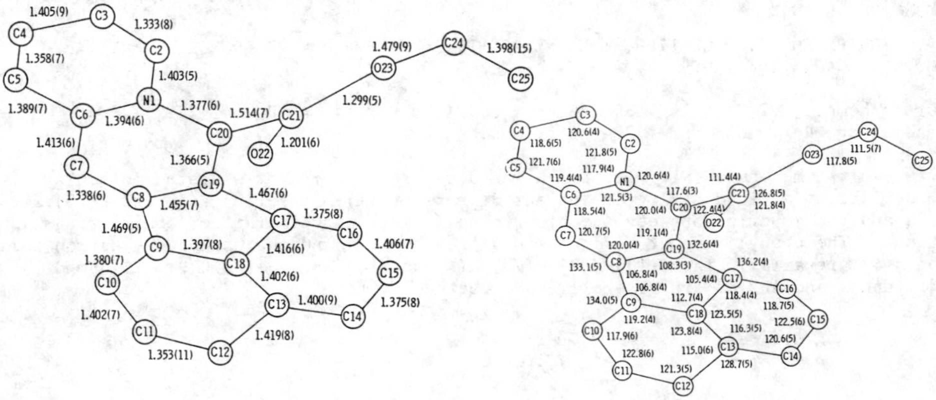

Fig. 1. Bond lengths and angles in the cation of $C_{22}H_{16}BrNO_2,1.5(H_2O)$.

2-(2-HYDROXY-2-METHYL-3-OXOBUTYL)-3-METHYLNAPHTHO[2,3-f]QUINOXALINE-7,12-DIONE
$C_{22}H_{18}N_2O_4$

M. BARON, S. GIORGI-RENAULT, J. RENAULT, P. MAILLIET, D. CARRE and J. ETIENNE, 1984. Can. J. Chem., 62, 526-530.

Triclinic, $P\bar{1}$, a = 9.091, b = 16.966, c = 12.375 Å, α = 100.75, β = 101.83, γ = 100.29°, D_m = 1.40, Z = 4. Mo radiation, R = 0.039 for 3027 reflexions.

The analysis, the structure and conformation as shown in Fig. 1. The overall conformation of the molecule is essentially planar and there is an intramolecular bifurcated hydrogen bond.

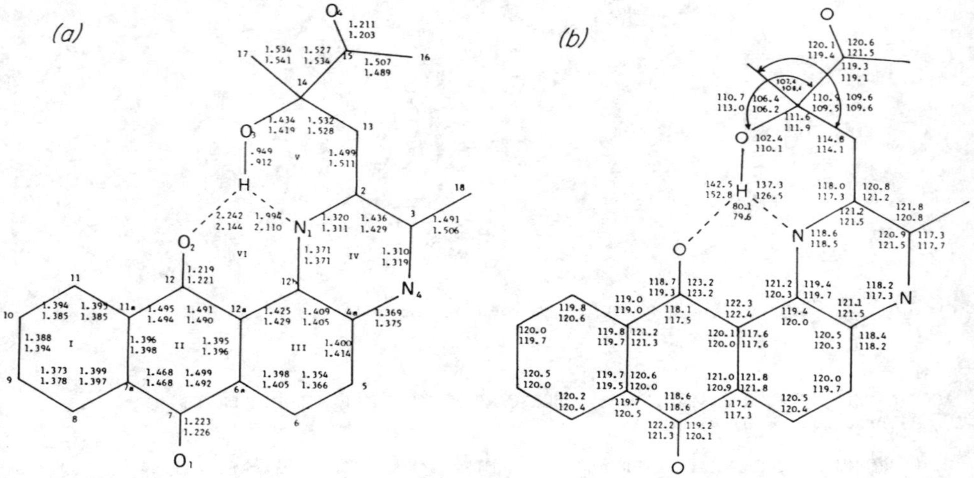

Fig. 1. Molecular dimensions (σ 0.005 Å, 0.3°) for $C_{22}H_{18}N_2O_4$.

(7S,11bS)-(+)-cis-1,3,4,6,7,11b-HEXAHYDRO-2-METHYL-7-PHENYL-2H,2H⁺-PYRAZINO[2,1-a]-ISOQUINOLINIUM HYDROGEN RR-(+)-TARTRATE METHANOL SOLVATE

$C_{23}H_{28}N_2O_6$,CH_4O $C_{19}H_{23}N_2^+$,$C_4H_5O_6^-$,CH_4O

R.C. GRIFFITH, R.J. GENTILE, W.L. ALBRECHT, G. HITE and M. RAPPOSCH, 1984. Acta Cryst., C40, 1103-1105.

Orthorhombic, $P2_12_12$, a = 39.646, b = 8.421, c = 7.149 Å, D_m = 1.279, Z = 4. Cu radiation, R = 0.038 for 1320 reflexions.

In the molecule (Fig. 1), which was designed by computer-assisted molecular modelling as a rigid analog of the antidepressant (+)-mianserin, C(1) and C(13) are cis and the junction between the piperidine (twist) and piperazine (chair) rings is trans. The phenyl ring at C(7), the hydrogen on N(2) and the unshared electron pair on N(5) are axial. The C(13)-C(7)-C(7a)-C(8) torsion angle is -70.9(5)°. Bond distances and angles are generally as expected.

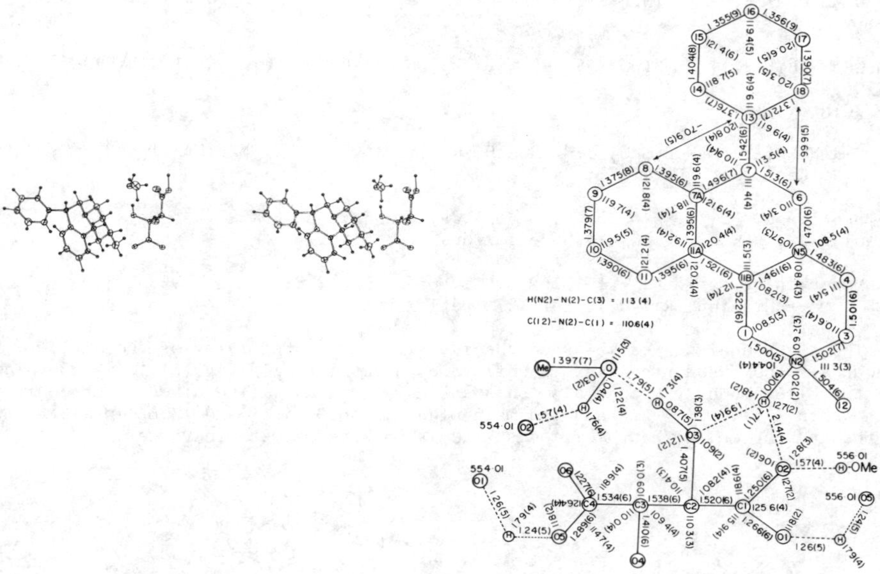

Fig. 1. $C_{23}H_{28}N_2O_6,CH_4O$: stereoscopic view and bond distances, angles, selected torsion angles and hydrogen-bonding scheme.

2a-(p-BROMOPHENYL)-4-METHYLTHIO-2-PHENOXY-1,2,2a,3-TETRAHYDROAZETO[1,2-a][1,5]BENZODIAZEPIN-1-ONE

$C_{24}H_{19}BrN_2O_2S$

M. SORIANO-GARCÍA, R.A. TOSCANO, E. CORTÉS and R. MARTÍNEZ, 1984. Acta Cryst., C40, 1460-1462.

Monoclinic, $P2_1/n$, a = 16.577, b = 7.838, c = 16.771 Å, β = 95.66°, Z = 4. Cu radiation, R = 0.043 for 2451 reflexions.

The diazepine ring has the boat conformation while the β-lactam ring is planar with the substituents at C(7), C(8) synperiplanar and those at C(8), C(9) synclinal. The angle between the planes of the benzene ring at N(6) and C(7) is 85.9(4)°. The C-Br bond length is 1.891(4) Å with Br 0.053(4) Å from the ring plane.

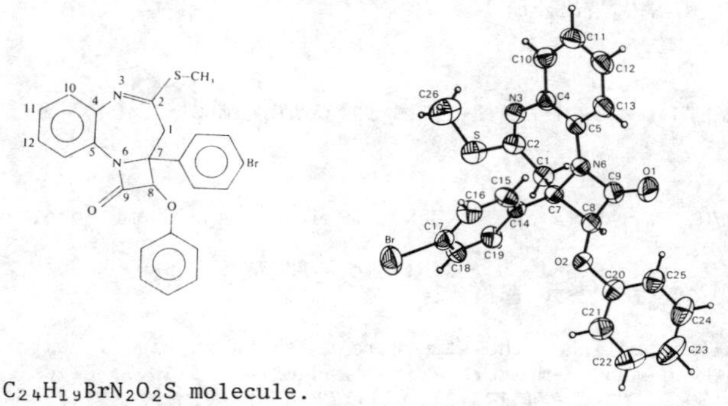

Fig. 1. The $C_{24}H_{19}BrN_2O_2S$ molecule.

2,12-DIMETHOXY-5,9-DIMETHYL-7H-(1)-BENZAZEPINO[1',2'.1,2]PYRROLO[5,4-b]QUINOLINE
$C_{24}H_{22}N_2O_2$ (I)

5-CHLOROMETHYL-2,11-DIMETHOXY-5,8-DIMETHYL-5,6-DIHYDROQUINO[1',2'.1,2]PYRROLO[5,4-b]-QUINOLINE
$C_{24}H_{23}ClN_2O_2$ (II)

M.B. STRINGER, V. CANDELORO, J.H. BOWIE, R.H. PRAGER, L.M. ENGELHARDT and A.H. WHITE, 1984. J. Chem. Soc. Perkin I, 2529-2534.

I. Monoclinic, $P2_1/c$, a = 20.49, b = 11.09, c = 8.72 Å, β = 97.8°, D_m = 1.23, Z = 4. Mo radiation, R = 0.096 for 1495 reflexions.

II. Monoclinic, $P2_1/n$, a = 15.316, b = 10.665, c = 12.859 Å, β = 99.51°, D_m = 1.29, Z = 4. Mo radiation, R = 0.10 for 1275 reflexions.

The X-ray analyses established the identity and connectivity of the title compounds. The fused ring systems defined by atoms 1 to 13 (Fig. 1) form good planes with the C(16) methoxy groups also substantially coplanar. The dihedral angle formed by the phenyl ring C(17-22) and the main plane is 46.9° in I and 13.6° in II. Bond lengths and angles, although of modest precision have acceptable values.

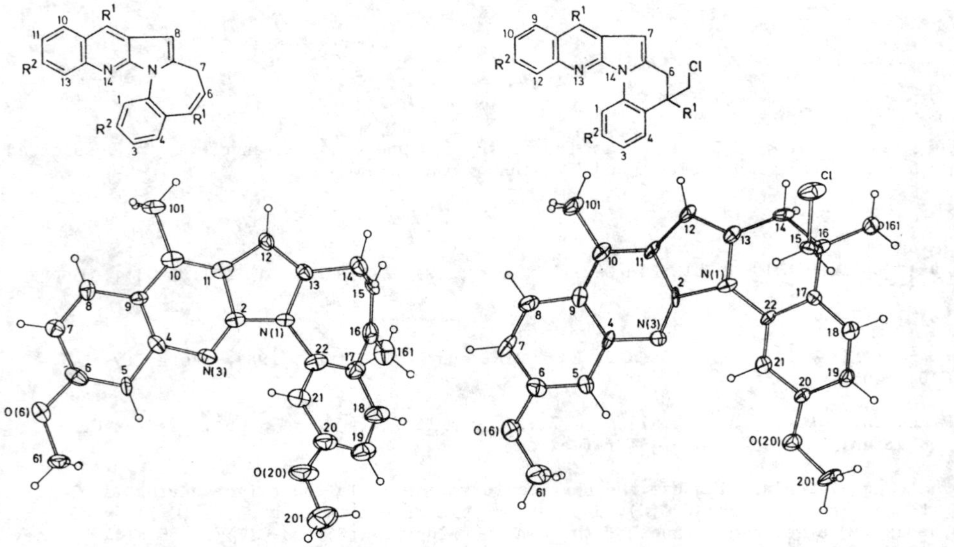

Fig. 1. The molecular skeletons and structures: $C_{24}H_{22}N_2O_2$ (left) and $C_{24}H_{23}ClN_2O_2$ (right).

2,2,6,6,9-PENTAMETHYL-4,4-DIPHENYL-1,7,9-TRIAZATRICYCLO[5.3.0^{3,5}]DECA-3(5)-ENE-8,10-DIONE
$C_{24}H_{25}N_3O_2$

W. ANDO, Y. HANYU, T. TAKATA and K. UENO, 1984. J. Am. Chem. Soc., 106, 2216-2217.

Monoclinic, $P2_1/c$, a = 10.381, b = 8.719, c = 23.717 Å, β = 98.30°, Z = 4. Cu radiation, R = 0.086 for 2670 reflexions.

The analysis establishes the structure as shown in Fig. 1, with an angle of 162.4° between the 3- and 6-membered rings and C2-C1-C3 49.9, C1-C2-C3 64.8, C1-C3-C2 65.3, C2-C3-C5 130.2, C3-C2-C4 129.2°. The C2-C3 distance is 1.286 Å.

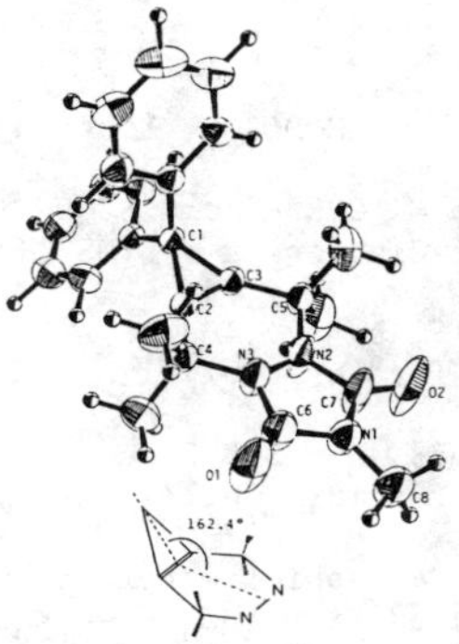

Fig. 1. A view of $C_{24}H_{25}N_3O_2$.

1-BENZYL-3,3-DIMETHYL-6'-NITROSPIRO([2H-1]-BENZOPYRAN-2',2-INDOLINE)
$C_{25}H_{22}N_2O_3$

K.Sh. KARAEV and N.G. FURMANOVA, 1984. J. Struct. Chem., 25, 168-170 [Zh. Strukt. Khim., 25(1), 185-188].

Triclinic, $P\bar{1}$, a = 13.534, b = 10.173, c = 8.374 Å, α = 105.66, β = 104.91, γ = 73.02°, Z = 2. Cu radiation, R = 0.043 for 2724 reflexions.

Bond lengths and angles (σ = 0.002-0.004 Å and 0.2-0.3°) are shown in Fig. 1. The plane of the indoline ring makes a dihedral angle of 86.6° with the plane of the benzopyran moiety. The five-membered ring has a flattened C(8)-envelope conformation.

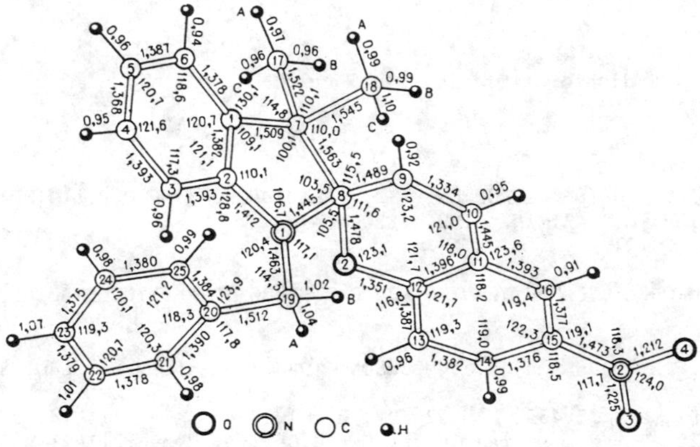

Fig. 1. Geometry of the 1-benzyl-3,3-dimethyl-6'-nitrospiro([2H-1]-benzopyran-2',2 -indoline molecule.

7-CHLORO-3-(p-CHLOROANILINO)-10-(p-CHLOROPHENYL)-2,10-DIHYDRO-2-(ISOPROPYLIMINO)-PHENAZINE (7-CHLOROCLOFAZIMINE)
$C_{27}H_{21}Cl_3N_4$ (I)

8-CHLOROCLOFAZIMINE
$C_{27}H_{21}Cl_3N_4$ (II)

4,9-DICHLOROCLOFAZIMINE
$C_{27}H_{20}Cl_4N_4$ (III)

M.B.H. BROOM, U. RYCHLEWSKA and D.J. HODGSON, 1984. Acta Cryst., C40, 1882-1887.

I. Triclinic, P$\bar{1}$, a = 12.152, b = 12.598, c = 8.699 Å, α = 91.68, β = 98.49, γ = 108.64°, D_m = 1.34, Z = 2. Mo radiation, R = 0.052 for 3097 reflexions.

II. Monoclinic, $P2_1/a$, a = 9.245, b = 24.403, c = 11.015 Å, β = 101.55°, D_m = 1.38, Z = 4. Mo radiation, R = 0.050 for 1897 reflexions.

III. Monoclinic, C2/c, a = 41.078, b = 5.575, c = 22.975 Å, β = 108.68°, D_m = 1.42, Z = 8. Mo radiation, R = 0.064 for 1619 reflexions.

All three compounds contain an approximately planar dihydrophenazine moiety; in II the maximum deviation of an atom from the least-squares plane is 0.054(3) Å; in I and III the two planar six-membered carbon rings intersect at 3.0 and 6.3° respectively. The p-chlorophenyl group at N(10) is nearly perpendicular to the phenazine plane (Fig. 1) with the p-chloroamilino group at C(3) inclined at about 30° to this plane. (This angle is 56.3° in III as a result of steric interactions).

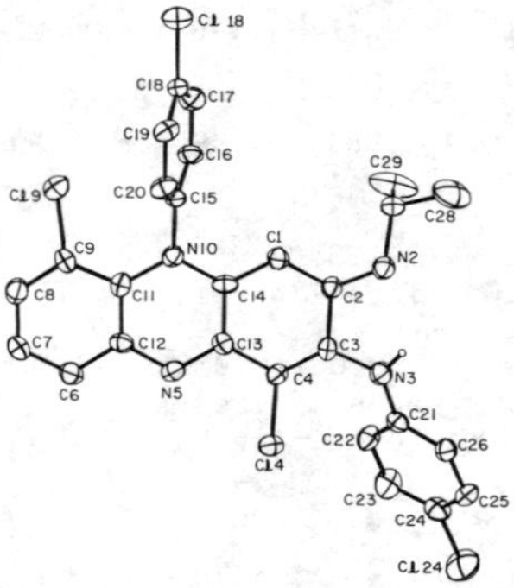

Fig. 1. The 4,9-dichloroclofazimine molecule.

3-(p-CHLOROANILINO)-10-(p-CHLOROPHENYL)-2,10-DIHYDRO-2-(ISOPROPYLIMINO)PHENAZINE DIMETHYLFORMAMIDE (CLOFAZIMINE DIMETHYLFORMAMIDE)
$C_{27}H_{22}Cl_2N_4, C_3H_7NO$ (I)

3-ANILINO-7-CHLORO-2-(CYCLOHEXYLIMINO)-2,10-DIHYDRO-10-PHENYLPHENAZINE
$C_{30}H_{27}ClN_4$ (II)

D.S. EGGLESTON, W.E. MARSH and D.J. HODGSON, 1984. Acta Cryst., C40, 288-292.

I. Triclinic, P$\bar{1}$, a = 12.435, b = 12.807, c = 10.424 Å, α = 111.74, β = 112.37, γ = 90.90°, D_m = 1.29, Z = 2. Mo radiation, R = 0.082 for 2939 reflexions.

II. Triclinic, P$\bar{1}$, a = 10.986, b = 24.369, c = 10.145 Å, α = 101.88, β = 96.84, γ = 88.53°, D_m = 1.19, Z = 4. Mo radiation, R = 0.077 for 3359 reflexions.

The structures of the clofazimine molecule and the two independent molecules of II are very similar (Fig. 1), approach to the potential active sites N2, N3 and N5 (in these antileprosy drugs) being unhindered in all cases. The torsion angle between the 10-phenyl [or 10-(p-chlorophenyl)] substituents are 83.3° (I) and 106.4, 92.4° (II).

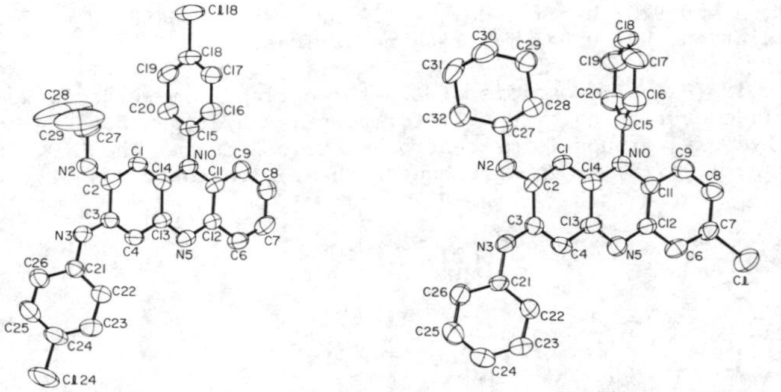

Fig. 1. Views of clofazimine (left) and one of the two independent molecules of (II) (right).

3-ALLYL-9-BENZOYL-2-METHYL-9-PHENYL-4,9-DIHYDROPYRAZOLO[5,1-b]QUINAZOLINE
$C_{27}H_{23}N_3O$ (I)

6,7-DIMETHYL-1,3,3-TRIPHENYL-1H-IMIDAZO[1,2-b]PYRAZOL-2(3H)-ONE
$C_{25}H_{21}N_3O$ (II)

A.L. RHEINGOLD, W.C. FULTZ, E.E. SCHWEIZER and K.J. LEE, 1984. Acta Cryst., C40, 687-690.

I. Orthorhombic, Pbca, a = 13.025, b = 15.991, c = 20.593 Å, Z = 8. Mo radiation, R = 0.0728 for 1671 reflexions.

II. Monoclinic, $P2_1/c$, a = 11.234, b = 7.021, c = 25.330 Å, β = 91.83°, Z = 4. Mo radiation, R = 0.0770 for 1383 reflexions.

Bond distances and angles in I and II are all within the expected ranges. The 1,2-diazacyclopentadiene rings in both structures are nearly planar (Fig. 1). In I the dihedral angle in the central six-membered ring related by a line joining N(4) and C(9) is 157.0(6)°. In II the dihedral angle relating the two five-membered rings is 8.5(6)°. The molecules in both I and II are well separated in the crystals.

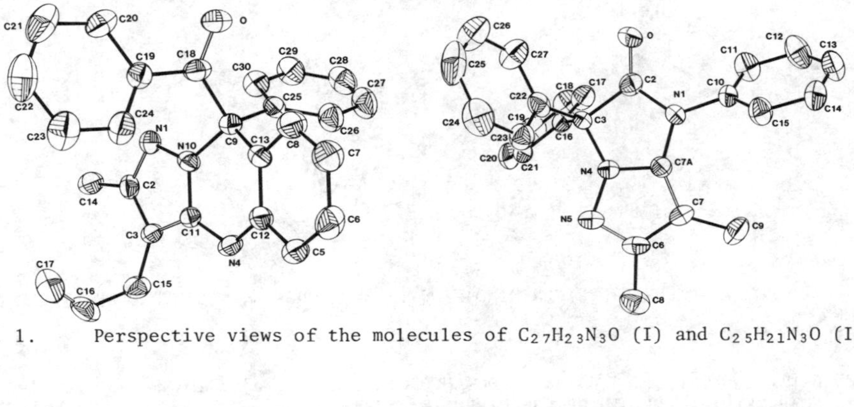

Fig. 1. Perspective views of the molecules of $C_{27}H_{23}N_3O$ (I) and $C_{25}H_{21}N_3O$ (II).

4,4a-DIHYDRO-9-METHYL-4a-[2,2,2-TRIFLUORO-1-(2-HYDROXY-9-METHYLCARBAZOL-3-YL)ETHYL]-CARBAZOL-2(3H)-ONE NITROMETHANE SOLVATE
$C_{28}H_{23}F_3N_2O_2,CH_3NO_2$

K. PROUT and J.M. MARÍN, 1984. Acta Cryst., C40, 1229-1231.

Triclinic, $P\bar{1}$, a = 9.920, b = 12.000, c = 12.390 Å, α = 73.93, β = 77.87, γ = 65.57°, Z = 2. Mo radiation, R = 0.043 for 1843 reflexions.

The molecule (Fig. 1) consists of two three-ring frameworks linked by a 2,2,2-trifluoroethylidene bridging group. Intermolecular hydrogen bonding between the carbonyl and hydroxyl groups forms centrosymmetric dimers. The C=O carbonyl bond is relatively long, 1.247(5) Å, in agreement with spectroscopic results.

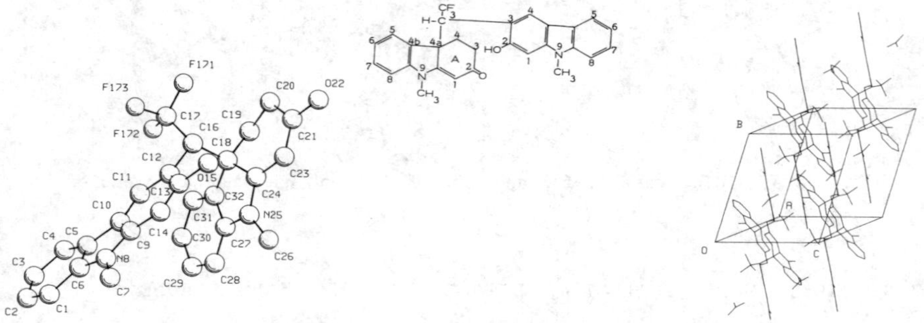

Fig. 1. $C_{28}H_{23}F_3N_2O_2,CH_3NO_2$: perspective view of the molecule and packing diagram of the unit cell in which the planar three-ring fragment of each molecule appears as a straight line.

cis-1-(BENZOYLOXY)-1,2,3,4-TETRAHYDRO-4-METHYL-1-PHENAZINEMETHANOL BENZOATE ESTER
$C_{28}H_{24}N_2O_4$

D.J. ABERHART, J. CLARDY, P.K. GHOSHAL, C.-H. HE and Q.-T. ZHENG, 1984. J. Org. Chem., 49, 2429-2433.

Monoclinic, $P2_1/n$, a = 13.985, b = 11.133, c = 14.385 Å, β = 92.90°, Z = 4. Cu radiation, R = 0.05 for 2822 reflexions.

The structure of this reaction product was determined to be as shown in Fig. 1.

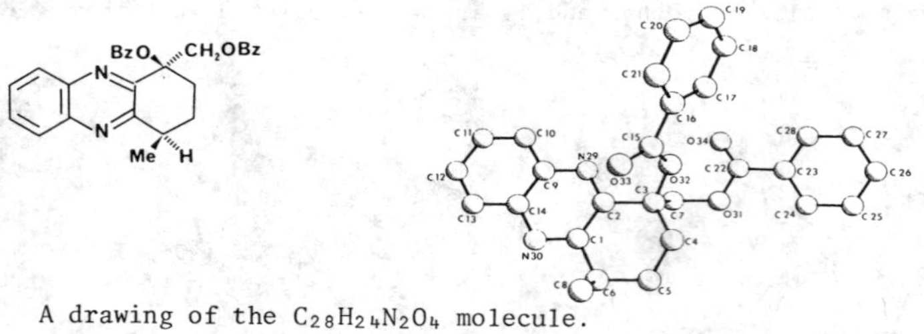

Fig. 1. A drawing of the $C_{28}H_{24}N_2O_4$ molecule.

5-[3-(4-CARBAMOYL-4-PIPERIDINOPIPERIDINO)PROPYL]-3-CHLORO-10,11-DIHYDRO-5H-DIBENZ[bf]-AZEPINE
$C_{28}H_{37}ClN_4O$ (I)

3-CHLORO-5-[3-(2-OXO-1,2,3,5,6,7,8,8a-OCTAHYDROIMIDAZO[1,2-a]PYRIDINE-3-SPIRO-4'-PIPERIDINO)PROPYL]-10,11-DIHYDRO-5H-DIBENZ[bf]AZEPINE
$C_{28}H_{35}ClN_4O$ (II)

5-[3-(2-OXO-1,2,3,5,6,7,8,8a-OCTAHYDROIMIDAZO[1,2-a]PYRIDINE-3-SPIRO-4'-PIPERIDINO)-PROPYL]-10,11-DIHYDRO-5H-DIBENZ[bf]AZEPINE
$C_{28}H_{36}N_4O$ (III)

I. UEDA and C. TASHIRO, 1984. Acta Cryst., C40, 422-426.

I. Monoclinic, C2/c, a = 27.81, b = 8.589, c = 22.14 Å, β = 102.82°, D_m = 1.23, Z = 8. Mo radiation, R = 0.0639 for 2566 reflexions.

II. Triclinic, P$\bar{1}$, a = 11.515, b = 11.837, c = 10.744 Å, α = 92.71, β = 112.73, γ = 68.76°, D_m = 1.26, Z = 2. Mo radiation, R = 0.0658 for 3114 reflexions.

III. Monoclinic, A2/a, a = 27.29, b = 10.254, c = 18.108 Å, β = 97.20°, D_m = 1.17, Z = 8. Mo radiation, R = 0.0620 for 2851 reflexions.

These three antischizophrenia drugs (Fig. 1) each form dimers by paired intermolecular O...H(N) hydrogen bonds at the carbamoyl moieties. The ethylene radical in the dibenzazepine ring is in plane with one benzene ring and is out of plane for the other. The Cl atom of (I) is connected to the out-of-plane benzene ring at the position meta to the N atom. That of (II) is connected to the in-plane benzene ring at the same position.

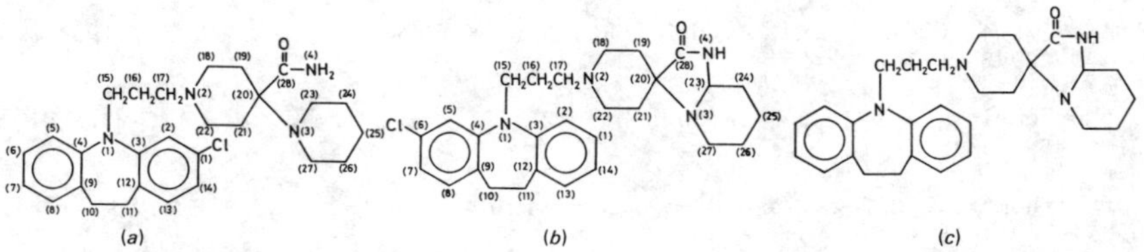

Fig. 1. Formulae for the three dibenz[bf]azepines; (a) I, (b) II, (c) III.

3-(p-CHLOROANILINO)-10-(p-CHLOROPHENYL)-2-CYCLOHEXYLIMINO-2,10-DIHYDROPHENAZINE
$C_{30}H_{26}Cl_2N_4$ (I)

3-(p-CHLOROANILINO)-10-(p-CHLOROPHENYL)-2,10-DIHYDRO-2-(4-METHYLCYCLOHEXYLIMINO)-PHENAZINE
$C_{31}H_{28}Cl_2N_4$ (II)

U. RYCHLEWSKA, M.B.H. BROOM and D.J. HODGSON, 1984. Acta Cryst., C40, 1004-1007.

I. Triclinic, P$\bar{1}$, a = 10.992, b = 12.191, c = 10.458 Å, α = 112.13, β = 91.57, γ = 88.58°, D_m = 1.29, Z = 2. Mo radiation, R = 0.068 for 1861 reflexions.

II. Triclinic, P$\bar{1}$, a = 12.161, b = 12.692, c = 10.448 Å, α = 111.60, β = 113.52, γ = 91.92°, D_m = 1.29, Z = 2. Mo radiation, R = 0.042 for 1784 reflexions.

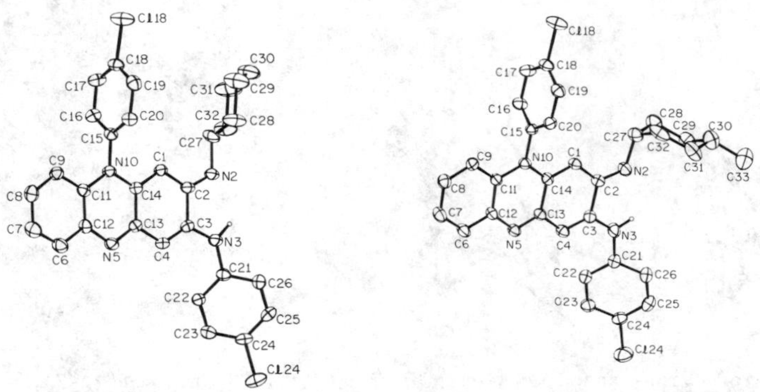

Fig. 1. Perspective views of molecules of $C_{30}H_{26}Cl_2N_4$ (I) and $C_{31}H_{28}Cl_2N_4$ (II).

The principal difference between the two structures (Fig. 1) is the orientation of the cyclohexyl ring at imino nitrogen atom N(2), which in I is equatorial and in II is axial. Consequently, approach to the presumed active sites at N(2) and N(3) is unhindered in I but hindered in II. The phenazine moieties in both compounds are planar and the aromatic rings at N(3) make interplanar angles with the phenazine rings of 31.8(4)° in I and 38.4(3)° in II respectively. As in similar molecules the C(3)-N(3) bond is shorter than the N(3)-C(21) bond: 1.375(4) and 1.407(4) Å in I and 1.369(4) and 1.405(4) Å in II. Weak intramolecular N(3)-H(3)...N(2) hydrogen bonds are probably present in both compounds.

5H-INDOLO[1,7-ab][1]BENZAZEPINE(6-6',7-7')DIMER
$C_{32}H_{22}N_2$

5H-INDOLO[1,7-ab][1]BENZAZEPINE(6-7',7-6')DIMER
$C_{32}H_{22}N_2$

R.B. BATES, K.A. CHRISTENSEN, A. HALLBERG, R.E. KLENCK and A.R. MARTIN, 1984. J. Org. Chem., 49, 2978-2981.

I. Orthorhombic, $P2_12_12_1$, a = 8.808, b = 12.197, c = 20.26 Å, Z = 4. Mo radiation, R = 0.051 for 1415 reflexions.

II. Orthorhombic, $P2_12_12_1$, a = 8.542, b = 14.404, c = 18.05 Å, Z = 4. Mo radiation, R = 0.051 for 1425 reflexions.

Solid phase irradiation of 5H-indolo[1,7-ab][1]benzazepine yields what is labelled a "mirror plane" syn dimer (I) whereas the major solution irradiation dimer is the "twofold-axis" syn dimer (II). Stereoviews of the dimers and details of dimensions are in Fig. 1. The compounds crystallize isomorphously. Both have puckered cyclobutane rings (torsion angles 13.6-14.3° in (I), 19.4-20.5° in (II)). The cyclobutane bonds connecting the two halves of the syn dimers are slightly lengthened (C-C 1.59 Å).

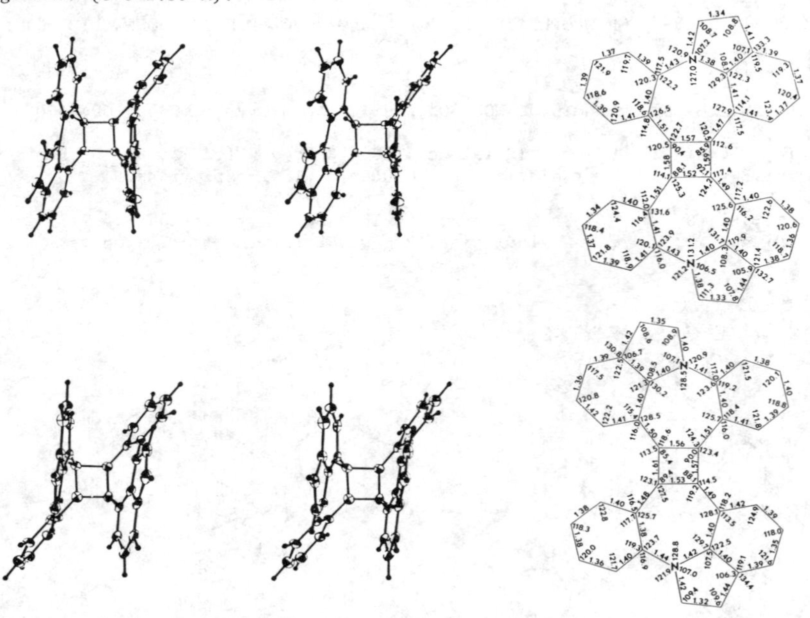

Fig. 1. Stereoviews and dimensions for the 6-6',7-7' (top) and 6-7',7-6' (bottom) $C_{32}H_{22}N_2$ molecules (σ 0.006-0.015 Å, 0.6-0.8°).

2,3,5,6-DIBENZO-4-(DICYANOMETHYLENE)-8,8,10-TRIS(METHOXYCARBONYL)-9-PHENYL-9-AZABICYCLO[5.3.0]DECA-2,5-DIENE
$C_{32}H_{25}N_3O_6$

D. FICHOU, F. TONNARD, L. TOUPET and R. CARRIE, 1984. Tetrahedron, 40, 5121-5133.

Orthorhombic, $P2_12_12_1$, a = 9.741, b = 15.636, c = 17.567 Å, Z = 4. R = 0.056 for three-dimensional data (at 130 K).

In the molecule (Fig. 1) the hydrogens at the junction of the five- and seven-membered rings are trans. The seven-membered ring has a boat conformation and the angle between the two aromatic ring planes is 70.3°.

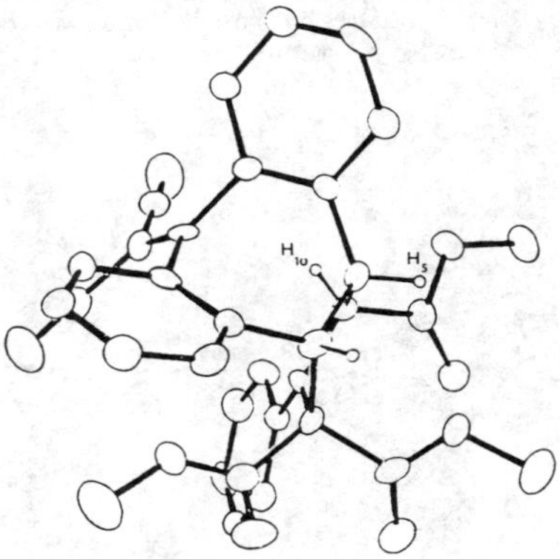

Fig. 1. A view of $C_{32}H_{25}N_3O_6$.

1-METHYL-1,3,5,7-TETRAAZAADAMANTAN-1-IUM BROMIDE HYDRATE
$C_7H_{15}BrN_4,H_2O$ $[(CH_2)_6N_4CH_3]Br,H_2O$

T.W.C. MAK, 1984. Inorg. Chem., 23, 620-622.

Monoclinic, $P2_1/c$, a = 6.840, b = 6.898, c = 22.597 Å, β = 94.91°, D_m = 1.584, Z = 4. Mo radiation, R = 0.050 for 1267 reflexions.

The most notable feature of the structure (Fig. 1) is that the water molecules form hydrogen bonds with the Br^- anions rather than the tertiary N atoms of the $[(CH_2)_6N_4CH_3]^+$ cation. The resulting cyclic $(H_2O,Br^-)_2$ entity occupies a site of $\bar{1}$ symmetry (O...Br 3.386(5), O...Br* 3.338(5), O-H 0.958(15) and 0.938(15), H...Br 2.463(15) and 2.511(15) Å, Br...O...Br 105.7(4), H-O-H 126(2)°). The geometry of the cation is as expected.

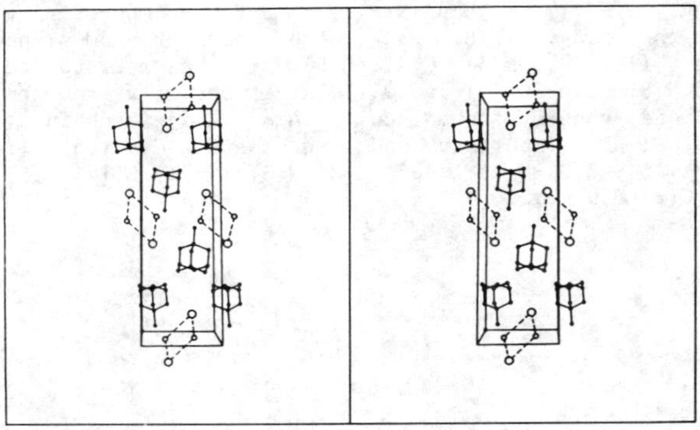

Fig. 1. The structure of $[(CH_2)_6N_4CH_3]Br,H_2O$.

5-AZATRICYCLO[4.3.1.$0^{3,8}$]DECAN-4-ONE HEMIHYDRATE
$C_9H_{13}NO,0.5H_2O$

P. KÁLAL, K. BLÁHA and V. LANGER, 1984. Acta Cryst., C40, 1242-1245.

Monoclinic, $P2_1/c$, a = 12.375, b = 12.695, c = 12.521 Å, β = 119.29°, D_m = 1.27, Z = 8. Cu radiation, R = 0.079 for 1494 reflexions.

In the unit cell there are two symmetric independent molecules of the lactam (Fig. 1). These molecules are of two conformational types: A, where the non-planarity of the amide group is caused mostly by a torsion around the C'-N bond, and B, in which there is a pyramidal arrangement of the bonds at N. In the crystal, molecules A are connected by inter-amide hydrogen bonds into cyclic dimers, while molecules B are connected indirectly by means of water molecules into chains.

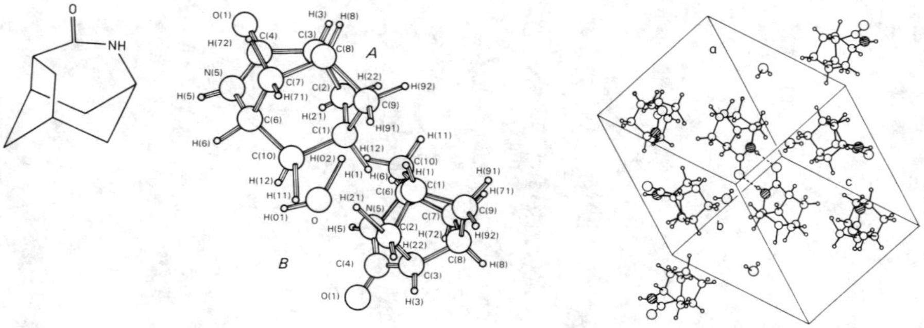

Fig. 1. $C_9H_{13}NO,0.5H_2O$: projection of the structure and a view of the crystal packing; broken lines indicate the hydrogen bonds forming the cyclic dimer.

(1R,4R,5R)-4-METHOXY-8-ACETYL-7,7-DIMETHYL-6-THIA-3,8-DIAZABICYCLO[3.2.1]OCTAN-2-ONE
$C_{10}H_{16}N_2O_3S$

B. KOJIĆ-PRODIĆ, Z. RUŽIĆ-TOROŠ, G. SNATZKE and J.J. HERAK, 1984. Croat. Chem. Acta, 57, 661-668.

Orthorhombic, $P2_12_12_1$, a = 15.857, b = 12.449, c = 6.120 Å, Z = 4. Mo radiation, R = 0.048 for 671 reflexions.

The diazabicyclo[3.2.1]octane ring is puckered, and the seven-membered ring with the sequence C(1), C(2), N(3), C(4), C(5), S(6), C(7) appears in a boat conformation (Fig. 1). C(1), C(7), S(6), C(5) and N(8) form a five-membered ring in an envelope conformation with the bridging N(8) lying out of the four atom plane by 0.638 Å. The six-membered ring C(1), C(2), N(3), C(4), C(5), N(8) has a distorted chair conformation with N(3) and N(8) being 0.131 and 0.774 Å respectively from the four atom plane. N-H...O hydrogen bonds connect the molecules in an infinite chain along c (N(3)-H...O(9) with N...O 2.881 Å). Important bond lengths are: C-S (both 1.849 Å), C(1)-N(8) 1.463, C(4)-N(3) 1.464, C(5)-N(8) 1.456, C(2)-N(3) 1.331, C(14)-N(8) 1.353, C(14)-O(15) 1.222, and C(2)-O(9) 1.236 Å.

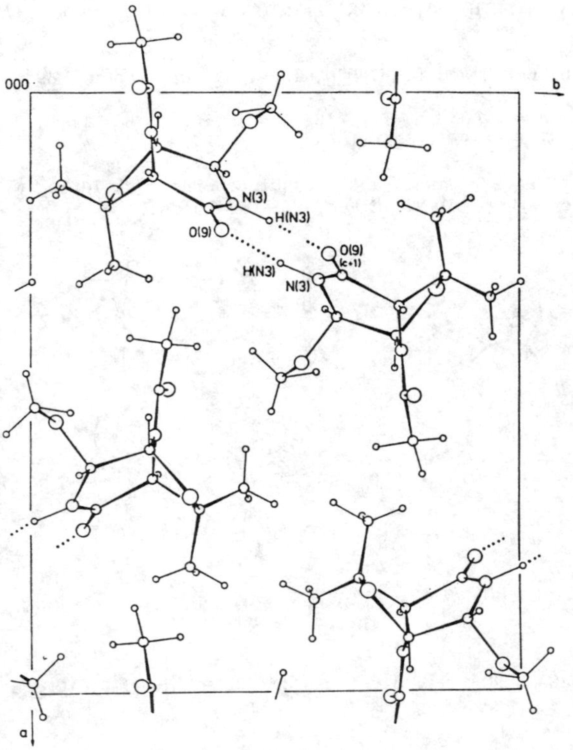

Fig. 1. A view of the crystal structure of $C_{10}H_{16}N_2O_3S$ along c.

1-AZA-8,9-DIOXO-7-OXA-6-PHENYLBICYCLO[4.2.1]NONANE
$C_{13}H_{13}NO_3$

W.J. BROUILLETTE and H.M. EINSPAHR, 1984. J. Org. Chem., 49, 5113-5116.

Monoclinic, C2/c, a = 25.383, b = 6.523, c = 18.945 Å, β = 133.63°, Z = 8. Cu radiation, R = 0.050 for 1823 reflexions.

The X-ray analysis establishes the structure of this bicyclic imide as shown in Fig. 1. The molecular dimensions are consistent with a slight decrease in resonance effects between N3 and the C4 carbonyl, while such effects with the C2 carbonyl appear to be undiminished.

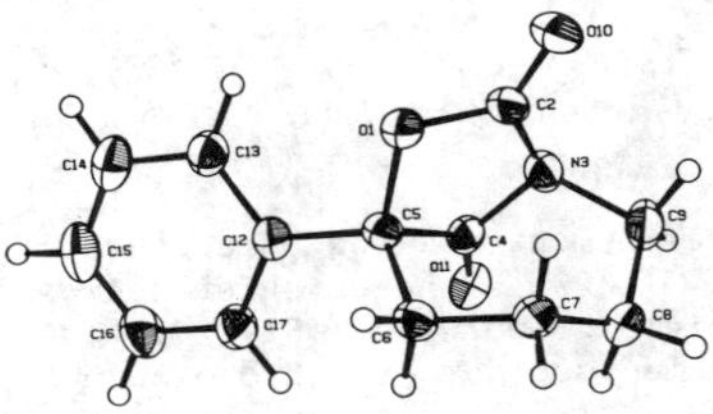

Fig. 1. A view of $C_{13}H_{13}NO_3$.

2,8:10,13-DIMETHANO-5-METHYL-3,5,7-TRIAZATRICYCLO$[7.4.0.0^{3,7}]$TRIDECANE-4,6-DIONE
$C_{13}H_{15}N_3O_2$

L.A. PAQUETTE, K.E. GREEN and L.-H. HSU, 1984. J. Org. Chem., 49, 3650-3652.

Monoclinic, $P2_1/n$, a = 6.603, b = 17.718, c = 10.766 Å, β = [?]°, Z = 4. Mo radiation, R = 0.037 for 1955 reflexions.

The molecule (Fig. 1) has an anti-sesquinorbornene skeleton, with a pseudo mirror plane bisecting the N1-N2, C7-C8, C12-C13 bonds.

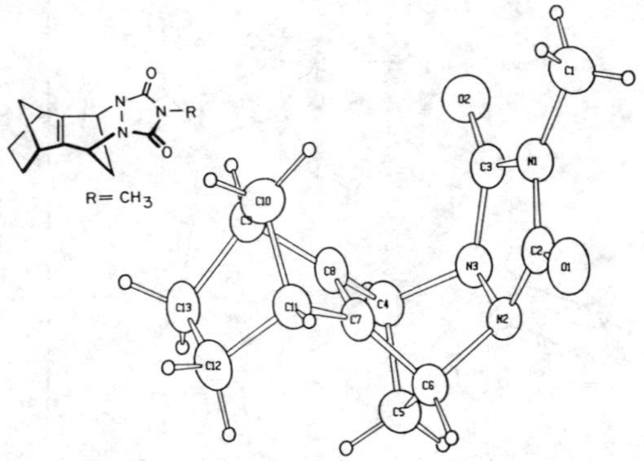

Fig. 1. A view of $C_{13}H_{15}N_3O_2$.

(t-5a,t-8a)-Δ^7-DODECAHYDRO-r-1,c-5,c-2,c-4-(NITRILOMETHENO)DICYCLOPENT[cd,g]INDOLE
$C_{14}H_{16}N_2$

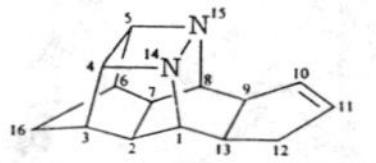

B. ALBERT, W. BERNING, C. BURSCHKA, S. HUNIG and F. PROKSCHY, 1984. Chem. Ber., 117, 1465-1475.

Orthorhombic, Pbnb, a = 6.135, b = 10.640, c = 31.58 Å, D_m = 1.28, Z = 8. Mo radiation, R = 0.051 for 855 reflexions.

In this strained molecule, the N-N distance is 1.550(4) Å; some C-N distances are: C4-N14 1.516(4), C5-N15 1.515(5) Å. The C1-C2-C3 and C6-C7-C8 angles are 103.4(3) and 102.6(3)° respectively.

BIS(QUINUCLIDIN-3-ONE) HYDROGEN PERCHLORATE
$C_{14}H_{23}ClN_2O_6$

D.J. JONES, I. BRACH and J. ROZIERE, 1984. J. Chem. Soc. Dalton, 1795-1800.

Monoclinic, C2/c, a = 16.169 (15.967), b = 7.341 (7.237), c = 14.363 (14.251) Å, β = 99.23 (98.58)°, Z = 4. Mo radiation, R = 0.086 (0.081) for 690 (1443) reflexions (at 298 (120) K).

The starting point for refinement was the parameters reported in 1 for the neutron diffraction analysis. At 298 K the perchlorate was severely disordered and was ultimately refined as a rigid group with a Cl-O distance of 1.37 Å. The quinuclidin-3-one units are held together by a hydrogen bond of 2.648(9) at 298 K and 2.629(4) Å at 120 K (Fig. 1). At 120 K this bond is more symmetrical than at 298 K probably due to bridge shortening. Indeed at room temperature the H atom is disordered as two half-atoms 0.68 Å apart, on either side of the centre of symmetry and off the N...N axis.

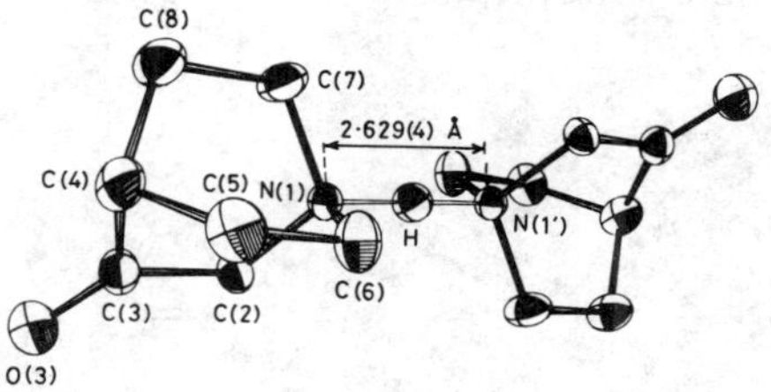

Fig. 1. $C_{14}H_{23}ClN_2O_6$: the bis(quinuclidin-3-one) cation.

1. J. Roziere, C. Belin and M.S. Lehmann, 1982. J. Chem. Soc. Chem. Commun., 388.

8-CHLORO-3-PHENYL-2,4,6-TRIAZATRICYCLO[7.2.1.0^{2,6}]DODECA-7,10-DIENE-3,5-DIONE
$C_{15}H_{12}ClN_3O_2$

K. PETERS, E.-M. PETERS, H.G. VON SCHNERING, W. ADAM and O. DE LUCCHI, 1984.
Z. Kristallogr., 168, 145-148.

Monoclinic, $P2_1/c$, a = 16.829, b = 6.868, c = 12.047 Å, β = 106.10°, Z = 4. Mo radiation, R = 0.042 for three-dimensional data.

The atom labels and the molecular structure are given in Fig. 1. As well as plane A(1,9,12), the structural fragments B(1,9,10,11), C(1,9,8,2), D(6,7,8,9,Cl), and E(2,3,O3,4,5,O5,6) are all planar within standard deviations. The dihedral angles are A/B 34.1°, A/C 72.4°, B/C 73.5°, C/E 26.1°, D/E 13.6°. The plane of the phenyl ring is orientated at an angle of 44.1° to the plane of the triazoledione ring.

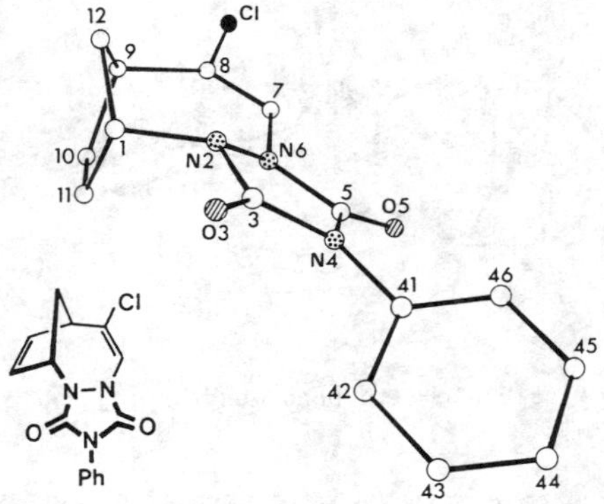

Fig. 1. The $C_{15}H_{12}ClN_3O_2$ structure.

(±)-2,3,4,5,6,6a-HEXAHYDRO-3-METHYL-8-HYDROXY-1H-4,11b-METHANOBENZOFURO[3,2-d]AZOCINE HYDROCHLORIDE
$C_{15}H_{20}ClNO_2$ $C_{15}H_{20}NO_2^+.Cl^-$

T.R. BURKE, A.E. JACOBSON, K.C. RICE and J.V. SILVERTON, 1984. J. Org. Chem., 49, 1051-1056.

Monoclinic, $P2_1/c$, a = 11.880, b = 10.674, c = 11.537 Å, β = 110.05°, D_m = 1.362, Z = 4. Cu radiation, R = 0.075 for 2792 reflexions.

The angle between the plane through atoms 1, 2, 4, 12 of the piperidine ring (Fig. 1) and the phenyl ring is 86°. It is thought that this locked conformation makes this molecule incapable of binding with opioid receptors.

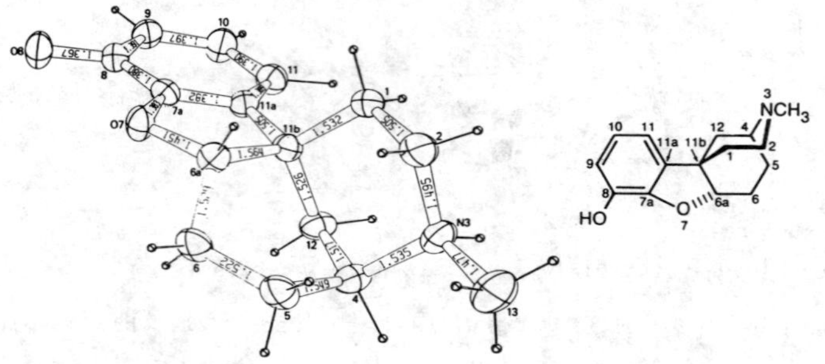

Fig. 1. A view of $C_{15}H_{20}ClNO_2$.

4-PHENYL-2,4,6-TRIAZAPENTACYCLO[5.4.2.$0^{2,6}$.$0^{8,10}$.$0^{9,11}$]TRIDEC-12-ENE-3,5-DIONE
$C_{16}H_{13}N_3O_2$

M. CHRISTL, R. LANG, W. REIMANN and H. IRNGARTINGER, 1984. Chem. Ber., 117, 959-965.

Orthorhombic, Pbca, a = 11.466, b = 15.123, c = 15.511 Å, Z = 8. Mo radiation, R = 0.052 for 1808 reflexions.

A view of the molecule with bond lengths is in Fig. 1. The dihedral angle for the bicyclo[1.1.0]butane group (the 2- and 4-positions of which are spanned by four atoms) is 122.7°.

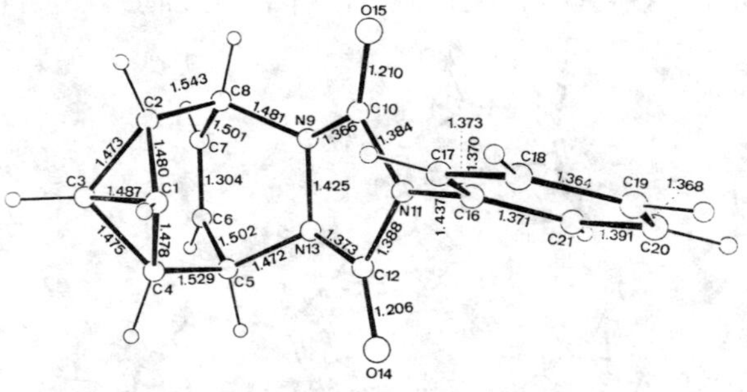

Fig. 1. The $C_{16}H_{13}N_3O_2$ molecule with bond lengths (Å).

rel-(5R,6R,9S,10S)-5,6,7,8,9,10-HEXAHYDRO-2-PHENYL-5,10:6,9-DIMETHANO-1H-[1,2,4]-TRIAZOLO[1,2-b]PHTHALAZINE-1,3(2H)-DIONE
$C_{18}H_{17}N_3O_2$

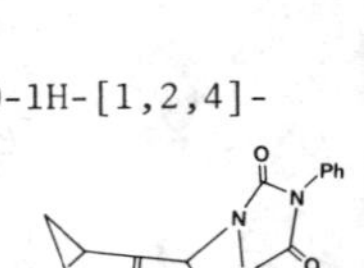

R.P. KASHYAP, W.H. WATSON, C. WU and P.D. BARTLETT, 1984. Acta Cryst., C40, 1851-1854.

Monoclinic, $P2_1/n$, a = 9.439, b = 19.454, c = 8.382 Å, β = 102.70°, Z = 4. Cu radiation, R = 0.035 for 1762 reflexions.

The central double bond in the anti-sesquinorbornene moiety is 1.327(3) Å with C atoms slightly pyramidalised leading to a 4.3(4)° deviation from planarity; the PTAD group is endo-oriented. The five-membered triazolo group has an envelope conformation with N-N 1.456(3) Å.

4,6,11,13-TETRAOXO-5,12-DIAZAPENTACYCLO[7.5.2.$0^{2,8}$.$0^{3,7}$.$0^{10,14}$]HEXADEC-15-ENE-5,12-DIYLDIMETHYLENE DIACETATE
$C_{20}H_{20}N_2O_8$

H.M. DEUTSCH, L. McGOWAN, D.G. VAN DERVEER, L.T. GELBAUM and L.H. ZALKOW, 1984. Acta Cryst., C40, 1925-1927.

Orthorhombic, $Pca2_1$, a = 16.646, b = 15.890, c = 7.222 Å, D_m = 1.444, Z = 4. Mo radiation, R = 0.065 for 2322 reflexions.

The two inside rings in the molecule (Fig. 1) have the exo-endo configuration and are nearly coplanar (dihedral angle 4.1°). The distance between the two nitrogen atoms is 7.79 Å.

Fig. 1. The $C_{20}H_{20}N_2O_8$ molecule.

1,1'-DI(4-PYRIDYL)-2,2',6,6'-BI(4-PIPERIDONE) DIHYDROCHLORIDE DIHYDRATE
$C_{20}H_{22}Cl_2N_4O_2,2H_2O$ $C_{20}H_{22}N_4O_2^{2+}.2Cl^-.2H_2O$

C.J. CHEER, J.P. COSGROVE and B.M. VITTIMBERGA, 1984. Acta Cryst., C40, 1474-1475.

Monoclinic, $P2_1/c$, a = 10.328, b = 12.174, c = 8.231 Å, β = 92.46°, D_m = 1.437, Z = 2. Mo radiation, R = 0.043 for 1513 reflexions.

The structure contains centrosymmetric dimers (Fig. 1). The piperidone ring has a distorted chair conformation flattened at the carbonyl group with its N atom 0.143(2) Å from the pyridine ring plane. The water molecule is hydrogen bonded to both the carbonyl group and the Cl^- ion. The Cl^- ion is also hydrogen bonded to N(1).

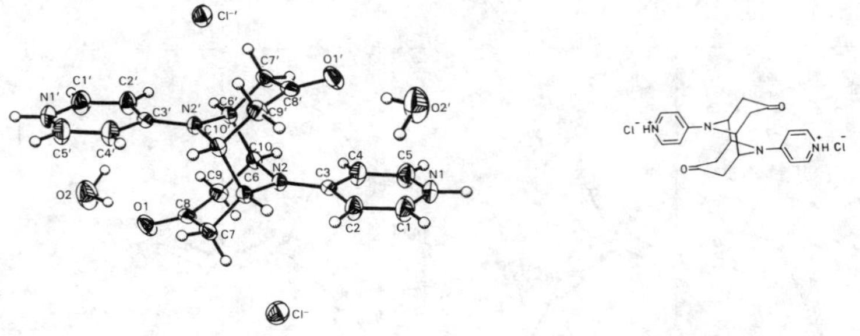

Fig. 1. The dimeric unit of $C_{20}H_{22}N_4O_2^{2+}.2Cl^-.2H_2O$.

7,16,21,26-TETRAOXA-1,4,10,13-TETRAAZATRICYCLO[11.5.5.$5^{4,10}$]OCTACOSANE
$C_{20}H_{40}N_4O_4$ (I)

7,16,21,26-TETRAOXA-1,4,10,13-TETRAAZATRICYCLO[11.5.5.$5^{4,10}$]OCTACOSANE DIHYDROCHLORIDE TETRAHYDRATE
$C_{20}H_{42}Cl_2N_4O_4,4H_2O$ (II) $C_{20}H_{40}N_4O_4,2HCl,4H_2O$

P. GROTH, 1984. Acta Chem. Scand., B38, 513-521.

I. Monoclinic, $P2_1/c$, a = 9.093, b = 12.940, c = 9.976 Å, β = 108.76°, D_m = 1.17, Z = 2. Mo radiation, R = 0.083 for 975 reflexions.

II. Orthorhombic, Fdd2, a = 39.72, b = 16.999, c = 8.386 Å, D_m = 1.25, Z = 8. Mo radiation, R = 0.056 for 771 reflexions.

Compound I has inversion symmetry and the twelve-membered ring adopts the diangular [2,10] conformation (Fig. 1). There appears to be disorder in atoms O(2), C(7) and C(8), and bond distances of 1.37 Å (O(2)-C(7)) and 1.47 Å (C(7)-C(8)) with a large O(2)-C(7)-C(8) angle of 124° bear this assumption out. Compound II has crystallographic C_2 symmetry and consists of two disordered twelve-membered rings labelled A and B (Fig. 1) which have different conformations. For A it is [3,9] and for B [3,3,3,3]. Occupancy factors are 0.7 for A and 0.3 for B. The Cl^- anion accepts three hydrogen bonds from the water molecules, Cl...O(W1) 3.20, Cl...O(W2') 3.25 and Cl...O(W2'') 3.27 Å.

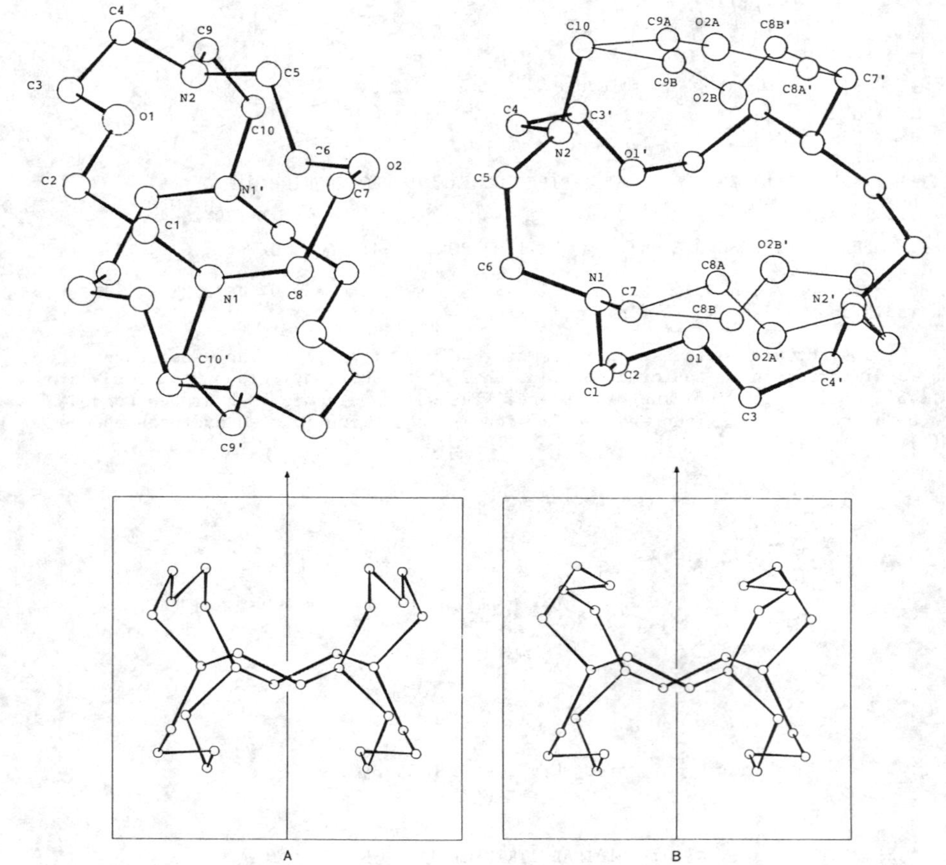

Fig. 1. Molecular structures of $C_{20}H_{40}N_4O_4$ (top left) and $C_{20}H_{42}Cl_2N_4O_4$ (top right), and the pure A and B ligands of II viewed roughly along the a axis.

2-AMINO-4-METHOXY-N-[8-(α-METHYLBENZYL)-8-AZABICYCLO[3.2.1]OCT-3-YL]PYRIMIDINE-5-CARBOXAMIDE

$C_{21}H_{27}N_5O_2$

F. DURANT, P. RENARD and G. EVRARD, 1984. Bull. Soc. Chim. Belg., 93, 925-926.

Orthorhombic, Pbca, a = 19.134, b = 24.227, c = 8.969 Å, Z = 8. Cu radiation, R = 0.051 for 786 reflexions.

The dihedral angle between the phenyl and amide mean planes (Fig. 1) is 23.2°. There is an intramolecular N-H...O hydrogen bond (N(16)...O(25) 2.79(1) Å). Pairs of intermolecular N-H...N hydrogen bonds link the molecules (N(27)...N(21') 3.07(1) Å).

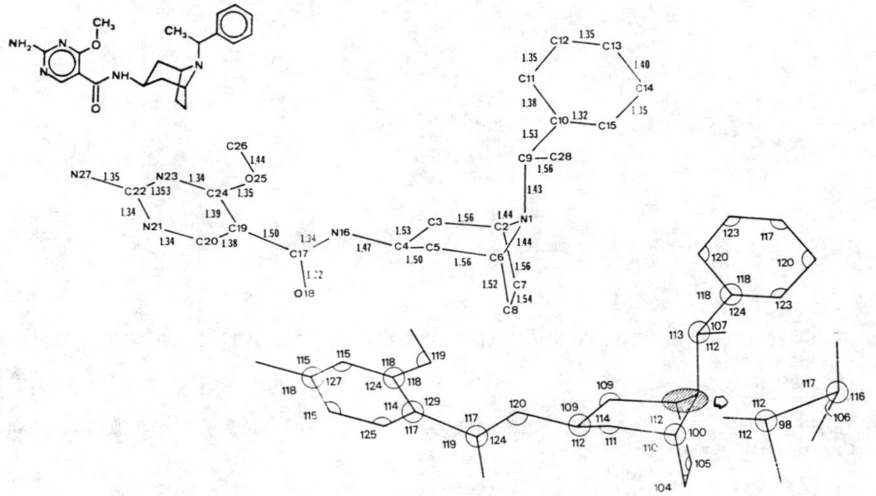

Fig. 1. The $C_{21}H_{27}N_5O_2$ molecule and dimensions.

exo-4-AMINO-5-BROMO-2-METHOXY-N[8-(2-CHLOROPHENYLMETHYL)-8-AZABICYCLO[3.2.1]OCT-3-YL]-BENZAMIDE

$C_{22}H_{25}BrClN_3O_2$

F. DURANT, P. RENARD, V. DE BEYS and G. EVRARD, 1984. Bull. Soc. Chim. Belg., 93, 923-924.

Monoclinic, $P2_1/c$, a = 7.673, b = 14.158, c = 19.993 Å, β = 92.31°, Z = 4. Cu radiation, R = 0.032 for 2704 reflexions.

The phenyl and amide groups (Fig. 1) are quasi coplanar (dihedral angle 6.3°). There is an intramolecular N-H...O hydrogen bond (N(16)...O(25) 2.697(6) Å) and an intermolecular one (N(27)...O(18') 2.832(7) Å).

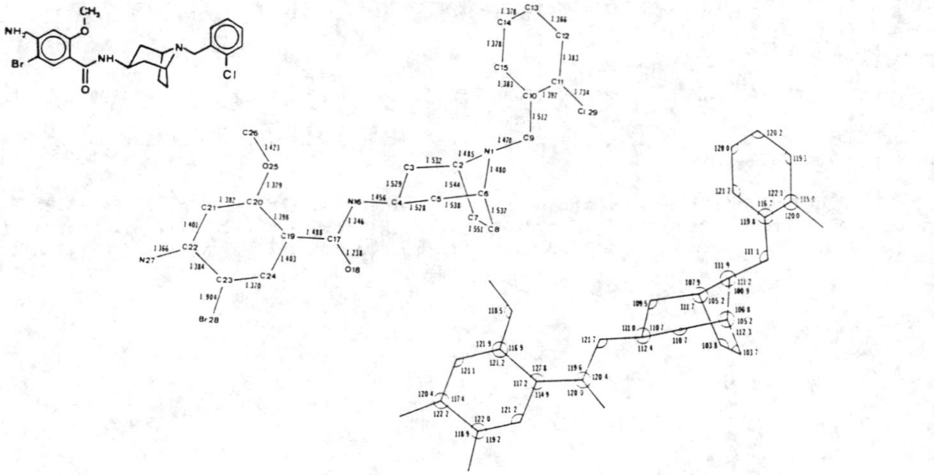

Fig. 1. The $C_{22}H_{25}BrClN_3O_2$ molecule and dimensions.

9-(m-HYDROXYPHENYL)-9α-METHOXY-3-METHYL-3-AZONIABICYCLO[3.3.1]NONANE p-TOLUENESULPHONATE MONOHYDRATE

$C_{23}H_{31}NO_5S,H_2O$ $C_{16}H_{24}NO_2{}^+,C_7H_7O_3S^-,H_2O$

G. HITE, P. SALVA, J.B. ANDERSON, M. RAPPOSCH, M. MANGION and M. FROIMOWITZ, 1984. Acta Cryst., C40, 850-853.

Monoclinic, $P2_1/c$, a = 10.006, b = 28.597, c = 8.352 Å, β = 106.29°, Dm = 1.305, Z = 4. Cu radiation, R = 0.065 for 2224 reflexions.

The crystal contains enantiomeric conformations (angles C(5)C(9)-C(10)C(11) ± 19.5(6); C(16)O(1)C(9)C(10) ± 54.6(5)°) but not the diastereomeric conformers generated by 180° rotation around the C(9)-C(10) bonds (Fig. 1). The chair-chair conformation of the bicyclic nucleus is flattened (N...C(7) 3.080(7) Å).

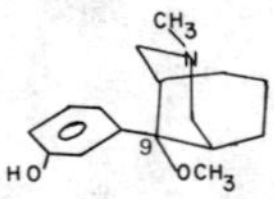

O(3)-S-O(5)	114.5(2)	N-C(4)-C(5)	114.9(4)
O(4)-S-C(18)	105.9(2)	H(C4e)-C(4)-H(C4a)	109 (4)
C(2)-N-H(N)	103 (3)	C(4)-C(5)-C(6)	113.9(4)
C(4)-N-C(17)	111.1(4)	C(9)-C(5)-H(C5)	107 (3)
C(2)-C(1)-H(C1)	102 (3)	C(5)-C(6)-C(7)	113.8(4)
C(8)-C(1)-C(9)	111.6(3)	H(C6e)-C(6)-H(C6a)	112 (4)
N-C(2)-H(C2a)	107 (3)	H(C7e)-C(7)-C(6)	110 (3)
C(1)-C(2)-H(C2e)	109 (3)	H(C7a)-C(7)-C(8)	110 (2)

H(C8e)-C(8)-C(1)	111 (3)
H(C8a)-C(8)-C(7)	108 (3)
O(1)-C(9)-C(5)	110.6(3)
C(1)-C(9)-C(10)	113.4(3)
H2(C17)-C(17)-N	107 (3)
H1(C17)-C(17)-H3(C17)	110 (4)
H1(C16)-C(16)-H2(C16)	91 (4)
O(1)-C(16)-H3(C16)	106 (3)

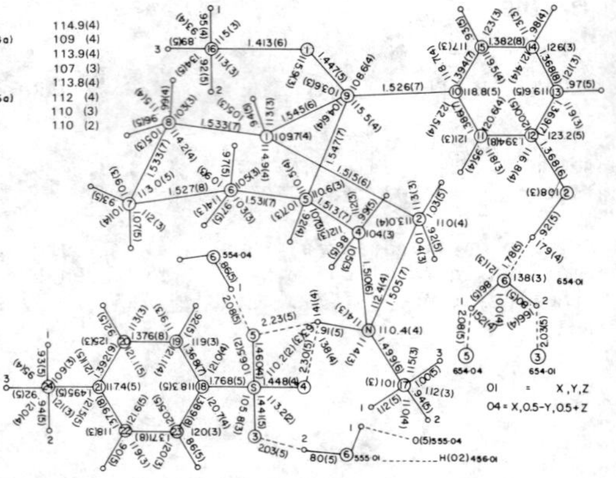

Fig. 1. $C_{23}H_{31}NO_5S,H_2O$: bond lengths and angles.

4-METHYL-1,2,4-TRIAZOLINE-3,5-DIONE ADAMANTYLIDENEADAMANTANE ADDUCT (1:1)
$C_{23}H_{31}N_3O_2$

C.-C. CHENG, C.A. SEYMOUR, M.A. PETTI, F.D. GREENE and J.F. BLOUNT, 1984. J. Org. Chem., 49, 2910-2916.

Triclinic, P$\bar{1}$, a = 9.306, b = 10.563, c = 10.553 Å, α = 84.61, β = 66.08, γ = 87.74°, Z = 2. Cu radiation, R = 0.045 for 2061 reflexions.

The adduct has the structure shown in Fig. 1. The C-C bond (C10-C20) connecting the two adamantyl units is unusually long (1.635 Å). Other bond lengths are N1-N2 1.399, N1-C1 1.354, N1-C10 1.514, N2-C20 1.500, N2-C2 1.343, N3-C1 1.399, N3-C2 1.393 Å.

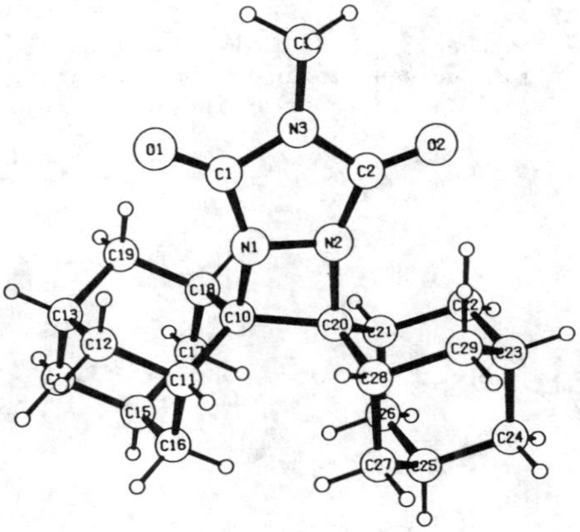

Fig. 1. The structure of $C_{23}H_{31}N_3O_2$.

N-(8-BENZYL-1αH,5αH-NORTROPAN-3β-YL)-2-ETHOXY-3-METHOXYBENZAMIDE HYDROCHLORIDE
$C_{24}H_{31}ClN_2O_3$

G. EVRARD, P. RENARD, V. DE BEYS and F. DURANT, 1984. Bull. Soc. Chim. Belg., 93, 417-418.

Monoclinic, $P2_1/c$, a = 16.243, b = 14.768, c = 9.923 Å, β = 105.07°, Z = 4. Cu radiation, R = 0.046 for 3046 reflexions.

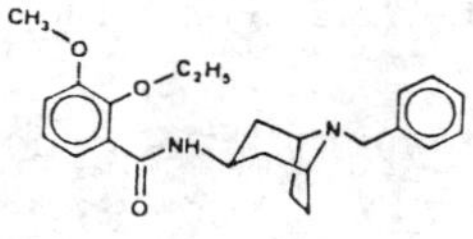

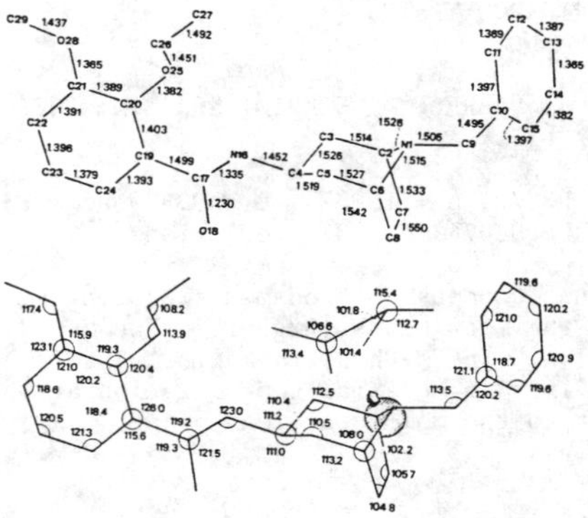

Fig. 1. Dimensions for $C_{24}H_{31}ClN_2O_3$.

In this molecule (Fig. 1), the amide and phenyl groups are quasi coplanar; there is also an intramolecular N-H...O hydrogen bond (N...O 2.719(4) Å). In the lattice there is an N(1)-H...Cl hydrogen bond N(1)...Cl 3.108(3) Å.

N-(8-BENZYL-1αH,5αH-NORTROPAN-3β-YL)-2,3,5-TRIMETHOXYBENZAMIDE MALEATE
$C_{28}H_{34}N_2O_8$ $C_{24}H_{32}N_2O_4^{2+},C_4H_2O_4^{2-}$

F. DURANT, P. RENARD and G. EVRARD, 1984. Bull. Soc. Chim. Belg., 93, 419-420.

Monoclinic, Cc, a = 14.813, b = 21.756, c = 8.829 Å, β = 105.56°, Z = 4. Cu radiation, R = 0.049 for 1625 reflexions.

The amide and trimethoxyphenyl groups (Fig. 1) are quasi coplanar (dihedral angle 11.0°) and are oriented almost perpendicular to the piperidine mean plane (dihedral angle with amide plane is 79.1°). There is an intramolecular N-H...O hydrogen bond (N(16)...O(25) 2.703(6) Å).

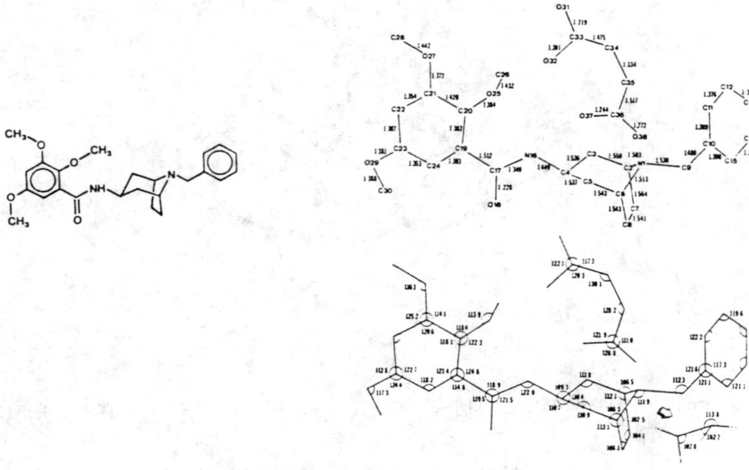

Fig. 1. Conformation and some dimensions for $C_{28}H_{34}N_2O_8$.

TRIMETHYL 8,13-DIPHENYL-13-AZATRICYCLO[8.2.1.0^{2,7}]-TRIDECA-2(7),3,5,11-TETRAENE-10,11,12-TRICARBOXYLATE
$C_{30}H_{27}NO_6$

R. MERCIER, J. VEBREL, M. SCHMIDT and G.M. SHELDRICK, 1984. Acta Cryst., C40, 1626-1628.

Monoclinic, $P2_1/c$, a = 15.074, b = 9.839, c = 18.545 Å, β = 104.39°, Z = 4. Cu radiation, R = 0.076 for 2070 reflexions.

The benzo condensed seven-membered ring has an envelope conformation with N 1.07 Å from the mean plane (Fig. 1). The N-C distances in this ring are 1.446(7) and 1.482(7) Å. The N-Ph distance is 1.392(7) Å. The pyrrolinic ring (with C=C 1.317(8) Å) lies in an axial configuration at the opposite side of the N-phenyl group with respect to the molecular mean plane.

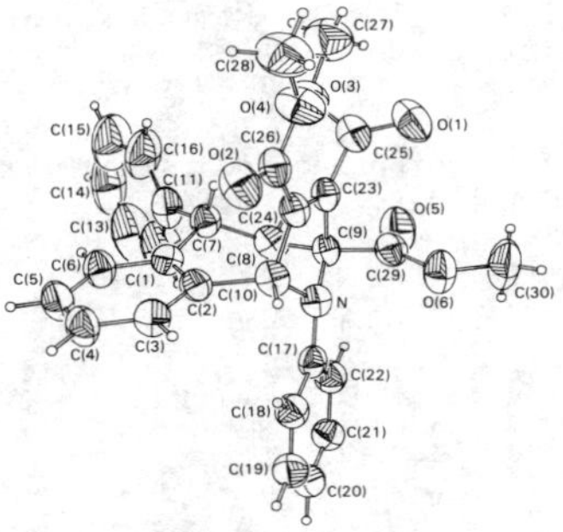

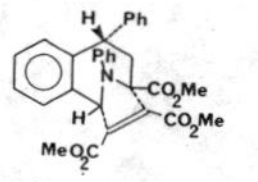

Fig. 1. The $C_{30}H_{27}NO_6$ molecule.

5,14-DIACETYL-4,5,8,9-TETRAHYDRO-2,7-DIPHENYL-4,8-O-BENZENO-3,9-IMINO-3H-3,5,6-BENZOTRIAZACYCLOUNDECINE ETHANOL SOLVATE

$C_{34}H_{28}N_4O_2,C_2H_6O$

K.R. MOTION, D.P. MUNRO, J.T. SHARP and M.D. WALKINSHAW, 1984. J. Chem. Soc. Perkin I, 2027-2033.

Monoclinic, $P2_1/c$, a = 15.924, b = 9.089, c = 21.319 Å, β = 109.35°, Z = 4. Mo radiation, R = 0.079 for 2802 reflexions.

The structure (Fig. 1) comprises two unsymmetrically coupled diazepine rings. One diazepine ring is slightly puckered with an angle of 8° between the best planes of atoms C(1)-C(13A)-C(9A)-C(9) and C(1)-C(2)-N(3)-N(14)-C(9) while the other is V-shaped with an angle of 120° between the best planes of atoms C(4)-C(16)-C(15)-C(8) and C(4)-N(5)-N(6)-C(7)-C(8). The angle between the benzene rings of the two monomer subunits is 72°. The two coupling bonds N(3)-C(4) (1.473(6)) and C(9)-C(8) (1.532(7) Å) are the longest C-N and C-C bond distances in the molecule. The N(5)-N(6) and N(3)-N(14) distances are 1.377(8) and 1.391(5) Å respectively. Atoms N(5) and N(14) show essentially planar trigonal geometry while N(3) adopts a more pyramidal coordination. The ethanol solvate molecules are grossly disordered.

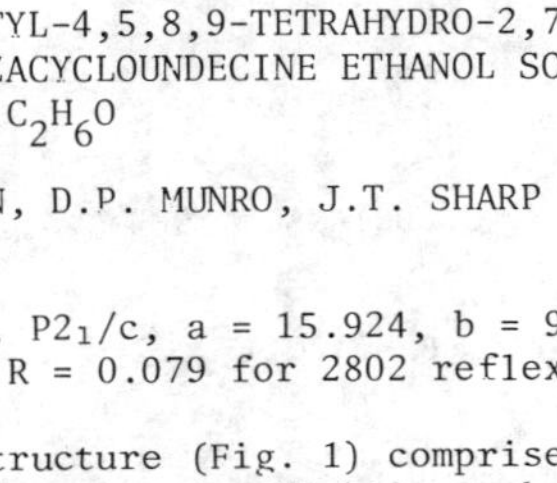

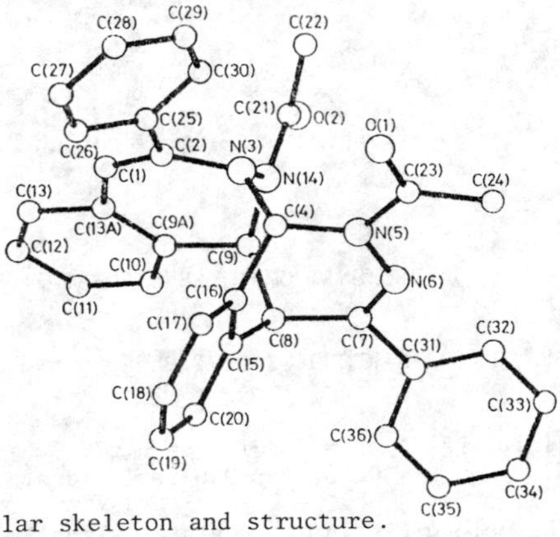

Fig. 1. $C_{34}H_{28}N_4O_2,C_2H_6O$: the molecular skeleton and structure.

OXETANE

C_3H_6O

P. LUGER and J. BUSCHMANN, 1984. J. Am. Chem. Soc., 106, 7118-7121.

Orthorhombic, Pnam, a = 8.620, b = 6.384, c = 6.038 Å, Z = 4. Mo radiation, R = 0.038 for 527 reflexions (at 90 K); a = 8.657, b = 6.401, c = 6.010 Å, Z = 4. Mo radiation, R = 0.041 for 606 reflexions (at 140 K).

The molecule lies across a crystallographic mirror plane and is non-planar (Fig. 1). The angle of pucker is 10.7(1)° at 90 K and 8.7(2) at 140 K.

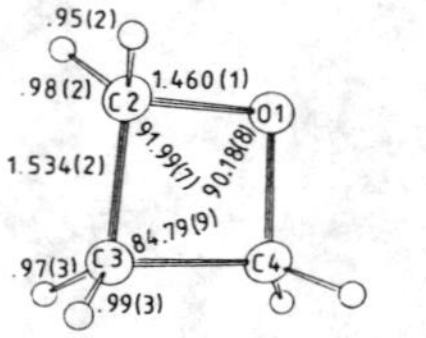

Fig. 1. Oxetane dimensions at 90 (left) and 140 K (right). (C4 is the mirror image of C2.)

DIBROMO-2,2-γ-BUTYROLACTONE
$C_4H_4Br_2O_2$

J.-P. LERE-PORTE, J. GALSOMIAS, J. PETRISSANS and M.-C. BRIANSO, 1984. J. Mol. Struct., 118, 277-292.

Monoclinic, $P2_1/c$, a = 6.792, b = 9.990, c = 11.111 Å, β = 120.17°, Z = 4. Cu radiation, R = 0.109 for 717 reflexions.

The molecule (Fig. 1) has a C(3) envelope conformation with dimensions shown in the figure.

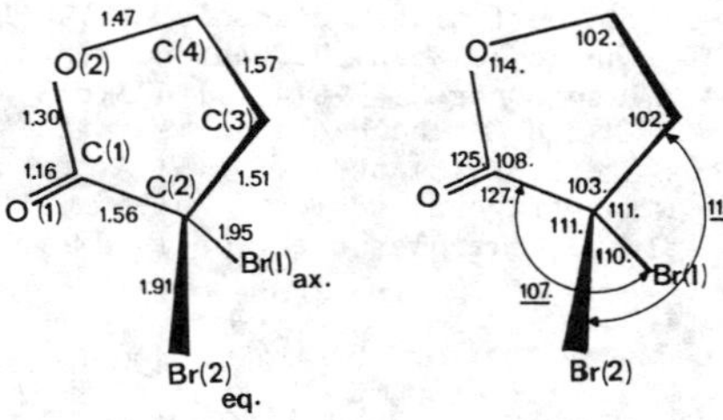

Fig. 1. Bond lengths and angles in $C_4H_4Br_2O_2$ (σ 0.02 Å, 2°).

(Z)-3-BROMO-5-(BROMONITROMETHYLENE)FURAN-2(5H)-ONE
$C_5HBr_2NO_4$

M.J. GRAY, M.P. HARTSHORN, B.R. PENFOLD and J. VAUGHAN, 1984. Aust. J. Chem., 37, 55-63.

Monoclinic, $P2_1/c$, a = 7.361, b = 6.678, c = 16.774 Å, β = 97.00°, D_m = 2.42, Z = 4. Mo radiation, R = 0.057 for 996 reflexions.

The molecule (Fig. 1) deviates slightly from planarity, with torsion angle O(61)-N(6)-C(6)-C(5) -177.0°.

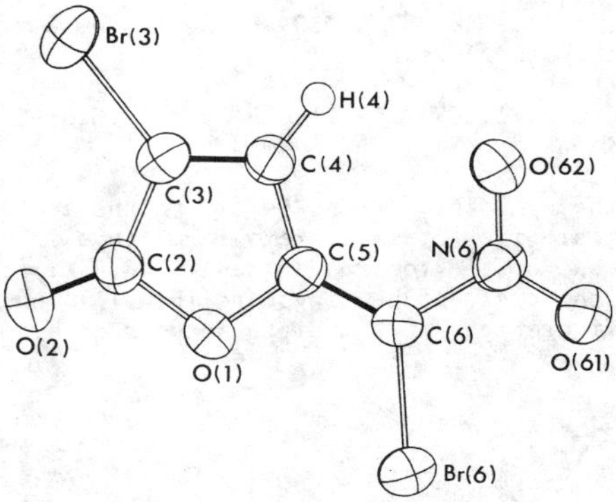

Fig. 1. The structure of $C_5HBr_2NO_4$.

1,2-BIS(FLUORODINITROMETHYL)ETHYLENE DIFORMATE
$C_6H_4F_2N_4O_{12}$ (I)

4,5-BIS(FLUORODINITROMETHYL)-1,3-DIOXOLAN-2-ONE
$C_5H_2F_2N_4O_{11}$ (II)

H.L. AMMON and S.K. BHATTACHARJEE, 1984. Acta Cryst., C40, 487-490.

I. Monoclinic, C2/c, a = 15.055, b = 9.3782, c = 11.714 Å, β = 129.058°, Z = 4. Cu radiation, R = 0.047 for 981 reflexions.

II. Monoclinic, $P2_1/c$, a = 6.283, b = 9.819, c = 18.679 Å, β = 96.00°, Z = 4. Cu radiation, R = 0.055 for 1509 reflexions.

The formate ester (I) lies on an inversion centre (Fig. 1) and has a conformation similar to what has been found previously in related molecules (1). The carbonate ester (II) has a quite different conformation; the two $CF(NO_2)_2$ substituents are cis and a ring twist is necessary to provide a staggered conformation about C(1)-C(3).

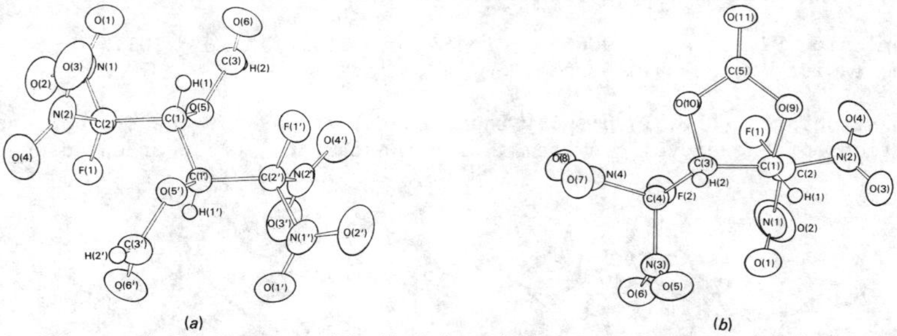

Fig. 1. Views of $C_6H_4F_2N_4O_{12}$ (I) (left) and $C_5H_2F_2N_4O_{11}$ (II) (right).

1. Structure Reports, 45B, 47; 49B,

(1SR,5SR,6SR)-4-OXO-3-OXABICYCLO[3.2.0]HEPTANE-6-CARBONITRILE
$C_7H_7NO_2$ (I)

(1SR,5SR,6RS)-4-OXO-3-OXABICYCLO[3.2.0]HEPTANE-6-CARBONITRILE
$C_7H_7NO_2$ (II)

F.W.B. EINSTEIN and A.C. WILLIS, 1984. Acta Cryst., C40, 1950-1952.

I. Monoclinic, $P2_1/n$, a = 11.134, b = 8.339, c = 6.826 Å, β = 92.05°, Z = 4. Mo radiation, R = 0.035 for 905 reflexions.

II. Orthorhombic, Pbca, a = 11.267, b = 11.095, c = 10.625 Å, Z = 8. Mo radiation, R = 0.035 for 773 reflexions.

In both molecules the cyclobutane ring is cis to the lactone (Fig. 1). All four rings are non-planar but are flatter in II (exo form) than in I (endo form); in the five-membered rings the sum of the torsion angles is 48.8 (I) and 32.0° (II) while the puckering angles in the cyclobutane rings are 15.4 (I) and 4.6° (II). The five-membered rings have configurations intermediate between envelope and half-chair.

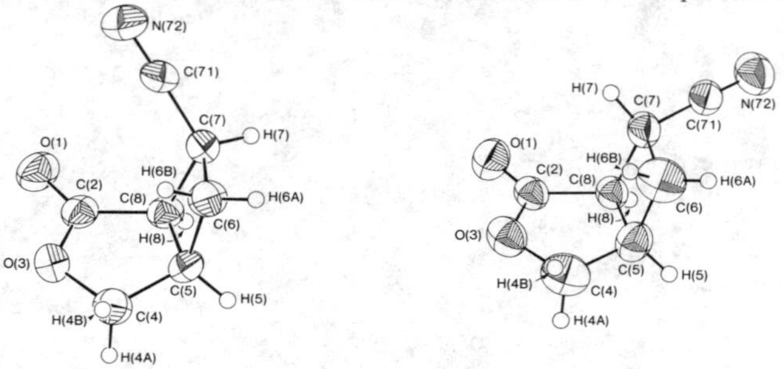

Fig. 1. The $C_7H_7NO_2$ molecules: endo (I) (left), and exo (II) (right).

trans-6,7-DIBROMO-8-OXABICYCLO[3.2.1]OCTAN-3-ONE
$C_7H_8Br_2O_2$ (I)

6-BROMO-8-OXABICYCLO[3.2.1]OCT-6-EN-3-ONE
$C_7H_7BrO_2$ (II)

M.M. OLMSTEAD, H. HOPE, V. SAMPATH and N.E. SCHORE, 1984. Acta Cryst., C40, 178-181.

I. Monoclinic, C2/c, a = 16.244, b = 12.500, c = 12.134 Å, β = 137.24°, Z = 8. Mo radiation, R = 0.051 for 1245 reflexions.

II. Monoclinic, $P2_1/c$, a = 7.900, b = 6.473, c = 13.980 Å, β = 101.55°, D_m = 1.9, Z = 4. Mo radiation, R = 0.048 for 1230 reflexions.

Both structures (Fig. 1) display chair conformations for the cyclohexane-like pyran portions of the bicyclic systems, with the carbonyl group oriented endo.

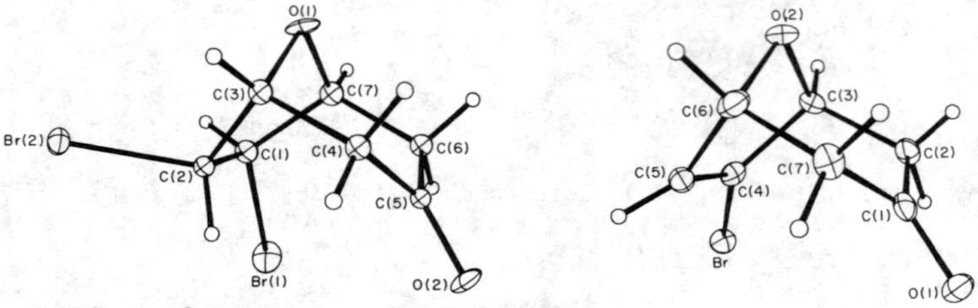

Fig. 1. Views of $C_7H_8Br_2O_2$ and $C_7H_7BrO_2$.

endo-7-HYDROXY-2,4-DIOXA-3-OXOBICYCLO[3.3.1]NONANE
$C_7H_{10}O_4$

W. WONG-NG and S.C. NYBURG, 1984. Can. J. Chem., 62, 1271-1274.

Orthorhombic, $Pna2_1$, a = 10.977, b = 6.489, c = 9.811 Å, Z = 4. Cu radiation, R = 0.044 for 591 reflexions.

The molecular conformation is shown in Fig. 1. Atom C(4) is +0.514(6) Å from the C(2),C(3),C(5),C(6) plane; the corresponding distance for C(1) is -0.729(7) Å. That C(4) is not displaced further is attributed to the presence of an intermolecular O-H...O hydrogen bond between O(1)-H and O(3) on an adjacent molecule.

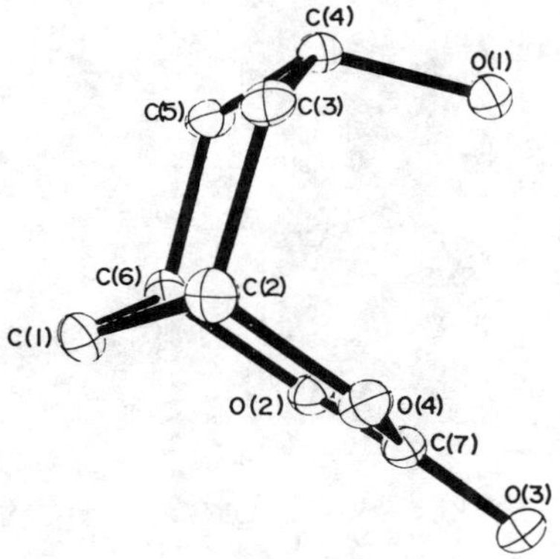

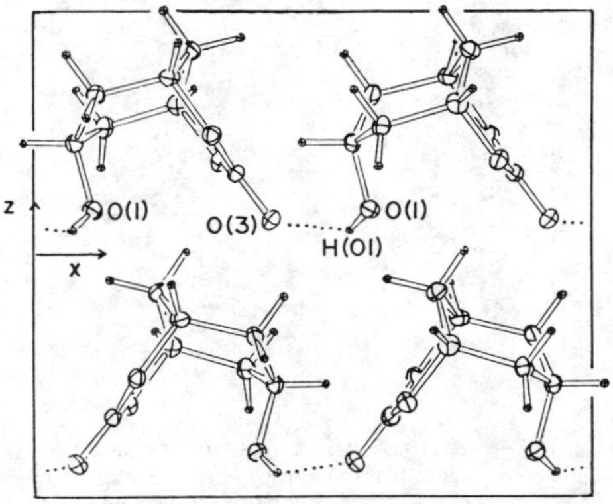

Fig. 1. A view of the $C_7H_{10}O_4$ molecule and the contents of one unit cell viewed along b.

2,5-DIMETHYL-2-HYDROPEROXY-5-METHOXY-2,5-DIHYDROFURAN
$C_7H_{12}O_4$

K. PETERS, E.-M. PETERS, H.G. VON SCHNERING, A GRIESBECK and K. GOLLNICK, 1984. Z. Kristallogr., 168, 153-155.

Monoclinic, $P2_1/n$, a = 7.224, b = 16.629, c = 7.067 Å, β = 106.45°, Z = 4. Mo radiation, R = 0.096 for three-dimensional data.

The structure (Fig. 1) is shown to be a cis-isomer. The dihydrofuran ring has an envelope conformation (O(1) 0.17 Å from the C2-C5 plane, dihedral angle of 11.7° between the planes C2-C5 and C2-O1-C5).

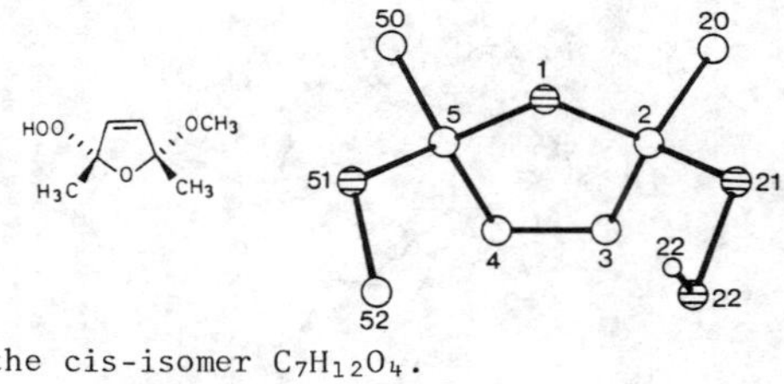

Fig. 1. View of the cis-isomer $C_7H_{12}O_4$.

(2S,3S)-TETRAHYDRO-3-ACETOXY-5-OXO-2,3-FURANDICARBOXYLIC ACID
$C_8H_6O_7$

D.E. ZACHARIAS, J.P. GLUSKER, R. GUTHRIE, A.C. SULLIVAN and P.A. SRERE, 1984. Acta Cryst., C40, 2100-2103.

Orthorhombic, $P2_12_12_1$, a = 9.495, b = 16.286, c = 5.841 Å, Z = 4. Cu radiation, R = 0.036 for 999 reflexions.

The structure (Fig. 1) is of interest because the compound is a derivative of citric anhydride and consists of two fused five-membered rings, one of which is an anhydride and the other a γ-lactone, with one carbon of the fusion bearing an acetoxy group. The molecule has bond lengths and angles comparable with those of other anhydrides. The packing is determined by interaction between carbonyl and ether oxygen atoms.

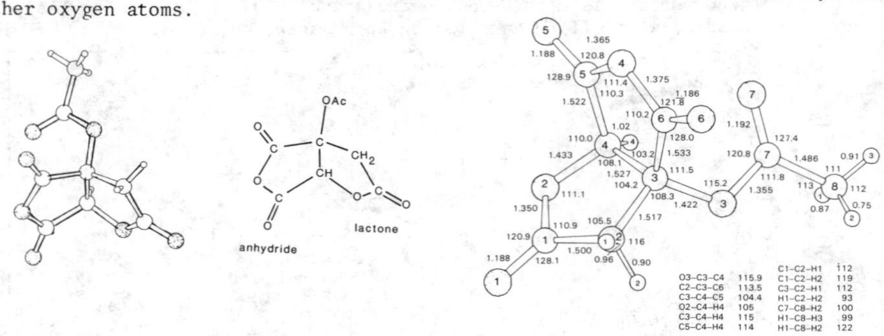

Fig. 1. $C_8H_6O_7$ with distances and angles (esd's are 0.002-0.003 Å for distances not involving H (0.03-0.06 if H is involved)) and 0.1-0.2° for angles (2-5° if H is involved).

3a,6a-DIMETHYL-2,3,3a,5,6,6a-HEXAHYDROFURO[3,2-b]FURAN-2,5-DIONE
$C_8H_{10}O_4$

M.R. CAIRA, W.H. WATSON, J. KAGAN and S.P. SINGH, 1984. Acta Cryst., C40, 498-500.

Orthorhombic, Fdd2, a = 13.055, b = 20.759, c = 6.080 Å, Z = 8. Cu radiation, R = 0.034 for 296 reflexions.

The molecule (Fig. 1) consists of two cis-fused γ-lactone rings with methyl groups attached to the bridgehead atoms. A crystallographic twofold axis passes through the midpoint of the bridgehead bond. The ring C-O distances are 1.343(5) and 1.463(5) Å and the lactone groups are planar. Each five-membered ring thus adopts an envelope form with the flap at the bridgehead atom. This conformation alleviates steric repulsion between the methyl groups.

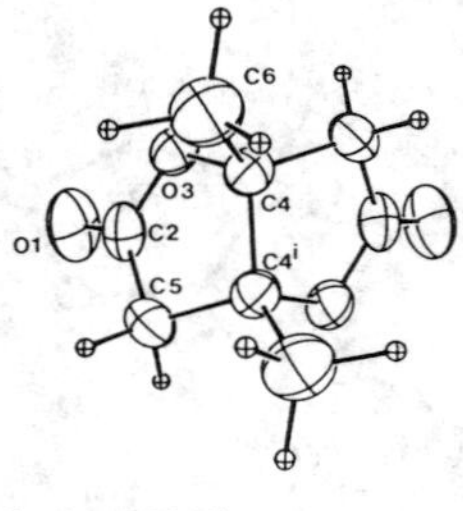

Fig. 1. A view of $C_8H_{10}O_4$ along [001].

2,4,6-TRIMETHYLPYRAN PERCHLORATE
$C_8H_{11}ClO_5$ (I)

BIS(2,6-DI-t-BUTYLPYRAN) BIS(μ_2-CHLORO)BIS(DICHLOROPALLADATE)
$C_{26}H_{42}Cl_6O_2Pd_2$ (II)

L.G. KUZ'MINA and Yu.T. Struchkov, 1983. Koord. Khim., 9, 705-710 [Sov. J. Coord. Chem., 9, 407-411].

I. Orthorhombic, Pbca, a = 10.052, b = 9.787, c = 19.64 Å, Z = 8. Mo radiation, R = 0.056 for 637 reflexions (at -120°C).

II. Monoclinic, C2/c, a = 12.950, b = 23.597, c = 11.620 Å, β = 103.77°, Z = 4. Mo radiation, R = 0.054 for 1460 reflexions.

The structures of I and II comprise discrete cations and anions (Fig. 1). The pyrylium cations in both I and II have acceptable dimensions. The perchlorate anion in I has a tetrahedral structure with Cl-O distances in the range 1.436(7)-1.47(1) Å. In II the hexachlorodipalladate anion is located in a special position on a twofold axis. The bridging Cl(1) and Cl(1') atoms are bonded symmetrically to the Pd and Pd' atoms (Pd-Cl(1) 2.312(3), Pd-Cl(1') 2.319(3) Å). The Pd-Cl (terminal) distances are somewhat shorter (Pd-Cl(2) 2.251(3), Pd-Cl(3) 2.264(3) Å). Coordination about the Pd atoms is square planar with a slight tetrahedral distortion.

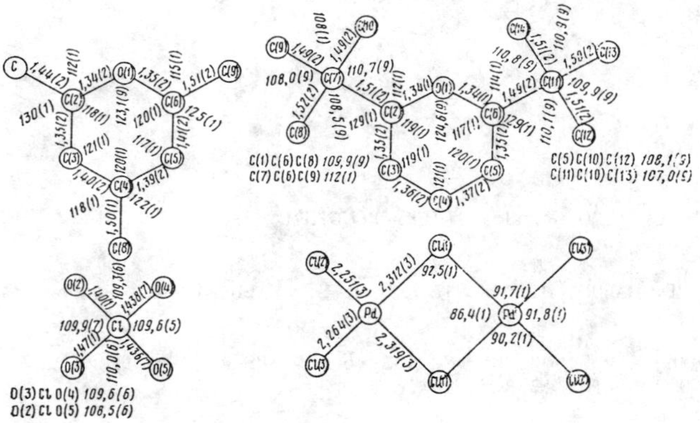

Fig. 1. Schematic representation of $C_8H_{11}ClO_5$ and $C_{26}H_{42}Cl_6O_2Pd_2$.

3-HYDROXY-3-(TRIFLUOROMETHYL)PHTHALIDE
$C_9H_5F_3O_3$

U.D.G. PRABHU, K.C. EAPEN and C. TAMBORSK, 1984. J. Org. Chem., 49, 2792-2795.

Orthorhombic, $P2_12_12_1$, a = 6.942, b = 8.047, c = 16.196 Å, Z = 4. Mo radiation, R = 0.164 for 1244 reflexions.

The phthalide and dimensions are shown in Fig. 1. The hydroxyl H atom is oriented so that it is equidistant from two F atoms of the adjacent trifluoromethyl group.

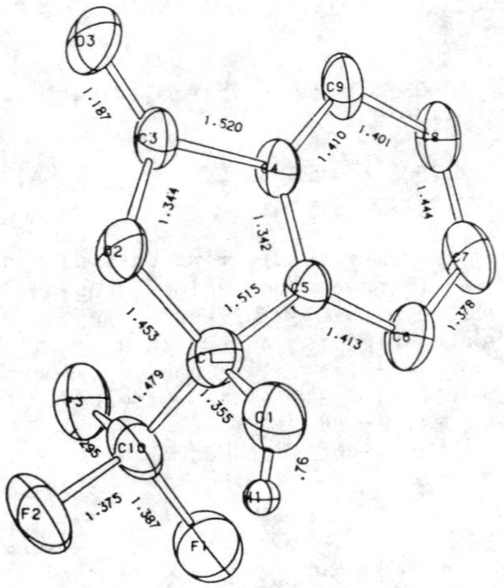

Fig. 1. A view of $C_9H_5F_3O_3$ and bond lengths.

1,4-DIMETHYL-5-exo-NITRO-6-endo-(TRICHLOROMETHYL)-7-OXABICYCLO[2.2.1]HEPT-2-ENE
$C_9H_{10}Cl_3NO_3$

T.M. BALTHAZUR, B. GAEDE, D.E. KORTE and H.-S. SHIEH, 1984. J. Org. Chem., 49, 4547-4549.

Monoclinic, $P2_1/n$, a = 7.002, b = 20.355, c = 9.389 Å, β = 106.86°, Z = 4. Cu radiation, R = 0.0788 for 1208 reflexions.

X-ray analysis established the structure of the molecule to be as shown in Fig. 1.

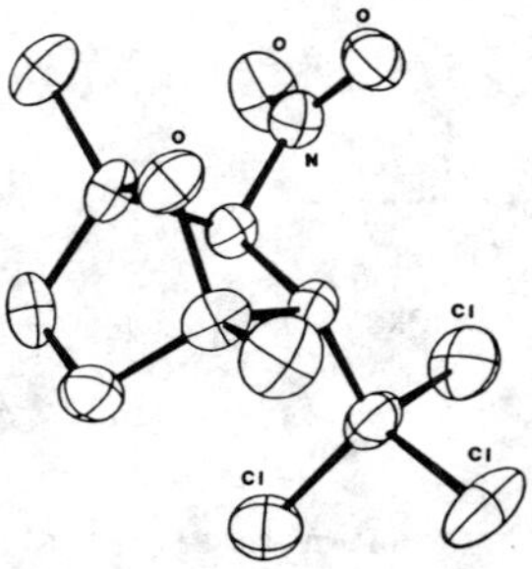

Fig. 1. A view of $C_9H_{10}Cl_3NO_3$.

3-(IMIDAZOLIDIN-2-YLIDENE)-4,5-DIMETHYL-4,5-DIHYDRO-2(3H)-FURANONE
$C_9H_{14}N_2O_2$

CHEN BENMING and HUANG ZHITANG, 1984. J. Struct. Chem., 3, 37-40.

Orthorhombic, $Pna2_1$, a = 10.698, b = 9.779, c = 8.872 Å, Z = 4. Mo radiation, R = 0.034 for 637 reflexions.

The lactone and imidazolidine ring planes form a 12.3° dihedral angle. The two methyl groups of the lactone are cis and have a C10-C5-C4-C11 torsion angle of

34.4°. The double bond linking the two rings is elongated (1.372(5) Å); other principal bond lengths are: C2-O12 1.229(4), C2-C3 1.424(6), C6-N9 1.357(5), C6-N9' 1.363(5), O1-C2 1.385(4), O1-C5 1.447(6) Å. An intramolecular N-H...O hydrogen bond links N9 and O12 (2.983 Å) and an intermolecular hydrogen bond links N9' to an adjacent O12 atom (N...O 2.917 Å).

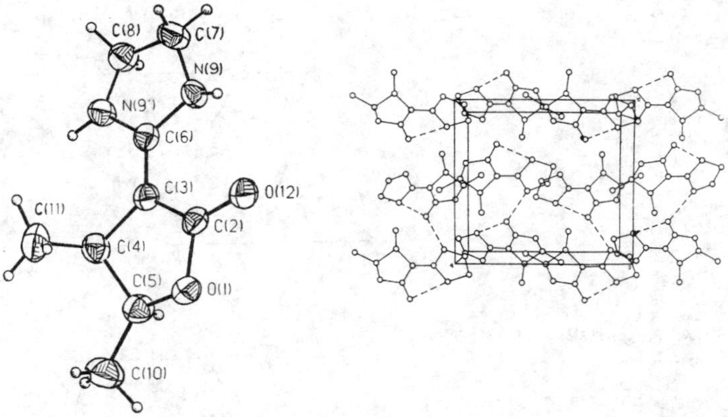

Fig. 1. A view of $C_9H_{14}N_2O_2$ and the packing arrangement.

3,4-METHYLENEDIOXYCINNAMIC ACID
$C_{10}H_8O_4$ (I)

3,4-DIMETHOXYCINNAMIC ACID
$C_{11}H_{12}O_4$ (II)

G.R. DESIRAJU, R. KAMALA, B.H. KUMARI and J.A.R.P. SARMA, 1984. J. Chem. Soc. Perkin II, 181-189.

I. Triclinic, $P\bar{1}$, a = 3.804, b = 10.502, c = 11.112 Å, α = 77.84, β = 84.26, γ = 80.17°, D_m = 1.54, Z = 2. Cu radiation, R = 0.063 for 742 reflexions.

II. Triclinic, $P\bar{1}$, a = 8.448, b = 15.072, c = 8.437 Å, α = 99.44, β = 94.71, γ = 101.59°, D_m = 1.30, Z = 4. Mo radiation, R = 0.107 for 725 reflexions.

Molecules of I exist as centrosymmetric hydrogen bonded dimers with the O-H...O distance 2.619(5) Å. The C(9)-O(1) and C(9)-O(2) distances and corresponding angles (Fig. 1) are indicative of partial disorder of the carboxy group. The molecular conformation can be defined in terms of three planes: (a) the phenyl ring, (b) the four-atom olefin linkage C(1),C(7),C(8),C(9) and (c) the four atom carboxy group. The ab, bc and ac dihedral angles are 4.56, 0.8 and 5.4° respectively, i.e. the overall structure is closely planar. In the crystal the hydrogen bonded dimers lie in sheets parallel to (-111). Within a sheet dimers are associated through C-H...O short contacts, e.g. H(8)...O(4) 2.48, H(5)...O(1) 2.88 Å. Sheets are stacked along (100) so that potentially photoreactive molecules are related by a translation along a with a spacing of 3.8 Å between reactive double bonds, suitable for topochemical reaction in the solid state. The molecular parameters of II are of limited accuracy. However the general trends are in accord with those found in the crystal structures of other cinnamic acids. As in I, disorder is also present in the carboxy group. In the crystal, pairs of pseudo-centrosymmetric molecules in the asymmetric unit are hydrogen bonded to form cyclic dimers. Nearest neighbours are related by a pseudo-translation and centres of inversion. Only half the molecules in the structure are potentially reactive: for each of these there are two inversion-related near neighbours. One of these is within photoreactive distance (4.04 Å) to give topochemical reaction on irradiation.

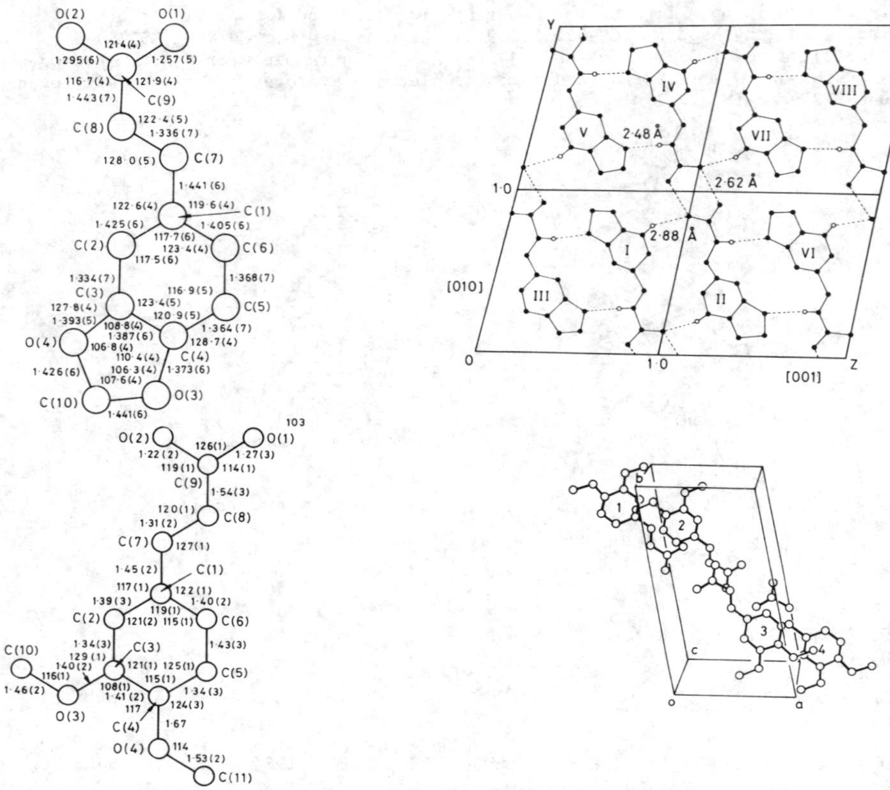

Fig. 1. $C_{10}H_8O_4$ (top): bond lengths, angles and projection of the crystal structure down [100]. $C_{11}H_{12}O_4$ (bottom): bond lengths and the four molecules in the unit cell.

r-2-ALLOPHANOYL-2-ALLYL-t-4-METHYLBUTYROLACTONE (VALOFAN)
$C_{10}H_{14}N_2O_4$

J. LAUTERWEIN, A.A. PINKERTON, H.H. WITTEKIND and B. TESTA, 1984. Helv. Chim. Acta, 67, 1662-1668.

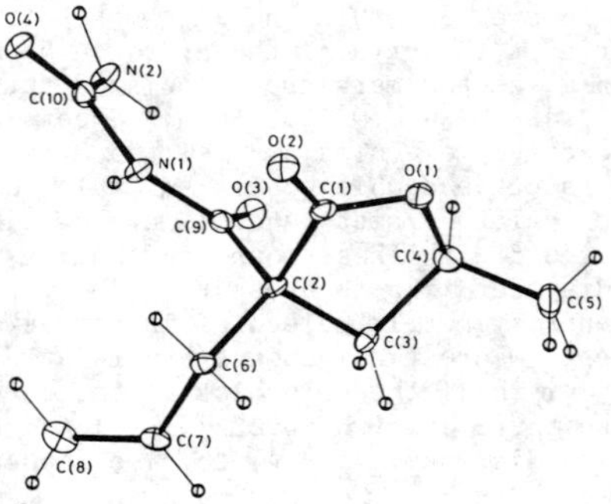

Fig. 1. A view of valofan.

Monoclinic, $P2_1/n$, a = 7.183, b = 15.517, c = 10.366 Å, β = 108.71°, Z = 4. Mo radiation, R = 0.053 for 863 reflexions.

The analysis establishes that the CH_3-group is cis to the allyl group and trans to the allophanoyl moiety (Fig. 1). The lactone ring is in an envelope conformation (C3 0.45 Å from the C1,C2,C4,O1 plane). There is an intramolecular N-H...O hydrogen bond (N2-H...O3 with H...O 1.98 Å and N-H...O 136°).

6-METHYL-2-PROPIONYLBENZOIC ACID HYDROXYLACTONE
$C_{11}H_{12}O_3$ (I)

3-METHYL-2-PROPIONYLBENZOIC ACID HYDROXYLACTONE
$C_{11}H_{12}O_3$ (II)

E. RUDINGER-ADLER, M. HOSTETTMANN, V. GRAMLICH and W. PETTER, 1984. Helv. Chim. Acta, 67, 743-747.

I. Monoclinic, $P2_1/c$, a = 7.513, b = 8.164, c = 16.150 Å, β = 93.78°, Z = 4. Mo radiation, R = 0.05 for 1258 reflexions.

II. Monoclinic, $P2_1$, a = 6.238, b = 10.321, c = 7.796 Å, β = 91.79°, Z = 2. Mo radiation, R = 0.038 for 864 reflexions.

The X-ray analysis established these structures as hydroxylactones as shown in Fig. 1. Molecules are linked O-H...O hydrogen bonds in the crystal lattice.

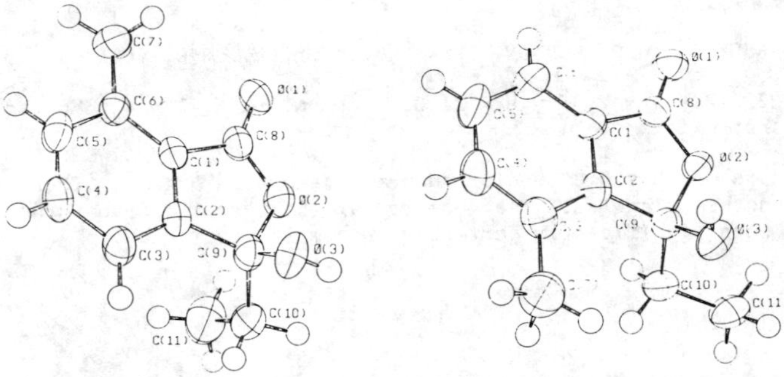

Fig. 1. Views of the 6-methyl and 3-methyl hydroxylactones.

(1S)-1α,4α,5β,5aβ,6β,8β,9β,9aβ-OCTAHYDRO-5,8-DIHYDROXY-1,4-EPOXY-6,9-METHANO-3-BENZOXEPIN-7-(2H)-ONE
$C_{11}H_{14}O_5$

M.G. ESSIG, T.T. STEVENSON, F. SHAFIZADEH, R.E. STENKAMP and L.H. JENSEN, 1984. J. Org. Chem., 45, 3652-3656

Tetragonal, $P4_3$, a = 12.9505, c = 6.3054 Å, Z = 4. Cu radiation, R = 0.023 for 732 reflexions.

In the molecule (Fig. 1) the pyranose ring has a conformation intermediate between a half-chair $^1H_0$ and a sofa E_0, with significant distortion toward a $^1C_4$ conformation. The cyclohexanone ring has a pronounced boat conformation. The anhydro ring adopts a twist conformation with some distortion towards an envelope. The cyclopentane and cyclopentanone rings both assume an ^{11}E conformation with some distortion towards twist forms $^{11}T_{10}$ and $^{11}T_7$, respectively. Molecules are linked about a 4_3 axis by O-H...O hydrogen bonds 2.679(4) and 2.650(4) Å.

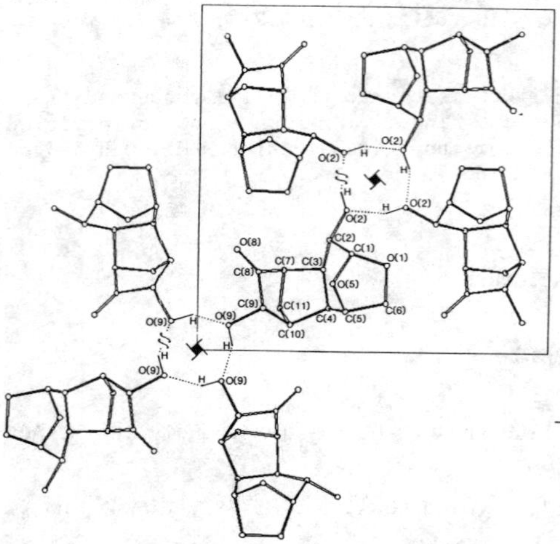

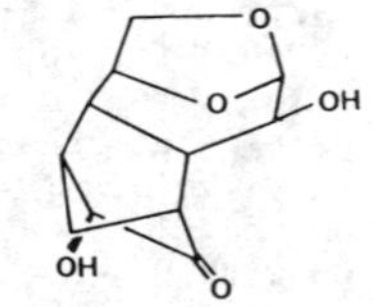

ring	Q (Å)	θ (deg)	φ (deg)	atom number 1	2	3
pyranose	0.629	135.3	195.3	O(5)	C(1)	C(2)
cyclohexanone	0.942	88.8	115.5	C(4)	C(3)	C(7)
anhydro	0.440		58.1	C(1)	O(5)	C(5)
cyclopentane	0.583		6.2	C(11)	C(10)	C(4)
cyclopentanone	0.602		1.1	C(11)	C(7)	C(8)

Fig. 1. The $C_{11}H_{14}O_5$ molecule and puckering parameters.

[2S-(2α,2aβ,4α,4aβ,5α,8α,8aβ,8bβ,10R*)]-2,2a,3,4,4a,5,6,8,8a,8b-DECAHYDRO-5,8-EPOXY-2,4-METHANO-1,7-DIOXACYCLOPENT[cd]AZULEN-10-OL MONOHYDRATE
$C_{11}H_{14}O_4,H_2O$

P. BHATE, J. GALLUCCI and D. HORTON, 1984. Acta Cryst., C40, 468-470.

Orthorhombic, $P2_12_12_1$, a = 7.832, b = 21.985, c = 6.115 Å, Z = 4. Mo radiation, R = 0.036 for 1241 reflexions.

The structure of this Diels-Adler adduct is shown in Fig. 1. There are no unusual bond lengths or angles. The geometry of the parent 1,6-anhydrohexapyranose system has not been significantly perturbed on complex formation.

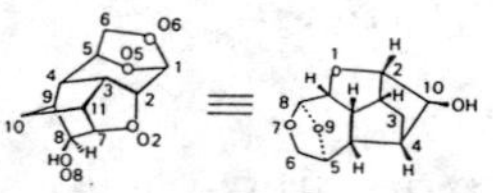

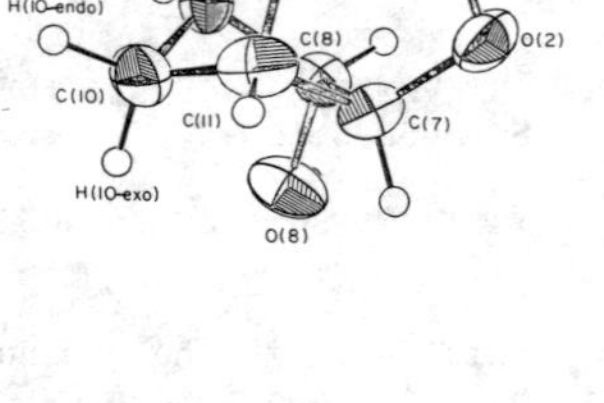

Fig. 1. A view of $C_{11}H_{14}O_4$.

[2R-(2α,2aβ,4α,4aβ,5β,6α,8aβ,8bβ,9R)]-DECAHYDRO-1,2,8-TRIHYDROXY-2,4-METHANO-1,7-DIOXACYCLOPENT[cd]AZULENE
$C_{11}H_{16}O_5$

M.G. ESSIG, F. SHAFIZADEH, T.G. COCHRAN and R.G. STENKAMP, 1984. Carbohydr. Res., 129, 55-61.

Monoclinic, $P2_1$, a = 10.545, b = 8.324, c = 5.973 Å, β = 83.53°, Z = 2. Cu radiation, R = 0.035 for 681 reflexions.

The septanose ring adopts a boat conformation (Fig. 1) with the furanoid and one cyclopentane ring in twist conformations and the other cyclopentane ring in an envelope conformation.

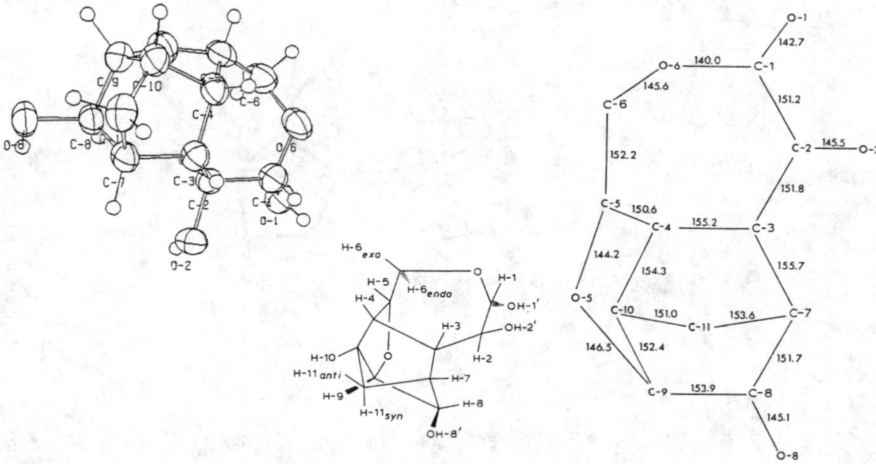

Fig. 1. A view of $C_{11}H_{16}O_5$ and bond lengths (pm).

4-(5-(2-IODO-1-HYDROXYETHYL)-5-METHYL-TETRAHYDRO-2-FURYL)-4-METHYL-1,3-DIOXOLAN-2-ONE
$C_{11}H_{17}IO_4$

P.G.M. WUTS, R. D'COSTA and W. BUTLER, 1984. J. Org. Chem., <u>49</u>, 2582-2588.

Monoclinic, $P2_1$, a = 7.557, b = 10.898, c = 9.169 Å, β = [?]°, Z = 2. Mo radiation, R = 0.048 for 967 reflexions.

The analysis established the structure as shown in Fig. 1.

Fig. 1. A view of $C_{11}H_{17}IO_4$.

(5R,7R,9R)-7-BROMO-5-NITRO-9-TRIMETHYLSILYL-2-OXATRICYCLO[4.2.1.0^{4,8}]NONANE
$C_{11}H_{18}BrNO_3Si$

J.P. MICHAEL, N.F. BLOM and J.C.A. BOEYENS, 1984. J. Chem. Soc. Perkin I, 1739-1743.

Triclinic, P$\bar{1}$, a = 6.70, b = 9.70, c = 11.36 Å, α = 93.31, β = 94.57, γ = 105.68°, D_m = 1.51, Z = 2. Mo radiation, R = 0.052 for 1905 reflexions.

The X-ray analysis established the structure of this tricyclic ether (Fig. 1). Bond lengths include: C(1)-O(2) 1.43(1), C(3)-O(2) 1.44(1), C(3)-C(4) 1.52(2), C(4)-C(5) 1.54(1), C(5)-C(6) 1.50(1), C(9)-C(1) 1.56(1), C(6)-C(7) 1.53(1), C(7)-C(8) 1.51(1), C(1)-C(8) 1.54(1), C(4)-C(8) 1.53(1), C(7)-Br(13) 1.959(8), C(9)-Si(14) 1.920(9) Å.

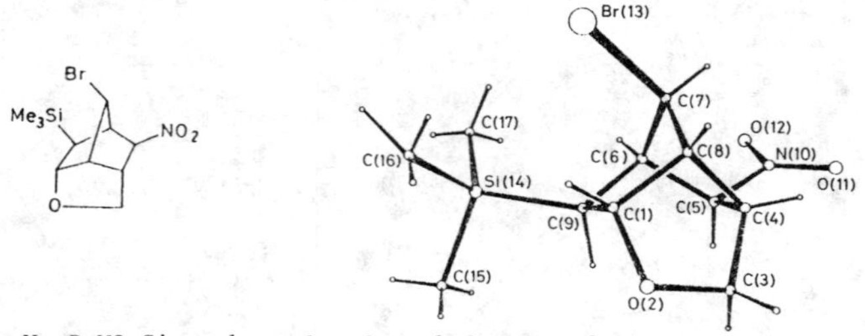

Fig. 1. $C_{11}H_{18}BrNO_3Si$: the molecular skeleton and structure.

DIBENZOFURAN
$C_{12}H_8O$

W.J. REPPART, J.C. GALLUCCI, A.P. LUNDSTEDT and R.E. GERKIN, 1984. Acta Cryst., C40, 1572-1576.

Orthorhombic, Pnma, a = 7.5154, b = 19.098, c = 5.7739 Å, Z = 4. Mo radiation, R = 0.039 for 1254 reflexions (at 169K).

The molecule has mirror symmetry with planar five- and six-membered rings; the dihedral angle between the benzene and furan rings is 1.5°. The mean C-C bond length in the six-membered rings is 1.392 Å (Fig. 1). The structure is disordered with nine percent of molecules adopting an alternative position related to the first by a 180° rotation about an axis passing through the centre of mass and perpendicular to the furan ring. Similar disorder has been found on reanalysis of room temperature data (1, 2). The presence of disorder (unreported in previous studies) has removed discrepancies reported earlier in C-O and C-C bond lengths.

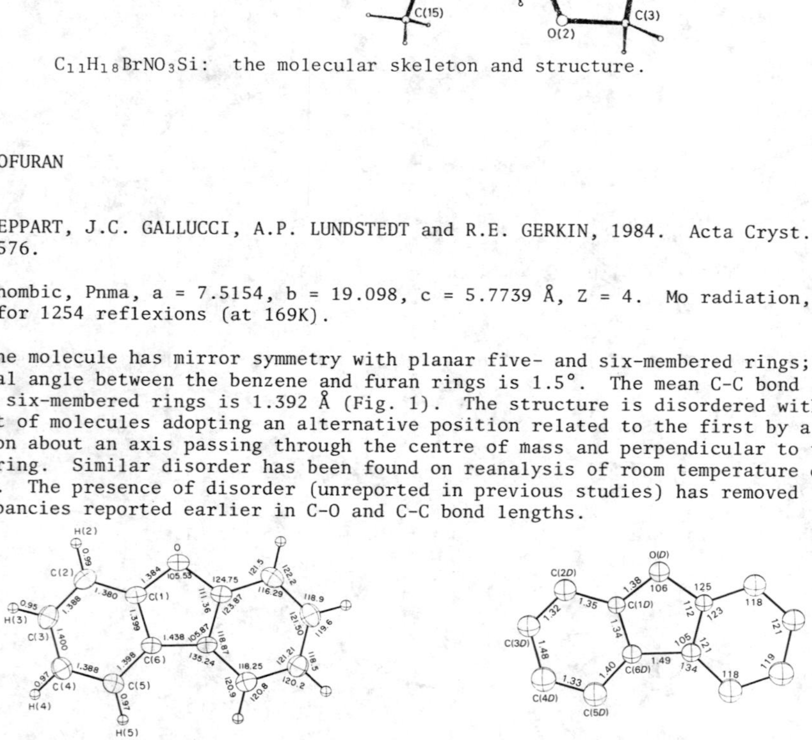

Fig. 1. Bond lengths and angles in the dibenzofuran molecules making up 91% (left) and 9% (right) of the structure.

1. Structure Reports, 38B, 332.
2. Ibid, 39B, 230.

syn-3,4-DIMETHOXY-1,2-EPOXY-1,2,3,4-TETRAHYDRONAPHTHALENE
$C_{12}H_{14}O_3$

C.L. KLEIN and E.D. STEVENS, 1984. Acta Cryst., C40, 315-317.

Orthorhombic, Pbca, a = 20.440, b = 13.383, c = 7.648 Å, D_m = 1.31, Z = 8. Mo radiation, R = 0.037 for 977 reflexions.

The molecule has the methoxy substituents bound axially to the ring system (Fig. 1). Bond lengths are as expected.

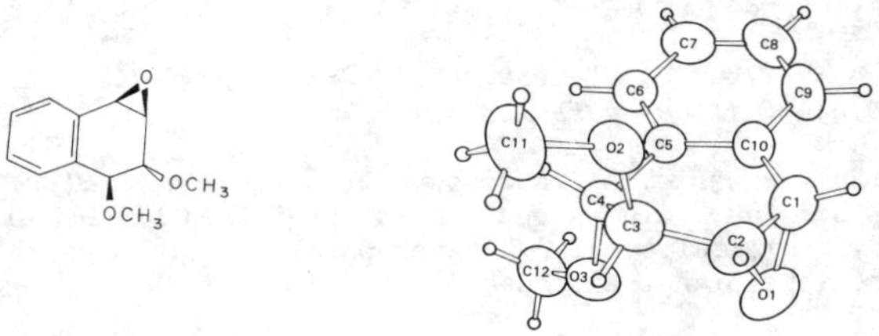

Fig. 1. The $C_{12}H_{14}O_3$ molecule.

AMMINETRIHYDROBORON 18-CROWN-6 METHANOL SOLVATE
$C_{12}H_{30}BNO_6,CH_4O$ (I)

AMMINETRIFLUOROBORON 18-CROWN-6 DICHLOROMETHANE SOLVATE
$C_{12}H_{27}BF_3NO_6,CH_2Cl_2$ (II)

H.M. COLQUHOUN, G. JONES, J.M. MAUD, J.F. STODDART and D.J. WILLIAMS, 1984. J. Chem. Soc. Dalton, 63-66.

I. Orthorhombic, $P2_1cn$, a = 8.416, b = 13.685, c = 16.923 Å, Z = 4. Cu radiation, R = 0.107 for 689 reflexions.

II. Orthorhombic, $P2_12_12_1$, a = 8.351, b = 14.715, c = 17.413 Å, Z = 4. Cu radiation, R = 0.048 for 1476 reflexions.

In both cases the 18-crown-6 molecule adopts the all-gauche conformation with pseudo D_{3d} symmetry. The N-H...O bonding is directly to the upper triangle of O atoms on the complexing face of the crown in both structures (Fig. 1). In I the N...O(1), N...O(7) and N...O(13) distances are 2.99, 3.01 and 3.02 Å respectively. The non-bonding N...O(4), N...O(10) and N...O(16) are 3.19, 3.25 and 3.18 Å respectively. The corresponding distances in II are 2.95, 2.92, 2.93, 3.10, 3.12 and 3.14 Å. The compounds are not isostructural due to packing arrangements with the different solvent molecules.

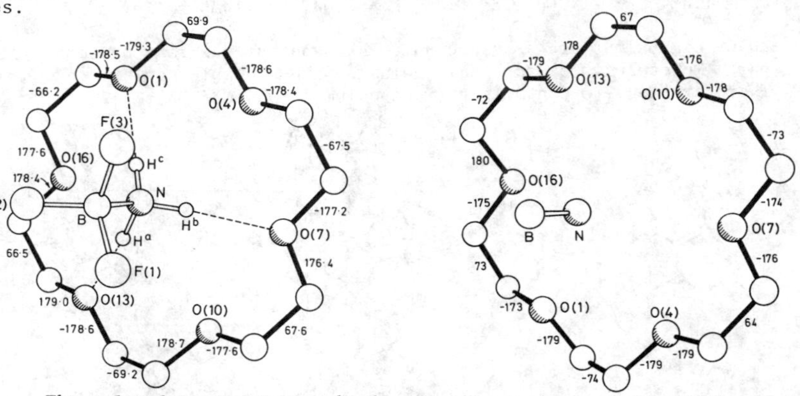

Fig. 1. The molecular structures showing torsion angles: $C_{12}H_{27}BF_3NO_6$ (left) and $C_{12}H_{30}BNO_6$ (right).

1-METHYL-2-NITRONAPHTHO[2,1-b]FURAN
$C_{13}H_9NO_3$

J.P. BIDEAU, G. BRAVIC and M. COTRAIT, 1984. Acta Cryst., C40, 1948-1950.

Monoclinic, $P2_1/n$, a = 7.703, b = 19.080, c = 7.255 Å, β = 97.17°, D_m = 1.44, Z = 4. Cu radiation, R = 0.044 for 1697 reflexions.

The naphthofuran ring (Fig. 1) is essentially planar with O(22) and O(23) 0.17 and -0.18 Å out of the plane. The structure is stabilised by CH...O interactions (2.50-2.63 Å) and by π orbital overlap between parallel molecules related by a symmetry centre; the interplanar distance is 3.42 Å.

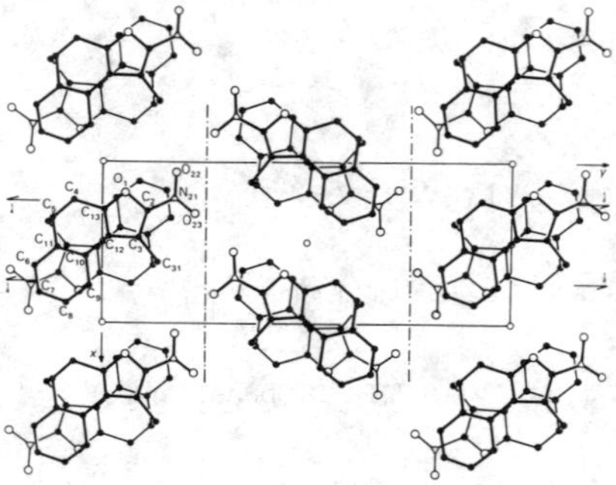

Fig. 1. Projection of the 1-methyl-2-nitronaphtho[2,1-b]furan structure along the z axis.

8-METHOXY-2-NITRO-NAPHTHO[2,3-b]FURAN
$C_{13}H_9NO_4$

J.-P. BIDEAU, G. GRAVIC, M. COTRAIT and C. COURSEILLE, 1984. Acta Cryst., C40, 322-324.

Triclinic, $P\bar{1}$, a = 7.175, b = 7.411, c = 11.340 Å, α = 89.10, β = 107.58, γ = 108.63°, D_m = 1.48, Z = 2. Cu radiation, R = 0.047 for 1435 reflexions.

Molecular dimensions are as expected. The naphthofuran ring system is close to being planar. Molecules are stacked in columns with a quasi-regular spacing of 3.35 Å. There are C-H...O contacts between columns (O...C 3.33, 3.49 Å).

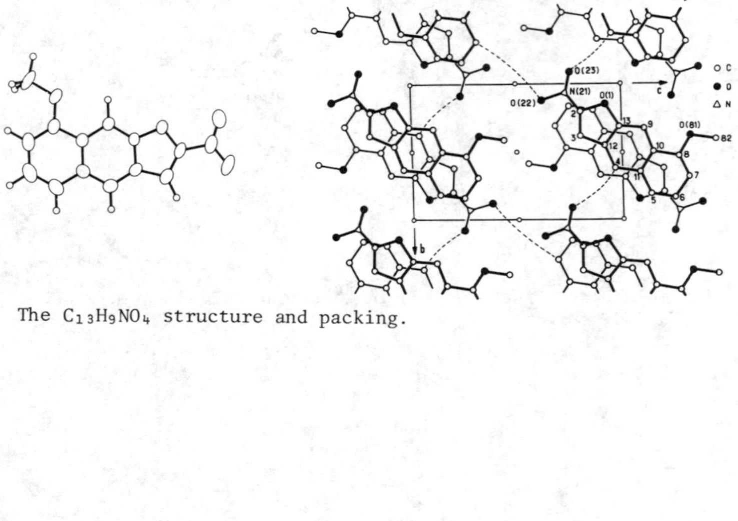

Fig. 1. The $C_{13}H_9NO_4$ structure and packing.

5-BENZYLIDENE-3-ACETYL-2,4-OXALANEDIONE
$C_{13}H_{10}O_4$

MAZHAR-UL-HAQUE, P. POLLET, J. AHMED and W. HORNE, 1984. J. Cryst. Spectrosc. Res., 14, 495-503.

Monoclinic, $P2_1/n$, a = 14.923, b = 4.165, c = 18.097 Å, β = 106.40°, Z = 4. Mo radiation, R = 0.048 for 1313 reflexions.

The five-membered oxolan ring and the phenyl ring are individually planar; the molecule as a whole approximates to a plane and this together with the bond lengths supports the view of a highly delocalized system. Molecules are linked to form centrosymmetric dimers by bifurcated O-H...O hydrogen bonds.

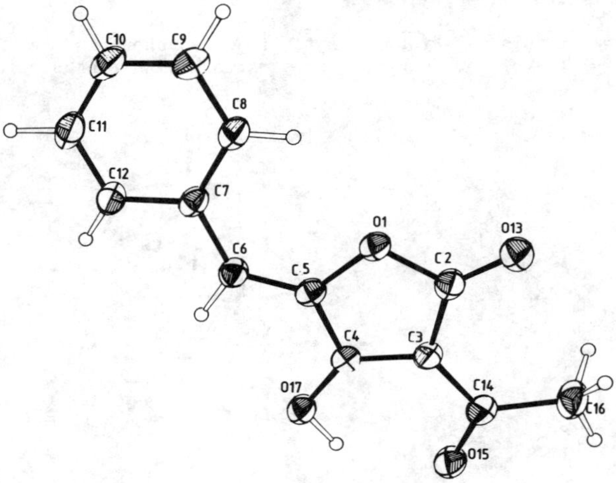

Fig. 1. A view of $C_{13}H_{10}O_4$.

5-ACETYL-5-METHYL-2-PHENYL-1,3-DIOXANE
$C_{13}H_{16}O_3$ (I)

5-ACETYL-5-METHYL-2-(2-CHLOROPHENYL)-1,3-DIOXANE
$C_{13}H_{15}ClO_3$ (II)

M. LAING, P. SOMMERVILLE, S.E. DREWES and M.W. DREWES, 1984. S. Afr. J. Chem., 37, 27-31.

I. Orthorhombic, $P2_12_12_1$, a = 6.020, b = 10.329, c = 18.612 Å, D_m = 1.26, Z = 4. Mo radiation, R = 0.070 for 781 reflexions.

II. Orthorhombic, $P2_12_12_1$, a = 9.684, b = 10.500, c = 12.601 Å, D_m = 1.32, Z = 4. Mo radiation, R = 0.072 for 1217 reflexions.

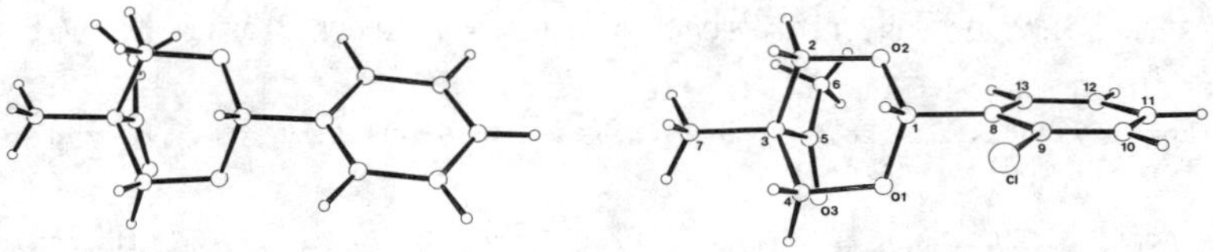

Fig. 1. Views of $C_{13}H_{16}O_3$ (left) and $C_{13}H_{15}ClO_3$ projected onto the O(1),O(2),C(2)-C(4) planes.

In both I and II, the dioxane rings are in chair conformation (Fig. 1) with the phenyl and methyl groups equatorial and the acetyl group axial. Incorporation of the o-chloro atom in II causes the dihedral angle between the phenyl ring plane and the O(1),O(2),C(2),C(4) plane to change from the 42° value in I to 69° in II.

(-)-2,6-DIOXO-6β-HYDROXY-1β-METHYLBICYCLO[3.3.1]NONYL-6-3'-PROPIONIC ACID LACTONE
$C_{13}H_{16}O_4$

B. CARTER, M. LAING and P. SOMMERVILLE, 1984. S. Afr. J. Chem., <u>37</u>, 53-56.

Orthorhombic, $P2_12_12_1$, a = 7.702, b = 11.190, c = 13.593 Å, Dm = 1.35, Z = 4. Mo radiation, R = 0.141 for 1194 reflexions.

The structure of this γ-lactone is established as shown in Fig. 1. The two six-membered rings have distorted chair conformations with ring B being somewhat flattened at C(3).

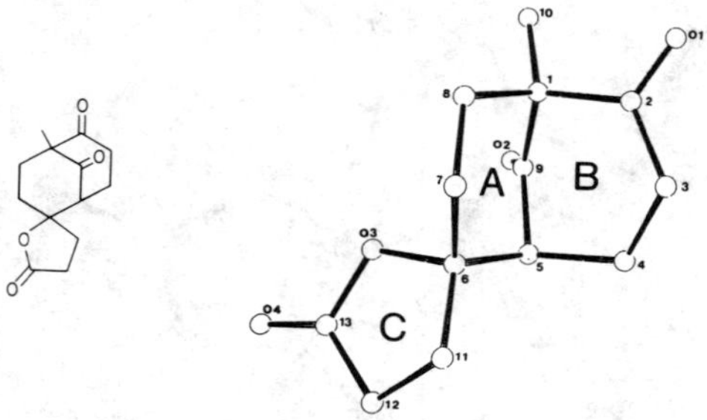

Fig. 1. The $C_{13}H_{16}O_4$ molecule.

(E)-1,2-BIS(2-HYDROXYPHENYL)ETHENE
$C_{14}H_{12}O_2$ (I)

(E)-1:2,5:6-DIBENZO-7,13-DIOXACYCLOTRIDECA-1,3,5-TRIENE
$C_{19}H_{20}O_2$ (II)

(E)-1:2,5:6,9:10-TRIBENZO-3,4-DIMETHYL-7,12-DIOXACYCLODODECA-1,3,5,9-TETRAENE
$C_{24}H_{22}O_2$ (III)

(Z)-1:2,5:6-DIBENZO-7,10,13-TRIOXACYCLOTRIDECA-1,3,5-TRIENE
$C_{18}H_{18}O_3$ (IV)

(Z)-1:2,5:6-DIBENZO-3,4-DIMETHYL-7,10,13-TRIOXACYCLOTRIDECA-1,3,5-TRIENE
$C_{20}H_{22}O_3$ (V)

(Z)-1:2,5:6,9:10-TRIBENZO-3,4-DIMETHYL-7,12-DIOXACYCLODODECA-1,3,5,9-TETRAENE
$C_{24}H_{22}O_2$ (VI)

J. TIRADO-RIVES, M.A. OLIVER, F.R. FRONCZEK and R.D. GANDOUR, 1984. J. Org. Chem., <u>49</u>, 1627-1634.

I. Orthorhombic, Pbca, a = 18.591, b = 11.392, c = 5.098 Å, Z = 4. Mo radiation, R = 0.044 for 514 reflexions.

II. Orthorhombic, Pbca, a = 22.369, b = 19.107, c = 7.211 Å, Z = 8. Mo radiation, R = 0.069 for 751 reflexions.

III. Orthorhombic, Pbca, a = 9.812, b = 22.150, c = 17.437 Å, Z = 8. Mo radiation, R = 0.101 for 1036 reflexions.

IV. Monoclinic, $P2_1$, a = 7.609, b = 7.956, c = 12.614 Å, β = 99.98°, Z = 2. Mo radiation, R = 0.050 for 1344 reflexions.

V. Orthorhombic, $P2_12_12_1$, a = 11.510, b = 18.511, c = 7.877 Å, Z = 4. Mo radiation, R = 0.042 for 1334 reflexions.

VI. Monoclinic, $P2_1/n$, a = 9.435, b = 17.329, c = 11.519 Å, β = 95.41°, Z = 4. Mo radiation, R = 0.062 for 896 reflexions.

The six molecules are shown in Fig. 1. Molecule I lies about an inversion centre and the central olefin moiety is strictly planar. The macrocycles II-VI all have different conformations determined by the E or Z geometry of the olefinic bonds. With these structures, and others, enough data have been made available to confirm previous criteria used to identify by NMR the E and Z isomers of diphenylethenes and butenes.

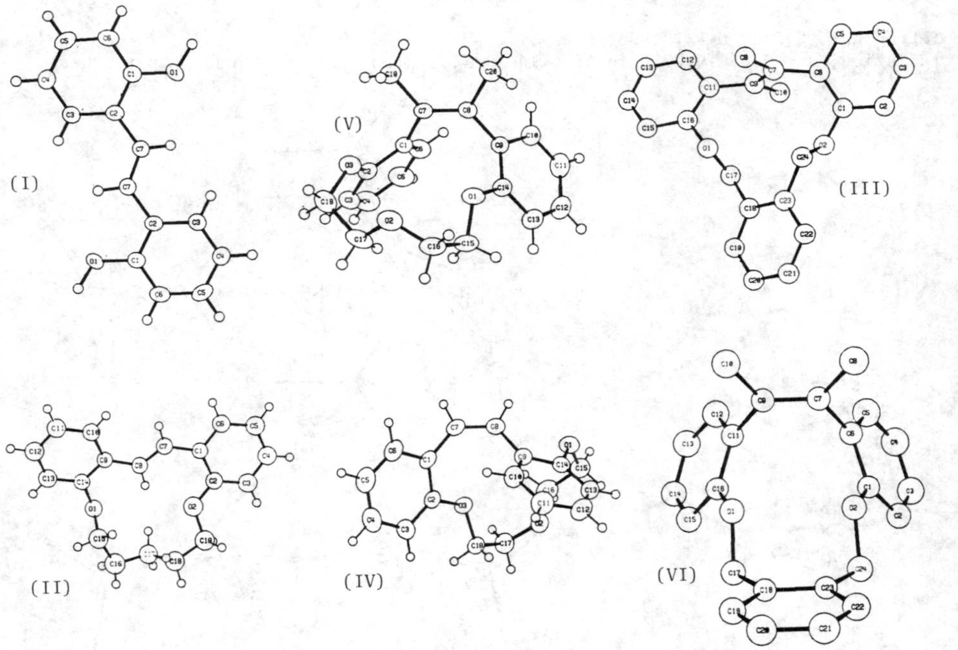

Fig. 1. Views of the six olefin systems showing their conformations.

3-ACETYL-5-(α-METHYLBENZYLIDENE)-2,4-OXOLANEDIONE
$C_{14}H_{12}O_4$

MAZHAR-UL-HAQUE, P. POLLET, J. AHMED and W. HORNE, 1984. Acta Cryst., C40, 1276-1278.

Monoclinic, $P2_1/n$, a = 4.763, b = 9.785, c = 24.955 Å, β = 92.79°, Z = 4. Mo radiation, R = 0.043 for 991 reflexions.

The molecule (Fig. 1) exists as the endo-enolic tautomer. It is essentially planar with the phenyl group cis to the lactone moiety. There is an intramolecular O-H...O hydrogen bond of 2.562(4) Å.

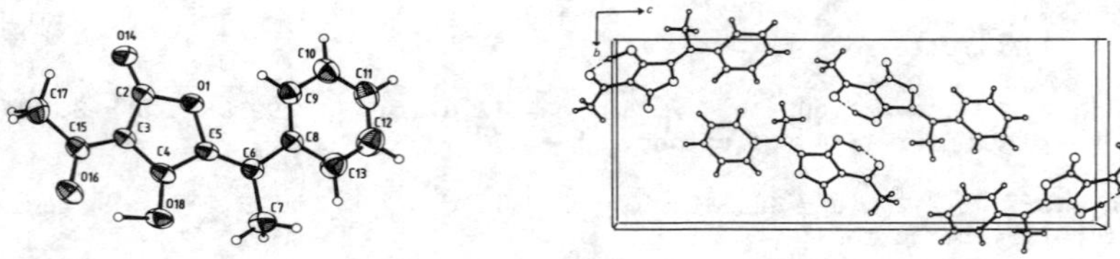

Fig. 1. $C_{14}H_{12}O_4$: perspective view of the molecule and a view of the unit cell.

[2]PARACYCLO[2](2,5)FURANOPHANE

$C_{14}H_{14}O$

F. HAMA, Y. KAI, N. YASUOKA and N. KASAI, 1984. Bull. Chem. Soc. Japan, 57, 3552-3556.

Tetragonal, $P\bar{4}2_1c$, a = 15.996(15.823), c = 8.374(8.330) Å, D_m = 1.23, Z = 8. Cu(Mo) radiation, R = 0.069(0.075) for 816(1461) reflexions (at 20°C and, in parentheses, at -150°C).

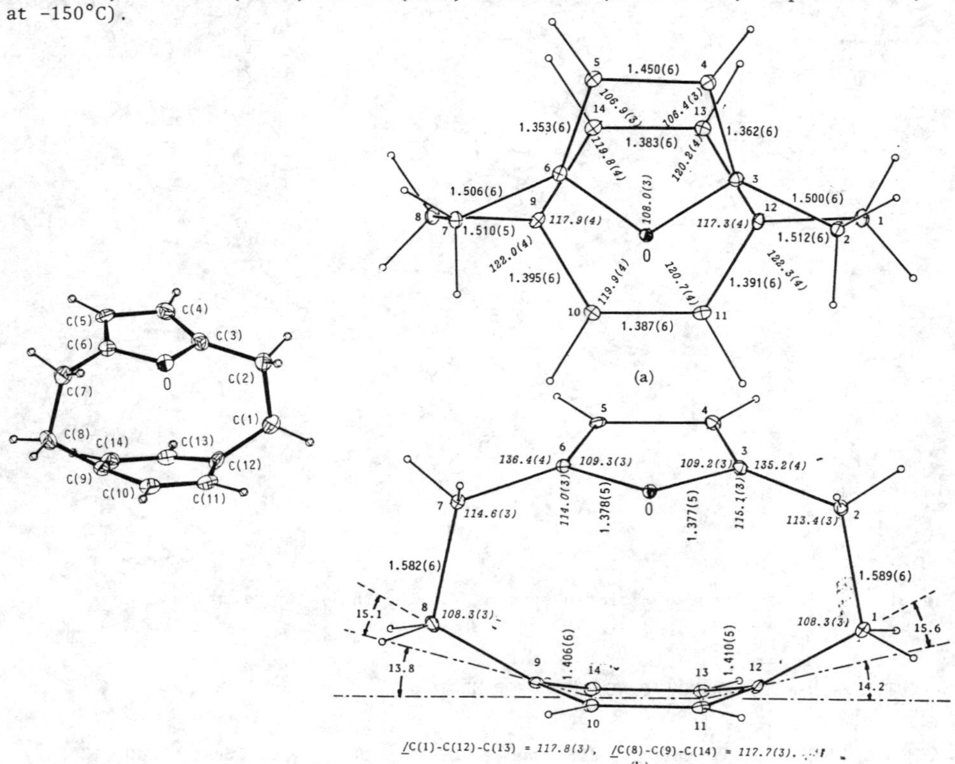

Fig. 1. Molecular structure of $C_{14}H_{14}O$ (at -150°C) (left), and bond lengths, bond angles and dihedral angles (right).

The molecule (Fig. 1) has approximate mirror symmetry and there are no essential differences between the structures observed at 20°C and -150°C. The furan ring takes an envelope form with the O atom away from the benzene ring, the dihedral angle between the planes made by the O, C(3) and C(6) atoms and the C(3), C(4), C(5) and C(6) being 3.3°. The benzene ring adopts a symmetrical boat form; the C(9) and

C(12) atoms deviate from the plane defined by C(10), C(11), C(13) and C(14) by 13.8 and 14.2° respectively (Fig. 1).

(2S,4R)-trans- AND (2R,4R)-cis-2-HYDROXY-2,4-DIMETHYL-3,4-DIHYDRO-2H,5H-PYRANO[3,2-c][1]BENZOPYRAN-5-ONE

$C_{14}H_{14}O_4$

E.J. VALENTE and V. SCHOMAKER, 1984. Acta Cryst., C40, 1068-1070.

Monoclinic, $P2_1$, a = 7.804, b = 18.248, c = 8.752 Å, β = 99.59°, Z = 4. Mo radiation, R = 0.066 for 1123 reflexions.

The crystal structure (Fig. 1) consists of diastereoisomers approximately inversion-related through a pseudocentre of symmetry at x = 0.241(4), z = 0.281(3) except the 4-methyl groups. The dihydropyran rings are half-chairs distorted towards the e,f-diplanar conformation. Like molecules are hydrogen-bonded between hydroxyl and carbonyl groups along a, O...O distances being 2.781(7) (trans) and 2.780(7) Å (cis).

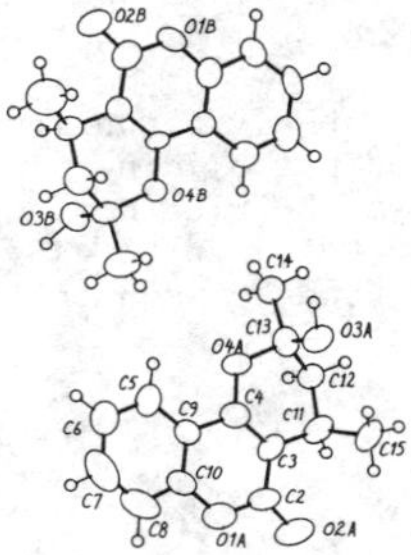

Fig. 1. $C_{14}H_{14}O_4$: views of the two molecules in projection along c, showing the near inversion symmetry.

2-endo-ACETYL-1,4-EPOXY-6,7-DIMETHOXY-1,2,3,4-TETRAHYDRONAPHTHALENE

$C_{14}H_{16}O_4$ (I)

2-endo-CARBOMETHOXY-1,4-EPOXY-6,7-DIMETHOXY-1,2,3,4-TETRAHYDRONAPHTHALENE

$C_{14}H_{16}O_5$ (II)

N.J. TAYLOR, B.A. KEAY and R. RODRIGO, 1984. Can. J. Chem., 62, 1099-1102.

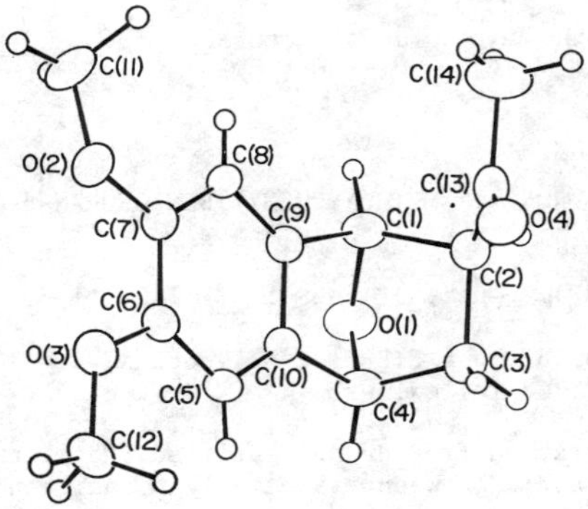

Fig. 1. A view of $C_{14}H_{16}O_4$ (I) (molecule II has the CH_3CO group of I replaced by CH_3CO_2).

I. Orthorhombic, $P2_12_12_1$, a = 5.524, b = 8.929, c = 25.479 Å, Z = 4. Mo radiation, R = 0.041 for 875 reflexions.

II. Monoclinic, $P2_1/c$, a = 10.878, b = 6.634, c = 18.224 Å, β = 92.76°, Z = 4. Mo radiation, R = 0.036 for 1327 reflexions.

Steric strain in these molecules (Fig. 1) is accommodated by bond angle distortions (e.g. C1-C9-C8 134.3° (I), 133.7° (II); C4-C10-C5 133.8° (I), 133.6° (II)). The C1-O4-C4 angles are 96.2° (I) and 96.3° (II). Bond lengths are as anticipated.

ETHYL (1SR,5RS,6SR,8RS)-6-ETHOXYCARBONYLAMINO-8-HYDROXYMETHYL-3,3-DIMETHYL-2,4-DIOXA-7-AZABICYCLO[3.3.0]OCTANE-7-CARBOXYLATE
$C_{14}H_{24}N_2O_7$

A.K. FORREST, R.R. SCHMIDT, G. HUTTNER and I. JIBRIL, 1984. J. Chem. Soc. Perkin I, 1981-1987.

Monoclinic, $P2_1/c$, a = 7.675, b = 24.063, c = 9.288 Å, β = 99.83°, Z = 4. Mo radiation, R = 0.066 for 1437 reflexions.

The X-ray analysis indicated an α-D-lyxo configuration for the molecule (Fig. 1). The conformation of the five-membered pyrrolidine ring (N(1)-C(7)-C(4)-C(5)-C(6)) is a distorted envelope with the pucker at the ring N atom. The endo- and exocyclic C-N bonds have lengths of 1.471(6) (C(6)-N(1)) and 1.446(7) Å (C(6)-N(2)) respectively. There is an intramolecular hydrogen bond between the alcohol and the N(1) amide carbonyl (O(5)-H...O(6) 1.65 Å).

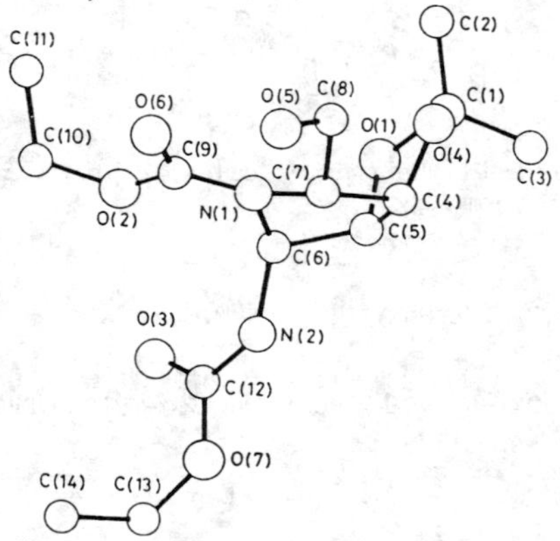

Fig. 1. $C_{14}H_{24}N_2O_7$: the molecular structure.

CENTROLOBOFURAN (2-(2'-HYDROXY-4'-METHOXYPHENYL)BENZOFURAN-6-OL)
$C_{15}H_{12}O_4$

L. JURD and R.Y. WONG, 1984. Aust. J. Chem., 37, 1127-1133.

Monoclinic, $P2_1/n$, a = 13.287, b = 5.287, c = 17.883 Å, β = 103.60°, Z = 4. Cu radiation, R = 0.049 for 1268 reflexions.

The compound was extracted from an unidentified Brazilian Centrolobium heartwood. There is a 5.3° dihedral angle between essentially planar benzofuran and methoxy phenyl rings (Fig. 1).

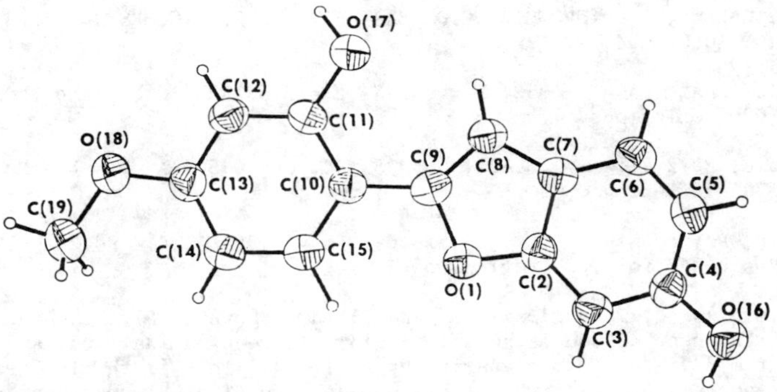

Fig. 1. The structure of centrolobofuran.

endo-N-METHYL-3,7-EPOXY-3-METHOXY-7-OXO-1,2-BENZO-1-CYCLOHEPTENE-4,5-DICARBOXIMIDE
$C_{15}H_{13}NO_5$ (I)

exo-N-METHYL-3,7-EPOXY-3-METHOXY-7-OXO-1,2-BENZO-1-CYCLOHEPTENE-4,5-DICARBOXIMIDE
$C_{15}H_{13}NO_5$ (II)

H. TAMURA, T. IBATA and K. OGAWA, 1984. Bull. Chem. Soc. Japan, 57, 926-931.

I. Orthorhombic, $P2_12_12_1$, a = 14.432, b = 11.874, c = 7.762 Å, D_m = 1.439, Z = 4. Cu radiation, R = 0.047 for 1160 reflexions.

II. Monoclinic, $P2_1/a$, a = 13.381, b = 12.825, c = 7.4793 Å, β = 92.13°, D_m = 1.485, Z = 4. Cu radiation, R = 0.067 for 1986 reflexions.

The corresponding bond distances (Fig. 1) and angles in the endo- and exo-adducts agree fairly well with each other. The difference between the O(1)-C(4) and O(1)-C(7) bond distances in the endo- and exo-adducts can be regarded as significant, and the intramolecular distances between O(1) and Me(15) are shorter than the van der Waals distance, i.e. they are 2.749 Å in the endo-adduct and 2.792 Å in the exo-adduct.

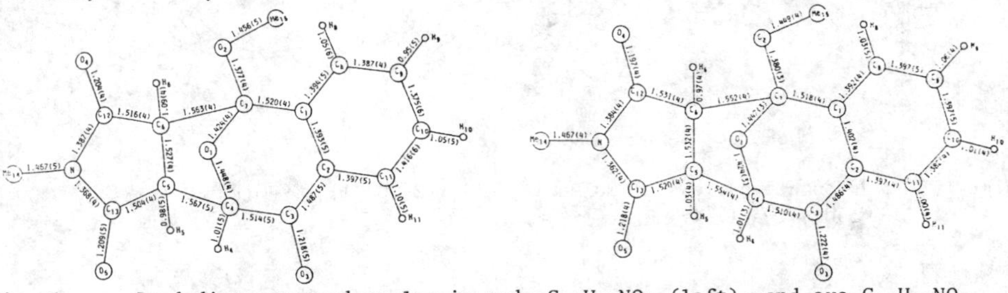

Fig. 1. Bond distances and angles in endo-$C_{15}H_{13}NO_5$ (left), and exo-$C_{15}H_{13}NO_5$ (right).

syn-OXABENZOSESQUINORBORNENE
$C_{15}H_{14}O$ (I)

syn-OXABENZOSESQUINORBORNENE DICHLOROCARBENE ADDUCT
$C_{16}H_{14}Cl_2O$ (II)

syn-OXABENZOSESQUINORBORNENE ANTHRANILIC ACID ADDUCT
$C_{23}H_{18}N_2O_3$ (III)

W.H. WATSON, J. GALLOY, D.A. GROSSIE, P.D. BARTLETT and G.L. COMBS JR, 1984. Acta Cryst., C40, 1050-1054.

I. Orthorhombic, $Pna2_1$, a = 21.552, b = 6.503, c = 7.652 Å, Z = 4. Mo radiation, R = 0.046 for 1026 reflexions.

II. Orthorhombic, $P2_12_12_1$, a = 13.821, b = 28.184, c = 6.561 Å, Z = 8. Mo radiation, R = 0.060 for 1795 reflexions.

III. Monoclinic, $P2_1/c$, a = 12.101, b = 18.658, c = 7.584 Å, β = 90.26°, Z = 4. Cu radiation, R = 0.043 for 1897 reflexions.

Perspective views of molecules I, II and III are shown in Fig. 1. In I the two carbon atoms of the central double bond (1.339(3) Å) are pyramidalised, and the double-bond system deviates from planarity by 22.1(2)°. This is the largest deviation yet reported for this type of distortion. In II dichlorocarbene has added to the exo-face of I. The two independent molecules have short intramolecular contacts, Cl...O 2.567(6) and 2.568(6) Å and Cl...H 2.05 and 2.14 Å. Carbon atoms forming the C(2)-C(3) bond in the phenyl ring are slightly pyramidalised but in a direction opposite to that normally observed in sesquinorbornene-type systems. This is attributed to an attractive interaction between the endo hydrogens H(6) and H(7) and the π system of C(2) and C(3). The four C(sp^2)...H distances in the two independent molecules range from 2.18 to 2.27 Å. Molecule III is formed by the reaction of I with anthranilic acid. By a yet not understood mechanism an N-formylindazolone moiety is formally added to the reactive double bond. The resulting tricyclic fragment is forced through intramolecular contacts to be planar which leads to extensive electron delocalisation. The C(2) and C(3) atoms of the phenyl ring are pyramidalised, and the associated π system deviates from planarity by 3.4(3)°.

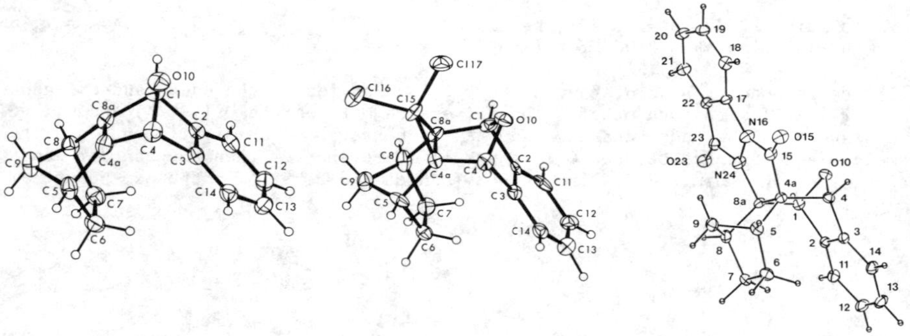

Fig. 1. Perspective views of molecules of $C_{15}H_{14}O$ (I), $C_{16}H_{14}Cl_2O$ (II) and $C_{23}H_{18}N_2O_3$ (III).

(1R,4R,5S,8S)-9,10-DIMETHYLIDENETRICYCLO[6.2.1.0^{2,7}]UNDEC-2(7)-ENE-4,5-DICARBOXYLIC ANHYDRIDE
$C_{15}H_{14}O_3$ (I)

(1R,4R,5S,8S)-11-ISOPROPYLIDENE-9,10-DIMETHYLIDENETRICYCLO[6.2.1.0^{2,7}]UNDEC-2(7)-ENE-4,5-DICARBOXYLIC ANHYDRIDE
$C_{18}H_{18}O_3$ (II)

(1R,4R,5S,8S)-9,10-DIMETHYLIDENETRICYCLO[6.2.2.0^{2,7}]DODEC-2(7)-ENE-4,5-DICARBOXYLIC ANHYDRIDE
$C_{16}H_{16}O_3$ (III)

(1R,4R,5S,8S)-TRICYCLO[6.2.2.0^{2,7}]DODECA-2(7),9-DIENE-4,5-DICARBOXYLIC ANHYDRIDE
$C_{14}H_{14}O_3$ (IV)

(4R,5S)-TRICYCLO[6.1.1.$0^{2,7}$]DEC-2(7)-ENE-4,5-DICARBOXYLIC ACID
$C_{12}H_{14}O_4$ (V)

A.A. PINKERTON, D. SCHWARZENBACH, J.-L. BIRBAUM, P.-A. CARRUPT, L. SCHWAGER and P. VOGEL, 1984. Helv. Chim. Acta, 67, 1136-1153.

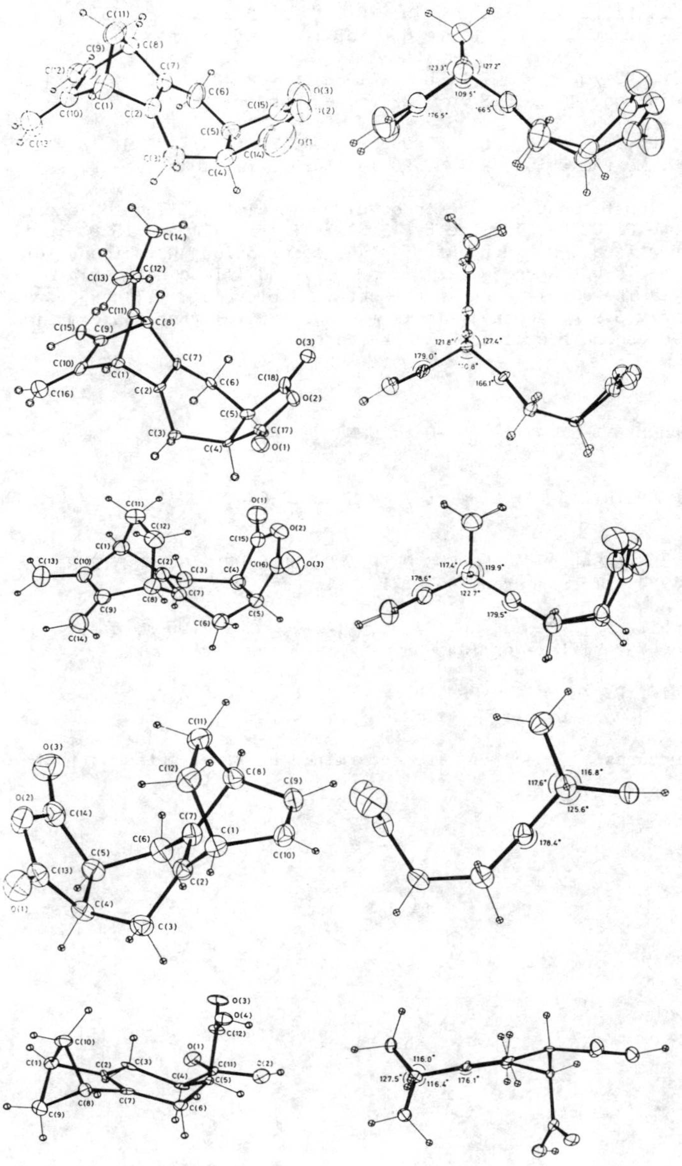

Fig. 1. Views (from top to bottom) of $C_{15}H_{14}O_3$ (I), $C_{18}H_{18}O_3$ (II), $C_{16}H_{16}O_3$ (III), $C_{14}H_{14}O_3$ (IV) and $C_{12}H_{14}O_4$ (V).

I. Monoclinic, $P2_1/a$, a = 10.915, b = 8.879, c = 13.517 Å, β = 106.43°, Z = 4. Mo radiation, R = 0.041 for 899 reflexions.

II. Orthorhombic, $Pbc2_1$, a = 6.871 (6.792), b = 13.240 (13.073), c = 32.199 (31.824) Å, Z = 8. Mo radiation, R = 0.059 (0.049) for 1807 (1035) reflexions. (Data in parentheses refer to 115 K.)

III. Triclinic, $P\bar{1}$, a = 6.552, b = 10.525, c = 11.042 Å, α = 115.11, β = 94.98, γ = 106.99°, Z = 2. Mo radiation, R = 0.036 for 1720 reflexions.

IV. Monoclinic, $P2_1/m$, a = 5.790, b = 11.508, c = 8.400 Å, β = 101.84°, D_m = 1.40, Z = 2. Mo radiation, R = 0.038 for 628 reflexions.

V. Triclinic, $P\bar{1}$, a = 6.683, b = 7.993, c = 10.739 Å, α = 75.91, β = 75.75, γ = 76.08°, Z = 2. Mo radiation, R = 0.090 for 1025 reflexions (at 120 K).

The alkyl substituents of the endocyclic bicyclo[2.2.1]hept-2-ene double bond deviate from the C1,C2,C3,C4 plane by 13.5° in I (Fig. 1) and 13.9° in II, leaning towards the endo face. No such out of plane deformations were found for III and IV. The exocyclic s-cis-butadiene moieties in I, II and III do not deviate significantly from planarity. The double bond in the strained bicyclo[2.1.1]hex-2-ene deviates from planarity by only 3.9°. An analysis is made of the relevant conformations of bicyclo-[2.2.1]hept-2-enes and bicyclo[2.2.2]oct-2-enes.

4-t-BUTYL-3-PHENYL-3-ETHAN-1-OL-2-OXETANONE
$C_{15}H_{20}O_3$ (I)

(2R*,3'R*,4'S*)-METHYL 2-(4'-t-BUTYL-3'-PHENYL-2'-OXETANON-3'-YL)-4-PENTENOATE
$C_{19}H_{24}O_4$ (II)

J. MULZER, P. DE LASALLE, A. CHUCHOLOWSKI, U. BLASCHEK, G. BRUNTRUP, I. JIBRIL and G. HUTTNER, 1984. Tetrahedron, 40, 2211-2218.

I. Orthorhombic, $P2_12_12_1$, a = 6.404, b = 14.22, c = 14.96 Å, Z = 4. Mo radiation, R = 0.0578 for 782 reflexions (at -35°C).

II. Monoclinic, $P2_1/c$, a = 9.059, b = 11.91, c = 18.86 Å, β = 119.14°, Z = 4. Mo radiation, R = 0.0586 for 1371 reflexions.

The structures of I and II were determined by X-ray methods; that for II is shown in Fig. 1.

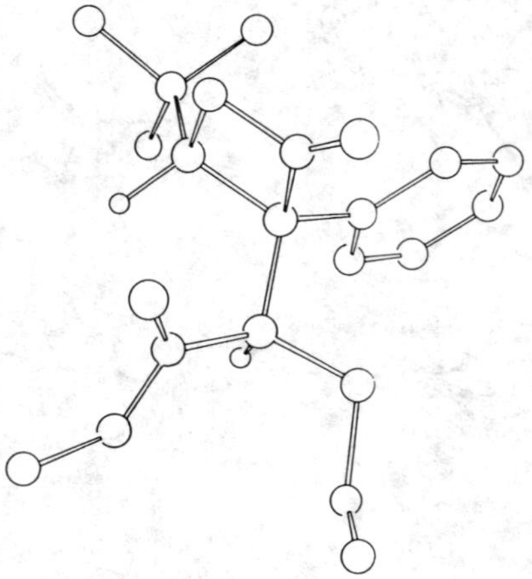

Fig. 1. A view of $C_{19}H_{24}O_4$.

[(6'RS,8'RS,2E),(6'RS,8'SR,2E)]-3-METHYL-3-(2',2',6'-TRIMETHYL-7'-OXABICYCLO[4.3.0]-NON-9'-EN-8'-YL)-2-PROPENAL
$C_{15}H_{22}O_2$

M. ACEMOGLU, R. PREWO, J.H. BIERI and C.H. EUGSTER, 1984. Helv. Chim. Acta, 67, 175-183.

Monoclinic, $P2_1/c$, a = 7.572, b = 14.894, c = 12.421 Å, β = 100.58°, Z = 4. Mo radiation, R = 0.090 for 4022 reflexions.

The five-membered ring (Fig. 1) has an envelope conformation and the six-membered ring has a deformed chair conformation.

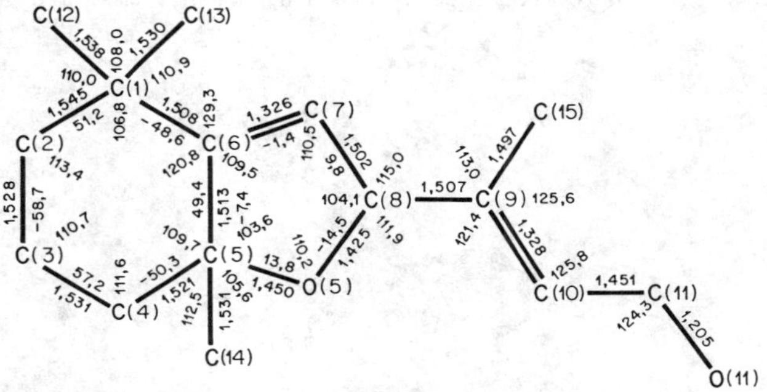

C(2),C(1),C(13) 109,1° C(4),C(5),O(5) 109,7°

C(6),C(1),C(12) 112,0° C(6),C(5),C(14) 115,1°

Fig. 1. Bond lengths, angles and torsion angles in $C_{15}H_{22}O_2$ (σ 0.002-0.003 Å, 0.1-0.2°).

7,8-BENZO-2,5-ETHENOBICYCLO[4.2.0]OCTANE-3,4-DICARBOXYLIC ANHYDRIDE
$C_{16}H_{12}O_3$

M.-E. GUNTHER, R. AYDIN, W. BUCHMEIER, B. ENGELEN and H. GUNTHER, 1984. Chem. Ber., 117, 1069-1076.

Rhombohedral, $R\bar{3}$, a = 30.66, c = 6.63 Å, Z = 18. Mo radiation, R = 0.041 for 679 reflexions.

The structure of this Diels-Alder adduct was determined to be as shown in Fig. 1. The interplanar angle between the aromatic and four-membered rings is 2.6°.

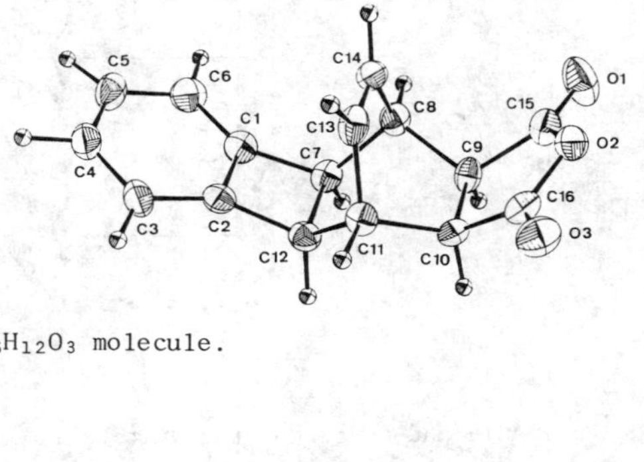

Fig. 1. The $C_{16}H_{12}O_3$ molecule.

4-BENZOYL-4-PHENYL-1,1-DIOXOLANE
$C_{16}H_{14}O_3$

A.Ya. VAINER, V.A. TAFEENKO, I.N. POLYAKOVA and K.M. DYUMAEV, 1984. J. Struct. Chem., 25, 818-820 [Zh. Strukt. Khim., 25(5), 163-165].

Monoclinic, $P2_1/c$, a = 9.168, b = 8.741, c = 16.876 Å, β = 104.44°, Z = 4. Mo radiation, R = 0.036 for 1472 reflexions.

The molecule is shown in Fig. 1. The five-membered dioxolane ring has a distorted O(1)-envelope conformation with O(1) displaced 0.573(2) Å from the mean plane of the remaining ring atoms (which are coplanar to within 0.023(3) Å). Bond lengths (σ = 0.002-0.003 Å for bonds between non-hydrogen atoms) are generally as expected.

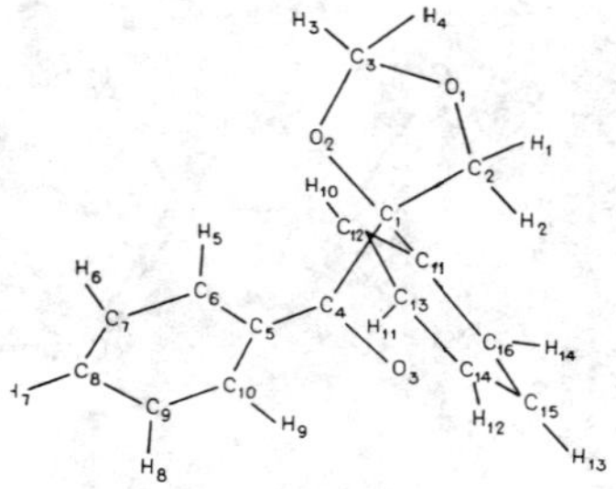

Fig. 1. The 4-benzoyl-4-phenyl-1,1-dioxolane molecule.

3-(N-DIETHYLAMINOETHYLTHIOCARBOXAMIDO)COUMARIN
$C_{16}H_{20}N_2O_2S$

N. RODIER, L. UZAN, P. REYNAUD and J.-D. BRION, 1984. Bull. Soc. Chim. Fr., I-317-320.

Triclinic, $P\bar{1}$, a = 6.757, b = 9.171, c = 13.133 Å, α = 97.49, β = 91.12, γ = 100.64°, Z = 2. Mo radiation, R = 0.055 for 1280 reflexions.

The structure (Fig. 1) contains layers of molecules parallel with $(10\bar{1})$; layers are linked by N-H...O hydrogen bonds (3.148 Å). In the molecule, the terminal ethyl groups are disordered; an intramolecular N-H...O hydrogen bond links N13 and O11 (N...O 2.665(4) Å).

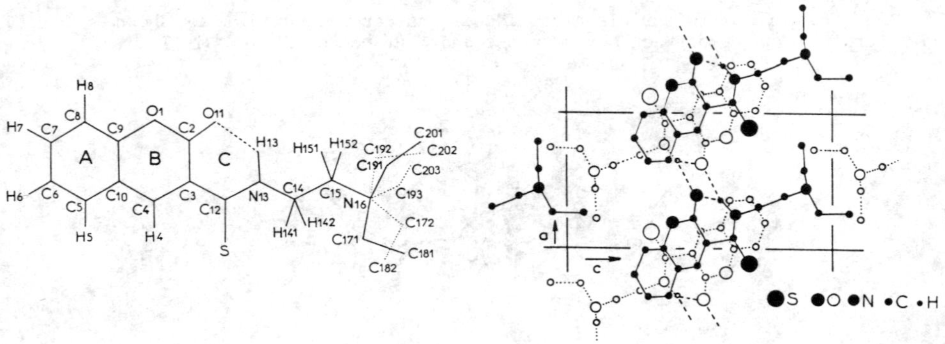

Fig. 1. The $C_{16}H_{20}N_2O_2S$ molecule and a view of the unit cell.

2-{8-HYDROXYBICYCLO[5.1.0]OCT-1(7)-EN-8-YL}-1-CYCLOHEPTENE-1-CARBOXYLIC ACID γ-LACTONE
$C_{16}H_{20}O_2$ (I)

2-{(1α,7α,8α)-8-HYDROXYBICYCLO[5.1.0]OCT-8-YL}-1-CYCLOHEPTENE-1-CARBOXYLIC ACID γ-LACTONE
$C_{16}H_{22}O_2$ (II)

C.M. WEEKS, W.L. DUAX, R.A. FINNEGAN, D.J. DELECKI and B. KOJIĆ-PRODIĆ, 1984. Acta Cryst., C40, 1376-1378.

I. Monoclinic, $P2_1/c$, a = 8.994, b = 14.955, c = 9.863 Å, β = 93.76°, Dm = 1.22, Z = 4. Cu radiation, R = 0.083 for 1080 reflexions.

II. Tetragonal, $P4_12_12$, a = 7.961, c = 42.919 Å, Dm = 1.23, Z = 8. Cu radiation, R = 0.052 for 1607 reflexions.

II is the hydrogenation product of I and the hydrogens added are on the same side of the cyclopropyl ring as the lactone group (Fig. 1). The conformations of the rings are almost identical in both molecules with the seven-membered rings in the chair form. There are no short intermolecular contacts.

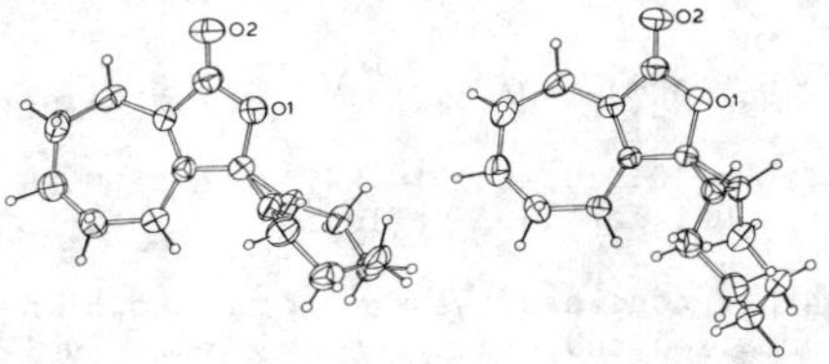

Fig. 1. The $C_{16}H_{20}O_2$ and $C_{16}H_{22}O_2$ molecules.

ETHYL (3R)-4,4,4-TRIFLUORO-3-[(1'S,4'R)-3'-OXO-4',7',7'-TRIMETHYL-2'-OXABICYCLO[2.1.1]-HEPTANE-1'-CARBONYLOXY]BUTANOATE
$C_{16}H_{21}F_3O_6$

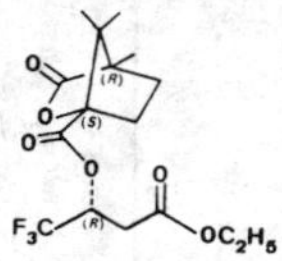

D. SEEBACH, P. RENAUD, W.B. SCHWEIZER, M.F. ZUGER and M.-J. BRIENNE, 1984. Helv. Chim. Acta, 67, 1843-1853.

Monoclinic, $P2_1$, a = 6.600, b = 9.329, c = 29.759 Å, β = 99.1°, Z = 4. Mo radiation, R = 0.050 for 2772 reflexions.

The analysis established the configuration as shown above.

(2S,8R)-8-METHYL-2-PHENYL-1,7-DIOXASPIRO[5.5]UNDECAN-4(R)-OL
$C_{16}H_{22}O_3$

A.M. DOHERTY, S.V. LEY, B. LYGO and D.J. WILLIAMS, 1984. J. Chem. Soc. Perkin I, 1371-1377.

Orthorhombic, F2dd, a = 8.749, b = 15.916, c = 42.316 Å, Z = 16. Cu radiation, R = 0.030 for 1045 reflexions.

The molecule has a spiro configuration with the C(2) phenyl and C(8) methyl equatorial and the C(4) hydroxyl axial (Fig. 1). There is an intramolecular hydrogen bond between O(12) and O(7) (2.77 Å). The structure is loosely packed with only one intermolecular contact less than 3.4 Å.

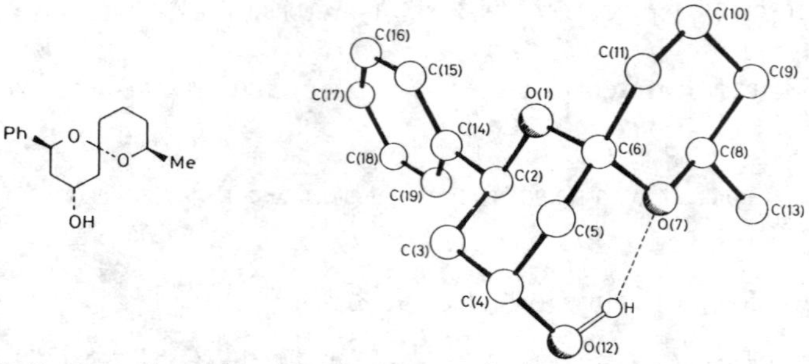

Fig. 1. $C_{16}H_{22}O_3$: the molecular skeleton and structure.

1,2,4,4,11,11-HEXAMETHYL-3,12-DIOXA-cis-anti-cis-TRICYCLO[6.4.0.0^{2,7}]DODECANE-6,9-DIONE
$C_{16}H_{24}O_4$

E. ANKLAM, W.A. KONIG and P. MARGARETHA, 1984. Z. Kristallogr., 169, 127-133.

Triclinic, P$\bar{1}$, a = 7.257, b = 10.410, c = 10.947 Å, α = 90.95, β = 96.70, γ = 99.87°, Z = 2. Cu radiation, R = 0.081 for 2392 reflexions.

This photodimer of 4-pyranone has the structure shown in Fig. 1. The cyclobutane ring has a puckered conformation and the pyranone rings have slightly distorted chair conformations with asymmetry parameters $\Delta C_s(8)$ 10.8°, $\Delta C_s(4)$ 10.5°.

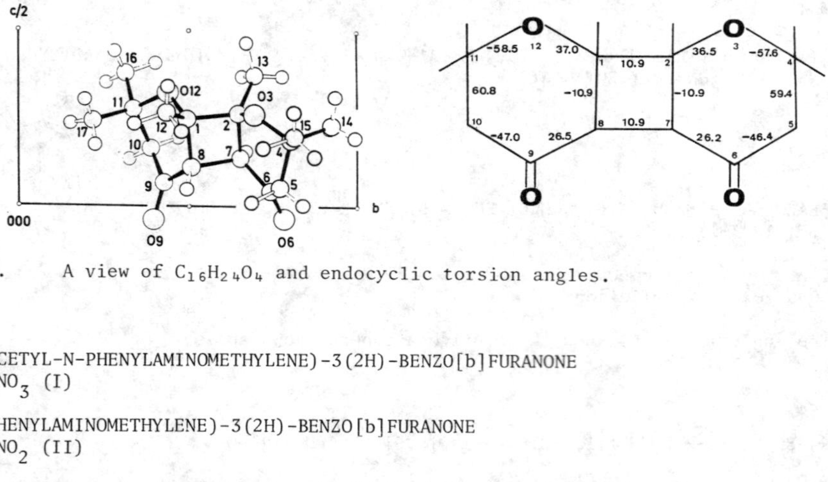

Fig. 1. A view of $C_{16}H_{24}O_4$ and endocyclic torsion angles.

2-(N-ACETYL-N-PHENYLAMINOMETHYLENE)-3(2H)-BENZO[b]FURANONE
$C_{17}H_{13}NO_3$ (I)

2-(N-PHENYLAMINOMETHYLENE)-3(2H)-BENZO[b]FURANONE
$C_{15}H_{11}NO_2$ (II)

S.M. ALDOSHIN, L.O. ATOVMYAN, O.A. D'YACHENKO, V.I. MINKIN, V.A. BREN', G.D. PALUI and Zh. V. BREN', 1984. J. Struct. Chem., 25, 93-98 [Zh. Strukt. Khim., 25(1), 106-112].

I. Monoclinic, $P2_1/n$, a = 15.047, b = 8.899, c = 10.532 Å, γ = 83.25°, Z = 4. Cu radiation, R = 0.058 for 835 reflexions.

II. Orthorhombic, $Pca2_1$, a = 26.262, b = 5.662, c = 7.676 Å, Z = 4. Cu radiation, R = 0.041 for 786 reflexions.

The structures of I and II are shown in Fig. 1. Both molecules have keto-amine structures with a Z-configuration about the C(1)=C(9) double bond. In I the N-phenyl substituent is rotated 87.2° out of the approximate plane of the remainder of the molecule, while molecules of II are essentially planar.

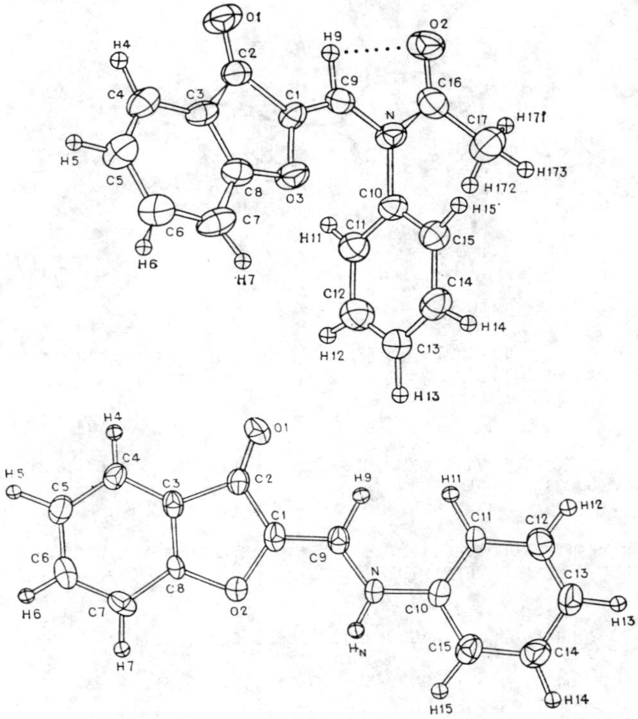

Fig. 1. The 2-(N-acetyl-N-phenylaminomethylene)-3(2H)-benzo[b]furanone (top) and 2-(N-phenylaminomethylene)-3(2H)-benzo[b]furanone (bottom) molecules.

(±)-trans-2,3-BIS(3-HYDROXYBENZYL)BUTAN-4-OLIDE (ENTEROLACTONE)
$C_{18}H_{18}O_4$

G. COOLEY, R.D. FARRANT, D.N. KIRK, S. PATEL, S. WYNN, M.J. BUCKINGHAM. G.E. HAWKES, M.B. HURSTHOUSE, A.M.R. GALAS, A.M. LAWSON and K.D.R. SETCHELL, 1984. J. Chem. Soc. Perkin II, 489-497.

Monoclinic, P2/n, a = 14.274, b = 10.065, c = 11.165 Å, β = 106.60°, Z = 4. Cu radiation, R = 0.048 for 2181 reflexions.

The X-ray analysis established the structure of enterolactone, a urinary lignan. Bond lengths and angles are normal. Principal ring torsion angles are shown in Fig. 1.

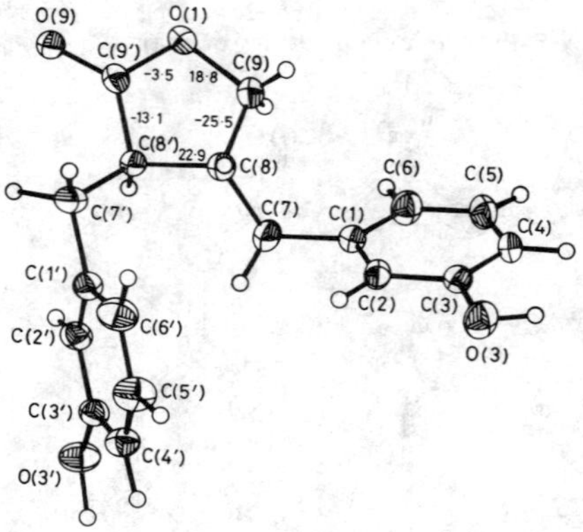

Fig. 1. $C_{18}H_{18}O_4$: a molecule of (8R,8'R) configuration showing principal ring torsion angles.

6-HYDROXY-5-[exo-7-(2-HYDROXY-5-OXOCYCLOPENTENYL)BICYCLO[4.1.0]-HEPT-7-YL]-2,2-DIMETHYL-4H-1,3-DIOXIN-4-ONE

$C_{18}H_{22}O_6$ (I)

6-HYDROXY-5-[exo-7-(2-HYDROXY-4,4-DIMETHYL-6-OXOCYCLOHEXENYL)BICYCLO[4.1.0]HEPT-7-YL]-1,3-DIMETHYL-2,4(1H,3H)PYRIMIDINEDIONE

$C_{21}H_{28}N_2O_5$ (II)

E. VILSMAIER, K. JOERG and G. MAAS, 1984. Chem. Ber., 117, 2947-2962.

I. Orthorhombic, $Pna2_1$, a = 11.003, b = 20.143, c = 7.611 Å, Z = 4. Mo radiation, R = 0.0451 for 1195 reflexions.

II. Triclinic, $P\bar{1}$, a = 9.323, b = 10.275, c = 11.814 Å, α = 99.86, β = 105.08, γ = 110.16°, Z = 2. Mo radiation, R = 0.048 for 2257 reflexions.

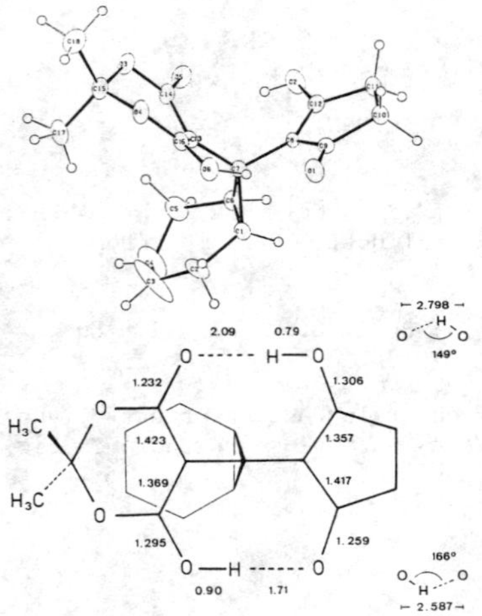

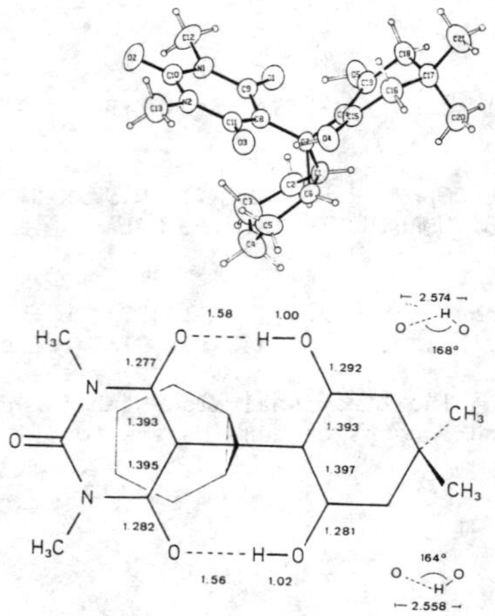

Fig. 1. The $C_{18}H_{22}O_6$ (I) and $C_{21}H_{28}N_2O_5$ (II) molecules and some details of dimensions.

The X-ray analyses show that in both (I) and (II) there are pairs of intramolecular nonsymmetric O-H...O hydrogen bonds. The β-diketonates connected by these hydrogen bonds have alternating bond order for (I) but non-alternating bond order for (II) (i.e. it exists as a dihydroxyallyl cation/enolate). Details of the structures and geometry are in Fig. 1.

8-ACETOXY-5-(5,5-DIMETHYL-1,3-DIOXACYCLOHEXAN-2-YL)-4,9-EPOXY-9-IODOMETHYL-2-METHOXY-3-OXABICYCLO[4.3.0]NONANE
$C_{18}H_{27}IO_7$

P. FLECKER, D. SCHOMBURG, P.W. THIES and E. WINTERFELDT, 1984. Tetrahedron, 40, 4843-4852.

Orthorhombic, $P2_12_12_1$, a = 14.102, b = 10.775, c = 26.789 Å, Z = 8. Cu radiation, R = 0.064 for 2424 reflexions.

No structural difference was found between the two independent molecules in the asymmetric unit; one of them is shown in Fig. 1. The dioxacyclohexane ring (C21-O26) has a chair conformation and that of the five-membered ring is between twist and envelope forms.

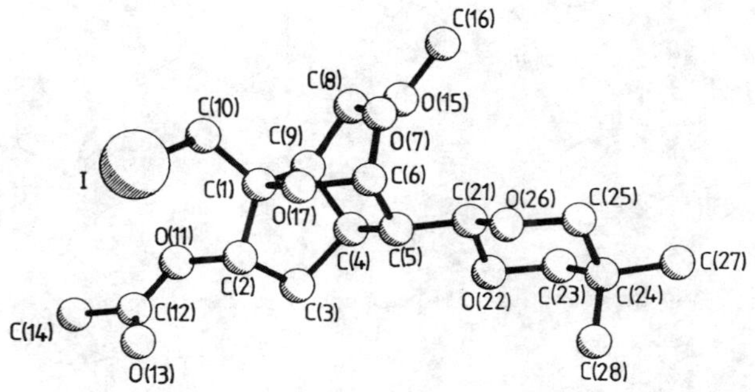

Fig. 1. A view of $C_{18}H_{27}IO_7$.

(cis-1,6,9,20-trans-12,17)-9,20-DIHYDROXY-7,11,18,22-TETRAOXA-TRICYCLO[$16.4.0.0^{12,17}$]-DOCOSANE (ISOMER A) (DICYCLOHEXANO-14-CROWN-4 3,14-DIOL)
$C_{18}H_{32}O_6$ (I)

(cis-1,6,9,12,17,20)-9,20-DIHYDROXY-7,11-18,22-TETRAOXATRICYCLO[$16.4.0.0^{12,17}$]DOCOSANE (ISOMER B) (DICYCLOHEXANO-14-CROWN-4 3,14-DIOL)
$C_{18}H_{32}O_6$ (II, III)

(cis-1,6,12,17-trans-9,20)-9,20-DIHYDROXY-7,11,18,22-TETRAOXATRICYCLO[$16.4.0.0^{12,17}$]-DOCOSANE (ISOMER C) (DICYCLOHEXANO-14-CROWN-4 3,14-DIOL)
$C_{18}H_{32}O_6$ (IV)

J.D. OWEN, J.A. BANDY, D.G. PARSONS, M.R. TRUTER and C.H.L. KENNARD, 1984. J. Chem. Soc. Perkin II, 129-137.

I. Monoclinic, $P2_1/n$, a = 12.010, b = 12.700, c = 13.290 Å, β = 112.78°, D_m = 1.219, Z = 4. Mo radiation, R = 0.041 for 2390 reflexions.

II, III. Monoclinic, $P2_1/c$, a = 8.999, 9.005, b = 17.580, 17.588, c = 12.682, 12.715 Å, β = 108.71, 108.79°, D_m = 1.202, Z = 4. Mo radiation, neutrons λ = 1.237 Å, R = 0.048, 0.064 for 3423, 1619 reflexions.

IV. Orthorhombic, $Pn2_1a$, a = 8.764, b = 18.298, c = 11.868 Å, D_m = 1.201, Z = 4. Mo radiation, R = 0.080 for 880 reflexions.

The four O atoms O(1), O(5), O(12), O(16) in each molecule (Fig. 1) are approximately coplanar. Isomer A has one cyclohexano ring on each side of this plane with the two hydroxy groups on the same side as the C(6)-C(11) ring. Isomer B has both cyclohexano rings and both hydroxy groups on the same side, while isomer C has both cyclohexano rings on one side and the hydroxy groups on the other. There are no significant differences for C-C, C-O and C-H distances from one isomer to the other. Isomer A has no molecular symmetry, while B and C have non-crystallographic twofold axes. The hydrogen bonding in each form correlates well with the i.r. spectra. In isomer B one hydroxy hydrogen does not take part in hydrogen bonding, the other gives a bond with an O...O distance of 2.925(2) Å (X-ray) or 2.935(10) Å (neutron). In isomers A and C each hydroxy hydrogen atom may occupy one of two sites yielding disordered hydrogen bonds of the type O-H...O or O...H-O; the O...O distances average 2.822(2) in A and 2.76(1) Å in C.

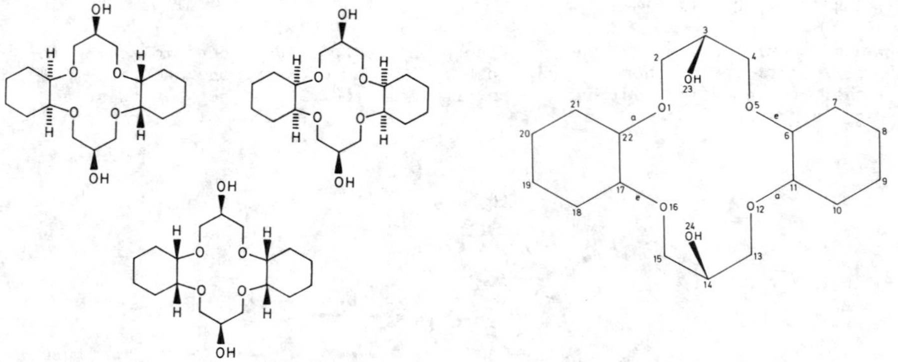

Fig. 1. $C_{18}H_{32}O_6$: the molecular skeletons and numbering scheme.

8-HYDROXY-1H-NAPHTHO[2,1,8-mna]XANTHEN-1-ONE
$C_{19}H_{10}O_3$ (I)

I.R. CASTLEDEN, L.M. ENGELHARDT, S.R. HALL and A.H. WHITE, 1984. Aust. J. Chem., 37, 1763-1768.

Monoclinic, $P2_1/c$, a = 5.128, b = 10.024, c = 27.45 Å, β = 107.73°, D_m = 1.51, Z = 4. Mo radiation, R = 0.056 for 892 reflexions, red needles with a green metallic lustre, brilliant fluorescence under UV, mp 346-348°. T = 295 K.

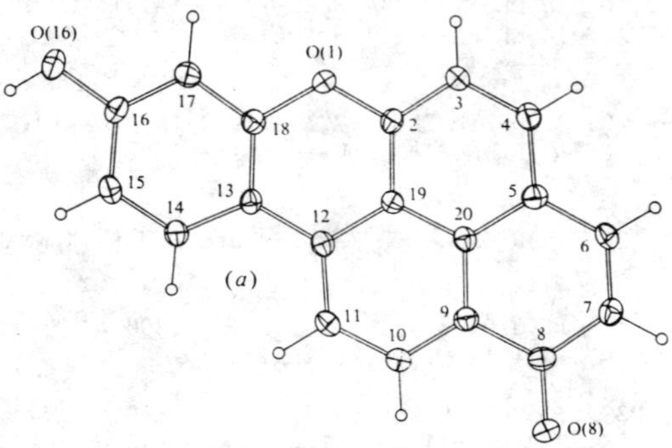

Fig. 1. The structure of $C_{19}H_{10}O_3$.

The molecule (Fig. 1) is planar with close intermolecular contacts (3.428-3.610 Å) with typical charge-transfer complex type packing. The cell contains an indeterminate amount of methanol.

cis-3-(p-CHLOROPHENYL)-10,10b-DIMETHYL-4a,7a-DIHYDRONAPHTHO[2,1-e][1,2,4]TRIOXIN-6(5H)-ONE
$C_{19}H_{17}ClO_4$

C.W. JEFFORD, D. JAGGI, S. KOHMOTO, J. BOUKOUVALAS and G. BERNARDINELLI, 1984. Helv. Chim. Acta, 67, 2254-2260.

Monoclinic, $P2_1/c$, a = 7.330, b = 6.193, c = 36.048 Å, β = 91.33°, Z = 4. Mo radiation, R = 0.091 for 930 reflexions.

The 1,2,4-trioxane ring has a chair conformation (Fig. 1), and is cis-fused to the benzocyclohexenone entity. The p-chlorophenyl and methyl substituents are in a 1,4-diequatorial arrangement.

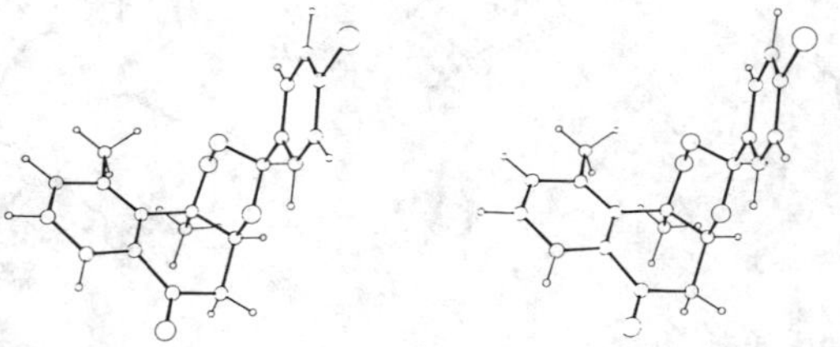

Fig. 1. A stereoview of $C_{19}H_{17}ClO_4$.

cis-3-(2,5-DIHYDROXYPHENYL)-5-HYDROXY-3-METHYL-2-(N-METHYLPROPIONAMIDO)-2,3-DIHYDRO-BENZO[b]FURAN ($^2H_6$)ACETONE
$C_{19}H_{21}NO_5,C_3D_6O$

Z. URBANCZYK-LIPKOWSKA, J.W. KRAJEWSKI, P. GLUZINSKI, L. KOZERSKI, G.D. ANDREETTI and G. BOCELLI, 1984. Acta Cryst., C40, 1750-1753.

Triclinic, $P\bar{1}$, a = 7.263, b = 15.272, c = 10.051 Å, α = 103.90, β = 81.23, γ = 88.53°, Z = 2. Cu radiation, R = 0.045 for 3039 reflexions.

In the compound (Fig. 1) the N-methylpropionamido and hydroquinone substituents situated at adjacent C atoms of the dihydrofuryl ring are almost eclipsed. The system of intermolecular hydrogen bonds in the crystal lattice consists of three types, two of which bond pairs of enantiomeric molecules into aggregates. The third type involves the acetone molecule. The aggregates form chains parallel to [011] in the crystal.

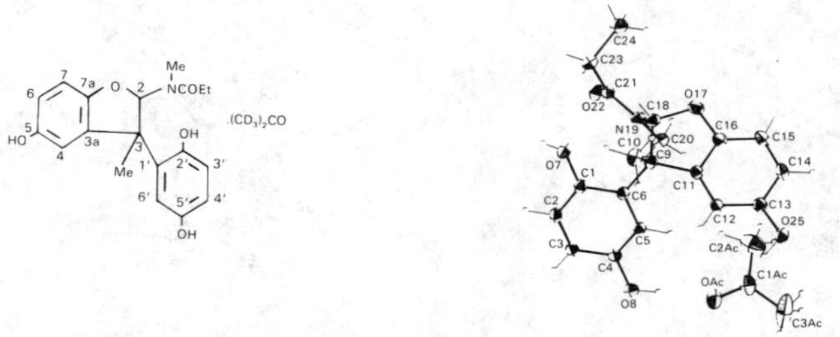

Fig. 1. The $C_{19}H_{21}NO_5.C_3D_6O$ structure.

9-(6,6-DIMETHYL-3-OXO-CYCLOHEX-1-EN-5-YL)-2-METHYL-8-OXO-2-VINYL-7-OXABICYCLO[4.3.0]-NONANE
$C_{19}H_{26}O_3$

R.Z. ANDRIAMIALISOA, M. FETIZON, I. HANNA, C. PASCARD and T. PRANGE, 1984. Tetrahedron, 40, 4285-4295.

Monoclinic, $P2_1/c$, a = 8.520, b = 13.063, c = 15.292 Å, β = 104.5°, Z = 4. R = 0.086.

In this molecule (a potential intermediate in taxane synthesis) (Fig. 1) the vinyl group at C(8) is in a axial disposition whereas the six-membered rings and the lactone display a cis junction (C(2)-C(3) : axial; C(4)-O(1) : equatorial). The cyclohexenone ring is linked via the C(1)-C(2) bond in equatorial position and is situated trans with respect to the vinyl at C(8). This leads to a unfavourable situation for cyclisation at C(11).

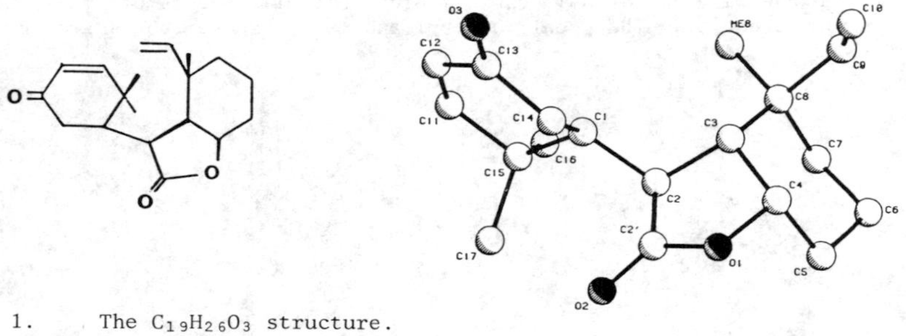

Fig. 1. The $C_{19}H_{26}O_3$ structure.

METHYL 3,3-DIMETHYL-7-PHENYLSULPHINYL-1,5-DIOXASPIRO[5.5]UNDECANE-9-CARBOXYLATE
$C_{19}H_{26}O_5S$

M. SORIANO-GARCIA, R.A. TOSCANO, G. GARCÍA, M.I. LARRAZA and I.H. SÁNCHEZ, 1984. Acta Cryst., C40, 887-889.

Monoclinic, $P2_1/a$, a = 11.094, b = 11.324, c = 15.298 Å, β = 96.22°, Z = 4. Cu radiation, R = 0.075 for 1496 reflexions.

In the molecule both the cyclohexane and 1,3-dioxane rings are in chair conformations (Fig. 1). The angle between the mean planes of the methoxycarbonyl group and the 1,3-dioxane ring is 93.8(5)° so that the molecule forms a spiral-like structure. The two C-S bonds are unequal with C(13)-S 1.794(6) and S-C(2) 1.830(6) Å. The packing in the crystal is entirely due to van der Waals forces.

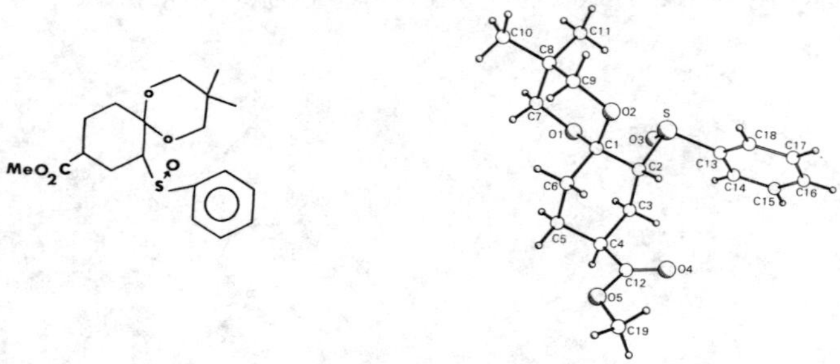

Fig. 1. $C_{19}H_{26}O_5S$: structural formula and perspective view of the molecule.

FLUORESCIN METHANOL SOLVATE
$C_{20}H_{12}O_5,CH_4O$

I.N. POLYAKOVA, Z.A. STARIKOVA, B.V. PARUSNIKOV, I.A. KRASAVIN, G.M. DOBRYAKOVA and B.V. ZHADANOV, 1984. J. Struct. Chem., 25, 752-757 [Zh. Strukt. Khim., 25(5), 98-104].

Triclinic, $P\bar{1}$, a = 7.835, b = 11.330, c = 11.759 Å, α = 115.86, β = 73.72, γ = 111.29°, Z = 2. Mo radiation, R = 0.043 for 2358 reflexions.

The fluorescin molecule has a lactone structure, stabilized by hydrogen bonds to the methanol molecule. The plane of the lactone ring is perpendicular to the tricyclic xanthene fragment. Bond lengths are shown in Fig. 1.

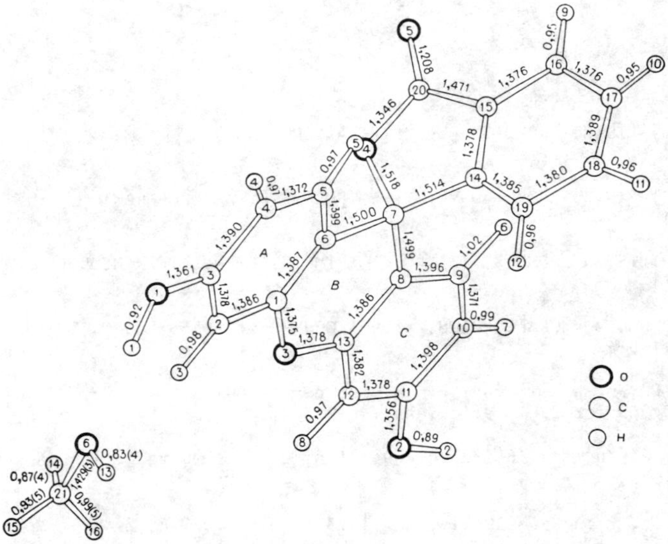

Fig. 1. Bond lengths for the fluorescin methanol molecular complex (σ = 0.002 Å for O-C, 0.003 Å for C-C, and 0.02-0.03 Å for O-H and C-H in the fluorescin molecule).

(6aS,8aS,14aS)-5,6,14,14a-TETRAHYDRO-4H,13H-6a,8a-EPOXYDINAPHTHO[1,8-cd:2',1'-f]-[1,2]DIOXEPIN
$C_{20}H_{18}O_3$

T.K. DOBBS, A.R. TAYLOR, J.A. BARNES, B.D. ISCIMENLER, E.M. HOLT and E.J. EISENBRAUN, 1984. J. Org. Chem., 49, 1030-1033.

Monoclinic, $P2_1/c$, a = 10.616, b = 9.607, c = 15.052 Å, β = 105.06°, D_m = 1.372, Z = 4. Mo radiation, R = 0.063 for 2385 reflexions.

The ozonide ring (Fig. 1) has an ether-oxygen envelope conformation, with O-O 1.471(3), peroxide C-O 1.443 and 1.460(4), ether C-O 1.417 and 1.417(4) Å.

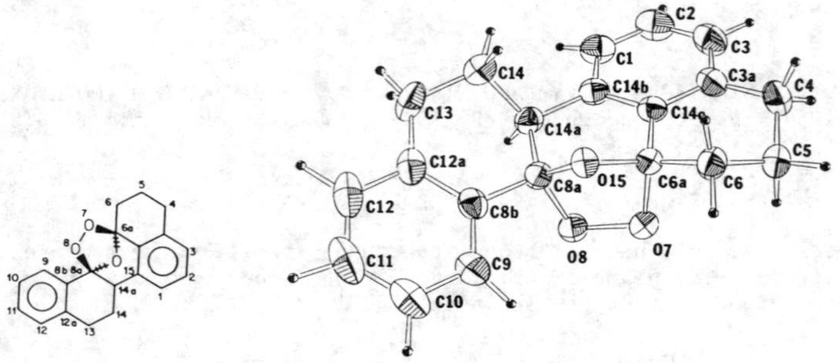

Fig. 1. A view of $C_{20}H_{18}O_3$.

(±)-2,2'-(μ2-OXO)-BIS(3α-BROMO-5β-PHENYLTETRAHYDROFURAN)
$C_{20}H_{20}Br_2O_3$ (I)

(±)-2,2'-(μ2-OXO)-BIS(3α-BROMO-4α-METHYL-5β-PHENYLTETRAHYDROFURAN)
$C_{22}H_{24}Br_2O_3$ (II)

(±)-2,2'-(μ2-OXO)BIS(3α-BROMO-5β-(2',4',6'-TRIMETHYLPHENYL)TETRAHYDROFURAN)
$C_{26}H_{32}Br_2O_3$ (III)

G. DANA, E. TOUBOUL, M. MORALES-RIOS, M. PHILOCHE-LEVISALLES and C. BOIS, 1984. Can. J. Chem., 62, 911-916.

I. Monoclinic, $P2_1/n$, a = 16.644, b = 6.089, c = 19.807 Å, β = 108.7°, Z = 4. Mo radiation, R = 0.065 for 2267 reflexions (at 233 K).

II. Tetragonal, $P4_2/n$, a = 15.135, c = 9.301 Å, Z = 4. Mo radiation, R = 0.052 for 1124 reflexions.

III. Orthorhombic, Pccn, a = 19.315, b = 14.311, c = 9.267 Å, Z = 4. Mo radiation, R = 0.052 for 1156 reflexions.

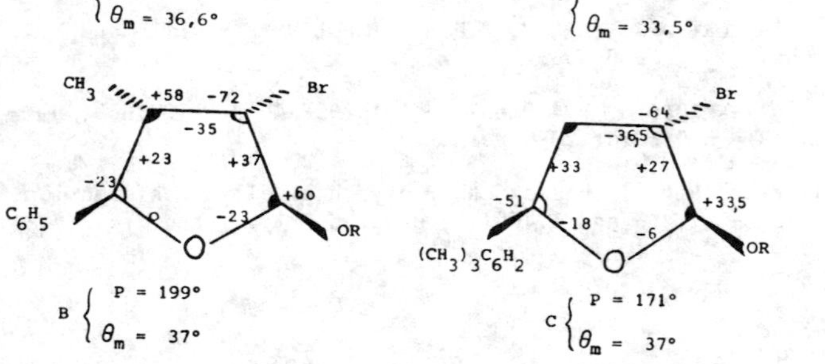

Fig. 1. The tetrahydrofuran ring conformations in I (A and A'), II (B) and III (C).

In the three molecules (Fig. 1) the tetrahydrofuran ring assumes a conformation characterized by polar groups in the 2-position (anomeric effect) and in the 3-position being diaxial and by the bulky aromatic group being equatorial. Molecules II and III have crystallographic twofold symmetry and molecule I has approximate twofold symmetry. Details of the five-membered ring conformation are in Fig. 1.

2,2-DIMETHYL-endo-5-(3-PHENYLTHIOMETHYLBICYCLO[2.2.1]HEPT-5'-EN-2'-YLIDENE)-1,3-DIOXAN-4,6-DIONE
$C_{20}H_{20}SO_4$ (I)

R.F.C. BROWN, K.J. COULSTON, F.W. EASTWOOD, B.M. GATEHOUSE, L.W. GUDDATT, M. PFENNINGER and I. RAINBOW, 1984. Aust. J. Chem., 37, 2509-2524.

Monoclinic, $P2_1/n$, a = 21.964, b = 12.824, c = 6.367 Å, β = 97.25°, D_m = 1.30, Z = 4. Cu radiation, R = 0.046 for 1778 reflexions. M.P. 85°C.

The molecule is shown in Fig. 1 and has S(1)-C(9) 1.764(4), S(1)-C(1)" 1.808(4) Å.

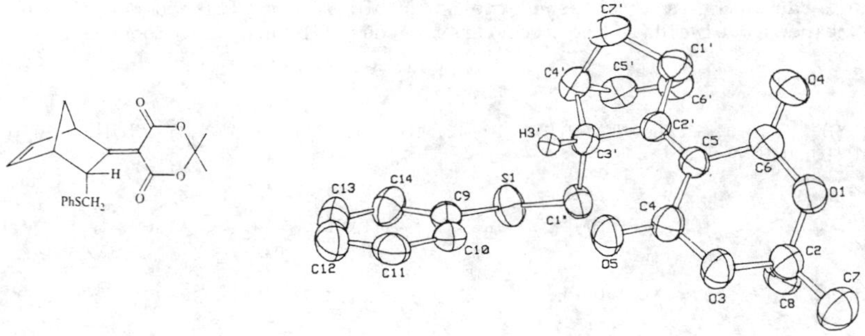

Fig. 1. The structure of $C_{20}H_{20}SO_4$.

cis-1,3,4b,6,8,9b-HEXAMETHYL-4b,9b-DIHYDROBENZO[3,2-b]BENZOFURAN
$C_{20}H_{22}O_2$

R. MERCIER, M. RAMAH, B. LAUDE, R. FAURE and H. LOISELEUR, 1984. Acta Cryst., C40, 1208-1210.

Triclinic, $P\bar{1}$, a = 8.656, b = 9.355, c = 10.901 Å, α = 101.26, β = 107.24, γ = 91.47°, Z = 2. Cu radiation, R = 0.041 for 1668 reflexions.

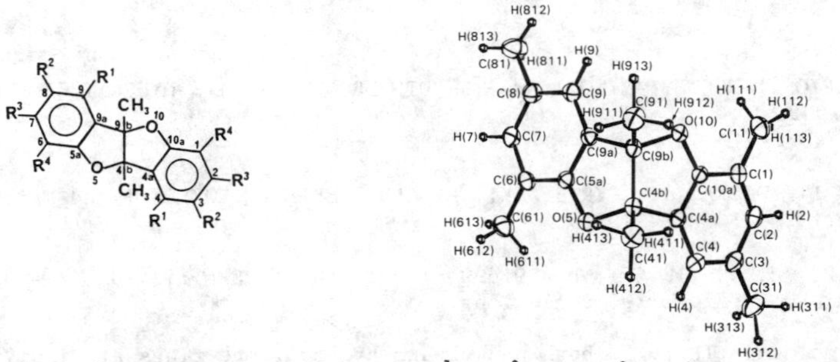

Fig. 1. $C_{20}H_{22}O_2$: structural formula ($R^1 = R^3 = H$, $R^2 = R^4 = CH_3$) and perspective view of the molecule.

In the molecule (Fig. 1) the methyl groups at positions 4b and 9b are in the cis configuration and the molecule shows a quasi-binary symmetry around an axis passing through the middle of the common C(4b)-C(9b) bond, whose length of 1.564(3) Å is in accord with those of equivalent bonds in other compounds.

3-ISOCHROMAN-1-YL-6-ISOPROPYL-2-METHOXYTROPONE
$C_{20}H_{22}O_3$

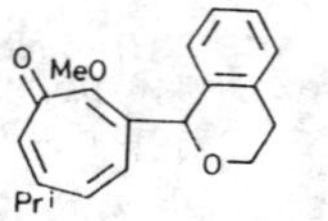

M. YAMATO, K. HASHIGAKI, N. KOKUBU and Y. NAKATO, 1984. J. Chem. Soc. Perkin I, 1301-1304.

Monoclinic, $P2_1/n$, a = 14.601, b = 13.427, c = 9.139 Å, β = 108.40°, Z = 4. Cu radiation, R not given, 2349 reflexions.

The X-ray analysis established the structure of the title compound. Knowledge of this helped to elucidate the structures of several related compounds.

3α-ETHENYL-1,2,3,6,7,11Bα-HEXAHYDRO-10-METHOXY-5H-3,5Aα-METHANONAPHTH[2,1-c]OXEPIN-12,6-ACETOLACTONE
$C_{20}H_{22}O_4$

K. SHISHIDO, T. SAITOH, K. FUKUMOTO and T. KAMETANI, 1984. J. Chem. Soc. Perkin I, 2139-2146.

Monoclinic, $P2_1/a$, a = 11.440, b = 13.537, c = 10.491 Å, β = 90.47°, Z = 4. R = 0.095.

The structure and stereochemistry (Fig. 1) of this pentacyclic model for the antitumour quassinoid bruceantin were unambiguously established by the X-ray analysis.

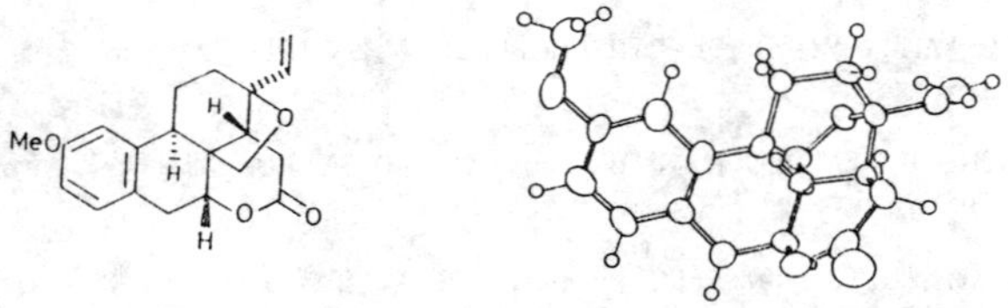

Fig. 1. $C_{20}H_{22}O_4$: the molecular skeleton and structure.

cis-2-(N-PROPIONYL-N-ETHYLAMINO)-3-(2',5'-DIHYDROXYPHENYL)-5-HYDROXY-3-METHYL-2,3-DIHYDROBENZOFURAN
$C_{20}H_{23}NO_5$

Z. URBANCZYK-LIPKOWSKA, J.W. KRAJEWSKI, P. GLUZINSKI, L. KOZERSKI and M. DOBLER, 1984. J. Mol. Struct., 125, 321-329.

Monoclinic, $P2_1/c$, a = 12.651, b = 9.699, c = 16.487 Å, β = 117.18°, Z = 4. Mo radiation, R = 0.047 for 1169 reflexions.

In the molecule (Fig. 1) both phenyl and hydroquinone rings are individually planar but the dihydrofuran ring has very flattened E_2 conformation with puckering parameters Q 0.1606 Å and φ 222°. The substituents at C(2) and C(3) are close to being eclipsed (torsion angle C(5)-C(3)-C(2)-N(20) -20.8(6)°). Bond lengths and angles are similar to those observed in related systems.

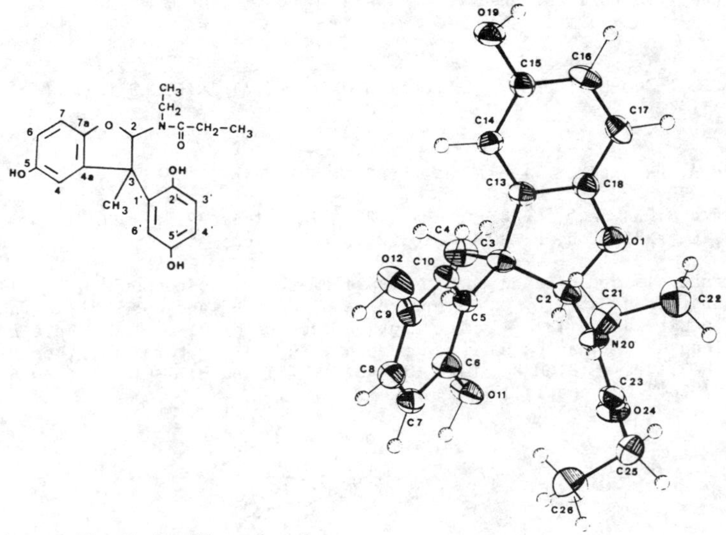

Fig. 1. The $C_{20}H_{23}NO_5$ molecule.

(1R*,9S*,10S*)-5,5,9-TRIMETHYLDECAHYDRO-1-NAPHTHYL-p-NITROBENZOATE
$C_{20}H_{27}NO_4$ (I)

(1R*,6S*,10R*)-5,5-DIMETHYL-11,12-DIOXATRICYCLO[8.2.1.0^{1,6}]TRIDECAN-10-OL
$C_{13}H_{22}O_3$ (II)

M.E. SCHELLER, P. MATHIES, W. PETTER and B. FREI, 1984. Helv. Chim. Acta, 67, 1748-1754.

I. Orthorhombic, Pbca, a = 36.476, b = 8.326, c = 12.235 Å, Z = 8. Mo radiation, R = 0.040 for 1519 reflexions.

II. Monoclinic, $P2_1/c$, a = 16.070, b = 6.948, c = 12.455 Å, β = 114.97°, Z = 4. Mo radiation, R = 0.040 for 1166 reflexions.

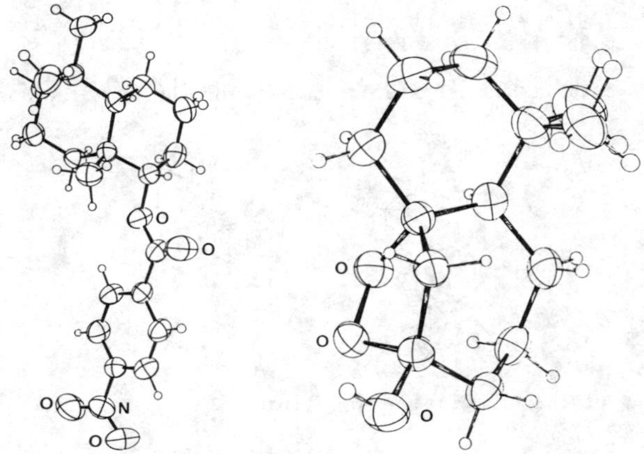

Fig. 1. Views of $C_{20}H_{27}NO_4$ (I) (left) and $C_{13}H_{22}O_3$ (II) (right).

X-ray analysis established that the two fused six-membered rings have a cis-ring function in I; the structure assigned to II was also confirmed.

METHYL (syn-7-ACETYL-4,9-DIISOPROPYL-1-METHYL-5-OXO-10-OXATETRACYCLO[4.4.0.$0^{2,4}$.$0^{3,8}$]-DECANE-9-CARBOXYLATE
$C_{20}H_{28}O_5$

C.P. FALSHAW and A. FRANKLINOS, 1984. J. Chem. Soc. Perkin I, 95-100.

Triclinic, P$\bar{1}$, a = 8.534, b = 10.441, c = 12.185 Å, α = 105.487, β = 96.46, γ = 111.30°, Z = 2. Cu radiation, R = 0.075 for 2249 reflexions.

The X-ray analysis established the structure of the title compound (Fig. 1). The isopropyl substituent at C(6) was found to be disordered and one methyl group was included in two positions (C(18)) with occupation factors of 0.5. These were not refined. Bond lengths and angles have acceptable values. The torsion angles H(2)-C(2)-C(3)-H(3), H(3)-C(3)-C(8)-H(8), H(8)-C(8)-C(7)-H(7) and H(7)-C(7)-C(6)-H(6) are 6, 25.5, 85.6 and 38.2° respectively.

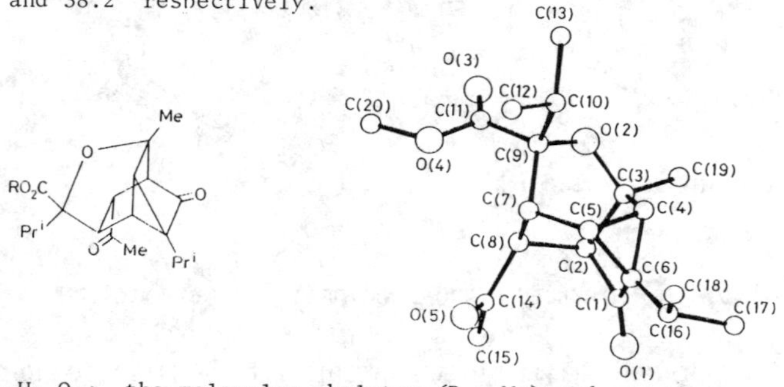

Fig. 1. $C_{20}H_{28}O_5$: the molecular skeleton (R = Me) and structure.

(3S-(3α,4aα,7β,7aβ,10β,11α,11aR*,12S*))-12-BROMODECAHYDRO-7,7a,10-TRIMETHYL-3-(1-METHYLETHYL)-4a,3,11-(EPOXYETHANYLYLIDENE)-1H,4aH-NAPHTHO[1,8a-b]PYRAN-13-ONE
$C_{21}H_{31}BrO_3$

S. DANISHEFSKY, P. HARRISON, M. SILVESTRI and B. SEGMULLER, 1984. J. Org. Chem., 49, 1319-1321.

Orthorhombic, $P2_12_12_1$, a = 7.248, b = 12.598, c = 20.971 Å, D_m = 1.427, [Z = 4]. Mo radiation, R = 0.038 for 1259 reflexions.

The analysis establishes the stereochemistry of the bromolactone as shown in Fig. 1.

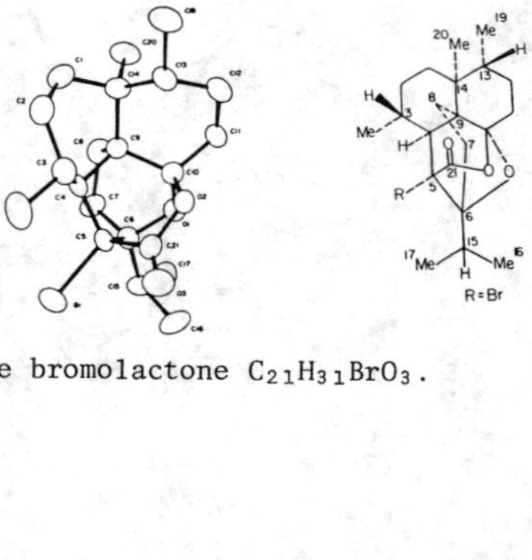

Fig. 1. A view of the bromolactone $C_{21}H_{31}BrO_3$.

STEGANE (3aS*,13aS*,8a,8bS*)-6,7,10,11-TETRAMETHOXY-3a,4,13,13a-TETRAHYDRODIBENZO-[4',5':6',7']CYCLOOCTA[1,2-c]FURAN-1(3H)-ONE
$C_{22}H_{24}O_6$ (I)

ISOSTEGANE (3aS*,13aS*,8a,8bR*)-6,7,10,11-TETRAMETHOXY-3a,4,13,13a-TETRAHYDRODIBENZO-[4',5':6',7']CYCLOOCTA[1,2-c]FURAN-1(3H)-ONE
$C_{22}H_{24}O_6$ (II)

R.C. CAMBIE, G.R. CLARK, P.A. CRAW, P.S. RUTLEDGE and P.D. WOODGATE, 1984. Aust. J. Chem., 37, 1775-1784.

I. Monoclinic, $P2_1$, a = 10.104, b = 9.622, c = 10.977 Å, β = 112.05°, D_m = 1.293, Z = 2. Mo radiation, R = 0.0438 for 1433 reflexions.

II. Monoclinic, $P2_1$, a = 8.749, b = 10.969, c = 10.781 Å, β = 108.61°, D_m = 1.302, Z = 2. Mo radiation, R = 0.0363 for 1050 reflexions.

Stegane and isostegane isomers (Fig. 1) differ in the nature of the molecular flexing at C(4) and C(13) and the relative twists of the biphenyl groups.

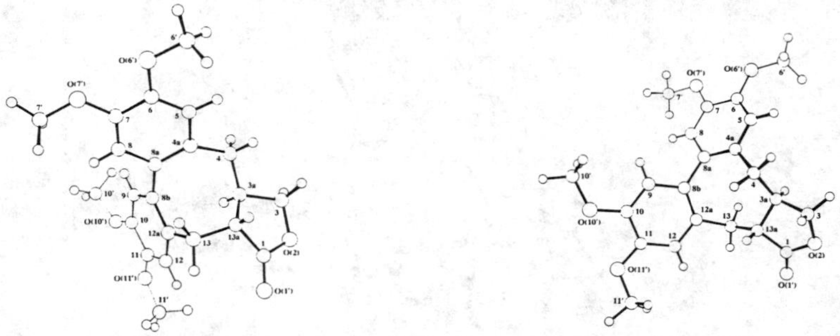

Fig. 1. The structures of stegane (left) and isostegane (right).

8'-OXO-(trans-1',6')-BICYCLO[4.4.0]DECAN-6'-YL-METHYL (trans-1,6-cis-1,9)-10-OXATRICYCLO[7.2.1.0^{1,6}]DODECAN-9-YL ETHER
$C_{22}H_{34}O_3$

D.J. CHADWICK and S.N. WHITTLETON, 1984. J. Chem. Res., (S) 398-399, (M) 3764-3785.

Triclinic, $P\bar{1}$, a = 7.476, b = 10.672, c = 12.746 Å, α = 78.61, β = 77.64, γ = 77.44°, Z = 2. Mo radiation, R = 0.030 for 2081 reflexions.

This decalin derivative (Fig. 1) has acetal and ketone functionalities in close proximity; the geometric features are interpreted as reflecting displacement from both termini of a γ-hydroxyketone-hemiacetal equilibrium towards the transition state. There is a remarkably short O...C=O intramolecular distance of 2.629(3) Å, and the O...C=O angle is 104.2(2)°. The endocyclic C-O bond is longer than the exocyclic one by 0.036(3) Å.

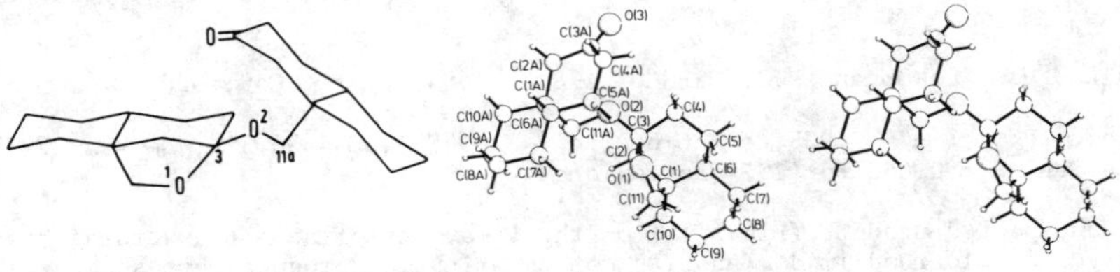

Fig. 1. $C_{22}H_{34}O_3$: skeleton and stereoview.

6-BENZOYL-5-HYDROXY-2,2,8,8-TETRAMETHYL-2H,8H-BENZO[1.2-b:3.4-b']DIPYRAN
$C_{23}H_{22}O_4$

X. SOLANS, M. FONT ALTABA, J.L. BRIANSÓ and J. GONZALEZ, 1984. Afinidad, 41, 373-374.

Orthorhombic, $Pn2_1a$, a = 20.941, b = 9.717, c = 9.687 Å, Z = 4. Mo radiation, R = 0.125 for 1030 reflexions.

The benzodipyrano moiety has an overall planar shape although the pyran-rings have boat-like conformation and the atoms of the benzene ring are planar (Fig. 1). The benzoyl substituent makes a dihedral angle of 15° with the best plane through the benzodipyrano moiety. An intramolecular hydrogen bond links the OH and carbonyl groups.

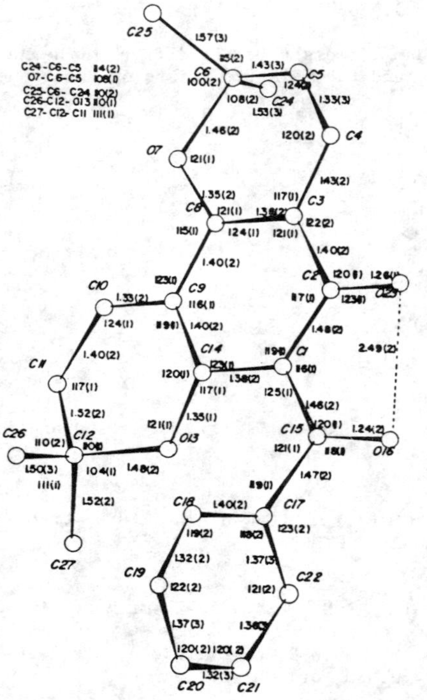

Fig. 1. Schematic diagram of the molecule $C_{23}H_{22}O_4$ showing bond lengths (Å) and angles (°).

1',3'-DICYCLOHEXYLSPIRO[2H-1-BENZOPYRAN-2,4'-IMIDAZOLIDINE]-2',4(3H),5'-TRIONE
$C_{23}H_{28}N_2O_4$

R.B. BATES, F.A. CAMOU and R.J. CREGGE, 1984. Acta Cryst., C40, 320-322.

Monoclinic, $P2_1/c$, a = 9.935, b = 6.669, c = 31.983 Å, β = 90.51°, D_m = 1.22, Z = 4. Mo radiation, R = 0.053 for 1934 reflexions.

An unexpected product (Fig. 1) from the treatment of an α,β-unsaturated acid with N,N'-dicyclohexylcarbodiimide (and its monochlorination product) are shown to be imidazolidine-2,4-diones. There are no unusual bond distances or bond angles.

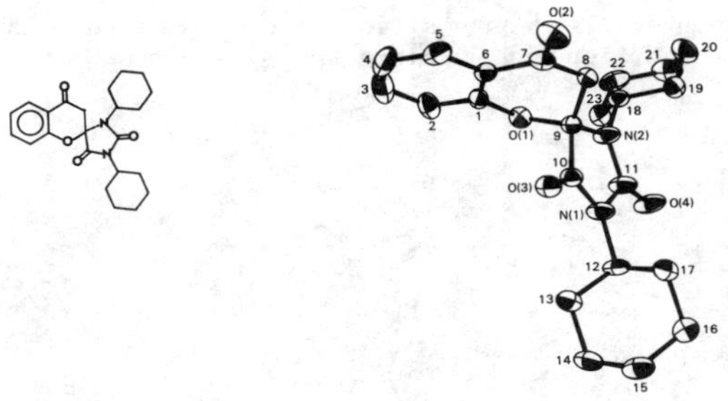

Fig. 1. The $C_{23}H_{28}N_2O_4$ molecule.

ANGELICOLIDE
$C_{24}H_{28}O_4$

S.K. BANERJEE, B.D. GUPTA, W.S. SHELDRICK and G. HOFLE, 1984. Justus Liebigs Ann. Chem., 888-893.

Monoclinic, C2/c, a = 25.275, b = 11.288, c = 16.846 Å, β = 115.18°, Z = 8. Cu radiation, R = 0.088 for 1558 reflexions.

Angelicolide is shown to be a synthetic dimer (Fig. 1) of butylidenephthalide. The propyl side chains at C12, C12', C13, C13' are disordered.

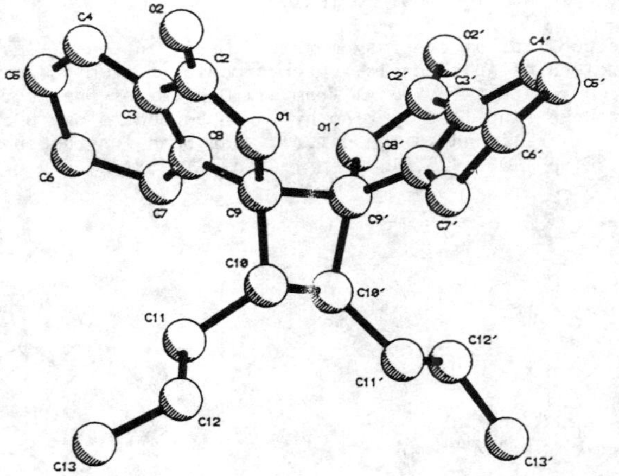

Fig. 1. A view of angelicolide.

BIS(HYDRONIUM DICHLOROPICRATE)-1,4,7,10,13,16-HEXAOXACYCLOOCTADECANE COMPLEX
$C_{24}H_{30}Cl_4N_6O_{22}$ $2[H_3O^+.C_6(NO_2)_3Cl_2O^-].(CH_2CH_2O)_6$

D. BRITTON, M.K. CHANTOONI JR, W.-J. WANG and I.M. KOLTHOFF, 1984. Acta Cryst., C40, 1584-1587.

Monoclinic, $P2_1/a$, a = 8.369, b = 25.538, c = 9.562 Å, β = 110.12°, Z = 2. Mo radiation, R = 0.056 for 1618 reflexions.

The complex (Fig. 1) is centrosymmetric with the two hydronium ions weakly hydrogen bonded to O atoms of the crown ether (O...O 2.803(7)-2.894(8) Å) and strongly hydrogen bonded to the phenolate O atom of a dichloropicrate anion (O...O 2.378(7) and 2.497(7) Å). The hydronium-ion positions are disordered. Bond lengths and angles are normal with torsion angles close to 69° about C-C and to 180° about C-O in the crown ether.

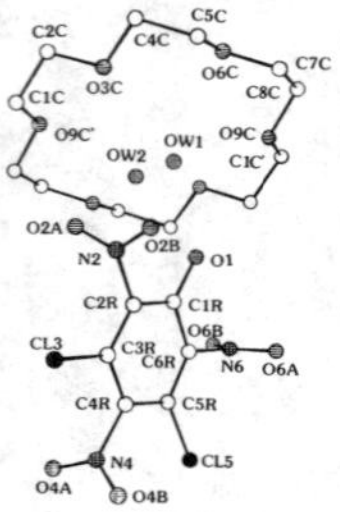

Fig. 1. One asymmetric unit plus the symmetry related remainder of the crown-ether ring in the $2[H_3O^+.C_6(NO_2)_3Cl_2O^-].(CH_2CH_2O)_6$ complex.

1,5,12,16,23,26,29,32-OCTAOXO[$10^{3,14}$][5.5]ORTHOCYCLOPHANE AMMONIUM IODIDE DIHYDRATE $C_{24}H_{34}INO_8,2(H_2O)$

J.A. BANDY, C.H.L. KENNARD, D.G. PARSONS and M.R. TRUTER, 1984. J. Chem. Soc. Perkin II, 309-315.

Monoclinic, C2/c, a = 22.734, b = 14.288, c = 17.514 Å, β = 100.46°, D_m = 1.480, Z = 8. Mo radiation, R = 0.062 for 3021 reflexions.

The structure consists of centrosymmetric dimeric cations (Fig. 1) with iodide anions and water molecules in channels. Each ammonium N atom is approximately coplanar with and equidistant (2.795 Å) from six O atoms of the macrocycle which is a pseudo benzo-18-crown-6 ring. It is also hydrogen bonded to one O atom (N...O 2.981(7) Å) of the other ligand in the dimer. Mean bond lengths in the cation are: C(sp^2)-O 1.375(8), C(sp^3)-O 1.425(8), C(sp^2)-C(sp^2) 1.388(22) and C(sp^3)-C(sp^3) 1.503(22) Å.

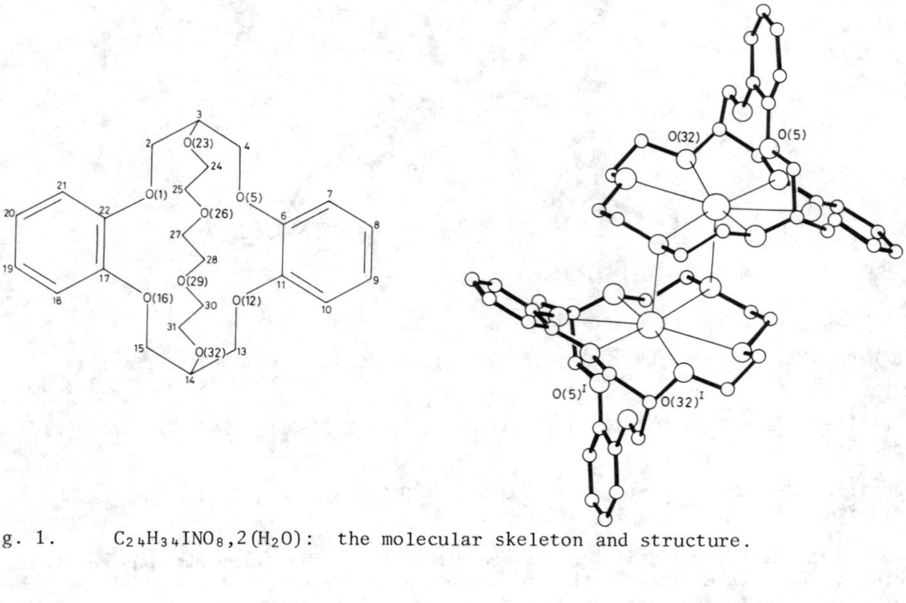

Fig. 1. $C_{24}H_{34}INO_8,2(H_2O)$: the molecular skeleton and structure.

2-(5-METHYL-2-FURYL)-3-KETO-L-GULONOLACTONE 3,6-HEMIKETAL SUCCINIC ANHYDRIDE (2:1 Complex)

$C_{26}H_{28}O_{17}$ $(C_{11}H_{12}O_7)_2.C_4H_4O_3$

G. FODOR, K. SUSSANGKARN, H. MATHELIER, R. ARNOLD, I. KARLE and C. GEORGE, 1984. J. Org. Chem., 49, 5064-5069.

Tetragonal, $P4_2$, a = 14.784, c = 6.235 Å, Z = 4. Mo radiation, R = 0.037 for 872 reflexions.

The asymmetric unit contains one molecule of the gulonolactone and half a succinic anhydride (which lies about a crystallographic twofold axis). The two fused five-membered rings (Fig. 1) of the gulonolactone are twisted about their common bond (torsion angles O8-C4-C3-C2 25.8°, O7-C3-C4-C5 27.5°, O8-C4-C3-O7 87.0°, C2-C3-C4-C5 -140.4°). The dihedral angle between the planes of the two rings is 108.5°. The methyl furyl ring is planar, with a dihedral angle of 90.5° with the plane of the lactone ring. The succinic anhydride ring is also planar (average deviation 0.004 Å). Molecules are linked by O-H...O hydrogen bonds, with O...O 2.74, 2.78 and 2.85 Å.

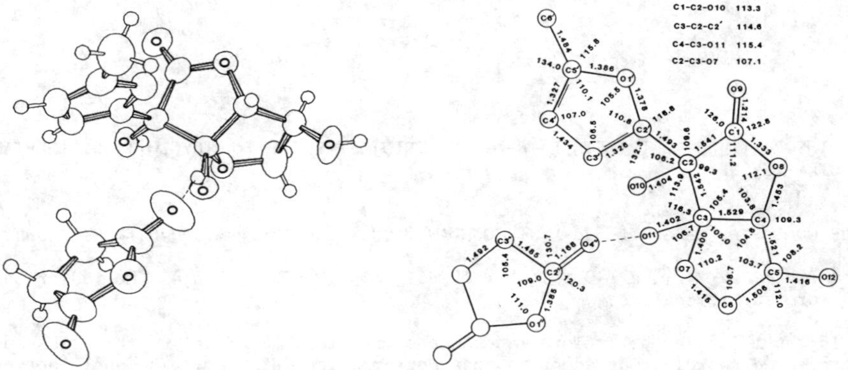

Fig. 1. Views of the $(C_{11}H_{12}O_7)_2.C_4H_4O_3$ complex and dimensions (σ 0.006 Å, 0.5°).

3,3-(p-N,N-DIMETHYLAMINOPHENYL)-6-N,N-DIMETHYLAMINOPHTHALIDE

$C_{26}H_{29}N_3O_2$

C.R. THEOCHARIS and W. JONES, 1984. J. Cryst. Spectrosc. Res., 14, 121-128.

Monoclinic, $P2_1/n$, a = 12.228, b = 17.385, c = 11.786 Å, Z = 4. Cu radiation, R = 0.046 for 2301 reflexions.

The molecule (Fig. 1) has a propeller shape with bond lengths and angles in the expected ranges.

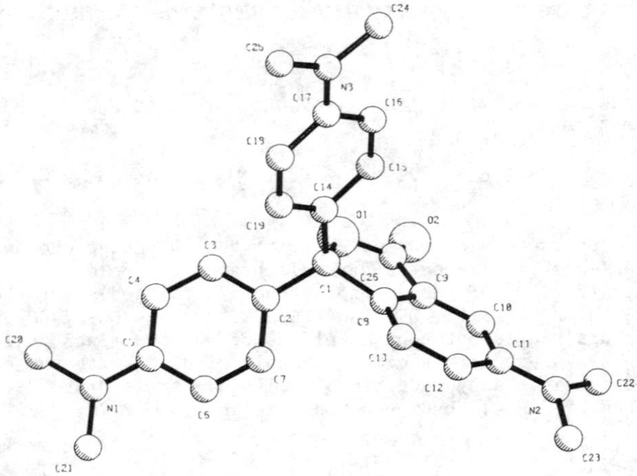

Fig. 1. A view of $C_{26}H_{29}N_3O_2$.

2,12-DIHYDROXY-4,8,12-TRIMETHYL-16-(PHENYLTHIO)METHYL-14-OXABICYCLO[11.3.1]HEPTADECA-4,8-DIEN-15-ONE
$C_{26}H_{36}O_4S$

L. LEBIODA, J.A. MARSHALL and M.J. COGHLAN, 1984. Acta Cryst., C40, 1191-1193.

Orthorhombic, $P2_12_12_1$, a = 9.897, b = 15.367, c = 15.633 Å, Z = 4. Mo radiation, R = 0.040 for 1202 reflexions.

In the molecule the cyclotetradecadiene ring is in a 'flat' conformation and the condensed caprolactone adopts a sofa conformation (Fig. 1). The angle between the average planes through these rings is 44.8(4)°. Bond distances and angles are normal, as are intermolecular contacts. Both hydroxyls form hydrogen bonds to the neighbouring molecule related by a twofold screw axis along a.

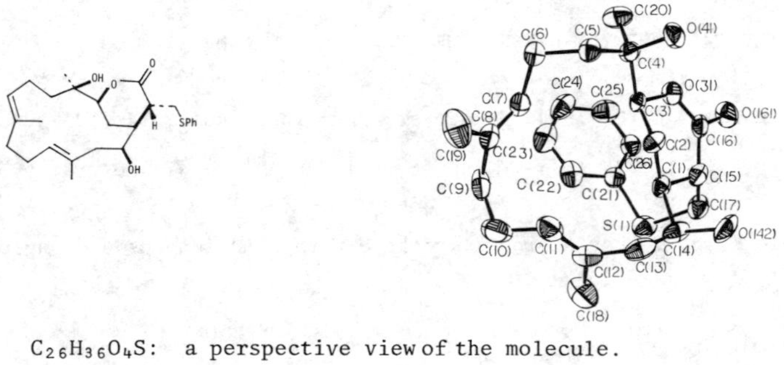

Fig. 1. $C_{26}H_{36}O_4S$: a perspective view of the molecule.

anti-DI-t-BUTYLBIS(BICYCLO[3.3.1]NON-9-YLIDENE)
$C_{26}H_{44}$ (I)

cis,cis-DI-t-BUTYLBIS(BICYCLO[3.3.1]NON-9-YLIDENE)DIOXETANE
$C_{26}H_{44}O_2$ (II)

Y. KABE, T. TAKATA, K. UENO and W. ANDO, 1984. J. Am. Chem. Soc., 106, 8174-8180.

I. Monoclinic, $P2_1/n$, a = 9.243, b = 10.760, c = 12.465 Å, β = 111.43°, Z = 2. Cu radiation, R = 0.084 for 1268 reflexions.

II. Orthorhombic, Pccn, a = 21.840, b = 8.428, c = 13.064 Å, Z = 4. Cu radiation, R = 0.088 for 1752 reflexions.

Compound (I) has crystallographic inversion symmetry and the central olefinic moiety is essentially planar (Fig. 1). Molecule II (Fig. 1) has crystallographic twofold symmetry. The central four-membered ring is non-planar with O-C-C-O 21.9° and O-O 1.507(5) Å.

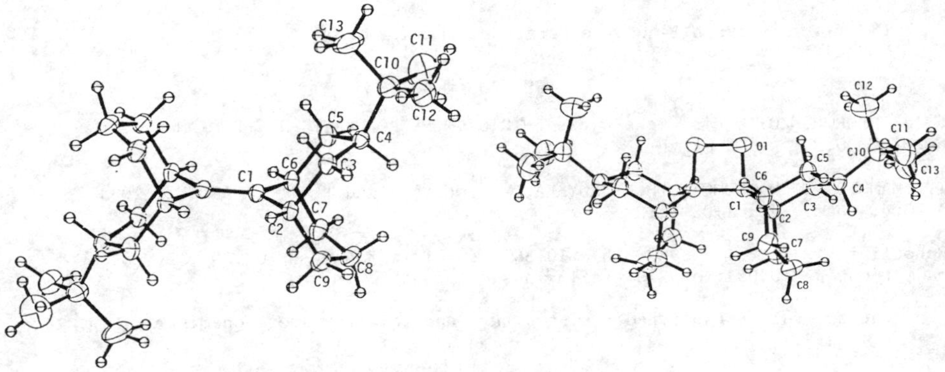

Fig. 1. Views of anti-$C_{26}H_{44}$ and cis,cis-$C_{26}H_{44}O_2$.

3-(2-{ETHYL-2-[4-(o-METHOXYPHENYL)-1-PIPERAZINYL]}ANILINO)-1(3H)-ISOBENZOFURANONE $C_{27}H_{29}N_3O_3$

A. CARPY, C. FENIOU, A. H'NAÏFI and J.-C. COLLETER, 1984. Acta Cryst., C40, 1970-1972.

Monoclinic, $P2_1/c$, a = 13.193, b = 9.902, c = 17.969 Å, β = 94.00°, Z = 4. Cu radiation, R = 0.058 for 2267 reflexions.

The molecule has a folded conformation (Fig. 1) with the four-membered carbon plane of the piperazine ring inclined at 17° to the benzofuranone ring. The methoxyphenyl and aniline rings are inclined at 80°. The piperazine ring has a chair conformation with the N atoms displaced 0.695 and 0.687(3) Å from the carbon atom plane. The C(20)-C(14)-C(13)-N(10) and C(15)-C(14)-C(13)-N(10) torsion angles are about 67 and -80° respectively. The N(10)...(centre of methoxyphenyl ring) distance is 5.66(1) Å.

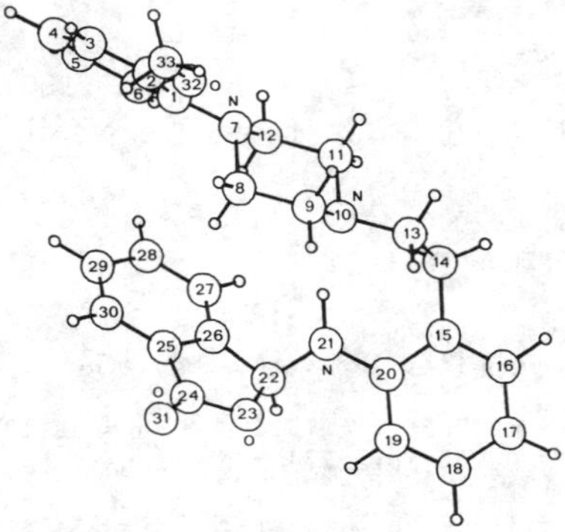

Fig. 1. The $C_{27}H_{29}N_3O_3$ molecule.

3-(N-BENZYL-N-METHYLCARBAMOYLAMINO)-2-(t-BUTYLCYANOACETYL)-5-(t-BUTYLCYANOMETHYLENE)-2-METHYL-DIHYDROFURAN
$C_{27}H_{34}N_4O_3$

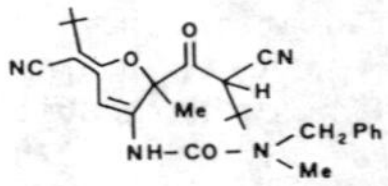

A. DONDONI, G. FANTIN, M. FOGAGNOLO, A. MASTELLARI, A. MEDICI and P. PEDRINI, 1984. J. Org. Chem., 49, 3478-3483.

Triclinic, P$\bar{1}$, a = 11.886, b = 12.054, c = 11.827 Å, α = 124.96, β = 90.60, γ = 70.02°, Z = 2. Cu radiation, R = 0.068 for 2618 reflexions.

The analysis established the structure shown above.

5-(p-CHLOROBENZOYL)-cis-2,cis-4,cis-6-TRIS(o-CHLOROPHENYL)-1,3-DIOXANE
$C_{29}H_{20}Cl_4O_3$

T. IMAMOTO, T. MATSUMOTO, H. YOKOYAMA, M. YOKOYAMA and K.-I. YAMAGUCHI, 1984. J. Org. Chem., 49, 1105-1110.

Monoclinic, C2/c, a = 36.870, b = 10.357, c = 14.437 Å, β = 107.09°, Z = 8. Cu radiation, R = 0.091 for 2751 reflexions.

The analysis established the structure and relative stereochemistry (Fig. 1).

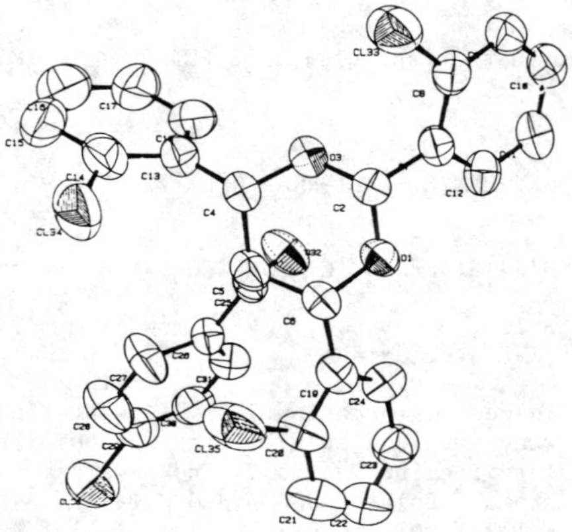

Fig. 1. A view of the dioxane derivative $C_{29}H_{20}Cl_4O_3$.

2-(N-METHYL-α-t-BUTYLACETAMIDO)-3-t-BUTYLCYCLOPROPANE-1-SPIRO-2'-(5',7'-DI-t-BUTYLBENZOFURAN-3'-ONE)
$C_{29}H_{45}NO_3$ (I)

CYCLOPENTANONE-4,6-DI(ISOBUTYL)BENZOFURAN-3-ONE
$C_{28}H_{40}O_3$ (II)

H.-D. BECKER, B.W. SKELTON and A.H. WHITE, 1984. Aust. J. Chem., 37, 2225-2233.

I. Triclinic, P$\bar{1}$, a = 15.101, b = 15.095, c = 13.469 Å, α = 95.10, β = 92.26, γ = 106.99°, Z = 4. Mo radiation, R = 0.055 for 3774 reflexions.

II. Triclinic, P$\bar{1}$, a = 19.27, b = 13.48, c = 10.813 Å, α = 83.14, β = 75.75, γ = 80.10°, Dm = 1.05, Z = 4. Mo radiation, R = 0.057 for 2018 reflexions.

There are two molecules in the asymmetric unit for both I and II. In I, there is pseudo-22 symmetry and in II a pseudo-glide relationship along b. Views of the molecules are in Fig. 1.

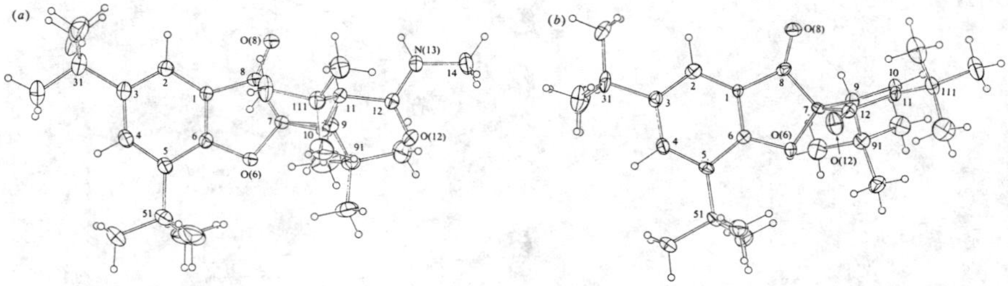

Fig. 1. The structure of $C_{29}H_{45}NO_3$ (I) (left) and $C_{28}H_{40}O_3$ (II) (right).

(±)-5α-[2'-HYDROXY-1'-METHOXYCARBONYL-2'-(3'',4'',5''-TRIMETHOXYPHENYL)ETHYL]-4α-(3'''-4'''-METHYLENEDIOXYPHENYL)-2α-PHENYL-1,3-DIOXANE
$C_{30}H_{32}O_{10}$

C. PICCINNI-LEOPARDI, J.P. DECLERCQ, G. GERMAIN, M. VAN MEERSSCHE, J. VAN DER EYCKEN, P. DE CLERCQ and M. VANDEWALLE, 1984. Bull. Soc. Chim. Belg., 93, 1047-1051.

Monoclinic, $P2_1/c$, a = 12.932, b = 9.670, c = 23.276 Å, β = 103.87°, Z = 4. Mo radiation, R = 0.056 for 2344 reflexions.

The dioxane ring (Fig. 1) has a nearly perfect chair conformation. The phenyl and methylenedioxyphenyl groups are equatorial and the substituted ethyl fragment is axial.

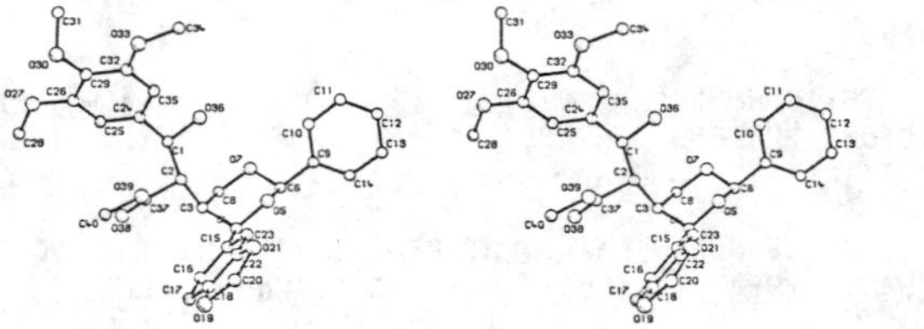

Fig. 1. $C_{30}H_{32}O_{10}$: a stereoview.

5-BENZYLIDENE-1,2,3,4,4a,5,6,7,8,9a-DECAHYDRO-4a-MORPHOLINO-9-PHENYLXANTHENE
$C_{30}H_{35}NO_2$ (I)

5-BENZYLIDENE-1,2,3,4,4a,5,6,7,8,9a-DECAHYDRO-9-PHENYL-4a-PIPERIDINOXANTHENE
$C_{31}H_{37}NO$ (II)

M.N. PONNUSWAMY and J. TROTTER, 1984. Acta Cryst., C40, 511-514.

I. Triclinic, $P\bar{1}$, a = 10.945, b = 13.545, c = 9.478 Å, α = 99.62, β = 114.36, γ = 73.02°, D_m = 1.160, Z = 2. Mo radiation, R = 0.059 for 2579 reflexions.

II. Triclinic, $P\bar{1}$, a = 10.879, b = 13.502, c = 9.587 Å, α = 99.12, β = 114.55, γ = 73.66°, D_m = 1.185, Z = 2. Mo radiation, R = 0.051 for 3022 reflexions.

The crystals of I (DBM) and II (DBP) are isomorphous. Both molecules have a similar configuration with a cis-fused B/C junction in the xanthene ring system (Fig. 1); this was not what had been inferred from earlier spectroscopic studies.

Rings A and B have half-chair conformations, ring C has a chair conformation as do the morpholino and piperidino rings.

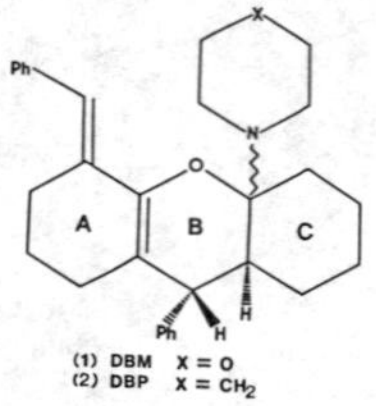

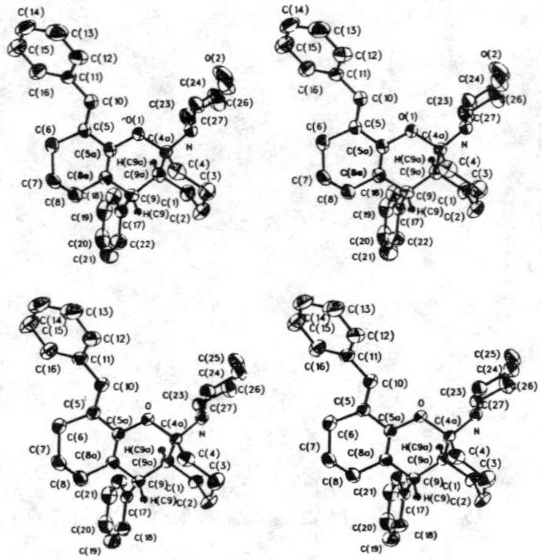

Fig. 1. Stereoviews of DBM (top) and DBP (bottom).

4,5-DIHYDRO-4-[(4-METHYLPHENYLCARBAMOYL)PHENYLMETHYLENE]-5-PHENOXY-5-PHENYLFURAN-2,3-DIONE
$C_{31}H_{23}NO_5$ (I)

3-(ACETOXYPHENYLMETHYLENE)-1-(4-METHYLPHENYL)-4-PHENYL-(3H,4H)-PYRROLE-2,5-DIONE
$C_{26}H_{21}NO_4$ (II)

4,5-DIHYDRO-5-(4-METHYLPHENYLCARBAMOYL)-4-[(2-NITROPHENYLAMINO)PHENYLMETHYLENE]-5-PHENYLFURAN-2,3-DIONE BENZENE SOLVATE
$C_{31}H_{23}N_3O_6, C_6H_6$ (III)

G. KOLLENZ, W. OTT, E. ZIEGLER, E.-M. PETERS, K. PETERS, H.G. VON SCHNERING, V. FORMACEK and H. QUAST, 1984. Justus Liebigs Ann. Chem., 1137-1164.

I. Monoclinic, C2/c, a = 26.380, b = 14.073, c = 14.465 Å, β = 111.43°, Z = 8. Mo radiation, R = 0.095 for 2846 reflexions.

II. Monoclinic, $P2_1/a$, a = 11.492, b = 20.652, c = 9.301 Å, β = 93.95°, Z = 4. Mo radiation, R = 0.072 for 3215 reflexions.

III. Triclinic, $P\bar{1}$, a = 12.554, b = 12.716, c = 10.952 Å, α = 100.13, β = 97.81, γ = 66.55°, Z = 2. Mo radiation, R = 0.090 for 5719 reflexions.

Details of conformation and geometry of these diones are shown in Fig. 1. The furan-dione rings in I and III are planar; the pyrrole-dione ring of II is almost planar.

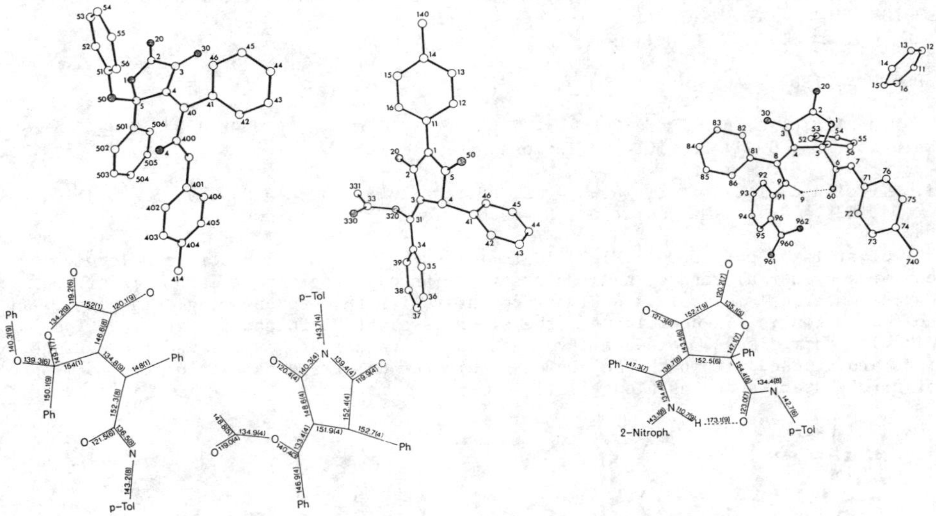

Fig. 1. Views and geometrical details (pm, °) for diones (I) (left), (II) (centre) and (III) (right).

2,3,11,12-BIS(1,2-ACENAPHTHO)-18-CROWN-6
$C_{32}H_{28}O_6$

A. MERZ, F. DIETL, R. TOMAHOGH, G. WEBER and G.M. SHELDRICK, 1984. Tetrahedron, 40, 665-671.

Orthorhombic, Pbca, a = 9.113, b = 14.706, c = 38.282 Å, Z = 8. Mo radiation, R = 0.078 for 2439 reflexions.

The molecule (Fig. 1) has an elongated structure, with the crown portion typical of free crown ethers. The acenaphthylene rings and adjacent oxygen atoms are planar. The molecule is essentially asymmetric but has approximate twofold symmetry (twofold axis perpendicular to and through the centre of the crown ring).

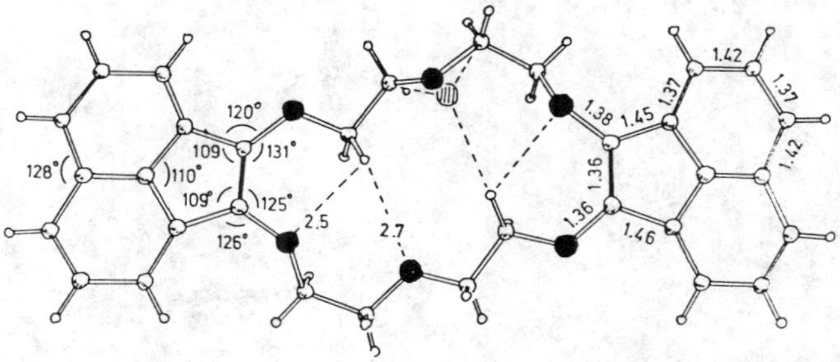

Fig. 1. The $C_{32}H_{28}O_6$ molecule with some averaged dimensions. One oxygen atom is disordered over two sites.

15,18,34,37-TETRAMETHOXY-1,4,7,10,13,20,23,26,29,32-DECAOXA[13.13]PARABENZENOPHANE
$C_{32}H_{48}O_{14}$ (I)

15,18,34,37-TETRAMETHOXY-1,4,7,10,13,20,23,26,29,32-DECAOXA[13.13]PARABENZENOPHANE - SODIUM THIOCYANATE WATER (1:2:2)
$C_{32}H_{48}O_{14}$,2(CNNaS),2(H_2O) (II)

J.D. OWEN, 1984. Acta Cryst., C40, 951-955.

I. Monoclinic, C2/c, a = 32.705, b = 7.637, c = 13.645 Å, β = 98.469°, Z = 4. Mo radiation, R = 0.044 for 1684 reflexions.

II. Triclinic, I$\bar{1}$, a = 13.178, b = 13.708, c = 23.567 Å, α = 83.054, β = 89.816, γ = 94.162°, Z = 4. Mo radiation, R = 0.089 for 2727 reflexions.

Crystals of compound I (Fig. 1) are formed by packing of discrete molecules separated by normal van der Waals contacts. The oxygen atoms of each half of the centrosymmetrical molecule are almost coplanar with the benzene ring. In complex II one sodium ion is six-coordinated, the other is eight-coordinated (Fig. 1). The molecules form dimers, joined by hydrogen-bonded waters and anions and the benzene rings are stacked within the molecule, within the dimer and between dimers forming infinitely stacked chains along [1$\bar{1}$1].

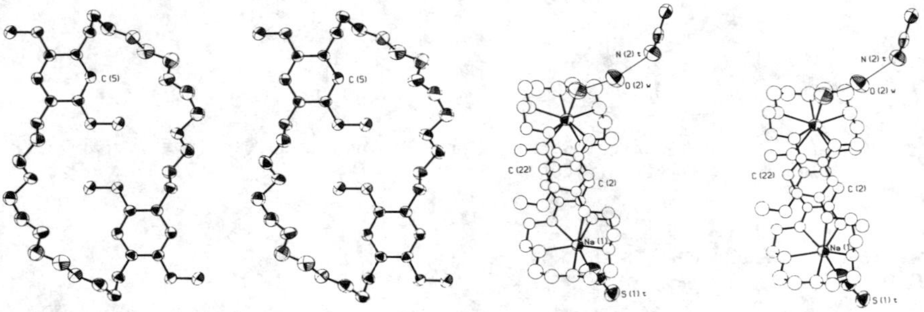

Fig. 1. Stereoscopic views of $C_{32}H_{48}O_{14}$ (I) and $C_{32}H_{48}O_{14}$,2(NaSCN,H_2O) (II).

cis-anti-cis-2,3,11,12-TETRAPHENYL-1,4,7,10,13,16-HEXAOXACYCLOOCTADECANE
$C_{36}H_{40}O_6$

E. BLASIUS, R.A. RAUSCH, G.D. ANDREETTI and J. REBIZANT, 1984. Chem. Ber., 117, 1113-1127.

Monoclinic, P2_1/n, a = 8.182, b = 20.035, c = 19.086 Å, β = 95.13°, Z = 4. Mo radiation, R = 0.042 for 1719 reflexions.

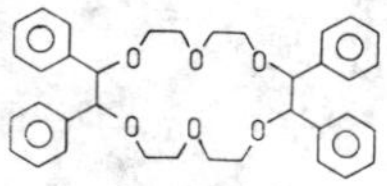

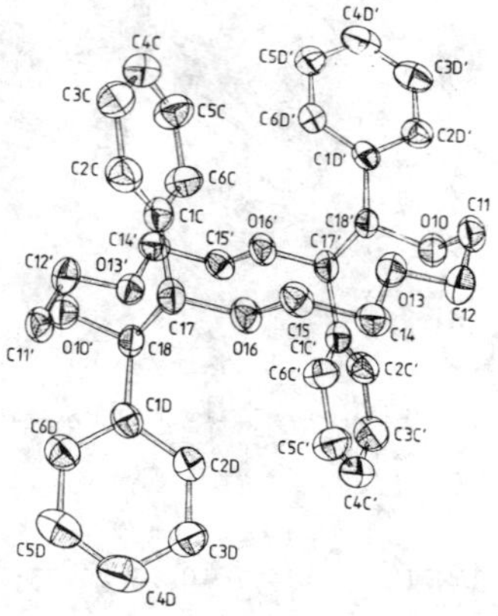

Fig. 1. The $C_{36}H_{40}O_6$ molecule.

There are two independent molecules in the system, each lying about a crystallographic inversion centre; one is shown in Fig. 1. The X-ray analysis establishes that the form examined is the cis-anti-cis isomer with the phenyl rings in a near ideal trans-conformation.

2,5,8,11-TETRA-t-BUTYL-peri-XANTHENOXANTHENE
$C_{36}H_{42}O_2$

H.P. SCHNEIDER, E. STREICH, K. SCHURR, N. PAULS, W. WINTER and A. RIEKER, 1984. Chem. Ber., 117, 2660-2674.

Monoclinic, $P2_1/c$, a = 11.291, b = 10.748, c = 12.193 Å, β = 92.83°, Z = 2. Cu radiation, R = 0.05 for 2321 reflexions.

The molecule (Fig. 1) lies on an inversion centre and has a planar carbon skeleton. The bond lengths are consistent with electron delocalization within the entire ring system.

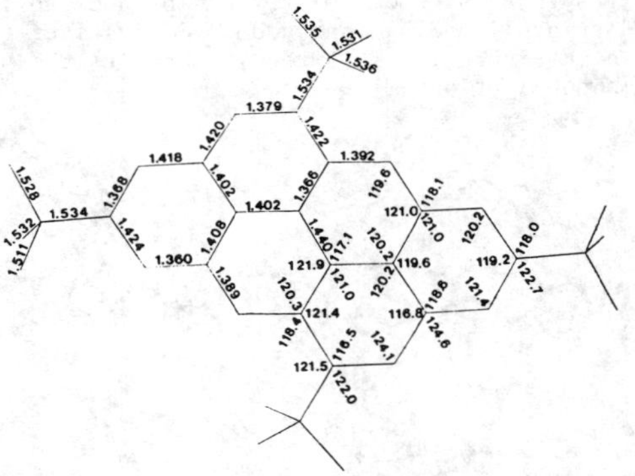

Fig. 1. Bond lengths and angles in $C_{36}H_{42}O_2$ (σ 0.003-0.004 Å, 0.1-0.2°).

BIS-(8α,13-EPOXY-15,14-BISNORLABD-12-ENE-12-YL)METHANE
$C_{37}H_{60}O_2$

A.I. YANOVSKII, S.V. LINDEMAN, R.G. GERR and Yu.T. STRUCHKOV, 1984. Kristallografiya, 29, 285-290 [Sov. Phys. Crystallogr., 29, 169-172].

Monoclinic, C2, a = 29.782, b = 6.920, c = 17.768 Å, β = 118.3°, Z = 4. Mo radiation, R = 0.045 for 4116 reflexions (at -120°C).

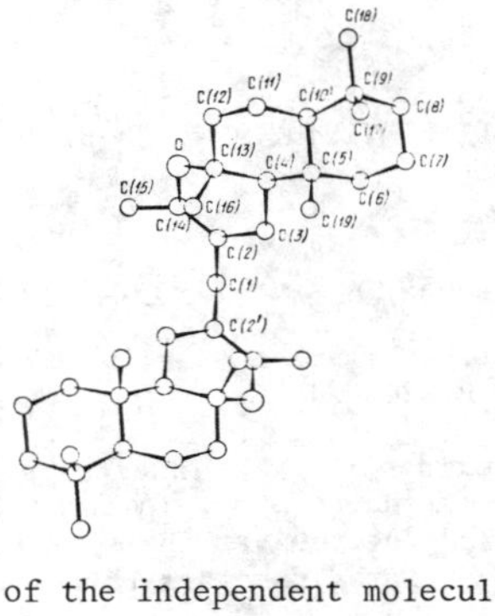

Fig. 1. Projection of one of the independent molecules of $C_{37}H_{60}O_2$ on the ac plane.

There are two independent molecules per asymmetric unit, each with twofold crystallographic symmetry (Fig. 1). These independent molecules have similar molecular geometry with mean bond lengths of 1.541 Å (Csp^3-Csp^3), 1.507 Å (Csp^3-Csp^2), 1.328 Å (Csp^2-Csp^2), and 1.391 Å (O-Csp^2). The bond angles at the bridging C(1) atom are 111.3 and 111.7° and at the O atom 115.4 and 115.6°. The conformation of the dihydropyran rings is intermediate between a 4,13 half chair and a 13-sofa. The conformation of the cyclohexane rings is a rather distorted chair.

(7R,8R)-7,8-EPOXY-6,19-(4'-PHENYL-1',2'-URAZOLO)-9,10-seco-5(10)-CHOLESTEN-3β-BENZOATE
$C_{42}H_{53}N_3O_5$

C. KRATKY, W. REISCHL and E. ZBIRAL, 1984. Monatsh., 115, 1453-1458.

Orthorhombic, $P2_12_12_1$, a = 7.402, b = 14.050, c = 35.48 Å, Z = 4. Mo radiation, R = 0.058 for 1824 reflexions (at 92 K).

Epoxidation of vitamin D_3 with meta-chloroperbenzoic acid yields exclusively the 7,8-monoxirane. Its 7R,8R stereochemistry has been established by X-ray analysis of the 4-phenyl-1,2,4-triazoline-3,5-dione adduct-derivative. The correct absolute configuration is shown in Figure 1. This result corrects the stereochemical assignment in previous publications (1).

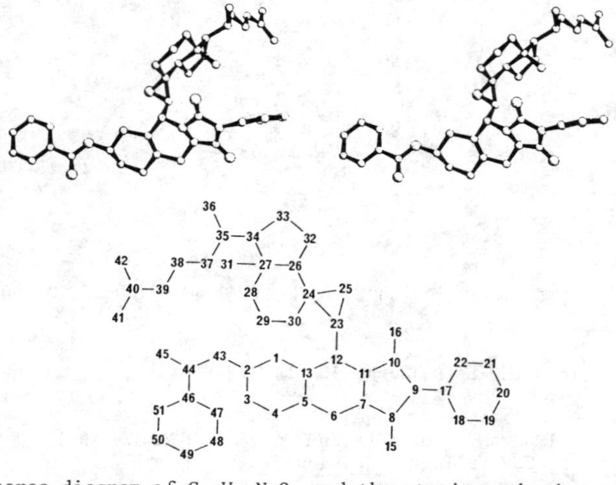

Fig. 1. Stereo diagram of $C_{42}H_{53}N_3O_5$ and the atomic numbering scheme.

1. Structure Reports, 49B, 392.

(±)-cis-1,3,6,8-TETRAKIS(1,1-DIMETHYLETHYL)-4b,9b-BIS(4-METHOXYPHENYL)-4b,9b-DIHYDROBENZOFURO[3,2-b]BENZOFURAN
$C_{44}H_{54}O_4$

R.Y. WONG and L. JURD, 1984. Aust. J. Chem., 37, 2593-2597.

Orthorhombic, $Pca2_1$, a = 20.617, b = 9.175, c = 19.717 Å, Z = 4. Cu radiation, R = 0.043 for 2419 reflexions. M.P. 243-245°C.

The molecule is a minor oxidant of the insect sterillant 2,4-bis(1,1-dimethylethyl)-6-[(4-methoxyphenyl)methyl]phenol. It is composed of a pair of planar dihydrobenzofuran moieties fused together at the side of their furan rings and substituted with a pair of cis related p-methoxyphenyl units (Fig. 1).

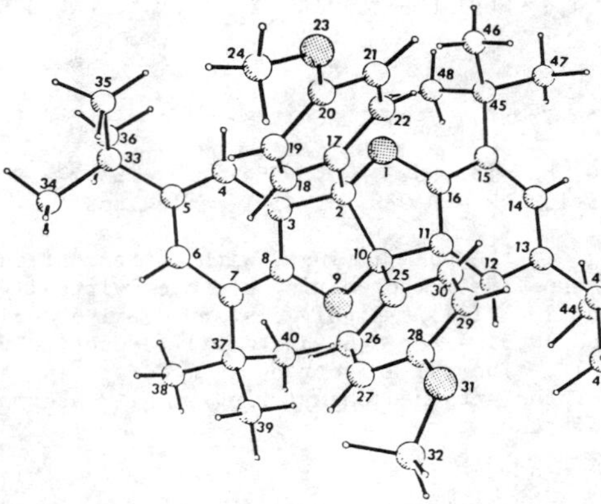

Fig. 1. The structure of $C_{44}H_{54}O_4$.

DIAMMONIUM BIS(DIBENZO-18-CROWN-6) TETRAISOTHIOCYANATOCOBALTATE(II) ACETONE SOLVATE

$C_{44}H_{56}CoN_6O_{12}S_4,2C_3H_6O$ $[NH_4(C_{20}H_{24}O_6)^+]_2[Co(NCS)_4]^{2-},2(CH_3)_2CO$

S.T. MALINOVSKII, Yu.A. SIMONOV, T.I. MALINOVSKII, V.E. ZUBAREVA, D.G. BATYR and I.M. REIBEL', 1984. Kristallografiya, 29, 466-470 [Sov. Phys. Crystallogr., 29, 279-281].

Monoclinic, C2/c, a = 20.498, b = 15.255, c = 19.158 Å, β = 95.2°, D_m = 1.29, Z = 4. Cu radiation, R = 0.069 for 3145 reflexions.

The complex cations (Fig. 1) are linked by hydrogen bonds to the acetone molecule (N(3)-H...O(7) 2.93 Å). The ammonium ion forms hydrogen bonds of the NH...O type and is at practically equal distances (2.87-2.96 Å) from the six ether O atoms. The N atom is 1.02 Å from the plane of the O atoms of the ring on the same side as that to which the benzene rings are turned. The anion has twofold symmetry with four SCN^- groups coordinating the metal ion in a tetrahedron. The bond angles at the Co atom lie between 104.3 and 112.0° and the Co-N distances are 1.970 and 1.952 Å.

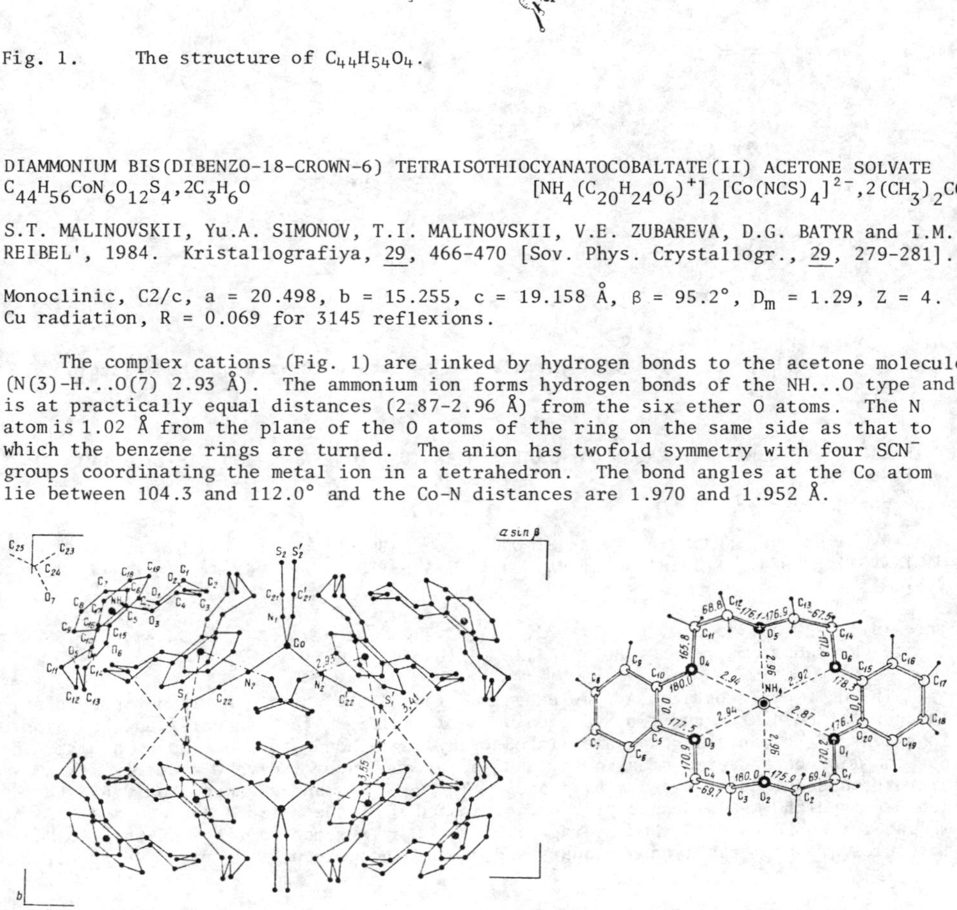

Fig. 1. Projection of the crystal structure of $C_{44}H_{56}CoN_6O_{12}S_4,2C_3H_6O$ on the xy plane (left), and structure of the complex cation showing torsion angles (°) (right).

4'5'-DIHYDRO-1',2,2',3,4,4,4'-HEPTAPHENYLSPIRO[CYCLOPENT-2-ENE-5,5'-NAPHTHO[2,1-b]-FURAN]-1-ONE

$C_{58}H_{40}O_2$

C.T. GRAINGER, 1984. Acta Cryst., C40, 1236-1239.

Triclinic, $P\bar{1}$, a = 12.030, b = 13.196, c = 13.754 Å, α = 93.27, β = 100.56, γ = 109.20°, Z = 2. Mo radiation, R = 0.039 for 4171 reflexions.

In the molecule (Fig. 1) the phenyl rings, with the exception of rings (10) and (11), the benzo ring and the five-membered ring are each virtually planar; ring (10) is somewhat distorted and ring (11) is highly distorted, with deviations from the plane as great as 0.33 and -0.31 Å for C(5) and C(9') respectively and torsional angles as large as 51.8(2)°. Thermal anisotropy is moderate in the central rings but significant in some of the attached phenyl rings. All intermolecular contacts are above van der Waals distances.

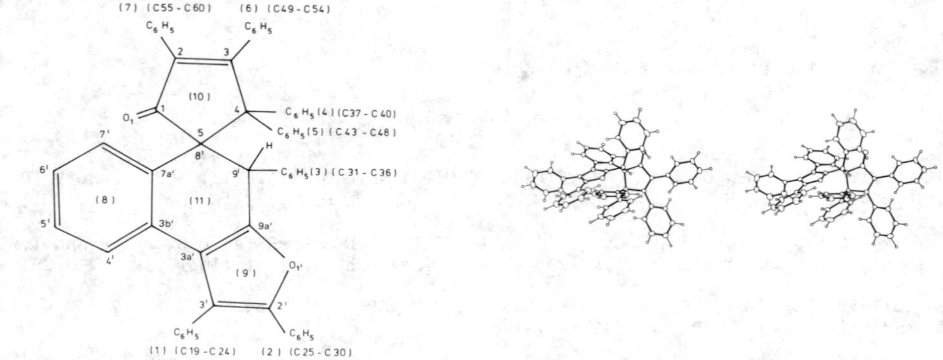

Fig. 1. $C_{58}H_{40}O_2$: structural formula and stereoscopic view of the molecule.

TETRAKIS(DICYCLOHEXYL-18-CROWN-6) HEXAKIS(NITRATO-O,O')CERATE(IV) BIS(OXONIUM) TRIHYDRATE

$C_{80}H_{150}CeN_6O_{44},3H_2O$

A.A. DVORKIN, N.F. KRASNOVA, Yu.A. SIMONOV, V.M. ABASHKIN, V.V. YAKSHIN and T.I. MALINOVSKII, 1984. Kristallografiya, 29, 471-475 [Sov. Phys. Crystallogr., 29, 282-285].

Triclinic, $P\bar{1}$, a = 15.510, b = 14.973, c = 13.456 Å, α = 71.53, β = 94.41, γ = 68.86°, Z = 1. Mo radiation, R = 0.080 for 5111 reflexions.

The unit cell contains a complex anion, four dicyclohexyl-18-crown-6 molecules, three water molecules and two hydroxonium cations (Fig. 1). The water and hydroxonium ions centre the macroring and are linked by hydrogen bonds to the ether O atoms. The Ce atom lies on a centre of symmetry and is surrounded by twelve O atoms from six nitro groups acting as bidentate ligands. The Ce-O distances range from 2.633 to 2.683 Å, and the N-O distances for uncoordinated O of the nitro group average 1.23 Å, whilst for coordinated O they average 1.26 Å. The 18-membered rings of the two DCH-6 molecules are appreciably non-planar, and one of the molecules is partly disordered.

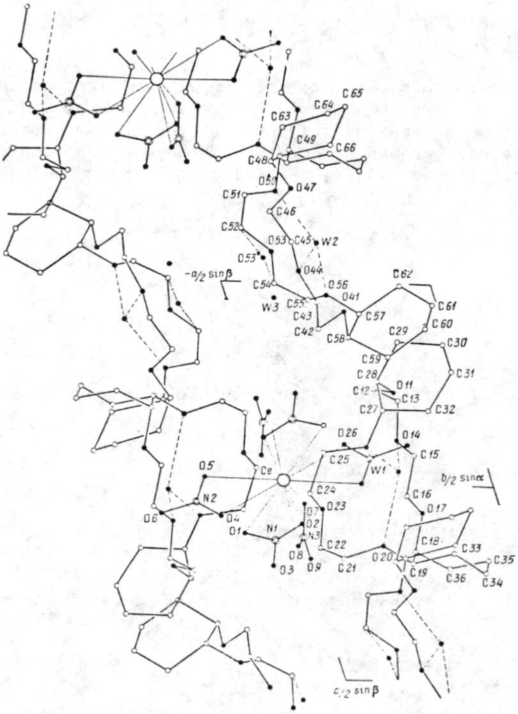

Fig. 1. Projection of the crystal structure of $C_{80}H_{150}CeN_6O_{44},3H_2O$ on xy plane.

1-(5-NITRO-1,3-THIAZOL-2-YL)-2-IMIDAZOLIDINONE (NIRIDAZOLE)
$C_6H_6N_4O_3S$

O.M. PEETERS, N.M. BLATON and C.J. DE RANTER, 1984. Acta Cryst., C40, 1748-1750.

Orthorhombic, $Pc2_1n$, a = 6.467, b = 6.533, c = 19.701 Å, D_m = 1.69, Z = 4. Cu radiation, R = 0.037 for 701 reflexions.

The non-H atoms of the molecule are nearly planar (Fig. 1) and the carbonyl moiety is cis to the S atom of the thiazolyl ring. The molecules are packed in parallel sheets perpendicular to the b axis and endless chains are formed in the c direction by a weak bifurcated hydrogen bond.

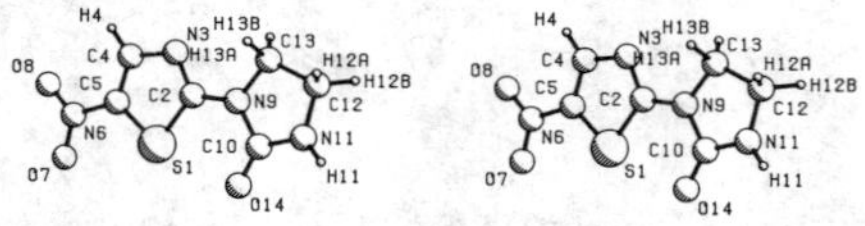

Fig. 1. A stereoview of $C_6H_6N_4O_3S$.

2-THIENYLMETHYLENEMALONONITRILE
$C_8H_4N_2S$

A.K. MUKHERJEE, M. MUKHERJEE, A. DE and S.P. BHATTACHARYYA, 1984. Acta Cryst., C40, 991-992.

Monoclinic, $P2_1/c$, a = 5.701, b = 18.613, c = 7.473 Å, β = 97.57°, Z = 4. Mo radiation, R = 0.050 for 929 reflexions.

In the crystal structure (Fig. 1) the two planar parts of the molecule, the thiophene ring and the dicyanovinyl moiety, are inclined at 3.6(5)° to one another and the vinylic H atom is trans with respect to the S atom. The electron-withdrawing properties of the cyanide groups affect two of the C-C bond lengths, C(3)-C(4) and C(4)-C(5), which have, respectively, partial single- and double-bond characters.

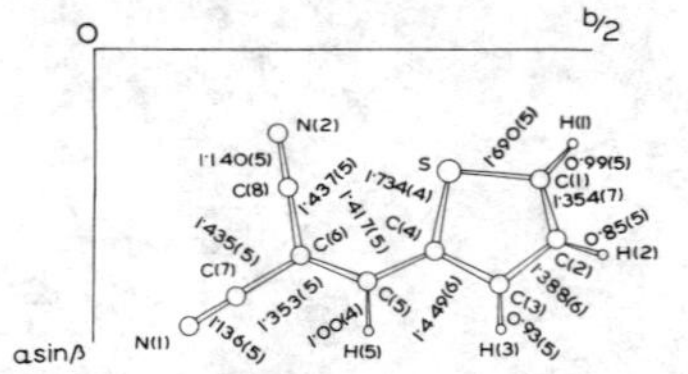

Fig. 1. $C_8H_4N_2S$: c-axis view of the molecule with bond lengths.

2-(1,3-DITHIOL-2-YLIDENE)-4-CYCLOPENTENE-1,3-DIONE
$C_8H_4O_2S_2$ (I)

2-(1,3-BENZODITHIOL-2-YLIDENE)CYCLOHEXANONE
$C_{13}H_{12}OS_2$ (II)

W.H. WATSON, J. GALLOY, D.A. GROSSIE and J. NAKAYAMA, 1984. Acta Cryst., C40, 190-193.

I. Orthorhombic, Pnma, a = 18.080, b = 11.102, c = 3.8904 Å, Z = 4. Cu radiation, R = 0.046 for 597 reflexions.

II. Monoclinic, $P2_1/c$, a = 6.459, b = 14.263, c = 13.514 Å, β = 112.16°, Z = 4. Cu radiation, R = 0.074 for 1429 reflexions.

Molecule I (Fig. 1) lies across a crystallographic mirror plane (C(1) and C(3) on the mirror) and is essentially planar; there is a small 2° fold along the S...S line. Compound II (Fig. 1) is planar except for atoms C(12)-C(14) and O(15). There are short intramolecular S...O contacts (2.944(3) Å in (I) and 3.064(8) Å in (II)).

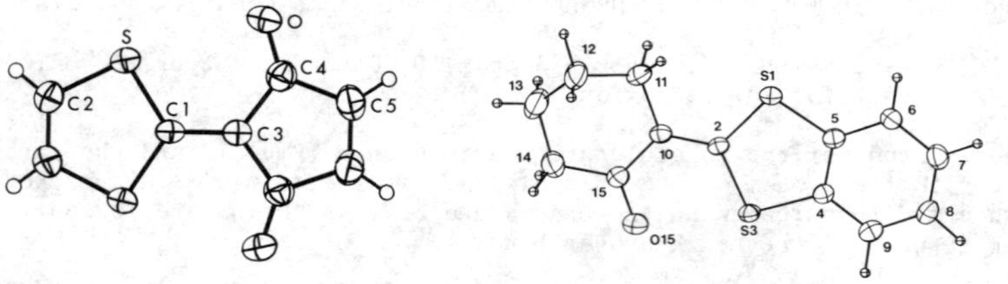

Fig. 1. Views of $C_8H_4O_2S_2$ and $C_{13}H_{12}OS_2$.

1-METHYL-exo,exo-3,5-DICHLORO-8-THIATRICYCLO[2.2.1.1^{2,6}]OCTANE TETRAFLUOROBORATE
$C_8H_{11}BCl_2F_4S$ $C_8H_{11}Cl_2S^+.BF_4^-$

A.L. RHEINGOLD, E.J. MILLER, R.L. DURNEY, R.L. CARET and J.J. UEBEL, 1984. Phosphorus and Sulfur, 21, 215-220.

Orthorhombic, Pbca, a = 9.269, b = 12.112, c = 21.418 Å, Z = 8. Mo radiation, R = 0.056 for 1289 reflexions.

The sulphur atom (Fig. 1) has pyramidal geometry, with C1-S 1.797(8), C2-S 1.844(5), C6-S 1.832(5) Å, external mean C-S-C 102.8(5)°, internal C-S-C 74.1(2)°.

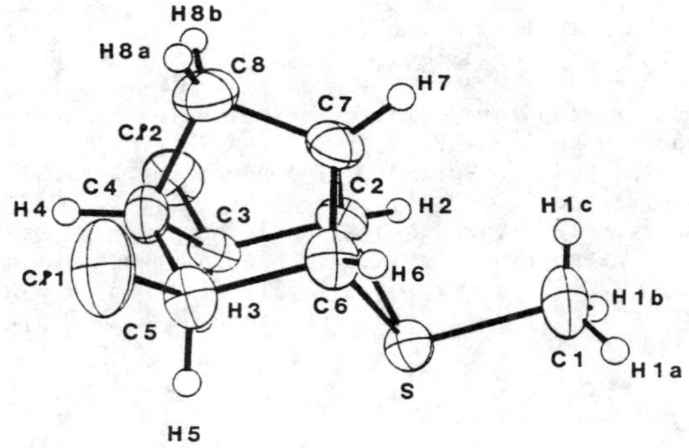

Fig. 1. The $C_8H_{11}Cl_2S^+$ cation.

(1R,2S,7R,8S)-1,2,7,8-TETRAHYDROXY-4,5,10,11-TETRATHIACYCLODODECANE HYDRATE
$C_8H_{16}O_4S_4,H_2O$ ($TTCD,H_2O$)

S. CAPASSO, L. MAZZARELLA, A. ZAGARI and A. IMMIRZI, 1984. J. Cryst. Spectrosc. Res., 14, 303-314.

Monoclinic, $P2_1/b$, a = 4.893, b = 15.110, c = 74.30 Å, β = 90.00°, D_m = 1.56, Z = 16. Cu radiation, R = 0.090 for 645 reflexions.

TTCD molecules (Fig. 1) sandwich water molecules, giving rise to two-molecule layers (bilayers) of $P2_1/b$ symmetry which gain remarkable rigidity because of their hydrogen bonds. In their turn, bilayers are piled according to four different structures having equivalent inter-bilayer H bonding and the same nonprimitive unit cell. The crystal is twinned by merohedry with the four structures in comparable amounts. The relationship of the solid-state conformations to the UV absorption spectrum in solution is also discussed.

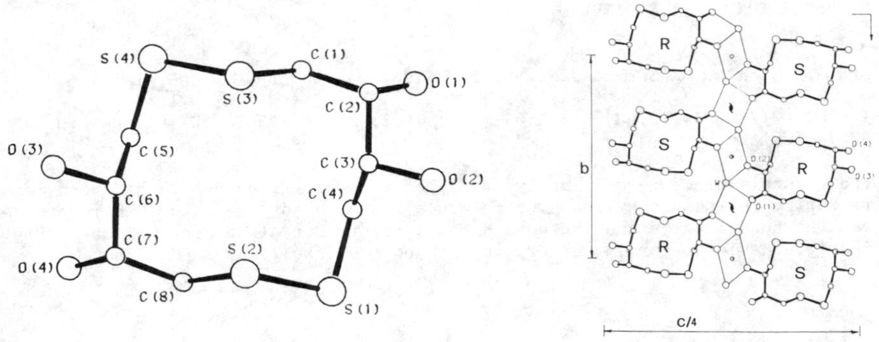

Fig. 1. A view of a TTCD molecule and a diagram showing a two-molecule layer constituting the basic structural unit of twinned monoclinic crystals.

(S)-α-(p-BROMOBENZENESULPHONAMIDO)-β-PROPIOTHIOLACTONE
$C_9H_8BrNO_3S_2$

I. MATIJAŠIĆ, G.D. ANDREETTI, P. SGARABOTTO, A. BEZJAK and D. FLEŠ, 1984. Croat. Chem. Acta, 57, 621-628.

Orthorhombic, $P2_12_12_1$, a = 10.125, b = 12.439, c = 9.304 Å, D_m = 1.83, Z = 4. Cu radiation, R = 0.045 for 2277 reflexions.

The crystal structure consists of discrete molecules connected along the a axis by intermolecular hydrogen bonds N-H...O of 2.919 Å (Fig. 1). The four-membered thietane ring is, as expected, non-planar. The puckering angles between the planes defined by S(1),C(1),C(2) and S(1),C(3),C(2) and by C(1),C(2),C(3) and C(1),S(1),C(3) are 8.9 and 8.4° respectively. The dihedral angle between the best planes of the β-propiothiolactone and benzene rings is 84.1°. Bond distances within the ring are S(1)-C(3) 1.808, S(1)-C(1) 1.762, C(1)-C(2) 1.521, C(2)-C(3) 1.540 Å, and distances outside the ring are: C(1)-O(1) 1.205, C(2)-N 1.438, N-S(2) 1.615, S(2)-O(2) 1.435, S(2)-O(3) 1.449 and S(2)-C(4) 1.767 Å.

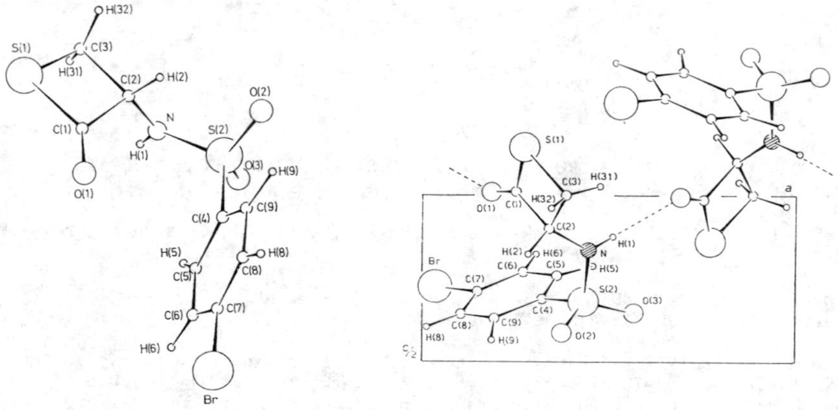

Fig. 1. Perspective drawing of $C_9H_8BrNO_3S_2$ and molecular packing along [010].

HEXACHLORONAPHTHALENE-1,8-DISULPHIDE
$C_{10}Cl_6S_2$

F.H. HERBSTEIN, R.E. MARSH and J.H. NOORDIK, 1984. Acta Cryst., C40, 1633-1634.

Monoclinic, $P2_1/n$, a = 10.68, b = 3.853, c = 15.31 Å, β = 93.3°, Z = 2. R = 0.083 for 485 reflexions.

The crystal structure originally described as space group Pn (1) has been more satisfactorily refined in $P2_1/n$. The molecule is now disordered about an inversion centre. The bond lengths and non-bonded contacts are normal and the molecule (Fig. 1) is slightly non-planar forming a very shallow boat or skullcap.

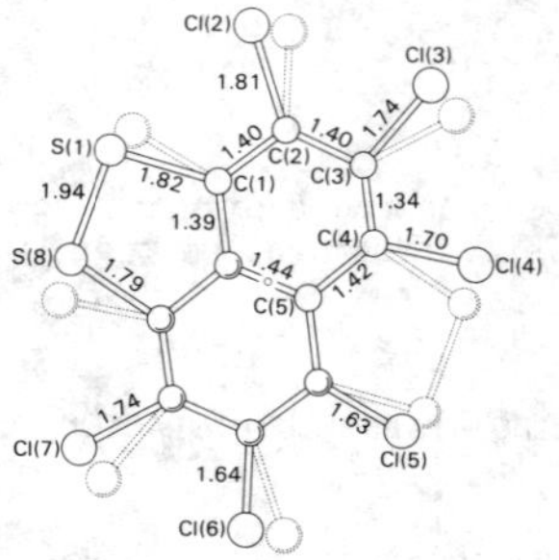

Fig. 1. The disordered structure of $C_{10}Cl_6S_2$.

1. Structure Reports, 49B, 400.

1,8:4,5-BIS(DISELENO)NAPHTHALENE (TETRASELENONAPHTHALENE)
$C_{10}H_4Se_4$

J.C. STARK, R. REED, L.A. ACAMPORA, D.J. SANDMAN, S. JANSEN, M.T. JONES and B.M. FOXMAN, 1984. Organometallics, 3, 732-735.

Monoclinic, $P2_1/c$, a = 7.722, b = 4.226, c = 15.305 Å, β = 102.15°, Z = 2. Mo radiation, R = 0.037 for 999 reflexions.

In the molecular structure the C atoms form a plane with no deviations >0.007 Å from the plane; however the Se atoms deviate markedly from this plane (Se(1) 0.101, Se(2) -0.049 Å). The C-Se-Se-C torsion angle is 4.4° (Fig. 1). The crystal structure consists of linear chains of TSeN stacked along b with interstack Se...Se contacts at 3.626 and 3.534 Å.

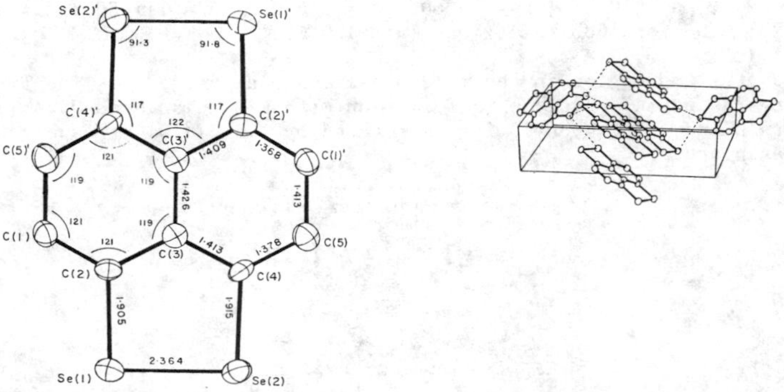

Fig. 1. Molecular structure of $C_{10}H_4Se_4$ showing selected bond lengths and angles (left), and a view of the packing showing short interstack Se...Se contacts (right).

2,4-BIS(CYANOMETHOXYCARBONYLMETHYLENE)-1,3-DITHIETANE
$C_{10}H_6N_2O_4S_2$

G. L'ABBE, P. VANGHELUWE, S. TOPPET, G.S.D. KING and L. VAN MEERVELT, 1984. Bull. Soc. Chim. Belg., 93, 405-415.

Monoclinic, $P2_1/c$, a = 13.196, b = 5.672, c = 8.590 Å, β = 107.72°, D_m = 1.53, Z = 2. Mo radiation, R = 0.059 for 566 reflexions.

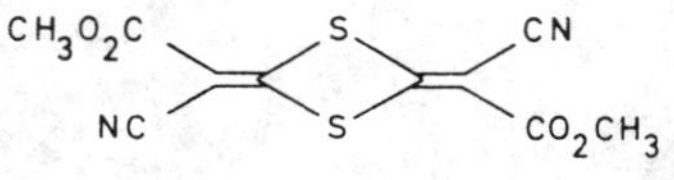

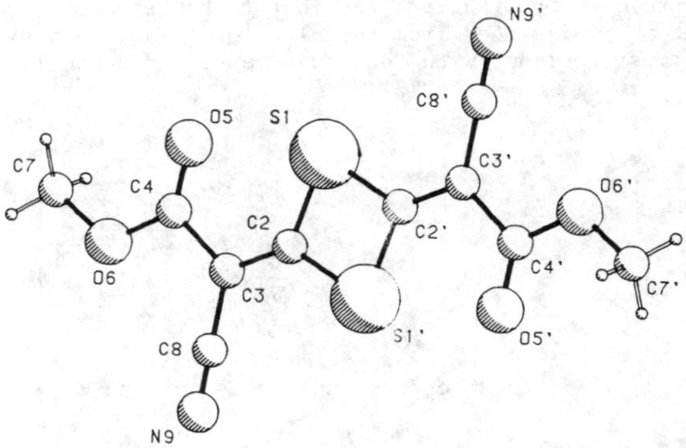

Fig. 1. A view of $C_{16}H_6N_2O_4S_2$.

The molecule (Fig. 1) lies about an inversion centre and consequently the 1,3-dithietane ring must be planar with trans-stereochemistry. The α,β-unsaturated system has an s-cis conformation and is coplanar with the ring, resulting in S(1)...O(5) being 2.74(4) Å. The two C-S bond lengths are different C(2)-S(1) 1.739(7), C(2)-S(1') 1.775(6) Å, but the exocyclic angles at C(2) are the same (130.7°).

3-PHENYL-2-THIOLENE-1,1-DIOXIDE

$C_{10}H_{10}O_2S$

G.G. RODE, T.E. BEZMENOVA, M.Yu. ANTIPIN, Yu.T. STRUCHKOV and G.V. KLIMUSHEVA, 1984. J. Struct. Chem., 25, 661-662 [Zh. Strukt. Khim., 25(4), 158-159].

Monoclinic, $P2_1/c$, a = 10.568, b = 10.602, c = 8.630 Å, β = 111.80°, Z = 4. Mo radiation, R = 0.03 for 1003 reflexions (at -120°C).

Bond lengths and angles in the molecule are shown in Fig. 1. The five-membered thiolene ring has a C(4)-envelope conformation with C(4) displaced 0.240(2) from the S,C(1-3) mean plane. The phenyl ring is rotated by 1.4(3)° with respect to the S,C(1-3) mean plane.

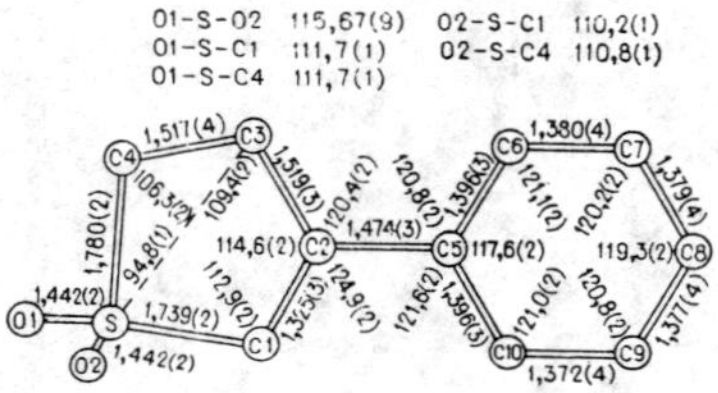

Fig. 1. Bond lengths and angles for the 3-phenyl-2-thiolene-1,1-dioxide molecule.

3-PHENYL-3-THIOLENE-1,1-DIOXIDE

$C_{10}H_{10}O_2S$

G.G. RODE, T.E. BEZMENOVA, M.Yu. ANTIPIN, Yu.T. STRUCHKOV, G.V. KLIMUSHEVA and A.P. POLISHCHUK, 1984. J. Struct. Chem., 25, 663-664 [Zh. Strukt. Khim., 25(4), 159-160].

Orthorhombic, $Pna2_1$, a = 9.615, b = 5.624, c = 16.58 Å, Z = 4. Mo radiation, R = 0.026 for 1006 reflexions (at -120°C).

Bond lengths and angles are shown in Fig. 1. The five-membered ring has an envelope conformation; the four carbon atoms are coplanar within 0.008 Å, and the S atom is displaced from the plane of these atoms by 0.3621(5) Å. The phenyl ring is almost coplanar with the above-mentioned plane (torsion angles about the C(2)-C(5) bond are -6.0(3) and -8.2(3)°).

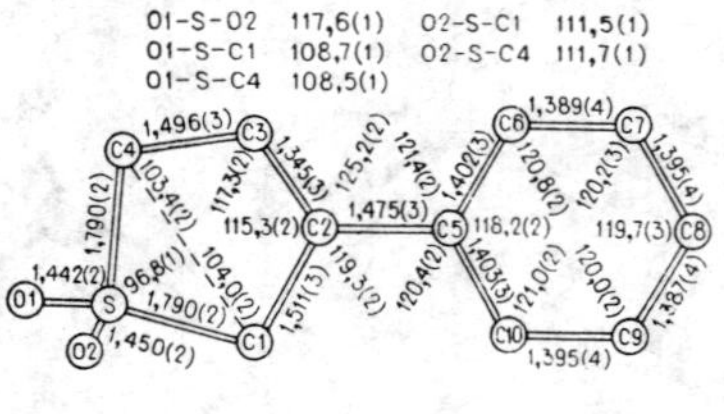

Fig. 1. Bond lengths and angles for the 3-phenyl-3-thiolene-1,1-dioxide molecule.

4-PHENYL-2-THIOLENE-1,1-DIOXIDE

$C_{10}H_{10}O_2S$

G.G. RODE, T.E. BEZMENOVA, M.Yu. ANTIPIN, Yu.T. STRUCHKOV, I.A. LITVINOV, A.I. YANOVSKII and G.V. KLIMUSHEVA, 1984. J. Struct. Chem., 25, 665-666 [Zh. Strukt. Khim., 25(4), 160-162].

Orthorhombic, $P2_12_12_1$, a = 8.466, b = 19.334, c = 5.7128 Å, Z = 4. Cu radiation, R = 0.054 for 813 reflexions.

Bond lengths and angles are shown in Fig. 1. The five-membered ring has a slightly distorted C(1)-envelope conformation; C(1) deviates 0.152(7) Å from the mean plane of the remaining ring atoms which are coplanar to within 0.006 Å. The angle between normals to the phenyl ring and the S,C(2-4) mean planes is 59.8(3)°.

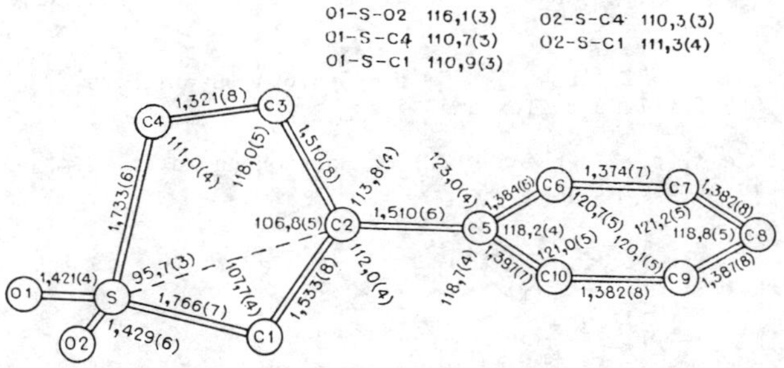

Fig. 1. Bond lengths and angles for the 4-phenyl-2-thiolene-1,1-dioxide molecule.

TETRAMETHYLTETRATHIAFULVALENE DIPERCHLORATE

$C_{10}H_{12}Cl_2O_8S_2$ $[C_{10}H_{12}S_2]^{2+}2ClO_4^-$

R.P. SHIBAEVA, 1984. Kristallografiya, 29, 480-483 [Sov. Phys. Crystallogr., 29, 288-289].

Monoclinic, $P2_1/n$, a = 12.385, b = 9.993, c = 7.004 Å, β = 101.73°, Z = 2. Mo radiation, R = 0.033 for 1130 reflexions.

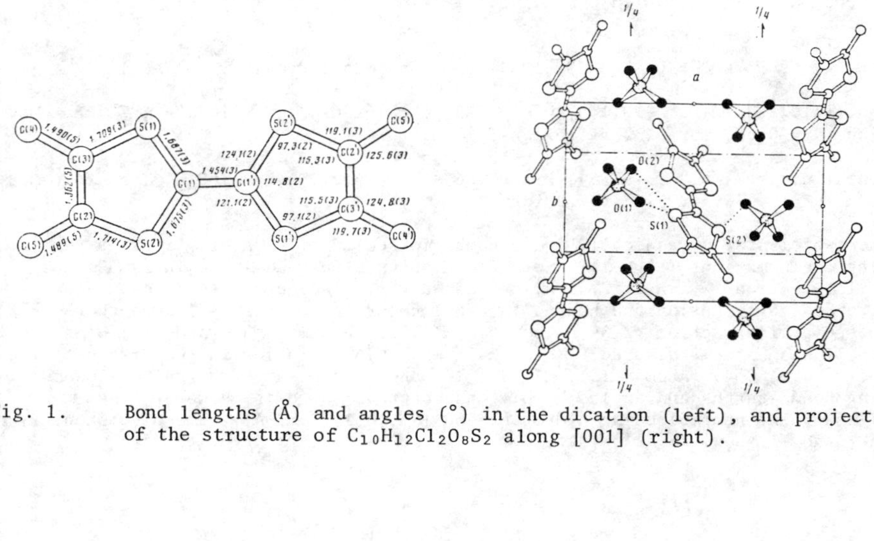

Fig. 1. Bond lengths (Å) and angles (°) in the dication (left), and projection of the structure of $C_{10}H_{12}Cl_2O_8S_2$ along [001] (right).

The dications lie on crystallographic centres of symmetry and are not quite planar (Fig. 1). The length of the central bond C(1)-C(1') (1.454 Å) is close to that of a single Csp^2-Csp^2 bond, and the angles S(1)-C(1)-C(1') and S(2)-C(1)-C(1') are unequal - two noteworthy features. The Cl-O distances in the anion are 1.441, 1.429, 1.425 and 1.411 Å with O-Cl-O angles ranging from 108.5 to 110.4°. The crystal structure consists of alternate anion and cation layers almost perpendicular to the a axis. There are no isolated conducting cation stacks. The short inter-ionic contacts are: S(1)...O(1') 3.02, S(1)...O(2') 3.21, and S(2)...O(1'') 3.29 Å.

S,S-TETRAMETHYLENE-N-(p-NITROPHENYL)SULPHILIMINE
$C_{10}H_{12}N_2O_2S$

S. TSUCHIYA, S.-I. MITOMO, M. SENO and H. MIYAMAE, 1984. J. Org. Chem., 49, 3556-3559.

Monoclinic, A2/a, a = 16.726, b = 25.877, c = 5.676 Å, β = 118.29°, Z = 8. Cu radiation, R = 0.055 for 1286 reflexions.

In the molecule (Fig. 1), the S-N1-C1-C2 torsion angle is -3.4° and the $SNC_6H_5NO_2$ moiety is almost planar.

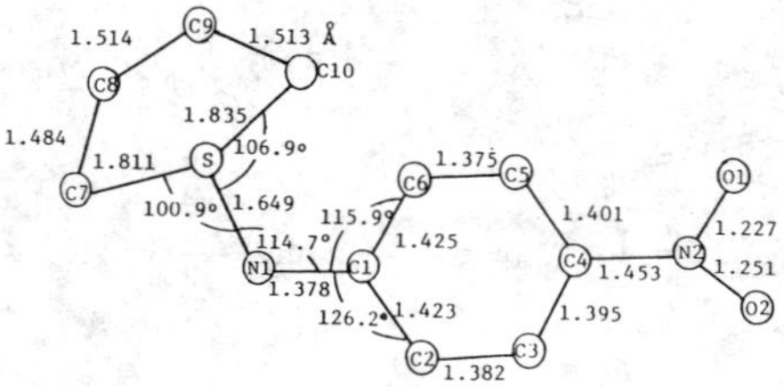

Fig. 1. Molecular dimensions for $C_{10}H_{12}N_2O_2S$.

2-(3-DIMETHYLAMINO-2-METHYLPROPYL)-1,3-DITHIANE-1,1,3,3-TETRAOXIDE
$C_{10}H_{21}NO_4S_2$ (I)

2-(2-MORPHOLINOETHYL)-1,3-DITHIONE-1,1,3,3-TETRAOXIDE
$C_{10}H_{19}NO_5S_2$ (II)

R.L. HARLOW, C. LI and M.P. SAMMES, 1984. J. Chem. Soc. Perkin I, 547-551.

I. Monoclinic, $P2_1/n$, a = 19.144, b = 9.372, c = 16.561 Å, β = 108.74°, Z = 8. Mo radiation, R = 0.052 for 3432 reflexions.

II. Monoclinic, $P2_1/n$, a = 9.534, b = 20.563, c = 6.936 Å, β = 99.81°, Z = 4. Mo radiation, R = 0.079 for 2168 reflexions.

The two crystallographically independent molecules of I (A, B) exist in two different conformations (Fig. 1). Molecule A shows a very definite intramolecular C-H...N interaction (N...H(1) 2.41 Å) and has a very disordered C(6) atom. B has a more extended conformation and has no intramolecular interaction although there is an intermolecular interaction (2.39 Å) between H(1) of this molecule and O(3) of A. Compound II has a badly disordered sidechain with C(6) existing in two distinct sites, occupancies 0.6, 0.4. Atoms C(5), C(7), C(10) and N have large thermal ellipsoids rendering bond lengths and angles highly inaccurate. There is, however, the suggestion of an intramolecular hydrogen bond (N...H(1) 2.57 Å). The disulphone rings of both I and II show slight twists.

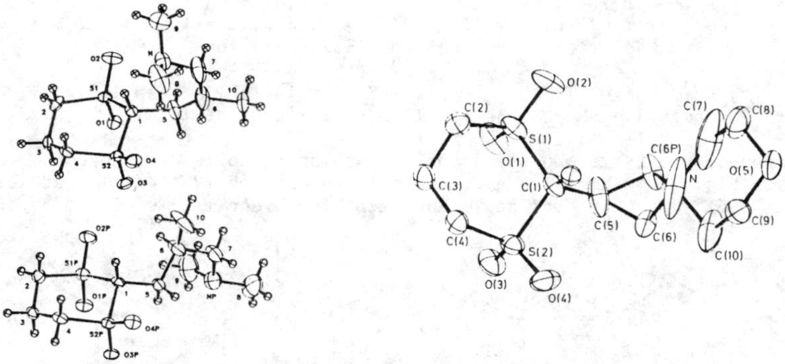

Fig. 1. The molecular structures: $C_{10}H_{21}NO_4S_2$ (left, A (top), B (bottom)) and $C_{10}H_{19}NO_5S_2$ (right).

THIANTHRENE 5,5,10,10-TETRAOXIDE
$C_{12}H_8O_4S_2$

V.M. LYNCH, S.B. LARSON, S.H. SIMONSEN and S.N. THOMAS, 1984. Acta Cryst., C40, 1378-1381.

Orthorhombic, Pbcn, a = 19.613, b = 12.161, c = 14.433 Å, D_m = 1.59, Z = 12. Mo radiation, R = 0.050 for 3077 reflexions (at 163K).

Within the unit cell, eight molecules are in general positions and four lie around twofold axes. In each molecule the SO_2 planes are almost at right angles to the corresponding CSC planes with the SO_2 groups coplanar. The S...S separation distance is 3.261(1) and 3.263(2) Å in each molecule. The angles between the two planes comprising one phenyl ring and the two S atoms are 138.01(5) and 144.01(5)° respectively in the two molecules.

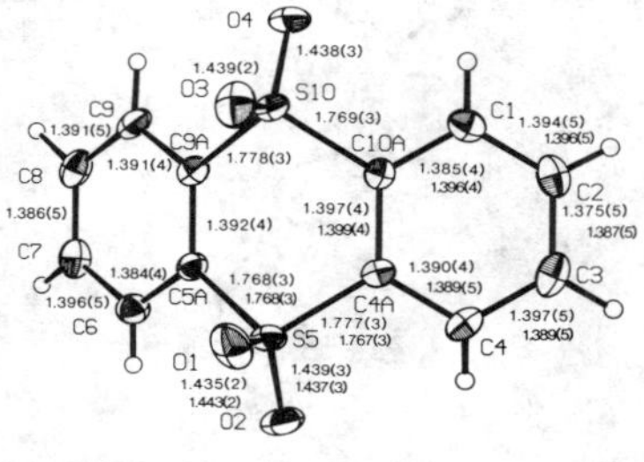

Fig. 1. Bond lengths in the thianthrene 5,5,10,10-tetraoxide molecule.

THIANTHRENE
$C_{12}H_8S_2$ (I)

1-AZATHIANTHRENE
$C_{11}H_7NS_2$ (II)

S.B. LARSON, S.H. SIMONSEN, G.E. MARTIN, K. SMITH and S. PUIG-TORRES, 1984. Acta Cryst., C40, 103-106.

I. At 163K. Monoclinic, $P2_1/c$, a = 11.8672, b = 6.0648, c = 14.3856 Å, β = 109.727°, Z = 4. Mo radiation, R = 0.034 for 2087 reflexions.

I. At 295K. Monoclinic, $P2_1/c$, a = 11.9559, b = 6.1578, c = 14.4959 Å, β = 109.964°, D_m = 1.44, Z = 4. Mo radiation, R = 0.043 for 1589 reflexions.

II. Monoclinic, $P2_1/c$, a = 11.7737, b = 6.0267, c = 14.1501 Å, β = 108.507°, D_m = 1.47, Z = 4. Mo radiation, R = 0.034 for 1958 reflexions (at 163K).

The compounds are isomorphous and the 1-azathianthrene is disordered. Bond lengths are in Fig. 1. Dihedral angles for thianthrene are 127.14(3)° (at 163K), 128.09(5)° at 295K. The dihedral angle in 1-azathianthrene is 130.37(3)° at 163K.

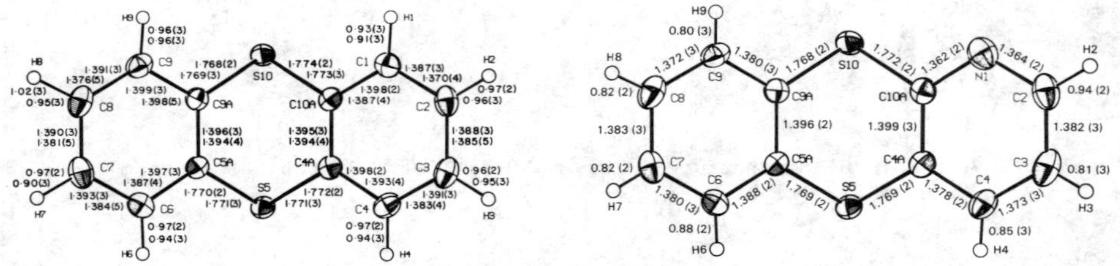

Fig. 1. Bond lengths in thianthrene (left) (upper numbers are 163K data, lower ones for 295K) and in 1-azathianthrene (right).

3,4-DICHLORO-6-SELENATRICYCLO[5.4.0.0^{2,5}]UNDECA-1(7),8,10-TRIEN-2-YL ACETATE
$C_{12}H_{10}Cl_2O_2Se$

F. BAERT, V. WARIN and P. GRANDCLAUDON, 1984. Acta Cryst., C40, 1372-1373.

Monoclinic, $P2_1/c$, a = 12.223, b = 21.631, c = 7.272 Å, β = 137.88°, Z = 4. Mo radiation, R = 0.043 for 2430 reflexions.

The chlorinated cyclobutane fragment has a trans configuration (Fig. 1). Se-C bond lengths are 1.917(7) and 1.965(7) Å and the C-Se-C angle is 89.6°.

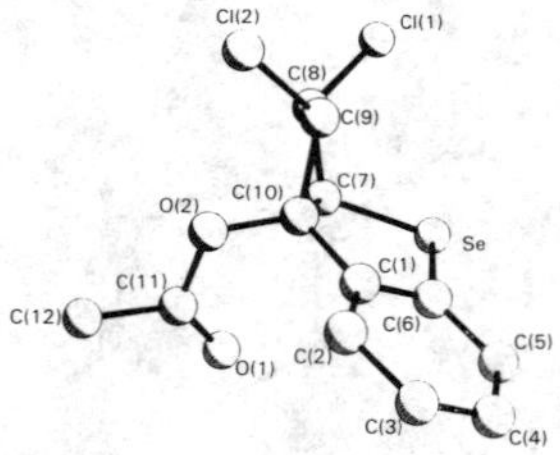

Fig. 1. The $C_{12}H_{10}Cl_2O_2Se$ molecule.

2-THIENYLAMMONIUM 5-HYDROXY-4-METHOXYCARBONYL-1-(2-THIENYL)-1,2,3-TRIAZOLE
$C_{14}H_{16}N_4O_3S_2$

P. MURRAY-RUST, J. McMANUS, S.P. LENNON, A.E.A. PORTER and J.A. RECHKA, 1984. J. Chem. Soc. Perkin I, 713-716.

Triclinic, $P\bar{1}$, a = 10.89, b = 8.41, c = 10.40 Å, α = 92.20, β = 106.39, γ = 111.40°, D_m = 1.394, Z = 2. Mo radiation, R = 0.057 for 1223 reflexions.

The structure consists of 2-thienylammonium cations hydrogen-bonded to the heterocyclic anions (Fig. 1). The heterocyclic anion is present in the enol form. Some bond distances are: C(35)-O(31) 1.261(8), N(31)-N(32) 1.378(7), N(32)-N(33) 1.312(7), N(31)-C(35) 1.382(8), C(34)-C(35) 1.416(9), C(11)-S(20) 1.723(7), C(14)-S(20) 1.699(8), C(21)-S(2) 1.700(8), C(24)-S(2) 1.684(9) Å. [C(31) in the figure in the paper should read C(34).]

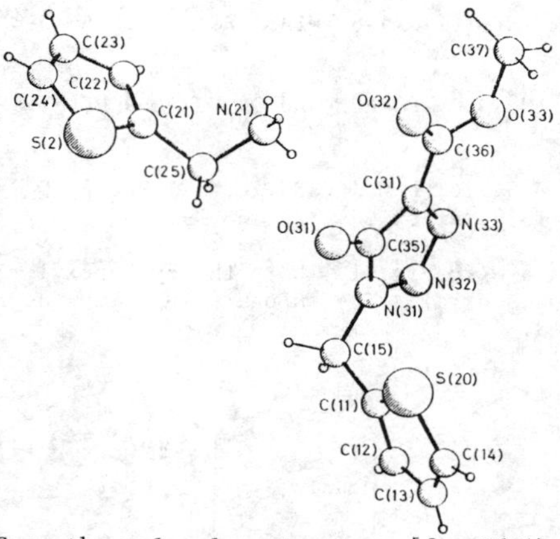

Fig. 1. $C_{14}H_{16}N_4O_3S_2$: the molecular structure [for C(31) read C(34)].

2-CYANO-3-(4-DIETHYLAMINOPHENYL)THIOACRYLAMIDE
$C_{14}H_{17}N_3S$ (I)

6-AMINO-3,5-DICYANO-3,4-DIHYDRO-2,4-BIS(3-METHOXYPHENYL)-2H-THIAPYRAN-3-CARBOTHIOAMIDE ETHANOL SOLVATE (1:1)
$C_{22}H_{20}N_4O_2S_2,C_2H_6O$ (II)

J.S.A. BRUNSKILL, A. DE, D.F. EWING and A.J. WELCH, 1984. Acta Cryst., C40, 493-498.

I. Monoclinic, $P2_1/n$, a = 10.7555, b = 12.3646, c = 10.8680 Å, β = 104.254°, D_m = 1.24, Z = 4. Cu radiation, R = 0.083 for 1866 reflexions.

II. Triclinic, $P\bar{1}$, a = 12.974, b = 12.645, c = 9.025 Å, α = 100.05, β = 69.03, γ = 118.25°, D_m = 1.31, Z = 2. Cu radiation, R = 0.102 for 1770 reflexions.

In (I) (Fig. 1) the aryl and thioamide groups occupy trans positions across the alkene double bond, whilst the diene fragment C=C-C=S has an s-trans geometry. The stereochemical consequences of conjugation in this molecule are successive twists of 8.3 and 18.9° about the single bonds linking aryl and alkene, and alkene and thioalkene, respectively. In (II) the thiapyran ring has a distorted half-chair conformation. The stereochemical pattern of the saturated residue C(3)C(4)C(5) is, respectively, $Ar_{eq}CN_{ax}Ar_{eq}$, and this has been rationalized on intramolecular steric grounds.

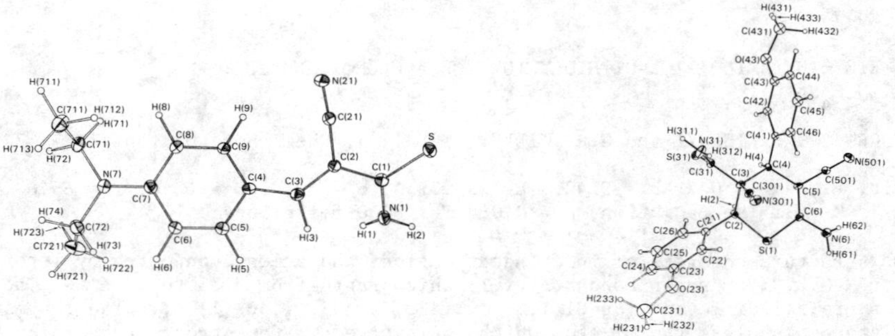

Fig. 1. Views of $C_{14}H_{17}N_3S$ (I) (left) and of $C_{22}H_{20}N_4O_2S_2$ (II) (right).

3-[(4-HYDROXY-4-THIANYL)METHYL]-1-OXA-8-THIASPIRO[4.5]DECAN-2-ONE
$C_{14}H_{22}O_3S_2$

N. SATYAMURTHY, K.D. BERLIN, D.R. POWELL and D. VAN DER HELM, 1984. Phosphorus and Sulfur, 19, 137-153.

Orthorhombic, $P2_12_12_1$, a = 5.916, b = 11.585, c = 21.986 Å, D_m = 1.287, Z = 4. Mo radiation, R = 0.045 for 1319 reflexions (at 138 K).

Molecular dimensions are in Fig. 1. Both thiacyclohexyl rings have chair confirmations with C-O bonds axial. The sulphur ring with the hydroxyl group is orientationally disordered.

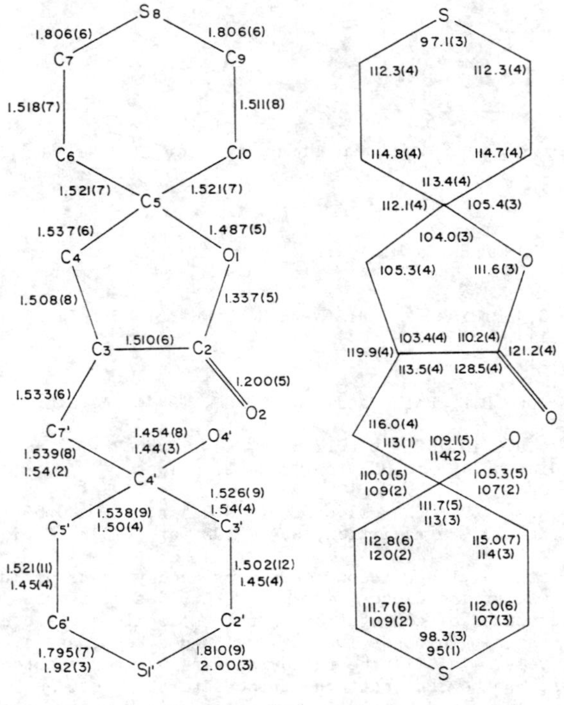

Fig. 1. Bond lengths and angles for $C_{14}H_{22}O_3S_2$.

BIS(3,5-BIS(ETHYLAMINO)-1,2-DITHIOLYLIUM) HEXACHLOROSTANNATE
$C_{14}H_{26}Cl_6N_4S_4Sn$

M.G.B. DREW, J.M. KISENYI and G.R. WILLEY, 1984. J. Chem. Soc. Dalton, 1727-1729.

Triclinic, $P\bar{1}$, a = 9.047, b = 9.429, c = 9.935 Å, α = 80.4, β = 63.0, γ = 65.6°, D_m = 1.71, Z = 1. Mo radiation, R = 0.064 for 3260 reflexions.

The structure consists of independent cations and anions connected by intermolecular N-H...Cl hydrogen bonds (Fig. 1) which range from 3.33 to 3.54 Å. The Sn-Cl distances are all different (Sn-Cl(1,2,3) 2.422, 2.405, 2.467(1) Å). The longest Sn-Cl distance involves the Cl atom which is involved in the strongest hydrogen bond. The 1,2-dithiolylium ring is planar with S(1)-S(2) 2.063(3) Å.

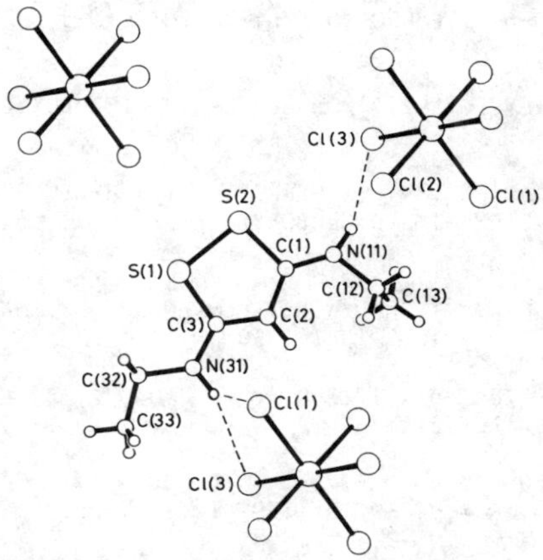

Fig. 1. $C_{14}H_{26}Cl_6N_4S_4Sn$: the environment of one cation.

1,4-DITHIA-8-SPIRO[4.5]DECYL p-TOLUENESULPHONATE
$C_{15}H_{20}O_3S_3$

A. DUBOURG, P. IRATCABAL, M.F. GRENIER-LOUSTALOT, R. ROQUES, J.P. DECLERCQ and G. GERMAIN, 1984. Acta Cryst., C40, 1221-1223.

Triclinic, P$\bar{1}$, a = 13.082, b = 19.595, c = 6.743 Å, α = 89.60, β = 104.68, γ = 81.84°, Z = 4. Mo radiation, R = 0.050 for 2592 reflexions.

In the molecule (Fig. 1) the geometrical modifications introduced by the spiro substitution on the cyclohexane ring are small and the cyclohexane ring has the usual chair conformation, with the p-toluenesulphonyloxy moiety in the equatorial position. The conformation of the dithiolane ring is nearer to a twist than to an envelope. The cohesion of the molecules in the crystal is influenced by hydrogen bonding.

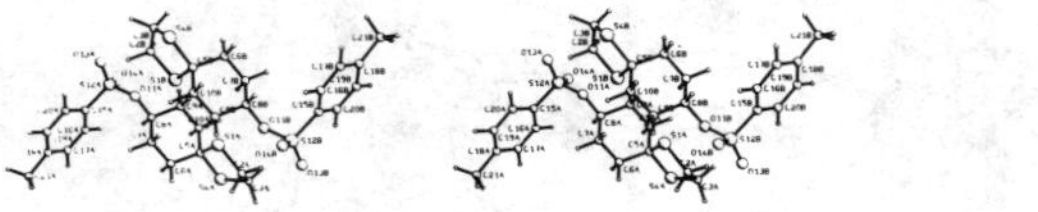

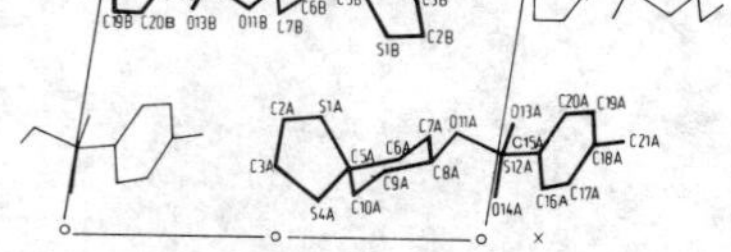

Fig. 1. $C_{15}H_{20}O_3S_3$: stereoscopic view of the molecules and projection of the unit-cell contents.

DIMETHYL cis,cis-7,8-DIHYDRO-6-(1-PYRROLIDINYL)-2H-THIOCIN-4,5-DICARBOXYLATE
$C_{15}H_{21}NO_4S$ (I)

DIMETHYL trans,cis-7,8-DIHYDRO-6-(1-PYRROLIDINYL)-2H-THIOCIN-4,5-DICARBOXYLATE
$C_{15}H_{21}NO_4S$ (II)

DIMETHYL cis,cis-3,8-DIHYDRO-6-(1-PYRROLIDINYL)-2H-THIOCIN-4,5-DICARBOXYLATE
$C_{15}H_{21}NO_4S$ (III)

DIMETHYL 8-(1-PIPERIDINYL)BICYCLO[5.2.0]NON-8-ENE-1,9-DICARBOXYLATE
$C_{18}H_{27}NO_4$ (IV)

DIMETHYL 8-[2,3-BIS(METHOXYCARBONYL)-4-(1-PYRROLIDINYL)-4-CYCLOOCTEN-1-YL]-3-(1-PYRROLIDINYL)-1,3-CYCLOOCTADIENE-1,2-DICARBOXYLATE
$C_{32}H_{46}N_2O_8$ (V)

D.N. REINHOUDT, W. VERBOOM, G.W. VISSER, W.P. TROMPENAARS, S. HARKEMA and G.J. VAN HUMMEL, 1984. J. Am. Chem. Soc., 106, 1341-1350.

I. Monoclinic, $P2_1/c$, a = 9.704, b = 17.260, c = 10.049 Å, β = 108.58°, Z = 4. Cu radiation, R = 0.050 for 2806 reflexions.

II. Triclinic, $P\bar{1}$, a = 13.600, b = 11.477, c = 10.546 Å, α = 74.66, β = 78.52, γ = 79.93°, Z = 4. Mo radiation, R = 0.059 for 3497 reflexions.

III. Triclinic, $P\bar{1}$, a = 12.028, b = 9.446, c = 7.027 Å, α = 100.92, β = 100.34, γ = 96.05°, Z = 2. Cu radiation, R = 0.044 for 2415 reflexions (at 218 K).

IV. Monoclinic, $P2_1/c$, a = 12.358, b = 10.885, c = 13.870 Å, β = 110.58°, Z = 4. Cu radiation, R = 0.045 for 2585 reflexions.

V. Monoclinic, $P2_1/n$, a = 13.718, b = 25.950, c = 8.922 Å, β = 102.11°, Z = 4. Cu radiation, R = 0.095 for 4369 reflexions (at 148 K).

The analyses established the structures and stereochemistry of these five cycloalkadienes (Fig. 1). In I the torsion angles for the C=C-C=C part of the eight-membered ring are -3.7° (2), -51.9° (2) and -17.4° (2), respectively. These values are indicative for a cis,cis-1,3-diene moiety with little conjugation between the double bonds. In II, the unit cell contains two crystallographically independent molecules with the same conformation. The mean torsion angles for the double bonds in the eight-membered ring are 142.5° and -27.8°, indicating the trans,cis conformation. The deviations of the torsion angles from the ideal values of 180° and 0° indicate the presence of severe strain in this molecule. Molecule III has C=C-C=C torsion angles -8.4(2), -58.8(2) and -1.1(2)°. In IV, the two rings are cis-fused, and in V the cyclooctadiene ring has the same conformation as found for I and II.

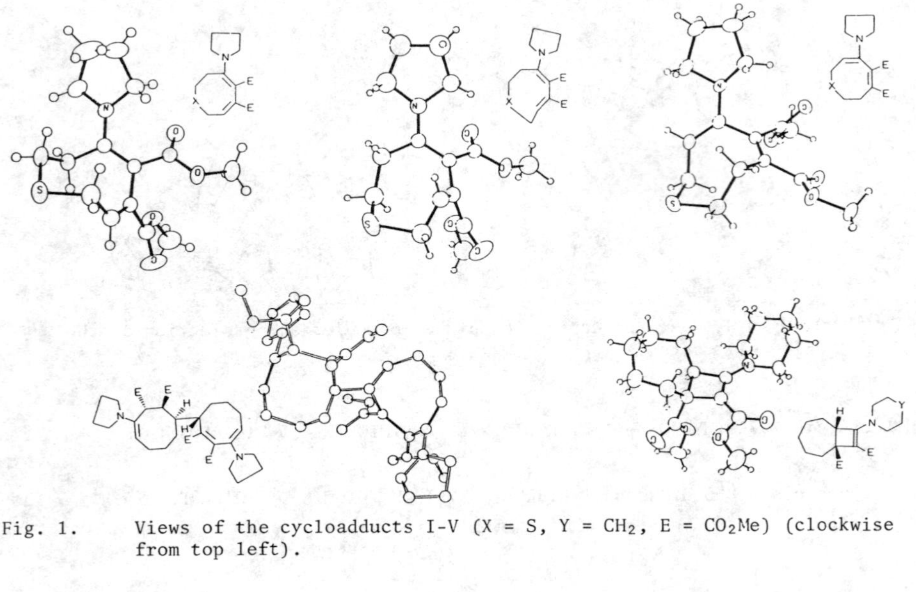

Fig. 1. Views of the cycloadducts I-V (X = S, Y = CH_2, E = CO_2Me) (clockwise from top left).

THIENO[3,2-e:4,5-e']BIS(BENZO[b]THIOPHENE) (TRITHIA[5]HETEROHELICENE)
$C_{16}H_8S_3$

H. NAKAGAWA, K. YAMADA, H. KAWAZURA and H. MIYAMAE, 1984. Acta Cryst., C40, 1039-1041.

Monoclinic, P2/c, a = 10.845, b = 8.420, c = 7.265 Å, β = 109.08°, D_m = 1.56, Z = 2.
Mo radiation, R = 0.050 for 1362 reflexions.

The molecule has typical helicenic geometry with crystallographic twofold symmetry (Fig. 1). Bond distances in the benzene ring vary from 1.423(4) Å for C(1)-C(2) in the inner core to 1.360(5) Å for C(6)-C(7) in the outer core, with the radial ones, C(2)-C(5) 1.409(4) and C(1)-C(8) 1.407(4) Å, being of medium length. The S-C distances are S(1)-C(8) 1.736(3), S(2)-C(4) 1.717(4) and S(2)-C(5) 1.738(3) Å. The dihedral angle between the terminal thiophene rings is 36.6(1)°. M and P enantiomers of the molecules are stacked alternately along c with nearly parallel overlap of the central thiophene moieties.

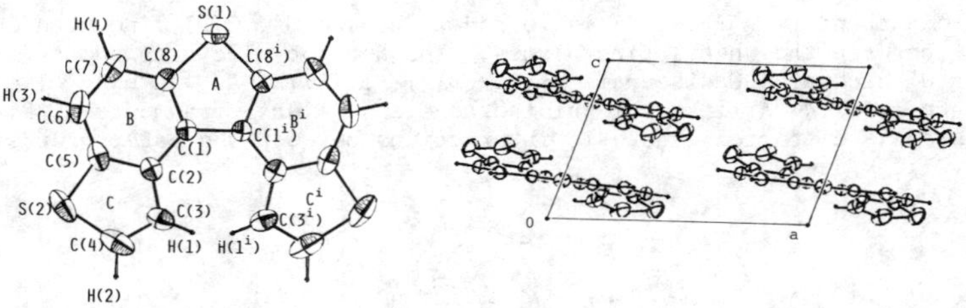

Fig. 1. Perspective view of the $C_{16}H_8S_3$ molecule and packing diagram of the molecules viewed down b.

(1α,4α,6α)-1-(PIPERIDIN-1-YL)-4-t-BUTYL-7-THIABICYCLO[4.2.0]OCTANE 7,7-DIOXIDE
$C_{16}H_{29}NO_2S$

S. BRUCKNER, F. GANAZZOLI, S. FATUTTA, G. PITACCO and E. VALENTIN, 1984. J. Heterocycl. Chem., 21, 1065-1067.

Monoclinic, $P2_1/c$, a = 9.079, b = 22.881, c = 8.468 Å, β = 108.42°, D_m = 1.20, Z = 4.
Mo radiation, R = 0.0494 for 1823 reflexions.

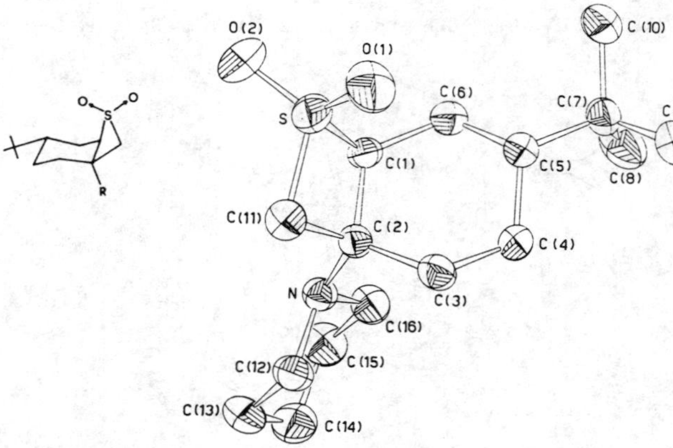

Fig. 1. A view of $C_{16}H_{29}NO_2S$.

The molecule is shown in Fig. 1. The four-membered ring is puckered, with a dihedral angle of 157.7° between the C1,S,C11 and C1,C2,C11 planes. The cyclohexane ring is considerably flattened along the ring junction (torsion angles C5-C6-C1-C2 -31°, C6-C1-C2-C3 23°). This flattening seems to have little effect on the direction of C1-S which is axial. In the four-membered ring C-S distances are significantly different (C1-S 1.805(3), C11-S 1.770(4)); the S-O distances are 1.444 and 1.435(3) Å.

DIMETHYL 9-THIOXANTHENYLIDENEMALONATE
$C_{18}H_{14}O_4S$

M.R. CAIRA, W.H. WATSON, A.L. TERNAY Jr and R. McKELLAR, 1984. Acta Cryst., C40, 1710-1712

Monoclinic, I2/c, a = 21.613, b = 7.536, c = 19.576 Å, β = 94.99°, Z = 8. Cu radiation, R = 0.044 for 1897 reflexions.

The thioxanthene ring (Fig. 1) is folded along the S...C(9) line with an angle of 137.6(6)° between the phenyl ring planes. The heterocyclic ring adopts a boat conformation with equal S-C distances of 1.765(2), 1.757(2) Å. An exocyclic double bond of length 1.347(4) Å links the thioxanthene and malonyl fragments. Planarity of the π system is distorted by steric interactions between the ester groups and atoms of the tricyclic system.

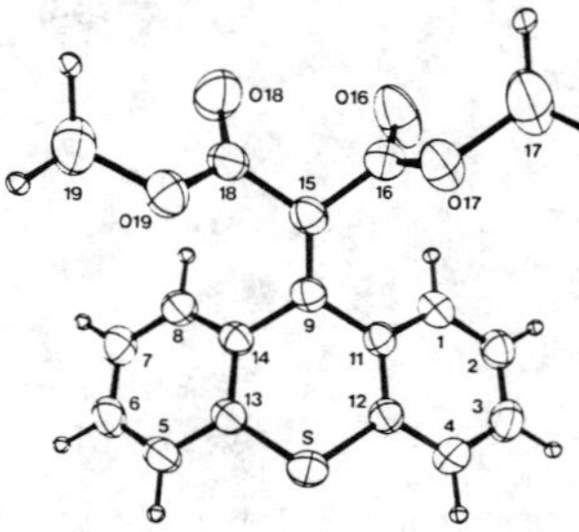

Fig. 1. A view of $C_{18}H_{14}O_4S$.

r-2-(DIPHENYLPHOSPHINOYL)-c-4,c-6-DIMETHYL-1,3-DITHIANE
$C_{18}H_{21}OPS_2$

E. JUARISTI, B.A. VALENZUELA, L. VALLE and A.T. McPHAIL, 1984. J. Org. Chem., 49, 3026-3027.

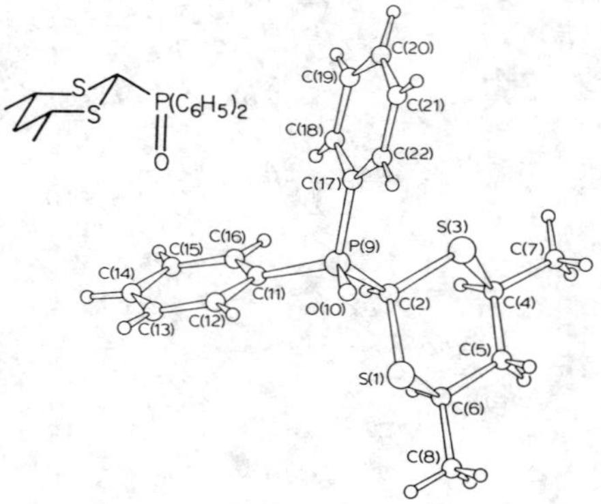

Fig. 1. The $C_{18}H_{21}OPS_2$ molecule.

Monoclinic, C2/c, a = 38.886, b = 5.773, c = 16.651 Å, β = 104.60°, Z = 8. Cu radiation, R = 0.048 for 2128 reflexions.

The 1,3-dithiane ring (Fig. 1) has a chair conformation with an equatorially oriented substituent at C(2); the diphenylphosphinoyl oxygen atom is gauche to both sulphur atoms but is asymmetrically disposed (S-C-P-O torsion angles are +32.9 and -91.5°) to accommodate non-bonded interactions between the bulky substituent and the ring atom.

(8R)-8-BENZYLOXY-6-(1,3-DIOXAN-2-YL)-1,4-DITHIASPIRO[4.4]NON-6-ENE $C_{18}H_{22}O_3S_2$

J.C. BARRIERE, A. CHIARONI, J. CLEOPHAX, S.D. GERO and C. RICHE, 1984. Acta Cryst., C40, 198-200.

Monoclinic, P2$_1$, a = 15.399, b = 7.785, c = 16.128 Å, β = 113.01°, Z = 4. Cu radiation, R = 0.067 for 1951 reflexions.

The two independent molecules in the asymmetric unit have the same chirality and similar conformations. The cyclopentene ring in one molecule has an envelope shape (C(5) 0.42 Å from the C(1)-C(4) plane); in the other molecule, the cyclopentene ring is planar. In both molecules the five-membered sulphur containing rings are in envelope conformation with C(9') at the flap.

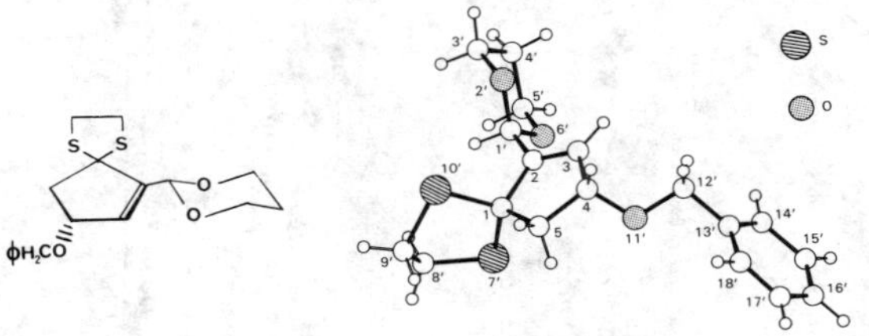

Fig. 1. The $C_{18}H_{22}O_3S_2$ molecule and a view of one of the two independent molecules.

9-METHYLTHIOXANTHENIUM-10-BIS(METHOXYCARBONYL)METHYLIDE 10-OXIDE $C_{19}H_{18}O_5S$

I. VICKOVIĆ, W.H. WATSON and A.L. TERNAY Jr., 1984. Acta Cryst., C40, 842-844.

Triclinic, P$\bar{1}$, a = 15.159, b = 8.672, c = 7.178 Å, α = 93.80, β = 101.12, γ = 104.91°, Z = 2. Cu radiation, R = 0.051 for 2016 reflexions.

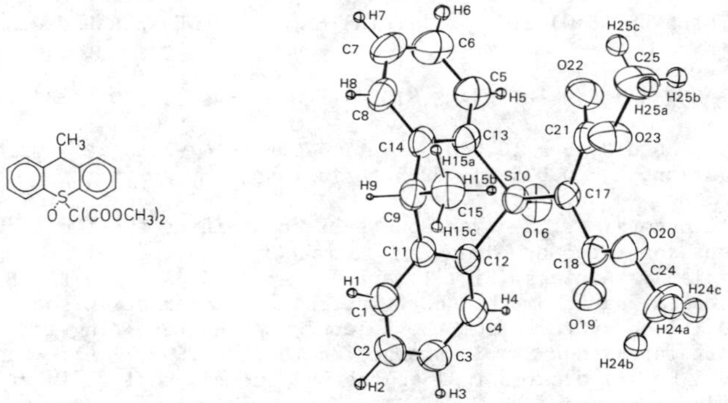

Fig. 1. $C_{19}H_{18}O_5S$: structural formula and perspective view of the molecule.

In the molecule (Fig. 1) the $^{+}S\text{-}C^{-}$ distance of 1.689(3) Å is shorter than the average distance of 1.718(8) Å observed for a series of sulphonium ylides. The two $^{+}S\text{-}C(sp^2)$ distances are 1.760(4) and 1.768(4) Å while the $^{+}S\text{-}O$ distance is 1.445(3) Å. The central ring in the tricyclic system is in a boat conformation with the methyl and carbanion moieties occupying axial sites. The C(17) carbanion carbon atom is 0.11(1) Å out of the plane of the attached atoms and towards the C(15) methyl group. An intramolecular interaction between the π cloud of the carbanion and the methyl hydrogen, C(17)...H(15b) 2.56 Å, may lead to a slight pyrimidalisation of C(17).

3,4;3',4'-BIS(ETHYLENEDITHIO)-2,2',5,5'-TETRATHIAFULVALENE HEXAFLUOROARSENATE (2:1)
$C_{20}H_{16}AsF_6S_{16}$ $(C_{10}H_8S_8)_2AsF_6$

P.C.W. LEUNG, M.A. BENO, G.S. BLACKMAN, B.R. COUGHLIN, C.A. MIDERSKI, W. JOSS, G.W. CRABTREE and J.M. WILLIAMS, 1984. Acta Cryst., C40, 1331-1334.

Monoclinic, A2/a, a = 14.890, b = 6.666, c = 35.3789 Å, β = 111.19°, Z = 4. Mo radiation, R = 0.055 for 1323 reflexions.

The structure (Fig. 1) consists of two-dimensional sheets parallel to bc of $(C_{10}S_8H_8)^{+}$ cations separated by AsF_6^{-} anions lying on inversion centres. Short S...S contacts (3.447(4) to 3.528(4) Å) are found normal to the molecular stacks but not along the stacking direction.

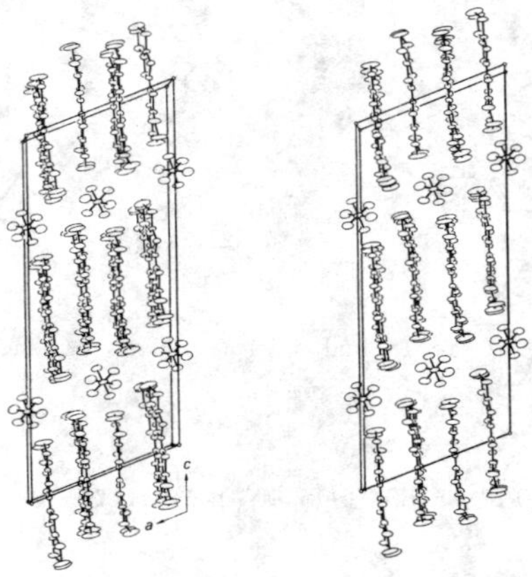

Fig. 1. Contents of the unit cell, viewed down b, of $(C_{10}H_8S_8)_2AsF_6$.

3,4;3'4'-BIS(ETHYLENEDITHIO)-2,2',5,5'-TETRATHIAFULVALENE TETRABROMOINDATE (2:1)
$C_{20}H_{16}Br_4InS_{16}$ $(C_{10}H_8S_8)_2InBr_4$

M.A. BENO, D.D. COX and J.M. WILLIAMS, 1984. Acta Cryst., C40, 1334-1337.

Triclinic, P$\bar{1}$, a = 6.618, b = 16.040, c = 17.470 Å, α = 95.29, β = 92.43, γ = 99.12°, Z = 2. Mo radiation, R = 0.095 for 2965 reflexions.

The structure consists of dimerised stacks containing the two independent $(C_{10}S_8H_8)$ cations separated by sheets of tetrahedral $InBr_4^{-}$ anions (mean In-Br 2.49 Å). Interplanar spacings (Fig. 1) are A...A' 3.43, A...B 3.87, B...B' 3.47 Å with molecules bowed away from the short interplanar interaction such that the S atoms deviate 0.3 Å on average from the molecular plane resulting in only one independent short intrastack S...S interaction (S...S 3.595(7) Å). Short S...S contacts (3.439-3.599 Å) occur normal to the molecular stacking direction involving the two independent cations.

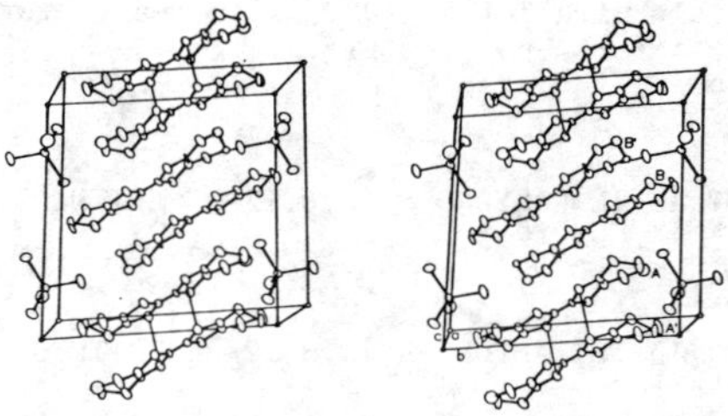

Fig. 1. The unit cell of $(C_{10}H_8S_8)_2InBr_4$ viewed down a.

BIS(ETHYLENEDITHIOLO)TETRATHIAFULVALENE TRIIODIDE

$C_{20}H_{16}I_3S_{16}$ $(C_{10}H_8S_8)_2{}^+I_3{}^-$

R.P. SHIBAEVA, V.F. KAMINSKII and V.K. BEL'SKII, 1984. Kristallografiya, 29, 1089-1093 [Sov. Phys. Crystallogr., 29, 638-640].

Triclinic, P$\bar{1}$, a = 6.609, b = 9.083, c = 15.267 Å, α = 85.63, β = 95.62, γ = 70.22°, Z = 1. Mo radiation, R = 0.031 for 2007 reflexions.

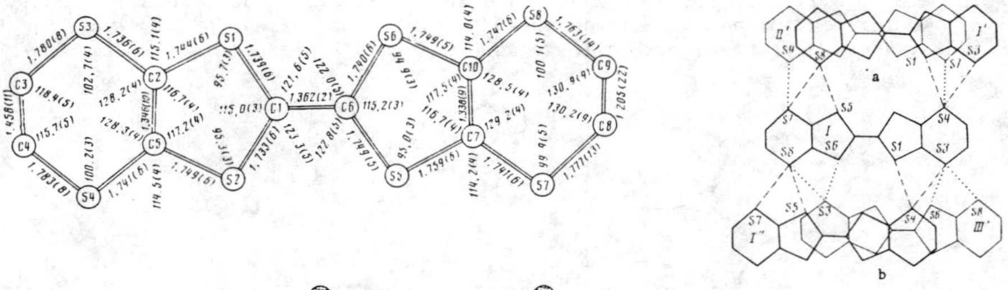

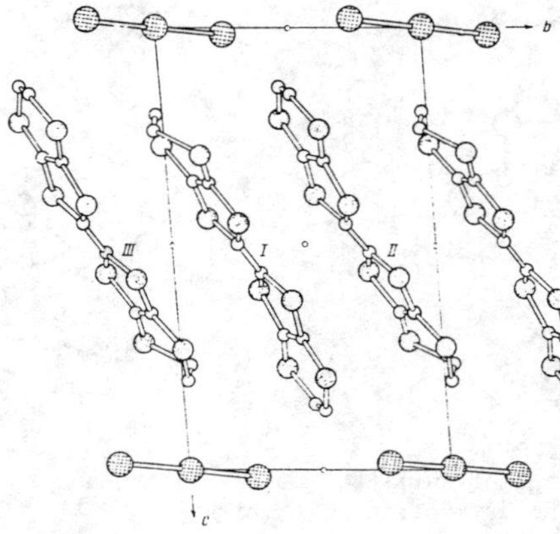

Fig. 1. Bond lengths (Å) and angles (°) in the cation radical (top left), mode of overlap of adjacent cations in a stack (S...S contacts from 3.568 to 3.688 Å) (top right), and projection of crystal structure along a direction in $C_{20}H_{16}I_3S_{16}$ (bottom).

The cation radical (Fig. 1) is not quite planar. Only its central fragment of two C atoms and four S atoms is actually planar. Additionally there is some positional disorder in one of the ethylene groups (C(8) and C(9)). The linear $I_3{}^-$ anions lie at centres of symmetry and the I-I distance is 2.909 Å. The shortest interanionic distance I...I is 4.204 Å. The cation radicals are packed in the crystal plane-to-plane approximately parallel to ($0\bar{2}2$) with mean interplanar distances of 3.75 Å (I to II) and 3.86 Å for I to III (Fig. 1).

anti-6,13-DIMETHYL-5,6,12,13-TETRAHYDRO-5,13:6,12-BISEPITHIODIBENZO[a,f]CYCLODECENE-7,14-DIONE
$C_{20}H_{16}O_2S_2$ (I)

syn-12,13-DIMETHYL-5,6,12,13-TETRAHYDRO-5,13:6,12-BISEPITHIODIBENZO[a,e]CYCLODECENE-11,14-DIONE
$C_{20}H_{16}O_2S_2$ (II)

O.H. JOHANSEN, P. GROTH and K. UNDHEIM, 1984. Acta Chem. Scand., B38, 617-622.

I. Monoclinic, $P2_1/n$, a = 6.633, b = 17.280, c = 7.488 Å, β = 103.51°, D_m = 1.38, Z = 2. Mo radiation, R = 0.035 for 1156 reflexions.

II. Orthorhombic, $P2_12_12_1$, a = 7.124, b = 16.45, c = 28.17 Å, D_m = 1.40, Z = 8. Mo radiation, R = 0.088 for 1439 reflexions.

For compound I the bond distances and angles do not differ significantly from those of earlier findings (1). Distances of note are: S-C(2) 1.816, S-C(4) 1.813, O(1)-C(1) 1.213, C(1)-C(2) 1.540, C(2)-C(4') 1.564, C(2)-C(3) 1.527 and C(4)-C(5) 1.505 Å. No information on bond distances and angles is supplied for II since estimated standard deviations are as large as 0.03 Å and 2° respectively (Fig. 1).

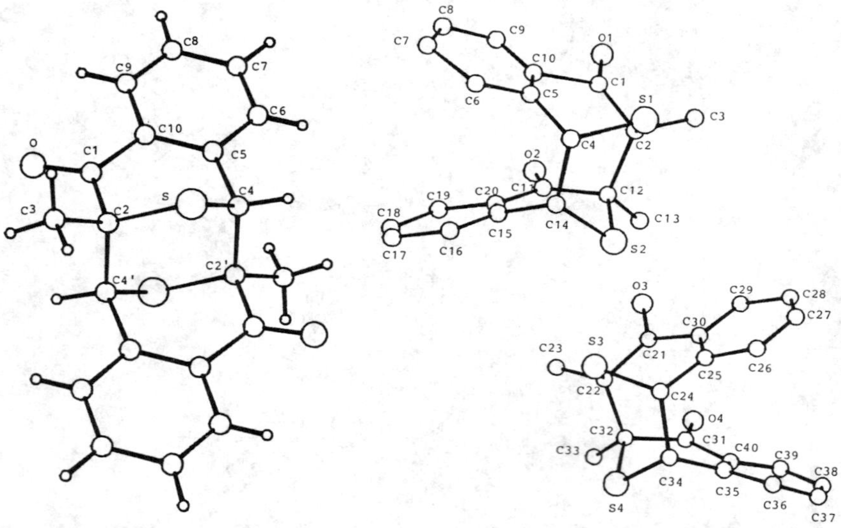

Fig. 1. Schematic drawings of anti-$C_{20}H_{16}O_2S_2$ (left), and the two syn-isomers of $C_{20}H_{16}O_2S_2$ in the asymmetric unit (right).

1. Structure Reports, 41B, 404, 413.

cis-10-METHYL-9-PHENYLTHIOXANTHENIUM TETRAFLUOROBORATE
$C_{20}H_{17}BF_4S$ $C_{20}H_{17}S^+.BF_4^-$

S. OHNO, H. SHIMIZU, T. KATAOKA, M. HORI and M. KIDO, 1984. J. Org. Chem., 49, 3151-3157.

Orthorhombic, $P2_12_12_1$, a = 9.976, b = 12.099, c = 14.700 Å, Z = 4. Mo radiation, R = 0.08 for 1130 reflexions.

The 9-phenyl and 10-methyl groups occupy axial positions (Fig. 1). The six-membered heterocyclic ring has a boat conformation, with S and C9 0.384(2) and 0.469(7) Å from the best plane. The dihedral angle between the two phenyl rings is 139.5°.

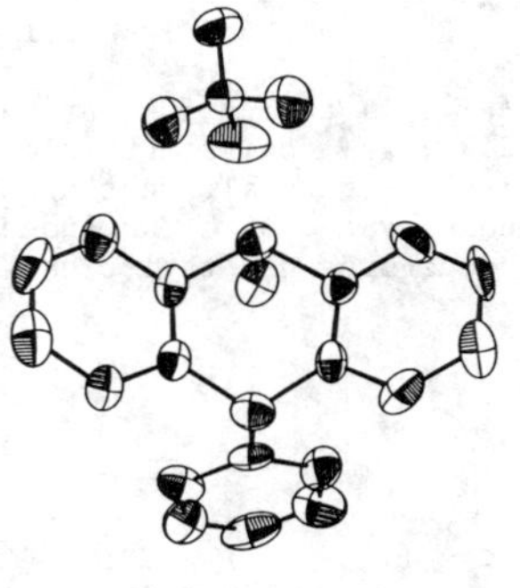

Fig. 1. A perspective view of $C_{20}H_{17}S^{+}.BF_4^{-}$.

2-ISOPROPYLIDENE-3-[1-(2-METHYL-5-PHENYL-3-THIENYL)ETHYLIDENE]SUCCINIC ANHYDRIDE (Z-ISOMER)
$C_{20}H_{18}O_3S$ (I)

2-ISOPROPYLIDENE-3-[1-(2-METHYL-5-PHENYL-3-THIENYL)ETHYLIDENE]SUCCINIC ANHYDRIDE (E-ISOMER)
$C_{20}H_{18}O_3S$ (II)

7,7a-DIHYDRO-4,7,7,7a-TETRAMETHYL-2-PHENYLBENZO[b]THIOPHENE-5,6-DICARBOXYLIC ANHYDRIDE
$C_{20}H_{18}O_3S$ (III)

M. KAFTORY, 1984. Acta Cryst., C40, 1015-1019.

I. Monoclinic, $P2_1$, a = 14.449, b = 7.619, c = 8.014 Å, β = 94.83°, Z = 2. Mo radiation, R = 0.063 for 1156 reflexions.

II. Monoclinic, $P2_1$, a = 14.295, b = 7.620, c = 8.317 Å, β = 104.21°, Z = 2. Mo radiation, R = 0.048 for 1500 reflexions.

III. Monoclinic, $P2_1/n$, a = 16.604, b = 7.490, c = 14.113 Å, β = 106.59°, Z = 4. Mo radiation, R = 0.073 for 1810 reflexions.

In the molecules (Fig. 1) S-C bond lengths are dramatically shorter in the thiophene rings in isomers I and II (1.712, 1.727, 1.713 and 1.726 Å) than the bonds in the thieno ring in III (1.853 and 1.764 Å); these differences are ascribed to the delocalisation of these bonds in the thiophene rings compared with the localised bonds in the thieno ring. The diene system in I and II is severely skewed. Only isomer II can undergo conrotatory photocyclisation to give III.

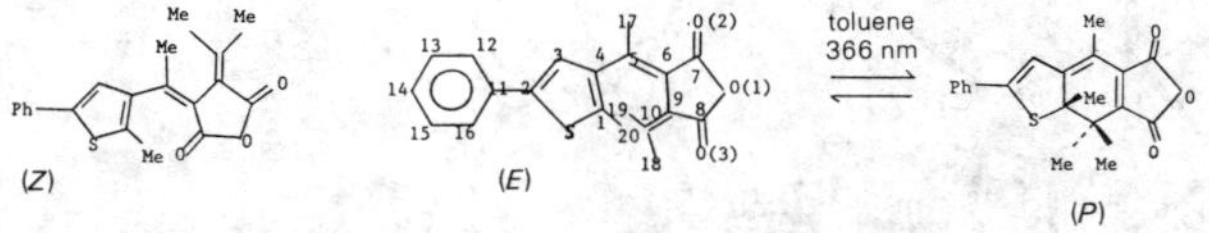

Fig. 1. Structures of the three $C_{20}H_{18}O_3S$ isomers.

DI(TETRAMETHYLTETRATHIAFULVALENIUM) PERCHLORATE
$C_{20}H_{24}ClO_4S_8$ $2C_{10}H_{12}S_4^{0.5+},ClO_4^{-}$

B. LIAUTARD, S. PEYTAVIN, G. BRUN, D. CHASSEAU, J.M. FARBE and L. GIRAL, 1984. Acta Cryst., C40, 1023-1026.

Triclinic, $P\bar{1}$, a = 7.115, b = 7.515, c = 12.992 Å, α = 84.41, β = 85.27, γ = 71.53°, D_m = 1.57, Z = 1. Cu radiation, R = 0.034 for 1068 reflexions.

In the crystal structure (Fig. 1) the cations are nearly planar, almost perpendicular to a (deviation 2.7°) and form stacks along a. They repeat by an inversion leading to overlap displacements alternating in the direction of the long molecular axis; they form a diadic system with two independent interplanar distances of 3.51 and 3.60 Å. The ClO_4^- ions present statistical disorder which maintains the centre of symmetry.

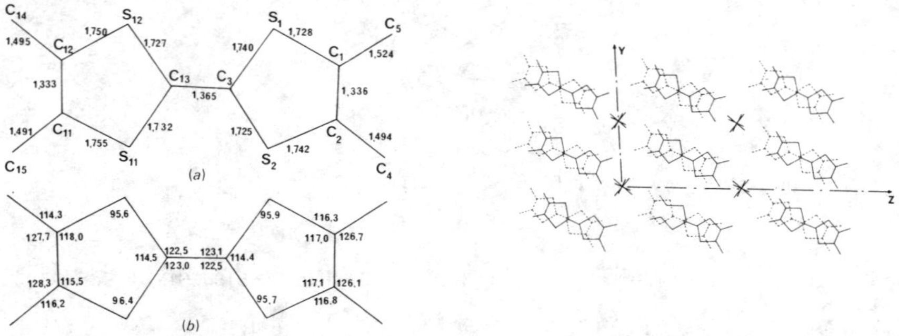

Fig. 1. $C_{20}H_{24}ClO_4S_8$: bond lengths and angles in the $C_{10}H_{12}S_4$ ion and a projection of the structure on the y 0 z plane.

(12E)-threo-12-CHLORO-3,4:5,6-DIBENZO-10-OXA-13-THIA[6.3.3]PROPELLA-3,5-DIENE-13-OXIDE
$C_{20}H_{19}ClO_2S$

K. WEINGES, W. KASEL, J. KLEIN, G. HUBERTUS, H. IRNGARTINGER, U. HUBER-PATZ and H. RODEWALD, 1984. Chem. Ber., 117, 966-981.

Orthorhombic, Pccn, a = 28.615, b = 15.192, c = 7.786 Å, Z = 8. Mo radiation, R = 0.051 for 1575 reflexions.

The structure of this propelladiene derivative was determined to be as shown in Fig. 1. Both five-membered rings are in envelope conformation. The 8-membered ring is severely twisted; the C2-C1-C8-C7 torsion angle is 25.9(6)°.

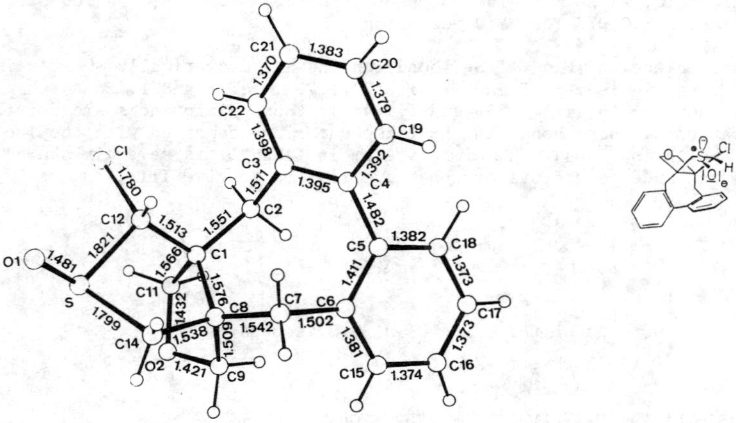

Fig. 1. The $C_{20}H_{19}ClO_2S$ molecule with bond lengths (σ 0.007 Å).

DI(4,4',5,5'-TETRAMETHYL-2,2'-BI-1,3-DISELENOLYLIDEN)IUM PERRHENATE
$C_{20}H_{24}O_4ReSe_8$ $C_{20}H_{24}Se_8^+,ReO_4^-$

G. RINDORF, H. SOLING and N. THORUP, 1984. Acta Cryst., C40, 1137-1139.

Triclinic, F$\bar{1}$, a = 14.295, b = 15.419, c = 26.649 Å, α = 83.96, β = 88.75, γ = 69.39°, Z = 8. Cu radiation, R = 0.052 for 4042 reflexions (at 120K).

The structure of the title compound has previously been determined at 295K and 120K (1). In the present study a better quality crystal was used in order to reveal the full superstructure at 120K. In the crystal structure (Fig. 1) the ReO_4^- anions are ordered, with Re-O 1.702(7), 1.707(8), 1.702(9) and 1.727(7) Å, and have almost perfect tetrahedral geometry; at room temperature these ions are disordered. Both the anions and the cations are displaced relative to their positions in the average structure previously reported. Bond distances and angles in the cation are the same as in the average structure.

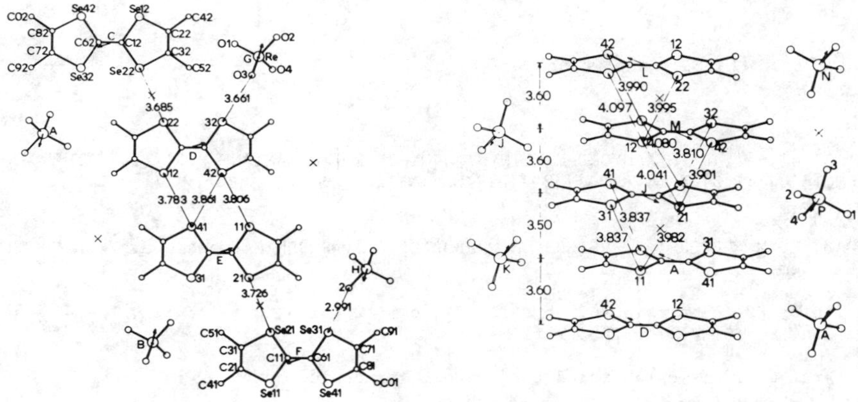

Fig. 1. $C_{20}H_{24}Se_8^+,ReO_4^-$: view of the structure along the a axis and side view of stacks (tilted 20°); centres of symmetry are marked with crosses.

1. Structure Reports, 49B, 417.

anti-5,7,9,14,16,18-HEXAMETHYL-1,3,10,12-TETRATHIA[3.3]METACYCLOPHANE
$C_{20}H_{24}S_4$

S. PAPPALARDO and F.R. FRONCZEK, 1984. J. Chem. Soc. Perkin II, 205-207.

Monoclinic, $P2_1/c$, a = 8.585, b = 15.189, c = 7.967 Å, β = 115.75°, Z = 2. Mo radiation, R = 0.036 for 1132 reflexions.

The molecular structure consists of discrete molecules which have a crystallographic centre of symmetry and thus the anti-conformation (Fig. 1). The aromatic rings are nearly planar, while C(7), C(9) and S(1) phenyl substituents are displaced from coplanarity by 0.224(3), 0.130(3) and -0.223(1) Å respectively. The averaged bond length values of C(Ar)-S (1.781), C(10)-S (1.810) and exocyclic C-C (1.511 Å) are normal. The dimensions in the benzene ring vary significantly, those in the most crowded parts of the molecule being longest (C(2)-C(3) 1.402(3)), those in the least crowded shortest (C(5)-C(6) 1.388(3) Å). Endocyclic C-C-C angles average 119.9°. The C(1)-S(1)-C(10) and C(10)-S(2)-C(3) bond angles are 100.7(1) and 105.9(1)° respectively.

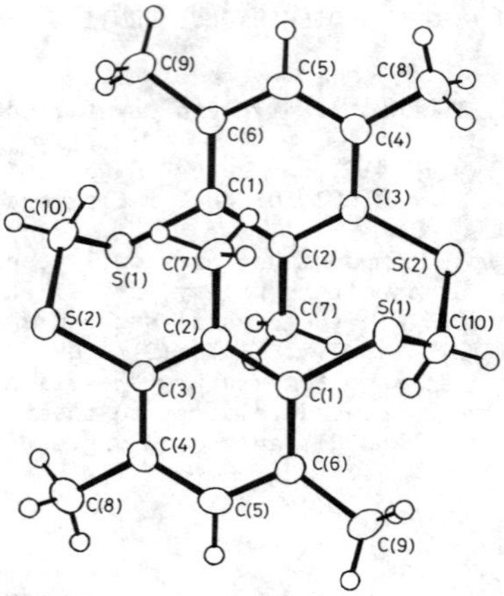

Fig. 1. $C_{20}H_{24}S_4$: the molecular structure.

1',2',3',4',4a',4b'-HEXAHYDRO-N-METHYL-7'-OXOSPIRO[ADAMANTANE-2,5'-(1,4-ETHENO-5H-THIENO[3',4':1,3]CYCLOPROPA[1,2-d]PYRIDAZINE)]-2',3'-DICARBOXIMIDE
$C_{21}H_{23}N_3O_3S$

W. ADAM, E.-M. PETERS, K. PETERS, H. REBOLLO, R.J. ROSENTHAL and H.G. VON SCHNERING, 1984. Chem. Ber., 117, 2393-2408.

Monoclinic, $P2_1/c$, a = 13.967, b = 11.777, c = 12.461 Å, β = 113.82°, Z = 4. Mo radiation, R = 0.041 for 2853 reflexions.

X-ray analysis established the urazole structure shown in Fig. 1.

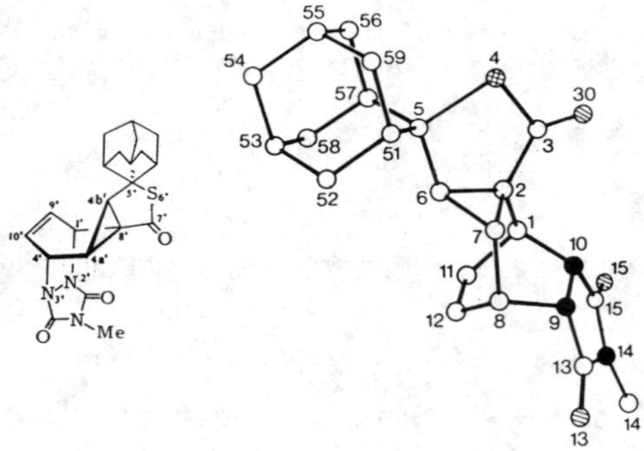

Fig. 1. A view of $C_{21}H_{23}N_3O_3S$.

8,9-BENZO-3,4-BIS(METHOXYCARBONYL)-7-(4,5-BIS(METHOXYCARBONYL)-1,3-DITHIACYCLOPENT-4-EN-2-YLIDENE)-2,5-DITHIABICYCLO[4.3.0]NONA-1(6),3-DIENE
$C_{22}H_{16}O_8S_4$

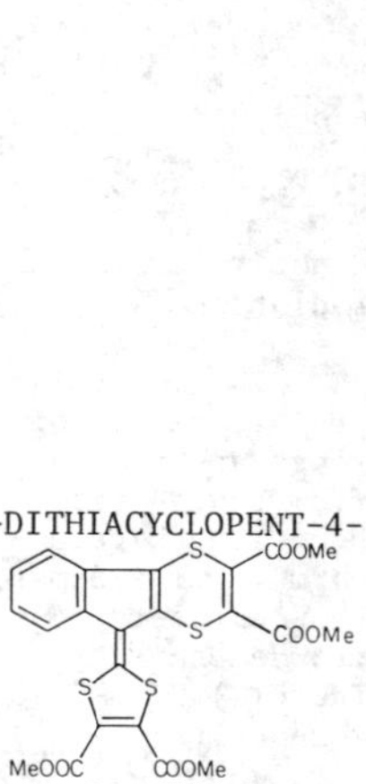

M.V. LAKSHMIKANTHAM, M.P. CAVA and P.J. CARROLL, 1984. J. Org. Chem., 49, 726-728.

Triclinic, $P\bar{1}$, a = 11.713, b = 13.154, c = 8.169 Å, α = 95.43, β = 98.13, γ = 64.93°, Z = 2. Mo radiation, R = 0.057 for 3094 reflexions.

The analysis established the structure shown above.

syn-9-PHENYL-2,11-DITHIA[3.3]METACYCLOPHANE
$C_{22}H_{20}S_2$

W. ANKER, K.A. BEVERIDGE, G.W. BUSHNELL and R.H. MITCHELL, 1984. Can. J. Chem., 62, 661-666.

Triclinic, $P\bar{1}$, a = 9.573, b = 12.010, c = 8.433 Å, α = 108.02, β = 95.95, γ = 102.25°, D_m = 1.313, Z = 2. Mo radiation, R = 0.072 for 2117 reflexions.

In the molecule (Fig. 1) the bridging S atoms differ in their conformations and the S...S distance is 6.391 Å. The metacyclophane rings have a distance of 3.57 Å between their centres and form a dihedral angle of 20.5°. The phenyl substituent forms an angle of 78.6° with the adjacent ring and has an ortho H-atom very close to the internal H-atom of the opposite ring. The phenyl substituted aromatic ring is non-planar and in a shallow boat form, the ring atoms C(11) (internal) and C(14) being 0.038(4) and 0.037(5) Å out-of-plane on the outside, the methylene C substituents are 0.212(5) and 0.199(5) Å out-of-plane on the inside. The first atom of the phenyl group is in the bowsprit position and is 0.274(4) Å above the mean plane on the outer side of the molecule. The opposite metacyclophane ring is less strained but in a similar boat form with the methylene C-atoms 0.092(4) and 0.067(6) Å to the inside. The mean bond angle at S is 103.5(2)°, the angles at C (methylene) range from 113.3(3)° to 118.0(3)°.

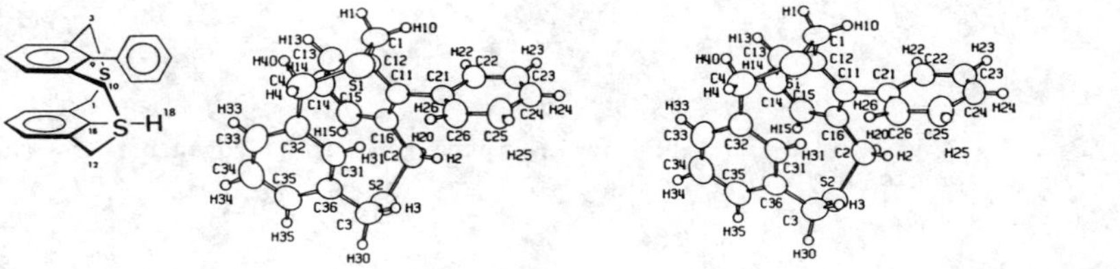

Fig. 1. The $C_{22}H_{20}S_2$ molecule and a stereoview.

3-[(2(Z)-BENZYLSULPHONYL)-2-PHENYLETHENYL]-4H-THIOCHROMAN-4-ONE
$C_{24}H_{18}O_3S_2$

C.H. CHEN, L.W. KELTS, H.R. LUSS and J.L. FOX, 1984. J. Org. Chem., 49, 5143-5147.

Monoclinic, $P2_1/n$, a = 16.930, b = 17.248, c = 6.974 Å, β = 100.51°, Z = 4. Mo radiation, R = 0.036 for 2830 reflexions.

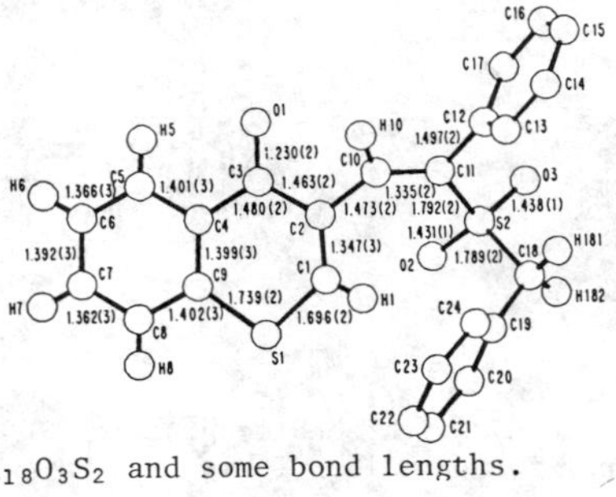

Fig. 1. A view of $C_{24}H_{18}O_3S_2$ and some bond lengths.

The molecule has the Z-configuration about the C10=C11 double bond (Fig. 1). Bond lengths in the thiochroman ring deviate slightly from expected values; in particular there is considerable localization of the bonds at C5-C6 and C7-C8. The bond lengths (Fig. 1) are also consistent with a significant contribution of a resonance form involving delocalization of the sulphur lone pair into the carbonyl.

15,30-DIHYDROXY-12,27-DIMETHYL-5,20-DIOXA-2,8,17,23-TETRATHIA[9,9]METACYCLOPHANE
$C_{26}H_{36}O_4S_4$

M. CANNAS, G. MARONGIU and T. PINTORI, 1984. Acta Cryst., C40, 431-433.

Orthorhombic, Pbca, a = 28.25, b = 9.14, c = 10.27 Å, D_m = 1.34, Z = 4. Mo radiation, R = 0.092 for 1248 reflexions.

The molecule (Fig. 1) is a centrosymmetric 24-membered macrocycle. There is intra-annular hydrogen bonding (O...O 2.75(2) Å).

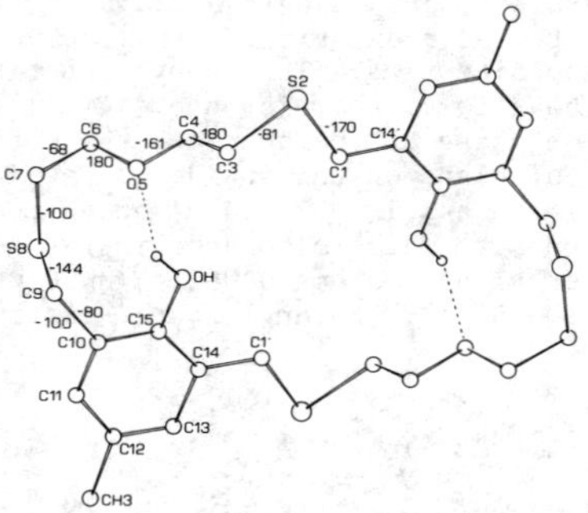

Fig. 1. A view of $C_{26}H_{36}O_4S_4$ showing intra-annular hydrogen bonding and torsion angles (σ ∿ 1°).

3-[2(Z)-(BENZYLTHIO)-2-PHENYLETHENYL]-2,6-DIPHENYL-4H-THIOPYRAN-4-ONE
$C_{32}H_{24}OS_2$

C.H. CHEN, G.A. REYNOLDS, D.L. SMITH and J.L. FOX, 1984. J. Org. Chem., 49, 5136-5143.

Triclinic, P$\bar{1}$, a = 10.381, b = 15.004, c = 9.433 Å, α = 109.67, β = 112.82, γ = 79.56°, Z = 2. Mo radiation, R = 0.077 for 2045 reflexions.

The X-ray analysis established that the molecule (Fig. 1) was the desired Z-isomer about the C18=C19 double bond. The bond lengths of the thiopyran ring are consistent with a 2:1 contribution to the ground state of a canonical form as shown in Fig. 1.

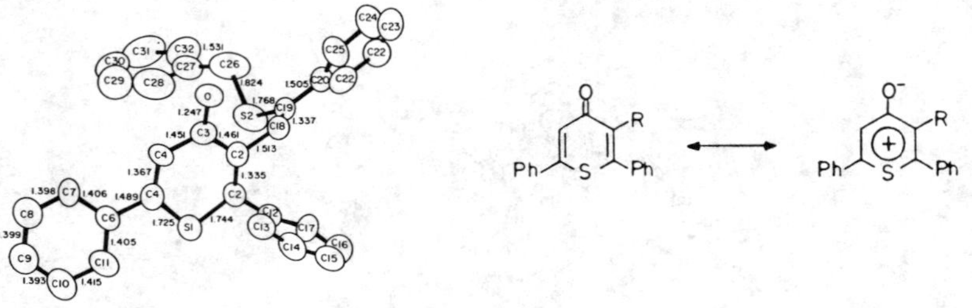

Fig. 1. The $C_{32}H_{24}OS_2$ molecule with bond lengths (σ 0.004-0.017 Å). Also shown are the two important canonical forms.

2,11,20,29-TETRATHIA[3.3.3.3]PARABENZENOPHANE
$C_{32}H_{32}S_4$

T.-L. CHAN and T.C.W. MAK, 1984. Acta Cryst., C40, 1452-1454.

Orthorhombic, Pccn, a = 20.607, b = 20.121, c = 6.622 Å, D_m = 1.316, Z = 4. Mo radiation, R = 0.045 for 2151 reflexions.

The molecule lies on a centre of symmetry with each benzene ring and its two methylene carbons coplanar (Fig. 1). The C2-C7 ring makes a dihedral angle of 26.4° with the (necessarily planar)S_4 system. The corresponding value for the C10-C15 ring is 45.9(4)°. The S-C bond lengths average 1.800(3) Å and the C-S-C angles 102.2°, indicating an absence of intermolecular steric strain.

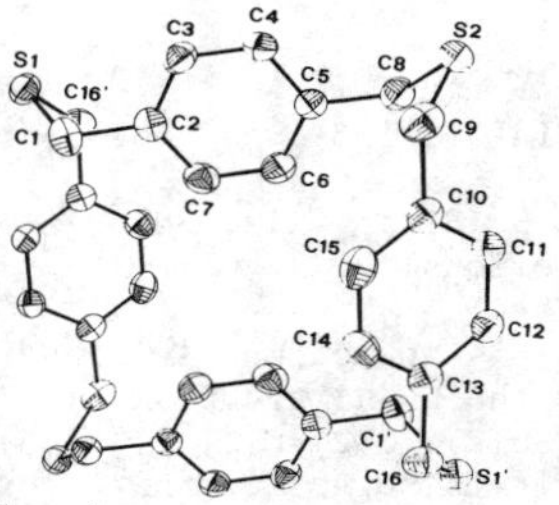

Fig. 1. The $C_{32}H_{32}S_4$ molecule.

TETRASELENATETRACENE HEXAIODODIMERCURATE
$C_{36}H_{16}Hg_2I_6Se_8$ $[(C_{18}H_8Se_4)_2^{+}][Hg_2I_6]^{2-}$

R.P. SHIBAEVA and V.F. KAMINSKII, 1984. Kristallografiya, 29, 606-609 [Sov. Phys. Crystallogr., 29, 361-363].

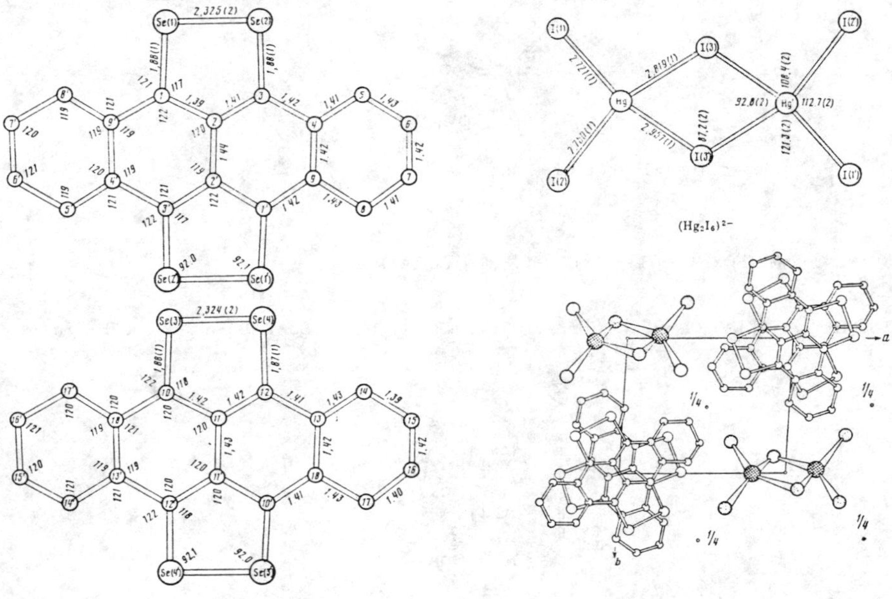

Fig. 1. Bond lengths (Å) and angles (°) in cation radicals (left), in the anion (top right), and one quarter of the unit cell of $C_{36}H_{16}Hg_2I_6Se_8$ projected along the c axis (bottom right).

Triclinic, $I\bar{1}$, a = 19.607, b = 16.252, c = 7.138 Å, α = 95.00, β = 78.28, γ = 93.41°, Z = 2. Mo radiation, R = 0.065 for 3312 reflexions.

The structure consists of two crystallographically independent cation radicals of $TSeT^{\ddagger}$ and similarly two centrosymmetric dimeric anions, composed of two tetrahedra with a common edge (Fig. 1). There are characteristic cation-radical stacks along the c axis with interplanar distances ($TSeT^{\ddagger}$ - TSeT•) of 3.57 Å. The shortest Se...Se distance in the stack is 4.15 Å. The shortest anion...cation contacts are I(1)...Se(1') 3.798, I(1)...Se(2') 3.662, I(2)-Se(3) 3.532, I(2)...Se(4) 3.625, I(2)...Se(2') 3.751 and I(2)...Se(4') 3.948 Å, where the prime signifies 0.5-x, 1.5-y, 0.5-z.

5-CHLOROMETHYL-2-OXAZOLIDINONE
$C_4H_6ClNO_2$

G. BOCELLI and M.-F. GRENIER-LOUSTALOT, 1984. J. Mol. Struct., 125, 11-17.

Monoclinic, $P2_1/c$, a = 9.257, b = 5.181, c = 12.620 Å, β = 110.75°, Z = 4. Cu radiation, R = 0.041 for 961 reflexions.

The oxazolidine ring (Fig. 1) is approximately planar (largest deviations C(2) -0.35(2) Å, C(3) 0.043(2) Å) with the CH_2Cl group axial. Main bond lengths are: C(1)-O(1) 1.215(3), C(1)-O(2) 1.350(2), O(2)-C(3) 1.448(3), C(1)-N 1.337(3), N-C(2) 1.440(3), C(2)-C(3) 1.540(3) Å. It is suggested that there is a certain degree of delocalization in the molecule. N-H...O hydrogen bonds (N...O 2.88(2) Å) link the molecules into dimers about an inversion centre.

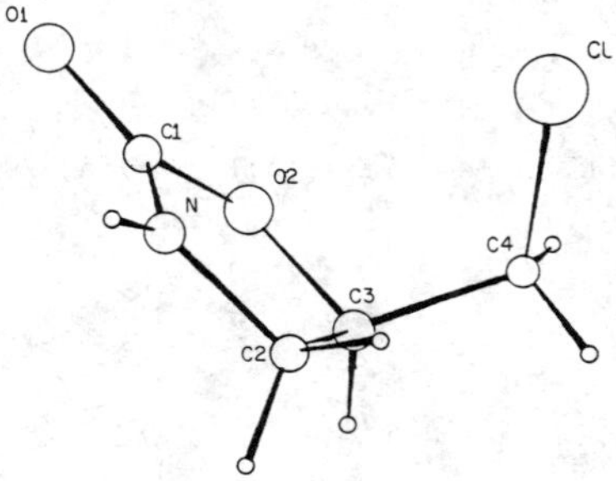

Fig. 1. A projection along a of $C_4H_6ClNO_2$.

4-CYANOIMINO-3-METHOXYMETHYLPERHYDRO-1,3,5-OXADIAZINE
$C_6H_{10}N_4O_2$

T. EBISUNO, M. TAKIMOTO and R. SHIBA, 1984. Bull. Chem. Soc. Japan, 57, 275-276.

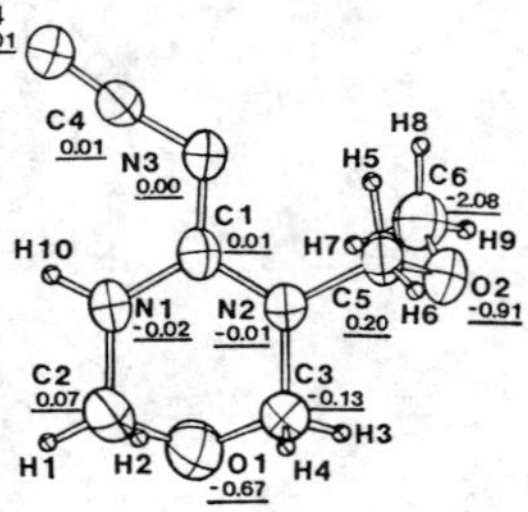

Fig. 1. The structure of $C_6H_{10}N_4O_2$ projected on the plane of the guanidine moiety (C(1),N(1),N(2),N(3)). Underscored numbers are heights in Å of the atoms above or below the plane.

Triclinic, P$\bar{1}$, a = 10.307, b = 8.166, c = 5.082 Å, α = 97.34, β = 91.98, γ = 107.13°, Z = 2. Mo radiation, R = 0.061 for 1651 reflexions.

One feature of the structure is a planar unit composed of four N atoms (N(1), N(2), N(3) and N(4)) and two C atoms (C(2) and C(3)) (Fig. 1). This planar structure is due to the triple bond character of the cyano group (C(4)-N(4) 1.144 Å) and to the resonated structure of the guanidine moiety. Other distances and angles of interest are: C(4)-N(3) 1.308, N(3)-C(1) 1.323, C(1)-N(2) 1.346, C(1)-N(1) 1.336, N(2)-C(5) 1.467, N(2)-C(3) 1.465, C(3)-O(1) 1.388, O(1)-C(2) 1.402 and N(1)-C(2) 1.451 Å; C(1)-N(3)-C(4) 118.5 and N(3)-C(4)-N(4) 174.5°.

1-(HYDROXYMETHYL)-2,1-BENZISOXAZOL-3(1H)-ONE
$C_8H_7NO_3$

W. WIERENGA, A.W. HARRISON, B.R. EVANS and C.G. CHIDESTER, 1984. J. Org. Chem., 49, 438-442.

Triclinic, P$\bar{1}$, a = 7.038, b = 7.718, c = 7.556 Å, α = 104.50, β = 83.11, γ = 115.10°, D_m = 1.45, Z = 2. Cu radiation, R = 0.035 for 1033 reflexions.

The nitrogen atom (Fig. 1) has a pyramidal configuration and is 0.12 Å out of the plane of the other atoms in the fused rings. Molecules are linked to form centrosymmetric dimers by O(9)-H...O(3) hydrogen bonds (O...O 2.757 Å).

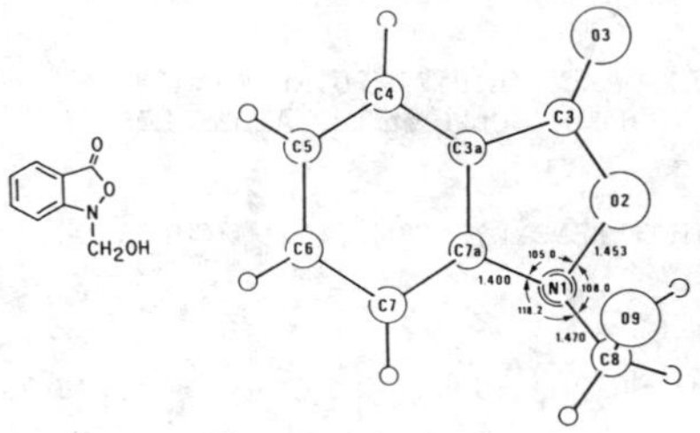

Fig. 1. A view of $C_8H_7NO_3$.

3-(p-ETHOXYPHENYL)SYDNONE
$C_{10}H_{10}N_2O_3$ (I)

3-(p-TOLYL)SYDNONE
$C_9H_8N_2O_2$ (II)

Y. WANG, P.L. LEE and M.Y. YEH, 1984. Acta Cryst., C40, 1226-1228.

I. Monoclinic, $P2_1/n$, a = 7.609, b = 7.350, c = 17.567 Å, β = 94.93°, D_m = 1.4, Z = 4. Mo radiation, R = 0.051 for 1391 reflexions.

II. Monoclinic, $P2_1/c$, a = 3.853, b = 11.819, c = 18.249 Å, β = 92.22°, D_m = 1.4, Z = 4. Mo radiation, R = 0.055 for 763 reflexions.

In both structures (Fig. 1) the bond distances of the sydnone ring are similar but N(1)-C(7) of I and II and C(7)-C(8) of I are significantly shorter than comparable bonds in other 3-phenylsydnone derivatives. The dihedral angles between the sydnone ring and the phenyl ring are quite different: 2° in I and 35° in II.

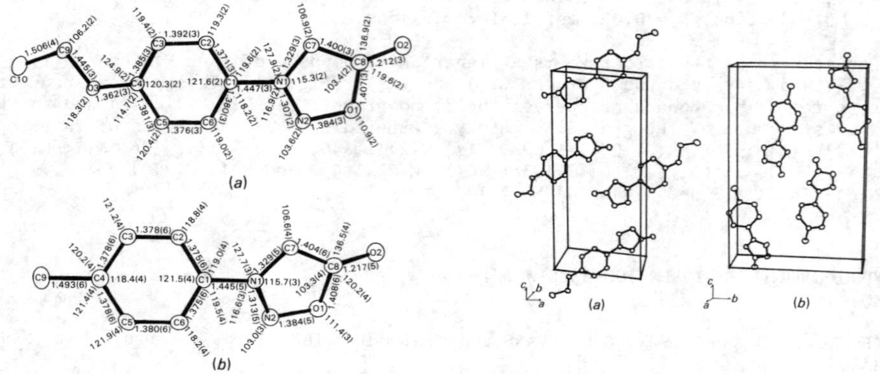

Fig. 1. Bond lengths and angles (left) and packing diagrams of the unit cells (right) in (a) $C_{10}H_{10}N_2O_3$ (I) and (b) $C_9H_8N_2O_2$ (II).

SODIUM 4-AMINO-N-(5-METHYL-3-ISOXAZOLYL)BENZENESULPHONAMIDE (SODIUM SULPHAMETHOXAZOLE)
$C_{10}H_{10}N_3NaO_3S$ (I)

2,4-DIAMINO-5-(3,4,5-TRIMETHOXYBENZYL)PYRIMIDINE 4-AMINO-N-(5-METHYL-3-ISOXAZOLYL)-BENZENESULPHONAMIDE (TRIMETHOPRIM SULPHAMETHOXAZOLE COMPLEX)
$C_{24}H_{29}N_7O_6S$ (II) $C_{14}H_{19}N_4O_3{}^+, C_{10}H_{10}N_3O_3S^-$

2,4-DIAMINO-5-(3,4,5-TRIMETHOXYBENZYL)PYRIMIDINIUM CHLORIDE (TRIMETHOPRIM HYDROCHLORIDE)
$C_{14}H_{19}ClN_4O_3$ (III)

H. NAKAI, M. TAKASUKA and M. SHIRO, 1984. J. Chem. Soc. Perkin II, 1459-1464.

I. Monoclinic, $P2_1/c$, a = 5.908, b = 17.434, c = 13.225 Å, β = 115.65°, Z = 4. Cu radiation, R = 0.027 for 2042 reflexions.

II. Orthorhombic, Pbca, a = 12.058, b = 24.479, c = 17.427 Å, Z = 8. Mo radiation, R = 0.057 for 3929 reflexions.

III. Monoclinic, $P2_1/c$, a = 22.052, b = 4.916, c = 14.256 Å, β = 91.39°, Z = 4. Cu radiation, R = 0.030 for 2578 reflexions.

Perspective views of the molecules along with bond lengths and angles are shown in Fig. 1. In II, the trimethoprim (TMP) and sulphamethoxazole (SMZ) hydrogen bond through N(1) (TMP) and N(7) (SMZ) (2.756 Å) and through N(7) (TMP) and N(2) (SMZ) (3.015 Å). These hydrogen bonds are stronger than any others in the structure of the complex. The proton in the N(1)...N(7) hydrogen bond is located closer to N(1) (TMP) than to N(7) (SMZ) to which it was originally bound. Protonation of TMP in III occurs at N(1) which is confirmed by the formation of the hydrogen bond N(1)...Cl. The N(7) atom of SMZ in I and II adopts a trigonal conformation. The N(7)-C(3) and N(7)-S distances of the anion in I (1.378(3), 1.573(3) Å) are short, indicating an ionising effect of SMZ on these bonds. The C(4)-C(5)-C(9)-C(10) and C(5)-C(9)-C(10)-C(11) torsion angles in TMP are -172.9(3), 90.6(4) and -156.4(1), 62.5(2)° for II and III respectively. The N(2)-C(3)-N(7)-S, C(3)-N(7)-S-C(8) and N(7)-S-C(8)-C(9) torsion angles in SMZ are 8.6(3), 67.8(2), -144.4(1) and 171.2(2), 90.4(2) and -65.2(3)° for I and II respectively.

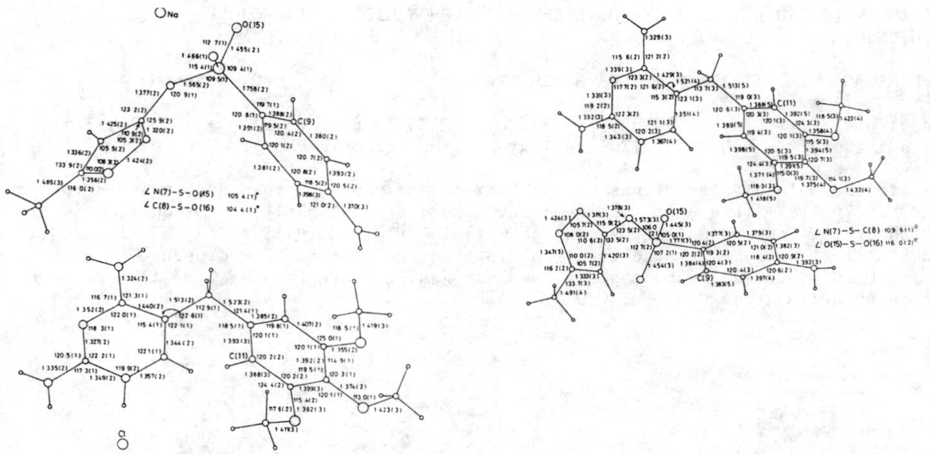

Fig. 1. The molecular skeleton and bond lengths and angles in $C_{10}H_{10}N_3NaO_3S$ (top left), in $C_{14}H_{19}ClN_4O_3$ (bottom left) and in $C_{24}H_{29}N_7O_6S$ (right).

4-AMINOPHENYL(N-MORPHOLYL)SULPHONE
$C_{10}H_{14}N_2O_3S$

G.V. GRIDUNOVA, S.V. LINDEMAN, V.E. SHKLOVER, Yu.T. STRUCHKOV and B.A. CHAYANOV, 1984. Kristallografiya, 29, 71-75 [Sov. Phys. Crystallogr., 29, 40-42].

Monoclinic, $P2_1$, a = 11.539, b = 7.825, c = 6.1571 Å, β = 91.32°, Z = 2. Cu radiation, R = 0.040 for 1251 reflexions.

The bond lengths and angles and torsion angles in the $>NSO_2-$ fragment of the molecule correspond to the geometry of most compounds in the class of arylsulphonyl-amides (1). The morpholine ring has the usual chair conformation (Fig. 1) whilst the benzene ring shows shortening of two bonds (C(8)-C(9) 1.352 and C(11)-C(12) 1.365 Å) resembling a quinoid form. The S atom has a distorted tetrahedral configuration with a marked increase in the angle O(1)-S-O(2) to 119.2°. The molecules are linked by weak hydrogen bonds (N(13)...O(2') 3.168, N(13)...O(4') 3.136 Å) into loose networks parallel to (011).

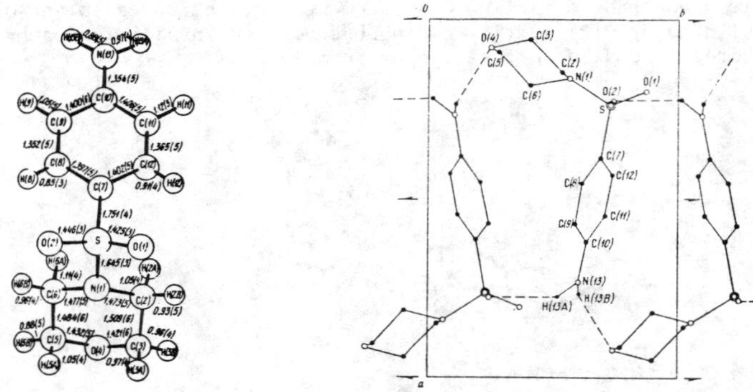

Fig. 1. Molecular structure of $C_{10}H_{14}N_2O_3S$ showing bond lengths (Å) (left), and projection of crystal structure on ab showing provisional hydrogen bonds (right).

1. Structure Reports, 42B, 812.

4-AMINO-N-(4,5-DIMETHYL-2-OXAZOLYL)BENZENESULPHONAMIDE (SULFAMOXOLE)
$C_{11}H_{13}N_3O_3S$

M. HARIDAS, R.K. TIWARI and T.P. SINGH, 1984. Acta Cryst., C40, 658-660.

Monoclinic, $P2_1/c$, a = 7.93, b = 12.88, c = 13.38 Å, β = 106.5°, D_m = 1.370, Z = 4. Cu radiation, R = 0.071 for 1297 reflexions (photographic data).

In the molecule the planes of the benzene and oxazole rings are inclined to each other at 76.7(8)° (Fig. 1). The geometry around S is slightly distorted from ideal tetrahedral. Some bond lengths are: S-C(1) 1.77(1), S-O(9) 1.49(1), S-O(10) 1.46(1) and S-N(11) 1.56(1) Å. There is a weak intermolecular N-H...O hydrogen bond with N(7)...O(10') 3.00(2) Å and a strong van der Waals interaction of 2.87(1) Å between the sulphonyl O atoms O(9) and O(9').

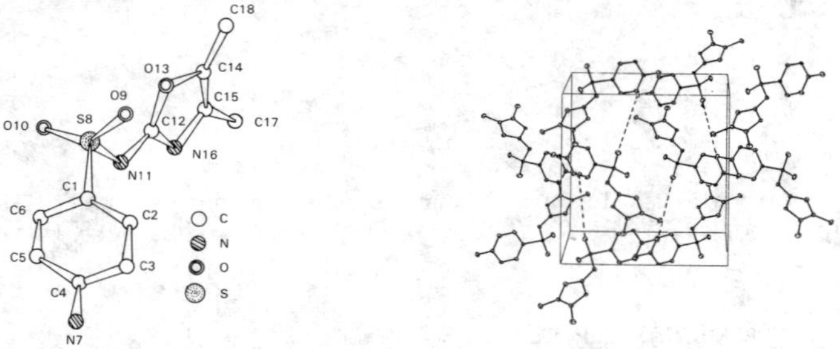

Fig. 1. $C_{11}H_{13}N_3O_3S$: views of the molecule and of the a-axial projection of the crystal structure; broken lines represent hydrogen bonds.

6-[1-(R,S)-HYDROXY-2-ISOPROPYLAMINOETHYL]-2,3-DIHYDRO-1,3-BENZOXAZOLE-2-ONE HYDROCHLORIDE
$C_{12}H_{17}ClN_2O_3$

G. MAIRESSE, J.C. BOIVIN, D.J. THOMAS, J.P. BONTE, D. LESIEUR and C. LESPAGNOL, 1984. Acta Cryst., C40, 1432-1434.

Monoclinic, $P2_1/c$, a = 9.562, b = 7.475, c = 18.843 Å, β = 94.8°, Z = 4. Mo radiation, R = 0.066 for 1588 reflexions.

The molecule is extended in two almost perpendicular planes (Fig. 1); the atoms of the bicyclic group show a maximum displacement from their mean plane of 0.03 Å. Molecules are linked by O(2)...N(15) hydrogen bonds and in addition each Cl ion is involved in hydrogen bonding to three separate molecules.

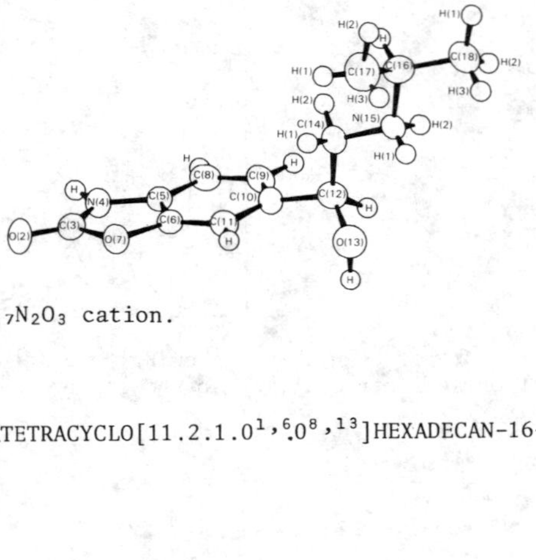

Fig. 1. The $C_{12}H_{17}N_2O_3$ cation.

14-OXA-7,15,16-TRIAZATETRACYCLO[11.2.1.0^{1,6}.0^{8,13}]HEXADECAN-16-OL
$C_{12}H_{21}N_3O_2$

H. SATO, S. IMAMURA, Y. KITANO, T. KANDA and T. ASHIDA, 1984. Bull. Chem. Soc. Japan, 57, 2162-2166.

Monoclinic, $P2_1/c$, a = 6.022, b = 19.245, c = 10.560 Å, β = 95.94°, D_m = 1.298, Z = 4. Cu radiation, R = 0.046 for 2022 reflexions.

The molecule has a fused ring structure (Fig. 1). The two outer cyclohexane rings are attached equatorially to the central six-membered heterocyclic ring and have chair conformations. The bonding within the three six-membered rings is normal with an average C-C bond length of 1.528 Å and C-N bond length of 1.471 Å. The bond angle C(1)-N(2)-C(12) is small at 100.1° due to the effect of strain where the five-membered ring is attached. The bond N(2)-O(2) is lengthened to 1.458 Å for the same reason. The five-membered ring has an envelope conformation with N(2) 0.610 Å out of the plane of the remaining four ring atoms. Bond lengths of note are: C(1)-O(1) 1.462, O(1)-N(3) 1.467, N(3)-C(12) 1.480, C(12)-N(2) 1.484, C(1)-N(2) 1.461, C(6)-N(1) 1.470 and N(1)-C(7) 1.467 Å. Each molecule is hydrogen-bonded to four neighbouring molecules. The hydrogen bond distances are: N(3)...N(1') 3.158 and O(2)...N(3') 2.807 Å, with the angles N(3)-H(N3)...N(1') 163°, and O(2)-H(O2)...N(3') 170°.

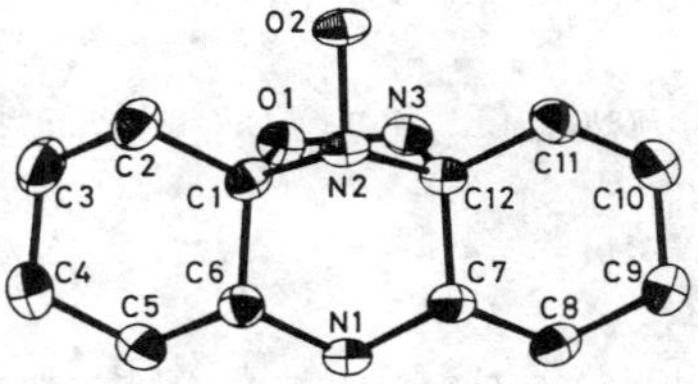

Fig. 1. Perspective view of the molecule $C_{12}H_{21}N_3O_2$.

BIS(N-ETHYLMORPHOLINIUM) TETRACHLOROMERCURATE(II)
$C_{12}H_{28}Cl_4HgN_2O_2$ (I) $[C_6H_{14}NO^+]_2HgCl_4^{2-}$

(N-AMMONIUMETHYLMORPHOLINIUM) TETRACHLOROMERCURATE(II)
$C_6H_{16}Cl_4HgN_2O$ (II) $[C_6H_{16}N_2O]^{2+}HgCl_4^{2-}$

I.M. VEZZOSI, A. BENEDETTI, A. ALBINATI, F. GANAZZOLI, F. CARIATI and L. PELLICCIARI, 1984. Inorg. Chim. Acta, 90, 1-7.

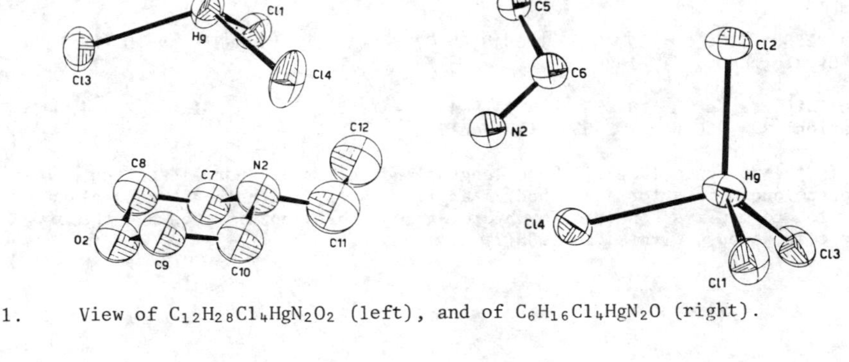

Fig. 1. View of $C_{12}H_{28}Cl_4HgN_2O_2$ (left), and of $C_6H_{16}Cl_4HgN_2O$ (right).

I. Orthorhombic, Pbca, a = 18.750, b = 14.363, c = 15.131 Å, Z = 8. Mo radiation, R = 0.073 for 1994 reflexions.

II. Monoclinic, $P2_1/c$, a = 6.657, b = 13.790, c = 14.404 Å, β = 91.48°, Z = 4. Mo radiation, R = 0.042 for 1787 reflexions.

Both compounds consist of discrete $HgCl_4^{2-}$ units hydrogen-bonded to the N atoms of the counter ions. In both cases the Hg(II) atom shows a distorted tetrahedral coordination. In I there is one long (2.559 Å) and three short (av. 2.456 Å) Hg-Cl distances whilst in II a more continuous range of Hg-Cl bond lengths (2.395-2.562 Å) is present. Hydrogen bond distances fall into two categories: those Cl...N distances greater than the sum of the van der Waals radii (3.30 Å) and those less than 3.30 Å. In both compounds the shortest Cl...N distances (3.14-3.16 Å) are those involving atoms Cl(1) and N(2). The morpholinium rings have the expected chair conformations (Fig. 1).

2-(2-CYANOETHYL)-cis-5-CYANO-3-PHENYLISOXAZOLIDINE
$C_{13}H_{13}N_3O$

R. GRIGG, M. JORDAN, A. TANGTHONGKUM, F.W.B. EINSTEIN and T. JONES, 1984. J. Chem. Soc. Perkin I, 47-57.

Monoclinic, $P2_1/c$, a = 15.090, b = 7.414, c = 12.721 Å, β = 119.77°, Dm = 1.17, Z = 4. Mo radiation, R = 0.034 for 1161 reflexions.

The X-ray analysis established the structure of this major isomer from the reaction of benzaldehyde oxime with acrylonitrile showing the cis-relationship between the phenyl ring and the 5-cyano group (Fig. 1). These groups are also cis to N(1). The five-membered heterocycle has an envelope conformation (flap angle 49.9°). Bond lengths and angles are normal.

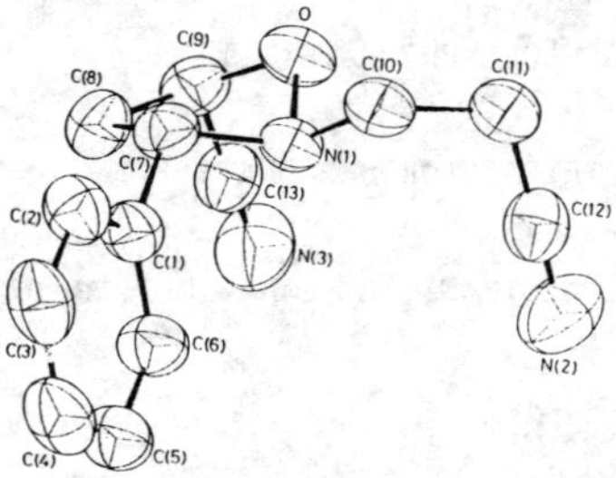

Fig. 1. $C_{13}H_{13}N_3O$: the molecular structure.

6-BENZOYL-2,3-DIHYDROBENZ-1,3-OXAZOLE-2-ONE
$C_{14}H_9NO_3$

G. MAIRESSE, J.C. BOIVIN, D.J. THOMAS, M.C. BERMANN, J.P. BONTE and D. LESIEUR, 1984. Acta Cryst., C40, 1019-1020.

Monoclinic, $P2_1/c$, a = 8.853, b = 14.944, c = 11.223 Å, β = 131.1°, Z = 4. Mo radiation, R = 0.046 for 2153 reflexions.

In the molecule (Fig. 1) bond lengths and angles are normal and the planar benzoxazolone moiety forms a dihedral angle of 54(1)° with the benzene plane. The molecules form pairs connected by hydrogen bonds between nitrogen and the oxygen of the cyclic ketone, with N...O 2.85(2) Å.

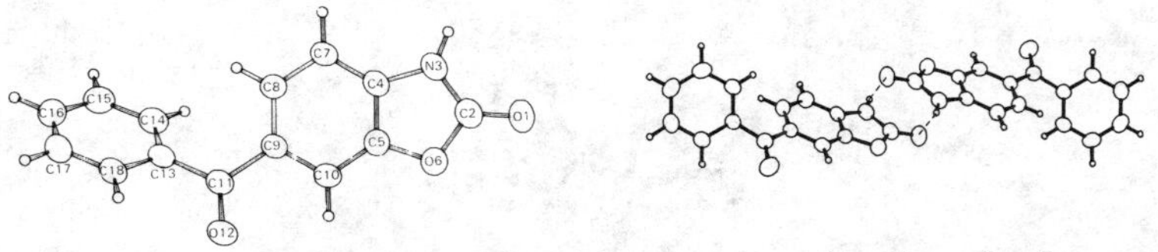

Fig. 1. $C_{14}H_9NO_3$: configuration of the molecule and a view showing a pair of molecules linked by N-H...O hydrogen bonds.

(4RS,5RS)-4,5-BIS(METHOXYCARBONYL)-3-[(4RS)-2,2,5,5-TETRAMETHYL-1,3-DIOXOLAN-4-YL]-Δ^2-ISOXAZOLINE

$C_{14}H_{21}NO_7$

R.H. JONES, G.C. ROBINSON and E.J. THOMAS, 1984. Tetrahedron, 40, 177-184.

Orthorhombic, Pbca, a = 7.642, b = 16.812, c = 24.822 Å, Z = 8. Cu radiation, R = 0.049 for 1860 reflexions.

Both five-membered rings (Fig. 1) have envelope conformations, with the 1,3-dioxolane ring being the more puckered. Molecular dimensions are as anticipated; the opening out of the C(3)-C(15)-C(14) angle to 116.7° is probably caused by steric crowding between C(15) and C(19) which have a 3.37 Å separation.

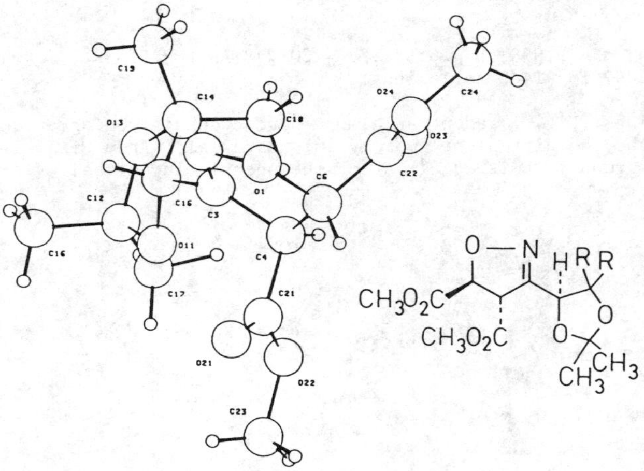

Fig. 1. The $C_{14}H_{21}NO_7$ molecule (R = CH_3).

5,5-DIPHENYL-1,3-OXAZOLIDINE-2,4-DIONE

$C_{15}H_{11}NO_3$

P.W. CODDING, 1984. Acta Cryst., C40, 2071-2074.

Monoclinic, $P2_1/c$, a = 12.473, b = 6.025, c = 16.371 Å, β = 90.57°, Z = 4. Mo radiation, R = 0.092 for 1779 reflexions at 173 K.

The angle between the two phenyl rings (Fig. 1) is 79.7(5)° and the angles between the oxazolidinedione ring and each of the two phenyl rings are 105.7(5) and 115.2(5)°; these values are similar to those found in the active anticonvulsant, diphenylhydantoin (1). The molecules form an intermolecular hydrogen bond between one carbonyl O atom and the amide N atom. Despite the different intermolecular environments of the two carbonyl groups, the C=O bond lengths are the same.

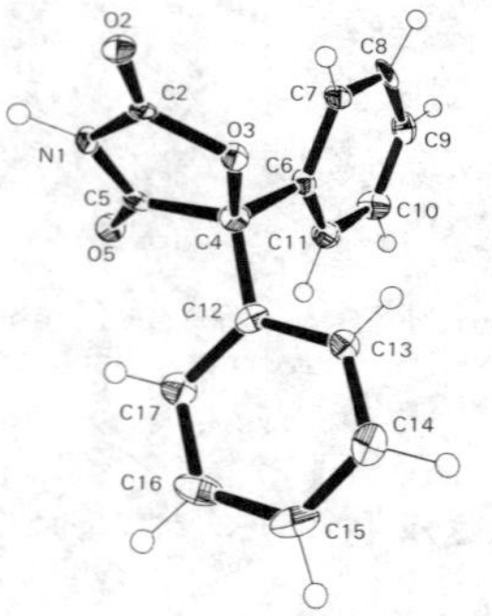

Fig. 1. A view of $C_{15}H_{11}NO_3$.

1. D. MASTROPAOLO, A. CAMERMAN and N. CAMERON, 1983. Mol. Pharmacol., 23, 273-277.

(r-4,t-2,c-5)-2-(p-NITROPHENYL)-3-METHYL-4,5-TETRAMETHYLENETETRAHYDRO-1,3-OXAZINE
$C_{15}H_{20}N_2O_3$

G. BERNATH, F. FULOP, A. KALMAN, G. ARGAY, P. SOHAR and I. PELCZER, 1984. Tetrahedron, 40, 3587-3593.

Monoclinic, $P2_1/n$, a = 9.958, b = 6.822, c = 20.748 Å, β = 91.39°, Z = 4. Mo radiation, R = 0.097 for 1445 reflexions.

The fused rings (Fig. 1) adopt an almost perfect chair conformation, the N(3)-Me group is axial, and the nitrophenyl moiety is equatorial. Atom C(13) is eclipsed with O(1). The nitro group is twisted 6° out of the phenyl ring plane.

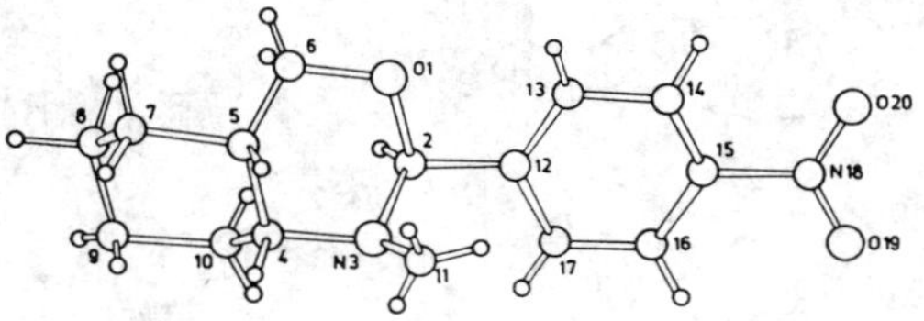

Fig. 1. A view of $C_{15}H_{20}N_2O_3$.

1-METHYL-2,4-DIAZA-6,7-BENZO-8-OXABICYCLO[3.3.1]NON-3-ENE
$C_{17}H_{16}N_2O$ (I)

1-PHENYL-2,4-DIAZA-6,7-BENZO-8-OXABICYCLO[3.3.1]NON-2-ENE
$C_{22}H_{18}N_2O$ (II)

A.L. WEISS and F. FROLOW, 1984. J. Org. Chem., 49, 3635-3637.

I. Monoclinic, C2/c, a = 24.660, b = 12.134, c = 10.139 Å, β = 114.26°, D_m = 1.270, Z = 8. Cu radiation, R = 0.042 for 2438 reflexions.

II. Monoclinic, $P2_1/a$, a = 8.969, b = 17.968, c = 10.595 Å, β = 102.16°, Dm = 1.289, Z = 4. Mo radiation, R = 0.068 for 1824 reflexions.

Views of the two molecules are in Fig. 1. Both have a half-chair conformation for the tetrahydropyrimidinic moieties, with the N-H, the C5 hydrogen and Me or Ph pseudoequatorial, and the C1-O pseudoxial.

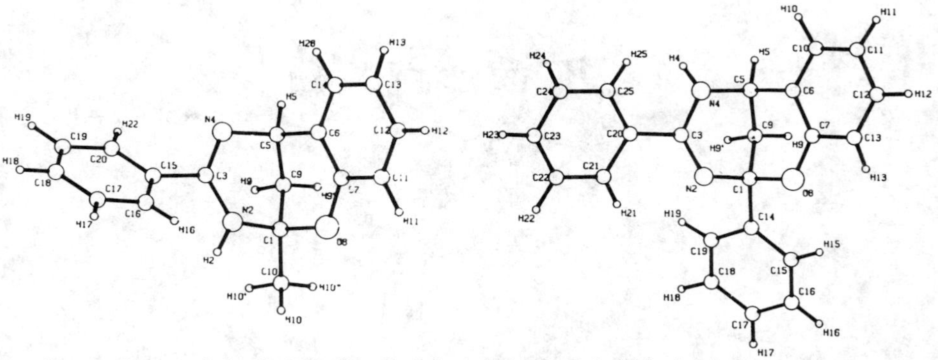

Fig. 1. Views of $C_{17}H_{16}N_2O$ (I) (left) and $C_{22}H_{18}N_2O$ (II) (right).

3-BENZYLOXYCARBONYL-1-[(N-(CHLOROACETYL)-N-METHOXYAMINO)METHYL]-2-OXA-3-AZABICYCLO-[2.2.2]OCT-5-ENE

$C_{18}H_{21}ClN_2O_5$

J.E. BALDWIN, P.D. BAILEY, G. GALLACHER, M. OTSUKA, K.A. SINGLETON, P.M. WALLACE, K. PROUT and W.M. WOLF, 1984. Tetrahedron, 40, 3695-3708.

Monoclinic, $P2_1/c$, a = 5.977, b = 19.945, c = 16.059 Å, β = 98.80°, Z = 4. Mo radiation, R = 0.039 for 1755 reflexions.

The bicyclic skeleton (Fig. 1) has slightly staggered D_3 pseudosymmetry, when viewed down the C(5)...C(8) axis. The configuration of both N(1) and N(2) is pyramidal (N atoms 0.260 and 0.356 Å from the appropriate plane).

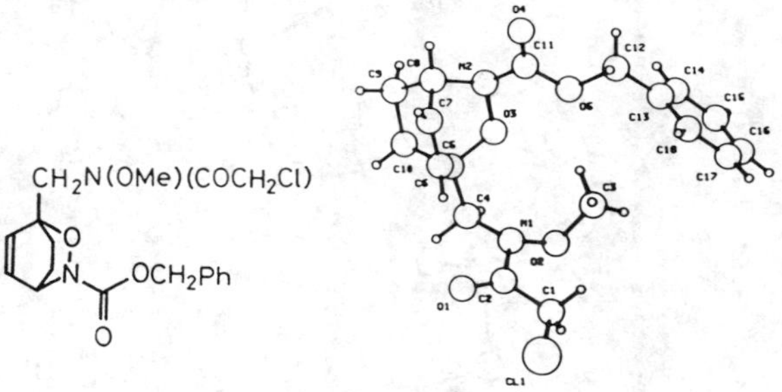

Fig. 1. A view of $C_{18}H_{21}ClN_2O_5$.

5'-MORPHOLINO-1'-(4-NITROPHENYL)-4',5'-DIHYDROSPIRO(BICYCLO[2.2.1]HEPT-2-ENE[5.4']-1',2',3'-TRIAZOLE)

$C_{18}H_{21}N_5O_3$

M.L. GELMI, D. POCAR, P. TRIMARCO, M. VALSECCHI, R. DESTRO and M. BELLABIO, 1984. Tetrahedron, 40, 4025-4032.

Monoclinic, $P2_1/c$, a = 6.590, b = 15.805, c = 16.541 Å, β = 92.53°, D_m = 1.37, Z = 4. Mo radiation, R = 0.040 for 1784 reflexions.

The molecule (Fig. 1) is highly strained. The C(1)-C(6) bond is lengthened to 1.582(3) Å. In the triazoline ring there is a 6° torsion angle at a formally double bond N(2)-N(3) leading to a distorted envelope conformation.

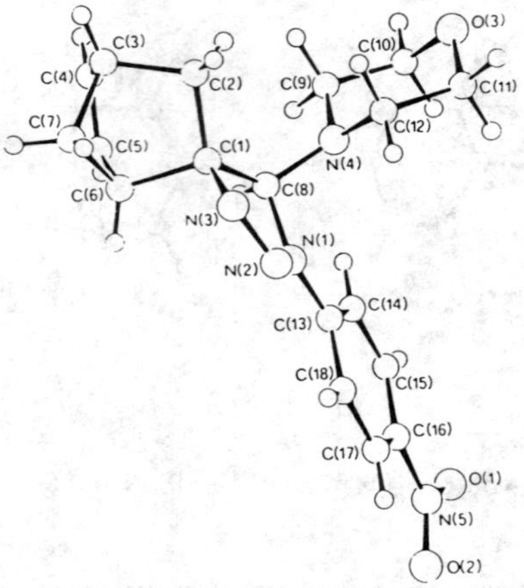

Fig. 1. A view of $C_{18}H_{21}N_5O_3$.

10,10'-ETHYLENEBIS(1,4,7-TRIOXA-10-AZACYCLODODECANE)
$C_{18}H_{36}N_2O_6$

P. GROTH, 1984. Acta Chem. Scand., A38, 342-344.

Monoclinic, $P2_1/n$, a = 10.662, b = 7.264, c = 12.926 Å, β = 92.68°, D_m = 1.22, Z = 2. Mo radiation, R = 0.027 for 1552 reflexions (at -130°C).

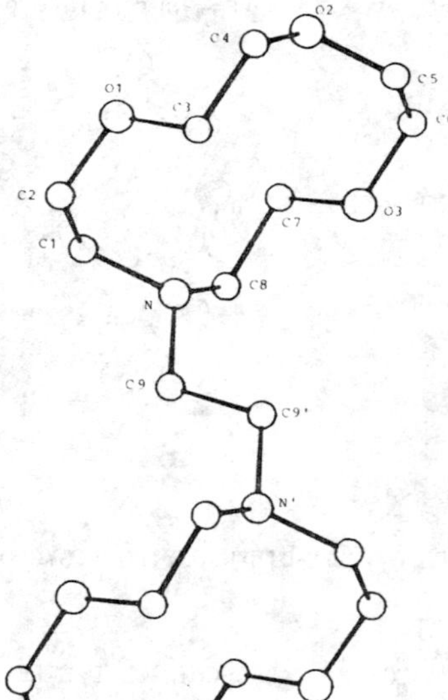

Fig. 1. Schematic drawing of the molecule $C_{18}H_{36}N_2O_6$.

The centrosymmetric molecule has its two twelve-membered rings in the biangular [6 6] conformation (Fig. 1). The bond distances of note are: O(1)-C(2) 1.425, O(2)-C(4) 1.431, O(3)-C(6) 1.424, N-C(9) 1.462, N-C(8) 1.465, C(1)-C(2) 1.516, C(5)-C(6) 1.507, O(1)-C(3) 1.431, O(2)-C(5) 1.428, O(3)-C(7) 1.435, N-C(1) 1.467, C(9)-C(9') 1.522, C(3)-C(4) 1.506 and C(7)-C(8) 1.510 Å. The N-C(9)-C(9')-N torsion angle is trans in the molecule.

9-METHYL-1-MORPHOLINO-4-PHENYL-2,3-OXAZABICYCLO[4.4.0]DEC-3-ENE (Two Isomers)
$C_{19}H_{26}N_2O_2$

M.T. COCCO, A. MACCIONI, A. PLUMITALLO, M. CANNAS and G. MARONGIU, 1984. Tetrahedron, 40, 2041-2045.

I. Triclinic, P$\bar{1}$, a = 10.159, b = 19.297, c = 9.561 Å, α = 89.08, β = 98.07, γ = 93.07°, D_m = 1.18, Z = 4. Cu radiation, R = 0.101 for 2371 reflexions.

II. Monoclinic, $P2_1/c$, a = 11.261, b = 13.312, c = 14.700 Å, β = 107.71°, D_m = 1.22, Z = 4. Mo radiation, R = 0.092 for 1816 reflexions.

X-ray analysis determined the structures of the two isomers as shown in Fig. 1. In I there are two independent molecules related by a pseudosymmetry element. In general, molecular dimensions agree well over all three molecules (of I and II). Slight conformation differences are present in the oxazine rings of I and II as shown by deviations from the O-N-C-C planes of the C(9) and C(15) atoms (mean deviations respectively 0.30, -0.42 Å in I, 0.24 and -0.50 Å in II).

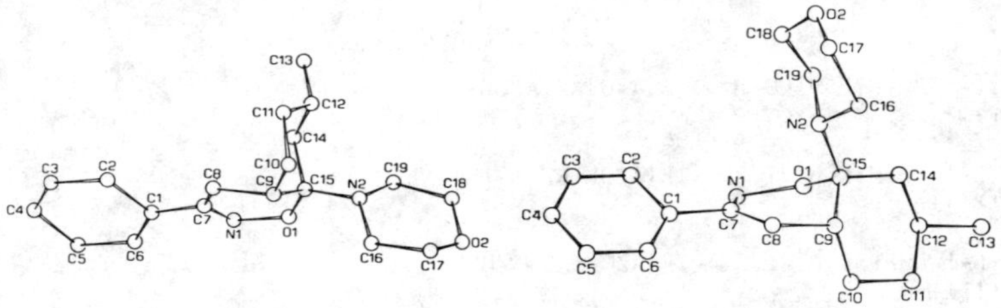

Fig. 1. Isomer I (left) and II (right) of $C_{19}H_{26}N_2O_2$.

14-BENZYL-7,12-DIMETHYL-1,4-DIOXA-7,12-DIAZACYCLOTETRADECANE-2,6,13-TRIONE
$C_{19}H_{26}N_2O_5$

Yu.G. GANIN, Yu.A. SIMONOV, N.G. LUK'YANENKO, Yu.A. POPKOV and T.A. SHIBANOVA, 1984. J. Struct. Chem., 25, 304-309 [Zh. Strukt. Khim. 25(2), 149-154].

Monoclinic, $P2_1/c$, a = 10.521, b = 16.094, c = 11.338 Å, α = 88.98°, Z = 4. Mo radiation, R = 0.061 for 1596 reflexions.

The fourteen-membered ring has a [3434] conformation (Fig. 1) if pseudocorners are included. The benzyl substituent occupies an equatorial position. One of the amide groups has a cis configuration while the other amide group and the ester group both have trans configurations. Torsion angles indicate some ring strain in the molecule.

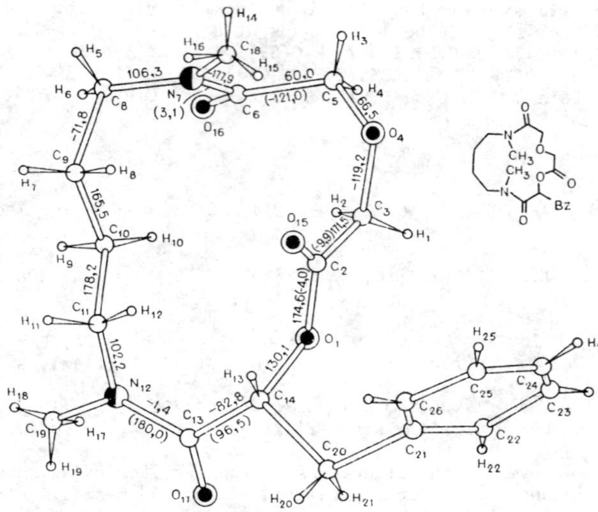

Fig. 1. Torsion angles in the fourteen-membered ring of 14-benzyl-7,12-dimethyl-1,4-dioxa-7,12-diazacyclotetradecane-2,6,13-trione.

5,5-DIPHENYL-2-PHENYLIMINO-Δ^3-1,3,4-OXADIAZOLINE
$C_{20}H_{15}N_3O$ (I)

3,3-DIMETHYL-1-PHENYL-Δ^1-1,2,4-TRIAZOLINE-5-THIONE
$C_{10}H_{11}N_3S$ (II)

3,3-DIMETHYL-1-PHENYL-Δ^1-1,2,4-TRIAZOLINE-5-ONE
$C_{10}H_{11}N_3O$ (III)

A.R. KATRITZKY, H.M. FAID-ALLAH, H. AGHABOZORG and G.J. PALENIK, 1984. Chem. Script., 23, 134-139.

I. Triclinic, P$\bar{1}$, a = 10.426, b = 9.688, c = 10.918 Å, α = 92.21, β = 116.86, γ = 119.56°, D_m = 1.28, Z = 2. Mo radiation, R = 0.085 for 1586 reflexions.

II. Monoclinic, $P2_1/n$, a = 14.195, b = 8.269, c = 9.087 Å, β = 98.48°, D_m = 1.29, Z = 4. Mo radiation, R = 0.037 for 1524 reflexions.

III. Monoclinic, $P2_1/n$, a = 8.384, b = 8.245, c = 14.335 Å, β = 103.30°, D_m = 1.30, Z = 4. Mo radiation, R = 0.048 for 1290 reflexions.

All three molecules consist of a planar five-membered ring which contains a tetrahedral C atom and an N-N double bond. The N=N distances (1.252, 1.244 and 1.244 Å in I, II and III respectively) are not significantly different but the remaining distances vary in the three compounds. The largest changes appear to occur in going from II to III where the only change is from S to O. The C(2)-O and C(2)-S distances are 1.201 and 1.630 Å respectively in III and II, similarly C(2)-N(3) 1.477 and 1.459 Å, N(4)-C(5) 1.479 and 1.468 Å, C(5)-C(6) 1.522 and 1.520 Å, C(5)-C(7) 1.509 and 1.507 Å, C(5)-N(1) 1.451 and 1.462 Å, N(1)-C(2) 1.344 and 1.331 Å, and N(1)-C(8) 1.430 and 1.434 Å. In I (Fig. 1) some distances of interest are: O(1)-C(2) 1.350, O(1)-C(5) 1.456, C(2)-N(3) 1.440, C(5)-N(4) 1.507, C(2)-N(6) 1.258, C(5)-N(4) 1.507, N(6)-C(31)(phenyl) 1.412 and N(3)-N(4) 1.252 Å.

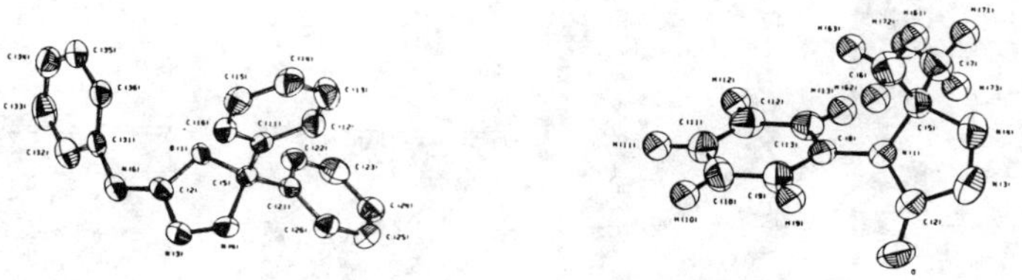

Fig. 1. Views of the molecular structure of $C_{20}H_{15}N_3O$ (left), and $C_{10}H_{11}N_3O$ (right).

3-(PHENYLSULPHONYL)-4-PHENYL-5-(METHOXYCARBOXYL)-3a,4,5,7a-TETRAHYDROPYRIDO[3,4-d]-ISOXAZOLE

$C_{20}H_{18}N_2O_5S$

S.K. DUBEY and E.E. KNAUS, 1984. J. Org. Chem., <u>49</u>, 123-125.

Triclinic, $P\bar{1}$, a = 7.830, b = 16.636, c = 7.596 Å, α = 90.77, β = 104.07, γ = 81.69°, Z = 2. Mo radiation, R = 0.091 for 697 reflexions.

The analysis establishes the geometry of this cycloadduct as shown in Fig. 1. The torsion angle C3-C3a-C4-CPh is 164.5°, indicating that the phenyl substituent at C4 and the 3a-3 carbon-carbon bond are axially disposed; the two hydrogen atoms at C4 and C3a are equatorial (torsion angle H3a-C3a-C4-H4 -70.5°).

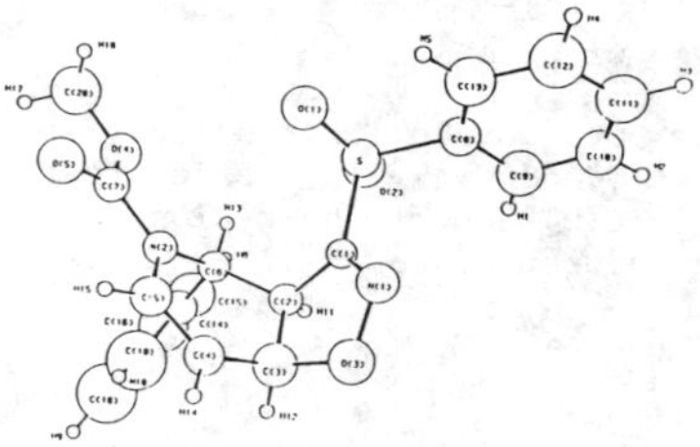

Fig. 1. A perspective view of $C_{20}H_{18}N_2O_5S$.

(5R*,1R*)-5-(1-METHYLMETHOXYMETHYL)-3-[4-(3-CYANOPHENYL-METHOXY)-PHENYL]OXAZOLIDIN-2-ONE

$C_{20}H_{20}N_2O_4$

F. DURANT, B. NORBERG, F. BUFKENS and G. EVRARD, 1984. Bull. Soc. Chim. Fr., II-183-186.

Monoclinic, C2/c, a = 35.199, b = 5.889, c = 20.314 Å, β = 122.40°, Z = 8. Cu radiation, R = 0.06 for 1633 reflexions.

Molecular dimensions and a stereoview of the crystal structure are in Fig. 1. The oxazolidinone moiety and adjacent phenyl ring are essentially coplanar.

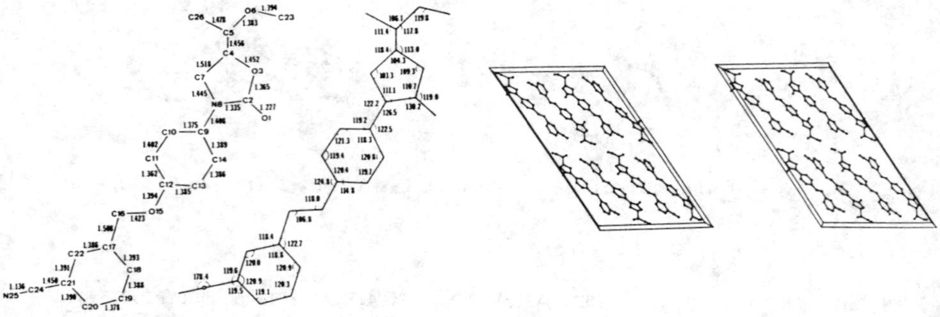

Fig. 1. Molecular dimensions for $C_{20}H_{20}N_2O_4$ and a stereoview of the crystal structure.

1,5-DIMORPHOLINO-5-METHYL-4-NITRO-3-PHENYLCYCLOPENTENE
$C_{20}H_{27}N_3O_4$

D. POCAR, P. TRIMARCO, R. DESTRO, E. ORTOLEVA and M. BALLABIO, 1984. Tetrahedron, 40, 3579-3585.

Orthorhombic, Pbca, a = 8.671, b = 18.768, c = 24.660 Å, D_m = 1.23, Z = 8. Mo radiation, R = 0.034 for 1644 reflexions.

In the molecule (Fig. 1) both morpholino rings are in chair form. The cyclopentene ring is overcrowded and C(4)-C(5) is lengthened to 1.578(2) Å. There is also a 5° twist about the C(1)=C(2) bond. The cyclopentene ring is in an envelope conformation (dihedral angle between planes C(1)C(2)C(3)C(5) and C(3)C(4)C(5) 30.8(1)°).

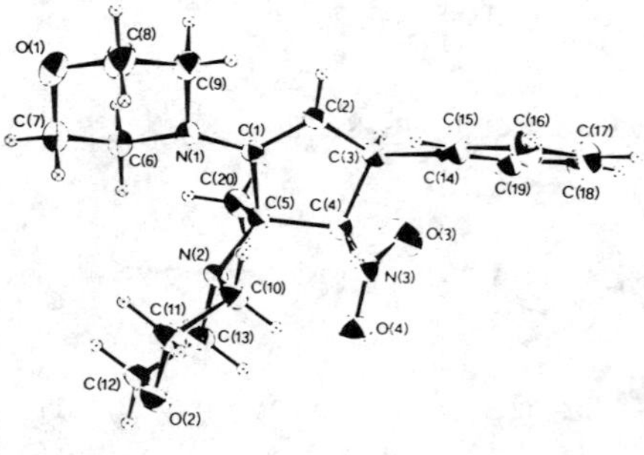

Fig. 1. A view of $C_{20}H_{27}N_3O_4$.

4,4-DIOXO-2,3-DIPHENYL-2,3-DIHYDROBENZO[b]THIENO[2,3-d]ISOXAZOLIDINE
$C_{21}H_{17}NO_3S$ (I)

4,4-DIOXO-3a-METHYL-2,3-DIPHENYL-2,3-DIHYDROBENZO[b]THIENO[2,3-d]ISOXAZOLIDINE
$C_{22}H_{19}NO_3S$ (II)

A. BENED, R. DURAND, D. PIOCH, P. GENESTE, C. GUIMON, G.P. GUILLOUZO, J.P. DECLERCQ, G. GERMAIN, P. BRIARD, J. RAMBAUD and R. ROQUES, 1984. J. Chem. Soc. Perkin II, 1-6.

I. Monoclinic, $P2_1/n$, a = 8.038, b = 14.880, c = 15.037 Å, β = 102.41°, D_m = 1.40, Z = 4. Mo radiation, R = 0.053 for 2021 reflexions.

II. Triclinic, P1, a = 11.735, b = 10.322, c = 9.210 Å, α = 63.27, β = 101.59, γ = 110.71°, D_m = 1.38, Z = 2. Mo radiation, R = 0.04 for 2021 reflexions.

The X-ray analyses established the structures of the title compounds (Fig. 1) and that in I 3a- and 3-H are trans to each other and 3a- and 8b-H are cis. The angles between rings A/B and B/C in I are 5.15 and 129° respectively.

Fig. 1. $C_{21}H_{17}NO_3S$: the molecular skeleton showing ring labelling and atom numbering. In I, $R^2 = R^3$ = H while in II, $R^2 = CH_3$, R^3 = H.

1,2,7,7a-TETRAHYDRO-1,7-DIMETHYL-2-METHYLIMINO-4,7a-DIPHENYLPYRROLO[2,3-d][1,3]OXAZINE-5,6-DIONE

$C_{21}H_{19}N_3O_3$ (I)

1,2,7,7a-TETRAHYDRO-1,7-DI-ISOPROPYL-2-ISOPROPYLIMINO-4,7a-DIPHENYLPYRROLO[2,3-d][1,3]-THIAZINE-5,6-DIONE

$C_{27}H_{31}N_3O_2S$ (II)

G. KOLLENZ, G. PENN, W. OTT, K. PETERS, E.-M. PETERS and H.G. VON SCHNERING, 1984. Chem. Ber., 117, 1310-1329.

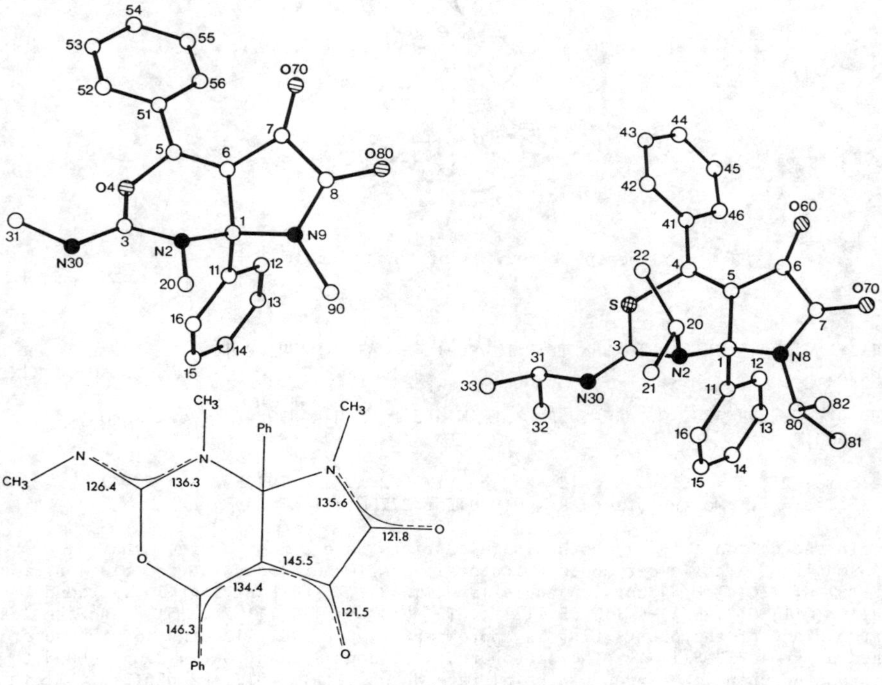

Fig. 1. A view of $C_{21}H_{19}N_3O_3$ and some bond lengths (left) and a view of $C_{27}H_{31}N_3O_2S$ (right).

I. Monoclinic, $P2_1/a$, a = 21.80, b = 8.895, c = 9.386 Å, β = 95.95°, Z = 4. Mo radiation, R = 0.043 for 2095 reflexions.

II. Triclinic, $P\bar{1}$, a = 12.221, b = 12.458, c = 9.702 Å, α = 96.89, β = 70.05, γ = 116.98°, Z = 2. Mo radiation, R = 0.042 for 3320 reflexions.

The structures of these reaction products are shown in Fig. 1. In I, there is some evidence for delocalisation. Some distances in II are: C3-S 1.825(2), S-C4 1.748(3), C1-N2 1.481(3), C1-N8 1.466(4), N2-C3 1.396(4), C3-N30 1.258(3), C7-N8 1.359(3), N30-C31 1.454(5) Å.

1,3,6-TRIMETHYL-4,5-DIPHENYL-9-OXA-2,7-DIAZATRICYCLO[3.3.1.$0^{2,4}$]NONAN-8-ONE
$C_{21}H_{22}N_2O_2$

R. FAURE, H. LOISELEUR, G. ALVERNNHE and A. LAURENT, 1984. Acta Cryst., C40, 1204-1205.

Monoclinic, $P2_1/c$, a = 11.669, b = 15.338, c = 13.596 Å, β = 133.75°, Z = 4. Cu radiation, R = 0.035 for 1834 reflexions.

The molecule (Fig. 1) results from the condensation of two molecules of 2-methyl-3-phenylazinine with one molecule of ethyl lactate and has a tricyclic structure. In spite of the presence of six asymmetric centres only one diastereoisomer, corresponding to the lowest steric hindrance, is formed. Bond lengths and angles are normal.

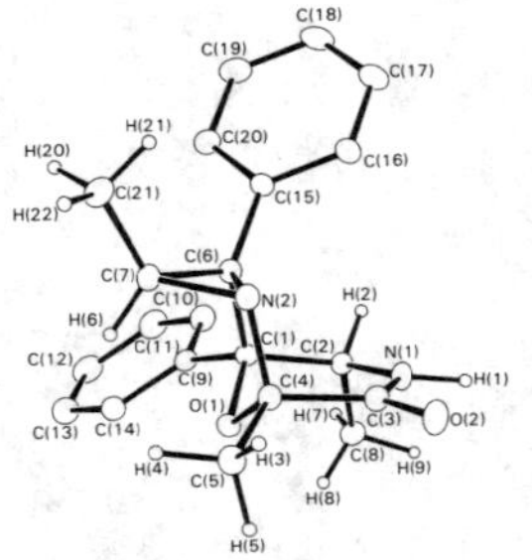

Fig. 1. $C_{21}H_{22}N_2O_2$: perspective view of the molecule.

N-BENZYL-cis-4,5-TETRAMETHYLENETETRAHYDRO-1,3-OXAZINIUM PICRATE
$C_{21}H_{24}N_4O_8$ $C_{15}H_{22}NO^+.C_6H_2N_3O_7^-$

F. FULOP, G. BERNATH, G. ARGAY, A. KALMAN and P. SOHAR, 1984. Tetrahedron, 40, 2053-2060.

Triclinic, $P\bar{1}$, a = 7.134, b = 12.830, c = 12.332 Å, α = 104.58, β = 87.41, γ = 104.37°, Z = 2. Mo radiation, R = 0.050 for 2720 reflexions.

In the cation (Fig. 1) both cis-fused rings have almost perfect chair conformations. The lowest endocyclic torsion angles are found at the cis junction. The H(5)-C(5)-C(4)-H(4) torsion angle is -47.3(9)°. The benzyl group is bound axially (O(1)-C(2)-N(3)-C(11) 65.2(2)°, C(5)-C(4)-N(3)-C(11) 71.3(3)°) and the least squares plane of the phenyl ring is approximately parallel with that of the heteroring (dihedral angle 17.8(1)°). The protonated N(3) donates a strong hydrogen-bond to the deprotonated phenolic oxygen of the picrate anion (N(3)...O(24) 2.676(2) Å, H(3)...O(24) 1.76 Å, NH...O 167.8(18)°). The ring distortion in the picrate anion is similar to that observed in other picrate moieties.

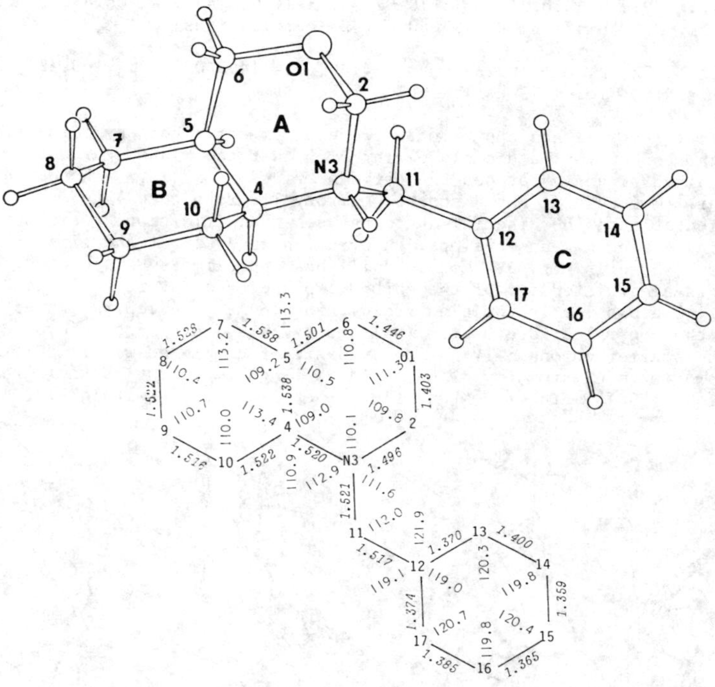

Fig. 1. The $C_{15}H_{22}NO^+$ cation, and bond lengths and angles (σ 0.002-0.004 Å, 0.2-0.4°).

SPIRO[1,3-DIOXOLANE-2,2'-(8,11,14,17,20-PENTAOXA-25,26-DIAZATRICYCLO[19.3.1.$1^{3,7}$]-HEXACOSA-1(25),3,5,7(26),21,23-HEXAENE)]
$C_{21}H_{26}N_2O_7$ (I)

SPIRO[1,3-DIOXOLANE-2,2'-(8,11,14-TRIOXA-20,21-DIAZATRICYCLO[13.3.1.$1^{3,7}$]ICOSA-1(20),-3,5,7(21),15,17-HEXAENE)]
$C_{17}H_{18}N_2O_5$ (II)

8,11,14,17,20,23,26-HEPTAOXA-31,32-DIAZATRICYCLO[25.3.1.$1^{3,7}$]DOTRIACONTA-1(31),3,5,-7(32),27,29-HEXAEN-2-ONE MONOHYDRATE
$C_{23}H_{30}N_2O_8,H_2O$ (III)

8,11,14,17,20,23,26-HEPTAOXA-31,32-DIAZATRICYCLO[25.3.1.$1^{3,7}$]DOTRIACONTA-1(31),3,5,-7(32),27,29-HEXAEN-2-ONE
$C_{19}H_{22}N_2O_6$ (IV)

G.R. NEWKOME, H.C.R. TAYLOR, F.R. FRONCZEK and T.J. DELORD, 1984. J. Org. Chem., 49, 2961-2971.

I. Monoclinic, $P2_1/c$, a = 9.051, b = 15.800, c = 14.662 Å, β = 96.85°, Z = 4. Cu radiation, R = 0.061 for 2411 reflexions.

II. Monoclinic, $P2_1/n$, a = 8.919, b = 16.636, c = 10.397 Å, β = 92.15°, Z = 4. Mo radiation, R = 0.033 for 1771 reflexions.

III. Triclinic, $P\bar{1}$, a = 10.265, b = 10.925, c = 11.633 Å, α = 106.41, β = 96.02, γ = 97.14°, Z = 2. Mo radiation, R = 0.060 for 2330 reflexions.

IV. Monoclinic, $P2_1/c$, a = 10.238, b = 13.140, c = 14.540 Å, β = 109.35°, Z = 4. Mo radiation, R = 0.044 for 1775 reflexions.

Molecule I (Fig. 1) has approximate twofold symmetry; both pyridine rings are arranged with respect to the dioxolane ring such that the N atoms are anti to the O atoms. The dihedral angle between the pyridine rings is 97.2°. The dioxolane ring is non-planar. Molecule II has a shorter macrocyclic ring than I and this affects the pyridine orientations; the dihedral angle between the pyridine rings is 66.8° and the dioxolane ring system has an envelope conformation with C16 at the flap. Coronands III and IV are shown in Fig. 2. Molecule III has approximate C2 symmetry; the pyridine rings are twisted out of the carbonyl plane by an average of 34.6° and the five ethereal O atoms lie in the plane to within 0.6 Å. The central water atom is 3.00 Å from the central bridge O atom (O5) and is 3.19-3.61 Å from the other ethereal O atoms. The smaller coronand IV has no evidence of complexed water; this macrocycle also has approximate C2 symmetry. The pyridine rings twist out of the carbonyl plane by an average of 30.1°. The two N and five ether O atoms are within 0.23 Å of their best plane.

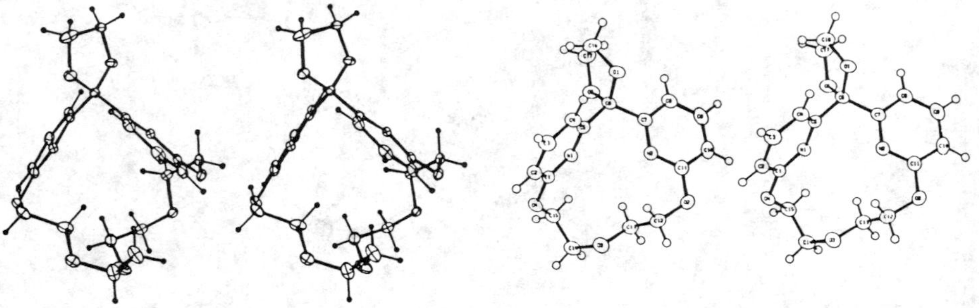

Fig. 1. Stereoviews of the spiro compounds (I) (left) and (II) (right).

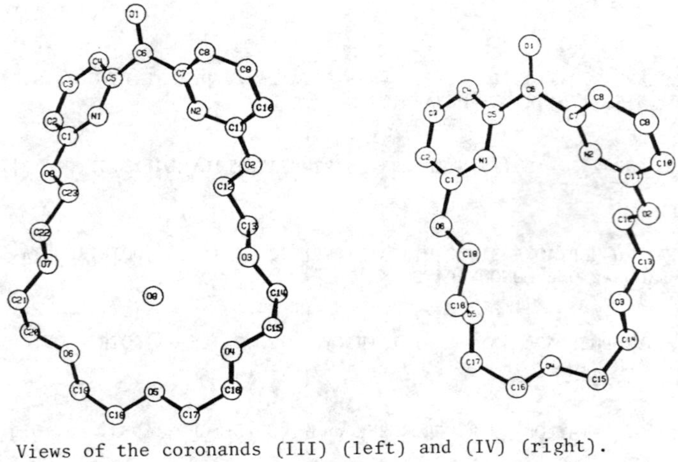

Fig. 2. Views of the coronands (III) (left) and (IV) (right).

2-BENZYL-1,7,10,13,19-PENTAOXA-4,16-DIAZACYCLOEICOSANE-3,17,21-TRIONE HYDRATE

$C_{21}H_{30}N_2O_8,H_2O$

Yu.G. GANIN, A.N. SOBOLEV, V.K. BEL'SKII, Yu.A. SIMONOV, N.G. LUK'YANENKO, Yu.A. POPKOV and T.I. MALINOVSKII, 1984. Kristallografiya, 29, 66-70 [Sov. Phys. Crystallogr., 29, 37-40].

Monoclinic, B2, a = 19.268, b = 14.652, c = 10.099 Å, γ = 54.77°, Z = 4. Cu radiation, R = 0.056 for 2296 reflexions.

The water molecule lies in the centre of the 21-membered macrocycle and is linked to it by three hydrogen bonds (Fig. 1), acting as an acceptor in interactions with the amide groups (N(4)...O(32) 3.265, N(16)...O(32) 3.006 Å) and also as a donor (O(32)...O(10) 3.041 Å). The water molecule is linked to another macrocycle by hydrogen bond (O(32)...O(23') 2.882 Å) thus forming chains along the c axis. The presence of hydrogen bonds between the water molecule and the ring donor atoms determines the conformation of the macrocycle and transforms the bonds O(1)-C(2), N(4)-C(5), O(13)-C(14) and C(15)-N(16) from the energetically favourable trans to gauche orientation. The benzyl radical has a pseudoaxial orientation and its plane makes an angle of 29° with the best plane through atoms O(1), N(4), O(7), O(10), O(13), N(16) and O(19).

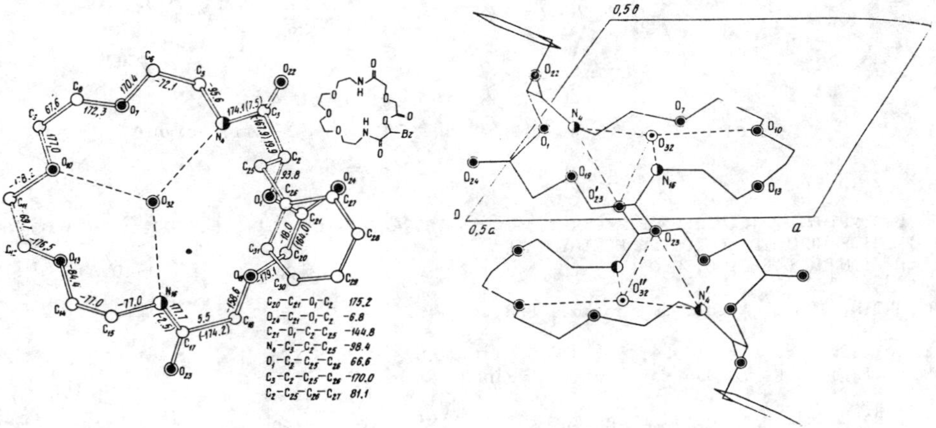

Fig. 1. Structure of the 21-membered macrocycle in $C_{21}H_{30}N_2O_8,H_2O$ showing torsion angles (left), and projection of the crystal structure on the xz plane (right).

4-METHOXYCARBONYLAMINO-5,7-DIMETHYL-1,8-DIPHENYL-2-OXA-4-AZABICYCLO[3.3.0]OCT-7-ENE-3,6-DIONE
$C_{22}H_{20}N_2O_5$ (I)

5-ETHOXYCARBONYL-3-METHOXY-6,8-DIMETHYL-1,9-DIPHENYL-2-OXA-4,5-DIAZABICYCLO[4.3.0]-NONA-3,8-DIEN-7-ONE
$C_{24}H_{24}N_2O_5$ (II)

D. MACKAY and D.D. McINTYRE, 1984. Can. J. Chem., 62, 355-360.

I. Orthorhombic, Pbca, a = 14.263, b = 16.229, c = 16.585 Å, Z = 8. Mo radiation, R = 0.076 for 1527 reflexions.

II. Triclinic, P$\bar{1}$, a = 8.736, b = 8.816, c = 15.093 Å, α = 96.80, β = 94.81, γ = 107.92°, Z = 2. Mo radiation, R = 0.077 for 2467 reflexions.

X-ray analysis establishes the structures of these molecules as shown in Fig. 1, with the ring junctions cis.

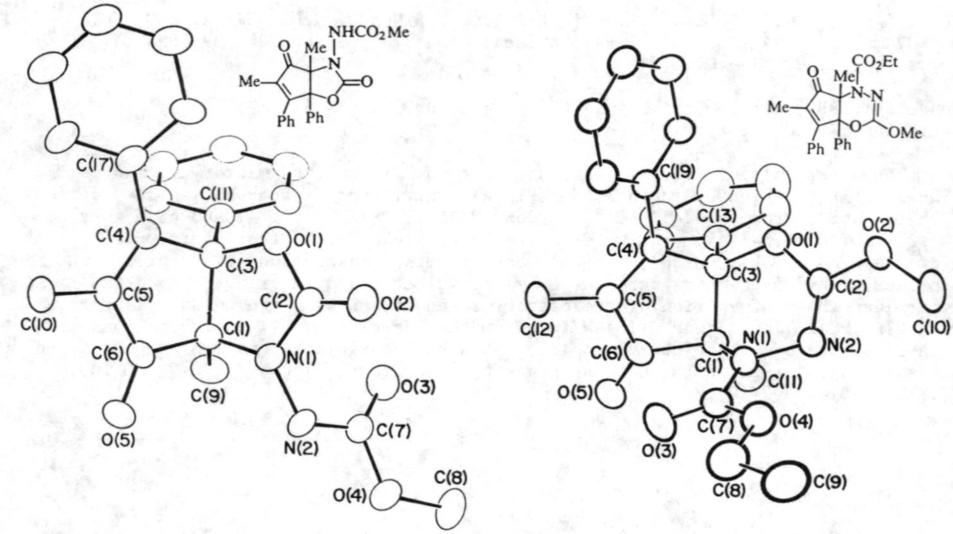

Fig. 1. Views of $C_{22}H_{20}N_2O_5$ (I) (left) and of $C_{24}H_{24}N_2O_5$ (II) (right).

1,10-DIMETHYL-7,16,21,26-TETRAOXA-1,4,10,13-TETRAAZATRICYCLO[11.5.5.5^{4,10}]OCTACOSANE DIIODIDE HEMIHYDROCHLORIDE HEMIHYDRATE

$C_{22}H_{44}I_2N_4O_4$,0.5HCl,0.5H_2O $[C_{22}H_{44}N_4O_4]^{2+}2I^-$,0.5HCl,0.5$H_2O$

P. GROTH, 1984. Acta Chem. Scand., B38, 179-182.

Monoclinic, Cc, a = 21.656, b = 8.410, c = 21.078 Å, β = 128.27°, D_m = 1.55, Z = 4. Mo radiation, R = 0.035 for 2261 reflexions (at -130°C).

Both twelve-membered rings adopt the [3 9] conformation (Fig. 1), and the torsion angles N(1)-C(10)-C(11)-N(4) (175°) and N(2)-C(22)-C(21)-N(3) (-175°) are anti. Distances of note are: O(1)-C(2) 1.44, O(1)-C(3) 1.43, O(2)-C(6) 1.38, O(2)-C(7) 1.41, N(1)-C(1) 1.48, N(1)-C(8) 1.52, N(2)-C(4) 1.41, N(2)-C(5) 1.55, O(3)-C(14) 1.40, O(3)-C(13) 1.45, O(4)-C(17) 1.44, O(4)-C(18) 1.51, N(3)-C(15) 1.61, N(3)-C(16) 1.51, N(4)-C(12) 1.57, N(4)-C(19) 1.38 Å. The water molecule is hydrogen-bonded to O(1) and O(3) (O(W)...O(1) 2.81, O(W)...O(3) 2.88 Å).

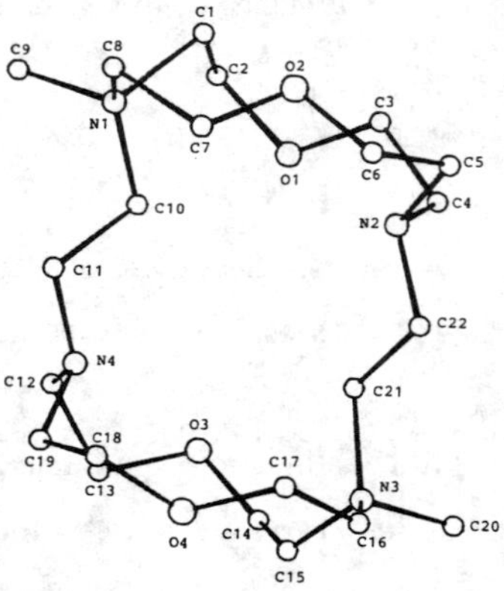

Fig. 1. Schematic drawing of the molecular cation in $C_{22}H_{44}I_2N_4O_4$.

Z-PHENYL-4-BENZHYDRYL-5-METHYL-4,5-DIHYDRO-1,3,4-OXADIAZIN-6-ONE
$C_{23}H_{20}N_2O_2$

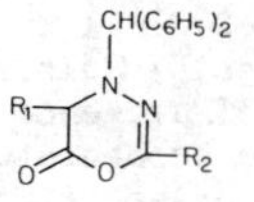

$R_1 = CH_3$; $R_2 = C_6H_5$

E.C. TAYLOR, H.M.L. DAVIES, W.T. LAVELL and N.D. JONES, 1984. J. Org. Chem., 49, 2204-2208.

Monoclinic, Pn, a = 9.353, b = 8.861, c = 11.638 Å, β = 101.03°, Z = 2. Cu radiation, R = 0.041 for 1352 reflexions.

The analysis established the structure of this rearrangement product.

(R,S)-3,3-DIPHENYL-2(1-MESITYLETHYL)OXAZIRIDINE
$C_{24}H_{25}NO$

D.R. BOYD, K.M. McCOMBE, T.A. HAMOR, W.B. JENNINGS and V.E. WILSON, 1984. J. Chem. Soc. Perkin II, 95-100.

Orthorhombic, $P2_12_12_1$, a = 8.843, b = 10.396, c = 21.687 Å, Z = 4. Mo radiation, R = 0.040 for 1242 reflexions.

The oxaziridine ring (Fig. 1) has the expected geometry with bond lengths of 1.511, 1.420 and 1.440 Å for the N-O, C-O and N-C bonds respectively. Bond angles at C, N and O are 63.8, 57.5 and 58.7° respectively. The arrangement about N(1)-C(19) is such that the N atom of the N-alkyl group is eclipsed by N(1)-C(21) of the oxaziridine ring. The torsion angles N(1),C(21),C(1),C(2), N(1),C(21),C(7),C(8) and N(1),C(19),C(13),C(14) are -94.2, -166.1 and 49.7° respectively. Mean bond lengths in the unsubstituted phenyl rings are 1.376(4) and 1.372(7) while that in the trisubstituted ring is 1.392(6) Å.

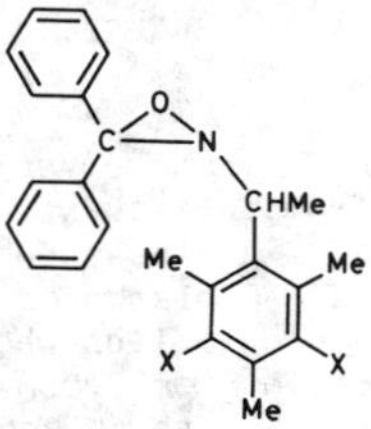

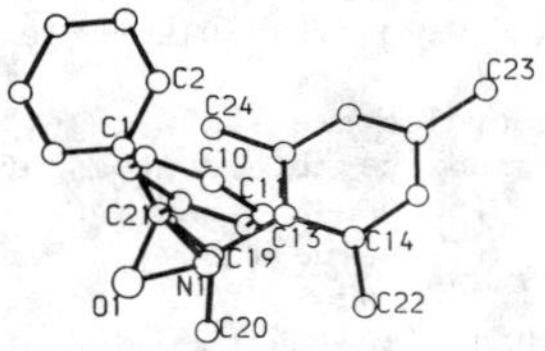

Fig. 1. $C_{24}H_{25}NO$: the molecular skeleton (X = H) and structure.

1,10-(6,6'-DICARBONYL-2,2'-AZOPYRIDINE-C(6),C(6')-1,10-DIAZA-4,7,14,17-TETRAOXACYCLO-OCTADECANE
$C_{24}H_{30}N_6O_6$ (I)

1,10-(3,3'-DICARBONYL-1,1'-AZOBENZENE)-1,10-DIAZA-4,7,14,17-TETRATHIACYCLOOCTADECANE
$C_{26}H_{32}N_4O_2S_4$ (II)

H.L. AMMON, S.K. BHATTACHARJEE, S. SHINKAI and Y. HONDA, 1984. J. Am. Chem. Soc., 106, 262-263.

I. Monoclinic, C2/c, a = 43.126, b = 7.9747, c = 14.367 Å, β = 93.45°, Z = 8. Cu radiation, R = 0.052 for 2936 reflexions.

II. Monoclinic, $P2_1/a$, a = 10.766, b = 23.552, c = 11.569 Å, β = 111.057°, Z = 4. Cu radiation, R = 0.076 for 2873 reflexions.

The molecules are shown in Fig. 1. The azoarene cap-crown ring angles are 82.5° in I and 79.3° in II. Bond lengths are normal in both compounds except for a possibly disordered part of the thiacrown ring in II. Two of the four crown O atoms in I are endodentate and two are exodentate, whereas all four crown S atoms in II are exodentate. Four of the eight C-S bonds in II adopt the anti conformation, unique among structurally characterized thiacrown ethers.

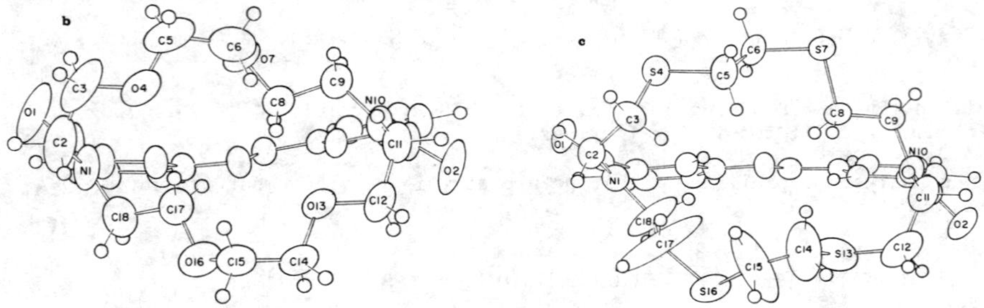

Fig. 1. Molecular structures of $C_{24}H_{30}N_6O_6$ (left) and $C_{26}H_{32}N_4O_2S_4$ (right).

7,16-DIPHENYL-1,4,10,13-TETRAOXA-7,16-DIAZACYCLOOCTADECANE
$C_{24}H_{34}N_2O_4$ (I)

7,16-BIS[2-(4-TOLUENESULPHONAMIDO)PHENYL]-1,4,10,13-TETRAOXA-7,16-DIAZACYCLOOCTADECANE
$C_{38}H_{48}N_4O_8S_2$ (II)

J.-P. DECLERCQ and E. SONVEAUX, 1984. Nouv. J. Chim., 8, 591-593.

I. Triclinic, P$\bar{1}$, a = 9.126, b = 11.771, c = 12.192 Å, α = 115.22, β = 111.26, γ = 78.39°, Z = 2. R = 0.053 for 2767 reflexions.

II. Triclinic, P$\bar{1}$, a = 11.638, b = 10.591, c = 8.716 Å, α = 67.82, β = 75.58, γ = 89.42°, Z = 1. R = 0.041 for 2019 reflexions.

The two macrocycles are shown in Fig. 1. In I there is no crystallographically required symmetry and the hetero atoms are displaced from their best mean plane by as much as 0.5 Å. The O-C-C-O and N-C-C-O torsion angles are near 60° except for the O(16)-C(17)-C(18)-N(1) value (180°). Mean C-C, C-N and C-O distances in the macrocycle are 1.511, 1.457 and 1.420 Å respectively. Molecule II lies about an inversion centre and is rather flat (maximum deviation of the hetero atoms from their mean plane is 0.09 Å). The O-C-C-O and N-C-C-O torsion angles are near to 60° except for the C(8)-C(9) bond (close to 180°). Mean C-C, C-N and C-O macrocycle distances are 1.497, 1.469 and 1.422 Å respectively. The conformation adopted by II has the tosylamide functions pointing toward the interior of the cavity; the H atom in N(31) is 2.10 Å from O(4) and 2.32 Å from N(1) (N(31)...O(4) 2.892, N(31)...N(1) 2.739 Å).

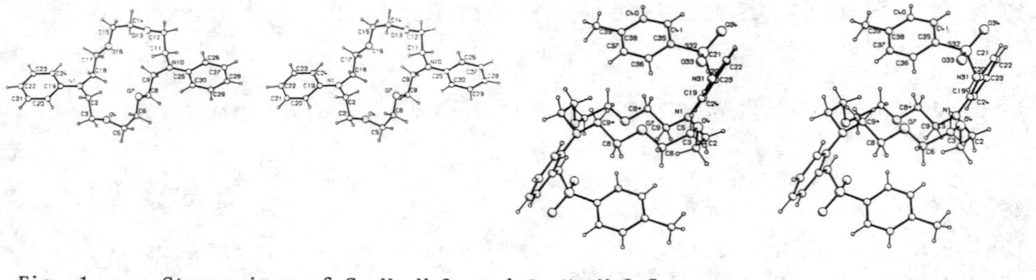

Fig. 1. Stereoviews of $C_{24}H_{34}N_2O_4$ and $C_{38}H_{48}N_4O_8S_2$.

7,19,30-TRIOXA-1,13-DIAZA-4,10,16,22,27,33-HEXAAZONIABICYCLO[11.11.11]PENTACONTANE AZIDE CHLORIDE TETRAKIS(HEXAFLUOROPHOSPHATE) TRIHYDRATE
$C_{24}H_{60}ClF_{24}N_{11}O_3P_4,3(H_2O)$ (I) $C_{24}H_{60}N_8O_3{}^{6+}.N_3{}^{-}.Cl^{-}.4(PF_6{}^{-}),3(H_2O)$

7,19,30-TRIOXA-1,13-DIAZA-4,10,16,22,27,33-HEXAAZONIABICYCLO[11.11.11]PENTACONTANE FLUORIDE PENTAPERCHLORATE HEMIHYDRATE
$C_{24}H_{60}Cl_5FN_8O_{23},0.5(H_2O)$ (II) $C_{24}H_{60}N_8O_3{}^{6+}.F^{-}.5(ClO_4{}^{-}),0.5(H_2O)$

7,19,30-TRIOXA-1,13-DIAZA-4,10,16,22,27,33-HEXAAZONIABICYCLO[11.11.11]PENTACONTANE HEXACHLORIDE
$C_{24}H_{60}Cl_6N_8O_3$ (III) $C_{24}H_{60}N_8O_3{}^{6+}.6Cl^{-}$

7,19,30-TRIOXA-1,13-DIAZA-4,10,16,22,27,33-HEXAAZONIABICYCLO[11.11.11]PENTACONTANE HEXABROMIDE
$C_{24}H_{60}Br_6N_8O_3$ (IV) $C_{24}H_{60}N_8O_3{}^{6+}.6Br^{-}$

B. DIETRICH, J. GUILHEM, J.-M. LEHN, C. PASCARD and E. SONVEAUX, 1984. Helv. Chim. Acta, 67, 91-104.

I. Monoclinic, $P2_1/n$, a = 15.236, b = 14.097, c = 24.410 Å, β = 90.67°, Z = 4. Cu radiation, R = 0.162 for 3240 reflexions.

II. Monoclinic, $P2_1/c$, a = 21.026, b = 12.611, c = 34.916 Å, β = 95.67°, Z = 8. Cu radiation, R = 0.198 for 3535 reflexions.

III. Trigonal, $P3_121$, a = 11.186, c = 24.801 Å, Z = 6. Cu radiation, R = 0.054 for 1651 reflexions.

IV. Trigonal, $P3_1$, a = 11.537, c = 25.413 Å, Z = 3. Mo radiation, R = 0.092 for 2249 reflexions.

In all four structures the hexaprotonated macrobicyclic ligand is in the endo,endo-conformation where the bridgehead N atoms are turned into the cavity (Fig. 1). In II, the location of the small F^- anion in one lobe of the bicyclic system distorts the ligand to achieve tetrahedral coordination of F^-. The Cl^- and Br^- complexes (III and IV) adopt the same general shape. The three bridges of the macrocycle form three rather large lobes not related by ternary symmetry. In III, two lobes are identical and related by a binary axis passing through the O atom of the third bridge. This crystallographic symmetry requirement is not found in IV. In the azide complex (I), the ligand is markedly expanded around the enclosed $N_3{}^-$ anion. The three lobes of the ligand form a trigonal arrangement centred on the N-N axis.

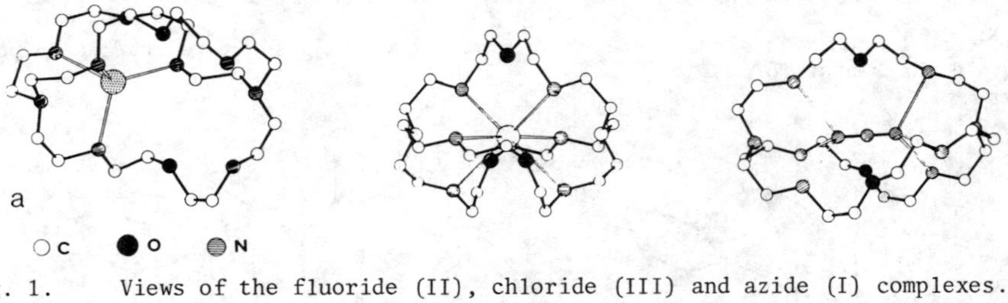

Fig. 1. Views of the fluoride (II), chloride (III) and azide (I) complexes.

3,7-BIS(N,N-DIETHYLAMINO)-10-BENZOYLPHENOXAZINE
$C_{27}H_{31}N_3O_2$

C.R. THEOCHARIS, 1984. Acta Cryst., C40, 1593-1596.

Triclinic, $P\bar{1}$, a = 8.667, b = 8.742, c = 16.382 Å, α = 87.34, β = 84.31, γ = 69.32°, Z = 2. Cu radiation, R = 0.048 for 3183 reflexions.

Within the molecule (Fig. 1), delocalisation of electrons occurs over the ring C(1)...C(6), the carbonyl group and the nitrogen lone pair (c.f. bond lengths C(1)-C(19) 1.488(5) and C(19)-N(3) 1.377(4) Å) although the torsion angle C(2)-C(1)-C(13)-N(3) is 43(1)°. The C(10)-N(1) and C(16)-N(2) bond lengths of 1.381(4) and 1.375(4) indicate some interaction between the N lone pairs and the adjacent rings. Dihedral angles between rings are C(1)...C(6)/C(7)...C(12) 60(1), C(1)...C(6)/C(13)...C(18) 30(1), C(7)...C(12)/C(13)...C(18) 33(1)°. The heterocyclic ring makes angles with attached benzene rings of 17(1) and 16(1)° and one of 44(1)° with C(1)...C(6). One ethyl group is disordered over two clearly distinguishable positions.

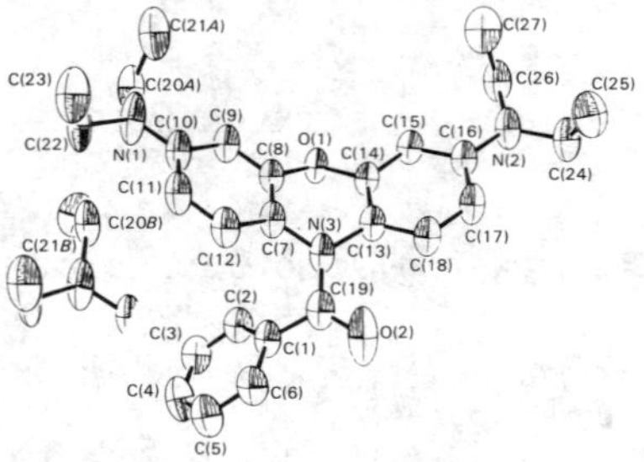

Fig. 1. The $C_{27}H_{31}N_3O_2$ molecule. (The insert shows the alternate position for the disordered ethyl group.)

1,5-BIS(ETHOXYCARBONYL)-7-METHYL-2,4-BIS(p-TOLYL)-3-OXA-7-AZABICYCLO[3.3.1]NONAN-9-ONE

$C_{28}H_{33}NO_6$ (I)

7-t-BUTYL-1,5-BIS(ETHOXYCARBONYL)-2,4-BIS(p-TOLYL)-3-OXA-7-AZABICYCLO[3.3.1]NONAN-9-ONE

$C_{31}H_{39}NO_6$ (II)

R. CAUJOLLE, A. LATTES, J. JAUD and J. GALY, 1984. Z. Kristallogr., 166, 81-95.

I. Triclinic, $P\bar{1}$, a = 10.605, b = 11.852, c = 12.189 Å, α = 109.20, β = 113.73, γ = 89.35°, Z = 2. Mo radiation, R = 0.051 for 2242 reflexions.

II. Triclinic, $P\bar{1}$, a = 10.273, b = 11.561, c = 12.605 Å, α = 101.65, β = 96.09 γ = 92.82°, Z = 2. Mo radiation, R = 0.054 for 3571 reflexions.

Compound I occurs in a chair-chair conformation whereas the more crowded compound II adopts a chair-boat form.

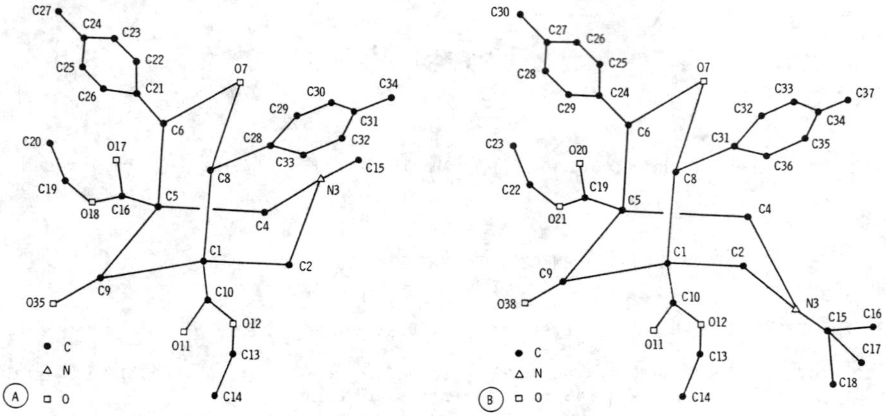

Fig. 1. Schematic drawings of (A) $C_{28}H_{33}NO_6$ and (B) $C_{31}H_{39}NO_6$.

7,16-DITOSYL-4,13-DIOXA-1,7,10,16-TETRAAZADICYCLO[8.8.2]EICOSANE
$C_{28}H_{42}N_4O_6S_2$

P. GROTH, 1984. Acta Chem. Scand., A38, 261-263.

Monoclinic, $P2_1/c$, a = 19.232, b = 5.327, c = 15.131 Å, β = 106.56°, D_m = 1.29, Z = 2. Mo radiation, R = 0.082 for 1037 reflexions.

The molecule has inversion symmetry (Fig. 1) and the twelve-membered ring in it adopts the [3,9] conformation. The bond distances and valency angles have normal values with the exception of the bond C(8)-C(9) which, at 1.34 Å, is somewhat short. In the macrocycle, C-C distances range from 1.52 to 1.56, N-C distances from 1.45 to 1.48, and bonds O(1)-C(4) and O(1)-C(5') are 1.42 Å each. In the tosyl group, bonds N(1)-S, S-O(2), S-O(3) and S-C(8) are respectively 1.61, 1.44, 1.44 and 1.77 Å. The two bridgehead N atoms (N(2) and N(2')) play the role as "corner" atoms in the twelve-membered ring (Fig. 1).

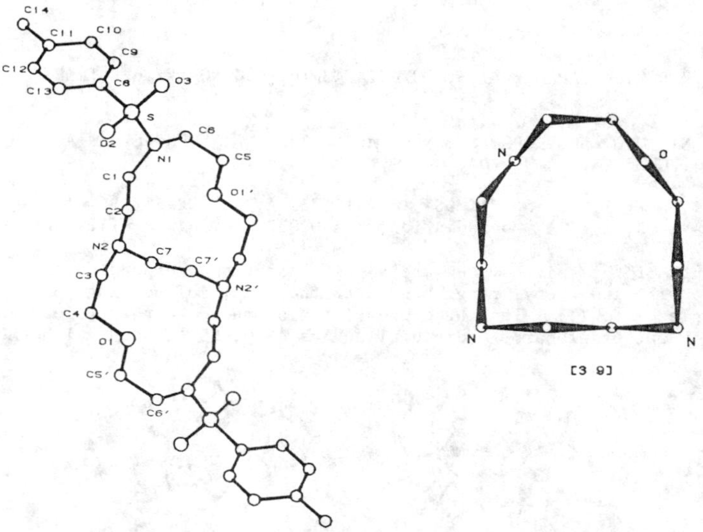

Fig. 1. Molecular structure of $C_{28}H_{42}N_4O_6S_2$ (left), and schematic representation of the [3,9] conformation (right).

5-BENZOYL-4-(p-CHLOROPHENYL)-2,2,5-TRIPHENYL-Δ^3-1,3-OXAZOLINE
$C_{34}H_{24}ClNO_2$

C. RUIZ-VALERO, E. GUTIERREZ-PUEBLA and A. MONGE, 1984. Acta Cryst., C40, 144-146.

Orthorhombic, Pbca, a = 16.354, b = 17.464, c = 18.599 Å, Z = 8. Mo radiation, R = 0.054 for 1171 reflexions.

The analysis of the molecule (Fig. 1) establishes that the reaction mechanism leading to its formation involved a 1,2-acyl migration. Molecular dimensions are normal.

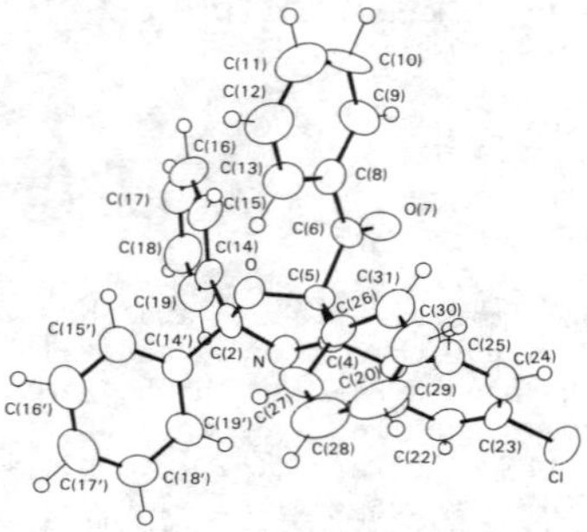

Fig. 1. A drawing of the $C_{34}H_{24}ClNO_2$ molecule.

6-METHYL-2,4,7,8,9-PENTAPHENYL-1,3,4-OXADIAZASPIRO[4.4]NONA-2,6,8-TRIENE
$C_{37}H_{28}N_2O$

E.G. TSATSARONI, N.G. ARGYROPOULOS, N.E. ALEXANDROU, A. HOUNDAS and A. TERZIS, 1984. J. Heterocycl. Chem., 21, 701-704.

Triclinic, $P\bar{1}$, a = 11.006, b = 12.681, c = 11.238 Å, α = 94.51, β = 104.10, γ = 112.03°, D_m = 1.23, Z = 2. Mo radiation, R = 0.0375 for 3148 reflexions.

Both rings of the spiro system are almost planar and perpendicular (interplanar angle 89.5°). The two phenyl rings bonded to the oxadiazoline ring form a planar unit (dihedral angles 3 to 7°). The three phenyl rings bonded to the cyclopentadiene ring are completely out of planarity (dihedral angles 45.0, 55.6, 50.0° with the C_5 plane.

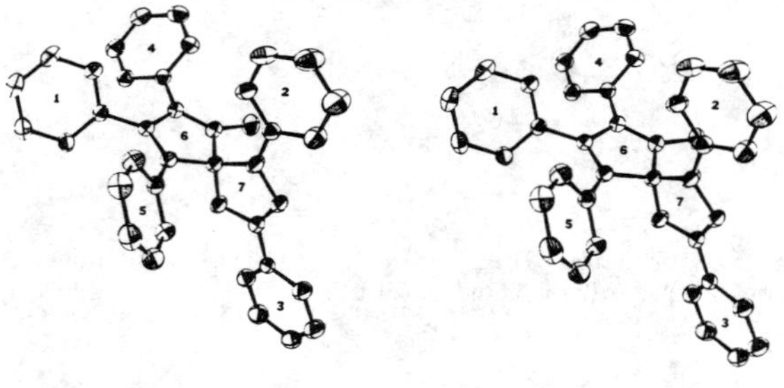

Fig. 1. A stereoview of $C_{37}H_{28}N_2O$.

5,10-DI-t-BUTYL-3,8,11,12-TETRAPHENYL-2,4,7,9-TETRAOXA-5,10-DIAZATETRACYCLO[4.4.2.-$0^{3,12}.0^{8,11}$]DODECANE
$C_{38}H_{40}N_2O_4$

H.G. AURICH, G. BAUM, W. MASSA, K.-D. MOGENDORF and M. SCHMIDT, 1984. Chem. Ber., 117, 2615-2621.

Orthorhombic, Pbcn, a = 9.565, b = 15.987, c = 21.018 Å, Z = 4. Mo radiation, R = 0.046 for 2531 reflexions.

The molecule (Fig. 1) has four fused five-membered rings and twofold crystallographic symmetry. The C2-C2'-C1'-O2-C3 ring is almost planar (maximum deviation 0.086 Å) whereas the C2-C3-O1-N1-C1 ring is quite puckered (O1 is 0.208 and N1 0.493 Å from the C1-C2-C3 plane).

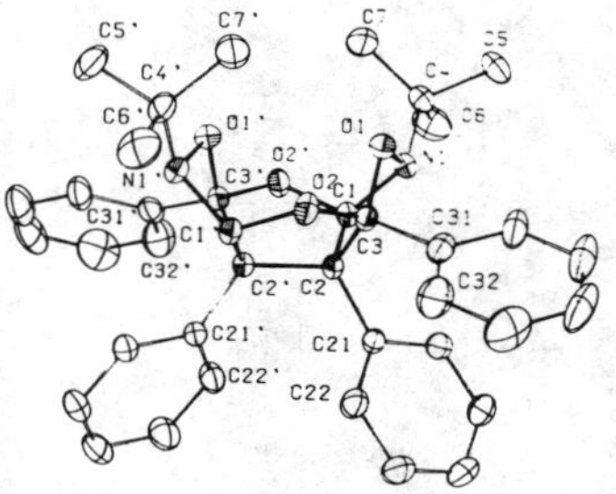

Fig. 1. A view of $C_{38}H_{40}N_2O_4$.

4,7,8,9-TETRAPHENYL-2-(4-NITROPHENYL)-6-ETHYL-1,3-OXAZASPIRO[4.4]NONA-3,6,8-TRIENE
$C_{39}H_{30}N_2O_3$

E.G. TSATSARONI, N.G. ARGYROPOULOS, N.E. ALEXANDROU, A. HOUNDAS and A. TERZIS, 1984. J. Heterocycl. Chem., 21, 1405-1408.

Triclinic, P$\bar{1}$, a = 15.438, b = 10.097, c = 10.621 Å, α = 109.64, β = 87.19, γ = 101.70°, D_m = 1.24, Z = 2. Mo radiation, R = 0.042 for 3018 reflexions.

The spiro-fused rings (Fig. 1) are almost perpendicular to one another (interplanar angle 85.5°). Phenyl ring (#2) is almost coplanar with oxazoline ring (#7) (dihedral angle 22.5°). Phenyl rings 1, 4 and 5 are rotated 34.9, 68.4 and 60.8° respectively out of the cyclopentadiene plane (#6).

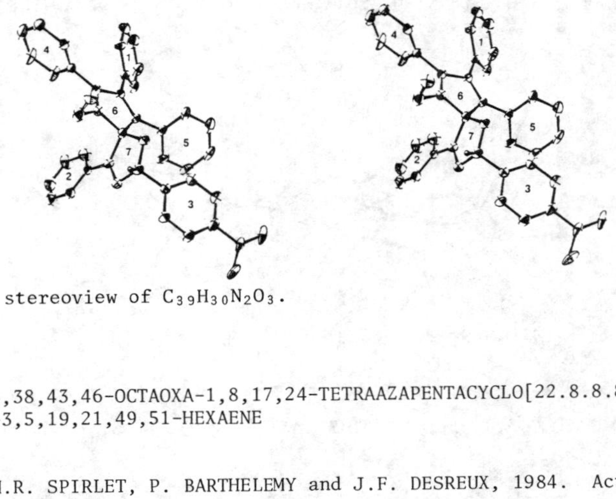

Fig. 1. A stereoview of $C_{39}H_{30}N_2O_3$.

11,14,27,30,35,38,43,46-OCTAOXA-1,8,17,24-TETRAAZAPENTACYCLO[22.8.8.8^{8,17}.2^{3,6}.2^{19,22}]-DOPENTACONTA-3,5,19,21,49,51-HEXAENE
$C_{40}H_{64}N_4O_8$

J. REBIZANT, M.R. SPIRLET, P. BARTHELEMY and J.F. DESREUX, 1984. Acta Cryst., C40, 484-486.

Triclinic, P$\bar{1}$, a = 8.823, b = 8.813, c = 13.859 Å, α = 87.40, β = 84.05, γ = 85.19°, Z = 1. Mo radiation, R = 0.048 for 2124 reflexions.

The molecule (Fig. 1) lies about an inversion centre and is a tricyclic cryptand with an internal cavity defined by two (parallel) 18-membered ($C_{12}H_{24}N_2O_4$) macrocyclic units bridged by two (parallel) phenyl rings.

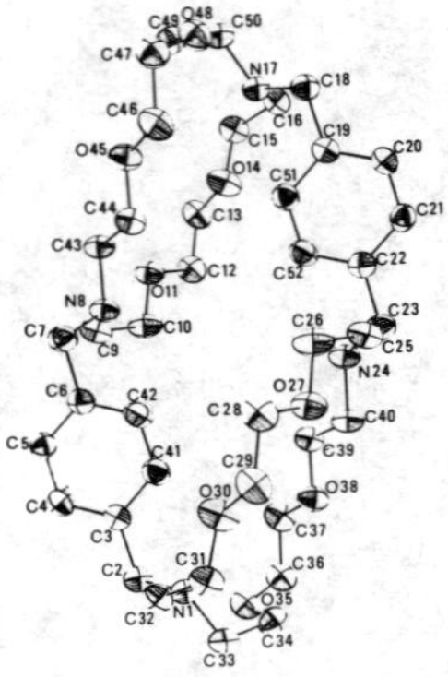

Fig. 1. A view of $C_{40}H_{64}N_4O_8$.

1,3-DICHLORO-5-TRIFLUOROMETHYL-1,3-DITHIA-2,4,6-TRIAZINE
$C_2Cl_2F_3N_3S_2$

H.-U. HOFS, G. HARTMANN, R. MEWS and G.M. SHELDRICK, 1984. Z. Naturforsch., B39, 1389-1392.

Orthorhombic, $P2_12_12_1$, a = 4.749, b = 11.015, c = 15.361 Å, Z = 4. Mo radiation, R = 0.056 for 1335 reflexions.

The molecule and principal bond lengths are shown in Fig. 1. The six-membered ring has a boat conformation with N1 0.370 Å and C1 0.053 Å from the S,S',N2,N2' plane. The chlorine substituents are cis.

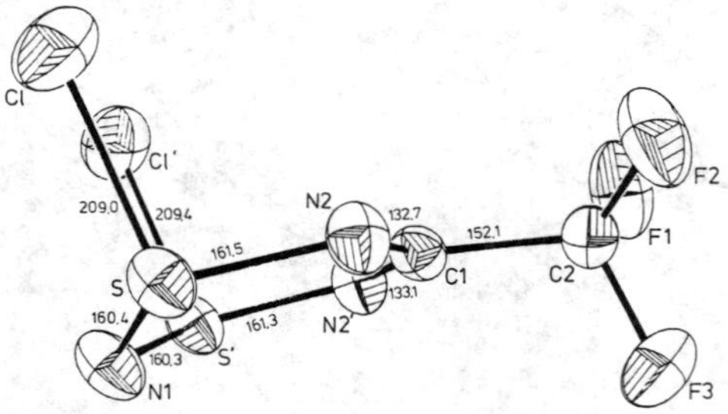

Fig. 1. A view of $C_2Cl_2F_3N_4S_2$ with bond lengths (pm).

4-METHYL[1,2,5]THIADIAZOLO[3,4-c][1,2,6]THIADIAZINE-7(6H)-ONE 5,5-DIOXIDE POTASSIUM SALT
$C_4H_3KN_4O_3S_2$

C. ESTEBAN-CALDERON, M. MARTINEZ-RIPOLL and S. GARCIA-BLANCO, 1984. Acta Cryst., C40, 80-82.

Monoclinic, C2/c, a = 25.999, b = 9.3396, c = 7.413 Å, β = 98.26°, Z = 8. Mo radiation, R = 0.026 for 1714 reflexions.

The five-membered ring is planar and the thiadiazine ring (Fig. 1) has an envelope conformation with the S atom at the flap, 0.171(4) Å out of plane. The K^+ ion is surrounded by seven oxygen atoms forming a very distorted laterally capped trigonal prism with K...O 2.72(4) to 3.10(1) Å.

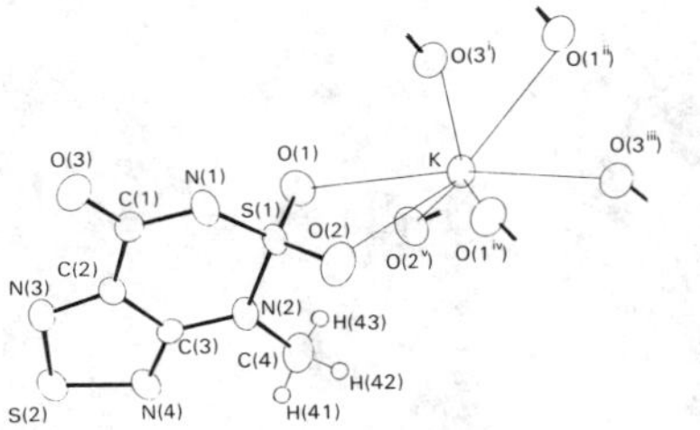

S(1)–O(1)	1·43 (2)	S(1)–O(2)	1·43 (1)
S(1)–N(1)	1·56 (1)	S(1)–N(2)	1·67 (2)
S(2)–N(3)	1·64 (3)	S(2)–N(4)	1·66 (1)
N(1)–C(1)	1·35 (1)	N(2)–C(3)	1·38 (1)
N(2)–C(4)	1·45 (1)	N(3)–C(2)	1·31 (1)
N(4)–C(3)	1·31 (2)	C(1)–C(2)	1·48 (2)
C(1)–O(3)	1·23 (1)	C(2)–C(3)	1·42 (1)
C(4)–H(41)	1·00	C(4)–H(42)	1·01
K–O(3)	2·72 (4)	K–O(3)	2·83 (1)
K–O(2)	2·88 (4)	K–O(1)	2·92 (1)
K–O(1)	3·00 (1)	K–O(2)	3·05 (2)
K–O(1)	3·10 (1)		

Fig. 1. The $(C_4H_3N_4O_3S_2)^-.K^+$ structure and bond lengths.

(Z)-2-(NITROMETHYLIDENE)THIAZOLIDINE
$C_4H_6N_2O_2S$ (I)

(E)-N-METHYL-2-(NITROMETHYLIDENE)PYRROLIDINE
$C_6H_{10}N_2O_2$ (II)

S. RAJAPPA, K. NAGARAJAN, K. VENKATESAN, N. KAMATH, V.M. PADMANABHAN, W. VON PHILIPSBORN, BAN CHIN CHEN and R. MULLER, 1984. Helv. Chim. Acta, 67, 1669-1680.

I. Monoclinic, C2/c, a = 8.821, b = 10.884, c = 12.870 Å, β = 95.6°, D_m = 1.40, Z = 8. Cu radiation, R = 0.071 for 841 reflexions.

II. Monoclinic, $P2_1/c$, a = 6.023, b = 15.851, c = 7.447 Å, β = 106.9°, D_m = 1.36, Z = 4. Cu radiation, R = 0.110 for 809 reflexions.

The analyses establish the Z- and E-configurations respectively for I and II. Both 5-membered rings have approximate C_2 symmetry (the twofold axis passing through C(2) and the mid-point of C(4)-C(5) (Fig. 1)). In I the exocyclic double bond length (C(2)=C(6)) is 1.405(5) Å (the corresponding bond in II is 1.357(8) Å).

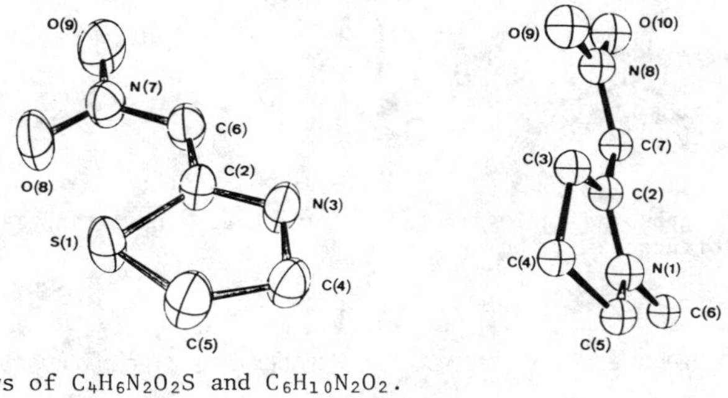

Fig. 1. Views of $C_4H_6N_2O_2S$ and $C_6H_{10}N_2O_2$.

METHYL 3-CYANO-1,2,4-THIADIAZOLE-5-CARBOXIMIDATE
$C_5H_4N_4OS$

H.W. ROESKY, H. HOFMANN, K. KELLER, W. PINKERT, P.G. JONES and G.M. SHELDRICK, 1984. Chem. Ber., 117, 2681-2685.

Monoclinic, $P2_1/c$, a = 5.903, b = 13.698, c = 9.237 Å, β = 98.21°, Z = 4. Mo radiation, R = 0.042 for 1266 reflexions.

The analysis establishes the structure as shown in Fig. 1. Apart from the methyl hydrogens, the molecule is planar. There is an intramolecular hydrogen bond (H(11)... N(4) 2.61 Å, N(11)...N(4) 2.94 Å). The S-N(2) is 1.646(2) Å.

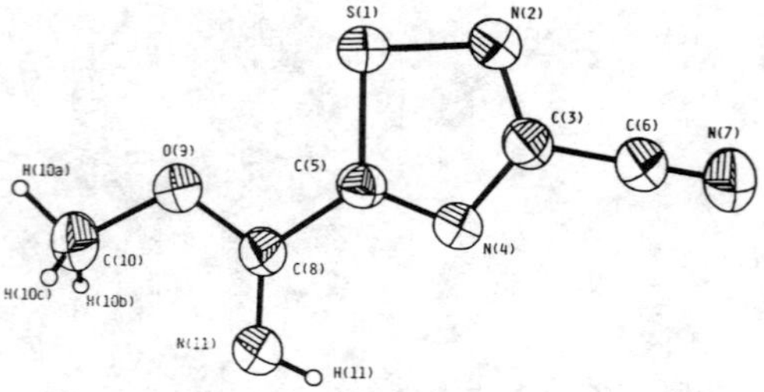

Fig. 1. The structure of $C_5H_4N_4OS$.

2-AMINOTHIAZOLIUM TRICHLOROACETATE

$C_5H_5Cl_3N_2O_2S$ $(C_3H_5N_2S)^+(C_2Cl_3O_2)^-$

L.G. KUZ'MINA and YU.T. STRUCHKOV, 1984. J. Struct. Chem., 25, 917-920 [Zh. Strukt. Khim., 25(6), 88-92].

Monoclinic, $P2_1/c$, a = 10.320, b = 8.381, c = 11.640 Å, β = 96.82°, Z = 4. Mo radiation, R = 0.037 for 2101 reflexions (at -120°C).

Bond lengths and angles are shown in Fig. 1. The cation exists in the imino form, geometry indicating delocalization of the positive charge from the exocyclic N atom to the N(1) and S atoms of the heterocycle. Extensive N-H...O hydrogen bonding forms a three-dimensional network in the solid.

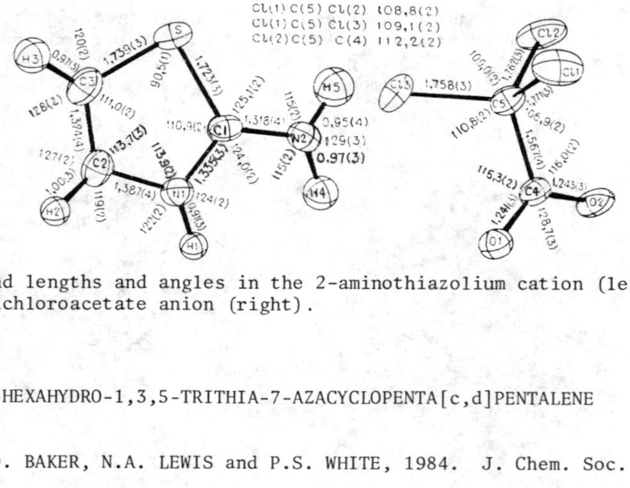

Fig. 1. Bond lengths and angles in the 2-aminothiazolium cation (left) and the trichloroacetate anion (right).

2,2a,4,4a,6,6a-HEXAHYDRO-1,3,5-TRITHIA-7-AZACYCLOPENTA[c,d]PENTALENE

$C_6H_9NS_3$

C.A. STEIN, A.D. BAKER, N.A. LEWIS and P.S. WHITE, 1984. J. Chem. Soc. Dalton, 2073-2075.

Rhombohedral, R3c, a = 7.721 Å, α = 105.14°, Z = 2. Mo radiation, R = 0.024 for 649 reflexions.

The X-ray analysis reveal a flattening of the molecule with the N atom being pulled slightly into the plane of the three S atoms (Fig. 1). The C-N-C angles (112.3(1)°) are slightly larger than the expected tetrahedral angles. The N-S (plane) distance is 0.952 Å. Bond distances are: S-C(1) 1.807(2), S-C(2) 1.822(2), N-C(2) 1.484(2), and C(1)-C(2) 1.523(4) Å.

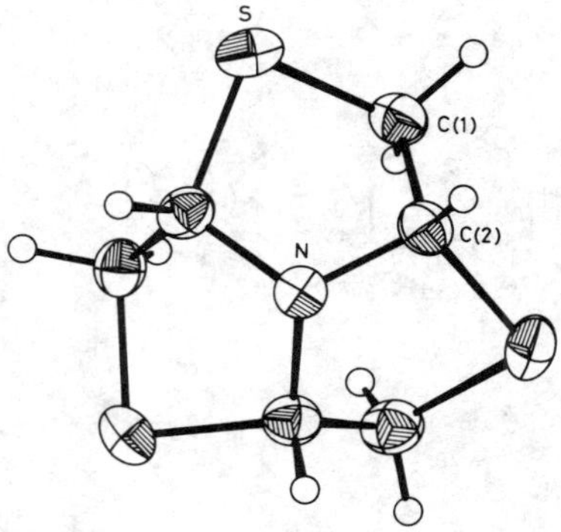

Fig. 1. $C_6H_9NS_3$: the molecular structure.

BIS(2-AMINO-4,5-DIHYDRO-3H$^+$-1,3-THIAZOLIUM) TETRACHLOROZINCATE(II)
$C_6H_{14}Cl_4N_4S_2Zn$ $2C_3H_7N_2S^+.ZnCl_4{}^{2-}$

M. KUBIAK and T. GLOWIAK, 1984. Acta Cryst., C40, 2039-2041.

Monoclinic, $P2_1/c$, a = 8.691, b = 12.856, c = 14.230 Å, β = 91.78°, D_m = 1.72, Z = 4. Mo radiation, R = 0.042 for 2551 reflexions.

The asymmetric unit consists of a $ZnCl_4^{2-}$ anion and two $C_3H_7N_2S^+$ (ATH^+) cations. There is no direct bonding between the metal atom and the ATH^+ cation. The distorted tetrahedral $ZnCl_4^{2-}$ anion is linked to ATH^+ cations through hydrogen bonding involving all H atoms attached to N. Both protonated ATH molecules have a similar conformation. Main bond lengths are Zn-Cl 2.243(1)-2.310(1) Å.

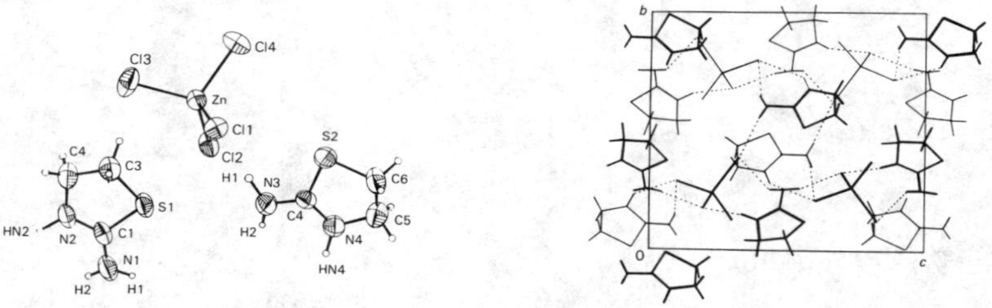

Fig. 1. The molecular geometry and crystal packing for $C_6H_{14}Cl_4N_4S_2Zn$.

3-METHYL-2(3H)-BENZOTHIAZOLONE
C_8H_7NOS

S. RUDD and G. BARANY, 1984. Acta Cryst., C40, 2118-2120.

Orthorhombic, Pbca, a = 16.941, b = 10.816, c = 8.424 Å, Z = 8. Mo radiation, R = 0.038 for 938 reflexions.

The five-membered heterocycle is flat and essentially coplanar with the fused benzene ring (Fig. 1). The C-N and C-S bonds of the thiazolone show partial double-bond character.

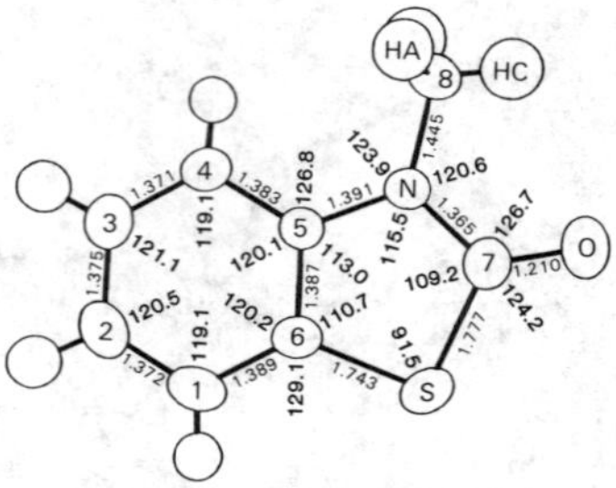

Fig. 1. Bond lengths (σ 0.003 Å) and angles (σ 0.2°) in C_8H_7NOS.

(3R)-cis-4-ACETYL-cis-3-METHYLCARBAMOYL-1,4-THIAZINANE 1-OXIDE
$C_8H_{14}N_2O_3S$

S.B. SANNI, M. DE WOLF, A.T.H. LENSTRA, C. VAN DER AUWERA and M.J. ANTEUNIS, 1984. Acta Cryst., C40, 1624-1626.

Orthorhombic, $P2_12_12_1$, a = 8.576, b = 15.845, c = 7.263 Å, Z = 4. Mo radiation, R = 0.047 for 1172 reflexions.

The six-membered ring has a distorted chair conformation with S=O and the methylcarbamoyl group axial and the N-acetyl group in an inclinal position (Fig. 1). The planes of the latter groups are inclined at 83.8(2)°. N(1) is 0.022(4) Å above the C(2)C(3)C(5) plane and C(7) is 0.015 Å out of the C(3)C(7)O(3)N(2) plane. The S=O bond length is 1.495(1) Å.

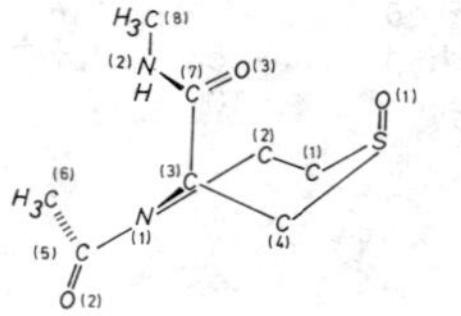

Fig. 1. The $C_8H_{14}N_2O_3S$ molecule.

3-PHENYLTHIAZOLIDINE-2-THIONE
$C_9H_9NS_2$ (I)

3-CYCLOHEXYLTHIAZOLIDINE-2-THIONE
$C_9H_{15}NS_2$ (II)

O. NUZZO, M. PIERROT, A. BALDY, M. CHANON and F. CHANON, 1984. Can. J. Chem., 62, 1410-1413.

I. Orthorhombic, $Pna2_1$, a = 11.777, b = 10.346, c = 7.668 Å, D_m = 1.40, Z = 4. Mo radiation, R = 0.027 for 1268 reflexions.

II. Orthorhombic, Pbca, a = 10.276, b = 10.916, c = 18.81 Å, D_m = 1.27, Z = 8. Mo radiation, R = 0.042 for 1144 reflexions.

The five-membered ring in I (Fig. 1) has a twist form, with torsion angle (ϕ) at C(4)-C(5) 23.2°. Compound II has the five-membered ring nearly planar; the conformation is a very flat C(4)C(5) out-of-plane half-chair with the C(4)-C(5) torsion angle 7.5°. In both compounds, the atoms S(1)C(2)S(2)N(3) are planar.

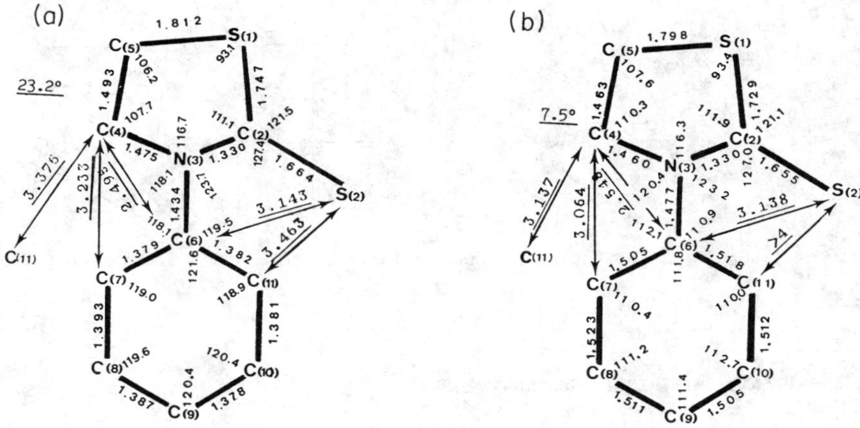

Fig. 1. Interatomic distances, angles and torsion angles in the thiazolidine-2-thiones (I) (left) and (II) (right).

3-METHYL-5-PHENYLIMINO-1,2,3-THIADIAZOLE
$C_9H_9N_3S$

V.A. KOZINSKIJ, O.V. ZELENSKAJA, S. BRUCKNER and L. MALPEZZI, 1984. J. Heterocycl. Chem., 21, 1889-1891.

Monoclinic, C2/c, a = 22.635, b = 5.544, c = 15.491 Å, β = 106.81°, Z = 8. Mo radiation, R = 0.044 for 1441 reflexions.

Molecular dimensions for this mesoionic system are in Fig. 1. The dihedral angle between the six- and five-membered ring planes is 13.5(3)°.

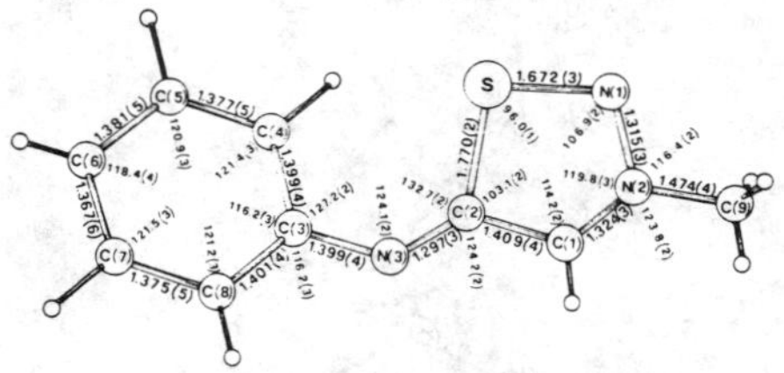

Fig. 1. A perspective view of $C_9H_9N_3S$ and dimensions.

NAPHTHO[2,3-c][1,2,5]THIADIAZOL-4,9-DIONE (β-MODIFICATION)
$C_{10}H_4N_2O_2S$

A. GIEREN, T. HÜBNER and V. LAMM, 1984. Acta Cryst., C40, 836-839.

Monoclinic, I2/a, a = 8.416, b = 10.952, c = 9.504 Å, β = 86.57°, D_m = 1.64, Z = 4. Cu radiation, R = 0.098 for 548 reflexions.

In the molecule the S-N and C-N bond lengths, 1.619(4) and 1.337(6) Å respectively, indicate a strong π interaction between the sulphur diimide group and the naphthoquinone system inducing a quasi-aromatic character of the 1,2,5-thiadiazole ring (Fig. 1). The molecular structure is not significantly different from that in the α-modification (1). In the crystal structure, the molecules form layers with co-planar molecular planes and also columnar stacks with an interplanar distance of 3.33 Å.

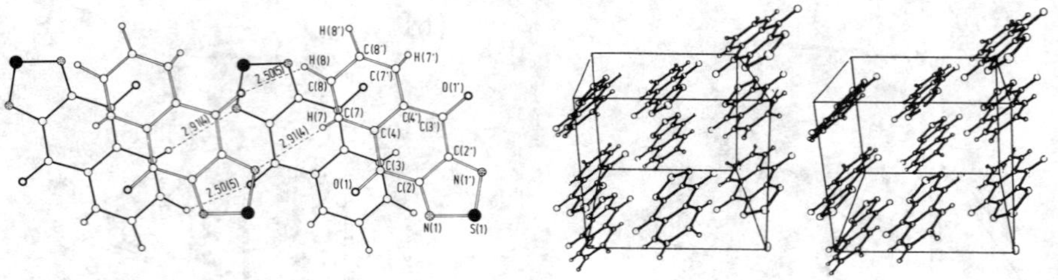

Fig. 1. $C_{10}H_4N_2O_2S$: a section of the crystal structure and a stereoscopic view.

1. Structure Reports, 49B, 470.

2-IMINO-N-PHENYL-3-THIAZOLIDINECARBOXAMIDE
$C_{10}H_{11}N_3OS$

C.R. RASMUSSEN, F.J. VILLANI Jr., E.A. GRIFFIN, D.S. MORRISON and R.A. OLOFSON, 1984. Acta Cryst., C40, 2120-2122.

Orthorhombic, $P2_12_12_1$, a = 7.550, b = 12.684, c = 22.095 Å, Z = 8. Mo radiation, R = 0.072 for 1625 reflexions.

The two independent molecules (Fig. 1) in the cell have very similar almost planar conformations. The geometry of the nearly flat molecule is determined by a hydrogen bond between the anilide NH and the imino nitrogen (imino NH not hydrogen-bonded to urea oxygen).

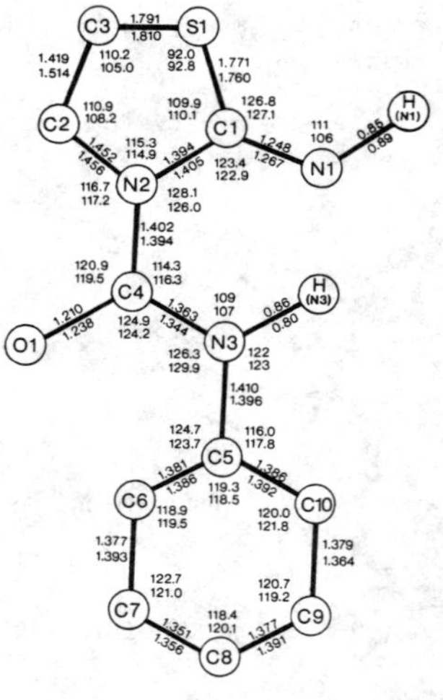

Fig. 1. Molecular dimensions for the two independent $C_{10}H_{11}N_3OS$ molecules.

ACENAPHTHYLENO[5,6-cd][1,2,6]THIADIAZINE
$C_{12}H_6N_2S$ (I)

2-METHYL-2H-ACENAPHTHYLENO[5,6-de]-1,2,3-TRIAZINE
$C_{13}H_9N_3$ (II)

NAPHTHO[1,8-cd]THIADIAZINE (β-FORM)
$C_{10}H_6N_2S$ (III)

A. GIEREN, V. LAMM, R.C. HADDON, M.L. KAPLAN, M.J. PERKINS and P. FLOWERDAY, 1984. Z. Naturforsch., B39, 975-984.

I. Triclinic, $P\bar{1}$, a = 16.327, b = 9.099, c = 7.541 Å, α = 115.23, β = 104.12, γ = 102.72°, D_m = 1.50, Z = 4. Cu radiation, R = 0.059 for 2740 reflexions.

II. Triclinic, $P\bar{1}$, a = 8.622, b = 8.874, c = 7.274 Å, α = 113.32, β = 101.17, γ = 81.34°, D_m = 1.35, Z = 2. Cu radiation, R = 0.052 for 1428 reflexions.

III. Tetragonal, $P4_1$, a = 7.696, c = 28.425 Å, D_m = 1.49, Z = 8. Mo radiation, R = 0.11 for 2051 reflexions.

Molecular dimensions for I and II are in Fig. 1. Structure III is disordered in the β-form. The data show that there is only a weak π-interaction between the SN_2 or N_3Me group and the acenaphthalene system.

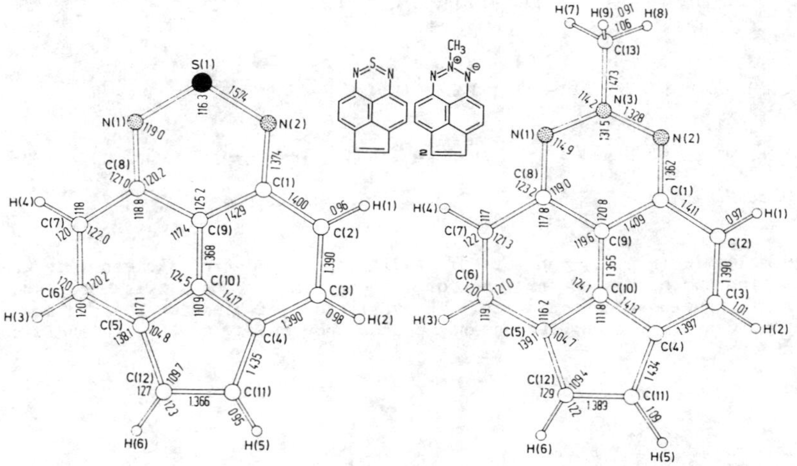

Fig. 1. Molecular dimensions for $C_{12}H_{16}N_2S$ (I) (left) and $C_{13}H_9N_3$ (II) (right).

1,4,10,13-TETRATHIA-7,16-DIAZACYCLOOCTADECANE
$C_{12}H_{26}N_2S_4$ (I)

7,16-BIS(4-CHLOROBENZOYL)-1,4,10,13-TETRATHIA-7,16-DIAZACYCLOOCTADECANE
$C_{26}H_{32}Cl_2N_2O_2S_4$ (II)

H.L. AMMON, K. CHANDRASEKHAR, S.K. BHATTACHARJEE, S. SHINKAI and Y. HONDA, 1984. Acta Cryst., C40, 2061-2064.

I. Monoclinic, $P2_1/c$, a = 5.200, b = 14.576, c = 11.203 Å, β = 100.95°, Z = 2. Cu radiation, R = 0.051 for 1017 reflexions.

II. Monoclinic, $P2_1/a$, a = 11.8146, b = 21.6043, c = 11.5440 Å, β = 94.255°, Z = 4. Cu radiation, R = 0.063 for 2764 reflexions.

Both molecules are shown in Fig. 1. Compound I has crystallographic inversion symmetry and mean C-S 1.811(5), mean C-C 1.496(6) Å. Disorder in II leads to some distorted distances; population averaged C-S and C-C distances are 1.792(7) and 1.502(9) Å respectively. Compound I has no exo- or endodentate S atoms and no anti C-S bonds, while II has two exodentate S atoms and three anti C-S bonds.

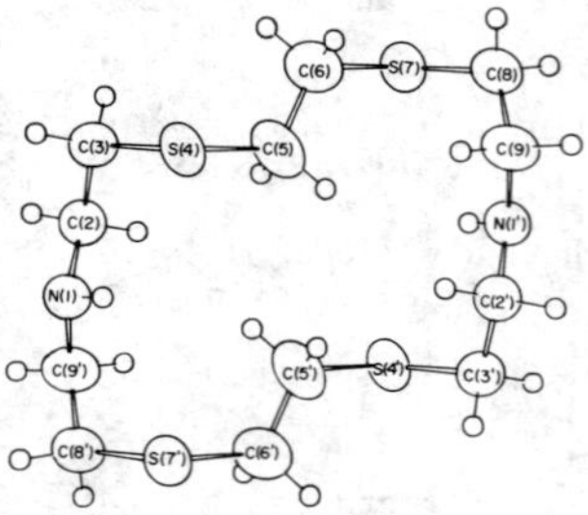
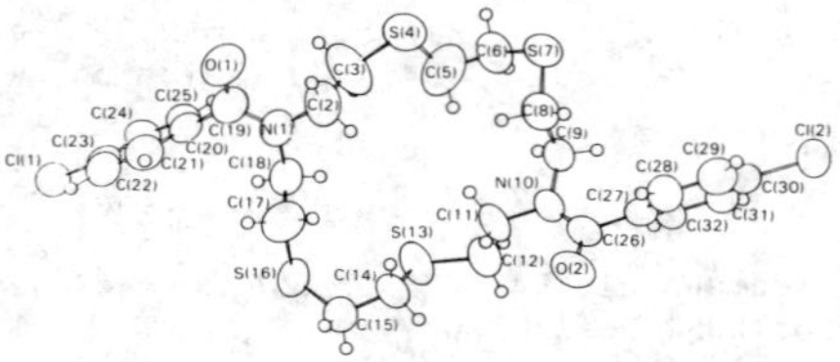

Fig. 1. Drawings normal to the crown-ring plane for $C_{12}H_{26}N_2S_4$ (I) (left) and $C_{26}H_{32}Cl_2N_2O_2S_4$ (II) (right).

METHYL 2(2H)-OXOPYRIMIDO[2,1-b][1,3]BENZOTHIAZOLE-4-ACETATE
$C_{13}H_{10}N_2O_3S$

S.B. SANNI, M. DE WOLF and A.T.H. LENSTRA, 1984. Acta Cryst., C40, 1724-1726.

Monoclinic, $P2_1/n$, a = 9.593, b = 9.213, c = 13.745 Å, β = 107.89°, Z = 4. Mo radiation, R = 0.035 for 1773 reflexions.

The molecule (Fig. 1) consists of a slightly warped ring skeleton, in which the thiazole ring approaches an envelope form, the pyrimido ring a twist-boat and the benzo ring a true-boat conformation. The variations of bond lengths and angles reflect the conflicting angular demands placed upon the system by the individual components.

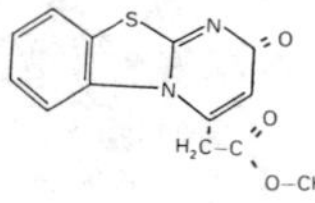
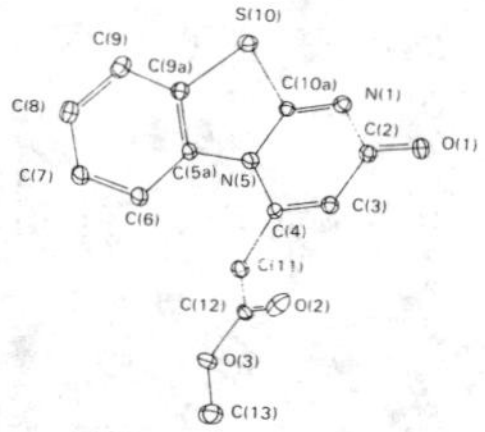

Fig. 1. The $C_{13}H_{10}N_2O_3S$ molecule.

2-(3,5-DICHLOROPHENYL-2-OXIDE)-4-METHYLTHIENO[2,3-a]-3,4,5,6-TETRAHYDROPYRIMIDINIUM
$C_{13}H_{12}Cl_2N_2OS$

S. RAJAPPA, B.G. ADVANI and G. KARTHA, 1984. J. Chem. Soc. Perkin I, 1631-1634.

Triclinic, P1, a = 7.608, b = 12.990, c = 15.917 Å, α = 67.02, β = 85.66, γ = 73.41°, Z = 2. Cu radiation, R = 0.058 for 3783 reflexions.

There are two crystallographically independent molecules (A, B) in the asymmetric unit. The molecules are essentially planar (Fig. 1) with a bond length of 1.435(3), 1.428(3) between the isothiazole and phenyl moieties. The S-N(1) distances are 1.730(2) and 1.725(2) Å. The S...O distance is 2.26 Å. The distances C(10)-C(13) (1.430(4), 1.388(4)), C(13)-C(11) (1.523(4), 1.424(4)) and C(11)-N(2) (1.470(3), 1.453(3) Å) are significantly different in molecules A and B making the saturated ring in B much flatter than that in A [C(3A)-C(4A) is given as 1.336 Å. This should probably be 1.366 Å.]

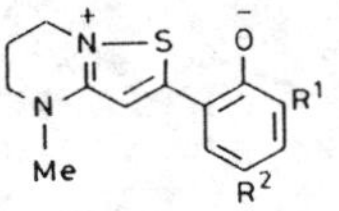

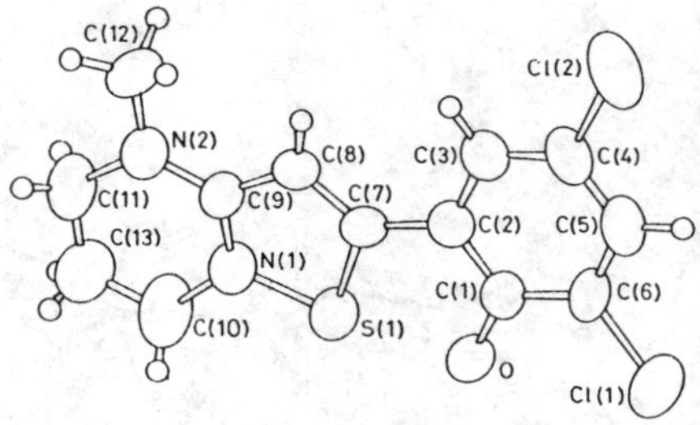

Fig. 1. $C_{13}H_{12}Cl_2N_2OS$: the molecular skeleton ($R^1 = R^2 = Cl$) and structure.

N-BENZYL-3-THIA-7-AZABICYCLO[3.3.1]NONAN-9,9-DIOL PERCHLORATE
$C_{14}H_{20}ClNO_6S$

B.R. BAILEY III, K.D. BERLIN and E.M. HOLT, 1984. Phosphorus and Sulfur, 20, 131-137.

Triclinic, P$\bar{1}$, a = 15.720, b = 17.425, c = 6.917 Å, α = 78.77, β = 90.35, γ = 62.13°, Z = 4. Mo radiation, R = 0.065 for 4635 reflexions.

The molecule is shown to be a chair-chair conformer in the solid state (Fig. 1) with little ring distortion near the gem diol group. Dimensions are unexceptional, with mean C-S 1.817. The N-H...S hydrogen bond distances in the two independent molecules are H...S 2.22(10) and 2.20(6) Å.

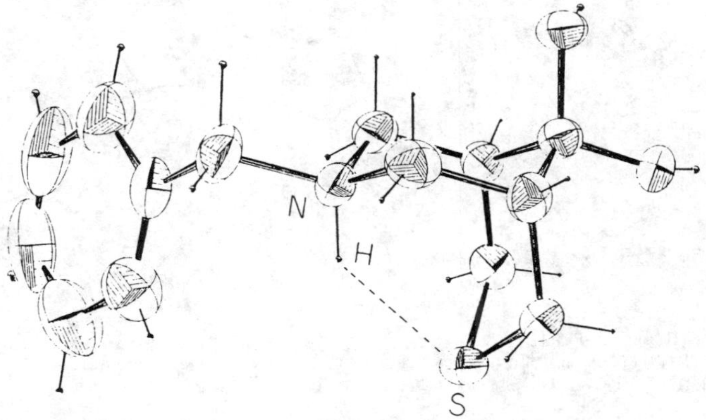

Fig. 1. A view of one of the $C_{14}H_{20}ClNO_6S$ molecules.

4-CROTONYL-10,10-DIMETHYL-4-AZA-3-THIATRICYCLO[5.2.1.0^{1,5}]DECANE-3,3-DIOXIDE
$C_{14}H_{21}NO_3S$

W. OPPOLZER, C. CHAPUIS and G. BERNARDINELLI, 1984. Helv. Chim. Acta, 67, 1397-1401.

Orthorhombic, $P2_12_12_1$, a = 7.410, b = 11.911, c = 16.228 Å, Z = 4. Mo radiation, R = 0.047 for 1131 reflexions.

The analysis establishes the structure shown in Fig. 1. Principal features include the synplanarity of the $C_\alpha C_\beta$-double bond with the carbonyl group which is anti to the SO_2-group; the N atom is slightly pyramidal.

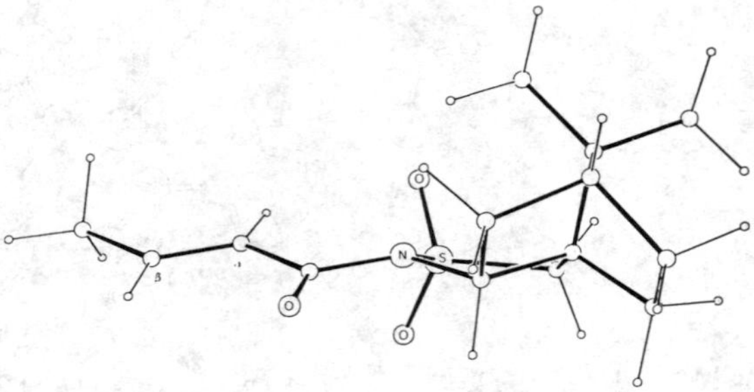

Fig. 1. A view of $C_{14}H_{21}NO_3S$.

3-HYDROXY-2-PHENYLTHIAZOLO[2,3-b]BENZOTHIAZOLIUM HYDROXIDE, INNER SALT
$C_{15}H_9NOS_2$

C.L. RASTON and A.H. WHITE, 1984. Aust. J. Chem., 37, 2583-2588.

Monoclinic, $P2_1/n$, a = 13.467, b = 9.426, c = 20.680 Å, β = 107.38°, D_m = 1.50, Z = 8. Mo radiation, R = 0.057 for 2006 reflexions.

There are two independent molecules in the unit cell (Fig. 1). Both molecules are substantially planar with only slight twists about the bond to their associated phenyl rings.

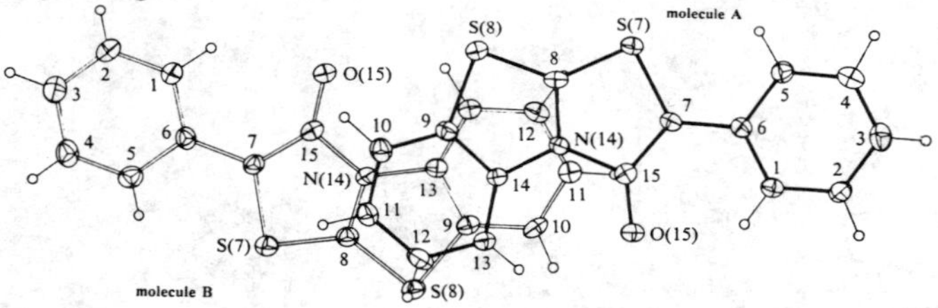

Fig. 1. The structure of $C_{15}H_9NOS_2$.

4-HYDROXY-2-METHYL-N-2-PYRIDYL-2H-1λ^6,2-BENZOTHIAZINE-3-CARBOXAMIDE 1,1-DIOXIDE MONOHYDRATE (PIROXICAM MONOHYDRATE)
$C_{15}H_{13}N_3O_4S,H_2O$

J. BORDNER, J.A. RICHARDS, P. WEEKS and E.B. WHIPPLE, 1984. Acta Cryst., C40, 989-990.

Triclinic, $P\bar{1}$, a = 12.721, b = 12.909, c = 10.481 Å, α = 99.31, β = 108.88, γ = 102.64°, D_m = 1.52, Z = 4. Cu radiation, R = 0.057 for 3110 reflexions.

The compound (Fig. 1) exists in a zwitterionic form, the enolic hydrogen on O(4B) having been transferred to pyridine N(13). Internal rotations of the neutral structure result in two intramolecular hydrogen bonds: N(11)-H...O(4B) (N...O 2.589(4) and 2.524(4) Å in the two independent molecules) and N(13)-H...O(10) (N...O 2.651(4) and 2.671(4) Å). The two independent molecules are hydrogen-bonded to water. A notable feature of the structure is the planarity of the side chain and the atoms of the thiazine ring.

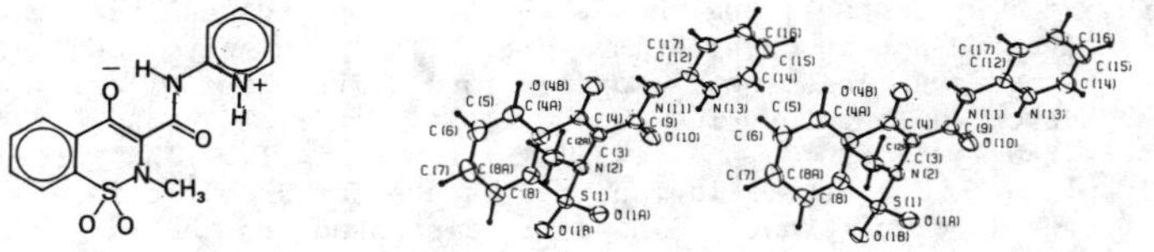

Fig. 1. $C_{15}H_{13}N_3O_4S,H_2O$: a stereoscopic view.

3-ETHYL-8-METHOXY-6-NITRO-2H-1-BENZOPYRAN-2-SPIRO-2'-(3'-METHYLTHIAZOLIDINE)
$C_{15}H_{18}N_2O_4S$

E. MILER-SRENGER and R. GUGLIELMETTI, 1984. Acta Cryst., C40, 2050-2053.

Triclinic, P$\bar{1}$, a = 6.973, b = 8.884, c = 13.438 Å, α = 84.48, β = 110.36, γ = 86.79°, Z = 2. Cu radiation, R = 0.048 for 2344 reflexions.

In the molecule (Fig. 1) the spiro carbon atom is involved in two long bonds to S and O. The thiazolidine ring is in an almost perfect half-chair conformation. The conformation of the pyran ring is 1,3-diplanar. Its puckering parameters and torsion energy increase through introduction of such bulky substituents in positions 3 and 8 of the benzopyran that cannot avoid short intramolecular contacts through steric accommodations. Molecules are piled along the b axis, alternately reversed in orientation and nearly equally spaced. Both enantiomers are present in the crystalline state as implied by the centrosymmetric space group.

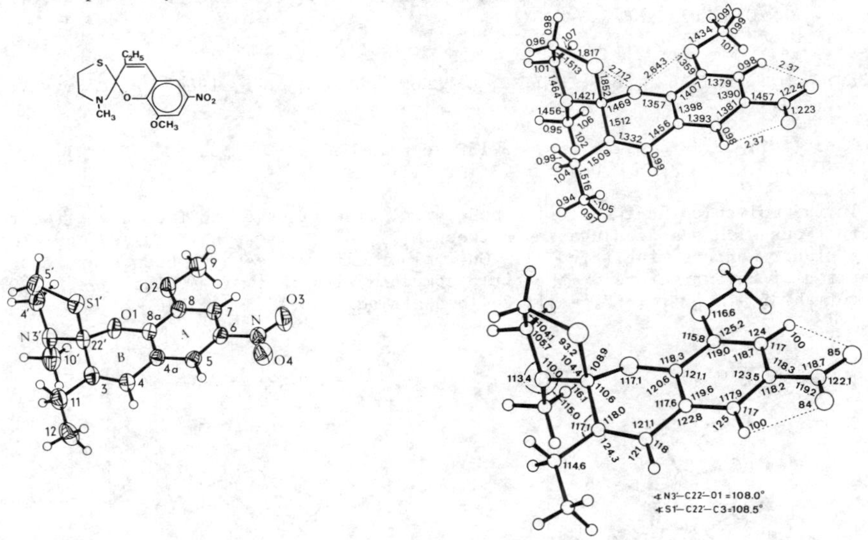

Fig. 1. Perspective view and molecular geometry details for $C_{15}H_{18}N_2O_4S$.

METHYL (Z)-2-METHOXY-3-METHYL-2-(N-METHYLANILINO)-4-OXOTHIAZOLIDIN-5-YLIDENE-ETHANOATE
$C_{15}H_{18}N_2O_4S$

A.P.K. ORRELL and J.D. WALLIS, 1984. J. Chem. Soc. Perkin II, 227-230.

Triclinic, P$\bar{1}$, a = 8.344, b = 10.045, c = 10.905 Å, α = 106.52, β = 109.99, γ = 97.10°, Z = 2. Mo radiation, R = 0.039 for 1981 reflexions.

The X-ray analysis established that the title compound was indeed a thiazolidinone derivative and not an open-chain isomer. The 2-methoxy group on the molecule (Fig. 1) is oriented such that the two oxygen lone pairs are antiperiplanar to the C-S and C-N bonds of the heterocycle. The C(2)-S(1), C(2)-O(1) and C(2)-N(2) distances are 1.880(2), 1.390(3) and 1.435(3) Å respectively. The heterocycle is approximately planar. The crystallographic evidence for anomeric interactions between oxygen or nitrogen lone pairs and C-S bonds is discussed and the latter appear to be stronger.

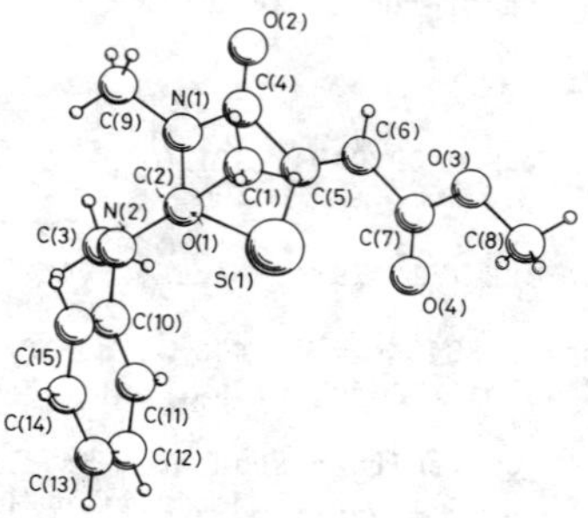

Fig. 1. $C_{15}H_{18}N_2O_4S$: the molecular structure.

(3aS)-7-CHLORO-4-(2-DIMETHYLAMINOETHYL)-8-METHYL-2,3,3a,4-TETRAHYDRO-1H-PYRROLO[2,1-c][1,2,4]BENZOTHIADIAZINE 5,5-DIOXIDE
$C_{15}H_{22}ClN_3O_2S$

R. OBERTI, M.T. BERNABEI, F. FORNI, R. CAMERONI and E. GALLI, 1984. Acta Cryst., C40, 882-884.

Monoclinic, $P2_1/c$, a = 10.226, b = 13.155, c = 12.722 Å, β = 100.59°, D_m = 1.35, Z = 4. Cu radiation, R = 0.064 for 1341 reflexions.

In the molecule the thiadiazine and pyrrolidine rings exhibit sofa and twist conformations with some conjugation between N(2), which is sp^2 hybridised, and the almost planar benzene ring (Fig. 1). Only minor differences have been found on comparison of the structure of the title compound with that of 7-chloro-4-ethyl-2,3,-3a,4-tetrahydro-1H-pyrrolo[2,1-c][1,2,4]benzothiadiazine 5,5-dioxide (1).

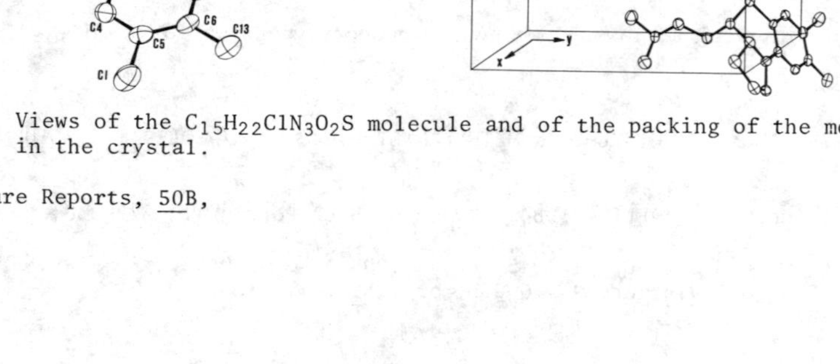

Fig. 1. Views of the $C_{15}H_{22}ClN_3O_2S$ molecule and of the packing of the molecules in the crystal.

1. Structure Reports, 50B,

(8aRS,9RS,10aSR)-10-ETHOXYCARBONYL-1,2,3,4,5,6,7,8,8a,10a-DECAHYDRO-10H-9-THIA-10-AZAPHENANTHRENE 9-OXIDE
$C_{15}H_{23}NO_3S$

Z. DAUTER, P. HANSON, C.D. REYNOLDS and W.A. STOCKBURN, 1984. Acta Cryst., C40, 521-523.

Monoclinic, P2₁/c, a = 5.264, b = 11.598, c = 25.694 Å, β = 104.36°, Z = 4. Cu radiation, R = 0.064 for 1518 reflexions.

The analysis establishes the relative configuration at S as shown in Fig. 1. The S-O bond (1.489(3) Å) is trans to the two C-H bonds on the heterocyclic ring. The S-N bond length is 1.689(3) Å.

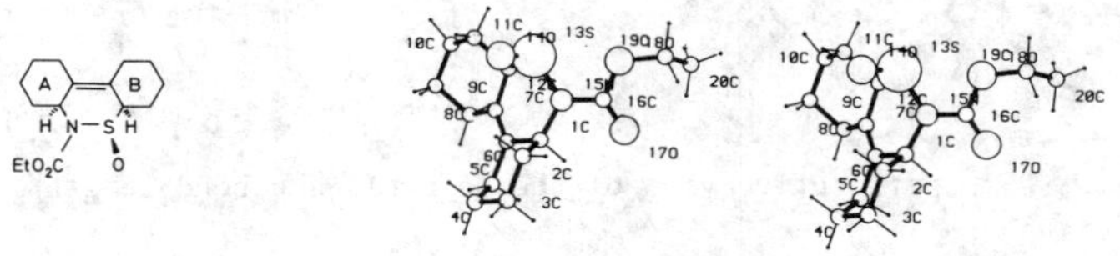

Fig. 1. The $C_{15}H_{23}NO_3S$ molecule and a stereoview.

10-(2'-PYRAZYL)PHENOTHIAZINE
$C_{16}H_{11}N_3S$

M.V. JOVANOVIC, E.R. BIEHL, R.D. ROSENSTEIN and S.S.C. CHU, 1984. J. Heterocycl. Chem., 21, 661-667.

Orthorhombic, Pbca, a = 10.633, b = 7.525, c = 32.984 Å, Z = 8. Cu radiation, R = 0.057 for 1944 reflexions.

A view of the molecule and dimensions are in Fig. 1. The plane of the pyrazine ring is nearly perpendicular to the S(5),N(10),C(2'),C(5') bisecting the phenothiazine molecule, with a dihedral angle of 88.2(1)°. The folding angle between the planes of the two benzo rings is 132.2°. The central ring is in a boat conformation with the pyrazyl group axial.

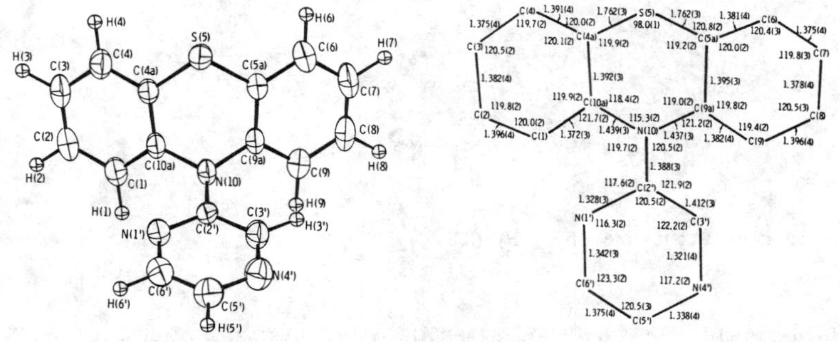

Fig. 1. The $C_{16}H_{11}N_3S$ molecule and dimensions.

1-ISOPROPYL-10-METHYLPHENOTHIAZINE
$C_{16}H_{17}NS$

S.S.C. CHU, S.V.L. NARAYANA and R.D. ROSENSTEIN, 1984. Acta Cryst., C40, 1281-1283.

Monoclinic, P2₁/c, a = 12.273, b = 7.998, c = 14.926 Å, β = 111.395°, Z = 4. Cu radiation, R = 0.047 for 2135 reflexions.

In the molecule (Fig. 1) the central ring of the phenothiazine ring system is in a boat conformation. The folding angle between the least-squares planes of the two benzo rings is 143.0(1)°. The 10-methyl group is in a quasi-axial conformation with respect to the central ring of the phenothiazine ring system due to the interaction with the 1-isopropyl group. Molecular packing in the unit cell is determined by van der Waals interactions.

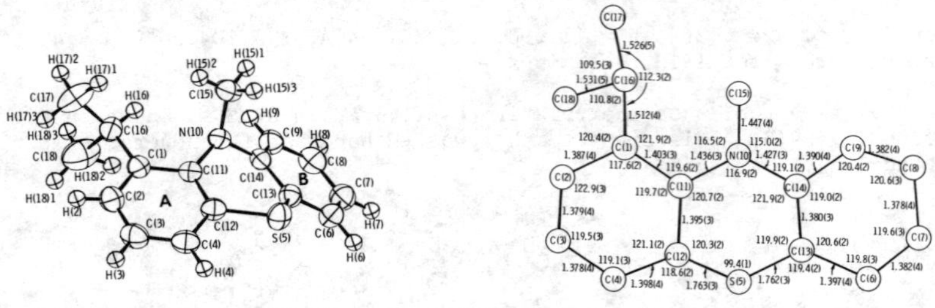

Fig. 1. $C_{16}H_{17}NS$: perspective view of the molecule and bond lengths and angles.

3,4-DIMETHYL-2,5-DIPHENYL-3,4-DIHYDRO-3a-THIA-1,3,4,6-TETRAAZAPENTALENE
$C_{17}H_{16}N_4S$

F. IWASAKI and K.-Y. AKIBA, 1984. Bull. Chem. Soc. Japan, 57, 2581-2583.

Orthorhombic, Pbcn, a = 14.533, b = 9.285, c = 11.305 Å, Z = 4. Mo radiation, R = 0.060 for 1107 reflexions.

The molecule (Fig. 1) has crystallographic twofold symmetry with the central S-C bond coincident with the twofold axis. The bonds S(6)-C(3) and S(6)-N(1) are 1.764 and 1.904 Å in length respectively, and other bond distances of interest are: N(1)-C(11) 1.460, N(1)-C(2) 1.316, C(3)-N(3) 1.336, C(2)-N(3) 1.354, and C(2)-C(21) 1.483 Å. Bond angles of interest are: C(3)-S(6)-N(1) 82.7, N(1)-S(6)-N(1') 165.5, S(6)-N(1)-C(2) 111.2, N(1)-C(2)-N(3) 116.3, C(2)-N(3)-C(3) 111.5, S(6)-C(3)-N(3) 117.9, and N(3)-C(3)-N(3') 124.1°. The dihedral angle between the planes of the central ring and the phenyl ring is 40.0°.

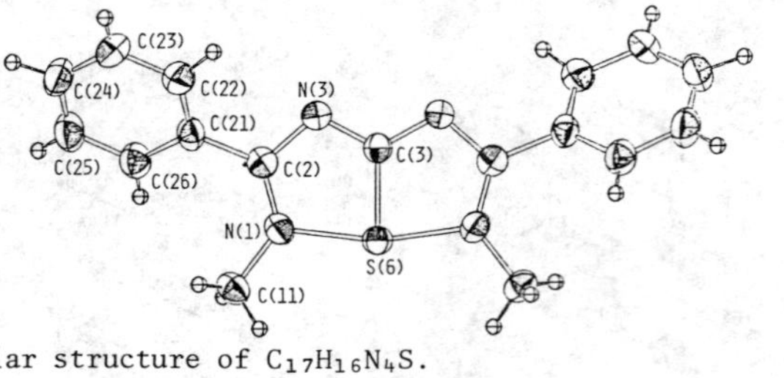

Fig. 1. Molecular structure of $C_{17}H_{16}N_4S$.

3-DIMETHYLIMINIO-3,4-DIHYDRO-7-METHOXY-2-(4-METHOXYPHENYL)-2H-1,2,4-BENZOTHIADIAZINE-1-OXIDE TETRAFLUOROBORATE
$C_{17}H_{20}BF_4N_3O_3S$

G. KRESZE, A. SCHWOBEL, A. HATJIISSAAK, K. ACKERMANN and T. MINAMI, 1984. Justus Liebigs Ann. Chem., 904-910.

Monoclinic, P2_1/c, a = 11.526, b = 12.997, c = 13.770 Å, β = 111.64°, Z = 4. Mo radiation, R = 0.046 for 2156 reflexions.

The BF_4^- ion is hydrogen bonded to N4-H4 in the solid state (Fig. 1) with N4...F3 2.953(4) Å. The thiadiazine ring has a boat form with S-O 1.463(2), S-C10

1.768(3), S-N2 1.736(3), N2-C21 1.451(4), N2-C3 1.363(4), C3-N4 1.339(4), N4-C5 1.407(4) Å. In the BF_4^- ion, the B-F bond involved in the hydrogen bond is longer (1.392(5) Å) than the other B-F bonds (1.353-1.364(5) Å).

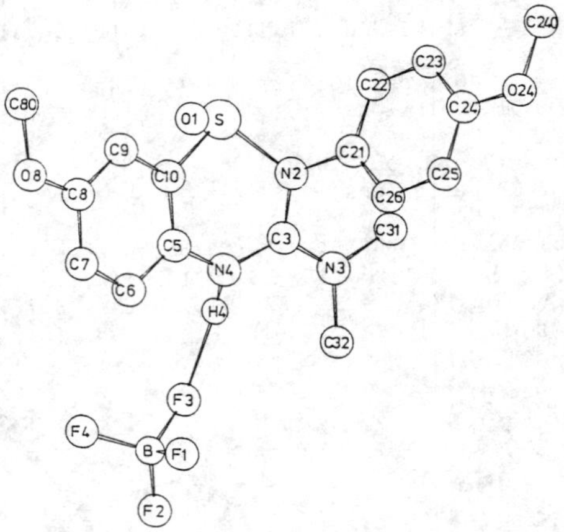

Fig. 1. A view of $C_{17}H_{20}BF_4N_3O_3$.

1-CHLOROPROMAZINE HYDROBROMIDE
$C_{17}H_{20}BrClN_2S$ $C_{17}H_{20}ClN_2S^+.Br^-$

A.R. MARTIN, A. HALLBERG, T.H. KRAMER, A. SVENSSON, R.B. BATES and R.B. ORTEGA, 1984. Acta Cryst., C40, 2113-2116.

Orthorhombic, Pbca, a = 10.166, b = 11.190, c = 32.93 Å, Z = 8. Mo radiation, R = 0.042 for 1227 reflexions.

This apparent antidepressant non-neuroleptic promazine analogue has a decidedly synclinal conformation, unlike the antipsychotics promazine and 2-chloropromazine. This result supports the view that a synclinal non-coplanar side-chain conformation is associated with the antidepressant activity of compounds of the tricyclic class and is also consistent with current conformational theories of antipsychotic action, which, in contrast, require an approximately coplanar (synperiplanar) side-chain conformation.

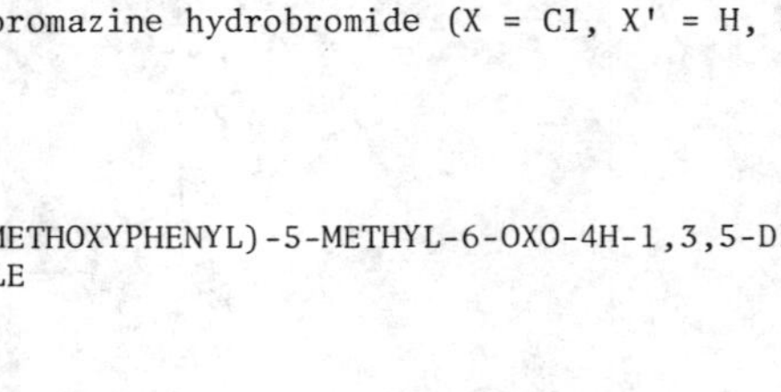

Fig. 1. 1-Chloropromazine hydrobromide (X = Cl, X' = H, X'' = Br) and an ORTEP plot.

2-(5,6-DIHYDRO-4-(4-METHOXYPHENYL)-5-METHYL-6-OXO-4H-1,3,5-DITHIAZIN-2-YLIDENE)-3,3-DIMETHYLBUTYRONITRILE
$C_{17}H_{20}N_2O_2S_2$ (I)

7-t-BUTYL-7-CYANO-3,3-DIMETHYL-2-METHOXYCARBONYL-8-THIOXO-4,6-DITHIA-1-AZABICYCLO-[3.3.0]OCTANE
$C_{14}H_{20}N_2O_2S_3$ (II)

E. SCHAUMANN, U. WRIEDE and G. ADIWIDJAJA, 1984. Chem. Ber., 117, 2205-2225.

I. Monoclinic, $P2_1/c$, a = 11.402, b = 10.482, c = 16.024 Å, β = 109.53°, Z = 4. Cu radiation, R = 0.064 for 2301 reflexions.

II. Monoclinic, $P2_1/c$, a = 12.112, b = 10.962, c = 13.896 Å, β = 109.38°, Z = 4. Cu radiation, R = 0.069 for 2099 reflexions.

The X-ray analyses established the structures shown in Fig. 1. Some dimensions are: for I, S1-C2 1.749(4), S1-C6 1.800(4), S3-C2 1.761(4), S3-C4 1.775(4), N5-C4 1.346(5), N5-C6 1.439(5) Å; for II, S4-C3 1.857(5), S4-C5 1.827(5), S6-C5 1.781(5), S6-C7 1.880(5), S81-C8 1.653(5), N1-C2 1.446(5), N1-C5 1.490(5), N1-C8 1.334(5) Å.

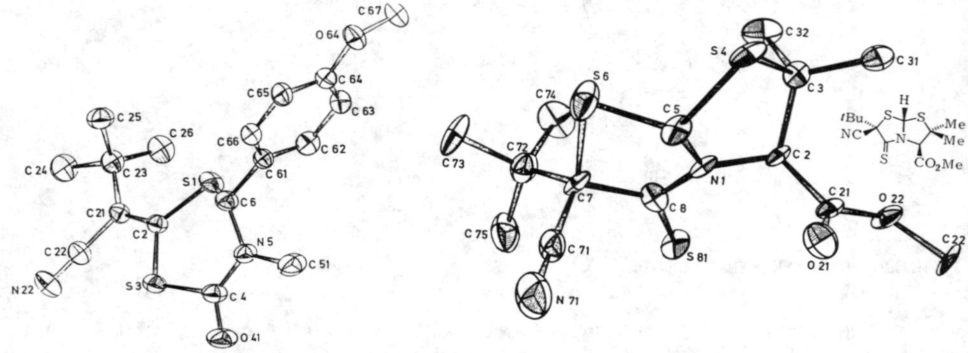

Fig. 1. Views of $C_{17}H_{20}N_2O_2S_2$ (left) and $C_{14}H_{20}N_2O_2S_3$ (right).

8-BENZYL-1-HYDROXY-11-METHYL-6,9-DIAZA-12-THIATRICYCLO[7.3.0.0^{2,6}]DODECANE-7,10-DIONE
$C_{17}H_{20}N_2O_3S$

G. ZANOTTI, F. PINNEN, G. LUCENTE, S. CERRINI, W. FEDELI and F. MAZZA, 1984. J. Chem. Soc. Perkin I, 1153-1157.

Orthorhombic, $P2_12_12_1$, a = 16.462, b = 15.763, c = 6.487 Å, Z = 4. Mo radiation, R = 0.11 for 1371 reflexions [crystal quality very poor].

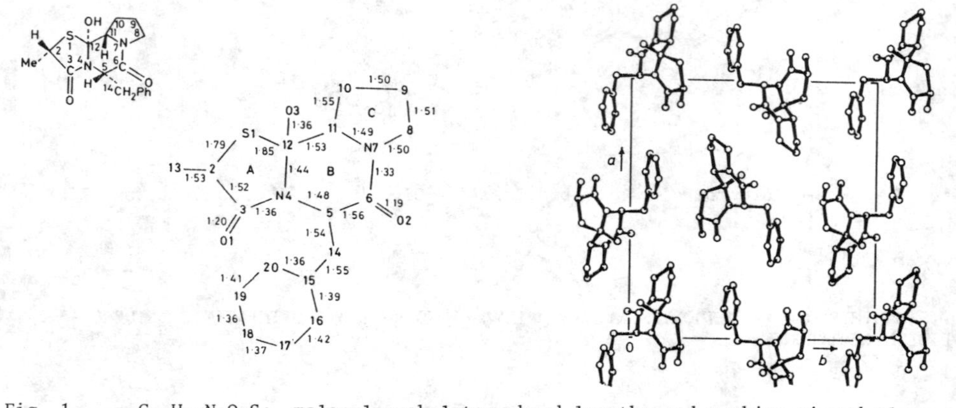

Fig. 1. $C_{17}H_{20}N_2O_3S$: molecular skeleton, bond lengths and packing viewed along c.

X-ray analysis established the structure of the title compound, a thiacyclol, (Fig. 1). The A/B ring junction is quasi-cis whereas the B/C ring junction is of the quasi-trans type. Ring A adopts an approximate C_S-C(12) envelope conformation, ring B a half-boat conformation with approximate C_S symmetry through C(6) and C(12), and ring C a C_S-C_β-endo conformation. The C(2)-C(3)-N(4)-C(5) and C(5)-C(6)-N(7)-C(11) torsion angles are 172.9 and -0.7° respectively. The benzylic side group adopts a conformation extended towards the phenylalanine N atom. Packing in the crystal is characterised by a hydrogen bond between O(2) of the phenylalanine residue and hydroxylic O(3) of a molecule translated along C(O(2)...O(3) 2.65(1) Å).

3β-BROMO-4β-HYDROXYMETHYL-4α-METHOXY-3α-METHYL-7β-PHENOXYACETAMIDOCEPHAM MONOHYDRATE
$C_{17}H_{21}BrN_2O_5S,H_2O$

A. BALSAMO, P. DOMIANO, B. MACCHIA, F. MACCHIA, E. ORLANDINI and A. ROSSELLO, 1984. J. Chem. Res., (S) 226-227, (M) 2163-2197.

Monoclinic, $P2_1$, a = 13.336, b = 6.720, c = 12.352 Å, β = 110.43°, Z = 2. Cu radiation, R = 0.057 for 1851 reflexions.

In the molecule (Fig. 1) atom N(5) is 0.115 Å from the plane of its three bonded C atoms. The tetrahydrothiazine portion of the molecule exists in a half-chair conformation and the C(3) Br atom has the same β-orientation of the phenoxyacetamido side chain, whereas the C(4) methoxy group is α-oriented and shows a trans diaxial relation with the Br atom.

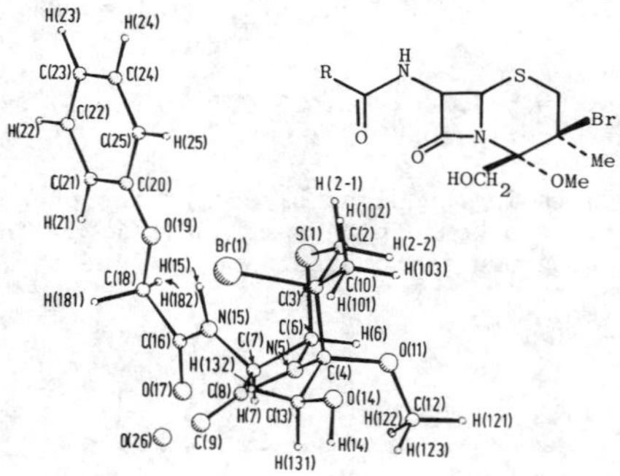

Fig. 1. A view of $C_{17}H_{21}BrN_2O_5S,H_2O$.

3,5-DICYANO-3,5-DI-t-BUTYL-7-THIA-1-AZABICYCLO[4,3,0]NON-8-ENE-2,4-DIONE
$C_{17}H_{21}N_3O_2S$ (I)

5,5-DICHLORO-2-DICHLOROMETHYLENE-6-ISOPROPYL-3-OXA-7-THIA-1-AZABICYCLO[4,3,0]NON-8-EN-4-ONE
$C_{10}H_9Cl_4NO_2S$ (II)

A. MEDICI, G. FANTIN, M. FOGAGNOLO, P. PEDRINI, A. DONDONI and G.D. ANDREETTI, 1984. J. Org. Chem., 49, 590-596.

I. Orthorhombic, Pbcn, a = 19.830, b = 13.123, c = 13.197 Å, Z = 8. Mo radiation, R = 0.074 for 1720 reflexions.

II. Monoclinic, $P2_1/n$, a = 14.264, b = 8.128, c = 12.356 Å, β = 106.76°, Z = 4. Cu radiation, R = 0.054 for 2230 reflexions.

The analyses established the structures of these cycloadducts to be as shown in Fig. 1.

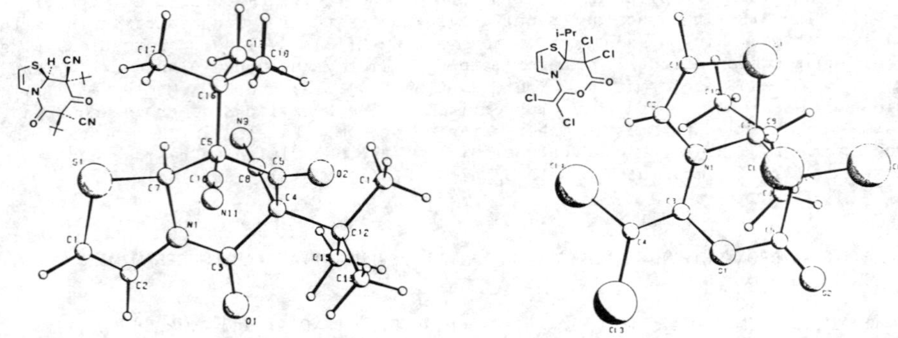

Fig. 1. Views of $C_{17}H_{21}N_3O_2S$ (left) and $C_{10}H_9Cl_4NO_2S$ (right).

2'-NITRO-10-PHENYLPHENOTHIAZINE
$C_{18}H_{12}N_2O_2S$

M.V. JOVANOVIC, E.R. BIEHL, P. DE MEESTER and S.S.C. CHU, 1984. J. Heterocycl. Chem., 21, 885-888.

Monoclinic, $P2_1/c$, a = 8.681, b = 7.340, c = 23.623 Å, β = 97.69°, Z = 4. Cu radiation, R = 0.039 for 1785 reflexions.

The 2-nitrophenyl group is in the plane bisecting the phenothiazine ring (Fig. 1). The folding angle of the two benzo-planes is 154.8(1)°. The central ring has a boat conformation. The nitro-group is rotated 55.6(1)° out of its phenyl-ring plane.

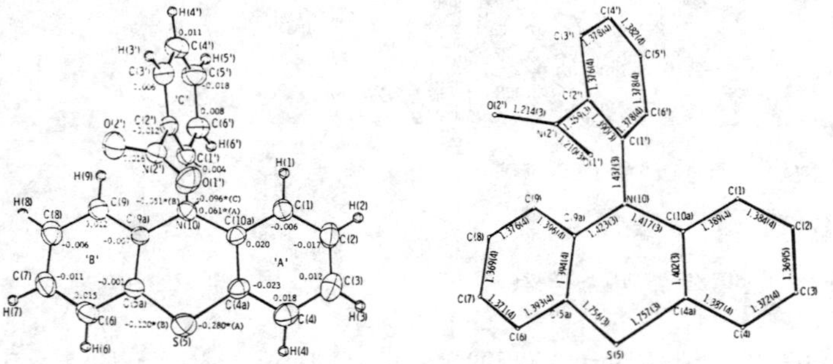

Fig. 1. A view of $C_{18}H_{12}N_2O_2S$ and bond lengths.

4'-NITRO-10-PHENYLPHENOTHIAZINE
$C_{18}H_{12}N_2O_2S$

M.V. JOVANOVIC, E.R. BIEHL, P. DE MEESTER and S.S.C. CHU, 1984. J. Heterocycl. Chem., 21, 1793-1800.

Monoclinic, $P2_1/c$, a = 7.269, b = 12.483, c = 34.498 Å, β = 107.00°, Z = 8. Cu radiation, R = 0.062 for 4031 reflexions.

The asymmetric unit contains two independent but similar molecules (Fig. 1). The central ring of the phenothiazine is in a boat conformation. The 10-aryl ring is perpendicular to the plane bisecting the dihedral angle of the phenothiazine nucleus. This geometry allows for resonance interaction between the lone pair of electrons on N(10) and the 4'-nitro-10-phenyl ring which influences both spectral properties and the structure of the tricyclic system.

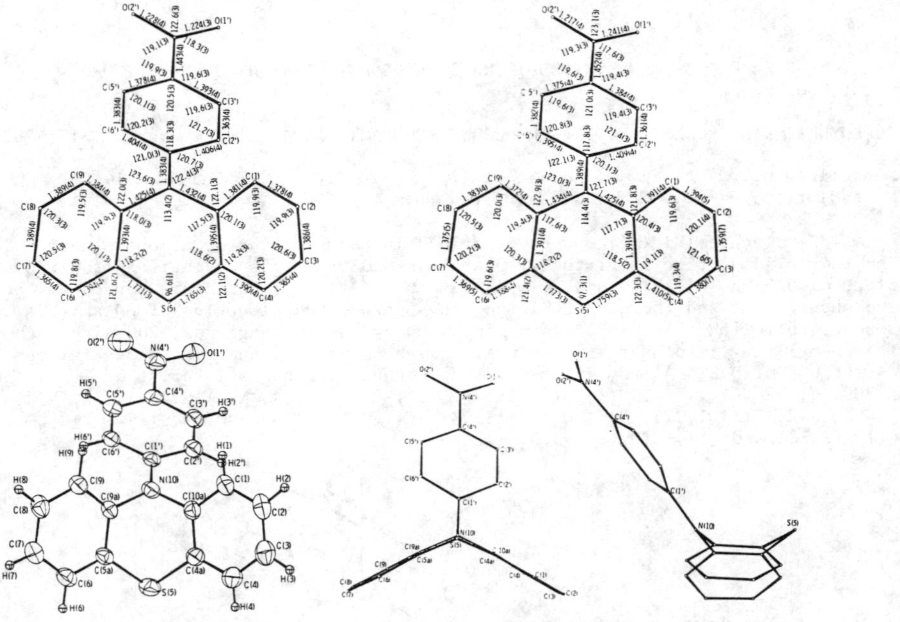

Fig. 1. Bond lengths and angles in $C_{18}H_{12}N_2O_2S$ and views of a molecule.

10-(2-METHOXYPHENYL)PYRIDO[3,2-b][1,4]BENZOTHIAZINE
$C_{18}H_{14}N_2OS$

P. DE MEESTER, S.S.C. CHU, M.V. JOVANOVIC and E.R. BIEHL, 1984. Acta Cryst., C40, 1753-1756.

Monoclinic, $P2_1/c$, a = 11.859, b = 8.121, c = 16.016 Å, β = 101.59°, Z = 4. Cu radiation, R = 0.050 for 1708 reflexions.

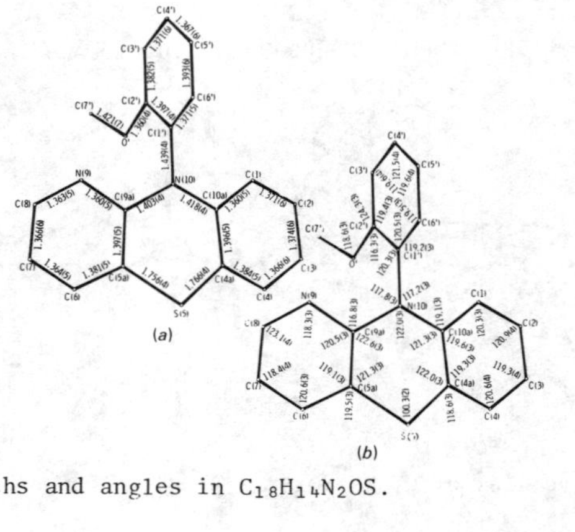

Fig. 1. Bond lengths and angles in $C_{18}H_{14}N_2OS$.

The central ring of the tricyclic ring system (Fig. 1) is in a boat conformation. The folding angle between the planes of the pyrido and benzo rings is 158.7(1)°. The plane of the 10-phenyl ring, almost perpendicular to the plane of the central ring (dihedral angle 95.4(1)°), nearly bisects the tricyclic ring system.

2-METHYL-3-(4-OXO-3-PHENYLTHIAZOLIDINE-2-YLIDENAMINO)-4(3H)-QUINAZOLINONE
$C_{18}H_{14}N_4O_2S$

S. BÜYÜKTIMKIN, S. ELZ, M. DRÄGER and W. SCHUNACK, 1984. Arch. Pharm., 317, 797-802.

Monoclinic, I2/c, a = 16.881, b = 10.470, c = 20.370 Å, β = 111.14°, D_m = 1.37, Z = 8. Mo radiation, R = 0.072 for 3129 reflexions.

The molecule (Fig. 1) can be described in terms of three, almost planar moieties. The thiazolidine ring is virtually planar (± 0.030 Å) with deviations from the best plane of 0.04 Å (C(4)), 0.004 Å (O(1)) and 0.13 Å (N(2)) for atoms bonded to the ring. The phenyl ring and the quinazolinone rings make dihedral angles of 80 and 82° respectively with the thiazolidine ring, and a dihedral angle of 120° with each other. The phenyl ring is planar (± 0.0060 Å) and the quinazolinone ring shows maximum deviations of 0.024 Å from its best plane. Bond lengths of interest are: S(1)-C(2) 1.799, S(1)-C(3) 1.744, N(1)-C(1) 1.378, N(1)-C(3) 1.384, N(1)-C(4) 1.437, N(2)-N(3) 1.426, N(2)-C(3) 1.271, N(4)-C(11) 1.293, N(4)-C(12) 1.386, O(1)-C(1) 1.210, O(2)-C(10) 1.221 and C(1)-C(2) 1.504 Å.

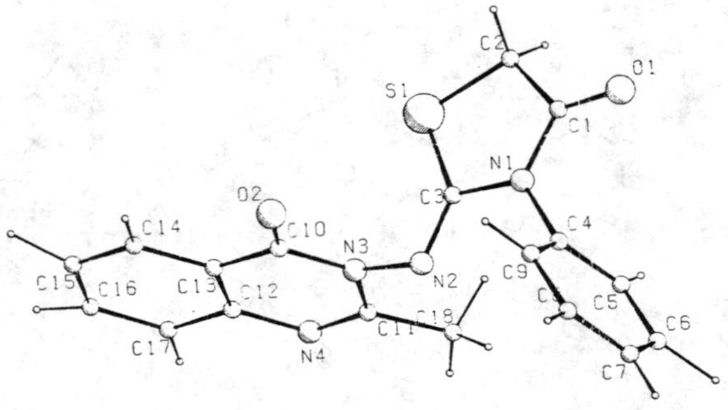

Fig. 1. Molecular structure of $C_{18}H_{14}N_4O_2S$.

4-HYDROXY-3,5-BIS(4-METHYLPHENYL)-1,4-THIAZINANE 1,1-DIOXIDE
$C_{18}H_{21}NO_3S$

H. YAMAMURA, D. KLEFFEL, J.F. BLOUNT, D. HUNKLER, L.J. TODARO and H.-H. OTTO, 1984. Justus Liebigs Ann. Chem., 1013-1023.

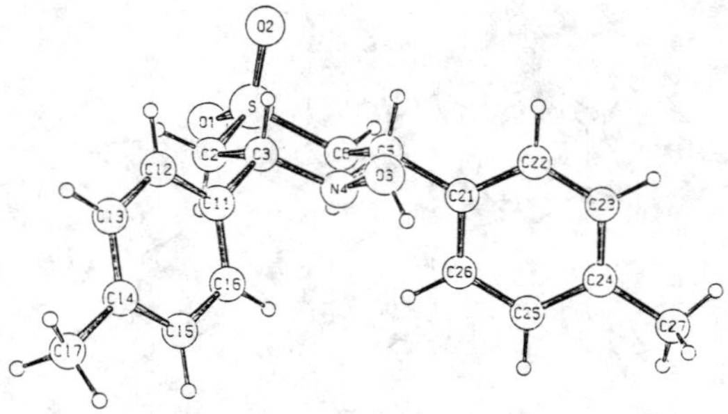

Fig. 1. A perspective view of $C_{18}H_{21}NO_3S$.

Monoclinic, $P2_1/n$, a = 11.041, b = 7.157, c = 22.694 Å, β = 100.44°, Z = 4. Cu radiation, R = 0.051 for 2124 reflexions.

The thiazinane ring has a chair conformation (Fig. 1) with the hydroxy group on N4 equatorial. Some dimensions are: S-O 1.440, 1.449(2), S-C 1.749 1.752(3), N4-O3 1.448(3), N4-C3 1.476(4), N4-C5 1.485(4) Å, O-S-O 117.9(1), C2-S-C6 99.7(1)°.

ETHYL 3-METHYLENE-2a-PHENYL-2a,3-DIHYDRO[1,2]DIAZETO[3,2-c][1,4]BENZOTHIAZINE-2-CARBOXYLATE
$C_{19}H_{16}N_2O_2S$

T. PILATI and M. SIMONETTA, 1984. Acta Cryst., C40, 1476-1477.

Monoclinic, $P2_1/c$, a = 14.677, b = 8.308, c = 16.008 Å, β = 116.79°, D_m = 1.275, Z = 4. Mo radiation, R = 0.045 for 2532 reflexions.

The heterocyclic six-membered ring has a conformation midway between E_2 and $^1S_2$ (Fig. 1). The four-membered ring is non-planar (Q_t = 0.0128(5) Å) and is very distorted, the lengths of the two diagonals being 2.091(3) and 1.963(2) Å. The C(4)-C(9) benzene ring deviates significantly from planarity (Q_t = 0.027(2) Å). Some bond lengths are N(1)-N(2) 1.422(3), N(2)-C(2) 1.510(2), N(1)-C(3) 1.304(3) Å.

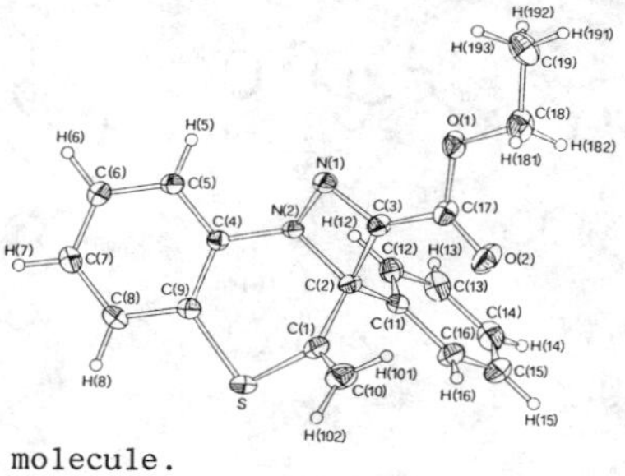

Fig. 1. The $C_{19}H_{16}N_2O_2S$ molecule.

4'-DIMETHYLAMINO-10-PHENYLPYRIDO[3,2-b]BENZOTHIAZINE
$C_{19}H_{17}N_3S$

M.V. JOVANOVIC, E.R. BIEHL, P. DE MEESTER and S.S.C. CHU, 1984. J. Heterocycl. Chem., 21, 1425-1429.

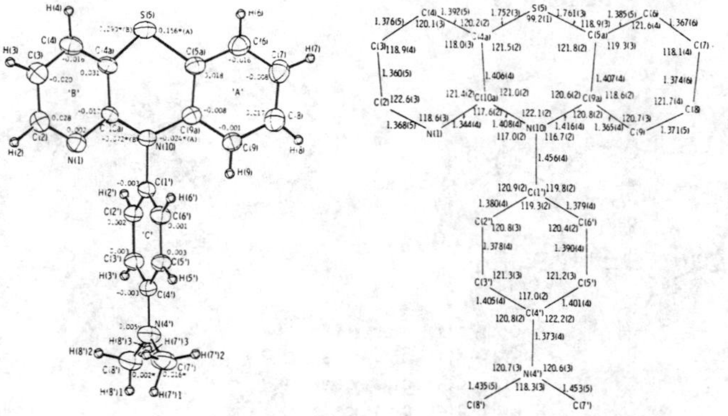

Fig. 1. A view of $C_{19}H_{17}N_3S$ and dimensions.

Monoclinic, P2₁/n, a = 8.580, b = 8.984, c = 20.880 Å, β = 98.42°, Z = 4. Cu radiation, R = 0.050 for 1865 reflexions.

The 10-aryl group is nearly parallel to the plane bisecting the pyridobenzothiazine ring (Fig. 1) (e.g. torsion angle C(9a)-N(10)-C(1')-C(2') 78.8(3)°). The central ring is in a boat conformation and the folding angle is 154.1(1)°.

trans-3a,11b-DIHYDRO-3-N-BUTYL-PHENANTHRO[9,10-d]THIAZOL-2-IMINE
$C_{19}H_{20}N_2S$

M. WEITZBERG, Z. AIZENSHTAT and J. BLUM, 1984. J. Heterocycl. Chem., 21, 1597-1601.

Monoclinic, I2/c, a = 21.908, b = 7.996, c = 19.286 Å, β = 106.84°, Z = 8. Mo radiation, R = 0.053 for 1857 reflexions.

The analysis established the structure and stereochemistry as shown in Fig. 1.

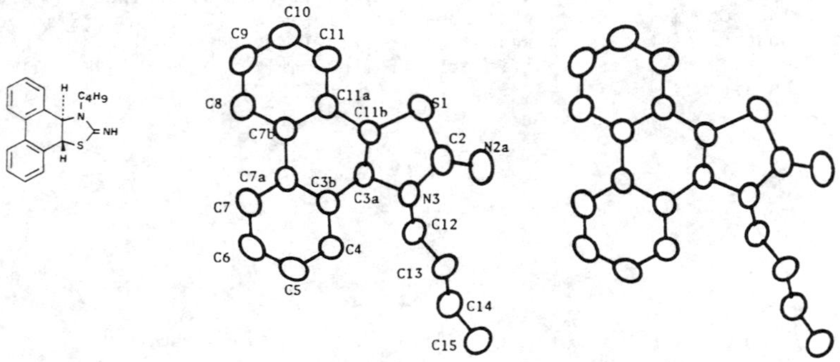

Fig. 1. The $C_{19}H_{20}N_2S$ molecule and a stereoview.

3-DIMETHYLAMINO-4,4-DIMETHYL-7-PHENYL-4,5,6,7-TETRAHYDRO-1,2,5,7-BENZOTHIATRIAZONIN-6-ONE 1,1-DIOXIDE
$C_{19}H_{22}N_4O_3S$

M. SCHLAPFER-DAHLER, R. PREWO, J.H. BIERI and H. HEIMGARTNER, 1984. Heterocycles, 22, 1667-1672.

Monoclinic, Cc, a = 16.724, b = 9.775, c = 12.125 Å, β = 109.82°, Z = 4. Mo radiation, R = 0.027 for 3469 reflexions (at -140°C).

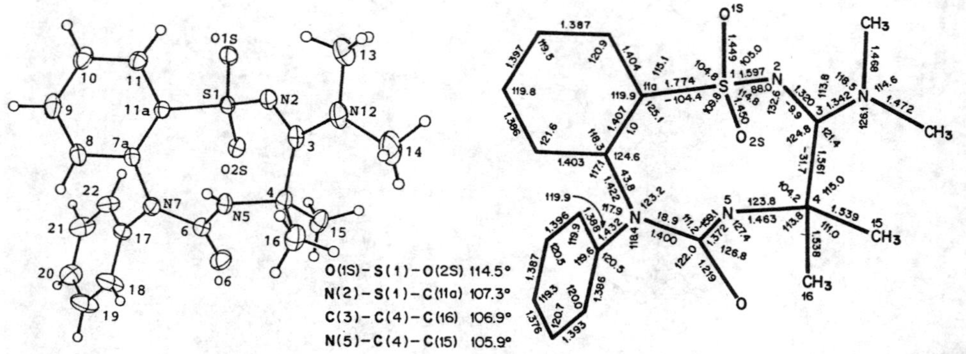

Fig. 1. The structure of $C_{19}H_{22}N_4O_3S$ with dimensions.

The structure of the compound is shown in Fig. 1 along with bond lengths, angles and torsion angles in the puckered nine-membered ring.

4-NITROBENZYL 3-BENZYLIDENE-7-OXO-4-THIA-1-AZABICYCLO[3.2.0]HEPTANE-2-CARBOXYLATE
$C_{20}H_{16}N_2O_5S$

J.L. DOUGLAS, A. MARTEL, G. CARON, M. MENARD, L. SILVEIRA and J. CLARDY, 1984. Can. J. Chem., 62, 2282-2286.

Triclinic, P$\bar{1}$, a = 5.529, b = 18.665, c = 13.481 Å, α = 129.04, β = 113.69, γ = 89.31°, Z = 2. Cu radiation, R = 0.056 for 1715 reflexions.

X-ray analysis established the structure shown in Fig. 1. The bridgehead nitrogen is 0.43 Å out of the plane of the three adjacent atoms.

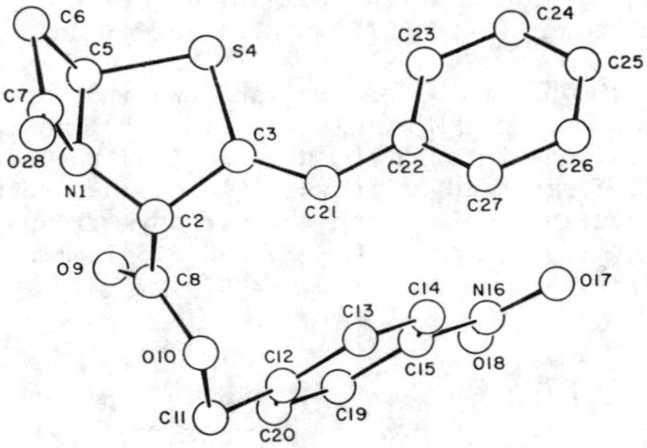

Fig. 1. A view of $C_{20}H_{16}N_2O_5S$.

2,2'-BIS(4-CHLOROPHENYL)-5,5'-DIMETHYL-[3,3'-BI-1,3-THIAZOLIDINE]-4.4'-DIONE
$C_{20}H_{18}Cl_2N_2O_2S_2$

G. BRUNO, G. BOMBIERI, A. DEL PRA, T. PREVITERA, M.G. VIGORITA and M. BASILE, 1984. Acta Cryst., C40, 671-674.

Monoclinic, $P2_1/n$, a = 13.857, b = 12.527, c = 12.199 Å, β = 94.58°, D_m = 1.41, Z = 4. Mo radiation, R = 0.069 for 1178 reflexions.

Fig. 1. $C_{20}H_{18}Cl_2N_2O_2S_2$: views down b of the molecule and of the crystal packing.

The molecular structure is characterised by the intramolecular interaction between the two parallel phenyl groups (mean distance 3.3 Å) and by the different conformations of the two thiazolidine rings, one, containing N(1), an envelope, and the other, containing N(2), a half-chair (Fig. 1). Corresponding bond distances and angles in these rings are in good agreement. The geometry of the bonds about N(1) and N(2) tends to be tetrahedral; the N(1)-N(2) bond length of 1.41(1) Å is normal. Packing is due mainly to van der Waals interactions.

DIPOTASSIUM SODIUM TRISACCHARINATE MONOHYDRATE
$C_{21}H_{12}K_2N_3NaO_9S_3,H_2O$ $K_2Na(C_7H_4NO_3S)_3.H_2O$

K.M.A. MALIK, S.Z. HAIDER, M.A. HOSSAIN and M.B. HURSTHOUSE, 1984. Acta Cryst., C40, 1696-1698.

Triclinic, P$\bar{1}$, a = 16.428, b = 12.122, c = 6.954 Å, α = 97.61, β = 77.36, γ = 110.09°, D_m = 1.72, Z = 2. Mo radiation, R = 0.054 for 4724 reflexions.

The structure (Fig. 1) consists of one Na^+ and two independent K^+ cations, three independent $C_7H_4NO_3S^-$ anions and one H_2O molecule, in a polymeric framework. The anions are bonded to the cations through the N as well as the carbonyl and sulphonyl O atoms. The H_2O molecule forms bridges between two Na^+ and between one Na^+ and one K^+ ion, and is also involved in H bonds with the carbonyl groups of two $C_7H_4NO_3S^-$ anions. Some interatomic distances are: Na-O 2.290-2.841, mean 2.438(2) Å; K-O 2.599-3.113, mean 2.818(2) Å; K-N 2.878-3.236, mean 3.001(2) Å.

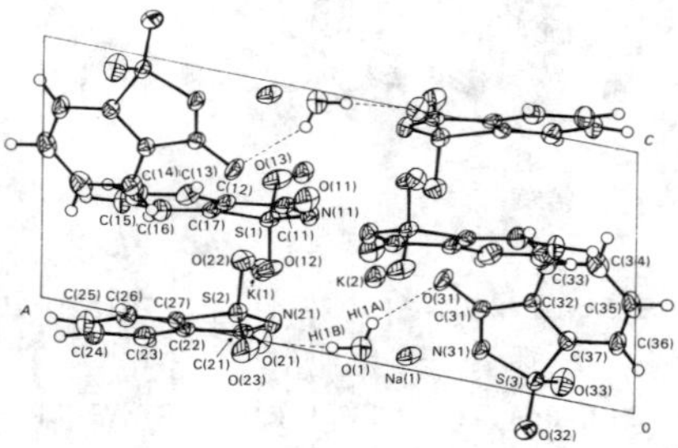

Fig. 1. The crystal structure of $K_2Na(C_7H_4NO_3S)_3.H_2O$ viewed along b.

4,4-DIMETHYL-2,9-DIPHENYL-7,7-BIS(TRIFLUOROMETHYL)-1,6-DITHIA-3,8-DIAZASPIRO[4.4]NONA-2,8-DIENE
$C_{21}H_{16}F_6N_2S_2$ (I)

cis-4,4-DIMETHYL-2,7,9-TRIPHENYL-1,6-DITHIA-3,8-DIAZASPIRO[4.4]NONA-2,7-DIENE
$C_{25}H_{22}N_2S_2$ (II)

trans-4,4-DIMETHYL-2,7,9-TRIPHENYL-1,6-DITHIA-3,8-DIAZASPIRO[4.4]NONA-2,7-DIENE
$C_{25}H_{22}N_2S_2$ (III)

T. BUCHEL, R. PREWO, J.H. BIERI and H. HEIMGARTNER, 1984. Helv. Chim. Acta, 67, 534-549.

I. Triclinic, P$\bar{1}$, a = 8.599, b = 10.113, c = 12.532 Å, α = 84.33, β = 77.46, γ = 89.54°, Z = 2. Mo radiation, R = 0.041 for 4862 reflexions (at -140°C).

II. Triclinic, P$\bar{1}$, a = 9.893, b = 10.068, c = 12.192 Å, α = 105.92, β = 98.03, γ = 105.73°, Z = 2. Mo radiation, R = 0.056 for 8577 reflexions.

III. Triclinic, P$\bar{1}$, a = 9.450, b = 10.435, c = 12.273 Å, α = 95.70, β = 108.07, γ = 110.80°, Z = 2. Mo radiation, R = 0.050 for 9184 reflexions.

The structures of these cycloadducts were established as shown in Fig. 1.

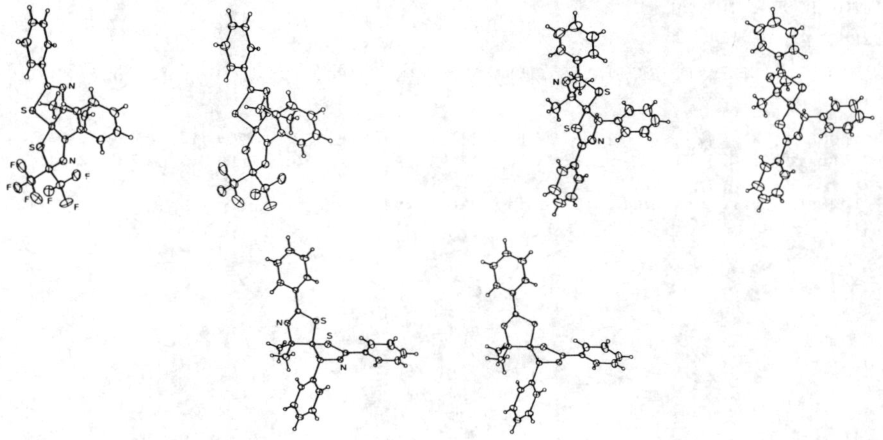

Fig. 1. Stereoviews of $C_{21}H_{16}F_6N_2S_2$ (top left), cis-$C_{25}H_{22}N_2S_2$ (top right) and trans- $C_{25}H_{22}N_2S_2$ (bottom).

(1Z)-17β-ACETOXY-5α-ANDROST-2-ENO[3,2-d][1,2,3]THIADIAZOLE-1-THIONE S-OXIDE
$C_{21}H_{28}N_2O_3S_2$

T.C. BRITTON, T.J. LOBL and C.G. CHIDESTER, 1984. J. Org. Chem., 49, 4773-4780.

Orthorhombic, $P2_12_12_1$, a = 8.143, b = 13.270, c = 18.838 Å, Z = 4. Cu radiation, R = 0.055 for 1647 reflexions (at -150°C).

The analysis reveals a close (2.66 Å) sulfine-O thiadiazole-S distance suggestive of a weakly attractive interaction (Fig. 1). The sulfine group is almost in the plane of the thiadiazole ring and the N1-S2...O1A angle is 169°. There is disorder of the O-C=O atoms O20, C21, O21 of the acetate group at C17.

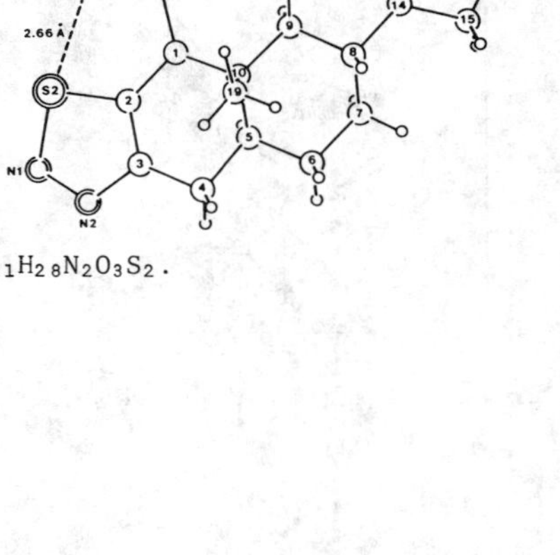

Fig. 1. A view of $C_{21}H_{28}N_2O_3S_2$.

(N-(2-(PIPERIDYLCARBONYLMETHYL)-5-NORBORNENE-3-YL)ACETYL)-4(R)-METHOXYCARBONYL-1,3-THIAZOLIDINE-2-THIONE
$C_{21}H_{28}N_2O_4S_2$

Y. NAGAO, T. INOUE, E. FUJITA, S. TERADA and M. SHIRO, 1984. Tetrahedron, 40, 1215-1223.

Orthorhombic, $P2_12_12_1$, a = 10.528, b = 34.061, c = 6.153 Å, Z = 4. Cu radiation, R = 0.046 for 1847 reflexions.

The structure and absolute configuration were determined to be as shown in Fig. 1.

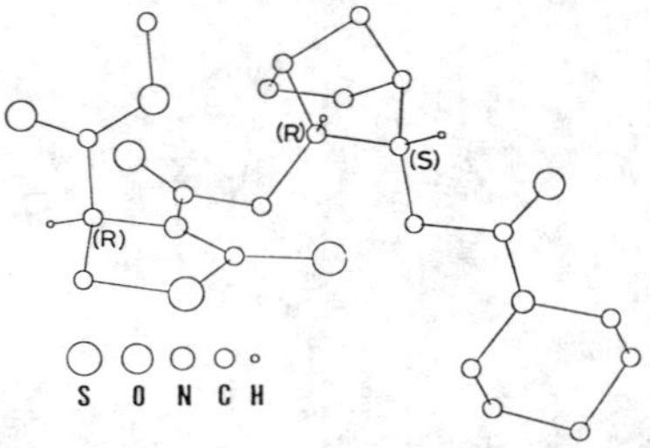

Fig. 1. The $C_{21}H_{28}N_2O_4S_2$ molecule showing the absolute configuration.

(2S,3S)-3-ACETOXY-5-(2-(DIMETHYLAMMONIO-ETHYL))-2-(4-METHOXYPHENYL)-2,3-DIHYDRO-1,5-BENZOTHIAZEPIN-4(5H)-ONE CHLORIDE (DILTHIAZEM HYDROCHLORIDE)
$C_{22}H_{27}ClN_2O_4$

B. KOJIC-PRODIC, Z. RUZIC-TOROS, V. SUNJIC, E. DECORTE and F. MOIMAS, 1984. Helv. Chim. Acta, 67, 916-926.

Orthorhombic, $P2_12_12_1$, a = 42.18, b = 9.079, c = 6.035 Å, Z = 4. Cu radiation, R = 0.052 for 2068 reflexions.

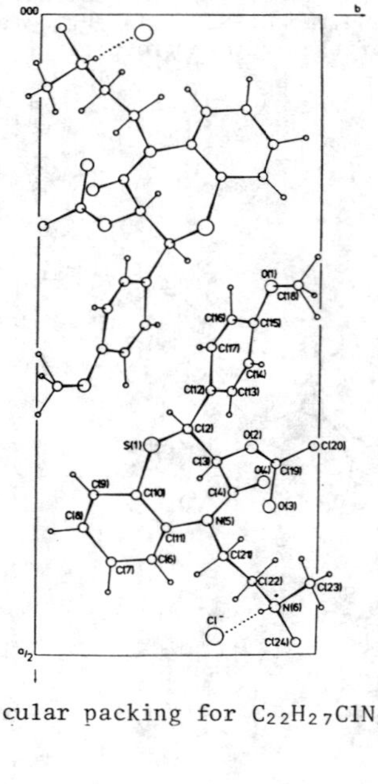

Fig. 1. Part of the molecular packing for $C_{22}H_{27}ClN_2O_4$.

The analysis determines the (2S,3S) absolute configuration and also shows that the seven-membered ring has a slightly twisted boat conformation in the solid state. The chloride ion is hydrogen bonded to N(6) (N...Cl 3.001(5) Å).

2,2,4,4-TETRAMETHYL-6,8-DIPHENYL-3-THIA-7-AZABICYCLO[3.3.1]NONAN-9-ONE
$C_{23}H_{27}NOS$

B.R. BAILEY III, K.D. BERLIN, D.R. POWELL and D. VAN DER HELM, 1984. Phosphorus and Sulfur, 21, 121-133.

Triclinic, P$\bar{1}$, a = 6.216, b = 10.507, c = 15.335 Å, α = 86.64, β = 86.66, γ = 98.51°, Z = 2. Mo radiation, R = 0.048 for 2762 reflexions (at 138 K).

The molecule (Fig. 1) adopts a chair-boat conformation. The gem-dimethyl groups at C(2) and C(4) add strain to the thiane ring. The N atom has pyramidal geometry with its hydrogen equatorial. There is significant flattening of the rings due to intramolecular interactions.

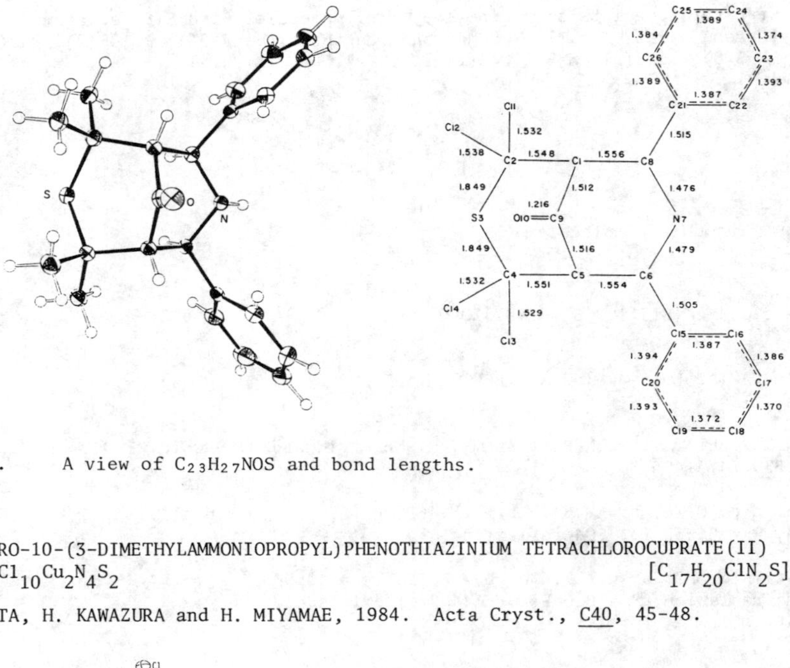

Fig. 1. A view of $C_{23}H_{27}NOS$ and bond lengths.

2-CHLORO-10-(3-DIMETHYLAMMONIOPROPYL)PHENOTHIAZINIUM TETRACHLOROCUPRATE(II)
$C_{34}H_{40}Cl_{10}Cu_2N_4S_2$ $[C_{17}H_{20}ClN_2S]_2[CuCl_4]_2$

A. OBATA, H. KAWAZURA and H. MIYAMAE, 1984. Acta Cryst., C40, 45-48.

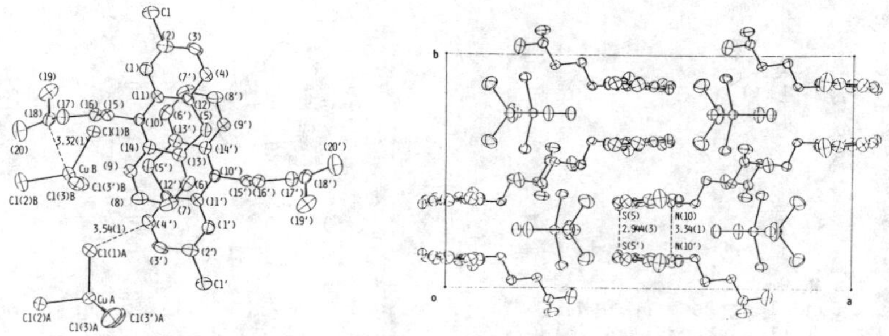

Fig. 1. Views of the overlap of chlorpromazine moieties.

Orthorhombic, Pnma, a = 23.745, b = 13.897, c = 12.580 Å, D_m = 1.622, Z = 4. Mo radiation, R = 0.057 for 2161 reflexions.

In the crystal (Fig. 1), to equilibrate the charge of the $CuCl_4^{2-}$, each chlorpromazine moiety needs a charge of 2+. Protonation of the N atom in the side chain accounts for 1+, and the other positive charge is spread over the phenothiazine ring resulting in the formation of a dimeric pair of chlorpromazines. In the pair of phenothiazine rings, the S...S' distance of 2.944(3) Å is extremely short suggesting some bonding interaction. The dimeric pairs stack one-dimensionally along b.

8-METHYL-1,2,3,4,5,6,7,8-HEPTATHIAPHOSPHOCAN-8-SULPHIDE
CH_3PS_8

T. HEINLEIN and K.-F. TEBBE, 1984. Acta Cryst., C40, 1596-1598.

Monoclinic, $P2_1/c$, a = 11.809, b = 7.957, c = 11.678 Å, β = 106.85°, Z = 4. Mo radiation, R = 0.033 for 1330 reflexions.

The structure consists of crown-like, ring molecules with S(8) equatorial and the methyl group axial (Fig. 1). Some bond lengths are: P-S(1) 2.105(2), P-S(7) 2.102(2), P-S(8) 1.938(2), S-S 2.043(2)-2.057(2), P-C 1.793(5) Å.

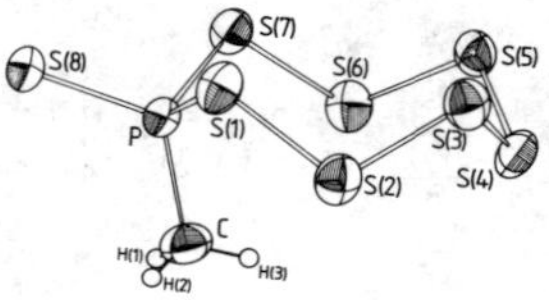

Fig. 1. The CH_3PS_8 molecule.

BENZO[1,2-c.3,4-c']BIS[1.2.5]THIADIAZOLE
$C_6H_2N_4S_2$ (I)

[1.2.5]SELENADIAZOLO[3,4-e]-2,1,3-BENZOTHIADIAZOLE
$C_6H_2N_4SSe$ (II)

A. GIEREN, H. BETZ, T. HUBNER, V. LAMM, R. NEIDLEIN and D. DROSTE, 1984. Z. Naturforsch., B39, 485-496.

I. Monoclinic, C2/c, a = 9.532, b = 11.469, c = 7.030 Å, β = 110.58°, Z = 4. Mo radiation, R = 0.032 for 894 reflexions.

II. Monoclinic, $P2_1/a$, a = 7.650, b = 7.729, c = 12.948 Å, β = 74.26°, D_m = 2.16, Z = 4. Mo radiation, R = 0.045 for 1440 reflexions.

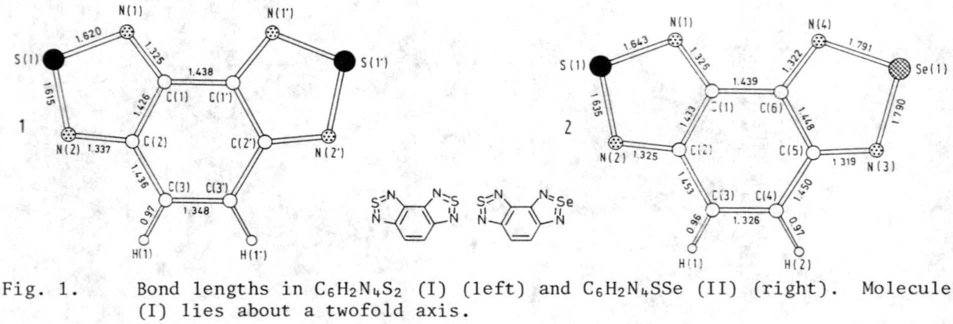

Fig. 1. Bond lengths in $C_6H_2N_4S_2$ (I) (left) and $C_6H_2N_4SSe$ (II) (right). Molecule (I) lies about a twofold axis.

The bond lengths (Fig. 1) of (I) and (II) indicate a quasi-aromatic-π-system in the [1.2.5]thiadiazole and selenadiazole rings, with changes to the aromatic character of the central ring. In the crystal structures, molecular layers and columnar stacks are formed. Quasi-polymer ribbon-like structures are established by short S...N (3.127 Å) and Se...N (2.930, 2.950 Å) contacts.

7,9-OXATHIABICYCLO[4.3.0]NON-1(6)-ENE 9,9-DIOXIDE
$C_7H_{10}O_3S$

E. BLOCK, M. ASLAM, R. IYER and J. HUTCHINSON, 1984. J. Org. Chem., 49, 3664-3666.

Orthorhombic, Pcmn, a = 7.385, b = 7.964, c = 13.419 Å, Z = 4. Mo radiation, R = 0.051 for 643 reflexions.

The molecule (Fig. 1) lies on a crystallographic mirror plane which passes through the atoms of the five-membered ring. Atom C4 of the six-membered ring is disordered over two sites on either side of the mirror plane. Some distances are S-O 1.430(3), S-C1 1.809(5), S-C2 1.727(5), C1-O1 1.414(6), O1-C7 1.373(6), C2-C7 1.330(6) Å.

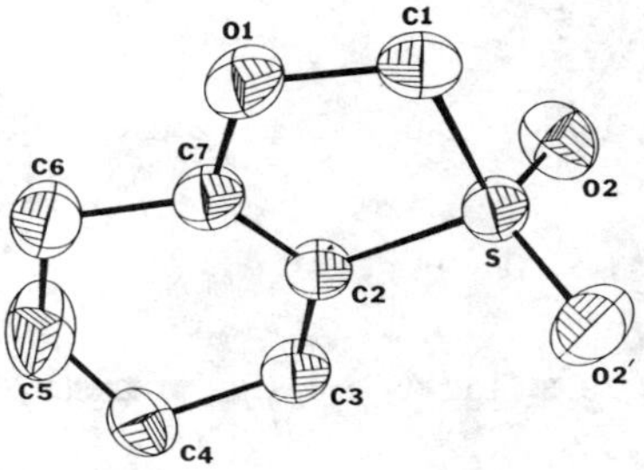

Fig. 1. A perspective view of $C_7H_{10}O_3S$.

4,8-DI-t-BUTYL-1,2,3,5,6,7-HEXASELENA-4,8-DIAZOCINE
$C_8H_{18}N_2Se_6$ (I)

3,5,8,10,13,15-HEXA-t-BUTYL-1,2,4,6,7,9,11,12,14-NONASELENA-3,5,8,10,13,15-HEXAAZA-CYCLOPENTADECANE
$C_{24}H_{54}N_6Se_9$ (II)

H.W. ROESKY, K.-L. WEBER and J.W. BATS, 1984. Chem. Ber., 117, 2686-2692.

I. Tetragonal, $P\bar{4}2_1m$, a = 9.135, c = 10.158 Å, Z = 2. Ag radiation, R = 0.050 for 635 reflexions.

II. Monoclinic, C2/c, a = 9.114, b = 22.541, c = 20.142 Å, β = 100.11°, Z = 4. Ag radiation, R = 0.093 for 1314 reflexions.

Molecule I has mm crystallographic symmetry and the conformation of the eight-membered ring (Fig. 1) is similar to that in sulphur. The 15-membered ring in II has twofold crystallographic symmetry and is stabilized by three short Se...Se contacts (Se...Se 3.355 and 3.386(4) Å).

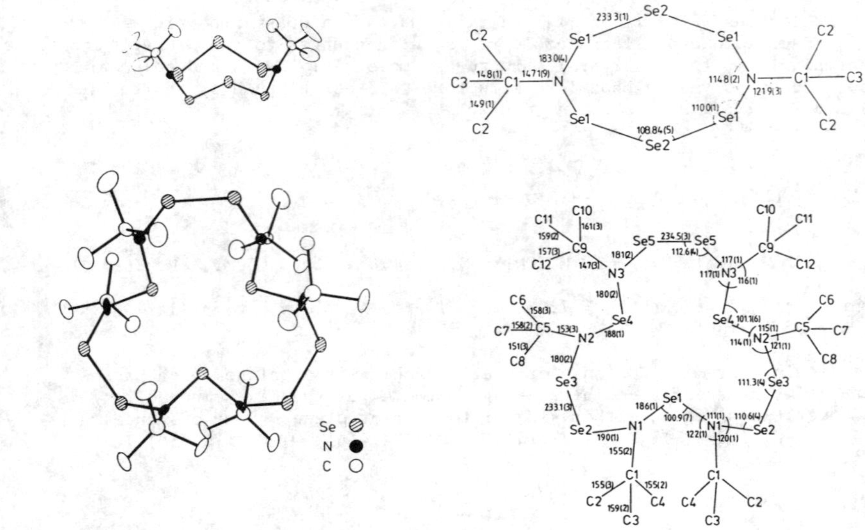

Fig. 1. Views and dimensions of $C_8H_{18}N_2Se_6$ and $C_{24}H_{54}N_6Se_9$.

1,3-DIAZAPHENOXATHIIN
$C_{10}H_6N_2OS$

S. PUIG-TORRES, G.E. MARTIN, S.B. LARSON and S.H. SIMONSEN, 1984. J. Heterocycl. Chem., 21, 995-1000.

Orthorhombic, Pbca, a = 5.754, b = 20.850, c = 14.207 Å, D_m = 1.526, Z = 8. Mo radiation, R = 0.0427 for 1492 reflexions (at -110°C).

The molecule and dimensions are shown in Fig. 1. The dihedral angle between the two wings of the molecule is 165.5(9)°.

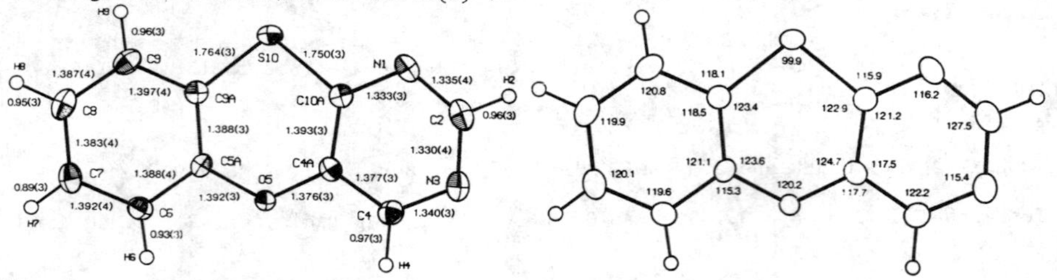

Fig. 1. Bond lengths and angles for 1,3-diazaphenoxathiin.

6,6,9,9-TETRAMETHYL-3,5-DIOXA-4-THIABICYCLO[6.1.0]NONANE-4-OXIDE
$C_{10}H_{18}O_3S$

B. CRAMMER, Z. GOLDSCHMIDT, R. IKAN and S. COHEN, 1984. J. Chem. Soc. Perkin I, 887-891.

Monoclinic, $P2_1/c$, a = 7.385, b = 8.756, c = 18.169 Å, β = 97.63°, Z = 4. Mo radiation, R = 0.041 for 1680 reflexions.

The X-ray analysis established that the macrocycle had a boat-chair conformation with the S=O group trans to the cyclopropane ring and in an equatorial position (Fig. 1). The S atom has pyramidal geometry. Some bond distances are: S-O(1) 1.448(2), S-O(2) 1.602(1), S-O(3) 1.615(1), C(6)-O(3) 1.467(3), C(1)-O(2) 1.493(2), C(3)-C(5) 1.520(2), C(3)-C(4) 1.508(2) and C(4)-C(5) 1.579(3) Å.

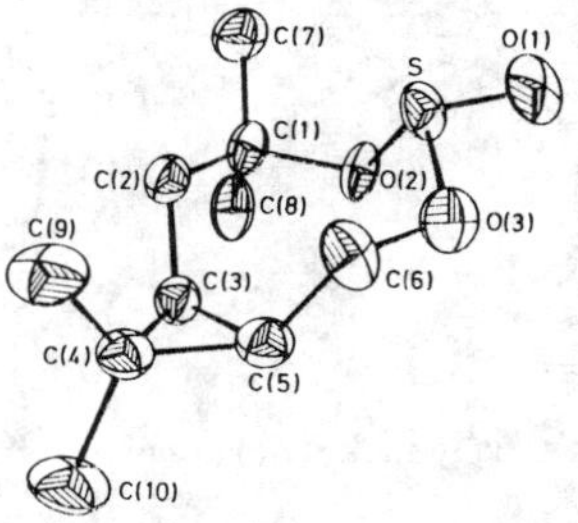

Fig. 1. $C_{10}H_{18}O_3S$: the molecular structure.

2-AZAPHENOXATHIIN 2-OXIDE
$C_{11}H_7NO_2S$

S.R. CALDWELL, J.C. TURLEY, G.E. MARTIN, C.A. DWIGGINS, M.B. HOSSAIN, J.V. MENDENHALL and D. VAN DER HELM, 1984. J. Heterocycl. Chem., 21, 449-454.

Monoclinic, $P2_1/n$, a = 17.50, b = 7.147, c = 16.044 Å, β = 110.60°, Z = 8. Cu radiation, R = 0.045 for 3831 reflexions (at -135°C).

There are two independent molecules (Fig. 1) with a slightly different fold angle between the two aromatic rings (160.8 and 151.3°) which persists even at room temperature.

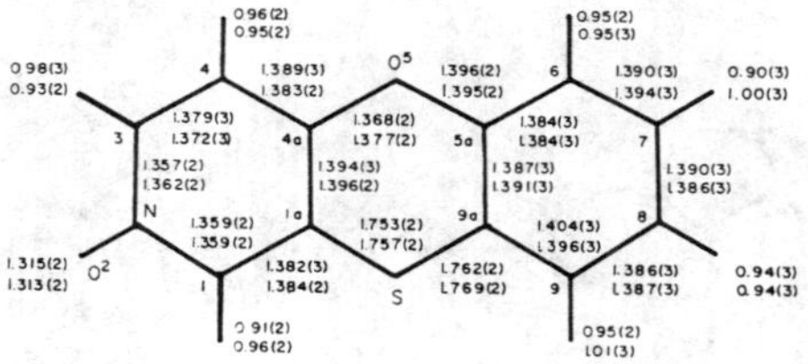

Fig. 1. Bond lengths for the two independent $C_{11}H_7NO_2S$ molecules.

3-BENZOYLIMINO-4-METHYL-1,2,4-OXATHIAZANE
$C_{11}H_{12}N_2O_2S$

C. COHEN-ADDAD, M.S. LEHMANN, P. BECKER, L. PARKANYI and A. KALMAN, 1984. J. Chem. Soc. Perkin II, 191-196.

Monoclinic, $P2_1/c$, a = 7.278, b = 10.978, c = 14.600 Å, β = 114.1°, Z = 4. Mo radiation, neutrons, R = 0.041, 0.037, for 2428, 3611 reflexions (at 122 K).

The central part of the molecule (Fig. 1) is planar and the benzoyl group forms an angle of 6.0(1)° with the S-C(3)-N(7)-C(8)-O(9) plane. There are no significant differences between the results from the X-ray and neutron experiments at 122 K. The S(2)...O(9) distance is 2.245(3) Å. This is shorter than the S...O distance found in 2-(2-chlorobenzoylimino)-1,3-thiazolidine (1). Simple MO theory is used to interpret the shortening of S...O distances in these compounds.

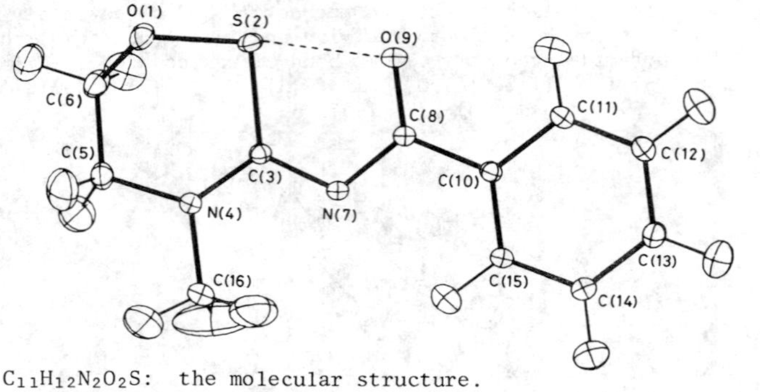

Fig. 1. $C_{11}H_{12}N_2O_2S$: the molecular structure.

1. Structure Reports, 46B, 400.

2-ACETYL-8-p-CHLOROPHENYL-4-METHYL-6-OXA-2,3,7-TRIAZA-1-PHOSPHABICYCLO[3.3.0]OCTA-3,7-DIENE
$C_{12}H_{11}ClN_3O_2P$

L. TOUPET, Y.Y.C. YEUNG LAM KO, F. DE SARLO and R. CARRIÉ, 1984. Acta Cryst., C40, 1034-1036.

Monoclinic, $P2_1/n$, a = 9.482, b = 8.855, c = 16.369 Å, β = 101.3°, Z = 4. Mo radiation, R = 0.047 for 1260 reflexions.

In the molecule (Fig. 1) the diazaphosphole and oxazaphosphole rings are in envelope conformations with a dihedral angle of 105(1)° between their mean planes. The phenyl group is 6.7(6)° out of the oxazaphosphole mean plane.

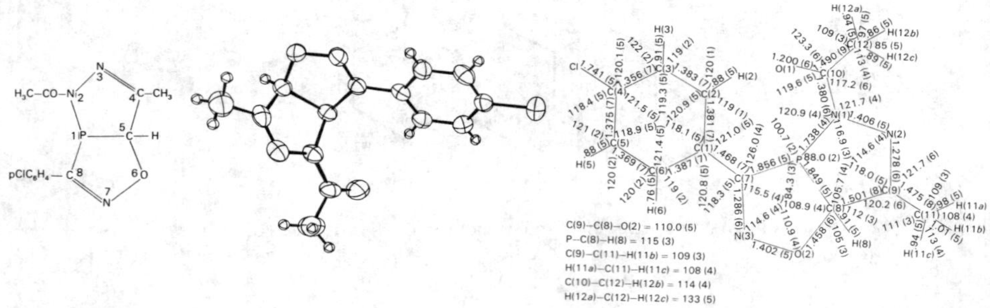

Fig. 1. Perspective view of the $C_{12}H_{11}ClN_3O_2P$ molecule and bond lengths and angles.

6-CHLORO-4-PHENYL-1,2,3-BENZOXATHIAZINE 2,2-DIOXIDE
$C_{13}H_8ClNO_3S$

S. MEHDI and B.R. RAO, 1984. Acta Cryst., C40, 1057-1059.

Orthorhombic, Pbca, a = 8.418, b = 14.708, c = 20.564 Å, D_m = 1.50, Z = 8. Mo radiation, R = 0.060 for 1272 reflexions.

In the molecule (Fig. 1) the oxathiazine ring exhibits a distorted half-chair conformation and the chlorophenyl and benzo groups are each planar, with a dihedral angle between these planes of -150.3(2)°. Bond lengths and angles are in agreement with expected values. There are no intermolecular contacts less than van der Waals distances.

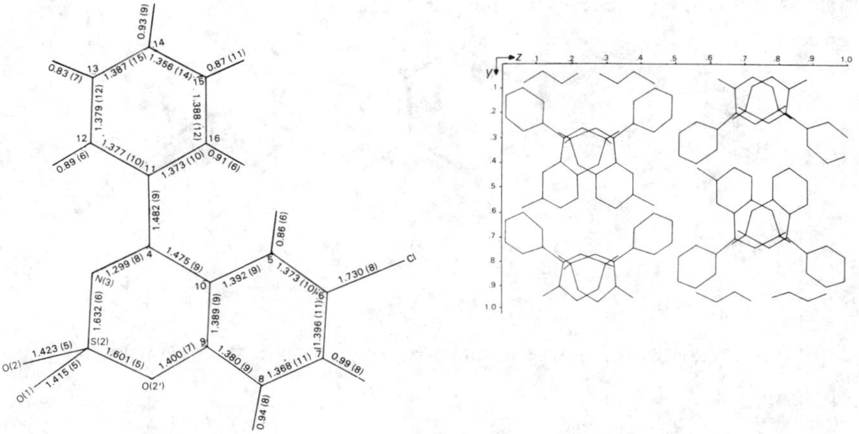

Fig. 1. $C_{13}H_8ClNO_3S$: bond distances in the molecule and packing of the molecule viewed along [100].

6,7,9-TRIMETHYL-4-AZAPHENOXATHIIN
$C_{14}H_{13}NOS$

V.M. LYNCH, S.H. SIMONSEN, G.E. MARTIN, S. PUIG-TORRES and K. SMITH, 1984. Acta Cryst., C40, 1483-1486.

Tetragonal, P4/mnc, a = b = 18.495, c = 6.805, Z = 8. Mo radiation, R = 0.036 for 1356 reflexions (at 163K).

With the exception of six methyl hydrogen atoms, the molecule (Fig. 1) is required by symmetry to be planar. The molecules are stacked perpendicular to c with the phenyl ring of one molecule lying above the pyridine ring of a neighbouring molecule. Some molecular dimensions are, S(10)-C(10a) 1.747(2), C(9a)-S(10) 1.761(2), C(5a)-O(5) 1.394(3), C(4a)-O(5) 1.372(3) Å, C(9a)-S(10)-C(10a) 101.00(10)°.

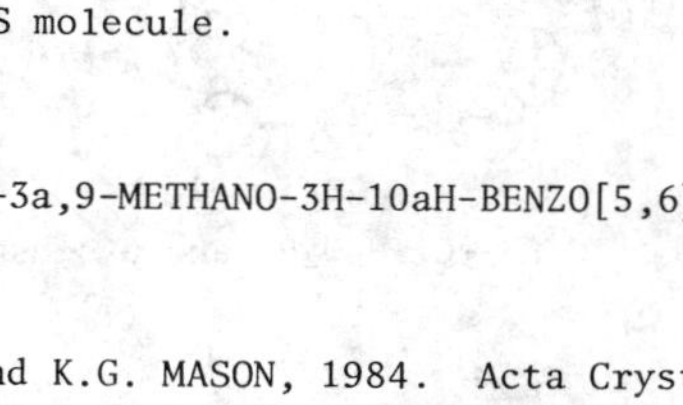

Fig. 1. The $C_{14}H_{13}NOS$ molecule.

9,10-DIHYDRO-10a-METHYL-3a,9-METHANO-3H-10aH-BENZO[5,6]CYCLOHEPT[1,2-d][1,2]OXATHIOL-4(3aH)-ONE 2,2-DIOXIDE
$C_{14}H_{14}O_4S$

D.S. BROWN, H. HEANEY and K.G. MASON, 1984. Acta Cryst., C40, 683-685.

Monoclinic, $P2_1/c$, a = 11.672, b = 7.717, c = 14.489 Å, β = 101.7°, Z = 4. Mo radiation, R = 0.074 for 1467 reflexions.

In the molecule, which is a saturated cyclic sultone (Fig. 1), bond lengths of interest are: mean acyclic S-O 1.417(4), cyclic S-O 1.559(3) and S-C 1.772(7) Å. There are no abnormal intermolecular contact distances.

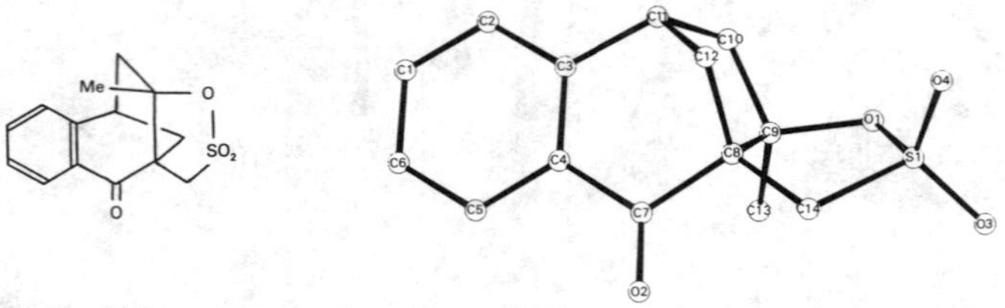

Fig. 1. $C_{14}H_{14}O_4S$: perspective view of the molecule.

trans-2-[BIS(2-CHLOROETHYL)AMINO]-4-p-TOLYL-TETRAHYDRO-2H-1,3,2-OXAZAPHOSPHORINANE-2-OXIDE
$C_{14}H_{21}Cl_2N_2O_2P$

L.-K. LIU, 1984. J. Chin. Chem. Soc., <u>31</u>, 125-129.

Monoclinic, $P2_1/c$, a = 11.636, b = 12.394, c = 11.819 Å, β = 93.9°, Z = 4. Mo radiation, R = 0.063 for 2524 reflexions.

The molecular structure (Fig. 1) clearly reveals a twisted chair-like structure with an equatorially-disposed tolyl substituent. More importantly the structure possesses a trans relationship between the equatorial tolyl and P=O moieties. The cyclophosphamide ring torsion angle O(1)-P-N(1)-C(4) is -36.8°, confirming the twisted chair conformation.

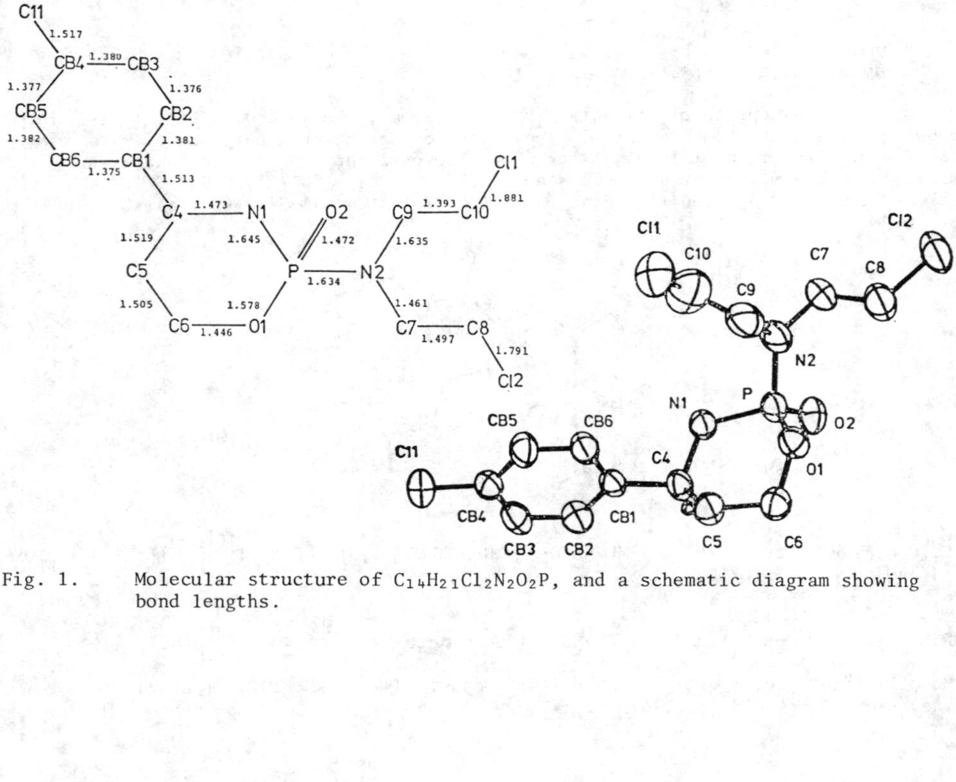

Fig. 1. Molecular structure of $C_{14}H_{21}Cl_2N_2O_2P$, and a schematic diagram showing bond lengths.

2-CHLORO-6-(p-CHLOROPHENYL)-3-(p-TOLYL)-3,4-DIHYDRO-1,3,2-OXAZAPHOSPHORINE 2-OXIDE
$C_{16}H_{14}Cl_2NO_2P$

K.R. ACHARYA, S.S. TAVALE, T.N. GURU ROW, S.D. SAHASRABUDHE and B.D. TILAK, 1984. Acta Cryst., C40, 1080-1082.

Monoclinic, $P2_1/a$, a = 10.141, b = 15.113, c = 10.891 Å, β = 93.22°, D_m = 1.41, Z = 4. Mo radiation, R = 0.067 for 1334 reflexions.

In the molecule (Fig. 1) the 1,3,2-oxazaphosphorine ring takes a half-chair conformation. The exocyclic chlorine and oxygen connected to phosphorus are in axial and equatorial positions respectively. The P-O(2) distance of 1.468(6) Å agrees with values found in other cyclophosphamide, isophosphamide and trophosphamide compounds. The packing of the molecules in the unit cell essentially involves van der Waals contacts.

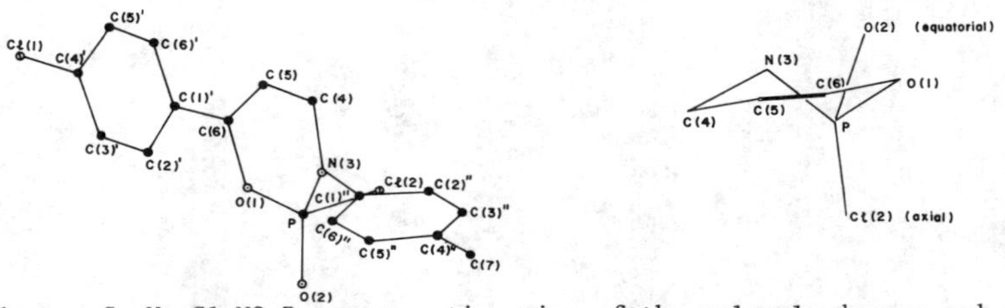

Fig. 1. $C_{16}H_{14}Cl_2NO_2P$: perspective view of the molecule down a and conformation of the oxazaphosphorine ring.

4,4-DIMETHYL-6,6-DIPHENYL-1,3,2-DIOXATHIANE-2-OXIDE
$C_{17}H_{18}O_3S$

G.H. PETIT, P. VAN NUFFEL, C. VAN ALSENOY, A.T.H. LENSTRA and H.J. GEISE, 1984. J. Am. Chem. Soc., 106, 6899-6904.

Monoclinic, C2/c, a = 17.239, b = 9.771, c = 18.390 Å, β = 107.74°, Z = 8. Mo radiation, R = 0.034 for 2147 reflexions (at 110 K).

The hetero ring has a boat-like conformation (Fig. 1) contrary to what was deduced from spectroscopic data. Bond lengths and some torsion angle data are also in Fig. 1. There is also an extensive review of 1,3,2-dioxathiane-2-oxide systems.

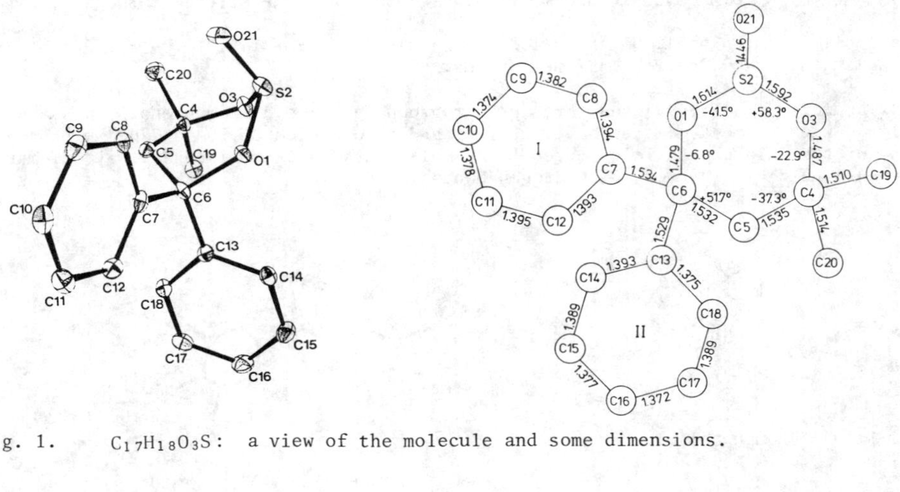

Fig. 1. $C_{17}H_{18}O_3S$: a view of the molecule and some dimensions.

3,4:12,13-DIBENZO-1-METHYL-1-PHOSPHA-2,5,8,11,14-PENTAOXA-1-THIOXOCYCLOTETRADECANE
$C_{17}H_{19}O_5PS$

S.T. MALINOVSKII, Yu.A. SIMONOV, T.N. KUDRYA, A.A. CHAIKOVSKAYA and T.I. MALINOVSKII, 1984. J. Struct. Chem., 25, 454-457 [Zh. Strukt. Khim., 25(3), 130-134].

Monoclinic, $P2_1/n$, a = 12.168, b = 17.256, c = 8.411 Å, γ = 92.39°, Z = 4. Mo radiation, R = 0.052 for 1354 reflexions.

Torsion angles in the 14-membered macrocycle are shown in Fig. 1. The oxygen atoms in positions 2, 5, 11 and 14 are coplanar (plane A) within 0.25 Å. The planes of the aromatic rings form dihedral angles of 15.5 and 70.5° with plane A. On the whole, the conformation of the macrocyclic ring can be represented as a chair. The O position 8, the P atom, and its substituents all lie on the same side of plane A. Bond lengths: P-S 1.916(3), P-O 1.589(7), P-C 1.786(9), mean $C(sp^3)$-$C(sp^3)$ 1.49(1), mean O-$C(sp^3)$ 1.42(1), and mean O-$C(sp^2)$ 1.38(1) Å.

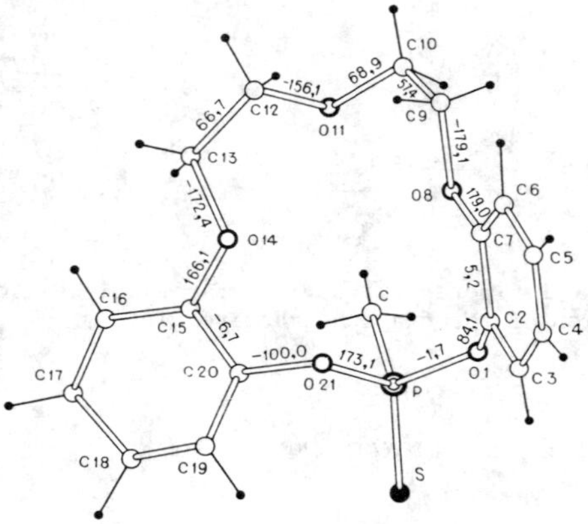

Fig. 1. The 3,4:12,13-Dibenzo-1-methyl-1-phospha-2,5,8,11,14-pentaoxa-1-thioxo-cyclotetradecane molecule.

3,3',5,5'-TETRABROMOPHENOLSULPHONPHTHALEIN (BROMPHENOL BLUE)
$C_{19}H_{10}Br_4O_5S$

B. VEERAPANDIAN, D.M. SALUNKE and M. VIJAYAN, 1984. Acta Cryst., C40, 500-502.

Monoclinic, C2/c, a = 31.003, b = 11.037, c = 21.183 Å, β = 143.7°, Dm = 2.06, Z = 8. Mo radiation, R = 0.061 for 1697 reflexions.

The molecule (Fig. 1) has three planar groupings; the dibromophenol rings are inclined with respect to each other at 73°, and make angles of 85° and 68° with the plane of the sulphonphthalein system. The molecules aggregate into helical columns which are held by a network of hydrogen bonds.

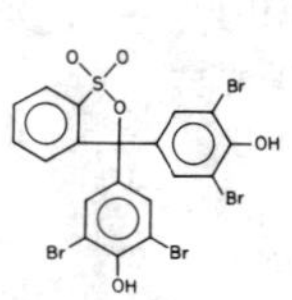

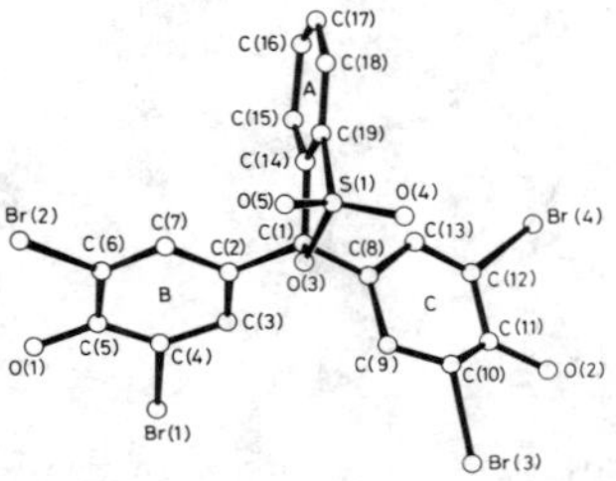

Fig. 1. Bromphenol blue.

1-METHYLISOGUANINE DIHYDRATE
$C_6H_7N_5O,2H_2O$

R.Y. WONG and R.J. NACHMAN, 1984. Acta Cryst., C40, 1409-1412.

Triclinic, $P\bar{1}$, a = 4.110, b = 9.454, c = 12.303 Å, α = 74.09, β = 84.52, γ = 79.96°, Z = 2. Cu radiation, R = 0.045 for 1043 reflexions.

1-methylisoguanine molecules in the N(12) amino form are linked across centres of symmetry by pairs of hydrogen bonds into infinite chains (Fig. 1). Each molecule is further hydrogen bonded to two different water molecules through the carbonyl and amino substituents on the pyrimidine ring. There is also hydrogen bonding between the water molecules. The purine ring is almost planar (maximum displacement from the mean plane is 0.017(5) Å) with O(10) displaced 0.054(6) Å from the ring plane. The endocyclic C-N bond lengths are in the range 1.301-1.430 Å (mean 1.36 Å).

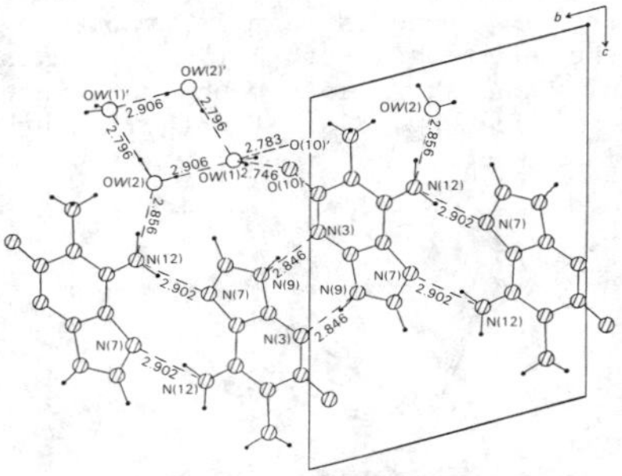

Fig. 1. Hydrogen bonding in a single layer of the 1-methylisoguanine dihydrate structure.

4,6-DIMETHYL-2-PYRIMIDONE HYDROCHLORIDE
$C_6H_9ClN_2O$ (I)

5,6-DIMETHYL-2-PYRIMIDONE HYDROCHLORIDE
$C_6H_9ClN_2O$ (II)

4,5,6-TRIMETHYL-2-PYRIMIDONE HYDROCHLORIDE
$C_7H_{11}ClN_2O$ (III)

1,4,6-TRIMETHYL-2-PYRIMIDONE HYDROCHLORIDE
$C_7H_{11}ClN_2O$ (IV)

1,5,6-TRIMETHYL-2-PYRIMIDONE HYDROCHLORIDE
$C_7H_{11}ClN_2O$ (V)

1,4,5,6-TETRAMETHYL-2-PYRIMIDONE HYDROCHLORIDE
$C_8H_{13}ClN_2O$ (VI)

T.W.S. LEE, S.J. RETTIG, R. STEWART and J. TROTTER, 1984. Can. J. Chem., 62, 1194-1202.

I. Orthorhombic, Pnma, a = 14.2523, b = 6.6674, c = 8.0192 Å, Z = 4. Mo radiation, R = 0.043 for 807 reflexions.

II. Monoclinic, $P2_1/c$, a = 6.7175, b = 15.8743, c = 7.3013 Å, β = 102.038°, Z = 4. Mo radiation, R = 0.031 for 1375 reflexions.

III. Orthorhombic, Pbcn, a = 11.4592, b = 14.593, c = 10.5894 Å, Z = 8. Mo radiation, R = 0.033 for 1103 reflexions.

IV. Monoclinic, $P2_1/n$, a = 6.8243, b = 7.2882, c = 16.7938 Å, β = 96.709°, Z = 4. Mo radiation, R = 0.033 for 1627 reflexions.

V. Monoclinic, $P2_1/c$, a = 10.4250, b = 8.3646, c = 10.2352 Å, β = 107.190°, Z = 4. Mo radiation, R = 0.031 for 1603 reflexions.

VI. Orthorhombic, $P2_12_12_1$, a = 6.2619, b = 9.7646, c = 15.6205 Å, Z = 4. Mo radiation, R = 0.026 for 941 reflexions.

The molecular geometry of the salts (Fig. 1) has been determined by X-ray crystallography with the view to determining whether the presence of strain in the pyrimidonium ring can account for the activating effect of adjacent methyl. The most reactive compound, the 1,5,6-trimethylpyrimidonium chloride (V), has a geometry that is consistent with high reactivity, viz. a non-planar ring and short H...H intermethyl distance. However, the 1,4,5,6-tetramethyl compound (VI), which also is non-planar and has an extremely short 5...6 intermethyl distance (1.99 Å), is not highly reactive; that is, the heightened strain is unable to overcome the deactivating inductive effect of the additional methyl group. It is concluded that deviations from ring planarity and short intermethyl distances are insufficient to account for the activation produced by adjacent methyl though they appear to be contributing factors to this effect.

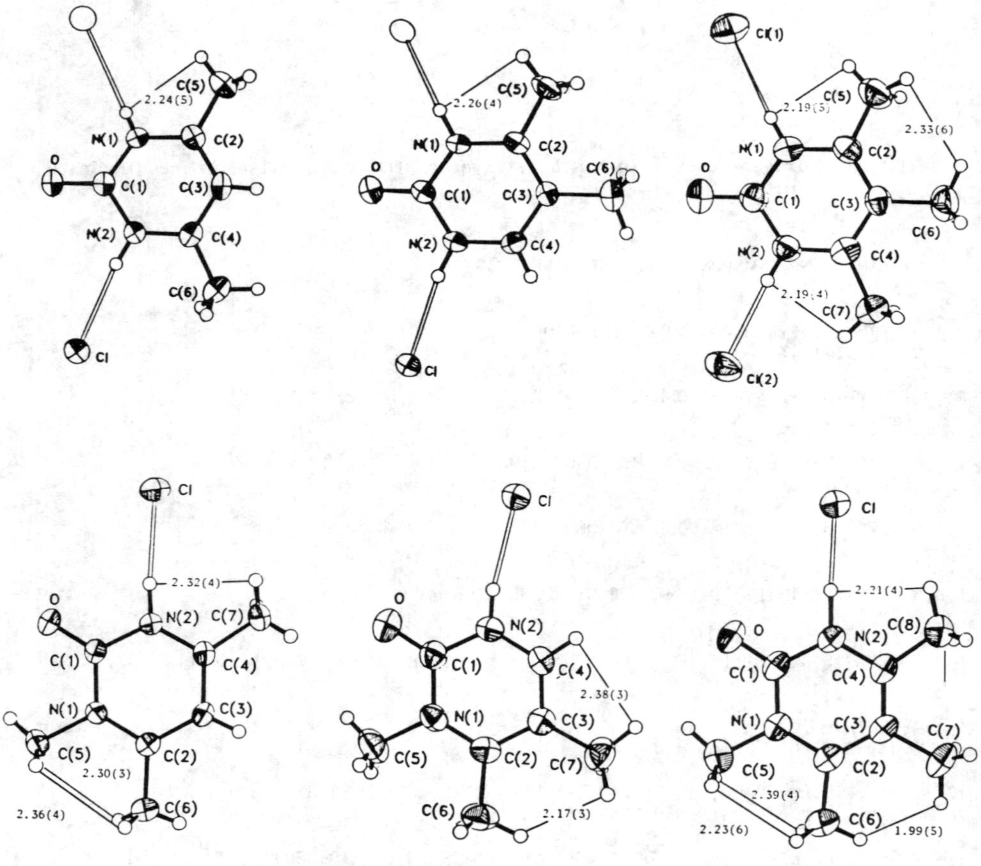

Fig. 1. Views of the pyrimidone salts I, II, III (top) and IV, V, VI (bottom).

1,2-DIHYDRO-1,3-DIMETHYL-2-OXO-PYRIMIDINIUM HYDROGEN SULPHATE
$C_6H_{10}N_2O_5S$ (I) $(C_6H_9N_2O)^+.(HSO_4^-)$

1,2-DIHYDRO-1,3-DIMETHYL-2-THIOXO-PYRIMIDINIUM PERCHLORATE
$C_6H_9ClN_2O_4S$ (II) $(C_6H_9N_2S)^+.(ClO_4^-)$

G. FERGUSON, B. KAITNER, D. LLOYD and H. McNAB, 1984. J. Chem. Res., (S) 184-185, (M) 1760-1790.

I. Monoclinic, $P2_1/n$, a = 14.411, b = 6.196, c = 10.523 Å, β = 94.93°, Z = 4. Mo radiation, R = 0.068 for 1622 reflexions.

II. Monoclinic, $P2_1/c$, a = 6.306, b = 14.192, c = 11.831 Å, β = 105.84°, Z = 4. Mo radiation, R = 0.065 for 1332 reflexions.

The molecular dimensions suggest that the electrons from the N atoms are delocalized largely into the vinamidinium system rather than into the oxo or thioxo groups. The six-membered ring of the oxodihydropyrimidinium cation of (I) is essentially planar, whereas that of the thioxodihydropyrimidinium cation (II) (Fig. 1) is deformed towards a boat conformation, with C(2) -0.047 and C(5) -0.023 Å from the N(1),N(3),C(4),C(6) plane. These small displacements for the cation of (II) correspond to flap angles of 3.6 and 1.9° at C(2) and C(5) respectively. The sulphur bonded to C(2) is -0.164 Å from the four-atom plane; this displacement presumably serves to minimize interaction with the vicinal methyl groups.

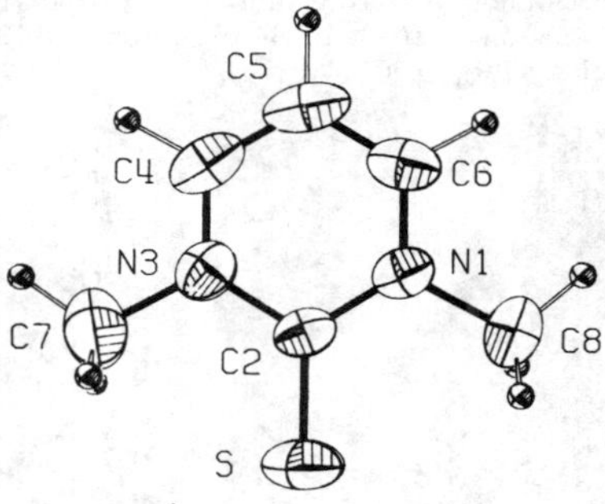

Fig. 1. A view of the $(C_6H_9N_2S)^+$ cation.

1-(2-CARBAMOYLETHYL)URACIL
$C_7H_9N_3O_3$

S. FUJITA, A. TAKENAKA and Y. SASADA, 1984. Acta Cryst., C40, 1605-1607.

Triclinic, $P\bar{1}$, a = 8.869, b = 11.318, c = 4.444 Å, α = 105.59, β = 99.75, γ = 103.87°, Dm = 1.51, Z = 2. Cu radiation, R = 0.040 for 1224 reflexions.

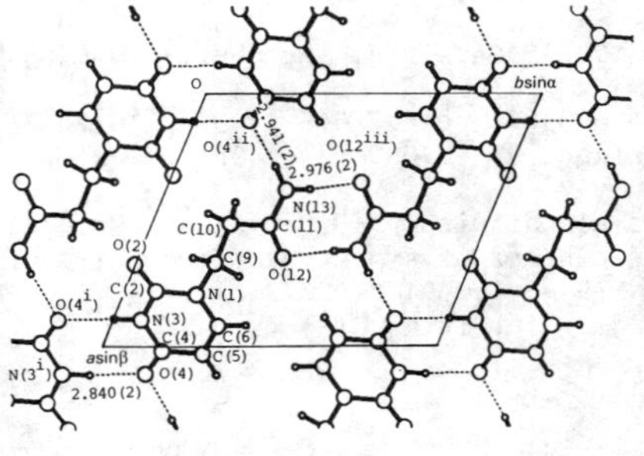

Fig. 1. The structure of $C_7H_9N_3O_3$ viewed down c.

The uracil moieties are united into pairs and thence into chains by pairs of N-H...O hydrogen bonds across inversion centres. Further hydrogen bonds cross link these chains (Fig. 1). The pyrimidine ring is planar within ±0.02 Å.

N(6),N(6)-DIMETHYLADENINE HYDROCHLORIDE MONOHYDRATE
$C_7H_{10}ClN_5,H_2O$ (I) $C_7H_{10}N_5{}^+Cl^-,H_2O$

N(6),N(6)-DIMETHYLADENINE DIHYDROCHLORIDE
$C_7H_{11}Cl_2N_5$ (II) $C_7H_{11}N_5{}^{2+}2Cl^-$

T. DAHL, 1984. Acta Chem. Scand., A38, 485-489.

I. Monoclinic, $P2_1/c$, a = 9.661, b = 14.307, c = 7.620 Å, β = 103.79°, D_m = 1.44, Z = 4. Cu radiation, R = 0.067 for 1434 reflexions.

II. Monoclinic, $P2_1/c$, a = 7.445, b = 19.675, c = 7.090 Å, β = 98.07°, D_m = 1.54, Z = 4. Cu radiation, R = 0.034 for 1778 reflexions.

The protonation of N(6),N(6)-dimethyladenine is at N(1) in I and at N(1) and N(7) in II, giving a mono- and a diprotonated ion with the H atoms bonded to the rings as observed for other adenine derivatives (Fig. 1). One part of each cation is planar: in I the planar part includes both rings apart from C(6), and for II the imidazole ring and C(6). In both compounds the ions and molecules are linked by hydrogen bonds in a nearly planar network, along (102) in I and along (10$\bar{2}$) in II. These planes are held together through stacks along [001].

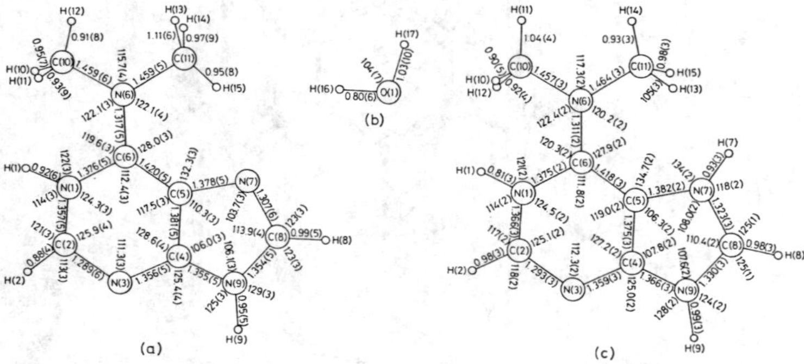

Fig. 1. Bond distances (Å) and angles (°) in $C_7H_{10}ClN_5$ (a) and the water molecule (b) in $C_7H_{10}ClN_5,H_2O$ and in $C_7H_{11}Cl_2N_5$ (c).

3-(1-CYTOSINYL)PROPIONAMIDE
$C_7H_{10}N_4O_2$

S. FUJITA, A. TAKENAKA and Y. SASADA, 1984. Acta Cryst., C40, 1467-1469.

Triclinic, $P\bar{1}$, a = 8.491, b = 8.906, c = 7.065 Å, α = 110.26, β = 95.33, γ = 118.15°, D_m = 1.45, Z = 2. Mo radiation, R = 0.059 for 1495 reflexions.

The pyrimidine ring is planar within ±0.019(3) Å. Molecular geometry is normal. Molecules are linked across centres of symmetry by two pairs of hydrogen bonds; N(4)-H...N(3) 2.988(3) and N(9)-H...O(9) 2.923(3) Å (Fig. 1). Further cross-linking is achieved by hydrogen bonding involving O(2).

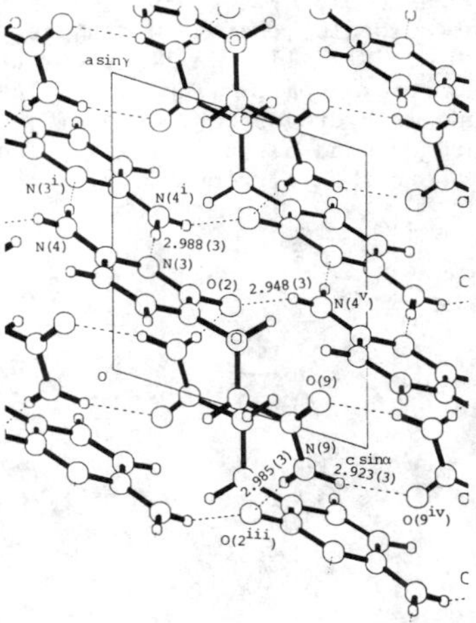

Fig. 1. Hydrogen bonding in the crystal structure of 3-(1-cytosinyl)propionamide.

2-ETHYLSULPHONYL-7-METHYL-5H-1,3,4-THIADIAZOLO[3,2-a]PYRIMIDIN-5-ONE
$C_8H_9N_3O_3S_2$

M. SUIKO, S. NAKATSU, T. KIYOSE and K. IMADA, 1984. Acta Cryst., C40, 1078-1080.

Orthorhombic, Pcab, a = 8.362, b = 11.368, c = 22.776 Å, D_m = 1.592, Z = 8. Mo radiation, R = 0.053 for 933 reflexions.

In the molecule (Fig. 1) the thiadiazole and pyrimidinone portions are both essentially planar and the sulphone part takes a staggered conformation around the fairly long S-C(aromatic) bond of 1.796(5) Å.

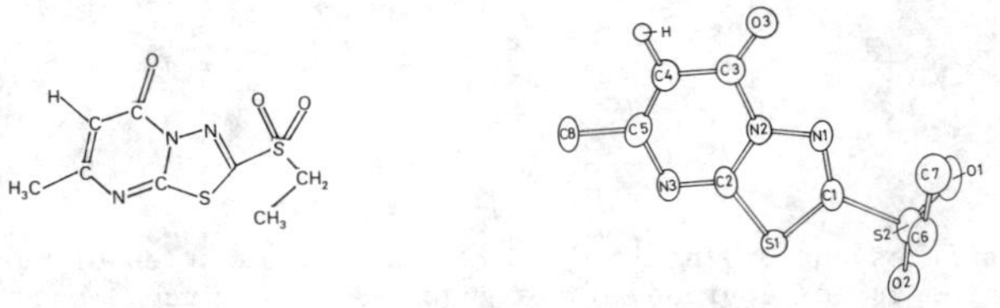

Fig. 1. $C_8H_9N_3O_3S_2$: structural formula and perspective view of molecule.

9-(2-CARBOXYETHYL)GUANINE
$C_8H_9N_5O_3$ (I)

9-(2-CARBOXYETHYL)GUANINE : 1-METHYLCYTOSINE COMPLEX
$C_{13}H_{16}N_8O_4$ (II) $C_8H_9N_5O_3$: $C_5H_7N_3O$

S. FUJITA, A. TAKENAKA and Y. SASADA, 1984. Bull. Chem. Soc. Japan, 57, 1707-1712.

I. Monoclinic, $P2_1/c$, a = 12.221, b = 11.022, c = 6.961 Å, β = 100.22°, D_m = 1.56, Z = 4. Cu radiation, R = 0.074 for 1126 reflexions.

II. Triclinic, $P\bar{1}$, a = 8.932, b = 12.370, c = 7.139 Å, α = 85.01, β = 99.21, γ = 84.64°, D_m = 1.49, Z = 2. Cu radiation, R = 0.044 for 1979 reflexions.

In I, the carboxyl group is double hydrogen-bonded with N(2) and N(3) of the guanine moiety of the adjacent molecule, the N...O distances being 2.939 Å for N(2)-H...O and 2.660 Å for N(3)...H-O (Fig. 1). In II, the guanine moiety and 1-methylcytosine form a Watson-Crick base pair. The carboxyl group takes an O=C-OH antiperiplanar conformation and is hydrogen-bonded to N(7) of guanine in the adjacent base pair. This hydrogen bond distance is very short, 2.634 Å through the unfavourable conformation of the carboxyl group.

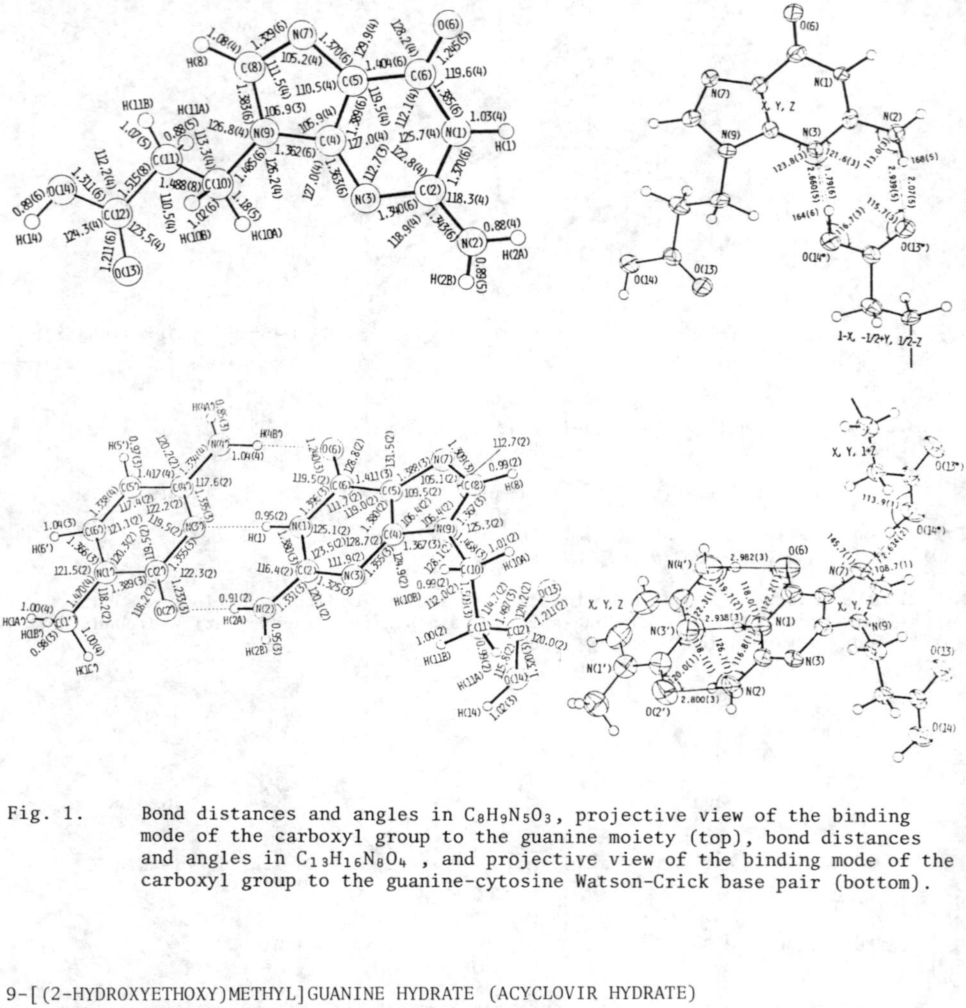

Fig. 1. Bond distances and angles in $C_8H_9N_5O_3$, projective view of the binding mode of the carboxyl group to the guanine moiety (top), bond distances and angles in $C_{13}H_{16}N_8O_4$, and projective view of the binding mode of the carboxyl group to the guanine-cytosine Watson-Crick base pair (bottom).

9-[(2-HYDROXYETHOXY)METHYL]GUANINE HYDRATE (ACYCLOVIR HYDRATE)
$C_8H_{11}N_5O_3$,0.67(H_2O)

G.I. BIRNBAUM, M. CYGLER and D. SHUGAR, 1984. Can. J. Chem., 62, 2646-2652.

Monoclinic, $P2_1/n$, a = 25.459, b = 11.282, c = 10.768 Å, β = 95.16°, Z = 12. Cu radiation, R = 0.053 for 3970 reflexions.

The asymmetric unit has three acyclovir and two water molecules. In two of the molecules the side chain is partially folded, while in the third one it is fully extended (Fig. 1). The glycosidic torsion angles are in the range 91.4-104.3°. There is an extensive hydrogen bond network in the crystal lattice. The conformational features are compared with those in other known acyclonucleosides. They are also examined in relation to the behavior of acyclonucleosides and acyclonucleotides in various enzymatic systems, including those related to antiviral activities.

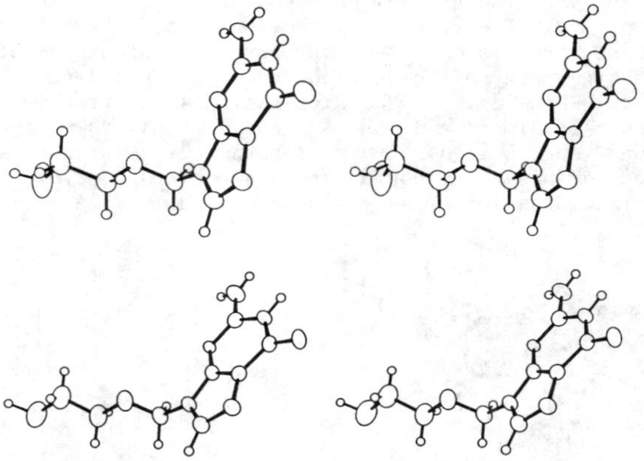

Fig. 1. Stereoviews of the two acyclovir conformers.

9-ETHYL-7-METHYLGUANINIUM IODIDE DIMETHYL SULFOXIDE SOLVATE
$C_8H_{12}IN_5O,C_2H_6OS$ $C_8H_{12}N_5O^+.I^-,C_2H_6OS$

T. ISHIDA, M. KATSUTA and M. INOUE, 1984. Acta Cryst., C40, 439-441.

Monoclinic, $P2_1/c$, a = 10.266, b = 6.976, c = 22.059 Å, β = 97.66°, D_m = 1.734, Z = 4. Cu radiation, R = 0.085 for 2659 reflexions.

The molecule (Fig. 1) has the structural features expected for an N(7) protonated guanine (N(7)-C(8) 1.31(1), C(8)-N(9) 1.33(1) Å). The crystal structure has extensive overlapping of centrosymmetrically related guaninium rings.

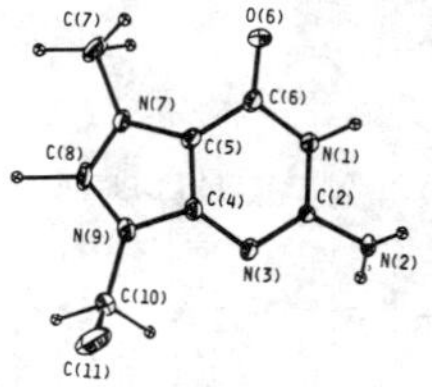

Fig. 1. A view of the 9-ethyl-7-methylguaninium cation.

DIPOTASSIUM cis-syn-THYMINE PHOTODIMER MONOHYDRATE
$C_{10}H_{10}K_2N_4O_4,H_2O$ (I)

SODIUM cis-syn-THYMINE PHOTODIMER MONOHYDRATE
$C_{10}H_{11}N_4NaO_4,H_2O$ (II)

C.H. WEI and J.R. EINSTEIN, 1984. Acta Cryst., B40, 271-279.

I. Monoclinic, Cc, a = 7.007, b = 25.97, c = 8.025 Å, β = 99.15°, Z = 4. Cu radiation, R = 0.101 for 523 reflexions.

II. Monoclinic, $P2_1/n$, a = 6.555, b = 29.010, c = 6.762 Å, β = 106.028°, D_m = 1.53, Z = 4. Cu radiation, R = 0.051 for 1006 reflexions.

The successful analysis of the structure was accomplished with the assumption that the crystal consists of the dimers intermingled with the monomer products of their X-ray cleavage and that the gradual photo-monomerization takes place as a linear function of exposure. The results of the analysis, in which monomers were treated as rigid groups, revealed detailed configurational features of the dimer (Fig. 1), including the geometry of the cyclobutane ring, coordination around the Na^+ ion, and intermolecular H-bonding. In the dipotassium salt structure (Fig. 1) although the coexistence of dimers and monomers was strongly suspected because of unusually long C-C bond lengths across the cyclobutane ring in what is evidently an average structure, monomer positions could not be located. However, the configuration of the cyclobutadipyrimidinate ion closely resembles that in the sodium-salt crystal.

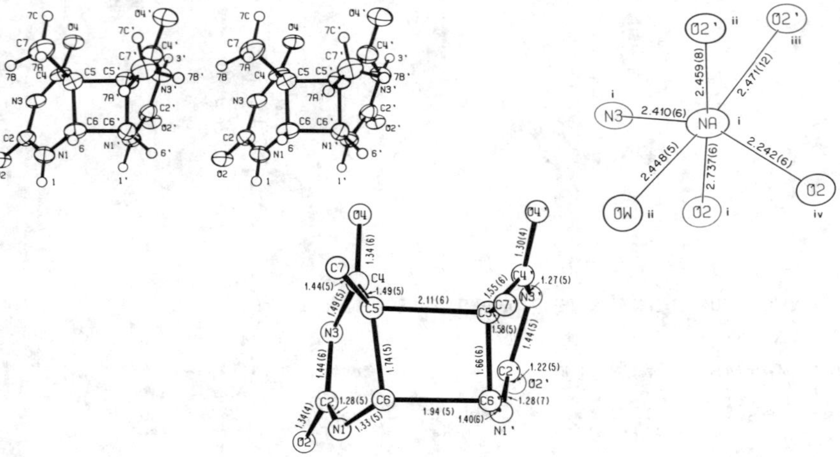

Fig. 1. A stereoview of the cyclobutadipyrimidinate ion in the sodium salt (II), the sodium ion coordination in (II) and (lower) the structure of the cyclobutadipyrimidinate ion in the dipotassium salt (I).

7-(5-HYDROXY-3-OXO-2,3-DIHYDRO-4-PYRAZOLYL)THEOPHYLLINE MONOHYDRATE

$C_{10}H_{10}N_6O_4,H_2O$

E. GAVUZZO, F. MAZZA, A. CAROTTI and G. CASINI, 1984. Acta Cryst., C40, 1231-1233.

Orthorhombic, $Pca2_1$, a = 19.049, b = 8.307, c = 7.869 Å, D_m = 1.58, Z = 4. Mo radiation, R = 0.058 for 1431 reflexions.

In the molecule (Fig. 1) the monocyclic nucleus is in the enolic form and the lengthening of its carbonyl bond may be favoured by its participation in a strong intermolecular hydrogen bond with the enolic group. The packing is governed by a network of hydrogen bonds in which, apart from the water molecule, all the acceptor and donor sites of the pyrazoline ring take part and by a short contact between methyl and carbonyl groups belonging to different molecules translated along the b direction.

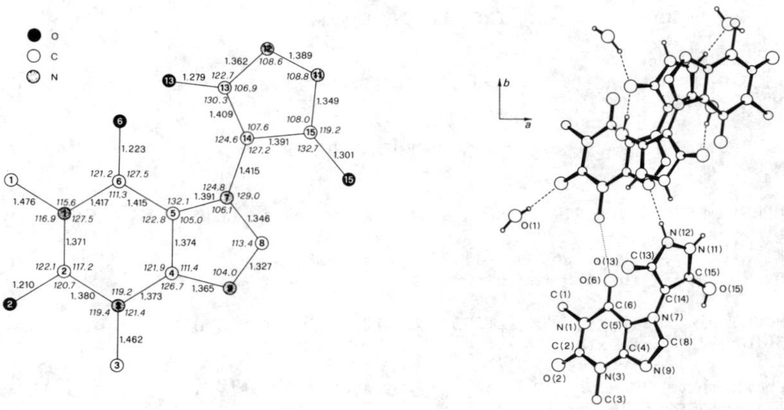

Fig. 1. $C_{10}H_{10}N_6O_4,H_2O$: bond lengths and angles in the molecule and a view along c of the molecular conformation of the compound showing the main interactions present in the crystal; dashed lines represent hydrogen bonds and the dotted line the methyl-carbonyl short contact.

7-(4,6-DIOXO-5-PYRIMIDINYL)THEOPHYLLINE DIHYDRATE
$C_{11}H_{10}N_6O_4,2H_2O$

E. GAVUZZO, F. MAZZA, A. TAMBURRINI, G. CASINI and A. CAROTTI, 1984. Acta Cryst., C40, 856-858.

Monoclinic, $P2_1/n$, a = 9.633, b = 12.794, c = 11.920 Å, β = 101.0°, Dm = 1.485, Z = 4. Mo radiation, R = 0.045 for 2894 reflexions.

In the molecule the planar dioxopyrimidine residue has a structure with a significant contribution of a mesomeric zwitterionic form; there is non-crystallographic mirror symmetry perpendicular to this ring through C(12) and C(15) (Fig. 1). The dihedral angle between the imidazole and pyrimidine rings is 67.3(6)°. The packing is governed by hydrogen bonding involving the CO and NH groups of the dioxopyrimidine moiety and the water molecules in the crystal.

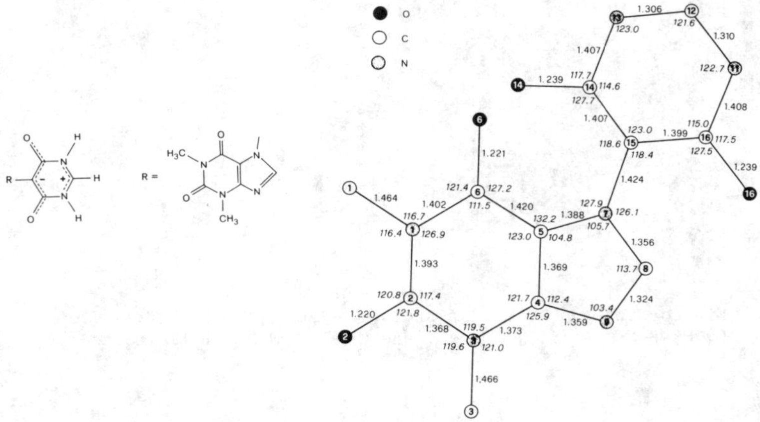

Fig. 1. $C_{11}H_{10}N_6O_4,2H_2O$: structural formula and bond lengths and angles.

5-FLUORO-1,3-DIMETHYLURACIL anti CYCLOBUTENE DIMER
$C_{12}H_{13}FN_4O_4$ (I)

5-FLUORO-1,3-DIMETHYLURACIL trans-anti CYCLOBUTANE DIMER
$C_{12}H_{14}F_2N_4O_4$ (II)

5-FLUORO-1,3-DIMETHYLURACIL cis-syn CYCLOBUTANE DIMER
$C_{12}H_{14}F_2N_4O_4$ (III)

J.L. FLIPPEN-ANDERSON and R. GILARDI, 1984. Acta Cryst., C40, 1957-1960.

I. Monoclinic, Pc, a = 8.456, b = 11.229, c = 7.386 Å, β = 112.55°, Z = 2. Cu radiation, R = 0.074 for 675 reflexions.

II. Orthorhombic, $P2_12_12_1$, a = 12.259, b = 12.842, c = 8.576 Å, Z = 4. Cu radiation, R = 0.053 for 1254 reflexions.

III. Orthorhombic, Pbca, a = 13.962, b = 25.418, c = 7.815 Å, Z = 8. Cu radiation, R = 0.061 for 1379 reflexions.

In I (Fig. 1) the double bond is delocalised over neighbouring C and N atoms. There is evidence of strain in the four-membered ring (e.g. a long C-C bond (1.60 Å), small internal angles (84.3 and 85.4°) and large external angles at the non-fluorinated corners (121.4, 141.9, 137.7°). There are several short C...O contacts in this structure (2.99-3.14 Å). In II (Fig. 1) the cyclobutyl ring is planar. The six-membered rings are planar to within 0.25 Å and are parallel with an approximate centre of symmetry at the middle of the cyclobutyl ring. In III (Fig. 1) the fluorine-substituted C-C bond is 0.017(7) Å shorter than the unsubstituted one with F...F 2.629(7) Å. The four-membered ring is twisted and the six-membered rings show significant deviations from planarity.

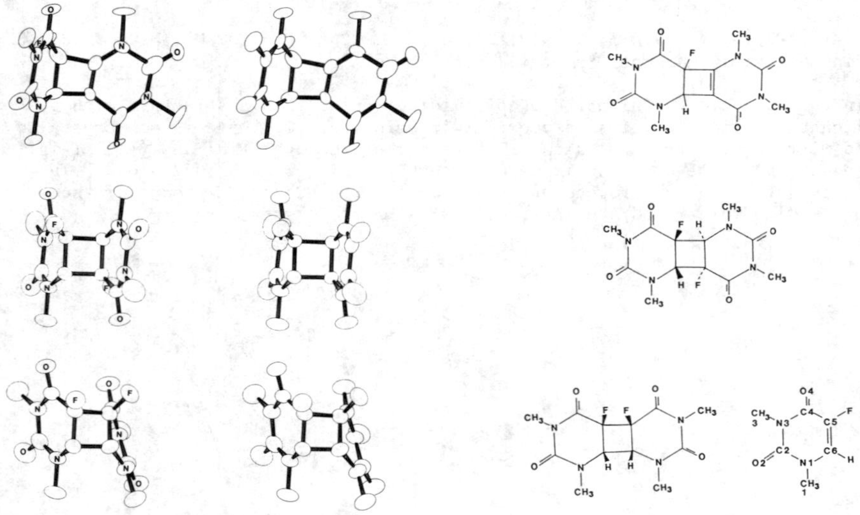

Fig. 1. Stereoviews and molecular skeletons for $C_{12}H_{13}FN_4O_4$ (I) (top), $C_{12}H_{14}F_2N_4O_4$ (II) (centre), and $C_{12}H_{14}F_2N_4O_4$ (III) (bottom).

3-(ADENIN-9-YL)-N-(2-SUCCINIMIDYL)PROPIONAMIDE
$C_{12}H_{13}N_7O_3$

M. TAKIMOTO, A. TAKENAKA and Y. SASADA, 1984. Bull. Chem. Soc. Japan, 57, 3070-3073.

Monoclinic, $P2_1/c$, a = 10.861, b = 12.766, c = 9.753 Å, β = 91.10°, D_m = 1.49, Z = 4. Cu radiation, R = 0.059 for 1875 reflexions.

The adenine moiety is planar, with a maximum displacement of 0.016 Å for N(9) from the best plane (Fig. 1). The succinimide moiety has a half-chair conformation and the dihedral angle between two amide planes (C(19),O(19),N(18)) and (C(17),O(17), N(18)) is 5.7°. The ureide moiety is hydrogen-bonded to adenine with NH...N(1) (2.903 Å) and O...HN(6) (2.955 Å) hydrogen bonds.

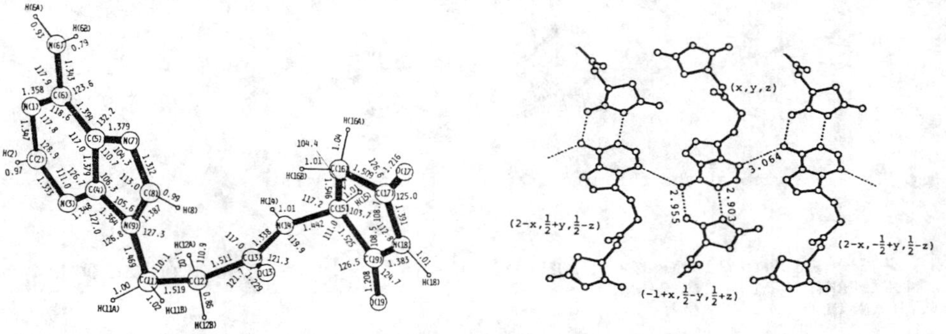

Fig. 1. $C_{12}H_{13}N_7O_3$: atomic numbering and bond lengths (Å) and angles (°), and hydrogen bonds in the molecular sheet.

4-AMINO-N-(4,6-DIMETHYL-2-PYRIMIDINYL)BENZENESULPHONAMIDE (SULFADIMIDINE)
$C_{12}H_{14}N_4O_2S$

R.K. TIWARI, M. HARIDAS and T.P. SINGH, 1984. Acta Cryst., C40, 655-657.

Monoclinic, $P2_1/c$, a = 9.27, b = 18.94, c = 7.46 Å, β = 97.3°, D_m = 1.465, Z = 4. Cu radiation, R = 0.065 for 1681 reflexions (photographic data).

In the molecule the mean planes of the benzene and pyrimidine rings are inclined to each other at 78.1(6)° (Fig. 1). The geometry around S is considerably distorted from the ideal tetrahedral geometry. Some bond lengths are: S-C(1) 1.765(9), S-O(9) 1.430(6), S-O(10) 1.426(8) and S-N(11) 1.632(7) Å. The structure is stabilised by a network of three-dimensional hydrogen bonds and van der Waals forces. This structure has already been reported elsewhere (1),

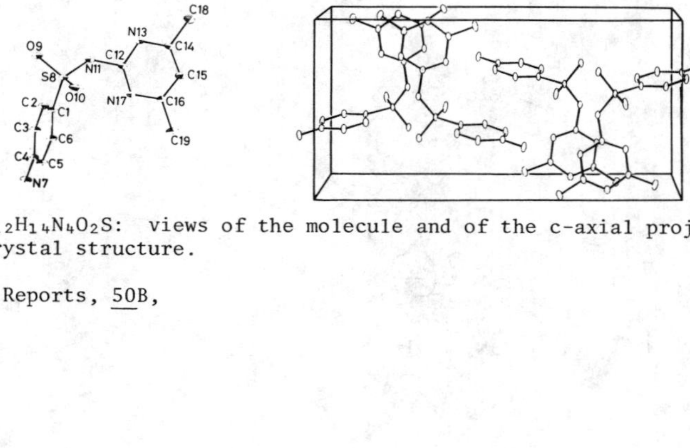

Fig. 1. $C_{12}H_{14}N_4O_2S$: views of the molecule and of the c-axial projection of the crystal structure.

1. Structure Reports, 50B,

THIAMINE NITRATE
$C_{12}H_{17}N_5O_4S$ $C_{12}H_{17}N_4OS^+.NO_3^-$

T. ISHIDA, A. TANAKA and M. INOUE, 1984. Acta Cryst., C40, 437-439.

Monoclinic, $P2_1/c$, a = 6.554, b = 12.275, c = 18.544 Å, β = 97.61°, D_m = 1.462, Z = 4. Cu radiation, R = 0.056 for 2515 reflexions.

The molecule (Fig. 1) has a neutral pyrimidine ring and the characteristic F conformation frequently observed in thiamine compounds with a neutral or cationic pyrimidine ring. This indicates the inherent stability of this F form for thiamine free of substituents on the thiazolium C(2) atom.

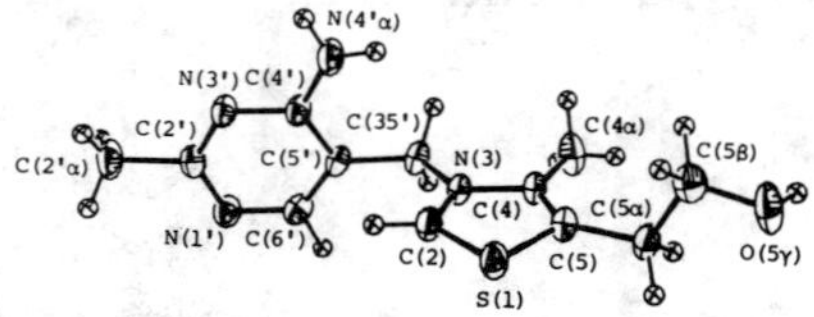

Fig. 1. A view of the thiamine cation in thiamine nitrate.

1-(2',3'-O-ISOPROPYLIDENE-β-D-RIBOFURANOSYL)-O(6),5'-CYCLO-7,7-DICHLOROCYCLOPROPA-[4,5]PERHYDROPYRIMIDIN-2-ONE
$C_{15}H_{18}Cl_2N_2O_5$

V.E. MARQUEZ, K.V.B. RAO, J.V. SILVERTON and J.A. KELLEY, 1984. J. Org. Chem., 49, 912-919.

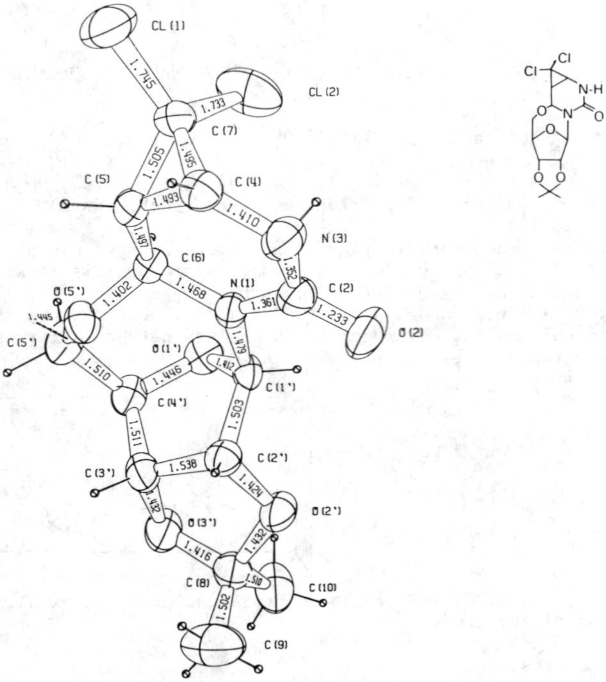

Fig. 1. A view of $C_{15}H_{18}Cl_2N_2O_5$ with bond lengths.

Orthorhombic, $P2_12_12_1$, a = 7.1547, b = 10.8571, c = 20.4126 Å, Z = 4. Cu radiation, R = 0.051 for 1861 reflexions.

Both five-membered rings (Fig. 1) have envelope conformations (O(1') and C(8) envelopes). The torsion angles in the six-membered ring fused to the cyclopropane ring are small (largest -11.4° at N(1)-C(6)). Bond lengths (Fig. 1) are generally as expected.

ADENINE N-METHYL-2-PYRROLIDONE (1:2)
$C_{15}H_{23}N_7O_2$ $C_5H_5N_5.2C_5H_9NO$

V. LANGER, K. HUML and J. ZACHOVA, 1984. Acta Cryst., C40, 2080-2082.

Monoclinic, $P2_1/c$, a = 9.907, b = 21.962, c = 7.994 Å, β = 100.73°, D_m = 1.275, Z = 4. Cu radiation, R = 0.057 for 1097 reflexions.

The adenine molecule (Fig. 1) is neutral and this feature is reflected in the geometry of the pyrimidine ring, in comparison with protonated molecules. There is an overlap of adenine with N-methyl-2-pyrrolidone but none between adenine molecules. A net of hydrogen bonds, forming adenine dimers, is found in the structure.

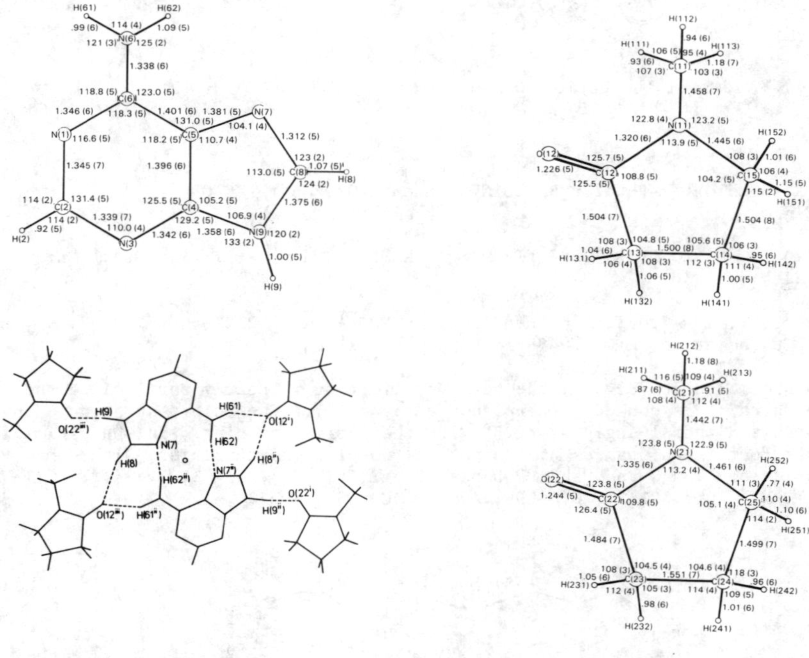

Fig. 1. Bond lengths and angles in the 1:2 complex adenine/N-methyl-2-pyrrolidone and the hydrogen bond scheme.

2,2-DIETHYL-1,2-DIHYDRO-4,6-DIMETHYLPYRIMIDINIUM PICRATE
$C_{16}H_{21}N_5O_7$ $(C_{10}H_{19}N_2)^+.(C_6H_2N_3O_7)^-$

G. FERGUSON, B. KAITNER, D. LLOYD and H. McNAB, 1984. J. Chem. Res., (S) 182-183, (M) 1738-1759.

Monoclinic, C2/c, a = 16.029, b = 19.780, c = 14.076 Å, β = 119.57°, Z = 8. Mo radiation, R = 0.078 for 1101 reflexions.

The six-membered ring of the cation (Fig. 1) adopts an envelope conformation with a slight boat distortion. The sp^3 C atom (C(22) in Fig. 1) is 0.223 Å from the plane of N(21),N(23),C(24),C(26), equivalent to a 15.0° bend at the flap; atom C(25) is only 0.038 Å from the same plane, equivalent to a 3.1° flap angle. The average bond lengths in the vinamidinium system are C•••N 1.32(1), C•••C 1.38(1) Å. The six-membered picrate ring is essentially planar; its bond lengths are consistent with major contributions from resonance forms in which the formal negative charge is not localized on the exocyclic oxygen O(1).

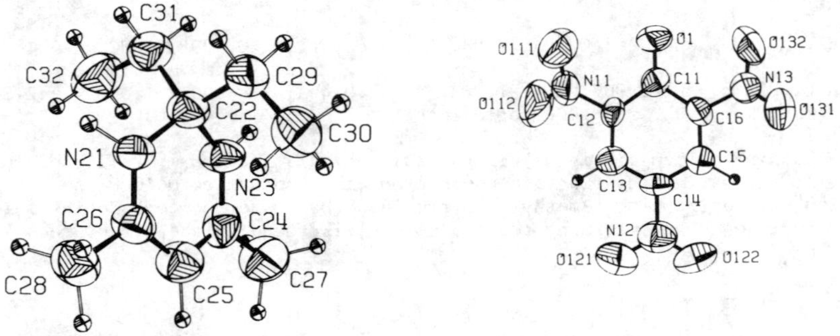

Fig. 1. Views of the cation and anion in $(C_{10}H_{19}N_2)^+.(C_6H_2N_3O_7)^-$.

5-HYDROXY-6-METHYL-2-PHENYLPYRIMIDIN-4-YL-p-CHLOROPHENYL KETONE
$C_{18}H_{13}ClN_2O_2$

G. TARZIA, G. PANZONE, L. ZERILLI, M. LANFRANCHI and G. PELIZZI, 1984. Gazz. Chim. Ital., 114, 431-436.

Triclinic, P$\bar{1}$, a = 7.065, b = 16.697, c = 6.909 Å, α = 92.05, β = 110.52, γ = 83.43°, Z = 2. Cu radiation, R = 0.0485 for 2509 reflexions.

In the molecule (Fig. 1) the heterocyclic ring is nearly parallel to the benzene ring to which it is bonded (interplanar angle 4.8°). The chloro-substituted benzene ring forms a dihedral angle of 31.7° with the pyrimidine ring. In the pyrimidine ring, all four C-N distances have a partial double bond character, with some alternation of "long" and "short" bonds (C8-N1 1.351(2), N1-C11 1.326(2), C11-N2 1.353(2), N2-C10 1.327(3), C10-C9 1.405(2), C9-C8 1.396(2) Å).

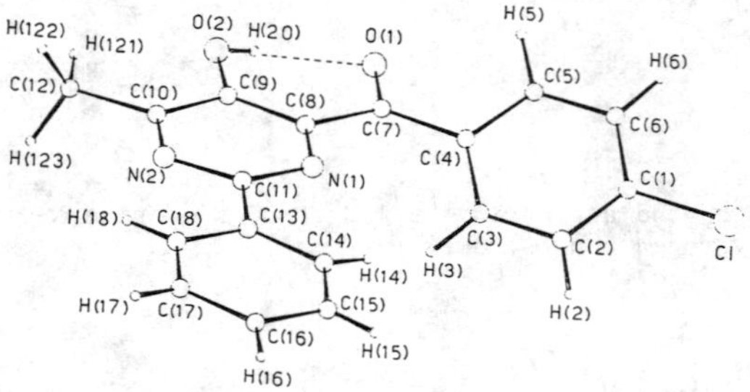

Fig. 1. A view of $C_{18}H_{13}ClN_2O_2$.

2,4-DIAMINO-5-(4-ISOPROPENYL-3,5-DIMETHOXYBENZYL)PYRIMIDINIUM ETHANESULPHONATE

$C_{18}H_{26}N_4O_5S$ (I) $C_{16}H_{21}N_4O_2{}^+,C_2H_5O_3S^-$

2,4-DIAMINO-5-(3,5-DIMETHOXY-4-METHOXYCARBONYLBENZYL)PYRIMIDINE

$C_{15}H_{18}N_4O_4$ (II)

V. CODY, 1984. Acta Cryst., C40, 1000-1004.

I. Triclinic, P$\bar{1}$, a = 9.400, b = 19.563, c = 5.419 Å, α = 93.29, β = 94.11, γ = 93.95°, Z = 2. Cu radiation, R = 0.048 for 3596 reflexions (at 257K).

II. Triclinic, P$\bar{1}$, a = 11.738, b = 18.204, c = 8.001 Å, α = 90.74, β = 107.32, γ = 99.94°, Z = 4. Cu radiation, R = 0.056 for 2975 reflexions (at 293K).

The conformation of I is antiskewed and both molecules in II are twist (Fig. 1). The 4'-isopropenyl group of I is perpendicular to the plane of the benzyl ring. The 4'-methoxycarbonyl group in molecule 2 of II is rotationally (67/33) disordered. In all three molecules the 3',5'-dimethoxy groups are coplanar with the benzyl ring and have their respective methyl carbons pointing away from the 4'-position.

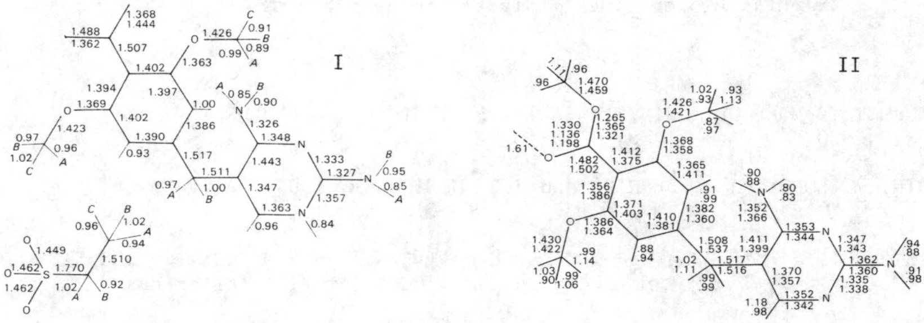

Fig. 1. Bond lengths in $C_{18}H_{26}N_4O_5S$ (I) and $C_{15}H_{18}N_4O_4$ (II). The second set of values on the 4' substituent in I refer to the disordered model. In II values are given for molecules 1 (top) and 2 (bottom); the first two sets of values at the 4'-methoxycarbonyl group are for the disordered model in molecule 1, the third set are for molecule 2.

2,4-DIAMINO-5-(3,4,5-TRIMETHOXYBENZYL)PYRIMIDINIUM BENZOATE (TRIMETHOPRIM MONO-BENZOATE)

$C_{21}H_{24}N_4O_5$ $C_{14}H_{19}N_4O_3{}^+,C_7H_5O_2{}^-$

G. GIUSEPPETTI, C. TADINI, G.P. BETTINETTI, F. GIORDANO and A. LA MANNA, 1984. Acta Cryst., C40, 650-653.

Monoclinic, $P2_1/n$, a = 11.295, b = 28.266, c = 6.543 Å, β = 100.97°, D_m = 1.36, Z = 4. Cu radiation, R - 0.039 for 1895 reflexions.

The compound contains the trimethoprim cation and benzoate anion (Fig. 1); bond lengths and angles within these ions are generally as expected. Two interionic hydrogen bonds N(1)-H(11)...O(11,1) and N(2)-H2(N2)...O(11,2), with N...O 2.578(3) and 2.813(3) Å respectively, lead to the formation of an eight-membered pseudo ring. The ion couples are linked to produce cyclic dimers via pairs of N(2)-H1(N2)... O(11,2) hydrogen bonds, with N...O 2.894(3) Å, across centres of symmetry.

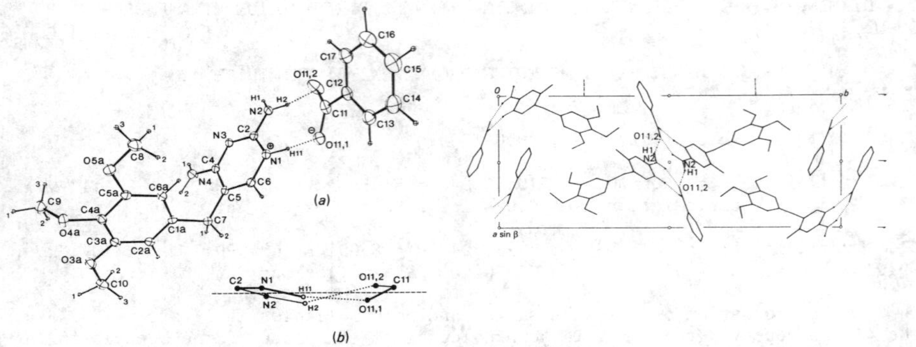

Fig. 1. $C_{14}H_{19}N_4O_3^+,C_7H_5O_2^-$: (a) the molecular conformation and (b) the eight-atom pseudo-ring system generated by the linkage between the ions (left) and the view along c showing the packing of the molecules in the crystal (right); broken lines indicate hydrogen bonds.

6-[CINNAMOYL(2-MORPHOLINOETHYL)AMINO]-1,3-DIMETHYLURACIL HYDRATE

$C_{21}H_{26}N_4O_4,H_2O$

V. WARIN, M. LOBRY, J.L. BERNIER and J.P. HENICHART, 1984. Acta Cryst., C40, 1428-1429.

Triclinic, P$\bar{1}$, a = 10.883, b = 10.465, c = 11.725 Å, α = 98.02, β = 123.54, γ = 93.62°, D_m = 1.17, Z = 2. Mo radiation, R = 0.05 for 2155 reflexions.

The uracyl and cinnamoyl planes are perpendicular, (Fig. 1), with the N(6) lone pair of electrons delocalised within the cinnamoyl π cloud. The closest contact to the water molecule is N(4)...O(W) at 2.77 Å. Bond lengths at N(6) are, C(6)-N(6) 1.423(3), C(7)-N(6) 1.368(6), C(16)-N(6) 1.480(6) Å.

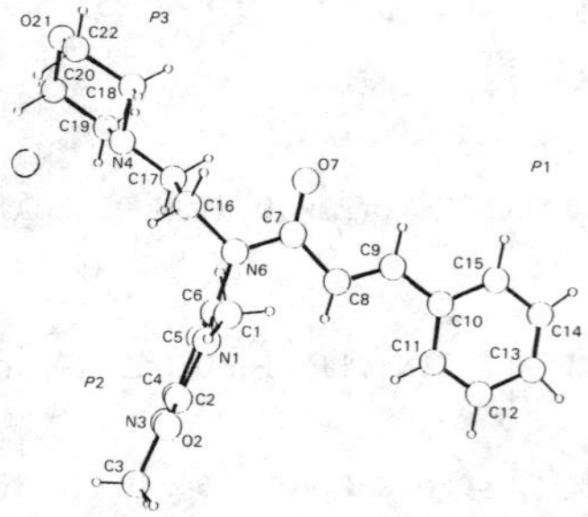

Fig. 1. The $C_{21}H_{26}N_4O_4$ molecule with a water molecule of crystallisation.

p-BROMOPHENACYLPHIDOPOLIN ACETONE SOLVATE

$C_{22}H_{18}BrN_5O_6,C_3H_6O$

S.W. AYER, R.J. ANDERSEN, C.-H. HE and J. CLARDY, 1984. J. Org. Chem., 49, 3869-3870.

Triclinic, P1, a = 9.549, b = 9.514, c = 15.212 Å, α = 72.69, β = 82.96, γ = 81.87°, Z = 2. Cu radiation, R = 0.0939 for 1255 reflexions.

A view of the molecule is given in Fig. 1; dimensions agree well with anticipated values.

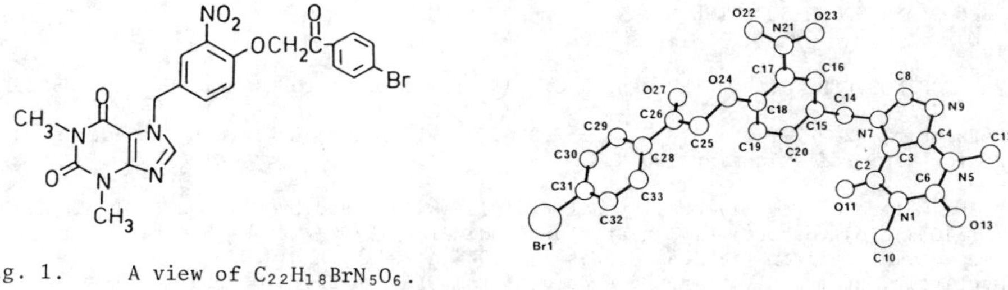

Fig. 1. A view of $C_{22}H_{18}BrN_5O_6$.

1,1,1-TRIFLUORO-3-[1,4,5,6-TETRAHYDRO-2-PHENYL-1-(2,4,6-TRIMETHYLPHENYL)-4-PYRIMIDINYLIDENE]-2-(TRIFLUOROMETHYL)-2-PROPANOL
$C_{23}H_{22}F_6N_2O$ (I)

1,6-DIHYDRO-2-PHENYL-α,α-BIS(TRIFLUOROMETHYL)-1-(2,4,6-TRIMETHYLPHENYL)-4-PYRIMIDINE ETHANOL
$C_{23}H_{22}F_6N_2O$ (II)

A. GIEREN, H. BETZ, G. WEBER, T. HÜBNER, K. BURGER, H. PARTSCHT and U. WASSMUTH, 1984. Chem. Zeit., 108, 215-217.

I. Monoclinic, $P2_1/c$, a = 12.223, b = 14.753, c = 12.291 Å, β = 92.16°, D_m = 1.36, Z = 4. Mo radiation, R = 0.067 for 3286 reflexions.

II. Triclinic, $P\bar{1}$, a = 9.140, b = 10.319, c = 12.297 Å, α = 99.14, β = 97.46, γ = 108.19°, D_m = 1.41, Z = 2. Mo radiation, R = 0.043 for 3247 reflexions.

The pyrimidine skeleton adopts an ideal half-boat conformation in I but in II the half-boat conformation is distorted towards a boat (Fig. 1). The overall arrangement of substituents is comparable in both compounds. Each of the two isomeric molecules is stabilized by one intramolecular O-H...N hydrogen bond (O(1)...N(3) 2.586 and 2.659 Å for I and II respectively, and O(1)-H(1)...N(3) 150 and 155° respectively.

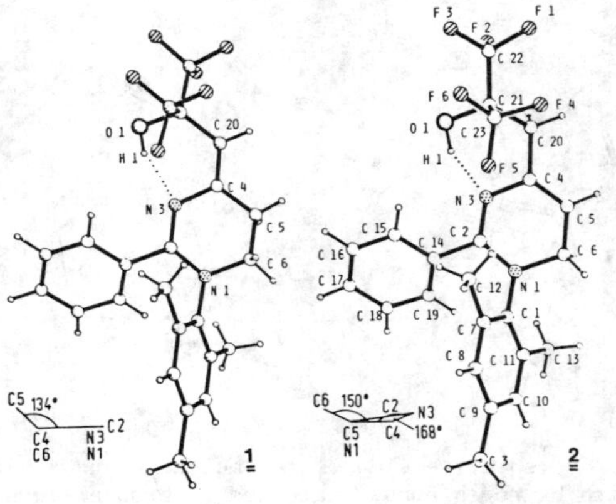

Fig. 1. Molecular structures of both isomers of $C_{23}H_{22}F_6N_2O$, and the conformations of the hydro-pyrimidine system.

1,4:3,6-DIANHYDRO-D-SORBITOL
$C_6H_{10}O_4$

H. VAN KONINGSVELD, J.A. PETERS and J.C. JANSEN, 1984. Acta Cryst., C40, 519-521.

Orthorhombic, $P2_12_12_1$, a = 5.313, b = 7.006, c = 17.174 Å, Z = 4. Mo radiation, R = 0.026 for 1263 reflexions at 100 K.

The molecule (Fig. 1) consists of two planes (defined by C(1)O(2)C(4)C(3) and by C(4)C(3)O(1)C(6) respectively) nearly perpendicular (87.0°) to each other. The maximum deviation from the mean planes is 0.004 Å (O(2)) and 0.006 Å (O(1)) respectively; atom O(4) is in the endo and atom O(3) in the exo position. Atoms C(2) and C(5) are displaced from the planes concerned by 0.58 and 0.59 Å respectively. There is evidence for the existence of a bifurcated hydrogen bond consisting of an intermolecular (O...O 2.771 Å) and an intramolecular (O...O 2.826 Å) hydrogen bond.

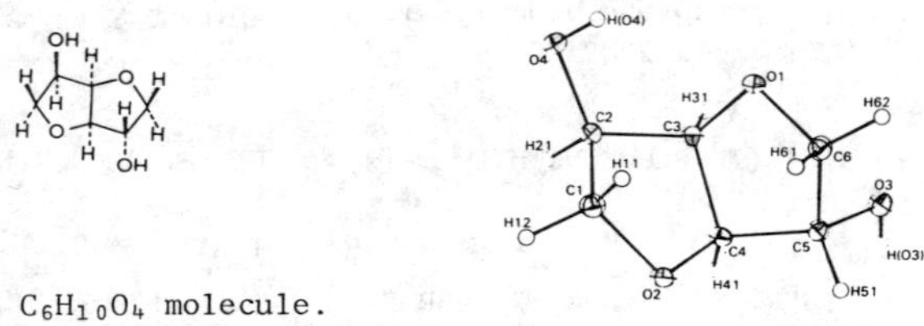

Fig. 1. The $C_6H_{10}O_4$ molecule.

4-DEOXY-4-FLUORO-β-D-FRUCTOPYRANOSE
$C_6H_{11}FO_5$

J. PODLAHOVÁ and J. LOUB, 1984. Acta Cryst., C40, 1284-1286.

Orthorhombic, $P2_12_12_1$, a = 7.787, b = 9.272, c = 10.224 Å, D_m = 1.615, Z = 4. Mo radiation, R = 0.043 for 1254 reflexions.

The molecular structure and conformation, $^2C_5(D)$ of the title compound (Fig. 1) are in accordance with literature data. The molecules are linked by hydrogen bonds, thus forming a three-dimensional network in the crystal.

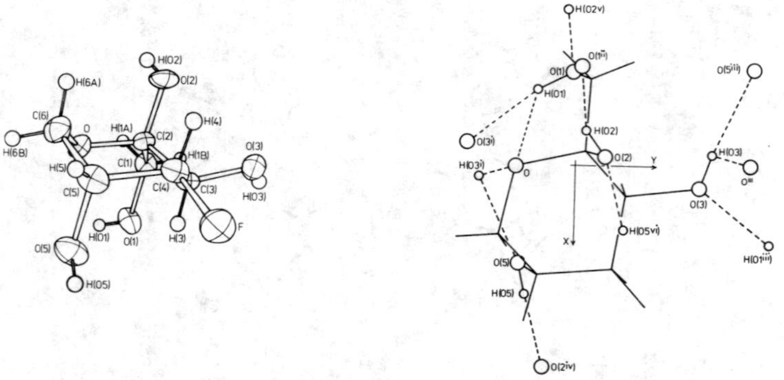

Fig. 1. $C_6H_{11}FO_5$: perspective view of the molecule and a view of the structure down z with hydrogen bonds indicated by broken lines.

2-DEOXY-β-D-lyxo-HEXOSE (2-DEOXY-β-D-GALACTOSE)
$C_6H_{12}O_5$

R. PULITI, C.A. MATTIA and G. BARONE, 1984. Carbohydr. Res., 135, 47-52.

Monoclinic, $P2_1$, a = 9.811, b = 6.953, c = 5.315 Å, β = 91.58°, Z = 2. Cu radiation, R = 0.032 for 800 reflexions.

The pyranose ring has near to ideal $^4C_1$ conformation (Fig. 1). The primary hydroxyl group is gauche (61.1°) with respect to the C5-O5 bond and trans (-178.0°) with respect to C5-C4. Molecular dimensions are normal. All hydroxyl groups are involved in hydrogen bonding to produce infinite helical chains around screw axes.

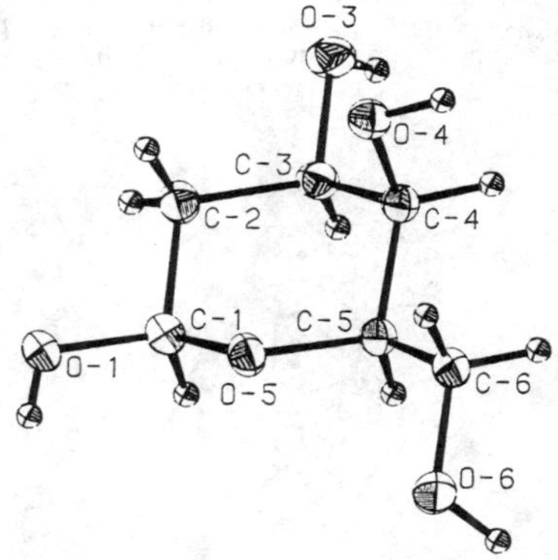

Fig. 1. A view of $C_6H_{12}O_5$.

β-D-ALLOSE
$C_6H_{12}O_6$

L.M.J. KROON-BATENBURG, P. VAN DER SLUIS and J.A. KANTERS, 1984. Acta Cryst., C40, 1863-1865.

Orthorhombic, $P2_12_12_1$, a = 4.918, b = 11.925, c = 12.805 Å, Z = 4. Mo radiation, R = 0.029 for 662 reflexions.

The molecule has the $^4C_1$ pyranose chair conformation; the conformation of the hydroxymethyl group is gauche-gauche. Each molecule is hydrogen-bonded to eight others (Fig. 1) with all five OH groups as proton donors and four OH groups and the ring O as proton acceptors; the anomeric O atom is the acceptor in two hydrogen bonds with H...O 2.45(5) and 1.96(5) Å.

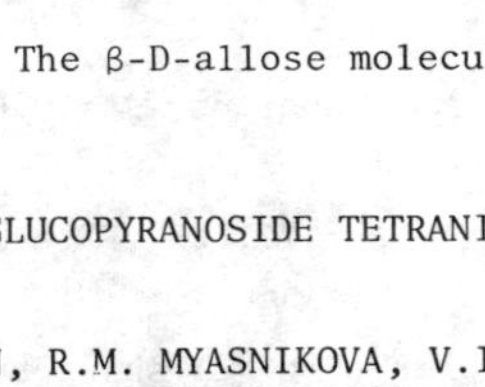

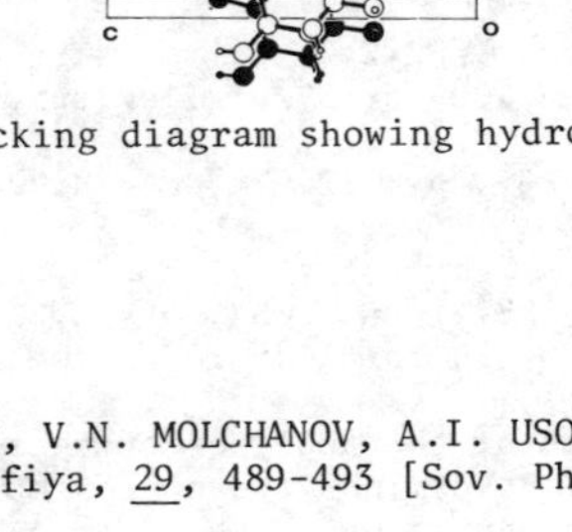

Fig. 1. The β-D-allose molecule and a packing diagram showing hydrogen bonds.

METHYL-β-D-GLUCOPYRANOSIDE TETRANITRATE
$C_7H_{10}N_4O_{14}$

A.V. NIKITIN, R.M. MYASNIKOVA, V.I. ANDRIANOV, V.N. MOLCHANOV, A.I. USOV, S.I. FIRGANG and A.U. PERTSIN, 1984. Kristallografiya, 29, 489-493 [Sov. Phys. Crystallogr. 29, 293-295].

Orthorhombic, $P2_12_12_1$, a = 15.503, b = 13.600, c = 7.052 Å, Z = 4. Mo radiation, R = 0.046 for 868 reflexions.

In the molecule (Fig. 1) the deviations of atoms C(1) and C(4) from the principal plane of the ring (through atoms C(2), C(3), C(5) and O(5)) are +0.67 and -0.65 Å respectively. Bond lengths of interest are: C(4)-C(5) 1.541, C(2)-O(2) 1.450, C(3)-O(3) 1.449 and C(1)-O(5) 1.417 Å, all the remaining parameters are considered typical for such a molecule. The nitrate groups are all practically planar and the deviations of the ether O atoms from the planes of the corresponding nitrate groups do not exceed 0.1 Å. Adjacent molecules have short contacts (3.01-3.02 Å) via the nitrate groups (Fig. 1).

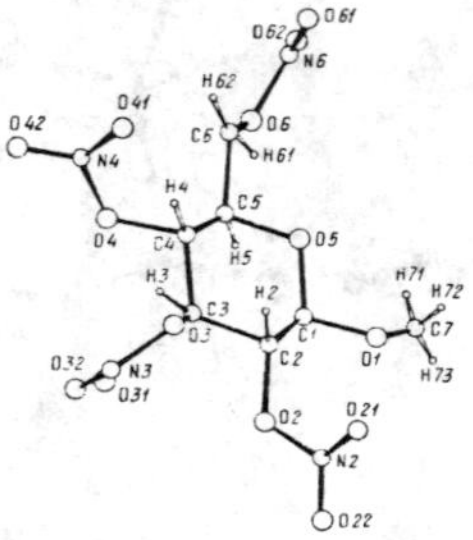

Fig. 1. Molecular structure of $C_7H_{10}N_4O_{14}$.

METHYL 3,6-ANHYDRO-β-D-GLUCOFURANOSIDE
$C_7H_{12}O_5$ (I)

METHYL 3,6-ANHYDRO-α-L-IDOFURANOSIDE
$C_7H_{12}O_5$ (II)

METHYL 3,6-ANHYDRO-β-D-MANNOFURANOSIDE
$C_7H_{12}O_5$ (III)

METHYL 3,6-ANHYDRO-α-D-GLUCOFURANOSIDE
$C_7H_{12}O_5$ (IV)

METHYL 3,6-ANHYDRO-α-D-MANNOFURANOSIDE
$C_7H_{12}O_5$ (V)

J. KOPF and P. KOLL, 1984. Carbohydr. Res., 135, 29-46.

I. Orthorhombic, $P2_12_12_1$, a = 6.829, b = 10.277, c = 11.629 Å, Z = 4. Mo radiation, R = 0.043 for 1421 reflexions.

II. Orthorhombic, $P2_12_12_1$, a = 6.639, b = 10.184, c = 11.700 Å, Z = 4. Mo radiation, R = 0.038 for 1787 reflexions.

III. Orthorhombic, $P2_12_12_1$, a = 7.475, b = 10.245, c = 10.976 Å, Z = 4. Mo radiation, R = 0.100 for 572 reflexions.

IV. Trigonal, $P3_1$, a = 12.974, c = 4.724 Å, Z = 3. Mo radiation, R = 0.051 for 1312 reflexions.

V. Monoclinic, $P2_1$, a = 7.054, b = 6.893, c = 8.590 Å, β = 107.23°, Z = 2. Mo radiation, R = 0.029 for 1176 reflexions.

The results of the analyses of these five molecules (Fig. 1) shows that the anomeric effect determines the conformation of the furanoid ring, which resulted in a quasi-axial orientation of the aglycon. Molecule II has an almost ideal E_2

conformation whereas I and III (having the same R-configuration at the anomeric centre, have furanoid conformations intermediate between E_2 and ${}^1T_2$. Molecules IV and V have an S configuration at C(1); IV has a ${}^0T_1$ conformation slightly distorted into ${}^0E$ and V is intermediate between ${}^2E$ and ${}^2T_1$. Except for V (which has C(6) exo) all compounds had a C(6) endo orientation.

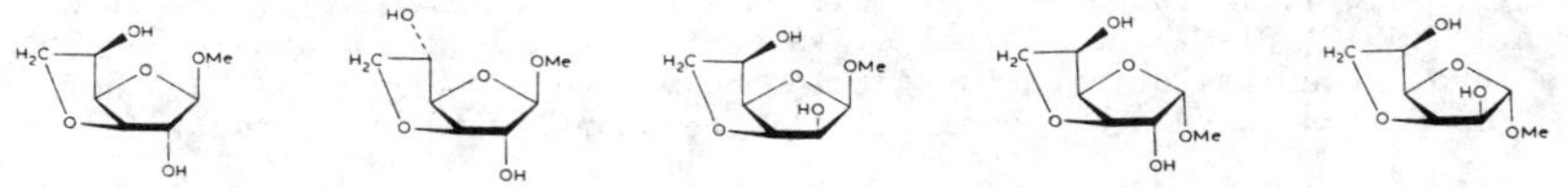

Fig. 1. The five isomers of $C_7H_{12}O_5$ (I - V from left to right).

6-O-ACETYL-2,3,4-TRIDEOXY-α-D,L-glycero-HEX-2-ENOPYRANOSE
$C_8H_{12}O_4$ (I)

3-O-(6-O-ACETYL-2,3,4-TRIDEOXY-α-L-glycero-HEX-2-ENOPYRANOSYL)-1,2,5,6-DI-O-ISOPROPYLIDENE-α-D-GLUCOFURANOSE
$C_{20}H_{30}O_9$ (II)

J.W. KRAJEWSKI, P. GLUZINSKI, Z. URBANCZYK-LIPKOWSKA, A. BANASZEK, K. VANGEHR and P. LUGER, 1984. Carbohydr. Res., 125, 203-216.

I. Monoclinic, $P2_1$, a = 21.123, b = 4.439, c = 10.085 Å, β = 110.22°, Z = 4. Cu radiation, R = 0.046 for 1065 reflexions.

II. Orthorhombic, $P2_12_12_1$, a = 22.110, b = 11.651, c = 8.658 Å, Z = 4. Cu radiation, R = 0.046 for 1884 reflexions.

Views of the molecules are in Fig. 1. The 2-enopyranose rings in (I) and (II) adopt half-chair conformations (H), and that in (II) is markedly deformed. The 1,2-dioxolane ring in (II) has an envelope (E) conformation, whereas the 5,6-dioxolane ring is dynamically disordered and can be represented by a conformational hybrid (E + P). The α-D-glucofuranose ring in (II) has a twist conformation (T). The glycoside bond in (II) is characterized by φ and ψ torsion angles of 47(2)° and 32(2)° respectively.

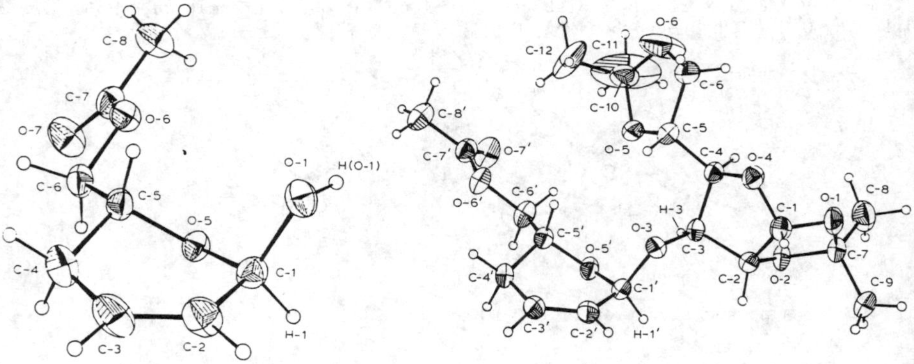

Fig. 1. Views of $C_8H_{12}O_4$ (I) (left) and of $C_{20}H_{30}O_9$ (II) (right).

ETHYL 2,3-DIDEOXY-α-D-erythro-HEX-2-ENOPYRANOSIDE
$C_8H_{14}O_4$

L.M. WINGERT, J.R. RUBLE and G.A. JEFFREY, 1984. Carbohydr. Res., 128, 1-10.

Orthorhombic, $P2_12_12_1$, a = 11.220, b = 18.387, c = 8.509 Å, D_m = 1.274, Z = 8. Cu radiation, R = 0.029 for 1992 reflexions (at 123 K).

There are two independent molecules in the asymmetric unit (Fig. 1) and both have the $^0H_5$ conformation. The alkenic bond is almost planar in one molecule (C1-C2-C3-C4 torsion angle +0.8°), in the other molecule this value is +3.7°. The glycosidic torsion angle (O5-C1-O1-C7) has normal exo-anomeric values of +71 and +64°. The primary alcohol group has orientation g/t in one molecule and g/g in the other. Both ring-bond (O5-C1) and glycosidic bond (C1-O1) are short, with distances 1.409-1.425 Å.

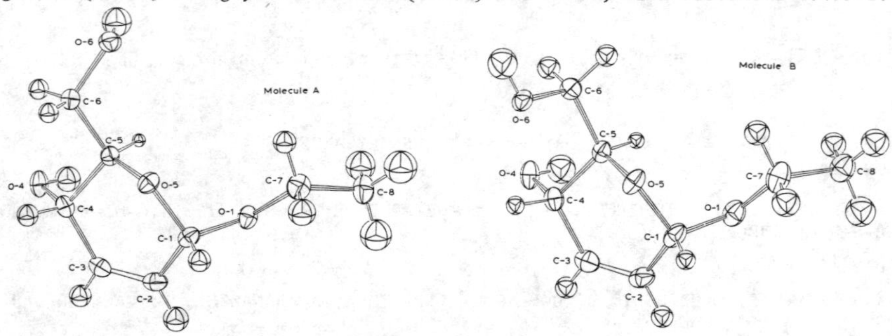

Fig. 1. The two $C_8H_{14}O_4$ molecules.

1,6-ANHYDRO-3,4-O-ISOPROPYLIDENE-β-D-GALACTOPYRANOSE
$C_9H_{14}O_5$

F.H. CANO, C. FOCES-FOCES, J. JIMINEZ-BARBERO and M. MARTIN-LOMAS, 1984. Carbohydr. Res., 127, 338-344.

Orthorhombic, $P2_12_12_1$, a = 17.7022, b = 5.4822, c = 9.7295 Å, Z = 4. Cu radiation, R = 0.040 for 860 reflexions.

The pyranoid ring (Fig. 1) has a $^1C_4$ conformation distorted towards an envelope at O5. The 1,6-anhydro has a distorted envelope (at O5) conformation. Molecules are linked by O-H...O hydrogen-bonding, O...O 2.916(4) Å.

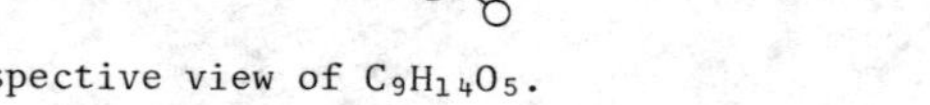

Fig. 1. A perspective view of $C_9H_{14}O_5$.

ISOPROPYL 2-AZIDO-2-DEOXY-β-D-GALACTOPYRANOSIDE
$C_9H_{17}N_3O_5$

N. DARBON, Y. ODDON, E. GUY, A.A. PAVIA, J.P. REBOUL and G. PEPE, 1984. Acta Cryst., C40, 292-294.

Orthorhombic, $P2_12_12_1$, a = 7.706, b = 8.275, c = 18.364 Å, D_m = 1.40, Z = 4. Mo radiation, R = 0.036 for 942 reflexions.

The conformation of the hydroxyl group about the C(5)-C(6) bond (Fig. 1) is trans-gauche. The conformational angles ϕ, ψ and ψ' are 157.8, 81.9 and 156.2° respectively.

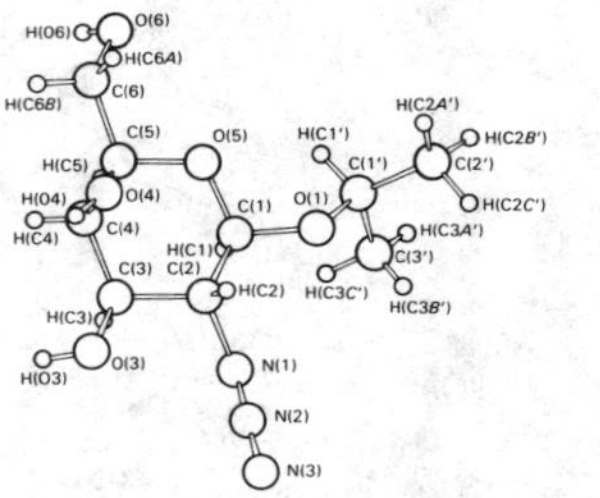

Fig. 1. A view of $C_9H_{17}N_3O_5$ with the numbering scheme.

1,2-S,O-(1R,2S-1-CYANO-2-METHYL-1,2-ETHANEDIYL)-1-THIO-β-D-GLUCOPYRANOSE (MEROSINIGRIN)

$C_{10}H_{15}NO_5S$

J.F. RICHARDSON and M.H. BENN, 1984. Can. J. Chem., 62, 1236-1237.

Triclinic, P1, a = 6.699, b = 9.347, c = 5.328 Å, α = 103.56, β = 109.14, γ = 86.71°, Z = 1. Mo radiation, R = 0.031 for 1401 reflexions.

The molecular structure (Fig. 1) consists of two fused six-membered rings, each in a chair conformation. The chirality at C(1) and C(2) is R and S, respectively, relative to the known chirality of the β-D-glucopyranose ring. The cyano function adopts an axial disposition. There is extensive intermolecular hydrogen bonding in the crystal.

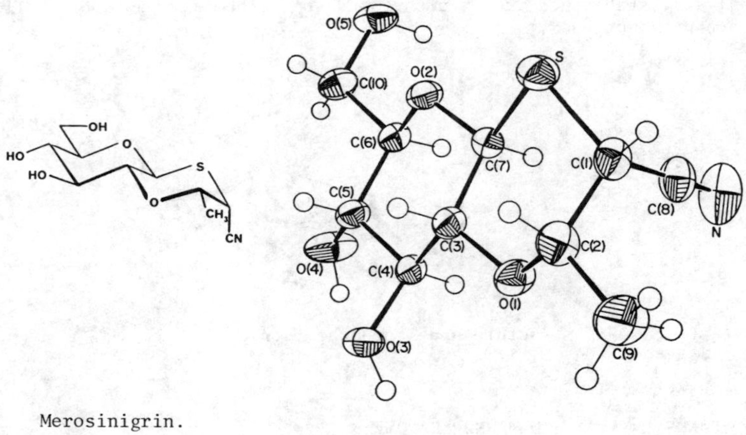

Fig. 1. Merosinigrin.

O-α-D-MANNOPYRANOSYL-(1-3)-L-THREONINE

$C_{10}H_{19}NO_8$

N. DARBON, Y. ODDON, J.-M. LACOMBE and A.A. PAVIA, 1984. Carbohydr. Res., 130, 55-64.

Monoclinic, P2₁, a = 7.732, b = 8.628, c = 9.939 Å, β = 112.45°, Z = 2. Cu radiation, R = 0.041 for 1182 reflexions.

The six-membered ring has a $^4C_1(D)$ conformation (Fig. 1). The threonine molecule has Hα and Hβ in a nearly gauche-gauche relationship.

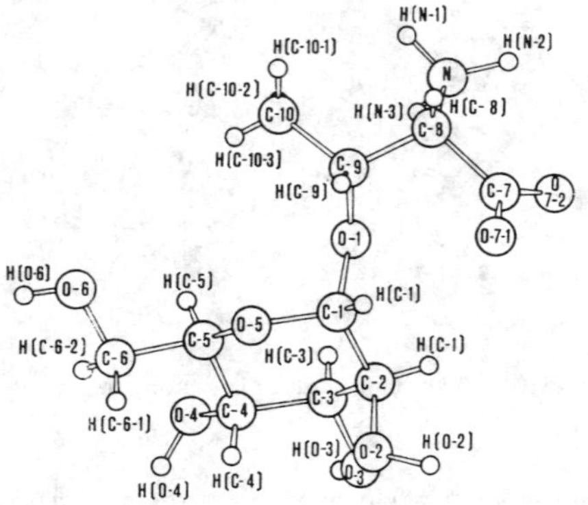

Fig. 1. A view of $C_{10}H_{19}NO_8$.

METHYL 3-C-ACETAMIDOMETHYL-3-CHLORO-2,3,6-TRIDEOXY-4-O-METHOXYMETHYL-α-L-arabino-HEXOPYRANOSIDE
$C_{12}H_{22}ClNO_5$

J.S. BRIMACOMBE and T.J.R. WEAKLEY, 1984. Carbohydr. Res., 126, 303-307.

Orthorhombic, $P2_12_12_1$, a = 13.135, b = 12.306, c = 9.392 Å, Z = 4. Cu radiation, R = 0.062 for 896 reflexions.

The analysis established the structure and stereochemistry as shown in Fig. 1. Dimensions are unexceptional.

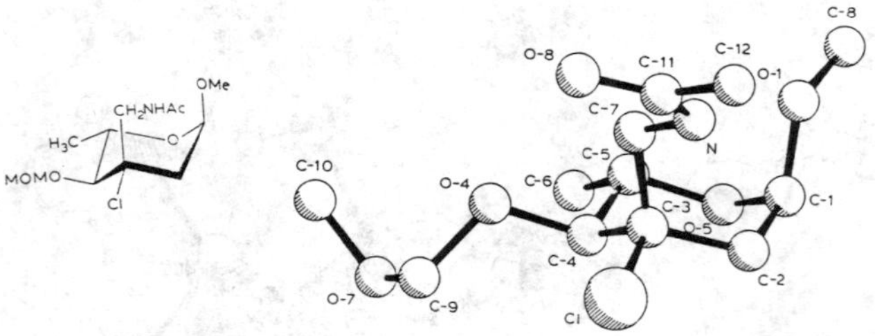

Fig. 1. $C_{12}H_{22}ClNO_5$: structure and stereochemistry.

O-β-D-MANNOPYRANOSYL-(1,4)-α-D-MANNOPYRANOSE
$C_{12}H_{22}O_{11}$

B. SHELDRICK, W. MACKIE and D. AKRIGG, 1984. Carbohydr. Res., 132, 1-6.

Orthorhombic, $P2_12_12_1$, a = 8.811, b = 7.485, c = 21.601 Å, Z = 4. Cu radiation, R = 0.0511 for 1024 reflexions.

The sugar residues have $^4C_1$ conformations (Fig. 1) and their relative orientation is stabilised by an intramolecular hydrogen bond O(13)-H...O(5). The mean C-C bond length is 1.518 Å. The average C-O (excluding C(1)-O(1)) bond length is 1.432; the anomeric bond C(1)-O(1) is 1.398 Å. There is a 68/32 mixture of α and β anomers in the crystal.

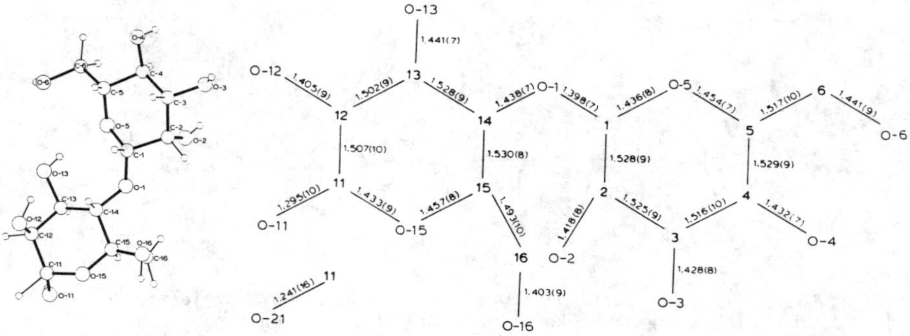

Fig. 1. A view of $C_{12}H_{22}O_{11}$ and bond lengths (Å).

4-O-β-D-GALACTOPYRANOSYL-α-D-GLUCOPYRANOSE MONOHYDRATE (α-LACTOSE MONOHYDRATE)
$C_{12}H_{22}O_{11},H_2O$

J.H. NOORDIK, P.T. BEURSKENS, P. BENNEMA, R.A. VISSER and R.O. GOULD, 1984.
Z. Kristallogr., 168, 59-65.

Monoclinic, $P2_1$, a = 7.937, b = 21.568, c = 4.815 Å, β = 109.77°, Z = 2. Cu radiation, R = 0.034 for 2627 reflexions.

A refinement with new data of a structure reported previously (1). The absolute configuration was established. The α-lactose molecules are oriented as shown in Fig. 1, with the glucose moieties in the direction of the -b axis.

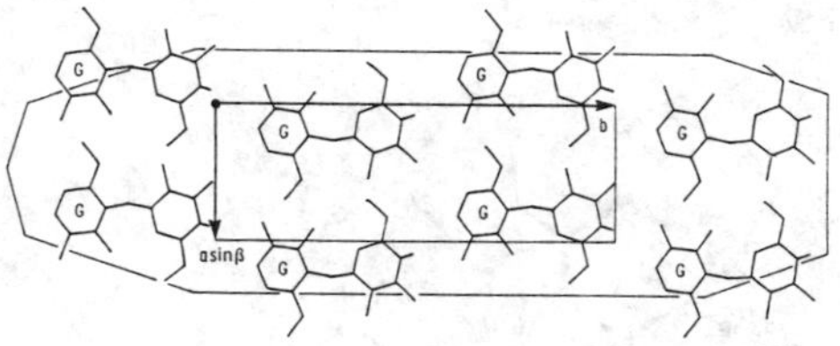

Fig. 1. A projection of the α-lactose structure down c.

1. Structure Reports, 37B, 244.

N-CYCLOHEXYL-D-GLUCONAMIDE
$C_{12}H_{23}NO_6$

N. DARBON, Y. ODDON, J.M. LACOMBE, E. DECOSTER, A.A. PAVIA and J.P. REBOUL, 1984.
Acta Cryst., C40, 1105-1107.

Monoclinic, $P2_1$, a = 14.654, b = 5.424, c = 8.610 Å, β = 97.67°, D_m = 1.37, Z = 2. Mo radiation, R = 0.050 for 1528 reflexions.

The gluconamide part of the molecule (Fig. 1) has a linear conformation with its long axis parallel to [100]. The C(1)-N bond distance is 1.342(8) Å, intermediate in length between a single and a double bond, whilst the C(7)-N bond distance of 1.463(9) Å corresponds to a single bond. The molecule contains one N-H...O intramolecular hydrogen bond, while three hydrogen bonds link the molecules to form the crystal.

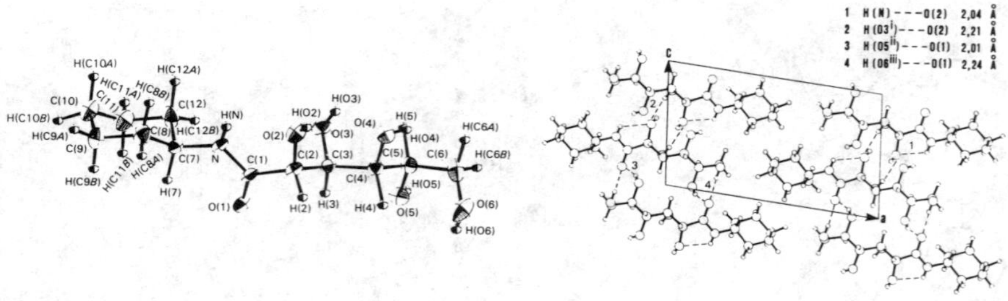

Fig. 1. $C_{12}H_{23}NO_6$: perspective view of the molecule and a projection of the unit cell along b; broken lines indicate hydrogen bonds.

SALICIN
$C_{13}H_{18}O_7$

K. UENO, 1984. Acta Cryst., C40, 1726-1728.

Orthorhombic, $P2_12_12_1$, a = 8.314, b = 21.169, c = 7.650 Å, D_m = 1.41, Z = 4. Cu radiation, R = 0.038 for 1552 reflexions.

In the molecule (Fig. 1) the torsion angle C(6)-C(1)-C(7)-O(7) is 37.1(4)°. The glucopyranose ring is in the usual C_1 chair conformation. The conformation of the C(6')-O(6') bond is trans-gauche with respect to the C(5')-C(4') and C(5')-O(5') bonds. The conformation angle of the glucosyl linkage (C(3)-C(2)-O(1')-C(1'):ϕ) is only 14.6(4)°, which is close to those in the other aromatic glucosides with an in-plane conformation. The bond angle at the glucosidic O atom is 117.8(2)°. The molecules are linked via hydrogen bonding to form strings along the twofold screw axes parallel to a and c.

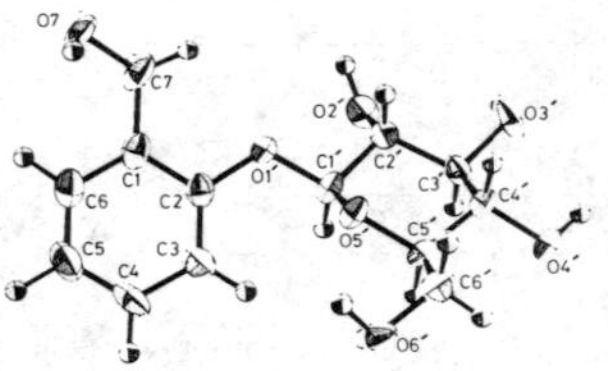

Fig. 1. A view of salicin.

METHYL 2,3-DIDEOXY-1',2',3',6'-TETRAHYDRO-4',5'-DIMETHYL-β-D-glycero-D-threo-BENZO[2,3]HEXOPYRANOSID-4-ULOSE
$C_{13}H_{20}O_4$

H. GNICHTEL, C. GUMPRECHT and P. LUGER, 1984. Justus Liebigs Ann. Chem., 1531-1544.

Orthorhombic, $P2_12_12_1$, a = 13.976, b = 9.865, c = 18.472 Å, Z = 8. Cu radiation, R = 0.044 for 2294 reflexions.

The two independent molecules differ in orientation of their CH_2OH group; one molecule is shown in Fig. 1. The two rings are trans-fused and the pyran ring has a $^1C_4$ conformation.

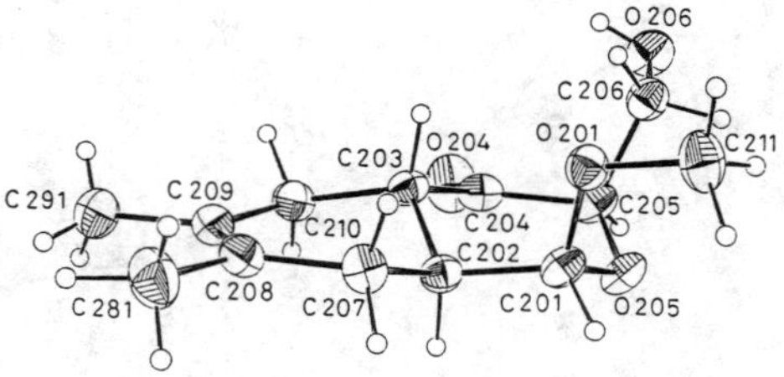

Fig. 1. A view of one molecule of $C_{13}H_{20}O_4$.

6-O-ACETYL-2,3,4-TRIDEOXY-4-C-ETHENYL-3-C-(METHOXYCARBONYLMETHYL)-α-D-lyxo-PYRANOSE
$C_{13}H_{20}O_6$

S. LÉGER, F. BELANGER-GARIEPY, S. HANESSIAN and F. BRISSE, 1984. Acta Cryst., C40, 1844-1846.

Monoclinic, C2, a = 21.935, b = 6.110, c = 11.002 Å, β = 107.01°, Z = 4. Cu radiation, R = 0.037 for 984 reflexions.

The pyranose ring has the $^4C_1$ conformation with endocyclic torsion angles averaging 54.4°. The primary acetate substituent is in the gt conformation, the ethenyl group is axial and the methoxycarbonyl substituent equatorial (Fig. 1). Hydrogen bonds between the anomeric OH and the ring O (O...O 2.823(3) Å) link the molecules into helices in the b direction.

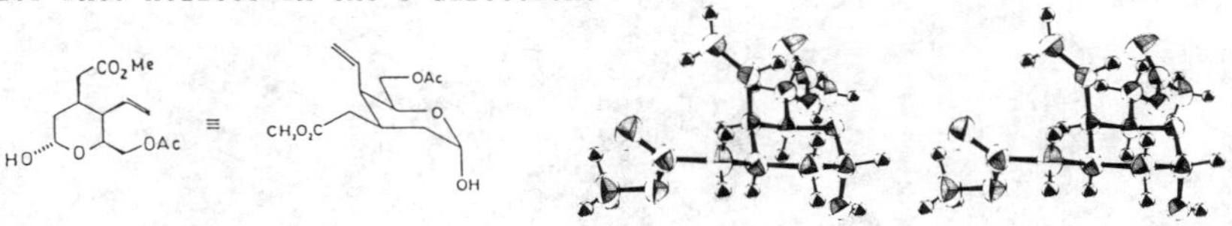

Fig. 1. The $C_{13}H_{20}O_6$ molecule and a stereoview.

METHYL α-THIOMALTOSIDE
$C_{13}H_{24}O_{10}S$

S. PEREZ and C. VERGELATI, 1984. Acta Cryst., B40, 294-299.

Monoclinic, $P2_1$, a = 14.196, b = 4.846, c = 12.410 Å, β = 110.12°, D_m = 1.54, Z = 2. Mo radiation, R = 0.044 for 4685 reflexions.

The two glucose residues (Fig. 1) have the $^4C_1$ conformation and are α(1 → 4) linked. The structural features at the glycosidic bridge are: C(1)-S(1) 1.826(1), C(4')-S(1) 1.828(1) Å, C(1)-S(1)-C(4') 100.3(1)°. Comparison with other sulphur-containing molecular segments has been performed. The conformational angles ϕ = O(5)-C(1)-S(1)-C(4') and ψ = C(1)-S(1)-C(4')-C(5') at the glycosidic linkage have the values 89 and -116.8° respectively. The observed conformational behaviour has been analysed with the aid of a conformational analysis investigation. The primary hydroxyl group in the non-reducing residue is in the gt conformation, whereas a gg orientation is observed for the corresponding hydroxyl in the reducing residue. There is no intramolecular hydrogen bond, and the structure is maintained by an extensive network of hydrogen bonds.

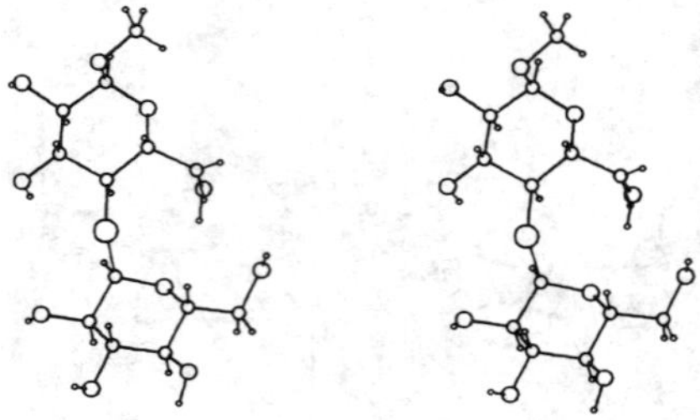

Fig. 1. A stereoview of methyl α-thiomaltoside.

3-ACETAMIDO-2,3,5,6-TETRADEOXY-5-FLUORO-4-NITROBENZOYL-D,L-ribo-HEXOFURANOSE
$C_{15}H_{17}FN_2O_6$

J.T. WELCH, B.-M. SVAHN, S. ESWARAKRISHNAN, J.P. HUTCHINSON and J. ZUBIETA, 1984. Carbohydr. Res., 132, 221-231.

Monoclinic, $P2_1/c$, a = 19.711, b = 8.601, c = 9.605 Å, β = 90.83°, Z = 4. Mo radiation, R = 0.074 for 920 reflexions.

The furanose ring adopts a conformation which has C(2) 0.5 Å below the least-squares plane of the other four atoms (Fig. 1). This results in the C(1) substituent being pseudo-axial and those at C(3) and C(4) being nearly equally above and below the plane respectively.

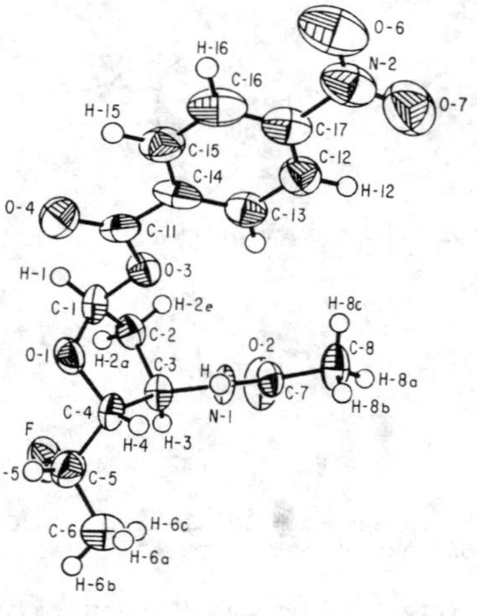

Fig. 1. A perspective view of $C_{15}H_{17}FN_2O_6$.

METHYL 2,3:4,6-BIS-O-(N-DIETHYLAMIDOTHIOPHOSPHATE)-α-D-MANNOPYRANOSIDE (diastereomers II and III)
$C_{15}H_{30}N_2O_6P_2S_2$

I.A. LITVINOV, Yu.T. STRUCHKOV, L.I. GURARII, L.S. LYAKISHEVA, E.T. MUKMENEV and B.A. ARBUZOV, 1984. J. Struct. Chem., 25, 448-453 [Zh. Strukt. Khim., 25(3), 124-129].

Diastereomer II: Monoclinic, $P2_1$, a = 15.564, b = 12.996, c = 11.407 Å, β = 94.83°, Z = 4. Cu radiation, R = 0.065 for 1881 reflexions.

Diastereomer III: Monoclinic, $P2_1$, a = 13.957, b = 7.1423, c = 11.371 Å, β = 96.52°, Z = 2. Cu radiation, R = 0.047 for 1517 reflexions.

The molecules (Fig. 1) are diastereomers differing in absolute configuration at the chiral P(2) atom. The configurations of the molecules correspond to trans-anti-cis linkages between the dioxaphosphorinane, pyranose, and dioxaphospholane rings. The conformations of the dioxaphospholane rings are ^3E envelopes, while those of the pyranose rings are $^3C_5$ chairs, and those of the dioxaphosphorinane rings are chairs.

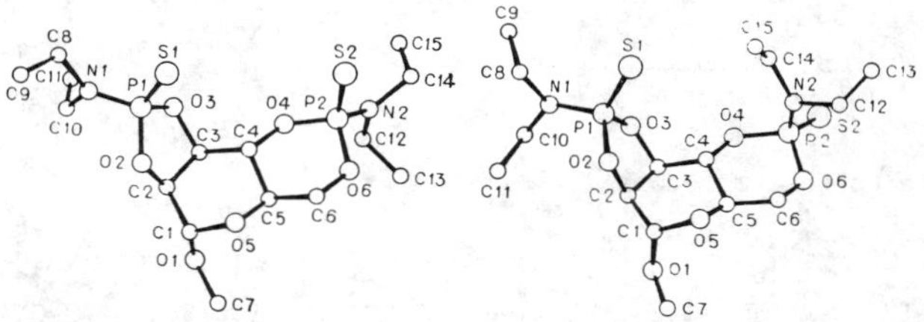

Fig. 1. The methyl 2,3:4,6-bis-O-(N-diethylamidothiophosphate)-α-D-mannopyranoside diastereomers (II, left, and III, right).

5,5'-ANHYDRO-5-BENZENESULPHONAMIDO-1-(2',3'-O-ISOPROPYLIDENE-β-D-RIBOFURANOSYL)-IMIDAZOLE-4-CARBONITRILE
$C_{18}H_{18}N_4O_5S$

S. SUGIO, K. KITAMURA, H. MIZUNO, A. WAKAHARA, M. IKEHARA and K.-I. TOMITA, 1984. Acta Cryst., C40, 1608-1610.

Orthorhombic, $P2_12_12_1$, a = 8.453, b = 20.905, c = 10.570 Å, D_m = 1.432, Z = 4. Cu radiation, R = 0.055 for 1316 reflexions.

In the molecule (Fig. 1) the glycosyl torsion angle is -107.2°, the furanose ring conformation is C(4')-endo-O(4')-exo ($^4T_0$) with P 244.0° and τ_m 22.8° and the seven-membered ring has the chair conformation. The cyano fragment is non-linear with the C(4)-C(6)-N(6) angle 175.9° and the C(6)-N(6) bond stacks parallel to the benzene ring at a distance of 3.22 Å. The structure is stabilised by van der Waals contacts only.

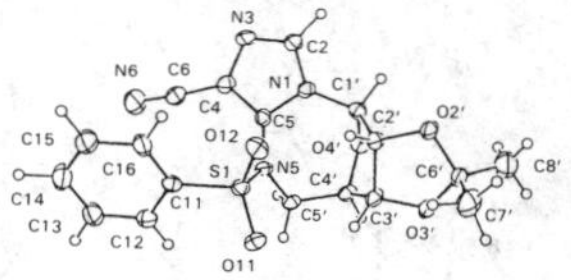

Fig. 1. The $C_{18}H_{18}N_4O_5S$ molecule.

6-KESTOSE MONOHYDRATE
$C_{18}H_{31}O_{16},H_2O$

V. FERRETTI, V. BERTOLASI, G. GILLI and C.A. ACCORSI, 1984. Acta Cryst., C40, 531-535.

Orthorhombic, $P2_12_12_1$, a = 8.718, b = 11.317, c = 23.386 Å, Z = 4. Cu radiation, R = 0.039 for 1381 reflexions.

The molecules (Fig. 1) are associated in the crystal by an extended network of hydrogen bonds where most O atoms are both donors and acceptors. Moreover, there are four bifurcated intramolecular hydrogen bonds. Ring conformation is discussed in terms of puckering-coordinate analysis and the conformation of the glucose-fructose linkage of the sucrose subunit in terms of the potential-energy map calculated in the atom-atom potential approximation. The conformations of the sugar rings are: glucose $^4C_1$, fructose-1 $^4T_3$, fructose-2 E_3.

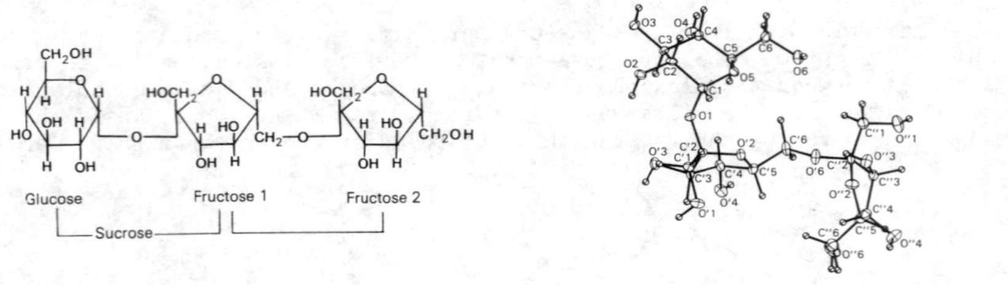

Fig. 1. The 6-kestose molecule.

2-PHENYL-(3,4,6-TRI-O-ACETYL-1,2-DIDEOXY-α-D-GLUCOPYRANO)-[2,1-d]-2-OXAZOLINE
$C_{19}H_{21}NO_8$

C. FOCES-FOCES, F.H. CANO, M. BERNABE, S. PENADES and M. MARTIN-LOMAS, 1984. Carbohydr. Res., 135, 1-11.

Orthorhombic, $P2_12_12$, a = 13.4160, b = 37.4931, c = 7.7886 Å, Z = 8. Cu radiation, R = 0.053 for 3213 reflexions.

There are two independent molecules in the asymmetric unit which although similar in overall conformation are significantly different. The pyranoid rings have conformations between $^0S_2$ and $D^{0,4}$. The five-membered rings also have slightly different conformations being near $_1T^2$ in one molecule and between $_1T^2$ and E_1 in the other; the overall puckering in both rings is quite low.

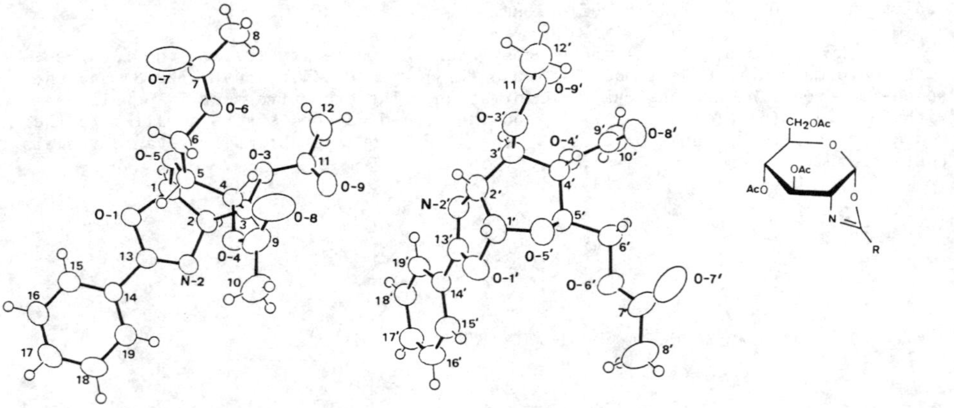

Fig. 1. Views of the two $C_{19}H_{21}NO_8$ molecules.

3-O-(6-O-ACETYL-2,3-ANHYDRO-4-DEOXY-α-L-ribo-HEXOPYRANOSYL)-1,2.5,6-DI-O-ISOPROPYLIDENE-α-D-GLUCOFURANOSE
$C_{20}H_{30}O_{10}$

J.W. KRAJEWSKI, P. GLUZINSKI, Z. URBANCZYK-LIPKOWSKA, A. BANASZEK and M. DOBLER, 1984. Carbohydr. Res., 134, 11-22.

Orthorhombic, $P2_12_12_1$, a = 8.790, b = 11.678, c = 21.457 Å, Z = 4. Mo radiation, R = 0.057 for 1275 reflexions.

In the molecule (Fig. 1) the 4-deoxy-2,3-pyranose ring has a sofa conformation ($_5E$), the 1,2-O-isopropylidene ring has a hybrid conformation (E + T) and the 5,6-O-isopropylidene and the α-D-glucofuranose rings have twist (T) conformations. The ϕ and ψ torsion angles for the glycosidic linkage are 54(4) and 29(4)° respectively.

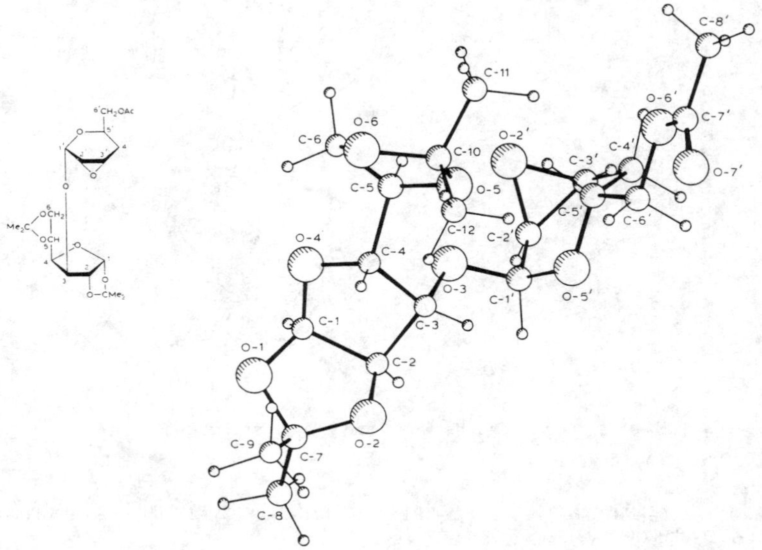

Fig. 1. A view of $C_{20}H_{30}O_{10}$.

3-O-(6-O-ACETYL-2,4-DIAZIDO-3-O-BENZYL-2,4-DIDEOXY-α-D-GLUCOPYRANOSYL)-1,6-ANHYDRO-2,4-DIAZIDO-2,4-DIDEOXY-β-D-GLUCOPYRANOSE
$C_{21}H_{24}N_{12}O_7$

M. STRUMPEL, H.-J. SCHMIDT, P. LUGER and H. PAULSEN, 1984. Carbohydr. Res., 125, 185-201.

Triclinic, P1, a = 8.893, b = 8.696, c = 9.995 Å, α = 105.83, β = 116.22, γ = 88.42°, Z = 1. Cu radiation, R = 0.063 for 2102 reflexions.

The anhydro-bridged system (Fig. 1) has a distorted $^1C_4$(D) conformation, in agreement with that of other anhydropyranoses so far investigated. The non-bridged ring adopts an almost perfect $^4C_1$(D) conformation. The anomeric effect is observed in both of the sugar-ring systems in terms of bond-length shortening. The disaccharide has an α-D-Glc-$^4C_1$-(1a→3e)-D-Glc-$^1C_4$ glycosidic linkage. No previous X-ray investigation of a compound of this type is known. The pyranoid rings are almost perpendicular to each other. The ϕ, ψ angles of the glycosidic linkage are +78.1(5) and -86.0(4)°. The magnitude of the ϕ value is in agreement with the predictions of the exo-anomeric effect. The glycosidic ϕ and ψ angles were very well reproduced by empirical calculations using the HSEA approach.

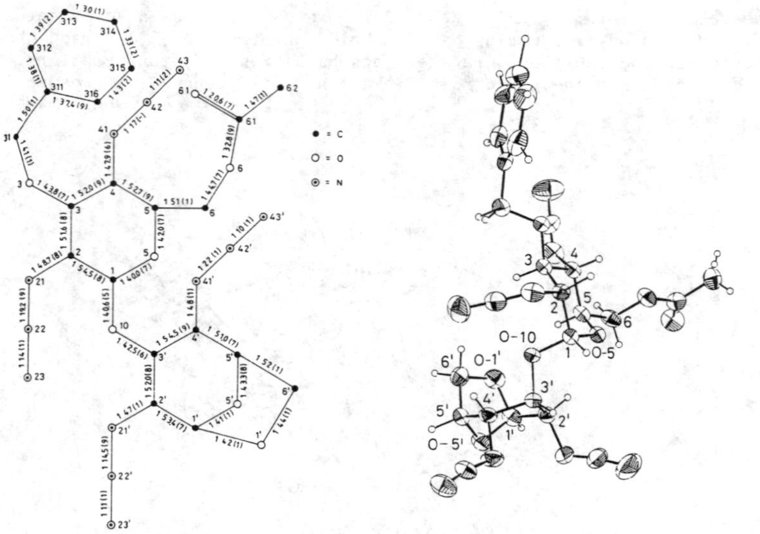

Fig. 1. $C_{21}H_{24}N_{12}O_7$: bond lengths (pm) and a general view.

2,3-DI-O-ACETYL-1,6-ANHYDRO-4-O-(3,4,6-TRI-O-ACETYL-2-DEOXY-2-FLUORO-α-D-GLUCO-PYRANOSYL)-β-D-GLUCOPYRANOSE
$C_{22}H_{29}FO_{14}$

J.G. SHELLING, D. DOLPHIN, P. WIRZ, R.E. COBBLEDICK and F.W.B. EINSTEIN, 1984. Carbohydr. Res., 132, 241-259.

Orthorhombic, $P2_12_12_1$, a = 8.839, b = 8.814, c = 32.721 Å, D_m = 1.39, Z = 4. Mo radiation, R = 0.042 for 1295 reflexions.

In the molecule (Fig. 1) the disaccharide linkage has the α-configuration. The anhydro bridge across the pyranose ring forms a seven-membered ring having a boat conformation; the bridge distorts the pyranose ring (the C(1)-O(1)-C(5) angle is 100.8(5)°).

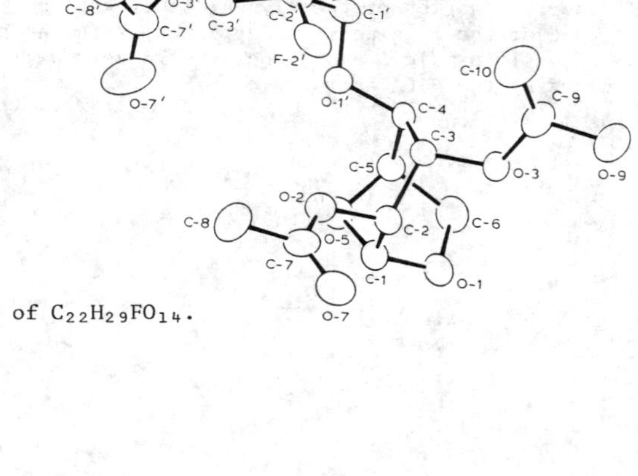

Fig. 1. A view of $C_{22}H_{29}FO_{14}$.

ETHYL 5,6,7,8,9-PENTAACETOXY-2-AMINO-4-NITROMETHYL-D-glycero-D-manno-2-NONENE-3-CARBOXYLATE
$C_{23}H_{34}N_2O_{14}$

R. VEGA, M.J. DIÁNEZ, A. LÓPEZ-CASTRO and R. MÁRQUEZ, 1984. Acta Cryst., C40, 1941-1944.

Orthorhombic, $P2_12_12_1$, a = 17.367, b = 18.176, c = 9.446 Å, D_m = 1.26, Z = 4. Cu radiation, R = 0.057 for 2303 reflexions.

The atoms C(1) to C(4) are coplanar within 0.018(6) Å; the acetoxy groups are all planar with the nitromethyl and exothycarbonyl groups almost planar. The configurations around the chiral centres C(8),C(7),C(6),C(5),C(4) are R,R,R,S,S. An intramolecular hydrogen bond (N...O 2.664(7) Å) connects N(21) and O(32) (Fig. 1).

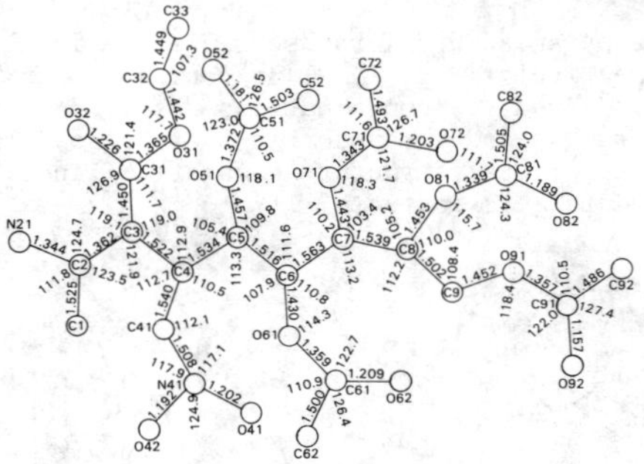

Fig. 1. Bond lengths and angles in the $C_{23}H_{34}N_2O_{14}$ molecule; (e.s.d.'s 0.009-0.012 Å and 0.5-0.8°).

METHYL 2,3,4-TRIDEOXY-2,4-DI-C-METHYL-6-O-TRIPHENYLMETHYL-α-D-lyxo-HEXOPYRANOSIDE
$C_{28}H_{32}O_3$

Y. CHAPLEUR, F. GERMAIN, A. AUBRY and D. BAYEUL, 1984. J. Carbohydr. Chem., 3, 443-459.

Orthorhombic, $P2_12_12_1$, a = 11.269, b = 12.405, c = 16.945 Å, Z = 4. Cu radiation, R = 0.039 for 1600 reflexions.

M. MORI, T. CHUMAN, T. FUJIMORI, K. KATO, A. OHKUBO and S. TODA, 1984. Carbohydr. Res., 127, 171-179.

Orthorhombic, $P2_12_12_1$, a = 11.268, b = 12.405, c = 16.967 Å, D_m = 1.168, Z = 4. Mo radiation, R = 0.065 for 2770 reflexions.

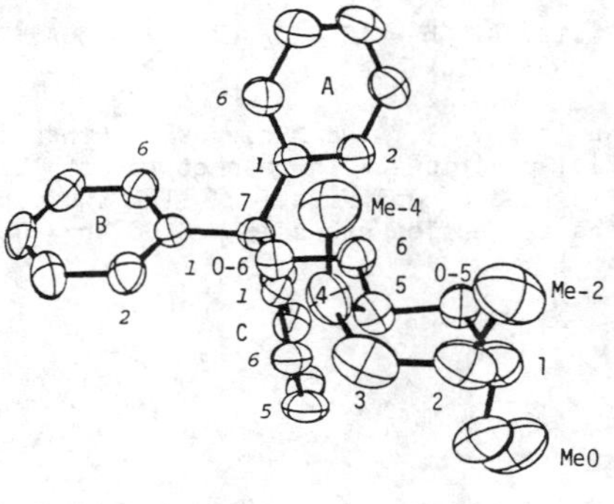

Fig. 1. A view of $C_{28}H_{32}O_3$.

The molecule is shown to adopt the $^4C_1$(D) conformation in the solid state (Fig. 1) with endocyclic torsion angles (starting at C1-C2) +48.3, -46.4, +50.5, -56.6, +64.7, -59.8(4)°. High proton NMR studies indicate that this is not the sole species in solution.

SUCROSE OCTAACETATE
$C_{28}H_{38}O_{19}$

J.D. OLIVER and L.C. STRICKLAND, 1984. Acta Cryst., C40, 820-824.

Orthorhombic, $P2_12_12_1$, a = 18.350, b = 21.441, c = 8.350 Å, Z = 4. Mo radiation, R = 0.053 for 2156 reflexions (at 173K).

In the molecule the pyranose and furanose rings are in chair ($^4C_1$) and twist ($^4T_1$) conformations respectively (Fig. 1). The conformations about the glycosidic bonds differ by 14 and 23° from the conformations observed in sucrose (1). The conformationns of the pyranose and furanose rings are also different from those observed for sucrose. The crystal structure consists of individual molecules separated by normal van der Waals contacts.

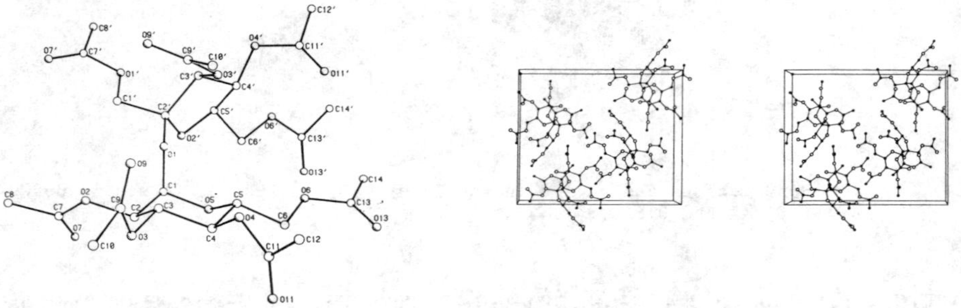

Fig. 1. $C_{28}H_{38}O_{19}$: perspective view of the molecule and stereoview of the molecular packing.

1. Structure Reports, 39B, 309.

N,N'-DI-β-D-TETRAACETATOGLUCOPYRANOSYL-PHENYLHYDRAZINE
$C_{34}H_{44}N_2O_{18}$

B.K. SADYBAKASOV, V.A. AFANES'EV, M.Yu. ANTIPIN and Yu.T. STRUCHKOV, 1984. J. Struct. Chem., 25, 930-934 [Zh. Strukt. Khim., 25(6), 104-108].

Orthorhombic, $P2_12_12_1$, a = 12.087, b = 13.835, c = 23.563 Å, D_m = 1.28, Z = 4. Mo radiation, R = 0.045 for 3560 reflexions.

The molecule is shown in Fig. 1. The N-glycoside bonds in this diglucopyranosyl derivative occupy equatorial positions with respect to both rings, resulting in an energetically favorable β-pyranose configuration. The two rings have the same $^4C_1$ conformation. Bond lengths and angles are as expected (σ = 0.004-0.006 Å and 0.2-0.4°).

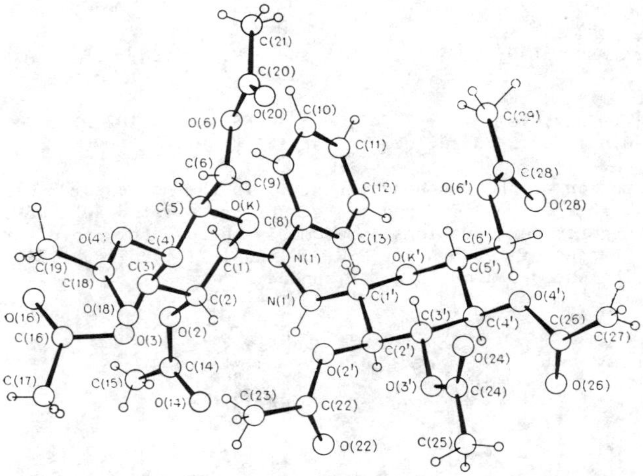

Fig. 1. The N,N'-di-β-D-tetraacetatoglucopyranosyl-phenylhydrazine molecule.

1-O-ACETYL-2,5,6-TRI-O-BENZOYL-4-HYDROXY-3,4-O-(α-SUCCINIMIDOBENZYLIDENE)-β-D-GALACTOFURANOSE

$C_{40}H_{33}NO_{13}$

R.J. FERRIER, S.R. HAINES, G.J. GAINSFORD and E.J. GABE, 1984. J. Chem. Soc. Perkin I, 1683-1687.

Orthorhombic, $P2_12_12_1$, a = 14.797, b = 23.671, c = 10.716 Å, Z = 4. Mo radiation, R = 0.048 for 1723 reflexions.

In the molecule (Fig. 1) the succinimide group occupies the exo-site and the furanose ring adopts a conformation close to $^3T_2$. Phenyl rings C(28-33) and C(35-40) showed some disorder.

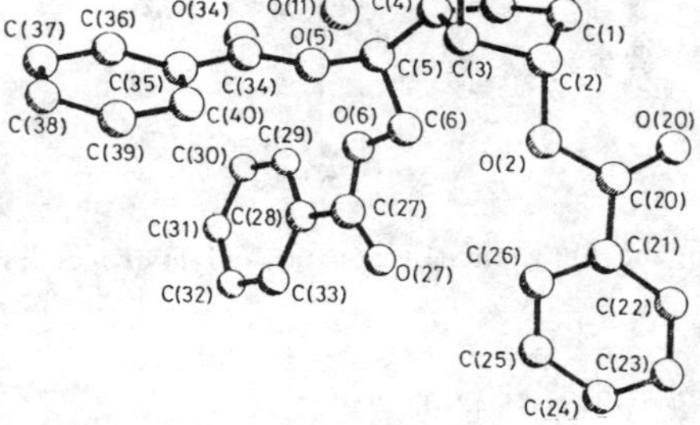

Fig. 1. $C_{40}H_{33}NO_{13}$: the molecular structure.

PHOSPHORYLETHANOLAMINE (2-AMINOETHYL PHOSPHATE)
$C_2H_8NO_4P$

H.-P. WEBER, R.K. McMULLAN, S. SWAMINATHAN and B.M. CRAVEN, 1984. Acta Cryst., B40, 506-511.

Monoclinic, $P2_1/c$, a = 9.015, b = 7.745, c = 8.788 Å, β = 102.51°, Z = 4. Neutron radiation, R = 0.041 for 2473 reflexions (at 122 K).

The conformation at the P-O bonds (Fig. 1) in the monoester H-O-P-O-C group is (+)-ac, (-)-sc. The terminal O-H and NH_3 groups are twisted 40° and 20° respectively from ideally staggered conformations, presumably because they form strong H bonds, having H...O distances of 1.535 Å for O-H...O and 1.713, 1.714, 1.747 Å for N-H...O interactions. The ester O atom is not H bonded.

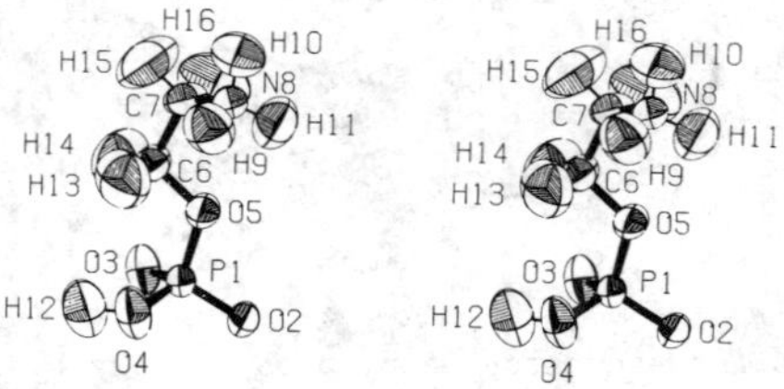

Fig. 1. A stereoview of 2-aminoethyl phosphate

ETHYLENEDIAMMONIUM HYDROGEN PHOSPHATE
$C_2H_{11}N_2O_4P$ $(C_2H_{10}N_2)^{2+}(HPO_4)^{2-}$

S.N. GOLUBEV and Yu.D. KONDRASHEV, 1984. J. Struct. Chem., 25, 471-473 [Zh. Strukt. Khim., 25(3), 150-153].

Monoclinic, $P2_1/b$, a = 7.500, b = 8.045, c = 11.806 Å, γ = 110.13°, D_m = 1.59, Z = 4. Mo radiation, R = 0.041 for 1619 reflexions.

The structure consists of tetrahedral $(HPO_4)^{2-}$ anions and planar $H_3NCH_2CH_2NH_3{}^{2+}$ cations in a trans conformation. An extensive system of hydrogen bonds with O-H...O 2.57 Å and N-H...O 2.71-2.90 Å creates anion layers in the ab plane which are connected to each other by layers of ethylenediammonium ions. Bond lengths for the ions (σ = 0.001-0.003 Å and 0.1° for non-hydrogen atoms, 0.05 Å and 2-4° involving hydrogen) are shown in Fig. 1.

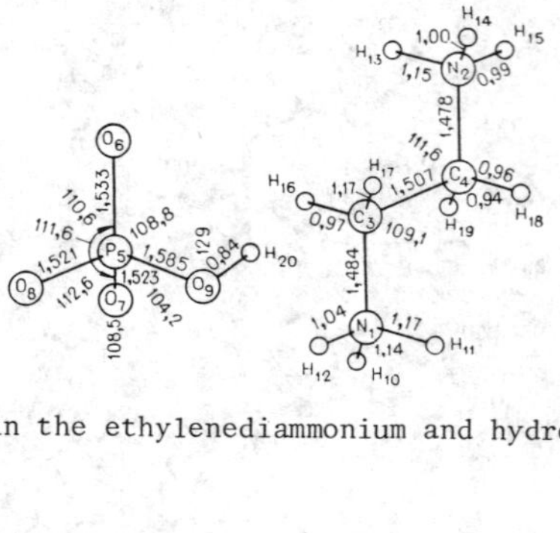

Fig. 1. Bond lengths in the ethylenediammonium and hydrogen phosphate ions.

GLUCOSE 1-PHOSPHATE HYDRATE DISODIUM SALT
$C_6H_{11}Na_2O_9P,3.5H_2O$ $2Na^+.C_6H_{11}O_9P^{2-}.3.5H_2O$

N. NARENDRA, T.P. SESHADRI and M.A. VISWAMITRA, 1984. Acta Cryst., C40, 1338-1340.

Monoclinic, C2, a = 8.429, b = 10.184, c = 16.570 Å, β = 99.18°, D_m = 1.73, Z = 4. Cu radiation, R = 0.069 for 1524 reflexions.

The pyranose ring has the $^4C_1$ chair conformation with C(4) and C(1) displaced -0.672(8) and 0.703(8) Å respectively from the C(2)-C(3)-C(5)-O(5) plane (Fig. 1). The conformation about C(5)-C(6) is gauche-gauche with torsion angles O(6)-C(6)-C(5)-O(5) -61.2(8) and O(6)-C(6)-C(5)-C(4) 58.2(9)°. Two sodium ions are six-coordinate and lie on twofold axes; the third sodium ion is five-coordinate. The structure is held together by hydrogen bonds but there is no intramolecular hydrogen bonding. Some bond lengths are: C(5)-O(5) 1.445(10), C(1)-O(5) 1.424(10), P-O(1) 1.641(6) Å.

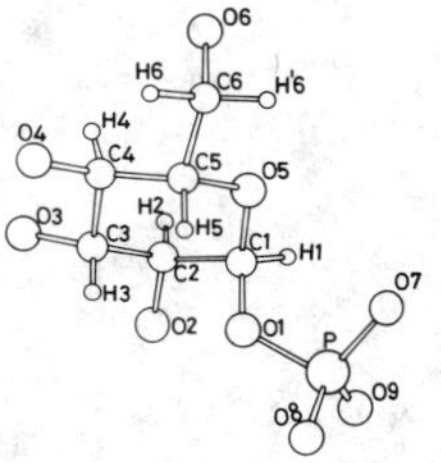

Fig. 1. The $C_6H_{11}O_9P^{2-}$ dianion.

CALCIUM 2'-DEOXYADENOSINE 5'-PHOSPHATE PENTAHYDRATE

$C_{10}H_{12}CaN_5O_6P,5H_2O$ $Ca^{2+},C_{10}H_{12}N_5O_6P^{2-},5H_2O$

T. SATO, 1984. Acta Cryst., C40, 738-740.

Orthorhombic, $P2_12_12_1$, a = 6.882, b = 39.761, c = 6.569 Å, Z = 4. Cu radiation, R = 0.024 for 1884 reflexions.

The molecule (Fig. 1) has a C(4)-exo-O(4')-endo ($^oT_4$) sugar associated with an anti glycosidic bond rotation, with torsion angle O(4')-C(1')-N(9)-C(8) 60.9(3)°. The conformations about the C(4')-C(5') and C(5')-O(5') bonds are trans-gauche, with torsion angle C(3')-C(4')-C(5')-O(5') -71.3(2)° and trans, with torsion angle C(4')-C(5')-O(5')-P 169.1(1)°, respectively. All available H atoms participate in intermolecular hydrogen bonds.

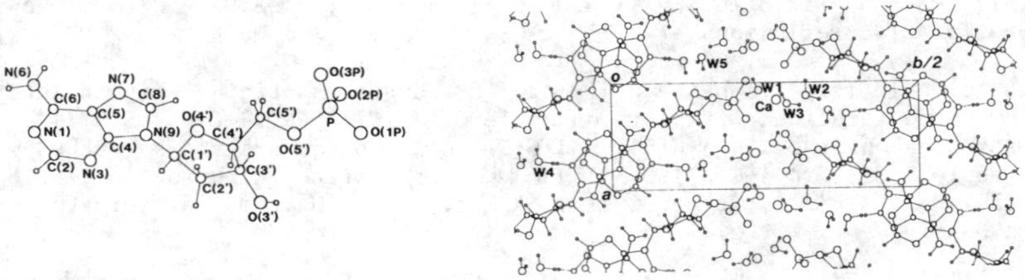

Fig. 1. $C_{10}H_{12}CaN_5O_6P,5H_2O$: a perspective view of the molecule and a projection of the structure viewed along the c axis.

5-IODO-2'-DEOXYURIDINE (3',5')-CYCLIC MONOPHOSPHATE P-O-METHYL ESTER

$C_{10}H_{12}IN_2O_7P$

J. BERES, P. SANDOR, A. KALMAN, T. KORITSANSZKY and L. OTVOS, 1984. Tetrahedron, 40, 2405-2414.

Orthorhombic, $P2_12_12_1$, a = 8.956, b = 10.520, c = 15.496 Å, Z = 4. Mo radiation, R = 0.043 for 1941 reflexions.

The P-methoxy group is axial to the dioxaphosphorinane ring (Fig. 1). The P-O(6) distance is 1.446 Å, other P-O distances are in the 1.560-1.586(6) Å range. The phosphorinane ring has a distorted chair conformation (shifted towards an envelope-like form). Asymmetry factors in the ribofuranose ring are fC_2 = 4.7 pm (fC_S = 7.4 pm). Molecules are linked by N(3)-H...O(6) intermolecular hydrogen bonds (N...O 2.77(1) Å).

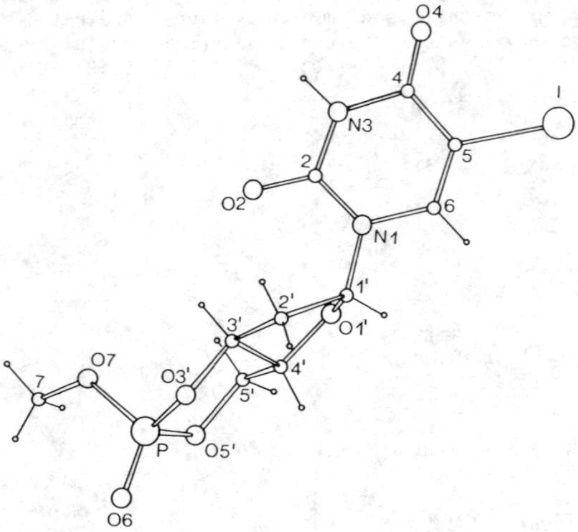

Fig. 1. A view of $C_{10}H_{12}IN_2O_7P$.

CALCIUM THYMIDINE 5'-PHOSPHATE DIHYDRATE

$C_{10}H_{13}CaN_2O_8P,2H_2O$ $Ca^{2+},C_{10}H_{13}N_2O_8P^{2-},2H_2O$

T. SATO, 1984. Acta Cryst., C40, 736-738.

Monoclinic, $P2_1$, a = 12.821, b = 11.050, c = 5.543 Å, β = 98.95°, Z = 2. Mo radiation, R = 0.030 for 2146 reflexions.

The molecule (Fig. 1) has a C(4')-exo (E_4) sugar associated with an anti glycosidic bond rotation, with torsion angle O(4')-C(1')-N(1)-C(6) 65.7(4)°. The conformations about the C(4')-C(5') and C(5')-O(5') bonds are trans-gauche, with torsion angle C(3')-C(4')-C(5')-O(5') -60.4(3)°, and trans, with torsion angle C(4')-C(5')-O(5')-P -153.5(2)°, respectively. All available H atoms participate in hydrogen bonding.

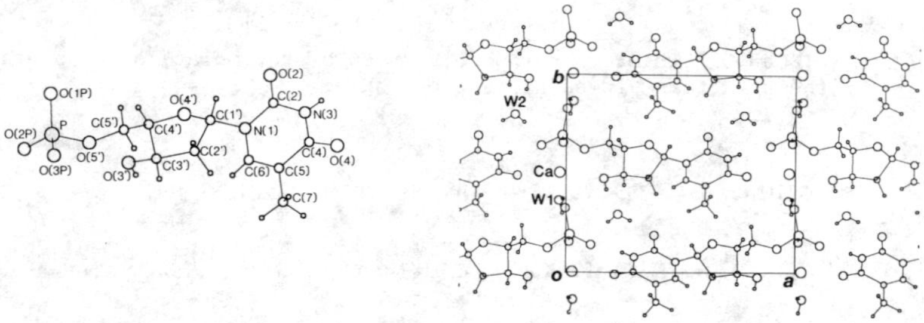

Fig. 1. $C_{10}H_{13}CaN_2O_8P,2H_2O$: a perspective view of the molecule and a projection of the structure viewed along the c axis.

O-p-TOLYL-O-α-DIETHYLAMINOMETHYL-β-CHLOROVINYLPHOSPHATE MONOHYDRATE
$C_{14}H_{21}ClNO_4P,H_2O$

M.Yu. ANTIPIN, A.N. CHERNEGA, T.V. KIM, E.I. KISELEVA, Yu.G. GOLOLOBOV and Yu.T. STRUCHKOV, 1984. J. Struct. Chem., 25, 463-466 [Zh. Strukt. Khim., 25(3) 141-144].

Monoclinic, Cc, a = 7.77, b = 27.48, c = 7.893 Å, β = 92.5°, Z = 4. Mo radiation, R = 0.047 for 2350 reflexions (at -120°C).

The molecule (Fig. 1) has a strong intramolecular hydrogen bond N-H...O-P (N...O 2.713(4) Å, N-H...O 160(4)°, P-O 1.491 Å) which stabilizes the betaine with the formation of a seven-membered non-planar ring. Bond lengths are shown in Fig. 1. Hydrogen bonding with the water molecule links the molecules in infinite chains.

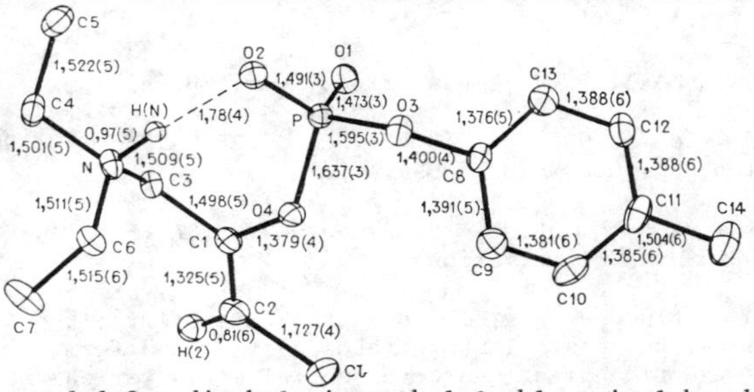

Fig. 1. The O-p-tolyl-O-α-diethylaminomethyl-β-chlorovinylphosphate molecule.

ADENOSINE 5'-O-(DIETHYL PHOSPHATE)
$C_{14}H_{22}N_5O_7P$

R.G. BRENNAN, N.S. KONDO and M. SUNDARALINGAM, 1984. J. Am. Chem. Soc., 106, 5671-5676.

Triclinic, P1, a = 6.799, b = 7.923, c = 9.003 Å, α = 86.86, β = 77.98, γ = 77.85°, D_m = 1.45, Z = 1. Cu radiation, R = 0.039 for 1917 reflexions.

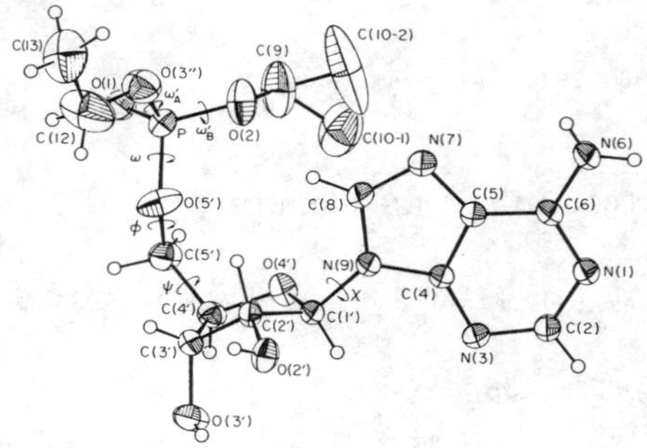

Fig. 1. A view of $C_{14}H_{22}N_5O_7P$.

In the molecule (Fig. 1) the D-ribofuranosyl ring (Fig. 1) is in a symmetrical twist conformation, C(2')-endo-C(1')-exo($^{2}_{1}T$), with pseudorotational parameters P 145.0(2)° and τ_m 39.4(2)°. The adenine base is anti (69.0(3)°) about the C(1')-N(9) glycosyl bond, and the conformation about the exocyclic bond C(4')-C(5') is the preferred gauche$^+$ (50.8(3)°). Hydrogen bonding is centred about the ribosyl hydroxyls and the N(6) amino group of the base. The molecular packing is dominated by intermolecular base-alkyl stacking and alkyl-alkyl van der Waals interactions. The methyl group of one of the ethoxy groups is two-site disordered. All four P-O bonds are significantly shorter than those of the nonalkylated nucleotides. The three combinations of phosphodiester linkages display the (g^-,t), (g^+,g^-), and (t,t) conformations which are different from the (g^-,g^-) conformation of right-handed polynucleotide helices. The lack of preference of the alkylated sugar phosphate backbone for the (g^-,g^-) phosphodiester conformation would tend to destack the bases and promote alkyl-base stacking. This will lead to sugar phosphate backbone configurations which would provide a mechanism for the biochemical and biological effects induced by phosphate alkylation.

DISODIUM URIDINE DIPHOSPHOGLUCOSE DIHYDRATE
$C_{15}H_{22}N_2NaO_{17}P_2,2H_2O$ (I)

DIPOTASSIUM GLUCOSE 1-PHOSPHATE DIHYDRATE
$C_6H_{11}K_2O_9P_2,2H_2O$ (II)

Y. SUGAWARA and H. IWASAKI, 1984. Acta Cryst., C40, 389-393.

I. Monoclinic, $P2_1$, a = 8.950, b = 18.724, c = 7.289 Å, β = 101.98°, D_m = 1.80, Z = 2. Mo radiation, R = 0.049 for 5895 reflexions.

II. Monoclinic, $P2_1$, a = 10.458, b = 9.027, c = 7.532 Å, β = 110.39°, Z = 2. Mo radiation, R = 0.029 for 3761 reflexions.

Uridine diphosphoglucose (Fig. 1) takes the extend form; the uracil is anti to the ribose and the pyrophosphate linkage takes a staggered form. O(4") and C(6"), which are known as reaction sites and/or essential sites for enzyme specificity are very close to the mean plane through the pyranose ring. The structure of glucose 1-phosphate (1) has been refined, details are also in Fig. 1.

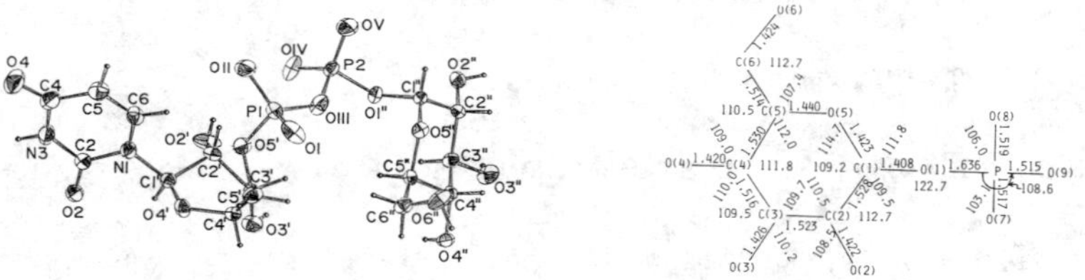

Fig. 1. The uridine diphosphoglucose structure, and dimensions of glucose 1-phosphate.

1. Structure Reports, 30B, 191.

BIS(CYCLOHEXYLAMMONIUM) PROPARGYL PHOSPHATE DIHYDRATE
$C_{15}H_{31}N_2O_4P,2H_2O$ $2C_6H_{14}N^+.C_3H_3O_4P^{2-},2H_2O$

P.G. JONES, G.M. SHELDRICK, A.J. KIRBY and K.W.Y. ABELL, 1984. Acta Cryst., C40, 547-549.

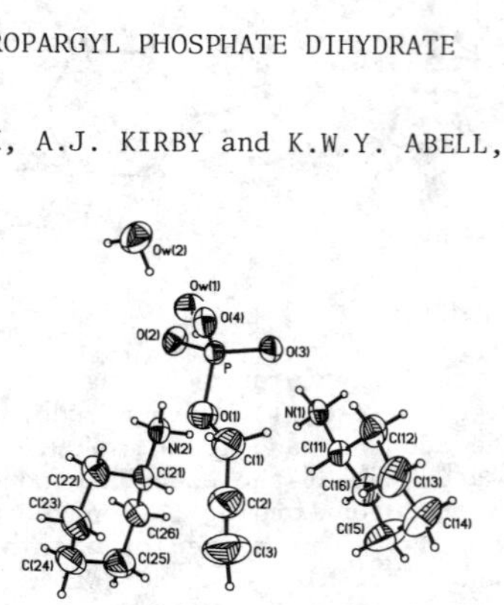

Fig. 1. A plot of the asymmetric unit for $2C_6H_{14}N^+.C_3H_3O_4P^{2-},2H_2O$.

Monoclinic, $P2_1/c$, a = 14.186, b = 6.355, c = 21.417 Å, β = 105.55°, Z = 4. Mo radiation, R = 0.080 for 1978 reflexions.

In the anion (Fig. 1) the P-OR bond length 1.626(4) Å, is appreciably longer than in related ester dianions with R = Me, Et. The mean P-OR bond length of all phosphate monoester dianions with R < 10% in the Cambridge Data Base was 1.614(4) Å. A complex hydrogen-bonding network links the hydrophilic residues in bands parallel to z.

BIS(CYCLOHEXYLAMMONIUM) 4-NITROPHENYL PHOSPHATE DIHYDRATE

$C_{18}H_{32}N_3O_6P,2(H_2O)$ $2C_6H_{14}N^+.C_6H_4NO_6P^{2-},2H_2O$

P.G. JONES, G.M. SHELDRICK, A.J. KIRBY and K.W.Y. ABELL, 1984. Acta Cryst., C40, 550-552.

Triclinic, $P\bar{1}$, a = 6.584, b = 12.138, c = 14.965 Å, α = 85.80, β = 87.28, γ = 81.45°, Z = 2. Mo radiation, R = 0.078 for 1953 reflexions.

In the anion (Fig. 1) the P-OR bond length 1.664(5) Å, is appreciably greater than in alkyl phosphate dianions. A system of hydrogen bonds links the hydrophilic residues in bands parallel to b at z = 0.50.

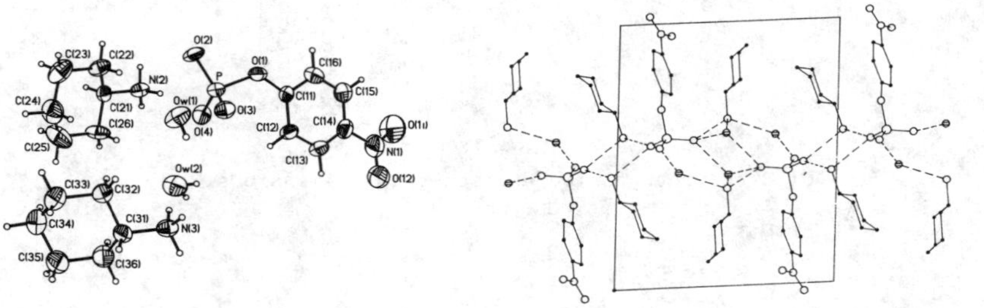

Fig. 1. The asymmetric unit of $2C_6H_{14}N^+.C_6H_4NO_6P^{2-},2H_2O$ and a projection of the structure along a.

8-CHLOROGUANOSINE DIHYDRATE

$C_{10}H_{12}ClN_5O_5,2(H_2O)$

G.I. BIRNBAUM, P. LASSOTA and D. SHUGAR, 1984. Biochemistry, 23, 5048-5053.

Orthorhombic, $P2_12_12_1$, a = 4.871, b = 12.040, c = 24.506 Å, Z = 4. Mo radiation, R = 0.031 for 1599 reflexions.

Molecular dimensions are shown in Fig. 1. The conformation about the glycosidic bond is syn with χ_{CN} -131.1°. The ribose ring has a C(2')-endo/C(1')-exo ($^2T_1$) pucker, and the gauche$^+$ conformation of the $-CH_2OH$ side chain is stabilized by an intramolecular O(5')-H...N(3) hydrogen bond (2.707 Å).

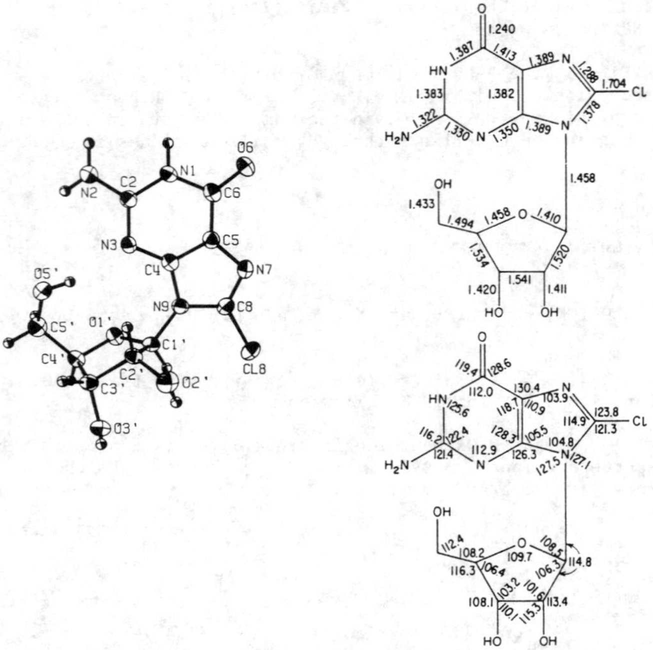

Fig. 1. A view of 8-chloroguanosine and molecular dimensions.

XANTHOSINE
$C_{10}H_{12}N_4O_6$

B. LESYNG, C. MARCK and W. SAENGER, 1984. Z. Naturforsch., C39, 720-724.

Orthorhombic, $P2_12_12_1$, a = 17.406, b = 12.378, c = 5.350 Å, Z = 4. Cu radiation, R = 0.039 for 984 reflexions.

The molecule is found in the syn conformation with χ -132.7° which is stabilized by an intramolecular hydrogen bond N(3)-H...O(5') (N...O 2.836 Å). The pseudorotation parameters for sugar puckering are P = 156° and τ_m = 38.7° corresponding to a C(2')-endo form. The crystal packing of this anhydrous form is quite different from that of the dihydrate (1), yet the molecular conformations are similar.

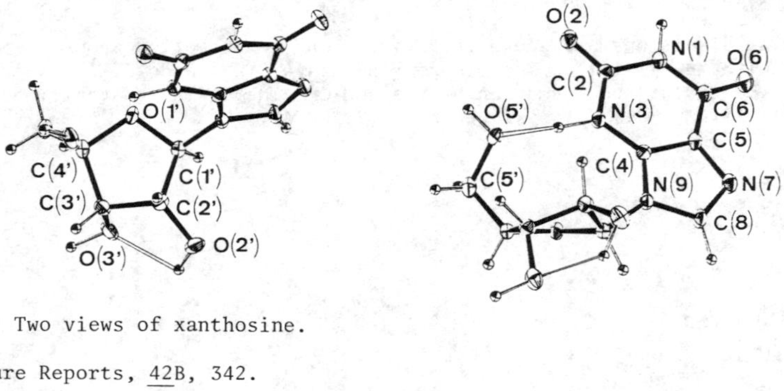

Fig. 1. Two views of xanthosine.

1. Structure Reports, 42B, 342.

2'-DEOXY-2'-FLUOROADENOSINE
$C_{10}H_{12}N_5O_3F$

K. MORISHITA, T. HAKOSHIMA, T. FUJIWARA, K.-I. TOMITA, T. KANEYASU, S. UESUGI and M. IKEHARA, 1984. Acta Cryst., C40, 434-436.

Monoclinic, $P2_1$, a = 4.791, b = 10.428, c = 11.507 Å, β = 102.06°, D_m = 1.576, Z = 2. Cu radiation, R = 0.066 for 1066 reflexions.

The molecule and dimensions are in Fig. 1. The base orientation around the glycosidic bond is anti and the conformation around the exocyclic C(4')-C(5') bond is gauche-trans. The sugar ring has C(3')-endo (^3E) puckering.

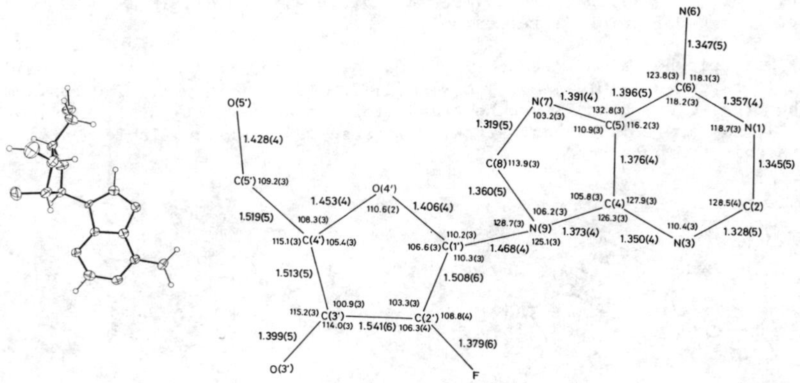

Fig. 1. The 2'-deoxy-2'-fluoroadenosine molecule and dimensions.

1-(5'-AMINO-5'-DEOXY-β-D-ALLOFURANURONOSYL)-4-FORMYL-4-IMIDAZOLIN-2-ONE TRIHYDRATE
$C_{10}H_{13}N_3O_7,3(H_2O)$ (I)

1-(5'-AMINO-5'-DEOXY-β-D-ALLOFURANURONOSYL)URACIL TRIHYDRATE
$C_{10}H_{13}N_3O_7,3(H_2O)$ (II)

R. RATHMANN, W.A. KONIG, H. SCHMALLE, G. CARLSSON, R. BOSCH, H. HAGENMAIER and W. WINTER, 1984. Justus Liebigs Ann. Chem., 1216-1229.

I. Orthorhombic, $P2_12_12_1$, a = 18.968, b = 14.297, c = 5.453 Å, Z = 4. Cu radiation, R = 0.067 for 2534 reflexions.

II. Orthorhombic, $P2_12_12_1$, a = 5.410, b = 14.254, c = 19.154 Å, Z = 4. Mo radiation, R = 0.042 for 1191 reflexions.

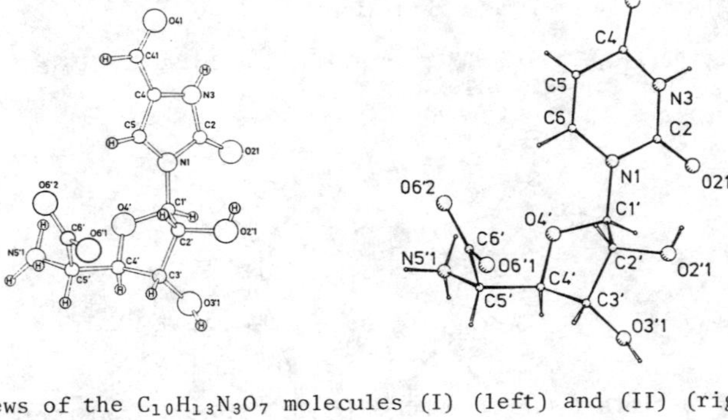

Fig. 1. Views of the $C_{10}H_{13}N_3O_7$ molecules (I) (left) and (II) (right).

Both compounds (Fig. 1) have the C-2'-endo-C-1'-exo conformation (${}_1^2T$) with gauche (g^+) position of the CO_2^- group, trans (t) position of the NH_3^+, and anti position of the base. NMR studies indicate only minor deviations from the conformation of the crystal.

2'-DEOXYADENOSINE
$C_{10}H_{13}N_5O_3$

T. SATO, 1984. Acta Cryst., C40, 880-882.

Monoclinic, $P2_1$, a = 11.298, b = 10.393, c = 4.819 Å, β = 101.51°, Z = 2. Mo radiation, R = 0.035 for 2241 reflexions.

The molecule has a C(3')-endo (^3E) sugar associated with an anti glycosidic bond rotation (torsion angle O(4')-C(1')-N(9)-C(8) 23.4(2)°) (Fig. 1). The conformation about the C(4')-C(5') bond is gauche-trans (torsion angle C(3')-C(4')-C(5')-O(5') 175.5(1)°). No appreciable base stacking is observed. All available H atoms participate in hydrogen bonds.

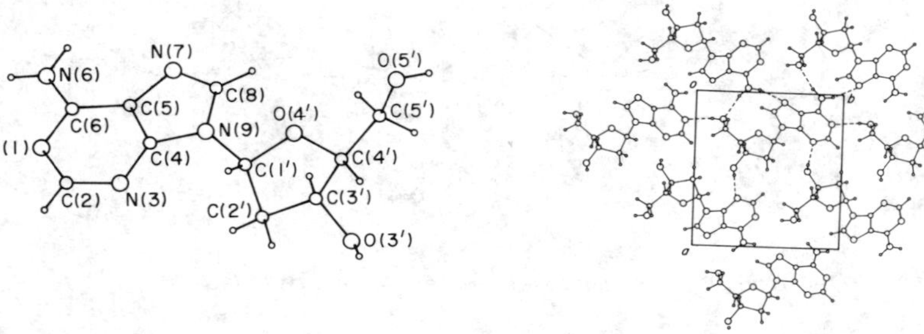

Fig. 1. A perspective view of the $C_{10}H_{13}N_5O_3$ molecule and a projection of structure along the c axis; hydrogen bonds are shown as broken lines.

1-β-D-RIBOFURANOSYLIMIDAZO[1,2-b]PYRAZOLE-7-CARBONITRILE
$C_{11}H_{12}N_4O_4$ (I)

5-β-D-RIBOFURANOSYLIMIDAZO[1,2-b]PYRAZOLE-7-CARBONITRILE
$C_{11}H_{12}N_4O_4$ (II)

S.G. WOOD, N.K. DALLEY, R.D. GEORGE, R.K. ROBINS and G.R. REVANKAR, 1984. J. Org. Chem., 49, 3534-3540.

I. Monoclinic, $P2_1$, a = 8.651, b = 7.031, c = 10.133 Å, β = 97.79°, D_m = 1.44, Z = 2. Mo radiation, R = 0.046 for 2254 reflexions.

II. Orthorhombic, $P2_12_12_1$, a = 7.507, b = 9.015, c = 16.907 Å, D_m = 1.53, Z = 4. Mo radiation, R = 0.050 for 822 reflexions.

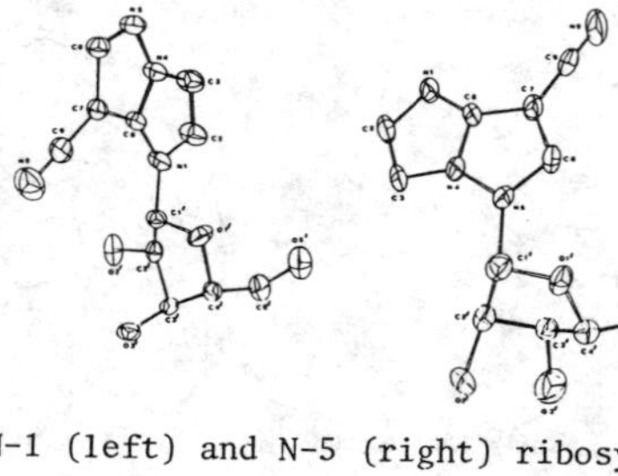

Fig. 1. Views of the N-1 (left) and N-5 (right) ribosyl isomers of $C_{11}H_{12}N_4O_4$.

The sugar conformations in the two isomers differ (Fig. 1). In (I) there is a $_3T^2$ conformation, with O5'-C5'-C4'-O1' and O5'-C5'-C4'-C3' torsion angles -63.2 and +62.7° respectively. Molecule (II) has a $^3T_2$ conformation, with torsion angles O5'-C5'-C4'-O1' and O5'-C5'-C4'-C3' -59.8 and +60.3° respectively. The glycoside torsion angles are O1'-C'-N1-C8 101.0° in (I) and O1'-C1'-N5-N4 -174.2° in (II). The glycosidic bond in (II) (1.505(6) Å) is considerably longer than that in (I) (1.451(2) Å). In both (I) and (II), molecules are linked by hydrogen bonds.

1,N^6-ETHENOADENOSINE HYDROCHLORIDE
$C_{12}H_{14}ClN_5O_4$

M. JASKOLSKI, 1984. J. Cryst. Spectrosc. Res., 14, 45-57.

Monoclinic, $P2_1$, a = 6.1540, b = 18.511, c = 6.7412 Å, β = 112.602°, Z = 2. Mo radiation, R = 0.045 for 2662 reflexions.

The ε-adenosine-H^+ moiety (Fig. 1) is not planar but has an S shape. The ribose is 2'-endo-3'-exo puckered, with a g^+ side chain; the glycosidic torsion angle is in the anti region. The structure contains a three-dimensional network of H bonds in which also C-H...N and C-H...Cl interactions seem to play an important role. There is no base stacking in the structure.

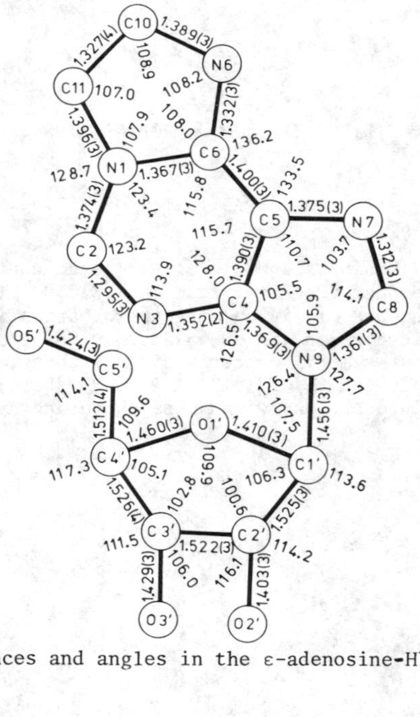

Fig. 1. Bond distances and angles in the ε-adenosine-H^+ ion.

6,3'-ANHYDRO-1-(4'-O-ACETYL-2'-DEOXY-β-D-XYLOPYRANOSYL)-6-METHYLURACIL HEMIMETHANOL
$C_{12}H_{14}N_2O_5,0.5(CH_4O)$

Y. YAMAGATA, T. FUJIWARA, K.-I. TOMITA, S. SHUTO and T. UEDA, 1984. Acta Cryst., C40, 543-545.

Monoclinic, C2, a = 14.050, b = 7.066, c = 13.211 Å, β = 93.18°, D_m = 1.430, Z = 4. Cu radiation, R = 0.052 for 1108 reflexions.

The bond distances and angles of the base moiety agree very well with standard values. The pyrimidine ring (Fig. 1) is essentially planar and the maximum atom deviation from the mean plane is 0.032(6) Å for C(4). The orientation of the base moiety is anti, with the torsion angle O(5')-C(1')-N(1)-C(6) 99.1(4)°. The pyranose ring shows a chair conformation, which is slightly distorted from the typical chair form found in sucrose; i.e. the mean deviation from the average absolute endocyclic torsion angle (55.6°) in sucrose is 5°, reflecting the strain caused by the cyclization between C(6) in the uracil moiety and C(3') in the pyranose ring. The newly formed six-membered heterocycle has an envelope conformation with C(2') displaced 0.772(6) Å from the plane through the remaining five atoms.

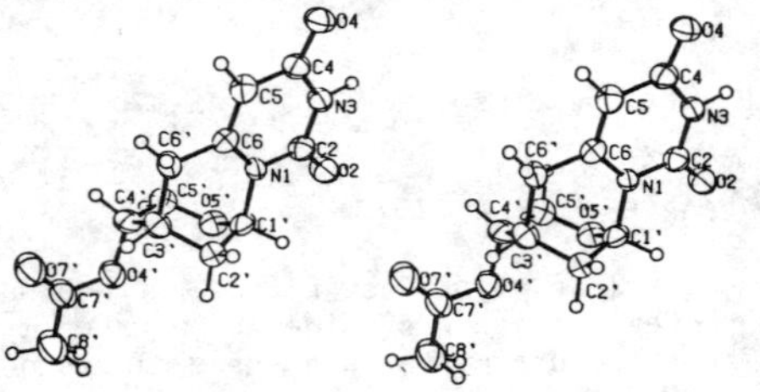

Fig. 1. A stereoview of $C_{12}H_{14}N_2O_5$.

5-(PROPYN-1-YL)-1-(β-D-ARABINOFURANOSYL)URACIL
$C_{12}H_{14}N_2O_6$

M. CYGLER, W.F. ANDERSON, J. GIZIEWICZ and M.J. ROBINS, 1984. Can. J. Chem., 62, 147-152.

Orthorhombic, $P2_12_12_1$, a = 4.925, b = 14.326, c = 17.454 Å, Z = 4. Cu radiation, R = 0.032 for 1155 reflexions.

The sugar ring (Fig. 1) exhibits an ^3E or a C(3')endo conformation with a pseudorotation angle P = 28.3° and puckering amplitude τ_m = 31.7°. The orientation of the $-CH_2OH$ side chain is g^+. The base is in an anti conformation with respect to the sugar ring, with a glycosidic torsion angle χ = 33.7°. Changes in the C(5)-C(6) and C(6)-N(1) bond lengths suggest some interaction of the propynyl group with the pyrimidine ring. The C≡C bond length of 1.121 Å is very short; its shortening could result from intermolecular interactions with the neighbouring pyrimidine ring and nearby oxygen atoms.

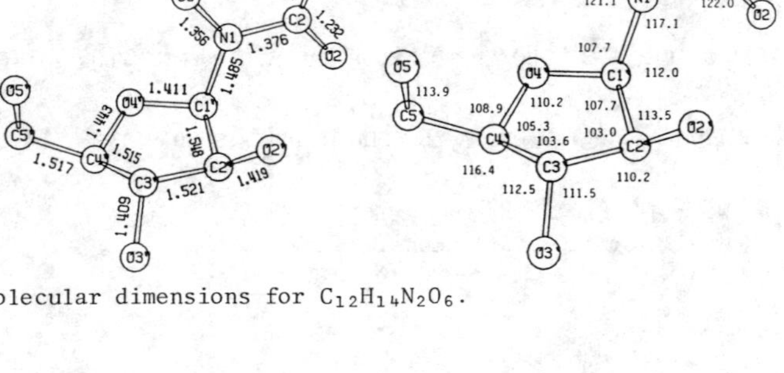

Fig. 1. Molecular dimensions for $C_{12}H_{14}N_2O_6$.

3',5'-DI-O-ACETYLTHYMIDINE
$C_{14}H_{18}N_2O_7$

C.C. WILSON, J.N. LOW, P. TOLLIN and H.R. WILSON. Acta Cryst., C40, 1712-1715.

Monoclinic, $P2_1$, a = 9.796, b = 18.758, c = 8.829 Å, β = 100.72°, Z = 4. Mo radiation, R = 0.086 for 2232 reflexions.

The two molecules in the asymmetric unit (Fig. 1) form a hydrogen-bonded base-paired dimer with both molecules in the anti conformation, and both displaying sugar pucker ^2E [C(2')-endo]. The conformation about C(4')-C(5') for both molecules is gauche-gauche (g^+). A close contact of O(3')1 with a neighbouring base ring occurs.

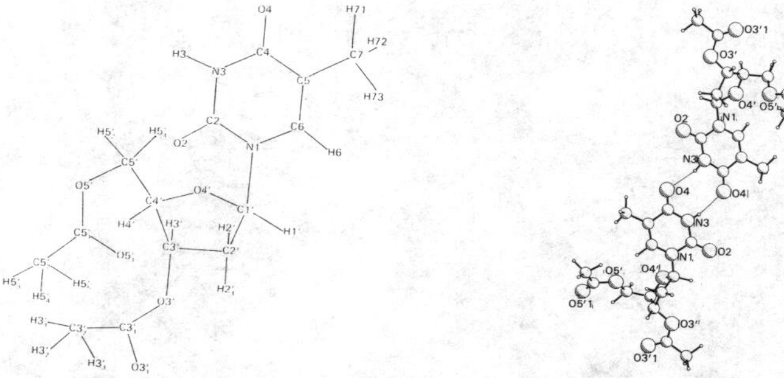

Fig. 1. The $C_{14}H_{18}N_2O_7$ numbering scheme and hydrogen-bonded dimer.

2',3',5'-TRI-O-ACETYLURIDINE
$C_{15}H_{18}N_2O_9$

J.N. LOW and C.C. WILSON, 1984. Acta Cryst., C40, 1030-1032.

Orthorhombic, $P2_12_12_1$, a = 7.493, b = 14.327, c = 16.380 Å, Z = 4. Mo radiation, R = 0.089 for 1712 reflexions.

In the molecule (Fig. 1) the sugar-ring pucker is $^3T_4$ (C(3')-endo/C(4')-eno), with P 46.5(6)°, and χ_{CN} (C(2)-N(1)-C(1')-O(4')) is 74.2(6)°, in the syn range. A close contact of 2.90(3) Å between acetyl oxygen O(5'1) and a neighbouring base ring is noted. The pyrimidine base ring is essentially planar.

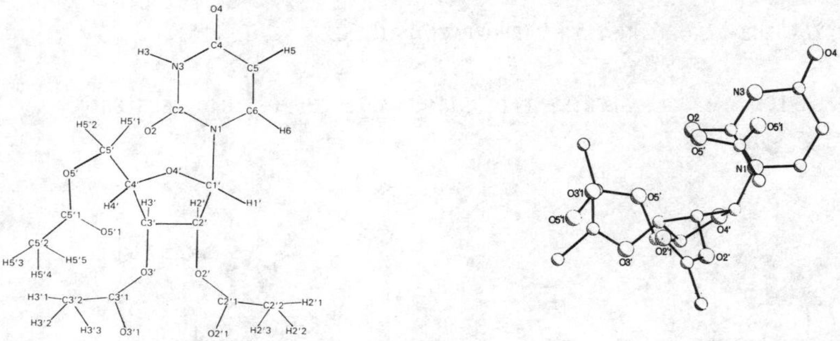

Fig. 1. $C_{15}H_{18}N_2O_9$: molecular structure and view of molecule perpendicular to the pyrimidine ring, showing stacking of O(5'1); in the latter view only 5'-acetyl atoms of the second molecule are shown, and H atoms are omitted for clarity.

6-(4-NITROBENZYL)THIOINOSINE

$C_{17}H_{17}N_5O_6S$

M. SORIANO-GARCIA, R. PARTHASARATHY, B. PAUL and A.R.P. PATERSON, 1984. Acta Cryst., C40, 1897-1901.

Monoclinic, $P2_1$, a = 14.718, b = 8.678, c = 7.269 Å, β = 93.55°, D_m = 1.49, Z = 2. Cu radiation, R = 0.042 for 2205 reflexions.

The six-substituent is distal to the imidazole ring (Fig. 1). The nucleoside has the syn conformation (ψ_{CN} -127.9(4)°) with an intramolecular hydrogen bond (O(5')...N(3) 2.813(5) Å). The ribose has the C(2')-endo (^2E) conformation (P 158.9(3), τ_m 38.0(2)°). The conformation across C(4')-C(5') is gauche-gauche. Molecules are linked with no overlap of base rings by O(2')-H...O(3'), 2.764(4) and by C(8)-H...O(5'), 3.244(4), C(2)-H...N(7), 3.463(4) Å hydrogen bonds.

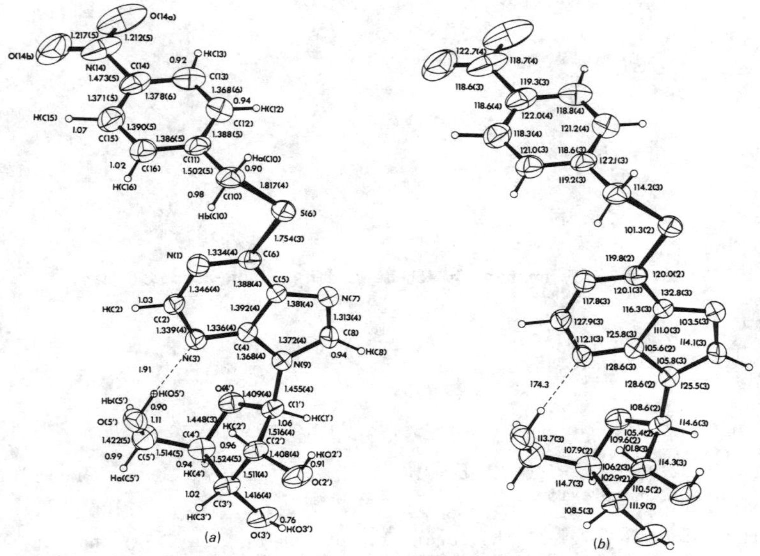

Fig. 1. Dimensions in 6-(4-nitrobenzyl)thioinosine.

2',3'-O-CYCLOHEXYLIDENE-4'-C-(2-METHYL-2-PROPENYL)URIDINE

$C_{19}H_{26}N_2O_6$

W.J. COOK, S.E. EALICK and J.A. SECRIST III, 1984. Acta Cryst., C40, 885-887.

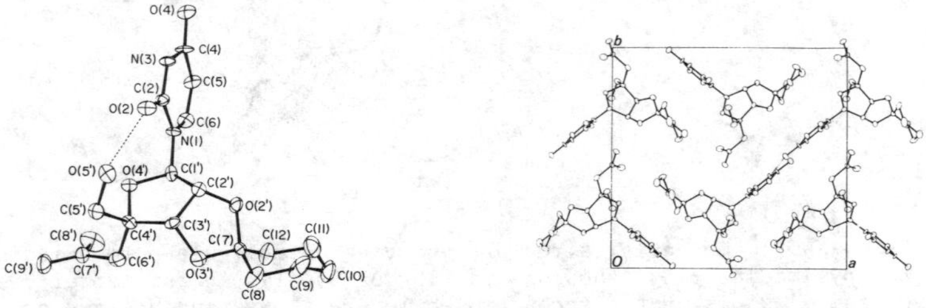

Fig. 1. $C_{19}H_{26}N_2O_6$: perspective view of the molecule and the molecular packing viewed along the c axis; the intermolecular N(3)...O(4') hydrogen bond is represented by a thin line.

Orthorhombic, $P2_12_12_1$, a = 16.785, b = 15.922, c = 7.115 Å, Z = 4. Cu radiation, R = 0.098 for 1107 reflexions.

The syn conformation of the base is stabilised by an unusual intramolecular O(5')...O(2) hydrogen bond of length 2.817(7) Å (Fig. 1). The ribose assumes an unusual conformation, O(4')-endo, C(1')-exo ($_1^oT$) with the phase angle of pseudorotation P 107(1)°. The structure is stabilised by one intermolecular hydrogen bond, N(3)...O(4') 2.884(7) Å. There is no base-pairing of the uracil bases.

5-CINNAMOYL-1,3-DIMETHYL-6-[(2-MORPHOLINOETHYL)AMINO]URACIL HYDROCHLORIDE
$C_{21}H_{27}ClN_4O_4$ $C_{21}H_{27}N_4O_4^+.Cl^-$

V. WARIN, F. BAERT, J.L. BERNIER and J.P. HENICHART, 1984. Acta Cryst., C40, 127-129.

Orthorhombic, Pbca, a = 24.635, b = 14.463, c = 12.147 Å, D_m = 1.22, Z = 8. Cu radiation, R = 0.049 for 1423 reflexions.

The introduction of a morpholinoethyl group at the N(6) position (Fig. 1) of this uracil derivative in order to increase the solubility, has deformed the planarity of the molecule and its length. This disruption has led to a higher toxicity for the molecule and a lower biological activity.

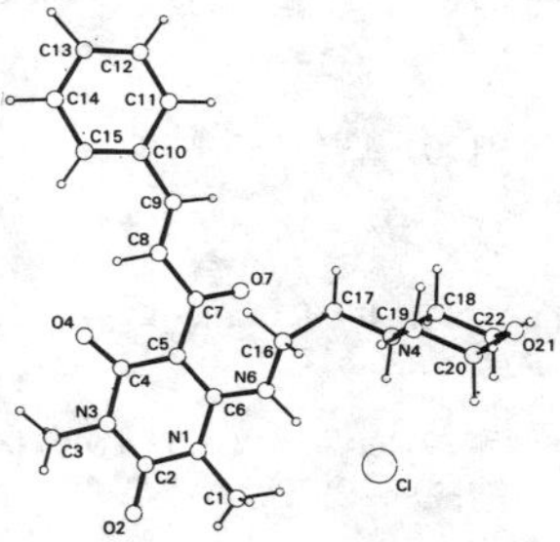

Fig. 1. A perspective view of $C_{21}H_{27}ClN_4O_4$.

O-2,O-3,O-5-TRIACETYL-6-ACETYLAMINO-1,6-DIDEOXY-1-α-(1,2,3,4-TETRAHYDRO-3-METHYL-2,4-DIOXO-1-PYRIMIDINYL)-4-THIO-L-glycero-L-IDOHEPTOFURANURONIC ACID METHYL ESTER S-((R)-OXIDE)
$C_{21}H_{27}N_3O_{12}S$

G. BENZ, L. BORN, M. BRIEDEN, R. GROSSER, J. KURZ, H. PAULSEN, V. SINNWELL and B. WEBER, 1984. Justus Liebigs Ann. Chem., 1408-1423.

Triclinic, P1, a = 7.3157, b = 8.9781, c = 11.8726 Å, α = 75.09, β = 79.93, γ = 69.80°, Z = 1. Cu radiation, R = 0.023 for 4806 reflexions.

The analysis establishes the absolute configuration shown in Fig. 1. The five-membered ring has an envelope conformation with C4' out of the plane of the other four atoms. The plane of the planar six-membered makes a dihedral angle of 89.6° with the planar 4-atom fragment of the five-membered ring.

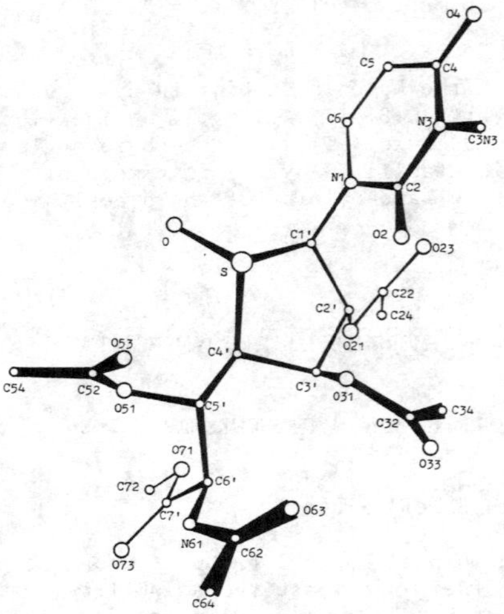

Fig. 1. A view of $C_{21}H_{27}N_3O_{12}S$.

5,2'-ANHYDRO-5-(p-TOLUENESULPHONAMIDO)-1-(3'-O-TOSYL-β-D-ARABINOFURANOSYL)IMIDAZOLE-4-CARBONITRILE
$C_{23}H_{22}N_4O_7S_2$

S. SUGIO, H. MIZUNO, K. KITAMURA, M. IKEHARA and K.-I. TOMITA, 1984. Acta Cryst., C40, 186-188.

Monoclinic, $P2_1$, a = 11.099, b = 11.015, c = 10.114 Å, β = 92.27°, D_m = 1.404, Z = 2. Mo radiation, R = 0.033 for 2764 reflexions.

In this molecule (Fig. 1) the glycosyl torsion angle χ C(2)-N(1)-C(1')-O(4') is -74.2(2)°. The sugar pucker is in a twisted form, C(4')-endo-C(3')-exo, with pseudorotation angle P 220.9°.

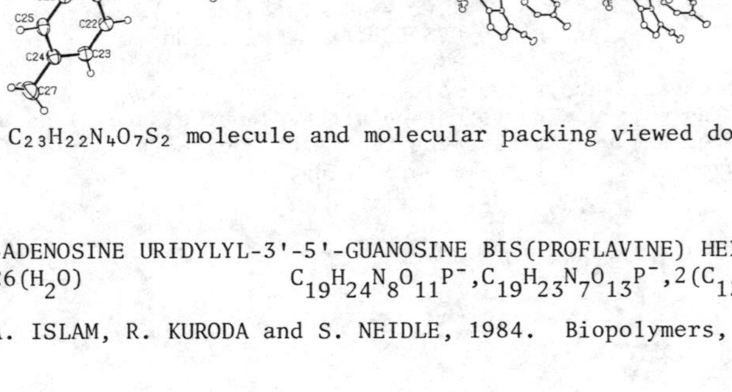

Fig. 1. The $C_{23}H_{22}N_4O_7S_2$ molecule and molecular packing viewed down b.

CYTIDYLYL-3'-5'-ADENOSINE URIDYLYL-3'-5'-GUANOSINE BIS(PROFLAVINE) HEXACOSAHYDRATE
$C_{64}H_{71}N_{21}O_{24}P_2,26(H_2O)$ $C_{19}H_{24}N_8O_{11}P^-,C_{19}H_{23}N_7O_{13}P^-,2(C_{13}H_{12}N_3^+),26(H_2O)$

A. AGGARWAL, S.A. ISLAM, R. KURODA and S. NEIDLE, 1984. Biopolymers, 23, 1025-1041.

Monoclinic, C2, a = 29.079, b = 13.560, c = 28.244 Å, β = 111.13°, Z = 8. Cu radiation, R = 0.200 for 1828 reflexions.

Planar proflavine molecules are intercalated between C...G and U...A Watson-Crick base pairs in a double-helical fragmentlike arrangement (Fig. 1). The sugar conformations at the 3'-ends of the strands are dissimilar. The separation of the two strands and the basic twist angle of 16° are indicative of sequence-dependent effects in the drug binding.

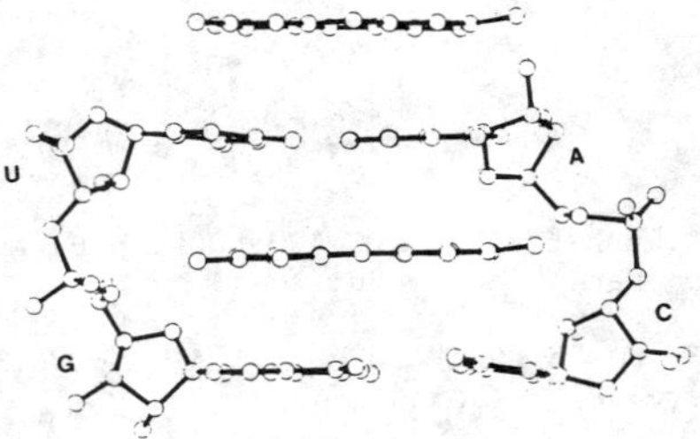

Fig. 1. A view of a proflavine-CpA-UpG complex.

GLYCINE CALCIUM DINITRATE DIHYDRATE

$C_2H_5CaN_3O_8,2(H_2O)$

S. NATARAJAN, K. RAVIKUMAR and S.S. RAJAN, 1984. Z. Kristallogr., 168, 75-82.

Orthorhombic, $P2_12_12_1$, a = 6.865, b = 13.250, c = 11.275 Å, D_m = 1.82, Z = 4. Cu radiation, R = 0.066 for 1005 reflexions.

The Ca coordination is distorted pentagonal bipyramidal, with Ca-O 2.366-2.583(7) Å (mean value 2.446 Å). The glycine exists as a zwitterion with normal dimensions. A hydrogen bonding scheme is proposed.

DL-SELENOMETHIONINE

$C_5H_{11}NO_2Se$

M. RAJESWARAN and R. PARTHASARATHY, 1984. Acta Cryst., C40, 647-650.

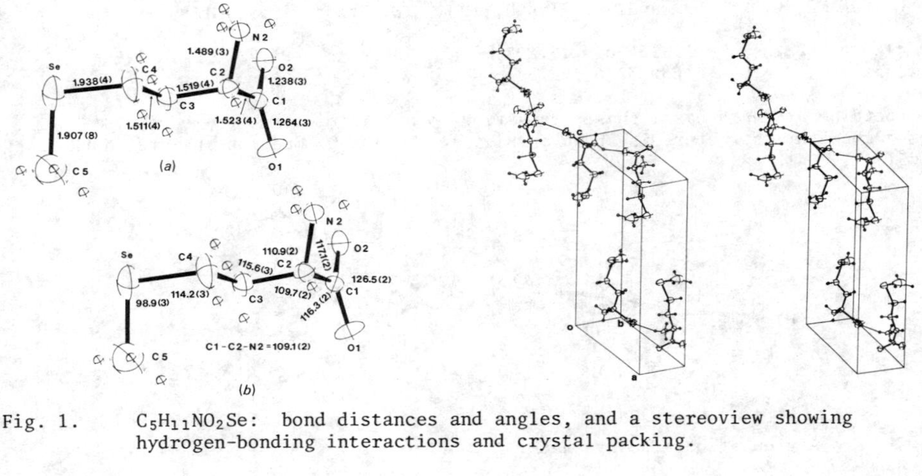

Fig. 1. $C_5H_{11}NO_2Se$: bond distances and angles, and a stereoview showing hydrogen-bonding interactions and crystal packing.

Monoclinic, $P2_1/a$, a = 9.893, b = 4.713, c = 17.082 Å, β = 101.63°, D_m = 1.67, Z = 4. Cu radiation, R = 0.079 for 1391 reflexions.

The crystal structure (Fig. 1) is isomorphous with that of the α-form of DL-methionine (1). Bond lengths and angles are generally as expected. The molecules are linked in the crystal by N-H...O hydrogen bonds; such $COO^-...H_3N^+$ hydrogen bonds are found in all amino acids. There are, however, no short contacts with the Se atom.

1. Structure Reports, 46B, 472.

D-β,β-DIMETHYLCYSTEIC ACID
$C_5H_{11}NO_5S$

A. CALVO, R. FAGGIANI, D.A. HARVEY, H.E. HOWARD-LOCK, W.F. KEAN and C.J.L. LOCK, 1984. J. Cryst. Spectrosc. Res., 14, 59-70.

Monoclinic, $P2_1$, a = 8.133, b = 8.094, c = 6.3138 Å, β = 96.36°, D_m = 1.58, Z = 2. Mo radiation, R = 0.025 for 1081 reflexions.

The molecule is in the zwitterion form, with the sulphonate group deprotonated, rather than the carboxyl group (Fig. 1). Bond lengths do not differ from those in similar structures. Molecules are linked in the crystal by a network of O-H...O and N-H...O hydrogen bonds.

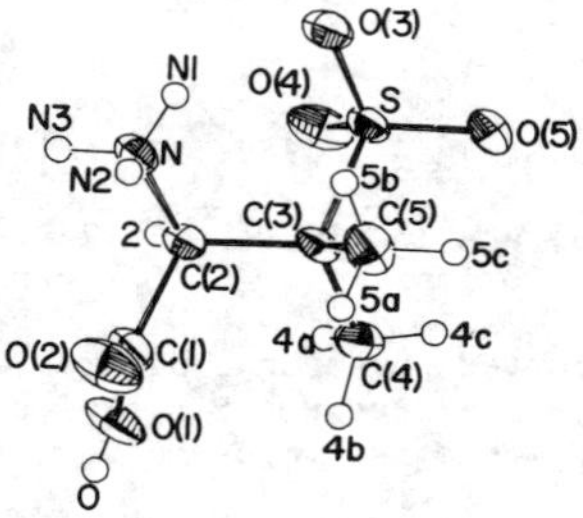

Fig. 1. A view of D-β,β-dimethylcysteic acid.

BETAINE ARSENATE (FERROELASTIC, PARAELECTRIC)
$C_5H_{14}AsNO_6$ $C_5H_{11}NO_2.H_3AsO_4$

W. SCHILDKAMP, G. SCHAFER and J. SPILKER, 1984. Z. Kristallogr., 168, 187-195.

Monoclinic, $P2_1/n$, a = 9.556, b = 12.897, c = 8.128 Å, β = 91.23°, D_m = 1.704, Z = 4. Mo radiation, R = 0.056 for 3206 reflexions.

Betaine arsenate has a phase transition to a ferroelectric phase at 119 K. The room temperature structure is shown in Fig. 1. The As-O bond lengths are in the range 1.675(2)-1.704(2) Å.

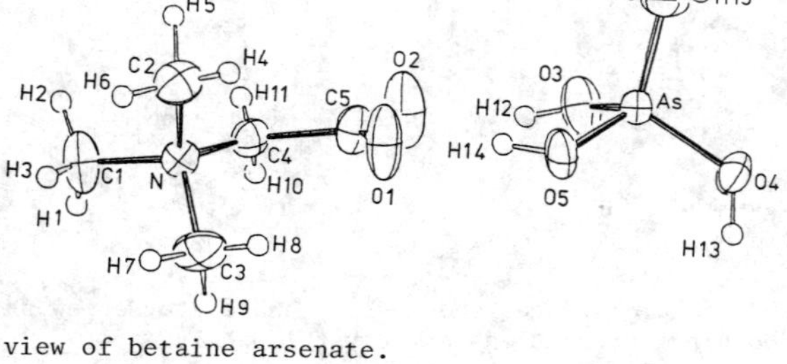

Fig. 1. A view of betaine arsenate.

BETAINE PHOSPHATE at 383 K
$C_5H_{14}NO_6P$ (I) $C_5H_{12}NO_2{}^+.H_2PO_4{}^-$

BETAINE PHOSPHATE at 295 K
$C_5H_{14}NO_6P$ (II) $C_5H_{12}NO_2{}^+.H_2PO_4{}^-$

BETAINE TRIDEUTEROPHOSPHATE at 294 K
$C_5H_{11}D_3NO_6P$ (III) $C_5H_{11}NO_2.D_3PO_4$

BETAINE TRIDEUTEROPHOSPHATE at 150 K
$C_5H_{11}D_3NO_6P$ (IV) $C_5H_{11}NO_2.D_3PO_4$

BETAINE TRIDEUTEROPHOSPHATE at 113 K
$C_5H_{11}D_3NO_6P$ (V)

W. SCHILDKAMP and J. SPILKER, 1984. Z. Kristallogr., 168, 159-171.

I. Monoclinic, $P2_1/m$, a = 11.114, b = 7.907, c = 6.543 Å, β = 119.77°, Z = 2. Mo radiation, R = 0.049 for 1272 reflexions.

II. Monoclinic, $P2_1/c$, a = 11.076, b = 7.852, c = 12.944 Å, β = 119.62°, Z = 4. Mo radiation, R = 0.056 for 1699 reflexions.

III. Monoclinic, $P2_1/c$, a = 11.079, b = 7.852, c = 12.922 Å, β = 119.51°, Z = 4. Mo radiation, R = 0.044 for 1267 reflexions.

IV. Monoclinic, $P2_1/c$, a = 11.066, b = 7.810, c = 12.766 Å, β = 119.41°, Z = 4. Mo radiation, R = 0.037 for 1513 reflexions.

V. Monoclinic, $P2_1/c$, a = 22.126, b = 7.798, c = 12.704 Å, β = 119.23°, Z = 8. Mo radiation, R = 0.053 for 2811 reflexions.

Betaine phosphate undergoes a phase transition from a paraelectric high temperature phase (space group $P2_1/m$; Z = 2) to an antiferrodistortive phase (space group $P2_1/c$; Z = 4) and another one to an antiferroelectric phase (space group $P2_1/c$; Z = 8). The high temperature phase is described by a disorder of the non-hydrogen atoms, which order in the antiferrodistortive phase. The hydrogen atoms of the phosphate group remain disordered. These hydrogen atoms induce the phase transition to the antiferroelectric phase. They can, however, not explain the diffraction intensities observed in this phase. A view of the structure at 150 K is in Fig. 1.

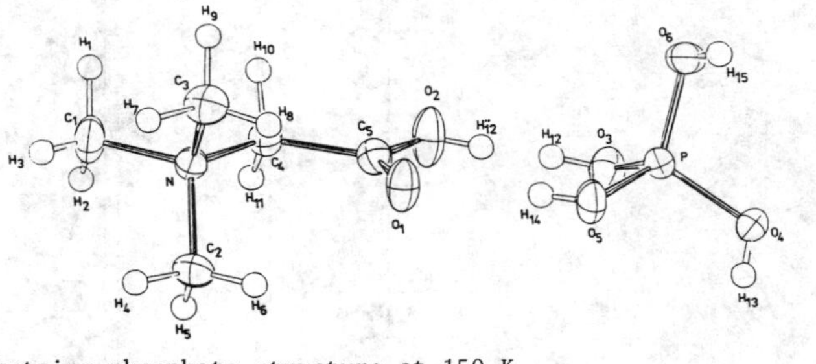

Fig. 1. The betaine phosphate structure at 150 K.

(R)-N-CHLOROACETYL-(+)-3,3,3-TRIFLUORO-2-AMINOISOBUTYRIC ACID
$C_6H_7ClF_3NO_3$

J.W. KELLER and C.S. DAY, 1984. Acta Cryst., C40, 1224-1226.

Monoclinic, $P2_1$, a = 10.258, b = 8.357, c = 5.749 Å, β = 96.45°, Z = 2. Mo radiation, R = 0.042 for 1407 reflexions.

The title compound is a heavy-atom derivative of (+)-3,3,3-trifluoro-2-amino-isobutyric acid and the absolute configuration is determined to be R. In the molecule (Fig. 1) bond lengths and angles are generally as expected; the backbone angle at C(2), C(1)-C(2)-N 111.0(2)°, is close to the usual value of 110°. Each molecule participates in two intermolecular hydrogen bonds: one with carboxylate as donor and chloroacetyl O of an adjacent molecule as acceptor and a second with N as donor and the carboxylate carbonyl O of a different adjacent molecule as acceptor.

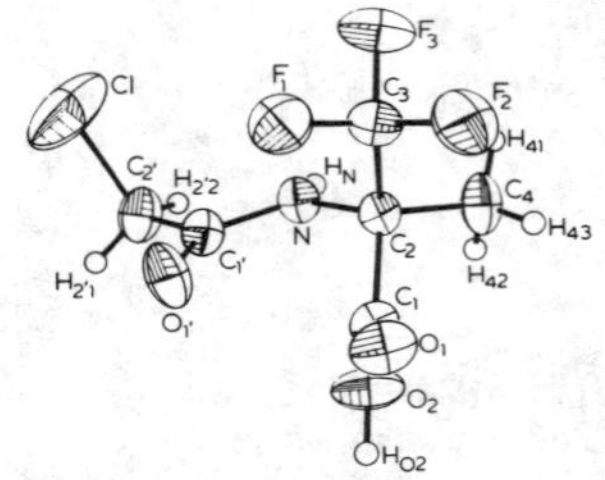

Fig. 1. $C_6H_7ClF_3NO_3$: perspective view of the molecule.

N-CARBOXY-L-VALINE ANHYDRIDE
$C_6H_9NO_3$

H. KANAZAWA, Y. OHASHI and Y. SASADA, 1984. Acta Cryst., C40, 1094-1096.

Orthorhombic, $P2_12_12_1$, a = 5.787, b = 22.740, c = 5.395 Å, D_m = 1.33, Z = 4. Mo radiation, R = 0.056 for 845 reflexions.

In the molecule (Fig. 1) the atoms of the five-membered ring and the exocyclic O atoms are planar. The conformation around the C(3)-C(4) bond is a stable staggered form. The molecules are linked by intermolecular hydrogen bonds between N(1)-H(1) and O(1) along the a axis; N...O is 2.888(4) Å. The N-carboxy anhydride rings are arranged in a layer sandwiched by hydrophobic side chains, as observed in the crystal of N-carboxy-L-leucine anhydride (1).

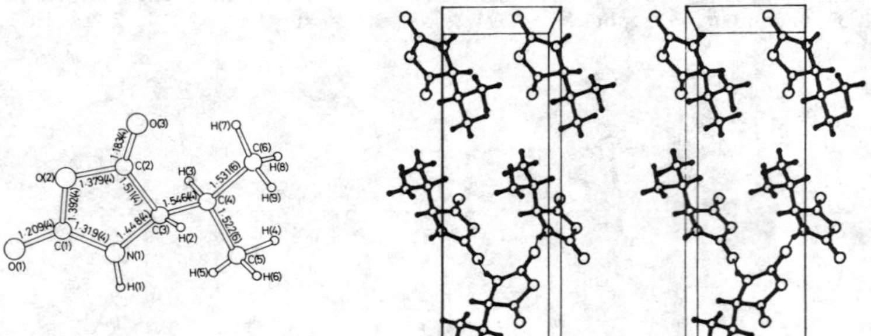

Fig. 1. $C_6H_9NO_3$: perspective view of the molecule showing bond distances and a stereoview of the crystal packing showing hydrogen bonding (fine lines).

1. Structure Reports, 44B, 434.

α-KETOGLUTARIC ACID SEMICARBAZONE
$C_6H_9N_3O_5$

B.Ya. ANTOSYAK, V.N. BIYUSHKIN and N.I. BELICHUK, 1984. Kristallografiya, 29, 722-725 [Sov. Phys. Crystallogr., 29, 427-429].

Tetragonal, $I4_1/a$, a = 16.093, b = 13.830 Å, Z = 16. Cu radiation, R = 0.045 for 1167 reflexions.

The atoms C(5), C(6), O(4), O(5) deviate from the mean plane of the other non-hydrogen atoms and are oriented almost perpendicular (89.3°) to it (Fig. 1). Interatomic distances are: C(1)-O(1) 1.247, C(1)-N(1) 1.366, C(1)-N(3) 1.330, N(1)-N(2) 1.360, N(2)-C(2) 1.281, C(2)-C(3) 1.485, C(3)-O(2) 1.313, C(3)-O(3) 1.193, C(2)-C(4) 1.524, C(4)-C(5) 1.521, C(5)-C(6) 1.498, C(6)-O(4) 1.206 and C(6)-O(5) 1.319 Å. The planar group N(2),C(2),C(3),C(4) makes an angle of 10.2° with the carbamide group (O(1),C(1),N(1),N(3)) and one of -3.7° with the group C(2),C(3),O(2),O(3). Each molecule forms ten hydrogen bonds with six adjacent molecules. Some donor...acceptor distances are: N(3)...O(3') 2.919, O(1)...O(2') 2.610, N(1)...O(4') 2.883, N(3)...O(3'') 3.079, O(5)...O(1') 2.679 Å.

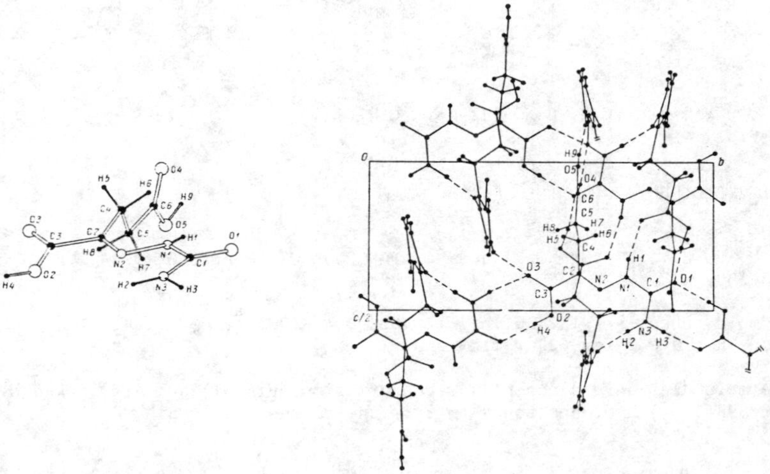

Fig. 1. Molecular structure of $C_6H_9N_3O_5$ (left), and projection of the crystal structure on (100) plane showing hydrogen bonds (right).

N-CHLOROACETYL-α-AMINOISOBUTYRIC ACID
$C_6H_{10}ClNO_3$ (I)

N-DICHLOROACETYL-α-AMINOISOBUTYRIC ACID
$C_6H_9Cl_2NO_3$ (II)

N-TRICHLOROACETYL-α-AMINOISOBUTYRIC ACID
$C_6H_8Cl_3NO_3$ (III)

G. VALLE, G.M. BONORA and C. TONIOLO, 1984. Gazz. Chim. Ital., 114, 481-486.

I. Orthorhombic, $P2_12_12_1$, a = 17.045, b = 8.798, c = 5.828 Å, D_m = 1.36, Z = 4. Mo radiation, R = 0.069 for 729 reflexions.

II. Monoclinic, $P2_1/n$, a = 11.137, b = 9.742, c = 9.171 Å, β = 111.2°, D_m = 1.52, Z = 4. Mo radiation, R = 0.051 for 1468 reflexions.

III. Monoclinic, $P2_1/c$, a = 15.775, b = 6.343, c = 10.092 Å, β = 95.7°, D_m = 1.63, Z = 4. Mo radiation, R = 0.100 for 1145 reflexions.

Views of I and III are in Fig. 1. The dihedral angles between the chloroacetamido and carboxylic acid groups are 74, 74 and 103° for I, II and III respectively. The secondary amide torsion angles are 177, 180 and -173° (trans conformation), and the ϕ,ψ_T angles are -51, -45; -55, -39; -60, -36° (for I, II and III respectively). In all three compounds the carboxyl group adopts a conformation with respect to the C(3)-N(1) bond between the antiplanar and anticlinal conformations, the N(1)-C(3)-C(6)-O(2) torsion angles being 137, 147 and 148° for I, II and III respectively.

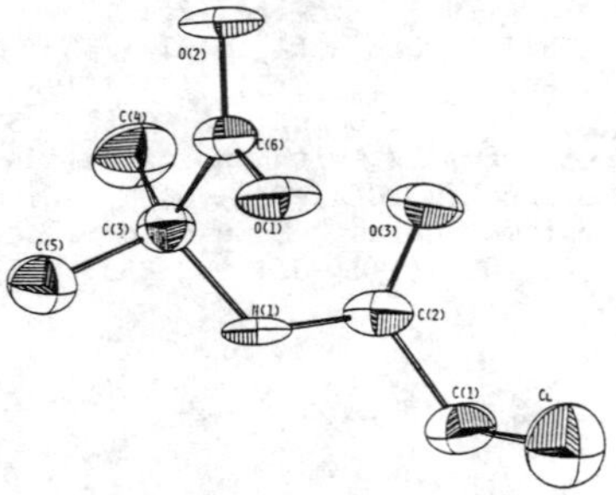

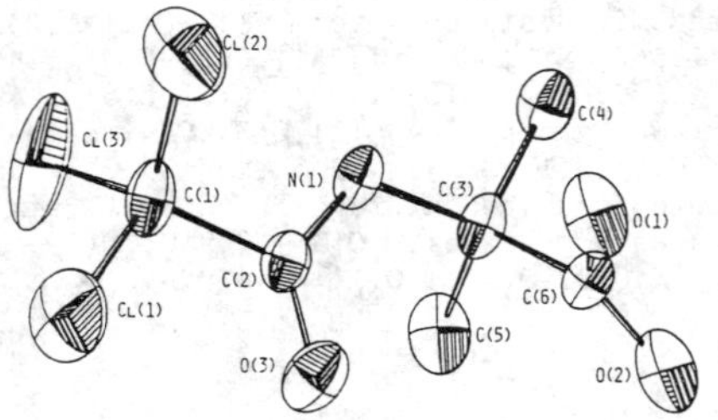

Fig. 1. Views of $C_6H_{10}ClNO_3$ (I) and $C_6H_8Cl_3NO_3$ (II).

N-ACETYL-2,3-DIDEHYDROPROLINE
$C_7H_9NO_3$

D. AJO, V. BUSETTI, G. GRANOZZI and M. LIAKOPOULOU-KYRIAKIDES, 1984. Acta Cryst., C40, 327-330.

Orthorhombic, Pbca, a = 13.455, b = 13.466, c = 8.069 Å, D_m = 1.41, Z = 8. Cu radiation, R = 0.042 for 625 reflexions.

The molecules have a non-planar conformation (Fig. 1) and are linked by O-H...O (O...O 2.598(5) Å) hydrogen bonds to form one-dimensional chains.

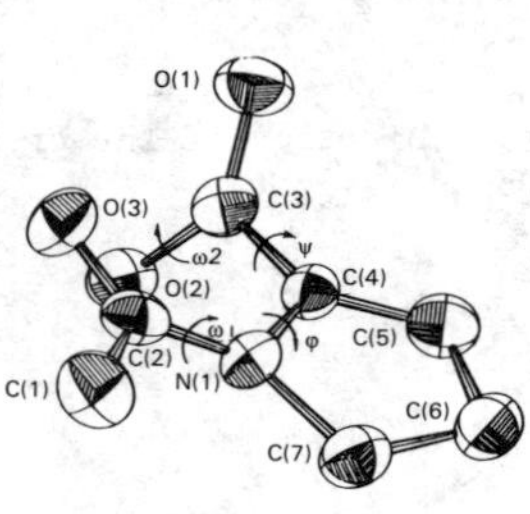

ω_1[C(1)–C(2)–N(1)–C(4)] = – 172·5 (5)°
φ [C(2)–N(1)–C(4)–C(3)] = – 26·3 (7)
ψ [N(1)–C(4)–C(3)–O(2)] = – 58·3 (6)
ω_2[C(4)–C(3)–O(2)–H(O2)] = – 160 (4).

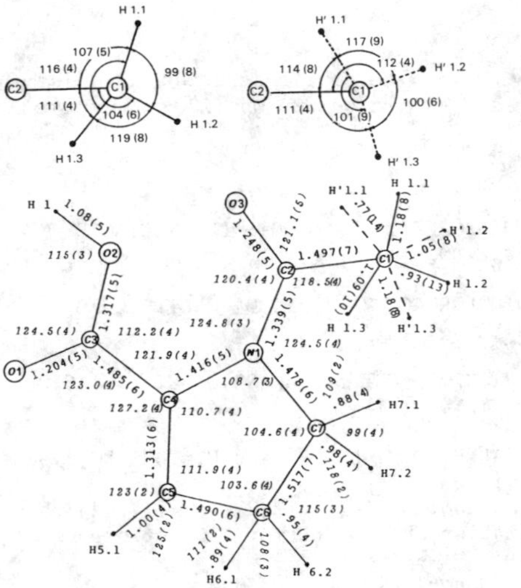

Fig. 1. The $C_7H_9NO_3$ molecule, with details of dimensions and conformation. The methyl hydrogens at C1 are disordered over two sites.

3-HYDROXY-α-METHYL-L-TYROSINE SESQUIHYDRATE
$C_{10}H_{13}NO_4,1.5H_2O$

A. NEUMAN, H. GILLIER, D. AVENEL and G. PERRET, 1984. Acta Cryst., C40, 2124-2126.

Monoclinic, $P2_1$, a = 14.30, b = 8.678, c = 8.911 Å, β = 96.57°, Z = 4. Cu radiation, R = 0.049 for 3402 reflexions.

The two independent zwitterions (Fig. 1) show the two gauche conformations of the N atom relative to the C(7)-C(8) bond, with the carboxylic group gauche in molecule (I) and trans in molecule (II). The angle C(1)-C(7)-C(8) is 115.5(3) and 116.8(3)° significantly greater than tetrahedral. The tyrosine and water molecules are linked by an extensive H-bond network.

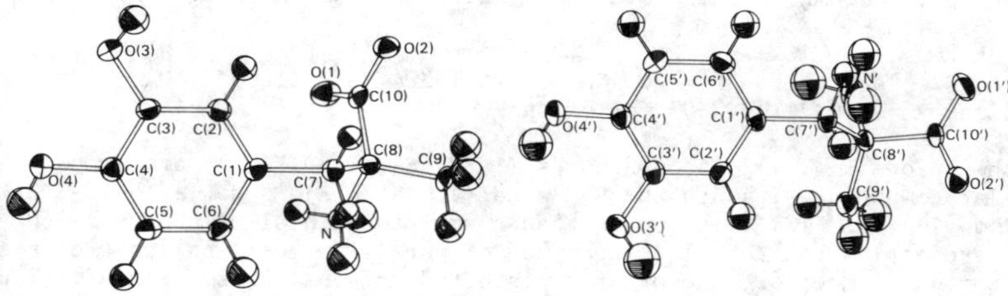

Fig. 1. Views of the two independent $C_{10}H_{13}NO_4$ zwitterions.

9-AMINOBICYCLO[3.3.1]NONANE-9-CARBOXYLIC ACID DIHYDRATE
$C_{10}H_{17}NO_2,2H_2O$

P. NARASIMHAN and K.K. CHACKO, 1984. Acta Cryst., C40, 1234-1236.

Monoclinic, $P2_1/c$, a = 13.582, b = 7.182, c = 11.936 Å, β = 105.5°, D_m = 1.35, Z = 4. Mo radiation, R = 0.068 for 987 reflexions.

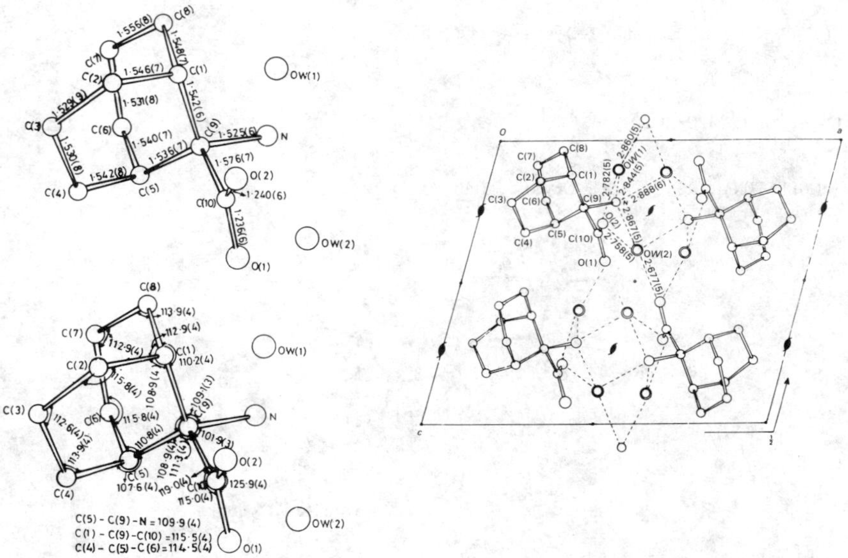

Fig. 1. $C_{10}H_{17}NO_2,2H_2O$: bond lengths and angles and the projection of the structure down the b axis.

In the molecule (Fig. 1) the bicyclo[3.3.1]nonane system exists in the twin-chair conformation, with the two cyclohexane rings as slightly flattened chairs. The structure is stabilised by a network of N-H...O and O-H...O hydrogen bonds involving the ionised carboxyl and amino groups and the water molecules. The structure is compared with that of the hydrobromide (1).

1. Structure Reports, 45B, 520.

DIPOTASSIUM 3,3,3',3'-TETRAMETHYLCYSTINATE TRIHYDRATE
$C_{10}H_{18}K_2N_2O_4S_2,3(H_2O)$

R. FAGGIANI, H.E. HOWARD-LOCK, C.J.L. LOCK, M.L. MARTINS and P.S. SMALLEY, 1984. Can. J. Chem., 62, 1127-1133.

Monoclinic, $P2_1$, a = 6.160, b = 26.473, c = 6.193 Å, β = 113.94°, D_m = 1.53, Z = 2. Mo radiation, R = 0.047 for 2303 reflexions.

In the anion (Fig. 1), the C-S bonds (1.877(6), 1.891(6) Å) are longer than in many similar compounds although the S-S bond (2.040(2) Å) is not. The C-S-S-C torsional angle (108.7(3)°) is larger than normal in dithiol compounds. Other distances and angles are normal. Two types of potassium coordination are present, one a distorted octahedron, the other a distorted trigonal prism. In addition to the ionic interactions, hydrogen bonds involving the water molecules are important in stabilizing the structure.

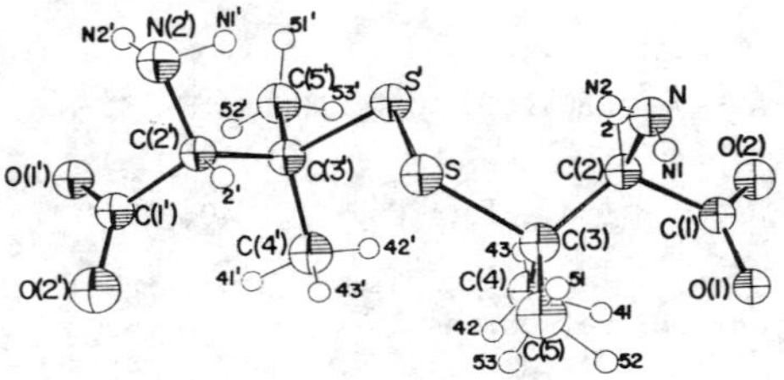

Fig. 1. The tetramethyl-D-cystinate anion.

L-VALYL-L-GLUTAMIC ACID
$C_{10}H_{18}N_2O_5$

D.S. EGGLESTON, 1984. Acta Cryst., C40, 1250-1252.

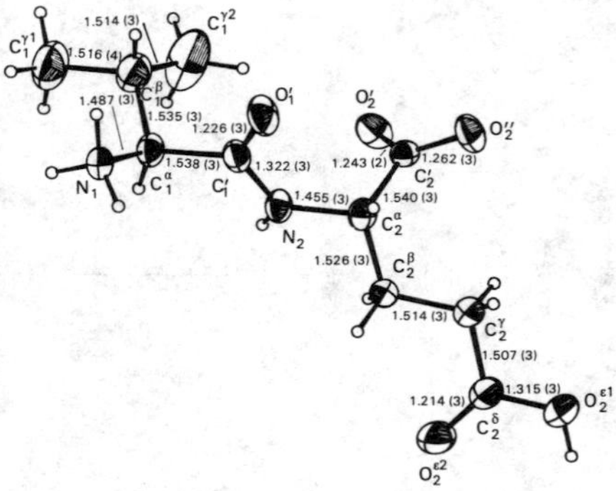

Fig. 1. $C_{10}H_{18}N_2O_5$: bond lengths in the molecule.

Orthorhombic, $P2_12_12$, a = 16.781, b = 13.827, c = 5.367 Å, D_m = 1.30, Z = 4. Mo radiation, R = 0.038 for 1374 reflexions.

The dipeptide (Fig. 1) crystallises as a zwitterion with the main-chain carboxyl ionised and the amino terminus protonated. The configuration of the peptide bond is trans with an ω torsion angle of 175°. The peptide backbone is in an extended conformation as is the glutamyl side chain. There is extensive intermolecular hydrogen bonding in the crystal lattice.

N-BENZOYLGLYCINE ETHYLDITHIO ESTER
$C_{11}H_{13}NOS_2$ (I)

N-(p-CHLOROBENZOYL)GLYCINE ETHYLDITHIO ESTER
$C_{11}H_{12}ClNOS_2$ (II)

K.I. VARUGHESE, A.C. STORER and P.R. CAREY, 1984. J. Am. Chem. Soc., 106, 8252-8257.

I. Triclinic, $P\bar{1}$, a = 8.900, b = 11.037, c = 13.539 Å, α = 85.91, β = 95.56, γ = 109.73°, D_m = 1.28, Z = 4. Cu radiation, R = 0.042 for 4427 reflexions.

II. Monoclinic, $P2_1/c$, a = 8.182, b = 9.312, c = 17.362 Å, β = 91.52°, Dm = 1.37, Z = 4. Cu radiation, R = 0.053 for 2326 reflexions.

Crystals of (I) contain two independent molecules in the asymmetric unit. Both (I) and (II) molecules (Fig. 1) assume a B-type conformation in which the glycinic N atom comes into close contact with the thiol S atom. (II) differs from (I) in the orientation of the end methyl group. The data are used to show that there is a strongly preferred line of approach of N to S in the N...S nonbonding contact and that this line represents the conformational pathway with maximum nitrogen-sulphur orbital interactions.

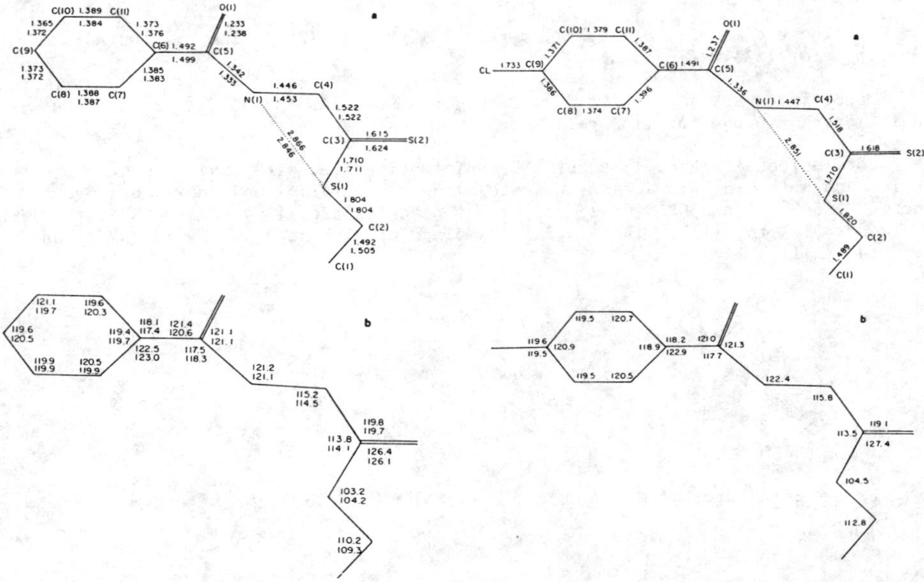

Fig. 1. Molecular dimensions for the two molecules of $C_{11}H_{13}NOS_2$ (I) (left) and for $C_{11}H_{12}ClNOS_2$ (II) (right).

ETHYL (Z)-N-ACETYLDEHYDROPHENYLALANINATE
$C_{13}H_{15}NO_3$

D. AJO, V. BUSETTI, J.C.J. OTTENHEIJM and R. PLATE, 1984. Acta Cryst., C40, 324-327.

Orthorhombic, Pbca, a = 49.869, b = 10.898, c = 9.544 Å, D_m = 1.18, Z = 16. Mo radiation, R = 0.095 for 1758 reflexions.

The two independent molecules in the asymmetric unit (Fig. 1) are approximate enantiomers (ϕ = 80.8° for molecule A and -64.0° for B). There is some disorder of the ethyl groups. The conformations of the two molecules are defined by the torsion angles given in Fig. 1.

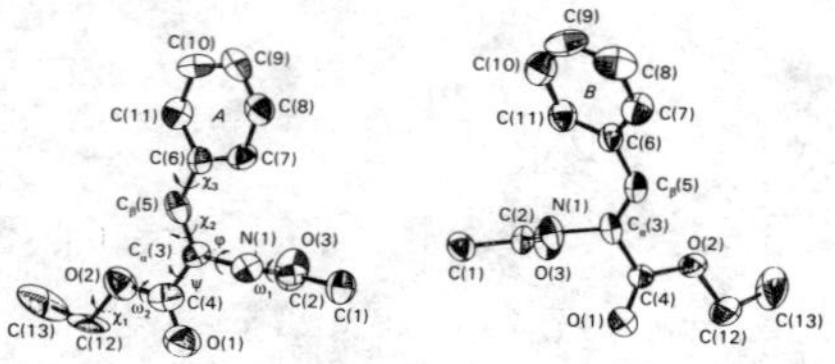

	Molecule *A*	Molecule *B*
ω_1 [C(1)–C(2)–N(1)–C(3)]	177·7 (6)	177·2 (5)
φ [C(2)–N(1)–C(3)–C(4)]	80·8 (7)	−64·0 (8)
ψ [N(1)–C(3)–C(4)–O(2)]	−177·4 (5)	158·5 (5)
ω_2 [C(3)–C(4)–O(2)–C(12)]	−177·6 (5)	−178·2 (5)
χ_1 [C(4)–O(2)–C(12)–C(13)]	−174·4 (6)	162·8 (5)
χ_2 [N(1)–C(3)–C(5)–C(6)]	0·(1)	−5·(1)
χ_3 [C(3)–C(5)–C(6)–C(7)]	−18·(1)	150·2 (7)

Fig. 1. The two independent $C_{13}H_{15}NO_3$ molecules and torsion angles.

N-(β-PHENYLPROPIONYL)GLYCINE ETHYLTHIOL ESTER
$C_{13}H_{17}NO_2S$ (I)

N-BENZOYLGLYCINE ETHYLTHIOL ESTER
$C_{11}H_{13}NO_2S$ (II)

C.P. HUBER, P.R. CAREY, SHI-CHUAN HSI, H. LEE and A.C. STORER, 1984. J. Am. Chem. Soc., 106, 8263-8268.

I. Orthorhombic, $Pca2_1$, a = 9.541, b = 12.231, c = 12.128 Å, Z = 4. Cu radiation, R = 0.037 for 1273 reflexions.

II. Monoclinic, Pc, a = 5.675, b = 10.439, c = 9.944 Å, β = 93.56°, Z = 2. Cu radiation, R = 0.050 for 1145 reflexions.

Both molecules (Fig. 1) adopt a B conformation i.e. with small N-C-C-S torsion angles and an N...S interaction and with amide and ester planes nearly orthogonal. The N-C4-C3-S (ψ) and C5-N-C4-C3 (ϕ') torsion angles are -15.4 and -84.8° for (I) and 5.4 and -88.7° for (II). The N...S intramolecular distances are 2.930(2) (I) and 2.879(4) Å (II).

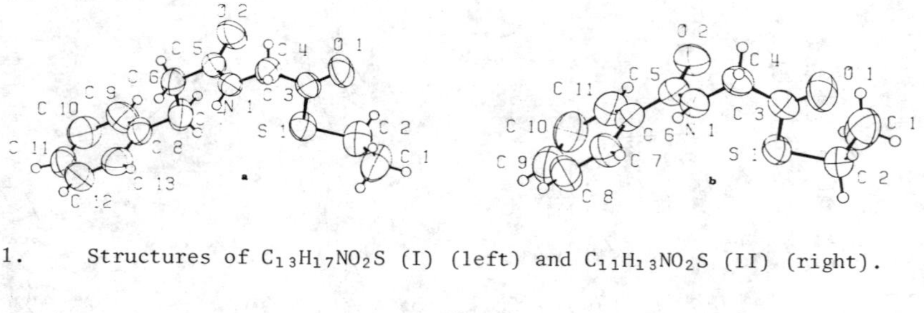

Fig. 1. Structures of $C_{13}H_{17}NO_2S$ (I) (left) and $C_{11}H_{13}NO_2S$ (II) (right).

cyclo(-L-PHENYLALANINE-L-PROLINE)
$C_{14}H_{16}N_2O_2$

F. MAZZA, G. LUCENTE, F. PINNEN and G. ZANOTTI, 1984. Acta Cryst., C40, 1974-1976.

Monoclinic, $P2_1$, a = 10.789, b = 10.061, c = 5.668 Å, β = 92.70°, Z = 2. Mo radiation, R = 0.038 for 1739 reflexions.

The diketopiperazine ring has a boat conformation (Fig. 1). The degree of folding (β) along the line joining the two C^{α} atoms is 50.7(1)° with the axial H atoms 2.37(6) Å apart. The pyrrolidine ring has the C_2-C^{β}_{endo}-C^{γ}_{exo} half-chair conformation with C^{β} and C^{γ} respectively 0.231(4) and -0.361(4) Å from the plane of the other three atoms. Screw-related molecules are linked by hydrogen bonds between the phenylalanine NH and CO groups.

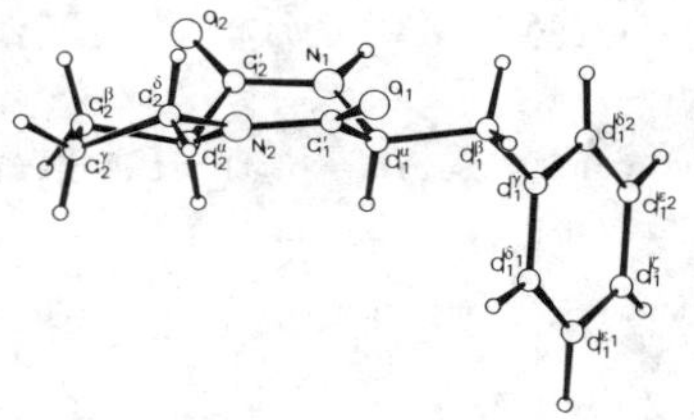

Fig. 1. The cyclo(-L-Phe-L-Pro) molecule.

(Z)-N-ACETYL-α,β-DIDEHYDROPHENYLALANYL-L-ALANINE HYDRATE
$C_{14}H_{16}N_2O_4$,0.33(H_2O)

V. BUSETTI, D. AJÒ and M. CASARIN, 1984. Acta Cryst., C40, 1245-1248.

Hexagonal, $P6_5$, a = 10.534, c = 23.753 Å, D_m = 1.230, Z = 6. Mo radiation, R = 0.052 for 950 reflexions.

In the crystal structure (Fig. 1) the molecules, which are linked by intermolecular hydrogen bonds, are arranged in helices along the z axis. The tendency of the phenyl ring to be planar with the C-C-N system causes a change to the expected bond angles (C5-C6-C7 117.0, C5-C6-C11 124.8, C3-C5-C6 132.2, N1-C3-C5 124.1(8)°).

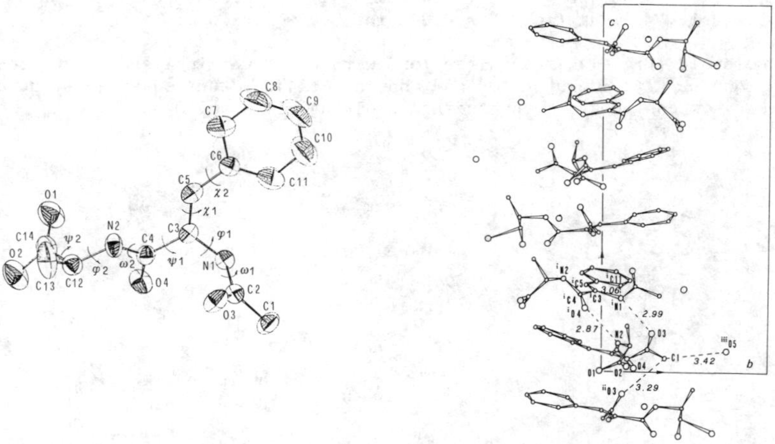

Fig. 1. $C_{14}H_{16}N_2O_4$,0.33(H_2O): perspective view of the molecule and the view along x of the crystal packing; broken lines indicate hydrogen bonds.

threo-N-ACETYL-3-FLUOROPHENYLALANINE ISOPROPYL ESTER
$C_{14}H_{18}FNO_3$ (I)

erythro-N-ACETYL-3-FLUOROPHENYLALANINE
$C_{11}H_{12}FNO_3$ (II)

T. TSUSHIMA, K. KAWADA, J. NISHIKAWA, T. SATO, K. TORI, T. TSUJI and S. MISAKI, 1984. J. Org. Chem., 49, 1163-1169.

I. Monoclinic, $P2_1/c$, a = 5.167, b = 22.089, c = 12.856 Å, β = 103.85°, Z = 4. Cu radiation, R = 0.041 for 1768 reflexions.

II. Orthorhombic, $P2_12_12_1$, a = 11.899, b = 16.714, c = 5.615 Å, Z = 4. Cu radiation, R = 0.044 for 1033 reflexions.

The analyses establish the configuration of the two derivatives as shown in Fig. 1.

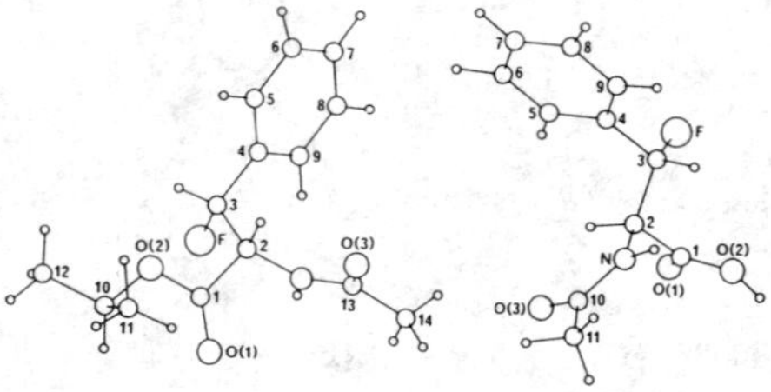

Fig. 1. Views of $C_{14}H_{18}FNO_3$ (I) (left) and of $C_{11}H_{12}FNO_3$ (II) (right).

L-TYROSYL-L-GLUTAMIC ACID MONOHYDRATE
$C_{14}H_{18}N_2O_6,H_2O$

J. PANDIT, T.P. SESHADRI and M.A. VISWAMITRA, 1984. Acta Cryst., C40, 169-170.

Triclinic, P1, a = 5.801, b = 7.977, c = 9.110 Å, α = 102.33, β = 97.92, γ = 109.82°, D_m = 1.45, Z = 1. Mo radiation, R = 0.046 for 990 reflexions.

The molecule (Fig. 1) is a zwitterion, with the N-terminal end protonated, the carboxyl group at C(4) deprotonated and that at C(14) protonated. The structure is stabilized by extensive hydrogen bonding involving all N and O atoms.

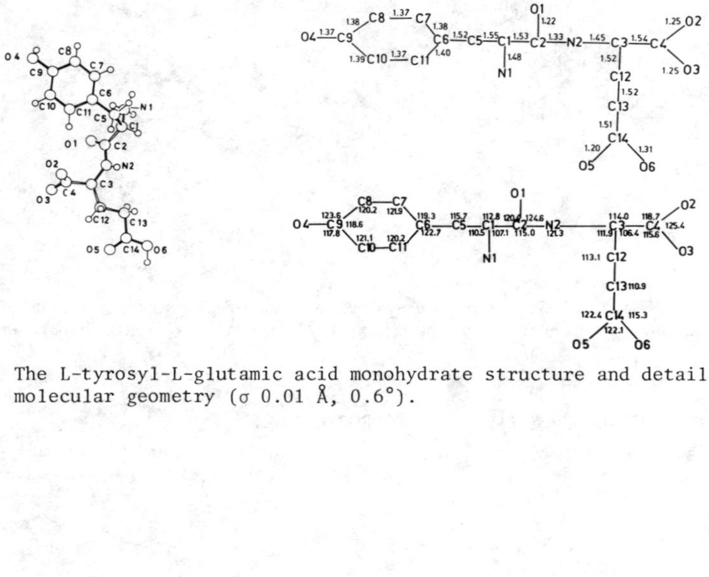

Fig. 1. The L-tyrosyl-L-glutamic acid monohydrate structure and details of molecular geometry (σ 0.01 Å, 0.6°).

L-TYROSYL-L-VALINE MONOHYDRATE
$C_{14}H_{20}N_2O_4,H_2O$

B. RAMAKRISHNAN, T.P. SESHADRI and M.A. VISWAMITRA, 1984. Acta Cryst., C40, 1248-1250.

Orthorhombic, $P2_12_12_1$, a = 5.629, b = 8.702, c = 31.007 Å, D_m = 1.31, Z = 4. Cu radiation, R = 0.071 for 1144 reflexions.

The molecule (Fig. 1) exists as a zwitterion with the N-terminal end protonated. The side-chain conformations are as expected with χ^{11}, χ^{12}, χ^{22}, χ^{21} 174(1), -60(1), -101(1) and 81(1)°, respectively. The crystal structure is stabilised by extensive hydrogen bonding involving N and O atoms.

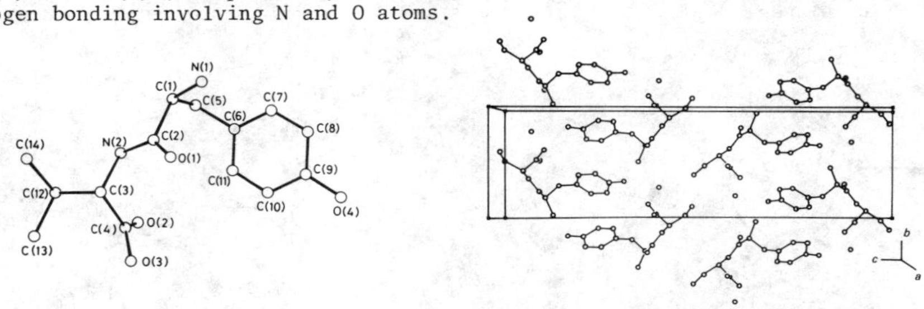

Fig. 1. $C_{14}H_{20}N_2O_4,H_2O$: perspective view of the molecule and view along a of the crystal packing.

cyclo(-L-LEUCYL-L-TYROSYL-) MONOHYDRATE
$C_{15}H_{20}N_2O_3,H_2O$

K. SUGUNA, S. RAMAKUMAR and K.D. KOPPLE, 1984. Acta Cryst., C40, 2053-2056.

Monoclinic, $P2_1$, a = 8.883, b = 6.105, c = 14.582 Å, β = 99.53°, D_m = 1.25, Z = 2. Cu radiation, R = 0.075 for 1147 reflexions.

In this structure (Fig. 1) one of the cis peptide units shows significant non-planarity with ω -11.7(10)°, the other is planar with ω -2.8(10)°. The diketo-piperazine ring takes a boat form. The leucyl side chain is in an extended conformation (χ^2 173.7(7)°); the tyrosyl residue shows a folded geometry (χ^1 59.5(7)°). χ^2 of the tyrosyl side chain is 98.4(8)°.

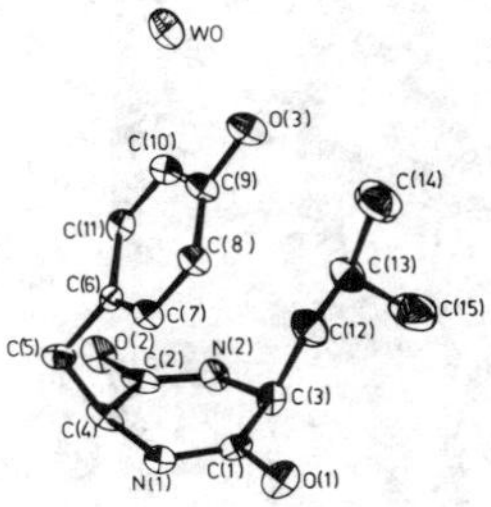

Fig. 1. A view of $C_{15}H_{20}N_2O_3,H_2O$.

1-ADAMANTYLOXYCARBONYL-α-AMINOISOBUTYRIC ACID
$C_{15}H_{23}NO_4$ (I)

2-p-BIPHENYLISOPROPYLOXYCARBONYL-L-ISOLEUCINE
$C_{22}H_{27}NO_4$ (II)

G. VALLE, G.M. BONORA and C. TONIOLO, 1984. Gazz. Chim. Ital., 114, 487-493.

I. Tetragonal, $P4_12_12$, a = 11.970, c = 21.458 Å, D_m = 1.21, Z = 8. Mo radiation, R = 0.065 for 1584 reflexions.

II. Orthorhombic, $P2_12_12_1$, a = 30.130, b = 9.170, c = 7.531 Å, D_m = 1.17, Z = 4. Mo radiation, R = 0.060 for 1425 reflexions.

Views of the two molecules are in Fig. 1. The conformation of the α-aminoisobutyric acid residue is folded while that of L-isoleucine is partially extended. The conformation of the -CONH- group of the secondary methane moiety is unusual; E (cis) in I and Z (trans) in II. The biphenyl moiety of II has a 34° dihedral angle between the two ring planes.

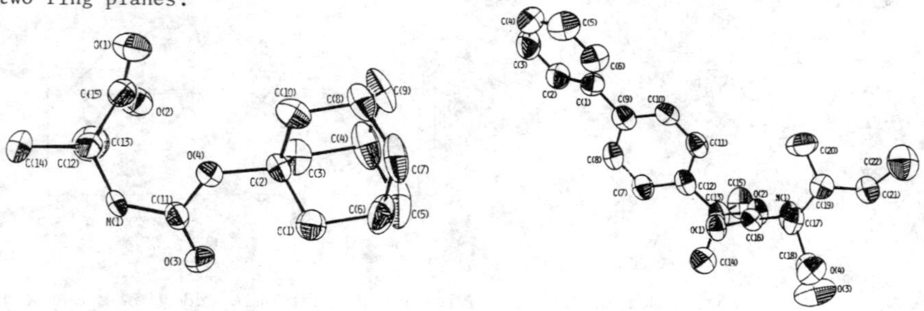

Fig. 1. Views of $C_{15}H_{23}NO_4$ (I) and $C_{22}H_{27}NO_4$ (II).

N-t-BUTOXYCARBONYL-L-PROLYL-D-VALINE
$C_{15}H_{26}N_2O_5$

R. BOSCH, H. SCHMITT, G. JUNG and W. WINTER, 1984. Acta Cryst., C40, 1096-1098.

Orthorhombic, $P2_12_12_1$, a = 9.426, b = 11.876, c = 16.085 Å, Z = 4. Cu radiation, R = 0.0493 for 3024 reflexions.

In the molecule (Fig. 1) the butoxycarbonyl-L-prolyl urethene bond adopts the cis conformation. For the proline ring system an envelope conformation is observed, with C(7) situated about 0.2 Å above the best plane through the ring atoms. Bond lengths and angles are normal. In the crystal there is intermolecular packing of polymeric hydrogen-bonded sheets of peptide molecules which have only van der Waals contacts with one another.

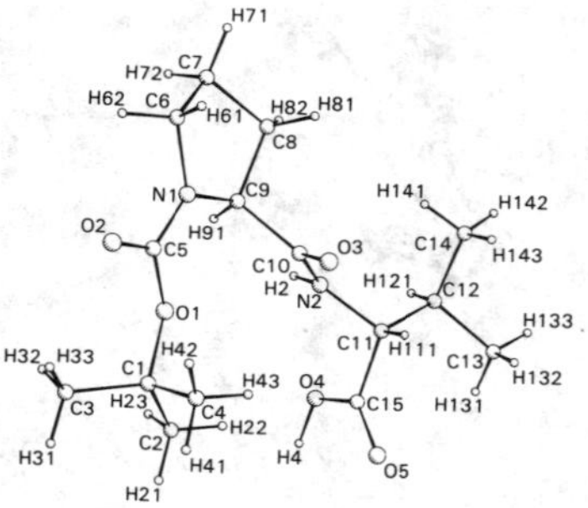

Fig. 1. $C_{15}H_{26}N_2O_5$: perspective view of the molecule.

N-(t-BUTYLOXYCARBONYL)-L-ALANYL-α-AMINOISOBUTYRYL-L-ALANININE METHYL ESTER
$C_{16}H_{29}N_3O_6$

R. BOSCH, G. JUNG, K.-P. VOGES and W. WINTER, 1984. Justus Liebigs Ann. Chem., 1117-1128.

Monoclinic, $P2_1$, a = 11.732, b = 6.013, c = 14.195 Å, β = 91.76°, Z = 2. Cu radiation, R = 0.047 for 3089 reflexions.

This peptide (Fig. 1) adopts a new type of β-turn with a very long distance of 3.621 between N3 and O2 which is too large to be considered a real intramolecular hydrogen bridge. It is possible that two strong intermolecular hydrogen bonds (N2...O4' 2.966 Å, N1...O3' 2.969 Å) are instrumental in causing the N3...O2 distance to be lengthened.

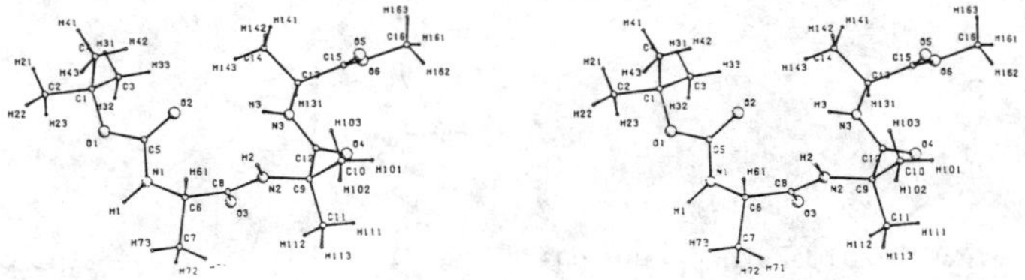

Fig. 1. A stereoview of $C_{16}H_{29}N_3O_6$.

4-O-(2'-ACETYLAMINO-2'-DEOXY-β-D-GLUCOPYRANOSYL)-4-HYDROXY-5-HYDROXYMETHYLPROLYL-TAURINE MONOHYDRATE
$C_{16}H_{29}N_3O_{11}S,H_2O$

S. SHINAGAWA, F. KASAHARA, Y. WADA, S. HARADA and M. ASAI, 1984. Tetrahedron, 40, 3465-3470.

Triclinic, P1, a = 8.620, b = 8.898, c = 8.100 Å, α = 111.86, β = 90.27, γ = 72.11°, Z = 1. Mo radiation, R = 0.039 for 1777 reflexions.

X-ray analysis establishes the structure as shown in Fig. 1.

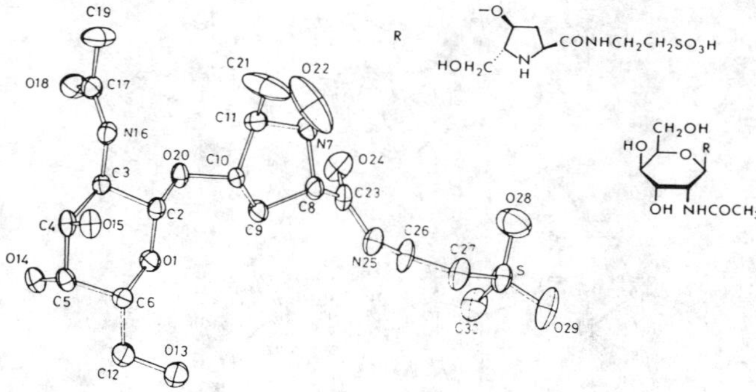

Fig. 1. The $C_{16}H_{29}N_3O_{11}S$ molecule.

N-ACETYL-DL-PHENYLALANINE p-NITROANILIDE
$C_{17}H_{17}N_3O_4$

B.K. SINHA, V. PATTABHI, S. RAGHUNATHAN and E.J. GABE, 1984. Acta Cryst., C40, 166-168.

Triclinic, P$\bar{1}$, a = 7.219, b = 9.122, c = 13.805 Å, α = 99.07, β = 92.55, γ = 111.41°, D_m = 1.307, Z = 2. Cu radiation, R = 0.054 for 1903 reflexions.

The peptide groups of this chymotripsin substrate (Fig. 1) are non-planar. Molecular dimensions are also in Fig. 1.

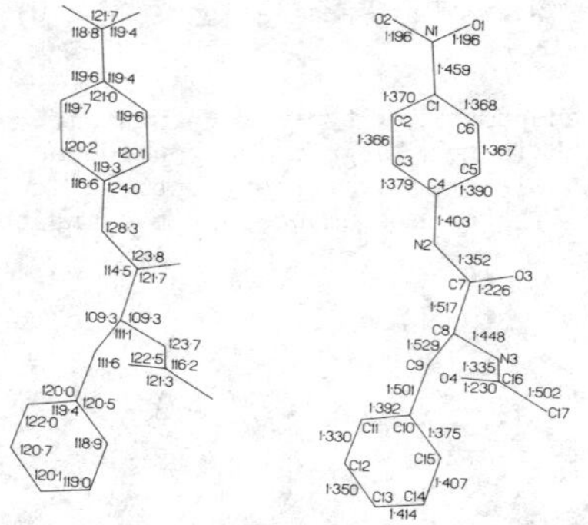

Fig. 1. Molecular dimensions (σ 0.003-0.0095 Å, 0.2-0.5°) for $C_{17}H_{17}N_3O_4$.

L-PYROGLUTAMYL-β-(2-THIENYL)-L-ALANYL-L-PROLINAMIDE
$C_{17}H_{22}N_4O_4S$

R. NORRESTAM, B. STENSLAND and S. CASTENSSON, 1984. Acta Chem. Scand., A38, 473-478.

Monoclinic, C2, a = 15.358, b = 10.830, c = 23.155 Å, β = 108.29°, Z = 8. Cu radiation, R = 0.115 for 1173 reflexions.

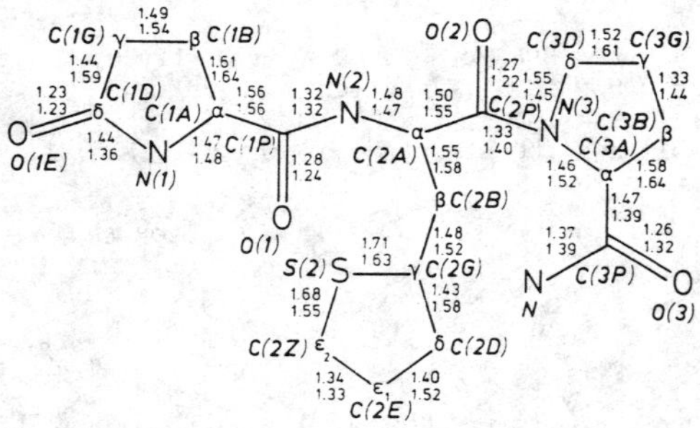

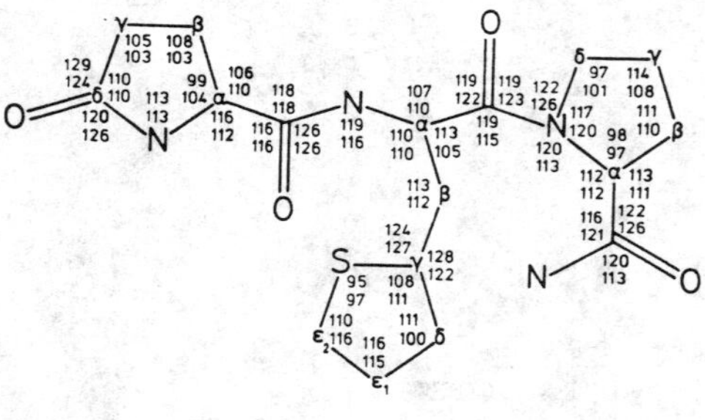

Fig. 1. Interatomic distances (Å) (top), and bond angles (°) (bottom) for the two independent molecules of $C_{17}H_{22}N_4O_4S$.

The structure is composed of hydrogen-bonded dimers in which the molecules are built up in antiparallel β-pleated sheet arrangements. Comparison of this structure with that of a different crystalline modification (1) reveals that the conformation properties of the present molecular structures deviate slightly from those previously determined (in the main chain backbone of the second residue) but nevertheless adopt approximately the same conformation (Fig. 1). The thiophene rings have rotational disorder and this, together with the conformational freedom of the pyrrolidine rings, is common to both structure determinations.

1. Structure Reports, 49B, 529.

N-(FLUOREN-9-YL-METHOXYCARBONYL)ALANINE MONOHYDRATE
$C_{18}H_{17}NO_4,H_2O$ (I)

N-(FLUOREN-9-YL-METHOXYCARBONYL)-α-AMINOISOBUTYRIC ACID
$C_{19}H_{19}NO_4$ (II)

G. VALLE, G.M. BONORA and C. TONIOLO, 1984. Can. J. Chem., 62, 2661-2666.

I. Orthorhombic, $P2_12_12_1$, a = 25.054, b = 9.226, c = 7.360 Å, D_m = 1.20, Z = 4. Mo radiation, R = 0.047 for 1326 reflexions.

II. Orthorhombic, Pbcn, a = 29.785, b = 12.119, c = 9.556 Å, D_m = 1.24, Z = 8. Mo radiation, R = 0.086 for 1124 reflexions.

The solid-state conformations of I and II are shown in Fig. 1. Bond lengths and angles are in accord with expected values. The dihedral angle between the planes of the two aromatic rings of the fluorene system is 3.35° for I and 3.47° for II. The dihedral angles between the plane of the urethane group and the planes of the fluorene system and the carboxylic acid group are 61.5° and 106.1°, respectively, for I and 89.2° and 73.1°, respectively, for II. The secondary urethane linkage in both compounds adopt one of its most common conformations (trans, trans, or b (2, 3)), the corresponding torsional angles being 178.1° and 179.1° for I, and ±168.2° and ∓173.6° for II.

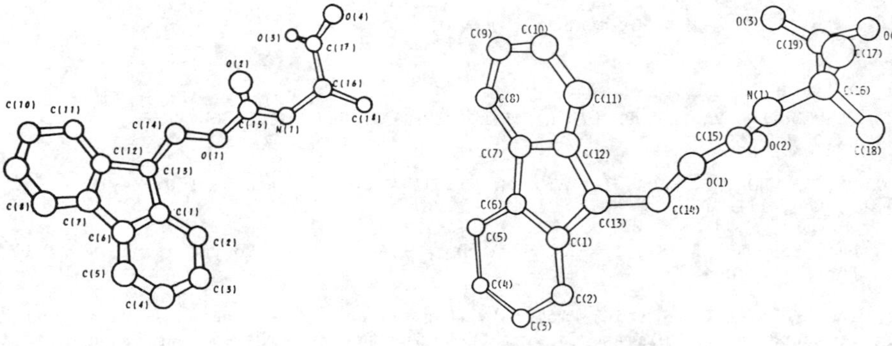

Fig. 1. Views of $C_{18}H_{17}NO_4$ (I) (left) and $C_{19}H_{19}NO_4$ (II) (right).

L-TYROSYL-L-TYROSINE DIHYDRATE
$C_{18}H_{20}N_2O_5,2H_2O$

M. COTRAIT, J.P. BIDEAU, G. BEURSKENS, W.P. BOSMAN and P.T. BEURSKENS, 1984. Acta Cryst., C40, 1412-1416.

Orthorhombic, $C222_1$, a = 12.118, b = 12.800, c = 24.523 Å, D_m = 1.34, Z = 8. Cu radiation, R = 0.037 for 1568 reflexions.

The dipeptide (Fig. 1) exists in the crystal as a zwitterion with the fully extended ββ conformation; χ_1 angles close to -60 and 180° and χ_2 to 60 and 120° respectively for the two residues. Both phenolic groups are planar (except for phenolic H) but C(3) and C(18) deviate from the mean planes through the phenyl groups. There is an extensive system of hydrogen bonds with water molecules involved as bridges between NH_3^+ and COO^- groups.

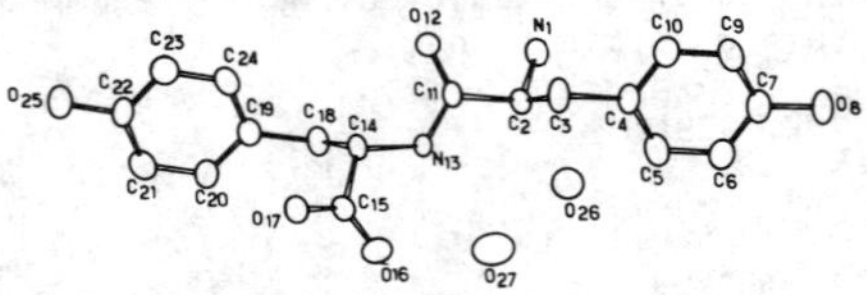

Fig. 1. The L-tyrosyl-L-tyrosine molecule.

1-{N-[1(S)-CARBOXY-3-PHENYLPROPYL]-L-ALANYL}-L-PROLINE TRIHYDRATE
$C_{18}H_{24}N_2O_5,3(H_2O)$

M.J. WYVRATT, E.W. TRISTRAM, T.J. IKELER, N.S. LOHR, H. JOSHUA, J.P. SPRINGER, B.H. ARISON and A.A. PATCHETT, 1984. J. Org. Chem., 49, 2816-2819.

Orthorhombic, $P2_12_12_1$, a = 9.798, b = 10.452, c = 19.819 Å, Z = 4. Cu radiation, R = 0.046 for 1360 reflexions.

The molecule (Fig. 1) has absolute configuration SSS.

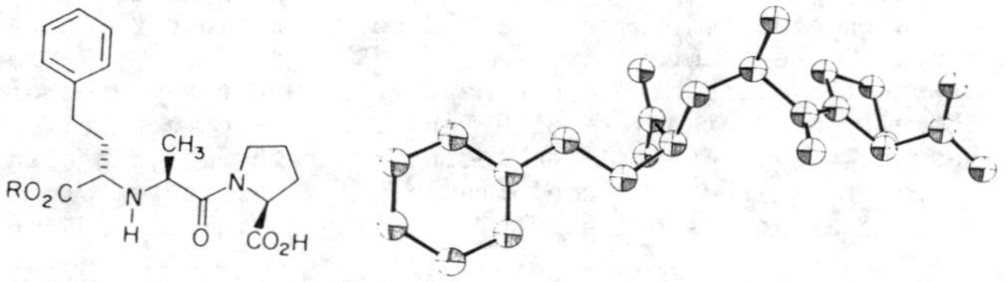

Fig. 1. The $C_{18}H_{24}N_2O_5$ molecule (R = H).

CYCLO(L-PROLYL-L-VALYL-L-PROLYL-L-VALYL) DIMETHYL SULPHOXIDE SOLVATE
$C_{20}H_{32}N_4O_4,C_2H_6OS$

I. UEDA, T. UEDA, I. SADA, T. KATO, M. MIKURIYA, S. KIDA and N. IZUMIYA, 1984. Acta Cryst., C40, 111-113.

Trigonal, $P3_1$, a = 9.584, c = 23.692 Å, D_m = 1.30, Z = 3. Mo radiation, R = 0.071 for 1866 reflexions.

The cyclo(-L-Pro-L-Val-)$_2$ molecule (Fig. 1) has a cis-trans-cis-trans backbone ring geometry and approximate C_2 symmetry. The twelve-membered ring has a distorted boat conformation. The Me_2SO molecule is disordered but linked by N-H...O hydrogen bonds to valyl moieties.

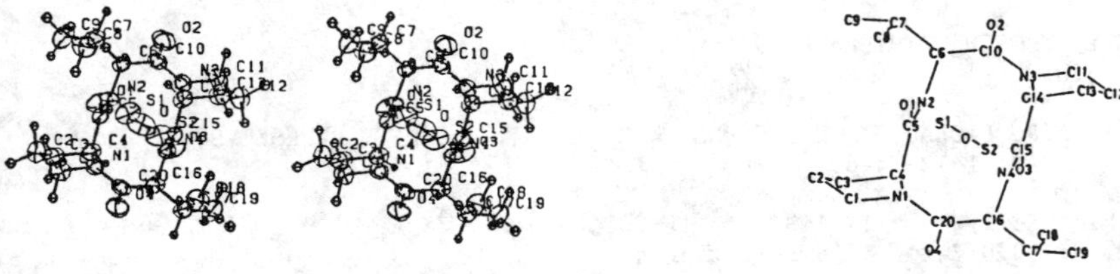

Fig. 1. A stereoview of $C_{20}H_{32}N_4O_4,C_2H_6OS$ and line drawing.

trans-ACETYLASPARAGINYLPROLYL-N-METHYLTYROSINAMIDE
$C_{21}H_{29}N_5O_6$ (I)

trans-ACETYLASPARAGINYLPROLYL-N-METHYLTYROSINAMIDE ETHANOL SOLVATE
$C_{21}H_{29}N_5O_6,C_2H_6O$ (II)

ACETYLTYROSINYLPROLYL-N-METHYLASPARAGINAMIDE DIHYDRATE
$C_{21}H_{29}N_5O_6,2(H_2O)$ (III)

ACETYLTYROSINYL-N-METHYLPROLINAMIDE
$C_{17}H_{23}N_3O_4,2(H_2O)$ (IV)

G.T. MONTELIONE, E. ARNOLD, Y.C. MEINWALD, E.R. STIMSON, J.B. DENTON, S.-G. HUANG, J. CLARDY and H.A. SCHERAGA, 1984. J. Am. Chem. Soc., 106, 7946-7958

I. Orthorhombic, $P2_12_12_1$, a = 17.587, b = 13.413, c = 9.912 Å, Z = 4. Cu radiation, R = 0.047 for 1653 reflexions.

II. Orthorhombic, $P2_12_12_1$, a = 25.770, b = 12.848, c = 7.476 Å, Z = 4. Mo radiation, R = 0.064 for 1291 reflexions.

III. Orthorhombic, $P2_12_12_1$, a = 26.490, b = 10.924, c = 8.598 Å, Z = 4. Cu radiation, R = 0.076 for 1191 reflexions.

IV. Orthorhombic, $P2_12_12_1$, a = 14.337, b = 13.821, c = 9.872 Å, Z = 4. Cu radiation, R = 0.062 for 1078 reflexions.

Views of the four peptides are in Fig. 1; they are of interest because they occur in ribonuclease A. Both (I) and (II) are characterised by a type I β-bond at Pro-Tyr with a trans-peptide bond at Asn-Pro and intramolecular hydrogen bonding involving both TyrNH and NHMe amide protons. Molecules (III) and (IV) have extended peptide backbone conformations with all trans-peptide bonds, and no intramolecular hydrogen bonds.

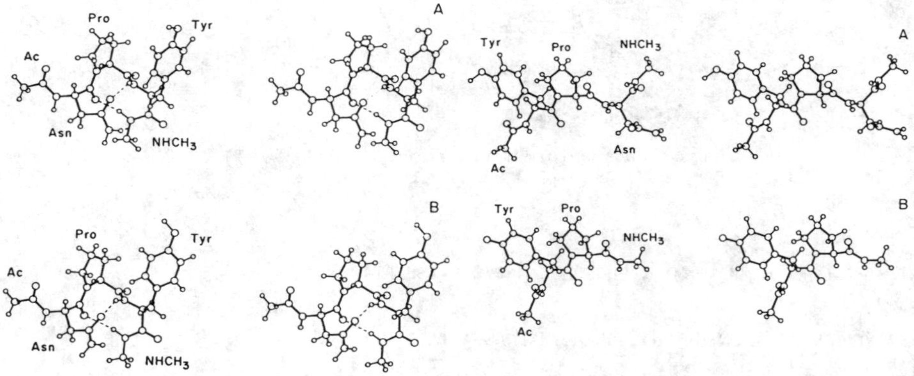

Fig. 1. Stereoviews of the peptides (I)(top left), (II) (lower left), (III) (top right) and (IV) (lower right).

(N-TRIFLUOROACETYL)-(N',N'-DIBENZYLHYDRAZINO)-α,α-DI-n-PROPYLGLYCINE
$C_{24}H_{30}F_3N_3O_2$ (I)

(N-TRIFLUOROACETYL-α,α-DI-n-PROPYLGLYCYL)-[(N',N'-DIBENZYLHYDRAZINO)-α,α-DI-n-PROPYLGLYCINE]
$C_{32}H_{45}F_3N_4O_3$ (II)

E. BENEDETTI, C. TONIOLO, P. HARDY, V. BARONE, A. BAVOSO, B. DI BLASIO, P. GRIMALDI, F. LELJ, V. PAVONE, C. PEDONE, G.M. BONORA and I. LINGHAM, 1984. J. Am. Chem. Soc., 106, 8146-8152.

I. Monoclinic, $P2_1/n$, a = 19.615, b = 12.284, c = 22.153 Å, β = 108.68°, D_m = 1.18, Z = 8. Cu radiation, R = 0.072 for 2720 reflexions.

II. Triclinic, $P\bar{1}$, a = 11.259, b = 12.729, c = 13.099 Å, α = 65.21, β = 98.31, γ = 101.91°, D_m = 1.18, Z = 2. Cu radiation, R = 0.076 for 2361 reflexions.

There are two independent molecules in (I). Bond lengths and conformations for (I) and (II) are shown in Fig. 1 and are largely unexceptional. The results of theoretical analysis of the conformations are in agreement with those derived from the X-ray work.

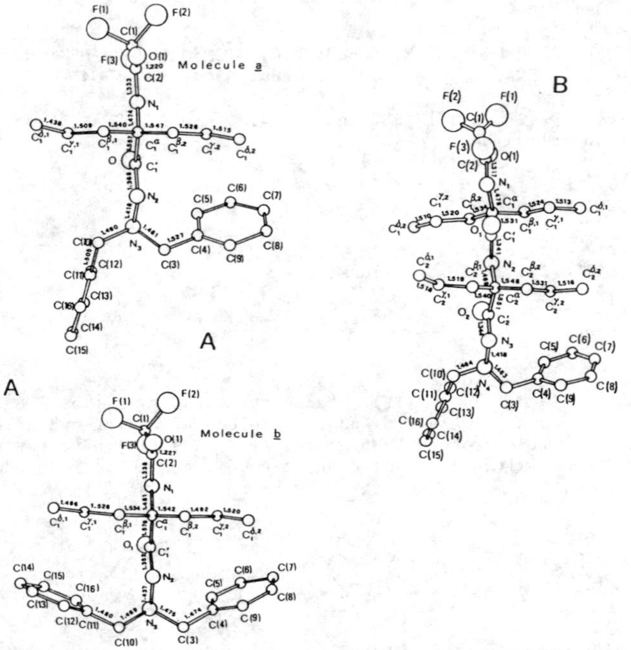

Fig. 1. Bond lengths and conformations for $C_{24}H_{30}F_3N_3O_2$ (I) (left) and $C_{32}H_{45}F_3N_4O_3$ (II) (right).

TRIS(CYCLOHEXYLAMMONIUM) N-PHOSPHONACETYL-L-ASPARTATE
$C_{24}H_{49}N_4O_8P$ (I)

BIS(CYCLOHEXYLAMMONIUM) N-CARBAMOYL-L-ASPARTATE
$C_{17}H_{34}N_4O_5$ (II)

G. ZANOTTI, H.L. MONACO and J. FOOTE, 1984. J. Am. Chem. Soc., 106, 7900-7904.

I. Orthorhombic, $P2_12_12_1$, a = 23.431, b = 6.749, c = 19.237 Å, Z = 4. Mo radiation, R = 0.090 for 1170 reflexions.

II. Monoclinic, C2, a = 26.238, b = 5.889, c = 15.416 Å, β = 116.1°. Z = 4. Mo radiation, R = 0.100 for 1325 reflexions.

Compound (I) is a potent inhibitor of the allosteric enzyme aspartate carbamoyltransferase and its structure is compared with those of the substrates carbamoyl-L-aspartate (II) and L-aspartate (1). (I) and (II) have very similar dimensions. Carbamoyl-L-aspartate and L-aspartate adopt an almost identical conformation; that of N-phosphonacetyl-L-aspartate has a different disposition of carboxyl groups.

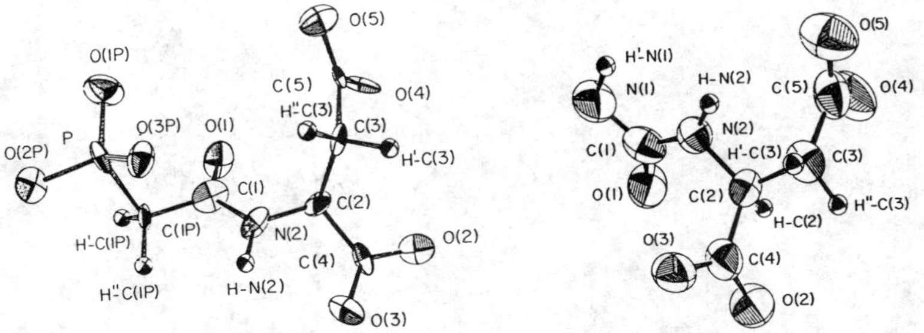

Fig. 1. Views of N-phosphonacetyl-L-aspartate (left) and N-carbamoyl-L-aspartate (right).

1. Structure Reports, 33B, 176.

CYCLO(L-PROLINE-L-PROLINE-GLYCINE-L-PROLINE-L-LEUCINE-GLYCINE) METHANOL MONOHYDRATE
$C_{25}H_{38}N_6O_6,CH_4O,H_2O$

T. NAKASHIMA, T. YAMANE, I. TANAKA and T. ASHIDA, 1984. Acta Cryst., C40, 171-174.

Orthorhombic, $P2_12_12_1$, a = 11.635, b = 17.716, c = 14.334 Å, D_m = 1.28, Z = 4. Cu radiation, R = 0.077 for 2439 reflexions.

The structure is characterized by a type-I β-turn with Pro-Leu at the corner in half of the molecule and two consecutive cis-peptide bonds, Gly-Pro and Pro-Pro, in the other half (Fig. 1). One of the cis-peptide bonds is severely distorted from planarity (ω -17.3°).

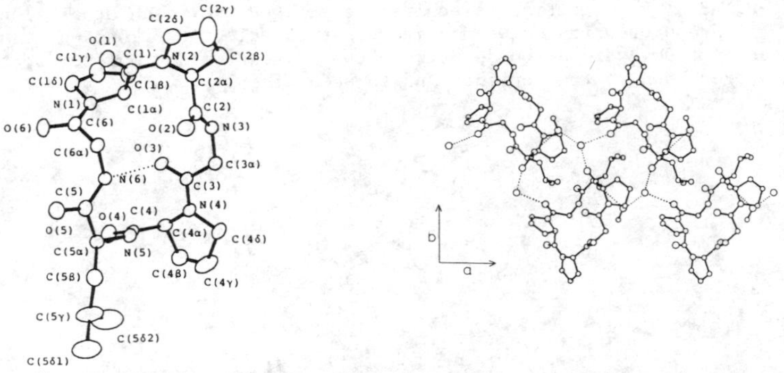

Fig. 1. The $C_{25}H_{38}N_6O_6$ molecule and a view of the crystal packing.

N-[2-(2-OXO-3-PHENOXY-4-PHENYLAZETIDIN-1-YL)PROPANOYL]-L-PROLINE t-BUTYL ESTER
$C_{27}H_{32}N_2O_5$

C.J. WHARTON, R. WRIGGLESWORTH and M. ROWE, 1984. J. Chem. Soc. Perkin I, 29-39.

Monoclinic, P2, a = 11.822, b = 6.464, c = 16.555 Å, β = 95.62°, Z = 2. Cu radiation, R = 0.078 for 1822 reflexions.

The X-ray analysis established the structure of this β-lactam peptide and the stereochemistry of the β-lactam ring (Fig. 1). Bond distances in this ring are: N(1)-C(1) 1.361(6), N(1)-C(3) 1.467(6), C(2)-C(3) 1.610(7), C(1)-C(2) 1.506(7) and C(1)-O(1) 1.212(6) Å.

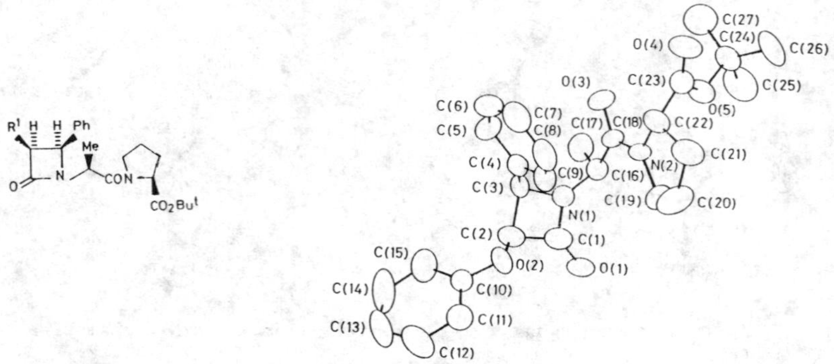

Fig. 1. $C_{27}H_{32}N_2O_5$: the molecular skeleton and structure.

cyclo(-L-PROLYL-L-VALYL-L-PHENYLALANYL-L-PHENYLALANYL-L-ALANYL-GLYCYL-) WATER ETHANOL (1:1:3) (CYCLOAMANIDE A - FORM II)
$C_{33}H_{42}N_6O_6, H_2O, 3(C_2H_6O)$

I.L. KARLE and C.C. CHIANG, 1984. Acta Cryst., C40, 1381-1386.

Orthorhombic, $P2_12_12_1$, a = 16.716, b = 24.007, c = 10.918 Å, Z = 4. Cu radiation, R = 0.072 for 4007 reflexions.

In both forms I (1) and II the peptide molecules have almost identical conformations and are isostructural. They have the unusual feature of a type II' hydrogen bond in the β-bend encompassing an LL sequence (L-phe-L-ala) but with conformational angles characteristic of a DL sequence (Fig. 1). The ϕ, ψ values are +60, -122 and -86, -5°. Peptide molecules are stacked over each other along c with N(5)-H...O(1) hydrogen bonds between adjacent stack molecules. One water molecule participates in four hydrogen bonds between two adjacent stack molecules; the ethanol O atoms hydrogen bond to the peptides and to each other.

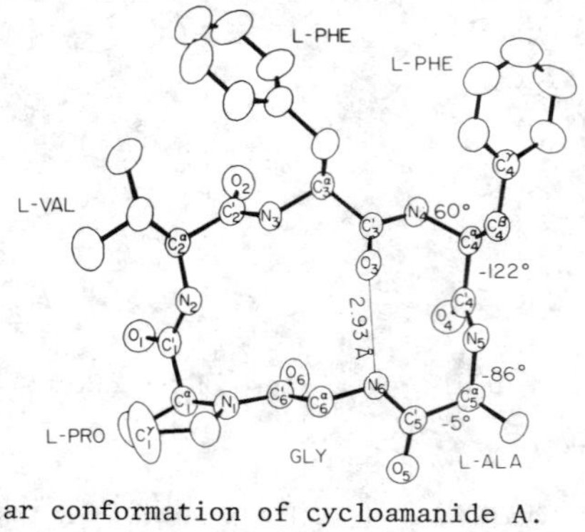

Fig. 1. The molecular conformation of cycloamanide A.

1. C.C. CHIANG, I.L. KARLE and Th. WIELAND, 1982. Int. J. Pept. Protein Res., 20, 414-420.

α-t-BUTYL N-(N-t-BUTOXYCARBONYL-O^5-BENZYL-α-L-GLUTAMYL)-O^5-BENZYL-L-GLUTAMATE
$C_{33}H_{44}N_2O_9$

D.S. EGGLESTON and D.J. HODGSON, 1984. Acta Cryst., C40, 1201-1204.

Orthorhombic, $P2_12_12_1$, a = 10.108, b = 11.528, c = 28.387 Å, D_m = 1.23, Z = 4. Mo radiation, R = 0.066 for 2066 reflexions.

The fully blocked peptide molecule adopts a β-sheet conformation with the side chains extended on opposite sides of the peptide backbone (Fig. 1). Bond lengths and angles within the peptide linkage are similar to averages compiled for this particular molecular fragment; however, there are exceptions involving C_1^{α}. The C_1^{α}-C_1^{β} bond of 1.556(8) Å is lengthened by 0.02 Å over the normally opserved sp^3-sp^3 C-C distance while the C_1^{α}-C_1^1 distance of 1.502(7) Å is 0.02 Å shorter than the normally observed value of 1.522 Å. The urethane and peptide bonds are in the trans configuration as are all the ester linkages. The side-chain conformations are unique compared to those of other structurally characterized glutamyl peptides. There appears to be weak hydrogen bonding in the crystals involving both available amide H atoms.

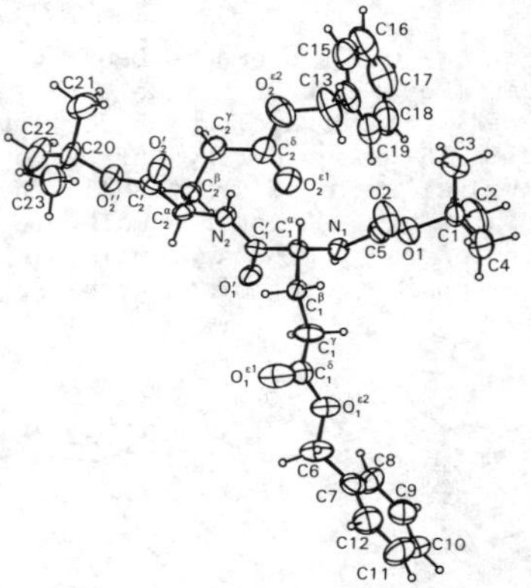

Fig. 1. $C_{33}H_{44}N_2O_9$: perspective view of the molecule.

CYCLOBIS(L-PHENYLALANYL-L-PROLYL-D-ALANYL) OCTAHYDRATE

$C_{34}H_{42}N_6O_6$,8(H_2O)

G. KARTHA, K.K. BHANDARY, K.D. KOPPLE, A. GO and PENG-PENG ZHU, 1984. J. Am. Chem. Soc., 106, 3844-3850.

Monoclinic, $P2_1$, a = 18.730, b = 9.239, c = 21.544 Å, β = 95.17°, Z = 4. Cu radiation, R = 0.059 for 7765 reflexions.

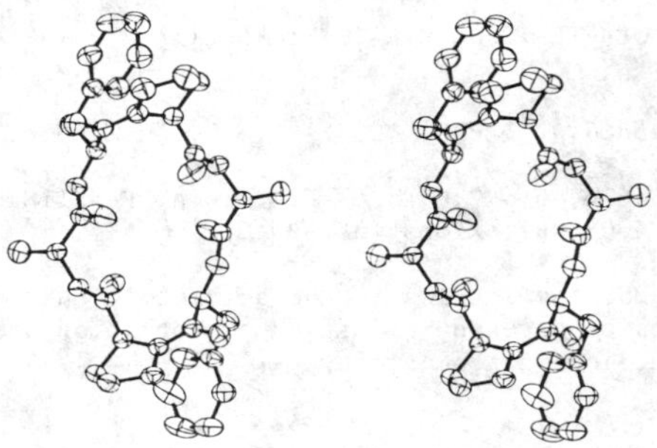

Fig. 1. A stereoview of one of the $C_{34}H_{42}N_6O_6$ molecules.

The two independent molecules in the asymmetric unit have nearly identical conformations (Fig. 1). Both molecules have near twofold symmetry. All the L-Phe-L-Pro peptide bonds are cis and they show considerable deviations from planarity with ω values -3 to -21°. In one molecule both proline rings are C^{γ}-endo, while in the other molecule one is endo and one is exo. There is an extensive hydrogen-bonded network involving the two molecules and the waters of hydration. It is suggested that the conformation in solution is the same as that in the crystal.

S-DEOXO[Ile[3]]AMANINAMIDE ETHANOL SOLVATE TRIHYDRATE
$C_{39}H_{53}N_{10}O_{10}S,2(C_2H_6O),3(H_2O)$

G. SHOHAM, D.C. REES, W.N. LIPSCOMB, G. ZANOTTI and T. WIELAND, 1984. J. Am. Chem. Soc., 106, 4606-4615.

Monoclinic, $P2_1$, a = 12.147, b = 11.250, c = 19.267 Å, β = 92.41°, Z = 2. Mo radiation, R = 0.065 for 3894 reflexions.

The octapeptide ring (Fig. 1) of 24 atoms is bent at the bridging points. The bridging segment of five atoms (containing the indole ring and the thioether group) participates in two 18-membered rings both of which display a fairly "flat" conformation, often observed in 18-membered rings of cyclohexapeptides. The planes of these two rings form a dihedral angle of about 80°. The indole ring lies approximately in the "plane" of ring 1 giving the molecule an overall distorted "T" shape. The three water molecules and two ethanol molecules, crystallized with the amatoxin, form a strong and extensive intermolecular hydrogen-bonding system.

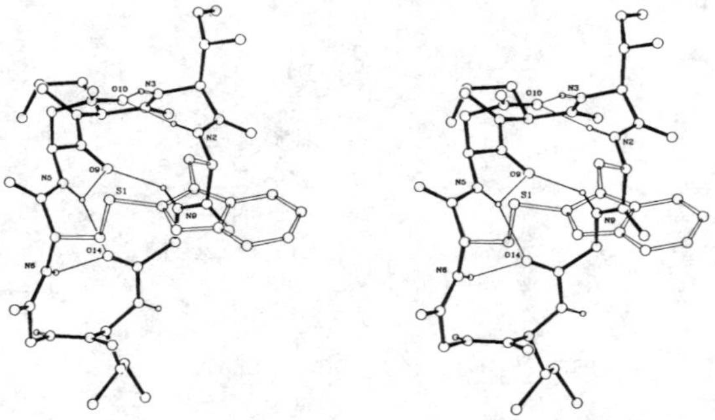

Fig. 1. A stereoview of the amatoxin $C_{39}H_{53}N_{10}O_{10}S$ showing the intramolecular hydrogen bonds.

(5,10,15,20-TETRAMETHYLISOBACTERIOCHLORINATO)NICKEL(II)
$C_{24}H_{24}N_4Ni$

M.P. SUH, P.N. SWEPSTON and J.A. IBERS, 1984. J. Am. Chem. Soc., 106, 5164-5171.

Monoclinic, $P2_1/n$, a = 23.656, b = 7.644, c = 24.014 Å, β = 118.94°, Z = 8. Cu radiation, R = 0.065 for 4008 reflexions (at -69°C).

There are two independent molecules in the asymmetric unit; selected bond lengths are in Fig. 1. The molecules have an S_4-ruffled conformation and a bond length pattern indicating a reduction in aromaticity.

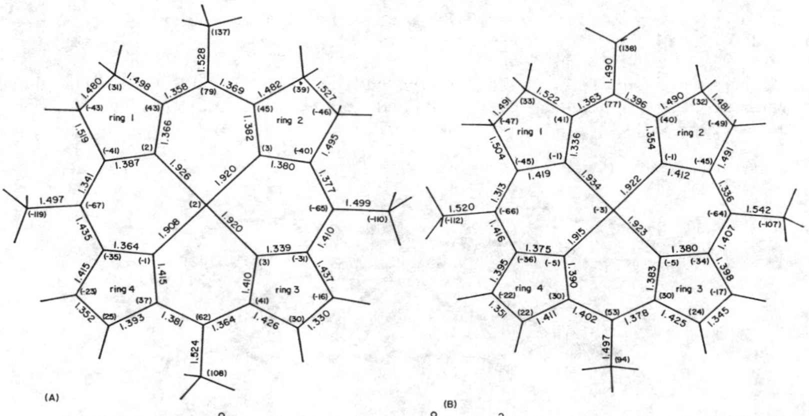

Fig. 1. Bond lengths (Å) and deviations (Å x 10^3) from the plane of the 24-atom macrocyclic skeleton for the two independent molecules.

catena-(μ-OXO)(HEMIPORPHYRAZINATO)IRON(IV)
$C_{26}H_{14}FeN_8O$ $[FeO(C_{26}H_{14}N_8)]_n$

W. HILLER, J. STRÄHLE, A. DATZ, M. HANACK, W.E. HATFIELD, L.W. ter HAAR and P. GÜTLICH, 1984. J. Am. Chem. Soc., 106, 329-335.

Monoclinic, P2/n, a = 16.090, b = 3.975, c = 16.070 Å, β = 92.97°, Z = 2. Mo radiation, R = 0.045 for 1459 reflexions (at 210 K).

Distorted $Fe(C_{26}H_{14}N_8)$ units stack along the b axis and are linked into polymeric, uniformly spaced, linear chains by axially bound O atoms. The O atoms are disordered into two crystallographically distinct sites lying about a central location which is normal to the C_2 axis collinear with the chain. The $[FeO(C_{26}H_{14}N_8)]$ units have exact C_2 symmetry (Fig. 1). Bond lengths and angles: Fe-O 1.97(3)-2.03(3), Fe-N 2.135(3) and 1.910(3) Å, Fe-O-Fe 158.2(7) and 170.5(7)°.

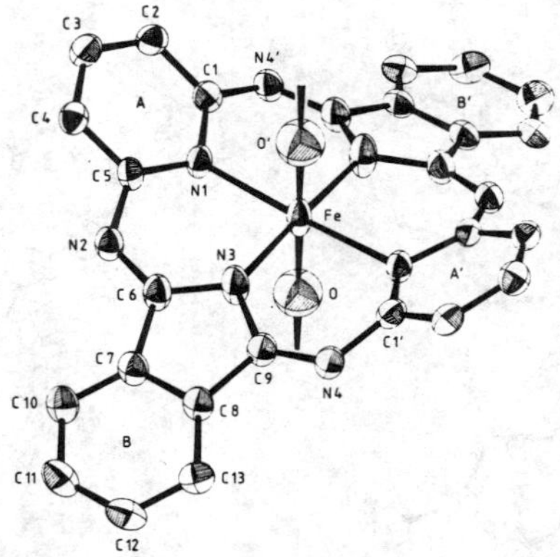

Fig. 1. The structural unit $[FeO(C_{26}H_{14}N_8)]$.

OXOPHTHALOCYANINATOMOLYBDENUM(IV)
$C_{32}H_{16}MoN_8O$

V. BORSCHEL and J. STRAHLE, 1984. Z. Naturforsch., B39, 1664-1667.

Monoclinic, $P2_1/c$, a = 13.441, b = 13.292, c = 14.041 Å, β = 103.80°, Z = 4. Mo radiation, R = 0.084 for 1409 reflexions.

The Mo atom has square-pyramidal coordination (Fig. 1) and is 0.69 Å above the phthalocyaninato ligand plane. The Mo-O bond length is 1.668(6) Å and Mo-N values are in the range 2.068-2.112(7) Å with O-Mn-N 107.8-110.9(6)° and N-Mo-N 82.2-85.1(6)°.

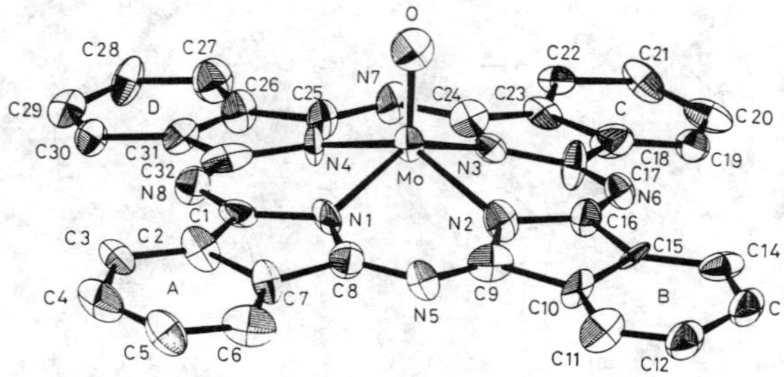

Fig. 1. The $C_{32}H_{16}MoN_8O$ structure.

CHLORO(PHTHALOCYANINATO)GALLIUM(III)
$C_{32}H_{20}ClGaN_8$ (I) $Ga(C_{32}H_{20}N_8)Cl$

CHLORO(PHTHALOCYANINATO)ALUMINUM(III)
$C_{32}H_{20}AlClN_8$ (II) $Al(C_{32}H_{20}N_8)Cl$

K.J. WYNNE, 1984. Inorg. Chem., 23, 4658-4663.

I. Triclinic, $P\bar{1}$, a = 9.301, b = 11.272, c = 13.143 Å, α = 105.46, β = 105.51, γ = 96.80°, Z = 2. Mo radiation, R = 0.036 for 3483 reflexions.

II. Triclinic, $P\bar{1}$, a = 13.776, b = 13.775, c = 14.059 Å, α = 98.36, β = 108.60, γ = 90.16°, Z = 4. Cu radiation, R = 0.137 for 9454 reflexions. One molecule is disordered.

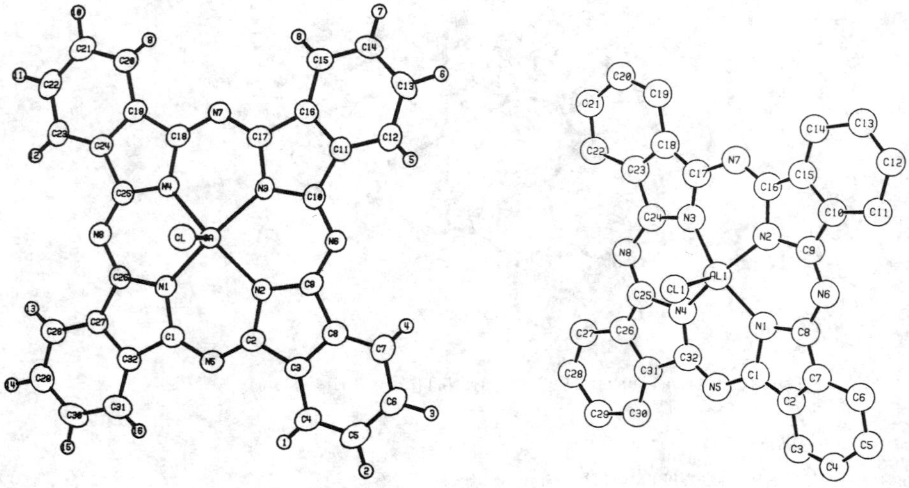

Fig. 1. The $Ga(C_{32}H_{20}N_8)Cl$ molecule (left) and the ordered $Al(C_{32}H_{20}N_8)Cl$ molecule (right).

The coordination about Ga (I) and Al (II) is square pyramidal with the Cl^- ligand apical (Ga-Cl 2.217(1), Ga-N 1.979(4)-1.988(4), Al-Cl 2.179(6), Al-N 1.961(12)-2.018(12) Å). Ga and Al are displaced 0.439(1) and 0.410(6) Å, respectively, from the basal planes. The phthalocyanine ring is essentially planar in I; packing gives an interleaved slipped-stacked array with an interplanar distance of 3.34 Å.

DEOXOPHYLLOERYTHROETIOPORPHYRINOXOVANADIUM(IV)
$C_{33}H_{35}Cl_3N_4OV$ VO(dpep),$CHCl_3$

S.A. MILLER, T.W. HAMBLEY and J.C. TAYLOR, 1984. Aust. J. Chem., 37, 761-766.

Monoclinic, $P2_1/c$, a = 12.912, b = 14.151, c = 18.404 Å, β = 70.34°, D_m = 1.37, Z = 4. Mo radiation, R = 0.077 for 2296 reflexions. Dark red prismatic crystals.

The V atom lies 0.51 Å above the least square plane containing the porphyrin ring system (V-O 1.582(3), V-N 2.055(5) Å, N-V-N 86.7(2)°) (Fig. 1).

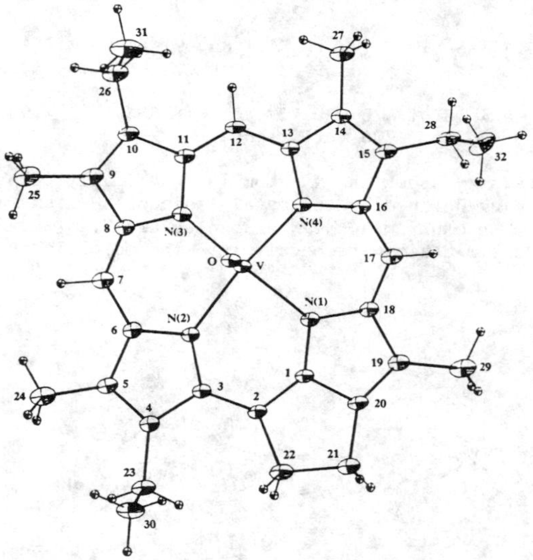

Fig. 1. The structure of $C_{33}H_{35}Cl_3N_4OV$.

(2,2,3,3,7,7,8,8,12,12,13,13,17,17,18,18-HEXADECAMETHYL-2,3,7,8,12,13,17,18-OCTAHYDRO-1H,23H-10,20-DIAZAPORPHINATO)COBALT(I)
$C_{34}H_{51}CoN_6$

L. WALDER, G. RYTZ, U. VOGELI, R. SCHEFFOLD and P. ENGEL, 1984. Helv. Chim. Acta, 67, 1801-1817.

Monoclinic, Pc, a = 13.250, b = 10.896, c = 13.037 Å, β = 119.15°, D_m = 1.17, Z = 2. Mo radiation, R = 0.103 for 1664 reflexions.

Dimensions are in Fig. 1. The Co atom is 0.025 Å from the best plane. The four inner N-atoms form a very flat tetrahedron with Co at the centre.

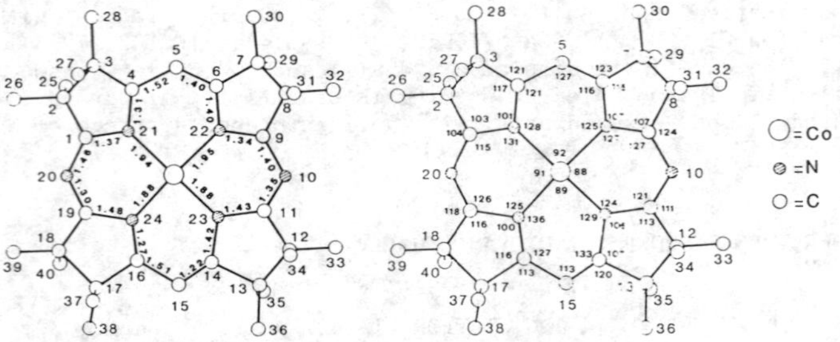

Fig. 1. A view of $C_{34}H_{51}CoN_6$ with bond lengths and angles.

BIS(PYRIDINE)(HEMIPORPHYRAZINATO)NICKEL(II)
$C_{36}H_{24}N_{10}Ni$ $Ni(C_{26}H_{14}N_8)(C_5H_5N)_2$

E. AGOSTINELLI, D. ATTANASIO, I. COLLAMATI and V. FARES, 1984. Inorg. Chem., 23, 1162-1165.

Monoclinic, $P2_1/c$, a = 16.418, b = 9.727, c = 18.632 Å, β = 94.59°, D_m = 1.47, Z = 4. Mo radiation, R = 0.053 for 1726 reflexions.

The stereochemistry about the Ni atom is rhombically compressed octahedral, the unique direction being defined by the metal-isoindole N atoms which are 1.97(1) Å from the Ni atom. The other mean Ni-N distances are 2.18(2) and 2.22(2) Å for macrocycle and free pyridinic N respectively. The molecule is shown in Fig. 1.

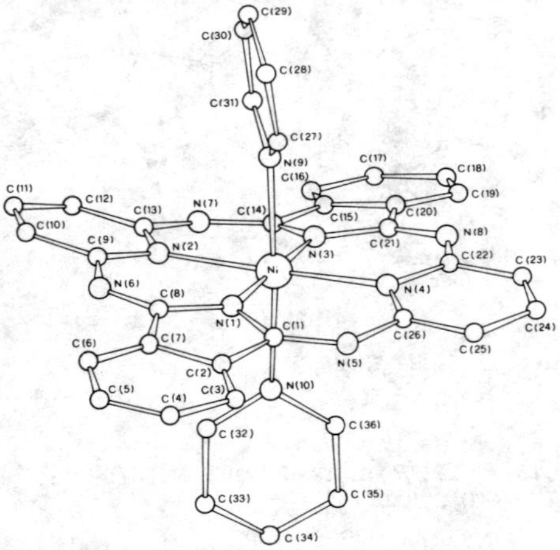

Fig. 1. The $Ni(C_{26}H_{14}N_8)(C_5H_5N)_2$ molecule.

2,12-DIETHYL-3,7,13,17-TETRAMETHYL-8,18-NONAMETHYLENE-21H,23H-PORPHYRIN METHANOL SOLVATE
$C_{37}H_{46}N_4,2CH_4O$

F.W.B. EINSTEIN and T. JONES, 1984. Acta Cryst., C40, 696-700.

Monoclinic, $P2_1/c$, a = 14.774, b = 13.823, c = 17.634 Å, β = 103.31°, Z = 4. Mo radiation, R = 0.036 for 2349 reflexions (at 173K).

The structure (Fig. 1) consists of a free-base porphyrin which exhibits considerable folding as the result of the nine-carbon 'strap' which spans diametrically opposed β-positions. Calculations indicate that the strain induced by this folding is distributed throughout the entire molecule. Bond distances and angles are generally as expected.

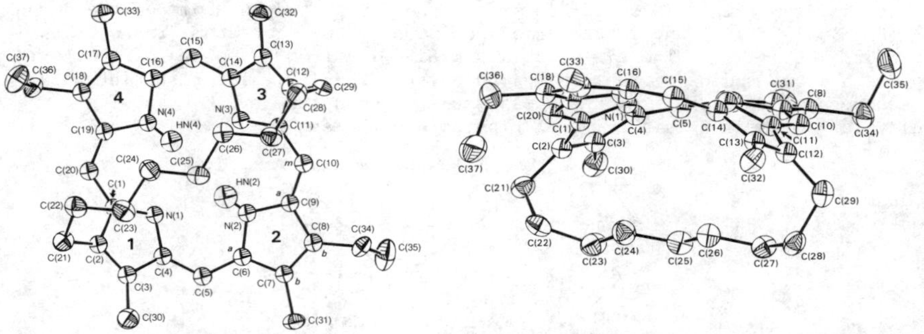

Fig. 1. $C_{37}H_{46}N_4,2CH_4O$: molecular structure of the porphyrin molecule and a side view of the molecule showing the degree of distortion of the core.

3,13-DIMETHYLENE-2,2',7,8,12,12',17,18-OCTAETHYLPORPHYRIN HEXANE DICHLOROMETHANE SOLVATE

$C_{38}H_{50}N_4,0.8(C_6H_{14}),0.2(CH_2Cl_2)$ (I)

3,13-DIMETHYLENE-2,2',7,8,12,12',17,18-OCTAETHYLPORPHYRIN

$C_{38}H_{50}N_4$ (II)

K.M. BARKIGIA, J. FAJER, C.K. CHANG and R. YOUNG, 1984. J. Am. Chem. Soc., 106, 6457-6459.

I. Monoclinic, $P2_1/n$, a = 11.244, b = 8.079, c = 21.402 Å, β = 96.54°, Z = 2. Cu radiation, R = 0.063 for 3129 reflexions (at 140 K).

II. Triclinic, $P\bar{1}$, a = 8.655, b = 13.492, c = 7.856 Å, α = 105.83, β = 104.43, γ = 97.23°, Z = 1. Mo radiation, R = 0.072 for 1235 reflexions.

The macrocycles in I and II lie about crystallographic inversion centres and are effectively planar. For (I) the largest deviations from the 24-atom core are at C(2) (-0.068(2) Å) and C(3) (+0.066(2) Å). The C1-C2-C3-C4 torsion angle is 5.3(2)° and C6-C7-C8-C9 is 0.2(2)°.

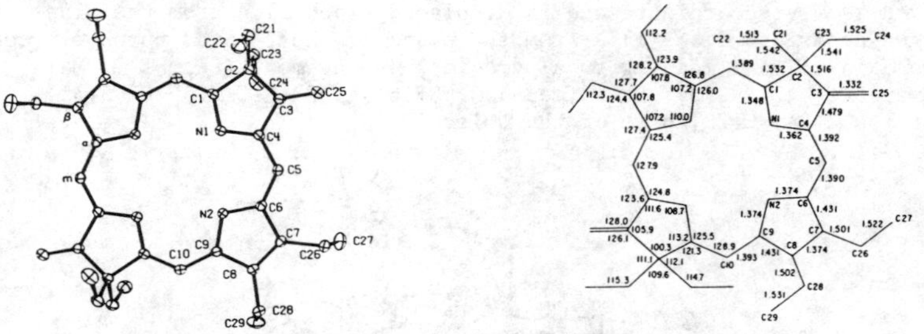

Fig. 1. The structure of $C_{38}H_{50}N_4$ and details of dimensions (at 140 K).

[2,3,7,8,12,13,17,18-OCTAETHYLPORPHINATOTIN(II)]TETRACARBONYLIRON

$C_{40}H_{44}FeN_4O_4Sn$

J.-M. BARBE, R. GUILARD, C. LECOMTE and R. GERARDIN, 1984. Polyhedron, 3, 889-894.

Triclinic, $P\bar{1}$, a = 12.253, b = 13.910, c = 15.087 Å, α = 59.53, β = 61.14, γ = 68.88°, Z = 2. Mo radiation, R = 0.0409 for 5965 reflexions.

The tin atom (Fig. 1) is coordinated by the four nitrogen atoms of the porphinato group (Sn-N 2.187(3) Å) and in the axial position by the iron atom (Sn-Fe 2.492(1) Å, Sn-Fe-C(52) 179.1(1)°). The tin atom lies 0.818(9) Å from the plane of the four nitrogens and the radius of the central hole is 2.028(3) Å. The resulting polyhedron is a square pyramid. The iron atom possesses a pseudo-symmetry C_{3v} and its coordination polyhedron is a trigonal bipyramid with mean Fe-C 1.754(5) Å.

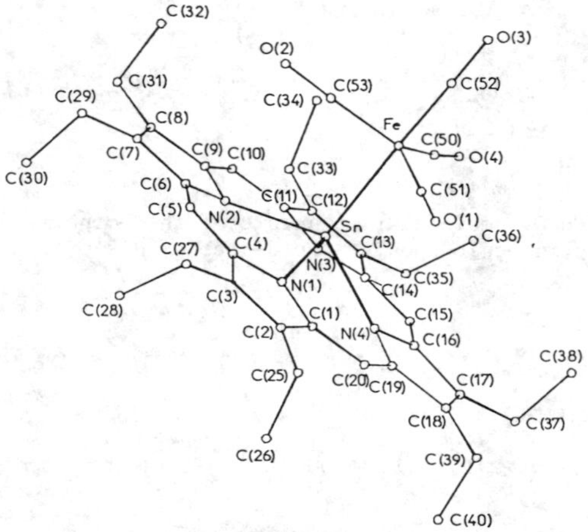

Fig. 1. A view of $C_{40}H_{44}FeN_4O_4Sn$.

(BENZENETHIOLATO)(2,3,7,8,12,13,17,18-OCTAETHYLPORPHINATO)IRON(III)

$C_{42}H_{49}FeN_4S$ $Fe(C_{36}H_{44}N_4)(C_6H_5S)$

K.M. MILLER and C.E. STROUSE, 1984. Acta Cryst., C40, 1324-1327.

Triclinic, $P\bar{1}$, a = 10.246, b = 13.040, c = 14.900, α = 107.82, β = 43.00, γ = 101.98°, D_m = 1.30, Z = 2. Mo radiation, R = 0.069 for 4676 reflexions.

The iron is five-coordinate and is displaced 0.466(1) Å from the plane of the four nitrogen atoms and 0.512(1) Å from the plane of the 24-atom porphyrin core towards the axial ligand leading to a pyramidal arrangement (Fig. 1). Main dimensions are Fe-S 2.299(3), Fe-N (mean) 2.057(6) Å, Fe-S-C(37) 102.5(3)°. There is disorder in the C(24)/C(28) part of the molecule.

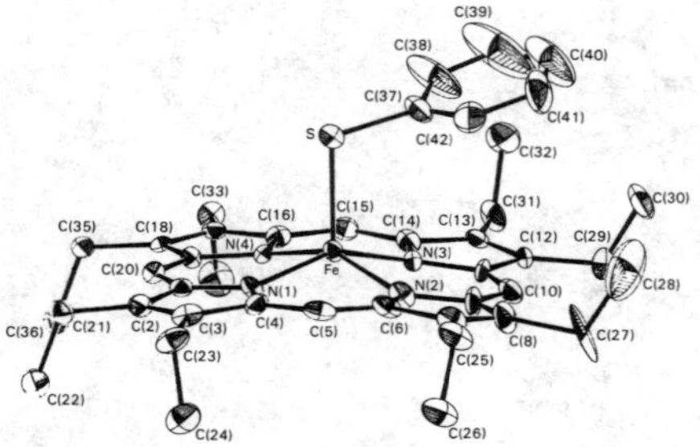

Fig. 1. The $Fe(C_{36}H_{44}N_4)(C_6H_5S)$ molecule (omitting H atoms).

TETRAPHENYLPORPHYRINOXOVANADIUM(IV)
$C_{44}H_{28}N_4OV$ (I) $VON_4C_{44}H_{28}$

ETIOPORPHYRINOXOVANADIUM(IV)
$C_{32}H_{36}N_4OV$ (II) $VON_4C_{32}H_{36}$

ETIOPORPHYRINOXOVANADIUM(IV) HEMIHYDROQUINONE
$C_{32}H_{36}N_4OV,0.5(C_6H_6O_2)$ (III) $C_{32}H_{36}N_4OV,0.5(C_6H_4(OH)_2)$

M.G.B. DREW, P.C.H. MITCHELL and C.E. SCOTT, 1984. Inorg. Chim. Acta, 82, 63-68.

I. Tetragonal, I4/m, a = 13.345, c = 9.745 Å, D_m = 1.31, Z = 2. Mo radiation, R = 0.099 for 556 reflexions.

II. Triclinic, P$\bar{1}$, a = 10.52, b = 11.70, c = 12.97 Å, α = 70.7, β = 113.8, γ = 108.8°, D_m = 1.34, Z = 2. Mo radiation, R = 0.086 for 2316 reflexions.

III. Orthorhombic, Pnaa, a = 17.72, b = 18.14, c = 18.57 Å, D_m = 1.28, Z = 8. Mo radiation, R = 0.082 for 1265 reflexions.

In all three structures, the V atom is five-coordinate being bonded to the four N atoms of the porphyrin and one terminal O atom (Fig. 1). The metal geometry is square pyramidal with V=O distances and N_4 plane-to-metal displacements of (I) 1.63, 0.53 Å; (II) 1.599, 0.49 Å; (III) 1.614, 0.51 Å. Compound I has crystallographic fourfold symmetry and a V-N(1) distance of 2.080 Å. In compounds II and III the V-N distances range from 2.049 to 2.086 Å and from 2.042 to 2.086 Å. In III, the molecules are linked by hydrogen bonds through the VO groups to the quinol OH groups (O...O 2.82 Å, V=O...O 159.3° and C-O...O 118.9°).

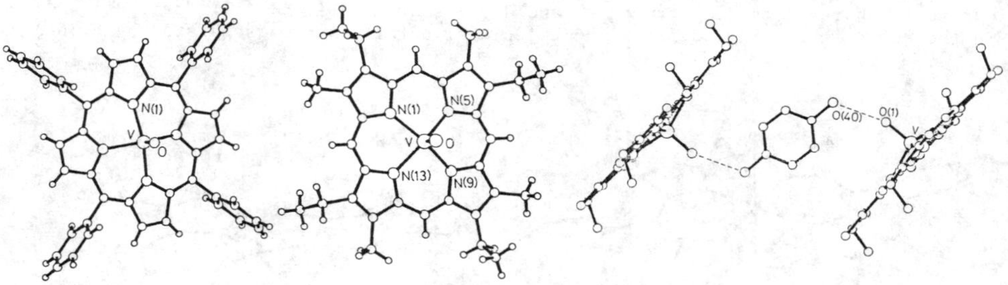

Fig. 1. Structure of $C_{44}H_{28}N_4OV$ (left), $C_{32}H_{36}N_4OV$ (centre), and $C_{32}H_{36}N_4OV,0.5C_6H_4(OH)_2$ (right).

(5,10,15,20-TETRAPHENYL-21H,23H-PORPHINATO)PLATINUM(II)
$C_{44}H_{28}N_4Pt$

A. HAZELL, 1984. Acta Cryst., C40, 751-753.

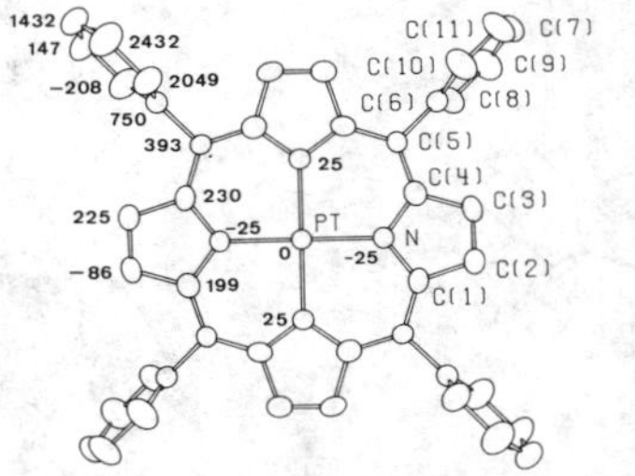

Fig. 1. $C_{44}H_{28}N_4Pt$: perspective view of the molecule. Deviations of the atoms from the (001) plane through Pt are shown (Å × 10^3).

Tetragonal, $I\bar{4}2d$, a = 15.073, c = 13.988 Å, Z = 4. Cu radiation, R = 0.012 for 1051 reflexions.

In the molecule the porphine nucleus is slightly ruffled, with N alternately 0.025(3) above and below the plane (Fig. 1). Pt-N is 2.008(2) Å. The pyrrole ring is tilted at an angle of 12.0° with the (001) plane and the phenyl group at an angle of 79.6°, the angle between these rings being 89.5°. The crystal is isomorphous with the corresponding Fe, Co, Ni, Cu and Pd complexes (1).

1. Structure Reports, 29, 617; 41B, 566; 42B, 378.

AQUANITROSYL(meso-TETRAPHENYLPORPHINATO)IRON(III) PERCHLORATE
$C_{44}H_{30}ClFeN_5O_6$ (I) $[Fe(TPP)(NO)(H_2O)]ClO_4$

NITROSYLOCTAETHYLPORPHINATOIRON(III) PERCHLORATE CHLOROFORM SOLVATE
$C_{36}H_{44}ClFeN_5O_5,CHCl_3$ (II) $[Fe(OEP)(NO)]ClO_4,CHCl_3$

W.R. SCHEIDT, Y.J. LEE and K. HATANO, 1984. J. Am. Chem. Soc., 106, 3191-3198.

I. Monoclinic, $P2_1/n$, a = 10.303, b = 8.124, c = 21.364 Å, β = 97.76°, D_m = 1.48, Z = 2. Mo radiation, R = 0.057 for 3999 reflexions (at 96 K).

II. Monoclinic, $P2_1/n$, a = 12.890, b = 20.363, c = 14.969 Å, β = 95.48°, D_m = 1.41, Z = 4. Mo radiation, R = 0.058 for 5956 reflexions (at 292 K).

Both I and II are low-spin ferric porphyrinates; II is the first five-coordinate one to be characterised. Structures and details are in Fig. 1. II forms a π-π dimer in the solid state (Fig. 2) with the two planar parallel cores separated by 3.36 Å.

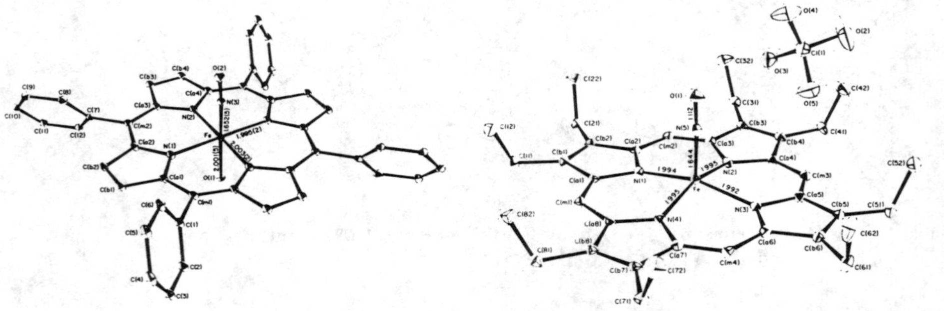

Fig. 1. Views and principal bond lengths in $C_{44}H_{30}ClFeN_5O_6$ (I) and $C_{36}H_{44}ClFeN_5O_5$ (II).

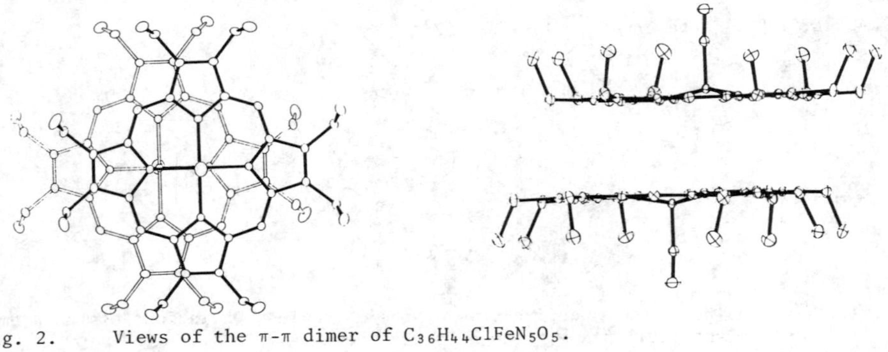

Fig. 2. Views of the π-π dimer of $C_{36}H_{44}ClFeN_5O_5$.

BIS(2-METHYLIMIDAZOLE)OCTAETHYLPORPHINATOIRON(III) PERCHLORATE
$C_{44}H_{56}ClFeN_8O_4$

D.K. GEIGER, Y.J. LEE and W.R. SCHEIDT, 1984. J. Am. Chem. Soc., 106, 6339-6343.

Triclinic, $P\bar{1}$, a = 10.274, b = 12.276, c = 9.038 Å, α = 91.60, β = 109.38, γ = 88.71°, D_m = 1.31, Z = 1. Mo radiation, R = 0.046 for 7692 reflexions.

The molecule lies on an inversion centre. Details of molecular dimensions and the structure are in Fig. 1. It is suggested that the state-dependent magnetic properties of this complex are a result of the axial ligand orientation. In the solid the projection of the 2-MeIm ligand plane onto the porphorinato core is within 22° of eclipsing $Fe-N_p$ bonds. Thus steric hindrance does not permit the short axial Fe-N bond required to achieve a low spin state. In solution this can be achieved.

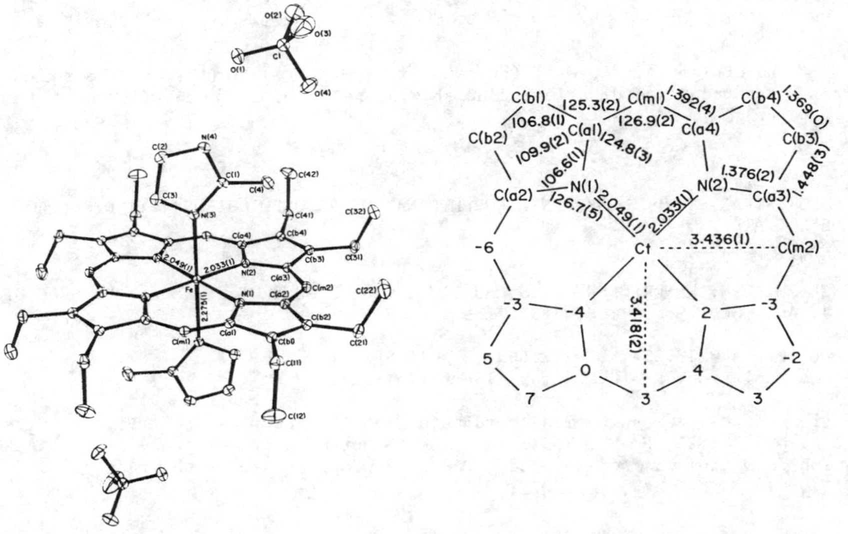

Fig. 1. $C_{44}H_{56}ClFeN_8O_4$: a view of the complex and mean values of dimensions of the porphinato core (also displayed are perpendicular displacements (Å × 100) of the atoms from the core plane).

FORMYLMETHYL(TETRAPHENYLPORPHINATO)COBALT(III)
$C_{46}H_{31}CoN_4O$ $[Co^{III}(tpp)CH_2CHO]$

H. MASUDA, T. TAGA, H. SUGIMOTO and M. MORI, 1984. J. Organometal. Chem., 273, 385-392.

Monoclinic, Pc, a = 13.132, b = 12.747, c = 13.443 Å, β = 129.83°, D_m = 1.37, Z = 2. Cu radiation, R = 0.057 for 2531 reflexions.

The Co(III) ion is coordinated by the four porphinato N atoms and one C atom of the formylmethyl group, and is displaced by 0.14 Å from the mean plane of the four porphinato N atoms towards the axial formylmethyl ligand.

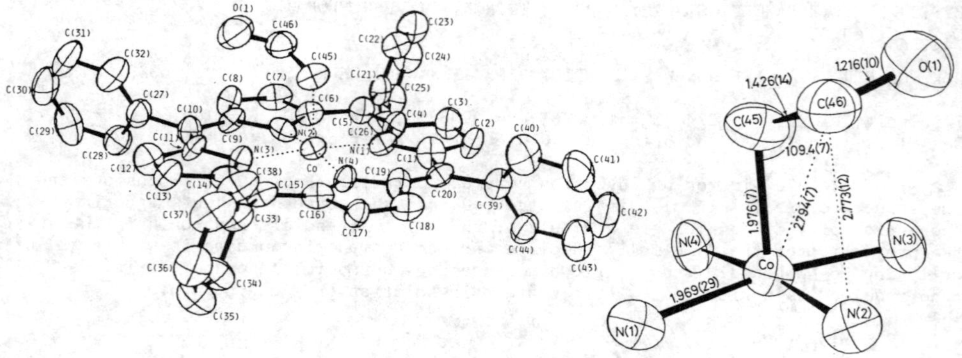

Fig. 1. Molecular structure of the complex $C_{46}H_{31}CoN_4O$ (left), and the square pyramidal coordination group showing selected distances (Å) and angles (°) (right).

5,10,15,20-TETRAKIS(4-METHOXYPHENYL)PORPHYRINATOHYDROSULPHIDOIRON(III) DICHLOROMETHANE SOLVATE
$C_{48}H_{37}FeN_4O_4S,CH_2Cl_2$

D.R. ENGLISH, D.N. HENDRICKSON, K.S. SUSLICK, C.W. EIGENBROT Jr. and W.R. SCHEIDT, 1984. J. Am. Chem. Soc., 106, 7258-7259.

Triclinic, P1, a = 12.354, b = 12.604, c = 15.557 Å, α = 69.28, β = 82.28, γ = 72.18°, Z = 2. Mo radiation, R = 0.073 for 6130 reflexions.

A view of this low-spin five-coordinate iron(III) complex is in Fig. 1, with average Fe-N 2.015(2), Fe-S 2.298(3) Å. The Fe out-of-plane distance is 0.33 Å. The hydrosulphide H atom was not located. The S-Fe-N angles are in the range 97.3-100.8(1)°.

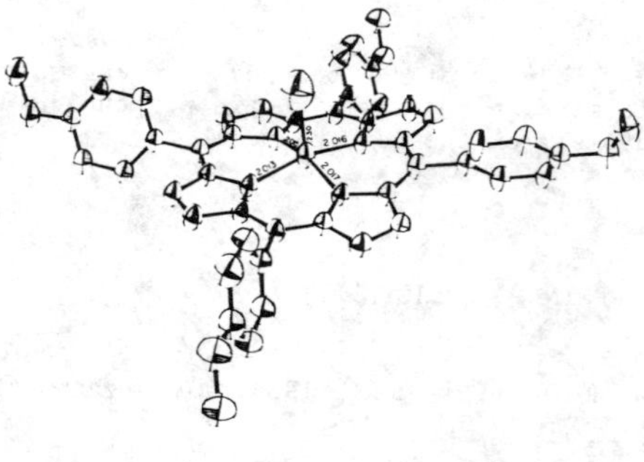

Fig. 1. A perspective view of $C_{48}H_{37}FeN_4O_4S$.

(2,3,5,6-TETRAFLUOROBENZENETHIOLATO)(TETRAPHENYLPORPHINATO)IRON(III)
$C_{50}H_{31}F_4FeN_4S$ $Fe(C_{44}H_{30}N_4)(SC_6HF_4)$

K.M. MILLER and C.E. STROUSE, 1984. Inorg. Chem., 23, 2395-2400.

Monoclinic, $P2_1/a$, a = 13.272, b = 12.685, c = 22.46 Å, β = 97.17°, Z = 4. Mo radiation, R = 0.078 for 1554 reflexions with k+l = 2n only (at 117 K).

The five-coordinate complex (Fig. 1) has square-pyramidal coordination geometry with Fe displaced 0.395(3) Å from the N_4 basal plane, Fe-S 2.298(3), Fe-N 2.02(2)-2.10(2) Å. The hkl, k+l = 2n+1, reflexions are observed to be diffuse and indicate a lack of long range order in the b and c directions. The disorder has been mimicked in a two-dimensional simulation of the crystal growth.

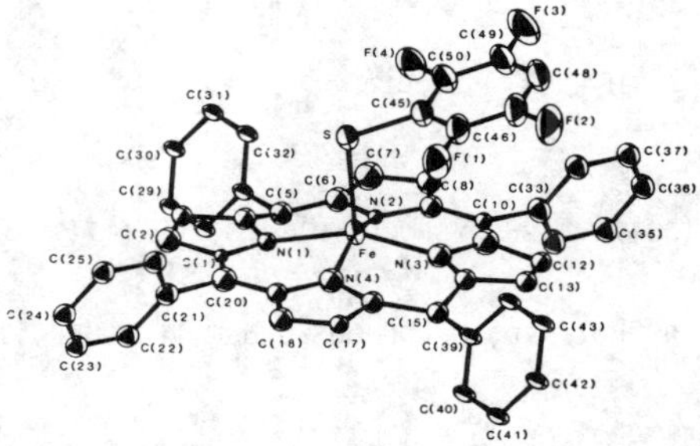

Fig. 1. The $Fe(C_{44}H_{30}N_4)(SC_6HF_4)$ molecule.

CHLORO(N-PHENYL-5,10,15,20-TETRAPHENYLPORPHINATO)ZINC(II)

$C_{50}H_{33}ClN_4Zn$ $ZnCl(C_{50}H_{33}N_4)$

D. KUILA, D.K. LAVALLEE, C.K. SCHAUER and O.P. ANDERSON, 1984. J. Am. Chem. Soc., 106, 448-450.

Orthorhombic, $P2_12_12_1$, a = 15.078, b = 15.272, c = 17.268 Å, D_m = 1.30, Z = 4. Mo radiation, R = 0.055 for 2928 reflexions.

The molecule is shown in Fig. 1. The N-phenyl group is poised so that one O-hydrogen atom is near the Zn atom (Zn-H 2.52 Å). The N-substituted pyrrole ring is canted 42.0° with respect to the plane of the three unsubstituted N atoms. Bond lengths: Zn-NPh 2.499, Zn-N (adjacent) 2.086 and 2.102, Zn-N (opposite) 2.015, and Zn-Cl 2.24 Å.

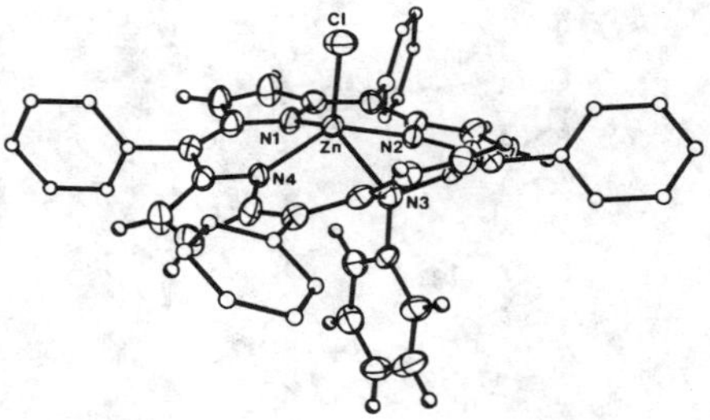

Fig. 1. The $ZnCl(C_{50}H_{33}N_4)$ molecule.

PHENYL(meso-TETRAPHENYLPORPHYRINATO)IRON(III) PENTANE SOLVATE

$C_{50}H_{33}FeN_4,C_5H_{12}$ $Fe(C_6H_5)(C_{44}H_{28}N_4).C_5H_{12}$

P. DOPPELT, 1984. Inorg. Chem., 23, 4009-4011.

Triclinic, $P\bar{1}$, a = 12.556, b = 16.866, c = 12.216 Å, α = 109.79, β = 114.91, γ = 72.30°, D_m = 1.23, Z = 2. Cu radiation, R = 0.060 for 4968 reflexions.

The complex (Fig. 1) is five-coordinate with mean Fe-N 1.962 Å and Fe-C 1.955(3) Å. The Fe(III) atom is displaced by 0.17 Å from the mean plane of the porphyrin core; this displacement is very short. In evidence $C_6H_5^-$ is a very strong field ligand, and this property is at the origin of the low-spin character of the compound.

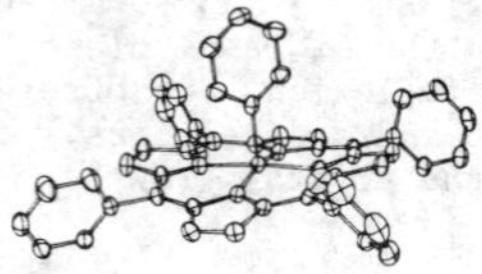

Fig. 1. The $Fe(C_6H_5)(C_{44}H_{28}N_4)$ molecule.

BENZENESULPHINATO(5,10,15,20-TETRAPHENYLPORPHYRINATO)IRON(III)
$C_{50}H_{33}FeN_4O_2S$

P. COCOLIOS, G. LAGRANGE, R. GUILARD, H. OUMOUS and C. LECOMTE, 1984. J. Chem. Soc. Dalton, 567-574.

Triclinic, P1, a = 11.01, b = 11.09, c = 11.37 Å, α = 107.2, β = 107.9, γ = 110.2°, Z = 1. Cu radiation, R = 0.097 for 1901 reflexions.

The coordination polyhedron of the Fe atom (Fig. 1) is of the 4:1 type in accord with O-sulphinato bonding. The Fe atom lies 0.45(2) Å from both the four-N atom and porphyrinato planes. The mean Fe-N distance is 2.05(2) and Fe-O is 1.92(1) Å. There are two different S-O bond distances in the sulphinato group (1.46(2), 1.56(2) Å). The shorter corresponds to S-O(2) which is not bound to Fe. Fe...O(2) is 3.66(2) and S-C(30) is 1.73(2) Å. There is a degree of packing disorder.

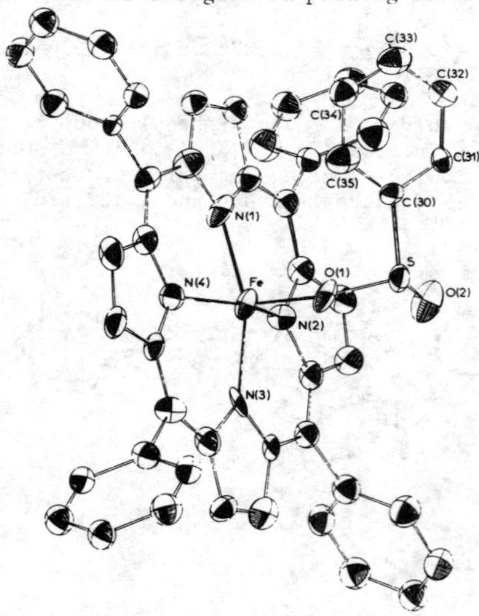

Fig. 1. $C_{50}H_{33}FeN_4O_2S$: a view of the molecule.

cis-DICARBONYLTETRA-p-TOLYLPORPHYRINATOMOLYBDENUM(II)
$C_{50}H_{36}MoN_4O_2$ (I) $Mo(CO)_2(TTP)$

trans-DIPYRIDINETETRA-p-TOLYLPORPHYRINATOMOLYBDENUM(II) PYRIDINE SOLVATE
$C_{58}H_{46}MoN_6,2(C_5H_5N)$ (II) $Mo(py)_2(TTP),2(py)$

J. COLIN, A. STRICH, M. SCHAPPACHER, B. CHEVRIER, A. VEILLARD and R. WEISS, 1984. Nouv. J. Chim., 8, 55-64.

I. Monoclinic, $P2_1/n$, a = 29.015, b = 15.081, c = 10.948 Å, β = 92.29°, Z = 4. Cu radiation, R = 0.091 for 1676 reflexions.

II. Monoclinic, $P2_1/n$, a = 18.808, b = 9.221, c = 16.944 Å, β = 110.99°, Z = 2. Cu radiation, R = 0.058 for 2997 reflexions.

In the cis isomer (I), (Fig. 1), the two carbonyl groups are in a staggered conformation with respect to the pyrrole N atoms and the coordination at Mo is nearly trigonal prismatic. The Mo atom is 0.64 Å from the N_4 plane towards the CO groups (Mo-N 2.11-2.16(1), Mo-C 1.84 and 1.87(2) Å). The trans-isomer (II) has the Mo atom on an inversion centre (with a planar MoN_4 unit, Mo-N(1) 2.067(3), Mo-N(2) 2.075(3), Mo-N(pyridine) 2.215(3) Å). The two pyridines (mutually eclipsed) are rotated about the Mo-N bond to make an angle of 34° with Mo-N(1) direction.

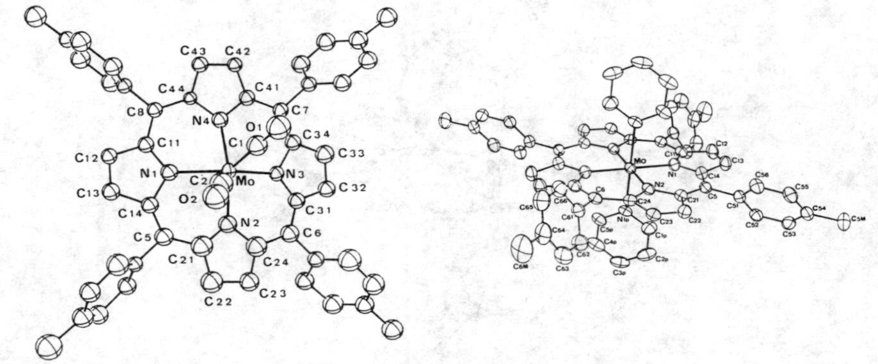

Fig. 1. Views of $Mo(CO)_2TTP$ and $Mo(py)_2TTP$.

ACETATO(5,10,15,20-TETRA-p-TOLYLPORPHYRINATO)IRON(III) ACETIC ACID SOLVATE
$C_{50}H_{39}FeN_4O_2,0.5(C_2H_4O_2)$

H. OUMOUS, C. LECOMTE, J. PROTAS, P. COCOLIOS and R. GUILARD, 1984. Polyhedron, 3, 651-659.

Monoclinic, I2/c, a = 24.464, b = 9.332, c = 37.174 Å, β = 90.49°, Z = 8. Cu radiation, R = 0.0584 for 5132 reflexions.

The iron atom is pentacoordinated by the four nitrogen atoms and one oxygen atom of the acetate group. It lies at 0.520(1) Å out of the porphinato plane and 0.485(1) Å out of the four nitrogen plane. Main bond lengths are Fe-O 1.898(4), Fe-N 2.051-2.073(3) Å.

BIS(1-METHYLIMIDAZOLE-N)TETRAPHENYLPORPHYRINATOMAGNESIUM(II)
$C_{52}H_{40}MgN_8$ (I) $Mg(C_{44}H_{28}N_4)(C_4H_6N_2)_2$

BIS(4-PICOLINE-N)TETRAPHENYLPORPHYRINATOMAGNESIUM(II)
$C_{56}H_{42}MgN_6$ (II) $Mg(C_{44}H_{28}N_4)(C_6H_7N)_2$

BIS(PIPERIDINE-N)TETRAPHENYLPORPHYRINATOMAGNESIUM(II)
$C_{54}H_{50}MgN_6$ (III) $Mg(C_{44}H_{28}N_4)(C_5H_{11}N)_2$

V. McKEE, C.C. ONG and G.A. RODLEY, 1984. Inorg. Chem., 23, 4242-4248.

I. Tetragonal, $P4_2/n$, a = 20.764, c = 9.659 Å, D_m = 1.28, Z = 4. Mo radiation, R = 0.074 for 1132 reflexions.

II. Triclinic, $P\bar{1}$, a = 10.146, b = 11.210, c = 11.643 Å, α = 65.63, β = 76.32, γ = 67.42°, D_m = 1.25, Z = 1. Mo radiation, R = 0.045 for 2059 reflexions.

III. Triclinic, $P\bar{1}$, a = 9.944, b = 11.463, c = 11.914 Å, α = 101.78, β = 104.59, γ = 115.60°, D_m = 1.23, Z = 1. Mo radiation, R = 0.059 for 2144 reflexions.

The three molecules (Fig. 1) are all centrosymmetric and have nearly planar porphyrin cores. The Mg(II) have 4+2 six-coordination with Mg-N(porphyrin) 2.082(6) and 2.074(6) Å in I, 2.072(3) and 2.069(2) Å in II, 2.074(3) and 2.071(3) Å in III. The Mg-N(axial) bond lengths are 2.297(8), 2.386(2), and 2.419(3) Å respectively for I-III.

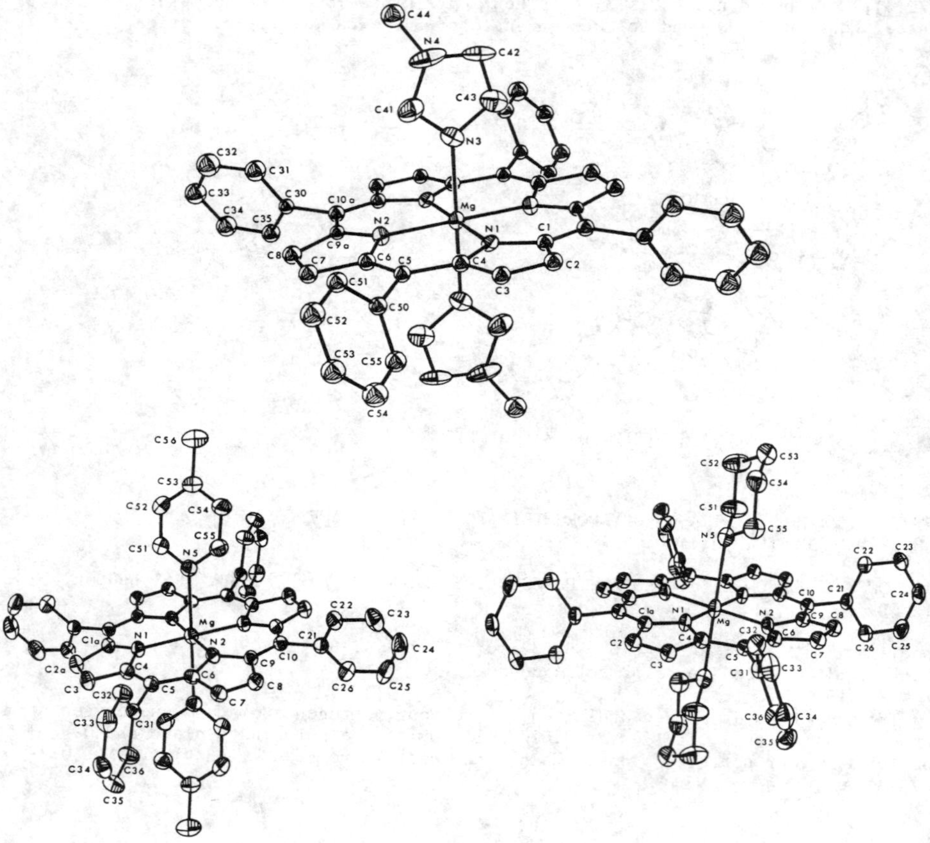

Fig. 1. The structures of $Mg(C_{44}H_{28}N_4)L_2$; L = $C_4H_6N_2$ (top), C_6H_7N (lower left), and $C_5H_{11}N$ (lower right).

BIS(DIMETHYLPHENYLPHOSPHINE)(2,3,7,8,12,13,17,18-OCTAETHYLPORPHYRINATO)VANADIUM(II)
$C_{52}H_{66}N_4P_2V$ (I)

BIS(TETRAHYDROFURAN)(2,3,7,8,12,13,17,18-OCTAETHYLPORPHYRINATO)VANADIUM(II)
$C_{44}H_{60}N_4O_2V$ (II)

H. OUMOUS, C. LECOMTE, J. PROTAS, J.L. PONCET, J.M. BARBE and R. GUILARD, 1984. J. Chem. Soc. Dalton, 2677-2682.

I. Orthorhombic, Pbca, a = 14.540, b = 21.431, c = 15.305 Å, Z = 4. Cu radiation, R = 0.045 for 1733 reflexions.

II. Triclinic, $P\bar{1}$, a = 10.078, b = 10.202, c = 10.815 Å, α = 75.54, β = 87.63, γ = 66.43°, Z = 1. Cu radiation, R = 0.073 for 2360 reflexions.

In both (I) and (II) the V atom lies on an inversion centre, has trans geometry and perfectly planar VN_4 moieties. Main bond lengths are V-P 2.523(1), V-N 2.046(4) Å for (I) (Fig. 1) and V-O 2.174(4), V-N 2.051(4) Å for (II).

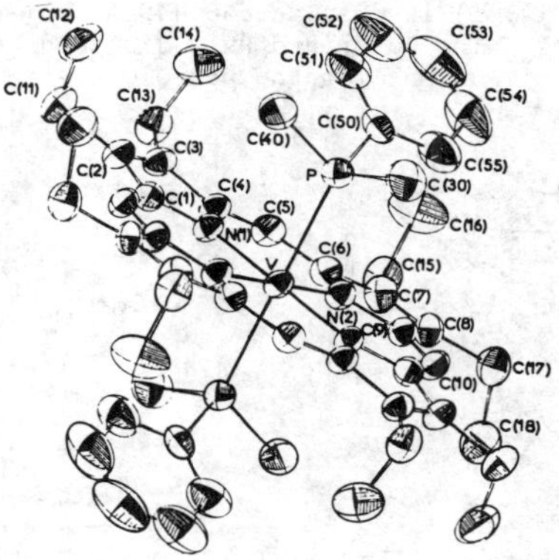

Fig. 1. A view of $C_{52}H_{66}N_4P_2V$.

(meso-TETRA-p-CHLOROPHENYLPORPHINATO)((2,2,6,6-TETRAMETHYL-1-PIPERIDYL)NITRENE)IRON(II)
$C_{53}H_{42}Cl_4FeN_6$

J.-P. MAHY, P. BATTIONI, D. MANSUY, J. FISHER, R. WEISS, J. MISPELTER, I. MORGENSTERN-BADARAU and P. GANS, 1984. J. Am. Chem. Soc., 106, 1699-1706.

Monoclinic, C2/c, a = 16.127, b = 17.969, c = 16.516 Å, β = 95.13°, D_m = 1.32, Z = 4. Cu radiation, R = 0.051 for 2215 reflexions.

The molecule (Fig. 1) has twofold crystallographic symmetry and an axial linear Fe-N-N moiety with Fe-N 1.809(4), N-N 1.232(5) Å. The average Fe-N(p) distance is 2.096 Å. The Fe atom is 0.49 Å from the four N(p) atom plane.

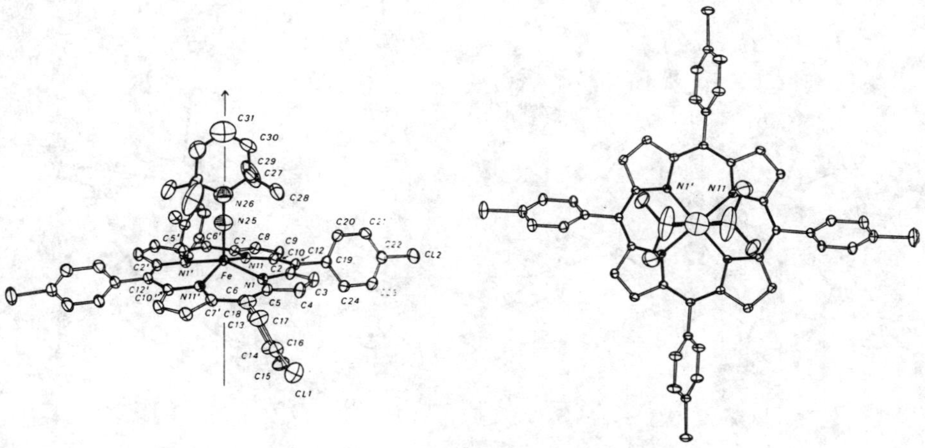

Fig. 1. Two views of the $C_{53}H_{42}Cl_4Fe_6N_6$ complex.

BROMOOCTAETHYLPORPHINATOTRIPHENYLPHOSPHINERUTHENIUM(III) HEXANE SOLVATE

$C_{54}H_{59}BrN_4PRu,C_6H_{14}$ Ru(OEP)(PPh_3)Br

B.R. JAMES, D. DOLPHIN, T.W. LEUNG, F.W.B. EINSTEIN and A.C. WILLIS, 1984. Can. J. Chem., 62, 1238-1245.

Monoclinic, $P2_1/c$, a = 13.037, b = 15.372, c = 29.577 Å, β = 115.19°, D_m = 1.32, Z = 4. Mo radiation, R = 0.076 for 2536 reflexions.

The ruthenium atom (Fig. 1) is displaced 0.049 Å from the plane of the pyrrole N atoms towards the phosphine. The Ru-Br and Ru-P distances are 2.552(2) and 2.412(5) Å respectively. Dimensions for the pyrrole system are also in Fig. 1.

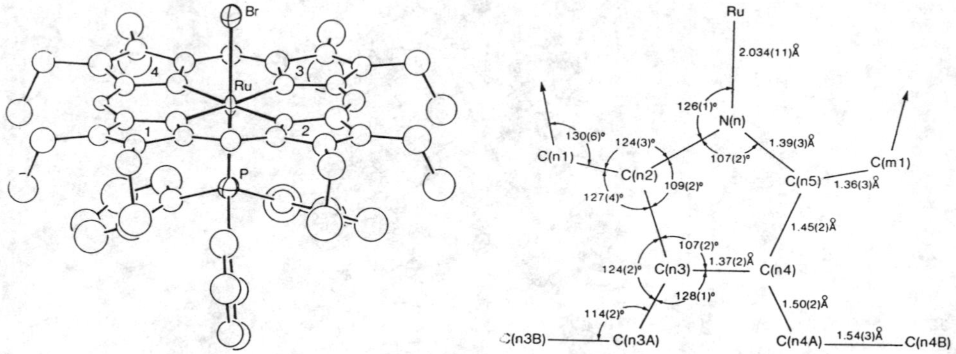

Fig. 1. A view of Ru(OEP)(PPh_3)Br and averaged geometry for the pyrrole rings.

DICARBONYL-((N^{22},N^{23}-(μ_2-2,2-BIS(p-CHLOROPHENYL)VINYLENE)-5,10,15,20-TETRAPHENYL-PORPHYRIN-C^6,N^{21},N^{22},N^{24}))RUTHENIUM DICHLOROMETHANE SOLVATE

$C_{60}H_{36}Cl_2N_4O_2Ru,2(CH_2Cl_2)$

YEE-WAI CHAN, F.E. WOOD, M.W. RENNER, H. HOPE and A.L. BALCH, 1984. J. Am. Chem. Soc., 106, 3380-3381.

Triclinic, $P\bar{1}$, a = 9.796, b = 12.944, c = 22.377 Å, α = 80.14, β = 89.41, γ = 70.25°, Z = 2. R = 0.075 for 7658 reflexions (at 140 K).

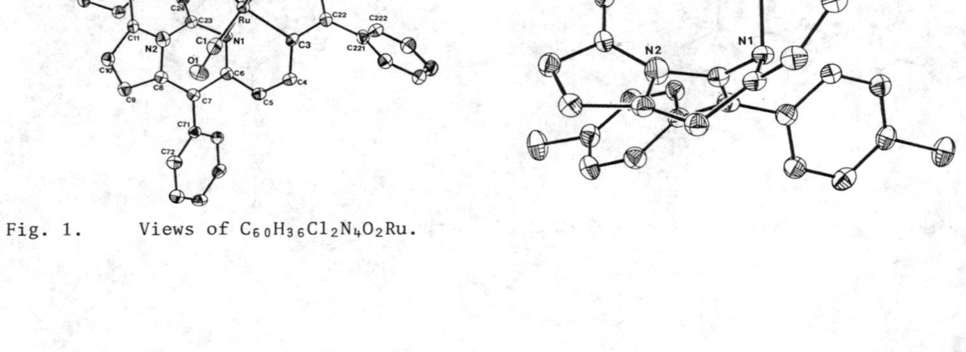

Fig. 1. Views of $C_{60}H_{36}Cl_2N_4O_2Ru$.

The Ru atom (Fig. 1) has inserted itself into a pyrrole C-N bond and the disrupted porphyrin is no longer planar. Some selected bond lengths are: Ru-C(1) 1.878(7), Ru-C(2) 1.864(6), Ru-C(3) 2.086(6), Ru-N(1) 2.206(5), Ru-N(3) 2.257(6), Ru-N(4) 2.068(5) Å.

N,N'-BIS(5,5-DIMETHYLCYCLOHEX-1-EN-1-OLATO-3-ON-2-YL-C,O)TETRAPHENYLPORPHYRINATO-IRON(II)
$C_{60}H_{48}FeN_4O_4$

D. MANSUY, J.-P. BATTIONI, I. AKHREM, D. DUPRE, J. FISCHER, R. WEISS and I. MORGENSTERN-BADARAU, 1984. J. Am. Chem. Soc., 106, 6112-6114.

Triclinic, P$\bar{1}$, a = 12.755, b = 15.963, c = 12.104 Å, α = 91.77, β = 105.46, γ = 78.71°, D_m = 1.32, Z = 2. R = 0.076 for 3255 reflexions.

The centrosymmetric hexacoordinate complex is shown in Fig. 1. The Fe-N(alkyl) bond lengths (2.242(5) Å) are longer than the Fe-N(p) values (2.082(5) Å). The FeN_4 core is planar and the angle between the N-alkylated pyrrole plane and the Fe-N4 plane is 50.3°.

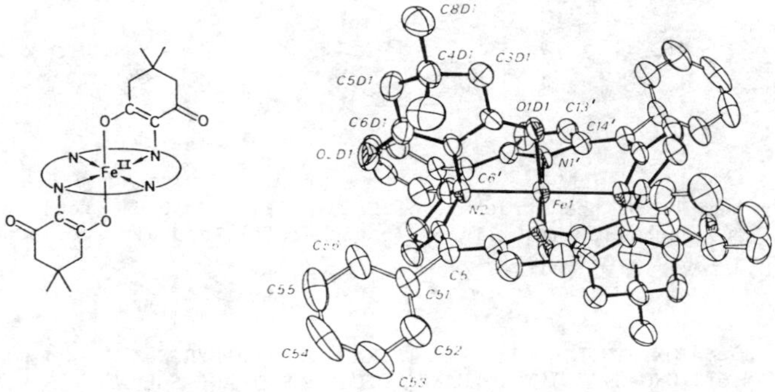

Fig. 1. A view of $C_{60}H_{48}FeN_4O_4$.

β-BIS(PHTHALOCYANINE)NEODYMIUM
$C_{64}H_{32}N_{16}Nd$ $Nd^{3+}[C_{32}H_{16}N_8]^{2-}[C_{32}H_{16}N_8]^{1-}$

A.N. DAROVSKIKH, A.K. TSYTSENKO, O.V. FRANK-KAMENETSKAYA, V.S. FUNDAMENSKII and P.N. MOSKALEV, 1984. Kristallografiya, 29, 455-461 [Sov. Phys. Crystallogr., 29, 273-276].

Monoclinic, C2/c, a = 19.010, b = 19.066, c = 15.538 Å, β = 116.1°, Z = 4. Mo radiation, R = 0.052 for 4015 reflexions.

The Nd atoms lie at the centres of distorted antiprisms formed by the pyrrole N atoms of both ligands (Fig. 1). The Nd-N bonds range from 2.46 to 2.48 Å and the base of the antiprism is a square with mean side N...N 2.803 Å. The sandwich molecules lie in layers parallel to (001) with the closest distance between C atoms in adjacent molecules 3.482 Å. The planes of the benzene and pyrrole rings of the ligand are inclined to the plane of the pyrrole N atoms at angles α_i (Fig. 1) equal to 7.6, 8.7, 12.6, 12.0, 6.8, 5.3, 5.6 and 5.4° for i = 2-9 respectively. The plane of the pyrrole N atoms N(5), N(6), N(7) and N(8) is practically parallel (within 1°) to the plane of the macro ring through atoms C(1), C(8), C(9), C(16), C(17), C(24), C(25) and C(32).

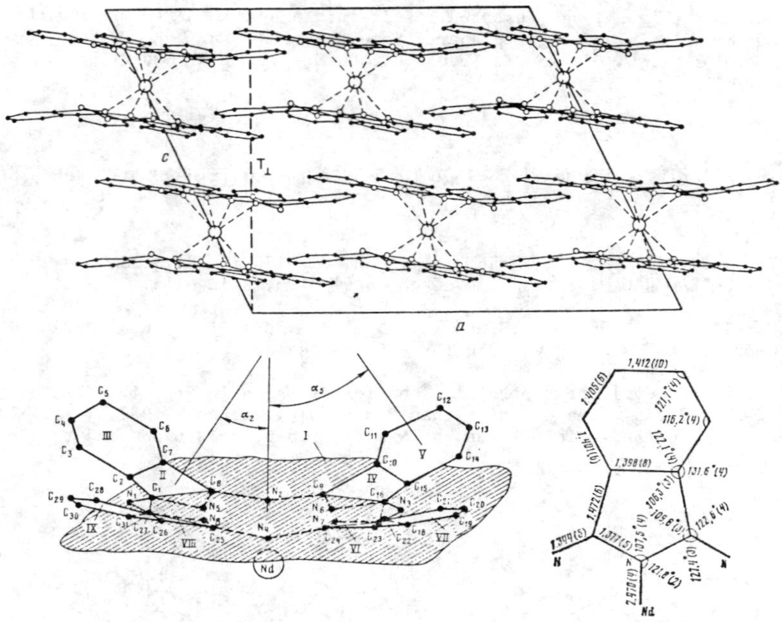

Fig. 1. Projection of the structure of $C_{64}H_{32}N_{16}Nd$ on (010) (top), conformation of the phthalocyanine ligand (bottom left), and averaged bond lengths in the ligand (bottom right).

[8,18-BIS(2-CARBOBENZOXYETHYL)-2,7,12,17-TETRAMETHYLPORPHYRIN-3,13]ADAMANTANE-1,3-[6(2-OXO-3-AZA),6(2-OXO-3-AZA)CYCLOPHANE] DICHLOROMETHANE SOLVATE
$C_{64}H_{72}N_6O_6,2(CH_2Cl_2)$

T.G. TRAYLOR, N. KOGA, L.A. DEARDURFF, P.N. SWEPSTON and J.A. IBERS, 1984. J. Am. Chem. Soc., 106, 5132-5143.

Triclinic, $P\bar{1}$, a = 15.858, b = 17.779, c = 12.970 Å, α = 109.62, β = 97.88, γ = 114.18°, Z = 2. Mo radiation, R = 0.065 for 5489 reflexions (at -152°C).

The porphyrin macrocycle (Fig. 1) is approximately planar (mean displacement from the 24-atom plane is 0.038 Å, maximum displacement 0.101(4) Å). The adamantane bridge has an offset orientation relative to the porphyrin plane and there is no free cavity between the strap and the porphyrin.

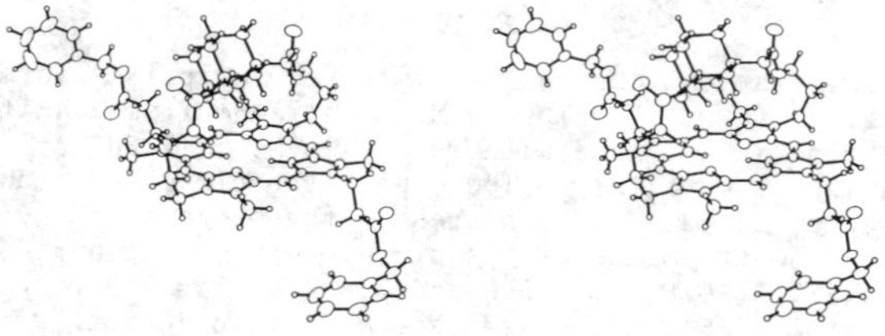

Fig. 1. A stereoview of $C_{64}H_{72}N_6O_6$.

CHLORO(meso-α,α,α,α-TETRAKIS(o-NICOTINAMIDOPHENYL)PORPHINATO)IRON(III) MONOHYDRATE CHLOROFORM SOLVATE
$C_{68}H_{44}ClFeN_{12}O_4,CHCl_3,H_2O$

M.J. GUNTER, G.M. McLAUGHLIN, K.J. BERRY, K.S. MURRAY, M. IRVING and P.E. CLARK, 1984. Inorg. Chem., 23, 283-300.

Monoclinic, $P2_1/c$, a = 14.739, b = 21.924, c = 19.524 Å, β = 101.03°, D_m = 1.37, Z = 4. Mo radiation, R = 0.104 for 5042 reflexions.

The structure consists of polymeric chains, with the Fe atom of one molecule coordinated to a pyridine N of the nicotinamide unit of a second molecule (Fig. 1). The Cl^- ion occupies the sixth coordination site, inside the "pocket" of the four nicotinamide groups. The Fe atom is displaced 0.109(1) Å from the mean plane of the porphyrin toward Cl^-. Long Fe-Cl (2.311(3) Å) and Fe-N_{py} (2.330(8) Å) distances and an average Fe-N_{porph} distance of 2.042(8) Å indicate an essentially high-spin Fe, accommodated by an S_4 ruffling of the porphyrin.

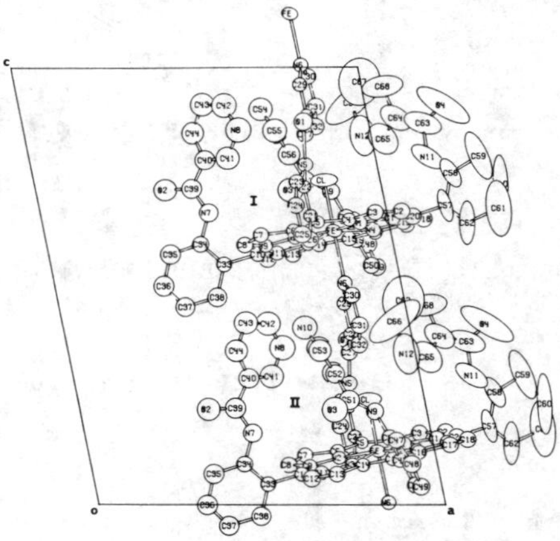

Fig. 1. Projection of two $C_{68}H_{44}ClFeN_{12}O_4$ molecules, showing the polymeric nature of the structure.

BIS(2,3,5,6-TETRAPHENYLTHIOLATO)(m-TERPHENYLPORPHYRINATO)IRON(III) (18-CROWN-6)SODIUM BENZENE SOLVATE
$C_{68}H_{54}F_8FeN_4NaO_6S_2,C_6H_6$

P. DOPPELT, J. FISCHER and R. WEISS, 1984. Croat. Chem. Acta, 57, 507-518.

Triclinic, $P\bar{1}$, a = 12.628, b = 21.594, c = 12.881 Å, α = 104.02, β = 98.26, γ = 76.40°, Z = 2. Cu radiation, R = 0.060 for 5674 reflexions.

The asymmetric unit contains two half independent anionic porphyrinates located on a crystallographic inversion centre, along with one Na crown-ether cation and one molecule of benzene located in general positions. In both anions, the metal and the four N atoms of the FeN_4S_2 octahedral coordination unit form square, necessarily planar entities with Fe(1)-N(av.)=Fe(2)-N(av.) 1.998 Å and $(N\text{-}Fe(1)\text{-}N)_{av.}=(N\text{-}Fe(2)\text{-}N)_{av.}$ 89.95°. The axial thiolato ligands eclipse two Fe-N bonds in both anions (Fig. 1). The mean value of the Fe-S bonds is 2.312 Å, and the Fe(1)-S(1)-CT(1) and Fe(2)-S(2)-CT(2) angles are 108.5 and 106.7° respectively. The phenyl rings have normal geometries, and the average bond lengths within the crown-ether cation are: C-O 1.401 Å, C-C 1.446 Å and the Na-O bond length range is from 2.364 to 2.80 Å.

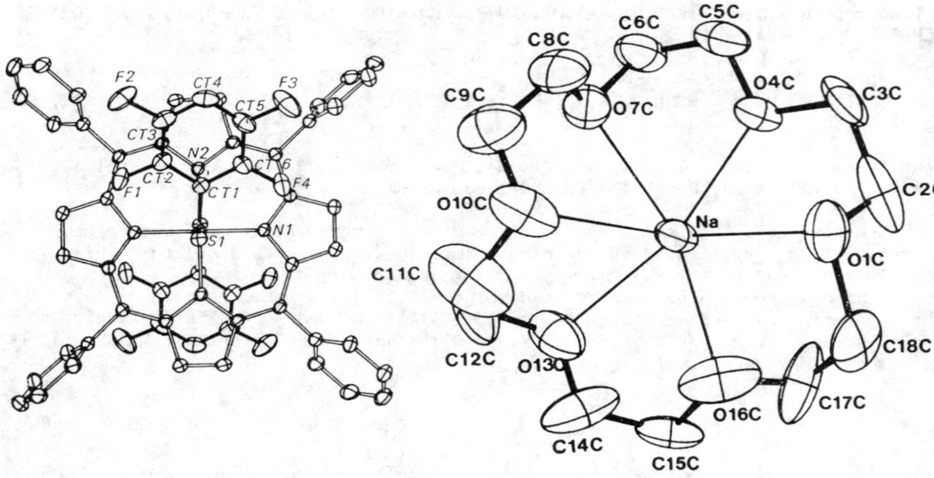

Fig. 1. Plot of one porphyrin unit in $C_{68}H_{54}F_8FeN_4NaO_6S_2$, and of the 18-crown-6 complex.

[α-(5,15).β-(10,20)-BIS{2,2'-[4,4'-(p-PHENYLENE)DIBUTOXY]DIPHENYL}PORPHYRIN]CHLORO-IRON(III) DICHLOROMETHANE SOLVATE

$C_{72}H_{64}ClFeN_4O_4,CH_2Cl_2$

L. RICARD, J. FISCHER, R. WEISS and M. MOMENTEAU, 1984. Nouv. J. Chim., 8, 639-642.

Monoclinic, $P2_1/n$, a = 17.040, b = 24.234, c = 15.238 Å, β = 99.88°, Z = 4. Cu radiation, R = 0.061 for 4668 reflexions.

The structure of this "basket-handle" porphyrin is shown in Fig. 1. The geometry about the Fe atom is typical for a FeCl(porphyrin) system with Fe-Cl 2.207(2), Fe-N 2.051-2.067(4) Å and Fe 0.46 Å above the mean N_4 plane. The porphyrin is slightly ruffled and the phenyl rings of both handles are in inequivalent off-set positions leaving the centre of the porphyrin ring largely exposed.

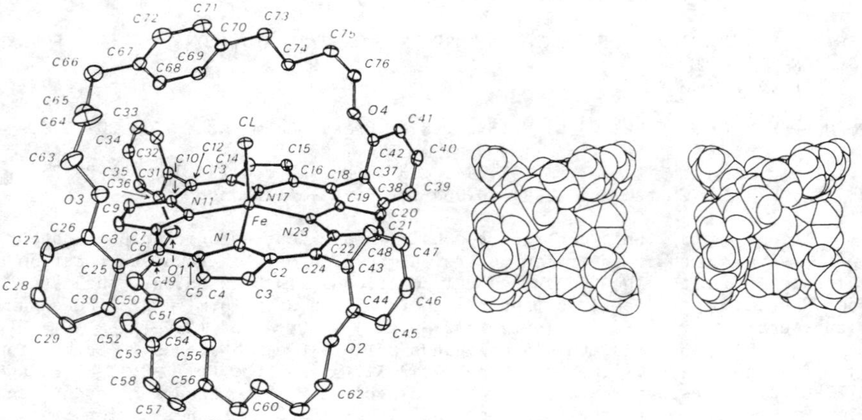

Fig. 1. A view of $C_{72}H_{64}ClFeN_4O_4$ and a stereo view of a space-filling diagram viewed perpendicular to the N_4 plane on the chlorine side.

(DIBENZO-18-CROWN-6)BIS(TETRAHYDROFURAN)SODIUM (meso-TETRAPHENYLPORPHINATO)IRON(I) TETRAHYDROFURAN SOLVATE

$C_{72}H_{68}FeN_4NaO_8,C_4H_8O$ (I) $[Na(C_{20}H_{24}O_6)(C_4H_8O)_2][Fe(C_{44}H_{28}N_4)].C_4H_8O$

BIS[TRIS(TETRAHYDROFURAN)SODIUM] (meso-TETRAPHENYLPORPHINATO)IRON(0)

$C_{68}H_{76}FeN_4Na_2O_6$ (II) $[Na(C_4H_8O)_3]_2[Fe(C_{44}H_{28}N_4)]$

T. MASHIKO, C.A. REED, K.J. HALLER and W.R. SCHEIDT, 1984. Inorg. Chem., 23, 3192-3196.

I. Monoclinic, $P2_1/m$, a = 12.796, b = 21.861, c = 12.108 Å, β = 103.47°, Z = 2. Mo radiation, R = 0.068 for 4160 reflexions.

II. Triclinic, $P\bar{1}$, a = 11.877, b = 13.242, c = 11.242 Å, α = 113.69, β = 110.01, γ = 77.55°, Z = 1. Mo radiation, R = 0.091 for 4904 reflexions.

The structure of I consists of centrosymmetric $[Fe(C_{44}H_{28}N_4)]^-$ anions and $[Na(C_{20}H_{24}O_6)(C_4H_8O)_2]^+$ cations (Fig. 1). The Fe atom has square planar coordination with mean Fe-N 1.980(6) Å. The Na is eight-coordinate with hexagonal bipyramidal geometry, the THF molecules being axial ($Na-O_{eq}$ 2.573(3)-2.745(3) Å, $Na-O_{ax}$ 2.299(4) and 2.279(4) Å). The structure of II consists of centrosymmetric ion-paired units (Fig. 1) in which two $[Na(THF)_3]^+$ cations interact with a pair of porphinato N atoms (Na-N 2.784(4), 2.825(4) Å) above and below the porphyrin plane. The Na-O distances are 2.311(4)-2.339(4) Å. The Fe coordination is square planar with mean Fe-N 1.968(1) Å. The porphinato core geometries in both structures reveal considerable π-radical anion character.

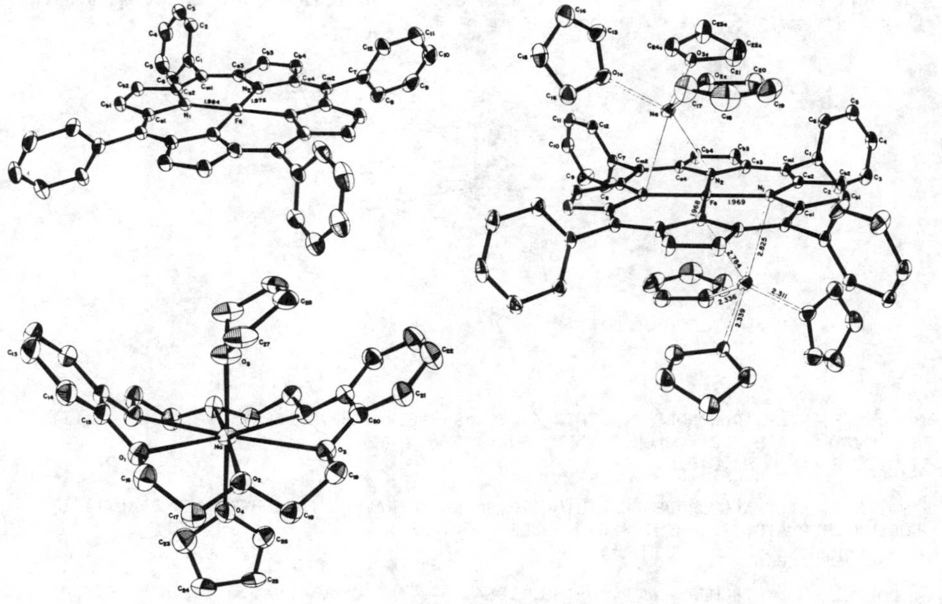

Fig. 1. The $[Fe(C_{44}H_{28}N_4)]^-$ anion (top left), the $[Na(C_{20}H_{24}O_6)(C_4H_8O)_2]^+$ cation (bottom left), and the $[Na(C_4H_8O)_3]_2[Fe(C_{44}H_{28}N_4)]$ complex (right).

OCTAETHYLPORPHINATOBIS(TRIPHENYLPHOSPHINE)RUTHENIUM(II)

$C_{72}H_{74}N_4P_2Ru$

S. ARIEL, D. DOLPHIN, G. DOMAZETIS, B.R. JAMES, T.W. LEUNG, S.J. RETTIG, J. TROTTER and G.M. WILLIAMS, 1984. Can. J. Chem., 62, 755-762.

Triclinic, $P\bar{1}$, a = 12.139, b = 12.120, c = 20.878 Å, α = 86.51, β = 86.25, γ = 79.90°, Z = 2. Mo radiation, R = 0.0454 for 5697 reflexions.

The asymmetric unit contains two independent half molecules with the Ru atoms lying on inversion centres. In each complex, the Ru atom has the expected octahedral coordination, with average Ru-P 2.428, average Ru-N 2.050 and average P-C 1.84 Å. Two of the ethyl groups in one complex are disordered.

BIS((OCTAETHYLPORPHYRINATO)RUTHENIUM(II)) PENTANE SOLVATE
$C_{72}H_{88}N_8Ru_2,2(C_5H_{12})$ $(Ru(OEP))_2,2(C_5H_{12})$

J.P. COLLMAN, C.E. BARNES, P.N. SWEPSTON and J.A. IBERS, 1984. J. Am. Chem. Soc., 106, 3500-3510.

Monoclinic, C2/c, a = 27.99, b = 27.25, c = 17.737 Å, β = 102.50°, Z = 8. Mo radiation, R = 0.042 for 7304 reflexions (at -103°C).

In the molecule (Fig. 1) the distance between the Ru atoms is 2.408(1) Å and the mean Ru-N distance is 2.050(5) Å. Each Ru atom is situated 0.30 Å out of a plane defined by four coordinating N atoms in the direction of the other Ru atom. The two porphyrin macrocycles are twisted 23.8(1)° with respect to each other. Both porphyrinato cores have domed-type distortions with average deviations from the least-squares planes of 0.07 and 0.10 Å and maximum deviations of 0.19 and 0.24 Å, respectively. X-ray and NMR results are consistent with a qualitative MO diagram that suggests a formal Ru-Ru bond order of 2.

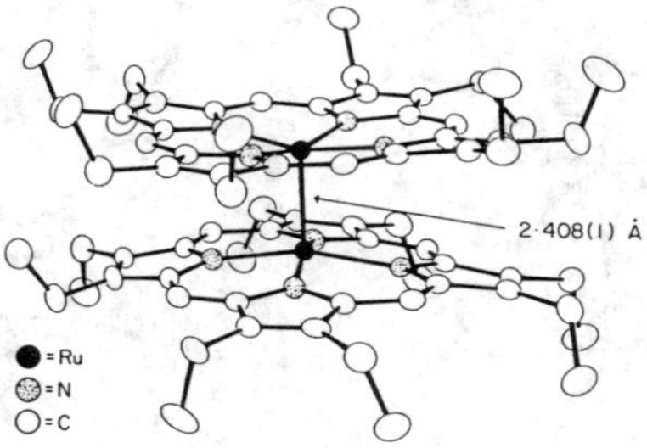

Fig. 1. A view of $(Ru(OEP))_2$.

BIS(2,3,5,6-TETRAFLUOROBENZENETHIOLATO)-meso-TETRAPHENYLPORPHYRINATOCOBALT(III) SODIUM CRYPTATE[222] CHLOROBENZENE SOLVATE
$C_{74}H_{66}CoF_8NaN_6O_6S_2,C_6H_5Cl$ (I)

BIS(2,3,5,6-TETRAFLUOROBENZENETHIOLATO)-meso-TETRAPHENYLPORPHYRINATOCOBALT(III) SODIUM CRYPTATE[222] HEMICHLOROBENZENE SOLVATE
$C_{74}H_{66}CoF_8NaN_6O_6S_2,0.5(C_6H_5Cl)$ (II)

P. DOPPELT, J. FISCHER and R. WEISS, 1984. J. Am. Chem. Soc., 106, 5188-5193.

I. Triclinic, $P\bar{1}$, a = 13.053, b = 25.034, c = 12.176 Å, α = 103.56, β = 110.57, γ = 82.03°, Z = 2. Cu radiation, R = 0.089 for 6317 reflexions.

II. Triclinic, $P\bar{1}$, a = 15.963, b = 16.007, c = 14.806 Å, α = 108.10, β = 99.31, γ = 91.41°, Z = 2. Cu radiation, R = 0.073 for 7312 reflexions.

In I there are two independent centrosymmetric complexes; II has no crystallographically imposed symmetry. Views of the bis(mercapto)cobalt(III) porphyrins are in Fig. 1. The average equatorial Co-Np distance is 1.977(4) in I and 1.973(4) Å in II. Co-S distances are 2.330 and 2.346(3) Å in I and 2.346 and 2.351(3) Å in II.

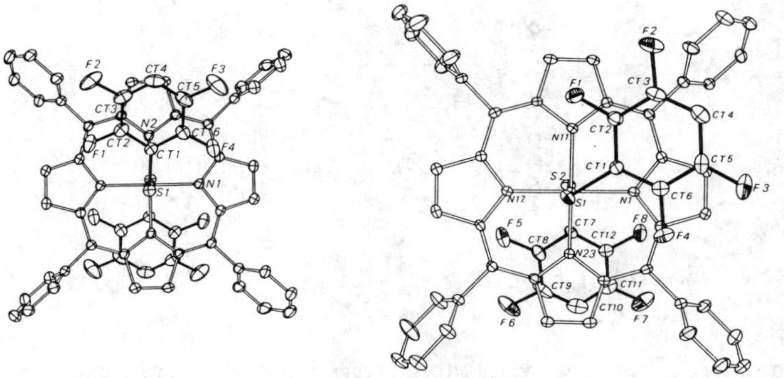

Fig. 1. Views of the bis(mercapto)cobalt(III) porphyrins (I) (left) and (II) (right).

μ-OXO-BIS[METHOXO(OCTAETHYLPORPHINATO)OSMIUM(IV)]
$C_{74}H_{94}N_8O_3Os_2$

H. MASUDA, T. TAGA, K. OSAKI, H. SUGIMOTO and M. MORI, 1984. Bull. Chem. Soc. Japan, 57, 2345-2351.

Monoclinic, P2/c, a = 24.171, b = 14.037, c = 24.401 Å, β = 119.68°, D_m = 1.40, Z = 4. Mo radiation, R = 0.094 for 3838 reflexions.

There are two independent dimers in the unit cell, each with a crystallographically-imposed twofold axis. The dimers are in an approximately orthogonal arrangement and have very similar conformations (Fig. 1). The main averaged bond distances are 2.033 Å for Os(IV)-N, 1.808 Å for $Os(IV)-O^{2-}$, and 1.997 Å for $Os(IV)-OCH_3$. The two porphinato cores bridged by a μ-oxo oxygen atom are rotated round the Os-O-Os bond by 23° from the eclipsed form. The porphine ring is planar to within 0.10 Å, and deviation of the Os atom from the porphinato plane of the four N atoms is 0.062 Å.

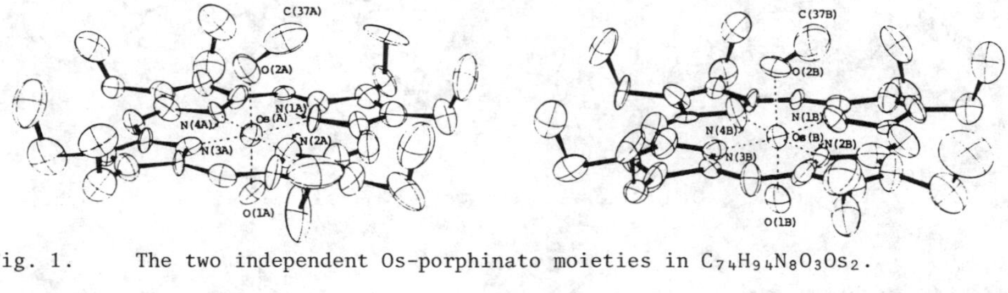

Fig. 1. The two independent Os-porphinato moieties in $C_{74}H_{94}N_8O_3Os_2$.

($μ_2$-OXO)BIS[(t-BUTYLDIMETHYLSILOXY)(PHTHALOCYANINATO-N,N',N'',N''')SILICON]
$C_{76}H_{62}N_{16}O_3Si_4$

E. CILIBERTO, K.A. DORIS, W.J. PIETRO, G.M. REISNER, D.E. ELLIS, I. FRAGALA, F.H. HERBSTEIN, M.A. TATNER and T.J. MARKS, 1984. J. Am. Chem. Soc., 106, 7748-7761.

Orthorhombic, Pbcn, a = 21.670, b = 13.274, c = 23.031 Å, D_m = 1.30, Z = 4. Mo radiation, R = 0.127 for 1975 reflexions.

The molecule has a cofacial (Pc)Si-O-Si(Pc) core of C_2 symmetry, virtually planar phthalocyanine rings, interplanar spacing (Si...Si) 3.32(1) Å, Si-O-Si 179(1)° and a ring-ring staggering of 36.6°. The groups capping the Si atoms are disordered.

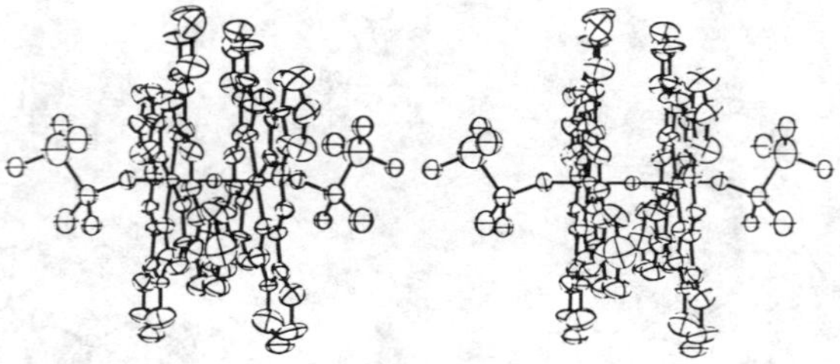

Fig. 1. A stereoview of $C_{76}H_{62}N_{16}O_3Si_4$.

(μ-SULPHATO)BIS[(meso-TETRAPHENYLPORPHINATO)IRON(III) CHLOROFORM SOLVATE MONOHYDRATE
$C_{88}H_{56}Fe_2N_8O_4S,CHCl_3,H_2O$

W.R. SCHEIDT, Y.J. LEE, T. BARTZCAR and K. HATANO, 1984. Inorg. Chem., 23, 2552-2554.

Trigonal, $R\bar{3}c$, a = 32.431, c = 36.355 Å, D_m = 1.42, Z = 18. Mo radiation, R = 0.088 for 3499 reflexions.

The SO_4 ligand bridges the two Fe atoms in this binuclear complex (Fig. 1). The complex has exact C_2 symmetry. The Fe atom is displaced 0.43 Å from the N_4 plane. The mean Fe-N distance is 2.062(8) Å, Fe-O 1.894(4) Å.

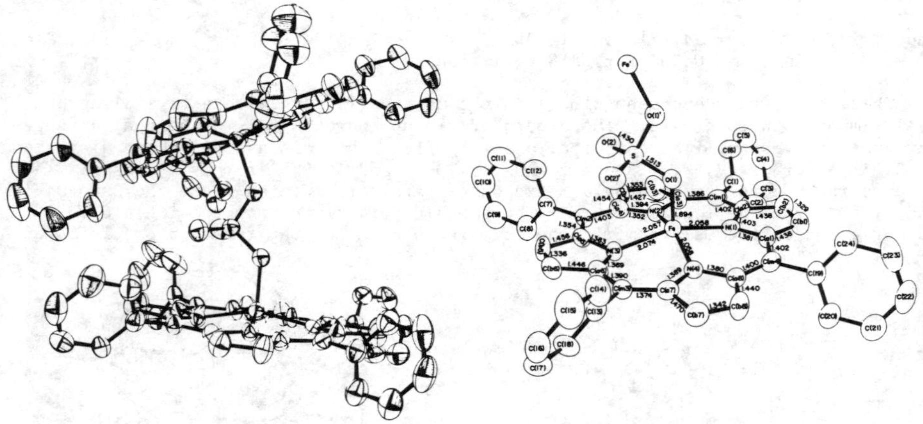

Fig. 1. The $C_{88}H_{56}Fe_2N_8O_4S$ molecule, geometry of the asymmetric unit at right.

BIS[(5-(2-PHENOLATO-(O)-10,15,20-TRITOLYLPORPHINATO))IRON(III)] BIS(TOLUENE) SOLVATE
$C_{94}H_{66}Fe_2N_8O_2,2C_7H_8$ $[(C_{47}H_{33}N_4O)Fe]_2,2CH_3C_6H_5$

H.M. GOFF, E.T. SHIMOMURA, Y.J. LEE and W.R. SCHEIDT, 1984. Inorg. Chem., 23, 315-321.

Orthorhombic, P2na, a = 16.199, b = 14.408, c = 18.622 Å, D_m = 1.24, Z = 2. Mo radiation, R = 0.049 for 4457 reflexions.

The dimeric molecule (Fig. 1) possesses exact C_2 symmetry. Each iron atom is coordinated to four porphyrin N atoms (mean 2.068(6) Å) and to the phenolate O atom from the C_2-related porphyrin Fe-O 1.847(2) Å. The Fe atom is displaced 0.46 Å toward the apical O atom from the N_4 mean plane.

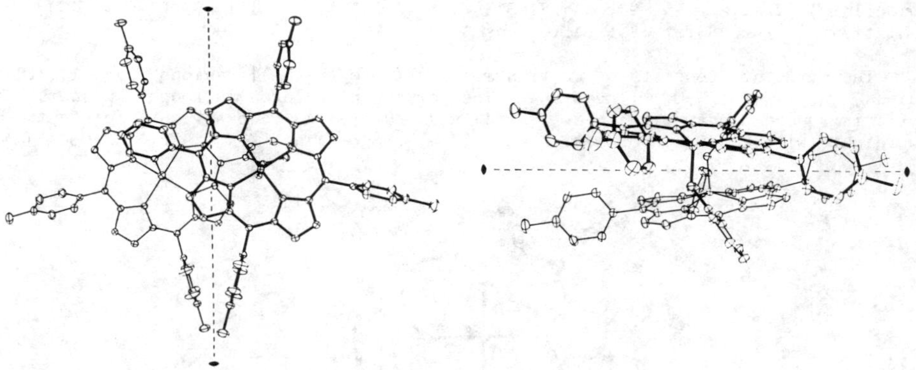

Fig. 1. Two views of $[(C_{47}H_{33}N_4O)Fe]_2$.

(μ_2-OXO)BIS[(p-METHYLPHENOXO)(meso-TETRAPHENYLPORPHYRINATO)RUTHENIUM(IV)]
$C_{102}H_{70}N_8O_3Ru_2$ (I)

ETHOXO(meso-TETRAPHENYLPORPHYRINATO)ETHANOLRUTHENIUM(III) ETHANOL SOLVATE
$C_{48}H_{39}N_4O_2Ru,2(C_2H_6O)$ (II)

J.P. COLLMAN, C.E. BARNES, P.J. BROTHERS, T.J. COLLINS, T. OZAWA, J.C. GALLUCCI and J.A. IBERS, 1984. J. Am. Chem. Soc., 106, 5151-5163.

I. Triclinic, $P\bar{1}$, a = 16.911, b = 19.802, c = 12.979 Å, α = 99.96, β = 104.31, γ = 77.53°, Z = 2. Mo radiation, R = 0.099 for 4416 reflexions (at -157°C).

II. Triclinic, $P\bar{1}$, a = 9.894, b = 12.946, c = 9.758 Å, α = 112.06, β = 94.12, γ = 71.85°, Z = 1. Mo radiation, R = 0.062 for 3526 reflexions (at -160°C).

Compound I (Fig. 1) has Ru-O(Ru) and Ru-O($p-OC_6H_4CH_3$) bond lengths of 1.789(11) and 1.944(11) Å respectively and Ru-O-Ru 177.8(7)°. Complex II has crystallographic inversion symmetry, with Ru-O 2.019(3) Å.

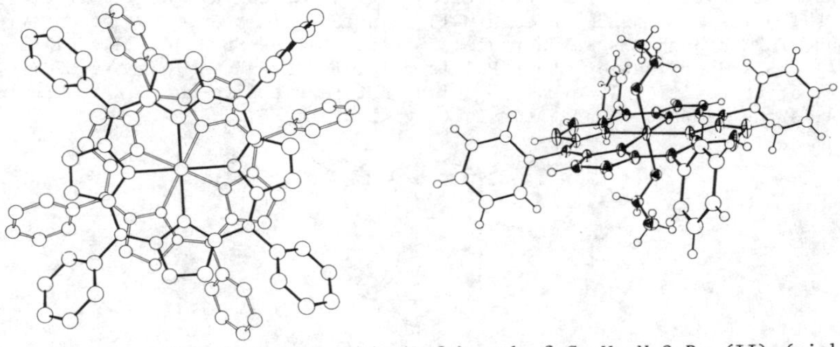

Fig. 1. Views of $C_{102}H_{70}N_8O_3Ru_2$ (I) (left) and of $C_{48}H_{39}N_4O_2Ru$ (II) (right).

TETRAKIS[(1,10-DIAZA-4,7,13,16,21,24-HEXAOXABICYCLO[8.8.8]HEXACOSANE)POTASSIUM] meso-(TETRAPHENYLPORPHYRINATO)COBALT(I) TRICHLORIDE MONOHYDRATE
$C_{116}H_{172}Cl_3CoK_4N_{12}O_{24},H_2O$ $[(C_{18}H_{36}N_2O_6)K]_4[Co(C_{44}H_{28}N_4)]Cl_3.H_2O$

P. DOPPELT, J. FISCHER and R. WEISS, 1984. Inorg. Chem., 23, 2958-2962.

Monoclinic, C2/c, a = 15.190, b = 21.313, c = 40.763 Å, β = 101.58°, Z = 4. Cu radiation, R = 0.068 for 4811 reflexions.

The structure consists of centrosymmetric $[Co(C_{44}H_{28}N_4)]^-$ anions (Fig. 1), Cl^- anions, and $[(C_{18}H_{36}N_2O_6)K]^+$ cations. The porphyrin skeleton is roughly planar. The Co(I) is square planar with Co-N 1.939(3) and 1.946(3) Å. The K^+ ions are eight coordinate with a N_2O_6 donor set, K-O 2.777(4)-2.846(4) and K-N 2.972(5)-3.060(5) Å.

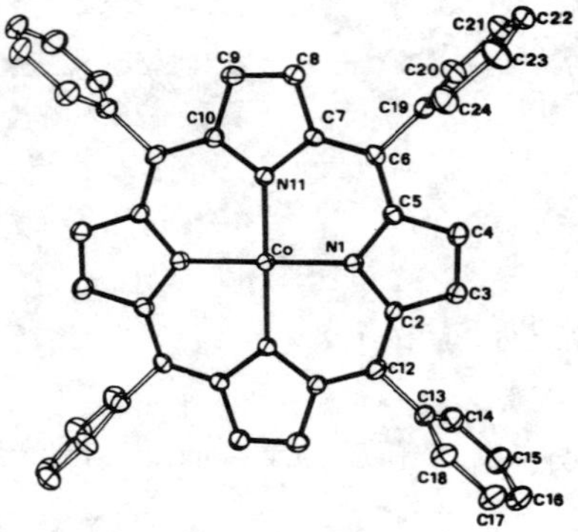

Fig. 1. The $[Co(C_{44}H_{28}N_4)]^-$ anion.

TETRA-N-BUTYLAMMONIUM (μ_2-cis-BIS(1,2-DICYANO-1,2-ETHYLENEDITHIOLATO)COPPER(II))BIS-meso-TETRA-p-CHLOROPHENYLPORPHYRINATO IRON(III)) cis-BIS(1,2-DICYANO-1,2-ETHYLENE-DITHIOLATO)COPPER(II) BENZENE SOLVATE
$C_{120}H_{84}Cl_8Cu_2Fe_2N_{17}S_8,3(C_6H_6)$

C.K. SCHAUER, K. AKABORI, C.M. ELLIOTT and O.P. ANDERSON, 1984. J. Am. Chem. Soc., 106, 1127-1128.

Triclinic, $P\bar{1}$, a = 13.378, b = 17.356, c = 28.918 Å, α = 104.05, β = 99.62, γ = 98.38°, Z = 2. Mo radiation, R = 0.063 for 1042 reflexions (at -130°C).

The analysis establishes the structure (Fig. 1) as containing a pair of $Fe(p\text{-}Cl_4TPP)^+$ units that sandwich a $Cu(MNT)_2{}^{2-}$ anion. This anion is bound strongly to both Fe1 and Fe2 through cis sulphur atoms S5 and S8 (Fe2-S8 2.444(2), Fe1-S5 2.549(2) Å). Six-coordination about Fe1 is completed by a weak bond to S2 (Fe1-S2 2.956(2) Å) of an additional $Cu(MNT)_2{}^-$ anion, but Fe2 is five-coordinate (Fe2-S3' 3.641(3) Å is the closest ligand atom contact trans to S8).

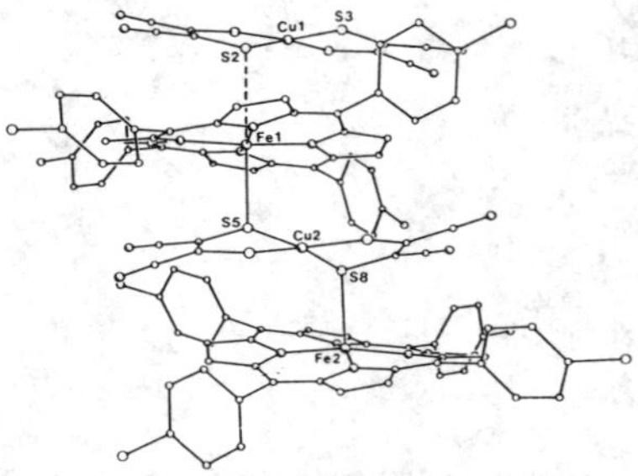

Fig. 1. A view of the metal-containing anionic unit in $C_{120}H_{84}Cl_8Cu_2Fe_2N_{17}S_8$.

METHYL FOSFOMYCIN SODIUM SALT
$C_4H_8NaO_4P$

A. PERALES, M. MARTINEZ-RIPOLL and J. FAYOS, 1984. Acta Cryst., C40, 357-359.

Monoclinic, $P2_1$, a = 11.098, b = 10.301, c = 6.448 Å, β = 96.67°, Z = 4. Cu radiation, R = 0.042 for 1238 reflexions.

The two crystallographically independent anions have identical absolute configurations but quite different conformations in the bond between phosphate and epoxy groups (Fig. 1). The Na^+ ions are five-coordinate. In both cases, four short Na-O bonds are formed with the O atoms of phosphate groups (Na-O 2.275-2.403(3) Å), and a longer Na-O bond involves the epoxy O atom (Na-O 2.612 and 2.666(3) Å).

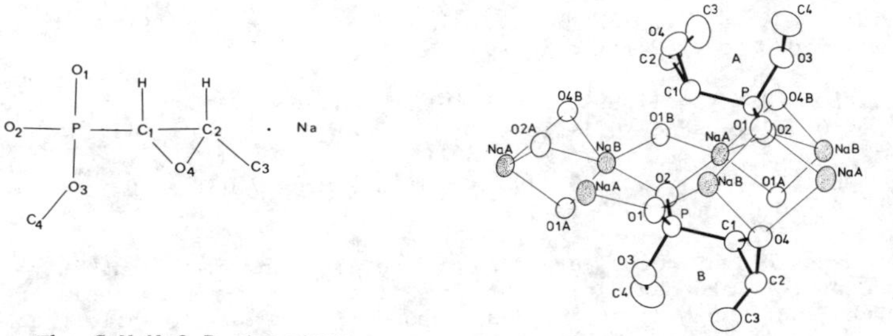

Fig. 1. The $C_4H_8NaO_4P$ structure.

REDUCTIOMYCIN
$C_{14}H_{15}NO_6$

J. DONOHUE, A.B. SMITH and P.J. CARROLL, 1984. J. Cryst. Spectrosc. Res., 14, 35-43.

Orthorhombic, $P2_12_12_1$, a = 14.701, b = 19.730, c = 4.821 Å, D_m = 1.41, Z = 4. Cu radiation, R = 0.091 for 969 reflexions.

The structure of reductiomycin reported previously [1] is shown to be in error; the imino group and heterocyclic oxygen of the previous study should be interchanged. The correct structure is shown in Fig. 1.

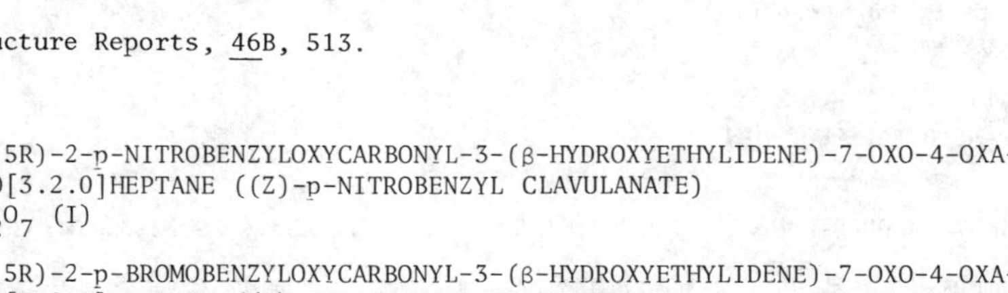

Fig. 1. The structure of reductiomycin.

1. Structure Reports, 46B, 513.

(Z)-(2R,5R)-2-p-NITROBENZYLOXYCARBONYL-3-(β-HYDROXYETHYLIDENE)-7-OXO-4-OXA-1-AZA-BICYCLO[3.2.0]HEPTANE ((Z)-p-NITROBENZYL CLAVULANATE)
$C_{15}H_{14}N_2O_7$ (I)

(Z)-(2R,5R)-2-p-BROMOBENZYLOXYCARBONYL-3-(β-HYDROXYETHYLIDENE)-7-OXO-4-OXA-1-AZA-BICYCLO[3.2.0]HEPTANE ((Z)-BROMOBENZYL CLAVULANATE)
$C_{15}H_{14}BrNO_5$ (II)

(E)-2-p-BROMOBENZYLOXYCARBONYL-3-(β-HYDROXYETHYLIDENE)-7-OXO-4-OXA-1-AZABICYCLO-[3.2.0]HEPTANE ((E)-p-BROMOBENZYL ISOCLAVULANATE)
$C_{15}H_{14}BrNO_5$ (III)

(1S,2aS,5aR,8aR,9aR)-TETRAHYDRO-1-HYDROXYMETHYL-2a-PHENYL-1H,7H-AZETO[2,1-b]OXETO-(3',2'.4,5)PYRANO[3,4-d]OXAZOLE-5,7(5aH)DIONE
$C_{16}H_{15}NO_6$ (IV)

A.G. BROWN, D.F. CORBETT, J. GOODACRE, J.B. HARBRIDGE, T.T. HOWARTH, R.J. PONSFORD, I. STIRLING and T.J. KING, 1984. J. Chem. Soc. Perkin I, 635-650.

I. Monoclinic, $P2_1$, a = 19.528, b = 7.795, c = 5.025 Å, β = 95.85°, D_m = 1.44, Z = 2. Mo radiation, R = 0.055 for 1438 reflexions.

II. Monoclinic, $P2_1$, a = 13.155, b = 4.763, c = 12.670 Å, β = 95.22°, Z = 2. Mo radiation, R = 0.088 for 762 reflexions.

III. Monoclinic, $P2_1$, a = 11.996, b = 5.113, c = 12.346 Å, β = 92.92°, Z = 2. Mo radiation, R = 0.077 for 833 reflexions.

IV. Orthorhombic, $P2_12_12_1$, a = 13.479, b = 15.227, c = 7.031 Å, D_m = 1.45, Z = 4. Cu radiation, R = 0.039 for 1583 reflexions.

'Streptomyces clavuligerus' produces the novel β-lactamase inhibitor clavulanic acid. Compounds I-IV are derivatives of this acid. The X-ray analysis of I established the configuration of the double bond (Fig. 1) as Z, while that of II, the p-bromobenzyl ester, though not of high accuracy, established the absolute configuration of clavulanic acid as (2R,5R). Prolonged hydrogenation of benzyl clavulanate resulted in formation of a mixture of two compounds one of which was (E)-p-bromobenzyl isoclavulanate (III) and irradiation of phenacyl isoclavulanate resulted in the formation of the oxetane IV. Molecular dimensions, though not of high accuracy in II and III, have acceptable values in all four compounds.

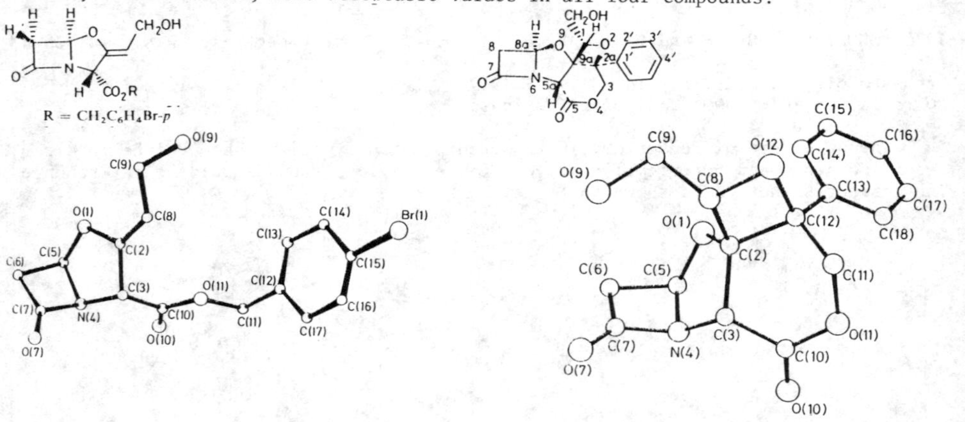

Fig. 1. The molecular skeletons and structures: $C_{15}H_{14}BrNO_5$ (left) and $C_{16}H_{15}NO_6$ (right).

(3Z,6E)-1-N-METHYLALBONOURSIN
$C_{16}H_{18}N_2O_2$

A.A. FREER, D. GARDNER and J.P. POYSER, 1984. J. Chem. Res., (S) 283, (M) 2601-2622.

Monoclinic, $P2_1/a$, a = 11.939, b = 7.933, c = 15.783 Å, β = 105.7°, Z = 4. Mo radiation, R = 0.060 for 1538 reflexions.

In this antibiotic (Fig. 1), there is delocalisation of the electron density around the piperazine-2,5-dione ring as seen in the longer C=O (1.233 Å) and shorter C-N (1.344 Å) bonds. The terminal methyl groups of the (E)-isobutylidene group are disordered.

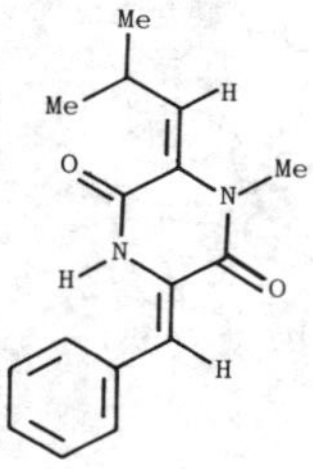

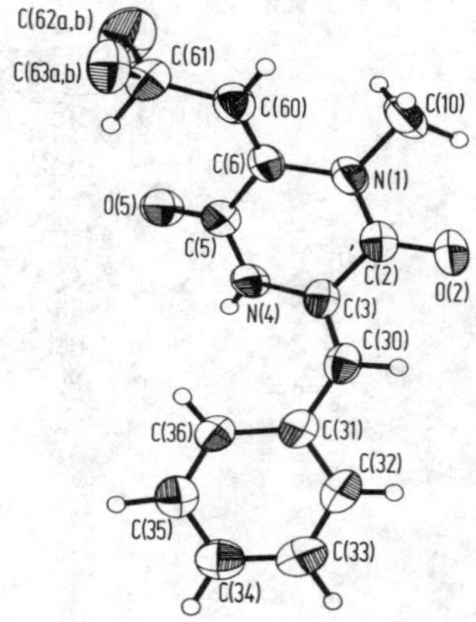

Fig. 1. $C_{16}H_{18}N_2O_2$: structure and stereochemistry.

PENTALENOLACTONE G METHYL ESTER
$C_{16}H_{18}O_6$

H. SETO, H. NOGUCHI, U. SANKAWA and Y. IITAKA, 1984. J. Antibiot., 37, 816-819.

Orthorhombic, $P2_12_12_1$, a = 10.893, b = 21.868, c = 6.482 Å, Z = 4. Cu radiation, R = 0.066 for 1743 reflexions.

The analysis establishes the correct stereochemistry for the compound as shown in Fig. 1.

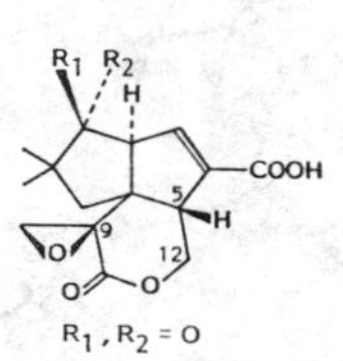

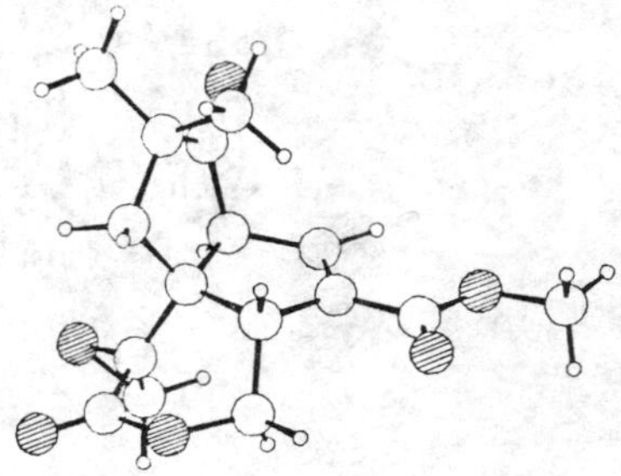

Fig. 1. Pentalenolactone G methyl ester.

METHYL 5,9-DIMETHOXY-2-METHYL-4H,6H-PYRANO[3,4-g]-1-BENZOPYRAN-4,6-DIONE-8-CARBOXYLIC ACID (DIMETHYL LATEROPYRONE)
$C_{17}H_{14}O_8$

G.W. BUSHNELL, YUN-LONG LI and G.A. POULTON, 1984. Can. J. Chem., 62, 2101-2106.

Monoclinic, $P2_1/c$, a = 9.663, b = 22.317, c = 7.747 Å, β = 113.03°, D_m = 1.477, Z = 4. Mo radiation, R = 0.054 for 2100 reflexions.

The molecule is shown in Fig. 1. The central ring is aromatic and the rings on either side are of the α- and γ-pyrone types. There is considerable overcrowding, the most striking feature is the lateral displacement of O(1) and O(11) away from the O(13) methoxy group. In- and out-of-plane distortions are rationalized in terms of relief from molecular overcrowding.

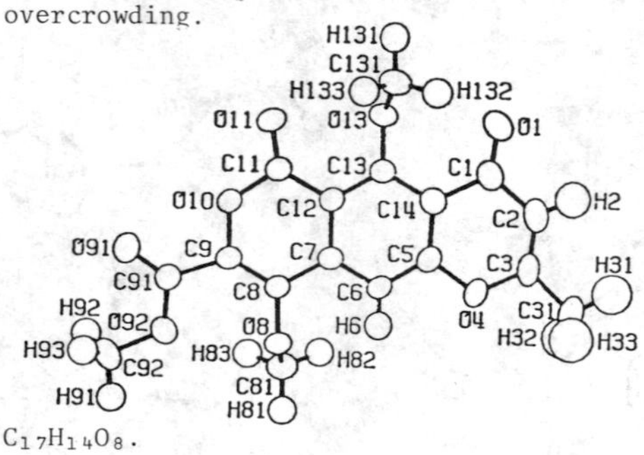

Fig. 1. A view of $C_{17}H_{14}O_8$.

RISHIRILIDE A MONOHYDRATE
$C_{21}H_{24}O_7,H_2O$

H. IWAKI, Y. NAKAYAMA, M. TAKAHASHI, S. UETSUKI, M. KIDO and Y. FUKUYAMA, 1984. J. Antibiot., 37, 1091-1093.

Orthorhombic, $P2_12_12_1$, a = 9.578, b = 11.573, c = 17.991 Å, Z = 4. Mo radiation, R = 0.042 for 1372 reflexions.

The analysis establishes the relative stereochemistry for rishirilide A to be as shown in Fig. 1.

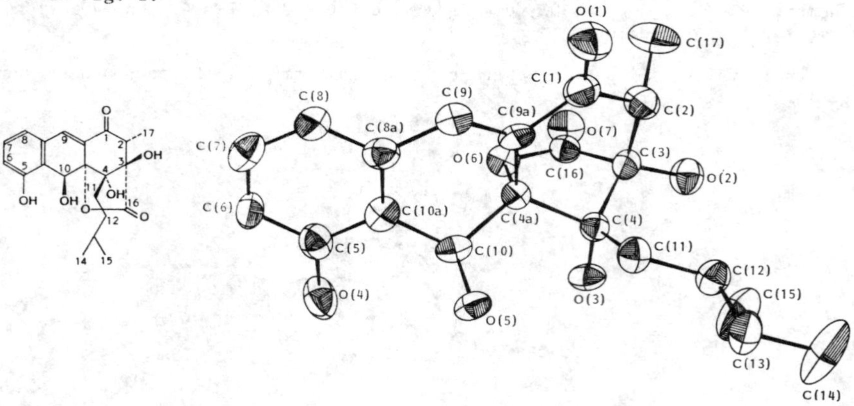

Fig. 1. The structure and relative stereochemistry of rishirilide A.

4,6,12-TRIMETHOXY-2-OXATRICYCLO[13.2.2.1^{3,7}]ICOSA-3,5,7(20),11,15,17,18-HEPTAEN-10-ONE (GARUGANIN-I)
$C_{22}H_{24}O_5$

V. PATTABHI, S. KRISHNASWAMI and E.J. GABE, 1984. Acta Cryst., C40, 832-834.

Monoclinic, $P2_1/a$, a = 9.3993, b = 15.0431, c = 13.7949 Å, β = 102.577°, D_m = 1.30, Z = 4. Mo radiation, R = 0.040 for 1962 reflexions.

Most of the bond lengths and angles in the molecule agree with the standard values (Fig. 1). The phenyl group including C(3) to C(7) and C(20) is planar, whilst that including C(15) to C(19) and C(1) shows significant non-planarity. The structure is stabilised by van der Waals forces.

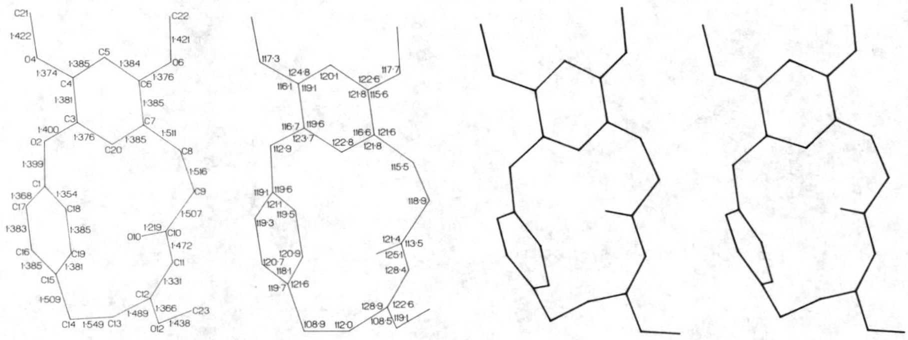

Fig. 1 $C_{22}H_{24}O_5$: bond lengths and angles and a stereoview of the molecule.

1,2-O-ISOPROPYLIDENE-3-O-METHYL-5,6-O-[2,3,4,6-TETRA-O-ACETYL-1-DESOXY-D-GLUCOPYRANOSYLIDENE-1-(1S)]-α-D-GLUCOFURANOSE
$C_{24}H_{34}O_{15}$

J.P. PRALY, G. DESCOTES, R. FAURE and H. LOISELEUR, 1984. Acta Cryst., C40, 1206-1207.

Orthorhombic, $P2_12_12_1$, a = 13.310, b = 14.0005, c = 15.626 Å, Z = 4. Cu radiation, R = 0.045 for 1434 reflexions.

The molecule (Fig. 1) exhibits a spiro ortho-ester linkage at the anomeric carbon which characterizes a new class of antibiotics (orthosomycines). There are no unusual interatomic distances or angles.

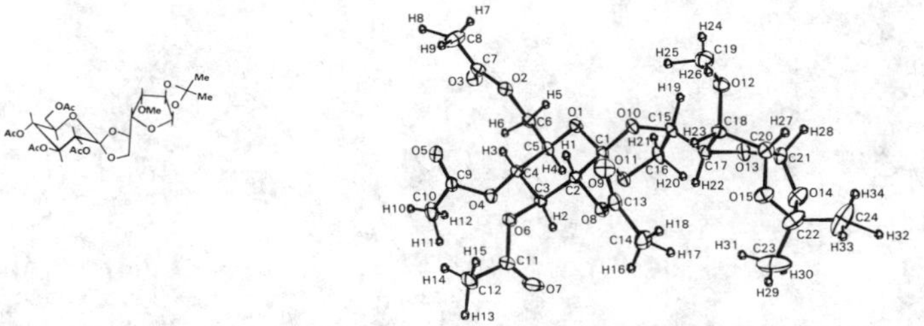

Fig. 1. $C_{24}H_{34}O_{15}$: structural formula and perspective view of the molecule.

MONACOLIN K
$C_{24}H_{36}O_5$

S. SATO, T. HATA, Y. TSUJITA, A. TERAHARA and C. TAMURA, 1984. Acta Cryst., C40, 195-198.

Orthorhombic, $P2_12_12_1$, a = 22.154, b = 17.321, c = 5.968 Å, Z = 4. Cu radiation, R = 0.088 for 1515 reflexions.

The analysis shows that the six-membered rings of the hexahydronaphthalene system contain a transoid conjugated diene system. Ring A (Fig. 1) has an approximate envelope conformation with C(5) at the flap, but distorted slightly into a half-chair form. The B and C rings adopt a cyclohexene-like half-chair conformation.

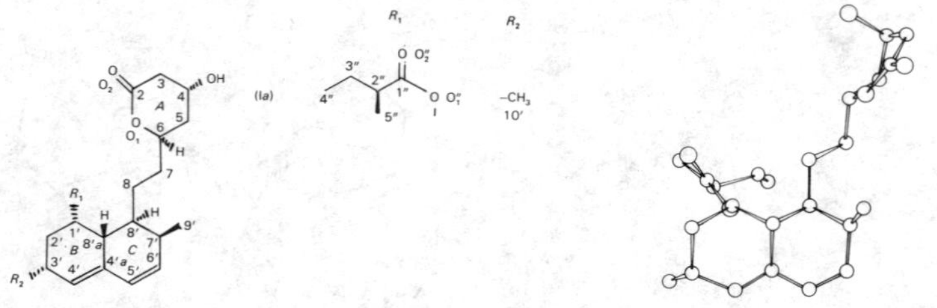

Fig. 1. The monacolin K structure.

3-N-p-BROMOBENZOYL-1-O-p-PHENYLBENZOYL-4-O-(2'-TETRAHYDROPYRANYL)-β-L-DUANOSAMINE
$C_{31}H_{32}BrNO_6$ (I)

1-O-p-BENZOYL-3-N-p-BROMOBENZOYL-4-O-(2'-TETRAHYDROPYRANYL)-α-L-DAUNOSAMINE
$C_{25}H_{28}BrNO_6$ (II)

H. UMEZAWA, Y. TAKAHASHI, T. TAKEUCHI, H. NAKAMURA, Y. IITAKA and K. TATSUTA, 1984. J. Antibiot., 37, 1094-1097.

I. Monoclinic, $P2_1$, a = 13.901, b = 11.042, c = 9.550 Å, β = 104.35°, Z = 2. Cu radiation, R = 0.0668 for 1740 reflexions.

II. Triclinic, P1, a = 9.749, b = 15.157, c = 8.853 Å, α = 93.64, β = 107.82, γ = 87.04°, Z = 2. Cu radiation, R = 0.0674 for 3781 reflexions.

The structures and absolute stereochemistries of (I) and the two independent molecules of (II) are shown in Fig. 1. For (I), the absolute configuration at C2' is R and in II, the C2' atom has the S-configuration.

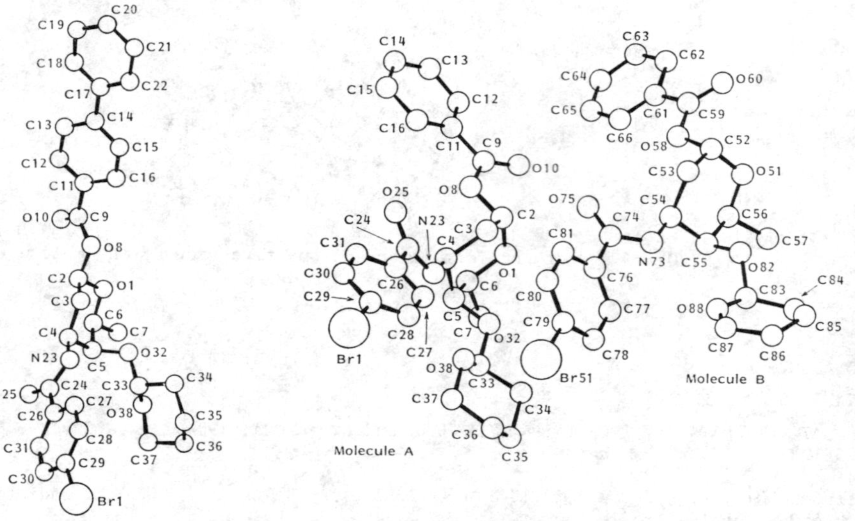

Fig. 1. Views of $C_{31}H_{32}BrNO_6$ and the two independent $C_{25}H_{28}BrNO_6$ molecules.

CYCLORIFAMYCIN SV
$C_{38}H_{49}NO_{11}$

S.K. AKORA and P. MAIN, 1984. J. Antibiot., 37, 178-181.

Monoclinic, $P2_1$, a = 20.347, b = 7.978, c = 23.364 Å, β = 100.6°, Z = 4. Cu radiation, R = 0.079 for 3923 reflexions.

There are two independent molecules in the asymmetric unit and they have similar geometries; a view of one molecule with averaged bond lengths is in Fig. 1. The double bonds at C16-C17, C18-C19 and C28-C29 have cis, trans and trans configurations respectively. The spatial arrangement of oxygens O1, O2, O9 and O10 is an important factor in binding rifamycins to DNA directed RNA polymerase.

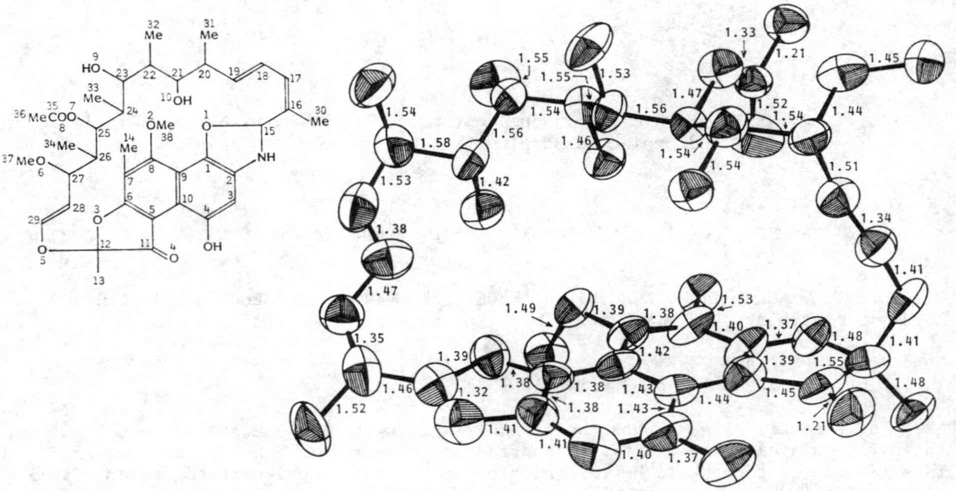

Fig. 1. Cyclorifamycin SV: a view of one molecule and averaged bond lengths.

OCTA-meso-HYI-VALINOMYCIN
$C_{40}H_{68}N_4O_{12}$

A.D. VASIL'EV, Yu.V. NEKRASOV, G.N. TISHCHENKO and V.I. SIMONOV, 1984. Kristallografiya, 29, 266-273 [Sov. Phys. Crystallogr., 29, 158-162].

Monoclinic, $P2_1/b$, a = 21.550, b = 20.306, c = 20.860 Å, γ = 90.71°, Z = 8. Cu radiation, R = 0.095 for 4271 reflexions.

In the crystal there are four conformationally different types of molecules: molecules A, their enantiomorphic molecules $\bar{A}$, molecules B, and their enantiomorphs $\bar{B}$ (Fig. 1). Geometrical parameters reveal that A and B are enantiomorphically identical with some precision because of the presence of a non-crystallographic pseudocentre of symmetry. The main differences between $\bar{A}$ and B involve the orientation of the side chains of the hydroxy acid residues. The conformation of the depsipeptide spines of the molecules is typically of the bracelet type, stabilised by weak intramolecular hydrogen bonds (N...O distances range from 3.08 to 3.23 Å). The carbonyl O atoms of the ester groups form a heavily distorted tetrahedron (with edges ranging from 2.9 to 6.7 Å) within the cavity of the molecule. Bond lengths and angles are normal and typical of known values for peptides.

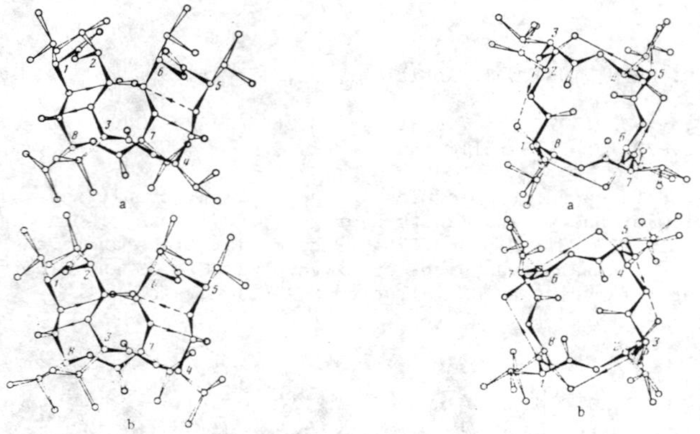

Fig. 1. Crystallographically independent molecules of $C_{40}H_{68}N_4O_{12}$ (a) Ā (b) B (left), and projected on plane through the N atoms (a) A (b) B (right).

4-DEOXY-3'-BROMOPYRIDO(1',2'-1,2)IMIDAZO[5,4-c]RIFAMYCIN-S DIHYDRATE
$C_{42}H_{46}BrN_3O_{11},2(H_2O)$

M. BRUFANI, L. CELLAI, S. CERRINI, W. FEDELI, E. MARCHI, A. SEGRE and A. VACIAGO, 1984. J. Antibiot., 37, 1623-1627.

Monoclinic, P2₁, a = 9.581, b = 14.508, c = 16.752 Å, β = 97.31°, Z = 2. Mo radiation, R = 0.089 for 3163 reflexions.

The molecule (Fig. 1) has the pyridoimidazo system in a mesomeric betaine form, the pyrido N atom being positively charged and the imidazo N being negatively charged. This feature is thought to be responsible for the pharmacokinetic behaviour of this class of compound. The mean plane of the ansa side chain makes an angle of 123° with that of the chromophore ring. The naphthoquinone and pyrido-imidazo systems are coplanar. The conformation is stabilized by three intramolecular O-H...O bonds (O1...O2 2.54, O8...O9 2.69, O9...O10 2.62 Å).

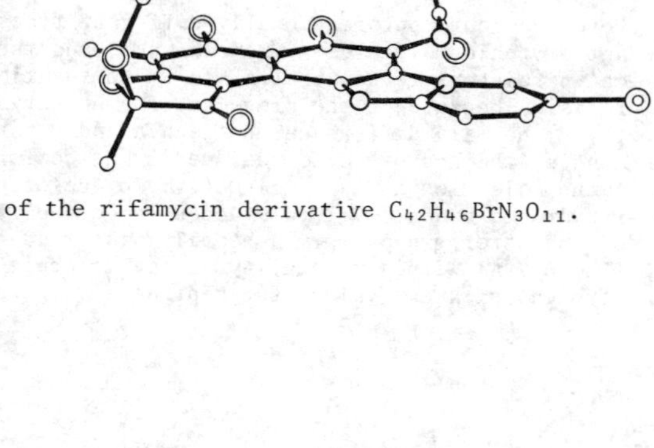

Fig. 1. A view of the rifamycin derivative $C_{42}H_{46}BrN_3O_{11}$.

25-O-METHYLNAPHTHOMYCIN A IMINOMETHYL ETHER HYDRATE
$C_{42}H_{50}ClNO_9,2H_2O$

W. KELLER-SCHIERLEIN, M. MEYER, L. CELLAI, S. CERRINI, D. LAMBA, A. SEGRE, W. FEDELI and M. BRUFANI, 1984. J. Antibiot., 37, 1357-1361.

Orthorhombic, $P2_12_12_1$, a = 12.565, b = 14.410, c = 24.239 Å, Z = 4. Cu radiation, R = 0.085 for 2898 reflexions.

A view of the molecule showing the absolute configuration is in Fig. 1. All chiral centres have the S configuration.

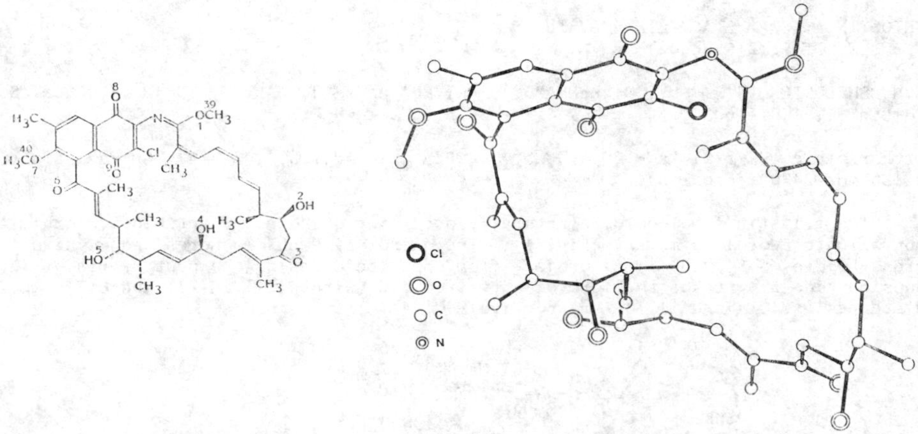

Fig. 1. The $C_{42}H_{50}ClNO_9$ molecule and absolute configuration.

SODIUM 6-DEMETHYLALBORIXIN DIMETHYLFORMAMIDE SOLVATE
$C_{47}H_{81}NaO_{14},0.85(C_3H_7NO)$

P. VAN ROEY, W.L. DUAX, P.D. STRONG and G.D. SMITH, 1984. Israel J. Chem., 24, 283-289.

Fig. 1. A space-filling model showing the conformation of sodium 6-demethylalborixin (O atoms are marked with concentric circles).

Orthorhombic, $P2_12_12_1$, a = 14.716, b = 25.392, c = 14.634 Å, Z = 4. Mo radiation, R = 0.094 for 1118 reflexions.

The ionophore wraps completely around a central cavity which is occupied by the cation which is thus isolated. The surface of the molecule is hydrophobic except for the O9 hydroxyl group which is exposed. The Na^+ ion is effectively five-coordinate. The structure is compared with the free acid and the Ag^+ and K^+ complexes. The similarity of all the conformations is thought to arise from the presence of intramolecular hydrogen bonding.

TRIOSTIN A ISOAMYL ACETATE SOLVATE
$C_{50}H_{62}N_{12}O_{12}S_2,2(C_7H_{14}O_2)$

G.M. SHELDRICK, J.J. GUY, O. KENNARD, V. RIVERA and M.J. WARING, 1984. J. Chem. Soc. Perkin II, 1601-1605.

Orthorhombic, $P2_12_12_1$, a = 20.940, b = 18.528, c = 18.795 Å, Z = 4. Cu radiation, R = 0.150 for 2450 reflexions.

The triostin A molecule, a naturally occurring quinoxaline antibiotic, exhibits approximate twofold symmetry (Fig. 1). The quinoxaline chromophores and the disulphide cross-bridge (S-S' 2.00(1) Å) project from opposite sides of the peptide ring. The depsipeptide ring takes the form of a twisted disk with C(2)...C(2') and C(7)...C(7') distances of 4.14 and 11.34 Å respectively.

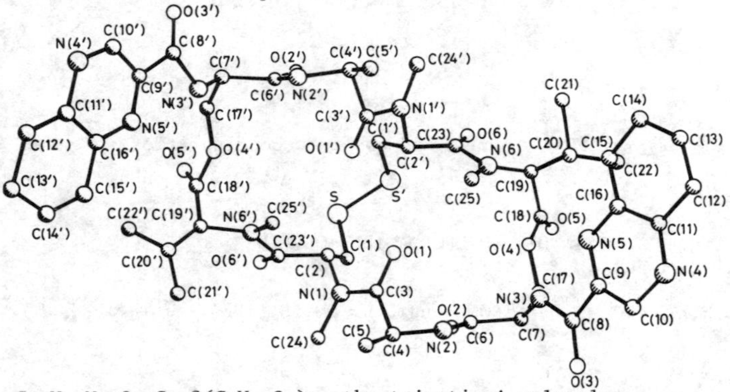

Fig. 1. $C_{50}H_{62}N_{12}O_{12}S_2,2(C_7H_{14}O_2)$: the triostin A molecule.

CONCANAMYCIN A DIACETATE
$C_{50}H_{79}NO_{16}$

J.M. WESTLEY, C.-M. LIU, L.H. SELLO, R.H. EVANS, N. TROUPE, J.F. BLOUNT, A.M. CHIU, L.J. TODARO and P.A. MILLER, 1984. J. Antibiot., 37, 1738-1741.

Monoclinic, $P2_1$, a = 13.375, b = 18.871, c = 11.317 Å, β = 106.39°, Z = 2. Cu radiation, R = 0.040 for 5026 reflexions.

The antibiotic has the structure and stereochemistry shown in Fig. 1.

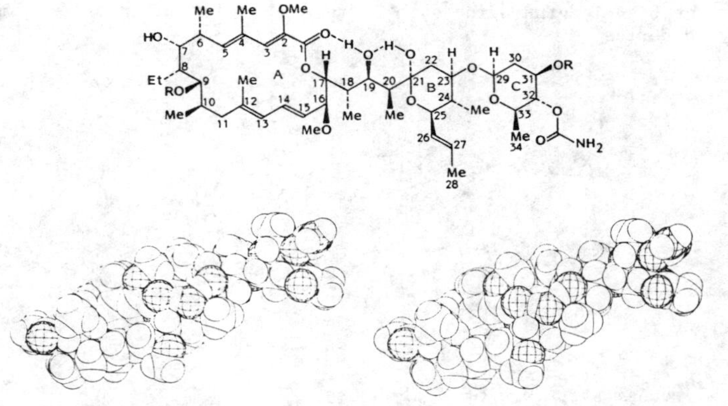

Fig. 1. Concanamycin A diacetate (R = MeCo) and a space-filling stereoview.

SALBOMYCIN DIHYDRATE
$C_{54}H_{88}O_{18},2H_2O$

E.F. PAULUS, L. VERTESY and G.M. SHELDRICK, 1984. Acta Cryst., C40, 700-703.

Monoclinic, $P2_1$, a = 9.770, b = 10.037, c = 30.890 Å, β = 92.58°, D_m = 1.17, Z = 2. Cu radiation, R = 0.094 for 3660 reflexions (at 140K).

The molecule consists of two almost identical sub-units and has 26 chiral C atoms (Fig. 1). The central region of the molecule is a macrolidic sixteen-membered ring with two lactone groups. There are two chains of deoxysaccharides attached to opposite atoms on the macrolidic ring. The molecule has a plate-like conformation with polar and nonpolar sides, which accounts for its action on biological membranes. The molecules are linked in the crystal by hydrogen bonds.

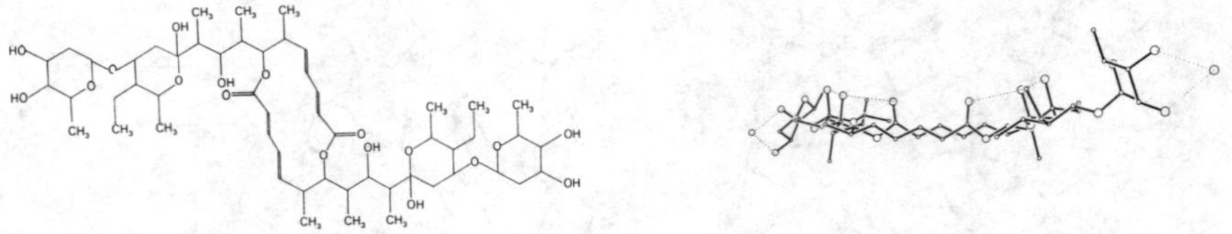

Fig. 1. $C_{54}H_{88}O_{18},2H_2O$: structural formula and a projection of the molecule, viewed parallel to a direction in the macrolidic ring.

METHYLALANYLVALINOMYCIN
$C_{55}H_{93}N_7O_{17}$

V.I. SMIRNOVA and G.N. TISHCHENKO, 1984. Kristallografiya, 29, 252-259 [Sov. Phys. Crystallogr., 29, 149-154].

Monoclinic, $P2_1$, a = 14.628, b = 22.276, c = 10.474 Å, β = 105.74°, Z = 2. Cu radiation, R = 0.067 for 3850 reflexions.

The main chain (macroring) of the molecule forms the surface of a very flat tube of oval cross section (Fig. 1) stabilised by four intramolecular hydrogen bonds involving carbonyl O atoms from amide groups (N...O distances 3.060, 3.088, 2.864 and 2.996 Å), and one involving the carbonyl O of an ester group (N(3)-H(3)...O(11),

N(3)...O(11) 3.022 Å). Both the amide and the ester groups have planar trans structures. The bond lengths agree with corresponding values in other compounds of the valinomycin series.

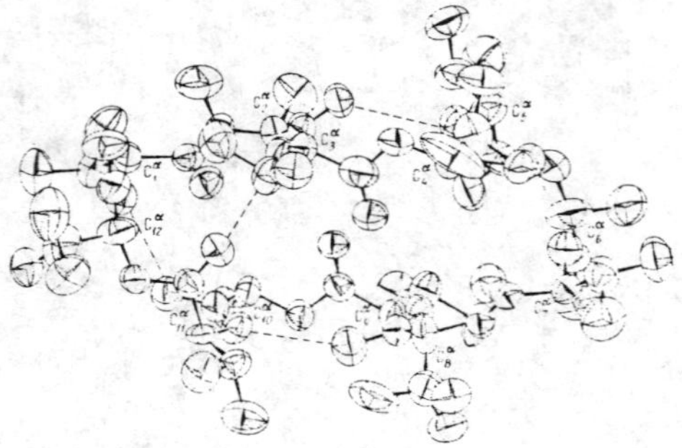

Fig. 1. Molecular structure of $C_{55}H_{93}N_7O_{17}$.

VALINOMYCIN CESIUM PICRATE
$C_{60}H_{92}CsN_9O_{25}$

L.K. STEINRAUF and K. FOLTING, 1984. Israel J. Chem., 24, 290-296.

Trigonal, $P3_1$, a = 14.028, c = 31.888 Å, Z = 3. Mo radiation, R = 0.111 for 4088 reflexions (at -170°C).

The Cs cation is too large for the valinomycin cavity which as a result is incompletely closed (Fig. 1) and the average Cs-O coordinating distance is 2.98(2) Å. A weak bond (Cs...O 3.28 Å) is formed with the picrate anion at the D-valyl end of the ligand, in agreement with a proposal that the cation enters and leaves the complex by the D-valyl end.

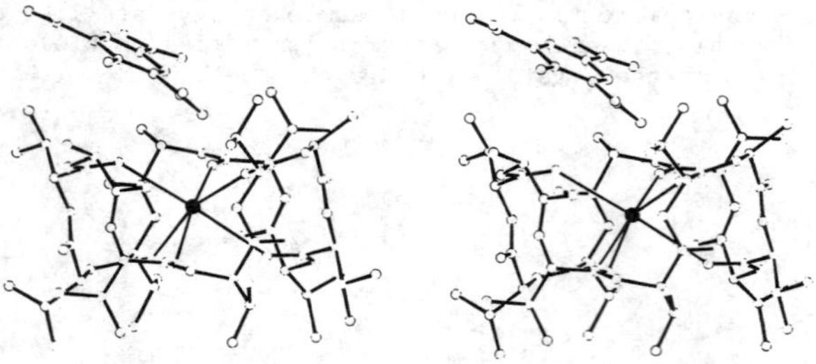

Fig. 1. A stereoview of valinomycin cesium picrate.

ZINC MAGNESIUM ZINCOPHORIN
$C_{66}H_{108}Mg_{0.5}O_{14}Zn_{0.5}$ $0.5(Zn^{2+}).0.5(Mg^{2+}).(C_{33}H_{59}O_7^{-})_2$

H.A. BROOKS, D. GARDNER, J.P. POYSER and T.J. KING, 1984. J. Antibiot., 37, 1501-1503.

Monoclinic, $P2_1$, a = 12.629, b = 12.259, c = 22.958 Å, β = 90.4°, Z = 2. Mo radiation, R = 0.1195 for 4400 reflexions.

The structure and absolute stereochemistry were determined to be as shown in Fig. 1 for this mixed Zn^{2+}/Mg^{2+} salt. Chelation to the cation is octahedrally through the carboxylate and hydroxyls at C11 and C13 of two molecules of zincophorin.

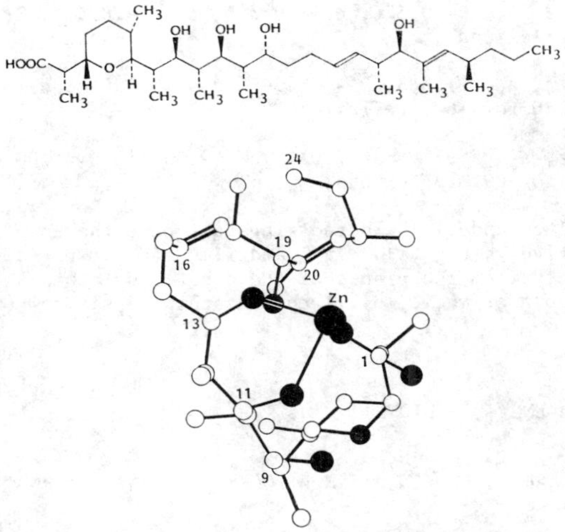

Fig. 1. The zincophorin antibiotic and a perspective drawing of half of the complex.

3,4-DINOR-2,5-SECO-9α-ESTRANE-1,5,17-TRIONE
$C_{16}H_{22}O_3$ (I)

D-HOMO-4-NORESTR-3(5)-ENE-2,17-DIONE
$C_{18}H_{24}O_2$ (II)

1,2,3,4-TETRANORESTR-9-ENE-5,17-DIONE
$C_{14}H_{18}O_2$ (III)

E.O. SCHLEMPER, B. NASSIM and P. CRABBÉ, 1984. Acta Cryst., C40, 1455-1458.

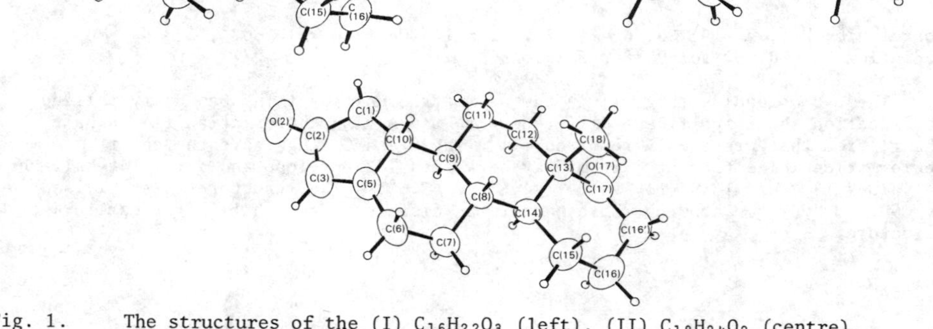

Fig. 1. The structures of the (I) $C_{16}H_{22}O_3$ (left), (II) $C_{18}H_{24}O_2$ (centre), (III) $C_{14}H_{18}O_2$ (right) molecules.

I. Orthorhombic, $P2_12_12_1$, a = 9.326, b = 10.355, c = 14.824 Å, Z = 4. Mo radiation, R = 0.044 for 1156 reflexions.

II. Monoclinic, $P2_1/n$, a = 11.383, b = 11.645, c = 11.859 Å, β = 107.83°, Z = 4. Mo radiation, R = 0.046 for 1605 reflexions.

III. Monoclinic, $P2_1$, a = 8.947, b = 7.031, c = 9.557 Å, β = 100.60°, Z = 2. Mo radiation, R = 0.041 for 997 reflexions.

The saturated five- and six-membered rings all have the envelope and chair conformations respectively (Fig. 1). In I and III, C(14) is displaced 0.621(3) and 0.629(3) Å respectively from the planes of the other four atoms. The unsaturated ring in II is almost planar while in III the unsaturated six-membered ring has the sofa conformation.

10β-HYDROPEROXYESTR-4-ENE-3,17-DIONE
$C_{18}H_{24}O_4$

D.F. COVEY, W.F. HOOD, D.D. BEUSEN and H.L. CARRELL, 1984. Biochemistry, 23, 5398-5406.

Orthorhombic, $P2_12_12_1$, a = 9.640, b = 20.099, c = 8.216 Å. Z = 4. Cu radiation, R = 0.040 for 1695 reflexions.

Views of the molecule are in Fig. 1. Bond lengths in 10β-hydroperoxyl group are C(10)-O(19a) 1.444(3), O(19a)-O(19b) 1.460(2) Å, with C(10)-O(19a)-O(19b) 109.0(1)°.

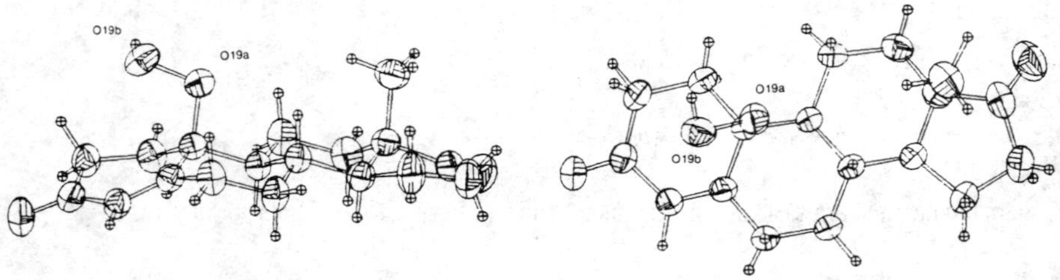

Fig. 1. Two views showing the conformation of $C_{18}H_{24}O_4$.

(+)-17β-HYDROXY-19-NOR-4-ANDROSTEN-3-ONE (19-NORTESTOSTERONE)
$C_{18}H_{26}O_2$

M.M. BHADBHADE and K. VENKATESAN, 1984. Acta Cryst., C40, 1905-1908.

Monoclinic, $P2_1$, a = 9.755, b = 11.467, c = 14.196 Å, β = 101.07°, Z = 4. Mo radiation, R = 0.060 for 2158 reflexions.

The two molecules (Fig. 1) in the asymmetric unit of this second crystalline modification show significant differences in the A-ring conformation (1α,2β-half-chair) from that previously reported (1). The B and C rings have the chair conformation while that of D is midway between a 13β-envelope and a 13β,14α-half-chair (Δ values 18.7 and 18.9°; ϕ_m values 44.8 and 47.6°). The structure is stabilised by O-H...O hydrogen bonding whose pattern is different from that in the earlier structure.

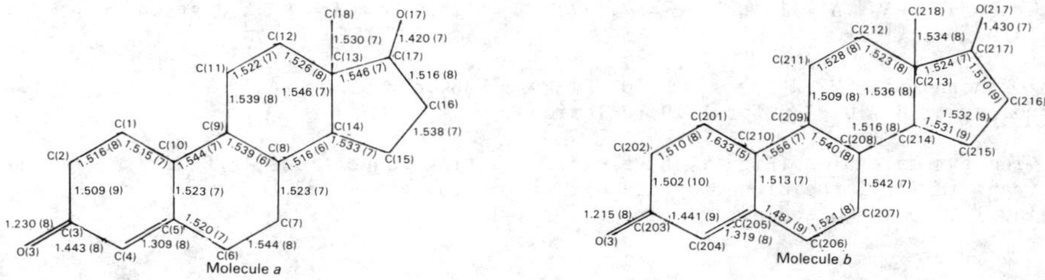

Fig. 1. Bond lengths (Å) in the two independent 19-nortestosterone molecules.

1. Structure Reports, 41B, 590.

5α-ANDROST-16-EN-3-ONE
$C_{19}H_{28}O$

P.J. COX and A.B. TURNER, 1984. Tetrahedron, 40, 3153-3158.

Orthorhombic, $P2_12_12_1$, a = 8.495, b = 10.057, c = 18.490 Å, Z = 4. Mo radiation, R = 0.044 for 1367 reflexions.

The X-ray dimensions (Fig. 1) are compared with those from molecular mechanics calculations. All except three distances agree to within 3σ. The largest difference involves C16=C17 (1.302(5) Å, X-ray; 1.336 Å molecular mechanics). The presence of the double bond in the D ring makes the α-face of the molecule less concave than usual. The axial methyl groups are significantly twisted; the pseudo torsion angle C19-C10-C13-C18 is 8.6(3)°.

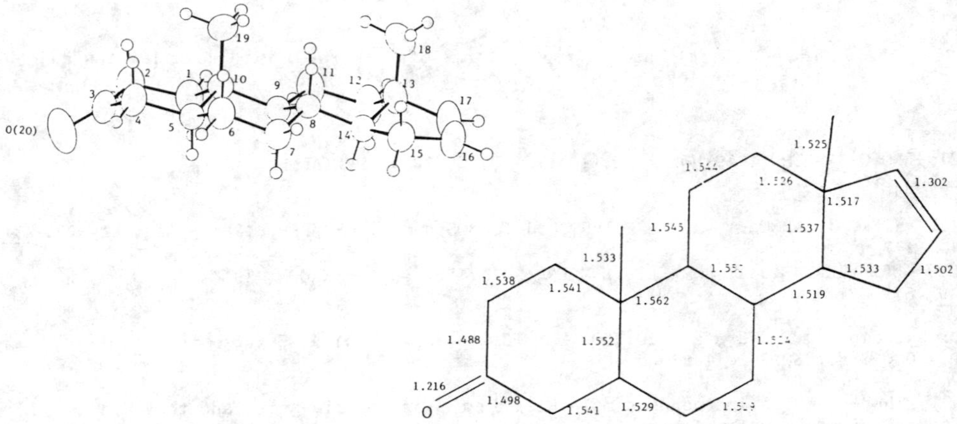

Fig. 1. A view of $C_{19}H_{28}O$ and bond lengths.

17β-HYDROXY-17α-METHYL-5β-ANDROSTANO-(2,3-C)-1',2',5'-OXADIAZOLE
$C_{19}H_{30}N_2O_2$ (I)

17β-HYDROXY-17α-METHYL-5α-ANDROSTANO-(2,3-C)-1',2',5'-OXADIAZOLE
$C_{19}H_{30}N_2O_2$ (II)

A.E. SHORA, R.A. PALMER, H. SINGH and D. PAUL, 1984. J. Cryst. Spectrosc. Res., 14, 315-332.

I. Trigonal, $P3_2$, a = 14.820, c = 7.177, D_m = 1.13, Z = 3. Cu radiation, R = 0.0376 for 1419 reflexions.

II. Orthorhombic, $P2_12_12_1$, a = 9.507, b = 18.528, c = 10.048 Å, D_m = 1.19, Z = 4. Cu radiation, R = 0.0770 for 1819 reflexions.

The two molecules (Fig. 1) have different ring connections, A/B cis in (I) and A/B trans in (II), the planar oxadiazole ring being cis-fused to ring A (which is strained) in both molecules.

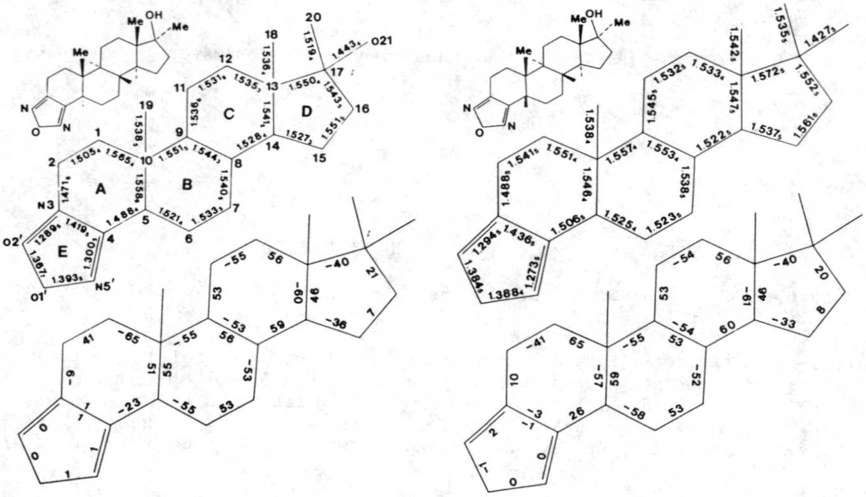

Fig. 1. The 5β (left) and 5α (right) forms of $C_{19}H_{30}N_2O_2$ with bond lengths and torsion angles.

21α-BROMO-3-METHOXY-19-NOR-1,3,5(10),17(20),20-PREGNAPENTAENE
$C_{21}H_{25}BrO$

A.L. SPEK, J. KROON, C.J. ELSEVIER, H.J.T. BOS and P. VERMEER, 1984. Acta Cryst., C40, 445-447.

C.J. ELSEVIER, H.J.T. BOS, P. VERMEER, A.L. SPEK and J. KROON, 1984. J. Org. Chem., 49, 379-381.

Orthorhombic, $P2_12_12_1$, a = 10.030, b = 13.286, c = 27.41 Å, Z = 8. Mo radiation, R = 0.067 for 3013 reflexions.

There are two independent molecules in the asymmetric unit, and they are structurally equivalent (Fig. 1). The Br atom is in the 21α position, which implies that Lowe-Brewster rules do not apply in this case.

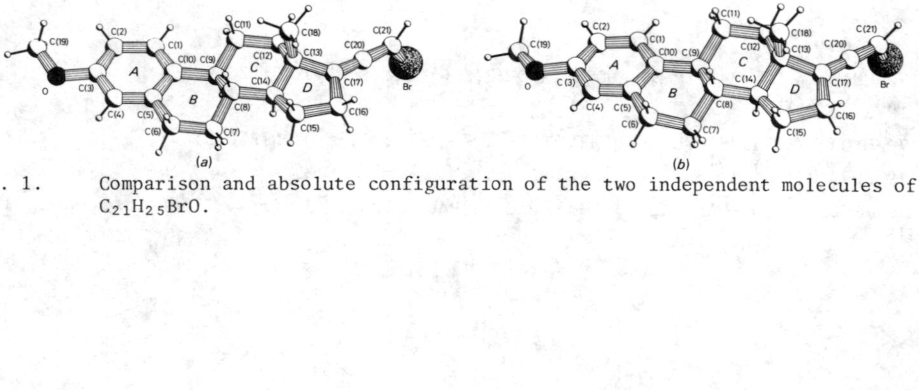

Fig. 1. Comparison and absolute configuration of the two independent molecules of $C_{21}H_{25}BrO$.

17-HYDROXY-7α-METHYL-19-NOR-17α-PREGN-5(10)-EN-20-YN-3-ONE
$C_{21}H_{28}O_2$

J.-P. DECLERCQ, M. VAN MEERSSCHE and F.J. ZEELEN, 1984. Rec. J.R. Neth. Chem. Soc., 103, 145-147.

Monoclinic, $P2_1$, a = 6.542, b = 41.213, c = 6.678 Å, β = 101.64°, Z = 4. Mo radiation, R = 0.042 for 2211 reflexions.

There are two independent molecules in the asymmetric unit and these have different ring A half-chair conformations (Fig. 1). The B-rings in both molecules are in half-chair forms with the 7α-methyl group axial, ring C is a chair form and ring D is a 13β-envelope. Molecules are linked by O-H...O hydrogen bonds (O...O 2.88 Å).

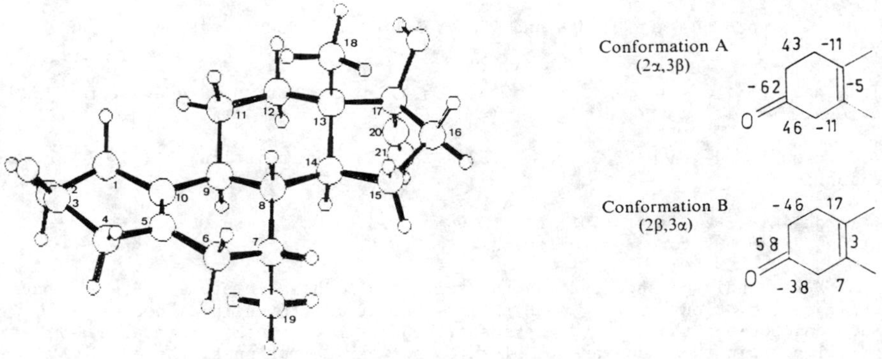

Fig. 1. A view of one of the $C_{21}H_{28}O_2$ molecules and details of the ring A torsion angles in the two independent molecules.

16α,17-EPOXY-4-PREGNENE-3,20-DIONE
$C_{21}H_{28}O_3$

K. GOUBITZ, H. SCHENK and F.J. ZEELEN, 1984. Steroids, 44, 153-158.

Orthorhombic, $P2_12_12_1$, a = 13.8635, b = 17.283, c = 7.424 Å, Z = 4. Cu radiation, R = 0.081 for 1710 reflexions.

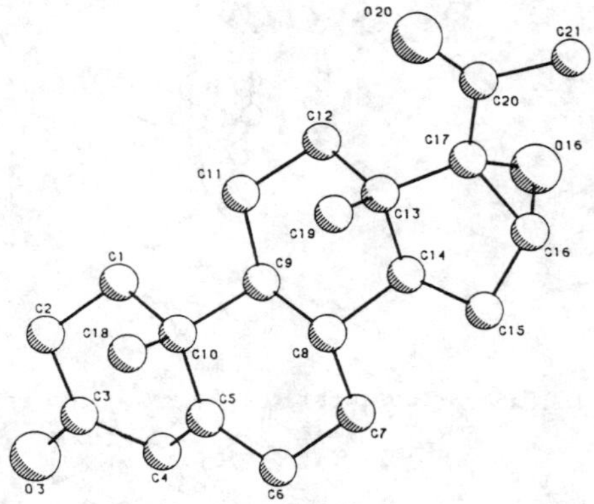

Fig. 1. A view of $C_{21}H_{28}O_3$.

The A-ring (Fig. 1) is in a 1α-sofa conformation, rings B and C are nearly perfect chairs and ring D is a 14α-envelope. The carbonyl group at C20 eclipses the C13-C17 bond (O20-C20-C17-C13 torsion angle 9°).

17β-HYDROXY-4,4,9-TRIMETHYL-9β,10α-ESTR-5-EN-3-ONE
$C_{21}H_{32}O_2$ (I)

2α-BROMO-17β-HYDROXY-4,4,9-TRIMETHYL-9β,10α-ESTR-5-EN-3-ONE
$C_{21}H_{31}BrO_2$ (II)

2β-BROMO-17β-t-BUTOXY-4,4,9-TRIMETHYL-9β,10α-ESTR-5-EN-3-ONE
$C_{25}H_{39}BrO_2$ (III)

J.R. BULL, R. PACHTER, A. TUINMAN and P.H. VAN ROOYEN, 1984. Tetrahedron, 40, 777-784.

I. Trigonal, $P3_1$, a = 14.67, c = 7.43 Å, Z = 3. Mo radiation, R = 0.052 for 1229 reflexions.

II. Tetragonal, $P4_3$, a = 23.35, c = 7.52 Å, Z = 8. Cu radiation, R = 0.084 for 2614 reflexions.

III. Monoclinic, $P2_1$, a = 14.08, b = 12.9, c = 14.54 Å, β = 98.96°, Z = 4. Cu radiation, R = 0.087 for 3461 reflexions.

The X-ray analyses show that in the solid state these molecules (Fig. 1) have a $^1C_4$ chair conformation for ring 4 which is somewhat flattened in II because of interactions between the 2α- and 4α- substituents. Each of the brominated derivatives II and III contain two independent molecules per asymmetric unit; these pairs of molecules differ slightly in conformation. Spectroscopic data are interpreted in terms of conformational equilibria for I and II.

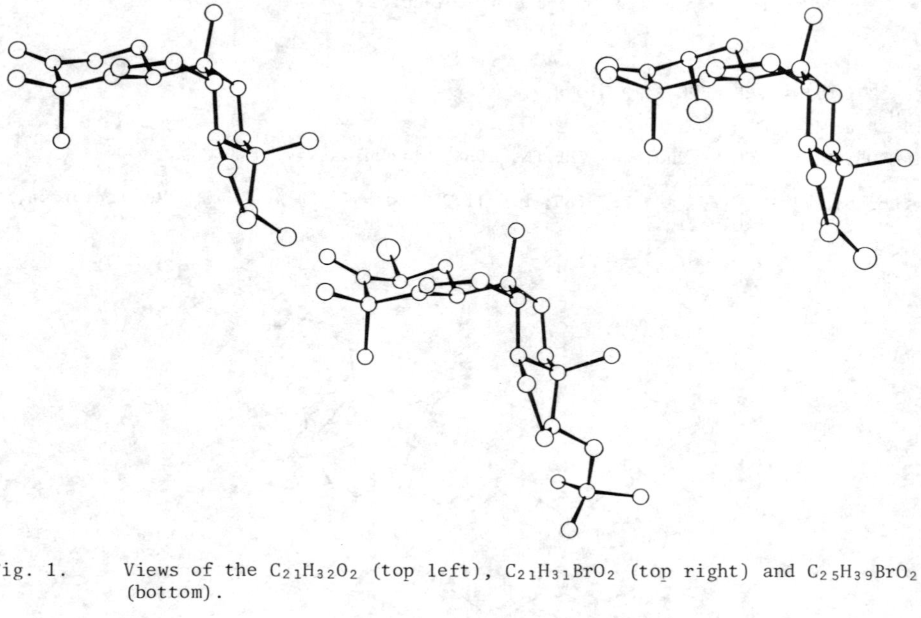

Fig. 1. Views of the $C_{21}H_{32}O_2$ (top left), $C_{21}H_{31}BrO_2$ (top right) and $C_{25}H_{39}BrO_2$ (bottom).

3β-ACETOXY-17β-HYDROXY-18-NOR-5α,17α-PREGNAN-20-ONE
$C_{22}H_{34}O_4$

C.P. HUBER and F.R. AHMED, 1984. Acta Cryst., C40, 703-706.

Monoclinic, C2, a = 14.746, b = 7.483, c = 36.916 Å, β = 97.40°, D_m = 1.188, Z = 8. Cu radiation, R = 0.049 for 3704 reflexions.

In the two independent molecules (Fig. 1) the hydroxyl group at C(17) is β, cis to the C(19) methyl group. Rings A, B and C show essentially identical chair conformations in the two molecules, while ring D, here in the uncommon C(14)-envelope form, shows some small conformational differences, which may be related to the intermolecular hydrogen-bonding pattern.

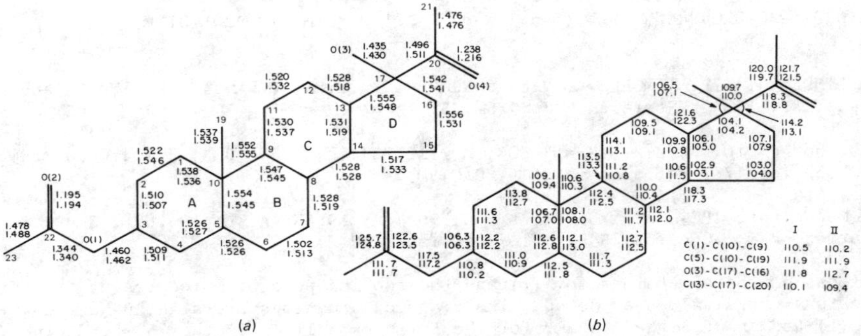

Fig. 1 $C_{22}H_{34}O_4$: (a) bond distances and (b) bond angles; values for molecule I are given above, those for molecule II below.

3β-HYDROXY-14,15β-EPOXY-5β,14β-CARD-20,22-ENOLIDE (DIGIREZIGENIN)
$C_{23}H_{32}O_4$

A. KÁLMÁN, G. ARGAY, B. RIBÁR, S. VLADIMIROV and D. ŽIVANOV-STAKIĆ, 1984. Croat. Chem. Acta, 57, 519-528.

Orthorhombic, $P2_12_12_1$, a = 7.288, b = 14.686, c = 18.480 Å, D_m = 1.25, Z = 4. Mo radiation, R = 0.059 for 1778 reflexions.

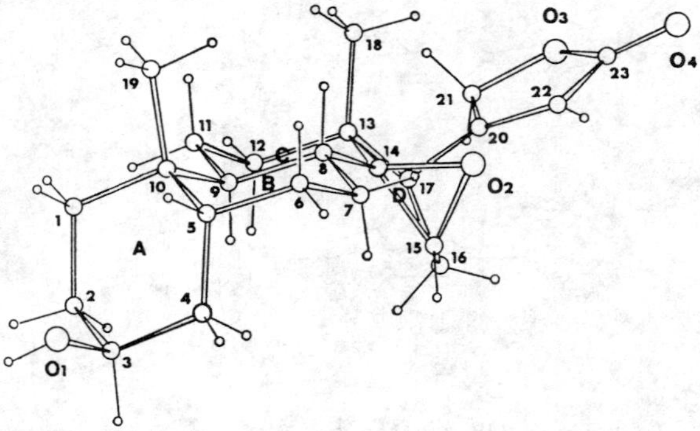

Fig. 1. Perspective view of $C_{23}H_{32}O_4$ showing numbering scheme.

The presence of the rigid 14,15β-epoxy ring alters considerably the shape of ring D (Fig. 1) relative to that in digitoxigenin (1). However, it has little effect upon the general features of the 14-iso-aethiocholane skeleton and the conformation of the γ-lactone ring. The amount of rotation about the C(17)-C(20) bond is hardly changed while the distance between the position of the carbonyl O atom relative to digitoxigenin is only 0.332 Å. Within the epoxy ring the distances O(2)-C(14), O(2)-C(15), and C(14)-C(15) are 1.452, 1.452 and 1.469 Å respectively, and the plane of this triangle makes a dihedral angle of 81.7° with the least-squares plane through ring D.

1. Structure Reports, 34B, 298.

3,20-DIOXOPREGN-4-ENE(16α,17α-D)-2',2'-DIMETHYLTHIAZOLIDINE
$C_{24}H_{35}NO_2S$ (I)

3β-HYDROXY-20-OXO-PREGN-5-ENE(16α,17α-D)-2',2'-DIMETHYLTHIAZOLIDINE
$C_{24}H_{37}NO_2S$ (II)

S.V. LINDEMAN, T.V. TIMOFEEVA, V.E. SHKLOVER, Yu.T. STRUCHKOV, A.M. TURUTA and A.V. KAMERNITZKY, 1984. Steroids, 43, 125-145.

I. Orthorhombic, $P2_12_12_1$, a = 7.338, b = 12.283, c = 23.648 Å, Z = 4. Cu radiation, R = 0.033 for 1737 reflexions (at -120°C).

II. Monoclinic, $P2_1$, a = 18.094, b = 6.356, c = 19.638 Å, β = 94.60°, Z = 4. Cu radiation, R = 0.066 for 2549 reflexions (at -120°C).

Most of the dimensions for (I) and the two independent molecules of (II) are close to average values expected. The ring conformations are a 1α,2β-half-chair for A in (I) and an 8β,9α-half-chair for the B rings in II; all other A, B, C rings are in chair conformations. Views of the molecules are in Fig. 1. Ring D in (I) and (IIA) is a 14α-envelope; ring D in (IIB) is intermediate between a 13β-envelope and a 13β,14α-half-chair. Ring E in (I) and (IIA) is a C22β-envelope whereas in IIB a Nβ,C(16)α-half-chair is found.

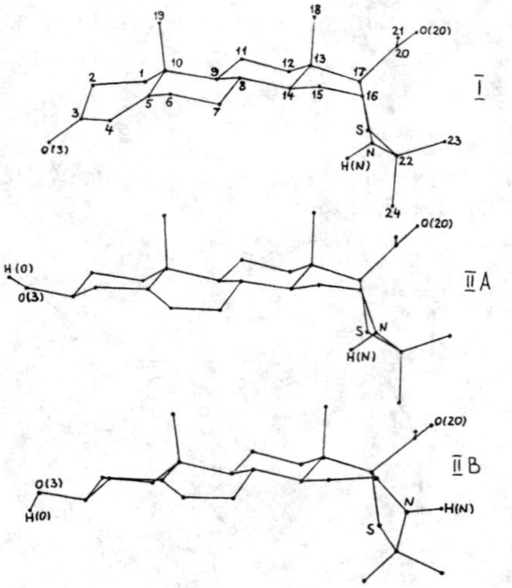

Fig. 1. Views of $C_{24}H_{35}NO_2S$ (I) and the two $C_{24}H_{37}NO_2S$ (II) molecules.

OLEANDRIGENIN MONOHYDRATE
$C_{25}H_{37}O_6,H_2O$

K. GO and G. KARTHA, 1984. Acta Cryst., C40, 1866-1869.

Monoclinic, $P2_1$, a = 17.878, b = 7.671, c = 19.531 Å, β = 108.26°, D_m = 1.241, Z = 4. Cu radiation, R = 0.055 for 4608 reflexions.

The conformation of the D ring (Fig. 1) in the two independent molecules is 14β-envelope in one and 13α,14β-half-chair in the other. The A, B, C rings have flattened chair conformations with the A and D rings cis with respect to B/C. The structure is stabilised by extensive hydrogen bonding. Some bond lengths are C(9)-C(10) 1.567(3) and 1.563(2), C(13)-C(17) 1.575(3) and 1.557(3) Å.

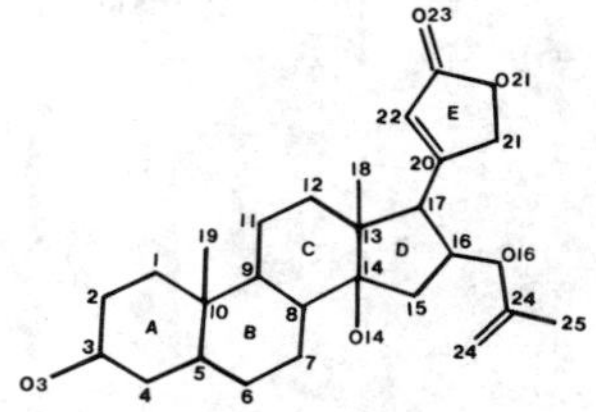

Fig. 1. The oleandrigenin molecule.

17aα-(DIETHYLCARBAMOYL)-4-METHYL-D-HOMO-4-AZA-5α-ANDROSTANE-3,17-DIONE HEMI-HYDRATE
$C_{25}H_{40}N_2O_3,0.5(H_2O)$

N.G. STEINBERG, G.H. RASMUSSEN, G.F. REYNOLDS, J.H. HIRSHFIELD and B.H. ARISON, 1984. J. Org. Chem., 49, 4731-4733.

Orthorhombic, $P2_12_12_1$, a = 8.242, b = 12.772, c = 23.112 Å, Z = 4. R = 0.048 for 1646 reflexions.

The analysis shows that the amide side chain is in the predicted axial configuration (Fig. 1).

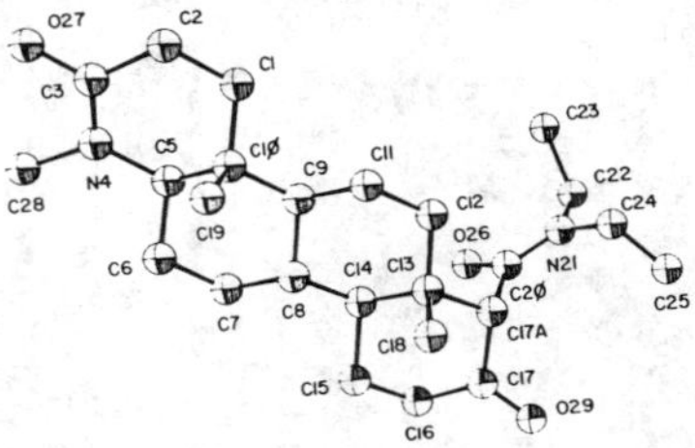

Fig. 1. A perspective view of $C_{25}H_{40}N_2O_3$.

17β-(2-HYDROXYETHYL)-3β-PYRROLIDINO-17a-AZA-D-HOMO-5-ANDROSTENE HYDRATE
$C_{25}H_{42}N_2O,H_2O$ (I)

17β-(2-HYDROXYETHYL)-3-PYRROLIDINO-17a-AZA-D-HOMO-5-ANDROSTENE (mixed 3α- and 3β-epimers)
$C_{25}H_{42}N_2O$ (II)

A.I. EL-SHORO, R.A. PALMER, H. SINGH, T.R. BHARDWAJ and D. PAUL, 1984. J. Cryst. Spectrosc. Res., 14, 89-115.

I. Orthorhombic, $P2_12_12_1$, a = 7.089, b = 11.502, c = 28.975 Å, Z = 4. Cu radiation, R = 0.056 for 1644 reflexions.

II. Triclinic, P1, a = 14.013, b = 12.572, c = 6.688 Å, α = 95.187, β = 103.491, γ = 86.210°, Z = 2. Mo radiation, R = 0.073 for 1731 reflexions.

There is one steroid molecule in the asymmetric unit of (I) (molecule A) and two (molecules B and C) in (II). Molecules A and B (Fig. 1) have similar 3β-pyrrolidino geometry (Fig. 1), differences in molecule C being related to the strain caused by the unusual 3α ring substituent which also produces a pronounced kink in the molecular backbone. The structures show hydrogen bonding but the pyrrolidine nitrogen N(31) in molecule C is congested and is unable to act as an acceptor. The hydroxyethyl side chain, important for activity, has a different conformation in the three molecules (t,g, t,-g, and t,t respectively). The mean C-C distances are 1.528, 1.537 and 1.533 Å (for A, B, C respectively). Average C-N distances are 1.484, 1.487 and 1.484 Å (for A, B, C respectively).

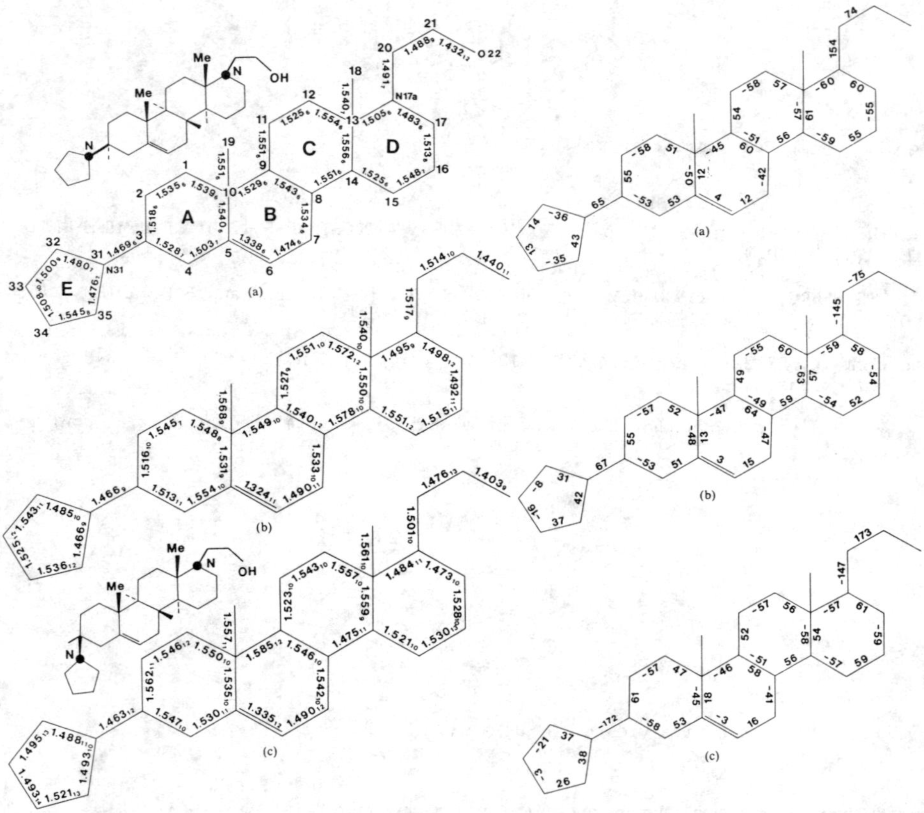

Fig. 1. Bond lengths, and torsion angles for molecules A, B and C of $C_{25}H_{42}N_2O$.

3β-ACETOXY-4,4-14α-TRIMETHYL-5α-PREGNA-7,9(11)-DIENE-20S-CARBOXYLIC ACID MONOHYDRATE
$C_{27}H_{40}O_4,H_2O$

N. CLAYDON, J.F. GROVE, M. POPLE and M.J. BEGLEY, 1984. J. Chem. Soc. Perkin I, 497-502.

Monoclinic, $P2_1$, a = 6.162, b = 31.110, c = 7.028 Å, β = 111.16°, Z = 2. Cu radiation, R = 0.059 for 990 reflexions.

The molecule (Fig. 1) has a pentanortriterpene structure with bond lengths, angles and torsion angles in the accepted ranges for this type of molecule. In the crystal, molecules are linked in layers perpendicular to the c axis by hydrogen bonds associated with the solvent water molecule (O(23)H...O(28) 2.61, O(28)H...O(22) 2.85, O(28)H...O(26) 2.84 Å).

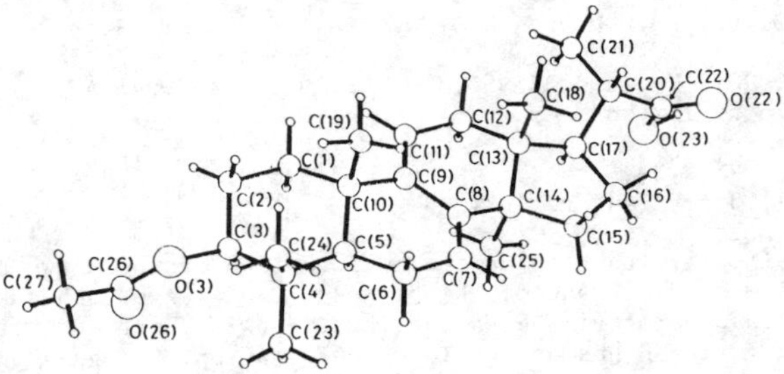

Fig. 1. $C_{27}H_{40}O_4,H_2O$: the molecular conformation.

(25R)-3β-HYDROXY-5α-SPIROSTAN-12-ONE MONOHYDRATE (HECOGENIN)
$C_{27}H_{42}O_4,H_2O$

M. SORIANO-GARCÍA, R.A. TOSCANO, I. LÓPEZ, A. HERNÁNDEZ and P. ENRÍQUEZ, 1984. Acta Cryst., C40, 2116-2118.

Orthorhombic, $P2_12_12_1$, a = 7.815, b = 9.582, c = 32.839 Å, Z = 4. Cu radiation, R = 0.044 for 1807 reflexions.

The six-membered rings (Fig. 1) have chair conformations. The substituents at C(3) and C(25) are equatorial. The five-membered rings adopt a conformation intermediate between half-chair and β-envelope. The structure consists of sheets of molecules parallel to the ac plane, linked by hydrogen bonds via a molecule of water.

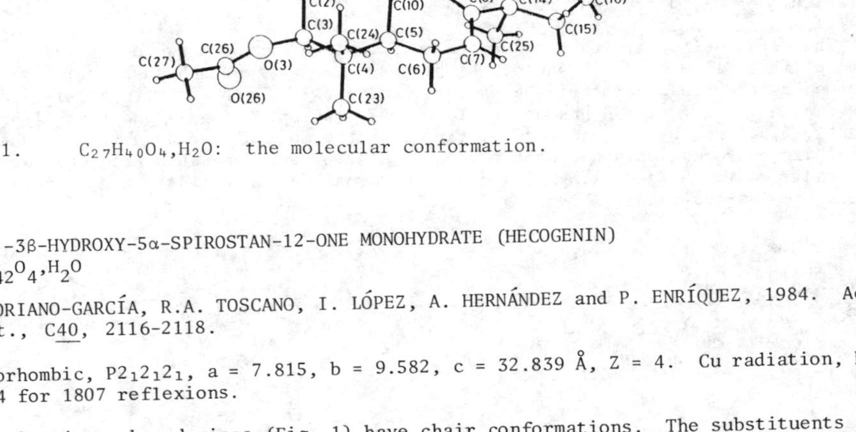

Fig. 1. $C_{27}H_{42}O_4,H_2O$.

(20R,22R)-20,22-EPOXY-3β-METHOXY-22-(PHENYLSULPHONYL)-23,24-BISNORCHOL-5-ENE
$C_{29}H_{40}O_4S$

M. ADAMCZYK, E.K. DOLENCE, D.S. WATT, M.R. CHRISTY, J.H. REIBENSPIES and O.P. ANDERSON, 1984. J. Org. Chem., 49, 1378-1382.

Monoclinic, $P2_1$, a = 9.454, b = 8.223, c = 16.993 Å, β = 100.43°, Z = 2. Mo radiation, R = 0.041 for 2259 reflexions.

The X-ray analysis established the 20R,22R stereochemistry for the molecule.

4α,5β-DIHYDROBENZO[4,5,6]CHOLEST-4-EN-5'(6'H)-ONE
$C_{30}H_{48}O$ (I)

4α,5α-DIHYDROBENZO[4,5,6]CHOLEST-4-EN-5'(6'H)-ONE
$C_{30}H_{48}O$ (II)

J.R. BULL, L.M. STEER and P.H. VAN ROOYEN, 1984. J. Chem. Soc. Perkin I, 397-404.

I. Orthorhombic, $P2_12_12_1$, a = 27.957, b = 12.091, c = 7.943 Å, Z = 4. Cu radiation, R = 0.068 for 2076 reflexions.

II. Orthorhombic, $P2_12_12_1$, a = 27.706, b = 12.608, c = 7.578 Å, Z = 4. Mo radiation, R = 0.110 for 1444 reflexions.

The effect of the fifth ring on the conformation of these two pentacyclic steroids (Fig. 1) was investigated. Rings A and B of I are both substantially deformed by the presence of ring E, A to a hybrid twist boat and B to a twist conformation. In I ring E has an envelope conformation. Rings A and B of II are slightly flattened by the presence of ring E in the molecule, ring A to a slight half-chair conformation and ring B more so. Ring E itself has a hybrid envelope-half-chair conformation. Rings C and D have normal conformations. Bond lengths and angles are unexceptional.

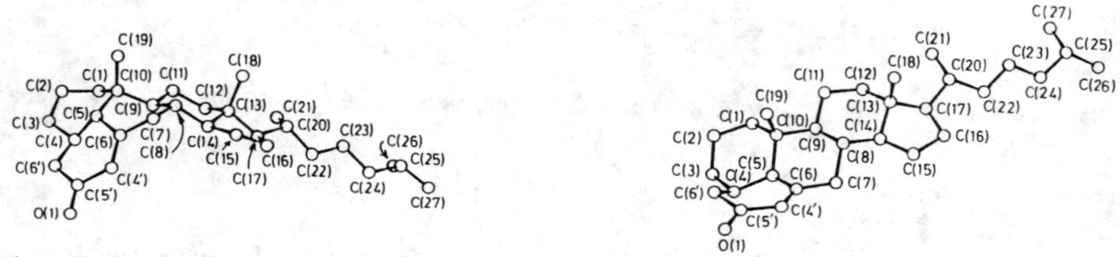

Fig. 1. $C_{30}H_{48}O$: isomer I (left) and isomer II (right).

D-1α,2β,3α,4β-TETRA(TRIMETHYLSILYL)-(10)-OESTRENE
$C_{30}H_{60}Si_4$

T.G. SHISHOVA, Z.I. KHAZHEEVA, V.I. SIMONOV, O.N. MINAILOVA and K.K. PIVNITSKII, 1984. Kristallografiya, 29, 274-278 [Sov. Phys. Crystallogr., 29, 162-165].

Orthorhombic, $P2_12_12_1$, a = 19.597, b = 17.960, c = 10.043 Å, Z = 4. Cu radiation, R = 0.045 for 2083 reflexions.

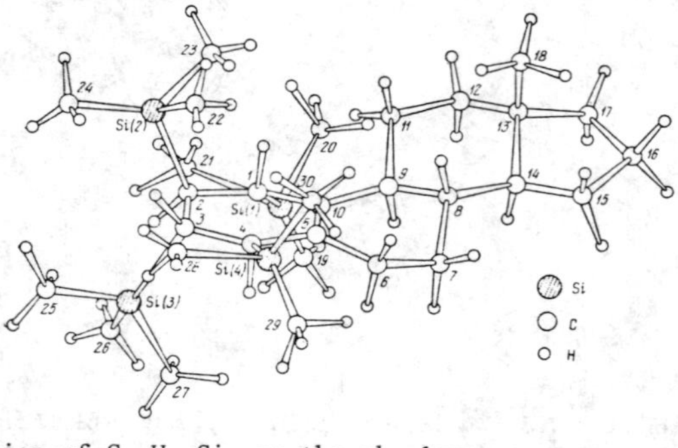

Fig. 1. Projection of $C_{30}H_{60}Si_4$ on the ab plane.

Ring A is markedly flattened (largest torsion angle 19.1°) and its conformation may be described as a flattened chair (Fig. 1). The deviations of atoms Si(1), Si(2), Si(3) and Si(4) from the best plane through atoms C(2), C(3), C(5) and C(10) are respectively -1.890, 1.492, -1.420 and 1.708 Å. Rings B and D have half-chair conformations whilst that of ring C is close to an ideal chair. The sum of the endocyclic angles in ring A is 716.6°, very close to 720°, the sum of the angles of a planar hexagon. The Si-C(ring) distances are 1.919, 1.870, 1.884 and 1.936 Å for Si(1) to Si(4) respectively. Bond C(5)-C(10) is, at 1.337 Å, a formal double bond. There are no hydrogen bonds in the crystal structure.

(3β,5β,14β)-3-[(2',6'-DIDEOXY-3',4'-o-(1'-METHYLETHYLIDENE)-β-D-ribo-HEXOPYRANOSYL)-OXY]-14-HYDROXY-CARD-20(22)-ENOLIDE
$C_{32}H_{48}O_7$ (I)

(3β,5β,14β,20E)-METHYL 14-HYDROXYPREGN-20-ENE-21-CARBOXYLATE MONOHYDRATE
$C_{23}H_{36}O_4,H_2O$ (II)

(3β,5β,14β,20E)-METHYL 3-[(2',6'-DIDEOXY-3',4'-o-(1'-METHYLETHYLIDENE)-β-D-ribo-HEXOPYRANOSYL)OXY]-14-HYDROXY-PREGN-20-ENE-21-CARBOXYLATE
$C_{32}H_{50}O_7$ (III)

(3β,5β,14β,20E)-METHYL 3-[(2',6'-DIDEOXY-β-D-ribo-HEXOPYRANOSYL)OXY]-14-HYDROXY-PREGN-20-ENE-21-CARBOXYLATE
$C_{29}H_{46}O_7$ (IV)

(3β,5β,14β,20E)-METHYL 14-HYDROXY-21-METHYLENEPREGN-21-CARBOXYLATE
$C_{24}H_{38}O_4$ (V)

(3β,5β,14β,20E)-METHYL 3-[(2',6'-DIDEOXY-3',4'-o-(1'-METHYLETHYLIDENE)-β-D-ribo-HEXOPYRANOSYL)OXY]-14-HYDROXY-21-METHYLENEPREGN-21-CARBOXYLATE
$C_{33}H_{52}O_7$ (VI)

3-[(2',6'-DIDEOXY-3',4'-o-(1'-METHYLETHYLIDENE)-β-D-ribo-HEXOPYRANOSYL)OXY](20S)-20,22-DIHYDRODIGITOXIGENIN
$C_{32}H_{50}O_7$ (VII)

D.C. ROHRER, M. KIHARA, T. DEFFO, H. RATHORE, K. AHMED, A.H.L. FROM and D.S. FULLERTON, 1984. J. Am. Chem. Soc., 106, 8269-8276.

I. Orthorhombic, $P2_12_12_1$, a = 16.025, b = 25.243, c = 7.637 Å, Z = 4. Mo radiation, R = 0.064 for 3074 reflexions.

II. Orthorhombic, $P2_12_12_1$, a = 16.5559, b = 17.999, c = 7.2666 Å. Z = 4. Cu radiation, R = 0.051 for 1675 reflexions.

III. Orthorhombic, $P2_12_12_1$, a = 11.758, b = 26.573, c = 10.131 Å, Z = 4. Cu radiation, R = 0.039 for 3418 reflexions.

IV. Orthorhombic, $P2_12_12_1$, a = 10.2354, b = 38.154, c = 7.3000 Å, Z = 4. Cu radiation, R = 0.044 for 3122 reflexions.

V. Monoclinic, $P2_1$, a = 10.628, b = 15.951, c = 6.418 Å, β = 102.12°, Z = 2. Mo radiation, R = 0.072 for 2523 reflexions.

VI. Monoclinic, $P2_1$, a = 17.621, b = 11.077, c = 8.115 Å, β = 98.60°, Z = 2. Mo radiation, R = 0.058 for 4073 reflexions.

VII. Triclinic, P1, a = 7.887, b = 15.350, c = 6.303 Å, α = 91.80, β = 99.84, γ = 93.53°, Z = 1. Cu radiation, R = 0.046 for 3000 reflexions.

The structural and conformational characteristics of a series of five cardenolide analogues, their digitoxose acetonide derivatives, and their digitoxoside derivatives are determined from analysis of crystal structure results and molecular mechanics calculations. The A, B, and C rings in the steroid backbone remain essentially conformationally invariant in all the structures, while the D rings show a high degree of flexibility. The conformational characteristics of the C17β side groups on the various analogues do not seem to be significantly affected by the nature of the

C3 substituent. The bonds linking the steroid to the sugar moieties show a surprisingly small range of rotational freedom. The C2'-C1'-O3-C3 torsion angles range only over 29.4° in 10 crystal structures, while the C1'-O3-C3-C2 torsion angles range over 110.1°.

RUBIDIUM DEOXYCHOLATE-WATER (3/10)
$C_{72}H_{117}O_{12}Rb_3,10H_2O$ $3(Rb^+,C_{24}H_{39}O_4^-),10H_2O$

A.R. CAMPANELLI, S. CANDELORO DE SANCTIS, E. GIGLIO and S. PETRICONI, 1984. Acta Cryst., C40, 631-635.

Monoclinic, $P2_1$, a = 19.834, b = 19.312, c = 11.586 Å, γ = 116.70°, D_m = 1.35, Z = 2. Cu radiation, R = 0.098 for 3057 reflexions.

In the crystal structure the molecules of rubidium deoxycholate form 2_1 helices mainly stabilised by hydrogen bonds and ion-ion and ion-dipole interactions (Fig. 1). The interior of the helix is filled with Rb^+ ions surrounded by water molecules. The helices have non-polar outer surfaces and pack together through van der Waals interactions.

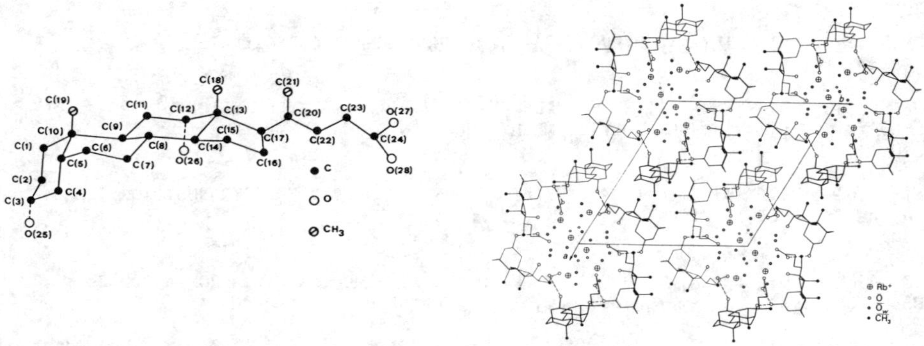

Fig. 1. $3(Rb^+,C_{24}H_{39}O_4^-),10H_2O$: perspective view of the deoxycholate anion and the c-axial projection of the crystal packing; broken lines indicate hydrogen bonds.

(1R,2S,4S,1'E)-2-BROMO-1-CHLORO-4-(2'-CHLOROETHENYL)-1-METHYL-5-METHYLENECYCLOHEXANE
$C_{10}H_{13}BrCl_2$ (I)

(1R,2S,4R,5R,1'E)-4-BROMO-1,2-DICHLORO-5-(2'-CHLOROETHENYL)-1,5-DIMETHYLCYCLOHEXANE
$C_{10}H_{14}BrCl_3$ (II)

R.J. CAPON, L.M. ENGELHARDT, E.L. GHISALBERTI, P.R. JEFFERIES, V.A. PATRICK and A.H. WHITE, 1984. Aust. J. Chem., 37, 537-544.

I. Orthorhombic, $P2_12_12_1$, a = 18.964, b = 9.820, c = 6.312 Å, Z = 4. Mo radiation, R = 0.034 for 1047 reflexions, T = 120 K.

II. Monoclinic, $P2_1$, a = 8.266, b = 11.679, c = 7.078 Å, β = 109.21°. Mo radiation, R = 0.044 for 1043 reflexions.

(I) is the major terpene metabolite of a red algae Plocamium mertensii. The cyclohexane ring is a pseudo chair with enlarged angles about C(1) (C(2)-C(1)-C(7) 113.6(5), C(6)-C(1)-C(7) 110.8(5)°). Enlarged angles are also associated with the double bond. Some distances are C(1)-Cl(1) 1.821(6), C(10)-Cl(10) 1.751(6), C(2)-Br(2) 1.976(6) Å. II is a metabolite of an unidentified Plocamium species, with the cyclohexane ring a pseudo chair with enlarged angles about C(1) (C(2)-C(1)-C(7)

115.3(7), C(6)-C(1)-C(7) 116.2(6)°), and for atoms associated with the double bond. Some distances are C(1)-Cl(1) 1.845(9), C(10)-Cl(10) 1.723(9), C(4)-Br(4) 1.962(9) Å, C(2)-Cl(2) 1.794(8) Å. Angle C(5)-C(6)-C(1) 116.4(6)° is enlarged and C(6)-C(5)-C(9) 104.2(6)° diminished. The molecules are shown in Fig. 1.

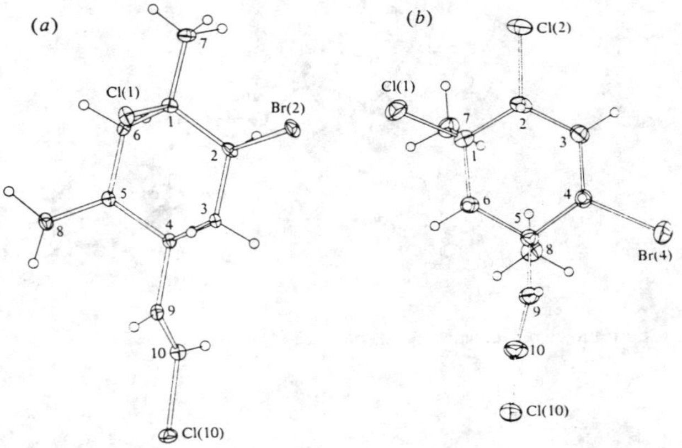

Fig. 1. The structures of (a) $C_{10}H_{13}BrCl_2$ and (b) $C_{10}H_{14}BrCl_3$.

1-(N-METHYLPHENYLSULPHONIMIDOYL)-2-[trans-2-(2-METHYL-1-PROPENYL)CYCLOPROPYL]-2-PROPANOL
$C_{17}H_{25}NO_2S$

M.R. BARBACHYN, C.R. JOHNSON and M.D. GLICK, 1984. J. Org. Chem., 49, 2746-2748.

Monoclinic, C2, a = 22.648, b = 6.619, c = 12.638 Å, β = 104.92°, D_m = 1.12, Z = [4]. Mo radiation, R = 0.079 for 918 reflexions.

The structure of this derivative of the monoterpene (-)-rothrockene is shown in Fig. 1.

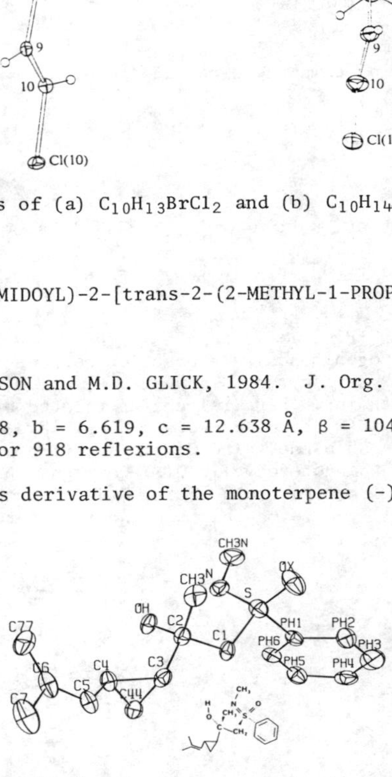

Fig. 1. The $C_{17}H_{25}NO_2S$ structure.

CAMPHANE-2-endo,3-exo-DIOL DIBENZOATE
$C_{24}H_{26}O_4$

S.J. ANGYAL, D.C. CRAIG and T.Q. TRAN, 1984. Aust. J. Chem., 37, 661-666.

Orthorhombic, $P2_12_12_1$, a = 18.883, b = 14.541, c = 7.900 Å, D_m = 1.17, Z = 4. Cu radiation, R = 0.068 for 1345 reflexions.

In 1979, doubt was cast on the correctness of previously assigned configurations for diastereoisomeric camphane-2,3-diols derived from (+)-camphor. However, an X-ray determination of a dibenzoate confirmed the structure of camphane-2-endo,3-exo-diol

(Fig. 1).

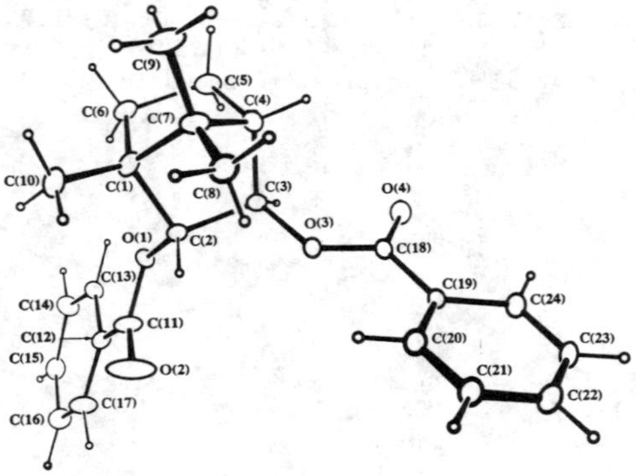

Fig. 1. The structure of camphane-2-endo,3-exo-diol.

(-)-EPICATECHIN
$C_{15}H_{14}O_6$

A.L. SPEK, B. KOJIC-PRODIC and R.P. LABADIE, 1984. Acta Cryst., C40, 2068-2071.

Orthorhombic, $P2_12_12_1$, a = 6.702, b = 13.273, c = 14.227 Å, Z = 4. Mo radiation, R = 0.052 for 623 reflexions.

The heterocyclic ring (Fig. 1) is in a half-chair conformation. All oxygen atoms are involved in hydrogen bonds. Interaction between the OH groups of dihydroxyphenyl and 3,5,7-trihydroxybenzopyran residues, O(5)-H...O(4), connects molecules along b to form infinite chains. Parallel chains related by 2_1 axes along b are connected through O(3)-H...O(2) bonds between 3,5,7-trihydroxybenzopyran residues and O(2)-H...O(5) bonds in the ab plane. The O(2) hydroxyl group also exhibits an intramolecular contact to the heterocyclic O(1) forming a bifurcated hydrogen bond with O(5) as does O(5)-H with O(6) (both of the 3,4-dihydroxyphenyl residue). A weak interaction, O(6)-H...O(1), completes this bonding in a three-dimensional network.

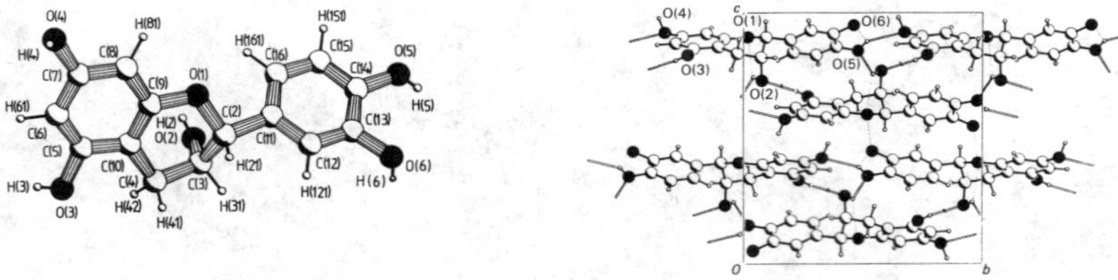

Fig. 1. The (-)-epicatechin molecule and a view of the structure down a.

DEHYDROELEMANSCHKUHRIOLIDE
$C_{15}H_{16}O_4$

G. DELGADO, H. HERNANDEZ and A. ROMO DE VIVAR, 1984. J. Org. Chem., 49, 2994-2996.

Orthorhombic, $P2_12_12_1$, a = 6.566, b = 13.843, c = 14.717 Å, Z = 4. Cu radiation, R = 0.0632 for 606 reflexions.

In the molecule (Fig. 1) the cyclohexane ring has a distorted chair form with C14 and O6 axially oriented and α to the plane of the ring; the vinyl groups are equatorial. The α-methylene γ-lactone is somewhat puckered and H7 is pseudoequatorial and α.

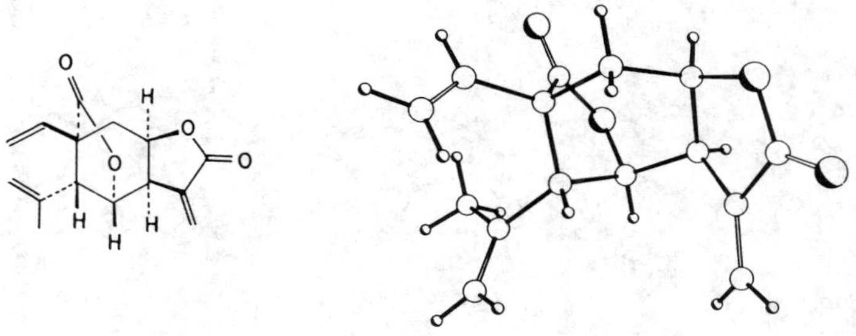

Fig. 1. Dehydroelemanschkuhriolide.

7α-HYDROXY-3-DESOXYZALUZANIN C

$C_{15}H_{18}O_3$

F.R. FRONCZEK, D. VARGAS, N.H. FISCHER and K. HOSTETTMANN, 1984. J. Nat. Prod., 47, 1036-1039.

Orthorhombic, $P2_12_12_1$, a = 14.611, b = 33.024, c = 5.420 Å, Z = 8. Mo radiation, R = 0.046 for 1536 reflexions.

The conformations of the two independent molecules differ only slightly; one is shown in Fig. 1. The molecule is shown to be a cis-fused guaianolide with an α-methylene-γ-lactone trans-fused at C6-C7 and a unique α-OH at C7. The seven-membered ring conformation is somewhat distorted from the ideal C_s (mirror) symmetry; the cyclopentane ring is in half-chair conformation. In one molecule the lactone ring is in a half-chair conformation (local twofold axis passing through C12 and bisecting C6-C7; in the other molecule the lactone ring is intermediate between half-chair (C12 on the axis) and envelope (with C7 at the flap). Molecules are linked into spirals by O-H...O hydrogen bonds.

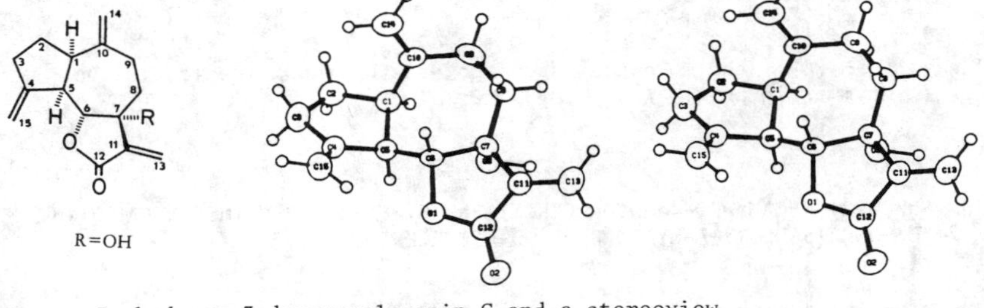

Fig. 1. 7α-hydroxy-3-desoxyzaluzanin C and a stereoview.

SUBCORDATOLIDE B

$C_{15}H_{18}O_4$

A.G. OBER, F.R. FRONCZEK and N.H. FISCHER, 1984. J. Nat. Prod., 47, 626-629.

Orthorhombic, $P2_12_12_1$, a = 8.756, b = 10.926, c = 14.020 Å, Z = 4. Mo radiation, R = 0.040 for 1026 reflexions.

The structure is established as shown in Fig. 1; an α-methylene-γ-lactone ring is trans-fused to the eudesmane skeleton at C7-C8. The cyclohexene ring has C_2

symmetry, the cyclohexane ring has a chair conformation and the non-planar lactone has close to a half-chair conformation.

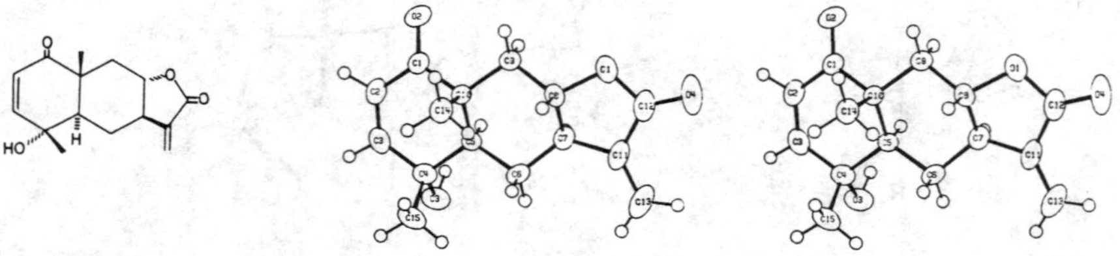

Fig. 1. Subcordatolide B and a stereoview.

5α-HYDROXY-10β-SELINA-1,4(15),7-TRIEN-6-ONE (STEMONOLONE)
$C_{15}H_{20}O_2$

A.P. PHADNIS, S.A. PATWARDHAN, A.S. GUPTA, K.R. ACHARYA, S.S. TAVALE and T.N. GUBUROW, 1984. J. Chem. Soc. Perkin I, 937-939.

Monoclinic, $P2_1$, a = 11.782, b = 5.961, c = 19.363 Å, β = 103.89°, Z = 4. Mo radiation, R = 0.077 for 1312 reflexions.

There are two molecules in the asymmetric unit of this sesquiterpene isolated from 'Pogostemon plectranthoides' (Fig. 1). Both six-membered rings adopt half-chair conformations. The angle between the mean-plane of each ring is 11.8° indicating a slightly folded overall conformation. The crystal is stabilised by intermolecular hydrogen bonds between similar molecules (O(1)...O(2) 3.01(1) Å).

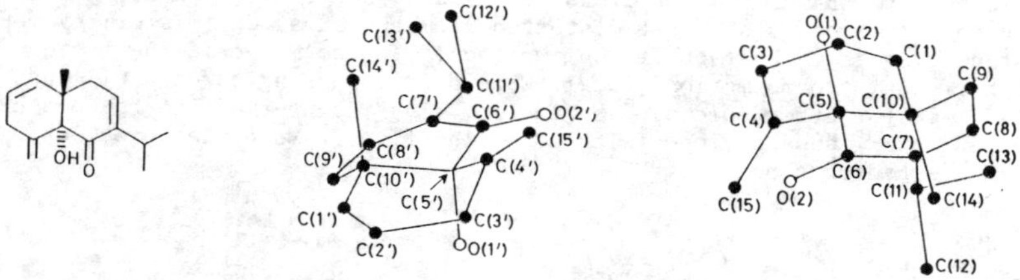

Fig. 1. $C_{15}H_{20}O_2$: the molecular skeleton and the two molecules in the asymmetric unit.

(3aR,4aS,8aR,9aR)-DECAHYDRO-4a-HYDROPEROXY-8a-METHYL-3,5-DIMETHYLENENAPHTHO[2,3-b]-FURAN-2(3H)-ONE (5-DEOXY-5-HYDROPEROXY-5-EPITELEKIN)
$C_{15}H_{20}O_4$ (I)

(3aR,4aS,8aR,9aR)-DECAHYDRO-4a-HYDROXY-8a-METHYL-3,5-DIMETHYLENENAPHTHO[2,3-b]FURAN-2(3H)-ONE (5-EPITELEKIN)
$C_{15}H_{20}O_3$ (II)

G. APPENDINO, M. CALLERI, G. CHIARI and D. VITERBO, 1984. J. Chem. Soc. Perkin II, 903-907.

I. Orthorhombic, $P2_12_12_1$, a = 8.165, b = 12.082, c = 13.991 Å, Z = 4. Mo radiation, R = 0.043 for 1160 reflexions.

II. Orthorhombic, $P2_12_12_1$, a = 9.677, b = 11.365, c = 12.109 Å, Z = 4. Cu radiation, R = 0.066 for 1188 reflexions.

In both compounds (Fig. 1) the cis-decalin system shows a chair-chair conformation of the steroid-like type in I, and the non-steroid-like type in II, the substituents at C(5) and C(10) being in a β-orientation. The γ-lactone ring, cis-fused at C(8) is puckered in both molecules and shows an envelope conformation. The O(3)-O(4) peroxy bond length in I is 1.464(3) Å. Other bond lengths and angles are normal. Packing of molecules in I and II is similar with spirals of intermolecular hydrogen bonds along the z-axis, O(4)...O(2') 2.815(2) (I), O(3)...O(2) 2.819(2) Å (II).

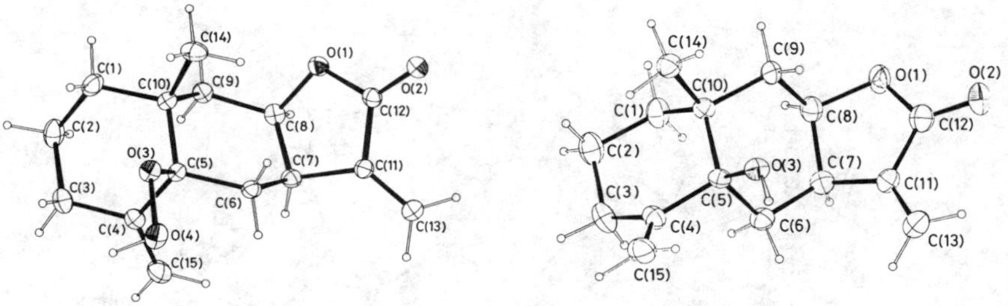

Fig. 1. The molecular structures: $C_{15}H_{20}O_4$ (left) and $C_{15}H_{20}O_3$ (right).

CARYOPHYLLEN-9β,10-OLIDE (PARVINOLIDE)
$C_{15}H_{22}O_2$

B. NANDA, S.A. PATWARDHAN, A.S. GUPTA, K.R. ACHARYA, N.N. DHANESHWAR, S.S. TAVALE and T.N.G. ROW, 1984. J. Chem. Res., (S) 394, (M) 3721-3742.

Orthorhombic, $P2_12_12_1$, a = 7.854, b = 9.783, c = 17.258 Å, Z = 4. Mo radiation, R = 0.076 for 1239 reflexions.

This sesquiterpene is shown to have the stereochemistry indicated in Fig. 1.

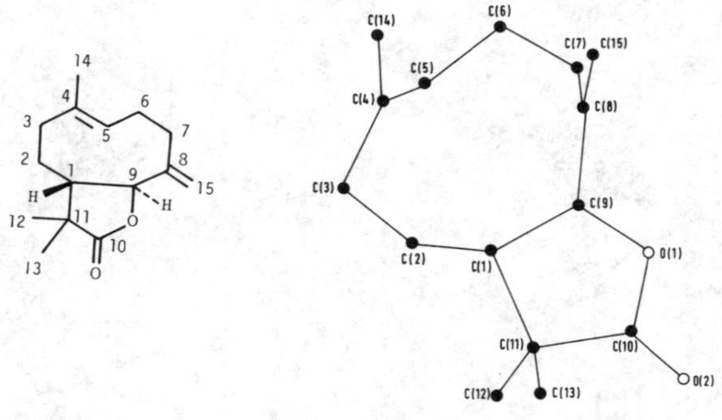

Fig. 1. Parvinolide: structure and stereochemistry.

LACTARORUFIN D
$C_{15}H_{22}O_4$

W.M. DANIEWSKI, A. WAWRZUN, M. DE BERNARDI, G. VIDARI, P. VITA-FINZI, G. FRONZA and G. GATTI, 1984. Tetrahedron, 40, 2757-2762.

Monoclinic, $P2_1$, a = 17.825, b = 6.091, c = 6.765 Å, β = 93.81°, Z = 2. Mo radiation, R = 0.0543 for 695 reflexions.

Lactarorufin D (Fig. 1) has a lactarane skeleton with the five-membered ring A fused to the seven-membered ring in a cis-manner. In ring C the C=O is at C-5 and in ring B CH_3-12, OH-4 and OH-8 are cis to the bridge-head protons H-2 and H-9. The carbons C-3, C-4, C-6, C-7, C-8 and C-2, C-3, C-8, C-9 lie approximately in two planes (maximum deviation from the least squares plane is 0.04 and 0.02 Å, respectively) forming a dihedral angle of 123.04° between each other. Thus the cycloheptene ring B exists in the hinge conformation with CH_3-12 and OH-8 in a quasi-equatorial and OH-4 in a quasi-axial orientation. The lactone ring C is essentially planar and the five-membered ring A has the ^{11}E conformation.

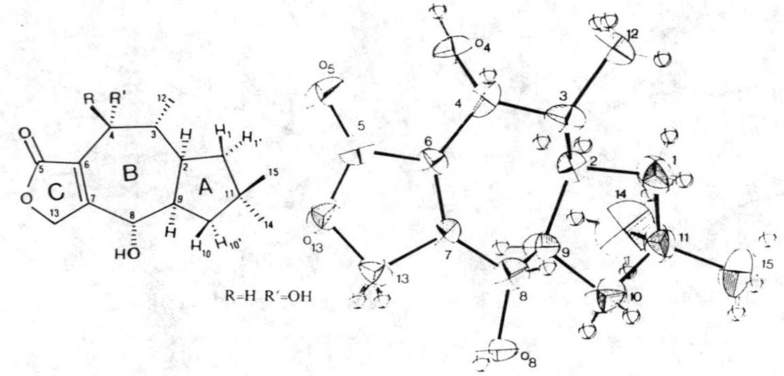

Fig. 1. Lactarorufin D.

(6R,7S)-6,11-CYCLOGERMACRA-1(10)E,4E-DIEN-14-OL ((+)-ISOBICYCLOGERMACRENOL) $C_{15}H_{24}O$

A. MATSUO, H. NOZAKI, N. KUBOTA, S. UTO and M. NAKAYAMA, 1984. J. Chem. Soc. Perkin I, 203-214.

Orthorhombic, $P2_12_12_1$, a = 6.457, b = 16.919, c = 25.936 Å, Z = 8. Mo radiation, R = 0.079 for 1911 reflexions.

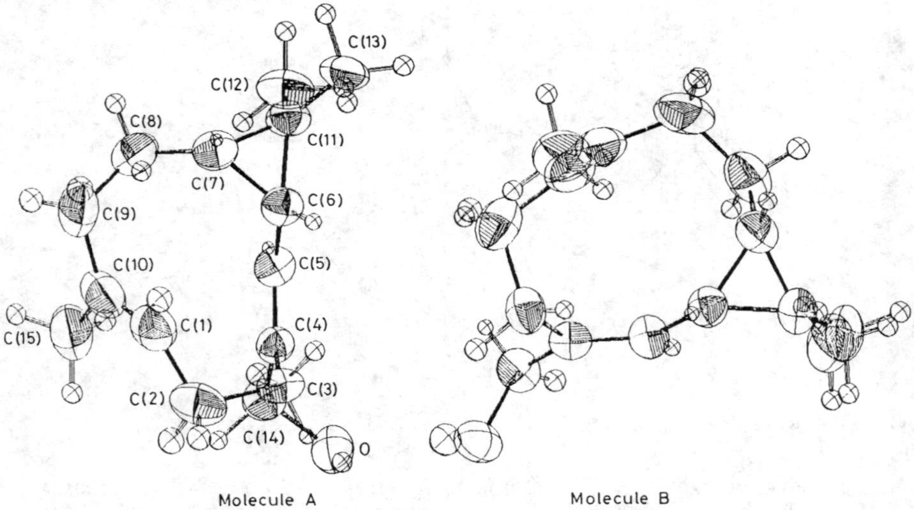

Fig. 1. $C_{15}H_{24}O$: the two independent molecules.

The structure of the title compound was established by the X-ray analysis. The two crystallographically independent molecules (A, B) in the asymmetric unit (Fig. 1) have the same chirality and almost identical conformation. The mean lengths of C-C bonds in the cyclodecadiene ring are $C(sp^3)$-$C(sp^3)$ 1.526 (1.526), $C(sp^3)$-$C(sp^2)$ 1.483 (1.495), $C(sp^2)$-$C(sp^2)$ 1.313 (1.314) Å for A (B). The mean bond length in the cyclopropane ring is 1.523 (1.574) Å. The C(1)-C(10) and C(4)-C(5) double bonds are trans and cis respectively. The ten-membered ring adopts a chair-twist conformation. In the crystal, two independent molecules in the asymmetric unit are linked by intermolecular hydrogen bonding through the hydroxy groups (OA...HOB 1.865 Å); this unit is further linked with others by the same bond around the 2_1 screw axis along the 'a' direction, i.e. the hydroxy groups tend to congregate in one region and the aliphatic rings in the other.

3,3,11-TRIMETHYL-7-METHYLENETRICYCLO[6.3.0.0^{2,4}]UNDECANE-5,11-DIOL
$C_{15}H_{24}O_2$

M. CREVOISIER, K.C. STEUDLE and H.B. BUERGI, 1984. Acta Cryst., C40, 979-980.

Orthorhombic, $C222_1$, a = 9.541, b = 15.115, c = 19.412 Å, D_m = 1.12, Z = 8. Mo radiation, R = 0.043 for 1092 reflexions.

In the molecule (Fig. 1) the five- and seven-membered carbocycles are trans fused; O(2) is cis relative to H(1) at C(1). The seven-membered ring shows the stable chair conformation. The C(1)-C(11) bond length of 1.566(4) Å is probably the result of the eclipsing of five non-hydrogen and one hydrogen substituents; the average length of the remaining four C-C bonds in the five-membered ring is 1.522(9) Å. The molecules are linked in the crystal by O-H...O hydrogen bonds with O(1)...O(2') 2.811(4) and O(2)...O(1'') 2.817(4) Å.

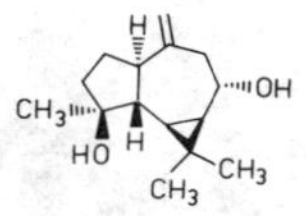

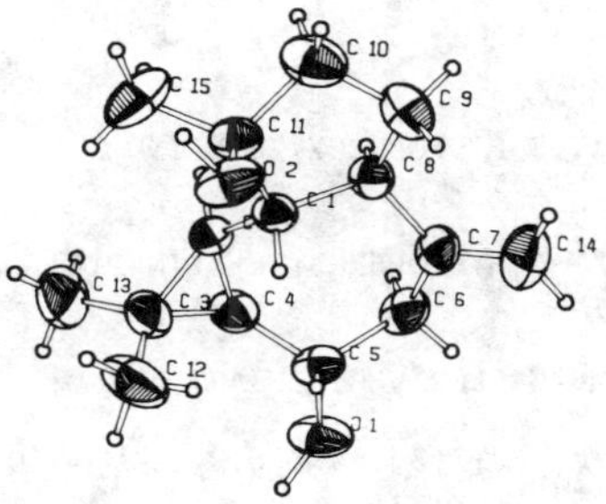

Fig. 1. $C_{15}H_{24}O_2$: a view of the molecule.

4-HYDROXY-10-EPI-EUDESM-11-ENE
$C_{15}H_{26}O$

R. BAKER, A.J. ORGAN, S.A. WALMSLEY, M. WEBSTER and A.M.R. GALAS, 1984. J. Chem. Res., (S) 138-139, (M) 1401-1429.

Monoclinic, A2, a = 20.105, b = 7.794, c = 21.423 Å, β = 122.68°, D_m = 0.99, Z = 8. Cu radiation, R = 0.0502 for 1584 reflexions.

The crystals contain two distinct but similar molecules differing only in the torsion angle of the isoprenyl group (Fig. 1). The cis-fused cyclohexane rings have chair conformation with an axial OH group.

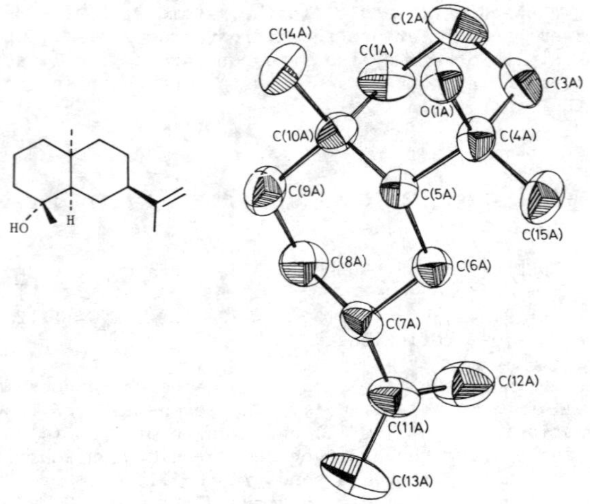

Fig. 1. $C_{15}H_{26}O$: a view of one molecule.

O-METHYL-5-DEOXY-5-HYDROPEROXYTELEKIN
$C_{16}H_{22}O_4$ (I)

O-METHYL-5-DEOXY-5-HYDROPEROXYTELEKIN PYRAZOLINE ADDUCT
$C_{17}H_{24}N_2O_4$ (II)

G. APPENDINO, M. CALLERI, G. CHIARI and D. VITERBO, 1984. Acta Cryst., C40, 97-100.

I. Orthorhombic, $P2_12_12_1$, a = 6.953, b = 10.108, c = 21.211 Å, Z = 4. Mo radiation, R = 0.040 for 1327 reflexions.

II. Monoclinic, $P2_1$, a = 6.278, b = 12.522, c = 10.885 Å, β = 95.68°, Z = 2. Mo radiation, R = 0.042 for 1106 reflexions.

In both compounds the trans-decalin system has chair-chair conformation and the cis-fused γ-lactone ring has a puckered conformation of the half-chair type with C(7) above and C(8) below the plane through O(1),C(11),C(12) (Fig. 1). In II, the pyrazoline ring is almost orthogonal to the lactone moiety. The peroxide group in II shows positional disorder. The O-O bond lengths are 1.478(2) Å (I) and 1.521(6), 1.505(6) Å in (II).

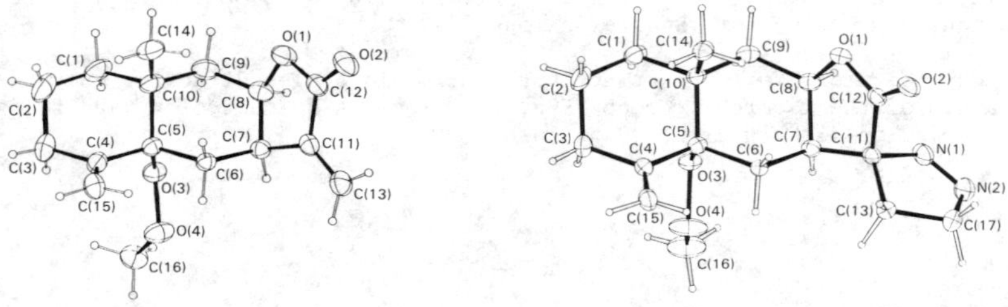

Fig. 1. The $C_{16}H_{22}O_4$ (I) (left) and $C_{17}H_{24}N_2O_4$ (II) (right) molecules.

ARTEMETHER
$C_{16}H_{26}O_5$ (I)

DIHYDROQINGHAOSU
$C_{15}H_{24}O_5$ (II)

XUAN-DE LUO, H.J.C. YEH, A. BROSSI, J.L. FLIPPEN-ANDERSON and R. GILARDI, 1984. Helv. Chim. Acta, 67, 1515-1522.

I. Monoclinic, $P2_1$, a = 9.868, b = 18.324, c = 10.172 Å, β = 117.55°, Z = 4. Mo radiation, R = 0.095 for 1432 reflexions.

II. Orthorhombic, $P2_12_12_1$, a = 5.659, b = 14.270, c = 18.954 Å, Z = 4. Mo radiation, R = 0.113 for 1182 reflexions.

Compound I is the methyl ether of II and both have the same relative conformation. The non-bridged six-membered rings are both in normal chair conformations while the six-membered ring formed by the O-O bond has a somewhat distorted boat conformation.

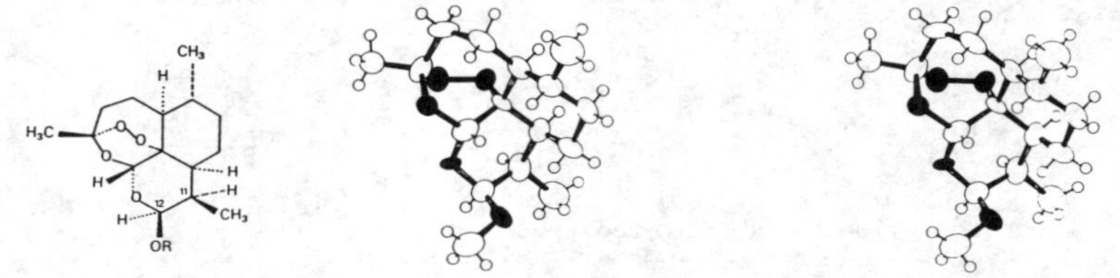

Fig. 1. The structure of artemether (I) (R = CH_3) and a stereoview. In II, R = H.

3-α-ACETOXY-7α,15-DIHYDROXY-12,13-EPOXYTRICHOTHEC-9-EN-8-ONE
$C_{17}H_{22}O_7$

R. GREENHAUGH, A.W. HANSON, J.D. MILLAR and A. TAYLOR, 1984. J. Agric. Food Chem., 32, 945-948.

Tetragonal, I4, a = 18.099, c = 10.631 Å, Z = 8. Cu radiation, R = 0.055 for 1269 reflexions.

The structure and relative stereochemistry are shown in Fig. 1. The molecules are held in the lattice by hydrogen bonds and this leads to the crystals containing channels parallel to the c-direction. The channel radius ranges from 3.35 Å to 4.25 Å.

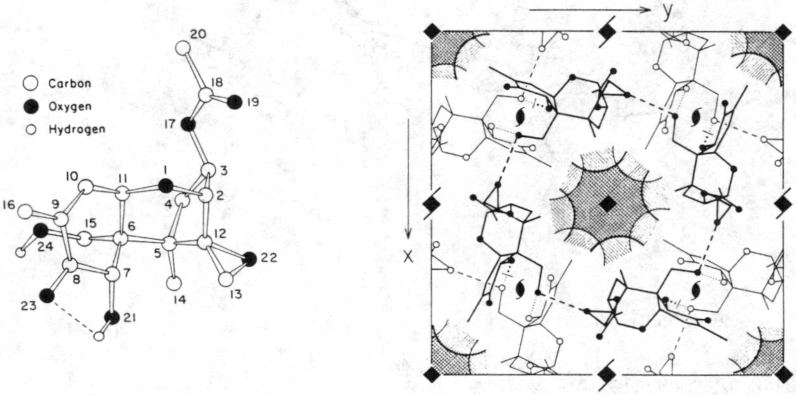

Fig. 1. A view of $C_{17}H_{22}O_7$ and the unit-cell contents projected along c.

SUBCORDATOLIDE C
$C_{19}H_{26}O_5$

A.G. OBER, F.R. FRONCZEK and N.H. FISCHER, 1984. J. Nat. Prod., 47, 920-923.

Orthorhombic, $P2_12_12_1$, a = 6.387, b = 9.980, c = 27.992 Å, Z = 4. Mo radiation, R = 0.041 for 1401 reflexions.

The α-methylene-γ-lactone ring is cis fused to the eudesmane skeleton at C7-C8 (Fig. 1). Both six-membered rings have chair conformations. The lactone ring has a half-chair conformation (with the local twofold axis passing through C12 and bisecting C7-C8).

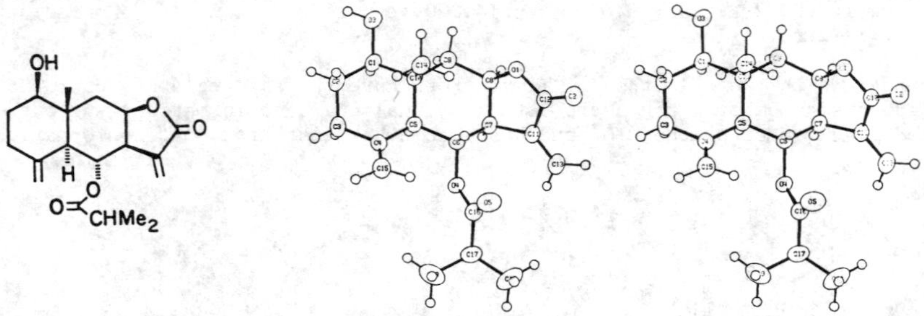

Fig. 1. Subcordatolide C and a stereoview.

8β-ANGELOYLOXY-9α-HYDROXYCALYCULATOLIDE
$C_{20}H_{24}O_7$

N.H. FISCHER, IHL-YOUNG LEE, F.R. FRONCZEK, G. CHIARI and L.E. URBATSCH, 1984. J. Nat. Prod., 47, 419-425.

Orthorhombic, $P2_12_12_1$, a = 14.466, b = 14.491, c = 8.883 Å, Z = 4. Cu radiation, R = 0.049 for 1763 reflexions.

The structure is established as shown in Fig. 1. The lactone ring is planar. There is disorder of the ester group at C(8) especially with terminal carbon C(20).

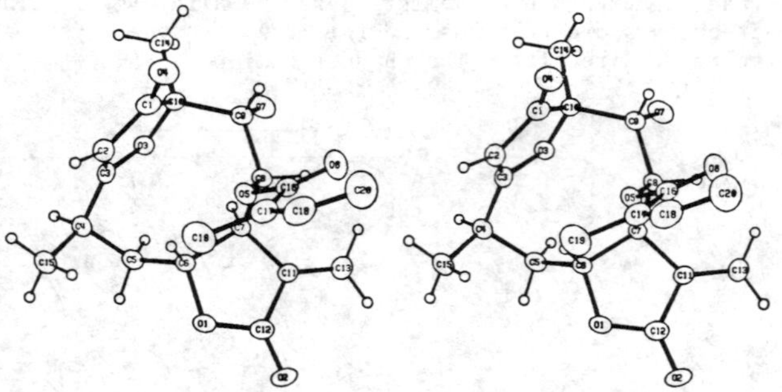

Fig. 1. A stereoview of $C_{20}H_{24}O_7$.

ACANTHODORAL (p-BROMOPHENYL)URETHANE
$C_{22}H_{30}BrNO_2$ (I)

ISOACANTHODORAL DINITROPHENYLHYDRAZONE
$C_{21}H_{32}N_4O_4$ (II)

S.W. AYER, R.J. ANDERSEN, C.-H. HE and J. CLARDY, 1984. J. Org. Chem., 49, 2653-2654.

I. Monoclinic, $P2_1$, a = 9.581, b = 6.406, c = 34.45 Å, β = 85.55°, Z = 2. Cu radiation, R = 0.078 for 2365 reflexions.

II. Monoclinic, $P2_1$, a = 6.027, b = 31.613, c = 8.910 Å, β = 80.89°.

The structures of these derivatives of sesquiterpenoid aldehydes are shown in Fig. 1.

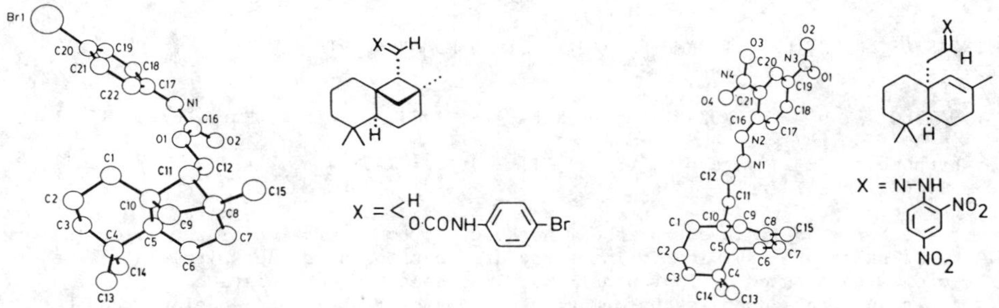

Fig. 1. The acanthodoral (left) and isoacanthadoral (right) derivatives.

3β-ACETOXY-1β-HYDROXY-4α,8α-DIANGELOYLOXYDAUC-5-ENE (ACETYLTINGITANOL) $C_{27}H_{40}O_7$

M. MISKI, A. ULUBELEN, T.J. MABRY, W.H. WATSON, I. VICKOVIC and M. HOLUB, 1984. Tetrahedron, 40, 5197-5201.

Orthorhombic, $P2_12_12_1$, a = 10.922, b = 23.074, c = 10.809 Å, Z = 4. Cu radiation, R = 0.048 for 2115 reflexions.

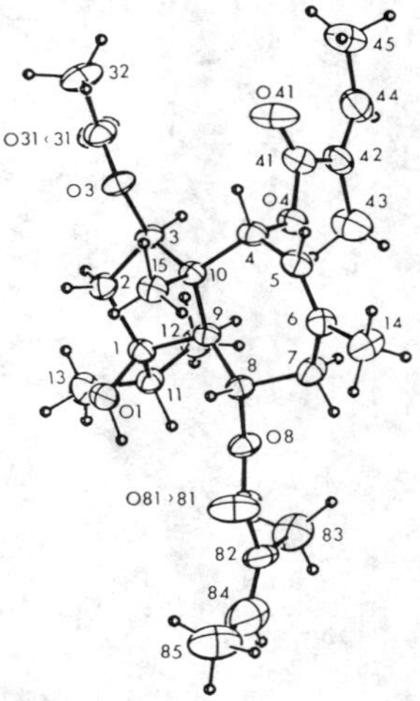

Fig. 1. An Ortep drawing of acetyltingitanol.

The five- and seven-membered rings (Fig. 1) are trans-fused with the five-membered ring in an envelope and the seven in a chair conformation. The conformation of the five-membered ring results in steric interactions between the substituents, O(3)...C(15) 2.768(7), O(3)...H(15a) 2.43(4) and O(1)...C(15) 3.171(7) Å. Each angeloyloxy moiety and the ring carbon to which it is attached is nearly planar. The two groups differ only in the orientation about the C-O bond with C(4) and O(41) being in a cis arrangement and O(8) and C(8) in a trans arrangement. The molecules are loosely packed in the crystal and the angeloyl groups exhibit high thermal motion.

(1R,4S,5S,6R,7S,10R)-VITRANE-5,14-DIYL DI-p-BROMOBENZOATE
$C_{29}H_{32}Br_2O_4$

A. MATSUO, S. UTO, H. NOZAKI and M. NAKAYAMA, 1984. J. Chem. Soc. Perkin I, 215-221.

Orthorhombic, $P2_12_12_1$, a = 9.568, b = 11.290, c = 24.814 Å, Z = 4. Mo radiation, R = 0.067 for 1170 reflexions.

The X-ray analysis established the absolute configuration of the molecule (Fig. 1), a plant growth inhibitor. The structure contains a unique spiro[4.5]decane skeleton with a fused cyclopropane ring. The mean C-C bond distance is 1.546 Å. The cyclopentane ring adopts a conformation between half chair and envelope and the cyclohexane ring adopts a deformed chair conformation. The cyclopropane ring is perpendicular to the cyclohexane ring.

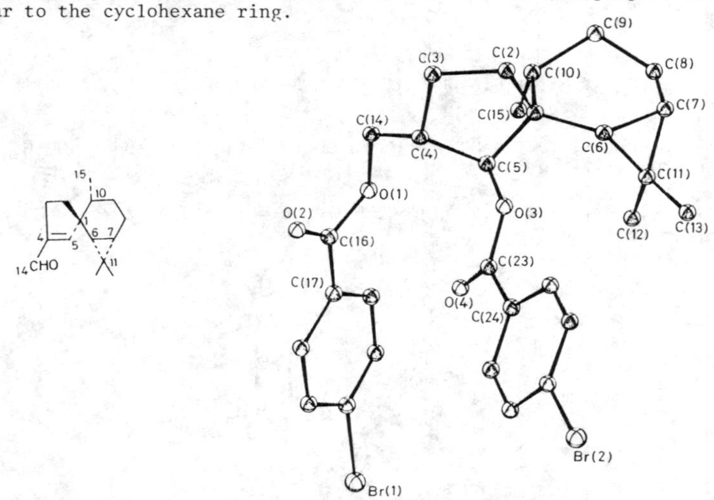

Fig. 1. $C_{29}H_{32}Br_2O_4$: the molecular skeleton and structure.

MORTONIN C
$C_{29}H_{32}O_8$

L. RODRIGUEZ-HAHN, M. MARTINEZ, M. MORA, C.R.E. DIAZ, M. SORIANO-GARCIA and H. CAMPANA, 1984. Tetrahedron, 40, 1743-1748.

Orthorhombic, $P2_12_12_1$, a = 11.372, b = 13.012, c = 18.396 Å, D_m = 1.315, Z = 4. Cu radiation, R = 0.056 for 2819 reflexions.

Analysis of Mortonin C (Fig. 1) was performed to confirm a previously assigned structure based on chemical and spectroscopic evidence.

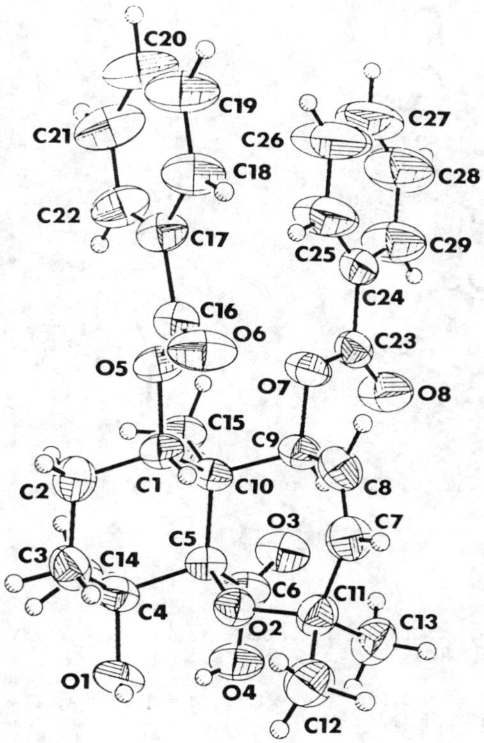

Fig. 1. A view of Mortonin C.

4A-HYDROXY-1-[2-(METHOXYCARBONYL)ETHYL]-1,2,5,5-TETRAMETHYLDECAHYDRONAPHTHALENE (12-METHOXYCARBONYLAMBLIOL C)
$C_{18}H_{32}O_3$ (I)

4A-HYDROXY-1-(2-CARBOXYETHYL)-1,2,5,5-TETRAMETHYLDECAHYDRONAPHTHALENE (12-CARBOXY-AMBLIOL B)
$C_{17}H_{30}O_3$ (II)

R.P. WALKER, R.M. ROSSER, D.J. FAULKNER, L.S. BASS, HE CUN-HENG and J. CLARDY, 1984. J. Org. Chem., 49, 5160-5163.

I. Monoclinic, $P2_1$, a = 7.678, b = 9.256, c = 12.345 Å, β = 102.01°, Z = 2. Mo radiation, R = 0.071 for 1278 reflexions.

II. Orthorhombic, $C222_1$, a = 7.944, b = 18.411, c = 50.325 Å, Z = 8. Cu radiation, R = 0.074 for 1689 reflexions.

The molecules are derivatives of ambliol-C and ambliol-B respectively. There are two independent molecules of (II) in the asymmetric unit and both have the same configuration and conformation (Fig. 1). The molecules differ principally in the orientation of the methyl group at C8 (axial in (I) and equatorial in (II)). The decalin rings are trans-fused with the bridgehead hydrogen at C10 in the α-configuration and the bridgehead hydroxyl at C5 in the β-configuration.

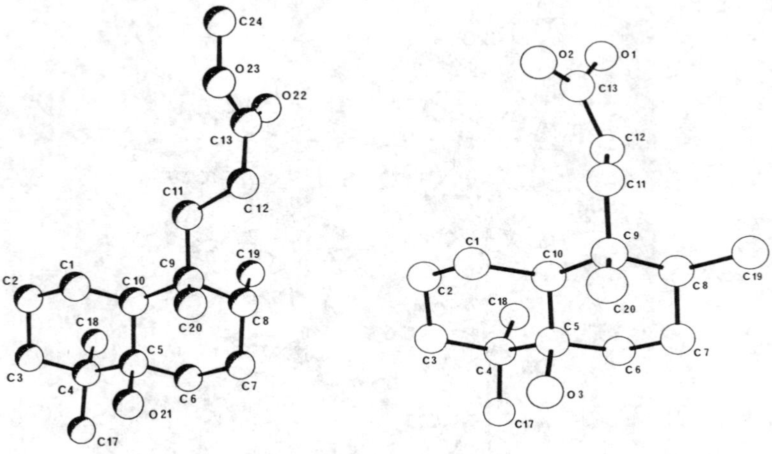

Fig. 1. Views of $C_{18}H_{32}O_3$ (I) (left) and $C_{17}H_{30}O_3$ (II) (right).

CROTONIN
$C_{19}H_{24}O_4$

J.F. BLOUNT, W.R. CHAN, J. CLARDY, P.S. MANCHAND and J.O. PEZZANITE, 1984. J. Chem. Res., (S) 114-115, (M) 1201-1221.

Orthorhombic, $P2_12_12_1$, a = 10.603, b = 12.019, c = 13.251 Å, Z = 4. Cu radiation, R = 0.041 for 1249 reflexions.

The structure and absolute stereochemistry of crotonin were determined as shown in Fig. 1.

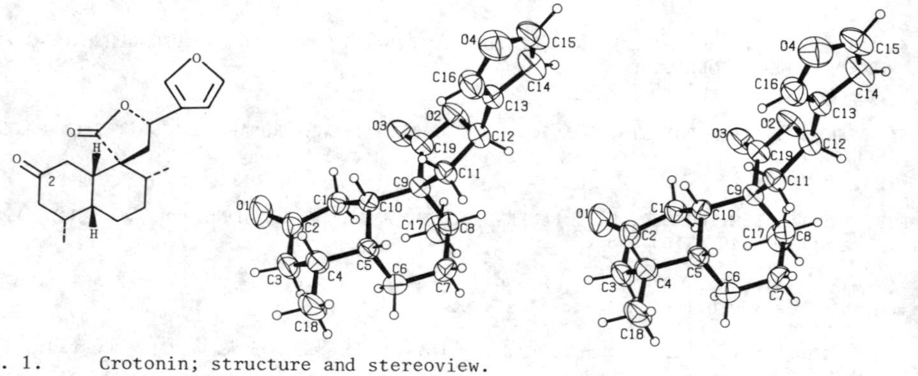

Fig. 1. Crotonin; structure and stereoview.

(4βH,6βH,11αH)-3β,10β-EPOXY-8β-ISOBUTYRYLOXY-1-OXOGERMACR-2-EN-6,12-OLIDE (TETRAHYDRO-ZEXBREVIN)
$C_{19}H_{26}O_6$

M. SORIANO-GARCÍA and R.A. TOSCANO, 1984. Acta Cryst., C40, 1425-1427.

Orthorhombic, $P2_12_12_1$, a = 9.060, b = 14.080, c = 14.628 Å, Z = 4. Cu radiation, R = 0.063 for 1208 reflexions.

The cyclodecene ring has a chair-boat conformation with the C(4) and C(10) methyl groups anti and is quasi-trans fused at C(6) and C(7) to a γ-lactone ring with a half-chair conformation (Fig. 1). The 3(2H)-furanone ring is a flattened envelope with O(6) as the flap. The isobutyrate group is almost perpendicular to the ten-membered ring with C=O syn to the C(8)-H bond. There is a short O(4)...H(6) intramolecular contact of 2.20 Å and the possibility of a weak C(7)-H(7)...O(6) interaction.

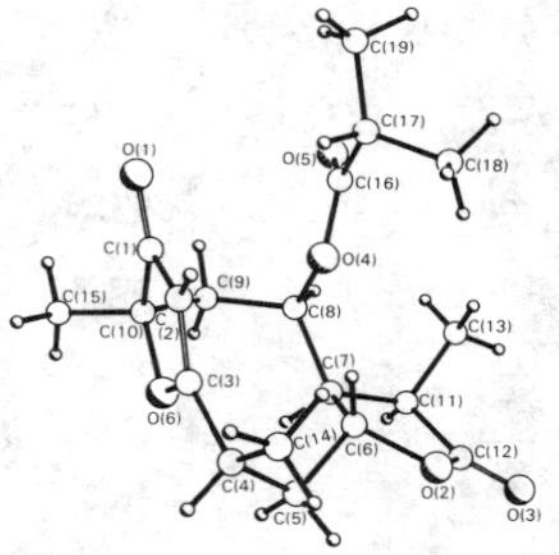

Fig. 1. The $C_{19}H_{26}O_6$ molecule.

PODOLACTONE C
$C_{20}H_{22}O_8S$

J.M. CASSADY, T.K. LIGHTNER, T.G. McCLOUD, J.A. HEMBREE, S.R. BYRN and C.-J. CHANG, 1984. J. Org. Chem., 49, 942-945.

Orthorhombic, $P2_12_12_1$, a = 7.597, b = 11.794, c = 22.330 Å, Z = 4. Cu radiation, R = 0.0582 for 1588 reflexions.

The analysis shows (Fig. 1) that the molecule has a 2,3-β-epoxide moiety.

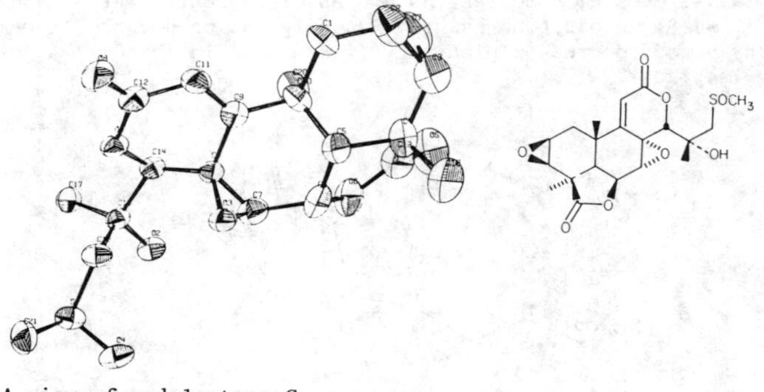

Fig. 1. A view of podolactone C.

ent-16α-CHLORO-10β-HYDROXY-13-METHYL-17,20-BISNOR-8,13-ISOGIBBERELLANE-7,19-DIOIC ACID 19-10β-LACTONE 7-METHYL ESTER
$C_{20}H_{27}ClO_4$ (I)

ent-16β-HYDROXY-17-NORGIBBERELL-2-ENE-7,19,20-TROIC ACID 7,19,20-TRIMETHYL ESTER
$C_{22}H_{30}O_7$ (II)

Z.J. DURI, J.R. HANSON and P.B. HITCHCOCK, 1984. J. Chem. Soc. Perkin I, 1099-1103.

I. Orthorhombic, $P2_12_12_1$, a = 9.032, b = 11.663, c = 17.910 Å, Z = 4. Mo radiation, R = 0.046.

II. Monoclinic, $P2_1$, a = 9.033, b = 13.197, c = 9.704 Å, β = 115.10°, Z = 2. Cu radiation, R = 0.042.

The X-ray analyses established the stereochemistries of the two molecules.

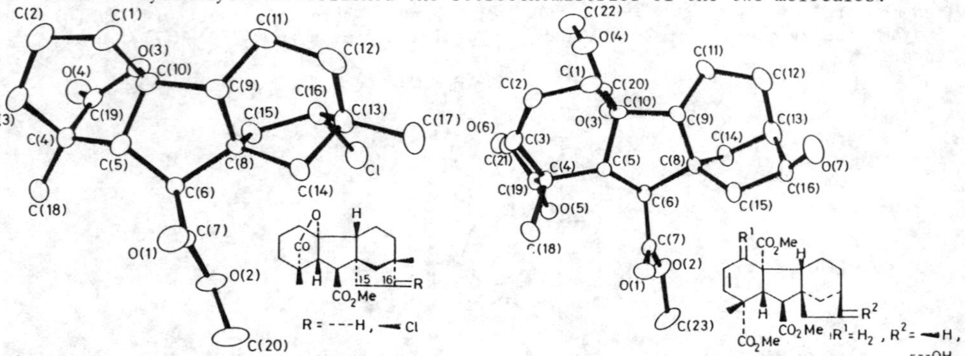

Fig. 1. $C_{20}H_{27}ClO_4$ (left) and $C_{22}H_{30}O_7$ (right): the molecular skeletons and structures.

12,13-BISEPIEUPALMERIN
$C_{20}H_{30}O_4$

Y. GOPICHAND, L.S. CIERESZKO, F.J. SCHMITZ, D. SWITZNER, A. RAHMAN, M.B. HOSSAIN and D. VAN DER HELM, 1984. J. Nat. Prod., 47, 607-614.

Monoclinic, $P2_1$, a = 6.1551, b = 8.3476, c = 18.202 Å, β = 96.11°, Z = 2. Cu radiation, R = 0.050 for 1969 reflexions (at -135°C).

The chemical structure, bond lengths and absolute configuration are shown in Fig. 1. The C(7)-C(8) double bond has the E-configuration and the 14-membered ring is characterised by a good degree of planarity (maximum deviation 0.77 Å for C(5)). The γ-lactone ring is planar.

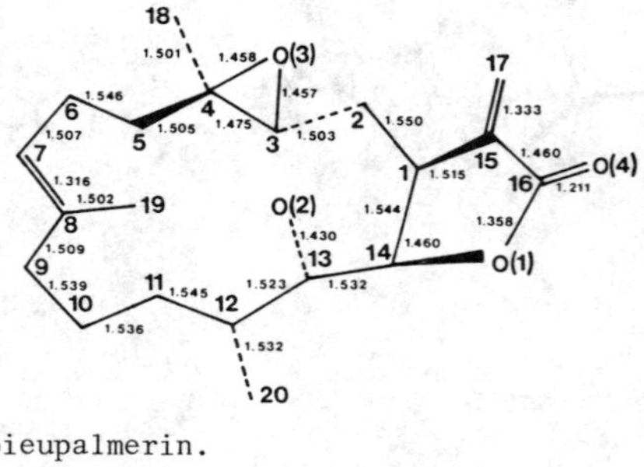

Fig. 1. 12,13-Bisepieupalmerin.

CALYCULONE A
$C_{20}H_{32}O_2$

S.A. LOOK, W. FENICAL, ZHENG QI TAI and J. CLARDY, 1984. J. Org. Chem., 49, 1417-1423.

Monoclinic, $P2_1$, a = 8.699, b = 10.771, c = 10.436 Å, β = 104.21°, Z = 2. Cu radiation, R = 0.044 for 1361 reflexions.

The analysis established the relative stereochemistry to be as shown in Fig. 1.

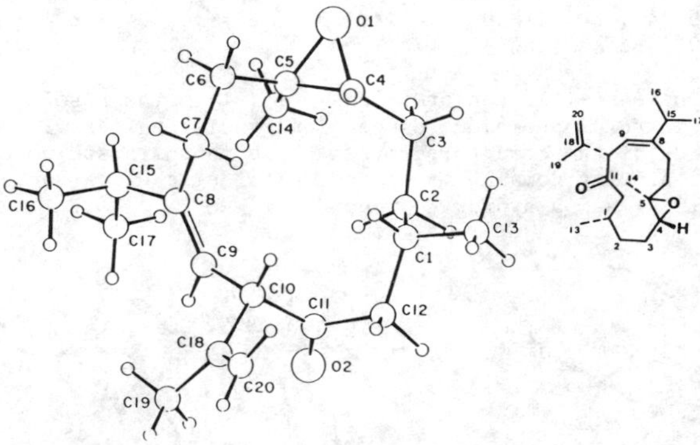

Fig. 1. Relative stereochemistry for calyculone A.

(-)-2(S),8(R)-DIHYDROXYVERRUCOSANE
$C_{20}H_{34}O_2$

I. KUBO, A. MATSUMOTO, K. HIROTSU, H. NAOKI and W.F. WOOD, 1984. J. Org. Chem., 49, 4644-4646.

Orthorhombic, $P2_12_12_1$, a = 18.515, b = 15.490, c = 6.327 Å, Z = 4. Mo radiation, R = 0.038 for 1451 reflexions.

Ring B (Fig. 1) has a 1,2 diplanar conformation (C1,C2,C3,C5,C6 coplanar, C7 0.73 Å from the plane). Ring C is in a slightly distorted chair form, and ring D has an envelope conformation with C10 as the flap.

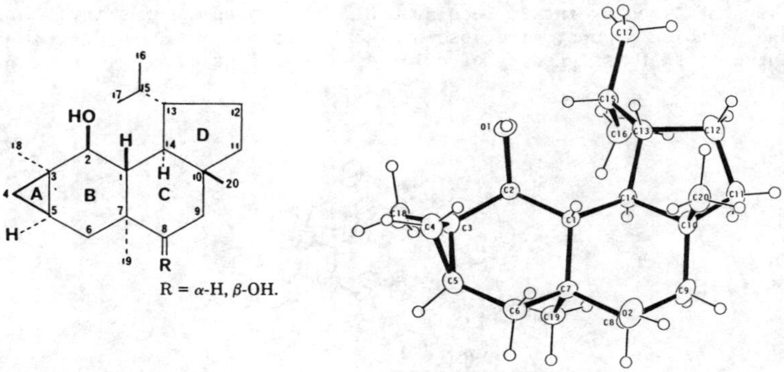

Fig. 1. A perspective drawing of $C_{20}H_{34}O_2$.

GRAYANOTOXIN III HYDRATE
$C_{20}H_{34}O_6$,2.5(H_2O) (I)

α-DIHYDROGRAYANOTOXIN II
$C_{20}H_{34}O_5$ (II)

P.W. CODDING, 1984. J. Am. Chem. Soc., 106, 7905-7909.

I. Monoclinic, C2, a = 21.941, b = 8.762, c = 18.504 Å, β = 143.50°, D_m = 1.23, Z = 4. Mo radiation, R = 0.068 for 2587 reflexions (at -100°C).

II. Monoclinic, $P2_1$, a = 9.749, b = 6.860, c = 14.670 Å, β = 107.64°, Z = 2. Cu radiation, R = 0.039 for 1932 reflexions.

The two grayanotoxins have similar conformations (Fig. 1). The absolute configuration determined by careful anomalous dispersion studies agrees with that deduced chemically. Features of the structures match those found in sodium channel neurotoxins. Two potential binding domains are found, a triangle of oxygen atoms (with O...O 2.8-2.9 and ∿4.9 Å) and a hydroxyl group.

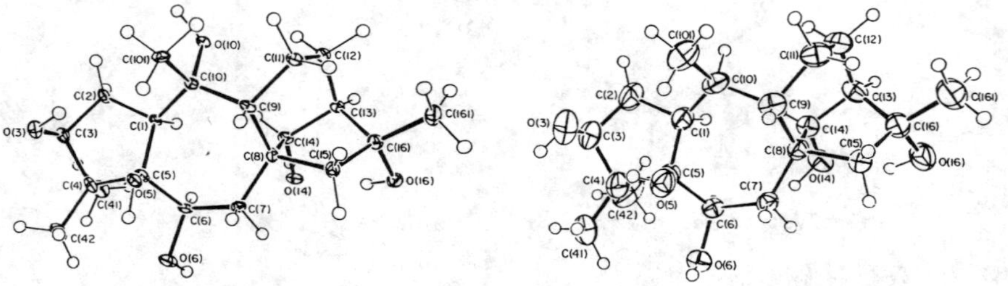

Fig. 1. Views of grayanotoxin III (left) and dihydrograyanotoxin II (right).

THUJAPLICATIN METHYL ETHER
$C_{21}H_{24}O_7$

M.R. CAIRA, W.H. WATSON and M. SILVA, 1984. Acta Cryst., C40, 299-301.

Monoclinic, $P2_1$, a = 11.715, b = 10.815, c = 7.411 Å, β = 92.66°, D_m = 1.365, Z = 2. Cu radiation, R = 0.033 for 1314 reflexions.

The lactone ring (Fig. 1) has a twist conformation (torsion angle C-C-C-O 27.6(3)°), with the substituent C atoms at the 2- and 3-position trans. The crystal structure is stabilized by intermolecular O-H...O hydrogen bonding involving the carbonyl and phenolic groups. The absolute configuration was also determined and is in agreement with that assigned by circular dichroism studies.

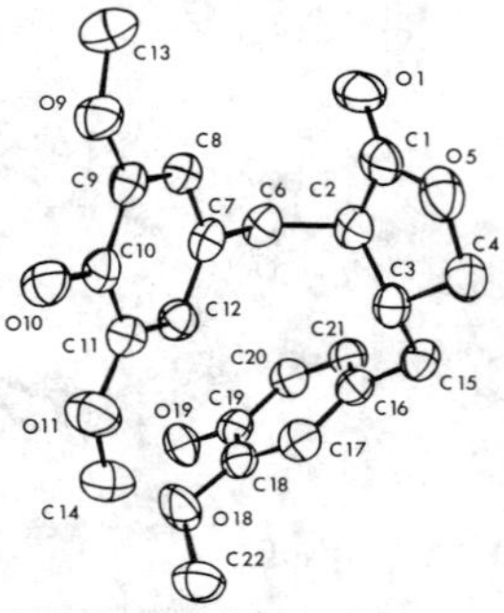

Fig. 1. A view of thujaplicatin methyl ether.

THURFERIN MONOHYDRATE
$C_{21}H_{34}O_4,H_2O$

I. VICKOVIC, W.H. WATSON, M. SILVA and M. BITTNER, 1984. Acta Cryst., C40, 1756-1758.

Monoclinic, $P2_1$, a = 7.131, b = 21.310, c = 7.033 Å, β = 110.93°, Z = 2. Cu radiation, R = 0.046 for 1257 reflexions.

The compound (Fig. 1) is a kaurane type diterpene which contains a methyl hemiacetal functionality at C(17). The A and B rings are in regular chair conformations while ring C (C(8)C(9)C(11)C(12)C(13)C(14)) is slightly distorted due to the fusion of the five-membered ring (a 1-3 ethylene bridge). The five-membered ring is in an envelope conformation with C(14) being the flap. The bond lengths and valence angles of the diterpene system are normal.

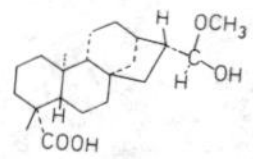

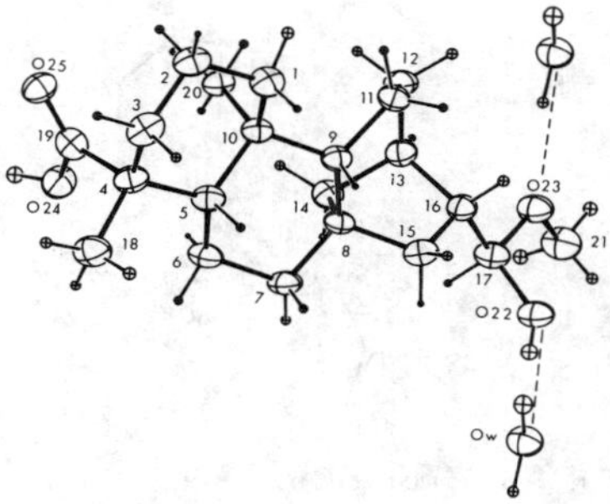

Fig. 1. The thurferin skeleton and an ORTEP plot.

KALIHINOL-A
$C_{22}H_{33}ClN_2O_2$

C.W.J. CHANG, A. PATRA, D.M. ROLL, P.J. SCHEUER, G.K. MATSUMOTO and J. CLARDY, 1984. J. Am. Chem. Soc., 106, 4644-4646.

Orthorhombic, $P2_12_12_1$, a = 6.899, b = 16.3897, c = 19.3295 Å, Z = 4. Cu radiation, R = 0.0643 for 1451 reflexions.

There are three six-membered rings in the structure (Fig. 1); two cyclohexane rings in the trans-decalin and a tetrahydropyran. The conformations of each of these rings is best described as a chair. There are two isonitriles: one in an axial position at C-14 and one equatorial at C-9. There is an equatorial chlorine substituent at C-2. In general bond distances and angles agree well with generally accepted values.

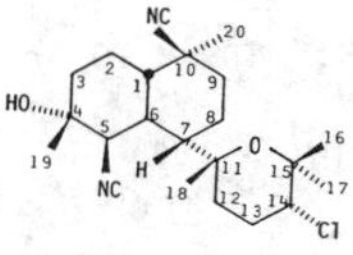

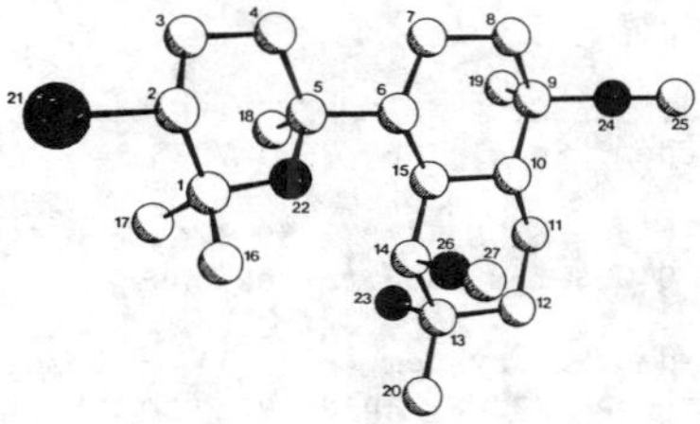

Fig. 1. Kalihinol-A (hydrogens omitted).

9-ACETOXYMETHYL-10-CHLORO-6,7-EPOXY-9-HYDROXY-5-(1-HYDROXY-1-METHYLETHYL)-2-METHYL-13-METHYLENE-TRICYCLO[8.4.0.0^{2,6}]TETRADECANE
$C_{22}H_{35}ClO_5$

XIA ZONGXIANG and ZHANG ZHIMING, 1984. J. Struct. Chem., 3, 29-32.

Orthorhombic, $P2_12_12_1$, a = 12.722, b = 14.726, c = 12.255 Å, Z = 4. Mo radiation, R = 0.052 for 2254 reflexions.

This acetylated derivative of a new diterpene from soft coral has the structure shown in Fig. 1.

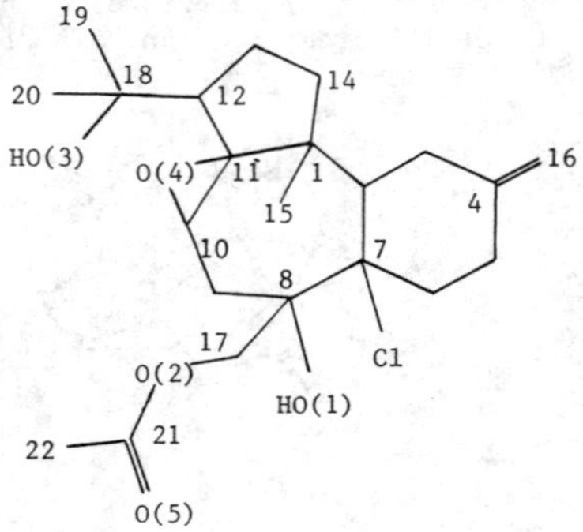

Fig. 1. The $C_{22}H_{35}ClO_5$ molecule.

DIVINORIN A (SALVINORIN A)
$C_{23}H_{28}O_8$

L.J. VALDES, W.M. BUTLER, G.M. HATFIELD, A.G. PAUL and M. KOREEDA, 1984. J. Org. Chem., 49, 4716-4720.

Orthorhombic, $P2_12_12_1$, a = 6.369, b = 11.366, c = 30.747 Å, Z = 4. Mo radiation, R = 0.087 for 1376 reflexions.

A redetermination of a structure reported previously (and called salvinorin A) (1).

1. Structure Reports, 49B, 668.

KALIHINOL-F
$C_{23}H_{33}N_3O_2$

A. PATRA, C.W.J. CHANG, P.J. SCHEUER, G.D. VAN DUYNE, G.K. MATSUMOTO and J. CLARDY, 1984. J. Am. Chem. Soc., 106, 7981-7983.

Monoclinic, $P2_1$, a = 6.5835, b = 35.676, c = 10.363 Å, β = 108.02°, Z = 2. Cu radiation, R = 0.078 for 2834 reflexions.

The trans-decalin system (Fig. 1) is described as two boats, with an axial isocyano substituent at C5, an axial hydroxy at C4 and an axial methyl at C10; the substituent at C7 is equatorial. The tetrahydrofuran ring is in envelope (C_s) form with C11 as the flap.

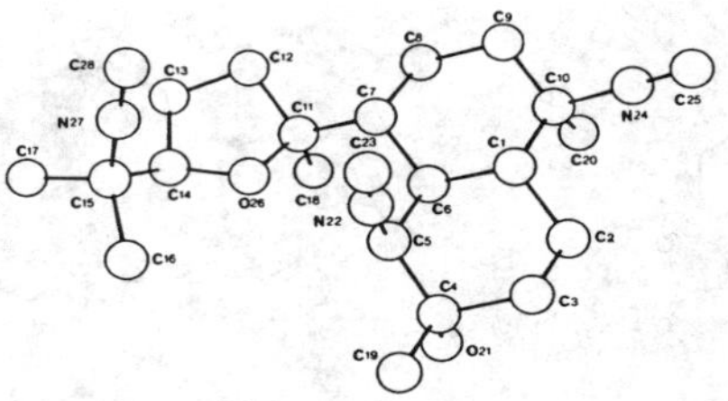

Fig. 1. A view of the relative stereochemistry of Kalihinol F with H atoms omitted.

MONTANIN C
$C_{24}H_{30}O_8$

J. FAYOS, F. FERNANDEZ-GADEA, C. PASCUAL, A. PERALES, F. PIOZZI, M. RICO, B. RODRIGUEZ and G. SAVONA, 1984. J. Org. Chem., 49, 1789-1793.

Monoclinic, $P2_1$, a = 16.748, b = 11.777, c = 12.634 Å, β = 111.78°, Z = 4. Cu radiation, R = 0.066 for 2368 reflexions.

There are two independent molecules in the asymmetric unit; both are chemically identical and have very similar conformations (Fig. 1) differing mainly in the orientation of the furan ring D.

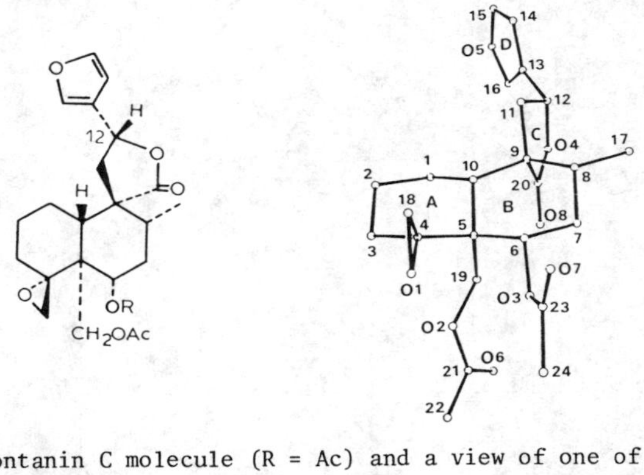

Fig. 1. The Montanin C molecule (R = Ac) and a view of one of the two independent molecules.

BRIANTHEN
$C_{24}H_{32}O_6$

J.H. CARDELLINA, T.R. JAMES, M.H.M. CHEN and J. CLARDY, 1984. J. Org. Chem., 49, 3398-3399.

Orthorhombic, $P2_12_12_1$, a = 8.783, b = 9.486, c = 28.231 Å, Z = 4. Cu radiation, R = 0.072 for 1212 reflexions.

The analysis established the structure and relative stereochemistry of brianthen (Fig. 1).

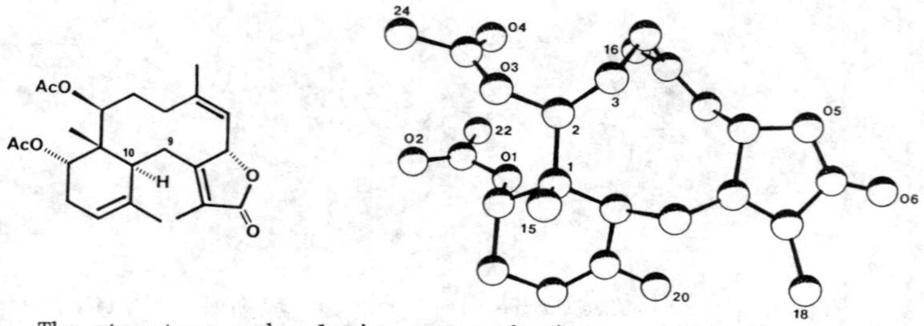

Fig. 1. The structure and relative stereochemistry of brianthen.

2-HYDROXY-1,1-DIMETHYLETHYLAMMONIUM (2E,1'S,3a'R,4'R,6'R,7a'S)-5-(7a'-HYDROXY-1',5',-5'-TRIMETHYLOCTAHYDRO-3a',6'-ETHANO-3a'H-INDEN-4'-YL)-2-METHYLPENT-2-ENOATE
$C_{24}H_{43}NO_4$

K.D. CROFT, E.L. GHISALBERTI, P.R. JEFFERIES, T.A. MORI, B.W. SKELTON and A.H. WHITE, 1984. Aust. J. Chem., 37, 785-793.

Trigonal, $P3_1$, a = 16.01, c = 8.150 Å, Z = 3. Mo radiation, R = 0.085 for 457 reflexions.

The absolute configuration of this eremane diterpene isolated from Eremophila cuneifolia Kraenzlin (Fig. 1) is taken to be that of (+)-isoeremolactone (1).

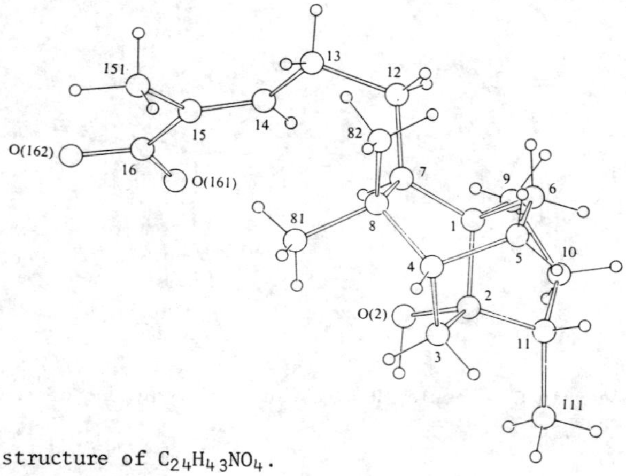

Fig. 1. The structure of $C_{24}H_{43}NO_4$.

1. Yow-Lam OH and E.N. MASLEN, 1966. Tetrahedron Lett., 3291.

(+)-8β-ACETOXY-12-(4-BROMOBENZOYLOXY)-13,14,15,16-TETRANORLABDANE
$C_{25}H_{35}BrO_4$

G. BERNARDINELLI, A. DUNAND, H.D. FLACK, K. YVON, W. GIERSCH and G. OHLOFF, 1984. Acta Cryst., C40, 1911-1914.

Orthorhombic, $P2_12_12_1$, a = 7.3547, b = 11.0209, c = 30.025 Å, Z = 4. Mo radiation, R = 0.046 for 2519 reflexions.

The absolute configuration of the molecule is shown in Fig. 1. Both, trans-fused, six-membered rings have the chair conformation.

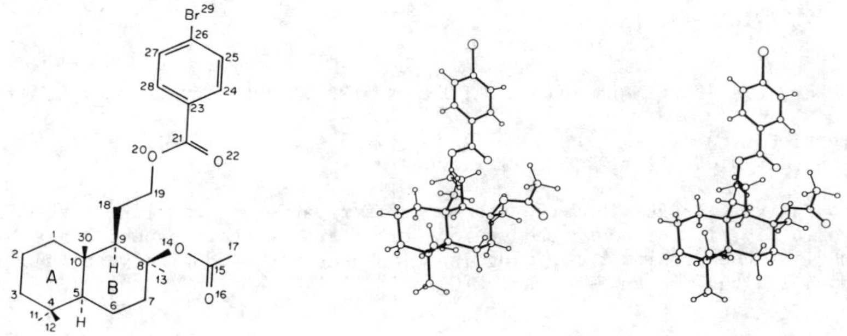

Fig. 1. The $C_{25}H_{35}BrO_4$ molecule and a stereoview.

WAIXENICIN-B
$C_{26}H_{36}O_8$

S.J. COVAL, P.J. SCHEUER, G.K. MATSUMOTO and J. CLARDY, 1984. Tetrahedron, 40, 3823-3828.

Monoclinic, $P2_1$, a = 8.545, b = 8.166, c = 18.721 Å, β = 94.26°, Z = 2. Cu radiation, R = 0.051 for 1853 reflexions.

The relative stereochemistry of this diterpene from coral is shown in Fig. 1.

Fig. 1. Waixenicin-B.

O-CINNAMOYLTAXICIN-I TRIACETATE
$C_{30}H_{42}O_{10}$

M.J. BEGLEY, E.A. FRECKNALL and G. PATTENDEN, 1984. Acta Cryst., C40, 1745-1747.

Monoclinic, $P2_1$, a = 9.151, b = 9.321, c = 19.539 Å, β = 91.06°, Z = 2. Cu radiation, R = 0.0478 for 2133 reflexions.

The analysis establishes the stereochemistry and conformation of the derivative as shown in Fig. 1. Although the various rings are in their expected stable conformations the strain at the ring junctions was revealed by several unusually large bond lengths and angles (max. C-C 1.593(7) Å).

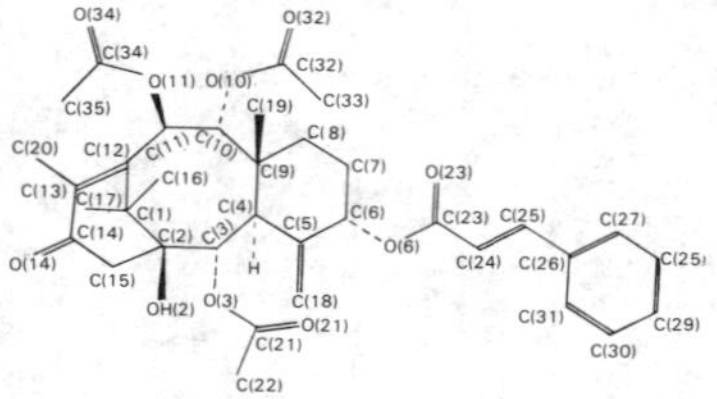

Fig. 1. The structure of O-cinnamoyltaxicin-I triacetate.

CHUANLIANSU TRIHYDRATE
$C_{30}H_{38}O_{11}, 3(H_2O)$

GUO FANG, HOU YONGGENG, ZHU NAIJUE and FU HENG, 1984. J. Struct. Chem., 3, 91-94.

Triclinic, P1, a = 7.396, b = 9.382, c = 13.671 Å, α = 115.92, β = 75.98, γ = 114.93°, D_m = 1.352, Z = 1. Mo radiation, R = 0.060 for 1897 reflexions.

In this triterpene derivative (Fig. 1), rings A and B are in distorted chair conformations, ring C has a twist boat and ring D an envelope form (C17 as flap). Dimensions are as anticipated. Molecules are linked through the water molecules by O-H...O hydrogen bonds.

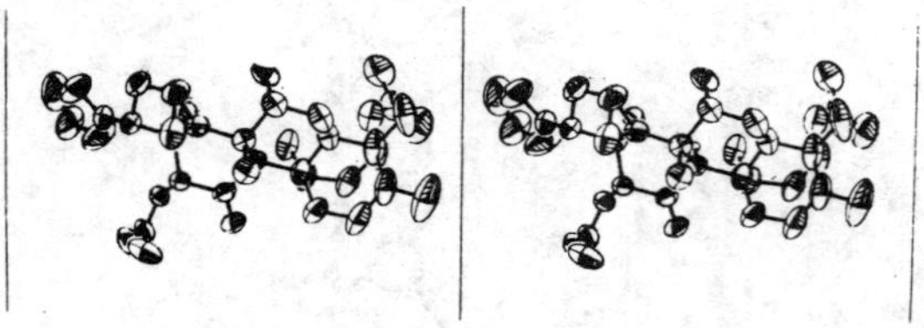

Fig. 1. A stereoview of chuanliansu.

3β,23-DIHYDROXY-LUPIN-20(29)-ENE-28-CARBOXYLIC ACID (ANEMOSUPOGENIN)
$C_{30}H_{48}O_4$

LIU SHIXIONG and HUANG JINLING, 1984. J. Struct. Chem., 3, 225-230.

Orthorhombic, $P2_12_12_1$, a = 12.595, b = 13.245, c = 15.923 Å, Z = 4. Mo radiation, R = 0.047 for 3195 reflexions.

In the molecule (Fig. 1) all four six-membered rings are in chair conformation. The molecules are linked by O-H...O hydrogen bonds.

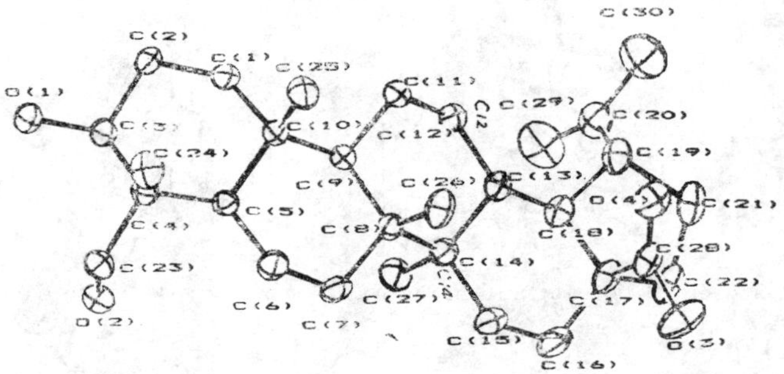

Fig. 1. A view of $C_{30}H_{48}O_4$.

19αH-LUPEOL METHYL ETHER
$C_{31}H_{52}O$

R.A. PAUPTIT, J.M. WATERS, D.D. ROWAN, G.B. RUSSELL, H.E. CONNOR and A.W. PURDIE, 1984. Aust. J. Chem., 37, 1341-1347.

Monoclinic, $P2_1$, a = 8.071, b = 14.5477, c = 11.6061 Å, β = 104.466°, Z = 2. Cu radiation, R = 0.054 for 1092 reflexions.

This compound was isolated from Chicnochloa bromoides (Hook. f.) Zotov.

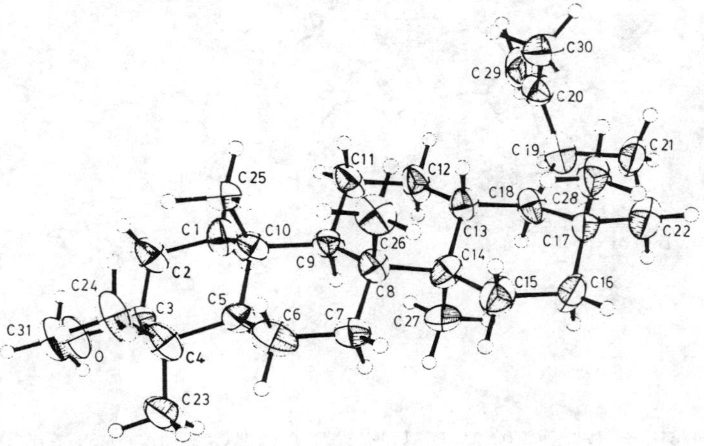

Fig. 1. The structure of 19-αH-lupeol methyl ether.

(20S,24R)-METHYL SHOREATE
$C_{31}H_{52}O_4$

D. LAVIE, F. FROLOW and H. MESHULAM, 1984. Tetrahedron, 40, 419-420.

Orthorhombic, $P2_12_12_1$, a = 7.818, b = 12.458, c = 29.366 Å, Z = 4. Mo radiation, R = 0.07 for 2531 reflexions.

The analysis establishes the configuration shown in Fig. 1.

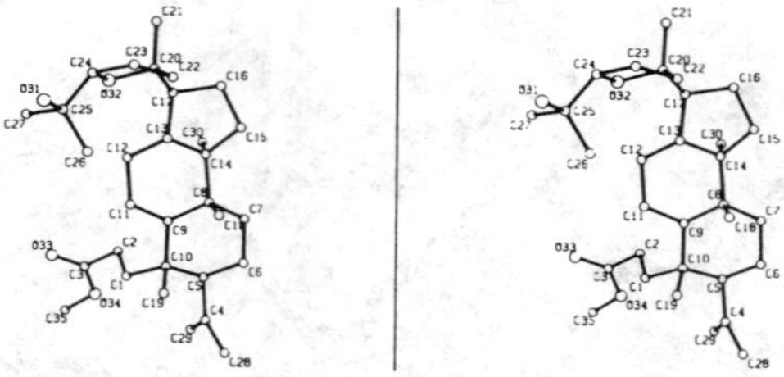

Fig. 1. A stereoview of methyl shoreate.

19-NOR-9β-METHYL-10α-2β,3α,16α,20,25-PENTAHYDROXY-Δ^5-LANOSTADIENE-11,22-DIONE-25-ACETATE

$C_{32}H_{50}O_8$

CHEN BENMING and JIANG DAZHI, 1984. J. Struct. Chem., 3, 73-78.

Monoclinic, $P2_1$, a = 13.099, b = 7.954, c = 15.446 Å, β = 108.97°, Z = 2. Mo radiation, R = 0.072 for 1726 reflexions.

The molecule (Fig. 1) has a bowed conformation with rings A and C in chair form, and rings B and D in half-chair form. Molecules are linked in the solid by O-H...O hydrogen bonds (2.798 and 2.995 Å).

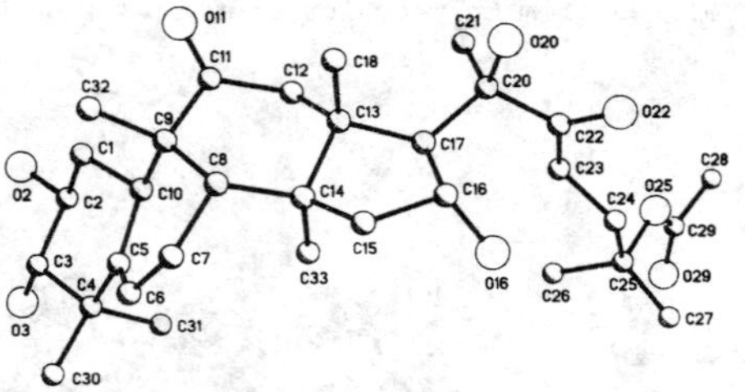

Fig. 1. A view of $C_{32}H_{50}O_8$.

5β,6β-EPOXYALNUSAN-3β-YL ACETATE

$C_{32}H_{52}O_3$

M. TORI, M. TAKAI, Y. MATSUMOTO, Y. MORIYAMA, T. TSUYUKI, T. TAKAHASHI, H. OHNISHI, A. ITAI and Y. IITAKA, 1984. Bull. Chem. Soc. Japan, 57, 2490-2495.

Orthorhombic, $P2_12_12_1$, a = 12.618, b = 32.776, c = 6.837 Å, Z = 4. Cu radiation, R = 0.047 for 2702 reflexions.

Unambiguous proof that the title compound exists as a β-epoxide has been obtained in this structure determination (Fig. 1).

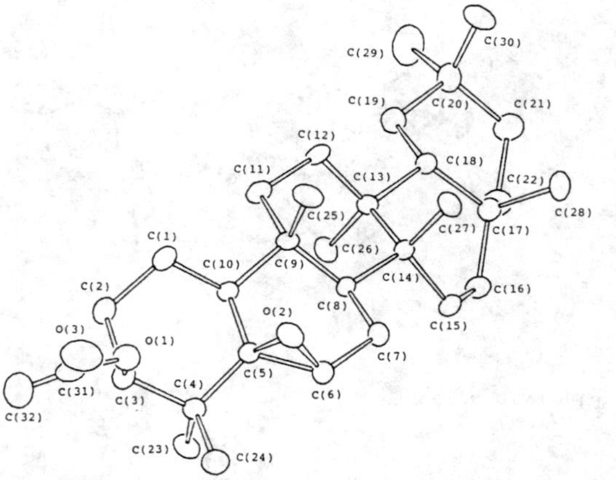

Fig. 1. Perspective view of the molecular structure of $C_{32}H_{52}O_3$.

D:C-FRIEDOOLEAN-8-ENE-3β,29-DIOL DIACETATE
$C_{34}H_{54}O_4$

W. KAMISAKO, K. ISOI, H. NAKAI and M. SHIRO, 1984. Acta Cryst., C40, 1013-1015.

Monoclinic, C2, a = 13.547, b = 6.385, c = 36.183 Å, β = 92.69°, Z = 4. Cu radiation, R = 0.038 for 2904 reflexions.

In the molecule (Fig. 1) the ring junctions A/B, C/D and D/E are respectively trans, trans and cis. The conformations of rings A-E are respectively chair, half-chair, half-chair, twist-boat and boat. The twist-boat form of ring D is considerably distorted to relieve methyl...methyl and methyl...H interactions. Bond lengths and angles are generally as expected.

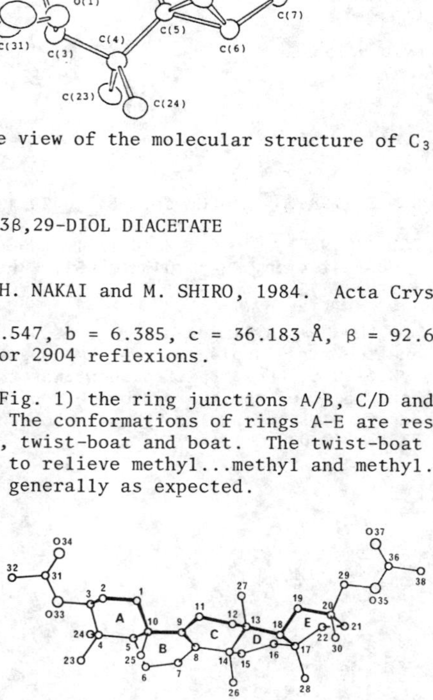

Fig. 1. $C_{34}H_{54}O_4$: perspective view of the molecule.

LUPEOL β-PHENYLPROPIONATE
$C_{39}H_{58}O_2$

P. CABALLERO, F.R. FRONCZEK, N.H. FISCHER, S. FERNANDEZ and E. HERNANDEZ, 1984. J. Nat. Prod., 47, 819-821.

Monoclinic, C2, a = 36.627, b = 11.260, c = 8.152 Å, β = 95.68°, Z = 4. Cu radiation, R = 0.069 for 1708 reflexions.

In the molecule (Fig. 1) all rings are trans-fused and the six-membered rings have chair conformation. The five-membered ring is puckered into a conformation resembling a half-chair (with C21 on the pseudo twofold axis).

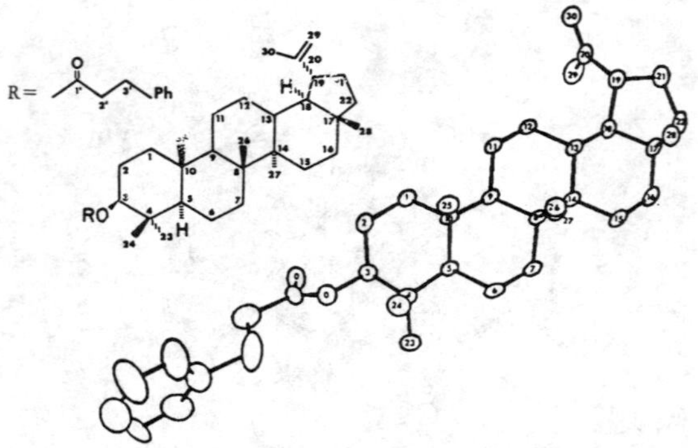

Fig. 1. Lupeol β-phenylpropionate.

PILOCARPINE HYDROCHLORIDE
$C_{11}H_{17}ClN_2O_2$

P.W. CODDING and M.N.G. JAMES, 1984. Acta Cryst., B40, 429-434.

Monoclinic, $P2_1$, a = 11.057 (10.965), b = 9.212 (9.177), c = 6.697 (6.507) Å, β = 110.05 (109.19°), Z = 2. Mo radiation, R = 0.043 (0.031) for 2751 (3864) reflexions; values in parentheses refer to 77 K.

In the cation (Fig. 1) the N-methylimidazole ring is protonated at the position of the secondary N atom and the ring is planar. The NMR spectrum of the compound was studied over the pH range 2.6-10.6; these spectra show that there is a delocalization of π electrons over the bonds between the two N atoms of the protonated imidazole ring. The Cl^- ion in the crystal structure is hydrogen-bonded to the protonated N atom at a distance of 3.035(2) (3.030(2)) Å.

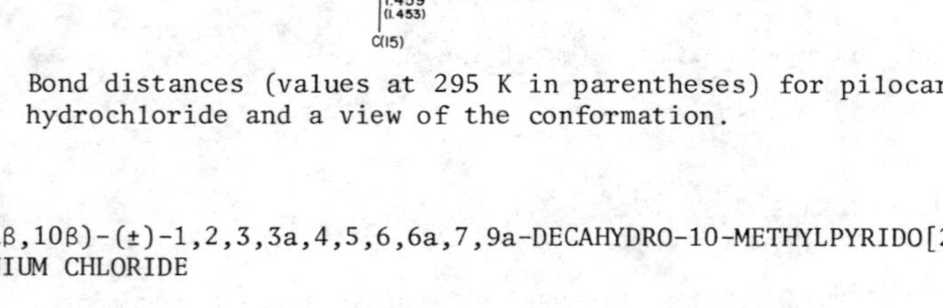

Fig. 1. Bond distances (values at 295 K in parentheses) for pilocarpine hydrochloride and a view of the conformation.

(3aα,6aβ,9aβ,10β)-(±)-1,2,3,3a,4,5,6,6a,7,9a-DECAHYDRO-10-METHYLPYRIDO[2,1,6-de]-QUINOLIZINIUM CHLORIDE
$C_{13}H_{22}ClN$ $C_{13}H_{22}N^+.Cl^-$

R.H. MUELLER, M.E. THOMPSON and R.M. DIPARDO, 1984. J. Org. Chem., 49, 2217-2231.

Monoclinic, $P2_1/n$, a = 16.158, b = 7.484, c = 20.466 Å, β = 95.26°, Z = 8. Mo radiation, R = 0.040 for 2530 reflexions.

There are two independent formula units in the asymmetric unit; one cation is shown in Fig. 1. In both cations, the double bond is disordered between C2=C3 and C3=C4.

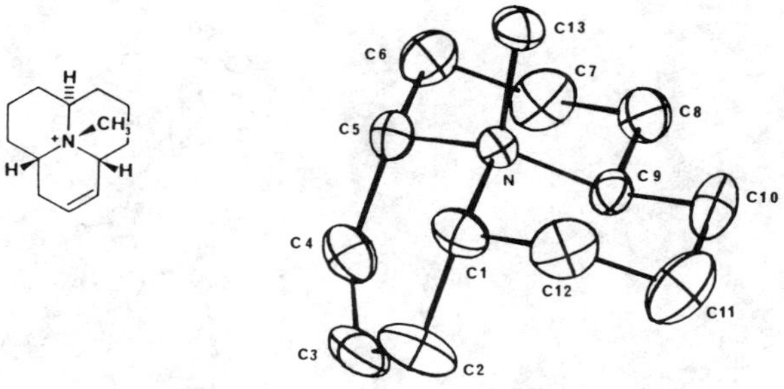

Fig. 1. The $C_{13}H_{22}N^+$ cation.

(±)-(1α,4aα,9aβ)-1,3,4,9a-TETRAHYDRO-2-METHYL-2H-1,4a-PROPANOBENZOFURO[2,3-c]PYRIDIN-8-OL

$C_{15}H_{19}NO_2$

T.R. BURKE Jr., A.E. JACOBSON, K.C. RICE and J.V. SILVERTON, 1984. J. Org. Chem., 49, 2508-2510.

Orthorhombic, $Pbc2_1$, a = 6.947, b = 12.896, c = 14.115 Å, D_m = 1.29, Z = 4. Cu radiation, R = 0.032 for 1337 reflexions.

In the molecule (Fig. 1) the phenyl ring plane makes an angle of 8° with the plane through the piperidine atoms C1,C3,C4,C9a. Angle C4-C4a-C9a is very small (97.9°) and C4a-C4b-C5 (133.4°) is unusually large. Molecules are linked by O-H...N hydrogen bonds (2.72 Å) along the twofold screw axis.

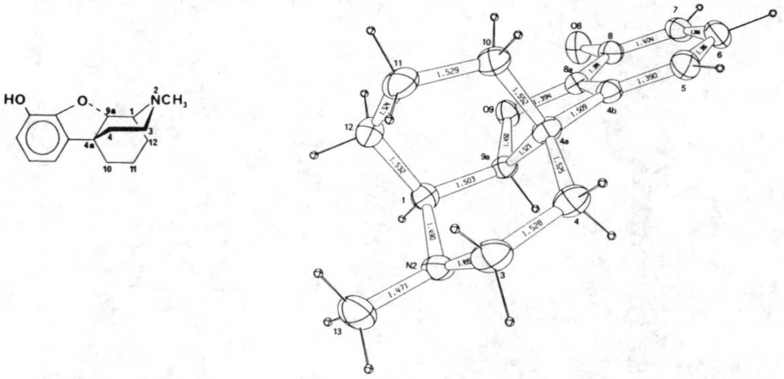

Fig. 1. A view of $C_{15}H_{19}NO_2$ and bond lengths.

DEHYDROMONOCROTALINE

$C_{16}H_{21}NO_6$

M.F. MACKAY, M. SADEK, C.C.J. CULVENOR and L.W. SMITH, 1984. Acta Cryst., C40, 473-476.

Orthorhombic, $P2_12_12_1$, a = 7.409, b = 10.551, c = 20.175 Å, Z = 4. Cu radiation, R = 0.055 for 1079 reflexions.

The conformation of the 11-membered ring (Fig. 1) is significantly different from that found in the parent monocrotaline alkaloid (1) as a consequence of the flattening of the pyrrolizidine ring.

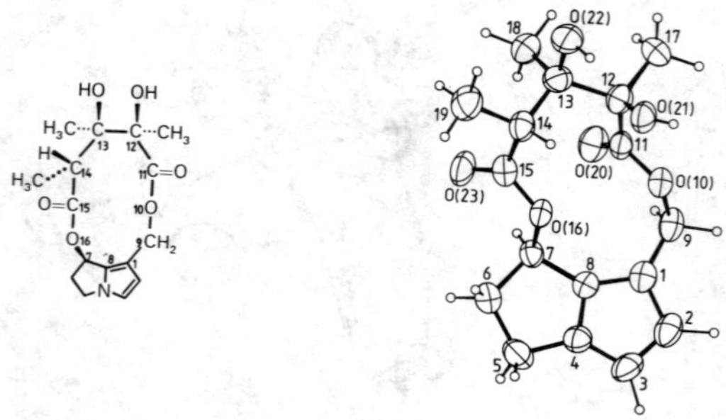

Fig. 1. Dehydromonocrotaline.

1. Structure Reports, 45B, 631.

CRISPATINE MONOHYDRATE
$C_{16}H_{23}NO_5,H_2O$

M.F. MACKAY, F. SADECK and C.C.J. CULVENOR, 1984. Acta Cryst., C40, 470-472.

Triclinic, P1, a = 5.670, b = 8.020, c = 9.943 Å, α = 91.25, β = 103.03, γ = 106.81°, Z = 1. Cu radiation, R = 0.037 for 1367 reflexions.

The conformation of the 11-membered macroring is shown in Fig. 1 and is similar to that found in related systems. The pyrrolizidine nucleus exhibits the usual exo-buckling of retronecine alkaloids with a pucker angle 47.5(6)° and an angle 123.1(7)° between the mean planes defined by the atoms C(5), N(4), C(8), C(7) and C(1), C(2), C(3), N(4), C(8). The atoms in the latter group are coplanar to within ±0.06(1) Å and C(9) lies within the plane.

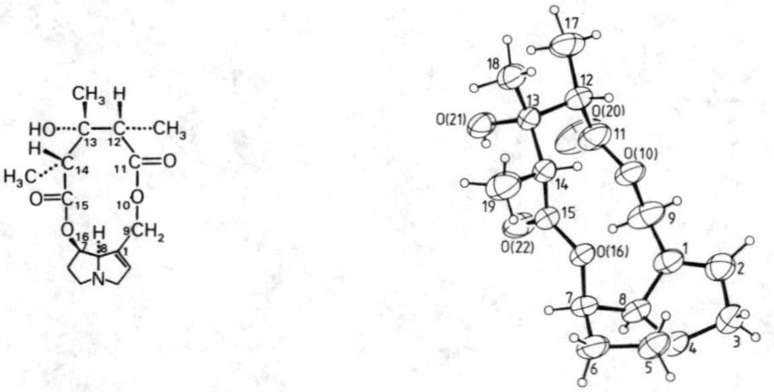

Fig. 1. The alkaloid crispatine.

1,2-DIHYDRO-1α,2α-EPOXYMONOCROTALINE
$C_{16}H_{23}NO_7$ (I)

1,2-DIHYDRO-1β,2β-EPOXYMONOCROTALINE MONOHYDRATE
$C_{16}H_{23}NO_7,H_2O$ (II)

M.F. MACKAY, M. SADEK and C.C.J. CULVENOR, 1984. Acta Cryst., C40, 2064-2068.

I. Orthorhombic, $P2_12_12_1$, a = 10.140, b = 11.691, c = 13.681 Å, Z = 4. Cu radiation, R = 0.059 for 1303 reflexions.

II. Orthorhombic, $P2_12_12_1$, a = 9.621, b = 12.416, c = 14.174 Å, Z = 4. Cu radiation, R = 0.037 for 1602 reflexions.

Apart from the pyrrolizidene nucleus, both epoxide molecules (Fig. 1) have conformations that closely resemble that of monocrotaline (1) and other pyrrolizidine alkaloids containing an eleven-membered macrocycle. The absolute configurations are assigned by comparison with other known structures.

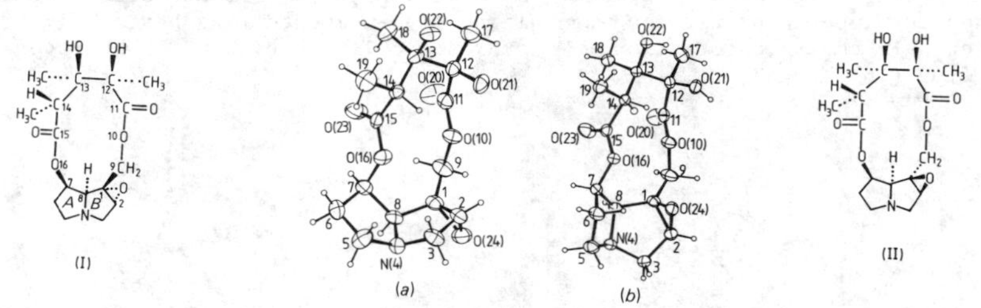

Fig. 1. Views of the α- (left) and β- (right) epoxymonocrotalines.

1. Structure Reports, 45B, 631.

(-)-NEFOPAM HYDROCHLORIDE MONOHYDRATE ((1R)-(-)-5-METHYL-1-PHENYL-1,3,4,6-TETRAHYDRO-5H-BENZO[f]-2,5-OXAZOCINE HYDROCHLORIDE MONOHYDRATE)

$C_{17}H_{19}ClNO,H_2O$ $C_{17}H_{19}NO^+Cl^-,H_2O$

L.K. HANSEN, A. HORDVIK and A.J. AASEN, 1984. Acta Chem. Scand., B38, 327-329.

Orthorhombic, $P2_12_12_1$, a = 8.507, b = 9.661, c = 19.781 Å, Z = 4. Cu radiation, R = 0.041 for 1919 reflexions.

The absolute configuration of the title compound has been determined and is seen (Fig. 1) to be the R-type. Bond lengths and angles in the molecule are normal, e.g. O(1)-C(2) 1.428, O(1)-C(13) 1.434, C(3)-N(4) 1.492, N(4)-C(5) 1.490, N(4)-C(6) 1.532 Å, C(3)-N(4)-C(6) 114.8, C(2)-O(1)-C(13) 114.8°. A Cl...H-N hydrogen bond is present in the crystal structure where the Cl...N distance is 3.06 Å and the Cl...H-N angle 158°.

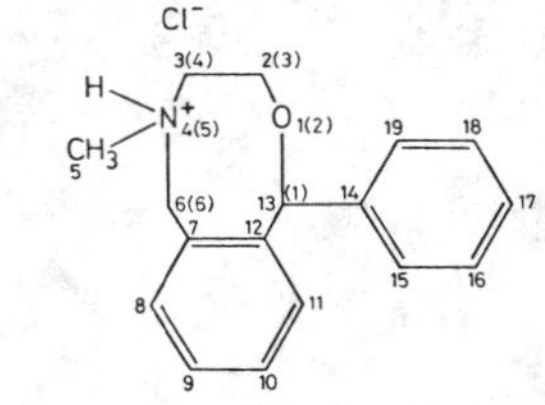

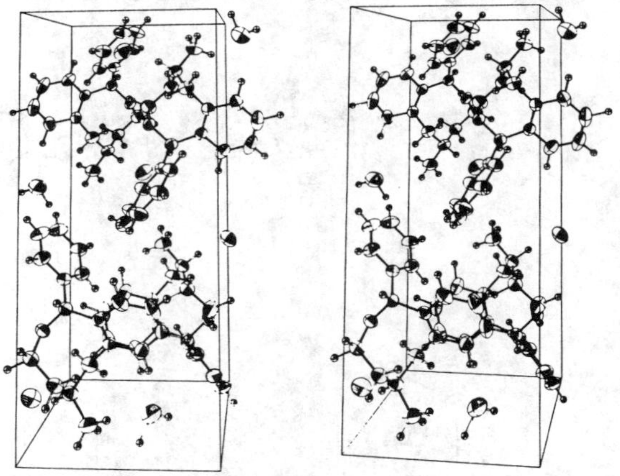

Fig. 1. Schematic drawing of $C_{17}H_{19}ClNO$ (left), and stereoscopic view of the unit cell (right).

HEMANTHAMINE

$C_{17}H_{19}NO_4$

W.H. WATSON, J. GALLOY and M. SILVA, 1984. Acta Cryst., C40, 156-158.

Orthorhombic, $P2_12_12_1$, a = 11.006, b = 14.022, c = 9.582 Å, Z = 4. Cu radiation, R = 0.070 for 1486 reflexions.

In this alkaloid (Fig. 1) the cyclohexene ring and the six-membered nitrogen-containing ring exist as 1,2-diplanar conformations; the five-membered ring is a flattened envelope. The nitrogen-containing five-membered ring has a slightly flattened half-chair conformation. Molecules are linked by O-H...N hydrogen bonds (O...N 2.693(16) Å).

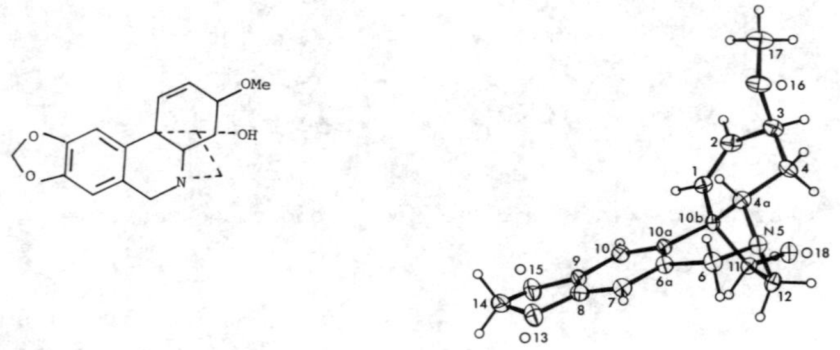

Fig. 1. Hemanthamine.

PIPLARTINE
$C_{17}H_{19}NO_5$

P.M. BOLL, J. HANSEN, O. SIMONSEN and N. THORUP, 1984. Tetrahedron, 40, 171-175.

Monoclinic, $P2_1/c$, Mo radiation, R = 0.047 for 1866 reflexions.

The analysis determined the structure shown in Fig. 1, with C3=C4 and C8=C9 as double bonds. The two outer methoxy groups are almost planar with the benzene ring and the central group is almost perpendicular to it.

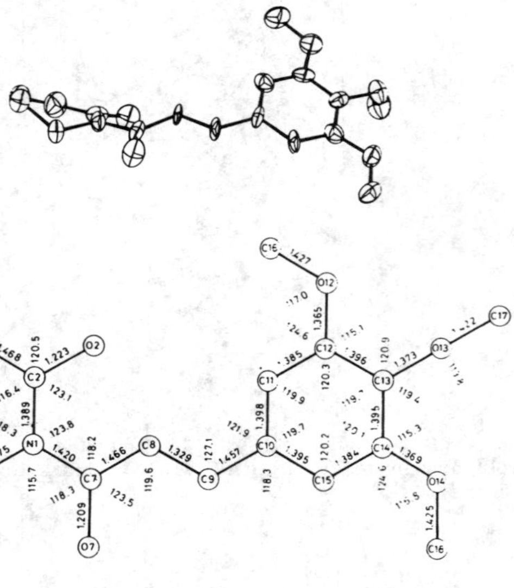

Fig. 1. The piplartine molecule and molecular dimensions (σ 0.002-0.03 Å, 0.1-0.2°).

8β-BROMO-6,7-DIDEHYDRO-4,5α-EPOXY-3-METHOXY-17-METHYLMORPHINAN
$C_{18}H_{20}BrNO_2$

H. VAN KONINGSVELD, T.S. LIE and L. MAAT, 1984. Acta Cryst., C40, 313-314.

Monoclinic, C2, a = 15.284, b = 7.999, c = 15.493 Å, β = 124.53°, Z = 4. Mo radiation, R = 0.062 for 1076 reflexions.

The analysis establishes that the Br atom is in the 8β-position of the C-ring of the morphinan skeleton (Fig. 1).

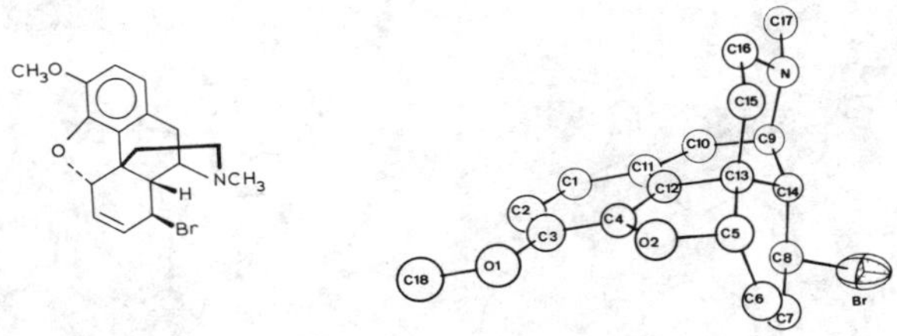

Fig. 1. The structure of $C_{18}H_{20}BrNO_2$.

N-PIPERIDYL-5-(2-METHOXY-4,5-METHYLENEDIOXYPHENYL)-trans-2-cis-4-PENTADIENAMIDE (PIPERX)
$C_{18}H_{21}NO_4$

K.A. WOODE, F.L. PHILLIPS, I. ADDAE-MENSAH, J.C.J. BART and S. CHAUDHURI, 1984. J. Nat. Prod., 47, 1024-1027.

Orthorhombic, $Pca2_1$, a = 16.907, b = 6.325, c = 15.007 Å, Z = 4. Cu radiation, R = 0.092 for 1403 reflexions.

In the piperx molecule (Fig. 1) the piperidine ring has a chair form with mean ring torsion angle 54.9°. The molecule is compared with the closely related trans-2-trans-4 isomer, wisanine (1),

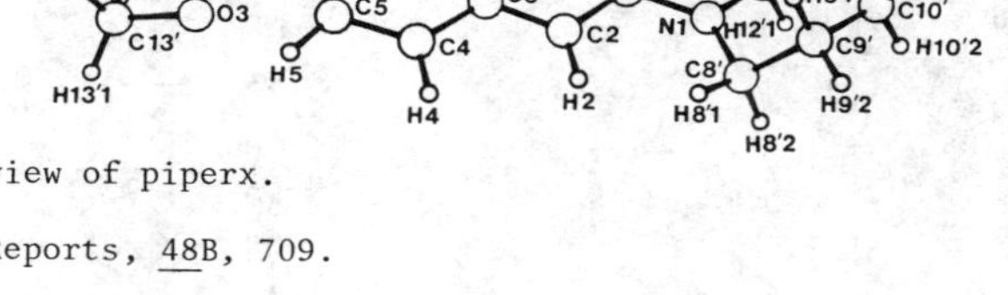

Fig. 1. A view of piperx.

1. Structure Reports, 48B, 709.

MORPHINE METHYL IODIDE MONOHYDRATE
$C_{18}H_{22}INO_3,H_2O$ (I) $(C_{18}H_{22}NO_3)^+.I^-.H_2O$

DI(MORPHINE) DIHYDROGENSULPHATE PENTAHYDRATE
$C_{34}H_{40}N_2O_{10}S,5H_2O$ (II) $2(C_{17}H_{20}NO_3)^+.SO_4^{2-}.5H_2O$

C. WONGWEIGHINTANA, E.M. HOLT and N. PURDIE, 1984. Acta Cryst., C40, 1486-1490.

I. Orthorhombic, $P2_12_12_1$, a = 9.574, b = 10.614, c = 17.549 Å, Z = 4. Mo radiation, R = 0.061 for 2575 reflexions.

II. Orthorhombic, $P22_12_1$, a = 8.463, b = 6.825, c = 30.71 Å, Z = 4. Mo radiation, R = 0.062 for 1754 reflexions.

In both structures, very similar molecular frameworks (Fig. 1) are linked by extensive hydrogen-bonding into clockwise spirals containing molecules related by the twofold screw axis along a. In (I) the hydrogen bonding involves phenolic hydrogens with oxygens of the water molecules while in (II) OH and NH hydrogen atoms and sulphate O atoms are involved.

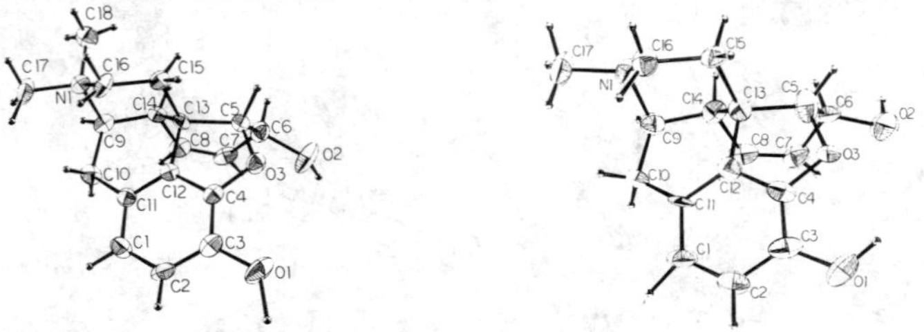

Fig. 1. The $C_{18}H_{22}NO_3^+$ and $C_{17}H_{20}NO_3^+$ cations.

(-)-(13Z,1aR,6bR,10R)-10-HYDROXY-13-ETHYLIDENE-11-METHYLENE-10-METHYL-1a,2,3,6b-TETRAHYDRO-5H-PYRROLIZINO[1a,6b,6a-b,c]-1,8-DIOXACYCLODODECANE-9,14-DIONE (SENECIPHYLLINE)
$C_{18}H_{23}NO_5$

H. WIEDENFELD, F. KNOCH, E. RÖDER and R. APPEL, 1984. Arch. Pharm., 317, 97-102.

Orthorhombic, $P2_12_12_1$, a = 8.828, b = 11.372, c = 17.185 Å, Z = 4. Mo radiation, R = 0.061 for 2591 reflexions.

The molecular structure and labelling scheme are shown in Fig. 1. The C-C bonds in the pyrrolizidine moiety are shorter than those in Retronecine (7β-hydroxy-1-hydroxymethyl-1,2-dehydropyrrolizidine) (C(1)-C(2) 1.302, C(2)-C(3) 1.502, C(1)-C(8) 1.501, C(5)-C(6) 1.502, C(6)-C(7) 1.525 and C(7)-C(8) 1.495 Å) but the three C-N bonds are longer (C(3)-N 1.468, C(5)-N 1.476 and C(8)-N 1.506 Å). This is responsible for the great stability of pyrrolizidine. The dihedral angle between planes C(1)-C(3)-C(8)-N and C(5)-C(7)-C(8)-N is 126.3°. The saturated five-membered ring has a half-chair conformation with a dihedral angle of 149.6° between planes C(5)-C(8)-N and C(6)-C(7)-C(8).

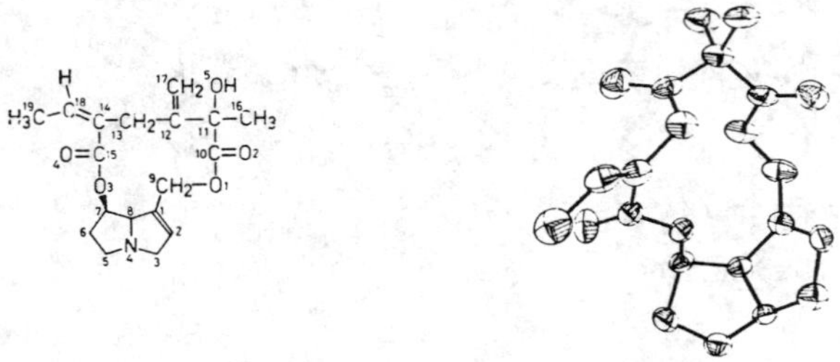

Fig. 1. Numbering scheme in $C_{18}H_{23}NO_3$, and perspective drawing of molecular structure.

GRANTIANINE
$C_{18}H_{23}NO_7$

H. STOECKLI-EVANS and D.J. ROBINS, 1984. Acta Cryst., C40, 1445-1449.

Monoclinic, $P2_1$, a = 12.93, b = 6.32, c = 10.78 Å, β = 102.73°, D_m = 1.40, Z = 2. Mo radiation, R = 0.059 for 1321 reflexions.

The carbonyl bonds of the ester groups are on opposite sides of the eleven-membered macrocycle (Fig. 1) and are inclined at 139.6(5)° to each other. The γ-butyrolactone ring is trans-fused with the methyl group at C(24) quasi-axial. The pyrrolizidine nucleus exists in the exo-puckered form with puckering angle 41.2(5)°. The shortest transannular distance is C(11)...O(16) 2.913(4) Å with O(10)...O(16) 3.119(6) Å. The C(9),O(10),C(11),C(12),O(18) group is planar within 0.01 Å with the OH at C(12) almost in this plane; there is an intramolecular hydrogen bond with O(19)...O(18) 2.611(5) Å.

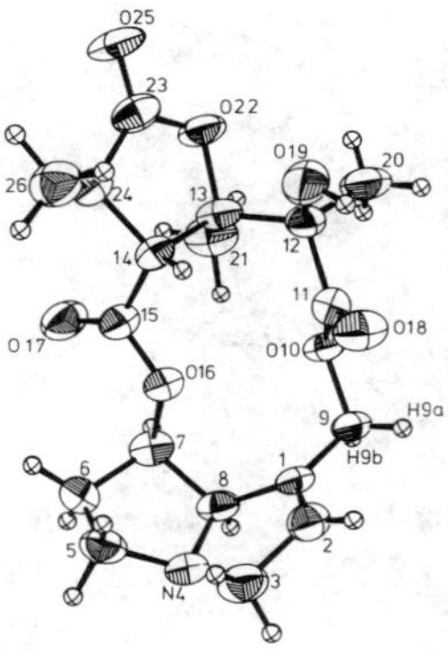

Fig. 1. The grantianine molecule.

FUNEBRINE
$C_{18}H_{24}N_2O_5$

R.F. RAFFAUF, T.M. ZENNIE, K.D. ONAN and P.W. Le QUESNE, 1984. J. Org. Chem., 49, 2714-2718.

Monoclinic, $P2_1$, a = 7.492, b = 8.052, c = 15.278 Å, β = 91.98°, Z = 2. Cu radiation, R = 0.057 for 1127 reflexions.

The two amino lactone fragments in funebrine (Fig. 1) have envelope conformations, with C10 and C20 0.46 and 0.54 Å respectively from the appropriate four-atom planes. Molecules are linked by O-H...O hydrogen bonds (O25...O11 2.81 Å).

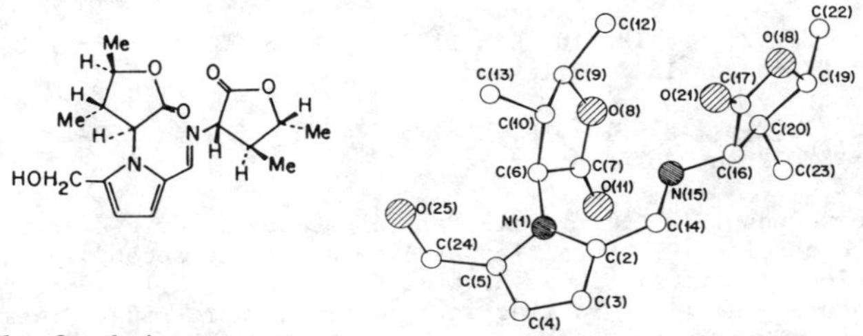

Fig. 1. The funebrine structure.

JACOBINE METHANOL SOLVATE
$C_{18}H_{25}NO_6,CH_4O$

D.C. ROHRER, J. KARCHESY and M. DEINZER, 1984. Acta Cryst., C40, 1449-1452.

Orthorhombic, $P2_12_12_1$, a = 10.9903, b = 20.530, c = 8.9550 Å, Z = 4. Cu radiation, R = 0.044 for 2373 reflexions (at 257K).

Bond lengths, angles and conformations are closely similar to those found in the unsolvated structure (1). The C(1),C(2),C(3),N(4),C(8) ring has the N(4) envelope conformation (ΔC_s(N4)0.8) while the N(4),C(5),C(6),C(7),C(8) ring has a C(6) envelope conformation (ΔC_s(C6)6.8) distorted towards a C(6)/C(7) half-chair conformation (ΔC_2(C6/C7)11.5). The two rings are pseudo-cis fused with a 41.4(2)° angle between their least-squares planes. Both OH groups are involved in hydrogen bonds with O(22)...O(M) 2.767(4) and O(22)...N(4) 2.819(2) Å.

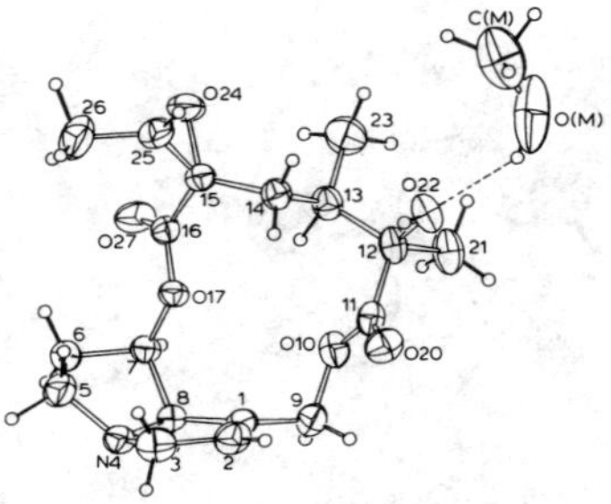

Fig. 1. The jacobine methanol solvate asymmetric unit.

1. Structure Reports, 44B, 532.

ANACROTINE ETHANOL WATER SOLVATE
$C_{18}H_{25}NO_6,0.5C_2H_6O,0.5H_2O$ (I)

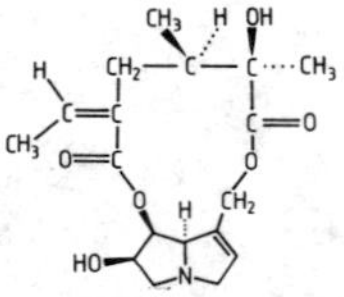

MADURENSINE ETHANOL WATER SOLVATE
$C_{18}H_{25}NO_6,0.5C_2H_6O,0.5H_2O$ (II)

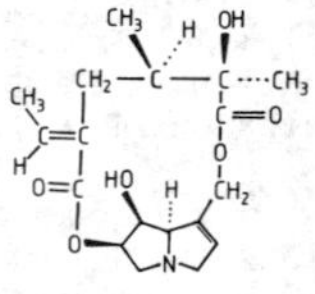

M.F. MACKAY, M. SADEK and C.C.J. CULVENOR, 1984. Acta Cryst., C40, 1073-1077.

I. Orthorhombic, $P2_12_12$, a = 19.810, b = 11.256, c = 8.931 Å, D_m = 1.28, Z = 4. Cu radiation, R = 0.068 for 1848 reflexions.

II. Orthorhombic, $P2_12_12_1$, a = 17.418, b = 16.074, c = 7.111 Å, D_m = 1.28, Z = 4. Cu radiation, R = 0.072 for 1676 reflexions.

The title compounds (Fig. 1) are diesters of stereoisomeric diprotic acid and the aminotriol crotanecine. The position of linkage of the secondary ester on the pyrrolizidine nucleus is at C(7) in anacrotine (I), forming a 12-membered macro-ring, and at C(6) in madurensine (II), forming a 13-membered macro-ring. This change in linkage has little effect on the general conformation of the macro-ring and the secondary ester group. In both crystals the molecules are linked by intermolecular hydrogen bonds.

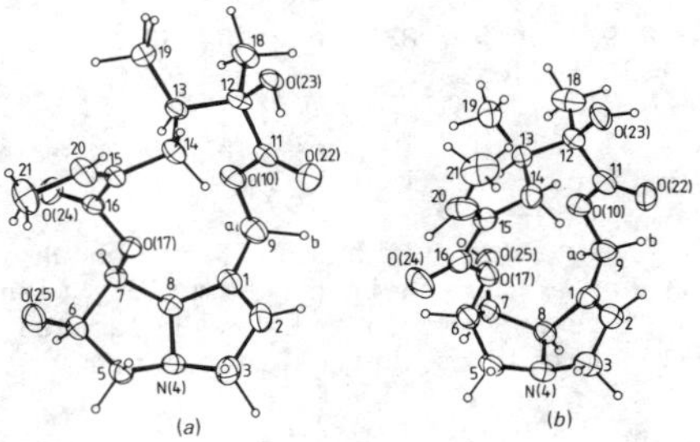

Fig. 1. Views of (a) anacrotine and (b) madurensine.

BULBOCAPNINE METHIODIDE
$C_{20}H_{22}INO_4$ $C_{20}H_{22}NO_4{}^+.I^-$

C.H. WEI, S.P. BASU, J.R. EINSTEIN and B.E. HINGERTY, 1984. Acta Cryst., C40, 1737-1740.

Monoclinic, $P2_1$, a = 10.446, b = 8.711, c = 10.317 Å, β = 97.778°, D_m = 1.66, Z = 2. Mo radiation, R = 0.028 for 2823 reflexions.

The alkaloid and its dimensions are in Fig. 1. Of the five fused rings, two six-membered rings form a biphenyl system with an angle of 30.2(1)° between their plane normals. The H atom of the hydroxyl group is hydrogen bonded to the I^- ion with O(25)...I^- 3.434(3) Å.

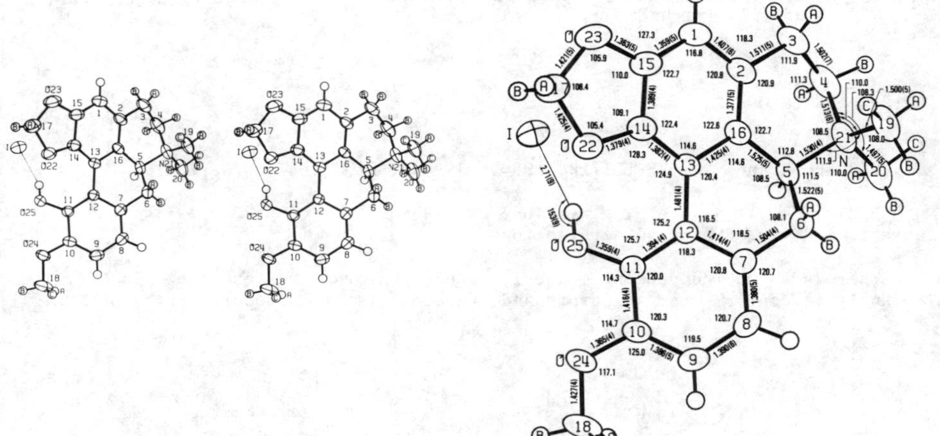

Fig. 1. A stereoview of the bulbocapnine methiodide structure and details of dimensions.

5,8-ETHANO-2-FORMYL-1-(p-METHOXYBENZYL)-1,2,3,4,5,6,7,8-OCTAHYDROISOQUINOLINE
$C_{20}H_{25}NO_2$ (I)

3,11c-ETHANO-10-METHOXY-6-METHYL-1,2,3,3a,11b,11c-HEXAHYDROAPORPHINE CHLORIDE HYDRATE
$C_{20}H_{28}ClNO,H_2O$ (II)

V.F. ROCHE, E.B. ROCHE, D.L. NAGEL and A.T. McPHAIL, 1984. J. Org. Chem., 49, 3881-3887.

I. Monoclinic, $P2_1$, a = 10.933, b = 7.885, c = 10.185 Å, β = 102.74°, Z = 2. Cu radiation, R = 0.043 for 1523 reflexions.

II. Monoclinic, C2/c, a = 37.622, b = 10.166, c = 22.416 Å, β = 112.88°, Z = 16. Cu radiation, R = 0.084 for 1603 reflexions.

The results of X-ray analyses of (I) and (II) are shown in Fig. 1. The azacyclohexene ring in (I) has a half-chair conformation, with the bulky benzyl substituent pseudoaxial. There are two independent molecules of II and in one the cyclohexene ring has a half-chair form whereas in the other it has a 1,3-diplanar form.

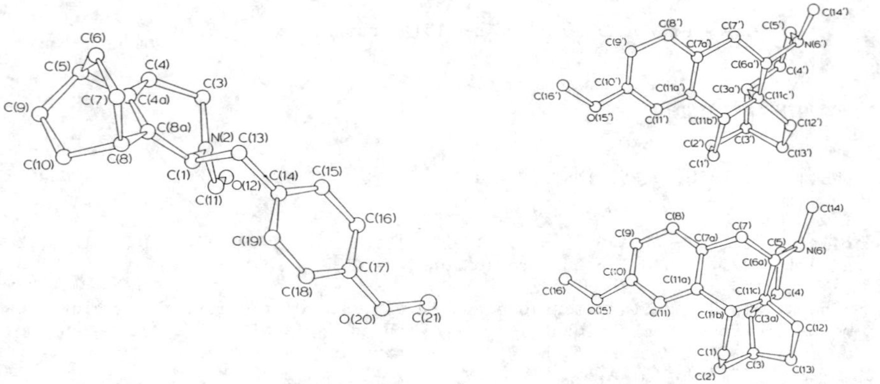

Fig. 1. Views of $C_{20}H_{25}NO_2$ (I) (left) and the two independent $C_{20}H_{28}NO^+$ cations of (II) (right).

HETISINE PERCHLORATE

$C_{20}H_{28}ClNO_7$ $C_{20}H_{28}NO_3^+,ClO_4^-$

K.I. VARUGHESE and S.W. PELLETIER, 1984. J. Nat. Prod., 47, 470-473.

Monoclinic, $P2_1$, a = 11.078, b = 9.463, c = 18.801 Å, β = 93.20°, Z = 4. Cu radiation, R = 0.069 for 4004 reflexions.

There are two independent molecules whose conformations are shown in Fig. 1. The analysis confirms the structure and stereochemistry as reported previously (1).

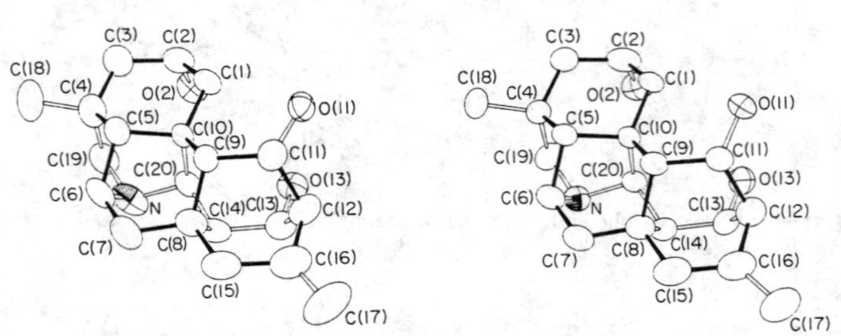

Fig. 1. Views of the two hetisine ions.

1. Structure Reports, 28, 552.

5-ETHYL-2'-HYDROXY-2-[(1-HYDROXYCYCLOPROPYL)METHYL]-9,9-DIMETHYL-6,7-BENZOMORPHAN HYDROCHLORIDE (BREMAZOCINE)

$C_{20}H_{30}ClNO_2$ $C_{20}H_{30}NO_2^{+}.Cl^{-}$

C.L. VERLINDE, N.M. BLATON, C.J. DE RANTER and O.M. PEETERS, 1984. Acta Cryst., C40, 1759-1761.

Monoclinic, $P2_1/n$, a = 15.326, b = 11.285, c = 11.381 Å, β = 108.17°, D_m = 1.24, Z = 4. Mo radiation, R = 0.053 for 1747 reflexions.

The (1-hydroxycyclopropyl)methyl side chain (Fig. 1) adopts a similar conformation to that in ketazocine (1), while the alcohol oxygen position and lone-pair orientation are those predicted for opiate kappa-agonist activity. The results of intramolecular potential-energy calculations are reported.

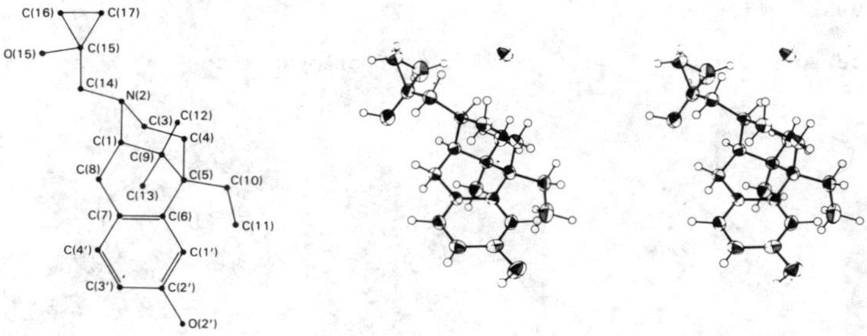

Fig. 1. The atomic numbering scheme and a stereoview of bremazocine hydrochloride.

1. Structure Reports, 50B.

(-)-PODOPETALINE and (-)-ORMOSANINE (Monoclinic and triclinic quasi-racemates)

$C_{20}H_{35}N_3, C_{20}H_{33}N_3$

W. WONG-NG, P.-T. CHENG and S.C. NYBURG, 1984. Acta Cryst., B40, 151-158.

I. Monoclinic, $P2_1$, a = 10.418, b = 19.634, c = 18.685 Å, β = 104.51°, Z = 4. Cu radiation, R = 0.085 for 4319 reflexions.

II. Triclinic, P1, a = 10.426, b = 19.574, c = 11.202 Å, α = 122.10, β = 105.25, γ = 89.50°, Z = 2. Cu radiation, R = 0.093 for 3386 reflexions.

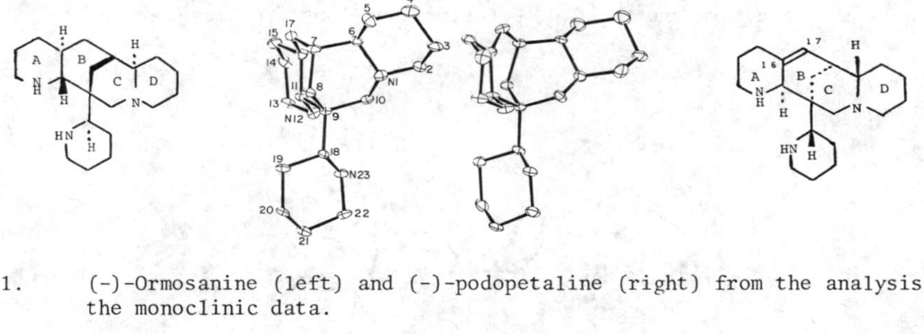

Fig. 1. (-)-Ormosanine (left) and (-)-podopetaline (right) from the analysis of the monoclinic data.

An extraction process yielded crystals containing the two alkaloids but repeated recrystallisation failed to resolve the two components. The crystals are typically triplets consisting of a monoclinic phase m and two identical, but differently oriented, triclinic phases t1 and t2. Three data sets were collected, two from a virtual twin from which most of t2 had been cut away and one from a virtually single m crystal. Both structures contain chirally opposite (and hence pseudo-symmetrically related) 1:1 ordered arrays of (-)-podopetaline and (-)-ormosanine (Fig. 1). Many of the bond lengths of the molecules in the t phase lay outside acceptable ranges. This could be caused by the t phase containing small, but significant amounts of the corresponding (+) isomers or other closely related alkaloid molecules of this series.

18-ACETYL-$\Delta^{17,18}$-DEHYDROALLOYOHIMBANE
$C_{21}H_{24}N_2O$

S. CHAO, F.-A. KUNNG, J.-M. GU, H.L. AMMON and P.S. MARIANO, 1984. J. Org. Chem., 49, 2708-2711.

Orthorhombic, Pbca, a = 10.648, b = 12.840, c = 25.29 Å, Z = 8. Cu radiation, R = 0.074 for 1938 reflexions.

The stereochemistry of this derivative of alloyohimbane is shown in Fig. 1.

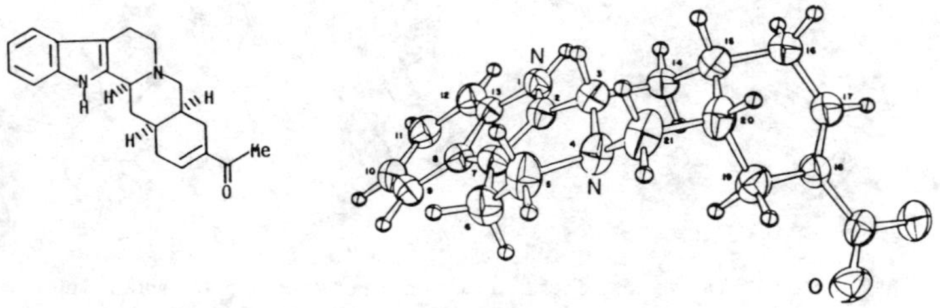

Fig. 1. A view of $C_{21}H_{24}N_2O$.

RAUBASINE HYDRATE
$C_{21}H_{24}N_2O_3,H_2O$

J.-P. DUBOST, J.-M. LEGER, M. GOURSOLLE, J.-C. COLLETER and A. CARPY, 1984. Acta Cryst., C40, 152-154.

Monoclinic, P2$_1$, a = 6.931, b = 8.438, c = 16.570 Å, β = 101.11°, Z = 2. Cu radiation, R = 0.044 for 2252 reflexions.

The molecule (Fig. 1) is quite rigid. Rings C and D have quasi-sofa conformation; ring E is in chair form and the C/D, D/E ring junctions are trans.

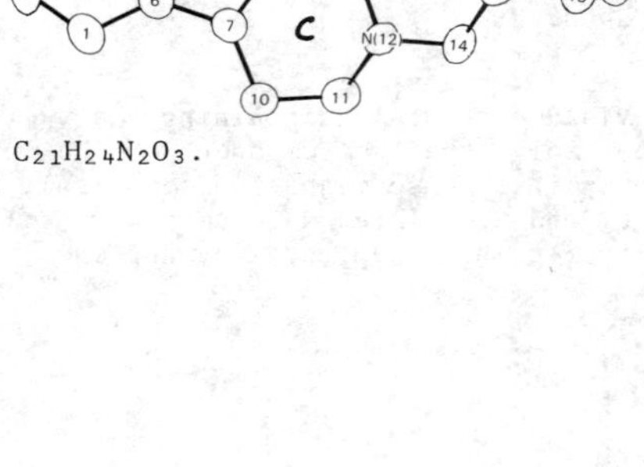

Fig. 1. A view of $C_{21}H_{24}N_2O_3$.

CATALINE
$C_{21}H_{25}NO_5$

I. FONSECA and S. GARCIA-BLANCO, 1984. Acta Cryst., C40, 176-178.

Orthorhombic, $P2_12_12_1$, a = 7.3558, b = 14.869, c = 17.410 Å, Z = 4. Cu radiation, R = 0.036 for 1822 reflexions.

The analysis confirms the absolute configuration shown in Fig. 1. Ring B has a half-chair conformation and ring C has a diplanar one.

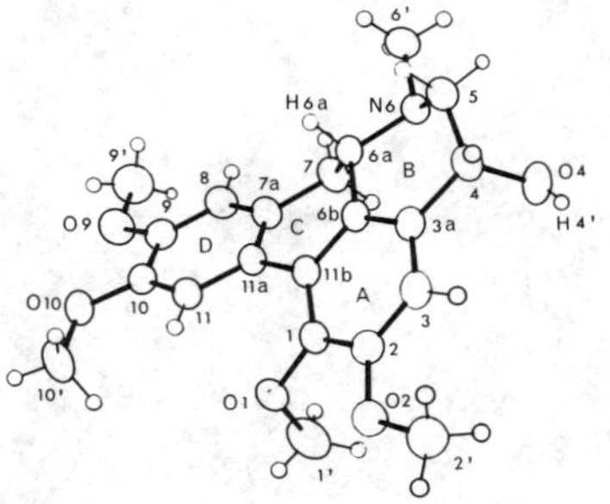

Fig. 1. A view of cataline.

3S,5S-N-METHYLDYSHOMERYTHRINIUM HEXAFLUOROPHOSPHATE
$C_{21}H_{28}F_6NO_4P$

A.J. ALADESANMI, C.J. KELLEY, J.D. LEARY and K.D. ONAN, 1984. J. Chem. Res., (S) 108-109, (M) 1001-1039.

Orthorhombic, $P2_12_12_1$, a = 12.593, b = 20.741, c = 8.584 Å, Z = 4. Cu radiation, R = 0.066 for 1676 reflexions.

The structure and absolute configuration are shown in Fig. 1. The seven-membered C-ring has a distorted twist-boat conformation, midway between a half-chair and a sofa form. Ring B adopts an envelope form with N9 as the flap.

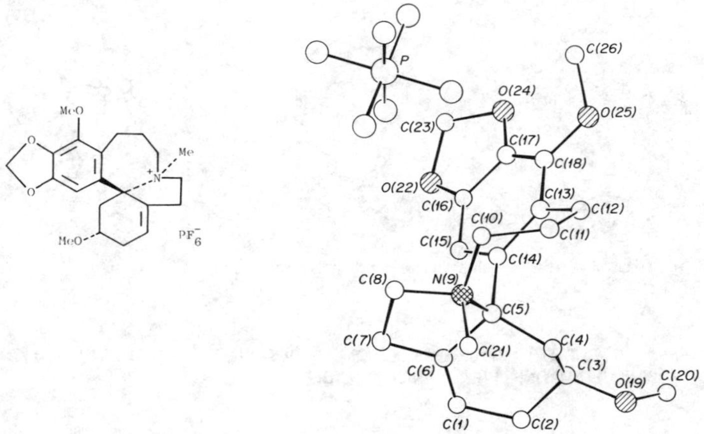

Fig. 1. The structure and absolute configuration of $C_{21}H_{28}F_6NO_4P$.

(7R,16R,20S,21S)-16-HYDROXY-N-METHYL-TETRAHYDROMELOSCINE METHIODIDE
$C_{21}H_{29}IN_2O_2$ $C_{21}H_{29}N_2O_2{}^+.I^-$

G. PALMISANO, B. DANIELI, G. LESMA, R. RNA, S. RIVA, F. DEMARTIN and N. MASCIOCCHI, 1984. J. Org. Chem., 49, 4138-4143.

Monoclinic, $P2_1$, a = 7.584, b = 12.841, c = 10.772 Å, β = 98.78°, Z = 2. Mo radiation, R = 0.030 for 2107 reflexions.

The X-ray analysis established the absolute configuration as shown in Fig. 1.

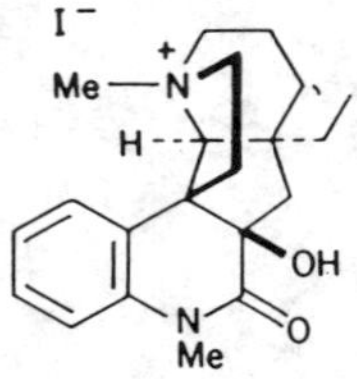

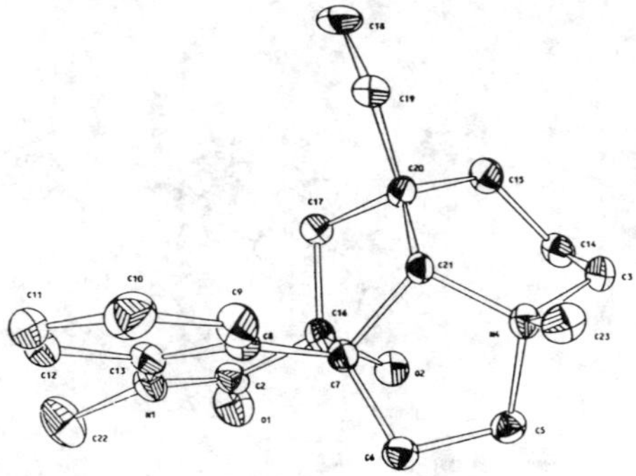

Fig. 1. A view showing the absolute configuration of the $C_{21}H_{29}N_2O_2{}^+$ cation.

(+)-DORONINE BENZENE SOLVATE
$C_{21}H_{30}ClNO_8,C_6H_6$

R.Y. WONG and J.N. ROITMAN, 1984. Acta Cryst., C40, 163-166.

Orthorhombic, $P2_12_12_1$, a = 10.873, b = 12.678, c = 20.511 Å, Z = 4. Cu radiation, R = 0.08 for 2162 reflexions.

The structure and absolute configuration are shown in Fig. 1. An intramolecular transannular interaction is postulated between the N and carbonyl C atoms of the eight-membered ring (N...C(8) 2.231 Å). The C(8)=O(23) bond length is 1.221 Å.

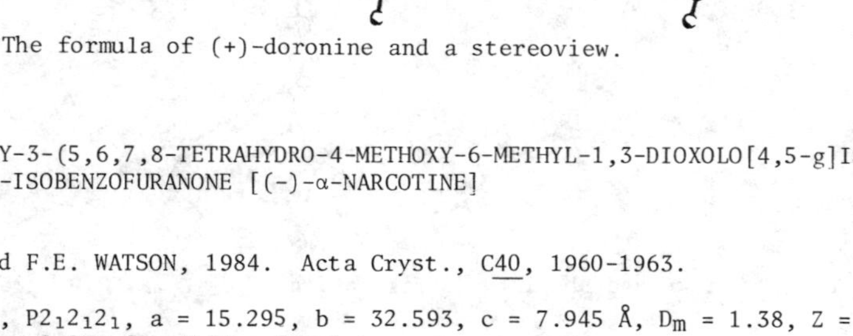

Fig. 1. The formula of (+)-doronine and a stereoview.

6,7-DIMETHOXY-3-(5,6,7,8-TETRAHYDRO-4-METHOXY-6-METHYL-1,3-DIOXOLO[4,5-g]ISOQUINOLIN-5-YL)-1(3H)-ISOBENZOFURANONE [(-)-α-NARCOTINE]
$C_{22}H_{23}NO_7$

D.S. MOSS and F.E. WATSON, 1984. Acta Cryst., C40, 1960-1963.

Orthorhombic, $P2_12_12_1$, a = 15.295, b = 32.593, c = 7.945 Å, D_m = 1.38, Z = 8. Cu radiation, R = 0.078 for 4166 reflexions.

Each of the two independent molecules consists of two planar moieties inclined at about 50° (Fig. 1). The bond joining the two moieties is axial with respect to the tetrahydropyridine ring; the N-methyl group is equatorial. The tetrahydropyridine rings have an approximate sofa conformation with C(3) the out-of-plane atom.

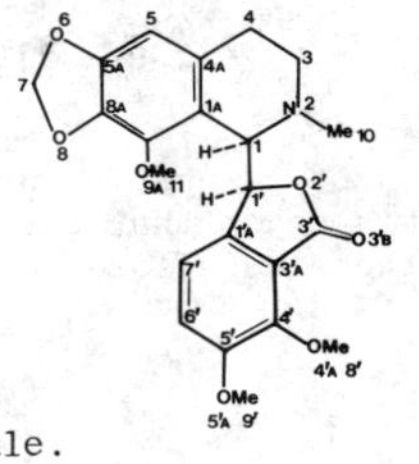

Fig. 1. The narcotine molecule.

ICAJINE
$C_{22}H_{24}N_2O_3$

A. MOSTAD, 1984. Acta Chem. Scand., B38, 821-827.

Orthorhombic, $P2_12_12_1$, a = 8.058, b = 14.237, c = 15.665 Å, Z = 4. Mo radiation, R = 0.057 for 1230 reflexions.

In icajine (Fig. 1), the five-membered ring (C(3), N(4), C(5), C(6), C(7)) exists in an envelope form with C(7) 0.7 Å out of the best plane through C(3), C(6), C(5) and N(4). The six-membered ring (C(3), C(14), C(15), C(20), C(21), N(4)) is in a slightly twisted boat form, atoms C(3), N(4), C(20), C(15) being planar within 0.05 Å and C(14) and C(21) being respectively 0.760 and 0.527 Å out of the plane on the same side. The N(4)...C(3) distance is 2.456 Å and the direction of the lone pair on N(4) deviates from the N(4)...C(3) direction by only 9.6° indicative of a transannular interaction. The molecular packing is governed entirely by van der Waals' forces.

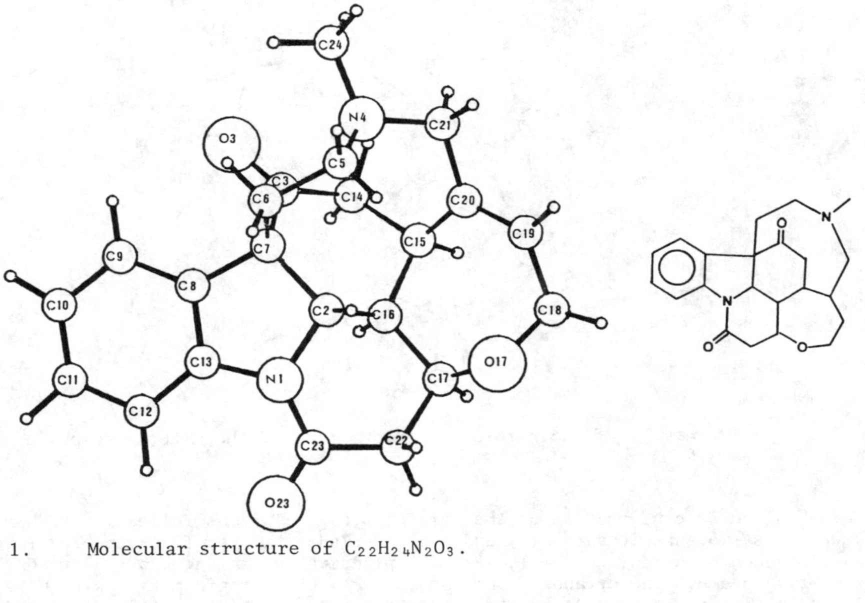

Fig. 1. Molecular structure of $C_{22}H_{24}N_2O_3$.

(-)-7α-ACETYL-4,5α-EPOXY-3-METHOXY-N-METHYL-6,14-ETHENOISOMORPHINAN
$C_{22}H_{25}NO_3$

H. VAN KONINGSVELD, L. MAAT and T.S. LIE, 1984. Acta Cryst., C40, 1082-1084.

Trigonal, $P3_2$, a = 10.899, c = 13.422 Å, Z = 3. Mo radiation, R = 0.039 for 2022 reflexions.

In the molecule (Fig. 1) the acetyl substituent is in the 7α position of the 6,14-ethenomorphinan skeleton. This means that the Diels-Alder reaction of methyl vinyl ketone with (-)-6-demethoxytheboine affords one isomer. There are no unusual bond lengths or angles.

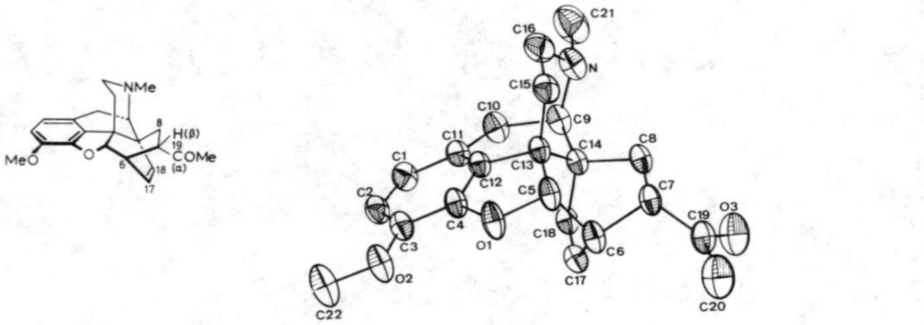

Fig. 1. $C_{22}H_{25}NO_3$: structural formula and perspective view of the molecule.

14-HYDROXY-STRYCHNOBRASILINE
$C_{22}H_{26}N_2O_4$

A. MOSTAD, 1984. Acta Chem. Scand.,B38, 381-386.

Orthorhombic, $P2_12_12_1$, a = 6.891, b = 9.021, c = 29.782 Å, Z = 4. Mo radiation, R = 0.065 for 1133 reflexions (at 121K).

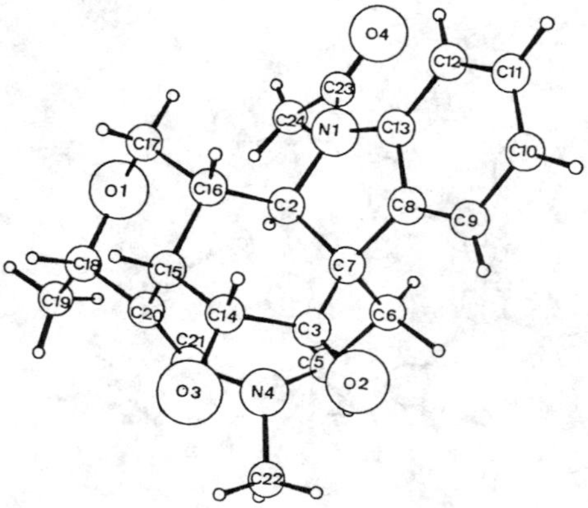

Fig. 1. Mirror image of the absolute configuration of the natural compound $C_{22}H_{26}N_2O_4$.

The molecular conformation of the compound (Fig. 1) is as follows: the benzene ring with its two substituents C(7) and N(1) is planar but the five-membered ring has an envelope conformation, the N(1),C(2),C(7) plane making an angle of 28° with the benzene ring plane. The pyranose ring has a normal chair conformation with C(17) and C(20) ca. 0.7 Å on either side of the best plane through C(15), C(16), C(18) and O(1).

The cyclohexane ring (C(2), C(7), C(3), C(14), C(15), C(16)) has a twisted chair conformation which, together with that of the interconnected nine-membered ring (N(4), C(5), C(6), C(7), C(3), C(14), C(15), C(20), C(21)), seems to be influenced by the transannular interaction between N(4) and C(3)=O. The N(4)...C(3) distance is 2.24 Å. The molecular packing appears to be governed by van der Waals' forces only as there are no intermolecular hydrogen bonds.

(+)-7β-ACETYL-3-METHOXY-N-METHYL-6,14-ETHENOMORPHINAN-4-OL
$C_{22}H_{27}NO_3$

H. VAN KONINGSVELD, T.S. LIE and L. MAAT, 1984. Acta Cryst., C40, 1963-1965.

Hexagonal, $P6_1$, a = 11.899, c = 22.478 Å, Z = 6. Mo radiation, R = 0.033 for 1707 reflexions.

The molecular structure is as shown in Fig. 1 with the acetyl substituent in the 7β position.

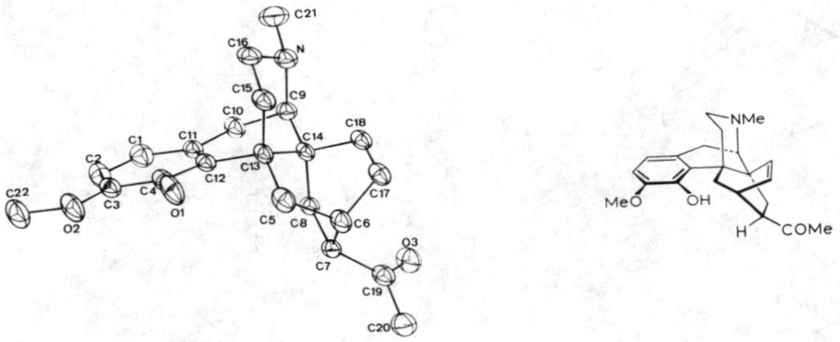

Fig. 1. The $C_{22}H_{27}NO_3$ molecule.

(+)-6-BENZYLSPARTEINE
$C_{22}H_{32}N_2$ (I)

(ETHYL)(BROMO)[(+)-6-BENZYLSPARTEINE]MAGNESIUM(II)
$C_{24}H_{37}BrMgN_2$ (II)

DIBROMO[(+)-6-BENZYLSPARTEINE]MAGNESIUM(II) TOLUENE SOLVATE
$C_{22}H_{32}Br_2MgN_2,C_7H_8$ (III)

H. KAGEYAMA, K. MIKI, Y. KAI, N. KASAI, Y. OKAMOTO and H. YUKI, 1984. Bull. Chem. Soc. Japan, 57, 1189-1196.

I. Orthorhombic, $P2_12_12_1$, a = 10.711, b = 25.868, c = 6.804 Å, Z = 4. Cu radiation, R = 0.055 for 1342 reflexions.

II. Orthorhombic, $P2_12_12_1$, a = 11.020, b = 23.660, c = 8.666 Å, Z = 4. Mo radiation, R = 0.169 for 1445 reflexions.

III. Orthorhombic, $P2_12_12_1$, a = 16.654, b = 19.723, c = 8.637 Å, Z = 4. Mo radiation, R = 0.118 for 1501 reflexions.

Molecule I is different from the others in terms of its overall conformation in that it takes a "one-boat" conformation. The six-membered ring C(7), C(8), C(9), C(11), N(1) and C(17) is in a boat form whereas all the other rings are in chair forms (Fig. 1).

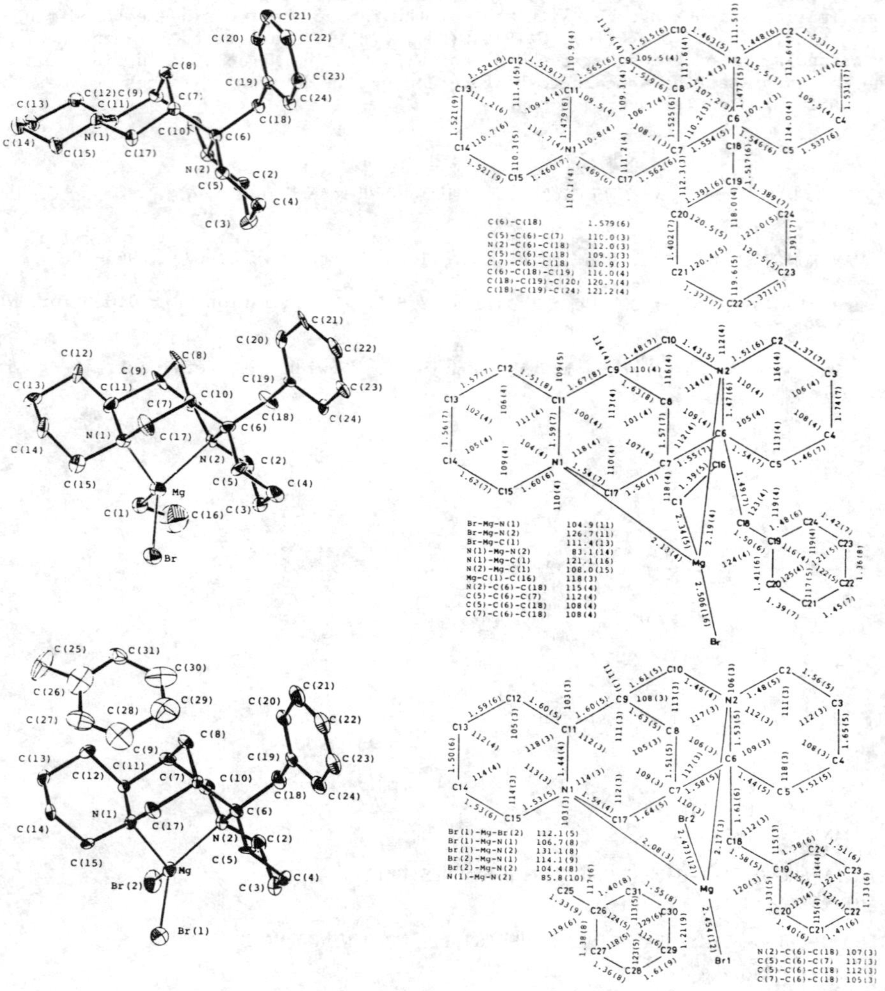

Fig. 1. Molecular structure and selected bond distances and angles for $C_{22}H_{32}N_2$ (top left and right), $C_{24}H_{37}BrMgN_2$ (centre left and right), and $C_{22}H_{32}Br_2MgN_2,C_7H_8$ (bottom left and right).

(±)-2α-ACETOXY-3β-BENZENESELENENYL-α-LYCORAN-5-ONE
$C_{24}H_{23}NO_5Se$

B. UMEZAWA, O. HOSHINO, S. SAWAKI, H. SASHIDA, K. MORI, Y. HAMADA, K. KOTER and Y. IITAKA, 1984. Tetrahedron, 40, 1783-1790.

Monoclinic, $P2_1/n$, a = 14.883, b = 24.151, c = 5.908 Å, β = 91.60°, Z = 4. Cu radiation, R = 0.068 for 3240 reflexions.

In the molecule (Fig. 1) the plane of the Ac group is perpendicular to the six-membered ring and bisecting the angle C(1)-C(2)-C(3). The phenyl group is situated at approximately trans position to one of the endocyclic bonds; torsion angles, C(3a)-C(3)-Se-C(14) and C(2)-C(3)-Se-C(14) are 162° and -74°, respectively, while those for the Ac group C(3)-C(2)-O(1)-C(12) and C(1)-C(2)-O(1)-C(12) are -138° and 105°. Ring C adopts a regular chair conformation and ring B is a distorted half-chair. The five-membered ring D takes an envelope form with the flap at C(3a).

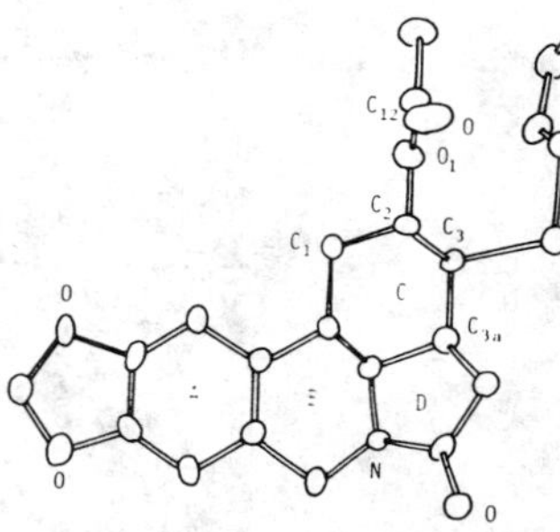

Fig. 1. A perspective drawing of $C_{24}H_{23}NO_5Se$.

BROWNIINE PERCHLORATE
$C_{25}H_{42}ClNO_{11}$ (I) $C_{25}H_{42}NO_7^+.ClO_4^-$

DICTYOCARPINE ACETONE SOLVATE
$C_{26}H_{39}NO_8,C_3H_6O$ (II)

S.W. PELLETIER and K.I. VARUGHESE, 1984. J. Nat. Prod., 47, 643-647.

I. Monoclinic, $P2_1$, a = 10.178, b = 10.649, c = 12.584 Å, β = 102.76°, Z = 2. Cu radiation, R = 0.078 for 2766 reflexions.

II. Orthorhombic, $P2_12_12_1$, a = 8.382, b = 17.568, c = 19.323 Å, Z = 4. Cu radiation, R = 0.085 for 2483 reflexions.

The two alkaloids are shown in Fig. 1. In I the A-ring has a boat conformation while in II it is in a chair conformation. The structural results require the configuration of the C(1)-methoxy group in all C_{19}-deterpenoid alkaloids related to lycoctonine to be revised from β to α.

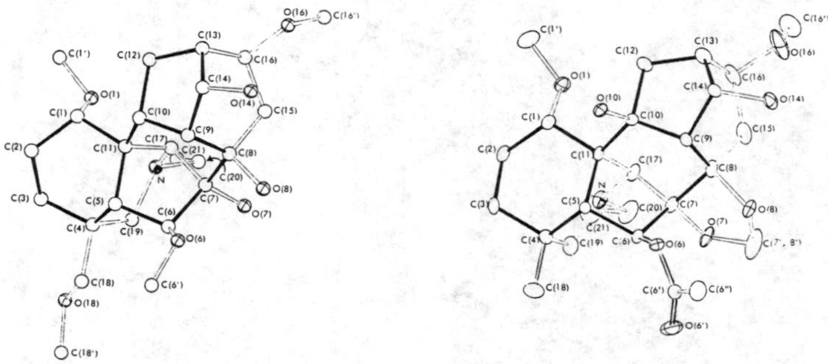

Fig. 1. Views of browniine (left) and dictyocarpine (right).

ZOANTHAMINE
$C_{30}H_{41}NO_5$

C.B. RAO, A.S.R. ANJANEYULA, N.S. SARMA, Y. VENKATATESWARLU, R.M. ROSSER, D.J. FAULKNER, M.H.M. CHEN and J. CLARDY, 1984. J. Am. Chem. Soc., 106, 7983-7984.

Orthorhombic, $P2_12_12_1$, a = 11.775, b = 13.88, c = 15.877 Å, Z = 4. Cu radiation, R = 0.058 for 1677 reflexions.

The analysis establishes zoanthamine to have the structure and relative stereochemistry shown in Fig. 1.

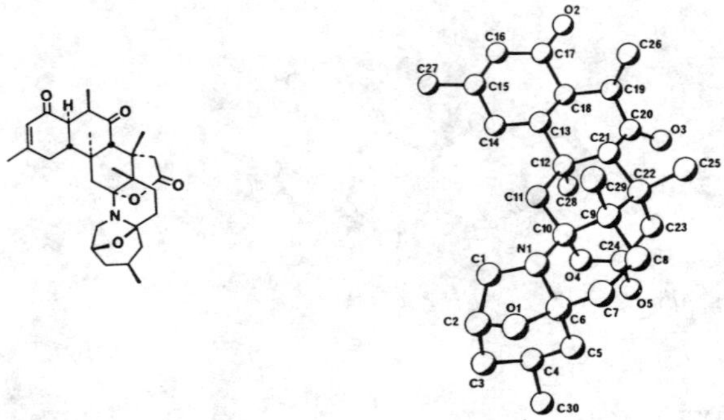

Fig. 1. Zoanthamine.

TETRAHYDRODEOXYSPIROLUCIDINE
$C_{30}H_{53}N_3O$

W.A. AYER, L.F. BALL, L.M. BROWNE, M. TORI, L.T.J. DELBAERE and A. SILVERBERG, 1984. Can. J. Chem., 62, 298-302.

Orthorhombic, $P2_12_12_1$, a = 14.482, b = 19.434, c = 10.507 Å, Z = 4. Cu radiation, R = 0.049 for 2177 reflexions.

The analysis establishes the structure and stereochemistry as shown in Fig. 1. All six-membered rings are in chair conformation.

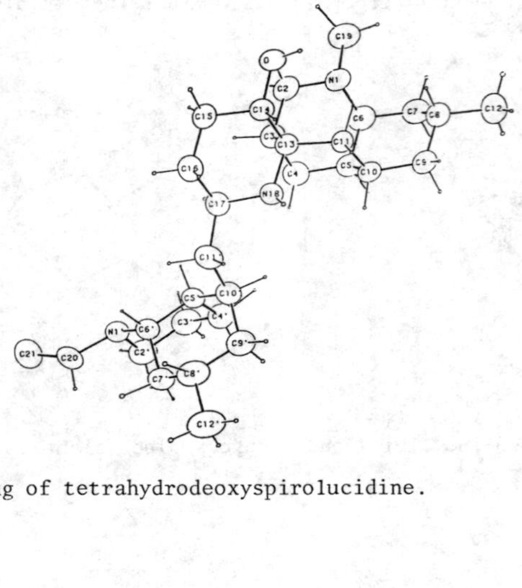

Fig. 1. A drawing of tetrahydrodeoxyspirolucidine.

2,3-BENZO-5,8-ETHENO-6-METHOXYCARBONYL-4-(p-METHOXYPHENYLSULPHONYL)-14-PHENYLTHIO-4,12-DIAZATETRACYCLO[6.6.1.0^{1,5}.0^{2,15}]PENTADECAN-13-ONE BENZENE SOLVATE
$C_{34}H_{32}N_2O_6S_2$,0.5(C_6H_6) (I)

2,3,4,5-TETRADEHYDRO-1-((p-METHOXYPHENYL)SULPHONYL)-11β-(PHENYLTHIO)-11α-(PROP-2-EN-1-YL)-20,21-DINORASPIDOSPERMIDIN-10-ONE
$C_{33}H_{30}N_2O_4S_2$ (II)

20,21-DIDEHYDRO-1-((p-METHOXYPHENYL)SULPHONYL)-11β-(PHENYLTHIO)-KOPSAN-10-ONE
$C_{33}H_{30}N_2O_4S_2$ (III)

P. MAGNUS, T. GALLAGHER, P. BROWN and J.C. HUFFMAN, 1984. J. Am. Chem. Soc., 106, 2105-2114.

I. Triclinic, P$\bar{1}$, a = 18.835, b = 11.486, c = 9.273 Å, α = 82.51, β = 62.51, γ = 62.49°, Z = 2. R = 0.068 for 4558 reflexions (at -164°C).

II. Monoclinic, P2_1/n, a = 15.064, b = 18.701, c = 9.851 Å, β = 98.49°, Z = 4. R = 0.091 for 2336 reflexions (at -163°C).

III. Monoclinic, P2_1/a, a = 17.386, b = 12.079, c = 12.754 Å, β = 95.21°, Z = 4. R = 0.053 for 2186 reflexions (at -163°C).

The structures of these three alkaloids (Fig. 1) were established by X-ray analyses.

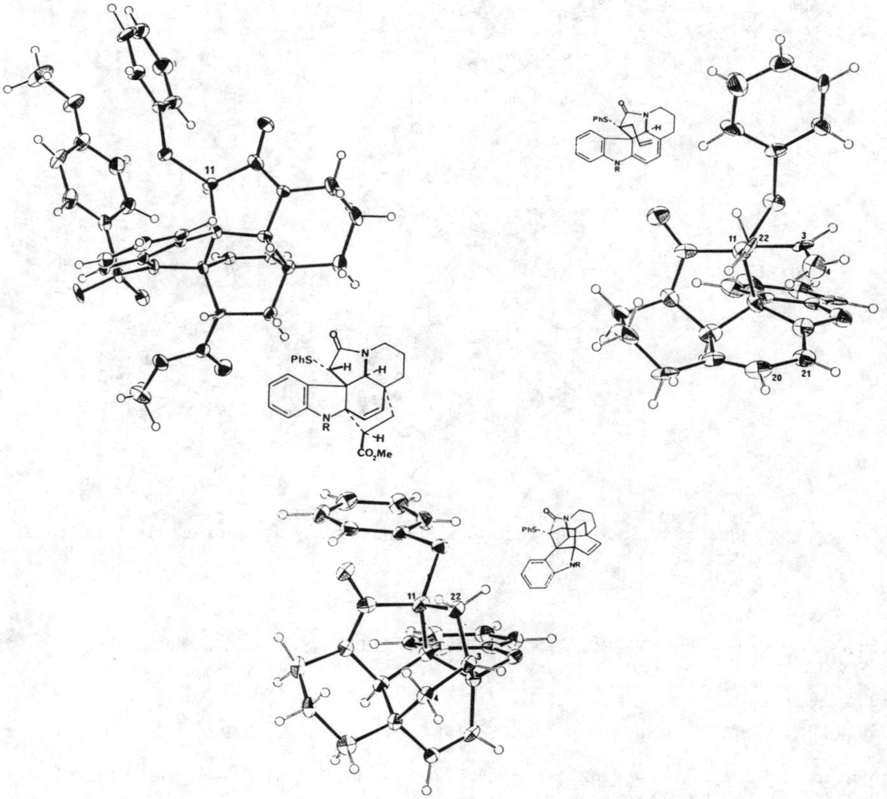

Fig. 1. Views of $C_{33}H_{32}N_2O_6S_2$ (I) (top left), $C_{33}H_{30}N_2O_4S_2$ (II) (right) and $C_{33}H_{30}N_2O_4S_2$ (III) (bottom left).

BISNORACRONYCINE
$C_{38}H_{34}N_2O_6$ (I)

ISONORACRONYCINE-NORACRONYCINE
$C_{38}H_{34}N_2O_6$ (II)

S. FUNAYAMA, G.A. CORDELL, H. WAGNER and H.L. LOTTER, 1984. J. Nat. Prod., 47, 143-154.

I. Triclinic, P1, a = 10.717, b = 12.479, c = 14.431 Å, α = 91.06, β = 103.53, γ = 110.77°, D_m = 1.284, Z = 2. R = 0.087 for 3033 reflexions.

II. Triclinic, P$\bar{1}$, a = 9.479, b = 10.443, c = 17.836 Å, α = 74.16, β = 79.39, γ = 63.04°, D_m = 1.318, Z = 2. R = 0.093 for 2686 reflexions.

The structures of these isomers were determined by X-ray methods. The connectivity and dimensions are in Fig. 1.

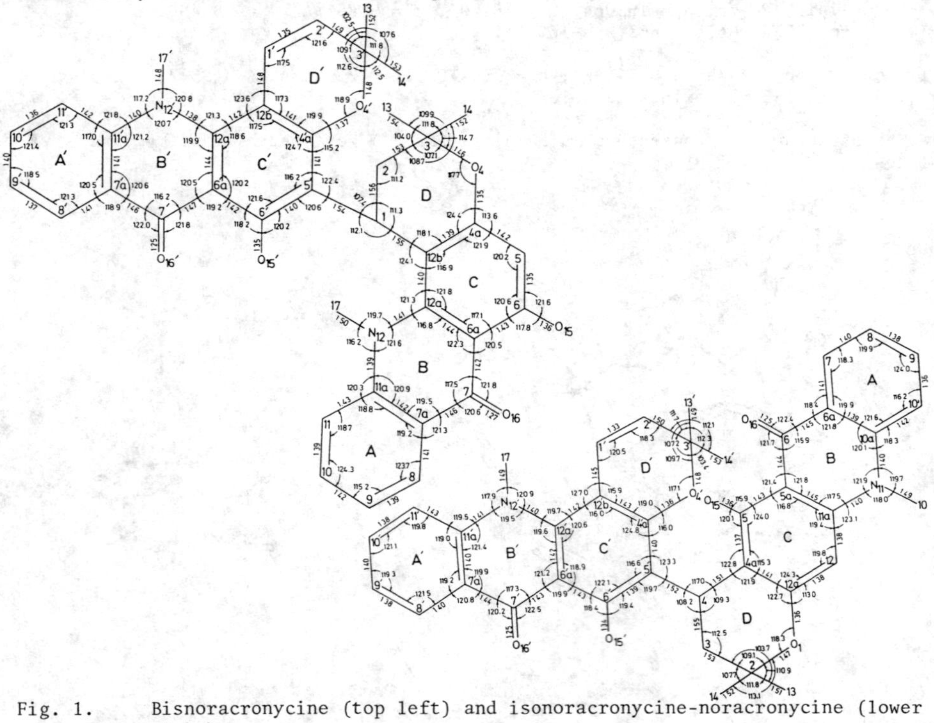

Fig. 1. Bisnoracronycine (top left) and isonoracronycine-noracronycine (lower right).

(5S,6R,8S,9S)-3-(1,2-EPOXYPROPYL)-5,6-DIHYDRO-5-HYDROXY-6-METHYLPYRAN-2-ONE (ASPYRONE)
$C_9H_{12}O_4$

M.J. GARSON, J. STAUNTON and P.G. JONES, 1984. J. Chem. Soc. Perkin I, 1021-1026.

Orthorhombic, $P2_12_12_1$, a = 4.427, b = 14.272, c = 14.968 Å, Z = 4. Mo radiation, R = 0.033 for 1131 reflexions.

Aspyrone (Fig. 1) is a polyketide metabolite from 'Aspergillus melleus'. The epoxy propyl sidechain is trans and the centres at C(5) and C(6) are defined as S and

R respectively. The configurations at C(8) and C(9) are both S. Some bond lengths in the molecule are: C(2)-O(1) 1.341(3), C(6)-O(1) 1.463(3), C(2)-O(2) 1.209(3), C(5)-O(5) 1.419(4), C(8)-C(9) 1.441(5), C(8)-O(8) 1.415(5), C(9)-O(8) 1.426(5), C(10)-C(9) 1.482(4) Å.

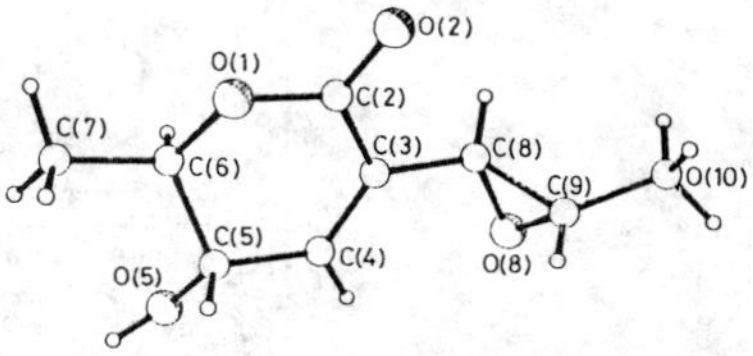

Fig. 1. $C_9H_{12}O_4$: perspective view of the molecule.

CITRININ
$C_{13}H_{14}O_5$

R. DESTRO and R.E. MARSH, 1984. J. Am. Chem. Soc., <u>106</u>, 7269-7271.

Orthorhombic, $P2_12_12_1$, a = 13.441, b = 7.291, c = 12.241 Å, D_m = 1.385, Z = 4. Mo radiation (at 290 K); a = 13.358, b = 7.226, c = 12.170 Å (at 147 K).

An uncertainty as to whether citrinin (Fig. 1) the p-quinonemethide formulation (1a) or the o-quinone form (1b) is resolved. At 147 K, difference maps show that an ordered form of (1a) is the only form present in the crystal structure. At 290 K difference maps are interpreted in favour of a 60/40 disordered model of the p-quinone form (1a).

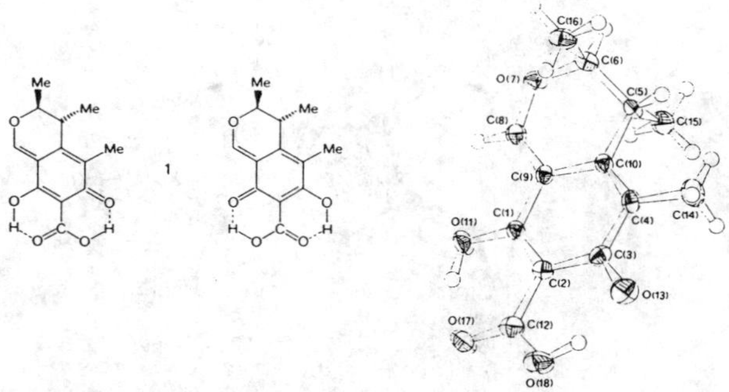

Fig. 1. Possible forms of citrinin and a view of the form at 147 K.

(-)-EPICATECHIN
$C_{15}H_{14}O_6$

F.R. FRONCZEK, G. GANNUCH, W.L. MATTICE, F.L. TOBIASON, J.L. BROEKER and R.W. HEMINGWAY, 1984. J. Chem. Soc. Perkin II, 1611-1616.

Orthorhombic, $P2_12_12_1$, a = 6.708, b = 13.291, c = 14.262 Å, Z = 4. Mo radiation, R = 0.041 for 1624 reflexions.

The aliphatic hydroxy group of this monomer unit of procyanidin polymers and three of the four phenolic groups participate in intermolecular hydrogen bond formation (Fig. 1). There are no intramolecular hydrogen bonds. The aromatic ring bonded to C(2) is approximately planar and is in an equatorial position while the hydroxy group bonded to C(3) adopts an axial position. The heterocyclic ring adopts a half-chair conformation.

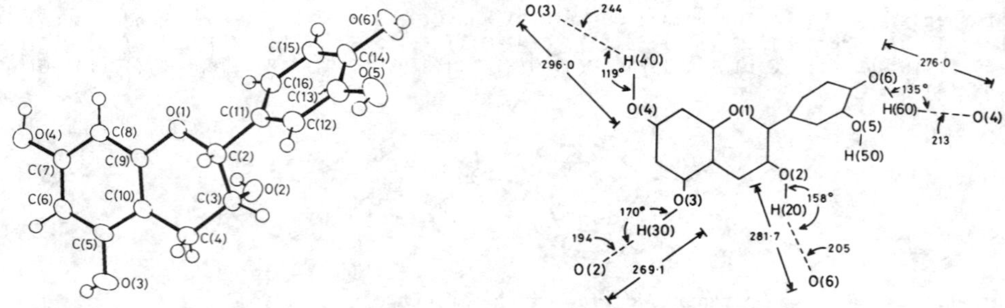

Fig. 1. $C_{15}H_{14}O_6$: the molecular structure and hydrogen bonding scheme.

3,4,6-TRIMETHOXYDIBENZOFURAN-2,7-DIOL (COTONEFURAN)
$C_{15}H_{14}O_6$

R.S. BURDEN, M.S. KEMP, C.W. WILTSHIRE and J.D. OWEN, 1984. J. Chem. Soc. Perkin I, 1445-1448.

Orthorhombic, $Pna2_1$, a = 9.870, b = 8.163, c = 16.836 Å, Z = 4. Mo radiation, R = 0.040 for 1685 reflexions.

The X-ray analysis established the structure of this anti-fungal compound, isolated from necrotic wood of a fungal-infected specimen of 'Cotoneaster lactea'. The molecule is essentially planar except for the methyl groups which are rotated 60° from the coplanar position on the same side of the molecule (Fig. 1). Each molecule is hydrogen bonded to four others (O...O 2.779(3)-2.901(3) Å) to form a three-dimensional network.

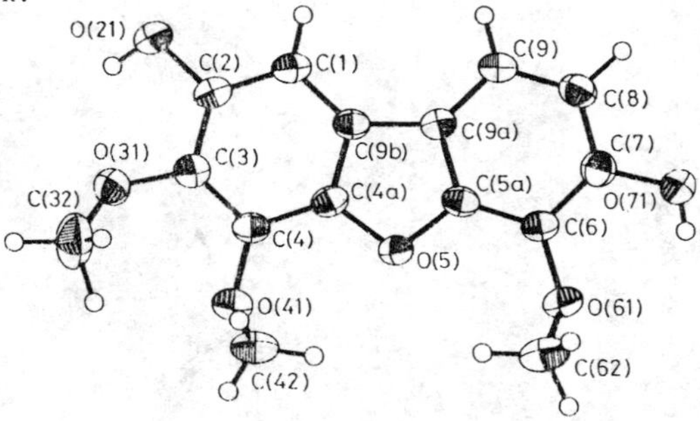

Fig. 1. $C_{15}H_{14}O_6$: the molecular structure.

SAMBUCINOL
$C_{15}H_{22}O_4$ (I)

SAMBUCOIN
$C_{15}H_{22}O_3$ (II)

P. MOHR, C. TAMM, W. ZURCHER and M. ZEHNDER, 1984. Helv. Chim. Acta, 67, 406-412.

I. Orthorhombic, $P2_12_12_1$, a = 6.609, b = 13.057, c = 15.531 Å, Z = 4, D_m = 1.32. Mo radiation, R = 0.076 for 1772 reflexions.

II. Orthorhombic, $P2_12_12_1$, a = 7.248, b = 11.918, c = 15.289 Å, Z = 4, D_m = 1.26. Mo radiation, R = 0.069 for 1745 reflexions.

The analysis established the structures of these mould metabolites as shown in Fig. 1.

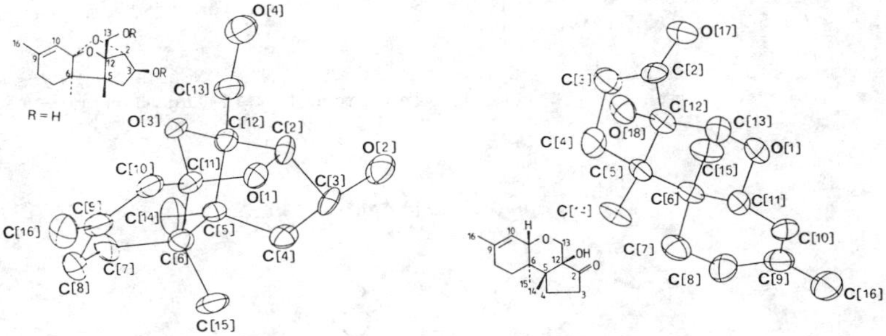

Fig. 1. Views of sambucinol (left) and sambucoin (right).

PSEUDOANISATIN
$C_{15}H_{22}O_6$ (I)

4-DEHYDROXY-7-DEOXO-7-HYDROXY-PSEUDOANISATIN MONOHYDRATE
$C_{15}H_{22}O_6,H_2O$ (II)

I. KOUNO, H. IRIE and N. KAWANO, 1984. J. Chem. Soc. Perkin I, 2511-2515.

I. Monoclinic, C2, a = 18.424, b = 7.828, c = 10.279 Å, β = 97.60°, D_m = 1.357, Z = 4. Cu radiation, R = 0.070 for 1141 reflexions.

II. Orthorhombic, $P2_12_12_1$, a = 19.229, b = 10.757, c = 7.440 Å, D_m = 1.360, Z = 4. Cu radiation, R = 0.050 for 1332 reflexions.

The X-ray analyses established the structure of pseudoanisatin (Fig. 1) which was isolated from Japanese star anise, 'Illicium anisatum'. The molecule has a ketonic group at C(7), a tertiary hydroxy group at C(4) and a lactone ring between the carboxy C(11) and the primary alcohdic C(14) group. Isomer II was prepared by refluxing I in methanol with potassium carbonate. The rearrangement of I into II involves steric inversions at C(3) and C(4) in addition to translactonisation into a stable γ-lactone ring and acetal ring formation between C(7) and C(4).

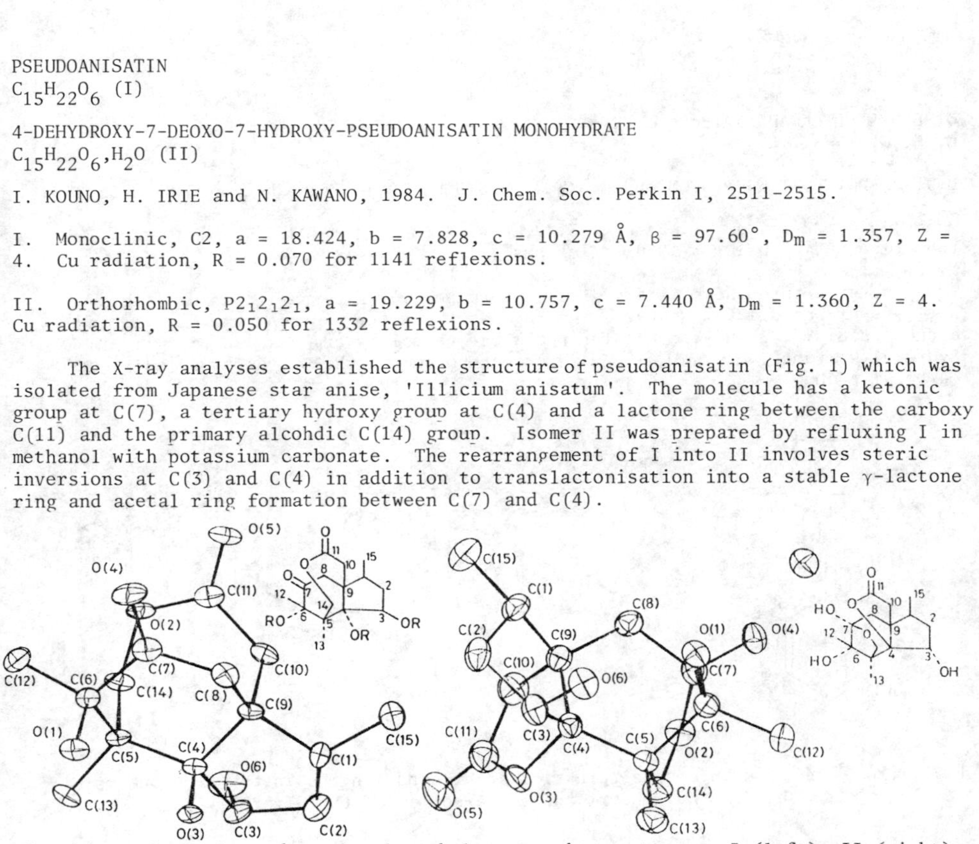

Fig. 1. $C_{15}H_{22}O_6$: the molecular skeletons and structures - I (left), II (right).

1,7-DIHYDROXY-3-METHOXY-2-METHYLDIBENZ[2,3-6,7]OXEPIN (PACHARIN)
$C_{16}H_{14}O_4$

A.S.R. ANJANEYULU, A.V.R. REDDY, D.S.K. REDDY, R.S. WARD, D. ADHIKESAVALU and T.S. CAMERON, 1984. Tetrahedron, 40, 4245-4252.

Monoclinic, P2/n, a = 17.170, b = 4.9154, c = 15.514 Å, β = 98.085°, Z = 4. Cu radiation, R = 0.056 for 1650 reflexions.

The structure of pacharin is established (Fig. 1). The oxepin ring has approximate C_2 symmetry with the C_2 axis passing through O(11) and the mid-point of C(5)-C(6).

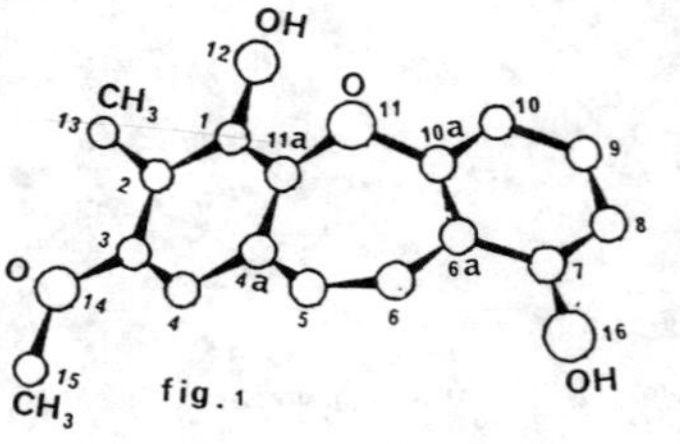

Fig. 1. Pacharin.

DIMETHYL ALTERNARIANATE
$C_{17}H_{16}O_8$

D.J. CHADWICK, I.W. EASTON and R.A.W. JOHNSTONE, 1984. Tetrahedron, 40, 2451-2455.

Monoclinic, $P2_1/c$, a = 8.42, b = 13.42, c = 14.64 Å, β = 104.7°, Z = 4. Cu radiation, R = 0.071 for 1970 reflexions.

The structure shown in Fig. 1 is determined for alternarian acid by this analysis of the dimethyl ester. The angle between the α-pyrone and phenyl ring planes is 81.5°. Bond lengths for the α-pyrone moiety are also in Fig. 1. The hydroxyl H atom is intramolecularly hydrogen-bonded to the O atom of an adjacent carboxyl group.

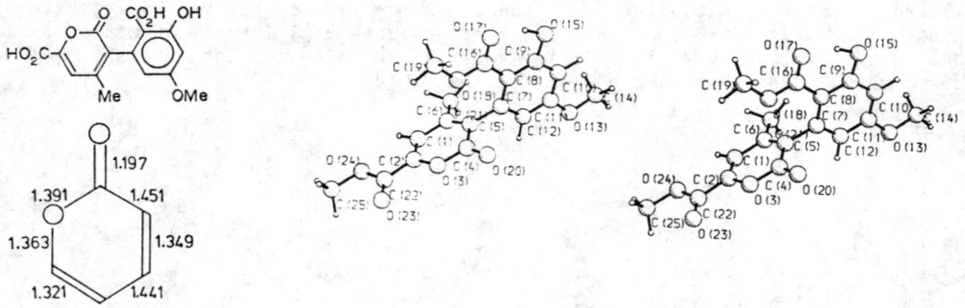

Fig. 1. The structure of alternarian acid, bond lengths in the α-pyrone moiety of the dimethyl ester, and a stereoview of the dimethyl ester.

17,18,19-TRISNORSTEMODAN-3,13-DIONE
$C_{17}H_{24}O_2$

S.V. EVANS and J. TROTTER, 1984. Can. J. Chem., 62, 1828-1830.

Monoclinic, $P2_1/c$, a = 10.395, b = 10.980, c = 13.181 Å, β = 113.55°, Dm = 1.26, Z = 4. Mo radiation, R = 0.051 for 1622 reflexions.

X-ray analysis establishes the structure shown in Fig. 1. Molecular dimensions are normal. Atom C(12) is disordered over two sites with 75/25% occupancy. In the major conformer, all six-membered rings have chair conformation; in the minor conformer, the ring containing C12 is in a shallow boat form. The five-membered ring has a half-chair conformation with C(9) and C(16) displaced in opposite directions from the C(8)C(15)C(14) plane.

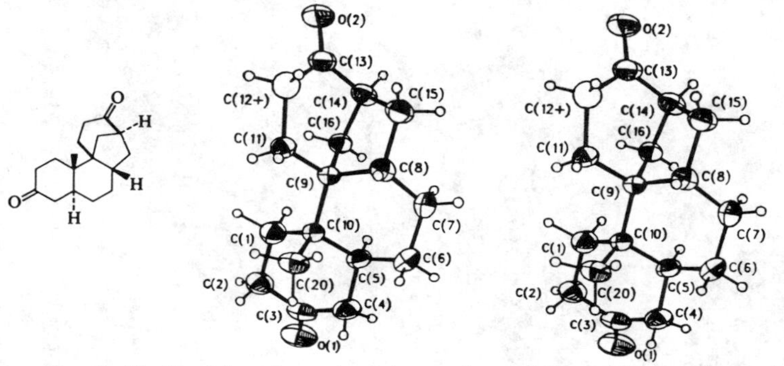

Fig. 1. The $C_{17}H_{24}O_2$ molecule and a stereoview of the major conformer.

NEODUNOL METHYL ETHER
$C_{18}H_{14}O_4$

R. SANDUJA, G.E. MARTIN, A.J. WEINHEIMER, M. ALAM, M.B. HOSSAIN and D. VAN DER HELM, 1984. J. Heterocycl. Chem., 21, 845-848.

Monoclinic, $P2_1$, a = 10.802, b = 13.218, c = 4.648 Å, β = 94.75°, Z = 2. Mo radiation, R = 0.032 for 1425 reflexions (at 138 K).

The analysis established the absolute configuration (Fig. 1). The molecule has a folded conformation such that all non-H atoms lie approximately in two planes (one through rings A, B and E, and the other through rings C and D) which are mutually inclined at 80°. The pyran ring is in a half-chair and the furan ring in a slightly distorted envelope conformation.

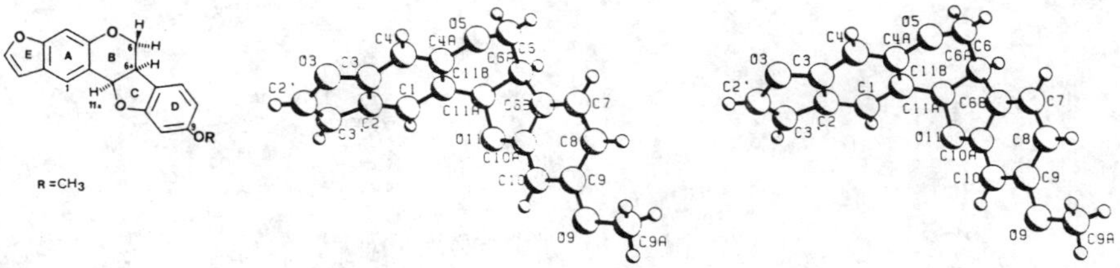

Fig. 1. Neodunol methyl ether (R = Me) and a stereoview.

CROTEPOXIDE
$C_{18}H_{18}O_8$

O. PANCHAROEN, V.A. PATRICK, V. REUTRAKUL, P. TUNTIWACHWUTTIKUL and A.H. WHITE, 1984. Aust. J. Chem., 37, 221-225.

Orthorhombic, $P2_12_12_1$, a = 23.53, b = 10.59, c = 6.99 Å, Z = 4. Mo radiation, R = 0.059 for 682 reflexions.

The bonds of the cyclohexane ring (Fig. 1) in the string C(1-4) are all 1.51 Å or greater, while those in string (C4-6,1) are between 1.46-1.47 Å including C(5)-C(6) which links two epoxide junctions.

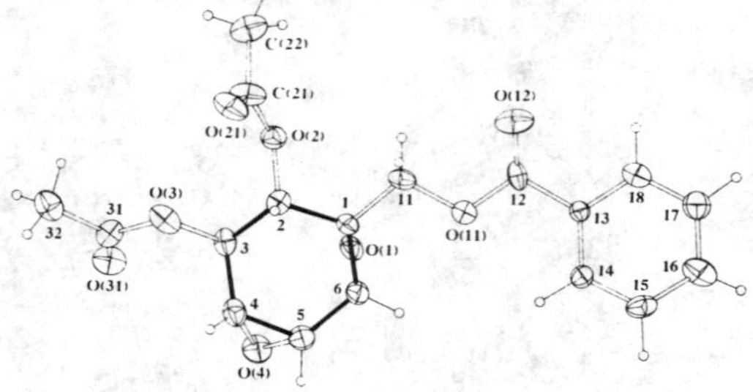

Fig. 1. The structure of crotepoxide.

7-DEACETAMIDOCOLCHICEINE
$C_{19}H_{20}O_5$

J. RIUS, E. MOLINS, C. MIRAVITLLES and A. BLADÉ-FONT, 1984. Acta Cryst., C40, 839-842.

Triclinic, P$\bar{1}$, a = 14.462, b = 8.185, c = 7.864 Å, α = 62.50, β = 82.18, γ = 78.34°, D_m = 1.34, Z = 2. Cu radiation, R = 0.060 for 2170 reflexions.

The troponoid ring exhibits an arrangement corresponding to an isocolchicine-like structure (Fig. 1). The benzene ring is almost planar while the troponoid ring has a skew-boat conformation and the other seven-membered ring a boat conformation. There is a possible intermolecular hydrogen bond between the O= and OH- groups on the troponoid ring but no other short intermolecular distances exist.

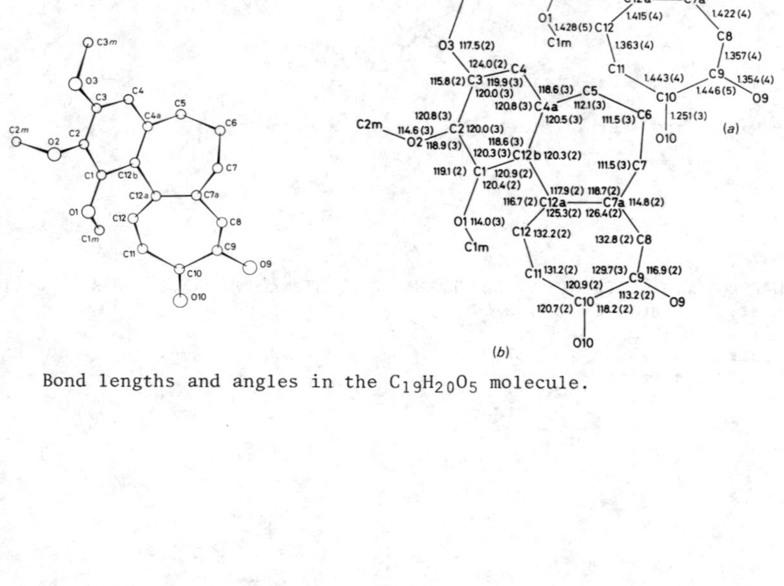

Fig. 1. Bond lengths and angles in the $C_{19}H_{20}O_5$ molecule.

(5R,10S)-11,20-DIHYDROXY-12,16-DIOXO-1,2-SECO-1-NOR-6(5→10)abeo-PICRASA-3,9(11)-DIEN-2,5-OLIDE (SHINJULACTONE B)
$C_{19}H_{22}O_7,H_2O$

T. FURUNO, M. ISHIBASHI, H. NAORA, T. MURAE, H. HIROTA, T. TSUYUKI, T. TAKAHASHI, A. ITAI and Y. IITAKA, 1984. Bull. Chem. Soc. Japan, 57, 2484-2489.

Orthorhombic, $P2_12_12_1$, a = 12.594, b = 19.016, c = 7.755 Å, Z = 4. Cu radiation, R = 0.039 for 2096 reflexions.

The absolute configuration of shinjulactone B has been established by X-ray crystallography to be as in Fig. 1.

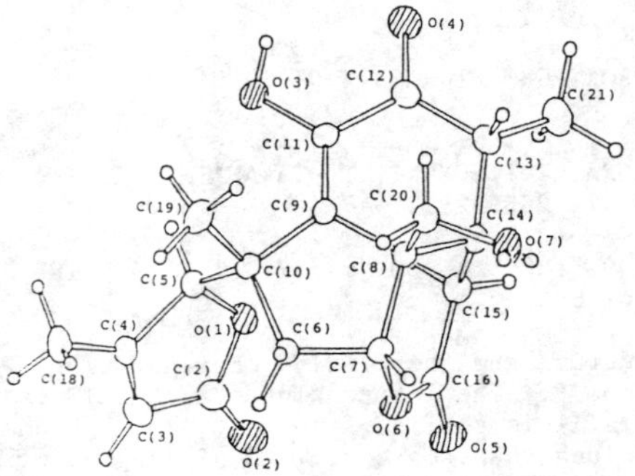

Fig. 1. Perspective view of $C_{19}H_{22}O_7$.

BERBERINE HYDROGEN SULPHATE
$C_{20}H_{19}NO_8S$ $(C_{20}H_{18}NO_4)^+.HSO_4^-$

B.E.A. ABADI, D.S. MOSS and R.A. PALMER, 1984. J. Cryst. Spectrosc. Res., 14, 269-281.

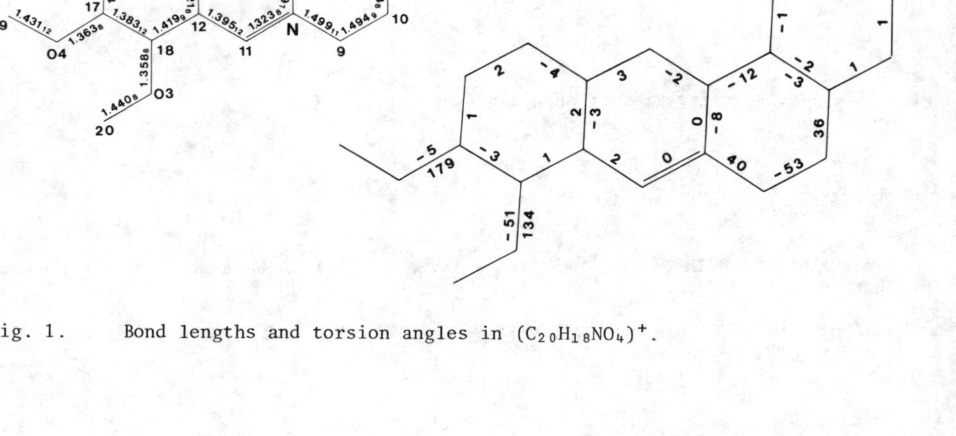

Fig. 1. Bond lengths and torsion angles in $(C_{20}H_{18}NO_4)^+$.

Triclinic, F1, a = 20.370, b = 7.435, c = 27.427 Å, α = 97.700, β = 116.222, γ = 85.456°, D_m = 1.556, Z = 8. Cu radiation, R = 0.070 for 1659 reflexions.

The ring system of the berberine molecule (Fig. 1) is approximately planar, with both $-OCH_3$ groups twisted slightly out of the plane. Heterocyclic ring B has a half-chair conformation with a psuedo-twofold axis bisecting the bonds C(9)-C(10) and C(7)-C(8), diagonal to the isoquinoline group AB. The five-membered ring E has an envelope conformation, the apical atom C(1) being out of plane. The HSO_4^- ion is a distorted tetrahedron, showing two short S-O distances of 1.418(5) Å, one intermediate of 1.489(5) Å, and a long 1.552(5) Å bond, and O-S-O bond angles ranging from 105.0(3) to 114.2(3)°. The hydrogen on O(6) was not located in the analysis.

1α,11-EPOXY-1β,20-DIHYDROXY-5βH-PICRASA-3,9(11)-DIENE-2,12,16-TRIONE DIHYDRATE
$C_{20}H_{22}O_7,2H_2O$

M. ISHIBASHI, S. YOSHIMURA, T. TSUYUKI, T. TAKAHASHI, A. ITAI and Y. IITAKA, 1984. Bull. Chem. Soc. Japan, 57, 2885-2892.

Monoclinic, $P2_1$, a = 8.392, b = 14.920, c = 7.581 Å, β = 105.3°, Z = 2. Cu radiation, R = 0.036 for 1883 reflexions.

The hemiacetal structure and the C(5β)-H configuration (Fig. 1) establish the structure of shinjulactone F. The interesting structural feature is the 5β-H picrasane skeleton with all-cis configurations at A/B, B/D and C/D ring junctures, resulting in a unique folded-shape molecule. Bond lengths of interest are: C(1)-C(2) 1.542, C(1)-C(10) 1.556, C(1)-O(1) 1.370, C(1)-O(3) 1.475, C(2)-C(3) 1.442, C(2)-O(2) 1.231, C(3)-C(4) 1.332, C(4)-C(5) 1.511, C(5)-C(6) 1.551, C(5)-C(10) 1.544, C(6)-C(7) 1.528, C(7)-C(8) 1.538, C(7)-O(6) 1.473, C(8)-C(9) 1.494, C(8)-C(14) 1.543, C(9)-C(10) 1.514, C(9)-C(11) 1.338, C(11)-C(12) 1.458, C(11)-O(3) 1.381, C(12)-C(13) 1.523, C(12)-O(4) 1.226, C(13)-C(14) 1.548, C(14)-C(15) 1.538, C(15)-C(16) 1.496, C(16)-O(5) 1.222, C(16)-O(6) 1.322, C(20)-O(7) 1.403 Å.

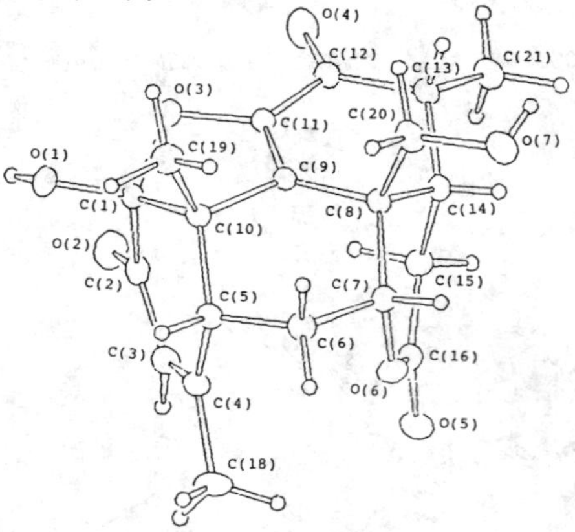

Fig. 1. Perspective view of shinjulactone F ($C_{20}H_{22}O_7$).

CORDIFOLIA-P2
$C_{20}H_{26}O_7$

M.J. BEGLEY, L. CROMBIE, W.M.L. CROMBIE, A.K. GATUMA and A. MARADUFU, 1984. J. Chem. Soc. Perkin I, 819-824.

Orthorhombic, $P2_12_12_1$, a = 6.959, b = 13.949, c = 19.762 Å, Z = 4. Cu radiation, R = 0.057 for 823 reflexions.

Cordifolia-P2 (Fig. 1) is one of four new compounds isolated from the insect antifeedant plant 'Erlangea cordifolia'. Bond lengths and angles are not of high accuracy. The ten-membered ring exhibits a different conformation from that of cordifene (1). For example the C(5)-C(6)-C(7)-C(8) torsion angle is 99° for cordifolia-P2 but 159° for cordifene. Only the torsion angles in the C(2)-C(1)-C(10)-C(9) portion of the molecule are comparable for the two compounds. Bond lengths also show marked differences with C(2)-C(3) (1.52(2)) and C(1)-C(2) (1.54(1)) being increased and C(4)-C(5) (1.48(1)) and C(5)-C(6) (1.49(2) Å) decreased compared with cordifene. The large bond angle at C(18) in (1) is observed here (C(17)-C(18)-C(19) 135.3(18)°). The angelate side-chain torsion angle is 40.6° (O(16)-C(16)-C(17)-C(18)). The five-membered ring adopts a flattened conformation.

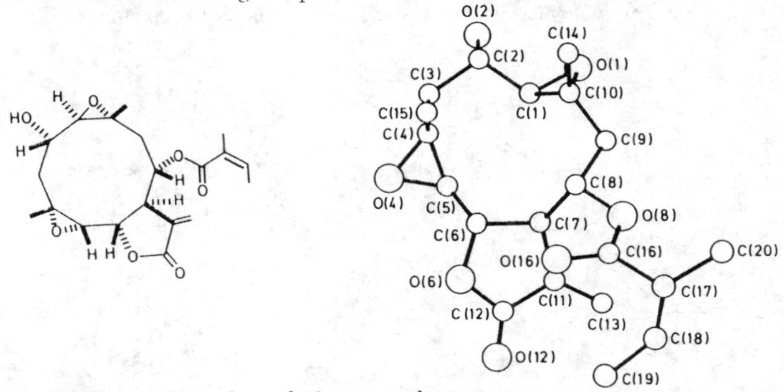

Fig. 1. $C_{20}H_{26}O_7$: molecular skeleton and structure.

1. Structure Reports, 48B, 724.

6-((1(S)-(3(S),4-DIHYDRO-8-HYDROXY-1-OXO-1H-2-BENZOPYRAN-3-YL)-3-METHYLBUTYL)AMINO)-4(S),5(S)-DIHYDROXY-6-OXO-3(S)-AMINOHEXANOIC ACID TETRAHYDRATE
$C_{20}H_{28}N_2O_8,4(H_2O)$

Y. SHIMOJIMA, H. HAYASHI, T. OOKA, M. SHIBUKAWA and Y. IITAKA, 1984. Tetrahedron, 40, 2519-2527.

Monoclinic, $P2_1$, a = 15.902, b = 10.161, c = 7.571 Å, β = 95.28°, Z = 2. Cu radiation, R = 0.054 for 2685 reflexions.

The structure determined by X-ray methods is shown in Fig. 1. All five asymmetric centres have the S-configuration.

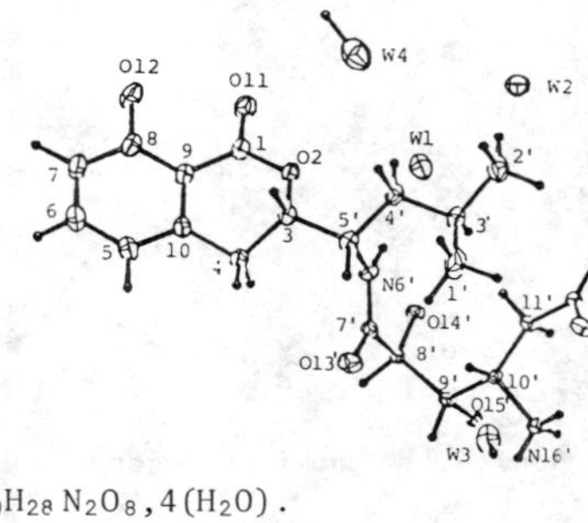

Fig. 1. A view of $C_{20}H_{28}N_2O_8,4(H_2O)$.

(±)-trans-CANNABITRIOL
$C_{21}H_{30}O_4$

A.T. McPHAIL, H.N. ELSOHLY, C.E. TURNER and M.A. ELSOHLY, 1984. J. Nat. Prod., 47, 138-142.

Monoclinic, $P2_1/c$, a = 14.993, b = 12.712, c = 10.603 Å, β = 96.05°, Z = 4. Cu radiation, R = 0.059 for 2535 reflexions.

The molecule is shown in Fig. 1. The cyclohexene ring approximates to a half-chair form; the 1,2-pyran ring adopts a 1,3-diplanar form. Molecules are linked by extensive O-H...O hydrogen bonding. Dimensions are as anticipated.

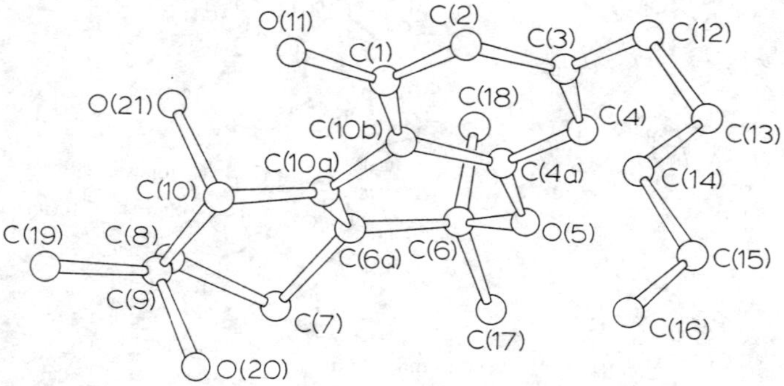

Fig. 1. A view of trans-cannabitriol.

(15R)-15-METHYL-12-epi-PROSTAGLANDIN F(2β)
$C_{21}H_{36}O_5$

K. WEINGES, W. HUBER, U. HUBER-PATZ, H. IRNGARTINGER, M. NIXDORF and H. RODEWALD, 1984. Justus Liebigs Ann. Chem., 761-772.

Monoclinic, $P2_1$, a = 9.491, b = 9.230, c = 24.970 Å, β = 94.20°, Z = 4. Mo radiation, R = 0.048 for 1562 reflexions (at -70°C).

There are two independent molecules (A and C) in the asymmetric unit. Both have the five-membered ring in a C12 envelope conformation (C12 0.70 Å (A), 0.68 Å (C) from the plane of the other four atoms). The double bond at C13-C14 has E configuration; that at C5=C6 is Z.

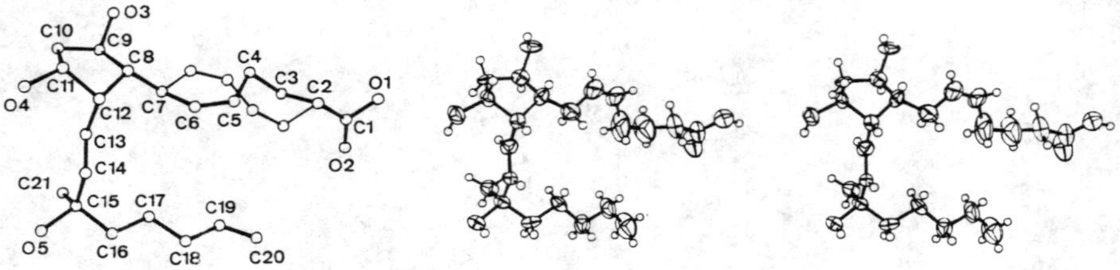

Fig. 1. A view of molecule A with the numbering scheme and a stereoview of molecule C of $C_{21}H_{36}O_5$.

(+)-12-BROMOISOSTEGANE
$C_{22}H_{21}BrO_7$

K. TOMIOKA, T. ISHIGURO, Y. IITAKA and K. KOGA, 1984. Tetrahedron, 40, 1303-1312.

Monoclinic, $P2_1$, a = 12.218, b = 23.107, c = 7.353 Å, β = 90.60°, Cu radiation, R = 0.056 for 2530 reflexions.

The analysis establishes the absolute stereochemistry of the molecule as shown in Fig. 1.

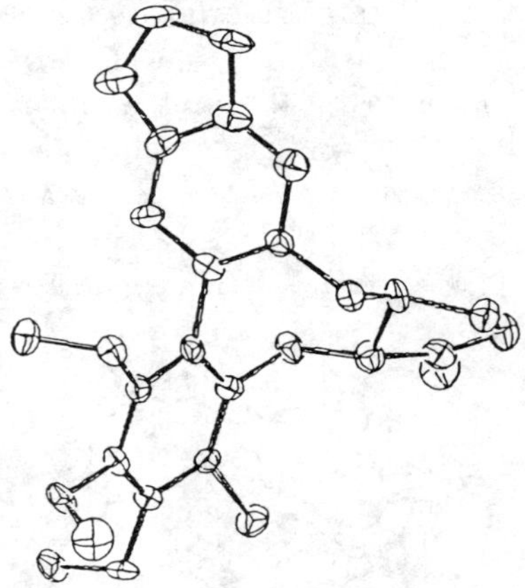

Fig. 1. The absolute stereochemistry of (+)-12-bromoisostegane.

TRICHORABDAL F ACETATE
$C_{22}H_{28}O_7$

R.P. KASHYAP, W.H. WATSON, D.A. GROSSIE, M. NODE, M. SAI, E. FUJITA and K. FUJI, 1984. Acta Cryst., C40, 515-517.

Orthorhombic, $P2_12_12_1$, a = 15.199, b = 17.334, c = 7.576 Å, Z = 4. Mo radiation, R = 0.040 for 914 reflexions.

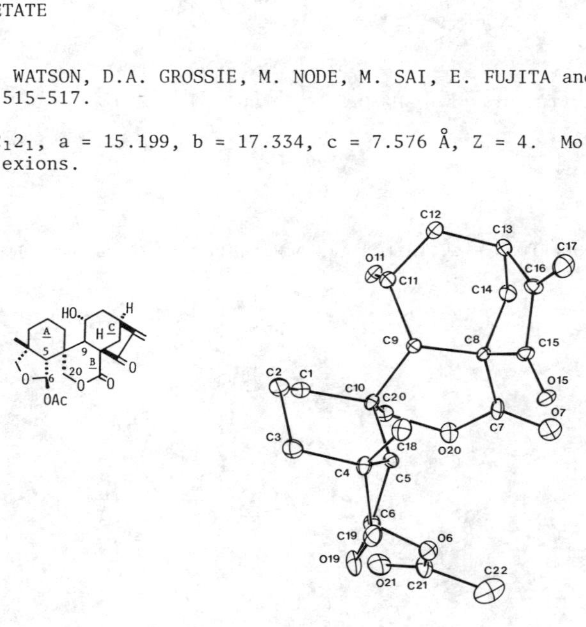

Fig. 1. Trichorabdal F acetate.

The trichorabdal skeleton (Fig. 1) contains spiro-fused bicyclic and tricyclic ring systems. The bicyclic system is composed of a cyclohexane ring in a chair conformation cis-fused to a five-membered hemiacetal in a half-chair conformation. The tricyclic ring system contains a cyclohexane ring in a skew-boat conformation, a six-membered lactone ring in a 1,3-diplanar conformation and a five-membered ketone in an envelope conformation with an exocyclic double bond.

(3aR,4S,6E,10E,14R,15R,15aS)-4-ACETOXY-14,15-EPOXY-6,10,14-TRIMETHYL-3-METHYLENE-3a,4,5,8,9,12,13,14,15,15a-DECAHYDROCYCLOTETRADECA[b]FURAN-2(3H)-ONE
$C_{22}H_{30}O_5$

B.F. BOWDEN, J.C. COLL, M.S.L. DE COSTA, M.F. MACKAY, M. MAHENDRAN, E.D. DE SILVA and R.H. WILLIS, 1984. Aust. J. Chem., 37, 545-552.

Orthorhombic, $P2_12_12_1$, a = 8.152, b = 9.915, c = 26.590 Å, Z = 4. Cu radiation, R = 0.054 for 1350 reflexions.

The compound (mp. 146°C) was extracted from Sri Lanka soft coral Lobophytum cristigalli.

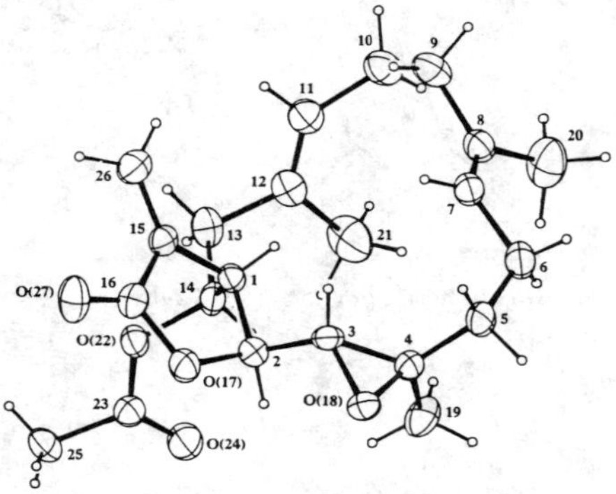

Fig. 1. The structure of $C_{22}H_{30}O_5$.

APLYSULPHURIN
$C_{22}H_{32}O_5$

P. KARUSO, B.W. SKELTON, W.C. TAYLOR and A.H. WHITE, 1984. Aust. J. Chem., 37, 1081-1093.

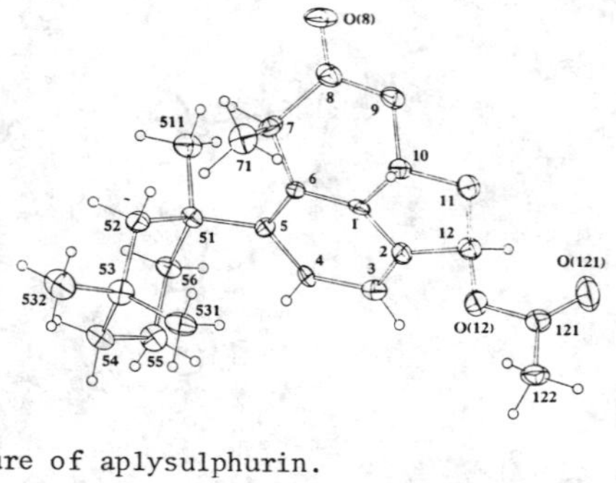

Fig. 1. The structure of aplysulphurin.

Orthorhombic, $P2_12_12_1$, a = 13.79, b = 12.493, c = 11.56 Å, Z = 4. Mo radiation, R = 0.051 for 785 reflexions.

The compound (Fig. 1) is isolated from Aplysilla sulphurea, an encrusting sea sponge, as a bright yellow compound which changes to blue-purple on collection and exposure to air.

7,11-EPOXY-12-HYDROXY-1,17-DIMETHOXY-16,17-seco-PICRASIN
$C_{22}H_{38}O_4$

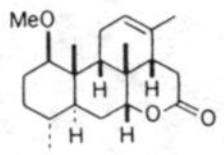

G. VIDARI, S. FERRINO and P.A. GRIECO, 1984. J. Am. Chem. Soc., 106, 3539-3548.

Monoclinic, $P2_1/n$, a = 8.522, b = 22.773, c = 10.190 Å, β = 111.32°, Z = 4. R = 0.066 for 1713 reflexions (at -180°C).

The analysis establishes the structure shown above.

DIACETYLTERFEROL
$C_{23}H_{20}O_5$

F. NAKAGAWA, S. TAKAHASHI, A. NAITO, S. SATO, S. IWABUCHI and C. TAMURA, 1984. J. Antibiot., 37, 10-12.

Monoclinic, Cc, a = 21.022, b = 11.656, c = 8.238 Å, β = 90.82°, Z = 4. Cu radiation, R = 0.059 for 1435 reflexions.

The structure of the diacetyl derivative of terferol was established to be as shown in Fig. 1.

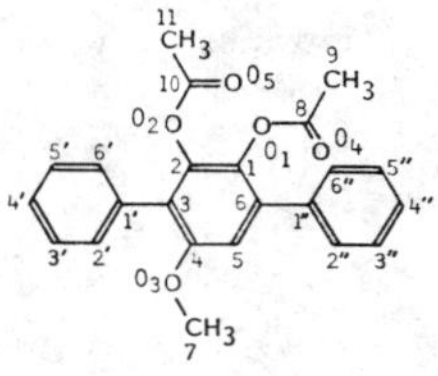

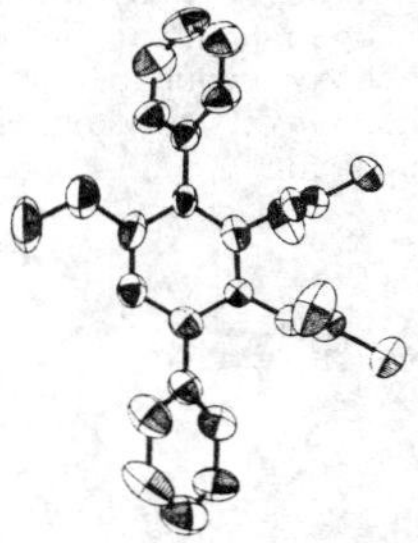

Fig. 1. Diacetylterferol.

3'-O-ACETYL-4'-O-BENZOYLKHELLACTONE
$C_{23}H_{20}O_7$

M. HAMBURGER, H. STOECKLI-EVANS and K. HOSTETTMANN, 1984. Helv. Chim. Acta, 67, 1729-1733.

Orthorhombic, $P2_12_12_1$, a = 9.06, b = 13.86, c = 16.42 Å, Z = 4. Mo radiation, R = 0.062 for 1348 reflexions.

The structure of this coumarin diester is established as shown in Fig. 1. The dihydropyranoside ring has a half-chair conformation. The acid substituents at C(13) and C(14) are cis (torsion angle O(18)-C(13)-C(14)-O(22) 48.5°). The coumarin moiety is planar.

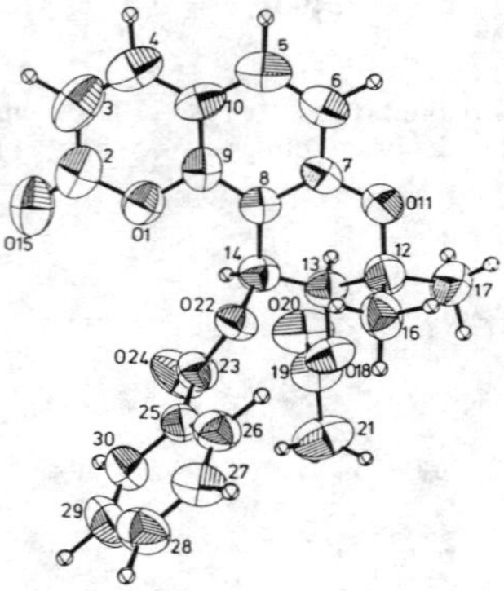

Fig. 1. A view of $C_{23}H_{20}O_7$.

(5S)-1,7-BIS(6-BROMO-3,4-DIMETHOXYPHENYL)-5-HYDROXYHEPTAN-3-ONE
$C_{23}H_{28}Br_2O_6$

S. OHTA, T. AOKI, T. HIRATA and T. SUGA, 1984. J. Chem. Soc. Perkin I, 1635-1642.

Monoclinic, $P2_1$, a = 4.720, b = 18.806, c = 13.651 Å, β = 98.65°, D_m = 1.52, Z = 2. Mo radiation, R = 0.054 for 1933 reflexions.

The absolute configuration of the molecule (Fig. 1) was determined to be S. The heptane chain is not quite planar with torsion angles of 139, -58 and -59° at C(2)-C(3)-C(4)-C(5), C(3)-C(4)-C(5)-C(6) and C(4)-C(5)-C(6)-C(7), respectively. There is intermolecular hydrogen bonding between the H atom of the 5-hydroxy group and the O atom of the 4''-methoxy group of an adjacent molecule. In this way, molecules build up along the b axis to form an infinite chain. Contacts between chains are van der Waals in nature.

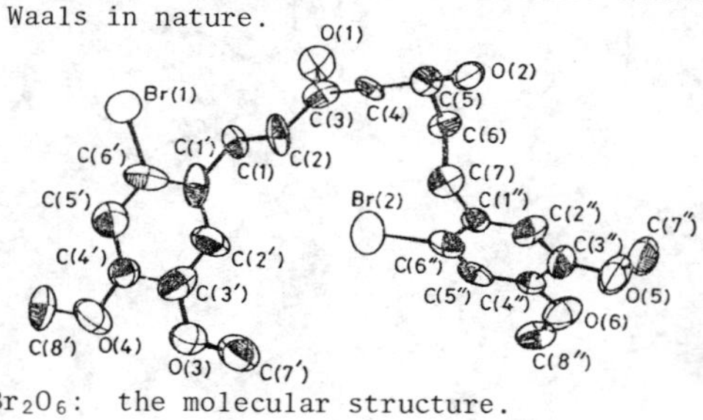

Fig. 1. $C_{23}H_{28}Br_2O_6$: the molecular structure.

TETRAMETHYLVIOLACEIN
$C_{24}H_{21}N_3O_3$

H. LAATSCH, R.H. THOMSON and P.J. COX, 1984. J. Chem. Soc. Perkin II, 1331-1339.

Monoclinic, $P2_1/n$, a = 14.253, b = 5.625, c = 29.658 Å, β = 124.71°, Z = 4. Mo radiation, R = 0.063 for 894 reflexions.

The individual rings in the molecule (Fig. 1) are planar, the angles between the indole ring systems and the pyrrole ring being 25.3 and -5.1° respectively. In-plane and out-of-plane distortions are greater about the C(8)-C(10) bond than about the C(13)-C(15) bond. High standard deviations associated with geometric parameters means it is not possible to draw firm conclusions regarding bond orders etc.

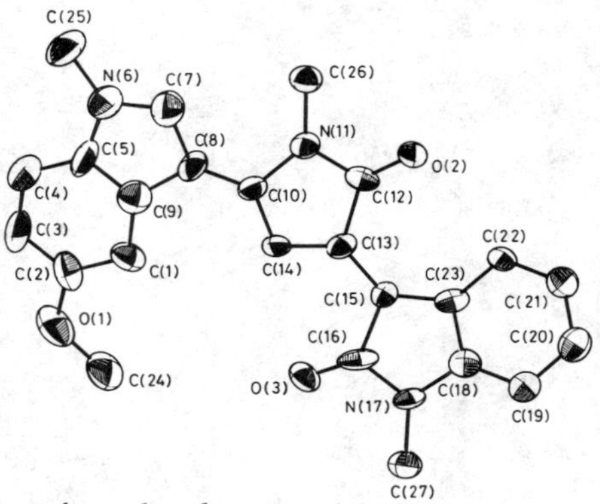

Fig. 1. $C_{24}H_{21}N_3O_3$: the molecular structure.

PHYLLANTHOCINDIOL METHYL ESTER
$C_{25}H_{32}O_8$

L.R. NASSIMBENI, M.L. NIVEN, G.M. CRAGG and G.R. PETTIT, 1984. Acta Cryst., C40, 146-149.

Monoclinic, $P2_1$, a = 9.891, b = 24.84, c = 9.687 Å, β = 94.82°, D_m = 1.21, Z = 4. Cu radiation, R = 0.072 for 3692 reflexions.

There are two independent molecules in the asymmetric unit (Fig. 1), both with similar folded conformations. Molecules are linked by a network of intra- and intermolecular hydrogen bonds. The configuration of the 7-OH group is established as axial (α) to the five-membered ring.

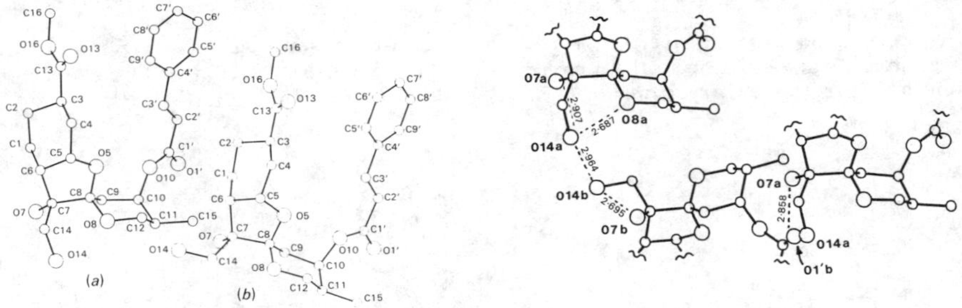

Fig. 1. A view of the two independent molecules of $C_{25}H_{32}O_8$ and details of hydrogen bonding.

ARENAROL DIACETATE
$C_{25}H_{34}O_4$

F.J. SCHMITZ, V. LAKSHMI, D.R. POWELL and D. VAN DER HELM, 1984. J. Org. Chem., 49, 241-244.

Orthorhombic, $P2_12_12_1$, a = 9.964, b = 6.81, c = 33.33 Å, D_m = 1.132, Z = 4. Cu radiation, R = 0.065 for 2268 reflexions (at 138 K).

The cis-fused rings (Fig. 1) are in chair conformations, with the exocyclic methylene at C4. The ring bearing the exocyclic methylene moiety is flattened around C(1) and C(2) because of steric interactions between C(2) and C(14) (C...C 3.242(7) Å).

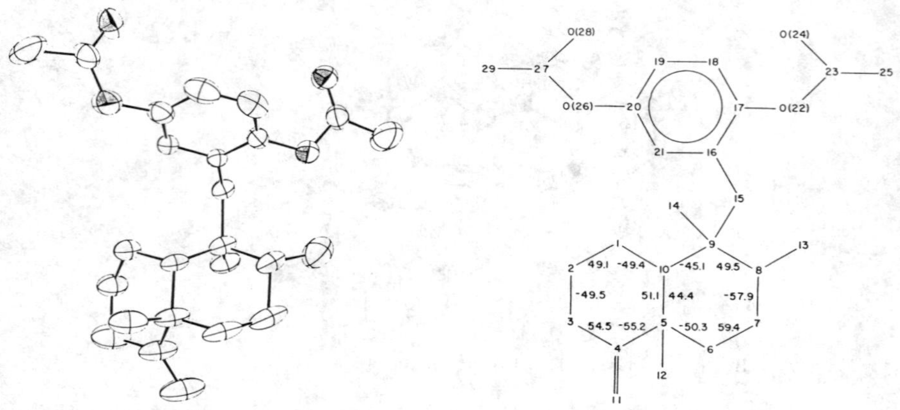

Fig. 1. A perspective view of the arenarol derivative and selected conformational angles.

PANDURATIN A (1'RS,2'SR,6'RS)-(2,6-DIHYDROXY-4-METHOXYPHENYL)[3'-METHYL-2'-(3''-METHYLBUT-2''-ENYL)-6'-PHENYLCYCLOHEX-3'-ENYL)METHANONE]
$C_{26}H_{30}O_4$ (I)

O. PANCHAROEN, V. REUTRAKUL, B.W. SKELTON, W.C. TAYLOR, P. TUNTIWACHWUTTIKUL and A.H. WHITE, 1984. Aust. J. Chem., 37, 2589-2592.

Monoclinic, $P2_1/n$, a = 10.140, b = 24.85, c = 9.415 Å, β = 90.69°, D_m = 1.14, Z = 4. Mo radiation, R = 0.055 for 1701 reflexions.

The compound was isolated from red rhizomes of Boesenbergia pandurata (Roxb.) used in Thai folk medicine. The proximity of C(2) to O(9) (2.803(5) Å) is presumably responsible for the enlargement of C(7)-C(8)-C(9) (126.3(4)°, Fig. 1).

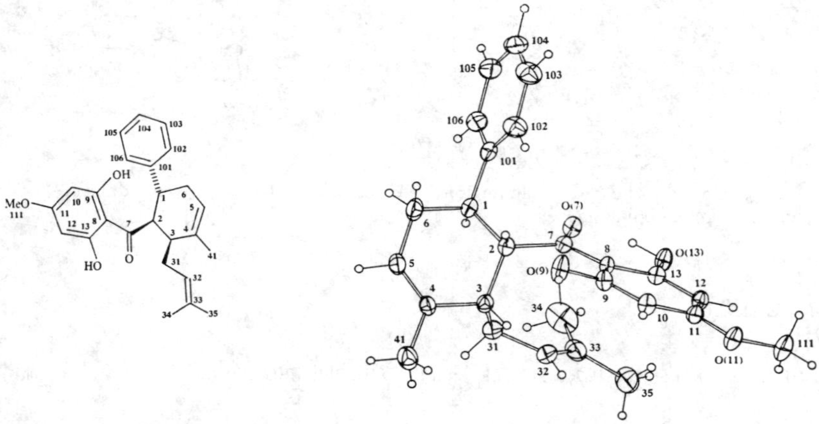

Fig. 1. The structure of panduratin A.

N[1]-{(4-BROMO-5,6,9,10,13,14,14a,15-OCTAHYDRO-1-HYDROXY-2,11-DIMETHOXY-3,12,16-TRIMETHYL-10,13-DIOXO-6,15-IMINO-7H-ISOQUINO[3,2-b][3]BENZAZOCIN-9-YL)METHYL}-ALANINAMIDE (SAFRACIN A BROMINATED PRODUCT)
$C_{28}H_{35}BrN_4O_6$

I. UEDA, S. KAWANO, Y. IKEDA, H. MATSUKI and T. OGAWA, 1984. Acta Cryst., C40, 1578-1580.

Orthorhombic, $P2_12_12_1$, a = 12.25, b = 17.773, c = 12.681 Å, D_m = 1.43, Z = 4. Mo radiation, R = 0.050 for 2299 reflexions.

The molecules (Fig. 1) are connected by O(5)-H...N(4) hydrogen bonds (O...N 2.866(8) Å) into chains along a. The dihedral angle between the terminal six-membered rings is 60°. The C(2)-C(10) bond length is 1.570(9) Å.

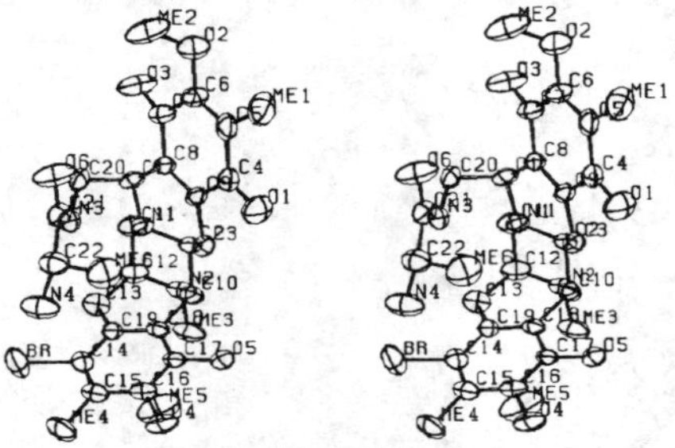

Fig. 1. The $C_{28}H_{35}BrN_4O_6$ molecule.

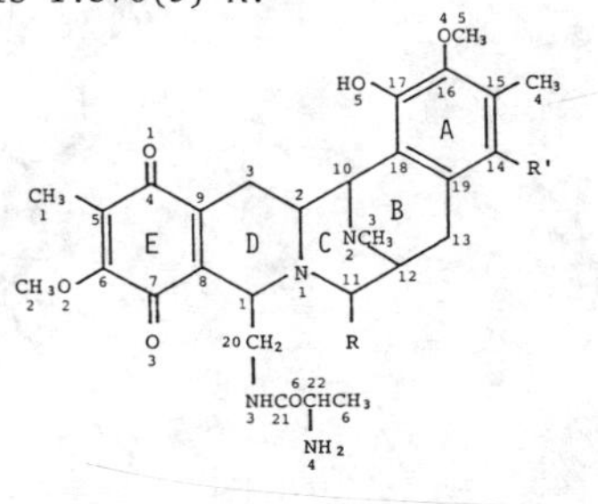

R=H, R'=H (Safracin A)
R=OH, R'=H (Safracin B)
R=H, R'=Br (Title compound)

PTAQUILOSIDE TETRAACETATE
$C_{28}H_{38}O_{12}$

S. OHBA, Y. SAITO, I. HIRONO, H. NIWA, M. OJIKA, K. WAKAMATSU and K. YAMADA, 1984. Acta Cryst., C40, 1877-1879.

Orthorhombic, $P2_12_12_1$, a = 15.900, b = 18.158, c = 10.252 Å, Z = 4. Mo radiation, R = 0.044 for 2520 reflexions.

The molecule (Fig. 1) has the cis-fused illudane skeleton with the cis-fused A/B rings inclined at 96.1(2)°. The allylic tertiary OH group forms a β-glycosidic linkage with D-(+)-glucose. The structure contains discrete molecules with intra-molecular hydrogen bonds linking O(1) and O(3). There is a short C(27)...O(3) contact of 2.957(5) Å.

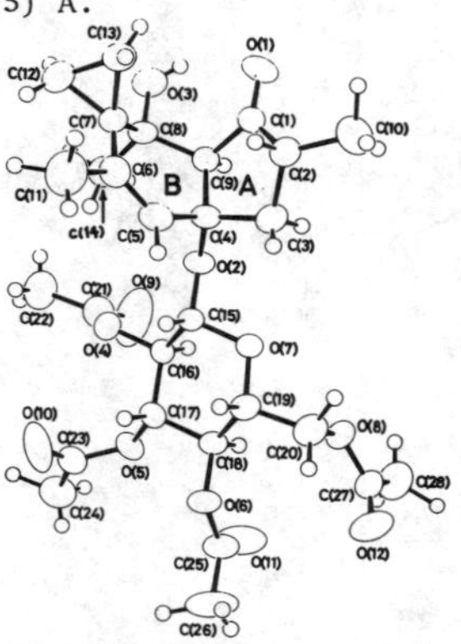

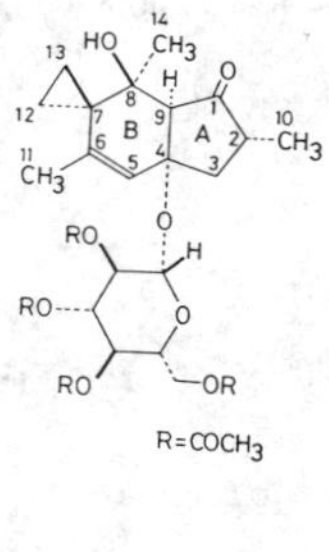

R=COCH3

Fig. 1. The ptaquiloside tetraacetate molecule.

BALEAROne

$C_{28}H_{40}O_5$

V. AMICO, F. CUNSOLO, M. PIATTELLI, G. RUBERTO and F.R. FRONCZEK, 1984. Tetrahedron, 40, 1721-1725.

Orthorhombic, $P2_12_12_1$, a = 10.048, b = 15.294, c = 17.251 Å, Z = 4. Mo radiation, R = 0.052 for 1566 reflexions.

The analysis establishes the relative configuration of balearone as shown in Fig. 1.

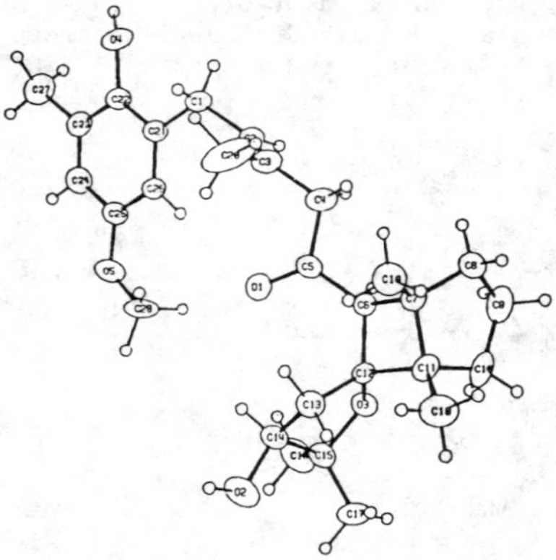

Fig. 1. The relative configuration of balearone.

PANICULIDE B BIS-p-BROMOBENZOATE

$C_{29}H_{26}Br_2O_7$

P. ANASTASIS, I. FREER, C. GILMORE, H. MACKIE, K. OVERTON, D. PICKEN and S. SWANSON, 1984. Can. J. Chem., 62, 2079-2088.

Monoclinic, $P2_1$, a = 14.326, b = 4.826, c = 19.974 Å, β = 102.27°, Z = 2. Mo radiation, R = 0.039 for 2101 reflexions.

The absolute configuration of paniculide B is established as shown in Fig. 1.

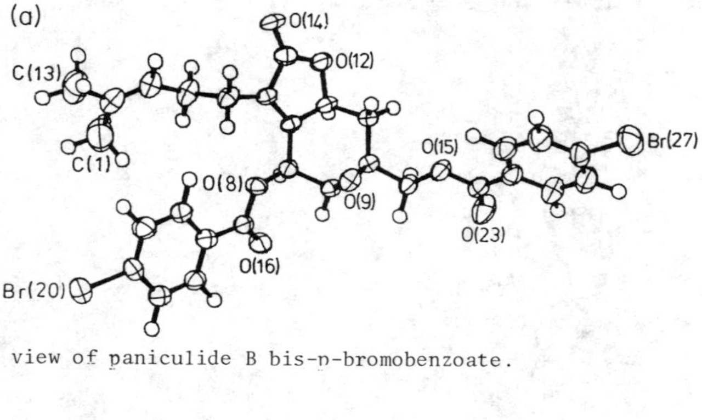

Fig. 1. A view of paniculide B bis-p-bromobenzoate.

MIOTOXIN A
$C_{29}H_{38}O_9$

G.G. HABERMEHL, L. BUSAM and J. STEGEMANN, 1984. Z. Naturforsch., C39, 212-216.

Orthorhombic, $P22_12_1$, a = 10.130, b = 10.184, c = 30.456 Å, Z = 4. Mo radiation, R = 0.079 for 2903 reflexions.

The molecule was shown to have the structure indicated in Fig. 1. The stereochemistry at C(6'), C(13') and C(4') is (R).

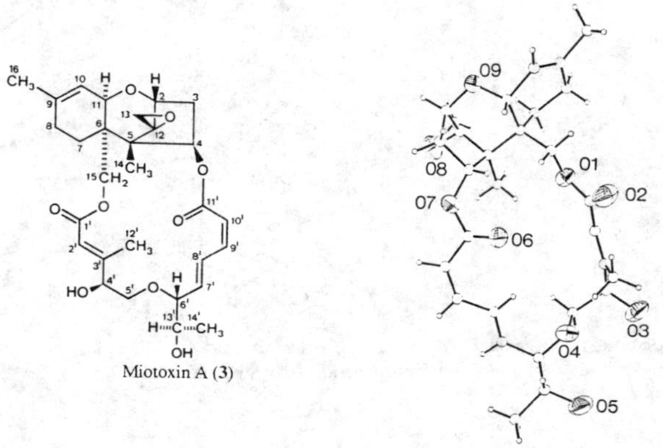

Fig. 1. $C_{29}H_{38}O_9$: structure and stereochemistry.

ROSEOTOXIN B BENZENE SOLVATE
$C_{30}H_{49}N_5O_7,C_6H_6$ (I)

ROSEOTOXIN B
$C_{30}H_{49}N_5O_7$ (II)

J.P. SPRINGER, R.J. COLE, J.W. DORNER, R.H. COX, J.L. RICHARD, C.L. BARNES and D. VAN DER HELM, 1984. J. Am. Chem. Soc., 106, 2388-2392.

I. Monoclinic, $P2_1$, a = 10.110, b = 10.590, c = 18.563 Å, β = 105.84°, Z = 2. Cu radiation, R = 0.089 for 2323 reflexions.

IIa. Monoclinic, $P2_1$, a = 9.516, b = 10.734, c = 16.495 Å, β = 101.72°, Z = 2. Cu radiation, R = 0.039 for 3537 reflexions (at 138 K).

IIb. a = 9.776, b = 10.785, c = 16.692 Å, β = 102.88°. Cu radiation, R = 0.092 for 2430 reflexions (at room temperature).

Crystals of I were from benzene, those of II were from ether/hexane. There are only minor differences in the three structure determinations. The roseotoxin molecule, details of torsion angles in the three analyses and bond length data from the low temperature study are in Fig. 1. The analyses correct previously reported amino acid and hydroxy acid sequences.

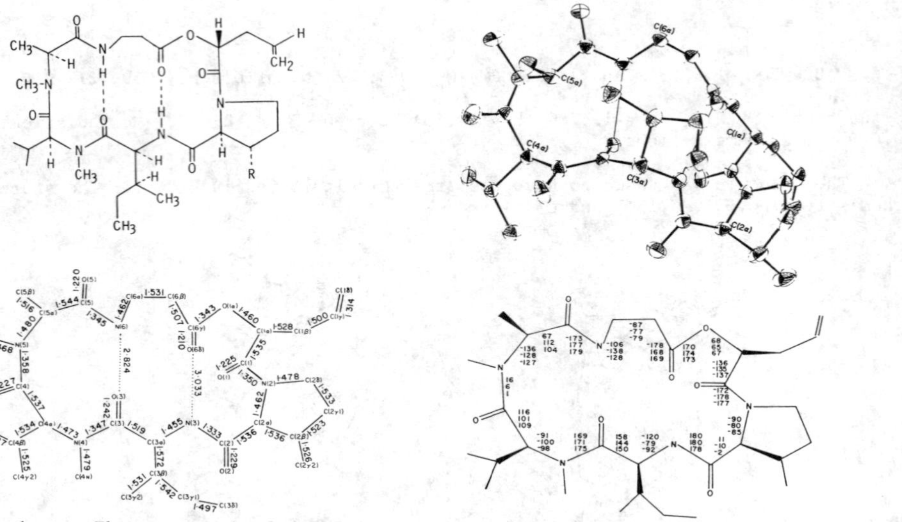

Fig. 1. The roseotoxin formula (R = CH_3), structure at 138 K, torsion angles (in the sequence I, IIa, IIb) and bond lengths at 138 K.

α-FLINDERCARPIN-2 METHYL(S)(ACETOXY){[3'S-(3α,3aα,4α,5β,6α,6aα,7α,10aα,11aα)]-5',6'-DIACETOXY-7'-(FURAN-3''-YL)-2',2',3a',6a',11a'-PENTAMETHYL-12'-METHYLENE-9'-OXO-2',3',3a',4',5',6',6a',9',10',11a'-DECAHYDRO-7'H-4',10a'-METHANO-10-a'H-FURO[2,3-b]-PYRANO-[3,4-g]OXOCIN-3'-YL} ACETATE

$C_{32}H_{40}O_{13}$,'3.1H_2O' (I)

β-FLINDERCARPIN-2 METHYL(S)(ACETOXY){[3'S-(3α,3aα,4α,5β,6α,6aα,7α,10aα,11aα)]-5',6'-DIACETOXY-7'-(FURAN-3''-YL)-2',2',3a',6a',11a'-PENTAMETHYL-12'-METHYLENE-9'-OXO-2',3',3a',4',5',6',6a',9',10',11a'-DECAHYDRO-7'H-4',10a'-METHANO-10-a'H-FURO[2,3-b]-PYRANO-[3,4-g]OXOCIN-3'-YL} ACETATE

$C_{32}H_{40}O_{13}$,'H_2O' (II)

G.W. BREEN, K. PICKER, C.L. RASTON, B.W. SKELTON, W.C. TAYLOR and A.H. WHITE, 1984. Aust. J. Chem., 37, 1461-1471.

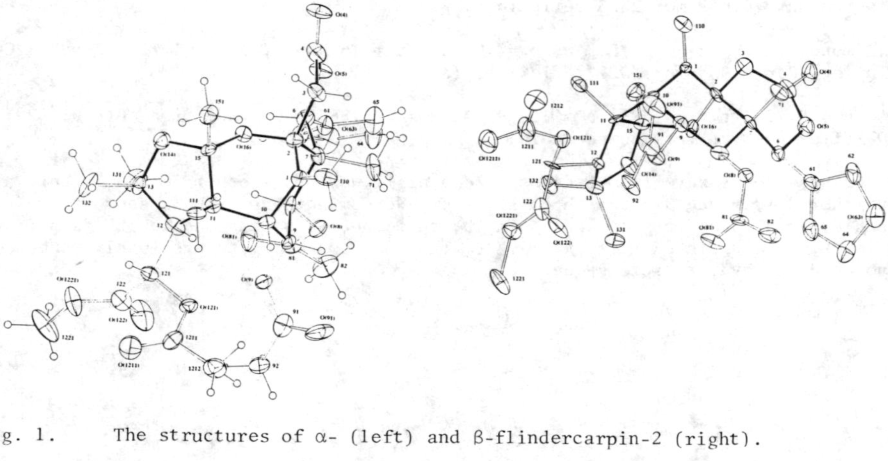

Fig. 1. The structures of α- (left) and β-flindercarpin-2 (right).

I. Orthorhombic, $P2_12_12_1$, a = 18.855, b = 14.693, c = 13.127 Å, Z = 4. Mo radiation, R = 0.072 for 1462 reflexions.

II. Orthorhombic, $P2_12_12_1$, a = 21.34, b = 14.921, c = 10.583 Å, Z = 4. Mo radiation, R = 0.10 for 1508 reflexions. A twinned form.

This compound was isolated from the bark of Flindersia laevicarpa C.T. White & Francis (Rutaceae).

TEDANOLIDE
$C_{32}H_{50}O_{11}$

F.J. SCHMITZ, S.P. GUNASEKERA, G. YALAMANCHILI, M.B. HOSSAIN and D. VAN DER HELM, 1984. J. Am. Chem. Soc., <u>106</u>, 7251-7252.

Orthorhombic, $P2_12_12_1$, a = 16.084, b = 29.850, c = 6.671 Å, Z = 4. Cu radiation, R = 0.045 for 2996 reflexions (at 138 K).

The structure and absolute stereochemistry of tedanolide are established (Fig. 1). The molecule is shaped like an elongated disk (length 9 Å, width 5 Å) with the hydroxyl at C2 and C13 directed to the interior of the ring. There is a possible intramolecular OH...O bond (O10...O9 2.824(4) Å.

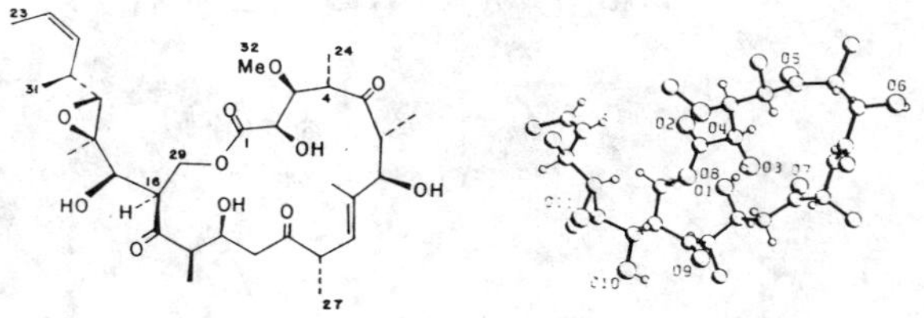

Fig. 1. Tedanolide.

ILIKONAPYRONE ACETONIDE
$C_{35}H_{52}O_7$

C.M. IRELAND, J.E. BISKUPIAK, G.J. HITE, J. RAPPOSCH, P.J. SCHEUER and J.R. RUBLE, 1984. J. Org. Chem., <u>49</u>, 559-561.

Orthorhombic, $P2_12_12_1$, a = 19.380, b = 23.031, c = 7.831 Å, Dm = 1.108, Z = 4. Cu radiation, R = 0.072 for 1697 reflexions.

The relative stereochemistry of the seven asymmetric centres was determined by this analysis (Fig. 1). It was not possible to determine the absolute configuration.

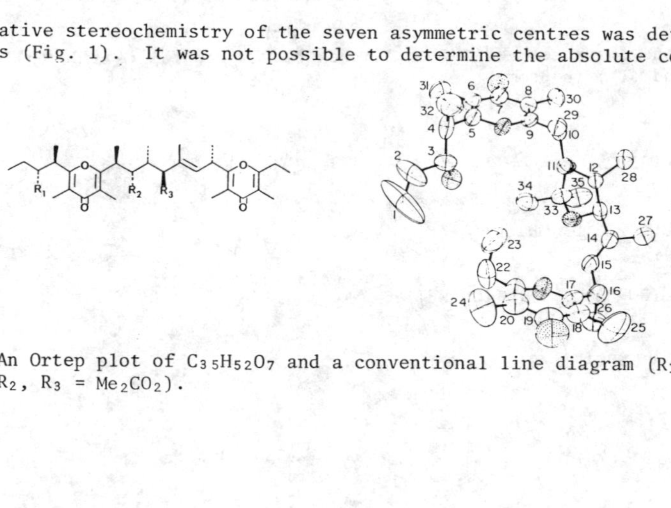

Fig. 1. An Ortep plot of $C_{35}H_{52}O_7$ and a conventional line diagram (R_1 = OH, R_2, R_3 = Me_2CO_2).

TRIETHYLENEDIAMINE - HYDROQUINONE
$C_{12}H_{18}N_2O_2$ (I)

TRIETHYLENEDIAMINE - PHENOL (1/2)
$C_{18}H_{23}N_2O$ (II)

T.C.W. MAK, W.-H. YIP and L. BOOK, 1984. J. Cryst. Spectrosc. Res., 14, 457-465.

I. Monoclinic, C2/c, a = 11.944, b = 9.491, c = 11.986 Å, β = 121.70°, D_m = 1.266, Z = 4. Mo radiation, R = 0.097 for 479 reflexions.

II. Monoclinic, $P2_1/c$, a = 12.987, b = 6.376, c = 21.350 Å, β = 106.94°, D_m = 1.175, Z = 4. Mo radiation, R = 0.092 for 1580 reflexions.

Both molecular components of I (Fig. 1) lie on twofold axes and are linked alternately by N...H-O hydrogen bonds to form infinite zig-zag chains. The structure of II corresponds to discrete hydrogen-bonded molecular aggregates. The triethylenediamine moieties in both adducts closely approximate idealized D_{3h} conformation.

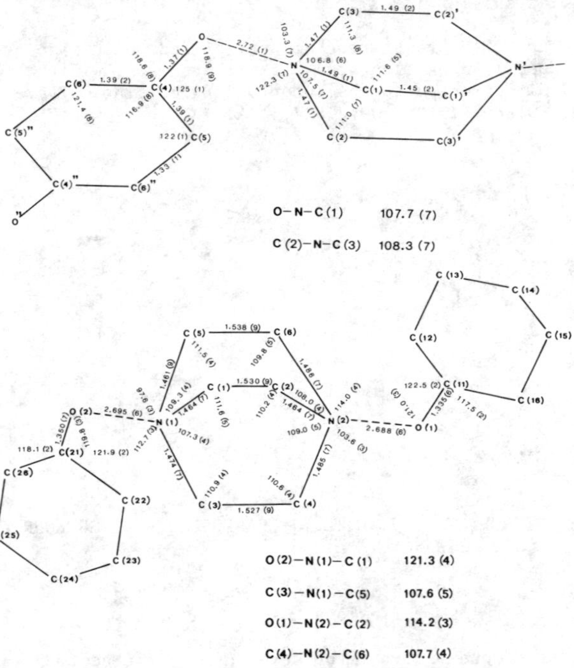

Fig. 1. Bond distances and angles in the triethylenediamine adducts (I) (top) and (II) (bottom).

DITHIZONE DIIODINE
$C_{13}H_{12}I_2N_4S$ (I)

ETHYLENETHIOUREA BIS(DIIODINE)
$C_3H_6I_4N_2S$ (II)

BIS(ETHYLENETHIOUREA) TRIS(DIIODINE)
$C_6H_{12}I_6N_4S_2$ (III)

(BIS(DITHIZONE) HEPTAKIS(DIIODINE))
$C_{26}H_{24}I_{14}N_8S_2$ (IV)

1-(1-IMIDAZOLIN-2-YL)-2-THIOXOIMIDAZOLIDINIUM TRIIODIDE (ETHYLENETHIOUREA DIIODINE) $C_9H_{17}I_5N_6S_2$ (V)

F.H. HERBSTEIN and W. SCHWOTZER, 1984. J. Am. Chem. Soc., 106, 2367-2373.

I. Monoclinic, $P2_1/n$, a = 12.233, b = 11.713, c = 12.030 Å, β = 102.8°, D_m = 2.01, Z = 4. Mo radiation, R = 0.059.

II. Triclinic, $P\bar{1}$, a = 10.276, b = 9.970, c = 6.792 Å, α = 105.1, β = 81.1, γ = 96.4°, Z = 2. Mo radiation, R = 0.041.

III. Triclinic, $P\bar{1}$, a = 9.362, b = 9.176, c = 7.434 Å, α = 116.9, β = 72.5, γ = 114.7°, Z = 1. Mo radiation, R = 0.063.

IV. Triclinic, $P\bar{1}$, a = 13.312, b = 10.637, c = 9.746 Å, α = 101.8, β = 87.1, γ = 109.4°, D_m = 2.85, Z = 1. Mo radiation, R = 0.086.

V. Monoclinic, $P2_1$, a = 11.102, b = 11.193, c = 9.551 Å, β = 110.0°, Z = 2. Mo radiation, R = 0.051.

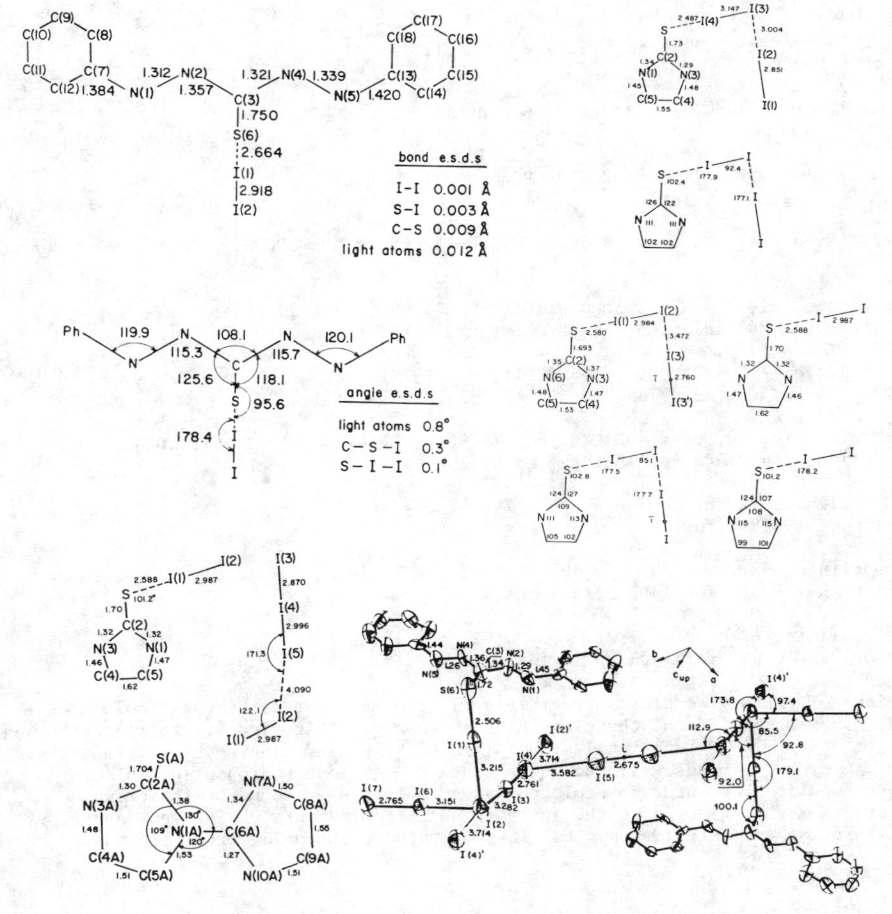

Fig. 1. Views of the thione - diiodine complexes I-V (clockwise from top left).

The structures of these complexes with dimensions are shown in Fig. 1. Compound I contains discrete moieties of the molecular formula with the I_2 fragment approximately normal to the dithizone fragment. Complex II also contains discrete moieties of the molecular formula arranged in the shape of a twisted U. Compound III has a centrosymmetric twisted-chain structure with three I_2 molecules held between two donor molecules. Complex IV has a centrosymmetric branched twisted chain of iodine molecules held at its ends by strong S...I interactions (2.506 Å). The salt-molecule complex V is intermediate between neutral structures and polyiodide salts and contains discrete ethylenethiourea-I_2 moieties together with cations and I_3^- ions of the other component.

FORMAMIDE - 18-CROWN-6 2:1 COMPLEX
$C_{14}H_{30}N_2O_8$ (I) $C_{12}H_{24}O_6.2(CHONH_2)$

N-METHYLTHIOUREA - 18-CROWN-6 1:1 COMPLEX
$C_{14}H_{30}N_2O_6S$ (II) $C_{12}H_{24}O_6.C_2H_6N_2S$

DITHIOOXAMIDE - 18-CROWN-6 2:1 COMPLEX
$C_{16}H_{32}N_4O_6S_4$ (III) $C_{12}H_{24}O_6.2(C_2H_4N_2S_2)$

3-NITROPHENOL - 18-CROWN-6 HYDRATE 2:1:2 COMPLEX
$C_{24}H_{38}N_2O_{14}$ (IV) $C_{12}H_{24}O_6.2(C_6H_5NO_3).2(H_2O)$

p-NITROBENZALDEHYDE OXIME - 18-CROWN-6 HYDRATE 2:1:2 COMPLEX
$C_{26}H_{40}N_4O_{14}$ (V) $C_{12}H_{24}O_6.2(C_7H_6N_2O_3).2(H_2O)$

2-GUANIDINOBENZIMIDAZOLE - DIAZA-18-CROWN-6 2:1 COMPLEX
$C_{28}H_{44}N_{12}O_4$ (VI) $C_{12}H_{26}N_2O_4.2(C_8H_9N_5)$

W.H. WATSON, J. GALLOY, D.A. GROSSIE, F. VÖGTLE and W.M. MÜLLER, 1984. J. Org. Chem., 49, 347-353.

I. Triclinic, $P\bar{1}$, a = 8.115, b = 8.194, c = 7.700 Å, α = 90.89, β = 111.22, γ = 98.34°, Z = 1. Mo radiation, R = 0.071 for 970 reflexions.

II. Triclinic, $P\bar{1}$, a = 8.903, b = 9.311, c = 15.772 Å, α = 120.69, β = 94.62, γ = 114.29°, Z = 2. Cu radiation, R = 0.068 for 2527 reflexions.

III. Monoclinic, $P2_1/c$, a = 8.620, b = 14.908, c = 11.366 Å, β = 121.02°, Z = 2. Mo radiation, R = 0.042 for 1441 reflexions.

IV. Monoclinic, $P2_1/c$, a = 12.47, b = 7.520, c = 15.90 Å, β = 102.50°, Z = 2. Mo radiation, R = 0.050 for 1170 reflexions.

V. Monoclinic, $P2_1/c$, a = 7.570, b = 7.875, c = 27.46 Å, β = 101.19°, Z = 2. Mo radiation, R = 0.072 for 1498 reflexions.

VI. Monoclinic, C2/c, a = 21.27, b = 8.191, c = 18.26 Å, β = 93.10°, Z = 4. Mo radiation, R = 0.069 for 1805 reflexions.

Views of the six complexes are in Fig. 1. In most cases the guest molecules approach from both sides of the plane of the crown ring, sterically fitting with XNH_3, XH_2 and $XH...H_2O$ functional groups (X = C, N, O) and binding by multiple NH_2...O and OH...O hydrogen bonds. If the guest contains only one hydrogen suitable for hydrogen bonding, one water molecule per guest molecule is bound by the crown receptor. N-Methylthiourea as the guest is able to bind in a chain-type manner by using its amino and methyl groups as links between neighbouring crown ligands.

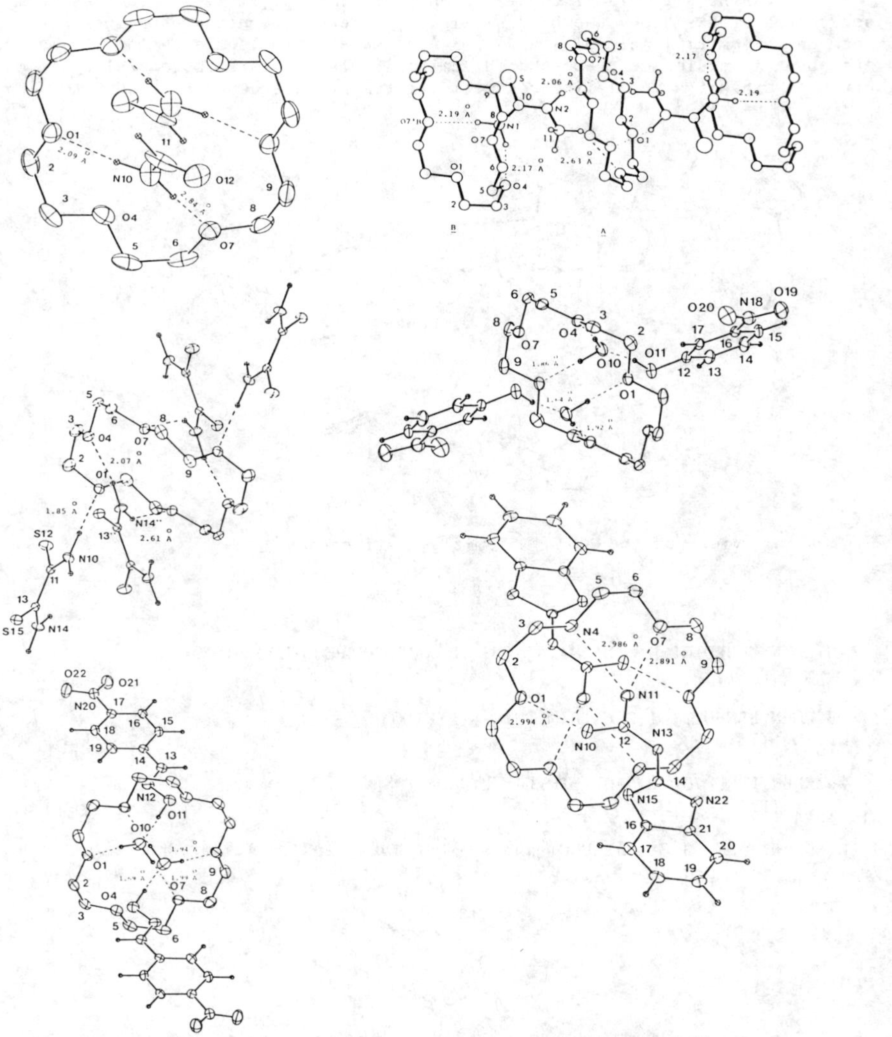

Fig. 1. Views of the 18-crown-6 complexes with formamide (top left), N-methylthiourea (top right), dithiooxamide (centre left), 3-nitrophenol water (centre right), p-nitro-benzaldehyde oxime water (bottom left) and 2-guanidinobenzimidazole (bottom right).

TETRATHIAFULVALENE BIS[BIS(4,5-DIMERCAPTO-1,3-DITHIOLE-2-THIONE)NICKEL(II)]

$C_{18}H_4Ni_2S_{24}$ $(TTF)[Ni(dmit)_2]_2$

M. BOUSSEAU, L. VALADE, M.-F. BRUNIQUEL, P. CASSOUX, M. GARBAUSKAS, L. INTERRANTE and J. KASPER, 1984. Nouv. J. Chim., 8, 3-6.

Monoclinic, C2/c, a = 46.22, b = 3.732, c = 22.86 Å, β = 119.29°, Z = 4. R = 0.0979 for 1475 reflexions.

The essentially planar donor TTF and acceptor $[Ni(dmit)_2]$ molecules form uniform segregated stacks along the b-axis (Fig. 1). Mean interplanar spacings are 3.55 Å between $[Ni(dmit)_2]$ molecules and 3.65 Å between TTF molecules. The TTF molecules lie about inversion centres, the $[Ni(dmit)_2]$ molecules are in general positions. The crystals exhibit a metal-like conductivity temperature dependence down to 4 K.

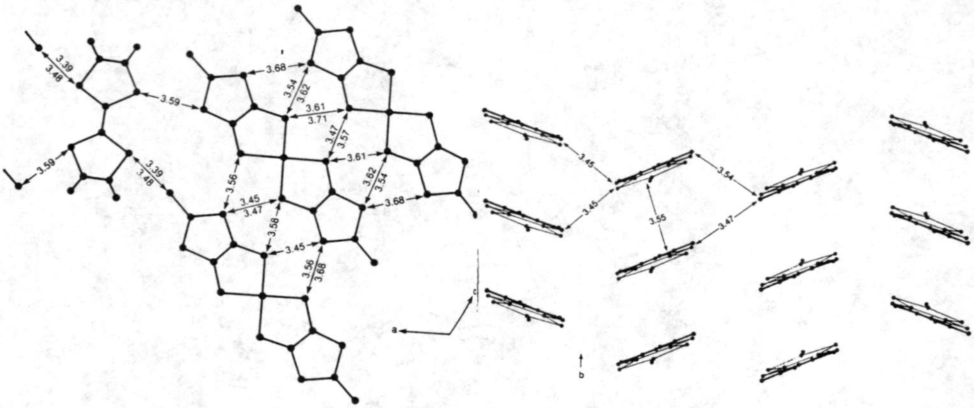

Fig. 1. Views of the $(TTF)[Ni(dmit)_2]_2$ structure.

1,8-BIPHENYLENEDIOL N,N,N',N',N'',N''-HEXAMETHYLPHOSPHORAMIDE
$C_{18}H_{30}N_3O_3P$ (I) $C_6H_{18}N_3OP,C_{12}H_8O_2$

1,8-BIPHENYLENEDIOL 1,2,6-TRIMETHYL-4-PYRIDONE
$C_{20}H_{19}NO_3$ (II) $C_8H_{11}NO,C_{12}H_8O_2$

1,8-BIPHENYLENEDIOL 2,6-DIMETHYL-γ-PYRONE
$C_{19}H_{16}O_4$ (III) $C_7H_8O_2,C_{12}H_8O_2$

J. HINE, K. AHN, J.C. GALLUCCI and S.-M. LINDEN, 1984. J. Am. Chem. Soc., 106, 7980-7981.

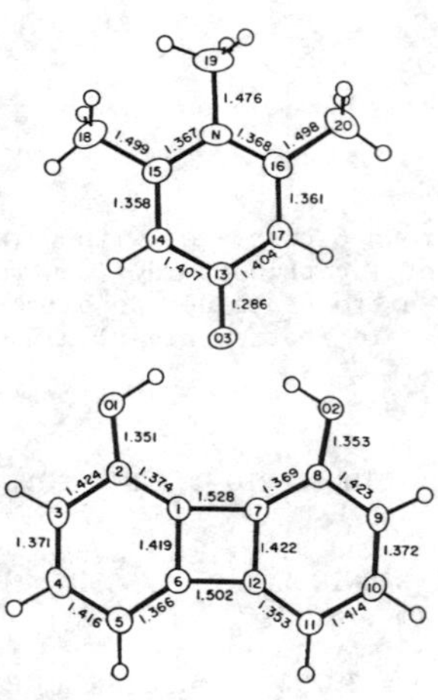

Fig. 1. A view of the $C_8H_{11}NO:C_{12}H_8O_2$ hydrogen bonded complex.

I. Monoclinic, $P2_1/n$, a = 8.527, b = 15.099, c = 15.354 Å, β = 95.17°, Z = 4. Mo radiation, R = 0.077 for 2351 reflexions (at -56°C).

II. Monoclinic, $P2_1/n$, a = 7.639, b = 22.130, c = 9.411 Å, β = 90.66°, Z = 4. Mo radiation, R = 0.048 for 2188 reflexions (at -123°C).

III. Triclinic, $P\bar{1}$, a = 8.983, b = 13.450, c = 6.926 Å, α = 85.15, β = 106.19, γ = 109.72°, Z = 2. Mo radiation, R = 0.048 for 2592 reflexions (at -125°C).

A view of complex (II) is in Fig. 1. The two components are almost coplanar (interplanar angle 3.2(1)°). The O-H...O distances are 2.545 and 2.548(3) Å, with O-H...O angles 177 and 174(3)°. For (I) and (III) the structures are similar, with corresponding dimensions 2.601, 2.613(5) Å, 168, 172(5)° for (I) and 2.585, 2.589(2) Å, 176, 178(3)° for (III).

BIS[BIS(ETHYLENEDITHIO)TETRATHIAFULVALENE] DIBROMOIODIDE (β phase)

$C_{20}H_{16}Br_2IS_{16}$ $(C_{10}H_8S_8)_2(IBr_2)$

J.M. WILLIAMS, H.H. WANG, M.A. BENO, T.J. EMGE, L.M. SOWA, P.T. COPPS, F. BEHROOZI, L.N. HALL, K.D. CARLSON and G.W. CRABTREE, 1984. Inorg. Chem., 23, 3839-3841.

Triclinic, $P\bar{1}$, a = 6.593, b = 8.975, c = 15.093 Å, α = 93.79, β = 94.97, γ = 110.54°, Z = 1. Mo radiation, R = 0.060 for 4416 reflexions.

The structure (Fig. 1) has the "corrugated-sheet" network of short interstack S...S interactions noted for the isostructural I_3^- analog (1). The shorter I-Br bonds vs. I-I bonds (2.702(1) vs. 2.912(1) Å) result in an average decrease of ∿0.02 Å in the interstack S...S distances.

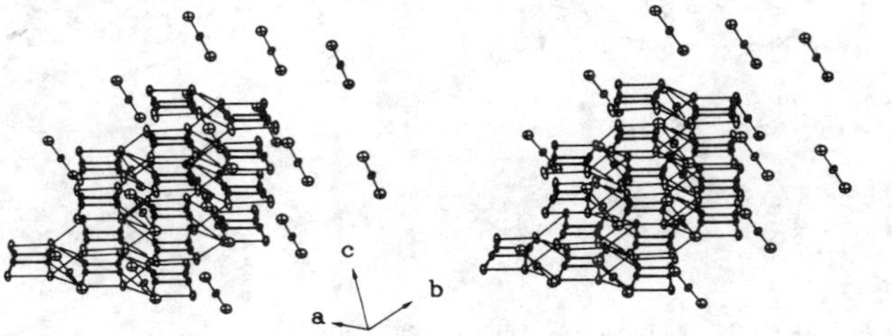

Fig. 1. Stereoview of the $(C_{10}H_8S_8)_2(IBr_2)$ structure.

1. Following Report.

BIS[BIS(ETHYLENEDITHIO)TETRATHIAFULVALENE] TRIIODIDE

$C_{20}H_{16}I_3S_{16}$ $(C_{10}H_8S_8)_2I_3$

J.M. WILLIAMS, T.J. EMGE, H.H. WANG, M.A. BENO, P.T. COPPS, L.N. HALL, K.D. CARLSON and G.W. CRABTREE, 1984. Inorg. Chem., 23, 2558-2560.

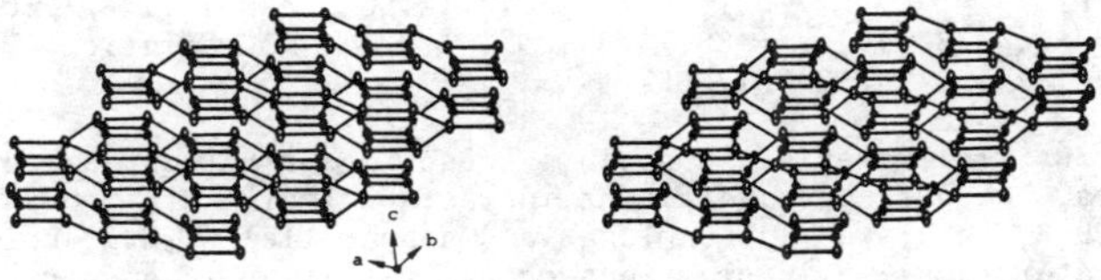

Fig. 1. Stereoview of the $(C_{10}H_8S_8)_2I_3$ structure showing interstack S-S distances < 3.6 Å.

Triclinic, $P\bar{1}$, a = 6.615, b = 9.100, c = 15.286 Å, α = 94.38, β = 95.59, γ = 109.78°, Z = 1. Mo radiation, R = 0.06 for 4142 reflexions.

The structure (Fig. 1) has a "corrugated sheet" structure based on interstack S-S interaction similar to that observed for the BrO_4^- and ReO_4^- derivatives (1).

1. This volume p. 692.

BIS[BIS(ETHYLENEDITHIO)TETRATHIOFULVALENE] TRIIODIDE
$C_{20}H_{16}I_3S_{16}$ $2(ET).I_3$

P.C.W. LEUNG, T.J. EMGE, M.A. BENO, H.H. WANG, J.M. WILLIAMS, V. PETRICEK and P. COPPENS, 1984. J. Am. Chem. Soc., 106, 7644-7646.

Triclinic, $P\bar{1}$, a = 6.561, b = 9.010, c = 15.173 Å, α = 95.09, β = 95.95, γ = 110.27°, Z = 1. Mo radiation, R = 0.060 for 4387 reflexions (at 125 K).

For details of the room temperature structure see (1). The average crystal structure at 125 K (Fig. 1) differs mainly from that at 298 K in that the intermolecular S...S distances are contracted by about 2%. Even at 125 K all but one of the interstack S...S distances exceed 3.60 Å. The displacement vectors for I_3^- ions and ET molecules are of different directions and magnitudes.

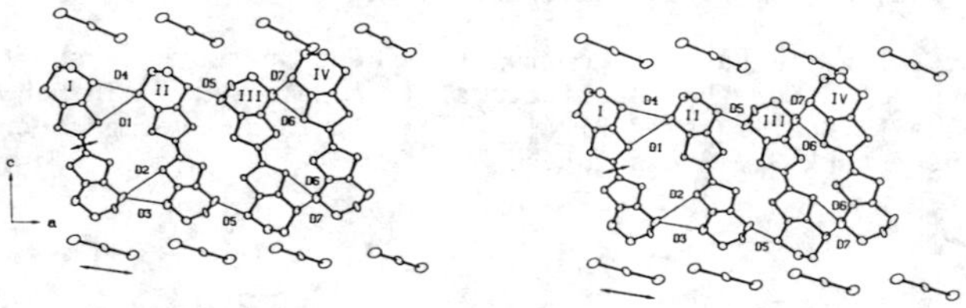

Fig. 1. A stereoview of $2(ET).I_3$ at 125 K.

1. Preceding Report.

BIS[BIS(ETHYLENEDITHIO)TETRATHIAFULVALENE] PERRHENATE
$C_{20}H_{16}O_4ReS_{16}$ (I) $(C_{10}H_8S_8)_2(ReO_4)$

BIS[BIS(ETHYLENEDITHIO)TETRATHIAFULVALENE] PERBROMATE
$C_{20}H_{16}BrO_4S_{16}$ (II) $(C_{10}H_8S_8)_2(BrO_4)$

J.M. WILLIAMS, M.A. BENO, H.H. WANG, P.E. REED, L.J. AZEVEDO and J.E. SCHIRBER, 1984. Inorg. Chem., 23, 1790-1792.

I. Triclinic, $P\bar{1}$, a = 7.798/7.668, b = 12.579/12.583, c = 17.102/17.043 Å, α = 73.37/73.27, β = 80.35/81.24, γ = 88.97/89.06°, Z = 2. Mo radiation, R = 0.041/0.032 for 4838/8065 reflexions (at 298/125 K).

II. Triclinic, $P\bar{1}$, a = 7.795/7.631, b = 12.613/12.564, c = 17.148/17.076 Å, α = 72.97/73.21, β = 80.44/80.49, γ = 88.74/89.09°, Z = 2. Mo radiation, R = 0.034/0.039 for 2993/8672 reflexions (at 298/125 K).

These isostructural materials contain a loose zigzag packing arrangement of $C_{10}H_8S_2$ molecules, however, no significant intrastack (columnar) S-S interactions occur even at 125 K, resulting in nonplanar and nonparallel $C_{10}H_8S_2$ moieties (Fig. 1). Numerous short interstack S-S interactions occur, resulting in an unusual side-by-side stacking of $C_{10}H_8S_2$ molecules. These materials constitute a different structural class of organic metals compound with bis(tetramethyltetraselenafulvalene)X systems.

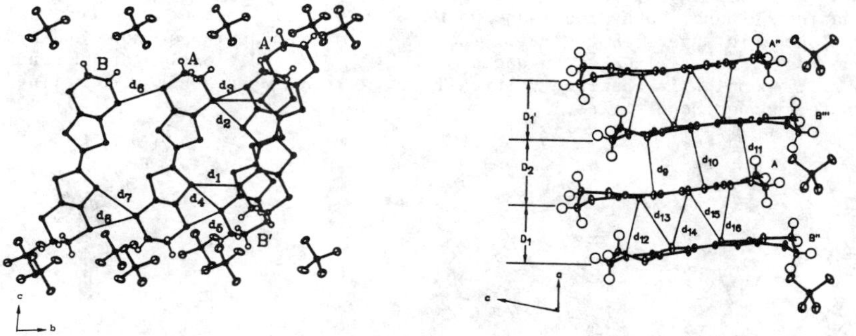

Fig. 1. Inter- (left) and intrastack (right) interactions in $(C_{10}H_8S_8)_2(BrO_4)$.

2,9-DIMETHYL-1,10-PHENANTHROLINE - RESORCINOL

$C_{20}H_{18}N_2O_2$ $C_{14}H_{12}N_2.C_6H_6O_2$

W.H. WATSON, J. GALLOY, F. VOGTLE and W.M. MULLER, 1984. Acta Cryst., C40, 200-202.

Triclinic, $P\bar{1}$, a = 7.461, b = 11.942, c = 14.392 Å, α = 139.25, β = 91.16, γ = 94.87°, Z = 2. Cu radiation, R = 0.078 for 1755 reflexions.

The crystal structure (Fig. 1) contains parallel sheets of 2,9-dimethyl-1,10-phenanthroline molecules which are almost perpendicular (81.7(4)°) to parallel sheets of resorcinol molecules. There is extensive O-H...N hydrogen bonding, with each phenanthroline bound from above and below by resorcinol molecules (O...N 2.925-3.116(4) Å).

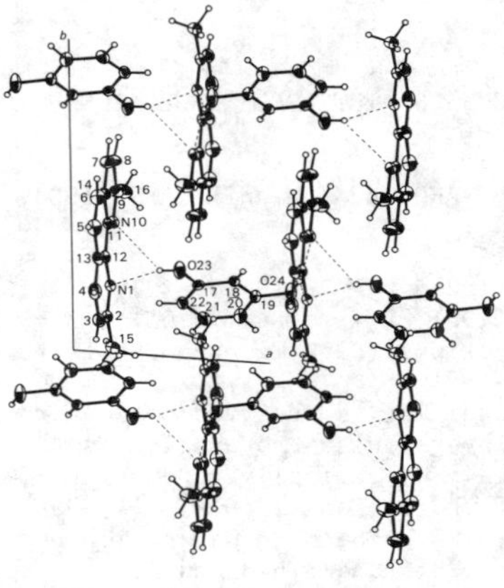

Fig. 1. A view of the packing of $C_{14}H_{12}N_2:C_6H_6O_2$.

2,5-DIMETHYLBENZOQUINONE BIS(HYDROQUINONE)

$C_{20}H_{20}O_6$ $C_8H_8O_2.2(C_6H_6O_2)$

A.O. PATIL, D.Y. CURTIN and I.C. PAUL, 1984. J. Am. Chem. Soc., 106, 4010-4015.

Triclinic, $P\bar{1}$, a = 7.683, b = 9.758, c = 6.050 Å, α = 103.62, β = 91.58, γ = 89.17°, Z = 1. Mo radiation, R = 0.039 for 1538 reflexions.

In this complex the basic structural unit is the hydrogen-bonded chain of alternating quinone and hydroquinone units (Fig. 1); all component molecules lie on crystallographic inversion centres. The planes of overlapping quinone and hydroquinone rings are inclined at an angle of 7.14°. The additional hydroquinone molecule in this 1:2 complex is inserted vertically to form hydrogen-bonded cross links between alternate hydroquinone strips.

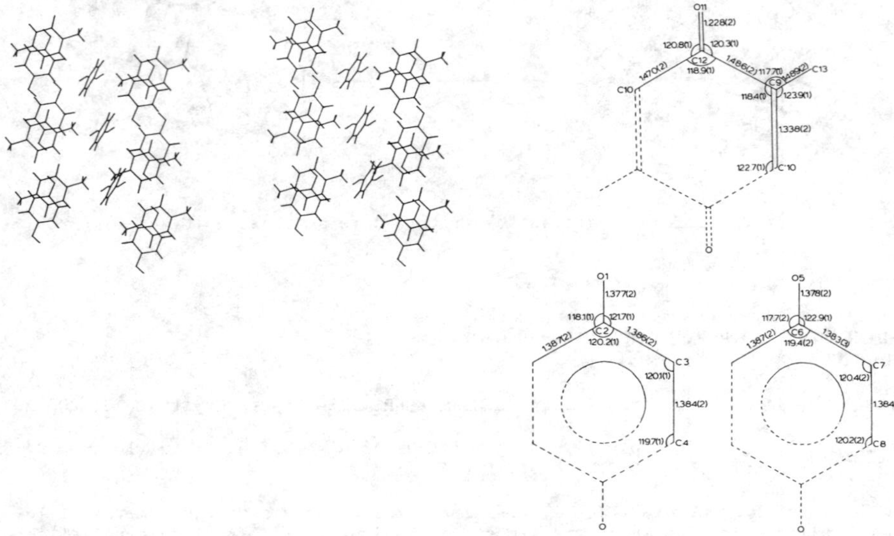

Fig. 1. A stereoview of the $C_8H_8O_2.2(C_6H_6O_2)$ complex and details of dimensions in the centrosymmetric molecules.

DI(3,3',4,4'-TETRAMETHYL-2,2',5,5'-TETRASELENAFULVALENIUM) PENTAFLUOROSILICATE

$C_{20}H_{24}F_5Se_8Si$ $(C_{10}H_{12}Se_4)_2SiF_5$

K. ERIKS, M.A. BENO, K. BECHGAARD and J.M. WILLIAMS, 1984. Acta Cryst., C40, 1715-1717.

Triclinic, $P\bar{1}$, a = 7.289 (7.122), b = 7.715 (7.682), c = 13.548 (13.495) Å, α = 83.53 (83.72), β = 86.20 (86.90), γ = 70.84 (70.03)°, Z = 1. Mo radiation, R = 0.051 (0.045) for 2951 (3969) reflexions (data in parentheses refer to 125 K).

No structural phase transition is observed between room temperature and 125 K. The SiF_5^- anion lies on an inversion centre and is disordered. The structure is isostructural with other highly electrically conducting $(TMTSF)_2X$ (X = monovalent anion) salts (1). A view of the cation is in Fig. 1; Se-C distances at 125 K are in the range 1.878(3) to 1.907(3) Å. The mean separation between alternate TMTSF donors in the columns is 3.560 Å at 125 K and 3.644 and 3.632 Å at 293 K.

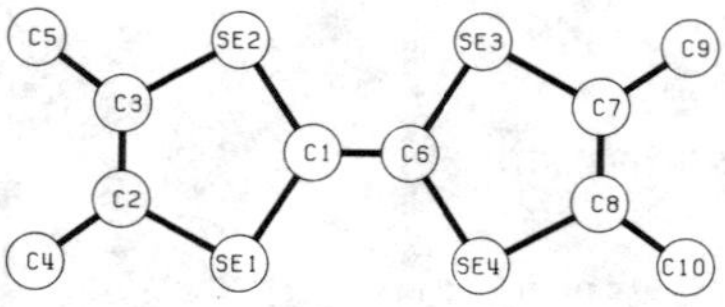

Fig. 1. A view of the TMTSF donor.

1. Structure Reports, 49B, 713.

BIS(TETRAMETHYLTETRATHIAFULVALENE) PERRHENATE

$C_{20}H_{24}O_4ReS_8$ $(C_{10}H_{12}S_4)_2{}^{+}ReO_4{}^{-}$

H. KOBAYASHI, A. KOBAYASHI, Y. SASAKI, G. SAITO and H. INOKUCHI, 1984. Bull. Chem. Soc. Japan, 57, 2025-2026.

Triclinic, $P\bar{1}$, a = 7.166, b = 7.621, c = 13.230 Å, α = 86.62, β = 95.56, γ = 108.20°, Z = 1. Mo radiation, R = 0.066.

The planar TMTTF molecules are stacked face-to-face in a zig-zag column along the a axis (Fig. 1). There are two independent interplanar distances (A...B 3.57, and A...C 3.59 Å). The shortest S...S distance (3.74 Å) along the column is longer than the van der Waals distance S...S (3.70 Å) but the shortest S...O contact is 3.12 Å which is slightly shorter than the corresponding van der Waals distance.

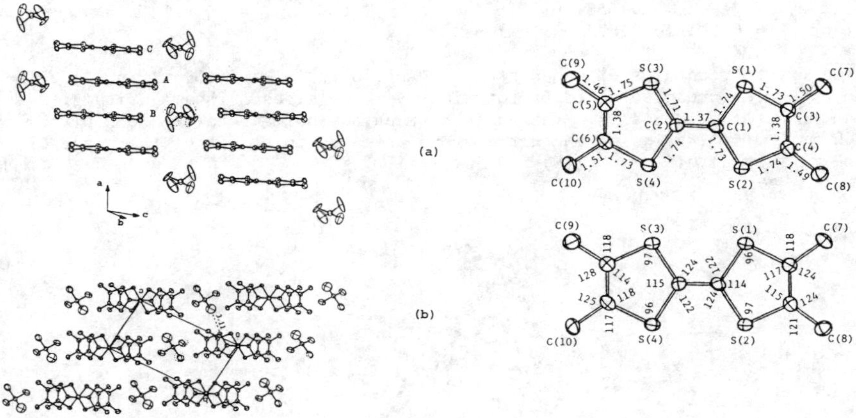

Fig. 1. Side view of stacking in $C_{20}H_{24}O_4ReS_8$ (tilted 10°) showing disorder in ReO_4^- anions, and view along a (left), bond lengths and angles in TMTTF (right).

1,4-DITHIINTETRACARBOXYLIC N,N'-DIMETHYLDIIMIDE - ACRIDINE (1:1)

$C_{23}H_{15}N_3O_4S_2$ $[C_{10}H_6N_2O_4S_2][C_{13}H_9N]$

Y. YAMAGUCHI and I. UEDA, 1984. Acta Cryst., C40, 113-115.

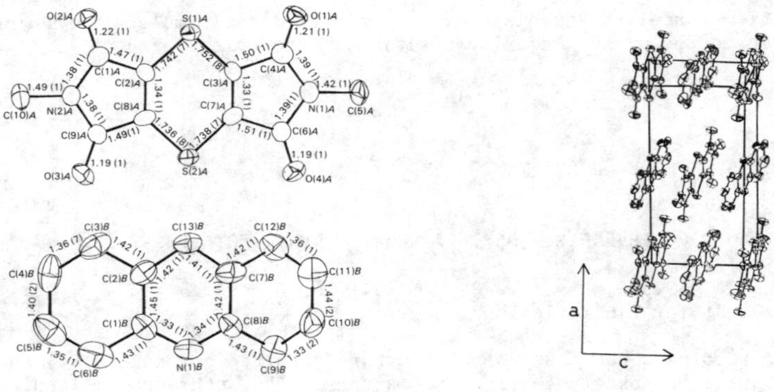

Fig. 1. Bond lengths in the $C_{10}H_6N_2O_4S_2$ and $C_{13}H_9N$ molecules and a view of the crystal structure projected along b.

Monoclinic, Pn, a = 13.701, b = 10.244, c = 7.208 Å, β = 92.20°, D_m = 1.515, Z = 2. Mo radiation, R = 0.039 for 1544 reflexions.

The component molecules of this complex have planar structures and the almost parallel molecules are alternately stacked along the c-axis (Fig. 1).

PYRENE 2,4,6-TRINITROANISOLE
$C_{23}H_{15}N_3O_7$ $C_{16}H_{10}:C_7H_5N_3O_7$

J.C. BARNES, J.A. CHUDEK, R. FOSTER, F. JARRETT, F. MACKIE, J. PATON and D.R. TWISELTON, 1984. Tetrahedron, 40, 1595-1601.

Monoclinic, $P2_1/c$, a = 10.633, b = 16.336, c = 11.683 Å, β = 94.62°, Z = 4. Cu radiation, R = 0.091 for 1572 reflexions.

The crystal contains extended stacks ADADA, parallel to (101) (Fig. 1). The pyrene molecules are disordered by rotation over two sites. The majority site has an occupancy factor of 0.544(2). The angle between the normals to the mean planes of A and D is 7.0(4)°, allowing the nitro groups of A to twist out of the plane of the benzene ring by 41.0(1.0), 2.7(5) and 20.7(8)°.

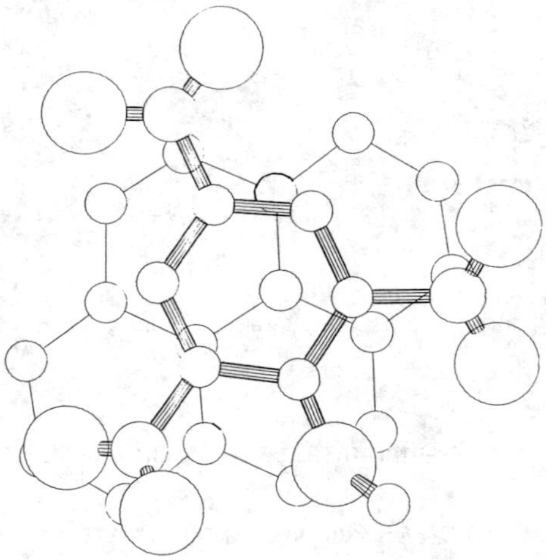

Fig. 1. A view normal to the plane of the trinitroanisole molecule showing the orientation of the majority position of the pyrene ring.

ANTHRACENE 1,2,4,5-TETRACYANOBENZENE COMPLEX
$C_{24}H_{12}N_4$ $C_{14}H_{10},C_{10}H_2N_4$

J.C.A. BOEYENS and D.C. LEVENDIS, 1984. J. Chem. Phys., 80, 2681-2688. .

Monoclinic, C2/m, a = 9.505, b = 12.748, c = 7.417 Å, β = 92.45°, Z = 2. Mo radiation, R = 0.058 for 646 reflexions.

This paper reports a new study of a structure reported previously (1) and uses the previously published observed structure factors. The anthracene molecule randomly occupies positions at ±6° away from the crystallographic mirror plane.

1. Structure Reports, 38B, 620; 46B, 610.

DIBENZO[b,d]TELLUROLE-7,7,8,8-TETRACYANO-p-QUINODIMETHANE
$C_{24}H_{12}N_4Te$

H.B. SINGH, W.R. McWHINNIE, R.F. ZIOLO and C.H.W. JONES, 1984. J. Chem. Soc. Dalton, 1267-1273.

Monoclinic, $P2_1/n$, a = 14.849, b = 8.172, c = 16.275 Å, β = 92.05°, Z = 4. Mo radiation, R = 0.041 for 4514 reflexions.

The crystal consists of quasi-hexagonal mixed stacks of alternating donor and acceptor molecules parallel to the a-axis. The dibenzotellurole and tetracyanoquinodimethane (TCNQ) molecules are essentially planar and dimensionally identical to the respective uncomplexed molecules. The TCNQ molecules lie on centres of symmetry with two unique halves in different molecules. This results in different orientations of TCNQ above and below the tellurium donor (Fig. 1). The mean interplanar spacing between the least squares planes of donor and acceptor molecules is 3.45 Å and dihedral angle 3.5° in one orientation and 3.51 Å and 5.7° in the other. The Te atom is involved in a short intermolecular contact to N(2) (3.216(6) Å). There is also an interstack Te...N(1) distance of 3.635(7) and a Te...C(14) distance of 3.746(6) Å.

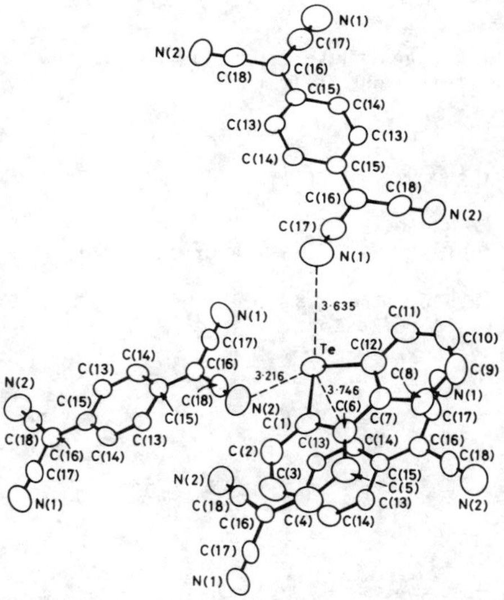

Fig. 1. $C_{24}H_{12}N_4Te$: coordination geometry and contacts about Te.

TETRATHIAFULVALENE BIS(NAPHTHO[2,3-c][1,2,5]THIADIAZOLE-4,9-DIONE)
$C_{26}H_{12}N_4O_4S_6$ (I) $C_6H_4S_4{:}2(C_{10}H_4N_2O_2S)$

TETRATHIAFULVALENE BENZO[1,2-c:4,5-c']BIS[1,2,5]THIADIAZOLE-4,8-DIONE
$C_{12}H_4N_4O_2S_6$ (II) $C_6H_4S_4{:}C_6N_4O_2S_2$

A. GIEREN, V. LAMM, T. HUBNER, M. RABBEN, R. NEIDLEIN and D. DROSTE, 1984. Chem. Ber., 117, 1940-1953.

I. Monoclinic, $P2_1/n$, a = 14.814, b = 11.108, c = 7.778 Å, γ = 81.19°, D_m = 1.70, Z = 2. Cu radiation, R = 0.069 for 1117 reflexions.

II. Monoclinic, $P2_1/n$, a = 7.376, b = 7.720, c = 13.705 Å, β = 101.14°, D_m = 1.88, Z = 2. Mo radiation, R = 0.039 for 1698 reflexions.

Both I and II are built up by mixed columnar stacks of donors and acceptors, forming a characteristic packing motive for organic D-A complexes with the properties of an electrical insulator. In I the tetrathiafulvalene lies about an inversion centre; in II both components lie about inversion centres. Dimensions are in Fig. 1.

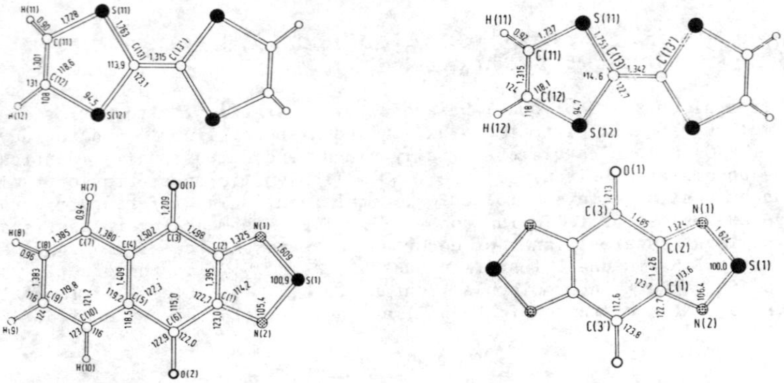

Fig. 1. Dimensions (averaged where appropriate) for the tetrathiafulvalene complexes I (left) and II (right).

PYRENE - 1,4,5,8-NAPHTHALENETETRONE (1:1)

$C_{26}H_{14}O_4$ $C_{16}H_{10}.C_{10}H_4O_4$

F.H. HERBSTEIN and G.M. REISNER, 1984. Acta Cryst., C40, 202-204.

Monoclinic, $P2_1/a$, a = 10.536, b = 12.877, c = 7.314 Å, β = 114.1°, Z = 2. Mo radiation, R = 0.057 for 878 reflexions.

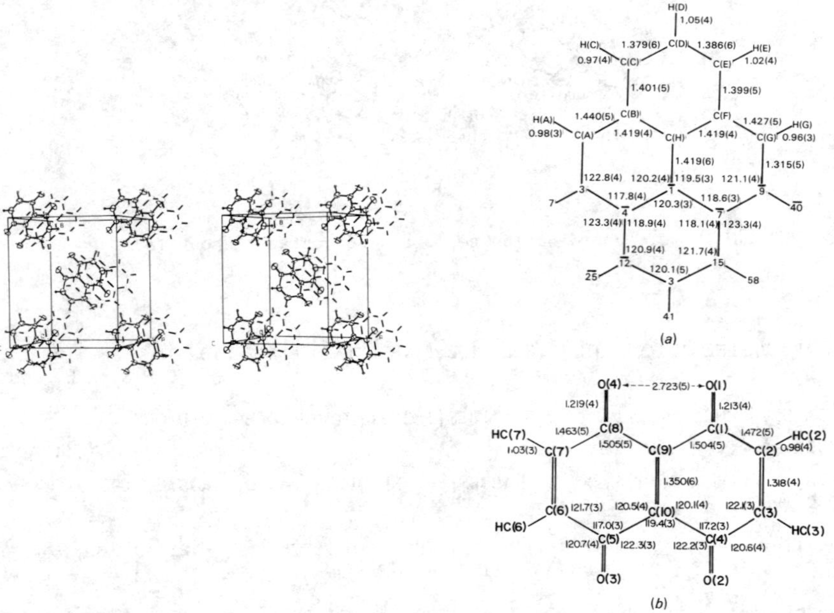

Fig. 1. A stereoview of the unit cell of $C_{16}H_{10}:C_{10}H_{14}O_4$ and details of molecular geometry.

Both component molecules (Fig. 1) lie on inversion centres. The packing has mixed stacking of donor and acceptor molecules, with the carbonyl groups aligned above six-membered rings of the pyrene molecule. The pyrene system is planar within experimental error; the naphthalenetetrone system is non-planar.

1,4,7,10,13,16-HEXAOXACYCLOOCTADECANE 2-(2-BENZIMIDAZOLYL)GUANIDINE (1:2 COMPLEX)
$C_{28}H_{42}N_{10}O_6$ $C_{12}H_{24}O_6,2C_8H_9N_5$

M.R. CAIRA, W.H. WATSON, F. VÖGTLE and W. MÜLLER, 1984. Acta Cryst., C40, 1047-1050.

Monoclinic, C2/c, a = 20.894, b = 8.399, c = 18.330 Å, β = 93.09°, Z = 4. Cu radiation, R = 0.049 for 1462 reflexions.

In the complex (Fig. 1) the 18-crown-6 molecule lies at an inversion centre and adopts the D_{3d} conformation. Two guest 2-(2-benzimidazolyl)guanidine molecules, related by the same centre of symmetry, bind to the crown ether via (guanidine)-N-H...O hydrogen bonds. Additional stabilisation is provided by intramolecular N-H...N, as well as intermolecular base-pairing N-H...N hydrogen bonding involving guest molecules.

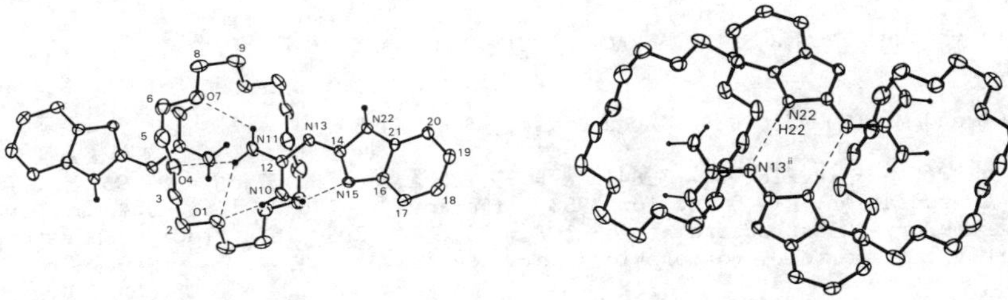

Fig. 1. $C_{12}H_{24}O_6,2C_8H_9N_5$: views of the title complex, with dashed lines representing the unique host-guest O...H-N and intramolecular guest N-H...N hydrogen bonds, and of the (010) projection, showing hydrogen bonding between diad-related guest molecules.

[2.2.2]PARACYCLOPHANE TETRACYANOETHYLENE
$C_{30}H_{24}N_4$ (I) $C_{24}H_{24},C_6N_4$

[2.2.2.2]PARACYCLOPHANE TETRACYANOETHYLENE
$C_{38}H_{32}N_4$ (II) $C_{32}H_{32},C_6N_4$

C. COHEN-ADDAD, M. LEBARS, A. RENAULT and P. BARET, 1984. Acta Cryst., C40, 1927-1931.

I. Triclinic, P$\bar{1}$, a = 9.057, b = 22.646, c = 6.830 Å, α = 104.85, β = 103.25, γ = 106.35°, Z = 2. Cu radiation, R = 0.085 for 2604 reflexions.

II. Monoclinic, C2/c, a = 27.298, b = 6.727, c = 17.156 Å, β = 105.25°, Z = 4. Cu radiation, R = 0.066 for 1438 reflexions.

In (I) (Fig. 1) only two of the three benzene rings which are related by a threefold, pseudo-axis of symmetry are involved in charge transfer with the centrosymmetric TCNE molecules. This leads to a one-dimensional charge transfer along the body diagonal. In (II) non-planar paracyclophane molecules lie on centres of symmetry as do the TCNE molecules (Fig. 1). Charge transfer is along a face diagonal in a sandwich arrangement.

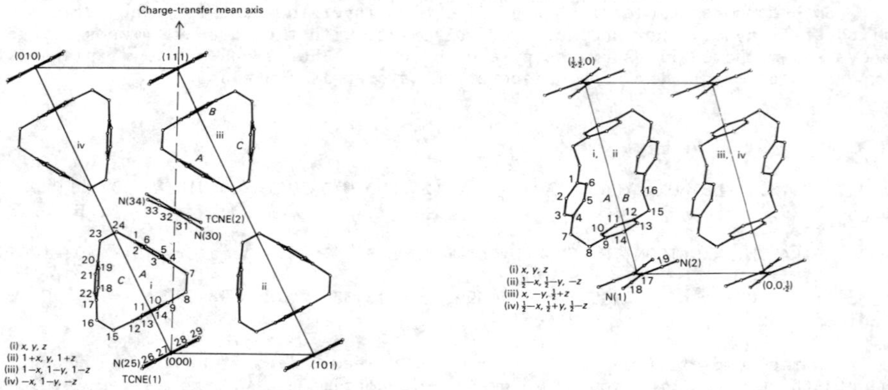

Fig. 1. Intermolecular arrangement in [2.2.2]paracyclophane-TCNE (left) and in [2.2.2.2]paracyclophane-TCNE (right).

TRIETHYLAMMONIUM - 7,7,8,8-TETRACYANO-p-QUINODIMETHANE
$C_{30}H_{24}N_9$

A. FILHOL and M. THOMAS, 1984. Acta Cryst., B40, 44-59.

Triclinic, P$\bar{1}$, a = 14.226, b = 7.843, c = 13.282 Å, α = 69.77, β = 83.93, γ = 73.55°, Z = 1. Cu radiation, R = 0.107 for 3953 reflexions at 110 K. [Cell data and analyses for the same crystal structure at 40, 108, 173, 234, 295, 298, "room temperature" and 345 K are also reported. For previous work see 1.]

The X-ray structure of TEA-$(TCNQ)_2$ has been measured at 110, 173, 234 and 345 K, which, in addition to data in the literature, enables a detailed description of the structural evolution of the compound with temperature. The charge transfer estimated from bond lengths indicates a partial localization (0.6 and 0.4 e) of the charges, on the two independent TCNQ molecules, in the temperature range considered. The unit-cell thermal expansion has been measured in the range 156 to 367 K. An anomaly at 200-220 K was found and the related structural changes are described. Since electrical-conductivity anomalies at 200-220 K have already been reported and sometimes attributed to a metal-insulator phase transition, an attempt is thus made to interpret the temperature dependence of the stacking distances in the TCNQ columns in terms of Peierls-like lattice distortions - namely dimerization and tetramerization.

1. Structure Reports, 46B, 613.

3-t-BUTYL-5-METHYLTHIO-1,2-DITHIOLYLIUM BIS(7,7,8,8-TETRACYANO-p-QUINODIMETHANE)
$C_{32}H_{21}N_8S_3$ $C_8H_{13}S_3{}^+, 2C_{12}H_4N_4{}^{0.5-}$

F. MATHIEU, 1984. Acta Cryst., C40, 1026-1029.

Monoclinic, Pc, a = 14.938, b = 6.472, c = 15.738 Å, β = 101.6°, D_m = 1.35, Z = 2. Mo radiation, R = 0.038 for 1445 reflexions.

In the crystal structure the 7,7,8,8-tetracyano-p-quinodimethane and 3-t-butyl-5-methylthio-1,2-dithiolylium ions form segregated stacks, with some weak interactions S...N and CH...N (Fig. 1). The stacking of the anions is diadic and the interplanar spacings between the anions in a column are quasi equal. The material is a semiconductor.

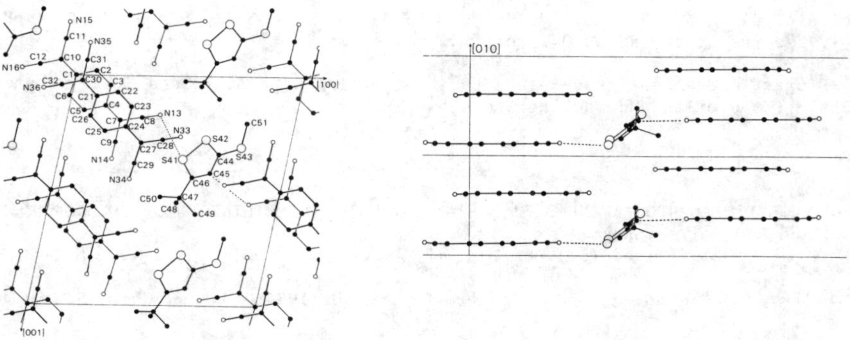

Fig. 1. $C_{32}H_{21}N_8S_3$: projections of the structure along (010) and on to the plane close to ($\bar{5}$06).

1-METHYLCINNOLINIUM BIS(7,7,8,8-TETRACYANOQUINODIMETHANIDE)

$C_{33}H_{17}N_{10}$ $(C_9H_9N_2)(C_{12}H_4N_4)_2$

QIAN MINXIE, ZHU NAIJUE and ZHU DAOBEN, 1984. J. Struct. Chem., 3, 83-86.

Triclinic, P1, a = 6.516, b = 7.715, c = 13.915 Å, α = 88.39, β = 84.29, γ = 78.75°, Z = 1. Mo radiation, R = 0.054 for 1230 reflexions.

Planar TCNQ ions stack along the a-direction with the molecular planes parallel to each other with 3.269 and 3.244 Å between ring planes. [It appears from calculations with the data from the paper that there is an inversion centre between the two TCNQ moieties and another at the centre of the cinnolinium ring system; the authors do not mention this - it would imply a disordered model for the cation. Similar observations apply to the following paper on the 2-methyl derivative.]

2-METHYLCINNOLINIUM BIS(7,7,8,8-TETRACYANOQUINODIMETHANIDE)

$C_{33}H_{17}N_{10}$ $(C_9H_9N_2)(C_{12}H_4N_4)_2$

QIAN MINXIE, ZHU NAIJUE and ZHU DAOBEN, 1984. J. Struct. Chem., 3, 79-82.

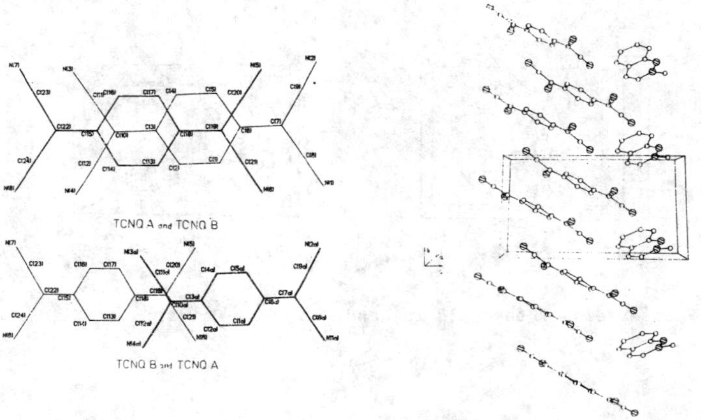

Fig. 1. The TCNQ overlap types and the crystal structure of $C_{33}H_{17}N_{10}$.

Triclinic, P1, a = 7.570, b = 7.599, c = 13.033 Å, α = 86.92, β = 80.92, γ = 69.39°, Z = 1. Mo radiation, R = 0.047 for 1736 reflexions.

Planar TCNQ anions stack along the b-direction with 3.212 and 3.321 Å between layers. The overlap and packing are shown in Fig. 1.

3,3'-DIMETHYLTHIACARBOCYANINE 3,3',5,5'-TETRAMETHYLTRIMETHINEOXONOL CHLOROFORM SOLVATE (Gold and Red Polymorphs)

$C_{34}H_{32}N_6O_6S_2$,$CHCl_3$ (I) and (II) $C_{19}H_{17}N_2S_2^+.C_{15}H_{15}N_4O_6^-$,$CHCl_3$

M.C. ETTER, R.B. KRESS, J. BERNSTEIN and D.J. CASH, 1984. J. Am. Chem. Soc., 106, 6921-6927.

I. Monoclinic, $P2_1/c$, a = 7.660, b = 23.950, c = 20.201 Å, β = 82.33°, Z = 4. Cu radiation, R = 0.096 for 1032 reflexions.

II. Monoclinic, $P2_1/a$, a = 14.528, b = 13.676, c = 18.548 Å, β = 90.54°, Z = 4. Cu radiation, R = 0.083 for 1610 reflexions.

Although the molecular geometries of the gold (I) and red (II) polymorphs are similar, they are packed quite differently (Fig. 1). In (I) nearest neighbour cyanine and oxonol ions are approximately coplanar, packed with their long molecular axes crossed and form stacks by translation along the a-direction. The red form (II) has nearest neighbour cyanine and oxonol ions which are not only coplanar but have their long molecular axes aligned nearly parallel to one another; stacks are formed along the a-glide direction.

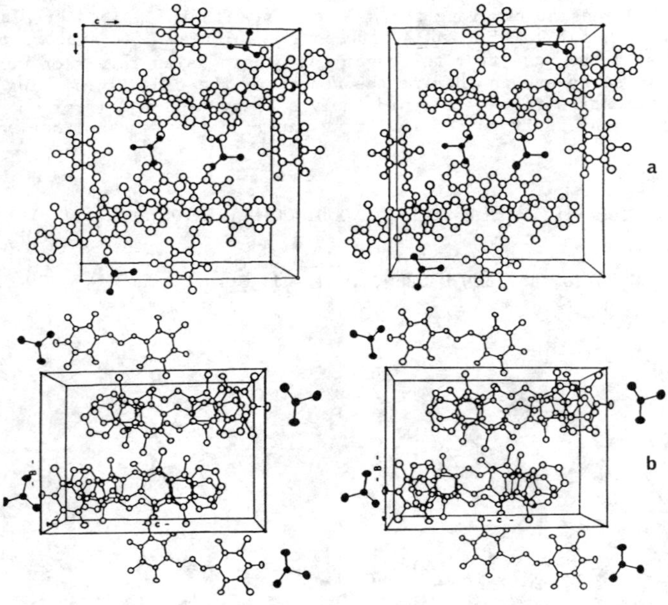

Fig. 1. Stereoviews of the gold (a) and red (b) polymorphs of $C_{34}H_{32}N_6O_6S_2$,$CHCl_3$.

N-(n-BUTYL)QUINOLINIUM 7,7,8,8-TETRACYANOQUINODIMETHANE
$C_{37}H_{24}N_9$ $(n-C_4H_9C_9H_7N)^+(C_{12}H_4N_4)_2^-$

G. MENCZEL, J. KÜRTI and R. LEHÖCZ, 1984. Acta Chim. Acad. Sci. Hung., 116, 335-344.

Triclinic, $P\bar{1}$, a = 7.75, b = 15.49, c = 13.49 Å, α = 72.9, β = 99.8, γ = 87.8°, Dm = 1.27, Z = 2. Cu radiation, R = 0.136 for 2367 reflexions.

The structure was solved using photographic data. The TCNQ moieties are stacked along c and form tetramers with a large perpendicular gap of 3.57 Å between neighbouring tetramers (B and $\bar{B}$). The (A) and (B) TCNQ rings are almost parallel, the dihedral angle between them is 1.4° (Fig. 1).

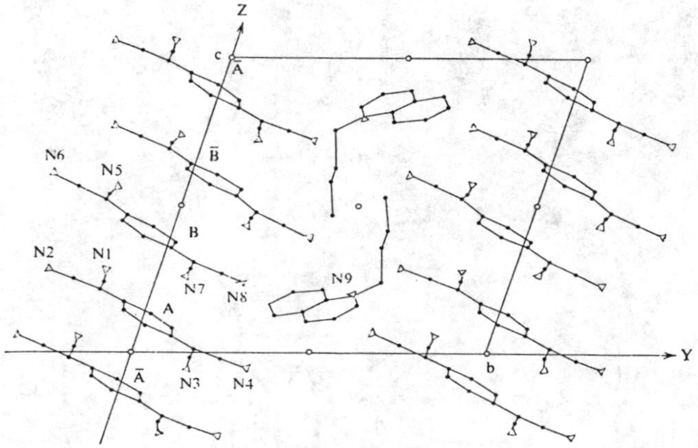

Fig. 1. The crystal structure of $C_{37}H_{24}N_9$ viewed down a.

5,5',12,12'-BIS(TRISELENO)BIS(NAPHTHACENE-6,11-QUINONE) : BENZONITRILE
$C_{43}H_{21}NO_4Se_6$ $(C_{18}H_8Se_3O_2)_2(C_6H_5CN)$

R.P. SHIBAEVA, V.F. KAMINSKII, E.E. KOSTYUCHENKO and E.B. YAGUBSKII, 1984. J. Organometal. Chem., 260, 293-299.

Monoclinic, I2/a, a = 14.263, b = 18.090, c = 14.314 Å, γ = 101.83°, Z = 4. Mo radiation, R = 0.079 for 1026 reflexions.

The benzonitrile molecules are in the inner cavity of the TSNQ molecules and are situated on a crystallographic twofold axis (Fig. 1). The two terminal rings in the tetracyclic carbon frame of the TSNQ molecule are almost planar whereas the central rings have the conformation of a flattened boat. The Se and O atoms are displaced from the plane through atoms C(2),C(7),C(9),C(11),C(16),C(18) by 0.59 and 0.75 Å (Se) and -0.36 and -0.27 Å (O). The five-membered groupings SeOCCC have envelope conformations with Se(1) and Se(2) deviating from the CCCO planes by 0.69 and 0.72 Å respectively.

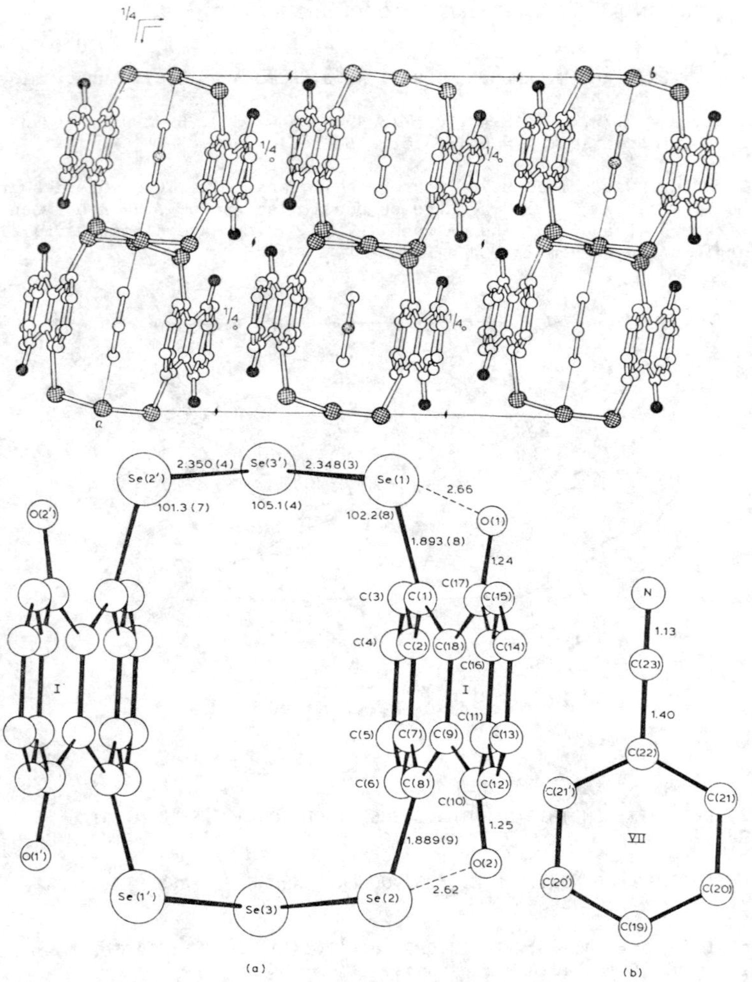

Fig. 1. Projection of the crystal structure of $C_{43}H_{21}NO_4Se_6$ along c, and atomic labelling with some interatomic distances (Å) and angles (°) in the component molecules.

BIS[N,N'-(μ_2-2,6-DIFORMYL-4-METHYLPHENOLATO)-1,3-DIAMINOPROPANE]DICOPPER BIS(7,7,8,8-TETRACYANOQUINODIMETHANIDE)

$C_{48}H_{34}Cu_2N_{12}O_2$ $[Cu_2L^{2+}].[TCNQ^-]_2$

P. LACROIX, O. KAHN, A. GLEIZES, L. VALADE and P. CASSOUX, 1984. Nouv. J. Chim., 8, 643-651.

Triclinic, P$\bar{1}$, a = 11.592, b = 14.232, c = 6.971 Å, α = 90.45, β = 98.52, γ = 112.84°, Z = 1. Mo radiation, R = 0.044 for 2201 reflexions.

The Cu_2L unit lies about an inversion centre and the TCNQ moiety is in a general position in the cell (Fig. 1). The crystal structure contains segregated stacks of Cu_2L and TCNQ entities parallel to the c-axis with angles of 26.5 and 60.0° (for TCNQ and Cu_2L respectively) for the normals to the appropriate mean molecular plane with the c-axis. The Cu coordination (Fig. 1) is completed by a TCNQ nitrogen (Cu-N 2.378(5) Å) to give each Cu square pyramidal coordination (with Cu 0.153 Å above the basal plane towards N). $TCNQ^-$ units are centrosymmetrically related (and their mean planes are parallel). They stack with alternating interplanar spacings of 3.15 and 3.25 Å.

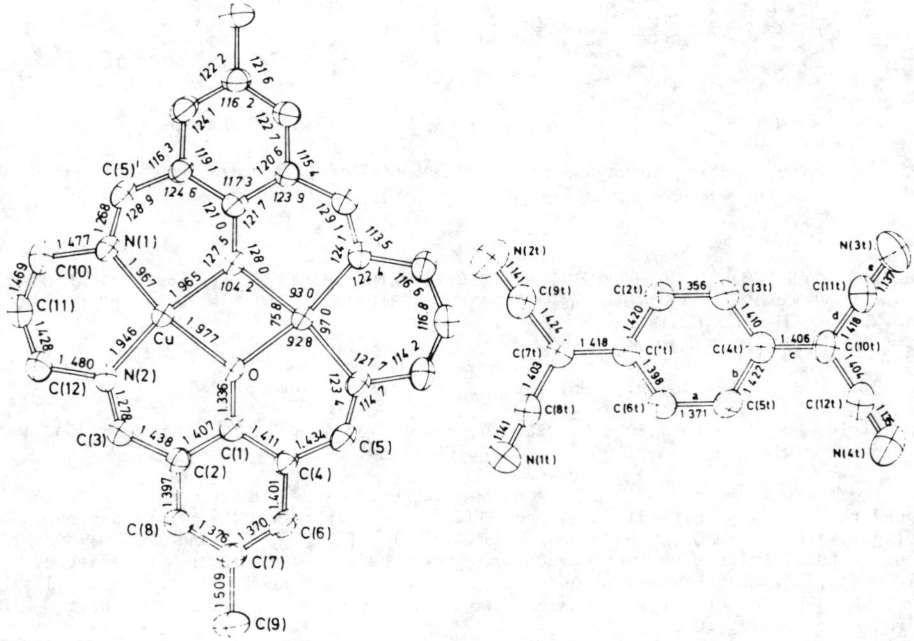

Fig. 1. The Cu_2L^{2+} and $TCNQ^-$ ions and bond lengths.

TETRAKIS(2,6-XYLYLISOCYANIDE)RHODIUM(I) 7,7,8,8-TETRACYANO-p-QUINODIMETHANE

$C_{48}H_{40}N_8Rh$ $[Rh(2,6\text{-}Me_2C_6H_3NC)_4]^+(TCNQ\cdot^-)$

Y. YAMAMOTO and H. YAMAZAKI, 1984. Inorg. Chim. Acta, 85, 83-87.

Monoclinic, $P2_1/n$, a = 15.211, b = 19.942, c = 14.072 Å, β = 98.35°, Z = 4. Mo radiation, R = 0.048 for 1229 reflexions.

The structure consists of mixed stacks of alternating two cations and two TCNQ molecules (Fig. 1). In the cation the Rh atom has slightly distorted square planar coordination geometry, the dihedral angle between planes Rh,C(10),C(20) and Rh,C(30),C(40) being 10.6°. The average Rh-C bond length is 1.96 Å and the N≡C triple bond lengths average 1.14 Å. Bond angles around Rh range from 87.5 to 92.4°. Phenyl ring 20 is tilted considerably (63.9°) with respect to the Rh,C(10),C(20) plane; the other rings have angles of 2-22° with this plane. The dihedral angles (Rh,C(10),C(20) or Rh,C(30),C(40) planes) with the quinoid ring plane are 16.4 and 10.9° respectively, and intermolecular distances between the Rh,C(30),C(40) plane and the TCNQ molecule range from 2.71 Å (N(50)) to 4.12 Å (N(52)). The distance between adjacent TCNQ planes is 3.1 Å.

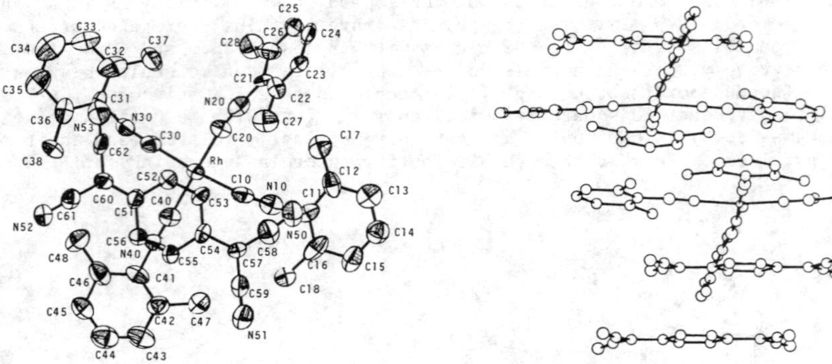

Fig. 1. The stacking arrangement of $C_{48}H_{40}N_8Rh$ (left), and the molecular structure of the complex (right).

HYDROBIS(10-METHYL-5,10-DIHYDROPHENARSAZINE 10-OXIDE) BIS(2,2'-(2,3,5,6-TETRAFLUORO-2,5-CYCLOHEXADIENE-1,4-DIYLIDENE)DIPROPANEDINITROLE) 10-METHYL-5,10-DIHYDROPHENARSAZINE 10-OXIDE ACETONITRILE

$C_{63}H_{37}As_3F_8N_{11}O_3,C_2H_3N$ $[(MPAO)_2H]^+[(TCNQF_4)_2]^-(MPAO)(AN)$

K. DIETZ, H.J. KELLER and D. WEHE, 1984. Acta Cryst., C40, 257-260.

Monoclinic, $P2_1/a$, a = 14.305, b = 25.420, c = 17.669 Å, β = 106.78°, D_m = 1.542, Z = 4. Mo radiation, R = 0.100 for 3357 reflexions.

The three 10-methyl-5,10-dihydrophenarsazine 10-oxide (MPAO) units are folded around the N-As line by 8(2), 3(2) and 11(2)°. Also the $C-(CN)_2$ groups are not coplanar with the $TCNQF_4$ ring planes. $[(MPAO)_2H]^+$, $[(TCNQF_4)_2]^-$ and $(MPAO)_2$ dimers occur in the lattice with intermolecular overlap between $[(MPAO)_2H]^+$ and adjacent $[(TCNQF_4)_2]^-$ dimers; however, these overlaps are not continued to other unit cells. The $[(MPAO)_2H]^+$ dimer (Fig. 1) has the two MPAO molecules connected by a short O-H-O hydrogen bridge (O...O 2.39 Å).

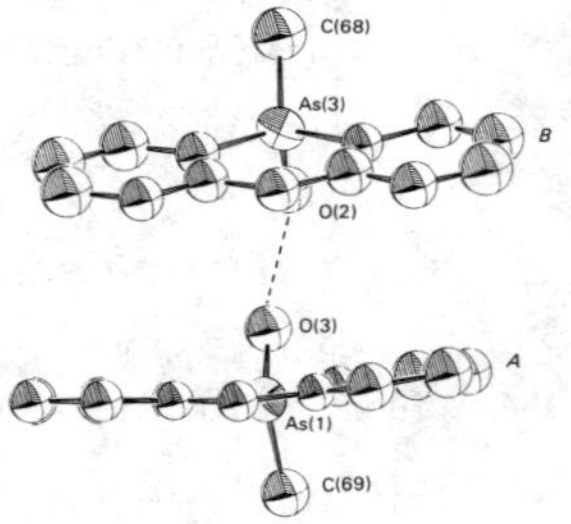

Fig. 1. A view of the $[(MPAO)_2H]^+$ dimer.

DODECASIL 1H PIPERIDINE DINITROGEN CLATHRATE

$C_5H_{11}N,34(SiO_2),5(N_2)$

H. GERKE and H. GIES, 1984. Z. Kristallogr., 166, 11-22.

Hexagonal, P6/mmm, a = 13.783, c = 11.190 Å. Mo radiation, R = 0.056 for 658 reflexions.

This clathrate has a silica host framework and guest molecules within three different types of cages (Fig. 1). Corner sharing SiO_4 tetrahedra form a 3-dimensional 4-connected net which is built up from hexagonal layers of pentagonododecahedra, the fundamental cage of the dodecasil series. The entrapped molecules are oriented within the cages such that van der Waals contacts between guest and host are optimized. In the cage, the mean Si-O-Si angle is 170.6° and the mean Si-O distance is 1.565 Å.

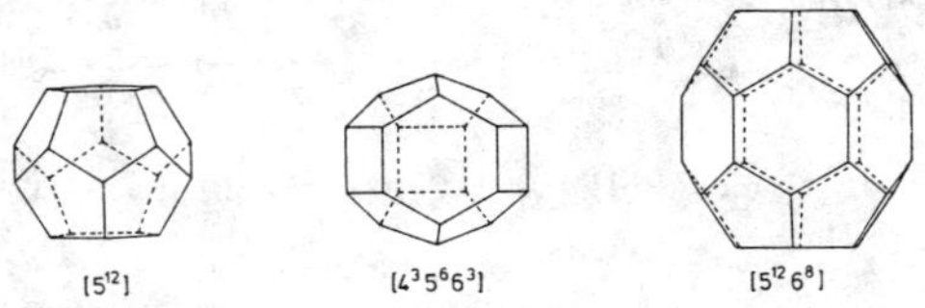

Fig. 1. Cage types present in dodecasil H.

1,4-DIAMINOBUTANECADMIUM(II) TETRACYANONICKELATE(II) 2,5-XYLIDINE CLATHRATE
$C_8H_{12}CdN_6Ni,C_8H_{11}N$

S.-I. NISHIKIORI and T. IWAMOTO, 1984. J. Inclusion Phenomena, 2, 341-349.

Monoclinic, $P2_1/m$, a = 9.795, b = 15.010, c = 7.125 Å, β = 105.56°, D_m = 1.62, Z = 2. Mo radiation, R = 0.052 for 2742 reflexions.

The Ni atom lies on an inversion centre and the Cd atom on a mirror plane; this clathrate has two dimensional extended wavy cyanometal layers bridged by 1,4-diaminobutane to give a three-dimensional host which provides a channel-like cavity for the guest molecule. Views are in Fig. 1. The 1,4-diaminobutane adopts a trans-cis conformation to make the packing more efficient.

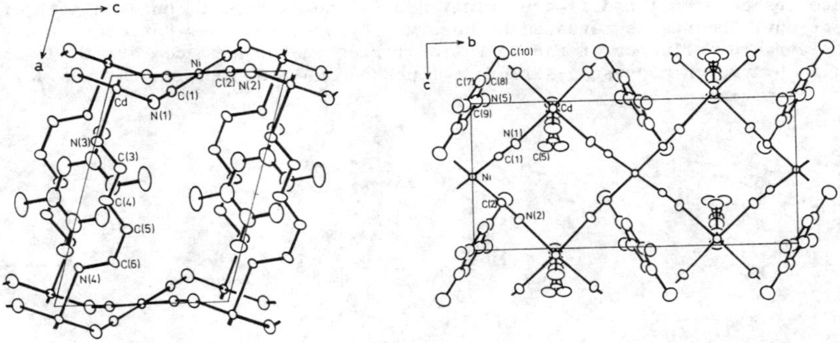

Fig. 1. Two views of the $C_8H_{12}CdN_6Ni,C_8H_{11}N$ clathrate.

1,6-DIAMINOHEXANECADMIUM(II) TETRACYANONICKELATE(II) o-TOLUIDINE CLATHRATE
$C_{10}H_{16}CdN_6Ni,C_7H_9N$

T. HASEGAWA, S.-I. NISHIKIORI and T. IWAMOTO, 1984. J. Inclusion Phenomena, 2, 351-357.

Monoclinic, P2/m, a = 9.541, b = 7.569, c = 7.199 Å, β = 100.3°, D_m = 1.61, Z = 1. Mo radiation, R = 0.036 for 1973 reflexions.

The Ni and Cd atoms lie at sites with 2/m symmetry and channel-like cavities are produced by the 1,6-diaminohexane molecules linking tetracyanonickelate moieties. In this clathrate (Fig. 1) the o-toluidine guest molecules are disordered over two sites.

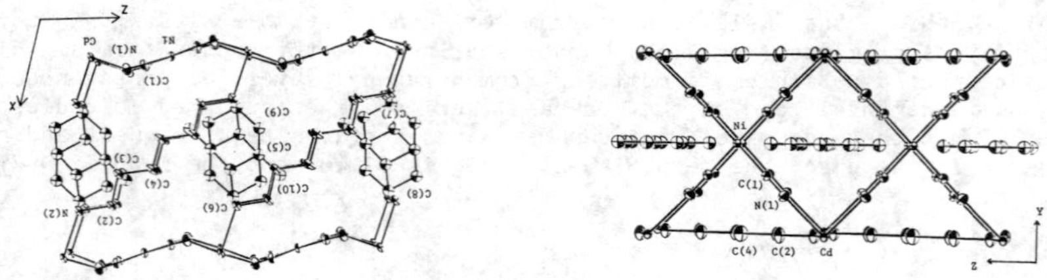

Fig. 1. Two views of the $C_{10}H_{16}CdN_6Ni,C_7H_9N$ clathrate.

2-HYDROXY-3,5-DINITRO-N'-(5-NITROFURFURYLIDENE)BENZOHYDRAZIDE - ACETONITRILE (1:1 COMPLEX)
$C_{12}H_7N_5O_9,C_2H_3N$ (I)

2-HYDROXY-3,5-DINITRO-N'-(5-NITROFURFURYLIDENE)BENZOHYDRAZIDE - ACETIC ACID (1:1 COMPLEX)
$C_{12}H_7N_5O_9,C_2H_4O_2$ (II)

I. GOLDBERG and Z. STEIN, 1984. Acta Cryst., C40, 666-671.

I. Monoclinic, $P2_1/c$, a = 12.977, b = 8.021, c = 16.830 Å, β = 106.95°, Z = 4. Mo radiation, R = 0.048 for 2062 reflexions.

II. Triclinic, $P\bar{1}$, a = 8.665, b = 8.696, c = 12.077 Å, α = 90.82, β = 99.82, γ = 105.15°, Z = 2. Mo radiation, R = 0.052 for 1690 reflexions.

In the crystal structures (Fig. 1) the solvent guest molecules are included in pseudo-channels formed between layers of the hydrazide molecules. Structural differences between I and II are mainly due to the different functionality of the solvent environments as revealed by observed hydrogen-bonding patterns. They reflect a considerable influence of the solvation forces on the host conformation, as a result of which polymorphic crystal-structure types are formed. Bond lengths and angles are generally normal.

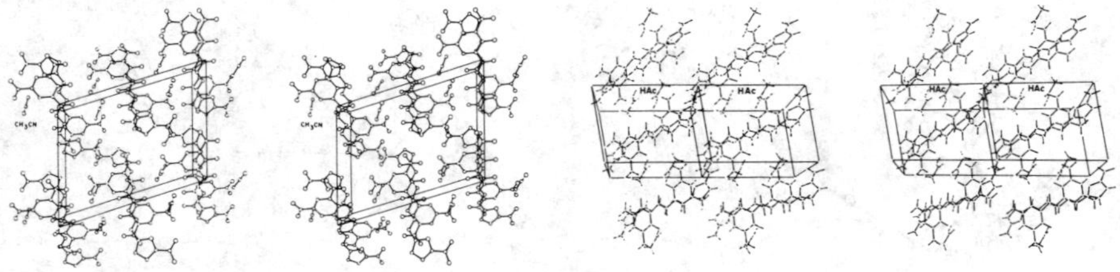

Fig. 1. Stereoviews of the inclusion structure formed with acetonitrile ($C_{12}H_7N_5O_9,C_2H_3N$) (I) and of the compound formed with acetic acid ($C_{12}H_7N_5O_9,C_2H_4O_2$) (II).

18-CROWN-6 TETRAKIS(THIOUREA)CLATHRATE
$C_{12}H_{24}O_6,4(CH_4N_2S)$

G. WEBER, 1984. J. Inclusion Phenomena, 1, 339-347.

Monoclinic, $P2_1$, a = 9.251, b = 19.285, c = 16.556 Å, β = 90.85°, Z = 4. Mo radiation, R = 0.086 for 4252 reflexions.

The two crystallographically independent adducts differ slightly in the way the thiourea molecules link to them via N-H...O hydrogen bonds, as shown in Fig. 1. In molecule A, all six crown O atoms serve as receptors; in B only four crown O atoms are involved. Despite these different hydrogen bond patterns, the macrocycle adopts the same "biangular[99]" conformation in both A and B.

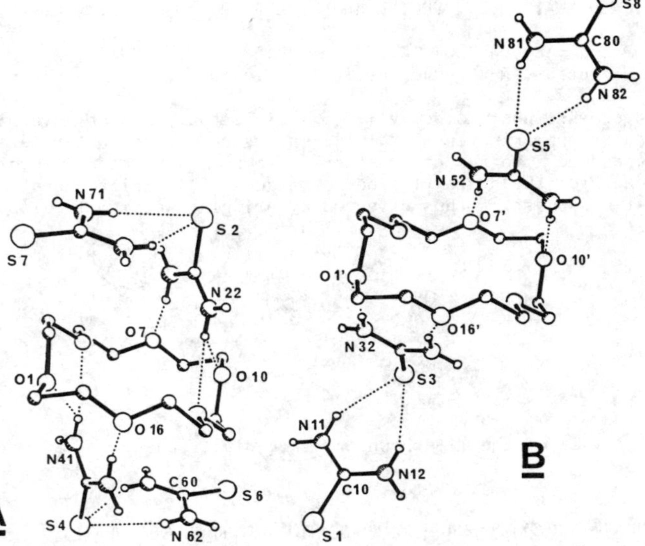

Fig. 1. Perspective views of the two 18-crown-6 tetrakis(thiourea) clathrates.

1,4,7,10,13,16-HEXAOXACYCLOOCTADECANE : N,N'-DIFORMOHYDRAZIDE 1:2

$C_{12}H_{24}O_6,2C_2H_4N_2O_2$ $C_{12}H_{24}O_6,2C_2H_4N_2O_2$

M.R. CAIRA, W.H. WATSON, F. VOGTLE and W. MULLER, 1984. Acta Cryst., C40, 136-138.

Monoclinic, $P2_1/n$, a = 14.176, b = 9.968, c = 8.121 Å, β = 103.34°, D_m = 1.29, Z = 2. Cu radiation, R = 0.044 for 1218 reflexions.

The 18-crown-6 molecule (Fig. 1) lies on an inversion centre and its conformation deviates only slightly from D_{3d} symmetry. The conformation of the guest, N,N'-diformohydrazide approximates C_2 symmetry with a torsion angle of 105.4(4)° around the N-N bond. Two inversion-related crown ether O atoms are each hydrogen-bonded to an N-H group of the guest molecule. Guest molecules are linked by two identical N-H...O=C hydrogen bonds which are related by another inversion centre.

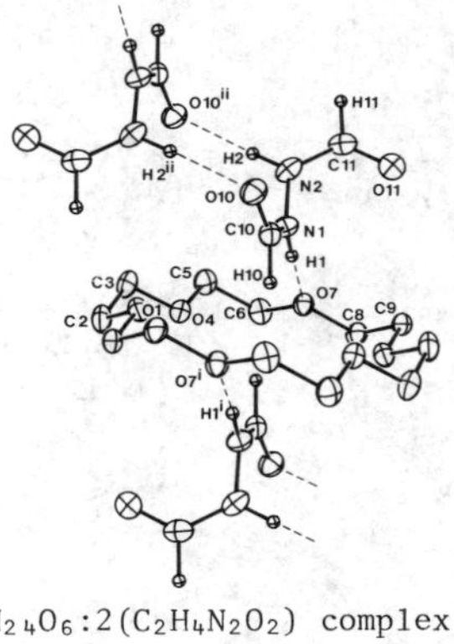

Fig. 1. A view of the $C_{12}H_{24}O_6:2(C_2H_4N_2O_2)$ complex.

(18-CROWN-6) - 2,4-DINITROPHENOL - WATER (1:2:2) COMPLEX
$C_{12}H_{24}O_6.2C_6H_4N_2O_5.2H_2O$

M.R. CAIRA, W.H. WATSON, F. VOGTLE and W. MULLER, 1984. Acta Cryst., C40, 491-493.

Triclinic, $P\bar{1}$, a = 8.064, b = 13.139, c = 7.682 Å, α = 98.53, β = 94.93, γ = 99.73°, Z = 1. Cu radiation, R = 0.049 for 1621 reflexions.

The 18-crown-6 host molecule (Fig. 1) is situated on an inversion centre and has pseudo-symmetry approximating D_{3d}. Two inversion-related water molecules each act as hydrogen-bond donors to two crown ether O atoms. Each water molecule accepts a hydrogen bond from the hydroxyl group of a guest 2,4-dinitrophenol molecule. Binding between host and guest is thus effected by bridging water molecules.

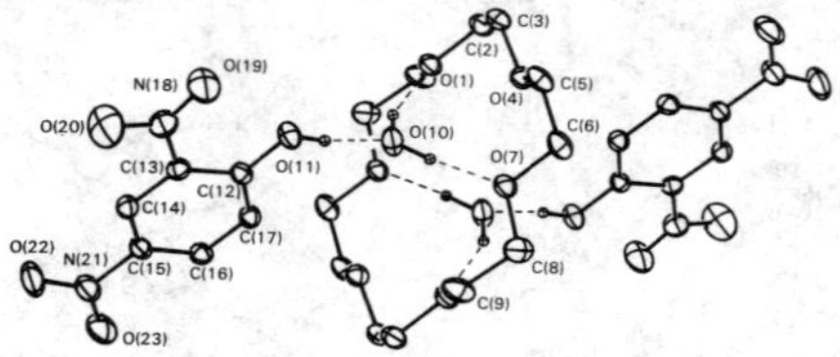

Fig. 1. A view of the (18-crown-6) - 2,4-dinitrophenyl - water (1:2:2) complex.

2,4-DICHLORO-5-CARBOXYBENZSULPHONIMIDE TETRAHYDRATE CLATHRATE
$C_{14}H_7Cl_4NO_8S_2,4H_2O$ (I)

2,4-DICHLORO-5-CARBOXYBENZSULPHONIMIDE ACETIC ACID CLATHRATE DIHYDRATE
$C_{14}H_7Cl_4NO_8S_2,C_2H_4O_2,2H_2O$ (II)

2,4-DICHLORO-5-CARBOXYBENZSULPHONIMIDE BIS(N,N-DIMETHYLACETAMIDE) CLATHRATE DIHYDRATE
$C_{14}H_7Cl_4NO_8S_2,2(C_4H_9NO),2(H_2O)$ (III)

I. GOLDBERG, 1984. J. Inclusion Phenomena, 1, 349-364.

I. Triclinic, $P\bar{1}$, a = 8.227, b = 8.964, c = 16.945 Å, α = 89.64, β = 97.51, γ = 114.28°, Z = 2. Mo radiation, R = 0.046 for 3583 reflexions.

II. Triclinic, $P\bar{1}$, a = 7.857, b = 11.379, c = 13.831 Å, α = 92.50, β = 101.21, γ = 101.12°, Z = 2. Mo radiation, R = 0.035 for 3970 reflexions.

III. Orthorhombic, $P2_12_12_1$, a = 14.838, b = 14.818, c = 14.500 Å, Z = 4. Mo radiation, R = 0.044 for 2552 reflexions.

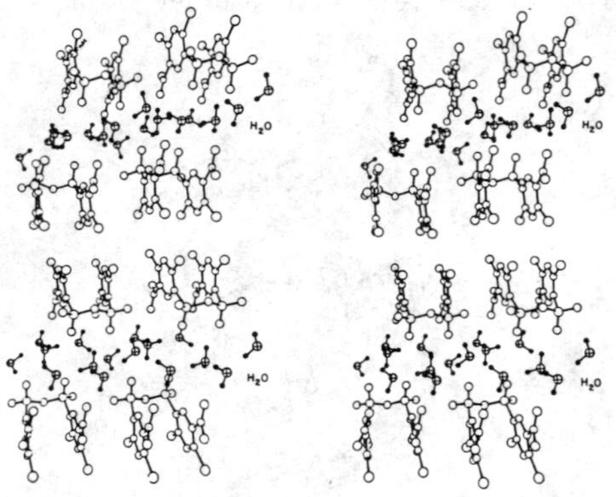

Fig. 1. A stereoview of the $C_{14}H_7Cl_4NO_8S_2,4H_2O$ clathrate.

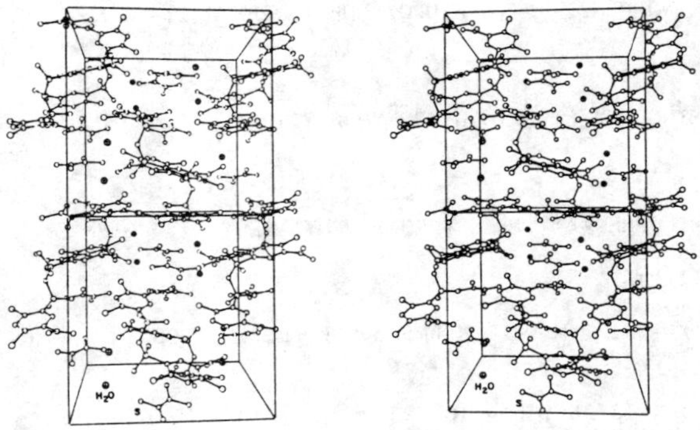

Fig. 2. A stereoview of the $C_{14}H_7Cl_4NO_8S_2,2(C_4H_9NO),2H_2O$ structure.

The water (I) and acetic acid (II) clathrates have layered structures consisting of alternating zones of host (with a folded conformation) and the solvent (Fig. 1). In the case of III a three-dimensional lattice of loosely packed host molecules (in an extended conformation) interspersed with solvent molecules is found (Fig. 2). The folded conformation for the host is induced in solvation environments which are good donors of hydrogen bonds; the extended conformation is favoured in a solvent which is a poor hydrogen bond donor.

N,N'-[5-t-BUTYL-1,3-PHENYLENE-BIS(METHYLENE)]BIS(TRIMETHYLAMMONIUM) DIIODIDE ETHANOL CLATHRATE

$C_{18}H_{34}I_2N_2,C_2H_6O$

H.-G. LOHR, H.-P. HOSEL, A. ENGEL, F. VOGTLE, W. SCHUH and H. PUFF, 1984. Chem. Ber., 117, 1487-1496.

Monoclinic, $P2_1/c$, a = 15.411, b = 12.023, c = 15.153 Å, β = 111.20°, Z = 4. Mo radiation, R = 0.081 for 4055 reflexions.

The X-ray analysis shows that one ethanol molecule is surrounded by the host molecules in a cage-type manner (Fig. 1).

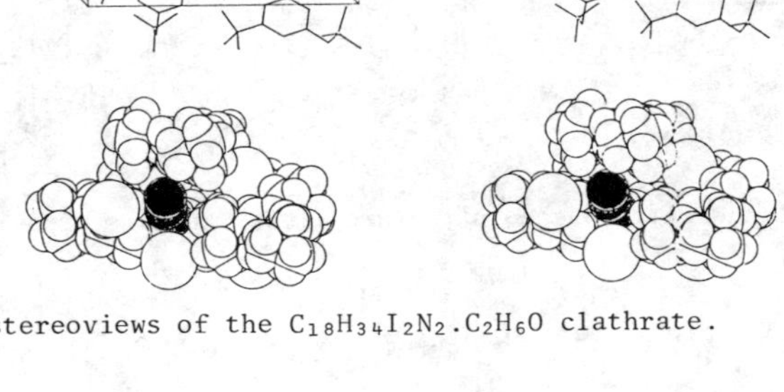

Fig. 1. Two stereoviews of the $C_{18}H_{34}I_2N_2.C_2H_6O$ clathrate.

E,E-1-(p-DIMETHYLAMINOPHENYL)-5-(o-HYDROXYPHENYL)PENTA-1,4-DIEN-3-ONE CHLOROFORM CLATHRATE

$C_{19}H_{19}NO_2$,0.4($CHCl_3$) (I) DHDK,0.4$CHCl_3$

E,E-1-(p-DIMETHYLAMINOPHENYL)-5-(o-HYDROXYPHENYL)PENTA-1,4-DIEN-3-ONE m-DINITROBENZENE CLATHRATE

$C_{19}H_{19}NO_2$,$C_6H_4N_2O_4$ (II)

E,E-1-(p-DIMETHYLAMINOPHENYL)-5-(o-HYDROXYPHENYL)PENTA-1,4-DIEN-3-ONE p-DIMETHYLAMINO-BENZALDEHYDE CLATHRATE

$C_{19}H_{19}NO_2$,$C_9H_{11}NO$ (III) DHDK.PDMB

F.H. HERBSTEIN, M. KAPON, G.M. REISNER and M.B. RUBIN, 1984. J. Inclusion Phenomena, 1, 233-250.

I. Triclinic, P$\bar{1}$, a = 12.086, b = 10.323, c = 8.015 Å, α = 94.58, β = 103.58, γ = 110.10°, D_m = 1.27, Z = 2. Mo radiation, R = 0.155 for 1535 reflexions.

II. Triclinic, P$\bar{1}$, a = 21.787, b = 13.850, c = 7.759 Å, α = 88.25, β = 84.70, γ = 88.86°, D_m = 1.31, Z = 4. Cu radiation, R = 0.071 for 6033 reflexions.

III. Monoclinic, $P2_1/n$, a = 22.383, b = 12.272, c = 8.917 Å, β = 92.99°, D_m = 1.21, Z = 4. Mo radiation, R = 0.103 for 2641 reflexions.

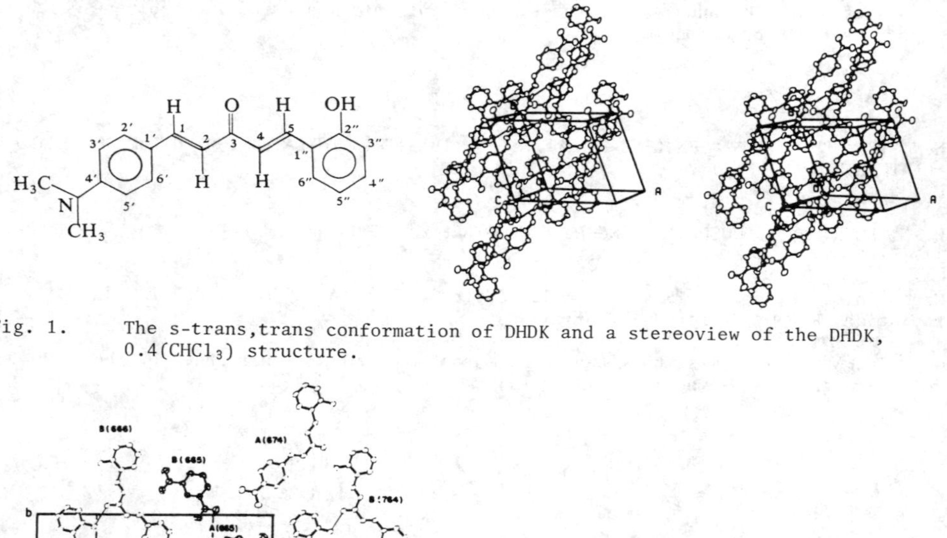

Fig. 1. The s-trans,trans conformation of DHDK and a stereoview of the DHDK, 0.4($CHCl_3$) structure.

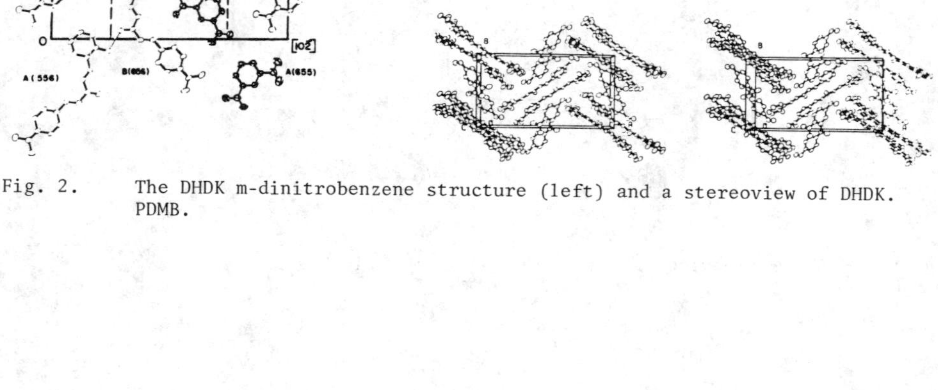

Fig. 2. The DHDK m-dinitrobenzene structure (left) and a stereoview of DHDK. PDMB.

The structure of I is characterised by centrosymmetrically related pairs of DHDK molecules linked by O-H...O hydrogen bonding between carbonyl and hydroxyl groups (Fig. 1). Approximately planar molecular pairs are arranged in sheets parallel to (112). The situation of the $CHCl_3$ molecule in its cavity is not defined. The DHDK molecule is in the s-trans,trans conformer. In II the DHDK molecule has an s-cis, trans conformation and the molecules are hydrogen bonded through carbonyl-hydroxyl interactions (O...O 2.64(1) Å) to form chains in the y-direction (Fig. 2). Both components lie in the molecular sheet in (201). The guest molecules are in sinuous channels between DHDK ribbons. Successive planes are mutually shifted so that the guest molecules are enclosed above and below by host molecules. In III, the DHDK molecules have s-trans,trans conformation and form the sides of a channel of approximately rectangular cross section with their long axes perpendicular to the channel axis (Fig. 2). The PDMB molecules are contained in the channels with their mean planes approximately perpendicular to those of the DHDK molecules. PDMB and DHDK molecules are linked by a hydrogen bond between a DHDK hydroxyl and a PDMB carbonyl (O...O 2.724(8) Å).

1,1'-BINAPHTHYL-2,2'-DICARBOXYLIC ACID BIS(METHANOL)CLATHRATE
$C_{22}H_{14}O_4,2(CH_4O)$ (I)

1,1'-BINAPHTHYL-2,2'-DICARBOXYLIC ACID BIS(ETHANOL)CLATHRATE
$C_{22}H_{14}O_4,2(C_2H_6O)$ (II)

1,1'-BINAPHTHYL-2,2'-DICARBOXYLIC ACID BIS(ISOPROPANOL)CLATHRATE
$C_{22}H_{14}O_4,2(C_3H_8O)$ (III)

1,1'-BINAPHTHYL-2,2'-DICARBOXYLIC ACID 2-BUTANOL CLATHRATE
$C_{22}H_{14}O_4,C_4H_{10}O$ (IV)

1,1'-BINAPHTHYL-2,2'-DICARBOXYLIC ACID ETHYLENE GLYCOL CLATHRATE
$C_{22}H_{14}O_4,C_2H_6O_2$ (V)

E. WEBER, I. CSOREGH, B. STENSLAND and M. CZUGLER, 1984. J. Am. Chem. Soc., 106, 3297-3306.

I. Monoclinic, $P2_1/n$, a = 15.642, b = 14.532, c = 9.292 Å, β = 95.14°, Z = 4. Mo radiation, R = 0.068 for 2308 reflexions.

II. Monoclinic, C2/c, a = 11.737, b = 14.522, c = 13.769 Å, β = 101.50°, Z = 4. Mo radiation, R = 0.072 for 842 reflexions.

III. Monoclinic, C2/c, a = 12.051, b = 14.776, c = 14.362 Å, β = 102.53°, Z = 4. Mo radiation, R = 0.075 for 707 reflexions.

IV. Monoclinic, $P2_1/n$, a = 12.009, b = 12.747, c = 14.982 Å, β = 105.52°, Z = 4. Mo radiation, R = 0.062 for 2281 reflexions.

V. Monoclinic, $P2_1/n$, a = 14.276, b = 9.533, c = 15.556 Å, β = 109.19°, Z = 4. Mo radiation, R = 0.050 for 2261 reflexions.

1,1'-Binaphthyl-2,2'-dicarboxylic acid is demonstrated as a novel type of clathrate host whose main mode of action is derived from a combination of the steric barrier and the coordinative interaction principle which constitutes the new "coordinatoclathrate" strategy. Inclusion properties of the crystal lattice are revealed for a variety of uncharged organic guest molecules (30 examples), ranging from OH-, to NH-, to CH-acidic compounds such as alcohols, carboxylic acids, amides, and nitriles to rather unpolar compounds like bromobenzene and toluene. Marked discrimination selectivities in the clathrate formation from solvent mixtures are found in regard to the group functionality, the substitution pattern, and the molecular size of the guest species, making accessible a simple process for chemical compound separation. The stoichiometries and the increased-temperature as well as the reduced-pressure stabilities of the various clathrates are discussed. The principles of structure of five different alcohol clathrates are determined (Fig. 1). In I, IV and

V the molecules lie in general positions; in II and III the binaphthyl moieties have twofold crystallographic symmetry. In all these cases hydroxyl groups of the host molecules were found to be intercalated via large pseudo-ring formation between the carboxyl functions of at least two host units of opposite chirality with a different mode of hydrogen bridging. Depending on the host:guest stoichiometry (1:1 or 1:2) and on the nature of the guest molecules, these entities consist of three, four, or eight moieties (COOH, OH). The direction of the strong and cooperative bonds is always homodromic. The shape and the size of the cleft formed in the matrix of cooperating host moieties are shown to vary, matching the specific needs of coordinating interactions (hydrogen bonding) and topological requirements (branching, e.g.) of the guest species.

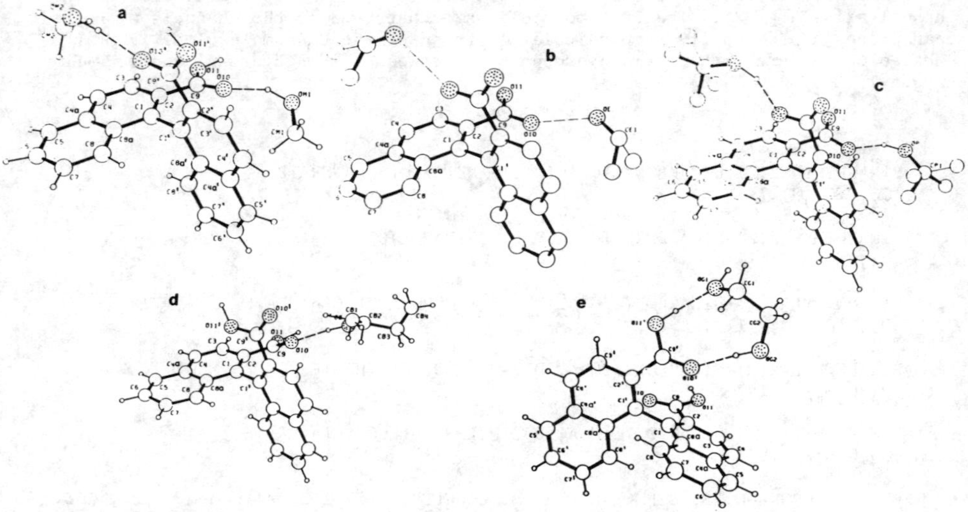

Fig. 1. Molecular structures of the five coordinatoclathrates.

1,4,7,14,23-PENTAOXA[7]ORTHOBENZENO[2](2,6)PYRIDINO[2]ORTHOBENZENOPHANE HYDRATED ETHANOL INCLUSION COMPOUND

$C_{23}H_{23}NO_5,C_2H_6O,0.12(H_2O)$ (I)

1,4,7,14,23-PENTAOXA[7]ORTHOBENZENO[2](2,6)PYRIDINO[2]ORTHOBENZENOPHANE SODIUM ISOTHIOCYANATE COMPLEX

$C_{24}H_{23}NNaO_5S$ (II) $C_{23}H_{23}NO_5,NaNCS$

G. WEBER, 1984. Acta Cryst., C40, 592-597.

I. Tetragonal, I4/m, a = 15.342, c = 19.397 Å, Z = 8. Mo radiation, R = 0.058 for 1403 reflexions.

II. Triclinic, $P\bar{1}$, a = 11.529, b = 14.667, c = 14.724 Å, α = 87.12, β = 81.11, γ = 68.58°, Z = 4. Mo radiation, R = 0.054 for 4900 reflexions.

In I the title crown ether possesses nearly the same molecular geometry and conformation as its xylylene analogue (1) (Fig. 1). In the lattice the fourfold axis is associated with channels along c which are occupied by disordered solvent molecules. The slight conformational changes in II (Fig. 1) imply weak cation-dipole interactions involving the six hetero-atoms; there are no unusual features in the packing of the discrete complex ion pairs.

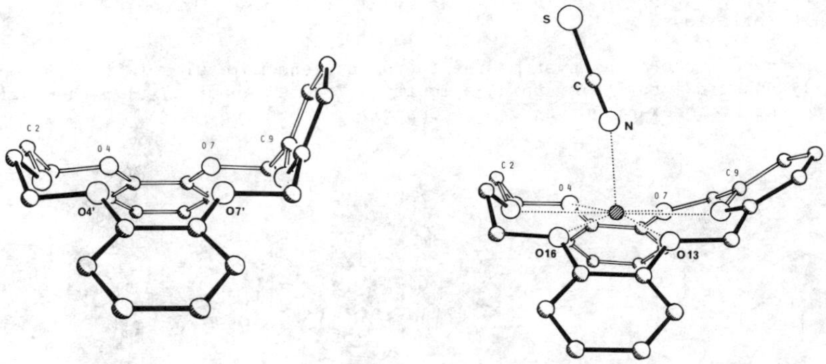

Fig. 1. Perspective views of the macrocycle in $C_{23}H_{23}NO_5,C_2H_5OH,0.12H_2O$ (I) (left) and in one of the two independent complexes of $C_{23}H_{23}NO_5$,NaNCS (II) (right); the other complex in II is similar.

1. Structure Reports, 50B,

TETRAPHENYLENE - CARBON TETRACHLORIDE (2/1)
$2C_{24}H_{16},CCl_4$

H.N.C. WONG, T.-Y. LAU and T.C.W. MAK, 1984. Acta Cryst., C40, 1721-1723.

Tetragonal, $P4_2/n$, a = 9.930, c = 18.948 Å, D_m = 1.355, Z = 2. Mo radiation, R = 0.097 for 1047 reflexions.

The $C_{24}H_{16}$ molecules have crystallographic twofold symmetry with the CCl_4 guest species disordered about a site with $\bar{4}$ symmetry and lying in a nearly spherical cavity of free diameter 7.2-7.4 Å (Fig. 1).

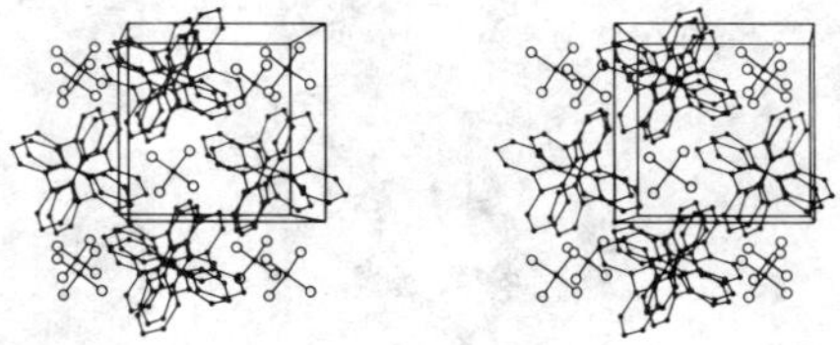

Fig. 1. A stereoview of the molecular packing of $2C_{24}H_{16}.CCl_4$.

BIS(TETRAPHENYLENE) BENZENE CLATHRATE
$2(C_{24}H_{16}),C_6H_6$ (I)

BIS(TETRAPHENYLENE) CYCLOHEXANE CLATHRATE
$2(C_{24}H_{16}),C_6H_{12}$ (II)

F.H. HERBSTEIN, T.C.W. MAK, G.M. REISNER and H.N.C. WONG, 1984. J. Inclusion Phenomena, 1, 301-308.

I. Tetragonal, $P4_2/n$, a = 10.069, c = 18.431 Å, Z = 2. Mo radiation, R = 0.128 for 926 reflexions.

II. Tetragonal, $P4_2/n$, a = 10.073, c = 18.712 Å, Z = 2. Mo radiation, R = 0.100 for 1180 reflexions.

In these isomorphous systems the tetraphenylene molecules (with C_2 symmetry) form a nearly spherical cavity of point symmetry $\bar{4}$ located about $\frac{1}{4},\frac{1}{4},\frac{1}{4}$. The enclosed benzene and cyclohexane guests are necessarily disordered.

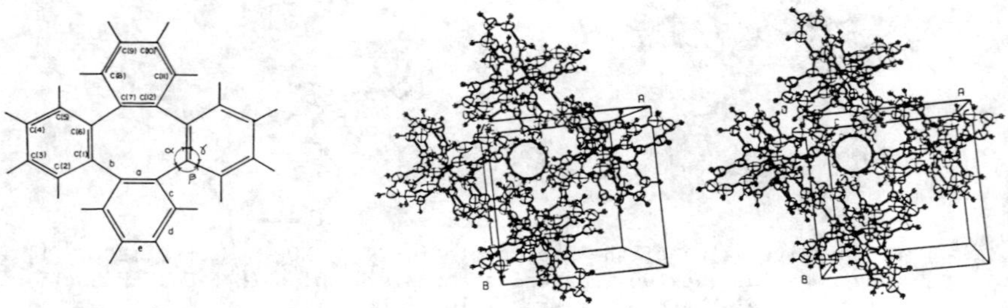

Fig. 1. The tetraphenylene molecule and a stereoview of the unit cell showing the cavity at $\frac{1}{4},\frac{1}{4},\frac{1}{4}$. The location of a disordered benzene is indicated.

DEOXYCHOLIC ACID DI-t-BUTYLTHIOKETONE CLATHRATE
$C_{24}H_{40}O_4,0.5(C_9H_{18}S)$

K. PADMANABHAN, K. VENKATESAN and V. RAMAMURTHY, 1984. Can. J. Chem., 62, 2025-2028.

Orthorhombic, $P2_12_12$, a = 13.933, b = 27.294, c = 7.285 Å, D_m = 1.130, Z = 4. Mo radiation, R = 0.101 for 1364 reflexions.

In this structure (Fig. 1) two halves of a bilayer are related by a 2_1 axis and are parallel to the a-axis. The bilayers are stacked along the b-direction and are related by the twofold rotation axis of the space group. There is a helical system of intermolecular (O-H...O) hydrogen bonds (O...O 2.644-2.755(14) Å). The guest molecule exhibits orientational disorder.

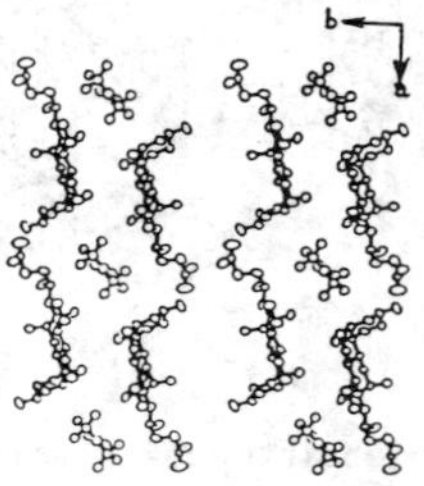

Fig. 1. The packing arrangement of the deoxycholic acid di-t-butylthioketone complex.

CALIX[4]ARENE ACETONE
$C_{28}H_{24}O_4,C_3H_6O$ (I)

TRIS(CALIX[4]ARENE) ACETONE
$(C_{28}H_{24}O_4)_3,C_3H_6O$ (II)

R. UNGARO, A. POCHINI, G.D. ANDREETTI and V. SANGERMANO, 1984. J. Chem. Soc. Perkin II, 1979-1985.

I. Orthorhombic, Pnma, a = 17.010, b = 14.127, c = 10.677 Å, Z = 4. Cu radiation, R = 0.090 for 1085 reflexions.

II. Hexagonal, $P6_3/m$, a = 14.543, c = 18.228 Å, γ = 120°, Z = 6. Mo radiation, R = 0.070 for 3940 reflexions.

The conformation of the calix[4]arene molecules (Fig. 1) in the two phases is different, mainly in the orientation of the phenyl rings with respect to the mean plane of the bridging methylene group (dihedral angles 115.3(1), 115.5(1), 137.2(1) in II, 123.0(3), 125.3(3)° in I). In II two opposite rings are pushed in the direction of the calix and two are pulled, giving the molecule a more asymmetric shape. This distortion has no effect on the intramolecular hydrogen-bonding system but does deform the angles at the bridging methylene carbon atoms which increase from 106.2(5), 108.2(8) and 107.6(5) in I to 112.7(3) and 113.0(3)° in II. In both crystals the molecule loses the fourfold symmetry, but they possess a plane of symmetry which goes through the methylene carbon atoms in I and through the O-C...C lines of two trans phenolic groups in II. In the crystal, I shows a molecular arrangement of an intercalatoclathrate type and II a tubulato-clathrate type of molecular arrangement.

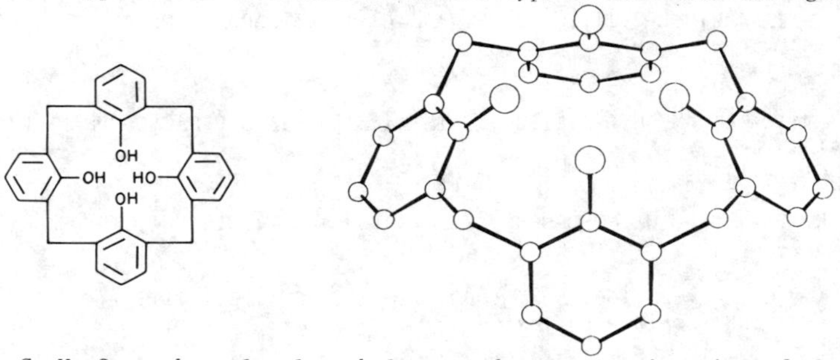

Fig. 1. $C_{28}H_{24}O_4$: the molecular skeleton and a perspective view of the calix[4]-arene molecule.

N,N',N'',N'''-TETRAMETHYL-2,11,20,29-TETRAAZA[3.3.3.3]PARACYCLOPHANE
$C_{36}H_{44}N_4$ (I)

N,N',N'',N'''-TETRAMETHYL-2,11,20,29-TETRAAZA[3.3.3.3]PARACYCLOPHANE CARBON DIOXIDE CLATHRATE
$C_{36}H_{44}N_4,CO_2$ (II)

N,N',N'',N'''-TETRAMETHYL-2,11,20,29-TETRAAZA[3.3.3.3]PARACYCLOPHANE ACETONITRILE CLATHRATE
$C_{36}H_{44}N_4,C_2H_3N$ (III)

K. HIROTSU, S. KAMITORI, T. HIGUCHI, I. TABUSHI, K. YAMAMURA and H. NONOGUCHI, 1984. J. Inclusion Phenomena, 2, 215-222.

I. Monoclinic, $P2_1/a$, a = 21.491, b = 15.123, c = 10.020 Å, β = 97.26°, Z = 4. R = 0.083 for 1269 reflexions.

II. Monoclinic, C2/c, a = 25.590, b = 5.420, c = 26.900 Å, β = 109.86°, Z = 4. Mo radiation, R = 0.076 for 1593 reflexions.

III. Monoclinic, C2/c, a = 25.544, b = 5.406, c = 26.948 Å, β = 109.90°, Z = 4. Mo radiation, R = 0.087 for 815 reflexions.

A view of the uncomplexed host molecule is in Fig. 1. In its clathrates (II and III), the host lies about a twofold crystallographic symmetry axis and the guest molecules lie along the twofold axis (Fig. 1). The parent macrocycle has a "square-box" structure and the size of the hydrophobic cavity is 6 Å deep ca. 4.6 × 4.6 at one end and ca. 6.4 × 6.4 at the wide end.

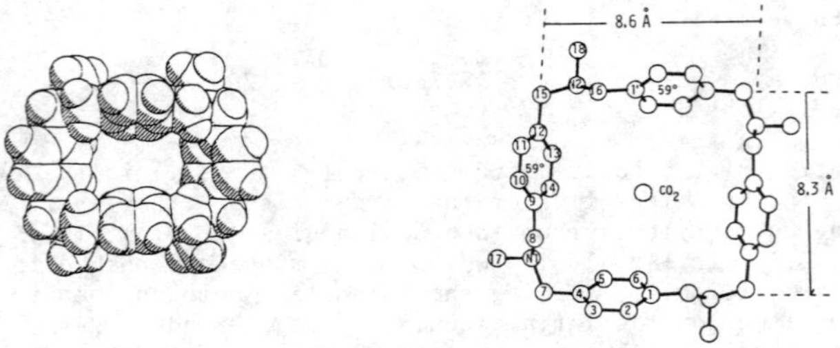

Fig. 1. View of uncomplexed $C_{36}H_{44}N_4$ (left) and its CO_2 complex (II) (right).

N,N',N'',N'''-TETRAMETHYL-2,11,20,29-TETRAAZA[3.3.3.3]PARACYCLOPHANE CHLOROFORM CLATHRATE
$C_{36}H_{44}N_4$,$CHCl_3$ (I)

rac-N,N',N'',N'''-TETRAMETHYL-2,11,20,29-TETRAAZA[3.3.3.3]PARACYCLOPHANE DICHLOROMETHANE CLATHRATE
$C_{36}H_{44}N_4$,CH_2Cl_2 (II)

I. TABUSHI, K. YAMAMURA, H. NONOGUCHI, K. HIROTSU and T. HIGUCHI, 1984. J. Am. Chem. Soc., 106, 2621-2625.

I. Monoclinic, C2, a = 25.166, b = 5.668, c = 13.438 Å, β = 111.31°, Z = 2. Mo radiation, R = 0.098 for 1219 reflexions.

II. Monoclinic, C2/c, a = 25.359, b = 5.485, c = 50.565 Å, β = 96.92°, Z = 8. Mo radiation, R = 0.079 for 2597 reflexions.

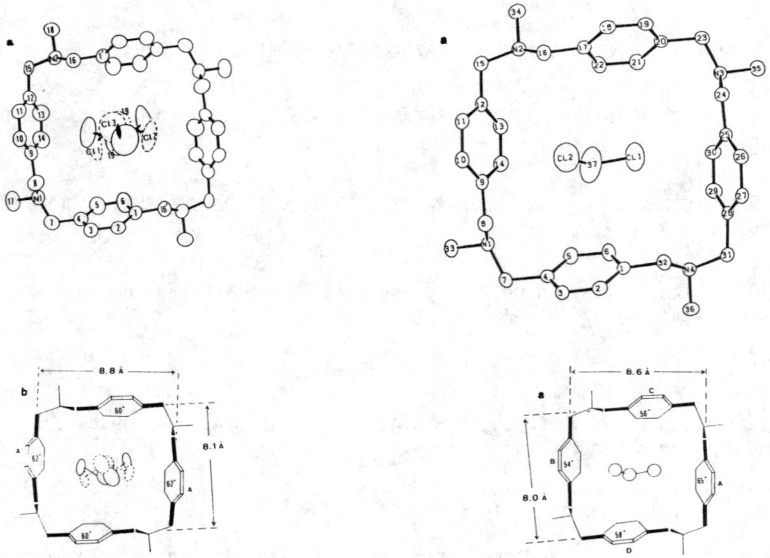

Fig. 1. Views of the $CHCl_3$ (left) and CH_2Cl_2 (right) clathrates.

Compound (I) has twofold crystallographic symmetry (Fig. 1) and the $CHCl_3$ molecule is disordered. The macrocycle has a "square box" shape. The walls are composed of four benzene rings that form angles of ca 60° with respect to a pseudo molecular plane defined by the four N atoms. This conformation is achieved by rotation of the benzene rings around the C(7)-C(4)-C(1)-C(16') or C(15)-C(12)-C(9)-C(8) axis. The complex is chiral. Compound (II) is optically inactive and the inclusions of the CH_2Cl_2 molecule in place of $CHCl_3$ causes small conformational changes in the host (Fig. 1).

N,N'-DITRITYLUREA PROPANAMIDE CLATHRATE
$C_{39}H_{32}N_2O,C_3H_7NO$ (I)

N,N'-DITRITYLUREA ETHYL N-ACETYLGLYCINATE CLATHRATE
$C_{39}H_{32}N_2O,C_6H_{11}NO_3$ (II)

BIS(N-TRITYLUREA) N,N-DIMETHYLFORMAMIDE CLATHRATE
$2(C_{20}H_{18}N_2O),C_3H_7NO$ (III)

I. GOLDBERG, L.-T.W. LIN and H. HART, 1984. J. Inclusion Phenomena, 2, 377-389.

I. Monoclinic, C2/c, a = 15.839, b = 9.088, c = 24.584 Å, β = 111.05°, Z = 4. Cu radiation, R = 0.070 for 2419 reflexions (at 115 K).

II. Triclinic, P$\bar{1}$, a = 9.010, b = 10.800, c = 19.810 Å, α = 105.29, β = 94.33, γ = 93.03°, Z = 2. Cu radiation, R = 0.084 for 4377 reflexions (at 115 K).

III. Monoclinic, Cc, a = 29.614, b = 8.906, c = 16.127 Å, β = 121.04°, Z = 4. Cu radiation, R = 0.069 for 2209 reflexions (at 115 K).

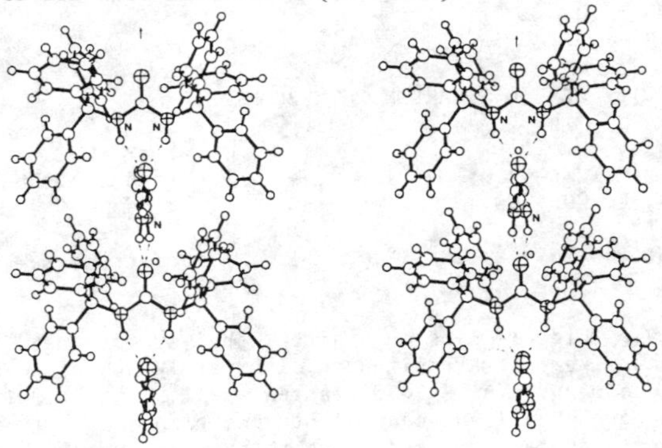

Fig. 1. A stereoview of the hydrogen bonding in $C_{39}H_{32}N_2O,C_3H_7NO$ (I).

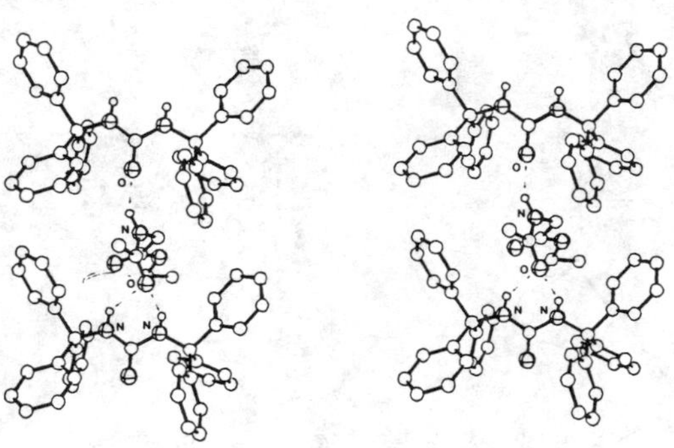

Fig. 2. A stereoview of the hydrogen bonding in $C_{39}H_{32}N_2O,C_6H_{11}NO_3$ (II).

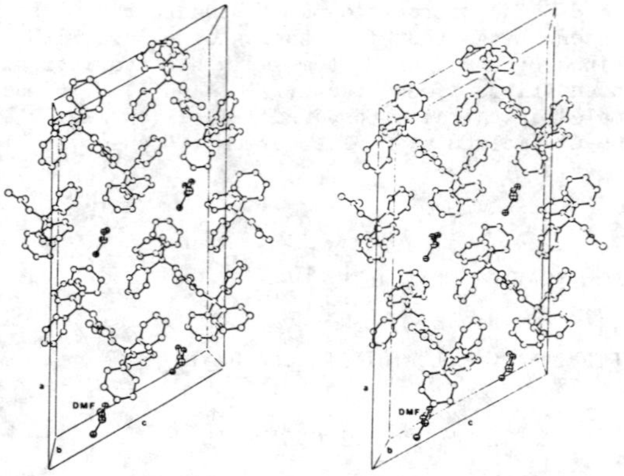

Fig. 3. A stereoview of the bis(N-tritylurea) N,N-dimethylformamide clathrate (III).

The structure of I is shown in Fig. 1; both host and guest have twofold crystallographic symmetry (guest disordered) and are linked by N-H...O hydrogen bonds. In II, the guest molecule lies between two host molecules linked by N-H...O hydrogen bonds as shown in Fig. 2. The structure of III has the guest molecule enclosed within lattice cavities formed by trityl groups of four neighbouring hosts (Fig. 3). The host molecules are linked by N-H...O hydrogen bonds along the b-direction; there is no H-bonding involving the guest molecules.

HEXAKIS(PHENYLSELENO)BENZENE BIS(CARBON TETRABROMIDE)
$C_{42}H_{30}Se_6,2(CBr_4)$

C.J. GILMORE, D.D. MACNICOL, P.R. MALLINSON, A. MURPHY and M.A. RUSSELL, 1984. J. Inclusion Phenomena, 1, 295-299.

Trigonal, $R\bar{3}$, a = 11.009, α = 82.28(1)°, Z = 1. Mo radiation, R = 0.049 for 1031 reflexions.

The host molecule (Fig. 1) lies on a site with $\bar{3}$ symmetry and crystallographically equivalent phenyl groups are situated alternately above and below the plane of the central benzene ring. The C-Se distances are 1.91 and 1.92(1) Å and the Se atom is 0.12 Å from both phenyl and central benzene planes. There are two CBr_4 molecules in each enclosed cage oriented such that a C-Br bond of each lies along the rhomb diagonal (Fig. 1).

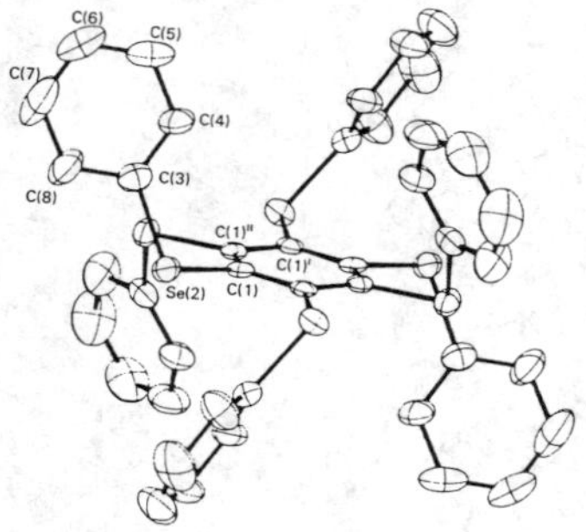

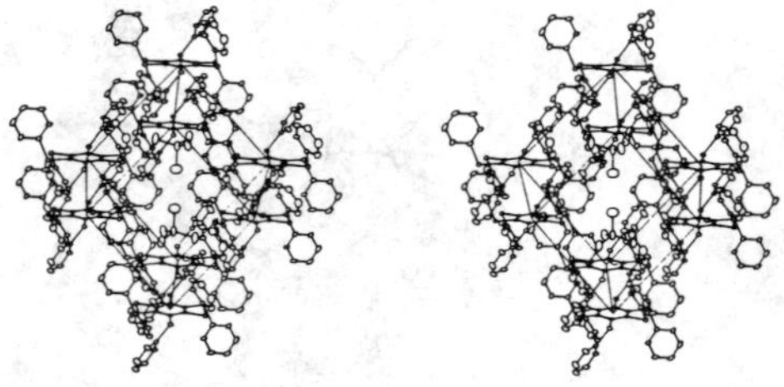

Fig. 1. The $C_{42}H_{30}Se_6$ molecule and a stereoview of the CBr_4 clathrate.

DEUTERO-β-CYCLODEXTRIN UNDECAKIS(DEUTERIUM OXIDE) CLATHRATE
$C_{42}H_{49}D_{21}O_{35},11(D_2O)$

C. BETZEL, W. SAENGER, B.E. HINGERTY and G.M. BROWN, 1984. J. Am. Chem. Soc., 106, 7545-7557

Monoclinic, $P2_1$, a = 21.261, b = 10.306, c = 15.123 Å, β = 112.3°, Z = 2. Neutron radiation, R = 0.068 for 6610 reflexions.

A view of the β-cyclodextrin molecule is given in Fig. 1. There are 53 H-bonds per asymmetric unit of which 35 are normal O-H...O and 18 are of the flip-flop type O-H...H-O, representing a statistical average O-H...O⇄O...H-O. The 11 water molecules are distributed over 16 positions, 8 in the cavity of β-CD (6.13 water molecules) and 8 in interstices (4.88 water molecules). Cavity waters form only 2 H-bonds to enclosing β-CD and 6 contacts to neighbouring β-CD. There positions are better occupied and they are better ordered than "outside" water in interstices, an effect probably due to the hydrophobic character of the cavity. Flip-flop O-H...H-O bonds are interconnected to produce more extended systems, with one infinite flip-flop chain running through the whole crystal structure. Flip-flop bonds are also observed between all seven intramolecular, interglucose O(2), O(3) hydroxyls, O(2)-H...H-O(3), and explain the unusual conformational stiffness of β-CD in solution. For details of previous work see 1. The α-form is described in (2).

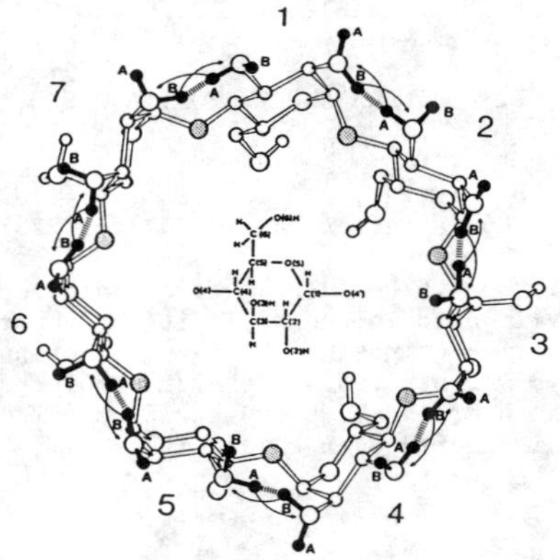

Fig. 1. The structure of β-cyclodextrin.

1. Structure Reports, 49B, 743.
2. Ibid, 46B, 448; 49B, 742.

β-CYCLODEXTRIN - HEXAMETHYLENETETRAMINE (1:1) COMPLEX HEXAHYDRATE
$C_{42}H_{70}O_{35},C_6H_{12}N_4,6H_2O$

K. HARATA, 1984. Bull. Chem. Soc. Japan, 57, 2596-2599.

Monoclinic, $P2_1$, a = 15.285, b = 10.345, c = 20.118 Å, β = 102.14°, Dm = 1.46, Z = 2. Cu radiation, R = 0.056 for 4369 reflexions.

The β-cyclodextrin molecule (Fig. 1) is slightly distorted from a regular heptagonal structure. These molecules are stacked along the crystallographic b axis which makes an angle of 46.7° with the plane through seven glycosidic O atoms. Water molecules occupy the intermolecular spaces between the β-cyclodextrin molecules whilst the guest hexamethylenetetramine molecule is located roughly at the centre of the β-cyclodextrin ring, in van der Waals contact with the interior surface of the host molecule. The orientation of two primary OH groups is statistically disordered. The secondary OH groups form intramolecular hydrogen bonds between adjacent glucose residues (O...O distances range from 2.73 to 3.00 Å), and two out of four N atoms of hexamethylenetetramine form N...H-O hydrogen bonds with water and adjacent β-cyclodextrin molecules (N...O distances 2.74 and 2.80 Å).

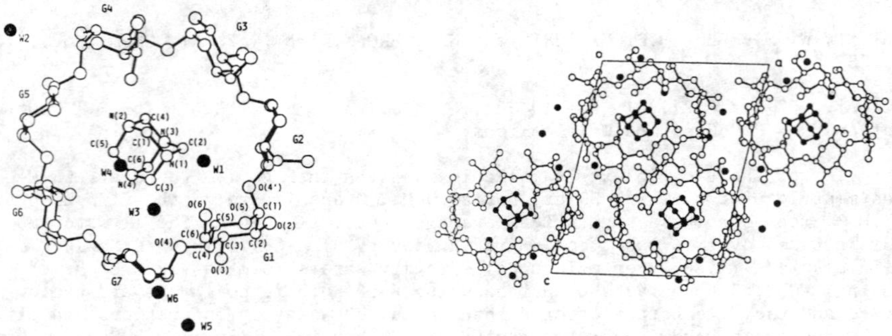

Fig. 1. The complex $C_{42}H_{70}O_{35},C_6H_{12}N_4,6H_2O$ seen from above the ring (left), and the crystal structure viewed along the b axis (right).

BIS(β-CYCLODEXTRIN ACETYLSALICYLIC ACID) SALICYLIC ACID CLATHRATE HYDRATE
$2(C_{42}H_{70}O_{35}),2(C_9H_8O_4),C_7H_6O_3,23.3(H_2O)$

F. NISHIOKA, I. NAKANISHI, T. FUJIWARA and K. TOMITA, 1984. J. Inclusion Phenomena, 2, 701-704.

Triclinic, P1, a = 19.777, b = 15.247, c = 15.475 Å, α = 102.63, β = 116.96, γ = 104.12°, D_m = 1.409, Z = 1. Cu radiation, R = 0.107 for 8733 reflexions.

The two β-cyclodextrin (β-CyD) molecules form a dimer unit which includes the two aspirin and one salicylic acid molecules in its hydrophobic cavity (Fig. 1). All fourteen glucopyranose moieties composing the β-CyD dimer have the $^4C_1$ conformation and form two circular β-CyD macrocycles. The hydrophobic benzene rings of the aspirin molecules protrude into the hydrophobic β-CyD cavities. The salicylic acid at the centre of the complex is disordered over three sites.

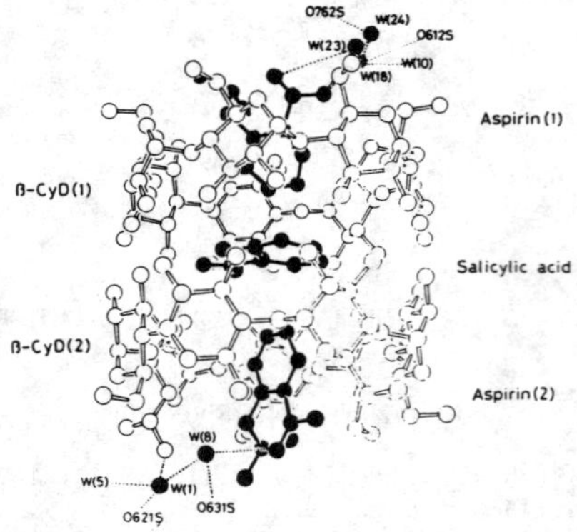

Fig. 1. The molecular structure of $(\beta\text{-CyD})_2(\text{aspirin})_2(\text{salicylic acid})$.

CUCURBITURIL p-XYLYLENEDIAMMONIUM DICHLORIDE DECAHYDRATE
$C_{44}H_{50}Cl_2N_{26}O_{12},10(H_2O)$ (I) $C_{36}H_{36}N_{24}O_{12}.C_8H_{14}N_2^{2+}.2Cl^-.10(H_2O)$

CUCURBITURIL BIS(CALCIUM BIS(HYDROGENSULPHATE)) TRIDECAHYDRATE
$C_{36}H_{40}Ca_2N_{24}O_{28}S_4,13(H_2O)$ (II) $C_{36}H_{36}N_{24}O_{12}.2Ca^{2+}.4(HSO_4^-).13(H_2O)$

W.A. FREEMAN, 1984. Acta Cryst., B40, 382-387.

I. Monoclinic, $P2_1/n$, a = 12.071, b = 15.757, c = 14.485 Å, β = 91.98°, D_m = 1.63, Z = 2. Mo radiation, R = 0.0624 for 3694 reflexions.

II. Triclinic, $P\bar{1}$, a = 11.948, b = 10.392, c = 15.669 Å, α = 92.15, β = 115.18, γ = 109.66°, D_m = 1.78, Z = 1. Mo radiation, R = 0.0643 for 3686 reflexions.

Cucurbituril is a hexagonal cage with an internal cavity of diameter ∿5.5 Å accessed by portals of diameter ∿4 Å. In the p-xylylenediammonium chloride adduct the p-xylylenediammonium ion is a guest inside the cucurbituril cage. The calcium bis(hydrogensulphate) adduct (Fig. 1) has a water molecule at the centre of the otherwise empty cage; Ca^{II} ions lie at the two cage portals and link adjoining cages by coordination to carbonyl O atoms. Crystal data are also reported for 11 other adducts of cucurbituril.

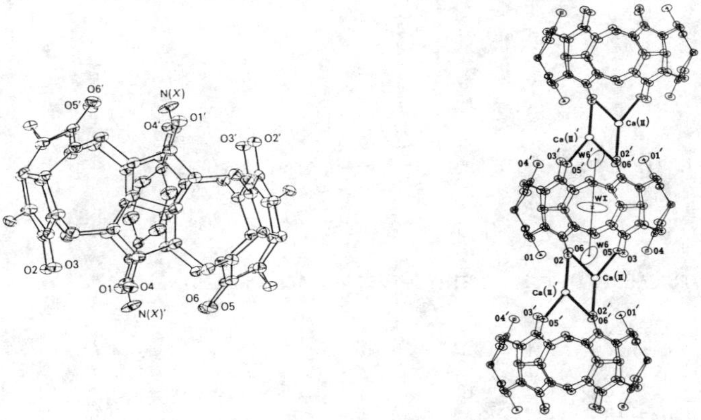

Fig. 1. Views of the p-xylylenediammonium (left) and calcium bis(hydrogensulphate) adducts of cucurbituril.

1,5-DIFLUORO-3,7-DIMETHYL-2,4,6,8,9-PENTAOXA-1,5-DIBORABICYCLO[3.3.1]NONADIENE
$C_4H_6B_2F_2O_5$

H. BINDER, W. MATHEIS, H.-J. DEISEROTH and HAN FU-SON, 1984. Z. Naturforsch., B39, 1717-1721.

Orthorhombic, Pmmn, a = 8.796, b = 9.772, c = 4.541 Å, Z = 2. Mo radiation, R = 0.086 for 356 reflexions.

The molecule has crystallographic mm symmetry. Molecular dimensions are in Fig. 1.

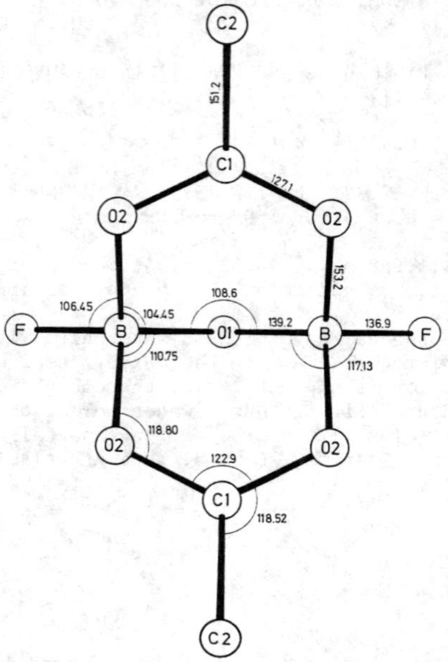

Fig. 1. The $C_4H_6B_2F_2O_5$ molecule and dimensions (pm and °).

2-CHLORO-1-(TRICHLOROGALLIUM)-1,3-DIMETHYL-1,3,2-DIAZABOROLIDINE
$C_4H_{10}BCl_4GaN_2$

K. ANTON, C. EURINGER and H. NOTH, 1984. Chem. Ber., 117, 1222-1234.

Monoclinic, $P2_1/c$, a = 7.244, b = 12.967, c = 12.652 Å, β = 95.36°, D_m = 1.70, Z = 4, Mo radiation, R = 0.047 for 1857 reflexions.

In the molecule (Fig. 1) there is a long (1.511(5) Å) B-N bond to chiral nitrogen N(1) and a shortening of the B-N(3) bond (1.356(6) Å). Other dimensions include Ga-Cl 2.149-2.156(1), Ga-N(1) 2.018(3), B-Cl(4) 1.736(4) Å.

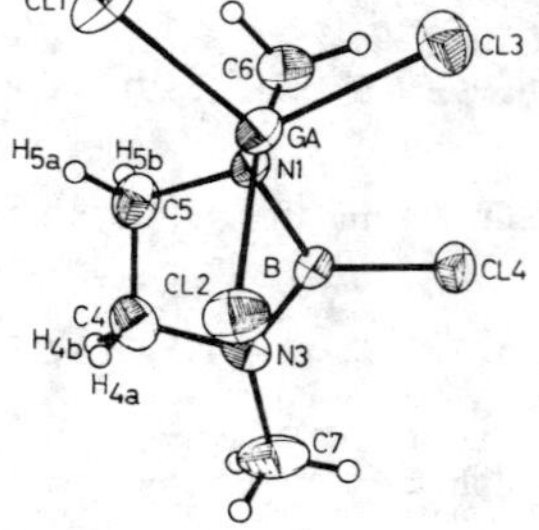

Fig. 1. A view of $C_4H_{10}BCl_4GaN_2$.

2-CHLORO-1,3-DIMETHYL-1-(TRICHLOROBORA)-1,3,2-DIAZABOROLIDINE
$C_4H_{10}B_2Cl_4N_2$

K. ANTON, H. NÖTH and H. POMMERENING, 1984. Chem. Ber., 117, 2479-2494.

Orthorhombic, $P2_12_12_1$, a = 6.827, b = 10.798, c = 29.622 Å, Z = 8. Mo radiation, R = 0.076 for 3843 reflexions.

There are two independent molecules in the asymmetric unit (Fig. 1). The five-membered rings have envelope conformations. In each molecule one B-N distance is shorter than the other two (B12-N11 1.545(7), B12-N13 1.352(8), B11-N11 1.600(7), B22-N21 1.527(8), B22-N23 1.320(9), B21-N21 1.575(10) Å).

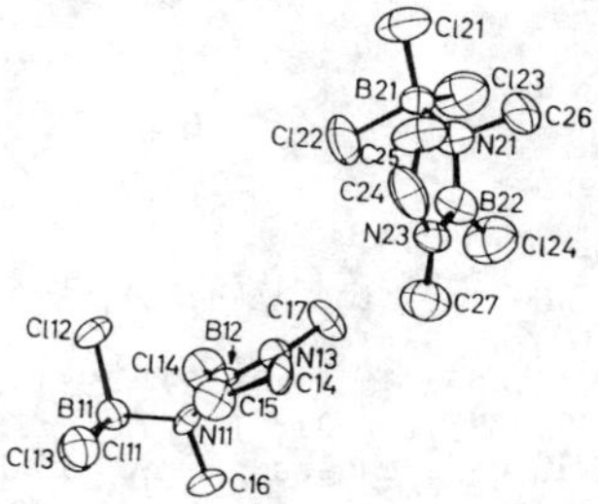

Fig. 1. A view of the two $C_4H_{10}B_2Cl_4N_2$ molecules.

8-DIMETHYLSULPHIDO-7,9-DICARBA-nido-UNDECABORANE(II)
$C_4H_{17}B_9S$

V. ŠUBRTOVÁ, C. NOVÁK, A. LÍNEK and J. HAŠEK, 1984. Acta Cryst., C40, 1955-1956.

Monoclinic, $P2_1/c$, a = 10.043, b = 8.936, c = 12.946 Å, β = 94.48°, Z = 4. Cu radiation. R = 0.071 for 1368 reflexions.

The structure consists of C_2B_9 nido carbaborane cages (Fig. 1) each cage having twelve nearest neighbours giving the appearance of distorted cubic close packing. Some bond lengths are B-S 1.886(4), S-C (mean) 1.790(6), B-C (mean) 1.65(3), B-B (mean) 1.78(2) Å; the longest B-B bond (B(10)-B(11) 1.841(6) Å) has a bridging hydrogen symmetrically bonded with B-H 1.28(3) Å and B-H-B 91(2)°.

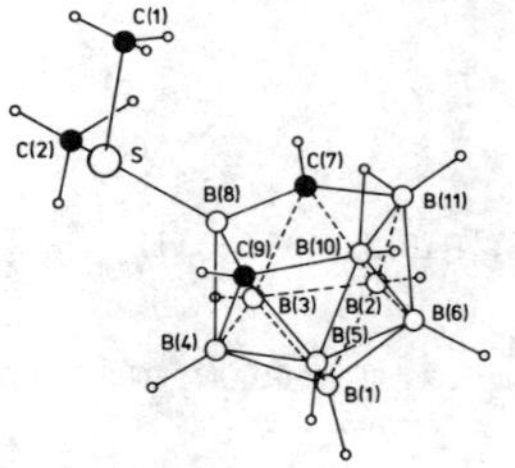

Fig. 1. The $C_4H_{17}B_9S$ cage structure.

1,1,4-TRIMETHYL-2,2,3-TRICHLORO-1-AZONIA-2-BORATA-3-BORA-4-AZA-CYCLOHEXANE
$C_5H_{13}B_2Cl_3N_2$

W. HAUBOLD, J. HREBICEK and G. SAWITZKI, 1984. Z. Naturforsch., B39, 1027-1031.

Monoclinic, $P2_1/c$, a = 6.851, b = 13.251, c = 11.609 Å, β = 91.59°, Z = 4. Mo radiation, R = 0.041 for 1973 reflexions (at 150 K).

In the molecule (Fig. 1) the angle sum at both B1 and N6 is 360° with bond lengths B1-B2 1.698(4), B1-N6 1.382(3), N6-C5 1.466(3), N6-C61 1.466(3), B2-N3 1.616(3), N3-C4 1.510(3), C4-C5 1.509(3), B1-Cl 1.815(3), B2-Cl 1.887 and 1.874(3) Å. The B1-N6 distance is close to that expected for a B=N moiety. The torsion angle C5-N6-B1-B2 is 8.4°.

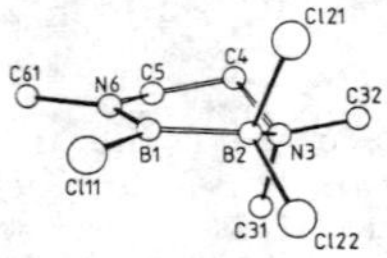

Fig. 1. A view of $C_5H_{13}B_2Cl_3N_2$.

BORON NITRILOTRIACETATE
$C_6H_6BNO_6$

E. MULLER and H.-B. BURGI, 1984. Helv. Chim. Acta, 67, 399-405.

Orthorhombic, $Pn2_1a$, a = 10.505 (10.468), b = 10.269 (10.193), c = 6.807 (6.754) Å, Z = 4. Mo radiation, R = 0.024 (0.037) for 818 (1296) reflexions measured at room temperature and at 110 K (values in parentheses are at 110 K).

The molecule (Fig. 1) has a non-crystallographic C_{3v} symmetry and a transannular dative B-N bond 1.620(3) Å (1.622(3) Å). The molecules pack tightly (density 1.804 g cm^{-3}) interacting via O...C contacts between carbonyl groups (2.92-3.05 Å). Corrections for rigid body motion were applied and yielded B-N distances of 1.624 and 1.623 Å (for room temperature and 110 K data respectively); the mean B-O distance is 1.450 Å.

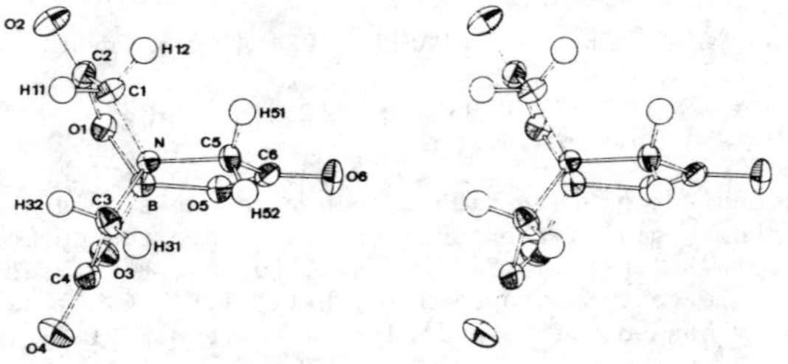

Fig. 1. A stereoview of $C_6H_6BNO_6$.

[BIS(TRIMETHYLPHOSPHINE)DIBORANE(4)]DICHLOROZINC(II)
$C_6H_{22}B_2Cl_2P_2Zn$ $ZnCl_2.B_2H_4.[P(CH_3)_3]_2$

S.A. SNOW, M. SHIMOI, C.D. OSTLER, B.K. THOMPSON, G. KODAMA and R.W. PARRY, 1984. Inorg. Chem., 23, 511-512.

Orthorhombic, $Pb2_1a$, a = 12.212, b = 13.700, c = 9.316 Å, Z = 4. Mo radiation, R = 0.031 for 1720 reflexions.

The molecule (Fig. 1) features a $B_2H_4.(PMe_3)_2$ adduct coordinated to Zn through two hydrogen bridge bonds. Bond lengths: Zn-Cl 2.213(1) and 2.208(1), Zn-H 1.87(5) and 1.90(4), B-B 1.814(6), B-H_b 1.08(4) and 1.20(4), B-H_t 0.99(5) and 1.11(5), B-P 1.912(4) and 1.910(4) Å.

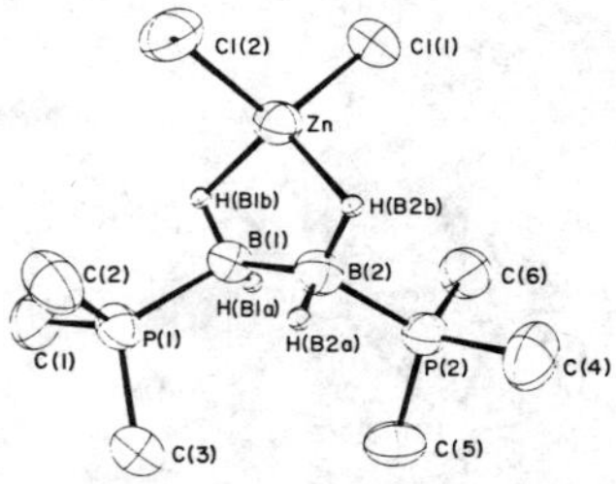

Fig. 1. The $ZnCl_2.B_2H_4.[P(CH_3)_3]_2$ complex.

(s-1,4',2,1')-3,3'-commo-BIS(8-METHOXY-1,2-DICARBA-3-NICKELA-closo-DODECABORANE(II))
$C_6H_{26}B_{18}NiO_2$

V. PETRICEK, K. MALY, A. PETRINA, K. BASE and A. LINEK, 1984. Z. Kristallogr., 166, 1-10.

Monoclinic, $P2_1/b$, a = 15.208, b = 13.762, c = 10.363 Å, γ = 117.25°, D_m = 1.318, Z = 4. Mo radiation, R = 0.066 for 2430 reflexions.

The structure consists (Fig. 1) of two dicarbollide ligands coordinated by the π-pentagonal faces to the Ni atom. The ligands adopt a staggered conformation with cages distorted to each other by an angle of 112°. Mean bond lengths are: Ni-C 2.037, Ni-B 2.144, C-C 1.637, C-B 1.696, B-B 1.786, O-B 1.381 Å.

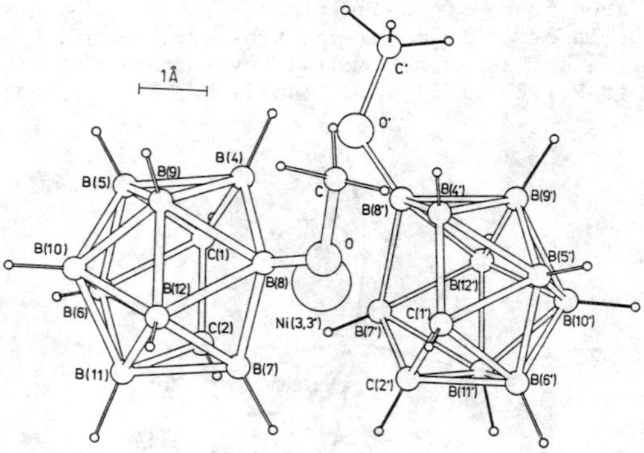

Fig. 1. A view of $C_6H_{26}B_{18}NiO_2$.

2-CARBONYL-2,2-BIS(TRIMETHYLPHOSPHINE)-nido-2-IRIDANONABORANE
$C_7H_{29}B_8IrOP_2$

J. BOULD, N.N. GREENWOOD and J.D. KENNEDY, 1984. J. Chem. Soc. Dalton, 2477-2485.

Monoclinic, $P2_1/n$, a = 9.160, b = 11.792, c = 16.953 Å, β = 93.81°, Z = 4. R = 0.032.

The IrB_8 cluster (Fig. 1) is based on that of a bicapped Archimedean square antiprism with an equatorial five-connected vertex missing and with the Ir atom on the open face. NMR spectroscopy indicated that each B atom had an H atom attached to it and that there were two B-H-B bridging hydrogen atoms. Within the cluster B-B distances are within normal ranges as are the Ir-P and Ir-CO distances. The Ir-B(3) and Ir-B(6) distances are 2.304(7) and 2.330(10) Å respectively. Ir-B(1) is much shorter at 2.185(8) Å. The hydrogen bridge distance Ir(2)-B(5) is much longer (2.500(6) Å).

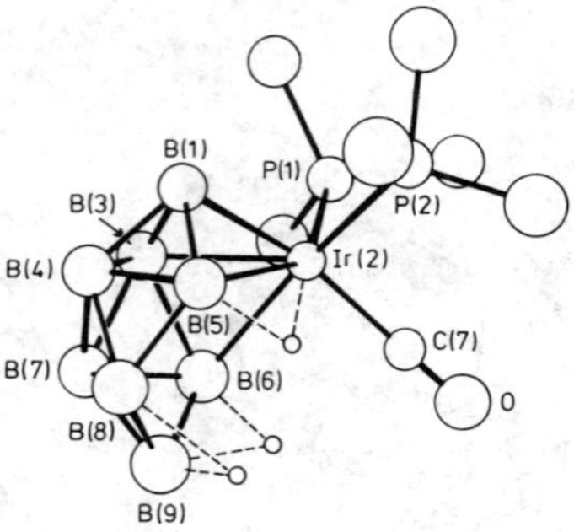

Fig. 1. $C_7H_{29}B_8IrOP_2$: the molecular structure.

arachno-1-CHLORO-4-CARBONYL-4-HYDRIDO-4,4-BIS(TRIMETHYLPHOSPHINE)-4-IRIDANONABORANE
$C_7H_{30}B_8ClIrOP_2$

J. BOULD, J.E. CROOK, N.N. GREENWOOD and J.D. KENNEDY, 1984. J. Chem. Soc. Dalton, 1903-1910.

Monoclinic, $P2_1/c$, a = 9.719, b = 24.239, c = 9.550 Å, β = 117.80°, Z = 4. Mo radiation, R = 0.034 for 2430 reflexions.

The gross structural geometry is similar to that of the arachno-nonaborane iso-B_9H_{15} with the metal atom in the four-position (Fig. 1). H atoms were not located but their positions can be inferred from NMR spectroscopy. B-B distances are within normal ranges. The Ir atom has an octahedral environment with Ir-B(1,5,9) 2.284(11), 2.261(10), 2.291(8), Ir-P(1,2) 2.334(2), 2.346(2) and Ir-C 1.934(10) Å.

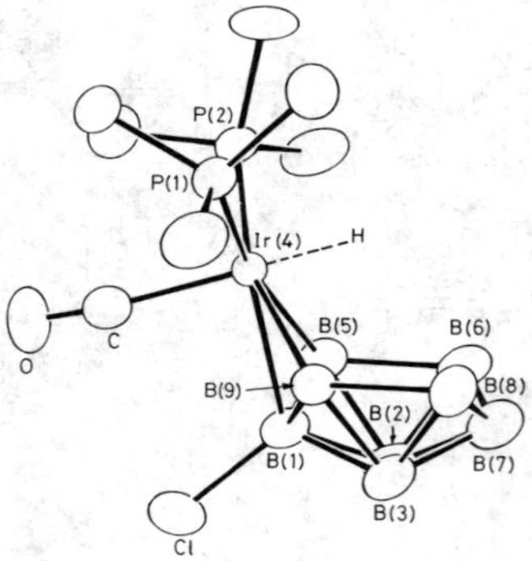

Fig. 1. $C_7H_{30}B_8ClIrOP_2$: the molecular structure minus hydrogen atoms.

t-BUTYL(t-BUTYLIMINO)BORANE
$C_8H_{18}BN$ (I)

TETRAKIS(t-BUTYL)-1,3,2,4-DIAZADIBORETIDINE
$C_{16}H_{36}B_2N_2$

P. PAETZOLD, C. VON PLOTHO, G. SCHMID, R. BOESE, B. SCHRADER, D. BOUGEARD, U. PFEIFFER, R. GLEITER and W. SCHAFER, 1984. Chem. Ber., 117, 1089-1102.

I. Orthorhombic, Pnma, a = 17.957, b = 10.014, c = 5.867 Å, Z = 4. Mo radiation, R = 0.054 for 657 reflexions (at -85°C).

II. Monoclinic, Pc, a = 9.179, b = 9.153, c = 11.951 Å, β = 110.46°, Z = 2. Mo radiation, R = 0.056 for 1980 reflexions (at -80°C).

Compound I lies on a crystallographic mirror plane (Fig. 1) and slowly dimerizes to yield compound II. In I, there is an essentially linear central C-B-N-C chain with B-N 1.258(4) Å. The four-membered ring in II is puckered because of over-crowding, with a 155° fold; the B-N distances are in the range 1.475(5)-1.500(4) Å.

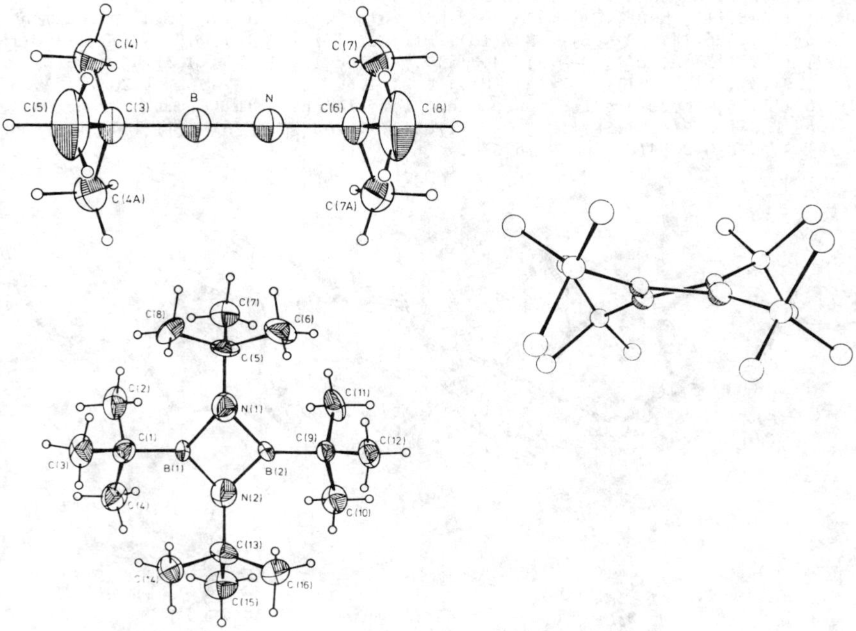

Fig. 1. The $C_8H_{18}BN$ molecule and its dimer.

3,3,4,7,7,8-HEXAMETHYL-2,6-DIOXONIA-4,8-DIAZA-3,7-DIBORATABICYCLO[3.3.0]OCTANE
$C_8H_{18}B_2N_2O_2$ (I)

2,2,3-TRIMETHYL-5-METHYLAMINO-1-OXONIA-3-AZA-2-BORATACYCLOPENTAN-4-ONE
$C_6H_{13}BN_2O_2$ (II)

2,2-DIMETHYL-3-PHENYL-5-PHENYLAMINO-1-OXONIA-3-AZA-2-BORATACYCLOPENTAN-4-ONE
$C_{16}H_{17}BN_2O_2$ (III)

5-ETHYLAMINO-4-ETHYLIMINIO-2,2-DIMETHYL-1-OXONIA-3-OXA-2-BORATA-4-IMINIOCYCLOPENTANE BROMIDE
$C_6H_{18}BBrN_2O_2$ (IV)

W. MARINGGELE, G.M. SHELDRICK, A. MELLER and M. NOLTEMEYER, 1984. Chem. Ber., 117 2112-2131.

I. Monoclinic, C2/c, a = 10.024, b = 6.885, c = 18.179 Å, β = 102.24°, Z = 4. Mo radiation, R = 0.057 for 364 reflexions.

II. Monoclinic, $P2_1/c$, a = 6.520, b = 15.583, c = 9.213 Å, β = 99.40°, Z = 4. Mo radiation, R = 0.086 for 506 reflexions.

III. Monoclinic, C2/c, a = 25.444, b = 5.865, c = 20.399 Å, β = 99.98°, Z = 8. Mo radiation, R = 0.053 for 2161 reflexions.

IV. Triclinic, P$\bar{1}$, a = 9.860, b = 12.270, c = 12.531 Å, α = 110.45, β = 102.52, γ = 90.98°, Z = 4. Mo radiation, R = 0.060 for 2854 reflexions.

Views of the four molecules are in Fig. 1. Molecule I lies on an inversion centre; principal dimensions are: O-B 1.610(9), C(1)-N 1.276(9), O(1)-C(1) 1.286(8), C(1)-C(1) 1.448(11), N-B 1.612(8) Å, O-B-N 97.6(5)°. In II the principal dimensions are: B-O(1) 1.606(14), B-N(2) 1.548(13), O(1)-C(1) 1.275(12), N(2)-C(2) 1.306(15), C(1)-C(2) 1.512(13) Å, O(1)-B-N(2) 94.8(8)°. Molecule III has O(2)-B 1.593(3), B-N(1) 1.591(3), N(1)-C(1) 1.324(2), C(1)-C(2) 1.507(3), C(2)-O(2) 1.263(2) Å, N(1)-B-O(2) 96.7(1)°. There are two independent formula units in IV, with principal dimensions B-O 1.577(9)-1.587(12), O-C 1.265(7)-1.268(9), C(3)-C(4) 1.493(9) and 1.484(10) Å, O-B-O 96.4(6) and 96.8(5)°.

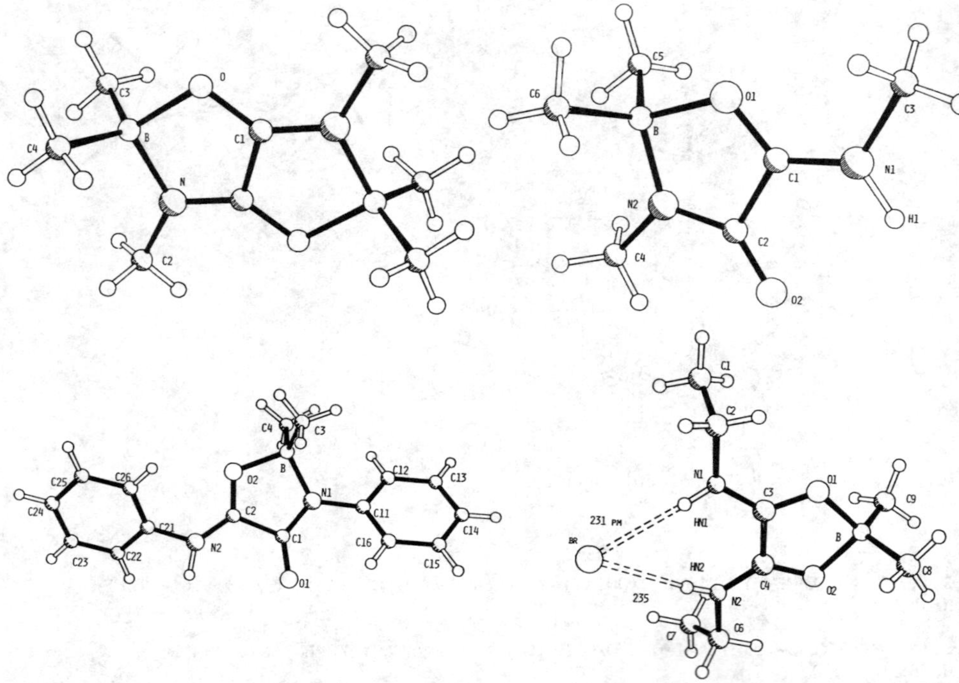

Fig. 1. Views of the boratacyclopentane derivatives I (top left), II (top right), III (lower left) and IV (lower right.

2-(1',3'-DIMETHYL-1',3',2'-DIAZABOROLIDIN-2'-YL)-1,3-DIMETHYL-2-(TRICHLOROGALLIUM)-1,3,2-DIAZABOROLIDINE
$C_8H_{20}B_2Cl_3GaN_4$

K. ANTON, H. NÖTH and H. POMMERENING, 1984. Chem. Ber., 117, 2495-2503.

Orthorhombic, Pbca, a = 13.2629, b = 14.439, c = 17.7533 Å, Dm = 1.42, Z = 8. Mo radiation, R = 0.076 for 1239 reflexions.

The molecular conformation is shown in Fig. 1. Principal dimensions include Ga-Cl 2.147(5)-2.155(5), Ga-N 1.999(11), B1-B2 1.705(23), N1-B1 1.536(19), N2-B1 1.369(20), N3-B2 1.400(19), N4-B2 1.404(20) Å. The dihedral angle between the two 5-rings is 72.8°,

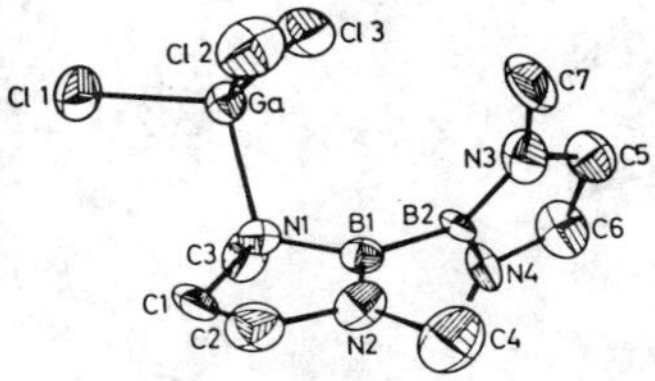

Fig. 1. A view of $C_8H_{20}B_2Cl_3GaN_4$ (without H atoms).

DIMERIC t-BUTYLAMINOBORON DIFLUORIDE
$C_8H_{20}B_2F_4N_2$ $[(CH_3)_3CNHBF_2]_2$

P.G. JONES, 1984. Acta Cryst., C40, 1465-1466.

Orthorhombic, Pbca, a = 7.615, b = 17.007, c = 9.706 Å, Z = 4. Mo radiation, R = 0.037 for 776 reflexions.

The structure consists of dimers lying across inversion centres (Fig. 1) with B-N 1.595(4) and B-F 1.355(4), 1.366(4) Å. The B_2N_2 rings are constrained to be exactly planar. The shortest intermolecular contacts are F(2)...C(2) 3.18(2) and F(1)...N(1) 3.29(2) Å.

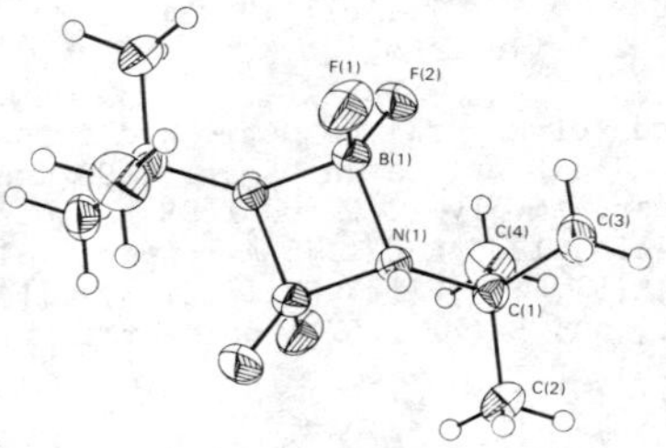

Fig. 1. The dimeric t-butylaminoboron difluoride molecule.

TETRAMETHYLAMMONIUM 2-(η^5-CYCLOPENTADIENYL)-1-CARBA-2-COBALTA-closo-NONAHYDRODECA-BORANE
$C_{10}H_{26}B_8CoN$ $C_6H_{14}B_8Co^-, C_4H_{12}N^+$

K.A. SOLNTSEV, L.A. BUTMAN, I.Yu. KUZNETSOV, N.T. KUZNETSOV, B. SHTIBR, Z. YANOUSHEK and K. BASE, 1983. Koord. Khim., 9, 993-997 [Sov. J. Coord. Chem., 9, 568-572].

Monoclinic, C2/c, a = 14.376, b = 15.318, c = 15.199 Å, β = 91.90°, D_m = 1.20, Z = 8. Mo radiation, R = 0.061 for 1453 reflexions.

The structure is composed of tetramethylammonium cations and complex anions (Fig. 1). The anion consists of a cyclopentadienyl ring and a distorted bicapped square antiprism, whose vertices are occupied by the Co(2) and C(1) atoms and eight B atoms. The mean Co-C(cyclopentadienyl) distance is 2.04(1) and the mean C-C distance within the cyclopentadienyl ring is 1.38(2) Å. The mean B-B bond lengths are 1.68(2), 1.81(2) and 1.78(4) Å for polar, tropical and equatorial distances respectively. The mean C-B distance in the polyhedron is 1.57(1) Å and the mean Co-B distance is 2.11(2) Å. The cation has the expected geometry.

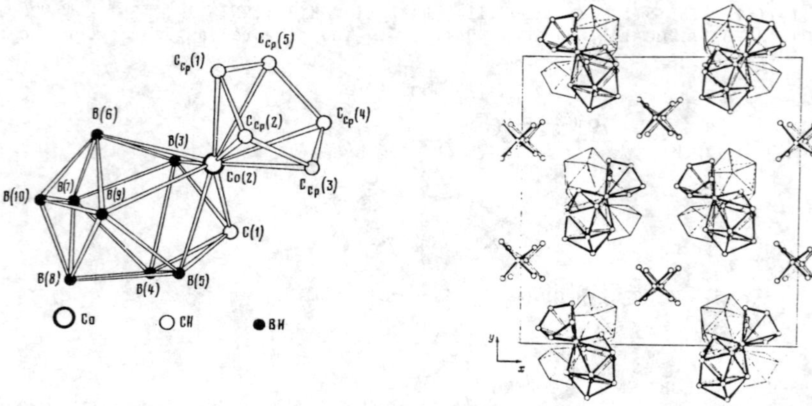

Fig. 1. $C_{10}H_{26}B_8CoN$: the complex anion (left) and a projection of the unit cell (right).

DIBROMOBORYL-1,2,3,4-TETRAHYDRO-1,10-PHENANTHROLINE

$C_{12}H_{11}BBr_2N_2$ $(C_{12}H_{11}N_2)BBr_2$

G. KLEBE and D. TRANQUI, 1984. Inorg. Chim. Acta, 81, 1-5.

Monoclinic, $P2_1/n$, a = 7.190, b = 16.274, c = 10.931 Å, β = 93.39°, Z = 4. Mo radiation, R_w = 0.053 for 2202 reflexions.

The B atom has distorted tetrahedral geometry and is part of a five-membered NBNCC heterocycle (Fig. 1). The B-Br distances are 1.99 and 2.07 Å and the B-N(1) and B-N(10) distances are respectively 1.49 and 1.58 Å. Distances N(1)-C(13), N(10)-C-(11) and C(11)-C(13) are 1.34, 1.33 and 1.45 Å respectively. The atoms of the aromatic system (N(10),C(9),C(8),C(7),C(12),C(11),C(13),C(14),C(5),C(6)) form a planar unit with Br(1) 1.62 Å above, and Br(2) 1.67 Å below this plane. B and N(1) lie in the plane, within experimental error.

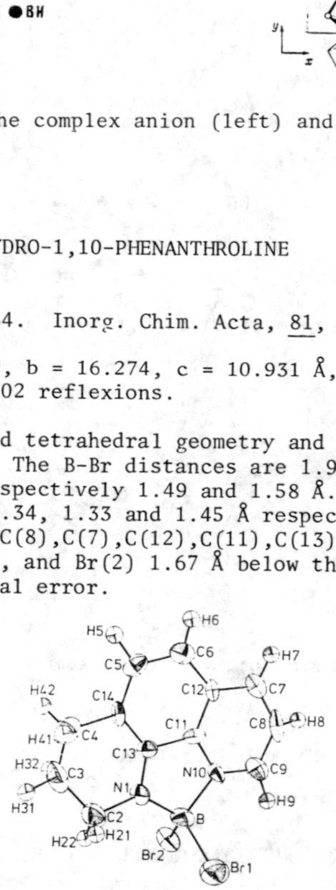

Fig. 1. Molecular structure of $C_{12}H_{11}BBr_2N_2$.

POTASSIUM DICITRATOBORATE DIHYDRATE

$C_{12}H_{12}BKO_{14},2H_2O$ $K[(C_6H_6O_7)_2B].2H_2O$

I.I. ZVIEDRE, V.S. FUNDAMENSKII, V.V. KRASNIKOV and G.P. KOLESNIKOVA, 1984. J. Struct. Chem., 25, 593-599 [Zh. Strukt. Khim., 25(4), 95-101].

Orthorhombic, Pbca, a = 10.137, b = 11.189, c = 33.326 Å, D_m = 1.63, Z = 8. Mo radiation, R = 0.052 for 1867 reflexions.

The structure consists of bicyclic $(C_6H_6O_7)_2B^-$ anions, potassium cations, and water molecules. Bond distances for the anion, containing tetrahedrally coordinated boron, are shown in Fig. 1. The geometry about the seven-coordinate potassium ion is also shown in Fig. 1.

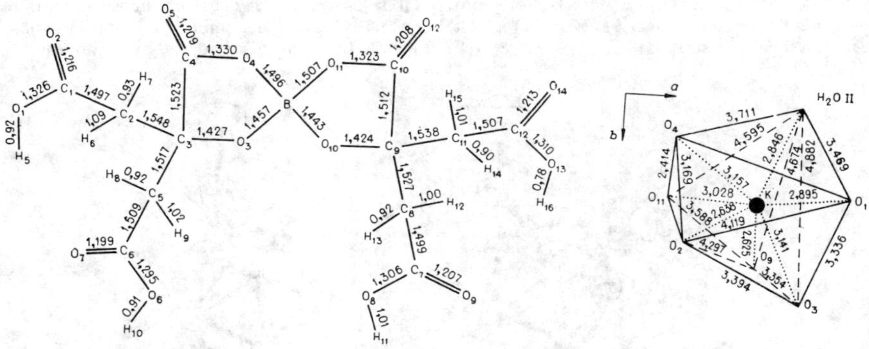

Fig. 1. Bond lengths in the dicitratoborate anion (σ = 0.005 Å or less for bonds between non hydrogen atoms) (left) and the coordination environment of the potassium cation (σ(K-O) = 0.004 Å).

4,4,5,5-TETRAMETHYL-1,3,2-DIOXABOROLAN-2-YL-4',4',5',5'-TETRAMETHYL-1',3',2'-DIOXABOROLANE

$C_{12}H_{24}B_2O_4$

H. NOTH, 1984. Z. Naturforsch., B39, 1463-1466.

Monoclinic, $P2_1/c$, a = 10.423, b = 7.625, c = 10.334 Å, β = 110.81°, Z = 2. Mo radiation, R = 0.0646 for 788 reflexions.

The molecule lies about an inversion centre (Fig. 1) and has close to D_{2h} point symmetry. The central B_2O_4 moiety is planar and the B atoms have sp^2 hybridisation. Main dimensions (corrected for libration) are B-B 1.716, B-O 1.328 and 1.329, C-O 1.442 and 1.445, C(1)-C(2) 1.562 Å, O-B-O 111.1(2), O-B-B 124.3 and 124.6(4), B-O-C 111.6 and 112.0(3)°.

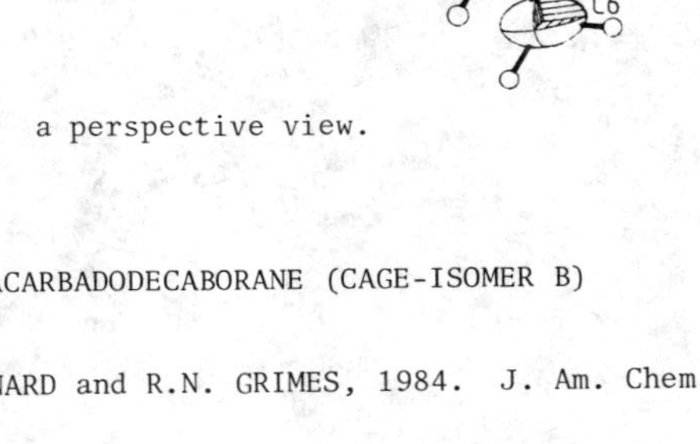

Fig. 1. $C_{12}H_{24}B_2O_4$: a perspective view.

2,3,7,8-TETRAETHYLTETRACARBADODECABORANE (CAGE-ISOMER B)

$C_{12}H_{28}B_8$

T.L. VENABLE, R.B. MAYNARD and R.N. GRIMES, 1984. J. Am. Chem. Soc., 106, 6187-6193.

Monoclinic, $P2_1/n$, a = 7.890, b = 17.815, c = 12.195 Å, β = 98.32°, Z = 4. Mo radiation, R = 0.010 for 1555 reflexions.

The structure (Fig. 1) consists of an open framework comprised of two pyramidal C_2B_4 units joined at their basal B-B edge. This is quite different from what was found for the tetramethyl isomer (1). The central C3-C7 distance is clearly non-bonding in the present structure (2.886(2) Å) but bonding in the methyl isomer (1.53(1) Å).

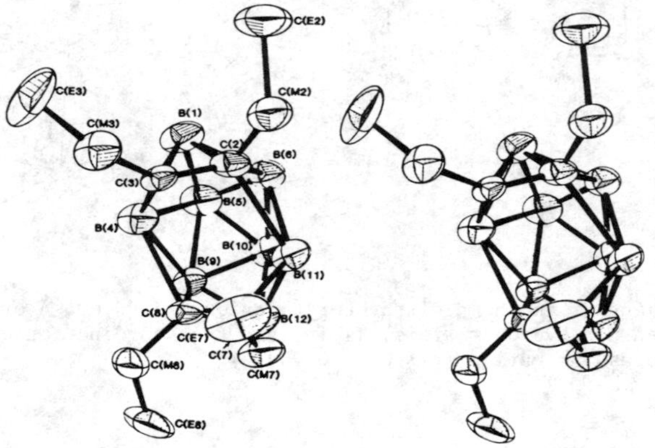

Fig. 1. A stereoview of $C_{12}H_{28}B_8$.

1. Structure Reports, 43B, 809.

1,7-DIMETHYL-3,5-DIPHENYL-2,4,6-TRIOXA-7-AZA-1-AZONIA-3-BORA-5-BORATABICYCLO[3.3.0]-OCTANE

$C_{15}H_{18}B_2N_2O_3$

W. KLIEGEL, S.J. RETTIG and J. TROTTER, 1984. Can. J. Chem., 62, 515-520.

Triclinic, $P\bar{1}$, a = 10.559, b = 15.156, c = 10.338 Å, α = 99.14, β = 89.90, γ = 96.73°, Z = 4. Mo radiation, R = 0.032 for 2396 reflexions.

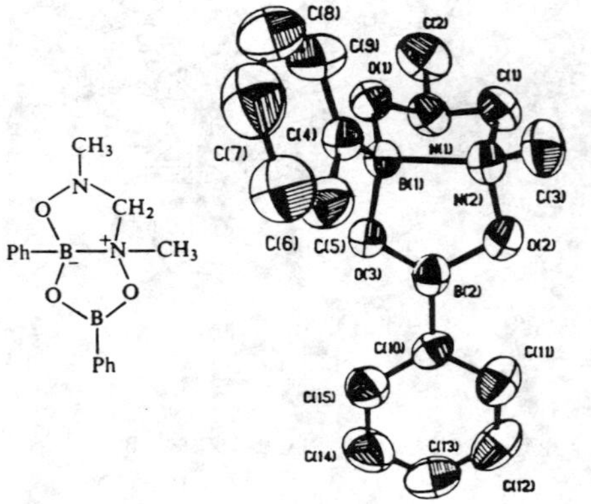

Fig. 1. A $C_{15}H_{18}B_2N_2O_3$ molecule.

The structure consists of two crystallographically independent, but virtually identical, molecules (Fig. 1) containing sterically lengthened N-B coordinative bonds (mean 1.740(3) Å). The structurally inequivalent (in the solid state) pairs of N-methyl and B-phenyl groups are observed by variable temperature ^{1}H NMR to be magnetically equivalent in solution, implying a rapid interconversion of "valence tautomers" at temperatures down to -60°C.

8-CHLORO-7,7-BIS(DIMETHYLPHENYLPHOSPHINE)-nido-7-PLATINAUNDECABORANE
$C_{16}H_{33}B_{10}ClP_2Pt$

J.E. CROOK, N.N. GREENWOOD, J.D. KENNEDY and W.S. MACDONALD, 1984. J. Chem. Soc. Dalton, 2487-2495.

Monoclinic, $P2_1/n$, a = 9.789, b = 2.674, c = 10.796 Å, β = 115.83°, Z = 4. Mo radiation, R = 0.041 for 1903 reflexions.

The X-ray analysis established that the molecule comprises an eleven vertex platinaborane cage (Fig. 1). The Pt-P distances are 2.325(3) and 2.354(3) and the B-Pt distances range from 2.206(12) to 2.342(13) Å. The B-B distances lie within the normal ranges. B-Cl is 1.841(13) Å.

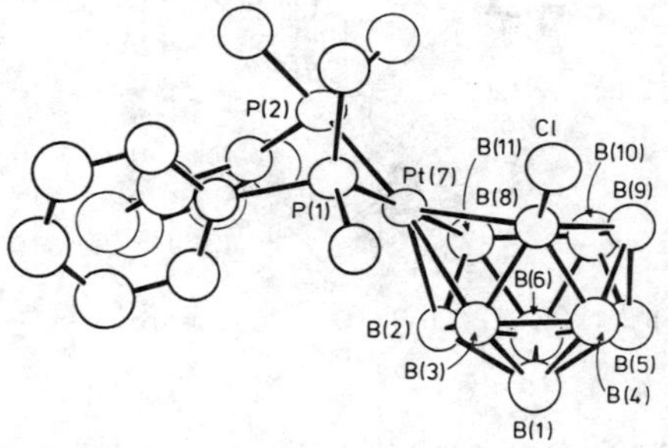

Fig. 1. $C_{16}H_{33}B_{10}ClP_2Pt$: the molecular structure.

1,3-DIMETHYL-2-(DIPHENYLAMINO)-1,3,2-DIAZABOROLIDINIUM(1+) TRIFLUOROMETHYLSULPHONATE
$C_{17}H_{21}BF_3N_3O_3S$ $[C_{16}H_{21}BN_3]^+[O_3SCF_3]^-$

C.K. NARULA and H. NÖTH, 1984. Inorg. Chem., 23, 4147-4152.

Triclinic, $P\bar{1}$, a = 9.256, b = 10.741, c = 11.482 Å, α = 68.69, β = 77.86, γ = 71.27°, Z = 2. Mo radiation, R = 0.056 for 3043 reflexions.

The structure is shown in Fig. 1. The boron atom of the cation is trigonal planar, asymmetrically bound to three N atoms. The B-N(2) bond to the tetracoordinated nitrogen is long (1.547(3) Å), that to the NPh_2 group normal (1.412(3) Å), and that to the ring atom N3 is short (1.386(3) Å).

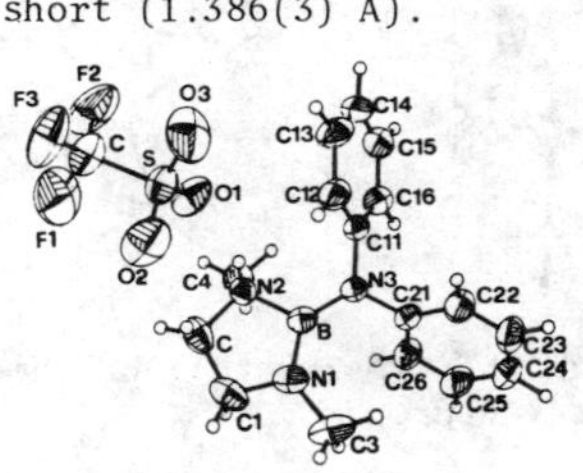

Fig. 1. The structure of $[C_{16}H_{21}BN_3]^+[O_3SCF_3]^-$.

1-CARBONYL-1,1-BIS(DIMETHYLPHENYLPHOSPHINE)-1-IRIDIAPENTABORANE

$C_{17}H_{31}B_4IrOP_2$ $[Ir(\eta^4\text{-}B_4H_9)(CO)\{P(CH_3)_2(C_6H_5)\}_2]$

S.K. BOOCOCK, M.A. TOFT, K.E. INKROTT, L.-Y. HSU, J.C. HUFFMAN, K. FOLTING and S.G. SHORE, 1984. Inorg. Chem., 23, 3084-3092.

Monoclinic, $P2_1/c$, a = 9.324, b = 23.162, c = 12.486 Å, β = 55.95°, Z = 4. Mo radiation, R = 0.054 for 3159 reflexions (at -162°C).

The molecule (Fig. 1) consists of an open five-atom cluster in which the $[Ir(CO)\{P(CH_3)_2(C_6H_5)\}_2]$ fragment occupies an apical vertex site and is bound to the four B atoms of the $[\eta^4\text{-}B_4H_9]$ ligand. Bond lengths: Ir-P 2.365(3) and 2.340(3), Ir-CO 1.911(11), Ir-B 2.147(14)-2.258(12), B-B 1.828(20)-1.855(19) Å.

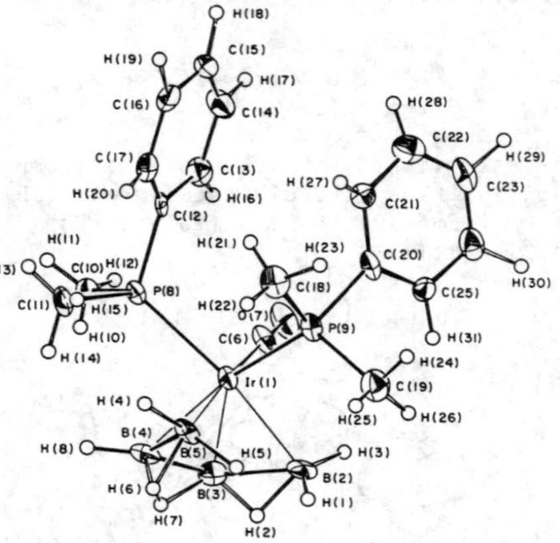

Fig. 1. The $[Ir(\eta^4\text{-}B_4H_9)(CO)\{P(CH_3)_2(C_6H_5)\}_2]$ molecule.

TRIS(Z-TRIMETHYLSILYLPROP-1-ENYL)BORANE

$C_{18}H_{39}BSi_3$ $B\{C(CHCH_3)Si(CH_3)_3\}_3$

N.S. HOSMANE, N.N. SIRMOKADAM, M.D. WALKINSHAW and E.A.V. EBSWORTH, 1984. J. Organometal. Chem., 270, 1-7.

Triclinic, $P\bar{1}$, a = 10.311, b = 13.693, c = 19.033 Å, α = 70.59, β = 85.43, γ = 79.63°, Z = 4. Mo radiation, R = 0.061 for 3422 reflexions.

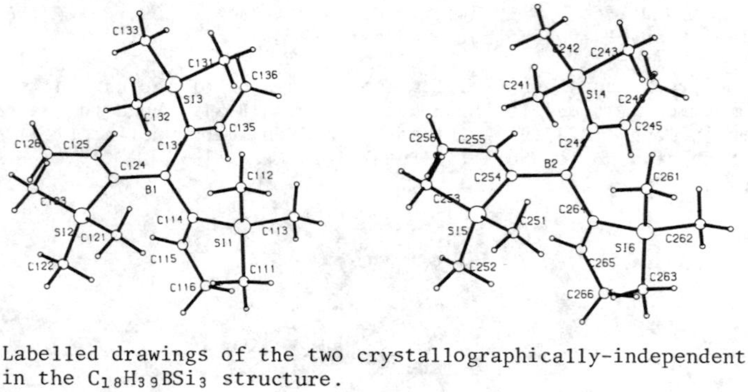

Fig. 1. Labelled drawings of the two crystallographically-independent molecules in the $C_{18}H_{39}BSi_3$ structure.

The two crystallographically-independent molecules in the asymmetric unit adopt almost identical conformations (Fig. 1). There are no statistically significant differences between any of the chemically (and conformationally) equivalent bond lengths, bond angles and torsion angles of the six independent trimethylsilylpropenyl side arms. Coordination about B is trigonal planar with an average C-B-C angle of 120.0°. The eighteen independent Si-Me bond lengths average 1.866 Å and the average Si-C(xy4) bond distance is 1.877 Å and is thus slightly, but not significantly, larger.

(SALICYLALDOXIMATO-O,N)DIPHENYLBORON
$C_{19}H_{16}BNO_2$

W. KLIEGEL, S.J. RETTIG and J. TROTTER, 1984. Can. J. Chem., 62, 1363-1368.

Triclinic, P$\bar{1}$, a = 10.125, b = 10.797, c = 14.760 Å, α = 89.05, β = 86.98, γ = 83.18°, Z = 4. Mo radiation, R = 0.048 for 4145 reflexions.

The structure consists of two crystallographically independent, but essentially identical, molecules linked by strong O-H...O hydrogen bonds (O...O 2.699(2) and 2.737(3) Å) to form continuous spirals along the a axis. The Ph_2B moiety is chelated by the phenolic oxygen and oxime nitrogen atoms of the ligand resulting in the formation of a non-planar six-membered chelate ring. Important libration-corrected mean bond lengths are O-B 1.516(1), N-B 1.609(2) and C-B 1.613(2) Å.

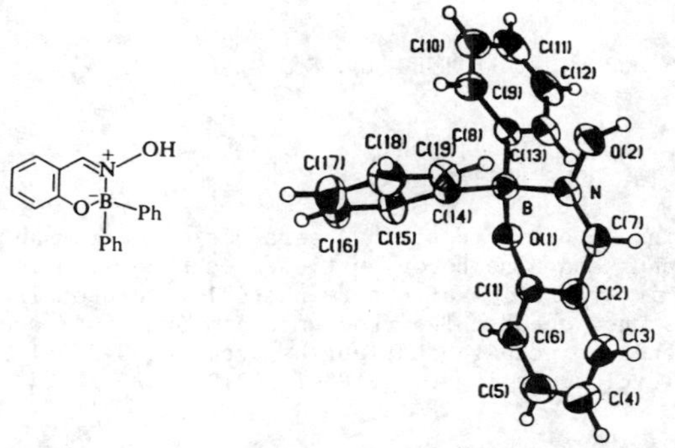

Fig. 1. The $C_{19}H_{16}BNO_2$ skeleton and one of the independent molecules.

4,4-DIMETHYL-2,5,5-TRIPHENYL-1,3-DIOXA-4-AZONIA-2-BORA-5-BORATACYCLOPENTANE
$C_{20}H_{21}B_2NO_2$ (I)

4,4,5,5-TETRAMETHYL-2,2-DIPHENYL-1,3-DIOXA-4-AZONIA-2-BORATACYCLOPENTANE
$C_{17}H_{22}BNO_2$ (II)

W. KLIEGEL, H.-W. MOTZKUS, S.J. RETTIG and J. TROTTER, 1984. Can. J. Chem., 62, 838-844.

I. Triclinic, P$\bar{1}$, a = 9.841, b = 14.722, c = 6.6701 Å, α = 86.79, β = 102.95, γ = 93.89°, Z = 2. Mo radiation, R = 0.032 for 1456 reflexions.

II. Monoclinic, Cc, a = 14.9164, b = 6.4732, c = 18.2248 Å, β = 117.263°, Z = 4. Mo radiation, R = 0.035 for 1649 reflexions.

The five-membered BOBON ring in I (Fig. 1) has a distorted B(2)-envelope conformation. The phenyl ring bonded to the trigonal-planar boron atom is twisted out of the B(1) coordination plane by 7.1°. The N(sp^3-Bsp^3) distance is unusually long (1.728(3) Å). The five-membered BOCNO ring in II has a distorted C(1)-envelope conformation in which both O(1) and B lie on the same side of the NCO(2) plane. The N(sp^3)-C(sp^3) bond is long (1.591(3) Å). In both I and II there is considerable steric crowding.

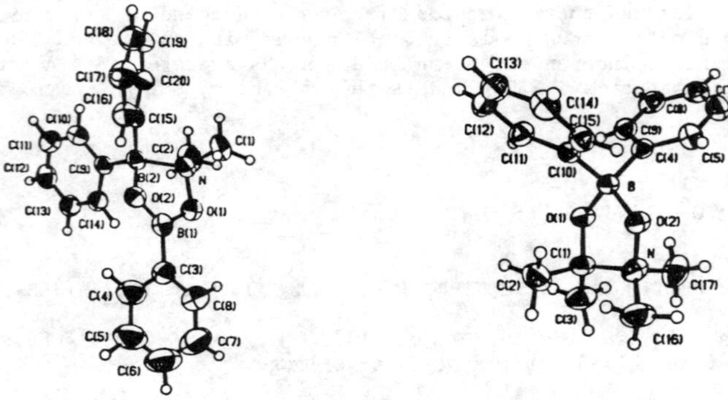

Fig. 1. Views of $C_{20}H_{21}B_2NO_2$ (I) (left) and $C_{17}H_{22}BNO_2$ (II) (right).

7-CHLORO-2,3-BIS(η-PENTAMETHYLCYCLOPENTADIENYL)-1-THIA-2,3-DIRHODA-closo-DODECA-BORANE(9)
$C_{20}H_{38}B_9ClRh_2S$

G. FERGUSON, M.F. HAWTHORNE, B. KAITNER and F.J. LALOR, 1984. Acta Cryst., C40, 1707-1709.

Monoclinic, $P2_1/c$, a = 8.724, b = 15.964, c = 20.454 Å, β = 106.77°, Z = 4. Mo radiation, R = 0.028 for 3955 reflexions.

The molecule (Fig. 1) can be described as a distorted icosahedron composed of two rhodium, one sulphur and nine boron vertices; each Rh atom is also coordinated by a C_5Me_5 ligand. The incorporation of a terminal chlorine atom into the structure (B-Cl 1.793(4) Å) was unexpected. The rhodium atoms occupy adjacent vertices with Rh-Rh 2.778(1) Å. Other principal bond lengths are: Rh-S 2.351 and 2.362(1), Rh-B 2.181-2.311(6), Rh-C(cyclopentadienyl) 2.183-2.295(5), S-B 2.014-2.104(7), B-B 1.737-1.859(7) Å.

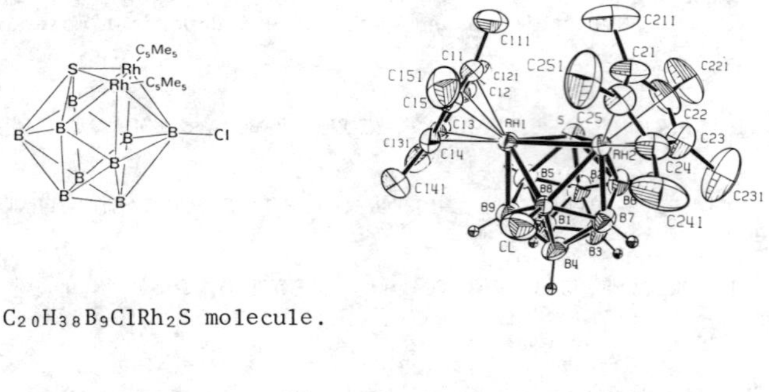

Fig. 1. The $C_{20}H_{38}B_9ClRh_2S$ molecule.

(DI-t-BUTYLMETHYLENEAMINO)DIPHENYLBORANE
$C_{21}H_{28}BN$

W. CLEGG, S.J. BRYAN and K. WADE, 1984. Acta Cryst., C40, 2082-2084.

Orthorhombic, Pbca, a = 11.534, b = 14.875, c = 22.909 Å, Z = 8. Mo radiation, R = 0.077 for 1638 reflexions.

This allene-like molecule (Fig. 1) has a dihedral angle of 94.4° between the C_2BN and NCC_2 planes. B-N and N-C bond lengths of 1.366(6) and 1.244(5) Å together with a B-N-C angle of 178.2(4)° indicate substantial π-contribution to both bonds.

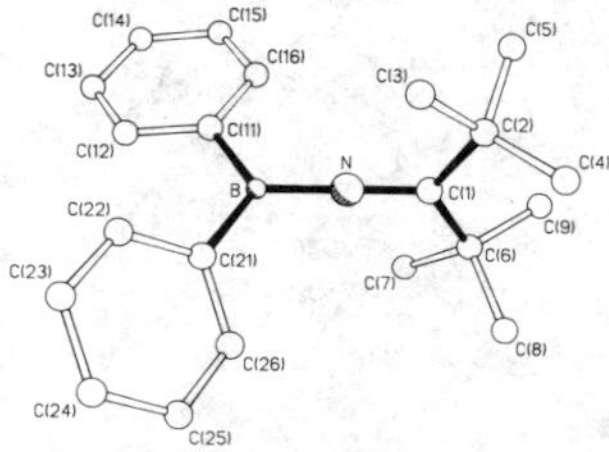

Fig. 1. The $C_{21}H_{28}BN$ structure, H atoms omitted.

1,3,5-TRI-t-BUTYL-2,4,6-TRIISOPROPYL-3,5-DIAZA-1-AZONIA-2,6-DIBORA-4-BORATA[2.2.0]-BICYCLOHEXANE
$C_{21}H_{48}B_3N_3$

P. PAETZOLD, C. VON PLOTHO, G. SCHMID and R. BOESE, 1984. Z. Naturforsch., B39, 1069-1075.

Monoclinic, $P2_1/c$, a = 17.899, b = 9.624, c = 16.068 Å, β = 114.24°, Z = 4. Mo radiation, R = 0.067 for 1954 reflexions.

This Dewar-borazine type molecule (Fig. 1) has two short peripheral B=N bonds (1.384(6) and 1.364(6) Å) and an extra long bridge bond (B-N 1.752(5) Å). The remaining four B-N bonds are in the range 1.548-1.564(5).

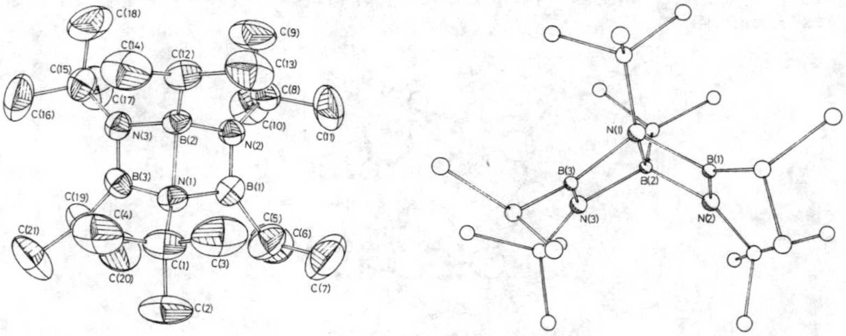

Fig. 1. Views of the $C_{21}H_{48}B_3N_3$ molecule.

BIS(BENZYL-t-BUTYLAMINO)BORANE
$C_{22}H_{33}BN_2$

H. NÖTH and S. WEBER, 1984. Chem. Ber., 117, 2504-2509.

Monoclinic, C2/c, a = 23.712, b = 5.9308, c = 30.075 Å, β = 91.21°, Z = 8. Mo radiation, R = 0.076 for 2841 reflexions.

The molecule (Fig. 1) has approximate twofold symmetry, with mean B-N 1.420(3) Å. The C_2N planes are twisted 23.8° from the HBN_2 plane.

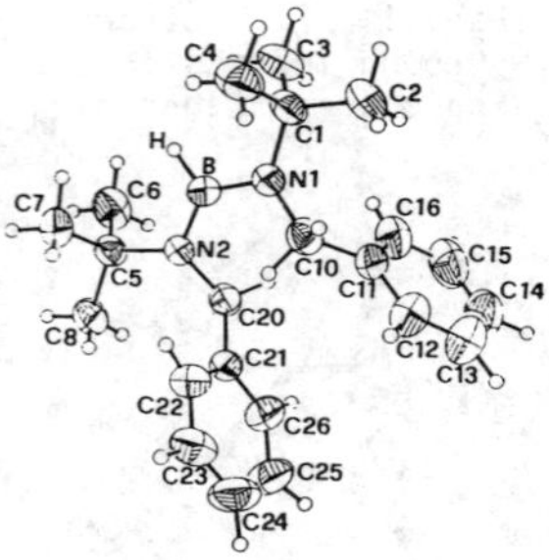

Fig. 1. A view of $C_{22}H_{33}BN_2$.

[TRIS(TRIMETHYLSILYL)METHYL]DIPHENYLBORANE

$C_{22}H_{37}BSi_3$ $(C_6H_5)_2BC(Si(CH_3)_3)_3$

C. EABORN, M.N.A. EL-KHELI, P.B. HITCHCOCK and J.D. SMITH, 1984. J. Organometal. Chem., 272, 1-9.

Triclinic, P$\bar{1}$, a = 9.436, b = 9.971, c = 13.893 Å, α = 89.33, β = 76.70, γ = 73.98°, Z = 2. Mo radiation, R = 0.037 for 2082 reflexions.

The coordination about the B atom is strictly planar but the $(Me_3Si)_3C$-B-C(Ph) angles (122.5 and 127.1°) are larger than the C(Ph)-B-C(Ph) angle (110.4°) (Fig. 1). The planes of the phenyl rings make angles of 91 and 47° respectively, with the BC_3 plane. The three B-C distances (mean 1.583 Å) are the same within the limits of error, and the mean B-C-Si angle is 110.3°. Within the $(Me_3Si)_3C$ group the C-$SiMe_3$ bonds (mean 1.934 Å) are significantly longer than the Si-Me bonds (mean 1.873 Å) and the C($(SiMe_3)_2$)-Si-C(Me) angles (mean 114°) substantially larger than the C(Me)-Si-C(Me) angles (mean 105°).

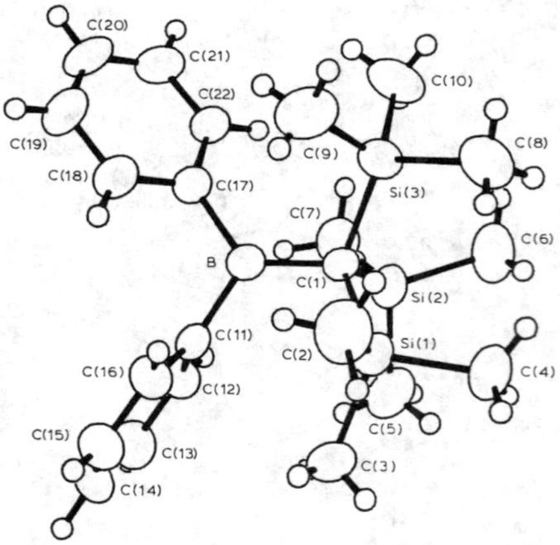

Fig. 1. Molecular structure of $C_{22}H_{37}BSi_3$.

3a-METHYL-2,4,9a-TRIPHENYL-1,2,3a,4,9,9a-HEXAHYDRO-1,3,9-TRIOXA-3a-AZONIA-2-BORA-9a-BORATACYCLOPENTA[b]NAPHTHALENE

$C_{26}H_{23}B_2NO_3$

W. KLIEGEL, D. NANNINGA, S.J. RETTIG and J. TROTTER, 1984. Can. J. Chem., 62, 845-849.

Monoclinic, $P2_1/n$, a = 11.8548, b = 22.7837, c = 8.5457 Å, β = 106.952°, Z = 4. Cu radiation, R = 0.040 for 2970 reflexions.

The backbone of the molecule (Fig. 1) consists of a system of three fused rings; the central six-membered C_3OBN ring has a distorted boat conformation. The five-membered heterocycle is cis-fused to the central ring at adjacent pseudo-equatorial positions, resulting in a marked folding of the fused-ring system, the angle between normals to the mean planes of the terminal rings being 71.9(2)°. The five-membered BOBON ring in 4b is significantly non-planar, having a flattened B(1)-envelope conformation. The $B(sp^3)-N(sp^3)$ bond length is very long (1.759(2) Å).

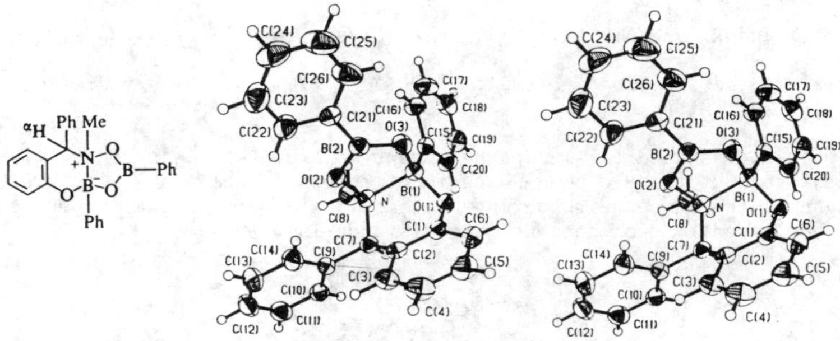

Fig. 1. The $C_{26}H_{23}B_2NO_3$ molecule and a stereoview.

1-METHYL-3,3-(n-BUTYLPROPIONATE-C,O)-3-TRIPHENYLPHOSPHINE-3-RHODA-1,2-DICARBADODECA-BORANE
$C_{28}H_{41}B_9O_2PRh$ (I)

2-CHLORO-2-TRIPHENYLPHOSPHINE-2-RHODA-1,7-DICARBADODECABORANE BENZENE SOLVATE
$C_{20}H_{26}B_9ClPRh,C_6H_6$ (II)

R.T. BAKER, M.S. DELANEY, R.E. KING III, C.B. KNOBLER, J.A. LONG, T.B. MARDER, T.E. PAXSON, R.G. TELLER and M.F. HAWTHORNE, 1984. J. Am. Chem. Soc., 106, 2965-2978.

I. Orthorhombic, $P2_12_12_1$, a = 24.578, b = 12.543, c = 10.377 Å, Z = 4. Mo radiation, R = 0.069 for 3159 reflexions.

II. Monoclinic, $P2_1/n$, a = 13.840, b = 17.000, c = 13.771 Å, β = 118.98°, Z = 4. Mo radiation, R = 0.039 for 4173 reflexions (at 119 K).

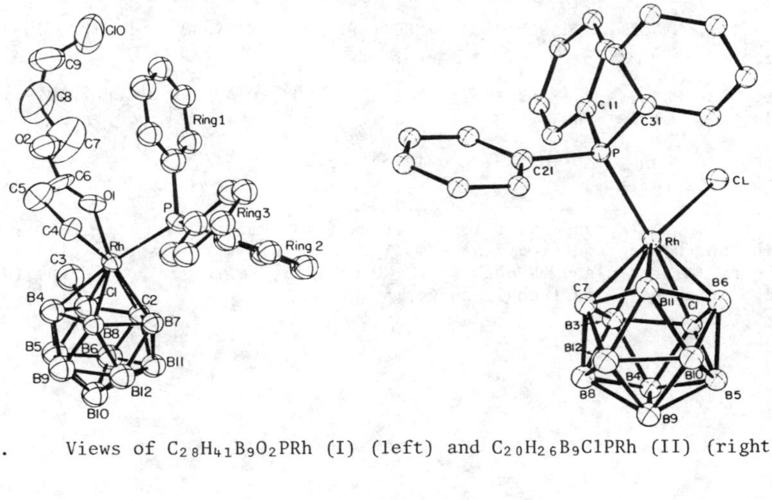

Fig. 1. Views of $C_{28}H_{41}B_9O_2PRh$ (I) (left) and $C_{20}H_{26}B_9ClPRh$ (II) (right).

In (I), the organic ligand is coordinated to Rh through both the oxygen atom O1 of the ester carbonyl group and alkyl carbon C4 (Rh-O1 2.185(8), Rh-C4 2.10(1) Å). The Rh atom has distorted octahedral geometry (assuming the cage occupies three coordination sites with Rh-P 2.323(3) Å). Molecule II (Fig. 1) shows no distortion of the 12-vertex polyhedron and has Rh-P 2.329(1), Rh-Cl 2.299(1) Å.

TRI-1-NAPHTHYLBORANE BENZENE (1/1)
$C_{30}H_{21}B,C_6H_6$

M.S. GUDIPATI and K. VENKATESAN, 1984. Acta Cryst., C40, 1920-1922.

Rhombohedral, R3, a = 8.710 Å, α = 91.10°, D_m = 1.17, Z = 1. Mo radiation, R = 0.046 for 481 reflexions.

The propeller-shaped tri-1-naphthylborane molecule has C_3 symmetry with a 41.5(7)° twist about the B-C bond which is of length 1.566(5) Å. B is 0.078 Å out of the plane of the three neighbouring C atoms in a direction away from the benzene ring; the boron-centroid of benzene ring distance is 4.39 Å.

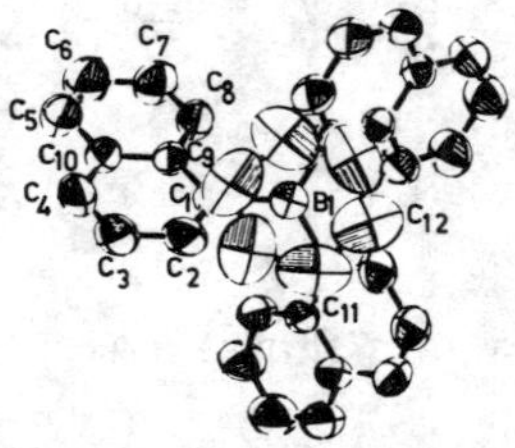

Fig. 1. The tri-1-naphthylborane benzene complex viewed down the threefold axis.

4,4,8,8-TETRAETHYL-2,2,6,6-TETRAPHENYL-1,3,5,7-TETRAOXA-2,6-DIPHOSPHA-4,8-DIBOROCIN CHLOROFORM SOLVATE
$C_{32}H_{40}B_2O_4P_2,CHCl_3$ (I)

4,8-BIS(CYCLOOCTANE-1,5-DIYL)-2,6-DIPHENYL-1,3,5,7-TETRAOXA-2,6-DIPHOSPHA-4,8-DIBOROCIN
$C_{28}H_{40}B_2O_4P_2$ (II)

L. SYNORADZKI, R. MYNOTT, J. ANBEI, C. KRUGER, YI-HUNG TSAY and R. KOSTER, 1984. Chem. Ber., 117, 2863-2874.

I. Orthorhombic, Pnna, a = 18.826, b = 21.393, c = 9.289 Å, Z = 4. Mo radiation, R = 0.075 for 2031 reflexions.

II. Orthorhombic, Pnma, a = 10.752, b = 18.491, c = 14.096 Å, Z = 4. Cu radiation, R = 0.070 for 1999 reflexions.

In (I) the molecule has crystallographic twofold symmetry and the $(POBO)_2$ eight-membered ring has a boat conformation (Fig. 1). Molecule (II) has crystallographic mirror symmetry (mirror plane through P1,P2,C9,C12,C13,C16 (Fig. 1)) and the $(POBO)_2$ eight-membered ring has a half-chair conformation.

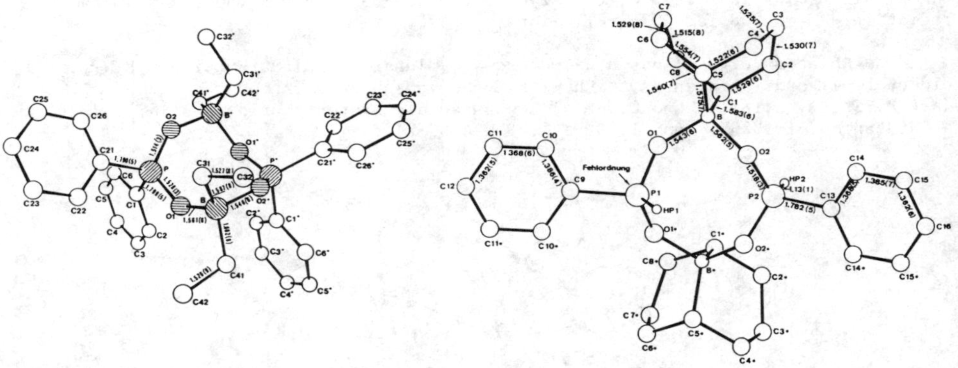

Fig. 1. Views with dimensions of the $(POBO)_2$ systems in (I) (left) and (II) (right).

[BIS(α-PYRIDYL)BIS(DIPHENYLBORANOXY)]METHANE
$C_{35}H_{28}B_2N_2O_2$

F. FLORENCIO, P. SMITH-VERDIER and S. GARCIA-BLANCO, 1984. Z. Kristallogr., 168, 173-178.

Triclinic, $P\bar{1}$, a = 10.415, b = 14.424, c = 10.289 Å, α = 99.92, β = 105.61, γ = 71.84°, Z = 2. Mo radiation, R = 0.060 for 3160 reflexions.

The molecule (Fig. 1) consists of two similar halves with both B atoms four-coordinate. One of the five-membered rings has an envelope conformation.

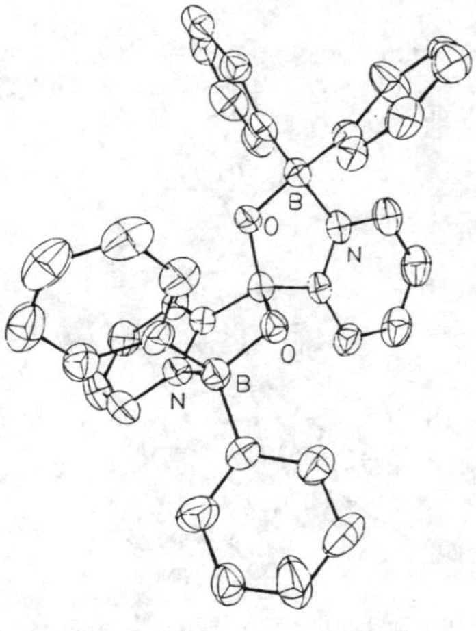

Fig. 1. A view of $C_{35}H_{28}B_2N_2O_2$.

9,9-BIS(TRIPHENYLPHOSPHINE)-7,8-DICARBA-9-RHODA-nido-UNDECABORANE(11)
$C_{38}H_{41}B_8P_2Rh$

P. LU, C.B. KNOBLER and M.F. HAWTHORNE, 1984. Acta Cryst., C4, 1704-1706.

Monoclinic, $P2_1/n$, a = 18.950, b = 12.057, c = 15.585 Å, β = 97.80°, Z = 4. Mo radiation, R = 0.038 for 5021 reflexions.

The Rh atom (Fig. 1) has a pseudo square-planar coordination with $C_2B_8H_{11}$ as a bidentate ligand. Distances to Rh within the icosahedral fragment are Rh-C 2.198(4), Rh-B 2.132(4), 2.181(4) and 2.238(4) Å. An acidic hydrogen bridges B(10) and B(11) at distances of 1.25(5) and 1.29(5) Å, B-H-B 95(3)°.

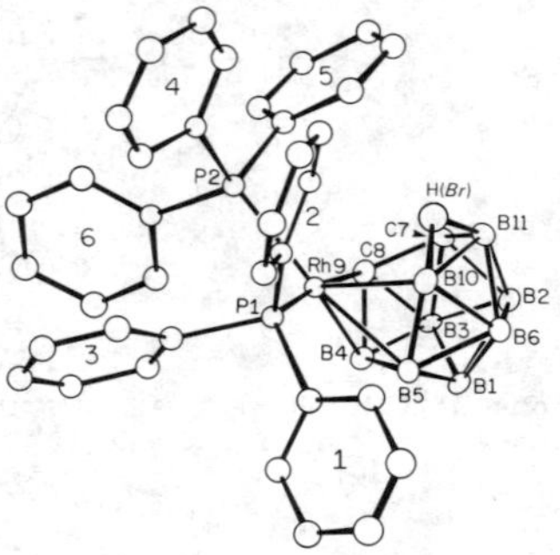

Fig. 1. A perspective view of $C_{38}H_{41}B_8P_2Rh$.

1,1-BIS(TRIPHENYLPHOSPHINE)-2,5-DIETHOXY-1-RUTHENAUNDECABORANE
$C_{40}H_{48}B_{10}O_2P_2Ru$

J.E. CROOK, M. ELRINGTON, N.N. GREENWOOD, J.D. KENNEDY and J.D. WOOLLINS, 1984. Polyhedron, 3, 901-904.

Monoclinic, $P2_1/n$, a = 13.090, b = 23.335, c = 13.902 Å, β = 93.63°, Z = 4. Mo radiation.

The molecule has the structure shown in Fig. 1 with Ru-P(1) 2.536(3), Ru-P(2) 2.511(3), Ru-B(2) 2.057(13), Ru-B(5) 2.052(13), Ru-B(3) 2.385(14), Ru-B(4) 2.415(13), Ru-B(6) 2.383(13), Ru-B(7) 2.408(13) Å, P-Ru-P 96.9(1)°.

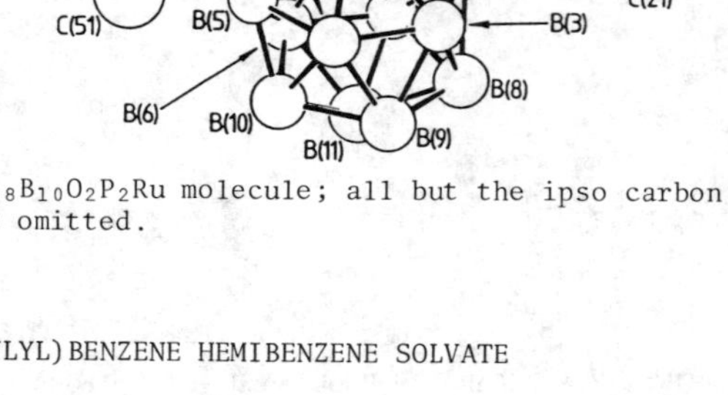

Fig. 1. The $C_{40}H_{48}B_{10}O_2P_2Ru$ molecule; all but the ipso carbon atoms of the phenyl rings are omitted.

1,3,5-TRIS(4-BIPHENYLYL)BENZENE HEMIBENZENE SOLVATE
$C_{42}H_{30}, 0.5C_6H_6$ (I)

1,3,5-TRIS[4-(C-o-CARBORANYLMETHYL)PHENYL]BENZENE CHLOROFORM SOLVATE
$C_{33}H_{54}B_{30}, CHCl_3$ (II)

S.V. LINDEMAN, V.E. SHKLOVER, Yu.T. STRUCHKOV, I.A. KHOTINA, T.M. SALYKHOVA, M.M. TEPLYAKOV and V.V. KORSHAK, 1984. Makromol. Chem., 185, 417-427.

I. Rhombohedral, $R\bar{3}c$, a = 20.168, c = 13.678 Å, Z = 6. Mo radiation, R = 0.063 for 563 reflexions.

II. Triclinic, $P\bar{1}$, a = 15.7528, b = 12.7004, c = 15.460 Å, α = 109.97, β = 111.92, γ = 100.69°, Z = 2. Mo radiation, R = 0.061 for 4240 reflexions.

In I, the molecule has 32 crystallographic symmetry (Fig. 1) with a propeller-like orientation of the biphenyl fragments with respect to the central ring (the rotation of the disubstituted benzene rings relative to the central one is 47.2° and relative to the terminal ones is 35.9°). The C-C bond lengths in the central benzene ring are 1.395 Å but in the ρ-phenylene fragments are slightly shortened (1.379-1.382 Å; average 1.381 Å), and they are still shorter in the terminal phenyl rings (av. 1.362 Å) due to thermal motion in the terminal molecular fragments. In II the disubstituted benzene rings are rotated with respect to the central one by different values (dihedral angles are 31.7, -30.7, and 52.5°). In all benzene rings of both molecules a decrease of the endocyclic bond angles at the ipso-atoms is observed which in II depends on the dihedral angles between the planes of bonded benzene rings. All three carboranylmethylphenyl fragments have the same conformation: the CH-group of the carborane skeleton "hangs" over the benzene ring plane. The C-C bond lengths in the central benzene ring average 1.394 Å, those in the p-phenylene rings 1.387 Å. In the terminal o-carboranyl groups the C-C bonds average 1.643 Å, the C-B bonds 1.699 Å and the B-B bonds 1.757 Å.

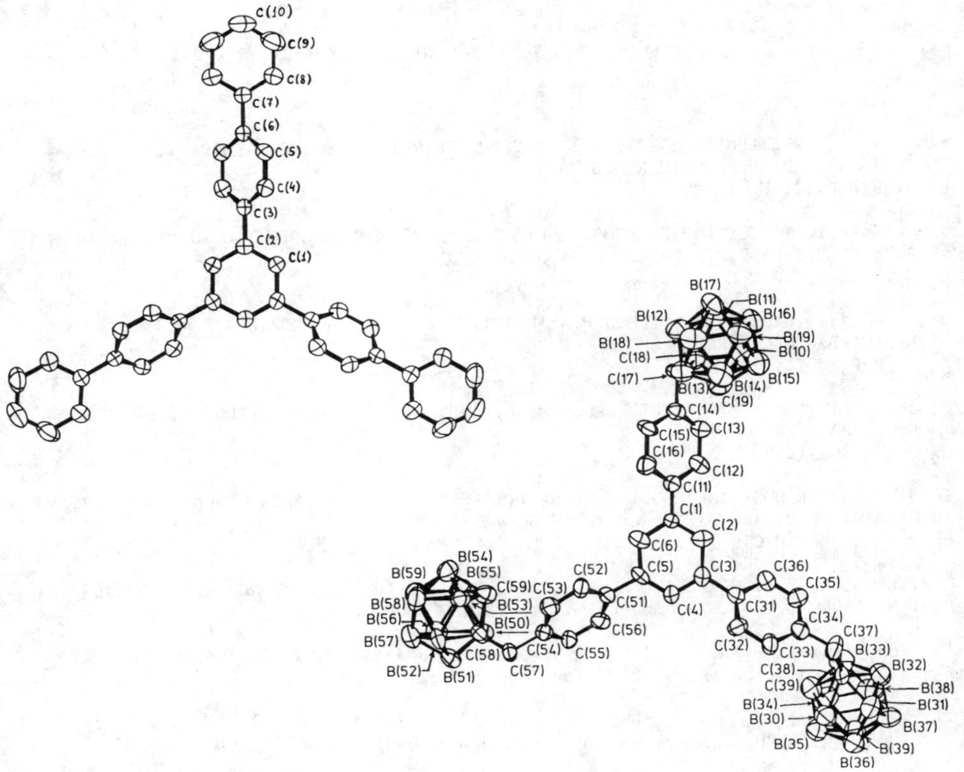

Fig. 1. Projection of the molecule $C_{42}H_{30}$ on the ab plane, and projection of the molecule $C_{33}H_{54}B_{30}$ on the plane of the central benzene ring.

closo-2,2-BIS(TRIPHENYLPHOSPHINE)-2-HYDRIDO-8-PHENYL-2,1,8-IRIDIADICARBADODECABORANE HEMI(DICHLOROMETHANE) SOLVATE

$C_{44}H_{46}B_9IrP_2,0.5(CH_2Cl_2)$ closo-2,2-$[P(C_6H_5)_3]_2$-2-H-8-(C_6H_5)-2,1,8-$IrC_2B_9H_{10},0.5CH_2Cl_2$

J.A. DOI, E.A. MIZUSAWA, C.B. KNOBLER and M.F. HAWTHORNE, 1984. Inorg. Chem., 23, 1482-1484.

Triclinic, P$\bar{1}$, a = 11.722, b = 19.994, c = 9.923 Å, α = 100.17, β = 107.59, γ = 86.40°, D_m = 1.48, Z = 2. Mo radiation, R = 0.050 for 5025 reflexions.

The most striking feature of the structure (Fig. 1) is the relative positions of the two carbon atoms of the C_2B_9 carborane ligand, which defines this complex as a member of the 2,1,8-isomer series. Important bond lengths: Ir-P 2.339(3) and 2.344(3), Ir-C 2.263(9), Ir-B 2.164(11)-2.224(12), C-B 1.667(15)-1.771(14), and B-B 1.732(18)-1.861(15) Å.

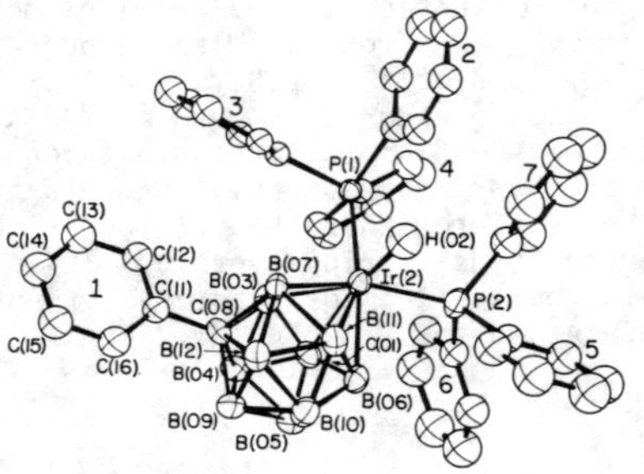

Fig. 1. The closo-2,2-$[P(C_6H_5)_3]_2$-2-H-8-(C_6H_5)-2,1,8-$IrC_2B_9H_{10}$ molecule.

closo-1,2-(o-XYLYLENE)-3,3-BIS(TRIPHENYLPHOSPHINE)-3-HYDRIDO-3-RHODA-1,2-DICARBA-DODECABORANE DICHLOROMETHANE SOLVATE

$C_{46}H_{48}B_9P_2Rh,2(CH_2Cl_2)$ (I)

exo-nido-6,10-(TRICYCLOHEXYLPHOSPHINETRIPHENYLPHOSPHINERHODIUM)-6,10-BIS(μ_2-HYDRIDO)-7,8-(o-XYLYLENE)-7,8-DICARBAUNDECABORANE BENZENE SOLVATE

$C_{46}H_{66}B_9P_2Rh,1.5(C_6H_6)$ (II)

exo-nido-4,9-(BIS(TRIPHENYLPHOSPHINE)RHODIUM)-4,9-BIS(μ_2-HYDRIDO)-7-METHYL-8-PHENYL-7,8-DICARBAUNDECABORANE

$C_{45}H_{48}B_9P_2Rh$ (III)

closo-1-METHYL-2,2-BIS(TRIETHYLPHOSPHINE)-2-HYDRIDO-8-PHENYL-2-RHODA-1,8-DICARBA-DODECABORANE

$C_{21}H_{48}B_9P_2Rh$ (IV)

TRIS(TRIPHENYLPHOSPHINE)RHODIUM nido-7-(1'-closo-1',2'-DICARBADODECABORANYL)-7,8-DICARBAUNDECABORANE DICHLOROMETHANE SOLVATE

$C_{58}H_{67}B_{19}P_3Rh,CH_2Cl_2$ (V) $[Rh(PPh_3)_3][C_2B_{10}H_{11}-C_2B_9H_{11}],CH_2Cl_2$

C.B. KNOBLER, T.B. MARDER, E.A. MIZUSAWA, R.G. TELLER, J.A. LONG, P.E. BEHNKEN and M.F. HAWTHORNE, 1984. J. Am. Chem. Soc., 106, 2900-3004.

I. Monoclinic, $P2_1/n$, a = 17.803, b = 21.425, c = 13.010 Å, β = 98.20°, D_m = 1.36, Z = 4. Mo radiation, R = 0.0597 for 2864 reflexions.

II. Triclinic, P$\bar{1}$, a = 11.565, b = 13.527, c = 18.313 Å, α = 91.90, β = 73.21, γ = 102.10°, D_m = 1.20, Z = 2. Mo radiation, R = 0.0561 for 5571 reflexions.

III. Monoclinic, C2/c, a = 10.732, b = 23.610, c = 17.247 Å, β = 91.32°, D_m = 1.24, Z = 4. Mo radiation, R = 0.0613 for 3104 reflexions.

IV. Monoclinic, $P2_1/n$, a = 12.905, b = 13.855, c = 16.201 Å, β = 91.46°, D_m = 1.28, Z = 4. Mo radiation, R = 0.039 for 3946 reflexions.

V. Monoclinic, $P2_1/c$, a = 14.819, b = 25.429, c = 17.578 Å, β = 98.65°, D_m = 1.257, Z = 4. Mo radiation, R = 0.0757 for 3319 reflexions.

Views of the five complexes are in Fig. 1. In I, both carborane C atoms are bridged by an o-xylylene group and the molecule has normal geometry. In II a $Rh(PPh_3)(PCy_3)$ moiety is attached to a nido C_2B_9 icosahedral fragment through a pair of three-centre two-electron M-H-B bonds and the Rh coordination is distorted square planar (with Rh-P1 2.247(2), Rh-P2 2.256(2), Rh-H6 1.84(5), Rh-H10 1.83(5) Å). Molecule III is disordered (eight molecules with half-occupancy) in the unit cell. Although molecule IV has icosahedral geometry resembling that of I, the bonding face is CB_4 while that in I is C_2B_3. Some dimensions for IV are: Rh-P1 2.357(1), Rh-P2 2.346(1), Rh-H 1.53(5) Å. The anion of V consists of a closo-$C_2B_{10}H_{11}$ icosahedron bonded to a nido-$C_2B_9H_{11}^-$ icosahedral fragment through carbon vertices. The $Rh(PPh_3)_3$ cation has unusual geometry as shown in Fig. 1.

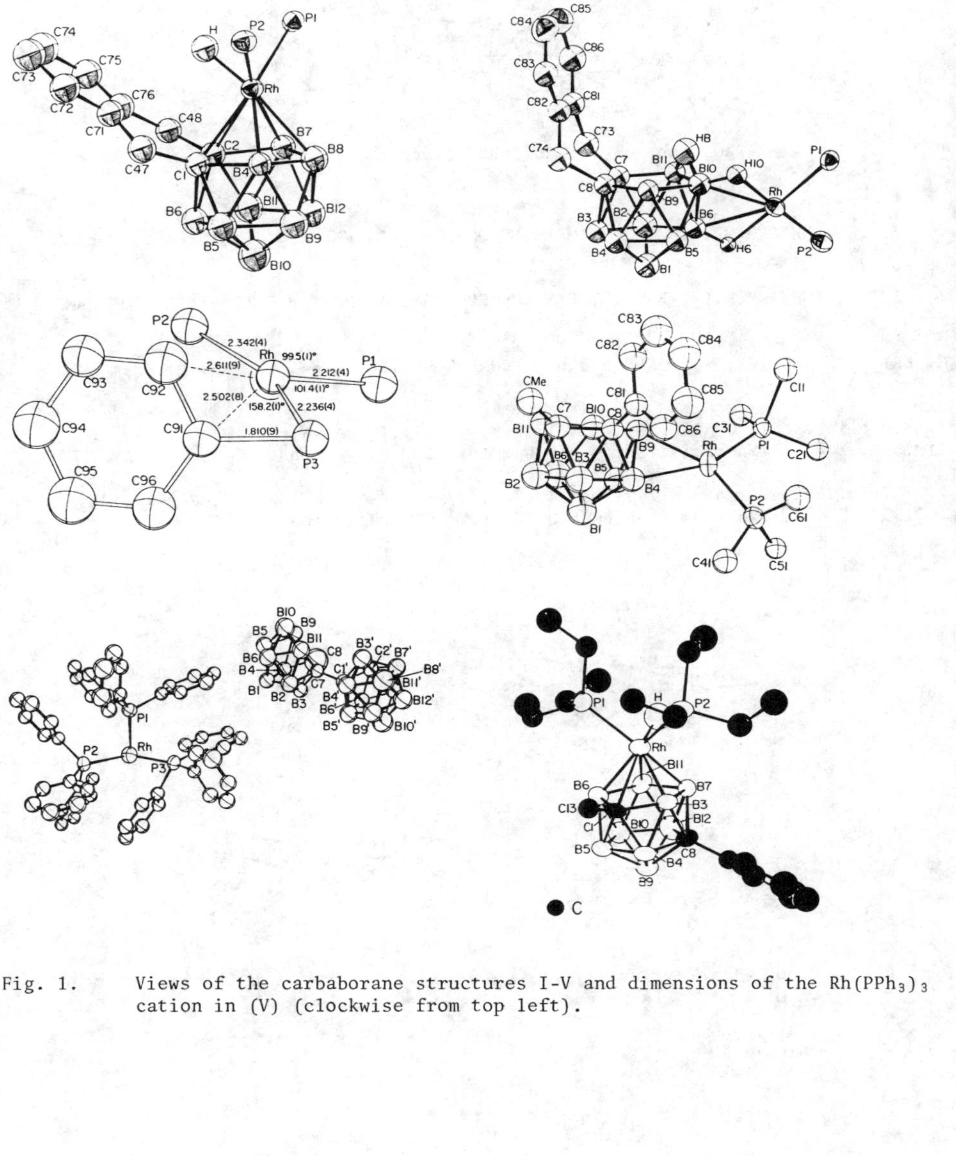

Fig. 1. Views of the carbaborane structures I-V and dimensions of the $Rh(PPh_3)_3$ cation in (V) (clockwise from top left).

exo,nido-(7,8-(μ_2-TRIMETHYLENE)-10-(n-BUTOXYCARBONYLETHYL)-7,8-DICARBAUNDECABORANE-B^4,B^9)BIS(TRIPHENYLPHOSPHINE)RHODIUM
$C_{48}H_{58}B_9O_2P_2Rh$

J.D. HEWES, C.W. KREIMENDAHL, T.B. MARDER and M.F. HAWTHORNE, 1984. J. Am. Chem. Soc., 106, 5757-5759.

Triclinic, P$\bar{1}$, a = 12.404, b = 13.826, c = 14.941 Å, α = 95.58, β = 101.92, γ = 109.35°, Z = 2. Mo radiation, R = 0.053 for 6220 reflexions (at 133 K).

The molecule (Fig. 1) has a Rh(1) bis(PPh_3) moiety attached to a nido C_2B_9 icosahedral fragment through a pair of three-centre two-electron Rh-H-B bonds. Some dimensions are: Rh-P(1) 2.225(1), Rh-P(2) 2.215(1), Rh-B(4) 2.355(6), Rh-B(9) 2.388(6), B(10)-C(91) 1.574(8) Å, P-Rh-P 95.83(5)°, B(9)-H(9)-Rh 97.4(4)°, B(4)-H-Rh 99.5(3)°.

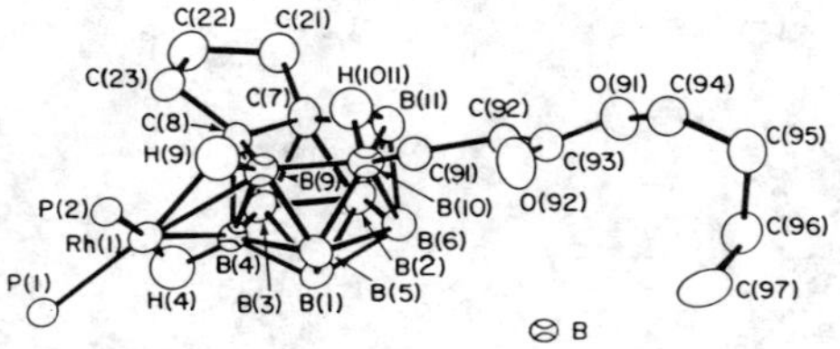

Fig. 1. $C_{48}H_{58}B_9O_2P_2Rh$: a view of the molecule.

1,1,3,3,5,5-HEXACHLORO-2-CHLORODIMETHYLPHOSPHORANYLIDENE-1,3,5-TRISILACYCLOHEXANE
$C_5H_{10}Cl_7PSi_3$ (I)

1,1,3,3,5,5-HEXAMETHYL-2-BROMODIMETHYLPHOSPHORANYLIDENE-1,3,5-TRISILACYCLOHEXANE
$C_{11}H_{28}BrPSi_3$ (II)

G. FRITZ, W. SCHICK, W. HONLE and H.G. VON SCHNERING, 1984. Z. Anorg. Chem., 511, 95-107.

I. Triclinic, P$\bar{1}$, a = 13.560, b = 17.808, c = 11.400 Å, α = 81.95, β = 90.12, γ = 110.39°, Z = 6. Mo radiation, R = 0.042 for 7299 reflexions.

II. Triclinic, P$\bar{1}$, a = 8.759, b = 18.261, c = 6.444 Å, α = 90.77, β = 108.40, γ = 107.35°, Z = 2. Mo radiation, R = 0.034 for 3076 reflexions.

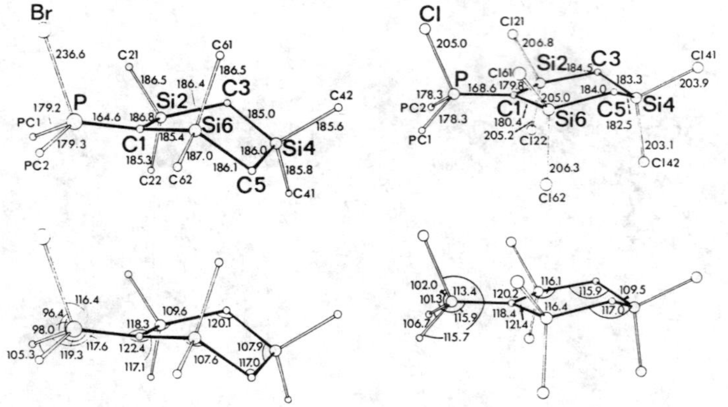

Fig. 1. Views of the ylids (I) (left) and (II) (right) with dimensions (mean values for I) in pm and °.

In I, the ylid part of the molecule (Fig. 1) is planar with shortened P-C 1.686 and Si-C 1.801 Å. In II the P-C distance is shorter (1.646 Å) but P-Br (2.366 Å) is enlarged; the ylidic C atom is 0.17 Å from the ylid plane. Molecule I has a flat chair conformation; II is in a boat form.

1-FLUOROSILATRANE
$C_6H_{12}FNO_3Si$ $FSi(OCH_2CH_2)_3N$

L. PÁRKÁNYI, P. HENSCEI, L. BIHÁTSI and T. MÜLLER, 1984. J. Organometal. Chem., 269, 1-9.

Monoclinic, Pc, a = 7.153, b = 7.177, c = 10.775 Å, β = 131.61°, Z = 2. Mo radiation, R = 0.043 for 1323 reflexions.

The silatrane skeleton can be regarded as a system of three five-membered rings of envelope conformation linked to each other by two annellation atoms (Si(1) and N(5)) with the C atoms attached to N as the envelope flaps (Fig. 1). These C atoms are disordered, with occupancy factors of 0.56 and 0.44 over the two sites. The N-Si bond length is 2.042 Å and the Si-F distance 1.622 Å. Other bond distances are: Si-O(2) 1.643, Si-O(8) 1.638, Si-O(9) 1.654, O(2)-C(3) 1.409, O(8)-C(7) 1.414 and O(9)-C(10) 1.408 Å. The N-C distances (primed and unprimed) are: 1.454, 1.465; 1.455, 1.454; 1.471 and 1.490 Å for C(4), C(6) and C(11) respectively.

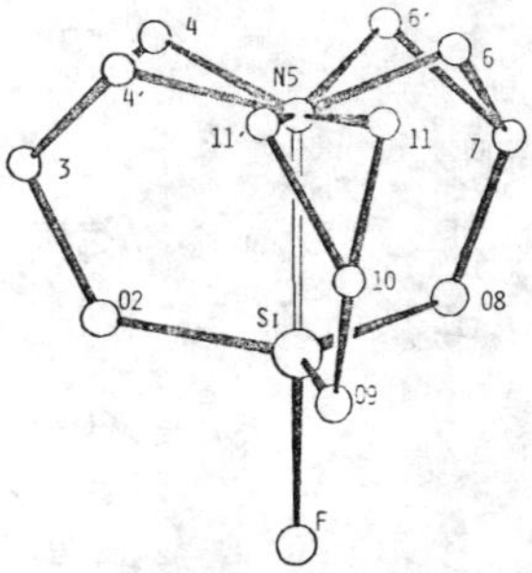

Fig. 1. Molecular diagram of $C_6H_{12}FNO_3Si$ showing disorder in the N-bonded C atoms.

2-[(METHYL(TRIFLUOROSILYL)AMINO)METHYL]PYRIDINE
$C_7H_9F_3N_2Si$

G. KLEBE, M. NIX and K. HENSEN, 1984. Chem. Ber., 117, 797-808.

Monoclinic, $P2_1/m$, a = 6.951, b = 7.137, c = 9.541 Å, β = 109.52°, Z = 2. Mo radiation, R = 0.056 for 941 reflexions.

The molecule (Fig. 1) has crystallographic mirror symmetry, with Si-F 1.603 and 1.621(5), Si-N1 1.974(6), Si-N2 1.701(10).

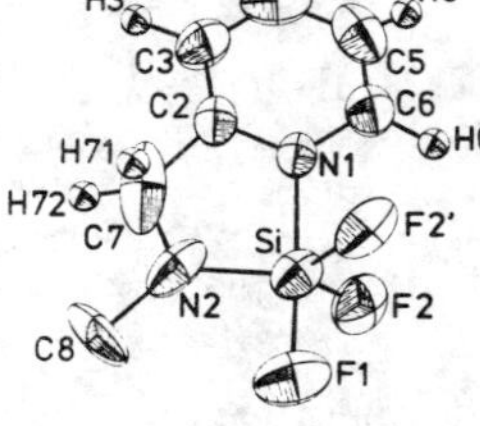

Fig. 1. A view of $C_7H_9F_3N_2Si$.

DI-t-BUTYLSILANEDIOL
$C_8H_{20}O_2Si$ (I)

1,1,5,5-TETRA-t-BUTYL-1,5-DIHYDROXY-3,3-DIMETHYL-1,3,5-TRISILOXANE
$C_{18}H_{44}O_4Si_3$ (II)

1,1,3,3,5-PENTA-t-BUTYL-5-FLUORO-1,3,5-CYCLOTRISILOXANE
$C_{20}H_{45}FO_3Si_3$ (III)

O. GRAALMANN, U. KLINGEBIEL, W. CLEGG, M. HAASE and G.M. SHELDRICK, 1984. Chem. Ber., 117, 2988-2997.

I. Orthorhombic, Ibam, a = 10.533, b = 12.775, c = 16.489 Å, Z = 8. Mo radiation, R = 0.098 for 776 reflexions.

II. Triclinic, $P\bar{1}$, a = 12.669, b = 14.328, c = 17.254 Å, α = 81.36, β = 71.19, γ = 66.97°, Z = 4. Mo radiation, R = 0.075 for 4826 reflexions.

III. Monoclinic, $P2_1/c$, a = 16.286, b = 11.541, c = 14.387 Å, β = 97.24°, Z = 4. Mo radiation, R = 0.066 for 3071 reflexions.

Molecule I (Fig. 1) lies on a mirror plane and infinite chains are formed by O-H...O hydrogen bonds (O...O 2.785 Å). The two independent molecules of II (Fig. 1) are linked by O-H...O hydrogen bonds (O...O 2.773 and 2.824 Å) In III, atom Si(3) lies 0.24 Å from the Si_2O_3 plane (Fig. 1). Si-O distances are 1.646(3) Å in I, 1.603-1.635(4) Å in II and 1.619-1.655(3) Å in III.

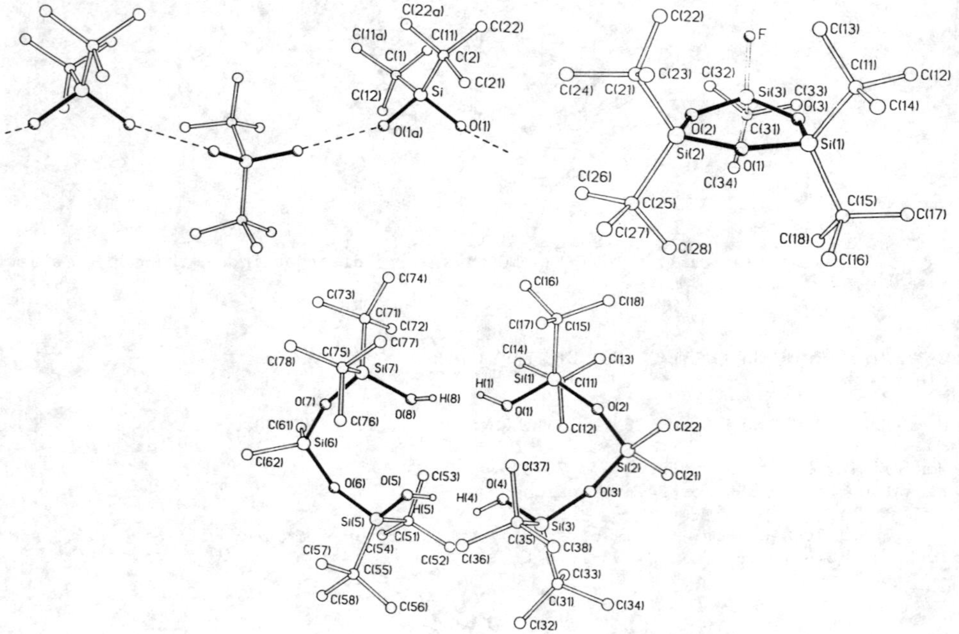

Fig. 1. Views of di-t-butylsilanediol (I) (top left), $C_{18}H_{44}O_4Si_3$ (II) (bottom) and the fluorocyclotrisiloxane (III) (top right).

3,9,10-TRIOXA-6-AZA-1,2-DISILABICYCLO[4.3.3]DODECANE
$C_9H_{21}NO_3Si_2$

J. GROBE, G. HENKEL, B. KREBS and N. VOULGARAKIS, 1984. Z. Naturforsch., B39, 341-351.

Monoclinic, $P2_1/c$, a = 7.088, b = 15.074, c = 12.701 Å, β = 104.96°, Z = 4. Mo radiation, R = 0.028 for 2582 reflexions (at -130°C).

There is a significant intramolecular N...Si interaction in this molecule (Fig. 1), with the coordination around Si(1) distorted from tetrahedral to trigonal bipyramidal and Si...N 2.768 Å. The Si(1)-Si(2) distance is 2.345(1), with Si-O in the range 1.652-1.657(1). Si-C 1.863-1.874(2) Å.

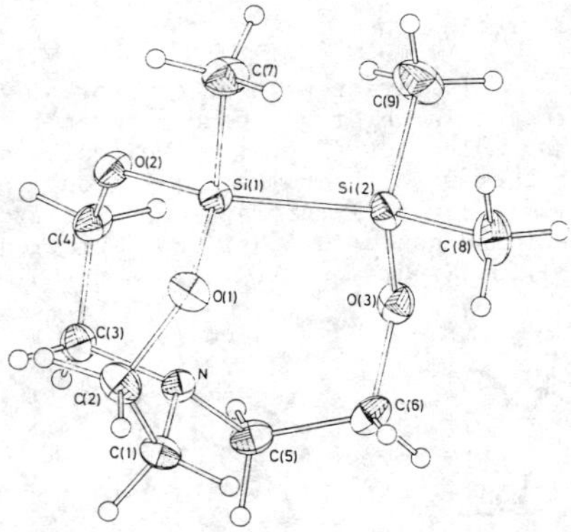

Fig. 1. A view of $C_9H_{21}NO_3Si_2$.

2,2,4,4,6,6-HEXAMETHYL-5-(HYDROXYDIMETHYLSILYL)-1,3-DIOXA-2,4,6-TRISILACYCLOHEXANE
$C_9H_{26}O_3Si_4$

C. EABORN, P.B. HITCHCOCK and P.D. LICKISS, 1984. J. Organometal. Chem., 264, 119-126.

Monoclinic, C2/c, a = 15.417, b = 19.688, c = 13.642 Å, β = 122.23°, Z = 8. Mo radiation, R = 0.068 for 1739 reflexions.

The six-membered ring has a flattened (and somewhat distorted) chair conformation (Fig. 1). The Si-C-Si angles around C(1) open to relieve the strain (the mean value is 114.9°), and the Si-C(1) bonds (1.886, 1.860, 1.884 Å) are significantly longer than the $Si-CH_3$ bonds (mean 1.834 Å). The Si-O-Si angles of 134.4 and 135.5° are considered normal. There is intermolecular hydrogen bonding, the OH group of one molecule interacting with the siloxane O(2) of another molecule; the O(3)-H distance is 1.22 Å and the H...O(2') distance 1.80 Å.

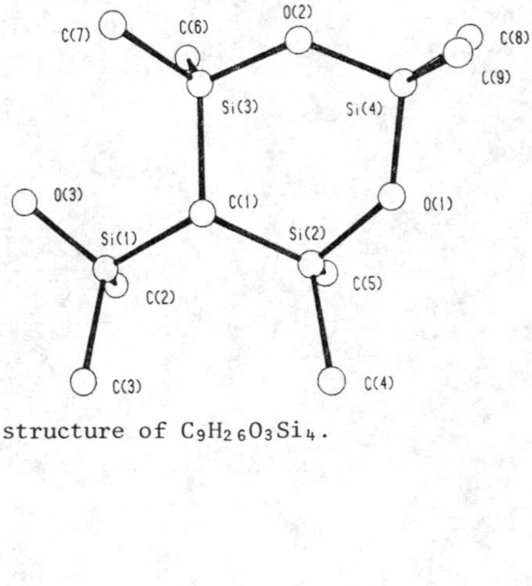

Fig. 1. Molecular structure of $C_9H_{26}O_3Si_4$.

2-TRIMETHYLSILYL-trans-2-METHOXYCYCLOHEXANOL
$C_{10}H_{22}O_2Si$

G. BERTI, P. CROTTI, F. MACCHIA and P. DOMIANO, 1984. J. Chem. Soc. Perkin II, 1987-1991.

Monoclinic, $P2_1/n$, a = 6.505, b = 17.104, c = 11.181 Å, β = 99.2°, Z = 4. Cu radiation, R = 0.045 for 1723 reflexions.

The molecule exists in a conformation with an equatorial trimethylsilyl group stabilised by a strong intermolecular hydrogen bond between the axial hydroxy and methoxy groups of neighbouring molecules (Fig. 1). Molecules of alternate chirality are thus stacked one upon another with the bulky Me_3Si group pointing in the opposite direction in order to have minimal mutual interaction. The cyclohexane ring is slightly flattened around C-Si and has a $^4C_1$ chair conformation. The C-Si bond distances range from 1.869(3) for Si-C(1) to 1.929(2) Å for Si-C(4).

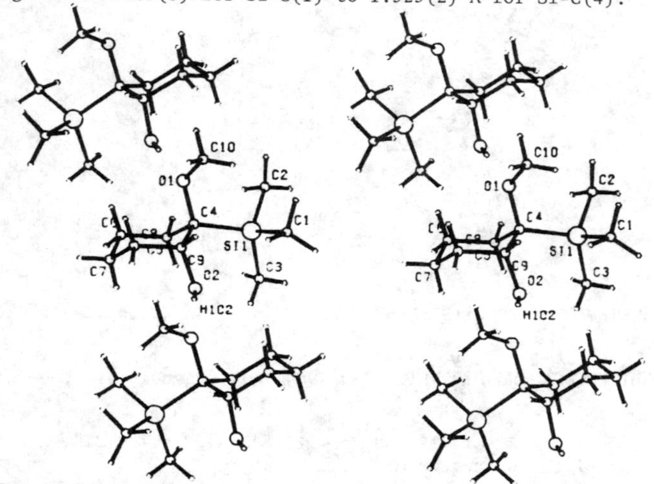

Fig. 1. $C_{10}H_{22}O_2Si$: stereoview of the packing viewed along the 010 axis.

1,1,3,3,5,5,7,7,9,9-DECAMETHYLCYCLOPENTA-λ^5-SIL-λ^4-AZANE
$C_{10}H_{45}N_5Si_5$ $(C_2H_9NSi)_5$

A.J. BLAKE, E.A.V. EBSWORTH and A.J. WELCH, 1984. Acta Cryst., C40, 895-897.

Monoclinic, $P2_1/m$, a = 6.408, b = 19.491, c = 9.467 Å, β = 90.79°, Z = 2. Mo radiation, R = 0.0504 for 2698 reflexions (at 118K).

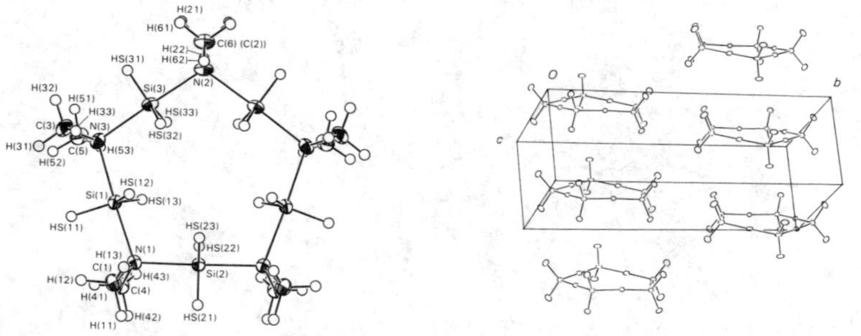

Fig. 1. A perspective view of the $(C_2H_9NSi)_5$ pentamer and the crystal packing diagram.

For previous work see 1. The unit cell contains two pentameric, ten-membered rings of alternating N and Si atoms (Fig. 1). Each ring lies across a crystallographic mirror plane. All Si-N distances (mean 1.976(4) Å) are equivalent and greater by ca. 0.26 Å than that found for the unassociated molecule in the gas phase (2). The coordination at N and Si is pyramidal and trigonal bipyramidal respectively. The crystal structure consists of essentially isolated rings packed parallel to each other.

1. Structure Reports, 32B, 301.
2. M.J. BARROW, E.A.V. EBSWORTH and M.M. HARDING, 1980, J. Chem. Soc. Dalton Trans., p. 1839.

9-METHYL-2,2,3,3-TETRAKIS(TRIFLUOROMETHYL)-1,4,6,12-TETRAOXA-9-AZA-5-SILASPIRO[4.7]-DODECANE

$C_{11}H_{11}F_{12}NO_4Si$

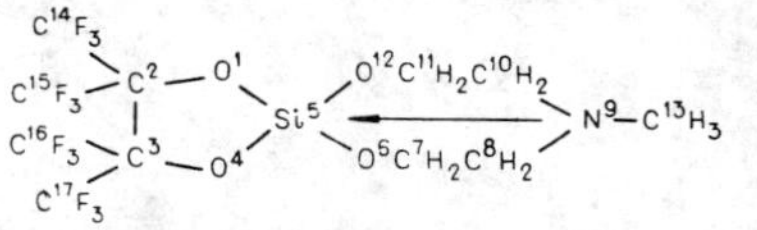

A.A. KEMME, J.J. BLEIDELIS, I.P. URTANE, G.I. ZELCHAN and E. LUKEVICS, 1984. J. Struct. Chem., 25, 147-149 [Zh. Strukt. Khim., 25(1), 165-168].

Orthorhombic, Pbca, a = 9.960, b = 14.493, c = 22.857 Å, Z = 8. Mo radiation, R = 0.083 for 1968 reflexions.

A short transannular Si-N interaction (2.032(7) Å) results in a trigonal bipyramidal arrangement about the silicon atom. The angle between axial ligands (N-Si-O) is 176.4(3)°. The axial Si-O distance is 1.731(6) Å and the equatorial Si-O distances range from 1.632(6) to 1.690(5) Å.

1,2,3,4-TETRAHYDRO-1-TRICHLOROSILYL-1,10-PHENANTHROLINE

$C_{12}H_{11}Cl_3N_2Si$

G. KLEBE, J.W. BATS and H. FUESS, 1984. J. Am. Chem. Soc., 106, 5202-5208.

Monoclinic, $P2_1/m$, a = 9.243, b = 7.076, c = 10.739 Å, β = 114.97°, Z = 2. Mo radiation, R = 0.055 for 3008 reflexions; neutron radiation (1.266 Å), R = 0.028 for 1428 reflexions; neutron radiation (0.894 Å), R = 0.030 for 1617 reflexions (all at 120 K).

Molecular dimensions are in Fig. 1. The molecule has two different Si-N bonds, the shorter being in the equatorial plane. The results of deformation density studies are reported.

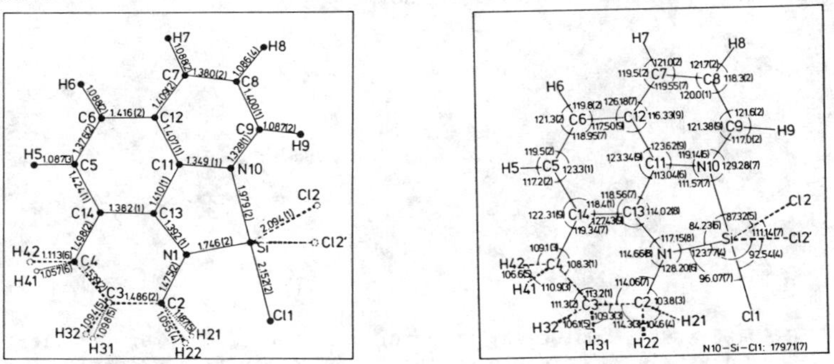

Fig. 1. Bond lengths and angles (neutron data) for $C_{12}H_{11}Cl_3N_2Si$.

cis-2,4-DICHLORO-2,4-BIS(TRIMETHYLSILYL)-1,1,3,3-TETRAMETHYL-1,3-DISILACYCLOBUTANE
$C_{12}H_{30}Cl_2Si_4$

G. FRITZ, J. THOMAS, K. PETERS, E.M. PETERS and H.G. VON SCHNERING, 1984. Z. Anorg. Chem., 514, 61-68.

Orthorhombic, Fdd2, a = 21.491, b = 22.292, c = 17.636 Å, Z = 16. Mo radiation, R = 0.041 for 1932 reflexions.

The disilacyclobutane ring is slightly folded (17.9°) (Fig. 1). The configurations at C2 and C4 approach sp^2 with Si-C-Si angles 87.1 and 87.3° within the ring and 124.6-125.5° outside the ring. The Cl atoms are cis with long C-Cl bonds (1.842 and 1.849(5) Å).

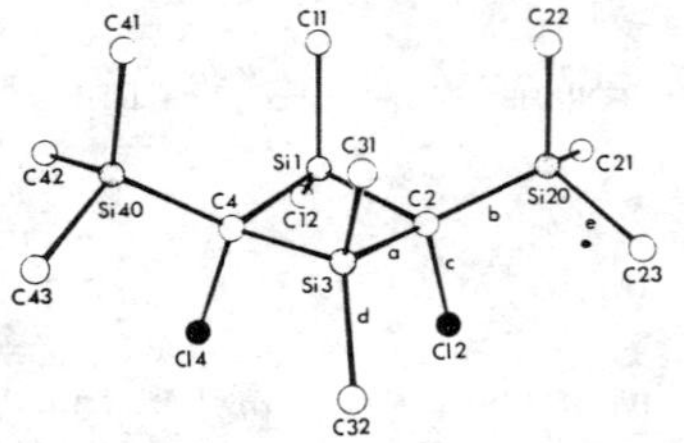

Fig. 1. A view of $C_{12}H_{30}Cl_2Si_4$.

cis-2,4,6-TRI-t-BUTYL-2,4,6-TRIFLUOROCYCLOTRISILAZANE
$C_{12}H_{30}F_3N_3Si_3$ (I)

trans-2,4,6-TRI-t-BUTYL-2,4,6-TRIFLUOROCYCLOTRISILAZANE
$C_{12}H_{30}F_3N_3Si_3$ (II)

W. CLEGG, G.M. SHELDRICK and D. STALKE, 1984. Acta Cryst., C40, 816-818.

I. Trigonal, $P\bar{3}$, a = 10.873, c = 10.040 Å, Z = 2. Mo radiation, R = 0.054 for 1232 reflexions.

II. Monoclinic, $P2_1/n$, a = 18.077, b = 10.569, c = 22.015 Å, β = 104.67°, Z = 8. Mo radiation, R = 0.071 for 4421 reflexions.

Molecule I has crystallographic threefold rotation symmetry and a virtually planar Si_3N_3 ring whilst the two crystallographically independent molecules of II lie in general positions and deviations from planarity of the ring are greater (Fig. 1). One molecule of II is disordered. Corresponding bond lengths and angles in I and II are similar; mean values are: Si-N 1.695(2) and 1.696(7), Si-F 1.602(2) and 1.600(5), Si-C 1.862(3) and 1.851(9) Å, N-Si-N 107.0(2) and 106.5(8), Si-N-Si 133.0(2) and 131.5(22)°, in I and II respectively.

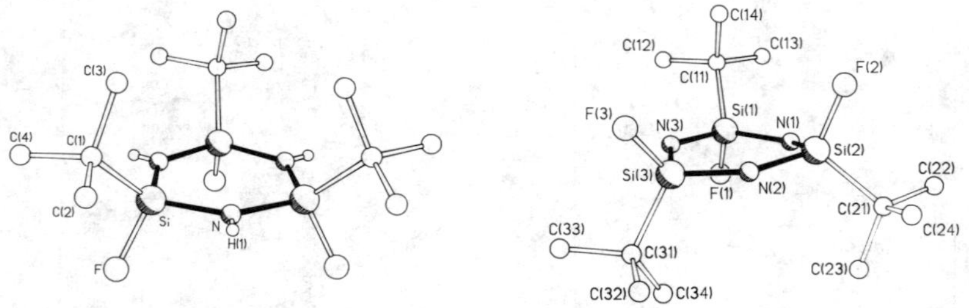

Fig. 1. $C_{12}H_{30}F_3N_3Si_3$: molecular structures of the cis isomer (I) (left) and of the undisordered molecule of the trans isomer (II) (right).

HEXAMETHYLENEDIAMMONIUM BIS[BIS(1,2-ETHYLENEDIOLATO)PHENYLSILICATE(IV)]
$C_{13}H_{22}NO_4Si$ (I) $[(C_2H_4O_2)_2SiPh]^-[NH_3(CH_2)_6NH_3]^{2+}{}_{0.5}$

TETRAETHYLAMMONIUM BIS(3,4,5,6-TETRACHLORO-1,2-BENZENEDIOLATO)PHENYLSILICATE(IV)
$C_{26}H_{25}Cl_8NO_4Si$ (II) $[NEt_4]^+[(Cl_4C_6O_2)_2SiPh]^-$

R.R. HOLMES, R.O. DAY, J.J. HARLAND, A.C. SAU and J.M. HOLMES, 1984. Organometallics, 3, 341-347.

I. Monoclinic, $P2_1/n$, a = 10.266, b = 12.344, c = 12.525 Å, β = 105.48°, Z = 4. Mo radiation, R = 0.048 for 2225 reflexions.

II. Monoclinic, $P2_1/n$, a = 8.502, b = 21.141, c = 17.406 Å, β = 90.29°, Z = 4. Mo radiation, R = 0.051 for 3542 reflexions.

The basic structural form assumed by both compounds is a rectangular pyramid although I is closer to a trigonal bipyramid (Fig. 1). This is shown by the angles O(2)-Si-O(2') and O(1)-Si-O(1') which are closer together in II (152.9 and 147.3° respectively) than in I (159.2 and 143.6° respectively). Also, the Si-O bond lengths reflect the greater rectangular pyramidal character in II: Si-C(7) (apical) 1.854, Si-O(1) 1.746, Si-O(1') 1.757, Si-O(2) 1.759 and Si-O(2') 1.753 Å (all basal). In I the corresponding distances are: Si-CP(1) 1.883, Si-O(1) 1.738, Si-(O1') 1.757 Å (equatorial); Si-O(2) 1.717 and Si-O(2') 1.685 Å (axial). There is hydrogen bonding in I between the amino H atoms and O atoms of the anions, O...H ranging from 1.83 to 1.91 Å.

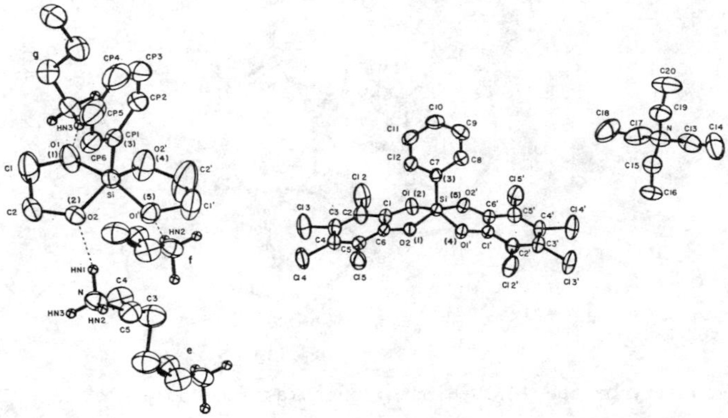

Fig. 1. Molecular structure of $C_{13}H_{22}NO_4Si$ showing three extra half-cations (left), and of $C_{26}H_{25}Cl_8NO_4Si$ (right).

DI-t-BUTYLFLUOROSILANOL PYRIDINE N-OXIDE
$C_{13}H_{24}FNO_2Si$ $C_8H_{19}FOSi,C_5H_5NO$

W. CLEGG, G.M. SHELDRICK, U. KLINGEBIEL, D. BENTMANN, G. HENKEL and B. KREBS, 1984. Acta Cryst., C40, 819-820.

Triclinic, $P\bar{1}$, a = 6.581, b = 9.168, c = 13.484 Å, α = 84.62, β = 76.22, γ = 83.63°, Z = 2. Mo radiation, R = 0.064 for 2944 reflexions (at 153K).

The silanol and pyridine N-oxide molecules are linked by a hydrogen bond with O...O 2.630(6) Å, Si-O...O 127.6(4) and N-O...O 117.2(4)° (Fig. 1). The electron-withdrawing fluoro substituent causes the Si-O bond to be much shorter than those in other hydrogen-bonded silanols (1). The adduct formation does not affect the normal tetrahedral coordination of Si.

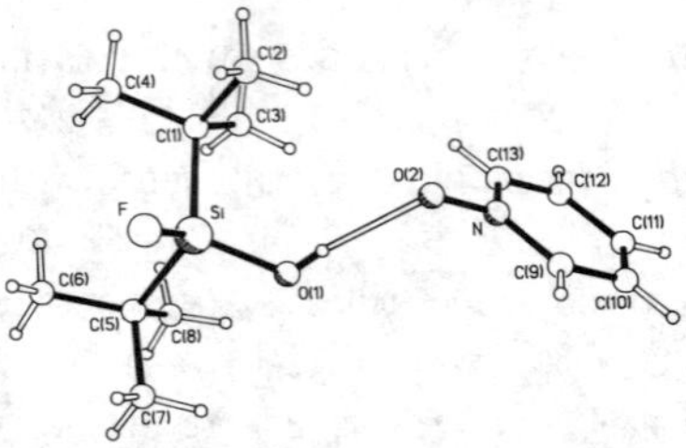

Fig. 1. A perspective view of the $C_8H_{19}FOSi,C_5H_5NO$ adduct.

1. Structure Reports, 50B.

10,10-DIMETHYLPHENOXASILANE
$C_{14}H_{14}OSi$ (I)

10,10-DIMETHYLPHENOTHIASILANE
$C_{14}H_{14}SSi$ (II)

O.A. D'YACHENKO, Yu.A. SOKOLOVA and L.O. ATOVMYAN, 1984. J. Struct. Chem., 25, 582-587 [Zh. Strukt. Khim. 25(4), 83-88].

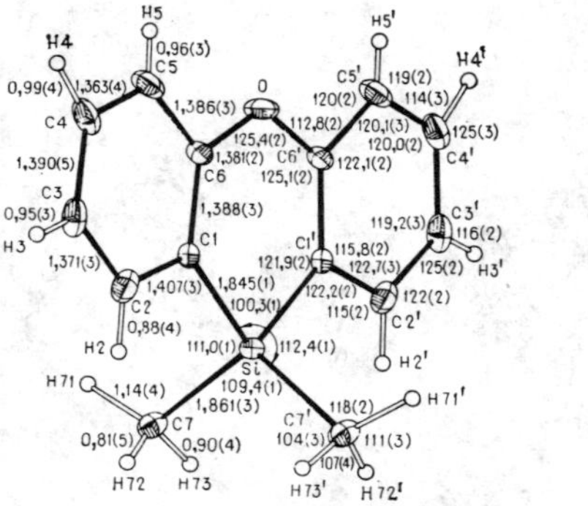

Fig. 1. Geometry of the 10,10-dimethylphenoxasilane molecule.

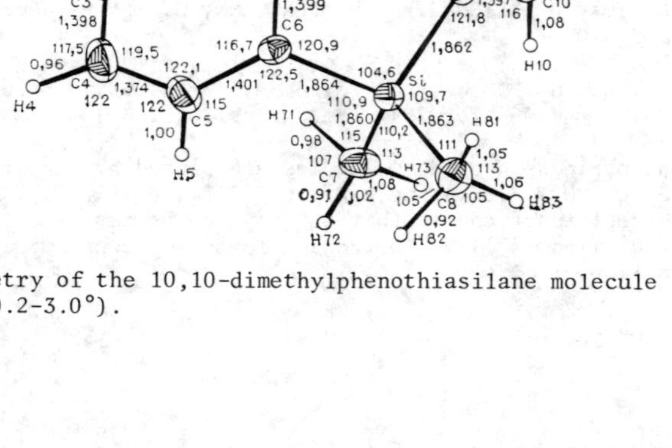

Fig. 2. Geometry of the 10,10-dimethylphenothiasilane molecule (σ = 0.004-0.06 Å and 0.2-3.0°).

I. Trigonal, $P\bar{3}c1$, a = 12.452, c = 14.379 Å, Z = 6. Cu radiation, R = 0.066 for 1023 reflexions.

II. Orthorhombic, Pbca, a = 30.850, b = 11.006, c = 7.590 Å, Z = 8. Cu radiation, R = 0.059 for 1651 reflexions.

Bond lengths and angles for I and II are shown in Figures 1 and 2, respectively. Molecule I has exact C_2 symmetry with the twofold axis passing through the O and Si atoms. The system of three fused rings is nearly planar in I while that in II deviates significantly from planarity (the dihedral angles between normals to the benzene rings being 175.6 and 148.0°, respectively).

2,8-DIBROMO-5,10-DIMETHYL-10-FLUORO-10,11-DIHYDRO-5H-DIBENZ[b,f][1,4]AZASILEPINE
$C_{15}H_{14}Br_2FNSi$

J.Y. COREY, E.R. COREY, V.H.T. CHANG, M.A. HAUSER, M.A. LEIBER, T.E. REINSEL and M.E. RIVA, 1984. Organometallics, 3, 1051-1060.

Monoclinic, $P2_1/c$, a = 8.869, b = 9.934, c = 18.392 Å, β = 98.75°, D_m = 1.71, Z = 4. Mo radiation, R = 0.042 for 1121 reflexions.

The molecule assumes a folded boat conformation with a dihedral angle between benzo group planes of 112° (Fig. 1). The sum of the bond angles about N(8) is 348.5°, and the C-Si-F angles are all smaller than the C-Si-C angles (mean values 106.5 and 112.3° respectively) due to the greater electronegativity of fluorine. Mean values for Br-C, Si-C, N-C and the benzo C-C distances are 1.904 Å [1.903 and 1.905 Å], 1.845 Å [range 1.822-1.862 Å], 1.446 Å [range 1.420-1.472 Å] and 1.384 Å [range 1.373-1.411 Å]; the Si-F bond length is 1.615 Å. All H atoms were located and refined. The mean C-H distance is 0.94 Å with a range of 0.78-1.03 Å.

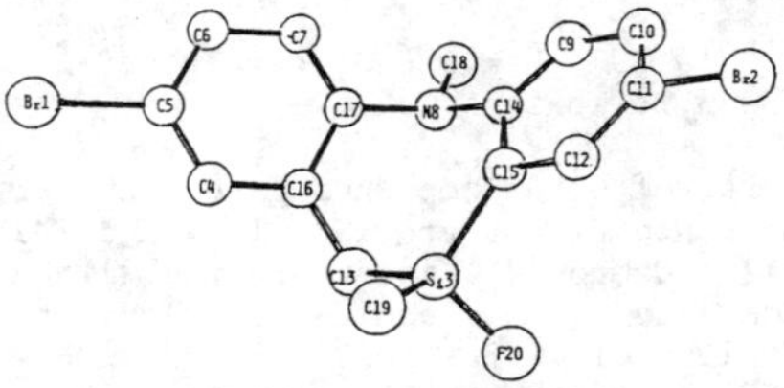

Fig. 1. Perspective view of the molecule $C_{15}H_{14}Br_2FNSi$.

13,13-DIMETHYL-8,13-DIHYDRO-5H-DIBENZO[d,g][1.2.6]-DITHIASILONINE
$C_{16}H_{14}S_2Si$

V.E. SHKLOVER, Yu.T. STRUCHKOV, O.G. RODIN, V.F. TRAVEN' and B.I. STEPANOV, 1984. J. Organometal. Chem., 266, 117-121.

Orthorhombic, Pbca, a = 16.5274, b = 22.272, c = 8.5940 Å, Z = 8. Cu radiation, R = 0.089 for 1802 reflexions.

The two intramolecular Si...S distances in the molecule (Fig. 1) are unequal (4.190 and 3.438 Å), the second of these indicating intramolecular transannular Si...S interaction which is confirmed by the distortion of the coordination polyhedron around the Si atom. The nine-membered heterocycle has a distorted BB conformation with an approximate symmetry plane passing through C(7) and the mid point of the C(9)-C(14) bond. It is folded along the S(2)...C(1) vector by 96.2°, though the C(1),Si,C(14),-C(9),C(8),S(2) and C(1),C(6),C(7),S(1),S(2) moieties are only approximately planar. The dihedral angle between the two phenyl rings is 97.1°.

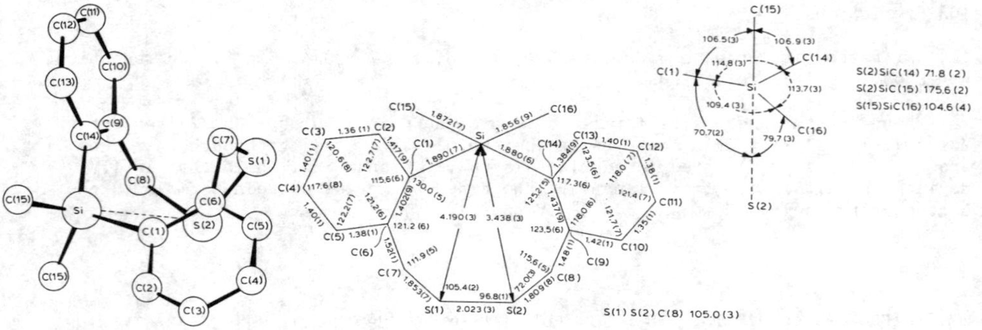

Fig. 1. Molecular structure of $C_{16}H_{14}S_2Si$, bond lengths (Å) and angles, and bond angles in the coordination polyhedron of the Si atom.

9,9,10,10-TETRAMETHYL-9,10-DIHYDRO-10-GERMA-9-SILAANTHRACENE
$C_{16}H_{20}GeSi$ (I)

9,9,10,10-TETRAMETHYL-9,10-DIHYDRO-10-STANNA-9-SILAANTHRACENE
$C_{16}H_{20}SiSn$ (II)

W.Z. McCARTHY, J.Y. COREY and E.R. COREY, 1984. Organometallics, 3, 255-263.

I. Orthorhombic, Pbca, a = 9.585, b = 16.200, c = 21.236 Å, D_m = 1.25, Z = 8. Mo radiation, R = 0.043 for 1512 reflexions.

II. Monoclinic, $P2_1/n$, a = 18.897, b = 9.093, c = 19.439 Å, β = 95.60°, Z = 8. Mo radiation, R = 0.051 for 3475 reflexions.

The tricyclic frameworks of both compounds have a butterfly conformation (Fig. 1). The dihedral angles of the benzo groups are 153° in I and 142 and 148° for the two independent molecules of II. Compound I is disordered with a 50/50 occupancy by Si and Ge of the heteroatom sites. The mean C-(Si,Ge) distance of 1.909 Å is intermediate between the values of 1.88 Å for C-Si and 1.95 Å for C-Ge. In II, the C-Si distances average 1.878 Å and the C-Sn distances 2.138 Å over the two molecules.

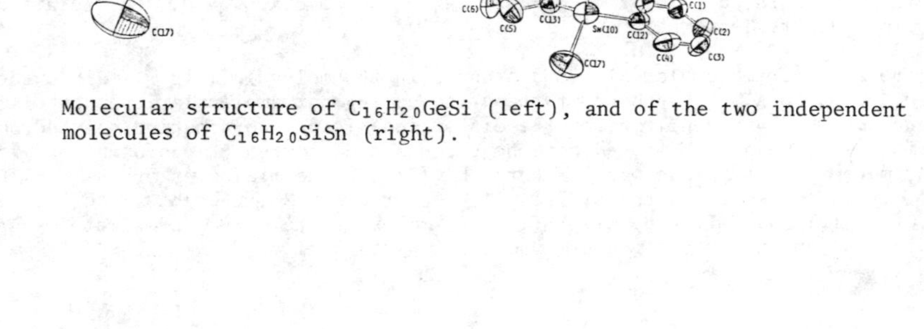

Fig. 1. Molecular structure of $C_{16}H_{20}GeSi$ (left), and of the two independent molecules of $C_{16}H_{20}SiSn$ (right).

TETRA-t-BUTOXY-1,3,2,4-DITHIADISILETANE
$C_{16}H_{36}O_4S_2Si_2$

W. WOJNOWSKI, K. PETERS, D. WEBER and H.G. VON SCHNERING, 1984. Z. Anorg. Chem., 519, 134-140

Orthorhombic, Pbca, a = 17.084, b = 15.608, c = 9.071 Å, Z = 4. Mo radiation, R = 0.070 for 1758 reflexions.

The molecule (Fig. 1) lies about a crystallographic inversion centre and the Si_2S_2 ring is consequently planar; principal dimensions are in the figure.

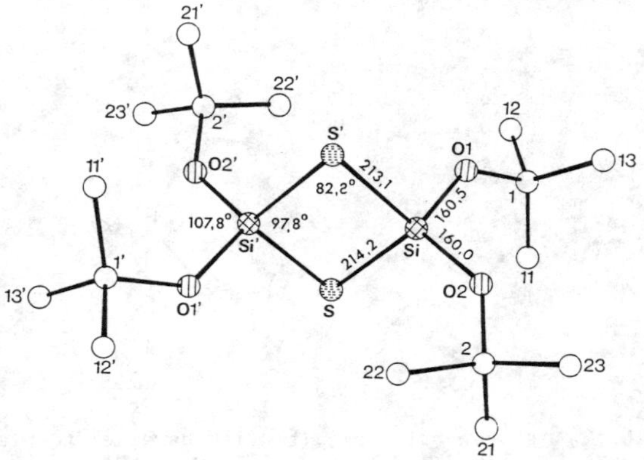

Fig. 1. A view of $C_{16}H_{36}O_4S_2Si_2$.

cis-1,4-BIS(TRIMETHYLSILYL)HEXAMETHYL-1,4-DISILACYCLOHEXA-2,5-DIENE
$C_{16}H_{36}Si_4$ (I)

trans-1,4-BIS(TRIMETHYLSILYL)HEXAMETHYL-1,4-DISILACYCLOHEXA-2,5-DIENE
$C_{16}H_{36}Si_4$ (II)

J.D. RICH, F. SHAFIEE, K.J. HALLER, S.G. HARSY and R. WEST, 1984. J. Organometal. Chem., 264, 61-78.

I. Triclinic, $P\bar{1}$, a = 9.955, b = 15.568, c = 7.951 Å, α = 104.43, β = 107.80, γ = 98.95°, Z = 2. Mo radiation, R = 0.040 for 3780 reflexions (-65°C).

II. Monoclinic, $P2_1/c$, a = 8.386, b = 9.526, c = 14.143 Å, β = 97.98°, Z = 2. Mo radiation, R = 0.033 for 2581 reflexions (-55°C).

The cis-isomer (I) exists in a slightly twisted boat conformation with the two bulky trimethylsilyl substituents located in pseudoaxial positions (Fig. 1). The trans-isomer (II) has crystallographic inversion symmetry and adopts a nearly-planar conformation, deformed slightly towards a chair form (Fig. 1). The Si-C bond lengths range from 1.868 to 1.884 Å in I (average 1.874 Å) and in II they range from 1.870 to 1.878 Å (average 1.873 Å). The Si-Si bond distances are 2.360 and 2.362 Å in I and II respectively. The Si-Si, Si-C and C-C bond distances are all within the normal ranges.

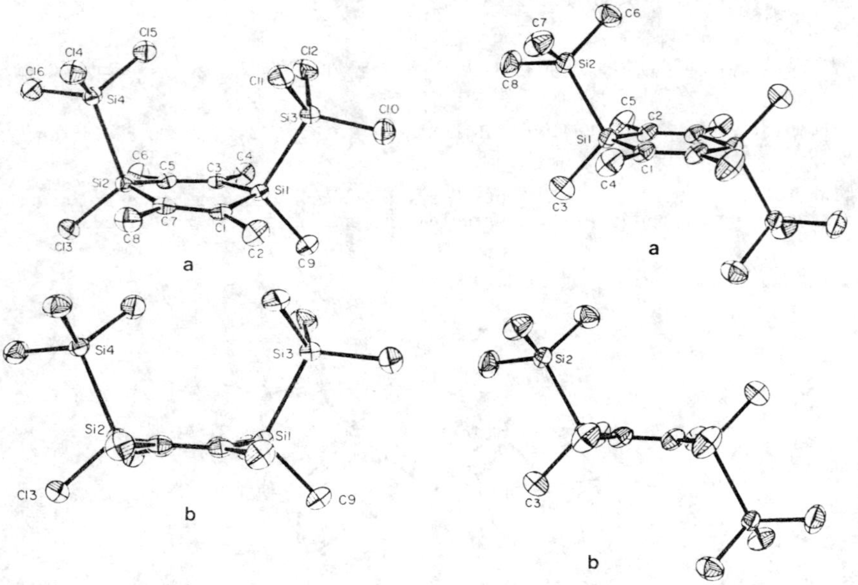

Fig. 1. Cis-$C_{16}H_{36}Si_4$ (a) general view, (b) view parallel to plane through the four olefinic C atoms (left), and trans-$C_{16}H_{36}Si_4$ (a) general view, (b) view parallel to plane through the four olefinic C atoms (right).

[2-(DIMETHYLAMINO)ETHYLTHIOMETHYL]DIPHENYLSILANOL
$C_{17}H_{23}NOSSi$

M. SÖDERHOLM, 1984. Acta Chem. Scand., B38, 303-307.

Monoclinic, $P2_1/c$, a = 6.121, b = 15.956, c = 18.348 Å, β = 93.09°, Z = 4. Mo radiation, R = 0.068 for 1925 reflexions.

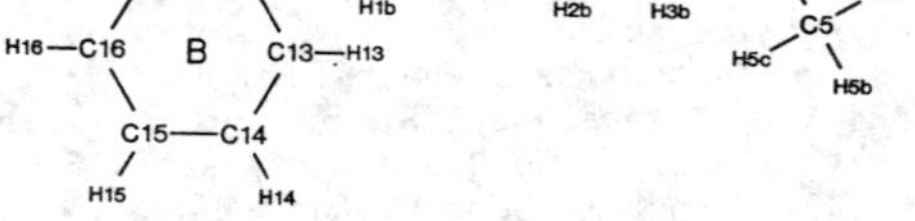

Fig. 1. Schematic diagram of the molecule $C_{17}H_{23}NOSSi$.

The phenyl rings have slight distortions in geometry, in particular the angles C(7)-C(6)-C(11) (117.4°) and C(17)-C(12)-C(13) (115.5°). This reduction of the endocyclic angle of the C atom attached to the Si atom is a common feature in such compounds. The dihedral angle between the phenyl rings is 86.9°. The aminoethylthiomethyl chain is maximally extended, with torsion angles of Si-C(1)-S-C(2) 178.1, C(1)-S-C(2)-C(3) -177.2, and S-C(2)-C(3)-N 168.0°. Centrosymmetrically-related molecules form dimers as they are linked by O-H...N hydrogen bonds. The O(1)...N distance is 2.744 Å. Distances of interest are: Si-O(1) 1.609, Si-C(1) 1.865, Si-C(6) 1.881, Si-C(2) 1.872, C(1)-S 1.785 and S-C(2) 1.784 Å.

10,10-DIHYDRO-10-(p-CHLOROPHENYL)PHENOTHIASILENE
$C_{18}H_{13}ClSSi$ (I)

10,10-DIHYDRO-10-(p-TOLYLPHENYL)PHENOTHIASILENE
$C_{19}H_{16}SSi$ (II)

10,10-DIHYDRO-10-(p-METHOXYPHENYL)PHENOTHIASILENE
$C_{19}H_{16}OSSi$ (III)

V.K. BEL'SKII, I.P. YAKOVLEV, Yu.S. FINOGENOV and V.O. REIKHSFEL'D, 1984. Kristallografiya, 29, 260-265 [Sov. Phys. Crystallogr., 29, 154-157].

I. Monoclinic, $P2_1/a$, a = 31.554, b = 10.810, c = 9.504 Å, γ = 91.88°, Z = 8. Mo radiation, R = 0.036 for 2494 reflexions.

II. Monoclinic, $P2_1/a$, a = 31.399, b = 10.979, c = 9.463 Å, γ = 91.78°, Z = 8. Mo radiation, R = 0.038 for 2273 reflexions.

III. Monoclinic, $P2_1/b$, a = 8.554, b = 17.059, c = 12.118 Å, γ = 108.28°, Z = 4. Mo radiation, R = 0.026 for 1031 reflexions.

The molecules all have a "butterfly" conformation with the following angles along the S...Si line: 143.1 and 143.8° in I, 144.2 and 141.6° in II (I and II are isostructural and have two molecules per asymmetric unit) and 133.0° in III, (Fig. 1). The central rings in all the molecules have boat conformations. The Si-C distances range from 1.844 to 1.862 Å, the Si-H distances from 1.41 to 1.48 Å, and the S-C distances from 1.766 to 1.782 Å. The endocyclic C-Si-C angles are all less than the usual tetrahedral angle, ranging from 103.1 to 104.6°.

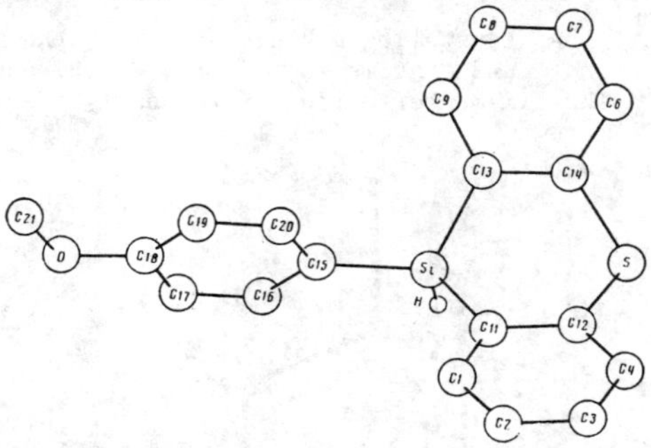

Fig. 1. Molecular conformation of $C_{19}H_{16}OSSi$.

(2-MORPHOLINOETHYL)DIPHENYLSILANOL
$C_{18}H_{23}NO_2Si$

M. SÖDERHOLM, 1984. Acta Chem. Scand., B38, 31-35.

Monoclinic, $P2_1/c$, a = 17.063, b = 10.863, c = 9.094 Å, β = 90.61°, Z = 4. Mo radiation, R = 0.054 for 1219 reflexions.

The morpholine ring has a chair conformation with atoms N and O(2) some -0.675 and +0.668 Å, respectively, from the least-squares plane through carbon atoms C(3) to C(6). The phenyl rings A and B (Fig. 1) are planar but the substitution of Si for H causes the endocyclic angles C(12)-C(7)-C(8) and C(18)-C(13)-C(14) to close up to values of 117.1 and 116.8° respectively. The angle between the phenyl ring planes is 99.0° and the angles between their planes and that through the morpholine ring are 83.1 and 58.6° for ring A and B respectively. The molecules are linked by O-H...N hydrogen bonds (O(1)...N' 2.776 Å) forming infinite chains.

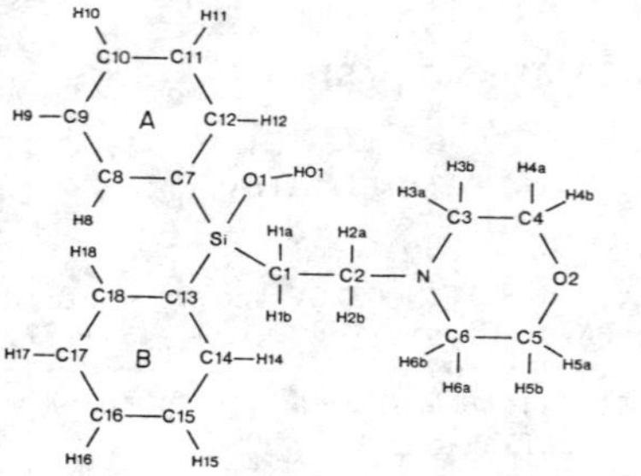

Fig. 1. Schematic drawing of the molecule $C_{18}H_{23}NO_2Si$.

1,3,5,7,9,11,13,15-OCTAMETHYL-1,3,5,7,9,11,13,15-OCTASILAHEPTACYCLO[7.7.1.$1^{3,15}$.$0^{2,7}$.-$0^{4,13}$.$0^{5,10}$.$0^{13,17}$]OCTADECANE

$C_{18}H_{40}Si_8$

G. FRITZ, K.-P. WORNS, K. PETERS, E.-M. PETERS and H.G. VON SCHNERING, 1984. Z. Anorg. Chem., 512, 126-130.

Orthorhombic, $Cmc2_1$, a = 13.545, b = 11.143, c = 17.072 Å, Z = 4. Mo radiation, R = 0.049 for 1115 reflexions.

The molecule (Fig. 1) has crystallographic m-symmetry, the mirror plane passing through Si atoms 1, 2, 3 and 6 and C atoms 10, 20, 50, 70, 31 and 61. The mean Si-C bond length is 1.886 Å. The six-membered rings form nearly ideal chair- and boat-conformations.

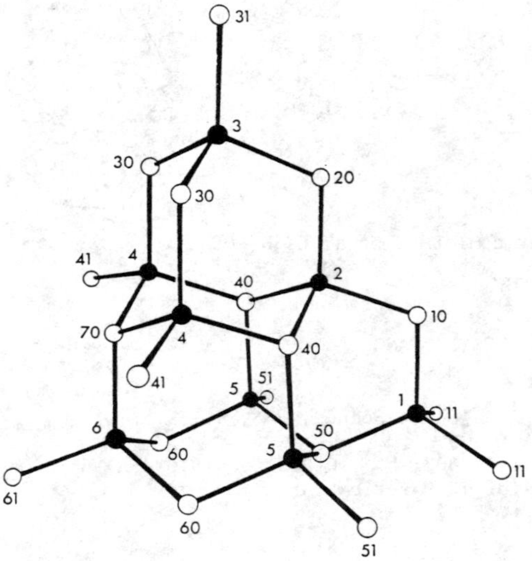

Fig. 1. A view of $C_{18}H_{40}Si_8$.

3,3,7,7-TETRA-t-BUTYL-1-FLUORO-5,5-DIMETHYL-1-BORA-3,5,7-TRISILA-2,4,6,8-TETRAOXA-CYCLOOCTANE
$C_{18}H_{42}BFO_4Si_3$

O. GRAALMANN, U. KLINGEBIEL, W. CLEGG, M. HAASE and G.M. SHELDRICK, 1984. Z. Anorg. Chem., 519, 87-95.

Triclinic, P$\bar{1}$, a = 9.387, b = 11.955, c = 13.401 Å, α = 71.78, β = 85.45, γ = 67.75, Z = 2. Mo radiation, R = 0.050 for 3966 reflexions.

In the molecule (Fig. 1), the planar Si_3O_4 unit makes a dihedral angle of 15.8° with the FBO_2 group. Some dimensions are: B-F 1.337(3), B-O 1.313 and 1.320(3), Si-O 1.612-1.635(2), Si-C 1.876-1.884(3) Å.

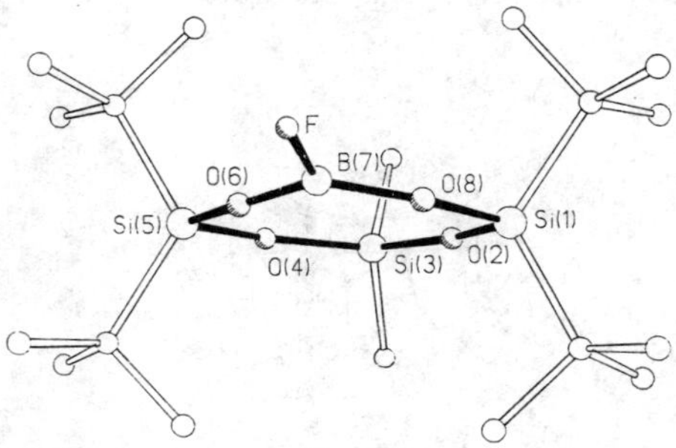

Fig. 1. A view of $C_{18}H_{42}BFO_4Si_3$.

HEXAKIS(N,N-DIMETHYLFORMAMIDE)SILICON(IV) TETRAIODIDE
$C_{18}H_{42}I_4N_6O_6Si$ $[Si.6(DMF)]I_4$

B. DEPPISCH, B. GLADROW and D. KUMMER, 1984. Z. Anorg. Chem., 519, 42-52.

Monoclinic, C2/c, a = 18.402, b = 16.843, c = 11.646 Å, β = 102.15°, Z = 4. Mo radiation, R = 0.041 for 1597 reflexions.

In the complex cation (Fig. 1), the Si atom lies on an inversion centre, with Si-O 1.754, 1.768 and 1.772(5) Å.

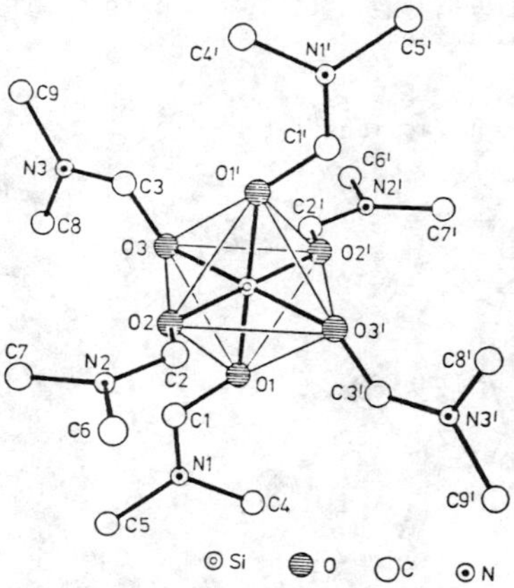

Fig. 1. A view of the $[Si.6DMF]^+$ ion.

CHLORO(METHYL)BIS(8-QUINOLINOLATO)SILICON(IV)
$C_{19}H_{15}ClN_2O_2Si$ $Si(CH_3)(C_9H_9NO)_2Cl$

G. KLEBE and D. TRAN QUI, 1984. Acta Cryst., C40, 476-479.

Triclinic, P$\bar{1}$, a = 8.085, b = 8.195, c = 15.040 Å, α = 76.14, β = 74.59, γ = 60.41°, Z = 2. Mo radiation, R = 0.049 for 2922 reflexions.

The Si atom is hexacoordinated through Lewis acid-base interaction (Fig. 1). The coordinative Si-N distance is 2.015(4) Å, the Si-O bond length 1.765(2) Å. The bidentate ligands are arranged in cis configuration along two edges of the octahedral polyhedron and methyl and chlorine are found at the remaining sites in equipopulation. Si represents a centre of chirality and two isomeric forms are present in the crystal, which correspond to the left- and right-hand arrangements.

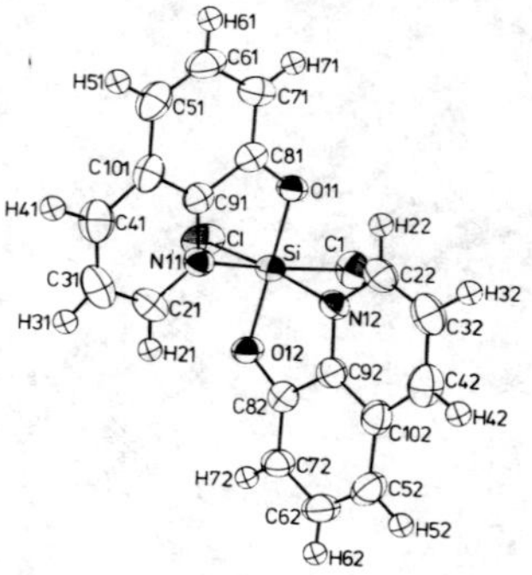

Fig. 1. A view of $C_{19}H_{15}ClN_2O_2Si$.

(η^5-CYCLOPENTADIENYL)DICARBONYL(FLUORODIPHENYLSILYL)IRON
$C_{19}H_{15}FFeO_2Si$ (I)

(η^5-CYCLOPENTADIENYL)DICARBONYL(TRICHLOROSILYL)IRON
$C_7H_5Cl_3FeO_2Si$ (II)

U. SCHUBERT, G. KRAFT and E. WALTHER, 1984. Z. Anorg. Chem., 519, 96-106.

I. Monoclinic, P2_1/c, a = 8.058, b = 14.175, c = 14.982 Å, β = 90.99°, Z = 4. Mo radiation, R = 0.037 for 2231 reflexions.

II. Monoclinic, P2_1/n, a = 8.17, b = 12.44, c = 11.42 Å, β = 99.9°, Z = 4. Mo radiation, R = 0.050 for 1780 reflexions (at -80°C).

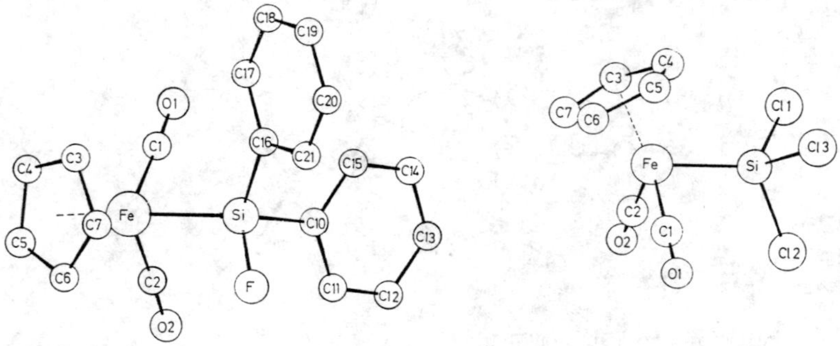

Fig. 1. Views of $C_{19}H_{15}FFeO_2Si$ (I) and $C_7H_5Cl_3FeO_2Si$ (II).

Views of (I) and (II) are in Fig. 1. In (I) there is a gauche conformation with respect to the Fe-Si bond; in (II) the silyl ligand is rotated 12° around the Fe-Si axis out of a fully staggered conformation. The Fe-Si bond length is 2.278(1) Å in (I) and 2.216(1) Å in (II).

(3-MORPHOLINOPROPYL)DIPHENYLSILANOL
$C_{19}H_{25}NO_2Si$

M. SÖDERHOLM, 1984. Acta Chem. Scand., B38, 95-99.

Triclinic, P$\bar{1}$, a = 15.073, b = 9.678, c = 6.770 Å, α = 99.04, β = 95.84, γ = 103.90°, Z = 2. Cu radiation, R = 0.072 for 3281 reflexions.

As in (2-morpholinoethyl)diphenylsilanol (1), the morpholine ring has a chair conformation. Atoms N and O(2) deviate -0.666 and +0.655 Å, respectively, from the least-squares plane through atoms C(4)-C(7). The planar phenyl rings are at 66.4° to each other and 86.0° (Ring A) and 29.6° (Ring B) to the plane through the morpholine ring (Fig. 1). Angles C(9)-C(8)-C(13) and C(15)-C(14)-C(19) in the phenyl rings are closed down to 116.3° because of the substitution of Si for H in the rings. The torsion angles of the Si-C(1)-C(2)-C(3) and C(1)-C(2)-C(3)-N fragments are 169.8 and 175.7° respectively. The molecules are linked by hydrogen bonds (O(1)-H(O1)...N) with an O...N distance of 2.840 Å, to form infinite chains.

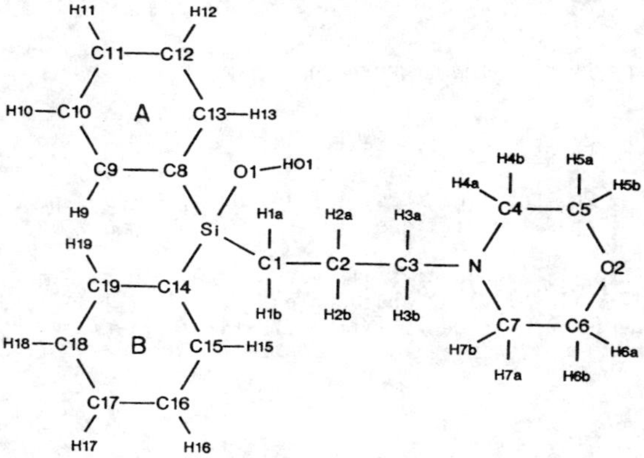

Fig. 1. Schematic drawing of the molecule $C_{19}H_{25}NO_2Si$.

1. This Volume, page 761.

BIS(TRIMETHYLSILYL)METHANIDEDIMETHYLSILYLTETRAHYDROFURANATE
$C_{19}H_{44}OSi_3$

N. WIBERG, G. WAGNER, G. MÜLLER and J. RIEDE, 1984. J. Organometal. Chem., 271, 381-391.

Orthorhombic, $P2_12_12_1$, a = 10.000, b = 15.386, c = 15.509 Å, Z = 4. Mo radiation, R = 0.052 for 1991 reflexions (-45°C).

The trigonal planar sp^2 geometry expected for both the Si(1) and C(1) atoms of the silaethene molecule is retained only for the latter. However, the short (1.747 Å) Si(1)-C(1) bond length together with the long Si-O dative bond from the THF molecule (Fig. 1) suggests that the adduct may be regarded as a zwitterion, the formal charges of which are located at O and C(1). Angles around C(1) are 126.7° (Si(1)-C(1)-Si(3)), 114.9° (Si(1)-C(1)-Si(2)), and 117.2°, while those around Si(1) vary between 96.5 and 124.1°. The Si-C(methyl) distances are 1.850 and 1.858 Å for

S(1), 1.879, 1.885 and 1.898 Å for Si(2), and 1.907 Å for Si(3). Bonds Si(3)-C(8) and Si(3)-C(12) are 1.939 and 1.927 Å respectively.

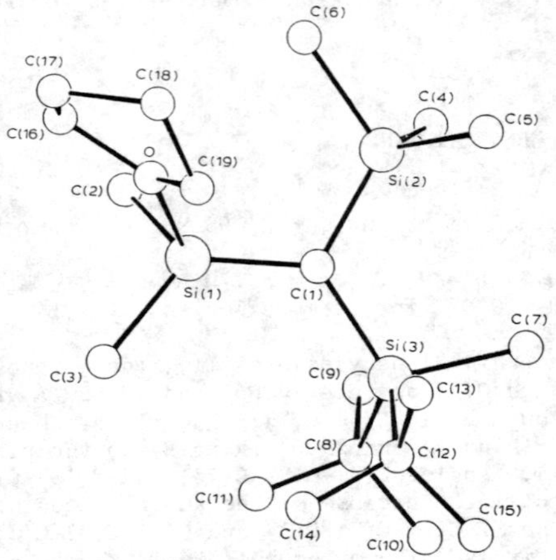

Fig. 1. Molecular structure of $C_{19}H_{44}OSi_3$.

1,1'-DIMETHYLSILANEDIYLBIS(ACETONE PHENYLHYDRAZONE)
$C_{20}H_{28}N_4Si$

W. CLEGG, 1984. Acta Cryst., C40, 529-531.

Monoclinic, C2/c, a = 20.788, b = 7.525, c = 13.200 Å, β = 102.62°, Z = 4. Mo radiation, R = 0.094 for 971 reflexions.

The molecule (Fig. 1) with a crystallographic twofold axis through Si contains essentially single N-N (1.451(2) Å) and double N=C (1.253(7) Å) bonds in the hydrazone fragment. Other dimensions include Si-N(1) 1.755(4), N(1)-C(5) 1.400(7), Si-C(1) 1.853(6) Å, N(1)-Si-N(1) 112.4(2)°.

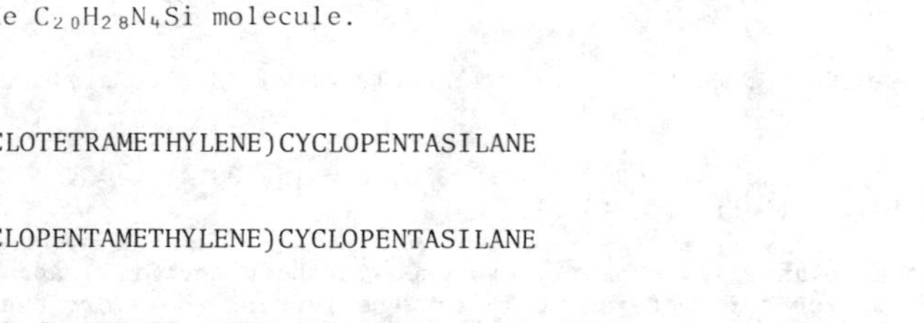

Fig. 1. The $C_{20}H_{28}N_4Si$ molecule.

PENTASPIRO(CYCLOTETRAMETHYLENE)CYCLOPENTASILANE
$C_{20}H_{40}Si_5$ (I)

PENTASPIRO(CYCLOPENTAMETHYLENE)CYCLOPENTASILANE
$C_{25}H_{50}Si_5$ (II)

C.W. CARLSON, K.J. HALLER, XING-HUA ZHANG and R. WEST, 1984. J. Am. Chem. Soc., 106, 5521-5531.

I. Monoclinic, $P2_1/c$, a = 15.246, b = 9.133, c = 18.452 Å, β = 107.59°, D_m = 1.04, Z = 4. Mo radiation, R = 0.035 for 4969 reflexions (at -85°C).

II. Monoclinic, $P2_1/c$, a = 13.310, b = 14.762, c = 14.968 Å, β = 103.39°, D_m = 1.10, Z = 4. Mo radiation, R = 0.029 for 5814 reflexions (at -90°C).

In (I) and (II) the Si_5 rings have a conformation intermediate between envelope and half-chair. The Si-Si-Si angles in (I) average 104.5(20)°; for (II) the corresponding value is 102.5(26)°. Mean Si-Si distances are 2.347(3) Å (I) and 2.359(7) Å (II). The silacyclopentane rings in (I) adopt a variety of conformations; the silacyclohexane rings in (II) are all in chair conformations.

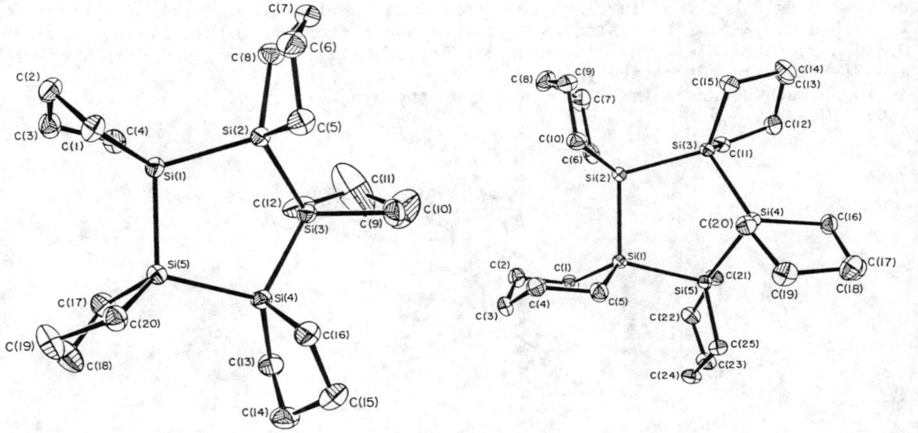

Fig. 1. Views of $C_{20}H_{40}Si_5$ and $C_{25}H_{50}Si_5$.

1,3-DI-t-BUTYL-2,2,4,4-TETRAISOPROPYLCYCLODISILAZANE
$C_{20}H_{46}N_2Si_2$

W. CLEGG, M. HAASE, G.M. SHELDRICK and N. VATER, 1984. Acta Cryst., C40, 871-873.

Monoclinic, $P2_1/n$, a = 8.842, b = 15.842, c = 9.266 Å, β = 115.89°, Z = 2. Mo radiation, R = 0.041 for 1670 reflexions.

The molecule has crystallographic inversion symmetry, with a central planar Si_2N_2 ring of dimensions Si-N 1.747(2) and 1.748(2) Å, Si-N-Si 93.9(1) and N-Si-N 86.1(1)° (Fig. 1). Similar ring geometry is found in other cyclodisilazanes (1).

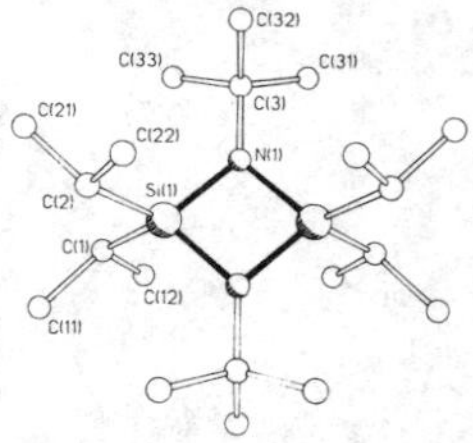

Fig. 1. $C_{20}H_{46}N_2Si_2$: molecular structure.

1. Structure Reports, 46B, 644.

10-(p-METHOXYPHENYL)-10-ETHOXY-10,10-DIHYDRO-10-SILA-9-THIAANTHRACENE
$C_{21}H_{20}O_2SSi$

V.K. BEL'SKII, I.P. YAKOVLEV, Yu. S. FINOGENOV and V.O. REIKHSFEL'D, 1984. Kristallografiya, 29, 613-614 [Sov. Phys. Crystallogr., 29, 365-366].

Monoclinic, I2/a, a = 14.439, b = 31.701, c = 8.421 Å, γ = 100.58°, Z = 8. Mo radiation, R = 0.031 for 1287 reflexions.

The molecule (Fig. 1) is non-planar, a characteristic feature of heterocyclic anthracenes. The phenyl rings A, C and D are individually planar, and the Si atom lies at distances of 0.153, 0.002, and 0.056 Å from these planes respectively. The central ring B has the flattened boat conformation with atoms C(11)-C(14) coplanar to within 0.018 Å. The distances of the Si and S atoms from this plane are 0.380 and 0.339 Å respectively. The distances S-C(14), S-C(12), Si-C(13), Si-C(11), Si-O(1) and Si-C(15) are respectively 1.759, 1.761, 1.855, 1.842, 1.636 and 1.862 Å. The angle between planes A and C (the angle of the bend along the Si...S vector) is 156.7°.

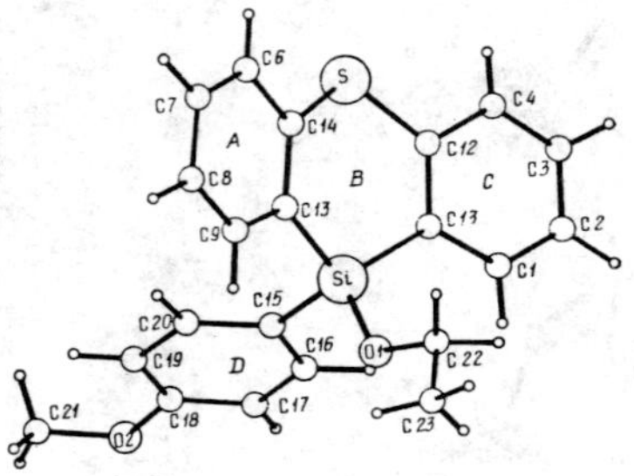

Fig. 1. Conformation of the molecule in $C_{21}H_{20}O_2SSi$.

1,2,4,5-TETRAAZA-3-PHOSPHA-6-SILACYCLOHEXANE
$C_{22}H_{27}N_4OPSSi$

U. ENGELHARDT, T. BUNGER and H. VIERTEL, 1984. J. Cryst. Spectrosc. Res., 14, 603-615.

Monoclinic, $P2_1/c$, a = 14.983, b = 8.267, c = 20.019 Å, β = 103.56°, Z = 4. Cu radiation, R = 0.073 for 1867 reflexions.

The six-membered ring adopts a twist conformation (Fig. 1). The methyl groups adopt positions between pseudo-equatorial and pseudo-axial. Bond distances are as expected with P(1)-S(1) 1.909(5), P(1)-O(1) 1.602(9), P-N 1.62 and 1.66, N-N 1.43(1), Si-N 1.73 and 1.74, Si-C 1.85(1) Å.

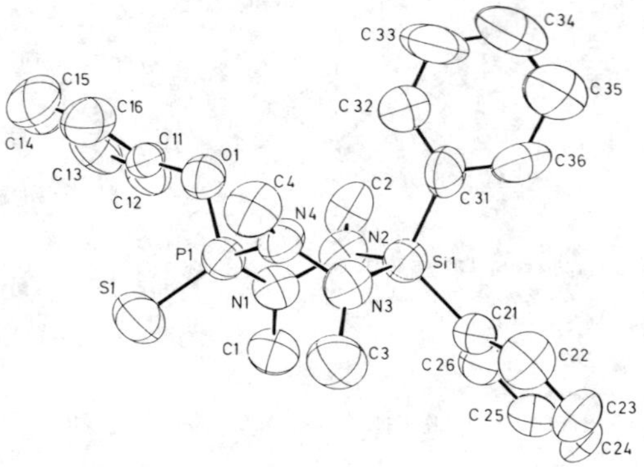

Fig. 1. A view of $C_{22}H_{27}N_4OPSSi$.

(DIMETHYLNITRATOSILYL)(METHYLDIPHENYLSILYL)BIS(TRIMETHYLSILYL)METHANE
$C_{22}H_{37}NO_3Si_4$

C. EABORN, P.B. HITCHCOCK, P.D. LICKISS, A. PIDCOCK AND K.D. SAFA, 1984. J. Chem. Soc. Dalton, 2015-2017

Triclinic, P$\bar{1}$, a = 9.693, b = 9.962, c = 15.094 Å, α = 105.64, β = 92.24, γ = 107.86°, Z = 2. Mo radiation, R = 0.055 for 1287 reflexions.

The molecular structure is shown in Fig. 1. Although the molecule is a highly sterically hindered species the Si-bound O atom of the nitrate group does not appear to be involved in any significant interaction with the Si atom of the $SiMePh_2$ group (O(1)-Si(1)-C(22)-Si(4) 44.5°, O(1)...Si(4) 3.20 Å). The O(1)-Si(1)-C(22) angle is 100.2(3)° whereas the mean of the C(22)-Si-C angles is 113.3°. The C(22)-Si(1) distance is 1.881 Å whereas the mean of the other C(22)-Si distances is 1.929 Å. The mean Si-Me bond distance is 1.861 Å. Distortion in the aromatic rings parallels that in related compounds.

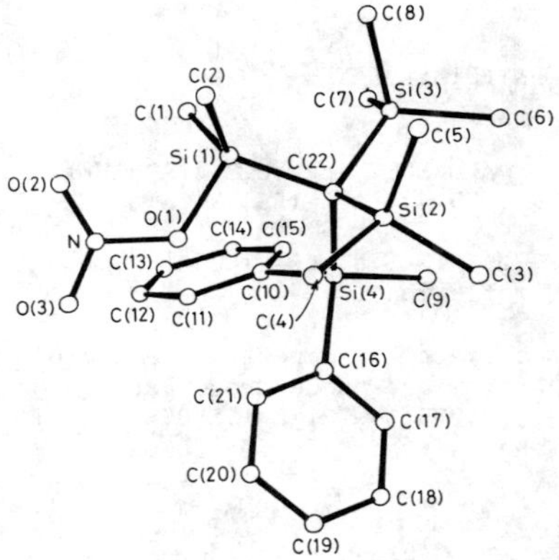

Fig. 1. $C_{22}H_{37}NO_3Si_4$: the molecular structure.

10,10-DIHYDRO-10,10-DIPHENYLPHENOXASILINE
$C_{24}H_{18}OSi$ (I) $O(C_6H_4)_2Si(C_6H_5)_2$

10,10-DIHYDRO-10,10-DIPHENYLPHENOTHIASILINE
$C_{24}H_{18}SSi$ (II) $S(C_6H_4)_2Si(C_6H_5)_2$

V.K. BEL'SKII, A.A. SIMONENKO, I.P. YAKOVLEV and V.O. REIKHSFEL'D, 1984. Kristallografiya, 29, 386-388 [Sov. Phys. Crystallogr., 29, 232-234].

I. Monoclinic, B2/b, a = 11.810, b = 22.380, c = 14.734 Å, γ = 105.1°, Z = 8. Mo radiation, R = 0.030 for 1214 reflexions.

II. Triclinic, P$\bar{1}$, a = 13.190, b = 8.804, c = 9.349 Å, α = 67.10, β = 73.56, γ = 89.99°, Z = 2. Mo radiation, R = 0.030 for 1507 reflexions.

In both I and II the central ring has a boat conformation which is flattened in I (Fig. 1). The conformation of the three-ring system is represented by the dihedral angle between the "halves" divided by the Si...X vector. In I this angle is 165.9° and in II it is 150.4°. The Si-C(exo) bond lengths are 1.868 and 1.864 Å in I, and 1.872 and 1.868 Å in II, while the Si-C(endo) distances are 1.849 and 1.855 Å in I and 1.856 and 1.861 Å in II. The O-C distances in I are 1.390 and 1.381 Å whilst the S-C distances in II are 1.778 and 1.771 Å.

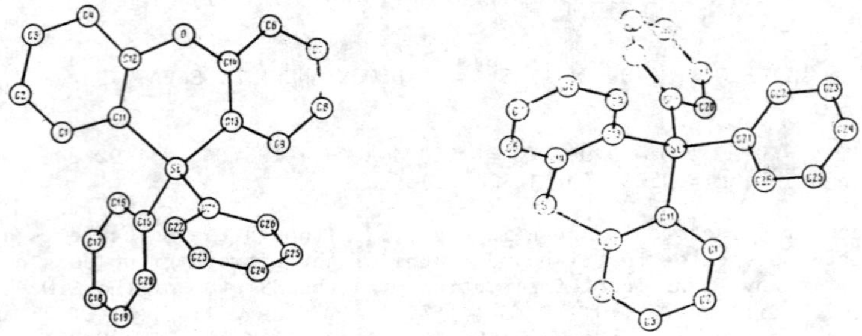

Fig. 1. Conformation of $C_{24}H_{18}OSi$ (left) and of $C_{24}H_{18}SSi$ (right).

1,3-DICHLORO-1,1,3,3-TETRAPHENYLDISILAZANE
$C_{24}H_{21}Cl_2NSi_2$

S.N. GURKOVA, A.I. GUSEV, N.V. ALEKSEEV and Yu.M. VAREZHKIN, 1984. J. Struct. Chem., 25, 655-658 [Zh. Strukt. Khim., 25(4), 153-155].

Monoclinic, C2/c, a = 21.792, b = 10.378, c = 10.546 Å, β = 110.61°, Z = 8. Mo radiation, R = 0.029 for 890 reflexions.

The molecule (Fig. 1) has exact C_2 symmetry, the twofold axis passing through the trigonal planar central N atom and the hydrogen attached to it. Bond lengths include: Si-N 1.702(2), Si-Cl 2.075(1), Si-C 1.850(4) and 1.851(4) Å. The Si atoms have slightly distorted tetrahedral coordination geometry.

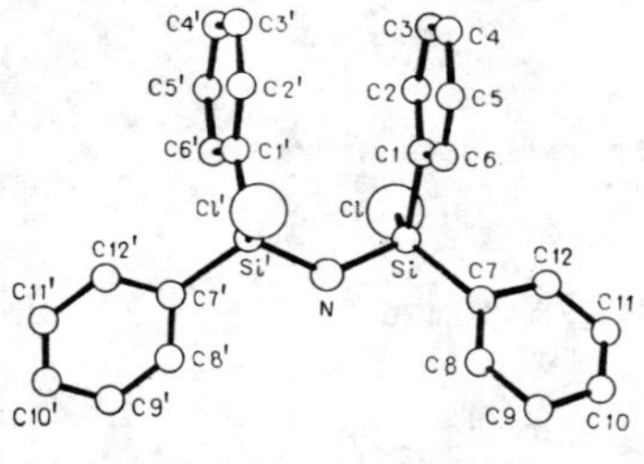

Fig. 1. The 1,3-dichloro-1,1,3,3-tetraphenyldisilazane molecule.

TRIPHENYLTRIVINYLDISILOXANE
$C_{24}H_{24}OSi_2$ $(C_2H_3)_3SiOSi(C_6H_5)_3$

A.I. GUSEV, M.G. LOS, S.D. VLASENKO, V.I. ZHUN and V.D. SHELUDYAKOV, 1984. J. Struct. Chem., 25, 154-156 [Zh. Strukt. Khim., 25(1), 172-175].

Rhombohedral, R$\bar{3}$, a = 10.415 Å, α = 90.23°, Z = 2. Mo radiation, R = 0.037 for 611 reflexions.

The molecule (Fig. 1) has exact threefold symmetry. The Si-O-Si linkage is linear. Structural parameters include: Si(1)-O 1.619(4), Si(2)-O 1.580(4), Si(1)-C(1) 1.852(4), Si(2)-C(7) 1.831(5) Å, O-Si(1)-C 108.5(1)°, O-Si(2)-C 110.3(1)°. The vinylic double bond C(7)=C(8) has a cis orientation relative to the Si(2)-O bond.

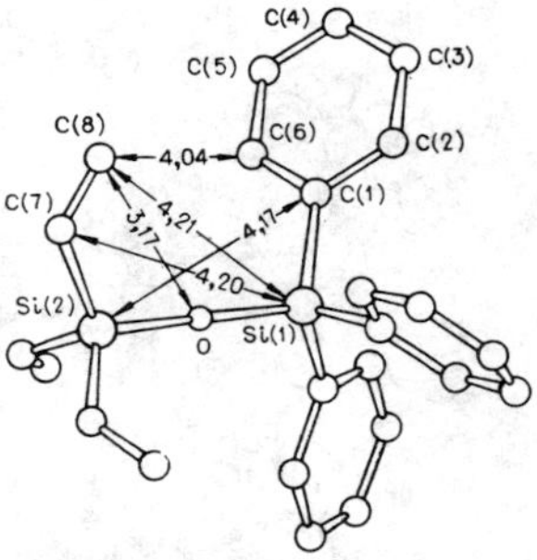

Fig. 1. The triphenyltrivinyldisiloxane molecule, showing important non-bonded distances.

2,2,4,4,6,6-HEXA-t-BUTYLCYCLOTRISILAZANE

$C_{24}H_{55}N_3Si_3$ $[(C_4H_9)_2SiNH]_3$

W. CLEGG, G.M. SHELDRICK and D. STALKE, 1984. Acta Cryst., C40, 433-434.

Rhombohedral, R$\bar{3}$c, a = 10.227, c = 49.051 Å (on hexagonal axes), Z = 6. Mo radiation, R = 0.060 for 736 reflexions.

The Si_3N_3 ring is strictly planar as a consequence of the crystallographic 32 (D_3) symmetry (Fig. 1), with Si-N 1.727(2), Si-C 1.924(3) Å.

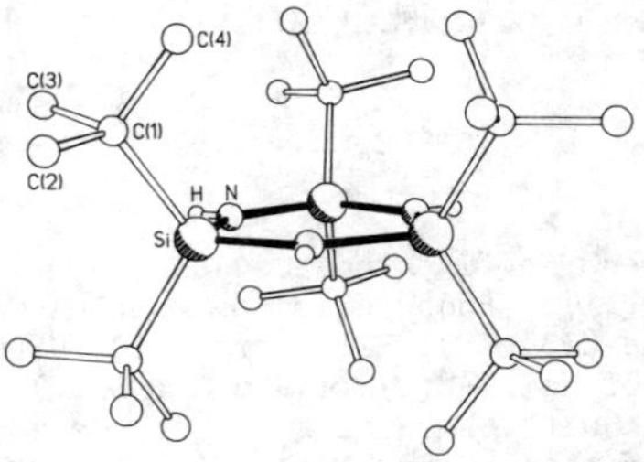

Fig. 1. The structure of $C_{24}H_{55}N_3Si_3$; t-butyl H atoms are omitted.

TRIS(PHENYLDIMETHYLSILYL)METHANE

$C_{25}H_{34}Si_3$ $[C_6H_5(CH_3)_2Si]_3CH$

C. EABORN, P.B. HITCHCOCK and P.D. LICKISS, 1984. J. Organometal. Chem., 269, 235-238.

Trigonal, P$\bar{3}$, a = 15.206, c = 6.625 Å, Z = 2. Mo radiation, R = 0.091 for 2463 reflexions.

The molecule has crystallographically-imposed threefold symmetry (Fig. 1). The central Si-C(1)-Si angles are 113.99° while the mean values of C(1)-Si-C and the other C-Si-C angles are 112.0 and 106.8° respectively. The Si-C(1) bond lengths are 1.895 Å and the remaining Si-C distances average 1.862 Å. There is significant distortion of the phenyl ring - the angle C(5)-C(4)-C(9) is 115.8° and the lengths of the C-C bonds to C(4) are longer (mean 1.392 Å) than those of the other ring C-C bonds (mean 1.363 Å).

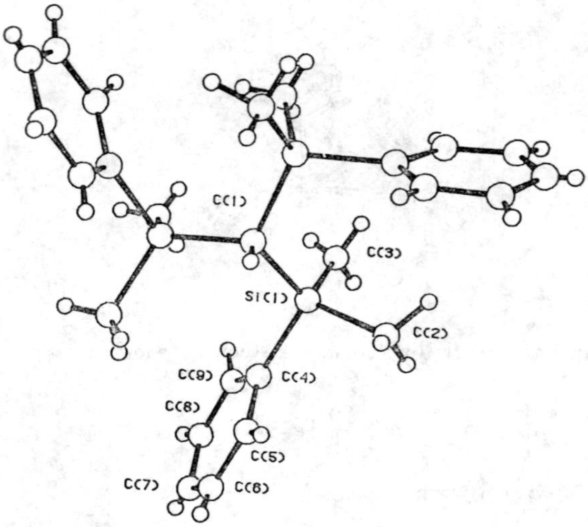

Fig. 1. Molecular structure of $C_{25}H_{34}Si_3$.

3-PHENYL-4-[PHENYL(TRIMETHYLSILYL)METHYLENE]-1,1,2-TRIS(TRIMETHYLSILYL)-1-SILACYCLOBUT-2-ENE
$C_{28}H_{46}Si_5$

M. ISHIKAWA, S. MATSUZAWA, K. HIROTSU, S. KAMITORI and T. HIGUCHI, 1984.
Organometallics, 3, 1930-1932.

Monoclinic, $P2_1/n$, a = 18.336, b = 16.002, c = 11.759 Å, β = 95.34°, Z = 4. R = 0.074 for 2280 reflexions.

The central four-membered ring is almost planar, a maximum deviation from the best plane being 0.01 Å (Fig. 1). Endocyclic bond angles C(2)-Si(1)-C(4), Si(1)-C(2)-C(3), C(2)-C(3)-C(4) and C(3)-C(4)-Si(1) are 74.0, 91.9, 106.6 and 87.5° respectively. Bond lengths of Si(1)-C(2) (1.906 Å), Si(1)-C(4) (1.915 Å), and C(3)-C(4) (1.500 Å) are consistent with those of single bonds while C(2)-C(3) (1.367 Å) is rather longer than that of the C=C double bond. The exocyclic C(4)-C(5) bond (1.334 Å) is in accord with the normal double bond length. The dihedral angles between the plane of the four-membered ring and the best plane through each of the C(10) phenyl and C(16) phenyl rings are 65.1 and 26.4° respectively.

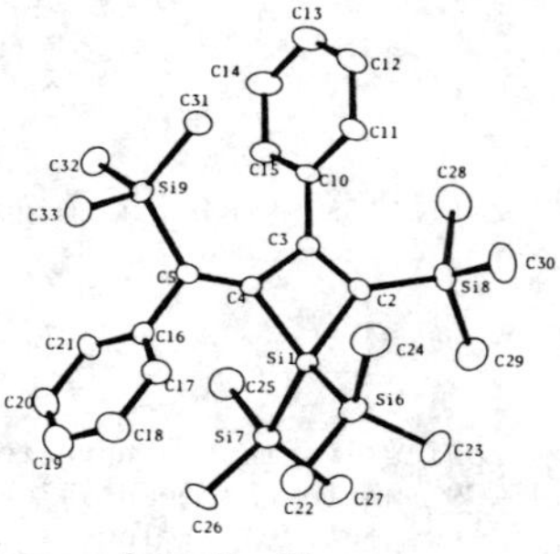

Fig. 1. Molecular structure of $C_{28}H_{46}Si_5$.

1-(t-BUTYLDIMETHYLSILYL)-3-(t-BUTYLMETHYL(METHYL(TRIMETHYLSILYL)AMINO)SILYL)-2-METHYL-2,4,4-TRIS(METHYL(TRIMETHYLSILYL)AMINO)CYCLODISILAZANE
$C_{28}H_{78}N_6Si_8$

W. CLEGG, U. KLINGEBIEL, G.M. SHELDRICK and D. STALKE, 1984. J. Organometal. Chem., 265, 17-25.

Monoclinic, $P2_1/n$, a = 12.458, b = 22.589, c = 16.376 Å, β = 102.33°, Z = 4. Mo radiation, R = 0.070 for 3377 reflexions.

The compound, a cyclodisilazane, consists of a central four-membered Si_2N_2 ring with four different substituents (Fig. 1). The bond distances within the ring are: Si(1)-N(1) 1.744, Si(1)-N(2) 1.756, Si(2)-N(1) 1.746 and Si(2)-N(2) 1.756 Å, and the angles are N(1)-Si(1)-N(2) 89.2, N(1)-Si(2)-N(2) 89.1, Si(1)-N(1)-Si(2) 91.2 and Si(1)-N(2)-Si(2) 90.5°. Elsewhere the Si-N bond distances range from 1.712 Å (Si(1)-N(4)) to 1.760 Å (Si(4)-N(4)) and the Si-C distances from 1.850 Å (Si(4)-C(42)) to 1.909 Å (Si(6)-C(62)).

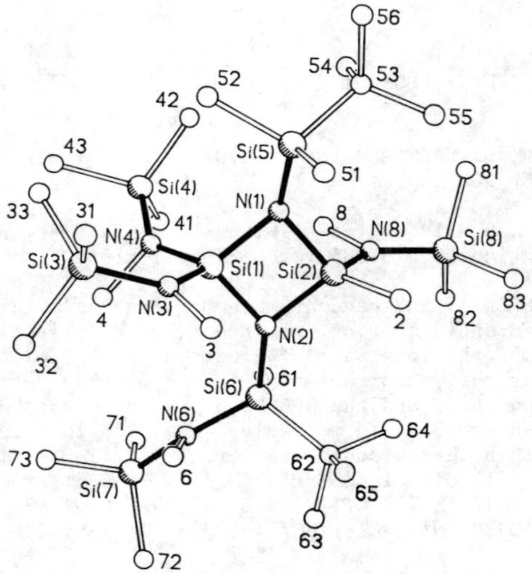

Fig. 1. Molecular structure of $C_{28}H_{78}N_6Si_8$. The Si-N bonds are filled in.

1-(1,10-PHENANTHROLINE)-3,3,3',3'-TETRAKIS(TRIFLUOROMETHYL)-1,1'-SPIROBI(3H-2,1-BENZOXASILOLE) TETRAHYDROFURAN SOLVATE
$C_{30}H_{16}F_{12}N_2O_2Si,2(C_4H_8O)$

W.B. FARNHAM and J.F. WHITNEY, 1984. J. Am. Chem. Soc., 106, 3992-3994.

Monoclinic, $P2_1/n$, a = 14.962, b = 25.322, c = 9.820 Å, β = 101.72°, Z = 4. Mo radiation, R = 0.097 for 5725 reflexions.

The complex (Fig. 1) adopts a slightly distorted octahedral framework, with Si-O1 1.772(7), Si-O2 1.772(7), Si-N1 2.051(9), Si-N2 2.089(9), Si-C21 1.930(10), Si-C27 1.914(10) Å, O-Si-O 91.7(3), N-Si-N 77.9(4), C-Si-C 99.4(4)°.

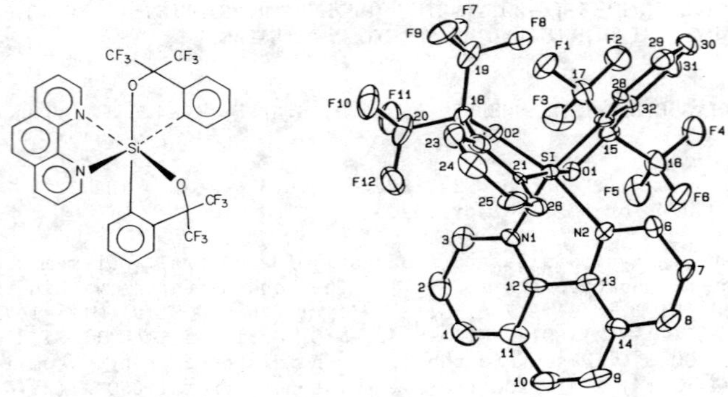

Fig. 1. The $C_{30}H_{16}F_{12}N_2O_2Si$ molecule.

[(TRIPHENYLSILYL)AMINO]DIPHENYLPHOSPHINE

$C_{30}H_{26}NPSi$ $(C_6H_5)_3SiN(H)P(C_6H_5)_2$

H. SCHMIDBAUR, A. SCHIER, S. LAUTESCHLÄGER, J. RIEDE and G. MÜLLER, 1984. Organometallics, 3, 1906-1909.

Triclinic, P$\bar{1}$, a = 9.914, b = 14.418, c = 19.558 Å, α = 110.48, β = 103.54, γ = 99.21°, Z = 4. Mo radiation, R = 0.083 for 5382 reflexions.

Both crystallographically-independent molecules have similar structures (Fig. 1), the most noteworthy differences arising from the relative orientations of the phenyl rings. The orientation of the lone pair of electrons at P (designated E1 and E2 in Fig. 1) was calculated as the vector pointing to the missing tetrahedral corner, and in both molecules the lone pair and the N-H bond are in a trans conformation. The torsional angles Si-N-P-E are 7.5° in molecule I and 5.5° in molecule II. The bond distances and angles within the molecule are normal: P(1)-N(1) 1.703, Si(1)-N(1) 1.737, P(1)-C(4) 1.840, P(1)-C(5) 1.838, P(2)-N(2) 1.705, Si(2)-N(2) 1.737, P(2)-C(9) 1.835, P(2)-C(10) 1.842, Si(1)-C(1) 1.868, Si(2)-C(6) 1.873 Å; P(1)-N(1)-Si(1) 125.5, P(2)-N(2)-Si(2) 123.0, N(1)-Si(1)-C(1) 102.6, N(2)-Si(2)-C(6) 106.3°.

Fig. 1. Perspective views of the independent molecules of $C_{30}H_{26}NPSi$.

HEXANEOPENTYLCYCLOTRISILANE

$C_{30}H_{66}Si_3$ $[(t\text{-}C_4H_9CH_2)_2Si]_3$

H. WATANABE, M. KATO, T. OKAWA, Y. NAGAI and M. GOTO, 1984. J. Organometal. Chem., 271, 225-233.

Orthorhombic, Pnna, a = 35.668, b = 16.963, c = 12.251 Å, Z = 8. Cu radiation, R = 0.069 for 851 reflexions.

There are two independent molecules per asymmetric unit, each with a twofold axis of molecular symmetry (Fig. 1). In both molecules the cyclotrisilane ring forms an isosceles triangle with Si(1) at the apex and the twofold axis passes through Si(1) and the midpoint of the Si(2)-Si(2') bond. The unique Si-Si bond lengths in molecules A and B are respectively 2.367 and 2.378 Å, and those of the two equivalent bonds are 2.386 and 2.414 Å respectively. The bond angles in the rings range from 59.0-60.5°. The Si-C bond lengths in the two molecules range from 1.89 to 1.94 Å.

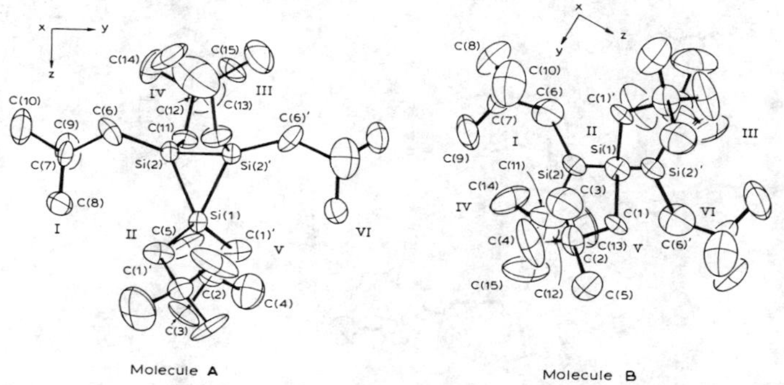

Fig. 1. Molecular structures of the two independent molecules A and B in $C_{30}H_{66}Si_3$.

PYRIDINIUM BIS(2,3-NAPHTHALENEDIOLATO)PHENYLSILICATE(IV)
$C_{31}H_{23}NO_4Si$ (I) $[C_5H_5NH]^+[(C_{10}H_6O_2)_2SiPh]^-$

PYRIDINIUM BIS(1,2-BENZENEDIOLATO)-1-NAPHTHYLSILICATE(IV)
$C_{27}H_{21}NO_2Si$ (II) $[C_5H_5NH]^+[(C_6H_4O_2)_2SiC_{10}H_7]^-$

PYRIDINIUM BIS(1,2-BENZENEDIOLATO)(2- AND 4-CHLOROPHENYL)SILICATE(IV)
$C_{23}H_{18}ClNO_4Si$ (III) $[C_5H_5NH]^+[(C_6H_4O_2)_2SiC_6H_4Cl]^-$ (2-Cl,0.85;5-Cl,0.15)

R.R. HOLMES, R.O. DAY, J.J. HARLAND and J.M. HOLMES, 1984. Organometallics, 3, 347-353.

I. Monoclinic, $P2_1/n$, a = 10.959, b = 20.332, c = 11.448 Å, β = 99.07°, Z = 4. Mo radiation, R = 0.044 for 2695 reflexions.

II. Monoclinic, Cc, a = 16.529, b = 17.603, c = 8.293 Å, β = 111.63°, Z = 4. Mo radiation, R = 0.036 for 1129 reflexions.

III. Monoclinic, $P2_1/n$, a = 9.397, b = 20.335, c = 11.067 Å, β = 102.82°, Z = 4. Mo radiation, R = 0.058 for 1313 reflexions.

The structure of I is nearly an ideal rectangular pyramid, II is displaced about halfway between a rectangular pyramid and trigonal bipyramid, and III is closest to a trigonal bipyramid (Fig. 1). The change in Si-O bond lengths is an indication of change in structure: the axial Si-O(1) and Si-O(1') distances in III are 1.820 and 1.751 Å, the equatorial distances 1.710, 1.698 and 1.888 Å. In I and II, the 'basal' Si-O distances are 1.750, 1.733, 1.734 and 1.759 Å, and 1.734, 1.772, 1.742 and 1.722 Å. The apical Si-O bonds in I and II are 1.871 and 1.875 Å respectively. There are hydrogen bonds between the pyridinium H atoms and O atoms in each of the anions ranging from 1.849 to 2.527 Å.

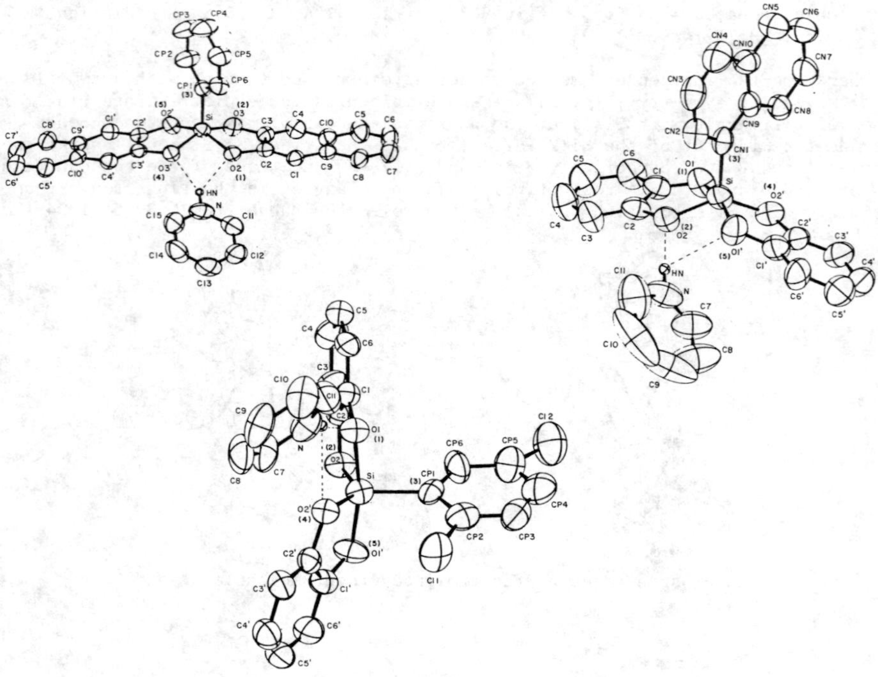

Fig. 1. Molecular structures of $C_{31}H_{23}NO_4Si$ (top left), $C_{27}H_{21}NO_4Si$ (top right), and $C_{23}H_{18}ClNO_4Si$ (bottom).

2,2,4,4,12,12,14,14-OCTAMETHYL-8,8,17,17-TETRAPHENYLDISPIRO[5.3.5.3]OCTAOXA-3,13-DIAZA-2,4,6,8,10,12,14,17-OCTASILAOCTADECANE

$C_{32}H_{46}N_2O_8Si_8$

I.L. DUBCHAK, V.E. SHKLOVER, Yu.T. STRUCHKOV, E.S. KHYNKU and A.A. ZHDANOV, 1984. J. Struct. Chem., 25, 574-578 [Zh. Strukt. Khim. 25(4), 74-78].

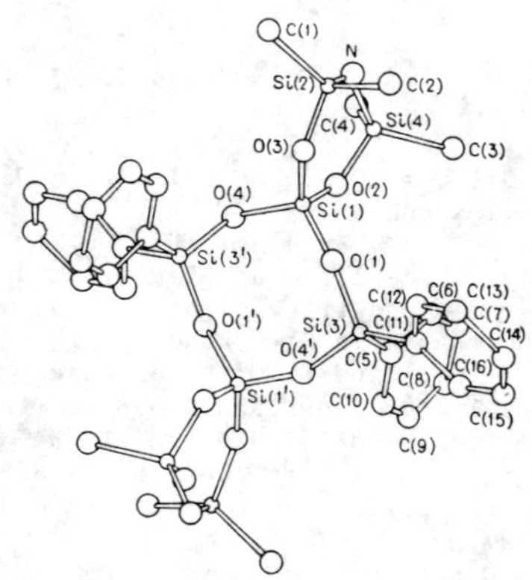

Fig. 1. The 2,2,4,4,12,12,14,14-octamethyl-8,8,17,17-tetraphenyldispiro[5.3.5.3]-octaoxa-3,13-diaza-2,4,6,8,10,12,14,17-octasilaoctadecane molecule.

Triclinic, $P\bar{1}$, a = 9.544, b = 10.925, c = 11.043 Å, α = 98.98, β = 102.67, γ = 68.65°, Z = 1. Mo radiation, R = 0.052 for 3695 reflexions (at -120°C).

The central tetrasiloxane ring of the centrosymmetric molecule (Fig. 1) has a chair conformation, and the six-membered trisilazaoxane rings have the sofa conformation. Bond lengths include: Si-O 1.557(3)-1.662(3), Si-N 1.687(3) and 1.748(3), Si-C 1.833(5)-1.919(4) Å.

1,6-BIS(2',2',6',6'-TETRAMETHYLPIPERIDINO)-5,10-DI-t-BUTYL-3,8-DIMETHYL-4,5,9,10-TETRAAZA-1,6-DISILATRICYCLO[5.3.0.0^{2,6}]DECA-3,8-DIENE
$C_{32}H_{62}N_6Si_2$ (I)

1,9-DI-t-BUTYL-3,7-DIMETHYL-1,2,8,9-TETRAAZA-5-SILASPIRO[4.4]NONA-2,7-DIENE
$C_{14}H_{28}N_4Si$ (II)

O. GRAALMANN, M. HESSE, U. KLINGEBIEL, W. CLEGG, M. HAASE and G.M. SHELDRICK, 1984. Z. Anorg. Chem., 514, 49-60.

I. Orthorhombic, Pbcn, a = 19.825, b = 18.574, c = 9.622 Å, Z = 4. Mo radiation, R = 0.079 for 1761 reflexions.

II. Monoclinic, C2/c, a = 12.928, b = 11.073, c = 12.264 Å, β = 101.39°, Z = 4. Mo radiation, R = 0.055 for 1445 reflexions.

Both compounds have twofold crystallographic symmetry. The central Si_2C_2 ring in I is planar and dimensions in both molecules are in the anticipated ranges. In the SiN_2C_2 ring of II, the N and C atoms are planar to within 0.008 Å and the Si atom is 0.172 Å from this plane.

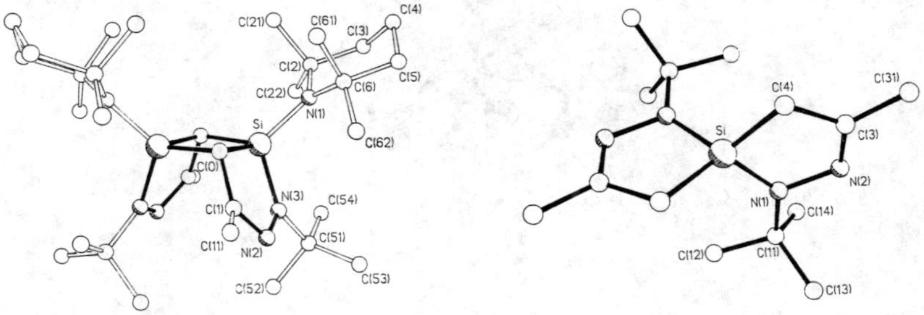

Fig. 1. Views of $C_{32}H_{62}N_6Si_2$ and $C_{14}H_{28}N_4Si$.

1,2,5,6,7,8,9,10-OCTA-t-BUTYL-1,2,5,6,7,8,9,10-OCTAPHOSPHA-3,4-DISILADISPIRO[2.0.4.2]-DECANE
$C_{32}H_{72}P_8Si_2$

K.-F. TEBBE and M. FEHÉR, 1984. Acta Cryst., C40, 1879-1882.

Monoclinic, C2/c, a = 35.296, b = 14.442, c = 18.501 Å, β = 105.37°, D_m = 1.11, Z = 8. Mo radiation, R = 0.047 for 3817 reflexions.

The P_4Si, P_2Si_2 and P_2Si rings are connected almost at right angles by means of the spiro Si atoms (Fig. 1). The endocyclic P-Si-P angles are in the range 59.1-107.5(1)°. Some bond lengths are, Si-Si 2.383(2), Si-P 2.233-2.317, P-P 2.197-2.210(2) Å.

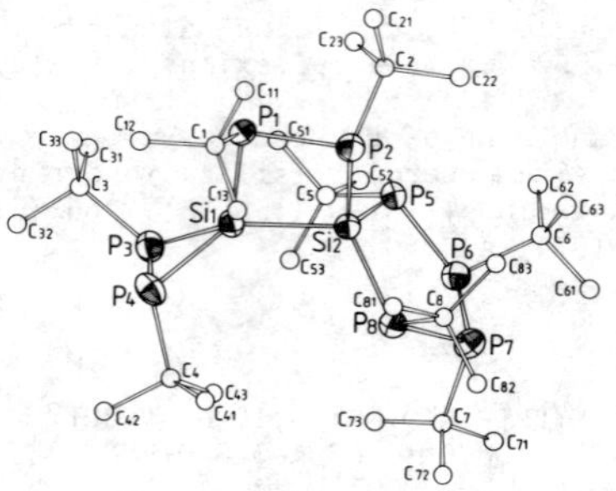

Fig. 1. The $C_{32}H_{72}P_8Si_2$ molecule.

TETRAMESITYLCYCLODISILOXANE TOLUENE SOLVATE
$C_{36}H_{44}O_2Si_2,C_7H_8$

M.J. FINK, K.J. HALLER, R. WEST and J. MICHL, 1984. J. Am. Chem. Soc., 106, 822-823.

Tetragonal, $I4_1/a$, a = 12.212, c = 52.425 Å, Z = 8. Mo radiation, R = 0.079 for 2292 reflexions (at -30°C).

The molecule (Fig. 1) has exact C_2 symmetry. The four-membered ring has a fold angle of 6°. The Si-O-Si angle is 86°, and the Si-Si distance is unusually short at 2.31 Å. Other distances: Si-O 1.66, 1.72 Å, Si-C 1.86, 1.87 Å, and O...O 2.47 Å.

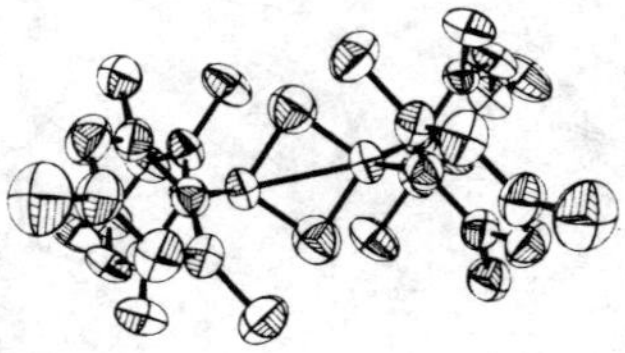

Fig. 1. The $C_{36}H_{44}O_2Si_2$ molecule.

TETRAMESITYLDISILENE
$C_{36}H_{44}Si_2,C_7H_8$ (I) $(C_6H_2(CH_3)_3)_2Si\text{-}Si(C_6H_2(CH_3)_3)_2,C_6H_5CH_3$

trans-1,2-DI-tert-BUTYL-1,2-DIMESITYLDISILENE
$C_{26}H_{40}Si_2$ (II) $Si\{C_6H_2(CH_3)_3(C_4H_9)\}_2Si$

M.J. FINK, M.J. MICHALCZYK, K.J. HALLER, R. WEST and J. MICHL, 1984. Organometallics, 3, 793-800.

I. Tetragonal, $I4_1/a$, a = 12.064, c = 52.506 Å, Z = 8. R = 0.068 for 3718 reflexions (at 188K).

II. Triclinic, $P\bar{1}$, a = 8.968, b = 10.377, c = 7.745 Å, α = 93.79, β = 114.19, γ = 98.63°, Z = 1. R = 0.065 for 2596 reflexions (at 248K).

Compound I has twofold crystallographic symmetry and the four aryl substituents are disposed about the Si-Si framework in roughly helical fashion in which two of the cis-related rings are nearly coplanar with the Si-Si bond axis (Fig. 1). The twist angle (half the difference of the angles subtended by the C(11)-Si-Si' and C(11')-Si-Si' planes (29.5°) and the C(21)-Si-Si and C(21')-Si-Si' planes (16.3°)) is 6.6°. The Si-Si' bond length is short at 2.160 Å and the independent Si-C bond lengths are

1.87 and 1.88 Å. Molecule II has crystallographic inversion symmetry midway between the two Si atoms. The structure is different to that of I in that the Si-C framework in II is planar. The mesityl rings are skewed nearly orthogonal to the disilene framework (dihedral angle is 88°). The Si-Si' bond length is even shorter than that in I (2.143 Å), and the Si-C bond distances to the tert-butyl and mesityl groups are 1.90 and 1.88 Å respectively.

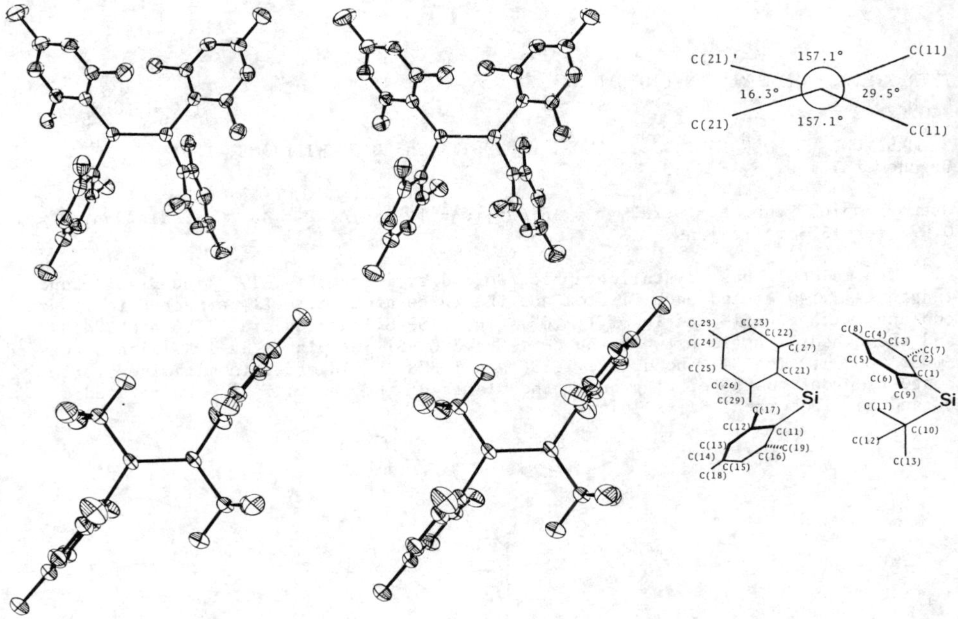

Fig. 1. Stereoview of $C_{36}H_{44}Si_2$ (top left), Newman projection of $C_{36}H_{44}Si_2$ along the Si-Si axis (top right), stereoview of $C_{26}H_{40}Si_2$ (bottom left), and numbering schemes for I and II (bottom right).

1,3-BIS(DIPHENYLHYDROXYSILYL)-2,2-DIMETHYL-4,4-DIPHENYLCYCLODISILAZANE $C_{38}H_{36}N_2O_2Si_4$

CHEN BENMING, XIE ZEMIN and WANG JINTING, 1984. J. Struct. Chem., 3, 113-117.

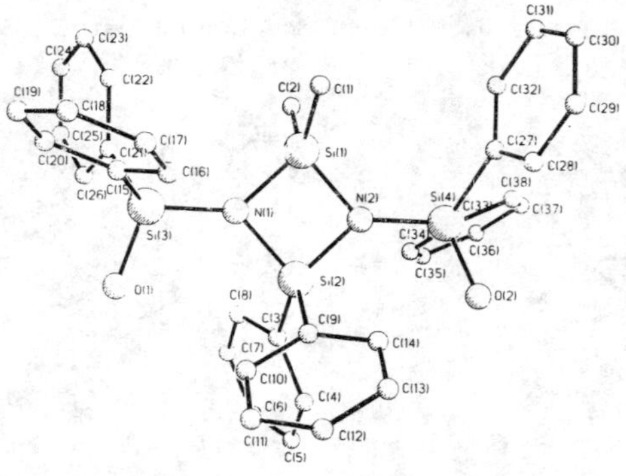

Fig. 1. A perspective view of $C_{38}H_{36}N_2O_2Si_4$.

Monoclinic, $P2_1/n$, a = 11.854, b = 10.136, c = 29.931 Å, β = 93.51°, Z = 4. Mo radiation, R = 0.072 for 3015 reflexions.

The central four-membered ring is nearly planar (deviations ±0.011 Å) and atoms N(1) and N(3) (Fig. 1) have planar geometry. Mean bond lengths are: Si-C(Me) 1.839, Si-C(phenyl) 1.847, Si-O 1.636, Si-N(endo) 1.744, Si-N(exo) 1.711 Å.

TETRAKIS(2,6-DIETHYLPHENYL)DISILENE

$C_{40}H_{52}Si_2$ $[\{C_6H_3(C_2H_5)_2\}_2Si]_2$

S. MASAMUNE, S. MURAKAMI, J.T. SNOW, H. TOBITA and D.J. WILLIAMS, 1984. Organometallics, 3, 333-334.

Orthorhombic, Pcnb, a = 9.787, b = 18.694, c = 19.289 Å, Z = 4. Cu radiation, R = 0.072 for 1575 reflexions.

The molecule has crystallographic twofold symmetry (Fig. 1). The Si=Si' bond length is 2.140 Å, and each Si atom and the three atoms directly attached to it are coplanar with a C-Si-C angle of 117.6°. The C-Si-Si' angles are 117.6 and 124.8°, and the dihedral angle between the C-Si-C and C'-Si'-C' planes is 10° (with a twisted Si=Si' bond). The Si-C bonds are 1.886 and 1.878 Å. In order to minimise steric congestion due to the ethyl groups, the planes of the phenyl rings are rotated by 62 and 64° from the C-Si-Si' plane.

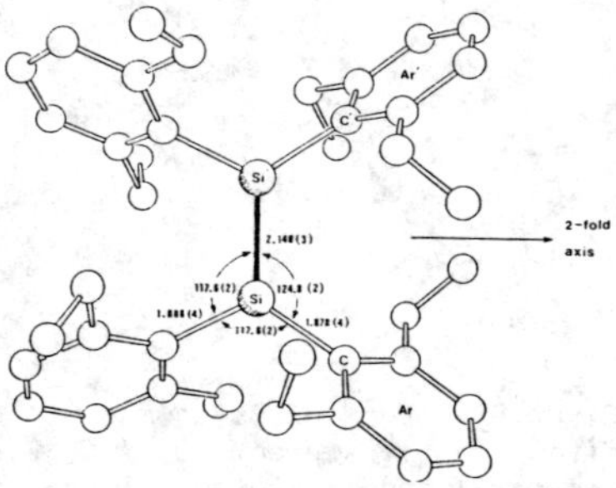

Fig. 1. Molecular structure of $C_{40}H_{52}Si_2$ showing selected bond lengths (Å) and angles (°).

DIFLUORO(METHYLAMINO)PHOSPHINE

CH_4F_2NP PF_2NHCH_3

A.J. BLAKE, E.A.V. EBSWORTH and A.J. WELCH, 1984. Acta Cryst., C40, 1260-1262.

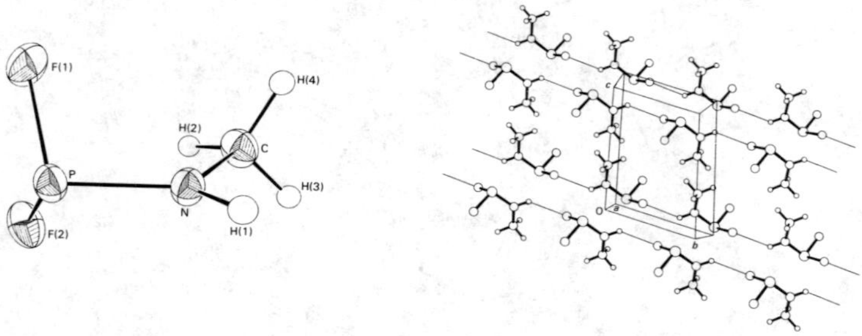

Fig. 1. CH_4F_2NP: perspective view of the molecule and crystal-packing diagram.

Triclinic, $P\bar{1}$, a = 4.9969, b = 5.4618, c = 8.2231 Å, α = 106.435, β = 108.088, γ = 76.780°, Z = 2. Mo radiation, R = 0.0397 for 1055 reflexions (at 97K).

In the molecule (Fig. 1) P has pyramidal geometry, N has planar geometry and there is a short P-N distance of 1.6174(12) Å. The P-F bond lengths are inequivalent, reflecting the involvement of F(1) in hydrogen bonding. The lone pair on P and the N p_z orbital are approximately orthogonal. In the crystal molecules are linked by hydrogen bonding into chains which run alternately parallel and antiparallel to the crystallographic b axis.

GUANIDINOMETHYLPHOSPHONIC ACID MONOHYDRATE
$C_2H_8N_3O_3P,H_2O$

W. SAWKA-DOBROWOLSKA, T. GLOWIAK and R. TYKA, 1984. Acta Cryst., C40, 174-176.

Monoclinic, C2/c, a = 13.222, b = 7.502, c = 13.936 Å, β = 104.17°, D_m = 1.70, Z = 8. Mo radiation, R = 0.042 for 1581 reflexions.

The molecule (Fig. 1) is a zwitterion; the C-N-C-P torsion angle is -95.9(3)°. Some bond lengths are: P-O(1) 1.579(2), P-O(2) 1.498(2), P-O(3) 1.507(2), N(1)-C(1) 1.455(3), N(1)-C(2) 1.323(3), C(2)-N(2) 1.341(3), C(2)-N(3) 1.328(3), P-C(1) 1.820(2) Å. Molecules are linked by N-H...O and O-H...O hydrogen bonds.

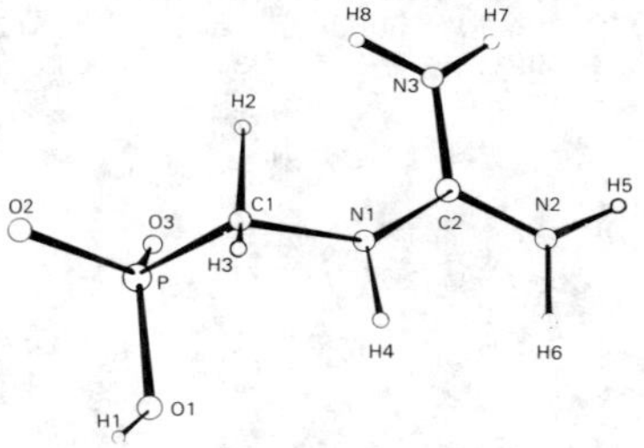

Fig. 1. A view of $C_2H_8N_3O_3P$.

DIMETHYLAMINOMETHANEDIPHOSPHONIC ACID
$C_3H_{11}NO_6P_2$

S. KULPE, I. SEIDEL and K. SZULZEWSKY, 1984. Crystal Res. & Technol., 19, 669-674.

Monoclinic, $P2_1/c$, a = 6.870, b = 10.824, c = 11.337 Å, β = 94.59°, Z = 4. Mo radiation, R = 0.033 for 2648 reflexions.

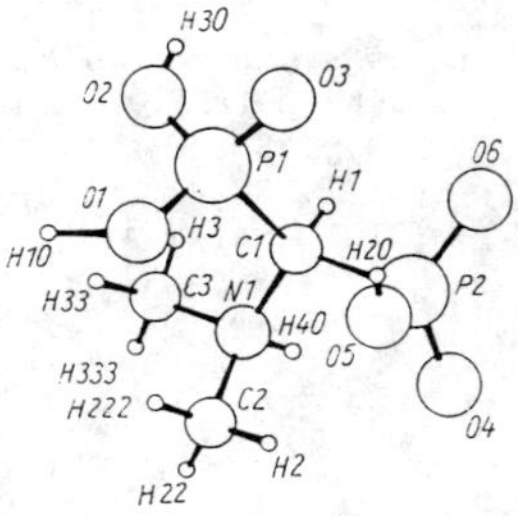

Fig. 1. The $C_3H_{11}NO_6P_2$ molecule.

1. H. Gross, B. Costisella, Th. Gnauck and L. Brennecke, 1976. J. Prakt. Chem., 318, 116.

This structure determination shows differences between it and an earlier determination (1). The N atom is shown to be bonded to C(1), C(2) and C(3), and also an H atom (H(40)) (Fig. 1). Two of the P-OH bonds are single (P(1)-O(2) 1.557(1), P(2)-O(5) 1.556(1) Å), two P=O bonds are found (P(1)-O(3) 1.495(1), P(2)-O(4) 1.481(1) Å) and two partially double-bonded P=O bonds are present (P(1)-O(1) 1.509(1), P(2)-O(6) 1.514(1) Å). Symmetrical O-H-O intermolecular hydrogen bonds exist in the crystal structure (O(1)--H(10)--O(6'), O...O 2.415 Å). Other bond lengths are: P(1)-C(1) 1.840(1), P(2)-C(1) 1.845(1), N(1)-C(1) 1.517(2), N(1)-C(2) 1.501(2), N(1)-C(3) 1.500(2) Å.

cis-3,6-DIETHOXY-1,2,4,5-TETRAAZA-3,6-DIPHOSPHACYCLOHEXANE-3,6-DISULPHIDE
$C_4H_{14}N_4O_2P_2S_2$

H. VIERTEL and U. ENGELHARDT, 1984. Acta Cryst., C40, 125-127.

Tetragonal, $P4_2/n$, a = 11.506, c = 9.278 Å, Z = 4. Cu radiation, R = 0.044 for 1014 reflexions.

The molecule has crystallographic twofold symmetry. The saturated six-membered ring (Fig. 1) has an almost ideal twist conformation with interplanar angles around ring bonds of +63.5(3) (PNNP), -32.1(3) and -28.7(3)° (NPNN). The PNN bond angles range from 116.4(3) to 118.5(3)°, the NPN bond angle is 102.6(2)°. Mean bond distances are P-N 1.643(4), N-N 1.409(5), P-S 1.925(2), P-O 1.579(3), O-C 1.470(6) Å.

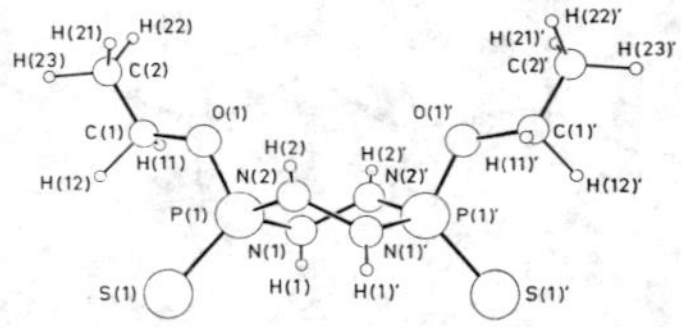

Fig. 1. A view of the $C_4H_{14}N_4O_2P_2S_2$ molecule.

2-DIMETHOXYPHOSPHORYL-1,3,5-TRITHIANE
$C_5H_{11}O_3PS_3$

M. MIKOLAJCZYK, P. BALCZEWSKI, K. WROBLEWSKI, J. KAROLAK-WOJCIECHOWSKA, A. MILLER, M.W. WIECZOREK, M.YU. ANTIPIN and YU.T. STRUCHKOV, 1984. Tetrahedron, 40, 4885-4892.

Monoclinic, $P2_1/c$, a = 5.5692, b = 16.4152, c = 11.6423 Å, β = 100.553°, Z = 4. Mo radiation, R = 0.035 for 1854 reflexions.

The six-membered ring (Fig. 1) has a chair conformation, slightly flattened at C1. The dimethoxyphosphoryl group is shown to be axial. Main bond lengths are P-O(1) 1.465(2), P-O(2) 1.572(2), P-O(3) 1.565(2), S(1)-C(1) 1.806(3), S(1)-C(2) 1.809(3), S(2)-C(2) 1.794(4), S(2)-C(3) 1.788(4), S(3)-C(1) 1.813(3), S(3)-C(3) 1.823(4), C(4)-O(2) 1.446(4), C(5)-O(3) 1.454(4) Å.

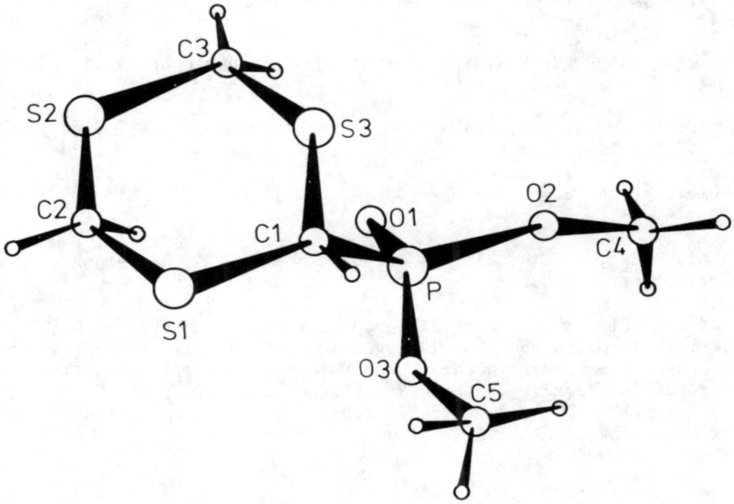

Fig. 1. The conformation of $C_5H_{11}O_3PS_3$ in the crystal.

N-(DICHLOROPHOSPHORYL)ANILINE
$C_6H_6Cl_2NOP$ (I)

3,7-DIPHENYL-2,2,4,4,6,6,8,8-OCTACHLORO-1,5-DIOXA-3,7-DIAZA-2,6-DIPHOSPHA-4,8-DIBORA-CYCLO-OCTA-1,5-DIENE
$C_{12}H_{10}B_2Cl_8N_2O_2P_2$ (II)

H. NOTH and W. STORCH, 1984. Chem. Ber., 117, 2140-2156.

I. Triclinic, P$\bar{1}$, a = 5.939, b = 8.873, c = 9.252 Å, α = 112.32, β = 93.94, γ = 101.73°, Z = 2. Mo radiation, R = 0.064 for 1539 reflexions.

II. Monoclinic, C2/c, a = 22.279, b = 6.294, c = 16.793 Å, β = 109.03°, Z = 4. Mo radiation, R = 0.034 for 1753 reflexions.

Both molecules are shown in Fig. 1. In I, the principal dimensions are: P-Cl 1.992 and 1.998(6), P-O 1.456(5), P-N 1.608(6), N-C 1.421(7) Å. Molecule II occurs as a centrosymmetric dimer, with the eight-membered ring having a chair-like conformation and P-Cl 1.965 and 1.979(1), P-O 1.499(2), P-N 1.598(2), N-C 1.474(3), N-B 1.560(4), B-Cl 1.823(4) and 1.844(3), B-O 1.503(4) Å.

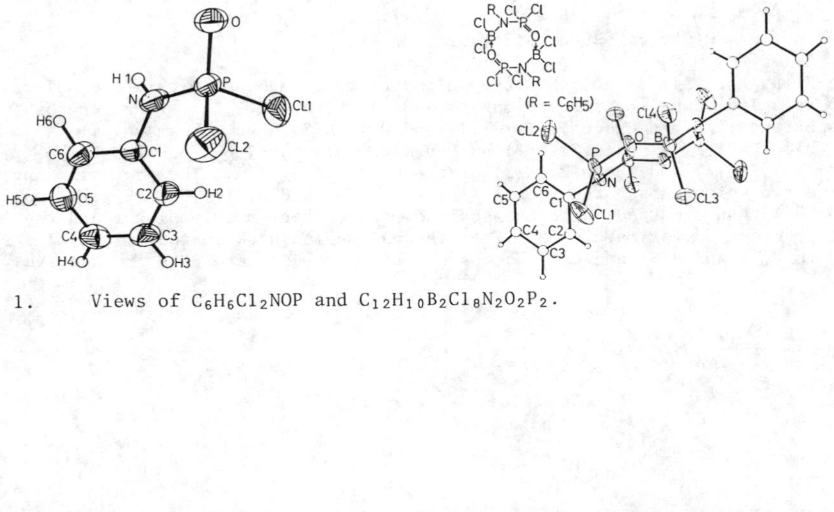

Fig. 1. Views of $C_6H_6Cl_2NOP$ and $C_{12}H_{10}B_2Cl_8N_2O_2P_2$.

4,5-DIMETHYL-2-(1,2,4-TRIAZOL-1-YL)-2-OXO-1,3,2-DIOXAPHOSPHOLE
$C_6H_8N_3O_3P$ (I)

4,5-DIMETHYL-2-(TETRAMETHYLGUANIDYL)-2-OXO-1,3,2-DIOXAPHOSPHOLE
$C_9H_{18}N_3O_3P$ (II)

F. RAMIREZ, J.S. RICCI Jr., H. OKAZAKI, J.F. MARECEK and M. LEWY, 1984. Phosphorus and Sulfur, 20, 279-299.

I. Orthorhombic, Pnma, a = 7.703, b = 9.074, c = 12.830 Å, D_m = 1.49, Z = 4. Cu radiation, R = 0.043.

II. Monoclinic, C2/c, a = 14.621, b = 7.246, c = 24.054 Å, β = 91.62°, D_m = 1.27, Z = 8. Cu radiation, R = 0.047.

Compound (I) (Fig. 1) has crystallographic mirror symmetry, with the methine carbon C(4) vicinal to the phosphoryl oxygen (C...O 3.038(2) Å); in (II) there is no required molecular symmetry. The dioxaphosphole rings in (I) and (II) are planar and have nearly identical geometry. The P-N distances are relatively short (1.680(2) Å (I), 1.591(2) Å (II)). The exocyclic P=O distances are 1.44 Å (I) and 1.46 Å (II). The P-O ring bond lengths are 1.58-1.61 Å.

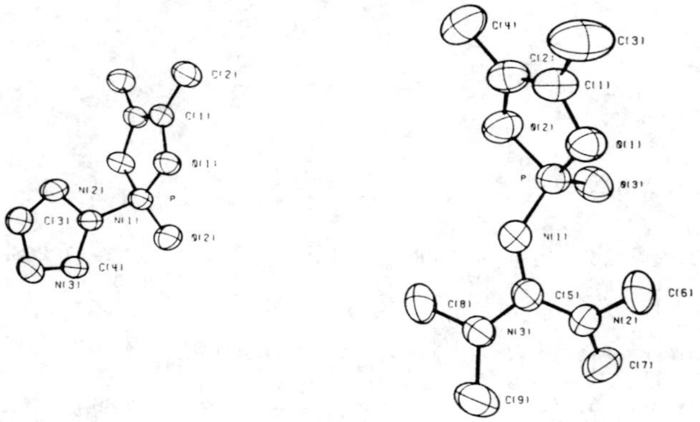

Fig. 1. Views of $C_6H_8N_3O_3P$ and $C_9H_{18}N_3O_3P$.

6,6-DICHLORO-2,2,4,4-DISPIRO(N-METHYLETHANOLAMINO)CYCLOTRIPHOSPHAZATRIENE
$C_6H_{14}Cl_2N_5O_2P_3$

V. CHANDRASEKHAR, S.S. KRISHNAMURTHY, H. MANOHAR, A.R.V. MURTHY, R.A. SHAW and M. WOODS, 1984. J. Chem. Soc. Dalton, 621-625.

Monoclinic, $P2_1/c$, a = 12.717, b = 9.677, c = 11.847 Å, β = 89.46°, D_m = 1.44, Z = 4. Mo radiation, R = 0.056 for 2968 reflexions.

The X-ray analysis established the spirocyclic arrangement within the molecule and also revealed the cis structure for this derivative (Fig. 1). Bond lengths in the phosphazene ring are unequal P(1)-N(1) and P(1)-N(3) have a mean value of 1.565 Å while P(2)-N(1) and P(3)-N(3) have a mean value of 1.602 Å. The P(3)-N(2) distance is 1.593(4) and the P(2)-N(2) distance is 1.565(4) Å. The exocyclic angles around the two spiro phosphorus atoms, O(1)-P(2)-N(4) and O(2)-P(3)-N(5) are smaller (95.7, 96.2°) than expected. The phosphazene ring is very nearly planar and the spirocyclic rings approximately so. The dihedral angles between the spirocyclic rings attached to P(2) and P(3) and the phosphazene ring are 89.1 and 89.9° respectively.

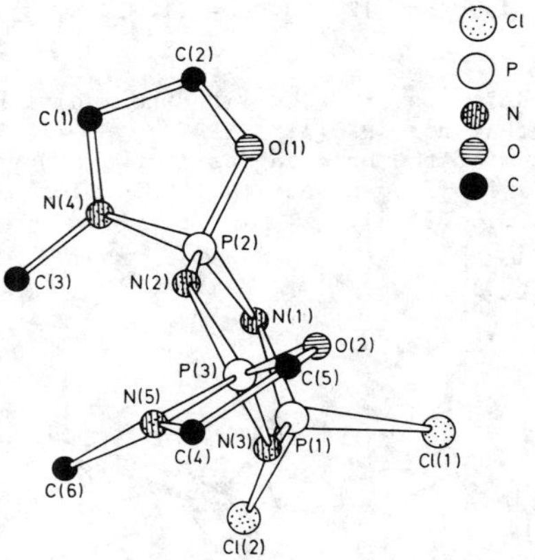

Fig. 1. $C_6H_{14}Cl_2N_5O_2P_3$: the molecular structure viewed down a.

2-DIMETHYLAMINO-5,5-DIMETHYL-1,3,2-DIOXAPHOSPHORINANE
$C_7H_{16}NO_2P$ (I)

2-PIPERIDYL-5,5-DIMETHYL-1,3,2-DIOXAPHOSPHORINANE
$C_{10}H_{20}NO_2P$ (II)

D.E. SCHIFF, J.W. RICHARDSON, Jr., R.A. JACOBSON, A.H. COWLEY, J. LASCH and J.G. VERKADE, 1984. Inorg. Chem., 23, 3373-3377.

I. Triclinic, $P\bar{1}$, a = 6.542, b = 11.731, c = 6.440 Å, α = 101.53, β = 83.52, γ = 90.96°, Z = 2. Mo radiation, R = 0.074 for 1220 reflexions (at -130°C).

II. Orthorhombic, Pbna, a = 8.464, b = 9.868, c = 27.832 Å, Z = 8. Mo radiation, R = 0.084 for 560 reflexions.

The six-membered C_3PO_2 rings in both structures have chair conformations with the R_2N groups in equatorial positions (Fig. 1). The stereochemistry about N is planar, this plane containing the lone pair on P. The P-N distances of 1.642(5) Å in I and 1.637(12) Å in II are indicative of N-Pπ bonding. The P-O distances are 1.640(4) and 1.652(4) Å in I, 1.675(16) and 1.623(18) Å in II.

Fig. 1. The molecules $C_7H_{16}NO_2P$ (left) and $C_{10}H_{20}NO_2P$ (right).

2-(DIMETHYLAMINO)-2-OXO-5,5-DIMETHYL-1,3,2-OXAZAPHOSPHORINANE
$C_7H_{17}N_2O_2P$

R.R. HOLMES, R.O. DAY, W.N. SETZER, A.E. SOPCHIK and W.G. BENTRUDE, 1984. J. Am. Chem. Soc., 106, 2353-2358.

Orthorhombic, Pbca, a = 7.409, b = 12.180, c = 21.912 Å, Z = 8. Mo radiation, R = 0.061 for 1055 reflexions.

The molecule has a chair conformation with the 2-dimethylamino substituent axial (Fig. 1). Some bond lengths are: P-O(1) 1.584(4), P-O(2) 1.465(4), P-N(3) 1.626(5), P-N(2) 1.644(5) Å. The sum of the bond angles at the exocyclic Me_2N group is 349.3° showing considerable deviation from planarity. Molecules are linked by N-H...O hydrogen bonds (HN3...O2' 2.03(7) Å).

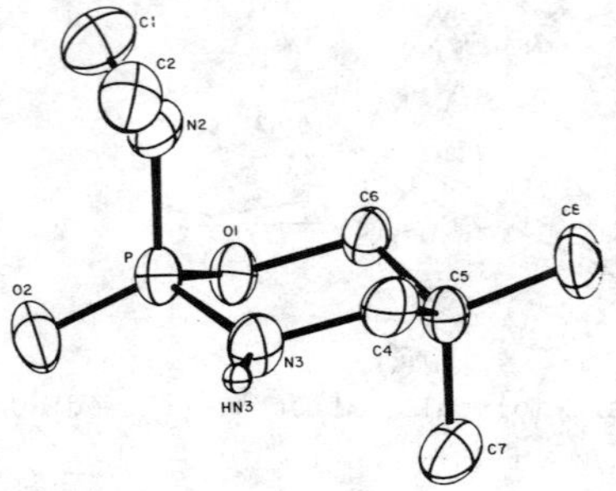

Fig. 1. A view of $C_7H_{17}N_2O_2P$.

cis-2-t-BUTYLAMINO-4-METHYL-1,3,2-DIOXAPHOSPHINANE-2-OXIDE
$C_8H_{18}NO_3P$

R.S. EDMUNDSON and T.J. KING, 1984. J. Chem. Soc. Perkin I, 1943-1947.

Monoclinic, $P2_1/n$, a = 11.304, b = 9.702, c = 11.386 Å, β = 116.12°, D_m = 1.17, Z = 4. Mo radiation, R = 0.097 for 823 reflexions. [Crystals poor in quality.]

The molecule (Fig. 1) possesses cis-geometry with the methyl and t-butylamino groups oriented equatorially. Bond lengths involving the P atom are: P(1)-O(1) 1.562(9), P(1)-O(2) 1.447(7), P(1)-O(3) 1.556(8), P(1)-N(1) 1.591(8) Å.

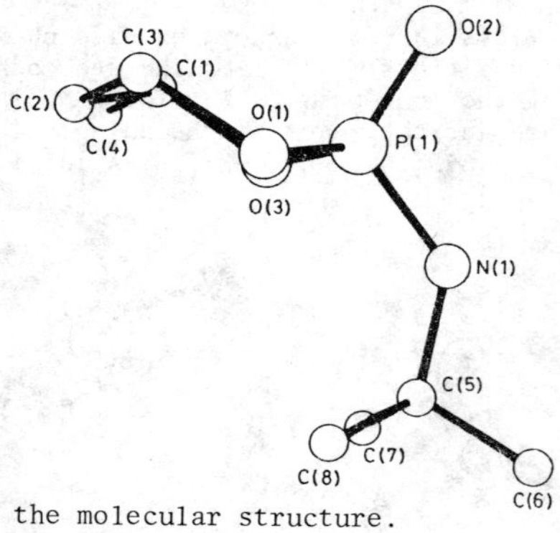

Fig. 1. $C_8H_{18}NO_3P$: the molecular structure.

BIS(DIETHYLDITHIOPHOSPHATO)SELENIUM(II)
$C_8H_{20}O_4P_2S_4Se$ $[Se(SP(S)(OC_2H_5)_2)_2]$

L.S. REFAAT, K. MAARTMANN-MOE and S. HUSEBYE, 1984. Acta Chem. Scand., A38, 303-307.

Triclinic, $P\bar{1}$, a = 8.364, b = 10.755, c = 12.378 Å, α = 63.28, β = 70.78, γ = 83.93°, D_m = 1.530, Z = 2. Mo radiation, R = 0.032 for 3623 reflexions (at -135°C).

Each Se atom is surrounded by four S atoms and the SeS_4 coordination sphere is roughly trapezoid planar with angles of S(1)-Se-S(2) 104.05, S(1')-Se-S(2') 92.92, S(1)-Se-S(1') 82.04 and S(2)-Se-S(2') 83.74° (Fig. 1). The bonds Se-S(1) and Se-S(2) are strong (2.209 and 2.210 Å respectively) but the remaining two are weak intermolecular bonds involving S atoms from two neighbouring molecules (Se-S(1'') 3.342, Se-S(2') 3.523 Å). There is a P-S and P=S bond in each ligand. For P-S, the values are P(1)-S(1) 2.097 and P(2)-S(2) 2.094 Å, and for P=S they are P(1)-S(3) 1.912 and P(2)-S(4) 1.918 Å. The P-O distances are P(1)-O(11) 1.565, P(1)-O(12) 1.580, P(2)-O(21) 1.574 and P(2)-O(22) 1.564 Å.

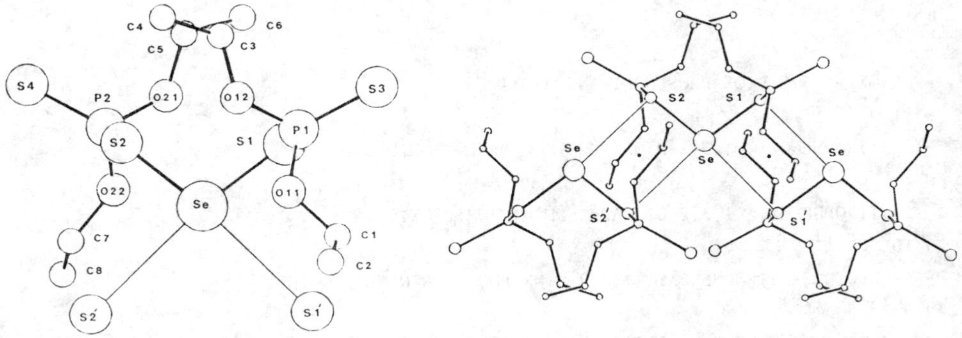

Fig. 1. The molecule of $C_8H_{20}O_4P_2S_4Se$ as seen along the normal to the Se,S(1),S(2) plane (left), and intermolecular interaction between one molecule and its two nearest neighbours (right).

1,1-BIS(TRIFLUOROMETHYL)-3,5-NORBORNADIENYLCYCLO-1-PHOSPHA-3,5-DITHIA-2,4,6-TRIAZENE $C_9H_8F_6N_3PS_2$ (I)

1,1,3,3-TETRAKIS(TRIFLUOROMETHYL)CYCLO-1λ^5,3λ^5-DIPHOSPHA-5,7-DITHIA-2,4,6,8-TETRAZENE $C_4F_{12}N_4P_2S_2$ (II)

H.W. ROESKY, J. LUCAS, M. NOLTEMEYER and G.M. SHELDRICK, 1984. Chem. Ber., 117, 1583-1590.

I. Monoclinic, $P2_1/c$, a = 11.392, b = 10.827, c = 12.112 Å, β = 115.60°, Z = 4. Mo radiation, R = 0.0815 for 1145 reflexions.

II. Triclinic, $P\bar{1}$, a = 7.328, b = 9.078, c = 12.602 Å, α = 69.72, β = 89.13, γ = 109.66°, Z = 2. Mo radiation, R = 0.0911 for 1451 reflexions.

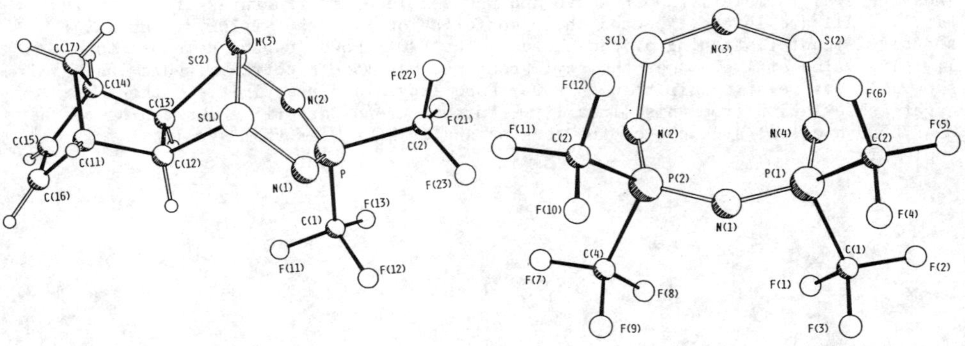

Fig. 1. Views of $C_9H_8F_6N_3PS_2$ (left) and $C_4F_{12}N_4P_2S_2$ (right).

Cycloaddition product I is shown in Fig. 1. Principal bond lengths include S(1)-N(1) 1.649(6), S(1)-N(3) 1.641(9), S(1)-C(12) 1.830(11), S(2)-N(2) 1.656(6), S(2)-N(3) 1.638(9), S(2)-C(13) 1.834(13), N(1)-P 1.574(7), N(2)-P 1.585(7), P-C(1) 1.853(11), P-C(2) 1.884(11), C(12)-C(13) 1.540(13) Å. Compound II has an eight-membered ring which has a folded conformation (there is a 48.5° angle between the normals to the planes through the N(2)-S(1)-N(3)-S(2)-N(4) and N(4)-P(1)-N(1)-P(2)-N(2) portions of the molecule). Main bond lengths are: P(1)-N(1) 1.550(7), P(1)-N(4) 1.583(6), P(2)-N(1) 1.568(7), P(2)-N(2) 1.575(6), S(1)-N(2) 1.607(8), S(1)-N(3) 1.557(9), S(2)-N(3) 1.568(8), S(2)-N(4) 1.603(8) Å.

2-PHENOXY-1,3,2-DIOXAPHOSPHORINANE 2-OXIDE
$C_9H_{11}O_4P$ (I)

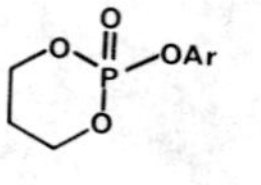

2-(3-NITROPHENOXY)-1,3,2-DIOXAPHOSPHORINANE 2-OXIDE
$C_9H_{10}NO_6P$ (II)

2-(2-NITROPHENOXY)-1,3,2-DIOXAPHOSPHORINANE 2-OXIDE
$C_9H_{10}NO_6P$ (III)

2-(4-CHLORO-2-NITROPHENOXY)-1,3,2-DIOXAPHOSPHORINANE 2-OXIDE
$C_9H_9ClNO_6P$ (IV)

2-(2,4-DINITROPHENOXY)-1,3,2-DIOXAPHOSPHORINANE 2-OXIDE
$C_9H_9N_2O_8P$ (V)

P.G. JONES, G.M. SHELDRICK, A.J. KIRBY and A.J. BRIGGS, 1984. Acta Cryst., C40, 1061-1065.

I. Orthorhombic, $Pna2_1$, a = 15.453, b = 11.005, c = 5.925 Å, Z = 4. Mo radiation, R = 0.045 for 2146 reflexions.

II. Orthorhombic, $P2_12_12_1$, a = 5.710, b = 9.042, c = 21.409 Å, Z = 4. Mo radiation, R = 0.037 for 2434 reflexions.

III. Triclinic, $P\bar{1}$, a = 7.394, b = 7.970, c = 9.966 Å, α = 86.68, β = 70.73, γ = 85.06°, Z = 2. Mo radiation, R = 0.047 for 2250 reflexions.

IV. Triclinic, $P\bar{1}$, a = 6.226, b = 9.586, c = 10.148 Å, α = 93.77, β = 92.51, γ = 100.96°, Z = 2. Mo radiation, R = 0.046 for 2110 reflexions.

V. Triclinic, $P\bar{1}$, a = 5.547, b = 10.382, c = 10.696 Å, α = 87.01, β = 89.63, γ = 79.90°, Z = 2. Mo radiation, R = 0.044 for 2425 reflexions.

The five compounds (I - V) are a homogeneous series of phosphate triesters whose general formula is shown above and whose molecular structures are shown in Fig. 1. All are in the expected chair conformation with the aryloxy group axial and significant flattening at phosphorus. The P-OAr bond length increases to a limiting value of 1.600 Å as the aryl group becomes more electron-withdrawing. I-IV show a linear relationship between P-OAr bond length and reactivity; V shows a negative deviation from this linearity. Further trends are (i) a shortening of the O-Ar bond and (ii) a change of the torsion angle about O-Ar from synclinal towards coplanar.

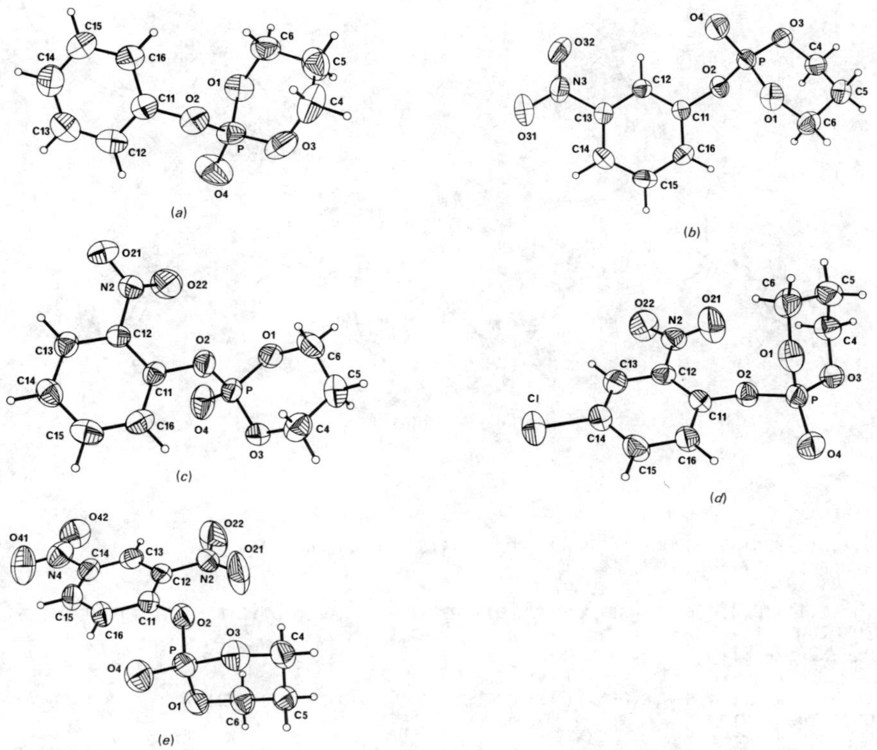

Fig. 1. Perspective views of (a) $C_9H_{11}O_4P$ (I), (b) $C_9H_{10}NO_6P$ (II), (c) $C_9H_{10}NO_6P$ (III), (d) $C_9H_9ClNO_6P$ (IV) and (e) $C_9H_9N_2O_8P$ (V).

2,4,6-TRIS(1,3-PROPYLENEDIAMINE-N,N')CYCLOTRIPHOSPHAZENE DIHYDRATE
$C_9H_{24}N_9P_3,2(H_2O)$

N. EL MURR, R. LAHANA, J.-F. LABARRE and J.-P. DECLERCQ, 1984. J. Mol. Struct., 117, 73-85.

Triclinic, $P\bar{1}$, a = 14.645, b = 11.518, c = 11.381 Å, α = 86.56, β = 103.58, γ = 103.92°, Z = 4. R = 0.094 for three-dimensional data.

There are two independent molecules which differ only in the conformation of one of the spiro loops (Fig. 1); these loops are not arranged in a propeller-like fashion. P-N bond lengths in the phosphazene ring are in the range 1.592-1.611(6) Å. The water molecules are essential for crystal stability and take part in a hydrogen-bonded network.

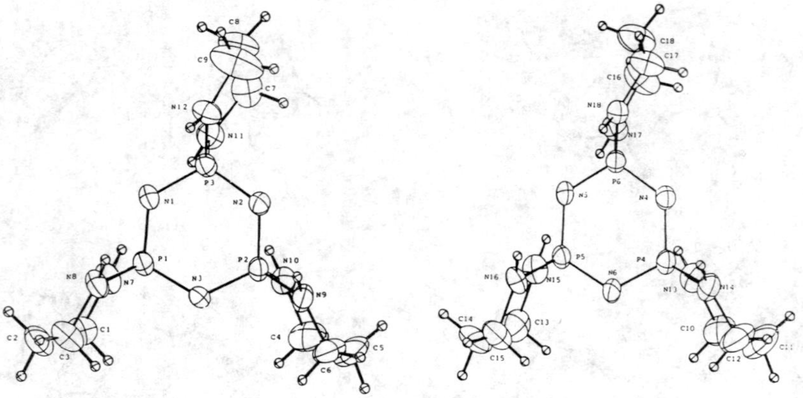

Fig. 1. Views of the two $C_9H_{24}N_9P_3$ molecules.

(1,3,3,5,5-PENTAFLUOROCYCLOTRIPHOSPHAZEN-1-YL)FERROCENE
$C_{10}H_9F_5FeN_3O_3$ (I)

(η^5,η^5-1,1'-(2,4,4,6-TETRAFLUOROCYCLOTRIPHOSPHAZENE-2,6-DIYL)BIS(CYCLOPENTADIENYL))-RUTHENIUM
$C_{10}H_8F_4N_3P_3Ru$ (II)

(η^5,η^5-1,1'-(2,4,4,6,8,8-HEXAFLUOROCYCLOTETRAPHOSPHAZENE-2,6-DIYL)BIS(CYCLOPENTA-DIENYL))RUTHENIUM
$C_{10}H_8F_6N_4P_4Ru$ (III)

(η^5,η^5-2,2'-(4,4,4',4',6,6,6',6'-OCTACHLOROBICYCLOTRIPHOSPHAZEN-2-YL)BIS(CYCLOPENTA-DIENYL))IRON
$C_{10}H_8Cl_8FeN_6P_6$ (IV)

H.R. ALLCOCK, K.D. LAVIN, G.H. RIDING, P.R. SUSZKO and R.R. WHITTLE, 1984. J. Am. Chem. Soc., 106, 2337-2347.

I. Monoclinic, C2/c, a = 19.946, b = 6.922, c = 22.340 Å, β = 111.19°, Z = 8. Mo radiation, R = 0.052 for 1938 reflexions.

II. Triclinic, P$\bar{1}$, a = 8.029, b = 8.286, c = 11.555 Å, α = 87.16, β = 87.91, γ = 61.83°, Z = 2. Mo radiation, R = 0.065 for 1958 reflexions.

III. Orthorhombic, Pbca, a = 15.534, b = 14.610, c = 14.430 Å, Z = 8. Mo radiation, R = 0.041 for 2005 reflexions.

IV. Monoclinic, $P2_1/n$, a = 11.064, b = 13.019, c = 18.196 Å, β = 102.55°, Z = 4. Mo radiation, R = 0.036 for 2590 reflexions.

In (I) the ferrocenyl group is oriented over one face of the phosphazene ring (Fig. 1) with a dihedral angle of 46.8° between organic and inorganic rings. The phosphazene ring in II is markedly distorted to accommodate the constraints imposed by the metallocene. The angle between cyclopentadienyl and phosphazene ring planes is nearly 90°. The mean P-C distance is 1.77(1) Å. The ruthenocyl unit of III is linked through P-C bonds (1.738(6) Å) to the 1,5-phosphorus atoms of the cyclotetraphosphazene ring which is in a boat conformation. Molecule IV has two non-planar phosphazene rings linked by a P-P bond (2.219(1) Å). The mean P-C distance is 1.762(6) Å. The dihedral angles between the cyclopentadienyl ring planes for I-IV are 1.4°, 5°, 4°, 5° respectively.

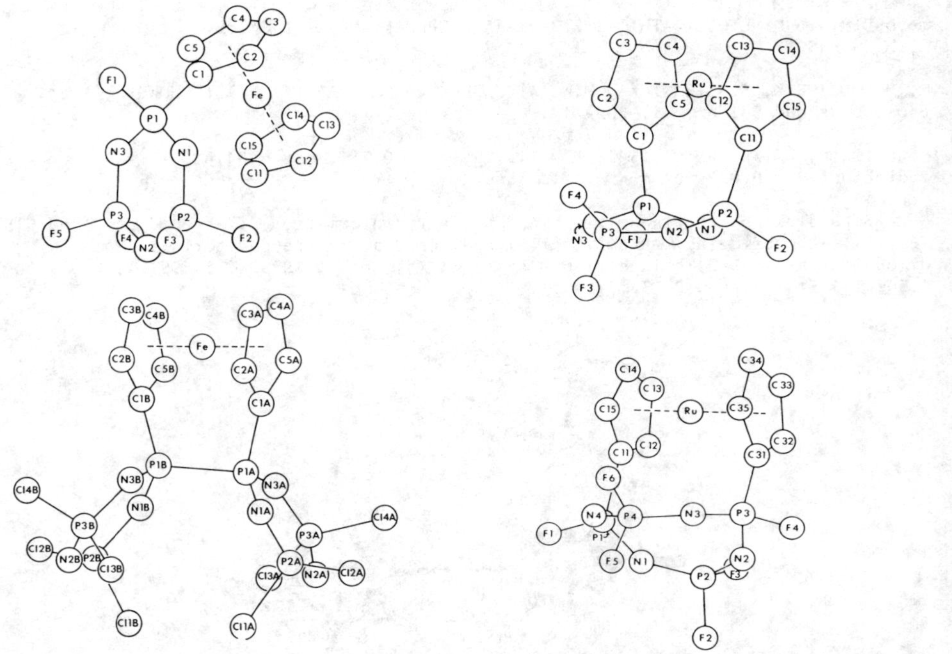

Fig. 1. Views of the metallocene cyclophosphazene structures I-IV (clockwise from top left).

p-DIMETHYLAMINOPHENYL-P,P-DIMETHYLPHOSPHINE SELENIDE

$C_{10}H_{16}NPSe$ $p\text{-}Me_2NC_6H_4P(Se)Me_2$

V.V. TKACHEV, L.O. ATOVMYAN, A.A. SHVETS and O.A. OSIPOV, 1984. J. Struct. Chem., 25, 142-144 [Zh. Strukt. Khim., 25(1) 161-162].

Orthorhombic, $P2_12_12_1$, a = 19.485, b = 7.910, c = 7.695 Å, Z = 4. Mo radiation, R = 0.076 for 870 reflexions.

The molecule, along with bond distances, is shown in Fig. 1. The angle between the planes of the phenyl ring and the C(1)PSe atoms is 20.5°.

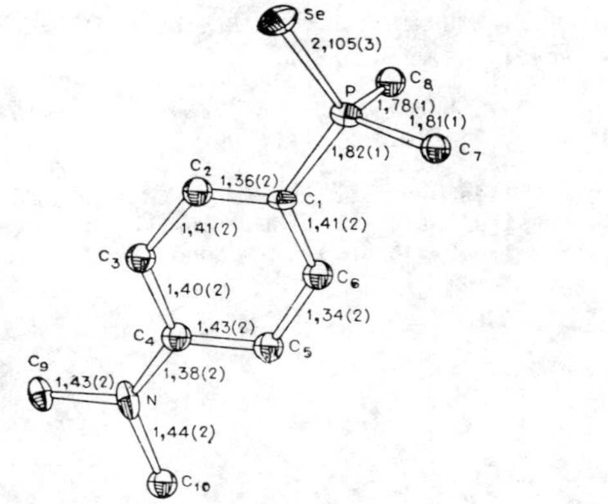

Fig. 1. The p-dimethylaminophenyl-P,P-dimethylphosphine selenide molecule.

meso-BIS(4,6-DIMETHYL-2-THIO-1,3,2-DIOXAPHOSPHORINANE-2-YL)OXIDE
$C_{10}H_{20}O_5P_2S_2$

M. MIKOLAJCZYK, B. ZIEMNICKA, M.W. WIECZOREK and J. KAROLAK-WOJCIECHOWSKA, 1984. Phosphorus and Sulfur, 21, 205-214.

Monoclinic, $P2_1/n$, a = 13.542, b = 11.175, c = 10.731 Å, β = 94.81°, Z = 4. Cu radiation, R = 0.072 for 1649 reflexions.

Both dioxaphosphorinane rings adopt chair conformation, flattened at P and C(5) (Fig. 1); the bridging O atom is axial and the S atoms are equatorial. Main dimensions are: P-O (bridge) 1.609 and 1.617(4), P-S 1.898 and 1.895(2), P-O (ring) 1.561-1.565(4) Å, P-O-P 131.7(2)°.

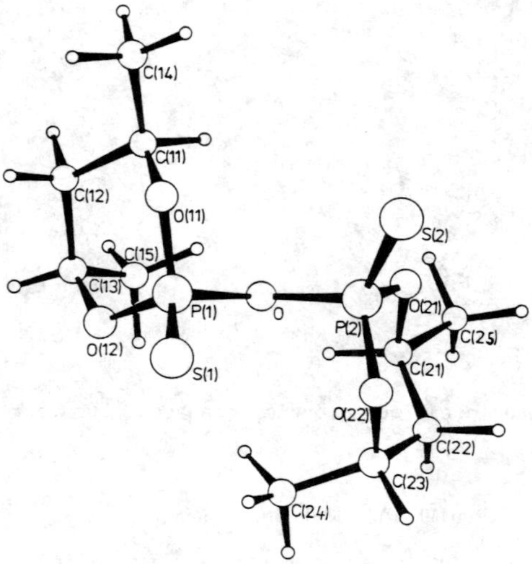

Fig. 1. A view of $C_{10}H_{20}O_5P_2S_2$.

(μ_2-TETRAMETHYLENE)-BIS[2-(1,3-PROPYLENEDIAMINE-N,N')-4,4,6,6-TETRACHLOROCYCLOTRIPHOSPHAZENE]
$C_{10}H_{22}Cl_8N_{10}P_6$

J.-F. LABARRE, G. GUERCH, F. SOURNIES, R. LAHANA, R. ENJALBERT and J. GALY, 1984. J. Mol. Struct., 116, 75-88.

Monoclinic, $P2_1/c$, a = 10.397, b = 7.869, c = 17.898 Å, β = 102.1°, D_m = 1.75, Z = 2. Mo radiation, R = 0.042 for 2814 reflexions.

The molecule (Fig. 1) lies about an inversion centre. The cyclophosphazene six-membered rings are quasi-planar; the other ring has a slight chair conformation. The N-P-N angles are 113.0-120.5(2)° while P-N bonds range from 1.554(3) to 1.634(3) Å.

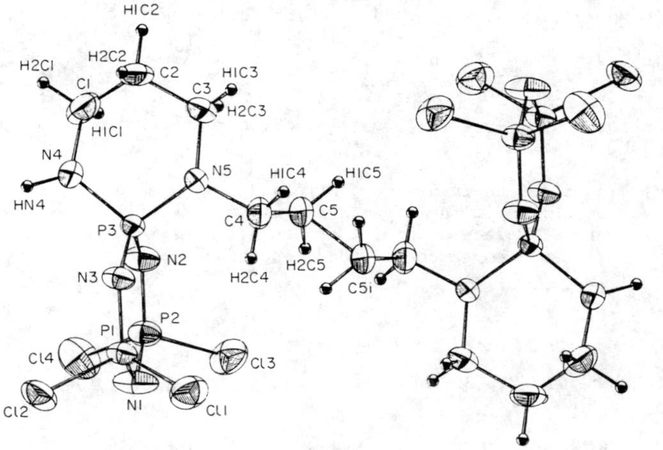

Fig. 1. A view of $C_{10}H_{22}Cl_8N_{10}P_6$.

DIMETHYLDI-t-BUTYLDIPHOSPHANE DISULPHIDE (meso form)
$C_{10}H_{24}P_2S_2$ (I)

DIMETHYLDI-t-BUTYLDIPHOSPHANE DISULPHIDE (racemic form)
$C_{10}H_{24}P_2S_2$ (II)

BIS(METHYL-t-BUTYLPHOSPHINE SULPHIDE)SULPHANE
$C_{10}H_{24}P_2S_3$ (III)

H. WUNDERLICH and H.-G. WUSSOW, 1984. Z. Naturforsch., B39, 1581-1585.

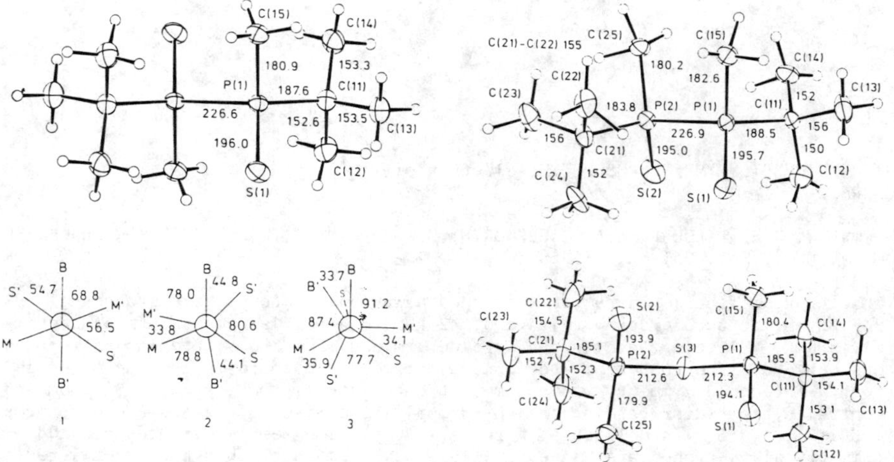

Fig. 1. Views of the disulphides I-III (clockwise from top left) and Newman projections along the P-P bonds. Bond lengths are in pm.

I. Triclinic, $P\bar{1}$, a = 6.765, b = 7.788, c = 8.060 Å, α = 73.39, β = 64.94, γ = 80.54°, Z = 1. Mo radiation, R = 0.033 for 2151 reflexions.

II. Monoclinic, $P2_1$, a = 6.582, b = 14.663, c = 8.463 Å, β = 110.27°, Z = 2. Mo radiation, R = 0.070 for 973 reflexions.

III. Orthorhombic, Pbca, a = 12.297, b = 13.432, c = 20.069 Å, Z = 8. Mo radiation, R = 0.049 for 2607 reflexions.

Molecule I lies about an inversion centre and is the meso form (Fig. 1). Molecule II is an enantiomeric form from a racemic mixture (with S atoms and methyls gauche t-butyls trans). Molecules of III are enantiomers in a racemic crystal (with S atoms and t-butyls close to gauche). Bond lengths are in Fig. 1. The P-S-P angle in III is 104.78(5)°.

(R,S,S,R)-CYCLAMPHOSPHANE-BIS(BORANE)

$C_{10}H_{27}B_2N_4P$ $C_{10}H_{21}N_4P.2BH_3$

J.-M. DUPART, A. GRAND, S. PACE and J.G. RIESS, 1984. Inorg. Chem., 23, 3776-3779.

Monoclinic, $P2_1/c$, a = 12.538, b = 9.835, c = 13.263 Å, β = 114.19°, Z = 4. Cu radiation, R = 0.066 for 1639 reflexions.

The molecule (Fig. 1) has an "open tautomeric form" structure. The BH_3 groups are coordinated to the P and N atoms of this tetracyclic tetraaminophosphorane and are oriented trans to one another. Important bond lengths: P-B 1.858(5), N-B 1.619(6), P-N 1.637(4), 1.653(4), and 1.686(3) Å.

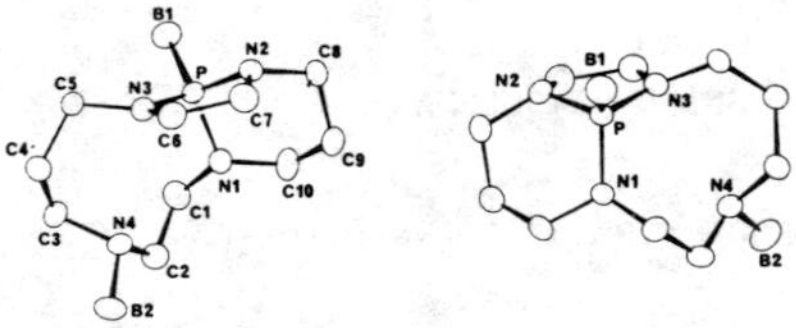

Fig. 1. Two views of $C_{10}H_{27}B_2N_4P$.

TRIS(TRIMETHYLSILYL)METHYLPHOSPHINE (electron diffraction)

$C_{10}H_{29}PSi_3$ $(Me_3Si)_3CPH_2$

A.H. COWLEY, J.E. KILDUFF, E.A.V. EBSWORTH, D.W.H. RANKIN, H.E. ROBERTSON and R. SEIP, 1984. J. Chem. Soc. Dalton, 689-693.

Electron diffraction data were recorded using nozzle-to-plate distances of 497.81 and 247.85 mm, an accelerating voltage of 42 kV and a nozzle temperature of approximately 390 K. Electron wavelengths were determined from the scattering patterns of gaseous benzene. The number of independent parameters used to define the molecule was fourteen. Threefold symmetry was assumed at each methyl group, at each Si atom and at the central C atom (disregarding the PH_2 group). The Si-C, Si-Me and P-C distances are 1.941(5), 1.883(2) and 1.808(9) Å respectively. The Me-Si-Me angles are 104.3(4), the trimethylsilyl groups are tilted 7.3(15) away from each other and are twisted 21.2(4)° away from the fully staggered conformation (Fig. 1).

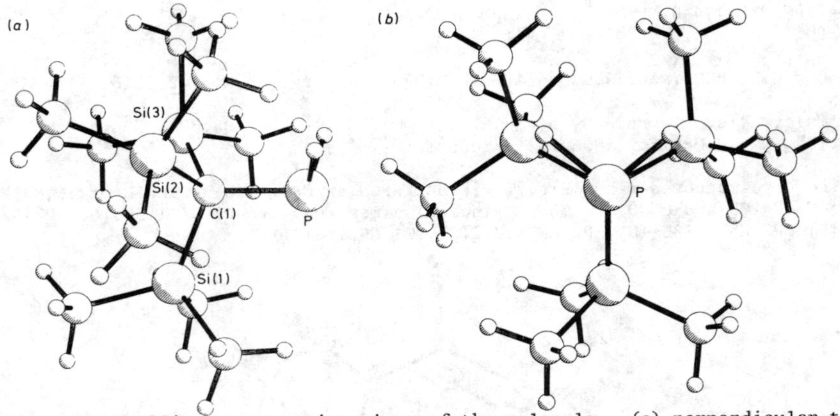

Fig. 1. $C_{10}H_{29}PSi_3$: perspective views of the molecule. (a) perpendicular to the C(1)-P bond, (b) viewed from P to the central C atom.

7-OXO-5-PHENYL-5λ⁵-PHOSPHA-1,4,6-TRIOXASPIRO[4.4]NONANE
$C_{11}H_{13}O_4P$

T. HIGUCHI and M. FUKUYO, 1984. Makromol. Chem., 185, 821-827.

Monoclinic, $P2_1/n$, a = 14.751, b = 5.577, c = 14.490 Å, β = 106.91°, Z = 4. Mo radiation, R = 0.068 for 1385 reflexions.

The geometry around the P atom is slightly distorted trigonal bipyramidal (Fig. 1). The two five-membered rings A and B are in apical-equatorial skeletal positions and the phenyl ring is in an equatorial position. The P atom deviates from the basal plane C(1),O(2),C(9) toward the O(3) atom by 0.084 Å. Both five-membered rings have envelope conformations.

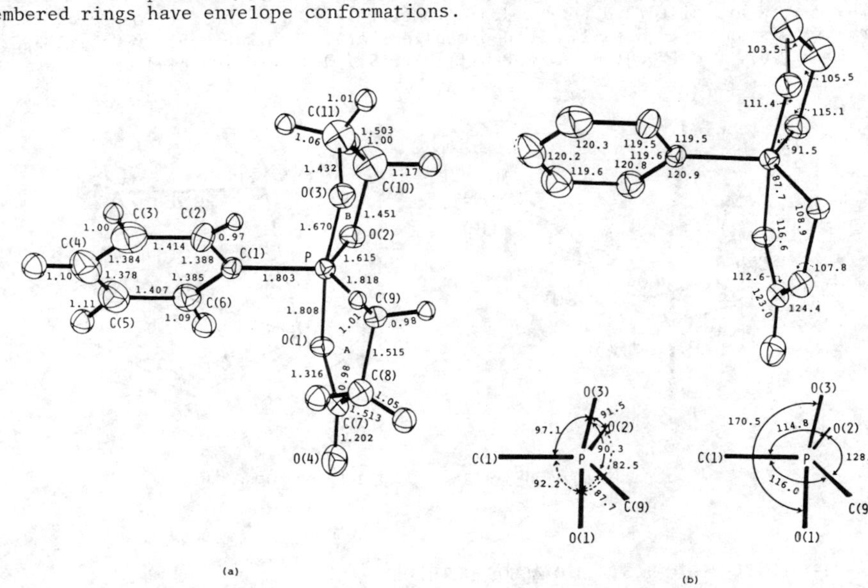

Fig. 1. Perspective view of the molecule $C_{11}H_{13}O_4P$ showing (a) bond lengths (Å), and (b) bond angles.

2-OXO-2-PHENYL-4,4-DIMETHYL-1,2-OXAPHOSPHOLANE
$C_{11}H_{15}PO_2$

G. SINGH, G.S. REDDY and J.C. CALABRESE, 1984. J. Org. Chem., 49, 5132-5136.

Monoclinic, C2/c, a = 26.688, b = 7.612, c = 11.305 Å, β = 98.88°, Z = 8. Mo radiation, R = 0.088 for 1096 reflexions (at -70°C).

The five-membered ring (Fig. 1) is puckered at C4, with an angle of 36.8° between planes C3-C4-C5 and C5-O1-P2-C3. Principal dimensions are P2-O1 1.599(6), P2-O2 1.472(5), P2-C3 1.785(10), P2-C8 1.772(7), O1-C5 1.411(10).

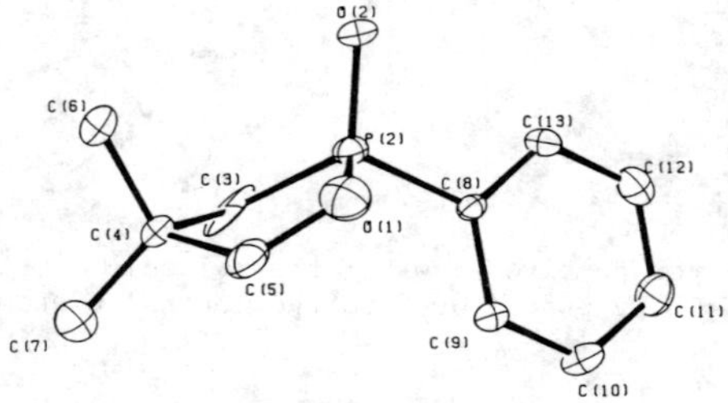

Fig. 1. A view of $C_{11}H_{15}PO_2$.

(2R,4S,5R)-2-THIOMETHYL-3,4-DIMETHYL-5-PHENYL-1,3,2-OXAZAPHOSPHOLIDINE-2-OXIDE
$C_{11}H_{16}NO_2PS$

T.J. BARTCZAK, Z. GALDECKI, M.Yu. ANTIPIN and Yu.T. STRUCHKOV, 1984. Phosphorus and Sulfur, 19, 11-16.

Orthorhombic, $P2_12_12_1$, a = 21.744, b = 8.409, c = 6.9704 Å, Dm = 1.338, Z = 8. Cu radiation, R = 0.113 for 960 reflexions.

The P atom has distorted tetrahedral geometry (Fig. 1); the five-membered ring conformation is shown in detail. Some dimensions are: P-S 2.059(5), P-O(1) 1.598(9), P-O(2) 1.491(10), P-N 1.669(10) Å, N-P-O(1) 96.8(5), O(2)-P-S 104.8(4)°.

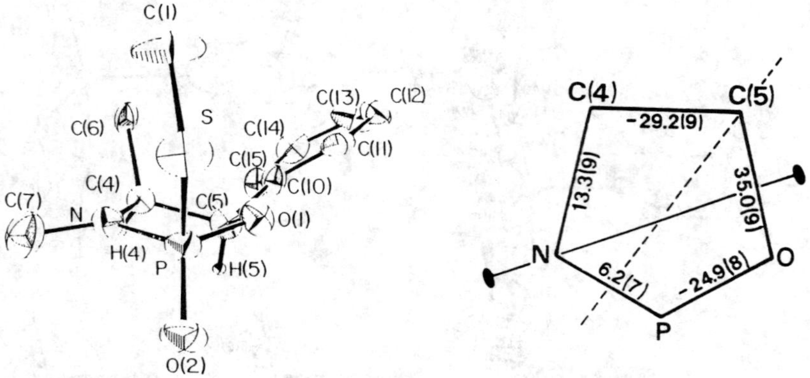

Fig. 1. A view of $C_{11}H_{16}NO_2PS$ and details of ring conformation.

cis-2,5-DI-t-BUTYL-2-OXO-1,3,2-DIOXAPHOSPHORINANE
$C_{11}H_{23}O_3P$

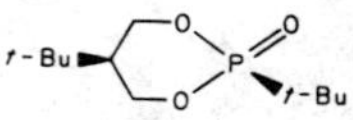

R.O. DAY, W.G. BENTRUDE, K.C. YEE, W.N. SETZER, J.A. DEITERS and R.R. HOLMES, 1984. J. Am. Chem. Soc., 106, 103-106.

Orthorhombic, $Pmc2_1$, a = 9.729, b = 11.634, c = 11.945 Å, Z = 4. Mo radiation, R = 0.042 for 974 reflexions.

The two independent molecules (Fig. 1), which have crystallographic C_s symmetry, are in the boat conformation with the t-butyl groups in cis positions. Bond lengths include: P=O 1.455(6) and 1.444(5), P-O 1.581(3) and 1.576(4), P-C 1.813(7) and 1.801(7) Å.

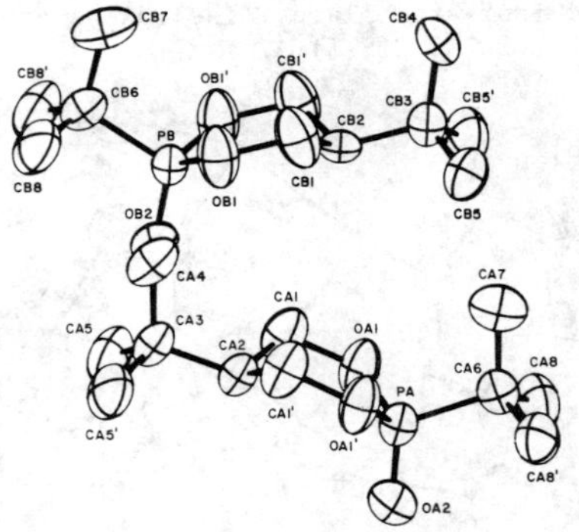

Fig. 1. The two independent molecules of $C_{11}H_{23}O_3P$.

2-(1,3-DIAMINOPROPANE-N,N')-4,4,6,6-TETRAKIS(AZARIDINYL)CYCLOTRIPHOSPHAZENE $C_{11}H_{24}N_9P_3$

J.-F. LABARRE, G. GUERCH, F. SOURNIES, F. SPREAFICO and S. FILIPPESCHI, 1984. J. Mol. Struct., 117, 59-72.

Monoclinic, $P2_1/a$, a = 18.369, b = 9.715, c = 9.722 Å, β = 94.79°, Z = 4. Mo radiation, R = 0.0470 for three-dimensional data.

The six-membered phosphazene ring is planar and there is considerable puckering of the spiro loop (Fig. 1). P-N bond lengths are in the range 1.571-1.690(4) Å.

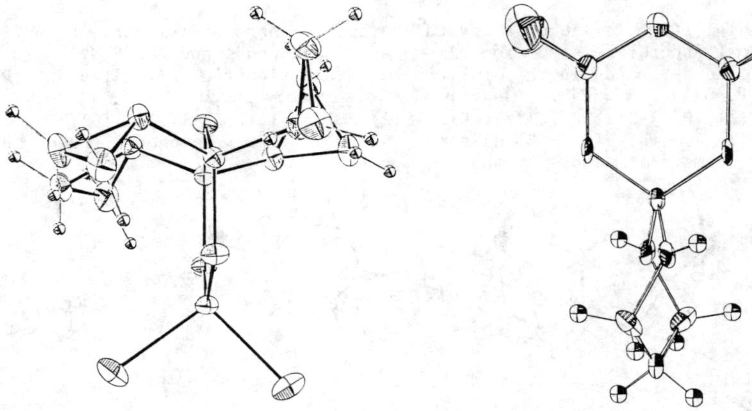

Fig. 1. Two views of $C_{11}H_{24}N_9P_3$.

1,2,5,6-TETRAPHOSPHA-3,4,7,8-DIBENZO-2,6-DIIODOBICYCLO[3.3.0]OCTANE
$C_{12}H_8I_2P_4$

H.-J. WORZ, H. PRITZKOW and H.P. LATSCHA, 1984. Z. Naturforsch., B39, 139-141.

Monoclinic, $P2_1/a$, a = 8.674, b = 21.540, c = 9.540 Å, β = 114.72°, Z = 4. Mo radiation, R = 0.058 for 3281 reflexions.

The molecule (Fig. 1) has approximate C_2 symmetry. The two five-membered rings have slightly distorted envelope forms with P2 and P3 as the flap atoms approximately 0.8 Å from the appropriate four-atom plane. The iodine atoms are axial, with P-I 2.493(1) and 2.490(2), P-P 2.207-2.215(3) Å.

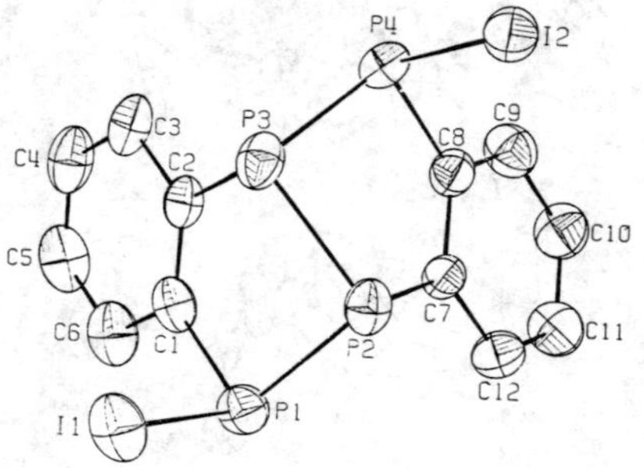

Fig. 1. $C_{12}H_8I_2P_4$: a perspective view.

2,4-DIPHENYL-2,4-DITHIOXO-1,3,2λ^5,4λ^5-DITHIADIPHOSPHETANE
$C_{12}H_{10}P_2S_4$ (I)

2,4,5-TRIPHENYL-2-THIOXO-1,3,2λ^5,4,5-DITHIATRIPHOSPHOLANE
$C_{18}H_{15}P_3S_3$ (II)

C. LENSCH, W. CLEGG and G.M. SHELDRICK, 1984. J. Chem. Soc. Dalton, 723-725.

I. Monoclinic, $P2_1/c$, a = 6.624, b = 14.438, c = 7.667 Å, β = 95.31°, Z = 2. Mo radiation, R = 0.060 for 880 reflexions. [Crystal twinned by reflexion in (100).]

II. Monoclinic, $P2_1$, a = 8.884, b = 13.713, c = 8.309 Å, β = 106.61°, Z = 2. Mo radiation, R = 0.051 for 1938 reflexions. [Crystal twinned by reflexion in (001).]

Molecules of I (Fig. 1) lie on a crystallographic centre of symmetry while II has no symmetry. In I, therefore, the exocyclic S atoms are trans and the P-S(2) distance is 1.920(2) Å. The P-S bonds in the P_2S_2 ring of I are significantly different (P-S(1) 2.108(2), P'-S(1) 2.133(2) Å). II contains a slightly puckered P_3S_2 ring with P(1)-P(2) 2.219(2), P(1)-S(1) 2.111(2), P(2)-S(2) 2.106(3), P(3)-S(1) 2.103(3) and P(3)-S(2) 2.121(2) Å. The P(3)-S(3) distance is 1.934(3) Å. The phenyl groups on the trivalent P atoms are mutually trans.

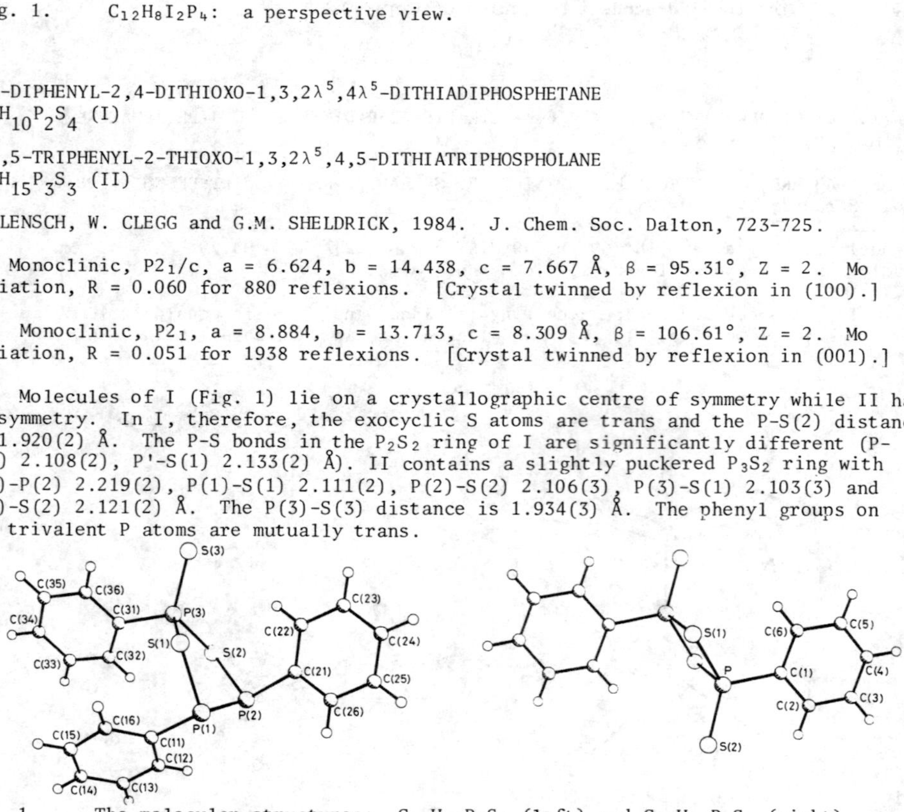

Fig. 1. The molecular structures: $C_{18}H_{15}P_3S_3$ (left) and $C_{12}H_{10}P_2S_4$ (right).

DIPHENYLTHIOPHOSPHINIC ACID
$C_{12}H_{11}OPS$ $C_6H_5P(=S)OH$

R. MATTES and D. RUHL, 1984. Acta Cryst., C40, 106-108.

Orthorhombic, Pbca, a = 22.62, b = 6.105, c = 16.562 Å, D_m = 1.34, Z = 8. Mo radiation, R = 0.049 for 2578 reflexions (at 140K).

The molecules (Fig. 1) form endless chains along the glide plane in the [010] direction through strong O-H...S hydrogen bonds (O...S 3.133(8) Å). Principal dimensions are P-O 1.582(2), P-S 1.956(1), P-C(11) 1.799(2), P-C(21) 1.805(3) Å, O-P-S 115.4(1), O-P-C 101.6(2) and 105.9(2), S-P-C 113.8(1) and 112.8(1), C-P-C 106.2(2)°.

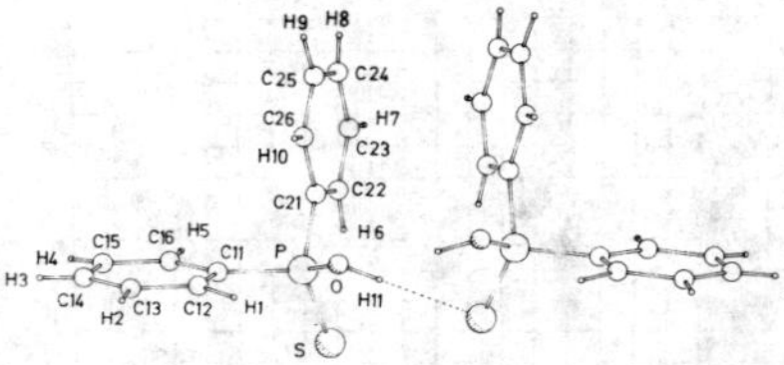

Fig. 1. A view of the hydrogen bonding in diphenylthiophosphinic acid.

6-PHENYL-1-OXA-6-PHOSPHASPIRO[2.5]OCTANE 6-SULPHIDE
$C_{12}H_{15}OPS$

N. SATYAMURTHY, K.D. BERLIN, M.B. HOSSAIN and D. VAN DER HELM, 1984. Phosphorus and Sulfur, 19, 113-129.

Orthorhombic, $Pna2_1$, a = 13.056, b = 14.268, c = 6.1522 Å, D_m = 1.341, Z = 4. Cu radiation, R = 0.031 for 1247 reflexions (at -135°C).

Molecular dimensions are shown in Fig. 1. The phosphorane ring has a slightly flattened chair conformation, with the phenyl-P bond equatorial and the P=S bond axial. The epoxide ring plane is virtually coincident with a pseudo-mirror plane through P(6),C(3),C(9),S(15). The phenyl group is rotated 28.2° out of this plane.

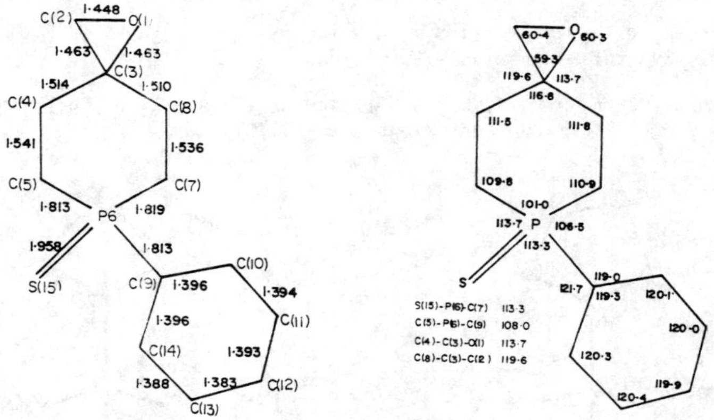

Fig. 1. Molecular dimensions for $C_{12}H_{15}OPS$.

5-AZA-2,8-DIOXO-3,7-DI-t-BUTYL-1-PHOSPHABICYCLO[3.3.0]OCTA-2,4,6-TRIENE
$C_{12}H_{20}NO_2P$

S.A. CULLEY and A.J. ARDUENGO III, 1984. J. Am. Chem. Soc., 106, 1164-1165.

Orthorhombic, $Pca2_1$, a = 11.493, b = 11.558, c = 11.154 Å, Z = 4. R = 0.068 for 526 reflexions.

The molecule (Fig. 1) is essentially planar except for t-butyl methyl groups. The principal dimensions are: P-O 1.843 and 1.805(10), P-N 1.68(1), C-O 1.34 and 1.34(2), C-N 1.38 and 1.39(2), C-C (ring) 1.32 and 1.35(2) Å, O-P-O 168.8(4), N-P-O 84.0 and 84.9(5)°.

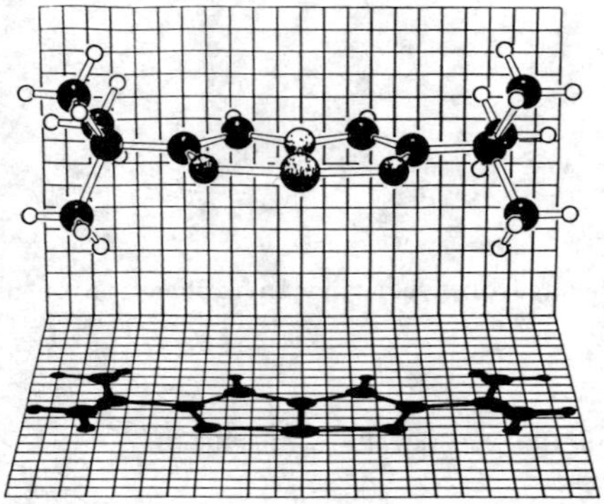

Fig. 1. A drawing of $C_{12}H_{20}NO_2P$.

TRIS(PYRROLIDINO)PHOSPHINE SELENIDE
$C_{12}H_{24}N_3PSe$ (I) $[(CH_2)_4N]_3PSe$

TRIS(PYRROLIDINO)PHOSPHINE TELLURIDE
$C_{12}H_{24}N_3PTe$ (II) $[(CH_2)_4N]_3PTe$

C. RØMMING, K. MAARTMANN-MOE and J. SONGSTAD, 1984. Acta Chem. Scand., A38, 349-357.

I. Orthorhombic, $P2_12_12_1$, a = 14.715, b = 16.021, c = 12.392 Å, Z = 8. Mo radiation, R = 0.066 for 4392 reflexions (at -150°C).

II. Monoclinic, $P2_1/n$, a = 13.983, b = 12.249, c = 8.856 Å, β = 91.31°, Z = 4. Mo radiation, R = 0.074 for 2290 reflexions (at -150°C).

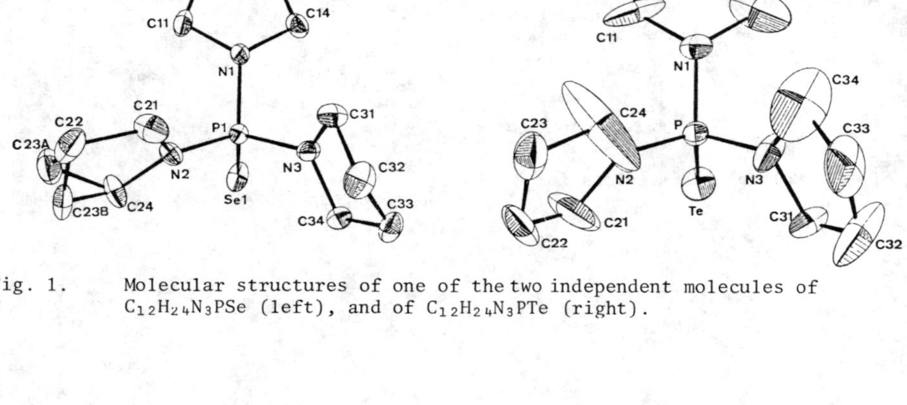

Fig. 1. Molecular structures of one of the two independent molecules of $C_{12}H_{24}N_3PSe$ (left), and of $C_{12}H_{24}N_3PTe$ (right).

In both compounds there exists disorder in one or more of the five-membered rings. This does not, however, cause any ambiguity with regard to the torsion angles about the P-N bonds. The conformation of the P-N bond is anti-periplanar for one amino group and synclinal for the two others in each molecule (Fig. 1). The P-Se bond length in I is 2.105 and 2.107 Å for the two non-equivalent molecules, while the P-Te bond length in II is 2.355 Å. The P-N bond lengths are in the range 1.633-1.660 for I and 1.632-1.653 for II.

1-t-BUTYL-2,2,3,4,4-PENTAMETHYLPHOSPHETANE 1-OXIDE
$C_{12}H_{25}OP$

J.A. CAMPBELL, C.N. CAUGHLAN, A. FITZGERALD, C. CAMPANA and S.E. CREMER, 1984. Acta Cryst., C40, 1918-1920.

Monoclinic, $P2_1/m$, a = 6.137, b = 12.181, c = 9.005 Å, β = 96.24°, D_m = 1.06, Z = 2. Mo radiation, R = 0.052 for 1138 reflexions.

The molecule (Fig. 1) has mirror symmetry with the C(3) methyl group trans to the t-butyl group on P. The four-membered ring is puckered, with an angle of 19.2(4)°, such that diaxial, cross-ring interactions are minimised.

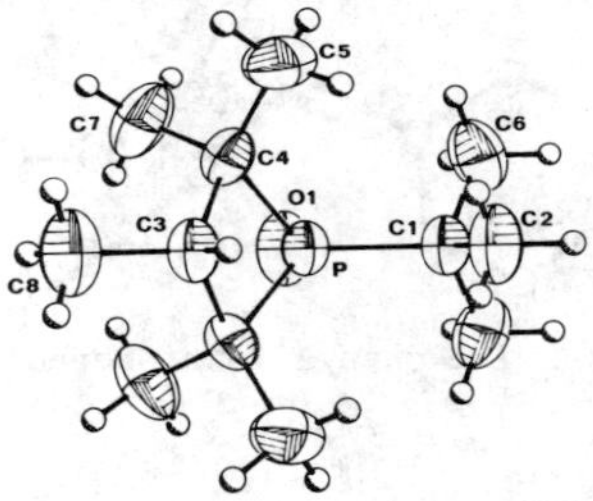

Fig. 1. The $C_{12}H_{25}OP$ molecule.

N_1,N_{11},N_4,N_8-BIS(μ-METHYLPHOSPHINO)-1,4,8,11-TETRAAZACYCLOTETRADECANE
$C_{12}H_{26}N_4P_2$ (I)

1,4,8,11-TETRAKIS(DIPHENYLPHOSPHINOMETHYL)-1,4,8,11-TETRAAZACYCLOTETRADECANE
$C_{62}H_{68}N_4P_4$ (II)

H. HOPE, M. VIGGIANO, B. MOEZZI and P.P. POWER, 1984. Inorg. Chem., 23, 2550-2552.

I. Monoclinic, $P2_1/n$, a = 6.519, b = 9.488, c = 11.730 Å, β = 90.71°, Z = 2. Mo radiation, R = 0.043 for 1426 reflexions.

II. Monoclinic, $P2_1/n$, a = 14.259, b = 9.289, c = 20.882 Å, β = 104.92°, Z = 2. Mo radiation, R = 0.045 for 4897 reflexions.

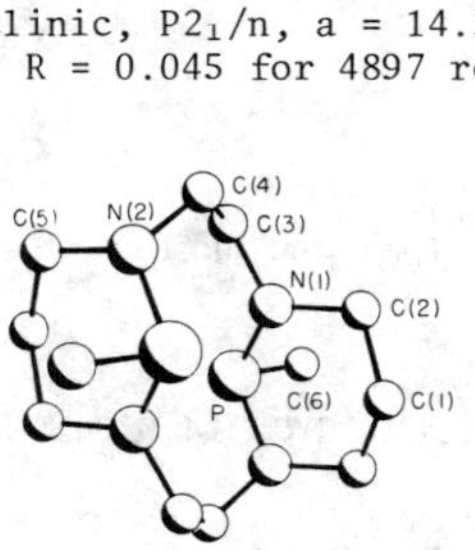

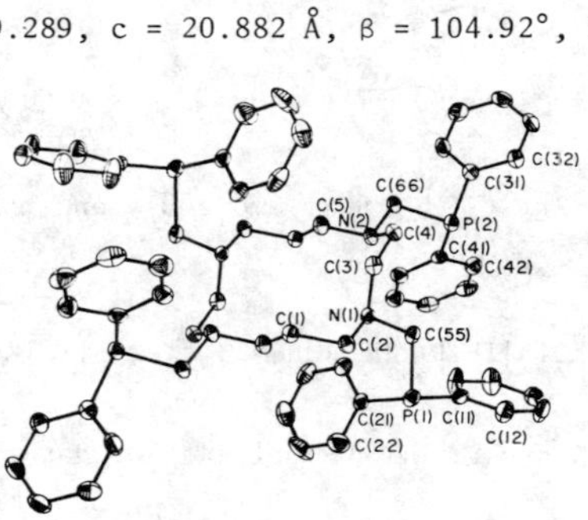

Fig. 1. The structures of $C_{12}H_{26}N_4P_2$ (left) and $C_{62}H_{68}N_4P_4$ (right).

The two centrosymmetric molecules are shown in Fig. 1. Compound I consists of two PN_2C_3 rings linked at the N's by a pair of $-CH_2CH_2-$ fragments resulting in a central 10-membered $P_2N_4C_4$ ring. The N coordination in I is planar with P-N 1.694(2) Å. II consists of a 14-membered $C_{10}N_4$ backbone with the pairs of $-CH_2PPh_2$ groups on N_1,N_4 and N_8,N_{11} on opposite sides of the ring. Bond lengths are normal.

trans-2,5-DIMETHYL-1-PHENYL-1-THIOXOPHOSPHORINAN-4-ONE (stereoisomer II)
$C_{13}H_{17}OPS$

A.I. YANOVSKII, A.E. SHALAMOV, Yu.T. STRUCHKOV, A.P. LOGUNOV, Yu.G. BOSYAKOV and O.V. AGASHKIN, 1984. J. Struct. Chem., 25, 578-581 [Zh. Strukt. Khim. 25(4), 79-82].

Orthorhombic, $P2_1ab$, a = 12.358, b = 13.876, c = 7.607 Å, Z = 4. Cu radiation, R = 0.031 for 1117 reflexions.

The six-membered heterocyclic ring in the molecule (Fig. 1) has a distorted chair conformation with equatorial methyl substituents and an axial phenyl group. Bond lengths include: S-P 1.958(1) and P-C 1.809(5)-1.826(6) Å.

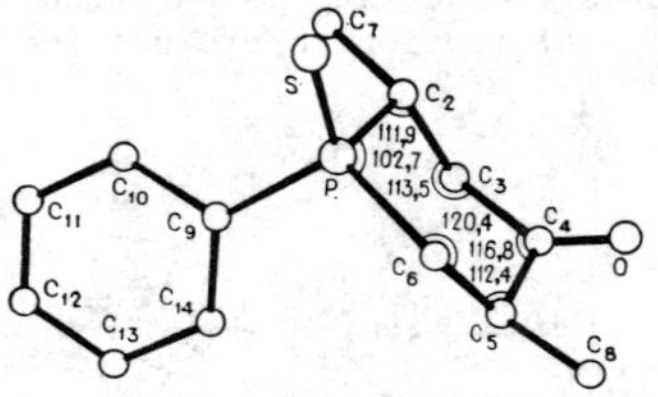

Fig. 1. The trans-2,5-dimethyl-1-phenyl-1-thioxophosphorinan-4-one molecule.

(BIS(TRIMETHYLSILYLSULPHANO)METHYLIDENE)PHENYLPHOSPHINE
$C_{13}H_{23}PS_2Si_2$

G. BECKER, O. MUNDT and G. UHL, 1984. Z. Anorg. Chem., 517, 89-102.

Monoclinic, $P2_1/c$, a = 13.864, b = 10.360, c = 12.817 Å, β = 101.23°, Z = 4. Mo radiation, R = 0.032 for 3173 reflexions (at -95°C).

A view of the molecule and details of dimensions are in Fig. 1.

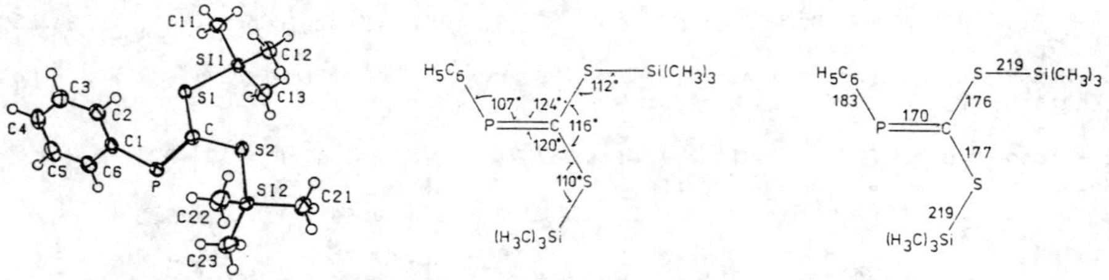

Fig. 1. The $C_{13}H_{23}PS_2Si_2$ molecule and bond lengths (pm) and angles.

6-METHYLBENZO[h]TETRACHLOROBENZO[c]-2,5,7-TRIOXA-1-PHOSPHABICYCLO[4.3.0]NONANE
$C_{14}H_7Cl_4O_3P$

H.-D. HAUSEN and G. WECKLER, 1984. Z. Naturforsch., B39, 628-631.

Monoclinic, $P2_1/c$, a = 8.499, b = 10.095, c = 18.594 Å, β = 94.03°, D_m = 1.6, Z = 4. Mo radiation, R = 0.07 for 1263 reflexions.

The molecule (Fig. 1) is shown to have a cis-configuration and an endo-conformation. The dihedral angle between the C31-C36,P,O3,C11 and P,C11,O1,O2 planes is 92.3° and between P,C11,O1,O2 and C21-C26,C11-C14,O1,O2 is 141.7°.

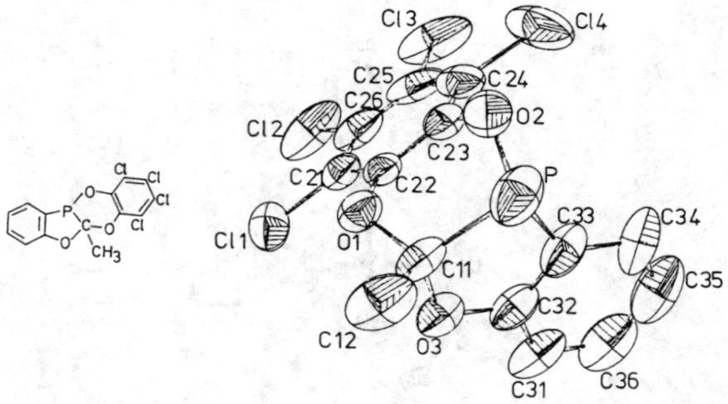

Fig. 1. $C_{14}H_7Cl_4O_3P$: a perspective view.

BIS(HEXAHYDRO-1-PYRIDAZINYL)THIOPHOSPHORIC ACID O-PHENYL ESTER (Two forms)
$C_{14}H_{23}N_4OPS$

U. ENGELHARDT and B. STROMBURG, 1984. Acta Cryst., C40, 441-445.

I. Orthorhombic, $P2_12_12_1$, a = 15.877, b = 11.843, c = 8.923 Å, Z = 4. Cu radiation, R = 0.033 for 1595 reflexions.

II. Triclinic, $P\bar{1}$, a = 15.551, b = 8.768, c = 6.446 Å, α = 93.63, β = 104.40, γ = 95.52°, Z = 2. Cu radiation, R = 0.034 for 2713 reflexions.

In both crystal structures, the molecules have almost identical geometries. The hexahydropyridazine rings have slightly distorted chair conformations with mean N-N distances of 1.437(4) Å and mean torsion angles for ring bonds of 56(4)°. Other mean bond distances: P-S 1.930(1), P-N 1.665(1) and 1.646(3), P-O 1.603(2), O-C 1.393 and N-C 1.465(3) Å.

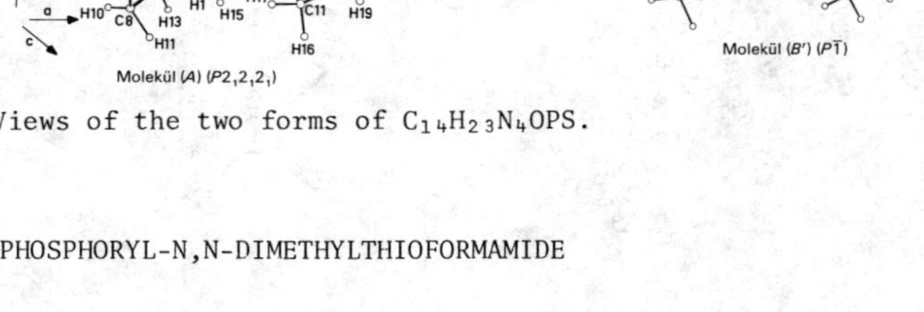

Fig. 1. Views of the two forms of $C_{14}H_{23}N_4OPS$.

DIPHENYLTHIOPHOSPHORYL-N,N-DIMETHYLTHIOFORMAMIDE
$C_{15}H_{16}NPS_2$

A. BRUNS, W. HILLER and U. KUNZE, 1984. Z. Naturforsch., B39, 14-18.

Monoclinic, C2/c, a = 17.915, b = 6.867, c = 25.913 Å, β = 93.10°, Z = 8. Mo radiation, R = 0.072 for 1226 reflexions.

The molecule has a planar thioamide moiety, with the P=S bond forming an angle of 53.4° to the plane (Fig. 1). Main dimensions are: P=S 1.950(2), P-C1 1.858(4), C1-S2 1.663(4), N-C1 1.337(5), N-C2 1.449(6), N-C3 1.470(6) Å. Angles at P involving S1 are 112.0-114.9(2)°; those not involving S1 are 104.6-106.0(2)°.

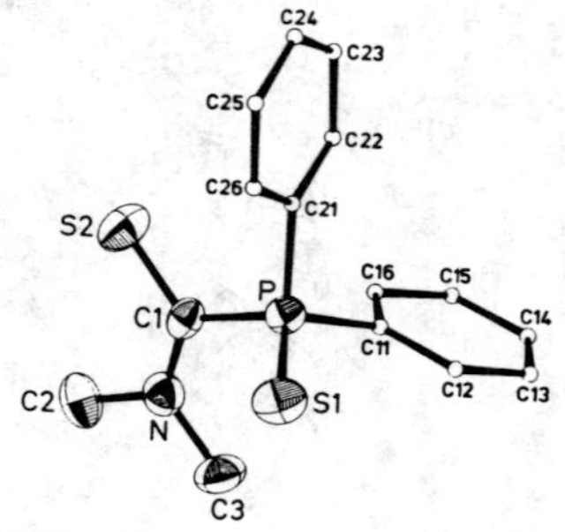

Fig. 1. A view of $C_{15}H_{16}NPS_2$.

1,4-DIMETHYL-2,3-DIPHENYL-1,4,2,3-DIAZADIPHOSPHOLIDINE-5-THIONE-2-SULPHIDE
$C_{15}H_{16}N_2P_2S_2$ (I)

N-(N,N'-DIMETHYLCARBAMIMIDOYL)-N,N'-DIMETHYLTHIOUREA HYDROCHLORIDE PYRIDINIUM CHLORIDE
$C_{11}H_{21}Cl_2N_5S$ (II) $C_5H_6N^+, C_6H_{15}N_4S^+, 2(Cl^-)$

H. RIFFEL, S. KLEEMANN, H. HESS and E. FLUCK, 1984. Z. Anorg. Chem., 508, 61-72.

I. Monoclinic, $P2_1/a$, a = 13.072, b = 9.824, c = 13.664 Å, β = 109.19°, Z = 4. Mo radiation, R = 0.039 for 3308 reflexions (at -130°C).

II. Triclinic, $P\bar{1}$, a = 8.187, b = 10.696, c = 10.936 Å, α = 109.98, β = 117.87, γ = 96.34°, Z = 2. Mo radiation, R = 0.028 for 4260 reflexions (at -130°C).

The five-membered ring of (I) is puckered (Fig. 1) with deviations ±0.11 Å from the best plane. Main bond lengths are: P1-S1 1.915(1), P1-P2 2.197(1), P1-N1 1.694(2), P1-C11 1.798(2), P2-N2 1.724(2), P2-C21 1.820(2), C3-S2 1.660(2), C3-N1 1.362(3), C3-N2 1.359(3), N1-C1 1.465(4) and N4-C2 1.466(4) Å. In II, the cation components are linked through chloride anions by N-H...Cl hydrogen bonds (Fig. 1) with H...Cl 2.24-2.32 Å. Principal bond lengths are: S-C1 1.681(1), C1-N1 1.334(2), N1-C2 1.455(2), C1-N2 1.378(2), N2-C3 1.473(2), N2-C4 1.417(2), C4-N4 1.310(2), N4-C6 1.460(2), C4-N3 1.316(2), N3-C5 1.457(2) Å.

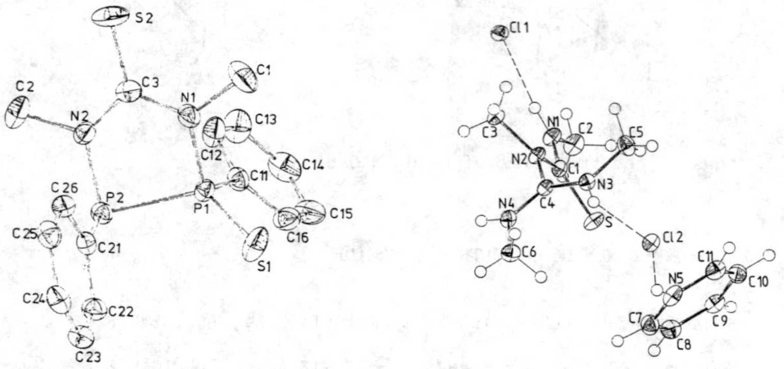

Fig. 1. Views of $C_{15}H_{16}N_2P_2S_2$ (I) (left) and of $C_{11}H_{21}Cl_2N_5S$ (II) (right).

DIPHENYL-3-HYDROXYPROPYLPHOSPHINE OXIDE
$C_{15}H_{17}O_2P$ $Ph_2P(O)(CH_2)_3OH$

V.V. TKACHEV and L.O. ATOVMYAN, 1984. J. Struct. Chem., 25, 117-122 [Zh. Strukt. Khim., 25(1), 135-141].

Monoclinic, $P2_1/b$, a = 16.653, b = 42.259, c = 5.931 Å, γ = 83.5°, Z = 12. Cu radiation, R = 0.048 for 2015 reflexions.

The asymmetric unit contains three crystallographically independent molecules. Molecules of each kind are linked in chains along c by P=O...H-O hydrogen bonds (O...H 2.21(1), 2.06(1), and 2.22(1) Å, see Fig. 1). Molecular geometry is as expected.

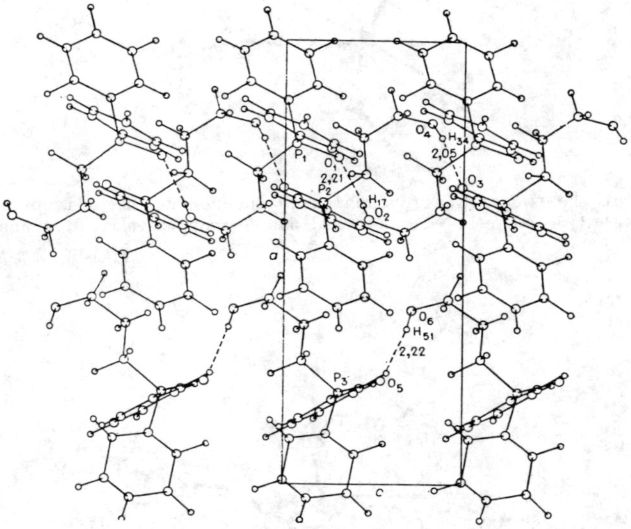

Fig. 1. The structure of diphenyl-3-hydroxypropylphosphine oxide.

5-(DIISOPROPYL)AMINO-6-METHYL-2,3-BENZO-1,4-DIOXA-6,9-DIAZA-5-PHOSPHA-SPIRO[4.4]NONANE
$C_{15}H_{26}N_3O_2P$ (I)

3,8-DI-n-BUTYL-2,7-BIS(o-PHENYLENEDIOXA)-1,3,6,8-TETRAAZA-2,7-DIPHOSPHATRICYCLO-[5.3.0.$0^{2,6}$]DECANE
$C_{24}H_{34}N_4O_4P_2$ (II)

A.N. CHERNEGA, M.Yu. ANTIPIN, Yu.T. STRUCHKOV, Yu.G. GOLOLOBOV, I.E. BOLDESKUL, N.I. GUSAR' and M.P. CHAUS, 1984. J. Struct. Chem., 25, 921-930 [Zh. Strukt. Khim., 25(6), 93-103].

I. Monoclinic, $P2_1/c$, a = 12.88, b = 19.66, c = 14.26 Å, β = 115.8°, Z = 8. Mo radiation, R = 0.065 for 3102 reflexions (at -120°C).

II. Monoclinic, $P2_1/n$, a = 8.3509, b = 14.407, c = 10.824 Å, β = 104.112°, Z = 2. Mo radiation, R = 0.037 for 2400 reflecions.

The molecules I and II are shown in Figs. 1 and 2, respectively. The binuclear complex II has exact inversion symmetry. In both structures the P atoms have distorted trigonal bipyramidal coordination geometry. The angles between axial ligands are 173.5 and 172.8° for the two independent molecules of I and 173.31° for II. In II the P atom is displaced 0.08 Å from the equatorial plane toward O(1). The distortions of the trigonal bipyramids toward square pyramidal along the Berry coordinate are 21.5% and 21.2% for I and 19.1% for II. Bond lengths: (for I) P-O(ax) 1.819(4) and 1.825(4), P-O(eq) 1.658(4) and 1.665(4), P-N(ax) 1.716(5) and 1.708(5), P-N(eq) 1.636(4)-1.659(4) Å; (for II) P-O(ax) 1.714(2), P-O(eq) 1.669(2), P-N(ax) 1.744(2), P-N(eq) 1.660(2) and 1.652(2) Å.

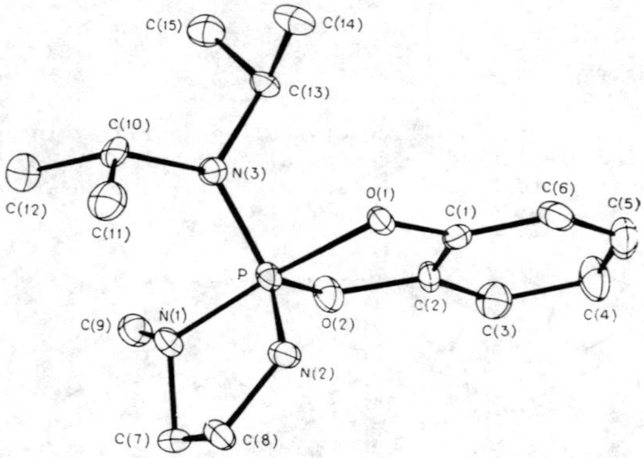

Fig. 1. One of the two crystallographically independent 5-(diisopropyl)amino-6-methyl-2,3-benzo-1,4-dioxa-6,9-diaza-5-phospha-spiro[4.4]nonane molecules.

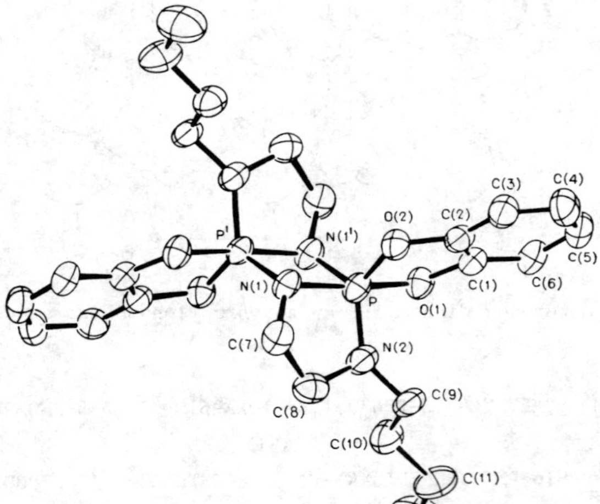

Fig. 2. The 3,8-di-n-butyl-2,7-bis(o-phenylenedioxa)-1,3,6,8-tetraaza-2,7-diphosphatricyclo[5.3.0.$0^{2,6}$]decane molecule.

1,2-DI-t-BUTYL-3-DIMETHYLAMINO-1-THIO-4-TRIMETHYLSILYLSULPHANO-1λ^5,2λ^3-DIPHOSPHET-3-ENE
$C_{15}H_{33}NP_2S_2Si$

G. BECKER, W. BECKER and G. UHL, 1984. Z. Anorg. Chem., 519, 31-41.

Orthorhombic, $P2_12_12_1$, a = 10.673, b = 10.771, c = 19.246 Å, Z = 4. Mo radiation, R = 0.038 for 2553 reflexions.

A view of the molecule and some dimensions are in Fig. 1. The four membered P_2C_2 ring is close to being planar (maximum deviation 0.045 Å).

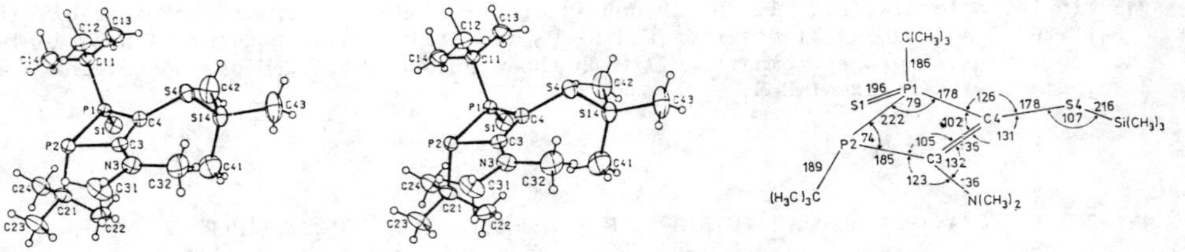

Fig. 1. A stereoview of $C_{15}H_{33}NP_2S_2Si$ and some dimensions (pm and °).

(1R,2S)DIHYDROGEN(2-BROMO-1-(4-METHOXYPHENYL)-3-PHENYL-3-OXO-1-PROPYL)PHOSPHONIC ACID ACETONE SOLVATE

$C_{16}H_{16}BrO_5P,C_3H_6O$

K.C. CALVO and J.M. BERG, 1984. J. Am. Chem. Soc., 106, 4202-4204.

Orthorhombic, $P2_12_12_1$, a = 6.038, b = 12.692, c = 27.000 Å, Z = 4. Mo radiation, R = 0.0537 for 1827 reflexions.

The analyses establishes the absolute stereochemistry of the crystal studied as the 1R,2S isomer (Fig. 1). Molecular dimensions are normal; the P-C1-C2-Br torsion angle is 162.4°.

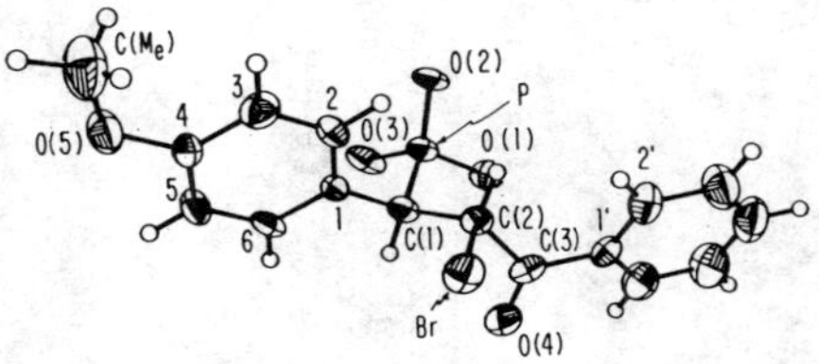

Fig. 1. A view of the (1R,2S)-isomer of $C_{16}H_{16}BrO_5P$.

2-METHYL-5,5-DIPHENYL-1,3,2λ^5-DIOXAPHOSPHORINANE 2-SULPHIDE

$C_{16}H_{17}O_2PS$

P. VAN NUFFEL, C. VAN ALSENOY, A.T.H. LENSTRA, H.J. GEISE and G.R. VAN DEN BERG, 1984. Acta Cryst., C40, 1263-1264.

Monoclinic, $P2_1/c$, a = 11.171, b = 11.174, c = 12.604 Å, β = 105.24°, Z = 4. Mo radiation, R = 0.060 for 2857 reflexions (at 150K).

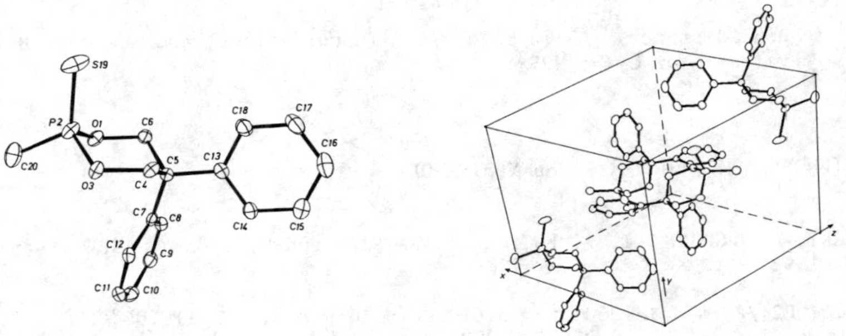

Fig. 1. $C_{16}H_{17}O_2PS$: conformation of the molecule and the contents of a unit cell.

In the molecule (Fig. 1) the thiophosphate ring adopts a chair conformation with P=S 1.926(1) Å in an axial position and P-CH_3 1.778(2) Å in an equatorial position; the geometrical features compare well with those of other 1,3,2-dioxaphosphorinane 2-sulphides with an axial P=S.

4-t-BUTYL-5-CYANO-2-PHENYL-1-THIOXO-7-THIA-1-PHOSPHABICYCLO[2.2.1]HEPT-2-ENE
$C_{16}H_{18}NPS_2$

T. UCHIDA, K. KOZAWA, T. SAITO and S. MOTOKI, 1984. Bull. Chem. Soc. Japan, 57, 1985-1989.

Triclinic, P$\bar{1}$, a = 12.731, b = 13.143, c = 12.483 Å, α = 102.39, β = 104.34, γ = 117.96°, Dm = 1.30, Z = 4. Mo radiation, R = 0.082 for 4495 reflexions.

The asymmetric unit contains two crystallographically-independent molecules (A and B), and differences in their molecular geometry are not highly significant (Fig. 1). Each of two five-membered rings takes an envelope conformation, having in common the P-S(2)-C(3) bridge. The average bond angle P-S(2)-C(3) is 82.5° indicating severe strain in the bicyclo skeleton. Molecules A and B are rotational isomers because the C(1)-C(61) bond is kept from free rotation by the steric hindrance of the ortho hydrogens H(62) and H(66).

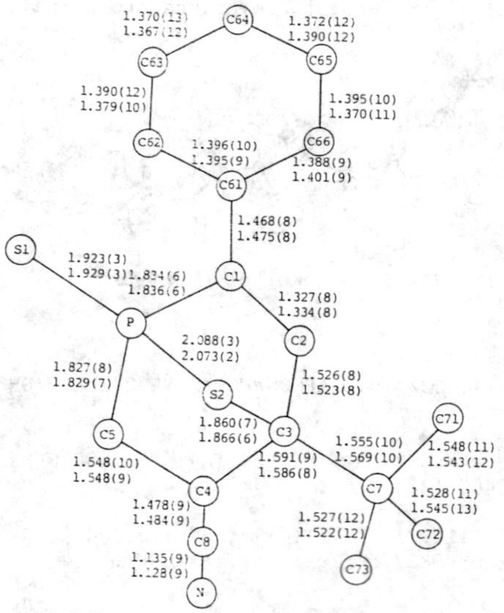

Fig. 1. Bond distances (Å) in molecule A (upper values) and molecule B (lower values) for $C_{16}H_{18}NPS_2$.

1-[ETHYL(PHENYL)PHOSPHINYL]-1-PHENYLETHANOL
$C_{16}H_{19}O_2P$

Z. GAŁDECKI, P. GROCHULSKI, B. LUCIAK, Z. WAWRZAK and W.L. DUAX, 1984. Acta Cryst., C40, 1197-1198.

Monoclinic, $P2_1/c$, a = 15.960, b = 5.614, c = 16.418 Å, β = 110.06°, D_m = 1.303, Z = 4. Mo radiation, R = 0.078 for 2973 reflexions.

There are no unusual bond distances and angles in the molecule (Fig. 1). The conformation about C(10)-P is such that the two phenyl groups are in antiperiplanar orientation. The molecules form chains in the crystal along the [010] direction by means of hydrogen bonding with O...O 2.758(3), H...O 1.94(6) Å and O...H-O 144(6)°.

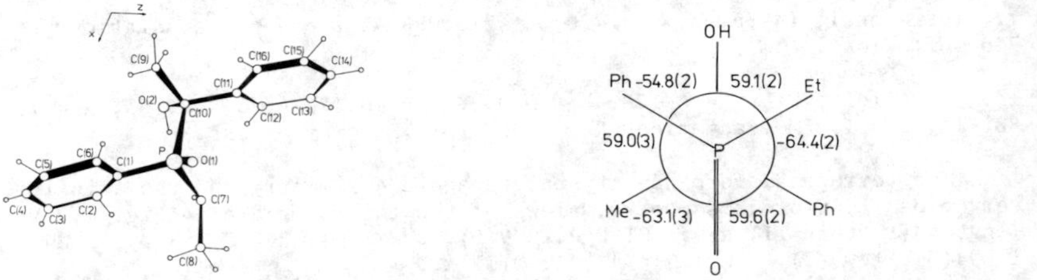

Fig. 1. $C_{16}H_{19}O_2P$: view of the molecule along [010] and Newman projections along C(10)-P showing torsion angles.

trans-3,6-DIPHENYL-1,2,4,5-TETRAMETHYL-1,2,4,5,3,6-TETRAAZADIPHOSPHACYCLOHEXAN-3,6-DISULPHIDE

$C_{16}H_{22}N_4P_2S_2$ (I)

trans-HEXAMETHYL-1,2,4,5,3,6-TETRAAZADIPHOSPHACYCLOHEXAN-3,6-DISULPHIDE

$C_6H_{18}N_4P_2S_2$ (II)

F. DIRSCHL and H. NOTH, 1984. Z. Naturforsch., B39, 269-274.

I. Monoclinic, C2/c, a = 21.443, b = 10.480, c = 18.552 Å, β = 109.99°, D_m = 1.32, Z = 8. Mo radiation, R = 0.035 for 3023 reflexions.

II. Triclinic, P$\bar{1}$, a = 6.949, b = 6.955, c = 7.242 Å, α = 69.43, β = 77.46, γ = 81.93°, Z = 1. Mo radiation, R = 0.029 for 1050 reflexions.

The P_2N_4 ring in (I) lies in a general position in the unit cell and adopts a twist conformation. In (II) the P_2N_4 ring lies about an inversion centre and has a chair conformation. The torsion angle data show that there is less interaction between lone pairs of electrons on adjacent N atoms in the chair conformation but stronger interaction to those bound to P. It is suggested that this accounts for N-P-N being 107° in II compared with 101° in (I).

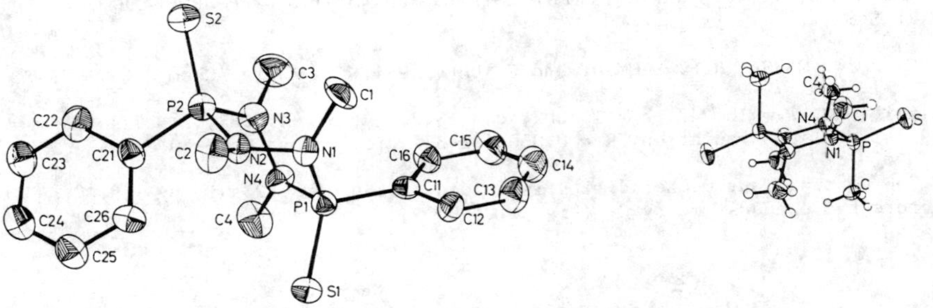

Fig. 1. Views of $C_{16}H_{22}N_4P_2S_2$ and $C_6H_{18}N_4P_2S_2$.

1,2,4,5-TETRA-t-BUTYL-1,2,4,5-TETRAPHOSPHA-3-SILASPIRO[2.2]PENTANE
$C_{16}H_{36}P_4Si$ (Tetragonal and Triclinic Forms)

K.-F. TEBBE and T. HEINLEIN, 1984. Z. Anorg. Chem., 515, 7-18.

I. Tetragonal, $I4_1/a$, a = 12.470, c = 15.055 Å, Z = 4. Mo radiation, R = 0.038 for 710 reflexions.

II. Triclinic, P1, a = 6.128, b = 9.963, c = 10.172 Å, α = 75.63, β = 72.38, γ = 88.71°, Z = 1. Mo radiation, R = 0.062 for 1561 reflexions.

The tetragonal form has crystallographic $\bar{4}$ symmetry and the triclinic form has very close to twofold symmetry; views of the two forms are in Fig. 1. Mean bond lengths (I, then II) are: Si-P 2.2009(9), 2.215(5), P-P 2.255(2), 2.242(5), P-C 1.894(3), 1.90(2), C-C 1.514(4), 1.52(3) Å.

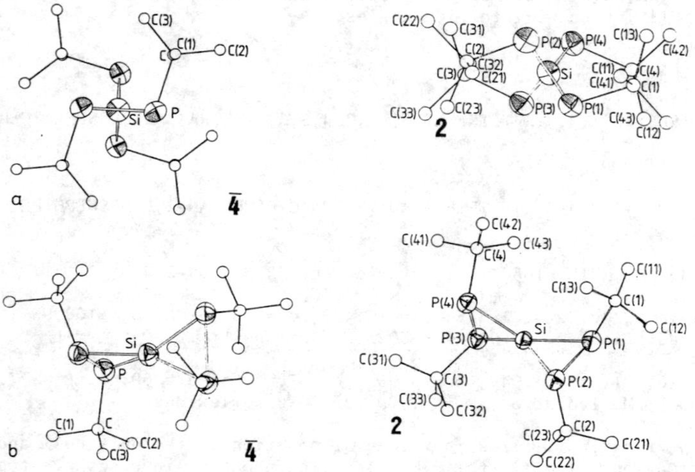

Fig. 1. Views of the two forms of $C_{16}H_{36}P_4Si$.

1,3-DIMETHYL-2,2,4,4-TETRAKIS(TRIMETHYLSILYLSULPHANO)-1,3-DIPHOSPHETANE
$C_{16}H_{42}P_2S_4Si_4$

G. BECKER, W. MASSA, R.E. SCHMIDT and G. UHL, 1984. Z. Anorg. Chem., 517, 75-88.

Triclinic, $P\bar{1}$, a = 10.247, b = 13.602, c = 13.263 Å, α = 117.85, β = 111.05, γ = 72.09°, Z = 2. Mo radiation, R = 0.030 for 3727 reflexions (at -80°C).

The central four-membered ring is folded at the P1...P2 axis by 149.1° and the structure has pseudosymmetry C_s (Fig. 1).

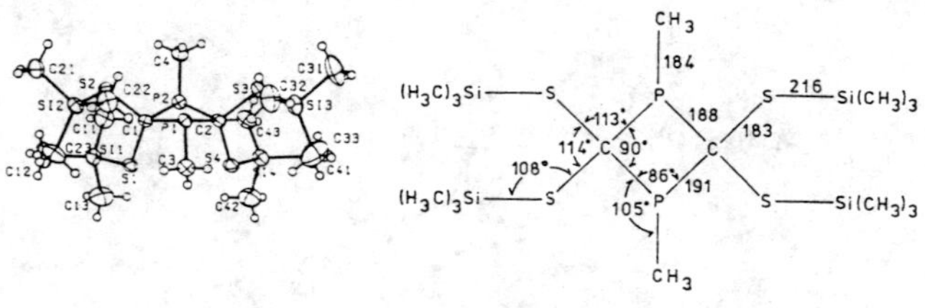

Fig. 1. The $C_{16}H_{42}P_2S_4Si_4$ molecule with averaged bond lengths (pm) and angles.

2-CHLORO-5-METHYL-6-PHENYL-3-(p-TOLYL)PERHYDRO-1,3,2λ^5-OXAZAPHOSPHORINE 2-OXIDE
$C_{17}H_{19}ClNO_2P$

K.R. ACHARYA, S.S. TAVALE and T.N.G. ROW, 1984. Acta Cryst., C40, 875-877.

Monoclinic, $P2_1/a$, a = 12.253, b = 11.650, c = 12.646 Å, β = 109.05°, D_m = 1.25, Z = 4. Mo radiation, R = 0.070 for 942 reflexions.

In the molecule the 1,3,2-oxazaphosphorine ring adopts a chair conformation (Fig. 1). The exocyclic O and Cl connected to P are in equatorial and axial positions respectively. Bond lengths and angles are normal. The packing of the molecules in the unit cell involves van der Waals contacts.

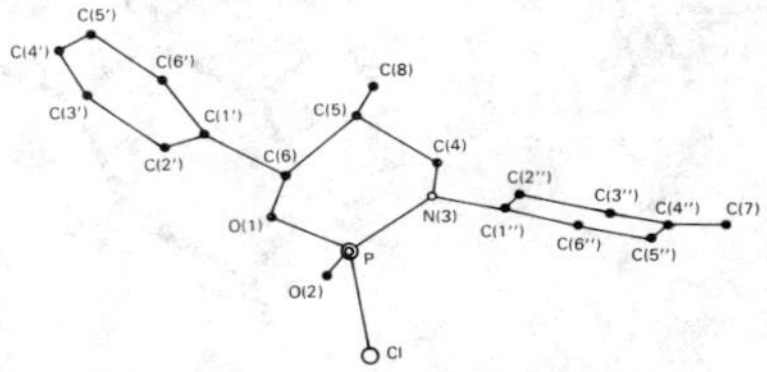

Fig. 1. A perspective view of the $C_{17}H_{19}ClNO_2P$ molecule down the a axis.

cis-2-OXO-2-(DIMETHYLAMINO)-3,5-DIPHENYL-1,3,2-OXAZAPHOSPHORINANE
$C_{17}H_{21}N_2O_2P$

W.G. BENTRUDE, R.O. DAY, J.M. HOLMES, G.S. QUIN, W.N. SETZER, A.E. SOPCHIK and R.R. HOLMES, 1984. J. Am. Chem. Soc., 106, 106-111.

Monoclinic, $P2_1/n$, a = 16.225, b = 6.149, c = 16.496 Å, β = 92.75°, Z = 4. Mo radiation, R = 0.047 for 1608 reflexions.

The molecule (Fig. 1) adopts a twist-boat conformation with the 5-phenyl and 2-dimethylamino substituents cis to one another in pseudoequatorial positions. Both the ring and dimethylamino N atoms have planar geometries. The ring is somewhat flattened at the phosphorus end. Bond lengths: P=O 1.457(3), P-O 1.584(3), P-N (ring) 1.656(3), $P-NMe_2$ 1.620(3) Å.

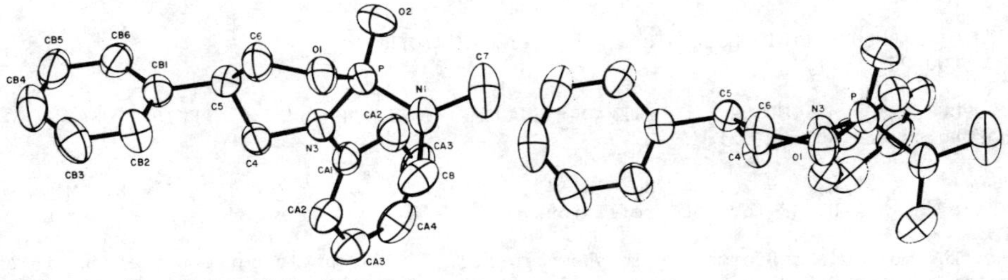

Fig. 1. Two views of $C_{17}H_{21}N_2O_2P$.

P,P-DIPHENYL-P-TETRAKIS(PERFLUOROMETHYL)ETHYLENE GLYCOL PHOSPHINATE
$C_{18}H_{11}F_{12}O_3P$

MAZHAR-UL-HAQUE, J. AHMED and W. HORNE, 1984. J. Cryst. Spectrosc. Res., 14, 169-177.

Triclinic, $P\bar{1}$, a = 8.796, b = 10.754, c = 12.271 Å, α = 68.56, β = 67.68, γ = 78.87°, Z = 2. Mo radiation, R = 0.048 for 2636 reflexions.

The phosphorus atom (Fig. 1) has tetrahedral geometry, with P-O(2) 1.611(2), P-O(22) 1.464(2) and P-C 1.784-1.785(3) Å. Molecules are linked to form centrosymmetric dimers by O-H...O hydrogen bonds (O(21)-H...O'(22), O...O 2.647(3) Å).

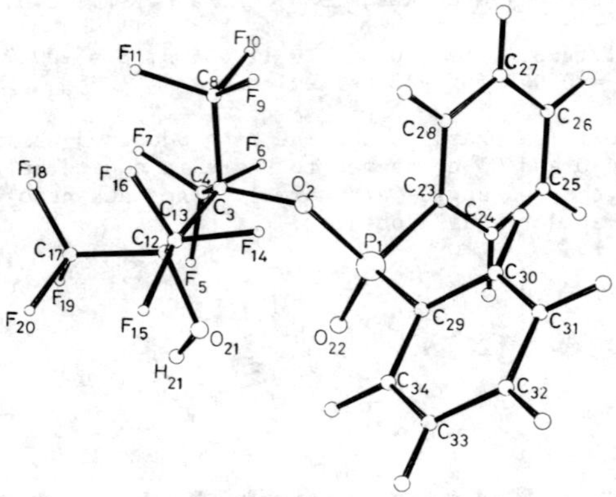

Fig. 1. A view of $C_{18}H_{11}F_{12}O_3P$.

(2-HYDROXYPHENYL)DIPHENYLPHOSPHONIUM BROMIDE
$C_{18}H_{16}BrOP$ $[Ph_2PH(C_6H_4OH)]^+.Br^-$

R. SCHMUTZLER, D. SCHOMBURG, R. BARTSCH and O. STELZER, 1984. Z. Naturforsch., B39, 1177-1184.

Monoclinic, $P2_1/n$, a = 9.631, b = 17.466, c = 10.229 Å, β = 110.46°, Z = 4. Cu radiation, R = 0.044 for 2324 reflexions.

The cation has tetrahedral geometry with P-C 1.783-1.790(4) Å, and takes part in strong hydrogen bonds with the bromide anion (P-H 1.42, H...Br 2.41 Å, P-H...Br 150°; O-H 1.00, H...Br 2.09 Å, O-H...Br 176°).

2-(O,O'-DIPHENYLDITHIOPHOSPHATE-S)-1,3-BENZOTHIAZOLE
$C_{19}H_{14}NO_2PS_3$

J.P. DECLERCQ, A. DUBOURG, G. ETEMAD-MOGHADAM, A. KLAEBE and J.J. PERIE, 1984. Phosphorus and Sulfur, 21, 31-37.

Monoclinic, C2/c, a = 20.238, b = 14.345, c = 15.472 Å, β = 120.11°, Z = 8. Mo radiation, R = 0.048 for 2067 reflexions.

The molecule conformation is shown in Fig. 1. The angle between the benzothiazole system and the P,S(2),S(3) plane is 95° with P-S(2) 1.900(2), P-S(3) 2.079(2), P-O 1.581(3) Å, S-P-S 109.8(1), O-P-O 99.4°.

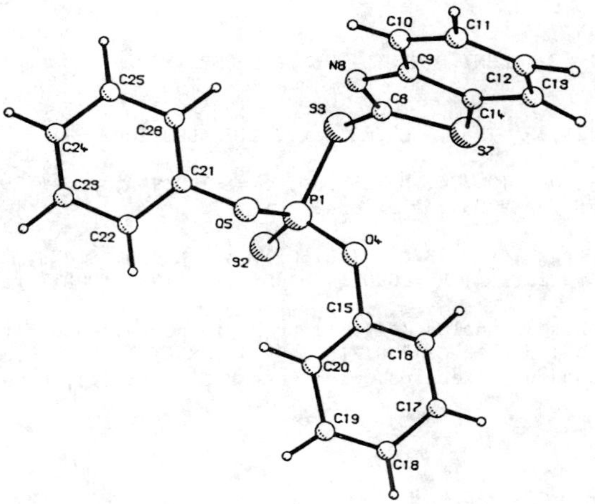

Fig. 1. A view of $C_{19}H_{14}NO_2PS_3$.

DICYCLOHEXYL(N-PHENYLTHIOCARBAMOYL)PHOSPHINE
$C_{19}H_{28}NPS$ (I) dcp

DIPHENYL(N-PHENYLTHIOCARBAMOYL)PHOSPHINE
$C_{19}H_{16}NPS$ (II) dpp

S.W. COWAN, B.F. HOSKINS and E.R.T. TIEKINK, 1984. Aust. J. Chem., 37, 1991-1998.

I. Monoclinic, $P2_1/c$, a = 20.541, b = 17.784, c = 10.2564 Å, β = 94.54°, D_m = 1.18, Z = 8. Cu radiation, R = 0.055 for 3119 reflexions.

II. Triclinic, $P\bar{1}$, a = 9.242, b = 9.994, c = 10.373 Å, α = 68.56, β = 71.21, γ = 86.00°, D_m = 1.26, Z = 2. Cu radiation, R = 0.061 for 2971 reflexions.

For I, the plane of the phenyl substituent of the isothiocyanate group has a dihedral angle of 18°, 32° with the P, C(1), S and N plane for the two discrete molecules in the asymmetric unit (Fig. 1). Structural analysis indicates substantial strain which possibly is the contributing factor enhancing this compound's reactivity compared to II. Comparable N-C(2)-C(3) angles for I and II are 121.1(6) and 124.4(3)°.

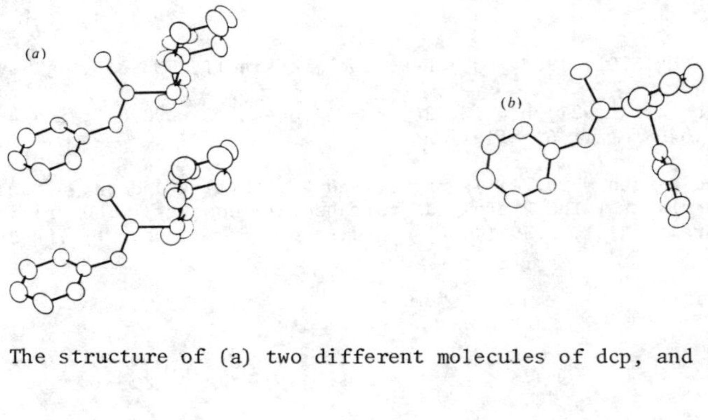

Fig. 1. The structure of (a) two different molecules of dcp, and (b) dpp.

1,2-DIPHENYL-1H-PHOSPHINDOLE
$C_{20}H_{15}P$ (I)

3-CHLORO-2-PHENYL-1-(2-(PHENYLETHYNYL)PHENYL)PHOSPHINDOLE
$C_{28}H_{18}ClP$ (II)

T. BUTTERS and W. WINTER, 1984. Chem. Ber., 117, 990-1002.

I. Monoclinic, $P2_1$, a = 10.474, b = 5.939, c = 13.019 Å, β = 112.55°, Z = 2. Mo radiation, R = 0.078 for 2293 reflexions.

II. Orthorhombic, $Pca2_1$, a = 18.873 (18.808), b = 8.499 (8.479), c = 27.51 (27.481) Å, Z = 8. Mo (Cu) radiation, R = 0.065 (0.051) for 1910 (3087) reflexions.

In these two phosphoindoles (Fig. 1) there is no evidence for shortening of the endocyclic P-C bonds (in I, P-C1 1.812(6), P-C4 1.836(6), P-C9 1.824(3) Å; in II there are two independent molecules, with mean P-C1 1.839(6), P-C4 1.810(7), P-C9 1.841(4) Å).

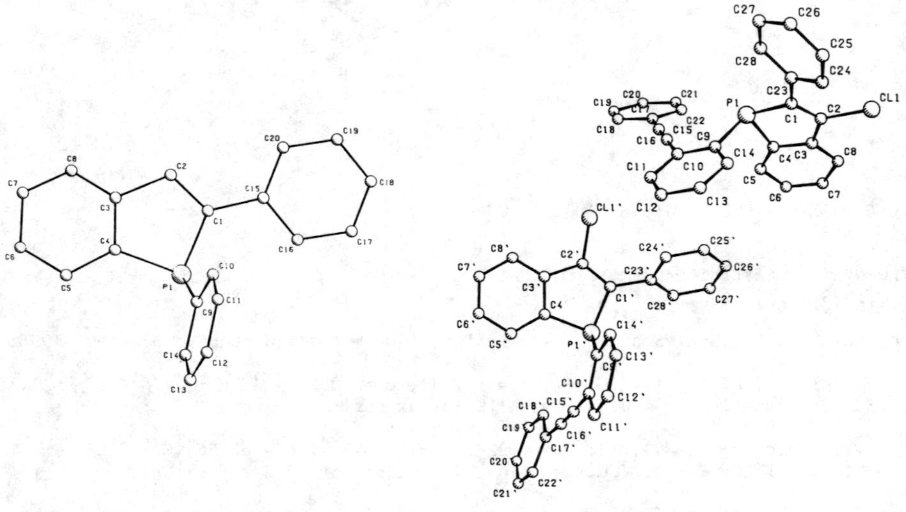

Fig. 1. A view of $C_{20}H_{15}P$ and the two independent molecules of $C_{28}H_{18}ClP$.

N,N-DIMETHYLAMINOTRIPHENYLPHOSPHONIUM TRIBROMIDE
$C_{20}H_{21}Br_3NP$ $(C_6H_5)_3PN(CH_3)_2Br_3$

L. CHICHE and H. CHRISTOL, 1984. J. Chem. Soc. Perkin II, 753-755.

Monoclinic, $P2_1/c$, a = 9.921, b = 15.426, c = 14.048 Å, β = 90.54°, Z = 4. Mo radiation, R = 0.046 for 1146 reflexions.

The P-N bond length (Fig. 1) is rather short 1.63(1) Å and the arrangement of bonds about N is planar. The P atom has tetrahedral geometry. The tribromide ion is asymmetric (Br-Br 2.470(4), 2.596(4) Å) and diverges from linearity (Br-Br-Br 174.4(1)°).

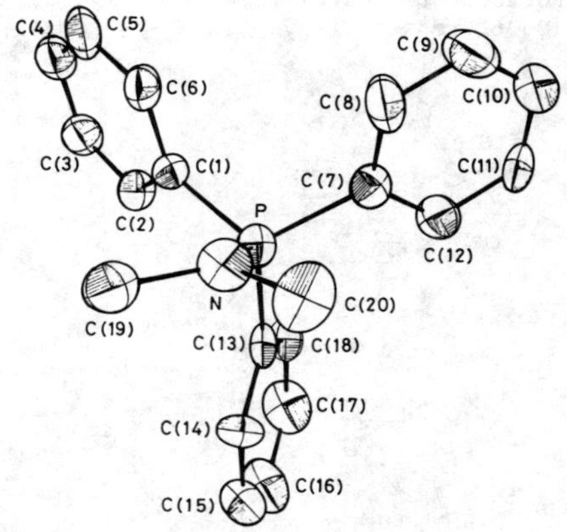

Fig. 1. $C_{20}H_{21}Br_3NP$: the structure of the cation.

CYCLOTETRA(N^3-n-PROPYL-1,3,2-DIAZAPHOSPHOLIDIN-1,2-DIYL)
$C_{20}H_{44}N_8P_4$

C. MALAVAUD, T. N'GANDO M'PONDO, L. LOPEZ, J. BARRANS and J.-P. LEGROS, 1984. Can. J. Chem., 62, 43-50.

Monoclinic, $P2_1/c$, a = 10.332, b = 12.288, c = 11.258 Å, β = 92.34°, D_m = 1.21, Z = 2. Mo radiation, R = 0.065 for 1537 reflexions.

The tetrameric molecule is shown in Fig. 1 and lies about a crystallographic inversion centre; the 8-membered ring has a twisted-chair conformation. The P-N bond lengths are in the range 1.671(6)-1.714(5) Å, with mean value 1.705(5) Å in the 8-membered ring.

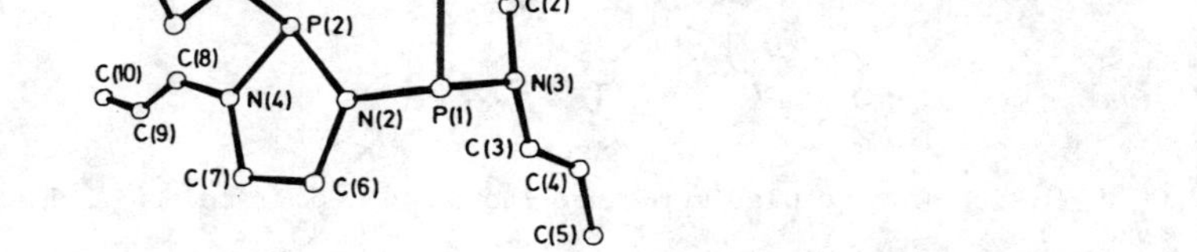

Fig. 1. A view of $C_{20}H_{44}N_8P_4$ and the 8-membered ring conformation.

TRIS(m-TRIFLUOROMETHYLPHENYL)PHOSPHINE SELENIDE
$C_{21}H_{12}F_9PSe$

D.W. ALLEN, L.A. MARCH and I.W. NOWELL, 1984. J. Chem. Soc. Dalton, 483-485.

Orthorhombic, Pbca, a = 31.822, b = 11.047, c = 12.426 Å, D_m = 1.68, Z = 8. Mo radiation, R = 0.054 for 1964 reflexions.

The P atom adopts distorted tetrahedral geometry (Fig. 1) with mean Se-P-C and C-P-C angles of 113.1 and 105.6° respectively. The P-Se bond length is 2.094(2) Å.

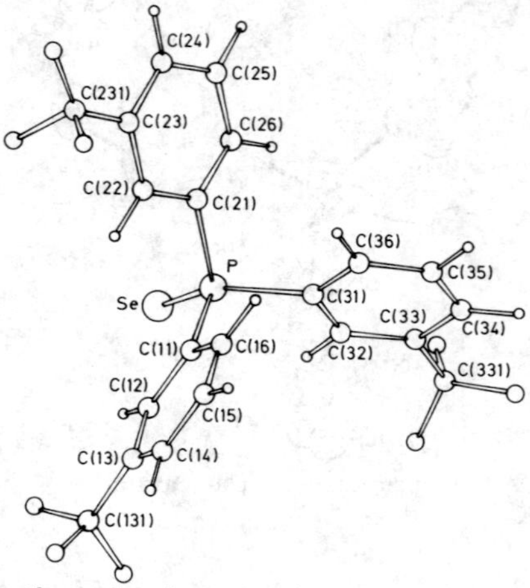

Fig. 1. $C_{21}H_{12}F_9PSe$: the molecular structure.

2-(1RS,2SR)-DIPHENYLPHOSPHINOYL-1-PHENYLPROPAN-1-OL
$C_{21}H_{21}O_2P$

A.D. BUSS, W.B. CRUSE, O. KENNARD and S. WARREN, 1984. J. Chem. Soc. Perkin I, 243-247.

Monoclinic, $P2_1/c$, a = 9.859, b = 17.218, c = 11.421 Å, β = 111.41°, D_m = 1.24, Z = 4. Cu radiation, R = 0.043 for 2642 reflexions.

The X-ray analysis established the structure and stereochemistry of the title compound (Fig. 1). Bond lengths and angles are unremarkable. The crystal is held together by van der Waals forces but without much stabilisation by stacking interactions between aromatic rings. An interesting feature is a P=O...H-O intramolecular hydrogen bond (O...O 2.738(4) Å).

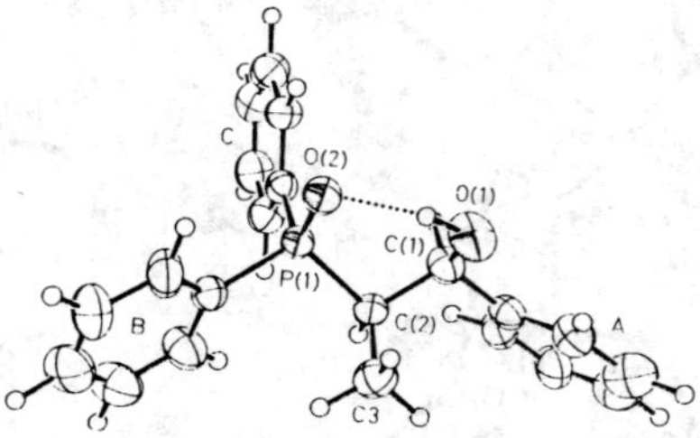

Fig. 1. $C_{21}H_{21}O_2P$: the molecular structure showing the intramolecular hydrogen bond.

2,4,6-TRIMETHYLPHENYL(DIPHENYLMETHYLENE)PHOSPHINE
$C_{22}H_{21}P$

T.A. VAN DER KNAAP, T.C. KLEBACH, F. VISSER, F. BICKELHAUPT, P. ROS, E.J. BAERENDS, C.H. STAM and M. KONIJN, 1984. Tetrahedron, 40, 765-776.

Monoclinic, $P2_1/c$, a = 11.938, b = 6.382, c = 23.66 Å, β = 90.81°, Z = 4. Cu radiation, R = 0.054 for 2420 reflexions.

In the molecule (Fig. 1), atoms P,C(1),C(2),C(8),C(14) are planar and the P=C bond length (P-C(1)) is 1.692(3) Å. The mesityl ring plane forms a dihedral angle of 71° with the P=C nodal plane; the phenyl groups at C(1) form smaller angles (42.9° and 36.6°) with the P=C nodal plane. Steric crowding causes P-C(1)-C(2) (127.2°) to be larger than P-C(1)-C(8) 116.2°.

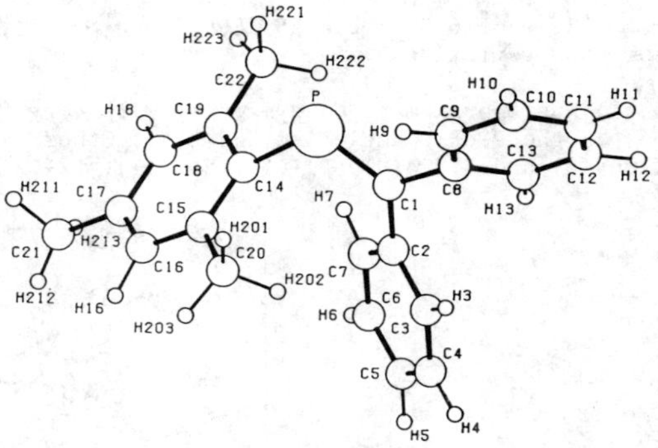

Fig. 1. A view of $C_{22}H_{21}P$.

2-DIMETHYLAMINO-1,3-DIPHENYL-2-PHENYLIMINO-1,3,2-DIAZAPHOSPHOLANE
$C_{22}H_{25}N_4P$

M.-R. MARRE, M. SANCHEZ, R. WOLF, J. JAUD and J. GALY, 1984. Can. J. Chem., 62, 2186-2191.

Monoclinic, $P2_1/c$, a = 10.243, b = 17.236, c = 11.992 Å, [β = 106.7°,] Dm = 1.22, Z = 4. Mo radiation, R = 0.056 for 1342 reflexions.

The geometry at the P atom (Fig. 1) is distorted tetrahedral with NPN angles varying from 91.8° (ring angle) to 121.7° for N(1)-P(2)-N(6). The diazophospholane ring has an envelope conformation where one of the 2 carbon atoms is at the tip. The three tricoordinated nitrogen atoms have a planar structure; the lone pairs on the two ring nitrogens have parallel orientation and both are orthogonal to the lone pair of the exocyclic nitrogen atom (N_7). The P-N bond lengths are relatively short, falling between 1.620 Å and 1.669 Å, while the P(2)=N(6) bond of 1.539(6) Å is among the shortest values reported.

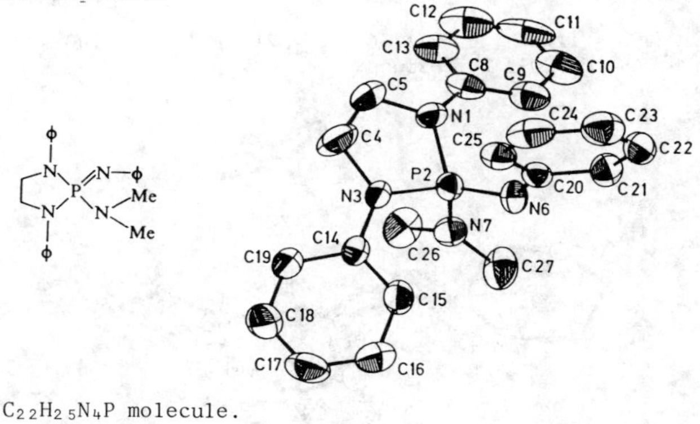

Fig. 1. The $C_{22}H_{25}N_4P$ molecule.

3-[BIS(4-CHLOROPHENYL)METHYLENE]-1,2-DI-t-BUTYL-1,2-DIPHOSPHIRANE n-PENTANE SOLVATE
$C_{22}H_{26}Cl_2P_2,0.5(C_5H_{12})$

M. BAUDLER, F. SAYKOWSKI, M. HINTZE, K.-F. TEBBE, T. HEINLEIN, A. VISSERS and M. FEHER, 1984. Chem. Ber., 117, 1542-1550.

Monoclinic, $P2_1/c$, a = 6.052, b = 16.378, c = 25.180 Å, β = 89.21°, Z = 4. Mo radiation, R = 0.067 for 1500 reflexions.

The molecule (Fig. 1) has approximate C_2 symmetry, with very small P-P-C3 angles (52.2 and 52.6(3)°); the P-C3-P angle is 75.2(3)°. Other dimensions include P1-P2 2.195(4), P-C3 1.804(8) and 1.794(9), P1-C1 1.861(11), P2-C2 1.883(10), C3-C4 1.353(11) Å.

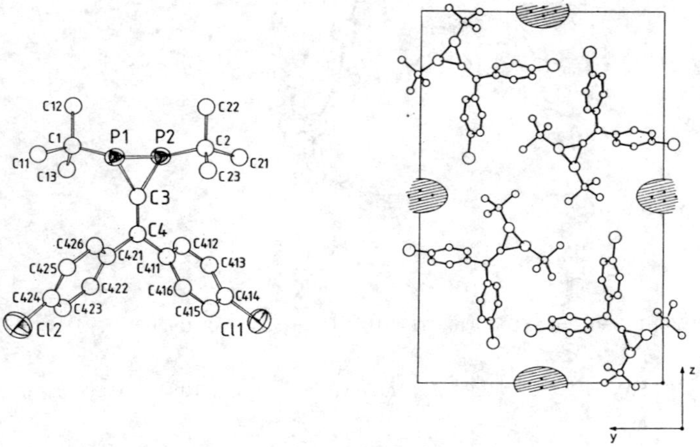

Fig. 1. The $C_{22}H_{26}Cl_2P_2$ molecule and details of the crystal packing; the n-pentane molecules lie in channels as shown in the shaded area.

N,N'-DICYCLOHEXYL-N-DINEOPENTOXYPHOSPHORYLTHIOUREA
$C_{23}H_{45}N_2O_3PS$

J. KAROLAK-WOJCIECHOWSKA, M. MIKOLAJCZYK, P. KIELBASINSKI and A. SUT, 1984. Phosphorus and Sulfur, 21, 221-225.

Monoclinic, $P2_1/c$, a = 11.225, b = 11.567, c = 21.423 Å, β = 100.92°, Z = 4. Cu radiation, R = 0.069 for 2954 reflexions.

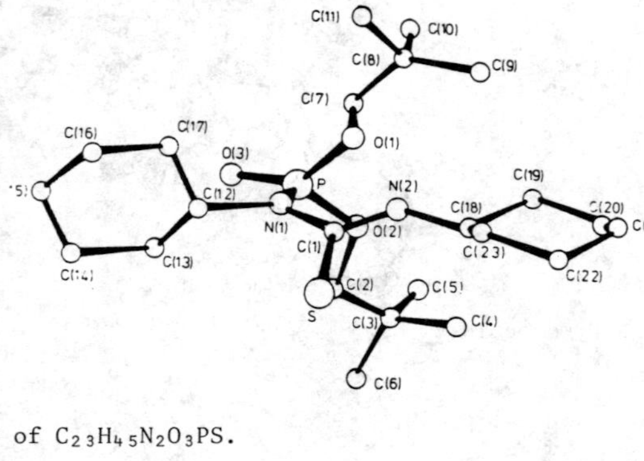

Fig. 1. A view of $C_{23}H_{45}N_2O_3PS$.

The cyclohexyl rings have chair conformations (Fig. 1) with N atoms equatorial. The P-N1-C1-S torsion angle is 125.7°. There are no hydrogen bonds in the crystal structure.

(8-CHLOROBENZOCYCLO-OCTEN-6-YL)-DIPHENYLPHOSPHINE OXIDE HEMIHYDRATE
$C_{24}H_{18}ClOP,0.5(H_2O)$

M. BOHSHAR, G. MAAS, H. HEYDT and M. REGITZ, 1984. Tetrahedron, 40, 5171-5176.

Monoclinic, $P2_1/n$, a = 18.140, b = 8.513, c = 26.873 Å, β = 108.3°, Z = 8. Mo radiation, R = 0.048 for 4069 reflexions.

The two independent molecules in the asymmetric unit differ in conformation; in one molecule (Fig. 1) the C(1)-C(2)-P-O(1) torsion angle is -11.6° (126.3° in the second molecule). The two cyclooctane systems have similar conformations. The P-O distances are 1.494 and 1.484(2) Å. The mean P-C2 distance is 1.814(3) Å. A centrosymmetrically related pair of molecules is linked by water molecules via O-H...O hydrogen bonds (O...O 1.860(4) Å).

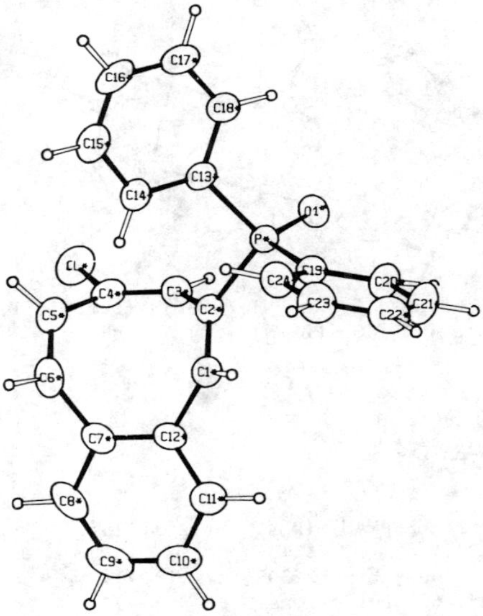

Fig. 1. A view of one of the $C_{24}H_{18}ClOP$ molecules.

1,5-DIBROMO-3,3,7,7-TETRAPHENYL-1,5-DITHIA-2,4,6,8-TETRAAZA-3,7-DI-λ^5-PHOSPHOCINE
$C_{24}H_{20}Br_2N_4P_2S_2$ (I) $1,5-(Ph_2PN)_2(NSBr)_2$

3,3,5,5,9,9,11,11-OCTAPHENYL-1,7-DITHIONIA-2,4,6,8,10,12-HEXAAZA-3,5,9,11-TETRA-PHOSPHACYCLODODECA-1,3,5,7,9,11-HEXAENE BIS(TRIBROMIDE) ACETONITRILE SOLVATE
$C_{48}H_{40}Br_6N_6P_4S_2,C_2H_3N$ (II) $[1,3,7,9-(Ph_2PN)_4(SN)_2][Br_3]_2,CH_3CN$

3,3,5,5,9,9,11,11-OCTAPHENYL-1,7-DITHIA-2,4,6,8,10,12-HEXAAZA-3,5,9,11-TETRAPHOSPHA-BICYCLO[5.5.0]DODECA-1,3,5,7,9,11-HEXAENE
$C_{48}H_{40}N_6P_4S_2$ (III) $1,3,7,9-(Ph_2PN)_4(SN)_2$

N. BURFORD, T. CHIVERS, M.N.S. RAO and J.F. RICHARDSON, 1984. Inorg. Chem., 23, 1946-1952.

I. Monoclinic, Pn, a = 11.476, b = 7.240, c = 16.914 Å, β = 106.42°, Z = 2. Mo radiation, R = 0.043 for 2588 reflexions.

II. Monoclinic, $P2_1/n$, a = 16.596, b = 13.978, c = 23.796 Å, β = 91.61°, Z = 4. Mo radiation, R = 0.051 for 3283 reflexions.

III. Triclinic, $P\bar{1}$, a = 9.191, b = 11.724, c = 21.409 Å, α = 79.29, β = 89.01, γ = 75.30°, Z = 2. Mo radiation, R = 0.046 for 6216 reflexions.

The three molecules are shown in Fig. 1. The structure of I consists of an eight-membered $P_2S_2N_4$ ring in a chair conformation with exocyclic Br substituents in a trans configuration (S-Br 2.454(7) and 2.440(8) Å). The average S-N and P-N distances are 1.53 and 1.62 Å, and the mean endocyclic bond angles at P, S, and N are 119.3, 114.0, and 139.3°, respectively. The neutral molecule III has a bicyclic structure containing a transannular S-S bond of 2.385(1) Å, resulting in fused seven-membered $P_2S_2N_3$ rings. The dication in II has a monocyclic structure with a 12-membered $P_4S_2N_6$ ring. The mean S-N, P-N(P-N-P), and P-N(P-N-S) bond lengths are 1.52, 1.58, and 1.66 Å in II, and 1.60, 1.58, and 1.62 Å in III. The absence of cross-ring bonding in III results in larger average intracyclic angles at N compared to those in II: PNP 137 vs. 133°, PNS 127 vs. 123°. The angles at P are closer to the tetrahedral value in II (113.7 vs. 119.6°).

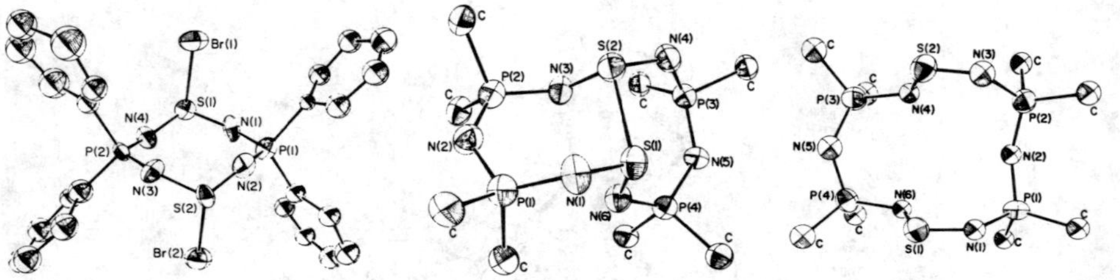

Fig. 1. The structures of 1,5-$(Ph_2PN)_2(NSBr)_2$ (left), $[1,3,7,9\text{-}(Ph_2PN)_4(SN)_2]^{2+}$ (middle), and 1,3,7,9-$(Ph_2PN)_4(SN)_2$ (right).

TETRAPHENYL IMIDODIPHOSPHATE
$C_{24}H_{21}NO_6P_2$

S. KULPE, I. SEIDEL and E. HERRMAN, 1984. Crystal Res. & Technol., 19, 661-668.

Monoclinic, $P2_1/c$, a = 12.009, b = 12.305, c = 18.993 Å, β = 122.1°, Z = 4. Mo radiation, R = 0.051 for 2690 reflexions.

The molecule exists in the phosphazene form with the two phosphoryl O atoms (O(3) and O(6)) in trans positions (Fig. 1). The imidophosphoryl group is approximately planar (deviations: O(3) 0.08, P(1) -0.09, N(1) 0.07, H(111) 0.07, P(2) -0.12, O(6) 0.06 Å). Some bond lengths are: P(1)-O(3) 1.448(4), P(2)-O(6) 1.474(5), P(1)-N(1) 1.653(6), P(2)-N(1) 1.639(6) Å. Pairs of molecules form dimers, connected by N(1)-H(111)...O(6') hydrogen bonds (N...O 2.747(7) Å, N-H...O 172.1°).

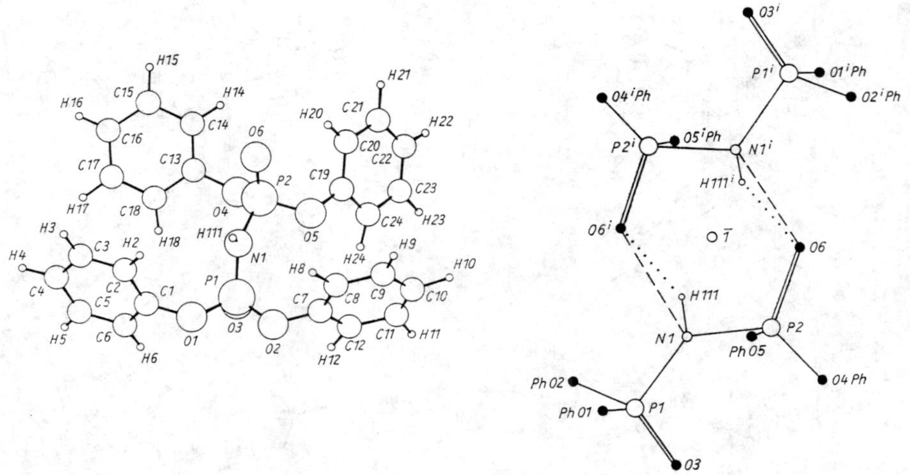

Fig. 1. The $C_{24}H_{21}NO_6P_2$ molecule and a view of a dimer pair.

BIS(DIPHENYLPHOSPHINO)AMINE
$C_{24}H_{21}NP_2$ (I)

BIS(P-BORANEDIPHENYLPHOSPHINO)AMINE
$C_{24}H_{27}B_2NP_2$ (II)

BIS(DIPHENOXYPHOSPHORYL)AMINE
$C_{24}H_{21}NO_6P_2$ (III)

DIETHYLPHOSPHINYLDIPHENOXYPHOSPHORYLAMINE
$C_{16}H_{21}NO_4P_2$ (IV)

H. NOTH and E. FLUCK, 1984. Z. Naturforsch., B39, 744-753.

I. Orthorhombic, Pccn, a = 8.922, b = 11.175, c = 20.970 Å, D_m = 1.21, Z = 4. Mo radiation, R = 0.059 for 1204 reflexions.

II. Monoclinic, $P2_1/n$, a = 9.870, b = 15.999, c = 14.860 Å, β = 90.78°, Z = 4. Mo radiation, R = 0.042 for 3122 reflexions.

III. Monoclinic, $P2_1/n$, a = 12.006, b = 12.305, c = 16.207 Å, β = 96.78°, Z = 4. Mo radiation, R = 0.040 for 2003 reflexions.

IV. Monoclinic, $P2_1/n$, a = 7.621, b = 14.966, c = 15.767 Å, β = 90.47°, D_m = 1.30, Z = 4. Mo radiation, R = 0.056 for 2459 reflexions.

The structures of these compounds having a P-NH-P unit are shown in Fig. 1. Molecule I lies on a twofold axis with P-N 1.692(2) Å, P-N-P 118.9(2)°. Molecule II has the BH_3 groups transoid and P-N 1.671 and 1.684(2), P-B 1.936 and 1.950(1) Å, P-N-P 132.6(1)°. In III and IV there is weak intermolecular hydrogen bonding to the O atom of a phosphoryl group. Dimensions in III are P-N 1.632 and 1.646(2) Å, P-N-P 130.9(1)°; in IV the corresponding values are 1.683, 1.630(3) Å and 129.7°.

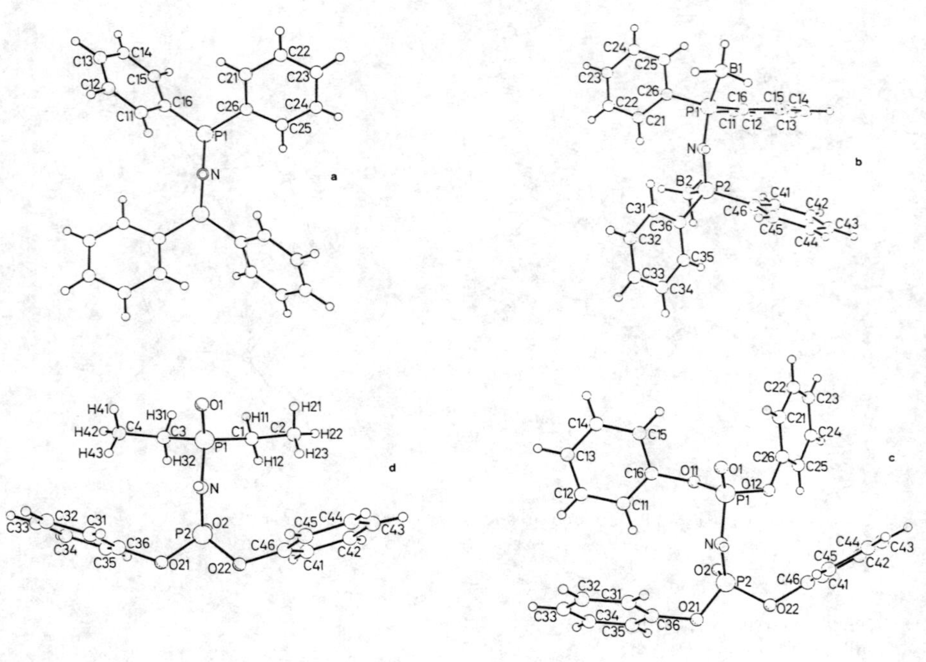

Fig. 1. Views of P-NH-P compounds I - IV (clockwise from top left).

2,2,4,4-TETRAKIS(DIISOPROPYLAMINO)-1,3,$2\lambda^5$,$4\lambda^5$-DIAZADIPHOSPHETE CHLOROFORM SOLVATE
$C_{24}H_{56}N_6P_2$,2($CHCl_3$)

A. BACEIREDO, G. BERTRAND, J.-P. MAJORAL, G. SICARD, J. JAUD and J. GALY, 1984. J. Am. Chem. Soc., 106, 6088-6089.

Monoclinic, $P2_1/n$, a = 10.723, b = 17.850, c = 11.143 Å, β = 112.2°, Z = 2. Mo radiation, R = 0.103 for 1587 reflexions.

The molecule lies about an inversion centre and the central four-membered ring is planar (Fig. 1), with P-N-P 85.0(5)°, P(1)-N(1) 1.651(8), P(1)-N'(1) 1.648(8), P(1)-N(2) 1.645(8), P(1)-N(3) 1.644(9) Å. Each N atom is trigonal planar and unsaturation in the system is strongly delocalized.

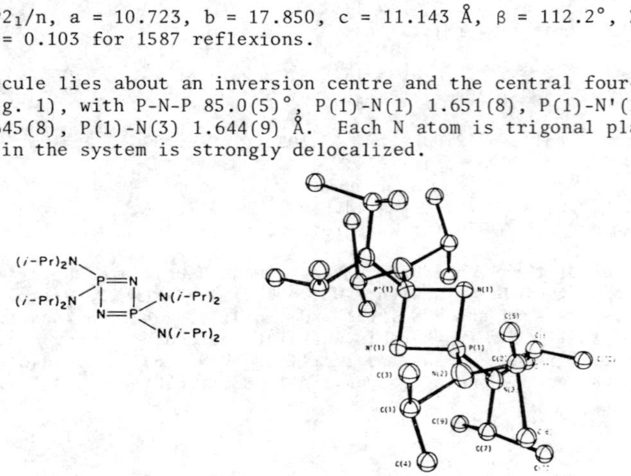

Fig. 1. The $C_{24}H_{56}N_6P_2$ structure.

(BENZOYLMETHYL)TRIPHENYLPHOSPHONIUM BROMIDE

$C_{26}H_{22}BrOP$ $[PhCOCH_2PPh_3]^+Br^-$

M.Yu. ANTIPIN and Yu.T. STRUCHKOV, 1984. J. Struct. Chem., 25, 106-110 [Zh. Strukt. Khim., 25(1), 122-127].

Orthorhombic, Pbca, a = 10.818, b = 18.359, c = 21.805 Å, Z = 8. Mo radiation, R = 0.063 for 1974 reflexions (at -120°C).

The structure contains an unusual C-H...Br hydrogen bond, formed by the bromide anion and a methylene hydrogen atom (C...Br 3.457 Å, H...Br 2.26 Å, C-H...Br 172°). Bond lengths are shown in Fig. 1.

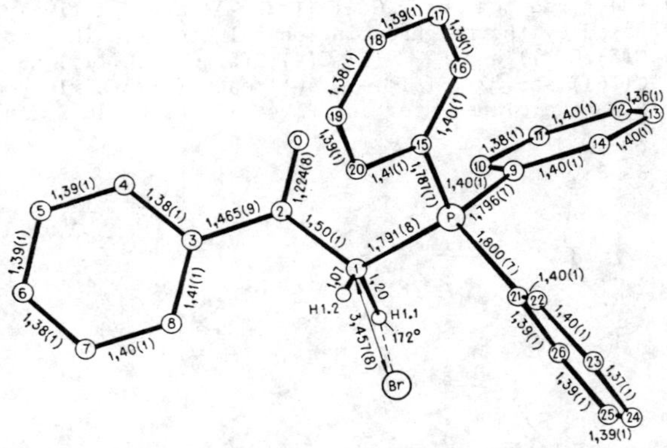

Fig. 1. The structure of (benzoylmethyl)triphenylphosphonium bromide.

1,1-BIS(DIPHENYLPHOSPHINO)ETHENE

$C_{26}H_{22}P_2$ $H_2C=C(PPh_2)_2$

H. SCHMIDBAUR, R. HERR and J. RIEDE, 1984. Chem. Ber., 117, 2322-2327.

Triclinic, P$\bar{1}$, a = 10.011, b = 11.209, c = 12.979 Å, α = 81.99, β = 120.50, γ = 119.98°, Z = 2. Mo radiation, R = 0.056 for 2084 reflexions.

The analysis determines the conformation in the solid state as shown in Fig. 1. The bond length data (C1=C2 1.327(6), C1-P1 1.829(4), C1-P2 1.838(4), P-C(ph) 1.819-1.838(5) Å) are consistent with no significant C=P interactions.

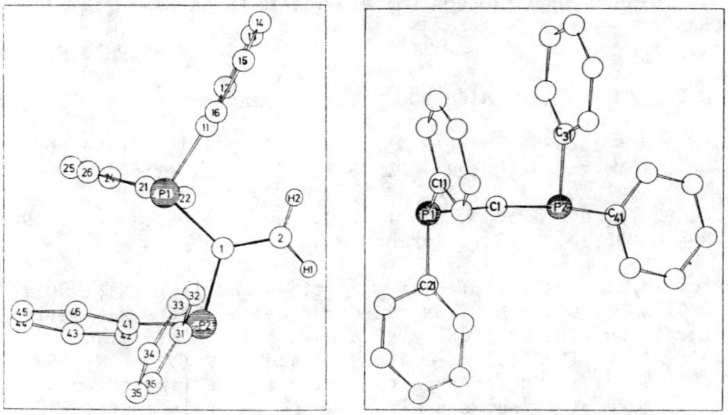

Fig. 1. Two views (in the ethylene plane and along the C1-C2 axis) for $C_{26}H_{22}P_2$.

3,8-DI-t-BUTYL-5,6-DIPHENYL-1-THIOXO-2,9-DITHIA-1-PHOSPHABICYCLO[4.3.0]NONA-3,7-DIENE
$C_{26}H_{31}PS_3$

T. UCHIDA, K. KOZAWA, T. SAITO and S. MOTOKI, 1984. Bull. Chem. Soc. Japan, 57, 1990-1993.

Orthorhombic, Pbca, a = 17.116, b = 19.798, c = 14.898 Å, D_m = 1.23, Z = 8. Mo radiation, R = 0.083 for 3318 reflexions.

The bicyclo skeleton has a shallow envelope thiaphosphacyclopentene ring and a boat form thiaphosphacyclohexene ring, which are cis-fused together via the bridging P-C(4) bond (one of the sides of the six-membered boat) (Fig. 1). The C(3)-C(4) bond distance of 1.594 Å and that of P-C(4) (1.895 Å) are relatively long and this is due to the strain caused by the neighbouring phenyl groups. The two ethylenic groups E1 (S(1),C(1),C(2),C(3),C(7)) and E2 (S(2),C(4),C(5),C(6),C(11)) and the two phenyl groups P1 (C(31)...C(36)) and P2 (C(41)...C(46)) each form fairly good planes. The dihedral angles between the planes are: E1/P1 44.2°, E2/P2 45.3° and P1/P2 48.9°.

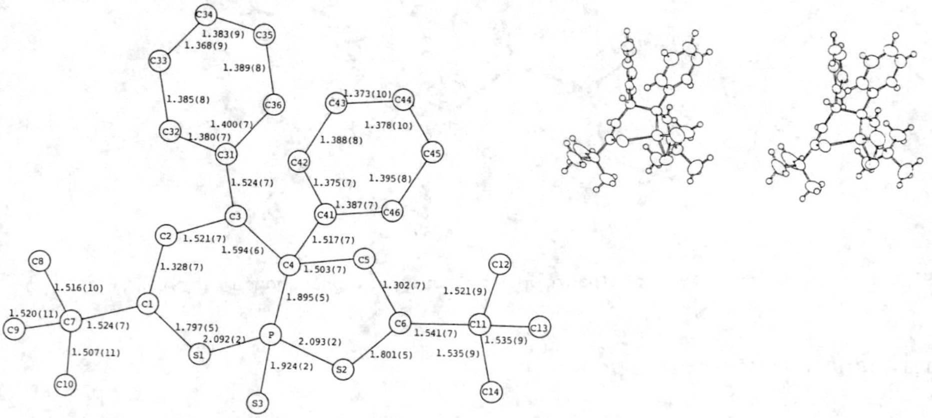

Fig. 1. Bond distances in $C_{26}H_{31}PS_3$, and stereoscopic view of the molecule.

DIETHYLAMMONIUM 1-OXO-3,3,5,5-TETRAPHENYL-1-THIA-2,4,6-TRIAZA-3,5-DIPHOSPHORINE-1-OXIDE
$C_{28}H_{32}N_4O_2P_2S$ (I) $Et_2NH_2^+[(Ph_2PN)_2(NSO_2)]^-$

1-OXO-1-(2,2,4,4-TETRAPHENYLDIPHOSPHAZENYL)-3,3,5,5-TETRAPHENYL-1-THIA-2,4,6-TRIAZA-3,5-DIPHOSPHORINE
$C_{48}H_{42}N_6OP_4S$ (II) $(Ph_2PN)_2(NS(O)NPPh_2NPPh_2NH_2)$

T. CHIVERS, M.N.S. RAO and J.F. RICHARDSON, 1984. Inorg. Chem., 23, 1531-1535.

I. Triclinic, $P\bar{1}$, a = 11.0853, b = 12.4189, c = 10.6956 Å, α = 96.008, β = 103.590, γ = 96.517°, Z = 2. Mo radiation, R = 0.043 for 2482 reflexions.

II. Monoclinic, $P2_1/c$, a = 17.284, b = 15.224, c = 18.320 Å, β = 114.906°, Z = 4. Mo radiation, R = 0.036 for 2625 reflexions.

The structure of I consists of $Et_2NH_2^+$ cations and $[(Ph_2PN)_2(NSO_2)]^-$ anions (Fig. 1) linked by N-H...O and N-H...N hydrogen bonds. The six-membered ring has a P-envelope conformation with P(1) ca. 0.45 Å out of the NSNPN plane. Bond lengths for I: S-N 1.613(3) and 1.537(3), S-O 1.419(3) and 1.399(3), P-N 1.582(3), 1.595(3), 1.605(3) and 1.611(2) Å. Bond length alternation in the ring may result from hydrogen bonding. The structure of II shows a mixed phosphazen-sulphanuric ring of the type $(Ph_2PN)_2(NS(O)R)$ (Fig. 1) where R is the diphosphazenyl chain $NPPh_2NPPh_2NH_2$. The ring adopts a chair conformation, with S and one N atom lying on opposite sides of the mean plane. An intramolecular N-H...O hydrogen bond is present. Bond lengths for II: S-N 1.580(4)-1.585(4), S-O 1.459(2), P-N 1.598(4)-1.614(3) (endocyclic), 1.561(4)-1.625(3) Å (exocyclic).

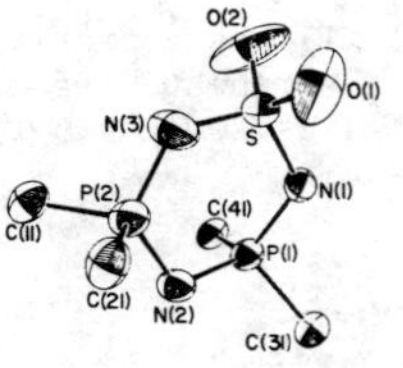

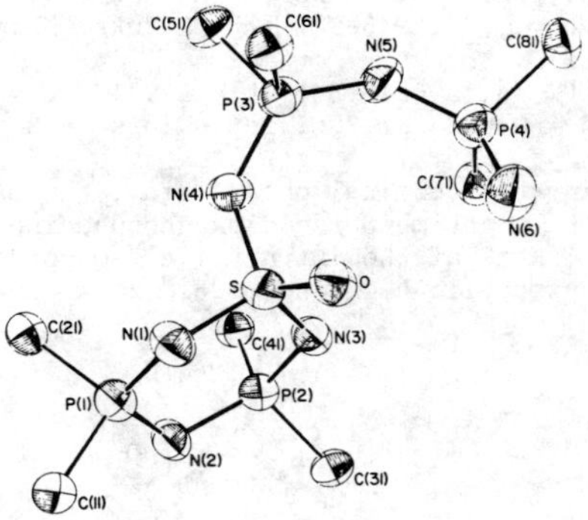

Fig. 1. The $[(Ph_2PN)_2(NSO_2)]^-$ anion (left) and the $(Ph_2PN)_2(NS(O)NPPh_2NPPh_2NH_2$ molecule (right).

2,4,4,6,6-PENTAKIS(AZIRIDINO)-2-(TRIPHENYLPHOSPHINEIMINO)CYCLOTRIPHOSPHAZENE $C_{28}H_{35}N_9P_4$

K.C.K. SWAMY, M.D. POOJARY, S.S. KRISHNAMURTHY and H. MANOHAR, 1984. Z. Naturforsch., B39, 615-617.

Triclinic, P$\bar{1}$, a = 9.336, b = 11.427, c = 14.922 Å, α = 100.47, β = 93.11, γ = 94.75°, Z = 2. Mo radiation, R = 0.056 for 4286 reflexions.

In the molecule (Fig. 1) the conformation of the $-NPPh_3$ group is defined by the N4-P1-N9-P4 torsion angle of 15°; this was not what was anticipated, and brings one of the phenyl groups in close proximity to the aziridino group geminal to the $-NPPh_3$ substituent. The phosphazine ring is planar to within 0.07 Å, with P-N 1.576-1.610(3), P-N(aziridino) 1.677-1.722(6) Å. The exocyclic P4-N9 (1.557(3) Å) and P1-N9 (1.590(3) Å) are indicative of π-bonding involving P(ring)-N(exo)-P(exo) units at the triphenylphosphazenyl end of the molecule.

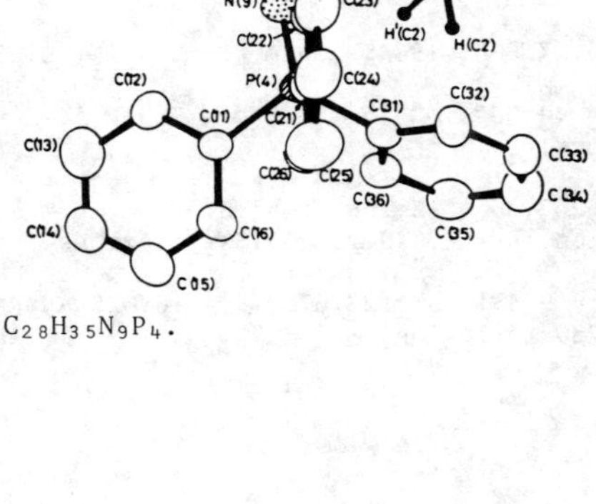

Fig. 1. A view of $C_{28}H_{35}N_9P_4$.

4',4''-DIMETHYL-6',6''-DI-t-BUTYL-2-PHENOXY-4,5,7,8-DIBENZO-6H-1,3,2-DIOXAPHOSPHOCINE
$C_{29}H_{35}O_3P$

I.A. LITVINOV, Yu.T. STRUCHKOV, B.A. ARBUZOV, R.P. ARSHINOVA and O.V. OVODOVA, 1984. J. Struct. Chem., 25, 943-947 [Zh. Strukt. Khim., 25(6), 118-122].

Monoclinic, $P2_1/a$, a = 15.846, b = 12.312, c = 15.325 Å, β = 118.283°, Z = 4. Mo radiation, R = 0.044 for 2961 reflexions.

Excluding the terminal phenyl ring, the molecule (Fig. 1) has approximate mirror symmetry. The eight-membered dioxaphosphocine ring has a chair-boat conformation. The average length of the intracyclic P-O bonds, 1.642(2) Å, is somewhat greater than that of the exocyclic P-O bond, 1.618(2) Å.

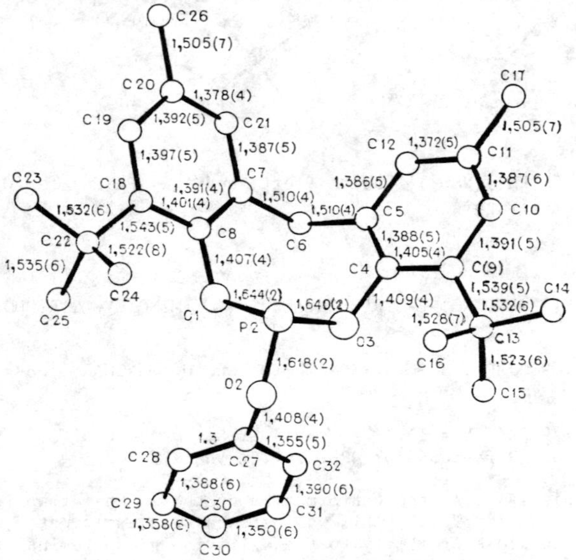

Fig. 1. Geometry of the 4',4''-dimethyl-6',6''-di-t-butyl-2-phenoxy-4,5,7,8-dibenzo-6H-1,3,2-dioxaphosphocine molecule.

(BIS(TRIMETHYLSILYL)METHYLENE)DIPHENYL(2,4,6-TRIMETHYLBENZOYL)PHOSPHORANE
$C_{29}H_{39}OPSi_2$ (I)

(BIS(TRIMETHYLSILYL)METHYLENE)((BIS(TRIMETHYLSILYL)METHYL)DIPHENYLPHOSPHORANYLIDENE-METHYL)DIPHENYLPHOSPHORANE
$C_{39}H_{58}P_2Si_4$ (II)

(DI-t-BUTYLPHOSPHINO)DIPHENYL(TRIMETHYLSILYLMETHYLENE)PHOSPHORANE
$C_{24}H_{38}P_2Si$ (III)

R. APPEL, G. HAUBRICH and F. KNOCH, 1984. Chem. Ber., 117, 2063-2075.

I. Orthorhombic, Fdd2, a = 44.032, b = 26.868, c = 9.936 Å, Z = 16. Mo radiation, R = 0.071 for 1840 reflexions.

II. Triclinic, $P\bar{1}$, a = 12.262, b = 13.165, c = 13.395 Å, α = 92.76, β = 90.70, γ = 106.10°, Z = 2. Mo radiation, R = 0.049 for 3813 reflexions.

III. Monoclinic, $P2_1/c$, a = 9.451, b = 22.674, c = 16.625 Å, β = 135.38°, Z = 4. Mo radiation, R = 0.047 for 2730 reflexions.

Views of the three molecules are in Fig. 1. In I, the P=C distance (P1-C1) is 1.694(9) Å and P1-C2 is 1.920(10) Å. C1-Si distances are 1.851 and 1.858(9) Å. In II, the P=C distances (P1-C2 and P2-C1) are 1.699(4) and 1.693(4) Å, with C2-Si 1.834 and 1.850(4) Å, and C3-Si 1.899 and 1.907(4) Å. The P1-C1 distance is 1.762(4) Å and the P1-C1-P2 angle is 135.2(2)°. In III, the principal dimensions are P1-P2 2.248(5), P1=C1 1.674(5), P2-C 1.900 and 1.901(5) Å, P1-C1-Si 131.7(2)°.

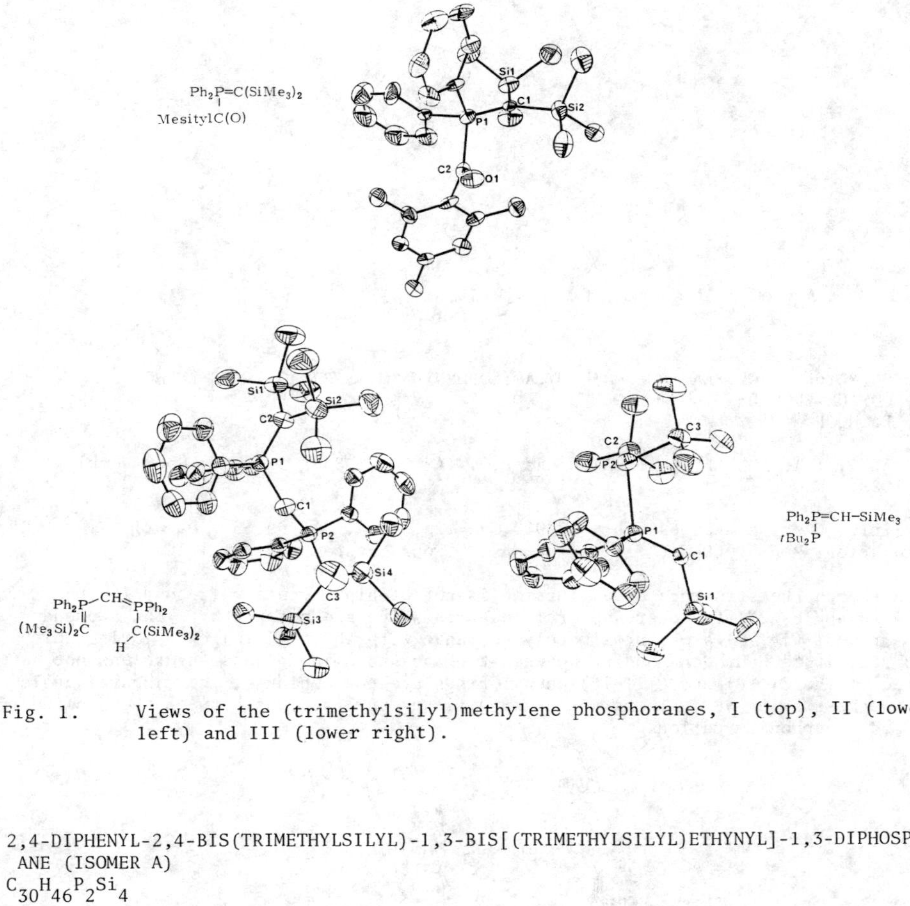

Fig. 1. Views of the (trimethylsilyl)methylene phosphoranes, I (top), II (lower left) and III (lower right).

2,4-DIPHENYL-2,4-BIS(TRIMETHYLSILYL)-1,3-BIS[(TRIMETHYLSILYL)ETHYNYL]-1,3-DIPHOSPHETANE (ISOMER A)
$C_{30}H_{46}P_2Si_4$

R. APPEL, C. CASSER and F. KNOCH, 1984. Chem. Ber., <u>117</u>, 2693-2702.

Monoclinic, $P2_1/n$, a = 7.032, b = 23.080, c = 11.268 Å, β = 103.22°, Z = 2. Mo radiation, R = 0.054 for 2844 reflexions.

The molecule (Fig. 1) lies about an inversion centre, with C(1)-P 1.887(2), C(1')-P 1.892(2), C(2)-P 1.773(2), C(2)-C(3) 1.191(4), C(3)-Si(1) 1.841(3), C(1)-Si(2) 1.924(2) Å, C(1)-P-C(1') 88.9(1)°.

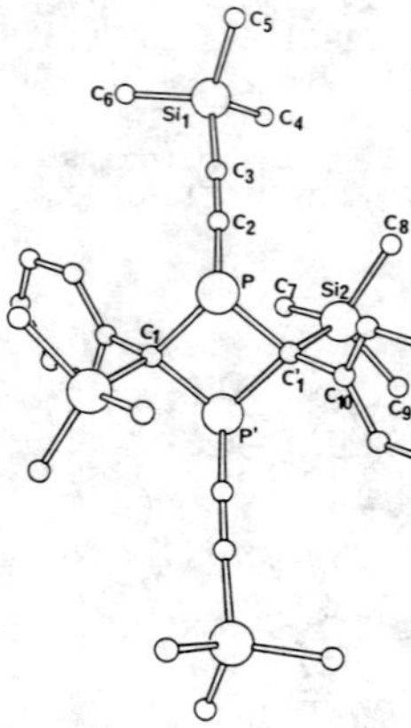

Fig. 1. A view of isomer A of $C_{30}H_{46}P_2Si_4$.

TRI-n-BUTYL(6-(2-ACETYLAMINO-4-DIETHYLAMINOPHENYLAZO)-3-BROMO-5-METHYLPHENYL)-PHOSPHONIUM BROMIDE
$C_{31}H_{49}Br_2N_4OP$

D.W. ALLEN, I.W. NOWELL, L.A. MARCH and B.F. TAYLOR, 1984. J. Chem. Soc. Perkin I, 2523-2528.

Monoclinic, C2/c, a = 25.695, b = 10.910, c = 25.363 Å, β = 90.63°, D_m = 1.27, Z = 8. Mo radiation, R = 0.069 for 710 reflexions. [Poor crystal quality.]

Although the structure determination is not of high accuracy (σ bond lengths 0.03 Å; σ angles 2°) there are several features which are noteworthy. The azo and amido groups (Fig. 1) are approximately coplanar with the C(11-16) phenyl ring which would facilitate hydrogen bonding between the azo and amido groups [hydrogens not located]. The C(1-6) and C(11-16) phenyl rings are not coplanar, the dihedral angle between these being 48°. There are no significant interactions between the bromide ions and phosphonium cations.

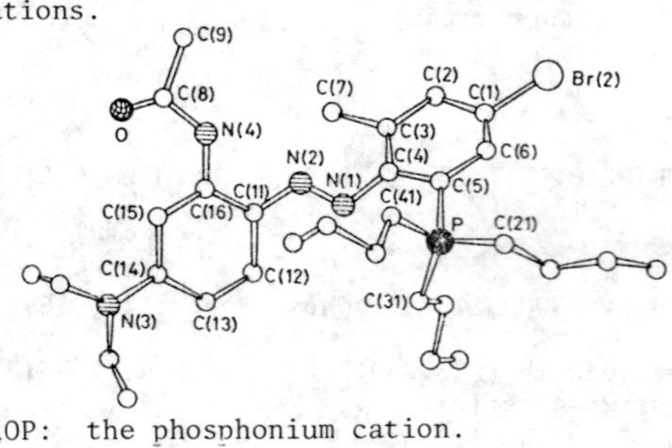

Fig. 1. $C_{31}H_{49}Br_2N_4OP$: the phosphonium cation.

1,2-DI-t-BUTYL-3,4-BIS(2-PHENYLETHENYL)-3,4-BIS(TRIMETHYLSILYLOXY)-1,2-DIPHOSPHETANE
$C_{32}H_{50}O_2P_2Si_2$

R. APPEL, F. KNOCH and H. KUNZE, 1984. Chem. Ber., <u>117</u>, 3151-3159.

Monoclinic, $P2_1/c$, a = 16.843, b = 10.329, c = 20.880 Å, β = 94.18°, Z = 4. Mo radiation, R = 0.054 for 3178 reflexions.

In this phosphetane (Fig. 1) the bond lengths in the four-membered ring are: P1-P2 2.198(1), P1-C1 1.929(3), P2-C2 1.922(3), C1-C2 1.582(5) Å and the P-P-C angles are 77.3 and 78.1(1)°, with P-C-C 95.2 and 94.6(2)°.

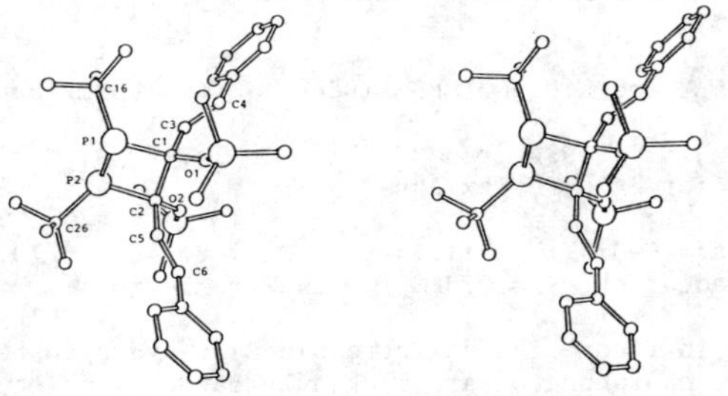

Fig. 1. A stereoview of $C_{32}H_{50}O_2P_2Si_2$.

FLUORENE-9-SPIRO-1-(3,4-DI-t-BUTYL)-3,4-DIPHOSPHETANE-2-SPIRO-9'-FLUORENE $C_{34}H_{34}P_2$ (I)

1,2-DI-t-BUTYL-1,2-DIPHOSPHORINANE-[6,5,4-1,m]-FLUORENE-SPIRO-9'-FLUORENE $C_{34}H_{34}P_2$ (II)

A.N. CHERNEGA, M.Yu. ANTIPIN, Yu.T. STRUCHKOV, V.P. KUKHAR', I.V. SHEVCHENKO, O.I. KOLODYAZHNYI and I.E. BOLDESKUL, 1984. J. Struct. Chem., 25, 280-287 [Zh. Strukt. Khim. 25(2), 122-130].

I. Monoclinic, $P2_1/c$, a = 14.033, b = 9.556, c = 20.612 Å, β = 95.42°, Z = 4. Cu radiation, R = 0.036 for 3363 reflexions.

II. Triclinic, $P\bar{1}$, a = 9.7084, b = 14.3189, c = 10.7600 Å, α = 102.26, β = 102.33, γ = 95.55°, Z = 2. Cu radiation, R = 0.050 for 4055 reflexions.

Bond lengths in the two molecules are shown in Fig. 1. Compounds I and II contain unusual diphosphorinane and diphosphetane rings with P-P bonds. The six-membered heterocycle in I is non-planar, having a distorted boat conformation. The four-membered heterocycle in II is non-planar with a fold angle (about P(2)...C(3)) of 49.9°. All P-C distances in the two molecules are significantly longer than usual and the C(1)-C(2) bond in II is long at 1.626(4) Å.

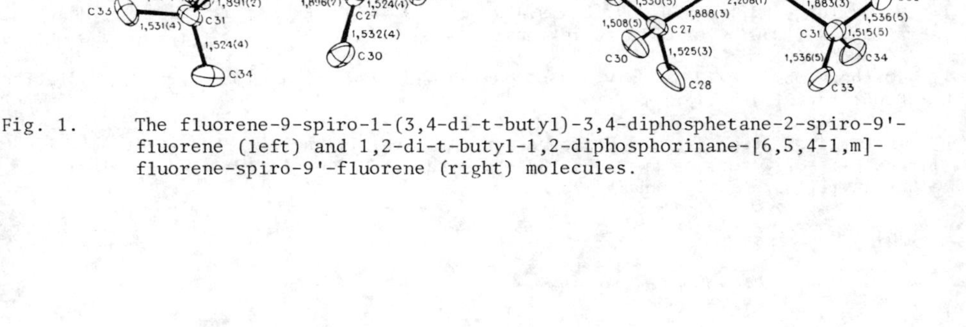

Fig. 1. The fluorene-9-spiro-1-(3,4-di-t-butyl)-3,4-diphosphetane-2-spiro-9'-fluorene (left) and 1,2-di-t-butyl-1,2-diphosphorinane-[6,5,4-1,m]-fluorene-spiro-9'-fluorene (right) molecules.

TETRAPHENYLPHOSPHONIUM 2-HYDROXYPHENYLPHENYLPHOSPHONATE CATECHOL
$C_{36}H_{30}O_4P_2,C_6H_6O_2$ (I)

PYRIDINIUM 2-HYDROXYPHENYLPHENYLPHOSPHONATE CATECHOL
$C_{17}H_{16}NO_4P,C_6H_6O_2$ (II)

C.A. POUTASSE, R.O. DAY and R.R. HOLMES, 1984. J. Am. Chem. Soc., 106, 3814-3820.

I. Monoclinic, $P2_1/n$, a = 9.996, b = 26.858, c = 14.160 Å, β = 109.57°, Z = 4. Mo radiation, R = 0.045 for 2470 reflexions.

II. Triclinic, $P\bar{1}$, a = 9.074, b = 11.149, c = 11.786 Å, α = 70.26, β = 87.54, γ = 73.72°, Z = 2. Mo radiation, R = 0.058 for 2422 reflexions.

The phosphonate in I contains an intramolecular hydrogen-bonded seven-membered ring (Fig. 1) and the phosphonates are linked by catechol molecules to form a chain. In II the pyridinium ring disrupts the intramolecular hydrogen bond and dimeric phosphonate units are linked by catechol molecules in a doubly hydrogen-bonded chain. A systematic classification of hydrogen bonding in phosphates is presented. It is noted that O-X lengths increase with decreasing proton acidity in the system P=O...H-X (X = O, N). Little variation in P=O bond lengths is apparent.

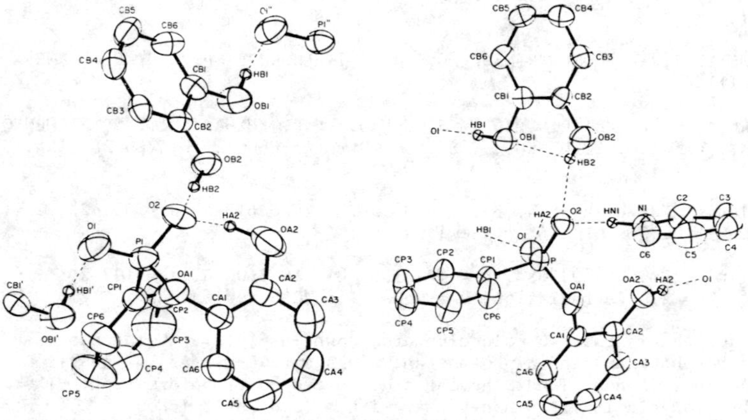

Fig. 1. Plots showing part of the hydrogen-bonded chains of the 2-hydroxyphenyl-phosphonate derivatives (I) (left) and (II) (right).

PHELLANPHOS DISULPHIDE
$C_{36}H_{38}P_2S_2$ (I)

NOPAPHOS DISULPHIDE
$C_{37}H_{38}P_2S_2$ (II)

O. SAMUEL, SHENG YONG ZHANG and H.B. KAGAN, 1984. Phosphorus and Sulfur, 21, 145-154.

I. Monoclinic, $P2_1$, a = 9.069, b = 9.439, c = 19.248 Å, β = 94.69°, Z = 2. Mo radiation, R = 0.036 for 2819 reflexions.

II. Orthorhombic, $P2_12_12_1$, a = 9.036, b = 20.052, c = 18.052 Å, Z = 4. Mo radiation, R = 0.061 for 2154 reflexions.

Details of structure and conformation for both (I) and (II) are in Fig. 1. In (I) the P-C1-C2-P torsion angle is 102.8°, with P...P 4.01 Å; the corresponding values in (II) are 131.7° and 4.275 Å.

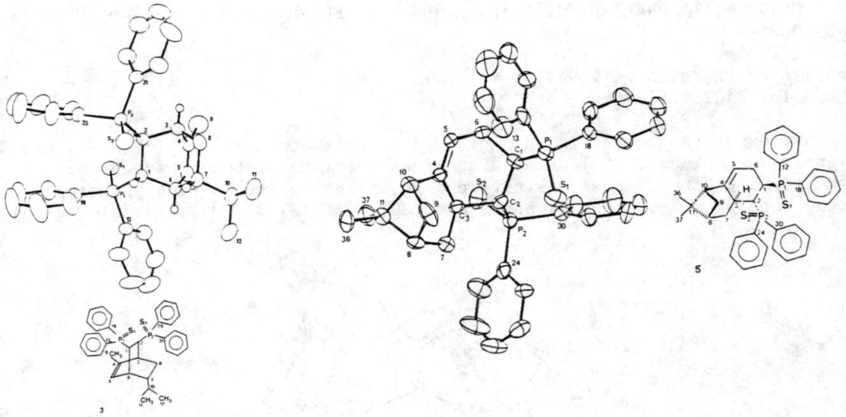

Fig. 1. Views of phellanphos and nopaphos disulphides.

CYCLOTETRA(MESITYLPHOSPHATHIANE)
$C_{36}H_{44}P_4S_4$ (I)

cyclo-DI(MESITYLTHIO-λ^5-PHOSPHATHIANE)
$C_{18}H_{22}P_2S_4$ (II)

C. LENSCH and G.M. SHELDRICK, 1984. J. Chem. Soc. Dalton, 2855-2857.

I. Monoclinic, Cc, a = 22.993, b = 11.679, c = 19.378 Å, β = 130.65°, Z = 4. Mo radiation, R = 0.050 for 3209 reflexions.

II. Monoclinic, $P2_1/c$, a = 8.901, b = 18.452, c = 6.413 Å, β = 98.74°, Z = 2. Mo radiation, R = 0.062 for 904 reflexions.

The eight-membered $(PS)_4$ ring of I adopts a crown conformation (Fig. 1) but is disordered over two components (78.7/21.3); the mean P-S distance is 2.127(1) Å. Molecule II lies about an inversion centre with exocyclic S atoms. The endocyclic P-S bonds are significantly different (2.093 and 2.125(1) Å). The exocyclic P-S distance is 1.917(1) Å.

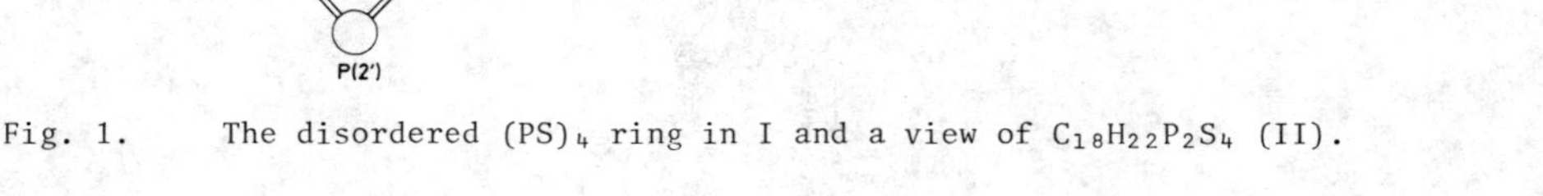

Fig. 1. The disordered $(PS)_4$ ring in I and a view of $C_{18}H_{22}P_2S_4$ (II).

[(4-PHENYLBENZOYL)(TRIPHENYLPHOSPHINE)METHYLENE]-o-TOLYLIODONIUM TETRAFLUOROBORATE
$C_{39}H_{31}BF_4IOP$ $(C_{39}H_{31}IOP).(BF_4)$

R.M. MORIARTY, I. PRAKASH, O. PRAKASH and W.A. FREEMAN, 1984. J. Am. Chem. Soc., 106, 6082-6083.

Monoclinic, $P2_1/n$, a = 11.944, b = 13.479, c = 21.424 Å, β = 91.07°, Z = 4. R = 0.0683 for 2102 reflexions.

A view of the cation is in Fig. 1. The five atoms I, C(1), P, C(2) and O are essentially coplanar. Molecular dimensions are interpreted in favour of a resonance form with a P=C bond and a positive charge on the iodine being a principal contributor to the bonding (quoted bond lengths are I-C(1) 2.053(13), P-C(1) 1.210(13) Å).

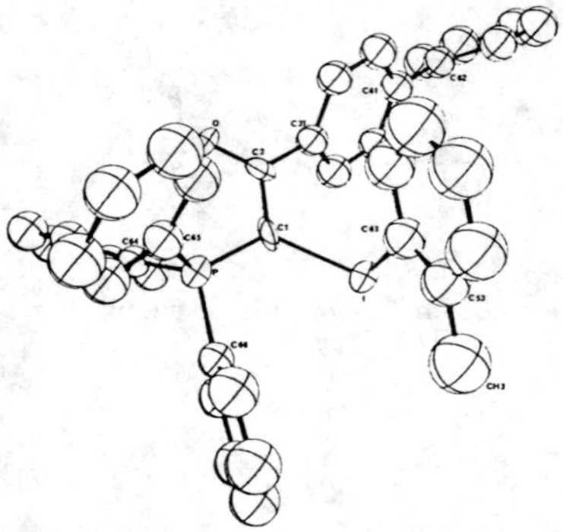

Fig. 1. A view of the $(C_{39}H_{31}IOP)^+$ cation.

SPIRO-(BIS(3,6-DI-t-BUTYLPHENYLENEDIOXY)-2'-HYDROXY-3',6'-DI-t-BUTYLPHENOXYPHOSPHANE)
$C_{42}H_{61}O_6P$

D.S. YUFIT, E.I. MATROSOV, N.V. MATROSOVA, D.I. LOBANOV, Yu.T. STRUCHKOV and M.I. KABACHNIK, 1984. J. Struct. Chem., 25, 568-573 [Zh. Strukt. Khim. 25(4), 68-73].

Triclinic, $P\bar{1}$, a = 14.483, b = 17.046, c = 17.998 Å, α = 97.37, β = 111.43, γ = 95.10°, Z = 2. Mo radiation, R = 0.062 for 8332 reflexions (at -120°C).

The molecule (Fig. 1) contains a five-coordinate, trigonal bipyramidal phosphorus atom with an additional weak interaction between the phosphorus atom and the phenolic hydroxyl oxygen atom. Principal interatomic distances are shown in Fig. 1.

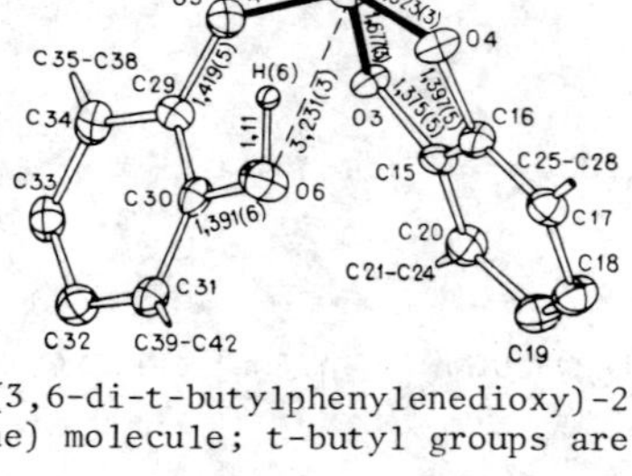

Fig. 1. The spiro-(bis(3,6-di-t-butylphenylenedioxy)-2'-hydroxy-3',6'-di-t-butylphenoxyphosphane) molecule; t-butyl groups are omitted for clarity.

TRIPHENYL(TRIPHENYLPHOSPHORANYLIDENEAMINO)PHOSPHONIUM PHENYLCYANOPHOSPHIDE

$C_{43}H_{35}N_2P_3$ $Ph_3PNPPh_3{}^+.PhPCN^-$

A. SCHMIDPETER, K.-H. ZIRZOW, G. BURGET, G. HUTTNER and I. JIBRIL, 1984. Chem. Ber., 117, 1695-1706.

Monoclinic, Cc, a = 15.36, b = 16.49, c = 15.01 Å, β = 111.32°, Z = 4. Mo radiation, R = 0.068 for 1753 reflexions (at -25°C).

The anion is almost planar (Fig. 1), with the C-P-C angle of 102°. The long P-C and short C-N bond lengths are consistent with the anion being "a cyanide complex of phenylphosphinidene".

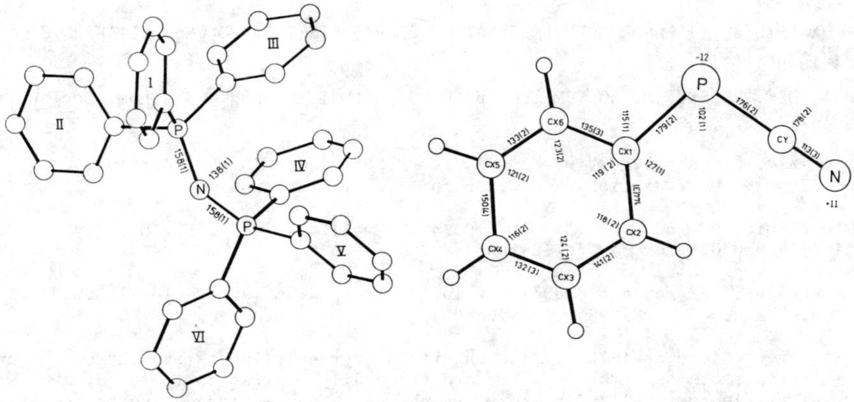

Fig. 1. Views of the cation and anion in $C_{43}H_{35}N_2P_3$.

DIBUTYLPHOSPHORANYLIDENEBIS(TRIPHENYLPHOSPHONIO)METHANE DIBROMIDE HYDROBROMIDE DICHLOROMETHANE SOLVATE

$C_{45}H_{50}Br_3P_3,CH_2Cl_2$ $(C_{45}H_{49}P_3)^{2+}2Br^-HBr.CH_2Cl_2$

M.Yu. ANTIPIN, Yu.T. STRUCHKOV, T.A. MASTRYUKOVA, I.M. ALADJEVA, I.V. LEONT'EVA and M.I. KABACHNIK, 1984. J. Struct. Chem., 25, 564-567 [Zh. Strukt. Khim., 25(4), 63-67].

Monoclinic, $P2_1/c$, a = 11.642, b = 30.465, c = 14.050 Å, β = 100.66°, Z = 4. Mo radiation, R = 0.115 for 1812 reflexions.

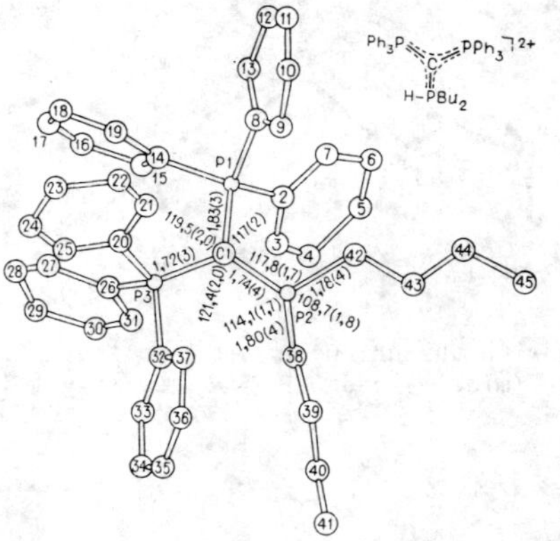

Fig. 1. The dibutylphosphoranylidenebis(triphenylphosphonio)methane dication.

The $(C_{45}H_{49}P_3)^{2+}$ dication (Fig. 1) occurs in the PH-ylide form with an sp^2-hybridized central carbon atom. The valence angles at the phosphorus atom connected to the butyl groups are close to tetrahedral. The structure also contains a weak P-H...Br interaction (P...Br 3.985(12), H...Br 2.75 Å, and P-H...Br 138°).

BIS(μ_2-OXO)BIS(HYDROXOBIS(TRIFLUOROMETHYL)ARSENIC)
$C_4H_2As_2F_{12}O_4$ (I)

BIS(μ_2-OXO)BIS(CHLOROHYDROXOTRIFLUOROMETHYLARSENIC)
$C_2H_2As_2Cl_2F_6O_4$ (II)

BIS(μ_2-OXO)BIS(μ_2-TRIFLUOROMETHYLARSENATO-O,O')BIS(BIS(TRIFLUOROMETHYL)ARSENIC)
$C_6H_2As_4F_{18}O_8$ (III)

R. BOHRA, H.W. ROESKY, M. NOLTEMEYER and G.M. SHELDRICK, 1984. J. Chem. Soc. Dalton, 2011-2014.

I. Triclinic, $P\bar{1}$, a = 5.842, b = 7.037, c = 9.586 Å, α = 103.91, β = 103.10, γ = 107.90°, Z = 1. Mo radiation, R = 0.036 for 1520 reflexions.

II. Monoclinic, $P2_1/n$, a = 5.543, b = 9.345, c = 11.496 Å, β = 92.70°, Z = 2. Mo radiation, R = 0.038 for 749 reflexions.

III. Monoclinic, $P2_1/n$, a = 8.803, b = 10.114, c = 12.567 Å, β = 104.17°, Z = 2. Mo radiation, R = 0.046 for 1328 reflexions.

All three molecules (Fig. 1) lie on crystallographic inversion centres. I and II have similar structures. The As atoms are trigonally bipyramidally coordinated with equatorial trifluoromethyl and hydroxy groups. The four-membered ring bridges equatorial and axial positions and the remaining axial position is occupied by a trifluoromethyl group in I and a Cl atom in II. The four-membered rings each involve long (2.003 in I, 1.937 Å in II) axial As-O bonds and short (1.713 in I, 1.718 Å in II) equatorial As-O bonds. The axial As-C bond in I is 2.034 while the equatorial As-C bonds are 1.998 and 2.008 Å in I and II respectively. In both molecules the equatorial As-C and As-OH bonds are bent out of the equatorial plane towards the long axial As-O bond. III possesses a novel cage structure involving octahedrally and tetrahedrally coordinated As atoms. The bond lengths in the $(AsO)_2$ ring are equal (1.831, 1.833 Å) whereas the As(1)-O-As(2) bridges are unsymmetrical, the four-coordinated As(2) having much shorter bonds (1.693, 1.686) than the six-coordinated As(1) (1.901, 1.905 Å).

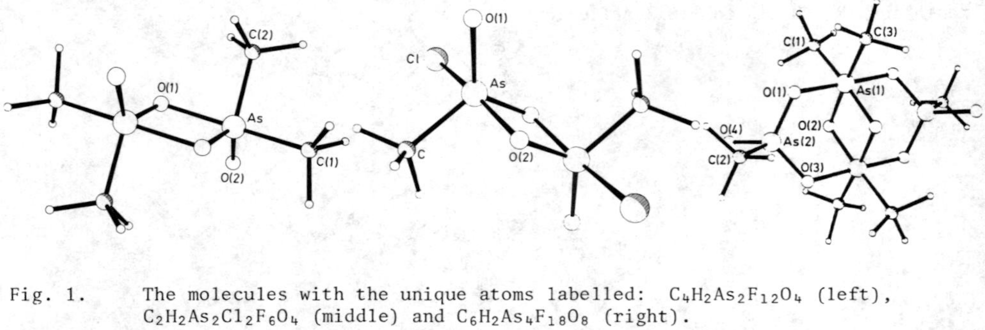

Fig. 1. The molecules with the unique atoms labelled: $C_4H_2As_2F_{12}O_4$ (left), $C_2H_2As_2Cl_2F_6O_4$ (middle) and $C_6H_2As_4F_{18}O_8$ (right).

1,3-DI-t-BUTYL-2,4-DICHLOROCYCLODIARSAZANE
$C_8H_{18}As_2Cl_2N_2$ $[C_4H_9NAsCl]_2$

R. BOHRA, H.W. ROESKY, M. NOLTEMEYER and G.M. SHELDRICK, 1984. Acta Cryst., C40, 1150-1152.

Orthorhombic, Pbca, a = 11.490, b = 16.093, c = 15.954 Å, Z = 8. Mo radiation, R = 0.036 for 1410 reflexions.

In the molecule (Fig. 1) the $[AsN(C)]_2$ unit is almost planar with the two Cl atoms lying 2.073(3) Å to the same side of this plane. Bond distances in the unit are: As(1)-N(1) 1.819(4), N(1)-As(2) 1.799(5), As(2)-N(2) 1.827(4) and N(2)-As(1) 1.826(5) Å. The two tert-butyl groups adopt different conformations, so that the approximate overall molecular symmetry is only m.

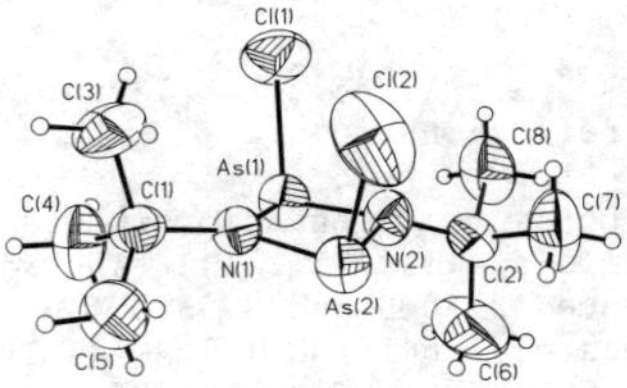

Fig. 1. Perspective view of the $[C_4H_9NAsCl]_2$ molecule.

3,7-DI-t-BUTYL-3H,7H-1λ⁴,5λ⁴,2,4,6,8,3,7-DITHIATETRAAZADIARSOCINE
$C_8H_{18}As_2N_4S_2$ (I)

BIS(DIPHENYLARSINO)SULPHUR DIIMIDE
$C_{24}H_{20}As_2N_2S$ (II)

A. GIEREN, H. BETZ, T. HUBNER, V. LAMM, M. HERBERHOLD and K. GULDNER, 1984. Z. Anorg. Chem., 513, 160-174.

I. Monoclinic, $P2_1/n$, a = 6.490, b = 10.145, c = 23.719 Å, β = 95.09°, D_m = 1.64, Z = 4. Mo radiation, R = 0.041 for 2691 reflexions.

II. Monoclinic, I2/c, a = 21.205, b = 5.904, c = 23.802 Å, β = 132.42°, D_m = 1.58, Z = 4. Mo radiation, R = 0.053 for 2327 reflexions.

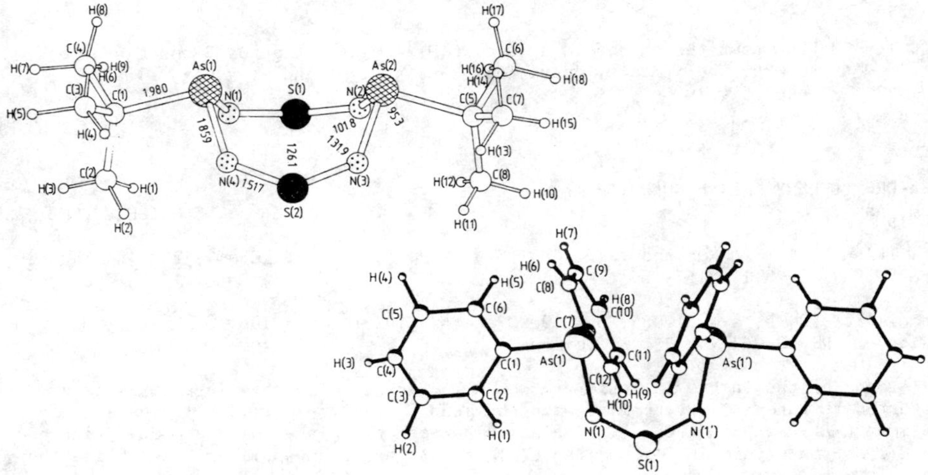

Fig. 1. Views of $C_8H_{18}As_2N_4S_2$ (I) (left) and $C_{24}H_{20}As_2N_2S$ (II) (right).

The eight-membered $As_2S_2N_4$ ring in I (Fig. 1) has a boat conformation with only small deviations from mm2 symmetry. The t-butyl groups are quasi-equatorial. Torsion angles up to 12° about the S=N bonds are explained by transannular As...As repulsion (As...As 3.683 Å). Molecule II has twofold crystallographic symmetry (Fig. 1) and an As...As separation of 3.379 Å. The mean S=N bond lengths in (I) (1.517(5) Å) and (II) (1.521(3) Å) are typical of sulphur diimides with little π-interaction and correspond to S^{IV}=N double bonds.

TRIS(O-ETHYLXANTHATO)ARSENIC(III)
$C_9H_{15}AsO_3S_6$ $As(C_2H_5OCS_2)_3$

B.F. HOSKINS, P.M. PIKO, E.R.T. TIEKINK and G. WINTER, 1984. Inorg. Chim. Acta, 84, L13-L14.

Trigonal (Hexagonal setting), R$\bar{3}$, a = 15.681, c = 12.909 Å, D_m = 1.57, Z = 6. Mo radiation, R = 0.024 for 1020 reflexions.

Crystals of the compound consist of enantiomorphous pairs of discrete molecules which are centred about a threefold axis (Fig. 1). The xanthate moiety bonds asymmetrically with As-S distances 2.3069 and 2.9560 Å. The C-S distances are also quite different, with magnitudes of 1.641 and 1.744 Å. The longer C-S distance is associated with the shorter As-S bond. The angles S(1)-As-S(2), As-S(1)-C(1), As-S(2)-C(1) and S(1)-C(1)-S(2) are respectively 67.56, 93.89, 74.93 and 123.1°. The coordination geometry about the As atom may be described as intermediate between trigonal prismatic and octahedral.

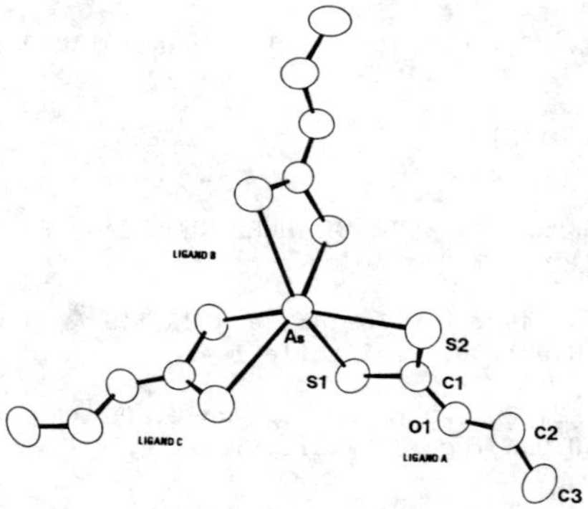

Fig. 1. View down the threefold axis of $C_9H_{15}AsO_3S_6$. Ligands B and C are related to A by the symmetry operations (-y, x-y, z) and (y-x, -x, z) respectively.

DI-n-PROPYLPHENYLARSINE SULPHIDE
$C_{12}H_{19}AsS$ $C_6H_5(CH_2CH_2CH_3)_2AsS$

V.V. TKACHEV, V.A. PEROV and Yu.F. GATILOV, 1984. J. Struct. Chem., 25, 144-146 [Zh. Strukt. Khim., 25(1), 163-165].

Monoclinic, $P2_1/b$, a = 11.085, b = 10.907, c = 12.023 Å, γ = 106.1°, Z = 4. Mo radiation, R = 0.064 for 787 reflexions.

Bond lengths in the molecule are shown in Fig. 1. The arsenic atom is in a distorted tetrahedral environment with the mean S-As-C angle (112.5°) greater than the mean C-As-C angle (106.3°). The angle between the planes of the phenyl ring and the C(1)AsS moiety is 16.7°, and the C(1)-As bond makes an angle of 5° with the phenyl mean plane.

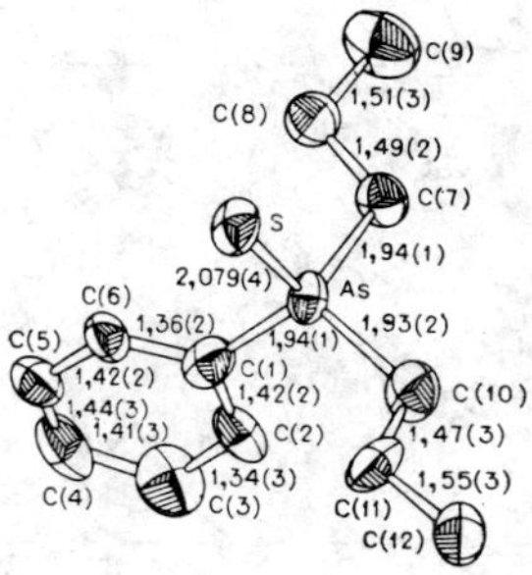

Fig. 1. Bond distances in the di-n-propylphenylarsine sulphide molecule.

TRIS(p-CHLOROPHENYL)ARSINOXIDE (MODIFICATION I)
$C_{18}H_{12}AsCl_3O$ (I) $(C_6H_4Cl)_3AsO$

TRIS(p-CHLOROPHENYL)ARSINOXIDE (MODIFICATION II)
$C_{18}H_{12}AsCl_3O$ (II) $(C_6H_4Cl)_3AsO$

TRIS(p-CHLOROPHENYL)ARSINSULPHIDE
$C_{18}H_{12}AsCl_3S$ (III) $(C_6H_4Cl)_3AsS$

V.K. BELSKY and V.E. ZAVODNIK, 1984. J. Organometal. Chem., 265, 159-165.

I. Triclinic, $P\bar{1}$, a = 11.969, b = 13.174, c = 15.366 Å, α = 91.11, β = 118.24, γ = 119.41°, Z = 4. Mo radiation, R = 0.036 for 3579 reflexions.

II. Hexagonal, $P6_3$, a = 14.276, c = 5.980 Å, Z = 2. Mo radiation, R = 0.034 for 470 reflexions.

III. Monoclinic, $P2_1/b$, a = 9.779, b = 18.174, c = 10.568 Å, γ = 90.49°, Z = 4. Mo radiation, R = 0.025 for 1657 reflexions.

The geometrical parameters of the molecules in polymorphs I and II are quite similar, with mean values for the bond distances and angles of: As=O 1.641, As-C 1.928, C-Cl 1.736 Å; C-As-C 107.3, C-As=O 111.6°. These features are essentially the same in III with As=S 2.074, mean As-C 1.934, C-Cl 1.748 Å, C-As-C 106.5 and C-As=S 112.3° so replacement of the O atom by S does not lead to significant changes in molecular geometry. However, the packing modes of I and II are quite different (Fig. 1), and the conformation of the molecules in I resembles that of III more than II. In I the independent molecules in the unit cell may be transformed into each other by a supersymmetrical operation which comprises rotation by 180° with subsequent shift. In the hexagonal modification II the molecules spiral round the 6_3 axes forming cylindrical channels along the z direction. The diameter of these channels is ca. 5 Å.

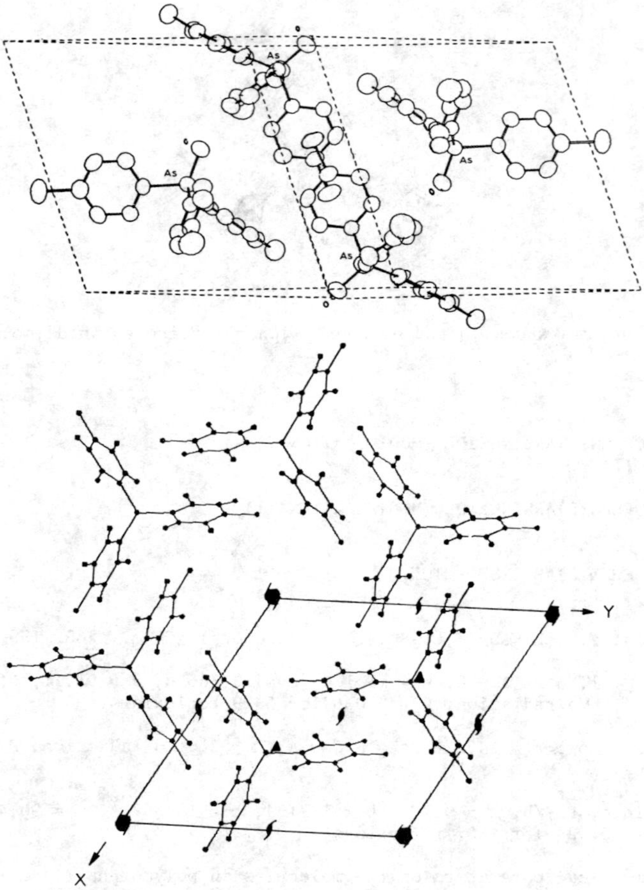

Fig. 1. Projection of $C_{18}H_{12}AsCl_3O$ (I) (top), and II (bottom).

5-METHYL-2,2,8,8-TETRAETHOXYCARBONYL-1,3,7-TRIARSATRICYCLO[3.3.1.$0^{3,7}$]NONANE
$C_{19}H_{29}As_3O_8$ (I)

5-METHYL-1,3,7-TRIARSA-2,8-DISELENATRICYCLO[3.3.1.$0^{3,7}$]NONANE
$C_5H_9As_3Se_2$ (II)

G. THIELE, H.W. ROTTER, M. LIETZ and J. ELLERMANN, 1984. Z. Naturforsch., B39, 1344-1349.

I. Triclinic, $P\bar{1}$, a = 7.693, b = 10.361, c = 15.992 Å, α = 98.24, β = 97.99, γ = 103.63°, Z = 2. Mo radiation, R = 0.0620 for 1821 reflexions.

II. Monoclinic, $P2_1/c$, a = 11.027, b = 7.275, c = 12.813 Å, β = 98.80°, Z = 4. Mo radiation, R = 0.0577 for 1065 reflexions.

The compounds are shown to be nor-heteroadamantanes (Fig. 1) with dimensions for (I) As-As 2.47(2), As-C 1.95-2.01(2) Å and for (II) As-As 2.483(3), As-Se 2.357-2.369(3), As-C 1.96-2.01(2) Å.

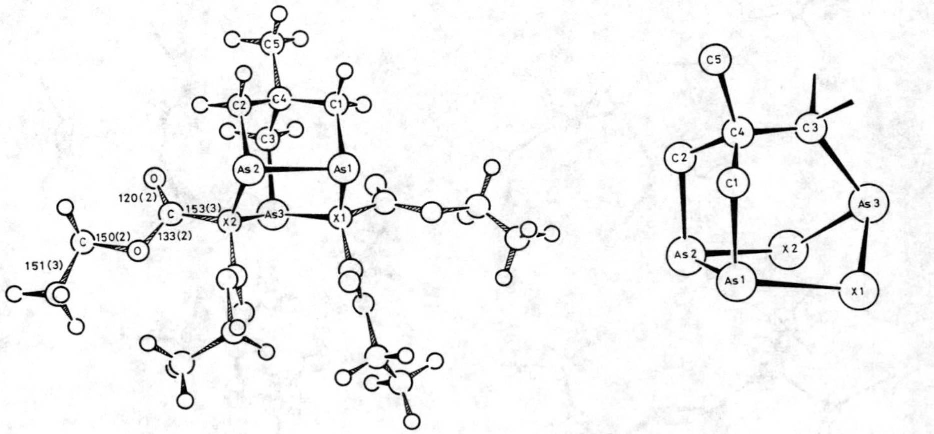

Fig. 1. Views of $C_{19}H_{29}As_3O_8$ (I) (left) X = C) and $C_5H_9As_3Se_2$ (II) (right) (X = Se).

BIS[TRIS(TRIMETHYLSILYL)METHYL]DIPHOSPHENE
$C_{20}H_{54}P_2Si_6$ (I) $(Me_3Si)_3CP{=}PC(SiMe_3)_3$

(P-2,4,6-TRI-t-BUTYLPHENYL)[As-BIS(TRIMETHYLSILYL)METHYL]PHOSPHAARSENE
$C_{25}H_{48}AsPSi_2$ (II) $[2,4,6\text{-}(t\text{-}Bu)_3C_6H_2]P{=}AsCH(SiMe_3)_2$

2,4,6-TRI-t-BUTYLPHENYL[BIS(TRIMETHYLSILYL)METHYL]DIARSENE
$C_{25}H_{48}As_2Si_2$ (III) $[2,4,6\text{-}(t\text{-}Bu)_3C_6H_2]As{=}AsCH(SiMe_3)_2$

A.H. COWLEY, J.E. KILDUFF, J.G. LASCH, S.K. MEHROTRA, N.C. NORMAN, M. PAKULSKI, B.R. WHITTLESEY, J.L. ATWOOD and W.E. HUNTER, 1984. Inorg. Chem., 23, 2582-2593.

I. Triclinic, $P\bar{1}$, a = 15.629, b = 9.155, c = 12.482 Å, α = 91.58, β = 111.42, γ = 90.02°, Z = 2. Mo radiation, R = 0.086 for 2046 reflexions.

II. Triclinic, $P\bar{1}$, a = 9.991, b = 10.452, c = 15.140 Å, α = 89.72, β = 85.03, γ = 77.72°, Z = 2. Mo radiation, R = 0.057 for 3862 reflexions.

III. Triclinic, $P\bar{1}$, a = 9.955, b = 10.393, c = 15.159 Å, α = 89.56, β = 85.39, γ = 78.31°, Z = 2. Mo radiation, R = 0.087 for 2508 reflexions.

Each of the three compounds (Fig. 1) contains a double bond between group 5A elements; P=P 2.014(6) and 2.004(6) Å in I (in which there are two independent centrosymmetric molecules), P=As 2.125(1) Å in II, and As=As 2.224(2) Å in III. The geometry about the double bond is trans-planar in each case with angles P-P-C 108.2(4) and 108.9(4)° in I, P-As-C 101.35(9) and As-P-C 96.37(9)° in II, As-As-C 99.9(3) and 93.6(3)° in III. Other bond lengths include: P-C 1.85(1) and 1.87(1) Å in I, As-C 1.991(3) and P-C 1.852(3) Å in II, As-C 1.976(10) (alkyl) and 1.946(10) Å (aryl) in III.

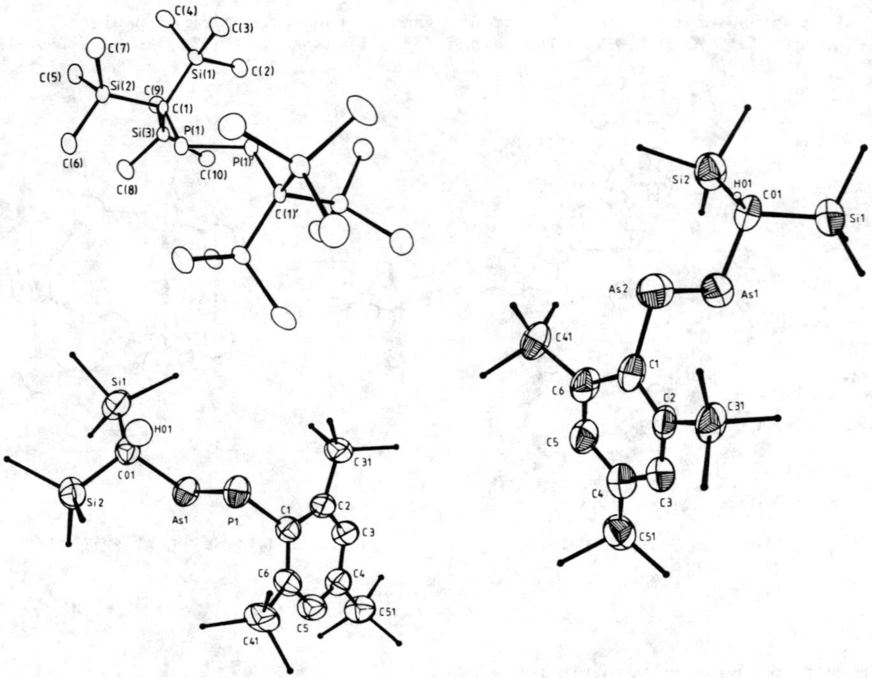

Fig. 1. The molecules $(Me_3Si)_3CP{=}PC(SiMe_3)_3$ (upper left), $[2,4,6\text{-}(t\text{-}Bu)_3C_6H_2]P{=}AsCH(SiMe_3)_2$ (lower left), and $[2,4,6\text{-}(t\text{-}Bu)_3C_6H_2]As{=}AsCH(SiMe_3)_2$ (right).

(R)-BENZYL(METHYL)(4-METHYLPHENYL)(NAPHTHALEN-1-YL)ARSONIUM BROMIDE
$C_{25}H_{24}AsBr$ (I)

(R)-BENZYL(METHYL)(4-METHYLPHENYL)(NAPHTHALEN-1-YL)ARSONIUM HEXAFLUOROPHOSPHATE
$C_{25}H_{24}AsF_6P$ (II)

D.G. ALLEN, C.L. RASTON, B.W. SKELTON, A.H. WHITE and S.B. WILD, 1984. Aust. J. Chem., 37, 1171-1181.

I. Orthorhombic, $P2_12_12_1$, a = 22.472, b = 15.724, c = 12.585 Å, Z = 8. Mo radiation, R = 0.031 for 1917 reflexions.

II. Orthorhombic, $P2_12_12_1$, a = 23.56, b = 16.40, c = 13.12 Å, D_m = 1.44, Z = 8. Mo radiation, R = 0.078 for 2256 reflexions.

I and II are isostructural with two molecules in the asymmetric unit (Fig. 1). These are conformational isomers with respect to the relative dispositions of 4-methylphenyl and naphthalen-1-yl substituents and have the same absolute configuration (R).

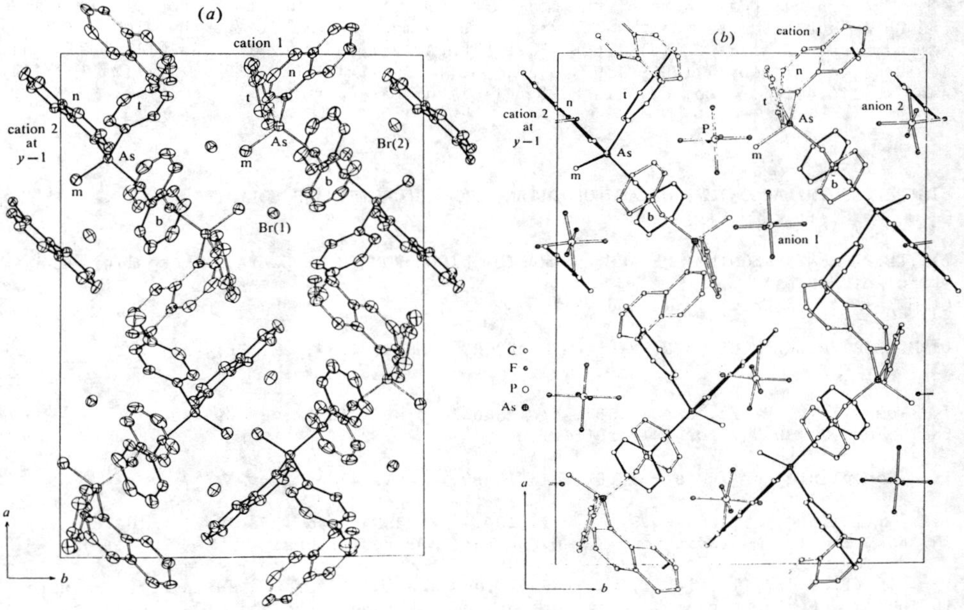

Fig. 1. Unit cell contents of (a) $C_{25}H_{24}AsBr$ and (b) $C_{25}H_{24}AsF_6P$.

TRIPHENYLPHOSPHONIUM-BIS(DIPHENYLARSINO)METHYLIDE
$C_{43}H_{35}As_2P$

H. SCHMIDBAUR, P. NUSSTEIN and G. MULLER, 1984. Z. Naturforsch., B39, 1456-1462.

Monoclinic, $P2_1/n$, a = 9.776, b = 18.668, c = 18.879 Å, β = 94.38°, Z = 4. Mo radiation, R = 0.046 for 3876 reflexions (at -35°C).

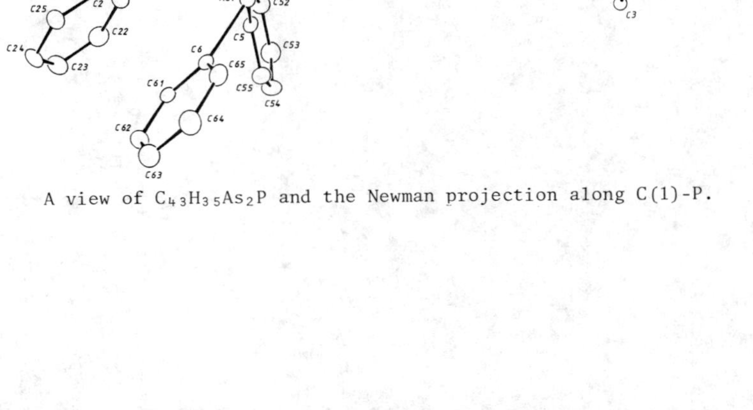

Fig. 1. A view of $C_{43}H_{35}As_2P$ and the Newman projection along C(1)-P.

The central $PCAs_2$ core of the molecule (Fig. 1) is found to be non-planar and the Ph_2As substituents adopt a cis/trans conformation relative to the ylidic P=C bond. Some dimensions are: P-C(1) 1.698(4), P-C(phenyl) 1.809-1.827(5), As-C(1) 1.936 and 1.938(4), As-C(phenyl) 1.959-1.974(5) Å, P-C(1)-As(1) 128.0(2), P-C(1)-As(2) 108.4(2), As-C-As 116.4(2)°. Atom C(1) is 0.28 Å from the PAs_2 plane.

BIS(TETRAPHENYLARSONIUM) HEXACHLOROTITANIUM DICHLOROMETHANE SOLVATE

$C_{48}H_{40}As_2Cl_6Ti,2(CH_2Cl_2)$ (I)

BIS(TETRAPHENYLARSONIUM) HEXACHLOROURANIUM DICHLOROMETHANE SOLVATE (Monoclinic and triclinic forms)

$C_{48}H_{40}As_2Cl_6U,2(CH_2Cl_2)$ (II and III)

U. MULLER, P. KLINGELHOFER, J. EICHER and R. BOHRER, 1984. Z. Kristallogr., 168, 121-131.

I. Monoclinic, $P2_1/c$, a = 17.587, b = 15.631, c = 19.190 Å, β = 98.58°, Z = 4. Mo radiation, R = 0.047 for 2417 reflexions.

II. Monoclinic, $P2_1/c$, a = 17.85, b = 15.89, c = 19.42 Å, β = 98.3°.

III. Triclinic, $P\bar{1}$, a = 10.52, b = 12.209, c = 22.96 Å, α = 98.29, β = 104.62, γ = 101.83°, Z = 1. Mo radiation, R = 0.039 for 2308 reflexions.

In (I) and (II) (Fig. 1) two $AsPh_4^+$ ions are paired around an inversion centre in such a way that three phenyl groups of one ion face three phenyl groups of the other ion in a staggered arrangement; the fourth phenyl group points away from the inversion centre. The $(AsPh_4^+)_2$ groupings are aligned to give parallel rows in which contacts between the fourth phenyl groups exist. Compound (III) (Fig. 1) has layers of $AsPh_4^+$ ions; within the layer, pairs of $AsPh_4^+$ ions have a similar arrangement as in (I). The MCl_5X^{2-} ions and CH_2Cl_2 molecules fill gaps between the cation arrangements; the CH_2Cl_2 molecules are associated to the anions via weak hydrogen bonds.

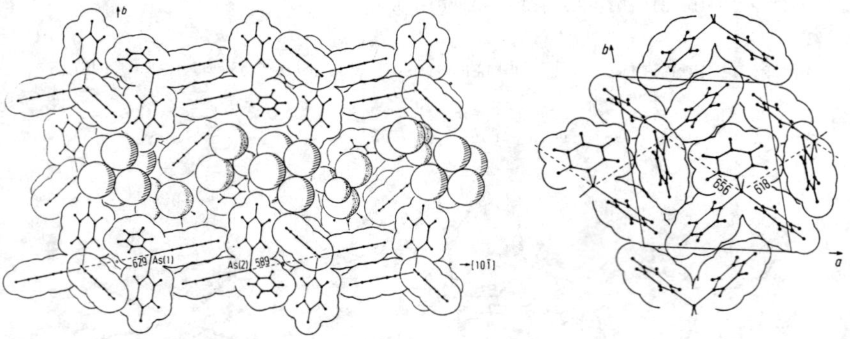

Fig. 1. Views of the packing of Ph_4As^+ ions in (I) and (II) (left) and in (III) (right).

RUBIDIUM (μ_2-OXALATO)PENTAFLUORODIANTIMONATE(III)
$C_2F_5O_4RbSb_2$

R.L. DAVIDOVICH, L.A. ZEMNUKHOVA, A.A. UDOVENKO and N.I. SIGULA, 1983. Koord. Khim., 9, 787-792 [Sov. J. Coord. Chem., 9, 461-466].

Orthorhombic, $P2_12_12_1$, a = 6.566, b = 22.776, c = 6.001 Å, D_m = 3.58, Z = 4. Mo radiation, R = 0.043 for 919 reflexions.

The structure is built up from dimeric $[Sb_2C_2O_4F_5]^-$ anions and Rb^+ cations (Fig. 1). The two Sb positions are crystallographically independent. The Sb(1) atoms have three F atoms (Sb(1)-F 1.93-2.01 Å) and one O atom from the $C_2O_4^{2-}$ moiety (Sb(1)-O(1') 2.51 Å) in their immediate environment. The immediate environment of Sb(2) consists of two F atoms (Sb(2)-F 1.96, 1.99 Å) and two O atoms from the oxalato group, which has bidentate bonding with the Sb atom (Sb-O 2.06, 2.27 Å). The coordination polyhedra of Sb(1) and Sb(2) are trigonal bipyramids with the lone pair of Sb at one of the equatorial vertices.

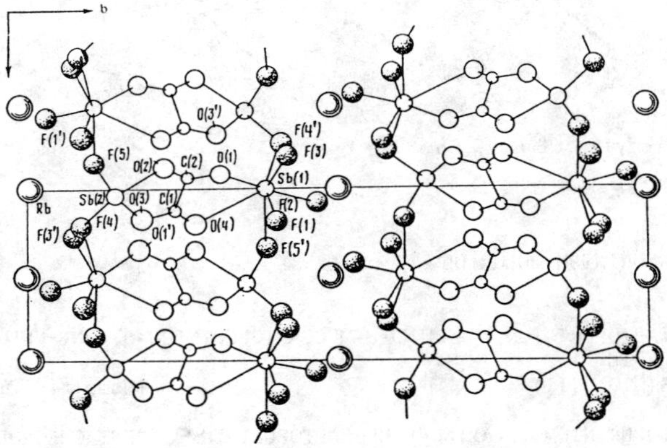

Fig. 1. $C_2F_5O_4RbSb_2$: projection of the structure onto the ab plane.

TETRAMETHYLDISTIBANE
$C_4H_{12}Sb_2$ $(CH_3)_4Sb_2$

A.J. ASHE III, E.G. LUDWIG, Jr., J. OLEKSYSZYN and J.C. HUFFMAN, 1984. Organometallics, 3, 337-338.

Orthorhombic, Pmnb, a = 11.314, b = 11.507, c = 6.507 Å, Z = 4. Mo radiation, R = 0.059 for 1769 reflexions (at -160°C).

The crystal packing structure of the title compound is shown in Figure 1. Six molecules are illustrated in two chains. The bond distances are as follows: Sb-Sb 2.862 Å and Sb-Sb ("intermolecular") 3.645 Å, and the mean Sb-C distance is 2.15 Å. The mean Sb-Sb-C and mean C-Sb-C angles are 93.6 and 93.5° respectively. The chains of Sb atoms are completely colinear with alternating intramolecular and short "intermolecular" separations. See also the following Report.

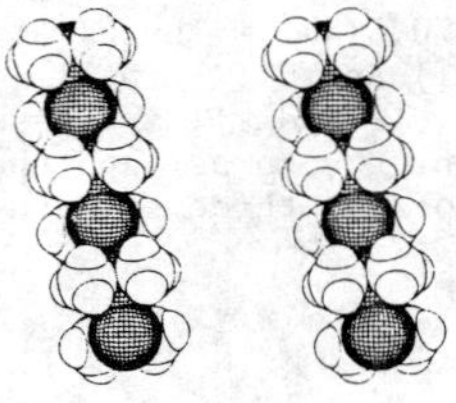

Fig. 1. Crystal packing drawing of two chains in $C_4H_{12}Sb_2$.

TETRAMETHYLDISTIBANE
$C_4H_{12}Sb_2$

O. MUNDT, H. RIFFEL, G. BECKER and A. SIMON, 1984. Z. Naturforsch., B39, 317-322.

Orthorhombic, Pnma, a = 11.591 (11.509), b = 11.577 (11.368), c = 6.539 (6.516) Å, Z = 4. Mo radiation, R = 0.049 (0.035) for 894 (1026) reflexions at -21°C (-139°C).

The molecule (Fig. 1) has crystallographic mirror symmetry with Sb-Sb 2.830(1) (-21°C) and 2.838(1) Å (-139°C). In the crystal lattice there are extended linear Sb chains with intermolecular contacts Sb...Sb 3.678(1) Å and Sb-Sb...Sb 179.2° (at -139°C). The Sb-C distances are in the range 2.118(9)-2.190(6) Å. See also the preceding Report.

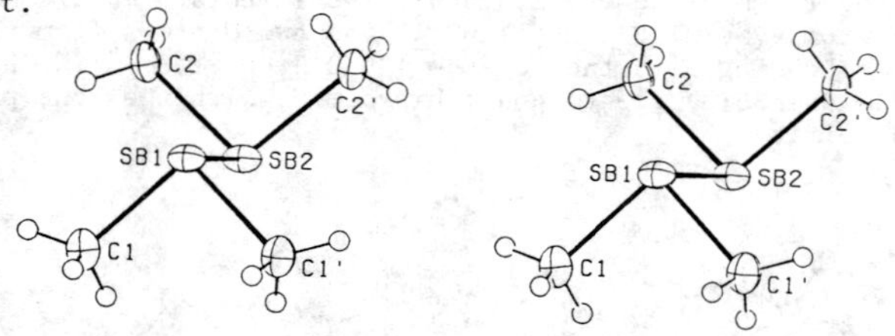

Fig. 1. A stereoview of $C_4H_{12}Sb_2$.

6-AMINO-8-MERCAPTOPURINIUM CHLORIDE
$C_5H_6ClN_5S$ (I) [Hamp]Cl

BIS(6-AMINO-8-MERCAPTOPURINIUM)(6-AMINO-8-MERCAPTOPURINIUM-S)PENTACHLOROBISMUTH CHLORIDE PENTAHYDRATE
$C_{15}H_{18}BiCl_6N_{15}S_3,5(H_2O)$ (II) $[Hamp]_2[BiCl_5(Hamp)]Cl.5H_2O$

(6-AMINO-8-MERCAPTOPURINIUM)(6-AMINO-8-MERCAPTOPURINIUM-S)PENTACHLOROBISMUTH DIHYDRATE
$C_{10}H_{12}BiCl_5N_{10}S_2,2(H_2O)$ (III) $[Hamp][BiCl_5(Hamp)].2H_2O$

L.P. BATTAGLIA and A. BONAMARTINI CORRADI, 1984. J. Chem. Soc. Dalton, 2401-2407.

I. Monoclinic, C2/c, a = 15.251, b = 7.982, c = 15.215 Å, β = 120.99°, D_m = 1.70, Z = 8. Mo radiation, R = 0.031 for 1040 reflexions.

II. Monoclinic, $P2_1$, a = 8.465, b = 27.669, c = 7.399 Å, β = 96.19°, D_m = 1.98, Z = 2. Mo radiation, R = 0.047 for 2532 reflexions.

III. Triclinic, $P\bar{1}$, a = 7.255, b = 9.260, c = 18.220 Å, α = 100.33, β = 91.86, γ = 84.84°, D_m = 2.10, Z = 2. Mo radiation, R = 0.038 for 3911 reflexions.

Compounds II and III comprise $[BiCl_5(Hamp)]^-$ complex anions, organic cations and water molecules and in addition in II Cl^- ions. In the anions the Bi atom is octahedrally surrounded by five Cl atoms and one ligand molecule coordinated via the S atom (Bi-S(1) 2.829(5), 2.935(3) Å for II, III). Bond lengths and angles involving the Cl atoms include: II - Bi-Cl 2.536(6)-2.825(6) Å, cis angles 86.2(2)-94.2(2), trans angles 172.8(2)-178.6(3)°; III - Bi-Cl 2.575(3)-2.883(3) Å, cis angles 83.2(1)-102.7(1), trans angles 166.8(4)-170.5(1)°. The Cl atoms involved in shorter bonds to Bi, Cl(1) and Cl(2) in II and Cl(1) and Cl(4) in III, are not involved in hydrogen bonding. The Hamp molecules in I, II and III are essentially planar. Only the molecular dimensions of I are of high accuracy and these have the expected values. Base stacking is present in each of the three structures (Fig. 1).

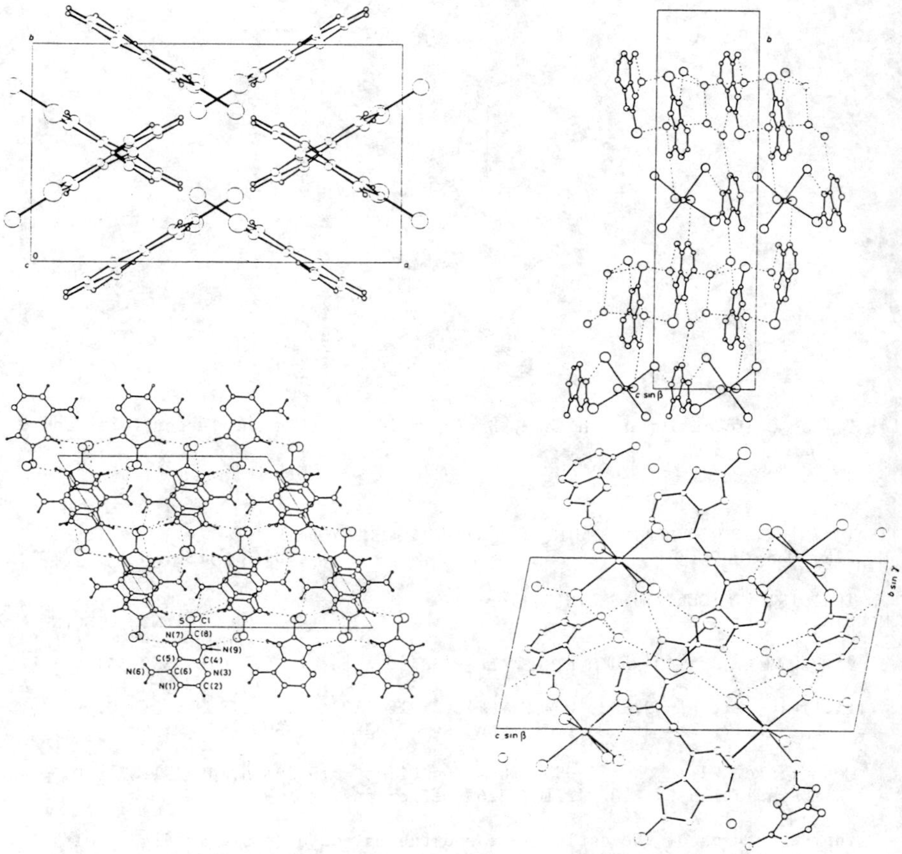

Fig. 1. $C_5H_6ClN_5S$ (left): stacking viewed along c (top) and the hydrogen bond system (bottom). Structure of $C_{15}H_{18}BiCl_6N_{15}S_3,5(H_2O)$ projected along a (top right) and of $C_{10}H_{12}BiCl_5N_{10}S_2,2(H_2O)$ (bottom right).

BIS(DIMETHYLDIBROMOSTIBORANO)METHANE

$C_5H_{14}Br_4Sb_2$ $[(CH_3)_2SbBr_2]_2CH_2$

W. SCHWARZ, W. KOLONDRA and J. WEIDLEIN, 1984. J. Organometal. Chem., 260, C1-C5.

Triclinic, $P\bar{1}$, a = 7.051, b = 9.536, c = 22.896 Å, α = 92.28, β = 94.16, γ = 109.65°, Z = 4. Mo radiation, R = 0.049 for 4561 reflexions.

There are two independent molecules (Fig. 1) per asymmetric unit and the major difference in molecular geometry between them is their twist angles. The twist angle, defined as the angle between planes C,Sb(1),C(11),C(12) and C,Sb(2),C(21),C(22), is 66.6° in molecule 1, and 72.8° in molecule 2. The Sb-Br and Sb-C (methyl) distances range from 2.603 to 2.655 Å and from 2.117 to 2.140 Å respectively. The $Sb-CH_2$ distances range from 2.111 to 2.137 Å and the Sb(1)-C-Sb(2) angles are 120.5° in both molecules.

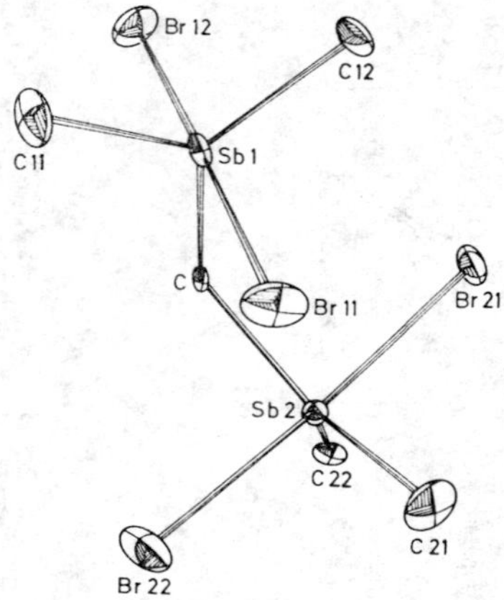

Fig. 1. Conformation of the $C_5H_{14}Br_4Sb_2$ molecule with the larger twist angle.

NITRATOPENTAKIS(THIOUREA)BISMUTH(III) NITRATE MONOHYDRATE
$C_5H_{20}BiN_{13}O_9S_5,H_2O$ (I) $[Bi(NO_3)\{SC(NH_2)_2\}_5](NO_3)_2,H_2O$

TRINITRATOTRIS(THIOUREA)BISMUTH(III)
$C_3H_{12}BiN_9O_9S_3$ (II) $[Bi(NO_3)_3\{SC(NH_2)_2\}3]$

G.B. JAMESON, E. BLASZO and H.R. OSWALD, 1984. Acta Cryst., C40, 350-354.

I. Triclinic, P$\bar{1}$, a = 9.292, b = 9.788, c = 15.729 Å, α = 118.55, β = 92.99, γ = 93.72°, D_m = 2.12, Z = 2. Mo radiation, R = 0.039 for 5300 reflexions.

II. Monoclinic, $P2_1/c$, a = 6.9556, b = 15.801, c = 15.561 Å, β = 91.87°, D_m = 2.422, Z = 4. Mo radiation, R = 0.042 for 3263 reflexions.

The structures of the cation in I and the molecule II are in Fig. 1 with principal bond lengths. Extensive hydrogen bonding and probable C...N π-donor - π-acceptor interactions occur for both structures.

Fig. 1. The structures of $[Bi(NO_3)\{SC(NH_2)_2\}_5]^{2+}$ and $[Bi(NO_3)_3\{SC(NH_2)_2\}_3]$ with principal bond lengths (Å).

DIISOPROPYLCYANAMIDE - PENTACHLOROANTIMONY
$C_7H_{14}Cl_5N_2Sb$

J.C. JOCHIMS, R. ABU-EL-HALAWA, L. ZSOLNAI and G. HUTTNER, 1984. Chem. Ber., 117, 1161-1177.

Orthorhombic, $P2_12_12_1$, a = 9.86, b = 12.07, c = 13.42 Å, Z = 4. Mo radiation, R = 0.038 for 1061 reflexions.

The molecule is neither a σ- nor a π-complex (Fig. 1). The C1-N1-Sb angle is 133° and torsion angle C3-N2-N1-Sb is 95°. Nitrogen N1 is not sp^2 hybridized.

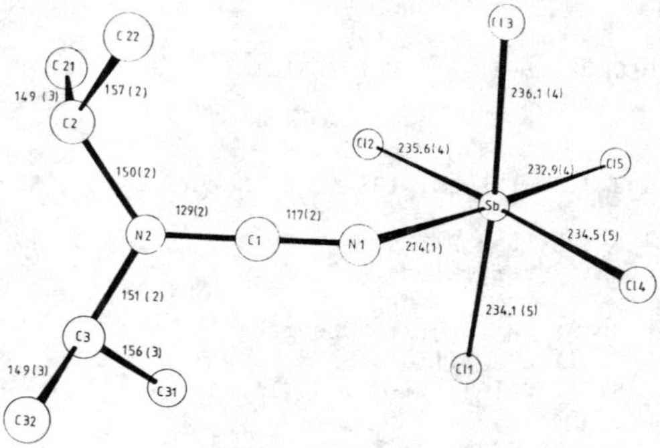

Fig. 1. The $C_7H_{14}Cl_5N_2Sb$ adduct and bond lengths (pm).

N,N'-DI-ISOPROPYLDITHIOOXAMIDE
$C_8H_{16}N_2S_2$ (I)

HEXACHLOROTRIS(N,N'-DI-ISOPROPYLDITHIOOXAMIDE-S,S')DIANTIMONY
$(C_{24}H_{48}Cl_6N_6S_6Sb_2)_n$ (II)

M.G.B. DREW, J.M. KISENYI, G.R. WILLEY and S.O. WANDIGA, 1984. J. Chem. Soc. Dalton, 1717-1721.

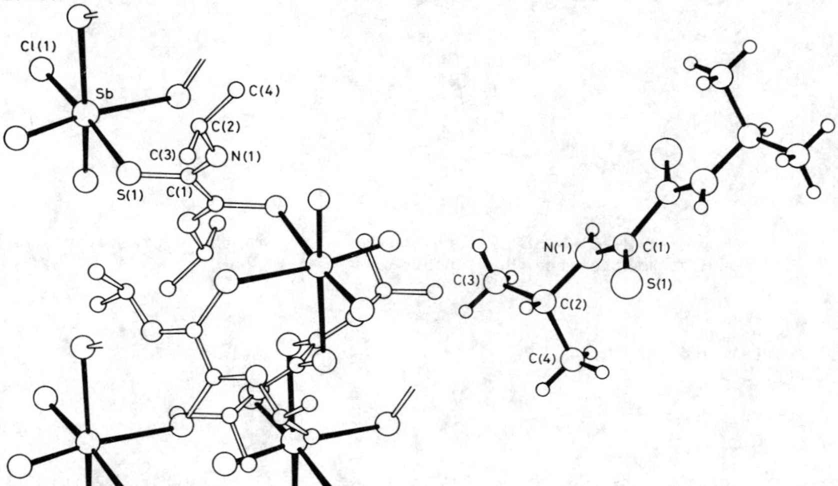

Fig. 1. The structure of polymeric $(C_{24}H_{48}Cl_6N_6S_6Sb_2)_n$ (left) and of $C_8H_{16}N_2S_2$ (right).

I. Monoclinic, $P2_1/c$, a = 6.226, b = 8.067, c = 11.187 Å, β = 91.0°, D_m = 1.21, Z = 2. Mo radiation, R = 0.077 for 635 reflexions.

II. Rhombohedral, $R\bar{3}$, a = 11.80 Å, α = 60°, Z = 2. Mo radiation, R = 0.068 for 1089 reflexions.

In II (Fig. 1), which is a three-dimensional polymer, the octahedrally-coordinated Sb atom lies on a threefold axis and is strongly bound to three chlorine atoms (2.413(2) Å) and also weakly bound to three sulphur atoms from different ligands (Sb-S 3.197(5) Å). In I there are two centrosymmetric ligands in the unit cell.

2,2,2,2,4,4,4,4-OCTACHLORO-1,3-BIS(TRIMETHYLSILYLISOCYANO)-1,3-DIAZA-2,4-DISTIBACYCLO-BUTANE
$C_8H_{18}Cl_8N_4Sb_2Si_2$

G. RAJCA, W. SCHWARZ and J. WEIDLEIN, 1984. Z. Naturforsch., B39, 1219-1223.

Monoclinic, $P2_1/c$, a = 6.383, b = 9.725, c = 20.346 Å, β = 92.17°, Z = 2. Mo radiation, R = 0.04 for 2362 reflexions.

The molecule lies about an inversion centre Fig. 1. The Sb atoms have octahedral coordination with Sb-N 2.111 and 2.136(5), Sb-Cl 2.331-2.351(2), N(1)-C(1) 1.272(8), C(1)-N2 1.151(9), N(2)-Si 1.786(6), Si-C 1.834-1.848(7) Å.

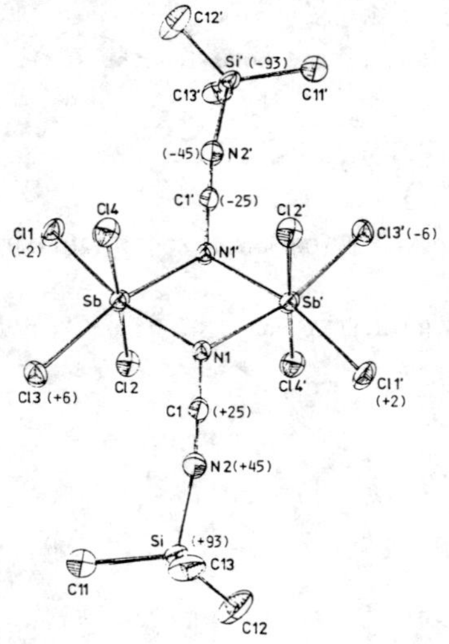

Fig. 1. A view of $C_8H_{18}Cl_8N_4Sb_2Si_2$; the numbers in parentheses are distances (pm) of the atoms from the Sb_2N_2 plane.

9-CHLORO-9-STIBAFLUORENE
$C_{12}H_8ClSb$ (I)

9-PHENYL-9-STIBAFLUORENE
$C_{18}H_{13}Sb$ (II)

V.K. BEL'SKII, 1984. J. Struct. Chem., 25, 965-967 [Zh. Strukt. Khim., 25(6), 138-140].

I. Orthorhombic, Pccn, a = 11.157, b = 13.590, c = 13.868 Å, Z = 8. Mo radiation, R = 0.027 for 1474 reflexions.

II. Monoclinic, $P2_1/n$, a = 5.172, b = 25.730, c = 10.526 Å, γ = 89.48°, Z = 4. Mo radiation, R = 0.044 for 1490 reflexions.

The structures of I and II are shown in Figs. 1 and 2, respectively (σ = 0.002-0.008 Å and 0.1-0.5° for I, 0.01-0.02 Å and 0.4-1.0° for II). The Sb atoms in both molecules have trigonal pyramidal coordination with sums of valence angles at Sb of 269.8 and 268.9° and displacements of the Sb atoms from the planes of the substituents of 1.285 and 1.240 Å, respectively for I and II.

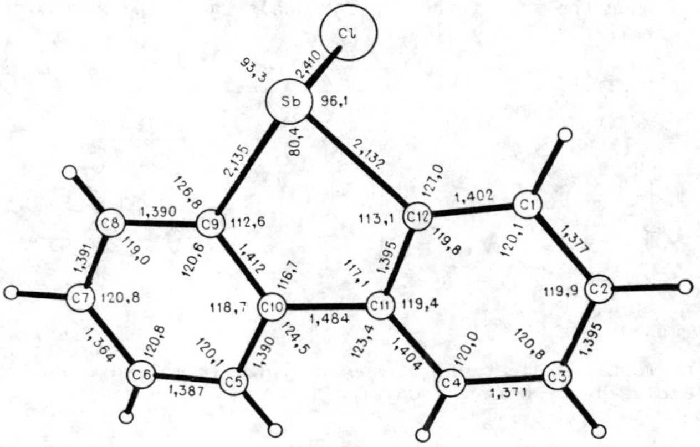

Fig. 1. The 9-chloro-9-stibafluorene molecule.

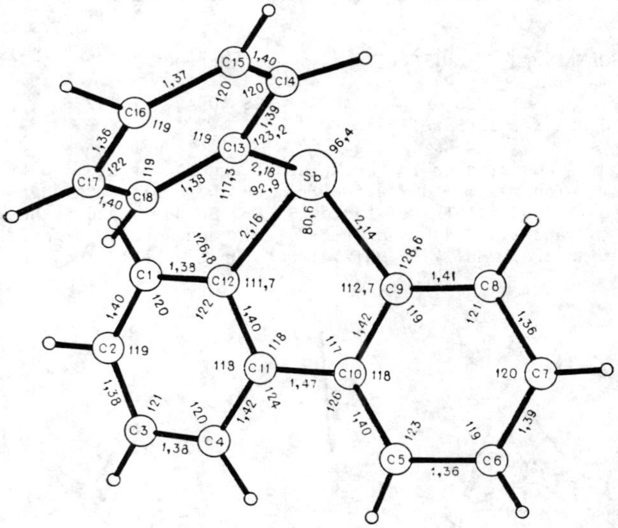

Fig. 2. The 9-phenyl-9-stibafluorene molecule.

TRIS(2-PYRIDINETHIOLATO)ANTIMONY(III)

$C_{15}H_{12}N_3S_3Sb$ $[Sb(C_5H_4NS)_3]$

A.P. BOZOPOULOS, S.C. KOKKOU, P.J. RENTZEPERIS and P. KARAGIANNIDIS, 1984. Acta Cryst., C40, 944-946.

Rhombohedral, $R\bar{3}c$, a = 12.568, c = 37.909 Å, D_m = 1.73, Z = 12. Mo radiation, R = 0.029 for 710 reflexions.

The complex molecule (Fig. 1) consists of three planar 2-pyridinethiol ligands surrounding Sb in C_3 symmetry with Sb-S bond distances of 2.472(2) and Sb...N intramolecular contacts of 2.830(7) Å. A gap in the Sb coordination can be attributed to a stereochemically active lone pair of electrons and the proximity of the rotation-equivalent molecule. The Sb...Sb' distance of 3.657(1) Å between two molecules is less than the sum of the van der Waals radii and a weak intermolecular interaction is very likely.

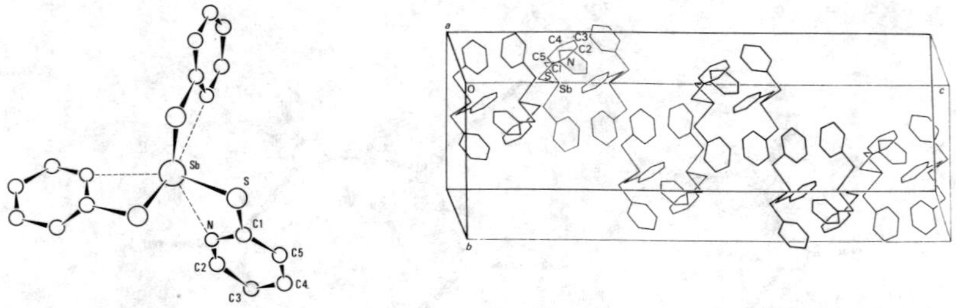

Fig. 1. Clinographic projections of the $Sb(C_5H_4NS)_3$ molecule and of the molecular packing in the unit cell.

(μ_2-BROMOSTIBINIDENE)TETRACARBONYLBIS(η^5-METHYLCYCLOPENTADIENYL)DIMANGANESE

$C_{16}H_{14}BrMn_2O_4Sb$ $[(\pi-CH_3C_5H_4)Mn(CO)_2]_2SbBr$

U. WEBER, L. ZSOLNAI and G. HUTTNER, 1984. J. Organometal. Chem., 260, 281-291.

Monoclinic, $P2_1/c$, a = 12.235, b = 7.144, c = 25.10 Å, β = 122.64°, Z = 4. Mo radiation, R = 0.051 for 1983 reflexions.

The Sb atom is coordinated to the Mn atoms and the Br atom in trigonal planar fashion (Fig. 1). The maximum deviation from the best plane through atoms Mn(1), Mn(2), Sb and Br is 0.07 Å. The distances Sb-Br, Sb-Mn(1) and Sb-Mn(2) are respectively 2.613, 2.407 and 2.402 Å. The Mn-C(Cp) distances range from 2.12 to 2.18 Å. The angles about Sb are: Mn(1)-Sb-Mn(2) 138.6, Mn(1)-Sb-Br 110.2 and Mn(2)-Sb-Br 110.8°.

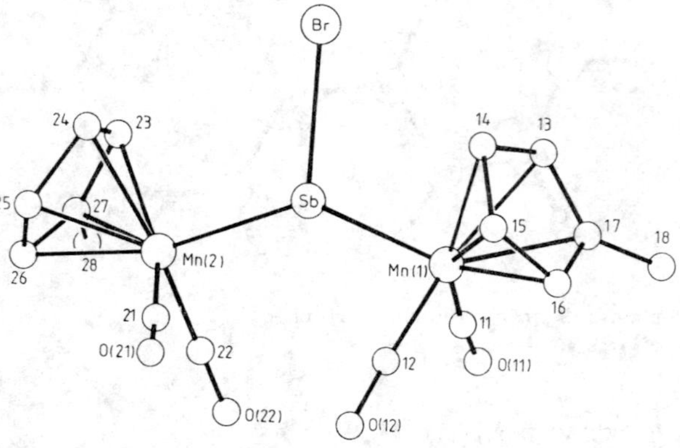

Fig. 1. Molecular structure of $C_{16}H_{14}BrMn_2O_4Sb$.

(μ_2-TETRACARBONYLIRON)BIS[BIS(TRIMETHYLSILYL)METHYL]DIANTIMONY
$C_{18}H_{38}FeO_4Sb_2Si_4$ (I)

[μ_2-BIS(TRIMETHYLSILYL)METHYLANTIMONY]BIS(TETRACARBONYLIRON)
$C_{15}H_{19}Fe_2O_8SbSi_2$ (II)

A.H. COWLEY, N.C. NORMAN and M. PAKULSKI, 1984. J. Am. Chem. Soc., 106, 6844-6845.

I. Monoclinic, $P2_1/n$, a = 19.958, b = 6.635, c = 24.408 Å, β = 103.51°, Z = 4. Mo radiation, R = 0.036 for 3884 reflexions.

II. Triclinic, $P\bar{1}$, a = 7.065, b = 9.179, c = 19.814 Å, α = 86.86, β = 85.23, γ = 70.04°, Z = 2. Mo radiation, R = 0.031 for 3283 reflexions.

Molecule (I) is a trans RE=ER complex with Sb-Sb 2.774(1) Å. In molecule (II) the Sb (Fig. 1) atom is pyramidal (angle sum 287.1(1)°) with Fe(1)-Sb 2.663(1), Fe(2)-Sb 2.641(1), Fe-Fe 2.801(1) Å, Fe-Sb-Fe 64.15(2)°. The Fe-Fe bond is weak.

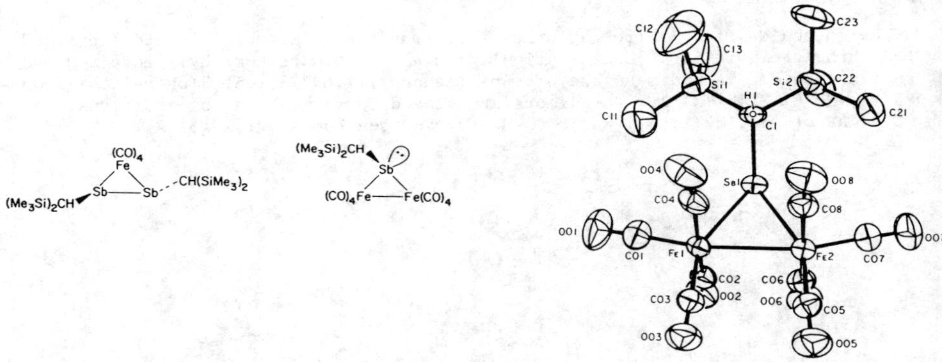

Fig. 1. Line drawings of (I) and (II) and a view of $C_{15}H_{19}Fe_2O_8SbSi_2$ (II).

ANTIMONY(III) MONOTHIOBENZOATE
$C_{21}H_{15}O_3S_3Sb$ $Sb(C_6H_5COS)_3$

R. MATTES and D. RÜHL, 1984. Inorg. Chim. Acta, 84, 125-127.

Rhombohedral, R3, a = 20.453, c = 4.197 Å, Z = 3. Mo radiation, R = 0.042 for 585 reflexions.

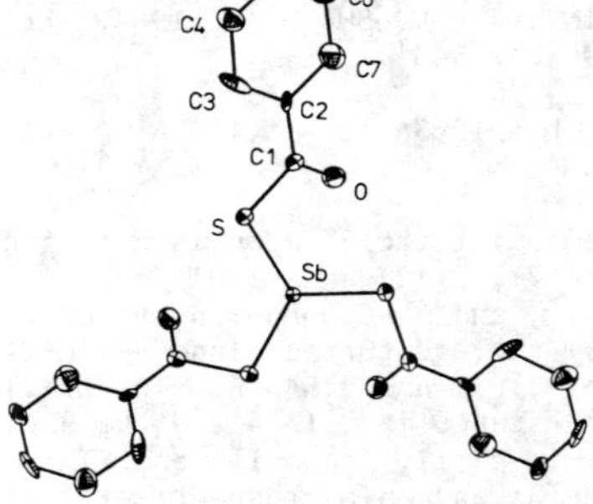

Fig. 1. Molecular structure of $C_{21}H_{15}O_3S_3Sb$ viewed along the hexagonal c axis.

Antimony, situated on the crystallographic threefold axis, is pyramidally coordinated to the S atoms of three ligands (Fig. 1). The Sb-S distance is 2.493 Å and the S-Sb-S angles are 91.1°. Secondary bonds between Sb and the O atoms occur with Sb-O distances of 2.802 Å causing the ligands to act as unsymmetrical chelates. The O-Sb-O angles are 120.0° meaning that the Sb atom and the three O atoms are virtually coplanar. Other distances are: C(1)-S 1.77, C(1)-O 1.23 and C(1)-C(2) 1.43 Å. Polymerization occurs along the c axis through three very weak Sb-S' bonds of 3.463 Å length.

BIS(ETHANOLMETHYL(μ_2-PHTHALATO-O,O')ANTIMONY)
$C_{22}H_{26}O_{10}Sb_2$

M. WIEBER, D. WIRTH and C. BURSCHKA, 1984. Z. Naturforsch., B39, 600-603.

Triclinic, P$\bar{1}$, a = 8.526, b = 9.102, c = 9.700 Å, α = 69.74, β = 77.57, γ = 62.54°, Z = 1. Mo radiation, R = 0.020 for 2054 reflexions.

The molecule (Fig. 1) lies about an inversion centre with the Sb atoms bridged by the phthalato moieties. Coordination at Sb also includes a methyl group and an ethanol molecule. Main dimensions are Sb-O(ethanol) 2.647(3), Sb-O(phthalato) 2.088-2.985(2), Sb-C1 2.107(4) Å. The dimers are linked to form chains by additional Sb-O3' contacts (3.132(2) Å), and by O-H...O hydrogen bonds (2.779(3) Å).

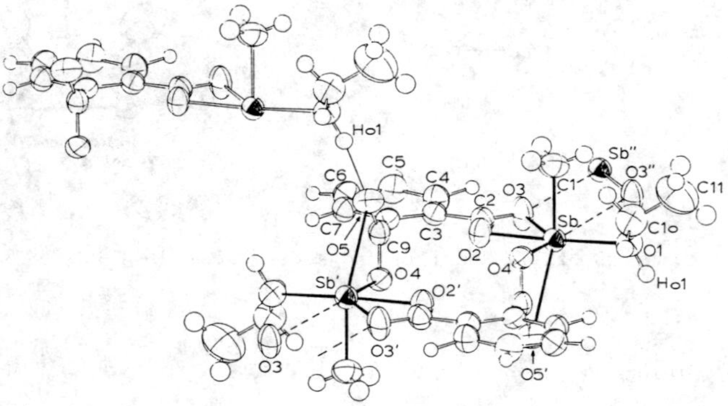

Fig. 1. A view of $C_{22}H_{26}O_{10}Sb_2$.

DIBROMOTRICYMANTRENYLANTIMONY
$C_{24}H_{12}Br_2Mn_3O_9Sb$ $(C_5H_4Mn(CO)_3)_3SbBr_2$

Yu.N. SAF'YANOV, É.A. KUZ'MIN and V.V. SHARUTIN, 1984. Kristallografiya, 29, 928-930 [Sov. Phys. Crystallogr., 29, 547-548].

Monoclinic, B2/b, a = 34.17, b = 19.36, c = 9.123 Å, γ = 111.9°, Z = 8. Mo radiation, R = 0.063 for 1637 reflexions.

The coordination polyhedron of the Sb atom is a trigonal bipyramid (Fig. 1). In the equatorial plane, atoms C(7), C(18) and C(29) form a triangle with C-Sb-C bond angles 121.5, 119.7 and 118.7°, and Sb-C bond lengths of 2.05, 2.09 and 2.10 Å. The Sb atom lies 0.03 Å away from the equatorial plane. Two Br atoms occupy the axial sites with Sb-Br distances of 2.617 and 2.545 Å. The Br(2)-Sb-Br(3) angle is 176.9°. The mean Mn-C(η^5-C_5H_4) bond distance is 2.13 Å and the angles between the Cp rings are 115.6, 63.4 and 60.4° (rings I and II, I and III and II and III where I, II and III are associated with Mn(4), Mn(5) and Mn(6) respectively).

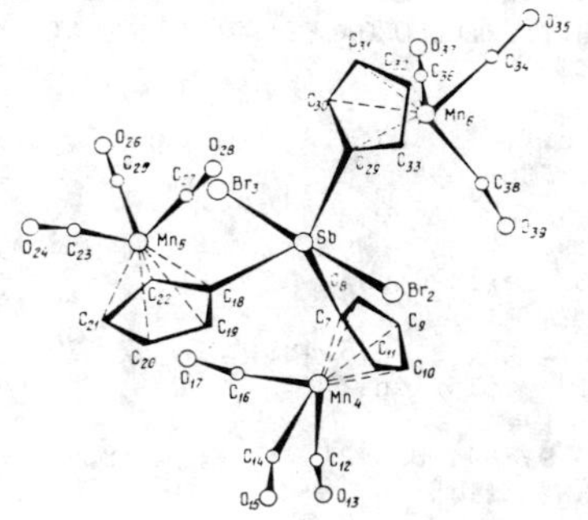

Fig. 1. General view of the molecule $C_{24}H_{12}Br_2Mn_3O_9Sb$.

TETRAPHENYLDIBISMUTH
$C_{24}H_{20}Bi_2$

F. CALDERAZZO, R. POLI and G. PELIZZI, 1984. J. Chem. Soc. Dalton, 2365-2369.

Triclinic, $P\bar{1}$, a = 7.865, b = 11.444, c = 6.257 Å, α = 105.24, β = 94.20, γ = 96.52°, Z = 1. Mo radiation, R = 0.056 for 1097 reflexions.

The molecule, which is constrained to lie on a crystallographic centre of symmetry, has a staggered trans conformation (Fig. 1). The environment of the Bi atom is approximately pyramidal and the Bi-Bi distance is 2.990(2) Å. The angles around Bi are close to 90°. The Bi-C distances are 2.28(2) and 2.26(2) Å. The phenyl rings are inclined at angles of 49.6 and 75.5° with respect to the C(1),C(7),Bi plane. Packing in the crystal is governed by van der Waals interactions.

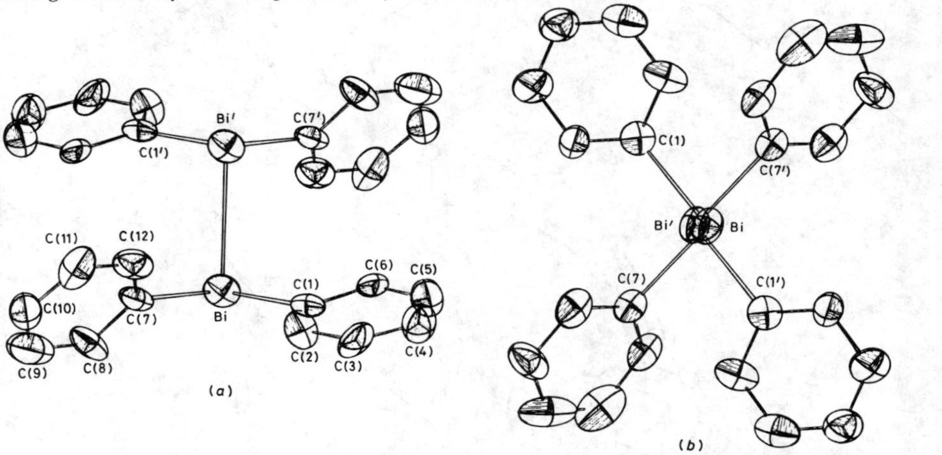

Fig. 1. $C_{24}H_{20}Bi_2$: views of the molecule. (a) Perpendicular to the Bi-Bi bond, (b) parallel to the Bi-Bi bond.

CHLORO(8-HYDROXYQUINOLINE-O)TRIPHENYLBISMUTH
$C_{27}H_{21}BiClNO$ (I)

TETRAPHENYLBISMUTH p-TOLUENESULPHONATE
$C_{31}H_{27}BiO_3S$ (II) $Ph_4Bi^+.MeC_6H_4SO_3^-$

TETRAPHENYLBISMUTH DIPHENYLBIS(TRIFLUOROACETATO)BISMUTHATE
$C_{40}H_{30}Bi_2F_6O_4$ (III) $Ph_4Bi^+.Ph_2Bi(O_2CCF_3)_2^-$

D.H.R. BARTON, B. CHARPIOT, E.T.H. DAU, W.B. MOTHERWELL, C. PASCARD and C. PICHON, 1984. Helv. Chim. Acta, 67, 586-599.

I. Monoclinic, $P2_1/n$, a = 19.908, b = 9.436, c = 13.088 Å, β = 106.602°, Z = 4. Mo radiation, R = 0.053 for 3087 reflexions.

II. Monoclinic, C2/c, a = 21.082, b = 15.488, c = 17.620 Å, β = 108.3°, Z = 8. Mo radiation, R = 0.050 for 2518 reflexions.

III. Monoclinic, $P2_1/c$, a = 9.884, b = 21.215, c = 18.556 Å, β = 95.1°, Z = 4. Mo radiation, R = 0.053 for 3259 reflexions.

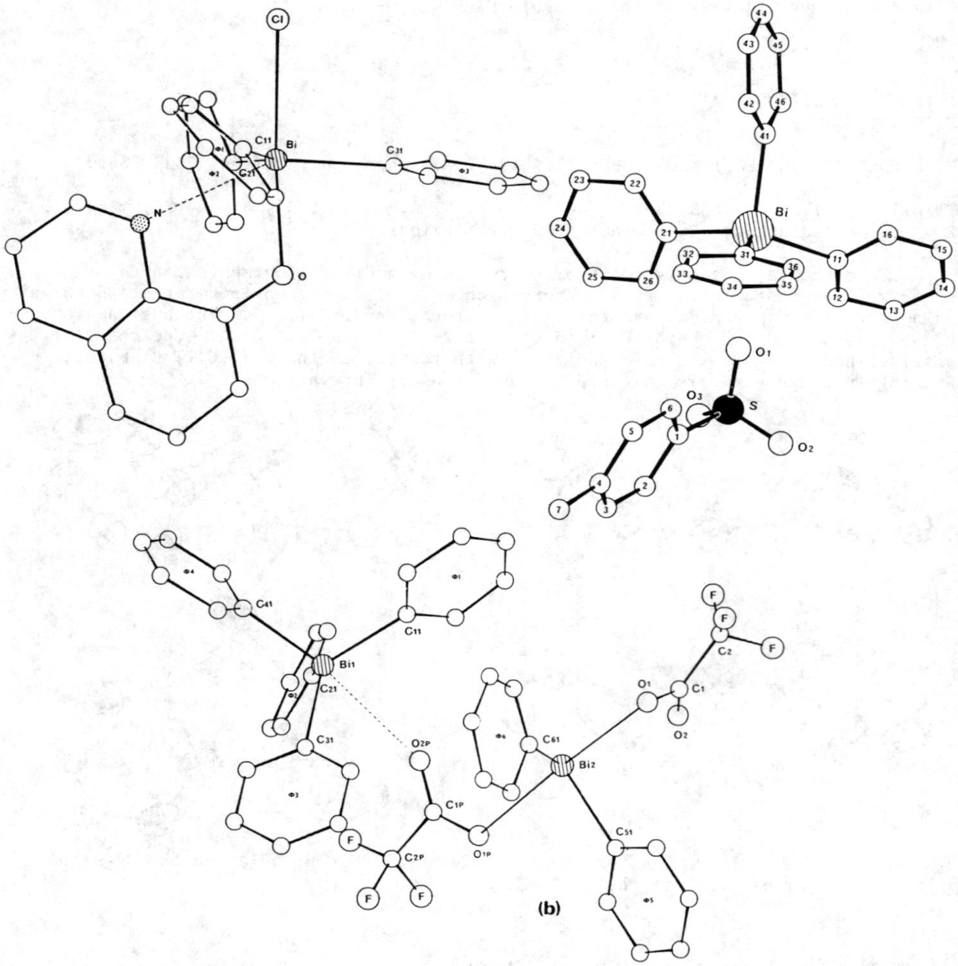

Fig. 1. Views of the three bismuth derivatives.

Molecule I has a trigonal bipyramidal structure (Fig. 1) with Bi-O 2.175(7), Bi-Cl 2.657(3), Bi-C 2.168-2.214(11) Å. The quinoline N atom is 2.81 Å from Bi and is directed towards it. In II there is considerable flattening of the bismuth tetrahedron because of the close approach (2.77(3) Å) of an oxygen atom from the tosylate anion. The Bi-C distances are in the range 2.17-2.24(3) Å and C-Bi-C angles range from 98 to 123(1)°. The O(1)...Bi-C(41) angle is 177°. Compound III has a distorted tetrahedral Ph_4Bi^+ cation (Bi-C 2.19-2.23(2) Å) and a trigonal bipyramidal anion with a stereochemically active lone pair. Some anion dimensions are Bi-O 2.38 and 2.41(2), Bi-C 2.24 and 2.26(2) Å, O-Bi-O 172, C-Bi-C 95°.

TRIS(DICYCLOHEXYLTHIOPHOSPHINATO)ANTIMONY(III)
$C_{36}H_{66}O_3P_3S_3Sb$ (I)

TRIS(DIPHENYLTHIOPHOSPHINATO)ANTIMONY(III)
$C_{36}H_{30}O_3P_3S_3Sb$ (II)

R. MATTES and D. RUHL, 1984. Z. Anorg. Chem., 508, 19-25.

I. Monoclinic, $P2_1/n$, a = 9.835, b = 21.412, c = 20.387 Å, β = 94.41°, Z = 4. Mo radiation, R = 0.060 for 5005 reflexions

II. Monoclinic, C2/c, a = 31.918, b = 17.723, c = 12.610 Å, β = 104.27°, Z = 8. Mo radiation, R = 0.045 for 5900 reflexions.

In both structures, the primary coordination is through oxygen (Fig. 1) with Sb-O 2.026(6)-2.031(6) in (I) and 2.037(3)-2.063(3) in (II). Sb...S distances in the mainly chelating ligands are in the range 3.001(1)-3.286(3) Å.

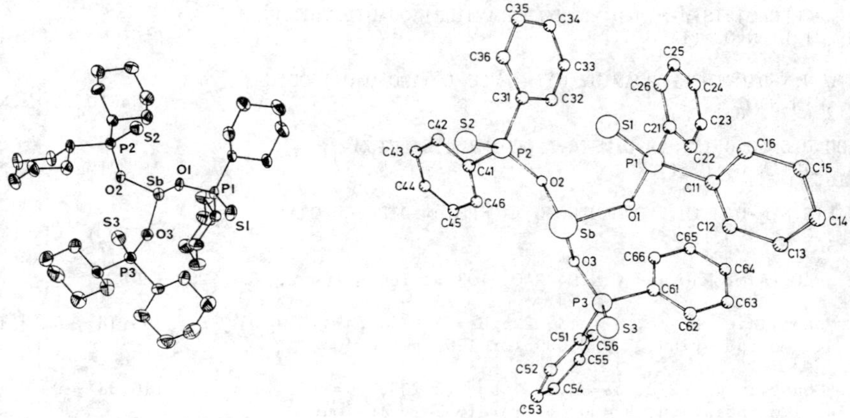

Fig. 1. Views of $C_{36}H_{66}O_3P_3S_3Sb$ and $C_{36}H_{30}O_3P_3S_3Sb$.

TRIPHENYLPHOSPHONIUMBIS(DIPHENYLSTIBINO)METHYLIDE
$C_{43}H_{35}PSb_2$ $(C_6H_5)_3PC(Sb(C_6H_5)_2)_2$

H. SCHMIDBAUR, B. MILEWSKI-MAHRLA, G. MÜLLER and C. KRÜGER, 1984. Organometallics, 3, 38-43.

Monoclinic, $P2_1/n$, a = 10.947, b = 14.514, c = 22.768 Å, β = 90.67°, Z = 4. Mo radiation, R = 0.030 for 4880 reflexions.

The molecule (Fig. 1) adopts the conformation with lone pairs of electrons at the Sb atoms oriented in the plane of the heavy atoms (P(1), Sb(1), Sb(2)) which also includes the ylidic C atom C(1). The deviation of C(1) from the plane of P(1), Sb(1), Sb(2) is 0.12 Å and the valency angles Sb(1)-C(1)-Sb(2), Sb(1)-C(1)-P(1) and Sb(2)-C(1)-P(1) are respectively 116.2, 130.5 and 111.5°. The P(1)-C(1) bond length of 1.692 Å indicates partial multiple bond character when compared to P(1)-C(2) (1.835 Å), P(1)-C(8) (1.825 Å), and P(1)-C(14) (1.825 Å). Bond distances involving the Sb atoms are: Sb(1)-C(1) 2.110, Sb(1)-C(20) 2.165, Sb(1)-C(26) 2.176, Sb(2)-C(1) 2.136, Sb(2)-C(32) 2.170, Sb(2)-C(38) 2.172 Å.

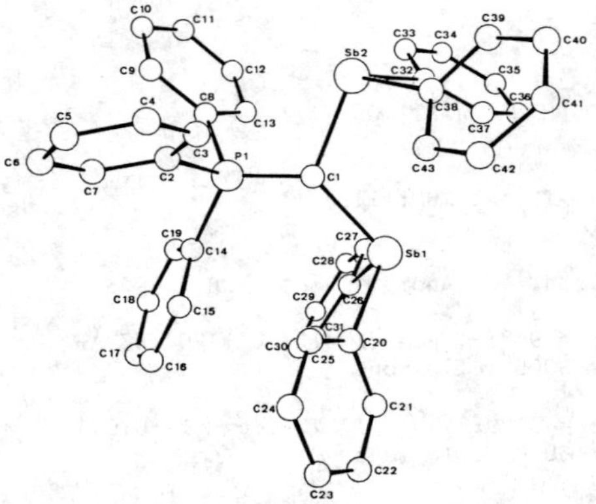

Fig. 1. Perspective view of the molecule $C_{43}H_{35}PSb_2$.

BIS(μ-CHLORO)BIS(μ-N,N-DIMETHYLFORMAMIDE)AQUADILITHIUM
$C_6H_{14}Cl_2Li_2N_2O_3$ (I) $[LiCl.Me_2NC(O)H]_2(H_2O)$

BIS(μ-N,N-DIMETHYLFORMAMIDE-O)(μ-PERCHLORATO-O,O')SODIUM
$C_6H_{12}ClN_2NaO_6$ (II) $NaClO_4.2[Me_2NC(O)H]$

DIAQUADICHLOROBIS(N,N-DIMETHYLFORMAMIDE)CALCIUM
$C_6H_{16}CaCl_2N_2O_4$ (III) $CaCl_2.2[Me_2NC(O)H],2H_2O$

HEXAKIS(N,N-DIMETHYLFORMAMIDE-O)MAGNESIUM DIPERCHLORATE
$C_{18}H_{36}Cl_2MgN_6O_{14}$ (IV) $Mg(ClO_4)_2.6[Me_2NC(O)H]$

C.P. RAO, A.M. RAO and C.N.R. RAO, 1984. Inorg. Chem., 23, 2080-2085.

I. Monoclinic, $P2_1/c$, a = 13.022, b = 5.978, c = 17.028 Å, β = 105.48°, D_m = 1.281, Z = 4. Mo radiation, R = 0.089 for 1581 reflexions.

II. Monoclinic, P2/c, a = 9.297, b = 10.203, c = 13.510 Å, β = 110.08°, D_m = 1.464, Z = 4. Mo radiation, R = 0.062 for 1663 reflexions.

III. Triclinic, $P\bar{1}$, a = 6.293, b = 6.944, c = 8.853 Å, α = 110.15, β = 105.60, γ = 95.34°, D_m = 1.382, Z = 1. Mo radiation, R = 0.055 for 1506 reflexions.

IV. Monoclinic, $P2_1/a$, a = 20.686, b = 10.962, c = 14.885 Å, β = 91.45°, D_m = 1.273, Z = 4. Mo radiation, R = 0.121 for 2900 reflexions.

The asymmetric units of I-IV are shown in Fig. 1. In I, each of the two crystallographically independent Li atoms is tetrahedrally coordinated, Li(1) by three O atoms and a Cl (Li-O 1.904(18)-1.984(19), Li-Cl 2.361(17) Å), and Li(2) by two O and two Cl atoms (Li-O 1.949(18) and 1.975(18), Li-Cl 2.347(17) and 2.341(17) Å). The structure I is polymeric, as is that of II. In II-IV the cations are all octahedrally coordinated. In II there are two independent Na^+ ions, one having C_i and the other C_2 symmetry. The Na-O distances are in the range 2.346(11)-2.578(9) Å. The structures III and IV both consist of discrete centrosymmetric complexes, there being two crystallographically independent moieties in the case of IV. The Ca^{2+} ion in III has O_4Cl_2 coordination with Ca-O 2.341(3) and 2.365(4), Ca-Cl 2.713(1) Å. In IV each Mg^{2+} is coordinated by six $Me_2NC(O)H$ ligands with Mg-O 2.006(19)-2.080(16) Å.

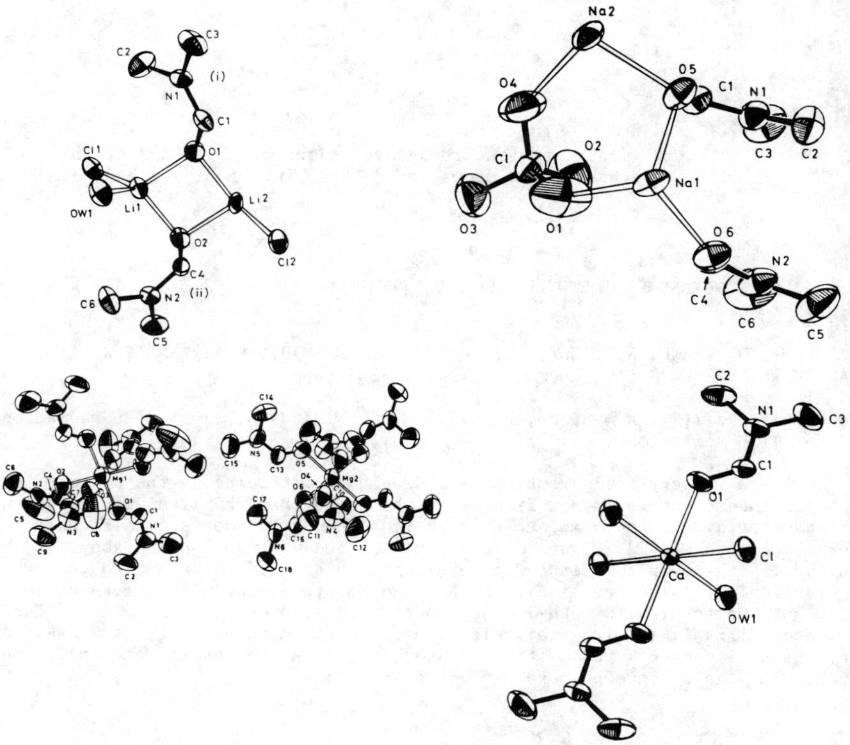

Fig. 1. Cation coordination in $[LiCl.Me_2NC(O)H]_2(H_2O)$ (top left), $NaClO_4.2[Me_2NC(O)H]$ (top right), $CaCl_2.2[Me_2NC(O)H]$ (bottom right), and $Mg(ClO_4)_2.6[Me_2NC(O)H]$ (bottom left).

[TETRAHYDROFURANTETRACARBOXYLATO(1-)]CAESIUM
$C_8H_7CsO_9$ (I)

HEXAAQUA[TETRAHYDROFURANTETRACARBOXYLATO(4-)]DICALCIUM DIHYDRATE
$C_8H_{16}Ca_2O_{15},2H_2O$ (II) $Ca_2(C_8H_4O_9)(H_2O)_6,2H_2O$

J.C. BARNES and J.D. PATON, 1984. Acta Cryst., C40, 1809-1812.

I. Monoclinic, $P2_1/c$, a = 8.842, b = 11.389, c = 11.359 Å, β = 100.82°, Z = 4. Mo radiation, R = 0.079 for 3114 reflexions.

II. Triclinic, $P\bar{1}$, a = 9.150, b = 10.512, c = 9.138 Å, α = 95.17, β = 91.48, γ = 77.67°, Z = 2. Cu radiation, R = 0.086 for 1409 reflexions.

The anions in both structures are of the trans, cis, trans isomer (Fig. 1); I contains three protonated carboxylate groups while in II all carboxylate groups are deprotonated. In I, eight-coordinate Cs atoms are linked into pairs by a square of O atoms with Cs...O 3.009-3.342 Å; in II, eight-coordinate Ca atoms are linked into sheets by μ-carboxylato and cross-ligand bridging. The furan O is coordinated in II but not in I. There is extensive hydrogen bonding involving the water molecules in II.

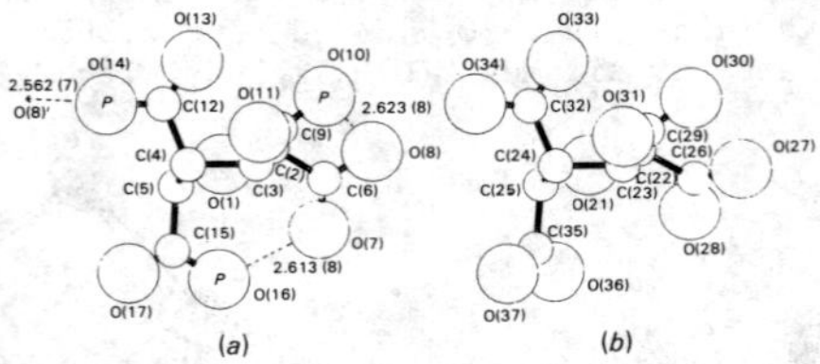

Fig. 1. The tetrahydrofurantetracarboxylato anions of I (a) and II (b). (P indicates protonated O atoms and dashed lines indicate hydrogen bonds).

(μ_2-σ^1,η^4-CYCLOHEXANONE DIMETHYLHYDRAZONE)LITHIUM
$(C_8H_{15}LiN_2)_n$

D.B. COLLUM, D. KAHNE, S.A. GUT, R.T. DEPUE, F. MOHAMADI, R.A. WANAT, J. CLARDY and G. VAN DUYNE, 1984. J. Am. Chem. Soc., 106, 4865-4869.

Orthorhombic, $P2_12_12_1$, a = 8.503, b = 11.247, c = 19.244 Å, Z = 8. Cu radiation, R = 0.043 for 1136 reflexions.

There are two crystallographically independent molecules in the asymmetric unit (Fig. 1). The structure is not dimeric in the sense that the lithiums and anions form an extended array. In each molecule the cyclohexene ring adopts a half-chair conformation with C1, C2, C3 and C6 being planar and C4 and C5 being above and below the plane. Lithium is coordinated to one anion in an η^4 fashion and to the other in an η^1 fashion. The four atoms C1, C2, N7, and N8 are planar with a mean deviation of 0.01 Å, but the terminal dimethylamino group is twisted out of this plane. The η^4 coordinated Li is 1.64 Å from this plane, and the η^1 coordinated Li is 0.5 Å from the plane. Each Li connects two of these planes, one in an η^1 fashion and one in an η^4 fashion. The angle between these two planes is 111°.

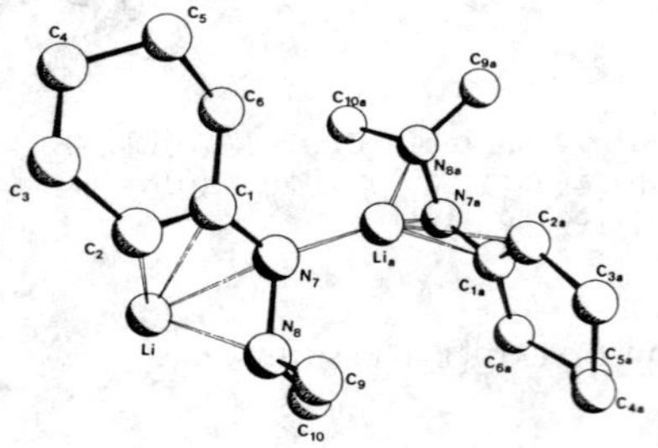

Fig. 1. The asymmetric unit of $(C_8H_{15}LiN_2)_n$.

BIS(1,2-DIMETHOXYETHANE-O,O')LITHIUM BROMIDE
$C_8H_{20}BrLiO_4$

R.D. ROGERS, R.V. BYNUM and J.L. ATWOOD, 1984. J. Cryst. Spectrosc. Res., 14, 29-34.

Monoclinic, C2/c, a = 13.518, b = 8.007, c = 12.252 Å, β = 98.08°, D_m = 1.35, Z = 4. Mo radiation, R = 0.033 for 489 reflexions.

The lithium and bromine atoms lie on a crystallographic twofold axis. The structure (Fig. 1) consists of discrete $LiBr-(CH_3OCH_2CH_2OCH_3)_2$ units. The Li-Br separation is 2.57(1) Å. The two DME molecules are coordinated to the lithium atom through the four oxygen atoms at an average Li...O separation of 2.09(3) Å.

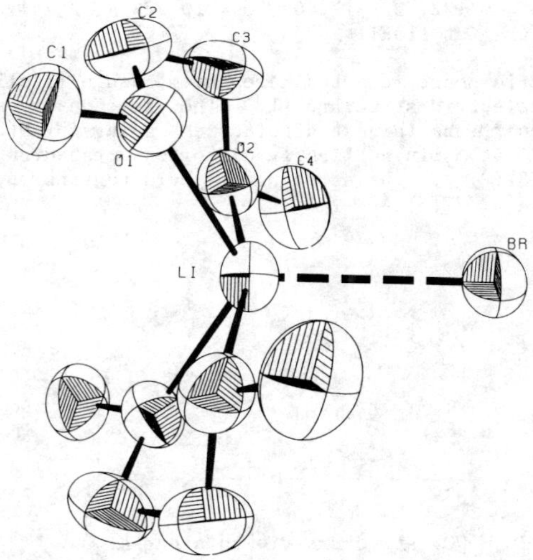

Fig. 1. A view of $C_8H_{20}BrLiO_4$.

BIS(1,2-DIMETHOXYETHANE)BIS(TETRAHYDROBORATE)CALCIUM
$C_8H_{28}B_2CaO_4$

E. HANECKER, J. MOLL and H. NOTH, 1984. Z. Naturforsch., B39, 424-430.

Monoclinic, $P2_1/c$, a = 8.884, b = 12.851, c = 14.711 Å, β = 101.85°, Z = 4. Mo radiation, R = 0.0507 for 2171 reflexions.

Each BH_4 group binds to the Ca atom via three H atoms as shown in Fig. 1, and the two ethers bind through their O atoms so that Ca is decacoordinated. Main dimensions are: Ca-O 2.415-2.465(2), Ca-H 2.35-2.58(3) Å, mean O-Ca-O 66.6(1)°.

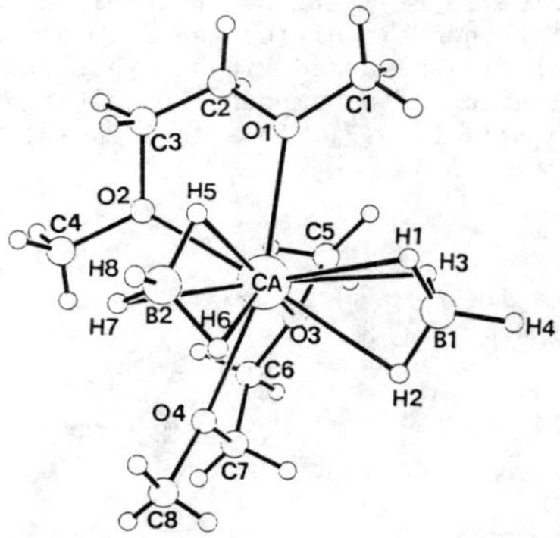

Fig. 1. A view of $C_8H_{28}B_2CaO_4$.

TRIS-SARCOSINE CALCIUM CHLORIDE
$C_9H_{21}CaCl_2N_3O_6$ $CaCl_2,3(CH_3NH_2^+CH_2COO^-)$

N. MISHIMA, K. ITOH and E. NAKAMURA, 1984. Acta Cryst., C40, 1824-1827.

Orthorhombic, $Pn2_1a$, a = 9.122, b = 17.408, c = 10.228 Å, Z = 4. Mo radiation, R = 0.036 for 5951 reflexions (at 118K).

In the ferroelectric phase reported here the molecule is slightly deformed from that found in the paraelectric structure (1). The skeleton of each sarcosine molecule is almost planar, the largest displacement being for N(2) at 0.10 Å. Ca^{2+} is coordinated by six O atoms in a slightly distorted octahedron (Fig. 1) with Ca...O in the range 2.294-2.387(2) Å. The structure is held together by N-H...Cl hydrogen bonds with N...Cl 3.178-3.227(3) Å.

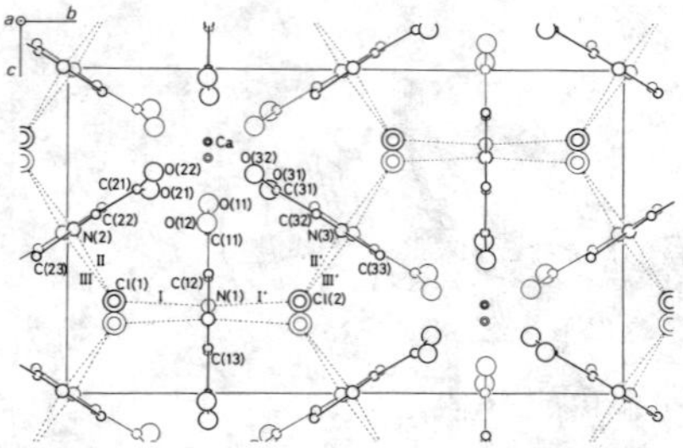

Fig. 1. The $CaCl_2,3C_3H_7NO_2$ structure viewed along a.

1. Structure Reports, 38B, 889.

BIS[μ-(2-DIMETHYLAMINO-N-METHYLETHYLAMIDO)-N',μ-N]BIS(HYDRIDOBERYLLIUM)

$C_{10}H_{28}Be_2N_4$ $Be_2(C_5H_{13}N_2)_2H_2$

N.A. BELL, G.E. COATES, M.L. SCHNEIDER and H.M.M. SHEARER, 1984. Acta Cryst., C40, 608-610.

Triclinic, $P\bar{1}$, a = 7.463, b = 6.061, c = 8.252 Å, α = 104.5, β = 95.90, γ = 96.80°, D_m = 1.05, Z = 1. Mo radiation, R = 0.048 for 958 reflexions.

The molecule (Fig. 1) is dimeric, with the two halves of the dimer being linked by a centrosymmetric four-membered Be_2N_2 ring. Be achieves a coordination number of 4 by formation of a dative N→Be bond involving the second N atom of the ligand resulting in the formation of five-membered rings. This represents the first example of the structural characterisation of a compound containing a terminal Be-H bond. Some bond distances are: Be-H(1) 1.39(2), Be-N(1) 1.746(3), Be-N(1') 1.747(3) and Be-N(2) 1.814(3) Å.

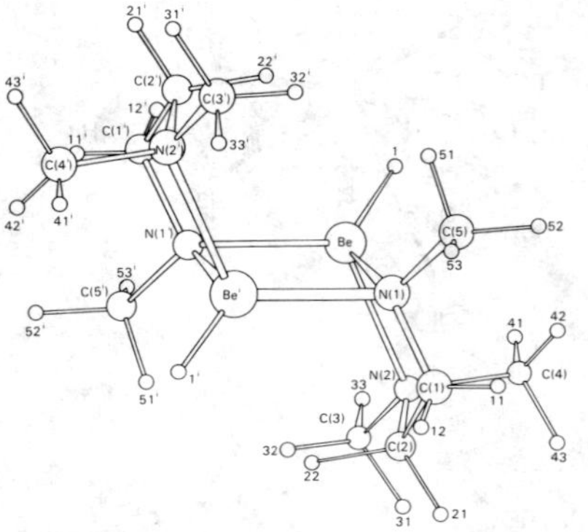

Fig. 1. Perspective view of the $Be_2(C_5H_{13}N_2)_2H_2$ molecule.

BIS(o-NITROPHENOLATO)BARIUM(II)
$C_{12}H_8BaN_2O_6$ $Ba(C_6H_4NO_3)_2$

J.A. KANTERS, W.J.J. SMEETS, A.J.M. DUISENBERG, K. VENKATASUBRAMANIAN and N.S. POONIA, 1984. Acta Cryst., C40, 1699-1701.

Monoclinic, C2/c, a = 27.23, b = 6.094, c = 7.736 Å, β = 97.20°, Z = 4. Mo radiation, R = 0.0356 for 1407 reflexions.

The Ba atom lies on a twofold axis and is coordinated to ten O atoms (Fig. 1) forming a distorted one-side-bicapped square antiprism; four of the coordinating atoms are phenolic. Eight of the ten O atoms are coordinated to two Ba^{2+} ions forming a two-dimensional network. Ba-O distances are in the range 2.711-3.117(5) Å. In the o-nitrophenolate moiety the plane of the nitro group is rotated 17.9° out of the phenyl ring plane.

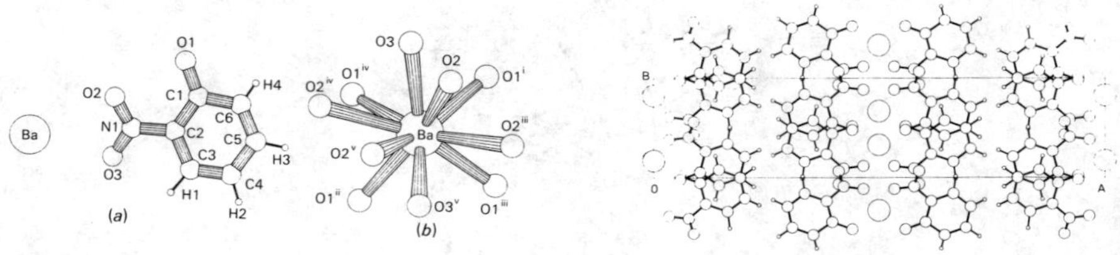

Fig. 1. Views of the asymmetric unit and the structure projected along c for $Ba(C_6H_4NO_3)_2$.

1-[(PENTAMETHYLDIETHYLENETRIAMINE-N,N',N'')LITHIO]-2-METHYL-1,2-DICARBA-closo-DODECABORANE
$C_{12}H_{36}B_{10}LiN_3$

W. CLEGG, D.A. BROWN, S.J. BRYAN and K. WADE, 1984. Polyhedron, 3, 307-311.

Monoclinic, $P2_1/m$, a = 8.242, b = 13.522, c = 10.799 Å, β = 108.75°, Z = 2. Mo radiation, R = 0.088 for 1128 reflexions.

The molecule (Fig. 1) has crystallographic mirror symmetry and is unusual in that it contains a lithium atom terminally attached by a formally single Li-C bond to the six-coordinate carbon atom of a carborane. Selected bond lengths are Li-C(6) 2.176(8), Li-N(1) 2.134(9), Li-N(2) 2.169(5), C(6)-C(7) 1.681(7), C(7)-C(8) 1.516(11), C(6)-B(1) 1.707(6), C(6)-B(2) 1.709(8) Å.

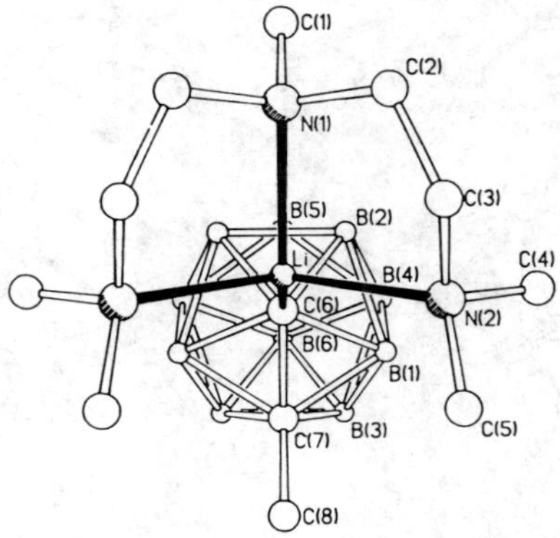

Fig. 1. The molecular structure of $C_{12}H_{36}B_{10}LiN_3$.

BENZO-14-CROWN-4 LITHIUM NITRATE

$C_{14}H_{20}LiNO_7$ $C_{14}H_{20}O_4 . LiNO_3$

E.M. HOLT, G.D. MALPASS, JR., R.G. GHIRARDELLI, R.A. PALMER and B. RUBIN, 1984. Acta Cryst., C40, 394-396.

Monoclinic, $P2_1/n$, a = 14.584, b = 9.074, c = 12.238 Å, β = 91.08°, Z = 4. Cu radiation, R = 0.081 for 1389 reflexions.

The Li atom has six-coordination (Fig. 1) via ligation to four oxygen atoms of the crown ether ring and two oxygens of a nitrate anion. Li-O distances are in the range 2.011-2.300(4) Å. The coordination is not octahedral principally because of the acute bite of the nitrate anion (O-Li-O 57.6(1)°).

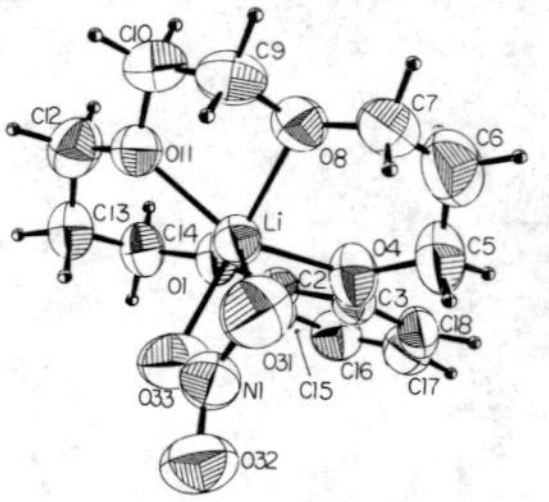

Fig. 1. A view of $C_{14}H_{20}O_4 . LiNO_3$.

TETRAMETHYLETHYLENEDIAMINE(TRIMETHYLSILYLCYCLOPENTADIENYL)LITHIUM

$C_{14}H_{29}LiN_2Si$ $[Li\{\eta\text{-}C_5H_4(SiMe_3)\}\{NMe_2(CH_2)_2NMe_2\}]$

M.F. LAPPERT, A. SINGH, L.M. ENGELHARDT and A.H. WHITE, 1984. J. Organometal. Chem., 262, 271-278.

Triclinic, $P\bar{1}$, a = 13.592, b = 8.618, c = 8.276 Å, α = 96.71, β = 100.23, γ = 101.17°, Z = 2. Mo radiation, R = 0.064 for 1319 reflexions.

The Li atom may be regarded as trigonal planar, with bonds to the two N atoms (2.142 and 2.119 Å) of the chelating TMEDA ligand, and bonds to the centroid of the planar η^5-cyclopentadienyl ring (1.928 Å) (Fig. 1). The Li-C bonds range from 2.257 to 2.286 Å (the carbon bearing the $SiMe_3$ substituent), and the cyclopentadienyl C-C bonds from 1.38 (the two bonds furthest removed from C-$SiMe_3$) to 1.421 Å. The Cp(centroid)-Li-N angles are 135.7 and 139.1° and the N(1)-Li-N(2) angle is 84.9°.

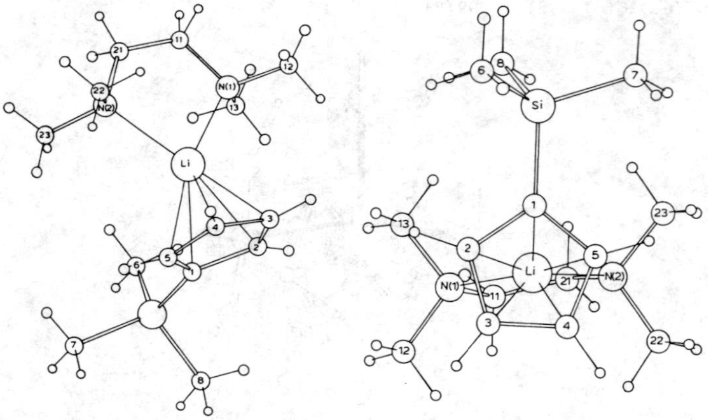

Fig. 1. Molecular structure of $C_{14}H_{29}LiN_2Si$.

(3,6,9,12,15,18,21-HEPTAMETHYL-1,4,7,10,13,16,19-HEPTAOXA-2,5,8,11,14,17,20-HEPTA-AZACYCLOHENEICOSA-2,5,8,11,14,17,20-HEPTAENE-$O^1,O^4,O^7,O^{10},O^{13},O^{16},O^{19}$)POTASSIUM THIOCYANATE

$C_{15}H_{21}KN_8O_7S$ $C_{14}H_{21}KN_7O_7^{+}.NCS^{-}$

G. CASINI, F. DE SARLO, D. DONATI, A. GUARNA and P. ORIOLI, 1984. J. Chem. Res., (S) 2-3, (M) 0201-0274.

Monoclinic, Pa, a = 24.236, b = 22.920, c = 8.315 Å, β = 91.48°, Z = 8. Mo radiation, R = 0.100 for 5713 reflexions.

The four independent complexes all have similar structures; one is shown in Fig. 1. The K atom is less than 0.1 Å from the plane of oxygen atoms (all of which are within 0.54 Å from the plane). The mean K-O distance is 2.82 Å. The K atoms are further coordinated by S and N atoms from the NCS^- ion (K-N 2.87, K-S 3.23 Å).

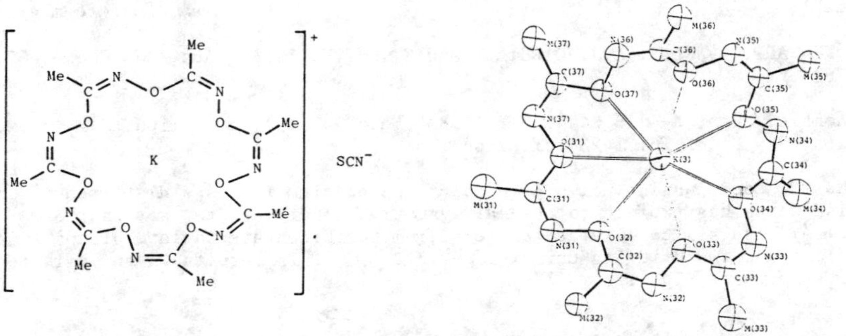

Fig. 1. The $C_{15}H_{21}KN_8O_7S$ structure and a view of one complex cation.

BARIUM (PHENYLTHIO)ACETATE

$C_{16}H_{14}BaO_4S_2$ (I) $[Ba(C_6H_5SCH_2COO)_2]_n$

POTASSIUM (PHENYLTHIO)ACETATE (PHENYLTHIO)ACETIC ACID

$C_{16}H_{15}KO_4S_2$ (II) $[K(C_6H_5SCH_2COO)(C_6H_5SCH_2COOH)]_n$

T.C.W. MAK, W.-H. YIP, G. SMITH, E.J. O'REILLY and C.H.L. KENNARD, 1984. Inorg. Chim. Acta, 88, 35-39.

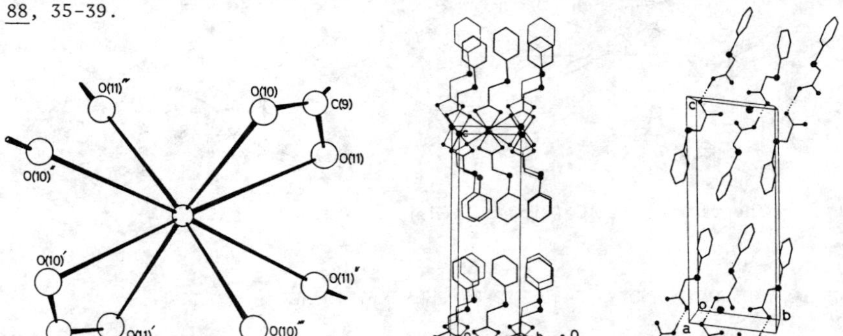

Fig. 1 Coordination polyhedron of $[Ba(PTA)_2]_n$ (left), and molecular packing for $[C_{16}H_{14}BaO_4S_2]_n$ (centre), and $[C_{16}H_{15}KO_4S_2]_n$ (right), both viewed perpendicular to bc.

I. Monoclinic, C2, a = 7.219, b = 5.894, c = 19.272 Å, β = 91.57°, D_m = 1.91, Z = 2. Mo radiation, R = 0.047 for 844 reflexions.

II. Triclinic, P$\bar{1}$, a = 6.272, b = 7.222, c = 18.459 Å, α = 97.23, β = 93.64, γ = 97.31°, D_m = 1.52, Z = 2. Mo radiation, R = 0.053 for 2628 reflexions.

The MO_8 coordination about Ba in (I) is distorted 'cubic' with four O's from two bidentate carboxylate groups (Ba-O 2.86, 2.97 Å] and four O's from adjacent carboxylates (Ba-O 2.68, 2.69 Å) forming an infinite two-dimensional polymer structure (Fig. 1). Compound II is polymeric and has bonded phenylthioacetate as well as adducted phenylthioacetic acid. There are seven K-O contacts ranging from 2.729 to 3.034 Å from both PTA and PTAH species in an irregular coordination.

catena-μ-[DIAQUABIS(3,6-DICHLORO-2-METHOXY-BENZOATO(O,O')]CALCIUM(II)
$C_{16}H_{14}CaCl_4O_8$ $[Ca(dicamba)_2(H_2)_2]_n$

C.H.L. KENNARD, B. KERR, E.J. O'REILLY and G. SMITH, 1984. Aust. J. Chem., 37, 1757-1761.

Orthorhombic, Pbcn, a = 15.8444, b = 16.833, c = 7.751 Å, D_m = 1.66, Z = 4. Mo radiation, R = 0.045 for 1023 reflexions.

The MO_8 coordination polyhedron about the calcium ion is a distorted Archimedes antiprism which has twofold rotational symmetry. There are two waters (Ca-O 2.414(3) Å), and six carboxylate oxygens from two bidentate chelate groups (Ca-O 2.516(3) Å) and two bridging carboxyls (2.372 Å). This results in an infinite linear polymer structure (Fig. 1).

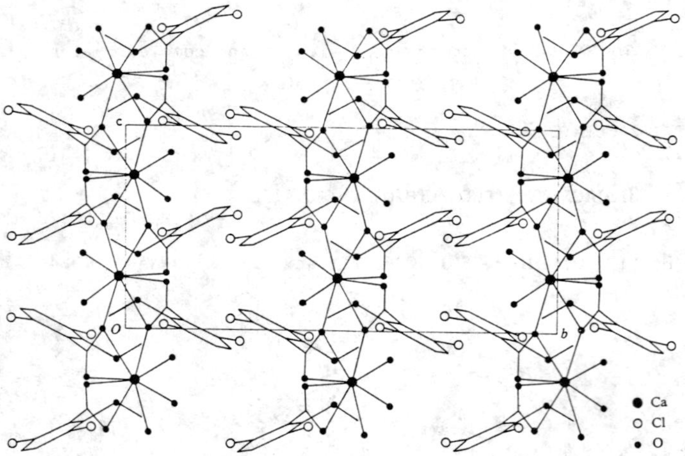

Fig. 1. The packing of $[Ca(dicamba)_2(H_2)_2]_n$.

1-(o-CARBOXYMETHOXYPHENOXY)-2-(o-HYDROXYPHENOXY)ETHANE
$C_{16}H_{16}O_6$ (I)

BIS[1-(o-CARBOXYMETHOXYPHENOXY)-2-(o-HYDROXYPHENOXY)ETHANE]POTASSIUM
$C_{32}H_{31}KO_{12}$ (II)

D.L. HUGHES and J.N. WINGFIELD, 1984. J. Chem. Soc. Dalton, 1187-1193.

I. Triclinic, P$\bar{1}$, a = 5.229, b = 8.759, c = 16.436 Å, α = 90.78, β = 96.33, γ = 105.06°, Z = 2. Mo radiation, R = 0.043 for 2079 reflexions.

II. Monoclinic, $P2_1/a$, a = 7.920, b = 14.993, c = 13.133 Å, β = 100.77°, Z = 2. Mo radiation, R = 0.13 for 621 reflexions.

I (Fig. 1) crystallises in an extended form with intermolecular hydrogen bonding around centres of symmetry linking the molecules in chains. The phenolic H atom is disordered, occupying two sites one of which is involved in a bifurcated hydrogen bond system. The O...O distances range from 2.618 to 3.177 Å. In II the two ligands of the anion are related by a centre of symmetry (O(21)...O(21') 2.45(3) Å). There are also hydrogen bonds between phenolic oxygen O(1) of one ligand and carbonyl oxygen O(22) of the other ligand (O...O 2.54 Å). The anions form sandwiches around the potassium ion in a pentagonal antiprismatic coordination to form a step-polymer. The K-O distances range from 2.71 to 2.87 Å.

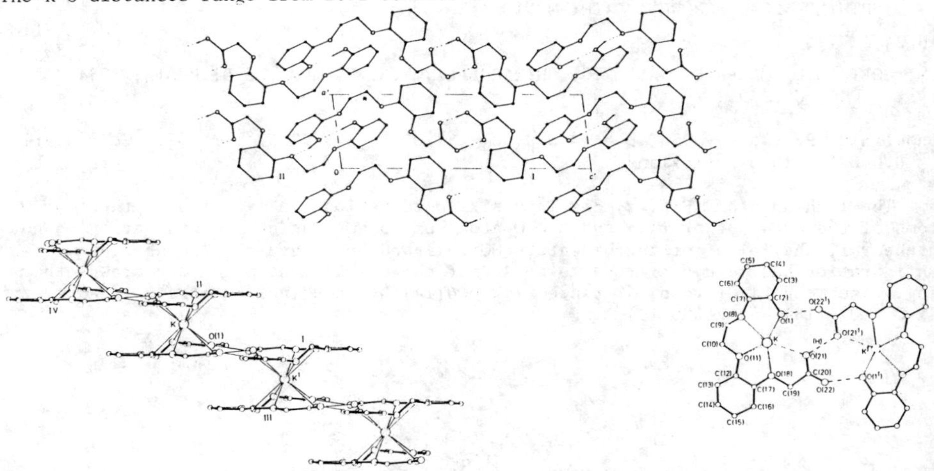

Fig. 1. $C_{16}H_{16}O_6$ (top), the packing in the unit cell and $C_{32}H_{31}KO_{12}$ (bottom), the step-polymer formation and the anion with its coordination potassium ions.

(μ-1,4-DIOXANE)(μ-DIPHENYLARSINO)SODIUM
$C_{16}H_{18}AsNaO_2$ $[(C_4H_8O_2)NaAs(C_6H_5)_2]_n$

A. BELFORTE, F. CALDERAZZO, A. MORVILLO, G. PELIZZI and D. VITALI, 1984. Inorg. Chem., 23, 1504-1508.

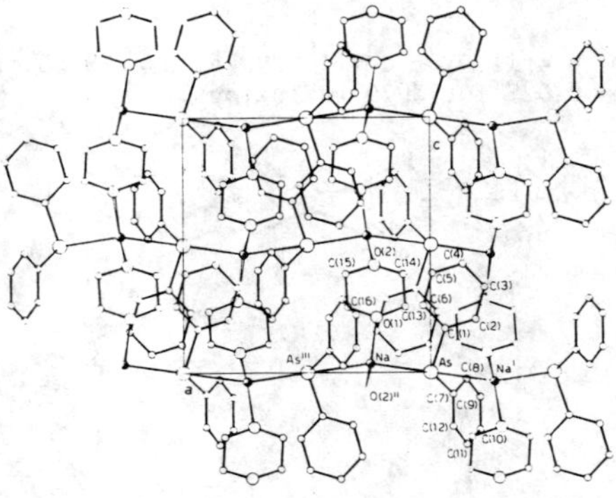

Fig. 1. The $[(C_4H_8O_2)NaAs(C_6H_5)_2]_n$ structure viewed along b.

Orthorhombic, $P2_12_12_1$, a = 10.280, b = 15.053, c = 10.735 Å, Z = 4. Cu radiation, R = 0.0435 for 946 reflexions.

The structure (Fig. 1) consists of an alternating sequence of dioxane-coordinated Na atoms and Ph_2As groups, giving rise to infinite -Na-As-Na-As- chains parallel to [100]. The chains are almost linear (173.6(1)°) at As and bent (121.5(1)°) at Na. An electron deficient As-Na bond is suggested. Bond lengths: As-Na 2.962(4) and 2.937(4), As-C 1.983(7) and 1.976(6), Na-O 2.320(11) and 2.307(10) Å. The geometry at As is trigonal bipyramidal with an equatorial lone pair.

6,7-DIMETHYLBENZO-14-CROWN-4 LITHIUM NITRATE

$C_{16}H_{24}LiNO_7$ $C_{16}H_{24}O_4.LiNO_3$

E.M. HOLT, G.D. MALPASS, JR., R.G. GHIRARDELLI, R.A. PALMER and B. RUBIN, 1984. Acta Cryst., C40, 396-398.

Monoclinic, $P2_1$, a = 7.835, b = 14.359, c = 8.207 Å, β = 97.37°, Z = 2. Cu radiation, R = 0.066 for 1076 reflexions.

The lithium atom (Fig. 1) displays six coordination via ligation to the four O atoms of the crown-ether ring and two O atoms of an NO_3^- anion. Coordination is not octahedral, the 'bite' of the bidentate NO_3^- ligand being acute. The absolute configuration has been confirmed to be R,R at the chiral C atoms of the crown-ether ring. The mean Li-O(crown) distance is 2.066(14) Å. Bonding to nitrate is asymmetric (Li-O 2.074(18) and 2.286(19) Å.

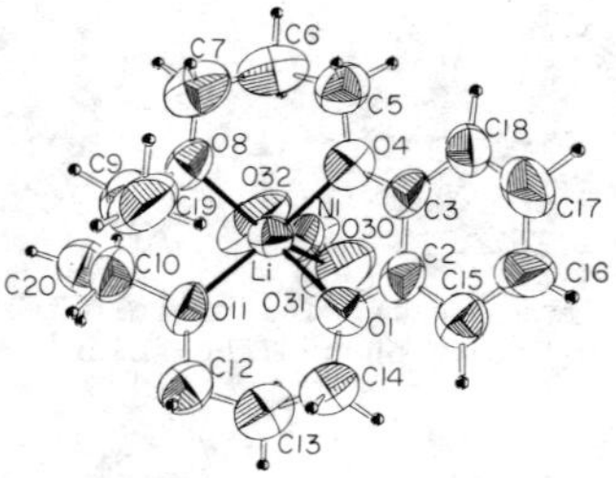

Fig. 1. A view of $C_{16}H_{24}O_4.LiNO_3$.

2,5,8,11,14,17,20-HEPTAOXAHENICOSANE BARIUM ISOTHIOCYANATE

$C_{16}H_{30}BaN_2O_7S_2$ $C_{14}H_{30}O_7.Ba(NCS)_2$

G. WEBER, F. HIRAYAMA, W. SAENGER and G.M. SHELDRICK, 1984. Acta Cryst., C40, 1570-1572.

Triclinic, $P\bar{1}$, a = 7.859, b = 11.010, c = 14.299 Å, α = 89.65, β = 78.39, γ = 78.50°, Z = 2. Mo radiation, R = 0.025 for 3979 reflexions.

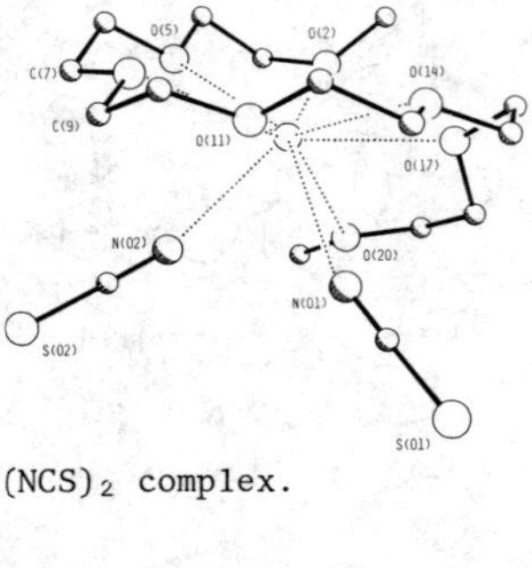

Fig. 1. The $C_{14}H_{30}O_7.Ba(NCS)_2$ complex.

The open-chain oligoether approximates to one turn of a helix leading to a sevenfold coordination of Ba^{2+} by all ether oxygens with mean Ba...O 2.91 Å. There are additional interactions with the two anions with mean $Ba^{2+}...N^-$ 2.83 Å. The O(2), O(5), O(8), O(11) atoms (Fig. 1) are coplanar within ±0.07 Å. There are no linkages between complex units.

BIS(μ_2-CHLORO)DICHLOROIRON(II)-TETRAKIS(TETRAHYDROFURAN)MAGNESIUM(II)
$C_{16}H_{32}Cl_4FeMgO_4$

P. SOBOTA, T. PLUZINSKI and T. LIS, 1984. Polyhedron, 3, 45-47.

Orthorhombic, Pbcn, a = 9.18, b = 16.30, c = 15.85 Å, D_m = 1.43, Z = 4. Mo radiation, R = 0.053 for 1194 reflexions.

Both metal atoms (Fig. 1) lie on a crystallographic twofold axis. The Fe...Mg distance (3.455(3) Å) rule out any metal-metal bond. The geometry around the Fe atom is distorted tetrahedral. The terminal Fe-Cl(2) bond lengths of 2.228(2) Å are shorter than the bridging Fe-Cl(1) distances of 2.377(2) Å. The Mg^{2+} ion has octahedral coordination with Mg-O(1) 2.073(5), Mg-O(2) 2.111(4) and Mg-Cl 2.501(2) Å.

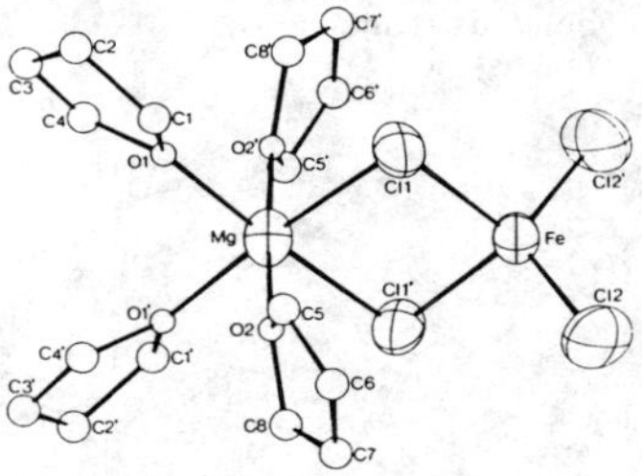

Fig. 1. The structure of $[FeCl_2(\mu\text{-}Cl)_2Mg(THF)_4]$.

TETRA-μ-t-BUTOXODICHLOROTRIBERYLLIUM
$C_{16}H_{36}Be_3Cl_2O_4$ $Be_3Cl_2(C_4H_9O)_4$

N.A. BELL, H.M.M. SHEARER and J. TWISS, 1984. Acta Cryst., C40, 610-613.

Orthorhombic, Cmcm, a = 13.91, b = 12.19, c = 13.71 Å, D_m = 1.12, Z = 4. Mo radiation, R = 0.054 for 720 reflexions.

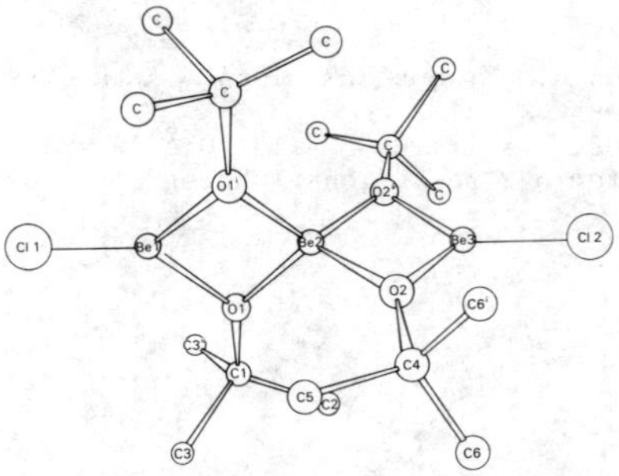

Fig. 1. Perspective view of the $Be_3Cl_2(C_4H_9O)_4$ molecule.

The molecule (Fig. 1) lies at the intersection of two perpendicular mirror planes in the unit cell. The two Cl and three Be atoms lie along the intersection of these mirror planes and are collinear. Adjacent Be atoms are linked by two bridging t-butoxy groups and the two $Be(\mu\text{-}OBu^t)_2Be$ units are at right angles to one another so that the central Be atom is four-coordinate and has an approximately tetrahedral distribution of surrounding O atoms. The terminal Be atoms are trigonal planar and involved in π-bonding to the surrounding O and Cl atoms. Some bond distances are: Be(1)-Cl(1) 1.87(1), Be(1)-O(1) 1.54(1), Be(2)-O(1) 1.64(1), Be(2)-O(2) 1.64(1), Be(3)-Cl(2) 1.89(1) and Be(3)-O(2) 1.54(1) Å.

DI-μ-t-BUTOXO-BIS[BROMO(DIETHYLETHER)BERYLLIUM]

$C_{16}H_{38}Be_2Br_2O_4$ $Be_2Br_2(C_4H_9O)_2(C_4H_{10}O)_2$

N.A. BELL, H.M.M. SHEARER and J. TWISS, 1984. Acta Cryst., C40, 605-607.

Monoclinic, $P2_1/n$, a = 9.035, b = 13.180, c = 9.805 Å, β = 96.4°, D_m = 1.35, Z = 2. Mo radiation, R = 0.0946 for 526 reflexions (photographic data).

The molecule (Fig. 1) contains a centrosymmetric Be_2O_2 four-membered ring in which each Be is four-coordinate and adopts a distorted tetrahedral arrangement. The t-butyl O atoms O(1) and O(1') are three-coordinate and adopt a trigonal arrangement as do the ether O(2) and O(2') attached to Be. The structure is therefore similar to the Mg analogue (1). Some bond distances are: Be-Br 2.19(3), Be-O(1) 1.60(4), Be-O(1') 1.63(3) and Be-O(2) 1.65(3) Å.

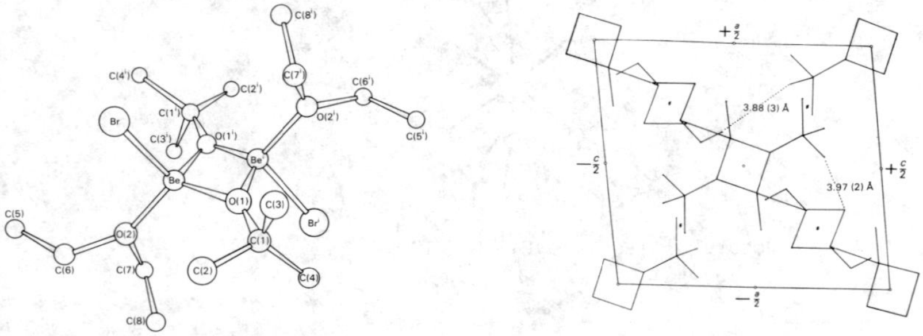

Fig. 1. $Be_2Br_2(C_4H_9O)_2(C_4H_{10}O)_2$: views of the molecule and of the projection of the crystal structure on (010).

1. Following report

DI-μ-t-BUTOXO-BIS[BROMO(DIETHYLETHER)MAGNESIUM]

$C_{16}H_{38}Br_2Mg_2O_4$ $Mg_2Br_2(C_4H_9O)_2(C_4H_{10}O)_2$

N.A. BELL, P.T. MOSELEY and H.M.M. SHEARER, 1984. Acta Cryst., C40, 602-604.

Monoclinic, $P2_1/c$, a = 9.68, b = 11.10, c = 15.10 Å, β = 129.1°, D_m = 1.30, Z = 2. Mo radiation, R = 0.0986 for 672 reflexions (photographic data).

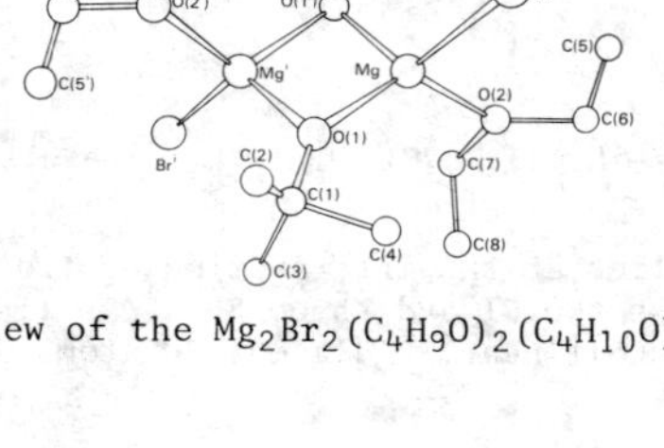

Fig. 1. Perspective view of the $Mg_2Br_2(C_4H_9O)_2(C_4H_{10}O)_2$ molecule.

The molecule (Fig. 1) is a dimer with a four-membered Mg_2O_2 ring linking the two halves of the dimer round a centre of symmetry. The Mg atoms of this ring are four-coordinate and the O atoms three-coordinate and trigonal with the central O and its three neighbours lying very accurately in a plane, suggesting that there is some degree of d_π-p_π interaction between the metal and t-butoxy O atoms. Some bond distances are: Mg-Br 2.435(7), Mg-O(1) 1.91(1), Mg-O(1') 1.91(1) and Mg-O(2) 2.01(1) Å.

BIS((μ_2-σ^2-PROPYNYL)PROPYNYLTRIMETHYLAMINEBERYLLIUM) BIS((μ_2-σ^1,η^2-PROPYNYL)PROPYNYLTRIMETHYLAMINEBERYLLIUM)

$C_{18}H_{30}Be_2N_2$ $(Me_3N)(MeC{\equiv}C)Be(\mu\text{-}C{\equiv}CMe)_2Be(C{\equiv}CMe)(NMe_3)$

N.A. BELL, I.W. NOWELL, G.E. COATES and H.M.M. SHEARER, 1984. J. Organometal. Chem., 273, 179-185.

Triclinic, P$\bar{1}$, a = 9.115, b = 13.519, c = 8.384 Å, α = 102.8, β = 99.0, γ = 94.8°, D_m = 0.90, Z = 2. Cu radiation, R = 0.074 for 1493 reflexions.

There are two independent centrosymmetric dimers (Fig. 1) in which the Be_2C_2 rings are crystallographically constrained to be planar. This leads to highly distorted tetrahedral coordination about Be, with bond angles ranging from 103.4 to 115.0° in dimer A, and from 96.2 to 121.1° in dimer B. The μ-alkynyl groups adopt quite different orientations with respect to the Be atoms in the two dimers. The Be-C(1), Be-C(1') and Be-C(2') distances in dimer A are 1.836, 1.904 and 2.812 Å, and 1.763, 2.042 and 2.538 Å in dimer B respectively. The angles Be-C(1)-C(2) and Be'-C(1)-C(2) are 154.9 and 128.5° in A and 168.6 and 100.2° in B. The resulting Be...Be distances are: 2.319 Å in A and 2.549 Å in B.

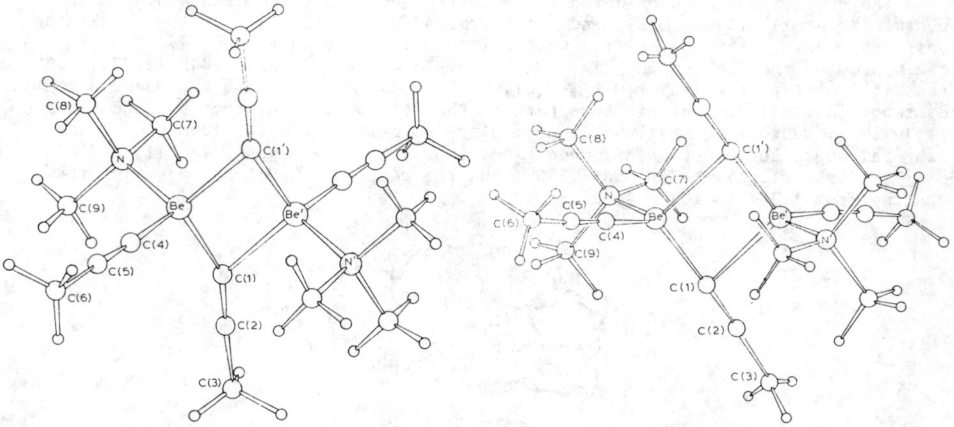

Fig. 1. Molecular structures of dimer A (left), and dimer B (right) in $C_{18}H_{30}Be_2N_2$.

BIS(N,N,N',N'-2,2-HEXAMETHYLPROPANEDIAMIDO)SODIUM(I) PERCHLORATE

$C_{18}H_{36}ClN_4NaO_8$ $[Na(C_9H_{18}N_2O_2)_2][ClO_4]$

S.F. LINCOLN, T.W. HAMBLEY, A.M. HOUNSLOW, N.J. MAEJI and M.R. SNOW, 1984. Aust. J. Chem., 37, 1363-1367.

Tetragonal, $P4_32_12$, a = 10.233, c = 24.540 Å, Z = 4. Mo radiation, R = 0.038 for 1080 reflexions.

The sodium atom is coordinated by four bridging $Me_2NCOCMe_2CONMe_2$ ligands (Fig. 1) and by a perchlorate. It has an intermediate coordination between a trigonal bipyramid and a square pyramid (Na-O 2.297(3), 2.325(3), 2.378(6) Å). Each bidentate ligand (Fig. 1), linked through the oxygen, acts as a bridge between two sodium atoms.

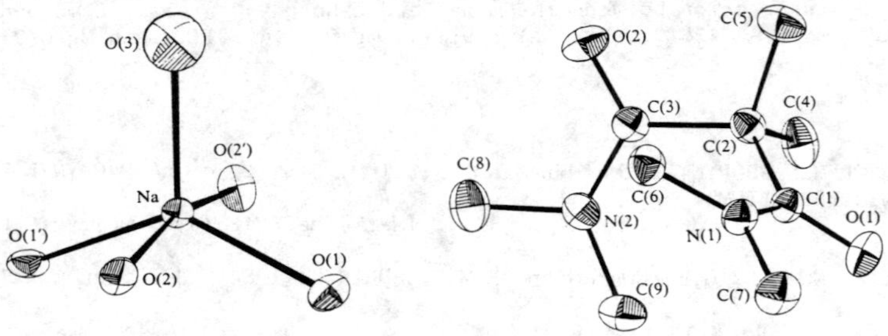

Fig. 1. The coordination sphere for sodium and a view of the ligand in $C_{18}H_{36}ClN_4NaO_8$.

1,4,7,10,13-PENTAOXA[13]FERROCENOPHANE SODIUM THIOCYANATE
$C_{19}H_{14}FeNNaO_5S$

S. AKABORI, Y. HABATA and M. SATO, 1984. Bull. Chem. Soc. Japan, 57, 68-72.

Orthorhombic, Cccn, a = 20.579, b = 25.622, c = 7.606 Å, Z = 8. Mo radiation, R = 0.100 for 2539 reflexions.

The complex can be considered in two parts: the roughly planar complexing unit containing atoms O(1) to O(5), and the ferrocene unit (Fig. 1). In the complexing unit the bonding from the Na^+ ion to the ligand is directed at the five O atoms and the N atom of the SCN^- anion. The Na^+...O distances are 2.439, 2.449, 2.474, 2.396 and 2.631 Å respectively for O(1), O(2), O(3), O(4) and O(5), and the Na^+...N (SCN^-) distance is 2.541 Å. The distance between the Fe atom and the incorporated Na^+ ion is 4.129 Å, indicating that the Fe atom plays no part in complex formation. The dihedral angle between the ferrocene rings is 3.30°. The bonds C(2)-O(1) and C(6)-O(5) are respectively 1.369 and 1.355 Å and the remaining C-O bonds have lengths ranging from 1.398 to 1.445 Å.

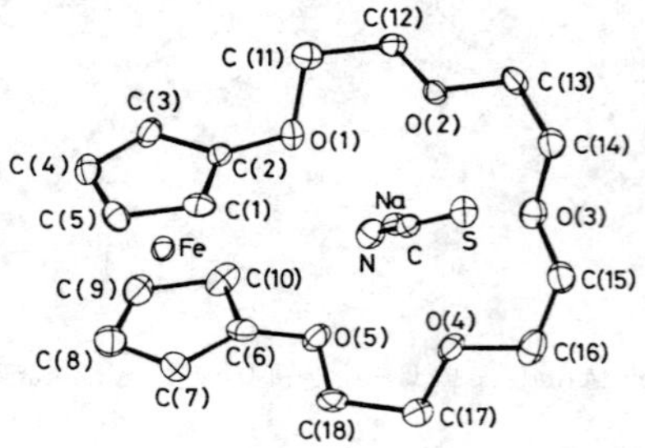

Fig. 1. Molecular structure of $C_{19}H_{14}FeNNaO_5S$.

10,10'-ETHYLENEBIS(1,4,7-TRIOXA-10-AZACYCLODODECANE)LITHIUM THIOCYANATE
$C_{19}H_{36}LiN_3O_6S$ $Li(C_{18}H_{36}N_2O_6)^+SCN^-$

P. GROTH, 1984. Acta Chem. Scand., A38, 337-341.

Monoclinic, $P2_1/c$, a = 12.222, b = 22.754, c = 16.482 Å, β = 100.02°, D_m = 1.24, Z = 8. Mo radiation, R = 0.042 for 5087 reflexions (at -130°C).

In the two independent molecules, the four 12-membered rings adopt three different conformations (Fig. 1). One from each complex has the most common [3 3 3 3] conformation and of the remaining two, one adopts the [3 9] conformation and the other [6 6] or free 12-crown-4. The coordination number of the Li^+ ion is uncertain. Although the Li^+...O distances (range 1.988 to 2.287 Å, average 2.168 Å) suggest hexacoordinated lithium, the mean value of the four Li^+...N contacts is much larger (2.418 Å).

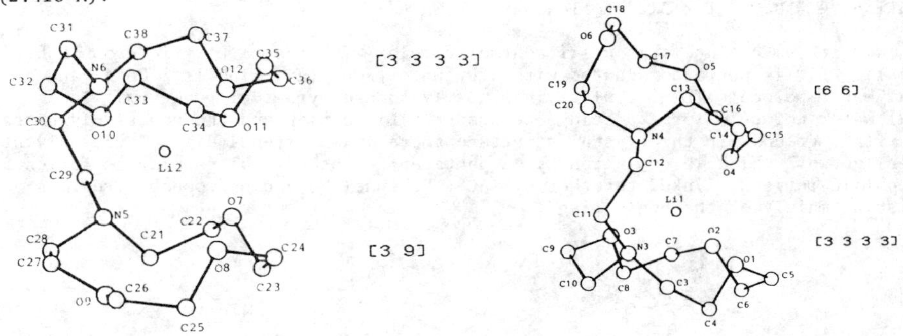

Fig. 1. Schematic drawing of the two independent cations in $C_{19}H_{36}LiN_3O_6S$.

SODIUM PICRATE BENZO-15-CROWN-5

$C_{20}H_{22}N_3NaO_{12}$ $Na^+.C_6H_2N_3O_7^-.C_{14}H_{20}O_5$

D.L. WARD, A.I. POPOV and N.S. POONIA, 1984. Acta Cryst., C40, 238-241.

Monoclinic, $P2_1/n$, a = 11.125, b = 13.555, c = 14.924 Å, β = 93.40°, Z = 4. Mo radiation, R = 0.061 for 2786 reflexions.

The Na^+ ion is coordinated to all five crown oxygen atoms (Na-O 2.400(3) to 2.501(3) Å) as well as to the phenoxide oxygen (Na-O 2.350(3) Å) and an ortho nitro oxygen (Na-O 2.510(3) Å). The Na^+ ion lies out of the crown cavity towards the picrate by 0.900(1) Å (Fig. 1).

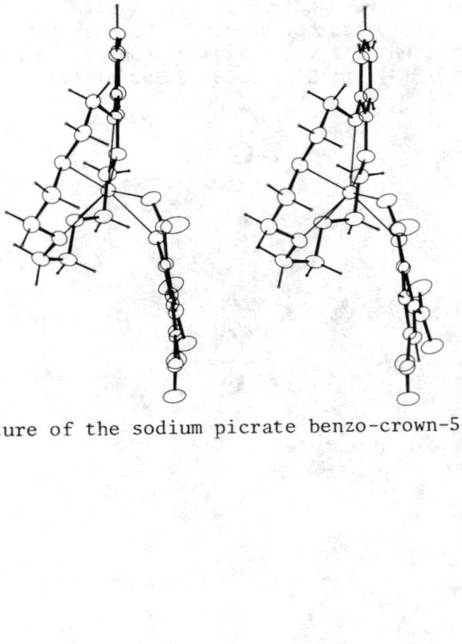

Fig. 1. The structure of the sodium picrate benzo-crown-5 complex (stereo view).

AQUA(sym-DIBENZO-14-CROWN-4-OXYACETATE)LITHIUM HYDRATE
$C_{20}H_{23}LiO_8,6.5(H_2O)$

G. SHOHAM, D.W. CHRISTIANSON, R.A. BARTSCH, G.S. HEO, U. OLSHER and W.N. LIPSCOMB, 1984. J. Am. Chem. Soc., 106, 1280-1285.

Monoclinic, C2/c, a = 20.779, b = 17.845, c = 13.903 Å, β = 98.19°, Z = 8. Cu radiation, R = 0.077 for 3124 reflexions.

The lithium cation forms a 1:1 neutral complex with the carboxylic crown ether (Fig. 1). Li^+ is pentacoordinated with 4 crown oxygens (av. Li-O is 2.06 Å) and an apical water molecule (Li-O 1.91 Å) in a nearly square pyramidal geometry. This apical water molecule forms a bridge between the Li^+ cation and the negatively charged carboxylate group. In the crystal structure there is an extensively ordered solvent (water) network which is dominated by O_5 pentagons. The crystal packing is formed of hydrophilic regions, linked together by water "channels", and hydrophobic regions, consisting mainly of the benzo rings.

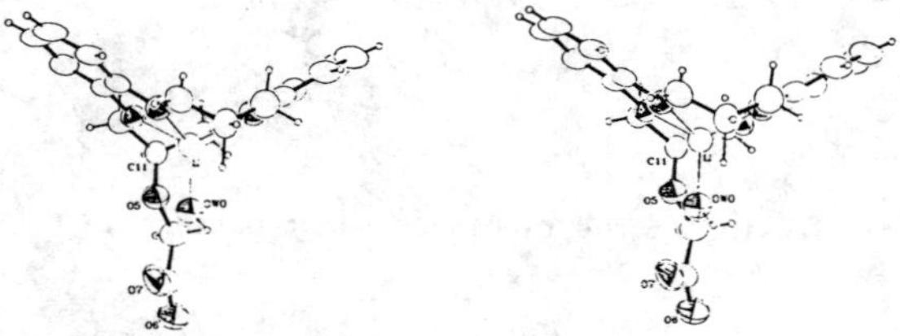

Fig. 1. A stereoview of $C_{20}H_{23}LiO_8$.

TETRAKIS[μ_3-(4-METHOXY-2-BUTYL)-O,μ_3-C^2]TETRALITHIUM
$C_{20}H_{44}Li_4O_4$ $Li_4(C_5H_{11}O)_4$

A.L. SPEK, A.J.M. DUISENBERG, G.W. KLUMPP and P.J. GEURINK, 1984. Acta Cryst., C40, 372-374.

Tetragonal, P$\bar{4}$b2, a = 12.389, c = 8.218 Å, Z = 2. Cu radiation, R = 0.046 for 664 reflexions.

The molecule is found as meso tetramers with $\bar{4}$ site symmetry (Fig. 1). The four Li atoms form a slightly distorted tetrahedron. To each face a 1-methoxybutane moiety is attached by four-centre bonding via chiral carbon C2. Some bond lengths: Li-Li 2.506, 2.474(7), Li-O 1.923(5), Li-C(2) 2.288-2.362(6) Å.

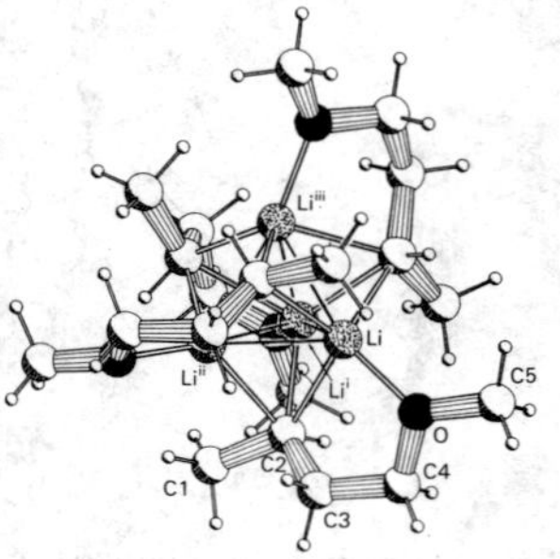

Fig. 1. The $Li_4(C_5H_{11}O)_4$ structure.

7,16,21,26-TETRAOXA-1,4,10,13-TETRAAZATRICYCLO[11.5.5.$5^{4,10}$]OCTACOSANE POTASSIUM THIOCYANATE

$C_{21}H_{44}KN_5O_4S$ $K(C_{20}H_{44}N_4O_4)^+SCN^-$

P. GROTH, 1984. Acta Chem. Scand., A38, 183-186.

Orthorhombic, Pbca, a = 11.834, b = 12.444, c = 34.38 Å, D_m = 1.26, Z = 8. Mo radiation, R = 0.078 for 1465 reflexions (at -130°C).

The cation of the complex consists of an eight-coordinated K^+ ion at the centre of three rings (Fig. 1). The conformation of the twelve-membered ring (C(1), C(2), O(2), C(3), C(4), N(2), C(5), C(6), O(1), C(7), C(8) and N(1)) may be described as a strained biangular [4 8] while the other ring (C(11), C(12), O(3), C(13), C(14), N(4), C(15), C(16), O(4), C(17), C(18) and N(3)) has the [3 3 3 3] conformation with approximate C_4 symmetry. Distances involving K^+ are: K-O(1) 2.84, K-O(5) 2.75, K-N(1) 2.76, K-N(5) 2.81, K-O(2) 2.69, K-O(4) 2.58, K-N(2) 2.81 and K-N(4) 2.74 Å. The K-O(4) distance is considered unusually short.

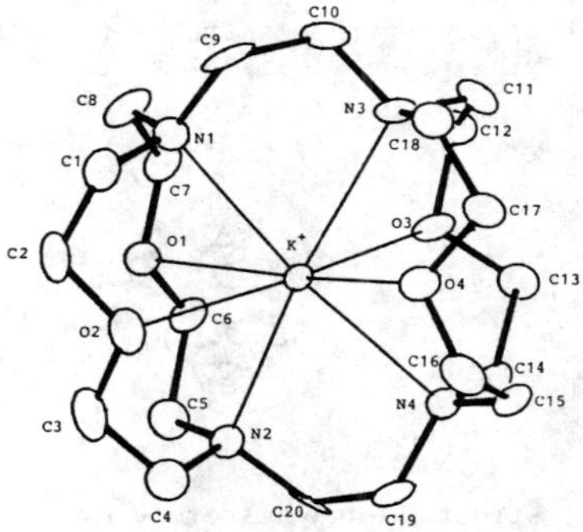

Fig. 1. Structure of the cation in $C_{21}H_{44}KN_5O_4S$.

AQUA(DICYCLOHEXYL-18-CROWN-6)DIISOTHIOCYANATOBARIUM

$C_{22}H_{38}BaN_2O_7S_2$

N.K. DALLEY, D. SYPHERD and R.D. GEORGE, 1984. J. Heterocycl. Chem., 21, 497-499.

Orthorhombic, Aba2, a = 21.239, b = 13.551, c = 9.635 Å, D_m = 1.53, Z = 4. Mo radiation, R = 0.033 for 725 reflexions.

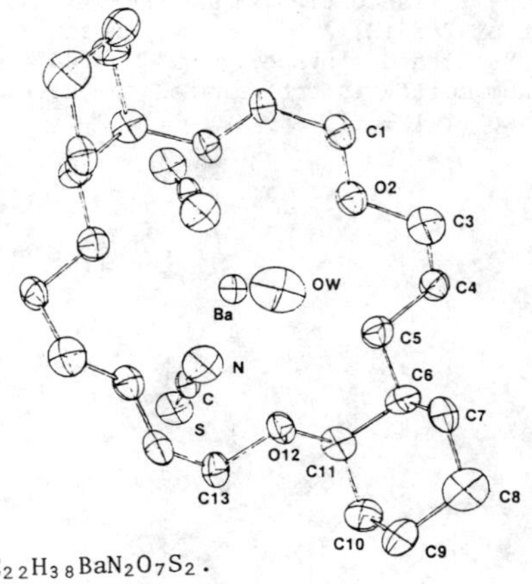

Fig. 1. A view of $C_{22}H_{38}BaN_2O_7S_2$.

The Ba atom lies on a twofold axis in the cavity of the macrocycle (Fig. 1) and space group symmetry requires this to be the cis-syn-cis isomer. The Ba coordination is ninefold as it interacts with six O atoms of the macrocycle (2.779-2.920(9) Å), two N atoms of the SCN^- groups (2.828(10) Å) and the water O atom (2.833(11) Å). In the macrocycle the average absolute values of the torsion angles about C-C and C-O bonds are 61(7) and 166(18)° respectively.

POTASSIUM 1,18-DIOXO-2,5,8,11,14,17-HEXAOXA[18](1,3)AZULENOPHANE THIOCYANATE

$C_{23}H_{26}KNO_8S$ $K^+.CNS^-.C_{22}H_{26}O_8$

H.-G. LOHR, F. VOGTLE, W. SCHUH and H. PUFF, 1984. Chem. Ber., 117, 2839-2849.

Monoclinic, C2/c, a = 29.245, b = 15.156, c = 15.011 Å, β = 109.5°, Z = 8. Mo radiation, R = 0.255 for 1253 reflexions.

The analysis shows the K^+ ion at the centre of the macrocycle (Fig. 1), with K...C2 3.66 Å. The azulene ring plane is at 45° to the macrocycle plane.

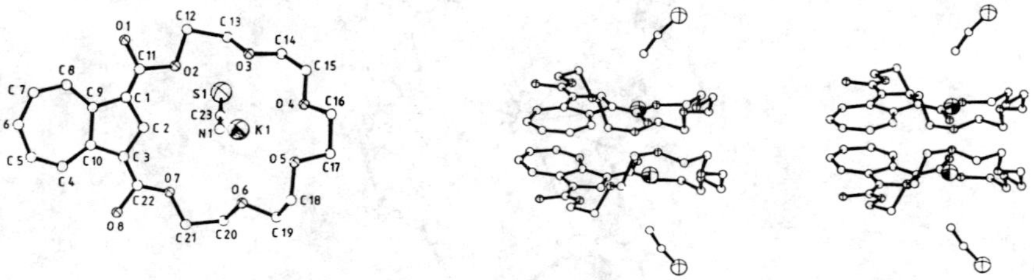

Fig. 1. The $C_{23}H_{26}KNO_8S$ structure and a stereoview.

POTASSIUM DIBENZO-18-CROWN-6 TRIMETHYLNITRATOALUMINATE BENZENE SOLVATE

$C_{23}H_{33}AlKNO_9,0.5(C_6H_6)$ $C_3H_9AlNO_3^-.C_{20}H_{24}KO_6^+,0.5(C_6H_6)$

R.D. ROGERS and J.L. ATWOOD, 1984. J. Cryst. Spectrosc. Res., 14, 1-11.

Monoclinic, $P2_1/c$, a = 11.804, b = 28.828, c = 9.118 Å, β = 96.61°, D_m = 1.23, Z = 4. Mo radiation, R = 0.045 for 983 reflexions.

The potassium atom (Fig. 1) is centred among the six oxygen atoms of the crown at an average K...O contact of 2.74(2) Å. Its environment is completed by an oxygen atom of a nitrate anion (2.88 Å) and a benzo group (> 3.44 Å) of a neighbouring crown ether. The trimethylaluminum moiety is coordinated to the nitrate anion by an oxygen atom at an Al-O bond distance of 1.92(1) Å.

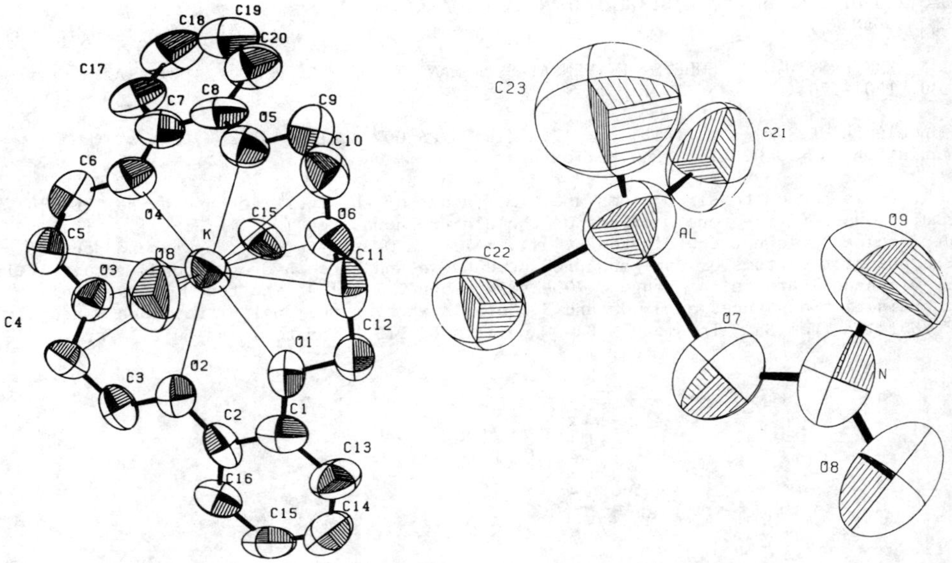

Fig. 1. The K^+ environment in $[C_{20}H_{24}KO_6]^+$ and a view of the $[AlMe_3NO_3]^-$ anion.

TETRAKIS(μ_2-CYANO)(TETRAKIS(PYRIDINE)BARIUM)MERCURY
$(C_{24}H_{20}BaHgN_8)_n$ $BaHg(CN)_4.4(C_5H_5N)$

K. BRODERSEN, I. BECK, R. BECK, H.U. HUMMEL and G. LIEHR, 1984. Z. Anorg. Chem., 516, 30-34.

Tetragonal, $I\bar{4}$, a = 11.194, c = 10.579 Å, D_m = 1.72, Z = 2. Ag radiation, R = 0.0397 for 1670 reflexions.

The Hg and Ba atoms lie on sites with $\bar{4}$ symmetry. The structure (Fig. 1) contains tetrahedral $Hg(CN)_4$ units and $Ba(NC)_4(py)_4$ moieties linked by Hg-C-N-Ba bridges, with Hg-C 2.198(8), Ba-N 2.921(7) and 2.906(7), C-N 1.124(9) Å, Hg-C-N 176.2(7), C-N-Ba 167.8(7), C-Hg-C 107.1(2) and 114.3(3)°.

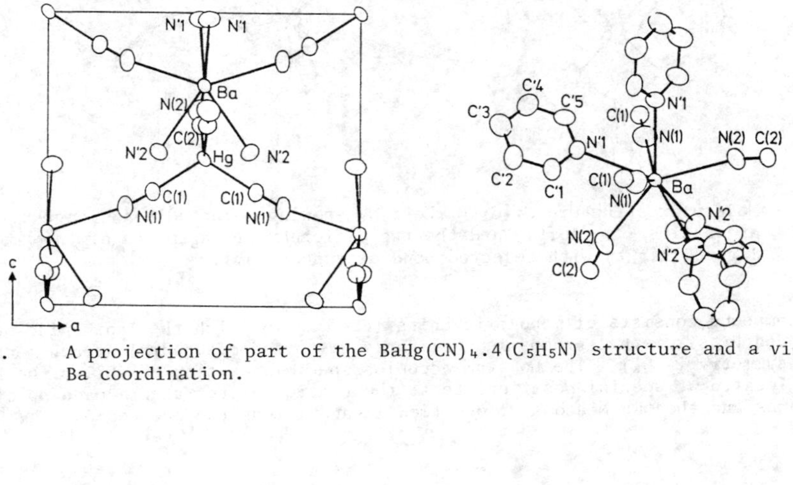

Fig. 1. A projection of part of the $BaHg(CN)_4.4(C_5H_5N)$ structure and a view of the Ba coordination.

BIS(2,4-DINITROPHENOLATO)BIS(TRIETHANOLAMINE)BARIUM(II)
$C_{24}H_{36}BaN_6O_{16}$ $Ba(C_6H_3N_2O_5)_2(C_6H_{15}NO_3)_2$

J.A. KANTERS, W.J.J. SMEETS, K. VENKATASUBRAMANIAN and N.S. POONIA, 1984. Acta Cryst., C40, 1701-1704.

Monoclinic, $P2_1/c$, a = 7.418, b = 17.884, c = 25.007 Å, β = 100.25°, Z = 4. Mo radiation, R = 0.039 for 5930 reflexions.

Ba is bonded to six O atoms and two N atoms of the two triethanolamines (tea's) and to one o-nitro-group O of one of the dinitrophenolates (dnp1) (Fig. 1). The dnp2 molecule is not coordinated at all. All tea O atoms are hydrogen-bond donors; the phenolic O atoms of dnp1 and dnp2 accept two and three hydrogen bonds respectively. Ba-O(ethanol) are in the range 2.703-2.894, mean 2.77(6) Å, the tea N atoms have coordination distances of 3.000 and 3.036(5) Å whereas the dnp1 nitro-group O atom is 2.785(4) Å from Ba. The mean N-Ba-O angle is 60.9(5)° for tea1 and 60(2)° for tea2.

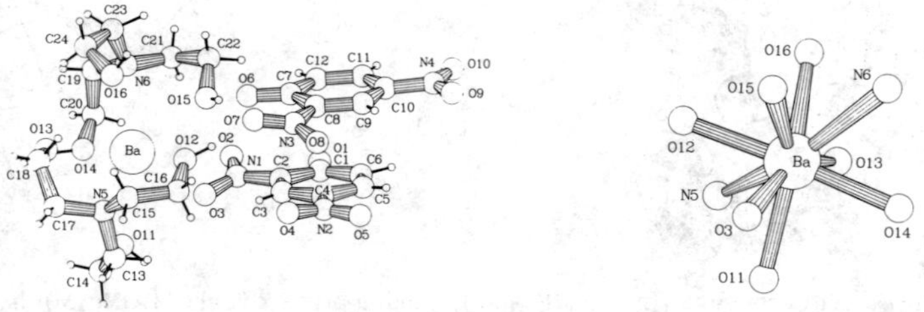

Fig. 1. View of the $C_{24}H_{36}BaN_6O_{16}$ complex and of the Ba coordination.

TRIS[(N,N,N',N'-TETRAMETHYLETHYLENEDIAMINE)LITHIUM] HEXAMETHYLHOLMATE(III)
$C_{24}H_{66}HoLi_3N_6$ $[Li(tmed)]_3[Ho(CH_3)_6]$

H. SCHUMANN, J. MÜLLER, N. BRUNCKS, H. LAUKE and J. PICKARDT, 1984. Organometallics, 3, 69-74.

Rhombohedral, $R\bar{3}c$, a = 12.807, α = 79.84°, D_m = 0.9, Z = 2. Mo radiation, R = 0.050 for 1535 reflexions (at 160K).

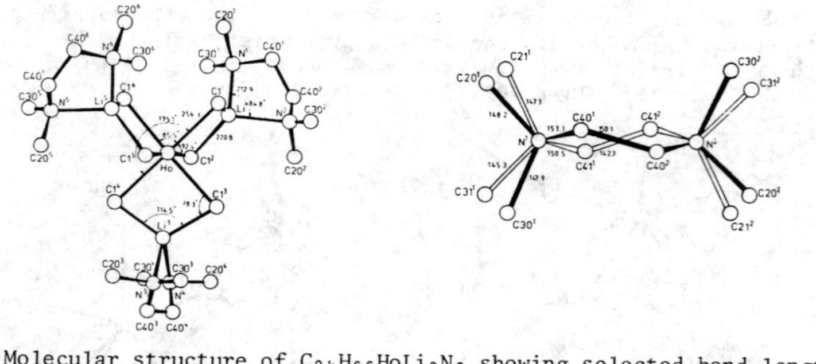

Fig. 1. Molecular structure of $C_{24}H_{66}HoLi_3N_6$ showing selected bond lengths (pm) and angles (°) (left), and the two different arrangements of the tmed ligand (right) with selected bond distances (pm).

The compound consists of monomeric units (Fig. 1) in which the lanthanide metal is surrounded by six methyl groups in a slightly distorted octahedral arrangement (the molecular symmetry is D_3). The Ho atom occupies special position 2a whilst the Li atoms are located at special positions 6e at the centres of tetrahedra made up of two methyl groups and the two N atoms of one tmed ligand. Thus the octahedron is connected to three tetrahedra in such a way that each tetrahedron shares one common edge with the octahedron. There is disorder in the tmed ligands (Fig. 1). The N atoms occupy fixed positions but the C atoms rotate around the Li-N axes; the rotation angle C(20)-N-C(21) is 19°.

AQUA(BENZO-CROWN-5)BIS(PICRATE-O,O')BARIUM

$C_{26}H_{26}BaN_6O_{20}$

K. VENKATASUBRAMANIAN, N.S. POONIA, K. CLINGER, S.R. ERNST and M.L. HACKERT, 1984. J. Inclusion Phenomena, 1, 319-327.

Monoclinic, $P2_1/c$, a = 11.43, b = 16.31, c = 17.38 Å, β = 92.265°, Dm = 1.73, Z = 4. Mo radiation, R = 0.049 for 5926 reflexions (at -32°C).

The Ba^{++} ion is 10-coordinate (Fig. 1) through all five crown O atoms (Ba-O 2.800 to 3.002 Å), the two bidentate picrates (Ba-O 2.642 and 2.666 Å, Ba-ONO 2.825 and 2.994 Å) and the water molecule (Ba-O 2.711 Å).

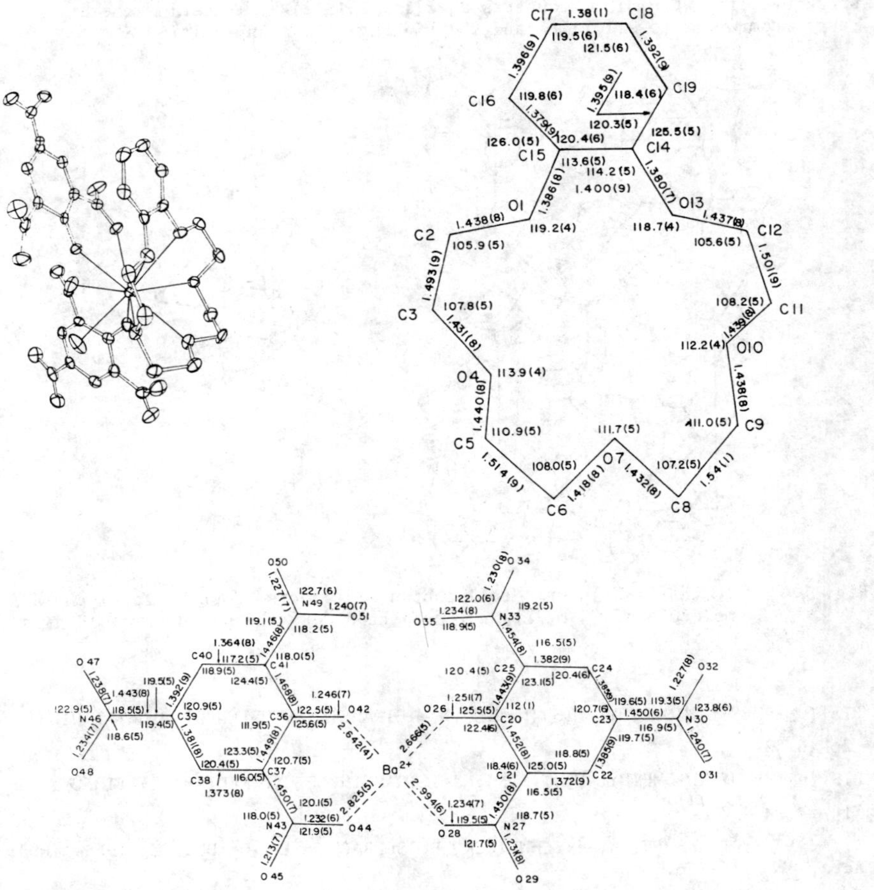

Fig. 1. A view of the $C_{26}H_{26}Ba_6O_{20}$ complex and details of dimensions.

α,α'-BIS[(N,N,N',N'-TETRAMETHYLETHYLENEDIAMINE)LITHIO]-2,2'-DIMETHYLBIPHENYL
$(C_{26}H_{44}Li_2N_4)_n$ (I)

α,α'-BIS[(N,N,N',N'-TETRAMETHYLETHYLENEDIAMINE)LITHIO]-2,2'-BIS(TRIMETHYLSILYLMETHYL)-BIPHENYL
$C_{32}H_{60}Li_2N_4Si_2$ (II)

L.M. ENGELHARDT, W.-P. LEUNG, C.L. RASTON, P. TWISS and A.H. WHITE, 1984. J. Chem. Soc. Dalton, 321-330.

I. Tetragonal, $P4_12_12$, a = 15.11, c = 28.11 Å, Z = 8. Mo radiation, R = 0.092 for 1043 reflexions.

II. Monoclinic, $P2_1/c$, a = 10.265, b = 39.297, c = 10.801 Å, β = 117.52°, Z = 4. Mo radiation, R = 0.066 for 2789 reflexions.

Compound I forms an infinite polymer (Fig. 1) generated by the 4_1 screw axis along c with a metallacycle lithium (Li-N 2.21(3), 2.10(3) Å, N-Li-N 83.8(10), 86.5(10)°) and a lithium bridging methylene group of different metallacycles (Li-C(mean) 2.30 Å, Li-C-Li 159.5, C-Li-C 106.5°). The dihedral angle between the mean planes of the phenyl groups of the biphenyl moiety is 73.7°. Compound II forms a contact ion-pair structure; Li...C(H)$SiMe_3$ contacts 2.22(1), 2.19(2) Å. The Li-N chelating distances range from 2.042 to 2.10(1) Å and the biphenyl twist angle is 54.3°.

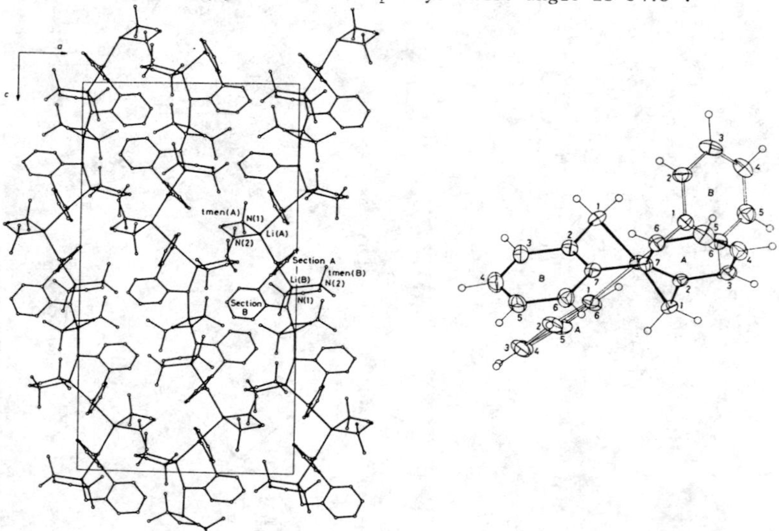

Fig. 1. $(C_{26}H_{44}Li_2N_4)_n$ unit cell contents projected down b (left) and a single molecule of $C_{32}H_{60}Li_2N_4Si_2$ projected down the pseudo-twofold axis (right).

BIS[(μ_2-PHENYLTHIOMETHYL-C,S)(N,N,N',N'-TETRAMETHYLETHYLENEDIAMINE)LITHIUM]
$C_{26}H_{46}Li_2N_4S_2$ (I)

BIS[(μ_2-METHYLTHIOMETHYL-C)(N,N,N',N'-TETRAMETHYLETHYLENEDIAMINE)LITHIUM]
$C_{16}H_{14}Li_2N_4S_2$ (II)

R. AMSTUTZ, T. LAUBE, W.B. SCHWEIZER, D. SEEBACH and J.D. DUNITZ, 1984. Helv. Chim. Acta, 67, 224-236.

I. Monoclinic, $P2_1/c$, a = 8.507, b = 14.336, c = 12.432 Å, β = 78.62°, Z = 2. Mo radiation, R = 0.034 for 2221 reflexions (at 95 K).

II. Monoclinic, $P2_1/c$, a = 10.254, b = 15.874, c = 15.144 Å, β = 74.45°, Z = 4. Mo radiation, R = 0.067 for 3703 reflexions (at 113 K).

Molecule I lies about an inversion centre (Fig. 1). The central ring has a chair form, with Li-S 2.555(3), Li-C(7) 2.131(4), Li-N(1) 2.148(4), Li-N(2) 2.133(3) Å, Li-C(7)-S 114.2(1), C(7)-S-Li 92.7(1), S-Li-C(7') 109.7(1)°. Molecule II has a four-membered Li_2C_2 ring (Fig. 1) with Li(1)-C(2) 2.256(6), Li(1)-C(21) 2.227(7), Li(11)-C(2) 2.236 (7), Li(11)-C(21) 2.241(7), Li-N 2.151-2.156(8) Å, Li-C-Li 66.5 and 66.9(4), C-Li-C 108.9 and 109.1(3)°. The four-membered ring is folded 22° about the Li...Li axis; it has approximate C_{2v} symmetry.

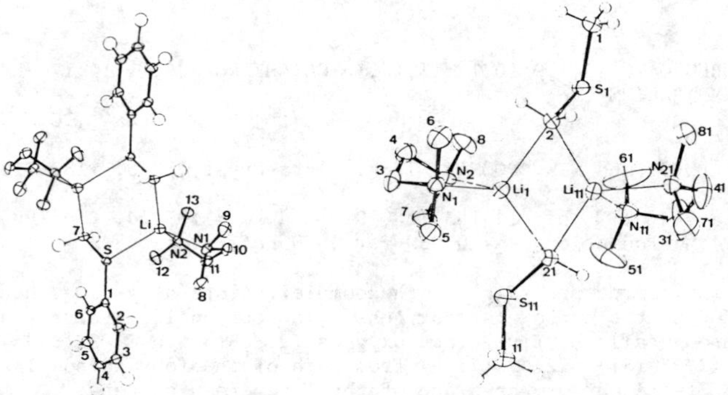

Fig. 1. Views of $C_{26}H_{46}Li_2N_4S_2$ (I) (left) and of $C_{16}H_{14}Li_2N_4S_2$ (II) (right).

BIS[(N,N,N',N'-TETRAMETHYLETHYLENEDIAMINE)LITHIUM]-α,α'-BIS(TRIMETHYLSILYL)-p-XYLENE
$C_{26}H_{56}Li_2N_4Si_2$

W.-P. LEUNG, C.L. RASTON, B.W. SKELTON and A.H. WHITE, 1984. J. Chem. Soc. Dalton, 1801-1807.

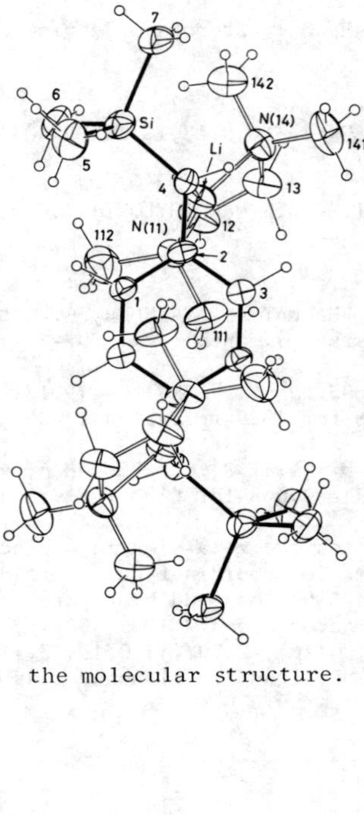

Fig. 1. $C_{26}H_{56}Li_2N_4Si_2$: the molecular structure.

Monoclinic, $P2_1/c$, a = 10.313, b = 11.176, c = 15.32 Å, β = 91.80°, Z = 2. Mo radiation, R = 0.057 for 780 reflexions.

The geometry of the aromatic fragment of the structure (Fig. 1) shows evidence of a dominant p-quinodimethanide contribution to the bonding with C(1)-C(2) 1.443(9), C(2)-C(3) 1.402(9) and C(1)-C(3') 1.376(10) Å. The C(2)-C(4) and C(4)-Si bond distances are 1.458(9) and 1.802(8) Å respectively. The mean Si-C(5,6,7) distance is 1.883 Å. The Li atom lies equidistant from N(11) and N(14) (2.06(1), 2.04(1) Å) with the LiN_2 plane approximately normal to the aromatic ring plane (dihedral angle 79.3°). The Li atom lies 2.10(1) and 2.38(1) Å from C(4) and C(2) respectively.

BIS(3,5-DINITROBENZOATO)(6,7,9,10,12,13,15,16-OCTAHYDRO-5,8,11,14,17-PENTAOXABENZO-CYCLOPENTADECENE)BARIUM

$C_{28}H_{26}BaN_4O_{17}$ $Ba^{2+},2C_7H_3N_2O_6^-,C_{14}H_{20}O_5$

D.L. WARD, A.I. POPOV and N.S. POONIA, 1984. Acta Cryst., C40, 1183-1186.

Triclinic, $P\bar{1}$, a = 13.694, b = 14.183, c = 9.781 Å, α = 105.64, β = 100.97, γ = 109.36°, Z = 2. Mo radiation, R = 0.039 for 3835 reflexions.

In the crystal structure (Fig. 1) the complex exists as a dimer because of bridging provided by the carboxylate oxygens. The cation is involved in a ninefold coordination through all the five crown oxygens (2.888(4) to 3.092(5) Å) and one anionic oxygen (2.639(5) to 2.694(4) Å) from each of the four carboxylate groups. The cation lies 1.877(1) Å out of the plane of the five crown oxygens.

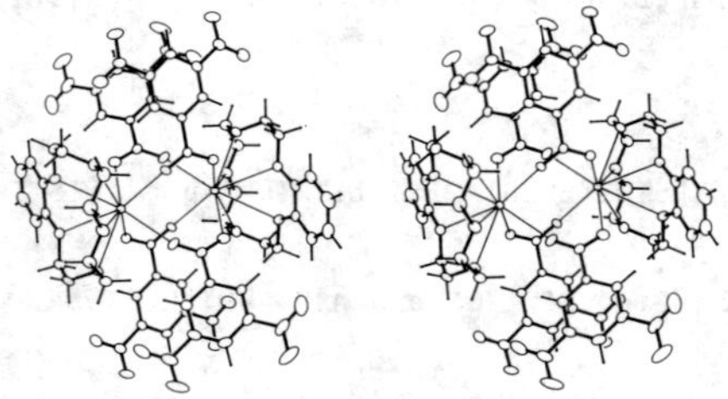

Fig. 1. $Ba^{2+},2C_7H_3N_2O_6^-,C_{14}H_{20}O_5$: stereoscopic view of the complex molecule.

DIISOTHIOCYANATO[1,2-BIS(PHENYLOXY)-3,6,9,12,15,18-HEXAOXAICOSANE]BARIUM HYDRATE

$C_{28}H_{38}BaN_2O_8S_2,0.37(H_2O)$ (I)

AQUAISOTHIOCYANATOTHIOCYANATO[1,20-BIS(NAPHTHYLOXY)-3,6,9,12,15,18-HEXAOXAICOSANE]-BARIUM

$C_{36}H_{44}BaN_2O_9S_2$ (II)

R. HILGENFELD, F. HIRAYAMA, V. BHAGWAT, W. SAENGER, G. WEBER, G.M. SHELDRICK, H. SIEGER and F. VOGTLE, 1984. Israel J. Chem., 24, 307-312.

I. Monoclinic, $P2_1/c$, a = 13.693, b = 13.375, c = 19.181 Å, β = 107.1°, D_m = 1.48, Z = 4. Cu radiation, R = 0.078 for 4829 reflexions.

II. Triclinic, $P\bar{1}$, a = 7.458, b = 11.758, c = 22.408 Å, α = 82.30, β = 85.21, γ = 85.94°, Z = 2. Mo radiation, R = 0.068 for 6103 reflexions.

The HPGE ligand in I encloses the metal ion in a discontinuous helix providing seven of its eight ether O atoms for coordination. Tenfold coordination of Ba^{2+} is completed by the N atoms of the SCN^- ions and by a water molecule (Fig. 1) with distances from Ba to the coordinated atoms O(1), 2.90(1); O(4), 2.83(1); O(7) 2.83(1); O(10), 2.83(1); O(13), 2.84(1); O(16), 2.79(1); O(19) 3.11(1); N(35), 2.77(1); N(38), 2.81(2); OW, 3.08(3) Å. The HGNE ligand in II behaves as a hexadentate system. The

Ba^{2+} ion is ninefold coordinated to six ether O atoms, to an H_2O molecule, to an SCN^- ion (via its N atom) and to a symmetry related SCN^- (via its S atom). The second SCN^- ion does not contact the Ba^{2+} but is hydrogen bonded to symmetry related water molecules. Distances from Ba to the coordinated atoms are O(4), 2.89(1); O(7), 2.85(1); O(10) 2.88(1); O(13), 2.81(1); O(16), 2.82(1); O(19), 2.82(1); N(43), 2.79(1); S(45), 3.31(1); OW, 2.75(1) Å.

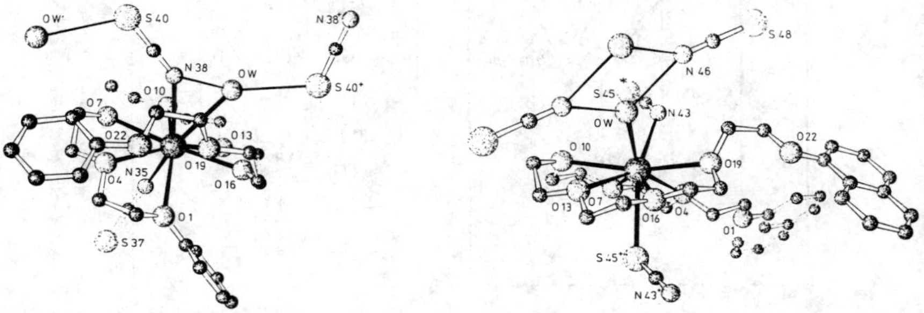

Fig. 1. The $Ba(NCS)_2$ complexes with HPGE (I) (left) and with HGNE (II) (right).

BIS(12-CROWN-4)LITHIUM DIPHENYLARSENIDE TETRAHYDROFURAN SOLVATE

$C_{28}H_{42}AsLiO_8,C_4H_8O$ (I) $[Li(12\text{-crown-}4)_2][AsPh_2]$.THF

BIS(12-CROWN-4)LITHIUM DIPHENYLPHOSPHIDE

$C_{28}H_{42}LiO_8P$ (II) $[Li(12\text{-crown-}4)_2][PPh_2]$

H. HOPE, M.M. OLMSTEAD, P.P. POWER and X. XU, 1984. J. Am. Chem. Soc., 106, 819-821.

I. Monoclinic, $P2_1/n$, a = 12.223, b = 23.155, c = 12.969 Å, β = 117.53°, Z = 4. Cu radiation, R = 0.065 for 3979 reflexions (at 140 K).

II. Triclinic, $P\bar{1}$, a = 8.931, b = 12.174, c = 14.902 Å, α = 90.54, β = 105.91, γ = 109.45°, Z = 2. Cu radiation, R = 0.074 for 4472 reflexions (at 140 K).

The structures (Fig. 1) consists of well-separated $Li(12\text{-crown-}4)_2{}^+$ cations and diaryl group 5 anions. The Li atoms are eight coordinate with a LiO_8 unit of approximate D_{4d} symmetry and Li-O ≈ 2.35 Å. Anion geometry is As-C 1.972(5) Å and C-As-C 108.6(2)° in I and P-C 1.811(4) and 1.804(4) Å, C-P-C 105.2(2)° in II.

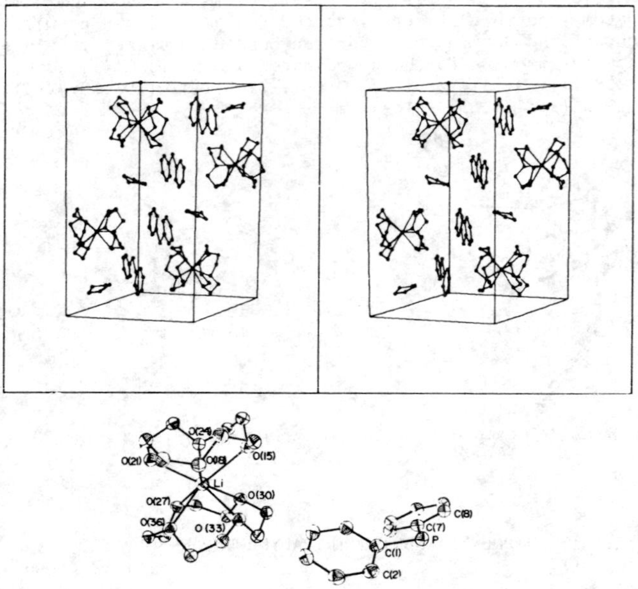

Fig. 1. Stereoview of the unit cell of $[Li(12\text{-crown-}4)_2][AsPh_2]$.THF (top) and the structure of $[Li(12\text{-crown-}4)_2][PPh_2]$ (bottom).

DIBENZO-30-CROWN-10:POTASSIUM THIOCYANATE (1:1) MONOHYDRATE
$C_{29}H_{40}KNO_{10}S,H_2O$ (I) $[K(C_{28}H_{40}O_{10})]^+.SCN^-.H_2O$

DIBENZO-30-CROWN-10:POTASSIUM THIOCYANATE (1:1)
$C_{29}H_{40}KNO_{10}S$ (II) $[K(C_{28}H_{40}O_{10})]^+.SCN^-$

J.D. OWEN, M.R. TRUTER and J.N. WINGFIELD, 1984. Acta Cryst., C40, 1515-1520.

I. Monoclinic, $P2_1/n$, a = 13.335, b = 12.386, c = 20.584 Å, β = 102.67°, Z = 4. Mo radiation, R = 0.129 for 3081 reflexions.

II. Monoclinic, C2/c, a = 18.749, b = 9.808, c = 17.681 Å, β = 98.64°, Z = 4. R = 0.053 for 2327 reflexions.

In I the cations have no symmetry (they lie on twofold axes in II) and have few torsion angles differing by more than 20° from those of the anhydrous form. There is some disorder in the aliphatic chains of the ligand in I similar to that in the isomorphous Rb complex (1). In I the K...O distances are significantly longer than in II (2.861(6)-3.106(7) against 2.857(2)-2.935(2) Å). Thiocyanate anions (disordered at N but not S) and water molecules are hydrogen bonded into chains along b in I. In II disordered thiocyanate ions occupy centres of symmetry. In II the benzene rings are almost parallel while in I the angle between the plane normals is 28.1(3)°. (The results for (II) are based on a reinterpretation of previous data (2).)

1. Structure Reports, 45B, 366.
2. Ibid, 46B, 343.

2,2-DIMETHYL-3-TRIMETHYLSILYL-2,3-DIHYDRO-2-SILAPHENALENE-1,3-DIYL BIS(N,N,N',N'-TETRAMETHYLETHYLENEDIAMINE)LITHIUM (α and β phases)
$C_{29}H_{54}Li_2N_4Si_2$ (I,II)

L.M. ENGELHARDT, R.I. PAPASERGIO, C.L. RASTON and A.H. WHITE, 1984. J. Chem. Soc. Dalton, 311-320.

I. Orthorhombic, $P2_1cn$, a = 16.224, b = 15.080, c = 13.828 Å, Z = 4. Mo radiation, R = 0.043 for 1261 reflexions.

II. Monoclinic, $P2_1/c$, a = 19.09, b = 9.05, c = 20.86 Å, β = 108.53°, Z = 4. Mo radiation, R = 0.20 for 2165 reflexions [large fragile plate crystal].

One formula unit of the α-form of the complex comprises the asymmetric unit in a structure which contains equal numbers of enantiomorphs by virtue of the presence of symmetry elements derived from mirror planes. The dianion of the structure (Fig. 1) is novel and comprises a planar naphthalene ring system fused through the 1,8-positions to a six-membered heterocycle which contains an Si atom at its apex and incorporating the immediate substituent atoms into the conjugated system as evidenced by the C(6)-C(7) distances (1.43(1), 1.40(1) Å). As in most naphthalene systems C(1)-C(2) is long (1.45(1)) and C(3)-C(4) short (1.35(1), 1.36(2) Å). C(5)-C(6) is 1.41(1), 1.43(1) and C(1)-C(6) 1.48(1) Å. Angles at C(5) are enlarged to 124.6(8), 125.5(8)°. The C(5) and C(7) atoms associate with the Li cations, Li(A) being 1.90 Å out of the plane of the aromatic system and Li(B) 1.97 Å in the opposite direction. The tetramethylethylenediamine groups are chelated to the Li atoms via N with Li-N 2.08(1) and 2.18(2) Å. The heterocyclic ring is folded with a dihedral angle of 27° between Si, C(7A), C(7B) and the associated C_5 plane. The β-form determination was of low precision and serves to define general atom positions only. Again the Li atoms are associated with the dianion.

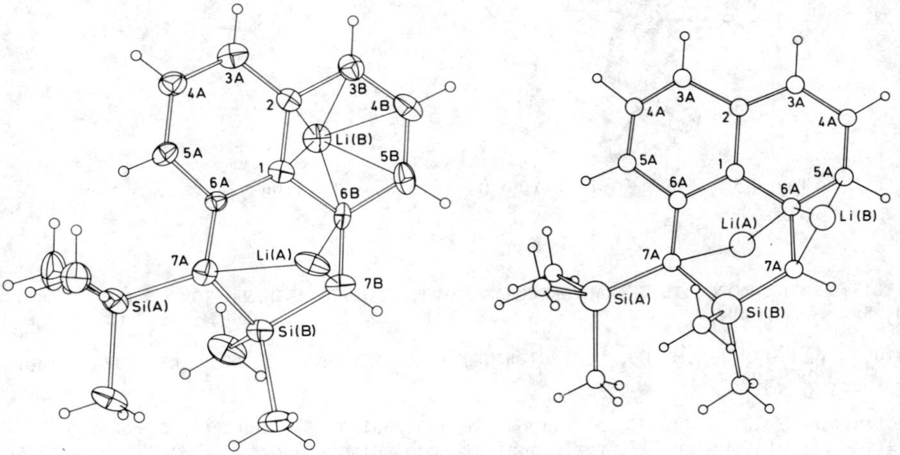

Fig. 1. $C_{29}H_{54}Li_2N_4Si_2$: the dianion and its associated Li cations, α-form (left) and β-form (right).

(DIBENZO-3-CROWN-12)BIS(HEXAFLUOROPHOSPHATOSODIUM)
$C_{32}H_{48}F_{12}Na_2O_{12}P_2$

J.M. MAUD, J.F. STODDART, H.M. COLQUHOUN and D.J. WILLIAMS, 1984. Polyhedron, 3, 675-679.

Monoclinic, $P2_1/n$, a = 14.189, b = 9.372, c = 16.750 Å, β = 108.98°, Z = 2. Cu radiation, R = 0.078 for 2662 reflexions.

The macrocycle (Fig. 1) possesses a crystallographic centre of symmetry and is twisted such that it presents two complexing faces at opposite ends of the molecule and each accommodates a bound $NaPF_6$ unit. Each sodium ion is seven coordinate, with bonding to five oxygen atoms at distances ranging from 2.419(3) to 2.544(2) Å, and the two fluorine atoms of a hexafluorophosphate anion at 2.424(4) and 2.528(4) Å. The sodium ion is displaced 0.45 Å from the mean plane of these five oxygen atoms towards the phosphorus atom.

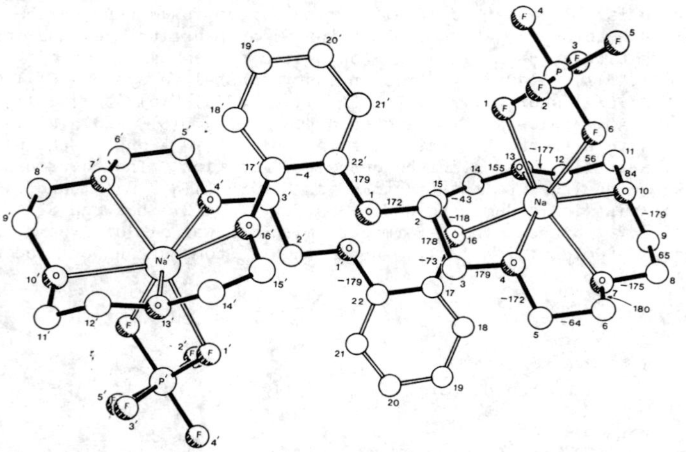

Fig. 1. The $C_{32}H_{48}F_{12}Na_2O_{12}P_2$ structure with some torsion angles.

TETRAKIS[BROMO(CHLORO)BIS(TETRAHYDROFURAN)MAGNESIUM]-HEXAHYDRIDOIRON
$C_{32}H_{70}Br_{3.5}Cl_{0.5}FeMg_4O_8$ $FeH_6Mg_4Br_{3.5}Cl_{0.5}(C_4H_8O)_8$

R. BAU, M.Y. CHIANG, D.M. HO, S.G. GIBBINS, T.J. EMGE and T.F. KOETZLE, 1984. Inorg. Chem., 23, 2823-2829.

Monoclinic, C2/c, a = 20.427, b = 11.623, c = 21.661 Å, β = 109.39°, Z = 4. Mo radiation, R = 0.049 for 1476 reflexions at room temperature. a = 20.808, b = 11.509, c = 21.009 Å, β = 116.47°, 1.16173 Å neutron radiation, $R(F^2)$ = 0.117 for 5951 reflexions at 80 K.

The molecule (Fig. 1) consists of an octahedral $[FeH_6]^{4-}$ core having exact C_2 symmetry with $[MgBr(C_4H_8O)_2]^+$ units situated over four of the eight faces of the octahedron. Each "Br" atom is actually a disorder of Br and Cl species to yield the formulation above. The mean Fe-H and Mg-H distances observed (1.609(2) and 2.045(18) Å, respectively) support the idea that the Fe-H bonds are largely "covalent" and the Mg-H bonds largely "ionic" in nature. The mean Mg-Br and Mg-O distances at 80 K are 2.483(4) and 2.077(9) Å respectively.

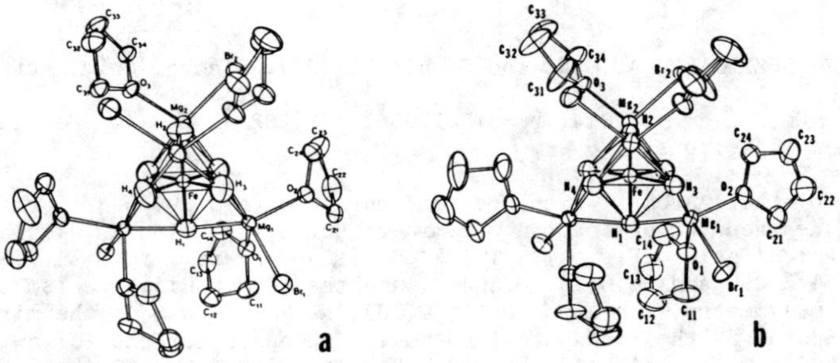

Fig. 1. The $FeH_6Mg_4Br_{3.5}Cl_{0.5}(C_4H_8O)_8$ structure at 80 K (a) and at room temperature (b).

BIS(μ_2-TRI-t-BUTYLMETHOXY)BIS(TRIMETHYLSILYL)AMINOMANGANESE(II)LITHIUM
$C_{32}H_{72}LiMnNO_2Si_2$ (I)

[BIS(μ_2-BROMO)(μ_2-TRI-t-BUTYLMETHOXY)(TETRAHYDROFURANLITHIUM)]MANGANESE(II)
$C_{34}H_{70}Br_2Li_2MnO_4$ (II)

B.D. MURRAY and P.P. POWER, 1984. J. Am. Chem. Soc., 106, 7011-7015.

I. Monoclinic, Cc, a = 21.243, b = 11.814, c = 20.334 Å, β = 133.60°, Z = 4. Mo radiation, R = 0.033 for 2558 reflexions (at 140 K).

II. Monoclinic, $P2_1/c$, a = 17.868, b = 8.563, c = 26.075 Å, β = 98.39°, Z = 4. Cu radiation, R = 0.088 for 3991 reflexions (at 140 K).

The geometry at manganese in (I) is distorted trigonal planar (Fig. 1) with Mn-O(1) 2.019(4), Mn-O(2) 1.980(4), Mn-N 2.001(3), Mn...Li 2.640(7), Li-O(1) 1.808(14), Li-O(2) 1.813(14) Å, O-Mn-O 86.4(2), O-Li-O 98.2(4), O-Mn-N 134.8 and 138.2(3)°. The Mn atom in (II) has a distorted tetrahedral environment (Fig. 1) with Mn-Br 2.553 and 2.576(2), Mn...Li(1) 2.650(19), Mn...Li(2) 3.36(2) Å. The geometry at Li(2) is also distorted tetrahedral, with Li(2)-O(7) 1.84(2) and Li(2)-1.91(2) Å, Br-Li(2)-Br 121.5(11)°.

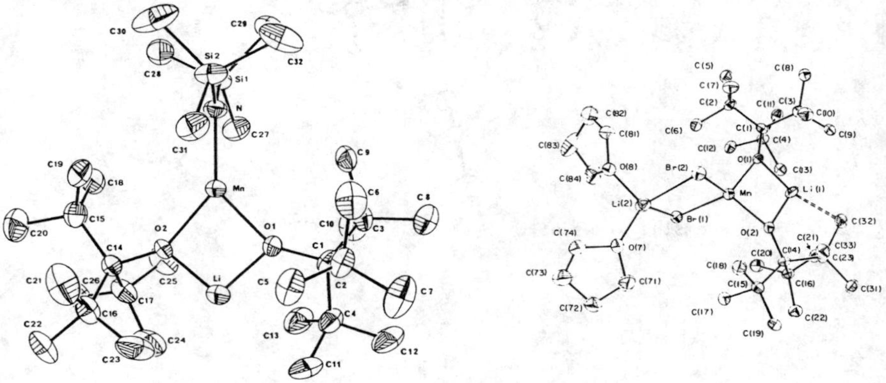

Fig. 1. Views of $C_{32}H_{72}LiMnNO_2Si_2$ and $C_{34}H_{70}Br_2Li_2MnO_4$.

2,3,11,12-(BIS-1,2-ACENAPHTHO)-18-CROWN-6 : POTASSIUM ISOTHIOCYANATE

$C_{33}H_{28}KNO_6S$ $C_{32}H_{28}O_6.KNCS$

G. WEBER, G.M. SHELDRICK, A. MERZ and F. DIETL, 1984. Inorg. Chim. Acta, 90, L1-L3.

Orthorhombic, $Cmc2_1$, a = 22.814, b = 15.170, c = 8.282 Å, Z = 4. Mo radiation, R = 0.083 for 1832 reflexions.

The six O(ether) atoms, which are coplanar to within ±0.35 Å, form the basal plane of the hexagonal coordination pyramid of potassium, with K^+ displaced by 0.73 Å towards the apical anion (Fig. 1). The K^+-O distances are: to O(1) 2.706, to O(4) 2.994, to O(7) 2.860 and to O(10) 2.804 Å, and the K^+...N distance is 2.727 Å. The molecule has m symmetry with atoms O(1), O(10), K, N, C and S on the mirror plane. The strain induced by the acenaphthylene nuclei accounts for the following torsion angles: C(3)-O(4)-C(5)-C(6) 137.2°, and C(5)-C(6)-O(7)-C(8) 130.6°.

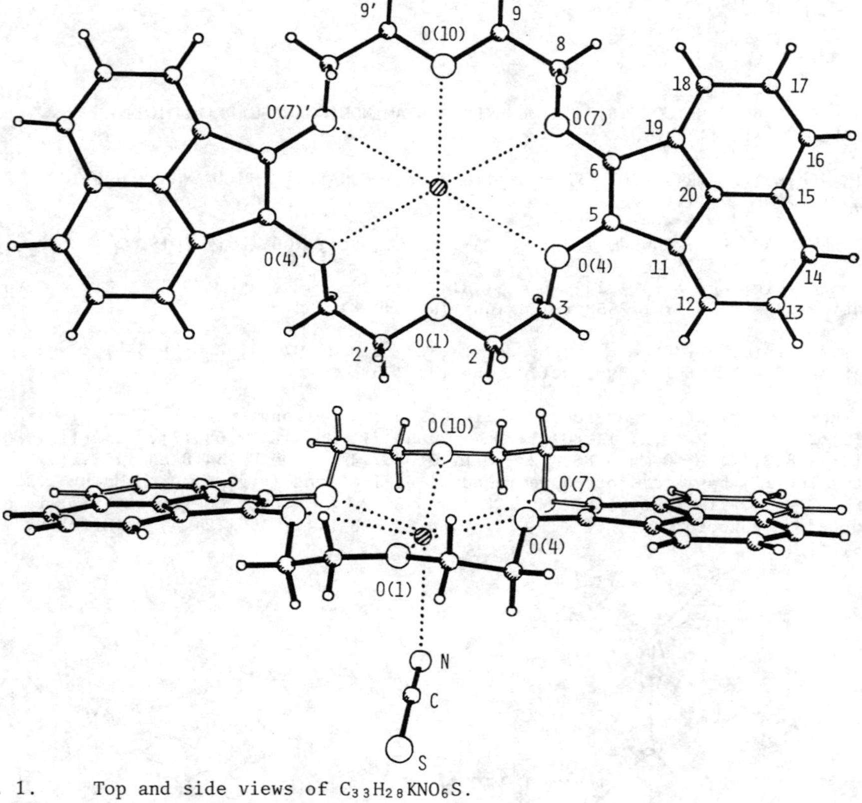

Fig. 1. Top and side views of $C_{33}H_{28}KNO_6S$.

DILITHIUM TRIBENZYLIDENEMETHANE BIS(TETRAMETHYLETHYLENEDIAMINE)
$C_{34}H_{50}Li_2N_4$

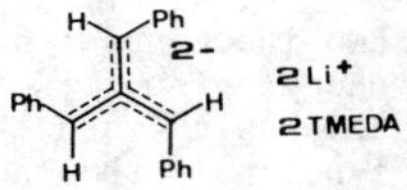

D. WILHELM, H. DIETRICH, T. CLARK, W. MAHDI, A.J. KOS and P. VON R. SCHLEYER, 1984. J. Am. Chem. Soc., 106, 7279-7280.

Triclinic, $P\bar{1}$, a = 8.627, b = 11.867, c = 17.026 Å, α = 72.51, β = 83.88, γ = 76.84°, Z = 2. Mo radiation, R = 0.152 for 3900 reflexions.

The structure (Fig. 1) has Li atoms above and below the tribenzylidenemethane plane with asymmetric Li-C contacts in the range 2.234(5)-2.784(6) Å (Fig. 1); one of the three C-C bonds at the central C atom is considerably shorter than the other two (1.388(4) Å compared with 1.433(3) and 1.460(3) Å). The Li coordination is completed by the tetramethylethylenediamine molecules, with Li-N 2.025-2.118 Å.

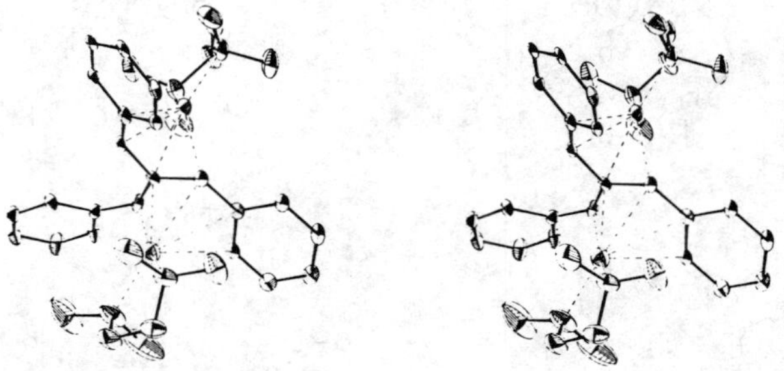

Fig. 1. A stereoview of $C_{34}H_{50}Li_2N_4$.

cis,anti,cis-2,3,11,12-TETRAPHENYL-18-CROWN-6
$C_{36}H_{40}O_6$ (I)

(cis,anti,cis-2,3,11,12-TETRAPHENYL-18-CROWN-6) SODIUM IODIDE DIHYDRATE DICHLOROMETHANE SOLVATE
$C_{36}H_{40}INaO_6,2H_2O,CH_2Cl_2$ (II)

trans,syn,trans-2,3,11,12-TETRAPHENYL-18-CROWN-6
$C_{36}H_{40}O_6$ (III)

(trans,syn,trans-2,3,11,12-TETRAPHENYL-18-CROWN-6) SODIUM IODIDE
$C_{36}H_{40}INaO_6$ (IV)

G. WEBER, G.M. SHELDRICK, T. BURGEMEISTER, F. DIETL, A. MANNSCHRECK and A. MERZ, 1984. Tetrahedron, 40, 855-863.

I. Monoclinic, $P2_1/n$, a = 8.193, b = 20.021, c = 19.102 Å, β = 95.12°, Z = 4. Mo radiation, R = 0.095 for 2353 reflexions.

II. Monoclinic, $P2_1/c$, a = 19.498, b = 19.662, c = 10.714 Å, β = 100.28°, Z = 4. Mo radiation, R = 0.072 for 3481 reflexions.

III. Monoclinic, $P2_1/c$, a = 12.546, b = 22.724, c = 11.684 Å, β = 108.57°, Z = 4. Mo radiation, R = 0.055 for 3013 reflexions.

IV. Triclinic, $P\bar{1}$, a = 8.986, b = 12.414, c = 16.059 Å, α = 89.58, β = 78.76, γ = 82.24°, Z = 2. Mo radiation, R = 0.043 for 5453 reflexions.

Views of the four molecules are in Fig. 1. In each of I, II and III there are two independent half molecules (or complex) per asymmetric unit. The cis-oriented phenyl groups in I are located at the anti-periplanar (ap) ethylene groups and are axial with respect to the crown ether ring. In III, the phenyl groups are at synclinal ethylene groups and appear synclinal and equatorial. Structure III is more rectangular in shape and less compact than I. The sodium iodide complex II is a "normal" crown complex with the Na^+ ion at the centre of the plane of six oxygen atoms and the phenyl groups in axial-equatorial positions. The Na^+ ion is further coordinated to two water molecules. Complex IV is quite different from II and has a seven-coordinated Na^+ ion. The ligand is strongly bent as shown in Fig. 1, with the $Na^+ \ldots I^-$ contact distance 3.166 Å.

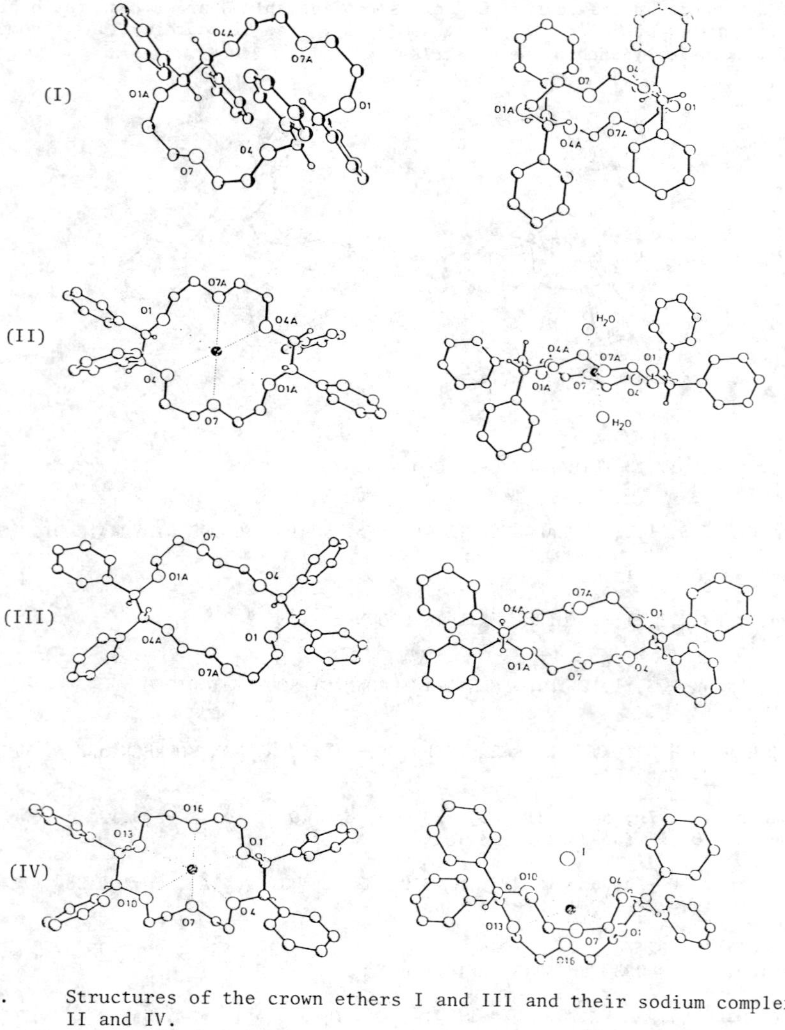

Fig. 1. Structures of the crown ethers I and III and their sodium complexes II and IV.

BIS((μ_2-2,6-DI-t-BUTYLPHENOLATO)TETRAHYDROFURAN-LITHIUM)
$C_{36}H_{58}Li_2O_4$ $\{Li[OC_6H_3\text{-}2,6\text{-}(Bu)_2]THF\}_2$

J.C. HUFFMAN, R.L. GEERTS and K.G. CAULTON, 1984. J. Cryst. Spectrosc. Res., 14, 541-547.

Orthorhombic, Pbca, a = 11.799, b = 18.758, c = 15.695 Å, Z = 4. Mo radiation, R = 0.062 for 1535 reflexions (at -160°C).

The molecule (Fig. 1) synthesized from phenol and LiH, is shown to exist in the solid as a centrosymmetric dimer with two bridging phenoxides. Terminal THF ligands (envelope conformation) complete trigonal planar coordination about Li. The THF ligand plane lies in the $Li_2(\mu\text{-}O)_2$ plane, while the phenyl ligands are perpendicular to this same Li_2O_2 plane. This explains the unusual ^1H NMR chemical shifts of the THF ligand. THF binds to Li as a dipole rather than using one of the two ether lone pairs. The Li-O distances are 1.828 and 1.868(7) Å in the four-membered ring and 1.900(7) Å (to THF). The Li...Li separation is 2.393(14) Å.

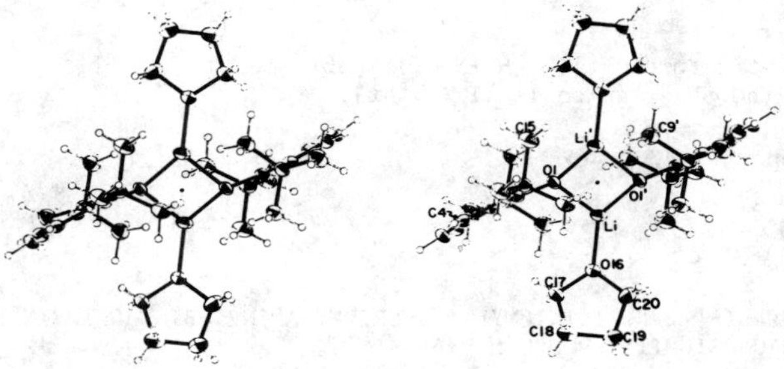

Fig. 1. A stereoview of the $C_{36}H_{58}Li_2O_4$ structure.

BIS[(4,7,13,16,21,24-HEXAOXO-1,10-DIAZABICYCLO[8.8.8]HEXACOSANE)POTASSIUM] TETRA-ANTIMONIDE(2-)
$C_{36}H_{72}K_2N_4O_{12}Sb_4$ (I) $[(C_{18}H_{36}N_2O_6)K]_2Sb_4$

TRIS[4,7,13,16,21,24-HEXAOXA-1,10-DIAZABICYCLO[8.8.8]HEXACOSANE)POTASSIUM] HEPTA-ANTIMONIDE(3-) BIS(ETHYLENEDIAMINE) SOLVATE
$C_{54}H_{108}K_3N_6O_{18}Sb_7,2C_2H_8N_2$ (II) $[C_{18}H_{36}N_2O_6)K]_3Sb_7,2H_2NCH_2CH_2NH_2$

S.C. CRITCHLOW and J.D. CORBETT, 1984. Inorg. Chem., 23, 770-774.

I. Triclinic, P$\bar{1}$, a = 11.555, b = 11.795, c = 11.067 Å, α = 97.60, β = 97.80, γ = 61.24°, Z = 1. Mo radiation, R = 0.081 for 3909 reflexions.

II. Triclinic, P$\bar{1}$, a = 14.589, b = 21.753, c = 13.857 Å, α = 94.46, β = 99.63, γ = 89.12°, Z = 2. Mo radiation, R = 0.075 for 7281 reflexions.

The structure of I and the Sb_4^{2-} (I) and Sb_7^{3-} (II) anions are shown in Fig. 1. I is isostructural with the Bi analog (1). Both structures consist of $[(C_{18}H_{36}N_2O_6)K]^+$ complex cations and polyantimonide anions. The centrosymmetric square planar Sb_4^{2-} anion in I has effective D_{4h} symmetry with Sb-Sb 2.749(1) and 2.751(1) Å and Sb-Sb-Sb 90.05(3) and 89.95(3)°. Structure II contains the Sb_7^{3-} anion with approximately C_{3v} symmetry with Sb-Sb 2.717(2)-2.785(2) Å for two-coordinate to three-coordinate bonds and 2.876(2)-2.906(2) Å for bonds between three-coordinate Sb atoms.

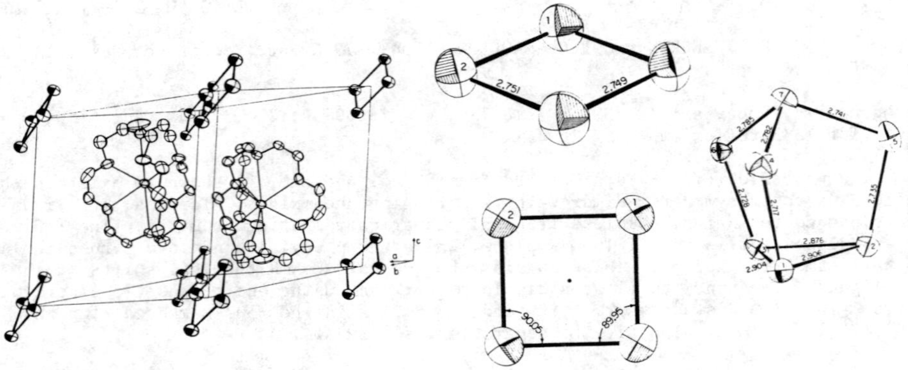

Fig. 1. The structure of $[(C_{18}H_{36}N_2O_6)K]_2Sb_4$ (left), the Sb_4^{2-} anion (middle), and the Sb_7^{3-} anion in II (right).

1. Structure Reports, 43B, 467.

HEXAAQUAMAGNESIUM BIS[(2-PYRIDINIUMYL)(2-PYRIDYL)AMINE] BIS(ADENOSINE-5'-TRI-PHOSPHORIC ACID-O,O',O'')MAGNESIUM(II) DODECAHYDRATE
$C_{40}H_{58}Mg_2N_{16}O_{32}P_6,12(H_2O)$ (I)

HEXAAQUACALCIUM BIS[(2-PYRIDINIUMYL)(2-PYRIDYL)AMINE] BIS(ADENOSINE-5'-TRI-PHOSPHORIC ACID-O,O',O'')CALCIUM(II) NONAHYDRATE
$C_{40}H_{58}Ca_2N_{16}O_{32}P_6,9(H_2O)$ (II)

R. CINI, M.C. BURLA, A. NUNZI, G.P. POLIDORI and P.F. ZANAZZI, 1984. J. Chem. Soc. Dalton, 2467-2476.

I,II. Orthorhombic, $C222_1$, a = 22.734, 22.965, b = 10.233, 10.154, c = 30.997, 32.390 Å, Z = 4. Mo radiation, R = 0.111, 0.124 for 1088, 1008 reflexions.

Compounds I and II are isostructural and contain complex anions, hexaquo cations, uncoordinated bis(2-pyridyl)amine cations and several free water molecules. In the anions the metal atoms (Mg,Ca) lie on a twofold axis with octahedral coordination geometry completed by the O atoms of the α, β and γ phosphate groups of two symmetry related adenosine 5'-triphosphate cations (Fig. 1). The mean Mg-O and Ca-O distances are 2.06(2) and 2.28(2) Å respectively. The coordination polyhedron in I is regular but that in II is distorted. The phosphates in both have a folded conformation. There are no bonding interactions between the metal ions and the adenine base. In the complex cations $M(H_2O)_6^+$ the metal atoms are located on twofold axes while the six coordinated water molecules form hydrogen bonds with the phosphate chains. The mean $Mg\text{-}OH_2$ and $Ca\text{-}OH_2$ distances are 2.12(6) and 2.40(7) Å respectively. The bis(2-pyridyl)amine cations do not coordinate to the metal atoms and are disordered around the twofold axis. Strong stacking interactions exist between these cations and the purine rings.

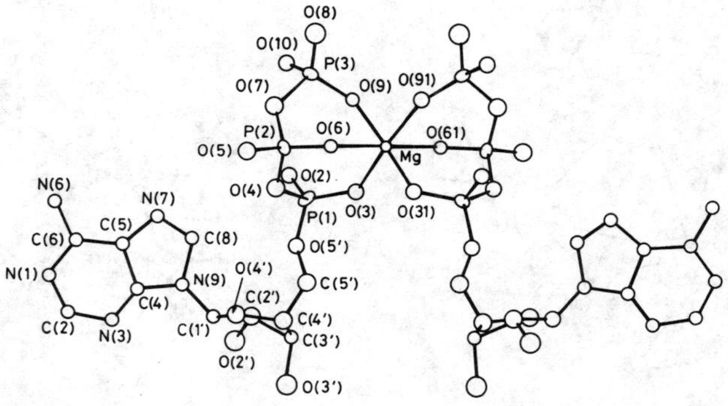

Fig. 1. $C_{40}H_{58}Mg_2N_{16}O_{32}P_6,12(H_2O)$: a view of the complex anion.

TETRAAQUABIS(1,10-PHENANTHROLINE)BARIUM(II) PERCHLORATE - BIS(1,10-PHENANTHROLINE)
$C_{48}H_{40}BaCl_2N_8O_{12}$ $[Ba(C_{12}H_8N_2)_2(H_2O)_4](ClO_4)_2.2C_{12}H_8N_2$

R.L. STANFIELD, S.R. ERNST, M.L. HACKERT, V. VIJAYWARGIA, K. VENKATASUBRAMANIAN and N.S. POONIA, 1984. Acta Cryst., C40, 1681-1683.

Triclinic, $P\bar{1}$, a = 7.026, b = 13.003, c = 13.361 Å, α = 75.51, β = 84.84, γ = 80.00°, D_m = 1.610, Z = 1. Cu radiation, R = 0.055 for 3955 reflexions.

A redetermination of a structure reported previously (1). Four N atoms from two o-phenanthrolines and four O atoms from water molecules are coordinated to Ba in a distorted cubic arrangement (Ba-O 2.778, 2.763(6), Ba-N 2.945, 2.940(6) Å).

1. Structure Reports, 43B, 379.

HEXAKIS(3-LITHIOMETHYL-1,1,2,2-TETRAMETHYLCYCLOPROPANE)
$C_{48}H_{90}Li_6$

A. MAERCKER, M. BSATA, W. BUCHMEIER and B. ENGELEN, 1984. Chem. Ber., 117, 2547-2554.

Rhombohedral, $R\bar{3}$, a = 18.48, c = 12.99 Å, Z = 18. Mo radiation, R = 0.037 for 628 reflexions.

In the crystal lattice (Fig. 1) there are hexameric clusters with the lithium atoms forming a trigonal antiprism with unoccupied triangular faces on the top and at the bottom while each of the six laterally triangular faces is coordinated to a 2,2,3,3-tetramethylcyclopropylmethyl ligand. The Li-Li distances (Fig. 1) are 2.976(12) and 2.462(10) Å. Li-C distances are in the range 2.123(8)-2.297(8) Å.

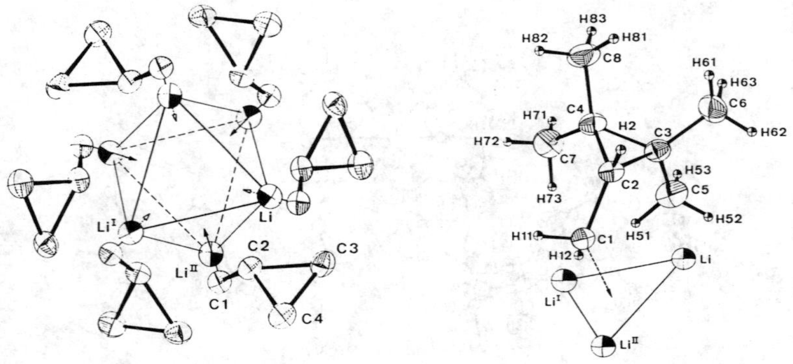

Fig. 1. Views of the $C_{48}H_{90}Li_6$ structure.

rel-(1S,2R,16S,18R,32S,33R,47S,49R)-3,6,9,12,15,19,22,25,28,31,34,37,40,43,46,50,53,-56,59,62-ICOSAOXAPENTACYCLO[31.29.1.$1^{2,32}$.$0^{16,49}$.$0^{18,47}$]DOHEXACONTANE POTASSIUM PICRATE (2:1)

$C_{56}H_{84}K_2N_6O_{34}$ $C_{44}H_{80}O_{20}.2[K(C_6H_2N_3O_7)]$

J.D. OWEN, 1984. Acta Cryst., C40, 246-248.

Monoclinic, $P2_1/c$, a = 9.170, b = 16.089, c = 22.846 Å, β = 92.797°, D_m = 1.43, Z = 2. Mo radiation, R = 0.064 for 3054 reflexions.

The complex cation lies about an inversion centre and wraps up the cations to the exclusion of the anions (Fig. 1). The two K^+ ions are each coordinated to 10 oxygen atoms of the ligand with K...O in the range 2.855(4) to 3.195(4) Å.

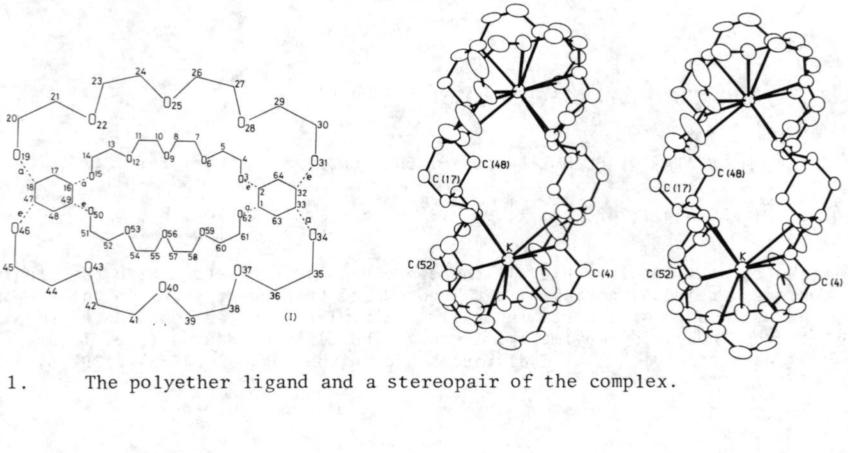

Fig. 1. The polyether ligand and a stereopair of the complex.

TRIS(1,10-PHENANTHROLINE)POTASSIUM TETRAPHENYLBORATE

$C_{60}H_{44}BKN_6$ $[K(C_{12}H_8N_2)_3][B(C_6H_5)_4]$

G. BOMBIERI, G. BRUNO, M.D. GRILLONE and G. POLIZZOTTI, 1984. Acta Cryst., C40, 2011-2014.

Triclinic, P$\bar{1}$, a = 10.455, b = 14.435, c = 18.209 Å, α = 103.5, β = 103.7, γ = 110.1°, Z = 2. Mo radiation, R = 0.074 for 3408 reflexions.

The structure (Fig. 1) consists of BPh_4^- anions and $[K(phen_3]_2{}^{2+}$ centrosymmetric dimeric cationic species (phen = 1,10-phenanthroline) in which the K atoms are eight coordinated by the N atoms of the phen molecules in a square-antiprismatic geometry. The K...K' contact distance in the dimer is 3.813(3) Å. The K-N bond distances range from 2.800(6) to 2.893(5) Å for the singly chelated phen molecules, while for the bridging phens these vary from 3.010(5) and 3.167(6) Å.

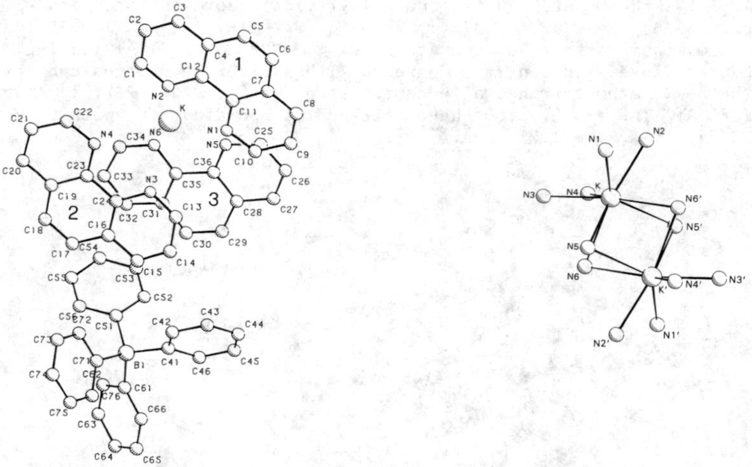

Fig. 1. View down a of the asymmetric unit for $C_{60}H_{44}BKN_6$ and the K coordination polyhedra.

BIS[LITHIUM TRIS(PHENYLETHYNYL)(N,N,N',N'-TETRAMETHYLETHYLENEDIAMINE)MAGNESIUM]
$C_{60}H_{62}Li_2Mg_2N_4$ (I)

(N,N,N',N'-TETRAMETHYLETHYLENEDIAMINE)LITHIUMBIS(μ_2-BENZYL)DIBENZYLMAGNESIUM BIS(N,N,-N',N'-TETRAMETHYLETHYLENEDIAMINE)LITHIUM
$C_{46}H_{76}Li_2MgN_6$ (II) $C_{12}H_{32}LiN_4^+.C_{34}H_{44}LiMgN_2^-$

B. SCHUBERT and E. WEISS, 1984. Chem. Ber., 117, 366-375.

I. Monoclinic, C2/c, a = 23.294, b = 11.030, c = 22.388 Å, β = 105.05°, Z = 4. Mo radiation, R = 0.076 for 2640 reflexions.

II. Monoclinic, P2$_1$/c, a = 18.164, b = 12.361, c = 23.668 Å, β = 109.62°, Z = 4. Mo radiation, R = 0.097 for 2301 reflexions.

Molecule I has the Mg atom in trigonal-bipyramidal coordination, with Mg-C(eq) 2.188(5) and 2.183(5), Mg-C(ax) 2.307(4), Mg-N(eq) 2.202(5), Mg-N(ax) 2.350(4) Å. The Li atom forms four contacts with C atoms (Li-C 2.257-2.342(8) Å) leading to dimer formation about an inversion centre. Compound II has discrete cations and anions (Fig. 1) with the Mg atom tetrahedrally coordinated (Mg-C 2.225-2.232(11) Å and Mg... Li(1) 2.719(18) Å). The two Li atoms have tetrahedral coordination, with Li-C 2.27(2), 2.23(2), Li-N 2.08-2.21(2) Å.

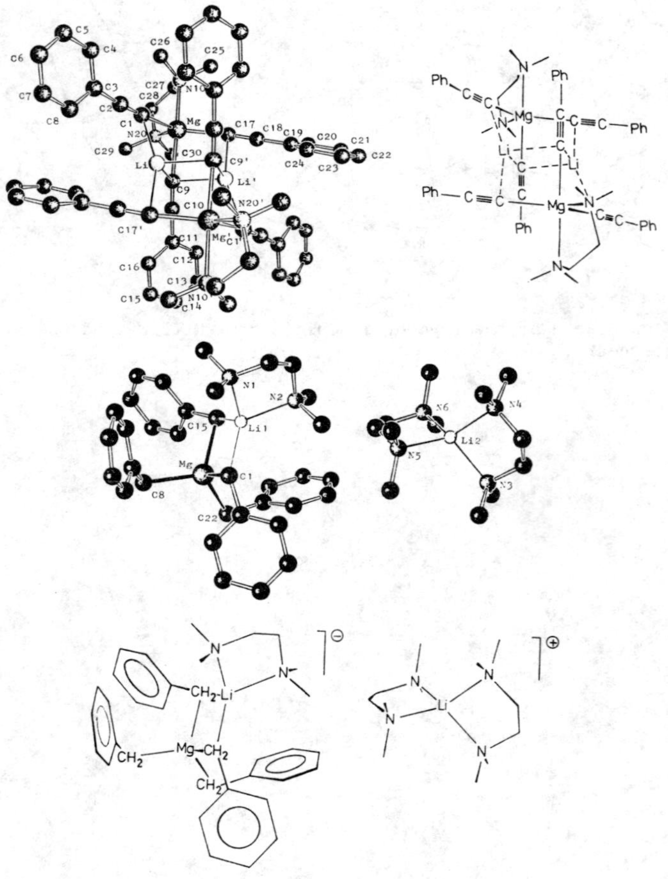

Fig. 1. Views of $(C_{30}H_{31}LiMgN_2)_2$ (I) (top) and $C_{46}H_{76}Li_2MgN_6$ (II) (bottom).